THE SYSTEM OF
MINERALOGY

By the late JAMES D. DANA

SYSTEM OF MINERALOGY. *Seventh Edition.*
Rewritten and enlarged by the late Charles Palache, the late Harry Berman, and Clifford Frondel.
Vol. I. 1944.
Vol. II. 1951.
Vol. III. In preparation.

MANUAL OF MINERALOGY. *Seventeenth Edition.*
Revised by Cornelius S. Hurlbut, Jr.

By the late EDWARD S. DANA

A TEXTBOOK OF MINERALOGY. *Fourth Edition.*
Revised by the late William E. Ford. 1932.

MINERALS AND HOW TO STUDY THEM. *Third Edition.*
Revised by Cornelius S. Hurlbut, Jr. 1949.

THE SYSTEM OF
MINERALOGY

of James Dwight Dana and Edward Salisbury Dana
Yale University 1837–1892

SEVENTH EDITION
Entirely Rewritten and Greatly Enlarged

By

CHARLES PALACHE
the late HARRY BERMAN
and CLIFFORD FRONDEL
Harvard University

VOLUME II

HALIDES, NITRATES, BORATES, CARBONATES,

SULFATES, PHOSPHATES, ARSENATES, TUNGSTATES,

MOLYBDATES, ETC.

JOHN WILEY & SONS
New York • Chichester • Brisbane • Toronto

COPYRIGHT, 1892
COPYRIGHT RENEWED, 1920
BY
EDWARD S. DANA

COPYRIGHT, 1944 AND 1951
BY
JOHN WILEY & SONS, INC.

All Rights Reserved

Reproduction or translation of any part of this work beyond that permitted by Sections 107 or 108 of the 1976 United States Copyright Act without the permission of the copyright owner is unlawful. Requests for permission or further information should be addressed to the Permissions Department, John Wiley & Sons, Inc.

SEVENTH EDITION

14

ISBN 0 471 19272 4

PRINTED IN THE UNITED STATES OF AMERICA

PREFACE

The preparation of the present volume of the seventh edition of Dana's *System of Mineralogy* was started in the summer of 1942, but after only a few months the two junior authors were drawn into services connected with World War II and the work was set aside. In August of 1944, Professor Berman, while traveling abroad, died in an accident. The science of mineralogy suffered a tragic loss, and the writing of these volumes a lasting setback with his death. The work was resumed in the fall of 1945 and, after many lengthy interruptions, was brought to completion in the winter of 1949–1950.

The preparation of the mineral descriptions has been with but rare exceptions entirely from original sources, although the references connected with the older synonomy have in general not been consulted. It must be acknowledged immediately, however, that essential guidance and aid have been obtained from the standard reference works in the field. Particular mention may be made here of Hintze's *Handbuch der Mineralogie* and of the earlier editions of this work, together with the *Referate* series of the *Neues Jahrbuch für Mineralogie* and the invaluable *Mineral Abstracts* published by the Mineralogical Society of Great Britain. As formerly, an effort was made to remedy recognized inadequacies or errors in existing descriptions by a first-hand examination of authentic specimens. The examination and organization of the literature and the actual writing of the manuscript were done by the present writer.

The general method of the work and the guiding framework of definitions and conventions have been outlined in the Introduction to Volume I, but some further remarks are pertinent. The concept of minerals as phases that vary serially in composition within natural limits, and that vary concomitantly therewith in properties, is implicit in this work. The natural unit of description is then the series. Although this unified treatment of the several species that may constitute a series is easily effected in the case of binary systems, the complexity of some polycomponent systems has made it advisable to subdivide the extended series into smaller, arbitrarily defined regions of composition and to describe these separately. This treatment has been found particularly desirable, although only with regard to the mechanics of presenting an adequate description, with certain common and important series such as the calcite group. At the same time, the separate descriptions are written so as to emphasize the continuity between the several species involved.

Of the multitude of new species names that have been introduced into the science during the years since the sixth edition of the *System*, comparatively few are based on adequate descriptions. Other new names, including many introduced to designate particular ranges of composition in series whose extent is already established by observation or by reasonable inference, do not appear to fill any real need. The disposition of these names has been the subject of much concern. Descriptive mineralogy is overburdened already by trivial names, and the interests of the science unquestionably would be served best by a reduction in size and simplification of its nomenclature. It is realized, however, that any extensive revision of nomenclature in order to be sustained by usage must come by the expression of mineralogists generally. In a few instances, nevertheless, the authors have acted summarily, and it is hoped wisely, to redefine established names (*vide* ankerite, cobaltocalcite, chalcocyanite, metastrengite, soda alum); they further have subscribed completely to the now very generally accepted adjectival scheme of varietal nomenclature. Thus, names introduced as new subdivisions of series in almost all instances have been relegated to the species synonymy, and their place has been taken by adjectival modifiers of the species name. This method of nomenclature has the advantages of elasticity, thus obviating the introduction of further names of closer meaning, and of telling something of the character of the substance at hand. Closer characterization of points or regions within a series, if desired, should be accomplished by a statement of the exact physical and chemical properties of the material. The systematization of nomenclature in this way, it may be noted, is a natural expression of the series concept of mineral constitution that has supplanted the older notion of species as constituting phases of essentially fixed composition.

The classification employed is essentially chemical in nature. Very few of the anisodesmic oxysalts have been analyzed structurally, and no unifying structural theme based on the anionic framework, such as obtains in the silicates, has yet been clearly set forth. The major classes are based on the nature of the predominant anion, and the several types within these classes are arranged according to the ratio of cations to anions. The arrangement of oxysalts containing in addition OH, O, or halogen is based on the ratio of the cationic elements to the anionic complexes only. This often permits a close grouping of minerals of similar properties and genesis that otherwise would be widely scattered among quite dissimilar species. In any event, this arrangement is necessitated by uncertainty as to the amount and role of OH and H_2O in most basic and hydrated oxysalts. In the borates and the vanadium oxysalts the nature of the anionic complex in most of the species is uncertain or unknown. These minerals have been classed temporarily for convenience of reference on the arbitrary ratio of cations to B or V. The system of numbering species is in conformity with and extends that followed in Volume I.

In the description of the structure cell, the symbols kX and $Å$ are employed to designate dimensions referred to the old and new x-ray wavelengths, respectively, following current convention.

Preface

The authors again wish to acknowledge the unstinting aid of their associates and colleagues in the science. The crystallographic and other computations for the most part were done by Miss Mary Mrose, whose devoted service has hastened the appearance of this volume. Professor C. W. Wolfe of Boston University computed a proportion of the crystallographic material during his connection with the project in 1945 and 1946. Most of the crystal figures were drawn by Professor Wolfe, and the rest by Dr. Howard T. Evans and others. Grateful acknowledgment is also made to Professor Adolf Pabst of the University of California (Berkeley) for assistance in the preparation of manuscript during the summer of 1948. The typing of the manuscript passed entirely through the capable hands of Mrs. Celia Lang. Gratitude also must be expressed, although space is lacking for specific acknowledgment, to the many people both here and abroad who have aided in giving counsel and in supplying factual material. The generous financial support provided by the Geological Society of America from its Penrose Fund for Volume I of this work was extended through a large supplementary grant in 1946 for the preparation of the present volume.

CLIFFORD FRONDEL

Cambridge, Massachusetts,
September, 1951

CONTENTS

HALIDES

Class 9 NORMAL ANHYDROUS AND HYDRATED HALIDES	3
Class 10 OXYHALIDES AND HYDROXYHALIDES	51
Class 11 HALIDE COMPLEXES. ALUMINO-FLUORIDES	91
Class 12 COMPOUND HALIDES	129

CARBONATES

Class 13 ACID CARBONATES	134
Class 14 ANHYDROUS NORMAL CARBONATES	141
Class 15 HYDRATED NORMAL CARBONATES	224
Class 16 CARBONATES CONTAINING HYDROXYL OR HALOGEN	244
Class 17 COMPOUND CARBONATES	294

NITRATES

Class 18 NORMAL ANHYDROUS AND HYDRATED NITRATES	300
Class 19 NITRATES CONTAINING HYDROXYL OR HALOGEN	308
Class 20 COMPOUND NITRATES	309

IODATES

Class 21 NORMAL ANHYDROUS AND HYDRATED IODATES	312
Class 22 IODATES CONTAINING HYDROXYL OR HALOGEN	315
Class 23 COMPOUND IODATES	318

BORATES

Class 24 ANHYDROUS NORMAL BORATES	321
Class 25 HYDRATED NORMAL BORATES	334
Class 26 BORATES CONTAINING HYDROXYL OR HALOGEN	369
Class 27 COMPOUND BORATES	385

SULFATES

Class 28 ANHYDROUS ACID AND NORMAL SULFATES	395
Class 29 HYDRATED ACID AND NORMAL SULFATES	436
Class 30 ANHYDROUS SULFATES CONTAINING HYDROXYL OR HALOGEN	541
Class 31 HYDRATED SULFATES CONTAINING HYDROXYL OR HALOGEN	572
Class 32 COMPOUND SULFATES	628

SELENATES AND TELLURATES; SELENITES AND TELLURITES

Class 33 SELENATES AND TELLURATES	635
Class 34 SELENITES AND TELLURITES	638

CHROMATES

Class 35 ANHYDROUS NORMAL CHROMATES	644
Class 36 COMPOUND CHROMATES	650

PHOSPHATES, ARSENATES, AND VANADATES

Class 37 ANHYDROUS ACID PHOSPHATES, ETC.	660
Class 38 ANHYDROUS NORMAL PHOSPHATES, ETC.	664
Class 39 HYDRATED ACID PHOSPHATES, ETC.	698
Class 40 HYDRATED NORMAL PHOSPHATES, ETC.	715
Class 41 ANHYDROUS PHOSPHATES, ETC., CONTAINING HYDROXYL OR HALOGEN	775
Class 42 HYDRATED PHOSPHATES, ETC., CONTAINING HYDROXYL OR HALOGEN	915
Class 43 COMPOUND PHOSPHATES, ETC.	1001

ANTIMONATES; ANTIMONITES AND ARSENITES

Class 44 ANTIMONATES	1017
Class 45 ACID AND NORMAL ANTIMONITES AND ARSENITES	1031
Class 46 BASIC OR HALOGEN-CONTAINING ANTIMONITES, ARSENITES	1036

VANADIUM OXYSALTS

Class 47 (ARRANGED ACCORDING TO CATION TO V RATIO)	1043

MOLYBDATES AND TUNGSTATES

Class 48 Normal Anhydrous Molybdates and Tungstates   1064

Class 49 Basic and Hydrated Molybdates and Tungstates   1091

ORGANIC COMPOUNDS

Class 50 Salts of Organic Acids   1099

Index   1109

HALIDES

All the minerals classed here contain a halogen as the sole or principal anionic constituent. A number of diverse types of structure are represented. They include isodesmic crystals such as halite and fluorite and a variety of more or less markedly anisodesmic structures. Among the latter are found examples of planar anionic groups, tetrahedral anions such as the $(BF_4)^-$ groups in the barite-type compound KBF_4, and octahedral anions such as the $(SiF_6)^=$ groups in K_2SiF_6 and the $(AlF_6)^{\equiv}$ groups in the alumino-fluorides. The latter compounds have been tentatively arranged according to the nature of the anionic framework. The other halides are classified solely on chemical grounds.

CLASS 9. Normal anhydrous and hydrated halides

Type 1. AX

9.1.1	Halite group	
9.1.1.1	Halite	NaCl
9.1.1.2	Sylvite	KCl
9.1.1.3	Villiaumite	NaF
9.1.1.4	Cerargyrite	AgCl
9.1.1.5	Bromyrite	AgBr
9.1.2	Salammoniac	NH_4Cl
9.1.3	Nantokite group	
9.1.3.1	Nantokite	CuCl
9.1.3.2	Miersite	(Ag,Cu)I
9.1.3.3	Marshite	CuI
9.1.4	Iodyrite	AgI
9.1.5	Calomel	HgCl

Type 2. AX_2

9.2.1	Fluorite	CaF_2
9.2.2	Sellaite	MgF_2
9.2.3	Lawrencite group	
9.2.3.1	Lawrencite	$FeCl_2$
9.2.3.2	Scacchite	$MnCl_2$
9.2.3.3	Chloromagnesite	$MgCl_2$
9.2.4	Hydrophilite	$CaCl_2$
9.2.5	Coccinite	HgI_2
9.2.6	Cotunnite	$PbCl_2$
9.2.7	Eriochalcite	$CuCl_2 \cdot 2H_2O$
9.2.8	Bischofite	$MgCl_2 \cdot 6H_2O$

Type 3. AX_3

9.3.1	Molysite	$FeCl_3$
9.3.2	Fluocerite	$(Ce,La,Nd)F_3$
9.3.3	Chloraluminite	$AlCl_3 \cdot 6H_2O$

HALIDES

CLASS 10. Oxyhalides and hydroxyhalides

Type 1. $A_m(O,OH)_pX_q$

10.1.1		Eglestonite	Hg_4OCl_2
10.1.2		Terlinguaite	Hg_2OCl
10.1.3		Lorettoite	$Pb_7O_6Cl_2$
10.1.4		Mendipite	$Pb_3O_2Cl_2$
10.1.5		Daviesite	—
10.1.6		Matlockite group	
10.1.6.1		Matlockite	$PbFCl$
10.1.6.2		Bismoclite	$BiOCl$
10.1.6.3		Daubréeite	$BiO(OH,Cl)$
10.1.7		Laurionite	$Pb(OH)Cl$
10.1.8		Paralaurionite	$Pb(OH)Cl$
10.1.9		Penfieldite	$Pb_2(OH)Cl_3$
10.1.10		Fiedlerite	$Pb_3(OH)_2Cl_4$
10.1.11		Atacamite group	
10.1.11.1		Atacamite	$Cu_2(OH)_3Cl$
10.1.11.2		Kempite	$Mn_2(OH)_3Cl$
10.1.12		Paratacamite	$Cu_2(OH)_3Cl$
10.1.13		Botallackite	Basic Cu chloride
10.1.14		Cadwaladerite	$Al(OH)_2Cl \cdot 4H_2O$

Type 2. $A_mB_n(O,OH)_pX_q$

10.2.1	Boleite	$Pb_9Cu_8Ag_3Cl_{21}(OH)_{16} \cdot 2H_2O$ (?)
10.2.2	Cumengite	$Pb_4Cu_4Cl_8(OH)_8 \cdot H_2O$ (?)
10.2.3	Pseudoboleite	$Pb_5Cu_4Cl_{10}(OH)_8 \cdot 2H_2O$ (?)
10.2.4	Percylite	$PbCuCl_2(OH)_2$
10.2.5	Diaboleite	$Pb_2CuCl_2(OH)_4$
10.2.6	Chloroxiphite	$PbCuO_2Cl_2(OH)_2$ (?)
10.2.7	Nocerite	$Ca_3Mg_3F_8O_2$
10.2.8	Koenenite	$Mg_5Al_2(OH)_{12}Cl_4$
10.2.9	Zirklerite	Al,Fe basic chloride
10.2.10	Kleinite	Hg,NH_4,Cl,SO_4
10.2.11	Mosesite	Hg,NH_4,Cl,SO_4

CLASS 11. Halide complexes. Alumino-fluorides

Type 1. $A_mBX_3 \cdot xH_2O$

11.1.1	Chlorocalcite	$KCaCl_3$
11.1.2	Carnallite	$KMgCl_3 \cdot 6H_2O$
11.1.3	Tachyhydrite	$CaMg_2Cl_6 \cdot 12H_2O$

Type 2. A_mBX_4

11.2.1	Pseudocotunnite	K_2PbCl_4
11.2.2	Avogadrite	$(K,Cs)BF_4$
11.2.3	Ferruccite	$NaBF_4$
11.2.4	Cryolithionite	$Na_3Li_3Al_2F_{12}$

Type 3. $A_mBX_4 \cdot xH_2O$

11.3.1	Douglasite	$K_2FeCl_4 \cdot 2H_2O$ (?)
11.3.2	Mitscherlichite	$K_2CuCl_4 \cdot 2H_2O$
11.3.3	Erythrosiderite series	
11.3.3.1	Erythrosiderite	$K_2FeCl_5 \cdot H_2O$
11.3.3.2	Kremersite	$(NH_4,K)_2FeCl_5 \cdot H_2O$

Type 4. A_mBX_6
11.4.1	Hieratite group	
11.4.1.1	Hieratite	K_2SiF_6
11.4.1.2	Cryptohalite	$(NH_4)_2SiF_6$
11.4.2	Malladrite group	
11.4.2.1	Malladrite	Na_2SiF_6
11.4.2.2	Bararite	$(NH_4)_2SiF_6$
11.4.3	Rinneite	NaK_3FeCl_6
11.4.4	Chlormanganokalite	K_4MnCl_6

Type 5. Alumino-fluorides (see below)

CLASS 12. Compound halides
Type 1.	Miscellaneous	
12.1.1	Creedite	$Ca_3Al_2F_4(OH,F)_6(SO_4) \cdot 2H_2O$
12.1.2	Arzrunite	Pb,Cu sulfate-chloride
12.1.3	Trudellite	$Al_{10}Cl_{12}(OH)_{12}(SO_4)_3 \cdot 30H_2O$
	Bandylite	[with borates]
	Teepleite	[with borates]

9 NORMAL ANHYDROUS AND HYDRATED HALIDES

TYPE 1. *AX*

9.1.1 HALITE GROUP

ISOMETRIC; HEXOCTAHEDRAL—$4/m\,\bar{3}\,2/m$

	a_0	G.	n(Na)
Halite, NaCl	5.627 kX	2.168	1.5443
Sylvite, KCl	6.277	1.993	1.4903
Villiaumite, NaF	4.619	2.79	1.3270
Cerargyrite series			
Cerargyrite, Ag(Cl,Br)	5.545	5.55	2.071
Bromyrite, Ag(Br,Cl)	5.755	6.47	2.253

The crystal structure of halite, NaCl, was the first structure to be analyzed by x-rays.[1] The halite structure-type is characteristic of a large number of AX compounds with a radius ratio between 0.41 and 0.73, and includes among minerals the halite group, periclase group, and galena group. The important members of the halite group are halite and sylvite. These minerals occur principally as extensive bedded deposits formed by the evaporation of oceanic waters. Villiaumite is a rare primary mineral found in the pegmatitic facies of nepheline-syenites. At ordinary temperatures the extent of substitutional solid solution between NaCl and KCl is extremely small.

The silver halides—cerargyrite, AgCl, and bromyrite, AgBr—belong to the halite structure-type but differ from the typically ionic alkali halides in that they possess an intermediate, ionic-covalent bond and show marked semi-metallic properties. Cerargyrite and bromyrite form a complete series of substitutional solid solutions, Ag(Cl,Br), the boundary between the two species being drawn at Cl:Br = 1:1. Both minerals are secondary in origin and occur in the oxidized zone of silver deposits.

Ref.

1. Bragg, W. L., *Proc. Royal Soc. London*, **89A**, 468 (1914).

9.1.1.1   **H A L I T E** [NaCl].   Common or Rock Salt.   Muriate of Soda, Sodium Chloride.   Kochsalz, Steinsalz, Bergsalz, *Germ.*   Soude muriatée, Chlorure de sodium, Sel gemme *Fr.*   Sal mare *Beudant* (1832).   Halites *Glocker* (290, 1847).   Sal gemma, Alite *Ital.* Sal gema, Sal marina *Span.*   Halite *Dana* (112, 1868).
   Martinsite *Karsten* (*J. pr. Chem.*, **36**, 127, 1845).   Natrikalite *Adam* (69, 1869).   Huantajayite *Raimondi* (in Domeyko, Min. Chili, 5th App., 1876); *Raimondi* (64, 1878).   β-halite *Paniche* (*Per. Min.*, **4**, 25, 1933).   Saltspar *Murzaev* (*C. r. ac. sc. U.R.S.S.*, **33**, 306, 1941).

C r y s t.   Isometric; hexoctahedral—$4/m\,\bar{3}\,2/m$.

Forms:[1]

$\quad\quad\quad\quad\quad$ *a* 001 $\quad\quad$ *d* 011 $\quad\quad$ *o* 111 $\quad\quad$ *e* 012

Structure cell.[2] Space group $Fm3m$. a_0 5.627 kX. Cell contents Na_4Cl_4.

Habit.[3] Usually cubic, rarely octahedral. The crystal faces often are cavernous and stepped (hopper-crystals). Embedded crystals are sometimes distorted into rhomboid or other shapes due to growth under directed pressure.[4] Also massive, coarsely granular to compact; rarely columnar, stalactitic or capillary.

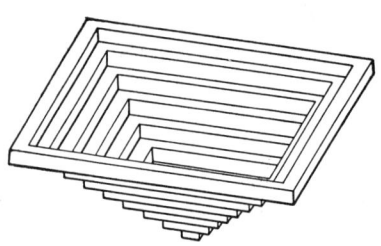

Twinning. Observed on {111} in artificial crystals.[5]

Etch figures.[6] Etch figures ordinarily conform to hexoctahedral symmetry; figures with gyroidal symmetry may be obtained from optically active solutions.

P h y s. Cleavage {001} perfect.[7] Fracture conchoidal. Rather brittle. H. 2.[8] G. 2.168; 2.165 (calc.), M.P. 804°.[9] Luster vitreous. Colorless; also white, yellow, red, blue, purple. The blue and purple tints usually are localized in growth bands or in irregular spots. The origin of the blue color has been the subject of much investigation,[10] and has been ascribed to the presence of free sodium atoms in the structure, and to other causes. The yellow and red tints are caused usually by mechanically enclosed clay or iron oxide. Streak colorless to white; relatively coarsely powdered blue or purple halite has a faint tint. Transparent. Taste saline. Highly diathermanous. Translation gliding [11] with: $T\{011\}$, $t[110]$; $T\{001\}$, $t[110]$; possibly with $T\{111\}$. Paramagnetic. Sometimes fluorescent.[19]

O p t. In transmitted light, colorless. Material with a relatively deep body color may show a faint corresponding tint in grains or sections under the microscope. Colored material may become weakly pleochroic after being subjected to pressure.[12] Usually isotropic, but may become weakly doubly refracting after mechanical deformation or heating.[13] Refractive indices.[14]

λ	C	D	F
n at 18°	1.54068	1.54432	1.55340

Temp.	14°	102°	177°	320°	425°
n 587.6	1.5446	1.5409	1.5372	1.5301	1.5246

Chem. Sodium chloride, NaCl. Na 39.34, Cl 60.66, total 100.00. Analyses of carefully purified material conform closely to the formula.[15] Often admixed with sand, clay, iron oxide, sylvite, kainite, gypsum, kieserite, etc. Minute cavities containing entrapped brine or gas are often present. Material containing calcium or magnesium chlorides may be highly deliquescent.

Tests. C.T. fuses, often with decrepitation. 35.7 grams of NaCl are soluble in 100 cc. water at 0°.

Occur. Halite occurs principally as extensive beds formed by the evaporation of enclosed bodies of salt water subsequently buried by other sedimentary deposits.[16] The salt beds occur in rocks of all geological ages and range from a few feet to over a thousand feet in thickness. In these deposits halite is commonly associated with gypsum, anhydrite, clay, polyhalite, sylvite, kieserite, kainite, carnallite, and a variety of other soluble salts. Mechanical deformation of stratified salt deposits may result in the local extrusion of massive plug-like bodies of salt, sometimes several miles in width, through overlying sediments as the so-called salt domes. Sodium chloride occurs in solution in salt springs that rise from salt beds and in the waters of the ocean, salt seas and lakes. Halite also is found as a sublimation product in volcanic regions and often occurs as an efflorescence in arid areas. Also found in hair-like or stalactitic forms on the walls of old mine workings. Only a very few of the many known localities of halite can be mentioned here.

In Europe, halite is found in a large deposit at Iletsk near Orenburg in southeastern Russia. In Galicia, in Tertiary rocks at Wieliczka, often as large crystals at Bochnia to the east and at Kalusz in large deposits. In the Austrian Alps in deposits with complex structure associated with clay, gypsum, anhydrite, polyhalite, glauberite, etc.; in Salzburg at Hallein, Hallstadt, and Ischl; in Styria at Aussee; in Tyrol at Hall near Innsbruck. From Switzerland at Bex in Vaud. Famous deposits of Permian age are found at Stassfurt, Saxony, and at the neighboring deposit of Leopoldshall in Anhalt, etc. Classical treatises on the chemistry of salt deposition have resulted from the study of these deposits. The lowest strata are composed of beds of halite with small included beds of anhydrite; followed by layers of halite with beds of polyhalite; halite with kieserite; lastly, smaller amounts of halite and much carnallite, also with large masses of chlorides and sulfates of potassium, magnesium, calcium, and sodium together with various double salts, the mixture being known as "Abraumsalz." Well-crystallized halite is found in Sicily at Girgenti, Racalmuto, Castrogiovanni, and elsewhere. In France large bodies occur in the neighborhood of Dax, in Landes, and from Vic and Dieuze in Lorraine. Near Cardona in Barcelona, Spain. At Northwich, Cheshire, England. Skeletal crystals containing large amounts of sand have been found in the Namib desert of Southwest Africa. Enormous deposits are found in the Salt Range in the Punjab, India. Halite also occurs in quantity in Algeria, Abyssinia, China, Peru, Colombia, Santo Domingo, and other places.

In the United States, large bedded deposits of salt occur in the Salina formation of the Upper Silurian over a wide area in central and western New York. Salt also occurs near Cleveland, Ohio, and extensive deposits are found in Michigan. In Louisiana in the southern portion of the state and in the neighborhood of Petite Anse Island. In Kansas, extensive beds occur near the base of the Triassic. Beautiful transparent masses associated with thenardite, glauberite, and mirabilite are found in the Verde Valley, Yavapai County, Arizona. In Nevada along the Virgin River in Clark County, and in distorted crystals from Humboldt County. From Borax Lake, San Bernardino County, California. In the Permian basin of southeastern New Mexico and adjacent parts of Texas. Obtained from the waters of Great Salt Lake, Utah. In Canada, an important deposit of salt lies along the eastern shores of Lake Huron in the counties of Bruce, Huron, and Lambton in Ontario.

Alter. Anhydrite, gypsum, polyhalite, celestite, dolomite, quartz, hematite, and pyrite have been found as pseudomorphs after halite; the removal of the salt crystals by their solution leaves a cavity which any mineral may then occupy. Cubic or hopper-shaped crystals may leave an impression of their form in shales or other sedimentary rocks,[17] and clay or other casts after these cavities are common.

Name. The name *halites* was originally given, from ἅλς, *sea*, and later was modified by J. D. Dana to halite.

HUANTAJAYITE. *Raimondi* in Domeyko (*Min. Chili*, 5th App., 1876); Raimondi (64, 1878). Lechedor *miner's name*.

Supposedly an argentian halite, (Na,Ag)Cl, but evidence of homogeneity of the natural material is lacking. Found as small cubes, usually aggregated into crusts; also fibrous. Cleavage {001}. Brittle, not sectile. H. 2. Color white, not altered by exposure to light. Isotropic; sometimes exhibits weak double refraction. Analysis gave AgCl 11, NaCl 89, total 100. Found at several mines in the Huantajaya district, Tarapacá, Chile, associated with calcite, embolite, cerargyrite, bromyrite, iodyrite, atacamite. The name refers to the locality; *lechedor* from the Spanish, *milk-giving*, alluding to the milky-white suspension of AgCl obtained when the material is dissolved in water.

Mixed crystals containing limited amounts of AgCl have been prepared from ammoniacal solutions of AgCl and NaCl at ordinary temperatures and in other ways, but the material is unstable and AgCl exsolves on standing.[18]

SYLVITE

Ref.

1. Goldschmidt (**8**, 80, 1922). Rare or uncertain:

015   014   013   035   023   034   045   112   122   233   123   345

Brauns, *Jb. Min.*, **1**, 113 (1889) lists as forms bordering etch pits: 0.2.21, 0.1.10, 019, 0.4.33, 018, 0.4.29, 017, 0.4.25, 016, 0.4.23, 0.2.11, 0.4.21, 015, 029, 014, 0.4.15, 027, 013.

2. The structure of halite was the first to be determined by x-rays, Bragg, *Proc. Royal Soc.*, London, **89A**, 468 (1914). Cell dimensions from Straumanis and Ievins, *Zs. Phys.*, **109**, 468 (1914), by powder method.

3. On habit variation in NaCl accompanying growth from solutions containing cosolutes see literature cited in Hintze (**1** [2B], 2101, 1911); Doelter (**4** [2], 1057, 1928); Frondel, *Am. Min.*, **25**, 91, 338 (1940).

4. Friedel, *Bull. soc. min.*, **53**, 122 (1930).

5. Löffler, *Jb. Min., Beil.-Bd.*, **68**, 125 (1934); Brauns (1889) lists and figures {1.20.20} as a twin plane appearing as striations on a cube.

6. See especially Sohncke, *Jb. Min.*, 938 (1875), *Ann. Phys.*, **157**, 329 (1876); Brauns, *Jb. Min.*, I, 113 (1889); Rosický, *Beitr. Kr. Min.*, **1**, 241 (1918); Herzfeld and Hettich, *Zs. Phys.*, **38**, 1 (1927), **40**, 327 (1927). On the polyhedral bodies obtained by solution of halite spheres see Nitschmann, Inaug. Diss., Breslau (1939); Neuhaus, *Zs. Kr.*, **68**, 15 (1928); Schnorr, *Zs. Kr.*, **68**, 1 (1928); Ernst, *Zs. Kr.*, **96**, 38 (1937).

7. On the existence of a {110} cleavage see Tertsch, *Zs. Kr.*, **81**, 264 (1932).

8. On the variation in hardness with direction see Exner, *Härte an Krystallflächen*, Wien (1873); also Cesàro, *Ann. Soc. géol. Belgique, Mém.*, **15**, 204 (1888), Müller, *Jb. Min.*, I, 191 (1907), Reis and Zimmermann, *Zs. Kr.*, **57**, 471 (1922).

9. *Int. Crit. Tab.*, **1**, 150 (1926).

10. Blue and purple tints have been produced in halite by exposure to cathode rays, x-rays, electric sparks, radium, by heating in sodium vapor, by exposure to light after having been subjected to pressure, and in other ways. For discussions of these, with reviews of the voluminous literature, see especially Doelter (**4** [2], 1105, 1928); Doelter, *Ak. Wien*, I, **138**, 113 (1929); Przibram, cited in *Min. Abs.*, **4**, 253 (1930); Przibram, *Kali*, **30**, 61 (1936), *Ak. Wien*, IIa, **143**, 489 (1934); Lorenz, *Fortschr. Min.*, **20**, 290 (1936); Friend and Allchin, *Nature*, **145**, 266 (1940), *Min. Mag.*, **25**, 584 (1940); Ekstein, *Bull. soc. min.*, **61**, 239 (1938); Steinmetz, *Jb. Min., Beil.-Bd.*, **65**, 119 (1932); Liermann and Rexner, *Naturwiss.*, **20**, 561, (1932); Kennard, Howell, and Jaeckel, *Am. Min.*, **22**, 65 (1937); Mott and Gurney, *Electronic Processes in Ionic Crystals*, Oxford (1940).

11. Cf. Buerger, *Am. Min.*, **15**, 180, 226 (1930); Tertsch, *Zs. Kr.*, **81**, 264 (1932).

12. Cf. Doelter (**4** [2], 1115, 1928); Cornu, *Jb. Min.*, I, 42 (1908); *Zbl. Min.*, 324 (1910).

13. Literature in Doelter (**4** [2], 1118, 1092, 1928); see also Paniche, *Per. Min.*, **4**, 25 (1933) on a supposedly dimorphous birefringent modification of halite.

14. Average values, Landolt-Börnstein (970, 1912). Temperature variation data from Marbach, Inaug. Diss., Leipzig (1913); *Zs. Kr.*, **56**, 212 (1921). For indices in the infrared see tabulation in Doelter (1928), Müller and Wetthauer, *Zs. Phys.*, **85**, 559 (1933).

15. For a tabulation of analyses of halite and of brines see Doelter (1928).

16. For a discussion with literature of the physical chemistry of formation of salt deposits see Hintze (**1** [2B], 2157–2173, 1911); Doelter (**4** [2], 1233–1397, 1928); Borchert, *Preuss. Geol. Landesanst., Arch. für Lagerst.*, **67** (1940).

17. Leith, *Univ. Toronto Stud., Geol. Ser.*, no. 46, 69, (1941).

18. Cornu, *Jb. Min.*, I, 22 (1908) and Hintze (**1** [2B], 2275, 1911). See also LeBlanc and Quenstädt, *Zs. phys. Chem.*, **150**, 321 (1930).

19. Murata and Smith, *Am. Min.*, **31**, 527 (1946).

9.1.1.2   **S Y L V I T E** [KCl].   Muriate of Potash (from Vesuvius) *Smithson (Ann. Phil.*, **6**, 258, 1823). Chloride of Potassium, Potassium Chloride. Chlorkalium *Germ.* Sylvine *Beudant* (**2**, 511, 1832). Sylviit *Glocker* (291, 1847). Hoevelit *Girard (Jb. Min.*, 568, 1863). Leopoldit *Zinken (Berg.- u. hütt. Ztg.*, **24**, 79, 1865), Reichardt (*Jb. Min.*, 331, 1866). Schätzellit and Hövellit (from Stassfurt) *Bischoff (Berg.- u. hütt. Ztg.*, **24**, 276, 1865; *Ann. Chem. Phys.*, **5**, 318, 324, 1865). Sylvite *Dana* (111, 1868). Chlornatrokalite *Johnston-Lavis (Nature*, **74**, 174, 1906).

8   HALIDES

C r y s t.   Isometric; hexoctahedral—$4/m\,\bar{3}\,2/m$.

Forms:[1]
a 001	k 025	w 148
d 011	m 113	z 247
o 111	n 112	s 123

Structure cell.[2] Space group $Fm3m$. a_0 6.277 ± 0.002 kX. Cell contents K_4Cl_4.

Habit.[3] Crystals cubic, less commonly cubo-octahedral or octahedral. Also massive, coarsely granular to compact; rarely columnar; as crusts.

Twinning. On {111}, observed only in artificial crystals.[4]

Etch figures.[5] Etch figures conforming to hexoctahedral symmetry are obtained from pure solutions, but figures with gyroidal symmetry may be obtained from optically active solutions.

Oriented growths. Hematite inclusions in sylvite have been observed with {0001} of the former parallel {001}, {111} or {011} of the latter.[17] Minute halite inclusions oriented in parallel position to the sylvite also have been noted.[18]

P h y s. Cleavage {001} perfect. Not as brittle as halite. Fracture uneven. H. 2. G. 1.993 ± 0.005;[6] 1.99 (calc.). M.P. $ca.$ 790°.[7] Luster vitreous. Colorless or white; also grayish, bluish, yellowish red or red; the red tints are usually due to included particles of hematite. Transparent. Soluble; taste much like that of halite, but somewhat bitter. Highly diathermanous.[8] Translation gliding[9] with T{011}, t[011] and also with T{001}, t[011].

O p t. In transmitted light, isotropic. Weak anomalous birefringence is sometimes observed, especially in mechanically deformed crystals, and in colored crystals this effect may be accompanied by distinct pleochroism.[10] Refractive indices[11] at 20°C.:

λ	A	B	C	D	E	F	G	H
n	1.48377	1.48597	1.48713	1.49031	1.49455	1.40830	1.50542	1.51061

C h e m. Potassium chloride, KCl. K 52.44, Cl 47.56, total 100.00. The Na reported in analyses[12] is due to admixture, in most instances of halite. Br,[13] possibly in substitution for Cl, and He, U,[14] have been reported in sylvite.

Tests. Easily fusible. Soluble in water (34.7 g. in 100 cc. H_2O at 20°). If small pieces are pressed between glasses, sylvite flattens like a piece of wax but halite breaks to a powder.

O c c u r. Sylvite occurs principally in basin-like bedded salt deposits of sedimentary origin but is much less common than halite. Associated with kieserite, kainite, carnallite, polyhalite, gypsum, anhydrite and especially halite. The name sylvinite has been applied to rocks composed of admixed halite and sylvite. Also found as a sublimation product in fumaroles (Vesuvius), and in the desert nitrate deposits of Peru and Chile.

Found abundantly in north Germany at Stassfurt and Leopoldshall, and at Sondershausen, Bleicherode and elsewhere in the southern Harz region;

at Lübtheen, Mecklenburg. At Kaluscz, Galicia. With halite in fumaroles on Vesuvius and ultimately derived by the action of HCl gas on potash-rich minerals in the lava; also on Etna, Sicily. With halite, soda niter, and various sulfates in the nitrate deposits of Tarapacá, Chile, and similarly in Peru. In the United States sylvite occurs in bedded salt deposits in the Permian basin of southeastern New Mexico and adjacent parts of Texas.[15]

Alter. Sometimes formed as a product of the dissolution of carnallite.

Name. The compound is the *Sal digestivus Sylvii*, or *salt of Sylvius*, whence the name.

CHLORNATROKALITE. *Johnston-Lavis* (*Nature*, **74**, 174, 1905). A name given to a supposed compound 6KCl·NaCl found at Vesuvius. Shown[16] to be a mixture of sylvite and halite.

Ref.

1. Goldschmidt (**8**, 108, 1922). Rare or uncertain forms:

027	114	144	158	458
012	227	9.11.11	127 *	234
047	3.3.10	1.4.20	249	356
045	558	1.4.12	124	8.9.12
119 *	223	139	256 *	12.15.16
117	334	149	235	

Starred forms by Kreutz, *Zs. Kr.*, **51**, 209 (1912).

2. The structural identity of KCl with NaCl was first shown by Bragg, *Proc. Royal Soc., London*, **89**, 448 (1913); cell dimensions from Strukturber. (**1**, 105, 1931).
3. On habit variation in artificially grown KCl crystals see literature in Hintze (**1** [2B], 2231, 1911) and Frondel, *Am. Min.*, **25**, 91, 338 (1940).
4. See Mügge, *Cbl. Min.*, 259 (1906); Johnsen, *Jb. Min., Beil.-Bd.*, **23**, 311 (1907) and Löffler, *Jb. Min., Beil.-Bd.*, **68**, 125 (1934).
5. For literature see *Zs. Kr., Ref. Bd.* **1**, 109, 1928; Lowry and Vernon, *Trans. Faraday Soc.*, **25**, 286 (1929); Royer, *Bull. soc. min.*, **53**, 350 (1930).
6. Range of best reported values.
7. Mellor (**2**, 534, 1922).
8. Magnus, *Ann. Phys.*, **134**, 302 (1868); **139**, 451, 456, 588, 592 (1870) and Knoblauch, *Ann. Phys.*, **136**, 66 (1869), **139**, 155 (1870).
9. Buerger, *Am. Min.*, **15**, 174 (1930).
10. Cf. Cornu, *Cbl. Min.*, 168 (1907), *Jb. Min.* I, 47 (1908); Brauns, *Die optischen Anom. der kryst.*, Leipzig, 163 (1891); Herzfeld and Lee, *Phys. Rev.*, **44**, 625 (1933).
11. Stefan, *Ak. Wien, Sitzber.*, **63**, 241 (1871) by prism method at 20°. Sprockhoff, *Jb. Min., Beil.-Bd.*, **18**, 132 (1904) gave $nD = 1.48965$ and Dufet, *Bull. soc. min.*, **14**, 443 (1891) gave $n(Na) = 1.490294$ at 20° by the prism method. Indices in the infrared range are given by Rubens and Snow, *Ann. Phys.*, **46**, 529 (1892) and in the ultraviolet range by Rubens and others, *Ann. Phys.*, **60**, 418, 724 (1897); Martens, *Ann. Phys.*, **8**, 459 (1902). The variation in refractive index with temperature for sodium light between 15°–42.7° is expressed $n = 1.49110 - 0.0000345T$ by Stefan (1871); for indices in the range 22°–320° see Marbach, Inaug. Diss., Leipzig, 1913, *Zs. Kr.*, **56**, 211 (1921).
12. See Doelter (**4** [2], 1142, 1928).
13. Cf. Boeke, *Zs. Kr.*, **45**, 386 (1908) and Carobbi, *Ann. R. Osservat. Vesuviano*, **1**, 27 (1925).
14. Strutt, *Proc. Royal Soc., London*, **81**, 278 (1908).
15. See Schaller and Henderson, *U. S. Geol. Sur., Bull. 833*, 74 (1932).
16. Spencer, *Min. Mag.*, **15**, 59 (1908).
17. Leonhardt and Tiemeyer, *Naturwiss.*, **26**, 410 (1938).
18. D'Ans and Kühn, *Kali*, **32**, 152 (1938).

9.1.1.3 VILLIAUMITE [NaF]. *Lacroix (C.R., 146, 213, 1908).*

Cryst. Isometric; hexoctahedral—$4/m\ \bar{3}\ 2/m$.[1]

Forms:[2]

         a 001 d 011 o 111

Structure cell. Space group $Fm3m$.[1] a_0 4.619 kX (artif.).[3] Cell contents Na_4F_4.

Habit. Massive, granular.

Phys. Cleavage {001} perfect. Brittle. H. 2–2½. G. 2.79; 2.76 (artif.);[4] 2.81 (calc.). M.P. 980°.[8] Luster vitreous. The natural material is carmine; pure artificial material is colorless. Streak white. Transparent. Translation gliding[5] with T{011}, t[011], and probably also T{001}, t[011].

Opt. In transmitted light artificial NaF is isotropic with nNa 1.3270. The natural mineral is carmine in color with weak anomalous birefringence; uniaxial negative $(-)$; strongly pleochroic with E = yellow and O = pink to deep carmine; nNa 1.328,[6] (French Guinea), 1.327 (Kola).[9] The natural material becomes colorless and isotropic when heated to 300°.[6]

Chem. Sodium fluoride, NaF. Analysis gave:

	Na	K	Ca	Mg	ZrO$_2$	F	Insol.	Total
1.	54.76					45.24		100.00
2.	53.4	tr.	1.2	tr.	1.5	44.2		100.3
3.	53.83	0.32				45.28	0.84	100.27

1. NaF. 2. French Guinea.[7] 3. Lovozero tundra, U.S.S.R.[9]

Tests. Easily fusible. Soluble in water (about 4 g. in 100 cc. H$_2$O at 20°).

Occur. On the islands of Los, French Guinea, filling small miarolytic cavities in nepheline syenite. Also found in alkalic rocks in the Lovozero and Khibina tundras, Kola peninsula, U.S.S.R.[9]

Name. After a French explorer, Villiaume, in whose collection of rocks from Guinea the mineral was first found.

Ref.

1. Inferred from the structural identity of NaF with NaCl as shown by Davey, *Phys. Rev.*, **18**, 102 (1921), *ibid.*, **21**, 143 (1923), and others. The mineral was originally considered to be tetragonal and pseudo-cubic, but later it was shown by Barth and Lunde, *Cbl. Min.*, 57 (1927), to be identical with artificial (isometric) NaF.
2. Observed only on artificial crystals. See Frondel, *Am. Min.*, **25**, 91, 338 (1940), on habit variation in artificially grown crystals.
3. Barth and Lunde (1927) by precision powder method; natural NaF gave a_0 4.614.
4. Lacroix (1908).
5. Buerger, *Am. Min.*, **15**, 64 (1930).
6. Barth and Lunde (1927), who ascribe the anomalous optical properties to strain and radiations probably of radioactive origin.
7. Pisani anal. in Lacroix (1908).
8. *Int. Crit. Tab.* **1**, 150 (1926).
9. Gerasimovsky, *C.R., ac. sc. U.R.S.S.*, **32**, 492 (1941).

CERARGYRITE-BROMYRITE SERIES

9.1.1.4 C E R A R G Y R I T E [AgCl]. Argentum cornu pellucido simile (from Marienberg) *Germ.* Hornfarbs-Silber *Gesner* (63, 1565). Argentum rude jecoris colore, lucem corneam habens (from Freiberg, etc.) *Fabricius* (*De Rebus Met.*, 1566). Glaserz, durchsichtig wie ein Horn in einer Lantern *Matthesius* (Sarept., 1585). Horn-Silfver, Minera argenti cornea, Argento sulphure et arsenico mineralisatum, Hornerz *Cronstedi* (159, 1758). Buttermilcherz, Buttermilchsilber (first mentioned early in seventeenth century). Hornerz *Werner* [in Emmerling (168, 1796); *Letzt. Min.-Syst.*, 18, 1817]. Hornsilber *Hausmann* (3, 1010, 1813). Silberhornerz *Leonhard* (208, 1821). Silber-Kerat *Breithaupt* (287, 1832). Silberhornspath, Chlorsilberspath, Silberspath *Glocker* (876, 1831; 606, 1839; 249, 1847). Kerargyre *Beudant* (**2**, 501, 1832). Cerargyrites *Breithaupt* (**2**, 315, 1841). Kerat *Haidinger* (506, 1845). Argyroceratite *Glocker* (249, 1847). Chlorargyrit *Weisbach* (37, 1875). Kerargyrite, Chlorargyrite *Prior* and *Spencer* (*Min. Mag.*, **13**, 174, 1902). Ostwaldit *Cornu* (*Zs. Chem. Ind. Kolloide*, **4**, 187, 1909). Chlorsilber, Hornsilber *Germ.* Chlorsilfver, Silfverhornmalm *Swed.* Horn Silver, Corneous Silver. Argent muriaté, Argent corné, Chlorure d'argent *Fr.* Cherargirio, Argent cornea *Ital.* Plata cornea *Span.*

Chlorobromure d'argent *Domeyko* (*Ann. mines*, **6**, 153, 1844), *Berthier* (*ibid.*, **2**, 540, 1842). Plata cornea verde *Domeyko* (202, 1845). Embolite *Breithaupt* (*Ann. Phys.*, **77**, 134, 1849). Chlorobromide of Silver. Chlorbromsilber. Microbromit *Breithaupt* (*B.H. Ztg.*, **18**, 449, 1859). Iodembolite pt. *Prior and Spencer* (*Min. Mag.*, **13**, 174, 1902).

9.1.1.5 B R O M Y R I T E [AgBr]. Bromure d'Argent, Plata verde *Mex.* (from Mexico and Huelgoet), *Berthier* (*Ann. mines*, **19**, 734, 742, 1841; *ibid.*, **2**, 256, 1842). Bromide of Silver; Bromic Silver. Bromsilber, Bromchlorsilber *Germ.* Bromsilber *Berthier* (*Ann. Phys.*, **54**, 586, 1841). Bromit *Haidinger* (506, 1845). Bromspath *Glocker* (250, 1847). Bromyrite *Dana* (93, 1854). Megabromit *Breithaupt* (*B.H. Ztg.*, **18**, 449, 1859). Bromargyrite *Rammelsberg* (196, 1860). Plata cornea amarilla melada *Domeyko* (214, 1860). Iodobromite *von Lasaulx* (*Jb. Min.*, 619, 1878); Jodbromchlorsilber, Chlorbromjodsilber, Jodobromit *Germ.* Chlorobromite *Jeremejew* (*Gornyi J.*, **3**, 263, 1887). Bromargyrite, Iodembolite pt., *Prior* and *Spencer* (*Min. Mag.*, **13**, 174, 1902). Orthobromite *Samojloff* (*Mat. Geol. Russ.*, **13**, 145, 1906).

C r y s t. Isometric; hexoctahedral—$4/m\,\bar{3}\,2/m$.

Forms:[1]

Cerargyrite:	a 001	d 011	o 111	k 114	n 112	p 122	ρ 144
Bromyrite:	a 001	d 011	o 111				

Structure cell.[2] Space group $Fm3m$. a_0 5.545 kX (AgCl); a_0 5.755 kX (AgBr). Cell contents $Ag_4(Cl,Br)_4$.

Habit. Crystals usually cubic, with {111} and {011} sometimes large and the other reported forms as small and rare modifying faces. Often in parallel or subparallel groups. Ordinarily massive: in crusts and waxy coatings, occasionally with a drusy surface; as wax- or horn-like masses; columnar or stalactitic; rarely fibrous.[3]

Twinning. On {111}. Less common for bromyrite.

P h y s. Fracture uneven to subconchoidal;[4] tough. Sectile and ductile. Very plastic. H. $2\frac{1}{2}$. G. 5.556,[5] 5.55 (calc. for AgCl); 6.474,[6] 6.50 (calc. for AgBr). Luster resinous to adamantine, horn-like. Colorless when pure and fresh, usually gray; also yellowish, greenish brown (*bromyrite*). An increase in iodine content deepens the color. On exposure to light [7] cerargyrite becomes violet-brown or purple, and bromyrite is relatively little altered. Transparent to translucent. M.P. 455° (AgCl), 434° (AgBr).[6]

Opt. In transmitted light, colorless to yellow or green. Isotropic. Mechanically deformed crystals may be weakly birefringent [8] with lamellar or undulose extinction. The iodine-rich members frequently show an increased anomalous birefringence. Refractive indices:

	n(Na)
Cerargyrite [9]	2.071
Bromyrite [10]	2.253

Chem. Silver chloride (*cerargyrite*) and bromide (*bromyrite*), Ag(Cl,Br). A complete series exists between Cl and Br.[11] The species names cerargyrite and bromyrite are here applied to material with Cl > Br and Br > Cl, respectively. Iodine is often present in substitution for (Cl,Br), but usually not in considerable amounts, and mercury has been reported.[12]

Anal.[13]

	1	2	3	4	5	6	7
Ag	75.26	75.27	67.28	66.91	65.16	65.49	61.40
Cl	24.74	24.73	14.36	13.20	10.71	11.17	8.81
Br			15.85	19.71	24.13	23.34	26.85
I			2.35	0.16			tr.
Rem.							2.99
Total	100.00	100.00	99.84	99.98	100.00	100.00	100.05
G.	5.55		5.82	5.66			

1. AgCl. 2. Cerargyrite. Chañarcillo, Chile.[14] 3, 4. Bromian cerargyrite. Broken Hill, N.S.W.[15] 5. Ag(Cl,Br), with Cl:Br = 1:1. 6. Chlorian bromyrite (*orthobromite*). Donetz Basin, Russia.[16] 7. Chlorian and mercuroan bromyrite. Chañarcillo, Chile.[17] Rem. is Hg.

	8	9	10	11	12	13
Ag	59.96	60.37	56.93	56.7	57.56	57.44
Cl	7.09	7.11	1.96	0.6		
Br	17.30	22.35	32.22	38.9	42.44	42.56
I	15.05	10.39	8.77	2.6		
Total	99.40	100.22	99.88	98.8	100.00	100.00
G.		6.17	6.31			6.50

8. Iodian bromyrite (*iodobromite*), I:(Br,Cl) = 1:3.5. Dernbach, Germany.[18] 9. Iodian bromyrite (*iodembolite*). Chañarcillo, Chile.[15] 10. Iodian bromyrite (*iodembolite*). Broken Hill, N.S.W.[15] 11. Bromyrite. Tombstone, Arizona.[19] 12. Bromyrite. San Onofre, Zacatecas, Mexico.[20] 13. AgBr.

Var. *Bromian Cerargyrite.* Embolite. Chlorobromure d'argent *Domeyko* (*Ann. mines*, **6**, 153, 1844); *Berthier* (*ibid.*, **2**, 540, 1842). Plata cornea verde *Domeyko* (202, 1845). Embolite *Breithaupt* (*Ann. Phys.*, **77**, 134, 1849). Chlorobromide of Silver. Chlorbromsilber. Microbromit *Breithaupt* (*B. H. Ztg.*, **18**, 449, 1859). Iodembolite pt. *Prior and Spencer* (*Min. Mag.*, **13**, 174, 1902). The varietal designation bromian cerargyrite is here given that portion of the series with Br:Cl > 1:2 and < 1:1. The color of cerargyrite takes on a greenish gray tint with increasing substitution of Br.

Chlorian Bromyrite. Megabromite *Breithaupt* (*B. H. Ztg.*, **18**, 449, 1859). Iodembolite pt. *Prior* and *Spencer* (*Min. Mag.*, **13**, 174, 1902). Embolite pt. *some authors.* Orthobromite *Samojlov* (*Mat. Geol. Russ.*, **23**, 145, 1906). Refers to that portion of the series with Cl:Br $> 1:2$ and $< 1:1$.

Iodian Bromyrite (and Cerargyrite). Iodobromite *von Lasaulx* (*Jb. Min.*, 619, 1878). Jodbromchlorsilber, Jodbromit, Chlorbromjodsilber *Germ.* Iodembolite pt. *Prior* and *Spencer* (*Min. Mag.*, **13**, 174, 1902). I substitutes from (Br,Cl) in the natural material to about I:(Br,Cl) = 1:3.5, especially in the bromine-rich part of the series. In artificial material [21] I enters to about I:(Br,Cl) = 7:3, and unstable artificial solid solutions may contain as much as 95 molecular per cent of AgI. The color of cerargyrite and bromyrite tends toward yellow and orange-yellow with increasing substitution of I.

Tests. In C.T. fuses without decomposition. A fragment of cerargyrite placed on a strip of zinc and moistened with a drop of water swells up, turns black, and finally is entirely reduced to metallic silver, which shows the metallic luster on being pressed with the point of a knife. Practically insoluble in water and dilute HCl, but soluble (bromyrite, with difficulty) in NH_4OH, KCN and $Na_2S_2O_3$ solution.

O c c u r. The minerals of the series are secondary and occur in the oxidized zone of silver deposits, especially in arid regions. Commonly associated with native silver, iodyrite, jarosite, wad, and limonite; also cerussite, atacamite, malachite, pyromorphite, wulfenite.

Cerargyrite occurs at Johanngeorgenstadt, Freiberg, Schneeberg, and other localities in Saxony and was obtained in great quantities during the early period of mining in these districts. An earthy variety (Buttermilk ore) containing admixed native silver and clay was early recognized (sixteenth century) at Andreasberg in the Harz. At Joachimsthal, Bohemia, and at Markirch, Alsace. In Russia, especially at Schlangenberg, in the Altai Mountains. The mineral also has been found, although in only minor quantities, at numerous localities in France, Spain, Italy, and England. Found in large amounts at many places in Chile, as at Caracoles in Antofagasta, the Huantajaya district in Tarapacá, and especially Chañarcillo in Atacama; bromian varieties are found at many of the localities. Also obtained at Cerro Negro, Argentina, at Potosi and other localities in Bolivia, and in Peru. In Mexico found in Zacatecas and elsewhere. Obtained in notable amounts in the Broken Hill district, New South Wales (*bromian* pt.).

In the United States, cerargyrite is found very widely in the western mining districts.[22] Notable localities include Leadville, Colorado; Lake Valley, near Silver City, New Mexico; Tombstone, Cochise County, Arizona; in the Rand district and in the Calico and Barstow mines, San Bernardino County, California. A mass of cerargyrite weighing over six tons was found at Treasure Hill, Nevada; also found in important amounts in Nevada in the Bullfrog district, Nye County, and in the Rochester district,

Humboldt County. In the Silver City district, Owyhee County, Idaho, especially in the Poorman mine, where masses weighing many pounds were obtained, in part well-crystallized in cubes and cubo-octahedrons up to half an inch across and as twinned cubes.

Bromyrite is found near Dernbach, Nassau, Germany (*iodobromite*) associated with iodyrite, carminite, and beudantite. A chlorian variety (*orthobromite*) occurs in the Donetz basin, in the Ukraine, Russia; also from Kotschkar, Orenburg. From Huelgoet, Finistère, France. Found abundantly at Chañarcillo, Chile (*megabromite, iodembolite*, pt.) associated with native silver, iodyrite, calcite. In Mexico in the San Onofre and Plateros districts, Zacatecas, and at Parilla, south of Durango. An iodian variety (*iodembolite*) occurs with native silver and iodyrite at Broken Hill, New South Wales. In the United States, bromyrite has been found at Tombstone, Arizona, and probably also occurs in that state in the Hechman mine, near Globe, Gila County. Many other occurrences of so-called silver bromides have been reported from the western United States, but the exact identity of the material is uncertain.

Alter. A common alteration product of native silver and of silver-bearing sulfides and sulfosalts. Frequently observed as pseudomorphs after native silver and, less commonly, after argentite, proustite, and pyrargyrite; also found altered to native silver. Noted as an alteration product of ancient silver coins.

Artif.[23] Obtained in minute crystals from ammoniacal solution, by heating precipitated AgCl in dilute HCl or concentrated AgNO$_3$ solution, by the reaction of HCl on metallic silver in a closed tube at 100°–150°, and in other ways.

Name. Cerargyrite, from κέρας, *horn*, and ἄργυρος, *silver*, in allusion to the horn-like luster of the mineral; the proper derivation is ceratargyrite, but contracted to cerargyrite. Embolite from ἐμβόλιον, an *intermediate*, because between the chloride and bromide of silver in composition. Bromyrite in allusion to the composition. The terms horn-silver and cerargyrite doubtless have been applied in the past to both cerargyrite and bromyrite, as here defined, and the true signification of some of the older varietal names is uncertain.

Ref.

1. Goldschmidt (**2**, 152, 1913). On bromyrite with a pyritohedral habit see Lovisato, *Acc. Linc., Rend.*, **7**, 246 (1898) and Trivelli and Sheppard, *Silver Bromide Grain of Phot. Emulsions*, N. Y. (1921).
2. Barth and Lunde, *Zs. phys. Chem.*, **122**, 293 (1926), *Norsk Geol. Tidsskr.*, **8**, 281 (1925) by the powder method on artificial material.
3. See Levi and Tabet, *Acc. Linc., Rend.*, [6], **19**, 723, (1934), on fibrous artificial AgCl and AgBr.
4. Traces of {111} cleavage in bromyrite are mentioned by von Lasaulx, *Jb. Min.*, 621 (1878), and {001} cleavage by Breithaupt, *Ann. Phys.*, **77**, 134 (1849), *B. H. Ztg.*, **18**, 449 (1859). The plasticity of the mineral makes the observation of cleavage difficult.
5. Prior in Prior and Spencer, *Min. Mag.*, **13**, 184 (1902), by pycnometer on material from Taltal, Chile, free from Br and I.
6. *Int. Crit. Tab.* **1**, 123 (1926).

7. On the photochemistry of the silver halides see Mott and Gurney, *Electronic Proc. in Ionic Crystals*, Oxford (1940), and Mellor (**3**, 408, 1923).
8. See Cornu, *Cbl. Min.*, 394 (1908) and Gaubert, *Bull. soc. min.*, **30**, 266 (1907).
9. Des Cloizeaux, *Bull. soc. min.*, **5**, 143 (1882), by prism method.
10. Average of prism determinations by Wernicke, *Ann. Phys.*, **142**, 565 (1871).
11. See Barth and Lunde (1926), Mellor (**3**, 418, 1923), Hintze (**1** [2B], 2284, 1912).
12. Dana (158, 1892), and analysis 7.
13. Additional analyses are listed under embolite and bromyrite in Hintze (**1** [2B], 2298, 1912) and Doelter (**4** [3], 70, 1929); also Chukhrov, *C. R. ac. sc. U.R.S.S.*, **27**, 697 (1940).
14. Field, *J. Chem. Soc., London*, **10**, 239 (1858).
15. Prior anal. in Prior and Spencer (1902).
16. Samojloff, *Mat. Geol. Russ.*, **23**, 145 (1906).
17. Moesta, *Vork. der Chlor-, Brom- und Jodverbind. des Silbers*, Marburg (1869).
18. von Lasaulx, *Jb. Min.*, 619 (1878).
19. Carrillo anal. in Rasor, *Am. Min.*, **23**, 157 (1938).
20. Berthier, *Ann. mines*, **2**, 526 (1842).
21. Barth and Lunde (1926); Prior and Spencer (1902).
22. Many localities are mentioned by Schrader, Stone, and Sanford, *U. S. Geol. Sur., Bull. 624* (1917); Galbraith, *Univ. Arizona, Bull. 149*, 24 (1941); Gianella, *Univ. Nevada, Bull., Geol. Ser.*, *36*, 56 (1941); Northrop, *Univ. New Mexico, Bull. 6*, no. 1, 103 (1942), Pabst, *Cal. State Div. Mines Bull. 113*, 88, (1938).
23. See Hintze (**1** [2B], 2297, 1912); Mellor (**3**, 390, 418, 1923); Trivelli and Sheppard (1921).

HYDROHALITE. *Hausmann* (1458, 1847). Natriumchloridhydrat.
Supposedly $NaCl \cdot 2H_2O$, found as a product of crystallization of sea water or of saline spring waters in extremely cold weather.[1] Originally reported from salt works at Hallein, Salzburg, Germany. Artificial $NaCl \cdot 2H_2O$ is probably monoclinic and is stable below *ca.* $-5°$.

Ref.

1. On reported occurrences in Siberia see Quenstedt (623, 1877) and Poljenov, *Jb. Min.*, II, 12 (1909).

SALAMMONIAC

Salammoniac, NH_4Cl, crystallizes in the CsCl structure-type and is the sole representative of this type in nature. Over $184.3°$ NH_4Cl crystallizes in the NaCl structure-type.

9.1.2 **S A L A M M O N I A C** [NH_4Cl]. Salammoniac, Sal ammoniacus *Agricola* (1546). Natürliches Salmiak (from Bucharia) *Model* (*Versuch über ein nat. Salmiak*, Leipzig, 1758). Muriate of Ammonia; Chloride of Ammonium. Salmiak *Germ.* Sel ammoniac, Ammoniaque muriatée *Fr.* Salmiac *Beudant* (1832). Chlorammonium *Rammelsberg* (41, 1855; **1**, 247, 1881). Chlorammonio *Scacchi* (Acc. Napoli, **6**, 28, 1875). Ammoniumchlorid *Groth* (**1**, 182, 1906). Sal-Ammoniac.

C r y s t. Isometric; gyroidal—4 3 2 (?).[1]

Forms:[2]

a 001	*o* 111	*μ* 114	*n* 112
d 011	*f* 013	*m* 113	*s* 123

Structure cell.[3] Space group $P\bar{4}3m$ (?). a_0 3.859 kX. Cell contents NH_4Cl.

Habit. Crystals usually trapezohedral {112}, sometimes modified by {113}; less commonly with gyroidal forms alone or in combination. Also dodecahedral or, rarely, cubic. The crystals often possess curved or stepped faces and may present a tetragonal or rhombohedral appearance due to suppression of faces and distortion. Ordinarily in skeletal or den-

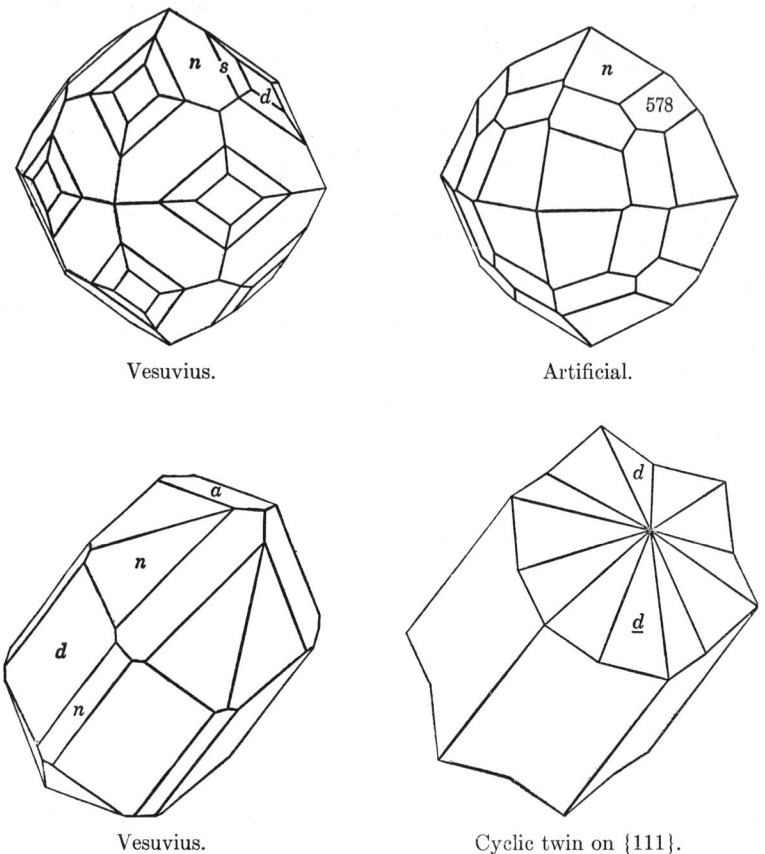

Vesuvius.   Artificial.

Vesuvius.   Cyclic twin on {111}.

dritic aggregates; also as crusts or stalactitic masses; fibrous; earthy or mealy.

Twinning. On {111}.[4]

Etch figures. Etch pits conforming to gyroidal symmetry have been observed.[5]

P h y s. Cleavage {111} imperfect. Fracture conchoidal. Between brittle and sectile, not easily powdered. Very plastic. G. 1.532;[12] 1.535 (calc.). Luster vitreous. Colorless or white when pure; grayish; yellow to brown in material admixed with ferric chloride. Taste salty, stinging. Transparent. Translation gliding [6] with $T\{011\}$, $t[001]$.

O p t. In transmitted light, colorless. Sometimes weakly birefringent, especially after mechanical deformation.[7] n 1.639 ± 0.001.[8]

Chem. Ammonium chloride, NH_4Cl. NH_4 33.72, Cl 66.28, total 100.00. Ferric chloride is sometimes present (apparently constituting an anomalous mixed crystal) [9] in amounts up to *ca.* 12 per cent $FeCl_3$ (Kilauea).[10]

Tests. In C.T. sublimes without fusion. 29.7 grams of NH_4Cl are soluble in 100 cc. H_2O at 0°.

Occur. Formed usually as a sublimation product about volcanic fumaroles, as at Vesuvius, Etna, Stromboli, Vulcano, Pelée, Hekla, etc. A relatively high ferrian variety was found at Kilauea. The mineral is found especially where the lava has flowed over soil and vegetation, from which the ammonia is presumed to have been derived. Also noted as a sublimation product in burning coal seams: at Duttweiler near Saarbrücken in the Saar, Germany; in mines of the Plauenschen Grund near Dresden; in the neighborhood of Saint-Étienne, Loire, France; at Newcastle, Northumberland, England, and at a number of similar occurrences. Here salammoniac is found associated with realgar, orpiment, sulfur, mascagnite, ammonia-alum. Salammoniac has been found in guano deposits on the islands of Guañape and Chincha, Peru, and in Tarapacá, Chile.

Artif. Crystals are easily obtained by sublimation or by crystallization from water solution. The habit and perfection of the crystals is markedly sensitive to the presence of metallic chlorides or other impurities in the crystallizing solution.[11]

Name. The ἅλς ἀμμωνιακός, *sal ammoniac,* of Dioscorides, Celsius, and Pliny, is common rock salt, dug in Egypt, near the oracle of Ammon. The name was afterwards transferred to the present compound, when subsequently manufactured in Egypt. Salammoniac is supposed to have been included by the ancients, with one or two other species, under the name of *nitrum,* which, according to Pliny, gave the test of ammonia when mingled with quicklime.

Ref.

1. Salammoniac is generally assigned to the gyroidal class on the basis of the morphological development and of etch symmetry (see literature in Groth, **1,** 182, 1906 and Hintze, **1** [2B], 2250, 1912). The x-ray diffraction effects are, however, compatible only with hextetrahedral symmetry (see Wyckoff, *Am. J. Sc.,* **3,** 177, 1922; *ibid.,* **4,** 469, 1922).
2. Goldschmidt (**8,** 2, 1922). Rare or uncertain: 012, 225, 349, 478, 023, 4.4.11, 578.
3. Wyckoff (1922). A modification stable above 184.3° described by Bartlett and Langmuir, *J. Am. Chem. Soc.,* **43,** 84 (1921), has a NaCl-type structure with a_0 6.533 kX and contains $(NH_4)_4Cl_4$.
4. Observed by Johnsen, *Jb. Min. Beil.-Bd.,* **33,** 311 (1907) and Groth (**1,** 184, 1906) on artificial crystals, and by Scacchi, *Acc. Napoli, Att.,* **6,** no. 9, 28 (1875) on crystals from Vesuvius.
5. Tschermak, *Min. Mitt.,* **4,** 531 (1882); see also Mügge, Jb. Min. I, 146 (1898).
6. Johnsen, *Jb. Min.* II, 149 (1902).
7. See Mügge (1898) on the nature of birefringent lamellae produced parallel {001} and other planes under pressure.
8. Merwin in Wyckoff (**4,** 469, 1922).

9. See Neuhaus, *Chem. Erde*, **5,** 554 (1930), for a summarizing account of the anomalous mixed crystals formed with $FeCl_3$.
10. Silliman in Dana (114, 1868).
11. See literature in Hintze (1912), also Gaubert, *Bull. soc. min.*, **38,** 149 (1915), Ehrlich, *Zs. anorg. Chem.*, **203,** 26 (1931).
12. Krickmeyer, *Zs. phys. Chem.*, **21,** 71 (1896).

9.1.3 NANTOKITE GROUP

ISOMETRIC; HEXTETRAHEDRAL—$\bar{4}3m$

	a_0	G.	n(Na)
Nantokite, CuCl	5.407 kX	4.136	1.930
Miersite, (Ag,Cu)I	6.491	5.68	
Marshite, CuI	6.047	5.68	2.346

The cuprous halides and miersite crystallize with the sphalerite type of structure, in which the metal atoms are tetrahedrally coordinated by halogen atoms. All the minerals are secondary in origin and occur in the oxidized zone of ore deposits, especially in arid regions.

9.1.3.1 **N A N T O K I T E** [CuCl]. Nantoquita *Sieveking, Domeyko* (2d App., 51, 1867; 3d App., 22, 1871). Nantokit *Breithaupt* (B.H. Ztg., **27,** 3, 1868; Jb. Min., 814, 1872).

C r y s t. Isometric; hextetrahedral—$\bar{4}\,3\,m$.
Structure cell.[1] Space group $F\bar{4}3m$. a_0 5.407 kX. Cell contents Cu_4Cl_4.
Habit. Massive, granular. Artificial crystals are tetrahedral.
P h y s. Cleavage {011}. Fracture conchoidal. H. $2\frac{1}{2}$. G. 4.136 (artif.);[6] 4.22 (calc.). Luster adamantine. Colorless to white when fresh; grayish to greenish. Streak white. Transparent.
O p t. In transmitted light, colorless. Isotropic; may exhibit anomalous birefringence. n 1.930 ± 0.005 (artif.).[2]
C h e m. Cuprous chloride, CuCl.
Anal.

	1	2	3
Cu	64.19	64.17	64.28
Cl	35.81	35.52	35.82
Total	100.00	99.69	100.10
G.	4.22	3.930	4.3±

1. CuCl. 2. Nantoko, Chile.[3] 3. Broken Hill, New South Wales.[4]

Tests. Easily fusible B.B. Gives off Cl when struck with a hammer. Slowly decomposed in water, yielding Cu_2O, $CuCl_2$, and Cu. Easily soluble in HCl, HNO_3, and NH_4OH.

O c c u r. Originally found at mines near Nantoko, Copiapó, Chile, associated with cuprite, native copper, atacamite, and hematite. At Broken Hill, New South Wales, with cuprite, native copper, cerussite, and hematite.

A l t e r. In air alters superficially to paratacamite.

NANTOKITE 19

A r t i f.[5] Obtained in small white tetrahedral crystals from solution in HCl, by the reaction of excess Cu with HCl or NaCl solution and by electrolysis of a solution of $CuCl_2$ in dilute HCl.

N a m e. From the locality at Nantoko, Chile.

Ref.

1. Barth and Lunde, *Norsk Geol. Tidsskr.*, **8**, 281 (1925), by powder method on artificial material. Space group from structural identity with sphalerite.
2. Larsen (114, 1921).
3. Sieveking in Domeyko (211, 1879).
4. Armstrong and Carmichael, *Proc. Royal Soc. New South Wales*, **28**, 96 (1894).
5. Mellor (**3**, 157, 1923).
6. Wyckoff and Posnjak, *J. Am. Chem. Soc.*, **44**, 30 (1922); a wide range of values, for the most part much lower than that here cited, is given in the literature for both natural and artificial material.

9.1.3.2 **M I E R S I T E** [AgI]. Spencer (*Nature*, **57**, 574, 1898; *Min. Mag.*, **13**, 41, 1901). α-AgI.

C r y s t. Isometric; hextetrahedral—$\bar{4}\,3\,m$.

Forms:

$a\,001 \qquad o\,111 \qquad -o\,\bar{1}11$

Structure cell.[1] Space group $F\bar{4}3m$. a_0 6.491 kX. Cell contents Ag_4I_4.

Habit. Tetrahedral, often with small faces of $\{001\}$; rarely cubo-octahedral by equal development of $\{001\}$, $\{111\}$ and $\{\bar{1}11\}$. $\{001\}$ striated parallel edges with the tetrahedral faces. The positive and negative tetrahedra show no difference in surface characters. As crusts and aggregates of indistinct crystals.

Twinning. On $\{111\}$, sometimes repeated.

P h y s. Cleavage $\{011\}$, perfect. Fracture conchoidal. Somewhat brittle (in contrast to iodyrite). H. $2\frac{1}{2}$. G. 5.64 (Broken Hill),[2] 5.680 (artif.);[3] 5.67 (calc.). Luster adamantine. Color and streak canary-yellow. Transparent.

O p t. In transmitted light, pale yellow. Isotropic; sometimes exhibits weak anomalous birefringence. n 2.20 $\pm$ 0.02.[5]

C h e m. Silver iodide, AgI. Cu substitutes for Ag, with Cu:Ag = 1:4 in the only reported analysis. The amount of Cu, however, varies in other specimens. A complete series[4] extends from AgI to CuI in artificial material. AgI exists in a number of polymorphic modifications (see *iodyrite*).

Anal.

	1	2
Cu		5.64
Ag	45.94	38.17
I	54.06	56.58
Total	100.00	100.39
G.	5.67	5.64

1. AgI. 2. Broken Hill.[6]

Tests. On heating, changes in color to bright orange-yellow and brick-red and fuses to a deep-red liquid; the color changes are reversed on cooling. Decomposed with precipitation of metallic Ag and Cu by means of Zn and H_2SO_4. Not affected by dilute HNO_3.

Occur. Found at Broken Hill, New South Wales, with malachite, cerussite, limonite, wad, cuprite. In part as oriented intergrowths with iodyrite, and as paramorphs in iodyrite.

Artif. Both miersite and iodyrite are obtained by metathical reaction in water solution at ordinary temperature; the formation of miersite is favored by an excess of Ag ions in the solution, and that of iodyrite by an excess of I ions.[7] Also from fusion by the inversion of iodyrite on cooling, and by powdering iodyrite.

Name. After Henry A. Miers (1858–1942), Professor of Mineralogy at Oxford.

Ref.
1. Aminoff, *Geol. För. Förh.*, **44**, 444 (1922), with analysis of the structure; cell dimensions of Barth and Lunde, *Zs. phys. Chem.*, **122**, 293 (1926).
2. Prior, *Min. Mag.*, **13**, 189 (1902), by pycnometer on material of analysis 2.
3. Kolkmeijer and van Hengel, *Zs. Kr.*, **88**, 317 (1934), by pycnometer on artificial material.
4. Barth and Lunde (1926). The observations of Spencer (1901) and Quercigh, *Acc. Linc., Rend.*, **23**, 441, 711, 825 (1914), suggest that material with less than *ca.* 15 mol per cent CuI in solid solution is unstable.
5. Larsen (110, 1921).
6. Prior (1902).
7. Kolkmeijer and van Hengel (1934).

9.1.3.3   **M A R S H I T E** [CuI]. Native copper iodide Marsh (*Proc. Royal Soc. New South Wales*, **26**, 326, 1892); Marshite Liversidge (*ibid.*, p. 328). Kupferjodür, Cuprojodid *Germ.*

Cryst.[1] Isometric; hextetrahedral—$\bar{4}\,3\,m$.

Forms:[2]

$a\,001$ $o\,111$ $-o\,\bar{1}11$ $m\,113$ $\beta\,223$

Structure cell.[3] Space group $F\bar{4}3m$. $a_0\,6.047\,kX$. Cell contents Cu_4I_4.

Habit. Usually tetrahedral, with modifying {001} or other forms; rarely cubo-octahedral through equal development of {001}, {111} and

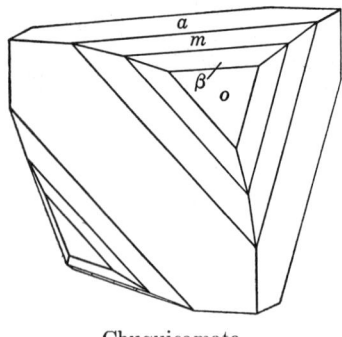
Chuquicamata.

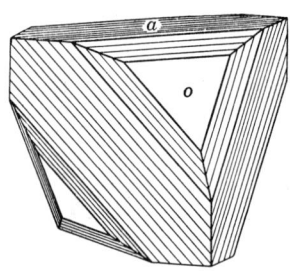
Chuquicamata.

{$\bar{1}11$}. {001} and {hhl} striated parallel edges, with the tetrahedra showing no difference in surface characters. As isolated crystals and crusts.
Twinning. On {111}, sometimes repeated.
Phys. Cleavage {011}, perfect. Fracture conchoidal. Rather brittle. H. $2\frac{1}{2}$. G. 5.68;[4] 5.60 (calc.). Luster adamantine. Colorless to pale shades of yellow when fresh, becoming pale salmon-red to dark brick-red on exposure. Streak yellow. Transparent. Fluoresces dark red in the ultraviolet.
Opt. In transmitted light, colorless. Isotropic; may exhibit anomalous birefringence. Refractive indices:

λ	Li	Na	Tl
n	2.313	2.346	2.385 (Broken Hill)[5]

The dispersion of the indices of refraction for the extreme visible ends of the spectrum of white light is about 0.45.

Chem. Cuprous iodide, CuI. Ag substitutes for Cu, to Ag:Cu = 1:46 in anal. 2. A complete series[6] is formed to AgI in artificial material. Cl apparently substitutes for I in small amounts (anal. 3); in artificial material a series extends at elevated temperatures from CuI to about 10 mol per cent $CuCl$.[7]

Anal.

	1	2	3
Cu	33.37	32.35	33.01
Ag		1.19	
I	66.63	65.85	66.67
Cl			0.33
Total	100.00	99.39	100.01
G.	5.60	5.59	5.68

1. AgI. 2. Broken Hill, New South Wales.[8] 3. Chuquicamata, Chile.[9]

Tests. Material free from Ag is blackened in cold dilute HNO_3 due to liberation of I. Not reduced by Zn and H_2SO_4. Argentian material behaves like miersite. Turns dark red and decrepitates when heated on a slide; on cooling the original color is regained. Decomposes before fusion.

Occur. Found originally at Broken Hill, New South Wales, associated with wad, cerussite, cuprite, native copper, and limonite; in part altered superficially to tenorite (?). Also at Chuquicamata, Chile, with atacamite, leightonite, and gypsum.

Artif. Obtained in tetrahedral crystals by the action of HI on metallic copper or chalcocite.[10]

Name. After C. W. Marsh, who first described the mineral.

Ref.

1. The mineral was described as tetragonal by Marsh (1892); the isometric hextetrahedral symmetry was first recognized by Miers, *Zs. Kr.*, **24**, 207 (1894).
2. Spencer, *Min. Mag.*, **13**, 38 (1901) and Jarrell, *Am. Min.*, **24**, 629 (1939).
3. Barth and Lunde, *Norsk Geol. Tidsskr.*, **8**, 281 (1925) by powder method on artificial material; space group from structural identity with sphalerite proved by Aminoff, *Geol. För. Förh.*, **44**, 444 (1922).
4. Jarrell (1939) by microbalance on material from Chuquicamata, Chile.

5. Spencer (1902) by prism method; see also Jarrell (1939).
6. Barth and Lunde, *Zs. phys. Chem.*, **122**, 293 (1926).
7. Mönkemeyer, *Jb. Min. Beil.-Bd.*, **22**, 41 (1906).
8. Prior, *Min. Mag.*, **13**, 189 (1902).
9. Gonyer anal. in Jarrell (1939).
10. Meusel, *Ber.*, **3**, 123 (1870).

CUPRO-IODARGYRITE. *Schulze* (*Chem.-Ztg.*, **16**, 1952, 1892; *Zs. Kr.*, **24**, 626, 1895).

A transparent sulfur-yellow mineral, somewhat harder and less sectile than iodyrite. Analysis gave: Cu 15.91, Ag 25.58, I 57.75, total 99.24. Found with atacamite in limestone at the San Augustin mine, Huantajaya, near Iquique, Chile, as an alteration of stromeyerite. Possibly an argentian marshite,[1] with Ag:Cu = 1:1.06.

Ref.

1. Suggested by Spencer, *Min. Mag.*, **13**, 38 (1901), to be an intermediate member of the marshite-miersite series.

9.1.4   I O D Y R I T E [AgI]. Iodure d'Argent *Vauquelin* (*Ann. chim. phys.*, **29**, 99, 1825); *Domeyko* (*Ann. mines*, **6**, 158, 1844). Plata cornea amarilla clara (?) *Domeyko* (205, 1845). Iodic Silver. Iodit *Haidinger* (506, 1845). Iodyrite *Dana* (95, 1854). Iodargyrit *Rammelsberg* (197, 1860). Iodsilber, Jodsilber, Jodin-silber *Germ.* Argent ioduré *Fr.* β-AgI.

C r y s t.[1] Hexagonal—P; dihexagonal-pyramidal—$6\,m\,m$.

$$a:c = 1:1.6408; \quad p_0:r_0 = 1.8946:1$$

Forms:[2]

Lower	Upper	ϕ	ρ	M	A_2
$\bar{c}$	c 0001	.....	0°00′	90°00′	90°00′
	m $10\bar{1}0$	30°00′	90 00	60 00	90 00
	a $11\bar{2}0$	0 00	90 00	90 00	60 00
	o $10\bar{1}2$	30 00	43 27	69 53½	90 00
$\bar{g}$	g $30\bar{3}4$	30 00	54 52	65 52	90 00
	r $70\bar{7}8$	30 00	58 54	64 39	90 00
$\bar{s}$	$15.0.\overline{15}.16$	30 00	60 37	64 10	90 00
$\bar{i}$	i $10\bar{1}1$	30 00	62 10½	63 45½	90 00
	t $70\bar{7}6$	30 00	65 39½	62 54	90 00
$\bar{f}$	f $30\bar{3}2$	30 00	70 37	61 51½	90 00
$\bar{u}$	u $20\bar{2}1$	30 00	75 13	61 05½	90 00
$\bar{w}$	w $90\bar{9}4$	30 00	76 48	60 52	90 00
	x $70\bar{7}2$	30 00	81 25½	60 22	90 00
$\bar{y}$	y $90\bar{9}2$	30 00	83 18½	60 13½	90 00
	z $33.0.\overline{33}.4$	30 00	86 20½	60 04	90 00

Structure cell.[3] Space group $C6mc$. a_0 4.58 kX, c_0 7.49 (at 19.5°); $a_0:c_0 = 1:1.637$. Cell contents Ag_2I_2.

Habit. Usually prismatic [0001]; also tabular {0001}. The usual development is distinctly hemimorphic. Common forms: $c\,m\,i\,\bar{i}$. Sometimes barrel-shaped, or with an irregular depression or extended cavity in {0001}. In parallel groups and rosettes. Also massive, lamellar or scaly {0001}.

IODYRITE 23

Twinning.[4] On {30$\bar{3}$4}, sometimes repeated as fourlings. Paramorphs of iodyrite after miersite show a mimetic repeated twinning with iodyrite [0001] parallel [111] of the original miersite.[5]

Etch figures.[6] Etch pits conforming to dihexagonal-pyramidal symmetry have been obtained.

Oriented growths. Upon AgBr (artif.) with iodyrite {0001} parallel AgBr {111}.[7]

P h y s. Cleavage {0001} perfect. Fracture conchoidal. Sectile. Flexible. H. 1$\frac{1}{2}$. G. 5.69;[8] 5.70 (calc.). M.P. 552°.[9] Luster resinous to

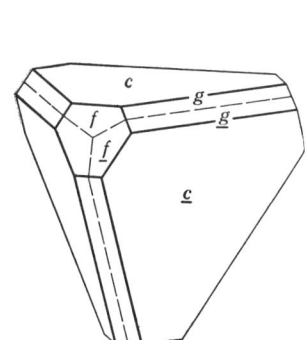
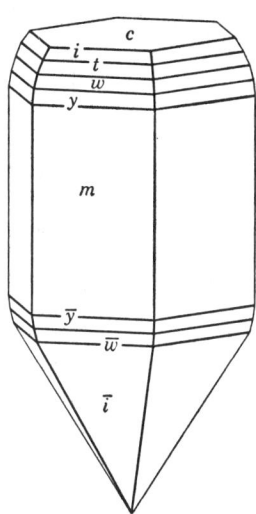

Broken Hill. Fourling on {30$\bar{3}$4}.   Tonopah.

adamantine; pearly on cleavage surfaces. Colorless when fresh, but becoming pale yellow on exposure to light and not changing in tint thereafter; also citron- and sulfur-yellow, greenish yellow, yellowish green, brownish, pearl-gray. Streak yellow, shining. Transparent. Not distinctly pyroelectric. On heating,[10] the axial ratio and specific gravity decrease regularly to 146°, where the substance inverts with absorption of heat to α-AgI.

O p t. In transmitted light, colorless. Uniaxial positive (+). Occasionally shows undulose anomalous birefringence, with small $2V$, due to mechanical deformation. Exhibits abnormal green interference colors. Refractive indices:[11] nO 2.21, nE 2.22. Dispersion very large.

C h e m. Silver iodide, AgI. Ag 45.94, I 54.06, total 100.00. The available analyses[12] conform very closely to the formula.

AgI exists in a number of polymorphous modifications:[13] a sphalerite-type modification stable below 137° (*miersite*), a wurtzite-type modification stable up to 146° (*iodyrite*), and an isometric modification, α-AgI, stable over 146°; a fourth modification has been reported[14] to exist at high pressures, and a rock-salt-type modification, containing, however, some Br in substitution for I, is known [15] (see *iodian bromyrite*). Mixtures of iodyrite and miersite usually are obtained by precipitation from solution at ordinary temperature or by sublimation.

Tests. In C.T. fuses and assumes a deep orange color, but resumes its yellow color on cooling. With Zn reacts like cerargyrite and bromyrite. Insoluble in water and in dilute HNO_3 but decomposes in hot concentrated HNO_3 or H_2SO_4. Soluble in concentrated KI solution and reprecipitates on dilution; insoluble in cold KCl or NaCl solution and slightly soluble in hot solution.

Occur. A secondary mineral, found in the oxidized zone of silver deposits. Associated principally with cerargyrite, bromyrite, iodian bromyrite, native silver, calcite, descloizite, vanadinite, pyromorphite, bismoclite, cerussite, limonite, wad.

In Germany at Dernbach, Nassau, with iodian bromyrite. At Hiendelaencina and Navacerrada, Spain; at Montmins, Allier, France. With malachite at Katanga, in the Belgian Congo. A notable locality at Chañarcillo, Atacama, Chile, associated with other silver halides; also at Caracoles in Antofagasta, and at Algodones, near Coquimbo. Originally found at Albarradón near Mazapil in Zacatecas, Mexico. Found abundantly at Broken Hill, New South Wales. At Dzhezkazgan, Karaganda district, Kazakhstan, U.S.S.R. In the United States, finely crystallized iodyrite was obtained at Tonopah, Nevada; also noted at Goldfield, Nevada, with bismoclite. At Lake Valley, Sierra County, New Mexico, associated with vanadates and calcite. At the Commonwealth mine, Pearce Hills, Cochise County, Arizona.

Alter. Found as paramorphs after miersite at Broken Hill. Also altered to native silver.

Artif.[16] Obtained in crystals from solution in HI, by the reaction of HgI_2, Ag, and KI solution in a closed tube at *ca*. 100°, by cooling a hot solution of HgI_2 and $AgNO_3$, and in other ways. Precipitates of AgI formed by metathical reaction in water solution at ordinary temperatures consist largely or entirely of iodyrite in the presence of an excess of I ions, and of miersite in the presence of an excess of Ag ions.

Name. In allusion to the composition.

Ref.

1. Angles of Kraus and Cook, *Am. J. Sc.*, **27**, 210 (1909); unit of Aminoff, *Geol. För. Förh.*, **44**, 444 (1922). Transformation: Kraus and Cook (1909) and Dana (160, 1892) to Aminoff (1922), 1000/0100/0010/0002.
2. Kraus and Cook (1909). Rare or uncertain: $10\bar{1}4$, $10\bar{1}4$. Also on artificial crystals: $10\bar{1}2$, $10\bar{1}3$, $30\bar{3}8$, $20\bar{2}5$, $9.9.\bar{1}8.40$. Schaller, *Am. Min.*, **26**, 654 (1941), lists for what are probably iodyrite crystals: $1.0.\bar{1}.16$, $1.0.\bar{1}.14$, $1.0.\bar{1}.12$, $1.0.\bar{1}.8$, $10\bar{1}6$, $30\bar{3}8$.
3. Aminoff (1922) and Helmholz, *J. Chem. Phys.*, **3**, 740 (1935); cell dimensions of Kolkmeijer and van Hengel, *Zs. Kr.*, **88**, 317 (1934).
4. On the structural interpretation of the twinning see Aminoff (1922).
5. Spencer, *Min. Mag.*, **13**, 45 (1901).
6. Kraus and Cook (1909).
7. Schwab, *Naturwiss.*, **31**, 322 (1943).
8. Kolkmeijer and van Hengel (1934) by pycnometer on artificial material.
9. Mönkemeyer, *Jb. Min., Beil.-Bd.*, **22**, 28 (1906).

CALOMEL 25

10. See literature in Hintze (**1** [2B], 2310, 1912); also Bloch and Möller, *Zs. phys. Chem.*, **152**, 245 (1930).
11. Larsen (92, 1921).
12. Doelter (**4** [3], 77, 1929), Chukhrov, *C. r. ac. sc. U.R.S.S.*, **27**, 697 (1940).
13. Bloch and Möller (1930), Strock, *Zs. phys. Chem.*, **25**, 441 (1934).
14. Tammann, *Zs. phys. Chem.*, **75**, 733 (1910).
15. Barth and Lunde, *Norsk Geol. Tidsskr.*, **8**, 293 (1925); *Zs. phys. Chem.*, **122**, 293 (1926).
16. Doelter (1929).

TOCORNALITE. *Domeyko* (*Min. Chili*, 2nd App., 41, 1867). Plata iodurada mercurial.

A name given to a supposed iodide of silver and mercury found with other ill-defined chloride-iodides of silver and mercury at Chañarcillo, Chile.[1] Granular massive. Color pale yellow, becoming darker on exposure. Streak yellow. An analysis of impure material gave Ag 33.80, Hg 3.90, I 41.77, siliceous residue 16.65, total 96.12; the loss is due to some water belonging with the residue, and probably some iodine. A mineral referred to tocornalite was found in the Proprietary mine, Broken Hill, New South Wales, as a deep yellow powder, blackening in sunlight, associated with iodyrite and cinnabar.[2] Named after S. F. Tocornal, former rector of Santiago University, Chile.

Ref.
1. See also Domeyko, *Min. Chili*, 3rd ed., 431 (1879).
2. Smith, Dept. of Mines, New South Wales, *Min. Res.*, no. **34**, 42 (1926).

9.1.5 **C A L O M E L** [HgCl]. Horn Mercury *Woulfe* (*Phil. Trans.*, 618, 1776). Mine de mercure cornée *Sage* (**2**, 61, 1777), *de Lisle* (**3**, 161, 1783). Turpeth *Suckow*. Quecksilber-Hornerz *Werner* (*Bergm. J.*, 381, 1789). Mercure muriaté *Haüy* (**3**, 447, 1801). Merkur-Hornerz *Breithaupt* (136, 1823). Chlormercur *Naumann* (338, 1828). Chlorquecksilber *von Kobell* (**2**, 93, 1830). Quecksilberhornspath *Glocker* (876, 1831). Calomel *Beudant* (**2**, 500, 1832). Chlormerkurspath, Merkurspath *Glocker* (605, 1839). Merkur-Kerat *Breithaupt* (**2**, 318, 1841). Quecksilberspath, Hydrargyrit *Glocker* (249, 1847). Horn Quicksilver. Kalomel, Chlorquecksilber, Chlormercur, Quecksilberchlorür, Quecksilberhornerz *Germ.* Mercure chloruré *Fr.* Calomelano *Ital.* Mercurio corneo *Ital.*

Bordosite (?) *Bertrand* (*Ann. mines*, **1**, 412, 1872).

C r y s t.[1] Tetragonal; ditetragonal-dipyramidal—$4/m\ 2/m\ 2/m$.

$$a:c = 1:2.4372; \quad p_0:r_0 = 2.4372:1$$

Forms:[2]

	ϕ	ρ	A	M
c 001	....	0°00′	90°00′	90°00′
a 010	0°00′	90 00	90 00	45 00
m 110	45 00	90 00	45 00	90 00
α 013	0 00	39 05½	90 00	63 31½
i 012	0 00	50 37½	90 00	56 52
r 011	0 00	67 41½	90 00	49 08½
γ 118	45 00	23 18½	73 45	90 00
t 114	45 00	40 45	62 30½	90 00
e 112	45 00	59 52½	52 17½	90 00
s 111	45 00	73 49½	47 13½	90 00
d 332	45 00	79 03	46 02	90 00
π 138	18 26	43 56	48 50½	71 55½
ρ 125	26 34	47 28	70 45½	76 31½
n 122	26 34	69 51	65 10½	72 44
v 233	33 41½	71 09	58 20	79 18

Less common forms:

o 021 p 031 j 1.1.24 z 116 ψ 121

Structure cell.[3] Space group $I4/mmm$. a_0 4.45 kX, c_0 10.89; $a_0:c_0$ = 1:2.448. Cell contents Hg_4Cl_4.

Habit. Variable: often tabular {001}; prismatic [001]; pyramidal; equant, especially in complex twins. Crystals often show complex development. As drusy crusts of minute crystals. Also massive, earthy. Common forms: $a\ r\ \alpha\ m\ c\ e\ i$.

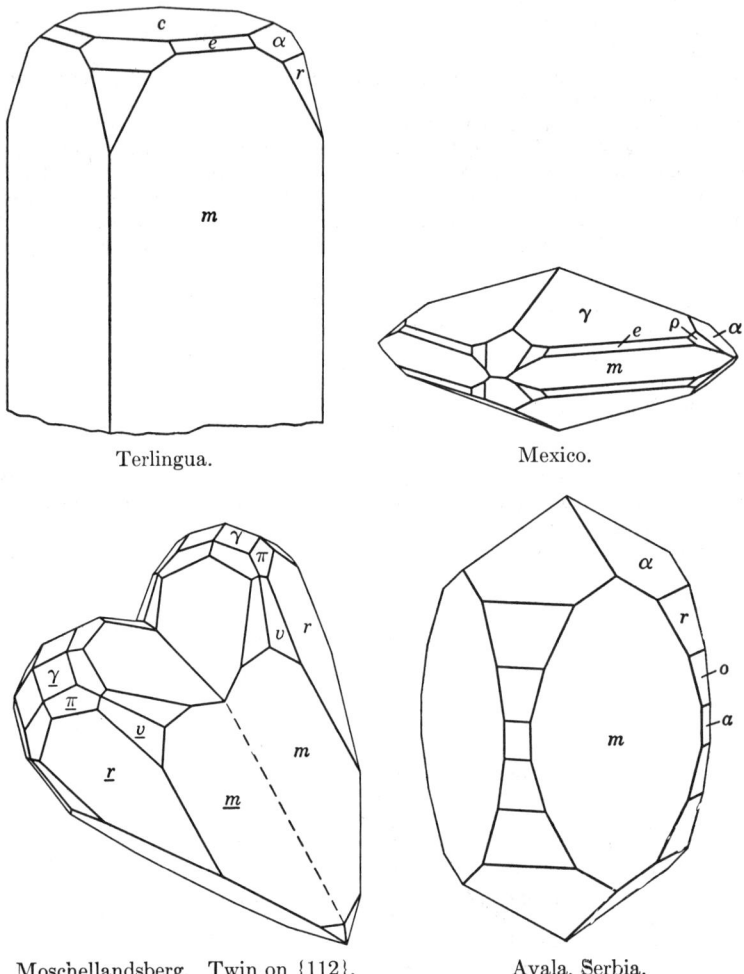

Terlingua.   Mexico.

Moschellandsberg. Twin on {112}.   Avala, Serbia.

Twinning. On {112}. As contact or interpenetration twins, often repeated and with irregular or concealed boundaries.

Phys. Cleavage {110} good but slightly uneven, {011} imperfect. Fracture conchoidal. Plastic; sectile. H. 1½. G. 7.15;[4] 7.23 (calc.). M.P. 302°.[4] Luster adamantine. Colorless, white, grayish and yellowish

CALOMEL 27

white, yellowish gray to ash-gray, brown. The color deepens on exposure to light. Streak pale yellowish white. Transparent. Non-conductor of electricity. Thermally positive.[5] Fluoresces brick-red in ultraviolet light.

O p t. In transmitted light, colorless. Mechanically deformed crystals may exhibit optical anomalies. Sometimes weakly pleochroic, with absorption $E > O$. Refractive indices:[6]

λ	Li	Na	Tl	
nO	1.956	1.973	1.991	Uniaxial positive (+).
nE	2.601	2.656	2.713	

C h e m. Mercurous chloride, HgCl. Hg 84.98, Cl 15.02, total 100.00. Modern analyses are lacking.

Tests. In the C.T. and on charcoal volatilizes without fusion, condensing as a white sublimate. Insoluble in water and HCl, but soluble in aqua regia; blackens when treated with alkalies.

O c c u r. A secondary mineral, formed by the alteration of cinnabar, mercurian silver, amalgam, mercurian tetrahedrite, selenian metacinnabar and other mercury-containing minerals. Associated with native mercury, eglestonite, terlinguaite, montroydite, cinnabar, calcite, limonite, clay.

Originally found at Moschellandsberg, in the Palatinate, Bavaria, Germany. At Avala, Serbia, as crusts upon cinnabar, barite, quartz. In Spain at Almadenejos in Ciudad Real province, Castile. In France at Montpellier, Hérault, with native mercury in a calcareous conglomerate. A notable occurrence at El Doktor, near Zimapan, Queretaro, Mexico, as an alteration of selenian metacinnabar. In the United States, calomel occurs at Terlingua, Brewster County, Texas, associated with native mercury, terlinguaite, montroydite, kleinite, eglestonite, calcite. Found with eglestonite, cinnabar, and native mercury near Jackport, Pike County, Arkansas, and in the Denio district, Oregon. With montroydite and eglestonite in a silicified serpentine near Redwood City, San Mateo County, California.

A r t i f.[7] Formed by the action of HCl or chlorides on metallic mercury or mercurous salts. Also obtained in crystals by sublimation, and from a solution of HgCl or HCl in mercurous nitrate.

N a m e. Calomel is an old term of uncertain origin and meaning, perhaps from καλος, *beautiful*, and μελι, *honey*, the taste being sweet, and the compound the *Mercurius dulcis* of early chemistry; or from καλός and μέλας, *black*, possibly in reference to black mercuric sulfide to which the name calomel was first applied. *Turpeth*, a word of Persian origin given to a preparation from the bark or root of a certain reedy plant, is properly applied to artificial basic mercuric sulfate.

BORDOSITE. *Bertrand (Ann. mines,* **1**, 412, 1872).

Yellow to red, turning dark on exposure. Found as an alteration of amalgam at Los Bordos, Atacama, Chile. Analysis gave AgCl 40.69, HgCl 59.31, after deducting 22.70 per cent HgO. Probably a mixture containing calomel, cerargyrite, and montroydite (?).

28   HALIDES

Ref.
1. Elements of Hillebrand and Schaller, *U. S. Geol. Sur., Bull. 405*, 159 (1909), the weighted average of four closely agreeing determinations, transformed to the body-centered structural cell of Mark and Steinbach, *Zs. Kr.*, **64**, 79 (1926). Transformations: Dana, Hillebrand, and Schaller to new, 110/1̄10/002; Mark and Steinbach to old, 1̄10/110/001.
2. Goldschmidt and Mauritz, *Zs. Kr.*, **44**, 393 (1908), who give a full discussion of the forms; and Hillebrand and Schaller (1909), who add new forms. Rare, uncertain, and vicinal forms:

l 130	ϵ 017	u 052	k 221	? 3.5.13
χ 7.11.0	h 014	Y 1.1.16	S 1.7.20	ϕ 354
g 570	y 059	δ 1.1.12	F 1.4.11	D 7.11.9
μ 340	x 058	q 1.1.10	λ 9.19.20	f 578
H 0.1.24	T 034	Φ 3.3.10	? 1.2.11	σ 7.9.20
ζ 019	K 053	β 558	B 123	

3. Mark and Steinbach (1926) by rotation and Laue methods on artificial (?) material.
4. *Int. Crit. Tab.*, **1**, 120 (1926).
5. Jannettaz, *Bull. soc. min.*, **15**, 138 (1892).
6. Dufet, *Bull. soc. min.*, **21**, 90 (1898), by prism method on artificial material; the values of Dufet are here rounded off to the third decimal place. The structural cause of the extreme birefringence of calomel has been discussed by Mark and Steinbach (1926) and Havighurst, *Am. J. Sc.*, **10**, 15 (1925).
7. Mellor (**4**, 796, 1923).

POTASSIUM FLUORIDE [KF]. *Carobbi (Att. acc. sc. Lett. Arti., Modena,* [5], **1**, 33, 1936).

Reported as small cubes admixed with mercallite, misenite, hieratite and an unidentified fluoride of Na, Ca, Mg, and Al in fumaroles in the crater of Vesuvius. Isotropic; index of refraction: 1.362. Artificial KF is isometric hexoctahedral with the rock-salt type structure and a_0 5.333 kX.[1] Cleavage {001}; G. 2.505;[3] 2.528 (calc.); colorless to white; n(Na) 1.3629.[2] Easily soluble in water; taste acrid and salty.

Ref.
1. Broch, Oftedal and Pabst, *Zs. phys. Chem.* (B), **3**, 209 (1929).
2. Wulff, *Zs. Kr.*, **77**, 84 (1931); see also Winchell (161, 1931).
3. Wulff (1931).

TYPE 2. AX_2

9.2.1 FLUORITE AND SELLAITE

	a_0	c_0	G.
Fluorite, CaF_2	5.452 kX		3.18
Sellaite, MgF_2	4.660	3.078	3.15

A number of fluorides and oxides of the AX_2 type crystallize with the fluorite type of structure, with 8 and 4 coordination for the cations and anions, respectively. All the cations are relatively large and afford values of the radius ratio within the theoretical limiting value of 0.73. The bonding is essentially ionic. Among the natural halides only fluorite itself is a representative of the structure-type. Uraninite (UO_2) and thorianite (ThO_2) belong in the type and baddeleyite (ZrO_2) has a distorted (monoclinic) fluorite-type structure. Digenite ($Cu_{2-x}S$) and berzelianite (Cu_2Se) have an anti-fluorite structure, as may have the isometric high-temperature polymorphs or disordered forms of other A_2X or ABX sulfides, selenides, and tellurides. Other AX_2 fluorides and oxides of relatively small cations crystallize in the rutile structure-type. The rutile-type structures theoretically obtain for values of the radius ratio between 0.414 and 0.73, and have 6 and 3 coordination for the cations and anions, respectively. Sellaite, MgF_2, is the only natural representative among halides. Rutile (TiO_2), cassiterite (SnO_2), pyrolusite (MnO_2), and plattnerite (PbO_2) crystallize in the type.

The principal type of compositional variation in fluorite is the substitution of Y''' for Ca''. This involves a concomitant substitution of F interstitially, in the vacant "holes" between six Ca ions at the cell centers, to provide valence compensation, and a partial series extends towards YF_3. The latter compound, not known in nature, has a structure similar to fluorite but with the vacant positions in sixfold coordination with Ca completely filled.

9.2.1 **F L U O R I T E** [CaF_2]. Fluores lapides gemmarum similis sed minus duri— qui ignis calore liquescunt [whence he derives the name]—Colores varii, jucundi, (1) rubri, (2) purpurei (vulgo amethysti), (3) candidi, (4) lutei, (5) cineracei, (6) subnigri, etc. [with mention also of its use as a flux in smelting] *Agricola* (458, 1529); Germ. Flusse *Agricola* (464, 1546). Fluores *de Boodt* (*Gemm. et lap. hist.*, 293, 1609; 575, 1647). Fluor mineralis Stolbergicus, Lithophosphorus Suhlensis *Woodward* (1728). Glas-Spat, Spatum vitreum *Wallerius* (64, 1747; 87, 1750). Fluss, Flussspat, Glasspat *Cronstedt* (93, 1758). Flussaures Kalk *Scheele* (*Ak. Handl. Stockholm*, 120, 1771). Calx fluorata *Bergmann* (1782). Spath fusible, Spath vitreux *de Lisle* (1772, 1783). Fluorite *Napione* (*Min.*, 373, 1797). Fluor Spar, Fluate of Lime, Fluoride of Calcium. Derbyshire Spar, Blue John *Vulg.* Chaux fluatée *Haüy* (**2**, 249, 1801). Fluorine *Beudant* (**2**, 517, 1832). Liparit *Glocker* (282, 1847). Fluorina, Spato fluore *Ital.* Espato fluor, Fluspat *Span.*

Chlorophane *De Grotthaus* (in Delamétherie (*J. Phys.*, 45, 398, 1794). Ratofkit *Fischer* (in John, *Chem. Unters.*, **6**, 232, 1812). Yttrocerit *Gahn* and *Berzelius* (*Afhandl. Fys. Kem. Min.*, **4**, 151, 1814). Yttrocererite *Leonhard* (573, 1826). Yttroflussspath, Flussyttrocalcit *Germ.* Yttrocalcit *Glocker* (283, 1847) [not yttrocalcit *Federov* (*Gyorni J.*, **3**, 254, 1905)]. Pyrosmaragd *Hausmann* (**2**, 1438, 1847). Stinkfluss *Hausmann* (1441, 1847). Antozonit *Schönbein* (*J. prakt. Chem.*, **83**, 95, 1861). Fluorbaryt *Hausmann* (1441, 1847). Gunnisonite *Clarke* and *Perry* (*Am. Chem. J.*, **4**, 140, 1882). Pseudonocerite, Pseudonocerina *Scacchi* (*Mem. acc. Napoli*, **2**, 1885). Bruiachite *Macadam* (*Min. Mag.*, **7**, 42, 1886).

C r y s t. Isometric; hexoctahedral—$4/m\,\overline{3}\,2/m$.

Forms: [1]

a 001	o 111	e 012	n 112	p 122	t 124	s 123
d 011	f 013	m 113	q 133	137	3.5.11	

Less common:

029	0.3.10	119	114	223	233	139
014	037	116	338	144	344	

Structure cell.[2] Space group $Fm3m$. a_0 5.452 ± 0.001 kX. Cell contents Ca_4F_8. The cell dimensions may increase slightly with increasing substitution of Y and Ce.[3]

Habit.[4] Simple crystals usually cubic; less frequently octahedral, rarely dodecahedral. Combinations of the cube with {111} are common; also with {011} or {011} and {111}; modifying faces of {0kl} are characteristic. Common forms: $a\,o\,d\,t\,m\,f$. {001} usually smooth and lustrous, {111} rough and dull. The crystals are sometimes distorted by unequal development of faces, as of {013}. Often markedly composite; sometimes minute

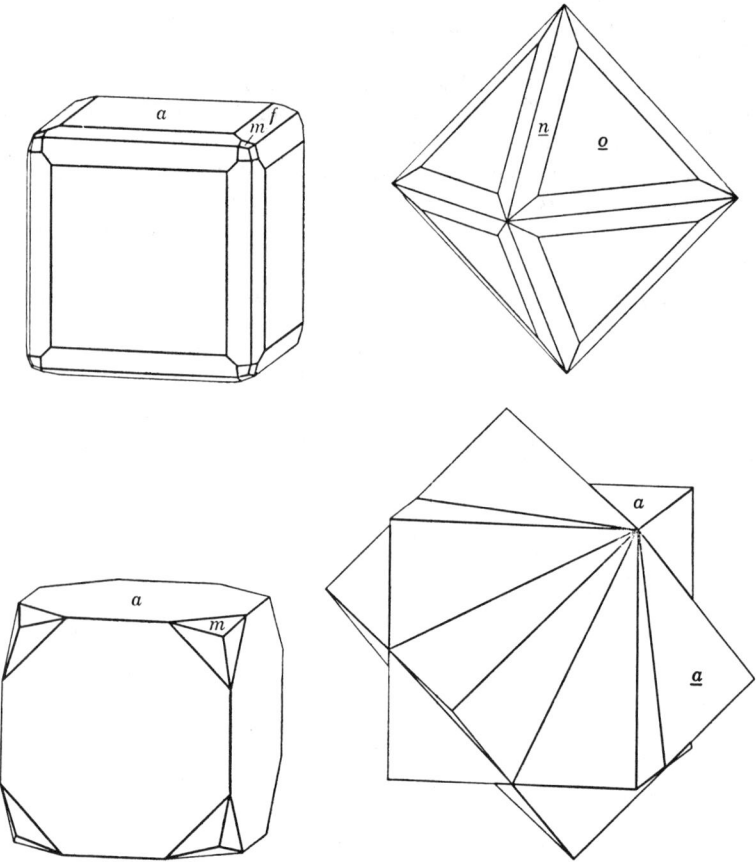

cubes aggregated together to form an octahedron, or as overgrowths of crystals upon the corners of an earlier formed crystal of different habit. Also massive: coarse to fine granular; compact, earthy; rarely columnar, fibrous or in globular aggregates.

Twinning.[5] On {111}, usually as penetration twins of cubes; the cube faces often are rectangularly striated [001] by combination with a vicinal tetrahexahedron. Rarely spinel twins of octahedra, which may be flattened {111} or have the two individuals unequally developed.

Oriented growths. The following types have been described: siderite upon fluorite with siderite [0001] parallel fluorite [111];[6] pyrite upon fluorite with parallel axes;[7] quartz upon fluorite;[8] scheelite included in fluorite.[8]

FLUORITE 31

Etch figures.[9] Pits principally bounded by {*hll*} and {*hhl*} are obtained from acid solvents; {0*kl*} from alkaline solvents.

P h y s. Cleavage {111}, perfect; also sometimes an indistinct parting or cleavage {011}. Brittle. Fracture flat-conchoidal to splintery or uneven. H. 4.[10] G. 3.180 ± 0.001;[11] 3.180 (calc.); 3.3–3.6 in yttrian and cerian material. M.P. 1360°.[12] Luster vitreous; glimmering to dull in massive varieties.

Colorless and water-clear when pure; commonly wine-yellow, green, greenish blue, violet-blue; also white, gray, yellow, sky blue, deep purple, bluish black, and brown; rarely rose-red, crimson-red, or pink. Some

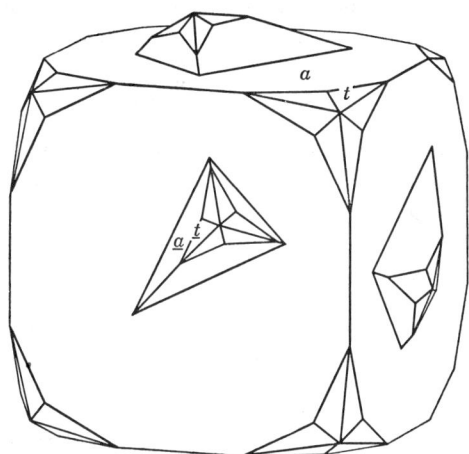

Alston Moor.

deeply colored crystals appear blue by reflected light and green by transmitted light. The color may be modified by various means, such as by heat, radium rays, cathode and x-rays, deuterons, ultraviolet light, pressure. The literature on the causes of the color of fluorite is extensive;[13] the color in part is secondary and due to photochemical processes. The color often is distributed[14] in zones parallel to the crystal faces or is restricted to particular growth segments. The pigmenting material also may be distributed irregularly, or be localized as pleochroic haloes[15] around radioactive inclusions. Massive columnar or fibrous material also may exhibit parallel bands of different color.

Some varieties show phosphorescence when heated gently[16] or after exposure to sunlight[17] or ultraviolet light.[18] The kind emitting a bright green light on heating has been called *chlorophane*. At times triboluminescent.[19] Usually strongly fluorescent under ultraviolet light[20] or cathode rays;[21] the fluorescence most often blue or violet, rarely red. Exhibits a difference of electric potential between the faces and angles of a cube, both under the action of heat and of light.[22] A non-conductor of electricity. Diamagnetic. Translation gliding [23] with $T\{001\}$ and probably $t[011]$. Linear thermal expansion from 20° to 100° = 0.16 per cent.[41]

Opt. In transmitted light, colorless to green, purple, etc., in thick grains, or in deeply pigmented material. Frequently exhibits very weak anomalous birefringence, especially in cleaved, cut, or pressed crystals.[24] The birefringence usually is distributed in lamellae parallel [001]. Indices:

n (AT 20°; ALL ± 0.0001) [25]

	B	Li	C	D	Tl	E	F
n	1.4320	1.4322	1.4325	1.43385	1.4353	1.4355	1.4370

The indices apparently do not vary significantly with the color of the crystals. An 8° to 10° rise in temperature causes a decrease of 0.0001 in the refractive index.

INDICES OF YTTRIAN FLUORITE

	Anal. 3 [26]	Anal. 4 [27]	Anal. 5 [27]
n_D	1.4572	1.4425	1.4483

Chem. Calcium fluoride, CaF_2. Y and Ce substitute [28] for Ca to the extent of (Y,Ce):Ca = 1:6 (anal. 3) and to (Ce,Y):Ca = 1:5.2 (anal. 7). In artificial material crystallized from melts,[29] Y substitutes for Ca to a limit of about Y:Ca = 1:1.87, and Ce to a limit of about Ce:Ca = 1:2.0. Y, Er, La, Sm, Dy, Pr, Nd, and other rare earths are common in fluorite in spectrographic amounts.[30]

Some specimens of fluorite, such as the dark violet-blue material from Wölsendorf, Bavaria (*antozonite; stinkfluss*), contain free fluorine and calcium released by radium rays and on grinding give an odor due to HF and ozone formed by reaction of the fluorine with water.[31] On heating, some fluorite yields a bituminous odor and a distillate of organic material and gas.[32] Minute cavities filled with gas, water, or an organic liquid are frequently present.

Anal.[33]

	1	2	3	4	5	6	7
Ca	51.33	51.24					
CaO			54.89				
CaF_2				89.07	83.87	45.81	50.00
Y_2O_3			17.35			17.39	8.10
Ce_2O_3			1.68			9.94	16.45
YF_3				10.59	16.88		
CeF_3				0.95			
F	48.67	48.29	[45.54]			41.64	[25.45]
Rem.		0.30	1.04			2.62	
Less O = F			19.17			17.53	
Total	100.00	99.83	[101.33]	100.61	100.75	99.87	[100.00]
G.		3.180	3.175	3.53–3.56	3.319	3.405	

1. CaF_2. 2. Corvara, Trentino, Italy. Rem. is MgO 0.03, Fe tr., SiO_2 0.05, ign. loss 0.22.[34] 3. Yttrian fluorite (*yttrofluorite*). Hundholmen, Norway. Rem. is alk. 0.15, H_2O −110° 0.22, ign. loss (includes some F) 0.67.[26] 4, 5. Yttrian fluorite, Hundholmen, Norway.[27] 6. Yttrian fluorite, Khibina tundra, Russia. Rem. is BaO 0.22, H_2O 2.40.[35] 7. Cerian fluorite (?) (*yttrocerite*), Fahlun, Sweden.[36]

Var. Yttrian. Yttrofluorite *Vogt* (*Cbl. Min.*, 274, 1911). Contains Y in substitution for Ca (anals. 3–6). Ce usually is present also; the variety is here restricted to material with Y > Ce. Massive, granular. Color yellow, brown, violet, blue. The specific gravity, hardness, and index of refraction are somewhat greater than in ordinary fluorite. The cleavage has been given variously as octahedral, cubical, or dodecahedral.[42]

Cerian. Yttrocerite *Gahn* and *Berzelius* (*Afhandl. Fys. Kem. Min.*, **4**, 151, 1814). Yttrocererite *Leonhard* (573, 1826). Yttria fluatée *Fr.* Fluate of Cerium and Yttria. Yttroflussspath, Flussyttrocalcit *Germ.* Yttrocalcit *Glocker* (283, 1847) [not Yttrocalcit *Federov* (*Gyorni J.*, **3**, 254, 1905)]. Contains Ce in substitution for Ca (anal. 7). Y also may be present; the variety is here restricted to material with Ce > Y. The evidence for the existence of this variety is not satisfactory, and most material formerly placed here properly belongs under yttrian fluorite.

Tests. In C.T. decrepitates and phosphoresces. B.B. in the forceps and on charcoal fuses to an enamel which reacts alkaline on test paper. With soda or gypsum fuses to a transparent bead, which becomes opaque on cooling. Decomposed by H_2SO_4; slightly soluble in hot HCl. Solubility in water 0.016 g. per liter at 18°.

Occur. Fluorite is a common mineral and is found in deposits of widely different character. It occurs most commonly as a vein mineral, either in deposits in which it is the chief constituent or as a gangue mineral associated with metallic ores, especially those of lead and silver. Often associated with quartz, calcite, dolomite, and barite. Also found in cavities in sedimentary rocks, such as dolomite and limestone, where it may be associated with celestite, anhydrite, gypsum, dolomite, sulfur, millerite, and as a cementing material in sandstones. Observed, sometimes in large amounts, as a hot spring deposit (Wagon Wheel Gap, Colorado). Fluorite is characteristic of certain so-called pneumatolytic deposits, especially greisens and high temperature veins carrying cassiterite, and is here associated with topaz, tourmaline, lepidolite, apatite, quartz. Less commonly found as a relatively late-crystallized, hydrothermal product in cavities or joints in granite and in other acid or alkalic igneous rocks. Often found finely crystallized in veins of the Alpine type. With calcite and minor apatite, hornblende, biotite, uraninite in vein-like bodies cutting syenite (Wilberforce, Ontario). Rarely noted as a secondary mineral in the oxidized zone of ore deposits. In pegmatites.[43]

Studies[37] of the crystal habit of fluorite suggest that the light-colored octahedral crystals are typical of relatively high temperatures of formation, whereas the more deeply colored cubic crystals are formed under low-temperature conditions. Dodecahedral crystals are characteristic of intermediate conditions.

Only a few of the many known localities can be mentioned in this place. Fluorite is widespread in the mining districts of Saxony, as at Freiberg, in the tin veins at Altenberg, and at Annaberg, Zschopau, and Marienberg. In the druses of the Striegau granite, Silesia. In several important deposits in the Harz, notably with barite, quartz, and sulfides in vein deposits at Stolberg; also found with silver ores at Andreasberg. In Baden at Münsterthal, in sandstone at Waldshut and Badenweiler, and in veins with quartz, chalcopyrite, barite at Schapbach. A deep purple variety which yields a pronounced odor of fluorine and ozone when struck (*antozonite*) occurs adjacent to uranium minerals at Wölsendorf in the Oberpfalz, Bavaria. Found in the lithium pegmatites of Epprechtstein in the Fichtelgebirge, Bavaria. In Bohemia in tin greisen at Zinnwald, and with apatite, topaz, and chalcopyrite at Schlaggenwald. Found in veins of the Alpine type at numerous localities in the Austrian, Italian, and French Alps, and in Switzerland. Rose-colored octahedral crystals are found at localities on the Aar massif, and elsewhere in the Alps. At Brienz in Berne, Switzerland. In Italy near Bolzano, Trentino, in a large vein deposit; near Brescia, in Lombardy; in druses in the granite of Baveno, Piedmont; in the volcanic tuffs of the Compania, Sicily. In France in large deposits in the neighborhood of Fréjus, Dept. Var; at numerous localities on the Mont Blanc massif; in veins in the neighborhood of Langeac and Barlet in Haute-Loire; at Roche Cornet near Saint-Jacques-d'Amburg in Puy-de-Dôme; near Autun, and with manganese ores at Romanèche, Saône-et-Loire; as a recent deposit in the hot springs at Plombières, Vosges.

A number of famous localities are in England. In Cumberland at Alston Moor and Cleator Moor; at Weardale in Durham; in Yorkshire; in the Beeralston mine and at Tavistock in Devonshire; at Castleton and other places in Derbyshire. A massive, blue fibrous or fine granular and often banded variety from Derbyshire, formerly much used for vases and other ornamental objects, is known as Blue John.[38] Fine specimens have been found in the copper and tin veins at Liskeard, St. Agnes, and other places in Cornwall. In Norway with barite, dolomite, and quartz in the silver veins of Kongsberg; near Dalen in Telemarken; highly yttrian fluorite (*yttrofluorite*) occurs in a pegmatite at Hundholmen on the Tyfs fiord in northern Norway, associated with fergusonite and other rare-earth minerals. Yttrian fluorite also occurs on the Khibina tundra, Kola peninsula, Russia, associated with calcian ancylite and natrolite in nepheline syenite pegmatite. Fluorite occurs in pipe-like deposits in dolomite of the Transvaal series in the Zeerust district, Transvaal, South Africa. In Tasmania with tin ores at Mt. Bischoff; in New South Wales with galena and pyrite in a vein deposit in Goulburn County, and in pipes and vugs in granite near Deepwater in Gough County. In Mexico mined at Monte Realejo near Guadalcazar in San Luis Potosi; at Guanajuato. With quartz, feldspar, siderite in the cryolite deposit at Ivigtut, Greenland. Cerian fluorite occurs at Finbo and Broddbo near Fahlun, Dalarne, Sweden (*yttrocerite*), with quartz, beryl, albite, and topaz.

In the United States, fluorite has been mined chiefly in Hardin (especially near Rosiclare) and Pope counties, Illinois, in adjacent areas in western Kentucky, and in Mercer and Woodford counties in central Kentucky. Here the mineral (sometimes of optical grade) occurs mostly in veins in Mississippian sedimentary rocks, associated with calcite, dolomite, barite, quartz, and subordinate amounts of galena, sphalerite, and other sulfides.

FLUORITE 35

Small deposits have been worked in Colorado at Jamestown and other places in Boulder County, at Wagon Wheel Gap in Mineral County, and elsewhere; in New Mexico in the Burro Mountain district, Grant County, near Deming in Luna County, and near Mesilla Park in Dona Ana County; near Carthage in Smith County, Tennessee; in California near Blythe, Riverside County, and in the Cave Canyon district, San Bernardino County; in Arizona in the Castle Dome district north of Yuma, and in the Serrita Mountains southwest of Tucson. A few other noteworthy occurrences of fluorite in the United States follow. At Trumbull, Fairfield County, Connecticut, with topaz and tungsten minerals. In large green masses, sometimes yielding cleavage octahedrons more than 21 cm. on an edge, in a vein at Westmoreland, Cheshire County, New Hampshire. In New York as large green cubes at Muscalonge Lake in Jefferson County, and at Macomb in St. Lawrence County. In pegmatite at the quarries around Amelia, Amelia County, Virginia. In Ohio as brown cubes with celestite at Clay Center, Ottawa County, and at Tiffin. In Colorado as octahedrons, sometimes twinned, in the pegmatites of the Mt. Antero region, Chaffee County; in drusy pegmatitic cavities in the granite of the Pikes Peak region; abundantly as a gangue mineral at Cripple Creek. In contact metamorphic copper deposits in the White Knob district, Custer County, Idaho. Yttrian fluorite occurs in several localities in the Franklin crystalline limestone in Orange County, New York, and Sussex County, New Jersey; yttrian material also has been reported from pegmatites at Mt. Mica, Paris, Maine, from Colorado, and from Worcester County, Massachusetts. In Canada, fluorite is found with calcite and apatite in vein-like bodies cutting syenite in Cardiff township, Haliburton County, Ontario. In crystals of optical quality in veins with barite, calcite, celestite, from Madoc, Hastings County, Ontario. With calcite, mica, apatite, hornblende, pyroxene, scapolite, and uraninite in veins near Wilberforce, Haliburton County, Ontario, and with amethyst at Thunder Bay, on Lake Superior, in Ontario. A large deposit at the Rock Candy mine, near Grand Forks, British Columbia, as veins in syenite, associated with barite, calcite, pyrite. At Little St. Lawrence Bay, Newfoundland.

Alter. Incrustation and substitution pseudomorphs of quartz and of chalcedony after fluorite are especially common.[39] The name *cubosilicite* has been given to the smalt-blue cubic chalcedony pseudomorphs from Tresztyan, Transylvania. *Guanabaquite* and *cubaite* are names that have been given to cubes of silica, apparently pseudomorphs after fluorite, from Cuba. Clay minerals and various iron and manganese oxides are often found as pseudomorphs after fluorite; also, less commonly, calcite, dolomite, siderite, sphalerite, cerussite, pyrite, marcasite, talc, hemimorphite, smithsonite, chlorite, feldspar, bertrandite (?). Some of these pseudomorphs are probably casts after embedded fluorite crystals that have been dissolved away, and others are the result of incrustation and later solution of the fluorite. Molds of fluorite crystals are often noted and express the ready solubility of CaF_2 in water, especially when carbonated. Fluorite has also been found as pseudomorphs after calcite, barite, galena, and as an alteration product of prosopite.

A r t i f.[40] Obtained in minute octahedral, cubic or cubo-octahedral crystals by various methods: the diffusion of ammonium fluoride into $CaCl_2$ solution; the evaporation of a solution of CaF_2 in HCl; the reaction and cooling of calcium fluosilicate and $CaCl_2$ solutions; the action of HF solution on calcium-containing glass. Obtained in relatively large single-crystal masses by the appropriately arranged crystallization of melts.

N a m e. From *fluere, to flow*, because it melts easily and in allusion to the use of the mineral as a flux. *Antozonite*, from the supposed presence of a compound, *antozone*, to which the odor of the Wölsendorf fluorite was ascribed. *Chlorophane* and *pyrosmaragd* in allusion to the green phosphorescence shown on heating. *Ratofkite*, from a locality on the Ratofka River, Russia, is a name given to an earthy impure fluorite. The term *fluorescence* is derived from the mineral name, because fluorite shows the effect so markedly.

Ref.

1. Goldschmidt (**4**, 8, 1918) and Holzgang, *Schweiz. min. Mitt.*, **10**, 374 (1930). Rare, uncertain, and vicinal forms:

0.1.40	0.5.12	118	155	168	2.7.10	4.7.16
0.1.35	049	2.2.15	299	3.10.25	146	3.8.14
0.1.32	035	229	3.11.11	2.6.15	3.10.16	5.9.19
0.1.10	023	227	277	157	4.13.21	4.7.15
019	079	337	255	3.4.20	125	9.14.32
018	045	334	355	3.14.20	135	237
016	078	667	477	2.9.13	5.12.24	7.10.24
015	1.1.19	778	577	126	249	5.7.17
027	1.1.17	188	677	2.3.11	259	3.4.10
0.3.11	1.1.12	177	2.6.25	2.5.11	4.9.17	234
025	1.1.10	166	128	2.7.11	134	345

2. Structure first analyzed by Bragg, *Royal Soc. London, Proc.*, **89**, 468 (1914).
3. Goldschmidt and Thomassen, *Vidensk. Skr., Kristiana*, no. 2, 40 (1923), give a_0 5.49 kX for an unanalyzed specimen of yttrofluorite from Hundholmen.
4. See Obenauer, *Jb. Min., Beil.-Bd.*, **66**, 89 (1932), on the relative frequency and combinations of the fluorite forms; also Bradistilov and Stranski, *Zs. Kr.*, **103**, 1 (1940).
5. For a discussion of the twinning in light of the crystal structure see Schaake, *Zs. Kr.*, **98**, 281 (1937).
6. Obenauer (1932).
7. Marx, Kastner's Arch., **5**, 306 (1825).
8. Mügge, *Jb. Min., Beil.-Bd.*, **16**, 351, 361 (1903).
9. See especially Becke, *Min. Mitt.*, **11**, 349 (1890). On the solution bodies from the sphere see Himmel, *Jb. Min., Beil.-Bd.*, **60**, 111 (1930); Kleber, *Jb. Min., Beil.-Bd.*, **65**, 557 (1932); Himmel and Kleber, *Jb. Min., Beil.-Bd.*, **69**, 42 (1934), **70**, 49 (1935).
10. See Exner, *Härte an Krystallflächen*, Wien, 31, 34 (1873) for sclerometric measurements of the vectorial hardness.
11. Merwin, *Am. J. Sc.*, **32**, 430 (1911).
12. *Int. Crit. Tab.*, **1**, 143 (1926).
13. On the origin of the color, see literature in Hintze (**1** [2B], 2384, 1912) and Doelter (**4** [3], 228, 1930); also Goebel, *Ak. Wien, Ber.*, **139**, 373 (1930), Hoffmann, *Chem. Erde*, **11**, 368, (1937), Iimori, *Sc. Pap. Inst. Phys. Chem. Res. Tokyo*, **20**, 189 (1933), Przibram, *Nature*, **141**, 970 (1938).
14. On the spatial distribution of the pigment within the crystals see especially Haberlandt and Schiener, *Zs. Kr.*, **90**, 193 (1935) and Steinmetz, *Zs. Kr.*, **61**, 380 (1925).

15. Brown, *Univ. Toronto Stud., Geol. Ser.*, no. 32, 51, (1932) and Ramdohr, *Jb. Min., Beil.-Bd.*, **67**, 63 (1933).
16. Rothschild, *Festschr. V. Goldschmidt*, 243 (1928); Northup and Lee, *J. Opt. Soc. Am.*, **30**, 206 (1940); Iwase, *Sc. Pap. Inst. Phys. Chem. Res. Tokyo*, **22**, 233 (1933), **23**, 22, 153, 212 (1934).
17. Brown, *Univ. Toronto Stud., Geol. Ser.*, no. 35, 19 (1933).
18. Iimori and Iwase, *Sc. Pap. Inst. Phys. Chem. Res., Tokyo*, **16**, 41 (1931).
19. See Wick, *J. Opt. Soc. Am.*, **27**, 275 (1937).
20. See literature in Doelter (1930) and Hintze (1913); also Haberlandt, *et al., Ak. Wien, Ber.*, **142**, 29, 235 (1933), **141**, 441 (1932), **143**, 151, 591 (1934), **144**, 77, 135 (1935).
21. Yoshimura, *Sc. Pap. Inst. Phys. Chem. Res. Tokyo*, **23**, 224 (1934).
22. Hankel, *Ann. Phys.*, **2**, 66 (1877), *ibid.*, **11**, 269 (1880); *Sächs. Ak. Wiss., Abh.*, **12**, 201 (1879); Schmidt, *Ann. Phys.*, **62**, 407 (1897).
23. Veit, *Jb. Min., Beil.-Bd.*, **45**, 121 (1922).
24. On the optic orientation and relation of the birefringent lamellae to the direction of pressure see Pockels, *Ann. Phys.*, **37**, 144, 372 (1889), Veit (1922); also literature in Hintze (1913).
25. An average by Merwin, *Am. J. Sc.*, **32**, 425 (1911), of selected published values. For measurements in the infrared see Matossi and Brix, *Zs. Phys.*, **92**, 303 (1934); Paschen, *Ann. Phys.*, **4**, 302 (1901); and in the ultraviolet Martens, *Ann. Phys.*, **6**, 616, (1901) and Tousey, *Phys. Rev.*, **50**, 1057 (1936). Fluorite has a very low transparency for large wavelengths.
26. Vogt, *Cbl. Min.*, 373 (1911).
27. Zambonini, *Riv. min.*, **45**, 148 (1915).
28. On the crystal chemistry of the substitution see Goldschmidt and Thomassen (1923) and Zintl, *et al., Zs. anorg. Chem.*, **240**, 150 (1939), **242**, 79 (1939).
29. Vogt, *Jb. Min.*, II, 9 (1914).
30. Humphreys, *Am. J. Sc.*, **19**, 202 (1905); Iwase (1933, 1934), Yoshimura (1934); Haberlandt, *Ak. Wien, Sitzber.*, **143**, 591 (1934), Bray, *Am. Min.*, **27**, 769 (1942).
31. See Hoffmann, *Chem. Erde*, **11**, 368 (1937) and Doelter (1930).
32. Garnet, *Trans. Chem. Soc. London*, **117**, 620 (1920); Blount and Sequeira, *Trans. Chem. Soc. London*, **115**, 705 (1919).
33. For additional analyses see Doelter (1930) under fluorite, yttrofluorite, and yttrocerite; also Morgante, *Per. Min.*, **10**, 379 (1939), Starik, *et al., C.r. ac. sc. U.R.S.S.*, **32**, 254 (1941), Fornaseri, *Per. Min.*, **16**, 129 (1947).
34. Morgante (1939).
35. Tschernik, *Bull. ac. sc. Russie* 81 (1923); *Jb. Min.*, II, 289 (1925).
36. Berzelius, *Afhandl. Phys. Kem. Min.*, **4**, 151 (1814); see also Rammelsberg, *Ber.*, **3**, 857 (1870).
37. Drugman, *Min. Mag.*, **23**, 137 (1932) and Obenauer, *Jb. Min., Beil.-Bd.*, **66**, 89 (1932).
38. For an account of the Derbyshire Blue John see Mawe, *Min. and Geol. of Derbyshire*, London, 69 (1802).
39. For a detailed description of some silica pseudomorphs see Murdoch, *Am. Min.*, **21**, 18 (1936).
40. Cf. Doelter (1930).
41. Cited in Birch, 30 (1942).
42. See summary in Hintze, **1** [2B], 2556 (1913).
43. Heinrich, *Am. Min.*, **33**, 64 (1948).

YTTRIUM CALCIUM FLUORIDE. Genth and Penfield (*Am. J. Sc.*, **44**, 386, 1892). A granular cleavable fluoride found with quartz, astrophyllite and danalite at West Cheyenne Canyon, El Paso County, Colorado. Color white, grayish white and reddish white. H. 4. G. 4.316. Analysis gave: CaO 19.41, $(Y,Er)_2O_3$ 47.58, CeO_2 0.83, $(La,Di)_2O_3$ 1.55, ign. loss 1.57 (total 70.94; F not det. and O = F not deducted). The ratio of Ca to rare earths is about 1:1. Possibly an yttrian fluorite or a calcian and yttrian tysonite.

9.2.2   S E L L A I T E [MgF_2].   Strüver (*Acc. Torino, Att.*, **4**, 35, 1868). Belonesite. Belonesia *Scacchi* (*Acc. Napoli, Mem.*, **1**, no. 5, announced September 8, 1883, published 1886). Zamboninite *Starraba* (*Soc. geol. ital., Boll.*, **48**, 259, 1930).

HALIDES

C r y s t. Tetragonal; ditetragonal-dipyramidal—$4/m\ 2/m\ 2/m$.

$$a:c = 1:0.6596;\ ^1 \qquad p_0:r_0 = 0.6596:1$$

Forms: [2]

	ϕ	ρ	A	$\overline{M}$
a 010	0°00′	90°00′	90°00′	45°00′
m 110	45 00	90 00	45 00	90 00
e 011	0 00	38 21½	90 00	67 05
p 111	45 00	43 00½	61 10	90 00

Less common:

120   112   551

Structure cell.[3] Space group $P4/mnm$. a_0 4.660 kX, c_0 3.078; $a_0:c_0$ = 1:0.660. Cell contents Mg_2F_4.

Twinning.[4] On {110}.

Habit. Stout prismatic to acicular [001]. Fibrous aggregates.

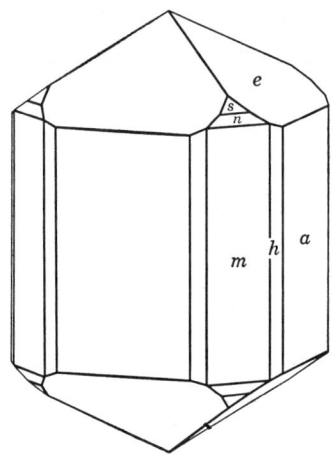

Gebroulaz, France.

P h y s. Cleavage {010} and {110} perfect. Fracture conchoidal. Brittle. H. 5. G. 3.15;[5] 3.08 (calc.). Luster vitreous. Colorless to white. Transparent. Warmed crystals show pale violet light when cleaved.

O p t. In transmitted light, colorless. Refractive indices (for Na):[6]

nO 1.378   Uniaxial positive (+).
nE 1.390

C h e m. Magnesium fluoride, MgF_2.

Anal.

	1	2
Mg	39.02	38.37
F	60.98	[61.63]
Total	100.00	[100.00]
G.	3.08	3.15

1. MgF_2. 2. Gebroulaz glacier, Savoy.[7]

Tests. B.B. in small fragments fuses with intumescence. Very slightly soluble in water. Decomposed by concentrated H_2SO_4, and by fusion in alkali carbonates.

O c c u r.[11] Originally found with native sulfur, fluorite, quartz, celestite in a bituminous dolomite-anhydrite rock in the moraine of the Gebroulaz glacier near Moutiers, north of Modane, Savoy, France. Found similarly associated at Bleicherode in the southern Harz, Germany. In Italy with native sulfur and gypsum in cavities in marble at Carrara; in ejected blocks on Vesuvius with anhydrite (*belonesite*); with fluorite as a fumarolic deposit on Etna, Sicily (*zamboninite*). Also reported from cassiterite veins at Nertschinsk, Urals, U.S.S.R., and from potash-salt deposits in Germany.

A r t i f.[8] Obtained by reaction of $MgSO_4$ and KF in water solution, by fusion of $MgCl_2$ with NaF and $NaCl$, and in other ways.

N a m e. After Quintino Sella (1827–1884), Italian mining engineer and mineralogist

---

BELONESITE. Belonesia *Scacchi (Acc. Napoli Mem.*, **1**, No. 5, announced September 8, 1883, published 1886). A supposed tetragonal molybdate of magnesium found as minute acicular crystals at Vesuvius. Identical with sellaite.[9]

ZAMBONINITE. *Starraba (Soc. geol. ital.*, **48**, 259, 1930.) A supposed orthorhombic mineral of the composition $CaF_2 \cdot 2MgF_2$. Radial fibrous. Friable. Found in extinct fumaroles on Etna, Sicily. Shown [10] to be a mixture of sellaite and fluorite. Named after Ferruccio Zambonini (1880–1932), Italian mineralogist and crystallographer.

Ref.

1. Sella, *Acc. Linc., Mem.*, **4**, 455 (1887).
2. Goldschmidt (**8**, 34, 1922), Pelloux, *Acc. Linc., Att.*, **28**, 284 (1919). Rare and vicinal forms:

230	031	221	233	373	5.16.1
065	558	255	576	792	5.17.1
052	334	122	494	4.11.1	791

3. Buckley and Vernon, *Phil. Mag.*, **49**, 945 (1925), by powder method.
4. Heidorn, *Cbl. Min.*, 356 (1932), on material from Bleicherode; also observed on artificial crystals.
5. Sella (1887) on material from Gebroulaz by pycnometer. Ferrari, *Acc. Linc., Rend.*, **1**, 664 (1925), obtained 3.148 on artificial material.
6. Sella (1887) by prism method on Gebroulaz material.
7. Sella (1887).
8. Mellor (**4**, 296, 1923).
9. Zambonini (82, 1935).
10. Ferrari and Curti, *Per. Min.*, **4**, 465 (1933), by x-ray (powder) and optical study.
11. On the occurrence and paragenesis of sellaite see Sahama, *Ann. Ac. Sc. Fennicae*, [A], Pt. III, no. 9, (1945).

9.2.3 LAWRENCITE GROUP

	a_0	c_0
Lawrencite, $FeCl_2$	3.57 kX	17.51
Scacchite, $MnCl_2$	3.68	17.48
Chloromagnesite, $MgCl_2$	3.58	17.59

These AX_2 substances crystallize in layer structures of the $CdCl_2$ type. Artificial crystals are hexagonal tablets on $\{0001\}$ with a perfect basal cleavage and are optically negative, in accordance with the structure. The substances occur in nature principally as sublimates in volcanic fumaroles and are difficult to preserve because of their highly deliquescent nature.

9.2.3.1 LAWRENCITE [FeCl$_2$]. *Jackson* (*Am. J. Sc.*, **48**, 146, 1845). Lawrencite *Daubrée* (*C.R.*, **84**, 69, 1877). Eisenchlorür *Germ.* Chlorure de fer *Fr.* Ferro muriato *Ital.*

C r y s t. Hexagonal—R; scalenohedral—$\bar{3}\ 2/m$ (?).
Structure cell.[1] Space group $R\bar{3}m$ (?). a_{rh} 6.19, α 33°33$\frac{1}{2}$'; a_0 3.57 kX, c_0 17.51; $a_0:c_0 = 1:4.898$. Hexagonal cell contents Fe$_3$Cl$_6$.
Habit. Thin hexagonal plates {0001} (artif.). The natural material is found only massive.
P h y s. Cleavage {0001} perfect (?).[2] Soft. G. 3.16 (artif.);[3] 3.22 (calc.). Color green to brown; fresh artificial material is white. Deliquescent.
O p t.[4] Uniaxial negative ($-$). $nO = 1.567 \pm 0.005$. Birefringence weak.
C h e m. Ferrous chloride, FeCl$_2$. Fe 44.05, Cl 55.95, total 100.00. Analyses show a considerable, but variable, content of Ni. It is not known whether this is due to impurities or to isomorphous substitution for Fe.[5]

Tests. B.B. easily fusible; volatilizes at higher temperatures. Readily soluble in water.

O c c u r. Found in fissures in iron meteorites.[6] Ferric chloride (cf. *molysite*) formed from lawrencite often exudes from the surface of the irons. Said to occur as a sublimation product at Vesuvius.[7] Found in the native iron of Ovifak, Greenland.
A l t e r. Deliquesces and alters by oxidation to ferric chloride.
A r t i f.[8] Formed by the action of HCl gas or of NH$_4$Cl on metallic iron at red heat, by the reaction of Cl on ferrous oxide, and in other ways. Minute crystals are obtained by sublimation.
N a m e. After J. Lawrence Smith (1818–1883), American chemist, mineralogist, and student of meteorites.

Ref.

1. Pauling, *Proc. Nat. Ac. Sc.*, **15**, 709 (1929) and Ferrari, Celeri, and Giorgio, *Acc. Linc., Rend.*, **9**, 782, (1929) by x-ray powder method on artificial material.
2. Inferred from the crystal structure.
3. Biltz and Birk, *Zs. anorg. Chem.*, **134**, 125 (1924).
4. Larsen (99, 1921) on artificial material. The value given for nO seems low, and the measurement possibly was made on partly hydrated material.
5. Jackson (1845) and Hayes, *Am. J. Sc.*, **48**, 153 (1845), on soluble extracts from the Claiborne, Alabama, iron and Buddhue, *Pop. Astronomy*, **48**, (1940), *Contr. Soc. Res. Meteorites*, **2**, 233 (1941) on a soluble extract from the Mt. Elden, Arizona, meteorite.
6. Smith, *Am. J. Sc.*, **19**, 159 (1855), **13**, 214 (1877).
7. Zambonini (84, 1935).
8. Mellor (**14**, 10, 1935).

9.2.3.2 SCACCHITE [MnCl$_2$]. Protochloruro di Manganese *Scacchi* (*Mem. Incend. Vesuv.*, 181, 1855). Scacchite *Adam* (70, 1869). [Not scacchite *Napoli* (*Acc. Napoli, Rend.*, 66, 1859) or sacchit [scacchite ?] *Nordenskiöld* (94, 1848)].

C r y s t. Hexagonal—R; scalenohedral—$\bar{3}\ 2/m$ (?).
Structure cell.[1] Space group $R\bar{3}m$ (?). a_{rh} 6.20, α 34°32'; a_0 3.68 kX, c_0 17.48, $a_0:c_0 = 1:4.478$. Cell contents MnCl$_2$ in the rhombohedral unit.

Phys. Cleavage {0001} perfect ?.[2] Soft. G. 2.98;[3] 3.04 (calc.). Color rose-red, becoming dirty red to brown on exposure; colorless when pure and fresh. Deliquescent.

Chem. Manganous chloride, $MnCl_2$. Mn 43.65, Cl 56.35, total 100.00. Analyses are lacking.[4]

Tests. B.B. easily fusible; volatilizes at higher temperatures. Readily soluble in water.

Occur. Found as a deliquescent salt associated with halite, sylvite, hydrophilite (?), lawrencite (?) and other salts in fumaroles on Vesuvius.

Artif.[5] Obtained by heating $MnCO_3$ in a current of HCl gas, by passing Cl over a mixture of MnO and C, and in other ways.

Name. After Arcangelo Scacchi (1810–1894), Italian mineralogist and Professor at the University of Naples.

Ref.
1. Pauling, *Proc. Nat. Ac. Sc.*, **15,** 709 (1929) and Ferrari, Celeri, and Giorgio, *Acc. Linc., Rend.*, **9,** 782 (1929) by powder method on artificial material.
2. Inferred from the crystal structure.
3. *Int. Crit. Tab.*, **1,** 127 (1926).
4. A complete series between artificial $FeCl_2$ and $MnCl_2$ can be formed by fusion.
5. Mellor (**12,** 348, 1932).

9.2.3.3 **Chloromagnesite** [$MgCl_2$]. Magnesia muriata *Monticelli* and *Covelli* (1825). Cloruro di Magnesio *Scacchi* (*Mem. Incend. Vesuv.*, 181, 1855). Cloromagnesite *Scacchi* (*Acc. Napoli, Att.*, **6,** no. 9, 43, 1875).

Anhydrous $MgCl_2$, reported as a sublimate in fumaroles on Vesuvius.[1] Associated with halite and sylvite. Artificial $MgCl_2$ is hexagonal—R, scalenohedral, and is isostructural with lawrencite and scacchite. Structure cell:[2] a_{rh} 6.22, α 33°30′; a_0 3.58 kX, c_0 17.59 with $a_0:c_0 = 1:4.908$; contains Mg_3Cl_6 in the hexagonal unit. G. 2.325 (artif.);[3] 2.44 (calc.). Colorless to white hexagonal plates. Uniaxial negative $(-)$; nO 1.675, nE 1.59.[4] Deliquescent, and very soluble in water.

Ref.
1. Cf. Zambonini (90, 1935).
2. Strukturber. (**1,** 743, 1913–1928); Pauling, *Proc. Nat. Ac. Sc.*, **15,** 709 (1929).
3. Hüttig, *Zs. anorg. Chem.*, **115,** 251 (1921).
4. Larsen (57, 1921).

9.2.4 **Hydrophilite.** *Hausmann* (**3,** 857, 1813). Chlorure de calcium *Beudant* (**2,** 512, 1832). Hydrophillite *wrong orthogr.* Chlorcalcium.

Hydrophilite was originally described as a hygroscopic calcium chloride from Lüneberg, Hanover, Germany, found as mealy crusts and crystalline particles in anhydrite, gypsum, and halite. Similar substances, loosely ascribed to $CaCl_2$, have been reported[1] from Etna on Sicily; as an efflorescence on soil in Tarapacá, Chile, and Chincha, Peru; in the crater of Barren Island, Bay of Bengal; and as a slimy exudation on sandstone on Guy's Cliff, Warwickshire, England. Definitive descriptions are lacking, however, and it is possible that most or all of these occurrences are of chlorocalcite (*q. v.*).

Artificial $CaCl_2$ is orthorhombic, pseudo-tetragonal, with $a_0 = 6.24$ kX, b_0 6.43, c_0 4.20, and G. 2.22, 2.17 (calc.).[2] Optically positive (+), biaxial, with $nX = 1.600$, $nY = 1.605$, $nZ = 1.613$ and $2V$ moderate.[3] Perfect prismatic cleavage, with complex polysynthetic twinning on a prism. A dimorphous form of unknown properties has been reported.[4]

Ref.

1. See Dana (161, 1892) and Hintze (**1** [2B], 2378, 1912).
2. van Bever and Nieuwenkamp, *Zs. Kr.*, **90**, 374 (1935). See also Menge, *Zs. anorg. Chem.*, **72**, 169 (1911), on the system $MgCl_2$-$CaCl_2$.
3. Slawson, *Am. Min.*, **14**, 160 (1929).
4. van Bever and Nieuwenkamp (1935); a supposedly isotropic modification reported by Larsen (89, 1921) apparently is a compound formed by reaction between $CaCl_2$ and organic index liquids.

9.2.5 **Coccinite.** Iodure de Mercure *Del Rio* (*Ann. mines*, **5**, 324, 1829); *Beudant* (**2**, 515, 1832). Coccinit *Haidinger* (572, 1845). Mercure ioduré *Fr.* Iodquecksilber, Quecksilberjodid *Germ.*

The name coccinite is held in a reserve status for mercuric iodide, which has been reported but not yet identified with certainty as occurring in nature. A bright scarlet mineral found as crusts of tiny cubes on limonite at the Broken Hill mine, New South Wales, has been referred here.[1] The original coccinite from Casas Viejas, Mexico, apparently is an oxychloride of mercury,[2] and the so-called coccinite from Zimapan and Culebras, Mexico, found as transparent red to yellow and yellowish green acicular, rhombic pyramids may be calomel.

Ref.

1. Moses, *Am. J. Sc.*, **12**, 98 (1901); see also Smith, *Dept. of Mines, New South Wales, Min. Res.*, no. **34**, 20, 42 (1926).
2. See also Burkart, *Jb. Min.*, 409 (1866).

9.2.6 **COTUNNITE** [$PbCl_2$]. Cotunnia *Monticelli* and *Covelli* (47, 1825). Cottunite *von Kobell* (**2**, 179, 1830). Lead chloride. Chlorblei *Germ.*

C r y s t. Orthorhombic; dipyramidal—$2/m$ $2/m$ $2/m$.

$$a:b:c = 0.8423:1:0.5013;\,^1 \qquad p_0:q_0:r_0 = 0.5951:0.5013:1$$

$$q_1:r_1:p_1 = 0.8423:1.6803:1; \qquad r_2:p_2:q_2 = 1.9948:1.1872:1$$

Forms:[2]

	ϕ	$\rho = C$	ϕ_1	$\rho_1 = A$	ϕ_2	$\rho_2 = B$
c 001	....	0°00′	0°00′	90°00′	90°00′	90°00′
b 010	0°00′	90 00	90 00	90 00	....	0 00
a 100	90 00	90 00	....	0 00	0 00	90 00
v 120	30 41½	90 00	90 00	59 18½	0 00	30 41½
m 110	49 53½	90 00	90 00	40 06½	0 00	49 53½
u 210	67 09½	90 00	90 00	22 50½	0 00	67 09½
e 011	0 00	26 37½	26 37½	90 00	90 00	63 22½
q 111	49 53½	37 53	26 37½	61 59	59 14½	66 41½
p 121	30 41½	49 23	45 04½	67 12½	59 14½	49 15

COTUNNITE

Structure cell.[3] Space group *Pnam*. a_0 7.67 kX, b_0 9.15, c_0 4.50; $a_0:b_0:c_0 = 0.838:1:0.492$. Cell contents Pb_4Cl_8.

Habit. Often more or less flattened {010} and elongated [001]. Sometimes doubly terminated with equivalent faces unequally developed. Re-entrant angles may be present in the prism zone due to oscillatory combination of {210} and {110}. Artificial crystals may tend toward a pyramidal habit. Also massive, granular.

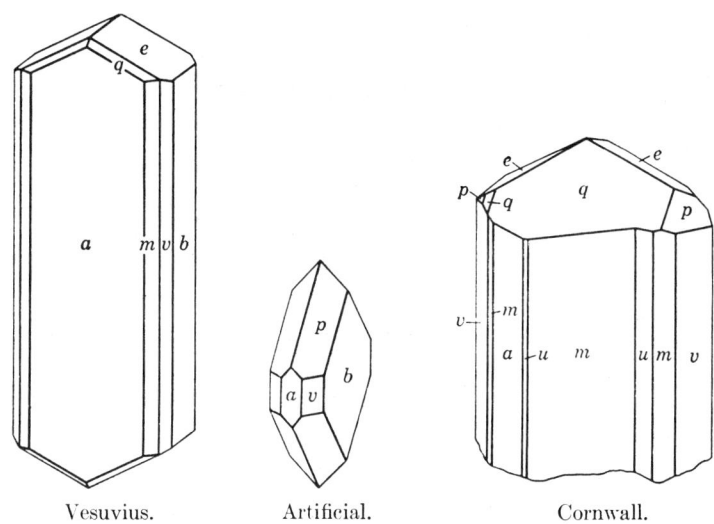

Vesuvius.   Artificial.   Cornwall.

Oriented growths. With galena.[13]

Twinning. On {120}, in artificial crystals.[4]

P h y s. Cleavage {010}, perfect. Fracture subconchoidal. Slightly sectile. H. $2\frac{1}{2}$.[5] G. 5.80 (artif.);[6] 5.81 (calc.). Colorless to white, also yellowish or greenish. Luster adamantine inclining to silky or pearly. Transparent to translucent.

O p t.[7] In transmitted light, colorless.

ORIENTATION		$n(Na)$	
X	b	2.199	Biaxial positive (+).
Y	a	2.217	2V 67°12′ (calc.).
Z	c	2.260	

C h e m. Lead chloride, $PbCl_2$. Pb 74.50, Cl 25.50, total 100.00. The two reported [8] analyses conform to the formula.

Tests. Easily fusible (M.P. 501°).[9] Soluble in water (1.08 g. $PbCl_2$ per 100 g. H_2O at 25°), dilute HCl or HNO_3, and in sodium acetate solution.

O c c u r. Cotunnite was originally found at Vesuvius as a product of sublimation.[10] The mineral also occurs associated with cerussite, anglesite, matlockite, and other secondary minerals as an alteration product of galena under arid, saline conditions. Here may be mentioned the occurrences in

Chile in the Sierra Gorda, near Caracoles in Antofagasta province and in the Cerro Challocollo in Tarapacá province; and in the La Pampa district of Pallasca province, Peru. Reported as small veinlets in chalcocite in the Bentley district, Mohave County, Arizona. Cotunnite also has been found as an alteration product of ancient leaden objects immersed in sea water.[11]

A l t e r. Found as alteration crusts upon or as complete pseudomorphs after galena.

A r t i f.[12] Easily obtained in crystals by evaporation or cooling of a water or HCl solution.

N a m e. After Domenico Cotugno (1736–1822), Professor of Anatomy at the University of Naples.

Ref.

1. Unit of Miller (1852); angles of Stöber, Ac. Belgique, Bull., **30** [3], 345 (1895); reoriented to make $c < a < b$. Transformations: Miller to new, 010/100/001; Stöber to new, 001/010/100; Dana (165, 1892) to new, 001/200/010; Goldschmidt (**2**, 193, 1913) and Lacroix (**4**, 889, 1910) to new 010/001/100.
2. Goldschmidt (1913). Rare or uncertain forms: 9.10.0, 211.
3. Bräkken and Harang, Zs. Kr., **68**, 123 (1928), by rotation and powder methods on artificial material. Transformation: Bräkken and Harang to new, 010/001/100.
4. Stöber (1895).
5. Frondel, priv. comm. (1947), on Vesuvius material.
6. Apparently the best representative value; higher values have been reported by Brügelmann, Ber., **17**, 2359 (1884) (5.88), and by Lorenz and Eitel, Zs. anorg. Chem., **91**, 48 (1915) (5.84).
7. Stöber (1895) by prism method on artificial crystals.
8. Cited in Hintze (**1** [2B], 2350, 1912).
9. Cf. Mellor (**7**, 709, 1927).
10. Cf. Zambonini (86, 1935); the material is strongly radioactive.
11. Lacroix (**4**, 890, 1910); Russell, Min. Mag., **14**, 64 (1920).
12. Mellor (1927).
13. Ramdohr, Abh. deutsch. Akad. Wiss. Berlin, no. 4, 1 (1947).

MELANOTHALLITE. Melanotallo A. Scacchi (Acc., Napoli Rend., **9**, 86, 1870).

Found as black scales associated with eriochalcite, chalcocyanite, euchlorin, and dolerophanite as a sublimation product in fumaroles on Vesuvius. Said to be soluble in water to an acid solution reacting for Cu and Cl. Composition [1] perhaps $CuCl(OH)$ or $CuCl_2$. On exposure rapidly changes to a green, more highly hydrated substance (hydromelanothallite).

Ref.

1. See E. Scacchi, Acc., Napoli Rend., **10**, 158 (1884), and Zambonini (111, 112, 1935).

9.2.7 **E R I O C H A L C I T E** [$CuCl_2 \cdot 2H_2O$]. Eriocalco Scacchi (Acc. Napoli, Rend., **23**, 158, 1884). Eriocalcite. Erythrocalcite Dana (174, 1892). Erythrochalcit Groth (52, 1898). Antofagastite Palache and Foshag (Am. Min., **23**, 85, 1938).

C r y s t.[1] Orthorhombic; dipyramidal—$2/m\ 2/m\ 2/m$.

$$a:b:c = 0.9177:1:0.4631; \qquad p_0:q_0:r_0 = 0.5046:0.4631:1$$

$$q_1:r_1:p_1 = 0.9177:1.9816:1; \qquad r_2:p_2:q_2 = 2.1594:1.0897:1$$

ERIOCHALCITE

Forms: [2]

	ϕ	$\rho = C$	ϕ_1	$\rho_1 = A$	ϕ_2	$\rho_2 = B$
c 001	.....	0°00′	0°00′	90°00′	90°00′	90°00′
b 010	0°00′	90 00	90 00	90 00	.....	0 00
a 100 *	90 00	90 00	.....	0 00	0 00	90 00
m 110	47 27½	90 00	90 00	42 32½	0 00	47 27½
t 520	69 50½	90 00	90 00	20 09½	0 00	69 50½
q 103 *	90 00	9 33	0 00	80 27	80 27	90 00
r 101 *	90 00	26 46½	0 00	63 13½	63 13½	90 00
s 301	90 00	56 33	0 00	33 27	33 27	90 00
p 111	47 27½	34 24½	24 51	65 24	63 13½	67 32½

* On artificial crystals

Structure cell.[3] Space group $Pbmn$. a_0 7.38 kX, b_0 8.04, c_0 3.72; $a_0:b_0:c_0 = 0.918:1:0.462$. Cell contents $Cu_2Cl_4 \cdot 4H_2O$.

Habit. Lichen-like aggregates of crystals elongated [001]; the center of these aggregates is occupied by small deeply grooved spire-like crystals, often bent or entirely recurved in the manner of gypsum "flowers" (Chile). Also as wool-like aggregates (Vesuvius).

Twinning.[4] On {021}, as nearly rectangular penetration twins.

Phys. Cleavage {110} perfect, {001} good. Fracture conchoidal. H. 2½. G. 2.47 (artif.)[5]; 2.55 (calc.). Luster vitreous. Color bluish green to greenish blue with at times a yellowish tinge. Transparent.

Opt.[6] In transmitted light, greenish blue.

ORIENTATION	n(Na)	PLEOCHROISM	
X b	1.646	Pale green	Biaxial positive (+).
Y c	1.685	Pale olive-green	$2V$ 75°.
Z a	1.745	Pale blue	$r < v$, strong.

Chem. Hydrated copper chloride, $CuCl_2 \cdot 2H_2O$.
Anal.

	1	2
Cu	37.22	36.89
Cl	41.52	40.68
H$_2$O	21.26	20.81
Rem.		1.34
Total	100.00	99.72
G.	2.55	2.4

1. $CuCl_2 \cdot 2H_2O$. 2. Mina Queténa, Antofagasta, Chile. Rem. is insol. 0.95, Fe$_2$O$_3$ 0.20, CaO 0.15, MgO 0.04.[7]

Tests. Easily fusible. In C.T. yields water and fuses easily, recrystallizing to a yellow or light brown anhydrous chloride. Easily soluble in water to a pale blue solution and in NH$_4$OH to an intense blue solution.

Occur. A secondary mineral, found with atacamite and bandylite at Mina Queténa, near Calama, Province of Antofagasta, Chile (*antofagastite*). Found originally in fumaroles during the eruption of 1869 at Vesuvius.

Artif. A well-known artificial product,[8] obtained as crystals from water or HCl solution.

Name. From ἔριον, *wool*, and χαλκός, *copper*, in allusion to the form of aggregation of the original Vesuvius material. The first definitive description of the mineral was made on Chilean material (*antofagastite*) in lack of knowledge of the original, ill-described material from Vesuvius. It was later shown [9] that the two minerals were identical, and the name eriochalcite has priority.

Ref.
 1. Palache and Foshag, *Am. Min.*, **23**, 85 (1938), on Antofagasta crystals; Marignac, *Mém. soc. phys. Genève*, **14**, 219, (1855), obtained almost identical results on artificial crystals.
 2. Palache and Foshag (1938); see Groth (**1**, 238, 1906) for artificial crystals.
 3. Harker, *Zs. Kr.*, **93**, 136 (1936), by oscillation and Laue methods on artificial crystals.
 4. Marignac (1855) on artificial crystals; also on natural material from Vesuvius (Frondel, priv. comm., 1947).
 5. Boedeker in Groth (1906).
 6. Palache and Foshag (1938) on Antofagasta crystals.
 7. Foshag anal. in Palache and Foshag (1938). Scacchi (1884) cited an analysis of a water extract of Vesuvius material which affords Cu:Cl = 1:2.
 8. Mellor (**3**, 168, 1923).
 9. Frondel, *Min. Mag.*, **29**, 34 (1950).

9.2.8 **BISCHOFITE** [$MgCl_2 \cdot 6H_2O$]. Ochsenius (Die Bildung der Salzlager, Halle, 126, 156, 1877—*Jahresber. Chem.*, 1284, 1285, 1877; *Zs. Kr.*, **1**, 414, 1877).

Cryst.[1] Monoclinic; prismatic—$2/m$.

$a:b:c = 0.3872:1:0.8543;\quad \beta\ 93°42';\quad p_0:q_0:r_0 = 0.6158:0.8525:1$

$r_2:p_2:q_2 = 1.1730:0.7223:1;\quad \mu\ 86°18';\quad p_0'\ 0.6171,\ q_0'\ 0.8543,\ x_0'\ 0.0647$

Forms:[1]

	ϕ	ρ	ϕ_2	$\rho_2 = B$	C	A
a 100	90°00′	90°00′	0°00′	90°00′	86°18′	......
m 110	35 50½	90 00	0 00	35 50½	87 50½	54°09½′
r $\bar{2}01$	−90 00	49 28	139 28	90 00	53 10	139 28
o 111	38 35½	47 32½	55 43	54 47	45 18½	62 36

Rare: c 001 s 201 u $\bar{1}11$

Habit. Crystalline granular and foliated, sometimes fibrous. Natural and artificial crystals are short prismatic [001].

Twinning. May exhibit secondary, polysynthetic twinning lamellae due to pressure.

Phys. Fracture conchoidal to uneven.[2] H. 1–2. G. 1.604.[3] Colorless to white. Luster vitreous to dull. Exhibits twin gliding with K_1 irrational, K_2 ($1\bar{1}1$), η [$1\bar{1}2$]; also translation gliding on {110}.[4] Deliquescent. Taste stinging and bitter.

Opt.[5] In transmitted light, colorless.

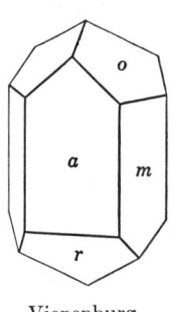

Vienenburg.

ORIENTATION	n(Na)	
X b	1.495	Biaxial positive (+).
$Y \wedge c\ 9\frac{1}{2}°$	1.507	$2V\ 79°24'$.
Z	1.528	$r > v$, weak.

MOLYSITE 47

Chem. Magnesium chloride hexahydrate, $MgCl_2 \cdot 6H_2O$. Artificial $MgBr_2 \cdot 6H_2O$ is isostructural and isomorphous [6] with $MgCl_2 \cdot 6H_2O$.
Anal.

	1	2
Mg	11.96	11.86
Cl	34.87	35.04
H_2O	53.17	[53.10]
Total	100.00	[100.00]

1. $MgCl_2 \cdot 6H_2O$. 2. Leopoldshall.[7]

Tests. Very fusible (M.P. 116.8°[8]). Easily soluble in alcohol and in water (35.36 g. $MgCl_2$ per 100 g. of saturated solution at 25°).

Occur.[9] Bischofite is a comparatively rare constituent of saline deposits (Stassfurt) where it occurs with carnallite and halite in the kieserite-rich zones. In part at least of secondary origin and formed by alteration of carnallite by water. Found in the salt deposits of northern Germany at Leopoldshall, Anhalt, and at Stassfurt and Vienenburg, Saxony.

Artif.[9] Obtained in crystals from water solution between $-3.4°$ and $116.8°$. At least six different hydrates of magnesium chloride are known.

Name. After the German mineral chemist and geologist Gustav Bischof (1792–1870).

Ref.

1. Crystallography based on the description of natural crystals from Vienenburg by Mügge, *Jb. Min.*, I, 91 (1906). Uncertain forms on artificial crystals: 130, 310, 221.
2. Przibylla, *Cbl. Min.*, 238 (1904) mentions three oblique cleavages, but these could not be verified on material from Stassfurt, Frondel, priv. comm. (1947), and may be partings along glide planes.
3. Dewar, *Chem. News*, **91**, 216 (1905); Przibylla (1904) gives 1.591.
4. Mügge (1906).
5. Görgey, *Min. Mitt.*, **29**, 200 (1910).
6. Boeke, *Zs. Kr.*, **45**, 355 (1908).
7. Koenig analysis in Ochsenius (1877).
8. Przibylla (1904).
9. On the physical chemistry of the formation of bischofite see van't Hoff, *Ozeanisch. Salzablag.*, Leipzig (1912, Arts. I, II, VII, XIII), and Doelter (4 [2], 1212, 1928).

TYPE 3. AX_3

9.3.1 MOLYSITE [$FeCl_3$]. Eisenchlorid *Hausmann* (1819; 1463, 1847). Molisite *Scacchi*. Molysite *Dana* (118, 1868). Stagmatite *Daubrée* (*C.R.*, **84**, 69, 1877).

Cryst.[1] Hexagonal—R; rhombohedral ($\bar{3}$) or trigonal-pyramidal (3).
Structure cell.[2] Space group $R\bar{3}$ or $R3$; a_{rh} 6.69, $\alpha\ 52°30'$; a_0 5.92 ± 0.02 kX, c_0 17.26 ± 0.02 with $a_0:c_0 = 1:2.915$. Cell contents Fe_2Cl_6 in the rhombohedral cell.

Habit. Tabular {0001} with a hexagonal outline (artif.). The natural material is massive, as coatings.

Phys. Cleavage {0001} perfect.[3] Soft. G. 2.90 (artif.);[4] 3.04 (calc.). M.P. 282°.[5] Color yellow to brownish red, purple-red; crystals are green by reflected light. Very deliquescent.

O p t.[6] Uniaxial negative (−). Birefringence very strong. Indices of refraction not known.

C h e m. Ferric chloride, $FeCl_3$. Fe 34.42, Cl 65.58, total 100.00. Analyses of natural material are lacking.

O c c u r. Found as a sublimation product in fumaroles on Vesuvius, Mt. Etna, Vulcano on the Lipari Islands, and Iceland. The mineral forms a brownish red or yellowish incrustation or film on the lava and soon yields by its decomposition a reddish brown deposit of iron oxide. Stated to exude as drops (*stagmatite*) on the surface of iron meteorites by oxidation of lawrencite, but this material doubtless is a hydrate.

A l t e r. Readily hydrolyzed by water to hydrous iron oxide.

A r t i f.[7] Formed by the action of Cl on heated metallic iron, whereby the ferric chloride volatilizes and condenses near by. Also obtained by passing a current of dry HCl gas over red-hot Fe_2O_3. In crystals by sublimation and by the slow cooling of melts.

N a m e. From μόλυσις, *a stain*, in allusion to its staining the lavas.

Ref.

1. Symmetry established by x-ray study of Wooster, *Zs. Kr.*, **83**, 35 (1932), on artificial material. Nordenskiöld, *Svenska Vet.-Akad. Handl. Stockholm, Bihang*, **2**, no. 2 (1874), described imperfect crystals from melts as hexagonal, with the forms {0001}, {10$\bar{1}$0}, {10$\bar{1}$1}, {30$\bar{3}$2}, {30$\bar{3}$4}, and $c = 1.235$; angular measurements are not cited. Transformation: Nordenskiöld to Wooster, 1$\bar{1}$00/01$\bar{1}$0/$\bar{1}$010/0002.
2. Wooster (1932) by Laue and Weissenberg methods.
3. Inferred from the crystal structure.
4. Biltz and Birk, *Zs. anorg. Chem.*, **134**, 125 (1924).
5. *Int. Crit. Tab.* **1**, 128 (1926).
6. Wooster (1932).
7. Mellor (**14**, 40, 1935).

9.3.2 **F L U O C E R I T E** [(Ce,La,Nd)F_3]. Neutralt flussspatssyradt *Berzelius* (**6**, 56, 1818). Neutrales flussaures Cerer, Flusscerium *Germ*. Neutral Fluate of Cerium. Cérium fluaté *Haüy* (**4**, 399, 1822). Flucérine *Beudant* (**2**, 519, 1832). Flusscerit *Glocker* (956, 1831; 595, 1839). Fluocerit *Haidinger* (500, 1845). Tysonite *Allen* and *Comstock* (*Am. J. Sc.*, **19**, 390, 1880).

C r y s t.[1] Hexagonal—P; dihexagonal-dipyramidal—$6/m\ 2/m\ 2/m$.

$$a:c = 1:1.0199; \qquad p_0:r_0 = 1.1777:1$$

Forms:[2]

	φ	ρ = c	M	A_2
c 0001	.....	0°00′	90°00′	90°00′
m 10$\bar{1}$0	30°00′	90 00	60 00	90 00
a 11$\bar{2}$0	0 00	90 00	90 00	60 00
r 30$\bar{3}$2	30 00	60 29	64 12½	90 00
p 11$\bar{2}$1	0 00	63 53	90 00	63 19½
s 22$\bar{4}$1	0 00	76 13½	90 00	60 57

Structure cell.[3] Space group $C6/mcm$. $a_0\ 7.124\ kX$, $c_0\ 7.280$; $a_0:c_0 = 1:1.022$. Cell contents $(Ce,La,Nd)_6F_{18}$.

FLUOCERITE

Habit. Crystals prismatic [0001] or tabular {0001}. Also massive, coarsely granular.

Phys. Cleavage {0001} distinct, {11$\bar{2}$0} indistinct. Fracture subconchoidal to uneven and splintery. Brittle. H. 4–5. G. 6.14 (Colorado); 5.93 (Broddbo);[4] 6.08 (calc. for Ce:La = 1:1); the G. decreases to 5.7 or less on alteration. Luster vitreous to resinous, somewhat pearly on cleavage surfaces. Color pale wax-yellow when fresh, changing to yellowish and reddish brown. Streak nearly white. Transparent to translucent.

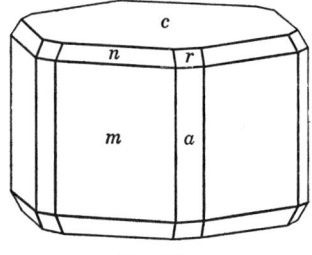

Broddbo.

Opt. In transmitted light, nearly colorless to pale pink. Often seen to be partly altered to material with high birefringence (bastnäsite).

	n(Colorado)[5]	n(Broddbo)[9]	
O	1.612	1.618	Uniaxial negative (−).
E	1.608	1.611	

Chem. A fluoride of cerium and lanthanum, (Ce,La)F_3. Y, Er, Nd, Yt also are present in small amounts. The reported analyses of the material from Colorado and Österby, Sweden, are not in satisfactory agreement probably due to partial alteration of the Österby material.

Anal.[12]

	Ce group	La group	Y group	F	O	H_2O	Rem.	Total	G.
1.	35.65	35.35		29.00				100.00	
2.	40.19	30.37		[29.44]				[100.00]	
3.	36.57	33.55		28.71			1.12	99.95	6.14
4.	39.53	30.82	3.19	19.49	4.43	1.78	1.50	100.74	5.70

1. (Ce,La)F_3 with Ce:La = 1:1. 2. Pikes Peak region, Colorado.[6] Atomic wt. of Ce, etc. = 141.2, of La, etc. = 138. 3. Pikes Peak region, Colorado.[7] Rem. is gangue. At. wt. of La, etc. = 139.7. Ce contains 13 per cent ThO_2? 4. Österby, Sweden.[11] Rem. is $CaCO_3$. Average of four.

Tests. B.B. infusible, darkening in color. Almost unaffected by HCl or HNO_3 but soluble in H_2SO_4.

Occur. Found with bastnäsite in Cheyenne Canyon in the Pikes Peak district, El Paso County, Colorado (*tysonite*).[8] Originally found (fluocerite) in pegmatite at Finbo and Broddbo in Dalarne, Sweden; also with gadolinite and allanite in pegmatite at Österby in Dalarne.

Alter. Commonly altered to bastnäsite, the two minerals oriented in parallel position.[10]

Ref.

1. Angles of Flink, *Ark. Kemi*, **3**, 1 (1910); unit and orientation of Oftedal, *Zs. phys. Chem.*, Abt. B, **13**, 190 (1931). Transformation: earlier settings to Oftedal, 1$\bar{1}$00/01$\bar{1}$0/$\bar{1}$010/0001. See also Koechlin, *Min. Mitt.*, **31**, 528 (1912).

2. Goldschmidt (**4**, 7, 1918).

3. Oftedal (1931); Zs. phys. Chem., Abt. B, **5**, 272 (1929).
4. Nordenskiöld, Ak. Stockholm, Öfv., **27**, 550 (1870).
5. Wolfe, priv. comm. (1945). Larsen (76, 1921) gives for partly altered Österby material a mean index of 1.615, birefringence about 0.002, and optically positive.
6. Allen and Comstock (1880).
7. Hillebrand, Am. J. Sc., **7**, 51 (1899).
8. On the identity of tysonite with fluocerite see Geijer, Geol. För. Förh., **43**, 19 (1921).
9. Geijer (1921).
10. On the structural interpretation of the orientation see Oftedal, Norsk Geol. Tidsskr., **12**, 459 (1931).
11. Weibull and Tedin, Geol. För. Förh., **8**, 496 (1886).
12. For additional analyses see Hintze (1 [2B], 2567, 1915).

9.3.3   **CHLORALUMINITE** [$AlCl_3 \cdot 6H_2O$]. Chloralluminio A. Scacchi (Acc. Napoli, Att., **6**, no. 9, 43, 68, 1874).   Chloraluminit vom Rath (Verh. naturhist. Ver. Rheinland., **34**, 152, 1877).   Chloralluminite Dana (165, 1892).

C r y s t.[1]   Hexagonal—R.

$a:c = 1:0.5356$; $\alpha\ 111°39'$; $p_0:r_0 = 0.6185:1$; $\lambda\ 48°59'$

Forms:

	ϕ	$\rho = C$	A_1	A_2
c 0001	111	....	0°00'	90°00'	90°00'
a 11$\bar{2}$0	10$\bar{1}$	0°00'	90 00	60 00	60 00
r 10$\bar{1}$1	100	30 00	31 44$\frac{1}{2}$	62 54	90 00

Habit.   In crystalline crusts and stalactites.   Natural crystals rhombohedral, artificial prismatic [0001].

P h y s.   Descriptive data are lacking.   Colorless to white or yellowish. Deliquescent.

O p t.[2]   In transmitted light, colorless.

nO 1.560   Uniaxial negative ($-$).
nE 1.507

C h e m.   Aluminum chloride hexahydrate, $AlCl_3 \cdot 6H_2O$.   Al 11.17, Cl 44.06, H_2O 44.77, total 100.00.   Analyses of natural material are lacking.

Tests.   Decomposes at a low temperature.   Very soluble in water.

O c c u r.   Formed with molysite and chloromagnesite in acid fumaroles on Vesuvius, Italy, during the eruption of April, 1872.   Later identified [3] as crusts of small crystals at the same locality in the eruption of 1906.

A r t i f.[4]   Obtained in crystals by evaporation of the water solution or by saturating a cold, concentrated HCl solution with HCl gas.

N a m e.   In allusion to the composition.

Ref.

1. Elements and forms of Gill in Dennis, Zs. anorg. Chem., **9**, 340 (1895), on artificial crystals of $AlCl_3 \cdot 6H_2O$. The identity of the artificial and natural material was shown by Lacroix, Bull. soc. min., **30**, 254 (1907). See also Andress and Carpenter, Zs. Kr., **87**, 446 (1934).
2. Dennis (1895); Takahashi, et al., J. Geol. Soc. Tokyo, **35**, 439 (1928).
3. Lacroix (1907).
4. Mellor (**5**, 314, 316, 1924).

10 OXYHALIDES AND HYDROXYHALIDES

TYPE 1. $A_m(O,OH)_pX_q$

10.1.1 EGLESTONITE [Hg_4OCl_2]. *Moses (Am. J. Sc., 16, 253, 1903).*

Cryst. Isometric; hexoctahedral—$4/m\,\overline{3}\,2/m$.

Forms: [1]

| a 001 | d 011 | o 111 | ϕ 116 | n 112 | r 233 | s 123 |

Less common or rare:

| f 013 | e 012 | ρ 144 | p 122 | i 189 | j 167 | w 156 |
| k 146 | F 126 | Σ 145 | v 4.7.11 | l 347 | M 234 | |

Structure cell.[2] Space group probably $Ia3d$. a_0 16.00 ± 0.04 kX. Cell contents $Hg_{96}O_{24}Cl_{48}$.

Habit. Dodecahedral, with striations parallel to the edges, and sometimes elongated on [001] into almost hair-like forms. Less commonly octahedral or cubic. Also massive, as crusts.

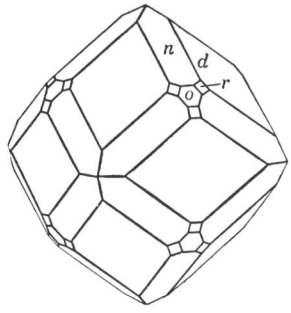
Terlingua.

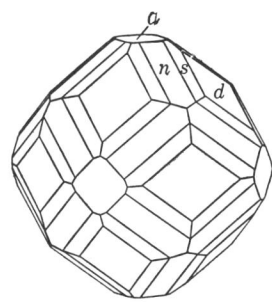
Terlingua.

Phys. Fracture uneven to conchoidal. Brittle. H. $2\frac{1}{2}$. G. 8.33,[3] 8.45;[4] 8.61 (calc.). Luster brilliant adamantine to resinous. Color yellow and orange-yellow to light brownish yellow and dark brownish, becoming dark brown and finally black on exposure to light. Streak yellow or greenish yellow, rapidly becoming black. Translucent.

Opt. In transmitted light, yellowish brown to brown. Isotropic, sometimes with weak anomalous birefringence. Refractive index:[5] 2.49 ± 0.02 (Li).

Chem. An oxychloride of mercury, Hg_4OCl_2.
Anal.[6]

	1	2
Hg	90.23	89.00
O	1.80	1.79
Cl	7.97	8.22
Total	100.00	99.01

1. Hg_4OCl_2. 2. Terlingua.[7] Average of three.

Tests. In C.T. decrepitates, becomes orange then black and finally dark red (HgO), and gives a white sublimate of calomel with mercury beyond. Blackened at once by hydrogen sulfide or ammonia. Decomposed by acids with separation of calomel.

Occur. From the mercury deposit at Terlingua, Brewster County, Texas, with calomel, calcite, native mercury, and, rarely, terlinguaite and montroydite. Commonly found upon calomel and has been derived by its oxidation.[8] Also noted in small amounts with cinnabar in Pike County, Arkansas,[9] and with cinnabar, native mercury, and calomel from a locality about five miles west of Palo Alto, San Mateo County, California.[10] In the Monarch cinnabar mine, Transvaal, South Africa.

Alter. Alters to a substance gray-black on the surface but yellow and dull within and containing both Cl and O much in excess of the ratios of eglestonite.[11]

Name. After Thomas Egleston (1832–1900), Professor of Mineralogy and Metallurgy in Columbia University.

Ref.
    1. Moses (1903); Hillebrand and Schaller, *U. S. Geol. Sur., Bull. 405* 143 (1909); Rogers, *Am. J. Sc.*, **32**, 48 (1911).
    2. Wolfe, priv. comm. (1945), by powder and Weissenberg methods on Terlingua crystals. Bird, *Am. Min.*, **17**, 541 (1932), found a simple cell with a_0 9.50 by the powder method, his powder spacing data and that of Wolfe not agreeing. Hedlik, *Experientia*, Basel, **4**, 66 (1948), gives a_0 8.02 Å and space group $Im3m$.
    3. Average of 8.309, 8.345 of Moses (1903), by pycnometer on Terlingua material.
    4. Wolfe (1945) by microbalance on Terlingua crystals.
    5. Larsen (71, 1921).
    6. Earlier analyses reported by McCord in Moses (1903) are less accurate and led to the erroneous formula $Hg_6O_2Cl_3$.
    7. Hillebrand analysis in Hillebrand and Schaller (1909).
    8. Broderick, *Econ. Geol.*, **11**, 645 (1916).
    9. Sohlberg, *Am. Min.*, **18**, 1 (1933).
    10. Rogers, *Am. J. Sc.*, **32**, 48 (1911).
    11. Hillebrand and Schaller (1909).

10.1.2 TERLINGUAITE [Hg_2OCl]. Moses (*Am. J. Sc.*, **16**, 255, 1903).

Cryst. Monoclinic; prismatic—$2/m$.

$a:b:c = 2.0245:1:1.6022;$ $\beta\ 105°37\frac{1}{2}';$ $p_0:q_0:r_0 = 0.7914:1.5430:1$

$r_2:p_2:q_2 = 0.6481:0.5129:1;$ $\mu\ 74°22\frac{1}{2}';$ $p_0'\ 0.8321,\ q_0'\ 1.6022,\ x_0'\ 0.2796$

TERLINGUAITE 53

Forms:[2]

		ϕ	ρ	ϕ_2	$\rho_2 = B$	C	A
c	001	90°00′	15°37½′	74°22½′	90°00′	......	74°22½′
a	100	90 00	90 00	0 00	90 00	74°22½′	......
m	110	27 26½	90 00	0 00	27 26½	82 52	62 33½
h	510	68 56½	90 00	0 00	68 56½	75 26½	21 03½
δ	012	19 14½	40 19	74 22½	52 21	37 39	77 41½
d	011	9 54	58 25	74 22½	32 57	57 03	81 34½
j	021	4 59	72 44	74 22½	17 57½	72 02½	85 14½
t	201	90 00	62 46½	27 13½	90 00	47 09	27 13½
x	$\bar{1}$02	−90 00	7 46	97 46	90 00	23 23½	97 46
Q	$\bar{2}$03	−90 00	15 23	105 23	90 00	31 00½	105 23
u	$\bar{1}$01	−90 00	28 55	118 55	90 00	44 32½	118 55
n	$\bar{2}$01	−90 00	54 09½	144 09½	90 00	69 47	144 09½
i	112	40 58½	46 41½	55 10½	56 40	37 54	61 30
p	111	34 45½	62 51	41 58½	43 01½	54 49½	59 31
Z	221	31 14½	75 03½	27 13½	34 18	67 28	59 55½

Also frequent:

b 010	L $\bar{4}$01	l $\bar{3}$34	X $\bar{3}$31	λ 511	I $\bar{1}$31
D 130	s 113	Δ $\bar{3}$32	T 312	α $\bar{3}$13	A $\bar{7}$12
f 310	g $\bar{1}$12	ψ $\bar{2}$21	π 211	O $\bar{3}$11	β $\bar{5}$11

Less common:

l: 710	Π $\bar{5}$06	ϵ 331	ϕ $\bar{4}$41	C $\bar{3}$14	N: $\bar{9}$15
B 023	y: $\bar{4}$03	ζ 441	σ $\bar{5}$51	Q $\bar{4}$25	J $\overline{15}$·1·6
G 104	N $\bar{3}$02	q: 551	θ 212	Υ $\bar{5}$15	t: $\bar{5}$12
y 101	x: $\bar{5}$03	G: 10·10·1	U: 733	Ω $\bar{2}$12	A: $\overline{11}$·1·4
p: 301	W $\bar{3}$01	γ $\bar{1}$14	Σ 421	D $\bar{5}$14	V $\bar{9}$13
n: 501	M: 335	o $\bar{1}$13	X: 621	E: $\bar{5}$34	ρ $\bar{3}$11
S $\bar{1}$04	χ 223	e $\bar{1}$11	Φ 821	Z: $\bar{4}$23	H $\bar{4}$11
z $\bar{1}$03	r 334	Γ $\bar{4}$43	K $\bar{1}$36	ξ $\bar{3}$12	ι $\bar{4}$21
P $\bar{3}$04	Y 332	L: $\bar{5}$52	E $\bar{3}$15	q $\bar{5}$13	

Structure cell.[3] Space group $C2/m$. a_0 11.63 kX, b_0 5.76, c_0 9.28; β 105°37′; $a_0:b_0:c_0 = 2.019:1:1.611$. Cell contents $Hg_{14}O_7Cl_7$.

Habit. Prismatic [010] and often somewhat flattened on {001}; rarely prismatic in a pyramidal direction; also equidimensional, or thick tabular {001}. Found also in powdery form, and as massive aggregates of imperfect crystals.

Phys. Cleavage {$\bar{1}$01} perfect. Brittle. H. 2½. G. 8.725;[4] 8.73 (calc.). Luster brilliant adamantine. Color of crystals sulfur-yellow to greenish yellow and rarely brown, of powder lemon-yellow, both becoming olive-green on exposure to light. Transparent to translucent.

Opt.[5] In transmitted light, pale olive-green. Slightly pleochroic in green and yellow.

	n(Li)	
X	2.35 ± 0.02	Biaxial negative (−).
Y	2.64 ± 0.02	$2V$ $20° \pm 2°$ (red).
Z	2.66 ± 0.02	$r < v$, extreme.
		Axial plane parallel b and inclined $-7°$ to c.

HALIDES

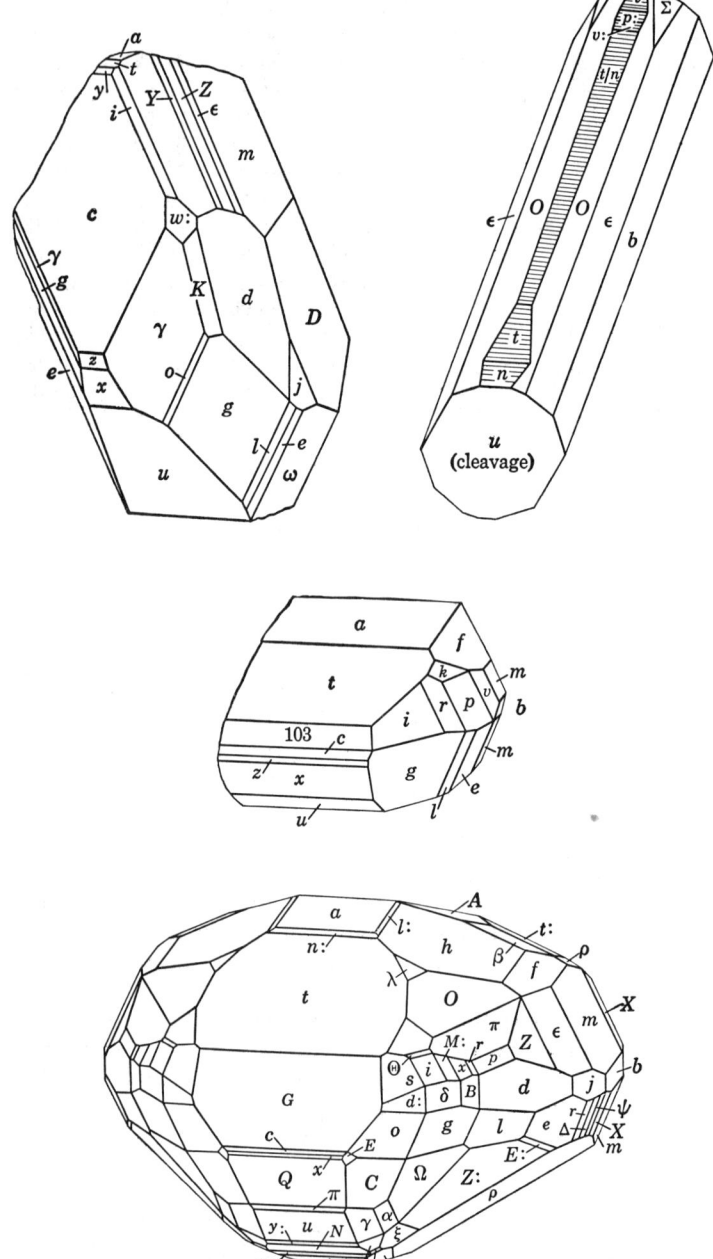

Chem. An oxychloride of mercury, Hg_2OCl. Contains both monovalent and divalent mercury.

Anal.

	1	2	3
Hg	88.63	88.24	88.61
O	3.53	3.47	3.75
Cl	7.84	7.89	7.83
Total	100.00	99.60	100.19
G.	8.73	8.725	

1. Hg_4OCl_2. 2. Terlingua.[6] 3. Terlingua.[7] Recalculated after deducting 3.14 per cent gangue. Analyses 2 and 3 are average of several.

Tests.[8] Heated rapidly in C.T. decrepitates violently and becomes red-brown. The resulting powder yields a sublimate of calomel and mercury with oxide of mercury remaining which on further heating is volatilized. Heated slowly decrepitates only slightly and yields at first a sublimate of mercury, then only calomel. Treated by hydrogen sulfide becomes black at once, with ammonia turns black slowly, differing in these reactions from kleinite. Eglestonite treated with ammonia becomes black at once. Decomposed with separation of calomel by acetic, hydrochloric, and nitric acids, hydrochloric acid producing no precipitate in the filtrate from the acetic acid solution in distinction to eglestonite.

Occur. Found with eglestonite, cinnabar, kleinite, montroydite, calomel, native mercury, at the mercury locality at Terlingua, Brewster County, Texas.

Artif.[9] A compound with the empirical composition of terlinguaite has been obtained by heating equivalent weights of mercuric oxide and calomel with water at 180°.

Name. From the locality.[10]

Ref.

1. Unit of Schaller in Hillebrand and Schaller, *U. S. Geol. Sur., Bull.* **405**, 83 (1909) and confirmed by x-ray study. The a and c axes of Schaller's A-centered cell are interchanged to conform with the convention of employing a C-centered cell. Transformation: Schaller to C-centered cell, $001/0\bar{1}0/100$; Moses (1903) to new orientation, $001/0\bar{1}0/300$. The figures are drawn in Schaller's orientation.

2. Schaller in Hillebrand and Schaller (1909). Rare or uncertain:

540	032	803	$\bar{1}2.0.13$	661	413	$\bar{1}8.7.27$
530	1.0.12	v: 401	$\bar{5}04$	881	k 433	$\bar{5}36$
210	109	$\bar{1}.0.12$	$\bar{7}01$	$\bar{2}25$	623	$\bar{4}33$
940	105	$\bar{1}07$	115	$\bar{2}23$	713	$\bar{1}8.2.9$
520	103	$\bar{1}05$	114	$\bar{2}5.25.33$	512	$\bar{6}12$
13.3.0 or 920	305	$\bar{2}05$	11.11.21	w 553	823	$\bar{1}1.1.3$
016	304	$\bar{1}0.0.21$	7.7.12	$\bar{7}76$	14.4.3	
d: 038	403	$\bar{8}.0.15$	443	236	$\bar{2}36$	
w: 025	r: 302	$\bar{3}05$	v 553	131	$\bar{4}18$	

3. Wolfe, priv. comm. (1945), by Weissenberg method.
4. Average of 8.728 and 8.723 by pycnometer method by Moses (1903).
5. Larsen (143, 1921) and Hillebrand and Schaller (1909).
6. McCord analysis in Moses (1903).
7. Hillebrand analysis in Hillebrand and Schaller (1909).
8. Hillebrand and Schaller (1909).
9. Fischer and von Wartenberg, *Chem.-Ztg.*, **29**, 308, (1905).

10. The name terlinguaite had been applied earlier by Turner, *Min. and Sc. Press*, July (1900), to several nondifferentiated mercury minerals from Terlingua. The species was first established by Moses (1903).

10.1.3 **LORETTOITE** [$Pb_7O_6Cl_2$?]. *Wells* and *Larsen* (*J. Wash. Ac. Sc.*, **6**, 669, 1916). Chubutite *Corti* (*Anales soc. quim. argentina*, **6**, 65, 1918).

Probably tetragonal. In compact masses of coarse fibers or blades, with a perfect (basal) cleavage parallel to the blades. H. $2\frac{1}{2}$–3. G. 7.39 (Loretto),[1] 7.65 (loc. unknown),[1] 7.95 (Argentina).[2] Color honey-yellow and yellow to reddish yellow. Streak pure yellow. Luster adamantine.

Opt.[3]

	n Li (Loretto)	n Li (loc. unknown)	
O	2.40 ± 0.02	2.35 ± 0.02	Uniaxial negative (−).
E	2.37 ± 0.02	2.33 ± 0.02	

Chem. An oxychloride of lead of uncertain formula, perhaps $Pb_7O_6Cl_2$.

Anal.

	1	2	3	4
PbO	82.80	83.72	80.62	84.90
$PbCl_2$	17.20	16.28	19.38	15.10
Total	100.00	100.00	100.00	100.00
G.		7.39	7.65	7.952

1. $Pb_7O_6Cl_2$. 2. Loretto, Tennessee.[4] Analysis recalculated to 100 after deducting MgO 0.56, CaO 0.48, ZnO 0.31, Al_2O_3 0.08, P_2O_5 0.11, CO_2 0.20, insol. 0.58, H_2O 0.03. 3. Locality unknown.[4] Values calculated from single determination of Cl (4.09 per cent). 4. Chubut, Argentina.[2] Analysis recalculated to 100 after deducting Fe_2O_3 0.19, Sb_2O_3 0.69, Al_2O_3 0.13, SiO_2 0.76.

Tests. Easily fusible (F = 1). Soluble in hot dilute nitric acid, more slowly in hydrochloric and sulfuric acid, and not appreciably soluble in water.

Occur. With cerussite and galena (?) at Loretto, Tennessee. An apparently identical mineral was found on a single specimen from an unknown locality,[5] and another mineral almost certainly identical with lorettoite was briefly described from the Cerro de las Coronas, Chubut, Argentina (*chubutite*). None of these materials can be said to be well-defined.

Artif. The compounds $Pb_7O_6Cl_2 \cdot 2H_2O$ and $Pb_8O_7Cl_2$ have been reported.[6]

Name. From the locality.

Ref.

1. Wells and Larsen (1916); the value is probably low.
2. Corti (1918).
3. Larsen in Wells and Larsen (1916).
4. Wells analysis in Wells and Larsen (1916).
5. Larsen and Wells (1916).
6. Mellor (**7**, 742, 1927).

10.1.4 **MENDIPITE** [$Pb_3O_2Cl_2$]. Lead-Ore, flaky and striated ... *Woodward* (**1**, Pt. 1, 214, 1729). Minera plumbi calciformis (?) *Cronstedt* (1758). Saltsyradt Bly (Salzsaures Blei) *Berzelius* (*Ak. Handl. Stockholm*, 184, 1823; *Edinburgh J. Sc.*,

MENDIPITE

1, 379, 1824). New ore of lead from Mendip, Peritomous Lead-baryte *Haidinger* (Mohs, **2**, 151, 1825). Muriate of Lead, Chloride of Lead. Plomb chloruré pt. *Fr.*, Kerasine pt. [rest phosgenite] *Beudant* (**2**, 502, 1832). Cerasite. Chlor-Spath *Breithaupt* (61, 1832). Berzélite *Lévy* (**2**, 448, 1837). Mendipite *Glocker* (604, 1839). Churchillite *Dufrénoy* (**3**, 280, 1856). Pseudomendipite *Rimann* (*Anal. soc. quim. argentina*, **6**, 326, 1918).

C r y s t.[1] Orthorhombic.

$$a:b = 0.8002:1$$

Forms:[2]

$c\ 001 \quad b\ 010 \quad a\ 100 \quad m\ 110 \quad d\ 101$

Structure cell.[3] a_0 9.50 kX, b_0 11.87, c_0 5.87; $a_0:b_0:c_0 = 0.8003:1:0.4952$. Cell contents $Pb_{12}O_8Cl_8$.

Habit. As fibrous or columnar masses, often radiated.

P h y s. Cleavage {110} perfect, {100} and {010} less perfect. Fracture conchoidal to uneven. H. $2\frac{1}{2}$. G. 7.24;[4] 7.22 (calc.). Colorless to white and gray, often with a tinge of yellow, red, or blue. Streak white. Transparent to feebly translucent. Luster pearly to silky on cleavage surfaces, resinous to adamantine on cross-fractures.

O p t.[5] In transmitted light, nearly colorless.

ORIENTATION		n	
X	a	2.24 ± 0.02	Biaxial positive (+).
Y	b	2.27 ± 0.02	$2V$ nearly 90°.
Z	c	2.31 ± 0.02	$r < v$, very strong.
			Elongation positive.

C h e m. An oxychloride of lead, $Pb_3O_2Cl_2$.

Anal.[6]

	1	2	3
Pb	85.79	85.87	85.69
O	4.42	4.53	[4.44]
Cl	9.79	9.35	9.87
Total	100.00	99.75	[100.00]
G.		7.22	7.24

1. $Pb_3O_2Cl_2$. 2. Higher Pitts, Mendip.[7] 3. Westphalia.[8]

Tests. In C.T., becomes yellow and decrepitates into fibers. Before these melt lead chloride vapor fills the tube and condenses to a white sublimate; the fibers then melt to a yellow liquid (lead oxide) which spreads within the tube. Soluble in dilute nitric acid.

O c c u r. Found in the Mendip Hills, Somersetshire, England, near Churchill and at Higher Pitts farm nine miles southeast of Churchill. It occurs as nodular masses in wad and is associated with cerussite, hydrocerussite, malachite, pyromorphite, calcite, chloroxiphite, diaboleite. Mendipite also occurs at the Kunibert mine, near Brilon, Westphalia, Germany. Reported from the Altai Mountains, Siberia.[9]

A r t i f.[10] Reported in the system $PbO-PbCl_2$; also by adding KOH solution to a suspension of $PbCl_2$ in water, or by shaking $Pb(OH)_2$ with a KCl solution.

N a m e. From the original locality.

Ref.

1. Partial axial ratio from measurements of Spencer, *Min. Mag.*, **20,** 67 (1923), on cleavage fragments of Higher Pitts material.
2. In Goldschmidt (**6,** 29, 1920).
3. Bannister, *Min. Mag.*, **23,** 596 (1934).
4. Spencer (1923) by pycnometer method on Mendip material; earlier reported values range from 7.0 to 7.4.
5. Larsen (108, 1921) by immersion method on material from Westphalia; see also Spencer (1923).
6. For earlier analyses see Spencer (1923).
7. Mountain analysis in Spencer (1923).
8. Schnabel, *Ann. Phys.*, **71,** 516 (1847).
9. Pilipenko, *Bull. Imp. Tomsk Univ.*, no. 63 (1915)—*Min. Abs.*, **2,** 111 (1923).
10. Mellor (**7,** 740, 1927).

10.1.5 DAVIESITE. Fletcher (*Min. Mag.*, **8,** 174, 1889).

Cryst.[1] Orthorhombic.

$$a:b:c = 0.7940:1:0.4778; \qquad p_0:q_0:r_0 = 0.6018:0.4778:1$$

$$q_1:r_1:p_1 = 0.7940:1.6618:1; \qquad r_2:p_2:q_2 = 2.0929:1.2594:1$$

Forms:

	ϕ	$\rho = C$	ϕ_1	$\rho_1 = A$	ϕ_2	$\rho_2 = B$
c 001	....	0°00′	0°00′	90°00′	90°00′	90°00′
b 010	0°00′	90 00	90 00	90 00	....	0 00
m 110	51 33	90 00	90 00	38 27	0 00	51 33
f 011	0 00	25 32½	25 32½	90 00	90 00	64 27½
g 031	0 00	55 06	55 06	90 00	90 00	34 54
d 101	90 00	31 02½	0 00	58 57½	58 57½	90 00
e 301	90 00	61 01	0 00	28 59	28 59	90 00

Also: h 051   v 221   s 121   t 211   r 251

Habit. Prismatic [001]; also thick tabular {010}, with {010} sometimes striated vertically.

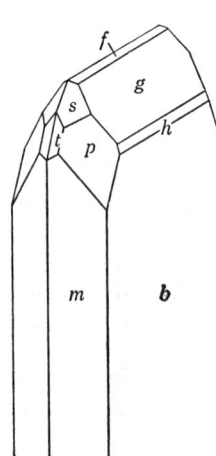

Phys. Cleavage possibly present on {011}. Fracture subconchoidal. G. > 3. Luster subadamantine. Colorless and transparent. Optically biaxial with $Y = b$, $Z = c$.

Chem. A lead oxychloride of unknown composition. Not affected by water. B. B. easily fusible. Readily soluble in HNO_3.

Occur. Found associated with caracolite and percylite in cavities in massive anglesite at the Mina Beatriz, Sierra Gorda, Atacama, Chile.

Name. After Thomas Davies (1837–1892), Assistant in the Department of Mineralogy of the British Museum.

Ref.

1. Angles and unit of Fletcher (1889); orientation of Goldschmidt (**3,** 30, 1916). Transformation: Fletcher to Goldschmidt, 010/100/001.

10.1.6 MATLOCKITE GROUP

TETRAGONAL; DITETRAGONAL-DIPYRAMIDAL—$4/m\ 2/m\ 2/m$

	a_0	c_0
Matlockite, PbFCl	4.09 kX	7.21
Bismoclite, BiOCl	3.89	7.37
Daubréeite, BiO(OH,Cl)	3.85	7.40 [OH:Cl = 1:1]

The members of the Matlockite Group together with the artificial compounds PbFBr, BiOBr, and BiOI are layer structures.[1] The structures are based on metal-oxygen or metal-fluorine layers in which the anions are in quadratic packing with the metal ions fitting into the interstices between the anions, half on each side of the layer. These layers are separated by double layers of Cl or (Cl,OH) each in quadratic packing which coordinate the metal ions of the first-mentioned layers. The perfect basal cleavage (breaking the double layer of Cl ions) and the optically negative character are consequences of the structural arrangement. It may be noted that BiOCl-BiO(OH,Cl) series is incomplete and that the end-member of the series, BiO(OH), is not known. The structure of laurionite, Pb(OH)Cl, and probably also that of its polymorph, paralaurionite, is near that of matlockite.

Ref.

1. Bannister and Hey, *Min. Mag.*, **23**, 587 (1934) and **24**, 49 (1935); Sillén, *Naturwiss.*, **22**, 309 (1942).

10.1.6.1 **M A T L O C K I T E** [PbFCl]. *Greg* (*Phil. Mag.*, **2**, 120, 1851); *Rammelsberg* (*Ann. Phys.*, **85**, 141, 1852).

C r y s t. Tetragonal; ditetragonal-dipyramidal—$4/m\ 2/m\ 2/m$.

$a:c = 1:1.763;$[1] $p_0:r_0 = 1.763:1$

Forms:[2]

	ϕ	ρ	A	$\overline{M}$
c 001	.....	0°00′	90°00′	90°00′
m 110	45°00′	90 00	45 00	90 00
e 011	0 00	60 26	90 00	52 02½
r 111	45 00	68 08½	48 59	90 00

Structure cell.[3] Space group $P4/nmm$. a_0 4.09 kX, c_0 7.21; $a_0:c_0 = 1:1.763$. Cell contents $Pb_2Cl_2F_2$.

Habit. Usually tabular {001}; also as stumpy pyramidal crystals with small prism faces. In subparallel aggregates of platy crystals, and as hemispherical, rosette-like groups. Massive, coarsely lamellar.

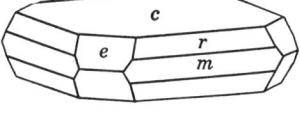

Cromford.

P h y s. Cleavage {001} perfect. Fracture uneven to subconchoidal. Thin cleavage flakes are brittle. H.2½–3. G. 7.12;[4] 7.16 (calc.). M.P. 601°.[5] Luster adamantine, inclining to pearly on {001}. Colorless or yellow to pale amber; greenish. Transparent.

O p t.[6] In transmitted light, colorless. Uniaxial negative (−). Sometimes biaxial with small $2V$ (due to subparallel growth or strain). Indices:

λ	656.3	589.3	535.0	486.1	425.0
nO	2.127	2.145	2.164	2.191	2.24
nE	1.994	2.006	2.021	2.039	2.07

C h e m. Lead chloride-fluoride, PbFCl. Formerly considered to be an oxychloride of lead, the early analysts [7] having overlooked the fluorine.

Anal.

	1	2	3
Pb	79.24	79.55	78.92
F	7.21	7.11	7.25
Cl	13.55	13.44	13.57
Total	100.00	100.10	99.67
G.	7.16	7.05	

1. PbFCl. 2. Cromford, Derbyshire.[8] 3. Mammoth mine, Arizona.[10]

Tests. B.B. easily fusible. Soluble in HNO_3. Decomposed by H_2SO_4 with formation of $PbSO_4$.

O c c u r. Originally found at Cromford near Matlock in Derbyshire, England, associated with phosgenite, anglesite, cerussite, galena, barite, sphalerite, and fluorite. Reported from Challacolla, Tarapacá, Chile, with percylite. Also reported from Laurium, Greece, with laurionite, fiedlerite, and georgiadesite in altered slag. In the United States at the Mammoth mine, Tiger, Arizona, as lamellar pieces several inches across the cleavage face, associated with diaboleite, cerussite, boleite, leadhillite.

A l t e r. To cerussite (Arizona).

A r t i f.[9] Obtained from melts of $PbCl_2$ and PbF_2 and by reaction in water solution.

N a m e. From the locality near Matlock, Derbyshire.

Ref.
1. Miller in Greg (1851).
2. Goldschmidt (**6**, 12, 1920).
3. Bannister, *Min. Mag.*, **23**, 587 (1934), by rotation and Laue methods on material from Cromford, with analysis of structure.
4. Frondel, priv. comm. (1942), by microbalance on material from Cromford.
5. Sandonnini, *Gazz. chim. ital.*, **41**, 144 (1911).
6. Bannister (1934); indices by prism method on Cromford material.
7. Smith anal. in Greg (1851), and Rammelsberg (1852). The true composition was first shown by Nieuwenkamp, *Zs. Kr.*, **86**, 470 (1933) who reported 6½ per cent fluorine.
8. Hey anal. in Bannister (1934).
9. Mellor (**7**, 732, 1927).
10. Gonyer, priv. comm. (1949).

10.1.6.2 **B I S M O C L I T E** [BiOCl]. Daubréeite (?) Means (*Am. J. Sc.*, **41**, 125, 1916). Bismoclite *Mountain* (*Min. Mag.*, **24**, 59, 1935).

10.1.6.3 **D A U B R É E I T E** [BiO(OH,Cl)]. Daubreite Domeyko (*C.R.*, **82**, 922, 1876; 297, 1879). Daubreeite.

C r y s t.[1] Tetragonal; ditetragonal-dipyramidal—$4/m\ 2/m\ 2/m$.

Structure cell. [2]

	Space group	a_0	c_0	c_0/a_0	Cell contents
Bismoclite	$P4/nmm$	3.89 ± 0.01 kX	7.37 ± 0.02	1.895	$Bi_2O_2Cl_2$
Daubréeite	$P4/nmm$	3.85 ± 0.01 kX	7.40 ± 0.02	1.922	$Bi_2O_2(OH,Cl)_2$ with OH:Cl $\sim$ 1:1

Habit. Found only massive; compact, earthy, columnar to platy-fibrous; as minute scaly crystals (*bismoclite*, Nevada). Artificial crystals of bismoclite are square plates {001} with a very flat vicinal pyramid.

P h y s. Cleavage {001} perfect. Very plastic. H. 2–2½. G. 7.717 (artif. BiOCl),[3] 7.70 (calc.); 7.56 (calc. for daubréeite with OH:Cl $\sim$ 1:1). Luster greasy to silky, pearly on cleavage surfaces; dull to earthy in massive material. Color creamy-white, grayish, yellowish brown. Transparent in small grains.

O p t. In transmitted light, colorless.

	nO	
Bismoclite (artif. BiOCl) [3]	2.15	Uniaxial negative ($-$).
Daubréeite (Bolivia) [4]	1.91 ± 0.01	

C h e m. Bismoclite is bismuth oxychloride, BiOCl. Daubréeite is bismuth oxy-hydroxychloride, BiO(OH,Cl). OH substitutes for Cl in artificial material, and a series extends to at least OH:Cl $\sim$ 1:1, to which ratio anal. 3, daubréeite, approximates. The compound BiO(OH) is not known.[5] The name daubréeite is here applied to that part of the presumed natural series with OH > Cl (although it is not certain whether the only reported analysis actually has OH > Cl), and bismoclite to material with OH < Cl.

Anal.

	1	2	3	4
Bi_2O_3	89.41	88.49	89.60	92.74
Cl	13.67	13.00	7.50	7.06
H_2O		0.87	3.84	1.79
Rem.		0.89	0.72	
	103.08	103.25	101.66	101.59
Less O = Cl	3.08	2.93	1.69	1.59
Total	100.00	100.32	99.97	100.00
G.	7.70		6.4–6.5	7.56

1. BiOCl. 2. Bismoclite. Namaqualand.[6] Rem. is Fe_2O_3 0.12, PbO tr., insol. 0.77. H_2O = $+110°$ 0.45, $-110°$ 0.42. 3. Daubréeite. Bolivia.[7] Rem. is Fe_2O_3. 4. BiO(OH,Cl) with OH:Cl = 1:1.

Tests. In C.T. may yield a small amount of acid moisture. On further heating turns yellow or orange and gives a white cloudy sublimate which collects into lemon-yellow drops; melts at red heat and on cooling solidifies into a lemon-yellow (*bismoclite*) or black (*daubréeite*) mass. Completely soluble in acids but re-precipitated on considerable dilution.

O c c u r.[5] Secondary minerals, formed by the alteration of bismuthinite or native bismuth. Bismoclite was first described from near Jackals Water, 18 miles NNE of Steinkopf, Namaqualand, South Africa, with bismutite in pegmatite. At Bygoo, New South Wales, as an alteration of bismuthinite in greisen. Bismoclite occurs in the Eagle and Blue Bell mines, Tintic district, Utah, with jarosite, alunite, cerussite, and bismutite. The so-called rhombohedral bismite from Goldfield, Nevada, associated with iodyrite, also is bismoclite. Daubréeite is found admixed with clay at the Constancia mine, Tazna, Bolivia.

A r t i f.[3] Bismoclite is obtained in minute crystals by the slow hydrolysis of $BiCl_3$ in HCl; $BiO(Cl,OH)$ by precipitation of $BiCl_3$ solution with ammonia.

N a m e. Bismoclite in allusion to the composition. Daubréeite after Gabriel Auguste Daubrée (1814–1896), French mineralogist and geologist.

Ref.

1. Symmetry established by x-ray study by Bannister, *Min. Mag.*, **24**, 49 (1935).
2. Bannister (1935) by rotation, oscillation, and powder methods on artificial bismoclite and on natural daubréeite.
3. Bannister (1935).
4. Larsen (66, 1921), on material presumed to be the same as anal. 3; birefringence about 0.01.
5. Cf. Frondel, *Am. Min.*, **28**, 526 (1943).
6. Mountain, *Min. Mag.*, **24**, 59 (1935).
7. Domeyko (1876).

10.1.7   L A U R I O N I T E [Pb(OH)Cl]. Koechlin (*Ann. naturhist. Hofmus. Wien*, **2**, 188, 1887; *Zs. Kr.*, **17**, 112, 1889).

C r y s t.   Orthorhombic; dipyramidal—$2/m\ 2/m\ 2/m$.

$a:b:c = 2.4084:1:1.7668;$[1] $p_0:q_0:r_0 = 0.7336:1.7668:1$

$q_1:r_1:p_1 = 2.4084:1.3631:1;$ $r_2:p_2:q_2 = 0.5660:0.4152:1$

Forms:[2]

	ϕ	$\rho = C$	ϕ_1	$\rho_1 = A$	ϕ_2	$\rho_2 = B$
c 001	.....	0°00′	0°00′	90°00′	90°00′	90°00′
a 100	90°00′	90 00	.....	0 00	0 00	90 00
d 110	22 33	90 00	90 00	67 27	0 00	22 33
k 101	90 00	36 15½	0 00	53 44½	53 44½	90 00
n 201	90 00	55 43½	0 00	34 16½	34 16½	90 00
q 211	39 42½	66 28	60 29½	54 09	34 16½	45 08½
r 311	51 14½	70 29½	60 29½	42 41½	24 26	53 50
p 812	73 14½	71 55½	41 27½	24 27	18 49	74 05½

Less common and rare:

310   102   302   111   312   411   511

Structure cell.[3] Space group $Pcmn$. a_0 9.7 kX, b_0 4.05, c_0 7.1; $a_0:b_0:c_0 = 2.39:1:1.75$. Cell contents $Pb_4(OH)_4Cl_4$. Related structurally to matlockite, PbFCl.

Habit. Elongated [010]; thick to thin tabular {100}. {100} sometimes striated [0$\bar{2}$1]. The important form {812} is generally replaced or accompanied by vicinal faces. Common forms: $a\ k\ d\ n\ p$.

LAURIONITE

Phys. Cleavage {101} distinct. Not brittle. H. 3–3½. G. 6.24 (artif.);[4] 6.14 (calc.). Luster adamantine, on {100} pearly. Colorless to white.

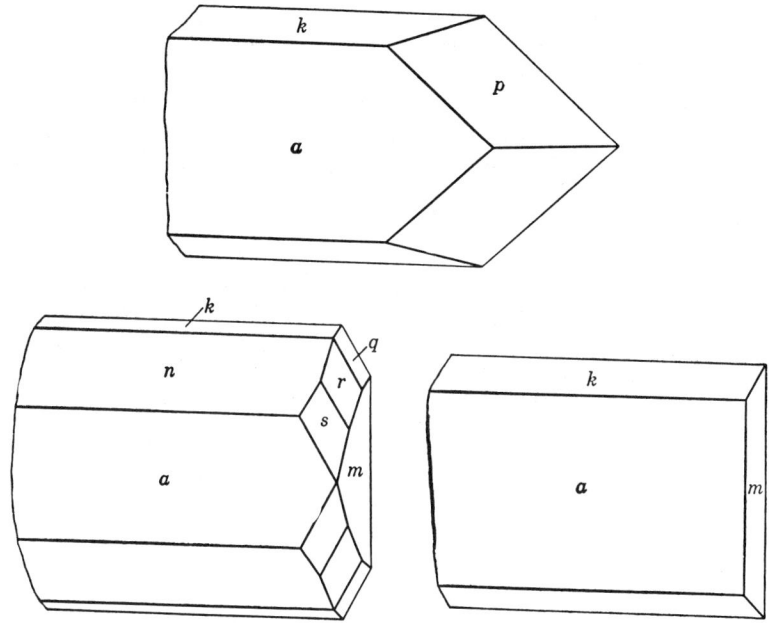

Opt. In transmitted light, colorless.

ORIENTATION		n [5]	n(Na) [6]	
X	c	2.08 ± 0.01	2.077	Biaxial, probably
Y	a		2.116	negative (−).
Z	b	2.16 ± 0.01	2.158	2V large.

Chem. Lead hydroxide-chloride, $Pb(OH)Cl$. Dimorphous with paralaurionite. The water is lost at about 142°.

Anal.

	1	2	3
Pb	79.80	79.38	79.86
O	3.08	3.17	[3.23]
Cl	13.65	13.77	13.57
H_2O	3.47	3.68	3.34
Total	100.00	100.00	[100.00]

1. $Pb(OH)Cl$. 2. Laurium.[7] 3. Artificial.[4]

Tests. B.B. easily fusible to a yellowish opaque bead. Soluble in nitric acid; very slightly soluble in cold water, more so in hot water.

Occur. Found in lead slags that have been exposed to the action of the sea at Laurium, Greece, associated with paralaurionite, penfieldite, fiedlerite, phosgenite, cerussite, anglesite, and other secondary lead minerals. Paralaurionite is the most common immediate associate. The

Laurium mines were worked extensively by the Athenians at the time of Pericles for their content of lead and especially silver. The ore, which contained principally galena, was largely smelted on the shore of the sea, in which the slags were dumped. Laurionite with paralaurionite, phosgenite, and anglesite also has been found in oxidized ore in the mine Wheal Rose, near Sithney, Cornwall.

Artif. Obtained in crystals by heating a saturated water solution of $PbCl_2$ in a sealed tube at 100°–130°. Also in cold solutions by the action of NaCl on excess lead acetate. The compounds Pb(OH)Br and Pb(OH)I also are known.

Name. From the locality.

Ref.

1. Palache, *Min. Mag.*, **23**, 573 (1934), combining the data of previous observers with new measurements on Laurium crystals. Position of Ktenas, *Bull. soc. min.*, **33**, 173 (1910). Transformations: Koechlin (1887) to Palache, $010/00\frac{1}{2}/100$; vom Rath, *Sitzber. niederrhein. Ges., Bonn*, 102, 149 (1887), to Palache, $100/00\frac{1}{8}/0\frac{2}{6}0$. Dana (171, 1892), Goldschmidt (**5**, 124, 1918), Goldsztaub, *C.R.*, **204**, 702 (1937), and Hintze (**1** [2B], 2632, 1915) use the position of Koechlin. The c-axis of Goldsztaub is one-half that of Koechlin, and the transformation to Palache is 010/001/100.
2. Goldschmidt (1918) and Palache (1934). Uncertain forms: 104, 203, 403, 601, 912, 10.1.2.
3. Goldsztaub (1937); *C.R.*, **208**, 1234 (1939); Brasseur, *Bull. soc. roy. sc. Liége*, **9**, 166 (1940).
4. de Schulten, *Bull. soc. min.*, **20**, 186 (1897).
5. Berman in Palache (1934).
6. Smith, *Min. Mag.*, **12**, 106 (1899).
7. Bettendorff in vom Rath (1887).

10.1.8 **PARALAURIONITE** [Pb(OH)Cl]. Smith (*Min. Mag.*, **12**, 102, 1899). Rafaëlit *Arzruni* and *Thaddéeff* (*Zs. K.*, **31**, 229, 1899).

Cryst.[1] Monoclinic; prismatic—$2/m$.

$a:b:c = 2.7052:1:1.8090;$ $\beta\ 117°12\frac{1}{2}';$[1] $p_0:q_0:r_0 = 0.6687:1.6088:1$

$r_2:p_2:q_2 = 0.6216:0.4157:1;$ $\mu\ 62°47\frac{1}{2}';$ $p_0'\ 0.7519,\ q_0'\ 1.8090,\ x_0'\ 0.5142$

Forms:[2]

		ϕ	ρ	ϕ_2	$\rho_2 = B$	C	A
c	001	90°00′	27°12½′	62°47½′	90°00′	.....	62°47½′
a	100	90 00	90 00	0 00	90 00	62°47½′	.....
m	110	22 34½	90 00	0 00	22 34½	79 53½	67 25½
n	310	51 16½	90 00	0 00	51 16½	69 06	38 43½
e	201	90 00	63 38½	26 21½	90 00	36 26	26 21½
g	$\overline{2}$03	90 00	0 44½	89 15½	90 00	26 28½	89 15½
d	$\overline{1}$01	−90 00	13 22½	103 22½	90 00	40 35	103 22½
h	$\overline{2}$01	−90 00	44 42	134 42	90 00	71 54½	134 42
k	$\overline{4}$01	−90 00	68 09	158 09	90 00	95 21½	158 09
l	$\overline{6}$01	−90 00	75 57½	165 57½	90 00	103 10	165 57½
j	$\overline{8}$01	−90 00	79 42	169 42	90 00	106 54½	169 42
p	111	34 59½	65 38	38 18	41 43½	52 43	58 30½

Less common:

b 010 f 401 $i\ \overline{5}$01 $O\ \overline{1}$12 $P\ \overline{1}$11 y 411 t 511 $S\ \overline{3}$11 $U\ \overline{7}$11

PARALAURIONITE

Structure cell.[3] Space group $C2_1/m$. a_0 10.77 kX, b_0 3.97, c_0 7.18; β 117°13'; $a_0:b_0:c_0 = 2.713:1:1.809$. Cell contents $Pb_4(OH)_4Cl_4$.

Habit. Thin tabular {100} or lath-like by elongation [001], the terminations of the plates or laths rectangular or wedge-shaped due to the development of pyramid faces at the corners. Commonest forms: $a \ m \ c \ p \ h$.

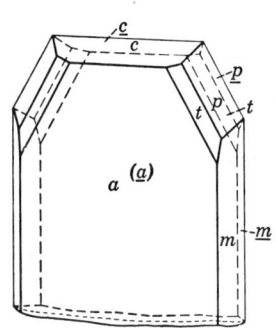
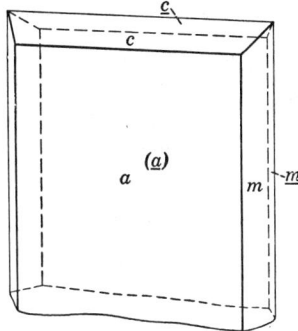

Laurium. Twinned on {100}.   Laurium. Twinned on {100}.

Twinning.[4] Twin plane {100}, very common, as symmetrical contact twins with composition face (100) and simulating orthorhombic holohedry.

Phys. Cleavage {001} perfect and easy. Not brittle. Soft. G. 6.15;[5] 6.28 (calc.). Luster sub-adamantine. Colorless to white, rarely pale greenish or violet. Transparent. Crystals bend readily about [010] due to translation gliding with T{100}, t[001].

Opt. In transmitted light, colorless. Violet-tinted crystals are pleochroic with absorption $Y > X, Z$. Plates on {100} show abnormal interference figures due to twinning.

Orientation	n [6]	
X	2.05 ± 0.01	Biaxial negative $(-)$.
$Y \ b$	2.15 ± 0.01	$2V$ medium to large.
$Z \wedge c \ +25°$	2.20 ± 0.01	$r < v$, strong.

Chem. A hydroxide-chloride of lead, $Pb(OH)Cl$. The water is lost at about 180°.

Anal.

	Pb	O	Cl	H$_2$O	Insol	Total	G.
1.	79.80	3.08	13.65	3.47		100.00	6.28
2.	78.1	[3.6]	14.9	3.4		[100.00]	6.05
3.	77.75	6.00	12.84	3.51	0.09	100.19	

1. Pb(OH)Cl. 2. Laurium.[7] 3. Mammoth mine, Arizona.[8]

Tests. B.B. easily fusible. Soluble in nitric acid.

Occur. A secondary mineral, found associated with laurionite in altered lead-slags at Laurium, Greece. Also with schwartzembergite from the San Rafael mine in the Sierra Gorda, Chile (*rafaëlite*),[9] and with phosgenite from the Wheal Rose near Sithney, Cornwall, England.[10] In the United States with leadhillite, matlockite, cerussite, hydrocerussite, diaboleite at the Mammoth mine, Tiger, Arizona.

Name. In allusion to the polymorphic relation to laurionite.

Ref.
1. Orientation and unit of Smith (1899); angles of Palache, *Min. Mag.*, **23**, 578 (1934). Transformation: Arzruni and Thaddéeff (1899) [rafaelite] to Smith, $2\overline{0}0/0\overline{2}0/001$.
2. Palache (1934); priv. comm. (1942). Rare or uncertain: 203, $\overline{3}01$, $\overline{1}12$, $\overline{5}11$.
3. Wolfe, priv. comm. (1945), by Weissenberg method.
4. See Palache (1934) and Ktenas, *Bull. soc. min.*, **33**, 173 (1910).
5. Berman, priv. comm. (1942).
6. Berman in Palache (1934).
7. Prior analysis in Smith (1899).
8. Gonyer, priv. comm. (1948).
9. Shown to be identical with paralaurionite by Smith, *Min. Mag.*, **12**, 183 (1899).
10. Russell, *Min. Mag.*, **21**, 221 (1927).

10.1.9 PENFIELDITE [$Pb_2(OH)Cl_3$]. Genth (*Am. J. Sc.*, **44**, 260, 1892).

Cryst.[1] Hexagonal—P; dihexagonal-dipyramidal—$6/m\ 2/m\ 2/m$ (?).

$$a:c = 1:0.7848; \qquad p_0:r_0 = 0.9062:1$$

Forms:[2]

	ϕ	$\rho = C$	M	A_2
c 0001	.....	0°00′	90°00′	90°00′
m 10$\overline{1}$0	30°00′	90 00	60 00	90 00
a 11$\overline{2}$0	0 00	90 00	90 00	60 00
l 21$\overline{3}$0	10 53½	90 00	79 06½	70 53½
o 10$\overline{1}$2	30 00	24 22½	78 05½	90 00
t 50$\overline{5}$1	30 00	77 33½	60 46½	90 00
u 11.0.$\overline{11}$.2	30 00	78 39½	60 38½	90 00
v 60$\overline{6}$1	30 00	79 34½	60 32½	90 00
w 80$\overline{8}$1	30 00	82 09	60 18½	90 00
x 11.0.$\overline{11}$.1	30 00	84 16	60 10	90 00

Less common and rare:

ϵ 31$\overline{4}$0 ρ 20$\overline{2}$9 τ 30$\overline{3}$7 ω 50$\overline{5}$8 q 11.0.$\overline{11}$.5 f 5.10.$\overline{15}$.1 e 6.6.$\overline{12}$.1
g 6.0.$\overline{6}$.11 σ 10$\overline{1}$3 ν 5.0.$\overline{5}$.12 θ 70$\overline{7}$9 s 90$\overline{9}$2 d 44$\overline{8}$7

Habit. Usually prismatic [0001] with $\{10\overline{1}0\}$ or steeply pyramidal; also tabular $\{0001\}$. Common forms $c\ m\ o\ u\ v$. Oscillatory combination is common in the zone $c\ m$. The crystals ordinarily are very small and grouped in parallel position.

Twinning.[2] (1) Twin axis [21$\overline{3}$0] with (0001) as composition face, the twinned crystals having [0001] in common but turned 21°47′ to each other. (2) Twin plane $\{41\overline{5}4\}$, the twinned crystals having c and c' almost at right angles.

Phys. Cleavage $\{0001\}$ distinct. Not brittle. G. 6.61,[2] 5.86.[3] Luster adamantine inclining toward greasy. Colorless

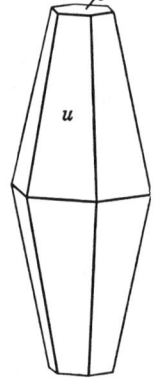

Laurium.   Sierra Gorda.

FIEDLERITE 67

and transparent; also white or tinted yellowish or bluish by impurities.

O p t. In transmitted light, colorless.

	n (Laurium) [6]	
O	2.13 ± 0.01	Uniaxial positive (+).
E	2.21 ± 0.01	

C h e m. A hydroxide-chloride of lead, $Pb_2(OH)Cl_3$.
Anal.

	1	2	3	4
Pb	77.06	78.25		76.55
Cl	19.78	18.55	17.94	19.82
O	1.49			1.47
H_2O	1.67			1.59
Insol.				0.14
Total	100.00	96.80		99.57

1. $Pb_2(OH)Cl_3$. 2. Laurium.[4] Tapering crystals. 3. Laurium.[4] Opaque crystals.
4. Laurium.[5]

Tests. B.B. decrepitates and affords a sublimate of $PbCl_2$. Decomposed by water, leaving a yellowish white oxychloride of lead. Soluble in dilute HNO_3.

O c c u r. Originally found at Laurium, Greece, where it has formed together with laurionite, fiedlerite, etc., by the action of sea water on ancient lead-slags. Also found with percylite in a mine on the western side of the Sierra Gorda about 10 km. from the Sierra Gorda station on the Antofagasta-La Paz railway, Antofagasta, Chile.

N a m e. After the American mineralogist and mineral chemist Samuel L. Penfield (1856–1905), of Yale University.

Ref.

1. Elements of Gordon, *Ac. Sc. Philadelphia, Not. Nat.*, no. 69 (1941) on Sierra Gorda crystals. Transformation: Penfield, *Am. J. Sc.*, **48**, 114 (1894), to Goldschmidt (260, 1897) and Gordon (1941), $\frac{1}{33}00/\frac{11}{33}00/i/0001$. Holohedral symmetry indicated by observations of Gordon.
2. Gordon (1941).
3. Berman, priv. comm. (1940).
4. Genth (1892).
5. Meier analysis in Gordon (1941).
6. Larsen (118, 1921).

10.1.10 **F I E D L E R I T E** [$Pb_3(OH)_2Cl_4$]. vom Rath (*Ber. niederrhein. Ges., Bonn*, 154, 1887; *Jb. Min.*, I, 388, 1889).

C r y s t.[1] Monoclinic; prismatic—$2/m$.

$a:b:c = 2.0488:1:0.8992;\quad \beta\ 102°30';\quad p_0:q_0:r_0 = 0.4389:0.8779:1$

$r_2:p_2:q_2 = 2.2785:0.9999:1;\quad \mu\ 77°30';\quad p_0'\ 0.4496,\ q_0'\ 0.8992,\ x_0'\ 0.2217$

68 HALIDES

Forms:[4]

	ϕ	ρ	ϕ_2	$\rho_2 = B$	C	A
c 001	90°00′	12°30′	77°30′	90°00′	.....	77°30′
a 100	90 00	90 00	0 00	90 00	77°30′	.....
l 120	14 02	90 00	0 00	14 02	86 59½	75 58
w 210	45 00	90 00	0 00	45 00	81 12	45 00
n 310	56 18½	90 00	0 00	56 18½	79 37½	33 41½
f 011	13 51	42 48	77 30	48 43	41 17	80 38½
d 201	90 00	48 15½	41 44½	90 00	35 45½	41 44½
x $\overline{2}$01	−90 00	34 07	124 07	90 00	46 37	124 07
t 111	36 44¼	48 17½	56 07½	53 15½	41 44	63 28½
p $\overline{1}$11	−14 13	42 51	102 50	48 45½	47 11	99 37
e 122	26 24½	45 07	65 56½	50 36½	40 47	71 38
r 211	51 15½	55 10	41 44½	59 05½	45 47½	50 11½
u 311	60 12½	61 04½	32 29½	64 13	50 27½	40 34½

Also:

M 520   g 015   y $\overline{4}$01   o 522   s 321   v $\overline{3}$22

Structure cell.[2] Space group $P2_1/a$. a_0 16.59 kX, b_0 8.00, c_0 7.19; β 102°12′; $a_0:b_0:c_0 = 2.074:1:0.900$. Cell contents $Pb_{12}(OH)_8Cl_{16}$.

Habit. Lath-like; tabular {100} and elongated [010] with a rectangular outline. The crystal faces ordinarily are of poor quality.

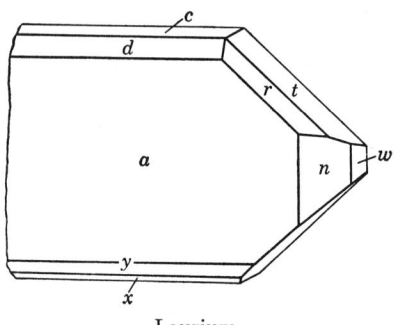

Laurium.

Twinning. On {100}, common.

Phys. Cleavage {100}, good. Not brittle. H. ∼ 3½. G. 5.88; 5.64 (calc.). Luster adamantine. Colorless to white. Transparent.

Opt.[3] In transmitted light, colorless.

Orientation	n	
X	1.98 ± 0.01	Biaxial negative (−).
Y ∧ c −34°	2.04 ± 0.01	2V large.
Z b	2.10 ± 0.01	$r < v$, perceptible.

Crystals resting on {100} may fail to show complete extinction due to twinning.

ATACAMITE 69

Chem. A hydroxide-chloride of lead, $Pb_3(OH)_2Cl_4$. Analysis gave:

	Pb	PbO	Cl	H$_2$O	Total
1.	51.97	27.99	17.78	2.26	100.00
2.	51.01	29.02	17.48	2.33	99.84

1. $Pb_3(OH)_2Cl_4$. 2. Laurium.[5]   Crystals superficially altered to cotunnite.

Tests. In C.T. decrepitates, loses water (at about 150°), becomes turbid, and fuses yielding a sublimate of lead chloride. Attacked by cold water and becomes turbid. Soluble in HNO$_3$ but less readily than laurionite.

Occur. Found at Laurium, Greece, where it has formed together with penfieldite, laurionite, etc., by the action of sea water on ancient lead-slags.

Name. After the Saxon Commissioner of Mines K. G. Fiedler (1791–1853), who was director of an exploratory expedition to the Laurium region in 1835.

Ref.

1. Elements of Palache, *Min. Mag.*, **23**, 573 (1934), representing the mean of new data and the earlier data of vom Rath (1887) and Smith, *Min. Mag.*, **12**, 102 (1899), transformed to the unit of the structure cell of Wolfe and Frondel, priv. comm. (1946). Transformations: Palache to new, 200/010/001; vom Rath to new, $\frac{1}{4}$00/010/001; Smith to new, $\frac{5}{4}$00/010/00$\frac{5}{4}$; Goldschmidt (4, 4, 1918) to new, 300/010/001.
2. Wolfe and Frondel (1946) by Weissenberg method.
3. Berman in Palache (1934).
4. Palache (1934).
5. de Schulten, *C.R.*, **140**, 315 (1905); Lacroix and de Schulten, *Bull. soc. min.*, **31**, 83 (1908).

10.1.11 ATACAMITE GROUP

10.1.11.1 **A T A C A M I T E** [$Cu_2(OH)_3Cl$]. Sable vert cuivreux du Pérou, Chaux cuivreuse unie à un peu d'acide muriatique et d'eau *Rochefoucauld*, *Baumé*, and *Fourcroy* [*Mém. ac. Paris*, 1786 (publ. in 1788)]; *Berthollet* (*ibid.*, 474, 1788). Kupfersand, Salzsaures Kupfer *Karsten* (46, 76, 1800). Cuivre muriaté *Haüy* (**3**, 561, 1801). Muriate of copper. Atacamit, Salzkupfererz *Blumenbach* (*Handb. Nat.*, 1805). Kupferhornerz, Atacamit *Ludwig* (**2**, 178, 1804). Smaragdochalcit *Hausmann* (1039, 1813). Chlorochalcit *Glocker* (845, 1831). Halochalzit, Hal-Chalzit *Breithaupt* (165, 1841). Remolinite *Brooke* and *Miller* (618, 1852). Marcylite pt. *Shepard* (*Am. J. Sc.*, **21**, 206, 1856; *Dana*, *Am. J. Sc.*, **24**, 122, 1857).

Cryst. Orthorhombic; dipyramidal—$2/m\ 2/m\ 2/m$.

$a:b:c = 0.6617:1:0.7535;$ [1] $p_0:q_0:r_0 = 1.1387:0.7535:1$

$q_1:r_1:p_1 = 0.6617:0.8782:1;$ $r_2:p_2:q_2 = 1.3271:1.5112:1$

HALIDES

Forms:[2]

		φ	ρ = C	φ₁	ρ₁ = A	φ₂	ρ₂ = B
c	001	....	0°00′	0°00′	90°00′	90°00′	90°00′
b	010	0°00′	90 00	90 00	90 00	....	0 00
γ	150	16 49	90 00	90 00	73 11	0 00	16 49
s	120	37 04½	90 00	90 00	52 55½	0 00	37 04½
l	230	45 13	90 00	90 00	44 47	0 00	45 13
m	110	56 30½	90 00	90 00	33 29½	0 00	56 30½
β	210	71 41½	90 00	90 00	28 18½	0 00	71 41½
e	011	0 00	37 00	37 00	90 00	90 00	53 00
g	031	0 00	66 08	66 08	90 00	90 00	23 52
B	061	0 00	77 31½	77 31½	90 00	90 00	12 28½
h	201	90 00	66 17½	0 00	23 42½	23 42½	90 00
r	111	56 30½	53 47	37 00	47 43	41 17½	63 34
q	221	56 30½	69 53½	56 26	38 27½	23 42½	58 47½
z	331	56 30½	76 17	66 08	35 53½	16 19	57 35
M	123	37 04½	32 11½	26 40½	71 16	69 13	64 50½
n	121	37 04½	62 06	56 26	57 48½	41 17½	45 09½
φ	131	26 44	68 26½	66 08	65 16	41 17½	33 50½

Less common and rare:

a 100	δ 270	λ 340	ψ 035	C 071	ρ 443	N 122	S 171
j 190	k 130	μ 790	d 023	D 091	σ 332	O 132	T 312
v 180	ε 250	t 560	i 0.10.9	E 0.18.1	H 551	ς 142	f 211
w 170	η 470	ν 320	ω 053	u 101	J 881	P 253	U 231
α 160	Δ 350	π 016	o 021	F 112	K 124	Q 141	V 241
x 140	κ 570	χ 013	A 051	G 223	L 154	R 151	y 321

Structure cell.[3] Space group $Pnam$. a_0 6.01 kX, b_0 9.13, c_0 6.84; $a_0:b_0:c_0 = 0.658:1:0.749$. Cell contents $Cu_8(OH)_{12}Cl_4$.

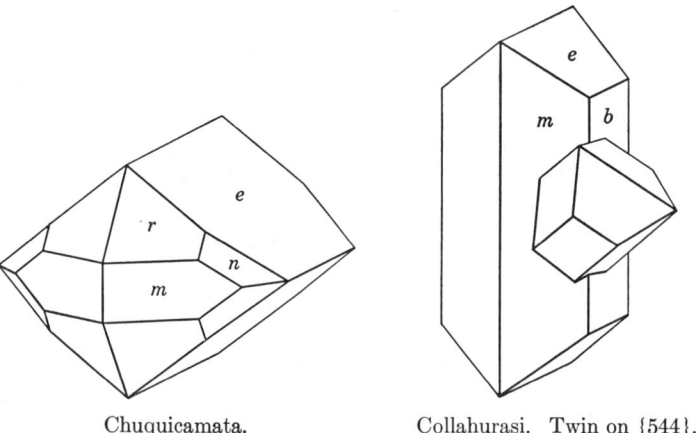

Chuquicamata.   Collahurasi. Twin on {544}.

Habit. Commonly in slender prismatic crystals [001] with bright terminal planes. {110} and other forms in the zone [001] striated parallel

ATACAMITE

[001]; {010} also striated parallel [$\bar{1}$01]. Also frequently tabular {010}; rarely pseudo-octahedral with equal development of {110} and {011}. Common forms: *b m s e r n*. In confused crystalline aggregates; also massive, fibrous or granular to compact; as sand.

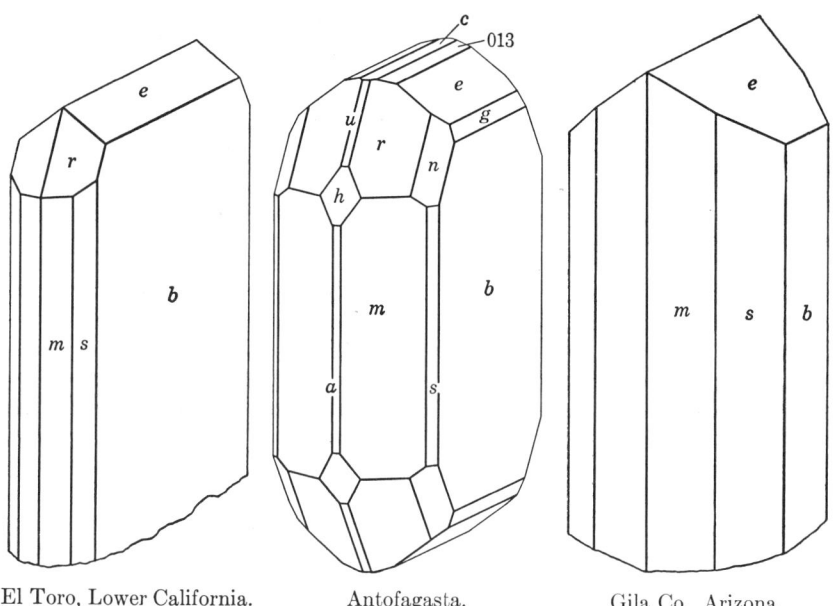

El Toro, Lower California.   Antofagasta.   Gila Co., Arizona.

Twinning. (1) Twin plane {110}, composition plane {110}; rare, forming doublets or triplets.[4] (2) In unusual, irrational twins that can be defined by a 120° rotation about [950] or a 180° rotation about [544]; rather common, as doublets, trillings, and complex groups, both contact and penetration.[5]

Phys. Cleavage {010}, perfect; {101} fair, easy to obtain on crystals tabular {010}. Fracture conchoidal. Brittle. H. 3–3½. G. 3.760 ± 0.015,[6] 3.776;[7] 3.756 (calc.). Luster adamantine to vitreous. Color bright green of various shades; dark emerald-green to blackish green. Streak apple-green. Transparent to translucent.

Opt. In transmitted light, various shades of green.

ORIENTATION		n (green)[8]	PLEOCHROISM	
X	b	1.831	Pale green	Biaxial negative (−).
Y	a	1.861	Yellow-green	2V 74°56′ (calc.).
Z	c	1.880	Grass-green	$r < v$, strong.

Chem. A basic chloride of copper, $Cu_2(OH)_3Cl$. No clear evidence of compositional variation is afforded by the numerous reported analyses. The Br analogue has been prepared artificially.[9]

Anal.[10]

	1	2	3	4	5	6
Cu	14.88	14.82	15.05	14.79	14.75	14.53
CuO	55.87	56.01	55.43	55.28	55.81	55.33
Cl	16.60	16.55	16.78	16.51	16.48	16.22
H_2O	12.65	12.69	[12.74]	12.42	13.09	13.15
Rem.				0.79		0.44
Total	100.00	100.07	[100.00]	99.79	100.13	99.67
G.	3.76				3.775	3.775

1. $Cu_2(OH)_3Cl$. 2. Collahurasi, Chile.[11] 3. Bimbouri, South Australia.[12] 4. Paposa, Chile.[13] Rem. is insol. 5. Antofagasta, Chile.[14] 6. Boleo, Lower California.[14] Rem. is CoO 0.21, CaO 0.23.

Tests. In C.T. gives off much water and forms a gray sublimate. Fusible. Easily soluble in acids.

Occur. A secondary mineral formed by the oxidation of other copper minerals, especially under arid, saline conditions. Commonly associated with malachite, cuprite, brochantite, chrysocolla, gypsum, limonite. Atacamite was originally found as sand at a not precisely known locality in the Atacama desert of northern Chile. The mineral occurs widely in the arid coastal belt of Chile and Peru. Notable localities in Chile include mines in the neighborhood of Paposa, Copiapó and Remolinos in Atacama province; at Collahurasi in Tarapacá province; and at Taltal, Tocopilla, and Chuquicamata in Antofagasta province. In Lower California at Boleo and El Toro. In South Australia at the Wallaroo, Moonta, and Cornwall (Kadina) mines on the Yorke peninsula; in the Burra district, and at Bimbowrie. As a fumarole product on Vesuvius and Etna, Italy. At the Turjinsk mines, Bogoslowsk district, U.S.S.R. Sparingly in Cornwall, at St. Just and elsewhere. In the United States, atacamite is found in small amounts in many mining districts in the western and southwestern states. In Arizona at Bisbee; at the United Verde mine, Jerome; at Clifton; at the Mammoth mine, Pinal County. In Utah with secondary copper arsenates in the Tintic district, Juab County.

Alter. Occurs altered to malachite (a change that has been duplicated artificially) [15] and to chrysocolla. Observed as an alteration product of ancient copper and bronze objects [16] and of nantokite and other copper minerals. Altered by KOH solution to hydrous cupric oxide.[17]

Artif.[18] Obtained in crystals by heating Cu_2O with ferric chloride solution in a sealed tube, by similarly treating a mixed solution of sodium chloride with basic copper nitrate, and in other ways.

Name. From the locality in Atacama province, northern Chile.

Ref.

1. Angles of Ungemach, *Bull. soc. min.*, **34,** 148 (1911); orientation of Mohs-Zippe (**2,** 177, 1839)—also used by Dana (172, 1892) and Goldschmidt (**1,** 117, 1913). Ungemach used the orientation of Lévy, **3,** 47 (1838); transformation: Lévy to Mohs-Zippe, 100/001/010.

2. Principally from Ungemach (1911), a complete survey of old observations with

many new data; a few added forms from Liffa and Tokody, *Cbl. Min.*, 183 (1926) and Buttgenbach, *Soc. géol. Belgique, Publ. Congo Belge*, **52**, C65 (1929). Uncertain forms:

7.13.0	13.19.0	890	773	992	388	5.14.9	273
7.12.0	450	092	552	12.12.1	3.11.8	4.11.7	17.20.9
13.21.0	670	0.30.1	772	2.11.9	275	385	762
580	780	117	441	297	4.13.9	3.90.5	891

3. Thoreau and Verhulst, *Ac. Belgique, Bull.* **24**, 716 (1938), by Laue and rotation methods; confirmed by Brasseur and Toussaint, *Bull. soc. roy. sc. Liége*, **11**, 555 (1942). See also Wells, *Acta Cryst.*, **2**, 175 (1949).
4. Law (1), noted only by Dana (121, 1868; 173, 1892), requires verification.
5. This remarkable twin law, first observed by Ford, *Am. J. Sc.*, **30**, 16 (1910), and verified by Ungemach (1911), was described by Ford as a rotation of 112°40′ about the normal to (011). The definition of the law as given by Friedel, *Bull. soc. min.*, **35**, 45 (1912) is used here.
6. Range of best values reported in literature, including new microbalance measurements on Australian crystals by Mrose, priv. comm. (1947).
7. Average of measurements by Ungemach (1911) by pycnometer method on Boleo and Antofagasta crystals.
8. Smith, *Min. Mag.*, **12**, 22 (1898), by prism method.
9. Dupont and Jansen, *Bull. soc. chim. Paris*, **9**, 193 (1893); the structural relation to atacamite is uncertain.
10. For additional analyses see Hintze (1 [2B], 2593, 1915); also Carobbi, *Acc. Napoli, Rend.*, **34** [3], 78 (1928) on impure material.
11. Ford, *Am. J. Sc.*, **30**, 16 (1910).
12. Mawson, *Trans. Roy. Soc., South Australia*, **30**, 67 (1906).
13. Keller, *Proc. Am. Phil. Soc.*, **47**, 79 (1908).
14. Ungemach (1911).
15. Tschermak, *Min. Mitt.*, 38 (1873).
16. Daubrée, *C.R.*, **81**, 182 (1875); Skinder; *Zs. Kr.*, **50**, 69 (1912); Frondel, *Min. Mag.*, **29**, 34 (1950).
17. Chumanov, *J. Russ. Phys. Chem. Soc.*, **47**, 1268 (1915).
18. Mellor (**3**, 178, 1923); see also Guillot and Geneslay, *Bull. soc. chim.*, 4 [5], 125 (1937); Hubbell, *J. Am. Chem. Soc.*, **59**, 215 (1937); Frondel (1950).

10.1.11.2 K E M P I T E [$Mn_2(OH)_3Cl$]. Rogers (*Am. J. Sc.*, **8**, 145, 1924).

C r y s t.[1] Orthorhombic; dipyramidal—$2/m\ 2/m\ 2/m$ (?).

$$a:b:c = 0.677:1:0.747; \qquad p_0:q_0:r_0 = 1.103:0.747:1$$

$$q_1:r_1:p_1 = 0.677:0.906:1; \qquad r_2:p_2:q_2 = 1.339:1.477:1$$

Forms:

	ϕ	$\rho = C$	ϕ_1	$\rho_1 = A$	ϕ_2	$\rho_2 = B$
b 010	0°00′	90°00′	90°00′	90°00′	....	0°00′
a 100	90 00	90 00	....	0 00	0°00′	90 00
m 110	55 53½	90 00	90 00	34 06½	0 00	55 53½
d 011	0 00	36 45½	36 45½	90 00	90 00	53 14½
t 121	36 26½	61 42	56 12	58 28	42 12	44 54

Habit. Prismatic [001]. {100} and {010} are usually absent or inconspicuous.

P h y s. H. about 3½. G. about 2.94. Color emerald-green.

O p t. In transmitted light, green.

Orientation		n (Orange filter)	
X	c	1.684 ± 0.001	Biaxial negative (−)
Y	b	1.695 ± 0.001	$2V$ 54°52′ (calc.)
Z	a	1.698 ± 0.001	

Chem. A basic chloride of manganese, $Mn_2(OH)_3Cl$. Isostructural with atacamite.[3]

Anal.

	1	2
Mn	55.95	50.59
Cl	18.06	16.41
O	12.22	[21.40]
H_2O	13.77	11.60
Total	100.00	[100.00]

1. $Mn_2(OH)_3Cl$. 2. California.[2]

Tests. In C.T. turns black and affords acid water. Soluble in dilute acid.

Occur. Found immediately associated with pyrochroite, hausmannite, and rhodochrosite in a boulder of manganese ore containing tephroite and psilomelane in Alum Rock Park, about five miles east of San José, Santa Clara County, California.

Name. After James Furman Kemp (1859–1926), mining geologist, and Professor of Geology at Columbia University.

Ref.
1. Doubly terminated crystals not observed.
2. Crook analysis in Rogers (1924).
3. Rogers, *Bull. Geol. Soc. Am.*, **60**, no. 12, pt. 2, 1944 (1949).

10.1.12 **P A R A T A C A M I T E** [$Cu_2(OH)_3Cl$]. Atelite, Atelina *A. Scacchi (Acc., Napoli, Att.*, **6**, no. 9, 22, 1873). Paratacamite *Smith (Min. Mag.*, **14**, 170, 1906).

Cryst.[1] Hexagonal—R; hexagonal-scalenohedral—$\bar{3}\,2/m$.

$a:c = 1:1.0248;$ $\alpha\ 96°23';$ $p_0:r_0 = 1.1833:1;$ $\lambda\ 82°49\frac{1}{2}'$

Forms:[2]

	ϕ	$\rho = C$	A_1	A_2
$c\ 0001$ 111	.....	0°00'	90°00'	90°00'
$a\ 11\bar{2}0$ $10\bar{1}$	0°00'	90 00	60 00	60 00
$w\ 20\bar{2}5$ 311	30 00	25 19½	68 15	90 00
$v\ 7.0.\bar{7}.13$ 922	30 00	32 30	62 16	90 00
$u\ 40\bar{4}7$ 511	30 00	34 04	60 59	90 00
$r\ 10\bar{1}1$ 100	30 00	49 48	48 35½	90 00
$e\ 01\bar{1}2$ 110	−30 00	30 36½	90 00	63 50
$f\ 02\bar{2}1$ $11\bar{1}$	−30 00	67 05½	90 00	37 05
$l\ 24\bar{6}1$ $31\bar{3}$	−10 53½	80 55½	71 08½	49 43½

Structure cell.[3] $a_{rh}\ 9.150\ kX$, $\alpha\ 96°28';$ $a_0\ 13.65\ kX$, $c_0\ 13.95;$ $a_0:c_0 = 1:1.022$. Cell contents in the rhombohedral unit $Cu_{16}(OH)_{24}Cl_8$.

Habit. Rhombohedral; common forms $r\,f\,e\,c$. Also granular massive, and as powdery incrustations produced by alteration; as microscopically fibrous or spherulitic crusts.

Twinning. On $\{10\bar{1}1\}$ common, sometimes polysynthetic.

P h y s. Cleavage {10$\bar{1}$1} good. Fracture conchoidal to uneven. H. 3. G. 3.74; 3.75 (calc.). Luster vitreous. Color green to dark green and greenish black. Streak green. Translucent to nearly opaque.

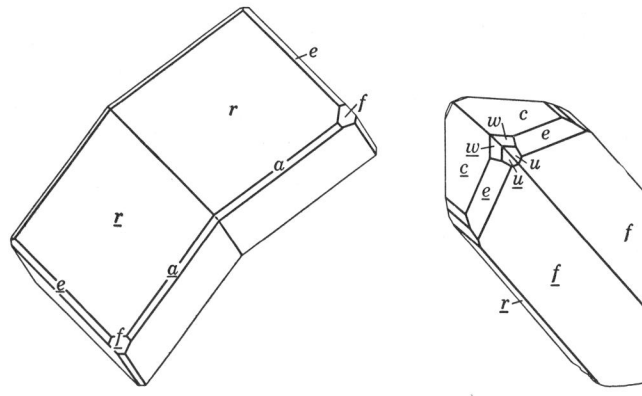

Sierra Gorda. Twin on {10$\bar{1}$1}.   Sierra Gorda. Twin on {10$\bar{1}$1}.

O p t. In transmitted light, green and not pleochroic.

	n (Sierra Gorda)[4]	n (Remolinos)[4]	
O	1.843	1.844	Uniaxial positive (+).
E	1.848	1.849	

Often biaxial, apparently due to strain, with 2V up to about 50° and $r > v$. The extinction often is undulant or patchy and the ultra-blue is commonly shown.

C h e m. A basic chloride of copper, $Cu_2(OH)_3Cl$. Dimorphous with atacamite. The thermal analysis curve differs from that of atacamite.[4]

Anal.

	1	2	3
Cu	14.88	14.60	14.27
CuO	55.87	55.98	56.10
Cl	16.60	16.29	15.97
H_2O	12.65	13.13	14.10
Total	100.00	[100.00]	100.44
G.	3.75	3.72	3.74

1. $Cu_2(OH)_3Cl$. 2. Remolinos, Chile.[5] H_2O includes H_2O+ 13.10, H_2O- 0.03. Recalculated to 100 after deducting 0.97 insol. 3. Sierra Gorda, Chile.[6]

Tests. Like atacamite.

O c c u r. A secondary mineral found associated with caracolite, schwartzembergite, and atacamite at the Generosa and Herminia mines, Sierra Gorda, Chile. Also from the Bolaco mine, San Cristóbal, and from Remolinos, in Chile. As an alteration of tenorite in fumaroles on Vesuvius (*atelite*).[7] In the Botallack mine, St. Just, Cornwall, England. Also as an alteration product of eriochalcite from Mina Queténa, Chile, and of nantokite from Nantoko, Chile, and Broken Hill, New South Wales; the latter alteration may occur by exposure to moist air after the specimens have been collected.

A r t i f.[4] Obtained as a fine-grained, hydrous precipitate by reaction of copper sheets or wire with salt water or solutions of various chlorides; also by exposure of cuprous chloride to a moist atmosphere. Found with atacamite and, rarely, botallackite as a constituent of the green alteration crusts and patina on ancient copper and bronze objects.

N a m e. In allusion to the dimorphous relation to atacamite.

Ref.

1. Smith (1906). Paratacamite was discredited by Ungemach, *Bull. soc. min.*, **34**, 148 (1911), as only twinned atacamite, but the species later was shown to be valid by Frondel, *Min. Mag.*, **29**, 34 (1950).
2. Smith (1906) on Sierra Gorda crystals.
3. Frondel (1950) by Weissenberg method on type crystals.
4. Frondel (1950).
5. Peck analysis in Frondel (1950).
6. Prior analysis in Smith (1906).
7. Shown to be identical with paratacamite by Frondel (1950).

TALLINGITE. *Church (J. Chem. Soc. London*, **18**, 77, 213, 1865).

Thin crusts with a minute globular structure. H. $\sim$ 3. G. $\sim$ 3.5. Color bright blue inclining to green. Streak white. A hydrated basic copper chloride of uncertain formula, approximately $Cu_5(OH)_8Cl_2 \cdot 4H_2O$. Analysis gave: CuO 53.57, Cu 10.11, Cl 11.33, H_2O 24.99, total 100.00. Found with paratacamite at the Botallack mine, Cornwall, England. Named after R. Talling, of Lostwithiel, Cornwall, by whom the mineral was collected.

10.1.13 Botallackite. *Church (J. Chem. Soc. London*, **18**, 212, 1865).

Orthorhombic (?). As crusts of tiny interlacing or columnar crystals. Color pale bluish green to green. Cleavable in one direction. G $\sim$ 3.6.

O p t.[1] In transmitted light, pale bluish green.

ORIENTATION	n	
$X \perp$ cleavage	1.775 ± 0.003	Biaxial positive (+).
Y	1.800 ± 0.005	$2V$ moderately large.
Z	1.846 ± 0.003	$r > v$, strong. Weakly pleochroic.

C h e m. A basic copper chloride of uncertain formula, perhaps $Cu_2(OH)_3Cl \cdot H_2O$.

Anal.

	CuO	Cl	H_2O		Cl = O	Total
1.	68.70	15.31	19.44	103.45	3.45	100.00
2.	66.25	14.51	22.60	103.36	3.27	100.09

1. $Cu_2(OH)_3Cl \cdot H_2O$. 2. Botallack mine.

O c c u r. A secondary mineral found with atacamite and paratacamite at the Botallack mine, St. Just, Cornwall, England. Also identified as a powdery greenish blue alteration of an ancient Egyptian bronze object. Other specimens described [2] from the Botallack mine and near-by mines have been referred to botallackite but probably represent atacamite, with which botallackite formerly was made identical.[3]

Ref.

1. Frondel, *Min. Mag.*, **29**, 34 (1950), on type material.
2. Maskelyne, *Proc. Roy. Soc. London*, **14**, 399 (1865), and Zepharovich, *Ak. Wien, Sitzber.*, **68**, 130 (1873).
3. The probable species validity of botallackite is indicated by the study of Frondel (1950), who gives x-ray powder spacing data.

HYDROMELANOTHALLITE [$Cu_2(OH)_2Cl_2 \cdot H_2O$?]. Idromelanotallite *Zambonini* (**57**, 1910). Mélanothallite (?) *Lacroix* (**30**, 223, 1907).

A green mineral formed by the alteration of melanothallite. Composition apparently $Cu_2(OH)_2Cl_2 \cdot H_2O$ from the analysis:[1] CuCl 57.37, CuO 31.39, H_2O [11.24], total [100.00]. Found with eriochalcite and other copper minerals in fumaroles on Vesuvius. Needs further study.[2]

Ref.

1. E. Scacchi, *Acc. Napoli, Rend.*, **10**, 158 (1884).
2. See Frondel, *Min. Mag.*, **29**, 34 (1950).

10.1.14 **C A D W A L A D E R I T E** [$Al(OH)_2Cl \cdot 4H_2O$]. *Gordon (Ac. Sc. Philadelphia, Not. Nat.*, no. 80, April, 1941).

Found as amorphous grains and small masses. Fracture conchoidal. G. 1.66. Luster vitreous. Color lemon-yellow. Transparent. Isotropic; with $n(Na)$ 1.513, variable.

Chem. A hydrated basic chloride of aluminum, $Al(OH)_2Cl \cdot 4H_2O$. Weakly hygroscopic.

Anal.

	1	2	3
Na_2O		1.85	
K_2O		0.90	
CaO		2.07	1.60
Al_2O_3	30.25	27.50	29.34
Cl	21.04	22.96	21.51
SO_3		0.82	
$H_2O + 110°$	53.46	24.99	26.27
$H_2O - 110°$		25.13	26.81
	104.75	106.22	105.53
Less O = Cl	4.75	5.18	5.53
Total	100.00	101.04	100.00

1. $Al(OH)_2Cl \cdot 4H_2O$. 2. Chile.[1] Original total given as 106.32. 3. Analysis 2 recalculated to 100 after deduction of SO_3 as $CaSO_4 \cdot 2H_2O$ and the Na and K as NaCl and KCl.

Occur. Found associated with halite in a sulfate deposit containing tamarugite, pickeringite, gypsum, and trudellite in Cerro Pintados, 80 km. southeast of Iquique, Tarapacá province, Chile.

Name. After Charles M. B. Cadwalader, President of the Academy of Natural Sciences of Philadelphia.

Needs further study.

Ref.

1. Pitman analysis in Gordon (1941).

78　　　　　　　　　　　　　　Halides

TYPE 2.　$A_m B_n(O,OH)_p X_q$

10.2.1　B O L E I T E [$Pb_9Cu_8Ag_3Cl_{21}(OH)_{16} \cdot 2H_2O$?]. *Mallard* and *Cumenge* (*C.R.*, **113**, 519, 1891). Argentopercylite *Schulze* (*Chem.-Ztg.*, **16**, 1953, 1892).

C r y s t.[1] Tetragonal; pseudo-isometric ?.

$$a:c = 1:3.996; \quad p_0:r_0 = 3.996:1$$

Forms:[2]

	ϕ	ρ	A	$\bar{M}$
c 001	.....	0°00′	90°00′	90°00′
m 110	45°00′	90 00	45 00	90 00
f 014	0 00	44 58	90 00	60 01
e 011	0 00	75 57	90 00	43 18½
o 114	45 00	54 32½	54 45	90 00

Structure cell.[3] Tetragonal. a_0 15.4 kX, c_0 62; $a_0:c_0 = 1:4.02$. Cell contents uncertain, perhaps $Pb_{108}Cu_{96}Ag_{36}Cl_{252}(OH)_{192} \cdot 24H_2O$.

Habit. Cubical, each crystal composed of three individuals with their c-axes parallel to a cubic axis and the basal planes of each individual forming the pseudo-cubic faces. Re-entrant angles are lacking. Also pseudo-cubo-octahedral and pseudo-dodecahedral. The crystals are often overgrown in parallel position by cumengite and pseudoboleite.

P h y s. Cleavage {001} perfect, {101} good, {100} poor. H. 3–3½. G. 5.05;[4] 5.05 (calc.). Luster weakly vitreous, pearly on the cleavages. Color deep Prussian blue, inclining to blackish blue on crystal faces. Streak blue with a greenish tint. Translucent.

O p t. In transmitted light, bluish green and not pleochroic.

	n (Boleo)[5]	
O	2.05 ± 0.02	Uniaxial negative (−).
E	2.03 ± 0.02	

Sections parallel to the pseudo-cubic faces show an isotropic core, due to the orientation and overlap of the several individuals, and a birefringent border.

C h e m. A hydroxide-chloride of lead, copper, and silver. The formula is uncertain, perhaps $Pb_9Cu_8Ag_3Cl_{21}(OH)_{16} \cdot 2H_2O$ [4] or simply $Pb(Cu,Ag)Cl_2(OH)_2 \cdot H_2O$.

Anal.[6]

	1	2	3	4	5
Pb	49.73	49.16	49.52	49.51	47.20
CuO	16.98	17.17	17.20	17.71	24.03
AgCl	11.47	12.03	11.16	11.51	10.96
Cl	17.02	17.04	17.28	17.13	10.79
H_2O	4.80	4.35	4.35	4.48	[7.02]
Insol.		0.21	0.25		
Total	100.00	99.96	99.76	100.34	[100.00]
G.					5.02

1. $Pb_9Cu_8Ag_3Cl_{21}(OH)_{16} \cdot 2H_2O$. 2. Boleo.[4] Inner isotropic zone. 3. Boleo.[4] Outer birefringent zone. 4. Boleo.[7] 5. Broken Hill, New South Wales.[9] A separate, direct determination of H_2O gave 6.39 per cent.

CUMENGITE

Tests. In C.T. loses water and fuses. Soluble in HNO_3. Not attacked by water.

Occur. A secondary mineral, found originally at Boleo near Santa Rosalia, Lower (Baja) California, Mexico, where it is associated with cumengite, pseudoboleite, anglesite, phosgenite, atacamite, cerussite, and gypsum in white clay. Crystals up to 2 cm. on edge have been found; but usually much smaller. Also noted in Chile at the San Augustin mine near Huantajaya and at Challocollo, Cerro Gordo, in Tarapacá. At the South mine, Broken Hill, New South Wales.

Artif.[8] Obtained as crystals by the action of $CuCl_2$ solution on precipitated Ag_2O and $Pb(OH)_2$.

Ref.
1. Elements of Friedel, *Bull. soc. min.*, **29**, 14 (1906), on cleavage surfaces of Boleo crystals. The true symmetry of boleite is uncertain and the mineral has been considered to be isometric by some authors—see Hadding, *Geol. För. Förh.*, **41**, 175 (1919); Gossner and Arm, *Zs. Kr.*, **72**, 202 (1929); Friedel, *Zs. Kr.*, **73**, 147 (1930); Gossner, *Zs. Kr.*, **75**, 365 (1930); Goldschmidt (**6**, 129, 1920). Gossner and Arm (1929) give a_0 15.4 kX in the isometric interpretation.
2. Mallard and Cumenge (1891); Friedel (1906).
3. Hocart, *Zs. Kr.*, **74**, 20 (1930).
4. Friedel (1906).
5. Larsen (49, 1921). Hadding (1919) gives 2.081, 2.087 for the refractive index in different parts of a zoned crystal.
6. Earlier analyses are listed in Hintze (**1** [2B], 2645, 1915).
7. Gossner and Arm (1929).
8. Friedel, *Bull. soc. min.*, **17**, 6 (1894).
9. Carmichael and Armstrong in Liversidge, *J. Roy. Soc. New South Wales*, 94, 98 (1894).

10.2.2 **CUMENGITE** [$Pb_4Cu_4Cl_8(OH)_8 \cdot H_2O$?]. *Mallard* (*Bill. soc. min.*, **16**, 184, 1893). Cumengeite. Not Cumengit *Kenngott* (29, 1853).

Cryst.[1] Tetragonal.

$$a:c = 1:1.625; \quad p_0:r_0 = 1.625:1$$

Forms:[2]

	ϕ	ρ	A	M
c 001	....	0°00'	90°00'	90°00'
m 110	45°00'	90 00	45 00	90 00
e 011	0 00	58 23½	90 00	52 58½

Structure cell.[3] Tetragonal. a_0 15.17 kX, c_0 24.71; $a_0:c_0 = 1:1.628$. Cell contents uncertain, probably $Pb_{40}Cu_{40}Cl_{80}(OH)_{80} \cdot 10H_2O$.

Habit. Octahedral, or cubo-octahedral with more or less equal development of *c m e*. Occurs as parallel overgrowths on crystals of boleite and pseudoboleite, sometimes completely enveloping them and giving regular groupings that simulate twins.[8]

Phys. Cleavage {101} good, {110} distinct, {001} poor. H. 2½ (less than boleite).[7] G. 4.67;[4] 4.60 (calc.). Luster weakly vitreous, not pearly on the cleavage surfaces. Color indigo-blue. Streak sky-blue. Translucent.

Opt.

	n [5]	$n(\lambda 510)$ [6]	DICHROISM	
O	2.026	2.041	Darker blue, with tint of green	Uniaxial negative (−).
E	1.965	1.926	Pure blue	

In transmitted light cumengite is a purer blue than the boleite and pseudoboleite with which it is intergrown.

Chem. A hydroxide-chloride of lead and copper, $Pb_4Cu_4Cl_8(OH)_8 \cdot H_2O$ or $PbCuCl_2(OH)_2$.

Anal.

	1	2	3	4
Pb	54.50	55.15	54.47	54.17
CuO	20.93	21.18	20.27	19.93
Cl	18.65	18.87	19.03	19.13
H_2O	5.92	4.80	5.90	6.19
Insol.			0.19	
Total	100.00	100.00	99.86	99.42

1. $Pb_4Cu_4Cl_8(OH)_8 \cdot H_2O$. 2. $PbCuCl_2(OH)_2$. 3. Boleo.[8] 4. Boleo.[9]

Tests. Soluble in HNO_3.

Occur. Found with boleite and pseudoboleite at Boleo near Santa Rosalia, Lower (Baja) California, Mexico.

Name. After Édouard Cumenge (1828–1902), French mining engineer.

Ref.

1. Elements of Friedel, *Bull. soc. min.*, **29**, 14 (1906). See also Hadding, *Geol. För. Förh.*, **41**, 175 (1919).
2. Friedel (1906) and Mallard (1893).
3. Gossner and Arm, *Zs. Kr.*, **72**, 202 (1929). Hocart, *Zs. Kr.*, **74**, 20 (1930), gives a_0 14.9 kX, c_0 24.15.
4. Friedel (1906). Berman, priv. comm. (1942), obtained 4.656 ± 0.015. Hadding (1919) gave 4.77, and Mallard (1893) 4.71.
5. Mallard (1893).
6. Hadding (1919).
7. Frondel, priv. comm. (1948).
8. Friedel (1906).
9. Gossner and Arm (1929).

10.2.3 PSEUDOBOLEITE [$Pb_5Cu_4Cl_{10}(OH)_8 \cdot 2H_2O$?]. Lacroix (*Bull. mus. hist. nat., Paris*, 39, 1895).

Cryst.[1] Tetragonal (pseudo-cubic).

$$a:c = 1:2.023; \quad p_0:r_0 = 2.023:1$$

Forms:[2]

	ϕ	ρ	A	M
c 001	....	0°00′	90°00′	90°00′
a 010	0°00′	90 00	90 00	45 00
m 110	45 00	90 00	45 00	90 00
e 011	0 00	63 41½	90 00	50 39½
o 112	45 00	55 02½	54 35	90 00

Structure cell.[2] Tetragonal. a_0 15.4 kX, c_0 31.2; $a_0:c_0 = 1:2.027$.
Cell contents uncertain, perhaps $Pb_{60}Cu_{48}Cl_{120}(OH)_{96} \cdot 24H_2O$.

Habit. Observed only in parallel growth upon boleite, with the {001} faces of the two minerals in common. Euhedral crystals of pseudoboleite may thus project from the several cube faces of boleite, forming re-entrant angles along the cube edges and simulating twins.[3]

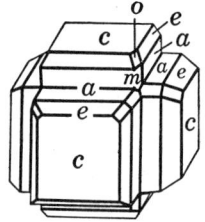

Phys. Cleavage {001} perfect, {101} nearly perfect. H. $2\frac{1}{2}$. G. 4.85; 4.89 (calc.). Luster pearly on the cleavages. Color indigo-blue, closely resembling boleite. Translucent.

Opt.

nO	2.03	Uniaxial negative $(-)$.
nE	2.00	Not dichroic.

Chem. A hydroxide-chloride of lead and copper. The only available analysis, made on a small sample admixed with boleite, suggests that the formula is $Pb_5Cu_4Cl_{10}(OH)_8 \cdot 2H_2O$.

Anal.

	Pb	CuO	AgCl	Cl	H$_2$O	Insol.	Total
1.	57.02	17.52		19.51	5.95		100.00
2.	53.5	16.5	1.6	20.2	5.5	0.8	98.1
3.	57.7	16.9		19.8	5.6		100.0

1. $Pb_5Cu_4Cl_{10}(OH)_8 \cdot 2H_2O$. 2. Boleo.[3] Analysis on 94 mg. 3. Analysis 2 recalculated after deducting the AgCl as boleite ($9PbCl_2 \cdot 8CuO \cdot 3AgCl \cdot 9H_2O$).

Tests. B.B. easily fusible. Soluble in HNO_3.

Occur. Found with boleite and cumengite at Boleo near Santa Rosalia, Lower California, Mexico.

Pseudoboleite and boleite have been considered to be identical but the evidence[4] favors the individual character and tetragonal symmetry of the former species.

Ref.

1. Elements and forms of Friedel, *Bull. soc. min.*, **29**, 14 (1906).
2. Hocart, *Zs. Kr.*, **74**, 21 (1930).
3. Friedel (1906).
4. See Gossner and Arm, *Zs. Kr.*, **72**, 202 (1929); Friedel, *Zs. Kr.*, **73**, 154 (1930); Gossner, *Zs. Kr.*, **75**, 365 (1930); Hocart (1930).

10.2.4 **PERCYLITE** [$PbCuCl_2(OH)_2$?]. Brooke (*Phil. Mag.*, **36**, 131, 1850).

Isometric? Crystals small, usually cubical or dodecahedral. Also massive. Forms: $a\{001\}$, $d\{011\}$, $o\{111\}$, $e\{012\}$. H. $2\frac{1}{2}$. G. not known. Luster vitreous. Color and streak sky-blue. Transparent. Optically isotropic,[1] with n 2.05 ± 0.01.

Chem. A hydroxide-chloride of lead and copper, $PbCuCl_2(OH)_2$?. The reported analyses were made on very impure material and the formula is uncertain.

Anal.

	Pb	Cu	Ag	Cl	H_2O	O	Rem.	Total
1.	55.15	16.92		18.87	4.80	4.26		100.00
2.	37.64	8.78	8.98	13.37	2.87	n.d.	24.37	96.01
3.	36.93	6.80	0.90	9.56	6.52	n.d.	33.80	94.51

1. $PbCuCl_2(OH)_2$. 2. South Africa.[6] Rem. is $PbSO_4$ 22.98, CO_2 1.39. 3. Tarapacá, Chile.[4] Rem. is Sb_2O_3 2.50, MnO 1.30, SiO_2 30.00.

Tests. In C.T. decrepitates, loses water, and fuses to a brownish melt. In boiling HNO_3 turns white and dissolves. On slight heating turns emerald-green but becomes blue again on cooling.

O c c u r. A secondary mineral, found associated with matlockite, caracolite, daviesite, cerargyrite, cerussite, anglesite, limonite. The original mineral was found with gold and limonite, and supposed to be from Sonora, Mexico. It has since been found in Namaqualand, South Africa,[3] and at localities in northern Chile notably the Beatriz mine in the Sierra Gorda, Antofagasta,[2] and in the Pica district in the Cerro de Challacolla, Tarapacá.[4]

A r t i f.[5] Obtained in crystals by the action of a solution of $CuCl_2$ on precipitated $Pb(OH)_2$.

N a m e. After John Percy (1817–1889), English metallurgist.

Ref.

1. Larsen (118, 1921) who gives nO 2.06 ± 0.01 and birefringence 0.02 for another specimen, locality and authenticity not known. Friedel, *Bull. soc. min.*, **15**, 96 (1892), found artificial octahedral crystals of the composition $PbCuCl_2(OH)_2$ to be strongly birefringent.
2. Fletcher, *Min. Mag.*, **8**, 171 (1889); Sandberger, *Jb. Min.*, II, 75 (1887); Websky, *Ak. Berlin, Sitzber.*, 1045 (1886).
3. Maskelyne and Flight, *J. Chem. Soc. London*, **25**, 1051 (1872).
4. Raimondi (174, 1878).
5. Friedel (1892).
6. Flight analysis in Maskelyne and Flight (1872).

10.2.5 D I A B O L E I T E [$Pb_2CuCl_2(OH)_4$]. *Spencer* and *Mountain* (*Min. Mag.*, **20**, 78, 1923).

C r y s t.[1] Tetragonal; ditetragonal-pyramidal—4 *m m*.

$$a:c = 1:0.9361; \quad p_0:r_0 = 0.9361:1$$

Forms:[1]

Lower	Upper	φ	ρ	A	M
c	c 001		0°00'	90°00'	90°00'
	a 010	0°00'	90 00	90 00	135 00
	m 110	45 00	90 00	45 00	90 00
$\bar{r}$	r 012	0 00	25 05	90 00	72 33½
$\bar{e}$	e 011	0 00	43 06½	90 00	61 49
$\bar{s}$	s 021	0 00	61 53½	90 00	51 25
$\bar{n}$	n 112	45 00	33 30	67 01½	90 00
	p 111	45 00	52 56	55 39	90 00

Structure cell.[2] Space group $P4mm$. a_0 5.83 ± 0.02 kX, c_0 5.46 ± 0.02; $a_0:c_0 = 1:0.937$. Cell contents $Pb_2Cu(OH)_4Cl_2$.

Diaboleite

Habit. Tabular {001}, with a square outline. The large base is negative; the positive pedion is either lacking or is very small with a more or less curved surface. Sometimes as subparallel aggregates of thin plates.

P h y s. Cleavage {001} perfect but not easy. Fracture conchoidal. Brittle. H. $2\frac{1}{2}$. G. 5.42 ± 0.01;[3] 5.48 (calc.). Color deep blue. Streak pale blue. Transparent.

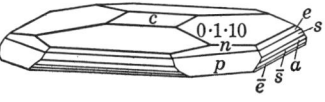

Mammoth mine, Arizona.

O p t.[4] In transmitted light, pale blue. May exhibit a slight distortion of the optic axis interference figure, perhaps due to subparallel growth.

	n	
O	1.98 ± 0.01	Uniaxial negative $(-)$.
E	1.85 ± 0.01	Absorption $O > E$.

C h e m. A hydroxide-chloride of lead and copper, $Pb_2CuCl_2(OH)_4$.

Anal.

	1	2	3
PbO	72.36	72.09	72.01
CuO	12.90	12.90	12.68
Cl	11.49	10.89	11.42
H$_2$O	5.84	6.14	6.03
Insol.			0.19
	102.59	102.02	102.33
Less O = Cl	2.59	2.46	2.57
Total	100.00	99.56	99.76
G.		5.48	5.42

1. $Pb_2CuCl_2(OH)_4$. 2. Mendip Hills.[5] 3. Mammoth mine.[6]

Tests. In the C.T. breaks into cleavage fragments with evolution of water followed by vaporization of lead chloride. The residue melts to a brown liquid which on cooling becomes a bright green glass. Completely soluble in nitric acid.

O c c u r. Found originally at Higher Pitts farm, Mendip Hills, Somerset, England, as minute crystals associated with mendipite, chloroxiphite, hydrocerussite, and cerussite in oxidized iron and manganese ores. Recently found well-crystallized in oxidized ore in the Collins vein of the Mammoth mine, Tiger, Arizona, associated with boleite, cerussite, hydrocerussite, dioptase, hemimorphite, and other secondary minerals.

N a m e. From διά, *apart* or *distinct from*, and boleite.

Ref.

1. Palache, *Am. Min.*, **26**, 605 (1941), on crystals from the Mammoth mine. Uncertain forms: 0.1.10, 029, 014, 02$\bar{3}$, 03$\bar{2}$, 04$\bar{1}$.
2. Wolfe in Palache (1941) by Weissenberg and Laue methods.
3. Berman in Palache (1941) by microbalance. The value given in Spencer and Mountain (1923), 6.412, apparently is in error.
4. Berman in Palache (1941).
5. Mountain anal. in Spencer and Mountain (1941).
6. Gonyer anal. in Palache (1941).

84 HALIDES

10.2.6 CHLOROXIPHITE [$Pb_3CuO_2(OH)_2Cl_2$ (?)]. *Spencer (Min. Mag., 20, 76, 1923).*

Cryst.[1] Monoclinic; prismatic—$2/m$.

$a:b:c = 1.805:1:1.138;$ $\quad \beta\, 97°11';$ $\quad p_0:q_0:r_0 = 0.631:1.129:1$

$r_2:p_2:q_2 = 0.886:0.559:1;$ $\quad \mu\, 82°49';$ $\quad p_0'\, 0.635,\, q_0'\, 1.138,\, x_0'\, 0.126$

Forms:[2]

	ϕ	ρ	ϕ_2	$\rho_2 = B$	C	A
c 001	90°00'	7°11'	82°49'	90°00'	.....	82°49'
a 100	90 00	90 00	0 00	90 00	82°49'	......
$D\,\bar{1}01$	−90 00	26 59½	116 59½	90 00	34 10½	116 59½

Structure cell.[3] Space group $P2_1/m$. $a_0\, 10.34\, kX$, $b_0\, 5.73$, $c_0\, 6.52$; $\beta\, 97°11'$; $a_0:b_0:c_0 = 1.804:1:1.138$. Cell contents $Pb_6Cu_2O_4(OH)_4Cl_4$.

Habit. In bladed crystals elongated [010] and flattened {$\bar{1}01$}, without measurable faces. In sub-parallel groupings. {$\bar{1}01$} roughly striated [010].

Phys. Cleavage {$\bar{1}01$} perfect, {100} distinct. Very brittle. H. 2½. G. 6.763,[4] 6.93;[5] 7.07 (calc.). Luster resinous to adamantine. Color dull olive-green or pistachio-green, not unlike that of epidote. Streak pale greenish yellow.

Opt.[6]

ORIENTATION	n	PLEOCHROISM	
$X \sim \perp \{\bar{1}01\}$	2.16		Biaxial negative (−).
Y	2.24	Brown	$2V \sim 70°$.
$Z\ b$	2.25	Green	$r > v$, medium to strong.

Chem. A basic oxychloride of copper and lead. The formula is uncertain, probably $Pb_3CuO_2(OH)_2Cl_2$.

Anal.

	1	2	3
Pb	75.61	75.34	74.10
Cu	7.73	8.71	8.37
Cl	8.63	7.19	8.97
O	5.84	6.38	5.79
H_2O	2.19	2.56	2.52
Total	100.00	100.18	99.75

1. $Pb_3CuO_2(OH)_2Cl_2$. 2,3. Higher Pitts mine.[7]

Tests. In C.T. gives off some water and decrepitates violently. Gives a sublimate of lead chloride and a fusible residue of a mixture of lead and copper oxides which solidifies to a green glass. Soluble in nitric acid.

Occur. Found in the Higher Pitts mine in the Mendip Hills, Somerset, England, as blade-like crystals embedded in mendipite. The largest crystals are 3 cm. long, 1 cm. across and 1 mm. thick. Diaboleite also occurs on the specimens, and hydrocerussite, cerussite, malachite, and wad are found at the locality.

Name. From χλωρός, *green*, and ξιφος, *a blade* or *straight sword*, in allusion to the habit of the crystals. Pronounced chloro-xiphite.

Ref.
1. Elements from x-ray Weissenberg study of Berman, priv. comm. (1942), on Spencer's material.
2. $\{\bar{1}01\}$ and $\{100\}$ from Spencer (1923), $\{001\}$ from Berman (1942). Transformation: Berman to Spencer, $10\bar{1}/010/001$, making $D \wedge c$ of Berman, $63°00\frac{1}{2}'$, equal to $a \wedge c$ of Spencer, $62\frac{3}{4}°$.
3. Berman (1942).
4. Mountain in Spencer (1923) by pycnometer.
5. Highest of five widely varying values measured by Berman (1942) on microbalance.
6. Berman (1942).
7. Mountain analysis in Spencer (1923).

10.2.7 **N O C E R I T E** [$Ca_3Mg_3F_8O_2$]. Nocerina *Scacchi* (*Acc. Linc., Trans.*, **5**, 270, 1881). Nocerin. Noceran. Nocerite *Dana* (174, 1892).

C r y s t.[1] Hexagonal—*P*.

$$a:c = 1:0.353; \quad p_0:r_0 = 0.403:1$$

Forms:[2]

	ϕ	ρ	$\overline{M}$	A_2
c 0001	.....	0°00'	90°00'	90°00'
m 10$\bar{1}$0	30°00'	90 00	60 00	90 00
a 11$\bar{2}$0	0 00	90 00	90 00	60 00

Structure cell.[3] Space group $C\bar{6}$, $C6$ or $C6/m$. a_0 8.84 kX, c_0 3.12; $a_0:c_0 = 1:0.353$. Cell contents $Ca_3Mg_3F_8O_2$.

Habit. Acicular [0001]; as fibrous aggregates.

P h y s. Data are lacking on cleavage and hardness. G. 2.96;[4] 2.947 (calc.). Luster vitreous. Colorless to white or brownish; rarely greenish. Transparent.

O p t. In transmitted light, colorless.

	$n(Na)_{brown}$[4]	$n(Na)_{colorless}$[4]	n[5]	
O	1.5084	1.5098	1.512	Uniaxial negative $(-)$.
E	1.4856	1.4855	1.487	Weakly dichroic in brown, with absorption $O > E$.

C h e m. An oxyfluoride of calcium and magnesium, probably $Ca_3Mg_3F_8O_2$. The reported variation in indices of refraction indicates a variation in the composition. The nature of this is not known but may involve a substitution of Na for Ca (anal. 2).

Anal.

	1	2	3
Ca	31.88	27.97	26.85
Na		2.19	2.47
Mg	19.34	18.18	17.47
F	40.30	40.08	37.55
O	8.48	8.30	11.40
H_2O		1.50	
Rem.		1.31	4.86
Total	100.00	99.53	100.60

1. $Ca_3Mg_3F_8O_2$. 2. Nocera, Italy.[4] Rem. is Li trace, K 0.69, Fe 0.59, Mn 0.03, Cl trace. 3. Nocera, Italy.[6] Rem. is Al 4.35, K 0.51. Average of two.

86 HALIDES

Tests. B.B. fusible with difficulty. Soluble in acids.

Occur. Found with fluorite, biotite, hornblende, and hematite in metamorphosed, embedded blocks of limestone in volcanic tuff between Nocera and Sarno, Campania, Italy.

Name. From the locality.

METANOCERITE. Metanocerin *Sandberger* (*Jb. Min.*, I, 221, 1892).
An ill-defined mineral found as small, rough white crystals associated with babingtonite, hornblende, and epidote at Arendal, Norway. H. $4\frac{1}{2}$. Reacts qualitatively for Ca, Mg, Na, F, and minor Al.

Ref.

1. Ratio from x-ray unit of Scherillo, *Per. Min.*, **9**, 229 (1938).
2. Zambonini, *Mem. serv. alla descr. cart. geol. d'Italia*, **7**, Pt. 2 (1919)—*Zs. Kr.*, **56**, 219 (1921), who also observed unmeasurable pyramids.
3. Scherillo (1938) by Laue and rotation methods. Unit cell (a_0 8.93 kX, c_0 3.12) and possible space groups confirmed by Frondel, priv. comm. (1942) by Weissenberg method.
4. Zambonini (1919).
5. Larsen (116, 1921).
6. Lederer analysis in Fischer, *Zs. Kr.*, **10**, 270 (1885); tests on other specimens indicate that the aluminum and alkalies are due to impurities.

10.2.8 **KOENENITE** [$Mg_5Al_2(OH)_{12}Cl_4$]. *Rinne* (*Cbl. Min.*, 493, 1902). Justit *Koechlin* (*Min. Mitt.*, **23**, 97, 1903).

Cryst. Probably hexagonal—*R*. Found as crusts of indistinct crystals, in part as very acute scalenohedra [1] and rhombohedra; as rosettes of intergrown tablets.

Phys. Cleavage {0001} perfect. Thin folia flexible. H. $\sim 1\frac{1}{2}$. G. 1.98.

Luster on cleavage surfaces pearly. Colorless when pure; usually pale yellow to deep red due to included scales of hematite. Transparent to translucent.

Opt.[2] In transmitted light, reddish brown to colorless.

	n	DICHROISM	
O	1.52	Red-brown	Uniaxial positive (+).
E	1.55	Colorless	

Chem. A hydroxide-chloride of magnesium and aluminum, probably $Mg_5Al_2(OH)_{12}Cl_4$. A small amount of Fe" apparently can substitute for Mg, and gives rise on oxidation to included hematite.

Anal.

	1	2	3
MgO	23.20	21.10	23.44
$MgCl_2$	36.51	35.70	36.85
Al_2O_3	19.59	17.79	18.25
H_2O	20.70	25.41	21.46
Total	100.00	100.00	100.00

1. $Mg_5Al_2(OH)_{12}Cl_4$. 2. Justus I mine, Hannover.[3] Recalc. to 100 after deduction of alkali chlorides 15.94, insol. 0.35, $MgCl_2$ and H_2O soluble in alcohol 1.87, 2.13. 3. Justus I mine.[4] Recalc. to 100 after deducting alkali chlorides 18.48, insol. 0.245.

Occur. Found with halite, sylvite, anhydrite, and carnallite in the potash mine Justus I near Volpriehausen in the Solling, Hannover, Germany. Also similarly at the Glückauf-Sarstedt mine, Hannover.[5]
Alter.[6] Decomposed by hot water or NH_4Cl solution with removal of Cl and Mg, leaving a porous scaly pseudomorph of $Al_2O_3 \cdot nH_2O$ with $n \sim 2$. On ignition, a pseudomorph of Al_2O_3 remains.
Name. After Adolph von Koenen (1837–1915), of Göttingen, geologist, who first found the mineral.

Ref.

1. Described by Rinne (1902) as making plane angles of 152° and 88° on {0001} sections and corresponding perhaps to {31̄41}.
2. Larsen and Berman (69, 1934).
3. Sundmacher analysis in Rinne (1902).
4. Buchholz analysis in Rinne (1902).
5. Erdmannsdörffer, *Cbl. Min.*, 449 (1913).
6. Rinne (1902); Erdmannsdörffer (1913).

10.2.9 Zirklerite. *Harbort* (*Kali*, **22**, 157, 1928).

Probably hexagonal—R. Found only massive, granular to fibrous. Cleavage rhombohedral. H. $\sim 3\frac{1}{2}$. G. ~ 2.6. Optically uniaxial positive $(+)$. Mean index of refraction 1.552; birefringence somewhat less than 0.009. Apparently a basic chloride of Al and Fe″, with minor Ca and Mg. Analysis gave, after deducting 31.26 insoluble, 5.08 anhydrite and 15.64 halite: $CaCl_2$ 2.47, $MgCl_2$ 6.83, $FeCl_2$ 57.20, Al_2O_3 12.29, H_2O 21.23, total 100.02. Found as the chief constituent of a rock containing halite, clay, anhydrite, quartz, and minor rinneite, dolomite, and chlorite occurring in breccia-like layers in halite and potash salts in the Adolfsglück mine at Hope, Hannover, Germany. Also reported from other potash deposits in northern Germany. Named after Dr. Zirkler, Director of the Aschersleben potash works. Needs verification.

10.2.10 KLEINITE. Yellow mercury mineral (No. 5) *Moses* (*Am. J. Sc.*, **16**, 253, 1903). [Mercurammonite] *Hillebrand* (*Science*, **22**, 844, 1905; *Am. J. Sc.*, **21**, 85, 1906; *J. Am. Chem. Soc.*, **28**, 122, 1906). Kleinite *Sachs* (*Ak. Berlin, Sitzber.*, 1091, 1905; *Cbl. Min.*, 200, 1906).

Cryst. Hexagonal (above 130° ?).[1]

$$a:c = 1:0.832;{}^2 \quad p_0:r_0 = 0.960:1$$

Forms:[3]

	ϕ	$\rho = C$	M	A_2
c 0001	.....	0°00′	90°00′	90°00′
m 10$\bar{1}$0	30°00′	90 00	60 00	90 00
a 11$\bar{2}$0	0 00	90 00	90 00	60 00
x 10$\bar{1}$1	30 00	43 50$\frac{1}{2}$	69 44$\frac{1}{2}$	90 00
p 20$\bar{2}$1	30 00	62 30	63 40$\frac{1}{2}$	90 00

Structure cell.[9] Space group probably $C6/mmc$. a_0 13.56 Å, c_0 11.13; $a_0:c_0 = 1:0.821$. Cell contents not known.

Habit. Short prismatic [0001] with variable development of $\{10\bar{1}1\}$; occasionally equi-dimensional. Commonest forms $m\ c\ p$.

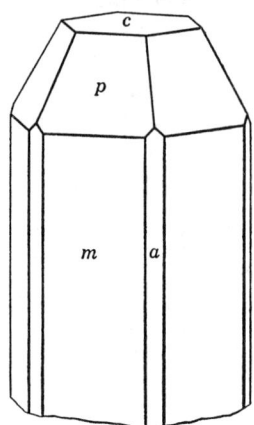

Phys. Cleavage $\{0001\}$ easy but uneven, $\{10\bar{1}0\}$ imperfect. Brittle. H. $3\frac{1}{2}$–4. G. ~ 8.0.[4] Luster adamantine to greasy. Color pale yellow and canary-yellow to orange, changing to reddish yellow or orange in daylight and reverting to the original color in darkness. Streak sulfur-yellow. Transparent to translucent.

Opt.[5] In transmitted light, yellow to colorless.

Ordinary Biaxial Form, below 130°			Uniaxial Form, above 130°		
nX	2.16	Biaxial negative (−)	nO	2.19	Uniaxial positive (+)
nY	2.18	$2V$ small to medium	nE	2.21	Isotropic over about 190°[6]
nZ	2.18	$r < v$, very strong			

Chem. A chloride-sulfate of divalent mercury and ammonia of unknown formula. An average of numerous partial analyses[7] gave: Hg 85.86, N 2.57, Cl 7.30, SO_4 3.10, H_2O 1.03, total 99.86.

Tests. In C.T. gives a little water, darkens in color, and at 260°–280° mercury and calomel sublime, leaving a nearly white residue; at higher temperatures another sublimate is formed. Soluble in warm HCl and HNO_3 without deposition of calomel. Soluble in ammonium bromide with evolution of ammonia.

Occur. Found at Terlingua, Brewster County, Texas, associated with gypsum, calcite, barite, terlinguaite, and the other mercury minerals of the locality.

Artif. Artificial mercury-ammonia chlorides and sulfates have been reported[8] but their relation to kleinite is not known.

Name. After Carl Klein (1842–1907), Professor of Mineralogy in the University of Berlin.

Ref.

1. The crystals apparently are inversion pseudomorphs after a high-temperature hexagonal form, as suggested by the disappearance at ca. 130° of optical anomalies observed at room temperature in basal sections and their reappearance on cooling.
2. Sachs, *Ak. Wiss. Berlin, Sitzber.*, 1091 (1905), with the c-axis halved to conform to the structure cell.
3. Hillebrand and Schaller, *U. S. Geol. Sur., Bull. 405*, 18 (1909).
4. Hillebrand and Schaller (1909) give values from 7.94 to 8.04, by pycnometer, but, with the possible exception of the high value, these are somewhat low due to contamination.
5. Indices from Larsen (96, 1921). The behavior of the optical anomalies on heating is described by Hillebrand and Schaller (1909).
6. Schaller, *U. S. Geol. Sur., Bull. 509*, 108 (1912).
7. Hillebrand in Hillebrand and Schaller (1909). See also Sachs, *Cbl. Min.*, 200 (1906), for additional analyses.
8. In Doelter (4 [3], 405, 1930) and Mellor (4, 1923).
9. Heritsch, *Min. Mitt.*, ser. 3, 1, 300 (1949).

10.2.11   **M O S E S I T E.** *Canfield, Hillebrand,* and *Schaller* (*Am. J. Sc.*, 30, 202, 1910).

C r y s t.   Isometric.[1]

Forms:[2]

a 001	o 111	d 011	ϕ 116	μ 114	n 112

Structure cell.[3] Face centered. $a_0 = 9.55\ kX$.

Habit. Usually octahedral; also cubo-octahedral or cubical.

Twinning. Twin plane {111} common; sometimes as repeated groups.

P h y s. Cleavage {111} imperfect. Fracture uneven. Brittle. H. $3\frac{1}{2}$. G. unknown. Luster adamantine. Color lemon-yellow to canary-yellow, becoming light olive-green on long exposure to light. Streak very pale yellow.

O p t. Isotropic. Sometimes weakly birefringent, becoming isotropic when heated over ca. 186°. Refractive index:[4] 2.065 ± 0.010 (Terlingua).

C h e m. A chloride-sulfate of monovalent mercury and ammonia. The formula is uncertain, perhaps $Hg_6(NH_3)_2Cl_2(SO_4)(OH)_4$.

Anal.

	1	2	3
Hg	81.73		83.0
NH_3	2.31		2.2
Cl	4.82	5.0	5.0
SO_4	6.52	3.5	7.0
H_2O	2.45		
O	2.17		
Total	100.00		97.2

1. $Hg_6(NH_3)_2Cl_2(SO_4)(OH)_4$. 2. Terlingua.[5] Analysis on 40 mg. 3. Fitting District, Nevada.[6] Analysis on 5.88 mg.

Tests. Heated slowly in C.T. turns dark reddish brown to almost black and then at higher temperature becomes white. Fumes of calomel are given off and condense in the tube, with globules of mercury formed beyond the sublimate. Heated rapidly in C.T. decrepitates violently, fuses, and volatilizes. In cold HCl changes slowly to a white substance which retains the original form of the grain.

Occur. Found very sparingly at Terlingua, Brewster County, Texas, associated with montroydite, calcite, and gypsum. Also noted [7] from the Clack Quicksilver mine in the Fitting District, northeast of Lovelock, Pershing County, Nevada, where the mineral occurs with native mercury in a cinnabar deposit.

Artif. A number of artificial mercury-ammonia compounds with chlorides and sulfates have been reported.[8]

Name. After Alfred J. Moses (1859–1920), Professor of Mineralogy at Columbia University, who first described several of the mercury minerals from Terlingua.

Ref.

1. Some crystals are weakly birefringent, becoming isotropic over *ca.* 186°, and may be pseudo-isometric inversion products—see Canfield, *et al.* (1910).
2. Canfield, *Sch. Mines Q.*, **34**, no. 3 (1913). $d\{011\}$ observed by Bird, *Am. Min.*, **17**, 541 (1932), on Nevada crystals.
3. Bird (1932) by powder method.
4. Larsen (113, 1921).
5. Hillebrand anal. in Canfield, *et al.* (1910); a little water is said to be present.
6. Smoot anal. in Bird (1932).
7. Bird (1932).
8. Mellor (**4,** 1923) and Doelter (4 [3], 405, 1930).

11 HALIDE COMPLEXES. ALUMINO-FLUORIDES

TYPE 1. $A_m BX_3 \cdot xH_2O$

11.1.1 CHLOROCALCITE [KCaCl$_3$]. Hydrophilite pt. *Hausmann* (**3**, 857, 1813). Chlorure de calcium pt. (?) *Beudant* (**2**, 512, 1832). Idroclorato di calce in piccola dose *Monticelli* and *Covelli* (*Storia di fenom. del Vesuvio avven. negli anni, 1821, etc., Naples, 1823*, p. 183). Chlorocalcite *Scacchi* (*Acc. Napoli, Rend.*, Oct. 12, 1872; *Acc. Napoli, Mem.*, Dec. 13, 1873). Baeumlerite *Renner* (*Cbl. Min.*, 106, 1912).

C r y s t.[1] Probably orthorhombic, and pseudo-isometric. In cube-like crystals, sometimes with octahedral or dodecahedral-like truncations; also prismatic with {100}, {010} and {011} or tabular with {001}, {110} and {100}.

P h y s. Cleavage cube-like, with one direction better than the other two.[1] H. $2\frac{1}{2}$–3.[2] Color white, sometimes stained violet. Transparent. Deliquescent. Taste bitter.

O p t.[3] Biaxial negative (−), with weak birefringence. The axial plane is perpendicular to a face of the pseudo-cube, with $nY \sim 1.52$. May exhibit polysynthetic twin lamellae parallel to the pseudo-cubical faces.

C h e m. A chloride of potassium and calcium, KCaCl$_3$.
Anal.[4]

	1	2
K	21.07	21.44
Na		0.17
Ca	21.60	21.40
Cl	57.33	57.56
Total	100.00	100.57

1. KCaCl$_3$. 2. Vesuvius.[4]

O c c u r. Found as a volcanic sublimate associated with sylvite, halite, and hematite on Vesuvius, Italy. Formed notably during the eruptions of 1822, 1872, and 1906. Also found intergrown with tachyhydrite and halite in the Desdemona potash mine in the Leinetal, Prussia (*baeumlerite*). Some or all of the localities given for the presumed mineral hydrophilite, CaCl$_2$, actually may be of this species.

A r t i f.[5] The compound KCaCl$_3$ has been observed in melts in the system KCl-CaCl$_2$. Also obtained as crystals by evaporation of solutions containing KCl and CaCl$_2$.

N a m e. The name chlorocalcite was originally given by Scacchi to a mineral found as a volcanic sublimate on Vesuvius that was presumed to be anhydrous calcium chloride. Earlier, the name hydrophilite had been applied by Hausmann to a mineral found in anhydrite at Lüneberg, Germany, also thought to be calcium chloride. The two names were then considered synonymous, although with much uncertainty about the existence and properties of the mineral, until Zambonini in 1910 showed that the Vesuvius mineral had the composition $KCaCl_3$. The name hydrophilite is here retained for the true $CaCl_2$, but there is no rigorous evidence that any of the reported occurrences actually are of that substance.

Ref.
1. Zambonini (50, 1910; 3, 1912; 99, 1935).
2. Renner, *Cbl. Min.*, 106 (1912), on material from the Desdemona potash mine (*baeumlerite*).
3. Renner (1912) and Zambonini (1935).
4. The only available analysis, by Zambonini (50, 1910), and cited in column 2, conforms closely to the formula given. A later analysis of material from the Desdemona mine is also said by Renner (1912) to conform to the formula but the numerical values are not given. The early partial analyses by Scacchi (1872) of material ascribed to $CaCl_2$ contained significant amounts of alkalies, although not directly reported, and, as appears from evidence of Zambonini (1910) and Lacroix, *Bull. soc. min.*, 30, 255 (1907), 31, 261 (1908), actually represent the present species.
5. Mellor (3, 719, 1923).

11.1.2 CARNALLITE [$KMgCl_3 \cdot 6H_2O$]. Rose (*Ann. Phys.*, 98, 161, 1856). Kalium Magnesium chlorid *Germ.*
   Almeraite (?) *Tomás* (*Treballs Inst. Catalana Hist. Nat. Barcelona*, 129, 1919–20). Sebkhainite *Berthon* (*L'Industrie minérale en Tunisie*, 176, 1922).

Cryst.[1] Orthorhombic; dipyramidal—$2/m\ 2/m\ 2/m$.

$a:b:c = 0.5930:1:1.3952; \qquad p_0:q_0:r_0 = 2.13528:1.3952:1$

$q_1:r_1:p_1 = 0.5930:0.4250:1; \qquad r_2:p_2:q_2 = 0.7167:1.6863:1$

Forms:[2]

	ϕ	$\rho = C$	ϕ_1	$\rho_1 = A$	ϕ_2	$\rho_2 = B$
c 001	....	0°00'	0°00'	90°00'	90°00'	90°00'
b 010	0°00'	90 00	90 00	90 00	....	0 00
m 110	59 20	90 00	90 00	30 40	0 00	59 20
d 023	0 00	42 55½	42 55½	90 00	90 00	47 04½
e 011	0 00	54 22	54 22	90 00	90 00	35 38
f 021	0 00	70 17	70 17	90 00	90 00	19 43
i 101	90 00	66 58½	0 00	23 01½	23 01½	90 00
s 113	59 20	42 21½	24 56½	54 35	51 53½	69 54
o 112	59 20	53 49½	34 54	46 01½	40 22	65 41
k 111	59 20	69 55	54 22	36 07	23 01½	61 22½

Structure cell.[3] Space group *Pban*. a_0 9.54 kX, b_0 16.02, c_0 22.52; $a_0:b_0:c_0 = 0.5955:1:1.406$. Cell contents $K_{12}Mg_{12}Cl_{36} \cdot 72H_2O$.

Habit. Crystals pseudo-hexagonal pyramidal in habit, due to the more or less equant development of pyramids and brachydomes; sometimes thick tabular with large {0001}. Ordinarily massive, granular.

Twinning.[4] Secondary twinning lamellae, with $K_1(110)$, $K_2(1\bar{3}0)$, can be developed by pressure.

Phys. No distinct cleavage. Fracture conchoidal. H. 2½. G. 1.602;[5] 1.598 (calc.). Luster greasy, dull to shining. Colorless to milk-white, often reddish with a metallic schiller due to enclosed oriented[6] scales of

CARNALLITE

hematite; rarely yellow or blue. Transparent to translucent. Deliquescent in a moist atmosphere. Taste bitter.

O p t. In transmitted light, colorless. May exhibit polysynthetic twinning.

ORIENTATION		n(Na) [7]	n (New Mexico) [8]	
X	c	1.4665	1.465	Biaxial positive (+).
Y	b	1.4753	1.474	$r < v$.
Z	a	1.4937	1.496	
$2V$		70°03′	66° (calc.)	

C h e m. A hydrated chloride of potassium and magnesium, $KMgCl_3 \cdot 6H_2O$. Br substitutes for Cl in natural material in amounts less than 0.5 weight per cent; also small amounts of Rb, Cs, NH_4, and Tl [9] substitute for K. Fe″ apparently substitutes in small amount for Mg, giving rise on oxidation and exsolution to enclosed oriented plates of hematite.[6] Small amounts of Na, Ca and SO_4 are sometimes reported, and are probably due to admixture. In artificial material, NH_4 and Br substitute for K to

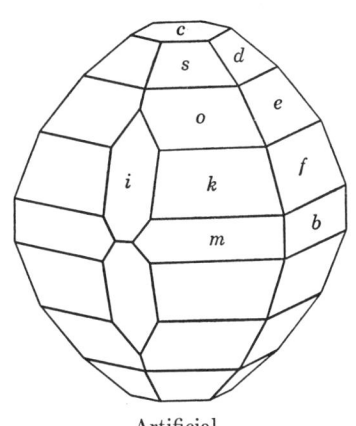

Artificial.

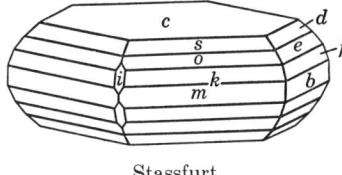

Stassfurt.

roughly 15 and 12 mol per cent, respectively; the series extend beyond to the end members but with the intervention of polymorphism.[10]

Anal.[15]

	1	2	3
K	14.07	13.51	14.07
Mg	8.75	8.80	8.80
Cl	38.28	38.16	38.32
H_2O	38.90	[39.53]	38.38
Rem.			0.16
Total	100.00	[100.00]	99.73
G.	1.598		

1. $KMgCl_3 \cdot 6H_2O$. 2. Königslutter, Germany.[11] 3. Eddy County, New Mexico.[12] Rem. is insol. 0.04, Na trace, Br 0.12.

Tests. B.B. fuses easily. Soluble in water (64.5 parts in 100 parts H_2O at 18.75°); the evaporated solution deposits sylvite and bischofite.

O c c u r. Carnallite is found associated with sylvite, halite, polyhalite, and especially kieserite in the upper layers of saline deposits of the marine type (Stassfurt). The mineral in part appears to have formed by the reaction of pre-existing saline minerals with liquors rich in potash. The principal occurrences are in the salt deposits of northern Germany, notably in the Stassfurt district, at Leopoldshall and Westeregeln, and at Beienrode in Upper Silesia. Also found at Kalusz, Galicia; at Maman, Azerbaijan; and in Barcelona and Lerida provinces, Catalonia, Spain. With sylvite at Mt. Dallol, on the plain of Danakil, Abyssinia, and with kainite and other potash salts at Ozinki, Saratov, U.S.S.R. Also found with halite and epsomite in salt pans in Tunisia.

Carnallite occurs in the United States in the Permian salt basin of southeastern New Mexico and adjacent parts of Texas.[14] Here it is found as granular masses and veinlets associated with polyhalite, sylvite, and halite in anhydrite. Also from Grand County, Utah, with sylvite and kieserite.

A r t i f.[13] Carnallite is stable between $-12°$ and $167.5°$ in the system $KCl-MgCl_2-H_2O$.

N a m e. After Rudolph von Carnall (1804–1874), Prussian mining engineer.

ALMERAITE. Tomás (*Treballs Inst. Catalana Hist. Nat. Barcelona*, 129, 1919–1920); Tomás and Folch (*Bull. Inst. Catalana Hist. Nat.*, **11**, 11, 1914). Almeraita, Almeraïta.

An ill-defined reddish granular material from the salt deposits at Suria, prov. Barcelona, Spain. An analysis is close to $KCl \cdot NaCl \cdot MgCl_2 \cdot H_2O$, but the substance may be only a mixture of carnallite with halite. Named after Dr. Jaume Almera, of Barcelona.

Ref.

1. Busz, *Jb. Min., Festband*, 126 (1907).
2. Goldschmidt (**2**, 105, 1913). Less common and rare:

120	032	118	126
012	072	114	136
043	103	334	132

3. Andress and Saffe, *Zs. Kr.*, **101**, 451 (1939), by rotation and Weissenberg methods on artificial crystals. An earlier study by Leonhardt, *Zs. Kr.*, **66**, 506 (1928), gave comparable cell dimensions but an erroneous space group, *Pnna*.
4. Cf. Doelter 4 [2], 1191 (1928) and Johnsen, *Jb. Min., Beil. Bd.*, **23**, 252 (1907).
5. Przibylla, *Cbl. Min.*, 236 (1904), by pycnometer method—rounded from 1.6018.
6. Cf. Ruff, *Jb. Min.*, II, 9 (1908), Johnsen, *Cbl. Min.*, 169 (1909), Boeke, *Jb. Min.*, I, 52 (1911). The hematite plates are for the most part oriented with {0001} parallel carnallite {001}, also parallel {110} or {100}.
7. Busz, *Jb. Min., Festband*, 127 (1907) by prism method on a crystal from Beienrode.
8. Schaller in Schaller and Henderson, *U. S. Geol. Sur., Bull.* **833**, 22 (1932).
9. Lindberg, *Am. Min.*, **31**, 486 (1946); Boeke, *Zs. Kr.*, **45**, 369, 384 (1908); Jander and Busch, *Zs. anorg Chem.*, **194**, 38 (1930); Biltz and Marcus, *Zs. anorg. Chem.*, **62**, 184 (1909); Erdmann, *J. pr. Chem.*, **86**, 377 (1862); Hammerbacher, *Ann. Chem. Pharm.*, **172**, 82 (1875); Schaller and Henderson (1932); Doelter, 4 [2], 1187 (1928).
10. Boeke (1908), *Cbl. Min.*, 710 (1908); Andress and Saffe (1939); see also Hintze, 1 [2B], 2373, 2374 (1912), and Doelter, 4 [2], 1200 (1928).
11. Kleinfeldt analysis in Bücking, *Ak. Berlin, Sitzber.*, 539 (1901).
12. Fahey analysis in Schaller and Henderson (1932).
13. van't Hoff and Meyerhoffer, *Ak. Berlin, Sitzber.*, 487 (1897).
14. Cf. Schaller and Henderson (1932), King, *J. Sed. Petrol.*, **16**, 14 (1946).
15. For additional analyses, see Doelter, 4 [2], 1185 (1928).

11.1.3 TACHYHYDRITE [$CaMg_2Cl_6 \cdot 12H_2O$]. Tachhydrit *Rammelsberg* (*Ann. Phys.*, **98**, 261, 1856). Tachydrite.

C r y s t. Hexagonal—*R*.

$$a:c = 1:1.76\ ^1$$

Forms:[2]

	ϕ	ρ	A_1	A_2
c 0001	....	0°00'	90°00'	90°00'
r 10$\bar{1}$1	30°00'	63 46$\frac{1}{2}$	39 01$\frac{1}{2}$	90 00

Habit. Massive; in rounded masses. Artificial crystals are rhombohedral with {0001}.

Twinning.[3] Sometimes exhibits twin lamellae, probably secondary and due to pressure.

P h y s. Cleavage {10$\bar{1}$1}, perfect. H. 2. G. 1.667.[4] Color wax- to honey-yellow; also colorless. Luster vitreous. Transparent. Very deliquescent. Taste sharp and bitter.

O p t.[5] In transmitted light, colorless to pale yellow. Not dichroic.

	n(Na)	
O	1.520	Uniaxial negative (−).
E	1.512	

C h e m. A hydrated chloride of calcium and magnesium, $CaMg_2Cl_6 \cdot 12H_2O$. Fe'', in amounts less than 0.1 weight per cent, apparently can substitute for Mg and gives rise on oxidation to the yellow color of the mineral.[6] Br can be made to substitute in part for Cl in artificial material.[10]

Anal.

	1	2	3	4
Ca	7.74	7.16	7.72	7.46
Mg	9.40	9.97	9.71	9.51
Cl	41.10	40.85	40.89	40.34
H_2O	41.76	42.50	42.20	[42.69]
Total	100.00	100.48	100.52	[100.00]

1. $CaMg_2Cl_6 \cdot 12H_2O$. 2. Stassfurt.[7] Average of two. 3. Krügershall. Saxony.[8] 4. Stassfurt.[9]

Tests. B.B. fuses superficially and decomposes with loss of H_2O and some Cl to an infusible material. Easily soluble in water (160.3 parts of tachyhydrite dissolving in 100 parts H_2O at 18.75°).

O c c u r. A rare mineral in potash-rich saline deposits of the oceanic type. Often associated with kainite and carnallite, from which it may have been derived;[11] also with sylvite, halite, kieserite, anhydrite, bischofite. Found in the Stassfurt district, Saxony; also at Krügershall-Teutschenthal, near Mansfield, Saxony.

A r t i f.[12] Obtained in crystals by cooling a hot solution of $MgCl_2$ containing a large excess of $CaCl_2$.

N a m e. From ταχύς, *quick*, and ὕδωρ, *water*, in allusion to its ready deliquescence.

Ref.

1. Jung, *Cbl. Min.*, 274 (1926), by contact measurements on cleavage fragments. Other values ranging from 1.72 to 2.11, all based on poor measurements, were reported by Groth (74, 1874), de Schulten, *C.R.*, **111**, 928 (1890) and Kling, *Cbl. Min.*, 14 (1915).
2. de Schulten (1890) on artificial crystals.
3. Noted by Des Cloizeaux (74, 1867), but not remarked by later observers.
4. Average of five determinations agreeing within ±0.004 cited in the literature.
5. Jung (1926) by immersion method; Kling (1915) gives nO 1.5215, nE 1.5128, by total reflection, but without statement of wavelength.
6. Kling, *Cbl. Min.*, 44 (1915).
7. Hammerbacher, Inaug. Diss., Erlangen, 84 (1874)—in Dana (178, 1892).
8. Jung (1926).
9. Rammelsberg (1856).
10. Boeke, *Zs. Kr.*, **45**, 354 (1908).
11. Cf. Doelter (4 [2], 1219, 1928).
12. de Schulten (1890); see also van't Hoff, *Ozean. Salzablag.*, Leipzig, 1912, Arts. IV, XIV, XL, XLIV, XLVI on the general physico-chemical relations.

TYPE 2. A_mBX_4

11.2.1 Pseudocotunnite [K_2PbCl_4 (?)]. Pseudocotunnia *Scacchi* (*Acc. Napoli, Att.*, 6, no. 9, 1873).
Mellonite (?) *Palmieri* (*Acc. Napoli, Rend.*, 92, 1873).

C r y s t.[1] Orthorhombic (?). Crystals are flattened on {010} or {100}, with indistinct {$hk0$} and {$0kl$} faces, and are elongated [001] into needle or lath-like shapes. Also found as warty aggregates and dendritic crusts.

P h y s. Data are lacking on the cleavage, hardness, and specific gravity. Colorless to white, but usually yellow or greenish yellow. Luster dull. Exhibits parallel extinction, with $Z = [001]$. Birefringence medium strong.

C h e m. Probably a potassium lead chloride, K_2PbCl_4.
Anal.[2]

	1	2	3
K	18.30	17.11	18.97
Na		1.53	
Ca		2.13	
Pb	48.50	43.00	47.67
Cl	33.20	36.23	33.36
Total	100.00	100.00	100.00

1. K_2PbCl_4. 2. Vesuvius. Also contains some F and SO_4 tr. 3. Analysis 2 recalculated to 100 after deduction of Na and Ca as chlorides.

Tests. Easily fusible. Completely soluble in warm water.

O c c u r. Found associated with cotunnite and tenorite in fumaroles of the eruptions of Vesuvius in 1872 and 1906.
A r t i f. Observed in melts in the system $PbCl_2$-KCl.[3]
N a m e. From a resemblance of the crystals to cotunnite. Mellonite after Macedonio Melloni (1798–1854), Italian physicist and director of a meteorological observatory on Vesuvius.

Ref.
1. Cf. Zambonini (103, 1935) on natural material and Lorenz and Ruckstuhl, *Zs. anorg. Chem.*, **51**, 70, 80 (1906), on artificial crystals.
2. For additional analyses see Scacchi (1873).
3. Cf. Lorenz and Ruckstuhl (1906) and Mellor (**7**, 729, 1927).

AVOGADRITE AND FERRUCCITE

ORTHORHOMBIC

Avogadrite, $(K,Cs)BF_4$

Ferruccite, $NaBF_4$

Avogadrite and artificial $CsBF_4$ and $TlBF_4$ are isostructural with barite. Ferruccite, with a much smaller cation, is isostructural with anhydrite.

11.2.2 AVOGADRITE [$(K,Cs)BF_4$]. Zambonini (*Acc. Linc., Rend.*, [6], **3**, 644, 1926).

C r y s t.[1] Orthorhombic.

$a:b:c = 1.5796:1:1.2830;$ $\quad p_0:q_0:r_0 = 0.8122:1.2830:1$

$q_1:r_1:p_1 = 1.5796:1.2312:1;$ $\quad r_2:p_2:q_2 = 0.7794:0.6331:1$

Forms:[2]

	ϕ	$\rho = C$	ϕ_1	$\rho_1 = A$	ϕ_2	$\rho_2 = B$
c 001	.....	0°00'	0°00'	90°00'	90°00'	90°00'
a 100	90°00'	90 00	.....	0 00	0 00	90 00
m 210	51 42	90 00	90 00	38 18	0 00	51 42
n 011	0 00	52 04	52 04	90 00	90 00	37 56
d 101	90 00	39 05	0 00	50 55	50 55	90 00
o 211	51 42	64 13	52 04	45 02½	31 37	56 04½

Habit. Minute crystals, tabular to platy {001} and sometimes elongated [010] or [100], resembling barite. As dense crusts.

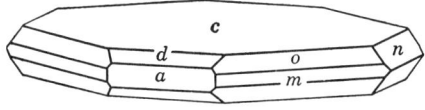

Vesuvius.

P h y s. G. 2.505 (for KBF_4); 2.617 (calc. for 18 per cent $CsBF_4$), 3.305 (for $CsBF_4$). Colorless to white; yellowish to reddish when impure. Taste bitter.

O p t.[3] In transmitted light, colorless.

Orientation		n (pure KBF_4)	
X	c	1.3239	Biaxial negative (−).
Y	b	1.3245	2V very large.
Z	a	1.3247	

Chem. Potassium, cesium fluoborate $(K,Cs)BF_4$. The Cs in natural material ranges from 0 to about 19 per cent $CsBF_4$. The corresponding K, Cs, Rb, and Tl salts probably are completely isomorphous. KBF_4 is isostructural with $KClO_4$, $KMnO_4$, and $BaSO_4$. An isometric high-temperature polymorph is known.[5] The available analyses [4] are on water extractions of admixed sublimates.

Tests. Easily fusible. Slightly soluble in water.

Occur. Found admixed with sassolite and other salts as a fumarolic incrustation on Vesuvius.

Artif.[6] Prepared by reaction of potassium salts with fluoboric acid, and by dissolving potassium carbonate and boric acid in hydrofluoric acid and evaporating.

Name. After Amedeo Avogadro (1776–1856), Professor of Physics in the University of Turin, Italy. Avogadro, known chiefly for his work in chemistry and physics, wrote the first complete treatise on crystallography in Italian.

Ref.

1. On artificial crystals of pure KBF_4 by Brugnatelli in Montemartini, *Acc. Linc., Rend.*, **3** [1], 339 (1894)—*Zs. Kr.*, **41**, 59 (1905). Partial x-ray data with b_0 5.68 kX and c_0 7.38 are given by Pesce, *Gazz. chim. Ital.*, **60**, 936 (1930). Crystal class not established with certainty. On the x-ray cell of $CsBF_4$ see Klinkenberg, *Rec. trav. chim. Pays-Bas*, **56**, 36 (1937). Transformation: Brugnatelli to new (barite unit), 200/010/001.
2. Artificial crystals; see Zambonini (93, 1935) for natural crystals.
3. Zambonini (93, 1935). Pure $CsBF_4$ has n(Na) 1.36.
4. Cf. Zambonini (1926); Zambonini and Coniglio, *Acc. Lincei, Rend.*, [6], **3**, 521 (1926); Carobbi, *Acc. Linc., Rend.*, [6], **4**, 382 (1926).
5. Finbak and Hassel, *Zs. phys. Chem.*, **32**, 433 (1936).
6. Mellor (**5**, 126, 1924).

11.2.3 **FERRUCCITE** [$NaBF_4$]. Carobbi (*Per. Min.*, **4**, 410, 1933).

Cryst. Orthorhombic. Artificial crystals [1] are tabular {001} or {010} with c{001}, a{100}, b{010}, and m{110}; $a \wedge m = 42°31'$, affording the partial ratio $a:b = 0.9169:1$.

Structure cell.[2] a_0 6.25 ± 0.02 kX, b_0 6.77 ± 0.01, c_0 6.82 ± 0.01; $a_0:b_0:c_0 = 0.922:1:1.007$. Cell contents $Na_4(BF_4)_4$.

Phys. Cleavage {100}, {010}, {001}. H. probably 3. G. 2.496; 2.511 (calc.). Colorless to white. Taste bitter and acid.

Opt. In transmitted light, colorless.

Orientation		n	
X	c	1.301	Biaxial positive (+).
Y	b	1.3012 (calc.)	$2V$ 11°25'.
Z	a	1.3068	

Chem. Sodium fluoborate, $NaBF_4$. The available analyses are on material much admixed with other salts. Ferruccite is isostructural with $NaClO_4$ and $KMnO_4$. Several high-temperature polymorphs are known.[3]

Tests. Easily fusible. Soluble in water.

Occur. Found as minute crystals in admixed fumarolic sublimates on Vesuvius. Obtained together with hieratite, avogadrite, sassolite, and malladrite by recrystallization of the sublimates from water solution.

Artif.[4] Formed in crystals by reaction of water solutions of sodium fluoride or sodium carbonate and fluoboric acid.

Name. After the Italian mineralogist Ferruccio Zambonini (1880–1932).

Ref.
1. Carobbi (1933).
2. Klinkenberg, *Rec. trav. chim. Pays-Bas*, **56**, 36 (1937); see also Bellanca, *Rend. soc. min. ital.*, **3**, 20 (1946).
3. Finbak and Hassel, *Zs. phys. Chem.*, **32**, 433 (1936).
4. Cf. Mellor (**5**, 126, 1924) and Carobbi (1933).

11.2.4 CRYOLITHIONITE [$Na_3Li_3Al_2F_{12}$]. Ussing (*Danske Vidensk. Selsk., Overs.*, **1**, 1904). Kryolithionite.

Cryst. Isometric; hexoctahedral—$4/m\,\bar{3}\,2/m$.

Forms:

$a\,001 \qquad d\,011 \qquad n\,112$

Structure cell.[1] Space group $Ia3d$. a_0 12.097 kX. Cell contents $Na_{24}Li_{24}Al_{16}F_{96}$. The structure is like that of garnet.

Habit. Dodecahedral, sometimes modified by {112} (Urals)[2]; also cubic (artif.).

Phys. Cleavage {011}, distinct. Fracture uneven to subconchoidal. Brittle. H. $2\frac{1}{2}$–3. G. 2.770 (Greenland)[3]; 2.772 (calc.). Luster vitreous. Colorless to white. Transparent. M.P. 710°.

Opt. In transmitted light, colorless. Isotropic.

	Li	Na	Tl
n	1.3382	1.3395	1.3408

Chem. Sodium lithium aluminum fluoride, $Na_3Li_3Al_2F_{12}$, or $Na_3Al_2(LiF_4)_3$—analogous to garnet.

Anal.

	1	2
Na	18.56	18.83
Li	5.60	5.35
Al	14.51	14.46
F	61.33	60.79
Ign.		0.36
Total	100.00	99.79

1. $Na_3Li_3Al_2F_{12}$. 2. Ivigtut.[4]

Tests. Easily fusible, more so than cryolite. In C.T. decrepitates violently and fuses to a colorless liquid. Decomposes at red heat, giving off fumes. Soluble in H_2SO_4, with evolution of HF. Slightly soluble in water (1 part in 1350 parts H_2O at 18°).

Occur. Found intimately associated with cryolite at Ivigtut, Greenland. The crystals of cryolithionite, which range up to 17 cm. in size, are often cloudy due to liquid and gaseous inclusions. Also found at the cryolite locality near Miask in the Ilmen Mountains, U.S.S.R.

Artif. Apparently deposited by evaporation of the water solution.

Name. From the lithium content and the resemblance to cryolite.

Ref.
1. Menzer, *Zs. Kr.*, **75**, 265 (1930), by powder method; the structural analogy to garnet was first suggested by Ussing (1904). The earlier x-ray study of Clausen, *Cbl. Min.*, 390 (1929), was found to be in error and withdrawn.
2. Bøggild, *Zs. Kr.*, **51**, 600 (1913).
3. Menzer (1930) by suspension method; Ussing (1904) gave 2.774–2.778.
4. Christensen analysis in Ussing (1904).

PROIDONITE. Proidonina Scacchi (*Acc. Napoli, Att.*, **6**, 65, 1873). Proidonite Dana (1868, App. III, 97, 1882).

A name applied to silicon tetrafluoride gas (SiF_4) emitted during the 1872 eruption of Vesuvius.

TYPE 3. $A_mBX_4 \cdot xH_2O$

11.3.1 Douglasite [$K_2FeCl_4 \cdot 2H_2O$?]. Ochsenius (*Steinsalzlager*, Halle, 94, 1877). Douglasite Precht (*Ber.*, **12**, 560, 1879; **13**, 2327, 1880). Eisenchlorürchlorkalium Germ.

Found as coarsely granular masses associated with carnallite, sylvite, and halite at Douglashall near Westeregeln, northwest of Stassfurt, Saxony. Color light green, altering on exposure to brownish red. Luster vitreous. Optically,[1] nearly uniaxial positive (+) with nO 1.488 ± 0.003, nE 1.500 ± 0.003.

Douglasite perhaps is $K_2FeCl_4 \cdot 2H_2O$, and identical with the artificial salt of this composition,[2] but definitive data are lacking.

Ref.
1. Larsen (69, 1921).
2. According to Schabus, *Ak. Wien, Sitzber.*, **4**, 475 (1850), the artificial salt is monoclinic prismatic, $a:b:c = 0.7367:1:0.5036$, β 104°46'; G. 2.162; cleavable $\{\bar{2}01\}$. See also Boeke, *Jb. Min.*, II, 44 (1909).

11.3.2 MITSCHERLICHITE [$K_2CuCl_4 \cdot 2H_2O$]. Zambonini and Carobbi (*Ann. R. Osservat. Vesuviano* [3], **2**, 7, 1925).

Cryst. Tetragonal; ditetragonal-dipyramidal—$4/m\ 2/m\ 2/m$ (?).

$$a:c = 1:1.0640;^1 \qquad p_0:r_0 = 1.0640:1$$

Forms:[2]

	ϕ	ρ	A	M
o 011	0°00'	46°46½'	90°00'	58°59'
m 110	45 00	90 00	45 00	90 00

Structure cell.[3] Space group probably $P4/mnm$. a_0 7.45 kX, c_0 7.88; $a_0:c_0 = 1:1.0575$. Cell contents $K_4Cu_2Cl_8 \cdot 4H_2O$.

Habit. Pyramidal $\{011\}$ or short prismatic $[001]$.

Twinning. On $\{011\}$ in artificial material.[7]

P h y s. H. $2\frac{1}{2}$. G. 2.418;[4] 2.41 (calc.). Luster vitreous. Color greenish blue. Transparent.
O p t.[5] In transmitted light, colorless.

	n(Na)	DICHROISM	
O	1.6365	Sky-blue	Uniaxial negative $(-)$.
E	1.6148	Grass-green	Absorption $O > E$.

C h e m. Potassium copper chloride dihydrate, $K_2CuCl_4 \cdot 2H_2O$. Analyses of natural material are lacking.[6]
O c c u r. Found in 1920 associated with sylvine, metavoltine, and gypsum as stalactites on the floor of the crater of Vesuvius, Italy.
A r t i f.[8] Obtained in crystals by cooling a warm concentrated water solution containing 2KCl and $CuCl_2$.
N a m e. After Eilhard Mitscherlich (1794–1863), German crystallographer and chemist, who first prepared the compound.

Ref.

1. Angles of Wyrouboff, *Bull. soc. min.*, **10**, 125 (1887) on artificial crystals; unit and orientation of Hendricks and Dickinson, *J. Am. Chem. Soc.*, **48**, 2149 (1927). On page 129 of Wyrouboff (1887) the angle 172°34′ apparently should be 172°54′ and is here so taken.
2. Observed on both natural and artificial crystals.
3. Hendricks and Dickinson (1927) on artificial material. See also Chrobak, *Zs. Kr.*, **88**, 35 (1934).
4. Zambonini and Carobbi (1925) on Vesuvius crystals; Wyrouboff (1887) gave 2.410 for artificial material.
5. Grailich, *Kryst.-opt. Unters.*, *Wien*, 86 (1858) on artificial material.
6. The natural material was shown by qualitative tests by Zambonini and Carobbi (1925) to contain only K, Cu, and H_2O and to be crystallographically identical with the artificial salt.
7. Wyrouboff (1887).
8. Mellor (**3**, 188, 1923).

ERYTHROSIDERITE SERIES

11.3.3.1 **E R Y T H R O S I D E R I T E** [$K_2FeCl_5 \cdot H_2O$]. Scacchi (*Acc. Napoli, Rend.*, 210, 1872; *Contr. min. incend. Vesuv.*, II, 42, 1874).
11.3.3.2 **K R E M E R S I T E** [$((NH_4),K)_2FeCl_5 \cdot H_2O$]. Eisenchlorid mit den Chlorkalium Kremers (*Ann. Phys.*, **84**, 79, 1851). Kremersit Kenngott (9, 1853).

C r y s t.[1] Orthorhombic; dipyramidal—$2/m\ 2/m\ 2/m$.

$$a:b:c = 1.3822:1:0.7178; \qquad p_0:q_0:r_0 = 0.5193:0.7178:1$$

$$q_1:r_1:p_1 = 1.3822:1.9256:1; \qquad r_2:p_2:q_2 = 1.3931:0.7235:1$$

Forms:[2]

	ϕ	$\rho = C$	ϕ_1	$\rho_1 = A$	ϕ_2	$\rho_2 = B$
b 010	0°00′	90°00′	90°00′	90°00′	.....	0°00′
a 100	90 00	90 00	90 00	0 00	0°00′	90 00
m 210	55 21	90 00	90 00	34 39	0 00	55 21
o 011	0 00	35 40	35 40	90 00	90 00	54 20
d 101	90 00	27 26½	0 00	62 33½	62 23½	90 00
e 201	90 00	46 05	0 00	43 55	43 55	90 00

Structure cell.[12] Space group *Pnma*.

	a_0	b_0	c_0	$a_0:b_0:c_0$
Erythrosiderite	13.75 Å	9.924	6.93	1.385:1:0.698
Kremersite	13.78	9.85	7.09	1.399:1:0.720

Cell contents $(K,NH_4)_8Fe_4Cl_{20}\cdot 4H_2O$.

Habit. Somewhat tabular {100} (*erythrosiderite*); also pseudo-octahedral (*kremersite*) with {011} and {210} in equal development.

Twinning. On [21$\bar{1}$] and [20$\bar{1}$] (artificial).[3]

Phys. Cleavage {210} and {011}, perfect.[4] G. 2.372 [artificial $K_2FeCl_5\cdot H_2O$];[5] 2.00 [artificial $(NH_4)_2FeCl_5\cdot H_2O$].[5] Color ruby-red to red and brownish red. Luster vitreous. Very deliquescent.

Opt.[6] In transmitted light, brown-red or yellowish.

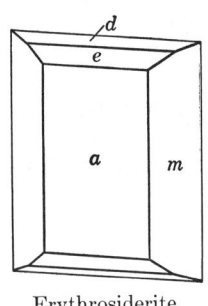
Erythrosiderite.

ORIENTATION		n (*erythrosiderite*)	
X	a	1.715	Biaxial positive (+).
Y	c	1.75	2V 62°.
Z	b	1.80	$r > v$, strong.

Chem. Erythrosiderite is a hydrated potassium ferric chloride, $K_2FeCl_5\cdot H_2O$, and kremersite in the only reported analysis contains ammonium substituting for slightly more than half the potassium, $[(NH_4)K]_2FeCl_5\cdot H_2O$. The pure ammonium compound has been prepared[7] and is isomorphous with the potassium compound. The names erythrosiderite and kremersite are here intended to apply to material in the series with $K > NH_4$ and $NH_4 > K$, respectively. Al apparently can substitute[8] for Fe.

Anal.

	1	2	3	4
K	23.74	24.21	12.07	
NH_4			6.17	12.56
Fe	16.96	16.81	16.89	19.44
Cl	53.83	53.30	55.15	61.73
H_2O	5.47	[5.68]	[9.56]	6.27
Rem.			0.16	
Total	100.00	[100.00]	[100.00]	100.00

1. $K_2FeCl_5\cdot H_2O$. 2. Erythrosiderite. Vesuvius.[9] 3. Kremersite. Vesuvius.[10] Rem. is Na. 4. $(NH_4)_2FeCl_5\cdot H_2O$.

Tests. Easily soluble in water.

Occur. Found in fumaroles on Vesuvius, Italy, associated with molysite and hematite; also reported from Mt. Etna. Erythrosiderite has been observed at Stassfurt, Germany, as an efflorescence on rinneite.

Artif.[11] The potassium salt can be prepared by passing hydrogen chloride into a concentrated solution of potassium ferric alum, and the ammonium salt by evaporation of a solution of ammonium chloride with a large excess of ferric chloride.

HIERATITE 103

N a m e. Erythrosiderite from ἐρυθρός, *red*, and σίδηρος, *iron*. Kremersite after the German chemist Peter Kremers (1827–?).

Ref.
1. Angles of Scacchi (1874) on Vesuvius erythrosiderite, unit of the structure cell; transformation: Scacchi to new, 200/010/001. Johnsen, *Jb. Min.* II, 93 (1903) gave $a:b:c = 1.3694:1:0.7023$ for artificial kremersite. Kremersite was early thought to be isometric because of the pseudo-octahedral habit but was shown to be isostructural with erythrosiderite (cf. Lacroix, *Bull. soc. min.*, **30**, 253, 1907; *C.R.*, **147**, 162, 1908).
2. Goldschmidt (**3**, 155, 1916). The forms on natural crystals and the artificial ammonium compound are the same.
3. Johnsen (1903).
4. Observed on artificial erythrosiderite by Gossner, *Zs. Kr.*, **40**, 73 (1905).
5. Gossner (1905).
6. Merwin in Larsen and Berman (135, 1934), and Lacroix (1907); Slavík, *Ac. sc. Bohême, Bull.*, **17**, 5 (1912) mentions weak pleochroism.
7. Cf. Johnsen (1903), Lacroix (1907), Mellor (**14**, 98, 1935).
8. Zambonini (107, 1935), in light-colored Vesuvius material.
9. Scacchi (1874).
10. Kremers (1851).
11. The ternary systems $KCl\text{-}FeCl_3\text{-}H_2O$ and $NH_4Cl\text{-}FeCl_3\text{-}H_2O$ are discussed by Mellor (1935).
12. Bellanca, *Per. Min.*, **17**, 1 (1948) by Weissenberg method on artificial crystals.

TYPE 4. $A_m BX_6$

HIERATITE AND MALLADRITE GROUPS

Hieratite group	Symmetry	a_0	c_0
Hieratite, K_2SiF_6	Isometric	8.168 kX	
Cryptohalite, $(NH_4)_2SiF_6$	Isometric	8.337	
Malladrite group			
Malladrite, Na_2SiF_6	Hexagonal		
Bararite, $(NH_4)_2SiF_6$	Hexagonal	5.76	4.77

The isometric polymorphs of these compounds are stable at higher temperatures. Their structures [1] are of the K_2PtCl_6 type and are based on cubic closest packing of the K and F ions with one-half of the octahedral holes occupied by Si atoms, giving rise to discrete $(SiF_6)^=$ ions. The corresponding Rb, Cs, and Tl salts are known artificially. Hieratite and cryptohalite doubtless can form a complete solid-solution series. The structure [1] of the malladrite group is more complex but again shows $(SiF_6)^=$ ions. All the species are formed in nature under fumarolic conditions.

Ref.
1. Ketelaar, *Zs. Kr.*, **92**, 155 (1935), Gossner and Kraus, *Zs. Kr.*, **88**, 223 (1934).

11.4.1 HIERATITE GROUP

11.4.1.1 H I E R A T I T E [K_2SiF_6]. Cossa (*Acc. Linc., Trans.*, **6**, 141, 1882; *Bull. soc. min.*, **5**, 61, 1882).

C r y s t. Isometric, hexoctahedral—$4/m\,\bar{3}\,2/m$.

Forms:
           a 001 o 111 d 011   (artif.)

104   HALIDES

Structure cell.[1]   Space group $Fm3m$. a_0 8.168 ± 0.007 kX.   Cell contents $K_8(SiF_6)_4$.

Habit.   Cubo-octahedral and octahedral.   As stalactitic concretions, spongy to dense in texture.

P h y s.   Cleavage {111}, perfect.   H. probably $2\frac{1}{2}$.   G. 2.665 (artif.);[2] 2.668 (calc.).   Colorless to white or gray.   Transparent.   Luster vitreous.

O p t.   In transmitted light, colorless.   Isotropic.   Refractive index (Na):[3] 1.340.

C h e m.   Potassium silicon fluoride, K_2SiF_6.   Isostructural with cryptohalite, $(NH_4)SiF_6$, but analytical evidence of compositional variation is lacking.   A hexagonal modification, isostructural with bararite, is known only as an artificial compound.

Anal.

	1	2
K	35.50	35.4
Si	12.74	12.7
F	51.76	51.9
Total	100.00	100.0

1. K_2SiF_6. 2. Vesuvius. Analysis on natural material, recrystallized from water solution.

Tests.   Soluble in cold water, more so in hot water.   Sublimes without residue.

O c c u r.   Found originally as a fumarolic deposit on Vulcano, one of the Lipari Islands, associated with sassolite, mirabilite, glauberite, selenian sulfur, realgar, and various alums.   Also found on Vesuvius, Italy, with avogadrite and malladrite.

A r t i f.[4]   Obtained in crystals by evaporating the water solution. Hieratite is isostructural with the corresponding salts of Rb, Cs, Tl, and NH_4, and with K_2PtCl_6.

N a m e.   Hieratite from *Hiera*, an ancient name for Vulcano.

Ref.

1. Ketelaar, *Zs. Kr.*, **92**, 155 (1935), by powder method on artificial material.
2. Stolba, *J. pr. Chem.*, **96**, 26 (1865); **102**, 1 (1867).
3. Zambonini and Carobbi, *Acc. Linc., Att.* [6], **4**, 171 (1926).
4. Cf. Mellor (**6**, 947, 1925).

11.4.1.2   **C R Y P T O H A L I T E**   [$(NH_4)_2SiF_6$].   Criptoalite Scacchi (*Acc. Napoli, Att.*, **6**, no. 9, 1873).

C r y s t.   Isometric; hexoctahedral—$4/m\ \overline{3}\ 2/m$.

Forms:

   a 001 o 111   (artif.)

Structure cell.[1]   Space group $Fm3m$; a_0 8.337 ± 0.007 kX.   Cell contents $(NH_4)_8(SiF_6)_4$.

Habit.   Cubo-octahedral and octahedral (artif.).   Massive, as lumps with a mammillary surface or crusts; arborescent.

MALLADRITE

Phys. Cleavage {111} perfect. H. $2\frac{1}{2}$. G. 2.004 (Barari)[2] 2.011 (artif.)[3] 2.029 (calc.). Colorless to white or gray. Transparent. Luster vitreous. Taste saline.

Opt. In transmitted light, colorless. Isotropic. Refractive index (Na): 1.369 ± 0.001 (Barari),[2] 1.3696 (artif.).[4]

Chem. Ammonium silicon fluoride, $(NH_4)_2SiF_6$. Dimorphous with bararite. Cryptohalite is isostructural with hieratite, K_2SiF_6, and probably isomorphous with that species but analytical evidence of compositional variation is lacking.

Anal.

	1	2
NH_4	20.25	20.43
Si	15.75	15.58
F	64.00	63.29
Rem.		0.58
Total	100.00	99.88
G.	2.028	2.004

1. $(NH_4)_2SiF_6$. 2. Barari, India.[2] Rem. is F 0.07 (excess over SiF_6), Cl tr., SO_4 0.06, H_2O 0.30, insol. 0.15.

Tests. Soluble in cold water, more so in hot. Sublimes without residue.

Occur. First recognized as a sublimate at Vesuvius, Italy, admixed with salammoniac and bararite. Also found as a sublimate over a burning coal seam at the Bararee colliery, Barari, Jharia coal field, India, associated with sulfur and bararite, and similarly at Libusín in the Kladno coal basin, Bohemia.

Artif.[5] Obtained in crystals from the water solution (cf. bararite).

Name. Cryptohalite from κρυπτός, *concealed*, and ἅλς, *salt*, in allusion to its intimate mixture with salammoniac in the original occurrence.

Ref.
1. Ketelaar, *Zs. Kr.*, **92**, 155 (1935) and Bozorth, *Am. Chem. Soc., J.*, **44**, 1066 (1922) on artificial material by powder, Laue and spectrometer measurements.
2. Christie, *Rec. Geol. Sur. India*, **59**, 233 (1926).
3. Gossner, *Zs. Kr.*, **38**, 147 (1903).
4. Topsøe, *Ann. chim. phys.*, **1**, 22 (1874).
5. Mellor (**6**, 945, 1925).

11.4.2 MALLADRITE GROUP

11.4.2.1 MALLADRITE $[Na_2SiF_6]$. *Zambonini* and *Carobbi* (*Acc. Linc., Rend.*, **4**, 173, 1926).

Cryst. Hexagonal.

$$a:c = 1:1.333 \text{ (artif.)}; \quad p_0:r_0 = 1.5392:1$$

Forms:

	φ	ρ = C	M	A_2
c 0001	.....	0°00'	90°00'	90°00'
m $10\bar{1}0$	30°00'	90 00	60 00	90 00
p $10\bar{1}1$	30 00	56 59$\frac{1}{2}$	65 12$\frac{1}{2}$	90 00

Habit. Prismatic [0001] with basal plane and pyramid.

P h y s. Observations on the cleavage and hardness are lacking. G. 2.714 (artif.).[1] Color pale rose (Vesuvius); white (artificial).

O p t. Basal sections of artificial crystals may show six anomalously birefringent sectors with the axial plane parallel to the prism faces and $X = c$.[2] Refractive indices:[3]

	n(Na)	
O	1.3125	Uniaxial negative $(-)$.
E	1.3089	

C h e m. Supposedly sodium silicon fluoride, Na_2SiF_6. Analyses of natural material are lacking, and the identification is based on the similarity in properties of admixed material re-crystallized from water solution with the artificial compound.[4]

Tests. Slightly soluble in cold water, and more so in hot.

O c c u r. Found as crusts on lava at Vesuvius, associated with salammoniac, avogadrite, hieratite, and ferruccite.

A r t i f.[5] Obtained by reaction of fluosilicic acid with NaOH, NaCl, or other sodium salts, and by reaction of a $NaHF_2$ solution on silica.

N a m e. After Alessandro Malladra, a director of the Vesuvius Observatory.

Ref.

1. Collins, *Chem. News*, **318**, 184 (1929); Stolba, *Zs. anal. Chem.*, **11**, 199 (1872) gives 2.7547.
2. Bertrand, *Bull. soc. min.*, **3**, 57 (1880).
3. Raiteri, *Acc. Linc., Rend.*, **31**, 115 (1922), on artificial crystals.
4. Cf. Zambonini and Carobbi (1926) and Carobbi, *Per. Min.*, **4**, 410 (1933).
5. Mellor (**6**, 947, 1925).

11.4.2.2   B A R A R I T E [$(NH_4)_2SiF_6$]. Criptoalite, Cryptohalite pt. *Scacchi* (*Acc. Napoli, Att.*, **6**, no. 9, 1873). Cryptohalite pt. *Christie* (*Rec. Geol. Sur. India*, **59**, 233, 1926). β-$(NH_4)_2SiF_6$.

C r y s t.[1] Hexagonal—P, scalenohedral—$\bar{3}\ 2/m$.

$$a:c = 1:0.8281\ (\text{artif.});^2 \qquad p_0:r_0 = 0.9562:1$$

Forms:[2]

	ϕ	$\rho = C$	M	A_2
c 0001	.....	0°00′	90°00′	90°00′
m 10$\bar{1}$0	30°00′	90 00	60 00	90 00
p 10$\bar{1}$1	30 00	43 43	69 47	90 00
x 20$\bar{2}$1	30 00	62 23½	63 42	90 00

Structure cell.[3] Space group $C\bar{3}m$. a_0 5.76 kX, c_0 4.77; $a_0:c_0 = 1:0.828$. Cell contents $(NH_4)_2SiF_6$.

Habit. Tabular {0001}, sometimes distorted by elongation $\perp c$. Also in arborescent and mammillary forms.

Twinning. Twins of paddle-wheel or dart-like form have been observed (Barari),[4] with the twin-plane inclined to {0001}.

Phys. Cleavage {0001} perfect. H. probably $2\frac{1}{2}$. G. $2.152,^2$ 2.144 (calc.). Color white. Luster vitreous. Taste saline.

Opt. In transmitted light, colorless.

$n(Na)^4$

O	1.406 ± 0.001	Uniaxial negative $(-)$.
E	1.391 ± 0.003	

Chem. Ammonium silicon fluoride, $(NH_4)_2SiF_6$. Dimorphous with cryptohalite. Chemical analyses of natural material are lacking.

Tests. Easily soluble in water. Sublimes without residue.

Occur. Found originally at Vesuvius, Italy, associated with sal-ammoniac and cryptohalite as a sublimation product. Also found as sublimed crusts with sulfur and cryptohalite on the surface of the ground above a burning coal seam in the Bararee colliery, Barari, Jharia coal field, India.[4]

Artif.[5] Formed in crystals from the water solution at temperatures below 5° C.; the isometric modification is formed at temperatures over 13°, and mixtures of the two phases may be obtained at intermediate temperatures.

Name. The name, here first given, refers to the Indian locality, Barari, from where the species was first fully described. The mineral was earlier recognized as a constituent of mixtures with cryptohalite but was not given an individual name.

Ref.

1. Gossner and Kraus, Zs. Kr., **88**, 223 (1934), by x-ray and morphological study of artificial crystals.
2. Gossner, Zs. Kr., **38**, 147 (1903), on artificial crystals.
3. Gossner and Kraus (1934).
4. Christie, Rec. Geol. Sur. India, **59**, 233 (1926).
5. See Mellor (**6**, 945, 1925) and Gossner (1903); the hexagonal modification was first prepared by Marignac, Ann. chim. [3], **60**, 301 (1860).

11.4.3 **RINNEITE** [NaK_3FeCl_6]. Boeke (Cbl. Min., 72, 1909; Jb. Min., II, 38, 1909).

Cryst.[1] Hexagonal—R; hexagonal-scalenohedral—$\bar{3}\,2/m$.

$a:c = 1:1.1532;^2 \quad \alpha\ 92°15'; \quad p_0:r_0 = 1.3316:1; \quad \lambda\ 57°22'$

Forms:[3]

		ϕ	$\rho = C$	A_1	A_2
c 0001	111	.....	0°00'	90°00'	90°00'
a 11$\bar{2}$0	10$\bar{1}$	0°00'	90 00	60 00	60 00
r 01$\bar{1}$2	110	-30 00	33 39$\frac{1}{2}$	90 00	61 19
R 10$\bar{1}$1	100	30 00	53 05$\frac{1}{2}$	46 10$\frac{1}{2}$	90 00

Structure cell.[4] Space group $R\bar{3}c$. a_{rh} 8.25, α 91°54'; a_0 11.86 kX, c_0 13.81; $a_0:c_0 = 1:1.164$. Cell contents $Na_2K_6Fe_2Cl_{12}$ in the rhombohedral unit.

Habit. Massive granular. Artificial crystals thick tabular $\{0001\}$ or short prismatic [0001].
Twinning. On $\{0001\}$ in artificial crystals.[5]
P h y s. Cleavage $\{11\bar{2}0\}$, good. Fracture splintery, conchoidal. H. 3. G. 2.347; 2.406 (calc.). Luster brilliant, often silky. Colorless when pure and fresh; usually rose, violet, or yellow, becoming brown on exposure. Taste astringent.
O p t.[6] In transmitted light, colorless to yellowish.

	$n(Li)$	$n(Na)$	$n(Tl)$	
O	1.5836	1.5886	1.5930	Uniaxial positive (+).
E	1.5842	1.5894	1.5939	

C h e m. A chloride of sodium, potassium, and divalent iron, NaK_3FeCl_6. The excess Na in analysis 3 probably is due to admixture. Isostructural with chlormanganokalite.

Anal.

	1	2	3	4
K	28.69	28.90	24.04	28.99
Na	5.62	5.61	10.70	5.69
Fe	13.66	13.94	11.58	13.57
Cl	52.03	51.87	52.92	52.02
Rem.		0.04		0.17
Total	100.00	100.36	99.24	100.44

1. NaK_3FeCl_6. 2. Wolkramshausen, Germany.[7] Average of two. Rem. is Br. 3. Hildesia, Germany.[8] 4. Hildesia.[9] Rem. is $MgSO_4 \cdot H_2O$.

Tests. Easily fusible in C.T. Soluble in water; decomposed by absolute alcohol with extraction of ferrous chloride.

O c c u r. Formed apparently as a secondary product in saline deposits of the oceanic type, associated with halite, sylvite, kieserite, and anhydrite. In Germany, at Wolkramshausen near Nordhausen, Saxony; at Hildesia near Diekholzen, at Salzdetfurth, and at Riedel, all in Hannover.

A r t i f.[10] Obtained in crystals from the system $FeCl_2$-$NaCl$-KCl-H_2O. Not redeposited from a water solution of the salt.

N a m e. After Friedrich Rinne (1863–1933), German crystallographer and petrographer.

Ref.

1. Point symmetry established by etch pits and form development, rhombohedral cell confirmed by x-ray study—see Cheng, Inaug. Diss., Leipzig (1929), *Jb. Min.*, I, 52 (1931), Boeke, *Jb. Min.*, II, 39 (1909), Schneider, *Cbl. Min.*, 504 (1909).
2. Angles of Boeke (1909) on artificial crystals; unit and orientation of Cheng (1929). Transformation: Boeke to Cheng, $0\bar{1}00/00\bar{1}0/\bar{1}000/0002$.
3. Boeke (1909) on artificial crystals; *a* and *r* only observed on natural crystals.
4. Cheng (1929) by Laue and rotation methods on natural material. See also Bellanca, *Per. Min.*, **16**, 199 (1948).
5. Cheng (1929).
6. Boeke (1909) by prism method on Wolkramshausen crystal.
7. Boeke (1909).
8. Schneider (1909).
9. Rinne and Kolb, *Cbl. Min.*, 337 (1911).
10. Boeke (1909); Mellor (**14**, 32, 1935).

11.4.4 CHLORMANGANOKALITE [K_4MnCl_6]. *Johnston-Lavis (Nature,* **74,** 103, 1906). *Lacroix (C.R.,* **142,** 1249, 1906; *Bull. soc. min.,* **30,** 219, 1907). *Johnston-Lavis* and *Spencer (Min. Mag.,* **15,** 54, 1908). Chloromanganokalite.

Cryst. Hexagonal—R.[1]

$a:c = 1:0.5797;$ $\alpha\,110°25';$ $p_0:r_0 = 0.6694:1;$ $\lambda\,57°36'$

Forms:[1]

		ϕ	$\rho = C$	A_1	A_2
$a\,11\bar{2}0$	$10\bar{1}$	0°00'	90°00'	60°00'	60°00'
$r\,10\bar{1}1$	100	30 00	33 48	61 12	90 00

Habit. Rhombohedrons, sometimes arranged in parallel groups.
Phys. Fracture conchoidal. Brittle. H. $2\frac{1}{2}$. G. 2.31. Luster vitreous. Color pale wine-yellow to lemon- or canary-yellow. Transparent. Uniaxial positive (+). Birefringence very low. Mean index 1.59.
Chem. A chloride of potassium and manganese, K_4MnCl_6.
Anal.

	1	2
KCl	70.32	69.42
$MnCl_2$	29.68	26.45
$MgCl_2$		0.16
Na_2SO_4		1.19
H_2O		1.52
Insol.		0.71
Total	100.00	99.45

1. K_4MnCl_6. 2. Vesuvius.[2]

Tests. In C.T. fuses readily to a clear, mobile, yellow liquid; on consolidation the mass is first white, and when cool reddish in color, changing again after some time to white. Readily soluble in water; a drop of solution evaporated on a glass slide deposits KCl and, on gently warming, $MnCl_2$, which disappears again as the slide cools in moist air.

Occur. Found with halite, sylvite, and hematite crystals lining cavities in blocks of scoria ejected from Vesuvius in the eruption of April, 1906; possibly also with chlorocalcite, halite, and sylvite on the lava of 1872.
Alter. On exposure to moist air the crystals rapidly become dull and deliquesce to a yellow liquid.
Artif.[3] Obtained at 75° from a saturated water solution of KCl and $MnCl_2$.
Name. In allusion to the composition.

Ref.
1. Spencer in Johnston-Lavis and Spencer (1908) shows the mineral to be hexagonal and probably rhombohedral and gives a measurement to the rhombohedron.
2. Johnston-Lavis anal. in Johnston-Lavis and Spencer (1908).
3. Süss, *Zs. Kr.,* **51,** 252 (1913).

TYPE 5. ALUMINO-FLUORIDES

The crystal structures of about half of the known alumino-fluorides have been worked out so far. In these structures, Al occurs in six-coordination with F as anionic $(AlF_6)^{\equiv}$ groups. These octahedral groups may occur either isolated or variously linked together by sharing of corners, analogous to the role played by (SiO_4) tetrahedra and (BO_3) triangles in determining the structure of the silicates and borates. The natural alumino-fluorides are here arranged in a tentative structural classification [1] based on the known or inferred type of anionic framework. The classification is restricted to compounds in which the central coordinating atom is Al, but it may be noted that a number of other halogen compounds of known structure, in which the central atom is Si, B, Fe''', or Mn''', and in part of different coordination, could be placed therein. An outline of the classification follows.

Type A. Isolated octahedra
11.5.1 Cryolite Na_3AlF_6
11.5.2 Elpasolite K_2NaAlF_6
11.5.3 Pachnolite $NaCaAlF_6 \cdot H_2O$
11.5.4 Thomsenolite $NaCaAlF_6 \cdot H_2O$
11.5.5 (?) Jarlite $NaSr_3Al_3F_{16}$ (hybrid type ?)

Type B. Chain structures
11.5.6 (?) Gearksutite $CaAl(OH)F_4 \cdot H_2O$

Type C. Sheet structures
11.5.7 Prosopite $CaAl_2(F,OH)_8$
11.5.8 Chiolite $Na_5Al_3F_{14}$ (interrupted sheet)

Type D. Network structures
11.5.9   Fluellite $AlF_3 \cdot H_2O$
11.5.10 Ralstonite $Na(Mg,Al)_6F_{12}(OH)_6 \cdot 3H_2O$
11.5.11 Weberite Na_2MgAlF_7 (interrupted network)

Ref.
1. Following essentially the earlier arrangements of Pabst, priv. contr. (1948), Ferguson, priv. contr. (1948), and Brosset, Doct. Diss., Stockholm (1942).

Type A. Isolated Octahedra

11.5.1 **C R Y O L I T E** [Na_3AlF_6]. Chryolith, Thonerde mit Flussäure *Abildgaard* (*Allg. J. Chem.*, **2**, 502, 1799); *d'Andrada* (*ibid.*, **4**, 37, 1800). Kryolith *Karsten* (28, 73, 1800); *Klaproth* (*J. Phys.*, **51**, 473, 1800; *Beitr.*, **3**, 207, 1802); *Vauqueline* (*Ann. Chim.*, **37**, 89, 1801). Alumine fluatée alkaline *Haüy* (**2**, 157, 1801). Eisstein *Glocker* (958, 1831).

C r y s t.[1] Monoclinic; prismatic—$2/m$.

$a:b:c = 0.9662:1:1.3883;$ $\beta\ 90°11';$ $p_0:q_0:r_0 = 1.4369:1.3883:1$

$r_2:p_2:q_2 = 0.7203:1.0350:1;$ $\mu\ 89°49'$ $p_0'\ 1.4369,\ q_0'\ 1.3883,\ x_0'\ 0.0032$

CRYOLITE

Forms:[2]

		φ	ρ	φ₂	ρ₂ = B	C	A
c	001	90°00′	0°11′	89°49′	90°00′	.....	89°49′
b	010	0 00	90 00	.....	0 00	90°00′	90 00
a	100	90 00	90 00	0 00	90 00	89 49	.....
m	110	45 59	90 00	0 00	45 59	89 52	44 01
r	011	0 08	54 14	89 49	35 46	54 14	89 53½
v	101	90 00	55 13½	34 46½	90 00	55 02½	34 46½
k	$\bar{1}$01	−90 00	55 06½	145 06½	90 00	55 17½	145 06½
p	111	46 03	63 26½	34 46½	51 37½	63 18½	49 55
s	121	27 25	72 16	34 46½	32 16	72 10½	63 59½

Less common:

A 012   D 105   B 102   C $\bar{1}$02   z 112   u $\bar{1}$12   F $\bar{2}$75

Structure cell.[3] Space group $P2_1/n$. a_0 5.39 kX, b_0 5.59, c_0 7.76; β 90°11′; $a_0:b_0:c_0$ = 0.9642:1:1.388. Cell contents $Na_6Al_2F_{12}$.

Habit. Usually massive, coarsely granular. Crystals ordinarily cuboidal with c, m, or modified by r, v, k; also short prismatic [001]. The {110} faces striated [1$\bar{1}$1], [1$\bar{1}\bar{1}$], [1$\bar{1}$0].

Twinning.[4] Very common. Often repeated or polysynthetic with several twin laws occurring together, and reflecting the pseudo-isometric symmetry of the cell cm. (1) By a 90° or 270° rotation on [110], penetration, common. (2) By a 180° rotation on [110], rhombic section (1$\bar{1}$0), repeated, less common. (3) By a 120° rotation on [021], composition surface irregular; common, especially in granular cryolite, as fine lamellae and probably always secondary. (4) By a 180° rotation on [$\bar{1}$11], rhombic section near (110), repeated; rare, never found in granular cryolite. (5) On (001) or by a 180° rotation on [100], composition plane (001).

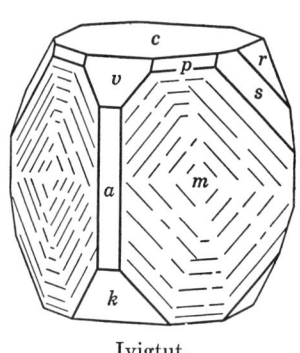

Ivigtut.

(6) On (100) or by a 180° rotation on [001], composition plane (100). (7) On (112), composition plane (112). (8) On ($\bar{1}$12), composition plane (112). (9) On (110), composition plane (110). (10) By a 180° rotation on [111], rhombic section near (1$\bar{1}$0). (11) On (211). Laws 5 to 10 are for lamellae in granular cryolite and are probably of secondary origin.

Phys. Cleavage none.[5] Parting on {001} and {110} to produce cuboidal forms. Fracture uneven. Brittle. H. 2½. G. 2.97 ± 0.01;[6] 2.963 (calc.). Luster vitreous to greasy, somewhat pearly on {001}. Colorless to white; also brownish, reddish, or brick-red, rarely black. Streak white. M.P. 1020°;[7] inverts at about 560° to an isometric form. Heat capacity,[7] for 0° 0.909 joule per gram, for 200°, 1.18. Specific heat[8] 0.2548. Weakly thermoluminescent. Artificial etch pits have been described.[9] Forms oriented growths with thenardite (artificial).[18]

O p t.[10] In transmitted light, colorless. May exhibit polysynthetic twin lamellae.

ORIENTATION	n(Na)	
X	1.3376	Biaxial positive (+).
Y b	1.3377	$2V$ 43°.
$Z \wedge c$ —44°	1.3387	$r < v$, horizontal.

The dispersion of the refractive indices is from 1.319 for λ = 477 $\mu\mu$ to 1.343 for λ = 662 $\mu\mu$.

C h e m. Sodium aluminum fluoride, Na_3AlF_6. Sometimes contains traces of Ca, Fe, Mn, or organic material in the darker colored samples. Al can substitute for Na, with a concomitant substitution of O for F, in artificial material.[11]

Anal.[12]

	1	2	3	4
Na	32.86	32.56	32.41	32.40
Al	12.85	13.07	13.01	12.81
F	54.29	54.15	54.28	53.55
Rem.				0.98
Total	100.00	99.78	99.70	99.74
G.	2.963			2.972

1. Na_3AlF_6. 2. Greenland. Massive.[13] 3. Greenland. Crystals.[13] 4. Colorado.[14] Rem. is Fe_2O_3 0.40, Ca 0.28, H_2O 0.30.

Tests. Fusible in small fragments in the flame of a candle. Affords acid water when heated in O.T. so that the flame enters the tube. On charcoal fuses easily to a clear bead which on cooling becomes opaque; after long blowing, the assay spreads out, the sodium fluoride is absorbed by the coal, an odor of fluorine is given off, and a crust of alumina remains. Soluble in sulfuric acid with evolution of HF. Slightly soluble in water (1 part in 2730 parts H_2O at 12°); easily soluble in $AlCl_3$ solution.

O c c u r. Found in a large deposit at Ivigtut in the Arksuk fiord, Frederikshaab district, West Greenland.[15] The cryolite occurs as a pegmatitic body in a granite stock intrusive into gneiss. The principal associated minerals include microcline, quartz, fluorite, chiolite, siderite, and topaz, together with smaller amounts of galena, molybdenite, arsenopyrite, pyrite, and other sulfides, cassiterite, wolframite, columbite, zircon, mica. A number of rare aluminum fluorides are found with the cryolite and have for the most part been derived by its alteration. These include pachnolite, thomsenolite, gearksutite, cryolithionite, weberite, jarlite, prosopite.

Cryolite also has been found in small amounts with topaz, chiolite, fluorite, and phenakite near Miask in the Ilmen Mountains, U.S.S.R. A minor occurrence is with fluorite near Sallent in the Pyrenees, Huesca province, Spain. Cryolite and its alteration products pachnolite, thomsenolite, prosopite, etc., occur in limited amounts with quartz and feldspar in pegmatitic veins at the southern base of Pikes Peak, El Paso County, Colorado, north and west of Saint Peter's Dome.

CRYOLITE 113

Alter. Cryolite alters easily to other aluminum fluorides, noted above. The change commonly begins by penetration along the parting directions and affords cellular or porous masses lined with crystals of the secondary minerals. Some of the alterations have been duplicated artificially.[16]

Artif.[17] Obtained by fusion of NaF and AlF_3, by adding NaF or NaCl to a water solution of AlF_3, and in other ways.

Name. From κρύος, *frost*, and λίθος, *stone*, hence meaning *ice-stone* in allusion to its appearance.

Ref.

1. The monoclinic character was first definitely established by Krenner, *Ber. Ungarn*, **1**, 151 (1883), and later verified by Bøggild, *Zs. Kr.*, **50**, 349 (1912). The mineral was earlier considered by Hagemann, *Am. J. Sc.*, **42**, 268 (1866), and others to be orthorhombic. Cryolite is pseudo-isometric and has been shown by Mügge, *Cbl. Min.*, 34 (1908), to invert to an isometric form at elevated temperatures. The ratio given is from Goldschmidt (**5**, 53, 1918).
2. Goldschmidt (**5**, 53, 1918). Rare and uncertain:

310	013	1̄14	1̄21	725
130	1̄03	1̄11	211	176
015	114	323	2̄11	

3. Náray-Szabó and Sasvári, *Mat. Termés. Ért.*, **57**, 664 (1938), *Zs. Kr.*, **99**, 27 (1938), showed that the earlier structure proposed by Menzer, *Cbl. Min.*, 378 (1928), was in error. The latter in a more recent paper, *Naturwiss.*, **26**, 236 (1938), gives the cell dimensions here used.
4. Summarized from an unpublished survey by J. D. H. Donnay, priv. comm. (1947). Bøggild, *Zs. Kr.*, **50**, 349 (1912), gives an elaborate study of cryolite twinning; see also Wallerant, *Bull. soc. min.*, **35**, 177 (1912), and Poduroff, *Mem. soc. russe min.*, **54**, 207 (1925). Law 1 is Baumhauer's law. Law 3, Bøggild's "new law," is here re-defined with rational elements. Law 10, recognized by Poduroff but not by Bøggild, may be "law d" of Cross and Hillebrand, *Am. J. Sc.*, **26**, 283 (1883). Law 11 was reported by Poduroff (1925) as probably very rare. To the list might be added, as possible twin laws, the following: (12) Twin axis [001], 90° rotation, a law very close to (9), (13) Twin axis [201], 120° rotation, a law very close to (8), (14) Twin axis [2̄01], 120° rotation, a law very close to (7). Bøggild did not attempt to differentiate these laws. It should be added that the twin laws given by various investigators are not universally accepted, and a completely satisfactory solution has not yet been reached.
5. The supposed cleavages on *c* (most perfect), *m*, and *k* were shown by Bøggild (1912) to be parting planes for the most part due to twinning.
6. Average of 5 new determinations on Greenland crystals on the microbalance, Rogers, priv. comm. (1942).
7. Cited in Birch (169, 231, 1942).
8. Joly, *Royal Soc. London, Proc.*, **41**, 250 (1887).
9. Baumhauer, *Zs. Kr.*, **11**, 139 (1885).
10. Cesàro and Mélon, *Ac. Belgique, Bull., Cl. Sc.*, 362 (1936).
11. Zintl and Morawietz, *Zs. anorg. Chem.*, **240**, 145 (1939) find about 11 per cent Al_2O_3 in excess, for which $x \sim .3$ in the formula $Na_{3-x}Al_{1+x}F_{2(3-x)}O_{2x}$. See also Doelter (**4** [3], 289, 1931), Tananaev and Lelchuk, *C.r. ac. sc. U.R.S.S.*, **41**, 114 (1943), Brosset, *Ark. Kemi*, 21A, no. 9 (1946).
12. For additional analyses see Doelter (**4** [3], 283, 1931).
13. Brandl analysis in Groth, *Zs. Kr.*, **7**, 386 (1882).
14. Hillebrand, *U. S. Geol. Sur., Bull.* **20**, 48 (1885).
15. Descriptions of the Ivigtut deposit are given by Doelter (**4** [3], 303, 1930), Bernard, *Min. Mag.*, **14**, 202 (1916), Boll, *Medd. Grønland*, **63** (1922), Baldauf, *Zs. pr. Geol.*, **18**, 432 (1910), and Bøggild, *Medd. Grønland*, 32 (1905).
16. Cf. Doelter (1930, p. 295).
17. Mellor (**5**, 304, 1924); Dixon and Scott, *Bull. Council Sc. Ind. Res. Australia*, no. 214 (1947).
18. Seifert, *Fortschr. Min.*, **26**, 116 (1947).

11.5.2 ELPASOLITE [K_2NaAlF_6]. Cross and Hillebrand (Am. J. Sc., **26**, 283, 1883; U. S. Geol. Sur., Bull. *20*, 57, 1885).

Cryst. Isometric; diploidal—$2/m\bar{3}$.

Forms:

                a 001 o 111

Structure cell.[1] Space group $Pa3$. a_0 8.093 kX. Cell contents $K_8Na_4Al_4F_{24}$.

Habit. Massive, sometimes as indistinct crystals with a suggestion of octahedral and cubic faces. Artificial crystals are cubo-octahedral.

Phys.[2] No cleavage. Fracture uneven. H. $2\frac{1}{2}$. G. 2.995; 3.015 (calc.). Luster weak vitreous to slightly greasy. Colorless. Transparent to translucent.

Opt.[2] In transmitted light, colorless. Isotropic. Refractive index: 1.376 ± 0.002.

Chem.[3] An alumino-fluoride of potassium and sodium, K_2NaAlF_6.

Anal.

	1	2
K	32.29	28.94
Na	9.50	9.90
Ca		0.72
Mg		0.22
Al	11.13	11.32
F	47.08	[47.90]
Total	100.00	[99.00]
G.	3.015	2.995

1. K_2NaAlF_6. 2. Colorado. The alkali determinations are approximations. F calculated in amount needed to satisfy the bases.

Occur. Found lining solution cavities in massive pachnolite in a cryolite-bearing pegmatite at St. Peter's Dome, at the southern base of Pikes Peak, El Paso County, Colorado.

Artif.[4] Observed in melts in the system K_3AlF_6-Na_3AlF_6. As crystals by evaporation of a solution containing KF, NaF, and $Al_2(SO_4)_3$.

Name. From the county in Colorado in which the mineral was found.

Ref.

1. Frondel, Am. Min., **33**, 84 (1948), by powder method on type material, confirming the earlier study by Menzer, Fortschr. Min., **17**, 61 (1932) on artificial material.
2. Frondel (1948) on type material.
3. Formula established by the demonstrated identity with artificial K_2NaAlF_6 (Frondel, 1948).
4. Náray-Szabó and György, Mat. Termés, Ért., **60**, 364 (1941) and Menzer (1932).

11.5.3 PACHNOLITE [$NaCaAlF_6 \cdot H_2O$]. Knop (Ann. Phys., **127**, 61, 1863). Pyroconite Wöhler (Ann. Phys., **180**, 231, 1875).

Cryst.[1] Monoclinic; prismatic—$2/m$ (?).

$a:b:c = 1.1665:1:1.5092$; β 90°20'; $p_0:q_0:r_0 = 1.2938:1.5092:1$

$r_2:p_2:q_2 = 0.6626:0.8573:1$; μ 89°40'; p_0' 1.2938, q_0' 1.5092, x_0' 0.0058

PACHNOLITE

Forms:[2]

		φ	ρ	φ₂	ρ₂ = B	C	A
c	001	90°00′	0°20′	89°40′	90°00′	....	89°40′
m	110	40 36½	90 00	0 00	40 36½	89°47′	49 23½
p	111	40 44	63 20½	37 34½	47 22½	63 07½	54 19½
q	221	40 40	75 53½	21 05	42 38½	75 40½	50 48
f	311	68 47	76 31	14 25½	69 23½	76 12	24 58½
r	1̄11	−40 29	63 15	142 10½	47 13	63 28	125 26

Less common:

s 554   t 553   v 331   x 551

Structure cell.[11] Space group $C2/c$ or Cc. a_0 12.12 kX, b_0 10.39, c_0 15.68; β 90°20′; $a_0:b_0:c_0 = 1.1665:1:1.5092$. Cell contents $Na_{16}Ca_{16}Al_{16}F_{96} \cdot 16H_2O$.

Habit. Prismatic [001]. Commonly acutely terminated; sometimes terminated by {001}. {110} striated parallel to intersection with {001}. Also massive, cleavable to granular, and as stalactitic masses with a dense to chalcedonic structure.

Twinning.[3] Twin plane {100} common, yielding crystals with an orthorhombic appearance.

Oriented growths. Growths have been described[4] of thomsenolite upon pachnolite, with thomsenolite (001) and (110) quasi-parallel pachnolite (110) and (001), in other instances with thomsenolite (110) and (1̄10) quasi-parallel pachnolite (110) and (001).

P h y s. Cleavage {001}, indistinct. Fracture uneven. Brittle. H. 3. G. 2.983;[5] 2.97 (calc.). Colorless to white. Luster vitreous. Transparent to translucent.

Greenland.

O p t.[6] In transmitted light, colorless.

ORIENTATION		n	
X	b	1.411	Biaxial positive (+).
Y		1.413	2V 76°, 2E 120°.
Z ∧ c	69°	1.420	$r < v$, weak; horizontal dispersion, strong.

C h e m. A hydrated alumino-fluoride of sodium and calcium, $NaCaAlF_6 \cdot H_2O$. Dimorphous with thomsenolite.

Anal.[10]

	1	2	3
Na	10.36	10.23	10.23
Ca	18.05	18.14	18.06
Al	12.14	12.50	12.14
F	51.34	51.54	51.33
H₂O	8.11	8.19	8.10
Total	100.00	100.60	99.86
G.		3.008	2.965

1. $NaCaAlF_6 \cdot H_2O$. 2. Greenland.[7] On crystals. 3. Colorado.[8] On crystals. Original total given as 99.88.

Tests. Easily fusible (more readily than cryolite). Heated in C.T. decrepitates, yields acid water, etches the glass, and gives a white deposit of silica. Easily soluble in sulfuric acid.

Occur. Found at Ivigtut, on the Arksuk fiord, Greenland, together with thomsenolite as alteration products of cryolite. Also described from a cryolite-pegmatite at St. Peter's Dome, near Pikes Peak, El Paso County, Colorado, associated with thomsenolite, elpasolite, and other secondary fluorides.

Artif.[9] A substance possibly identical with pachnolite (or thomsenolite) has been obtained by reaction of powdered cryolite with $CaCl_2$ solution.

Name. Pachnolite is from πάχνη, *frost*, and λίθος, *stone*, in allusion to its appearance. Pyroconite from πῦρ, *fire*, and κονία, *powder*, because it falls to pieces when ignited B.B.

Ref.

1. Elements of the structure cell, corresponding to the morphological unit of Knop (1863), Groth, *Zs. Kr.*, **7**, 457 (1883), and others (summarized by Ferguson, *Trans. Roy. Soc. Canada, Sect.* **4**, **40**, 11 (1946)). Berman, priv. comm. (1942), obtained $a:b:c = 1.160:1:1.509$, $\beta\ 90°40'$, on a Colorado crystal. Doubly terminated crystals are not known.
2. Goldschmidt (**6**, 113, 1920); rare or uncertain: 100, 101, 011 (latter form from Berman (1942)).
3. For descriptions see Des Cloizeaux, *Bull. soc. min.*, **5**, 310 (1882); Groth (1883); Cross and Hillebrand, *U. S. Geol. Sur., Bull.* **20**, 49 (1885).
4. Bøggild (1913), from Ivigtut; see also Ferguson (1946).
5. Average of the values 2.965, 3.008, and 2.976 reported, respectively, by Cross and Hillebrand (1885), Koenig, *Ac. Sc. Philadelphia, Proc.*, 42 (1876), and Bøggild, *Zs. Kr.*, **51**, 601 (1913); lower values, down to 2.92, have been reported but are questionable.
6. Larsen and Berman (**95**, 1934). Earlier, partial optical descriptions by Des Cloizeaux (1882); Groth (1883); Cross and Hillebrand (1885); Krenner, *Zs. Kr.*, **10**, 528 (1885).
7. Koenig (1876).
8. Hillebrand in Cross and Hillebrand (1885).
9. Lemberg, *Zs. deutsch. geol. Ges.*, **28**, 620 (1876), Noellner, *idem*, **33**, 139 (1881).
10. For earlier analyses see Dana (179, 1892). Some of these may have been on thomsenolite and not pachnolite.
11. Ferguson (1946) by Weissenberg method on Ivigtut crystals.

11.5.4 THOMSENOLITE [$NaCaAlF_6 \cdot H_2O$]. Dimetric Pachnolite *Hagemann* (*Am. J. Sc.*, **42**, 93, 1866). Thomsenolite *Dana* (129, 1868). Hagemannite pt. *Shepard* (*Am. J. Sc.*, **42**, 246, 1866).

Cryst.[1] Monoclinic; prismatic—$2/m$.

$a:b:c = 1.0127:1:2.9273;\quad \beta\ 96°27';\quad p_0:q_0:r_0 = 2.8906:2.9088:1$

$r_2:p_2:q_2 = 0.3438:0.9938:1;\quad \mu\ 83°33';\quad p_0'\ 2.9090,\ q_0'\ 2.9273,\ x_0'\ 0.1131$

Forms:[2]

		ϕ	ρ	ϕ_2	$\rho_2 = B$	C	A
c	001	90°00'	6°27'	83°33'	90°00'	.....	83°33'
m	110	44 49	90 00	0 00	44 49	85°27½'	45 11
t	$\bar{1}03$	−90 00	40 35	130 35	90 00	47 02	130 35
x	$\bar{1}02$	−90 00	53 18	143 18	90 00	59 45	143 18
p	111	45 54½	76 38	18 18½	47 24	72 02½	45 40
q	$\bar{1}13$	−41 17	52 24	130 35	53 27½	56 48	121 31
r	$\bar{2}23$	−43 06	69 29	151 18	46 51½	73 57½	129 47½
s	$\bar{1}11$	−43 41	76 07½	160 19	45 24½	80 37	132 06½

Structure cell.[9] Space group $P2_1/c$. a_0 5.57 kX, b_0 5.50, c_0 16.10; β 96°27′; $a_0:b_0:c_0 = 1.0127:1:2.9273$. Cell contents $Na_4Ca_4Al_4F_{24} \cdot 4H_2O$.

Habit. Usually prismatic [001]. Crystals often cubic in aspect with equal development of {001} and {110}, and grouped in parallel position. Also tabular {001}. {110} and the terminal prisms strongly striated parallel the intersection with {001}. {$h0l$} often curved. No twinning observed. Also found as opaline or chalcedonic crusts and stalactitic masses.

Oriented growths. Observed incrusting pachnolite (which see) in oriented position.

P h y s. Cleavage {001} perfect, with a pearly luster; {110} distinct. Fracture uneven. Brittle. H. 2. G. 2.981;[3] 2.99 (calc.). Colorless to white, sometimes with a brownish or reddish tint due to inclusions or coatings of iron oxide. Luster vitreous, on {001} somewhat pearly. Transparent to translucent.

O p t.[4] In transmitted light, colorless.

ORIENTATION		n	
$X \wedge c$	−52°	1.4072	Biaxial negative (−).
Y		1.4136	$2V\ 50°,\ 2E\ 73°$.
Z	b	1.4150	$r < v$, weak.

C h e m. A hydrated alumino-fluoride of sodium and calcium, $NaCaAlF_6 \cdot H_2O$. Dimorphous with pachnolite. Some analyses suggest a small amount of substitution of (OH) for F.

Anal.[5]

	1	2	3
Na	10.36	10.10	10.43
Ca	18.05	16.79	17.22
Al	12.14	13.74	13.26
F	51.34	50.37	50.61
H$_2$O	8.11	9.00	8.42
Total	100.00	100.00	99.94
G.		2.937	

1. $NaCaAlF_6 \cdot H_2O$. 2. Greenland.[6] 3. Greenland.[7]

Tests. Same as for pachnolite.

O c c u r. Originally described from the cryolite deposit at Ivigtut, Greenland, where it occurs with pachnolite, ralstonite, and fluorite. Also at Miask, in the Ilmen Mountains, Urals, U.S.S.R., with chiolite and cryolithionite in a cryolite-pegmatite at St. Peter's Dome, near Pikes Peak, El Paso County, Colorado.

A r t i f. A product obtained by reaction of powdered cryolite with $CaCl_2$ solution may be identical with thomsenolite or pachnolite.[8]

N a m e. After Julius Thomsen (1826–?), Danish physical chemist and founder of the Greenland cryolite industry, who first noticed the species.

HAGEMANNITE. *Shepard* (Am. J. Sc., **42**, 246, 1866).
Yellow, jasper-like crusts from Ivigtut, shown[10] to be a variable mixture of thomsenolite and ralstonite pigmented by iron oxide.

Ref.
1. Elements of the structure cell. Transformation: Nordenskiöld, *Geol. För. Förh.*, **2,** 81 (1874), Dana (180, 1892), Goldschmidt (**8,** 130, 1922) to new, 100/010/003. The morphological measurements, made on crystals of poor quality, vary widely—see Ferguson, *Trans. Roy. Soc. Canada, Sect. 4,* **40,** 11 (1946).
2. Goldschmidt (1922). Uncertain forms: 401, $\bar{4}01$, 443, 221, 883, 16.16.1, $\overline{10}.10.27$, 779, 889, 443, $\overline{10}.10.9$, 221, $\bar{4}41$, 661, 881.
3. Average of values 2.982 and 2.979 by Bøggild, *Zs. Kr.,* **51,** 601 (1913) and Groth, *Chem. Kryst.,* **1,** 461 (1906), respectively; lower values have been reported and are of questionable worth.
4. Bøggild (1913), by refractometer; for earlier partial optical descriptions see Des Cloizeaux, *Bull. soc. min.,* **5,** 316 (1882), and Kressner (1885).
5. For additional analyses see Dana (180, 1892).
6. Koenig, *Ac. Sc. Philadelphia, Proc.,* 42 (1876).
7. Brandl analysis in Groth, *Zs. Kr.,* **7,** 470 (1882).
8. Lemberg, *Zs. deutsch. geol. Ges.,* **28,** 620 (1876) and Noellner, *idem,* **33,** 139 (1881).
9. Ferguson (1946) by Weissenberg method.
10. Frondel, *Am. Min.,* **33,** 84 (1948), and Ferguson (1946).

11.5.5   JARLITE [$NaSr_3Al_3F_{16}$]. Bøgvad (*Medd. Grønland,* 92, no. 8, 11 pp., 1933). Meta-jarlite Bøgvad (*idem,* p. 7).

Cryst.[1] Monoclinic; prismatic—$2/m$ (?).

$a:b:c = 1.478:1:0.669;$ $\beta\ 101°49';$ $p_0:q_0:r_0 = 0.453:0.655:1$

$r_2:p_2:q_2 = 1.527:0.691:1;$ $\mu\ 78°11';$ $p_0'\ 0.462,\ q_0'\ 0.669,\ x_0'\ 0.209$

Forms:

	ϕ	ρ	ϕ_2	$\rho_2 = B$	C	A
$c\ 001$	90°00′	11°49′	78°11′	90°00′	.....	78°11′
$b\ 010$	0 00	90 00	.....	0 00	90°00′	90 00
$w\ 021$	8 52½	53 33½	78 11	37 22	52 38	82 52
$r\ 201$	90 00	48 34	41 26	90 00	36 45	41 26
$D\ \bar{4}01$	−90 00	58 36½	148 36½	90 00	70 25½	148 36½

Structure cell.[2] Space group $C2/m$, $C2$ or Cm. $a_0\ 15.99\ kX$, $b_0\ 10.82$, $c_0\ 7.24$; $\beta\ 101°49'$; $a_0:b_0:c_0 = 1.478:1:0.669$. Cell contents uncertain.

Habit. In minute crystals, tabular {100} and elongated [010], usually grouped in fan-like or spherulitic aggregates. Also massive, in part with a radial columnar structure.

Phys. H. 4–4½.[3] G. 3.93 (anal. 2), 3.78 (anal. 3). Colorless to white or gray. Luster vitreous.

Opt.[4] In transmitted light, colorless.

Orientation		$n(Na)$	
$X \wedge c$	−6° ± 2°	1.429	Biaxial negative (−)
Y	b	1.433	or positive (+).
$Z \wedge c$	+84° ± 2°	1.436	$2V\ 90° ± 10°$.

Chem. A fluoride of sodium, strontium, and aluminum of uncertain formula.[5] Small amounts of OH substitute for F. The material of analysis 2 contains relatively large amounts of Ca, Mg, and Ba in substitution for Sr.

GEARKSUTITE

Anal.

	Na	Mg	Ca	Ba	Sr	Al	F	H$_2$O+	H$_2$O−	Rem.	Total
1.	3.23	0.90	0.55	0.99	35.60	12.16	43.23	2.91	0.08	0.25	99.90
2.	3.54	1.38	3.20	2.25	28.70	12.49	45.50	2.14	0.08	0.35	99.63

1. Ivigtut.[6] Rem. is Li 0.08, Fe 0.17.   2. Ivigtut (*meta-jarlite*).[6] Rem. is Li 0.04, Fe 0.31.

Tests. B.B. melts easily with effervescence. Soluble in AlCl$_3$ solution.

Occur. Found with cryolite, chiolite, fluorite, topaz, and thomsenolite at Ivigtut, Greenland. The mineral in part occurs with thomsenolite as drusy crystals in cavities in partially dissolved cryolite.

Name. After Mr. C. F. Jarl, an official in the Danish cryolite industry.

META-JARLITE. *Bøgvad* (*Medd. Grønland*, 92, no. 8, 7, 1933).

A supposedly dimorphous form of jarlite, but shown [4] to be only a variety containing some Mg, Ca, and Ba in substitution for Sr (analysis 2).

Ref.

1. Elements of the structure cell of Brosset, Doctoral Diss., Stockholm (1942) and confirmed by Ferguson, *Am. Min.*, **34**, 383 (1949). The transformation from the morphological cell of Bøgvad (1933) is uncertain due to poor crystal measurements, but probably is 004/060/301.
2. Brosset (1948); cell dimensions of Ferguson (1949).
3. On material of analysis 2, the so-called meta-jarlite; the hardness of the crystallized material of analysis 1 could not be determined accurately.
4. Ferguson (1949).
5. The formula NaSr$_3$Al$_3$F$_{16}$ originally given cannot be reconciled with the x-ray data, and the analyses presumably are in error.
6. Blix analysis in Bøgvad (1933).

Type B. Chain Structures

11.5.6   GEARKSUTITE [CaAl(OH)F$_4$·H$_2$O]. *Hagemann* in Dana (130, 1868). Evigtokite *Flight* (*J. Chem. Soc.*, **43**, 140, 1883). Paragearksutite *Smolyaninov* and *Isakov* (Belyankin Jubilee vol., Ac. Sc. U.S.S.R., 145, 1946).

Cryst. Probably monoclinic.[1] Massive or nodular with an earthy, kaolin-like to chalky aspect, but consisting of very minute colorless needles with inclined extinction.

Phys. Hardness 2 (of the aggregates). G. 2.768 (Wagon Wheel Gap).[2] Color white. Luster dull.

Opt.[3] In transmitted light, colorless.

Orientation		n	
X	b	1.448 ± 0.003	Biaxial negative (−).
Y		1.454 ± 0.003	2V moderate.
Z		1.456 ± 0.003	Z ∧ elong., large.

Chem. An alumino-fluoride of calcium, CaAl(OH)F$_4$·H$_2$O. The near constancy of the ratio (OH):F = 1:4 in material from four different localities indicates the structural non-equivalence of these anions. OH is usually present in slight excess, however, as in analyses 5 and 6, and then substitutes for F. The water is lost at about 300°.[9]

Anal. [4]

	1	2	3	4	5	6
Na		0.10	0.04	0.06	0.20	
K		0.04	0.07	0.06	0.05	
Ca	22.51	22.30	22.41	22.13	22.15	21.45
Al	15.14	15.20	15.11	15.38	15.09	16.42
F	42.68	42.07	41.00	41.26	40.20	35.25
O	4.49	[4.83]	5.09	4.88	5.42	8.40
$H_2O + 110°$	15.18	15.46	15.20	15.88 } 15.52		16.19
$H_2O - 110°$			0.44	0.12 }		0.59
Insol.				0.23	0.96	2.12
Total	100.00	[100.00]	99.36	100.00	99.59	100.42
G.			2.768		2.72	

1. $CaAl(OH)F_4 \cdot H_2O$. 2. Pikes Peak, Colorado.[5] 3. Wagon Wheel Gap, Colorado.[6]
4. Gingin, West Australia.[7] 5. Hot Springs, Virginia.[8] Insoluble includes Fe_2O_3 0.03, K_2O 0.01 (from glauconite). 6. Transbaikalia (*paragearksutite*).[12] Insol. includes Fe_2O_3 0.56, SiO_2 1.56. Mean n, 1.454.

Tests. B.B. fuses easily to a white enamel. Gently heated in the C.T. gives off neutral water, but when strongly heated attacks the glass. Easily soluble in dilute acid.

O c c u r. Found originally in the cryolite deposit at Ivigtut, Greenland, associated with thomsenolite as one of the last formed of the secondary fluorides.[11] At Miask, Urals, U.S.S.R., with chiolite and other fluorides, phenakite, and lithia mica in a topaz- and cryolite-bearing pegmatite, and in Transbaikalia (*paragearksutite*). As a late-formed secondary fluoride associated with pachnolite, thomsenolite, fluorite, etc., in a cryolite-bearing pegmatite in the Pikes Peak granite at St. Peter's dome, El Paso County, Colorado. Found as nodular masses formed in situ in a glauconitic and phosphatic Cretaceous sandstone at Gingin, Western Australia. Gearksutite also has been found as a hydrothermal alteration product of the rhyolite wall rock of a fluorite-barite vein near Wagon Wheel Gap, Colorado, and by the action of hot spring water charged with calcium fluoride on a clay bed in limestone between Hot Springs and Warm Springs, Virginia.

A r t i f.[10] Hydrous fluorides of Ca and Al have been prepared but their relation to gearksutite is not known.

N a m e. From γῆ, *earth*, and *arksutite* (= chiolite) which it was thought to resemble in composition, in allusion to the earthy aspect of the mineral.

Ref.

1. Inferred from optical data of Larsen (78, 1921) and others.
2. Larsen and Wells, *Proc. Nat. Ac. Sc.*, **2**, 360, (1916); Simpson, *Min. Mag.*, **19**, 23 (1920), gives 2.72 by the pycnometer on Gingin material.
3. Larsen (78, 1921) on Ivigtut material.
4. For earlier analyses see Dana (130, 1868; 181, 1892).
5. Hillebrand anal. in Cross and Hillebrand, *U. S. Geol. Sur., Bull. 20*, 59 (1885).
6. Wells anal. in Larsen and Wells (1916).
7. Bowley anal. in Simpson, *Min. Mag.*, **19**, 33 (1920).
8. Henderson, *Am. Min.*, **14**, 281 (1929).
9. For dehydration data see Simpson (1920).
10. Mellor (**5**, 308, 1924).
11. For a description see Bøggild, *Min. Groenland*, Copenhagen, 1905, p. 127.
12. Smolyaninov and Isakov (1946).

Type C. Sheet Structures

11.5.7   PROSOPITE [CaAl$_2$(F,OH)$_8$]. Speckstein, in regelmässigen Kryst. im Eisenglanz eingewachsen [Altenberg] *Charpentier* (Beob. über die Lagerst. der Erze, Leipzig, 32, 1799). "Speckstein-Krystalle" *Breithaupt* (Über die Aechtheit der Kryst., Freiberg, 29, 1815). Prosopit *Scheerer* (*Ann. Phys.*, **90**, 315, 1853; **92**, 612, 1854; **101**, 361, 1857).

C r y s t. Monoclinic; prismatic—$2/m$.

$a:b:c = 0.6030:1:0.6584;$ $\beta\ 94°52';$ [1] $p_0:q_0:r_0 = 1.0919:0.6560:1$

$r_2:p_2:q_2 = 1.5243:1.6644:1;$ $\mu\ 85°08';$ $p_0'\ 1.0959,\ q_0'\ 0.6584,\ x_0'\ 0.0851$

Forms: [2]

	ϕ	ρ	ϕ_2	$\rho_2 = B$	C	A
b 010	0°00′	90°00′	....	0°00′	90°00′	90°00′
m 110	59 00	90 00	0°00′	59 00	85 50	31 00
μ 021	3 42	52 50½	85 08	37 19	52 41	87 03
z 111	60 51½	53 31	40 15½	66 57	49 18	45 23½
t $\bar{2}21$	−59 57½	69 11	156 17½	62 06	73 25	144 01

Less common:

h 011	j 041	l 061	X $\bar{7}73$
i 073	k 051	Z $\bar{1}11$	y 131

Structure cell.[3] Space group $C2/c$. a_0 6.69 kX, b_0 11.11, c_0 7.32 (all ±0.03); β 95°00′ ± 15′; $a_0:b_0:c_0 = 0.602:1:0.659$. Cell contents Ca$_4Al_8(F,OH)_{32}$.

Habit. More or less tabular {010}. Also massive, granular to powdery.

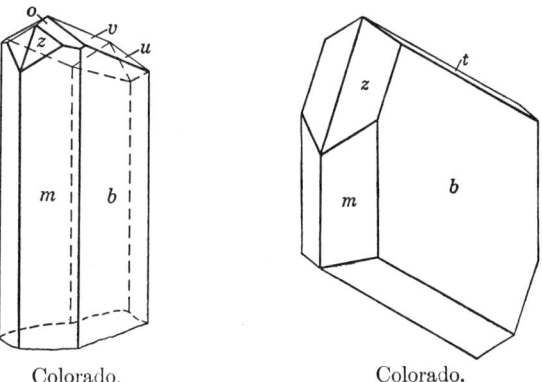

Colorado.   Colorado.

P h y s. Cleavage {111} perfect. Fracture uneven to conchoidal. Brittle. H. 4½. G. 2.894 (Altenberg),[4] 2.880 (Colorado);[5] 2.898 (calc.). Luster vitreous, weak. Colorless to grayish or white. The pseudomorphs are dull in luster and yellowish to grayish white in color. Transparent in fresh material.

O p t.[6] In transmitted light, colorless and transparent in fresh material.

Orientation		n	
X		1.501 ± 0.003	Biaxial positive (+).
Y	b	1.503 ± 0.003	$2V\ 62°45'$ (Na).
$Z \wedge c$	$-35°$	1.510 ± 0.003	$r > v$, strong.

C h e m. A basic alumino-fluoride of calcium, $CaAl_2(F,OH)_8$. The F and OH substitute mutually over only a narrow range (see analyses), and these anions may be structurally non-equivalent, as in gearksutite, at least in part.

Anal.[7]

	1	2	3	4
Na		0.33	0.48	
Ca	16.84	16.19	17.28	16.85
Mg		0.11	0.17	
Al	22.66	23.37	22.02	22.74
F	31.92	35.01	33.18	29.95
H$_2$O	15.14	12.41	13.46	16.12
O	13.44	[12.58]	[13.41]	[14.34]
Total	100.00	[100.00]	[100.00]	[100.00]
G.	2.898		2.880	2.87
F:OH	1:1	1.34:1	1.17:1	1.13:1

1. $CaAl_2(F,OH)_8$ with F:OH = 1:1. 2. Altenberg.[8] 3. St. Peter's Dome, Colorado.[9] Average of five analyses, some partial. 4. Dugway district, Utah.[10] Recalculated after deducting K 0.12, Na 0.32, Cu 0.17, and fluorite.

Tests. In C.T. affords water and silicon fluoride. Decomposed by sulfuric acid. The water goes off above about 260°.[9]

O c c u r. Found originally in greisen at Altenberg, Saxony, associated with hematite, fluorite, and siderite; the crystals are more or less altered into kaolin or fluorite or to a mixture of both minerals. Later found in tin veins and greisen at Schlaggenwald, Bohemia, with quartz, fluorite, cassiterite, apatite, sphalerite, mica, and kaolin; these crystals also are pseudomorphs, composed of admixed apatite, fluorite, and minor kaolin. At Mt. Bischoff, Tasmania,[11] as powdery masses admixed with kaolin and tourmaline and formed apparently as an alteration product of a topaz-bearing porphyry. In the United States, unaltered prosopite has been found in the cryolite-pegmatite on St. Peter's Dome, near Pikes Peak, El Paso County, Colorado, as crystals associated with pachnolite, fluorite, quartz, and astrophyllite. Also found in the Dugway district, Tooele County, Utah, with quartz, fluorite, and pyrite.

A l t e r. To mixtures of fluorite with kaolin or apatite (see above).

N a m e. From προσωπεῖον, *a mask*, in allusion to the deceptive (pseudomorphous) character of the mineral.

Ref.

1. Palache, priv. comm. (1936), from measurements on new material from St. Peter's Dome, Colorado. Transformation: Dana (178, 1892) and Goldschmidt (**6**, 176, 1920) to Palache, 001/010/½00 (making the lattice C-centered instead of A-centered, with a halved as indicated by the x-ray study of Berman and Wolfe (1941)). Earlier made triclinic by Des Cloizeaux, *Nouvelles Recherches*, 190 (1867), but regarded as monoclinic by Groth, *Zs. Kr.*, **7**, 487 (1883).

CHIOLITE

2. Goldschmidt (1920) and Palache (1936).
3. Berman and Wolfe, priv. comm. (1941), by Weissenberg method on Colorado material.
4. Scheerer (1857).
5. Cross and Hillebrand, *U. S. Geol. Sur., Bull. 20*, 62 (1885).
6. Indices from Larsen (123, 1921) on Altenberg crystals; optic angle from Des Cloizeaux (1867) on Altenberg crystals, who also gives nY = 1.502 (Na), 1.500 (red), 1.506 (blue).
7. For an early partial analysis on Altenberg material see Scheerer (1857).
8. Brandl in Groth (1883).
9. Hillebrand in Cross and Hillebrand (1885).
10. Hillebrand, *Am. J. Sc.*, **7**, 54 (1899).
11. Petterd (140, 1910).

11.5.8 CHIOLITE [$Na_5Al_3F_{14}$]. Chiolith (from Miask) *Hermann* and *Auerbach* (*J. pr. Chem.*, **37**, 188, 1846). Arksutite *Hagemann* (*Am. J. Sc.*, **42**, 94, 1866). Chiolith (from Miask) *von Wörth* and *Chodnev* (*Verh. min. Ges.*, 1845–46, 208, 216, 1846). Chodneffite *Dana* (234, 1850). Chodnewite. Nipholith *Naumann* (219, 1864).

C r y s t.[1] Tetragonal; ditetragonal-dipyramidal—$4/m\ 2/m\ 2/m$.

$$a:c = 1:1.4734; \quad p_0:r_0 = 1.4734:1$$

Forms:[2]

	ϕ	ρ	A	M
c 001	....	0°00′	90°00′	90°00′
n 114	45°00′	27 31	70 56	90 00
o 011	0 00	55 50	90 00	54 11½

Structure cell.[3] Space group $P4/mnc$. a_0 7.005 kX, c_0 10.39; $a_0:c_0$ = 1:1.483. Cell contents $Na_{10}Al_6F_{28}$.

Habit. Dipyramidal {011}, but distinct crystals rare and very small. Usually massive granular, resembling cryolite.

Twinning. On {011}, sometimes distorted into prismatic shapes.

P h y s. Cleavage {001} perfect, {011} distinct. H. 3½–4. G. 2.998;[4] 2.989 (calc.). Luster vitreous, pearly on the {001} cleavage. Snow-white to nearly colorless. Translucent to nearly transparent.

O p t.[5] In transmitted light, colorless.

nO	1.3486	Uniaxial negative (−).
nE	1.3424	

C h e m. An alumino-fluoride of sodium, $Na_5Al_3F_{14}$. Some early analyses diverge from this composition, probably due to admixture.[6]

Anal.

	1	2	3	4
Na	24.89	24.97	24.49	24.79
Al	17.52	17.66	17.68	17.54
F	57.59	57.30	[57.74]	57.81
Rem.			0.11	0.23
Total	100.00	99.93	[100.02]	100.37
G.	2.989		2.994	

1. $Na_5Al_3F_{14}$. 2. Miask, Urals.[7] 3. Ivigtut, Greenland.[8] Rem. is Mg. 4. Ivigtut.[9] Rem. is H_2O −110°.

Tests. Like cryolite, but somewhat more fusible.

Occur. Originally described from Miask, in the Ilmen Mountains, Urals, U.S.S.R., associated with topaz, phenakite, fluorite, cryolithionite, thomsenolite in a cryolite-pegmatite. Also at Ivigtut, Greenland, in part as very large masses embedded in cryolite.

A r t i f.[10] Obtained from fusions of the mixed Na and Al fluorides.

N a m e. From χιών, *snow*, and λίθος, *stone*, in allusion to its appearance and similarity to cryolite (= ice-stone).

Ref.

1. Point symmetry inferred from apparent dipyramidal development of an unindexed $\{hkl\}$ form figured by Koksharov (4, 389, 1862). Ratio from angles of Koksharov (1862) on Miask crystals; unit from x-ray studies of Brosset and others (ref. 3). Transformation; Koksharov to new, $\overline{1}10/110/002$.
2. Goldschmidt (2, 138, 1913); $x\{017\}$ doubtful. Bøggild *Zs. Kr.*, **51**, 598 (1912), gives also $\{034\}$ and $\{054\}$; and Koksharov (1862) gives $z\{hkl\}$.
3. Space group of Brosset, *Zs. anorg. Chem.*, **238**, 201 (1938)—*Chem. Abs.*, **37**, 13 (1943); cell dimensions of Clausen, *Zs. Kr.*, **95**, 394 (1936). See also Caglioti and Giacomello, *Naturwiss.*, **26**, 317 (1938), who give the space group $P4/mmm$.
4. From values 3.005 (Ivigtut) and 2.995 (Miask) by Bøggild (1912) and 2.994 (Miask) by Nordenskiöld, *Geol. För. Förh.*, **8**, 172 (1886).
5. Bøggild (1912) by refractometer on Ivigtut crystals.
6. Material from Miask approximating to the formula Na_2AlF_5 from analyses by Chodney (1846) and Rammelsberg, *Ann. Phys.*, **74**, 314 (1848), cited in Dana (128, 1868), was separated from chiolite under the name chodneffite by Dana (234, 1850; 128, 1868), but the impure nature of the material and probable identity with chiolite was shown by Groth, *Zs. Kr.*, **7**, 478 (1883). Groth's formula for chiolite, $Na_5Al_3F_{14}$, is confirmed by the analysis of Blix in Clausen (1936) and by the x-ray studies.
7. Brandl in Groth (1883).
8. Lindström in Nordenskiöld (1886).
9. Blix in Clausen (1936).
10. Brosset (in *Chem. Abs.*, **37**, 13, 1943); Puschin and Baskow, *Zs. anorg. Chem.*, **81**, 351 (1913); Fedotieff and Iljinsky, *Zs. anorg. Chem.*, **80**, 123 (1913); Michel-Lévy and Wyart, *Bull. soc. min.*, **70**, 168 (1948).

Type D. Network Structures

11.5.9 F L U E L L I T E [$AlF_3 \cdot H_2O$]. *Lévy* (*Ann. Phil.*, **8**, 242, 1824). Fluate of Alumine, Fluoride of Aluminum.

Pleysteinite *Groth* (*Zs. pr. Geol.*, **24**, 190, 1916). Kreuzbergite *Laubmann* and *Steinmetz* (*Zs. Kr.*, **55**, 551, 1920).

C r y s t. Orthorhombic; dipyramidal—$2/m\ 2/m\ 2/m$.

$a:b:c = 0.5325:1:0.4098;$ [1] $p_0:q_0:r_0 = 0.7696:0.4098:1$

$q_1:r_1:p_1 = 0.5325:1.2994:1;$ $r_2:p_2:q_2 = 2.4402:1.8780:1$

Forms:[1]

	ϕ	$\rho = c$	ϕ_1	$\rho_1 = A$	ϕ_2	$\rho_2 = B$
b 010	0°00'	90°00'	90°00'	90°00'	....	0°00'
m 110	61 58	90 00	90 00	28 02	0°00'	61 58
r 111	61 58	41 05	22 17	54 32½	52 25	72 00½

Structure cell.[2] Space group probably $Fddd$. a_0 11.40 kX, b_0 21.14, c_0 8.52; $a_0:b_0:c_0 = 0.539:1:0.403$. Cell contents $Al_{28}F_{84} \cdot 28H_2O$ (?).

Fluellite

Habit. Crystals dipyramidal {111}, often modified by {010}.
P h y s. Cleavage {010} and {111}, indistinct.[3] H. 3. G. 2.17 (Stenna Gwyn [4]), 2.139 (Pleystein [5]); 2.29 (calc. for 28(AlF$_3 \cdot$H$_2$O)). Luster vitreous. Colorless to white; sometimes faint yellow. Transparent.

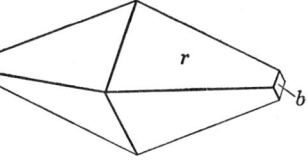

Cornwall.

O p t. In transmitted light, colorless.

ORIENTATION		n(Stenna Gwyn)[6]	n(Pleystein)[6]	n(Hagendorf)[7]	
X	a	1.473	1.489	1.490	Biaxial positive (+).
Y	c	1.490	1.495	1.496	2V very large.
Z	b	1.511	1.506	1.509	$r > v$, perceptible to rather strong.

C h e m. Hydrated aluminum fluoride, AlF$_3 \cdot$H$_2$O. The F probably is in part substituted for by (OH).

Anal.

	1	2
Na		0.58
Al	26.44	27.62
F	55.89	56.25
H$_2$O	17.67	[15.55]
Total	100.00	[100.00]
G.	2.29	2.17

1. AlF$_3 \cdot$H$_2$O. 2. Stenna Gwyn, Cornwall.[8]

O c c u r. First found at Stenna Gwyn, near St. Austell, Cornwall, in minute crystals on quartz with fluorite, arsenopyrite, torbernite, tavistockite, and wavellite (?). Later described (as *kreuzbergite*) [7] from several localities: with phosphosiderite and strengite in pegmatite at the Kreuzberg, Pleystein, in the Oberpfalz, Bavaria; also in the same region in pegmatite at Hagendorf; and as an alteration of triplite at Königswart, near Marienbad, Bohemia. Possibly occurs with amblygonite at Montaña de Cáceres, Estremadura, Spain.[9]

A r t i f. Several hydrates of aluminum fluoride including AlF$_3 \cdot$H$_2$O have been reported [10] but the relations to fluellite are not known.

N a m e. Fluellite in allusion to the composition. Kreuzbergite from the locality.

KREUZBERGITE. Laubmann and Steinmetz (*Zs. Kr.*, **55**, 551, 1920). Pleysteinite Groth (*Zs. pr. Geol.*, **24**, 190, 1916).
A supposed phosphate of aluminum later shown [7] to be identical with fluellite.

Ref.

1. Angles and unit of Miller (607, 1852), in orientation making $c < a < b$; transformation: Miller to new orientation, 010/001/100. Laubmann and Steinmetz, *Zs. Kr.*, **55**, 523 (1920), obtained, on kreuzbergite, 0.5261:1:0.3938 when transformed; transformation: Laubmann and Steinmetz to new, 001/010/100. Scholz and Strunz, *Cbl. Min.*, 133 (1940), who established the identity of kreuzbergite and fluellite, obtained 0.5343:1:0.4127 on kreuzbergite from Hagendorf.

2. Wolfe, priv. comm. (1945), by Weissenberg method on Cornwall crystals.
3. Laubmann and Steinmetz (1920) gave a cleavage {001} but this was later corrected by Sellner, *Zs. Kr.*, **59,** 504 (1924), to {010}.
4. Groth, *Zs. Kr.*, **7,** 484 (1882), by suspension method.
5. Laubmann and Steinmetz (1920).
6. Larsen and Berman (98, 1934).
7. Cf. Scholz and Strunz (1940).
8. Brandl analysis in Groth (1882).
9. Tenne and Calderón (142, 1902).
10. Mellor (**5,** 300, 1924).

11.5.10   **R A L S T O N I T E** [∼Na(Mg$_1$Al$_5$)$_6$F$_{12}$(OH)$_6$·3H$_2$O]. *Brush* (*Am. J. Sc.*, **2,** 30, 1871).

C r y s t.   Isometric; hexoctahedral—$4/m\,\bar{3}\,2/m$.

Forms:

$$a\,001 \qquad o\,111$$

Structure cell.[1] Space group $Fd3m$. a_0 9.87 kX. Cell contents Na$_{2.8}$(Mg$_{2.8}$Al$_{13.2}$)$_{16}$(F$_{31}$OH$_{17}$)$_{48}$·7H$_2$O (analysis 2), not all structure positions being occupied. The structure is of the pyrochlore type, A$_2$B$_2$O$_6$(OH,F).

Habit.   Octahedral, often with small {001}; less frequently cubooctahedral, or cubic with small {111}.

P h y s.   Cleavage {111} imperfect. Fracture uneven. Brittle. H. 4½. G. 2.56–2.62, varying with composition; 2.56 (calc. for anal. 2). Colorless to white, milky; often yellow on the surface due to iron oxide. Luster vitreous. Transparent to translucent.

O p t.   In transmitted light, colorless. Often shows weak anomalous birefringence, the crystal being divided into eight uniaxial (or biaxial) sectors or otherwise.[2]

$$n(\text{Na})^3 \qquad 1.399 \pm 0.001 \qquad \text{Isotropic}$$

C h e m.   A fluoride of sodium, magnesium, and aluminum. The ideal formula is A$_2$B$_2$F$_6$(OH,F), but the actual formula departs from this due to coupled substitutions and omissions in the structure.[4] Analysis 2 approximates to Na(Mg$_1$Al$_5$)$_6$F$_{12}$(OH)$_6$·3H$_2$O. The (OH) and part of the F may be driven off without destruction of the structure by heating to ca. 365°.[4]

Anal.[5]

	1	2	3
Na	5.05	4.27	4.06
Mg	3.90	4.39	4.30
Al	23.06	24.25	23.81
F	57.68	39.91	40.26
OH	[1.19]	[19.46]	18.02
H$_2$O	9.56	8.43	9.55
Rem.		0.15	
Total	[100.44]	[100.86]	100.00
G.		2.560	2.56

1. Ivigtut.[6] Recalculated after deduction of 8.51 per cent thomsenolite. Value for OH in 1 and 2 calculated to fill valence requirements.   2. Ivigtut.[7] Rem. is CaO 0.03, K 0.12.   3. NaMgAl$_5$F$_{12}$(OH)$_6$·3H$_2$O.

Tests. In C.T. whitens, yields acid water, then a copious white sublimate which etches the tube. B.B. does not fuse. Soluble with effervescence in a Na_2CO_3 bead. Decomposed by HF with evolution of HF.

O c c u r. Found with thomsenolite in solution cavities in cryolite at Ivigtut, Greenland. Also observed with pachnolite and thomsenolite in altered cryolite at St. Peter's Dome, near Pikes Peak, Colorado. Said to occur at Tanokamiyama, Omi, Japan.[8]

N a m e. After the Rev. J. Grier Ralston, of Norristown, Pennsylvania, who first observed the mineral.

Ref.
1. Pabst, *Am. Min.*, **24**, 566 (1939), by powder method on Greenland material, with analysis of the structure. See also Ferguson, *Am. Min.*, **34**, 383 (1949).
2. Cf. Brauns, *Die opt. Anom. der Kryst.*, Leipzig, 242 (1891), Bøggild, *Medd. Grønland*, **50**, 121 (1912), Bertrand, *Bull. soc. min.*, **4**, 34 (1881).
3. Moyd in Gordon, *Ac. Sc. Philadelphia, Not. Nat.*, no. 11 (1939); Bøggild (1912) gives 1.4267 by total reflection. Mrose, priv. comm. (1947), found $nX = 1.411$ in anomalously biaxial ($-$) material.
4. See discussion by Pabst (1939).
5. See additional analysis by Nordenskiöld and Penfield cited in Penfield and Harper, *Am. J. Sc.*, **32**, 380 (1886).
6. Brandl, *Ann. Chem.*, **213**, 1 (1882).
7. Penfield and Harper (1886).
8. Jimbo, *J. Sc. Coll. Imp. Univ. Tokyo*, **11**, 234 (1899).

UNNAMED MINERAL. *Naboko (Ac. sc. Leningrad., C. r.*, **33**, 140, 1941).
Isometric. As microcrystalline crusts and pulverulent. Color light yellow. Under the microscope isotropic with n 1.383. The grains are turbid due to minute inclusions. The composition approximates to the formula $NaCaMgAl_3F_{14} \cdot 4H_2O$. The formula may also be written, analogous to ralstonite,[1] as $(Ca,Na,K)_{7.72}(Al,Mg)_{16}F_{48}(OH)_{3.61} \cdot 13.7H_2O$. Analysis gave:[2] Na_2O 4.62, K_2O 1.07, CaO 11.15, MgO 8.65, Fe_2O_3 2.04, Al_2O_3 28.26, SiO_2 1.80, H_2O- 1.86, H_2O+ 11.60, F 43.40, Cl 0.81, O = F, Cl = 18.35, total 96.91. Decomposed by concentrated HCl. Found incrusting lava near the Bilyukai crater of the Klyuchevsky volcano, U.S.S.R.

Ref.
1. Pabst, *Am. Min.*, **24**, 566 (1939).
2. Nekrassova in Naboko (1941).

UNNAMED MINERAL. *Carobbi (Ac. sc. Lett. Arti, Modena, Att.*, [5], **1**, 33, 1936).
A supposed fluoride of Na, Ca, Mg, and Al found as amorphous stalactitic masses in a cavernous fumarole on Vesuvius. Isotropic; n 1.406. Analysis gave: K 5.23, Na 6.10, Ca 4.42, Mg 4.83, Fe 0.16, Al 15.05, Si 0.58, SO_4 14.87, F 28.73, Cl 2.45, H 0.20, OH 12.41, H_2O 4.50, total 99.53. The whitish exterior portion of the stalactites contains admixed mercallite, hieratite, potassium fluoride, and possibly $NaHSO_4$.

11.5.11 W E B E R I T E [Na_2MgAlF_7]. *Bøgvad (Medd. Grønland*, 119, no. 7, 1938).

Orthorhombic. Found as irregular grains and masses, in part as inclusions in cryolite.

Structure cell.[1] Space group *Immm*, *I*222, *I*$2_12_12_1$, or *I*2*mm*. a_0 7.29 kX, b_0 7.05, c_0 9.97; $a_0:b_0:c_0 = 1.034:1:1.414$. Cell contents $Na_8Mg_4Al_4F_{28}$.

P h y s. Cleavage {101} poor, {010} indistinct. Fracture uneven. H. $3\frac{1}{2}$. G. 2.96 or slightly higher; 2.97 (calc.). Luster vitreous. Color light gray. Streak white. Translucent.

O p t.[2] In transmitted light, colorless.

		n(Na)	
X	a	1.346	Biaxial positive (+).
Y	b	1.348	$2V\ 83° \pm 3°$ (meas.).
Z	c	1.350	

C h e m. An alumino-fluoride of sodium and magnesium, Na_2MgAlF_7. A small amount of K apparently substitutes for Na in the only reported analysis.

Anal.

	1	2
Na	19.97	19.08
Mg	10.56	10.43
Al	11.71	11.65
F	57.76	57.58
Rem.		1.80
Total	100.00	100.54

1. Na_2MgAlF_7. 2. Ivigtut.[3] Rem. is Fe 0.37, Ca 0.08, K 1.19, insol. 0.16.

Tests. B.B. turns white and swells up but does not fuse. In the C.T. gives acid vapors and a small white sublimate. Slightly soluble in water and easily soluble in $AlCl_3$ solution.

O c c u r. Found at Ivigtut, Greenland, associated with cryolite, fluorite, chiolite, topaz, green mica, and pyrite.

N a m e. After Theobald Weber, one of the founders of the cryolite industry in Denmark.

Ref.

1. Brosset, Doct. Diss., Stockholm (1942)—*Chem. Abs.*, **37**, 13 (1943). Ferguson, *Am. Min.*, **34**, 383 (1949), confirms the cell dimensions but finds the space group to be *Ibmm* or *Ibm*2. Byström, *Ark. Kemi*, 18B, no. 10 (1944), gives a structure based on *I*2*mm*, with another setting.
2. Optical orientation from Ferguson (1949).
3. Buchwald analysis in Bøgvad (1938).

12 COMPOUND HALIDES

TYPE 1. MISCELLANEOUS

12.1.1 **C R E E D I T E** [$Ca_3Al_2F_4(OH,F)_6(SO_4) \cdot 2H_2O$]. *Larsen* and *Wells* (*Proc. Nat. Ac. Sc.*, **2**, 360, 1916).

Cryst. Monoclinic; prismatic—$2/m$.

$a:b:c = 1.6199:1:1.1597$; $\beta\ 94°29\frac{1}{2}'$;[1] $p_0:q_0:r_0 = 0.7159:1.1561:1$

$r_2:p_2:q_2 = 0.8650:0.6192:1$; $\mu\ 85°30\frac{1}{2}'$; $p_0'\ 0.7181$, $q_0'\ 1.1597$, $x_0'\ 0.0786$

Forms:[2]

	ϕ	ρ	ϕ_2	$\rho_2 = B$	C	A
c 001	90°00′	4°29½′	85°30½′	90°00′	. . . . .	85°30½′
a 100	90 00	90 00	0 00	90 00	85°30½′	. . . . .
m 110	31 47½	90 00	0 00	31 47½	87 38	58 12½
δ 101	90 00	38 32½	51 27½	90 00	34 03	51 27½
l 201	90 00	56 34	33 26	90 00	52 04	33 26
i $\bar{1}01$	−90 00	32 36	122 36	90 00	37 05½	122 36
p 111	34 29½	54 36	51 27½	47 47½	52 08½	62 31
d 221	33 09	70 09	33 26	38 03	67 44½	59 03
v 331	32 41½	76 24	24 07½	35 07	74 00½	58 20
n $\bar{1}11$	−28 52½	52 56½	122 36	45 40	52 56½	112 40

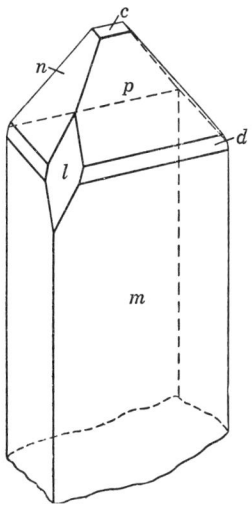

Wagon Wheel Gap.

Structure cell.[3] Space group $C2_1/c$. $a_0\ 13.88$ kX, $b_0\ 8.56$, $c_0\ 9.98$; $\beta\ 94°24'$; $a_0:b_0:c_0 = 1.621:1:1.166$. Cell contents $Ca_{12}Al_8F_{16}(OH,F)_{24}(SO_4)_4 \cdot 8H_2O$.

Habit. Crystals short prismatic to acicular [001]; usually with $\{111\}$ and $\{\bar{1}11\}$ or $\{111\}$ alone as the dominant terminal forms. Also as radiated aggregates resembling wavellite in structure, as drusy wart-like masses, and as embedded grains and crystals. Commonest forms: $m\ c\ p\ n$.

Phys. Cleavage $\{100\}$ perfect. Fracture conchoidal. Brittle. H. 4. G. 2.713,[5] 2.730;[4] 2.715 (calc.). Luster vitreous. Colorless to white; rarely purple. Transparent.

O p t. In transmitted light, colorless.

ORIENTATION		n (Creede)[4]	n (Granite)[5]	
X		1.461 ± 0.001	1.462	Biaxial negative (–).
Y	b	1.478 ± 0.001	1.478	$2V$ $64°22'$ (Na)[4].
$Z \wedge c$	$42°30'$ [5]	1.485 ± 0.001	1.483	$r > v$, strong.

C h e m. A basic hydrated sulfate-fluoride of calcium and aluminum, $Ca_3Al_2F_4(OH,F)_6(SO_4) \cdot 2H_2O$.

Anal.

	1	2	3
Ca	24.57	23.98	23.72
Al	11.02	11.58	11.74
SO$_4$	19.63	18.32	19.10
F	25.85	[30.35]	30.30
O	5.44	[3.97]	[4.36]
H$_2$O	13.49	11.80	10.78
Total	100.00	[100.00]	[100.00]
G.	2.715	2.730	2.713

1. $Ca_3Al_2F_4(OH,F)_6(SO_4) \cdot 2H_2O$, with F:OH = 2:1. 2. Wagon Wheel Gap, Colorado.[6] H$_2$O includes H$_2$O $-$ 110° 0.72. 3. Wagon Wheel Gap, Colorado.[5]

Tests. B.B. intumesces and fuses to a white enamel. Slowly soluble in acids.

O c c u r. Found with barite embedded in kaolinite in the upper portions of a fluorite-barite vein near Wagon Wheel Gap, Creede quadrangle, Mineral County, Colorado. Gearksutite occurs in the same deposit. Later found with clay and wad in the oxidized zone of gold-bearing fluorite-quartz veins at Granite, near Tonopah, Nye County, Nevada. In the tin veins of Colquiri, Bolivia.

N a m e. From the occurrence in the Creede quadrangle, Colorado.

Ref.

1. Foshag, *U. S. Nat. Mus., Proc.*, **59**, 419 (1922).
2. Foshag (1922); also *Am. Min.*, **17**, 75 (1932).
3. Wolfe and Frondel, priv. comm. (1945), by Weissenberg method on type material.
4. Larsen and Wells (1916).
5. Foshag (1932).
6. Wells anal. in Larsen and Wells (1916).

12.1.2 Arzrunite. *Arzruni* and *Thaddéeff* (*Zs. Kr.*, **31**, 230, 1899).

Apparently orthorhombic and pseudo-hexagonal with $a:b:c = 0.577:1:0.416$.[1] Found as druses of minute prismatic crystals with $\{110\}$, $\{010\}$, $\{001\}$, and small modifying faces of $\{111\}$ and $\{021\}$. Color blue or bluish green. Optically biaxial, with parallel extinction; an optic axis is nearly perpendicular $\{110\}$. Strongly pleochroic, blue and nearly colorless parallel the horizontal axes. Composition uncertain,[2] perhaps a sulfate-chloride of copper and lead. Found with lanarkite and daviesite (?) at the mine Buena Esperanza, Challacolla, Tarapacá province, Chile. Named after Andreas Arzruni (1847–1898), Professor of Mineralogy at Aachen, who first recognized the mineral.

Ref.

1. From (110) ∧ (1$\bar{1}$0) = 60°, (1$\underline{1}$1) ∧ (001) = 39°37'; other measured angles are (010) ∧ (021) = 50°28', (111) ∧ (1$\bar{1}$1) = 37°28'. Orthorhombic symmetry is inferred from the pleochroism observed.
2. Two partial analyses by Thaddéeff on very impure material are given in the original description and in Dana (App. II, 9, 1909).

12.1.3   **Trudellite** [$Al_{10}Cl_{12}(OH)_{12}(SO_4)_3 \cdot 30H_2O$ (?)].   Gordon (*Proc. Ac. Sc. Philadelphia*, **77**, 317, 1926).

C r y s t.[1]   Hexagonal—R (?).
Habit.   As compact masses.
P h y s.   Cleavage rhombohedral.   H. $2\frac{1}{2}$.   G. 1.93.   Luster vitreous. Color amber yellow.   Transparent.   Taste astringent.
O p t.   In transmitted light, pale yellow.

nO 1.560 ± 0.005   Uniaxial negative (−).
nE 1.495 ± 0.005

C h e m.   A hydrated basic sulfate-chloride of aluminum, $Al_{10}Cl_{12}(OH)_{12}(SO_4)_3 \cdot 30H_2O$.   The water content is uncertain.   Very deliquescent.
Anal.

	1	2
Na_2O		1.58
CaO		1.56
Fe_2O_3		1.00
Al_2O_3	29.50	25.67
SO_3	13.91	13.60
Cl	24.62	24.42
H_2O	37.54	36.60
Rem.		1.23
	105.57	105.66
Less O = Cl	5.57	5.49
Total	100.00	100.17

1. $Al_{10}Cl_{12}(OH)_{12}(SO_4)_3 \cdot 30H_2O$.   2. Chile.[2]   Rem. is SiO_2 0.57, MgO 0.66.

Tests.   B.B. infusible.   In the C.T. yields much acid water.   Easily soluble in water with precipitation of some of the aluminum as hydrous oxide.

O c c u r.   Found associated with gypsum, anhydrite, and pickeringite in the Cerro Pintados, 80 km. southeast of Iquique, Tarapacá province, Chile.

N a m e.   After Harry W. Trudell, amateur mineralogist of Philadelphia.

Ref.

1. The symmetry is trigonal, as indicated by the rhombohedral cleavage and the occurrence of microscopical crystals with a rhombohedral outline, but the lattice type is not certain.
2. Shannon anal. in Gordon (1926).

CARBONATES

The carbonates comprise a family of anisodesmic oxysalts in which the fundamental anionic unit of structure is the $(CO_3)^=$ group. In this group a carbon atom lies at the center of an equilateral triangle of three oxygen atoms. The carbon-oxygen bonds are essentially covalent, while the bonding of these units to the cationic elements is essentially ionic. Among the natural carbonates whose structures are known may be mentioned the members of the important Calcite and Aragonite groups. The structures of most of the hydrated and basic carbonates are not known.

CLASS 13. Acid carbonates
Type 1. Miscellaneous
13.1.1 Nahcolite $NaHCO_3$
13.1.2 Kalicinite $KHCO_3$
13.1.3 Teschemacherite $(NH_4)HCO_3$
13.1.4 Trona $Na_3H(CO_3)_2 \cdot 2H_2O$

CLASS 14. Anhydrous normal carbonates
Type 1. $A(XO_3)$
14.1.1 Calcite group
14.1.1.1 Calcite $CaCO_3$
14.1.1.2 Magnesite $MgCO_3$
14.1.1.3 Siderite $FeCO_3$
14.1.1.4 Rhodochrosite $MnCO_3$
14.1.1.5 Cobaltocalcite $CoCO_3$
14.1.1.6 Smithsonite $ZnCO_3$
14.1.1.7 Otavite $CdCO_3$
14.1.2 Vaterite $CaCO_3$
14.1.3 Aragonite group
14.1.3.1 Aragonite $CaCO_3$
14.1.3.2 Witherite $BaCO_3$
14.1.3.3 Strontianite $SrCO_3$
14.1.3.4 Cerussite $PbCO_3$

Type 2. $AB(XO_3)_2$
14.2.1 Dolomite group
14.2.1.1 Dolomite $CaMg(CO_3)_2$
14.2.1.2 Ankerite $Ca(Fe,Mg)(CO_3)_2$
14.2.1.3 Kutnahorite $Ca(Mn,Mg)(CO_3)_2$
14.2.2 Alstonite $CaBa(CO_3)_2$
14.2.3 Barytocalcite $CaBa(CO_3)_2$

Type 3. Miscellaneous
14.3.1 Fairchildite $K_2Ca(CO_3)_2$
14.3.2 Shortite $Na_2Ca_2(CO_3)_3$

CLASSIFICATION

CLASS 15. Hydrated normal carbonates
Type 1. $A(XO_3) \cdot xH_2O$
15.1.1	Thermonatrite	$Na_2CO_3 \cdot H_2O$
15.1.2	Nesquehonite	$MgCO_3 \cdot 3H_2O$
15.1.3	Trihydrocalcite	$CaCO_3 \cdot 3H_2O$
15.1.4	Pentahydrocalcite	$CaCO_3 \cdot 5H_2O$
15.1.5	Lansfordite	$MgCO_3 \cdot 5H_2O$
15.1.6	Natron	$Na_2CO_3 \cdot 10H_2O$

Type 2. Miscellaneous
15.2.1	Buetschliite	$K_6Ca_2(CO_3)_5 \cdot 6H_2O$
15.2.2	Pirssonite	$Na_2Ca(CO_3)_2 \cdot 2H_2O$
15.2.3	Gaylussite	$Na_2Ca(CO_3)_2 \cdot 5H_2O$
15.2.4	Schroeckingerite	$NaCa_3(UO_2)(CO_3)_3(SO_4)F \cdot 10H_2O$
15.2.5	Voglite	U,Cu,Ca carbonate
15.2.6	Bayleyite	$Mg_2(UO_2)(CO_3)_3 \cdot 18H_2O$
15.2.7	Swartzite	$CaMg(UO_2)(CO_3)_3 \cdot 12H_2O$
15.2.8	Andersonite	$Na_2Ca(UO_2)(CO_3)_3 \cdot 6H_2O$
15.2.9	Liebigite	$Ca_2U(CO_3)_4 \cdot 10H_2O$
15.2.10	Lanthanite	$(La,Ce)_2(CO_3)_2 \cdot 8H_2O$

CLASS 16. Carbonates containing hydroxyl or halogen
Type 1. $A_m(XO_3)_pZ_q$
16.1.1	Loseyite	$(Mn,Zn)_7(CO_3)_2(OH)_{10}$
16.1.2	Zaratite	$Ni_3(CO_3)(OH)_4 \cdot 4H_2O$
16.1.3	Hydrozincite	$Zn_5(CO_3)_2(OH)_6$
16.1.4	Aurichalcite	$(Zn,Cu)_5(CO_3)_2(OH)_6$
16.1.5	Rosasite	$(Cu,Zn)_2(CO_3)(OH)_2$
16.1.6	Malachite	$Cu_2(CO_3)(OH)_2$
16.1.7	Phosgenite	$Pb_2(CO_3)Cl_2$
16.1.8	Bismutite	$(BiO)_2(CO_3)$
16.1.9	Waltherite	Bi basic carbonate
16.1.10	Artinite	$Mg_2(CO_3)(OH)_2 \cdot 3H_2O$
16.1.11	Azurite	$Cu_3(CO_3)_2(OH)_2$
16.1.12	Hydrocerussite	$Pb_3(CO_3)_2(OH)_2$
16.1.13	Hydromagnesite	$Mg_4(CO_3)_3(OH)_2 \cdot 3H_2O$
16.1.14	Rutherfordine	$(UO_2)(CO_3)$ (?)
16.1.15	Sharpite	$(UO_2)_6(CO_3)_5(OH)_2 \cdot 6H_2O$ (?)

Type 2. $A_mB_n(XO_3)_pZ_q$
16.2.1	Dawsonite	$NaAl(CO_3)(OH)_2$
16.2.2	Northupite	$Na_3Mg(CO_3)_2Cl$
16.2.3	Dundasite	$PbAl_2(CO_3)_2(OH)_4 \cdot 4H_2O$
16.2.4	Alumohydrocalcite	$CaAl_2(CO_3)_2(OH) \cdot 2H_2O$ (?)
16.2.5	Beyerite	$Ca(BiO)_2(CO_3)_2$
16.2.6	Parisite	$Ce_2Ca(CO_3)_3F_2$
16.2.7	Cordylite	$Ce_2Ba(CO_3)_3F_2$
16.2.8	Synchisite	$CeCa(CO_3)_2F$
16.2.9	Bastnäsite	$Ce(CO_3)F$
16.2.10	Ancylite	Sr,Ca,Ce carbonate

CLASS 17. Compound carbonates
Type 1. Miscellaneous
17.1.1	Tychite	$Na_6Mg_2(CO_3)_4(SO_4)$
17.1.2	Bradleyite	$Na_3Mg(CO_3)(PO_4)$
17.1.3	Leadhillite	$Pb_4(CO_3)_2(OH)_2(SO_4)$
17.1.4	Susannite	$Pb_4(CO_3)_2(OH)_2(SO_4)$
	Caledonite	[with sulfates]

13 ACID CARBONATES

TYPE 1. MISCELLANEOUS

13.1.1   NAHCOLITE [$NaHCO_3$]. Thermokalite pt. *Johnston-Lavis* (Ms, ca. 1889; in Bannister, *Min. Mag.*, **22**, 53, 1928). Nahcolite *Bannister* (*Min. Mag.*, **22**, 53, 1928). Sodium bicarbonate.

C r y s t.[1] Monoclinic; prismatic—$2/m$.

$a:b:c = 0.7645:1:0.3582;$ $\beta\,93°19';$ [1] $p_0:q_0:r_0 = 0.4685:0.3576:1$

$r_2:p_2:q_2 = 2.7964:1.3102:1;$ $\mu\,86°41';$ $p_0'\,0.4693,\, q_0'\,0.3582,\, x_0'\,0.0580$

Forms: [2]

	ϕ	ρ	ϕ_2	$\rho_2 = B$	C	A
b 010	0°00′	90°00′	.....	0°00′	90°00′	90°00′
n 120	33 13½	90 00	0°00′	33 13½	88 10	56 46½
m 110	52 39	90 00	0 00	52 39	87 22	37 21
r 101	90 00	27 48	62 12	90 00	24 29	62 12
s $\bar{1}01$	−90 00	22 21½	112 21½	90 00	25 40½	112 21½
o 111	55 48½	32 31	62 12	72 25	29 49½	63 36

Structure cell.[3] Space group $P2_1/n$. $a_0\,7.51 \pm 0.04$ kX, $b_0\,9.70 \pm 0.04$, $c_0\,3.53 \pm 0.03$; $\beta\,93°19';$ $a_0:b_0:c_0 = 0.774:1:0.364.$ Cell contents $Na_4H_4(CO_3)_4$.

Habit. Prismatic [001] with {110} and {010} dominant. Usually terminated by {101} and minor {111}, also by {111} and {$\bar{1}$01} in equal development. As friable crystal aggregates and porous masses.

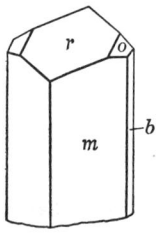
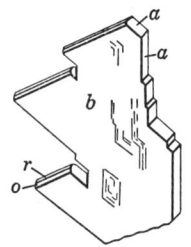

Searles Lake.   Searles Lake. Twin on {101}. After Foshag.

Twinning.[4] Very common, with {101} as twin and composition plane. Both contact and penetration types, often as reticulated groups, and flattened {010}.

P h y s. Cleavage {101} perfect, {111} good, {100} distinct. Fracture conchoidal. H. 2½. G. 2.21; [7] 2.16 (calc.). Luster vitreous, inclining toward resinous on the cleavage. Colorless to white, sometimes gray or buff from impurities. Streak colorless. Transparent.

NAHCOLITE

O p t. In transmitted light, colorless.

ORIENTATION n (Searles Lake) [5] n (artif.) [6]

$X \wedge c$ 27°	1.377	1.378	Biaxial negative $(-)$.
Y b	1.503	1.500	$2V \sim 75°$.
Z	1.583	1.584	$r < v$, weak. An optic axis is nearly perpendicular to the $\{101\}$ cleavage.

C h e m. Sodium bicarbonate, $NaHCO_3$.
Anal.

	1	2
Na_2O	36.90	36.74
CO_2	52.38	51.15
H_2O	10.72	10.76
Rem.		1.18
Total	100.00	99.83

1. $NaHCO_3$. 2. Searles Lake, California.[8] Rem. is CaO 0.20, $(Fe,Al)_2O_3$ 0.16, insol. 0.82.

Tests. Easily soluble in water, yielding a solution alkaline to phenolphthalein and to methyl red. The aqueous solution on heating gives off CO_2. Also soluble in glycerine. B.B. fuses quietly, yielding a white crystalline bead.

O c c u r. Found admixed with trona, thermonatrite, and thenardite as an efflorescence in an old Roman conduit from hot springs at Stufe de Nerone, in the Phlegrean Fields, near Naples, Italy. Also apparently admixed with thenardite and halite in a lava tunnel on Vesuvius. In these occurrences the mineral presumably has formed by the action of CO_2 and water vapor upon thermonatrite or trona.[9] Reported from Little Mogadi dry lake, 40 km. south of Mogadi Lake, in British East Africa, where the mineral may have formed by the action of CO_2 on trona in the playa muds. Occurs in large quantities in the central salt body of Searles Lake, San Bernardino County, California. The mineral forms thin beds associated with gaylussite, thenardite, burkeite, northupite, borax, halite; also disseminated in gaylussite or northupite marls free from other saline minerals. Nahcolite occurs as concretions up to five feet in diameter in oil shale of the Eocene Green River formation at Anvil Points, near Rifle, Colorado.[10] The mineral probably also occurs with trona in the soda lakes of Egypt.[11]

A l t e r. Observed as alteration rims around crystals of thermonatrite, and as fibrous pseudomorphs after gaylussite.

A r t i f.[12] Sodium bicarbonate is formed under many conditions when powdered and moist solid sodium carbonate or one of its hydrates is exposed to carbon dioxide. In crystals by the slow reaction of sulfuric acid and sodium carbonate solution, and by slow evaporation of the water solution.

N a m e. In allusion to the composition, from NaHCO-lite.

Ref.

1. Schabus, *Best. Krystallgest., Preiss. Ak. Wien*, 97 (1855)—Groth (**2**, 191, 1908) on artificial material. Foshag, *Am. Min.*, **25**, 769 (1940), obtained similar results on inferior crystals from Searles Lake.

2. Foshag (1940). Schabus (1855) in addition observed $i\{121\}$ and $w\{\bar{1}11\}$ on artificial material.
3. Zachariasen, *J. Chem. Phys.*, **1**, 634 (1933) by the Laue and rotation methods with analysis of the structure.
4. Foshag (1940) and Schabus (1855).
5. Average of measurements by Foshag and Bandy in Foshag (1940).
6. Average of measurements by Larsen (134, 1921) and Merwin in Winchell (198, 1931).
7. Average of the values 2.220, 2.206, and 2.192 from Stolba, *J. pr. Chem.*, **97**, 503 (1866), Schröder, *Ber.*, **11**, 2018 (1878), and Playfair and Joule, *Mem. Chem. Soc.*, **2**, 401 (1845).
8. Foshag (1940).
9. The conditions of formation of nahcolite are discussed by Bannister (1928) and Foshag (1940).
10. Ertle, *Am. Min.*, **32**, 117 (1947).
11. See discussion in Bannister (1928).
12. Mellor (**2**, 773, 1922).

13.1.2 KALICINITE [$KHCO_3$]. Kalicine Pisani (*C.R.*, **60**, 918, 1865). Kalicit Landero (255, 1888). Kalicinit Groth (57, 1898).

Cryst.[1] Monoclinic; prismatic—$2/m$.

$a:b:c = 2.6770:1:0.6558;$ $\beta\ 103°25';$ $p_0:q_0:r_0 = 0.2450:0.6379:1$

Forms:[2]

c 001 a 100 m 110 r 201 o $\bar{4}01$

Structure cell.[3] Space group $P2_1/a$. a_0 15.01 kX, b_0 5.69, c_0 3.68; $\beta\ 104°30';$ $a_0:b_0:c_0 = 2.64:1:0.647$. Cell contents $K_4(HCO_3)_4$.

Habit. As fine-crystalline aggregates. Artificial crystals are short prismatic [010].

Phys. Cleavage $\{100\}$, $\{001\}$, $\{101\}$. Soft. G. 2.168 (artif.);[4] 2.17 (calc.). Colorless to white or yellowish. Transparent.

Opt.[5] In transmitted light, colorless.

Orientation	$n(Na)$	
$X \wedge c$ +30°	1.380	Biaxial negative (−).
Y b	1.482	$2V\ 81\frac{1}{2}°$.
Z	1.578	

Chem. Potassium bicarbonate, $KHCO_3$. The only available analysis is on very impure material.[6]

Tests. On heating affords K_2CO_3 and fuses easily. Soluble in water (33.3 g. per 100 g. H_2O at 20°C.).

Occur. Found as a result of recent decomposition under a dead tree at Chypis, Canton Wallis, Switzerland. Said to occur associated with trona in Hungary.[7]

Artif. Obtained in crystals by passing CO_2 into a concentrated solution of K_2CO_3, and in other ways.[8]

Name. From *kalium, potassium*, in allusion to the composition.

Ref.

1. Orientation and angles of Brooke, *Ann. Phil.*, **22**, 42 (1823), x-ray unit of Dhar, *Curr. Sc.*, **4**, 867 (1936)—*SB* IV, 158 (1936); all data on artificial crystals. Transformation: Brooke to Dhar, $100/010/00\frac{1}{2}$.

2. Cf. Groth (**2**, 191, 1908).
3. Dhar (1936).
4. Average of values 2.158, 2.180, and 2.167 cited in Groth (1908).
5. Merwin, *Int. Crit. Tab.*, **7**, 27 (1930).
6. Pisani (1865).
7. Ochsenius, *Zs. pr. Geol.*, 198 (1893).
8. Cf. Mellor (**2**, 774, 1922).

13.1.3 **T E S C H E M A C H E R I T E** [$(NH_4)HCO_3$]. Bicarbonate of Ammonia
Teschemacher (*Phil. Mag.*, **28**, 548, 1846). Teschemacherite *Dana* (705, 1868). Ammonium acid carbonate; Ammonium hydrocarbonate.

C r y s t.[1] Orthorhombic; dipyramidal—$2/m\ 2/m\ 2/m$.

$a:b:c = 0.6726:1:0.7996;$ $p_0:q_0:r_0 = 1.1888:0.7996:1$

$q_1:r_1:p_1 = 0.6726:0.8412:1;$ $r_2:p_2:q_2 = 1.2506:1.4868:1$

Forms:[2]

| c 001 | b 010 | a 100 | m 110 | q 012 | r 102 |

Structure cell.[3] Space group $Pccn$. a_0 7.29 kX, b_0 10.79, c_0 8.76; $a_0:b_0:c_0 = 0.676:1:0.812$. Cell contents $(NH_4)_8(HCO_3)_8$.

Habit. As compact crystalline masses. Artificial crystals are short prismatic [001].

P h y s. Cleavage {110} perfect. Brittle. H. $1\frac{1}{2}$. G. 1.57 (artif.);[7] 1.51 (calc.). Colorless to white or yellowish. Transparent.

O p t.[4] In transmitted light, colorless.

Orientation		n(Na)	
X	a	1.4227	Biaxial negative ($-$).
Y	b	1.5358	$2V$ 41°38′ (calc.).
Z	c	1.5545	$r < v$, weak.

C h e m. Ammonium bicarbonate, $(NH_4)HCO_3$. The reported analyses[5] agree closely with the formula given. Slowly decomposes in moist air.

Tests. In the C.T. volatilized and decomposed. Soluble in water and, with effervescence, in dilute acids.

O c c u r. Found in guano deposits at Saldanha Bay, Cape Province, Africa. Found similarly on the Cincha and Guañape Islands, Peru, and on the west coast of the Patagonia region, South America.

A r t i f.[6] Obtained in crystals by cooling a hot concentrated solution of the sesquicarbonate in a closed vessel, by treating a solution of the normal carbonate or sesquicarbonate with CO_2, and in other ways.

N a m e. After Frederick Edward Teschemacher (1791–1863), English chemist, who first described the species.

Ref.

1. Angles and orientation of Rose, *Ann. Phys.*, **46**, 401 (1839); x-ray unit of Mooney, *Phys. Rev.*, **39**, 861 (1932); all data on artificial crystals.
2. Goldschmidt (**8**, 123, 1922), on artificial crystals.
3. Mooney (1932).
4. Lang, *Ak. Wien, Sitzber.*, **45** [II], 112 (1862).

5. Cited in Hintze (**1** [3A], 2751, 1916).
6. Mellor (**2**, 787, 1922).
7. Schiff, *Ann. Phys.*, **107**, 64 (1858). Ulex, *Ann. Phys.*, **61**, 44 (1847), gave 1.45 for the natural material.

13.1.4 **T R O N A** [$Na_3H(CO_3)_2 \cdot 2H_2O$]. Alkali orientale impurum terrestre pt. *Wallerius* (174, 1747). Trona *Bagge* (*Ak. Handl. Stockholm*, **35**, 140, 1773). Natrum von Tripole, Stralige Natrum *Klaproth* (**3**, 83, 1802). Urao *Boussingault* (*Ann. mines*, **12**, 278, 1826).

C r y s t. Monoclinic; prismatic—$2/m$.

$a:b:c = 2.8459:1:2.9696;$ $\beta\ 102°37';$ [1] $p_0:q_0:r_0 = 1.0435:2.8979:1$

$r_2:p_2:q_2 = 0.3451:0.3601:1;$ $\mu\ 77°23';$ $p_0'\ 1.0693, q_0'\ 2.9696, x_0'\ 0.2238$

Forms: [2]

	ϕ	ρ	ϕ_2	$\rho_2 = B$	C	A
c 001	90°00'	12°37'	77°23'	90°00'	.....	77°23'
a 100	90 00	90 00	0 00	90 00	77°23'	.....
ρ 304	90 00	45 43½	44 16½	90 00	33 06½	44 16½
e 101	90 00	52 17	37 43	90 00	39 40	37 43
s $\overline{3}02$	−90 00	54 04½	144 04½	90 00	66 41½	144 04½
p 111	23 32	72 50½	37 43	28 49½	68 09	67 34½
o $\overline{1}11$	−15 53½	72 03	130 13	23 48	75 53½	105 26
r 211	38 30	75 14	22 56½	40 49½	67 39	52 59

Habit. Elongated [010], also flattened {001}. {$h0l$} faces striated parallel [010]. Well-formed natural crystals are very rare. Often fibrous or columnar massive.

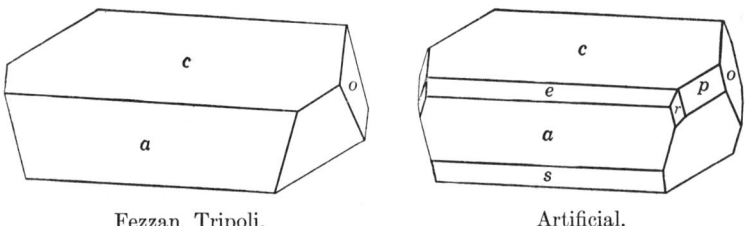

Fezzan, Tripoli.   Artificial.

P h y s. Cleavage {100} perfect, {$\overline{1}11$} and {001} in traces. Fracture uneven to subconchoidal. H. 2½–3. G. 2.14.[3] Luster vitreous, glistening. Colorless; usually gray or yellowish white. Translucent. Taste alkaline. Not altered by exposure to dry air.

O p t. In transmitted light, colorless.

Orientation [4]	n [5]	
X b	1.412	Biaxial negative (−).
Y	1.492	$2V\ 76°16'$ (Na).[4]
$Z \wedge c$ 83°	1.540	$r < v$, rather strong.
		Z is nearly normal to the perfect {100} cleavage.

Chem. A hydrated sodium acid carbonate, $Na_3H(CO_3)_2 \cdot 2H_2O$. Numerous analyses of trona have been reported.[6] A number of these apparently represent mixtures variously with natron, thermonatrite, nahcolite, halite, and sulfates.

Anal.

	1	2	3
Na_2O	41.14	41.22	41.00
CO_2	38.94	37.98	38.13
H_2O	19.92	19.13	20.07
Cl		0.86	0.19
SO_3		0.70	0.70
Rem.		0.08	0.02
	100.00	99.97	100.11
O = Cl		0.19	0.04
Total	100.00	99.78	100.07
G.		2.12	2.147

1. $Na_3H(CO_3)_2 \cdot 2H_2O$. 2. Lake Mandara, Fezzan, Egypt.[7] Rem. is CaO 0.02, insol. 0.06. May contain nahcolite. 3. Owens Lake, California.[8] Rem. is insol. 0.02. Average of several analyses.

Tests. In C.T. yields water and CO_2. Soluble in water, and effervesces in acids.

Occur.[9] Trona is found in the deposits of various saline lakes, or is produced by the evaporation of their waters, and also occurs as an efflorescence on the soil in arid regions. Commonly associated minerals include natron, thermonatrite, halite, glauberite, thenardite, mirabilite, gypsum. The oldest known deposits of natural sodium carbonates are those of the lower Nile Valley, near Memphis. Also from the Oasis of Bilma, in the eastern Sahara Desert, and at numerous places in the province of Fezzan, Tripoli. Also in the Sudan as efflorescences, in the neighborhood of Lake Chad, and in the region of Kilimanjaro, Tanganyika. At Lake Magadi, Kenya. Trona and other sodium carbonates and sulfates occur widely in soda lakes in Armenia, Persia, and the alkali deserts of Mongolia and Tibet. In the Szegedin district, Hungary, and as crusts on lava on Vesuvius, Italy. As a bottom deposit in a lake at Lagunillas, Venezuela. In the United States, trona occurs widely in alkali lake deposits in the western states. Found in soda lakes south of Laramie, Wyoming; also with shortite, northupite, bradleyite, and pirssonite in oil shale of the Green River formation in Sweetwater County, Wyoming. Found in considerable quantities in Nevada, in the Soda Lakes near Fallon, in Butterfield Marsh and Double Springs Marsh and other saline lakes and playas. Trona occurs abundantly in California with borax, hanksite, thenardite, and other salts at Searles Lake, San Bernardino County; also at Borax Lake, Lake County; and as layers along the shores of Owens Lake, Inyo County.

Artif.[10] Obtained in crystals by reaction of sodium bicarbonate and the decahydrated carbonate in sodium chloride solution, and in other ways. Trona crystallizes from solution at all temperatures up to 195°.

Name. Trona is the reduced form of the Arabic name for the native salt, *natrūn*. The origin of the name *urao* is unknown.

Ref.

1. Zepharovich, *Zs. Kr.*, **13**, 135 (1887) on artificial crystals. Ayres, *Am. J. Sc.*, **38**, 65 (1889) gives $a:b:c = 2.8426:1:2.9494$, $\beta\ 76°31'$ on artificial crystals.
2. Goldschmidt (**9**, 11, 1923). Rare or uncertain: 407, $\bar{1}.0.18$, $\bar{2}.0.13$, $\bar{2}09$, $\bar{1}02$.
3. From the values 2.14, 2.147, and 2.14 of Walther, *Am. Min.*, **7**, 86 (1922), Chatard, *Am. J. Sc.*, **38**, 59 (1889), and Reinitzer in Zepharovich (1887). Slightly lower values have been reported by others.
4. Zepharovich (1887).
5. Larsen and Berman (151, 1934), an average apparently of data earlier reported by Larsen (146, 1921).
6. See Doelter (**1**, 186, 1911) and Hintze (**1** [3A], 2759, 2770, 1916).
7. Piazza, *Per. Min.*, **2**, 17 (1931).
8. Chatard, *Am. J. Sc.*, **38**, 59 (1889).
9. On the occurrence and chemistry of formation of trona see Wegscheider in Doelter (**1**, 141, 1911), Hintze (**1** [3A], 2759, 1916), Chatard, *U. S. Geol. Sur., Bull. 60*, 27 (1890).
10. Mellor (**2**, 777, 1922). See also Waldeck, Lynn, and Hill, *J. Am. Chem. Soc.*, **56**, 43 (1934), Hill and Bacon (*ibid.*, **49**, 2487, 1927), on the system $NaHCO_3$-Na_2CO_3-H_2O.

14 ANHYDROUS NORMAL CARBONATES

TYPE 1. $A(XO_3)$

14.1.1 CALCITE GROUP

HEXAGONAL—R; HEXAGONAL-SCALENOHEDRAL—$\bar{3}\ 2/m$

	$a:c$
Calcite, $CaCO_3$	1:0.8543
Magnesite, $MgCO_3$	1:0.8112
Siderite, $FeCO_3$	1:0.8184
Rhodochrosite, $MnCO_3$	1:0.8183
Cobaltocalcite, $CoCO_3$	1:0.810
Smithsonite, $ZnCO_3$	1:0.8063
Otavite, $CdCO_3$	1:0.836

Most oxysalts of the AXO_3 type fall into either the calcite or aragonite structure-types. Structures of the former type are taken by nitrates and carbonates of relatively small cations, including Li, Na, Mg, Ca, Fe", Zn, Co, and Cd, and by certain metaborates. The large cations, including K, Ba, Sr, and Pb, crystallize in the aragonite structure-type. A few compounds near the critical radius ratio of 0.73 between the two types are polymorphous. These include $CaCO_3$ and KNO_3, the higher-temperature polymorphs crystallizing in the calcite structure-type. The nitrates of very large cations such as Tl and Rb crystallize in a third structure-type ($RbNO_3$) which is not represented among minerals. The structure [1] of calcite and its relatives is ionic and resembles that of NaCl, the place of the spherical Cl ions being taken by planar $(CO_3)''$ or $(NO_3)'$ groups. These groups are arranged in parallel position in the crystal with their threefold symmetry axes oriented along [0001], corresponding to a cube-diagonal or [111] direction in the (distorted) NaCl structure. The extreme birefringence and optically negative character of the members of the calcite structure-type are due to this arrangement of the anion.

The members of the Calcite Group enter into a wide range of substitutional solid solutions. The limits of substitution in the polycomponent system and their temperature dependency are not clearly defined by the available analyses of natural material. These analyses are essentially of chance origin and do not adequately or purposefully sample the range of physical and chemical environments in nature. A better understanding could be reached by laboratory investigations in which temperature, pressure, and the mutual access of the cations concerned in the formation of this structure-type can be controlled. These remarks apply generally to the interpretation of mineral series from natural-history evidence.

In the *binary* series between the members of the Calcite Group, the available evidence indicates that complete series exist between $CaCO_3$-$MnCO_3$, $FeCO_3$-$MgCO_3$, $FeCO_3$-$MnCO_3$, and probably between $ZnCO_3$-$MnCO_3$. Extensive series with small central gaps apparently extend between $MgCO_3$-$MnCO_3$ and $CaCO_3$-$FeCO_3$, and less extensive series between $ZnCO_3$-$CaCO_3$, $ZnCO_3$-$FeCO_3$, $ZnCO_3$-$MgCO_3$ and $CaCO_3$-$MgCO_3$. These relations are about what would be predicted from the relative sizes of the ions involved, except that the series of $ZnCO_3$ with $FeCO_3$ and $MgCO_3$ should be complete. Ba, Sr, and Pb substitute in only small amounts for Ca in calcite; these large ions ordinarily enter into aragonite-type structures.

Ref.

1. Bragg, W. L., *Proc. Roy. Soc. London*, **89A**, 468 (1914).

14.1.1.1   C A L C I T E [$CaCO_3$]. Marmor Marble pt., χάλξ, Calx *Pliny* (*Nat. Hist.*, 53, 36). Lapis calcarius. Saxum calcis (*Calx* in Latin meaning burnt lime], Kalchstein, Marmelstein *Agricola* (*De Nat. Foss.*, 320, 1546; Interpr., 468, 1546). Crystallus talcosa seu Islandica [Iceland Spar] *Bartholinus* (*Phil. Trans.*, 67, 1670; *Exper. cryst. Islandicae*, Hafniae, 1670). Kalksten [and as syn. or var., Calcareus lapis, Marmor fusareum, Spatum pellucidum objecta duplicans (Doppelstein), Cristallus islandica, etc.] *Wallerius* (1747; 53, 60, 67, 77, 1750). Spatig Kalksten, Kalkspat *Cronstedt* (13, 1758). Kalk, Kalkspath, Kalkstein *Germ.* Calx aerata [with erroneous analysis] *Bergmann* (1774; **1**, 24, 1780). Spath Calcaire *Romé de Lisle* (**1**, 490, 1783). Chaux carbonatée *Haüy* (**1**, 23, 1801). Calc Spar; Calcareous Spar; Limestone; Carbonate of Lime; Calcium carbonate. Chaux carbonatée *Fr.* Calcit *Haidinger* (498, 1845). Caliza, Espato caliza *Span.*
    Doppelstein, Doppelspat, Perlmutterspat, Schieferspat, Schaumerde, Bergmilch, Kreide, Stinkstein, Kanonenspat, Schweinszähne, Rogenstein kristallisierter Sandstein [from Fontainebleu] *Emmerling* (**1**, 430, 1793; **3**, 397, 1797). Erbsenstein pt. Pisolite *Werner*. Bergmehl. Montmilch. Farina fossilis *Bruckmann*. Papierspath. Schieferspath *Hofmann* (*Bergm. J.*, 188, 1789). Argentin *Kirwan* (**1**, 104, 1794). Fontainebleu Limestone [Lassonne, *Mém. ac. Paris*, 1775], Chaux carbonatée quartzifère *Haüy* (1801). Slate Spar. Aphrit *Karsten* (50, 1808). Schaumspath *Freiesleben*. Schaumerde, Schaumkalk *Werner*. Silvery Chalk *Kirwan*. Écume de Terre *Haüy* (**1**, 430, 1822). Satin Spar pt.; Faserkalk, Atlasspath *Germ.* Dog-tooth Spar, Nailhead Spar. Prunnerite *Esmark* (*Jb. Min.*, 71, 1830). Hislopite *Haughton* (*Phil. Mag.*, **17**, 16, 1859). Wurfelstein. Reichit *Breithaupt* (**24**, 311, 1865). Lublinite *Morozewicz* (*Kosmos*, **32**, 487, 1907). Patagosite *Meunier* (*C.R. somm. soc. géol. France*, 84, 1917). Drewite *Field* (*Carnegie Inst. Washington, Year Book*, **18**, 197, 1919). Capreite *Bellini* (*Boll. soc. geol. Ital.*, **40**, 228, 1921). Elatolite *Fersman* (*C. r. ac. sc. Russie*, 59, 1922). Vaterite-A *Gibson, Wyckoff*, and *Merwin* (*Am. J. Sc.*, **10**, 325, 1925). Protocalcite *Balogh* (*Erdélyi Múz., Kolozsvár*, **42**, 147, 1937).
    Neotyp *Breithaupt* (**2**, 313, 1841). Rhombohedral barytocalcite. Plumbocalcite *Johnston* (*Edinburgh Phil. J.*, **6**, 79, 1829). Strontianocalcite *Genth* (*Proc. Ac. Sc. Philadelphia*, **6**, 114, 1852). Manganocalcite pt., Spartait *Breithaupt* (*B. H. Ztg.*, **17**, 53, 1858). Calcimangite *Shepard* (*Am. J. Sc.*, **39**, 175, 1865). Baricalcite *Dana* (269, 1892). Ferrocalcite *Dana* (269, 1892). Zincocalcite *Dana* (269, 1892). Cobaltocalcite *Millosevich* (*Rend. Acc. Linc., Roma*, **19**, 92, 1910).

C r y s t.[1] Hexagonal—R; hexagonal-scalenohedral—$\bar{3}\,2/m$.

$a\!:\!c = 1\!:\!0.8543;$ $\alpha\,101°55';$ $p_0\!:\!r_0 = 0.9865\!:\!1;$ $\lambda\,74°55'$

CALCITE

Forms: [2]

			ϕ	$\rho = C$	A_1	A_2
c	0001	111	.....	0°00′	90°00′	90°00′
m	10$\bar{1}$0	2$\bar{1}\bar{1}$	30°00′	90 00	30 00	90 00
a	11$\bar{2}$0	10$\bar{1}$	0 00	90 00	60 00	60 00
p·	10$\bar{1}$1	100	30 00	44 36½	52 32½	90 00
m·	40$\bar{4}$1	3$\bar{1}\bar{1}$	30 00	75 47	32 55	90 00
δ·	01$\bar{1}$2	110	−30 00	26 15½	90 00	67 28½
φ·	02$\bar{2}$1	11$\bar{1}$	−30 00	63 07½	90 00	39 25½
π·	08$\bar{8}$1	33$\bar{5}$	−30 00	82 46½	90 00	30 46½
γ	8.8.$\bar{16}$.3	91$\bar{7}$	0 00	77 37½	60 46	60 46
T:	43$\bar{7}$1	40$\bar{3}$	4 43	80 32½	55 49½	65 05
P:	32$\bar{5}$1	30$\bar{2}$	6 35	76 54½	54 30½	67 14
N:	53$\bar{8}$2	50$\bar{3}$	8 13	73 51	53 33	69 07
t:	21$\bar{3}$4	310	10 53½	33 07½	69 02½	79 42
K:	21$\bar{3}$1	20$\bar{1}$	10 53½	69 02	52 19	72 12
w:	31$\bar{4}$5	410	16 06	35 25½	65 18½	82 00
b:	35$\bar{8}$4	52$\bar{3}$	−8 13	59 55	71 16	57 38½
𝔍:	4.8.$\bar{12}$.5	73$\bar{5}$	−10 53½	64 24½	72 50	53 48½
𝔭:	13$\bar{4}$1	21$\bar{2}$	−16 06	74 18	76 38	46 04½
𝕰:	2.8.$\bar{10}$.3	53$\bar{5}$	−19 06½	71 38½	79 40	44 09½

Less common:

ζ	31$\bar{4}$0	7$\bar{2}$5	ω·	0.11.$\bar{11}$.4	55$\bar{6}$	H:	31$\bar{4}$2	30$\bar{1}$
K·	50$\bar{5}$2	41$\bar{1}$	Δ·	07$\bar{7}$2	33$\bar{4}$	𝔍:	62$\bar{8}$1	5$\bar{1}$3
n·	50$\bar{5}$1	11.$\bar{4}$.$\bar{4}$	θ·	04$\bar{4}$1	55$\bar{7}$	f:	7.2.$\bar{9}$.11	920
q·	70$\bar{7}$1	5$\bar{2}$2	Ξ·	05$\bar{5}$1	22$\bar{3}$	G:	72$\bar{9}$5	70$\bar{2}$
r·	10.0.$\bar{10}$.1	7$\bar{3}$3	Σ·	0.11.$\bar{11}$.1	44$\bar{7}$	e:	41$\bar{5}$6	510
s·	13.0.$\bar{13}$.1	9$\bar{4}$4	Φ·	0.14.$\bar{14}$.1	55$\bar{9}$	F:	41$\bar{5}$3	40$\bar{1}$
t·	16.0.$\bar{16}$.1	11.$\bar{5}$.$\bar{5}$	Ψ·	0.16.$\bar{17}$.1	6.6.$\bar{11}$	q:	51$\bar{6}$7	610
η·	04$\bar{4}$5	33$\bar{1}$	π	11$\bar{2}$3	210	E:	51$\bar{6}$4	50$\bar{1}$
κ·	01$\bar{1}$1	22$\bar{1}$	λ	22$\bar{4}$3	31$\bar{1}$	Ξ:	14.2.$\bar{16}$.3	11.$\bar{3}$.$\bar{5}$
υ·	05$\bar{5}$4	33$\bar{2}$	α	44$\bar{8}$3	51$\bar{3}$	z:	12$\bar{3}$5	320
ξ·	04$\bar{4}$3	77$\bar{5}$	X:	7.6.$\bar{13}$.1	70$\bar{6}$	Θ	12$\bar{3}$1	52$\bar{4}$
π·	07$\bar{7}$5	44$\bar{3}$	W:	13.11.$\bar{24}$.2	13.0.$\bar{11}$	𝔮:	24$\bar{6}$1	31$\bar{3}$
ρ·	03$\bar{3}$2	55$\bar{4}$	V:	6.5.$\bar{11}$.1	60$\bar{5}$	𝕰:	25$\bar{7}$3	42$\bar{3}$
σ·	0.11.$\bar{11}$.7	66$\bar{5}$	U:	54$\bar{9}$1	50$\bar{4}$	𝔇:	4.12.$\bar{16}$.7	95$\bar{7}$
τ·	0.13.$\bar{13}$.8	77$\bar{6}$	O:	8.5.$\bar{13}$.3	80$\bar{5}$	𝕮:	27$\bar{9}$4	53$\bar{4}$
ψ·	05$\bar{5}$2	77$\bar{8}$	M:	7.4.$\bar{11}$.3	70$\bar{4}$	𝕏:	4.16.$\bar{20}$.3	9.5.$\bar{11}$

Structure cell.[3] Space group $R\bar{3}c$. a_0 4.98 kX, c_0 17.02; $a_0:c_0 = 1:3.418$; a_{rh} 6.36, α 46°06′. Cell contents $Ca_2(CO_3)_2$ in the rhombohedral unit.

Habit. *A. Single-Crystals.* The crystal habit of calcite is extremely varied and only a few of the more common face developments can be mentioned: (1) Prismatic [0001] with {10$\bar{1}$0} dominant, {11$\bar{2}$0} less common, and terminated by {0001}, {10$\bar{1}$2}, {21$\bar{3}$1}, or otherwise. (2) Thin to thick tabular {0001} and bounded laterally by {10$\bar{1}$0} or steep rhombohedrons. (3) Flat rhombohedral, ordinarily with {01$\bar{1}$2} and passing by rounding or distortion into lenticular shapes. (4) Rhombohedral, obtuse to acute and sometimes cuboid, the forms {02$\bar{2}$1}, {40$\bar{4}$1}, {05$\bar{5}$4}, {03$\bar{3}$2} relatively common alone or in combination. (5) Scalenohedral, with {21$\bar{3}$1} particularly common, {31$\bar{5}$1} less so, or in combination with other forms and sometimes of remarkable complexity; also in acute forms (dogtooth spar). (6) Rarely dipyramidal, with {8.8.$\overline{16}$.3} usually dominant. The commonest forms are $v\ e\ m\ f\ c\ M\ y$. The unit rhombohedron {10$\bar{1}$1} is rare, especially alone, and is usually rough and dull in luster; the dihexagonal prisms also are rare. {0001} is often rough and dull, sometimes with a pearly luster, and the subjacent parts of the crystal may be milky white and translucent with an indistinct parting {00$\underline{0}$1}. Faces other than {10$\bar{1}$1} in the zone [01$\bar{1}$1] are often striated [0$\bar{1}$11] and terminal planes, such as {01$\bar{1}$2}, may be rounded over in this direction; striations in other zones also are common. The identification of the positive and negative sextants in calcite can be established by reference to the {10$\bar{1}$1} cleavage.

B. *Parallel and Subparallel Aggregates.* Parallel growths of more or less merged-together individual crystals are frequently found. Crystals of lenticular habit may thus be stacked vertically with the free lateral edges forming coplanar surfaces. Very commonly, crystals of different periods of formation and of unlike habit may be overgrown in parallel position. Rhombohedral crystals may thus cap crystals of scalenohedral or other acute habits (nailhead spar); or the earlier formed crystals may be entirely enclosed and rendered visible by films of impurities. Occasionally as subparallel aggregates, in plumose, rosette-like, or other shapes; rarely as curved, composite crystals such as are typical of dolomite. Platy types of subparallel growths grade into lamellar aggregates with a pearly luster, the lamellae more or less undulating (*argentine*), and these pass into scaly varieties, the latter soft and often silvery white in color. Also fibrous, coarse to fine, and then with a silky luster; columnar.

C. *Massive Types.* Often massive, from coarse granular to impalpable; also stalactitic, nodular, tuberose, coralloidal and as other imitative shapes. Earthy to mealy crusts occur as deposits in limestone caves or as an efflorescence on calcareous soils (*rock-milk, rock-meal*). The latter types are usually soft and friable, also powdery to flocculent; this material is in part composed of microcrystals elongated along a rhombohedral edge and hence giving inclined extinction in polarized light (*lublinite*).[4] The massive calcite rocks are not properly considered mineralogical varieties. They principally include: limestones, chalk, marls, oolites, and other sedimentary deposits of variable structure, grain size, and degree of purity; the metamorphic equivalents of these rocks, usually relatively coarsegrained; and deposits from calcareous springs and streams (*calc-sinter, tufa, travertine, onyx*) or from percolating waters as stalactites and stalagmites in limestone caverns. Concretionary forms, variously oolitic, piso-

CALCITE

litic, or pearl-like, occur as sea deposits, about hot springs or in pools in caves; these ordinarily are concentrically banded, with a radially fibrous structure. Calcite, with aragonite, also occurs as a biochemical deposit as the nacreous matter of shells, pearls, etc.

Twinning. Four twin laws are known with certainty. (1) Twin plane $\{0001\}$; common, with $\{0001\}$ as the composition surface. Re-entrant angles are present about the equator of the crystal except when bounded laterally by $\{10\bar{1}0\}$, the twinning then being revealed by cleavage or by the apparent horizontal plane of symmetry. (2) Twin plane $\{01\bar{1}2\}$; very common, with $\{01\bar{1}2\}$ as composition face. Sometimes repeated, and in simple growth twins often accompanied by a distortion through the extension of certain faces.[5] Also lamellar and produced by pressure either naturally, as in marble, or artificially.[6] (3) Twin plane $\{10\bar{1}1\}$; uncommon, with composition surface $\{10\bar{1}1\}$. Sometimes repeated, and in simple growth twins often heart-shaped or accompanied by distortion of shape (which generally results in suppressing the re-entrant angle). The twinned individuals have their c-axes nearly at right angles and also have a cleavage plane in common. (4) Twin plane $\{02\bar{2}1\}$;[53] rare, with composition surface $\{02\bar{2}1\}$.

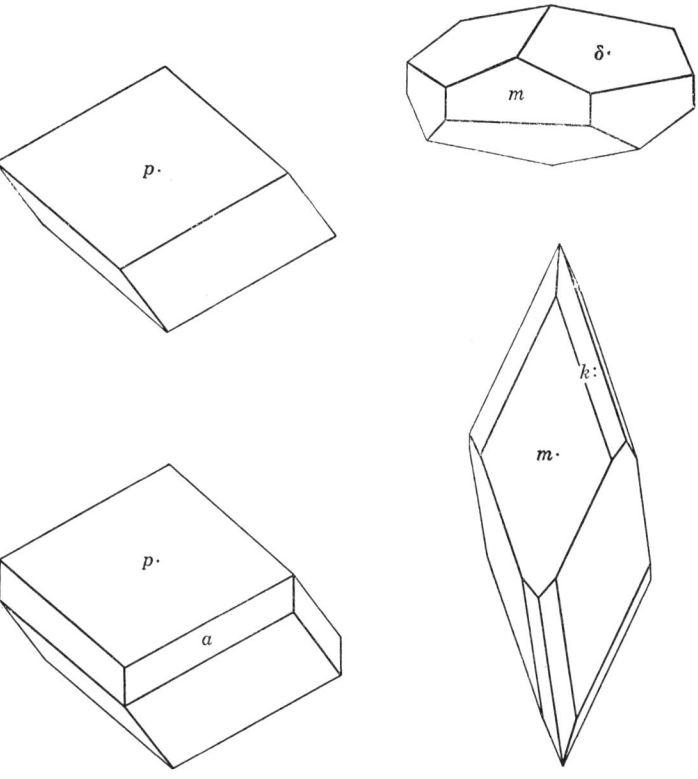

CARBONATES

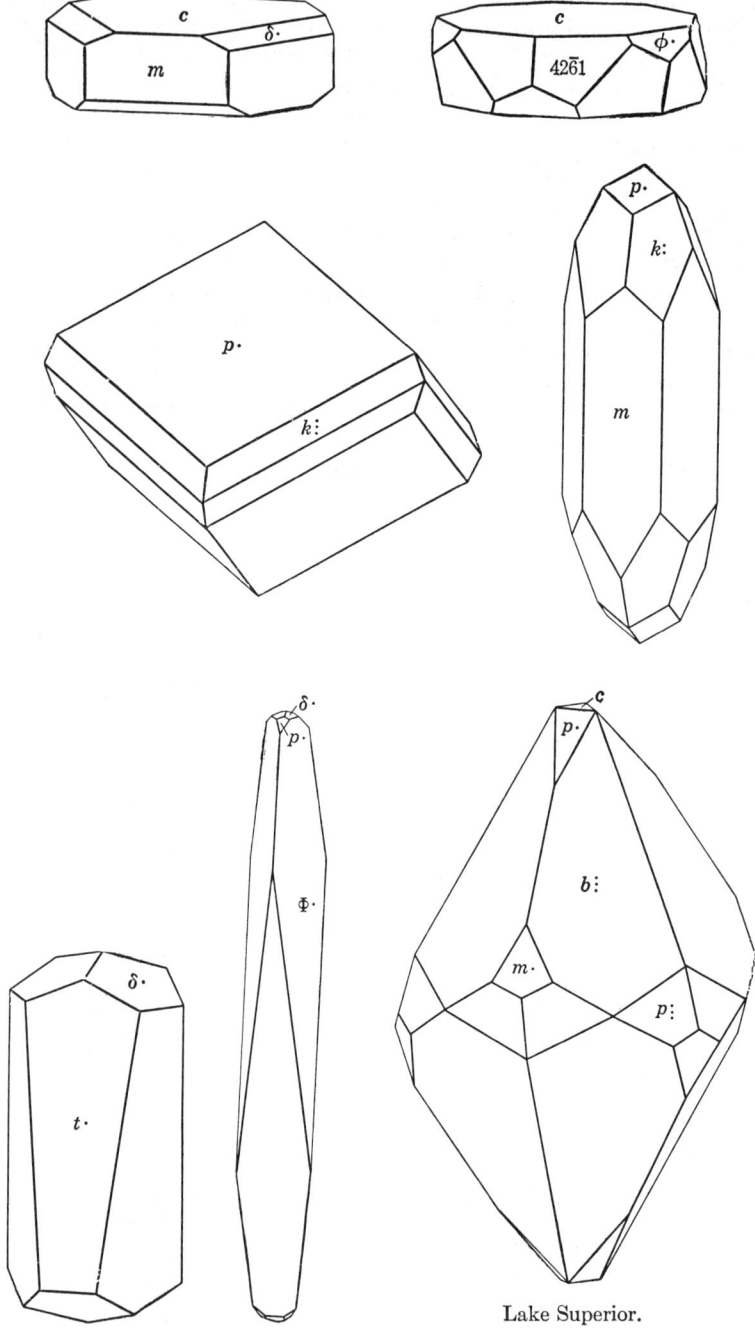

Lake Superior.

Calcite

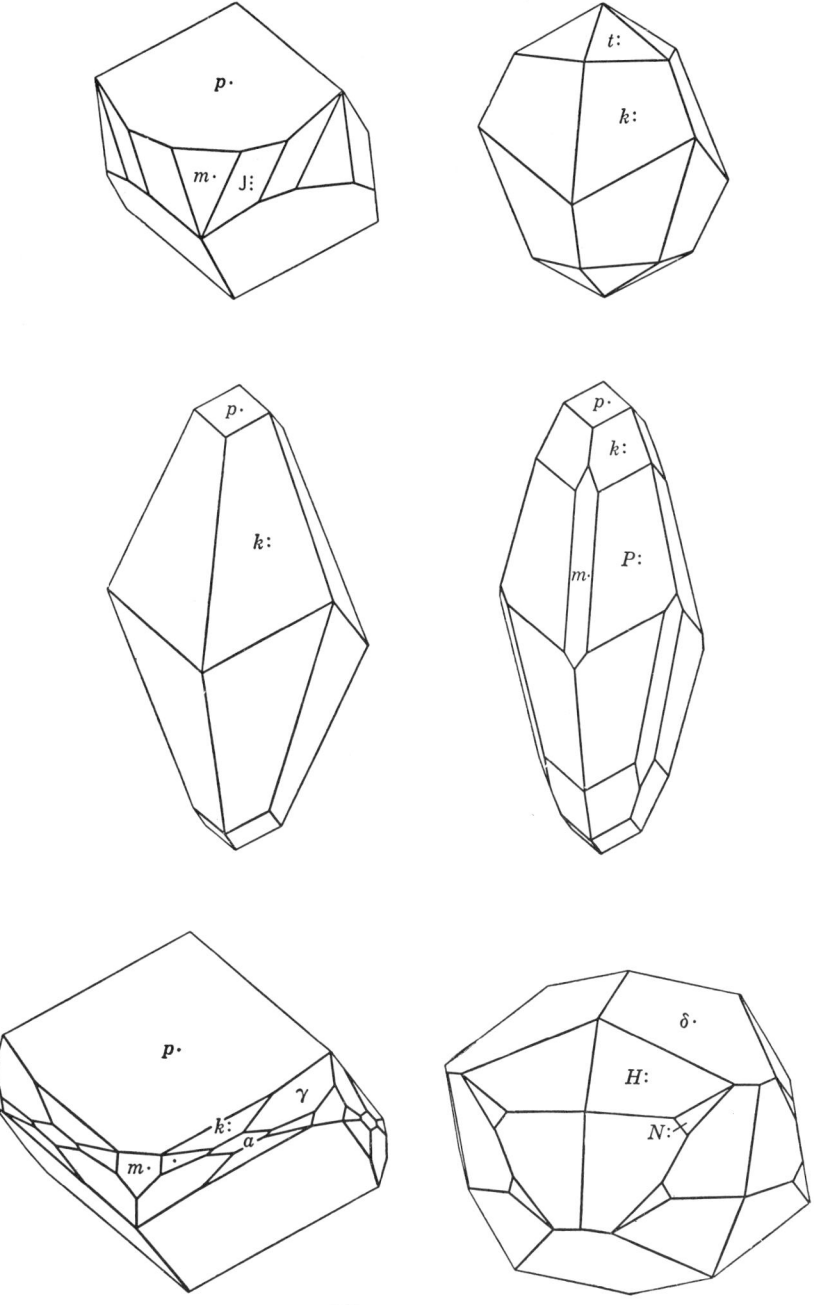

Union Springs, N. Y. With {19.10.$\overline{29}$.6}.   Rondout, N. Y. With {10.3.$\overline{13}$.2}.

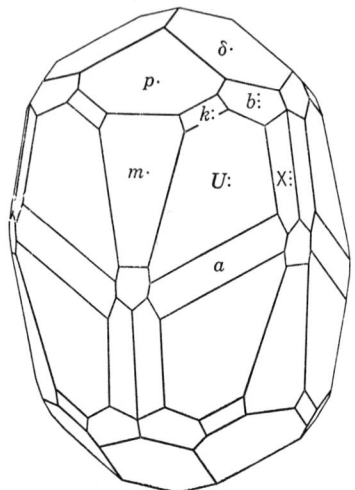

Lake Superior. With {18.0.$\overline{18}$.1}.

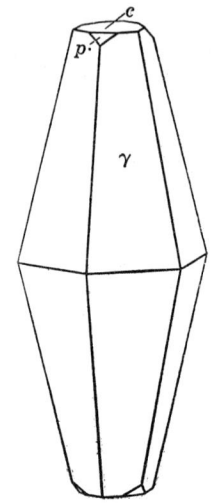

Kelley's Island, Ohio.

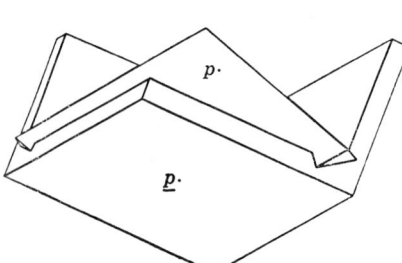

Rossie, N. Y. Twin on {0001}.

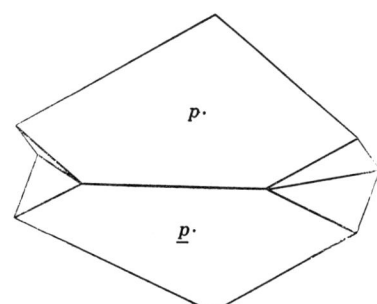

Twin on {0001}.

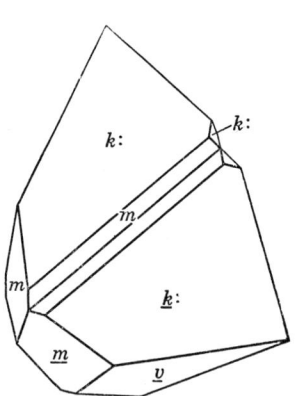

Twin on {10$\overline{1}$1}.

Calcite

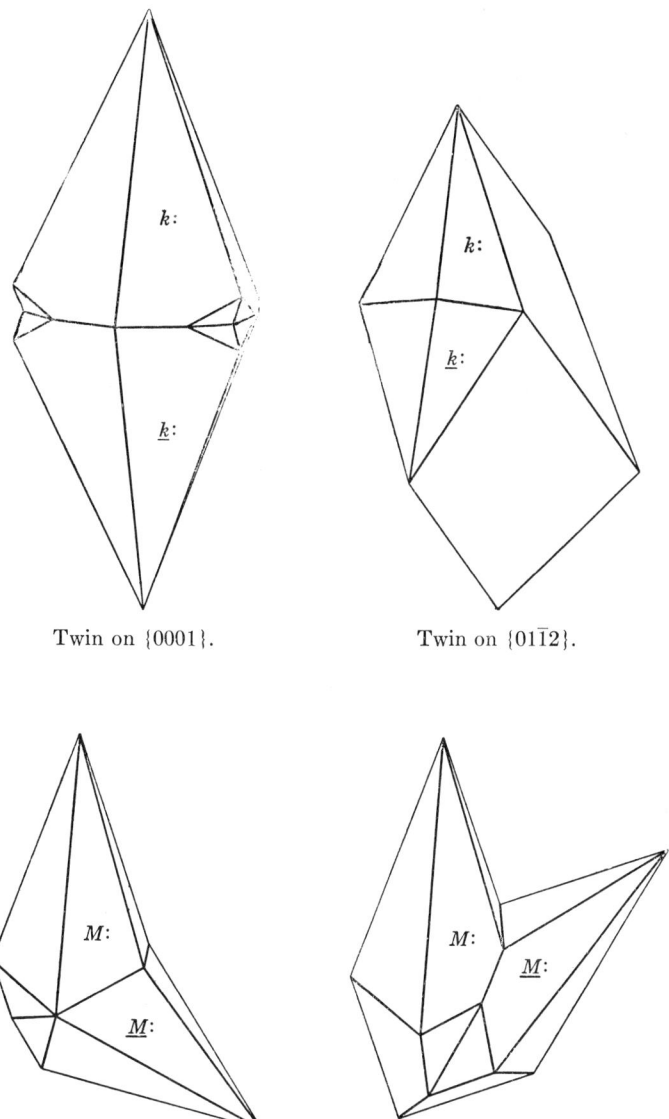

Twin on {0001}.

Twin on {01$\bar{1}$2}.

Twin on {01$\bar{1}$2}.

Twin on {022$\bar{1}$}.

Oriented growths. Often found overgrowing crystals of other members of the calcite group and of dolomite with the crystal axes oriented in parallel position; calcite similarly is found overgrown by these species. Also found in oriented position upon quartz,[7] with calcite {01$\bar{1}$2} parallel quartz {10$\bar{1}$1}. Artificial oriented growths of sodium nitrate upon calcite are well-known;[8] also of various alkali halides.[24]

Etch figures. Natural and artificial etch figures[9] and solution bodies[10] derived from spheres or cylinders conform to hexagonal-scalenohedral symmetry, $\bar{3}\,2/m$. Optically active solvents[11] produce dissymmetric etch figures on those faces which have as their sole element of symmetry one perpendicular plane.

Phys. Cleavage {10$\bar{1}$1} perfect. Also ready parting along twin lamellae on {01$\bar{1}$2}, and sometimes on {0001}. Fracture conchoidal, but difficult to produce. Brittle. H. 3, but varying somewhat on different planes and in different directions thereon;[12] relatively soft ($\sim 2\frac{3}{4}$) on {0001} and relatively hard ($\sim 3\frac{1}{4}$) on {10$\bar{1}$0}, especially in the vertical direction. G. 2.7102 ± 0.0002 at 20° for pure $CaCO_3$; the G. varies essentially linearly with variation in composition toward the several isostructural end-members; G. ~ 2.96 for Ca:Fe = 4:1, ~ 3.21 for Ca:Mn = 1:1. Luster vitreous; sometimes somewhat pearly or iridescent on cleavage surfaces, or pearly on {0001} in white or turbid crystals, and tending toward greasy on conchoidal fracture-surfaces. Colorless and transparent or white when pure. The color varies widely with isomorphous substitution or with mechanical admixture: yellow, brown, reddish pink, violet, pale blue, lavender, greenish, gray, black; bluish or greenish due to finely disseminated pyrite, greenish to chlorite or other ferrous iron compounds, reddish brown to hematite or cuprite, green to malachite, etc. Streak white to grayish. Twin gliding[15] with $K_1(01\bar{1}2)$, $K_2(0\bar{1}11)$; and translation gliding[16] with $T\{01\bar{1}2\}$, $t[12\bar{3}1]$. The lines of intersection of sets of gliding planes may be marked by minute linear cavities—hollow canals[13]—and these may give rise to asterism in transmitted light. Percussion figure[14] on {10$\bar{1}$1} triangular and striated horizontally, on {0001} three-rayed.

Luminescence. Many specimens are fluorescent and often also phosphorescent when excited by ultraviolet radiation, x-rays, cathode rays, or sunlight. Often also thermoluminescent (without prior artificial irradiation). The fluorescent color varies with the kind and amount of activating impurities, such as hydrocarbons, rare earths, or Mn.[17] Irradiation with x-rays or cathode rays may be accompanied by coloration of the crystals, which is destroyed as with some of the natural colors by gentle heating. The luminescence and coloration may vary in different growth zones or face loci in a single crystal.[18] Sometimes triboluminescent.

Opt. In transmitted light, colorless. Uniaxial negative (−). Sometimes abnormally biaxial[19] with uneven extinction and $2V$ up to 5°–10°, rarely to 30°, and generally due to mechanical deformation; the feature often is accompanied by pressure-twin lamellae and parting on {01$\bar{1}$2}.

CALCITE

Absorption $O > E$. Indices:[20]

	nO	nE	
n for $\lambda = D_2 = 588.99$ at $20°$	1.65838	1.48645	(Iceland)

INDICES AT 15° FOR DIFFERENT WAVELENGTHS [21]

λ	nO	nE	λ	nO	nE
795.0	1.64886	1.48216	330.28	1.70515	1.50745
768.24	1.64974	1.48255	303.42	1.71958	1.51364
706.56	1.65207	1.48353	274.86	1.74150	1.52266
656.30	1.65439	1.48456	257.31	1.76050	1.53012
589.31	1.65835	1.48639	244.58	1.77966	1.53731
560.71	1.66045	1.48734	231.29	1.80239	1.54550
527.01	1.66341	1.48870	226.51	1.81303	1.54914
486.14	1.66783	1.49074	219.44	1.83079	1.55512
434.06	1.67551	1.49424	214.44	1.84582	1.55992
396.16	1.68329	1.49777	209.88	1.86081	....
361.06	1.69316	1.50223			

INDICES AT DIFFERENT TEMPERATURES [22]

O index

Temp.	670.6	643.7	588.8	537.7	508.4	479.9
22°	1.6538	1.6551	1.6585	1.6624	1.6653	1.6687
53.2	1.6538	1.6552	1.6585	1.6625	1.6654	1.6687
88.9	1.6538	1.6552	1.6585	1.6626	1.6655	1.6688
208.2	1.6541	1.6554	1.6588	1.6628	1.6658	1.6691
256.2	1.6542	1.6555	1.6589	1.6629	1.6659	1.6692
330	1.6543	1.6556	1.6589	1.6630	1.6660	1.6694

E index

Temp.	670.6	643.7	588.8	537.7	508.4	479.9
22°	1.4844	1.4850	1.4865	1.4884	1.4897	1.4912
53.2	1.4848	1.4854	1.4869	1.4887	1.4900	1.4916
88.9	1.4852	1.4858	1.4873	1.4891	1.4905	1.4921
208.2	1.4866	1.4871	1.4887	1.4906	1.4920	1.4935
256.2	1.4871	1.4877	1.4893	1.4911	1.4925	1.4941
330	1.4880	1.4886	1.4901	1.4920	1.4935	1.4950

VARIATION IN INDICES WITH COMPOSITION (calc.) [50]

	$CaCO_3$ (pure)	Ca:Mn = 2:1	Ca:Mn = 1:1	Ca:Fe = 4:1
nO	1.658	1.711	1.737	1.701
nE	1.486	1.523	1.542	1.515

Chem. Calcium carbonate, $CaCO_3$. Calcite is the thermodynamically stable form of calcium carbonate at all pressures and temperatures so far investigated. Two metastable polymorphs are known, aragonite and vaterite (also known as *μ-calcite* or *vaterite-B*). These polymorphs change irreversibly to calcite on being heated. A fourth polymorph of calcium carbonate (α-$CaCO_3$; *elatolite*) stable at high temperatures and pressures has been reported but has not been verified.[23]

Calcite enters into a wide range of compositional variation through the substitution of other divalent cations for Ca. In these instances, the name calcite is applied as the species designation when Ca is the predominant cation in atomic per cent regardless of the total summation of the other substituting elements. The mineral as ordinarily found, however, is essentially pure and in some occurrences contains other elements in only spectrographic amounts. A complete series extends to rhodochrosite through the substitution of Mn'', and an at least partial series extends towards siderite, smithsonite, and cobaltocalcite by the entrance of Fe'', Zn, and Co, respectively. The cations Ba, Sr, and Pb sometimes substitute for Ca in significant, although relatively small, amounts; the pure carbonates of these elements are orthorhombic, like aragonite, and they do not form rhombohedral polymorphs isostructural with calcite. Mg substitutes for Ca only in very small amounts under ordinary conditions, although a more extended series toward dolomite may exist at high temperatures. Many other elements have been found in traces in calcite, including Cu, Al, Ni, V, Cr, Mo, and rare earths. Organic matter, which may be driven off on slight heating, is sometimes present.

The limits of variation in the indices of refraction, specific gravity, unit cell dimensions, and crystal angles accompanying variation in composition are not fully established, especially in the polycomponent systems. The variation in these properties may be taken to a first approximation as linear between the end-members in the two-component systems at least. The properties as here recorded refer to the essentially pure end-members of the calcite group; see further, however, under the compositional varieties of the several species.

Anal.[43]

	1	2	3	4	5	6	7
CaO	56.03	55.74	54.41	48.82	34.04	22.15	55.38
MgO		0.11	0.27		7.28	2.72	0.58
FeO		0.04	0.15		13.05	0.29	
MnO		0.04	0.42	6.21	1.71	16.67	
CO_2	43.97	43.95	43.55	42.62	43.84	42.08	43.69
Rem.		0.07	1.27	1.59	0.24	0.08	0.30
Total	100.00	99.95	100.07	99.24	100.16	100.47	100.00
G.					2.934	3.05?	

1. $CaCO_3$. 2. Joplin, Missouri.[44] Rem. is ZnO 0.01, Ce_2O_3 0.007, $(La,Sm,Di)_2O_3$ 0.012, $(Y,Er)_2O_3$ 0.013, SiO_2 0.032, traces of Na, Sn, Al, Cr, NH_3, Cl, S, P. 3. Elba.[45] Rem. is CoO. 4. Långban, Sweden.[46] Rem. is BaO 1.59. 5. Toggiano, Modena, Italy.[47] Rem. is H_2O. 6. Franklin, New Jersey.[48] Rem. is insol. 7. Marble (rock). Carrara, Italy.[49] Rem. is Na_2O 0.01, Al_2O_3 0.05, Fe_2O_3 0.06, $(NH_4)_2O$ 0.01, Cl 0.04, SO_3 0.02, P_2O_5 0.09, loss 0.02, traces of N, SiO_2, organics.

Var. The varieties of calcite are mainly of two kinds, those depending on differences in crystal shape or in the presence of mechanically enclosed impurities, and those depending on variation in chemical composition through isomorphous substitution.

CALCITE 153

A. VARIETIES BASED UPON CRYSTALLIZATION AND ACCIDENTAL IMPURITIES

1. *Ordinary.* In crystals and cleavable masses, the crystals varying very widely in habit as already noted.

2. *Fibrous.* Fine fibrous, with a silky luster (*satin spar*). Resembles fibrous gypsum, which is also called satin spar, but much harder than gypsum and effervesces with acids. The fiber elongation ordinarily is oblique to the c-axis, sometimes parallel.

3. *Lamellar.* Argentine *Kirwan* (**1,** 104, 1794). Schieferspath *Hofmann* (*Bergm. J.*, 188, 1789). Slate Spar. Aphrite pt. A lamellar to foliated calcite, the layers more or less undulating and usually flattened on {0001}. Luster pearly; color white, grayish, yellowish, or reddish. Passes into scaly types.

4. *Sand-calcite.* Siliceous Calcite. Fontainebleu limestone *Lassonne* (1775). Chaux carbonatée quartzifére *Haüy* (1801). Contains admixed sand grains up to 50 or 60 per cent, or more. Ordinarily found in sandstones or loosely consolidated sand rocks with a calcareous cement, the form of the crystals apparent only on weathered surfaces. The calcite may later be leached out, affording pseudomorphs of sand after calcite. The sand-calcite of Fontainebleu near Paris is well-known; also from South Dakota and many other localities.[25]

5. *Concretionary.* As rounded concretions in aggregates or composing rock masses and resembling the roe of fish (oolites). Also in larger, more distinct forms resembling a pea in size and shape (pisolites; Erbsenstein, *Germ.*; Cave-pearls [52]) also rarely externally in the form of a regular pentagonal dodecahedron,[26] or crudely tetrahedral. The concretionary varieties ordinarily have a distinct concentric layering or banding and are radially fibrous; also with a nucleus of foreign material. Similar shapes are also found with aragonite.

6. *Fetid calcite.* Stinkkalk. Contains liquid inclusions that afford H_2S on fracture.

B. VARIETIES BASED UPON CHEMICAL COMPOSITION

1. *Manganoan.* Manganocalcite. Mangankalkspat, Kalkmanganspat *Germ.* Spartaite *Breithaupt* (*B. H. Ztg.*, **17,** 53, 1858). Calcimangite *Shepard* (1865). Contains Mn'' in substitution for Ca, and a probably complete series extends to rhodochrosite. Other cations, especially Fe'', may substitute concomitantly for Ca. Color usually pinkish in high manganoan material; becomes black on exposure. The indices of refraction and specific gravity increase and the cell dimensions decrease regularly with increasing content of Mn. The reddish-yellow fluorescence in ultraviolet light of manganoan calcite rapidly increases in intensity up to about 3.5 per cent $MnCO_3$ and then gradually decreases.[38] Manganoan calcite occurs at Franklin and Sterling Hill in New Jersey, Vester Silfberg and Långban in Sweden, and at many other localities.

2. *Ferroan.* Ferrocalcite. Contains Fe″ in substitution for Ca, often together with Mn″ in significant amounts. The series extends toward siderite at least up to Fe:(Ca, etc.) = 1:4.5 (analysis 5) but a gap appears to exist between this region and the calcian varieties of siderite. Color white, yellowish, or brownish and turning brown on exposure. Amounts of Fe up to approximately 1 per cent FeO are commonly present.

3. *Zincian.* Zincocalcite. Contains Zn in substitution for Ca up to at least about 4 per cent ZnO.[27] Zincian material is rare.

4. *Cobaltian.* Cobaltocalcite (of older authors, not cobaltocalcite = $CoCO_3$ as here applied). Contains Co in substitution for Ca up to at least about 2 per cent CoO.[28] Color rose or rose-red. Often in curved or spheroidal aggregates. Turns black on exposure. Rare.

5. *Plumbian.* Plumbocalcite Johnston (*Edinburgh Phil. J.*, **6**, 79, 1829). Contains Pb in substitution for Ca up to at least Pb:Ca = 1:8.5. Material[40] with 1.9 per cent $PbCO_3$ has G. 2.73 and nO 1.6668, nE 1.4904. Relatively highly plumbian material is white or yellowish in color, opaque, and has a pearly or silky luster. Found especially at Leadhills and Wanlockhead in Scotland[29] and Bleiberg, Carinthia.[30] Highly plumbian material may be an oriented intergrowth of cerussite and plumbian calcite.[41]

6. *Barian.* Baricalcite. Neotyp Breithaupt (**2,** 313, 1841), Barytocalcite pt. Rhombohedral barytocalcite. Contains small amounts of Ba in substitution for Ca. A grayish white, unanalyzed type from Cumberland, England, occurs in rhombohedrons $\{02\bar{2}1\}$ with rr' = 74°57′ and G. 2.82–2.84. A barian calcite found with hedyphane at Långban, Sweden,[31] has Ba:Ca $\sim$ 1:1; it occurs as white granular masses with rhombohedral cleavage and G. 3.46. Barian calcite has been obtained artificially.[31]

7. *Strontian.* Strontianocalcite Genth (*Proc. Ac. Sc. Philadelphia*, **6,** 114, 1852). Contains Sr in substitution for Ca. Unanalyzed material from Girgenti, Sicily, occurs in coarsely fibrous globules whose surface consists of the terminations of acute rhombohedrons; color white; opaque. Also obtained artificially.

8. *Magnesian.* Dolomitic calcite pt. Contains small amounts of Mg in substitution for Ca. The maximum solubility of Mg in calcite appears to be small;[32] the material of analysis 5 has Mg:(Ca, etc.) = 1:4.5 and this probably is near the limit. Natural intergrowths of calcite and dolomite suggesting an origin by exsolution indicate that the limit of solubility may be larger at high temperature. The so-called magnesian limestones and magnesian marbles are mechanical mixtures of calcite and dolomite.

Tests. B.B. infusible, but becoming white and opaque (CaO) and glowing strongly. Dissociation temperature[33] for one atm. CO_2 about 900°. Fragments dissolve with brisk effervescence in cold acids. Solubility in CO_2-free water: 0.0145 g. $CaCO_3$ per liter at 17°, 0.0375 g. at 100°. The solubility is markedly greater in water containing dissolved CO_2 and also is increased by the presence of certain dissolved salts such as NaCl, KCl, Na_2SO_4. On staining tests for the distinction from aragonite see further under that species.

CALCITE 155

Occur. Calcite in its various forms is one of the commonest and most widely distributed of all minerals. It is an important rock-forming mineral in the sedimentary cycle and occurs in great thicknesses in relatively pure form as chalks and limestones and also in lesser amounts as cementing material in other sedimentary rocks; also in the metamorphic equivalents of these rocks. Calcite is a widespread although very minor constituent of the altered facies of igneous rocks, especially those of basic composition, where it often is formed by the decomposition of primary calcic silicates, and is frequently observed as a late hydrothermal deposit in veinlets or cavities in rocks of basic types. Also reported to be of magmatic crystallization in some alkalic rocks.[51] Calcite is deposited from lime-bearing carbonated waters under a wide range of geologic conditions and with many varying associations. It is very common in many kinds of mineral veins, formed either from meteoric waters or ascendant hydrothermal solutions, and is then typically associated with sulfides, quartz, barite, fluorite, dolomite, siderite, and other rhombohedral carbonates. Also in caves in calcareous rocks as crystalline incrustations and as stalactites and stalagmites; as a deposition in the form of travertine or calc-sinter from springs and streams; as the petrifying material of fossil animal and plant remains, and as the skeletal material of many living organisms. A few of the many known localities that have afforded finely crystallized specimens are mentioned beyond.[34]

Large rhombohedral crystals of optical quality, often tufted with stilbite, have long been obtained from a cavernous zone in basalt near Helgustadir on the Eskefiord, Iceland [35] (from which derives the name Iceland spar); one rhombohedral crystal measured 20 by $6\frac{1}{2}$ feet. In England, famous localities are at Alston Moor, Egremont, and Frizington in Cumberland, at Weardale in Durham, the Stank mine near Furness in Lancashire, and Liskeard and elsewhere in Cornwall. In Germany in the veins of Andreasberg in the Harz, at Freiberg, Schneeberg, Niederrabenstein, and Tharandt in Saxony, at Brilon, Westphalia, and at Oberscheld and Limburg, Hesse. At Příbram, Bohemia. A notable occurrence of travertine is at Tivoli near Rome, Italy. In France a long-known occurrence of sand-calcite at Fontainebleu near Paris, and as large tabular crystals in the veins of Chalanches near Allemont. As large pyramidal crystals at Rhisnes, Belgium. With zeolites at various places on the Faroe Islands. In Norway notably at Kongsberg, and in Sweden at Visby and Långban. Found in Russia, especially at Kara Dag and near Baidar-Tores in the Crimea, on the Chalyk river in the northern Caucasus, and on the Lower Tunguska river, Siberia. As tufa and pisolites in the hot springs of Hamman Meskoutine near Guelma in Algeria; in the Kenhardt district, Cape Province, South Africa. Simple crystals and twins of great variety and beauty come from Guanajuato, Mexico.

In the United States, well-crystallized calcite is found with zeolites in diabase at Great Notch, Upper Montclair, West Paterson, Bergen Hill and near-by localities in New Jersey. In New York as fine specimens at

Rossie in St. Lawrence County, in large crystals at Sterlingbush in Lewis County, and near Oxbow in Jefferson County. In Ohio on Kelley's Island, Lake Erie, in pyramidal crystals. In the Michigan copper deposits as limpid crystals of complex form, often enclosing scales or wires of native copper. At Warsaw, Illinois, with quartz as drusy crusts lining geodes. In Missouri at Joplin, Granby, and other points in the lead-zinc region in the southwestern part of the state, the large crystals usually scalenohedral in habit and of a brownish-yellow color. Also in the lead-zinc region of southern Wisconsin as at Mifflin, Mineral Point, and Galena. A well-known occurrence of sand-calcite is at Devils Hill in the Bad Lands of Washington County, South Dakota. At Bisbee, Arizona, as crystals and stalactites sometimes colored green by malachite inclusions. At Magdalena, New Mexico, and in the mercury deposit at Terlingua, Texas. Several giant crystals, each weighing upwards of 25 tons, in part of optical quality, were mined at the Iceberg claim near the Harding mine in Taos County, New Mexico.[36] Optical calcite also has been mined at Deer Creek near Livingston, Montana, and at Palm Wash near Thermal, San Diego County, California. In Canada in the apatite-phlogopite deposits of Ottawa County, Quebec, and similarly in Renfrew, Lanark, and Frontenac counties, Ontario.

A r t i f.[37] Calcite and its polymorphs aragonite and vaterite may all be formed by precipitation from solution, the modification obtained depending largely on the temperature and the presence of other substances in the solution. Below roughly 30° the precipitate is usually calcite, unless sulfates or salts of Sr, Ba, Pb, or Mg are present (see further under aragonite), and at very low temperatures hydrates may be obtained. The other modifications of $CaCO_3$ generally are converted into calcite quite rapidly when in contact with the mother solution. Macroscopic crystals of calcite have been variously obtained: by slow diffusion of a soluble carbonate into a gel containing a calcium salt; by heating gelatinous precipitated $CaCO_3$ together with water and ammonium chloride or sulfate or a soluble calcium salt in a closed tube at elevated temperatures; by slowly removing CO_2 from a solution of calcium bicarbonate, especially in the presence of potassium or sodium silicate; by the dissolution and slow interdiffusion of solid $CaCO_3 \cdot 6H_2O$ and $K_2CO_3 \cdot 1\frac{1}{2}H_2O$ in water, etc. Crystals composed of the unit rhombohedron $\{10\bar{1}1\}$ are often obtained in such experiments, although rare in nature, and the habit can be varied by addition of foreign substances to the solution.

A l t e r. Calcite by virtue of its relatively high chemical reactivity and ready solubility in carbonated waters is very commonly found replaced by or altered to other species. By far the commonest of these are the incrustation or substitution pseudomorphs of quartz (including its chalcedonic and opaline varieties) after calcite. Other common pseudomorphs after calcite are of limonite, hematite, goethite, wad, pyrolusite, and manganite; also, less commonly, dolomite, siderite, rhodochrosite, smithsonite, hemimorphite, cerussite, malachite, chrysocolla, leadhillite; and rarely talc, serpentine, hausmannite, apatite, prehnite, chlorite, aragonite (by incrustation), fluorite, gypsum, barite, minium, galena, pyrite, marcasite, sepiolite, dioptase, shattuckite, copper, azurite, and others. Calcite has been found pseudomorphous after aragonite, gypsum, barite, fluorite, celestite, sulfur, cerussite, thenardite, glauberite, gaylussite, anhydrite, quartz, apophyllite, and other minerals and often is found admixed among the complex alteration products of plagioclase and other calcic minerals. As the petrifying material of wood [39] and other organic remains. On the so-called glendonite, jarrowite, and thinolite pseudomorphs see beyond under *unidentified calcite pseudomorphs*.

Name and History. The name calcite, given by Haidinger in 1845 and supplanting the older name calcspar (Kalkspat, Kalkstein, *Germ.*), is derived from the Greek root χάλξ, meaning to reduce to a powder by heat and early applied properly to burnt lime or, more generally, to a base or calx (oxide) formed by ignition. The phenomenon of double refraction was discovered in cleavage fragments of calcite from Iceland and was described by Huygens in 1678 and Bartholinus in 1669. The study of calcite also led Malus in 1808 to the discovery of the polarization of light. The first chemical analysis of calcite, reported in 1774 by Bergmann, was erroneous (55 per cent calcis purae, 34 per cent aëris fixi, 11 per cent aquae), and the true composition was later established in 1804 by Buchholz (56 per cent Kalkerde, 44 per cent Kohlensaüre). The mineral by virtue of the size, abundance, and perfection of its crystals, and its remarkable complexity of habit—about 700 different forms have been reported—has always attracted the attention of crystallographers. The first angular measurements on calcite, made with the aid of Carangeot's contact goniometer, itself the first instrument devised for the purpose, were reported by Romé de Lisle in 1783. These observations were greatly extended by Haüy in 1781–1801, who from a study of the form development and cleavage of calcite developed a theory of crystal structure therefrom and established the foundations of geometrical crystallography. The crystallography of the mineral was further described by Bournon in 1808, in an important memoir containing 704 crystal drawings, and this was followed by the extensive publications of Lévy (1837, with 346 figures) and Zippe (1851, 700 figures). In the more recent literature, particular mention may be made of the morphological studies by Irby (1878), Rogers (1901), Whitlock (1910), and Palache (1898, 1943). The crystal structure of calcite, one of the earliest to be worked out by x-rays, was described by W. L. Bragg in 1914, who later derived (1924) in the first work of its kind, the optical properties of both calcite and its polymorph aragonite in terms of the atomic structure. The identity in composition and difference in crystal form of calcite and aragonite, early recognized by Haüy, was explained structurally by Mitscherlich in 1821 as one of the few instances of polymorphism then discovered. Investigations of the extensive variation in composition shown by calcite and other members of the group and the concomitant variation in crystallographic and physical properties have played an important part in the development of the theory of isomorphism. The variation in interfacial angle with composition among isostructural compounds was first established by Wollaston in 1812 in the calcite group, a discovery that antedated the formal description of isomorphism by Mitscherlich in 1820. The optical goniometer developed by Wollaston for the purpose, the first reflecting goniometer ever constructed, was later copied and used by Mitscherlich (1825) in his discovery, in calcite, of the variation of crystal angles as a function of temperature. The phenomenon of twin gliding or pressure twinning was first recognized by Brewster in 1815, in calcite, and gliding was first effected experimentally by Reusch in 1867 in the same species. The ($20\bar{2}2$) interplanar spacing of calcite, taken as 3.02945 kX at 18°, has been used as a relative standard for x-ray wavelengths (Siegbahn, 1925).

ELATOLITE *Fersman (C. r. ac. sc. Russie*, **59**, 1922; *Bull. ac. sc. Russie*, **17**, 251, 1923).

A name given to a substance represented by hollow triangular cavities resembling fir trees found in the nepheline-syenites of the Khibina and Lovozero tundras on the Kola peninsula, U.S.S.R. Supposed to have been a polymorph of calcium carbonate (α-$CaCO_2$) [23] stable at very high temperatures, later leached out. The cavities have been suggested [42] to be molds after skeletal crystals of villiaumite.

Ref.

1. The fundamental angle $rr' = 74°55'$ was first given by Wollaston, *Phil. Trans.*, **385** (1802), 159 (1812); values differing therefrom by 30″ or less were obtained by Malus (1810), Mitscherlich (1823), Breithaupt (1828), vom Rath (1867), and others. Hastings, *Am. J. Sc.*, **35**, 68 (1888) gives the probably superior value 74°54.93′ at 20° on Iceland spar. The angle varies appreciably with variation both in composition and in temperature. Rinne, *Cbl. Min.*, 706 (1914) gave 74°44′11″ at −165° and 75°53′31″ at 596°; see also V. M. Goldschmidt, *Zs. Kr.*, **51**, 21 (1912). See Rosenholtz and Smith, *Am. Min.*, **34**, 846 (1949) on the linear thermal expansion. The morphological unit cell here used is that of long-standing convention and has c one-fourth of the true structural unit; transformation: morphology to structure, 1000/0010/0004.

2. A review with literature of the 328 established, 296 uncertain, and 64 discredited forms reported for calcite is given by Palache, *Calcite: An Angle Table and Critical List*, Contr. No. 259, Cambridge, 27 pp., 1943 (privately printed) [copies obtainable by loan or purchase from Department of Mineralogy, Harvard University, Cambridge, Massachusetts]. Space is lacking to list all the forms here. The present angle table includes only forms described from more than 25 localities, and the list of less common forms, those from 10 to 20 localities. For morphological summaries see also Goldschmidt (**2**, 5, 1913), Whitlock, *New York State Mus. Mem.*, **13** (1910), Rogers, *Sch. Mines Q.*, **22**, 429 (1901), Irby, Inaug. Diss. Bonn (1878), and *Zs. Kr.*, **3**, 610 (1879).

3. Bragg, *Proc. Roy. Soc. London*, **89A**, 246 (1914), Wyckoff, *Am. J. Sc.*, **50**, 317 (1920). On the cell dimensions see Compton, *et al.*, *Phys. Rev.*, **25**, 618 (1925), Bearden, *Phys. Rev.*, **38**, 1389 (1931), Lipson, *Nature*, **151**, 250 (1943). On the variation in size with Mn content see Krieger, *Am. Min.*, **15**, 23 (1930).

4. Erroneously (and repeatedly) described as a distinct species; see Pacák, *Časop. Národ. Musea, Praha*, **102**, 148 (1928), Quercigh, *Riv. min.*, **44**, 65 (1916), Mizgier, *Zs. Kr.*, **70**, 160 (1929). Thugutt, *Arch. Min. Soc. Sc. Varsovie*, **5**, 97 (1929), Balogh, *Min. Abs.*, **7**, 515 (1940), Ulrich, *Věda Přír., Praha*, **19**, 45 (1938), Lang, *Cbl. Min.*, 298 (1915) and *Jb. Min., Beil.-Bd.*, **38**, 121 (1914).

5. See Becke, *Min. Mitt.*, **10**, 135 (1889); Kreutz, *Min. Mitt.*, **24**, 323 (1905) and **28**, 490 (1909); Tertsch, *Trachten der Kristalle*, Berlin, 83 (1926).

6. See Dove, *Ann. Phys.*, **110**, 286 (1860); Reusch, *Ann. Phys.*, **132**, 441, (1867), **136**, 130, 135, 632 (1869), **147**, 307 (1872); Baumhauer, *Zs. Kr.*, **3**, 588 (1879); Mügge, *Jb. Min.* I, **32**, 81 (1883); Liebisch, *Zs. Kr.*, **17**, 305 (1889); Voigt, *Zs. Kr.*, **19**, 485

CALCITE 159

(1891); Rinne, *Jb. Min.*, I, 160 (1903); Mügge, *Jb. Min.*, I, 119 (1898); Bell, *Am. Min.*, **26**, 247 (1941); Gogoberidze and Ananiaschwili, *Phys. Zs. Sowjetun.*, **7**, 547 (1935).
    7. Cf. Mügge, *Jb. Min., Beil.-Bd.*, **16**, 370 (1903).
    8. Finch and Whitmore, *Trans. Faraday Soc.*, **34**, 640 (1938).
    9. In particular see Baumhauer, *Ann. Phys.*, **138**, 563 (1869), **139**, 349 (1870), **140**, 271 (1870); Ebner, *Ak. Wien, Sitzber.*, **89**, 368 (1884), **91**, 760 (1885); Gaubert, *Bull. soc. min.*, **24**, 326 (1901); Ichikawa, *Am. J. Sc.*, **42**, 113 (1916), **48**, 124 (1919); Holzner, *Festschr. V. Goldschmidt*, Heidelberg, 163 (1928); Honess, *Am. J. Sc.*, **45**, 201 (1918). On burn-figures see Rinne, *Festschr. V. Goldschmidt*, Heidelberg, 213 (1928).
    10. Goldschmidt, *Zs. Kr.*, **38**, 656 (1904); Goldschmidt and Wright, *Jb. Min., Beil.-Bd.*, **17**, 363 (1903), **18**, 335 (1904); Goldschmidt and Porter, *Beitr. Kr. Min.*, **2**, 138 (1924); Homberg, *Geol. För. Förh.*, **12**, 617 (1890).
    11. Royer, *C. R.*, **202**, 429 (1936); Himmel and Kleber, *Jb. Min., Beil.-Bd.*, **72**, 347 (1937); Honess and Jones, *Bull. Geol. Soc. Am.*, **48**, 667 (1937); Nowacki, *Zs. Kr.*, **102**, 217 (1940).
    12. Pfaff, *Ak. München, Sitzber.*, **55**, 372 (1883); Exner, *Härte der Krystallfl.*, Vienna, 148 (1873); Holmquist, *Geol. För. Förh.*, **33**, 281 (1911); Tertsch, *Zs. Kr.*, **89**, 541 (1934).
    13. Rose, *Ak. Berlin, Abh.*, **23**, 57 (1868), Holmquist, *Ark. Kem.*, **12**, no. 10 (1936); Grishchinsky, *Mém. soc. nat. Kiev.*, **24**, 171 (1915); Graham, *Min. Mag.*, **18**, 252 (1918).
    14. Rose (1868); Mügge, *Jb. Min.*, I, 71 (1898); Tertsch, *Jb. Min.*, **76**, 291 (1940).
    15. Mügge, *Jb. Min.*, I, 32 (1883); Fairbairn and Hawkes, *Am. J. Sc.*, **239**, 617 (1941).
    16. Mügge (1898).
    17. Yoshimura, *Sci. Pap. Inst. Phys. Chem. Res., Tokyo*, **23**, 224 (1934); Iimori, *ibid.*, **21**, 220 (1933); Hata, *ibid.*, **20**, 163 (1933); Headden, *Am. J. Sc.*, **5**, 314 (1923), **6**, 247 (1923), **8**, 509 (1924); Pisani, *Bull. soc. min.*, **38**, 24 (1915); Fonda, *J. Phys. Chem.*, **44**, 435 (1940); Haberlandt, *Chem. Erde*, **13**, 221 (1940); Northup, *Rocks and Minerals*, **15**, 147 (1940); Gliszczynski and Stoicovici, *Zs. Kr.*, **98**, 344 (1937); Brown, *Univ. Toronto Stud., Geol. Ser.*, no. 36, 45 (1934); Haberlandt, *Wiener Chem. Ztg.*, **47**, 80 (1944).
    18. Frondel, Newhouse, and Jarrell, *Am. Min.*, **27**, 726 (1942).
    19. See Breithaupt, *Ann. Phys.*, **121**, 328, *Jb. Min.*, 341 (1860); Madelung, *Zs. Kr.*, **7**, 73 (1882); Gillson, *Am. Min.*, **12**, 357 (1927); Walker and Parsons, *Univ. Toronto Stud., Geol. Ser.*, No. 20, 14 (1925).
    20. Hastings, *Am. J. Sc.*, **35**, 60 (1888) on Iceland spar, with superior control of experimental conditions. Dufet, *Bull. soc. min.*, **16**, 157 (1893) gives almost identical values for $D_1D_2/2 = 589.31$ at $20°$ for Iceland spar. Most precision measurements reported in the literature (see Hintze 1 [3A], 2839, 1916) are within $O = 1.65842 \pm .00005$, $E = 1.48641 \pm .0007$ for Na at $20°$. Aside from ordinary experimental controls, the nature and condition of the polished surface, traces of impurities (Sr, Fe, etc.), and lineage structure are important factors in measurements beyond the fourth decimal. Much studied material was not analyzed and was only presumed to be "pure."
    21. Gifford, *Proc. Roy. Soc. London*, **70**, 336 (1902) on Iceland spar; the data are given to seven decimals and are said to be accurate to five and are here given without rounding off.
    22. Offret, *Bull. soc. min.*, **13**, 580 (1890) on Iceland spar; the data are given to six decimals and are here rounded off to four. Gifford (1902) gives the change in index for D as $O = +.00000479$, and $E = +.00001447$ for $1°C$. increase.
    23. Boeke, *Jb. Min.*, I, 91 (1912); Smith and Adams, *J. Am. Chem. Soc.*, **45**, 1167 (1923). Two high-pressure polymorphs have been reported by Bridgman, *Am. J. Sc.*, **237**, 7 (1939).
    24. Royer, *C.R.*, **205**, 1418 (1937).
    25. Rogers and Reed, *Am. Min.*, **11**, 23 (1926), Wanless, *Am. Min.*, **7**, 83 (1922), Stöber, *Chem. Erde*, **6**, 357 (1931).
    26. On the origin of similar shapes in aragonite pisolites see Fukuchi in Wada, *Min. of Japan*, no. 4, 133 (1912); Royer, *C.R.*, **208**, 1591 (1939).
    27. Gibbs analysis in Rammelsberg (3rd Suppl., 62 (1847)) on Olkucz, Poland, crystals, with ZnO 4.07.
    28. Hacquaert, *Natuurw. Tijdschr.*, **7**, 100 (1926) on Katanga crystal with CoO 1.96; see also analysis 3.
    29. See Collie, *J. Chem. Soc. London*, **55**, 91 (1899), and Lacroix, *Bull. soc. min.*, **8**, 36 (1885), with analyses showing up to 9.47 per cent $PbCO_3$.
    30. Höfer (44, 1870), with analyses by Schöffel showing up to 23.75 per cent $PbCO_3$.
    31. Lundström, *Geol. För. Förh.*, **3**, 291 (1877); on artificial material see Bourgeois,

Bull. soc. min., **12**, 464 (1889), Vater, *Zs. Kr.*, **21**, 462 (1893), Faivre, *C.R.*, **222**, 227 (1946).

32. Ford, *Trans. Conn. Ac. Arts Sc.*, **22**, 211 (1917), Foote and Bradley, *Am. J. Sc.*, **37**, 339 (1914).
33. See Johnston, *J. Am. Chem. Soc.*, **32**, 938 (1910); **30**, 1357 (1908).
34. Many additional localities are listed by Hintze (1926).
35. Eiriksson, *Trans. Inst. Min. Eng. London*, **39**, 36 (1920), Kašpar, *O islandském vápenci*, Prague, 48 pp. (1940)—*Min. Abs.*, **9**, 212 (1945).
36. Kelley, *Am. Min.*, **25**, 357 (1940).
37. The extensive literature on the synthesis of calcite and its polymorphs is reviewed by Hintze (1926), Mellor (**3**, 814, 1923), and Johnston, Merwin, and Williamson, *Am. J. Sc.*, **41**, 473 (1916); for more recent observations see also Morse and Donnay, *Bull. soc. min.*, **54**, 19 (1931), Copisarow, *J. Chem. Soc. London, Trans.*, **123**, 785 (1923), Reitemeier and Buehrer, *J. Phys. Chem.*, **44**, 535, 552 (1940).
38. See Brown (1934) and Fonda (1940).
39. Cf. Greenland, *Econ. Geol.*, **13**, 116 (1918), Wherry, *Proc. U. S. Nat. Mus.*, **53**, 227 (1917), Fettke, *Am. Min.*, **10**, 109 (1924).
40. Gaubert, *Bull. soc. min.*, **42**, 114 (1919).
41. See Siegl, *Min. Mitt.*, **48**, 286 (1936), and Seifert, *Zs. Kr.*, **100**, 120 (1938).
42. Gerasimovsky, *C. r. ac. sc. U.R.S.S.*, **32**, 492 (1941).
43. For additional analyses see Hintze (**1**, 3A, 2954, 1926); also Walker and Parsons (1925), Gagarin, *Ann. géol. pénin. Balkan.*, **13**, 72 (1936), Hacquaert (1926), Headden (1924), Harada, *J. Fac. Sc. Hokkaido Univ.*, **3**, no. 3, 4, 221 (1936), *ibid.*, **7**, no. 2, 143 (1948), Rost, *Rozpr. České Ak.*, **54**, no. 11 (1944).
44. Headden, *Am. J. Sc.*, **21**, 301 (1906).
45. Millosevich, *Acc. Linc., Rend.*, **19**, 91 (1910).
46. Sjögren, *Geol. För. Förh.*, **4**, 111 (1875).
47. Gallitelli, *Atti soc. nat. mat. Modena*, **8**, 86 (1929).
48. Roepper, *Am. J. Sc.*, **50**, 37 (1870).
49. Pollacci, *Gazz. chim. ital.*, **32**, 83 (1902).
50. On the correlation of indices and other properties with composition see Ford (1916); Sundius, *Min. Mitt.*, **38**, 175 (1925); von Philipsborn, *Jb. Min., Beil.-Bd.*, **64**, 187 (1931); Winchell (pt. 2, 104, 1951); Krieger (1930); Wayland, *Am. Min.*, **27**, 641 (1942); Gaubert (1919); Brown (1934).
51. See Brauns, *Cbl. Min.*, 1 (1926); Walker, *Univ. Toronto Stud., Geol. Ser.*, no. 20, 14 (1925).
52. Hess, *Proc. U. S. Nat. Mus.*, **76**, Art. 16 (1929); Mackin and Coombs, *J. Geol.*, **53**, 58 (1945), Baker and Frostick, *J. Sed. Petrol.*, **17**, 39 (1947).
53. Parker and Diehl, *Schweiz. Min., Mitt.*, **25**, 341 (1945); Drugman, *Bull. soc. min.*, **70**, 331 (1948).

UNIDENTIFIED CALCITE PSEUDOMORPHS. Thinolite *King* (*U. S. Geol. Sur., Rept. Geol. 40th Parallel*, **1**, 508, 1878). Jarrowite *Browell* (*Trans. Thyneside Nat. Club*, **5**, 103, 1860–62). Glendonite *David* and *Taylor* (*Rec. Geol. Sur. New South Wales*, **8**, 161, 1905). Pseudogaylussite *van Calker* (*Zs. Kr.*, **28**, 556, 1897); Barleycorn [Gerstenkörner] pseudomorphs.

Certain types of calcite pseudomorphs of more or less characteristic form and of wide distribution have been distinguished by given names, the identity of the original mineral or minerals not being known with certainty. The name thinolite has been given to acute pyramidal crystals approximating to tetragonal symmetry, found as extensive tufa deposits in northwestern Nevada and apparently representing a shore deposit of the Quaternary glacial lake, Lake Lahontan. It forms layers up to 60 feet thick of interlaced crystals of a light brown to yellowish or gray color that are solid or skeletonized and then affording a rectangular rib-work on the cross section. Also found about Mono Lake, California. The original

Calcite

mineral has been thought to be gaylussite or a calcium chlorocarbonate.[1]

The original so-called barleycorn pseudomorphs were from Obersdorf near Sangerhausen, Thuringia. They occurred as rounded, acute pyramidal single-crystals or as interpenetrating groups embedded in clay (see figures). Color gray, yellowish to brownish, the exterior usually smooth and hard and the interior cavernous with loosely coherent calcite grains. The original mineral has been thought to be gaylussite (from which the familiar name pseudo-gaylussite); also celestite, anhydrite, glauberite. Very similar pseudomorphs have been described [2] from Holland and a number of other

places. The so-called jarrowite pseudomorphs [3] found in river clays at Jarrow on Tyne, Durham, England, and at Cardross on the Clyde, near Glasgow, Scotland, are apparently the same as pseudo-gaylussite. These pseudomorphs usually are of an elongated or prismatic form with rounded pyramidal terminations and laterally grooved or rough; color brown and reddish brown. Calcite pseudomorphs of this general nature, the so-called glendonites,[4] are found widely distributed in Permo-Carboniferous mudstones in New South Wales and Queensland, Australia. These pseudomorphs apparently are after glauberite, as may be the well-known pineapple-like aggregates of pseudomorphous crystals consisting of opal from White Cliffs, New South Wales.

Ref.

1. See Dana, *U. S. Geol. Sur., Bull.* **12**, 25 (1884); Russell, *U. S. Geol. Sur., Monog.* **11**, 194 (1885); van Calker, *Zs. Kr.*, **28**, 556 (1897).
2. Cf. Calker (1897), who reviews the literature; also Dana (1884) and Hintze (**1** [3A], 2796, 1916).
3. See Trechmann, *Zs. Kr.*, **35**, 283 (1901).
4. Whitehouse, *Proc. Roy. Soc. Queensland*, **44**, 153 (1933); Raggatt, *J. Roy. Soc. New South Wales*, **71**, 336 (1938); Hodge-Smith, *Australian Mus. Mag.*, **6**, 337 (1938).

14.1.1.2 MAGNESITE [MgCO$_3$]. Kohlensaurer Talkerde *Mitchell* and *Lampadius* (**3**, 241, 1800) [first analysis]. Reine Talkerde, Talcum carbonatum *Werner* (Ludwig, **2**, 154, 1803). Magnesite pt. *Brongniart* (**1**, 489, 1807). Magnesit *Karsten* (48, 92, 1808). Carbonate of Magnesia. Magnésie carbonatée *Fr.* Kohlensaure Talkerde, Talkspath *Germ.* Baudisserite *Delamétherie* (*J. phys.*, **62**, 360, 1806; **2**, 1812); Baldisserite. Giobertite *Beudant* (410, 1824). Breunnerite *Haidinger* (**1**, 411, 1825); Breunerite. Walmstedtite *Leonhard* (297, 1826). Brown Spar pt. Pinolit *Rumpf* (*Min. Mitt.*, 265, 1873).

Mesitinspath pt. *Breithaupt* (*Ann. Phys.*, **11**, 170, 1827). Mesitit, Mesitin *Breithaupt* (*Ann. Phys.*, **70**, 148, 1847). Pistomesit *Breithaupt* (*Ann. Phys.*, **70**, 146, 1847).

Cryst.[1] Hexagonal—R; hexagonal-scalenohedral—$\bar{3}\,2/m$.

$a:c = 1:0.8112;$ $\alpha\ 103°18\frac{1}{2}';$ $p_0:r_0 = 0.9367:1,\ \lambda\ 72°36'$

Forms:[2]

		ϕ	$\rho = C$	A_1	A_2
c	0001	111	.....	0°00'	90°00'	90°00'
m	10$\bar{1}$0	2$\bar{1}\bar{1}$	30°00'	90 00	30 00	90 00
a	11$\bar{2}$0	10$\bar{1}$	0 00	90 00	60 00	60 00
r	10$\bar{1}$1	100	30 00	43 07$\frac{1}{2}$	53 42	90 00
f	02$\bar{2}$1	11$\bar{1}$	−30 00	61 54$\frac{1}{2}$	90 00	40 11
v	21$\bar{3}$1	20$\bar{1}$	10 53$\frac{1}{2}$	68 01$\frac{1}{2}$	52 37	72 20

Structure cell.[3] Space group $R\bar{3}c$.

Anal.	a_0	c_0	$a_0:c_0$	a_{rh}	α	G	nO	nE
2	4.584 kX	14.92	1:3.255	5.635	48°00'	3.031	1.712	1.518
3	4.599	14.95	1:3.2518	5.648	48°03'	3.134	1.719	1.517

Cell contents Mg$_2$(CO$_3$)$_2$ in the rhombohedral unit.

Habit. Distinct crystals are rare, usually rhombohedral $\{10\bar{1}1\}$, also $\{01\bar{1}2\}$; rarely prismatic [0001] with $\{11\bar{2}0\}$ and $\{0001\}$ or tabular $\{0001\}$; rarely scalenohedral. Magnesite is commonly massive, coarse to fine-granular or very compact and porcelaneous; earthy to somewhat chalky; lamellar or coarsely fibrous.

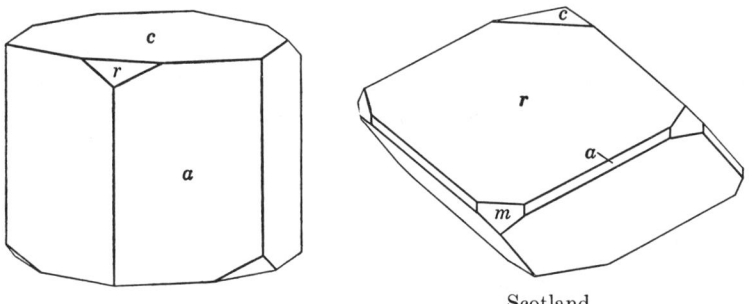

Scotland.

Twinning. Not known with certainty.[4]

Etch figures. Etch pits conform to hexagonal-scalenohedral symmetry, but sometimes are abnormal.[5]

MAGNESITE

Phys. Cleavage $\{10\bar{1}1\}$ perfect. Fracture conchoidal. Brittle. H. $3\frac{3}{4}$–$4\frac{1}{4}$. G. 3.00 ± 0.02 (pure $MgCO_3$); the G. varies essentially linearly toward the end-members with the substitution of Fe, etc., for Mg; G. $\sim$ 3.48 for Mg:Fe = 1:1; compact types with low Fe have G. $\sim$ 2.9. Luster vitreous. Colorless and transparent in pure crystals; white, grayish white, yellowish to brown. Streak nearly white. Transparent to subtranslucent.

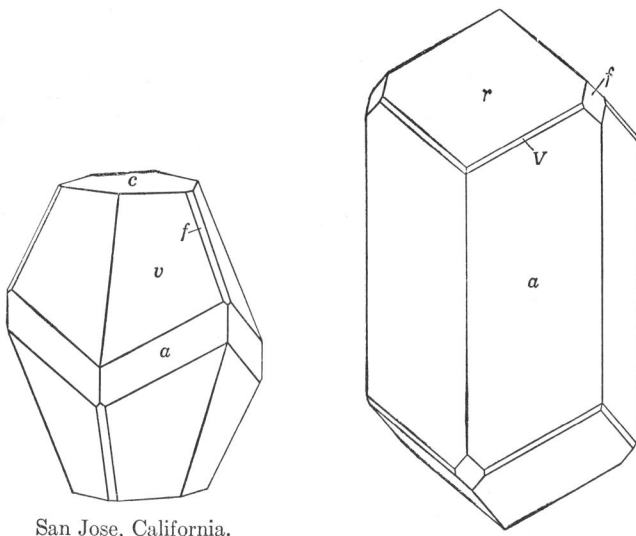

San Jose, California.

Translation gliding:[6] $T\{0001\}$, $t[10\bar{1}1]$. Sometimes shows a greenish or bluish fluorescence and phosphorescence in ultraviolet radiation; also triboluminescent.

Opt. In transmitted light, colorless. Uniaxial negative $(-)$. The indices of refraction[7] vary essentially linearly toward the end-members accompanying the substitution of Fe, etc., for Mg.

	$MgCO_3$ (pure)	Mg:Fe = 1:1
nO	1.700	1.788
nE	1.509	

The cobaltian material of analysis 7 is dichroic, with E violet-red, O flesh-red.

Chem. Magnesium carbonate, $MgCO_3$. Fe'' substitutes for Mg and a complete series extends to siderite. Mn and Ca substitute for Mg up to about Mn:(Mg, etc.) = 1:9.3 (analysis 7) and Ca:(Mg, etc.) = 1:8.9 (analysis 4), and gaps exist in the series toward rhodochrosite and calcite, respectively. Co substitutes for Mg with Co:(Mg,Fe) = 1:15 in analysis 7. Ni is sometimes present in very small amounts. The compact massive types of magnesite often contain SiO_2 (opal) or magnesium silicates as impurities.

Anal.[8]

	1	2	3	4	5	6	7	8
CaO		0.86	nil	6.41		0.43		1.85
MgO	47.81	44.98	41.62	39.17	47.53	46.62	33.41	17.39
FeO		2.41	7.22	2.95		0.56	6.50	36.38
MnO		0.03	0.90			0.12	7.50	2.03
CO_2	52.19	51.56	50.68	49.70	51.45	51.93	46.77	42.56
Rem.				2.13	0.72		5.43	
Total	100.00	99.84	100.42	100.36	99.70	99.66	99.61	100.21
G.		3.031	3.091			3.015		3.44

1. $MgCO_3$. 2. Sunk, Styria.[3] Original total given as 99.85. 3. Ochsenkogel, Styria.[3] 4. Häuselberg, Carinthia.[9] Rem. is insol. Original total given as 100.00. 5. Baudissero, Piedmont.[10] Rem. is Al_2O_3 0.06, Fe_2O_3 0.05, H_2O+ 0.48, H_2O- 0.13. Original total given as 100.17. 6. Eguas, Bahia, Brazil.[11] 7. Eiserfeld, Westphalia.[12] Rem. is CoO 5.12, H_2O 0.31. 8. Magnesian siderite. Saarbrücken, Germany.[13]

Var.

Ferroan. Breunnerite *Haidinger* (1825). Walmstedtite *Leonhard* (1826). Brown Spar pt. Mesitit, Mesitin, Mesitinspath pt. *Breithaupt* (1827, 1847). Pistomesit *Breithaupt* (1847). Contains Fe'' in substitution for Mg, a complete series extending to siderite. Color white, yellowish, brownish, rarely black and bituminous; often becoming brown on exposure.

Manganoan. Contains Mn in substitution for Mg up to at least Mn: (Mg, etc.) = 1:9.3 (analysis 7).

Calcian. Contains Ca in substitution for Mg up to at least Ca:(Mg, etc.) = 1:8.9 (analysis 4).

Tests. B.B. resembles calcite and dolomite. Slightly acted on by cold acids, readily soluble with effervescence in warm HCl. Solubility in CO_2-free water at 100° = 0.063 g. $MgCO_3$ per liter; the solubility increases with the presence of NaCl, Na_2SO_4, or CO_2.

Occur. Magnesite deposits are formed in various ways: through the alteration of rocks rich in magnesium such as serpentine, peridotite, and dunite by carbonated waters and then often cryptocrystalline and admixed with opaline silica; as crystalline beds associated with talc-, chlorite-, and mica-schists and of metamorphic origin; as a replacement of calcite rocks by magnesium-containing solutions, dolomite being formed as an intermediate product; as a sedimentary deposit. Rarely as a gangue mineral in hydrothermal ore-veins, as a cavity mineral in igneous flow-rocks, and in oceanic salt deposits. Also as a primary mineral in igneous rock [17] and in meteorites.

Notable localities for magnesite include the following. In chlorite-schist in the Pfitschtal and at Greiner in the Zillertal, Tyrol, in rock salt at Hall, Tyrol, and at Kraubat, Veitsch, Häuselberg, Sunk, and other places in Styria, Austria; a large deposit at Radenthein, Carinthia, Austria.[19] At numerous localities in Com. Gömör, Czechoslovakia. In the Zlatibore region, Yugoslavia. In Greece on the island of Eubœa, and in Macedonia. At Traversella and Baudissero in Piedmont, and in the Livorno hills, Tuscany, Italy. In Spain especially at Reinosa in Santander province, and in the Sierra de Gádor, Almeria province. At Snarum, Norway, and in Norbotten, Sweden. Also found near Salem, Madras, India; from Djebel

Hadifa and elsewhere in Algeria; in the Transvaal, South Africa, in Rhodesia, and at Katanga in the Belgian Congo. Large clear crystals occur in pegmatite at Bom Jesus dos Meiras, Brazil. Huge deposits of stratified, coarsely crystalline magnesite, dolomite, and marble extend on the Liao-Tung peninsula and in Sheng-king province in Manchuria; also near Taikwayodo, Korea.

In the United States, deposits of magnesite in serpentinous rocks extend along the Coast Range in California from Mendocino County to the south of Los Angeles and on the western side of the Sierra Nevada from Placer County to Kern County. Considerable bodies of sedimentary magnesite have been found in the Muddy Valley district, Clark County, Nevada, and at Bissell in Kern County, California.[15] Large deposits of hydrothermal origin in dolomite occur in the Paradise Range, Nevada.[18] In New Mexico in the South Canyon district, and Target Range Canyon in Dona Ana County, north of Lordsburg in Grant County, and in the Carlsbad potash district in Eddy County. Numerous minor localities for the most part with serpentine or dolomite rocks are known in the eastern United States. Found in Canada in Argenteuil County, Quebec, and on the Black River and Yukon River in the Yukon.

Alter. Pseudomorphs have been found of talc and rumpfite after magnesite and of limonite after ferroan magnesite. Magnesite has been found as incrustation pseudomorphs after calcite.

Artif.[14] Obtained as microscopic crystals by heating $CaCO_3$ with $MgCl_2$ or $MgSO_4$ solution, or precipitated $MgCO_3$ with carbonated water in a closed tube at 160°–200°. Precipitation at ordinary temperature and pressure gives either nesquehonite, $MgCO_3 \cdot 3H_2O$, or a basic carbonate.

Name. Magnesite in allusion to the composition. The name was early applied to a group of magnesium minerals, including the carbonate and the silicate, sepiolite, and was restricted to the carbonate by Karsten in 1808. The name giobertite, given in 1824 by Beudant after the Italian chemist, G. A. Giobert (1761–1834), is used to a large extent in France but lacks priority. Mesitite, from μεσίτης, *a go-between*, and pistomesite, from πίσος, *true*, and μέσον, *the middle*, were applied by Breithaupt and later by Dana and others as subspecies designations for supposed compounds between magnesite and siderite in composition; also by some [16] for intermediate ranges of composition in this continuous series. Both names are here abandoned.

Ref.

1. Fundamental angle of Koksharov (**7**, 181, 1875) on Tyrol crystals with G. 3.118; conventional unit, in which c is one-fourth of the unit of the structure cell. Transformation: morphology to structure, 1000/0100/0010/0004. The crystal angles vary appreciably with variation in composition—see Hintze (**1** [3A], 3113, 1926).
2. Goldschmidt (**5**, 175, 1918); Bücking, *Kali*, **5**, 221 (1911); Dobbel, *Am. Min.*, **8**, 223 (1923); Pardillo, *Treballs mus. cienc. nat. Barcelona*, **9**, 5 (1924). Rare or uncertain: $10\bar{1}2$, $40\bar{4}1$, $30.0.\overline{30}.1$, $05\bar{5}1$, $0.11.\overline{11}.1$, $0.17.\overline{17}.1$, $0.31.\overline{31}.1$, $16.4.\overline{20}.3$, $4.8.\overline{12}.5$.
3. Schoklitsch, *Zs. Kr.*, **90**, 433 (1935).
4. See Johnsen, *Jb. Min.*, II, 133 (1902).
5. Honess, *Am. Min.*, **45**, 210 (1918).
6. Johnsen, *Jb. Min.*, II, 142 (1902).

7. See Schoklitsch (1935); Wayland, *Am. Min.*, **27**, 614 (1942); Gaubert, *Bull. soc. min.*, **42**, 88 (1919) and *C.R.*, **164**, 46 (1917); Niggli, *Zs. Kr.*, **56**, 230 (1921), Ford, *Trans. Conn. Ac. Arts Sc.*, **22**, 211 (1917).
8. Numerous additional analyses are tabulated by Hintze (1926); Doelter (**1**, 220, 1911); Schoklitsch (1935); Lonsdale, *Am. Min.*, **15**, 238 (1930); Rogers, *Am. Min.*, **8**, 138 (1923); Koch and Zombory, *Földt. Közl.*, **64**, 160 (1934); Niinomy, *Econ. Geol.*, **20**, 25 (1925); Pavlovitch, *Bull. soc. min.*, **54**, 95 (1931); Petrascheck, *Fortschr. Min.*, **20**, 77 (1936); Du Rietz, *Geol. För. Förh.*, **57**, 133 (1935); Lacroix, *Le volcan . . . de la Réunion*, Paris (1936).
9. Ratz analysis in Redlich and Cornu, *Zs. pr. Geol.*, **16**, 145 (1908).
10. Fenoglio and Sanero, *Per. Min.*, **12**, 83 (1941).
11. Fornaseri, *Rend. soc. min. ital.*, **1**, 60 (1941).
12. Johnsen, *Cbl. Min.*, 13 (1903).
13. Weiss, *Jb. preuss. geol. Landesanst.*, 113 (1885).
14. Mellor (**4**, 350, 1923); Hintze (1926).
15. On the magnesite deposits of California, see Bradley, *Calif. Bur. Mines, Bull.* **79** (1925), and Murdoch and Webb, *Calif. Div. Mines, Bull.* **136**, 196 (1948).
16. Cf. Crook, *Trans. Ceramic Soc.*, **18**, 67 (1919).
17. Barth, *Norsk Geol. Tidsskr.*, **9**, 271 (1927), and Lacroix, *C.R.*, **213**, 261 (1941).
18. See Faust and Callaghan, *Bull. Geol. Soc. Am.*, **59**, 11 (1948).
19. Awerzger and Angel, *Radex-Rundschau*, **5**, 91 (1948).

14.1.1.3 **S I D E R I T E** [$FeCO_3$]. ? Vena ferri jecoris colore optima *Germ.* Stahelreich Eisen *Gesner* (90, 1565). Spatformig Jernmalm, Minera ferri alba spathiformis *Wallerius* (256, 1747). Järn med Kalkjord förenadt *Germ.* Stahlstein *Cronstedt* (29, 1758). Ferrum cum magnesio et terra calcarea acido aero mineralisatum *Bergmann* (**2**, 184, 1780). Spathiger Eisen, Spatheisenstein *Germ.* Fer spatique *de Lisle* (**3**, 281, 1783). Calcareous or Sparry Iron Ore *Kirwan*. Spathic Iron, Spathose Iron. Brown Spar pt. Steel Ore. Carbonate of Iron. Fer carbonaté, Mine d'acier *Fr.* Kohlensaures Eisen, Eisenkalk *Germ.* Eisenspath *Hausmann* (951, 952, 1813). Spherosiderite *Hausmann* (1070, 1813; 1847; 1853). Siderose *Beudant* (**2**, 346, 1832). Junckérite *Dufrénoy* (*Ann. chim. phys.*, **56**, 198, 1834); *Breithaupt* (*Ann. Phys.*, **58**, 278, 1843). Siderit *Haidinger* (499, 1845). Chalybit *Glocker* (241, 1847). Spatheisenstein.

Oligonspath *Breithaupt* (**2**, 235, 1841) = Oligonit *Hausmann* (1362, 1847). Thomäit *Meyer* (*Jb. Min.*, 200, 1845). Siderodot *Breithaupt* (*Ber. Mitt. Freund. Wiss. Wien*, **1**, 6, 1847). Sideroplesit *Breithaupt* (*B. H. Ztg.*, **17**, 54, 1858). Pelosiderit *Naumann-Zirkel* (457, 1885). Thoneisenstein = Clay Iron Ore pt., Clay Ironstone. Black-band Ore.

Manganosphärit *Busz* (*Jb. Min.*, II, 129, 1901). Kobalt-oligonspat, Cobaltosphärosiderit *Reissner* (*Zbl. Min.*, 170, 1935). Manganosiderite pt.

C r y s t.[1] Hexagonal—R; hexagonal-scalenohedral—$\bar{3}\ 2/m$.

$a:c = 1:0.8184;$ $\alpha\ 103°05\tfrac{1}{2}';$ $p_0:r_0 = 0.9450:1;$ $\lambda\ 73°00'$

Forms:[2]

		ϕ	$\rho = C$	A_1	A_2
c	0001	111	.....	$0°00'$	$90°00'$	$90°00'$
m	$10\bar{1}0$	$2\bar{1}\bar{1}$	$30°00'$	$90\ 00$	$30\ 00$	$90\ 00$
a	$11\bar{2}0$	$10\bar{1}$	$0\ 00$	$90\ 00$	$60\ 00$	$60\ 00$
r	$10\bar{1}1$	100	$30\ 00$	$43\ 23$	$53\ 30$	$90\ 00$
M	$40\bar{4}1$	$3\bar{1}\bar{1}$	$30\ 00$	$75\ 11$	$33\ 09$	$90\ 00$
e	$01\bar{1}2$	110	$-30\ 00$	$25\ 17\tfrac{1}{2}$	$90\ 00$	$68\ 17$
f	$02\bar{2}1$	$11\bar{1}$	$-30\ 00$	$62\ 07$	$90\ 00$	$40\ 03$
s	$05\bar{5}1$	$22\bar{3}$	$-30\ 00$	$78\ 03$	$90\ 00$	$32\ 05$
y	$32\bar{5}1$	$30\bar{2}$	$6\ 35$	$76\ 21\tfrac{1}{2}$	$54\ 36\tfrac{1}{2}$	$67\ 17$
v	$21\bar{3}1$	$20\bar{1}$	$10\ 53\tfrac{1}{2}$	$68\ 12$	$52\ 34$	$72\ 18\tfrac{1}{2}$
β	$24\bar{6}1$	$31\bar{3}$	$-10\ 53\tfrac{1}{2}$	$78\ 41\tfrac{1}{2}$	$71\ 11\tfrac{1}{2}$	$50\ 04$

Siderite

Structure cell.[3] Space group $R\bar{3}c$.

Anal.	a_0	c_0	$a_0:c_0$	a_{rh}	α	G.	nO	nE
~95%FeCO$_3$	4.71 kX	15.43	1:3.276	5.82	47°46′			
Anal. 8	4.677	15.267	1:3.264	5.761	47 54	3.71	1.844	1.606

Cell contents Fe$_2$(CO$_3$)$_2$ in the rhombohedral unit.

Habit. Crystals commonly rhombohedral, with $\{10\bar{1}1\}$ or, less frequently, $\{01\bar{1}2\}$; also with large $\{02\bar{2}1\}$ or $\{40\bar{4}1\}$ and modifying faces;

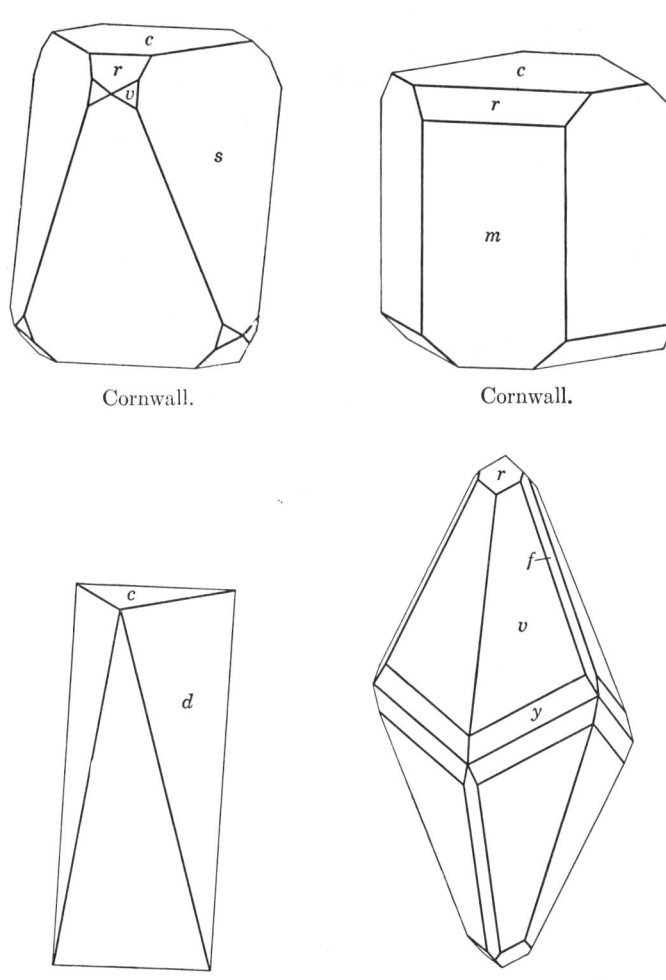

Cornwall.   Cornwall.

Cornwall.   Frostburg, Maryland.

also thin to thick tabular $\{0001\}$; prismatic [0001] with $\{11\bar{2}0\}$; scalenohedral. The crystal faces are often curved or composite. Often massive, coarse to fine granular. In botryoidal and globular forms, with usually a coarse fibrous structure; oolitic; earthy or stony and impure from admixture with clay or silica.

Twinning. (1) On $\{01\bar{1}2\}$, lamellar, uncommon. (2) On $\{0001\}$, rare.[20]

Oriented growths. Calcite and $LiNO_3$ (artificially) [4] upon siderite, the crystal axes parallel.

Etching. Etch pits are normally conformable with hexagonal-scalenohedral symmetry, but sometimes are distorted into asymmetric forms.[5]

P h y s. Cleavage $\{10\bar{1}1\}$ perfect. Fracture uneven to conchoidal. Brittle. H. $3\frac{3}{4}$–$4\frac{1}{4}$. G. 3.96 ± 0.01 (pure $FeCO_3$);[6] the G. decreases essentially linearly toward the end-members with substitution of Mn, Mg, Ca for Fe; G. $\sim$3.48 for Mg:Fe = 1:1, $\sim$3.83 for Fe:Mn = 1:1. Luster vitreous, inclining to pearly or silky in some aggregated types. Color yellowish brown and grayish brown to brown and reddish brown; also ash-gray, yellowish gray, greenish gray, pale green and sometimes white; the dark-brown and blackish-brown colors are due to partial oxidation. Sometimes tarnished iridescent. Streak white. Translucent to subtranslucent. Translation gliding:[7] $T\{0001\}$, $t[10\bar{1}0]$.

O p t. In transmitted light, colorless to yellow and yellow-brown. Uniaxial negative $(-)$. The indices of refraction [8] decrease essentially linearly toward the end-members with the substitution of Mn, Mg, Ca for Fe″.

	$FeCO_3$ (pure)	Mn:Fe = 1:1	Mg:Fe = 1:1	Ca:Fe = 1:2
nO	1.875	1.845	1.788	1.803
nE	1.633			

DISPERSION [9]

(± 0.001; analysis 3)

	Li	Na	Tl
nO	1.8647	1.8728	1.8804
nE	1.6295	1.6331	1.6366

C h e m. Ferrous carbonate, $FeCO_3$. Mn″ and Mg substitute for Fe″ and complete series extend to rhodochrosite and magnesite. Ca substitutes to only a limited extent for Fe″, up to about Ca:(Fe,Mg) = 1:3.4 (analysis 7), and a large gap exists in the binary series to calcite. Co can substitute for Fe″, with Co:(Fe, etc.) = 1:6.7 in analysis 5. Zn apparently substitutes for Fe″ in amounts up to about 2 weight per cent ZnO; it generally is absent.

Anal.[10]

	1	2	3	4	5	6	7	8
CaO		tr.	0.10	0.53	2.43	5.58	11.72	0.39
MgO		1.05	0.135	1.71	10.07	2.29	6.31	8.80
MnO		1.30	1.12	3.42	11.90	23.86		4.02
FeO	62.01	58.98	61.08	55.71	25.10	28.93	39.39	46.32
CO_2	37.99	38.07	38.19	38.35	41.04	38.79	[38.52]	40.73
Rem.		0.59			9.50		4.06	
Total	100.00	99.99	100.62	99.72	100.04	99.45	[100.00]	100.26
G.			3.938	3.879				3.713

1. $FeCO_3$. 2. Wölch, Carinthia.[11] Rem. is insol. 3. Camborne, Cornwall.[12] 4. Felsöbánya, Hungary.[13] 5. Locality unknown.[14] Rem. is CoO 9.10, ZnO 0.40. Microanalysis. 6. Felsöbánya, Hungary.[15] 7. Nezdenice, Moravia.[16] Rem. is insol. 8. Gollrad, Styria.[17]

Siderite

Var.

Manganoan. Oligonspath *Breithaupt* (1841); Oligonit *Hausmann* (1847). Mangano-sphärit *Busz* (1901). Contains Mn in substitution for Fe, a complete series extending to rhodochrosite. Color often somewhat pinkish or reddish, especially in highly manganoan material. Much siderite contains at least several weight per cent of MnO.

Magnesian. Sideroplesit *Breithaupt* (1858). Contains Mg in substitution for Fe, a complete series extending to magnesite. Mg ordinarily is present in amounts of several per cent MgO or more.

Calcian. Siderodot *Breithaupt* (1847). Contains Ca in substitution for Fe up to a maximum of about Ca:(Fe, etc.) = 1:3.4 (analysis 7), a gap in the series then extending to ferroan calcite with Fe:Ca $\sim$ 1:4.5. Ca is ordinarily present in siderite in amounts less than about three per cent CaO.

Cobaltoan. Kobalt-oligonspath, Cobalto-Sphärosiderit *Reissner* (1935). Contains Co in substitution for Fe with Co:(Fe, etc.) = 1:6.7 in analysis 5. Co ordinarily is absent.

Tests. In C.T. loses CO_2, blackens, and becomes magnetic. B.B. blackens and fuses at $4\frac{1}{2}$. Slowly soluble in cold acid, rapidly in hot. When heated, siderite loses CO_2 and forms FeO which may then oxidize to hematite, maghemite, or magnetite, depending on the availability of oxygen.[18] Solubility in water at 5°–8° at 2 atm. CO_2 pressure = 6.191 g. $FeCO_3$ per liter.

O c c u r. Found in important amounts as a bedded sedimentary deposit associated with shale, clay, or coal-seams. Usually massive fine-grained or compact, sometimes concretionary or with a spherulitic microstructure, and including the so-called clay ironstones and black-band (bituminous) ores. The deposits are often formed by biogenic processes in oxygen-deficient environments. Siderite also commonly occurs as a subordinate, primary gangue mineral in hydrothermal metallic veins, and sometimes as the predominant mineral in veins of considerable size (Westphalia). Also in globular-fibrous forms in cavities in basaltic igneous rocks; as a late hydrothermal mineral in some pegmatites; in metamorphosed sedimentary deposits; as a replacement of limestones by iron solutions of igneous origin; as a bog deposit, etc. Siderite occurrences are very common and of world-wide distribution; a few localities are mentioned below.

Important replacement and metamorphic deposits are found in Austria, as at Erzberg near Eisenerz in Styria and at Hüttenberg-Lölling in Carinthia. In Germany in the silver veins at Freiberg, at Neudorf and Stolberg in the Harz, in the Siegen region of Westphalia, at Lobenstein, Hirschberg, and Kamsdorf in Thuringia. Finely crystallized at Brosso and Traversella in Piedmont, Italy, and Allevard, Isère, France. In the Val Tavetsch, Grisons, Switzerland. Clay iron ores occur near Radom in central Poland. At Bilbao, Spain. Found well-crystallized in the tin veins of Cornwall, notably at Camborne, Redruth, St. Austell, Lostwithiel, Bodmin; also at Tavistock, Devonshire; black-band ores occur in Somerset, Durham, Northumberland, Yorkshire, etc. On the Kertsch and Taman peninsulas, U.S.S.R. At Djebel Ouenza, Constantine, Algeria. At Chorolque and Tatasi in Bolivia, and the Morro Velho mine, Minas Geraes, Brazil. In large cleavable masses with cryolite at Ivigtut, Greenland.

In the United States, siderite occurs in a large vein deposit at Mine Hill, Roxbury, Connecticut. At Napanoch, Ulster County, New York; and in talcose slate at Plymouth, Windsor County, Vermont. Formerly mined together with limonite at many points in eastern Pennsylvania; also found as black-band ore and as clay ironstone in the Pennsylvanian coal measures in Pennsylvania and similarly in Illinois, Indiana, Ohio, Kentucky, West Virginia, etc. In Cretaceous clays in the Coastal Plain region of Maryland. Globular-fibrous types are found in diabase at Weehawken, New Jersey, and Spokane, Washington. Siderite is an important gangue mineral in the silver-lead veins of Idaho and occurs especially in Coeur d'Alene district, Shoshone County, and in the Hailey and Elkhorn districts, Blaine County. In Arizona in the Bisbee district, Cochise County, and in the Cerbat Range, Mohave County. Abundantly at Leadville, Lake County, and Red Cliff, Eagle County, in Colorado. In the iron ranges of the northern peninsula, Michigan. Found in Canada in the Slocan district, British Columbia; at the Acadia mines, Londonderry, Nova Scotia; near Keno City in the Yukon.

Alter. Alteration pseudomorphs of limonite (usually identifiable as *goethite*) after siderite are very common; well-known localities are at Traversella, Italy, Eisenerz in Styria, and Pikes Peak, Colorado. Less common are pseudomorphs of hematite, magnetite, quartz, cacoxenite, and chlorite after siderite. Also pseudomorphs, usually by incrustation, after calcite, dolomite, fluorite, barite, gypsum, galena, bismuthinite.

Artif.[19] Obtained in crystals by heating $CaCO_3$ with $FeCl_2$ or $FeSO_4$ solution in a closed tube at 130°–200°; also by heating mixed solutions of $FeSO_4$ and $NaHCO_3$ with excess CO_2, and by reaction of $(NH_4)_2CO_3$ and $FeCl_2$ at red heat.

Name. Siderite from σίδηρος, *iron*, the name being shortened by Haidinger from the *spherosiderite* of Hausmann as originally applied to the globular variety. Oligonite, from ὀλίγος, *small*, because of a lower gravity than ordinary siderite due to the presence of much Mn; the name has been applied to an intermediate range of the siderite-rhodochrosite series. Chalybite, from χάλυψ, *steel*, because the mineral contains iron and carbon. This name has been much used in England but lacks priority over siderite.

Ref.

1. Fundamental angle of Wollaston, *Phil. Trans.*, 159 (1812), on crystals from Wheal Maudlin, Cornwall; conventional unit, in which c is one-fourth that of the structure cell. Transformation: morphology to structure, 1000/0100/0010/0004.

2. Goldschmidt (**3**, 107, 1916); the form development is discussed by de Klerk, *Beitr. Kr. Min.*, **3**, 85 (1926). Rare or uncertain:

$g\ 10\bar{1}2$	$k\ 50\bar{5}2$	$w\ 07\bar{7}3$	$d\ 08\bar{8}1$	$51\bar{6}7$
$z\ 30\bar{3}4$	$30\bar{3}1$	$03\bar{3}1$	$\vartheta\ 22\bar{4}3$	$3.7.\overline{10}.2$
$l\ 70\bar{7}5$	$h\ 03\bar{3}2$	$07\bar{7}2$	$\alpha\ 44\bar{8}3$	$26\bar{8}1$

3. Wyckoff, *Am. J. Sc.*, **50**, 317 (1920), and Schoklitsch, *Zs. Kr.*, **90**, 433 (1935).
4. Kreutz, *Min. Mag.*, **15**, 232 (1909).
5. See Honess, *Am. J. Sc.*, **45**, 201 (1918).
6. From graphical summary of available data.
7. Johnsen, *Jb. Min.*, II, 133 (1902).
8. See Wayland, *Am. Min.*, **27**, 614 (1942), Fornaseri, *Rend. soc. min. ital.*, **1**, 60 (1941), Ford, *Trans. Conn. Ac. Arts Sc.*, **22**, 211 (1917), Gaubert, *Bull. soc. min.*, **42**, 88 (1917), Niggli, *Zs. Kr.*, **56**, 224 (1921), von Philipsborn, *Jb. Min., Beil.-Bd.*, **64**, 187 (1931).
9. Hutchinson, *Min. Mag.*, **13**, 209 (1903) (averaged).
10. Numerous additional analyses are tabulated by Hintze (**1** [3A], 3187, 1926); Doelter (**1**, 418, 1911); Wahlstrom, *Am. Min.*, **20**, 377 (1935); Sundius, *Geol. För. Förh.*, **47**, 269 (1925); Schoklitsch (1935), Kratochvíl, *Věst. Stát. Geol. Úst. Českoslov. Rep.*, **18**, 157 (1943), Smythe and Dunham, *Min. Mag.*, **28**, 53 (1947).
11. Brunlechner, *Jb. Nat. Land.-Mus. Kärnten*, **17**, 1 (1885).
12. Hutchinson (1903).
13. Koch and Zombory, *Földt. Közl.*, **65**, 18 (1935).
14. Reissner, *Zbl. Min.*, 170 (1935).
15. Dittrich analysis in Schroeckinger, *Verhl. geol. Reichsanst. Wien.* 114 (1877).
16. Rosický, *Festschr. V. Goldschmidt*, 229, 1928.
17. Schoklitsch (1935).
18. Thermal analysis curves of siderite are discussed by Rowland and Jonas, *Am. Min.*, **34**, 550 (1949).
19. Mellor (**14**, 357, 1935), Hintze (1926), Biltz, *Zs. anorg. Chem.*, **220**, 312 (1934).
20. Spencer, *Min. Mag.*, **14**, 343 (1907).

14.1.1.4 **R H O D O C H R O S I T E** [$MnCO_3$]. Magnesium acido aëro mineralisatum *Bergmann* (1782) [without descr. or loc.]. Rother Braunsteinerz, Rothspath, Magnesium ochraceum rubrum, Oxide de manganèse couleur de rose, pt. *of latter part of 18th Cent.* [it being confused with rhodonite]. Luftsaures Braunsteinerz pt. *Lenz* (2, 1794). Manganèse oxydé carbonaté *Haüy* (111, 1809). Dichter Rothstein pt. *Hausmann* (302, 1813). Rhodochrosit, Kohlensaures Magnesium oxydul [from an anal. of a Kapnik spec. by Lampadius, *Pr. Chem. Abh.*, **3**, 239, 1800] *Hausmann* (1081, 1813). Carbonate of Manganese. Manganspath *Werner*. Dialogite *Jasche* (in Germor, *J. Chemie u. Phys.*, **26**, 119, 1819) = Blättrige Rothmanganerz *Jasche* (*Kl. Min. Schrift.*, 4, 1817). Diallogite [wrong orthogr.]. Rosenspath, Himbeerspath *Breithaupt* (67, 68, 1832; 228, 229, 1841). Manganocalcit *Breithaupt* (*Ann. Phys.*, **69**, 429, 1846). Manganosiderit *Bayer* (*Verhl. Ver. Brünn*, **12**, 1873). Rodocrosite *Ital.* Zincorhodochrosite. Viellaurite pt., Torrensite pt., Lacroixite pt., Schokoladenstein pt., Huelvite pt. *Lienau* (*Chem.-Ztg.*, **23**, 418, 1899). Zincorhodochrosite *Manasse* (*Mem. Soc. Tosc.*, **27**, 76, 1911). Ponite *Butureanu* (Ann. sc. univ. Jassy, **7**, 185, 1912).

C r y s t.[1] Hexagonal—R; hexagonal-scalenohedral—$\bar{3}\,2/m$.

$a{:}c = 1{:}0.8183$; $\alpha\ 103°04\tfrac{1}{2}'$; $p_0{:}r_0 = 0.9449{:}1$; $\lambda\ 72°59\tfrac{1}{2}'$

Forms:[2]

			ϕ	$\rho = C$	A_1	A_2
c	0001	111	.....	0°00′	90°00′	90°00′
m	10$\bar{1}$0	2$\bar{1}\bar{1}$	30°00′	90 00	30 00	90 00
a	11$\bar{2}$0	10$\bar{1}$	0 00	90 00	60 00	60 00
r	10$\bar{1}$1	100	30 00	43 22$\tfrac{1}{2}$	53 30	90 00
M	40$\bar{4}$1	3$\bar{1}\bar{1}$	30 00	75 11	33 09	90 00
e	01$\bar{1}$2	110	−30 00	25 17$\tfrac{1}{2}$	90 00	68 17$\tfrac{1}{2}$
f	02$\bar{2}$1	11$\bar{1}$	−30 00	62 07	90 00	40 03
s	05$\bar{5}$1	22$\bar{3}$	−30 00	78 03	90 00	32 05
y	32$\bar{5}$1	30$\bar{2}$	6 35	76 21	54 36$\tfrac{1}{2}$	67 17
v	21$\bar{3}$1	20$\bar{1}$	10 53$\tfrac{1}{2}$	68 12	52 34	72 18$\tfrac{1}{2}$

Structure cell.[3] Space group $R\bar{3}c$. a_{rh} 5.84, α 47°46' [on material of analysis 2]; a_0 4.73 kX, c_0 15.51; $a_0:c_0 = 1:3.279$. Cell contents $Mn_2(CO_3)_2$ in the rhombohedral unit.

Habit. Distinct crystals are not common: rhombohedral $\{10\bar{1}1\}$ or, less frequently, $\{01\bar{1}2\}$ and then often rounded and composite; rarely scalenohedral; thick tabular $\{0001\}$; prismatic $[0001]$. $\{11\bar{2}0\}$ and $\{21\bar{3}1\}$ often striated $[1\bar{1}0\bar{1}]$. Massive, coarsely granular to compact; columnar; incrusting. Also globular and botryoidal.

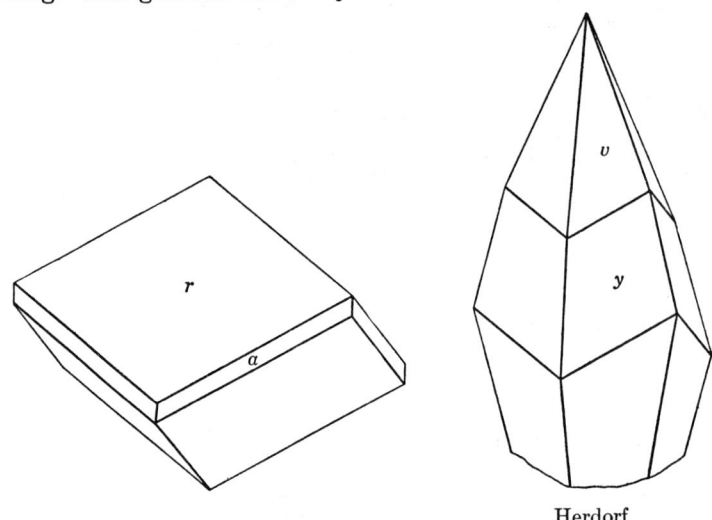

Herdorf.

Twinning.[4] On $\{01\bar{1}2\}$, lamellar, rare.

Etching.[5] Conforms to hexagonal-scalenohedral symmetry.

Oriented growths.[18] Parallel growths have been observed of rhodochrosite with dolomite and $LiNO_3$ (artificial).

P h y s. Cleavage $\{10\bar{1}1\}$ perfect. Sometimes parting $\{01\bar{1}2\}$. Fracture uneven to conchoidal. Brittle. H. $3\frac{1}{2}$–4. G. 3.70 (pure $MnCO_3$);[6] the G. varies essentially linearly toward the respective end-members in material containing Fe'', Ca, Mg, or Zn in substitution for Mn. Luster vitreous, inclining to pearly in some aggregated types. Color various shades of pink, rose and rose-red, sometimes red; yellowish gray, fawn-colored, brown. Streak white. Subtransparent to translucent. Translation gliding[7] with $T\{0001\}$, $t[10\bar{1}0]$.

O p t. In transmitted light, colorless to pale rose. Uniaxial negative ($-$). The indices of refraction[6] vary essentially linearly toward the end-members in the substitution of Fe'', Ca, Mg (and Zn) for Mn. Deep red varieties may be faintly dichroic with absorption $O > E$.

	$MnCO_3$ (pure)	Mn:Fe = 1:1	Mn:Ca = 1:1	Mn:Mg = 2:1
nO	1.816	1.845	1.737	1.777
nE	1.597			

RHODOCHROSITE 173

Chem. Manganese carbonate, $MnCO_3$. Fe'' and Ca substitute for Mn, and complete series extend to siderite and calcite, respectively. Mg substitutes for Mn to a limited extent, up to about $Mg:(Mn,Fe) = 1:1.93$ (analysis 5), and a gap apparently exists in the binary series to magnesite. Zn substitutes for Mn up to at least $Zn:(Mn,Ca) = 1:1.24$ (analysis 6), and Co for Mn in amounts [9] up to at least several weight per cent CoO. Cd also may be present (analysis 3). Essentially pure $MnCO_3$ is rare, and either or both Fe'' and Ca are commonly present in amounts of several weight per cent or more.

Anal.[8]

	1	2	3	4	5	6	7	8	9
CaO		0.28	0.09	0.51		2.10	1.1	19.40	
MgO		0.33	1.08	tr.	12.98	0.05	4.5	3.18	
FeO		1.16	0.73	0.77	4.63	0.66	0.06	2.87	26.18
MnO	61.71	59.11	59.24	60.87	39.63	30.17	54.6	36.44	35.28
CO_2	38.29	37.89	38.36	38.26	41.63	36.60	39.74	38.11	37.98
Rem.			0.82	1.11		0.74	31.03		
Total	100.00	99.59	100.61	100.41	99.61	100.61	100.00	100.00	99.44
G.			3.71	3.70	3.57				3.722

1. $MnCO_3$. 2. John Reed mine, Alicante, Colo.[9] Rem. is insol. 3. Rákosbánya, Czechoslovakia.[10] Rem. is CdO 0.96, ZnO 0.15, PbO and CuO tr. 4. Ljubija, Yugoslavia.[11] 5. Hambach, Nassau, Germany.[12] Rem. is SiO_2. 6. Rosseto, Elba.[13] Rem. is ZnO. 7. Philipsburg, Montana.[14] 8. Kaso mine, Japan.[15] Recalculated to 100 after deducting insol. 12.86, SiO_2 4.37, Al_2O_3 1.08, H_2O 0.95. 9. Leadville, Colorado.[16]

Var.

Calcian. Manganocalcite pt. Contains Ca in substitution for Mn, a complete series extending to calcite. The G. and the indices of refraction decrease with increasing content of Ca. Highly calcian material is rare. Color usually pale pink or gray. See also kutnahorite.

Ferroan. Manganosiderite Bayer (1873). Ponite Butureanu (1912). Contains Fe'' on substitution for Mn, a complete series extending to siderite. The G. and indices of refraction increase with increasing content of Fe''. Color fawn or brownish in highly ferroan material.

Magnesian. Contains Mg in substitution for Mn up to at least $Mg:Mn = 1:1.24$.

Zincian. Contains Zn in substitution for Mn, with $Zn:Mn = 1:1.2$ in analysis 6. Zn usually is absent.

Tests. B.B., turns gray, brown, or black but is infusible. Soluble with effervescence in warm acids. On heating, begins to dissociate at about 300° with the formation of CO_2 and MnO, the latter later oxidizing to hausmannite.[19] Solubility in CO_2-free water at 25° = 0.065 g. $MnCO_3$ per liter; the solubility increases with the presence of CO_2.

Occur. Commonly found as a primary gangue mineral in moderate- to low-temperature hydrothermal veins, especially of silver, lead, zinc, and copper ores, where it is associated with calcite, siderite, dolomite, fluorite, barite, quartz, manganite, alabandite, tetrahedrite, sphalerite, etc. Also in high-temperature metasomatic (or metamorphic) deposits associated with rhodonite, friedelite, garnet, braunite, alabandite, hausmannite, tephroite. As a secondary mineral in residual deposits of iron or manganese oxides; also in manganese deposits of sedimentary origin (and in their metamorphosed equivalents); and as a late hydrothermal mineral in pegmatites, especially those carrying lithiophilite.

A well-known locality is at Kapnik, Roumania; also at Nagyág, Macskamezö, and Offenbánya. In the Ljubija district, Yugoslavia. Found in Germany at Freiberg, Saxony; in the Siegen district, Westphalia; at Elbingerode in the Harz; Bockenrod in the Odenwald, Hesse; near Horhausen in the Rhineland. In France, with rhodonite, friedelite at Vielle Aure, Hautes-Pyrénées, and at Las Cabesses near Saint-Girons, Ariège. At Moët-Fontaine in the Ardennes, Belgium. In western Merionethshire, England. In the Polunochny district in the northern Urals, and at Mazul near Achinsk in the Krasnoyarsk region, Siberia. In the Queluz district and at Miguel Burnier, Minas Geraes, Brazil, and at Casapalca, Peru. In Catamarca and la Rioja provinces, Argentina. In the United States, rhodochrosite occurs abundantly at Butte, Montana, also at Philipsburg. In Utah at Park City, Marysvale, Bingham, and other localities. In Colorado, finely crystallized rhodochrosite of a beautiful rose-red color is found in the John Reed and other mines at Alicante near Leadville in Lake County; also near Lake City, Hinsdale County, at Alma in Park County, the Rico district in Dolores County, Silverton in Ouray County, etc. At Austin, Tonopah, and other districts in Nevada. In California in the Tesla district, Alameda County, with bementite in southern Trinity County, at the Buckeye mine in Stanislaus County. Found at Batesville, Arkansas, and the Sevier district, Tennessee. In pegmatite at Poland, Maine. At Placentia Bay, Newfoundland.

A l t e r. On exposure rhodochrosite turns brown or black superficially and ultimately may alter completely to manganite or pyrolusite. Found as an alteration product of alabandite, and as incrustation pseudomorphs after calcite, dolomite, galena, and barite; also pseudomorphs of quartz after rhodochrosite.

A r t i f.[17] As microscopic crystals by heating $CaCO_3$ with $MnCl_2$ or $MnSO_4$ solution in a closed tube at 150°–200° or by heating precipitated $MnCO_3$ in carbonated water.

N a m e. From ῥόδον, *a rose*, and χρῶσις, *color*. Dialogite from διαλογή, *doubt*, from the early uncertainty as to composition.

Ref.

1. Angles of Sansoni, *Zs. Kr.*, **5**, 250 (1881), on Horhausen, Germany, crystals with 1.14 FeO; conventional unit, in which c is one-fourth that of the structure cell. Transformation: morphology to structure, 1000/0100/0010/0004. The crystal angles of rhodochrosite and siderite are identical within the ordinary limits of error; the variation in angle with substitution of Ca, Mg, Zn, etc. is not established.
2. Goldschmidt (**5**, 198, 1918); Koch, *Ann. Hist.-Nat. Mus. Nat. Hungar.*, **21**, 67 (1924). Rare or uncertain: 0772, 21$\bar{3}$4, 8.4.$\overline{12}$.1.
3. Wyckoff cited in Strukturber., I, 316 (1913–26).
4. Lacroix (**3**, 623, 1910).
5. Honess, *Am. J. Sc.*, **45**, 201 (1918).
6. Wayland, *Am. Min.*, **27**, 614 (1942), who gives a chart showing the variation in G. and indices of refraction with substitution of Fe, Ca, and Mg for Mn. Data on the

variation in properties with composition also are given by Fornaseri, *Rend. soc. min. ital.*, **1**, 60 (1941), Otto, *Min. Mitt.*, **47**, 89 (1935), Ford, *Trans. Conn. Acad. Arts Sc.*, **22**, 211 (1917), Gaubert, *Bull. soc. min.*, **42**, 97 (1919), Niggli, *Zs. Kr.*, **56**, 231 (1921).
  7. Veit, *Jb. Min., Beil.-Bd.*, **45**, 121 (1921).
  8. Additional analyses are summarized by Hintze (**1** [3A], 3216, 1927) and Wayland (1942); also Dolar-Mantuani, *Zs. Kr.*, **98**, 181 (1937), Ham and Oakes, *Econ. Geol.*, **39**, 412 (1944).
  9. Wherry and Larsen, *J. Washington Ac. Sc.*, **7**, 365 (1917).
  10. Zsivny, *Zs. Kr.*, **65**, 728 (1927).
  11. Barić and Tućan, *Ann. Geol. Pen. Balkan.*, Belgrade, **8**, 129 (1925).
  12. Manchot and Lorenz, *Zs. anorg. Chem.*, **134**, 297 (1924).
  13. Manasse, *Atti soc. tosc., Mem., Pisa*, **27**, 76 (1911).
  14. Fahey anal. cited in Wayland (1942).
  15. Yosimura, *J. Fac. Sc. Hokkaido Imp. Univ.*, **4**, 361 (1939).
  16. Mayo and O'Leary, *Am. Min.*, **19**, 304 (1934).
  17. Mellor (**12**, 432, 1932); Hintze (1927); Biltz, *Zs. anorg. Chem.*, **220**, 312 (1934).
  18. Kreutz, *Min. Mag.*, **15**, 232 (1909).
  19. Thermal analyses are described by Kulp, Wright, and Holmes, *Am. Min.*, **34**, 195 (1949).

14.1.1.5 **C O B A L T O C A L C I T E** [$CoCO_3$]. Sphäro-kobaltit *Weisbach (Berg.-Hütt. Jb.*, 53, 1877). Kobaltspath *Germ.* Sphérocobaltite *Fr.* Sphaerocobaltite.

C r y s t.[1] Hexagonal—R; hexagonal-scalenohedral—$\bar{3}\ 2/m$.

$a:c = 1:0.81;\quad \alpha\ 103°21';\quad p_0:r_0 = 0.94:1;\quad \lambda\ 72°32'$

Forms:[2]

c	0001	111	$m\cdot$	$40\bar{4}1$	$3\bar{1}\bar{1}$	w	$07\bar{7}3$	10.10.11
m	$10\bar{1}0$	$2\bar{1}\bar{1}$	$\delta\cdot$	$01\bar{1}2$	110	$\Xi\cdot$	$05\bar{5}1$	$22\bar{3}$
$p\cdot$	$10\bar{1}1$	100	$\phi\cdot$	$02\bar{2}1$	$11\bar{1}$	$\pi\cdot$	$08\bar{8}1$	$33\bar{5}$

Structure cell.[3] Space group $R\bar{3}c$. a_{rh} 5.72, $\alpha\ 48°14'$; a_0 4.67 kX, c_0 15.13; $a_0:c_0 = 1:3.24$. Cell contents $Co_2(CO_3)_2$ in the rhombohedral unit.

Habit. Crystals rare. As small spherical masses, with a crystalline surface and concentric and radiated structure; as crusts.

P h y s. Presumably cleavable on $\{10\bar{1}1\}$. H. 4. G. 4.13; 4.11 (calc.). Luster vitreous. Color rose-red, altering superficially to gray, brown or velvet-black. Streak peach-blossom-red. Translucent to subtranslucent.

O p t.[4]

	n (Boleo)	DICHROISM	
O	1.855 ± 0.005	Violet-red	Uniaxial negative $(-)$.
E	$1.60\ \pm 0.01$	Rose-red	

Chem. Cobalt carbonate, $CoCO_3$. Small amounts of Ni, Fe″, Ca, and Cu (?) occur in substitution for Co.

Anal.

	CoO	NiO	FeO	CaO	CuO	CO_2	Rem.	Total
1.	63.00					37.00		100.00
2.	58.86			1.80		34.65	4.63	99.94
3.	59.68		0.90	0.18	2.87	[36.12]	0.25	[100.00]
4.	55.72	1.21		3.12	1.95	35.87	1.57	99.44

1. $CoCO_3$. 2. Schneeberg.[5] Rem. is Fe_2O_3 3.41, H_2O 1.22. Oxidized ferroan material? 3. Libiola, Italy.[6] Rem. is H_2O. 4. Valle del Neva, Italy.[7] Rem. is Fe_2O_3 0.27, H_2O 1.30, MnO and MgO tr.

Tests. B.B. and in C.T. becomes black. Slowly soluble in cold HCl; rapidly soluble with effervescence in hot acids.

O c c u r. Originally found at Schneeberg, Saxony, with roselite, erythrite, and annabergite in the Co-Ni veins. From Libiola near Casarza and Valle del Neva, in Liguria, Italy. In dolomite at the Étoile du Congo mine and elsewhere in the Katanga district, Belgian Congo,[9] and in the Jervois Range in Central Australia. From Boleo, Lower California.

A l t e r. To stainierite.

A r t i f.[8] As microscopic rhombohedral crystals by heating $CoCl_2$ with carbonates at 150° in a closed tube. Several different hydrates of $CoCO_3$ are known.

N a m e. Sphaerocobaltite in allusion to the spheroidal shape and the composition. This name is unsatisfactory because of the suggested relation to cobaltite and the use of the mode of aggregation as a species rather than as a varietal designation and is here replaced by the name cobaltocalcite.

Ref.

1. Elements of the structure cell of Baccaredda, *Acc. Linc., Rend.*, **16**, 248 (1932), but with *c* quartered to conform to the morphological unit used for the calcite group.
2. Pelloux, *Ann. museo civico stor. nat. Genoa*, **52**, 269 (1927), who states that the angles are close to those of siderite.
3. Baccaredda (1932) [analysis 4]. Ferrari and Colla, *Acc. Linc., Rend.*, **10**, 594 (1929), obtained a cell with a_{rh} 5.91 kX, α 103°21' on artificial material.
4. Larsen (135, 1921).
5. Weisbach (1877).
6. Ferro, *Atti soc. sc. Genoa*, **10**, 264 (1899).
7. Baccaredda (1932).
8. de Sénarmont, *Ann. Chem. Pharm.*, **80**, 216 (1851); *C.R.*, **32**, 409 (1851); and Mellor (**14**, 808, 1935).
9. Buttgenbach, *Les Min. de Belgique et du Congo belge*, Liége, 236 (1947).

REMINGTONITE. Booth (*Am. J. Sc.*, **14**, 48, 1852).

Supposedly a cobalt carbonate found as earthy, rose-colored incrustations on serpentine at a copper mine at Finksburg, Carroll County, Maryland. Apparently a serpentine-like mineral stained with cobalt and not a valid species.[1]

Ref.

1. Shannon, *Am. Min.*, **9**, 208, (1924).

14.1.1.6 **S M I T H S O N I T E** [$ZnCO_3$]. Calamine pt. Galmei pt. Zincum acido aëro mineralisatum *Bergmann* (144, 1782; **2**, 209, 1780). Zinc carbonaté *Brongniart* (47, 1827). Zinkspath, Kohlengalmei *Germ.* Carbonate of Zinc. Smithsonite *Beudant* (**2**, 354, 1832). Zinkspath, Kapnit (or Capnit), Zinkischer Carbonspat *Breithaupt* (241, 236, 1841). Calamine *Miller* (589, 1852); *Dana* (211, 1837; 263, 1844). Smithsonite *Dana* (447, 1854). Zinkglas. Dry-bone *U. S. Miners term.* Bonamite (Sterrett, *U. S. Geol. Sur. Min. Res.*, **2**, 805, 1908). Orthorhombic Zinc Carbonate (?) *Griffiths* and *Dreyfus* (*Chem. News*, **54**, 67, 1886).

Monheimite *Kenngott* (23, 1853). Zinkeisenspath, Eisenzinkspath *Germ.* Cadmiumzinkspat *Blum* (1858). Herrerite *Del Rio* (*Am. J. Sc.*, **18**, 193, 1830).

C r y s t.[1] Hexagonal—R; hexagonal-scalenohedral—$\bar{3}\ 2/m$.

$a:c = 1:0.8063;$ $\alpha\ 103°28';$ $p_0:r_0 = 0.9311:1;$ $\lambda\ 72°20'$

Smithsonite

Forms:[2]

			φ	ρ = C	A_1	A_2
c	0001	111	.....	0°00′	90°00′	90°00′
m	10$\bar{1}$0	2$\bar{1}\bar{1}$	30°00′	90 00	30 00	90 00
a	11$\bar{2}$0	10$\bar{1}$	0 00	90 00	60 00	60 00
r	10$\bar{1}$1	100	30 00	42 57½	53 49½	90 00
e	01$\bar{1}$2	110	−30 00	24 58	90 00	68 34
f	02$\bar{2}$1	11$\bar{1}$	−30 00	61 46	90 00	40 16
v	21$\bar{3}$1	20$\bar{1}$	10 53½	67 54½	52 39½	72 20½

Structure cell.[3] Space group $R\bar{3}c$. a_{rh} 5.669, α 48°26′; a_0 4.65 kX, c_0 14.95; $a_0:c_0 = 1:3.215$. Cell contents $Zn_2(CO_3)_2$ in the rhombohedral unit.

Habit. Rarely well-crystallized: rhombohedral {10$\bar{1}$1} or less commonly {02$\bar{2}$1}, the faces generally curved and rough or composite; rarely scalenohedral. Ordinarily botryoidal, reniform, or stalactitic; as crystalline incrustations; coarsely granular to compact massive; earthy and friable.

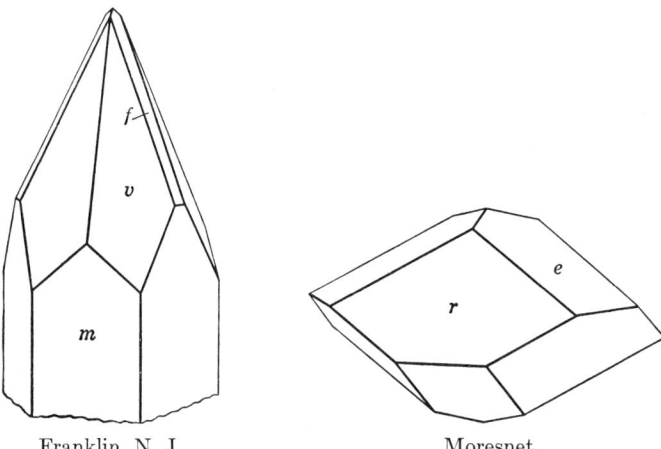

Franklin, N. J.   Moresnet.

As porous to cellular or cavernous masses, often with drusy surfaces in the cavities (*dry-bone*).

Twinning. Not observed.

Oriented growths. Upon calcite, with parallel axes.[4] Oriented pseudomorphs of ZnO are formed by thermal dissociation.[5]

Etch figures. Conform to hexagonal-scalenohedral symmetry.[7]

P h y s. Cleavage {10$\bar{1}$1}, not quite perfect. Fracture uneven to imperfectly conchoidal. Brittle. H. 4–4½. G. 4.43 ± 0.01 for pure $ZnCO_3$, 4.43 (calc.); the G. varies widely, 4.0–4.45, with the composition and mode of aggregation. Luster vitreous, sometimes pearly on crystal faces. Color often grayish white to dark gray, greenish or brownish white; also green to apple-green, bluish green, blue, bluish gray, yellow, brown, white; rarely colorless and transparent. Usually translucent. Streak white. Translation gliding[8] with $T\{0001\}$, $t[10\bar{1}0]$. Some specimens fluoresce greenish or bluish white in ultraviolet radiation.

O p t. In transmitted light, colorless or very faintly tinted. Indices of refraction:

BROKEN HILL, RHODESIA (analysis 2)[9]

λ	Li	C	Na	Tl	F	G
nO	1.842	1.843	1.848	1.855	1.862	1.874
nE	1.619	1.619	1.621	1.624	1.627	1.633

The material of analysis 3 has nE 1.6176.

C h e m. Carbonate of zinc, $ZnCO_3$. Fe'' is often present in substitution for Zn; also, less frequently and in smaller amounts, Co, Cu, Mn'', Ca, Cd, Mg, Pb. The limits of these substitutions are discussed below under *Var.* Essentially pure $ZnCO_3$ is rare. Ge and In have been found in traces.[11]

Anal.[10]

	1	2	3	4	5	6
CaO		0.27	0.35	0.44		0.12
MgO		0.45	0.04		7.22	0.22
CuO				3.48	1.65	
FeO		0.58		tr.	0.33	0.59
MnO		0.01		tr.	3.36	
PbO				tr.		
CdO				0.16		2.70
ZnO	64.90	63.18	64.55	60.97	39.02	62.06
CO_2	35.10	35.35	35.25	35.12	36.94	33.90
Rem.		0.04	0.07		11.65	0.40
Total	100.00	99.88	100.26	100.17	100.17	99.99
G.	4.43	4.398		4.41	3.874	

1. $ZnCO_3$. 2. Broken Hill, Rhodesia.[13] Rem. is H_2O −. 3. Aachen, Germany.[14] Rem. is insol. 4. Kelly, New Mexico.[15] 5. Boleo, Lower California.[16] Rem. is CoO 10.25, H_2O 1.29, Cl 0.11. 6. Laurium, Greece.[16] Rem. is Al_2O_3 0.02, S 0.19, SiO_2 0.19.

	7	8	9	10	11
CaO	tr.	0.94	12.74	1.27	2.9
MgO		1.86			
CuO	6.09				
FeO	tr.	1.98	17.40	22.61	33.0
MnO		9.25		2.14	1.4
PbO	0.98				
CdO	0.50				
ZnO	57.99	48.30	39.15	36.27	26.3
CO_2	34.22	35.83	30.36	35.80	37.3
Rem.		0.76	0.35	0.41	
Total	99.78	98.92	100.00	98.50	100.9
G.	4.40	3.98			4.00

7. Zyryanovsky mine, Altai, U.S.S.R.[17] 8. Herrenberg, Germany.[18] Rem. is SiO_2 0.20, H_2O 0.56. 9. Marien mine, Silesia.[19] Rem. is SiO_2. 10. Altenberg, Germany.[20] Rem. is hemimorphite. 11. Zincian siderite. Altenberg, Germany.[20] Original sum given as 100.24.

Var. *Ferroan.* Monheimite *Kenngott* (1853). Zinkeisenspath, Eisenzinkspath *Germ.* Capnit *Breithaupt* (1841). Contains Fe'' in substitution for Zn. The series extends up to at least Fe:(Zn, etc.) = 1:1.59 (analysis

10) and possibly to Fe:(Zn, etc.) = 1:0.85 (analysis 11). There is apparently a gap in the binary series between this material and siderite. Amounts of FeO up to about 4 weight per cent are commonly present in smithsonite.

Calcian. Material with Ca:(Zn, etc.) = 1:3.18 (analysis 9) has been reported, but Ca is ordinarily present in amounts of about 2 weight per cent CaO or less.

Cobaltian. Material with Co:(Zn, etc.) = 1:5.3 (analysis 5) has been described from Boleo but Co is usually absent. Color pink.

Cuprian. Herrerite *Del Rio* (1830). Contains Cu in substitution for Zn up to at least Cu:Zn = 1:9.3 (analysis 7). Color apple-green to dark green and bluish green.

Manganoan. Contains Mn″ in substitution for Zn up to at least Mn:(Zn, etc.) = 1:5.2 (analysis 8). Mn usually is present in amounts of only a few weight per cent or is absent. The occurrence of a zincian rhodochrosite [6] with Zn:Mn = 1:1.2 suggests that this series may be complete.

Cadmian. Material with Cd:Zn = 1:36 (analysis 6) has been described from Laurium, but Cd is ordinarily absent or present in amounts less than about one weight per cent CdO. Color sometimes bright yellow (but then due to admixed greenockite) [12] and constituting in part the "turkey-fat" ore of miners.

Magnesian. Mg substitutes for Zn with Mg:(Zn, etc.) = 1:3.8 in the material of analysis 5, but Mg ordinarily is absent or present in amounts of less than about one weight per cent.

Plumbian. Pb occasionally is found in substitution for Zn in amounts up to about one weight per cent PbO.

Tests. B.B. infusible. Soluble in acids with effervescence.

O c c u r. A secondary mineral found chiefly in the oxidized zone of ore deposits, or as a replacement of calcareous rocks adjacent thereto, and derived by the alteration of primary zinc minerals, especially sphalerite. Often associated with hemimorphite; also with cerussite, malachite, azurite, anglesite, pyromorphite, mimetite, aurichalcite, willemite, hydrozincite, etc. Smithsonite has not been described as a primary hydrothermal vein mineral although it appears that it would be stable under such conditions. A few of the many known localities are mentioned below.

At Laurium, Greece, abundantly in a variety of colors. Large deposits occur in Silesia, especially at Tarnowitz and Beuthen, and in the district about Aachen in the Rhineland, Germany, including Altenberg, Schmalgraf, and Moresnet in Belgium; also at Wiesloch, Baden, near Iserlohn in Westphalia, and at Bleiberg in Carinthia, Austria. Abundant in the lead-zinc deposits of Sardinia, Italy, as in the Iglesias district, and in Santander and Asturias provinces in Spain. In France in the region of Alais, Gard, and at St. Laurent du Minier in Hérault. In Africa found at numerous places in Algeria and Tunis; also from the Broken Hill mine in Rhodesia and in fine specimens from Tsumeb in South West Africa.

In the United States, smithsonite is found in Pennsylvania at Friedensville, Lehigh County, and at Bamford in Lancaster County. In Wisconsin at Mineral Point, Iowa County, and at Shullsberg and elsewhere in Lafayette County. In Missouri, especially in the Granby and Joplin districts; in the Dubuque district, Iowa; at Batesville in Searcy County and in Marion County, Arkansas. At Austinville, Virginia, and in numerous small deposits in Tennessee, Kentucky, Illinois. As large deposits at Monarch and Leadville, Colorado; in the Tintic and the Ophir districts, Utah; in the Elkhorn district, Montana. Large masses of a fine green color are found at Kelly, Magdalena County, New Mexico; also in the Fierro-Hanover district. At Goodsprings, Nevada, and in the Wood River district, Blaine County, Idaho. In the Cerro Gordo district, Inyo County, California.

A l t e r. Incrustation and substitution pseudomorphs after calcite crystals are very common. Also found pseudomorphous after dolomite, anglesite, and by incrustation after galena and sphalerite. Pseudomorphs after smithsonite include quartz, limonite, pyrolusite, and tarbuttite. Occasionally found as the petrifying material of fossil shells (after calcite).

A r t i f.[21] Obtained as microscopic crystals by reaction of $CaCO_3$ with $ZnCl_2$ or $ZnSO_4$ solution, especially when heated in a closed tube; also by heating metallic zinc with carbonated water in a closed tube, and by long exposure (17 years) of a cleavage fragment of calcite in a solution of $ZnSO_4$ at room temperature.

N a m e. After James Smithson (1754–1829), who founded the Smithsonian Institution in Washington. The name calamine formerly was used for the species in England.

Ref.

1. Fundamental angle of Wollaston [orig. ref. not known]; conventional unit, with c one-fourth of the unit of the structure cell. Transformation: morphology to structure, 1000/0100/0010/0004. The variation in angle with composition is not established.
2. Goldschmidt (**9**, 117, 1923); Palache, *Am. Min.*, **13**, 321 (1928); Buttgenbach, *Bull. soc. min.*, **29**, 190 (1906). Rare or uncertain: $40\bar{4}5$, $40\bar{4}1$, $01\bar{1}5$, $0.17.\overline{17}.7$, $07\bar{7}2$, $05\bar{5}1$.
3. Goldschmidt and Hauptmann, *Nach. Ges. Göttingen, Math.-phys. Kl.*, 53 (1932) on Tsumeb crystals.
4. Bergt, *Isis*, **1**, 20 (1903).
5. Rose, *C.R.*, **208**, 1914 (1939); Mehmet and Valensi, *Bull. soc. chim. France*, **2**, 1295 (1935); Rose, *Bull. soc. min.*, **71**, 15 (1948).
6. Manasse, *Mem. Soc. Tosc.*, **27**, 76 (1911).
7. Honess, *Am. J. Sc.*, **45**, 217 (1918).
8. Johnsen and Veit, *Cbl. Min.*, 265 (1918).
9. Mountain, *Min. Mag.*, **21**, 51 (1926) by prism method and here rounded to three figures. Palache (1928) and Neuhaus, *Ber. Freiberger Geol. Ges.*, no. 20, 39 (1944) cite additional data.
10. Additional analyses are tabulated by Hintze (**1** [3A], 3243, 1927) and Doelter (**1**, 443, 1911).
11. Müller, *Ind. Eng. Chem.*, **16**, 604 (1924); Tanner, *Chem. News*, **30**, 141 (1874).
12. Schaller and Fairchild, *Am. Min.*, **23**, 894 (1938).
13. Mountain (1926).
14. Ortloff, *Zs. phys. Chem.*, **19**, 214 (1896).
15. Headden, *Am. Min.*, **10**, 18 (1925).
16. Christomanos, *C.R.*, **123**, 62 (1896).
17. Pilipenko, *Bull. Imp. Tomsk Univ.*, 763 (1915).

18. Monheim, *J. pr. Chem.*, **49**, 382 (1850).
19. Gellhorn in Kenngott (42, 1853).
20. Monheim, *Jb. Min.*, 705 (1851).
21. Mellor (4, 642, 1923), Hintze (1927), Hüttig *et al.*, *Ak. Wien, Sitzber.*, **147**, Abt. II, 107 (1938).

14.1.1.7   **O T A V I T E** [$CdCO_3$] *Schneider* (*Cbl. Min.*, 388, 1906).

C r y s t. Hexagonal—R; hexagonal-scalenohedral—$\bar{3}\,2/m$. Found as crusts of minute rhombohedral crystals, $\{10\bar{1}1\}$. Artificial material [1] has $a:c = 1:0.8363$ and a_{rh} 6.11, α 47°24'; a_0 4.91 kX, c_0 16.24; $a_0:c_0 = 1:3.309$. Cell contents $Cd_2(CO_3)_2$ in the rhombohedral unit. Luster brilliant, adamantine. Color white to yellow-brown and reddish. Not brittle. G. 4.96 (artif.);[2] 5.03 (calc.). Chemically,[3] cadmium carbonate, $CdCO_3$, with CdO 74.5, CO_2 25.5, total 100.0. From Tsumeb, near Otavi, South West Africa, as a secondary mineral associated with smithsonite (in part as oriented growths thereon), cerussite, olivenite, pyromorphite, azurite, malachite.

Ref.

1. The minute, unmeasurable natural crystals, originally supposed to be a basic cadmium carbonate, were shown to be identical with artificial $CdCO_3$ and to be isostructural with calcite by Ramdohr and Strunz, *Zbl. Min.*, 97 (1941A). Morphological elements of de Schulten, *Bull. soc. min.*, **20**, 196 (1897), cell dimensions of Zachariasen, *Norske Vidensk.-Akad. Skr. Oslo*, no. 4 (1928).
2. de Schulten (1897); the morphological unit has c one-fourth the structure cell.
3. A partial analysis by Wölbling, in Schneider (1906), gave 70.2 per cent CdO.

14.1.2   **V A T E R I T E** [$CaCO_3$]. Vaterite pt., Vater's third modification of $CaCO_3$ pt., *Meigen* (*Verh. Ges. deutsch. Naturfor., Ärzte, Königsberg*, **2**, 124, 1911). μ-$CaCO_3$ *Johnston, Merwin*, and *Williamson* (*Am. J. Sc.*, **41**, 473, 1916). Vaterite-B *Gibson, Wyckoff*, and *Merwin* (*Am. J. Sc.*, **10**, 325, 1925). μ-Calcite.

Hexagonal. Artificial preparations are microscopic hexagonal plates or lens-shaped skeletal crystals resembling snowflakes; also spherulitic with a radial fibrous structure. G. 2.645 (calc.). Optically [1] uniaxial positive (+), with nO 1.550, nE 1.640–1.650.

Structure cell.[2] Hexagonal. a_0 4.120 kX, c_0 8.556; $a_0:c_0 = 1:2.077$. Cell contents $Ca_2(CO_3)_2$.

C h e m. Calcium carbonate, $CaCO_3$. Polymorphous with calcite and aragonite. When boiled in water, vaterite is converted to aragonite or to calcite, and, when boiled in NaCl solution, to calcite. Dry crystals are stable in air but convert to calcite when heated to about 440°.

A r t i f.[3] Obtained as microscopic crystals or as spherulitic growths, usually admixed with calcite or aragonite, by allowing a gelatinous precipitate of calcium carbonate to crystallize at 5° in the presence of a large excess of potassium carbonate solution, by slowly adding a 0.1 molar solution of calcium chloride into potassium carbonate solution, and in other ways.

O c c u r. Vaterite has not yet been identified in nature but may occur in pisolites or certain biogenic deposits, as in the nacreous matter of living shells. The name vaterite, here used in place of μ-$CaCO_3$ and vaterite-B, in the earlier literature of the subject [4] included a type of supposedly polymorphous $CaCO_3$ (*vaterite-A*) later found to be fine-grained or spherulitic calcite or aragonite.

Ref.

1. See Johnston, *et al.* (1916); Gibson, *et al.* (1925); Yoshimura, *Jap. J. Geol. Geog.*, **7**, 3 (1930); Donnay, *Soc. géol. Belgique, Bull.*, **59**, 215 (1936).
2. Heide, *Cbl. Min.*, 198 (1925).
3. See Gibson, *et al.* (1925), Lucas, *Bull. soc. min.*, **70**, 185 (1948), Stolkowski, *C.R.*, **225**, 312 (1947), Faivre, *C.R.*, **222**, 140 (1946).
4. For summary see Hintze (**1** [3B], 2883, 1926), Doelter (**1**, 113, 1911), Gibson, *et al.* (1925).

14.1.3 ARAGONITE GROUP

ORTHORHOMBIC; DIPYRAMIDAL—$2/m\ 2/m\ 2/m$

	a_0	b_0	c_0
Aragonite, $CaCO_3$	4.94 kX	7.94	5.72
Witherite, $BaCO_3$	5.252	8.828	6.544
Strontianite, $SrCO_3$	5.118	8.404	6.082
Cerussite, $PbCO_3$	5.173	8.480	6.130
Alstonite, $CaBa(CO_3)_2$	4.99	8.77	6.11

The members of the aragonite group are orthorhombic in crystallization. The cations are arranged in the structure [1] nearly in the manner of hexagonal closest-packing, giving rise to marked pseudo-hexagonal symmetry in the crystals as shown by their angles and by the mimetic twinning characteristic of the members of the group. $CaCO_3$ is polymorphous, since the Ca ion is of such size as to place it in the critical range of the radius ratio between the calcite and aragonite structure types, but calcite-type structures are not formed by the carbonates of the relatively large Sr, Ba, and Pb ions. There is, however, a limited range of solubility of Sr, Ba, and Pb in calcite. Only partial substitutional solid solutions extend between the members of the aragonite group at ordinary temperatures, although a complete series extends between $SrCO_3$ and $BaCO_3$ at elevated temperatures.

Alstonite, $CaBa(CO_3)_2$, appears to be a superstructure in the aragonite structure type based on a 1:1 ordering of the cations and is then analogous to the position of dolomite in the calcite-magnesite series. Barytocalcite is a monoclinic polymorph of alstonite.

Ref.

1. Bragg, W. L., *Proc. Roy. Soc. London*, **105A**, 16 (1924).

14.1.3.1 **A R A G O N I T E** [$CaCO_3$]. Spath calcaire crist. en prismes hexagones dont les deux bouts sont striés du centre à la circonference, id. dont les deux bouts sont lisses [from Spain] *Davila* (**2**, 50, 52, 1767). Arragonischer Apatit *Werner* (*Bergm. J.*, **1**, 95, 1788), *Klaproth* (*ibid.*, **1**, 299, 1788; *Crell's Ann.*, **1**, 387, 1788) [making it carbonate of lime]. Arragonischer Kalkspath *Werner* (*Bergm. J.*, **2**, 74, 1790) [after Klaproth's analysis]. Spath calcaire des limites entre l'Aragon et Valence en Espagne

ARAGONITE

Born (**2**, 320, 1790). Arragonit *Emmerling* (**2**, 684, 1802). Spathum prismaticum in igne lucem spargens *Gmelin* (*Linnaei Syst. Nat.*, 13th ed., **3**, 92, 1793). Arragon Spar [var. of Calc Spar] *Kirwan* (**1**, 87, 1794). Arragonit *Werner* (*Estner's Min.*, **2**, 1039, 1796). Excentrischer Kalkstein *Karsten* (34, 74, 1800). Arragonite [first made distinct from Calc Spar through crystallography] *Haüy* (**2**, 1801), *Brochant* (**1**, 576, 1800). Iglit [from Iglo, Transylvania] *Esmark* (*Bergm. J.*, **3**, 99, 1798). Igloit. Nadelstein *Lenz*. Erbsenstein pt., Faserkalk pt., Schallenkalk pt., Sprudelstein *Germ*. Pisolite. Chimborazite *Clarke* (*Ann. Phil.*, **2**, 57, 147, 1821). Stalactites Flos Ferri, Marmoreus ramulosus *Linnaeus* (183, 1768). Stalagmites coralloides *Wallerius* (**2**, 388, 1778). Coralloidal Aragonite. Chaux carbonatée coralloides *Haüy* (**2**, 1801). Eisenblüthe pt. *Werner*. Flos Ferri. Flowers of Iron. Zeiringit *Pantz* (*Leonh. Tasch.*, **5**, 372, 1811). Tarnovizit *Breithaupt* (252, 1841). Tarnovicit *Haidinger* (1845). Tarnowitzite. Mossottite *Luca* (*Nuovo Cimento*, **7**, 453, 1858). Oserskit *Breithaupt* (*B.-H. Ztg.*, **17**, 54, 1858). Erzbergit Hatle (*Mitt. naturw. Ver. Steiermark*, 1892). Plumbo-aragonite *Collie* (*J. Chem. Soc. London*, **55**, 95, 1889). Ktypeïte *Lacroix* (*C.R.*, **126**, 602, 1898). Ctypeite. Conchit Kelly (*Ak. München, Sitzber.*, **30**, 187, 1900). Pelagosit *Moser* (*Min. Mitt.*, **1**, 174, 1878). Nicholsonite *Butler* (*Econ. Geol.*, **8**, 8, 1913).

Cryst. Orthorhombic; dipyramidal—$2/m\ 2/m\ 2/m$.

$a:b:c = 0.6224:1:0.7206;$ [1] $p_0:q_0:r_0 = 1.1578:0.7206:1$

$q_1:r_1:p_1 = 0.6224:0.8637:1;$ $r_2:p_2:q_2 = 1.3877:1.6067:1$

Forms: [2]

		ϕ	$\rho = C$	ϕ_1	$\rho_1 = A$	ϕ_2	$\rho_2 = B$
c	001	....	0°00′	0°00′	90°00′	90°00′	90°00′
b	010	0°00′	90 00	90 00	90 00	....	0 00
m	110	58 06	90 00	90 00	31 54	0 00	58 06
x	012	0 00	19 49	19 49	90 00	90 00	70 11
k	011	0 00	35 46½	35 46½	90 00	90 00	54 13½
i	021	0 00	55 14½	55 14½	90 00	90 00	34 45½
v	031	0 00	65 10½	65 10½	90 00	90 00	24 49½
h	041	0 00	70 52	70 52	90 00	90 00	19 08
e	051	0 00	74 29½	74 29½	90 00	90 00	15 30½
q	061	0 00	76 58½	76 58½	90 00	90 00	13 01½
λ	091	0 00	81 14	81 14	90 00	90 00	8 46
F	0.11.1	0 00	82 48½	82 48½	90 00	90 00	7 11½
ρ	0.20.1	0 00	86 02	86 02	90 00	90 00	3 58
η	0.24.1	0 00	86 41½	86 41½	90 00	90 00	3 18½
u	101	90 00	49 11	0 00	40 49	40 49	90 00
o	112	58 06	34 17½	19 49	61 25½	59 56	72 41
p	111	58 06	53 45	35 44½	46 47½	40 49	64 46½
ι	661	58 06	83 02	76 59½	32 34½	8 11½	58 22
σ	991	58 06	85 20½	81 14	32 12	5 29	58 13
π	24.24.1	58 06	88 15	86 41½	31 56½	2 03½	58 07
n	122	38 46½	42 45	35 46½	64 50½	59 56	58 03
E	132	28 10½	50 48	47 13½	68 32½	59 56	46 54½
s	121	38 46½	61 35½	55 14½	56 34½	40 49	46 42½

Less common:

A 350	β 0.13.2	O 0.18.1	d 102	g: 21.21.1	S 231
M 570	χ 671	P 0.19.1	f 201	Γ 158	v: 271
α 540	ν 081	Q 0.21.1	ω 13.13.2	H 125	Z 572
B 970	F: 0.17.2	R 0.45.2	ψ 771	h: 133	L 341
N: 430	J 0.10.1	T 0.26.1	γ 881	t 243	w: 431
D 530	j 0.12.1	U 0.27.1	Θ 10.10.1	Δ 151	
Ξ 073	ε 0.13.1	V 0.30.1	J: 11.11.1	G: 413	
Ω 052	ϑ 0.14.1	W 0.32.1	G 13.13.1	λ: 342	
C 072	μ 0.16.1	X 0.35.1	δ 14.14.1	s: 352	
N 092	K 0.17.1	Y 0.40.1	Ψ 20.20.1	Σ 362	

Structure cell.[3] Space group $Pmcn$. a_0 4.94 kX, b_0 7.94, c_0 5.72; $a_0:b_0:c_0 = 0.622:1:0.720$. Cell contents $Ca_4(CO_3)_4$.

Habit. Untwinned crystals are very rare. Usually short to long prismatic [001], and sometimes flattened somewhat on {010}. Also acicular or chisel-shaped and characterized by the presence of steep domes or pyramids; less frequently pyramidal, or thick tabular {001} with {110} and {010}. Pseudo-hexagonal symmetry is often marked. Sometimes in columnar aggregates and crusts composed of straight or divergent fibers; as radiating or stellate groups of acicular crystals. In coralloidal, reniform, or globular shapes; pisolitic, with a radial fibrous and concentrically zoned structure; stalactitic; as laminated, fibrous crusts.

Twinning.[4] Twin plane {110}, extremely common. Usually repeated with composition plane {110} resulting in pseudohexagonal aggregates both of the contact and penetration types. Also as thin polysynthetic lamellae producing fine striations parallel [001] or a sixfold, feather-like series of striations on {001}; lamellar twinning often is revealed by optical or etch tests on seemingly untwinned crystals. Twinning has been doubtfully observed on {103}.[5]

Oriented growths. A mutual orientation has been reported in certain calcite pseudomorphs after aragonite.[6] Also in aragonite pseudomorphs after gypsum, with aragonite {010}[001] parallel gypsum {010}[001].[7]

Etch figures. The etch symmetry conforms to the orthorhombic dipyramidal class, but anomalous figures have been obtained with optically active solvents.[8]

Phys. Cleavage {010} distinct; also {110} and {011} barely noticeable. Fracture subconchoidal. Brittle. H. $3\frac{1}{2}$–4. G. 2.947 ± 0.002;[9] 2.944 (calc.). The G. increases with increasing substitution of Pb to about 3.0 in material with ca. 5 weight per cent PbO. The apparent G. may be greatly diminished in fibrous and massive material. Luster vitreous, inclining to resinous on surfaces of fracture. Colorless to white; also gray, yellowish, blue, bluish green, green, pale to deep violet, rose-red. Sometimes exhibits an hourglass-like coloration or fibrosity subjacent to {001}.[10] Streak uncolored. Transparent to translucent. Translation gliding [11] with T{010}, t[100]; also twin gliding [12] with K_1(110), $K_2(1\bar{3}0)$, η_2 310. Elastic coefficients (in 10^{-13} cm.^{2}/dyne):[13] s_{11} 6.97, s_{22} 13.2, s_{33} 12.2, s_{44} 24.3,

ARAGONITE

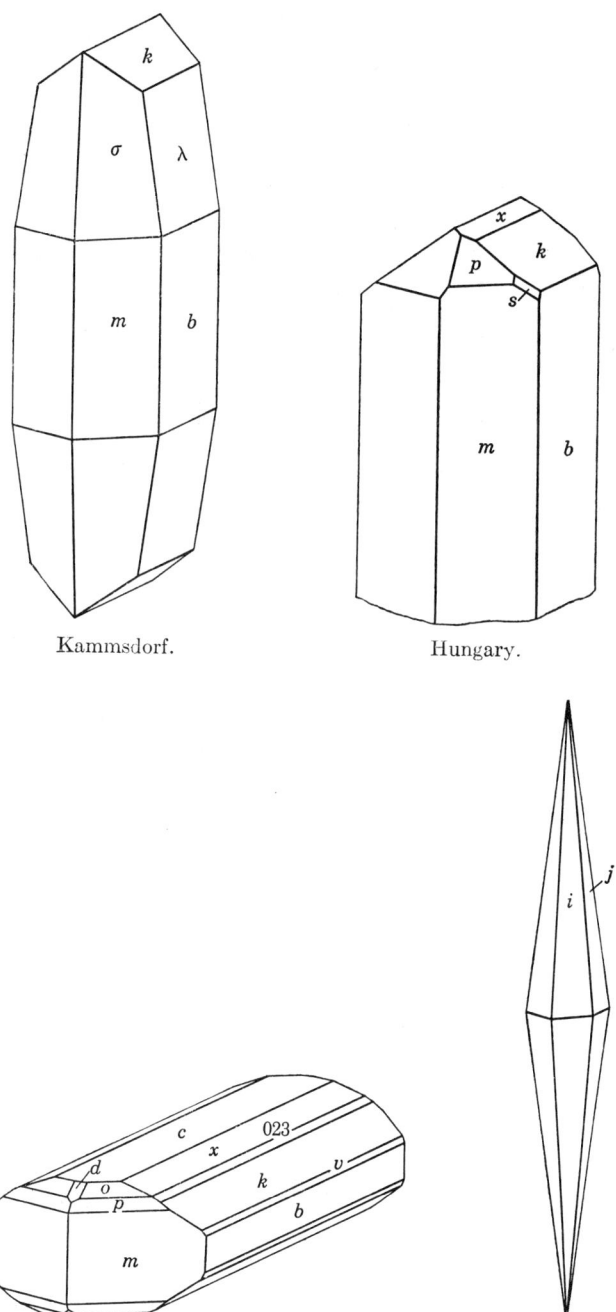

Kammsdorf.

Hungary.

Vesuvius.

Framont

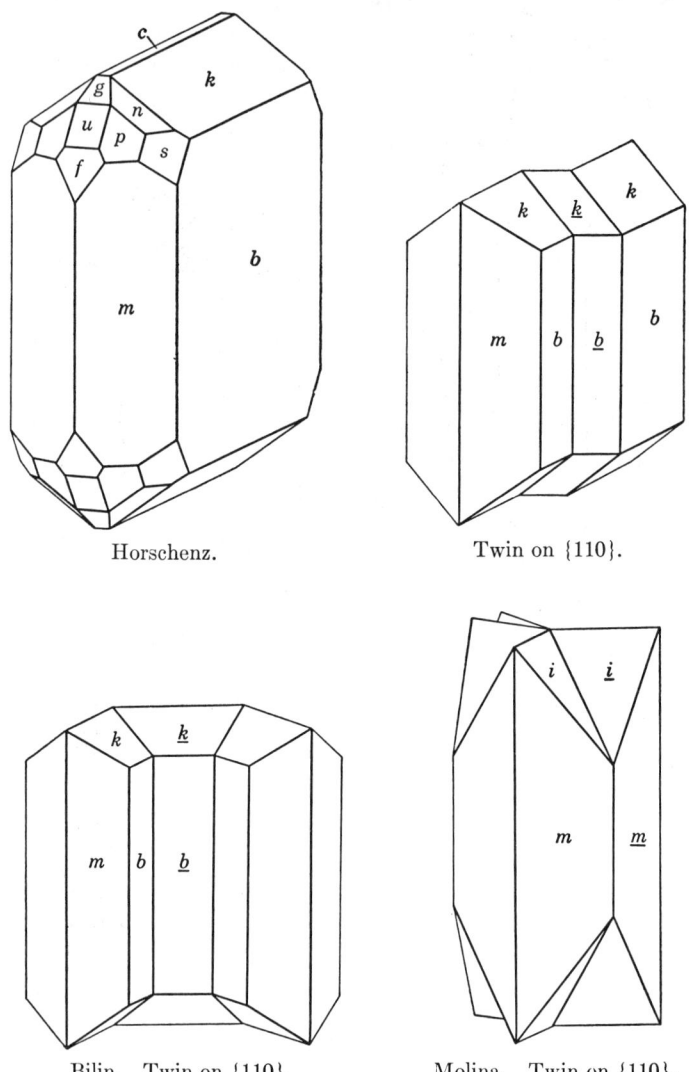

Horschenz.

Twin on {110}.

Bilin. Twin on {110}.

Molina. Twin on {110}.

s_{55} 39.0, s_{66} 23.5, s_{23} −2.37, s_{31} +0.43, s_{12} −3.04. Linear thermal expansion [14] at 450°: [100] 0.38, [010] 0.86, [001] 1.66 per cent. Aragonite is fluorescent [30] under excitation by electron beams, x-rays, and ultraviolet light, the intensity and color varying among different specimens; usually pale rose, yellow or rarely bluish, with a greenish phosphorescence in long-wave ultraviolet, and yellowish in short-wave ultraviolet. Also thermoluminescent. Aragonite has absorption maxima in the infrared [31] at approximately 6.5 μ, 11.5μ, and 14.5μ.

O p t. In transmitted light, colorless. Basal sections between crossed nicols commonly exhibit a sectoral distribution of twinned individuals. Conical refraction is clearly seen in thick, suitably oriented sections.[15] Indices of refraction:[16]

ORIENTATION		(687,B)	(656,C)	(589,D)	(527,E)	(486,F)	
nX	c	1.5273	1.5279	1.5300	1.5325	1.5346	Biaxial negative ($-$).
nY	a	1.6758	1.6772	1.6810	1.6858	1.6900	$r < v$, weak.
nZ	b	1.6801	1.6815	1.6854	1.6904	1.6947	
$2E$ (calc.)		30°19′	30°23′	30°39′	30°57′	31°14′	
$2V$ (calc.)		17°57′	17°59′	18°05′	18°13′	18°20′	

INDICES FOR $\lambda = 587.57$ AT DIFFERENT TEMPERATURES [17]

T:	20°	120°	220°	320°
nX	1.5306	1.5288	1.5276	1.5263
nY	1.6807	1.6783	1.6758	1.6731
nZ	1.6852	1.6826	1.6798	1.6768
$2V$ (calc.)	18°15′	17°55$\frac{1}{2}$′	17°24$\frac{1}{2}$′	16°49$\frac{1}{2}$′

The variation in optical properties with composition is not clearly defined. Plumbian material from Tsumeb (analysis 5) [18] has for $\lambda = 579$: nX 1.5397, nY 1.6950, nZ 1.7026, and $2V$ 23°06′ (calc.). A strontian aragonite from New Zealand [19] (analysis 7) has nX 1.527, nZ 1.676, $2V$ very small. Data are lacking on zincian aragonite.

C h e m. Calcium carbonate, $CaCO_3$. Polymorphous with calcite and vaterite. Aragonite is unstable relative to calcite at atmospheric pressure at all temperatures down to ordinary, but it may be the stable form below roughly $-100°$ or when foreign elements are present in solid solution.[28] On heating [20] in dry air aragonite begins to invert to calcite at about 400°, the rate of inversion increasing with temperature. The transformation is much more rapid in contact with water or solutions containing dissolved $CaCO_3$ and may then take place as low as room temperature.

Aragonite exhibits a considerable range of compositional variation. Sr is often present in substitution for Ca in small amounts, ordinarily of the order of a few weight per cent of SrO, with a maximum substitution in natural material in the neighborhood of Sr:Ca = 1:25 (anal. 7).[29] Pb also is found to substitute for Ca. The amount of substitution usually is in the range from 2 to 6 weight per cent PbO, but the series apparently extends in natural material up to at least Pb:Ca = 1:12 (anal. 6). Zn has been reported [32] in substitution for Ca in amounts up to 10 weight per cent ZnO (?) but this variety is rare. Ba apparently does not substitute in analytically significant amounts for Ca. Mg has been reported in small amounts in a few analyses,[21] in at least one instance due to admixture, and the validity of this substitution is uncertain. CuO, FeO, Fe_2O_3, As_2O_3, P_2O_5, and SiO_2 have been reported in small amounts in some analyses and are probably due to admixture.

Anal.[22]

	1	2	3	4	5	6	7	8
CaO	56.03	54.81	55.96	54.67	51.96	45.77	52.30	52.80
SrO		0.30					3.87	
PbO				0.67	5.23	15.08		
ZnO				0.89				3.07
CO_2	43.97	44.22	43.95	43.57	42.24	38.98	42.62	43.06
Rem.		0.96	0.16			0.20	1.42	0.53
Total	100.00	100.29	100.07	99.80	99.43	100.03	100.21	99.46
G.	2.944		2.936	3.13	3.015	3.15	2.91	

1. $CaCO_3$. 2. Sarajevo, Yugoslavia.[23] Rem. is $(Al,Fe)_2O_3$ 0.32, MgO 0.12, H_2O 0.22, insol. 0.30, Na_2O tr. Fibrous hot-spring deposit. 3. Matsushiro, Japan.[24] Rem. is MgO 0.03, insol. 0.13. 4. Tarnowitz, Silesia.[25] 5. Plumbian aragonite, Tsumeb, South West Africa.[26] 6. Plumbian aragonite. Postenje, Serbia.[27] Rem. is insol. 7. Strontian aragonite. Otago, New Zealand.[29] Rem. is SiO_2 and insol. with BaO tr. 8. Friendensville, Pennsylvania.[41] Rem. is insol.

Tests. B.B., whitens and falls to pieces (calcite), but does not fuse. Varieties containing Sr impart a more intensely red color to the flame than does lime. Easily soluble with effervescence in dilute acids.

Powdered aragonite when boiled with cobalt nitrate solution takes on a lilac-rose tint due to precipitation of a basic cobalt carbonate (the so-called Meigen reaction).[33] The reaction takes place much more rapidly than with calcite and has been used as a means of distinguishing these species. The distinction can be emphasized by further treatment with ammonium sulfide solution, aragonite becoming black due to the formation of cobalt sulfide and calcite turning only slightly gray.[34] A differential staining also can be effected by treating the powdered grains with dilute solutions first of silver nitrate and then of potassium chromate, the aragonite grains becoming coated with orange-red silver chromate.[35] A more sensitive and reliable test [36] is to place the carbonate in a solution containing manganous sulfate and silver sulfate. Aragonite becomes distinctly gray in a minute or so and soon thereafter turns black; calcite reacts very much more slowly, becoming only slightly gray in an hour or so. The test can be applied to thin sections under the microscope. The order of reactivity of the common carbonates to this test is: aragonite, strontianite, witherite, smithsonite, cerussite, dolomite, calcite, siderite, magnesite. Staining tests in general are not reliable with material that is very fine-grained or that contains organic matter.

Var. 1. *Ordinary.* Varieties based on the several modes of aggregation of the mineral are not here recognized. Found variously in simple or twinned crystals, the latter much the more common. Also coralloidal, in groupings of delicate interlacing and coalescing stems of a snow-white color and looking a little like coral; often called *Flos-ferri* (Eisenblüte, *Germ.*) Sometimes stalactitic or stalagmitic, either compact or fibrous in structure; as snow-white scaly masses pseudomorphous after gypsum; columnar massive, and resembling onyx; pisolitic (Erbsenstein, *Germ.*); etc.

2. *Plumbian.* Tarnowitzite. Tarnovizit *Breithaupt* (252, 1841). Tarnovicit *Haidinger* (1845). Plumbo-aragonite *Collie* (*J. Chem. Soc. London*, **55**, 95, 1889). Contains Pb and sometimes also Zn in substitution for Ca. The series extends up to at least Pb:Ca = 1:12, but highly plumbian material is rare. The crystals generally are zoned, and may be developed complexly morphologically with steep vicinal forms. Increase in content of Pb is accompanied by an increase in G. and in indices of refraction. Found particularly in the oxidized zone of ore deposits and then associated

with cerussite, malachite, etc., as at Tarnowitz, Silesia, and Tsumeb, Africa.

3. *Strontian.* Zeiringit Pantz (*Leonh. Tasch.*, **5,** 372, 1811). Mossottite Luca (*Nuovo Cimento,* **7,** 453, 1858). Sr is often present in small amounts in substitution for Ca, particularly in material from hot springs and cave deposits and, in part, from occurrences in gypsum and clay. The maximum range of substitution is uncertain, apparently near Sr:Ca = 1:25. The correlation between content of Sr and the gravity, optical properties, and crystallography is not well-defined.

4. *Zincian.* Nicholsonite Butler (*Econ. Geol.*, **8,** 8, 1913). Contains Zn in substitution for Ca. The variety is rare and not well-defined.

O c c u r. Aragonite is much less widespread and abundant than calcite. It is formed under a much narrower range of physico-chemical conditions than calcite and being metastable relative to calcite is often converted to that species accompanying changes in environment. In all of its occurrences aragonite is a low-temperature, near-surface deposit. It occurs principally: (1) as crystals, pisolites, or sinter deposits from hot springs and geysers and in their channelways, and as stalactites or helictites in caves in calcareous rocks; in these occurrences the depositing waters commonly are found to contain sulfates or Sr, Mg salts in solution. Also as oolites in recent sea-bottom deposits; (2) as disseminated crystals or masses in gypsum and clay beds; (3) with limonite and siderite in deposits of iron ores, often as coralloidal aggregates; (4) in the oxidized zone of ore deposits associated with limonite, calcite, malachite, smithsonite, cerussite, and other secondary metallic minerals; (5) with calcite, dolomite, hydromagnesite, brucite, and other magnesium minerals in veins and cavities in serpentine and altered basic igneous rock, and with calcite and sometimes zeolites in basalts and other flow rocks; (6) with celestite in sulfur deposits of the Sicilian type; (7) as the nacreous layer or skeletal part of living or fossil lower organisms, particularly those of recent origin. Aragonite, in part with calcite, also constitutes the substance of pearls.[37] Brief mention of some of the more important localities follows.

Aragonite occurs in Germany in cavities in basaltic rocks in the Kaiserstuhl, Baden; as pseudomorphs after gypsum (*schaumkalk*) at Wiederstädt, Eisleben, Ilfeld and elsewhere in the Harz; at Kamsdorf, near Saalfeld, Thuringia; a well-crystallized plumbian variety (*tarnowitzite*) occurs with lead ores in dolomite at Tarnowitz, Silesia; also in fine crystals at Werfen and Leogang in Salzburg. As a recent deposit in hot springs at Rohitsch-Sauerbrunn, Styria; also found abundantly as beautiful coralloidal masses (*flos ferri; eisenblüte*) and as crystals at Erzberg near Eisenerz and in other siderite deposits in Styria, and found similarly at Hüttenberg-Lölling and elsewhere in Carinthia. Pisolites and fine-fibrous sinter deposits are found in the hot springs at Carsbad, Bohemia (ktypeïte, pt.).[38] In fine crystals in veins in basalt from the Spitzberg near Horschenz (usually given as Bilin), Bohemia. In Hungary in good crystals at Dognácska and Herrengrund, and in cavities in basalt at Korlát and Fülek, Com. Nógrád.

Groups of large and fine crystals in part altered to calcite are found associated with celestite and sulfur at Racalmuto, Girgenti, and Cianciana in Sicily. On Monte Somma, Italy, in erratic blocks of leucite-tephrite with apatite and zeolites.

In France [39] found in deposits of gypsum, rock salt, and clay in the region of Dax, Pouillon, and Bastennes in Landes, and similarly in the region of Bayonne-Briscous in Basses-Pyrenées; the aragonite forms tabular to stout prismatic, pseudohexagonal crystals and the occurrences are like those of Aragon, Spain. Also found in veins and crevices in volcanic rocks and as hot spring deposits in the Auvergne, Puy-de-Dôme. Aragonite originally was recognized in specimens from Aragon, Spain, where it is found abundantly at Molina and numerous near-by localities in Guadalajara province. The mineral occurs as pseudohexagonal prisms and tablets embedded in a reddish gypsum-bearing clay. In England at Alston Moor, Cleator Moor, and Frizington in Cumberland; a plumbian variety occurs at Leadhills, Scotland. Both plumbian and zincian varieties are found well-crystallized at Tsumeb, South West Africa, associated with secondary lead and zinc minerals. Found in Japan as cone-shaped deposits at geyser and hot spring orifices at Kuriyama, Yuzawa, Shimotsuke province; [40] as pisolitic or cauliflower-like masses in pools of sulfurous hot-spring waters on the Takasegawa river, Shinano province; as nodular aggregates in clay and gypsum at Matsushiro, Iwami province. A strontian variety occurs in a vein cutting schist near Alexandra, Central Otago, New Zealand.

In the United States, aragonite occurs as fibrous crusts with hydromagnesite on serpentine at Hoboken, New Jersey; sparingly as coralloidal masses with gypsum in cavities in dolomite rock at Lockport, New York; at Lancaster and Wood's chrome mine, Lancaster County, Pennsylvania. Fine crystals associated with calcite and cerussite were found in the Magdalena district, Socorro County, New Mexico; also found as hexagonal tablets near Nara Visa, Quay County, and near Lake Arthur, Chaves County; as coralloidal masses in the Organ district, Dona Ana County. Fine coralloidal masses were found in limestone caves at Bisbee, Cochise County, Arizona; also as stalactitic masses in the Turquoise district, Dragoon Mountains, Cochise County, and the Hiltano district, Empire Mountains, Pima County; at Dubuque, Iowa, and Mine-la-Motte, Missouri. Near Fort Collins, Larimer County, Colorado, in clay; with zeolites in basalt on Table Mountain, near Golden, Clear Creek County; a zincian variety occurs with secondary zinc ores at Leadville, Colorado (*nicholsonite*), and has been reported from the Tintic district, Utah. In California, concretions of aragonite occur with gypsum in sand at Kettleman Hills, Kern County; also with opal in lava altered by hot springs at Sulfur Banks, Lake County. As hexagonal tablets in Big Horn County, Wyoming.

Alter. Aragonite very commonly alters to calcite, and complete pseudomorphs (paramorphs) after crystals of aragonite are often found.[42] Such specimens have been found, among other localities, at Vesuvius;

Aragon, Spain; Baroda, Kansas; and notably at Fort Collins, Larimer County, Colorado. Crystals from the latter locality are hexagonal tablets (twins) on {001}, sometimes with smaller crystals irregularly intergrown toward the center to give a rosette-like appearance, and range up to an inch in thickness and three inches across; they may be further altered into quartz. More or less hollow incrustation pseudomorphs of calcite after aragonite, thin septa of calcite often marking the composition surfaces of the twins, have been found at Girgenti and other localities in Sicily, at Herrengrund, Hungary, in Cumberland, England, and elsewhere. In these instances there has been solution and redeposition of $CaCO_3$ and not a direct structural inversion. Another but less common type of alteration is that of gypsum to aragonite, the pearly white, opaque and somewhat porous pseudomorphs (Schaumkalk, *Germ.*) being composed of minute, oriented scales and plates of aragonite.[43] Paramorphs of aragonite after calcite have been reported but are very doubtful.[44] Fine pseudomorphs of copper after aragonite have been found at Corocoro, Peru;[45] also pseudomorphs of deweylite after aragonite, dolomite after aragonite,[49] and, doubtfully, of aragonite after celestite.

A r t i f.[46] Aragonite has been synthesized from solutions, usually by metathical reaction of calcium salts with alkali carbonates. Calcite also may be obtained, the crystallization of aragonite being favored by the presence of small amounts of Ba, Sr, Mg, or Pb salts or of calcium sulfate in the solution and by relatively high temperatures—over about 30° to 70°—with calcite forming in the lower range. Rapid precipitation and relatively high concentrations of reactants also tend to favor aragonite. The influence of Sr, Mg, calcium sulfates, and other cosolutes in inducing the formation of aragonite is well illustrated by the natural occurrence and association of the mineral.

N a m e. From the original locality in Aragon province, Spain. Attention was first directed toward the Spanish mineral in the latter part of the eighteenth century. After being confused with apatite it was recognized as an individual substance by Werner in 1790 on the basis of an analysis by Klaproth proving it carbonate of lime. The distinction between it and calcite was not clearly recognized until Haüy showed it to differ in crystallography, gravity, and cleavage from that species, although he did not explain the identity in composition. The differences were ascribed by Stromeyer in 1813 on the basis of analyses to a small content of impurities (strontium), which he considered could influence the crystallization of a substance with which they were admixed. Strontium later was found to be not necessarily present, but the structural nature of the relation to calcite did not become apparent until the discovery of polymorphism by Mitscherlich in 1822. The transformation of aragonite to calcite on heating was first recognized in 1827 by Haidinger.

Conchite, from κονχη, *shell*, was a name given to a supposed new polymorph of $CaCO_3$ occurring in certain shells later shown[47] to be aragonite. Ktypeïte, from κτυπέω, in allusion to the marked decrepitation on heating, is a name given to the substance of certain fibrous pisolites in the belief that it represented a new polymorph but which is very probably identical with aragonite.[48]

Ref.
1. Koksharov (**6**, 261, 1870) on Bilin material, in very close agreement with the measurements of most other observers. The relation between the crystallography and compositional variation in aragonite is not well-defined. A plumbian aragonite (analysis 5) from Tsumeb described by O'Daniel, Zs. Kr., **74**, 333 (1930) had $a:b:c$ = 0.6212:1:0.7188; a_0 4.97 kX, b_0 8.06, c_0 5.72; $a_0:b_0:c_0$ = 0.617:1:0.710; G. 3.015; and indices as given under **O p t.** See also Stevanović, Ann. géol. pénin. Balkan., **7**, 85 (1922)—Min. Abs., **2**, 116 (1923), and Seifert, Zs. Kr., **100**, 120 (1938). The significance to be attached to data on compositionally variant aragonite is rendered uncertain by the zoning present apparently invariably in the crystals, successive zones differing markedly in composition.
2. Goldschmidt (**1**, 90, 1913). The supposedly new forms reported by Shannon, U. S. Nat. Mus., Bull. *131*, 237 (1926) are due to an incorrect orientation; transformation: Shannon to Koksharov, 1$\bar{1}$0/310/002. On the relation between morphology and structure see Rottenbach, Inaug. Diss., Bonn (1937), and Chudoba and Rottenbach, Zbl. Min., 261 (1938); on form statistics and recent bibliography see Kleber, Jb. Min., Beil.-Bd., **75**, 465 (1940); also Barca, Mus. Nac. Cien. Nat. Trab. Madrid, Ser. Geol., no. 24 (1919) on Spanish aragonite; Buttgenbach, Ann. Soc. géol. Belgique, **42**, 93 (1919); Vendl, Hist. Nat. Mus. Natl. Hung., **24**, 223 (1926); Obenauer, Jb. Min., Beil.-Bd., **64**, 437 (1931); Reichert, Földt. Közl., **62**, 196 (1932); Heritsch, Zbl. Min., 33 (1936); Tokody, Hist. Nat. Mus. Natl. Hung., **31**, 171 (1937); Yugovics, Földt. Közl., **71**, 23 (1941).

Rare, uncertain or vicinal forms:

100	50.59.0	0.11.10	0.23.7	0.60.1	6.7.21	512	16.17.3
1.28.0	50.57.0	076	0.24.7	304	2.19.7	8.11.3	40.42.7
1.19.0	9.10.0	0.11.9	0.11.3	114	123	11.15.4	24.33.4
1.15.0	13.14.0	043	0.13.3	332	215	14.20.5	12.17.2
1.14.0	20.21.0	0.11.8	0.19.4	331	255	28.40.9	48.49.7
1.10.0	25.24.0	032	0.16.3	441	9.2.16	10.12.3	25.27.2
190	17.16.0	053	0.15.2	551	2.60.3	11.14.3	16.22.1
160	760	0.19.11	0.19.2	21.21.2	425	15.11.4	24.25.1
120	25.21.0	0.11.6	0.37.2	12.12.1	212	19.27.5	
13.19.0	650	0.15.8	0.22.1	45.45.2	25.27.24	461	
11.15.0	850	0.19.10	0.23.1	30.30.1	312	9.12.2	
25.34.0	0.1.12	0.42.19	0.25.1	45.45.1	17.25.8	32.35.7	
25.32.0	013	0.23.10	0.29.1	126	9.13.4	23.31.5	
450	023	0.23.9	0.42.1	3.2.12	7.10.3	104.112.21	
9.11.0	0.20.19	0.16.5	0.48.1	124	12.17.5	561	

3. Bragg, Roy. Soc. London, Proc., **105**, 16 (1924); Wyckoff, Am. J. Sc., **9**, 145 (1925). See also ref. 1.
4. See in particular de Sénarmont, Ann. chim. phys., **41** [3], 61 (1854); Schrauf, Ak. Wien, Sitzber., **62** [2], 734 (1870); Leydolt, Ak. Wien, Sitzber., **19**, 10 (1856).
5. Bauer, Jb. Min., I, 72 (1886); Stevanović (1922).
6. See Mügge, Jb. Min., Beil.-Bd., **16**, 379 (1903); the calcite in such instances is found to be unoriented.
7. Köhler, Chemie Erde, **6**, 257 (1931); Rose, Ann. Phys., **97**, 161 (1856); Mügge (1903, p. 406).
8. Kleber, Jb. Min., Beil.-Bd., **75**, 465 (1940); Krüger, Ak. Leipzig, Sitzber., **74**, 253 (1922); Beckenkamp, Zs. Kr., **14**, 375 (1888); Leydolt (1856); Tomkeieff, Min. Mag., **20**, 408, 1925).
9. From best values reported in the literature.
10. See Beckenkamp, Zs. Kr., **32**, 25 (1900).
11. Veit, Jb. Min., Beil.-Bd., **45**, 121 (1922).
12. Mügge, Jb. Min., Beil.-Bd., **14**, 246 (1901).
13. Voigt, Lehrb. der Kristallphys., Berlin (1928).
14. Kôzu and Kani, Proc. Imp. Ac. Tokyo, **10**, 222 (1934).

15. Melmore, *Nature*, **150**, 382 (1942); Kalkowsky, *Zs. Kr.*, **9**, 497 (1884).
16. Indices measured by total reflection method by Mülheim, *Zs. Kr.*, **14**, 229 (1888) on an unanalyzed but probably pure crystal from Bilin; the data are here rounded off to four decimal places. Closely agreeing values are given by Offret, *Bull. soc. min.*, **13**, 582 (1890); see also Yamaguchi, *J. Geol. Soc. Tokyo*, **34**, 159, (1927); Hintze (**1** [3A], 2974, 1926). On the relation between optical properties and crystal structure see Bragg, *Proc. Royal Soc. London*, **105A**, 370 (1924).
17. Wünscher in Hintze (1926).
18. O'Daniel, *Zs. Kr.*, **74**, 333 (1930). See also Siegl, *Min. Mitt.*, **48**, 286 (1936).
19. Hutton, *Trans. Royal Soc. New Zealand*, **66**, 35 (1936).
20. Shoji, *Zs. Kr.*, **84**, 74 (1932); Leitmeier and Feigl, *Min. Mitt.*, **45**, 447 (1934); Kleber, *Jb. Min., Beil.-Bd.*, **75**, 465 (1940); Faust, *Am. Min.*, **35**, 207 (1950).
21. See Traube, Inaug. Diss., Greifswald (1884); Breithaupt, *B. H. Ztg.*, **24**, 319 (1865); Jeremejev, *Bull. Acad. Imp. Sc. St. Petersburg*, **7**, no. 1 (1897).
22. For additional analyses see Hintze (**1** [3A], 3019, 1926); Doelter (**1**, 337, 1911).
23. Gagarin, *Ann. géol. pénin. Balkan*, **13**, 72 (1936).
24. Yamaguchi (1927).
25. Traube, *Zs. Kr.*, **15**, 410 (1889).
26. O'Daniel (1930).
27. Stevanović (1922).
28. The relations between aragonite and calcite are discussed fully by Johnson, Merwin, and Williamson, *Am. J. Sc.*, **41**, 473 (1916); see further under **A r t i f**.
29. Hutton (1936). Up to 2.9 per cent SrO was reported in old analyses of aragonite from Molina, Spain, by Stromeyer, *Ann. Phys.* (1813)—cited in Hintze (1926, p. 3020), and 4.69 per cent SrO was reported in material from Gerfalco, Tuscany, by Luca, *Nuovo Cimento*, **7**, 453 (1858), but the value of these particular observations is questionable. Numerous other analyses, however, show SrO up to about 2.0 per cent.
30. Frondel, priv. comm. (1947); also Haberlandt, *Chem. Erde*, **13**, 212 (1940); Krejci-Graff, *Zs. Kr.*, **88**, 260 (1934); Köhler and Leitmeier, *Zs. Kr.*, **87**, 146 (1934); Keilhack, *Zs. deutsche geol. Ges.*, **50**, 131 (1898).
31. Schäfer and Schubert, *Ann. Phys.*, **50**, 283 (1916).
32. Zincian aragonite from Leadville, Colorado, is said by Butler, *Econ. Geol.*, **8**, 8 (1913) to contain up to 10 per cent "zinc" but analyses are not cited; see also analyses 4 and 8.
33. Meigen, *Cbl. Min.*, 577 (1901); Wyrouboff, *Bull. soc. min.*, **24**, 371 (1901); Panebianco, *Riv. min.*, **28**, 5 (1902); for a critical discussion see Johnston, Merwin, and Williamson, *Am. J. Sc.*, **41**, 473 (1916); Bray, *J. Roy. Soc. New South Wales*, **78**, 113 (1945); Hugi, *Schweiz. min. Mitt.*, **25**, 114 (1945).
34. Quercigh, *Acc. Linc., Rend.*, **24**, [5], 1231 (1915).
35. Thugutt, *Cbl. Min.*, 786 (1910); Lemberg, *Zs. deutsche geol. Ges.*, **44**, 232 (1892).
36. Leitmeier and Feigl, *Min. Mitt.*, **45**, 447 (1934).
37. See Alexander, *Am. J. Sc.*, **238**, 366 (1940); *Science*, **93**, 110 (1941); *Gemmologist*, **10**, 93 (1941).
38. For descriptions see Vater, *Zs. Kr.*, **35**, 149 (1902), and Lacroix, *C.R.*, **126**, 602 (1898).
39. See Lacroix (**3**, 670, 1909).
40. Watanabe, *Beitr. Min. Japan*, no. 5, 237 (1915).
41. Roepper analysis in Genth, *Second Geol. Sur., Penn. Rpt.*, 163 (1875).
42. For descriptions see Rose, *Ak. Berlin, Abh.* 64 (1856); *Ann. Phys.*, **91**, 147 (1853); Lasaulx, *Jb. Min.*, 505 (1879); Scharff, *Jb. Min.*, 31 (1861); Bauer, *Jb. Min.*, I, 79 (1886).
43. See Rose, *Ann. Phys.*, **97**, 161 (1856); Geinitz, *Jb. Min.*, 449 (1876); Vater, *Zs. Kr.*, **31**, 575 (1899), Wetzel, *Jb. Min.*, II, 70 (1910), and reference 7.
44. See Sandberger, *Ann. Phys.*, **129**, 472 (1866); Rose, *Ak. Berlin, Abh.* 66 (1856); Bauer, *Jb. Min.*, I, 12 (1890).
45. Domeyko, *Ann. mines*, **18** [7], 531 (1881); Forbes, *Quart. J. Geol. Soc.*, **17**, 45 (1862).
46. For reviews of the considerable literature on the synthesis of aragonite see Hintze (1926, p. 2986); Doelter (**1**, 346, 1911); Mellor (**3**, 817, 1923); Johnston, Merwin, and Williamson, *Am. J. Sc.*, **41**, 473 (1916); also for recent observations Reitemeier and Buehrer, *J. Phys. Chem.*, **44**, 535, 552 (1940); Fonda, *J. Phys. Chem.*, **44**, 435 (1940); Copisarow, *J. Chem. Soc. London, Trans.*, **123**, 785 (1923).
47. Brauns, *Cbl. Min.*, 134 (1901); Vater, *Zs. Kr.*, **35**, 149, (1902).
48. See Johnston, Merwin, and Williamson (1916); Vater (1902); Zavaritzky, *Doklady Ac. Sc. U.S.S.R.*, **63**, 725 (1948).
49. Andrews and Schaller, *Am. Min.*, **27**, 135 (1942).

14.1.3.2 **W I T H E R I T E** [BaCO₃]. Terra ponderosa aerata *Withering* (29, 1783; *Phil. Trans.*, 293, 1784). Witherit *Werner* (*Bergm. J.*, **2**, 225, 1790). Aerated Barytes *Watt* (*Mem. Manchester Soc.*, **3**, 599, 1790). Barolite *Kirwan* (**1**, 134, 1794). Kohlensaurer Baryt *Germ.* Baryte carbonatée *Fr.* Sulphato-Carbonate of Baryta *Thomson* (1836).

C r y s t. Orthorhombic; dipyramidal—$2/m\ 2/m\ 2/m$.

$a:b:c = 0.6032:1:0.7302$; [1] $p_0:q_0:r_0 = 1.2105:0.7302:1$

$q_1:r_1:p_1 = 0.6032:0.8261:1$; $r_2:p_2:q_2 = 1.3695:1.6578:1$

Forms: [2]

	ϕ	$\rho = C$	ϕ_1	$\rho_1 = A$	ϕ_2	$\rho_2 = B$
c 001	.....	0°00′	0°00′	90°00′	90°00′	90°00′
b 010	0°00′	90 00	90 00	90 00	.....	0 00
m 110	58 54	90 00	90 00	31 06	0 00	58 54
x 012	0 00	20 03½	20 03½	90 00	90 00	69 56½
k 011	0 00	36 08	36 08	90 00	90 00	53 52
i 021	0 00	55 36	55 36	90 00	90 00	34 24
l 031	0 00	65 27½	65 27½	90 00	90 00	24 32½
n 041	0 00	71 06	71 06	90 00	90 00	18 54
F 114	58 54	19 28	10 21	73 25	73 09½	80 05½
o 112	58 54	35 15	20 03½	60 22½	58 49	72 39
p 111	58 54	54 43½	36 08	45 39	39 33½	65 03½

Structure cell.[3] Space group *Pmcn*. a_0 5.252 kX, b_0 8.828, c_0 6.544; $a_0:b_0:c_0 = 0.595:1:0.741$. Cell contents $Ba_4(CO_3)_4$.

Habit. Crystals always repetitively twinned on {110} yielding pseudohexagonal dipyramids; also short prismatic [001] or tabular to lenticular

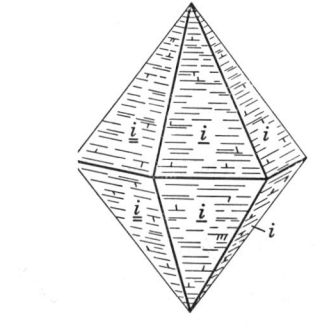

New Brancepeth, Durham.

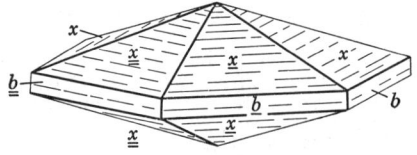

New Brancepeth, Durham.

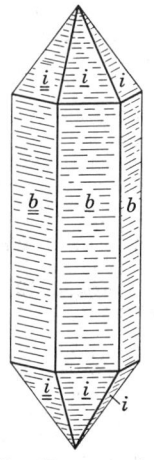

New Brancepeth, Durham.

WITHERITE

with a convex base. The morphology often is complex and obscure. Faces usually rough and horizontally striated. Also in globular, tuberose, and botryoidal forms; structure columnar, granular, or coarse fibrous.

Twinning. Twin plane {110} universal.

Oriented growths.[4] Barite upon witherite, with barite [010]{102} parallel witherite [100]{011} and {031}; also otherwise oriented.

Phys. Cleavage {010} distinct, {110} imperfect, probably also on {110} and {112}.[4] Fracture uneven. H. 3-3$\frac{1}{2}$. G. 4.291 ± 0.002;[5] 4.29 (calc.). Luster vitreous, inclining to resinous on fracture surfaces. Colorless to milky, white, or grayish; also weakly tinted yellow, brown, or green. Streak white. Transparent to translucent. Fluorescent and phosphorescent in x-rays, electron beams, and ultraviolet light; also thermoluminescent.[6]

Opt. In transmitted light, colorless.

ORIENTATION		n(Na)[7]	
X	c	1.529	Biaxial negative (−).
Y	b	1.676	2V 16°, 2E 26$\frac{1}{2}$°.
Z	a	1.677	$r > v$, very weak.

Chem. Barium carbonate, $BaCO_3$. BaO 77.70, CO_2 22.30, total 100.00. Inverts on heating under CO_2 pressure to a hexagonal (?) and then isometric modification.[8] Sr and Ca are sometimes present in substitution for Ba, in amounts less than 1 weight per cent oxide. Modern analyses on pure material are few.[9] A complete series with $SrCO_3$ and a partial series up to Ca:Ba ∼ 1:2 exist in artificial preparations.[10]

Tests. B.B. fuses at 2. Soluble in dilute HCl.

Occur. Witherite is the most common barium mineral after barite, but it is not of frequent occurrence. It occurs in low-temperature hydrothermal veins associated with barite and galena. Witherite occurs in Austria at Leogang in Salzburg and at Peggau north of Graz, Styria. In Germany at the Himmelsfürst mine, Freiberg, and the Maximilian mine at Andreasberg, in the Harz, Saxony. At Příbram, Bohemia. In France at Château Thinières, Beaulieu, Cantal, with barite in a lead vein; at Lamothe in coal measures and at Brioude, Haute-Loire. Important commercial deposits are found in the north of England, where the mineral occurs principally with barite, together with lesser amounts of calcite, galena, and other sulfides, and, rarely, alstonite and barytocalcite, in veins traversing Carboniferous rocks. Here found particularly at Alston Moor, Cumberland (the original locality); at the Settlingstones mine near Fourstones and at Fallowfield near Hexham in Northumberland; at the Morrison mine, Annfield Plain, Durham. As radial aggregates at the Tsubaki silver mine, Ugo province, Japan. In Russia in barite veins of the Arkhyz and Djalankol deposits in the North Caucasus, and in the Kopets Mountains, Karakala region, Turkmenia.

In the United States, witherite has been found associated with barite near Lexington, Kentucky, and filling cavities in limestone at Many Glacier National Park, Montana. As large crystals from Rosiclare, Illinois. With barite in veins near El Portal, Mariposa County, and on Beegum Creek, near Platina, Shasta County, California. As a gangue mineral in lead ores in the Castle Dome district, Yuma County, Arizona. Also reported from a silver-bearing vein in Gillies township, Thunder Bay district, Ontario, Canada.

A r t i f.[11] In crystals by fusion of precipitated $BaCO_3$ in KCl or NaCl, by heating $BaCO_3$ with ammonium carbonate or chloride and water in a closed tube at 150°–180° and by long digestion of the precipitate formed by reaction of $KHCO_3$ and $BaCl_2$.

N a m e. After William Withering (1741–1799), English physician and mineralogist, who first called attention to the mineral.

Ref.

1. Des Cloizeaux (**2**, 75, 1874).
2. Goldschmidt (**9**, 80, 1923). Rare or uncertain: 130, 014, 118, 332, 221.
3. Wilson, *Phys. Rev.*, **31**, 305 (1928), by powder method on artificial $BaCO_3$. See also Colby and LaCoste, *Zs. Kr.*, **90**, 1 (1935).
4. Mügge, *Jb. Min., Beil.-Bd.*, **16**, 399 (1903).
5. From value 4.289 of Madelung and Fuchs, *Ann. Phys.*, **65**, 289 (1921), and new determinations by Mrose, priv. comm., (1947), by microbalance; some reported values are slightly lower, whereas Groth (**2**, 209, 1908) and Dana (284, 1892) give the ranges 4.29–4.35 and 4.28–4.37.
6. On the luminescence phenomena see Jackson, *J. Chem. Soc. London*, **65**, 734 (1894), Bary, *C.R.*, **130**, 776 (1900), Kunz and Baskerville, *Science*, 769 (1903), Köhler and Leitmeier, *Zs. Kr.*, **87**, 146 (1934). Different specimens fluoresce deep rose or dull orange in both long- and short-wave ultraviolet (Frondel, priv. comm., 1947).
7. Mallard, *Bull. soc. min.*, **18**, 8 (1895).
8. Boeke, *Mitt. naturfor. Ges. Halle*, **3**, 1 (1913), *Zs. anorg. Chem.*, **50**, 244 (1906); Samojloff, *Cbl. Min.*, 161 (1915); see also Hackspill and Wolf, *C.R.*, **204**, 1820 (1937).
9. For existing analyses see Hintze (**1** [3A], 3045, 1926); also Szebellédy, Inaug. Diss., Budapest (1926)—*Min. Abs.*, **3**, 261 (1927); Sidorenko, *C.r. ac. sc. U.R.S.S.*, **55**, 149 (1947).
10. Cork and Gerhard, *Am. Min.*, **16**, 71 (1931); Faivre, *C.R.*, **222**, 227 (1946).
11. Cf. Doelter (**1**, 499, 1911), Cohen and von Hengel, *Zs. phys. Chem.*, **161**, 179 (1932).

14.1.3.3 **S T R O N T I A N I T E** [$SrCO_3$]. Strontianite *Sulzer* (*Lichtenberg's Mag.*, **7**, 3, 68; *Bergm. J.*, **1**, 5, 433, 1791). Strontian *Werner*. Strontianit, Kohlensaure Strontianerde *Klaproth* (*Crell's Ann.*, **2**, 189, 1793; **1**, 99, 1794; *Beitr.*, **1**, 268). Mineral from Strontian, Strontian Spar [not Strontites = strontium oxide] *Hope* (*Edinburgh Trans.*, **4**, 3, 1798 (read Nov., 1793)). Strontiane carbonatée *Fr.*

Emmonite, Calcareo-carbonate of Strontian *Thomson* (*Rec. Gen. Sc.*, **3**, 415, 1836). Calciostrontianit *Cathrein* (*Zs. Kr.*, **14**, 366, 1888). Barystrontianite, Stromnite *Traill* (*Edinburgh Phil. J.*, **1**, 380, 1819).

C r y s t. Orthorhombic; dipyramidal—$2/m\ 2/m\ 2/m$.

$a:b:c = 0.6090:1:0.7239$;[1] $p_0:q_0:r_0 = 1.1887:0.7239:1$

$q_1:r_1:p_1 = 0.6090:0.8413:1$; $r_2:p_2:q_2 = 1.3814:1.6420:1$

STRONTIANITE

Forms: [2]

	ϕ	$\rho = C$	ϕ_1	$\rho_1 = A$	ϕ_2	$\rho_2 = B$
c 001	.....	0°00'	0°00'	90°00'	90°00'	90°00'
b 010	0°00'	90 00	90 00	90 00	.....	0 00
m 110	58 39½	90 00	90 00	31 20½	0 00	58 39½
e 012	0 00	19 54	19 54	90 00	90 00	70 06
δ 023	0 00	25 46	25 46	90 00	90 00	64 14
k 011	0 00	35 54	35 54	90 00	90 00	54 06
i 021	0 00	55 22	55 22	90 00	90 00	34 38
z 041	0 00	70 56½	70 56½	90 00	90 00	19 03½
q 061	0 00	77 02	77 02	90 00	90 00	12 58
t 102	90 00	30 43½	0 00	59 16½	59 16½	90 00
ε 113	58 39½	24 53½	13 34	68 56	68 23	77 21½
o 112	58 39½	34 50	19 54	60 48	59 16½	72 43
p 111	58 39½	54 18	35 54	46 05	40 04½	65 00½
φ 331	58 39½	76 32	65 16½	33 50	15 40	59 37

Structure cell.[3] Space group $Pmcn$. a_0 5.118 kX, b_0 8.404, c_0 6.082; $a_0:b_0:c_0 = 0.609:1:0.724$. Cell contents $Sr_4(CO_3)_4$.

Habit. Crystals short to long prismatic [001], often acicular or acute spear-shaped. Steep pyramidal forms are frequently found in highly calcian material. Often pseudo-hexagonal in aspect, due to equal development of $\{110\}$ and $\{010\}$ or of $\{hhl\}$ and $\{0.2h.l\}$. $\{110\}$ and $\{010\}$ striated horizontally, the steep $\{hhl\}$ and $\{0kl\}$ forms sometimes rounded. Also massive, columnar to fibrous; granular; in rounded masses.

Twinning. Very common. Twin plane $\{110\}$, usually as contact-, rarely penetration-twins; also repeated, as trillings, fourlings, or polysynthetic giving enclosed twin lamellae.

Phys. Cleavage $\{110\}$ nearly perfect, $\{021\}$ poor, $\{010\}$ in traces. Fracture uneven to subconchoidal. Brittle. H. 3½. G. 3.76 ± 0.02 (est. for pure $SrCO_3$), 3.72 (material with ca. 3.3 per cent CaO), 3.66 ± 0.02 (est. for material with ca. 7.3 per cent CaO); 3.722 (calc. for $SrCO_3$). Luster vitreous, inclining to resinous on uneven fracture surfaces. Colorless to gray, yellowish or greenish; also yellowish brown and reddish. Transparent to translucent. Fluorescent and phosphorescent in x-rays, electron beams, and ultraviolet light; sometimes thermoluminescent.[4]

Opt.[7] In transmitted light, colorless.

ORIENTATION	$n(Li)$	$n(Na)$	$n(Tl)$	
X c	1.5181	1.5199	1.5219	Biaxial negative (−).
Y b	1.6624	1.6666	1.6704	$r < v$, weak.
Z a	1.6640	1.6685	1.6728	
2V (calc.)	7°03'	7°07'	7°16'	

Chem. Strontium carbonate, $SrCO_3$. More or less Ca is almost always present in substitution for Sr, up to Ca:Sr $\sim$ 1:4.5 in natural material.[14] A complete series to $BaCO_3$ exists in artificial material.[5] On heating inverts to a hexagonal modification.[6]

198   Carbonates

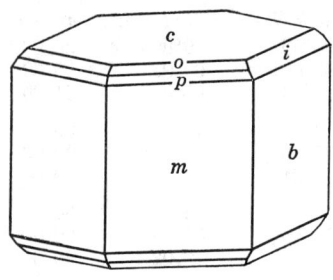

Strontian, England.

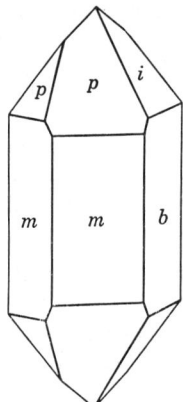

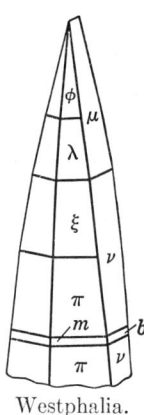

Westphalia.

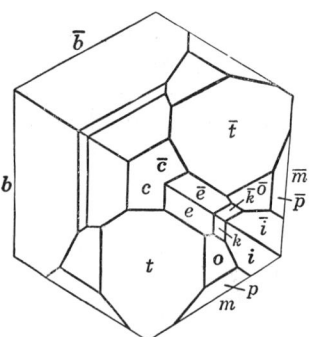

Clausthal. Twin on {110}. Projection on {001}.

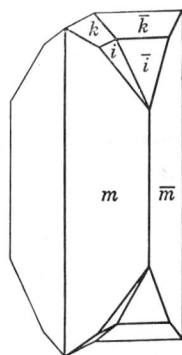

Strontian, England. Twin on {110}.

Anal.[3]

	1	2	3	4	5	6
CaO		2.70	3.14	4.05	6.10	7.36
SrO	70.19	66.31	66.09	65.01	62.62	60.99
CO_2	29.81	30.35	30.54	30.84	31.39	31.68
Rem.		0.17		0.06		
Total	100.00	99.53	99.77	99.96	100.11	100.03
G.			3.728		3.628	

1. $SrCO_3$. 2. Strontian, Scotland.[9] Rem. is BaO. 3. Drensteinfurt, Westphalia.[10]
4. Strontian, Scotland.[12] Rem. is SiO_2. 5. Albersloh, Westphalia.[10] 6. Brixlegg, Tyrol[11] (*calciostrontianite*).

Tests. B.B. swells up, throws out minute sprouts, but fuses only on thin edges. With soda on charcoal fuses to a clear glass and is absorbed by the coal with the exception of any lime present. Soluble in dilute HCl.

O c c u r. A low-temperature hydrothermal mineral, found associated with barite, celestite, and calcite in veins in limestone or marl, less frequently as a gangue mineral in sulfide veins. Also as geodes or concretionary masses in limestone and clay. Commercially important deposits occur in the marls of Westphalia, Germany. The principal localities are in the region of Hamm and Münster, notably at Drensteinfurt, Ascheberg, and Ahlen. Also in Germany in metallic veins at Clausthal, Grund, and St. Andreasberg in the Harz and at Bräunsdorf near Freiberg in Saxony; with zeolites at Oberschaffhausen in the Kaiserstuhl, Baden. In Austria at Leogang in Salzburg and at Brixlegg in Tyrol. Found originally in Scotland with barite, calcite, galena, harmotome, and rarely brewsterite in veins in gneiss at Strontian, Argyllshire. With celestite in a lead-silver deposit in the Sierra Mojada district, Coahuila, Mexico, and with celestite, gypsum, and phosphatic nodules in clay at Trichy, India.

In the United States found in geodes and thin veins with celestite and calcite in limestone at Schoharie, Schoharie County, New York, and at numerous other but minor occurrences. Early reported from Massachusetts (from New York?) (*emmonite*). With aragonite in limestone near Mt. Union, Mifflin County, Pennsylvania. In the Tijeras Canyon district, Bernalillo, New Mexico. As large deposits in limestone in the Strontium Hills, north of Barstow, San Bernardino County, California; also near Shoshone, Inyo County, and in the Genessee Valley, Plumas County. In Texas on Mt. Bonnell, near Austin, Travis County. Also with celestite in the calcite cap rock of salt domes on the Gulf coast of Texas and Louisiana. Near La Conner, Skagit County, Washington. In Canada in veins in limestone at Nepean, Carleton County, Ontario; and in the Cariboo District, British Columbia.

A l t e r. To celestite, analogous to the alteration of witherite to barite; also found as an alteration of celestite.

A r t i f.[13] In crystals by fusion of $SrCO_3$ with NaCl or KCl, and by heating precipitated $SrCO_3$ with water and ammonium chloride in a closed tube at 150°–180°. As spherulites by diffusion of Na_2CO_3 solution into gelatin containing $SrCl_2$.

Name. From the town in Scotland where the mineral was first found. The element strontium was discovered in this material in 1790.

Ref.
1. Naumann (from Dana (285, 1892), original reference unknown). The variation in angles with Ca content is not clearly defined, but appears to be very small.
2. Goldschmidt (**8**, 91, 1922). Rare or uncertain:

032	μ 071	0.12.1	445	11.11.4	881	π 24.24.1	325
031	081	0.24.1	332	λ 441	10.10.1	36.36.1
051	ν 0.11.1	115	221	ξ 661	12.12.1	40.40.1

3. Wilson, *Phys. Rev.*, **31**, 305 (1928), by powder method on artificial material.
4. See Jackson, *J. Chem. Soc. London*, **65**, 734 (1894); Bary, *C.R.*, **130**, 776 (1900); Kunz and Baskerville, *Science*, 769 (1903); Köhler and Leitmeier, *Zs. Kr.*, **87**, 146 (1934); Krejci-Graf, *Zs. Kr.*, **88**, 260 (1934). Fluoresces variously rose or greenish yellow in short-wave and rose or pale yellow in long-wave ultraviolet (Frondel, priv. comm., 1947).
5. Cork and Gerhard, *Am. Min.*, **16**, 71 (1931).
6. Boeke, *Mitt. naturfor. Ges. Halle*, **3**, 10 (1913).
7. Beykirch, *Jb. Min., Beil.-Bd.*, **13**, 427 (1900) by prism method on Walstedde, Westphalia, crystals with 5.95 per cent $CaCO_3$. Also older measurements on unanalyzed material by Buchrucker, *Zs. Kr.*, **19**, 146 (1891); Zirngibl, *Zs. Kr.*, **27**, 543 (1897); and Mallard, *Bull. soc. min.*, **18**, 12 (1895), with values slightly lower.
8. For additional analyses see Hintze (**1** [3A], 3034, 1926) and Doelter (**1**, 481, 1911).
9. Macadam, *Min. Mag.*, **6**, 173 (1885).
10. Beykirch (1900).
11. Cathrein, *Zs. Kr.*, **14**, 366 (1888).
12. Szebellédy, Inaug. Diss., Budapest (1926)—*Min. Abs.*, **3**, 261 (1927).
13. Doelter (**1**, 488, 1911); Biltz, *Zs. anorg. Chem.*, **220**, 312 (1934); Morse and Donnay, *Am. Min.*, **18**, 66 (1933).
14. Material containing much more Ca has been reported (see Beykirch (1900) and Dana (286, 1892; 699, 1868)), but probably due to admixed calcite.

14.1.3.4   **CERUSSITE** [$PbCO_3$].   Ψιρύθιον *Theophrastus*, etc.   Cerussa *Pliny*, *Ovid*, *Vitruvius*, *Agricola*, etc. (but only the artificial material).   Cerussa nativa ex agro Vicentino *Gesner* (85, 1565). Blyspath (= Bleispath *Germ.*). Minera Plumbi spathacea *Wallerius* (295, 1747). Plumbum terrestre vel lapideum, minera spathiformi, alba vel grisea *Wallerius*. Plomb spathique *Wallerius* (**1**, 536, 1753). Plumbum spathosum *Linné* (1768). Bly-Spat, Spatum Plumbi (the hard); Bly-Ochra, Cerussa nativa (the pulverulent) *Cronstedt* (1758). Plumbum acido aero mineralisatum *Bergmann* (**2**, 426, 1780). Weissbleierz *Werner, Emmerling, Reuss*. Plombe blanche *Fr.* White Lead Ore. Kohlensaures Blei *Germ.* Carbonate of Lead. Plomb carbonaté *Fr.* Céruse *Beudant* (**2**, 363, 1832). Cerussit *Haidinger* (503, 1845). Iglésiasite (Zinc-Bleispath *Kersten*) *Huot* (618, 1841). Cerusite.

Cryst. Orthorhombic; dipyramidal—$2/m\ 2/m\ 2/m$.

$a:b:c = 0.6100:1:0.7230;$ [1] $p_0:q_0:r_0 = 1.1852:0.7230:1$

$q_1:r_1:p_1 = 0.6100:0.8437:1;$ $r_2:p_2:q_2 = 1.3831:1.6393:1$

CERUSSITE

Forms: [2]

	ϕ	$\rho = C$	ϕ_1	$\rho_1 = A$	ϕ_2	$\rho_2 = B$
c 001	.....	0°00′	0°00′	90°00′	90°00′	90°00′
b 010	0°00′	90 00	90 00	90 00	.....	0 00
a 100	90 00	90 00	.....	0 00	0 00	90 00
r 130	28 39	90 00	90 00	90 00	0 00	28 39
m 110	58 37	90 00	90 00	90 00	0 00	58 37
x 012	0 00	19 52½	19 52½	0 00	90 00	70 07½
k 011	0 00	35 52	35 52	61 21	90 00	54 08
i 021	0 00	55 20	55 20	31 23	90 00	34 40
v 031	0 00	65 15	65 15	90 00	90 00	24 45
z 041	0 00	70 55½	70 55½	90 00	90 00	19 04½
n 051	0 00	74 32	74 32	90 00	90 00	15 28
y 102	90 00	30 39	0 00	59 21	59 21	90 00
g 113	58 37	24 50	13 33	68 59½	68 26½	77 22
o 112	58 37	34 46	19 52½	60 52	59 21	72 43½
p 111	58 37	54 14	35 52	46 09½	40 09½	65 00
s 121	39 20½	61 51½	55 20	56 01	40 09½	47 00
w 211	73 02½	68 01½	35 52	27 30	22 52½	74 18

Less common:

G 150	I 025	B 095	u 071	A 304	ε 331
d: 380	q 023	R 052	ζ 081	e 101	μ 324
χ 120	e: 087	C 072	n: 091	π 302	K 354
Γ 350	f: 076	D 0.11.2	g: 0.10.1	l 201	φ 131
f 530	S 032	t: 061	J 0.12.1	h 114	η 352
γ 013	F 053	M 0.13.2	h: 0.14.1	τ 221	O 241

Structure cell.[3] Space group $Pmcn$. a_0 5.173 kX, b_0 8.480, c_0 6.130; $a_0:b_0:c_0 = 0.610:1:0.723$. Cell contents $Pb_4(CO_3)_4$.

Habit. Extremely varied; dominant forms usually $m\ b\ p\ i\ x\ k\ r$. Simple crystals often tabular {010} and elongated [001] or [100]; also equant, or dipyramidal and then pseudo-hexagonal; rarely acicular [001] or very thin tabular {001}. {010} and {0kl} usually striated [100]; {111} often striated [1$\bar{1}$0] or [11$\bar{2}$]. The crystals commonly grouped in clusters or reticular aggregates (see under Twinning). Also massive, granular to dense and compact; sometimes stalactitic, or pulverulent to earthy. Rarely fibrous.

Twinning. Almost universal. Most common on {110}, either as twin lamellae or as contact types producing stellate pseudo-hexagonal groups or reticulated aggregates. Less common on {130}, chiefly as contact twins with a heart-shaped outline. Both laws sometimes occur in the same aggregate.

P h y s. Cleavage {110} and {021} distinct; {010} and {012} in traces. Fracture conchoidal. Very brittle. H. 3–3½. G. 6.55 ± 0.02;[4] 6.558 (calc.). Luster adamantine, inclining to vitreous, resinous, or pearly; sometimes submetallic if colors are dark due to surface films. Colorless to white and gray or smoky; sometimes dark gray to black due to included particles of sulfides, manganese oxide, or carbonaceous material; or blue to green due to included copper compounds. Streak colorless to white. Transparent to subtranslucent. Fluorescent in shades of yellow in x-rays and long-wave ultraviolet light.[25]

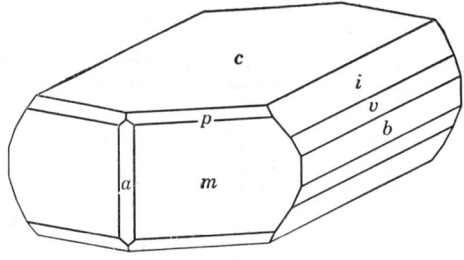

Dognácska.

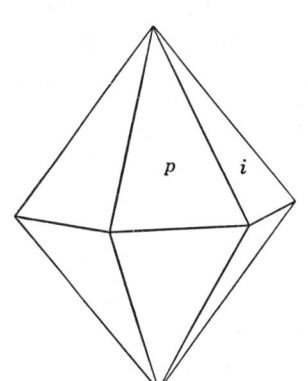

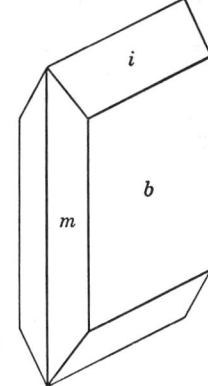

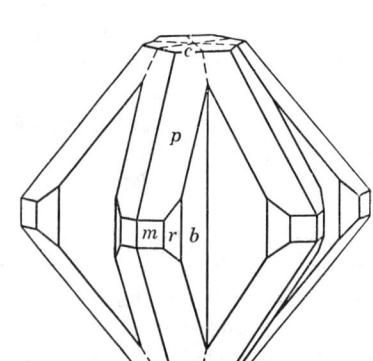

Cyclic twin on {110}.

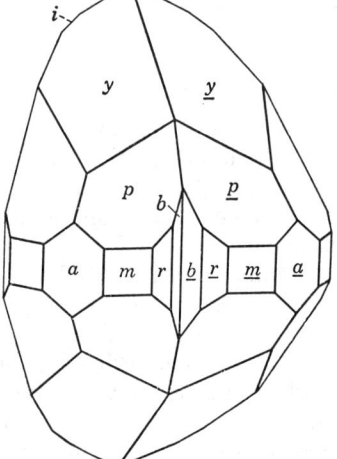

Friedrichssegen. Twin on {110}.

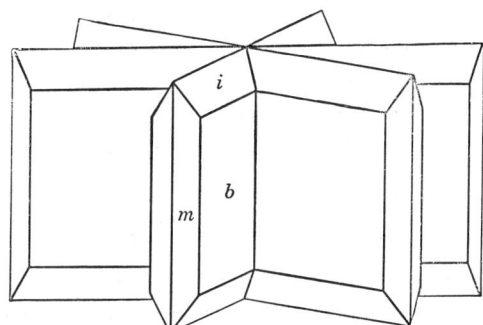

Cyclic twin on {110}.

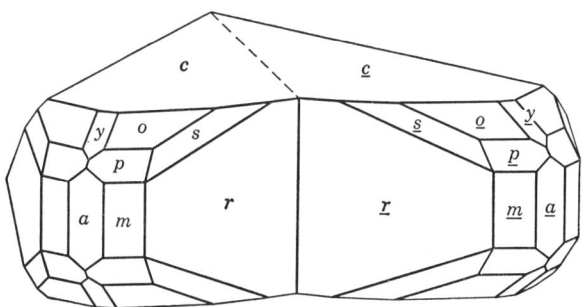

Tsumeb. Twin on {130}.

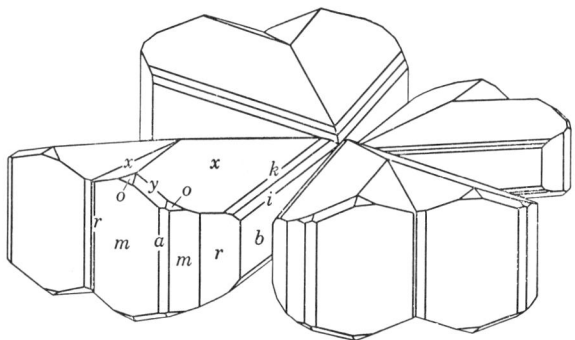

Mammoth mine, Arizona. Twinned on {110} and {130}

Opt.[5] In transmitted light, colorless.

ORIENTATION	$n(636.4)$	$n(578.2)$	$n(512.9)$	n(Na)[6]	
X c	1.7928	1.8026	1.8141	1.8036	Biaxial negative (−).
Y b	2.0598	2.0742	2.0918	2.0765	$r > v$, strong.
Z a	2.0618	2.0762	2.0934	2.0786	
2E (calc.)	18°45′	17°50′	16°01′	18°43½′	
2V (calc.)	9°04½′	8°34′	7°38′	9°00′	

Temp.:	25°	60°	80°	100°	160°	180°
2E	17°50′	18°28′	19°10′	20°15′	22°28′	22°45′

Chem. Lead carbonate, $PbCO_3$. Cerussite generally does not show a significant amount of isomorphous substitution by other elements. Sr has been observed substituting for Pb, with Sr:Pb = 1:10.5 (anal. 5). Small amounts of Zn apparently can also substitute for Pb with Zn:Pb = 1:8.4 (anal. 8), but the validity of this variety is uncertain. On heating,[7] cerussite breaks down to a basic carbonate at *ca.* 300° which in turn decomposes to PbO at *ca.* 500° (?).

Anal.[8]

	1	2	3	4	5	6	7	8	9
PbO	83.53	82.69	82.67	83.21	79.59	83.27	76.82	78.65	83.37
CaO		0.24	0.27	0.04					
SrO					3.15				
ZnO		0.08					4.56	3.41	
CO_2	16.47	16.57	16.27	16.40	17.02	16.64	17.74	[17.94]	16.52
Rem.		0.39	0.83	0.19		0.24			0.11
Total	100.00	99.97	100.04	99.84	99.76	100.15	99.12	[100.00]	100.00
G.					6.409		6.187	5.9	6.54

1. $PbCO_3$. 2. Pelsöc-Ardo, Czechoslovakia.[9] Rem. is FeO 0.09, MgO 0.06, H_2O 0.03, insol. 0.21. 3. Rézbánya, Hungary.[9] Rem. is FeO 0.11, MgO 0.02, H_2O 0.18, CuO 0.32, insol. 0.20. 4. Dognácska, Hungary.[9] Rem. is FeO 0.06, H_2O 0.01, insol. 0.12. 5. Isle, Colorado.[10] 6. Tsumeb, Africa.[11] Rem. is insol. 7. Iglesias, Sardinia [12] (*iglesiasite*). Perhaps a mixture. 8. Radzionkau, Silesia.[13] 9. Byelousovsky mine, Altai, U.S.S.R.[24] Rem. is FeO.

Tests. In C.T. decrepitates, loses CO_2, turns first yellow and at a higher temperature dark red, but becomes yellow again on cooling. B.B. fuses very easily. Soluble with effervescence in dilute HNO_3.

Occur. Cerussite is a common secondary lead mineral, found typically in the upper oxidized portion of ore deposits where it occurs with anglesite, limonite, and, less frequently, pyromorphite, phosgenite, malachite, smithsonite, and other secondary minerals containing lead, zinc, or copper. A few of the more important or interesting localities follow. In Germany [14] notably at Badenweiler, Baden; in the Friedrichssegen mine at Ems, Hesse-Nassau; at Johanngeorgenstadt, Saxony; and at Tarnowitz and Beuthen in Silesia. In fine crystals at Mies, Bohemia, and Rézbánya, Comitat Bihar, Hungary.[15] With anglesite and phosgenite in fine specimens at numerous lead deposits in Sardinia, notably Monteponi and Montevecchio near Iglesias and at Rosas. In France [16] notably at Poullaouen and Huelgoat, Brittany; at Pontgibaud, Puy-de-Dôme, Plateau

Central; at La Croix-aux-Mines, Vosges. Abundantly in Spain [17] in the Sierra de Cartagena, Murcia; in the Linares region, Andalusia; in the Sierra de Gador, Almeria. Formerly found well-crystallized together with leadhillite, caledonite, linarite, etc., at Leadhills, Lanarkshire, Scotland. In Russia in the Nerchinsk district, Transbaikalia, and at Berezovsk in the Urals. In Tunis at Sidi-Amor-ben-Salem. In exceptionally fine and large crystallizations at Tsumeb, South West Africa, associated with malachite, smithsonite, azurite, anglesite; also at Broken Hill, Rhodesia, with zinc phosphates; at Mindouli in the Belgian Congo. Another notable locality, especially for twinned reticular aggregates, is at Broken Hill, New South Wales.[18] In New Caledonia at Mine Mérétrice on the Diahot River.

In the United States,[19] cerussite was found in fine specimens in the Wheatley mines, Phoenixville, Pennsylvania. Found sparsely with lead ores at Granby, Frederickstown, and Joplin, Missouri, and at Mineral Point and other lead-zinc deposits in southern Wisconsin. An abundant ore at Leadville, Colorado, and found also at the Terrible mine, Isle, Custer County, and other localities in the state. In the Tintic district, Utah; in the Summit and other mining districts, Stevens County, Washington; at Esterbrook, Albany County, Wyoming; abundantly in the Elkhorn and Wickes districts, Jefferson County, Montana. At Galena, Lawrence County, South Dakota. An important ore mineral in the silver-lead deposits of Idaho,[20] notably in Shoshone County; also in the Bayhorse district, Custer County; Lemhi County; the Dome district, Fremont County; and in the Coeur d'Alene district, at Wardner and elsewhere. In New Mexico, notably in the Organ district, Dona Ana County (in part as very large heart-shaped twins), and in the Magdalena district, Socorro County. Abundant at Eureka and Goodsprings, Nevada. In Arizona in fine specimens at the Mammoth mine, Pinal County; at Tombstone, Cochise County; at the Red Cloud mine, Yuma County, and numerous other localities. Cerussite occurs in California in the Cerro Gordo and other silver-lead districts of Inyo County, and in the Silver Reef district and elsewhere in San Bernardino County. In Canada at Moyie and the H.B. mine at Salmo, in British Columbia.

Alter. Cerussite is very commonly found as an alteration product of galena, occurring as crystalline crusts or dense masses with a concentrically banded structure. A core of unaltered galena may be present and often is separated from the investing cerussite by an intermediate zone of anglesite. More or less complete pseudomorphs after crystals of anglesite are common, and cerussite also occurs as alteration pseudomorphs after phosgenite, leadhillite, caledonite, hydrocerussite, bournonite, linarite, pyromorphite, vanadinite, lanarkite (?). Incrustation or substitution pseudomorphs have been noted after calcite and sphalerite and, more doubtfully, after barite, fluorite, quartz, and pyrite. Pseudomorphs of other minerals after crystals of cerussite also have been observed, notably of pyromorphite, minium, and malachite, and less frequently of quartz, galena, limonite, calcite, siderite, malachite, phosgenite, dolomite, and chrysocolla. Thin films of galena are sometimes noted upon cerussite crystals.[21] Cerussite has been found as an alteration of ancient coins and various leaden metallic objects.[22]

A r t i f.[23] Easily obtained by reaction of solution of lead salts and alkali carbonates, and otherwise. In crystals by slow inter-diffusion of solutions of lead nitrate and potassium carbonate; by heating lead formate, or a mixture of precipitated lead carbonate with NH_4Cl, with water in a closed tube; by the long-continued action of sodium bicarbonate solutions on fragments of galena; and by heating a $PbCl_2$ and urea solution at 180°.

N a m e. There are frequent allusions to artificial lead carbonate in writings at least as early as 400 B.C., it being termed Ψιρίθιον by the Greeks and *cerussa*, whence the modern name, by the Romans. The normal and basic carbonates, the latter (artificial white lead) now known as hydrocerussite, were not early distinguished.

Ref.

1. Koksharov (**6**, 100, 1870), conforming to the structure cell.
2. Goldschmidt (**2**, 107, 1913). Also Dübigk, *Jb. Min., Beil.-Bd.*, **36**, 214 (1913); Ledoux and Walker, *Ottawa Nat.*, **32**, 7 (1918); Buttgenbach, *Bull. soc. min.*, **43**, 24 (1920); Stevanovic, *Ann. géol. pénin. Balkan.*, **7**, 85 (1922); Billows, *Att. accad. Veneto*, **14** [3], 89 (1923); Maier, *Zs. Kr.*, **58**, 75 (1923); Barthoux, *Bull. soc. min.*, **47**, 36 (1924); Shannon, *U. S. Nat. Mus., Bull.* **131**, 240 (1926); Hintze (**1** [3A], 3045, 1926); Tokody, *Zs. Kr.*, **63**, 385 (1926); *ibid.*, **96**, 325 (1937); O'Daniel, *Zs. Kr.*, **74**, 333 (1930); Laskiewicz, *Arch. Min. Soc. Warsaw*, **7**, 147 (1931); Garrido, *Bol. soc. espan. hist. nat.*, **34**, 301 (1934); Amaral, *Min. e met., Rio de Janeiro*, **13**, 59 (1948). Rottenbach, Inaug. Diss., Bonn (1937) discusses the relation of the morphology to the structure. Rare, uncertain or vicinal forms:

1.31.0	3.10.0	016	073	0.29.1	443	133	233	362
1.23.0	8.25.0	029	083	0.33.1	332	173	243	623
1.22.0	7.19.0	014	092	0.37.1	995	5.8.14	293	251
1.17.0	7.17.0	034	0.14.3	0.50.1	441	3.12.8	394	562
1.12.0	7.15.0	0.17.18	0.25.4	0.63.1	14.14.1	285	546	833
1.10.0	230	0.11.10	0.29.4	0.250.1	4.86.45	377	313	311
190	540	054	0.11.1	105	1.15.8	346	323	351
180	750	043	0.13.1	104	125	122	3.10.3	11.13.2
170	210	085	0.15.1	103	124	132	161	11.13.1
2.11.0	310	074	0.17.1	203	134	1.35.2	171	1.21.0
290	28.1.0	094	0.22.1	607	154	10.3.18	322	160
270	0.1.28	0.25.11	0.25.1	223	174	395	342	10.23.0

3. Space group of Colby and La Coste, *Zs. Kr.*, **84**, 300 (1933); cell dimensions of Lindsay and Hoyt, *Zs. Kr.*, **100**, 360 (1938).
4. From best reported values on crystals; massive material and crystals containing Sr or Zn have lower G.
5. Dübigk (1913), average of prism measurements on Tsumeb crystals (anal. 6). Earlier measurements by Schrauf, *Sitzber. Ak. Wien*, **42**, 120 (1860); Ohm, *Jb. Min., Beil.-Bd.*, **13**, 31 (1899). On the relation of birefringence to wavelength see also Sève, *Bull. soc. min.*, **46**, 34 (1923); on the variation of optic angle with temperature, Des Cloizeaux (49, 1867) and Panichi, *Acc. Linc., Mem.*, **4** [5a], 419 (1902), the mineral becoming uniaxial at −119° with $r < v$ and $Y = a$ at lower temperatures.
6. Negri, *Riv. min.*, **4**, 53 (1889).
7. Friedrich, *Cbl. Min.*, 621 (1912); Joulin, *Bull. soc. chim. phys.*, **19** [2], 345 (1873); Colson, *C.R.*, **140**, 865 (1905); Mügge, *Jb. Min., Beil.-Bd.*, **14**, 259 (1901).

8. For additional analyses see Doelter (**1**, 510, 1911), Vavrinecz, *Zs. Kr.*, **89**, 521 (1934).
9. Vavrinecz (1934).
10. Warren, *Am. J. Sc.*, **16**, 337 (1903).
11. Dübigk (1913).
12. Karsten, *J. Chemie u. Phys.*, **45**, 365 (1832).
13. Traube, *Zs. deutsche geol. Ges.*, **46**, 50 (1894).
14. See further in Hintze (1926).
15. Additional localities in Tokody (1926, 1937).
16. Additional localities in Lacroix (**3**, 699, 1909).
17. Additional localities in Calderón (**2**, 96, 1910), and Garrido (1934).
18. Smith, *New South Wales Dept. Mines, Min. Res.*, no. 34, 93 (1926).
19. See further in Schrader, *U. S. Geol. Sur., Bull. 624* (1917).
20. See further in Shannon, *U. S. Nat. Mus., Bull. 131*, 240 (1926).
21. Hobbs, *Am. J. Sc.*, **50**, 121 (1895); Boutwell, *U. S. Geol. Sur., Prof. Pap. 77*, 111 (1912).
22. Lacroix (**3**, 727, 1909); Brown, *Am. J. Sc.*, **32**, 377 (1886); Fletcher, *Min. Mag.*, **7**, 187 (1887); Rogers, *Am. Geol.*, **31**, 45 (1903); de Luca, *C.R.*, **84**, 1457 (1877).
23. Mellor (**7**, 830, 1927); Biltz, *Zs. anorg. Chem.*, **220**, 312 (1934).
24. Pilipenko, *Bull. Imp. Tomsk Univ.*, no. 63 (1915), *Min. Abs.*, **2**, 111 (1923).
25. See Tokody, *Magyar Mat. Termés. Ért.*, **61**, 1116 (1942).

TYPE 2. $AB(XO_3)_2$

14.2.1 DOLOMITE GROUP

HEXAGONAL—R; RHOMBOHEDRAL—$\bar{3}$

	a_0	c_0
Dolomite, $CaMg(CO_3)_2$	4.832 kX	15.92
Ankerite, $CaFe(CO_3)_2$		
Kutnahorite, $CaMn(CO_3)_2$		

The members of the dolomite group have a structure [1] similar to that of calcite. In calcite all the Ca ions are structurally equivalent with a sequence ... Ca-Ca-Ca ... along the threefold symmetry axes of the unit cell. In dolomite, however, alternate Ca positions are occupied by Mg, these becoming structurally non-equivalent and the crystal point-symmetry being lowered from $\bar{3}\ 2/m$ to $\bar{3}$. The relation between hematite (Fe_2O_3) and ilmenite ($FeTiO_3$) is entirely analogous. The dolomite structure is essentially an ordered sorting appearing in the calcite-magnesite random-substitution series when $Ca:Mg \sim 1:1$. The formation of the structure and the limited extent of the calcite-magnesite series itself are primarily a consequence of the relatively large dissimilarity in size of the Ca and Mg ions. Ankerite is stable only when considerable Mg is present in substitution in the Fe positions, the essentially pure compound $CaFe(CO_3)_2$ not being known, and the existence of kutnahorite, $CaMn(CO_3)_2$, is known only in a single questionable instance as a variety containing much Mg in substitution for Mn. The seemingly diminished stability of these compounds may be a consequence of the increasing similarity in size of Mg, Fe, and Mn with Ca.

Ref.
1. Wyckoff and Merwin, *Am. J. Sc.*, **8**, 447 (1924).

14.2.1.1 D O L O M I T E [Ca(Mg,Fe,Mn)(CO$_3$)$_2$]. Marmor pt. *Arduino (Osserv. chim., Venezia*, 1779). Pierres calcaires très-peu effervescentes avec les acides *Dolomieu (J. phys.*, **39**, 1, 1791). Dolomie *Saussure (J. phys.*, **40**, 161, 1792; *Voy. Alpes*, § 1929, 1796). Dolomite *Kirwan* (**1**, 111, 1794). Bitterspath, Rhomboidalspath, Kohlensauerter Kalkerde, Bittersalzerde [with anal.] *Klaproth (Schrift. Nat. Freund. Berlin*, **5**, 51, 1784; *Beitr.*, **1**, 300, 1795; also *Beitr.*, **3**, 297, 1802, **4**, 204, 236, 1807, **5**, 103, 1810, **6**, 323, 1815). Spath magnésien *Delamétherie* (**1**, 207, 1792). Miemit *Klaproth (Beitr.*, **3**, 292, 1802) [discov. at Miemo by D. Thomson in 1791 and labeled by him as Magnesian spar]. Rautenspath pt. *Werner* (1800; Ludwig's Werner, **1**, 51, 154, 1803). Chaux carbonatée, magnesifère pt.; Chaux carbonatée aluminifère [from Saussure's anal.] *Haüy* (1801). Bitterkalk pt. *Hausmann* (960, 1813). Perlspath pt., Rauhkalk, Kalktalkspath *Germ.* Pearl Spar pt., Brown Spar pt., Rhomb Spar pt., Magnesian Limestone. Spath perlé *Fr.*
    Conites, Flintkalk *Retzius (Min.*, 1795). Conite *Schumacher (Verzeichn.*, 20, 1801). Konit *Germ.* Gurhofian *Karsten (Mag. Nat. Freund. Berlin*, **1**, 4, 257, 1807; 50, 1808). Gurhofite. Tharandit *Freiesleben (Geognost. Arbeit*, **5**, 212, 1820). Lucullan *Kenngott (Min. Forsch.*, 43, 1850). Ridolphite *Des Cloizeaux*. Taraspit *von John (Verh. geol. Reichsanst. Wien*, **68**, 1899). Leesbergite *Blum (Ann. géol. belgique*, **34**, B 118, 1908). Teruelite *Maestre (Anales de Minas*, **3**, 264, 1845). Normaldolomite. Gajit *Tućan (Cbl. Min.*, 313, 1911). Codazzite *Codazzi (Biblioteca de Musso Nac. Bogotá*, 94, 1927—*Am. Min.*, **13**, 570, 1928). Magnesio-dolomite *Winchell* (pt. 2, 75, 1927).
    Ankerit [applied as ferroan dolomite] *Haidinger (Min. Mohs*, **1**, 411, 1825; **2**, 100, 1825); Paratomes Kalk-Haloid *Mohs* (**1**, 536, 1822; **2**, 116, 1824). Rohwand, Wandstein *Styrian Miners*. Dolomite pt. Brown Spar and Pearl Spar pt. Tautoklin *Breithaupt* (70, 1832; 20, 1830). Brossit *Hirzel (Zs. Pharm.*, **2**, 24, 1850). Normalankerit, Parankerit, Normal-parankerit *Bořický (Min. Mitt.*, **47**, 1876). Manganankerite *Koiké (J. Jap. Assoc. Min. Petr. Ec. Geol.*, **14**, 216, 1935). Mangandolomit *Eisenhuth (Zs. Kr.*, **35**, 582, 1902). Greinerite *Boldyrev (Kurs. optsat. min. Leningrad*, pt. 2, 168, 1928—*Min. Mag.*, **24**, 611, 1937). Plumbodolomit *Siegl (Min. Mitt.*, **48**, 288, 1936). Iron-dolomite, pt.

14.2.1.2 A N K E R I T E [Ca(Fe'',Mg,Mn)(CO$_3$)$_2$]. Ferrodolomite *Winchell* (pt. 2, 75, 1927). Eisendolomit.

C r y s t.[1] Hexagonal—R; rhombohedral—$\bar{3}$.

$a{:}c = 1{:}0.8322$; $\alpha\ 102°38'$; $p_0{:}r_0 = 0.9609{:}1$; $\lambda\ 73°45'$

Forms: [2]

			ϕ	$\rho = C$	A_1	A_2
c	0001	111	.....	0°00'	90°00'	90°00'
a	11$\bar{2}$0	10$\bar{1}$	0°00'	90 00	60 00	60 00
m	10$\bar{1}$0	2$\bar{1}\bar{1}$	30 00	90 00	30 00	90 00
r	10$\bar{1}$1	100	30 00	43 51½	53 07½	90 00
M	40$\bar{4}$1	3$\bar{1}\bar{1}$	30 00	75 25	33 03½	90 00
e	01$\bar{1}$2	110	−30 00	25 39½	90 00	67 58½
l	04$\bar{4}$5	33$\bar{1}$	−30 00	37 33	90 00	58 08½
f	02$\bar{2}$1	11$\bar{1}$	−30 00	62 30½	90 00	39 48
d	08$\bar{8}$1	33$\bar{5}$	−30 00	82 35½	90 00	30 49
v	21$\bar{3}$1	20$\bar{1}$	10 53½	68 31½	52 28	72 16

Less common:

Q 16.0.$\overline{16}$.1	11.5.5	α 44$\bar{8}$3	51$\bar{3}$	j 32$\bar{5}$1	30$\bar{2}$
h 03$\bar{3}$2	55$\bar{4}$	γ 33$\bar{6}$1	10.1.$\bar{8}$	k 8.4.$\overline{12}$.1	7$\overline{15}$

Dolomite–Ankerite

Structure cell.[3] Space group $R\bar{3}$.

Anal. no.	a_0	c_0	$a_0:c_0$	a_{rh}
3 (Fe:Mg = 1:32)	4.832 kX	15.92	1:3.294	5.995
10 (Fe:Mg = 1:1.1)	4.822	16.11	1:3.342	6.050

Anal. no.	α	G. meas.	G. calc.	nO	nE
3 (Fe:Mg = 1:32)	47°32′	2.87	2.85	1.690	1.505
10 (Fe:Mg = 1:1.1)	46°58′	3.01	3.02	1.728	1.531

Cell contents $Ca(Mg,Fe)(CO_3)_2$ in the rhombohedral unit.

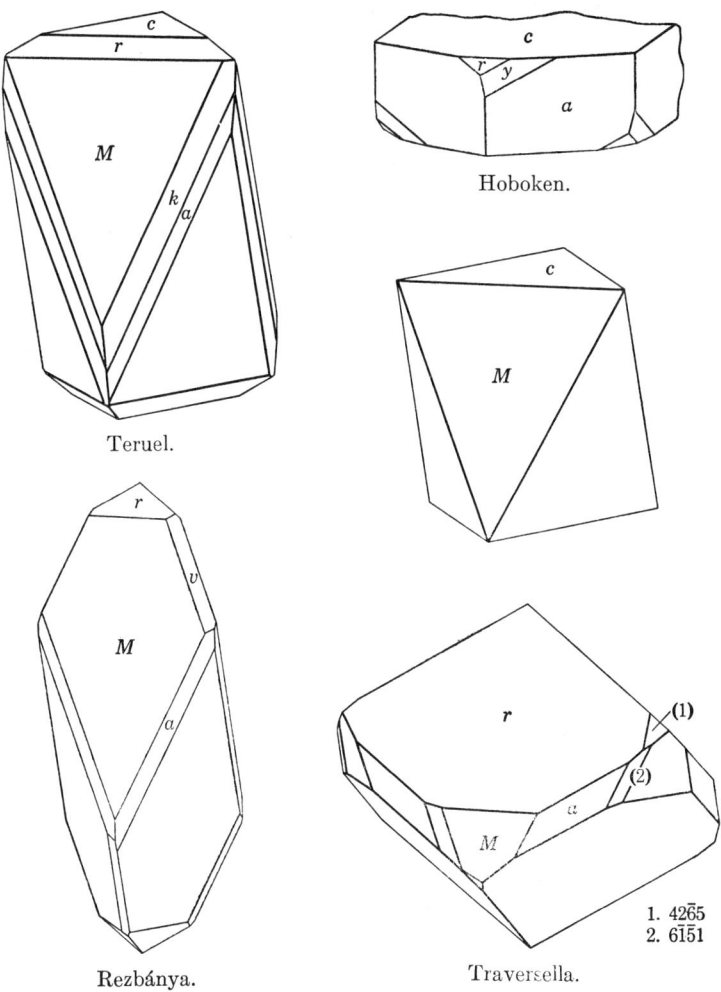

Teruel.

Hoboken.

Rezbánya.

Traversella.

1. $42\bar{6}5$
2. $6\bar{1}51$

Habit. Commonly rhombohedral with $\{10\bar{1}1\}$ or, less frequently, $\{40\bar{4}1\}$ dominant. Also prismatic with $\{11\bar{2}0\}$ and terminated by rhombohedrons, tabular {0001} with {11$\bar{2}$0}, or octahedral with {0001} and {10$\bar{1}$1} equally developed. The presence of second- and third-order rhombohedrons, usually as small modifying faces, is characteristic. Common forms: $r\,c\,M\,a$. {10$\bar{1}$1} often is striated horizontally; also commonly curved or made up of sub-individuals and passing into saddle-shaped forms. Massive, coarse to fine granular; columnar; sometimes porcelaneous and subtranslucent with a conchoidal fracture, occasionally opaline; in imitative shapes; rarely fibrous or pisolitic. Dolomite is an important sedimentary or metamorphic rock-forming mineral and then is compact massive, like ordinary limestone, or granular to saccharoidal and constituting many of the kinds of ornamental and other marbles.

Twinning.[4] (1) Twin plane {0001}, common, with re-entrant angles about the middle edges. (2) Twin plane {10$\bar{1}$0}, which also is a plane of symmetry for the twin; common. (3) Twin plane {11$\bar{2}$0}, common, as complementary twins simulating holohedral symmetry; also as double twins by combination of this law with twins on {10$\bar{1}$0} or {0001}. (4) Twin plane {10$\bar{1}$1}, rare. (5) Twin plane {02$\bar{2}$1} as lamellae, noted especially in the grains of dolomite marble;[14] said to be produced artificially by pressure but less readily than in calcite.

Etching.[5] Etch figures correspond to rhombohedral symmetry ($\bar{3}$).

Oriented growths.[6] Growths with parallel axes of calcite upon dolomite or, more commonly, of dolomite upon calcite have often been described; also dolomite upon rhodochrosite and siderite. Dolomite also occurs in oriented growths with antigorite (?), and with chlorite ({0001} and {001} parallel). Artificial growths of sodium nitrate upon dolomite have been obtained.

P h y s. Cleavage {10$\bar{1}$1} perfect. Sometimes parting in lamellar twins on {02$\bar{2}$1}. Fracture subconchoidal. Brittle. H. 3$\frac{1}{2}$–4.[15] G. 2.85 ± 0.01 for pure $CaMg(CO_3)_2$. The G. increases as Fe'' or Mn'' substitute for Mg, and is ∼3.02 in material with Mg:Fe = 1:1. Luster vitreous to pearly. Colorless and transparent or white in pure dolomite but sometimes gray or greenish, becoming yellowish brown or brown with increasing content of Fe''; also pink, rose or flesh-red, especially in varieties containing much Mn''. Ordinarily translucent to subtranslucent. Ferroan dolomite and ankerite on weathering turn darker brown or reddish in color. Shows twin gliding[7] on {02$\bar{2}$1}, and translation gliding with T{0001}, t[10$\bar{1}$0]. Some varieties are fluorescent in ultraviolet radiation; also triboluminescent.

O p t. In transmitted light, colorless. Uniaxial negative (−); rarely biaxial with very small $2V$ due to strain. The indices of refraction[8] and the birefringence increase with increasing substitution of Fe'' or Mn'' for Mg.

	% MgO	% FeO	Mg:Fe	nO(Na)	nE(Na)	Ref.
Dolomite	22.21	0.08		1.6799	1.5013	9
Dolomite	21.86	0	0	1.679	1.500	11
Ferroan dolomite	16.91	7.54	4:1	1.696		11
Ferroan dolomite	13.79	12.29	2:1	1.707		11
Ankerite	10.07	17.94	1:1	1.721		11
Ankerite	6.54	23.32	1:2	1.735		11
Ankerite		40.76	0	1.764		11

DISPERSION

				nO	nE	
Ferroan dolomite	21.06	1.19	Li	1.6784	1.5015	10
			Na	1.6830	1.5034	
			Tl	1.6870	1.5050	
Ferroan dolomite	17.58	6.68	Li	1.6931	1.5116	10
			Na	1.6983	1.5133	
			Tl	1.7031	1.5149	

The highly manganoan material of analysis 5, with Ca:Mg:Mn:Fe = 2:3.9:3.6:1 has nO 1.7005, nE 1.5148 (for Na).

C h e m. Carbonate of calcium and magnesium, $CaMg(CO_3)_2$. Fe'' and to a less extent Mn'', Co'', Pb, and Zn substitute for Mg; the limits of these substitutions are discussed beyond under Varieties. Fe'' and Mn'' apparently also can substitute in part for Ca. Ca also can substitute for Mg (these cations are structurally non-equivalent in dolomite) up to a maximum [12] of about Ca:Mg = 1:5 in the Mg positions, and Mg similarly can substitute for Ca up to a maximum of about Mg:Ca = 1:20 in the Ca positions. Material showing Mg in substitution for Ca or vice versa in amounts considerably in excess of these ratios has been reported, but the analyzed material has been shown or is presumed to have been impure. In some instances the mixtures appear to represent the breakdown of a more extended solid solution stable at higher temperatures. The interpretation of some analyses of other compositional varieties also is complicated by uncertainty as to the identity of the mineral, whether dolomite or a member of the calcite group.

In the case of Fe'', a series extends from the end-member $CaMg(CO_3)_2$ up to $Ca(Fe'',Mg)(CO_3)_2$ with Mg:(Fe,Mn) = 1:2.6 (analysis 14) at least. This series [36] is here arbitrarily divided at Mg:Fe = 1:1; material with Mg > Fe is referred to the species dolomite and material with Fe > Mg to the *species* ankerite. In former usage, the name ankerite included material now termed ferroan dolomite. The nomenclature is complicated by the fact that Fe'' and Mn'' apparently may substitute for both Ca and

Mg (analyses 5, 11); the application of the names dolomite and ankerite should depend on the ratio of Mg to Fe" in the Mg positions.

In the case of Mn", a probably continuous series extends up to Ca(Mg,Mn")(CO$_3$)$_2$ with Mn:(Mg,Fe) = 1:3.2 (analysis 9) at least. Dolomite containing more than a few weight per cent of MnO is rare, and considerable amounts of Fe" usually are present together with Mn". Material has been doubtfully reported in this series with Mn > Mg (see *kutnahorite*).

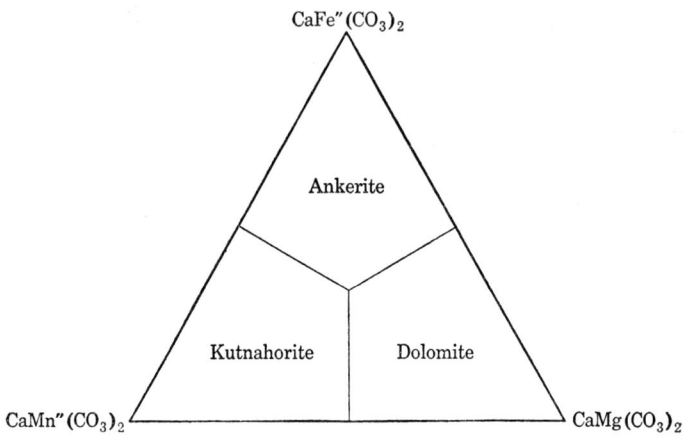

The boundaries between the Mg, Fe", and Mn" members of the dolomite group are taken as in the accompanying figure.

Anal.[13]

	1	2	3	4	5	6	7
CaO	30.41	29.90	30.19	30.83	10.48	43.71	30.83
MgO	21.86	22.21	21.07	21.06	14.58	9.81	16.85
FeO		0.08	1.17	1.19	6.59		2.56
MnO		nil			23.41		3.75
CO$_2$	47.73	47.77	47.52	46.76	45.59	44.23	46.05
Rem.				0.06	0.16	2.19	
Total	100.00	99.96	99.95	99.90	100.81	99.94	100.04
G.	2.85	2.882	2.87		2.96		

1. CaMg(CO$_3$)$_2$. 2. Binnenthal, Switzerland.[16] 3. Sunk, Styria, Austria.[17] 4. Greiner, Zillertal, Switzerland.[18] Rem. is insol. 5. Manganoan dolomite. Greiner.[18] Rem. is insol. 6. Calcian dolomite. Selmeczbánya, Hungary.[19] Rem. is Al$_2$O$_3$ 0.14, Fe$_2$O$_3$ 0.22, insol. 1.83. 7. Ferroan dolomite. Greiner and Traversella.[18]

	8	9	10	11	12	13	14
CaO	29.56	27.92	28.10	13.68	28.01	27.84	27.1
MgO	17.58	10.35	10.49	14.47	10.07	9.53	5.4
FeO	6.68	9.13	16.43	26.00	17.95	18.77	23.5
MnO		8.60	0.75				1.5
CO$_2$	45.64	44.25	44.09	42.32	43.97	43.86	42.5
Rem.		0.26		3.84			
Total	99.46	100.51	99.86	100.31	100.00	100.00	100.0
G.			3.01		3.0		

8. Ferroan dolomite. Greiner and Traversella.[18] 9. Manganoan dolomite. Budô, Japan.[20] Rem. is insol. 10. Ferroan dolomite. Erzberg, Styria.[21] 11. Ferroan dolomite. Felsöbánya.[19] Rem. is Al_2O_3 0.31, insol. 3.53. 12. $Ca(Mg,Fe'')(CO_3)_2$ with Mg:Fe = 1:1. 13. Magnesian ankerite. Erzberg, Styria.[22] 14. Ankerite. Locality unknown.[23]

Var.

DOLOMITE

Cobaltian. A reddish variety from Příbram, Bohemia,[27] contains 5.17 per cent CoO.

Plumbian. Plumbodolomite *Siegl* (1936). A plumbian variety has been reported [28] from Kreuth, Carinthia.

Manganoan. Greinerite *Boldyrev* (1928). Mangandolomite *Eisenhuth* (1902). Contains Mn'' in substitution for Mg, usually together with Fe''. The series extends up to Mn:(Mg,Fe) = 1:3.2 (analysis 9) and probably beyond Mn:Mg = 1:1 (*kutnahorite*). The variation in G. and indices of refraction with Mn content is not established but probably closely parallels the Fe''–Mg series.

Ferroan. Ankerite *older authors*. Tautoklin *Breithaupt* (1832). Brossite *Hirzel* (1850). Normal-ankerit, Parankerit, Normalparankerit *Bořický* (1876). Contains Fe'' in substitution for Mg, a series extending from essentially pure dolomite to ankerite with at least Mg:(Fe,Mn) = 1:2.6 (analysis 14). In some highly ferroan (and manganoan) types the substitution apparently is in part in the Ca positions (analysis 5).

Calcian. Magnesian. Contains Ca in substitution for Mg or of Mg for Ca in excess of the normal 1:1 ratio. The variation in properties accompanying these substitutions is not established. The calcian variety is much more common, and the substitution of greater extent, than the magnesian.

ANKERITE

Manganoan. Contains Mn'' in substitution for Fe''

Tests. Dolomite is infusible B.B. and glows brightly; the moistened residue, as with calcite, is strongly alkaline. Ankerite and ferroan dolomite heated B.B. darken in color and become magnetic. Fragments of dolomite placed in cold acids, unlike calcite, are only very slowly dissolved, but the finely powdered mineral dissolves readily with effervescence in warm acids. Numerous staining tests based on the difference in rate of solution of calcite and dolomite have been devised for the distinction of these species; in one method,[29] the material to be tested is placed in a 1 N solution of copper nitrate for about five hours and is then immersed briefly in strong ammonia water, whereby a deep blue color is produced on calcite. On heating [30] in air, dolomite dissociates to $CaCO_3 + MgO + CO_2$ or at a higher temperature to $CaO + MgO + CO_2$.

O c c u r. Massive dolomite rocks are found as extensive strata in many parts of the world. These rocks are believed to be of secondary origin and transformed from ordinary limestone, coral, or marble by the action of solutions containing magnesium. Most dolomite rocks are mixtures in varying proportions of dolomite proper and calcite. In high-grade thermal metamorphism the dolomite of such rocks may be dissociated [34] into a mixture of calcite and periclase (which may later alter into brucite or hydromagnesite), and siliceous types of dolomitic rocks also may afford diopside and other magnesium silicates.

In the localities listed below, dolomite rocks are excluded and reference is made only to outstanding occurrences of crystallized dolomite.[31] In this form, dolomite occurs in hydrothermal veins where it is associated with fluorite, barite, calcite, siderite, quartz, and various metallic ores; also in cavities or veins in limestone or dolomite rocks together with calcite, sellaite, celestite, gypsum, quartz; in quartz geodes in sedimentary rocks; frequently as veins or embedded crystals in serpentine and talcose rocks and in the altered facies of basic, magnesian igneous rocks in general. In Italy, fine crystals have been found at Brosso and especially Traversella in Piedmont, and in the Pfitschtal in Trentino and Miemo in Tuscany. At numerous localities in Switzerland, notably in dolomite, at Lengenbach in the Binnenthal, Valais, at Tarasp in Grisons, and in rock salt at Bex. In France, as a gangue mineral in veins at Markirch in Alsace, near Vieillevigne, Loire-Inférieure, from the Gebroulaz glacier and elsewhere in Savoy. In Germany in the mines of Freiberg and Schneeberg in Saxony; in Austria at Leogang in Salzburg and at Hall in Tyrol. At Rezbánya and Kapnik in Roumania. In England, dolomite crystals occur at Frizington and elsewhere in Cumberland and in metallic veins in Cornwall. As large clear crystals near Djelfa in Algeria. In Brazil at the Morro Velho gold mine near Ouro Preto, Minas Geraes, in pegmatitic veins at Bom Jesus do Meiras, Bahia, and in crystals of optical quality at Brumada, Bahia. Dolomite, in part supposedly containing a small amount of rare earths (codazzite), occurs in the emerald mines of the Muzo district, Colombia. In Mexico notably at Guanajuato in the silver veins.

In the United States, dolomite occurs as large rhombohedral crystals in talc at Roxbury, Vermont; also at Chester with magnetite, pyrite, and talc in chlorite rocks. In New York with gypsum, celestite, etc., in cavities in dolomite rock at Rochester, Lockport, and Niagara Falls; also with magnetite, chondrodite, serpentine at the Tilly Foster iron mine near Brewster, Putnam County. In serpentine near Hoboken, New Jersey, and on Staten Island, New York City. Fine crystals occur with rutile and muscovite in pegmatitic veins at Stony Point and elsewhere in Alexander County, North Carolina. In the quartz geodes of Keokuk, Lee County, Iowa. Abundantly as a gangue mineral in the lead-zinc deposits of the Joplin district, Missouri, and in similar deposits elsewhere in the midcontinent. In the Marquette iron district, Michigan, and at Warsaw, Hancock County, Illinois. Found as a gangue mineral in veins in many western mining districts.

Ferroan dolomite is rather common and has the same mode of occurrence and association as ordinary dolomite, although highly ferroan types do not occur on a rock-forming scale. It is frequently associated with iron ores. Found in the province of Styria, Austria, with siderite, especially at Erzberg near Eisenerz. At Traversella in Piedmont, Italy. At Csetnek,

Com. Gömör, Czechoslovakia, and at Kutnahora (Kuttenberg) in Bohemia; at Vaskö, Hungary. In the Northumberland coal field, England. In the United States found with hematite at Antwerp, Jefferson County, New York; at the Wheatley mines, Lancaster County, and Doan's mine, Bucks County, in Pennsylvania. A gangue mineral in the gold-quartz veins of the Mother Lode mines of Mariposa, Nevada, and Tuolumne Counties, California; also at Carson Hill, Calaveras County, and near Coulterville, Mariposa County. In Nova Scotia in the iron ores near Londonderry, Colchester County. In the gold veins of the Porcupine district, Ontario.

Ankerite (with Fe > Mg) occurs at Erzberg, Styria, Austria,[21] and with siderite at Csetnek, Com. Gömör, Czechoslovakia (*manganoan ankerite*).[24] At Hollópatak [25] and Bindtbánya,[35] Szepes, Hungary; and at Zaječov near Romorau in Bohemia [26] (Fe:Mg = 1:1). In Algeria at Oued Allelah. In England in the Northumberland coal field and in the ironstone deposits of the northern Pennines. In the United States ankerite probably occurs at the Antwerp iron mine, Jefferson County, New York. In Idaho in the Coeur d'Alene region, in Shoshone County,[33] in sulfide veins.

A r t i f. The synthesis of dolomite has been the subject of much investigation,[32] particularly with reference to the origin of dolomite rocks in nature. In general, it appears that dolomite does not crystallize directly at ordinary temperatures and pressures from solutions containing Ca, Mg, CO_2, or, in addition, other ions such as Cl, SO_4, or Na. Under such conditions fine-grained precipitates, sometimes gelatinous or spherulitic, are obtained which consist apparently of admixed calcite and magnesite, although hydrous basic carbonates of Ca and Mg have in some instances been claimed in addition. Undoubted dolomite has been obtained at ordinary temperatures from solutions of Ca and Mg carbonates under confining pressures upwards of 10 atm. of CO_2 and the yield and size of the dolomite crystals obtained increases with increasing CO_2 pressure, temperature, and concentration. Also similarly by the reaction of solid calcite, aragonite or vaterite with solutions of magnesium chloride or sulfate, the principal reaction products comprising dolomite, calcite, and magnesite. Many of the reported syntheses are prejudiced by lack of rigorous identification of the reaction products.

A l t e r. Dolomite and ferroan dolomite are very frequently found as pseudomorphs after calcite, with the dolomite generally oriented in parallel position to the original calcite. Also after aragonite and rarely (principally as incrustation pseudomorphs) after cerussite, barite, fluorite. Alteration pseudomorphs of dolomite after chondrodite occur at Tilly Foster, New York. Cubical casts of dolomite after halite molds occur in some sedimentary rocks. Calcite rocks such as limestone or marble often are more or less completely replaced by dolomite on a geologic scale variously by the action of sea water during diagenesis of the sediment or by the action of magnesia-containing circulating meteoric waters or ascendent hydrothermal solutions. A number of different minerals have been found as incrustation or substitution pseudomorphs after dolomite crystals or

CARBONATES

masses. These include siderite, calcite, smithsonite, quartz, talc, and limonite among the more common examples; also, less frequently, hematite, pyrite, malachite, azurite, magnetite, cinnabar, sphalerite, pyrolusite, marcasite, and serpentine.

Name. After Déodat Dolomieu (1750–1801), French engineer and mineralogist, who described some of the marked characteristics of the rock in 1791 and noted its occurrence in nature. Ankerite after M. J. Anker (1772–1843), Styrian mineralogist.

Ref.

1. Angles of Wollaston, *Phil. Trans.*, 159 (1812) on dolomite from unstated localities. The conventional morphological unit here used has c one-fourth that of the structure cell; dolomite, unlike calcite, however, could be described as satisfactorily by reference to the structure cell. Transformation: morphological to structural unit, $1000/0100/0010/0004$. The angles vary only slightly with varying composition, and the physical quality of the crystals is of more consequence in goniometric work. The tetartohedral (rhombohedral-$\bar{3}$) symmetry was first established by Tschermak, *Min. Mitt.*, **4**, 102 (1881). On the variation in angle with temperature see Rinne, *Cbl. Min.*, 705 (1914).

2. Goldschmidt (**3**, 65, 1916); Koller, *Jb. Min., Beil.-Bd.*, **42**, 457 (1918); Ungemach, *Zs. Kr.*, **58**, 161 (1923). There is no systematic treatment of the form development of ferroan dolomite or of ankerite. Rare or uncertain:

$21\bar{3}0$	$30\bar{3}2$	$11.8.\bar{19}.3$	$5\bar{1}43$	$4.8.\overline{12}.5$
$3\bar{1}20$	$30\bar{3}1$	$10.7.\overline{17}.13$	$6\bar{1}51$	$24\bar{6}1$
$11.2.\overline{13}.0$	$0.1.\bar{1}.10$	$24.16.\overline{40}.5$	$32.\bar{4}.\overline{28}.33$	$4.8.\overline{12}.1$
$4.0.\bar{4}.19$	$03\bar{3}5$	$53\bar{8}2$	$14.\bar{1}.\overline{13}.12$	$5.10.\overline{15}.1$
$10\bar{1}4$	$03\bar{3}2$	$42\bar{6}5$	$10.11.\overline{21}.8$	$8.16.\overline{24}.1$
$20\bar{2}5$	$44\bar{8}9$	$10.5.\overline{15}.2$	$6.7.\overline{13}.5$	$20.40.\overline{60}.1$
$40\bar{4}7$	$8.8.\overline{16}.3$	$20.4.\overline{24}.1$	$45\bar{9}2$	$25\bar{7}3$
$20\bar{2}3$	$6\bar{3}31$	$9.1.\overline{10}.2$	$6.8.\overline{14}.1$	$8.20.\overline{28}.3$
$7.0.\bar{7}.10$	$8\bar{4}49$	$19.\bar{8}.\overline{11}.21$	$8.12.\overline{20}.5$	$4.12.\overline{16}.1$
$8.0.\bar{8}.11$	$8\bar{4}43$	$21.\bar{8}.\overline{13}.23$	$23\bar{5}2$	$3.14.\overline{17}.1$
$30\bar{3}4$	$16.\bar{8}.\bar{8}.3$	$11.\bar{4}.\bar{7}.12$	$5.8.\overline{13}.3$	$1.14.\overline{15}.2$
$10.0.\overline{10}.13$	$7.6.\overline{13}.1$	$12.\bar{4}.\bar{8}.13$	$12.20.\overline{32}.13$	$37\bar{4}8$
$11.0.\overline{11}.14$	$6.5.\overline{11}.1$	$12.\bar{4}.\bar{8}.1$	$6.10.\overline{16}.1$	$\bar{8}.20.\overline{12}.5$
$40\bar{4}5$	$8.6.\overline{14}.3$	$13.\bar{4}.\bar{9}.14$	$16.28.\overline{44}.3$	$26\bar{4}7$
$14.0.\overline{14}.17$	$4\bar{3}71$	$17.\bar{4}.\overline{13}.18$	$5.9.\overline{14}.2$	$\bar{4}.16.\overline{12}.1$

3. Schoklitsch, *Zs. Kr.*, **90**, 433 (1935), by powder method. See also Wyckoff and Merwin, *Am. J. Sc.*, **8**, 447 (1924); Garrabos, *Bull. soc. min.*, **49**, 110 (1926); Halla, *Ak. Wien, Sitzber.*, **139**, 683 (1930).

4. Detailed descriptions are given by Sella, *Studii sulla min. Sarda*, Torino, 1856; Tschermak, *Min. Mitt.*, **4**, 109 (1881); Becke, *Min. Mitt.*, **11**, 224 (1890), **10**, 138 (1889); Koller (1918).

5. For summary see Koller (1918); also Royer, *C.R.*, **202**, 429 (1936); Honess, *Am. Min.*, **2**, 57 (1917).

6. Cf. Mügge, *Jb. Min., Beil.-Bd.*, **16**, 374 (1903); de Klerk and Goldschmidt, *Min. Mitt.*, **38**, 159 (1925).

7. Cf. Fairbairn and Hawkes, *Am. J. Sc.*, **239**, 617 (1941); Johnsen, *Jb. Min.*, II, 133 (1902); Tschermak, *Min. Mitt.*, **4**, 111 (1882).

8. On the variation with composition see Gaubert, *Bull. soc. min.*, **42**, 88 (1919); Koller (1918); Niggli, *Zs. Kr.*, **56**, 230 (1921); Eisenhuth, *Zs. Kr.*, **35**, 582 (1902); Schoklitsch (1935); Hawkes and Smythe, *Min. Mag.*, **24**, 65 (1935).

9. Koller (1918) on material of analysis 2.

10. Eisenhuth (1902) on material of analyses 4, 8.

11. From a graphical tabulation of available data.
12. Ford, *Trans. Conn. Ac. Arts Sc.*, **22**, 213 (1917); also Rocza, *Zbl. Min.*, 229 (1926) and analysis 6.
13. For tabulations of analyses see Hintze (**1** [3A], 3333, 1927), Doelter (**1**, 360, 1911); also Harding, et al., *Chem. News*, **121**, 50 (1920), Zsivny, *Zs. Kr.*, **65**, 728 (1927), Strobentz, *Földt Közl.*, **55**, 49 (1926), Koch and Zombory, *Földt. Közl.*, **64**, 160 (1934), Zsivny, *Ann. Hist.-Nat. Mus. Nat. Hungar.*, **24**, 423 (1926), Du Rietz, *Geol. För. Förh.*, **57**, 133 (1935), Koritnig and Ehrlich, *Zbl. Min.*, **41** (1940), Hawkes and Smythe (1935), Vavrinecz, *Magyar Chem. Foly.*, **38**, 140 (1932), Ulke, *Am. Min.*, **18**, 312 (1933), Johansson, *Geol. För. Förh.*, **70**, 349 (1948), Smythe and Dunham, *Min. Mag.*, **28**, 53 (1947).
14. See Fairbairn and Hawkes (1941).
15. On the variation in hardness with direction see Tertsch, *Zs. Kr.*, **92**, 39 (1935).
16. Pooth analysis in Koller (1918).
17. Meixner analysis in Schoklitsch (1935).
18. Eisenhuth (1902).
19. Strobentz (1926).
20. Koiké, *J. Jap. Assoc. Min. Petr. Ec. Geol.*, **14**, 216 (1935).
21. Schoklitsch (1935).
22. Rocza (1926).
23. Cited in Larsen and Berman (229, 1934).
24. Zsivny (1926).
25. Koch and Zombory (1934).
26. Bořický, *Min. Mitt.*, **47**, (1876).
27. Gibbs, *Ann. Phys.*, **71**, 361 (1847).
28. Siegl, *Min. Mitt.*, **48**, 288 (1936).
29. Rodgers, *Am. J. Sc.*, **238**, 788 (1940).
30. See Faust, *Am. Min.*, **34**, 789 (1949); Garnett, *Min. Mag.*, **20**, 54 (1923); Mitchell, *J. Chem. Soc. London*, **123**, 1055 (1923); Bäckström, *ibid.*, **125**, 430 (1924); Kani, *J. Geol. Soc.*, *Tokyo*, **35**, 279 (1928); Eitel, *Jb. Min., Beil.-Bd.*, **51**, 477 (1925).
31. A lengthy list of localities is given by Hintze (1927).
32. For summaries see Hintze (1927), Doelter (**1**, 360, 1911), Mellor (**4**, 372, 1923).
33. Shannon (226, 1926).
34. See Faust, *Am. Min.*, **34**, 789 (1949).
35. Tokody, *Mat. Termés. Ért.*, **50**, 650 (1934).
36. See also the discussion of Smythe and Dunham, *Min. Mag.*, **28**, 53 (1947).

14.2.1.3 **Kutnahorite.** *Bukowsky (Anz. II! Congress böhm. Naturfor. und Ärzte, Prag., p. 293, 1901).* Mangandolomit *Germ.* Manganodolomite. Kutnohorite, Kutnohorrite.

A name proposed for members of the dolomite group with Mn > Mg or Fe. The original material from Kutnahora or Kutná Hora (Kuttenberg), Bohemia, was found as granular to coarse cleavable masses with a white to pale rose color and G. 3.0. Analysis gave:

	CaO	MgO	FeO	MnO	CO_2	Total
1.	28.08	10.09		17.76	44.07	100.00
2.	24.66	5.18	4.27	23.76	42.62	100.49
3.	29.74	6.89	8.68	11.42	43.27	100.00

1. $Ca(Mn'',Mg)(CO_3)_2$ with Mn:Mg = 1:1. 2. Kutnahora. 3. Manganoan dolomite. Kutnahora.

The identification of this material as a member of the dolomite group is uncertain since crystallographic data are lacking. The approach of an analysis to the dolomite ratio, Ca:(Mn,Mg,Fe) = 1:1, is insufficient evidence in itself since this ratio might be fortuitously reached in the calcite group.

14.2.2   A L S T O N I T E [CaBa(CO$_3$)$_2$]. Barytocalcite *Johnston* (*Phil. Mag.*, **6**, 1, 1835; **10**, 373, 1837). Bicalcareocarbonate of Barytes [from a wrong analysis] *Thomson* (*Rec. Gen. Sc.*, **1**, 373, 1835). Bromlite *Thomson* (*Phil. Mag.*, **11**, 45, 48, 1837). Alstonite *Breithaupt* (**2**, 255, 1841).

C r y s t.   Orthorhombic; dipyramidal—$2/m\ 2/m\ 2/m$.

$a:b:c = 0.5827:1:0.7195;$ [1] $p_0:q_0:r_0 = 1.2348:0.7195:1$

$q_1:r_1:p_1 = 0.5827:0.8099:1;$ $r_2:p_2:q_2 = 1.3899:1.7161:1$

Forms: [2]

	ϕ	$\rho = C$	ϕ_1	$\rho_1 = A$	ϕ_2	$\rho_2 = B$
b 010	0°00′	90°00′	90°00′	90°00′	.....	0°00′
m 110	59 46½	90 00	90 00	30 13½	0°00′	59 46½
k 011	0 00	35 44	35 44	90 00	90 00	54 16
i 021	0 00	55 12	55 12	90 00	90 00	34 48
p 111	59 46½	55 01	35 44	44 56	39 00	65 38½
h 221	59 46½	70 43	55 12	35 21½	22 02½	61 37½

Structure cell. [3]   Orthorhombic. a_0 4.99 kX, b_0 8.77, c_0 6.11; $a_0:b_0:c_0 = 0.569:1:0.697$. Cell contents Ca$_2Ba_2$(CO$_3$)$_4$.

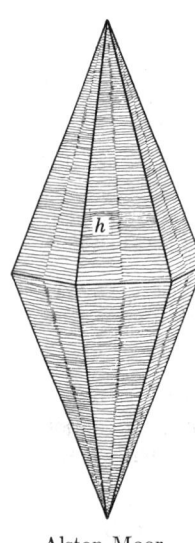

Alston Moor.

Habit. [4]   Crystals are pseudo-dihexagonal dipyramids formed by repeated twinning of individuals bounded by {111} and {021}; also as acute dipyramids bounded by {221}. Pseudo-hexagonal prisms result from twinned individuals bounded by {010} and {110}. The apparent dipyramidal faces are strongly striated horizontally and are divided vertically by a medial, slightly reentrant twinning line.

Twinning. [4]   Twin planes {110} and {130}. Untwinned crystals are rare.

P h y s.   Cleavage {110} imperfect. Fracture uneven. H. 4–4½. G. 3.707 ± 0.004 (Alston, anal. 2), [5] 3.67 (New Brancepeth, anal. 3), [6] varying presumably with the Sr content; 3.67 (calc.). Luster vitreous. Colorless to snow-white; also grayish, pale cream, pink to pale rose-red (and bleaching on exposure). Streak white. Transparent to translucent. Weakly fluorescent in yellow in long-wave ultraviolet radiation. [14]

O p t.   In transmitted light, colorless.

Orientation	n (Alston) [7]	n (artificial) [8]	
X c	1.526	1.525	Biaxial negative (−).
Y a	1.671		$r > v$, weak.
Z b	1.672	1.68	
$2V$ (meas.)	6°±	7°±	

C h e m.   A carbonate of calcium and barium, CaBa(CO$_3$)$_2$. Dimorphous with barytocalcite. The correct structural formulation of this compound is uncertain, [9] especially with regard to the Sr sometimes present in solid solution. The departure from the ratio Ca:Ba = 1:1 in some older analyses, suggesting the formulation $(Ca,Ba)CO_3$, or $Ca(Ba,Ca)(CO_3)_2$, may be due to analytical error. Small amounts of Mn($\sim$0.2 MnO) are generally present in both alstonite and barytocalcite.

Anal.[10]

	1	2	3
CaO	18.85	17.60	18.0
SrO		4.25	none
BaO	51.56	48.54	52.3
CO_2	29.59	29.41	[29.1]
Rem.			0.26
Total	100.00	99.80	[99.66]
G.	3.67	3.707	3.60

1. $CaBa(CO_3)_2$. 2. Alston Moor.[11] 3. New Brancepeth.[12] Rem. is MnO 0.06, insol. 0.2. CO_2 calculated to satisfy bases.

Tests. B.B. fuses on thin edges. Soluble in dilute HCl.

O c c u r. Found associated with calcite, barite, and witherite as a low-temperature hydrothermal deposit. Found at the Brownley Hill lead and zinc mine near Alston, Cumberland, England; in the Fallowfield lead mine near Hexham in Northumberland. Also in a vein at New Brancepeth, near Durham.[12]

A r t i f.[13] Obtained as tabular crystals from melts of $CaCO_3$, $BaCO_3$, and NaBr.

N a m e. From the locality near Alston. The older name bromlite, used by Dana in the sixth edition of this work, is derived from an erroneous spelling, Bromley, of the original locality at Brownley Hill mine.

Ref.

1. Kreutz, *Bull. Ac. Sc. Cracow*, 771 (1909), who discusses the earlier work.
2. Goldschmidt (**1**, 9, 1913).
3. Gossner and Mussgnug, *Cbl. Min.*, 220 (1930); transformation: Gossner to Kreutz, 010/100/001. No analysis or locality is given for the crystal used.
4. The habit, twinning, and pseudo-hexagonal symmetry of alstonite have been discussed at length by Kreutz (1909) and Gossner and Mussgnug (1930) who summarize the earlier work.
5. Kreutz (1909) by pycnometer; the values 3.718 and 3.706 also have been reported by Thomson (1837) and Johnston (1837).
6. Spencer, *Min. Mag.*, **15**, 302 (1910), by pycnometer.
7. Kreutz (1909), who gives $2E = 11°29'$, $12°44'$ (Na); Des Cloizeaux (**2**, 79, 1874) gave $2E = 9°50'$ (red, at 17°), $11°10'$ (red, at 141.5°).
8. Bellanca, *Per. Min.*, **12**, 127 (1941).
9. Alstonite is here considered to be a double salt, related to aragonite and witherite in the manner of dolomite to calcite and magnesite. This interpretation is suggested by the crystallographic relations and by thermal analysis of the system $BaCO_3$–$CaCO_3$ by Bellanca (1941), but it is not supported by all analyses of the mineral, notably the old and probably inferior analyses of Becker, *Zs. Kr.*, **12**, 222 (1886), in which Sr was not determined. The Sr sometimes present is presumed to substitute for Ba, but in the best available analysis (No. 2) (and in a nearly identical and very early analysis by Johnston (1837)) the ratios are $Ca:Ba:Sr:CO_2 = 7.6:7.7:1:16.3$. The problem is further discussed in detail by Kreutz (1909), Hintze (**1** [3A], 3096, 1926), and Gossner and Mussgnug (1930).
10. For earlier analyses see Dana (284, 1892) and (698, 1868).
11. Kreutz (1909).
12. Spencer (1910).
13. Bellanca (1941), Boeke, *Mitt. naturfor. Ges. Halle*, **3**, no. 3 (1913).
14. Frondel, priv. comm. (1947).

14.2.3 **BARYTOCALCITE** [CaBa(CO$_3$)$_2$]. Brooke (Ann. Phil., **8**, 114, 1824).

Cryst. Monoclinic; prismatic—$2/m$.

$a:b:c = 1.5534:1:1.2509;$ $\quad \beta\ 106°08';$ [1] $\quad p_0:q_0:r_0 = 0.8053:1.2016:1$

$r_2:p_2:q_2 = 0.8322:0.6701:1;$ $\quad \mu\ 73°52';$ $\quad p_0'\ 0.8383,\ q_0'\ 1.2509,\ x_0'\ 0.2893$

Forms: [2]

		ϕ	ρ	ϕ_2	$\rho_2 = B$	C	A
c	001	90°00′	16°08′	73°52′	90°00′	.....	73°52′
a	100	90 00	90 00	0 00	90 00	73°52′	.....
m	110	33 50	90 00	0 00	33 50	81 06	56 10
g	210	53 16½	90 00	0 00	53 16½	77 08	36 43½
o	101	90 00	48 26	41 34	90 00	32 18	41 34
x	$\bar{1}11$	−23 42	53 47½	118 46	42 22	61 29½	108 55½
y	$\bar{2}52$	−9 57½	72 31	118 46	20 02½	75 57	99 29½
ρ	$\bar{1}31$	−8 19½	75 13½	118 46	16 54½	78 06½	98 03

Structure cell. [3] Space group $P2_1$ or $P2_1/m$. $a_0\ 8.15\ kX$, $b_0\ 5.22$, $c_0\ 6.58; \beta\ 106°08'; a_0:b_0:c_0 = 1.561:1:1.261$. Cell contents Ca$_2Ba_2$(CO$_3$)$_4$.

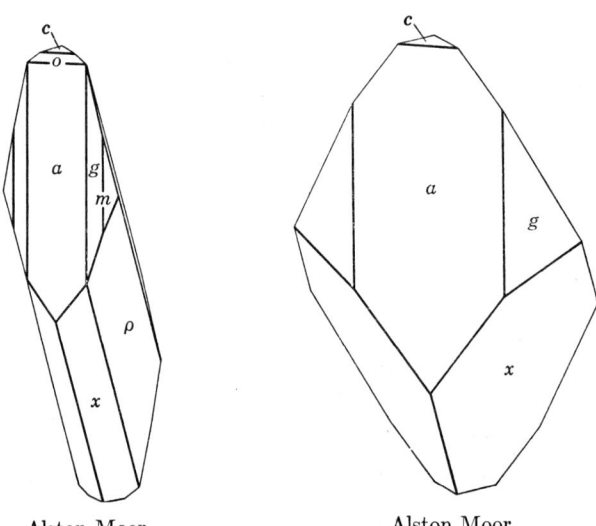

Alston Moor.   Alston Moor.

Habit. Usually short to long prismatic [101], also short prismatic [001] or equant. {100} striated [001]; and {$\bar{1}11$}, {$\bar{2}52$}, {$\bar{1}31$} striated [101]. Sometimes massive.

Oriented growths. [4] Barite upon barytocalcite, with barite {001}[010] parallel barytocalcite {001}[010]. Also calcite upon barytocalcite, with calcite ($10\bar{1}1$), ($\bar{1}101$), and ($01\bar{1}1$) approximately parallel barytocalcite (101), ($\bar{1}11$), ($\bar{1}\bar{1}1$).

Phys. Cleavage {210} perfect, {001} imperfect. Fracture uneven to subconchoidal. Brittle. H. 4. G. 3.66,[5] 3.71;[6] 3.65 (calc.). Luster vitreous, inclining to resinous. Colorless to white, grayish, greenish, or yellowish. Streak white. Transparent to translucent. Translation gliding on {100} (?).[6] Weakly fluorescent in ultraviolet light.[11]

Opt. In transmitted light, colorless.

Orientation[7]		n[8]	
$X \wedge c$	64°	1.525	Biaxial negative (−).
$Y \wedge c$	−26°	1.684	$2V$ 15°, $2E$ 25°.
$Z\ b$		1.686	$r > v$, slight; horizontal dispersion not apparent.

Chem. A carbonate of calcium and barium, $CaBa(CO_3)_2$. BaO 51.56, CaO 18.85, CO_2 29.59, total 100.00. Dimorphous with alstonite. The reported analyses [9] conform closely to the formula.

Tests. B.B. fusible at a high temperature on thin edges. Soluble in dilute hydrochloric acid.

Occur. Found as crystals up to 2 inches in length and as cleavable masses with barite, calcite, and fluorite in veins in limestone at Alston Moor, Cumberland, England. Quartz pseudomorphs after barytocalcite have been reported from Mies in Bohemia and Badenweiler, Baden, Germany, but the unaltered mineral was not found; at the Himmelsfürst mine at Freiberg, Saxony. A mineral apparently near barytocalcite has been reported [10] from Långban, Sweden, associated with hedyphane and hausmannite.

Alter. Pseudomorphs of barite and quartz after barytocalcite have been found.

Artif. The dimorphous modification alstonite has been prepared artificially, but not barytocalcite.

Name. In allusion to the composition.

Ref.

1. Orientation and angles of Brooke (1824), unit of structure cell of Gossner and Mussgnug, *Cbl. Min.*, 321 (1930). Transformation: Brooke to G. and M., 100/0½0/001; Goldschmidt (**1**, 171, 1913) to G. and M., 102/010/$\overline{1}$00.
2. Goldschmidt (1913).
3. Gossner and Mussgnug (1930), who consider the space group to be $P2_1$ (sphenoidal) from structural considerations although the form development indicates prismatic symmetry making the space group $P2_1/m$.
4. Mügge, *Jb. Min., Beil.-Bd.*, **16**, 385, 406 (1903); also sodium nitrate oriented upon barytocalcite described by Kreutz, *Min. Mag.*, **15**, 232 (1909); *Zs. Kr.*, **48**, 185 (1910).
5. Value of Children, *Ann. Phil.*, **7**, 275 (1824).
6. Kreutz, *Min. Mag.*, **15**, 232 (1909).
7. Des Cloizeaux (**2**, 80, 1874), who also gives $2E$ = 24°53′ (red, 17°) and 25°38′ (red, 170.8°).
8. Mallard, *Bull. soc. min.*, **18**, 10 (1895).
9. Cited in Dana (289, 1892; 702, 1868); also Kolbeck, *Min. Mitt.*, **54**, 373 (1942).
10. Sjögren, *Geol. För. Förh.*, **3**, 289 (1878); Des Cloizeaux, *Bull. soc. min.*, **4**, 95 (1881).
11. Pale yellowish in both long- and short-wave radiation, Frondel, priv. comm. (1947).

TYPE 3. MISCELLANEOUS

14.3.1 FAIRCHILDITE [$K_2Ca(CO_3)_2$]. *Milton* and *Axelrod* (*Am. Min.*, **32**, 607, 1947).

Hexagonal. As microscopic hexagonal plates flattened on {0001}. Cleavage {0001}, good. Uniaxial negative $(-)$;[1] nO 1.530, nE 1.48.

Chem. Potassium calcium carbonate, $K_2Ca(CO_3)_2$. The reported analyses[2] of natural material are on bulk clinkers containing in addition to fairchildite considerable amounts of buetschliite, calcite, and other impurities.

Occur. Found with buetschliite and calcite as clinkers formed by the fusion of wood ash in partly burned trees, especially hemlock and fir, at numerous localities in the western United States. Fairchildite slowly hydrates on exposure to the atmosphere, forming buetschliite which in turn may be decomposed and leached leaving calcite.

Artif.[2] Obtained in hexagonal tablets by treating calcite with strong solutions of KOH (35 per cent) or K_2CO_3 (59 per cent at 19°); buetschliite forms from less concentrated solutions. Also obtained from melts in the system K_2CO_3-$CaCO_3$; pure $K_2Ca(CO_3)_2$ melts at 815° in air. The artificial compound apparently forms isomorphous series with analogous compounds containing Na, Ba, or Sr.

Name. After John G. Fairchild, analytical chemist, of the U. S. Geological Survey.

Ref.
1. Niggli, *Zs. anorg. Chem.*, **98**, 241 (1916), on artificial material.
2. Summarized by Milton and Axelrod (1947).

14.3.2 SHORTITE [$Na_2Ca_2(CO_3)_3$]. *Fahey* (*Am. Min.*, **24**, 514, 1939).

Cryst.[1] Orthorhombic; pyramidal—$m\ m\ 2$.

$$a:b:c = 0.455:1:0.648; \quad p_0:q_0:r_0 = 1.425:0.648:1$$

$$q_1:r_1:p_1 = 0.455:0.702:1; \quad r_2:p_2:q_2 = 1.543:2.199:1$$

Forms:[1]

Lower	Upper	ϕ	$\rho = C$	ϕ_1	$\rho_1 = A$	ϕ_2	$\rho_2 = B$
c 001		....	0°00′	0°00′	90°00′	90°00′	90°00′
	a 100	90°00′	90 00	....	0 00	0 00	90 00
	e 011	0 00	56 30	56 30	90 00	90 00	33 30
	p 111	65 33	57 26	32 56½	39 54	35 03½	69 35

Structure cell.[2] Space group $Amm2$. a_0 4.98 kX, b_0 10.97, c_0 7.10; $a_0:b_0:c_0 = 0.454:1:0.647$. Cell contents $Na_4Ca_4(CO_3)_6$.

Habit. Wedge-shaped crystals, either tabular {100} or equant to short prismatic [100]. {011} striated [100] and in oscillatory combination with {00$\bar{1}$}.

P h y s. Cleavage {010}, distinct. Fracture conchoidal. H. 3. G. 2.60;[3] 2.605 (calc.). Luster vitreous. Colorless to pale yellow. Transparent. Strongly pyroelectric. Fluoresces light amber in ultraviolet light.
O p t. In transmitted light, colorless.

Orientation		n [4]	
X	c	1.531 ± 0.002	Biaxial negative $(-)$.
Y	a	1.555 ± 0.002	$2V\ 75°$ (calc.).
Z	b	1.570 ± 0.002	$r < v$, moderate.

C h e m. A carbonate of sodium and calcium, $Na_2Ca_2(CO_3)_3$. Shortite inverts at about 200° and melts incongruently at about 600° C.
Anal.

	1	2
Na_2O	20.25	19.91
CaO	36.63	36.34
MgO		0.04
CO_2	43.12	42.90
Insol.		0.66
Total	100.00	99.85
G.	2.605	2.60

1. $Na_2Ca_2(CO_3)_3$. 2. Wyoming.

Tests. Decomposed by water with separation of $CaCO_3$.

O c c u r. Found with calcite and pyrite embedded in an Eocene montmorillonite-type clay from an oil well about 20 miles west of Green River, Sweetwater County, Wyoming. Trona occurs in the same formation.

A r t i f. Not known artificially, although the compound $Na_2Ca(CO_3)_2$ has been prepared.

N a m e. After Maxwell N. Short (1889–   ), Professor of Mineralogy at the University of Arizona.

Ref.

1. Schaller in Fahey (1939).
2. Richmond, *Am. Min.*, **26**, 288, 629 (1941), by Weissenberg method on type material. See also Wickman, *Arkiv. Min. Geol.*, **1**, 95 (1949) for structure determination.
3. Richmond (1941) by microbalance; Fahey (1939) gives 2.629 by pycnometer on probably impure material.
4. Fahey (1939) by immersion method.

15 HYDRATED NORMAL CARBONATES

TYPE 1. $A(XO_3) \cdot xH_2O$

15.1.1 THERMONATRITE [$Na_2CO_3 \cdot H_2O$]. Νιτρον and Nitrum pt. *Vetruvius*. Natron, Alkali orientale impurum terrestre, Jordblandadt Alkaliskt-salt *Wallerius* (174, 1747). Natürliches mineralisches Alkali *Werner*. Prismatisches Natronsalz *Mohs* (Mohs-Haidinger, **2**, 29, 1825). Thermonatrit *Haidinger* (487, 1845). Thermonitrit *Hausmann* (**2**, 1411, 1847). Soude carbonatée prismatique *Dufrénoy* (**2**, 220, 1856).

C r y s t.[1] Orthorhombic; dipyramidal—$2/m\ 2/m\ 2/m$.

$a:b:c = 0.6047:1:0.4892;\qquad p_0:q_0:r_0 = 0.8090:0.4892:1$

$q_1:r_1:p_1 = 0.6047:1.2361:1;\qquad r_2:p_2:q_2 = 2.0442:1.6537:1$

Forms:[2] (artificial crystals).

c 001	b 010	a 100	μ 120	g 011	u 021	e 201	p 111

Structure cell.[3] Space group $Pmmm$. a_0 6.44 kX, b_0 10.72, c_0 5.24; $a_0:b_0:c_0 = 0.601:1:0.489$. Cell contents $Na_8(CO_3)_4 \cdot 4H_2O$.

Habit. Usually as a crust or efflorescence. Artificial crystals thin-lamellar {001}, also tabular {010}; elongated [100].

P h y s. Cleavage {100}, difficult. Somewhat sectile. H. $1–1\frac{1}{2}$. G. 2.255 (artif.);[4] 2.259 (calc.). Luster vitreous. Colorless to white; grayish, yellowish. Transparent. Taste alkaline.

O p t.[5] In transmitted light, colorless.

Orientation	n (artif.)[5]	n (artif.)[6]	
X b	1.420	1.420	Biaxial negative ($-$).
Y c	1.506	1.509	$2V$ 48°; $2E$ 75°.
Z a	1.524	1.525	$r < v$, weak.

C h e m. Sodium carbonate monohydrate, $Na_2CO_3 \cdot H_2O$. Na_2O 50.03, CO_2 35.45, H_2O 14.52, total 100.00. The reported analyses[7] are for the most part on very impure material.

Tests. Gently heated loses water and falls to a powder of Na_2CO_3. Easily fusible. Soluble in water (50.3 g. $Na_2CO_3 \cdot H_2O$ in 100 g. H_2O at 31.8°), affording an alkaline solution.

O c c u r. Found as an efflorescence on the soil in arid regions and as a deposit from saline lakes. Often associated with natron and trona, and formed by the partial dehydration of these species; also deposited directly from solution. Typical localities include the soda lakes of Szegedin and Debreczin in Hungary, and as salt crusts with trona in the desert region of the Sudan, Egypt. As a bedded deposit[9] in the Gorodki oil field near

Chusovsk in the Kama region, U.S.S.R. As a fumarole product, presumably resulting from the hydration of Na_2CO_3, on Vesuvius. Thermonatrite has been reported from Death Valley and the San Joaquin valley, California, as surface efflorescences, and it probably occurs at Owens Lake, Borax Lake, Searles Lake, and other saline lakes affording trona and natron, both in California and other western states.

A r t i f.[8] Thermonatrite is the stable phase from Na_2CO_3 solutions at temperatures between 32° and 112°; stable at lower temperatures when K_2CO_3, $NaHCO_3$, NaCl, or certain other salts are present. Formed by the dehydration of higher hydrates at the appropriate vapor pressure.

N a m e. From θερμός, *heat*, and *natron*, because it results from the drying out of natron, $Na_2CO_3 \cdot 10H_2O$, on increase of temperature.

Ref.
1. Angles of Marignac, *Ann. mines*, **12**, 55 (1857), x-ray unit of Colby and Harper, *Zs. Kr.*, **89**, 191 (1934); new orientation with $c < a < b$. Transformation: Marignac to new, 010/200/001; Colby and Harper to new, 010/100/001. The crystallographic data given by Goldschmidt (**8**, 129, 1922) apparently do not refer to thermonatrite.
2. Marignac (1857) on artificial crystals.
3. Colby and Harper (1934) by oscillation method on artificial crystals.
4. Pabst, *Am. Min.*, **15**, 72 (1930) by pycnometer. Colby and Harper (1934) give 2.25; values earlier given in the literature were 1.5–1.6.
5. Larsen (143, 1921).
6. Merwin cited in Winchell (200, 1931).
7. Cited in Hintze (**1** [3A], 2776, 1930).
8. Mellor (**2**, 751, 1922); Pabst (1930); Svatik and Stevens, *Am. Min.*, **17**, 538 (1932); Bury and Redd, *J. Chem. Soc. London*, 1160 (1933).
9. Aprodova, *C.r. ac. sc. U.R.S.S.*, **48**, 274 (1945).

15.1.2 **N E S Q U E H O N I T E** [$MgCO_3 \cdot 3H_2O$]. *Genth* and *Penfield* (*Am. J. Sc.*, **39**, 121, 1890; *Zs. Kr.*, **17**, 561, 1890).

C r y s t.[1] Orthorhombic; dipyramidal—$2/m\ 2/m\ 2/m$.

$a:b:c = 0.6444:1:0.4515$;[2] $p_0:q_0:r_0 = 0.7007:0.4515:1$

$q_1:r_1:p_1 = 0.6444:1.4272:1$; $r_2:p_2:q_2 = 2.1248:1.5518:1$

Forms:[3]

	φ	ρ = C	φ₁	ρ₁ = A	φ₂	ρ₂ = B
c 001	....	0°00′	0°00′	90°00′	90°00′	90°00′
b 010	0°00′	90 00	90 00	90 00	....	0 00
m 110	57 12	90 00	90 00	32 48	0 00	57 12
d 011	0 00	24 18	24 18	90 00	90 00	65 42

Structure cell.[4] Space group $Pmmm$. a_0 7.68 kX, b_0 11.93, c_0 5.39; $a_0:b_0:c_0 = 0.644:1:0.452$. Cell contents $Mg_4(CO_3)_4 \cdot 12H_2O$.

Habit. Prismatic [001], with {110} deeply striated [001]. In radiating tufts of acicular crystals, and as flat-radial coatings; also botryoidal, and as felted aggregates pseudomorphous after lansfordite.

P h y s. Cleavage {110} perfect, {001} less so. Fracture splintery to fibrous. H. $2\frac{1}{2}$. G. 1.852;[5] 1.856 (calc.). Luster vitreous or slightly greasy. Colorless to white. Transparent to translucent.

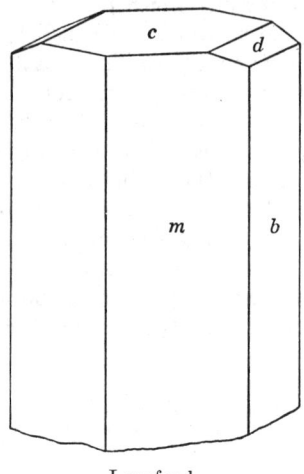

Lansford.

O p t.[6] In transmitted light, colorless.

ORIENTATION		n(Na)	
X	a	1.417 (calc.)	Biaxial negative $(-)$.
Y	c	1.503 ± 0.001	$2V\ 53°\ 2E\ 84°$ (Na).
Z	b	1.527 ± 0.001	$r < v$, weak.

C h e m. Magnesium carbonate trihydrate, $MgCO_3 \cdot 3H_2O$. The water is lost in two stages, the last at ca. 380° with some CO_2 lost simultaneously.[7] The formula may perhaps be better written $Mg(HCO_3)(OH) \cdot 2H_2O$.

Anal.

	1	2	3	4	5	6	7
MgO	29.14	29.22	29.24	28.52	29.14	29.20	29.09
CO_2	31.80	30.22	31.85	[31.09]	31.90	31.69	31.63
H_2O	39.06	40.32	39.11	40.03	39.51	39.30	39.21
Total	100.00	99.76	100.20	[99.64]	100.55	100.19	99.93
G.		1.856	1.852	1.854	1.834	1.824	1.842

1. $MgCO_3 \cdot 3H_2O$. 2. Lansford, Penn.[8] Average of four. 3. La Mure, France.[9] 4. Deposit from Donati spring, Rohitsch, Austria.[10] 5. Franscia, Val Lanterna, Italy.[11] 6. Viu, Val di Lanzo, Italy.[12] 7. Cogne, Val d'Aosta, Italy.[13]

Tests. Infusible. Easily soluble in dilute acids with effervescence. Very slightly soluble in water, but much more so in the presence of CO_2.

O c c u r. Nesquehonite was originally found in an anthracite coal mine at Nesquehoning near Lansford, Carbon County, Pennsylvania. When found it comprised the base of stalactites and incrustations, the remainder of which consisted of lansfordite from which it had been formed; on later exposure the entire mass became converted to white, chalky nesquehonite. Later observed in an old coal mine tunnel at La Mure, Isère, France. Also found lining crevices in serpentine rock at localities in northern Italy: Cogne, in the Val d'Aosta, and Viu in the Val di Lanzo, Piedmont; the

Franscia asbestos mine in the Val Lanterna, and in the Val Malenco, Lombardy. In serpentine at Kraubat, Styria, and near Rain, at Taufers in the southern Tyrol. As a deposit from the mineral springs at Rohitsch-Sauerbrunn, Austria. In all its occurrences nesquehonite, like lansfordite, appears to be a recent product formed under essentially atmospheric conditions of temperature and pressure.

A l t e r. Observed as a natural dehydration product of lansfordite (which see).

A r t i f.[14] Nesquehonite can be obtained as crystals by reaction of soluble carbonates with a Mg salt or by allowing CO_2 to escape slowly from a solution of magnesium bicarbonate. The precipitate formed in cold solutions contains some lansfordite, and in hot solutions some basic carbonate.

N a m e. From the original locality.

Ref.
1. Point symmetry established by Fenoglio, *Per. Min.*, **6**, 1 (1935).
2. Fenoglio (1935) on crystals from Cogne, Italy, and confirmed by his x-ray study. The earlier ratio of Genth and Penfield (1890) was based on crystals of inferior quality.
3. Genth and Penfield (1890).
4. Fenoglio (1935) by Laue and rotation methods on Cogne crystals.
5. Genth and Penfield (1890) on Lansford crystals. The values 1.854 and 1.842 were reported by Leitmeier, *Zs. Kr.*, **47**, 108 (1910), and Fenoglio (1935) on Rohitsch and Cogne crystals.
6. Fenoglio (1935), who shows the value calculated for X by Genth and Penfield, *Zs. Kr.*, **17**, 561 (1890), to be in error.
7. For dehydration studies see Beck, Doctoral Thesis, Harvard Univ. (1946); Davis, *J. Soc. Chem. Ind.*, **25**, 788 (1906); and Hintze (1 [3A], 3473, 1929).
8. Genth analysis in Genth and Penfield (1890).
9. Friedel, *Bull. soc. min.*, **14**, 62 (1891).
10. Leitmeier (1910).
11. Artini, *Acc. Linc., Rend.*, **30**, 153 (1921).
12. Fenoglio, *Acc. Linc., Rend.*, **11**, 310 (1930).
13. Fenoglio (1935).
14. von Knorre, *Zs. anorg. Chem.*, **34**, 260 (1903); Leitmeier, *Jb. Min., Beil.-Bd.*, **40**, 655 (1916); Hepburn, *J. Chem. Soc. London*, 96 (1940); also Mellor (**4**, 355, 1923).

15.1.3 Trihydrocalcite [$CaCO_3 \cdot 3H_2O$]. Kalkschaum (?) *Jurkiewicz*, 1872, mentioned in Krischtokovich (*Ann. géol. min. Russie*, [8], **3**, 124, 1906). Hydrocalcite *Kosman* (*Zs. deutsch. geol. Ges.*, **44**, 159, 1892). Bergmilch pt. (?). Trihydrocalcite *Tschirwinsky* (*Ann. géol. min. Russie*, [8], **8**, 238, 1910). Subhydrocalcite *Copisarow* (*J. Chem. Soc. London*, **123**, 785, 1923).

Found as a soft, white pulpy substance in a limestone cave at Wolmsdorff, Glatz, Silesia (*hydrocalcite*). A probably identical substance was found [1] as white mold-like or felty coatings in crevices in marl in the neighborhood of New Alexandria, Lublin region, Poland (U.S.S.R.). Under the microscope this material appeared as fibers with an extinction angle of 40° to 50° and high birefringence. G. ~ 2.63(?). Analyses of the above materials approximate to $CaCO_3 \cdot 3H_2O$; both lose water readily and afford distorted needle-like rhombohedral crystals of calcite (cf. *"lublinite"*). Further work is needed to establish the nature of this material and its relations to $CaCO_3 \cdot 5H_2O$ or $CaCO_3 \cdot 6H_2O$.

Ref.
1. Ivanoff, *Ann. géol. min. Russie*, [8], **1**, 23 (1905); see also Krischtokovich, *ibid.*, [8], **3**, 124 (1906), Tschirwinsky, *ibid.*, [8], **8**, 238 (1910), and Doelter (**1**, 356, 1912).

15.1.4 Pentahydrocalcite [$CaCO_3 \cdot 6H_2O$ (?)].

Fünffach gewässerter kohlensaurer kalk *Scheerer* (*Ann. Phys.*, **68**, 382, 1846). Hydroconite *Hausmann* (**2**, 1405, 1847). Pentahydrocalcite *Tschirwinsky* (*Ann. géol. min. Russie*, [8], **8**, 238, 1910).

Colorless crystals close in composition to $CaCO_3 \cdot 5H_2O$ have been observed as a recent formation in a water pipe [1] and as acute rhombohedrons (with G. 1.75) deposited upon wood in a brook near Oslo, Norway.[2] Artificial material approaching this composition has been described,[3] but without definitive crystallographic and physical data, and some of the observations may have been made on mixtures. The only well-defined hydrate of calcium carbonate is the hexahydrate, $CaCO_3 \cdot 6H_2O$. This substance [4] is monoclinic, and stout prismatic to tabular {001} in habit. G. 1.77. Optically negative ($-$), with nX 1.460, nY 1.535, nZ 1.545; $2V$ 38°; $Z \wedge c$ 17° $\pm$ 2° (Na). The hexahydrate changes rapidly into calcite and water at temperatures down to at least 0°; a monohydrate may form as an intermediate product, according to the conditions of formation.

Ref.
1. Salm-Horstmar, *Ann. Phys.*, **35**, 515 (1835).
2. Scheerer (1846).
3. Cf. Copisarow, *J. Chem. Soc. London*, **123**, 785 (1923), Tschirwinsky in Doelter (**1**, 356, 1912), and Mellor (**5**, 822, 1923).
4. Johnston, Merwin, and Williamson, *Am. J. Sc.*, **41**, 492 (1916); Vetter, *Zs. Kr.*, **48**, 72 (1910); Mackenzie, *J. Chem. Soc. London, Trans.*, **123**, 2409 (1923); Kraus and Schriever, *Zs. anorg. Chem.*, **188**, 259 (1930).

15.1.5 **L A N S F O R D I T E** [$MgCO_3 \cdot 5H_2O$].

Genth (*Zs. Kr.*, **14**, 255, 1888); Genth and Penfield (*Am. J. Sc.*, **39**, 121, 1890).

C r y s t.[1] Monoclinic; prismatic—$2/m$.

$a:b:c = 1.6446:1:0.9567;$ $\beta\ 101°46';$ [2] $p_0:q_0:r_0 = 0.5817:0.9366:1$

$r_2:p_2:q_2 = 1.0677:0.6211:1;$ $\mu\ 78°14';$ $p_0'\ 0.5942,\ q_0'\ 0.9567,\ x_0'\ 0.2083$

Forms: [3]

		ϕ	ρ	ϕ_2	$\rho_2 = B$	C	A
c	001	90°00′	11°46′	78°14′	90°00′	.....	78°14′
a	100	90 00	90 00	0 00	90 00	78°14′	.....
m	110	31 50½	90 00	0 00	31 50½	83 49½	58 09½
h	210	51 10	90 00	0 00	51 10	80 51½	38 50
r	011	12 17	44 23½	78 14	46 52½	43 07½	81 26½
o	$\bar{1}01$	$-90\ 00$	21 06	111 06	90 00	32 52	111 06
p	111	39 59½	51 18½	51 15	53 16½	44 25	59 53½
e	$\bar{1}11$	$-21\ 58$	45 53½	111 06	48 15	51 12	105 35

Less common:

$l\ 120$ $n\ \bar{2}01$ $q\ 211$ $s\ \bar{1}21$ $z\ \bar{2}11$ $f\ \bar{3}11$

LANSFORDITE

Structure cell.[4] Space group perhaps $P2_1/m$. a_0 12.48 kX, b_0 7.55, c_0 7.34; [β 101°46′]; $a_0:b_0:c_0 = 1.653:1:0.972$. Cell contents $Mg_4(CO_3)_4 \cdot 20H_2O$.

Habit. In stalactitic forms, elongated [001], bounded at the free extremity with faces (Lansford). Also as minute crystals, short prismatic [001].

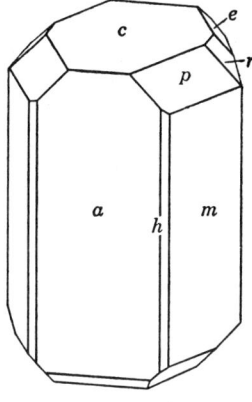

Cogne, Italy.

Phys. Cleavage {001} perfect, {100} less perfect.[5] H. $2\frac{1}{2}$. G. 1.692;[6] 1.700 (calc.). Luster of unaltered material vitreous. Colorless to white and translucent and resembling paraffin when unaltered but speedily becoming dull and opaque on exposure. Effloresces readily, affording the trihydrate nesquehonite.

Opt. In transmitted light, colorless.

ORIENTATION [8]		n (Atlin) [7]	n (Cogne) [8]	
X	b	1.465 ± 0.001	1.456	Biaxial positive (+).
Y	⊥ {100}	1.468 ± 0.001	1.469	
Z	c	1.507 ± 0.001	1.508	
2V		59°30′	59°48′	

Chem. Magnesium carbonate pentahydrate, $MgCO_3 \cdot 5H_2O$. The water is lost completely at about 110°.[9]

Anal.

	1	2	3
MgO	23.12	23.27	22.80
CO_2	25.23	25.09	25.43
H_2O	51.65	51.75	51.77
Total	100.00	100.11	100.00
G.	1.700	1.694	

1. $MgCO_3 \cdot 5H_2O$. 2. Cogne, Italy.[8] 3. Artificial.[10]

Tests. Infusible. Soluble in dilute acids with effervescence.

Occur. First found as small stalactites attached to the carbonaceous shale forming the roof of a gallery in the anthracite mine at Nesquehoning near Lansford, Carbon County, Pennsylvania. The lansfordite was associated with nesquehonite, and, when the specimens were exposed to the air, the former converted entirely to the latter. Also found as small crystals in a deposit of hydromagnesite at Atlin, British Columbia, Canada, and with nesquehonite in serpentinous rock at Cogne, Val d'Aosta, Piedmont, Italy.

A r t i f.[9] Precipitated admixed with nesquehonite when alkali carbonates are added to magnesium chloride solutions in the cold.

N a m e. From the original locality.

Ref.
1. Made triclinic by Genth and Penfield (1890) from study of the natural pseudomorphs and later shown to be monoclinic by Cesàro, *Ac. Belgique, Cl. sc., Bull.*, **4,** 234 (1910) from artificial crystals. Transformation: Genth and Penfield to Cesàro, $310/\overline{1}10/002$. Point symmetry confirmed by Fenoglio, *Per. Min.*, **4,** 443 (1933), from morphology and etch figures.
2. La Forge, priv. comm. (1933), recomputed from measurements of Genth and Penfield (1890); Cesàro (1910); Leitmeier, *Zs. Kr.*, **47,** 104 (1910); Fenoglio (1933).
3. Genth and Penfield (1890); Fenoglio (1933); Poitevin, *Am. Min.*, **9,** 225 (1924). Rare or uncertain: 010, 0.10.1, 102, 101, 302, 201, 9.11.11, 323, 321.
4. Fenoglio (1933) by Laue method; the data indicate only a primitive lattice with no glide, and the space group here given is inferred from the morphology.
5. Leitmeier (1910) and Fenoglio (1933).
6. Keeley in Genth and Penfield (1890); Fenoglio (1933) gives 1.694 on Cogne crystals.
7. Poitevin (1924).
8. Fenoglio (1933).
9. Fenoglio (1933) and Leitmeier (1910).
10. Cesàro (1910).

15.1.6 N A T R O N [$Na_2CO_3 \cdot 10H_2O$]. Νιτρον, Nitrum, pt. *of the Ancients*. Cristaux d'alkali fixe minéral *Romé de Lisle* (I, 149, 1783). Soude carbonatée, pt. *Haüy* (2, 374, 1801). Subcarbonate of Soda *Brooke* (*Ann. Phil.*, **6,** 287, 1823). Hemiprismatisches Natronsalz *Mohs* (2, 27, 1824) and *Zippe* (2, 29, 1839). Soda *Hausmann* (832, 1813). Natrit *Weisbach* (7, 1875). Soda. Carbonate of Soda. Sodium Carbonate. Sodium carbonate decahydrate. Soude carbonatée.

C r y s t.[1] Monoclinic; prismatic—$2/m$.

$a:b:c = 1.4828:1:1.4004;$ $\beta\, 121°08';$ $p_0:q_0:r_0 = 0.9444:1.2692:1$

$r_2:p_2:q_2 = 0.7879:0.7441:1;$ $\mu\, 58°52';$ $p_0'\, 1.1033,\, q_0'\, 1.4004,\, x_0'\, 0.6040$

Forms:[2] (on artificial crystals).

| c 001 | b 010 | a 100 | m 110 | e 011 | s $\overline{1}$01 | p $\overline{1}$12 |

Habit. Artificial crystals more or less tabular {010}, with varying development of the [100] and [001] zones. In nature found as crystalline, granular, or columnar crusts, as mealy coatings and as an efflorescence.

Twinning. Twin plane {001}.

P h y s. Cleavage {001} distinct, {010} imperfect, {110} in traces. Fracture conchoidal. Brittle. H. $1-1\frac{1}{2}$. G. 1.478.[3] Luster of crystals vitreous. Colorless to white, sometimes gray or yellow due to impurities. Taste alkaline. M.P. 34.5°. Effloresces rapidly in dry air, affording the monohydrate, thermonatrite.

O p t.[4] In transmitted light, colorless.

ORIENTATION	n	
X b	1.405 ± 0.003	Biaxial negative ($-$).
Y	1.425 ± 0.003	$2V$ large.
$Z \wedge c \sim 41°$	1.440 ± 0.003	$r > v$, perceptible; crossed.

C h e m. Sodium carbonate decahydrate, $Na_2CO_3 \cdot 10H_2O$. Na_2O 21.66, CO_2 15.38, H_2O 62.96, total 100.00. Analyses[5] of pure material conform closely to the formula.

Tests. Melts at a low temperature in its water of crystallization with the separation of thermonatrite. Easily soluble in water (40 g. Na_2CO_3 in 100 g. H_2O at 30.3°), the solution reacting alkaline and effervescing in acids.

O c c u r. The decahydrate, natron, crystallizes from sodium carbonate solutions only at comparatively low temperatures (under $ca.$ 32°), these being lowered still further by the presence of other dissolved substances. The substance occurs in nature most commonly in solution, as in the soda lakes of Wady Natrum, Egypt; Ragtown, Nevada; Borax Lake and Owens Lake, California. Many of the reported [6] natural occurrences of natron or "soda" are not certainly of this species and may represent the trona or thermonatrite, the latter readily forming from natron in dry air at ordinary temperatures and higher. Material definitely identified as the decahydrate, however, has been found in the soda-lake deposits of the Debreczin and Szegedin regions, Hungary, as an efflorescence on lava on Vesuvius and Etna, as a deposit from the soda lakes of Ragtown, Nevada, and Clinton, Lilloet district, British Columbia, in the salt lakes of the eastern Gobi desert, Mongolia, and at other localities. Usually associated with thermonatrite, trona, gaylussite, and calcite.

N a m e. See trona and niter.

Ref.

1. Angles of Mohs (**2**, 34, 1824); orientation and unit of Lévy (**1**, 324, 1837). Dana (301, 1892), Goldschmidt (**3**, 139, 1891), and Hintze (**1** [3A], 2778, 1916) have followed Lévy. Groth (**2**, 197, 1908), and others have chosen another orientation. Transformation: Groth to Lévy, 001/010/$\overline{1}$01.
2. Goldschmidt (**8**, 63, 1922).
3. Schröder, *Ber.*, **12**, 119 (1879). Schiff, *Ann. Phys.*, **108**, 334 (1858), gave 1.475, and other values have been reported down to 1.423.
4. Larsen (115, 1921).
5. For analyses see Hintze (1916) and Walker and Parsons, *Univ. Toronto Stud., Geol. Ser.*, **24**, 15 (1927).
6. Numerous localities are listed in Hintze (1916); see also Chatard, *U. S. Geol. Sur., Bull. 60*, 27 (1890), on the U. S. occurrences.

TYPE 2. MISCELLANEOUS

15.2.1 **B U E T S C H L I I T E** [$K_6Ca_2(CO_3)_5 \cdot 6H_2O$]. *Milton* and *Axelrod* (*Am. Min.*, **32**, 607, 1947).

Probably hexagonal. As microscopic, barrel-shaped crystals elongated [0001]. Uniaxial negative ($-$); nO 1.595, nE 1.455.

Chem. A hydrated carbonate of potassium and calcium, $K_6Ca_2(CO_3)_5 \cdot 6H_2O$. The reported analyses [1] of natural material are on clinkers containing fairchildite, calcite, and other material in addition to buetschliite.

Occur. Found with calcite as a product of hydration of fairchildite in clinkers formed by the fusion of wood-ash in partly burned trees. Further slaking of the clinkers may decompose the buetschliite leaving a residue of calcite.

Artif.[1] As crystals by reaction of calcite with relatively dilute solutions of KOH or K_2CO_3; also by hydration of fairchildite.

Name. After Otto Buetschli (1848–1920), Professor of Zoology at Heidelberg, who studied the double carbonates of potassium and calcium.

Ref.
1. Summarized by Milton and Axelrod (1947).

15.2.2 **P I R S S O N I T E** [$Na_2Ca(CO_3)_2 \cdot 2H_2O$]. Pratt (Am. J. Sc., **2**, 126, 1896).

Cryst. Orthorhombic; rhombic-pyramidal—$m\,m\,2$.

$a:b:c = 0.5662:1:0.3019;$ $\qquad p_0:q_0:r_0 = 0.5332:0.3019:1$

$q_1:r_1:p_1 = 0.5662:1.8755:1;$ $\qquad r_2:p_2:q_2 = 3.3124:1.7662:1$

Forms:

Lower	Upper	ϕ	$\rho = C$	ϕ_1	$\rho_1 = A$	ϕ_2	$\rho_2 = B$
	b 010	0°00′	90°00′	90°00′	90°00′		0°00′
	m 110	60 29	90 00	90 00	29 31	0°00′	60 29
$\bar{p}$	p 111	60 29	31 30	16 48	62 57½	61 56	75 05
	e 131	30 29	46 25½	42 10	68 26	61 56	51 22
	x 311	79 18½	58 26	16 48	33 09	32 01	80 54½

Structure cell.[4] Space group $Fdd2$. a_0 11.32 Å, b_0 20.06, c_0 6.00; $a_0:b_0:c_0 = 0.5643:1:0.2991$. Cell contents $Na_{16}Ca_8(CO_3)_{16} \cdot 16H_2O$.

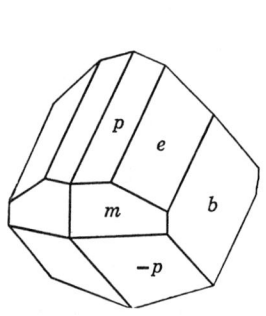
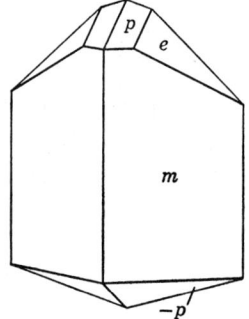

Borax Lake, Calif. Borax Lake, Calif.

Habit. Short prismatic [001] with variable upper termination; also tabular {010}, or pyramidal with e and $\bar{p}$.

PIRSSONITE 233

P h y s. Fracture conchoidal. Brittle. H. 3–3½. G. 2.352; 2.37 (calc.). Luster vitreous. Colorless to white; sometimes grayish due to inclusions. Pyroelectric, the end of the [001] axis terminated by {131} becoming negatively electrified on cooling. Transparent.

O p t.[1] In transmitted light, colorless.

ORIENTATION	n(Li)	n(Na)	n(Tl)	
X a		1.5043		Biaxial positive (+).
Y c	1.5056	1.5095	1.5115	2V increases slightly with
Z b	1.5710	1.5751	1.5789	increasing temperature.
2V (meas.)	31°11½′	31°26′	31°27′	$r < v$, slight.

C h e m. A hydrated carbonate of sodium and calcium, $Na_2Ca(CO_3)_2 \cdot 2H_2O$.

Anal.

	1	2
Na_2O	25.61	25.70
CaO	23.16	23.38
CO_2	36.35	36.07
H_2O	14.88	14.73
Rem.		0.57
Total	100.00	100.45

1. $Na_2Ca(CO_3)_2 \cdot 2H_2O$. 2. Borax Lake, California. Rem. is K_2O 0.15, Al_2O_3 0.13, SiO_2 0.29. Average of two.

Tests. B.B. decrepitates and fuses at about 2–2½ to an alkaline mass. Soluble in cold dilute acids with effervescence.

O c c u r. Found with northupite, tychite, and gaylussite in drill cuttings from clay beds in Borax Lake, San Bernardino County, California. Later recognized in the Otjiwalundo Salt Pan, about 250 kilometers north of Otavi, South West Africa, where it occurs associated with thenardite and trona.[2] Also found with shortite, bradleyite, northupite, gaylussite, and trona in drill cores from oil shale near Green River, Sweetwater County, Wyoming.

A r t i f.[3] Observed in the system Na_2CO_3-$CaCO_3$-H_2O; pirssonite cannot form below 37° C. or in a solution containing less than 21.7 per cent of Na_2CO_3.

N a m e. After Louis Valentine Pirsson (1860–1919), petrographer and mineralogist, for many years Professor of Physical Geology at Yale University.

Ref.

1. Pratt (1896); indices for Li and Tl by prism method, for Na by total reflection. Foshag, *Am. Min.*, **18**, 431 (1933), obtained almost identical values by the immersion method on material from Otjiwalundo.
2. Foshag (1933).
3. Bury and Redd, *J. Chem. Soc. London*, 1160 (1933).
4. Evans, *Am. Min.*, **33**, 261 (1948), by Weissenberg method on Searles Lake crystal.

15.2.3 **GAYLUSSITE** [$Na_2Ca(CO_3)_2 \cdot 5H_2O$]. Gay-Lussite *Boussingault* (*Ann. chim. phys.*, **31**, 270, 1826). Natrocalcit *Glocker* (673, 1839), *Hartmann* (Minéralogie, **2**, 271, 1843). Natrium-calciumcarbonat-Pentahydrat. Gay-Lussite. Gaylussacite.

Cryst. Monoclinic; prismatic—$2/m$.

$a:b:c = 1.4897:1:1.4441;$ $\beta\, 101°33';$ [1] $p_0:q_0:r_0 = 0.9694:1.4149:1$

$r_2:p_2:q_2 = 0.7068:0.6852:1;$ $\mu\, 78°27';$ $p_0'\, 0.9894,\, q_0'\, 1.4441,\, x_0'\, 0.2044$

Forms: [2]

	ϕ	ρ	ϕ_2	$\rho_2 = B$	C	A
c 001	90°00′	11°33′	78°27′	90°00′	.....	78°27′
b 010	0 00	90 00	.....	0 00	90°00′	90 00
a 100	90 00	90 00	0 00	90 00	78 27	.....
m 110	34 25	90 00	0 00	34 25	83 30	55 35
e 011	8 03½	55 34	78 27	35 15	54 45	83 22
s $\bar{1}01$	−90 00	38 08	128 08	90 00	49 41	128 08
r $\bar{1}12$	−21 54	37 53½	106 11½	55 15½	43 20	103 14½

Habit. Often elongated [100]; also flattened wedge-shaped with dominant {110}, {011}. Surfaces usually rough, with {011} striated [1$\bar{1}$1]. Common forms: $m\ e\ c\ r$.

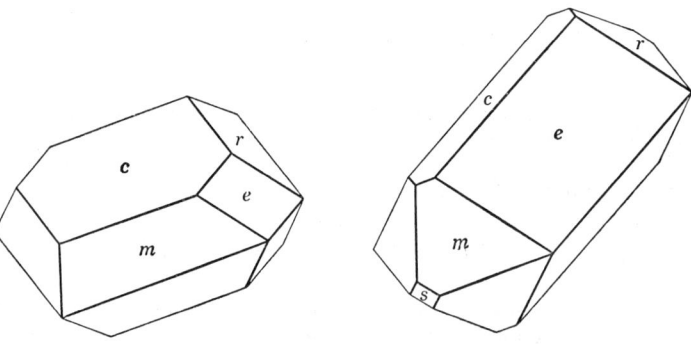

Wyoming.   Borax Lake, Calif.

Phys. Cleavage {110} perfect, {001} rather difficult. Fracture conchoidal. Very brittle. H. $2\frac{1}{2}$–3. G. 1.991.[3] Luster vitreous. Colorless to yellowish white, grayish white, and white. Streak uncolored to grayish white. Transparent to translucent. Effloresces slowly in dry air.

Opt.[4] In transmitted light, colorless.

ORIENTATION		n(Na)	
X	b	1.4435	Biaxial negative (−).
Y		1.5156	$2V\, 34°,\, 2E\, 52°$ (Na.)
$Z \wedge c$	−15°(Na)	1.5233	$r < v$ strong, crossed.

Chem. A hydrated carbonate of sodium and calcium, $Na_2Ca(CO_3)_2 \cdot 5H_2O$. The water is lost at *ca.* 100°.[5]

Anal.[6]

	1	2
Na_2O	20.93	20.40
CaO	18.94	19.02
CO_2	29.72	30.02
H_2O	30.41	30.47
Rem.		0.29
Total	100.00	100.20
G.		1.991

1. $Na_2Ca(CO_3)_2 \cdot 5H_2O$. 2. Taboos-nor, Gobi Desert.[7] Rem. is insol. 0.16, SiO_2 0.08, Fe_2O_3 0.03, MgO 0.01, MnO 0.01.

Tests. Easily fusible to a white enamel. C.T. decrepitates, loses water easily, and turns white. Soluble in acids with effervescence. Slightly soluble in water; the dehydrated powder is decomposed by water, sodium carbonate being extracted and calcium carbonate left as a residue.

Occur. Found as a deposit from soda lakes, where it occurs with natron, thermonatrite, trona, pirssonite, calcite, borax. Originally found at Lagunillas, near Merida, Venezuela, where its crystals are found in a bed of clay in a small alkali lake. In the Taboos-nor (Salt Lake) in the eastern Gobi desert, Mongolia. In the United States found at several sinks in the Carson desert near Ragtown, Nevada. With northupite and pirssonite in trona at Borax Lake, Lake County, at Mono Lake, Mono County, and at Searles Lake (Borax Lake), San Bernardino County, in California. From Independence Rock in the Sweetwater Valley, Wyoming. Other localities have been reported on the basis of calcite pseudomorphs supposedly after gaylussite.

Alter. Gaylussite alters readily to calcite. Not all the calcite pseudomorphs referred to gaylussite on the basis of crystal habit, however, are known definitely to be after this species (see pseudo-gaylussite, under calcite).

Artif.[8] Gaylussite is a stable phase between 15° and 40° in the system Na_2CO_3-$CaCO_3$-H_2O.

Name. After L. J. Gay-Lussac (1778–1850), French chemist.

Ref.

1. Angles of Phillips, *Phil. Mag.*, **1**, 263 (1827); unit of Dana (301, 1892) and all later authors. Transformation: Phillips to Dana, 100/010/002.
2. Goldschmidt (**4**, 29, 1918).
3. Average of the values 1.992, 1.990, 1.991 reported by Pratt, *Am. J. Sc.*, **2**, 130 (1896), Niinomy, *Econ. Geol.*, **25**, 758 (1930), and Phillips (1827) on Borax Lake, Lagunilla, and Gobi crystals. Mrose, priv. comm. (1947), obtained 1.986 on Searles Lake crystals of poor quality. The range sometimes given, 1.93–1.95, is too low.
4. Pratt (1896) by total reflection. See also Des Cloizeaux, *Ann. mines*, **14**, 400 (1858); Niinomy (1925); and Arzruni, *Zs. Kr.*, **6**, 30 (1881).
5. Bütschli, *J. pr. Chem.*, **75**, 556 (1907), and de Schulten, *C.R.*, **123**, 1023 (1896).
6. For earlier analyses see Hintze (**1** [3A], 2796, 1916).
7. Okada analysis in Niinomy (1925).
8. Mellor (**3**, 845, 1923); Bury and Redd, *J. Chem. Soc. London*, 1160 (1933); Hintze (1916).

15.2.4 SCHROECKINGERITE [$NaCa_3(UO_2)(CO_3)_3(SO_4)F \cdot 10H_2O$].
Schröckingerite *Schrauf* (*Min. Mitt.*, 137, 1873). Schröckeringite. Dakeite *Larsen and Gonyer* (*Am. Min.*, **22**, 561, 1937).

Hexagonal? As clusters or globular aggregates of scales flattened {0001}, sometimes with a six-sided outline and interfacial angles of about 60°.[1] Cleavage {0001} perfect. Laminae flexible. H. $2\frac{1}{2}$. G. 2.51. Luster weakly vitreous, somewhat pearly on {0001}. Color greenish yellow. Transparent. Strongly fluorescent in yellowish green in ultraviolet light.

Opt.

	n (Wyoming)[2]	nNa (Joachimsthal)[3]	
O	1.542	1.539–1.545	Uniaxial (?) negative (−).
E	1.489	1.496	

Usually biaxial with small and variable 2V, 0°–25°.

Chem. A hydrated fluo-carbonate-sulfate of sodium, calcium, and uranium, $NaCa_3(UO_2)(CO_3)_3(SO_4)F \cdot 10H_2O$. The H_2O is largely or entirely lost at about 75°. Analyses gave:[4]

	Na_2O	K_2O	MgO	CaO	UO_3	CO_2	SO_3	H_2O	F	insol.	Total
1.	3.49			18.93	32.19	14.86	9.01	20.28	2.14		100.00
										(less O = F =	0.90)
2.	3.19	0.23	0.63	18.44	31.28	14.67	9.24	20.20	2.09	1.53	100.62
										(less O = F =	0.88)
3.	3.63			18.14	31.44	14.20	9.17	20.15	2.15	1.03	99.01
										(less O = F =	0.90)

1. $NaCa_3(UO_2)(CO_3)_3(SO_4)F \cdot 10 H_2O$. 2. Arizona.[5] 3. Wyoming.[6]

Tests. Soluble in water and in acids. In C.T. yields neutral water.

Occur. Originally found as an alteration product of uraninite at Joachimsthal, Bohemia; reported from Johanngeorgenstadt, Saxony. In the United States found as a near-surface deposit from meteoric waters as small concretions in a gypsum-containing clay and in partly indurated arkose near Wamsutter, Wyoming (*dakeite*).[8] Also from the Hillside mine, Yavapai County, Arizona, associated with bayleyite, swartzite, andersonite, and gypsum as an efflorescence on tunnel walls.

Name. After J. von Schroeckinger who found and later described[7] the occurrence at Joachimsthal. Dakeite after H. C. Dake of Portland, Oregon.

Ref.

1. Originally interpreted as orthorhombic by Schrauf (1873), with (010) (110) $58\frac{1}{2}°$, but the data have little value; see also Nováček, *Am. Min.*, **24**, 317 (1939).
2. Larsen and Gonyer (1937).
3. Nováček (1939).
4. Earlier analyses by Nováček (1939), Larsen and Gonyer (1937), and Schrauf (1873) did not report F.
5. Grimaldi analysis in Axelrod, *et al.*, *Am. Min.*, **36**, 1, (1951).
6. Sherwood analysis in Jaffe, *et al.*, *Am. Min.*, **33**, 152 (1948).
7. Schroeckinger, *Geol. Reichsanst.*, *Wien*, *Verh.*, 66 (1875).
8. The identity of dakeite with schroeckingerite was proved by Nováček (1939).

15.2.5 Voglite [$Ca_2CuU(CO_3)_5 \cdot 6H_2O$ (?)]. Uran-Kalk-Kupfer-Carbonat *Vogl* (*Jb. Geol. Reichsanst. Wien*, 4, 222, 1853). Voglit *Haidinger* (*ibid.*, p. 223).

Triclinic. As coatings of scales. The scales are rhomboidal with an interior acute angle of 75°–80°. Luster pearly. Color emerald-green to bright grass-green. Streak pale green.

O p t.[1]

ORIENTATION	n	PLEOCHROISM	
X	1.541	Deep bluish green	Biaxial positive (+).
Y	1.547	Deep bluish green	$2V$ 60° (Na).
Z	1.564	Pale yellowish	$r < v$, very strong.

X is nearly normal to the rhomboidal plates and Z makes an angle of about 33° with the longer side in the acute angle. When turned on edge the plates show lamellar twinning with small extinction angles.

C h e m. A hydrated carbonate of calcium, copper, and uranium. The formula is uncertain, perhaps $Ca_2CuU(CO_3)_5 \cdot 6H_2O$. Analysis gave:

	CaO	CuO	UO_2	CO_2	H_2O	Total
1.	14.20	10.07	34.19	27.86	13.68	100.00
2.	14.09	8.40	37.00	26.41	13.90	99.80

1. $Ca_2CuU(CO_3)_5 \cdot 6H_2O$.   2. Joachimsthal.[2]

Tests. B.B. infusible. In C.T. blackens and yields water. Soluble in acids with effervescence.

O c c u r. Found with liebigite as an alteration of uraninite in the Elias mine near Joachimsthal, Bohemia.

N a m e. After J. F. Vogl.

Ref.

1. Larsen (154, 1921). Schrauf, *Min. Mitt.*, 138 (1873), gave the extinction angle as 36°, and the acute angle of the rhomboids as 78°–80°; Larsen (1921) gives the latter angle as ~75°, and Haidinger in Vogl (1853) as ~80°.
2. Lindacker analysis in Vogl (1853).

UNNAMED MINERAL. *Antipov* (*Vh. min. Ges.*, **38**, 38, 1900).
Found as scales with calcite in the Utsch-Kirtan pass, Ferghana, U.S.S.R. G. 3.35. An analysis on material admixed with calcite and iron oxide gave: CaO 1.35, CuO 9.71, Fe_2O_3 1.12, U_2O_3 35.45, CO_2 10.88, P_2O_5 tr., H_2O 42.13, total 100.64.

15.2.6 B A Y L E Y I T E [$Mg_2(UO_2)(CO_3)_3 \cdot 18H_2O$]. *Axelrod, Grimaldi, Milton,* and *Murata* (*Am. Min.*, **36**, 1, 1951).

Monoclinic. As minute, short-prismatic crystals; also in acicular, divergent groups (artificial). Color sulfur-yellow. G. 2.05; 2.06 (calc.). Fluoresces feebly in ultraviolet light.

Structure cell.[1] Space group $P2_1/a$. a_0 26.65 ± 0.05 Å, b_0 15.3 ± 0.05, c_0 6.53 ± 0.02; $a_0:b_0:c_0 = 1.742:1:0.4268$; β 93°04′. Cell contents $Mg_8(UO_2)_4(CO_3)_{12} \cdot 72H_2O$.

Opt.

	n	PLEOCHROISM	
X	1.455	Pinkish	Biaxial negative (−).
Y	1.490	Pale yellow	$2V$ 30° (calc.).
Z	1.500	Pale yellow	Inclined extinction.

Chem. A hydrated carbonate of magnesium and uranium, $Mg_2(UO_2)(CO_3)_3 \cdot 18H_2O$. Analysis gave:

	MgO	UO_3	CO_2	H_2O	Rem.	Total
1.	9.80	34.76	16.04	39.40		100.00
2.	9.76	35.28	16.72	37.91	0.33	100.00

1. $Mg_2(UO_2)(CO_3)_3 \cdot 18H_2O$. 2. Arizona.[2] Rem. is Na_2O 0.21, K_2O 0.10, SO_3 0.02. Analysis recalculated after deducting 8.45 per cent gypsum and 0.45 insol.

Tests. Soluble in water.

Occur. Found with schroeckingerite, andersonite, swartzite, and gypsum as a coating on mine walls in the Hillside mine, Yavapai County, Arizona.

Alter. On exposure rapidly breaks down into a lower hydrate (with nX 1.502, nZ 1.551, and fairly strong green fluorescence).

Artif. Obtained as groups of acicular crystals by reaction of solutions of uranyl nitrate, magnesium nitrate, and potassium carbonate.

Name. After William S. Bayley (1861–1943), American mineralogist and geologist.

Ref.

1. Axelrod in Axelrod, et al. (1951) by Weissenberg method.
2. Grimaldi analysis; a second, comparable analysis is cited.

15.2.7 **SWARTZITE** [$CaMg(UO_2)(CO_3)_3 \cdot 12H_2O$]. *Axelrod, Grimaldi, Milton,* and *Murata* (*Am. Min.*, **36**, 1, 1951).

Monoclinic. As clusters of tiny prismatic crystals. Color green. G. 2.3; 2.32 (calc.). Fluoresces bright green in ultraviolet light.

Structure cell.[1] Space group $P2_1/m$ or $P2_1$. a_0 11.12 ± 0.05 Å, b_0 14.72 ± 0.05, c_0 6.47 ± 0.02; β 99°26′; $a_0:b_0:c_0 = 0.7554:1:0.4395$. Cell contents $Ca_2Mg_2(UO_2)_2(CO_3)_6 \cdot 24H_2O$.

Opt.

Orientation	n	Pleochroism	
X	1.465	Colorless	Biaxial negative (−).
Y	1.51	Yellow	$2V$ 40° (calc.).
Z	1.540	Yellow	

Chem. A hydrated carbonate of calcium, magnesium, and uranium, $CaMg(UO_2)(CO_3)_3 \cdot 12H_2O$. Analysis gave:[2]

	MgO	CaO	UO_3	CO_2	H_2O	Rem.	Total
1.	5.52	7.67	39.15	18.07	29.59		100.00
2.	5.47	7.32	38.85	17.92	29.69	0.75	100.00

1. $CaMg(UO_2)(CO_3)_3 \cdot 12H_2O$. 2. Arizona. Rem. is Na_2O 0.26, K_2O 0.49. Recalculated after deducting 4.26 per cent gypsum, 0.30 insol.

Tests. Soluble in water.

Occur. Found with gypsum, schroeckingerite, bayleyite, and andersonite as an efflorescence on walls of the Hillside mine, Yavapai County, Arizona.

Alter. On exposure breaks down to a feebly fluorescent lower hydrate.
Artif. Obtained as crystals by seeding an evaporating solution of potassium carbonate and the nitrates of calcium, magnesium, and uranium.
Name. After Charles K. Swartz (1861–1949), of Johns Hopkins University, geologist and mineralogist.

Ref.
1. Axelrod in Axelrod, *et al.* (1951) by Weissenberg method.
2. Grimaldi analysis in Axelrod, *et al.* (1951).

15.2.8   **ANDERSONITE** [$Na_2Ca(UO_2)(CO_3)_3 \cdot 6H_2O$]. *Axelrod, Grimaldi, Milton*, and *Murata* (*Am. Min.*, **36**, 1, 1951).

Hexagonal—*R*. As clusters of minute pseudo-cubic crystals. Color bright green. G. 2.8; 2.86 (calc.). Fluoresces bright green in ultraviolet light. Stable on exposure.

Structure cell.[1] Space group $R\bar{3}$ or $R3$. a_{rh} 13.11 ± 0.02, α 86°56'; a_0 18.04 ± 0.05 Å, c_0 23.90 ± 0.05; $a_0:c_0 = 1:1.325$. Cell contents in the rhombohedral unit $Na_{12}Ca_6(UO_2)_6(CO_3)_{18} \cdot 36H_2O$.

Opt.

	n	Dichroism	
O	1.520	Colorless	Uniaxial positive (+).
E	1.540	Pale yellow	

Chem. A hydrated carbonate of sodium, calcium, and uranium, $Na_2Ca(UO_2)(CO_3)_3 \cdot 6H_2O$. Analyses gave:[2]

	Na_2O	MgO	CaO	UO_3	CO_2	H_2O	Total
1.	9.62		8.71	44.40	20.49	16.78	100.00
2.	9.61		8.80	44.27	20.61	16.50	99.79
3.	9.6	0.5	8.1	44.9	20.3	[16.6]	[100.00]

1. $Na_2Ca(UO_2)(CO_3)_3 \cdot 6H_2O$. 2. Artificial. 3. Arizona. Microanalysis on 3.5 mg. Recalculated after deducting 3.4 per cent gypsum. Mg determined spectrographically.

Tests. Soluble in water.

Occur. Found with gypsum, schroeckingerite, bayleyite, and swartzite as an efflorescence on walls of the Hillside mine, Yavapai County, Arizona.

Artif. Obtained as crystals by evaporating a solution containing potassium carbonate and the nitrates of sodium, calcium, and uranium.

Name. After Charles A. Anderson, geologist of the U. S. Geological Survey, who first observed and collected andersonite and its associated minerals.

Ref.
1. Axelrod in Axelrod, *et al.* (1951) by the Weissenberg method.
2. Grimaldi (3) and Eiland (2) analyses in Axelrod, *et al.* (1951).

15.2.9 LIEBIGITE [$Ca_2U(CO_3)_4 \cdot 10H_2O$]. Smith (Am. J. Sc., **5**, 336, 1848; **11**, 259, 1851). Kalk-Uran-carbonate Vogl (Geol. Reichsanst., Wien, Jb., **4**, 221, 1853). Flutherite Weisbach (48, 1875). Uranothallite Schrauf (Zs. Kr., **6**, 410, 1882).

Cryst.[1] Orthorhombic.

$a:b:c = 0.952:1:0.786;$ $p_0:q_0:r_0 = 0.826:0.786:1$

$q_1:r_1:p_1 = 0.952:1.211:1;$ $r_2:p_2:q_2 = 1.272:1.050:1$

Forms:

	ϕ	$\rho = C$	ϕ_1	$\rho_1 = A$	ϕ_2	$\rho_2 = B$
c 001	.....	0°00′	0°00′	90°00′	90°00′	90°00′
b 010	0°00′	90 00	90 00	90 00	.....	0 00
a 100	90 00	90 00	.....	0 00	0 00	90 00
n 230	35 01	90 00	90 00	54 59	0 00	35 01
m 110	46 25½	90 00	90 00	43 34½	0 00	46 25½
o 210	64 33½	90 00	90 00	25 26½	0 00	64 33½
d 011	0 00	38 10	38 10	90 00	90 00	51 50
p 111	46 25½	48 45	38 10	57 00	50 26½	58 47
r 121	27 43	60 37	57 32½	66 05½	50 26½	39 31½
q 141	14 02	73 38	73 09½	76 32½	50 26½	21 26
t 311	72 24	68 57½	38 10	27 10½	21 58½	73 36½

Doubtful: x 787   u 343   s 232   y 8.15.8

Structure cell.[2] Space group Bbam or Bba. a_0 16.71 Å, b_0 17.55, c_0 13.79; $a_0:b_0:c_0 = 0.952:1:0.786$. Cell contents $Ca_{16}U_8(CO_3)_{32} \cdot 80H_2O$.

Habit. Crystals equant or short prismatic [001], usually indistinct with rounded edges and convex or vicinal faces. Commonly as granular or scaly aggregates and thin crusts; also botryoidal.

Phys. Cleavage {100}. H. 2½–3. G. 2.41; 2.43 (calc.). Luster vitreous, slightly pearly on the cleavage. Color siskin-green to yellowish green. Transparent to translucent. Fluoresces green in long- and short-wave ultraviolet light.

Opt.[3]

Orientation		n	Pleochroism	
X	a	1.497	Nearly colorless	Biaxial positive (+).
Y		1.502	Pale yellowish green	2V 40°.
Z		1.539	Pale yellowish green	$r > v$, moderate.

Chem. A hydrated carbonate of uranium, $Ca_2U(CO_3)_4 \cdot 10H_2O$. The water is entirely lost below 200°.[4]

Anal.

	1	2	3	4
CaO	15.19	15.56	16.42	16.28
FeO				2.48
UO_2	36.57	37.11	36.29	35.45
CO_2	23.84	23.87	22.95	23.13
H_2O	24.40	23.35	23.72	22.44
Total	100.00	99.89	99.38	99.78
G.	2.43			2.14–2.15

1. $Ca_2U(CO_3)_4 \cdot 10H_2O$. 2. Joachimsthal.[5] Average of three. 3. Joachimsthal.[6] Analysis on 0.15 grams. 4. Joachimsthal.[7] Analysis on 0.151 grams.

Tests. B.B. infusible. Soluble with effervescence in H_2SO_4, leaving a white deposit. The solution in H_2SO_4 and HCl is green, in HNO_3 yellow.

Occur. A secondary mineral, in some instances found associated with schroeckingerite and β-uranophane. Found at various mines in the Joachimsthal district, Bohemia (*uranothallite*).[8] Also from Schneeberg, Saxony, and the Wheal Basset, Redruth, Cornwall, England.[9] From Schmiedeberg, Silesia, and Eisleben, Thuringia.[11] Reported [10] from Johanngeorgenstadt, Saxony. Liebigite was originally from near Adrianople, Turkey.

Alter. To an unidentified yellow ocherous substance.

Name. After the German chemist Justus Liebig (1803–1873).

Ref.

1. Elements and forms of Brezina, *Ann. nat. Hofmus. Wien*, **5**, 495 (1890).
2. Evans in Evans and Frondel, *Am. Min.*, **35**, 251 (1950), by precession method on Joachimsthal material.
3. Indices average of numerous measurements in Evans and Frondel (1950).
4. Dehydration data are given by Schrauf, *Zs. Kr.*, **6**, 410 (1882), and von Foullon in Brezina (1890).
5. Lindacker analyses in Vogl, *Gangverh. Mineralreich. Joachimsthal, Teplitz* (1856).
6. Schrauf (1882).
7. von Foullon in Brezina (1890).
8. Shown to be identical with liebigite by Evans and Frondel (1950).
9. Evans and Frondel (1950).
10. Smith (1851).
11. Meixner, *Zbl. Min.*, 145 (1940).

15.2.10 **LANTHANITE** [$(La,Ce)_2(CO_3)_3 \cdot 8H_2O$]. Kohlensaures Cereroxydul *Berzelius* (*Zs. für Min.*, **2**, 209, 1825; *Ak. Stockholm, Handl.*, 134, 1824). Kohlensaures Ceroxydul *Hisinger* (*Afh. Min. Geog. Schwed.*, 144, 1826). Carbonate of Cerium *Dana* (206, 1837). Cérium carbonaté *Dufrénoy* (**2**, 377, 1845). Carbocérine *Beudant* (**2**, 354, 1832). Hydrocerit *Hartmann* (*Min.*, **2**, 816, 1843). Lanthanit *Haidinger* (500, 1845). Hydrolanthanit *Glocker* (248, 1847). Crystallized Carbonate of Lanthanum *Blake* (*Am. J. Sc.*, **16**, 228, 1853).

Cryst.[1] Orthorhombic; dipyramidal—$2/m\ 2/m\ 2/m$.

$a:b:c = 0.5568:1:0.5629;\qquad p_0:q_0:r_0 = 0.9463:0.5269:1$

$q_1:r_1:p_1 = 0.5568:1.0567:1;\qquad r_2:p_2:q_2 = 1.8979:1.7960:1$

Forms:[2]

	ϕ	$\rho = C$	ϕ_1	$\rho_1 = A$	ϕ_2	$\rho_2 = B$
c 001	....	0°00'	0°00'	90°00'	90°00'	90°00'
b 010	0°00'	90 00	90 00	90 00	....	0 00
a 100	90 00	90 00	....	0 00	0 00	90 00
d 101	90 00	43 25	0 00	46 35	46 35	90 00
o 121	41 55½	54 46½	46 30	56 55	46 35	52 34

Structure cell.[3] Orthorhombic; space group uncertain. $a_0\ 9.50\ kX$, $b_0\ 17.1$, $c_0\ 9.00$; $a_0:b_0:c_0 = 0.556:1:0.526$. Cell contents $(La,Ce)_8(CO_3)_{12} \cdot 32H_2O$.

Habit. Platy to thick tabular {010}, sometimes lath-like by extension [001]. Also fine granular to earthy; as scales.

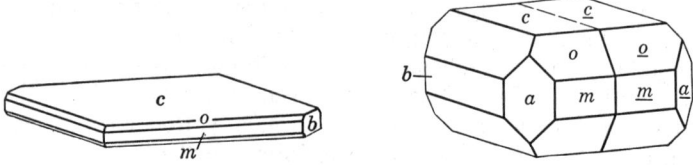

Twinning.[4] Twin and composition plane {101}.
P h y s. Cleavage {010}, micaceous. Not brittle. H. $2\frac{1}{2}$–3. G. 2.69–2.74 (Bastnaes),[5] 2.61 (Bethlehem);[6] 2.73 (calc. for M.W. = 140 of rare earths). Luster pearly. Colorless to white, pink, yellowish. Transparent.
O p t. In transmitted light, colorless.

ORIENTATION		n(Na), Bethlehem[7]	
X	b	1.52 ± 0.01	Biaxial negative (−).
Y	c	1.587 ± 0.003	$2V \sim 63°$(Na).
Z	a	1.613 ± 0.003	$r < v$, weak.

C h e m. A hydrated carbonate essentially of lanthanum and cerium, $(La,Ce)_2(CO_3)_3 \cdot 8H_2O$. Small amounts of Y and other rare earths are present.
Anal.

	1	2	3	4
$(La,Di)_2O_3$	28.54	28.34		
Ce_2O_3	25.67	25.52	55.03	54.95
Y_2O_3		0.79		
CO_2	21.89	21.95	21.95	21.08
H_2O	23.90	23.40	24.21	[23.97]
Insol.		0.13		
Total	100.00	100.13	101.19	[100.00]
G.		2.69–2.74	2.84	2.605

1. $(La,Ce)_2(CO_3)_3 \cdot 8H_2O$ with La:Ce = 1.12:1. 2. Bastnaes, Sweden.[5] 3. Bethlehem, Pennsylvania.[8] 4. Bethlehem.[6]

Tests. C.T. yields water. B.B. whitens but does not fuse. Soluble in acids. Decomposed into a white powder by boiling in water.

O c c u r. Originally found coating cerite at the Bastnaes mine near Riddarhyttan, Vastmanland, Sweden. Rarely at Bethlehem, Lehigh County, Pennsylvania, in oxidized zinc ores. Reported as an alteration product of allanite in pegmatite at Baringer Hill, Llano County, Texas, and similarly in a magnetite deposit at the Sanford mine, Moriah, Essex County, New York.
A r t i f.[9] Obtained as a gelatinous precipitate, which crystallizes on standing, by reaction of La and Ce hydroxide with $NaHCO_3$ solution saturated with CO_2.
N a m e. From the composition.

Ref.
1. Angles from average of values of von Lang, *Phil. Mag.*, **25,** 43 (1863), on Bethlehem crystals and Flink, *Ark. Kemi*, **3,** 165 (1910), on Bastnaes crystals; unit and orientation of the structure cell. Transformation: von Lang and Flink to new, 010/002/100.
2. Goldschmidt (**5,** 119, 1918).
3. Wolfe and Frondel, priv. comm. (1947), on crystals from Bethlehem.
4. Flink (1910).
5. Lindström, *Geol. För. Förh.*, **32,** 214 (1910).
6. Genth, *Am. J. Sc.*, **23,** 425 (1857).
7. Larsen (98, 1921); also Des Cloizeaux (**2,** 177, 1874).
8. Smith, *Am. J. Sc.*, **18,** 378 (1854).
9. Mellor (**5,** 664, 1924).

16 CARBONATES CONTAINING HYDROXYL OR HALOGEN

TYPE 1. $A_m(XO_3)_pZ_q$

16.1.1 LOSEYITE $[(Mn,Zn)_7(CO_3)_2(OH)_{10}]$. Bauer and Berman (Am. Min., 14, 150, 1929).

Cryst.[1] Monoclinic; prismatic—$2/m$.

$a:b:c = 0.70:1:0.62;$ $\beta\ 94°30';$ $p_0:q_0:r_0 = 0.88:0.62:1$

$r_2:p_2:q_2 = 1.62:1.43:1;$ $\mu\ 85°30';$ $p_0'\ 0.89;$ $q_0'\ 0.62;$ $x_0'\ 0.079$

Forms:[2]

c 001	s 130	e 011	d 101	p $\bar{1}36$	r $\bar{7}.14.2$

Structure cell.[3] Space group $A2/a$. $a_0\ 16.2 \pm 0.1$ kX, $b_0\ 5.55 \pm 0.05,$ $c_0\ 14.92 \pm 0.05;$ $\beta\ 95°24';$ $a_0:b_0:c_0 = 2.91:1:2.69.$ Cell contents $(Mn,Zn)_{28}(CO_3)_8(OH)_{40}$.

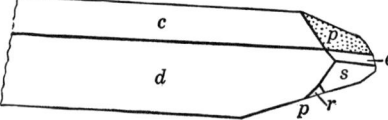

Habit. As subparallel aggregates and radiating bundles of lath-like crystals elongated [010].

Phys. Cleavage not observed. H. about 3. G. 3.27; 3.37 (calc. for Mn:Zn:Mg = 5:4:1). Color bluish white; also brownish. Transparent.

Opt. In transmitted light, colorless.

Orientation		n	
X		1.637	Biaxial positive (+).
Y	b	1.648	$2V\ 64°$.
Z		1.676	$r > v$, weak.

Chem. A basic carbonate of manganese and zinc, $(Mn,Zn)_7(CO_3)_2(OH)_{10}$. Small amounts of Mg are present in isomorphous substitution, with Mn:Zn:Mg = 5:4:1 in the only reported analysis.

Anal.

	1	2
MnO	36.27	34.94
FeO		0.64
ZnO	33.29	32.77
MgO	4.42	3.42
CO$_2$	12.86	12.59
H$_2$O	13.16	13.83
Rem.		1.19
Total	100.00	99.38
G.		3.27

1. $(Mn,Zn,Mg)_7(CO_3)_2(OH)_{10}$, with Mn:Zn:Mg = 5:4:1. 2. Franklin, New Jersey.[4] Rem. is Mn$_2$O$_3$ 1.03, insol. 0.16.

Occur. Found with altered pyrochroite, sussexite, chlorophoenicite, and calcite in small veinlets in massive ore at Franklin, Sussex County, New Jersey.

Name. After Samuel R. Losey (1833 ?–1906 ?), a native of Franklin, New Jersey, and long a collector of the minerals occurring there.

Ref.
1. Monoclinic symmetry proven by Bauer and Berman (1929); crystal class established by x-ray study by Wolfe and Frondel, priv. comm. (1946). Neither the morphological nor x-ray elements are satisfactory due to the very poor quality of the available crystals. The x-ray elements are $p_0:q_0:r_0 = 0.924:2.68:1$; $r_2:p_2:q_2 = 0.373:0.345:1$; p_0' 0.929; q_0' 2.69; x_0' 0.095.
2. Bauer and Berman (1929). No angle table is given because of the poor correlation between the morphological and the x-ray measurements.
3. Wolfe and Frondel, priv. comm. (1946), by Weissenberg method on type material. Transformation: Bauer-Berman to Wolfe-Frondel, 400/010/004.
4. Bauer analysis in Bauer and Berman (1929).

WISERITE. *Haidinger* (493, 1845).

As fibrous, asbestiform masses. Luster silky. Color white to brownish or reddish. H. $\sim 2\frac{1}{2}$. Optically uniaxial negative $(-)$, with $nO \sim 1.74$ (colorless), nE 1.66–1.67 (bright yellow-orange). Composition approximately $(Mn'', Mg)_2Mn_8'''(CO_3)(OH)_2O_{12} \cdot 8H_2O$. The Mn''' probably is due to partial oxidation of original Mn''. Analysis gave: MnO 9.38, MgO 3.09, Mn_2O_3 64.23, Fe_2O_3 0.13, CO_2 5.26, H_2O+ 16.17, H_2O 1.10, insol. 0.11, total 99.47 (original total given as 100.00). Found with pyrochroite, rhodochrosite, and hausmannite at Gonzen, Switzerland. Long considered to be altered pyrochroite but recently shown [1] to be a distinct species.

Ref.
1. Epprecht, *Beitr. Geol. Schweiz., Geotechn. Ser. Lief.*, **24**, (1946); *Schweiz. min. Mitt.*, **26**, 19 (1946).

16.1.2 **Z A R A T I T E** [$Ni_3(CO_3)(OH)_4 \cdot 4H_2O$]. Hydrate of Nickel *Silliman, Jr.* (*Am. J. Sc.*, **3**, 407, 1847). Emerald Nickel *Silliman, Jr.* (*Am. J. Sc.*, **6**, 248, 1848). Nickel Smaragd *Germ.* Texasite *Kenngott* (18, 1853). Carbonato hidratado de Niquel *Casares* (A. M. Alcibar in *Min. Revista* of Madrid, 304, 1850); Zaratita *Casares* (*ibid.*, 176, March, 1851). Zamtit *wrong orthogr.*

Probably isometric.[1] Found as incrustations, often small stalactitic or minute mammillary, with a microcrystalline structure. Also massive, compact; and as thin, single-crystal oriented films on millerite.[2]

Structure cell.[3] Simple cubic. a_0 6.15 kX. Cell contents $Ni_3(CO_3)(OH)_4 \cdot 4H_2O$.

Phys. Fracture (of aggregates) conchoidal. Brittle. H. $3\frac{1}{2}$. G. 2.57–2.69, 2.649 (Lilaz);[4] 2.67 (calc.). Luster vitreous to greasy. Color emerald-green. Streak paler. Transparent to translucent.

Opt. In transmitted light, green. Isotropic; in part weakly birefringent, perhaps due to strain. Often banded, with variable indices, and consisting of isotropic portions and spherulitic microcrystals. Indices of isotropic material: 1.56–1.61 (Lancaster),[5] 1.565 (Heazlewood).[6] The crystalline fibers have parallel extinction and positive elongation; pleochroism weak, with the fiber length emerald-green and perpendicular thereto yellowish green; indices variable, 1.559–1.566 (Lancaster), 1.589 (Heazlewood). It is not certain if the isotropic and fibrous components are the same material.

Chem. A hydrated basic carbonate of nickel, $Ni_3(CO_3)(OH)_4 \cdot 4H_2O$. Mg apparently can substitute in part for Ni.[7] Modern analyses are lacking.

Anal.

	1	2	3	4
NiO	59.56	58.81	56.82	59.30
MgO		tr.	1.68	
CO_2	11.70	11.69	11.63	11.42
H_2O	28.74	29.50	29.87	29.08
Total	100.00	100.00	100.00	99.80

1. $Ni_3(CO_3)(OH)_4 \cdot 4H_2O$. 2. Texas, Pennsylvania.[8] 3. Texas, Pennsylvania.[9] 4. Artificial zaratite.[10]

Tests. In C.T. yields water and CO_2, leaving a grayish black magnetic residue. B.B. infusible. Easily soluble with effervescence in heated dilute HCl.

Occur. A secondary mineral, often found with chromite, pentlandite, and pyrrhotite in basic igneous rocks and serpentine. Typically associated with dolomite, aragonite, calcite, hydromagnesite, genthite, brucite. Also observed as an alteration of millerite and of meteoric iron. Most of the reported localities are mentioned below.

Occurs near Cape Ortegal, Galicia, Spain, as an incrustation upon magnetite containing a nickel sulfide. With magnetite and nickelian pyrrhotite in the Stubachtal, Tyrol, and with millerite and chromite near Kraubat, Styria. As an alteration of millerite in geodes and sideritic septaria at Rapice, Dubi, and other localities in central Bohemia.[6] In Italy at Lilaz, Cogne, Piedmont, in serpentine with aragonite and nesquehonite. From Swinaness, Unst, Shetland Islands, in serpentine. Reported from nickel deposits in ultrabasic rocks in the Khalilova massif near Orsk in the southern Urals and from other localities in Russia. In Tasmania, as an alteration of heazlewoodite (Ni_2S_3) at Heazlewood and also on the Whyte river. From Dun Mountain, Nelson, New Zealand, and Broken Hill, New South Wales. With chromite at Nuasahi, Keonjhar State, India. Occurs at Igdlokunguak, Greenland. In the United States, found with chromite at Low's mine and Wood's mine in Lancaster County, and at West Nottingham, Chester County, Pennsylvania.

Artif. Formed by the action of a dilute solution of nickel chloride on nesquehonite.[10]

Name. After Señor Zarate of Spain.

Ref.

1. Indicated by the x-ray powder study of Fenoglio, *Per. Min.*, **5**, 33 (1934). Optically zaratite is isotropic or shows weak (anomalous?) birefringence. The mineral in some instances occurs as hardened gel masses and the isotropic character may then be due to lack of crystallinity or to aggregate polarization.
2. Cf. Slavík, *Am. Min.*, **11**, 279 (1926).
3. Fenoglio (1934) by powder method on Lilaz and Piedmont material.
4. Fenoglio (1931) by pycnometer.
5. Larsen (158, 1921).
6. Slavík (1926); *Časopis Národ. Musea, Praha*, **99**, 112, (1925).
7. Casares (1850, 1851) in Spanish material; also analysis 3.
8. Silliman, *Am. J. Sc.*, **6**, 248 (1848).
9. Smith and Brush, *Am. J. Sc.*, **16**, 52 (1853).
10. Fenoglio, *Per. Min.*, **5**, 265 (1934).

16.1.3 HYDROZINCITE [$Zn_5(OH)_6(CO_3)_2$].

Calamine *Smithson* (*Phil. Trans.*, 12, 1803). Zinkblüthe *Karsten* (70, 99, 1808). Hydro-carbonate of zinc. Earthy calamine. Zinconise *Beudant* (**2**, 357, 1832). Zinc Bloom. Hydrozinkit *Kenngott* (1853). Marionite *Elderhorst* (*Geol. Rept. Arkansas*, 153, 1858). Cegamit *Weisbach* (36, 1875).

C r y s t. Monoclinic.[1] Found massive, earthy and porous to compact; as incrustations, sometimes concentrically banded with a fine-fibrous radial structure; as dense, gel- or agate-like masses; sometimes stalactitic, reniform, or pisolitic. Minute lath- or blade-like crystals flattened {100}, and elongated [001], often tapering to a sharp point, may be observed projecting from massive aggregates.

Structure cell.[2] Space group $C2/m$, Cm, or $C2$. a_0 13.452 kX, b_0 6.307, c_0 5.357; β 95°30′; $a_0:b_0:c_0 = 2.133:1:0.849$. Cell contents $Zn_{10}(OH)_{12}(CO_3)_4$.

P h y s. Cleavage {100}, perfect. Very brittle. H. 2–2½ (of aggregates). G. ~4.0,[3] mostly 3.5–3.8 in massive material. Luster dull to earthy; also silky to shining in relatively coarsely crystallized material; of crystals pearly. Color pure white to gray; also yellowish, brownish, pinkish, pale lilac. Streak dull to shining. Fluoresces pale blue or lilac in ultraviolet light.

O p t. In transmitted light, colorless. Often too minutely fibrous for satisfactory optical study; in part spherulitic, with fiber elongation positive or negative.

ORIENTATION	n (Narlarla)[4]	n (Bou Thaleb)[5]	n (Goodsprings)[6]	
X b	1.635	1.640	1.650	Biaxial negative (−).
Y		1.736		$2V$ 40° (calc.).
Z	1.745(Z′)	1.750	1.740 (?)	$r < v$, rather strong.
Z ∧ c [7]	40°	Mod.	0° (?)	

C h e m. A carbonate-hydroxide of zinc, $Zn_5(OH)_6(CO_3)_2$. Some analyses diverge from this ratio,[9] perhaps due to admixture, and the correct structural formulation is uncertain. Cu is not present in significant amounts (compare *auricalcite*). On heating, the H_2O and CO_2 are lost beginning at about 230°, leaving ZnO (*zincite*).[14] The gel-like varieties contain a considerable amount of nonessential, adsorbed and capillary water.

Anal.[8]

	1	2	3	4	5	6
ZnO	74.12	74.67	73.72	74.20	74.15	73.66
CO_2	16.03	[16.41]	[15.47]	15.93	16.17	14.87
H_2O	9.85	8.92	10.81	9.83	9.50	11.62
Rem.					0.10	
Total	100.00	[100.00]	[100.00]	99.96	99.92	100.15

1. $Zn_5(OH)_6(CO_3)_2$. 2. Goodsprings, Nevada.[10] Average of two. 3. Sardinia.[10] 4. Buggerru, Sardinia.[11] 5. Buggerru, Sardinia.[12] Rem. is insol. 6. Laurium, Greece.[13] Average of two

Tests. In C.T. yields water. B.B. infusible. Easily soluble in acids.

Occur. A secondary mineral, formed in the oxidized zone of ore deposits by the alteration of sphalerite. Associated with smithsonite, but much less common than that species; also with hemimorphite, calcite, cerussite, aurichalcite, limonite. The more important localities include the following: Bleiberg, Carinthia, Austria; Raibl, Jugoslavia; near Brilon, Westphalia. In Italy at Auronzo in Venetia, and at Monte Malfidano near Iglesias and elsewhere in Sardinia. Found in great quantities at the Dolores mine, Udias valley, and at Comillas and Picos de Europa in Santander province, Spain.[15] At Quarsenis and Bou Thaleb, Algeria. From Llanidloes, Montgomery, Wales. With cerussite, limonite, and sphalerite at Narlarla, West Kimberley district, Western Australia. In the United States, hydrozincite has been found at Friedensville, Lehigh County, Pennsylvania; at Linden, Iowa County, Wisconsin; in concentric and contorted laminae and botryoidal crusts in Marion County, Arkansas. At Joplin, Missouri, and in the Galena district, Cherokee County, Kansas, with hemimorphite and smithsonite. In the Magdalena district, Socorro County, New Mexico, and in the Tintic district, Juab County, Utah. Rarely in the oxidized zinc ores at Leadville, Colorado. With hemimorphite and willemite in the Cerro Gordo mine, Inyo County, California. Abundantly at Goodsprings, Nevada.

Alter. Found as an alteration product of sphalerite, hemimorphite, and smithsonite; also as pseudomorphs after dolomite.

Artif. A number of different basic zinc carbonates have been reported formed by the action of alkali carbonates on soluble zinc salts, but only the compound $5ZnO \cdot 2CO_2 \cdot 4H_2O$ (agreeing with hydrozincite except for $1H_2O$) has been validated.[16] Also formed by exposing metallic zinc to moist air.

Name. In allusion to the composition.

Ref.

1. Symmetry indicated by Weissenberg x-ray study of Ramsdell, *Am. Min.*, **32**, 207 (1947), and by the optical properties. Morphological data are lacking.
2. Ramsdell (1947) on Goodsprings, Nevada, material. Suggested to be isostructural with aurichalcite by Ford and Bradley, *Am. J. Sc.*, **42**, 59 (1916). X-ray powder study by Frondel, priv. comm. (1940), of material from the following localities gave identical patterns and indicated that the mineral is not isostructural with aurichalcite (a fact also established by Ramsdell) or rosasite: Goodsprings, Nevada; Lafatsch, Bavaria; Carinthia; Cumillas, Spain; Auronzo, Italy; Motrico, Spain; Bou Thaleb, Algeria. The agate- and gel-like material gives few and weak diffraction effects.
3. Ramsdell (1947), in Clerici solution.
4. Prider, *Min. Mag.*, **26**, 60 (1941).
5. Larsen (90, 1921), who also gives rough measurements of nX 1.65, nY 1.73, elongation positive, rarely negative, on Malfidano, Sardinia, material; and nX 1.63, nY 1.73, elongation negative, on Tintic, Utah, material.
6. Ford and Bradley (1916); their Z index may be Z', with $Z \wedge c$ 70°.
7. Also given $Z < c$ 13° by Larsen and Berman (194, 1934). Ulrich, *Příroda, Brno*, **23**, 387 (1930), gives parallel extinction with negative elongation, nX 1.628, for material from Nová Ves, Moravia; nX 1.645, nZ 1.681, elongation negative, from Mežica; nX 1.64, nZ 1.658, elongation negative, Bleiberg.
8. For additional analyses see Hintze (1 [3A], 3354 (1929); Prider (1941); Ulrich, *Příroda, Brno*, **23**, 387 (1930).
9. See discussion by Perrier, *Atti soc. ital. sc. nat.*, **54**, 188 (1916), and Ulrich (1930).
10. Bradley anal. in Ford and Bradley (1916).
11. Perrier (1916).

12. Lauro, *Per. Min.*, **9**, 120 (1938).
13. Cabolet anal. in Kraut, *Zs. anorg. Chem.*, **13**, 8 (1897).
14. Beck (Doctoral Thesis, Harvard, 1946) by differential thermal analysis of Cumillas material.
15. Additional localities cited by Calderon (**2**, 107, 1910).
16. Mikusch, *Zs. anorg. Chem.*, **56**, 371 (1908); Kraut, *Zs. anorg. Chem.*, **13**, 8 (1897); Perrier (1916).

16.1.4 AURICHALCITE [$(Zn,Cu)_5(OH)_6(CO_3)_2$].

Calamine verdâtre (containing "une bonne quantié de cuivre"), Mine de Laiton [= Brass-ore] *Patrin* (*J. phys.*, **33**, 81, 1788). Mine de Laiton de Pise en Toscane, Aurichalcum of the ancients? *Sage* (*J. phys.*, **38**, 155, 1791). Messingblüthe *Germ*. Kupferzinkblüthe. Aurichalcit *Böttger* (*Ann. Phys.*, **48**, 495, 1839). Buratite *Delesse* (*Ann. chim. phys.*, **18**, 478, 1846). Orichalcit *Glocker* (230, 1847). Messingblüthe *Risse* (*Ver. Rheinl. Corr.-Bl.*, **22**, 95, 1865). Risséite, Messingite *Adam* (26, 1869).

Orthorhombic.[1] Found as delicate acicular or lath-like crystals elongated [001] and forming tufted, feathery, or plumose incrustations; rarely columnar, laminated, or granular. The crystals are flattened {010} and striated or furrowed [001] with a series of steep {$h0l$} forms affording wedge-shaped terminations.[2]

Structure cell.[3] Orthorhombic, A-centered cell (?). a_0 6.40 (?) kX, b_0 27.78 (?), c_0 5.25; $a_0:b_0:c_0 = 0.2346:1:0.1924$. Cell contents $(Zn,Cu)_{20}(OH)_{24}(CO_3)_8$ (?).

Phys. Cleavage {010} perfect. Fragile. H. 1–2. G. 3.64;[4] 4.23 (calc. for Zn:Cu = 2.65:1). Luster silky to pearly. Color pale green to greenish blue and sky blue. Transparent.

Opt. In transmitted light, colorless to pale shades of blue or green.

Orientation		n(Tintic)[5]	n(Utah)[6]	n(Bisbee)[6]	
X	a	1.654	1.658	1.661	Biaxial negative (−).
Y	b	1.740	1.746	1.749	2V very small.
					$r < v$, strong.
Z	c	1.743	1.751	1.756	Weakly pleochroic, with X colorless, Y and Z blue-green.

Chem. A carbonate-hydroxide of zinc and copper, $(Zn,Cu)_5(OH)_6(CO_3)_2$. Zn and Cu substitute mutually over a considerable range, the limits of which are uncertain; Cu:Zn = 1:3.16 in analysis 6 and Cu:Zn = 1:1.57 in analysis 4. The Ca reported in some analyses apparently is due to admixture.[7]

Anal.[8]

	1	2	3	4	5	6	7
CaO		0.36	0.46	0.22			
CuO	19.92	19.87	19.91	28.40	25.82	17.94	20.37
ZnO	54.08	54.01	54.77	45.67	46.73	58.12	53.41
CO_2	16.11	16.22	16.22	16.06	16.48	12.55	16.41
H_2O	9.89	9.93	8.50	9.98	11.14	9.44	9.90
Rem.					0.44	1.96	0.20
Total	100.00	100.39	99.86	100.33	100.61	100.01	100.29
G.		3.63			3.64		3.27

1. $(Zn,Cu)_5(OH)_6(CO_3)_2$ with Cu:Zn = 1:2.65. 2. Utah.[9] 3. Torreon, Chihuahua, Mexico.[10] With SO_3 tr. 4. Wanlockhead, Scotland.[11] 5. Zolotuschinsky, Urals.[12] Rem. is FeO. 6. Province Nagato, Japan.[13] Rem. is Fe_2O_3 1.46, insol. 0.22, H_2O − 0.28, MgO and SO_3 tr. 7. Campiglia, Italy.[14] Rem. is insol.

Tests. B.B. infusible. In C.T. loses water and turns black. Soluble in acids and in ammonia.

Occur. A secondary mineral found in the oxidized zone of copper and zinc deposits associated with malachite, azurite, cuprite, smithsonite, hemimorphite, hydrozincite, rosasite, limonite. A restricted list of localities follows.

In Russia near Loktevsk on the Alei river and elsewhere in the Altai district, Tomsk, Siberia. At Laurium, Greece, and at Moravicza and Rézbánya, Roumania. In Italy at Campiglia, Tuscany, and at Monteponi near Iglesias, Sardinia. Found in France at Chessy near Lyon, Rhône. In the zinc mines of the province of Santander, Spain. In England at Matlock in Derbyshire and at various localities in Cumberland; in Scotland at Leadhills, Lanark. In Africa at Mindouli, Belgian Congo, and at Tsumeb, South West Africa. In Province Nagato, Japan.[13] At Torreon and other localities in Chihuahua, Mexico. In the United States in notable specimens from Bisbee, Cochise County, Arizona; Cottonwood Canyon, Salt Lake County, Utah; and the Magdalena district, Socorro County, New Mexico. From Goodsprings and numerous minor localities in Nevada. At the Cerro Gordo mine and the Defiance mine in the Darwin district, Inyo County, California. In Utah at Ophir and elsewhere in Toole County, and in the Tintic district, Juab County.

Artif. Reported but not verified.[15]

Name.[16] Probably from ὀρειχαλκος, *mountain brass*.

Ref.

1. Aurichalcite has been variously classed on the basis of optical or inconclusive morphological evidence as monoclinic (see Kato, *J. Geol. Soc., Tokyo*, **20**, 9 (1913); Belar, *Zs. Kr.*, **17**, 113 (1889); and D'Achiardi, *Att. soc. tosc.*, **16**, 3 (1897)); triclinic (Navarro, *Bol. soc. espan. hist. nat.*, 117 (1908)—*Zs. Kr.*, **49**, 297 (1911); or orthorhombic (Ledoux, *J. Washington Ac. Sc.*, **7**, 361 (1917), Buttgenbach, *Mém. soc. roy. sc. Liége* [3], **12**, 3 (1924); and others). The x-ray study of Frondel and Wolfe, priv. comm. (1947), indicates orthorhombic symmetry. Transformation: Ledoux (1917) and Cesàro, *Mém. soc. roy. sc. Belg.*, **53**, 1 (1897) to structure cell, 010/100/001. Belar (1889) to structure cell, unchanged.
2. Angular measurements made under the microscope are reported by Ledoux (1917), Belar (1889) and others: $(h0l) \wedge [001] = 6°, 10°, 14°, 19°, 23°, 27°, 30°, 44°30', 52°30', 81°$.
3. Frondel and Wolfe (1947) by Weissenberg method on crystals of poor quality from the Kelly mine, Magdalena, New Mexico.
4. The highest and probably the best of the reported values, which range down to 3.2.
5. Larsen (43, 1921).
6. Mrose, priv. comm. (1947).
7. See Delesse, *Ann. chim. phys.*, **18**, 478 (1846) and Lacroix (739, 1909).
8. Hintze (**1** [3A], 3398, 1929) cites additional analyses, some of which probably do not represent this species.
9. Penfield, *Am. J. Sc.*, **41**, 106 (1891).
10. Collins, *Min. Mag.*, **10**, 15 (1892).
11. Heddle (**1**, 146, 1901).
12. Pilipenko, *Min. Abs.*, **2**, 109 (1925).
13. Kato, *Geol. Soc. Tokyo, J.*, **20**, 9 (1913).
14. Lauro, *Per. Min.*, **9**, 105 (1938).
15. Doelter (**1**, 476, 1911).
16. For a discussion of the origin of the name see Dana (298, 1892).

16.1.5 **R O S A S I T E** [$(Cu,Zn)_2(OH)_2(CO_3)$]. *Lovisato* (*Acc. Linc., Rend.*, **17**, 723, 1908). Paraurichalzit-I *Biehl* (Inaug. Diss. Münster, 1919—*Min. Abs.*, **1**, 202, 1921). Paraurichalcite-I.

In mammillary, botryoidal or warty crusts with a fibrous to spherulitic structure. Apparently monoclinic, pseudo-orthorhombic, with a_0 9.40 kX, b_0 12.30, c_0 3.43 (analysis 3 ?), a_0 9.40 kX, b_0 12.25, c_0 3.35 (analysis 4 ?) and $\beta \sim 90°$.[1] Cell contents $(Cu,Zn)_8(OH)_8(CO_3)_4$.

P h y s. Cleavable in two directions at right angles and appearing under the microscope as rectangular, platy, or lath-like grains. Brittle. H. $\sim 4\frac{1}{2}$. G. 4.0–4.2. Color green to bluish green and sky blue.

O p t. In transmitted light, colorless to pale blue.

ORIENTATION	n (Sardinia)[2]	n (Sardinia)[3]	n (Sardinia, anal. 4)[4]	
X elong.	1.672	1.688	1.708	Biaxial negative (−). 2V very small.
$Y \perp a$ clv.	1.83			$r < v$, strong. Extinction parallel.[5]
Z	1.83	1.831	1.823	Weakly pleochroic, with X colorless, Y and Z pale blue.

C h e m. A carbonate-hydroxide of copper and zinc, $(Cu,Zn)_2(OH)_2(CO_3)$. Cu and Zn substitute mutually over a small range at least, with Zn:Cu $\sim 2:3$ in analysis 2 and $\sim 1:2$ in analysis 4. Rosasite and the probably related minerals cuproplumbite and paraurichalcite-II, which see, may represent highly zincian malachite.

Anal.

	1	2	3	4	5
CuO	41.15	41.58	41.11	47.26	47.10
ZnO	30.99	28.96	29.21	24.46	24.49
CO_2	19.77	20.18	20.08	19.43	18.61
H_2O	8.09	8.58	8.24	8.28	10.26
Rem.		0.97	1.13	0.87	0.27
Total	100.00	100.27	99.77	100.30	100.73

1. $(Cu,Zn)_2(OH)_2(CO_3)$ with Zn:Cu = 1:1.47. 2. Sardinia.[6] Rem. is PbO 0.23, NiO 0.04, MgO 0.21, Fe_2O_3 0.31, insol. 0.18. 3. Sardinia.[4] Rem. is PbO 0.14, NiO tr., MgO 0.29, Fe_2O_3 0.42, insol. 0.28. 4. Sardinia.[7] Rem. is PbO tr., MgO 0.15, Fe_2O_3 0.39, insol. 0.33. 5. Tsumeb (paraurichalcite-I).[8] Rem. is Fe_2O_3.

Tests. In C.T. affords a little water and turns black. B.B. fusible. Soluble in acids.

O c c u r. A secondary mineral found in oxidized zinc-copper-lead ores. Originally from the Rosas mine, at Sulcis, Sardinia, associated with brochantite, malachite, aurichalcite, greenockite, and siderite. From Tsumeb, South West Africa, with malachite, smithsonite, hydrozincite (paraurichalcite-I).[9] Also from the United States at the Jack Pot Claim near Wellington and from Majuba Hill, in Nevada.[10] Reported from the Kyzyl-Espe lead-zinc deposit in northeastern Turkestan,[11] and from the Tombstone district, Cochise County, Arizona. With aurichalcite at Kelly, New Mexico.[10]

N a m e. From the locality.

Ref.
1. Lauro, *Per. Min.*, **9**, 105 (1938), by powder method from presumed structural analogy to malachite.
2. Barth and Berman, *Chem. Erde*, **5**, 22 (1930).
3. Mrose, priv. comm. (1948).
4. Lauro, *Per. Min.*, **8**, 151 (1937).
5. Lauro (1937) states that rare grains have inclined extinction of 20°–22° (analysis 4).
6. Perrier, *Acc. Linc., Rend.*, **30** [5], 119 (1921). Also an earlier, probably erroneous analysis by Rimatori in Lovisato (1908).
7. Lauro (1938).
8. Biehl (1919), who gives additional analyses.
9. Shown to be identical with rosasite by Lauro (1937, 1938).
10. Frondel, priv. comm. (1949), by x-ray powder study.
11. Sumin, *C. r. ac. sc. U.R.S.S.*, **31**, 779 (1941).

PARAURICHALCITE-II. Biehl (*Beitr. zur Kenntn. der Min. der Erzlagerst. von Tsumeb. Inaug. Diss. Münster*, 1919—*Min. Abs.*, **1**, 202, 1921; Busz, *Sitzber. naturhist. Ver. preuss. Rheinl.*, Bonn, 16B, 1923). Cuprozincite Biehl (1919).

Names given to ill-defined botryoidal or bead-like crusts and earthy materials formed by the action of zinc-bearing solutions on malachite at Tsumeb, South West Africa. The substances closely resemble malachite in appearance, structure, and color and sometimes contain unaltered cores of that mineral. One type of paraurichalcite has been shown [1] to be identical with rosasite. Another type (analysis 1) approximates to the formula $(Cu,Zn)_2(OH)_2(CO_3)$ with $Cu:Zn \sim 5:3$; G. 4.137, H. 4. Cuprozincite (analysis 2) approximates to $(Cu,Zn)_2(OH)_2(CO_3)$ with $Cu:Zn \sim 9:2$, G. 4.104, H. 3. This substance apparently is monoclinic, twinned on $\{100\}$, with cleavages $\{001\}$ and $\{010\}$. Optically biaxial with inclined extinction against the twin plane of 21°–23°; also (001) cleavage $\wedge$ (100) twin plane = 61°. These substances need further study, and may prove to be identical with rosasite or to be highly zincian varieties of malachite. Analyses gave: [2]

	FeO	CuO	ZnO	CO_2	H_2O	Total
1.		43.58	28.22	18.42	9.53	99.75
2.	0.56	58.58	12.71	19.74	7.61	99.20

Ref.
1. Lauro, *Per. Min.*, **8**, 151 (1937); **9**, 105 (1938).
2. Biehl (1919), who gives additional analyses.

SCHUILINGITE. Vaes (*Soc. géol. Belgique, Bull.*, **70**, B233, 1947).

Orthorhombic? Crusts of minute acicular crystals elongated [001] and bounded by a prism (with mm' 60°–70°) and terminal domes. Cleavage $\{110\}$ perfect, $\{100\}$ poor. Color azure blue. Luster adamantine. Transparent. Biaxial negative ($-$) with indices of refraction between 1.74 and 1.93; $2V$ about 60°; and $X = a$, $Y = c$, $Z = b$. Weakly pleochroic. Apparently a carbonate of lead and copper. Locality not given, but presumably the Belgian Congo. Named after H. J. Schuiling, geologist. Needs verification.

BLEIMALACHITE. Glinka and Antipov (*Vh. Min. Ges.*, St. Petersburg, 468, 1901 (?)); Glinka (*Cbl. Min.*, 281, 1901).

Acicular, monoclinic crystals with cleavage in three directions. Twinned. Strongly pleochroic in yellow and green with high indices of refraction and birefringence, $2E$ about 80°. Analysis is said to correspond to $Cu_3Pb(OH)_2(CO_3)_3$. Found in druses in the Syrjanovsk mine, Altai district, Urals, U.S.S.R. Needs verification.

16.1.6 MALACHITE [$Cu_2(OH)_2(CO_3)$]. Χρυσόκολλα pt. Theophrastus, Dioscorides, etc. Ψευδὴς Σμάραγδος [False Emerald of Copper Mines] pt. *Theophrastus*. Chrysocolla, Molochites pt. *Pliny, Agricola*. Berggrün *Germ*. Molochit *Agricola* (*Interpr.*, 1546). Ærugo nativa, Viride montanum pt., Koppargrön, Bärggrönt pt., Malachit *Wallerius* (278, 279, 1747). Cuivre carbonaté vert *L'Abbé Fontana* (*J. phys.*, **2**, 509, 1778) [proving the existence of a green carbonate]. Green Carbonate of

MALACHITE

Copper. Green Malachite. Mountain Green pt. Berggrün pt. *Germ.* Atlaserz [fibrous var.] *Germ.* Rame carbonato verde, Verdi di monte *Ital.* Malaquita *Span.* Mysorin *Thomson* (**1**, 601, 1836). Lime-Malachite, Calco-malachite, Kalk-malachit Zincken (*B. H. Ztg.*, **1**, 1842).

Cryst. Monoclinic.[1]

$a:b:c = 0.7914:1:0.2691$; $\beta\ 98°44'$;[2] $p_0:q_0:r_0 = 0.3400:0.2660:1$

$r_2:p_2:q_2 = 3.7607:1.2784:1$; $\mu\ 81°16'$; $p_0'\ 0.3440,\ q_0'\ 0.2691,\ x_0'\ 0.1536$

Forms:[3]

		ϕ	ρ	ϕ_2	$\rho_2 = B$	C	A
c	001	90°00′	8°44′	81°16′	90°00′	.....	81°16′
b	010	0 00	90 00	.....	0 00	90°00′	90 00
a	100	90 00	90 00	0 00	90 00	81 16	.....
m	110	51 58	90 00	0 00	51 58	83 08	38 02
p	$\overline{2}01$	−90 00	28 07	118 07	90 00	36 51	118 07

Less common:

$x\ \overline{1}02$ $\gamma\ 011$ $\epsilon\ 111$

Structure cell.[4] Space group $P2_1/a$. $a_0\ 9.49 \pm 0.02$ Å, $b_0\ 12.00 \pm 0.02$, $c_0\ 3.24 \pm 0.02$; $\beta\ 98°42'$; $a_0:b_0:c_0 = 0.790:1:0.270$. Cell contents $Cu_8(OH)_8(CO_3)_4$.

Habit. Crystals are very rare; usually short or long prismatic [001] to needle-like and grouped in tufts and rosettes. Untwinned crystals are practically unknown. The form is seldom distinct, and the faces often uneven or rounded; the prismatic faces are striated [001], and the terminal faces striated [010]. Commonly massive or incrusting, with surface mammillary, botryoidal or tuberose; internally radial or divergent fibrous on a coarse to very fine scale, and banded in color. Also in delicate fibrous aggregates, sheaf-like or compact; and in concentrically banded stalactites, sometimes alternating with azurite.

Twinning. Twin plane {100} very common, often as penetration twins, sometimes polysynthetic. Twin axis [201] also apparently common.

Phys. Cleavage {$\overline{2}01$} perfect, {010} fair. Fracture of compact massive types subconchoidal to uneven. H. $3\frac{1}{2}$–4. G. 4.05 ± 0.02 (crystals);[5] 4.00 (calc.); the gravity of the massive, fibrous material ranges down to about 3.6. Luster of crystals adamantine, inclining toward vitreous; of fibrous varieties more or less silky or velvety; often dull and earthy. Color bright green, the crystals inclining toward dark green and blackish green. Streak pale green. Translucent to opaque.

Twin on {100}.

Opt.[6] In transmitted light, green to yellowish green.

Orientation		n	Pleochroism	
$X \wedge c$	$23\frac{1}{2}°$	1.655 ± 0.003	Nearly colorless	Biaxial negative (−).
Y	b	1.875 ± 0.003	Yellowish green	$2V\ 43° \pm 2°$ (Na), meas.
Z		1.909 ± 0.003	Deep green	$r < v$, rather strong.

Chem. A basic carbonate of copper, $Cu_2(OH)_2(CO_3)$. A small amount of Zn substitutes for Cu in analysis 3; but the composition ordinarily does not vary significantly. There is no proof that reported highly zincian material (see *rosasite*) actually is of this species. The water is lost at *ca*. 315°, lower than in azurite, leaving tenorite.[7]

Anal.[8]

	1	2	3	4	5
CuO	71.95	71.84	71.31	71.99	72.03
ZnO			0.45		
CO_2	19.90	19.95	19.87	19.68	20.04
H_2O	8.15	8.21	8.36	8.22	8.09
Rem.			0.04	0.09	
Total	100.00	100.00	100.03	99.98	100.16
G.	4.10			4.07	

1. $Cu_2(OH)_2(CO_3)$. 2. Chessy, France.[9] 3. Chessy, France.[10] Rem. is Fe_2O_3, with PbO and insol. in traces. 4. Eiserfeld, Germany.[11] Rem. is FeO. 5. Rio Marina, Elba.[15]

Tests. C.T. blackens and yields water. B.B. fuses at 2. Easily soluble in dilute acids. Very slightly soluble in water containing CO_2.

Occur. Malachite is a minor ore of copper, occurring as a secondary mineral in the upper oxidized zone of copper deposits. Found typically with azurite, but more abundant and widespread than that species. Associated also with cuprite, tenorite, limonite, wad, calcite, chalcedony, chrysocolla and, less commonly, atacamite, brochantite and other secondary copper, lead or zinc minerals. Only the more important localities can be mentioned here.[12] It has been found in large amounts and in notable quality in Siberia at the Demidoff copper mine at Nizhne-Tagilsk, at Ekaterinburg, and at other localities. One banded mass measured at top 9 by 18 feet and the portion uncovered contained at least a half million pounds of the pure mineral. Much of this material, mostly obtained in the early 1800's, was cut into slabs and used for table tops and other ornamental purposes. Found in small amounts at various mines in Cornwall; in France at Chessy, near Lyon, and with secondary arsenates and phosphates in crevices in Keuper sandstone at Cap Garonne, near Hyères, Maures. As small crystals at Betzdorf and Horhausen, Rhenish Prussia, Germany. From Rudabánya and Moldawa in the Banat, Roumania. A notable occurrence as banded masses, stalactites, and fine pseudomorphs after azurite at Tsumeb, South West Africa; also at Bwana Mkubwa in northern Rhodesia; in Namaqualand, Cape Province; and Katanga, Belgian Congo. In South Australia in the Burra district and at Wallaroo; in New South Wales at Broken Hill and Cobar.

In the United States, malachite is widespread in the western mining districts. Found abundantly in fine masses and as acicular crystals at the Copper Queen mine and other mines at Bisbee, Cochise County, Arizona; also at Morenci, Greenleaf County, in banded forms with azurite, and in the Globe district, Gila County, with chalcedony and chrysocolla; in the Sierrita Mountains, Pima County. With brochantite as pseudomorphs after azurite at Good Springs, Nevada. Found in New Mexico especially in the Magdalena district, Socorro County, and the Fierro-Hanover and Santa Rita districts in Grant County. From the Tintic district, Utah. At the Jones mine, Fritz Island, Berks County, Pennsylvania, and in the copper mines at Ducktown, Tennessee.

A l t e r. Frequently observed as pseudomorphs after azurite (which see). Also as alteration pseudomorphs after cuprite (fine examples coming from Chessy, France), and less commonly after atacamite, brochantite, chalcopyrite, tetrahedrite, chalcophyllite, gypsum, libethenite, calcite, sphalerite, cerussite, pyrite. Malachite is rarely found altered to azurite or cuprite.

A r t i f.[13] Very fine-grained green precipitates of hydrated copper carbonates ascribed to malachite have been obtained by the action of alkali carbonates on solutions of cupric salts, by heating soluble cupric salts in contact with calcite in sealed tubes at 150°–225°, by passing CO_2 into suspended cupric hydroxide, and in other ways. Crystals have been formed by heating these precipitates with water and ammonium nitrate or urea in sealed tubes at *ca.* 140°; also by allowing a solution of basic cupric carbonate in carbonic acid to stand at ordinary temperature for some time, and by action of a dilute cupric sulfate solution saturated with CO_2 on marble. Also formed on a copper anode by electrolysis of soluble carbonates. Banded precipitates resembling natural banded malachite can be prepared by diffusing copper salts into silica gel containing a dissolved carbonate.[14]

N a m e. From μαλαχή, *mallows*, in allusion to the color.

Ref.

1. No definitive work on the symmetry class has been done.
2. Angles of Lang, *Phil. Mag.*, **25**, 432 (1863), **28**, 502 (1864). Unit and orientation of structure cell. Transformations: Goldschmidt (**5**, 187, 1918) to new, $\overline{1}0\frac{1}{3}/0\overline{1}0/00\frac{2}{3}$; Dana (294, 1892) to new, $\overline{1}0\frac{4}{3}/0\overline{1}0/00\frac{2}{3}$.
3. Goldschmidt (1918). Rare or uncertain: 130, 104, 101, $\overline{1}08$, $\overline{7}08$, 134, 232, 562, $\overline{1}68$, $\overline{1}98$, $\overline{1}65$, $\overline{8}.15.16$, $\overline{1}22$.
4. Ramsdell and Wolfe, *Am. Min.*, **35**, 119 (1950), by Weissenberg method on crystals from Horhausen. See also Brasseur, *Zs. Kr.*, **82**, 111 (1932), who assumed an incorrect β value and gave the wrong space group.
5. Range of best reported values, including new determinations on single-crystals (4.028, average) by Mrose, priv. comm. (1947).
6. Larsen (103, 1921); orientation from Des Cloizeaux (**2**, 185, 1874).
7. Beck, Doctoral Thesis, Harvard Univ. (1946). The synthetic and some natural materials contain a small amount of water over the requirements of the formula, that is lost below 200° C. (Binder, *C.R.*, **204**, 1200 (1937). The formula has been written $Cu_8(OH)_8(CO_3)_4 \cdot H_2O$ by Guillot and Geneslay, *C.R.*, **202**, 136 (1936).
8. For additional analyses see Hintze (**1** [3A], 3368, 1929).
9. Berthier analysis in Gonnard, *Min. du Rhône et de la Loire*, Paris, 82 (1906).
10. Perrier, *Acc. Linc., Att.*, **30** [5], 309 (1921).
11. Haege (Inaug. Diss., Jena 1888).
12. Additional localities are listed by Hintze (1929), Schrader, *et al., U. S. Geol. Sur., Bull., 624* (1917) and, for the western United States, by Pabst, *Calif. Div. Mines, Bull.* 113 (1938); Galbraith, *Arizona Bur. Mines, Geol. Ser., Bull. 149* (1914), Northrop, *Univ. New Mexico Bull., Geol. Ser.*, **6**, no. 1 (1942), Shannon, *U. S. Nat. Mus., Bull. 131* (1926).
13. Mellor (**3**, 270, 1923); Binder, *C.R.*, **204**, 1200 (1937); Hepburn, *J. Chem. Soc.*

London, **127**, 1007 (1925); Hémar, C.R., **204**, 1739 (1937); Guillot and Geneslay, Bull. soc. chim. France, 4 [5], 125 (1937).
14. Hartman et al., J. Chem. Ed., **11**, 346 (1934).
15. Lauro, Per. Min., **9**, 105 (1938).

16.1.7 PHOSGENITE [$Pb_2(CO_3)Cl_2$].

Hornblei *Karsten* (78, 1800). Salzsaures Bleierze *Klaproth* (**3**, 141, 1802). Corneous Lead *Jameson* (1804). Bleihornerz, Chlorbleispath *Germ*. Plomb carbonaté muriatifère, Plomb chloro-carbonaté, Plomb corné *Fr*. Phosgen-spath *Breithaupt* (61, 1832). Kerasine *Beudant* (**2**, 502, 1832). Phosgenite *Breithaupt* (**2**, 183, 1841). Galenoceratite, Bleikerat *Glocker* (248, 1847). Cromfordite *Greg* and *Lettsom* (421, 1858).

Cryst.[1] Tetragonal; trapezohedral—4 2 2.

$$a:c = 1:1.0889;\ ^2 \quad p_0:r_0 = 1.0889:1$$

Forms:[3]

		φ	ρ	A	M
c	001	....	0°00′	90°00′	90°00′
a	010	0°00′	90 00	90 00	45 00
m	110	45 00	90 00	45 00	90 00
u	120	26 34	90 00	63 26	71 34
e	011	0 00	47 26	90 00	58 37
o	021	0 00	65 20	90 00	50 01
x	111	45 00	57 00	53 37½	90 00
s	121	26 34	67 40½	65 33½	28 39

Less common:

    η 012   f 023   A 041   w 221   t 552   h 141   g 162

Structure cell.[4] Space group uncertain. a_0 8.139 Å, c_0 8.856; $a_0:c_0 = 1:1.0881$. Cell contents $Pb_8(CO_3)_4Cl_8$.

Habit. Short prismatic to prismatic [001]; rarely thick tabular {001}. Usually terminated with large {001}, sometimes with {111} dominant. Also massive, granular.

Oriented growths.[5] Cerussite upon phosgenite, with cerussite {100} [010] parallel phosgenite {001} [120].

Phys. Cleavage {001} and {110} distinct, {010} less distinct. Fracture conchoidal. Rather sectile. H. 2–3, varying noticeably with direction.[6] G. 6.133,[7] 6.134 (artif.);[8] 6.136 (calc.). Luster adamantine. Color yellowish white to yellowish brown, pale brown or smoky brown; also colorless, white, pale rose, gray, yellowish gray, and greenish. Streak white. Transparent to translucent. Translation gliding[9] with T{110}, t[001], the crystals being easily flexed about directions perpendicular [001]. Shows percussion figure[9] on {001} with the principal rays parallel [110]. Weakly fluorescent in shades of yellow in long-wave ultraviolet light and in cathode rays and x-rays.

Opt. In transmitted light, colorless. In thick sections very weakly pleochroic with O reddish and E greenish.[10] Sometimes weakly biaxial, noticeably so after mechanical deformation.

	n(Na)[11]	
O	2.1181	Uniaxial positive (+).
E	2.1446	

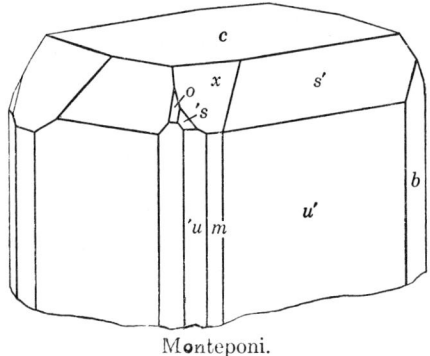

Monteponi.

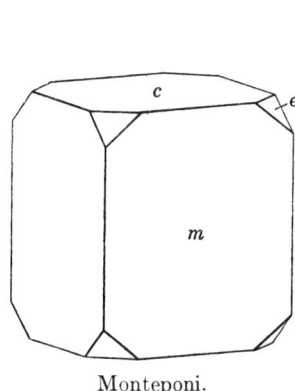

Monteponi.

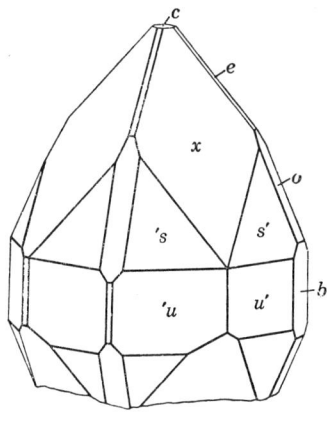

Monteponi.

Chem. A chlorocarbonate of lead, $Pb_2(CO_3)Cl_2$. Phosgenite is isostructural and isomorphous with the (artificial) compound $Pb_2(CO_3)Br_2$.
Anal.[12]

	1	2	3	4
PbO	81.86	80.84	81.73	81.78
Cl	13.00	13.08	13.06	12.91
CO_2	8.07	9.16	7.91	8.13
Rem.			0.01	
Total	102.93	103.08	102.71	102.82
O = Cl	2.93	2.95	2.95	2.91
	100.00	100.13	99.76	99.91
G.	6.136	6.12	6.05	6.24

1. $Pb_2(CO_3)Cl_2$. 2. Sidi Amor ben Salem.[13] 3. Monteponi.[14] Rem. is Ag. 4. Altai district, Siberia.[15]

Tests. B.B. fuses readily to a yellow globule, which on cooling becomes white and crystalline. At high temperatures CO_2 is driven off from the melt and Pb_2OCl_2 remains. Soluble with effervescence in dilute HNO_3.

Occur. Phosgenite is a secondary mineral formed by the alteration under surface conditions of galena and other lead minerals. Commonly associated with cerussite and anglesite. Found in fine crystals up to five inches across at Monteponi and Montevecchio near Iglesias and at Gibbas near Cagliari in Sardinia. At Laurium, Greece, where it has formed with laurionite and other lead minerals by the action of sea water upon ancient lead slags. Also as an alteration of ancient metallic objects immersed in sea water near Mahdia, Tunis,[16] and of leaden pipe in the hot springs at Bourbonne-les-Bains, France.[17] At Sidi Amor ben Salem, Tunis, and at Tsumeb, South West Africa. Phosgenite occurs abundantly in crystals and masses in large part altered to cerussite at Tarnowitz, Upper Silesia, Poland. In large crystals with matlockite at Matlock and nearby Cromford, Derbyshire, England. At the Comet mine, Dundas, Tasmania, and at Broken Hill, New South Wales. From the Syzjanov mine, Altai district, Siberia. In a vein in travertine at Salar del Plomo, Argentina.[19] In the United States, phosgenite has been found sparingly at the old Southampton lead mine, Hampshire County, Massachusetts. Abundantly in large masses in part altered to cerussite at the Terrible mine, Isle, Custer County, Colorado. With diaboleite at the Mammoth mine, Pinal County, Arizona. In the Organ district, Dona Ana County, New Mexico, and the Silver Sprout mine, Inyo County, California.

Alter. Phosgenite alters readily to cerussite, and pseudomorphs of this species after crystals and masses of phosgenite have been found at Tarnowitz and other localities. It is found experimentally that moisture and carbon dioxide transform phosgenite into the normal carbonate. Cold water slowly decomposes the mineral by extracting lead chloride.

Artif.[18] Phosgenite has been obtained in crystals by heating $PbCl_2$, $PbCO_3$, and H_2O in a sealed tube at 180°, by heating $Pb(OH)_2$ with phosgene in a sealed tube at 175°, by passing CO_2 over a solution of $PbCl_2$, and by the action of CO_2 on $Pb(OH)Cl$. Also by reaction of $PbCl_2$ and $KHCO_3$ in ethanol.

Name. From *phosgen*, a name for carbonyl chloride, $COCl_2$, because the mineral contains carbon, oxygen, and chlorine.

Ref.

1. Crystal class established from morphological evidence by Goldschmidt (**6**, 150, 1920), *Zs. Kr.*, **21**, 321 (1893), *ibid.*, **23**, 139 (1894), and Koch, *Jb. Min., Beil.-Bd.*, **59**, 97 (1929). Artificial etch figures, however, conform to holohedral symmetry—see Anders, *Ak. Leipzig, Ber.*, **73**, 117 (1921). The weak piezoelectricity reported by Onorato, *Per. Min.*, **5**, 37 (1934), could not be verified by the cathode-ray oscilloscope method by Frondel, priv. comm. (1947), contradictory signals probably due to contact potentials being observed.

2. Goldschmidt, *Zs. Kr.*, **21**, 321 (1893).

3. Goldschmidt (1920) and Koch (1929), the latter giving a critical discussion of the literature. Rare or uncertain forms:

150	230	116	332	129	122	131	7.13.3
130	12.7.0	114	11.11.9	127	485	362	
580	013	113	881	125	232	241	

4. Sillén and Pettersson, *Norsk Geol. Tidsskr.*, **24**, 79 (1945), *Ark. Kemi*, **21**, 1 (1945), by Weissenberg and powder methods on Monteponi and Laurium crystals.

The x-ray diffraction effects indicate a Laue symmetry of $4/mmm$ with systematic extinctions in $h0l$ when h is odd. The b glide thus indicated is not consistent with any space group in the crystal class, 4 2 2, indicated by the morphology. The space group in the holohedral class would be $P4/mbm$. The earlier x-ray study of Onorato (1934) is erroneous. See also Oftedal, *Norsk Geol. Tidsskr.*, **24**, 79 (1945).
  5. Tacconi, *Ist. Lombardo, Rend.*, **44**, 991 (1911).
  6. Mügge, *Jb. Min.*, I, 43 (1914); on {110} the hardness is less parallel [001] than perpendicular thereto.
  7. Average of numerous closely agreeing measurements on Monteponi crystals by Mrose, priv. comm. (1947), by microbalance.
  8. de Schulten, *Bull. soc. min.*, **20**, 191 (1897).
  9. Taricco, *Acc. Linc., Att.*, **19**, 278 (1910), and Mügge, *Jb. Min.*, I, 43 (1914).
  10. Sohncke, *Ak. München, Ber.*, **26**, 97 (1896).
  11. Smith, *Min. Mag.*, **12**, 107 (1909), by prism method on Laurium crystals.
  12. For earlier analyses see Dana (703, 1868).
  13. Mélon analysis in Buttgenbach, *Livre jubil., Soc. géol. Belgique* (1925).
  14. Rodolico, *Acc. Linc., Att.*, **8**, 171 (1928).
  15. Pilipenko, *Nachr. Tomsk. Univ.*, 14 (1906)–*Jb. Min.*, II, 368 (1909).
  16. Lacroix, *C.R.*, **151**, 276 (1910).
  17. Lacroix (3, 779, 1909).
  18. Mellor (7, 852, 1927), Sillén and Pettersson (1945).
  19. Angelelli and Valvano, *Rev. soc. geol. Argentina*, **1**, 257 (1946).

BISMUTITE AND BEYERITE

	a_0	c_0
Bismutite, $(BiO)_2(CO_3)$	3.859 kX	13.658
Beyerite, $Ca(BiO)_2(CO_3)_2$	3.78	21.77

Both of these tetragonal species have layer structures, as evidenced by their platy habit and basal cleavage. The structures [1] are based on bismuth-oxygen layers in which the oxygen atoms are in sheet-like quadratic packing with the Bi atoms fitting into the interstices, half on each side of the oxygen sheet. Identical $(XO)_n$ layers are found in a large family of bismuth and lead oxyhalides. In bismutite and beyerite the $(BiO)_n$ layers are separated by layers composed of (CO_3) groups, the individual (CO_3) groups having their planes of symmetry vertical and rotating probably in (110) and (1$\bar{1}$0) planes. In beyerite two such (CO_3) layers centrally coordinated by Ca are present between the $(BiO)_n$ layers.

Ref.
  1. Lagercrantz and Sillén, *Ark. Kemi*, **25**, 1 (1948).

16.1.8 **B I S M U T I T E** [$(BiO)_2(CO_3)$]. Luftsaures Wismuth *Beyer*, 1805. Arsenikwismuth *Werner* (56, 1817). Bismutit *Breithaupt* (*Ann. Phys.*, **53**, 627, 1841). Kohlensaures Wismuthoxyd, Wismuthspath *Germ*. Bismuthite. Carbonate of Bismuth.
  Bismutosphärit *Weisbach* (*Jb. Berg.-hütt. Sachsen*, **49**, 1877). Bismutosphaerite. Normannite *Weisbach*, 1877. Hydrobismutite, Basobismutite *Nenadkevich* (*Bull. Ac. Sc., Petrograd*, **11** [6], 448, 454, 1917). Boksputite *Mountain* (*Min. Mag.*, **24**, 62, 1935).

C r y s t. Tetragonal.
Structure cell.[13] Tetragonal, body-centered, with Laue symmetry $4/mmm$. a_0 3.859 kX, c_0 13.658; $a_0:c_0 = 1:3.541$. Cell contents $(BiO)_4(CO_3)_2$.

Habit. As pulverulent to dense earthy masses, as opaline crusts, and as radially fibrous crusts or spheroidal aggregates; rarely as scaly to lamellar aggregates. Bismutite often possesses a prismatic or other pseudomorphous structure.

P h y s. A more or less distinct cleavage on {001} has been observed in microscopic grains.[1] H. variable, depending on the state of aggregation; mostly $2\frac{1}{2}$–$3\frac{1}{2}$.[2] G. variable, 6.1–7.7, mostly 6.7–7.4; 8.28 (calc.), 8.15 (artif.). Color usually straw-yellow to brownish yellow; also yellowish white, greenish gray, pale green, gray, brown; sometimes bright bluish green or blue, due to included copper minerals, or deep gray to black, especially in pseudomorphs after bismuthinite. Luster vitreous, grading to pearly in coarse fibrous and scaly material; also dull to earthy. Streak gray. Transparent in small grains.

O p t.[3] In transmitted light, colorless or faintly tinted in shades of yellow, green, etc. Ordinarily cryptocrystalline and indistinctly polarizing. The mean index is variable, due to the content of non-essential water, and ranges from about 2.12 to above 2.30. Birefringence moderate. Fibrous material usually has parallel extinction and positive elongation.

C h e m. Bismuth subcarbonate, $(BiO)_2(CO_3)$. Ordinarily from 1 to 3.5 per cent of water is present, but this is non-essential.[1] On heating, the CO_2 is lost at about 290° and α-Bi_2O_3 remains. The Pb and small amounts of Cu, Fe, and Ca sometimes reported are believed due to admixture.

Anal.[4]

	1	2	3	4	5	6	7
Bi_2O_3	91.37	90.00	91.68	92.07	91.64	91.82	88.65
CO_2	8.63	6.56	8.29	8.01	8.03	7.54	9.53
H_2O		3.44		0.90	0.47	0.94	0.08
Rem.			tr.		0.42		1.64
Total	100.00	100.00	99.97	100.98	100.56	100.30	99.90
G.		7.67	7.64		7.42	6.83	

1. $(BiO)_2(CO_3)$. 2. Chesterfield, South Carolina.[5] 3. Guanajuato, Mexico.[6] Rem. is Fe_2O_3, SiO_2. 4. Willimantic, Connecticut.[7] 5. Willimantic, Connecticut.[8] Rem. is SO_3 0.34, insol. 0.08, Fe_2O_3 tr. 6. Portland, Connecticut.[8] 7. Ampangabe, Madagascar.[12] Rem. is CaO 1.20, PbO 0.44.

Tests. B.B. fuses readily. Soluble with effervescence in acids.

O c c u r.[9] A rather common secondary mineral, formed by the alteration under superficial conditions of bismuthinite, native bismuth, and other primary bismuth minerals. Bismutite occurs particularly in the oxidized portions of veins of the Co-Ni-Ag-Bi type, hypothermal veins and pegmatites carrying bismuthinite and native bismuth, and the Bolivian bismuth- and tin-rich sulfide veins. Found associated with bismite, sillenite, malachite, beyerite, cerussite, pucherite, mixite, arsenobismite, limonite, and other oxidation products.

Bismutite was originally described from Ullersreuth, Voigtland, Germany, but a bismuth carbonate, later shown to be bismutite, was known to occur at Schneeberg as early as 1805. Other well-known European localities include Neustädtel (*bismutosphaerite*), Johanngeorgenstadt, and Aul in Saxony; Meymac, Corrèze, France; Cornwall, England. Found at Schorl Mountain in Transbaikalia (*hydro-bismutite, basobismutite*) and as an alteration of aikinite at Berezov, Urals, U.S.S.R. In pegmatite at Ampangabe, Madagascar, and at Boksput, in Gordonia, Cape Province, South Africa (*boksputite*). In South America, bismutite occurs at numerous localities in Bolivia, notably in the Tazna district; in Brazil at São Jose de Bryamba and elsewhere in Minas Geraes; at Huarancaca, Peru. Found at the El Carmen mine and other localities in Durango, and at Guanajuato, in Mexico. In the pegmatite pipes at Kingsgate, New South Wales. Only a few of the numerous localities in the United States can be mentioned here. In pegmatite at Willimantic and Portland, Connecticut; at Casher's Valley, Jackson County, and in Gaston County, North Carolina. In New Mexico in pegmatite in the Petaca district, Rio Arriba County, in the San Andreas Mountains west of Tularosa, Otero County, in the Organ district, Dona Ana County, at Eagle Station, Sierra County, and in the Hachita district, Hidalgo County. In Nevada near the Bagdad copper mine, west of Hillside, Yavapai County in the Hualpai Mountains east of Yucca, Mohave County, and in Maricopa County. Found sparingly at Leadville, Colorado, as an alteration of bismuth sulfosalts; also at Salida, Chaffee County, at Telluride, San Miguel County, and in pegmatites in Fremont and Park Counties. With bismuth arsenates and bismoclite in the Tintic district, Utah. With beyerite in pegmatite at Pala, San Diego County, California.

Alter. Bismutite is a typical alteration product of bismuthinite, native bismuth, aikinite, tetradymite, and other bismuth minerals and often is found encrusting or as pseudomorphs after crystals of such species.

Artif.[10] Obtained as a voluminous, fine-grained precipitate containing much adsorbed and capillary water by reaction of a solution of bismuth nitrate with alkali or ammonium carbonates. The fresh precipitate gives a faint and diffuse x-ray pattern which becomes sharp after aging or slight heating of the material.

Name. Bismutite, hydrobismutite, and basobismutite in allusion to the composition. Boksputite from the locality. Bismutosphaerite in allusion to the spherical form of aggregation of the original material.

BISMUTOSPHÄRIT. *Weisbach (Jb. Berg.-hütt. Sachsen,* **49**, 1877). Bismutosphaerite.
Described as a species with the composition Bi_2CO_5 and distinct from bismutite, in the belief that the latter mineral was a hydrated carbonate, perhaps $Bi_2CO_5 \cdot H_2O$. The identity of the two substances was later proved [1] by x-ray and thermal analysis. The original bismutosphaerite was from Neustädtel, near Schneeberg, Saxony, where it occurred as fibrous spheroidal masses associated with beyerite

BOKSPUTITE. *Mountain (Min. Mag.,* **24**, 62, 1935).
A supposed carbonate of lead and bismuth, $6PbO \cdot Bi_2O_3 \cdot 3CO_2$, from Boksput, Gordonia, Cape Province, South Africa. Shown [11] to be a mixture of bismutite with a lead oxide and an unidentified mineral.

HYDROBISMUTITE, BASOBISMUTITE. *Nenadkevich (Bull. Ac. Sc., Petrograd,* **11** [6], 448, 454, 1917).

Hydrous bismuth carbonates from Schorl Mountain, Transbaikalia, U.S.S.R., identical with bismutite.[1]

Ref.

1. Frondel, *Am. Min.*, **28**, 521 (1943).
2. Dana (307, 1892) gives $4-4\frac{1}{2}$ for bismutite, and values up to $5\frac{1}{2}$ have been reported. Frondel (1943) reports a maximum of $3\frac{1}{2}$.
3. Frondel (1943); the optical data given by Larsen (49, 1921) for "bismutosphaerite" refers to beyerite.
4. For additional analyses see Hintze (1 [3A], 3404, 3406, 1929).
5. Rammelsberg, *Ann. Phys.*, **76**, 564 (1849).
6. Winkler in Weisbach, *Jb. Min.*, II, 254 (1882).
7. Sperry in Wells, *Am. J. Sc.*, **34**, 271 (1887).
8. Wells (1887).
9. A list of 46 localities proven by x-ray study is given by Frondel (1943).
10. Cf. Mellor (**9**, 703, 1929) and Frondel (1943).
11. Heinrich (*Am. Min.*, **32**, 365, 1947).
12. Raoult analysis in Lacroix (**1**, 297, 1922).
13. Lagercrantz and Sillén, *Ark. Kemi*, **25**, 1 (1948).

16.1.9 Waltherite. Vogl (*Gangverhältn. und Mineralreich. Joachimsthal*, Teplitz, 169, 1857). Walthérite.

An ill-defined mineral found as doubly terminated prismatic crystals with a prism angle of approximately $116\frac{1}{2}°$.[1] The crystals have a monoclinic aspect but are too roughly terminated to permit measurement. The value of c_0 is 5.42 ± 0.05 kX.[2]

P h y s. Cleavage $\{001\}$ good; also $\{110\}$ and $\{010\}$.[3] H. about 4.[4] G. 5.32.[5] Luster vitreous to subresinous. Color siskin green to olive-drab and brownish green, varying within the same crystal. Streak yellowish. Transparent in small grains.

O p t.[6] In transmitted light, pale brown or green. The brown and green materials have unlike optical properties but identical x-ray powder patterns. The brown material is biaxial negative $(-)$ with $2V$ about $75°$ and strong birefringence; $Y = b = 1.91\pm$ and $X \wedge c = 16°; r < v$; weakly pleochroic in brown with absorption $X = Y < Z$. Most of the greenish material has higher indices and stronger birefringence, with $2V$ up to $90°$ and in part optically positive $(+)$ with $r > v$.

C h e m.[7] Apparently a hydrated carbonate of bismuth, possibly with uranium as an essential constituent.

Tests. Soluble in dilute acids with effervescence.

O c c u r. Found associated with torbernite, chrysocolla, malachite, tellurite, and bismutite in the Elias mine, Joachimsthal, Bohemia.

N a m e. Perhaps after Walther, a mining official of Vienna.[8]

Needs further study.

Ref.

1. Bertrand, *Bull. soc. min.*, **4**, 58 (1881), and Frondel, *Am. Min.*, **28**, 521 (1943).
2. Frondel (1943).
3. Bertrand (1881); Vogl (1857) mentions only $\{010\}$.
4. Frondel (1943); Vogl (1857) gives 3.

5. Frondel (1943) by microbalance; Vogl (1857) gives 3.8–4.0.
6. The optical data here given, from Frondel (1943), do not agree entirely with those of Bertrand (1881).
7. Lindacker in Vogl (1857) as cited by Zepharovich (2, 341, 1873) gives the major constituents as UO_2, Bi_2O_3, TeO_2, CO_2, and H_2O with minor CuO and Fe_2O_3.
8. Chester (285, 1896).

16.1.10 A R T I N I T E [$Mg_2(CO_3)(OH)_2 \cdot 3H_2O$]. *Brugnatelli* (*Ist. Lombardo, Rend.*, 35, 869, 1902; 36, 824, 1903).

C r y s t.[1] Monoclinic (class unknown).

$a:b:c = 5.268:1:1.968;$ $\beta\ 98°56';$ $p_0:q_0:r_0 = 0.374:1.944:1$

$r_2:p_2:q_2 = 0.514:0.192:1;$ $\mu\ 81°04';$ $p_0'\ 0.378,\ q_0'\ 1.968,\ x_0'\ 0.157$

Forms:[2]

		ϕ	ρ	ϕ_2	$\rho_2 = B$	C	A
c	001	90°00'	8°56'	81°04'	90°00'	.....	81°04'
a	100	90 00	90 00	0 00	90 00	81°04'	......
F	$\overline{2}$01	−90 00	30 55½	120 55½	90 00	39 51½	120 55½

Structure cell.[3] Space group possibly $C2/m$. a_0 16.54 Å, b_0 3.14, c_0 6.18; [$\beta\ 98°56'$]; $a_0:b_0:c_0 = 5.27:1:0.197$. Cell contents $Mg_4(CO_3)_2(OH)_4 \cdot 6H_2O$.

Habit. As crusts of acicular crystals, elongated [010]. Also as botryoidal masses of silky fibers, as spherical aggregates of radiating fibers, and as cross-fiber veinlets.

P h y s. Cleavage {100} perfect, {001} good. Brittle. H. 2½. G. 2.02 ± 0.01,[4] 2.027;[5] 2.047 (calc.). Color and streak white. Luster of fibrous aggregates silky or satiny, of single crystals vitreous. Transparent.

O p t. In transmitted light, colorless.

Orientation		nNa (Nevada [6])	
X		1.488 ± 0.001	Biaxial negative (−).
Y	b	1.534 ± 0.001	$2V\ 70°$ (meas.).[12]
Z ∧ c	30°	1.556 ± 0.001	Optic axis nearly ⊥ {100}.

C h e m. A hydrated basic carbonate of magnesium, $Mg_2(CO_3)(OH)_2 \cdot 3H_2O$.[7]

Anal.

	1	2	3	4	5	6
MgO	41.00	41.81	41.04	41.12	40.97	41.34
CO_2	22.37	22.82	22.21	22.16	22.36	22.37
H_2O	36.63	35.46	36.64	36.54	36.62	[36.29]
Total	100.00	100.09	99.89	99.82	99.95	[100.00]
G.	2.047		2.027	2.022	2.02	2.028

1. $Mg_2(CO_3)(OH)_2 \cdot 3H_2O$. 2. Luning, Nevada.[8] 3. Cogne, Val d'Aosta, Italy.[9] 4. Viu, Val di Lanzo, Italy.[10] 5. Monte Ramazzo, Italy.[11] Average of two. 6. Franscia, Val Lanterna, Italy.[14]

Tests. B.B. whitens but does not fuse. Easily soluble with effervescence in cold acids.

O c c u r. A low-temperature hydrothermal mineral commonly found as veinlets or crusts along fracture surfaces in serpentinized ultrabasic rocks. Associated with hydromagnesite, brucite, aragonite, calcite, dolomite, magnesite, pyroaurite, and chrysotile. Found originally in asbestos deposits at Val Brutta, and Franscia in the Val Lanterna, Lombardy, Italy. Found similarly in Italy at Emarese and Cogne in the Val d'Aosta, near Torre Santa Maria in the Val Malenco, at Fubina in the Val di Lanzo, and at Monte Ramazzo, near Borzoli, Liguria. Occurs with hydromagnesite in serpentine at Kraubat, Styria, and Lojane in southern Serbia. In the United States artinite occurs at Hoboken, New Jersey, associated with brucite and hydromagnesite. At Luning, Nevada, as cross-fiber veinlets with hydromagnesite in massive brucite. Also found with pyroaurite along joint planes in serpentine near Eden Mills, Vermont.

A r t i f. Not known artificially.[13]

N a m e. After Ettore Artini (1866–1928), Professor of Mineralogy at the University of Milan.

Ref.

1. Elements derived from x-ray study of Hurlbut, *Am. Min.*, **31**, 365 (1946).
2. Brugnatelli (1903). Forms were not indexed, but angular values are simply related to structural cell with good agreement. Meixner's measurements, *Zbl. Min.*, Abt. A, 5 (1938), indicating orthorhombic symmetry are probably the results of twinning.
3. Hurlbut (1946) by Weissenberg method on material from Nevada. The data agree closely with those of Heritsch, *Zbl. Min.*, Abt. A., 25 (1940). The lattice is base-centered; since crystal class is unknown space group may be $C2$, Cm, or $C2/m$.
4. Range of best reported values.
5. Fenoglio, *Per. Min.*, **7**, 47 (1936) by suspension method. Brugnatelli (1902) obtained 2.028.
6. Hurlbut (1946); nearly identical values are given by Larsen (42, 1921) and Lincio, *Acc. Linc., Att.*, **11**, 420 (1930), on Val Lanterna and Mt. Ramazzo material.
7. Differential thermal analysis data are given by Beck in Hurlbut (1946).
8. Gonyer anal. in Hurlbut (1946).
9. Fenoglio (1936).
10. Fenoglio, *Soc. geol. ital., Boll.*, **46**, 13 (1927).
11. Lincio (1930).
12. Clar in Meixner, *Zbl. Min.*, 5 (1938).
13. Cf. Leitmeier, *Jb. Min., Beil.-Bd.*, **40**, 662 (1916).
14. Brugnatelli (1902).

GAJITE. *Tućan* (*Cbl. Min.*, 312, 1911).

As dense, snow-white masses resembling some magnesite. H. $3\frac{1}{2}$. G. 2.619. Under the microscope, grains show a rhombohedral cleavage and polysynthetic twin lamellae. Optically uniaxial negative (−), with strong birefringence. Composition apparently $(Ca,Mg)_7(CO_3)_4(OH)_4O$. Analysis gave (average of two): CaO 37.08, MgO 23.85, CO_2 32.34, H_2O 6.67, total 99.94. Easily soluble with effervescence in acids. In C.T. affords alkaline water. Found near Plešce in the Gorski-Kotar district, Croatia. Name after Ljudevit Gaj (1809–1872), a Croatian political leader.

Needs verification.

16.1.11 A Z U R I T E $[Cu_3(OH)_2(CO_3)_2]$. Cæruleum, Lapis armenius pt. *Pliny* (33, 57). Cæruleum *Germ.* Lasur, Berglasur pt. *Agricola.* Koppar-Lazur, Cuprum lazureum, Cæruleum montanum *Wallerius* (280, 1747). Bleu de montagne, Cuivre azuré *French transl. of Wallerius* (**1**, 506, 1753). Kupferlasur *Werner.* Bergblau *Germ. Abbé Fontana* (*J. phys.*, **2**, 1778) [with analysis making it a carbonate]. Unächter Lasurstein *Stütz* (*Einricht. Nat. Wien*, **49**, 1798). Blue carbonate of copper, Blue malachite. Chessy copper. Blue Spar. Azure Copper Ore. Cuivre carbonaté bleu *Fr.* Azurite *Beudant* (417, 1824). Lasur *Haidinger* (508, 1845). Chessylite *Brooke* and *Miller* (594, 1852). Zinkazurit *Breithaupt* (*B. H. Ztg.*, **11**, 101, 1852). Lasurit

AZURITE

von Kobell (32, 1853). Azzurrite. Rame carbonato azzurro, Bleu di Monte *Ital.* Azurita, Cobre azul *Span.*

C r y s t.[1] Monoclinic; prismatic—$2/m$.

$a:b:c = 0.8565:1:1.7688$; $\beta\,92°25'$; $p_0:q_0:r_0 = 2.0652:1.7672:1$

$r_2:p_2:q_2 = 0.5659:1.1686:1$; $\mu\,87°35'$; $p_0'\,2.0670,\,q_0'\,1.7688,\,x_0'\,0.0422$

Forms:[2]

	ϕ	ρ	ϕ_2	$\rho_2 = B$	C	A
c 001	90°00'	2°25'	87°35'	90°00'	.....	87°35'
b 010	0 00	90 00	.....	0 00	90°00'	90 00
a 100	90 00	90 00	0 00	90 00	87 35	.....
w 120	30 18	90 00	0 00	30 18	88 47	59 22
m 110	49 26½	90 00	0 00	49 26½	88 10	40 33½
l 013	4 05½	30 35½	87 35	59 30	30 30	87 55
f 012	2 44	41 31½	87 35	48 32	41 28	88 11½
p 011	1 22	60 31½	87 35	29 30	60 30	88 48½
σ 102	90 00	47 05½	42 55	90 00	44 40½	42 54½
ϕ 101	90 00	64 38	25 22	90 00	62 13	25 22
n $\bar{1}$04	−90 00	25 23	115 23	90 00	27 48	115 23
θ $\bar{1}$02	−90 00	44 45	134 45	90 00	47 10	134 45
η $\bar{3}$04	−90 00	56 27	146 27	90 00	58 52	146 27
v $\bar{1}$01	−90 00	63 43	153 43	90 00	66 08	153 43
P 113	51 07	43 12½	53 49½	64 33	41 21	57 47½
s 112	50 34½	54 19	42 55	58 56½	52 28	51 08½
h 111	50 01	70 02	25 22	52 51	68 11½	43 56
x $\bar{1}$12	−48 15½	53 02	134 45	57 52	54 51	126 36
k $\bar{1}$11	−48 51½	69 36	153 43	51 56	71 25½	134 54
e $\bar{1}$25	−27 41	38 37½	110 22	56 26½	50 16½	106 51½
d $\bar{1}$23	−28 44½	53 22	122 53½	45 17	54 33½	112 42
λ $\bar{1}$93	−6 57	79 24½	122 53½	12 39	79 42½	96 50
R $\bar{1}$21	−29 47	76 13	153 43	32 33	77 25½	118 50½

Less common forms:
i 320	j 025	μ $\bar{1}$.0.10	A $\bar{1}$06	B $\bar{5}$08	u $\bar{1}$13	ν 122	y $\bar{2}$12
g 210	ζ $\bar{1}$04	D $\bar{1}$08	I $\bar{1}$05	z: $\bar{5}$06	δ $\bar{1}$23	ς: $\bar{1}$38	
q 015	r $\bar{1}$.0.16	F $\bar{1}$07	T $\bar{2}$05	ψ $\bar{3}$02	q: 214	α $\bar{1}$22	

Structure cell.[3] Space group $P2_1/c$. $a_0\,4.96\,kX$, $b_0\,5.83$, $c_0\,10.27$; $a_0:b_0:c_0 = 0.851:1:1.762$. Cell contents $Cu_6(OH)_4(CO_3)_4$.

Habit.[4] Crystals are varied in habit and often are highly modified. Often tabular {001} or, less commonly, tabular {102} or {$\bar{1}$02}. Also short prismatic [001], with {110} or {100} usually prominent; short prismatic [010]. Sometimes equant, or rhombohedral in aspect. The faces are commonly slightly undulating, with {001} striated [100] and {100} striated [010]. The crystals occasionally are markedly composite or occur in subparallel groupings and then grade into rhomboidal, lenticular, or crudely spherical aggregates. Less frequently massive, or presenting stalactitic or imitative shapes with a columnar or coarse radial structure; also dull and earthy.

Carbonates

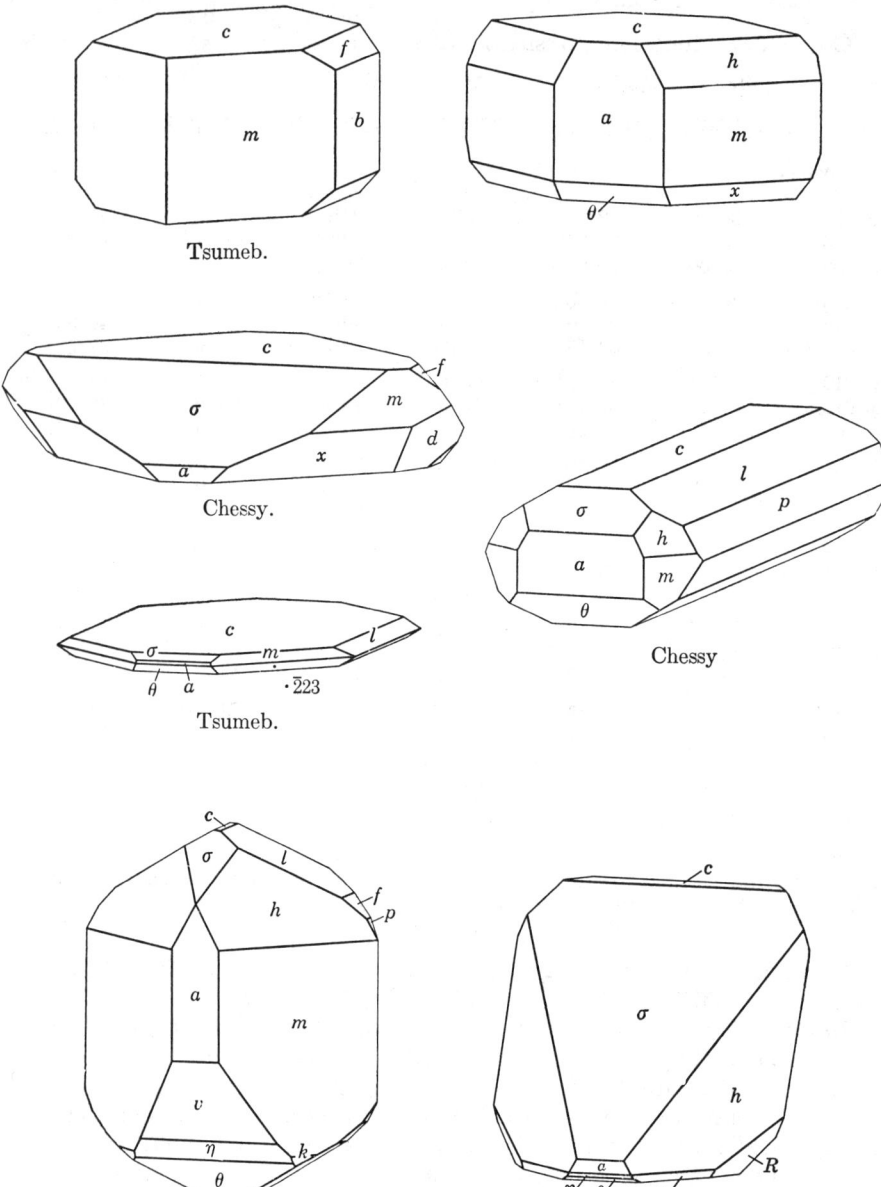

Tsumeb.

Chessy.

Chessy

Tsumeb.

Tsumeb.

Bisbee.

AZURITE

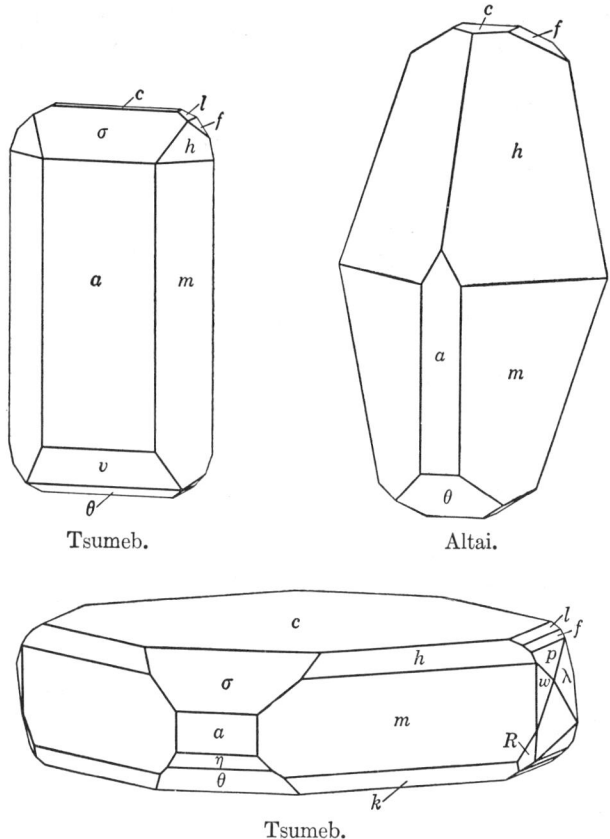

Tsumeb. Altai.

Tsumeb.

Twinning. Rare. (1) Twin plane $\{\bar{1}01\}$. (2) Twin plane $\{\bar{1}02\}$. (3) Twin plane $\{001\}$.

Phys. Cleavage $\{011\}$ perfect but interrupted; $\{100\}$ fair; $\{110\}$ in traces. Fracture conchoidal. Brittle. H. $3\frac{1}{2}$–4. G. 3.773 ± 0.003;[5] 3.834 (calc.). Luster vitreous, almost adamantine. Color various shades of azure-blue, passing into berlin-blue; also very dark blue, and in some massive or earthy types, light blue. Streak blue, lighter than the color. Transparent to subtranslucent.

Opt. In transmitted light, pale blue. Pleochroic in blue, with absorption $Z > Y > X$.

Orientation		n(Na)[6]	n[7]	n[7]	
X	b	1.730 ± 0.002	1.730 ± 0.005	1.730 ± 0.005	Biaxial positive (+).
Y		1.758 ± 0.003	1.754 ± 0.005	1.755 ± 0.005	$2V$ $67°$.
$Z \wedge c$	$-12°36'$	1.838 ± 0.003	1.836 ± 0.005	1.835 ± 0.005	$r > v$, rather strong.

Chem. A basic carbonate of copper, $Cu_3(OH)_2(CO_3)_2$. CuO 69.24, CO_2 25.53, H_2O 5.23, total 100.00. Modern analyses are lacking.[8] The available data do not suggest any significant compositional variation in the mineral. The water is lost on heating to ca. 410°, leaving tenorite.[9]

Tests. B.B. turns black and melts easily. Soluble in dilute acids, ammonia, and hot concentrated solutions of $NaHCO_3$. Slowly decomposed with removal of CO_2 by boiling water but not by cold water.

Occur. Azurite is a secondary mineral found in the upper oxidized portions of ore deposits. It has been formed chiefly by the action of carbonated waters on other copper minerals or of copper sulfate or chloride solutions on limestone. Azurite is almost invariably associated with malachite, although less frequent in occurrence and less abundant than that mineral. Also commonly associated with limonite, wad, cuprite, tenorite, calcite, chalcocite, chrysocolla, and, less often, with brochantite, antlerite, atacamite, and other secondary copper minerals including native copper. Secondary lead and zinc minerals ordinarily are not immediately associated. Space permits mention of only a few of the more important or notable localities.[10]

Azurite occurs in small amounts although in fine crystals in Siberia at Beresov and at Tomsk and Solotuschinsk in the Altai; at Laurium, Greece; at Moldawa and Rudabánya, in the Banat, Hungary; at Montes Chauves, near Kielce, Poland. A notable occurrence at Chessy,[11] near Lyon, Rhône, France (whence the name *chessylite*). On Sardinia, Italy, at Rosas and near Alghero. Crystals of unusual size and beauty have been found abundantly at Tsumeb, Africa, where the mineral is associated with olivenite, mimetite, mottramite, anglesite, cerussite, and malachite, the last in part as fine pseudomorphs after azurite.[12] In New South Wales at Broken Hill, Condobolin, and Cobar; at Mungana near Chillagoe, Queensland; in South Australia at Moonta, at Wallaroo and near Adelaide. Notable localities in the United States are few, the chief being at Bisbee, Arizona, where spectacular crystals and aggregated groups have been obtained formerly, at Morenci, Arizona, and at Kelly, New Mexico. Also from Arizona[13] in the San Xavier district, Santa Rita Mountains and the Black Hills district, Yavapai County. In New Mexico[14] in the Georgetown district, altered to pseudomorphs of copper and malachite, and in the Fierro-Hanover district, both in Grant County, and in the Magdalena district, Socorro County. Numerous minor occurrences are found in California[15] and other western states.

Alter. Alteration pseudomorphs of malachite after azurite are common, and fine examples have been found at Chessy, Bisbee, Tsumeb, and other localities; the reverse change, azurite after malachite, is rare. Pseudomorphs also have been found of copper after azurite,[16] and of azurite after cuprite, cerussite, tetrahedrite.

Artif.[17] Obtained in crystals by heating to low temperatures solutions of cupric nitrate or sulfate with calcite in a closed tube. Also formed by reaction of cold solutions of cupric nitrate and sodium bicarbonate with calcite. Malachite (which see) is more readily synthesized and apparently from somewhat more alkaline solutions and at higher temperatures.

Name. In allusion to the color. Both azure and lazure (lazurite, lazulite) ultimately derive from the Persian *lazhward, blue color*. Azurite was the most important blue pigment in the ancient wall paintings of the East and was used more widely than ultramarine in European painting from the fifteenth to the middle of the seventeenth century.

Ref.

1. Orientation and unit of Schrauf, *Ak. Wien, Sitzber.*, **64**, 123, (1871), and Atlas (1872); angles of Palache and Lewis, *Am. Min.*, **12**, 99 (1927). Schrauf's elements, in accordance with the structure cell of Brasseur, *Soc. roy. sc. Liége, Mem.* **18**, 1 (1933), *Zs. Kr.*, **82**, 195 (1932), has c doubled in comparison to the elements of Goldschmidt (5, 85, 1918), Dana (295, 1892), Hintze (1 [3A], 3369, 1929), and most other authors. Transformation: Goldschmidt, Palache and Lewis, Dana, *et al.* to Schrauf, 100/010/002.

2. Goldschmidt (1918); Stecher, *Jb. Min., Beil.-Bd.*, 59A, 159 (1929); Brasseur, *Zs. Kr.*, **77**, 177 (1931). Rare and vicinal forms:

150	032	$\bar{3}.0.28$	$\bar{7}.0.10$	$\bar{3}34$	376	$\bar{1}44$	$\bar{3}64$
130	021	$\bar{1}09$	$\bar{7}08$	$\bar{2}21$	9.12.16	$\bar{1}.10.4$	$\bar{3}74$
230	052	$\bar{3}.0.22$	$\overline{15}.0.16$	$\bar{7}72$	345	$\bar{5}.2.16$	$\bar{3}.12.4$
530	031	$\bar{5}.0.36$	$\overline{13}.0.12$	1.2.10	568	$\bar{2}36$	$\bar{5}76$
310	108	$\bar{3}.0.20$	$\bar{7}06$	3.8.24	456	$\bar{4}.7.12$	$\bar{3}43$
0.1.20	106	$\bar{2}.0.13$	$\overline{19}.0.16$	147	263	$\bar{4}.9.12$	$\bar{1}31$
0.1.16	105	$\bar{2}.0.11$	$\bar{5}04$	156	324	$\bar{2}76$	$\bar{1}51$
0.1.12	3.0.14	$\bar{7}.0.30$	$\bar{7}04$	218	364	$\bar{1}53$	$\bar{7}46$
0.1.10	103	$\bar{1}03$	114	124	212	$\bar{5}.2.10$	$\bar{6}55$
018	304	$\bar{5}.0.14$	335	134	232	$\bar{2}14$	$\bar{3}22$
017	506	$\bar{3}08$	223	1.11.4	121	$\bar{5}.3.10$	$\bar{3}52$
016	9.0.10	$\overline{13}.0.32$	334	257	322	$\bar{3}26$	$\bar{4}12$
014	302	$\bar{5}.0.12$	556	8.1.16	351	$\bar{2}34$	$\bar{2}41$
038	$\bar{1}.0.14$	$\overline{13}.0.30$	221	7.2.14	$\bar{1}.2.10$	$\bar{3}56$	$\bar{2}51$
037	$\bar{1}.0.12$	$\bar{9}.0.16$	$\bar{1}15$	316	$\bar{1}.3.15$	$\bar{1}32$	$\bar{3}41$
0.7.12	$\bar{1}.0.11$	$\bar{3}05$	$\bar{1}14$	5.2.10	$\bar{1}26$	$\bar{1}42$	
034	$\bar{2}.0.19$	$\bar{2}03$	$\bar{2}27$	234	$\bar{1}24$	$\bar{3}24$	

3. Brasseur (1932) by rotation and spectrometer methods with determination of the structure; an experimental β value is not given, and the morphological value is apparently employed.

4. Monographic studies of the form development of azurite have been contributed by Palache and Lewis (1927), *Am. Min.*, **27**, 334 (1942), Stecher (1929), Brasseur (1931), and Zedlitz, *Zs. Kr.*, **71**, 1 (1929), in the more recent literature.

5. From numerous new measurements by microbalance by Mrose and Wolfe, priv. comm. (1947), and the value 3.770 of Schröder, *Jb. Min.*, 712 (1874).

6. Merwin, *J. Washington Ac. Sc.*, **4**, 253 (1914) on Broken Hill and Butte crystals; also $nX_{671} = 1.719$, $nX_{486} = 1.756$.

7. Larsen (43, 1921) on Broken Hill and Rochester district, Nevada, crystals.

8. For older analyses see Dana (715, 1868).

9. Beck, Doctoral Thesis, Harvard Univ. (1946); much lower decomposition temperatures were reported by Rose, *Ann. Phys.*, **84**, 484 (1851).

10. Many additional localities are listed by Hintze (1929) and, for the United States, by Schrader, *et al.*, *U. S. Geol. Sur., Bull. 624* (1917).

11. Cf. Lacroix (3, 756, 1909).

12. Cf. Palache and Lewis (1927) and Zedlitz (1929).

13. Cf. Galbraith, *Arizona Bur. Mines, Geol. Ser., Bull. 149* (1941).

14. Cf. Northrop, *Univ. New Mexico Bull., Geol. Ser.*, **6**, no. 1 (1942).

15. Cf. Murdoch and Webb, *Calif. Division of Mines, Bull. 136* (1948).

16. Yeates, *Am. J. Sc.*, **38**, 405 (1889).

17. Mellor (**3**, 274, 1923); Hintze (1929); Hémar, *C.R.*, **204**, 1739 (1937).

16.1.12 HYDROCERUSSITE [$Pb_3(CO_3)_2(OH)_2$]. Nordenskiöld (Geol. För. Förh., **3**, 381, 1877). Hydrocerusite. Plumbonacrite Heddle (Min. Mag., **8**, 201, 1889).

Cryst.[1] Hexagonal—P (?).

$$a:c = 1:2.5984; \quad p_0:r_0 = 3.0003:1$$

Forms:

	ϕ	$\rho = C$	M	A_2
c 0001	.....	0°00'	90°00'	90°00'
m 10$\bar{1}$0	30°00'	90 00	60 00	90 00
o 10$\bar{1}$4	30 00	36 52½	72 32½	90 00
p 10$\bar{1}$2	30 00	56 19	65 25	90 00
q 10$\bar{1}$1	30 00	71 34	61 41	90 00
r 20$\bar{2}$1	30 00	80 32½	60 27	90 00

Structure cell.[2] a_0 8.97 kX, c_0 23.8; $a_0:c_0 = 1:2.66$. Cell contents $Pb_9(CO_3)_6(OH)_6$.

Habit. As thin scales with a hexagonal outline and flattened {0001}; also thick tabular {0001}, or steep pyramidal {$h0\bar{h}1$} with {0001} sometimes very small. Some crystals, not certainly of this species, show a rhombohedral development.[7]

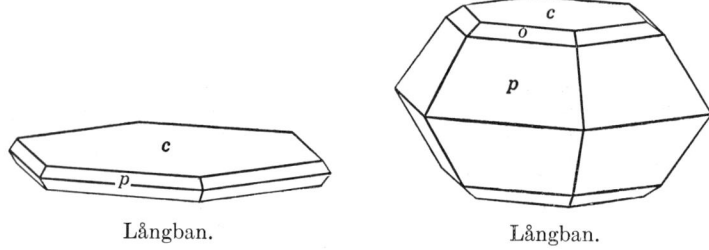

Långban.   Långban.

Phys. Cleavage {0001} perfect. Brittle. H. 3½. G. 6.80 (Mendip);[3] 6.94 (calc.). Luster adamantine, pearly on {0001}. Colorless to white; gray, sometimes faintly greenish. Transparent to translucent.

Opt.[4] In transmitted light, colorless.

    nO 2.09   Uniaxial negative (−).
    nE 1.94

Chem. A basic carbonate of lead, $Pb_3(CO_3)_2(OH)_2$. Small amounts of Cl substitute for (OH).

Anal.

	PbO	CO_2	H_2O	Cl	O = Cl	Total	G.
1.	86.33	11.35	2.32			100.00	6.94
2.	86.52	11.21	2.23	0.27	0.06	100.17	6.80
3.	86.43	11.32	2.00	0.32	0.07	100.00	6.786

1. $Pb_3(CO_3)_2(OH)_2$. 2,3. Mendip Hills.[5]

Tests. B.B. easily fusible. In C.T. decrepitates and becomes yellow-brown. Soluble in acids with effervescence.

Occur. A secondary mineral found associated with leadhillite, matlockite, cerussite, mendipite, paralaurionite. Originally found at Långban, Sweden, as an alteration product of native lead. Found in Scotland in cavities in galena at Wanlockhead, Dumfries, and from Leadhills, Lanark; also abundantly in the Mendip Hills, Somerset, England. Reported as occurring as an alteration product of the ancient lead slags of Laurium, Greece, and as occurring at the Rederovski mine, Altai Mountains, U.S.S.R. In the United States found as large, fine crystals at the Mammoth mine, Tiger, Arizona.

Alter. Found as alteration zones around mendipite, and also partly or completely altered to cerussite.

Artif.[6] Obtained in the system $PbCO_3$-PbO-H_2O. Also reported by heating powdered wulfenite in sodium carbonate solution and by heating $PbCl_2$ and urea solution at 180°. Hydrocerussite is one of the chief components of commercial white lead.

Name. In allusion to the composition, a basic (hydrated) carbonate of lead.

Ref.

1. Angles of Palache, priv. comm. (1946), on crystals from the Mammoth mine, Arizona; unit of the x-ray cell of Wolfe, priv. comm. (1946), on Mammoth crystals. Discordant angles have been earlier reported by Flink, *Bull. Geol. Inst. Upsala*, **5**, 94 (1901), and Aminoff, *Geol. För. Förh.*, **48**, 44 (1926), on Långban crystals with c halved, and by Spencer, *Min. Mag.*, **20**, 80 (1925), on Mendip Hills crystals. Forms of Flink (1901), Aminoff (1926), and Palache (1946).
2. Wolfe (1946).
3. Spencer, *Min. Mag.*, **20**, 80 (1925); lower values have been reported.
4. Larsen (202, 1921) on artificial material.
5. Mountain analysis in Spencer (1925), who cites earlier analyses.
6. Mellor (**7**, 836, 1927) and Biltz, *Zs. anorg. Chem.*, **220**, 312 (1934).
7. Shown by some crystals from the Mendip Hills (Spencer, 1925) and from the Mammoth mine, Arizona (Palache, 1949).

16.1.13 **HYDROMAGNESITE** [$Mg_4(OH)_2(CO_3)_3 \cdot 3H_2O$]. *Wachmeister* (*Ak. H. Stockholm*, **18**, 1827). Hydromagnesite *von Kobell* (*J. pr. Chem.*, **4**, 80, 1835). Hydrocarbonate of Magnesia.
  Lancasterite pt. *Silliman, Jr.* (*Am. J. Sc.*, **9**, 216, 1850). Baudisserite pt. *Delamétherie* (**2**, 1812). Predazzite pt. *Petzholdt* (*Beitr. Geogn. Tyrol*, 194, 1843). Pencatite pt. *Roth* (*Zs. deutsch. geol. Ges.*, **3**, 140, 143, 1851). Hibbertite (?) *Heddle* (*Min. Mag.*, **2**, 24, 1878). Hydrogiobertite *Scacchi* (*Acc. Napoli, Rend.*, **24**, 310, 1885). Hydrodolomite *Dana* (213, 1850). Hydromagnesite (?) *von Kobell* (*J. pr. Chem.*, **36**, 304, 1845). Kalkmagnesit (?) *Hausmann* (1404, 1847). Hydromanganocalcit *Hartmann* (*Min. Nachr.*, 299, 1850). Hydromagnocalcit pt. Hydrodolomit *Rammelsberg*. Hydronickelmagnesite *Shepard* (*Am. J. Sc.*, **6**, 250, 1848). Pennite (?) *Hermann* (*J. pr. Chem.*, **47**, 13, 1849).

Cryst.[1] Monoclinic.

$a:b:c = 1.1209:1:0.9473;$ $\beta\ 113°32';$ $p_0:q_0:r_0 = 0.8451:0.8685:1$

$r_2:p_2:q_2 = 1.1514:0.9731:1;$ $\mu\ 66°28';$ $p_0'\ 0.9218,\ q_0'\ 0.9473,\ x_0'\ 0.4355$

Forms:[2]

		φ	ρ = C	φ₂	ρ₂ = B	C	A
c	001	90°00'	23°32'	66°28'	90°00'	0°00'	66°28'
a	100	90 00	90 00	0 00	90 00	66 28	0 00
m	110	44 13	90 00	0 00	44 13	73 50	45 47
y	011	24 41½	46 11½	66 28	49 01½	40 58½	72 27½
g	021	12 56½	62 46½	66 28	29 55½	60 40½	78 30½
t	041	6 33½	75 18½	66 28	16 03½	73 56½	83 39½
o	1̄11	−27 10½	46 48	115 56	49 34½	60 21	109 26½
p	1̄21	−14 23½	62 55½	115 56	30 24½	70 47½	102 47½

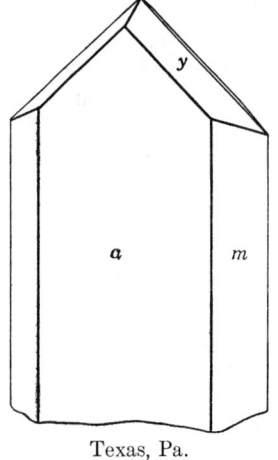

Texas, Pa.

Structure cell.[3] (In orthorhombic interpretation.) Space group $Pmmm$. a_0 9.32 kX, b_0 8.98, c_0 8.42; $a_0:b_0:c_0$ = 1.0378:1:0.9376. Cell contents uncertain.

Habit. Crystals small, as tufts, rosettes, or crusts of acicular or bladed crystals elongated [001] and flattened {100}. Also massive, chalky, or mealy.

Twinning.[4] Very common, with twin plane {100} and composition face {100}; polysynthetic.

Phys. Cleavage {010} perfect; also parting (?) on {100}. Brittle. H. 3½. G. 2.236.[5] Luster of crystals vitreous, of aggregates earthy to silky or pearly. Colorless to white. Transparent.

Opt. In transmitted light, colorless.

Orientation[6]	n(Na)[6]	n[7]	n[8]	
X ∧ c 47°09'	1.523 ± 0.003		1.524	Biaxial positive (+).
Y	1.527 ± 0.003	1.529	1.528	2V moderate
Z b	1.545 ± 0.001	1.543	1.544	

Chem. A hydrated carbonate-hydroxide of magnesium, $Mg_4(OH)_2$-$(CO_3)_3 \cdot 3H_2O$. Not all analyses adhere closely to this formula, perhaps due to admixture. Very small amounts of Ca apparently substitute for Mg.[14] On heating,[9] the water is lost in two stages with the loss of CO_2 beginning at ca. 485°, intermediate oxy-carbonates being formed, and MgO finally remaining.

Anal.[10]

	1	2	3	4	5	6	7	8
MgO	44.14	44.02	44.16	43.52	43.16	41.30	43.11	42.82
CO_2	36.14	35.85	35.16	36.72	37.10	36.71	37.09	37.36
H_2O	19.72	19.99	20.30	19.53	19.65	19.32	19.73	19.26
Rem.			0.25			2.89		0.26
Total	100.00	99.86	99.87	99.77	99.91	100.22	99.93	99.70
G.		2.14	2.152	2.236				

1. $Mg_4(OH)_2(CO_3)_3 \cdot 3H_2O$. 2. Val di Lanzo, Lombardy.[11] 3. Alameda County, California.[12] Rem. is CaO 0.20, insol. 0.05. 4. Texas, Pennsylvania.[3] 5. Emarese, Piedmont.[3] 6. Texas, Pennsylvania.[13] Rem. is CaO 1.47, FeO 1.28, SiO_2 0.14, MnO tr. 7. Gorance valley, Serbia.[15] 8. Paradise Range, Nye County, Nevada.[16] Rem. is CaO 0.04, Fe_2O_3 0.09, insol. 0.07, H_2O− 0.06.

Tests. In C.T. loses water and CO_2. B.B. whitens but infusible. Soluble with effervescence in acids.

O c c u r. Hydromagnesite occurs principally as a low temperature hydrothermal mineral in veinlets in serpentine and altered magnesium-rich igneous rocks; sometimes probably formed by weathering processes. Associated with opal, calcite, dolomite, aragonite, brucite, magnesite, artinite, pyroaurite, deweylite. Also found as an alteration product of brucite in periclase marbles (*predazzite*). Found as crusts in serpentine at Kraubat, Styria, Austria;[17] with magnesite in serpentine between Eibental and Tissovicza, Roumania, and at Hrubschütz north of Kromau in Moravia. From the Gorance valley and elsewhere in chromite mines in the serpentine district of southern Serbia.[17] From Limburg near Sasbach, Baden. As a constituent of altered periclase marble at Predazzo, in the southern Tyrol, Italy; also with calcite as nodular masses formed by the alteration of blocks of periclase marble on Monte Somma (*hydrodolomite*); and as a hydrothermal alteration of inclusions of dolomitic marble in pipernoid tuff at Fiano in the Campagnia, near Naples; also in serpentine at Emarese, Val d'Aosta, Piedmont, and Viu, Val di Lanzo, Lombardy. In the Shabani serpentine area, Belingwe district, Rhodesia.

In the United States, hydromagnesite occurs with artinite, aragonite, brucite, and dolomite as acicular crusts and earthy masses in veinlets and crevices in serpentine at Hoboken, New Jersey; found similarly at Wood's and Low's chrome mines near Texas, Lancaster County, Pennsylvania. Found in serpentine rock at numerous localities in California, notably in the Sulfur Creek area, Colusa County, and with magnesite on Larious Creek, on Sampson Peak, San Benito County. Abundantly with brucite and artinite at Luning, Nevada. With magnesite in deposits in valley bottoms in the Cariboo, Atlin, and Kamloops districts, British Columbia.[18]

A l t e r. Frequently observed as an alteration product of brucite, which in some instances may itself be an alteration of periclase as in predazzite rock. Also observed altered to deweylite and serpentine, and found as an alteration of dolomite.

A r t i f.[19] A number of supposedly distinct basic magnesium carbonates have been prepared by precipitating solutions of magnesium sulfate, chloride, or nitrate with alkali carbonates. Several of these at least are identical with hydromagnesite.

N a m e. In allusion to the composition.

HYDROGIOBERTITE. Scacchi (*Acc. Napoli, Rend.*, **24**, 310, 1885). Hydrodolomite *Dana* (213, 1850).
Material from Vesuvius (the type locality) and from Chiles Valley, Napa County, California,[21] has been shown to be a mixture of hydromagnesite and calcite.[20]

Ref.

1. The crystals appear to be monoclinic, although pseudo-orthorhombic, from the morphological studies of Rogers, *Am. J. Sc.*, **6**, 37 (1923), and Palache, priv. comm. (1939). The x-ray Laue study of Fenoglio, *Per. Min.*, **7**, 257 (1936), on {100} plates indicates orthorhombic symmetry, but, since the crystals are usually intimately twinned

on {100}, as shown by Rogers and others, the results are open to some question. The morphological elements here given are based on the angles of Palache.
2. Palache (1939) and Rogers (1923).
3. Fenoglio (1936).
4. See Rogers (1923).
5. Fenoglio (1936) on Texas, Pennsylvania, crystal; reported values range from 2.14 to 2.32.
6. Rogers (1923) on Alameda, California, crystals studied morphologically. Inclined extinction also is given by Weinschenk, Zs. Kr., 27, 570 (1897), but not by Meixner, Zbl. Min., 8 (1938), Fenoglio (1936), and others.
7. Average of selected determinations by various workers cited in Fenoglio (1936).
8. Kisselev, Ann. Inst. Mines Leningrad, 11, 59 (1938); Min. Abs., 9, 267 (1946), on Khalilovo, Urals, crystal.
9. Beck (Doctoral Thesis, Harvard, 1946); see also Caillère, Bull. soc. min., 66, 55 (1943).
10. For older analyses see Hintze (1 [3A], 3522, 1926).
11. Fenoglio, Boll. soc. geol. ital., 46, 13 (1927).
12. Boynton analysis in Rogers (1923).
13. Levi and Ghiron, Gazz. chim. ital., 62, 218 (1932).
14. See Kramm, Proc. Am. Phil. Soc., 49, 344 (1910), also analyses 3 and 6.
15. Meixner, Cbl. Min., 363 (1937).
16. Milton analysis in Callaghan, Univ. Nevada Bull., 27, 7 (1933).
17. The paragenesis of hydromagnesite in these occurrences has been described by Meixner, Zbl. Min., 8 (1938), 363 (1937).
18. Cummings, Bull. Brit. Columbia Dept. Mines, no. 4 (1940).
19. See Mellor (4, 364, 1923); Levi, Ann. chim. appl., Rome, 8, 265 (1924), Giorn. chim. indust. appl., no. 5, 224 (1930); Menzel and Brückner, Zs. Elektrochem., 36, 63 (1930); Leitmeier, Jb. Min., Beil.-Bd., 40, 655 (1916).
20. Caillère, Bull. soc. min., 66, 55 (1943); Larsen (89, 1921); Brugnatelli, Zs. Kr., 31, 54 (1899); Zambonini (95, 1910); Frondel, priv. comm. (1940).
21. Wells, Am. J. Sc., 30, 189 (1910).

GIORGIOSITE. *Lacroix (Bull. soc. min., 28, 198, 1905; C.R., 140, 1308, 1905).*

An imperfectly described substance found as spherulitic growths in white powdery crusts containing sodium chloride, sodium sulfate, and magnesium carbonate formed on lava of 1866 at Alphroëssa, Santorin.[1] The fibers have parallel extinction with positive elongation and nZ between 1.48 and 1.51. Supposedly a basic carbonate of magnesium. Not identical with hydromagnesite.[2]

Ref.

1. Fouqué, *Santorin et ses éruptions*, Paris, 211 (1879).
2. Caillère, *Bull. soc. min.*, 66, 54 (1943). See also Zambonini, *Jb. Min.*, II, 134 (1922).

16.1.14 Rutherfordine [$(UO_2)(CO_3)$]. *Marckwald (Cbl. Min., 761, 1906).*

As dense to pulverulent masses composed of minute, matted fibers. Color yellow. G. 4.82. Optically biaxial,[1] with nX 1.72 ± 0.01, nZ 1.80 ± 0.01.
Chem. Uranyl carbonate, $(UO_2)(CO_3)$. Analysis gave:

	PbO	CaO	FeO	UO_3	CO_2	H_2O	Insol.	Total
1.				86.68	13.32			100.00
2.	1.0	1.1	0.8	83.8	12.1	0.7	0.8	100.3

1. $(UO_2)(CO_3)$. 2. Morogoro.

Soluble in acids. The CO_2 is lost on heating over 300°. Found as pseudomorphs after uraninite in pegmatites in the Uruguru mountains, Morogoro district, Tanganyika, East Africa. Named after the English atomic physicist Ernest Rutherford (1871–1937). Most reputed specimens of rutherfordine are mixtures containing uranium silicates and oxides, and the existence of the species requires verification.

Ref.

1. Larsen (129, 1921).

16.1.15   Sharpite [$(UO_2)_6(CO_3)_5(OH)_2 \cdot 7H_2O$ (?)]. *Mélon (Bull. séan. inst. roy. colon. Belge*, **9**, 333, 1938).

Orthorhombic ? As crusts of thin radiating fibers. H. $\sim 2\frac{1}{2}$. G. > 3.3. Color greenish yellow.

Opt.

Orientation	n	Pleochroism	
X	1.633	Brownish	Biaxial positive (+).
$Y \perp$ laths			
Z elong.	~ 1.72	Greenish yellow	

Chem. A hydrated basic carbonate of uranium, perhaps $(UO_2)_6(CO_3)_5(OH)_2 \cdot 7H_2O$. The water and CO_2 are lost below 325°. Analysis gave:

	CaO	UO_3	CO_2	H_2O	Total
1.		82.50	10.57	6.93	100.00
2.	2.70	81.04	10.30	6.81	100.85
3.		84.38	8.53	7.09	100.00

1. $(UO_2)_6(CO_3)_5(OH)_2 \cdot 7H_2O$. 2. Kasolo. 3. Analysis 2 recalculated to 100 after deducting $CaCO_3$ 4.81.

Tests. Soluble with effervescence in dilute acids.

Occur. Found with uranophane, curite, and becquerelite at Chinkolobwe, Katanga, in the Belgian Congo.

Name. After Major R. R. Sharp, who discovered the Chinkolobwe uranium deposit in 1915.

DIDERICHITE. *Vaes (Bull. soc. belge géol.*, **70**, B 212, 1947).

An ill-defined mineral, apparently a hydrated carbonate of uranium, found at Chinkolobwe, Katanga, Belgian Congo. Orthorhombic. Yellow-green fibrous crusts. Biaxial positive, with $nX \sim 1.725$, $nY \sim 1.728$, nZ between 1.728 and 1.74, and $2V$ large. Named after Norbert Diderich, mining engineer.

STUDTITE. *Vaes (Bull. soc. belge géol.*, **70**, B 212, 1947).

An ill-defined mineral, apparently a hydrated carbonate of lead and uranium, found at Chinkolobwe, Katanga, Belgian Congo. Orthorhombic. As crusts of flexible, fibrous crystals. Color yellow. Biaxial negative (−), with nX 1.545, nY 1.555, nZ 1.68 (parallel elongation), and $2V$ large. Named after F. E. Studt, geologist.

TENGERITE. Kolsyrad Ytterjord *Svanberg* and *Tenger (Årsberätt.*, **16**, 206, 1838). Tengerite *Dana* (710, 1868). Ytterspath *Germ.* Carbonyttrine *Adam* (24, 1869).

As thin coatings, earthy and pulverulent to crystalline; also small mamillary with a radial fibrous structure. Luster dull, chalky. Color white. Supposed to be a yttrium carbonate, but analyses of material from the original locality at Ytterby are lacking. Found as thin coatings resulting from the alteration of gadolinite at Ytterby, northeast of Stockholm, Sweden.[1] Material supposedly tengerite from this locality was optically positive (+), with $2V$ large, $nX = 1.555$, $nY = 1.585$, and X parallel to the elongation.[2]

Also reported to occur as an alteration of thalenite at Åskagen, Vermland, Sweden,[3] and perhaps at Kragerö as an alteration of hellandite and at Hundholmen, Norway.[4]
A white mineral found as globular radiated incrustations as an alteration of gadolinite at Baringer Hill, Burnett County, Texas, may belong here (analysis 1).[5] Another mineral found as an alteration of yttrialite at Iisaka, Japan, also has been classed with tengerite (analysis 2).[6] This mineral occurred as a white powder or globular concretions with a minute scaly structure and pearly luster; $nX = 1.622$, $nY = 1.642$; G. = 3.12. The analysis is close to $CaY_3(OH)_3(CO_3)_4 \cdot 3H_2O$.

	Y_2O_3, etc.	Ce_2O_3, etc.	CaO	BeO	ThO_2	Fe_2O_3
1.	40.8	7.0		9.7		4.0
2.	49.0	1.3	8.3	0.7	0.3	tr.

	CO_2	PbO	SiO_2	H_2O+	H_2O-	Total
1.	19.6		0.4	14.1	3.2	100.0
2.	28.7	0.2	0.8	10.1	0.3	99.7

1. Baringer Hill, Texas.[7] Includes MgO, alkalies and loss, 1.2. Analysis recalculated after deducting gangue, 7.0. M.W. of Y_2O_3, etc. = 226; of Ce_2O_3, etc. 335. 2. Iisaka, Japan.[6]

Ref.

1. See also Flink, *Ark. Kemi*, **3**, 166 (1910).
2. Larsen (142, 1921).
3. Sjögren, *Geol. För. Förh.*, **28**, 96 (1906).
4. Brögger, *Vidensk. Skr. Oslo, Mat.-Nat. Kl.*, no. 6, 23 (1906), and Vogt, *ibid.*, no. 1, 23 (1922).
5. Genth, *Am. J. Sc.*, **38**, 199 (1889); Hidden and Mackintosh, *Am. J. Sc.*, **38**, 474 (1889); Hidden, *Am. J. Sc.*, **19**, 425 (1905).
6. Iimori, *Sci. Pap. Inst. Phys. Chem. Res.*, Tokyo, **34**, 832 (1938).
7. Hillebrand analysis in Hidden (1905).

TYPE 2. $A_m B_n (XO_3)_p Z_q$

16.2.1 **D A W S O N I T E** [$NaAl(CO_3)(OH)_2$]. Harrington (*Can. Nat.*, **7**, 305, 1874).

C r y s t.[1] Orthorhombic; dipyramidal—$2/m\ 2/m\ 2/m$.

$a:b:c = 0.6475:1:0.5339$;[2] $p_0:q_0:r_0 = 0.8246:0.5339:1$

$q_1:r_1:p_1 = 0.6475:1.2128:1$; $r_2:p_2:q_2 = 1.8730:1.5444:1$

Forms:[3]

	ϕ	$\rho = C$	ϕ_1	$\rho_1 = A$	ϕ_2	$\rho_2 = B$
c 001	.....	0°00′	0°00′	90°00′	90°00′	90°00′
b 010	0°00′	90 00	90 00	90 00	.....	0 00
a 100	90 00	90 00	.....	0 00	0 00	90 00
m 110	57 04½	90 00	90 00	32 55½	0 00	57 04½
D 011	0 00	28 06	28 06	90 00	90 00	61 54

Structure cell.[15] Space group Ima. a_0 6.72 Å, b_0 10.34, c_0 5.56; $a_0:b_0:c_0 = 0.650:1:0.538$. Cell contents $Na_4Al_4(CO_3)_4(OH)_8$.

Habit. As thin incrustations or rosettes of bladed to acicular crystals, and as tufts of fine needles. Elongated [001].

P h y s. Cleavage $\{110\}$ perfect. H. 3. G. 2.44.[4] Luster vitreous; silky in fine aggregates. Colorless to white. Streak colorless. Transparent.

Dawsonite

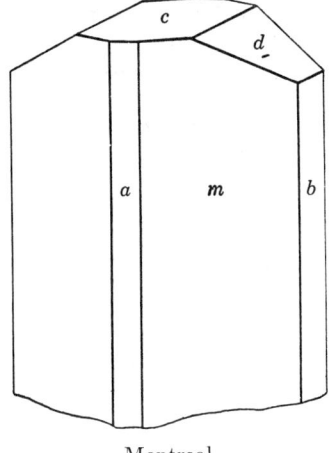

Montreal.

O p t.[7] In transmitted light, colorless.

Orientation	n (Montreal)[5]	n (Albania)[6]	
X a	1.466	1.462	Biaxial negative (−).
Y c	1.542	1.537	$2V$ 76°46′ (Na); [5] $2E$ 146°27′.
Z b	1.596	1.589	$r < v$, not marked.

C h e m. A basic carbonate of sodium and aluminum, $NaAl(CO_3)(OH)_2$. On heating,[8] the OH is lost at 300°–320°.

Anal.[9]

	1	2	3	4	5
Na_2O	21.53	21.81	21.05	21.98	[23.24]
Al_2O_3	35.40	36.01	35.91	34.90	34.86
CO_2	30.56	30.57	28.45	30.30	29.87
H_2O	12.51	11.61	11.05	12.70	11.93
Rem.			1.06		
Total	100.00	100.00	97.52	99.88	
G.		2.44			

1. $NaAl(CO_3)(OH)_2$. 2. Montreal.[10] Recalculated to 100 after deduction of CaO 1.59 as calcite. 3. Algeria.[11] Rem. is CaO 1.06, MgO tr. 4. Albania.[12] 5. Albania.[13] Analysis on 43 mg.

Tests. B.B. swells but does not fuse. Soluble in acids with effervescence.

O c c u r. A low-temperature hydrothermal mineral probably formed by the decomposition of aluminous silicates. Originally found as coatings on joint surfaces of a feldspathic dike cutting the Trenton limestone near McGill University, Montreal, Canada, and associated with calcite, dolomite, pyrite, and some galena and wad. Later found at Pian Castagnaio and Santa Fiora near Mte. Amiata, province of Siena, Tuscany, associated with dolomite, calcite, pyrite, fluorite, and cinnabar. Also east of Tenès in Alger, Algeria, with barite in an argillaceous sandstone, and from Komana, Drin valley, northern Albania, with calcite and quartz.

278   CARBONATES

Artif.[14] Obtained as a chalky powder by reaction of solutions of sodium aluminate and sodium bicarbonate, with excess CO_2. Also reported formed in other ways, but the identity with dawsonite has not been proved.

Name. After John William Dawson (1820–1899), Canadian geologist and Principal of McGill University.

Ref.

1. Graham, *Roy. Soc. Canada, Trans.*, **2** [3], 165 (1908), gives this symmetry class, although the crystals are not doubly terminated and general forms are lacking. The symmetry was earlier said by Harrington (1874), Des Cloizeaux, *Bull. soc. min.*, **1**, 8 (1878), and others to be probably monoclinic.
2. Graham (1908).
3. Graham (1908). Rare or uncertain: 130, 230, 210, 101.
4. Identical values reported by Graham (1908) and Bader, *Jb. Min., Beil.-Bd.*, **74**, 449 (1938), on material from Montreal and Albania, respectively.
5. Graham (1908).
6. Pelloux, *Per. Min.*, **3**, 15 (1932).
7. Larsen (66, 1921) gives indices for material from Siena, Italy, doubtfully referred to dawsonite, which differ markedly from the data here given.
8. Bader (1938).
9. For earlier analyses see Dana (300, 1892).
10. Graham (1908).
11. Curie and Flamand, *Ann. fac. sc., Marseille*, **2**, 50 (1889)—in Lacroix (3, 776, 1901).
12. Pelloux (1932).
13. Wiedmann analysis in Bader (1938).
14. See Bader (1938); Doelter (**1**, 203, 1916); Bader and Esch, *Zs. Elektrochem.*, **50**, 266 (1944).
15. Lauro, *Atti. acc. naz. Ital.*, ser. 7, **3**, 146 (1941), by Weissenberg method on Komana crystals.

16.2.2   NORTHUPITE [$Na_3MgCl(CO_3)_2$]. Foote (*Am. J. Sc.*, **50**, 480, 1895); Pratt (*Am. J. Sc.*, **2**, 123, 1896). Northrupite *wrong spelling*.

Cryst.[1] Isometric; diploidal—$2/m\,\bar{3}$.

Forms:

o 111

Structure cell. Space group $Fd3$. a_0 13.99 kX;[3] 14.13 kX.[2] Cell contents $Na_{48}Mg_{16}Cl_{16}(CO_3)_{32}$.

Habit. Octahedral. The crystals sometimes contain symmetrically arranged inclusions of clay.

Phys. Fracture conchoidal. Brittle. H. $3\frac{1}{2}$–4. G. 2.380;[4] 2.40 (calc. for a_0 13.99). Luster vitreous. Colorless; also pale yellow to gray and brown due to included clay or organic matter. Transparent.

Opt. In transmitted light, colorless. Isotropic, but sometimes exhibits anomalous birefringence distributed in sectors.[5] Indices:[6]

λ	Li	Na	Tl
n	1.5117	1.5144	1.5180

Chem. A chloride-carbonate of sodium and magnesium, $Na_3MgCl(CO_3)_2$. Northupite is isostructural with tychite, $Na_6Mg_2(SO_4)(CO_3)_4$, and a complete series with that substance has been obtained in artificial material.[7]

Anal.

	1	2
Na_2O	37.38	36.99
MgO	16.21	16.08
Cl	14.25	14.10
CO_2	35.38	35.12
Rem.		1.02
Total	103.22	103.31
$O = Cl$	3.22	3.16
	100.00	100.15
G.	2.40	2.380

1. $Na_3MgCl(CO_3)_2$. 2. Searles Lake.[8] Rem. is SO_3 0.08, H_2O 0.72, insol. 0.22. Average of two.

Tests. B.B. fuses at 1 with frothing to a white alkaline mass. Decrepitates violently in C.T. Easily soluble in dilute acids with effervescence. Decomposed by hot water with separation of magnesium carbonate.

O c c u r. Found as octahedral crystals embedded in clay sediments in Searles (Borax) Lake, San Bernardino County, California. Associated with tychite and pirssonite. Also found with shortite, bradleyite, trona, pirssonite, and gaylussite in drill cores from oil shale near Green River, Sweetwater County, Wyoming.

A r t i f.[9] Obtained as a gelatinous precipitate, which crystallizes on continued heating, by reaction of a $MgCl_2$ solution with a mixed solution of Na_2CO_3 and $NaCl$. The Br analogue also has been prepared. (G. 2.67; n 1.515; a_0 14.17 kX).

N a m e. After Mr. C. H. Northup of San Jose, California, who found the first specimens.

Ref.

1. Crystal class established by x-ray study by Gossner and Koch, *Zs. Kr.*, **80**, 458 (1931).
2. Gossner and Koch (1931) with determination of space group.
3. Shiba and Watanabé, *C.R.*, **193**, 1421 (1931), with determination of structure.
4. Pratt (1896) by suspension method on natural crystals. The values 2.377 and 2.366 have been reported on artificial crystals.
5. Gossner and Koch (1931).
6. Pratt (1896) by prism method on natural crystals.
7. de Schulten, *C.R.*, **143**, 403 (1906).
8. Pratt (1896).
9. de Schulten, (1906), *Bull. soc. min.*, **19**, 164 (1896), and Watanabé, *Sci. Papers Inst. Phys. Chem. Res., Tokyo*, **21**, 35 (1933).

16.2.3 **D U N D A S I T E** $[PbAl_2(CO_3)_2(OH)_4 \cdot 2H_2O]$. Petterd (*Minerals of Tasmania*, Hobart, 66, 1893).

Found as small spherical aggregates of radiating crystals, elongated [001], and as matted or felted crusts.

P h y s. Cleavage {010}, perfect, in microscopic grains.[2] H. 2. G. about 3.25.[1] Luster vitreous to silky. Color white. Transparent.

O p t.[2] In transmitted light, transparent.

Orientation		n	
X	a	1.603 ± 0.003	Biaxial negative $(-)$.
Y	b	1.716 ± 0.003	$2V$ very large.
Z	c	1.742 ± 0.003	

280   CARBONATES

Chem. A hydrated basic carbonate of lead and aluminum, $PbAl_2$-$(CO_3)_2(OH)_4 \cdot 2H_2O$. The Fe_2O_3 reported is due to admixture.

Anal.

	1	2	3
PbO	46.00	41.86	43.20
Al_2O_3	21.01	26.06	21.39
Fe_2O_3		5.50	1.61
CO_2	18.14 ⎱	28.08	16.45
H_2O	14.85 ⎰		15.01
Insol.			1.80
Total	100.00	101.50	99.46

1. $PbAl_2(CO_3)_2(OH)_4 \cdot 2H_2O$. 2. Dundas, Tasmania.[3] 3. Wales.[4] H_2O includes H_2O at 100° 1.41 per cent.

Tests. B.B. infusible. Soluble with effervescence in acids.

Occur. Found originally associated with crocoite, pyromorphite (?), and limonite in gossan in the Adelaide Proprietary mine, Dundas, Tasmania. Also with cerussite and gibbsite at the Hercules mine, Mt. Read, Tasmania. Later found [5] with cerussite and allophane at mines near Trefriw, Carnarvonshire, Wales, and at the Clements lead mine, near Maam, County Galway, Ireland; with cerussite, hemimorphite, and greenockite at the Mill Close mine, Wensley, Derbyshire; at the Wheal Rose, Sithney, and with allophane at the Port Quin antimony mine, in Cornwall.

Name. From the occurrence in Dundas, Tasmania.

Ref.

1. Prior, *Min. Mag.*, **14**, 167 (1906), by suspension method on material from Dundas and Wales.
2. Frondel, priv. comm. (1947), on Dundas material.
3. Pascoe analysis in Petterd (1893).
4. Prior (1906).
5. Cf. Prior (1906); Russell, *Min. Mag.*, **16**, 272 (1912); *ibid.*, **27**, 1 (1944).

16.2.4 Alumohydrocalcite $[CaAl_2(CO_3)_2(OH)_4 \cdot 2H_2O\ ?]$. *Bilibine (Mem. soc. russe min.,* **55**, 243, 1926).

Probably monoclinic. Found as chalky masses with a microscopically radial-fibrous, spherulitic structure. Cleavage {100} perfect, {010} imperfect. Brittle. H. $2\frac{1}{2}$. G. 2.231. Color chalky white to pale blue, rarely violet, light yellow, or gray.

Opt. In transmitted light, colorless.

ORIENTATION		n	
X	b	1.485	Biaxial negative (−).
Y		1.553	$2V$ 50°–55°.
Z		1.570	Fibers show extinction angles up to 7°–10°.

Chem. A hydrated basic carbonate of calcium and aluminum, probably $CaAl_2(CO_3)_2(OH)_4 \cdot 2H_2O$. The water is lost in several stages on heating.

Anal.

	1	2	3	4
CaO	17.63	15.46	16.28	15.62
Al_2O_3	32.05	28.60 ⎫	30.00	31.77
Fe_2O_3		0.45 ⎭		
CO_2	27.67	25.20 ⎫		
H_2O+	22.65	26.40 ⎬	51.82	51.88
H_2O-		2.48 ⎭		
Rem.		1.76	2.40	0.27
Total	100.00	100.35	100.50	99.54

1. $CaAl_2(CO_3)_2(OH)_4 \cdot 2H_2O$. 2. Siberia. Rem. is FeO 0.35, SiO_2 0.67, P_2O_5 0.74. 3,4. Siberia. Rem. is SiO_2.

Tests. Infusible. Easily soluble in acids, and decomposed by boiling water with separation of calcium carbonate and hydrous aluminum oxide.

Occur. Found in the Khakassy District, Siberia, with allophane, wad, limonite, volborthite, cuprite, malachite, and native copper. Apparently produced by the action of carbonated waters on allophane.

Name. In allusion to the composition.

16.2.5 **BEYERITE** [$Ca(BiO)_2(CO_3)_2$]. *Frondel (Am. Min.,* **28**, 532, 1943).

Cryst.[1] Tetragonal; ditetragonal-dipyramidal—$4/m\ 2/m\ 2/m$.

$$a:c = 1:5.759; \quad p_0:r_0 = 5.759:1$$

Forms:

	ϕ	ρ	A	M
c 001	.....	0°00′	90°00′	90°00′
o 111	45°00′	83 00	45 25½	90 00

Structure cell.[2] Space group $I4/mmm$. a_0 3.78 ± 0.01 kX, c_0 21.77 ± 0.05; $a_0:c_0 = 1:5.579$. Cell contents $Ca_2Bi_4(CO_3)_4O_4$.

Habit. Thin rectangular plates flattened {001}. Also as compact earthy masses.

Phys. Fracture conchoidal (crystals). H. 2–3. G. 6.56 (Schneeberg crystals), less in massive, porous material; 6.47 (calc.). Luster vitreous in crystals. Color bright yellow to lemon-yellow; also, in massive material, yellowish white to grayish green and gray.

Opt. In transmitted light, pale yellow to colorless. Some crystals have an anomalous biaxial character with very small $2V$. Not pleochroic.

n (Schneeberg)[3]

O	2.13 ± 0.02	Uniaxial negative (−).
E	1.99 ± 0.02	

Chem. A carbonate of bismuth and calcium, $Ca(BiO)_2(CO_3)_2$. A small amount of Pb substitutes for Ca in the reported analyses:

	CaO	PbO	Bi_2O_3	CO_2	Rem.	Total
1.	9.19		76.38	14.43		100.00
2.	8.85	1.73	73.65	13.59	2.01	99.83
3.	7.44	1.25	76.61	11.70	2.89	99.89

1. $Ca(BiO)_2(CO_3)_2$. 2. Mica Lode, Colorado.[4] Rem. is CuO 1.10, MnO 0.12, insol. 0.79. 3. Meyers Ranch, Colorado.[4] Rem. is CuO 0.25, Fe_2O_3 0.84, H_2O 0.96, insol. 0.84. Contains some bismutite.

Tests. Easily soluble with effervescence in acids.

Occur. A secondary mineral found associated with bismutite. Found as tiny yellow platy crystals scattered over bismutite and chalcedonic quartz at Schneeberg, Saxony. Also as greenish gray compact masses at the Stewart mine, Pala, California, where it occurs as an alteration product of bismuthinite (?) in pegmatite. Also found massive with bismutite as an alteration of primary bismuth minerals in the Mica Lode and School Section pegmatites in the Eight Mile Park district, near Canon City, Fremont County and the Meyers Ranch pegmatite in Park County, Colorado.[5]

Artif.[6] Obtained as thin plates by heating bismuth nitrate, calcium carbonate and urea in dilute HNO_3 in a bomb at 220°, and in other ways.

Name. After Adolph Beyer (1743–1805), a mining engineer and mineralogist of Schneeberg, Saxony, who in 1805 recognized the occurrence of a bismuth carbonate [bismutite] in nature.

Ref.
1. The x-ray elements are used here in place of the morphological data, obtained on crystals of inferior quality (average ρ for $\{111\}$ = 81°32').
2. Frondel (1943) by Weissenberg method on Schneeberg crystals. Lagercrantz and Sillén, *Ark. Kemi*, **25**, 1 (1948), give a_0 3.759 kX, c_0 21.646 for artificial material and describe the structure.
3. Frondel (1943). Larsen (49, 1921) gives nO 2.13, nE 1.99 for platy Schneeberg crystals labeled bismutosphaerite that doubtless are of this species.
4. Gonyer analysis in Heinrich, *Am. Min.*, **32**, 660 (1947).
5. Heinrich (1947).
6. Lagercrantz and Sillén (1948).

16.2.6 **PARISITE** [$(Ce,La)_2Ca(CO_3)_3F_2$]. Musite *Medici-Spada*, 1835. Parisite *Medici-Spada* in Bunsen (*Ann. Chemie*, 53, 147, 1845).

Cryst.[1] Hexagonal—P; ditrigonal-dipyramidal—$\bar{6}\,m\,2$ (?).

$$a:c = 1:3.8736; \qquad p_0:r_0 = 4.4729:1$$

Forms:[2]

	ϕ	$\rho = C$	M	A_1	A_2
c 0001	.....	0°00'	90°00'	90°00'	90°00'
a 11$\bar{2}$0	0°00'	90 00	90 00	60 00	60 00
B 2.0.$\bar{2}$.15	30 00	30 48½	75 09½	63 40	90 00
C 3.0.$\bar{3}$.22	30 00	31 23	74 54½	63 11½	90 00
k 10$\bar{1}$1	30 00	77 24	60 47½	32 18½	90 00
R 20$\bar{2}$1	30 00	83 37	60 12	30 36½	90 00
T 50$\bar{5}$2	30 00	84 53½	60 08	30 23½	90 00
δ 0.3.$\bar{3}$.20	−30 00	33 51½	106 10½	90 00	61 09
ϵ 0.9.$\bar{9}$.50	−30 00	38 50	108 16½	90 00	57 06
μ 01$\bar{1}$1	−30 00	77 24	119 12½	90 00	32 18½
π 05$\bar{5}$4	−30 00	79 51½	119 33½	90 00	31 31
ϕ 07$\bar{7}$4	−30 00	82 43	119 44	90 00	30 47½
i 11$\bar{2}$6	0 00	52 14½	90 00	66 43	66 43
n 5.5.$\overline{10}$.24	0 00	58 13	90 00	64 51	64 51
t 22$\bar{4}$9	0 00	59 51	90 00	64 23	64 23
r 11$\bar{2}$3	0 00	68 50	90 00	62 12½	62 12½
w 22$\bar{4}$3	0 00	79 02½	90 00	60 36	60 36

Less common:

$E\ 5.0.\bar{5}.32$	$L\ 50\bar{5}8$	$V\ 30\bar{3}1$	$\gamma\ 01\bar{1}7$
$H\ 10\bar{1}5$	$h\ 30\bar{3}4$	$\beta\ 0.2.\bar{2}.15$	$l\ 11.11.\overline{22}.54$

Structure cell.[3] Space group $C\bar{6}c$ (?). $a_0\ 7.091\ kX$, $c_0\ 27.93$; $a_0:c_0 = 1:3.939$. Cell contents $Ce_{12}Ca_6(CO_3)_{18}F_{12}$.

Habit. Crystals usually acute double hexagonal pyramids; sometimes prismatic in appearance due to oscillatory combination of steep pyramids,

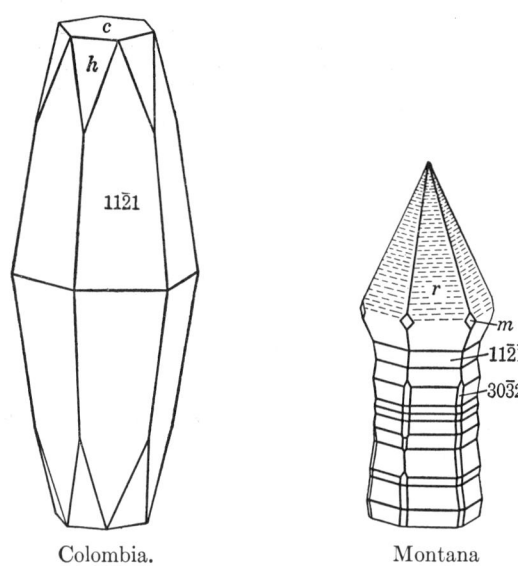

Colombia.   Montana

true prism faces being lacking or very small. Also trigonal (or rhombohedral ?) in development. Elongated crystals sometimes have locally enlarged portions or have sceptre-like terminations. The lateral faces commonly are striated or deeply grooved. {0001} very commonly present as small faces, sometimes large; other common forms: $a\ i\ T\ \delta\ \epsilon\ \pi\ r\ t$.

P h y s. A distinct to perfect parting or cleavage on {0001} is sometimes present apparently due to alteration. Fracture subconchoidal to splintery. Brittle. H. $4\frac{1}{2}$. G. 4.36 ± 0.03; 4.38 (calc.). Luster vitreous to resinous, on the {0001} parting pearly. Color brownish yellow; also brown, waxyellow, grayish yellow. Streak yellowish white. Transparent to translucent.

O p t. Uniaxial positive (+). In transmitted light, colorless to yellow. Weakly pleochroic with absorption $E > O$.

	n (Muzo)[4]			n (Quincy)[5]	
λ	453	589	667	Na	Dichroism[5]
O	1.6841	1.6717	1.6679	1.676	Light yellow
E	1.7882	1.7712	1.7657	1.757	Golden yellow

Chem. A fluo-carbonate of calcium and the cerium group of rare earths, $(Ce,La)_2Ca(CO_3)_3F_2$.

Anal.[6]

	1	2	3	4	5
CaO	10.44	10.98	10.70	11.40	3.78
Ce_2O_3	30.56	26.14	30.67	30.94	21.88
$(La,Di)_2O_3$	30.33	28.46	29.74	27.31	33.11
Y_2O_3		tr.	tr.	tr.	7.86
CO_2	24.58	22.93	24.22	24.16	23.69
F	7.07	5.90	6.82	6.56	6.37
Rem.		8.07	0.50	1.84	5.01
	102.98	102.48	102.65	102.21	101.70
O = F	2.98	2.48	2.87	2.76	2.68
Total	100.00	[100.00]	99.78	99.35	99.02
G.		4.128	4.302	4.32	4.396

1. $(Ce,La)_2Ca(CO_3)_3F_2$ with Ce:La = 1:1. 2. Ravalli County, Montana.[9] Rem. is Fe_2O_3 0.80, Na_2O 0.69, K_2O 0.19, H_2O 0.26, gangue 6.13 (by diff.). 3. Muzo, Colombia.[9] Rem. is Fe_2O_3 0.20, Na_2O 0.20, K_2O 0.10. 4. Quincy, Mass.[5] Rem. is Fe_2O_3 0.32, SrO tr., Na_2O 0.30, K_2O 0.20, H_2O tr., gangue 1.02. 5. Yttrian parisite ? Mukden, Manchuria.[7] Rem. is ThO_2 tr., Fe_2O_3 0.28, MnO tr., Na_2O 1.97, K_2O 0.31, H_2O 2.45. Analysis on zoned and perhaps altered crystals.

Tests. B.B. infusible; glows. Soluble in hot, strong acids. In C.T. loses CO_2 and turns lighter in color.

Occur. Originally found in the emerald deposits in the Muzo district, about 100 km. north of Bogota, Colombia. The parisite is associated with pyrite, marcasite, cerian dolomite, albite, calcite, quartz, gypsum, and emerald in veinlets and pockets in intensely folded, carbonaceous shale beds. Reported as occurring with weibyeite and eudidymite on the island Övre-Arö in the Langesund fiord, Norway, in an alkali-pegmatite in nepheline-syenite; also from Hundholmen in northern Norway. At Montorfano, Lombardy, Italy, in veinlets in riebeckite-granite. With bastnäsite and tscheffkinite from Ifasina near Torendrika, Madagascar. A yttrian parisite (?) has been found with fluorite, zircon, and pyrite in a granite-pegmatite boulder at Mukden, Manchuria. In the United States, a notable occurrence at the Ballou and Fallon Brothers quarries at Quincy, Massachusetts. The mineral occurs here with aegirite, microcline, riebeckite, fluorite, ilmenite, and anatase in pegmatite pipes in a riebeckite-aegirite-granite. With pyrite in an altered trachyte (?) near Pyrites, Ravalli County, Montana.

Alter. Heated in air parisite loses CO_2 and affords oriented, single-crystal pseudomorphs.[8] Some natural parisite crystals appear to be altered, perhaps by hydration. Other crystals show zones or cores of different color or optical properties which may be due to alteration or to parallel growths of parisite with a related mineral.

Name. After J. J. Paris, proprietor of the mine at Muzo where the mineral was discovered and whence, in 1835, specimens were sent to the Italian collector Medici-Spada.

CORDYLITE

Ref.
1. Point symmetry and lattice type indicated by the x-ray study of Oftedal, *Zs. Kr.*, **79**, 437 (1931). The usual development of the crystals is trigonal. Palache and Warren, *Am. J. Sc.*, **31**, 533 (1911), *Zs. Kr.*, **49**, 332 (1911), and other authors have suggested a rhombohedral lattice, but many of the forms do not conform to the rhombohedral centering rule of $h + i + l = 3n$. Further study of the crystallography of parisite and its related minerals is desirable. The angles and orientation here used are those of Palache and Warren (1911) on Quincy crystals, in the unit of the x-ray cell of Oftedal (1931). Transformation: P. and W. to Oftedal, 1000/0100/0010/0002. Somewhat discordant angles have been reported by others—see Cesàro, *Soc. roy. sci. Liége, Mém.*, **12**, no. 1 (1924). See also Ungemach, *Zs. Kr.*, **91**, 1 (1935), for a discussion of the morphology of parisite and its relations to synchisite.
2. Goldschmidt (**6**, 120, 1920). The angle table is for both hexagonal and rhombohedral lattice modes. Palache and Warren (1911) give frequency data for the forms. Goldschmidt's form no. 44, π, should be $-\frac{5}{2}$. Rare or uncertain:

$10\bar{1}0$	$10\bar{1}2$	$03\bar{3}8$	$11\bar{2}9$	$3.3.\bar{6}.2$
$10\bar{1}8$	$20\bar{2}3$	$03\bar{3}5$	$2.2.\bar{4}.15$	$2.1.\bar{3}.11$
$3.0.\bar{3}.20$	$11.0.\overline{11}.12$	$05\bar{5}8$	$11\bar{2}4$	$51\bar{6}4$
$5.0.\bar{5}.32$	$11.0.\overline{11}.8$	$03\bar{3}2$	$5.5.\overline{10}.18$	
$10\bar{1}6$	$30\bar{3}2$	$05\bar{5}3$	$5.5.\overline{10}.12$	
$5.0.\bar{5}.26$	$50\bar{5}3$	$02\bar{2}1$	$11\bar{2}2$	
$10\bar{1}4$	$0.3.\bar{3}.16$	$05\bar{5}2$	$33\bar{6}4$	
$4.0.\bar{4}.11$	$01\bar{1}4$	$03\bar{3}1$	$5.5.\overline{10}.6$	
$7.0.\bar{7}.16$	$0.5.\bar{5}.16$	$1.1.\bar{2}.10$	$o\ 11\bar{2}1$	

3. Oftedal (1931) on crystals from Muzo and Quincy.
4. Quercigh, *Acc. Linc., Rend.*, **21**, 581 (1912), by prism method.
5. Warren in Palache and Warren (1911).
6. Additional analyses cited by Hintze (**1** [3A], 3440, 1929).
7. Tschernik, *Verh. russ. min. Ges.*, **44**, 507 (1906)—*Jb. Min.*, II, 336 (1909).
8. Aminoff, *Geol. För. Förh.*, **42**, 291 (1920).
9. Warren in Penfield and Warren, *Am. J. Sc.*, **8**, 21 (1899).

YTTROPARISITE *Nefedov* (*C. r. ac. sc. U.R.S.S.*, **32**, 361, 1941—*Min. Abs.*, **8**, 279, 1942).

Hexagonal. Structure cell dimensions: a_0 4.008 ± 0.03 kX, c_0 4.469 ± 0.04; $a_0:c_0 =$ 1:1.117. Optically uniaxial positive (+) with nO 1.643, nE 1.755. Composition near parisite but contains much yttria. Found as inclusions in fluorite in pegmatite in the Adun-Cholon district, Transbaikalia, U.S.S.R.

16.2.7 **C O R D Y L I T E** [(Ce,La)$_2$Ba(CO$_3$)$_3$F$_2$]. *Flink* (*Medd. Grønland*, **14**, 236, 1898; **24**, 42, 1901). Barium-Parisite. Kordylit *Germ.*

C r y s t.[1] Hexagonal—P.

$$a:c = 1:1.1288; \quad p_0:r_0 = 1.3034:1$$

Forms: [2]

	ϕ	ρ	M	A_2
c 0001	.....	0°00′	90°00′	90°00′
m $10\bar{1}0$	30°00′	90 00	60 00	90 00
p $40\bar{4}5$	30 00	46 12	68 51	90 00
q $10\bar{1}1$	30 00	52 30½	66 37½	90 00
r $20\bar{2}1$	30 00	69 01	62 10½	90 00
s $40\bar{4}1$	30 00	79 08½	60 35½	90 00

Structure cell.[3] Space group not known. a_0 4.35 kX, c_0 22.8; $a_0:c_0$ = 1:5.24. Cell contents $(Ce,La)_4Ba_2(CO_3)_6F_4$.

Habit. Short prismatic [0001] with hexagonal dipyramidal terminations. Dominant forms $m\ p\ q$. Crystals tiny, often with enlarged terminations on a slender prism and sceptre-like.

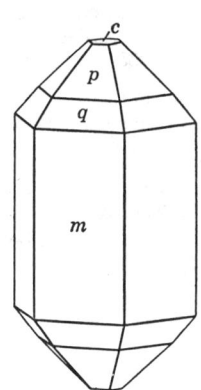
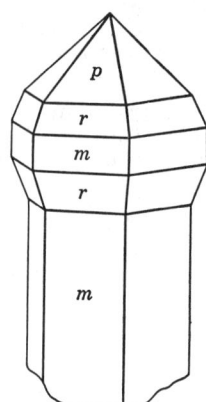

P h y s. Cleavage {0001} distinct, but perhaps a parting due to alteration. Fracture conchoidal. Rather brittle. H. $4\frac{1}{2}$. G. 4.31 (probably low); 5.61 (calc.). Luster greasy to adamantine, on {0001} pearly. Colorless to wax-yellow and transparent when fresh, but often ocher-yellow and dull due to surface alterations.

O p t.[4] In transmitted light, colorless to pale yellow. Weakly dichroic.

	n	Dichroism	
O	1.764	Greenish yellow	Uniaxial negative $(-)$.
E	1.577	Brownish yellow	

Basal sections sometimes show concentric hexagonal zones of colorless and cloudy or opaque material together with radial sets of striae.

C h e m. A fluo-carbonate of barium and the cerium group of rare earths, $(Ce,La)_2Ba(CO_3)_3F_2$. Chemically the barium analogue of parisite.

Anal.

	CaO	BaO	FeO	Ce_2O_3	$(La,Di)_2O_3$	CO_2	F	Rem.	Total	G.
1.		24.17		25.87	25.68	20.81	5.99		100.00	
2.	1.91	17.30	1.43	23.72	25.67	23.47	[4.87]	3.68	[100.00]	4.31

1. $(Ce,La)_2Ba(CO_3)_3F_2$, with Ce:La = 1:1. Total less F = O. 2. Narsarsuk.[5] Rem. is ThO_2 0.30, Y_2O_3 tr., H_2O 0.80, insol. 2.58. Total less O = F = 2.05.

Tests. B.B. decrepitates strongly but does not fuse. Easily soluble in acids.

O c c u r. Found at Narsarsuk in the Julianehaab district, Greenland, associated with aegirite, ancylite, synchisite, and neptunite in pegmatitic veins in nepheline-syenite.

N a m e. From κορδύ, *club*, in allusion to the shape of the crystals.

Ref.
1. Angles of Flink (1901), unit of Bøggild, *Medd. Grønland*, **33**, 101 (1906). Transformation: Flink to Bøggild, $1000/0100/0010/000\frac{1}{3}$.
2. Flink (1901).
3. Oftedal, *Zs. Kr.*, **79**, 437 (1931), from an oscillation photograph; the values given may represent a pseudo-cell. Transformation, Oftedal to Bøggild: $1\bar{1}00/01\bar{1}0/\bar{1}010/000\frac{3}{8}$.
4. Bøggild (1906) by prism method.
5. Mauzelius analysis in Flink (1901).

16.2.8 SYNCHISITE [(Ce,La)Ca(CO$_3$)$_2$F]. Parisite *Nordenskiöld* (*Geol. För. Förh.*, **16**, 338, 1894). Synchisite *Flink* (*Bull. Geol. Inst. Upsala*, **5**, 81, 1901; *Medd. Grønland*, **24**, 29, 1901). Synchysite.

Cryst.[1] Hexagonal—P; ditrigonal-dipyramidal—$\bar{6}\ m\ 2$ (?).

$$a:c = 1:3.8736; \quad p_0:r_0 = 4.4729:1$$

Forms:[2]

c 0001	o 10$\bar{1}$2	t 11$\bar{2}$9	v 3$\bar{3}$68	α 3$\bar{3}$64
m 10$\bar{1}$0	z 30$\bar{3}$2	q 11$\bar{2}$4	p 11$\bar{2}$2	γ 3$\bar{3}$62
a 11$\bar{2}$0	i 1.1.$\bar{2}$.10	r 11$\bar{2}$3	s 22$\bar{4}$3	

Structure cell.[3] Space group $C\bar{6}c2$ (?). a_0 7.091 kX, c_0 18.20; $a_0:c_0 = 1:2.538$. Cell contents Ce$_6$Ca$_6$(CO$_3$)$_{12}$F$_6$.

Habit. Acute rhombohedral with small {0001}, or thick tabular {0001} and hemimorphic (ditrigonal-pyramidal). Crystals of these two habits

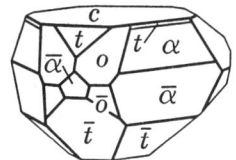

sometimes occur intergrown with parallel axes and may represent separate species. {0001} smooth and brilliant, lateral faces striated horizontally.

Twinning.[4] On {0001}, common; also lamellar.

Phys. Sometimes a parting or cleavage on {0001} due to alteration. Fracture subconchoidal to splintery. Brittle. H. $4\frac{1}{2}$. G. 3.90; 3.88 (calc.). Luster vitreous to greasy or sub-adamantine. Color wax-yellow to brown, also grayish yellow. Translucent.

Opt. Uniaxial positive (+). In transmitted light, colorless to pale yellow. Weakly pleochroic, with absorption $E > O$.

	n (Narsarsuk)[5]			n (Narsarsuk)[6]
λ	453	589	667	Na
O	1.6836	1.6730	1.6691	1.6742
E	1.7879	1.7690	1.7633	1.7701

Chem. A fluo-carbonate of calcium and the cerium rare earths, apparently (Ce,La)Ca(CO$_3$)$_2$F. Analyses 2 and 3 probably were made on somewhat altered and perhaps admixed material, and analysis 4 may represent parisite and not the true synchisite.

Anal.

	1	2	3	4
CaO	17.60	17.13	16.63	11.96
Ce_2O_3	25.75	28.14	21.98 ⎫	60.95
$(La,Di)_2O_3$	25.57	22.88	28.67 ⎭	
Y_2O_3		1.23	1.18	
CO_2	27.63	26.54	25.99	
F	5.96	5.82	5.04	
Rem.		0.31	2.51	
	102.51	102.05	102.00	
O = F	2.51	2.45	2.12	
Total	100.00	99.60	99.88	
G.			3.90	

1. $(Ce,La)Ca(CO_3)_2F$, with Ce:La = 1:1. 2. Narsarsuk.[6] Rem. is Na_2O 0.19, K_2O 0.12. 3. Narsarsuk.[7] Rem. is ThO_2 (?) 0.30, FeO (?) 0.11, H_2O 2.10. 4. Narsarsuk.[8]

Tests. B.B. infusible; glows. Relatively easily soluble in acids.

O c c u r. Found in pegmatitic segregations in syenite at Narsarsuk in the Julianehaab district, Greenland, associated with astrophyllite, aegirite, microcline, catapleiite, albite, neptunite, elpidite, epididymite, cordylite, fluorite, rhodochrosite. Also reported [9] with parisite at Quincy, Massachusetts, and Ravalli County, Montana; at the Zaaiplaats tin mine in the Potgietersrust district, Transvaal, South Africa (G. 3.90, nO 1.644, nE 1.744); [10] and with gadolinite, adularia, xenotime, etc., in an Alpine vein at Piz Blas, Val Nalps, Graubünden, Switzerland.[11]

A l t e r.[12] The Narsarsuk crystals are often more or less altered, becoming grayish white in color with reduced hardness and losing their transparency and luster, perhaps with the formation of bastnaesite or calcite. Some crystals, more or less hydrated, show a central core of apparently altered material with relatively dark colored, transparent terminations (parisite ?, anal. 4) that are anhydrous. The smaller crystals generally are homogeneous.

N a m e. From σύγχυσις, *confounding*, in allusion to its being originally mistaken for parisite.

Ref.

1. Point symmetry and lattice type indicated by the x-ray study of Oftedal, *Zs. Kr.*, **79**, 437 (1931). The morphological symmetry was earlier classed as rhombohedral by Flink (1901) and as trigonal pyramidal by Bøggild, *Medd. Grønland*, **33**, 99 (1907). Further study is needed to resolve the obvious inconsistencies between the x-ray crystallography and the morphological symmetries and twinning and to establish the morphological elements and forms. Apparently several different minerals, one of which seems to be parisite, have been confused under the name. The Narsarsuk crystals do not afford good morphological measurements, and both Flink (1901) and Bøggild (1907) referred their forms to the elements obtained by Vrba, *Zs. Kr.*, **15**, 210 (1889), on parisite from Muzo; Nordenskiöld (1894) earlier had concluded that the crystals were identical with parasite, and this view was later taken by Palache and Warren, *Am. J. Sc.*, **31**, 533 (1911), and Quercigh, *Acc. Linc., Rend.*, **21**, 581 (1912). The forms are here referred for convenience to the elements of parisite (which see) in its new unit using Palache and Warren's (1911) angles. Transformation; Flink (1901), Bøggild (1907), and Vrba (1889) to the present parisite elements, $01\bar{1}0/\bar{1}100/\bar{1}010/0002$. These elements do not correspond to the x-ray cell of Oftedal (1931) on synchisite, whose value for c is

two-thirds that here given. Transformation: Flink, Bøggild, and Vrba to Oftedal, $01\bar{1}0/\bar{1}100/\bar{1}010/000\frac{4}{3}$.
2. Flink (1901) and Bøggild (1907), transformed to the elements here given. Bøggild's upper and lower and right and left forms are not here distinguished. For angles see the parisite angle table.
3. Oftedal (1931).
4. Flink (1901), Nordenskiöld (1894).
5. Quercigh, *Acc. Linc., Rend.*, **21,** 581 (1912). The measurements may have been made on parisite and not true synchisite.
6. Flink (1901).
7. Mauzelius in Flink (1901).
8. Quercigh (1912).
9. Oftedal (1931).
10. Söhnge, *Trans. Geol. Soc. South Africa*, **47,** 176 (1944).
11. Parker, et al., *Schweiz. min. Mitt.*, **19,** 293 (1939).
12. See Flink (1901), Quercigh (1912), Oftedal (1931).

16.2.9   **B A S T N A E S I T E** [(Ce,La)(CO$_3$)F]. Basiskfluor-cerium *Hisinger* (*Ak. Stockholm, Öfv.*, 189, 1838). Bastnäsite *Huot* (**1,** 296, 1841). Hamartite *Nordenskiöld* (*Ak. Stockholm, Öfv.*, **25,** 399, 1868).

Basisk flussspatssyradt Cerium *Berzelius* (*Afhandl. Fys. Chem. Min.*, 5, 64, 1818). Basisches Fluorcerium. Basic fluocerine. Basicérine *Beudant* (**2,** 520, 1832). Fluocerine *Hausmann* (**2,** 1447, 1847). Hydrofluocerite. Cérium hydrofluaté *Dufrénoy* (**2,** 382, 1845). Sub-fluate of cerium *Phillips* (267, 1867).

Kischtim-Parisit *Karavayev* (*Bull. Ac. Imp. Sc. St. Petersburg*, **4,** 401, 1861; *J. prakt. Chem.*, 85, 442, 1862); *Koksharov* (4, 40, 1862). Kischtimite *Brush* (*Am. J. Sc.*, 35, 427, 1863). Kyshtymo-parisite.

C r y s t.[1] Hexagonal—P; ditrigonal-dipyramidal—$\bar{6}\ m\ 2$ (?).

$$a:c = 1:1.3598; \quad p_0:r_0 = 1.5702:1$$

Forms:[2]

	ϕ	ρ	M	A_2
$c\ 0001$	.....	0°00'	90°00'	90°00'
$m\ 10\bar{1}0$	30°00'	90 00	60 00	90 00
$a\ 11\bar{2}0$	60 00	90 00	90 00	60 00
$t\ 10\bar{1}3$	30 00	27 37½	76 35½	90 00
$p\ 10\bar{1}2$	30 00	38 08	72 01	90 00
$q\ 10\bar{1}1$	30 00	57 30	65 03½	90 00
$s\ 11\bar{2}2$	0 00	53 40	90 00	66 15

Structure cell.[3] Space group $C\bar{6}c2$. a_0 7.094 kX, c_0 9.718; $a_0:c_0 = 1:1.370$. Cell contents (Ce,La)$_6$(CO$_3$)$_6$F$_6$.

Habit. Tabular $\{0001\}$, usually with $\{10\bar{1}0\}$ alone. The crystals generally deeply grooved horizontally due to oscillatory combination of $\{10\bar{1}1\}$ and $\{10\bar{1}0\}$, whereby they may resemble a pile of thin plates. Also as large anhedral masses; granular.

Oriented growths.[4] Bastnaesite often forms oriented overgrowths and alteration pseudomorphs after fluocerite, the crystal axes of the two minerals oriented in parallel position.

P h y s. Cleavage $\{10\bar{1}0\}$ indistinct; often a distinct to perfect parting $\{0001\}$ of secondary origin. Fracture uneven. Brittle. H. 4–4½. G. 4.9–5.2; 4.99;[5] 5.12 (calc. for Ce(CO$_3$)F). Luster vitreous to greasy, pearly on the parting surfaces. Color wax-yellow to reddish brown. Transparent to translucent.

O p t.[6] In transmitted light, colorless to pale yellow. Faintly pleochroic, with absorption $E > O$.

	n (Bastnäs)	n (Pikes Peak)	
O	1.722	1.717	Uniaxial positive (+).
E	1.823	1.818	

C h e m. A fluo-carbonate of the cerium group of rare earths, $(Ce,La,Di)(CO_3)F$.

Anal.[7]

	1	2	3	4
Ce_2O_3	37.55	37.71	40.50 ⎫	75.84
$(La,Di)_2O_3$	37.27	36.29	36.30 ⎭	
CO_2	20.15	20.03	20.20	19.55
F	8.69	7.83	6.23	2.24
H_2O		0.08		1.83
Rem.		0.40	0.60	1.57
	103.66	102.34	103.83	101.03
$O = F$	3.66	3.30	2.61	0.95
Total	100.00	99.04	101.22	100.08
G.		5.12	4.948	4.746

1. $(Ce,La)(CO_3)F$, with $Ce:La = 1:1$. 2. Cheyenne Mountain, near Pikes Peak, Colorado.[8] Rem. is Na_2O 0.18, Fe_2O_3 0.22. Ce_2O_3 includes 0.1 ThO_2?; $(La,Di)_2O_3$ probably low, at. wt. 141. 3. Madagascar.[9] Rem. is P_2O_5. 4. Kychtym district, Urals [10] (*kischtimite?*). Rem. is CaO 0.60, MnO tr., SiO_2 0.97, P_2O_5 tr. (nO 1.723, nE 1.825).

Tests. B.B. infusible, becoming white and opaque. Soluble in strong, hot acids.

O c c u r. Originally found at Bastnäs in the Riddarhyttan district, Västmanland, Sweden, with allanite, cerite, fluocerite in narrow bands in a contact metamorphic amphibole-skarn.[12] As an alteration of fluocerite at Finbo near Fahlun, Sweden. In the Kychtym district in the Urals, U.S.S.R., with allanite, cerite, britholite, and törnebohmite as pebbles in the gold placers of the Motchaline-log, a tributary of the Borzovka River, and ultimately derived from a contact zone of alkali-syenites. In Madagascar in the region of Torendrika-Ifasina with glaucophane and tscheffkinite in a contact zone with pegmatitic facies of an alkali-granite. Also from the Belgian Congo, and reported from the Potgietersrust tin field, Transvaal, South Africa. In the United States found in Colorado at Jamestown with cerite, fluorite, allanite, törnebohmite in a contact zone, and with fluocerite in pegmatite in the Pikes Peak granite at Cheyenne Mountain and elsewhere in the region of Pikes Peak, El Paso County. Occurs with fluorite in a brecciated contact zone in the Gallinas Mountains, Lincoln County, New Mexico.[11]

A l t e r. Found as an alteration product of fluocerite and of tscheffkinite, sometimes as complete pseudomorphs. Alters to lanthanite (?).

Ref.

1. Point symmetry from space group considerations and observed piezoelectric character—see Oftedal, *Zs. Kr.*, **78**, 462 (1931). Angles and orientation of Koechlin,

Min. Mitt., **31**, 525 (1912), on Pikes Peak crystals, unit of the structure cell of Oftedal (1931). Transformation: Koechlin to Oftedal, 1000/0100/0010/0002; Lacroix, *Bull. soc. min.*, **35**, 108 (1912), to Oftedal, 1000/0100/0010/000$\frac{2}{5}$.
  2. Koechlin (1912), Lacroix (1912; **1**, 297, 1922).
  3. Oftedal (1931) on crystals from West Cheyenne Canyon, Colorado.
  4. See Koechlin (1912); Oftedal, *Norsk Geol. Tidsskr.*, **12**, 459 (1931); Lacroix (1912).
  5. Value obtained by Fleischer in Glass and Smalley, *Am. Min.*, **30**, 601 (1945), on Gallinas material and by Wolfe, priv. comm. (1946), on Colorado material and probably best representative of the mineral. The wide reported range probably includes altered material.
  6. See summary by Glass and Smalley (1945).
  7. Older analyses are summarized by Hintze (**1** [3A], 3418, 1929) and Glass and Smalley (1945).
  8. Hillebrand, *Am. J. Sc.*, **7**, 51 (1899).
  9. Pisani analysis in Lacroix (1912).
  10. Silberminz, *C. r. ac. sc. U.R.S.S.*, ser. A, 55 (1929). The earlier analysis of kischtimite by Karavayev (1861) shows F 6.35 per cent [see Flink in Doelter (**1**, 536, 1912)].
  11. Glass and Smalley (1945).
  12. Geijer, *Sver. Geol. Undersök., Årsbok*, **14**, no. 6 (1920).

BEIYINITE.   Ho (*Bull. Geol. Soc. China*, **14**, 279, 1935).

A name provisionally given to a greenish yellow mineral found as minute grains with a perfect basal cleavage. Presumed to be a carbonate of the rare earths from spectrographic analysis of the enclosing rock. H. $\sim 4\frac{1}{2}$. G. 4.83. Uniaxial negative ($-$), with $nO = 1.717$, $nE = 1.791$ (for Na). Feebly pleochroic. Insoluble in HCl, but soluble with difficulty in hot concentrated H_2SO_4 with effervescence. Infusible. Found associated with grains of magnetite, pyrite, barite, and oborite in fluorite veins in an iron ore deposit at Beiyin-Obo, north of Paoto, Suiyuan, Inner Mongolia. The mineral resembles bastnaesite.

OBORITE.   Ho (*Bull. Geol. Soc. China*, **14**, 279, 1935).

Found with beiyinite (which see) as minute, greenish yellow grains with a rhombohedral (?) cleavage. Perhaps hexagonal, with $a:c = 1:1.07$, from the angle 51°12′ measured optically from c to the $\{10\bar{1}1\}$ cleavage. H. $\sim 4\frac{1}{2}$. G. ~ 4.83. Tests as for beiyinite, but less soluble in H_2SO_4. Presumed to be a carbonate of the rare earths.

16.2.10   **ANCYLITE**   [$(Ce,La)_4(Sr,Ca)_3(CO_3)_7(OH)_4 \cdot 3H_2O$]. Ankylit *Flink* (*Medd. Grønland*, **14**, 235, 1899; **24**, 49, 1901). Ansilite. Calcio-ancylite *Tschernik* (*Bull. ac. sc. Russie*, [6], **17**, 81, 1923).

   CALCIO-ANCYLITE   [$(Ce,La)_4(Ca,Sr)_3(CO_3)_7(OH)_4 \cdot 3H_2O$]. Eisenkalkankylit, Mangankalkankylit *Tschernik* (*Verh. min. Ges. St. Petersburg*, **41**, 43, 1904).

C r y s t.[1]   Orthorhombic; dipyramidal—$2/m\ 2/m\ 2/m$.

$$a:b:c = 0.571:1:0.534; \quad p_0:q_0:r_0 = 0.935:0.534:1$$

$$q_1:r_1:p_1 = 0.571:1.069:1; \quad r_2:p_2:q_2 = 1.873:1.751:1$$

Forms:[2]

	ϕ	$\rho = C$	ϕ_1	$\rho_1 = A$	ϕ_2	$\rho_2 = B$
d 120	41°12′	90°00′	90°00′	48°48′	0°00′	41°12′
e 111	60 16	47 07	28 07	50 29	46 55½	68 41½

Habit.   Pseudo-octahedral, the faces curved and often dull; short prismatic [001]. As groups and crusts of small rounded crystals.

P h y s.   No cleavage. Fracture splintery, rather tough. Somewhat brittle. H. 4–4½. G. 3.95 (*ancylite*). Luster vitreous on faces, greasy on fracture surfaces. Color pale yellow with tinge of orange; also yellowish brown to brown, gray. Streak white. Translucent to subtranslucent.

O p t.[3] In transmitted light colorless, but sometimes clouded by inclusions (Narsarsuk).

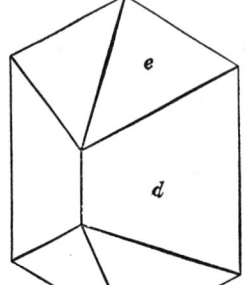

Narsarsuk.

ORIENTATION		n_{Hg} yellow (ancylite, Narsarsuk)	
X	a	1.625	Biaxial negative (−).
Y	b	1.700	2V medium large.
Z	c	1.735	

C h e m. A hydrated basic carbonate of strontium, calcium, and the cerium group of rare earths. Analysis 2 is close to the formula $(Ce,La)_4(Sr,Ca)_3(CO_3)_7(OH)_4 \cdot 3H_2O$, but the materials of analyses 3, 4, and 5 afford somewhat different ratios. Sr, Ca, Fe″, and Mn″ substitute mutually. The species names ancylite and calcio-ancylite are applied to the halves of the series with Sr > Ca and Ca > Sr, respectively.

Anal.

	1	2	3	4	5	6
CaO		1.52	4.36	13.01	12.83	13.78
SrO	22.81	21.03	12.11			
BaO			1.59	tr.	tr.	
FeO		0.35		5.36		
MnO		tr.			5.55	
Y_2O_3		tr.	0.74	tr.	tr.	
Ce_2O_3	24.08	22.22 }	47.27	44.58	35.61	26.90
$(La,Di)_2O_3$	23.90	24.04 }		5.42	14.21	26.70
CO_2	22.60	23.28	28.38	23.78	23.70	25.24
H_2O	6.61	6.52	5.55	6.97	6.94	7.38
Insol.		0.80			0.22	
Total	100.00	99.76	100.00	99.12	99.06	100.00
G.		3.95	3.82	4.298	3.962	

1. $(Ce,La)_4Sr_3(CO_3)_7(OH)_4 \cdot 3H_2O$, with Ce:La = 1:1. 2. Ancylite. Narsarsuk.[4] Rem. is ThO_2 0.20, F tr., insol. 0.60. 3. Calcian ancylite. Khibina tundra, U.S.S.R.[5] Recalculated to 100 after deducting natrolite. 4. Ferroan calcio-ancylite. Russia (*eisenkalkankylit*).[6] 5. Manganoan calcio-ancylite. Russia (*mangankalkankylit*).[6] 6. $(Ce,La)_4Ca_3(CO_3)_7(OH)_4 \cdot 3H_2O$, with Ce:La = 1:1.

Tests. B.B. infusible. Easily soluble in acids. In C.T. loses water freely

O c c u r. Ancylite occurs in druses in pegmatitic veinlets in nepheline-syenite at Narsarsuk in the Julianehaab district, Greenland, associated with aegirite, albite, microcline, zircon, synchisite, cordylite, eudidymite. A calcian ancylite occurs similarly in nepheline-syenite pegmatite in the Khibina tundra, Kola peninsula, U.S.S.R., associated with natrolite and yttrocerite. Calcio-ancylite is known only from an unstated locality in western Russia, where it was found in a granitic boulder.

N a m e. From ανκιλος, *curved*, in allusion to the rounded and distorted character of the crystals. The name calcio-ancylite has been used by some authors as a name for calcian varieties of ancylite but is here applied as a species designation for material with Ca > Sr.

Ref.
1. Gordon, *Proc. Ac. Sc. Philadelphia*, **77,** 1 (1925) from approximate measurements on rounded ancylite crystals from Narsarsuk.
2. Gordon (1925); Flink (1899) on ancylite.
3. Gordon (1925). Larsen (39, 1921) gives $nX = 1.865$ (variable), $2V \sim 0°$, and optically positive for a mineral labeled ancylite but probably not that species. Flink (1899) also states the mineral to be positive. The optical orientation is doubtful.
4. Mauzelius analysis in Flink (1901).
5. Tschernik (1923).
6. Tschernik (1904).

AMBATOARINITE. *Lacroix (Bull. soc. min.*, **38,** 265, 1915).

Probably orthorhombic. In microscopic crystals, often with parallel axes in skeletal groups. Color pink to reddish or black (due to ferruginous impurities). Optically negative $(-)$, with $nX > 1.658$ and a birefringence of about 0.08. Axial plane {010}, with $X = c$. In thin section the crystals show a prism angle near 95°. Material separated from celestite by solution in cold dilute HCl gave on analysis: Ce_2O_3 34.1, $(La,Di)_2O_3$ 22.7, SrO 17.8, CO_2 25.4, total 100.0 (recalculated to 100 from original total of 18.96). This is close to the formula $Sr_5(Ce,La,Di)_{10}(CO_3)_{17}O_3$.

Found at Ambatoarina, near Ambositra, Madagascar, enclosed in plates of celestite in veins of manganoan calcite and quartz in metamorphosed limestone. Associated minerals are aegirite, biotite, microcline, galena, and abundant monazite.

Needs investigation, particularly in relation to ancylite.

WEIBYEITE. *Brögger (Zs. Kr.*, **16,** 650, 1890).

An incompletely described mineral that may be related to or identical with ancylite. In minute orthorhombic crystals with a habit and angles near those of zircon. Observed forms: {111}, {110}, {540}?, {201} or {021}, with $a:b:c = 1.00:1:0.64$. Colorless internally but covered with a thin yellow ocher-like crust that penetrates inward to some extent. Biaxial negative $(-)$ with $Z = c$ and $2E \sim 110°$. A partial analysis made on impure material indicates that the substance is a fluo-carbonate of Ce and La with minor Ca and Sr. Found with eudidymite, natrolite, and analcite in pegmatitic veins in nepheline-syenite on the island Övre-Arö in the Langesund fiord, Norway. A similar mineral has been reported [1] as occurring very sparingly as yellowish to brownish red bipyramidal crystals, with G. 3.19, in cavities in granite at Baveno, Italy. Named after the Norwegian mineralogist P. C. Weibye.

Ref.
1. Artini, *Acc. Linc., Rend.*, ser. 5, **24,** 313 (1915).

17 COMPOUND CARBONATES

TYPE 1. MISCELLANEOUS

17.1.1 **T Y C H I T E** [$Na_6Mg_2(SO_4)(CO_3)_4$]. Penfield and Jamieson (Am. J. Sc., **20**, 217, 1905).

C r y s t.[1] Isometric; diploidal—$2/m\ \bar{3}$.

Forms:

o 111

Structure cell.[2] Space group $Fd3$. a_0 13.87 kX. Cell contents $Na_{48}Mg_{16}(SO_4)_8(CO_3)_{32}$.

P h y s. Fracture conchoidal. Brittle. H. $3\frac{1}{2}$–4. G. 2.456;[3] 2.588 (artif.);[4] 2.59 (calc.). Luster vitreous. Color white. Transparent.

O p t. In transmitted light, colorless. Isotropic. $n(Na)$:[5] 1.508; 1.510 (artif.).

C h e m. A sulfate-carbonate of sodium and magnesium, $Na_6Mg_2(SO_4)(CO_3)_4$. A complete series with northupite has been found in artificial material.[6]

Anal.

	1	2
Na_2O	35.58	35.57
MgO	15.42	15.80
SO_3	15.32	15.07
CO_2	33.68	33.50
Total	100.00	99.94
G.	2.59	2.456

1. $Na_6Mg_2(SO_4)(CO_3)_4$. 2. Borax Lake. Average of two.

Tests. Easily fusible. Not decomposed by hot water. Soluble in dilute acids.

O c c u r. Found as octahedral crystals associated with northupite, gaylussite, thenardite, schairerite, and pirssonite in well cuttings from a clay bed at Borax Lake, San Bernardino County, California.

A r t i f.[7] Obtained as a gelatinous precipitate, which crystallizes on long-continued heating, by reaction of a $MgSO_4$ solution with a mixed solution of Na_2CO_3 and Na_2SO_4. The chromate analogue also has been prepared (G. 2.506; n 1.555).

N a m e. Tychite from τυχή, luck or chance, in allusion to the circumstance that of two crystals found in a stock of about 5000 northupite crystals the first and one of the last ten proved to be tychite.

Ref.

1. Crystal class from isostructural and isomorphous relation to northupite; cf. Gossner and Koch, Zs. Kr., **80**, 458 (1931); Shiba and Watanabé, C.R., **193**, 1421 (1931); and de Schulten, C.R., **143**, 403 (1906).
2. Shiba and Watanabé (1931) by powder method on artificial crystals; space group from relation to northupite.

LEADHILLITE 295

3. Pratt in Penfield and Jamieson (1905) by suspension method.
4. Penfield and Jamieson (1905) by suspension method. Shiba and Watanabé (1931) give 2.490 for artificial crystals.
5. Penfield and Jamieson (1905) by prism method; Shiba and Watanabé (1931) also give 1.510 for artificial crystals.
6. de Schulten (1906).
7. Penfield and Jamieson (1905) and de Schulten (1906).

17.1.2 **B R A D L E Y I T E** [$Na_3Mg(PO_4)(CO_3)$]. *Fahey* (*Am. Min.*, **26**, 646, 1941).

Found as extremely fine-grained masses. Cleavage and hardness not determinable. G. 2.734.[1] Color light gray; probably colorless or white when free from admixed clay. Indices of refraction: $nX \sim 1.49$, $nY \sim 1.56$; mean index ~ 1.525.[1]

C h e m. A phosphate-carbonate of sodium and magnesium, $Na_3Mg(PO_4)(CO_3)$.

Anal.

	1	2
Na_2O	37.45	37.57
MgO	16.24	15.44
P_2O_5	28.59	26.34
CO_2	17.72	18.39
Rem.		2.26
Total	100.00	100.00

1. $Na_3Mg(PO_4)(CO_3)$. 2. Wyoming. Analysis recalculated to 100 after deduction of clay 14.46 and CaO 0.36 as shortite. Rem. is Fe_2O_3 0.62, Al_2O_3 0.29, SO_3 0.55, Cl 0.42, SiO_2 0.02, $H_2O-110°$ 0.36.

Tests. Slowly decomposed by cold water, Na_2CO_3 going into solution. Easily soluble in dilute HCl.

O c c u r. Found as a one-inch thick layer admixed with shortite and montmorillonite in a core drill from oil shale in the Green River formation, about 20 miles west of Green River, Sweetwater County, Wyoming. Other closely associated minerals include trona, pirssonite, northupite, gaylussite, alstonite, dolomite, quartz, and pyrite.

N a m e. After Wilmot H. Bradley, geologist of the U. S. Geological Survey, who had carried on geologic investigations in the Green River area.

Ref.

1. Calculated from a value obtained on a sample admixed with clay. Brasseur, *Bull. soc. roy. sc. Liége*, **15**, 527 (1946), calculates nX 1.477, nZ 1.540, $2V$ small, optically negative.

17.1.3 **L E A D H I L L I T E** [$Pb_4(SO_4)(CO_3)_2(OH)_2$]. Plomb carbonaté rhomboidal *Bournon* (343, 1817). Sulphato-tricarbonate of Lead pt. *Brooke* (*Edinburgh Phil. J.*, **3**, 117, 1820). Leadhillite *Beudant* (**2**, 366, 1832). Bleisulphotricarbonat, Ternärbleierz *Weiss*. Plomb sulfato-tricarbonaté *Dufrénoy* (3, 152, 1847). Psimythit *Glocker* (256, 1847). Maxite *Laspeyres* (*Jb. Min.*, 407, 508, 1872; 292, 1873). Schwefelkohlensaures Blei *Koksharov* (**1**, 76, 1853).

C r y s t.[1] Monoclinic; prismatic—$2/m$.

$a:b:c = 0.4371:1:0.5561$; $\beta\ 90°29\frac{1}{2}'$; $p_0:q_0:r_0 = 1.2722:0.5561:1$

$r_2:p_2:q_2 = 1.7982:2.2877:1$; $\mu\ 89°30\frac{1}{2}'$; $p_0'\ 1.2722$, $q_0'\ 0.5561$, $x_0'\ 0.0086$

296 CARBONATES

Forms:[2]

		φ	ρ	φ2	ρ2 = B	C	A
c	001	90°00′	0°29½′	89°30½′	90°00′	.....	89°30½′
a	100	90 00	90 00	0 00	90 00	89°30½′	.....
m	140	29 46	90 00	0 00	29 46	89 45½	60 14
l	120	48 50½	90 00	0 00	48 50½	89 38	41 09½
d	110	66 23½	90 00	0 00	66 23½	89 33	23 36½
g	021	0 26½	48 02½	89 30½	41 57½	48 02½	89 40
φ	041	0 13½	65 47½	89 30½	24 12½	65 47½	89 48
w	101	90 00	52 01	37 59	90 00	51 31½	37 59
u	201	90 00	68 36½	21 23½	90 00	68 07	21 23½
e	2̄01	−90 00	68 28½	158 28½	90 00	68 58	158 28½
s	111	66 32	54 23½	37 59	71 06½	53 56½	41 46½
q	1̄11	−66 15	54 05	141 38½	70 57½	54 32	137 50½
x	141	29 56	68 43	37 59	36 09	68 28	62 17½
Q	1̄32	−36 57	46 13½	122 06½	54 45½	46 31½	115 43½
v	1̄42	−29 26	51 56	122 06½	46 42½	52 10½	112 45½
r	1̄41	−29 36	68 39	141 38½	35 55	68 53½	117 23½

Less common:

b	010	h	031	E	2̄03	μ	1̄12	n	1̄71	p	1̄21
j	210	y	401	f	1̄01	t	142	k	121	o	1̄31
L	130	α	011	δ	112	β	143	θ	131	γ	321
ν	012	Γ	032	λ	1̄13	A	1̄51	ω	211	p	2̄11
χ	023	i	203	ζ	221	G	1̄61	W	2̄31	R	2̄81

Structure cell.[3] Space group $P2_1/a$. a_0 9.07 ± 0.04 kX, b_0 11.55 ± 0.03, c_0 20.70 ± 0.05; [β 90°29½′]; $a_0:b_0:c_0 = 0.438:1:0.558$. Cell contents $Pb_{32}(SO_4)_8(CO_3)_{16}(OH)_{16}$.

Habit. The crystals usually are markedly pseudo-hexagonal and of two main habits: thick to thin tabular {001} with a hexagonal outline,

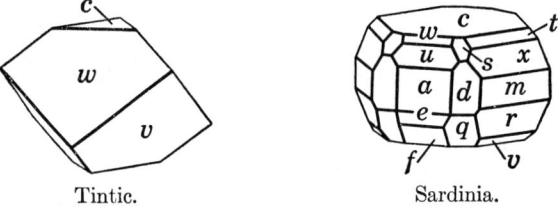

Tintic.   Sardinia.

and rhombohedral through the equal development of a positive orthodome with a negative pyramid of about the same inclination to the vertical (such as w with v or u with r); sometimes prismatic [001] or equant. Also massive, granular. Commonest forms: $c\ a\ m$ (80 per cent or more of all combinations); also $w\ e\ q\ v\ x\ r$ (50 per cent or more). The zone [101] often largely developed and striated or curved.

Twinning.[4] (1) Twin plane {140}, very common; as contact twins of the aragonite type, and also as irregular interpenetration twins or as lamellar twins with the composition plane parallel to a face of {1̄42}.

(2) Twin plane {340}, common, as lamellae; also as inversion twins by heating the crystal. (3) Twin axis [1$\bar{4}$0], rare, as contact twins and as relatively broad, indistinct lamellae with irrational composition surfaces inclined at about 55° to (001). Twin laws 1 and 2 give rise to pseudo-hexagonal groupings and can also be defined as rotations of 180° about [430] or [410]; these directions are pseudo-binary axes of the pseudo-hexagonal cell in which [001] is the pseudo-hexagonal axis [(140) ∧ (100) 60°14', (340) ∧ (010) 59°46']. Twinning also has been doubtfully reported on {101}, {010}, and {001}.

Phys. Cleavage {001} perfect, easy. Fracture conchoidal, scarcely observable. Rather sectile. H. 2$\frac{1}{2}$–3. G. 6.55; 6.57 (calc.). Luster resinous to adamantine; on {001} pearly. Colorless to white, passing into gray, pale green, pale bluish green, pale blue or yellowish. Transparent to translucent. Sometimes shows yellow fluorescence in ultraviolet light. Translation gliding on {001}; also twin gliding with $K_1(340)$, $\sigma_2[140]$; $K_2(3\bar{4}0)$, $\sigma_1[140]$.[6]

Opt.[5] In transmitted light, colorless.

Orientation		n	
$X \wedge c$	$-5\frac{1}{2}°$	1.87 ± 0.01	Biaxial negative (−).
Y		2.00 ± 0.01	$r < v$ strong, horizontal.
Z	b	2.01 ± 0.01	

$2E$ is approximately 20° ($2V \sim 10°$) at room temperature (for Na) and decreases with increasing temperature,[7] the mineral becoming uniaxial negative at about 125° and so remaining at higher temperatures. Concordant values are not obtained on the cooling cycle, and the temperature of uniaxiality (inversion?) of some crystals is upwards of 250°.

Chem. A basic sulfate-carbonate of lead, $Pb_4(SO_4)(CO_3)_2(OH)_2$. The water is lost at about 200° and the CO_2 at strong red heat. The reported analyses[8] of pure material conform closely to the formula. A cuprian variety with CuO 2.94 per cent has been reported.[10]

	PbO	SO_3	CO_2	H_2O	Total
1.	82.75	7.42	8.16	1.67	100.00
2.	82.44	7.33	8.14	1.68	99.59

1. $Pb_4(SO_4)(CO_3)_2(OH)_2$. 2. Granby, Missouri.[9]

Tests. B.B. intumesces, fuses at 1$\frac{1}{2}$, and turns yellow but becomes white on cooling. In C.T. affords water. Soluble with effervescence in nitric acid leaving a residue of lead sulfate. Exfoliates in hot water.

Occur. A secondary mineral, found in the oxidized zone of lead deposits associated with cerussite, anglesite, lanarkite, caledonite, linarite, pyromorphite. Originally found at Leadhills, Lanarkshire, Scotland, especially in the Susanna mine; also at Wanlockhead in Dumfriesshire, Scotland. Among the other known localities may be mentioned Red Gill, Cumberland, and Matlock, Derbyshire, in England; Laquorre, Arriège, France; the Mala-Calzetta lead mine near Iglesias, and the Tiny mine in the Oriddo valley, Sardinia. At Bleiberg, Carinthia, Austria.

Also at Nertschinsk and Beresovsk in the Urals, U.S.S.R. In Africa in fine crystals from Tsumeb, South West Africa, and Djebel Ressas, Tunis. From the Toroku mine, Miyazaki prefecture, Japan. From Dundas and the Victoria Magnet mine, Whyte River, Tasmania. In the United States, leadhillite occurs well-crystallized in the Tintic District, Utah. Also at Searchlight, Lincoln County, Nevada, and from the Lookout and Caledonia mines in Shoshone County, Idaho. In Arizona in notable specimens from the Mammoth mine at Tiger, Pinal County, and from the Cole mine, Bisbee, Cochise County. At the Cerro Gordo mine, Inyo County, California. With lanarkite at Leadville, Colorado. Found near Granby and Joplin, Missouri. Reported from the Morgan mine, Spartanburg district, South Carolina.

Alter. Leadhillite has been observed as pseudomorphs after galena and calcite, and cerussite and calcite have been found pseudomorphous after leadhillite. Sometimes formed by the alteration of lanarkite, and then followed by cerussite.

Artif.[11] Obtained by heating $PbSO_4$ and $PbCO_3$ with water in a closed tube at 180°.

Ref.

1. Angles and orientation of Palache and LaForge, *Proc. Am. Ac. Arts Sc.*, **44**, 435 (1909), *Zs. Kr.*, **48**, 129 (1910), on crystals from the Tintic District, Utah; unit of the structure cell. Transformations: P. and LaF. (1909) and Goldschmidt (**2**, 301, 1890; **5**, 133, 1918) to new, 100/020/001; Dana (921, 1892), Artini, *Giorn. Min.*, **1**, 1 (1890), Laspeyres, *Zs. Kr.*, **1**, 193 (1877), Hintze (**1** [3B], 4243, 1929) to new, 100/040/001.
2. Palache and LaForge (1909), Palache, priv. comm. (1948). Uncommon and uncertain forms:

340	061	$\bar{2}21$	$\bar{1}43$	162	263	$\bar{1}91$	$\bar{2}51$
043	102	123	122	$\bar{1}22$	151	$\bar{3}12$	$\bar{2}61$
0.10.3	403	133	132	$\bar{1}52$	161	241	
051	302	153	152	$\bar{2}33$	$\bar{1}81$	$\bar{2}41$	

3. Wolfe, priv. comm. (1946), by Weissenberg method on crystals from Mammoth mine, Arizona. There is a marked pseudo-cell with b_0 quartered.
4. See Palache and LaForge (1909), Artini (1890), Mügge, *Jb. Min., Beil.-Bd.*, **14**, 259 (1901).
5. Larsen (99, 1921).
6. Mügge (1901).
7. See Hintze, *Ann. Phys.*, **152**, 256 (1874); Pirsson and Wells, *Am. J. Sc.*, **48**, 219 (1894); Mügge (1901); Palache and LaForge (1909).
8. For summary see Hintze (1929).
9. Pirsson and Wells (1894).
10. Yosimura, *J. Fac. Sc. Hokkaido Univ.*, **4**, 453 (1939).
11. Groth (**2**, 444, 1908).

17.1.4 Susannite. Sulphato-tricarbonate of Lead pt. Brooke (*Edinburgh New Phil. J.*, **3**, 117, 138, 1827). Suzannit *Haidinger* (505, 1845).

Supposedly a rhombohedral species dimorphous with leadhillite. Originally from the Susanna mine, Leadhills, Scotland, as acute rhombohedral crystals with interfacial angles of $72\frac{1}{2}°$ and $107\frac{1}{2}°$ and with modifying faces of the basal pinacoid or a hexagonal prism. Colorless to greenish or yellowish. Optically uniaxial. Also reported from Moldawa, Hungary, Nertschinsk in Siberia, and the Mammoth mine, Tiger, Arizona. Susannite has been regarded as only leadhillite of pseudo-rhombohedral habit,[1] but it may represent a primary occurrence of the uniaxial polymorph of leadhillite obtained by heating that species. Conclusive evidence of this is lacking.

Ref.

1. Kenngott, *Jb. Min.*, 319 (1868).

NITRATES

The nitrates are anisodesmic structures and contain discrete $(NO_3)^-$ groups. These are of the same shape as the $(CO_3)^=$ group and of only slightly smaller size. The nitrates are very similar to carbonates, although in general less stable due to increased polarization of the coordinated oxygen atoms by the pentavalent N atom. The natural nitrates are few in number, relatively soluble, and with the exception of soda-niter are of rare occurrence. Soda-niter, which is isostructural with calcite, occurs abundantly in some arid regions as in South America along the Chilean coastal region.

CLASS 18. Normal anhydrous and hydrated nitrates
Type 1. $A(XO_3)$
  18.1.1 Soda-niter $Na(NO_3)$
  18.1.2 Niter $K(NO_3)$
  18.1.3 Ammonia-niter $NH_4(NO_3)$

Type 2. $A(XO_3)_2$
  18.2.1 Nitrobarite $Ba(NO_3)_2$
  18.2.2 Nitrocalcite $Ca(NO_3)_2 \cdot 4H_2O$
  18.2.3 Nitromagnesite $Mg(NO_3)_2 \cdot 6H_2O$

CLASS 19. Nitrates containing hydroxyl or halogen
Type 1. Miscellaneous
  19.1.1 Gerhardtite $Cu_2(NO_3)(OH)_3$
          Buttgenbachite   [with connellite]

CLASS 20. Compound nitrates
Type 1. Miscellaneous
  20.1.1 Darapskite $Na_3(NO_3)(SO_4) \cdot H_2O$

18 NORMAL ANHYDROUS AND HYDRATED NITRATES

TYPE 1. $A(XO_3)$

18.1.1 **SODA-NITER** [NaNO$_3$]. Soude nitratée native *M. de Rivero* (*Ann. mines*, **6**, 596, 1821). Salpetersaures Natron *Mohs* (**2**, 671, 1824). Zootinsalz *Breithaupt* (171, 1823). Natronsalpeter *Leonhard* (246, 1826), *Naumann* (262, 1828). Natron-Nitrat *Breithaupt* (27, 1832). Nitratin *Haidinger* (488, 1845). Natronitrite *Weisbach* (8, 1875). Nitrate of Soda. Soda Nitre. Soda Niter. Cubic Niter. Nitratite. Niter cubique. Chilisalpeter, Salpetersaures Natron *Germ*. Nitro, Salitre sodico, Caliche *Span*.

SODA-NITER 301

C r y s t.[1] Hexagonal—R; scalenohedral—$\bar{3}\ 2/m$.

$a:c = 1:0.8276$; $\alpha\ 102°46\frac{1}{2}'$; $p_0:r_0 = 0.9556:1$; $\lambda\ 73°30'$

Forms:[2]
$c\ 0001$ $a\ 11\bar{2}0$ $r\ 10\bar{1}1$ $e\ 01\bar{1}2$ $f\ 02\bar{2}1$

Structure cell.[3] Space group $R\bar{3}c$. a_{rh} 6.32, α 47°15'; a_0 5.07, c_0 16.81; $a_0:c_0 = 1:3.316$. Cell contents $Na_2(NO_3)_2$ in the rhombohedral unit. Isostructural with calcite.

Habit. Crystals rhombohedral $\{10\bar{1}1\}$, rarely with modifying faces.[4] Usually massive granular; as an incrustation.

Twinning.[5] (1) on $\{01\bar{1}2\}$, often elongated along the common rhombohedron edge; (2) on $\{0001\}$, as penetration twins; (3) on $\{02\bar{2}1\}$, sometimes as aggregates of three or six individuals; (4) on $\{10\bar{1}1\}$, rare. Also twin gliding[6] with K_1 $\{01\bar{1}2\}$, K_2 $\{0\bar{1}11\}$, as in calcite. Exhibits percussion figures.[13]

Oriented growths. Readily forms overgrowths upon calcite,[7] the crystal axes of the substances ordinarily parallel; also oriented upon muscovite,[8] dolomite,[9] barytocalcite.[10] Alkali halides[11] and certain phenols[12] also orient upon $NaNO_3$ rhombohedra.

P h y s. Cleavage $\{10\bar{1}1\}$ perfect; $\{01\bar{1}2\}$ and $\{0001\}$ also reported[5] as imperfect. Fracture conchoidal, but seldom observable. Rather sectile. H. $1\frac{1}{2}$–2. G. 2.24–2.29; G. 2.266;[14] 2.25 (calc.). Luster vitreous. Colorless; also white and, when tinted by impurities, reddish brown, gray, lemon-yellow. Transparent. Taste cooling. Diamagnetic, positive. Thermal conduction greatest parallel [0001]. M.P. 306.8° C.

O p t.[15] Colorless in transmitted light. Uniaxial negative (−). Index on cleavage, $E' = 1.467$.

$\lambda(17°)$	B	D	E	H
nO	1.5793	1.5874	1.5954	1.6260
nE	1.3346	1.3361	1.3374	1.3440

C h e m. Sodium nitrate, $NaNO_3$. The numerous reported analyses[16] of soda-niter and of crude nitrate rocks are only of technical interest. In artificial crystals,[16] Li, K, Tl, and NH_4 substitute for Na only in traces; Ag substitutes for Na and (ClO_3) and (BrO_3) for (NO_3) in considerable amount.

Tests. Deflagrates on charcoal with less violence than niter. Slightly deliquescent, and easily soluble in water (91.5 grams $NaNO_3$ in 100 grams H_2O at 25° C.).

O c c u r. A water-soluble salt found widely in arid regions as a surficial impregnation or as an efflorescence upon soil or in sheltered places. Usually associated with niter, nitrocalcite, gypsum, epsomite, mirabilite, halite. The principal deposits of soda-niter occur in a belt roughly 450 miles long and 10 to 50 miles wide along the eastern slopes of the coast range in the deserts of northern Chile. The deposits consist of a layer varying from a few inches up to a few feet in thickness with a shallow overburden of sand and rock fragments. The crude soda-niter, known as caliche, contains a variable admixture of other salts, notably bloedite, anhydrite, gypsum, polyhalite, halite, glauberite, and darapskite together with minor amounts of various iodates, chromates, and borates. The iodates lautarite and dietzeite occur in the so-called *azufrado* or *caliche jaune* and afford iodine as an important by-product of the refining operation. Chilean nitrate had a monopoly of the world's fertilizer market for many years but the industry was forced into a subordinate position with the development during World War I and subsequent years of synthetic processes for fixed nitrogen. The origin of the Chilean deposits [17] is not yet clearly understood but the principal controls have been physiographic and climatic factors.

Small deposits of soda-niter similar to those in Chile occur in Bolivia, Peru, North Africa, Egypt, U.S.S.R., India, and western United States.[18] Found with atacamite and bloedite in the upper portion of the oxidized zone and in the surrounding alluvium at Chuquicamata, Chile. In California, soda-niter and niter occur in Inyo and San Bernardino Counties along the Amargosa River and in Death Valley; also in the Calico borate district, San Bernardino County, and in the San Joaquin Valley near Tulare, Tulare County. In Nevada at the Niter Buttes about 25 miles southeast of Lovelock. In New Mexico in Hidalgo County. Near Homedale, Owyhee County, Utah, with niter and sulfates in veinlets in rhyolite; on Aqua Fria Mountain, Brewster County, Texas, with niter in veinlets in trachyte.

A r t i f. Easily obtained in crystals by evaporation of the water solution.

N a m e. From the composition, the sodium analogue of niter, KNO_3. See further under niter.

Ref.

1. Brooke, *Ann. Phil.*, **21**, 452 (1823).
2. Goldschmidt (**6**, 78, 1920).
3. Wyckoff, *Phys. Rev.*, **16**, 149 (1920), by Laue method on artificial crystals. Transformation: morphology to Wyckoff, 1000/0100/0010/0004.
4. On habit variation in artificial crystals due to impurities in the crystallizing solution see France and Wolfe, *J. Phys. Chem.*, **45**, 395 (1941). On the growth of large single crystals of $NaNO_3$ see Stöber, *Zs. Kr.*, **61**, 315 (1925), and West, *J. Opt. Soc. Am.* **35**, 26 (1945).
5. Wulff, *Ak. Berlin, Sitzber.*, 137 (1896), 717 (1895).
6. Johnsen, *Jahrb. Radioakt. und Elektronik*, **11**, 246 (1914).
7. Finch and Whitmore, *Trans. Faraday Soc.*, **34**, 640 (1939); Barker, *Trans. Chem. Soc., London*, **89**, 1123 (1906); Settele, *Jb. Min., Beil.-Bd.*, **61**, 227 (1930).
8. Mügge, *Cbl. Min.*, 355 (1902), *Jb. Min., Beil.-Bd.*, **16**, 389 (1903).
9. Tschermak, *Min. Mitt.*, **4**, 118 (1882).
10. Kreutz, *Min. Mag.*, **15**, 233 (1909).
11. Royer, *C.R.*, **205**, 1418 (1937); Heintze, *Zs. Kr.*, **97**, 241 (1937).
12. Willems, *Zs. Kr.*, **105**, 53 (1943).
13. Mügge, *Jb. Min.*, I, 123 (1898).
14. Haigh, *Am. Chem. Soc., J.*, **34**, 1137 (1912); this value is very close to the average of the best modern determinations.
15. Schrauf, *Ak. Wien, Sitzber.*, **41**, 787 (1860), by prism method, average of three sets. Merwin, *Int. Crit. Tables* (**7**, 26, 1930), gives $nO = 1.5848$ for Na.
16. Cf. Hintze (**1** [3A], 2677, 1916) and Doelter (**3** [1], 264, 1913).
17. Cf. Whitehead, *Econ. Geol.*, **15**, 187 (1920); Singewald and Miller, *Econ. Geol.*, **11**, 103 (1916); Wetzel, *Chemie Erde*, **3**, 385, 411 (1928).
18. For U. S. occurrences see Mansfield and Boardman, *U. S. Geol. Sur., Bull. 838* (1932).

NITER 303

18.1.2 NITER [KNO$_3$]. Nitrum, pt. (rest alkaline carbonate) *Ancient writings.*
Sal petrae, *about 12th–13th Cent.* Nitre. Saltpetre; Saltpeter. Salpeter, Salpetersaures Kali, *Germ.* Natürliche salpeter *Werner.* Salpeter *Reuss* (pt. **2, 3,** 21, 1803), *Hausmann* (**3,** 849, 1813). Nitre, Nitrate of Potash *Philips* (137, 1818). Kalisalpeter *Leonhard* (247, 1826), *Naumann* (261, 1828), *Hausmann* (1416, 1847). Kalinitrat *Breithaupt* (27, 1832; **2,** 94, 1847). Niter *Dana* (871, 1892). Potasse nitratée *Fr.* Salitre *Span.*

C r y s t.[1] Orthorhombic; dipyramidal—$2/m\ 2/m\ 2/m$.

$a:b:c = 0.5910:1:0.7010;\qquad p_0:q_0:r_0 = 1.1861:0.7010:1$

$q_1:r_1:p_1 = 0.5910:0.8431:1;\qquad r_2:p_2:q_2 = 1.4265:1.6920:1$

Forms:[2]

| c 001 | b 010 | a 100 | m 110 | x 012 | k 011 | i 021 | p 111 |

Structure cell.[3] Space group $Pmcn$; a_0 5.42 kX, b_0 9.17, c_0 6.45; $a_0:b_0:c_0 = 0.591:1:0.701$. Cell contents $K_4(NO_3)_4$. Isostructural with aragonite.

Habit. Generally in thin crusts, silky tufts, and delicate acicular crystallizations; also massive, granular, or columnar; earthy; mealy. Artificial crystals are prismatic [001], with {110}, {010} and usually {011}.

Twinning. On {110} common, often as pseudo-hexagonal groupings, analogous to aragonite.

P h y s. Cleavage {011} nearly perfect, {010} fairly good; {110} imperfect. Fracture subconchoidal to uneven. Brittle. H. 2. G. 2.109 ± 0.002;[4] 2.08 (calc.). Luster vitreous. Transparent. Color and streak colorless to white; sometimes gray, or tinted by mechanically admixed impurities. Twin gliding:[5] $K_1(110)$, $K_2(1\bar{3}0)$, as in aragonite.

O p t.

ORIENTATION		n[8]	n[6]	n[7]	
X	c	1.3320	1.332	1.3346	Biaxial negative (−).
Y	a	1.5038	1.504	1.5056	$r < v$, weak.
Z	b	1.5042	1.504	1.5064	
$2V$		5°01′ (calc.)	7° (meas.)	7°12′ (calc.)	

C h e m. Potassium nitrate, KNO_3. The reported analyses are only of technical interest. Na and Li do not substitute for K; but Rb and to a less extent Cs, Tl, and NH_4 substitute for K in artificial crystals.[9]

Tests. Deflagrates vividly on charcoal and detonates with combustible substances. Easily soluble in water (37.3 grams per 100 grams water at 25°). Taste saline and cooling.

O c c u r. Found commonly, although usually in small amounts, as a surface efflorescence in arid regions and in caves or other dry, sheltered places. Occurs as a surface crust or surficial impregnation in soils rich in organic material and may form abundantly during hot weather after rains. Usually associated with soda-niter, epsomite, nitrocalcite, and gypsum. The formation of the mineral is often due to the action of certain bacteria on nitrogenous or animal matter. The calcium salt, nitrocalcite, usually forms in place of niter on calcareous soils or rocks. Niter formerly was manufactured by heaping soil rich in humus together with animal excrement and wood ashes under sheds for protection from the rain; the niter, or nitrocalcite if plaster or other calcareous material is present, effloresces on the windward side of the pile and is regularly scraped off and purified by re-crystallization.

Niter occurs associated with soda-niter in the desert regions of northern Chile and is found variously as an efflorescence on soils and rocks in Spain, Italy, Egypt, Arabia, Persia, India, U.S.S.R., and western United States.[15] Found in sheltered recesses and as an impregnation in shale in the Prieska and Hay districts, Cape Province, Africa,[10] the potash being afforded by the shales and the nitrate by bacterial action on organic remains. With borax in the playa deposits of Cochabamba, Bolivia. Formerly obtained in some abundance from limestone caves in Tennessee, Kentucky, Alabama, and Ohio and used for the manufacture of gunpowder during the War of 1812 and the Civil War. Occurs near Dubois, Idaho, in a cave in lava and derived from a surface deposit of bird guano. A massive granular deposit of over 125 tons was mined from a cave near Lava Station, Socorro County, New Mexico.[11] Found with soda-niter along the Amargosa River and in Death Valley, Inyo and San Bernardino Counties, California.

A r t i f. Easily obtained in crystals from water solution. Niter inverts at 129° C. to a rhombohedral phase isostructural with soda-niter,[12] and several other polymorphs stable at high pressures have been reported.[13]

N a m e. The alkaline salt extracted by water from the ashes of vegetable matter was known to the ancients as *neter, nether* (Hebraic). This name, ultimately derived from *natar*, a substance that effervesces, corresponds to the *nitron* of the Greeks and the *nitrum* of the Romans. It was variously applied to the saline materials obtained from the trona deposits and salt lakes of Egypt[14] and from saline efflorescences on walls and soils, as well as from plant ashes, and included both alkaline carbonates and nitrates. Saltpeter, derived from *sal petrae, salt of the rocks*, in allusion to the mode of occurrence, appears to have been recognized as distinct from ordinary nitrum in the twelfth or thirteenth century with the introduction of gunpowder into Europe. The term *natron* then came to be applied to the native sodium carbonate, and nitrum itself, translated *nitre* or *niter*, came to be used as a synonym for saltpeter. A clear distinction between the nitrates of sodium and potassium was not reached, however, until the seventeenth century.

Ref.

1. Miller, *Phil. Mag.*, **17**, 38 (1840).
2. Goldschmidt (**5**, 3, 1918). Rare or uncertain: 210, 025, 041, 114, 112, 221.
3. Edwards, *Zs. Kr.*, **80**, 154 (1930), on artificial crystals.
4. From 2.111 Gossner, *Zs. Kr.*, **38**, 143 (1903), 2.109 Haigh, *J. Am. Chem. Soc.*, **34**, 1137 (1912), 2.109 Retgers, *Zs. phys. Chem.*, **3**, 313 (1889), and 2.106 Günther, abstr. in *Zs. Kr.*, **50**, 92 (1912), on artificial crystals.
5. Johnsen, *Jahrb. Radioakt. und Elektron.*, **11**, 246 (1914).

6. Larsen and Berman (152, 1934) by immersion method.
7. Schrauf, *Ak. Wien, Sitzber.*, **41**, 790 (1860); average of measurements at 15° and 17° C.
8. Merwin, *Int. Crit. Tables* (**7**, 27, 1930).
9. Cf. Mellor (**2**, 808 et seq., 1922) and Hintze (**1** [3A], 2713, 1930).
10. Frood and Hall, *Union South Africa Geol. Sur., Mem.* **14** (1919).
11. Northrop, *Univ. New Mexico Bull., Geol. Ser.*, **6**, no. 1, 227 (1942).
12. Mellor (1922).
13. Barth, *Zs. phys. Chem.*, **43** [B], 448 (1939), and Bridgman, *Proc. Am. Ac. Arts Sc.*, **51**, 597 (1916).
14. Gibbs, *Ann. Sc. London*, **3**, 213 (1938).
15. Cf. Mansfield and Boardman, *U. S. Geol. Sur., Bull. 838* (1932).

18.1.3 A M M O N I A - N I T E R [NH_4NO_3]. Nitrammite *Shepard* (1857 ?); *Am. J. Sc.*, **24**, 124, 1857). Ammoniak-salpeter.

Ammonia-niter was said to occur in the earth of the Nicojack Cave, Tennessee, but there has been no later verification of its occurrence in nature. NH_4NO_3 has one high-pressure and five low-pressure polymorphs.[1] The stable form at ordinary conditions ($-18°$ to $32.2°$) is orthorhombic dipyramidal, space group $Pmmn$, a_0 4.93 kX, b_0 5.44, c_0 5.73. Biaxial negative $(-)$[2] with $nX = b = 1.413$, $nY = a = 1.611$, $nZ = c = 1.637$, $2V$ 35°. G. 1.72. Very soluble in water.

Ref.

1. Hendricks, Posnjak, and Merwin, *J. Am. Chem. Soc.*, **54**, 2766 (1932) and West, *J. Am. Chem. Soc.*, **54**, 2256 (1932).
2. Optical data from Winchell (209, 1931).

TYPE 2. $A(XO_3)_2$

18.2.1 N I T R O B A R I T E [$Ba(NO_3)_2$]. Barytsalpeter, Salpetersaures Baryt, Nitrobaryt *Germ.* Nitrobarite *Lewis* (*Am. Nat.*, **16**, 78, 1882).

C r y s t. Isometric; diploidal—$2/m\ \overline{3}$.

Forms:[1]

a 100	$-o$ $\overline{1}11$	e' 012	m 113	$-n$ $\overline{1}12$	$-t$ $\overline{1}24$
d 011	f' 013	$'e$ 102	M 338	$-p$ $\overline{1}22$	V 135
o 111	$'f$ 103	$-w$ $\overline{1}15$	n 112	t 124	

Structure cell.[2] Space group $Pa3$. a_0 8.11 kX. Cell contents $Ba_4(NO_3)_8$.

Habit. Natural crystals octahedral; the habit of artificial crystals varies widely with the conditions of crystallization.[3]

Twinning. On the spinel law, {111}, in both natural and artificial crystals.

P h y s. No cleavage. G. 3.250 ± 0.005;[4] 3.24 (calc.). Transparent. Colorless.

O p t. In transmitted light, colorless. Isotropic. Anomalous birefringence may be developed in sectors. n 1.5714 ± 0.003 (for Na).[4]

C h e m. Barium nitrate, $Ba(NO_3)_2$. Analyses of natural material are lacking.

Tests. Easily soluble in water.

Occur. Known[5] only from one specimen from Chile, presumably from the nitrate deposits, but the exact locality not known.

Artif. A well-known artificial salt, easily crystallized from water solution.

Name. In allusion to the composition.

Ref.

1. Forms other than {111} and {1̄11} for artificial crystals only; see literature in Groth (**2**, 104, 1908) and Goldschmidt (**1**, 173, 1913).
2. Vegard, *Zs. Phys.*, **9**, 395 (1922).
3. Cf. Buckley, *Zs. Kr.*, **76**, 147 (1930).
4. From best reported values on artificial material.
5. Groth, *Zs. Kr.*, **6**, 195 (1881).

18.2.2 **NITROCALCITE** [$Ca(NO_3)_2 \cdot 4H_2O$]. Nitre à base calcaire *de Lisle* (**1**, 360, 1783). Kalksalpeter *Hausmann* (858, 1813). Nitrate de chaux *Beudant* (421, 1824; **2**, 383, 1832). Nitrate of lime. Calcium nitrate. Nitrocalcite *Shepard* (**2**, 84, 1835). Calcinitre *Huot* (**2**, 430, 1841).

Cryst.[1] Monoclinic; prismatic—$2/m$.

$a:b:c = 1.5839:1:0.6876$; $\beta\, 98°06'$; $p_0:q_0:r_0 = 0.4341:0.6807:1$

$r_2:p_2:q_2 = 1.4690:0.6377:1$; $\mu\, 81°54'$; $p_0'\, 0.4385,\ q_0'\, 0.6876,\ x_0'\, 0.1423$

Forms:

| b 010 | a 100 | m 110 | n 210 | q 011 | r 101 | ρ 1̄01 | o 111 | ω 1̄11 |

In efflorescent silken tufts and masses. Artificial crystals are long prismatic [001]. Cleavable. Soft. G. 1.90. Transparent. Color white or gray.

Opt.[2] In transmitted light, colorless.

	n (artificial)	
X	1.465 ± 0.003	Biaxial negative ($-$).
Y	1.498 ± 0.003	$2V\, 50° \pm 2°$.
Z	1.504 ± 0.003	Dispersion not perceptible.
		X is $\perp$ to a cleavage.

Chem. Calcium nitrate tetrahydrate, $Ca(NO_3)_2 \cdot 4H_2O$. Modern analyses on natural material are lacking. The identification as the tetrahydrate is based on the known stability of this salt.

Tests. Easily soluble in water. Taste sharp and bitter.

Occur. A water-soluble salt said to occur with niter and nitromagnesite as an efflorescence in limestone caves or upon calcareous rocks and soil. The reported localities need verification. In Hungary, in the southern part of the nitrate district from Szegedin to Titel, and in Spain, in Catalonia and Aragon provinces as a surface deposit on alkaline and calcareous soils. Similarly found at various localities in France.[3] Reported from the Mammoth Cave and other limestone caverns in Kentucky, and from the Wyandotte Cave, Indiana. During the War of 1812 the nitrocalcite from the Kentucky localities is said to have been concentrated by leaching and then converted to niter for use in gunpowder by filtration through wood ashes. Also reported in fissures in limestone along the Gila River two miles above Winkelman, Arizona; and with niter near Adobe Creek, 23 miles south of Animas station, New Mexico. Reported from niter beds in the lower end of Death Valley, San Bernardino County, California.

A r t i f. A well-known artificial salt, obtained in crystals from water solution.

N a m e. In allusion to the composition.

Ref.
1. Marignac, *Ann. mines*, **9**, 31 (1856), on artificial material.
2. Larsen (116, 1921) on artificial material.
3. Cf. Lacroix (**3**, 412, 1901) for a discussion of origin.

18.2.3 **N I T R O M A G N E S I T E** [$Mg(NO_3)_2 \cdot 6H_2O$]. Magnesiasalpeter *Naumann* (263, 1828). Nitrate de magnésie *Beudant* (**2**, 384, 1832). Nitromagnesite *Shepard* (**2**, 85, 1835). Magnesinitre *Huot* (**2**, 431, 1841). Magnésie nitratée. Magnesia salpeter. Magnesium nitrate.

C r y s t.[1] Monoclinic; prismatic—$2/m$.

$a:b:c = 0.5191:1:0.9698;$ $\beta\ 92°56';$ $p_0:q_0:r_0 = 1.8682:0.9685:1$

$r_2:p_2:q_2 = 1.0325:1.9289:1;$ $\mu\ 87°04';$ $p_0'\ 1.8707, q_0'\ 0.9698, x_0'\ 0.0512$

Forms:

c 001 m 110 k 012 q 011

Habit. In flocculent to earthy efflorescences Artificial crystals are long prismatic [001] with {001} and {110} dominant.

P h y s. Cleavage {110} perfect. G. 1.46. Luster vitreous. Transparent. Colorless to white.

O p t.[2]

	n (Kentucky)	
X	1.34 ± 0.01	Biaxial negative ($-$).
Y	1.506 ± 0.003	$2V\ 5° \pm 1°$.
Z	1.506 ± 0.003	$r < v$, perceptible.

C h e m. Hydrated magnesium nitrate, $Mg(NO_3)_2 \cdot 6H_2O$. Analyses of natural material are lacking.

Tests. Easily soluble in water. Taste bitter. M.P. 90°.

O c c u r. Reported to occur as an efflorescence associated with nitrocalcite in limestone caves but definitely identified [2] only from Madison County, Kentucky. Also reported as an efflorescence in marl from Baume and Salina, Jura, France.[3]

A r t i f. A well-known artificial salt, easily obtained in crystals from water solution.

N a m e. In allusion to the composition.

Ref.
1. Marignac, *Ann. mines*, **9**, 31 (1856).
2. Larsen (116, 1921).
3. Ogérien in Lacroix (**3**, 414, 1901).

19 NITRATES CONTAINING HYDROXYL OR HALOGEN

TYPE 1. MISCELLANEOUS

19.1.1 GERHARDTITE [$Cu_2(NO_3)(OH)_3$]. *Wells* and *Penfield* (*Am. J. Sc.*, 30, 50, 1885).

Cryst.[1] Orthorhombic; dipyramidal—$2/m\ 2/m\ 2/m$.

$a:b:c = 0.9206:1:1.1498;$ [1] $p_0:q_0:r_0 = 1.2490:1.1498:1$

$q_1:r_1:p_1 = 0.9206:0.8007:1;$ $r_2:p_2:q_2 = 0.8697:1.0862:1$

Forms:[2]

	ϕ	$\rho = C$	ϕ_1	$\rho_1 = A$	ϕ_2	$\rho_2 = B$
c 001	....	0°00′	0°00′	90°00′	90°00′	90°00′
m 110	47°22′	90 00	90 00	42 38	0 00	47 22
z 201	90 00	68 11	0 00	21 49	21 49	90 00
y 112	47 22	40 19½	29 53½	61 34	58 01	64 00
w 223	47 22	48 32	37 28	56 32½	50 13	59 29½
p 111	47 22	59 30	48 59	50 39½	38 41	54 18
s 221	47 22	73 35	66 29½	45 06½	21 49	49 29

Habit. Crystals thick tabular {001}, the pyramid zone strongly striated and the faces often in oscillatory combination.

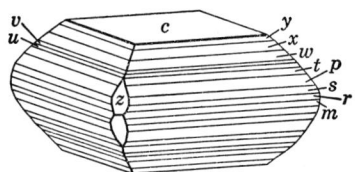

Phys. Cleavage {001} perfect, {100} good. H. 2. G. 3.43 (Arizona), 3.40 (Katanga).[3] Crystals flexible, with separation along {100}. Transparent. Color dark green to emerald-green. Streak light green.

Opt.[4]

Orientation		n	Pleochroism	
X	a	1.703	Green	Biaxial positive (+).
Y	b	1.713	Green	2V large.
Z	c	1.722	Blue	$r < v$, very strong.

Chem. Basic copper nitrate, $Cu_2(NO_3)(OH)_3$.

Anal.

	1	2	3
CuO	67.10	66.26	66.38
N_2O_5	22.77	[22.25]	22.76
H_2O	10.13	11.49	11.26
Total	100.00	[100.00]	100.40

1. $Cu_2(NO_3)(OH)_3$.   2,3. Jerome, Arizona.[5]

Tests. In C.T. gives nitrous fumes and acid water. Insoluble in water, soluble in dilute acids.

Occur. A secondary mineral, originally found associated with malachite and atacamite in cavities in massive cuprite at the United Verde mine, Jerome, Arizona. Doubtfully reported [6] as botryoidal films with atacamite (?) in Chase Creek Cañon, near Metcalf in the Clifton-Morenci district, Arizona. Also found in cavities in cuprite at Likasi, Katanga district, Belgian Congo.

Artif. Prepared artificially by heating a solution of the normal nitrate, by boiling solutions of cupric nitrate with sodium acetate or other soluble salts of organic acids, and in other ways.[7] An artificial dimorphous, monoclinic phase also is known.[5]

Name. After Charles Gerhardt, the chemist who first prepared the artificial compound.

Ref.

1. Thoreau, *Soc. géol. Belgique, Ann.*, **46**, B 285 (1923). Wells and Penfield (1885) give $a:b:c = 0.9218:1:1.1562$.
2. Thoreau (1923) and Wells and Penfield (1885). Rare or uncertain: 7.7.18, 11.11.20, 13.13.20, 7.7.10, 334, 778, 25.25.26, 441, 551.
3. Thoreau (1923).
4. Indices from Larsen and Berman (129, 190, 1934); optical sign, orientation, and dispersion from Thoreau (1923) on Katanga crystals and Frondel, priv. comm. (1946), on artificial crystals. Larsen and Berman give $r > v$; Wells and Penfield (1885) state it optically negative $(-)$.
5. Wells and Penfield (1885).
6. Lindgren and Hillebrand, *Am. J. Sc.*, **18**, 448 (1904).
7. Mellor (**3**, 284, 1923).

20 COMPOUND NITRATES

TYPE 1. MISCELLANEOUS

20.1.1   **D A R A P S K I T E** [$Na_3(NO_3)(SO_4) \cdot H_2O$]. Dietze (*Zs. Kr.*, **19**, 445, 1891).

Cryst.[1] Monoclinic; prismatic—$2/m$.

$a:b:c = 1.5258:1:0.7514;$ $\beta\ 102°55';$ $p_0:q_0:r_0 = 0.4925:0.7324:1$

$r_2:p_2:q_2 = 1.3654:0.6724:1;$ $\mu\ 77°05';$ $p_0'\ 0.5052,\ q_0'\ 0.7514,\ x_0'\ 0.2293$

Forms:

		ϕ	ρ	ϕ_2	$\rho_2 = B$	C	A
c	001	90°00'	12°55'	77°05'	90°00'	.....	77°05'
b	010	0 00	90 00	.....	0 00	90°00'	90 00
a	100	90 00	90 00	0 00	90 00	77 05	......
m	110	33 55	90 00	0 00	33 55	82 50	56 05
q	011	16 58	38 09	77 05	53 46½	36 13½	79 36½
r	101	90 00	36 18	53 42	90 00	23 23	53 42
e	302	90 00	44 37½	45 22½	90 00	31 42½	45 22½
n	1̄01	−90 00	15 25½	105 25½	90 00	28 20½	105 25½
d	2̄01	−90 00	37 59½	127 59½	90 00	50 54½	127 59½
o	111	44 21	46 25	53 42	58 48	38 16	59 34½
s	1̄11	−20 09½	38 40½	105 25½	54 05	44 32	102 26½
v	121	26 03	59 07½	53 42	39 32½	54 14	67 51½

Habit. Tabular {100}, and pseudo-tetragonal [100].

Twinning.[2] Frequently twinned on {100}, with re-entrant angles on the edges and composition face {100}; sometimes polysynthetic.

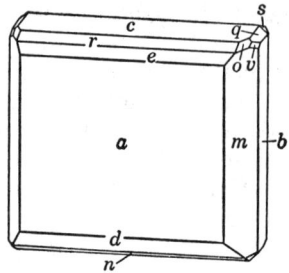

P h y s. Cleavage {010} perfect; also a perfect cleavage or parting {100}.[3] H. $2\frac{1}{2}$. G. 2.20.[4] Transparent. Colorless.

O p t.[5]

Orientation	n	
X b	1.391 ± 0.005	Biaxial negative (−).
Y	1.481 ± 0.003	$2V$ 27° ± 1°.
Z ∧ c 12°	1.486 ± 0.003	$r > v$, rather strong.

C h e m. A hydrated sodium nitrate-sulfate, $Na_3(NO_3)(SO_4) \cdot H_2O$.

Anal.

	1	2
Na_2O	37.94	38.27
N_2O_5	22.04	22.26
SO_3	32.67	32.88
H_2O	7.35	7.30
Total	100.00	100.71

1. $Na_3(NO_3)(SO_4) \cdot H_2O$.   2. Oficina Lautaro, Chile.[6]

Tests. Easily soluble in water.

O c c u r.[7] Found abundantly in certain of the Chilean nitrate deposits, especially those rich in sulfates, associated with soda-niter, bloedite, halite, anhydrite. The darapskite often occurs in cavities or crevices in the massive nitrate rock or as a surficial concentration. Not all the localities at which it has been found in Chile are exactly specified or described. Originally from the Pampa del Toro, Oficina Lautaro, near Taltal, in the Atacama desert. From the Oficina Maria Elena, Tocopilla, the Oficina Progresso, Iquique, and other localities in Antofagasta and Atacama provinces, Chile. Also found associated with kroehnkite and bloedite in veins up to 6 inches wide at Chuquicamata, Chile. Reported from Death Valley, San Bernardino County, California, associated with soda-niter and niter.

A r t i f.[8] Obtained in crystals from the system Na_2SO_4-$NaNO_3$-H_2O over a temperature range from 13.5° to somewhat over 50° C.

N a m e. After L. Darapsky, of Santiago, Chile.

Ref.

1. Osann, *Zs. Kr.*, **23**, 584 (1894).
2. Cf. Osann (1894).
3. Cf. Larsen (66, 1921).
4. From 2.203 on natural crystals, Osann (1894), and 2.197 on artificial crystals, de Schulten, *Bull. soc. min.*, **19**, 161 (1896).
5. Larsen (66, 1921); Osann (1894) gives $X = b$, $Y \sim c$, $Z \sim \perp \{100\}$.
6. Dietze (1891).
7. See also Wetzel, *Chem. Erde*, **3**, 385, 411 (1928); Semper and Michels, *Zs. Berg.-Hütt.- und Salinenwesen*, **52**, 359 (1904).
8. See Foote, *Am. J. Sc.*, **9**, 441 (1925); also Hintze (**1** [3B], 4553, 1930).

NITROGLAUBERITE. Schwartzenberg, Domeyko (*Min. Chili*, 3rd app., 46, 1871).

A white, fibrous substance found in caliche in the Cerro Reventon, near Paposo, Antofagasta, Chile. Supposed to be $2Na_2SO_4 \cdot 6NaNO_3 \cdot 3H_2O$, but no phases other than darapskite, $Na_2SO_4 \cdot NaNO_3 \cdot H_2O$, have been found [1] in the system Na_2SO_4-$NaNO_3$-H_2O. The substance repeatedly has been suggested to be a mixture of darapskite and soda-niter. A specimen from Atacama, Chile, of uncertain authenticity was in stout laths elongated X, tabular $\perp Y$ with a cleavage $\perp Z$; optically negative $(-)$ with parallel extinction and $r < v$ (rather strong); $nX = 1.418$, $nY = 1.500$, $nZ = 1.543$ (all ± 0.003).[2]

Ref.

1. Foote, *Am. J. Sc.*, **9**, 441 (1925).
2. Larsen (116, 1921).

IODATES

Among the oxy-acids of the halogens only a few iodates and an iodate-chromate are known to occur in nature. The structure of none of these is known. Salesite apparently has a structure of the olivine type.[1] The compounds are classed separately rather than with the nitrates or carbonates since the structures are probably of the multiple isodesmic type, with I in six-coordination with oxygen. The natural iodates are rare and occur only along the arid western coasts of South America.

Ref.
1. Strunz, *Zs. Kr.*, **103**, 359 (1941).

CLASS 21. Normal anhydrous and hydrated iodates
Type 1. $A(XO_3)_2 \cdot xH_2O$
    21.1.1 Lautarite $Ca(IO_3)_2$
    21.1.2 Bellingerite $Cu(IO_3)_2 \cdot \tfrac{2}{3}H_2O$

CLASS 22. Iodates containing hydroxyl or halogen
Type 1. Miscellaneous
    22.1.1 Salesite $Cu(IO_3)(OH)$
    22.1.2 Schwartzembergite $Pb_5(IO_3)Cl_3O_3$

CLASS 23. Compound iodates
    23.1.1 Dietzeite $Ca_2(IO_3)_2(CrO_4)$

21 NORMAL ANHYDROUS AND HYDRATED IODATES

TYPE 1. $A(XO_3)_2 \cdot xH_2O$

21.1.1 LAUTARITE $[Ca(IO_3)_2]$. *Dietze* (*Zs. Kr.*, **19**, 447, 1891).

Cryst.[1] Monoclinic; prismatic—$2/m$.

$a:b:c = 0.6331:1:0.6462;$ $\beta\ 106°22';$ $p_0:q_0:r_0 = 1.0207:0.6200:1$

$r_2:p_2:q_2 = 1.6129:1.6462:1;$ $\mu\ 73°38';$ $p_0'\ 1.0638,\ q_0'\ 0.6462,\ x_0'\ 0.2937$

Forms:

	ϕ	ρ	ϕ_2	$\rho_2 = B$	C	A
c 001	90°00′	16°22′	73°38′	90°00′	.....	73°38′
b 010	0 00	90 00	.....	0 00	90°00′	90 00
l 120	39 27½	90 00	0 00	39 27½	79 41	50 32½
m 110	58 43½	90 00	0 00	58 43½	76 03½	31 16½
q 011	24 26½	35 22	73 38	58 12	31 48	76 08½
r 101	90 00	53 37	36 23	90 00	37 15	36 23
n $\bar{1}01$	−90 00	37 36	127 36	90 00	53 58	127 36

***Structure cell.*[2]** Space group $P2_1/c$. a_0 7.18 kX, b_0 11.38, c_0 7.32; β 106°22′; $a_0:b_0:c_0 = 0.631:1:0.643$. Cell contents $Ca_4(IO_3)_8$.

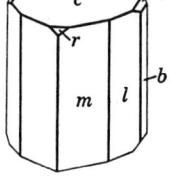

Habit. Short prismatic [001]; faces in zone [001] rounded and striated [001]. Often radially arranged or in stellate aggregates.

Phys. Cleavage {011} good, {100} and {110} in traces. H. $3\frac{1}{2}$–4. G. 4.59; 4.48 (calc.). Colorless; also yellowish, due to impurities which may be zonally arranged. Transparent.

Opt.[3]

ORIENTATION	n	
$X \wedge c$ +25°	1.792 ± 0.003	Biaxial positive (+).
Y b	1.840 ± 0.003	$2V$ nearly 90°.
Z	1.888 ± 0.003	$r > v$, moderate.

Chem. Calcium iodate, $Ca(IO_3)_2$.

Anal.

	1	2
CaO	14.38	14.95
I_2O_5	85.62	85.04
Total	100.00	99.99
G.	4.48	4.59

1. $Ca(IO_3)_2$. 2. Pampa del Pique III, Chile.[4] I_2O_5 calculated from average of two determinations as I (64.70, 64.62).

Tests. Slightly soluble in water and easily soluble in HCl with evolution of Cl.

Occur. Found as crystals coating fractures or embedded in gypsum bands in the nitrate deposits (*caliche*) of the Pampa del Pique III, Oficina Lautaro, and neighboring Pampas, in the Atacama desert, Antofagasta Province, Chile.

Artif.[5] Obtained in crystals by fusion of $Ca(IO_3)_2 \cdot H_2O$ in $NaNO_3$.

Name. From the locality.

Ref.

1. Osann, *Zs. Kr.*, **23**, 586 (1894).
2. Gossner and Mussgnug, *Zs. Kr.*, **75**, 410 (1930).
3. Larsen and Berman (140, 1934).
4. Dietze (1891).
5. de Schulten, *Bull. soc. min.*, **21**, 144 (1898).

21.1.2 **BELLINGERITE** [$3Cu(IO_3)_2 \cdot 2H_2O$]. Berman and Wolfe (*Am. Min.*, **25**, 505, 1940).

Cryst. Triclinic; pinacoidal—$\bar{1}$.

$a:b:c = 0.9264:1:1.0149$; α 105°06′, β 96°57$\frac{1}{2}$′, γ 92°55′

$p_0:q_0:r_0 = 1.0591:1.0088:1$; λ 74°23$\frac{1}{2}$′, μ 81°59′, ν 85°03′

p_0' 1.1078, q_0' 1.0552; x_0' 0.1225, y_0' 0.2815

Forms:

		φ	ρ	A	B	C
c	001	23°26′	17°03½′	81°59′	74°23½′	....
b	010	0 00	90 00	85 03	....	74°23½′
a	100	85 03	90 00	....	85 03	81 59
k	120	26 36	90 00	58 27	26 36	72 58
n	210	60 33	90 00	24 30	60 33	76 28½
N	2$\bar{1}$0	111 22½	90 00	26 19½	111 22½	89 24
w	011	5 14	53 19	81 51½	37 00½	37 23
V	0$\bar{1}$2	153 00½	15 06½	84 23½	103 25½	29 02
Ξ	$\bar{1}$12	−124 23½	27 29½	113 42	105 07	42 49½
π	$\bar{1}$11	−131 32½	52 39½	129 40½	121 49	68 22½
Φ	$\bar{2}$11	−114 50	66 28½	149 34	112 39	79 35

Less common forms:

M 1$\bar{1}$0	W 0$\bar{1}$1	d 102	E $\bar{1}$01	S 1$\bar{2}$2	τ $\bar{1}$21
K 1$\bar{2}$0	X 0$\bar{2}$1	e 101	p 111	σ $\bar{1}$22	φ $\bar{2}$11
x 021	Y 0$\bar{3}$1	D $\bar{1}$02	P 1$\bar{1}$1	T 1$\bar{2}$1	

Structure cell. Space group $P\bar{1}$. a_0 7.22 kX, b_0 7.82, c_0 7.92; $a_0:b_0:c_0$ = 0.9235:1:1.0128. Cell contents $Cu_3(IO_3)_6 \cdot 2H_2O$.

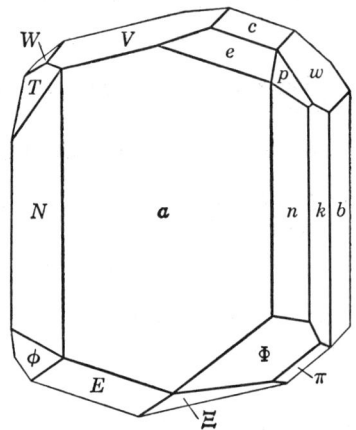

Habit. Prismatic [001] and somewhat tabular {100}. {100} striated parallel [001].

Twinning. On {$\bar{1}$01}, with or without re-entrant angles.

Phys. Fracture subconchoidal. Brittle. H. 4. G. 4.89 ± 0.01, 4.932 (calc.). Color light green. Streak very pale green.

Opt. In transmitted light, pale bluish green in color.

	Orientation			Pleochroism	
	φ	ρ	n (589)		
X	−70°	70°	1.890	Light bluish green	Biaxial positive (+).
Y	175°	38°	1.90	Light bluish green	2V medium.
Z	34°	59°	1.99	Blue-green	$r > v$, strong. Absorption $Z > X, Y$.

Chem. A hydrated copper iodate, $3Cu(IO_3)_2 \cdot 2H_2O$.
Anal.

	1	2
CuO	18.70	18.65
I_2O_5	78.47	77.55
H_2O	2.83	3.22
Total	100.00	99.42
G.	4.92	4.89

1. $3Cu(IO_3)_2 \cdot 2H_2O$.   2. Chuquicamata, Chile.[1]

Tests. In C.T. gives water with purple fumes of iodine which crystallizes on the walls of the tube. Slightly soluble in hot water and easily soluble in dilute HCl.

Occur. A secondary mineral found at Chuquicamata, Chile, as tiny isolated crystals associated with leightonite and gypsum. Also as veinlets in altered granitic rock.

Artif. A salt apparently identical with bellingerite has been prepared artificially.[2]

Name. After Herman C. Bellinger (1867–1940), metallurgist.

Ref.

1. Gonyer anal. in Berman and Wolfe (1940).
2. Granger and de Schulten, *Bull. soc. min.*, **27**, 137 (1904).

22 IODATES CONTAINING HYDROXYL OR HALOGEN

TYPE 1. MISCELLANEOUS

22.1.1   SALESITE [$Cu(IO_3)(OH)$]. *Palache* and *Jarrell* (*Am. Min.*, **24**, 388, 1939).

Cryst. Orthorhombic; dipyramidal—$2/m\ 2/m\ 2/m$.

$a:b:c = 0.4442:1:0.6241;\qquad p_0:q_0:r_0 = 1.4050:0.6241:1$

$q_1:r_1:p_1 = 0.4442:0.7117:1;\qquad r_2:p_2:q_2 = 1.6023:2.2512:1$

Forms:

		ϕ	$\rho = C$	ϕ_1	$\rho_1 = A$	ϕ_2	$\rho_2 = B$
c	001	.....	0°00′	0°00′	90°00′	90°00′	90°00′
b	010	0°00′	90 00	90 00	90 00	.....	0 00
n	130	36 53	90 00	90 00	53 07	0 00	36 53
m	110	66 03	90 00	90 00	23 57	0 00	66 03
e	023	0 00	22 35½	22 35½	90 00	90 00	67 24½
d	011	0 00	35 58	35 58	90 00	90 00	54 02
p	111	66 03	56 57½	31 58	40 00	35 26½	70 06½
(?) r	552	66 03	75 25	57 20½	27 49	15 53½	66 52

Structure cell.[1] Space group $Pbnm$. $a_0\ 4.78\ kX$, $b_0\ 10.77$, $c_0\ 6.70$; $a_0:b_0:c_0 = 0.444:1:0.622$. Cell contents $Cu_4(IO_3)_4(OH)_4$.

Habit. Stout prismatic [001] with pyramidal terminations.

IODATES

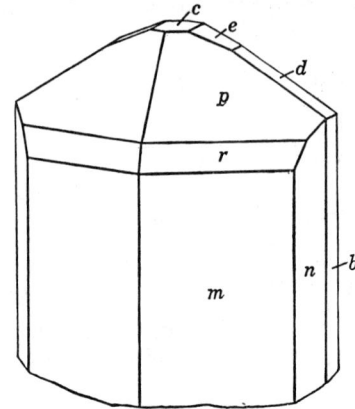

Phys. Cleavage {110} perfect. H. 3. G. 4.77 ± 0.05; 4.888 (calc.). Luster vitreous. Color bluish green, very similar to caledonite. Transparent.

Opt. In transmitted light, bluish green in color.

ORIENTATION		n	PLEOCHROISM	
X	a	1.786 ± 0.005	Colorless	Biaxial negative (−).
Y	c	2.070 ± 0.01	Light bluish green	$2V$ 0° to 5°.
Z	b	2.075 ± 0.01	Bluish green	(Uniaxial for blue).
				$r > v$, extreme.

Chem. A basic copper iodate, $Cu(IO_3)(OH)$.

Anal.

	1	2
Na_2O		0.59
CuO	31.14	30.62
I_2O_5	65.33	64.79
H_2O	3.53	3.68
Total	100.00	99.68
G.	4.89	4.77

1. $Cu(IO_3)(OH)$. 2. Chuquicamata, Chile.[2]

Tests. In C.T. decrepitates and gives water and copious fumes of iodine which crystallize on the walls of the tube. Insoluble in water but easily soluble in HNO_3.

Occur. Found at Chuquicamata, Chile, in oxidized ore.

Artif.[3] A basic copper iodate probably dimorphous with salesite is known.

Name. After Reno H. Sales (1876–), Chief Geologist of the Anaconda Copper Mining Company.

Ref.

1. Richmond in Palache and Jarrell (1939) by Weissenberg method. Recorded space group is $Pcmn$ which is incorrect for the chosen orientation.
2. Gonyer anal. in Palache and Jarrell (1939).
3. See Granger and de Schulten, *Bull. soc. min.*, **27**, 137 (1904), and Palache and Jarrell (1939).

22.1.2 **SCHWARTZEMBERGITE** [$Pb_5(IO_3)Cl_3O_3$]. Oxychloroïodure de plomb (from Atacama) *Domeyko* (*Ann. mines*, **5**, 453, 1864); Plomo oxichloro-ioduro *Domeyko* (319, 705, 1879). Bleioxychlorojodür. Schwartzembergite *Dana* (120, 1868). Plumbiodite *Adam* (67, 1869).

Cryst.[1] Tetragonal or pseudo-tetragonal.
Habit. In rounded, flat-pyramidal crystals. Also as compact to earthy crusts and masses.
Phys. Cleavage {001} distinct. H. 2 to $2\frac{1}{2}$. G. 7.39.[2] Color honey-yellow to reddish brown; straw-yellow, lemon-yellow. Streak straw-yellow. Luster adamantine.
Opt. Crystals from the San Rafael mine, Chile, are optically biaxial and pseudo-tetragonal; the crystals are divided into variously oriented biaxial sectors with variable $2V$.[3] Indices:[4]

	n(Li)	
X	2.25 ± 0.02	Biaxial negative (−).
Y	2.35 ± 0.02	2V small.
Z	2.36 ± 0.02	Dispersion not noticed.

Chem. A lead iodate-oxychloride, $Pb_5(IO_3)Cl_3O_3$ or $Pb(IO_3)_2 \cdot 3Pb_3O_2Cl_2$.
Anal.

	1	2
$PbCl_2$	30.55	31.17
$Pb(IO_3)_2$	20.40	18.95
PbO	49.05	48.29
CaO		0.67
SO_3		0.47
Total	100.00	99.55
G.		7.39

1. $Pb_5(IO_3)Cl_3O_3$. 2. San Rafael mine, Sierra Gorda, Chile.[5]

Tests. Easily fusible. Soluble in HCl with evolution of Cl.

Occur. Originally found with cerussite and anglesite as a crust upon galena at Cachinal between Paposo and Taltal in the Atacama desert, Chile. Later described from the San Rafael mine, Sierra Gorda, Caracoles district, Chile, associated with percylite, paralaurionite, and gypsum. Also reported [6] from several other localities in northern Chile, including the silver mines of Huantajaya near Iquique and at Palestina about 50 miles east of the port of Antofagasta, but the identity of these minerals is uncertain.

Name. After Dr. Schwartzemberg, an assayer at Copiapó, Chile, who first drew attention to the mineral.

Ref.

1. Approximate measurements by Smith and Prior, *Min. Mag.*, **16**, 77 (1911), of rounded pyramidal crystals (anal. 2) indicate tetragonal symmetry, with $p\{101\}$, $q\{441\}$ and unidentified second-order pyramids; $c \wedge q \sim 67°41'$ and $a:c = 1:0.430$. The optical characters suggest that these crystals are inversion pseudomorphs. Crystals not certainly of the same mineral earlier were said to be rhombohedral by Liebe, *Jb. Min.*, 159 (1867).

2. Smith and Prior (1911), by pycnometer on material of analysis 2. Earlier reported values, made on massive, impure material, are much lower.
3. Cf. Smith and Prior (1911); crystals from San Rafael, not certainly of the same mineral, earlier were said to be uniaxial negative (−) by Bertrand, *Bull. soc. min.*, **4**, 87 (1881).
4. Larsen (132, 1921) on material from the San Rafael mine.
5. Prior anal. in Smith and Prior (1911). Earlier analyses by Liebe (1867) and Domeyko (1864) were made on very impure material.
6. Cf. Domeyko (319, 1879).

23 COMPOUND IODATES

23.1.1 **DIETZEITE** [$Ca_2(IO_3)_2(CrO_4)$]. Jodchromate *Dietze* (*Zs. Kr.*, **19**, 449, 1891). Dietzeit *Osann* (*Zs. Kr.*, **23**, 588, 1894).

Cryst.[1] Monoclinic; prismatic—$2/m$.

$a:b:c = 1.3826:1:1.9030$; $\beta\ 106°32'$; $p_0:q_0:r_0 = 1.3764:1.8243:1$

$r_2:p_2:q_2 = 0.5481:0.7545:1$; $\mu\ 73°28'$; $p_0'\ 1.4358,\ q_0'\ 1.9030,\ x_0'\ 0.2968$

Forms:

	ϕ	ρ	ϕ_2	$\rho_2 = B$	C	A
c 001	90°00′	16°32′	73°28′	90°00′	.....	73°28′
b 010	0 00	90 00	.....	0 00	90°00′	90 00
a 100	90 00	90 00	0 00	90 00	73 28	.....
m 110	37 02	90 00	0 00	37 02	80 07½	52 58
l 210	56 28	90 00	0 00	56 28	76 16½	33 32
r 1̄02	−90 00	22 50	112 50	90 00	39 22	112 50
s 1̄13	−15 59½	33 25	100 18	58 02	40 48	98 44
o 1̄11	−30 54	65 43½	138 43	38 32	74 52½	117 55

Structure cell.[2] Space group probably $P2_1/c$. $a_0\ 10.16\ kX$, $b_0\ 7.30$, $c_0\ 14.03$; $\beta\ 106°32'$; $a_0:b_0:c_0 = 1.392:1:1.922$. Cell contents $Ca_8(IO_3)_8(CrO_4)_4$.

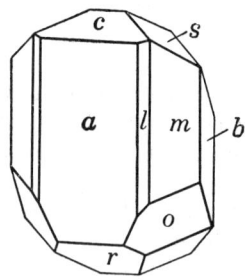

Habit. Crystals tabular {100} and elongated [001]. Usually in fibrous crusts or columnar.

Phys. Cleavage {100} incomplete. Fracture conchoidal. H. 3½. G. 3.698,[3] 3.617;[2] 3.61 (calc.). Color deep golden yellow. Transparent.

O p t.[4]

Orientation	n	
X	1.825	Biaxial negative $(-)$.
$Y\ b$	1.842	$2V\ 86°$.
$Z \wedge c\ 6°$	1.857	$r < v$, very strong; marked inclined dispersion.

C h e m.[5] A calcium iodate-chromate, $Ca_2(IO_3)_2(CrO_4)$.
Anal.

	1	2	3	4
CaO	20.54	22.01	21.50	21.10
I_2O_5	61.14	58.12	58.10	58.89
CrO_3	18.32	19.00	19.90	20.28
Total	100.00	99.13	99.50	100.27

1. $Ca_2(IO_3)_2(CrO_4)$. 2,3,4. Atacama desert, Chile.[6] Total for anal. 4 given as 100.37 in original.

Tests. Slowly soluble in cold water and more readily in hot water; the solution deposits $Ca(IO_3)_2 \cdot 6H_2O$ on cooling.

O c c u r. Originally found in nitrate deposits in the same general region that afforded lautarite, $Ca(IO_3)_2$, in the Atacama desert, Antofagasta province, Chile. Later reported [7] with lopezite, tarapacaite, and ulexite in nitrate rock in the Oficina Maria Elena, near Tocopilla, Chile.

N a m e. After August Dietze (died 1893 ?), chemist, who first described the mineral.

Ref.

1. Angles and orientation of Osann (1894); unit of Gossner and Mussgnug, *Zs. Kr.*, **75**, 410 (1930). Transformation: Osann to G. and M., 100/010/002.
2. Gossner and Mussgnug (1930).
3. Osann (1894) by pycnometer using CCl_4.
4. Larsen (68, 1921); see also Osann (1894).
5. The formula here given is indicated by the x-ray data and the structural relations to lautarite and crocoite, Gossner and Mussgnug (1930). The formula $7Ca(IO_3)_2 \cdot 8CaCrO_4$, given originally, is somewhat closer to the analyses.
6. Dietze (1891).
7. Bandy, *Am. Min.*, **22**, 929 (1937).

BORATES

In oxysalts boron occurs in triangular coordination with oxygen. The structure of borates thus is mesodesmic,[1] since the oxygen coordination is equal to the charge of the boron, and the possibility exists of forming indefinitely extending anionic groups by linkage of (BO_3) units analogous to the linkage of (SiO_4) units in the silicates. Indefinitely extending $(BO_2)^-$ chains are found in the structure of the artificial compound CaB_2O_4. In potassium metaborate, $K_3B_3O_6$, not known as a mineral, closed links of $(B_3O_6)^\equiv$ are found, and discrete (BO_3) triangles occur in the structure of hambergite, $Be_2(BO_3)(OH)$. Boron also occurs in fourfold coordination with oxygen in some borosilicates. Unfortunately, the crystal structures of most natural borates are still unknown, and a systematic classification [1] based on the nature of the anionic framework cannot yet be formulated. The present classification is based solely on the ratio of boron to the total number of cations.

Ref.
1. See Hendricks, *Washington Ac. Sc., J.*, **34**, 241 (1944).

CLASS 24. Anhydrous borates

24.1.1	Ludwigite series	
24.1.1.1	Ludwigite	$(Mg,Fe'')_2Fe'''BO_5$
24.1.1.2	Paigeite	$(Fe'',Mg)_2Fe'''BO_5$
24.1.2	Pinakiolite	$Mg_3Mn''Mn_2'''B_2O_{10}$
24.1.3	Hulsite	$(Fe'',Ca,Mg)_4(Fe''',Sn'''')_2B_2O_{10}$ (?)
24.1.4	Warwickite	$(Mg,Fe)_3TiB_2O_8$
24.1.5	Kotoite	$Mg_3(BO_3)_2$
24.1.6	Rhodizite	$NaKLi_4Al_4Be_3B_{10}O_{27}$ (?)
24.1.7	Jeremejevite	$AlBO_3$
24.1.8	Nordenskiöldine	$CaSn(BO_3)_2$

CLASS 25. Hydrated borates

25.1.1	Pinnoite	$Mg(BO_2)_2 \cdot 3H_2O$
25.1.2	Kernite	$Na_2B_4O_7 \cdot 4H_2O$
25.1.3	Tincalconite	$Na_2B_4O_7 \cdot 5H_2O$
25.1.4	Borax	$Na_2B_4O_7 \cdot 10H_2O$
25.1.5	Priceite	$Ca_4B_{10}O_{19} \cdot 7H_2O$ (?)
25.1.6	Probertite	$NaCaB_5O_9 \cdot 5H_2O$
25.1.7	Ulexite	$NaCaB_5O_9 \cdot 8H_2O$
25.1.8	Veatchite	$Sr_3B_{16}O_{27} \cdot 5H_2O$ (?)
25.1.9	Colemanite	$Ca_2B_6O_{11} \cdot 5H_2O$
25.1.10	Hydroboracite	$CaMgB_6O_{11} \cdot 6H_2O$
25.1.11	Inderborite	$CaMgB_6O_{11} \cdot 11H_2O$
25.1.12	Meyerhofferite	$Ca_2B_6O_{11} \cdot 7H_2O$
25.1.13	Inyoite	$Ca_2B_6O_{11} \cdot 13H_2O$
25.1.14	Kurnakovite	$Mg_2B_6O_{11} \cdot 13H_2O$
25.1.15	Inderite	$Mg_2B_6O_{11} \cdot 15H_2O$

CLASS 25. Hydrated borates—*Continued.*

25.1.16	Howlite	$Ca_2SiB_5O_9(OH)_5$
25.1.17	Bakerite	$Ca_4B_4(BO_4)(SiO_4)_3(OH)_3H_2O$
25.1.18	Paternoite	$MgB_8O_{13} \cdot 4H_2O$
25.1.19	Ginorite	$Ca_2B_{14}O_{23} \cdot 8H_2O$
25.1.20	Larderellite	$(NH_4)_2B_{10}O_{16} \cdot 5H_2O$ (?)
25.1.21	Ammonioborite	$(NH_4)_2B_{10}O_{16} \cdot 5H_2O$ (?)
25.1.22	Kaliborite	$KMg_2B_{11}O_{19} \cdot 9H_2O$

CLASS 26. Borates containing hydroxyl or halogen

26.1.1	Fluoborite	$Mg_3(BO_3)(F,OH)_3$
26.1.2	Hambergite	$Be_2(BO_3)(OH)$
26.1.3	Teepleite	$Na_2B_2O_4 \cdot 2NaCl \cdot 4H_2O$
26.1.4	Bandylite	$CuB_2O_4 \cdot CuCl_2 \cdot 4H_2O$
26.1.5	Sussexite series	
26.1.5.1	Sussexite	$(Mn,Zn)(BO_2)(OH)$
26.1.5.2	Szaibelyite	$Mg(BO_2)(OH)$
26.1.6	Roweite	$(Mn,Mg,Zn)Ca(BO_2)_2(OH)_2$
26.1.7	Boracite	$Mg_3B_7O_{13}Cl$
26.1.8	Hilgardite	$Ca_8(B_6O_{11})_3Cl_4 \cdot 4H_2O$
26.1.9	Parahilgardite	$Ca_8(B_6O_{11})_3Cl_4 \cdot 4H_2O$

CLASS 27. Compound borates

27.1.1	Luenebergite	$Mg_3B_2(OH)_6(PO_4)_2 \cdot 6H_2O$
27.1.2	Cahnite	$Ca_2B(OH)_4(AsO_4)$
27.1.3	Sulfoborite	$Mg_6H_4(BO_3)_4(SO_4)_2 \cdot 7H_2O$
27.1.4	Seamanite	$Mn_3(PO_4)(BO_3) \cdot 3H_2O$

24 ANHYDROUS BORATES

LUDWIGITE AND PINAKIOLITE

ORTHORHOMBIC

	$a:b:c$
Ludwigite, $(Mg,Fe'')_2Fe'''BO_5$	$0.6595:1:?$
Paigeite, $(Fe'',Mg)_2Fe'''BO_5$	
Pinakiolite, $Mg_3Mn''Mn_2'''B_2O_{10}$	$0.8339:1:0.5881$
Hulsite, $(Fe'',Ca,Mg)_4(Fe''',Sn'''')_2B_2O_{10}$ (?)	$0.550\ :1:?$

Ludwigite and paigeite are isostructural [1] and a complete series extends between them by mutual substitution of Mg and Fe''. Pinakiolite, although similar in formula, appears to be unrelated. The ill-defined mineral hulsite seems to be distinct from both pinakiolite and ludwigite-paigeite.

Ref.

1. W. T. Schaller, priv. comm. (1949).

LUDWIGITE SERIES

24.1.1.1 **L U D W I G I T E** [$Mg,Fe)_2Fe'''BO_5$]. *Tschermak* (*Min. Mitt.*, 59, 1874). Magnesioludwigite *Butler* and *Schaller* (*Wash. Ac. Sc., J.*, **7**, 29, 1917). Collbranite *Higgins* (*Econ. Geol.*, **13**, 19, 1918).

24.1.1.2 **P A I G E I T E** [(Fe,Mg)$_2$Fe'''BO$_5$]. *Knopf* and *Schaller* (*Am. J. Sc.*, **25**, 323, 1908). Ferroludwigite *Butler and Schaller* (*Wash. Ac. Sc. J.*, **7**, 29, 1917). Vonsenite *Eakle* (*Am. Min.*, **5**, 141, 1920).

C r y s t.[1] Orthorhombic.

$$a:b:c = 0.6595:1:?$$

Forms:[2]

	ϕ	$\rho = C$	ϕ_1	$\rho_1 = A$	ϕ_2	$\rho_2 = B$
c 001	.....	0°00'	0°00'	90°00'	90°00'	90°00'
a 100	90°00'	90 00	.....	0 00	0 00	90 00
l 140	20 45½	90 00	90 00	69 14½	0 00	20 45½
z 130	26 49	90 00	90 00	63 11	0 00	26 49
g 120	37 10	90 00	90 00	52 50	0 00	37 10
f 230	45 18½	90 00	90 00	44 41½	0 00	45 18½
m 110	56 35½	90 00	90 00	33 24½	0 00	56 35½
n 210	71 45	90 00	90 00	18 15	0 00	71 45
x 310	77 36	90 00	90 00	12 24	0 00	77 36

Habit. As fibrous masses, the fibers radiating or short and interwoven and then grading into dense, microcrystalline felted aggregates with a faint silky luster; also as embedded sheaf-like aggregates or as rosettes of needle-like crystals; sometimes granular. Rarely as ill-formed single-crystals prismatic [001] with a rhombic cross section but without terminal faces.

P h y s. Cleavage not observed.[3] Tough upon fracture. H. 5. G. 4.7 (essentially pure *paigeite* end-member)[4] to 3.6 (essentially pure *ludwigite* end-member).[5] Luster silky on fresh fracture in the fibrous types. Color coal-black to greenish black in paigeite and inclining toward dark green in ludwigite. Streak black to blackish green. Opaque except in small grains of highly magnesian ludwigite.

O p t.[6] Deeply colored and strongly pleochroic; paigeite high in iron is opaque.

Orientation	Ludwigite n (Hungary) anal. 4	Ludwigite, low Fe'' n (Mountain Lake mine, Utah)	Pleochroism	
X	1.85 ± 0.01	1.83 ± 0.01	Dark green	Biaxial positive
Y	1.85 ± 0.01	1.83 ± 0.01	Dark green	(+).
Z c	2.02 ± 0.02	1.97 ± 0.01	Dark reddish brown to opaque	$r > v$, extreme.
2V	Small	Very small		

C h e m. A borate of trivalent iron, divalent iron and magnesium, (Mg,Fe)$_2$Fe'''BO$_5$. The names ludwigite and paigeite are given to material with Mg > Fe'' and Fe'' > Mg, respectively, and a complete substitutional series probably extends to the end-members. Al substitutes for Fe''' in small amounts, with Al:Fe = 1:8.9 in analysis 4, and Mn and Ca substitute for (Mg,Fe''). Small amounts of non-essential water have been reported in some analyses. Sn apparently substitutes for Fe''' in some material.[16]

LUDWIGITE-PAIGEITE

Anal.[7]

	1	2	3	4	5	6	7	8
FeO		5.14	10.40	15.84	28.8	39.75	44.48	55.62
MgO	41.29	36.42	34.54	28.88	13.89	10.71	1.44	
Fe_2O_3	40.88	35.90	32.49	35.67	38.3	34.82	16.72	30.90
Al_2O_3		2.08	2.32		nil			
B_2O_3	17.83	14.59	16.80	[17.02]	13.55	14.12	9.83	13.48
H_2O+		2.28	1.42	0.82			2.03	
H_2O-		0.03		0.51	0.07			
Rem.		3.99	2.62	1.26	5.31		15.65	
Total	100.00	100.43	100.59	[100.00]	99.92	99.40	90.15	100.00
G.					4.29	4.21	4.71	

1. $Mg_2Fe'''BO_5$.   2. Ludwigite. Lemhi County, Idaho.[8] Rem. is SiO_2 0.90, CuO 2.87, S 0.22, MnO and CaO tr. Contains admixed bornite.   3. Ludwigite. Hol Kol mine, Korea.[9] Rem. is SiO_2 0.40, CaO 1.86, MnO 0.36.   4. Ludwigite. Moravicza, Hungary.[10] Rem. is SiO_2 0.36, CO_2 0.90.   5. Paigeite. Chersky range, Siberia.[11] Rem. is Na_2O 0.36, K_2O 0.01, CaO 0.87, MnO 0.20, TiO_2 0.24, SiO_2 0.35, F 0.07, ign. loss 3.21.   6. Paigeite. Riverside, California (*vonsenite*).[12]   7. Paigeite. Brooks Mountain, Alaska.[13] Rem. is admixed vesuvianite and arsenopyrite.   8. $Fe''_2Fe'''BO_5$.

Tests. B.B. fusible with difficulty to a black, magnetic slag. Heated in air turns red. Slowly soluble in acids. Paigeite has F 3.

O c c u r. The members of the ludwigite-paigeite series are high-temperature minerals found in contact metamorphic deposits associated with magnetite, diopside, forsterite, szaibelyite. Ludwigite was originally from Moravicza in the Banat, Hungary; also reported from Vaskö. From Yauli Province, Peru. Found with szaibelyite, fluoborite, and members of the chondrodite group in the Hol Kol mine, Suan, Korea,[15] and at Norberg and a number of other magnetite skarn deposits in Sweden.[14] In the United States ludwigite occurs at the Mountain Lake mine and other localities in the Big and Little Cottonwood districts, Utah. From Pioche, Lincoln County, Nevada, and the Texas district in Lemhi County, Idaho. In Montana at Philipsburg, Granite County, and in Colorado Gulch near Helena, Lewis and Clark County. Paigeite was originally from Brooks Mountain on the Seward Peninsula, Alaska. Also from Riverside, Riverside County, California (*vonsenite*) and from the northwest part of the Chersky range in Yakutia, Siberia.

A l t e r. To limonite.

A r t i f.[17] Reported by fusion of Fe_2O_3, MgO, and B_2O_3.

N a m e. Ludwigite after Ernst Ludwig (1842- ?), Professor of Chemistry at the University of Vienna. Paigeite after Sidney Paige (1880-), geologist, of the U. S. Geological Survey. Vonsenite after M. Vonsen, mineral collector of Petaluma, California.

Ref.

1. Partial ratio from angles (110) $\wedge$ ($\bar{1}$10) = 90°40' of Mallard, *Bull. soc. min.*, **11**, 305 (1888), on thin fibers from Moravicza, Hungary. Mallard's angles and forms are not consistent; his prism angle (110) $\wedge$ (140) of 10°56' is taken as a misprint for 30°56' as suggested by Groth, *Zs. Kr.*, **15**, 650 (1889), and accepted by Goldschmidt (224, 1897), and his form (110) is made (230), with (230) $\wedge$ ($\bar{2}$30) = 90°38' calc., as indicated by the rhomboidal rather than square outline of the crystals (Schaller, priv. comm.,

1948). Transformation: Eakle, *Am. Min.*, **5,** 141 (1920), on so-called vonsenite [= paigeite] to the new cell, $0\frac{1}{2}0/100/001$; his partial ratio in the new cell is $a:b = 0.6615:1$.

2. Mallard (1888) [z, f, n, x], Eakle (1920) [a, l, g, n, x], and Schaller (1948) on Utah material. Less common and rare:

d 470	j 760	p 760	t 750	w 530
e 350	k 9.10.0	q 650	u 10.7.0	
y 340	Y 21.22.0	r 540	h 320	
i 560	O 980	s 430	V 850	

3. Knopf and Schaller (1908) mention an apparent imperfect cleavage of unstated orientation.
4. Value of Knopf and Schaller (1908) after adjustment for impurities present in material of analysis 7.
5. Switzer in Knopf, *Am. Min.*, **27,** 824 (1942).
6. Larsen (102, 1921) who gives additional data on unanalyzed material.
7. Additional analyses are given by Harada, *J. Fac. Sc. Hokkaido Univ.*, **4,** 165 (1938); Edwards, *Am. J. Sc.*, **7,** 486 (1924), Schaller, *Am. J. Sc.*, **30,** 146 (1910); and Dana (877, 1892). Butler and Schaller (1917) describe material (*magnesioludwigite*) from the Mountain Lake mine, Big Cottonwood Canyon, Utah, with 2.55 per cent FeO.
8. Shannon, *Proc. U. S. Nat. Mus.*, **59,** 667 (1921).
9. Shannon, *Am. Min.*, **6,** 86 (1921).
10. Schaller (1910).
11. Vakar, Knipovich, and Schafpanovsky, *Mem. soc. russe min.*, **63,** 381 (1934).
12. Eakle (1920).
13. Knopf and Schaller (1908), Schaller, *Am. J. Sc.*, **29,** 543 (1910). Shown by x-ray study to be isostructural with ludwigite—Schaller, priv. comm. (1949).
14. Geijer, *Geol. För. Förh.*, **61,** 19 (1939).
15. Watanabe, *Min. Mitt.*, **50,** 441 (1939).
16. Schaller, *Am. J. Sc.*, **29,** 543 (1910).
17. Ebelmen in Mallard (1888).

24.1.2 **PINAKIOLITE** [$Mg_3Mn''Mn_2'''B_2O_{10}$]. Pinakiolith *Flink* (*Zs. Kr.*, **18,** 361, 1890).

Cryst.[1] Orthorhombic.

$$a:b:c = 0.8339:1:0.5881; \qquad p_0:q_0:r_0 = 0.7052:0.5881:1$$

$$q_1:r_1:p_1 = 0.8339:1.4180:1; \qquad r_2:p_2:q_2 = 1.7004:1.1992:1$$

Forms:

	ϕ	$\rho = C$	ϕ_1	$\rho_1 = A$	ϕ_2	$\rho_2 = B$
b 010	0°00′	90°00′	90°00′	90°00′	....	0°00′
x 310	74 28	90 00	90 00	15 32	0°00′	74 28
w 011	0 00	30 27½	30 27½	90 00	90 00	59 32½

Habit. Thin tablets {010} with a rectangular outline. Rarely short prismatic [001]. The crystals are often bent or broken.

Twinning. On {011} common, both as contact and cruciform interpenetration types.

Phys. Cleavage {010} good. Very brittle. H. 6. G. 3.88. Luster metallic, {010} usually brilliant, and the prism faces dull. Color pure black. Streak brownish gray. Opaque.

Opt. In transmitted light, deep reddish brown.

ORIENTATION		n^2	PLEOCHROISM	
X	b	1.908 ± 0.005	Deep reddish brown	Biaxial negative (−).
Y	c	2.05 ± 0.01	Nearly opaque	$2V$ 32° ± 1° (meas.).
Z	a	2.065 ± 0.01	Reddish yellow	$r < v$ (?), moderate.

PINAKIOLITE

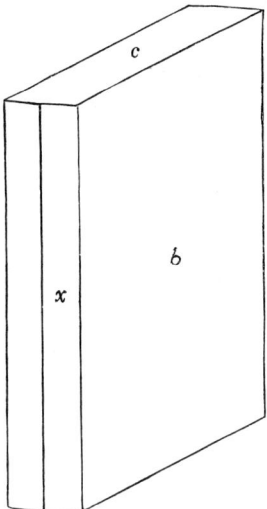

Chem. A borate of magnesium, divalent manganese and trivalent manganese, $Mg_3Mn''Mn_2'''B_2O_{10}$ (analysis 2).

Anal.

	1	2	3
CaO		1.12	1.35
PbO		0.78	1.22
MgO	28.84	29.30	22.36
MnO	16.91	15.70	16.36
Mn_2O_3	37.64	34.93	34.04
Fe_3O_4		2.12	10.52 (Fe_2O_3)
B_2O_3	16.61	16.05	13.92
Rem.			0.78
Total	100.00	100.00	100.55
G.		3.88	3.935

1. $Mg_3Mn''Mn_2'''B_2O_{10}$. 2. Långban. Recalculated to 100 after deducting SiO_2 1.21, H_2O 0.47. Direct determination of available oxygen gave 4.34 per cent. 3. Pinakiolite (?). Långban.[3] Rem. is SiO_2.

Tests. B.B. fuses with difficulty to a black, non-magnetic slag. On strong heating, the brown powdered mineral turns black. Soluble in concentrated HCl with evolution of Cl.

Occur. At Långban, Sweden, in bands in granular dolomite with hausmannite, tephroite, berzeliite, and manganophyllite.

Name. From πῐνάκιον, *a small tablet*, and λίθος, *stone*, in allusion to the thin tabular habit.

Ref.
1. Flink (1890). {011} occurs only as a twin plane, with (011) ∧ (0$\bar{1}$1) = 60°55', and the $b:c$ ratio is taken therefrom. See Brögger, *Zs. Kr.*, **18**, 376 (1890), on the morphological relations to manganite, ludwigite, etc.
2. Larsen, 120 (1921).
3. Bäckström, *Geol. För. Förh.*, **17**, 257 (1895). Described as prismatic crystals (prism angle = 68°36½') without terminal faces. Cleavage not observed and not twinned, but otherwise like pinakiolite.

24.1.3   Hulsite [(Fe″,Ca,Mg)$_4$(Fe‴,Sn⁗)$_2$B$_2$O$_{10}$ (?)]. *Knopf* and *Schaller* (*Am. J. Sc.*, **25**, 323, 1908).

Orthorhombic (?). As small crystals, rectangular in shape, with uneven and dull faces; also as tabular masses. $a:b:c = 0.550:1:?$, from cleavage faces (110) ∧ (1̄10) = 57°38′. Forms: $a\{100\}$, $b\{010\}$, $c\{001\}$, $m\{110\}$.
Twinning.   On [001], one individual rotated 120° from the other.
P h y s.   Cleavage {110} good. H. ~3. G. 4.28. Luster submetallic to vitreous. Color black.
C h e m.   A borate of trivalent iron, divalent iron, magnesium, and calcium, perhaps (Fe″,Ca,Mg)$_4$(Fe‴,Sn⁗)$_2$B$_2$O$_{10}$. Sn⁗ substitutes for Fe‴, with Sn:Fe = 1:4 in the cited analysis:[1] MgO 4.29, CaO 9.11, FeO 27.71, Fe$_2$O$_3$ 15.21, SnO$_2$ 7.07, B$_2$O$_3$ 9.20, insol. 18.63, H$_2$O (with some CO$_2$, SiO$_2$, Al$_2$O$_3$) [8.78], total [100.00].

Tests.   Fuses quietly to a black slag. Easily soluble in HCl or HF, less so in other acids.

O c c u r.   Found with vesuvianite, garnet, magnetite, and diopside in contact metamorphosed limestone at Brooks Mountain, Seward Peninsula, Alaska.
A l t e r.   To limonite.
N a m e.   After Alfred Hulse Brooks of the U. S. Geological Survey.

Ref.

1. Average of several partial analyses of impure material by Schaller, *Am. J. Sc.*, **29**, 543 (1910).

24.1.4   W A R W I C K I T E   [(Mg,Fe)$_3$Ti(BO$_4$)$_2$]. *Shepard* (*Am. J. Sc.*, **34**, 313, 1838; **36**, 85, 1839). Enceladite *Hunt* (*Am. J. Sc.*, **2**, 30, 1846; **11**, 352, 1851).

C r y s t.   Orthorhombic.

$$a:b:c = 0.972:1:0.318$$

Forms:[1]

		φ	ρ
b	010	0°00′	90°00′
a	100	90 00	90 00
g	130	18 50	90 00
m	110	45 40	90 00
h	310	71 57½	90 00

Structure cell.[8]   Space group *Pnam*. a_0 9.20 Å, b_0 9.45, c_0 3.01; $a_0:b_0:c_0 = 0.972:1:0.318$. Cell contents (Mg,Fe)$_6Ti_2B_4O_{16}$.
Habit.   In slender prismatic crystals with rounded terminations.
P h y s.   Cleavage {100}, perfect; not easily observed in altered material. Fracture uneven. Brittle. H. 3½–4. G. 3.35 ± 0.01.[2] Luster often nearly dull; of cleavage surface submetallic; pearly to subvitreous when fresh. Color dark hair-brown to dull black, sometimes a copper-red tinge on cleavage surface. Streak bluish black. Transparent in very small grains.

O p t. In transmitted light, reddish brown in color. Indices somewhat variable.

ORIENTATION[3]		n[4]	PLEOCHROISM[3]	
X	c	1.806 ± 0.005	Yellow brown	Biaxial positive (+).
Y	b	1.809 ± 0.005	Reddish brown	2V small and variable.
Z	a	1.830 ± 0.005	Cinnamon-brown	Absorption $X > Y > Z$.

C h e m. A titano-borate of magnesium and iron, probably $(Mg,Fe)_3Ti(BO_4)_2$.

Anal.[7]

	1	2	3	4
MgO	38.89	36.80	35.71	38.63
FeO	7.70	7.02	9.15	8.07
Fe_2O_3			4.76	
Al_2O_3		2.21	2.91	
SiO_2		1.00	1.39	1.56
TiO_2	28.54	23.82	24.86	27.87
B_2O_3	24.87	27.80	21.29	23.87
Total	100.00	98.65	100.07	100.00
G.		3.362	3.342	

1. $(Mg,Fe)_3Ti(BO_4)_2$, with Mg:Fe = 9:1. 2. Warwick, New York.[5] Spinel present as impurity. 3,4. Warwick, New York.[6] Anal. 3 average of two. Anal. 4 represents analysis 3 recalculated to 100 after deduction of Al_2O_3 and Fe_2O_3 as spinel and magnetite, respectively.

Tests. Yields a little water, apparently due to alteration. B.B. infusible. Decomposed by H_2SO_4.

O c c u r. Found in crystalline limestone (Franklin formation) about $2\frac{1}{2}$ miles southwest of Edenville, near the town of Warwick, Orange County, New York. Associated with chondrodite, blue and black spinel, graphite, magnetite, ilmenite, diopside, and pseudomorphous grains and masses of serpentine.

A l t e r. Alters on weathering to a hydrated, relatively soft, friable to pulverulent material. Color dull black to ocher-yellow.

N a m e. From the locality. Enceladite after Enceladus, one of the Titans of ancient mythology.

Ref.

1. Angles and forms of Des Cloizeaux (**2,** 16, 1874).
2. Range of best reported values; cf. Dana (881, 1892) and Bradley, *Am. J. Sc.,* **27,** 179 (1909).
3. Lacroix, *Bull. soc. min.,* **7,** 74 (1886).
4. Larsen (156, 1921).
5. Smith, *Am. J. Sc.,* **8,** 432 (1874).
6. Bradley (1909).
7. In earlier analyses by Shepard (1838, 1839) and Hunt (1846, 1851), made on altered and impure material, the boron was overlooked.
8. Takéuchi, Watanabé and Ito, *Acta Cryst.,* **3,** 98 (1950).

24.1.5 **K O T O I T E** [$Mg_3(BO_3)_2$]. *Watanabe* (*Min. Mitt.*, **50**, 441, 1939; *ibid.*, **51**, 162, 1939; *Fortschr. Min.*, **23**, clxvi, 1939).

C r y s t. Orthorhombic.[1]

$a:b:c = 0.6412:1:0.5494$;[2] $\quad p_0:q_0:r_0 = 1.1671:0.5494:1$

Forms:[2]

| m 110 | q 011 | r 101 | t 403 |

Structure cell.[3] a_0 5.41 kX, b_0 8.42, c_0 4.51; $a_0:b_0:c_0 = 0.643:1:0.536$. Cell contents $Mg_6(BO_3)_4$.

Habit. Massive granular, and as disseminated grains.

Twinning. Twin plane and composition plane {101}, polysynthetic.

P h y s. Cleavage {110}, perfect, with parting {101}. H. $6\frac{1}{2}$. G. 3.10; 3.06 (calc.). Colorless and transparent. Luster vitreous. M.P. about 1340° C. Exhibits translation gliding, gliding elements unknown.

O p t. In transmitted light, colorless. Biaxial positive (+). Dispersion, $r > v$.

		Hol		(ALL INDICES FOR Na)	
ORIENTATION		Kol[4]	Rézbánya[4]	Artificial[5]	Artificial[6]
nX	a	1.652	1.652	1.6514	1.6527
nY	b	1.653	1.653	1.6521 (calc.)	1.6537 (calc.)
nZ	c	1.673	1.674	1.6725	1.6748
2V		21° (obs.)		22° (obs.)	24°30′ (calc.)

C h e m. Magnesium orthoborate, $Mg_3(BO_3)_2$.

Anal.

	1	2
MgO	63.46	62.78
B_2O_3	36.54	35.20
Rem.		2.62
Total	100.00	100.60
G.		3.10

1. $Mg_3(BO_3)_2$. 2. Hol Kol, Korea. Rem. is SiO_2 1.32, Al_2O_3 0.26, Fe_2O_3 0.20, FeO 0.61, CaO 0.18, H_2O + 110° 0.05. Forsterite and spinel present as impurities.

Tests. B.B. infusible. Easily soluble in warm HCl or H_2SO_4.

O c c u r. Found abundantly at the Hol Kol mine, Suan, Korea, associated with forsterite, clinohumite, ludwigite, spinel, fluoborite, and szaibelyite in the contact zone of a granite intrusion into dolomite. Also found in the interior of nodules of szaibelyite in marble at Rézbánya, Hungary.

A l t e r. Under hydrothermal conditions alters to szaibelyite.

A r t i f.[7] Obtained in crystals by direct fusion of MgO and B_2O_3. Isostructural with artificial $Mn_3(BO_3)_2$ and $Co_3(BO_3)_2$.

N a m e. After Bundjirô Kotô (1856–1935), Japanese geologist and petrographer.

Ref.

1. Symmetry established by optical evidence and the identity with artificial $Mg_3(BO_3)_2$ as proved by x-ray and optical study.

2. Ratio and forms of Mallard, *C.R.*, **105**, 1260 (1887); *Ann. mines*, **12** [8], 427 (1887)— *Zs. Kr.*, **15**, 650 (1889) on artificial crystals.

3. Watanabe, *Fortschr. Min.*, **23,** clxvii (1939), by unstated method on artificial (?) crystals.
4. Watanabe, *Min. Mitt.*, **50,** 441 (1939), by immersion method.
5. Watanabe (1939) by prism method.
6. Mallard (1887) by prism method.
7. Cf. Mellor (**5,** 96, 1924).

24.1.6 **R H O D I Z I T E** [$NaKLi_4Al_4Be_3B_{10}O_{27}$]. Rhodozit *Rose (Ann. Phys.,* **33,** 253, 1834; **39,** 321, 1836). Rhodicit *Hausmann.*

C r y s t. Isometric; hextetrahedral—$\bar{4}\,3\,m$.

Forms: [1]

$$a\,001 \quad d\,011 \quad o\,111 \quad -o\,\bar{1}11$$

Structure cell.[2] Space group $P\bar{4}3m$. a_0 7.303 ± 0.025 kX. Cell contents $NaKLi_4Al_4Be_3B_{10}O_{27}$.

Habit. Dodecahedral, with o smooth and shining and d often uneven; also tetrahedral.

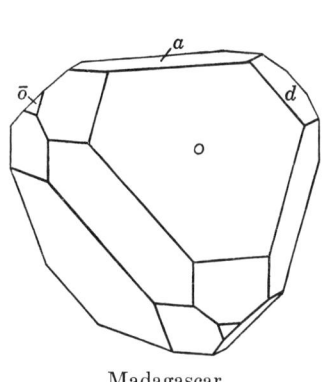

Madagascar.

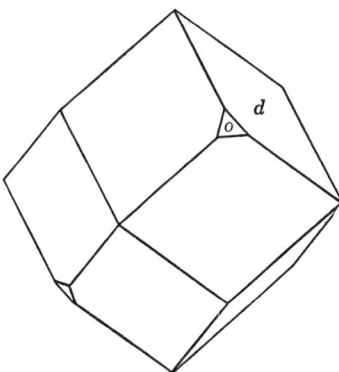

Madagascar.

P h y s. Cleavage {111} and {$\bar{1}11$}, difficult. Fracture conchoidal. H. 8. G. 3.38 (Urals); 3.305, 3.344 (Madagascar); 3.24 (calc.). Colorless to white, also grayish or yellowish white. Luster vitreous, inclining to adamantine. Transparent to translucent. Pyroelectric, the angles replaced by (111) the antilogous pole.[6]

O p t.[3] In transmitted light, colorless.

n(Li)	n(Na)	n(Tl)
1.6895	1.6935	1.6965

Exhibits weak anomalous double refraction distributed in sectors [4] with Z (+) inclined about 10° to [110]. Does not become isotropic on heating to bright redness.[5]

C h e m. A complex borate of beryllium, aluminum, and alkalies, probably $NaKLi_4Al_4Be_3B_{10}O_{27}$. Rb and Cs substitute for K, and OH may be present as an essential constituent.

Anal.

	1	2	3	4
Li_2O	7.81		7.30	0.68
Na_2O	4.05	1.62	3.30	1.78
K_2O				1.41
Rb_2O	6.16 }	12.00 }	5.90	2.29
Cs_2O				3.47
Al_2O_3	26.65	41.40	30.50	27.40
BeO	9.81		10.10	14.93
B_2O_3	45.52	33.93	40.60	[43.33]
Rem.		6.45	1.81	4.71
Total	100.00	95.40	99.51	[100.00]
G.		3.38	3.305	3.344

1. $NaKLi_4Al_4Be_3B_{10}O_{27}$. 2. Urals.[7] Rem. is CaO 0.74, MgO 0.82, FeO 1.93, ign. loss 2.96. The Be has been determined as Al_2O_3 and the analysis probably is otherwise unreliable. 3. Antandrokomby, Madagascar.[8] Rem. is SiO_2 1.36, ign. loss 0.45. Contains a little spodumene. 4. Sahatany valley, Madagascar.[9] Rem. is SiO_2 3.18, ign. loss 1.42, MgO 0.11.

Tests. B.B. fusible with difficulty on the edges to a white opaque glass, tingeing the flame at first green, then green below and red above, and finally red throughout. With borax and microcosmic salt fuses to a transparent glass. Insoluble in acids.

Occur. Originally found as minute crystals on red tourmaline from Sarapulsk and Schaitansk, both near Mursinsk north of Sverdlovsk (Ekaterinburg) in the Ural Mountains, U.S.S.R. Also found as crystals up to 2 cm. in size with spodumene and red tourmaline in pegmatite at Antandrokomby near Mt. Bity and at Manjaka in the valley of Sahatany.

Name. From $ροδίζειν$, *to be rose-colored*, because it tinges the blowpipe flame red.

Ref.

1. Goldschmidt (**7**, 123, 1922).
2. Strunz, Doctoral Dissert. (1938), and *Naturwiss.*, **31**, No. 5 (1943), by powder method on Schaitansk material. Cell contents calculated from analysis 3 of Madagascar material.
3. Indices of Duparc, Wunder, and Sabot, *Bull. soc. min.*, **34**, 131 (1911), by prism method on material of analysis 4.
4. Bertrand, *Bull. soc. min.*, **5**, 31, 72 (1882), and Klein, *Jb. Min.*, I, 65, (1891).
5. Klein (1891).
6. Ries and Rose, *Ann. Phys.*, **59**, 353 (1843).
7. Damour, *Bull. soc. min.*, **5**, 98 (1882).
8. Pisani, *Bull. soc. min.*, **33**, 37 (1910).
9. Duparc, Wunder, and Sabot, *Bull. soc. min.*, **34**, 136 (1911).

24.1.7 **JEREMEJEVITE** [$AlBO_3$]. Jeremeiewit *Damour* (*Bull. soc. min.*, **6**, 20, 1883). Eichwaldit *Websky* (*Ak. Berlin, Sitzber.*, 671, 1883; *Jb. Min.*, I, 1, 1884). Jeremejewite, Jeremejeffite, Yeremeyevite, Eremeyevite.

Hexagonal prisms elongated [0001], with rounded, irregular, or indented terminations and the prism faces modified by vicinal planes. The crystals are composite and consist of an optically uniaxial outer zone which is hexagonal and to which the name jeremejevite is applied. Within, is an optically biaxial core divided into six sectors to which the name eichwaldite is given. The terminal planes referred to the eichwaldite project at the summit beyond the jeremejevite. The morphological measurements based on the reflections from the several parts of the rounded and composite terminations must be regarded as more or less uncertain. The outer zone has been described [1] as perhaps hexagonal-dipyramidal, $6/m$, with $a\{11\bar{2}0\}$, $e\{21\bar{3}0\}$, $n\{10\bar{1}4\}$, $f\{10\bar{1}3\}$, $d\{10\bar{1}1\}$, $q\{70\bar{7}5\}$, $g\{41\bar{5}3\}$, and several vicinal third-order pyramids. The reported angles afford the ratio $a:c = 1:0.6836$ and are not consistent with the cell found by x-ray study. For the eichwaldite has been deduced $a:b:c = 0.5523:1:0.5434$ with the forms $x\{104\}$, $p\{101\}$, and $y\{136\}$ in an orthorhombic interpretation.

Structure cell.[2] Space group $C6_3/m$ (?). a_0 8.57 $\pm$ 0.02 kX, c_0 8.17 $\pm$ 0.02; $a_0:c_0 = 1:0.953$. Cell contents $Al_{12}B_{12}O_{36}$. The biaxial parts apparently are monoclinic, with the space group $C2_1/m$ (?).

P h y s. No cleavage. Fracture conchoidal. H. $6\frac{1}{2}$. G. 3.28,[3] in both uniaxial and biaxial parts; 3.27 (calc.). Luster vitreous. Colorless to pale yellowish brown. Piezoelectric.[4]

O p t.[5] In transmitted light, colorless.

Orientation	n(Na) (uniaxial part)	n(Na) (biaxial part)	n(Na) (biaxial part)
E or X	1.640	1.640	1.640
Y		1.653	
O or Z	1.653	1.653	1.653
$2V$	0°	$\sim$10°	$\sim$50°

In polarized light,[6] basal sections consist of an outer uniaxial part A; a narrow biaxial negative zone B divided into six sectors, each with variable $2E$ of 6° to 35° and the axial plane perpendicular to the edge BC; an inner biaxial negative zone C divided into six sectors, each with $2E$ 52° and the axial plane perpendicular to the line dividing the angle between the sectors of zone B; and a narrow central core, which may or may not be present, which is uniaxial. Increase of temperature does not change the optical characters. Pressure applied normal to [0001] makes zone A biaxial with the axial plane perpendicular to the direction of pressure. With both zones B and C the axial angle can be increased or diminished to zero according as the pressure is applied normal or parallel to the axial plane.

C h e m. Aluminum borate, $AlBO_3$. Fe''' substitutes for Al with Fe:Al = 1:21 in the only reported analysis. The analysis was made on a mixed sample of the uniaxial and biaxial parts, and the distribution of the Fe_2O_3 and K_2O therein is not known; the biaxial parts (*eichwaldite*) have been considered [7] to be relatively rich in K_2O. Spectrographic analysis [8] of the biaxial material showed the following elements present in trace amounts: Si > Pb, Cu > Fe, Mg, Ca, Na, Mn, Ga > K, Ti, Zr, Sn, Cr, V, Ag, Ba, Be, Tl, Bi.

Anal.

	1	2
K_2O		0.70
Al_2O_3	59.41	55.03
Fe_2O_3		4.08
B_2O_3	40.59	[40.19]
Total	100.00	[100.00]
G.	3.27	3.28

1. $AlBO_3$. 2. Nertschinsk.[3]

Tests. B.B. turns white and opaque, colors the flame green, but does not fuse. After ignition easily soluble in hot H_2SO_4 or concentrated KOH.

Occur. Found as a few single-crystals up to several inches in length associated with orthoclase and quartz on Mt. Soktuj, a northerly extension of the Adun-Chilon Range in Dauria, in the Nertschinsk district, eastern Siberia. The crystals were found loose in granitic debris under the turf.

Artif. Several different artificial aluminum borates including $AlBO_3$ have been reported.[9]

Name. After Pavel Vladimirovitch Jeremejev (1830–1899), Russian mineralogist and engineer. Eichwaldite after J. I. Eichwald, a director of the Nertschinsk mines, who found the first specimens.

Ref.
1. Websky (1883, 1884).
2. Strunz, Inaug. Diss. Univ. Berlin (1938); comparable cell dimensions are given by Gossner and Kraus, *Cbl. Min.*, 348, (1934). The space group $C6_3/m$ and the subgroup $C2_1/m$ given by Strunz are not consistent with the observed piezoelectricity.
3. Damour, *Bull. soc. min.*, **6**, 20 (1883). Mrose, priv. comm. (1949), obtained 3.27 on natural crystals.
4. Frondel, priv. comm. (1948), by the oscilloscope method on a sawn plate. Gossner and Kraus (1934) obtained no effect by the Giebe and Scheibe method.
5. Frondel, priv. comm. (1948), on natural material.
6. Klein, *Jb. Min.*, I, 84 (1891).
7. Strunz (1938).
8. Harrison, priv. comm. (1948), on material of Frondel (1948).
9. Mellor (**5**, 102, 1924); Baumann and Moore, *J. Am. Ceramic Soc.*, **25**, 391 (1942); Michel-Lévy, *C.R.*, **228**, 1814 (1949).

24.1.8 NORDENSKIÖLDINE [$CaSn(BO_3)_2$]. Brögger (*Geol. För. Förh.*, **9**, 255, 1887; *Zs. Kr.*, **16**, 61, 1890).

Cryst.[1] Hexagonal—R; rhombohedral—$\bar{3}$.

$a:c = 1:0.8221;$ [2] $\alpha\ 102°57\frac{1}{2}';$ $p_0:r_0 = 0.9493:1;$ $\lambda\ 73°12'$

Forms:[3]

		ϕ	$\rho = C$	A_1	A_2
c 0001	111	....	0°00'	90°00'	90°00'
a 11$\bar{2}$0	10$\bar{1}$	0°00'	90 00	60 00	60 00
r 10$\bar{1}$1	100	30 00	43 30$\frac{1}{2}$	53 24	90 00

Structure cell.[4] Space group $R\bar{3}$. a_{rh} 6.001, $\alpha\ 47°41\frac{1}{2}';$ a_0 4.852 kX, c_0 15.92; $a_0:c_0 = 1:3.282$. Cell contents $CaSn(BO_3)_2$ in the rhombohedral unit. Isostructural with dolomite.

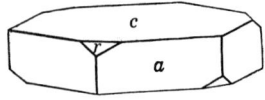

Norway.

Habit. Thin to thick tabular {0001}. Also as thick, lens-like crystals and as subparallel growths.

Oriented growths.[5] Overgrowths with parallel axes are found of calcite and siderite upon nordenskiöldine; also cassiterite, with [001] parallel nordenskiöldine {0001} [10$\bar{1}$1] or [11$\bar{2}$0].

Phys. Cleavage {0001} perfect; {10$\bar{1}$1} indistinct (seen in thin sections). Fracture conchoidal. Brittle. H. 5$\frac{1}{2}$–6. G. 4.20; [6] 4.22 (calc.).

Colorless; also sulfur-, lemon-, or wine-yellow. Luster vitreous, somewhat pearly on {0001}. Transparent.

O p t.[5] In transmitted light, colorless.

nO 1.778   Uniaxial negative $(-)$
nE 1.660

C h e m. A borate of calcium and tin, $CaSn(BO_3)_2$.
Anal.

	1	2
CaO	20.29	20.45
SnO_2	54.52	53.75
B_2O_3	25.19	[23.18]
Rem.		2.62
Total	100.00	[100.00]
G.	2.22	2.20

1. $CaSn(BO_3)_2$. 2. Arö, Norway.[6] Rem. is ZrO_2(?) 0.90, ign. loss and B_2O_3 (?) 1.72.

Tests. B.B., when strongly heated, sinters but does not fuse. Imperfectly decomposed by hydrochloric acid.

O c c u r. Found very sparingly with melinophanite, homilite, zircon, feldspar, molybdenite, cancrinite, analcime, etc. in an alkaline pegmatite on the island Arö, in the Langesund fiord, southern Norway. Later recognized [7] in an ore pipe in marble near a granite contact at Arandis, South West Africa, associated with tourmaline, cassiterite, calcite, siderite, stannite, chalcopyrite, and pyrrhotite.

A l t e r. To cassiterite.

N a m e. After the Swedish mineralogist and explorer N. A. E. Nordenskiöld (1832–1901).

Ref.

1. Brögger (1890) gives the crystal class as hexagonal-scalenohedral—$\bar{3}\ 2/m$. Ramdohr, *Jb. Min., Beil.-Bd.*, **68,** 288 (1934), describes etch pits on {0001} which fix the symmetry as rhombohedral—$\bar{3}$.
2. Brögger (1890), rounded from 0.82207. Brögger's unit, one-fourth that of the structural cell, is retained because of the structural similarity to dolomite where the quartered cell is used for morphological purposes. Transformation: Brögger to structure, 1000/0100/0010/0004.
3. Brögger (1890). Ramdohr (1934) lists {$10\bar{1}4$} as probable.
4. Ehrenberg and Ramdohr, *Jb. Min., Beil.-Bd.*, **69,** 1 (1934), by powder method. Space group here assigned on basis of isostructural relation to dolomite.
5. Ramdohr (1934) on Arandis material.
6. Cleve in Brögger (1890) on Arö material.
7. Ramdohr (1934).

25 HYDRATED BORATES

25.1.1 PINNOITE [$Mg(BO_2)_2 \cdot 3H_2O$]. Staute (*Ber. deutsche chemische Gesellschaft*, **17**, 1584, 1884).

C r y s t.[1] Tetragonal; dipyramidal—$4/m$.

$$a:c = 1:1.0761; \quad p_0:r_0 = 1.0761:1$$

Forms:

	ϕ	ρ	A	M
a 110	45°00′	90°00′	45°00′	90°00′
o 011	0 00	47 06	90 00	58 48
d 112	45 00	37 16	64 39	90 00
z 122	26 34	50 16	69 53	75 55

Structure cell.[7] Space group $P4_2/m$. a_0 7.617 Å, c_0 8.190; $a_0:c_0 = 1:1.075$. Cell contents $Mg_4(BO_2)_8 \cdot 12H_2O$.

Habit. Rarely in distinct crystals, short prismatic [001]. Usually crystalline and fine granular to faintly fibrous; in nodules with radiated fibrous structure and a crystalline surface.

P h y s. Fracture uneven. H. $3\frac{1}{2}$. G. 2.27;[2] 2.29 (calc.). Luster vitreous. Color sulfur-, straw-, or greenish yellow; sometimes pistachio-green. Translucent.

O p t. In transmitted light, yellow.

	$n(Na)$ [3]	
O	1.565	Uniaxial positive (+).
E	1.575	

C h e m. Hydrated magnesium metaborate, $Mg(BO_2)_2 \cdot 3H_2O$.

Anal.

	1	2	3	4
MgO	24.58	24.45	24.19	24.07
Fe		0.15	0.23	0.21
Cl		0.18	0.40	0.37
B_2O_3	42.46	42.50	[42.68]	[42.85]
H_2O	32.96	32.85	32.50	32.50
Total	100.00	100.13	[100.00]	[100.00]

1. $Mg(BO_2)_2 \cdot 3H_2O$. 2. Stassfurt. Average of several analyses.[4] 3,4. Stassfurt.[5] Anal. 3, dense yellow material; anal. 4, grayish yellow, granular aggregate.

Tests. B.B. decrepitates, turns white, and fuses to a dull white mass. Soluble in dilute acids. Dissolves in boiling water with the formation of an alkaline solution and a flocculent precipitate, the latter re-dissolving as the solution cools.

O c c u r. Originally found in the salt deposit at Stassfurt, Saxony, where it occurs in the upper kainite layers, associated with earthy boracite

KERNITE 335

from which it probably has been formed by reaction with residual borate-rich liquors. Also reported from Aschersleben, Saxony and Leopoldshall, Anhalt, Germany.

A r t i f.[6] Prepared by mixing 100 g. borax dissolved in 450 cc. H_2O with 53 g. of $MgCl_2 \cdot 6H_2O$ in 50 cc. H_2O, warming, and then adding 70 g. of the latter salt in 60 cc. H_2O. The solution is now concentrated by evaporation, seeded with pinnoite, and kept at 100°. Needle-like crystals of pinnoite appear after a few days.

N a m e. After Chief Councillor of Mines Pinno, of Halle.

Ref.

1. Angles of Luedecke, *Zs. Nat. Halle*, **58**, 645 (1885), setting of Goldschmidt (**6**, 155, 1920) conforming to the structure cell. Transformation: Goldschmidt to Luedecke, 110/$\bar{1}$10/001.
2. Staute (1884); checked by Mrose, priv. comm. (1946), by microbalance on Stassfurt material. Boeke, *Cbl. Min.*, 531 (1910), obtained 2.292 and Luedecke (1885), 2.373.
3. Boeke (1910); checked by Mrose, priv. comm. (1946).
4. Staute (1884).
5. Strohmeyer anal. in Staute (1884).
6. van't Hoff and Bruni, *Ak. Berlin, Sitzber.*, 805 (1902).
7. Stadler, *Min. Mag.*, **28**, 26 (1947), by oscillation and powder methods on Stassfurt material.

25.1.2 **K E R N I T E** [$Na_2B_4O_7 \cdot 4H_2O$]. Schaller (*Am. Min.*, **12**, 24, 1927; *U. S. Geol. Sur., Prof. Pap.* **158**, 137, 1930). Rasorite Palmer (*Eng. Mining J.*, **123**, 494, 1927).

C r y s t.[1] Monoclinic; prismatic—$2/m$.

$a:b:c = 1.6989:1:0.7615;$ $\beta\ 108°52';$ $p_0:q_0:r_0 = 0.4482:0.7206:1$

$r_2:p_2:q_2 = 1.3877:0.6620:1;$ $\mu\ 71°08';$ $p_0'\ 0.4737,\ q_0'\ 0.7615,\ x_0'\ 0.3417$

Forms:[2]

	ϕ	ρ	ϕ_2	$\rho_2 = B$	C	A
c 001	90°00'	18°52'	71°08'	90°00'	.....	71°08'
a 100	90 00	90 00	0 00	90 00	71°08'	.....
e 110	31 53	90 00	0 00	31 53	80 10	9 50
d 201	90 00	52 12	37 48	90 00	33 20	37 48
$H\ \bar{1}0.0.7$	−90 00	18 31½	108 31½	90 00	37 23½	108 31½
$G\ \bar{8}05$	−90 00	22 36	112 36	90 00	41 28	112 36
$E\ \bar{12}.0.7$	−90 00	25 11½	115 11½	90 00	44 03½	115 11½
$D\ \bar{2}01$	−90 00	31 12	121 12	90 00	50 04	121 12
$B\ \bar{12}.0.5$	−90 00	38 29½	128 29½	90 00	57 21½	128 29½

Less common and rare:

b 010	f 11.2.0	r 807	l 502	$R\ \bar{1}02$	$M\ \bar{6}07$	$F\ \bar{5}03$
i 120	u 102	q 403	k 301	$Q\ \bar{2}03$	$L\ \bar{1}01$	$C\ \bar{7}03$
h 230	t 203	p 302	j 501	$P\ \bar{3}04$	$K\ \bar{8}07$	$A\ \bar{1}0.0.3$
g 210	s 101	n 16.0.9	$S\ \bar{1}03$	$N\ \bar{4}05$	$J\ \bar{4}03$	

Structure cell.[3] Space group $P2/c$; $a_0\ 15.52\ kX$, $b_0\ 9.14$, $c_0\ 6.96$; [$\beta\ 108°52'$]; $a_0:b_0:c_0 = 1.6980:1:0.7615$. Cell contents $Na_8B_{16}O_{28} \cdot 16H_2O$.

Habit.[4] Nearly equant, usually slightly elongated [100] and heavily striated [010], with irregular form development. Commonest forms $c\,a\,e\,D$ with either c or D dominant. Sometimes wedge-shaped, or rounded due to repetition of faces. Also as cleavable masses, with a fibrous structure simulated by development of excellent cleavages. Cleavage fragments commonly bent or warped around [001]. Massive.

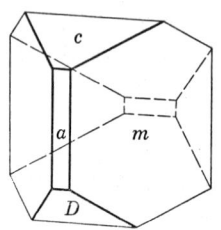

Kramer.

Twinning.[11] On {110}.

P h y s. Cleavage {100} perfect, {001} somewhat less so, {$\bar{2}$01} fair. H. $2\frac{1}{2}$, varying slightly on different faces and in different directions. G. 1.908;[5] 1.93 (calc.). Luster vitreous, slightly satiny on fibrous cleavage surface. Colorless; usually white and opaque due to a surface film of tincalconite. Transparent. Streak white. Thin cleavage fragments parallel {100} are somewhat flexible and elastic.

O p t.

Orientation		n [6]	
$X \wedge c$	$+38\frac{1}{2}°$	1.454	Biaxial negative (−).
$Y \wedge c$	$-51\frac{1}{2}°$	1.472	$2E$ 142° $2V$ 80°.
$Z\ b$		1.488	$r > v$, distinct.

C h e m. A hydrated sodium borate, $Na_2B_4O_7 \cdot 4H_2O$.
Anal.[7]

	1	2	3
Na_2O	22.66	22.63	22.65
B_2O_3	51.02	50.76	[50.80]
H_2O	26.32	26.50	26.55
Total	100.00	99.89	[100.00]
G.		1.93	1.908

1. $Na_2B_4O_7 \cdot 4H_2O$. 2. Kern County, California. 3. Artificial.

Tests. B.B. swells and readily fuses to an opaque white cauliflower mass, which on further heating becomes clear (borax glass). Slowly soluble in cold water. Readily soluble in hot water and acids.

O c c u r. In the Kramer borate district, Kern County, California, associated with minor amounts of borax, ulexite, tincalconite, and probertite, and forming a deposit a hundred feet thick in the clay-shales. The kernite is abundant as crystals 2 to 3 feet thick. The largest crystal noted measured 8 by 3 feet. Also as veins and irregular masses in the clay. The deposit apparently was formed by the melting and dehydration in situ of a buried lake deposit of borax heated by a local intrusion of igneous rock.

A l t e r. In contact with borax, and possibly other borates, kernite alters by dehydration to opaque white tincalconite, $Na_2B_4O_7 \cdot 5H_2O$. Isolated crystals do not alter. On heating to 100°–120°, kernite alters to $Na_2B_4O_7 \cdot 2H_2O$ (*metakernite*), and in a moist atmosphere it yields the decahydrate; these changes are not reversible.[8]

Artif. Formed by repeated heating of borax at 150° in contact with its water of crystallization.[9] Also by heating $Na_2B_4O_7$ with 4.5–6 mols H_2O in a closed tube at 130°–135°.[8]

Name. After the county in which it occurs. The mineral had been earlier called rasorite [10] for Mr. C. M. Rasor, an engineer of the Pacific Coast Borax Company, but the name kernite has published priority.

Ref.
1. First reported as orthorhombic (Schaller, 1927). Elements by Schaller (1930); unit and orientation by Garrido, *Zs. Kr.*, **82**, 468 (1932), who confirmed the monoclinic symmetry by Laue photographs and obtained the cell dimensions: a_0 15.65, b_0 9.07, c_0 7.01 by the rotation method. Transformation: Schaller to Garrido, $002/0\bar{2}0/100$. The unequal form development along [100] suggests a lower symmetry class, but etchings do not definitely indicate any deviation from holohedral symmetry (Schaller, 1930).
2. Schaller (1930). Vicinal:

$\bar{8}.0.15$	$\bar{14}.0.23$	$\bar{5}06$	$\bar{16}.0.13$	$\bar{32}.0.21$	$\bar{13}.0.7$
$\bar{16}.0.29$	$\bar{11}.0.17$	$\bar{12}.0.13$	$\bar{22}.0.17$	$\bar{14}.0.9$	$\bar{40}.0.21$
$\bar{10}.0.17$	$\bar{12}.0.17$	$\bar{26}.0.27$	$\bar{46}.0.31$	$\bar{20}.0.11$	$\bar{44}.0.21$

On artificial crystals: Schaller (1930), $c\ a\ e\ d\ D\ f$ and the new form $\{443\}$; Menzel and Schulz, *Zs. anorg. Chem.*, **245**, 157 (1940), $c\ b\ a\ e\ g\ d\ D$ and the new forms: 780, 760, 650, 970, 750, 13.0.1, $\bar{1}11$, 5.10.2, 14.12.13, $\bar{10}.18.9$, $\bar{14}.16.15$, $\bar{52}.46.43$.
3. Minder, *Zs. Kr.*, **92**, 301 (1935), from rotation photographs of artificial crystals. Transformation: Minder to Garrido, $001/0\bar{1}0/100$. See also Garrido (1932).
4. Artificial crystals prepared by Wells (in Schaller, 1930) and Menzel and Schulz (1940) were elongated [010] with $c\ a\ e$ dominant.
5. Schaller (1930). The average of 2 determinations, 1.911 and 1.904, obtained by suspending clear fragments in diluted bromoform.
6. Average of 3 determinations in close agreement, in Schaller (1930). Artificial kernite gave nX 1.455, nY 1.472, nZ 1.487.
7. Schaller (1930).
8. Menzel, Schulz and Deckert, *Naturwiss.*, **23**, 832 (1935); Menzel and Schulz (1940).
9. R. C. Wells in Schaller (1930).
10. Palmer, *Eng. Mining J.*, **123**, 494 (1927).
11. See Amorós, *Inst. invest. Geol. Barcelona*, no. 7 (1947).

25.1.3 **TINCALCONITE** [$Na_2B_4O_7 \cdot 5H_2O$]. Octahedral borax *Payen* (*J. chim. méd.*, **3**, 594, 1827). Tincalconite *Shepard* (*Bull. soc. min.*, **1**, 144, 1878). Mohavite *Schaller* (*Min. Mag.*, **23**, 634, 1934).

Cryst. Hexagonal—R; probably rhombohedral—$\bar{3}$.[1]

$a:c = 1:1.87;$[2] $\alpha\ 72°02';$ $p_0:r_0 = 2.16:1;$ $\lambda\ 103°38'$

Forms:[2]

		ϕ	$\rho = C$	A_1	A_2
$c\ 0001$	111	.....	0°00'	90°00'	90°00'
$r\ 10\bar{1}1$	100	30°00'	65 11	38 11	90 00
$d\ 01\bar{1}2$	110	$-30\ 00$	47 14	90 00	50 $31\frac{1}{2}$

Structure cell.[3] Space group probably $R\bar{3}$. a_{rh} 9.56, $\alpha\ 71°42';$ a_0 11.20 kX, c_0 21.13; $a_0:c_0 = 1:1.887$ Cell contents $Na_6B_{12}O_{21} \cdot 15H_2O$ in the rhombohedral unit.

Habit. Known in nature only as a fine-grained powder. Artificial crystals are pseudo-cubic, the equal development of the two commonest forms, r and c, giving an octahedron-like habit.

P h y s. Natural tincalconite is a dull white powder and unsuitable for the determination of its physical properties. Artificial: fracture hackly, occasionally poor conchoidal. G. 1.880; [4] 1.89 (calc.). Luster vitreous. Colorless. Transparent.

O p t. Colorless in transmitted light. Usually so fine grained as to appear almost opaque under the microscope.

n (artificial [4])
O	1.461	Uniaxial positive (+).
E	1.474	

C h e m. Hydrated sodium tetraborate, $Na_2B_4O_7 \cdot 5H_2O$.
Anal.

	1	2	3	4	5	6
Na_2O	21.29	20.80	20.67	21.28	20.72	21.40
B_2O_3	47.80	47.52	48.67	47.26	47.19	47.26
H_2O	30.91	30.59	30.81	30.78	30.93	31.01
Insol.		0.48		0.37	0.90	0.25
Total	100.00	99.39	100.15	99.69	99.74	99.92

1. $Na_2B_4O_7 \cdot 5H_2O$. 2, 3, 4. Kramer district, California.[4] Anal. 2 and 3 on hydration pseudomorphs after kernite; 3 recalculated after deduction of 3.43 per cent water-insoluble impurities; anal. 4 on a pseudomorph after borax. 5. Artificial dehydration pseudomorph after a borax crystal from Searles Lake, California.[4]

Tests. Like borax.

O c c u r. Found with kernite and borax in the Kramer district, Kern County, California.[4] From Searles Lake, San Bernardino County, California.[6] In a dry atmosphere borax quickly dehydrates to a dull-white, fine-grained mass of tincalconite. This species no doubt occurs commonly in the borax deposits of the Mohave desert in southern California, and elsewhere in the United States and abroad.

A r t i f.[5] Tincalconite is the stable form of $Na_2B_4O_7 \cdot nH_2O$ above 60° or above about 35° in saturated salt solutions. It can be made in crystals by boiling a solution of borax until crystallization ensues, by cooling a hot saturated solution to not below 60°, and by melting borax in its water of crystallization in either a closed or open tube.

N a m e. Tincalconite from *tincal*, the Oriental name of borax, and κονία, *powder*, in allusion to the pulverulent form. The name was first given by Shepard in 1878 to material from California, later shown [4] to be identical with the mineral from the Kramer district, California, and with artificial material. The octahedron-like combination of six rhombohedral and two basal faces of crystals precipitated from a hot borax solution suggested isometric symmetry to Payen, who first described the compound in 1827, and resulted in the name "octahedral borax."

Ref.
1. The rhombohedral symmetry was first recognized by Arzruni, *Ann. Phys.*, **158**, 250 (1876). The point symmetry was established by Minder, *Zs. Kr.*, **92**, 301 (1935), by the morphology and etching of artificial crystals. See also Pabst and Sawyer, *Am. Min.*, **33**, 472 (1948).
2. Arzruni (1876) on artificial crystals.
3. Minder (1935) by rotation method on artificial crystals.
4. Schaller, *U. S. Geol. Sur., Prof. Paper 158*, 137 (1930).
5. Cf. Mellor (**5**, 70, 1924).
6. Pabst and Sawyer (1948).

25.1.4 **B O R A X** [$Na_2B_4O_7 \cdot 10H_2O$]. Tinkál or Tincal of India. Tinkar. Būraq *Arabic;* Būrak *Persian.* Baurach, Albaurach, Bauracia *Alchem.* Chrysocolla (ex nitro confecta), Borras *Agricola (De re metallica, 453, 1556).* Borax *Wallerius* (**1**, 346, 1753). Soude boratée *Haüy* (**2**, 366, 1801). Borate of soda *Phillips* (141, 1818). Prismatic borax-salt *Haidinger* (**2**, 52, 1825).

C r y s t. Monoclinic; prismatic—$2/m$.

$a:b:c = 1.0995:1:1.1264;$ $\beta\ 106°35';$ [1] $p_0:q_0:r_0 = 1.0245:1.0795:1$

$r_2:p_2:q_2 = 0.9263:0.9490:1;$ $\mu\ 73°25';$ $p_0'\ 1.0689, q_0'\ 1.1264, x_0'\ 0.2978$

Forms: [2]

	ϕ	ρ	ϕ_2	$\rho_2 = B$	C	A
c 001	90°00'	16°35'	73°25'	90°00'	.....	73°25'
b 010	0 00	90 00	.....	0 00	90°00'	90 00
a 100	90 00	90 00	0 00	90 00	73 25	0 00
m 110	43 30	90 00	0 00	43 30	78 40	46 30
s 021	7 24	66 36½	73 25	24 48	65 32	83 12½
o $\bar{1}12$	−22 48	31 25½	103 19	61 16½	40 31	101 39
z $\bar{1}11$	−34 23½	53 46½	127 38	48 16	64 08	117 06½

Structure cell.[3] Space group probably $C2/c$. a_0 11.82 kX, b_0 10.61, c_0 12.30; β 106°35'; $a_0:b_0:c_0 = 1.114:1:1.159$. Cell contents $Na_8B_{16}O_{28} \cdot 40H_2O$.

Habit. Very similar to pyroxene in angles and habit. Usually short prismatic [001] and often somewhat tabular {100}. Prominent zones [001] and [110]. Commonly malformed through abnormal development of part of one or more of these zones. Commonest forms $a\ m\ c\ o\ z\ b$. The faces m, o, z often striated parallel [110].

Twinning.[4] Twin plane {100}, rare.

P h y s. Cleavage [5] {100} perfect, {110} less so, {010} in traces. Fracture conchoidal. Rather brittle. H. 2–2½. G. 1.715 ± 0.005; [5] 1.70 (calc.). Colorless to white, also grayish, bluish, or greenish. Luster vitreous to resinous; sometimes earthy. Streak white. Translucent to opaque. Taste sweetish alkaline, feeble. Diamagnetic.

Opt. In transmitted light, colorless.

ORIENTATION [7]		n(Na) [6]	
X	b	1.4466	Biaxial negative ($-$).
Y		1.4687	2V 39°58′ (calc.).
Z $\wedge$ c	$-55°35'$(Na)	1.4717	$r > v$, strong, crossed [9]

Sections perpendicular to an optic axis exhibit bluish gray to pinchbeck-brown interference colors.

Chem. A hydrated sodium borate, $Na_2B_4O_7 \cdot 10H_2O$. Na_2O 16.26, B_2O_3 36.51, H_2O 47.23, total 100.00. The few reported [8] analyses of pure natural crystals conform closely to the formula.

Tests. B.B. intumesces and fuses to a colorless glass (borax glass). Soluble in water, yielding a faintly alkaline solution. Solubility in grams $Na_2B_4O_7$ per 100 grams H_2O at various temperatures: 10°, 1.60; 30°, 3.86; 60°, 19.0; 90°, 41.0.

Occur. Borax occurs in the evaporated deposits and muds of saline lakes and playas, where it is associated with halite, trona, ulexite, thenardite, aphthitalite, glauberite, gypsum, calcite, soda-niter, gaylussite, hanksite, and various rare sulfates and carbonates of sodium. Borax also occurs as an efflorescence on the soil in arid regions and in solution in hot springs or as a deposit therefrom.

Borax has been obtained since very early times from the salt lakes of Kashmir and Tibet; it was brought to Europe in the crude state under the name of *tincal* and there purified. The deposits are found in the Ladakh district of Kashmir and to the east in Tibet around Rudok. Other deposits occur to the north of Lhasa at the lakes of Tengri-Nor and Bul Tso, and to the south at the lake of Yamdok Tso. Small deposits of borax occur variously in India, Russia, and Persia. In the United States, important commercial deposits have been worked in California. The first discovery of borax in this state was made in 1856 at Borax Lake,[10] near Clear Lake, in Lake County, and fine large crystals were obtained from the mud of the lake bottom. The largest deposit is at Searles Lake, a pan-like depression about five by ten miles in size in northern San Bernardino County. Here borax occurs with trona, hanksite, thenardite, and various rare sodium salts, including pirssonite, sulfohalite, schairerite, burkeite, tychite, and northupite. Borax also has been mined in California at Furnace Creek and Resting Springs in Death Valley, Inyo County, where the famous 20-mule teams hauled the ore to the railroad at Mojave. Masses of borax the size of an ordinary room were found in the kernite deposit in the Kramer district, Kern County. Borax also occurs with ulexite and other saline minerals in Rhodes Marsh, Teels Marsh, and other playa deposits in Esmeralda County, Nevada, and has been reported from Alkali Flat, west of White Sands, Dona Ana County, New Mexico.

Artif. Easily obtained in large crystals by evaporation of the water solution. The pentahydrate, tincalconite, also known as octahedral borax, is formed by crystallization from solution at temperatures over 60° C.

Priceite

The conditions of formation in the presence of other salts and its stability range have been studied in detail.[11]

Alter. Effloresces in dry air and yields soft, white pseudomorphs.

Name. Named borax from the Arabic *bauraq*, *būraq*, white, which also included the niter and natron of the ancients. The name tincal, from Malay *tinkal*, Persian *tinkār*, is given in the Orient to borax in its crude, native state.

Ref.

1. Angles of Naumann (in Mohs, **2**, 54, 1839) on natural crystals (?); unit of Miller (Brooke and Miller, 604, 1852). Transformation: Naumann, also Dana (886, 1892), to Miller, 100/010/002.
2. Goldschmidt (**1**, 214, 1913). Rare: $h\{750\}$, $u\{\bar{1}01\}$.
3. Minder, *Zs. Kr.*, **92**, 301 (1935), from rotation photographs of artificial crystals.
4. Rammelsberg, *Handb. der kryst. Chemie*, 171 (1855). See also Hahn, *Archiv der Pharmacie*, **99** (2), 146 (1859).
5. Range of best reported values; see also Mellor (**5**, 71, 1924).
6. Average of concordant measurements by the prism method on artificial crystals by Dufet, *Bull. soc. min.*, **10**, 218 (1887); Kohlrausch, *Ann. Phys.*, **4**, 30 (1878); and Tschermak, *Ak. Wien, Sitzber.*, **57**, II, 641 (1868).
7. Tschermak (1868).
8. Hintze (**1** [4A], 155, 1921).
9. Cf. Becke, *Ak. Wien, Math.-nat. Kl., Anz.*, 2 (1921).
10. An interesting early account of this deposit and of other deposits in California and Nevada is given by Hanks, *Calif. State Mining Bureau, Third Ann. Rept.*, Pt. II (1883).
11. Cf. Teeple, *The Industrial Development of Searles Lake Brines, N. Y.* (1929); van't Hoff and Blasdale, *Ak. Berlin, Sitzber.*, 1086 (1905); dehydration data in McIntosh and Matthews, *Am. Min.*, **33**, 747 (1948).

25.1.5 **PRICEITE** [$Ca_4B_{10}O_{19} \cdot 7H_2O$?]. Cryptomorphite *Chase* (*Am. J. Sc.*, **5**, 287, 1873). Priceite *Silliman* (*Am. J. Sc.*, **6**, 128, 1873). Pandermite *vom Rath* (*Ber. Niederhein. Ges.*, 193, 1877).

Triclinic ?[1] As nodules or irregular masses, soft and chalky to hard, compact and tough. Under the microscope[1] composed of shreds and small platy grains with occasional rhombic outline having an angle of 58°.

Phys. Fracture earthy to conchoidal in the hard compact variety. H. 3–3½, less in earthy material. G. 2.42.[2] Luster earthy. Color white.

Opt. In transmitted light, colorless.

	n (Oregon [1])	n (Panderma [1])	n (Furnace Creek [3])
X	1.572 ± 0.003	1.573 ± 0.003	1.571
Y	1.591 ± 0.003	1.591 ± 0.003	1.590
Z	1.594 ± 0.003	1.593 ± 0.003	1.593
2V	42°56′ (calc.)	32° ± 2° (meas.)	42°56′ (calc.)

Biaxial negative (−). $r < v$, rather strong. In rhomboid crystals lying on the flat face, Y' makes an angle of about 14° with the bisectrix of the acute angle of the rhombs and X makes a considerable angle with the normal to the plates. Turned on edge the plates show an extinction angle, Z' to the elongation, of 25° ± 2°.

Chem. A hydrated calcium borate of uncertain formula, probably $Ca_4B_{10}O_{19} \cdot 7H_2O$.

Anal.

	1	2	3	4	5	6	7
CaO	32.11	29.96	29.80	32.38	31.37	31.73	32.0
B_2O_3	49.84	[47.04]	[45.20]	[48.50]	[49.34]	[50.01]	50.1
H_2O	18.05	22.75	25.00	18.29	18.29	18.29	17.9
Rem.		0.25		0.93	1.00	0.97	
Total	100.00	[100.00]	[100.00]	[100.10]	[100.00]	[101.00]	100.00
G.					2.262–2.298		2.48

1. $Ca_4B_{10}O_{19} \cdot 7H_2O$. 2, 3. Curry County, Oregon.[4] Rem. in anal. 2 is $(Na,K)_2O$. The analyses were made on average samples from commercial shipments. 4, 5, 6. Curry County, Oregon.[5] The rem. is $NaCl + (Al,Fe)_2O_3$. Sum of anals. 4 and 6 as given in original. 7. "Persia." [6]

	8	9	10	11	12	13	14
CaO	29.33	32.15	32.16	32.30	32.42	31.7	32.20
B_2O_3	[54.59]	48.44	48.63	49.92	49.92	49.8	49.03
H_2O	15.45	19.42	19.40	18.10	18.20	18.4	17.86
Rem.	0.63						1.16
Total	[100.00]	100.01	100.19	100.32	100.54	99.9	100.25
G.							2.43

8. Panderma (*pandermite*).[7] Rem. is MgO 0.15, Fe_2O_3 0.30, K_2O 0.18. 9. Curry County, Oregon.[8] 10. Panderma (*pandermite*).[8] 11, 12. Panderma (*pandermite*).[9] 13. Panderma (*pandermite*).[10] 14. Furnace Creek, Inyo County, California.[3] Rem. is SiO_2 0.58, Al_2O_3 0.20, $H_2O(-)$ 0.38.

Tests. Fuses at red heat. Insoluble in water but easily soluble in acids.

O c c u r. Originally found on the seacoast five miles north of Chetco, Curry County, Oregon, both compact and as roundish masses up to several hundred pounds in weight with aragonite in what appears to be a hot-spring deposit. Also found in the United States with colemanite and gypsum as nodules in shale in Furnace Creek wash, Death Valley, Inyo County, California. Occurs (*pandermite*) at Sultan Tschair, in Brussa province, 70 kilometers south of the port of Panderma on the Sea of Marmora, Anatolia. The locality is sometimes described as being on the Chinar San, a small stream tributary to the Rhyndarcus (Susighirlig) River. The mineral is found as nodules and masses up to a ton in weight underlying beds of gypsum and clay and probably was formed in solfataric lagoons.[11] Some of the so-called pandermite from this locality apparently is a different mineral (see below). Also reported from the "boundary of Persia," perhaps the same locality given above.

A l t e r. To colemanite and calcite; the latter change can be imitated experimentally by the action of sodium carbonate or bicarbonate solutions [11] on priceite.

A r t i f.[12] Products apparently but not certainly identical with priceite have been obtained by the action of alkaline borax solutions on gypsum and by treating solid $NaCaB_5O_9 \cdot 8H_2O$ (*ulexite*) or $Ca_2B_6O_{11} \cdot 7H_2O$ with a boiling saturated solution of NaCl and KCl. The compound $Ca_4B_{10}O_{19} \cdot 3\frac{1}{2}H_2O$ possibly exists.[15]

PROBERTITE

N a m e. After Thomas Price, metallurgist, of San Francisco, who first analyzed the mineral. Chase called the mineral cryptomorphite in the mistaken belief that it was identical with the cryptomorphite of How from Nova Scotia (= *ulexite* ?). Pandermite from the locality. The identity of pandermite with priceite has been established by chemical and optical study.[13]

PANDERMITE. *Linck* (*Cbl. Min.*, 193, 1923). Monoclinic,[14] with $a:b = 0.555:1$, $\beta\ 110°$. Cleavages: {001} perfect, {110} good, {010} imperfect. G. 2.433. Optically negative $(-)$, with nX 1.582, nY 1.592, nZ 1.606, $2V$ 35°. The five reported analyses apparently were on impure material. From Sultan Tschair, Anatolia. The properties are unlike those established for pandermite (= *priceite*) by several independent workers but are close to those of colemanite.

Ref.

1. Larsen, *Am. Min.*, **2**, 1 (1917), on the basis of an optical study.
2. Average of the following new determinations on the microbalance by Mrose, priv. comm. (1946): 2.44, Furnace Creek; 2.42, Panderma; 2.40, Oregon; and the value 2.43, Furnace Creek, by Foshag, *Am. Min.*, **9**, 11 (1924).
3. Foshag (1924).
4. Price anal. in Chase (1873).
5. Silliman (1873).
6. Pisani (1875, p. 215).
7. Muck anal. in vom Rath (1877).
8. Whitfield, *Am. J. Sc.*, **34**, 283 (1887).
9. Schultze anal. in Kraut, *Zs. anal. Chem.*, **36**, 165 (1897).
10. van't Hoff, *Ak. Berlin, Sitzber.*, 572 (1906).
11. Cf. Linck, *Cbl. Min.*, 193 (1923).
12. Cf. Gutbier, Hüttig, and Linck, *Zs. Elektrochem.*, **32**, 79 (1926), Linck (1923), and van't Hoff (1906).
13. Larsen (1917) and Whitfield (1887). The x-ray powder patterns of material from the three known localities are identical (Frondel, priv. comm., 1946).
14. Symmetry, elements, and cleavage from a study of grains under the microscope.
15. Cf. the dehydration data of Gutbier, Hüttig, and Linck (1926).

25.1.6   **P R O B E R T I T E** [$NaCaB_5O_9 \cdot 5H_2O$]. *Eakle* (*Am. Min.*, **14**, 427, 1929). Kramerite *Schaller* (*U. S. Geol. Sur., Prof. Paper 158*, 137, 1929).

C r y s t. Monoclinic;[1] prismatic—$2/m$.

$a:b:c = 1.1051:1:0.5237$; $\beta\ 107°44'$;[2] $p_0:q_0:r_0 = 0.4740:0.4989:1$

$r_2:p_2:q_2 = 2.0046:0.9501:1$; $\mu\ 72°16'$; $p_0'\ 0.4976,\ q_0'\ 0.5237,\ x_0'\ 0.3197$

Forms:[3]

	ϕ	ρ	ϕ_2	$\rho_2 = B$	C	A
b 010	0°00′	90°00′	.....	0°00′	90°00′	90°00′
a 100	90 00	90 00	0°00′	90 00	72 16	.....
m 110	43 32	90 00	0 00	43 32	77 53½	12 06½
e 011	31 24	31 32	72 16	63 29½	26 30½	74 11
t 101	90 00	39 15½	50 44½	90 00	21 31½	50 44½
d $\bar{1}$01	−90 00	10 05	100 05	90 00	27 49	100 05
p 111	57 21	44 09	50 44½	67 55½	30 27	54 05½
o $\bar{1}$11	−18 46	28 57	100 05	62 43½	38 11½	98 57½

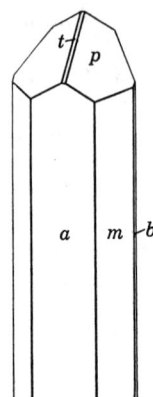

Habit. Commonly in rosettes or radial groups of needles or laths up to 30 mm. in length; spherulitic; as tough, compactly reticulated aggregates. Rarely as single crystals. Acicular [001], usually slightly sometimes considerably flattened {100}; rarely flattened {110}. Faces $b\ a\ m$ generally striated [001] and rounded. Commonest forms: $a\ m\ b\ e$.

Structure cell.[4] Space group $P2_1/n$. a_0 13.88 Å, b_0 12.56, c_0 6.609; β 107°40′; $a_0:b_0:c_0 = 1.1053:1:0.5263$. Cell contents $Na_2Ca_2B_{10}O_{18} \cdot 10H_2O$.

Twinning. Doubtfully observed in crushed fragments under the microscope.[3]

Phys. Cleavage {110}, perfect. Brittle. H. $3\frac{1}{2}$.[5] G. 2.14; 2.13 (calc.). Colorless and transparent. Luster vitreous.

Opt. In transmitted light, colorless.

ORIENTATION	n (Kramer, Cal.[3])	n (Kramer, Cal.[6])	n (Ryan, Cal.[5])	
X	1.514	1.515	1.517	Biaxial positive (+).
$Y\ b$	1.524	1.525	1.525	$2V$ 73° (meas.),[3] $2E$ 126°.
$Z \wedge c$ 12°–13°	1.543	1.544	1.544	$r > v$.

Chem. A hydrated borate of sodium and calcium, $NaCaB_5O_9 \cdot 5H_2O$. **Anal.**[7]

	1	2	3	4
Na_2O	8.83	8.53	8.12	9.00
CaO	15.98	15.45	15.42	15.88
B_2O_3	49.56	50.44	[50.73]	49.10
H_2O	25.63	25.58	25.73	25.64
Rem.				0.64
Total	100.00	100.00	[100.00]	100.26
G.		2.141		2.135

1. $NaCaB_5O_9 \cdot 5H_2O$. 2, 3. Kramer district, California.[3] Recalculated to 100 after deduction of 2.73 insol. (anal. 2) and 0.84 insol. (anal. 3). Anal. 2 average of 3 analyses; anal. 3 on material washed with water to remove any traces of borax. 4. Ryan district, California.[5] Rem. is MgO 0.06, (Fe,Al)$_2$O$_3$ 0.38, insol. 0.20.

Tests. In C.T., decrepitates, loses water, turns opaque white, and, after swelling slightly, fuses imperfectly to a clinker-like mass. B.B. fuses readily to a clear bubbly glass. Easily soluble in dilute acids but only partly decomposed in cold or hot water, Na being leached relative to Ca.

Occur. Found originally in the kernite deposit in the Kramer borate district in southeastern Kern County, California. Later described from the Ryan district, Inyo County, associated with colemanite and ulexite, and from Lang, Los Angeles County, where it is altered in part to ulexite.

Artif.[8] Obtained by heating a mixture of two parts of ulexite and one of borax to slightly over 60°; also by heating a similar mixture in a saturated solution of NaCl at about 90°.

ULEXITE   345

Name. After Frank H. Probert, Dean of the Mining College, University of California.

Ref.
1. Eakle (1929).
2. Schaller (1930), calculated from the average measured angles: ϕ_{110} 43°32′, ϕ_{011} 31°24′, ρ_{011} 31°32′.
3. Schaller (1930).
4. Barnes, *Am. Min.*, **34**, 19 (1949); **34**, 611 (1949).
5. Foshag, *Am. Min.*, **16**, 338 (1931).
6. Ross in Schaller (1930).
7. An earlier reported analysis, in Eakle (1929), apparently is considerably in error.
8. van't Hoff, *Preuss. Ak. Wiss., Sitzber.*, 303 (1907), and Schaller (1930).

25.1.7   ULEXITE [NaCaB$_5$O$_9 \cdot$8H$_2$O]. Boronatro-calcite *Ulex* (*Ann. Chem. Pharm.*, **70**, 49, 1849). Natronkalk-borat. Ulexite *Dana* (695, 1850). Natronborocalcite. Tinkalzit *Kletzinsky* (*Polyt. Centr.*, 1384, 1859). Tiza *South America*. Bornatrocalcit.
Cryptomorphite *How* (*Am. J. Sc.*, **32**, 9, 1861; *Min. Mag.*, **1**, 257, 1877). Franklandite *Reynolds* (*Phil. Mag.*, **3**, 284, 1877). Hydrous borate of lime *Hayes* (*Am. J. Sc.*, **46**, 377, 1844; **47**, 215, 1844). Borocalcite. Hydroborocalcite *Hausmann* (1429, 1847). Hayesine *Dana* (217, 1850). Hayesite. Hayesinite.

Cryst.[1] Triclinic; probably pinacoidal—$\bar{1}$.

$a:b:c = 0.6855:1:0.5191; \quad \alpha\ 90°16′, \beta\ 109°08′, \gamma\ 105°07′$

$p_0:q_0:r_0 = 0.7852:0.5080:1; \quad \lambda\ 84°20\tfrac{1}{2}′, \mu\ 70°05\tfrac{1}{2}′, \nu\ 73°53\tfrac{1}{2}′$

$p_0'\ 0.8351,\ q_0'\ 0.5403;\ x_0'\ 0.3466,\ y_0'\ 0.1048$

Forms:[2]

		φ	ρ	A	B	C
c	001	73°10′	19°54½′	70°05½′	84°20½′	.....
b	010	0 00	90 00	73 53½	.....	84°20½′
a	100	73 53½	90 00	.....	73 53½	70 05½
m	110	45 05½	90 00	27 48	46 05½	72 21
M	1$\bar{1}$0	111 02	90 00	37 08½	111 02	74 23
s	0$\bar{1}$1	141 29½	29 06	79 19	112 22½	28 09
o	$\bar{1}$01	−105 35	25 19	115 15	96 33	45 13½
t	$\bar{1}$11	−47 47½	31 36	105 58½	69 23	45 09½
p	$\bar{1}$11	−145 41	38 57	118 59	121 17	55 38
d	$\bar{1}$21	−25 32½	46 35	96 50½	49 03	52 29½

Structure cell.[3] a_0 8.71 kX, b_0 12.72, c_0 6.69; $a_0:b_0:c_0 = 0.686:1:0.526$. Cell contents Na$_2Ca_2B_{10}O_{18} \cdot$16H$_2$O.

Habit. In small nodular, rounded, or lens-like masses, sometimes loose in texture ("cotton-balls"), and consisting of capillary or acicular crystals; the fibers may be radially arranged, or randomly oriented toward the centers of the masses and arranged in parallel position at the periphery. Also as botryoidal crusts of randomly oriented fibers, or as compact veins with a parallel fibrous structure. Rarely as distinct crystals, greatly elongated [001].

Twinning.[4] Polysynthetic, common in massive aggregates and crushed fragments and perhaps in part due to mechanical deformation. A number of twin laws have been reported: {010} and {100} are fairly well-established; {3$\bar{4}$0} or {2$\bar{3}$0} less certain; also additional twin planes bearing no very simple relation to the axes but approaching such planes as {011}, {$\bar{2}$31}, and {0$\bar{7}$1}.

P h y s. Cleavage {010} perfect, {1$\bar{1}$0} good, {110} poor.[4] The needles have a ready cross fracture but too uneven to be considered a cleavage. Brittle. H. $2\frac{1}{2}$.[5] G. 1.955;[6] 2.00 (calc.). Color of aggregates pure white; individual crystals are colorless. Luster of aggregates silky or satiny, of crystals vitreous. Tasteless.

O p t.

	Orientation [7]		n (Kramer) [7]	n (Kramer) [8]	n (Inder) [14]	
	ϕ	ρ				
X	$11\frac{1}{2}°$	$81°$	1.491	1.493	1.496	Biaxial positive (+).
Y	$-100°$	$21\frac{1}{2}°$	1.504	1.506	1.505	Elongation generally
Z	$107°$	$70°$	1.520	1.519	1.519	positive, some-
$Y \wedge c$			$21\frac{1}{2}°$	Variable, about 20°	$2°$–$4°$	times negative.
$2V$			$73° \pm 1°$	Large	$78°$	

C h e m. A hydrated borate of sodium and calcium, $NaCaB_5O_9 \cdot 8H_2O$. The reported water content sometimes is less than required by this formula, due to loss by dehydration and grinding.[9]

Anal.[17]

	1	2	3	4	5
Na_2O	7.65	7.78	[7.05]	7.09	6.08
CaO	13.85	13.92	14.14	14.06	14.12
B_2O_3	42.95	43.07	43.13	42.94	41.99
H_2O	35.55	35.34	35.68	35.54	36.95
Rem.				0.10	0.57
Total	100.00	100.11	[100.00]	99.73	99.71
G.	2.00	1.955		1.963	1.91

1. $NaCaB_5O_9 \cdot 8H_2O$. 2. Suckow shaft, Kramer district, California.[10] 3. Lang, California.[11] 4. Kramer district, California.[12] Rem. is insol. 5. Inder, Kazakhstan, U.S.S.R.[13] Rem. is 0.07 $(Al,Fe)_2O_3$, 0.14 K_2O, 0.25 insol., 0.11 MgO.

Tests. B.B. fuses at 1 with intumescence to a clear blebby glass. In the C.T. yields water. Slightly decomposed in cold water and more so in hot water, with loss of soda to the solution.

O c c u r. Ulexite occurs typically in arid regions, as in the Mohave desert of southeastern California and western Nevada, where it is found in salt playas and desiccated saline lakes. The boron appears to have been derived by the leaching of sediments and pyroclastic rocks by meteoric waters. The principal associated minerals include borax, halite, glauberite, trona, mirabilite, soda-niter; also colemanite and other calcium borates, especially in bedded deposits and apparently largely as a reworking product of ulexite. Among the typical playa or salt marsh de-

ULEXITE

posits, so-called because they may be covered by a few inches of water during the rainy season, may be mentioned Columbus Marsh, Rhodes Marsh, and Teel's Marsh in Esmeralda County, Nevada. In Death Valley and Saline Valley in Inyo County and the Kramer district in San Bernardino County, California, ulexite occurs both in playas and with colemanite and other borates in bedded Tertiary sediments. Ulexite is widely distributed in the Chilean nitrate region, notably in the playas of Iquique, Tarapacá province, where it is associated with soda-niter, halite, glauberite, pickeringite, etc. Also found in playa deposits in Jujuy Province, northern Argentina, and in Santiago del Estero and adjoining provinces in west central Argentina. In Dept. Arequipa and elsewhere in Peru. Ulexite occurs in gypsum deposits in the Maritime Provinces, Canada, as at Windsor, Nova Scotia, and Hillsborough, New Brunswick, associated with anhydrite, glauberite, and borates. Found in borate deposits near Inder Lake, western Kazakhstan, U.S.S.R.

Alter. Occurs altered to gypsum and colemanite.[12]

Artif.[15] Obtained as a mass of capillary crystals by the action of a cold saturated solution of borax on calcium chloride. Also by seeding a solution of 110 g. $CaB_2O_4 \cdot 6H_2O$, 40 g. boric acid, 100 g. borax, 450 g. calcium chloride in $2\frac{1}{2}$ liters of water. Ulexite is stable up to about 60°, breaking down in water to sodium borate and priceite or colemanite, depending on the other salts present and the character of the inoculation.[16] A mixture of 2 parts ulexite and 1 part borax heated in a saturated solution of NaCl at about 90° deposits kramerite.[12]

Name. After the German chemist George Ludwig Ulex (1811–1883), who first gave a correct analysis of the species.

Several ill-defined borates have been described and later presumed or proved to be identical with ulexite. Cryptomorphite of How (1861) from Windsor, Nova Scotia, very probably is a mixture of ulexite with gypsum or anhydrite. Franklandite of Reynolds (1877) from Tarapacá, Chile, has been shown [16] to be only impure ulexite. Hayesine of Hayes (1844; Dana, 217, 1850) from Iquique, Chile, supposedly $CaB_4O_7 \cdot 6H_2O$, also is ulexite [18] as is some of the material later ascribed by others [19] to hayesine or to the supposed species bechilite (*borocalcite*).

Ref.

1. Murdoch, *Am. Min.*, **25**, 754 (1940), on Kramer crystals.
2. Murdoch (1940). Less common and rare:

l 270	r 530	R $3\bar{2}0$	z $2\bar{3}2$
h 350	L $7\bar{2}0$	K $2\bar{3}0$	x $\bar{5}32$
k 230	N $3\bar{1}0$	H $1\bar{2}0$	q $\bar{2}51$

3. Murdoch (1940) from an oscillation photograph around [001], using the morphological axial angles.
4. Cf. Murdoch (1940).
5. Murdoch (1940); Foshag, *Am. Min.*, **3**, 35 (1918), gives $3\frac{1}{2}$ on compact material from Lang, California. The value 1 is usually given and refers to the cohesion of loose aggregates.
6. From values 1.955 ± 0.001 by microbalance, in Murdoch (1940), and 1.9555 on artificial material by de Schulten, *C.R.*, **132**, 1576 (1901). Schaller (1930) gives 1.963.

7. Murdoch (1940). The indices are close to those reported by Larsen (148, 1921) for material from various localities. The extinction angle $Y \wedge c$ is usually reported as between 20°–24°, but it is given as 0° or nearly so by Buttgenbach, *Soc. géol. Belgique, Ann.*, **28**, Mem. 99 (1900–01), Des Cloizeaux (**2**, 10, 1874), and Godlevsky (ref. 13), presumably due to use of a reference direction other than c.
8. Schaller (1930).
9. Cf. Walker, *Univ. Toronto Stud., Geol. Ser.*, no. 12, 54 (1921), who gives a temperature dehydration curve.
10. Gonyer anal. in Murdoch (1940).
11. Foshag (1918).
12. Schaller (1930).
13. Godlevsky, *Mem. soc. russe min.*, **66** [2], 315 (1937)—*Min. Abs.*, **7**, 122 (1938).
14. Godlevsky (1937); Boky, *Bull. ac. sc., U.R.S.S., Cl. sc.*, 871 (1937), gives very similar values.
15. Cf. Mellor (**5**, 94, 1924).
16. van't Hoff, *Preuss. Ak. Wiss., Sitzber.*, 303 (1907).
17. Numerous earlier analyses are cited in Hintze (**1**, [4A], 156, 1933).
18. Cf. Dana (599, 1868) and Larsen (148, 1921).
19. Cf. Hintze (**1** [4A], 165, 1933).

25.1.8 **V E A T C H I T E** [$Sr_3B_{16}O_{27} \cdot 5H_2O$ (?)]. Switzer (*Am. Min.*, **23**, 409, 1938).

Cryst. Monoclinic.

$a:b:c = 0.163:1:0.998;$ $\beta\ 121°02';$ [1] $p_0:q_0:r_0 = 6.1227:0.8552:1$

$r_2:p_2:q_2 = 1.1694:7.1598:1;$ $\mu\ 58°58';$ $p_0'\ 7.1454,\ q_0'\ 0.9980,\ x_0'\ 0.6017$

Forms: [2]

	ϕ	ρ	ϕ_2	$\rho_2 = B$	C	A
b 010	0°00′	90°00′	.....	0°00′	90°00′	58°58′
s 180	41 49½	90 00	0°00′	41 49½	69 53½	48 10½
t 160	50 02	90 00	0 00	50 02	66 43½	39 58
k 140	60 48½	90 00	0 00	60 48½	63 15	29 11½
n 120	74 23½	90 00	0 00	74 23½	60 14	15 36½
l 310	87 20	90 00	0 00	87 20	59 00½	2 40
f 104	67 28½	33 05	58 58	77 56	12 04	59 43½
q 013	61 04	34 30½	58 58	74 05½	15 54½	60 16½
h 023	42 07½	41 53½	58 58	60 19	29 41	63 23½
d 011	31 05	49 22	58 58	49 28	40 32	66 56

Structure cell. [3] $a_0\ 6.72\ kX,\ b_0\ 41.26,\ c_0\ 41.20;\ [\beta\ 121°02'];\ a_0:b_0:c_0 = 0.163:1:0.998$. Cell contents uncertain, perhaps $Sr_{48}B_{256}O_{432} \cdot 80H_2O$.

Habit. Platy, flattened on {010} and elongated with [001], terraced along [001] and [100]; slender prismatic to fibrous. As groups of divergent plates and cross-fiber veins.

Phys. Cleavage {010} perfect and easy; {001} imperfect. H. 2. G. 2.69,[4] 2.58[5] (meas.); 2.59 (calc.). Luster vitreous, pearly on cleavage, silky in fibrous masses. Colorless. Transparent to white.

Opt.

		n(Na)	
$X \wedge c$	+52°	1.551	Biaxial positive (+).
Y	b	1.553	$2V$ 37°.
$Z \wedge c$	−38°	1.621	$r > v$, perceptible.

Chem. A hydrated strontium borate of uncertain formula, perhaps $Sr_3B_{16}O_{27} \cdot 5H_2O$. A small amount of Ca substitutes for Sr.

Anal.

	1	2	3
CaO		1.7	1.3
SrO	32.45	29.5	29.8
B_2O_3	58.16	57.3	57.7
H_2O	9.39	n.d.	9.6
Insol.		1.0	
Total	100.00	89.5	98.4

1. $Sr_3B_{16}O_{27} \cdot 5H_2O$. 2, 3. Lang.[6]

Tests. B.B. fuses easily with intumescence to an opaque white bead. In C.T. yields a moderate amount of water, with an acid reaction.

Occur. Found as cross-fiber veins cutting limestone and howlite and as crystals deposited upon colemanite in a borate deposit at Lang, Los Angeles County, California.

Name. After John A. Veatch, who was the first to detect the presence of borates in the mineral waters of California.[7]

Ref.

1. Murdoch, *Am. Min.*, **24**, 130 (1939), on platy crystals from Lang. Murdoch accepted the ratios of the cell sides determined by Switzer, *Am. Min.*, **23**, 409 (1938), but replaced Switzer's approximate β by the goniometric angle 121°02′. There is no evidence for the crystal class.
2. Murdoch (1939). Rare and uncertain:

| 170 | 130 | 560 | 320 | 017 | 027 | 056 | $\bar{1}66$ |
| 150 | 230 | 110 | 0.1.30 | 016 | 035 | 054 | $\bar{1}.6.12$ |

3. Switzer (1938), by rotation and Weissenberg photographs. The space group was not determined owing to uncertainties of indexing.
4. Switzer (1938), by suspension in bromoform.
5. Murdoch (1939), by suspension in Clerici solution.
6. Brannock analysis in Switzer, *Am. Min.*, **35**, 90 (1950); it is shown here that the original interpretation as a calcium borate, $Ca_2B_6O_{11} \cdot 2H_2O$, was erroneous.
7. Cf. letter from Veatch to H. G. Hanks, *California State Mining Bur.*, 3rd Ann. Rpt. (1883).

25.1.9 COLEMANITE $[Ca_2B_6O_{11} \cdot 5H_2O]$. Colemanite *Neuschwander*, in *Hanks* (3rd Rep. Min. California, 86, 1883). Neocolemanite *Eakle* (*Univ. California, Dept. Geol., Bull.*, **6**, 179, 1911). Borspar.

Cryst. Monoclinic; prismatic—$2/m$.[1]

$a:b:c = 0.7769:1:0.5430;$ $\quad \beta\, 110°07';$[2] $\quad p_0:q_0:r_0 = 0.6989:0.5099:1$

$r_2:p_2:q_2 = 1.9612:1.3707:1;$ $\quad \mu\, 69°53';$ $\quad p_0'\, 0.7443,\, q_0'\, 0.5430,\, x_0'\, 0.3663$

350 BORATES

Forms:[3]

		φ	ρ	φ₂	ρ₂ = B	C	A
c	001	90°00′	20°07′	69°53′	90°00′	.....	69°53′
b	010	0 00	90 00		0 00	90°00′	90 00
a	100	90 00	90 00	0 00	90 00	69 53	.....
z	120	34 25½	90 00	0 00	34 25½	78 47½	55 34½
m	110	53 53½	90 00	0 00	53 53½	73 52	36 06½
t	210	69 57½	90 00	0 00	69 57½	71 09	20 02½
κ	011	34 00	33 13½	69 53	62 59	27 01	72 09½
α	021	18 38½	48 53½	69 53	44 26½	45 33½	76 04
h	$\bar{2}$01	−90 00	48 18	138 18	90 00	68 25	138 18
W	$\bar{3}$01	−90 00	61 49	151 49	90 00	81 56	151 49
β	111	63 56½	51 02	42 00	70 02	33 49½	45 42
y	$\bar{1}$11	−34 50½	33 29½	110 42½	63 04½	47 34	108 22½
v	$\bar{2}$21	−45 56½	57 22	138 18	54 09	72 39	127 14½
d	$\bar{1}$21	−19 11½	48 59½	110 42½	44 33	57 56	104 21½
o	$\bar{2}$11	−64 11	51 16	138 18	70 08½	69 45½	134 36½

Less common:

H 130	ρ 301	ψ $\bar{4}$0!	q $\bar{3}$31	p 142	n 141	Q $\bar{2}$41	γ $\bar{3}$21
l 310	i $\bar{1}$01	U $\bar{6}$01	P $\bar{1}$23	η 232	r $\bar{2}$32	φ 522	s $\bar{3}$41
V 101	E $\bar{3}$02	f $\bar{8}$01	μ 165	e 121	x $\bar{1}$31	k 311	B $\bar{4}$11
λ 201	g $\bar{5}$02	σ 331	u 164	ω 131	ε $\bar{2}$31	Θ $\bar{3}$11	Z $\bar{6}$21

Structure cell.[4] Space group probably $P2/m$. a_0 8.72 ± 0.02 kX, b_0 11.29 ± 0.02, c_0 6.06 ± 0.04; [β 110°07′, morph.]; $a_0:b_0:c_0$ = 0.772:1:0.537. Cell contents $Ca_4B_{12}O_{22} \cdot 10H_2O$.

Habit. Equant; commonly short prismatic [001] with large (110), large to small (001) and complex terminations; pseudo-rhombohedral with large

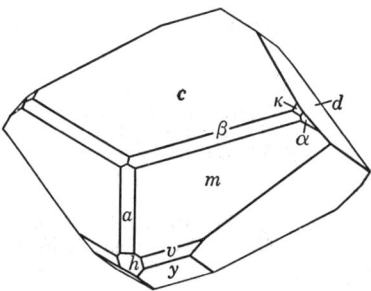

Death Valley.

(110) and ($\bar{3}$01); pseudo-octahedral with large ($\bar{2}$21) and (011). Also massive, cleavable to granular and compact; as crudely spherulitic aggregates.

Phys. Cleavage {010} perfect, {001} distinct. Fracture uneven to subconchoidal. H. 4½. G. 2.423 ± 0.005;[5] 2.422 (calc.). Luster vitreous to adamantine, brilliant. Colorless, also milky white, yellowish white, gray or muddy. Transparent to translucent.

COLEMANITE 351

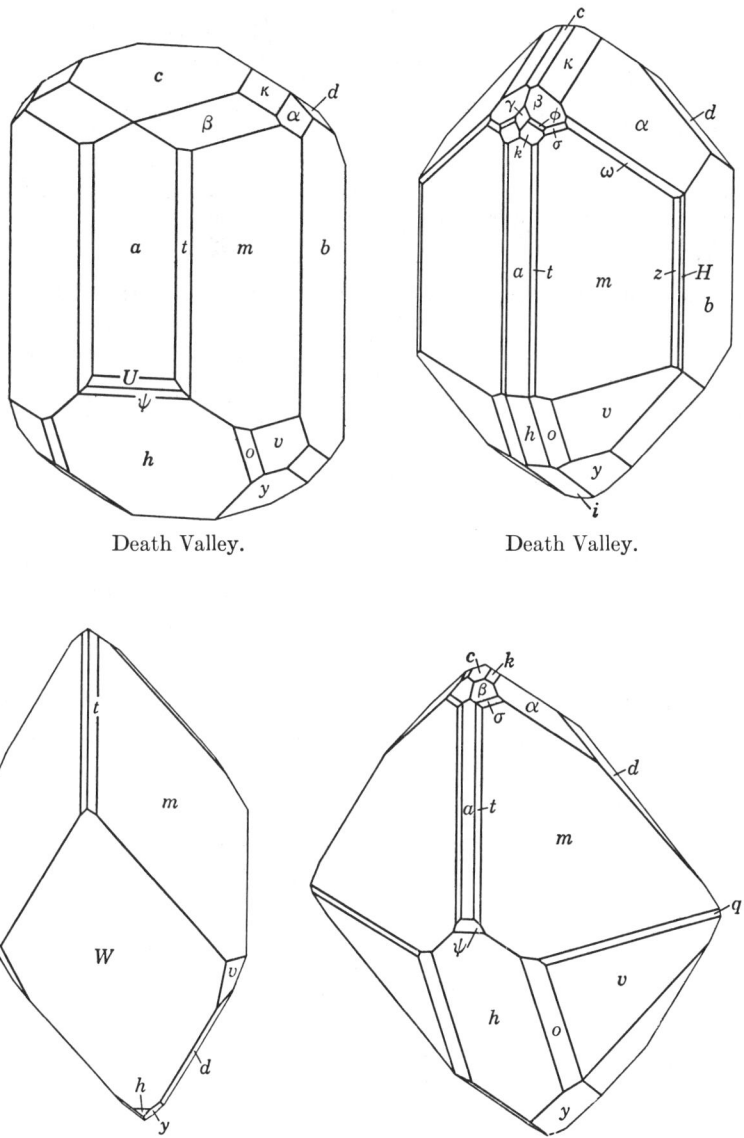

Death Valley.

Death Valley.

Death Valley.

Death Valley.

O p t.[6] In transmitted light, colorless.

ORIENTATION		n(Na)	
X	b	1.5863	Biaxial positive (+).
$Y \wedge c$	$-6°$	1.5920	$2V\ 55°,\ 2E\ 95°.$
$Z \wedge c$	$+84°$	1.6140	$r > v$, very slight.

C h e m. A hydrated calcium borate, $Ca_2B_6O_{11} \cdot 5H_2O$.[7]

Anal.[8]

	1	2	3	4	5	6	7
CaO	27.28	27.56	27.76	27.31	27.38	27.21	27.37
B_2O_3	50.81	50.96	49.45	50.70	49.59	50.93	49.92
H_2O	21.91	21.70	22.48	21.87	22.68	21.85	22.73
Rem.				0.10	0.72		0.43
Total	100.00	100.22	99.69	99.98	100.37	99.99	100.45
G.	2.422		2.423				2.44

1. $Ca_2B_6O_{11} \cdot 5H_2O$. 2. California. CaO average of four determinations.[9] 3. Lang, California (*neocolemanite*).[10] 4. Death Valley, California.[11] Rem. is MgO. 5. Death Valley, California.[11] Average of two; rem. is MgO 0.26, SiO_2 0.46. 6. Gila River, Arizona.[12] 7. Inder, U.S.S.R.[13] Rem. is MgO 0.10, Al_2O_3 0.06, Fe_2O_3 0.01, $(Na,K)_2O$ 0.10, insol. 0.07, SO_3 0.09.

Tests. B.B. decrepitates, exfoliates, sinters and fuses imperfectly. Soluble in hot HCl, with separation of boric acid on cooling. Solubility in water about 1 part in 1100 at 20°–25.[14]

O c c u r. First discovered in Death Valley, Inyo County, California, where large deposits occur along Furnace Creek in the Amargosa Range and near Ryan. Also found in California in the Calico district, near Yerma, San Bernardino County; at Lang, Los Angeles County (*neocolemanite*); with kernite in the Kramer district, Kern County; and on Frazier Mountain in Ventura County. Found at White Basin and Callville Wash in the Muddy Mountains, Clark County, Nevada. In these deposits the colemanite occurs principally as geodal masses in bedded Tertiary sediments and appears to have been formed by the action of meteoric waters on ulexite and borax formed originally in playa deposits and later buried by sedimentation.[15] The colemanite is associated with howlite, ulexite, gypsum, calcite, and celestite. Found with bitumen in a fossil egg from the Gila River, Arizona.[12] Colemanite has been reported from the borate zone of the playa Salinas Grandes, Jujuy Province, Argentina, but ordinarily it does not occur in such deposits. Also found in gypsum rock and clay in the Inder region, Kazakhstan, U.S.S.R., and formed principally by the replacement of hydroboracite.

Alter. Observed formed from ulexite and hydroboracite and found as pseudomorphs after inyoite.[18] Also observed altered to calcite.

Artif.[16] Obtained by heating to 70° a mixture of 140 cc. H_2O with 50 g. NaCl, 4 g. ulexite, and 0.4 g. H_3BO_3.

Name. After William T. Coleman, a founder of the California borax industry and owner of the mine where the mineral was first found.

NEOCOLEMANITE. *Eakle* (*Univ. California, Dept. Geol., Bull.*, **6,** 179, 1911). A supposed new species, later shown [17] to be colemanite.

Ref.
1. The crystal morphology is typically holohedral. Etch figures, studied in detail by Baumhauer, *Zs. Kr.*, **30**, 97 (1899), and McNairn, *Trans. Roy. Can. Inst.*, **11**, 231 (1917), frequently suggest, but do not prove, lower symmetry.
2. Eakle, *Univ. Calif., Dept. Geol., Bull.*, **3**, 31 (1902), on crystals from the Calico district, California.
3. Eakle (1902); Boky, *Bull. ac. sc. U.R.S.S., Cl. sci. mat. nat., Ser. chim.*, 871 (1937) —*Min. Abs.*, **7**, 123 (1940). Doubtful forms: 370, 10.19.0, 19.19.6, 771, 711, 731, 10.1.1, $\bar{1}82$, $\bar{4}12$, $\bar{7}21$. The supposed new forms of neocolemanite of Eakle (1911) and Goldschmidt (**2**, 176, 1913), namely 230, 221, $\bar{2}23$, $\bar{6}61$, 231, $\bar{2}63$, prove to be known forms of colemanite when transformed by the appropriate formula: $\bar{1}00/010/\tfrac{1}{2}01$.
4. Nikolsky, *C. r. ac. sc. U.R.S.S.*, **28**, 59 (1940)—*Min. Abs.*, **9**, 44 (1944); comparable dimensions are given by Dér, *Zs. Kr.*, **103**, 431 (1941).
5. Range of best values.
6. Indices by prism method, Mülheims, *Zs. Kr.*, **14**, 230 (1888). Comparable data are given by Larsen and Berman (1934), Godlevsky, *Mem. soc. russe min.*, **66** [2], 353 (1937), and others.
7. For infrared absorption data see Van Arkel and Fritzius, *Rec. trav. chim. Pay-Bas*, **50**, 1035 (1931).
8. For additional analyses see Hintze (**1** [4A], 180, 1922).
9. Ley anal. in Kraut, *Zs. anal. Chem.*, **36**, 165 (1897).
10. Eakle (1911).
11. Whitfield, *Am. J. Sc.*, **34**, 281 (1887).
12. Morgan and Tallmon, *Am. J. Sc.*, **18**, 363 (1904).
13. Boldyreva anal. in Godlevsky (1937).
14. Hicks in Gale, *U. S. Geol. Sur., Prof. Pap. 85A*, 7 (1913).
15. Cf. Foshag, *Econ. Geol.*, **16**, 199 (1921).
16. van't Hoff, *Ak. Berlin, Sitzber.*, 566, 689 (1906).
17. Hutchinson, *Min. Mag.*, **16**, 239 (1912).
18. Rogers, *Am. Min.*, **4**, 135 (1919).

25.1.10 **H Y D R O B O R A C I T E** [$CaMgB_6O_{11} \cdot 6H_2O$]. *Hess* (*Ann. Phys.*, **31**, 49, 1834).

C r y s t. Monoclinic.

$a:b:c = 1.7650:1:1.2330;$ $\quad \beta\, 102°39';$ [1] $\quad p_0:q_0:r_0 = 0.6986:1.2031:1$

$r_2:p_2:q_2 = 0.8312:0.5807:1;$ $\quad \mu\, 77°21';$ $\quad p_0'\, 0.7160,\, q_0'\, 1.2330,\, x_0'\, 0.2244$

Forms: [2]

	ϕ	ρ	ϕ_2	$\rho_2 = B$	C	A
b 010	0°00′	90°00′	.....	0°00′	90°00′	90°00′
a 100	90 00	90 00	0°00′	90 00	77 21	.....
m 110	30 08½	90 00	0 00	30 08½	83 41	59 51½
e 011	10 19	51 24½	77 21	39 44	50 16	81 57
p 111	37 20	57 11	46 45½	48 04	50 10½	59 21½

Less common:

c 001 l 130 h 120 k 310 n 410 j 810 v 012 d 102 r 112 s 343 t $\bar{2}11$

Habit. Elongated [001] and flattened {010}. Lamellar-fibrous, radiating or columnar, resembling fibrous and foliated gypsum. Compact and fine-grained.

P h y s. Cleavage {010} perfect; also {100}.[3] H. 2 (crystals) to 3 (compact masses). G. 2.167.[4] Luster vitreous, in fibrous masses silky. Clear colorless to white. Transparent.

O p t.[5]

ORIENTATION		n	
$X \wedge c$	33°	1.520–1.523	Biaxial positive (+).
Y	b	1.534–1.535	$2V$ 60°–66°.
$Z \wedge c$	57°	1.569–1.571	$r < v$, perceptible.

C h e m. A hydrated borate of calcium and magnesium, $CaMgB_6O_{11} \cdot 6H_2O$.

Anal.

	1	2	3	4	5	6
CaO	13.57	13.52	14.06	14.96	13.63	13.86
MgO	9.75	10.57	10.14	9.88	9.84	9.93
Fe_2O_3			0.12	0.30		
B_2O_3	50.53	[49.58]	47.71	46.79	49.04	49.22
CO_2			tr.	1.37	1.65	
H_2O	26.15	26.33	27.37	25.59	25.45	26.59
Rem.			0.23	0.58	0.86	0.28
Total	100.00	[100.00]	99.63	99.47	100.47	99.88
G.		1.9–2.08	2.168			2.14

1. $CaMgB_6O_{11} \cdot 6H_2O$. 2. Caucasus. Average of two analyses.[6] 3. Ryan, California. Rem. is SiO_2.[7] 4. Inder, Siberia. Rem. is 0.36 SiO_2, 0.22 Al_2O_3. Anal. on colorless needles.[8] 5. Inder, Siberia. Rem. is 0.33 SiO_2, 0.38 $(Al,Fe)_2O_3$, 0.15 alkalies. Anal. on compact fine-grained masses.[8] 6. Inder, Siberia. Rem. is 0.06 $(Al,Fe)_2O_3$, 0.12 $(K,Na)_2O$, 0.10 insol.[9]

Ryan, California.

Tests. B.B. fuses easily to a clear glass. In the C.T. affords water with an acid reaction. Readily soluble in acids. Nearly insoluble in cold water, but partly dissolved by prolonged boiling in water, the solution giving a faint alkaline reaction.

O c c u r. Originally described in a specimen from an unknown locality in the Caucasus. Reported[10] to occur with halite at Stassfurt, Germany. Later found with colemanite and calcite near Ryan, Inyo County, California, and with colemanite, inyoite, inderite, ulexite, and other borates in the gypsum capping of a salt dome and in the underlying salt beds near Inder salt lake, western Kazakhstan, U.S.S.R.

A r t i f. Not known artificially.[11]

N a m e. In allusion to the composition.

Ref.

1. Schaller, *Festsch. Goldschmidt*, Heidelberg, 256 (1928), on crystals from Ryan, Inyo County, California. On crystals from Inder, Kazakhstan, Boky, *Bull. ac. sc. U.R.S.S., Cl. sc. mat. nat., Ser. chim.*, 871 (1937)—*Min. Abs.*, **7**, 123 (1938), got $a:b:c = 1.762:1:1.241$, β 102°39'.
2. Schaller (1928); Boky (1937) noted $c\ b\ a\ m\ p$.
3. Noted by Larsen (88, 1921) and Boky (1937), but not found by Schaller (1928).
4. Schaller (1928) on crystals from Ryan. Erdmann, *Chemie u. Industrie der Kalisalze*, Berlin, 1907, p. 12, gives 2.168 for material from the Caucasus.
5. From concordant observations by Schaller (1928), Ross in Schaller (1928), for Ryan, California; and by Boldyreva, *Mater. Cent. Sc. Geol. Prosp. Inst.*, no. 1, 62 (1936)—*Min. Abs.*, **6**, 336 (1936); Boldyreva in Godlevsky, *Mem. soc. russe min.*, **66**, 315 (1937)—*Min. Abs.*, **7**, 122 (1938) for Inder, Kazakhstan. Larsen (1921) noted nX 1.517, nY 1.534, nZ 1.565 for Ryan, California.
6. Hess (1834).
7. Schaller (1928).

8. Boldyreva (1936).
9. Boldyreva in Godlevsky (1937).
10. Cf. Erdmann (1907) and Lottner, *Zs. deutsch. geol. Ges.*, **17**, 430 (1865).
11. Cf. van't Hoff, *Unters. ozean. Salzablag.*, **2**, 69 (1909).

25.1.11   I N D E R B O R I T E [$CaMgB_6O_{11} \cdot 11H_2O$]. *Gorshkov* (*C. r. ac. sc. U.R.S.S.* **33**, 254, 1941). Metahydroboracite *Ikornikova* and *Godlevsky* (*C. r. ac. sc. U.R.S.S.*, **33**, 257, 1941).

C r y s t.[1] Monoclinic; prismatic—$2/m$.

$a:b:c = 1.6346:1:1.3173;$ $\quad \beta\, 90°48';$ $\quad p_0:q_0:r_0 = 0.8059:1.3172:1$

$r_2:p_2:q_2 = 0.7592:0.6118:1;$ $\quad \mu\, 89°12';$ $\quad p_0'\, 0.8060,\, q_0'\, 1.3173,\, x_0'\, 0.0140$

Forms:

	ϕ	ρ	ϕ_2	$\rho_2 = B$	C	A
c 001	90°00′	0°48′	89°12′	90°00′	.....	89°12′
a 100	90 00	90 00	0 00	90 00	89°12′	.....
m 110	31 27½	90 00	0 00	31 27½	89 35	58 32½
p 111	31 54	57 12	50 39	44 28½	56 47	63 37½
x 221	31 41	72 06	31 35½	35 55½	22 21½	60 01
r $\bar{1}12$	−30 34	37 25	111 15½	58 27½	37 49½	108 00
q $\bar{1}11$	−31 01	56 57	128 23	44 05	57 22	115 35½
y $\bar{2}21$	−31 14½	72 01	147 57½	35 35	72 26	119 33½

Habit. As coarsely crystalline aggregates and as well-developed crystals up to 2 cm. in size.

P h y s. Cleavage {100} good. Fracture conchoidal. H. 3½. G. 2.00. Luster vitreous. Colorless to white. Transparent.

O p t.

Orientation[2]	$n(Na)$[2]	n[3]	
$X \wedge c$ 2½°	1.483 ± 0.002	1.496	Biaxial negative (−).
Y	1.512 ± 0.002	1.521	
Z b	1.530 ± 0.002	1.54	
$2V$	77°	80°–86°	

C h e m. A hydrated borate of calcium and magnesium, $CaMgB_6O_{11} \cdot 11H_2O$.

Anal.

	1	2	3
CaO	11.14	11.27	11.16
MgO	8.01	8.00	8.01
B_2O_3	41.49	41.70	40.90
H_2O	39.36	39.48	39.54
Rem.			0.03
Total	100.00	100.45	99.64
G.		1.93	2.004

1. $CaMgB_6O_{11} \cdot 11H_2O$.   2. Inder Lake.[3] H_2O is ign. loss.   3. Inder Lake.[4] Rem. is insol. 0.01, R_2O_3 0.02. F nil. H_2O is ign. loss.

Tests. B.B. cracks and fuses to a colorless glass. Slowly soluble in cold water, quickly soluble in hot HCl.

Occur. Found in the borate deposits of Inder Lake, Kazakhstan, U.S.S.R. Associated minerals are inyoite, colemanite, ulexite, szaibelyite.
Name. From the locality and the composition.

Ref.
1. Elements and forms of Gorshkov (1941). Comparable elements (with c doubled) are given by Ikornikova and Godlevsky (1941).
2. Ikornikova and Godlevsky (1941).
3. Gorshkov (1941).
4. Tikhomirova analysis in Ikornikova and Godlevsky (1941).

25.1.12   **MEYERHOFFERITE** [$Ca_2B_6O_{11} \cdot 7H_2O$]. *Schaller (J. Washington Ac. Sc.*, **4**, 355, 1914; *U. S. Geol. Sur., Bull. 610*, 41, 1916).

Cryst. Triclinic;[1] pinacoidal—$\bar{1}$.

$$a:b:c = 0.7904:1:0.7763; \quad \alpha\ 90°41',\ \beta\ 101°51',\ \gamma\ 86°44'\ ^2$$

$$p_0:q_0:r_0 = 0.9837:0.7610:1; \quad \lambda\ 89°59',\ \mu\ 78°10\tfrac{1}{2}',\ \nu\ 93°11\tfrac{1}{2}'$$

$$p_0' = 1.0051,\ q_0' = 0.7775,\ x_0' = 0.2098,\ y_0' = 0.0003$$

Forms:[2]

		φ	ρ	A	B	C	Z
c	001	89°55′	11°51′	78°10½′	89°59′	.....	78°09′
b	010	0 00	90 00	93 11½	.....	89°59′	....
a	100	93 11½	90 00	.....	93 11½	78 10½	0 00
m	110	54 17	90 00	38 54½	54 17	80 23½	0 00
M	1$\bar{1}$0	129 43	90 00	36 31½	129 43	80 55½	0 00
y	101	92 38	50 32	39 28	92 02	38 42	39 29½
t	$\bar{1}$01	−85 56½	38 30½	128 30	92 31½	50 20	128 26½
u	$\bar{1}$11	−43 35½	49 01	123 23	56 51	57 39	128 26½
p	$\bar{1}\bar{1}$1	−132 15½	47 00	120 52½	119 27½	56 13½	128 26½

Less common:
k 370 A 350 n 520 s 310 r 810 h $\bar{3}$10 w 4$\bar{3}$0 v 3$\bar{5}$0 f $\bar{6}$05 g $\bar{5}$04 o 111

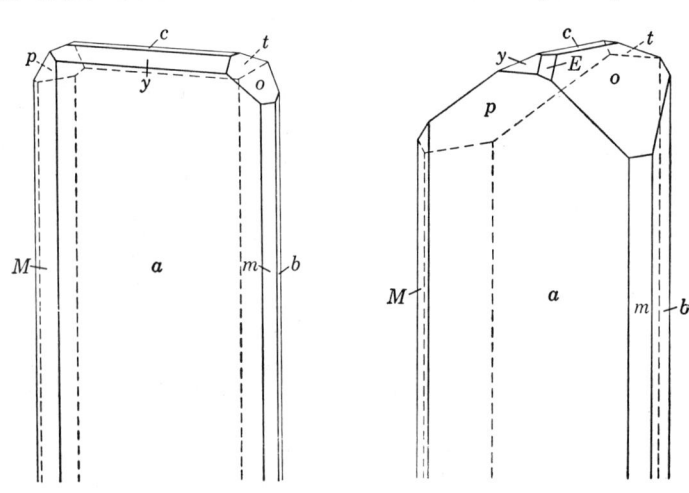

MEYERHOFFERITE

Structure cell.[3] Space group $P\bar{1}$. a_0 6.60 kX, b_0 8.33, c_0 6.48; α 91°00′, β 101°31′, γ 86°55′; $a_0:b_0:c_0 = 0.792:1:0.778$. Cell contents $Ca_2B_6O_{11} \cdot 7H_2O$.

Habit. Elongated [001] and generally flattened (100), with $a\ m\ M\ b\ p\ y\ t$; o usually replaced by a rounded surface. Also fibrous.

Phys. Cleavage {010} perfect; {100} and {1$\bar{1}$0} in traces. H. 2. G. 2.120 (meas.),[4] 2.116 (calc.). Luster vitreous in crystals, silky in fibrous masses. Colorless to white. Transparent to translucent.

Opt.[5]

	ϕ	ρ	n(Na)	
X	165°00′	62°00′	1.500	Biaxial negative (−).
Y	45 30	47 00	1.535	2V 78°.
Z	−83 00	55 00	1.560	Dispersion not observed.

On (010) $X' \wedge c\ 30°$; on (100) $Z' \wedge c\ 25°$

Chem. A hydrated calcium borate, $Ca_2B_6O_{11} \cdot 7H_2O$.

Anal.

	1	2	3
CaO	25.08	25.45	25.6
B_2O_3	46.72	46.40	45.6
$H_2O(-110°)$		1.01	0.3
$H_2O(+110°)$	28.20	27.75	28.5
Total	100.00	100.61	100.00

1. $Ca_2B_6O_{11} \cdot 7H_2O$. 2. Average of three analyses on opaque fibrous mass. 3. Colorless transparent crystal.

Tests. B.B. fuses readily with intumescence to an opaque enamel. In C.T. fuses and yields abundant water. Easily soluble in acids.

Occur. Found largely as an alteration product of inyoite in the colemanite deposits of the Mt. Blanco district, Furnace Creek, near Death Valley, Inyo County, California. The meyerhofferite occurs as pseudomorphs after inyoite with a fibrous internal structure, as small transparent colorless crystals on the surface of these pseudomorphs, and as masses of interlaced glassy crystals or fibrous aggregates embedded in clay.

Artif.[6] Obtained in crystals by heating $Ca_2B_6O_{11} \cdot 9H_2O$ in a 3 per cent boric acid solution at 100° C.

Name. After Wilhelm Meyerhoffer, German chemist, who worked for many years with J. H. van't Hoff on the composition and origin of the saline minerals and who synthesized the mineral here described.

Ref.

1. Schaller, *U. S. Geol. Sur., Bull. 610*, 41 (1916); Palache, *Am. Min.*, **23**, 644 (1938). The development of the {$hk0$} forms indicates pinacoidal symmetry.
2. Palache (1938) combining his observations with those of Schaller (1916) and using the setting which gives $c < a < b$, α and β obtuse. Transformation: Schaller to Palache, 100/0$\bar{1}$0/001.
3. Switzer, in Palache (1938), from rotation and Weissenberg photographs on crystals rotated about [001], [100], [010].
4. In agreement with van't Hoff and Meyerhoffer, *Preuss. Ak. Wiss., Sitzber.* 689 (1906), on artificial $Ca_2B_6O_{11} \cdot 7H_2O$.

5. Orientation and 2V by Pough, in Palache (1938); n(Na) by Larsen, in Schaller (1916); extinction angles from Schaller (1916).
6. van't Hoff and Meyerhoffer, *Ann. Chemie*, **351**, 100 (1907).

25.1.13   I N Y O I T E [$Ca_2B_6O_{11} \cdot 13H_2O$].   *Schaller (U. S. Geol. Sur., Bull. 610, 35, 1916).*

C r y s t.   Monoclinic; prismatic—$2/m$.

$a:b:c = 0.8833:1:0.6950;$ $\beta\ 114°01';$ [1] $p_0:q_0:r_0 = 0.7868:0.6348:1$

$r_2:p_2:q_2 = 1.5752:1.2394:1;$ $\mu\ 65°59';$ $p_0'\ 0.8614,\ q_0'\ 0.6950,\ x_0'\ 0.4456$

Forms: [2]

		ϕ	ρ	ϕ_2	$\rho_2 = B$	C	A
c	001	90°00′	24°01′	65°59′	90°00′	.....	65°59′
b	010	0 00	90 00	.....	0 00	90°00′	90 00
a	100	90 00	90 00	0 00	90 00	65 59	.....
z	120	31 47	90 00	0 00	31 47	77 37½	58 13
m	110	51 06	90 00	0 00	51 06	71 32	38 54
k	011	32 40	39 32½	65 59	57 35½	32 24½	69 54
v	021	17 46½	55 35	65 59	38 13½	51 46½	75 25
i	$\overline{1}$01	−90 00	22 34½	112 34½	90 00	46 35½	112 34½
h	$\overline{2}$01	−90 00	51 56½	141 56½	90 00	75 57½	141 56½
p	111	62 00	55 57½	37 25½	67 06½	35 59½	42 58½
y	$\overline{1}$11	−30 53½	39 00	112 34½	57 18½	54 40	108 51

Less common or rare: $l\ 112$ $g\ \overline{3}12$

Habit.   Short prismatic [001] to tabular {001}, with {110} and {001} dominant. Also massive granular, and as coarse spherulitic aggregates.

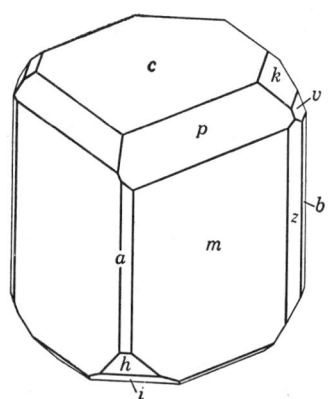

New Brunswick.

P h y s.   Cleavage {001} good; also {010}.[3] Fracture irregular. Brittle. H. 2. G. 1.875.[4] Luster vitreous. Colorless and transparent, becoming white and cloudy on partial dehydration.

INYOITE 359

Opt. In transmitted light, colorless.

ORIENTATION		n (Inyo County)[5]	n (Inder)[6]	n (Hillsborough)[8]	n (artif.)[7]	
$X \wedge c$	$+37°$	1.495	1.492		1.491	Biaxial negative ($-$).
Y	b	1.51	1.505	1.501	1.505	$r < v$, slight.
$Z \wedge c$	$-53°$	1.520	1.516		1.518	
$2V$		$70°$	$84°$	$80°33'$		

Chem. A hydrated calcium borate, $Ca_2B_6O_{11} \cdot 13H_2O$. $10H_2O$ are lost below $110°$ C.[9]

Anal.

	1	2	3	4	5
CaO	20.20	20.5	20.42	20.34	20.80
B_2O_3	37.62	[37.2]	37.44	36.57	36.26
$H_2O + 110°$	42.18	26.1	9.46	43.00	42.31
$H_2O - 110°$		16.2	32.46		
Rem.			0.55	0.28	0.43
Total	100.00	[100.00]	100.33	100.19	99.80
G.		1.875	1.885	1.82	

1. $Ca_2B_6O_{11} \cdot 13H_2O$. 2. Inyo County, California.[10] 3. Hillsborough, New Brunswick.[8] Rem. is SO_3 0.55, present as gypsum. 4. Inder, U.S.S.R.[11] Rem. is MgO 0.08, $(Al,Fe)_2O_3$ 0.07, $(Na,K)_2O$ 0.09, insol. 0.04. 5. Inder, U.S.S.R.[12] Rem. is SiO_2 0.11, $(Al,Fe)_2O_3$ 0.15, MgO tr., $(Na,K)_2O$ 0.17.

Tests. B.B., a fragment first whitens and crumbles, then fuses very easily (1) with intumescence to an opaque white bead. Soluble in hot water and dilute acids.

Occur. Found originally in the Mount Blanco borate district, near Death Valley, Inyo County, California, associated with meyerhofferite, colemanite, and priceite. As drusy crusts lining cavities in massive gypsum at Hillsborough, Albert County, New Brunswick. Found abundantly in the borate deposits of the Inder region, western Kazakhstan, U.S.S.R., as granular masses or spherulitic aggregates in clay and as drusy crusts along fissures or in geodes.

Alter. Alters by partial dehydration to meyerhofferite, the latter often forming white, opaque pseudomorphs with a silky-fibrous internal structure after the inyoite. Colemanite also occurs as dehydration pseudomorphs after inyoite.

Artif.[7] Obtained by long standing of a solution of borax, $CaCl_2$ and $MgCl_2 \cdot 6H_2O$.

Name. From the locality in Inyo County, California.

Ref.

1. Poitevin and Ellsworth, *Geol. Sur. Canada, Bull.*, 32 (1921), on good crystals from New Brunswick; very similar elements were obtained by Palache on crystals from Nova Scotia (priv. comm.), by Boky, *Bull. ac. sc. U.R.S.S., Cl. sc. mat. nat., Ser. chim.*, 871 (1937), on crystals from Inder, and by Murdoch and Webb, *Am. Min.*, 25, 549 (1940), on transparent crystals from Inyo County. The original elements of Schaller (1916) were based on contact measurements on dull crystals from Inyo County.
2. Poitevin and Ellsworth (1921); Boky (1937).
3. Cleavage {010} noted only by Boky (1937).
4. Schaller (1916) by suspension method on Inyo material; Poitevin and Ellsworth (1921) give 1.885 on material from Hillsborough containing gypsum.
5. Larsen in Schaller (1916).

6. Boldyreva in Godlevsky, *Mem. soc. russe min.*, **66**, 315 (1937).
7. Nikolaev and Chelishcheva, *C. r. ac. sc. U.R.S.S.*, **18**, 431 (1938).
8. Poitevin and Ellsworth (1921).
9. Schaller (1916), Poitevin and Ellsworth (1921), and Chelishcheva, *C. r. ac. sc. U.R.S.S.*, **18**, 508 (1940), give dehydration data.
10. Schaller (1916).
11. Boldyreva anal. in Godlevsky (1937).
12. Boldyreva, *Mater. Cent. Sc. Geol. Prosp. Inst.*, 1936, no. 1, 62—*Min. Abs.*, **6**, 336 (1936).

25.1.14 **KURNAKOVITE** [$Mg_2B_6O_{11} \cdot 13H_2O$]. *Godlevsky (C. r. ac. sc. U.R.S.S.*, **28**, 638, 1940).

Found as dense granular aggregates. Probably monoclinic, with an indistinct cleavage {010}. Sometimes twinned. H. 3. G. 1.85. Color white. Luster vitreous.

Opt. In transmitted light, colorless.

Orientation		n	
X		1.489 ± 0.002	Biaxial negative $(-)$.
Y	b	1.510 ± 0.002	$2V = 80°$.
Z		$1.525 \pm 0.00\underline{2}$	An optic axis is nearly $\perp$ {001}.

Chem. A hydrated magnesium borate, probably $Mg_2B_6O_{11} \cdot 13H_2O$. Dehydrates at *ca.* 81°–105°.

Anal.

	1	2
MgO	15.39	15.46
B_2O_3	39.89	37.58
H_2O	44.72	47.09
Rem.		0.60
Total	100.00	100.73
G.		1.85

1. $Mg_2B_6O_{11} \cdot 13H_2O$. 2. Inder.[1] Rem. is CaO 0.16, R_2O_3 0.20, SiO_2 0.10, F_2 0.14.

Tests. B.B. fuses to an enamel. Insoluble in water but soluble in warm acids.

Occur. Found in the borate deposit at Inder, Kazakhstan, U.S.S.R., in part associated with szaibelyite.

Name. After N. S. Kurnakov (1860–1941), Russian mineralogist and chemist.

Ref.

1. Egorova anal. in Godlevsky (1940).

25.1.15 **INDERITE** [$Mg_2B_6O_{11} \cdot 15H_2O$]. *Boldyreva* and *Egorova (Mat. Centr. Sc. Geol. Prosp. Inst. U.S.S.R., Gen. Ser.*, no. 2, 1937); *Boldyreva (Mem. soc. russe min.*, **66**, 651, 1937).

Cryst. Triclinic.[1]

Structure cell.[1] Space group $\bar{1}$ or 1. a_0 8.14 Å, b_0 10.47, c_0 6.33; α 96°56½′, β 106°28′, γ 106°03′; $a_0:b_0:c_0 = 0.768:1:0.604$. Cell contents $Mg_2B_6O_{11} \cdot 15H_2O$.

Habit. Faceted crystals unknown. Tabular with two pinacoidal cleavages at 66°20′ (California); acicular, in reniform nodular aggregates (Inder); prismatic (artif.).

INDERITE 361

Phys.[2] Cleavage {010} perfect; {1$\bar{1}$0} good. H. 3. G. 1.860; 1.903 (calc.). Luster vitreous to pearly on cleavage planes, dull and greasy on irregular surfaces. Crystals colorless and transparent, massive aggregates white to pink.

Opt. In transmitted light, colorless.

ORIENTATION	n(Na), United States [2]	n (Inder) [3]	
$X \sim b$	1.488 ± 0.002	1.488	Biaxial negative $(-)$.
Y	1.508 ± 0.002	1.505–1.506	$r > v$, weak.
Z	1.515 ± 0.002		
$2V$	$63° \pm 3°$	Large	
$Z \wedge c$	$-22°$	5°–7°	

Chem. A hydrated magnesium borate, $Mg_2B_6O_{11} \cdot 15H_2O$.

Anal.

	1	2	3	4	5
MgO	14.41	14.70	14.34	14.65	14.36
B_2O_3	37.32	36.41	35.60	36.20	37.31
H_2O	48.27	48.12	48.20	48.20	48.88
Rem.		0.86	2.62	0.97	
Total	100.00	100.09	100.76	100.02	100.55
G.		1.87	1.86		

1. $Mg_2B_6O_{11} \cdot 15H_2O$. 2. United States.[4] Rem. is CaO 0.32, insol. 0.54. 3. Inder.[5] Rem. is SiO_2 0.71, Al_2O_3 0.10, Fe_2O_3 0.23, CaO 1.02, $(K,Na)_2O$ 0.18, CO_2 0.38. 4. Inder.[5] Rem. is SiO_2 0.13, Al_2O_3 0.02, Fe_2O_3 0.32, CaO 0.16, $(K,Na)_2O$ 0.17, CO_2 0.17. 5. Artificial inderite.[6]

Tests. B.B. fuses at about 600° C. to an opaque white enamel. In the C.T. intumesces and affords much water with a weakly alkaline reaction. Insoluble in water but easily soluble in warm dilute HCl.

Occur. Originally found as small nodules associated with hydroboracite in red clay near Inder Lake, western Kazakhstan, U.S.S.R. Later described from an unstated locality [California] in the western United States.

Artif.[6] Obtained as acicular crystals from a mixed solution of borax and $MgSO_4 \cdot 7H_2O$ at 35° C.

Name. From the locality.

Ref.

1. Heinrich, *Am. Min.*, **31**, 71 (1946) by Weissenberg method on material from the United States. There is no evidence for the crystal class. The identity of this specimen with the Inder material is indicated by x-ray, optical, and chemical evidence although not with absolute certainty.
2. Heinrich (1946) for material from the U. S. Boldyreva (1937) gives H. 1 and G. 1.79 for aggregates (Inder).
3. Boldyreva (1937); see also Heinrich (1946). Artificial inderite is reported to be optically positive, $Z \wedge c = 6°$, nY ($= X$?) 1.487–1.489, nZ ($= Y$?) 1.505, $2V = 60°$ or $< 60°$, G. = 1.78, by Feigelson, Grushvitsky, and Korobochkina, *C. r. ac. sc. U.R.S.S.*, **22**, 242 (1939), Nikolaev and Chelishcheva, *ibid.*, **28**, 127 (1940). Artificial inderite prepared by Heinrich (1946) gives an x-ray powder pattern identical with that of the natural material from the U. S.
4. Gonyer anal. in Heinrich (1946).
5. Egorova anal. in Boldyreva (1937).
6. Feigelson, Grushvitsky, and Korobochkina (1939); Nikolaev and Chelishcheva (1940).

25.1.16 HOWLITE [$Ca_2SiB_5O_9(OH)_5$]. Silicoborocalcite *How* (*Phil. Mag.*, **35,** 32, 1868). Howlite *Dana* (598, 1868).
Winkworthite *How* (*Phil. Mag.*, **41,** 270, 1871).

Probably monoclinic.[1] As compact nodular masses, internally dense and structureless and resembling unglazed porcelain; sometimes chalk-like, earthy, scaly, or with a slaty structure.

Phys. Fracture of porcelaneous types nearly even and smooth. H. $3\frac{1}{2}$; often less. G. 2.53–2.59; 2.59 (Windsor).[2] Luster subvitreous, glimmering. Color white. Translucent in thin splinters.

Opt.[3] In transmitted light, colorless.

ORIENTATION		n (Windsor)	n (Ryan)	
X	b	1.586 ± 0.003	1.583 ± 0.003	Biaxial negative (–).
Y		1.598 ± 0.003	1.596 ± 0.003	$2V$ large.
$Z \wedge c(?)$	$44°$	1.605 ± 0.003	1.605 ± 0.003	

Chem. A hydrated calcium silico-borate, $Ca_2SiB_5O_9(OH)_5$.

Anal.

	1	2	3	4	5	6	7
CaO	28.66	28.45	28.44	27.94	28.69	28.26	29.22
SiO_2	15.34	15.50	15.33	15.33	15.25	14.81	15.31
B_2O_3	44.49	44.38	43.78	44.52	[44.22]	45.56	44.32
H_2O	11.51	11.58	11.39	11.55	11.84	11.37	11.44
Rem.		0.09	1.06	0.66			
Total	100.00	100.00	100.00	100.00	[100.00]	100.00	100.29
G.				2.59	2.55	2.531	

1. $Ca_2SiB_5O_9(OH)_5$. 2. Daggett, San Bernardino County, California.[4] Soft, scale-like crystals. Rem. is Na_2O and MgO. 3. Daggett, California.[4] Hard, dense material. Rem. is Na_2O and MgO. 4. Windsor, Nova Scotia.[5] Average of two analyses recalculated to 100 after deducting 4.32 per cent gypsum. Rem. is Na_2O 0.53, K_2O 0.13. 5. Windsor, Nova Scotia.[6] Recalc. after deducting gypsum. Average of three analyses. 6. Lang, Los Angeles County, California.[7] Total given as 100.38 in original. 7. Daggett, California.[8]

Tests. Ignited in the C.T., yields water reacting for boron with turmeric paper. Easily soluble in dilute acids, yielding gelatinous silica on evaporation.

Occur. Found originally as small nodules embedded in anhydrite or gypsum and associated with ulexite near Windsor, Hants County, Nova Scotia; also found at Wentworth (formerly Winkworth; *winkworthite*) and elsewhere in the same region. Found abundantly as nodules, sometimes ranging up to several hundred pounds in weight, in California in the colemanite deposit at Lang, Los Angeles County; also with ulexite and bakerite in the Mohave desert near Daggett, San Bernardino County, and in veins in Gower Gulch, Inyo County. Howlite is easily confused with both priceite and bakerite.

Name. After Henry How (? –1879), chemist, geologist, and mineralogist, of Nova Scotia, who first described the species.

Ref.

1. Larsen (87, 1921) from optical study of microscopic, pointed crystals tabular {100} or {001}, with narrow {010}, from Windsor, Nova Scotia. Earlier described as probably orthorhombic by Penfield and Sperry, *Am. J. Sc.*, **34,** 220 (1887), from similar evidence.

2. Penfield and Sperry (1887) by pycnometer; the variation in G. is due to the mode of aggregation. Porcelaneous material from Lang has G. 2.58 (Mrose, priv. comm., 1946).
3. Larsen (1921).
4. Giles, *Min. Mag.*, 13, 353 (1903).
5. Penfield and Sperry (1887).
6. How (1868).
7. Eakle, *Univ. Calif. Geol. Dept., Bull.* 6, 179 (1911).
8. Lawson anal. in Pabst (1938).

25.1.17 **BAKERITE**. Giles (*Min. Mag.*, 13, 353, 1903).

As nodules and veins. Dense and microcrystalline, resembling unglazed porcelain. Color white. H. $4\frac{1}{2}$. G. 2.88.[1] Optically [1] biaxial with moderate birefringence and a mean index of refraction of 1.642; in part spherulitic.

Chem. A basic calcium borate-silicate, probably $Ca_4B_4(BO_4)(SiO_4)_3(OH)_3H_2O$. This formula is based on the structural resemblance to datolite, herderite, and homilite indicated by x-ray powder diffraction study.[1] The reported analysis is in close agreement with this interpretation.

Datolite $Ca_4B_4(SiO_4)(SiO_4)_3(OH)_4 = CaB(SiO_4)(OH)$
Bakerite $Ca_4B_4(BO_4)(SiO_4)_3(OH)_3H_2O$
Herderite $Ca_4Be_4(PO_4)(PO_4)_3(OH)_4 = CaBe(PO_4)(OH)$
Homilite $(CaFe)_4B_4(SiO_4)(SiO_4)_3(OH)_4 = (CaFe)B(SiO_4)(OH)$ (?)

Anal.

	CaO	B_2O_3	SiO_2	H_2O	$(Al,Fe)_2O_3$	Total
1.	35.97	27.92	28.89	7.22		100.00
2.	35.05	27.30	28.25	8.46	0.94	100.00

1. $Ca_4B_4(BO_4)(SiO_4)_3(OH)_3H_2O$. 2. California.[2]

Occur. Originally found associated with howlite at Borate in the Calico Mountains, 9 miles east of Yermo, San Bernardino County, California. Also found abundantly with natrolite and thomsonite as veinlets and lenses in an altered volcanic rock in upper Baker Canyon in the Black Mountains to the west of Furnace Creek wash in Death Valley, Inyo County, California. Named after R. C. Baker, who found the mineral. Some of the dense types of datolite may be this mineral.

Ref.
1. Frondel, priv. comm. (1947).
2. Giles (1903).

25.1.18 **PATERNOITE** [$MgB_8O_{13} \cdot 4H_2O$]. Millosevich (*Acc. Linc., Rend.*, [5], 29, 286, 1920).

Found as fine-granular nodular masses. Under the microscope appears as rhombic scales with plane angles of about 62°, the acute angle sometimes truncated to give a hexagonal outline. Cleavable. G. 2.11. Color white. Transparent.

Opt. In transmitted light, colorless. Extinction parallel to the diagonals of the rhombic plates, with a mean index of refraction between 1.47 and 1.48.[1]

Chem. A hydrated borate of magnesium, $MgB_8O_{13} \cdot 4H_2O$.

Anal.

	1	2	3
Na_2O		0.36	
K_2O		1.08	
MgO	10.31	10.93	10.67
B_2O_3	71.26	66.02	71.66
SO_3		1.06	
Cl		2.35	
H_2O	18.43	19.16	17.67
Total	100.00	100.96	100.00

1. $MgB_8O_{13} \cdot 4H_2O$. 2. Calascibetta. 3. Analysis 2 recalculated to 100 after deduction of K and Cl, Na and SO_3, as carnallite and bloedite.

Tests. Partially soluble in water, yielding an alkaline solution. Easily soluble in acids.

Occur. Found as small nodules in bloedite beds in the salt deposits at Monte Sambuco, Calascibetta, Sicily.

Artif. The compound $MgB_8O_{13} \cdot 3H_2O$ has been reported.[2]

Name. After Emanuele Paterno (1847– ?), Italian chemist.

Ref.

1. Optical data given for so-called paternoite by Barth and Berman, *Chem. Erde*, **5**, 29 (1930), probably refer to kaliborite (Schaller, priv. comm., 1940); x-ray powder study of their specimen does not check the pattern of paternoite given by Millosevich, *Per. Min.*, **1**, 214 (1930), although exact comparison is not possible (Frondel, priv. comm., 1947).
2. Cf. Mellor (**5**, 99, 1924).

25.1.19 **GINORITE** [$Ca_2B_{14}O_{23} \cdot 8H_2O$]. *D'Achiardi* (*Per. Min.*, **5**, 22, 1934).

Monoclinic?[1] Flat tablets {010}. As dense masses. Cleavage flakes are lozenges flattened {010} with an angle of $\sim 78°$.

Phys. Cleavage {010}. H. $3\frac{1}{2}$. G. 2.09. Color white. Transparent in small grains.

Opt.

Orientation		n	
$X \wedge$ elong. (c?)	51°	1.517	Biaxial positive (+).
Y	b	1.524 (calc.)	$2V$ 42° ± 2°.
Z		1.577	Inclined dispersion.

Chem. A hydrated calcium borate of uncertain formula, probably $Ca_2B_{14}O_{23} \cdot 8H_2O$.

Anal.

	1	2	3
CaO	15.08	16.00	15.40
B_2O_3	65.54	64.06	63.00
H_2O	19.38	19.27	19.40
Rem.		1.22	2.20
Total	100.00	100.55	100.00

1. $Ca_2B_{14}O_{23} \cdot 8H_2O$. 2. Tuscany.[2] Rem. is insol. 1.07, CO_2 0.15. 3. Tuscany.[3] Rem. is insol. and loss.

Occur. Found with calcite in veins in sandstone at Sasso Pisano, Tuscany, Italy.

Name. After Piero Ginori Conti, of Florence, a leader in the development of the Tuscan borax industry.

Ref.
1. From optical evidence.
2. Rossoni anal. in D'Achiardi (1934).
3. Gallori anal. in D'Achiardi (1934).

25.1.20 **LARDERELLITE** [$(NH_4)_2B_{10}O_{16} \cdot 5H_2O$ (?)]. *Mascagni* (Viagg. Tosc., 3, 1806), *D'Achiardi* (**1**, 258, 1872). Larderellite *Bechi* (*Am. J. Sc.*, **17**, 129, 1854).
        Borate de Chaux *Beudant* (**2**, 249, 1832). Hayesine ? *Bechi* (*Am. J. Sc.*, **17**, 129, 1854). Bechilite *Dana* (597, 1868). Hydrous Borate of Lime. Borocalcite *Alger-Phillips* (318, 1844), *Groth* (38, 1874). Hayesine ? *Forbes* (*Phil. Mag.*, **25**, 113, 1863).

Probably monoclinic.[1] In microscopic rhomboidal tablets flattened {100}. The acute plane angle of the rhombs is ~68°.

Phys. Cleavage {001} ? and {010} ? perfect. Hardness and specific gravity not known. Color white; sometimes yellowish due to admixed impurities. Tasteless.

Opt. In transmitted light, colorless.

Orientation	n^2	
X b	1.493 ± 0.001	Biaxial positive (+).
Y	1.509 ± 0.001	2V 58°.
$Z \wedge c \sim 15°$	1.561 ± 0.001	$r < v$.

Chem. A hydrated ammonium borate, probably $(NH_4)_2B_{10}O_{16} \cdot 5H_2O$. Dimorphous with ammonioborite.[3]

Anal.

	1	2	3	4
$(NH_4)_2O$	10.62	9.87	9.78	10.04
B_2O_3	71.01	71.70	72.42	71.21
H_2O	18.37	[18.43]	[17.80]	[18.75]
Total	100.00	[100.00]	[100.00]	[100.00]

1. $(NH_4)_2B_{10}O_{16} \cdot 5H_2O$. 2,3. Tuscan lagoons.[4] 4. Tuscan lagoons.[5] Average of two.

Tests. B.B. fuses easily to a colorless glass. Decomposed by hot water.

Occur. Found associated with sassolite and ammonioborite in the boric acid fumaroles of Tuscany, Italy.

Artif. Not known artificially. The compound $(NH_4)_2B_{10}O_{16} \cdot 8H_2O$ has been formed, however, from cold water solutions.[6]

Name. After F. de Larderell, a proprietor of the Tuscan borax industry.

BECHILITE. Borate de Chaux *Beudant* (**2**, 249, 1832). Hayesine ? *Bechi* (*Am. J. Sc.*, **17**, 129, 1854). Bechilite *Dana* (597, 1868). Hydrous Borate of Lime. Borocalcite *Alger-Phillips* (318, 1844), *Groth* (38, 1874). Hayesine ? *Forbes* (*Phil. Mag.*, **25**, 113, 1863).
        Supposedly $CaB_4O_7 \cdot 4H_2O$, found as an incrustation in the boric acid lagoons of Tuscany, Italy. A compound of this composition has not been found to exist,[7] however, and specimens labeled bechilite recently examined [8] are mixtures in varying proportions of larderellite, ammonioborite, sassolite, and gypsum.

Ref.

1. Des Cloizeaux (**2**, 9, 1874) and Larsen (98, 1921) from optical study of microscopic crystals. See also D'Achiardi, *Acc. Linc., Rend.*, **9**, 342 (1900), and Schaller, *Am. Min.*, **18**, 485 (1933).
2. Ross in Schaller (1933); Schaller (1933) gives $Z \wedge c \sim 24°$.
3. The early and presumably erroneous analyses of Bechi, *Am. J. Sc.*, **17**, 129 (1854), afford the formula $(NH_4)_2B_8O_{15} \cdot 4H_2O$. On the present formula and disproved isostructural relation to paternoite see D'Achiardi, *Per. Min.*, **1**, 208 (1930); **3**, 36 (1932), and Millosevich, *Per. Min.*, **1**, 214 (1930).
4. D'Achiardi (1900).
5. Bonatti anal. in D'Achiardi (1930).
6. Cf. Schaller (1933).
7. Cf. van't Hoff, *Ak. Berlin, Sitzber.*, 652 (1907), Hintze (**1** [4A], 165, 1933), and Schaller (1933).
8. Cf. Larsen (98, 1921), D'Achiardi, *Per. Min.*, **3**, 1 (1932), and Schaller (1933).

25.1.21 A M M O N I O B O R I T E [$(NH_4)_2B_{10}O_{16} \cdot 5H_2O(?)$]. *Schaller* (*Am. Min.*, **16**, 114, 1931; *ibid.*, **18**, 480, 1933).

Found as compact, fine-grained granular masses. In part as microscopic monoclinic or triclinic crystals, platy in habit and often grouped in parallel position. Cleavage apparently lacking. Hardness and specific gravity not known. Color white.

O p t. In transmitted light, colorless.

	n [1]	
X	1.470	Biaxial positive (+).
Y	1.487	$2V\ 60° \pm$.
Z	1.540	$r < v$, slight.

Z makes an angle of $7°$–$13°$ with the direction of elongation of the plates, and X is $\perp$ to the plane of flattening.

C h e m. A hydrated ammonium borate, probably $(NH_4)_2B_{10}O_{16} \cdot 5H_2O$ and dimorphous with larderellite.

Anal.

	1	2
$(NH_4)_2O$	10.62	9.8
B_2O_3	71.01	74.2
H_2O	18.37	[16.0]
Total	100.00	[100.0]

1. $(NH_4)_2B_{10}O_{16} \cdot 5H_2O$. 2. Larderello, Italy.

Tests. Slowly but completely soluble in cold water. B.B. fuses easily to a colorless glass.

O c c u r. Found admixed with sassolite and larderellite, in part apparently as an alteration product of the latter species, in the boric acid lagoon at Larderello, Tuscany, Italy.

A r t i f.[2] By slowly evaporating at 95° a water solution of $(NH_4)_2O \cdot 5B_2O_3 \cdot 8H_2O$.

N a m e.[3] In allusion to the composition.

Ref.

1. Average of measurements by Schaller (1933) on admixed specimens variously labeled bechilite and boussingaultite.
2. Schaller, priv. comm. (1946) and Sborgi and Meccacci, *Acc. Linc., Att.*, **25**, 455 (1916).

3. Ammonioborite appears from the observations of Schaller (1933), Larsen (1921), and D'Achiardi, *Per. Min.*, **3**, 36 (1932), to be a constituent of the discredited mixture earlier described under the name bechilite, but it seems preferable to reject the latter name entirely.

25.1.22 **KALIBORITE** [$KMg_2B_{11}O_{19} \cdot 9H_2O$]. Feit (*Chem.-Ztg.*, **13**, 1188, 1889; *J. Chem. Soc. London*, **58**, 341, 1890).
  Heintzite Luedecke (*Zs. Nat. Halle*, **62**, 354, 1889; *Zs. Kr.*, **18**, 481, 1890); Hintzeite *Milch* (*Zs. Kr.*, **18**, 478, 1890).

Cryst. Monoclinic; prismatic—$2/m$.

$a:b:c = 2.1937:1:1.7338;\quad \beta\ 99°48';$[1] $\quad p_0:q_0:r_0 = 0.7904:1.7085:1$

$r_2:p_2:q_2 = 0.5853:0.4370:1;\quad \mu\ 80°12';\quad p_0'\ 0.8021,\ q_0'\ 1.7338,\ x_0'\ 0.1727$

Forms:[2]

		ϕ	ρ	ϕ_2	$\rho_2 = B$	C	A
a	100	90°00′	90°00′	0°00′	90°00′	80°12′	0°00′
m	110	24 49½	90 00	0 00	24 49½	85 54	65 10½
x	$\bar{1}01$	−90 00	32 11	122 11	90 00	41 59	122 11
n	111	29 21	63 18½	45 44	38 51	58 51½	64 02
r	311	56 05½	72 09½	21 12	57 55	64 08	37 49
o	$\bar{1}12$	−14 45½	41 52½	102 52	49 48	45 11	99 47½

Habit. In small crystals, often aggregated and with the faces unsymmetrically developed. Dominant forms: $a\ n\ m$. Also massive, granular.

Phys. Cleavage $\{001\}$ and $\{\bar{1}01\}$ perfect, $\{100\}$ good. H. 4–4½. G. 2.128.[3] Luster vitreous. Colorless to white; also reddish brown. Transparent.

Opt.[4] In transmitted light, colorless.

Orientation	n(Na) Neustassfurt[4]	n (Inder)[5]	
X	1.5081	1.508	Biaxial positive (+).
$Y\ b$	1.5255	1.527	$2V\ 80°38'$ (obs.).
$Z \wedge c\ 64°30'$	1.5500	1.549	Stassfurt.

Chem. A hydrated borate of potassium and magnesium, $KMg_2B_{11}O_{19} \cdot 9H_2O$.

Anal.

	1	2	3	4
K_2O	7.00	6.48	8.14	7.39
MgO	11.98	12.06	13.80	12.23
B_2O_3	56.92	[57.46]	52.39	60.53
H_2O	24.10	24.00	23.83	19.85
Rem.			0.74	
Total	100.00	[100.00]	98.90	100.00
G.		2.05	2.127	2.129

1. $KMg_2B_{11}O_{19} \cdot 9H_2O$. 2. Schmidtmannshall[6] (*kaliborite*). After deduction of 1–2 per cent NaCl. 2. Leopoldshall.[7] Rem. is Na 0.39, Cl 0.35; a duplicate B_2O_3 determination gave 51.88. 4. Leopoldshall.[8] Average of two; a duplicate B_2O_3 determination gave 59.27.

Tests. B.B. fuses very easily. Slightly soluble in water, yielding an alkaline solution. Easily soluble in acids.

Occur. Found with boracite and pinnoite in the upper kainite layers at Schmidtmannshall near Aschersleben, Saxony; also with pinnoite and kainite at Neustassfurt,[4] and with pinnoite at Leopoldshall near Stassfurt, in Saxony, Germany. Kaliborite occurs also with anhydrite and halite in the Inder salt deposits, Kazakhstan, U.S.S.R.[5]

Artif.[9] Obtained by treating $MgB_6O_{10} \cdot 7H_2O$ with a solution containing boric acid and potassium hydroxide.

Alter.[10] Kaliborite can be produced artificially from pinnoite by the action of solutions saturated with kainite and is slowly converted into pinnoite in water at 100°.

Name. Kaliborite in allusion to the composition. Hintzeite after the German mineralogist Carl A. F. Hintze (1851–1916). Heintzite after Wilhelm Heinrich Heintz (1817–1880), Professor of Chemistry in the University of Halle. Heintzite (and hintzeite, the two names being proposed simultaneously for the mineral) was later shown [11] to be identical with kaliborite.

Ref.

1. Unit and orientation of Milch, *Zs. Kr.*, **18**, 478 (1890). Transformation: Luedecke, *Zs. Kr.*, **18**, 481 (1890), to Milch, $20\bar{1}/0\bar{1}0/001$. Luedecke's transformed elements become $a:b:c = 2.2149:1:1.7572; \beta\ 99°48'$.
2. Milch (1890). $c\{001\}$ observed only as cleavage.
3. Average of values 2.127 and 2.129 given by Milch (1890) and Luedecke (1890).
4. Boeke, *Cbl. Min.*, 531 (1910); indices by total reflection method. Indices close to these were obtained by Barth and Berman, *Chem. Erde*, **5**, 29 (1930), on material erroneously labeled paternoite.
5. Godlevsky, *Mem. soc. russe min.*, [2], **67**, 258 (1938); Yarzhemsky, *C. r. ac. sc. U.R.S.S.*, **47**, 642 (1945).
6. Feit (1889).
7. Baurath analysis in Milch (1890).
8. Luedecke (1890).
9. Mellor (**5**, 99, 1924).
10. Cf. van't Hoff (1902) and Loewe, *Zs. pr. Geol.*, **11**, 355 (1903).
11. Feit, *Chem.-Ztg.*, **15**, 115 (1891).

26 BORATES CONTAINING HYDROXYL OR HALOGEN

26.1.1 FLUOBORITE [$Mg_3(BO_3)(F,OH)_3$]. Geijer (Geol. För. Förh., **48**, 84, 1926; Sver. Geol. Undersök., Ser. C, no. 343, Yearbook 20, no. 4, 26, 1927).

C r y s t. Hexagonal.
Structure cell.[1] a_0 9.05 kX, c_0 3.09; $a_0:c_0$ = 1:0.341. Cell contents $Mg_6(BO_3)_2(F,OH)_6$.

Habit. As acicular hexagonal prisms without measurable end-faces, sometimes arranged in fan-shaped or stellate groups; also as fluffy, felted aggregates.

P h y s. Cleavage [0001], indistinct.[2] H. probably $3\frac{1}{2}$. G. 2.98 ($Mg_3(BO_3)F_3$) to 2.85 ($Mg_3(BO_3)(OH)_3$).[3] Colorless to white.

O p t.[3] In transmitted light, colorless.

	$n(Mg_3(BO_3)F_3)$	$n(Mg_3(BO_3)(F,OH))$ with F:OH = 1:1	$n(Mg_3(BO_3)(OH)_3)$	
O	1.502	1.577	1.579	Uniaxial
E	1.487	1.522	1.532	negative (−).

C h e m. A fluoborate of magnesium, $Mg_3(BO_3)(F,OH)_3$. OH substitutes for F up to at least F:OH = 5:12 (anal. 2), and a complete series probably extends between the F and OH end-members. Small amounts of Zn, Mn'', and Ca apparently substitute for Mg.

Anal.

	1	2	3	4	5
MgO	66.17	61.65	60.07	62.02	64.07
B_2O_3	19.05	17.90	17.25	17.67	18.44
H_2O	14.78	10.78	5.22	3.23	
F		9.30	17.60	20.94	30.19
Rem.		3.57	6.76	3.91	
	100.00	103.20	106.90	107.77	112.70
Less F = O		3.92	7.41	8.82	12.70
Total	100.00	99.28	99.49	98.95	100.00
G.		2.89	2.92	2.95	

1. $Mg_3(BO_3)(OH)_3$. 2. Tallgruvan, Sweden.[4] Rem. is SiO_2 0.45, Al_2O_3 0.90, Fe_2O_3 0.81, MnO 0.05, CO_2 1.36. 3. Sterling Hill, N. J.[5] Rem. is ZnO 2.41, MnO 1.93, CaO 1.19, CO_2 1.23. Contains about 3 per cent of manganoan calcite. 4. Selibin, Malaya.[6] Rem. is SiO_2 0.88, Al_2O_3 0.92, Fe_2O_3 0.36, FeO 1.16, CaO 0.56. 5. $Mg_3(BO_3)F_3$.

O c c u r. Found originally at the Tallgruvan mine, east of Kallmora, Norberg, Sweden, associated with ludwigite and chondrodite in a contact metasomatic magnetite deposit. At Sterling Hill, New Jersey, as a hydrothermal mineral associated with mooreite, willemite, fluorite, hydrozincite, and pyrochroite in secondary veinlets cutting the massive franklinite ore.

Also found in the Beatrice mine at Selibin in the Federated Malay States associated with phlogopite, fluorite, and tremolite in a cassiterite pipe in dolomite, and in the Hol Kol mine, Korea,[7] associated with kotoite, ludwigite, forsterite, and magnetite in a contact deposit in marble. Probably occurs with szaibelyite in a contact deposit near Pioche, Lincoln County, Nevada.

Name. In allusion to the composition.

Ref.
1. Aminoff in Geijer (1927) by rotation method on crystal from Tallgruvan (anal. 2). See also Takéuchi, *Acta Cryst.*, **3**, 208 (1950) for structure analysis.
2. Johnston and Tilley, *Geol. Mag.*, **77**, 141 (1940), on material from Selibin.
3. Schaller, *Am. Min.*, **27**, 467 (1942); data for the end-members by extrapolation.
4. Bygden anal. in Geijer (1927).
5. Bauer anal. in Bauer and Berman, *Am. Min.*, **14**, 165 (1929).
6. Johnston anal. in Johnston and Tilley (1940).
7. Watanabe, *Min. Mitt.*, **50**, 441 (1939).

26.1.2   H A M B E R G I T E [$Be_2(OH)(BO_3)$]. *Brögger* (*Zs. Kr.*, **16**, 65, 1890).

Cryst.[1] Orthorhombic; dipyramidal—$2/m\ 2/m\ 2/m$.

$a:b:c = 0.8023:1:0.3634;$ [2] $p_0:q_0:r_0 = 0.4529:0.3634:1$

$q_1:r_1:p_1 = 0.8023:2.2078:1;$ $r_2:p_2:q_2 = 2.7518:1.2464:1$

Forms:[3]

	ϕ	$\rho = C$	ϕ_1	$\rho_1 = A$	ϕ_2	$\rho_2 = B$
c 001	....	0°00′	0°00′	90°00′	90°00′	90°00′
b 010	0°00′	90 00	90 00	90 00	....	0 00
a 100	90 00	90 00	....	0 00	0 00	90 00
m 110	51 15½	90 00	90 00	38 44½	0 00	51 15½
n 210	68 08½	90 00	90 00	21 51½	0 00	68 08½
l 410	78 39½	90 00	90 00	11 20½	0 00	78 39½
e 021	0 00	36 00½	36 00½	90 00	90 00	53 59½
d 102	90 00	12 45½	0 00	77 14½	77 14½	90 00
r 111	51 15½	30 08½	19 58½	66 56½	65 38	71 41
v 441	51 15½	66 42½	55 28½	44 14½	28 54	54 55
q 243	31 55½	29 43½	25 51	74 48	73 12	65 07
y 121	31 55½	40 34½	36 30	69 53	65 38	56 29½
w 131	22 33½	49 44	47 28½	72 58½	65 38	45 12

Less common:

k 230   z 310   p 221   t 122   s 211   u 241   x 361

Structure cell.[4] Space group *Pbca*. $a_0\ 9.73 \pm 0.01\ kX$, $b_0\ 12.18 \pm 0.02$, $c_0\ 4.42 \pm 0.01$; $a_0:b_0:c_0 = 0.799:1:0.363$. Cell contents $Be_{16}(OH)_8(BO_3)_8$.

Habit. Prismatic [001], often flattened {100}. Usually in large crystals; with {100} striated [001], and the terminal faces dull and etched. Common forms: *m e b a n*.

Twinning.[5] Twin plane and composition plane {110}.

HAMBERGITE

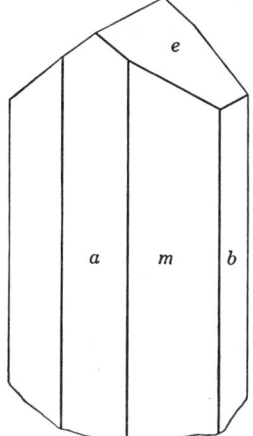

Phys. Cleavage {010} perfect, {100} good. Brittle. H. $7\frac{1}{2}$. G. 2.359; [6] 2.365 (calc.). Luster vitreous. Colorless, inclining toward grayish white or yellowish. Transparent.

Opt. In transmitted light, colorless.

ORIENTATION		n(Na) Madagascar [7]	n(Na) Sweden [8]	
X	a	1.5536	1.5595	Biaxial positive (+).
Y	b	1.5873	1.5908	$r > v$, weak.[12]
Z	c	1.6278	1.6311	
2V		87°1′	87°7′	

Chem. A basic borate of beryllium, $Be_2(OH)(BO_3)$. The water is lost only at a high temperature.

Anal.

	1	2	3	4
BeO	53.31	53.25	54.80	52.40
B_2O_3	37.09	[36.72]	35.10	[37.39]
H_2O	9.60	10.03	10.95	10.21
Total	100.00	[100.00]	100.85	[100.00]
G.	2.365	2.347		2.36

1. $Be_2(OH)(BO_3)$. 2. Sweden.[9] BeO average of three. 3. Imalo.[10] With $(Na,K)_2O$ trace. 4. Kashmir.[11]

Tests. B.B. decrepitates, becomes opaque due to loss of water, but does not fuse. Soluble in HF.

Occur. First found with barkevikite, feldspar, biotite, altered sodalite, and, rarely, zircon, fluorite, and analcime in syenite pegmatite near Helgaråen at the entrance to Langesund fiord, southern Norway. Later discovered in alkali-rich pegmatites in Madagascar, associated with danburite, beryl, and spodumene, at Imalo near Mania and at Maharitra on Mt. Bity. From gem gravels in Kashmir, India, and probably derived from granite pegmatites.

Name. After Axel Hamberg (1863–1933), Swedish mineralogist and geographer, who drew attention to the first specimens.

Ref.
1. Dipyramidal symmetry is indicated but not proved by the available evidence.
2. Orientation of Brögger, *Zs. Kr.*, **16**, 65 (1890); angles of Goldschmidt and Müller, *Zs. Kr.*, **48**, 473 (1910); unit of Zachariasen, *Zs. Kr.*, **76**, 289 (1931). Transformation: Brögger and Goldschmidt to Zachariasen, $100/010/00\tfrac{1}{2}$.
3. Goldschmidt (**4**, 106, 1918) and Burton, *Geol. Sur. India, Rec.*, **43**, 168 (1913). Uncertain: 661, 641.
4. Zachariasen (1931) by the Laue and oscillation methods on a natural crystal from Madagascar, with determination of the structure.
5. Drugman and Goldschmidt, *Zs. Kr.*, **50**, 596 (1912); see also Burton (1913).
6. Wolfe, priv. comm. (1946), by microbalance; Burton (1913) and Lacroix, *Bull. soc. min.*, **32**, 320 (1909), give 2.36 and Brögger (1890), 2.347.
7. Average of measurements by the prism method by Müller in Goldschmidt and Müller, *Zs. Kr.*, **48**, 473 (1910), and by total reflection by Gaubert in Lacroix, *Bull. soc. min.*, **32**, 320 (1909). Payne, *Gemmologist, London*, **9**, 33 (1939) gives data on the dispersion.
8. Brögger (1890) by the prism method.
9. Bäckström analysis in Brögger (1890).
10. Pisani analysis in Lacroix, *Bull. soc. min.*, **33**, 49 (1910).
11. Burton (1913).
12. Cf. Goldschmidt and Müller (1910).

26.1.3 TEEPLEITE [$Na_2B_2O_4 \cdot 2NaCl \cdot 4H_2O$]. Gale, Foshag and Vonsen (*Am. Min.*, **24**, 48, 1939).

Cryst. Tetragonal; ditetragonal-dipyramidal—$4/m\ 2/m\ 2/m$.

$$a:c = 1:0.6690;\ ^1 \qquad p_0:r_0 = 0.6690:1$$

	ϕ	ρ	A	M
c 001	.....	0°00′	90°00′	90°00′
m 110	45°00′	90 00	45 00	90 00
e 011	0 00	33 47	90 00	66 51

Structure cell.[2] Space group $P4/nmm$. a_0 7.27 kX, c_0 4.84; $a_0:c_0 = 1:0.666$. Cell contents $Na_4B_2Cl_2O_4 \cdot 4H_2O$.

Habit. Tabular {001}. The crystals are usually rounded into cushion- or lens-like shapes and are often aggregated into interpenetrating groups.

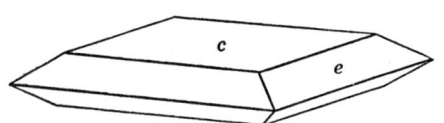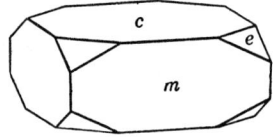

Phys. Fracture irregular to subconchoidal. Very brittle. H. $3-3\tfrac{1}{2}$. G. 2.076; 2.07 (calc.). Luster glassy to somewhat greasy, becoming dull upon exposure. Colorless to white; pale buff in material tinted by organic matter. Transparent.

Opt. In transmitted light, colorless.

nO	1.519	Uniaxial negative $(-)$.
nE	1.503	

BANDYLITE

Chem. A hydrated borate-chloride of sodium, $Na_2B_2O_4 \cdot 2NaCl \cdot 4H_2O$.

Anal.

	1	2
Na	28.70	28.93
Cl	22.15	18.72
B_2O_4	26.70	22.05
CO_3		9.42
H_2O	22.45	20.48
Rem.		0.23
Total	100.00	99.83
G.	2.07	2.076

1. $Na_2B_2O_4 \cdot 2NaCl \cdot 4H_2O$. 2. Borax Lake, California. Rem. is Ca 0.08, insol. 0.15. The Ca and CO_3 in the analysis were ascribed to admixed calcite and trona.[3]

Tests. B.B. fuses easily to a white enamel-like crystalline bead, coloring the flame yellow with a momentary flash of green. Easily soluble in water to an alkaline solution.

Occur. Found with halite and trona in pools of brine remaining after a period of extreme desiccation of Borax Lake, Lake County, California, during the late summer of 1934. The teepleite appears in part to have formed by the reaction of earlier-formed halite with the borate-rich mother liquor, as the composition of the brine changed, giving rise to pseudomorphs of teepleite after halite.

Artif. Known as an artificial product.[4]

Name. After John E. Teeple, chemist, in recognition of his contributions to the chemistry of Searles Lake.

Ref.

1. Gale, Foshag, and Vonsen (1939). {110} on artificial crystals only.
2. Switzer in Gale, Foshag, and Vonsen (1939) by the Weissenberg method on artificial crystals.
3. Foshag anal. in Gale, Foshag, and Vonsen (1939).
4. Teeple, *The Industrial Development of Searles Lake Brines*, New York, 130 (1929).

26.1.4 **BANDYLITE** [$CuB_2O_4 \cdot CuCl_2 \cdot 4H_2O$]. Palache and Foshag (*Am. Min.*, **23**, 85, 1938).

Cryst. Tetragonal; ditetragonal-dipyramidal—$4/m\ 2/m\ 2/m$.

$a:c = 1:0.9070;\qquad p_0:r_0 = 0.9070:1$

Forms:

	ϕ	ρ	A	M
c 001	....	0°00′	90°00′	90°00′
a 010	0°00′	90 00	90 00	45 00
m 110	45 00	90 00	45 00	90 00
d 012	0 00	24 23½	90 00	73 01½
e 023	0 00	31 09½	90 00	68 32½
f 011	0 00	42 12½	90 00	61 38
g 021	0 00	61 08	90 00	51 44½
o 112	45 00	32 40½	67 33½	90 00
p 111	45 00	52 03½	56 06½	90 00
q 221	45 00	68 42	48 47½	90 00

Structure cell.[1] Space group $P4/nmm$. a_0 6.13 kX, c_0 5.54; $a_0:c_0$ = 1:0.904. Cell contents $Cu_2B_2Cl_2O_4 \cdot 4H_2O$.

Habit. Tabular {001} or equant with large {001}, {201}, and {110}, with quadrangular terraces on {001}. Also as subparallel clusters spread out over a surface into irregular button-like or lichen-like groups, with the individual crystals flattened on the base or on a pyramidal face.

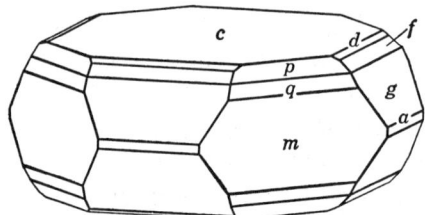

Phys. Cleavage {001}, perfect. Very flexible. H. $2\frac{1}{2}$. G. 2.810;[2] 2.81 (calc.). Luster vitreous, on cleavage surfaces pearly. Color deep blue with greenish lights. Streak pale blue.

Opt. Blue in transmitted light.

	n(Na)[3]	PLEOCHROISM	
O	1.691	Deep cendre-blue	Uniaxial negative (−).
E	1.641	Pale greenish yellow	

Chem. A hydrated copper borate-chloride $CuB_2O_4 \cdot CuCl_2 \cdot 4H_2O$.

Anal.

	1	2
Cu	35.74	34.94
Cl	19.94	19.47
B_2O_4	24.08	23.35
H_2O	20.24	19.60
Rem.		2.74
Total	100.00	100.10
G.		2.810

1. $CuB_2O_4 \cdot CuCl_2 \cdot 4H_2O$. 2. Mina Quetena, Chile. Rem. is insol. 1.84, CaO 0.05, MgO 0.05, Na_2O 0.40, Fe_2O_3 0.35, SO_3 0.05. The analyzed sample probably contained a small amount of admixed eriochalcite and atacamite.[4]

Tests. C.T. yields water and a yellow crystalline sublimate; in the O.T. yields water and a yellow crystalline sublimate that oxidizes black. Decomposed by water, leaving a residue of verdigris-green copper borate. Easily soluble in NH_4OH to an intense blue solution; more slowly soluble in HCl or HNO_3.

Occur. A secondary mineral found with atacamite and eriochalcite at Mina Quetena, near Calama, Chile.

Name. After Mark C. Bandy, mining engineer, who collected the mineral.

Ref.
1. Berman in Palache and Foshag (1938), by Weissenberg method.
2. Berman in Palache and Foshag (1938), by pycnometer.
3. Berman in Palache and Foshag (1938).
4. Foshag analysis in Palache and Foshag (1938).

SUSSEXITE SERIES

ORTHORHOMBIC

Sussexite, $(Mn,Mg)(BO_2)(OH)$
Szaibelyite, $(Mg,Mn)(BO_2)(OH)$

	a_0	b_0	c_0
Roweite, $(Mn,Mg,Zn)Ca(BO_2)_2(OH)_2$	8.27 kX	9.01	6.62

Sussexite and szaibelyite are isostructural and form a probably complete series by mutual substitution of Mg and Mn. Roweite, allied in formula, is not a member of the group.

26.1.5.1 **S U S S E X I T E** [$(Mn,Mg)(BO_2)(OH)$]. *Brush* (*Am. J. Sc.*, **46**, 140, 240, 1868).

26.1.5.2 **S Z A I B E L Y I T E** [$(Mg,Mn)(BO_2)(OH)$]. Sjajbélit *Peters* (*Ak. Wien Sitzber*, **44**, 143, 1861). Szaibelyite *Dana* (594, 1868). Boromagnesit *Groth* (38, 1874). Szajbelyite. Ascharite *Feit* (*Chem.-Ztg.*, **15**, 327, 1891). Camsellite *Ellsworth* and *Poitevin* (*Trans. Roy. Soc. Canada*, **15**, 1, 1921). Magnesiosussexite *Gruner* (*Am. Min.*, **17**, 509, 1932). Beta-ascharite *Godlevsky* (*Mem. soc. russe min.*, **66**, 315, 345, 1937). Alpha-ascharite *Godlevsky* (*Mem. soc. russe min.*, **66**, 315, 345, 1937).

Probably orthorhombic. As veinlets or masses with a felted or matted fibrous structure; also as parallel-fibrous or transverse-fibrous veinlets; as embedded nodules, sometimes dense or chalky. Under the microscope appears as minute laths or flattened fibers.

P h y s. Fibers not flexible. H. 3–3$\frac{1}{2}$. G. 3.30 (sussexite end-member) to 2.62 (szaibelyite end-member).[1] Luster silky to dull or earthy. Color white to buff or straw-yellow. Streak white.

O p t.[1] In transmitted light, colorless.

ORIENTATION	n (sussexite end-member)	n (szaibelyite end-member)	
X elongation	1.670	1.575	Biaxial negative (−)
Y	1.728	1.646	$2V \sim 25°$.
$Z \perp$ flattening	1.732	1.650	$r > v$.
			Parallel extinction.

C h e m. A basic borate of divalent manganese and magnesium, $(Mn,Mg)(BO_2)(OH)$. Mg and Mn substitute mutually, and a probably complete series extends between the Mg and Mn end-members. The names sussexite and szaibelyite are applied to the halves of the series with Mn > Mg and Mg > Mn, respectively. Small amounts of Fe″ and Zn substitute for (Mn,Mg), with Zn:(Mn,Mg) = 1:20 in analysis 4. The large amounts of SiO_2 reported in some analyses are due to admixture.[1] No essential water is lost below 500°.[1]

Anal.

	1	2	3	4	5	6
CaO		2.03	0.35	0.10	0.05	
MgO		9.56	14.57	16.29	29.32	45.24
MnO	61.81	49.40	40.42	37.58	23.48	1.09
FeO		0.16				1.28
B_2O_3	30.34	30.52	33.24	[33.16]	36.18	40.40
H_2O	7.85	8.33	8.91	7.90	10.40	10.81
SiO_2			0.89	0.50	0.43	
Rem.			0.98	4.47	0.34	1.18
Total	100.00	100.00	99.36	[100.00]	100.20	100.00
G.			3.0–3.1		2.83	

1. $Mn(BO_2)(OH)$. 2. Sussexite. Franklin, New Jersey.[2] Recalculated to 100 after deducting 4.5 per cent willemite. 3. Sussexite. Chicagon mine, Michigan.[3] Rem. is insol. 0.69, $(Fe,Al)_2O_3$ 0.29. 4. Sussexite, Franklin, New Jersey.[4] Rem. is Fe_2O_3 0.60, ZnO 3.87, H_2O includes H_2O- 0.10. 5. Szaibelyite. Gogebic range, Michigan (*magnesiosussexite*).[5] Rem. is $(Fe,Al)_2O_3$ 0.34. H_2O includes H_2O- 0.22. 6. Szaibelyite. Douglas Lake, B.C. (*camsellite*).[6] Rem. is Fe_2O_3 0.85, Al_2O_3 0.29, Alk. 0.04. Recalculated to 100 after deducting chrysotile and dolomite.

	7	8	9	10	11	12
CaO			nil		1.26	
MgO	48.84	47.12	46.70	49.44	46.72	47.92
MnO			0.02			
B_2O_3	40.49	40.17	40.85	34.60	[31.22]	41.38
H_2O	10.67	12.00	11.27	12.37	11.13	10.70
SiO_2			0.20		4.83	
Rem.		0.39	1.55	3.40	4.84	
Total	100.00	99.68	100.59	99.81	[100.00]	100.00
G.	2.6		2.64	3.0	2.76	

7. Szaibelyite. Bolinas Bay, California.[7] Recalculated to 100 after deducting admixed serpentine. 8. Szaibelyite. Schmidtmannshall (*ascharite*).[8] Rem. is R_2O_3 0.29, insol. 0.10. 9. Szaibelyite. Inder salt lake, U.S.S.R. (*beta-ascharite*).[9] Rem. is Al_2O_3 0.16, Fe_2O_3 0.13, Na_2O 0.20, K_2O 0.26, Cl 0.11, SO_3 0.69. 10. Szaibelyite. Rézbánya, Hungary.[10] Made on impure material. Rem. is Cl 0.20, Fe_2O_3 3.20. 11. Szaibelyite. Pioche, Nevada.[11] Rem. is Al_2O_3 0.63, Fe_2O_3 4.21. H_2O includes H_2O- 1.26. Analysis on 0.05 g. of impure material. 12. $Mg(BO_2)(OH)$.

Tests. B.B. easily fusible, material high in Mn affording a nearly black enamel. In C.T. yields water. Slowly soluble in acids.

Occur. Sussexite was found originally in hydrothermal veinlets cutting massive franklinite ore at Franklin, Sussex County, New Jersey, and was associated with pyrochroite, rhodochrosite, willemite, leucophoenicite. Much of what is labeled sussexite from Franklin in collections is a form of tremolite asbestos cemented by calcite or a mixture of fibrous calcite and zincite ("calcozincite"). Sussexite also occurs with seamanite as veinlets in hematite and cherty gangue at the Chicagon mine, Iron County, Michigan. Szaibelyite was originally found as nodular masses in crystalline limestone at Rézbánya, Hungary. Also found with boracite, sylvite, halite, kainite in salt deposits at Schmidtmannshall and Vienenberg, near Aschersleben, and later at Neu-Stassfurt, Saxony, Germany (*ascharite*).[12] Abundantly in the borate deposits near Inder Lake, western Kazakhstan, U.S.S.R. (*β-ascharite*).[12] With fluoborite and magnesioludwigite at the Hol Kol mine, Suan, Korea. With ludwigite at Norberg and other magnetite skarn deposits in Sweden.[13] With dolomite and chrysotile in serpentine near Douglas Lake, Nicola District, British Columbia (*camsellite*).[12] In the United States in serpentine at Bolinas Bay, Marin County, California. With magnesioludwigite and fluoborite (?) at Blind Mountain near Pioche, Lincoln County, Nevada. Manganoan szaibelyite occurs as veinlets in the iron ore of the Eureka mine, Gogebic range, Michigan (*magnesiosussexite*).[12]

Alter. As an alteration of colemanite, inyoite, and hydroboracite (Inder).

Name. Sussexite from the locality in Sussex County, New Jersey. Szaibelyite after Stephan Szaibely [Sjájbely] (1777–1855), mine surveyor of Rézbánya, Hungary, who first collected the mineral. Ascharite from the old Latin name, Ascharia, of the province in which the mineral is found in Germany. Camsellite after Charles Camsell, of the Geological Survey of Canada.

Ref.
1. Schaller, *Am. Min.*, **27**, 467 (1942).
2. Bauer analysis in Palache, *Am. Min.*, **13**, 323 (1928). Earlier analyses of sussexite from Franklin are given by Brush (1868) and Penfield and Sperry, *Am. J. Sc.*, **36**, 323 (1888).
3. Slawson, *Am. Min.*, **19**, 575 (1934).
4. Poitevin and Ellsworth, *Am. Min.*, **9**, 188 (1924).
5. Kameda analysis in Gruner (1932).
6. Ellsworth and Poitevin (1921).
7. Eakle, *Am. Min.*, **10**, 100 (1925), as recalculated by Schaller (1942).
8. Schaller (1942). Additional analyses of the German ascharite are given by van't Hoff, *Unters. ozean. Salzablag.*, **2**, 60 (1909), and Feit (1891).
9. Godlevsky (1937); see also Godlevsky and Egorova (1936), and Boky (1937).
10. Stromeyer, *Ak. Wien, Sitzber.*, **47**, 347 (1863), who cites another analysis.
11. Gillson and Shannon, *Am. Min.*, **10**, 137 (1925).
12. Material from the following localities has been shown to be isostructural or identical by x-ray powder study: (1) Rézbánya, (2) Gogebic mine, Michigan, (3) Marin County, California, (4) near Pioche, Nevada, (5) Franklin, New Jersey, (6) British Columbia, (7) Germany, (8) Inder Lake, by the following: Frondel, priv. comm. (1948) [1, 2, 3, 4, 5]; Richmond in Schaller (1942) [5, 6, 7]; Watanabe, *Min. Mitt.*, **50**, 454 (1939) [1, 3, 7]; Gruner (1932) [2, 5, 6]; Agafonova and Isküll (1939) [6, 8]. On optical and chemical evidence of these relations, see also Schaller (1942).
13. Geijer, *Geol. För. Förh.*, **61**, 19 (1930).

26.1.6 **ROWEITE** [(Mn,Mg,Zn)Ca(BO$_2$)$_2$(OH)$_2$]. Berman and Gonyer (*Am. Min.*, **22**, 301, 1937).

Orthorhombic.[1] As rough lath-shaped crystals, flattened {010} and elongated [001], without measurable terminations.

Structure cell.[3] a_0 8.27 ± 0.01 kX, b_0 9.01 ± 0.01, c_0 6.62 ± 0.02; $a_0:b_0:c_0 = 0.916:1:0.735$. Cell contents (Mn,Mg,Zn)$_4$Ca$_4$(BO$_2$)$_8$(OH)$_8$.

Phys. Cleavage {101}, poor. The crystals are brittle and break with an even fracture across the elongation. H. ~5. G. 2.92 ± 0.02;[2] 2.85 (calc.). Color light brown. Transparent.

Opt. In transmitted light, colorless.

Orientation		n	
X	a	1.648 ± 0.003	Biaxial negative (−).
Y	c	1.660 ± 0.003	2V 15°.
Z	b	1.663 ± 0.003	$r < v$, strong.

Chem. A basic borate of calcium and manganese, (Mn,Mg,Zn)-Ca(BO$_2$)$_2$(OH)$_2$. Minor amounts of Mg and Zn substitute for Mn in the only reported analysis.

Anal.

	1	2
CaO	26.33	25.55
MgO	1.58	1.67
ZnO	3.18	3.15
MnO	27.75	28.48
B$_2$O$_3$	32.70	32.59
H$_2$O	8.46	8.56
Total	100.00	100.00
G.		2.92

1. (Mn,Mg,Zn)Ca(BO$_2$)$_2$(OH)$_2$, with Mn:Mg:Zn = 10:1:1. 2. Franklin, New Jersey.[4] Recalc. to 100 after deduction of 0.84 insol.

Tests. Fuses at 1 to a black glass. In the C.T. turns brown, gives off water and after long heating melts to a glass. Easily soluble in dilute HCl.

Occur. Found at Franklin, New Jersey, associated with thomsonite and willemite in veinlets cutting massive franklinite-zincite ore. Only one specimen is known.

Name. After Mr. George Rowe of Franklin, for many years Mine Captain and a collector of the local minerals.

Ref.

1. Established by optical and single-crystal Weissenberg and Laue x-ray study. The crystals were too rough to permit an accurate morphological study.
2. By suspension in heavy liquids.
3. Berman in Berman and Gonyer (1937) by Weissenberg method.
4. Gonyer anal. in Berman and Gonyer (1937).

26.1.7 **BORACITE** [Mg$_3$B$_7$O$_{13}$Cl]. Würfelstein. Kubische Quarzkrystalle (from Lüneburg) *Lasius* (*Crell's Ann.*, **2**, 333, 1787). Lüneburger Sedativ-Spath *Westrumb* (*Kl. phys.-ch. Abh.*, **3**, 167, 1789). Borazit *Werner* (*Bergm. J.*, 393, 1789; 234, 1790). Quartz cubique, Borate magnésiocalcaire *Westrumb* (*Ann. chim.*, [1], **2**, 101, 1801). Boracite *Kirwan* (**1**, 155, 1810). Magnésie boratée *Haüy* (**2**, 56, 1822). Borate of Magnesia. Parasit *Volger* (*Ann. Phys.*, **92**, 77, 1854). Stassfurtit *Rose* (*Ann. Phys.*, **97**, 632, 1856). Alpha-boracite, Beta-boracite *Mehmel* (*Zs. Kr.*, **87**, 239, 1934; **88**, 1, 1934).
Eisenstassfurtit *Huyssen* (*Jb. Min.*, 329, 1865). Huyssenite, Iron-boracite *Dana* (596, 799, 1868). Eisenboracit.

Cryst. Above 265°,[1] isometric hextetrahedral—4 $\bar{3}$ m. At ordinary temperatures, in paramorphic isometric crystals,[2] orthorhombic pyramidal —$mm2$ with $a:b:c = 1:1:0.7072$.

Forms:[3] (Isometric)

| a 001 | d 011 | o 111 | $-o$ $\bar{1}$11 | $-n$ $\bar{1}$12 | V 135 |

Less common and rare:

| η 0.3.17 | h 014 | e 012 | τ $\bar{1}$.16.16 | Γ 255 | n 112 | u 134 |
| ϑ 0.3.13 | f 013 | l 035 | σ $\bar{1}$88 | p 122 | ρ 144 | |

BORACITE

Structure cell. The isometric high-temperature form [4] has the probable space group $F\bar{4}3c$, with a_0 12.07 kX (colorless; at 265°), increasing with content of Fe. Cell contents $Mg_{24}B_{56}O_{104}Cl_8$.

The inverted ordinary-temperature orthorhombic form [5] has the space group Cmm, with a_0 16.97 kX, b_0 16.97, c_0 12.10 (colorless); a_0 17.11 kX, b_0 17.11, c_0 12.10 (greenish; ferroan); $a_0:b_0:c_0 = 1:1:0.707$. Cell contents $Mg_{48}B_{112}O_{208}Cl_{16}$.

Habit. Commonly as isolated, embedded crystals; less often in groups, or with small individuals projecting from the face-centers of larger single-crystals. Often cubic; also dodecahedral, tetrahedral; pseudo-octahedral or cubo-octahedral with {111} usually larger and brighter than {$\bar{1}11$}. Faces often pitted due to etching. Also as fine-granular and fibrous to plumose aggregates.

Twinning.[6] As penetration twins on {111} (growth twins); rare.

P h y s. Cleavage none.[7] Fracture conchoidal to uneven. H. 7–7½. G. 2.91–2.97,[8] 2.93 ± 0.02 [9] (colorless crystals); 2.97–3.10 (green; ferroan); 2.97 (calc.). Luster vitreous, inclining toward adamantine. Colorless to white; also gray, yellow, bluish green, green and dark green (ferroan). Streak white. Transparent to translucent. Strongly piezoelectric and pyroelectric.[10] The [111] axes of the paramorphs are polar, with the faces of the larger and brighter tetrahedron o{111}—the antilogous pole—becoming positively charged on cooling or compression.

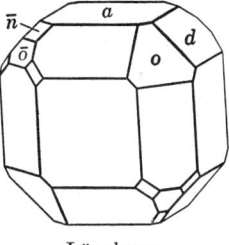

Lüneburg.

O p t. In transmitted light, colorless.

	(Isometric modification) [11]		(Orthorhombic modification)			
				Locality unknown [12]	Yorkshire [13]	
λ	501.6	587.6	nX	1.6622	1.658	Biaxial positive (+).
n at 290°	1.6776	1.6714	nY	1.6670	1.662	$2V$ 82½°.
n at 502°	1.6796	1.6741	nZ	1.6730	1.668	

C h e m. A magnesium chloroborate,[14] $Mg_3B_7O_{13}Cl$. Sometimes contains Fe″ in substitution for Mg, up to Fe″:Mg = 1:1 in an old analysis.[15] The fibrous and plumose varieties usually contain a small amount of non-essential water.

Anal.[16]

	1	2	3	4
FeO		tr.	1.09	6.91
MgO	25.71	26.39	26.38	21.59
MgCl$_2$	12.14	12.35	12.17	11.91
B$_2$O$_3$	62.15	[61.26]	59.68	59.77
Ign. loss			0.55	0.05
Total	100.00	[100.00]	99.87	100.23

1. $Mg_3B_7O_{13}Cl$. 2. Eime, Germany.[17] Average of two. 3. Lüneberg, Germany.[18] 4. Solvay, Germany.[18] Analyses 3 and 4 each average of two.

Tests. B.B. fuses at 2 with intumescence to a bead, clear and yellowish while hot and white and crystalline when cold. Slowly but completely soluble in HCl. Very slowly decomposed by water.

Var. Ferroan. Eisenstassfurtit *Huyssen* (*Jb. Min.*, 329, 1865). Huyssenite, Iron-boracite *Dana* (586, 799, 1868). Eisenboracit. (Contains Fe'' in substitution for Mg. Color greenish, with increased G. and unit cell dimensions. The temperature of inversion to the isometric form also is increased.

O c c u r. Found in bedded sedimentary deposits of anhydrite, gypsum, and halite; and in potash deposits of the oceanic type. Occurs as crystals and masses in the carnallite and kainite layers at Stassfurt, Douglashall and Leopoldshall, in Saxony, Germany. In rock salt at Solvayhall near Bernburg and at Hildesia near Hildesheim, Saxony. Also embedded in anhydrite and gypsum at Kalkberg and Schildstein near Lüneberg, Hannover, at Eime, and at Hohenfels near Sehnde. From Segeberg near Kiel, Schleswig-Holstein. At Luneville, La Meurthe, France. From a boring in Permian salt and anhydrite at Aislaby, Yorkshire. In the United States, boracite has been found with hilgardite, parahilgardite, magnesite, danburite, and anhydrite in the water-insoluble residue of rock salt from the Choctaw salt dome, Iberville Parish, Louisiana. Probably in none of the natural occurrences has boracite formed at temperatures near or over 265°.

A l t e r. Observed as pseudomorphs after quartz at Douglashall.[19]

A r t i f.[20] Obtained in crystals by slowly cooling a fusion of MgO and H_3BO_3 in a great excess of $MgCl_2$ and NaCl. Also by heating 1 part $Na_2B_4O_7$ with 2 parts $MgCl_2$ with H_2O at *ca.* 280° in a pressure bomb; and by the action at a high temperature of $MgCl_2$ vapor on ulexite. Zn, Cd, Pb, Co, Ni, Cu and Fe compounds apparently isostructural with boracite have been synthesized, as have the corresponding bromoborates and iodoborates.

N a m e. In allusion to the composition.

Ref.

1. Although crystals exhibit hextetrahedral symmetry, colorless boracite is isometric only above 265° when it becomes isotropic (Mallard, *Bull. soc. min.*, **6**, 122, 1883; **5**, 214, 1882. Klein, *Jb. Min.*, I, 235, 1884) and shows no pyroelectricity (Friedel and Curie, *Bull. soc. min.*, **6**, 191, 1883). According to Rinne, *Jb. Min.*, II, 108 (1900), the birefringence of green, ferroan boracite is considerably reduced at 285° but does not disappear even at 400°.
2. The non-isometric symmetry of boracite at ordinary temperatures was first recognized by Brewster, *Edinburgh Phil. J.*, **5**, 217 (1821), who observed the anomalous birefringence; he considered the symmetry rhombohedral, pseudo-cubic. Biot, *Ac. sc. Paris, Mém. 18*, 667 (1842), established the biaxial character of the crystals. Mallard, *Ann. mines*, [7], **10**, 93, (1876), found that certain crystals cut parallel (111) showed optically the triangular faces irregularly divided into three segments, joined along lines from the middle of the sides to the face-center. He concluded that such crystals were composed of 12 orthorhombic individuals with each having a dodecahedral face for base and projecting inward to the crystal center, any two adjacent individuals being related by twinning across some (110) plane of the isometric form. A geometrical consideration of their packing gave the ratio $a:b:c = 1:0.7072:1$, related to the struc-

tural ratio of Gruner, *Am. Min.*, **13**, 481 (1928), *Am. J. Sc.*, **17**, 453 (1929), by the transformation $100/00\bar{1}/010$. In Gruner's setting [010] and the polar axis [001] are, respectively, the long and short diagonals of the dodecahedral faces with [100] normal. See also the confirmatory observations of Rinne (1900), Feng, *Neue Beobacht. an Boraciten*, Rostock (1930), and Shôji, *Sc. Rpt. Tôhoku Univ.*, [1], **26**, 86 (1937).

Baumhauer, *Zs. Kr.*, **3**, 337 (1879), concluded on the basis of etching that six orthorhombic individuals with a and b parallel to the cube face diagonals and c parallel to the cube crystal axes composed the aggregate, giving $a:b:c = 1:1:0.7072$. Klein, *Jb. Min.*, II, 209 (1880), after a study of many crystals stated that crystals with cubic or dodecahedral habit were segmented as suggested by Mallard, those with tetrahedral habit as by Baumhauer. A detailed study by Mehmel, *Zs. Kr.*, **87**, 239 (1934), **88**, 1 (1934), of variously oriented sections at different temperatures shows almost universal polysynthetic twinning and relationships usually more complicated than pictured by Mallard and Baumhauer. Twinning accompanies the inversion from the isometric to the orthorhombic form, which invariably proceeds in directions normal to (110) planes resulting in Mallard's twinned aggregate under favorable conditions.

3. Goldschmidt (**1**, 211, 1913). Uncertain forms:

0.1.12   3.20.60 $\bar{1}16$ 126   6.7.21   236

4. Mehmel, *Zs. Kr.*, **88**, 1 (1934).

5. Gruner (1928; 1929) by the Seeman-Schiebold oscillation method on sections cut from large dodecahedrons from Hannover. Colorless crystals from Westeregeln, G. 2.92–2.97; green from Solvay, G. 2.97–3.10 (not analyzed). Rinne, *Zs. Kr.*, **60**, 67 (1924), obtained a_0 12.10 kX (isometric) on crystals from Eime by the powder method. See Mehmel (1934) for further measurements on material from various localities and at different temperatures; also Hocart, *Bull. soc. min.*, **57**, 62 (1934).

The orthorhombic modification is pseudo-isometric; [100] and [010] are the [110] and [$\bar{1}$10] directions, respectively, in the isometric orientation; [001] orthorhombic = 0 01] isometric. Hocart (1934) confirmed the orthorhombic space group but found close approximation to face-centered cubic; see also Shôji (1937).

6. Schrauf, *Atlas*, Pl. 36, fig. 6 (1873); also Schultze, *Jb. Min.*, 844 (1871), and Bücking, *Ak. Berlin, Sitzber*, **28**, 533 (1895).

7. Sometimes observed in traces parallel {111}, {$\bar{1}$11}, but probably due to parting along twinned surfaces.

8. The wide reported range in G. apparently is significant, but the nature of the accompanying variation in composition is not known.

9. Range of most frequently reported values, including new measurements on Stassfurt crystals by Mrose, priv. comm. (1946).

10. The pyroelectricity of boracite was first noted by Haüy in 1791 and later was described in detail by Hankel, *Ann. Phys.*, **50**, 471 (1840), **61**, 282 (1844); Mack, *Zs. Kr.*, **8**, 503 (1884); Friedel and Curie, *Bull. soc. min.*, **6**, 191 (1883); the piezoelectricity was first recognized by J. and P. Curie, *Bull. soc. min.*, **3**, 90 (1880); see also Cady, *Piezoelectricity*, New York, 230 (1946).

11. Marbach, Inaug. Diss., Leipzig (1913)—in Hintze (**1**, [4A], 113, 1921).

12. Mallard, *Bull. soc. min.*, **6**, 129 (1883), by prism method.

13. Guppy, *Min. Mag.*, **27**, 51 (1944).

14. See Mehmel (1934) and Gruner (1928) for a discussion of the long-standing problem of the composition of boracite. The observed variation in G. and the indices of refraction suggests that the composition varies, aside from the substitution of Fe, but the mechanism of this is not known. Mehmel (1934) suggests a series toward $Mg_7B_{16}O_{30}Cl_2$.

15. Huyssen, *Jb. Min.*, 329 (1865), with FeO 23.9. Bücking, *Zs. Kr.*, **15**, 574 (1889), cites material with FeO 7.9; see also analysis 4.

16. For earlier analyses see Doelter (**3** [2], 418, 1922); also Guppy (1944).

17. Ward analysis in Gruner, *Am. Min.*, **13**, 481 (1928).

18. Mehmel (1934).

19. Ochsenius, *Jb. Min.*, I, 271 (1889).

20. Mellor (**5**, 137, 139, 1924).

26.1.8 **H I L G A R D I T E** [$Ca_8(B_6O_{11})_3Cl_4 \cdot 4H_2O$]. *Hurlbut* and *Taylor* (*Am. Min.*, **22**, 1052, 1937).

Cryst. Monoclinic; domatic—*m*.

$a:b:c = 1.0147:1:0.5585$; $\beta\ 90°00'$; $p_0:q_0:r_0 = 0.5504:0.5585:1$

$r_2:p_2:q_2 = 1.7905:0.9855:1$; $\mu\ 90°00'$; $p_0'\ 0.5504,\ q_0'\ 0.5585,\ x_0'\ 0.00$

Forms:

		ϕ	$\rho = C$	ϕ_2	$\rho_2 = B$	A
b	010	0°00'	90°00'	......	0°00'	90°00'
M	$\bar{1}$10	−44 35	90 00	180°00'	135 25	134 35
E	0$\bar{1}\bar{1}$	180 00	150 49	−90 00	119 11	90 00
p	111	44 35	38 06	61 10½	63 55½	64 20
Q	1$\bar{1}\bar{1}$	135 25	141 54	−61 10½	116 04½	64 20
R	2$\bar{1}\bar{1}$	116 54	129 00½	−42 15	110 35	46 08
s	$\bar{3}$11	−71 18½	60 10½	148 48	73 52	145 15

Less common and rare:

   $K\ \bar{1}30$ $L\ \bar{2}30$ $q\ \bar{1}11$ $r\ \bar{2}11$

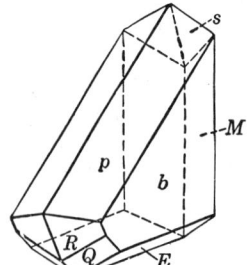

Structure cell.[1] Space group Pc or Pm. a_0 11.35 kX, b_0 11.12, c_0 6.20; [β 90°00']; $a_0:b_0:c_0 = 1.027:1:0.559$. Cell contents $Ca_8B_{18}O_{33}Cl_4 \cdot 4H_2O$.

Habit. Tabular {010}, with a marked hemimorphic aspect.

Oriented growths. Occurs in oriented intergrowth with parahilgardite (which see).

Phys. Cleavage {010} perfect, {100} also perfect but less easy to produce. H. 5. G. 2.71; 2.65 (calc.). Colorless and transparent. Luster vitreous.

Opt. In transmitted light, colorless.

ORIENTATION		n(Na)	
$X \wedge c$	88½°	1.630 ± 0.002	Biaxial positive (+).
$Y\ \ b$		1.636 ± 0.002	$2V\ \ 35°$.
$Z \wedge c$	1½°	1.664 ± 0.002	$r > v$.

Chem. A hydrated chlor-borate of calcium, $Ca_8(B_6O_{11})_3Cl_4 \cdot 4H_2O$.

Anal.

	1	2
CaO	35.67	35.14
B_2O_3	49.86	50.22
H_2O	5.73	6.44
Cl	11.28	10.59
Total	102.54	102.39
Less O = Cl	2.54	2.39
Total	100.00	100.00

1. $Ca_8(B_6O_{11})_3Cl_4 \cdot 4H_2O$. 2. Louisiana.[2] Recalculated to 100 after deduction of 1.89 per cent insoluble.

Tests. B.B. fuses at 2 to a white globule. In C.T. easily loses water with a strong acid reaction.

Occur. Found among the water-insoluble residues [3] of rock salt from the Choctaw salt dome, Iberville Parish, Louisiana. Associated minerals include parahilgardite, anhydrite, boracite, danburite, magnesite, dolomite, calcite, quartz, pyrite, hauerite.

Name. After Eugene Woldemar Hilgard (1833–1916), geologist, one of the first to describe the saline deposits of Louisiana.

Ref.
1. Hurlbut in Hurlbut and Taylor (1937) by Weissenberg and Laue methods.
2. Gonyer anal. in Hurlbut and Taylor (1937).
3. For a general description of these residues see Taylor, *Bull. Am. Assoc. Petr. Geol.*, **21**, 1268 (1937).

26.1.9 **PARAHILGARDITE** $[Ca_8(B_6O_{11})_3Cl_4 \cdot 4H_2O]$. Hurlbut (*Am. Min.*, **23**, 765, 1938).

Cryst.[1] Triclinic; pedial—1.

$a:b:c = 0.5045:1:0.2783;$ $\alpha\ 90°00',\ \beta\ 90°00',\ \gamma\ 91°12'$

$p_0:q_0:r_0 = 0.5518:0.2781:1;$ $\lambda\ 90°00',\ \mu\ 90°00',\ \nu\ 88°48'$

$p_0'\ 0.5518,\ q_0'\ 0.2784,\ x_0'\ 0.00,\ y_0'\ 0.00$

Forms:[2]

	ϕ	ρ	A	B	C
$b\ 0\bar{1}0$	180°00′	90°00′	91°12′	180°00′	90°00′
$a\ 100$	88 48	90 00	0 00	88 48	90 00
$\bar{a}\ \bar{1}00$	− 91 12	90 00	180 00	91 12	90 00
$j\ \bar{1}60$	− 18 23½	90 00	107 11½	18 23½	90 00
$k\ \bar{1}20$	− 45 20	90 00	134 08	45 20	90 00
$l\ \bar{1}20$	−135 51	90 00	135 21	135 51	90 00
$n\ \bar{1}60$	−161 50½	90 00	109 21½	161 50½	90 00
$p\ \bar{1}21$	−135 51	141 37	116 12½	116 27½	141 37
$r\ \bar{1}6\bar{1}$	− 18 23½	119 46	104 52	34 33	119 46
$s\ \bar{3}42$	−124 41½	134 49	126 16	113 48½	134 49
$t\ \bar{3}21$	−109 40	60 21½	145 31½	107 00½	60 21½

Structure cell.[3] Space group P1. a_0 11.24 kX, b_0 22.28, c_0 6.20; $a_0:b_0:c_0 = 0.5045:1:0.2783$; $\alpha\ 90°00'$, $\beta\ 90°00'$, $\gamma\ 91°12'$. Cell contents $Ca_{16}(B_6O_{11})_6Cl_8 \cdot 8H_2O$.

Habit. Found only as oriented intergrowths with hilgardite. The intergrowths consist of two partly merged individual crystals of parahilgardite, one right-handed and one left-handed and apparently in twin position, attached by the positive end of [100] to ($\bar{1}$00) of a single hilgardite crystal. Parahilgardite {100} [001] parallels hilgardite {100} [001], with the {010} cleavages of the two minerals inclined at 1° 12′. Common forms: $b\ p\ t$,

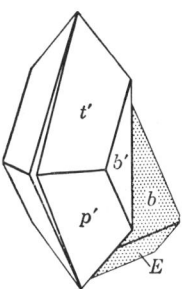

Intergrowth with hilgardite.

Phys. Cleavage {010} and {100}, perfect. H. 5. G. 2.71; 2.67 (calc.) Colorless and transparent. Luster vitreous.
Opt. In transmitted light, colorless.

	ORIENTATION		n(Na)	
	ϕ	ρ		
X	156°	84°	1.630	Biaxial positive (+).
Y	66°	89°	1.636	2V 35°.
Z	0°	5°	1.664	$r > v$.

In (001) sections, hilgardite has parallel extinction and parahilgardite an extinction angle of about 20°.

Chem. A hydrated chlor-borate of calcium, $Ca_8(B_6O_{11})_3Cl_4 \cdot 4H_2O$. Dimorphous with hilgardite.

Anal.

	1	2
CaO	35.67	35.13
B_2O_3	49.86	50.26
H_2O	5.73	6.23
Cl	11.28	10.82
Total	102.54	102.44
Less O = Cl	2.54	2.44
	100.00	100.00

1. $Ca_8(B_6O_{11})_3Cl_4 \cdot 4H_2O$. 2. Louisiana.[4] Recalculated to 100 after deduction of 2.21 insol.

Tests. Like hilgardite.

Occur. As oriented overgrowths upon hilgardite crystals in the water-insoluble residues of rock salt from the Choctaw salt dome, Iberville Parish, Louisiana.

Name. In allusion to the close relation to hilgardite.

Ref.

1. Point symmetry established by a positive test for piezoelectricity, using the electronic method of Noyes and Pierce, *J. Acoust. Soc. Am.*, **9**, 205 (1938), in conjunction with other evidence proving the triclinic nature of the crystals. The x-ray elements, here given, are superior to those based on morphological measurements due to the poor surface quality of the crystals.
2. Angle table calculated from the x-ray elements.
3. By Weissenberg method.
4. Gonyer anal. in Hurlbut (1938).

      JOHACHIDOLITE. Iwase and Saito (*Sci. Papers Inst. Phys. Chem. Res., Tokyo*, **39**, 300, 1942).

A partly described mineral found as grains and lamellar masses in nepheline dikes cutting limestone in the Johachido district, Kenkyohokudo prefecture, Korea. Colorless and transparent. H. 6½-7. G. slightly below 3.4. Optically biaxial negative (−) with nX 1.715, nY 1.720, nZ 1.726 (for D) with 2V 72° (calc.) and $r > v$ strong. Chemically a hydrous fluoborate of sodium, calcium, and aluminum. Analysis gave: Na_2O 8.27, CaO 24.77, MnO 0.23, Fe_2O_3 0.09, Al_2O_3 28.34, SiO_2 0.34, P_2O_5 0.03, B_2O_3 24.21, Cl none, F 12.21, H_2O- 0.07, H_2O+ 6.52, total 105.08 (less O = F 5.14 = 99.94). Fluoresces an intense blue in ultraviolet light owing to traces of rare earths.

27 COMPOUND BORATES

27.1.1 L U E N E B U R G I T E [$Mg_3B_2(OH)_6(PO_4)_2 \cdot 6H_2O$]. *Nöllner (Ak. München, Sitzber., 291, 1870).*

Probably monoclinic.[1] In flattened masses and nodules with fine fibrous to earthy structure. Also as minute pseudo-hexagonal tablets.

P h y s. Under the microscope exhibits a prismatic cleavage with an angle of about 73°.[2] H. ~2. G. 2.05. Color white to brownish white.

O p t.

ORIENTATION	n (Lüneburg [3])	n (New Mexico [4])	
X	1.520	1.522	Biaxial negative (−).
Y b?	1.54	1.541	$2V$ moderate.
Z	1.545	1.548	
$2V$		62° (calc)	

C h e m. A hydrated basic borate-phosphate of magnesium, $Mg_3B_2(OH)_6(PO_4)_2 \cdot 6H_2O$. The water content is uncertain.[5]

Anal.

	1	2	3
MgO	24.44	25.3	25.13
CaO			0.15
B_2O_3	14.07	12.7	12.9
P_2O_5	28.71	29.8	29.61
H_2O	32.78	32.2	32.16
Total	100.00	100.0	99.95

1. $Mg_3B_2(OH)_6(PO_4)_2 \cdot 6H_2O$. 2. Lüneburg.[6] Stated to lose about 0.7 BF_3 on ignition. 3. Lüneburg.[7]

O c c u r. Found with boracite in a gypsum-bearing marl at Lüneburg, Hannover, Germany. Also found associated with halite, sylvite, polyhalite, and clay in the Permian salt basin of southeastern New Mexico and adjacent parts of Texas.[8] A probably identical mineral has been described from a guano deposit at Mejillones, Peru.[9]

N a m e. From the original locality.

Ref.

1. Mügge in Biltz and Marcus, *Zs. anorg. Chem.*, **77**, 124 (1912), and Larsen (103, 1921) from optical evidence.
2. Mügge in Biltz and Marcus (1912).
3. Larsen (1921).
4. Henderson in Schaller and Henderson, *U. S. Geol. Sur., Bull. 833*, 47 (1932).
5. Dehydration data are given in Biltz and Marcus (1912).
6. Nöllner (1870).
7. Biltz and Marcus (1912).
8. Schaller and Henderson (1932).
9. Domeyko, *C.R.*, **90**, 544 (1880).

27.1.2 C A H N I T E [Ca$_2$B(OH)$_4$(AsO$_4$)]. *Palache* and *Bauer* (*Am. Min.*, **12**, 149, 1927).

C r y s t.[1] Tetragonal; disphenoidal—$\bar{4}$.

$a{:}c = 1{:}0.615; \quad p_0{:}r_0 = 0.615{:}1$

Forms:[2]

	φ	ρ = C	A	$\bar{M}$	M
a 010	0°00′	90°00′	90°00′	45°00′	45°00′
p 111	45 00	41 01	62 21	90 00	48 59
o $\bar{1}$11	−45 00	41 01	117 39	48 59	90 00
's 311	71 34	62 47½	32 28	113 26	37 18

Habit. Single untwinned crystals, of pseudo-tetrahedral habit, are rare. Commonly twinned on {110}, the individuals interpenetrating symmetri-

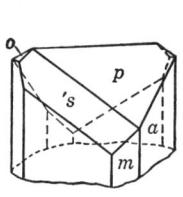

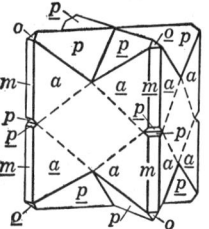

Idealized twin.

cally. Twins with well-developed sphenoid faces have a characteristic cross-like appearance.

P h y s. Cleavage {110}, perfect. Brittle. H. 3. G. 3.156. Colorless to white. Transparent. Luster vitreous.

O p t.[3] In transmitted light, colorless.

nO 1.662   Uniaxial positive (+).
nE 1.663   Dispersion strong; exhibits abnormal interference colors.

C h e m. A basic borate-arsenate of calcium, Ca$_2$B(OH)$_4$(AsO$_4$).
Anal.

	1	2	3	4
CaO	37.64	38.27	37.13	37.62
B$_2$O$_3$	11.69	10.14	11.64	11.86
As$_2$O$_5$	38.57	36.79	37.47	38.05
H$_2$O	12.10	11.75	11.78	12.42
Rem.		1.39	1.58	
Total	100.00	98.34	99.60	99.95

1. Ca$_2$B(OH)$_4$(AsO$_4$). 2,3,4. Franklin, New Jersey.[4] Total of analysis 2 originally given as 99.99. Analyses 2 and 3 on impure 0.1- and 0.26-gram samples; analysis 4 on a pure 0.5-gram sample. Rem. in no. 2 is PbO 1.15, MgO 0.24; in no. 3, ZnO.

Tests. B.B. fuses at about 3. In C.T. yields water and becomes opaque but does not fuse. Easily soluble in dilute HCl.

SULFOBORITE

Occur. Found in the United States at Franklin, New Jersey, associated with hedyphane, barite, pyrochroite, willemite, rhodonite, and datolite in cavities in axinite veinlets. The mineralization is associated with pegmatites cutting the main ore-body. Also found associated only with rhodonite, or implanted on garnet crystals lining drusy cavities in franklinite ore.

Name. After Lazard Cahn (1865–1940), mineral collector and dealer, who first recognized the species here described.

Ref.
1. Point symmetry established by the development of {311}. Cf. Palache, *Am. Min.*, **26**, 429 (1941).
2. Palache and Bauer (1927); Palache (1941). The larger brighter {hhl} sphenoid is taken as positive {111}.
3. Berman in Palache and Bauer (1927).
4. Bauer analyses in Palache and Bauer (1927).

27.1.3 **S U L F O B O R I T E** [$Mg_6H_4(BO_3)_4(SO_4)_2 \cdot 7H_2O$]. Sulfoborit *Naupert* and *Wense* (*Ber.*, **26**, 874, 1893); *Bücking* (*Ak. Berlin, Sitzber.*, 967, 1893). Sulphoborite.

Cryst.[1] Orthorhombic; pyramidal—$m\,m\,2$.

$a:b:c = 0.6196:1:0.8100$;[2] $p_0:q_0:r_0 = 1.3072:0.8100:1$

$q_1:r_1:p_1 = 0.6196:0.7649:1$; $r_2:p_2:q_2 = 1.2346:1.6139:1$

Forms:[2]

	ϕ	$\rho = C$	ϕ_1	$\rho_1 = A$	ϕ_2	$\rho_2 = B$
c 001	.....	0°00′	0°00′	90°00′	90°00′	90°00′
b 010	0°00′	90 00	90 00	90 00	.....	0 00
m 110	58 13	90 00	90 00	31 47	0 00	58 13
r 101	90 00	52 35	0 00	37 25	37 25	90 00
o 111	58 13	56 58	39 00½	44 33	37 25	63 48

Habit. Prismatic [001], of varying habit.

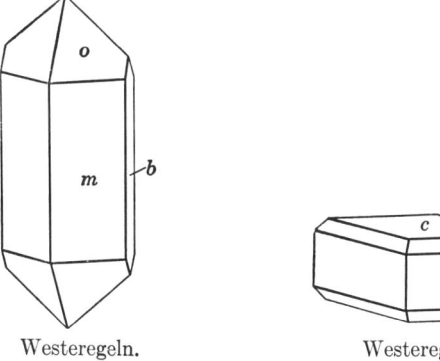

Westeregeln.   Westeregeln.

Phys. Cleavage {110} good, {001} fair. Brittle. H. 4–4½. G. 2.38–2.45; 2.440.[3] Colorless and transparent when pure and fresh; sometimes reddish from admixed iron oxide.

Opt. In transmitted light, colorless.

ORIENTATION		n(Na) [4]	n [5]	
X	c	1.5272	1.527	Biaxial negative $(-)$.
Y	b	1.5362	1.540	
Z	a	1.5443 (calc.)	1.544	
2V		79°36′ (meas.)	70°±	

Chem. A hydrated acid borate-sulfate of magnesium, $Mg_6H_4(BO_3)_4(SO_4)_2 \cdot 7H_2O$.

Anal.

	1	2	3
MgO	34.39	32.91	33.48
SO$_3$	22.76	21.95	22.46
B$_2$O$_3$	19.80	[23.64]	19.79
H$_2$O	23.05	21.50	23.53
Rem.			0.43
Total	100.00	[100.00]	99.69
G.			2.440

1. $Mg_6H_4(BO_3)_4(SO_4)_2 \cdot 7H_2O$. 2. Westeregeln.[6] 3. Westeregeln.[7] Average of five partial analyses; rem. is Fe$_2$O$_3$ 0.11, insol. 0.32. Water, determined by ignition, includes 0.10 lost between 110°–170°.

Tests. Fuses with intumescence. Soluble in acids, and decomposed by water.

Occur. Found sparingly in the insoluble residue of carnallite, associated with anhydrite, boracite, celestite, kieserite, etc., in the salt deposit at Westeregeln, Saxony. Later found similarly near Wittmar on the Asse, Brunswick, Germany, and perhaps widely distributed in small amounts in the salt deposits of northern Germany.

Artif. Efforts to prepare sulfoborite by reaction of magnesium sulfate and borax in water or a solution of KCl at 150°–160° were unsuccessful.[7]

Name. In allusion to the composition.

Ref.

1. Crystal class established by Bücking, *Ak. Berlin*, 967 (1893), *Zs. Kr.*, **36,** 156 (1902), on morphological evidence.
2. Bücking (1893).
3. Thaddéeff, *Zs. Kr.*, **28,** 264 (1897), by pycnometer on clear crystals from Westeregeln.
4. Bücking (1893) by prism method.
5. Winchell (120, 1933).
6. Naupert and Wense (1893).
7. Thaddéeff (1897).

27.1.4 **SEAMANITE** [$Mn_3(PO_4)(BO_3) \cdot 3H_2O$]. Kraus, Seaman, and *Slawson* (*Am. Min.*, **15,** 220, 1930).

Cryst.[1] Orthorhombic; dipyramidal—$2/m \; 2/m \; 2/m$.

$a:b:c = 0.5195:1:0.4508; \qquad p_0:q_0:r_0 = 0.8678:0.4508:1$

$q_1:r_1:p_1 = 0.5195:1.1524:1; \qquad r_2:p_2:q_2 = 2.2183:1.9442:1$

SEAMANITE

Forms:[2]

	ϕ	$\rho = C$	ϕ_1	$\rho_1 = A$	ϕ_2	$\rho_2 = B$
n 120	43°54½′	90°00′	90°00′	46°05½′	0°00′	43°54½′
m 110	62 33	90 00	90 00	27 27	0 00	62 33
o 111	62 33	44 21½	24 16	51 39	49 03	71 12

Structure cell.[3] Pbnm. a_0 7.83 ± 0.02 Å, b_0 15.14 ± 0.02, c_0 6.71 ± 0.02; $a_0:b_0:c_0 = 0.517:1:0.443$. Contains $Mn_{12}(PO_4)_4(BO_3)_4 \cdot 12H_2O$.

Habit. Acicular [001].

P h y s. Cleavage {001} distinct. Brittle. H. 4. G. 3.08;[4] 3.09 (calc.). Color pale yellow to wine-yellow. Transparent.

O p t. In transmitted light, colorless.

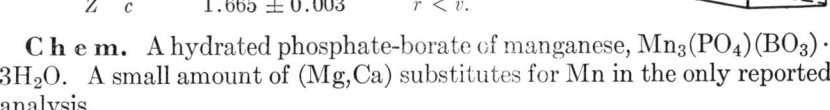

ORIENTATION		n	
X	a	1.640 ± 0.003	Biaxial positive (+).
Y	b	1.663 ± 0.003	2V 40°.
Z	c	1.665 ± 0.003	$r < v$.

C h e m. A hydrated phosphate-borate of manganese, $Mn_3(PO_4)(BO_3) \cdot 3H_2O$. A small amount of (Mg,Ca) substitutes for Mn in the only reported analysis.

Anal.

	1	2
MnO	57.10	56.42
MgO		⎱ 1.33
CaO		⎰
FeO		0.11
P_2O_5	19.05	15.94
B_2O_3	9.34	9.94
H_2O	14.51	14.57
Total	100.00	98.31
G.	3.08	3.09

1. $Mn_3(PO_4)(BO_3) \cdot 3H_2O$. 2. Michigan.[5] Average of four analyses. F not determined. The P_2O_5 and B_2O_3 probably are somewhat in error.

Tests. B.B. fuses readily to a dark slag. Soluble in cold dilute acid.

O c c u r. Found with sussexite, calcite and a manganese oxide in the Chicagon mine, near Iron River, Iron County, Michigan.

N a m e. After Arthur E. Seaman, Professor of Geology and Mineralogy in the Michigan College of Mining and Technology, who first drew attention to the mineral.

Ref.

1. Symmetry established by etching and x-ray data of McConnell and Pondrom, *Am. Min.*, **26**, 446 (1941). A morphological resemblance to reddingite is remarked by Kraus, Seaman, and Slawson (1930) but the two minerals appear to be structurally distinct—cf. McConnell and Pondrom (1941).
2. Kraus, *et al.* (1930).
3. McConnell and Pondrom (1941) by the Weissenberg method.
4. McConnell and Pondrom (1941) by the suspension method; Kraus, *et al.* (1930) give 3.128.
5. Slawson analysis in Kraus, *et al.* (1930).

SULFATES

The very numerous anisodesmic oxysalts containing anionic groups of the general type $(XO_4)^n$, and which constitute the bulk of the minerals described in this volume, fall into four main types: I. Compounds containing pentavalent X atoms in tetrahedral coordination with oxygen and yielding symmetrical $(XO_4)^{\equiv}$ groups. In these, X is either P, As, or V. There is commonly a partial or complete substitutional series between P and As and between As and V, less commonly between P and V. The natural phosphates, arsenates, and vanadates are here described as one broad family. II. Compounds containing hexavalent X atoms in tetrahedral coordination with oxygen and yielding symmetrical $(XO_4)^{=}$ groups. In these compounds, X is either S, Cr, Se, or Te, giving rise to the sulfates, chromates, selenates, and tellurates. These oxysalts are here described separately (the tellurates being included with the selenates), largely as a matter of convenience of reference and in part because of their dissimilar geochemical relations. The sulfates, chromates, and selenates are well known to form partial or complete substitutional series in artificial preparations. Substitutional series are lacking largely between natural sulfates and phosphates, although the sizes of the (XO_4) groups are very similar, because a concomitant cationic or other substitution must be made in order to effect valence compensation. III. Compounds containing relatively large hexavalent X atoms in a distorted tetrahedral coordination with oxygen. These comprise the molybdates and tungstates. They form partial or complete substitutional series, although not with the (XO_4) compounds of types I and II, and are here described jointly. IV. Compounds with planar (XO_4) groups, analogous to the $(PtCl_4)$ groups in K_2PtCl_4. Natural oxysalts with this structure are not known.

Among the sulfates, the crystal structures are now known of many of the more important groups, including the Barite Group, Alum Group, Mascagnite Group, and the extensive Alunite Group and its analogues. Numerous important species still remain to be investigated even among the simpler structures, however, and the constitution of the numerous hydrated and basic sulfates is not understood in even elementary regards.

Classification

Class 28. Anhydrous acid and normal sulfates
  Type 1. Anhydrous acid sulfates
    28.1.1 Mercallite — $KHSO_4$
    28.1.2 Misenite — $K_8H_6(SO_4)_7$
    28.1.3 Letovicite — $(NH_4)_3H(SO_4)_2$
  Type 2. A_2XO_4
    28.2.1 Mascagnite group
      28.2.1.1 Mascagnite — $(NH_4)_2SO_4$
      28.2.1.2 Arcanite — K_2SO_4
      28.2.1.3 Taylorite — $(K,NH_4)_2SO_4$ (?)
    28.2.2 Aphthitalite — $(K,Na)_3Na(SO_4)_2$
    28.2.3 Palmierite — $(K,Na)_2Pb(SO_4)_2$
    28.2.4 Thenardite — Na_2SO_4
  Type 3. AXO_4
    28.3.1 Barite group
      28.3.1.1 Barite — $BaSO_4$
      28.3.1.2 Celestite — $SrSO_4$
      28.3.1.3 Anglesite — $PbSO_4$
    28.3.2 Anhydrite — $CaSO_4$
    28.3.3 Chalcocyanite — $CuSO_4$
  Type 4. $A_mB_n(XO_4)_p$
    28.4.1 Vanthoffite — $Na_6Mg(SO_4)_4$
    28.4.2 Glauberite — $Na_2Ca(SO_4)_2$
    28.4.3 Langbeinite group
      28.4.3.1 Langbeinite — $K_2Mg_2(SO_4)_3$
      28.4.3.2 Manganolangbeinite — $K_2Mn_2(SO_4)_3$

Class 29. Hydrated acid and normal sulfates
  Type 1. Hydrated acid sulfates
    29.1.1 Rhomboclase — $FeH(SO_4)_2 \cdot 4H_2O$
    29.1.2 Minasragrite — $(VO)_2H_2(SO_4)_3 \cdot 15H_2O$
  Type 2. $A_2(XO_4) \cdot xH_2O$
    29.2.1 Lecontite — $Na(NH_4,K)(SO_4) \cdot 2H_2O$
    29.2.2 Mirabilite — $Na_2SO_4 \cdot 10H_2O$
  Type 3. $A_2B(XO_4)_2 \cdot xH_2O$
    29.3.1 Syngenite — $K_2Ca(SO_4)_2 \cdot H_2O$
    29.3.2 Koktaite — $(NH_4)_2Ca(SO_4)_2 \cdot H_2O$
    29.3.3 Kroehnkite — $Na_2Cu(SO_4)_2 \cdot 2H_2O$
    29.3.4 Loeweite — $Na_2Mg(SO_4)_2 \cdot 2\frac{1}{2}H_2O$
    29.3.5 Bloedite group
      29.3.5.1 Bloedite — $Na_2Mg(SO_4)_2 \cdot 4H_2O$
      29.3.5.2 Leonite — $K_2Mg(SO_4)_2 \cdot 4H_2O$
    29.3.6 Wattevilleite — $Na_2Ca(SO_4)_2 \cdot 4H_2O$ (?)
    29.3.7 Picromerite group
      29.3.7.1 Picromerite — $K_2Mg(SO_4)_2 \cdot 6H_2O$
      29.3.7.2 Cyanochroite — $K_2Cu(SO_4)_2 \cdot 6H_2O$
      29.3.7.3 Boussingaultite — $(NH_4)_2Mg(SO_4)_2 \cdot 6H_2O$
  Type 4. $A_mB_n(XO_4)_p \cdot xH_2O$, with $(m+n):p < 3:2$ and $> 1:1$
    29.4.1 Ferrinatrite — $Na_3Fe(SO_4)_3 \cdot 3H_2O$
    29.4.2 Polyhalite — $K_2Ca_2Mg(SO_4)_4 \cdot 2H_2O$
    29.4.3 Leightonite — $K_2Ca_2Cu(SO_4)_4 \cdot 2H_2O$

Type 5. $AB(XO_4)_2 \cdot x\mathrm{H_2O}$

29.5.1	Krausite	$KFe(SO_4)_2 \cdot H_2O$
29.5.2	Voltaite	$(K,Fe'')_3Fe'''(SO_4)_3 \cdot 4H_2O$ (?)
29.5.3	Tamarugite group	
29.5.3.1	Tamarugite	$NaAl(SO_4)_2 \cdot 6H_2O$
29.5.3.2	Amarillite	$NaFe(SO_4)_2 \cdot 6H_2O$
29.5.4	Mendozite group	
29.5.4.1	Mendozite	$NaAl(SO_4)_2 \cdot 11H_2O$
29.5.4.2	Kalinite	$KAl(SO_4)_2 \cdot 11H_2O$
29.5.5	Alum group	
29.5.5.1	Potash Alum	$KAl(SO_4)_2 \cdot 12H_2O$
29.5.5.2	Soda Alum	$NaAl(SO_4)_2 \cdot 12H_2O$
29.5.5.3	Ammonia Alum	$(NH_4)Al(SO_4)_2 \cdot 12H_2O$

Type 6. $A(XO_4) \cdot x\mathrm{H_2O}$

29.6.1	Bassanite	$2CaSO_4 \cdot H_2O$
29.6.2	Kieserite group	
29.6.2.1	Kieserite	$MgSO_4 \cdot H_2O$
29.6.2.2	Szomolnokite	$FeSO_4 \cdot H_2O$
29.6.2.3	Szmikite	$MnSO_4 \cdot H_2O$
29.6.3	Gypsum	$CaSO_4 \cdot 2H_2O$
29.6.4	Ilesite	$MnSO_4 \cdot 4H_2O$ (?)
29.6.5	Chalcanthite group	
29.6.5.1	Chalcanthite	$CuSO_4 \cdot 5H_2O$
29.6.5.2	Siderotil	$FeSO_4 \cdot 5H_2O$
29.6.5.3	Pentahydrite	$MgSO_4 \cdot 5H_2O$
29.6.6	Hexahydrite group	
29.6.6.1	Hexahydrite	$MgSO_4 \cdot 6H_2O$
29.6.6.2	Bianchite	$ZnSO_4 \cdot 6H_2O$
29.6.7	Retgersite	$NiSO_4 \cdot 6H_2O$
29.6.8	Melanterite group	
29.6.8.1	Melanterite	$FeSO_4 \cdot 7H_2O$
29.6.8.2	Pisanite	$(Fe,Cu)SO_4 \cdot 7H_2O$
29.6.8.3	Kirovite	$(Fe,Mg)SO_4 \cdot 7H_2O$
29.6.8.4	Boothite	$CuSO_4 \cdot 7H_2O$
29.6.8.5	Bieberite	$CoSO_4 \cdot 7H_2O$
29.6.8.6	Mallardite	$MnSO_4 \cdot 7H_2O$
29.6.9	Epsomite group	
29.6.9.1	Epsomite	$MgSO_4 \cdot 7H_2O$
29.6.9.2	Goslarite	$ZnSO_4 \cdot 7H_2O$
29.6.9.3	Morenosite	$NiSO_4 \cdot 7H_2O$
29.6.9.4	Tauriscite	$FeSO_4 \cdot 7H_2O$

Type 7. $A_2B(XO_4)_4 \cdot x\mathrm{H_2O}$

29.7.1	Ransomite	$Cu(Fe,Al)_2(SO_4)_4 \cdot 7H_2O$
29.7.2	Roemerite	$Fe''Fe_2'''(SO_4)_4 \cdot 14H_2O$
29.7.3	Halotrichite group	
29.7.3.1	Pickeringite	$MgAl_2(SO_4)_4 \cdot 22H_2O$
29.7.3.2	Halotrichite	$Fe''Al_2(SO_4)_4 \cdot 22H_2O$
29.7.3.3	Apjohnite	$Mn''Al_2(SO_4)_4 \cdot 22H_2O$
29.7.3.4	Dietrichite	$ZnAl_2(SO_4)_4 \cdot 22H_2O$
29.7.3.5	Bilinite	$Fe''Fe_2'''(SO_4)_4 \cdot 22H_2O$
29.7.3.6	Redingtonite	$(Fe'',Mn,Ni)(Cr,Al)_2(SO_4)_4 \cdot 22H_2O$ (?)

CLASSIFICATION

Type 8. $A_2(XO_4)_3 \cdot xH_2O$
- 29.8.1 Lausenite $Fe_2(SO_4)_3 \cdot 6H_2O$
- 29.8.2 Kornelite $Fe_2(SO_4)_3 \cdot 7H_2O$
- 29.8.3 Coquimbite $Fe_2(SO_4)_3 \cdot 9H_2O$
- 29.8.4 Paracoquimbite $Fe_2(SO_4)_3 \cdot 9H_2O$
- 29.8.5 Quenstedtite $Fe_2(SO_4)_3 \cdot 10H_2O$
- 29.8.6 Alunogen $Al_2(SO_4)_3 \cdot 18H_2O$

CLASS 30. Anhydrous sulfates containing hydroxyl or halogen

Type 1. $A_m(XO_4)_pZ_q$, with $m:p > 2:1$
- 30.1.1 Brochantite $Cu_4(SO_4)(OH)_6$
- 30.1.2 Antlerite $Cu_3(SO_4)(OH)_4$
- 30.1.3 Caracolite Na,Pb chloride-sulfate
- 30.1.4 Chlorothionite $K_2Cu(SO_4)Cl_2$
- 30.1.5 Schairerite $Na_3(SO_4)(F,Cl)$
- 30.1.6 Sulfohalite $Na_6ClF(SO_4)_2$

Type 2. $A_2(XO_4)Z_q$
- 30.2.1 Lanarkite $Pb_2(SO_4)O$
- 30.2.2 Dolerophanite $Cu_2(SO_4)O$
- 30.2.3 Linarite $PbCu(SO_4)(OH)_2$
- 30.2.4 Alunite group
- 30.2.4.1 Alunite $KAl_3(SO_4)_2(OH)_6$
- 30.2.4.2 Natroalunite $NaAl_3(SO_4)_2(OH)_6$
- 30.2.4.3 Jarosite $KFe_3(SO_4)_2(OH)_6$
- 30.2.4.4 Ammoniojarosite $(NH_4)Fe_3(SO_4)_2(OH)_6$
- 30.2.4.5 Natrojarosite $NaFe_3(SO_4)_2(OH)_6$
- 30.2.4.6 Argentojarosite $AgFe_3(SO_4)_2(OH)_6$
- 30.2.4.7 Carphosiderite $(H_2O)Fe_3(SO_4)_2[(OH)_5 \cdot H_2O]$
- 30.2.4.8 Beaverite $Pb(Cu,Fe,Al)_3(SO_4)_2(OH)_6$
- 30.2.4.9 Plumbojarosite $PbFe_6(SO_4)_4(OH)_{12}$
- 30.2.5 Euchlorin K,Na,Cu basic sulfate

CLASS 31. Hydrated sulfates containing hydroxyl or halogen

Type 1. $A_mB_n(XO_4)_pZ_q \cdot xH_2O$, with $(m+n):p > 4:1$
- 31.1.1 Connellite group
- 31.1.1.1 Connellite $Cu_{19}(SO_4)(OH)_{32}Cl_4 \cdot 3H_2O$
- 31.1.1.2 Buttgenbachite $Cu_{19}(NO_3)_2(OH)_{32}Cl_4 \cdot 3H_2O$
- 31.1.2 Glaucocerinite $Zn_{13}Al_8Cu_7(SO_4)_2(OH)_{60} \cdot 4H_2O$ (?)
- 31.1.3 Mooreite $(Mg,Mn,Zn)_8(SO_4)(OH)_{14} \cdot 4H_2O$
- 31.1.4 Torreyite $(Mg,Mn,Zn)_7(SO_4)(OH)_{12} \cdot 4H_2O$
- 31.1.5 Spangolite $Cu_6Al(SO_4)(OH)_{12}Cl \cdot 3H_2O$
- 31.1.6 Cyanotrichite $Cu_4Al_2(SO_4)(OH)_{12} \cdot 2H_2O$
- 31.1.7 Zincaluminite $Zn_3Al_3(SO_4)(OH)_{13} \cdot 2\tfrac{1}{2}H_2O$
- 31.1.8 Woodwardite $Cu_4Al_2(SO_4)(OH)_{12} \cdot 2-4H_2O$ (?)
- 31.1.9 Chalcoalumite $CuAl_4(SO_4)(OH)_{12} \cdot 3H_2O$
- 31.1.10 Uranopilite $(UO_2)_6(SO_4)(OH)_{10} \cdot 12H_2O$
- 31.1.11 Meta-uranopilite $(UO_2)_6(SO_4)(OH)_{10} \cdot 5H_2O$

Type 2. $A_4(XO_4)Z_q \cdot xH_2O$
- 31.2.1 Klebelsbergite Basic Sb sulfate
- 31.2.2 Langite $Cu_4(SO_4)(OH)_6 \cdot H_2O$ (?)
- 31.2.3 Felsöbányaite $Al_4(SO_4)(OH)_{10} \cdot 5H_2O$ (?)
- 31.2.4 Basaluminite $Al_4(SO_4)(OH)_{10} \cdot 5H_2O$
- 31.2.5 Hydrobasaluminite $Al_4(SO_4)(OH)_{10} \cdot 36H_2O$ (?)
- 31.2.6 Glockerite $Fe_4(SO_4)(OH)_{10} \cdot nH_2O$ (?)

394   SULFATES

Type 3. $A_mB_n(XO_4)_pZ_q \cdot xH_2O$, with $(m+n):p$ from 5:2 to 3:1

31.3.1	Kamarezite	$Cu_3(SO_4)(OH)_4 \cdot 6H_2O$ (?)
31.3.2	Ettringite	$Ca_6Al_2(SO_4)_3(OH)_{12} \cdot 26H_2O$
31.3.3	Devillite	$Cu_4Ca(SO_4)_2(OH)_6 \cdot 3H_2O$
31.3.4	Arnimite	$Cu_5(SO_4)_2(OH)_6 \cdot 3H_2O$ (?)
31.3.5	Serpierite	$(Zn,Cu,Ca)_5(SO_4)_2(OH)_6 \cdot 3H_2O$ (?)

Type 4. $(AB)_z(XO_4)Z_q \cdot xH_2O$

31.4.1	Kainite	$KMg(SO_4)Cl \cdot 3H_2O$
31.4.2	Ungemachite	$Na_9K_3Fe(SO_4)_6(OH)_3 \cdot 9H_2O$
31.4.3	Clino-ungemachite	
31.4.4	Zippeite	$(UO_2)_2(SO_4)(OH)_2 \cdot 4H_2O$
31.4.5	Aluminite	$Al_2(SO_4)(OH)_4 \cdot 7H_2O$

Type 5. $A_3(XO_4)_2Z_q \cdot xH_2O$

31.5.1	Natrochalcite	$NaCu_2(SO_4)_2(OH) \cdot H_2O$
31.5.2	Metasideronatrite	$Na_4Fe_2(SO_4)_4(OH)_2 \cdot 3H_2O$
31.5.3	Sideronatrite	$Na_2Fe(SO_4)_2(OH) \cdot 3H_2O$
31.5.4	Johannite	$Cu(UO_2)_2(SO_4)_2(OH)_2 \cdot 6H_2O$
31.5.5	Vernadskite	$Cu_4(SO_4)_3(OH)_2 \cdot 4H_2O$

Type 6. $A(XO_4)Z_q \cdot xH_2O$

31.6.1	Metahohmannite	$Fe_2(SO_4)_2(OH)_2 \cdot 3H_2O$
31.6.2	Butlerite	$Fe(SO_4)(OH) \cdot 2H_2O$
31.6.3	Parabutlerite	$Fe(SO_4)(OH) \cdot 2H_2O$
31.6.4	Amarantite	$Fe(SO_4)(OH) \cdot 3H_2O$
31.6.5	Hohmannite	$Fe_2(SO_4)_2(OH)_2 \cdot 7H_2O$
31.6.6	Fibroferrite	$Fe(SO_4)(OH) \cdot 5H_2O$ (?)
31.6.7	Botryogen	$MgFe(SO_4)_2(OH) \cdot 7H_2O$
31.6.8	Guildite	$Cu_3Fe_4(SO_4)_7(OH)_4 \cdot 15H_2O$
31.6.9	Metavoltine	$(K,Na,Fe)_5Fe_3'''(SO_4)_6(OH)_2 \cdot 9H_2O$ (?)
31.6.10	Slavikite	$Na_2Fe_{10}(SO_4)_{13}(OH)_6 \cdot 63H_2O$ (?)
31.6.11	Copiapite group	
31.6.11.1	Copiapite	$Fe''Fe_4'''(SO_4)_6(OH)_2 \cdot 20H_2O$
31.6.11.2	Magnesiocopiapite	$MgFe_4'''(SO_4)_6(OH)_2 \cdot 20H_2O$
31.6.11.3	Cuprocopiapite	$CuFe_4'''(SO_4)_6(OH)_2 \cdot 20H_2O$

CLASS 32. Compound sulfates

Type 1. Miscellaneous

32.1.1	Hanksite	$Na_{22}K(SO_4)_9(CO_3)_2Cl$
32.1.2	Caledonite	$Cu_2Pb_5(SO_4)_3(CO_3)(OH)_6$
32.1.3	Wherryite	$Pb_4Cu(CO_3)(SO_4)_2(OH,Cl)O$ (?)
32.1.4	Burkeite	$Na_6(SO_4)_2(CO_3)$
	Arzrunite	[with halides]
	Trudellite	[with halides]
	Creedite	[with halides]
	Tychite	[with carbonates]
	Sulfoborite	[with borates]
	Darapskite	[with nitrates]
	Chalcophyllite	[with arsenates]
	Ardeallite	[with phosphates]
	Lindackerite	[with phosphates]
	Beudantite group	[with phosphates]

MERCALLITE 395

28 ANHYDROUS ACID AND NORMAL SULFATES

TYPE 1. ANHYDROUS ACID SULFATES

28.1.1   M E R C A L L I T E [KHSO$_4$].   Carobbi (Acc. Linc., Rend., **21**, 385, 1935).

C r y s t.[1]   Orthorhombic; dipyramidal—$2/m\ 2/m\ 2/m$.

$a:b:c = 0.8609:1:1.9344;\quad p_0:q_0:r_0 = 2.2470:1.9344:1$

Forms:[2]

c 001	a 100	m 110	k 012	u 021	r 101	o 111	y 211
b 010	l 120	n 210	q 011	s 102	x 113	z 121	

Habit.   As stalactites composed of minute tabular crystals. Artificial crystals are tabular {001} and sometimes elongated [100].

P h y s.   No cleavage. G. 2.310; 2.322 (artif.).[3] Luster vitreous. Colorless when pure; also sky-blue (Vesuvius), presumably due to an admixed copper salt. Transparent. Taste acid.

O p t.[4]   In transmitted light, colorless.

Orientation		n	
X	b	1.445	Biaxial positive (+).
Y	c	1.460 (calc.)	$2V\ 56°$.
Z	a	1.491	$r < v$, weak.

C h e m.   Potassium acid sulfate, KHSO$_4$.
Anal.

	1	2
K	28.70	21.99
HSO$_4$	71.30	65.68
Rem.		12.08
Total	100.00	99.75
G.		2.307–2.310

1. KHSO$_4$. 2. Vesuvius. Rem. is SO$_4$ 4.25, Na 3.67, Cu 0.77, Ca 0.25, Al 0.12, H$_2$O (over CaCl$_2$) 1.81, H$_2$O (at 110°) 0.64, insol. 0.57. The sample contained 6 to 7 per cent of misenite and Na$_2$SO$_4$.

Tests.   Easily fusible. Soluble in water to an acid solution.

O c c u r.[5]   Found as stalactites in fumaroles in the crater of Vesuvius, admixed with misenite, hieratite, potassium fluoride, and an unidentified fluoride of Na, Ca, Mg, and Al.

A r t i f.   A well-known artificial product, easily obtained in crystals from water solution.

N a m e.   After Giuseppe Mercalli (1850–1914), a former director of the Vesuvius Observatory.

Ref.

1. Elements of Marignac, Ann. mines, [5], **9**, 6 (1856), Oeuvres, **1**, 436 (1856), on artificial crystals; see also Groth (**2**, 313, 1908).
2. Artificial crystals; cf. Groth (**2**, 313, 1908).
3. Gossner, Zs. Kr., **39**, 381 (1904).
4. Carobbi (1935) and Lang, Ak. Wien, Sitzber., **51**, 95 (1858).
5. See also Carobbi, Acc. Litt. Arti, Modena, Att., [5], **1**, 33 (1936).

28.1.2 MISENITE [$6KHSO_4 \cdot K_2SO_4$]. Misénite *Scacchi* (Mem. geol. sulla Campania, 98, 1849; *Acc. Napoli, Rend.*, 332, 1849; *C. R.*, **31**, 263, 1850).

Cryst.[1] Monoclinic.

$a:b:c = 3.2196:1:2.1842;$ $\quad \beta\, 102°05';$ $\quad p_0:q_0:r_0 = 0.6784:2.1358:1$

$r_2:p_2:q_2 = 0.4682:0.3176:1;$ $\quad \mu\, 77°55';$ $\quad p_0'\, 0.6938,\, q_0'\, 2.1842,\, x_0'\, 0.2141$

Forms:[2]

		ϕ	ρ	ϕ_2	$\rho_2 = B$	C	A
c	001	90°00′	12°05′	77°55′	90°00′	.....	77°55′
b	010	0 00	90 00	.....	0 00	90°00′	90 00
m	110	17 37½	90 00	0 00	17 37½	86 22	72 22½
l	430	22 57	90 00	0 00	22 57	84 55	67 03
e	403	90 00	48 43½	41 16½	90 00	36 38½	41 16½
D	$\bar{1}01$	−90 00	25 37½	115 37½	90 00	37 42½	115 37½

Habit. As fibrous masses and aggregates of needle- or lath-like crystals elongated [100] and flattened {001}. {001} striated [100].

Phys. Cleavage {010} distinct. G. 2.32.[3] Luster pearly to silky. Colorless (artificial) to grayish white (natural). Transparent to translucent. Taste bitter and acid.

Opt. In transmitted light, colorless.

Orientation		n[4]	
$X \wedge c$	+29°	1.475 ± 0.003	Biaxial positive (+).
Y		1.480 ± 0.003	$2V$ large.
Z	b	1.487 ± 0.003	

Chem. An acid sulfate of potassium, $K_8H_6(SO_4)_7$ or $6KHSO_4 \cdot K_2SO_4$.

Anal.

	1	2
K_2O	38.01	38.32
SO_3	56.54	56.45
H_2O	5.45	5.23
Total	100.00	100.00

1. $6KHSO_4 \cdot K_2SO_4$. 2. Cape Miseno.[5] With Al_2O_3 tr. Analysis recalculated to 100 after deducting an unstated amount of hygroscopic water. Titration gave 29.2 per cent H_2SO_4, with 29.7 required for the formula.

Tests. Easily fusible. Soluble in water to an acid solution.

Occur. Found with potash alum, alunogen, and arcanite (?) as a transient efflorescence in a fumarole (Grotto del Solfo) on Cape Miseno near Naples, Italy.

Artif. Obtained as crystals from water solution containing $H_2SO_4 : K_2SO_4 = 2.18:1$. Readily breaks down into $3KHSO_4 \cdot K_2SO_4$ and $KHSO_4$.

Name. From the locality.

Ref.

1. Wyrouboff, *Bull. soc. min.*, **7**, 5 (1884), on poor artificial crystals supposedly of a monoclinic polymorph of $KHSO_4$ to which the natural material also was ascribed; this substance was later shown by Stortenbecker, *Rec. trav. chim. Pays-Bas*, **21**, 399 (1902),

to have the composition $6KHSO_4 \cdot K_2SO_4$ and by Zambonini, *Acc. Napoli, Rend.*, 328 (1907), to be identical with the natural mineral.
  2. Wyrouboff (1884).
  3. Stortenbecker (1902) on artificial crystals. Zambonini (1907) gave 2.299–2.312 on natural material.
  4. Larsen (111, 1921), who gives $Z \wedge$ elong. $= 33°$. It is not certain if these data actually refer to misenite.
  5. Zambonini (1917). Earlier reported analyses by Scacchi (1849) were on impure material.

28.1.3   **L E T O V I C I T E** [$(NH_4)_3H(SO_4)_2$]. Sekanina (*Zs. Kr.*, **83**, 117, 1932).

C r y s t.[1] Monoclinic; prismatic—$2/m$.

$a:b:c = 1.7390:1:2.6474$; $\beta\ 102°06'$; $p_0:q_0:r_0 = 1.5224:2.5886:1$

$r_2:p_2:q_2 = 0.3863:0.5881:1$; $\mu\ 77°54'$; $p_0'\ 1.5570, q_0'\ 2.6474, x_0'\ 0.2144$

Forms:

   c 001 a 100 m 110 q 011 $e\ \bar{1}01$ x 113 o 111 $w\ \bar{1}11$

Habit. As minute pseudo-hexagonal plates on {001}; also as granular masses. Artificial crystals are thick tabular {001}. Lamellar twinning is common.

P h y s. Cleavage {001} distinct. Fracture uneven. G. 1.83 (artif.).[2] Colorless to white. Transparent.

O p t.[3] In transmitted light, colorless.

ORIENTATION		n (artif.)	n (Na; Kladno)[4]	
$X \wedge c$	$78°$	1.499	1.501	Biaxial negative ($-$).
$Y \wedge c$	$-12°$		1.516	$2V\ 75°$ (calc.).
$Z\ b$		1.526	1.525	

C h e m. Triammonium acid sulfate, $(NH_4)_3H(SO_4)_2$.

Anal.

	1	2
NH_3	20.67	20.19
SO_3	64.76	63.92
H_2O	14.57	[15.89]
Total	100.00	[100.00]
G.		1.804

1. $(NH_4)_3H(SO_4)_2$.   2. Kladno, Bohemia.[4]

Tests. Volatile. Easily soluble in water.

O c c u r. At Letovice, Moravia, with sulfur and formed during the burning of waste-heaps of a coal mine. Later noted as skeletal crystals formed similarly at a coal mine near Kladno, Bohemia.

N a m e. From the locality.

Ref.
  1. Elements of Marignac, *Ann. mines*, **12**, 523 (1857), on artificial crystals; see also Groth (**2**, 317, 1908).
  2. Gossner, *Zs. Kr.*, **38**, 158 (1903); Sekanina (1932) gives 1.81.
  3. Wyrouboff, *Bull. soc. min.*, **3**, 209 (1880), and Sekanina (1932) on artificial crystals.
  4. Rost, *Ak. Česká Roz.*, **47**, no. 11 (1937)—*Jb. Min.*, I, 360 (1938).

TYPE 2. A_2XO_4

28.2.1 MASCAGNITE GROUP

ORTHORHOMBIC; DIPYRAMIDAL—$2/m\ 2/m\ 2/m$

	a_0	b_0	c_0
Mascagnite, $(NH_4)_2SO_4$	$5.98\ kX$	10.62	7.78
Arcanite, K_2SO_4	5.731	10.008	7.424

Taylorite, $(K,NH_4)_2SO_4$ (?)

Compounds of the A_2XO_4 type fall principally into the spinel, phenakite, olivine, chrysoberyl, Na_2SO_4, and K_2SO_4 structure-types. The latter structure-type [1] is taken principally by relatively large cations and includes the Mascagnite Group, tarapacaite (K_2CrO_4), and a number of other anisodesmic artificial sulfates, chromates, and selenates principally of K, Rb, Cs, and (NH_4). The ill-defined mineral taylorite probably is an intermediate member of the mascagnite-arcanite series. Decrease in size of the cation brings about the formation of crystals of the Na_2SO_4 (*thenardite*) structure-type, and still further decrease results in the mesodesmic and isodesmic types mentioned above.

Ref.

1. Ogg, *Phil. Mag.*, **5**, 354 (1928).

28.2.1.1 **M A S C A G N I T E** [$(NH_4)_2SO_4$]. Sal ammoniacum secretum Glauberi, Alkali volatil vitreolé *Anc. chem.* Sale ammoniacale composto d'acido sulfureo *Mascagni* (*Dei Lagoni del Senese. . . .*, Sienna, 36, 1779). Sel ammoniac vitriolique *Sage* (**1**, 62, 1777). Maskagnin *Karsten* (40, 75, 1800). Ammoniaque sulfatée *Haüy* (**2**, 220, 1822). Sulfate of Ammonia. Schwefelsaures Ammoniak *Germ.* Mascagnine.

C r y s t.[1] Orthorhombic; dipyramidal—$2/m\ 2/m\ 2/m$.

$a:b:c = 0.5642:1:0.7309;\qquad p_0:q_0:r_0 = 1.2955:0.7309:1$

Forms:[2]

| c 001 | a 100 | m 110 | w 034 | v 021 | o 111 |
| b 010 | f 130 | x 012 | u 011 | q 112 | |

Structure cell.[3] Space group $Pmcn$. $a_0\ 5.98\ kX$, $b_0\ 10.62$, $c_0\ 7.78$; $a_0:b_0:c_0 = 0.563:1:0.733$. Cell contents $(NH_4)_8(SO_4)_4$.

Habit. Artificial crystals equant to short prismatic [001] and then often flattened $\{010\}$; also prismatic [100]; rarely flattened $\{100\}$. Well-formed natural crystals are rare; usually found as mealy crusts and stalactitic forms.

Twinning. On $\{110\}$ common, often repeated to give a pseudo-hexagonal habit; also polysynthetic.

P h y s. Cleavage $\{001\}$, good. Fracture uneven. Slightly sectile. H. $2-2\frac{1}{2}$. G. 1.768;[4] 1.765 (calc.). Luster vitreous to dull. Colorless when pure; also gray, yellowish gray, lemon-yellow. Transparent in large single-crystals, often turbid due to inclusions or open cavities; translucent to opaque in dendritic or aggregated forms. Taste sharp and bitter. Slightly hygroscopic. Twin gliding [5] with $K_1(110)$, $K_2(1\bar{3}0)$.

O p t.[6] In transmitted light, colorless.

Orientation	n(Li)	n(Na)	n(G)	
X c	1.5177	1.5202	1.5318	Biaxial positive (+).
Y b	1.5199	1.5230	1.5340	$r > v$, feeble.
Z a	1.5297	1.5330	1.5445	
$2V$ (meas.)	52°18′	52°12′		

C h e m. Ammonium sulfate, $(NH_4)_2SO_4$. Complete series exist in artificial material with the corresponding K, Rb, Tl, and Cs salts.[7] Analyses of natural material are lacking.

Tests. Heated in air melts and decomposes at a low temperature. Soluble in water (73.8 g. in 100 g. H_2O at 8.2°).

O c c u r. Found with sal-ammoniac as a sublimation product in fumaroles on Vesuvius and Etna, and with sassolite as a crystallization product from the solfataras of Tuscany, Italy. With ammonia-alum, gypsum, and native sulfur around fumaroles on the volcano Nyamlagira in the Belgian Congo. Also found as a sublimation product from burning coal seams, as at Commentry, France, and at Bradley, Staffordshire, England, and Arniston, Midlothian, Scotland; similarly from Kladno, Czechoslovakia. In the guano deposits of the Chincha and Guañape Islands. At the Geysers, Sonoma County, California, with boussingaultite and ammonia-alum.

A r t i f. Obtained as crystals by the slow evaporation of a water solution.

N a m e. After Paolo Mascagni (1755–1815), Professor of Anatomy in the University of Siena, who first described the natural salt.

Ref.

1. Mitscherlich, *Ann. Phys.*, **18,** 169 (1830), on artificial crystals. See also Tutton, *Zs. Kr.*, **38,** 602 (1904).
2. Goldschmidt (**6,** 10, 1920) and Tutton (1904).
3. Taylor and Boyer, *Mem. Manchester Lit. Phil. Soc.*, **72,** 125 (1928). Ogg, *Phil. Mag.*, **5,** 354 (1928) gives a_0 5.951 kX, b_0 10.560, c_0 7.729; $a_0:b_0:c_0 = 0.564:1:0.732$; G. 1.795 (calc.).
4. Tutton (1904), with a discussion of earlier values.
5. Fischer, *Jb. Min., Beil.-Bd.*, **32,** 1 (1911).
6. Tutton (1904) by prism method, who also gives indices of refraction at 80°.
7. See Groth (**2,** 322, 1908).

28.2.1.2 A R C A N I T E [K_2SO_4]. *Haidinger* (492, 1845). Glaserite *Hausmann* (1137, 1847). Schwefelsaures Kali, Kalisulphat *Germ.* Potasse sulfatée *Fr.* Sulfate of Potash.

Orthorhombic dipyramidal. The following data refer to artificial crystals: $a:b:c = 0.5727:1:0.7418$ (morphology).[1] Space group[2] *Pmcn*; a_0 5.731 kX, b_0 10.008, c_0 7.424; $a_0:b_0:c_0 = 0.573:1:0.742$. Often twinned on {110}, sometimes as cyclic, pseudo-hexagonal groupings. Cleavage {010} and {001} good. G. 2.663;[1] 2.70 (calc.). Colorless to white. Biaxial positive (+), with indices of refraction[1] (for Na): $nX = b = 1.4935$, $nY = a = 1.4947$, $nZ = c = 1.4973$; $2V = 67°20′$; $r > v$ moderate.

A mineral found in a pine railroad tie in the Santa Ana mine, Trabuco Canyon, Orange County, California, has been shown [3] to belong to this species. Found as thin tablets which are cyclic twins of six individuals bounded by {001}, {011}, and {112}.

Ref.
1. Tutton, *Zs. Kr.*, **24**, 5 (1895).
2. Ogg, *Phil. Mag.*, **5**, 354 (1928).
3. Eakle, *Bull. Dept. Geol. Univ. California*, **5**, 232 (1908), and Frondel, *Am. Min.*, **35**, 596 (1950).

28.2.1.3 **Taylorite.** Glascerite [i.e. glaserite] *Taylor* (*Proc. Ac. Sc. Philadelphia*, 309, 1859). Taylorite *Dana* (614, 1868).

As compact lumps or concretions with a crystalline structure. H. 2. Color yellowish white. Taste pungent and bitter. Analysis gave:

	$(NH_4)_2O$	Na_2O	K_2O	SO_3	Organic	Total
1.	5.37	1.68	43.45	48.40	tr.	98.90
2.	5.10	46.49		48.30	tr.	99.89

The composition is close to $5K_2SO_4 \cdot (NH_4)_2SO_4$ or $(K,NH_4)_2SO_4$. B.B. blackens and then turns white, fusing with difficulty. Soluble in water and in acids. Found in the guano beds of the Chincha Islands, Peru. Named after W. J. Taylor (1833–1864) of Philadelphia, mineral chemist.

Taylorite may be an ammonian variety of arcanite, since mascagnite and arcanite are isostructural and form a complete series, or if the Na content is higher than appears it may be an ammonian variety of aphthitalite (which see).

28.2.2 **APHTHITALITE** [$(K,Na)_3Na(SO_4)_2$]. Vesuvian Salt *Smithson* (*Phil. Trans.*, 256, 1813). Aphthalose *Beudant* (**2**, 477, 1832). Aphthitalite *Shepard* (**1**, 36, 1835). Aftalosa, Aftalosio, Solfato potassico *Ital.* Schwefelkalisalz *Glocker* (676, 1839). Arcanite pt., *some authors.* Glaserite *Hausmann* (**2**, 1137, 1847).

C r y s t. Hexagonal—P; scalenohedral—$\bar{3}\ 2/m$.[1]

$$a:c = 1:1.2879;\ ^2 \qquad p_0:r_0 = 1.4871:1$$

Forms: [3]

	ϕ	ρ	M	A_2
c 0001	.....	0°00′	90°00′	90°00′
m 10$\bar{1}$0	30°00′	90 00	60 00	90 00
e 10$\bar{1}$2	30 00	36 38	72 38½	90 00
$-e$ 01$\bar{1}$2	−30 00	36 38	107 21½	58 53
r 10$\bar{1}$1	30 00	56 05	65 29	90 00
$-r$ 01$\bar{1}$1	−30 00	56 05	114 31	44 03½

Rare or uncertain:

| a 11$\bar{2}$0 | $-g$ 01$\bar{1}$4 | s 11$\bar{2}$1 | x 02$\bar{2}$1 | 20$\bar{2}$1 |

Structure cell.[4] Space group $C\bar{3}m$. a_0 5.65 kX, c_0 7.29; $a_0:c_0 = 1:1.290$. Cell contents $K_3Na(SO_4)_2$.

APHTHITALITE

Habit. Usually thin to thick tabular {0001} with pronounced trigonal development. Also in distorted pseudo-orthorhombic shapes. In twinned groups on {11$\bar{2}$0} resembling the pseudo-hexagonal twins of aragonite but optically uniaxial throughout. In bladed aggregates; imperfectly mammillary, and in crusts; massive. The habit of artificial crystals varies markedly with the composition of the solution.

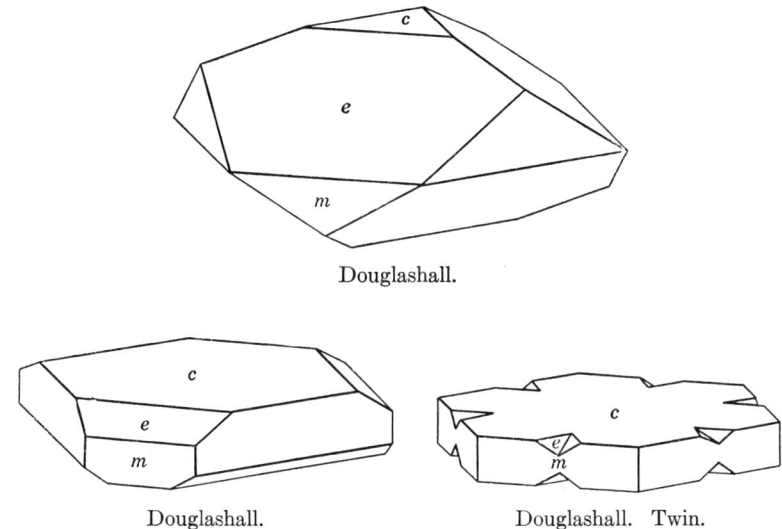

Douglashall.

Douglashall.   Douglashall. Twin.

Twinning. On {0001} or {11$\bar{2}$0}, repeated, yielding tabular, aragonite-like shapes.

P h y s. Cleavage {10$\bar{1}$0} fair, {0001} poor. Fracture conchoidal to uneven. Brittle. H. 3.[5] G. 2.656 (Douglashall),[6] 2.71 (Vesuvius),[7] 2.698 (artif.);[8] 2.72 (calc.). Luster vitreous inclining toward resinous. Color white, rarely colorless; also gray, blue or greenish, or reddish due to included iron oxide. Transparent to translucent or opaque. Taste saline and bitter.

O p t. In transmitted light, colorless.

$K_2SO_4:Na_2SO_4$	2.46:1 [6]	2.12:1 [9]	0.51:1 [10]	(blue, Vesuvius) [11]	
nO	1.491	1.490	1.487	1.4929	Uniaxial positive (+).
nE	1.499	1.496	1.492	1.5001	Sometimes weakly biaxial with small $2V$.

C h e m. A sulfate of potassium and sodium, $(K,Na)_3Na(SO_4)_2$. Na can substitute for K, the extent of the solid solution depending on the conditions of formation (see below under *Artif.*). Small amounts of Pb and Cu are present in some natural material (Vesuvius), and cuprian aphthitalite has been synthesized.[12] (NH$_4$) also can substitute for K (see further under *Var.*).

Anal.[18]

	Na₂O	K₂O	PbO	CuO	SO₃	Rem.	Total	G.
1.	9.32	42.51			48.17		100.00	2.721
2.	9.6	35.9			43.0	11.5	100.00	
3.	7.98	39.10			44.75	7.94	99.77	
4.	10.42	40.89			46.35	2.22	99.88	
5.	22.76	23.72		0.46	51.50	1.43	99.87	
6.	11.61	38.38	0.08	1.03	48.55	0.27	99.92	
7.	9.65	36.82	2.31	2.20	47.89	1.21	100.08	2.7

1. $K_3Na(SO_4)_2$. 2. Douglashall near Westeregeln.[13] Rem. is NaCl 10.1, H_2O 1.0, insol. 0.4. 3. Well G75, Searles Lake, California.[14] Rem. is NaCl 7.84, H_2O 0.10. 4. Eddy County, New Mexico.[15] Rem. is NaCl 2.16, insol. 0.06. 5. Aphthitalite with slightly bluish tinge. Kilauea, Hawaii, T.H.[16] Rem. is CaO 0.39, Cl 0.03, H_2O 0.25, insol. 0.76. 6. Sky-blue aphthitalite, Vesuvius, collected 1870. Rem. is CaO 0.02, Fe 0.11, insol. 0.14. 7. Sky-blue aphthitalite, Vesuvius.[17] Rem. is Fe_2O_3 0.70, H_2O 0.51.

Tests. B.B. easily fusible. Soluble in water and in acids.

Var. Ammonian. An ammonian variety of aphthitalite occurs in phosphate deposits on the Guañape Islands, Peru.[21] Analysis gave: $(NH_4)_2O$ 5.68, K_2O 33.87, Na_2O 9.65, SO_3 48.62, P_2O_5 0.70, H_2O at 95° 0.21, organic n.d., total 98.73, corresponding to $(K,NH_4)_3Na(SO_4)_2$ with $K:(NH_4) = 3.3:1$. G. 2.51. a_0 5.67 kX, c_0 7.34; $a_0:c_0 = 1:1.290$. nO 1.498, nE 1.503. As fibrous crusts with a yellowish white color.

Occur. Aphthitalite has two widely differing modes of occurrence. It is found as an incrustation in fumaroles of volcanoes and as a constituent of both oceanic and lacustrine salt deposits. First noted at Vesuvius where it has formed with thenardite, jarosite, sylvite, and hematite in fumaroles after various eruptions. Also at Etna and Kilauea. Occurs as small crystals enclosed in bedded bloedite at Douglashall near Stassfurt. Also found at Westeregeln and elsewhere in the German potash deposits. With syngenite and mirabilite at Stebnik and as crystals embedded in picromerite at Kalusz, Galicia. In the United States with octahedral halite on borax in the lacustrine salt deposit of Searles Lake, San Bernardino County, California. Very sparingly with halite in the Carlsbad potash district, Eddy County, New Mexico.

Artif. In the dry system Na_2SO_4-K_2SO_4 solid solutions of the end-members form from the melt in any proportion. Below 470° these invert to another solid solution series over the composition range 44 to 75 mol. per cent K_2SO_4 which remains stable to low temperatures though presumably with some contraction of the stability limit on the low K_2SO_4 end of the series. This lower temperature series corresponds to aphthitalite. The high K_2SO_4 end of the series corresponds to what is sometimes taken to be the double salt $K_3Na(SO_4)_2$. This inverts to a high-temperature solid solution of the same composition at 431°.[19] The limits of homogeneity of crystals grown from aqueous solution are much less, being set at 71 and 75 mol per cent K_2SO_4 at 25°C.[20] In conformity with these experimental results is the observation that high-sodium aphthitalites are found as fumarole products, whereas those of salt deposits mostly lie near the high-K_2SO_4 limit of the series. Some high-sodium aphthitalites of Vesuvius are found to be bluish and opalescent due to inclusions of a finely dispersed phase formed by exsolution. They can be homogenized by annealing at 425°.

Name. Aphthalose from ἄφθιτος, *unalterable*, and ἅλς, *salt*, in allusion to its stability in air. Glaserite after the seventeenth century chemist Christoph Glaser, the salt having been early called *Sal polychrestum Glaseri*. The name glaserite, although it lacks priority, has generally been used for the mineral in the German potash deposits.

Ref.

1. Gossner, *Jb. Min., Beil.-Bd.*, **57,** 89 (1928), from x-ray and morphological evidence.
2. Bücking, *Zs. Kr.*, **15,** 561 (1889).
3. Bücking (1889); {2021} from Koechlin, *Cbl. Min.*, 256 (1934); Goldschmidt (**4,** 41, 1918).
4. Gossner (1928) by rotation method probably on artificial material. See also Laskiewicz, *Arch. Min. Soc. Warsaw*, **10,** 117 (1934) who gives a_0 5.635 kX, c_0 7.27; and Bellanca, *Ric. sci. e ricostr.*, **16,** 1254 (1946); *Per. Min.*, **14,** 67 (1943).
5. Foshag, *Am. J. Sc.*, **49,** 367 (1920). Values from $2\frac{3}{4}$ to $3\frac{1}{2}$ have been reported.
6. Bücking (1889).
7. Breithaupt (**2,** 106, 1841) and Zambonini (133, 1935).
8. Nacken, *Jb. Min., Beil.-Bd.*, **24,** 21 (1907), cites G. as a function of solid solution of Na for K.
9. Foshag (1920).
10. Washington and Merwin, *Am. Min.*, **6,** 121 (1921).
11. Zambonini, *Boll. com. geol. Ital.*, 48 [3] (1921).
12. Minguzzi, *Per. Min.*, **9,** 359 (1938).
13. Recalculated from Bücking (1889).
14. Recalculated from Foshag (1920).
15. Fahey analysis in Wells, *U. S. Geol. Sur., Bull. 878*, 120 (1937).
16. Washington and Merwin (1921).
17. Zambonini (1921).
18. For additional analyses see Hintze (**1** [3B], 3696 (1929), Minguzzi (1938), and Bianchini, *Acc. Napoli, Rend.*, [4] **7,** 43 (1937).
19. Nacken (1907) and *Cbl. Min.*, 262 (1910). See also Perrier and Bellanca, *Per. Min.*, **11,** 163 (1940), and Bellanca, *Per. Min.*, **13,** 21 (1942).
20. Druzhinin, *Ac. Sci. Leningrad, Bull.*, Ser. Chim., 1141 (1938).
21. Frondel, *Am. Min.*, **35,** 596 (1950).

28.2.3 **Palmierite** [$(K,Na)_2Pb(SO_4)_2$]. Lacroix (*C.R.*, **144,** 1397, 1907).

Cryst. Hexagonal—R; scalenohedral—$\bar{3}\ 2/m$.[1]

$a:c = 1:3.761$;[2] $\alpha\ 42°28\frac{1}{2}'$; $p_0:r_0 = 4.343:1$; $\lambda\ 64°53'$

Forms:[2]

		ϕ	$\rho = C$	A_1	A_2
c	0001	111	.....	$0°00'$	$90°00'$	$90°00'$
m	$10\bar{1}0$	$2\bar{1}\bar{1}$	$30°00'$	$90\ 00$	$30\ 00$	$90\ 00$
s	$10\bar{1}2$	311	$30\ 00$	$65\ 16\frac{1}{2}$	$38\ 07\frac{1}{2}$	$90\ 00$
r	$10\bar{1}1$	100	$30\ 00$	$77\ 02$	$32\ 26\frac{1}{2}$	$90\ 00$
p	$01\bar{1}5$	221	$-30\ 00$	$40\ 58\frac{1}{2}$	$90\ 00$	$55\ 23\frac{1}{2}$
t	$01\bar{1}3$	441	$-30\ 00$	$55\ 22$	$90\ 00$	$44\ 33$
v	$01\bar{1}2$	110	$-30\ 00$	$65\ 16\frac{1}{2}$	$90\ 00$	$38\ 07\frac{1}{2}$

Structure cell.[6] Space group $R\bar{3}m$. a_0 5.58 kX, c_0 20.668; $a_0:c_0 =$ 1:3.704. Cell contents in the rhombohedral unit (a_{rh} 7.650 kX, α 42°51′) $K_2Pb(SO_4)_2$.

Habit. Microscopic, micaceous plates {0001}, with a hexagonal outline.
P h y s. G. 4.5 (artificial);[2] 4.24 (calc.). Luster pearly on base, vitreous otherwise. Colorless or white.
O p t.[2] In transmitted light, colorless.

        nO 1.712   Uniaxial negative (−).

C h e m.[3] A sulfate of potassium and lead with some substitution of sodium for potassium, $(K,Na)_2Pb(SO_4)_2$.
Anal.

	Na_2O	K_2O	PbO	SO_3	Rem.	Total
1.		19.73	46.74	33.53		100.00
2.	2.60	9.10	40.65	21.80	2.64	76.79

1. $K_2Pb(SO_4)_2$. 2. Vesuvius.[4] Rem. is NaCl.

A r t i f. Can be crystallized from a fusion of its constituent sulfates.[5]

O c c u r. Found with aphthitalite in fumarole deposits formed after the eruptions of 1906 and 1919 at Vesuvius, Italy.

N a m e. After Luigi Palmieri (1807–1896), Italian scientist and director of the observatory on Vesuvius.

Ref.
1. Crystals show trigonal, not rhombohedral, development; the crystal class is assumed without adequate evidence on the basis of a possible similarity to aphthitalite.
2. Zambonini, *Boll. com. geol. Ital.*, **48** [3] (1921). The c/a ratio is 3 times that of aphthitalite.
3. See Lacroix, *Bull. soc. min.*, **30**, 234 (1907) and *Bull. soc. min.*, **31**, 261 (1908), as well as Cesàro, *Ac. Belgique, Bull.*, [5], **20**, 683 (1934), for a discussion of the formula.
4. Pisani in Lacroix (1907).
5. Zambonini (1921). The system K_2SO_4-Na_2SO_4-$PbSO_4$ has been studied by Perrier and Bellanca, *Soc. it. per. il progr. d. sc.*, 24, **3**, 302 (1936); *Per. Min.*, **11**, 163 (1940).
6. Bellanca, *Per. Min.*, **15**, 5 (1946).

28.2.4 T H E N A R D I T E [Na_2SO_4]. Casaseca (*Ann. chim. phys.*, **32**, 308, 1826; *Ann. Phil.*, **12**, 313, 1826). Anhydrisches Natronsulfat Breithaupt (33, 1832). Pyrotechnite Scacchi (*Mem. Incend. Vesuv.*, Naples, 1855). β-Na_2SO_4. Thénardite.

C r y s t. Orthorhombic; dipyramidal—$2/m\ 2/m\ 2/m$.

$a:b:c = 0.7984:1:0.4771;$[1] $p_0:q_0:r_0 = 0.5976:0.4771:1$

$q_1:r_1:p_1 = 0.7984:1.6734:1;$ $r_2:p_2:q_2 = 2.0960:1.2525:1$

Forms:[2]

	ϕ	$\rho = C$	ϕ_1	$\rho_1 = A$	ϕ_2	$\rho_2 = B$
b 010	0°00′	90°00′	90°00′	90°00′		0°00′
r 011	0 00	25 30½	25 30½	90 00	90°00′	64 29½
d 101	90 00	30 55	0 00	59 05	59 05	90 00
o 111	51 24	37 24½	25 30½	61 39½	59 08	67 44
s 311	75 06	61 40½	25 30½	31 43	29 09	76 55

Less common or rare:
    a 100 m 110 t 061 ? u 301 v 131

Structure cell.[3] Space group $Fddd$. a_0 9.75 kX, b_0 12.29, c_0 5.85; $a_0:b_0:c_0 = 0.793:1:0.476$. Cell contents $Na_{16}(SO_4)_8$.

Habit. Dipyramidal $\{111\}$ with minor truncations; tabular $\{010\}$ with this form rough and striated and $\{101\}$ present as large faces; rarely prismatic $[100]$. Crystals several inches or so in size are not uncommon. Also as pulverulent crusts and efflorescences.

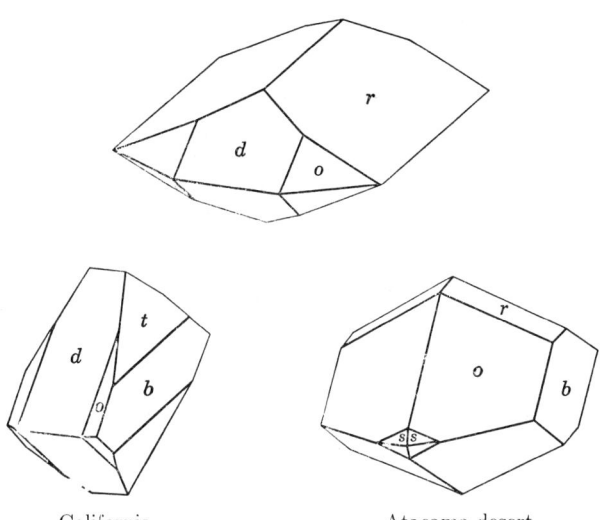

California.   Atacama desert.

Twinning. On $\{110\}$, common, producing cruciform groups. Also on $\{011\}$, with a habit somewhat similar to the butterfly twins of gypsum.

Phys. Cleavage $\{010\}$, perfect, $\{101\}$ fair, $\{100\}$ incomplete. Not very brittle. Fracture uneven to hackly. H. $2\frac{1}{2}$–3. G. 2.664;[5] 2.674 (calc.). Luster vitreous, inclining toward resinous. Colorless when pure; also grayish white, yellowish, or yellow-brown or reddish. Transparent to translucent. Taste faintly salty. M.P. 883°.

Opt. In transmitted light, colorless.

ORIENTATION[4]		$n(Na)$[6]	$n(Na)$[7]	$n(Na)$[12]	
X	c	1.471	1.464	1.4667	Biaxial positive (+).
Y	b	1.477	1.474	1.4729	$2V$ 82°35' (Na; meas.).[8]
Z	a	1.484	1.485	1.4812	

Chem. Sodium sulfate, Na_2SO_4. Analyses of pure crystals conform closely to the formula; most reported analyses[10] show small amounts variously of K_2O, MgO, CaO, Cl, and H_2O due to admixture. Five different polymorphs of Na_2SO_4 are known.[9]

Occur. Lakes and playas containing sodium sulfate are found widely in the arid parts of western United States and in Canada, Siberia, northern Africa, Central Asia, and other regions. The deposits from these waters include thenardite (especially from warm solutions and relatively concentrated brines), mirabilite, bloedite, glauberite, epsomite, gypsum, natron, halite, and, less commonly, borates. Thenardite occurs widely, although not as commercially important deposits, as an efflorescence on the soil of arid regions and in caliche, where it may be associated with sodaniter. The mineral also occurs mixed with alkali halides and sulfates as a deposit around fumaroles and as an incrustation on recent lavas.

Originally described from the salt lake of Espartinas near Aranjuez in the province of Madrid, Spain; also found as extensive bedded deposits in Miocene sediments in the neighborhood of San Martin de la Vega in the same province. In Russia as deposits from the salt lakes of the lower Volga region; at Lake Balkash in Semipalatinsk, Siberia, and in the Yenisseisk region; at Shermakha northwest of Baku in Azerbaijan. As coatings on lava and as a fumarolic deposit on both Vesuvius and Etna, Italy, and on Kilauea. In Africa, thenardite is found in soda lakes and salt pans at a number of localities in the Libyan and Sahara deserts, notably at the Natron Lakes northwest of Cairo, in the salt pans of the Bilma Oasis, and Eguéi in the French Sudan. Found at numerous localities in arid regions along the west coast of South America, notably in Chile near Tarapacá northeast of Iquique, at Caracoles in Antofagasta province and from Aguas Blancas near Copiapó in Atacama.

In the United States, found in California at Searles Lake, San Bernardino County, in layers several feet thick, at Soda Lake on the Carrizo Plain, San Luis Obispo County, in the Funeral Range and the dry depressions of Death Valley, Inyo County, and near Bertram in the Salton Sink, Imperial County. In Arizona as extensive beds in the salt deposits near Camp Verde, Yavapai County. Many occurrences are known in Nevada, notably at Rhodes Marsh, Esmeralda County. Found in the sodium sulfate lakes of western Canada.[11]

Alter. Pseudomorphs of calcite or less often quartz after thenardite occur in the phonolitic tuff of the hill Rosenegg, southern Würtemberg, Germany. Incrustation pseudomorphs of thenardite after halite and, in artificial material, of thenardite after mirabilite have been described.

Artif.[13] Thenardite crystallizes from pure water solution at temperatures over 32.4°.

Name. After the French chemist Louis Jacques Thénard (1777–1857).

Ref.

1. Angles and unit of Baerwald, *Zs. Kr.*, **6**, 36 (1882), on Aguas Blancas crystals in the orientation of the structure cell. Transformation: Baerwald to new, 001/010/100; Dana (895, 1892) and Goldschmidt (**8**, 126, 1922) to new, 010/001/100.
2. Goldschmidt (1922).
3. Zachariasen and Ziegler, *Zs. Kr.*, **81**, 92 (1932), by oscillation method with structure determination.
4. Baerwald (1882).
5. Kracek and Gibson, *J. Phys. Chem.*, **33**, 1304 (1929), on artificial material.
6. Görgey, *Min. Mitt.*, **29**, 202 (1910), on San Bernardino County, California, crystals.
7. Larsen (143, 1921) on Searles Lake crystals.
8. Baerwald (1882); see also Des Cloizeaux, *Ann. mines*, **11**, 318 (1857).
9. See Kracek, *J. Phys. Chem.*, **33**, 1281, 1304 (1929).

10. See tabulation in Hintze (1 [3B], 2679, 1929).
11. See Cole: *Canada, Mines Branch, Ottawa*, no. 646 (1926).
12. Spencer, *Min. Mag.*, **27**, 29 (1944).
13. See Hill and Wills, *J. Am. Chem. Soc.*, **60**, 1647 (1938), and Jänecke in Doelter (4 [2], 35, 1926) for details on solubility and equilibrium relations in the presence of other salts.

MAKITE. Abich (*J. pr. Chem.*, **38**, 9, 1846; *Ac. sc., St. Pétersburg, Bull. Cl. phys.-math.*, **5**, 121, 1847). As fibrous crusts from salt lakes at Güsgündag near Mt. Ararat, Armenia. The identity of the material is uncertain, and it may be a mixture of thenardite with sodium carbonate or possibly represent a sulfate-rich member of a series containing burkeite.[1]

Ref.

1. Wegscheider, *Min. Mitt.*, **39**, 316 (1928), and Ramsdell, *Am. Min.*, **24**, 109 (1939).

METATHENARDITE. Lacroix (*Bull. soc. min.*, **23**, 65, 1905). A name given to a high-temperature, hexagonal (?) polymorph of Na_2SO_4 found in fumaroles on Mt. Pelée, Martinique. On cooling and standing the transparent crystals soon break down to white, opaque paramorphs of thenardite.

TYPE 3. AXO_4

28.3.1 BARITE GROUP

ORTHORHOMBIC; DIPYRAMIDAL—$2/m\ 2/m\ 2/m$

	a_0	b_0	c_0
Barite, $BaSO_4$	$8.85\,kX$	5.43	7.13
Celestite, $SrSO_4$	8.36	5.36	6.84
Anglesite, $PbSO_4$	8.45	5.38	6.93

Among $A(XO_4)$ compounds, the barite structure-type [1] includes a number of oxysalts with tetrahedral anions and cations of relatively large size, the latter falling into twelve-coordination with oxygen. Among these salts are numerous alkali and alkaline-earth chlorates, manganates, selenates, chromates, and sulfates. The only natural representatives of the type are the members of the Barite Group, all relatively insoluble, and avogadrite, KBF_4. Ferrucite, $NaBF_4$, is isostructural with anhydrite and not barite. Anhydrite, $CaSO_4$, differs in structure from barite, due primarily to the relatively small size of the Ca ion, and is not a member of the structure-type. With increasing size of the X ion, a transition occurs to the scheelite structure-type, and the substances are less markedly anisodesmic.

Barite and celestite form a probably complete solid-solution series, the boundary between the two species being placed at Ba:Sr = 1:1, but most material is near one or the other end of the series. Anglesite ordinarily is essentially pure although Ba can substitute for Pb to a limited extent. Ca substitutes for Ba and Sr in barite and celestite to a very limited extent.

Ref.

1. James and Wood, *Proc. Roy. Soc. London*, **109A**, 598 (1925).

SULFATES

28.3.1.1 BARITE [BaSO$_4$]. Lapis Bononiensis, Litheosphorus *F. Licetus*, Utini, 1640; *Mentzel* (in *Misc. Ac. Nat. Cur.*, 1673, 1674, and *Lap. Bon. in obscuro lucens*, 1675). Lysesten, Bononiensisksten, Gypsum irregulare, lamellosum, etc. *Wallerius* (56, 1747); Marmor metallicum, Spatum tessulare *Wallerius* (58, 1747). Gypsum spatosum pt., Marmor metallicum, Spatum Bononiense, Tungspat *Cronstedt* (21, 1758); Terra calcarea phlogisto et acido vitrioli mixta, Leswersten, Lapis hepaticus *Cronstedt* (25, 1758). Gypsum ponderosum *von Born* (**1**, 14, 1772). Spath pesant ou séléniteux *de Lisle* (1772; 1783). Heavy Spar; Bolognian Spar. Cauk, Calk, Cawk *Derbyshire miners*, Withering (*Phil. Trans.*, 1784). Schwerspath *Werner*. Spathum ponderosum = Terra ponderosa vitriolata *Bergmann* (1782). Sulphate of Baryta. Baryte sulfatée *Fr.* Schwefelsaures Baryt *Germ.* Stangenspath *Werner*. Strahlbaryt. Baroselenite *Kirwan* (**1**, 136, 1794). Barytite *Delamétherie* (**2**, 8, 1797). Baryt *Karsten* (38, 75, 1800). Baryte *Haüy* (**2**, 1801). Barytine *Beudant* (441, 1824). Barytes. Michel-lévyte *Lacroix* (*C.R.*, **108**, 1126, 1889). Schwerspath *Germ.* Tungspat *Swed.* Spato pesato *Ital.* Baritina *Ital.*, *Span.* Barite *Dana* (616, 1868).

Hepatit *Karsten* (38, 75, 1800) = Lapis hepaticus *Cronstedt* (25, 1758) = Terr. pond. vit. petroleo imbuta *Bergmann* (1782) = Leberstein pt. *Germ.* = Fetid Heavy Spar. Calcareobarite *Thomson* (**1**, 105, 1836). Allomorphit *Breithaupt* (*J. pr. Chem.*, **15**, 322, 1838). Calstronbarite *Shepard* (*Am. J. Sc.*, **34**, 161, 1838). Barytocölestin *Waltershausen* (*Ann. Phys.*, **94**, 137, 1855). Barytocelestite. Celestobarite *Dana* (617, 1868). Schoharite *Eaton* (in Macneven, *Atomic Theory Chym.*, N.Y., 19, 1819); Schoarite [erron. sp.] *Adam* (62, 1869). Dréelite *Dufrénoy* (*Ann. chim. phys.*, **60**, 102, 1835); Dreeit *Glocker* (261, 1847); Dreelite. Leedsite pt. *Dana*, *Thomson* (704, 1850). Hokutolite *Jimbō* (*Am. J. Sc.*, **35**, 464, 1913); Okamoto (in Wada's *Min. of Japan*, no. 4, 178, 1912); Angleso-barite *Hayakawa* and *Nakano* (*Zs. anorg. Chem.*, **78**, 183, 1912).

Cryst.[1] Orthorhombic; dipyramidal—$2/m\ 2/m\ 2/m$.

$$a:b:c = 1.6304:1:1.3136; \qquad p_0:q_0:r_0 = 0.8057:1.3136:1$$

$$q_1:r_1:p_1 = 1.6304:1.2412:1; \qquad r_2:p_2:q_2 = 0.7613:0.6133:1$$

Forms:[2]

		ϕ	$\rho = C$	ϕ_1	$\rho_1 = A$	ϕ_2	$\rho_2 = B$
c	001	....	0°00′	0°00′	90°00′	90°00′	90°00′
b	010	0°00′	90 00	90 00	90 00	....	0 00
a	100	90 00	90 00	.....	0 00	0 00	90 00
x	230	22 14½	90 00	90 00	67 45½	0 00	22 14½
n	110	31 31½	90 00	90 00	58 28½	0 00	31 31½
m	210	50 49	90 00	90 00	39 11	0 00	50 49
η	310	61 28½	90 00	90 00	28 31½	0 00	61 28½
λ	410	67 49½	90 00	90 00	22 10½	0 00	67 49½
ϕ	012	0 00	33 18	33 18	90 00	90 00	56 42
o	011	0 00	52 43	52 43	90 00	90 00	37 17
i	021	0 00	69 09½	69 09½	90 00	90 00	20 50½
w	103	90 00	15 02	0 00	74 58	74 58	90 00
l	102	90 00	21 56½	0 00	68 03½	68 03½	90 00
g	203	90 00	28 14½	0 00	61 45½	61 45½	90 00
d	101	90 00	38 51½	0 00	51 08½	51 08½	90 00
u	201	90 00	58 10½	0 00	31 49½	31 49½	90 00
μ	112	31 31½	37 37	33 18	71 23½	68 03½	58 39
y	111	31 31½	57 01	52 43	63 59	51 08½	44 21
q	214	50 49	27 28	18 11	69 03½	68 03½	73 03½
γ	212	50 49	46 06½	33 18	56 02½	51 08½	62 54½
S	121	17 03	70 00	69 09½	74 00½	51 08½	26 03
z	211	50 49	64 18½	52 43	45 41½	31 49½	55 17½

Less common:

E 250	τ 810	t 032	A 504	P 322	K 216	f 213	ζ 613	ρ 411
L 120	e 018	Ω 031	O 403	F 2.1.20	M 133	X 233	θ 412	ψ 421
B 670	j 035	W 104	C 805	G 146	N 276	Z 243	ν 231	R 423
h 520	Y 023	σ 205	D 301	H 219	v 215	α 414	ξ 512	s 232
Π 10.3.0	k 034	κ 405	U 401	I 218	Q 122	δ 312	π 612	
β 610	r 054	V 607	p 113	J 217	T 254	ϵ 814	Γ 311	

Structure cell.[3] Space group $Pnma$. a_0 8.85 kX, b_0 5.43, c_0 7.13; $a_0:b_0:c_0 = 1.630:1:1.313$. Cell contents $Ba_4(SO_4)_4$.

Habit.[9] Commonly well-crystallized. Usually thin to thick tabular {001} and then often bounded by {210} alone or in combination with

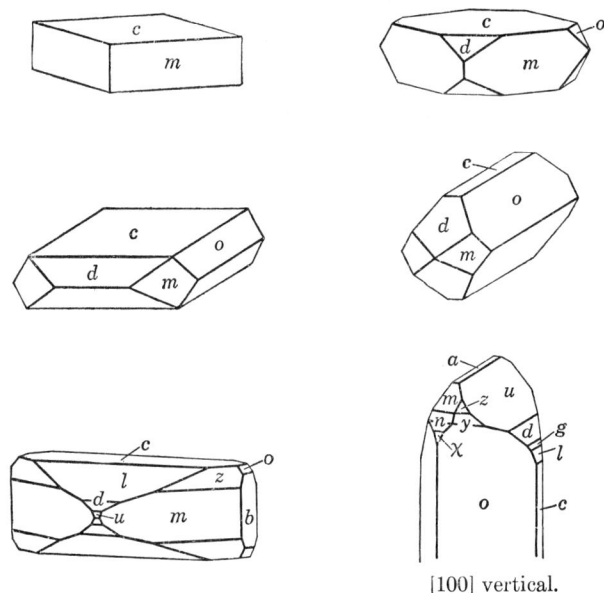

[100] vertical.

{101}, {011}, or other forms; also tabular {001} and extended [100] or [010]. Less frequently short to long prismatic [001] (the so-called *wolnyn* habit); prismatic [100] or [010]; equant. Some of the forms occasionally are only partly developed, as rarely with celestite and anglesite, so as to suggest merohedral symmetry although the true symmetry is holohedral.[4] Common forms: *c m o d z b a u l*. The crystals usually broad or stout; sometimes as aggregates of tabular crystals whose edges project at the surface into crest-like forms; as concretionary, rosette-like aggregates of tabular crystals (desert-roses).[10]

Also found massive, granular to compact and cryptocrystalline; coarsely laminated, the laminae convergent and often curved; in globular or nodular concretions, fibrous to columnar within; columnar to fibrous, either parallel or radiated; stalactitic and stalagmitic; earthy.

410   SULFATES

Twinning. Instances of true growth twins are not well authenticated.[5] Twin gliding of secondary origin produced by pressure is frequent in massive material (see beyond).

Etch figures.[6] The natural and artificial etch symmetry conforms to holohedral symmetry, although sometimes anomalous.

Oriented growths.[7] With anglesite, witherite, barytocalcite, and calcite; also with $KMnO_4$ and $KClO_4$ (artificial).

P h y s. Cleavage {001} perfect, {210} less perfect, {010} usually imperfect to indistinct but sometimes equal in quality to {210}. Fracture uneven. Brittle. H.[13] $3-3\frac{1}{2}$. G. 4.50, less in aggregated varieties and varying also with composition; 4.50 (calc.). Luster vitreous, inclining to resinous; sometimes pearly on {001} or on other forms. Colorless to white, also inclining to yellow, brown, dark brown, reddish, gray, rarely greenish or blue. Also pigmented by inclusions, as of hematite, sulfides, organic matter. The color may be distributed in growth zones or face loci and may alter on heating or exposure to sunlight or other radiation.[18] Streak white. Transparent to subtranslucent. Occasionally fetid when rubbed. Some varieties are fluorescent in ultraviolet light; also often thermoluminescent and phosphorescent (the *lapis Bononiensis* or "Bologna phosphorus"). Diamagnetic. Translation gliding:[8] $T\{001\}$, $t[100]$; $T\{011\}$, $t[011]$; $T\{101\}$, $t[010]$; $T\{010\}$, $t[100]$; $T\{110\}$, $t[1\bar{1}0]$ (the last three only at elevated temperatures); twin gliding: $K_1(110)$, $K_2(1\bar{1}0)$.

O p t. In transmitted light, colorless or faintly tinted yellow, brown, green, blue, etc. (see below). The variation in optical properties with composition is not clearly defined.

Orientation		$n(Na)$ [11]	
X	c	1.6362	Biaxial positive (+).
Y	b	1.6373	$r > v$ weak.
Z	a	1.6482	

Indices and $2V$ at Different Wavelengths [12]

λ	B	C	D	F	G
nX	1.6326	1.6336	1.6363	1.6427	1.6481
nY	1.6337	1.6346	1.6373	1.6440	1.6494
nZ	1.6446	1.6456	1.6484	1.6551	1.6607
$2V$	36°26′	36°37′	37°02′	37°45′	38°23′

Most colored and some colorless crystals are weakly pleochroic. The absorption usually is $Z > Y > X$, but variable with the color and locality and presumably dependent on the nature of the pigmenting material. Some observations follow:

Body color	X	Y	Z
Brown	Straw-yellow	Wine-yellow	Violet
Yellow	Pale yellow-brown	Yellow brown	Brown
Green	Nearly colorless	Pale green	Amethystine
Blue green	Bluish violet	Bluish green	Violet

C h e m. Barium sulfate, $BaSO_4$. Sr substitutes for Ba, and a complete series [14] probably extends to celestite, $SrSO_4$; the boundary between the two species is placed at Ba:Sr = 1:1. Ca substitutes for Ba to only

BARITE   411

a limited extent, with an apparent maximum substitution of about Ca:Ba = 1:12; anhydrite, $CaSO_4$, is not isostructural with barite. Pb substitutes for Ba up to at least Pb:Ba = 1:4 (analysis 8). Barite containing more than a few atomic per cent of either Sr or Ca is uncommon, and Pb is generally absent. (NH_4) has been reported in barite;[15] and liquid-filled cavities affording CO_2 or H_2S are frequently present. (MnO_4), (SeO_4), and (CrO_4) have been obtained in substitution for (SO_4) in artificial material.[16] Hg is sometimes present in traces.[17] On heating,[19] barite inverts to a monoclinic (?) polymorph at 1149° and melts at about 1580°.

Anal.[20]

	1	2	3	4	5	6	7	8
BaO	65.70	65.35	65.06	60.30	61.58	48.6	32.14	48.95
SrO		n.d.	0.17	1.38	2.70	14.7	28.27	
CaO		0.05	0.25	1.89	tr.		0.26	
PbO								17.78
SO_3	34.30	34.21	34.51	35.69	34.44	36.2	38.97	32.24
Rem.			0.11		0.91		0.45	0.60
Total	100.00	99.61	100.10	99.26	99.63	99.5	100.09	99.57
G.	4.50				4.389	3.9		4.62

1. $BaSO_4$. 2. Svárov, Bohemia.[21] 3. Racalmuto, Sicily.[22] Rem. is CO_2 tr., ign. loss 0.11. 4. Würzburg, Germany.[23] 5. Marano, Italy.[24] Rem. is Fe_2O_3 0.60, H_2O 0.31. 6. Strontian barite. Clifton, England.[25] 7. Greiner, Tirol.[26] Rem. is MgO 0.10, $(Fe,Al)_2O_3$ 0.16, SiO_2 0.19, properly barian celestite, with Ba:Sr = 1:1.02. 8. Plumbian barite. Shibukuro, Japan.[27] Rem. is ign. loss 0.60, Fe tr.

Var. Strontian. Barytocölestin *Waltershausen* (1855). Celestobarite *Dana* (1868). Contains Sr in substitution for Ba, a complete series probably extending to celestite with accompanying variation in optical and other properties.

Calcian. Calcareobarite *Thomson* (1836). Contains Ca in small amounts in substitution for Ba, with Ca:Ba = 1:12 in analysis 4. Material with larger amounts of Ca has been reported but is not well authenticated.[29]

Plumbian. Hokutolite *Jimbō, Okamoto* (1912). Angleso-barite *Hayakawa* and *Nakana* (1912). Contains Pb in substitution for Ba, a series[28] extending toward anglesite up to at least Pb:Ba = 1:4

Tests. B.B. decrepitates and fuses with difficulty (F = 3). Insoluble in acids. Solubility in water: 0.0024 g. $BaSO_4$ per liter of saturated solution at 20°; the solubility is much greater in solutions of salts or acids.

O c c u r.[31] Barite is the most common mineral containing barium. It occurs as a gangue mineral in hydrothermal metalliferous veins, especially those formed at moderate or low temperatures, and is frequently associated with fluorite, calcite, siderite, dolomite, quartz, galena, manganite, stibnite, etc. Also widely distributed in limestones and other sedimentary rocks as veins and lenses, cavity fillings, or replacement deposits and formed by solutions either of hypogene origin or of the meteoric circulation. Barite occurs in important amounts in residual clay deposits resulting from the weathering of limestones and other rocks. Also found in hematite deposits, with manganese oxides in lacustrine and marine deposits, in cavities in basic igneous flow rocks or intrusives, as the petrifying material of fossils, as a hot-spring deposit, etc. A few localities are mentioned below.[30]

Found at many places in Germany, notably Freiberg and Marienberg in Saxony; at Sieber, Claustal, and Lauterberg in the Harz; as a large deposit at Meggen in Westphalia. In the veins of Felsöbánya and Kapnik in Roumania and at Mies and Schemnitz in Czechoslovakia. At Kitzbühel in the Tyrol, Austria. As bituminous, radial-fibrous concretions with gypsum in marl at Mt. Paterno near Bologna, Italy. In France, especially in the Limagne district in Puy-de-Dôme. Large deposits occur in Castile and Andalusia, Spain. Many well-known localities are in England: the Dufton lead mines in Westmorland; at Alston Moor, Frizington, Cleator Moor, and elsewhere in Cumberland associated with calcite and hematite; in the veins of Cornwall; in Northumberland; as veinings in the septaria of Durham. Plumbian varieties occur as hot-spring deposits at Hokuto on Taiwan Island and at Shibukuro in Japan. In Africa in a hematite deposit in northeastern Namaqualand and similarly at Khenifra in French Morocco, and in the Gwelo district of southern Rhodesia. Highly strontian barite occurs at Clifton near Bristol, England; in the Binnenthal, Switzerland; at Görzig, Anhalt, Germany; at Grosskogel near Schwaz in the Tyrol and at Werfen in Salzburg, Austria.

In the United States found in Connecticut at Cheshire, New Haven County, in veins in sandstone; in New York at De Kalb in St. Lawrence County, at Pillar Point and near Chaumont in Jefferson County, with strontianite in limestone near Schoharie in Schoharie County. In veins on the west end of Isle Royale, Michigan. Commercial deposits occur in veins and residual deposits [32] in Paleozoic limestones in the southern Appalachian and central states, especially Georgia, Missouri, Tennessee, Kentucky, Virginia, North Carolina, Alabama. Common in the lead ores of Missouri and Wisconsin. Yellow-brown and brown crystals with vicinal faces line cavities in concretions and fossils in the Bad Lands of South Dakota. "Desert-roses" and concretionary crystals containing much sand occur at Norman, Cleveland County, and elsewhere in central Oklahoma; at Salina, Kansas. In New Mexico near Barton and elsewhere in Bernalillo County, and in various districts in Dona Ana, Socorro, and Grant Counties; in Nevada mined at Carson City, Battle Mountain, Cherry Creek, and other places; in the Castle Dome district, Yuma County, and elsewhere in Arizona. Near Hailey in Blaine County, Idaho. In California, large deposits are found in southeastern Tulare County, at Liberty Hill near Alta and the Spanish mine near Washington in Nevada County, near El Portal and Jerseydale in Mariposa County, in Inyo County, etc. Barite occurs in Canada in Nova Scotia in the Londonderry mines and along Bass and East rivers in Colchester County; in Ontario in the Timiskaming District, in Neebing township, and on Jarvis, McKellar, and Pic islands in Lake Superior in the Thunder Bay District, and a strontian variety at Galetta in Carleton County, and at Madoc; in Ottawa County, Quebec.

BARITE 413

Artif.[33] Measurable crystals have been obtained by slow interdiffusion of dilute solutions of $BaCl_2$ and sulfates, by long digestion of precipitated $BaSO_4$ in dilute HCl in closed tubes at elevated temperatures, and from fusions of $BaSO_4$ with $BaCl_2$, Na_2SO_4, or NaCl-KCl.

Alter. Pseudomorphs of quartz and chalcedony after barite, both by replacement and by incrustation, are very common, as are hollow molds of barite in quartz and other minerals. Witherite is a common alteration product of barite. Other minerals observed as forming pseudomorphs of various types after barite are siderite, calcite, rhodochrosite, dolomite, magnesite, cerussite, fluorite, limonite, plumbogummite, pyrite, marcasite, chalcopyrite, talc, witherite, malachite, wad. Barite itself occurs as pseudomorphs after barytocalcite, witherite, and calcite, and the mineral often occurs as the petrifying material (after calcite?) of fossil remains.

Name. Barite, βάρος, *weight*, and barytes from βάρυς, *heavy*.

Ref.

1. Angles of Helmhacker, *Ak. Wien, Denkschr.*, **32**, Pt. 2, 1 (1872), on Svárov, Bohemia, crystal; unit of the structure cell; conventional orientation of Dana (899, 1892), Hintze (**1** [3B], 3782, 1929), Goldschmidt (**1**, 140, 1913), Helmhacker (1872), *et alii*. Transformation: conventional to new, 200/010/001. The angles vary considerably, in part doubtless due to isomorphous substitution but principally, as with celestite, because of lineage and other crystal imperfections—see Rosický, *Ac. sc. Bohême, Bull.*, **13** (1908). Extended morphological studies of barite are given by Samoiloff, *Bull. soc. nat. Moscou*, **16**, 105 (1902); Helmhacker (1872); Henglein, *Jb. Min., Beil.-Bd.*, **32**, 71 (1911); Haas, *Jb. Min., Beil.-Bd.*, **67**, 217 (1933); Niggli, *Zs. Kr.*, **59**, 266 (1924). The variation in angle with content of Sr or Ca is not well established.

2. Haas (1933); see also Goldschmidt (1913); Tokody, *Ak. Magyar, Értes.*, **54**, 650 (1936); Erdélyi, *Földt. Közl.*, **69**, 290 (1939); Kašpar (in *Min. Abs.*, **7**, 336, 1939); Buttgenbach, *Soc. géol. Belgique, Ann.*, **55**, 165 (1932); Zeller, *Földt. Közl.*, **53**, 139 (1924); Bobkova, *Publ. Foc. Sc. Univ. Masaryk*, no. 211 (1935); Franco, *Bol. fac. fil. Cienc. Let. Univ. São Paulo*, no. 10, 75 (1938); Tavora, *Estud. Brasil. de Geol.*, **1**, 47 (1946). Rare or uncertain:

150	20.1.0	0.15.11	8.0.11	2.1.26	237	475	723
270	0.1.31	075	304	2.1.25	257	637	42.21.17
490	0.1.20	0.10.7	10.0.13	2.1.24	267	6.10.7	30.5.12
350	0.1.16	085	19.0.24	1.3.12	277	6.21.7	18.9.7
8.11.0	0.1.12	053	809	1.8.12	236	14.17.16	24.4.9
340	015	083	20.0.21	1.11.11	123	12.6.13	843
10.13.0	014	0.31.10	807	1.10.10	296	434	23.3.8
450	027	0.10.3	302	2.23.20	4.5.11	252	14.7.5
560	013	041	503	2.1.19	22.5.55	14.20.13	20.1.7
20.23.0	025	051	20.0.11	199	255	635	18.1.6
14.13.0	0.5.12	0.11.2	23.0.12	2.19.18	265	20.10.17	913
650	047	071	12.0.5	2.1.17	285	655	12.1.4
14.11.0	057	0.10.1	803	188	4.15.9	14.5.11	30.3.10
430	045	1.0.40	44.0.15	2.15.15	386	413	632
10.7.0	056	1.0.25	10.0.3	2.1.14	132	453	321
850	065	1.0.22	18.0.5	167	8.4.15	715	28.2.9
72.35.0	089	1.0.20	36.0.7	177	6.22.11	634	112.8.35
12.5.0	0.20.19	1.0.15	1.1.22	2.1.11	427	322	56.7.16

18.7.0	0.17.16	2.0.25	1.1.20	2.12.11	467	38.19.17	110.11.30
830	0.14.13	1.0.11	114	4.21.21	4.13.7	855	22.3.6
14.5.0	0.10.9	1.0.10	223	2.1.10	325	865	511
20.7.0	098	108	556	259	476	14.1.8	631
720	087	2.0.13	774	269	253	12.1.6	14.7.2
26.7.0	0.13.11	105	331	279	263	14.2.7	841
510	065	209	16.16.5	299	314	241	10.5.1
810	0.11.9	207	2.1.46	258	415	251	12.1.1
10.1.0	0.14.11	4.0.13	1.8.16	134	44.20.55	72.8.33	40.20.3
		4.0.11					
12.1.0	097	5.0.16	4.2.63	144	425	16.2.7	
14.1.0	0.17.13	6.0.11	2.1.27	298	8.7.10	16.8.7	
18.1.0	043	7.0.10	2.25.27	184	455	56.7.24	

3. James and Wood, *Proc. Roy. Soc. London*, **109A**, 598 (1925); also Basche and Mark, *Zs. Kr.*, **64**, 1 (1926).
4. See Samoiloff (1902); Hankel, *Sächs. Ges. Wiss., Abh.*, **10**, 281 (1874) [pyroelectricity]; Valentin, *Zs. Kr.*, **15**, 576 (1889), and etch symmetry (ref. 6).
5. See Gonnard, *Bull. soc. min.*, **13**, 354 (1890), for a description of a twin-like intergrowth.
6. Samoiloff (1902); Valentin (1889); Beckenkamp, *Zs. Kr.*, **28**, 69 (1897); Vogt, *Norsk. geol. Tidsskr.*, **1**, 3 (1908); and others.
7. Mügge, *Jb. Min., Beil.-Bd.*, **16**, 399 (1903); Maier, *Zs. Kr.*, **58**, 75 (1923); Ungemach, *Bull. soc. min.*, **31**, 92 (1908); Barker, *Zs. Kr.*, **45**, 25 (1908).
8. Heide, *Zs. Kr.*, **78**, 257 (1931); Veit, *Jb. Min., Beil.-Bd.*, **45**, 121 (1922); Mügge, *Jb. Min.*, I, 71 (1898); Bauer, *Jb. Min.*, I, 37 (1885).
9. For studies of the habit in relation to genesis, see Braun, *Jb. Min., Beil.-Bd.*, **65**, 173 (1932); Kalb and Koch, *Zs. Kr.*, **78**, 169 (1931); Buschendorf, *Zs. Kr.*, **81**, 38 (1932). Also (artificial) Ardagh, *et al.*, *J. Chem. Soc. London*, **53**, 1035 (1934); Popoff and Neuman, *Ind. Eng. Chem.*, **2**, 45 (1930). On vicinal faces see Kalb, *Zs. Kr.*, **81**, 342 (1932); **74**, 469 (1930).
10. Tarr, *Am. Min.*, **18**, 260 (1933); Pogue, *Proc. U. S. Nat. Mus.*, **38**, 17 (1910).
11. Average of eleven concordant (selected) sets of measurements on single-crystals reported in the literature (room temperature).
12. Kolb, *Zs. Kr.*, **49**, 14 (1911), for the Frauenhofer lines, who also gives data on the variation in indices and crystal angles with temperature.
13. On the variation in hardness with direction see Tertsch, *Zs. Kr.*, **95**, 296 (1936); Jannetaz and Goldberg, *Zs. Kr.*, **28**, 103 (1897).
14. Grahmann, *Jb. Min.*, 1 (1920); Collie, *Min. Mag.*, **2**, 220 (1879); Bruce and Light, *Am. Min.*, **12**, 396 (1927).
15. Luedeking and Wheeler, *Am. J. Sc.*, **42**, 495 (1891).
16. Wagner, *Zs. phys. Chem.*, **2**, 27 (1929), *Zs. angew. Chem.*, **44**, 665 (1931); Grimm, Peters, and Wolff, *Zs. anorg. Chem.*, **236**, 57 (1938).
17. Saukov, *C. r. ac. sc. U.R.S.S.*, **22**, 254 (1939).
18. Kolaczkowska, *Arch. min. soc. Varsovie*, **12**, 181 (1936).
19. Grahmann (1920; *Zs. anorg. Chem.*, **81**, 257, 1915); Masuda, *Proc. Imp. Ac. Tokyo*, **8**, 436 (1932).
20. For additional analyses see Hintze (1929); Doelter (4 [2], 227, 1927); Buschendorf (1932); Howland, *Am. Min.*, **21**, 584 (1936); Russell, *Min. Mag.*, **24**, 318 (1934).
21. Farský analysis in Helmhacker (1872).
22. Ruiz, *Acc. Linc., Rend.*, 3 [6], 342 (1926).
23. Sandberger, *Jb. Min.*, 383 (1875).
24. Gallitelli, *Atti soc. nat. mat. Modena*, **8**, 86 (1929).
25. Collie (1879).
26. Zepharovich, *Ak. Wien, Sitzber.*, **57**, 740 (1886).
27. Ōhashi, *Min. Mag.*, **19**, 73 (1920).
28. See Okamoto in Wada (no. 4, 178, 1912) and Ōhashi (1920).
29. See Vojtěch, *Časopis min. geol.*, **1**, 5 (1923)—*Jb. Min.*, I, 102 (1928); Dufrénoy, *Ann. chim. phys.*, **60**, 102 (1835); Lacroix, *Bull. soc. min.*, **8**, 435 (1885).
30. An extended list of localities is given by Hintze (1925).

CELESTITE

31. On the geochemistry of barite see Trener, *Jb. geol. Reichsanst. Wien*, **58**, 387 (1908); Palmquist, *Kungl. Fysiograf. Sällsk. Lund, Förh.*, **8**, 1, 21 (1939); Chukhrov, *Bull. ac. sc. U.R.S.S., Sér. Géol.*, no. 3, 562 (1937); Englehardt, *Chem. Erde*, **10**, 187 (1936).
32. Tarr, *Econ. Geol.*, **14**, 46 (1919).
33. See Mellor (3, 763, 1923), Hintze (1929).

28.3.1.2   CELESTITE [$SrSO_4$].   Fasriger Schwerspath (from Pennsylvania) Schütz (*Beschr. Nordamer. Foss.*, 12, Leipzig, 1791). Schwefelsaurer Strontianit aus Pennsylvanien Klaproth (*Beitr.*, **2**, 92, 1797). Strontiane sulfatée (from Sicily) [after Vauquelin's analysis] Dolomieu (*J. phys.*, **46**, 203, 1798). Cœlestin Werner (1798); Lenz (233, 1800); Karsten (54, 95, 1808). Sicilianite Lenz (233, 1800). Schützit Gerhard, Karsten (36, 75, 1800). Eschwegite Lévy (**1**, 224, 1837). Zölestin Germ., pt. Calciocelestine Wicke (*Arch. Pharm.*, **152**, 32, 1860).

Barytosulfate of Strontian Thomson (**1**, 111, 1836). Barytocölestin Glocker (634, 1839). Barytocelestite. Calciocelestite Dana (620, 1868).

C r y s t.[1]   Orthorhombic; dipyramidal—$2/m\ 2/m\ 2/m$.

$a:b:c = 1.5616:1:1.2823;$ $\quad p_0:q_0:r_0 = 0.8211:1.2823:1$

$q_1:r_1:p_1 = 1.5616:1.2178:1;$ $\quad r_2:p_2:q_2 = 0.7798:0.6404:1$

Forms:[2]

		ϕ	$\rho = C$	ϕ_1	$\rho_1 = A$	ϕ_2	$\rho_2 = B$
c	001	....	0°00′	0°00′	90°00′	90°00′	90°00′
b	010	0°00′	90 00	90 00	90 00	....	0 00
a	100	90 00	90 00	....	0 00	0 00	90 00
n	110	32 38	90 00	90 00	57 22	0 00	32 38
m	210	52 01	90 00	90 00	37 59	0 00	52 01
h	012	0 00	32 40	32 40	90 00	90 00	57 20
o	011	0 00	52 03	52 03	90 00	90 00	37 57
l	102	90 00	22 19	0 00	67 41	67 41	90 00
d	101	90 00	39 23½	0 00	50 36½	50 36½	90 00
θ	112	32 38	37 17	32 40	70 56	67 41	59 19½
y	111	32 38	56 42½	52 03	63 12½	50 36½	45 15½
χ	122	17 45	53 24	52 03	75 50	67 41	40 08
f	213	52 01	34 46	23 08½	63 17	61 18	69 27
ψ	233	23 07	54 21	52 03	71 23½	61 18	41 38½
x	232	23 07	64 26½	62 32	69 15½	50 36½	33 55½
z	211	52 01	64 21½	52 03	44 43	31 20½	56 18

Structure cell.[3]   Space group *Pnma*. $a_0\ 8.36\ kX$, $b_0\ 5.36$, $c_0\ 6.84$; $a_0:b_0:c_0 = 1.560:1:1.276$.   Cell contents $Sr_4(SO_4)_4$.

Twinning.   Very rare, if existent.   Reported[5] on {210}, {101}, and other planes.

Habit.[4]   The development of the crystals is variable: usually thin to thick tabular {001}, commonly with large {210}; tabular {001} and elongated [100] into lath-like shapes; or elongated [100] with equant cross section. Also, less frequently: equant by development of {001}, {011},

{101} or otherwise; pyramidal {122};[14] elongated [010] or [001]; tabular {100}, {100} commonly striated [001]. Common forms: *c a m o l d z y*. Also as fibrous veinlets or nodules, parallel or radiated; massive granular; as rounded, lenticular crystals[14] or aggregates; rarely lamellar; earthy, impure with clay or calcite.

P h y s. Cleavage {001} perfect, {210} good, {010} poor, reported[6] on {011}; the cleavages in general are better than in anglesite but not so good as in barite. Fracture uneven. H. 3–3½. G. 3.97 ± 0.01 (for essentially pure $SrSO_4$); 3.96 (calc.). Luster vitreous, inclining to pearly on the cleavages. Colorless to pale blue,[8] also white, reddish, greenish, or brownish. The blue color often is unequally distributed in growth zones or face

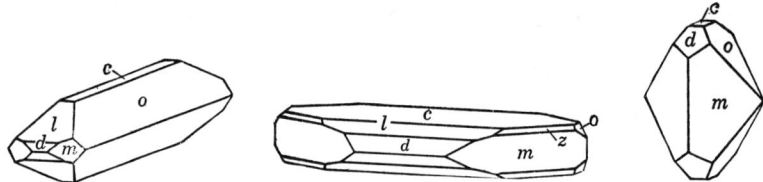

loci; also white or turbid zones similarly distributed. Transparent to translucent. Twin gliding and translation gliding[7] as in barite. Occasionally fluorescent and thermoluminescent; the fluorescence sometimes is due to inclusions of organic matter.

Etch figures.[9] Etch pits and solution bodies conform to holohedral symmetry.

Oriented growths. With barite, $KMnO_4$, and $KClO_4$ (artificially).[10]

O p t. In transmitted light, colorless or faintly tinted. Blue crystals are weakly pleochroic in shades of indigo- and lavender-blue, bluish green or violet.

Orientation		n(Na)[11]	
X	c	1.6215	Biaxial positive (+).
Y	b	1.6237	$r < v$.
Z	a	1.6308	

Indices and 2V at Different Wavelengths[12]

λ	B	C	D	F	G
nX	1.6180	1.6189	1.6215	1.6273	1.6322
nY	1.6198	1.6207	1.6232	1.6292	1.6341
nZ	1.6266	1.6279	1.6307	1.6367	1.6418
2V	49°40′	49°50′	50°25′	51°36′	52°44′

C h e m. Strontium sulfate, $SrSO_4$. Ba substitutes for Sr, and a probably complete series extends to barite.[13] Ca substitutes for Sr in limited amounts, with an apparent maximum at about Ca:Sr = 1:11.[22] Both Ba and Ca are ordinarily present in amounts ranging up to only a few atomic per cent. $SrSO_4$ inverts[23] at about 1152° to hexagonal α-$SrSO_4$ and melts at about 1605°.

Anal.[15]

	1	2	3	4	5	6	7	8
SrO	56.42	56.20	54.67	55.30	53.76	45.66	39.51	32.83
CaO		0.40	0.46	0.08	0.46	1.30		
BaO			0.78	0.89	1.29	8.47	20.26	26.18
SO_3	43.58	43.40	43.04	43.42	43.59	42.09	41.09	39.04
Rem.			0.59		0.49	1.03	tr.	
Total	100.00	100.00	99.54	99.69	100.59	98.55	100.86	98.05
G.		3.96		3.984	3.887	3.979	4.123	4.188

1. $SrSO_4$. 2. Djebel Kebbouch, Tunis.[16] 3. Trichy, India.[17] Rem. is SiO_2 0 32, Fe_2O_3 0.15, P_2O_5 tr., loss 0.12. 4. Darabana, Bessarabia.[18] 5. Maybee, Michigan.[19] Rem. is MgO 0.13, R_2O_3 0.14, SiO_2 0.22. Average of two. 6. Caramanico, Italy.[20] Rem. is Fe_2O_3 1.03, Al_2O_3 tr. 7. Ontario.[21] $(Al,Fe)_2O_3$ 0.005. 8. Lansdowne, Ontario.[21]

Tests. B.B. frequently decrepitates and fuses at 3 to a white pearl. Slowly soluble in hot concentrated acids or alkali carbonate solutions. Slightly soluble in water (0.0132 g. $SrSO_4$ in 100 cc. sat. soln. at 20°); the solubility is greater than barite but less than anhydrite.

Var. Barian. Barytosulfate of Strontian *Thomson* (1836). Barytocölestin *Glocker* (1839). Barytocelestite. Contains Ba in substitution for Sr, a complete series probably extending to barite, $BaSO_4$. The names celestite and barite are applied to the halves of the series with Sr > Ba and Ba > Sr, respectively. Increase in the content of Ba is accompanied by an increase in the gravity and indices of refraction, and by variation in crystal angles, but the precise relations are not well-defined.

Calcian. Calciocelestite *Dana* (1868). Contains Ca in substitution for Sr in limited amounts. The concomitant variation in physical properties is not well-defined.

O c c u r.[24] Celestite occurs chiefly in sedimentary rocks: in bedded deposits of gypsum, anhydrite and rock salt, where it is often associated with sulfur; in cavities and veins and disseminated in limestone and dolomite, where it may be associated with crystallized strontianite, gypsum, calcite, dolomite, fluorite; disseminated in shale, marls, and sandstone; in minor amounts in deposits of potash salts and of borates. The mineral variously has formed as a direct deposit from sea water, during diagenesis, or in favorable horizons as replacement deposits formed by traversing meteoric waters whose strontium content has been derived from adjacent sedimentary formations. Celestite also occurs as a primary mineral in hydrothermal veins, and as a cavity- or vein-filling mineral in basic eruptive rocks. A greatly abbreviated list of localities follows.

Found at numerous places in Germany: notably in sediments at Giershagen, Helmscheid (Waldeck) and Obergembeck in Westphalia; at Scharfenberg near Meissen, Saxony, in veins; in Mesozoic marls at Lüneberg, Hannover; at Dornburg near Jena in Thuringia; in limestone in the region of Würzburg, Bavaria. In Austria at Leogang, Salzburg, and at Bex, Vaud, Switzerland. Found in Italy at Montecchio Maggiore, Venetia, with zeolites in basalt and in tuff. Celestite occurs abundantly and in splendid groups of crystals associated with sulfur, aragonite, and gypsum at Girgenti and the near-by localities of Caltanissetta, Cianciana, etc., in Sicily. At many places in France, especially in the region of Wassy, Haute-Marne, and at Condorcet, Drôme. In Egypt southeast of Cairo as large crystals at Mokattam in Wadi-el-Tih. At Djebel Mezzouna and elsewhere in Cretaceous marls in southern Tunisia. Large deposits occur in England in the neighborhood of Bristol, Gloucestershire, at Clifton, Yate, Wickwar, and Durdham Down. In basaltic rocks at St. Andrews, Fifeshire, and elsewhere in Scotland. Widespread in Permian dolomite-anhydrite deposits in eastern European Russia, in sulfur deposits in the middle Volga region, and in Turkestan. As a gangue mineral in veins at Sierra Mojada, Mexico.[31]

In the United States, celestite occurs in cavities in the Lockport dolomite in western New York and disseminated in shales and clayey or dolomitic limestones in the region of Syracuse; [25] also in the lead deposit at Rossie, St. Lawrence County. At Bellwood (Bell's Mills), Blair County, Pennsylvania, as fibrous veinlets with dolomite and anhydrite; this was the original celestite taken to Europe by Schütz and named by Werner after an analysis by Klaproth. Hundreds of tons of finely crystallized celestite, some individuals 18 inches in length, were obtained from a cave in dolomite at Put-in-Bay on South Bass Island in Lake Erie, Ohio. In Michigan in dolomite at Drummond in Chippewa County and at Maybee in Monroe County. Fine specimens occur with fluorite at Clay Center, Ohio,[27] in cavernous dolomite. In Texas as large crystals at Lampasas, Lampasas County, and in the Mount Bonnell district near Austin in Travis County. In limestone in Buffalo Cove, Fentress County, Tennessee.[33] In California [26] found together with gypsum and halite in extensive lake-bed deposits along the northeast margin of the Avawatz Mountains in San Bernardino County; also in this county as massive beds, 10 to 20 feet thick, northwest of Ludlow, and as a large deposit on the southwest margin of Bristol Dry Lake, south of Amboy; also in the colemanite ores of Borate in the Calico Hills. A sedimentary deposit with gypsum occurs 15 miles south of Gila Bend, Maricopa County, Arizona. In veins in a fault zone in dunite at La Conner, Fidalgo Island, Skagit County, Washington.[28] Found in Canada at Lansdowne, Leeds County, and at Calabogie and near Kingston [29] in Ontario.

A r t i f.[30] Obtained in measurable crystals by methods analogous to those employed with barite.

A l t e r. Strontianite frequently occurs as an alteration product of celestite.[31] Calcite, quartz, chalcedony, barite, and sulfur have been found pseudomorphous after this species (see also under calcite). Hollow molds after celestite occur in dolomite and other sedimentary rocks, and these may later be filled by halite, calcite, or other material. Celestite has been found as pseudomorphs after gypsum [32] and as the petrifying material of fossils.

N a m e. From *cœlestis, celestial*, in allusion to the faint shade of blue often present.

CELESTITE 419

Ref.
1. Unit of the structure cell of James and Wood, *Proc. Roy. Soc. London*, **109A**, 598 (1925); conventional orientation of Dana (905, 1892); Hintze (**1** [3B], 3905, 1929), Goldschmidt (**2**, 163, 1913), *et alii;* angles are average of many good readings on crystals from various localities (Palache, priv. comm., 1947). Transformation: Dana (1892) to structure cell, 200/010/001. The reported angles vary appreciably due to isomorphous substitution of Ba and to composite structure—see Auerbach, *Ak. Wien, Sitzber., Abt.* I, **59**, 549 (1869); Arzruni and Thaddéeff, *Zs. Kr.*, **25**, 38 (1895); Hintze (1929).
2. Goldschmidt (1913); Kreutz, *Ak. Krakau, Abh.*, **55**, 1 (1915); Ranfaldi, *Acc. Linc., Att.*, **31**, 430, 468, 506 (1922); Di Franco, *Ist. Min. Vulc. Univ. Catania, Mem.*, **19** (1918); Giuşcă, *Ac. Roumaine, Sect. Sc., Bull.*, **9**, 25 (1924); Quercigh, *Acc. Linc., Rend.*, **33**, 262 (1924); Ruiz, *ibid.*, p. 267; Szadecsky-Kardoss, *Földt. Közl.*, **53**, 94 (1924); Thibault, *Am. Min.*, **20**, 147 (1935). For reviews of the morphology, see Henglein, *Cbl. Min.*, 692 (1911); Surganoff, *Bull. soc. nat. Moscou*, **18**, 435 (1905); Lacroix (**4**, 103, 1910). On vicinal faces see Kreutz (1915), Giuşcă (1924), Thibault (1935). Rare or uncertain:

450	0.1.16	023	209	904	217	437	14.5.14	512
430	0.1.14	067	104	14.0.5	277	477	838	311
850	0.1.12	087	103	401	287	243	212	632
12.5.0	0.1.10	021	4.0.11	221	123	253	565	10.2.3
14.5.0	019	0.15.2	205	1.8.12	133	314	121	421
310	018	2.0.17	407	2.24.23	276	627	836	531
10.3.0	017	107	203	2.19.19	215	8.1.10	453	
18.5.0	016	2.0.13	12.0.11	188	235	516	312	
410	015	2.0.11	14.0.15	155	255	415	412	
20.3.0	0.3.10	6.0.31	22.0.23	2.11.10	4.10.9	475	18.5.9	
810	013	105	302	219	214	414	231	
0.1.20	0.15.22	6.0.29	201	269	254	313	18.5.8	

3. James and Wood (1925). Also Basche and Mark, *Zs. Kr.*, **64**, 1 (1926).
4. On the habit and the relation of habit to structure and genesis see Schilly, *Jb. Min., Beil.-Bd.*, **67**, 323 (1933).
5. Di Franco (1918); Hintze (1929).
6. Negri, *Rev. min.*, **1**, 33 (1887), and Bettanini, *Zs. Kr.*, **14**, 507 (1888).
7. See Heide, *Zs. Kr.*, **78**, 257 (1931).
8. On the behavior and origin of the blue color see Friend and Allchin, *Nature*, **144**, 633 (1939); Dravert, *Ann. géol. min. Russie*, **17**, 75 (1916); Hintze (1929).
9. See Kemter, *Ak. Sächs., Verh.*, **72**, 56 (1921); Samoiloff, *Zs. Kr.*, **45**, 113 (1908); Honess (124, 1927).
10. Gaubert, *Bull. soc. min.*, **32**, 139 (1909); Barker, *Zs. Kr.*, **45**, 14 (1908).
11. Average of numerous single-crystal measurements in the literature.
12. Kolb, *Zs. Kr.*, **49**, 14 (1911), who also gives the temperature variation of indices and crystal angles.
13. See also Grahmann, *Jb.Min.*, I, 1(1920).
14. Williams, *Am. J. Sc.*, **39**, 183(1890).
15. For tabulations of additional analyses see Hintze (1929), Doelter (**4** [2], 205, 1927); see also Miropolsky, *C. r. ac. sc. U.R.S.S.*, **33**, 64 (1941); *ibid.*, **34**, 114 (1942); Solignac, *Bull. soc. min.*, **54**, 64 (1931); Serdyuchenko, *C. r. ac. sc. U.R.S.S.*, **60**, 433 (1948).
16. Pisani analysis in Termier, *Bull. soc. min.*, **25**, 173 (1902).
17. Jayaraman, *Quart. J. Indian Inst. Sc.*, **3**, 11 (1940).
18. Sidorenko, *Mém. soc. nat. Nouv. Russie*, **27**, 1 (1905)—*Jb. Min.*, II, 377 (1907).
19. Kraus and Hunt, *Am. J. Sc.*, **21**, 237 (1906).
20. Onorato, *Acc. Linc., Att.*, **33**, 259 (1924).
21. Volney, *J. Am. Chem. Soc.*, **21**, 386 (1898).
22. Wicke, *Arch. Pharm.*, **102**, 32 (1860), with 8.31 $CaSO_4$, on material from Wassel, Hannover.
23. Grahmann (1920).
24. A lengthy list of localities is given in Hintze (1929).
25. Kraus, *Am. J. Sc.*, **18**, 30 (1904); *ibid.*, **19**, 286 (1905).

26. See Murdoch and Webb, *Calif. State Div. Mines, Bull.* **136**, 87 (1948).
27. Morrison, *Am. Min.*, **21**, 780 (1935).
28. Landes, *Am. Min.*, **14**, 408 (1929).
29. Fairbairn, *Am. Min.*, **14**, 286 (1929).
30. See Hintze (1929); Mellor (**3**, 763, 1923); Lambert and Hume-Rothery, *J. Chem. Soc. London*, 2637 (1926).
31. See Krieger, *Am. Min.*, **18**, 345 (1933).
32. Linck, *Chem. Erde*, **2**, 481 (1926); Mellis, *Bull. comm. géol. Finlande*, no. 140, 239 (1947).
33. Kesler, *Econ. Geol.*, **39**, 287 (1944).

28.3.1.3 ANGLESITE [PbSO$_4$].

Vitriol de Plomb *Monnet* (*Syst. Min.*, 371, 1779). Plumbum acido vitriolico mineralisatum *Bergmann* (116, 1782). Lead mineralized by vitriolic acid *Withering* (trans. of Bergmann, 1783). Vitriol de Plomb (from Andalusia) *Proust* (*J. Phys.*, **30**, 394, 1787). Vitriolo nativo de plomo *Span. miners name*. Bleiglas (from Harz) *Lasius* (*Beob. Harzgeb.*, **2**, 355, 1789). Nat. Bleivitriol *Karsten* (24, 1791). Lead vitriol. Sulphate of Lead. Vitriolbleierz *Germ.* Plomb sulfaté *Fr.* Anglesite *Beudant* (**2**, 459, 1832). Sardinian (*B.-H. Ztg.*, **24**, 320, 1865; **25**, 194, 1866). Bouglisite *Cumenge* (in Lacroix, *Bull. mus. d'hist. nat.*, 42, 1892). Weisbachit *Kolbeck* (Plattner's Probierk. m.d. Lötr., 7th ed., Leipzig, 241, 253, 1907; Hlawatsch, *Ann. naturhist. Hofmus. Wien*, **38**, 19, 1925). Barytoanglesite *Ramdohr* (*Abh. deutsch. Ak. Wiss. Berlin*, no. 4, 1, 1947).

Cryst.[1] Orthorhombic; dipyramidal—$2/m\ 2/m\ 2/m$.

$$a:b:c = 1.5703:1:1.2894; \quad p_0:q_0:r_0 = 0.8211:1.2894:1$$

$$q_1:r_1:p_1 = 1.5703:1.2179:1; \quad r_2:p_2:q_2 = 0.7756:0.6368:1$$

Forms:[2]

	ϕ	$\rho = C$	ϕ_1	$\rho_1 = A$	ϕ_2	$\rho_2 = B$
c 001	.....	0°00′	0°00′	90°00′	90°00′	90°00′
b 010	0°00′	90 00	90 00	90 00	....	0 00
a 100	90 00	90 00	.....	0 00	0 00	90 00
n 110	32 29½	90 00	90 00	57 30½	0 00	32 29½
m 210	51 51½	90 00	90 00	38 08½	0 00	51 51½
ϕ 012	0 00	32 48½	32 48½	90 00	90 00	57 11½
o 011	0 00	52 12½	52 12½	90 00	90 00	37 47½
l 102	90 00	22 19	0 00	67 41	67 41	90 00
d 101	90 00	39 23½	0 00	50 36½	50 36½	90 00
y 111	32 29½	56 48½	52 12½	63 17½	50 36½	45 06
t 221	32 29½	71 53½	68 48½	59 18	31 20½	36 42½
r 212	51 51½	46 14	32 48½	55 23½	50 36½	63 31
p 312	62 22	54 16	32 48½	44 00½	39 04½	67 53
z 211	51 51½	64 24½	52 12½	44 49	31 20½	56 09

Less common:

k 230	λ 410	μ 112	π 255	x 122	s 232	τ 421
h 320	θ 021	w 114	f 214	ψ 233	ξ 121	

Structure cell.[3] Space group *Pnma*. a_0 8.45 kX, b_0 5.38, c_0 6.93; $a_0:b_0:c_0 = 1.571:1:1.265$. Cell contents Pb$_4$(SO$_4$)$_4$.

Habit.[4] Often well-crystallized, in various habits: thin to thick tabular {001}, often with {210}, {101} and rhomboidal in outline, and sometimes also extended [100] or [010]; prismatic [001] with large {210} and vertically striated; prismatic [100], with large {011}; stout prismatic [010], with {101}, {102}; tabular {100}; equant, or pyramidal with {111}, {211} or otherwise. {100} and {210} often striated [001]. Commonest forms: *m d y c z l n*. Commonly massive, granular to compact; nodular; stalactitic; often massive with concentric banding and enclosing an unaltered core of galena.

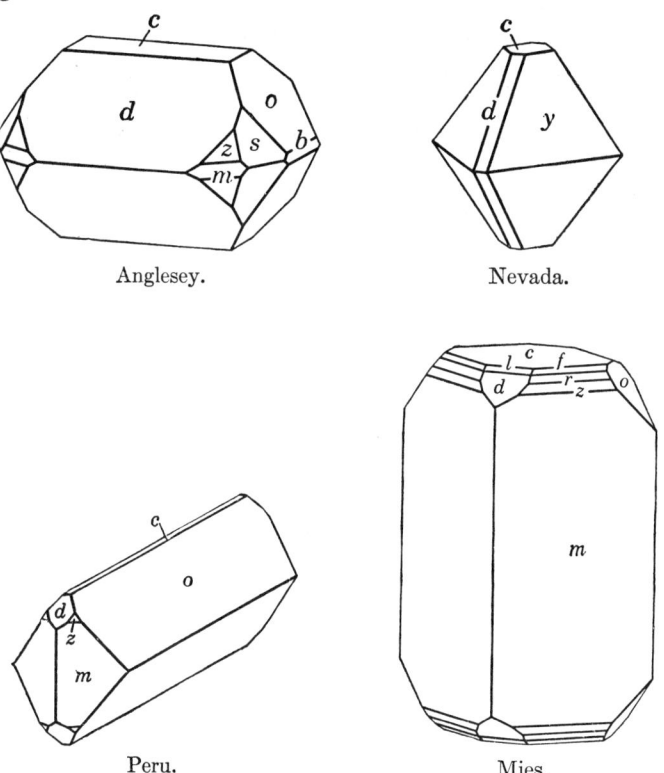

Anglesey.   Nevada.

Peru.   Mies.

Twinning. Undoubted instances of twinning are not known.

Oriented growths.[5] Upon barite, in parallel position; also with $KClO_4$ and $KMnO_4$ (artificial), and with galena.[15]

Etch figures.[6] Etch pits conform to holohedral symmetry.

P h y s. Cleavage {001} good, {210} distinct, {010} in traces. Fracture conchoidal. Brittle. H. $2\frac{1}{2}$–3. G. 6.38 ± 0.01 for crystals, less in massive types; 6.36 (calc.). Luster adamantine, inclining to resinous and vitreous in some specimens. Colorless to white, and often tinged gray, yellow, green, and sometimes blue. Streak colorless. Transparent to opaque. Translation gliding and twin gliding [7] as in barite. Often fluoresces yellow in ultraviolet light.

Opt. In transmitted light, colorless.

Orientation		$n(Na)$ [8]	
X	c	1.8771	Biaxial positive (+).
Y	b	1.8826	2V 75°24′ (Na, at 20°).[10]
Z	a	1.8937	

Indices and 2V at Different Wavelengths [9] at 20°

λ	B	C	D_2	F	G
nX	1.8682	1.8707	1.8781	1.8965	1.9128
nY	1.8734	1.8761	1.8832	1.9020	1.9183
nZ	1.8842	1.8869	1.8947	1.9137	1.9306
2V	68°07′		68°28′	68°42′	68°42′

Chem. Lead sulfate, $PbSO_4$. The reported analyses [11] of anglesite are old and unsatisfactory. A barian variety has been reported.[15] On heating,[12] anglesite inverts at about 864° to a monoclinic polymorph and decomposes between 900°–1000°.

Tests. B.B. decrepitates and fuses in a candle flame (F. $1\frac{1}{2}$). Soluble with difficulty in nitric acid and more readily in ammonium citrate or acetate. Fragments placed in ammonium carbonate solution are coated white ($PbCO_3$) and in ammonium sulfide turn black and lustrous (PbS). Solubility in water: 0.0427 g. $PbSO_4$ per liter H_2O at 20°.

Occur. Anglesite is a common secondary mineral, and usually is formed by the oxidation of galena. The formation of anglesite is accompanied and followed by that of cerussite. Other associated minerals are mimetite, pyromorphite, sulfur, lanarkite, leadhillite, linarite, wulfenite, gypsum, massicot, silver halides. Only a few of the many known localities can be given here.

In Germany at Müsen, Wissen, and Littfeld near Siegen in Westphalia; at Badenweiler and the Pfingstsegen mine near Schönau in Baden. At Bleiberg in Carinthia, Austria. In France at Huelgoat and Poullaouen, Brittany. On the island of Sardinia at Montevecchio and finely crystallized from Monteponi. In Russia, especially in the Nertschinsk region, Siberia. Fine specimens have been found at Matlock and Cromford in Derbyshire, England; also at Pary's mine on the island of Anglesey, Wales (the original locality); and in Scotland at Leadhills, Lanark, and at Wanlockhead, Dumfries. In Africa found as exceptional crystals at Sidi-Amor-ben-Salem in Tunis and at Tsumeb in South West Africa. At Broken Hill, New South Wales, and Dundas, Tasmania. From the Mérétrice mine in New Caledonia. In Mexico abundant in massive form at Sierra Mojada, Coahuila, and at Potosi, Chihuahua; crystallized in sulfur at Los Lamentos, Chihuahua. At Boleo, Lower California (*bouglisite*).

In the United States, anglesite was formerly found in good specimens at the Wheatley mine, Phoenixville, Chester County, Pennsylvania. Found in very minor amounts in the lead-zinc deposits in Missouri. In Utah in the Park City district, Summit County, and crystallized in the Tintic district, Juab County. As fine specimens in the Coeur d'Alene district in Shoshone County, Idaho.[13] In Arizona in the Castle Dome district, Yuma

County, and the Banner district in Gila County; at Goodsprings and Eureka, Nevada. In California in the Cerro Gordo district, Inyo County.

A r t i f.[14] Obtained as distinct crystals by the action over a period of years of copper sulfate-sodium chloride solution on fragments of galena, and similarly by the action of lead nitrate solution on pyrite. Also in crystals by heating precipitated lead sulfate with dilute HCl in a closed tube at 150°, by very slow precipitation by dilute H_2SO_4 of a dilute solution of lead chloride made acid with HCl, and in other ways.

A l t e r. Alteration pseudomorphs of anglesite after galena are common, as are pseudomorphs or alteration crusts of cerussite after anglesite. The alterations are found in all stages of development, and masses of compact anglesite with concentric color banding may pass inwardly into fresh cores of galena and outwardly in turn be altered to or veined by cerussite. Cavities in altering galena are often lined by fine crystals of anglesite, sometimes also with sulfur crystals. Less common are pseudomorphs of galena, mimetite, and smithsonite after anglesite, and of anglesite after palmierite and cerussite.

N a m e. From the locality in Anglesey, Wales.

Ref.

1. Angles of Koksharov (**1**, 34, 1853) on Monteponi crystals; unit of Ungemach, *Zs. Kr.*, **58**, 163 (1923), corresponding to the structure cell; conventional orientation of Dana (907, 1892), Hintze (**1** [3B], 3962, 1929), Goldschmidt (**1**, 41, 1913), *et alii*. Transformation: K. to new, 002/100/010; Dana, *et al*. to new, 200/010/001. Closely agreeing angles have been reported by others.

2. Goldschmidt (1913); Tacconi, *Rend. real. ist. lombardo*, **44**, 986 (1911); Cesàro, *Soc. géol. Belgique*, *Mém.*, **39**, 239 (1912); Dürrfeld, *Zs. Kr.*, **50**, 585 (1912); Kraus and Peck, *Jb. Min.*, II, 17 (1916); Shannon, *Am. J. Sc.*, **47**, 287 (1919) and *U. S. Nat. Mus., Bull.*, **131**, 444 (1926); Himmel and Schroeder, *Cbl. Min.*, 114 (1935); Billows, *Atti. Accad. Ven.-Trent.-Istr.*, *Padova*, **14**, 82 (1924); Saldanha, *Bol. Univ. São Paulo*, **8**, no. 1 (1938).

Rare or uncertain:

470	610	041	203	4.5.25	495	412	14.9.2
540	810	051	403	166	414	231	16.9.2
430	18.1.0	1.0.12	21.0.1	134	313	30.15.11	851
14.9.0	0.1.16	1.0.11	115	287	343	50.25.17	951
850	0.1.14	108	227	216	131	321	20.11.2
740	0.1.10	2.0.15	113	123	12.1.11	652	10.6.1
950	018	107	225	133	10.9.8	28.14.9	11.6.1
64.33.0	016	104	223	235	615	14.11.4	12.7.1
20.9.0	0.2.11	4.0.15	334	4.15.7	524	431	14.8.1
520	029	207	1.5.10	326	433	531	14.10.1
830	013	103	2.13.18	436	322	16.8.3	16.10.1
20.7.0	035	205	12.7.165	213	674	16.11.3	18.12.1
310	023	409	4.28.27	243	30.8.19	631	
720	045	12.0.25	24.13.156	8.9.10	835	641	
510	031	12.0.23	2.11.13	455	523	741	

3. James and Wood, *Proc. Roy. Soc. London*, **109A**, 598 (1925); also Basche and Mark, *Zs. Kr.*, **64**, 1 (1926).

4. See Lang, *Ak. Wien, Sitzber.*, **36**, 241 (1859); Hermann, *Zs. Kr.*, **39**, 463 (1904); Niggli, *Zs. Kr.*, **59**, 266 (1923).

5. Maier, *Zs. Kr.*, **58**, 89 (1923); Barker, *Zs. Kr.*, **45**, 14 (1908).

6. Samoiloff, Zs. Kr., **45**, 122 (1904); Kruse, Jb. Min., Beil.-Bd., **27**, 541 (1909); Honess (143, 1926).
7. Heide, Zs. Kr., **78**, 257 (1931).
8. Average of numerous single-crystal measurements reported in the literature (at 20°).
9. Kolb, Zs. Kr., **49**, 14 (1911) on Monteponi crystal, who gives data on variation in indices and crystal angles with temperature. See also Kruse (1909), Ehringhaus and Rose, Zs. Kr., **58**, 460 (1923).
10. Arzruni, Zs. Kr., **1**, 186 (1877).
11. See Hintze (1929); Doelter (4 [2], 624, 1927).
12. Jaeger and Germs, Zs. anorg. Chem., **119**, 150 (1921); Grahmann, Zs. anorg. Chem., **81**, 257 (1913).
13. Shannon (1926).
14. Hintze (1929); Mellor (**7**, 803, 1927).
15. Ramdohr, Abh. deutsch. Ak. Wiss. Berlin, no. 4, 1 (1947), describes anglesite with possibly 7 or more per cent $BaSO_4$, with $nY = a$, $nZ = c$, and a_0 8.55 Å, b_0 5.41, c_0 6.95.

28.3.2 **A N H Y D R I T E** [$CaSO_4$]. Muriazit, Salzsaurer Kalk (from Hall, Tyrol) Abbé Poda (in Fichtel, 228, 1794). Würfelspath Werner (1800; Ludwig's Werner, **1**, 51, 166, 1803) = Cube Spar. Soude muriatée gypsifère (of Hall) [from Klaproth's analysis in Beiträge, **1**, 307 (1795)] Haüy (**2**, 1801). Chaux sulfatée anhydre (from Bex) Vauquelin (in Haüy, **4**, 1801). Anhydrit Werner (1803); Ludwig's Werner, **2**, 212, 1804). Würfelgyps Ludwig (**2**, 169, 1804). Anhydrous Sulfate of Lime, Anhydrous Gypsum. Karstenit Hausmann (880, 1813). β-$CaSO_4$. Metanhydrit Sommerfeldt (Jb. Min., I, 139, 1907; Cbl. Min., 22, 189, 1909).

Gekrösstein (from Bothnia and Wieliczka) Werner; Tripe Stone English; Pierre de tripes Fr. = Anhydrit Klaproth (Beitr., **4**, 231, 1807). Pierre de Vulpino; Marmor Bardiglio di Bergamo; Bardiglione; Chaux sulfatée quartzifère Vauquelin (in Haüy, **4**, 251, 1801); Siliceous Anhydrous Gypsum. Kieselgyps, Vulpinit Ludwig (**2**, 170, 1804).

C r y s t.[1] Orthorhombic; dipyramidal—$2/m\ 2/m\ 2/m$.

$a:b:c = 0.9992:1:0.8925;$ $p_0:q_0:r_0 = 0.8932:0.8925:1$

$q_1:r_1:p_1 = 0.9992:1.1196:1;$ $r_2:p_2:q_2 = 1.1204:1.0008:1$

Forms:[2]

	ϕ	$\rho = C$	ϕ_1	$\rho_1 = A$	ϕ_2	$\rho_2 = B$
c 001	....	0°00'	0°00'	90°00'	90°00'	90°00'
b 010	0°00'	90 00	90 00	90 00	....	0 00
a 100	90 00	90 00	.....	0 00	0 00	90 00
m 110	45 01½	90 00	90 00	44 58½	0 00	45 01½
r 011	0 00	41 45	41 45	90 00	90 00	48 15
l 054	0 00	48 07½	48 07½	90 00	90 00	41 52½
q 032	0 00	53 14½	53 14½	90 00	90 00	36 45½
u 021	0 00	60 44½	60 44½	90 00	90 00	29 15½
e 052	0 00	65 51½	65 51½	90 00	90 00	24 08½
v 031	0 00	69 31	69 31	90 00	90 00	20 29
t 041	0 00	74 21	74 21	90 00	90 00	15 39
w 051	0 00	77 22	77 22	90 00	90 00	12 38
o 111	45 01½	51 37½	41 45	56 19½	48 13½	56 21
n 211	63 27	63 24	41 45	36 51	29 14½	66 26
f 311	71 32½	70 28	41 45	26 37½	20 30½	72 38

Less common and rare:

d 120	τ 450	ρ 210	z 015	i 012	x 043	φ 061	p 511
α 230	320	σ 310	013	γ 035	g 053	302	
340	μ 530	017	h 025	k 034	β 095	301	

ANHYDRITE

Structure cell.[3] Space group $Amma$. a_0 6.94 kX, b_0 6.97, c_0 6.20; $a_0:b_0:c_0 = 0.996:1:0.890$. Cell contents $Ca_4(SO_4)_4$. Not isostructural with the barite group.

Habit. Crystals not common: equant or nearly so with large pinacoidal faces; also thick tabular on {010}, {100} or {001}; elongated [100] or [001].

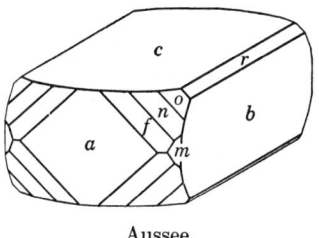
Aussee.

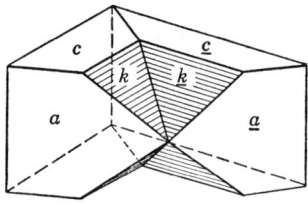
Santorin. Twin on {120}.

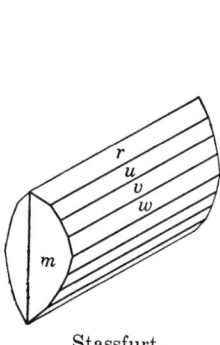
Stassfurt.

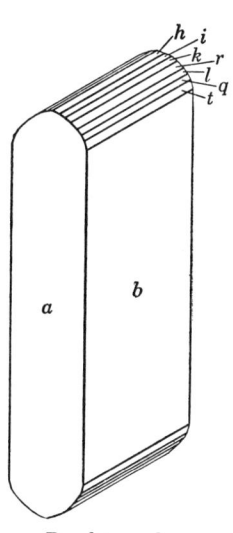
Berchtesgaden.

Usually massive (fragments then resembling an isometric mineral with cubic cleavage); fine granular to scaly granular; fibrous, either parallel, radiated or plumose and often curved; in contorted concretionary forms (*tripe-stone*).

Twinning. (1) Twin plane {011} as contact twins, and as polysynthetic lamellae which may be produced by pressure or heating;[4] (2) Twin plane {120}, as contact twins, rare.

Phys. Cleavage[5] {010} perfect, {100} nearly perfect, {001} good to imperfect. Fracture uneven to splintery. Brittle. H. $3\frac{1}{2}$. G. 2.98; 3.00 (calc.). Luster on {010} pearly, on {001} vitreous to somewhat greasy, on {100} vitreous; massive varieties vitreous to pearly. Colorless to bluish or violet [6] and transparent when in perfect crystals; also white, mauve, rose, brownish, reddish; often gray to dark gray due to impurities such as clay and then turbid or translucent. Streak white or grayish white. Translation gliding:[7] $T\{010\}$ $t[100]$, $T\{120\}$ $t[001]$?; twin gliding: $K_1(011)$ $K_2(0\bar{1}1)$.

Etch figures. Etch pits and solution bodies conform to dipyramidal symmetry.[20]

Opt. In transmitted light, colorless. The violet-tinted crystals are pleochroic (observable by the unaided eye in thick pieces) with X colorless to very pale yellow or rose, Y pale violet or rose, Z violet; absorption [14] $Z > Y > X$.

Orientation	n(Na) [8]	
X b	1.5698	Biaxial positive (+).
Y a	1.5754	$2V$ 43°41′ (Na)
Z c	1.6136	

Indices and 2V at Different Wavelengths [9]

λ	B	C	c	F	G
nX	1.5669	1.5677	1.5746	1.5751	1.5798
nY	1.5724	1.5731	1.5804	1.5811	1.5858
nZ	1.6103	1.6112	1.6192	1.6199	1.6250
$2V$	43°13′	43°18′	43°53′	43°57′	44°13′

Chem. Calcium sulfate, $CaSO_4$. The composition of anhydrite apparently varies only very slightly through isomorphous substitution for Ca. Sr is sometimes reported in amounts less than 1 per cent SrO, and Ba occurs in even smaller amounts. The MgO, Al_2O_3, SiO_2, etc., reported is believed due to admixture, and H_2O (or ignition loss) when present doubtless is due to partial hydration to gypsum. On heating, anhydrite inverts at 1193° to α-$CaSO_4$, isostructural and isomorphous with α-$BaSO_4$ and α-$SrSO_4$, and melts at about 1450°.[24] In saturated aqueous solutions the transition temperature [15] of gypsum to anhydrite is 42°.

Anal.[10]

	CaO	SrO	BaO	SO_3	H_2O	Rem.	Total	G.
1.	41.19			58.81			100.00	2.98
2.	40.47	0.71	0.05	58.94	0.40		100.57	2.97
3.	41.46			57.96		0.24	99.66	
4.	42.09	0.06	0.04	57.79		0.02	100.00	2.981

1. $CaSO_4$. 2. Kalgoorlie, Western Australia.[11] 3. Kanô mine, Ugo, Japan.[12] Rem. is Al_2O_3 0 05, FeS_2 0.02, CO_2 0.17. Sr, Zn, Cu nil. 4. Stromboli, Italy.[13] Rem. is Fe_2O_3, SiO_2 tr.

Tests. B.B. fuses at 3 to a white enamel. Soluble in acids. Solubility in water (in grams $CaSO_4$ per 100 g. sat. soln. in contact with anhydrite): 0.201 at 45°, 0.0650 at 100°, 0.0076 at 200°.

Occur. Anhydrite is an important rock-forming mineral and occurs in sedimentary terranes with gypsum, limestone, dolomite, and salt beds. Anhydrite is deposited directly by evaporation of sea water [15] at temperatures close to or above 42° or, at lower temperatures, at salinities about

4.8 times the ordinary salinity. At lower salinities, below 42°, gypsum is deposited. Sedimentary anhydrite deposits in part appear to be derived by the conversion of earlier formed gypsum; and anhydrite beds may later be converted to gypsum near the outcrops by meteoric waters. Anhydrite also occurs as an accessory mineral in sedimentary rocks, especially in rock salt, or occurs in crevices in such rocks and is then associated with celestite, dolomite, calcite, gypsum, quartz. Also found less commonly as a hypogene gangue mineral in metalliferous veins,[18] and with prehnite, quartz, glauberite, and zeolites in cavities in igneous flow-rocks. Also found in caliche and as a fumarole deposit. A few of the very large number of known localities are mentioned below.

Anhydrite was first found in the salt deposit at Hall near Innsbruck, Tyrol; also found in Austria in salt beds at Ischl in Salzburg, Aussee in Styria and at Hallein. In Switzerland at Bex, Vaud, and in the Simplon Tunnel in Valais. In France found at Arnave in Ariège and at numerous places in the gypsum and rock-salt deposits of the Pyrenees. Found in Germany as fine crystals in kieserite at Stassfurt and similarly at Douglashall and Leopoldshall; at Wathlingen and Lüneburg in Hannover; at Berchtesgaden in Bavaria. From the salt beds of Wieliczka in Krakow, Poland. In the ore of the Hanaoka and Kanô mines, Ugo, Japan. In the salt and gypsum deposits of the Salt Range, Punjab, India.

In the United States anhydrite occurs in druses in the Lockport dolomite in western New York; also in the zinc mines at Balmat, St. Lawrence County, and in the magnetite deposit at Lyon Mountain, Clinton County. As crystalline masses and as pseudomorphs (molds) with zeolites in cavities in Triassic diabase at Great Notch, West Paterson, Montclair, and other places in New Jersey; similarly at Westfield, Greenfield, and Holyoke in western Massachusetts. With gypsum, halite, and potash salts in the Carlsbad district, Eddy County, New Mexico, and elsewhere in the Permian Basin of Texas-New Mexico.[22] In the cap rock of salt domes in Texas and Louisiana.[23] In Canada in bedded deposits in Nova Scotia and as a replacement of Grenville marble on Calumet Island, Quebec.

Alter. Anhydrite alters readily by hydration to gypsum; the alteration, which often takes place on a geologic scale, is accompanied by an increase in volume with resulting mechanical deformation of the rock. Pseudomorphs after anhydrite, for the most part by incrustation, have been reported of quartz, siderite, dolomite, calcite (also effected artificially),[19] thaumasite, gypsum, and marcasite; also pseudomorphs of anhydrite after calcite and gypsum. Hollow rectangular molds after anhydrite crystals are very common in the zeolite-containing cavities in diabasic igneous rocks (New Jersey;[16] Westfield, Massachusetts [17]); the incrusting mineral usually is prehnite or quartz. Rectangular gashes or molds known or suspected to be after anhydrite also occur in some sulfide veins and in crevices and veins of the Alpine type.

Artif.[21] Obtained in crystals by slowly cooling fusions of $CaSO_4$ with $CaCl_2$, $BaCl_2$, or $NaCl$ or of $CaCl_2$ with K_2SO_4; also by heating gypsum with NaCl or CaCl₂ solution in a closed tube, and in other ways. Fine-grained anhydrite is obtained at ordinary temperatures by precipitating $CaSO_4$ from solutions containing much $MgCl_2$ or $CaCl_2$ or from concentrated acid solutions; gypsum is obtained at lower concentrations (and temperatures).

N a m e. From ἄν-υδρος, *without water*. Anhydrite was first recognized as a separate species by Nicolaus Poda von Neuhaus (1723–1798), Jesuit, although erroneously as a hydrated chloride of calcium (whence the name *muriacite*) and the true composition was first established by the analysis of Klaproth in 1795. Anhydrite rock probably was included among the translucent building stones termed phengites by Pliny.

Ref.

1. Angles and unit of Hessenberg, *Senck. Ges. Frankfurt, Abh.*, **8**, 1 (1872); orientation of the structure cell [with $c < a < b$]. Transformation: Dana (910, 1892), Goldschmidt (**1**, 56, 1913; **1**, 211, 1886), Hintze (**1** [3B], 3735, 1929), Basche and Mark, *Zs. Kr.*, **64**, 22 (1926), to new, 010/001/100; Hessenberg (1872) and Schrauf, *Ak. Wien, Sitzber.*, **39**, 887 (1860) to new, 010/100/001. Kolb, *Zs. Kr.*, **49**, 14 (1911), gives data on the variation of angle with temperature.
2. Goldschmidt (1913).
3. Basche and Mark (1926). See also Dickson and Binks, *Phil. Mag.*, **2**, 114 (1926); Wasastjerna, *ibid.*, 994.
4. Mügge, *Jb. Min.*, II, 258 (1883).
5. On the ease of cleavage see Tertsch, *Zs. Kr.*, **87**, 326 (1934).
6. On the violet and blue color see Kinoshita, *J. Geol. Soc. Tokyo*, **32**, 9 (1925); Przibram, *Kali*, **30**, 61 (1936).
7. Mügge, *Jb. Min.*, I, 71 (1898), and Veit, *Jb. Min., Beil.-Bd.*, **45**, 133 (1922).
8. Average of single-crystal measurements reported by Danker, *Jb. Min., Beil.-Bd.*, **4**, 272 (1886), Kolb, *Zs. Kr.*, **49**, 25 (1911), Mülheims, *Zs. Kr.*, **14**, 228 (1888), Zimányi, *Zs. Kr.*, **22**, 341 (1893).
9. Kolb (1911) for Fraunhofer lines, who also gives data on variation in indices with temperature.
10. For a tabulation of additional analyses see Hintze (**1** [3B], 3780, 1929).
11. Simpson, *J. Roy. Soc. Western Australia*, **16**, 25 (1930).
12. Kinoshita, *J. Geol. Soc. Tokyo*, **32**, 9 (1925).
13. Ponte, *Acc. Linc., Att.*, **26** [5], 348 (1917).
14. Absorption measurements are given by Berek and Strieder, *Zs. Kr.*, **86**, 212 (1933).
15. See Posnjak, *Am. J. Sc.*, **238**, 559 (1940), Engelhardt, *Chem. Erde*, **15**, 424 (1945).
16. Schaller, *U. S. Geol. Sur., Bull. 832*, (1932).
17. Emerson, *Am. J. Sc.*, **42**, 233 (1916).
18. Zimmer, *Am. Min.*, **32**, 647 (1947); Butler, *Econ. Geol.*, **14**, 581 (1919).
19. Vater, *Zs. Kr.*, **31**, 571 (1899).
20. Burkhardt, *Zs. Kr.*, **50**, 209 (1912); Preiswerk, *Jb. Min.*, I, 33 (1905).
21. Mellor (**3**, 763, 1923); Hintze (1929); Doelter (**4** [2], 196, 1927); Chassevent, *C.R.*, **194**, 786 (1932). On systems containing $CaSO_4$ and H_2O see Hill and Wills, *J. Am. Chem. Soc.*, **60**, 1647 (1938); Conley, *et al.*, *J. Phys. Chem.*, **42**, 587 (1938); Ramsdell and Partridge, *Am. Min.*, **14**, 59 (1929).
22. Schaller and Henderson, *U. S. Geol. Sur., Bull. 833*, 14 (1932).
23. Barnes, *Am. Min.*, **18**, 335 (1933), and Gulf Coast Oil Fields Sympos., *Am. Assoc. Pet. Geol.*, 119 (1936).
24. Grahmann, *Jb. Min.*, I, 1 (1920).

ZINKOSITE [$ZnSO_4$]. *Breithaupt* (*B.-H. Ztg.*, **11**, 100, 1852). Almagrerite.

Anhydrous $ZnSO_4$ has been reported from the mine of Barranco Jaroso in the Sierra Almagrera near Aquilas, Murcia, Spain. G. 4.33 (nat.), 3.86 (calc.). Definitive descriptive data are lacking, and the occurrence is very doubtful. Artificial $ZnSO_4$ is orthorhombic [1] with a_0 8.58 kX, b_0 6.74, c_0 4.76. Optically [2] negative (−), $2V$ medium to small, nX 1.658, nY 1.669, nZ 1.670.

Ref.
1. See Schiff, Zs. Kr., **87**, 379 (1934), on structure cell; de Schulten, C.R., **107**, 405 (1888), on morphology.
2. Larsen (159, 1921).

28.3.3 **C H A L C O C Y A N I T E** [$CuSO_4$]. Idrociano Scacchi (Acc. Napoli, Att., 5, 26, 1873). Hydrocyan. Hydrocyanite. Hydrokyanite.

C r y s t.[1] Orthorhombic; dipyramidal—$2/m\ 2/m\ 2/m$.

$a:b:c = 0.7971:1:1.1300;\qquad p_0:q_0:r_0 = 1.4176:1.1300:1$

$q_1:r_1:p_1 = 0.7971:0.7054:1;\qquad r_2:p_2:q_2 = 0.8850:1.2545:1$

Forms:[2]

	ϕ	$\rho = C$	ϕ_1	$\rho_1 = A$	ϕ_2	$\rho_2 = B$
b 010	0°00′	90°00′	90°00′	90°00′	....	0°00′
m 110	51 26½	90 00	90 00	38 33½	0°00′	51 26½
k 120	32 02½	90 00	90 00	57 57½	0 00	32 02½
u 102	90 00	35 19½	0 00	54 40½	54 40½	90 00
e 012	0 00	29 28	29 28	90 00	90 00	60 32
d 011	0 00	48 29½	48 29½	90 00	90 00	41 30½
μ 112	51 26½	42 11	29 28	58 19	54 40½	65 15

Less common:
c 001   p 232   n 212

Habit. Variable; often tabular {010} and slightly elongated [001].

P h y s. Cleavage not observed. H. 3½. G. 3.65 ± 0.05.[3] Pure artificial crystals are colorless; natural crystals are pale green, brownish or yellowish, also sky-blue. Transparent to translucent.

O p t. In transmitted light, colorless.

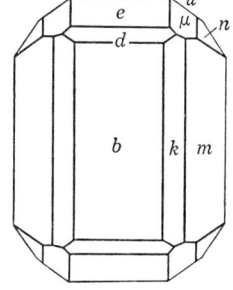

ORIENTATION		n(Na)[4]	
X	b	1.724 ± 0.003	Biaxial negative (−).
Y	a	1.733 ± 0.003	2V large.
Z	c	1.739 ± 0.003	$r > v$, extreme.

C h e m. Anhydrous cupric sulfate, $CuSO_4$.
Anal.

	1	2
CuO	49.85	49.47
SO_3	50.15	50.30
Total	100.00	99.77

1. $CuSO_4$. 2. Vesuvius.[2]

O c c u r. Found with dolerophanite, melanothallite, eriochalcite, and euchlorin as a sublimate in fumaroles of the eruption of 1868 of Vesuvius. Also observed[5] during the eruptions of 1880 and 1895.

A l t e r. Chalcocyanite is very hygroscopic, and readily soluble in water. On exposure to moist air the crystals rapidly turn blue, owing to the formation of the pentahydrate, and may ultimately fall to a powder of chalcanthite.

N a m e. From χαλκός, *copper*, and κύανος, *azure-blue*, in allusion to the composition and color change noted above. The original name hydrocyanite is objectionable for an anhydrous substance.

Ref.
1. Angles and unit of Scacchi (1873); orientation of Dana (912, 1892) to show the relation to the barite group.
2. Scacchi (1873).
3. Range of best reported values.
4. Posnjak and Tunell, *Am. J. Sc.*, **18**, 27 (1929).
5. Zambonini (158, 1935).

TYPE 4. $A_m B_n (XO_4)_p$

28.4.1 VANTHOFFITE [$Na_6Mg(SO_4)_4$]. Kubierschky (*Ak. Berlin, Ber.*, **21**, 404, 1902).

Found only massive, in anhedral grains or bedded aggregates.

P h y s. Fracture uneven to flat conchoidal. Friable. H. $3\frac{1}{2}$. G. 2.694.[1] Luster vitreous to pearly. Colorless. Transparent.

O p t.[1] In transmitted light, colorless.

	n (Hall, Tyrol)	
X	1.485	Biaxial negative (−).
Y	1.488	$2V\ 84°$ (meas.).
Z	1.489	$r < v$, weak.

C h e m. A sulfate of sodium and magnesium, $Na_6Mg(SO_4)_4$.

Anal.

	Na_2O	MgO	SO_3	Total	G.
1.	34.03	7.38	58.59	100.00	
2.	33.64	7.37	58.54	99.55	2.694

1. $Na_6Mg(SO_4)_4$. 2. Hall, Tyrol.[1]

O c c u r. Found only in oceanic salt deposits. First found in the Wilhelmshall potash mine near Halberstadt. With loeweite and langbeinite in halite at the Berlepsch mine near Stassfurt, and elsewhere in the German potash deposits. Sparingly in bloedite at Hall, Tyrol.

A r t i f. Can be made by heating a compressed mixture of the constituent sulfates at 80°.[2] Crystallizes from an aqueous solution above 57° at which temperature the solution is also in equilibrium with bloedite and thenardite.[3] From a solution containing the ions Na, K, Mg, SO_4, Cl and saturated with NaCl, corresponding to the brines of the oceanic salt deposits, it may crystallize above 46°.[4]

N a m e. After the Dutch physical chemist Jacobus Henricus van't Hoff (1852–1911), whose studies on equilibria in salt solutions were fundamental to an understanding of the formation of vanthoffite and other minerals of the oceanic salt deposits.

Ref.
1. Görgey, *Min. Mitt.*, **28**, 334 (1909).
2. Ide, *Kali*, **29**, 83 (1935)—*Min. Abs.*, **6**, 351 (1936).
3. *Int. Crit. Tables*, **4**, 350 (1928).
4. As determined by van't Hoff and co-workers and summarized in Doelter, **4** [2], 93 (1926).

28.4.2 **G L A U B E R I T E** [$Na_2Ca(SO_4)_2$]. *Brongniart* (*J. Mines*, **23**, 5, 1808).
Brongniartin *Leonhard* (*Naturges. des Mineralreiches*, 159, 1825; 270, 1826).

C r y s t. Monoclinic; prismatic—$2/m$.

$a:b:c = 1.2200:1:1.0275;$ $\beta\, 112°11';$ [1] $p_0:q_0:r_0 = 0.8422:0.9515:1$

$r_2:p_2:q_2 = 1.0510:0.8851:1;$ $\mu\, 67°49';$ $p_0'\, 0.9095,\, q_0'\, 1.0275,\, x_0'\, 0.40775$

Forms: [2]

		ϕ	ρ	ϕ_2	$\rho_2 = B$	C	A
c	001	90°00'	22°11'	67°49'	90°00'	.....	67°49'
a	100	90 00	90 00	0 00	90 00	67°49'	.....
m	110	41 31	90 00	0 00	41 31	75 30½	48 29
s	111	52 03	59 06	37 12	58 09	43 01½	47 25
n	$\overline{1}11$	−26 02	48 50	116 39	47 26	61 00	109 17½
x	$\overline{3}31$	−48 29	72 07½	156 41½	50 53	89 07	135 26½
e	$\overline{3}11$	−66 07½	68 30	156 41½	67 52½	88 57	148 18

Less common:

$\delta\, 112$ $v\, \overline{1}13$ $u\, \overline{1}12$

Structure cell.[3] Space group $C2/c$. $a_0\, 9.99\, kX$, $b_0\, 8.19$, $c_0\, 8.41$;
$[\beta\, 112°11'];$ $a_0:b_0:c_0 = 1.220:1:1.027$. Cell contents $Na_8Ca_4(SO_4)_8$.

Habit. Variable: tabular $\{001\}$, with $\{110\}$ sometimes lacking; prismatic by extension $[\overline{1}01]$ with large $\{111\}$; dipyramidal through combinations of $\{111\}$ and $\{110\}$; prismatic $[001]$ with large $\{110\}$. $\{001\}$ and

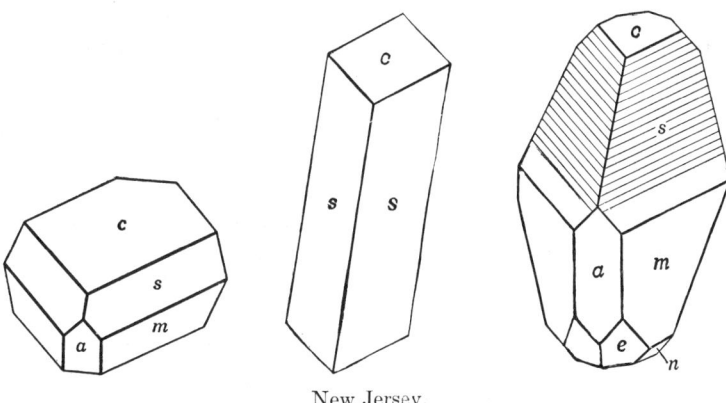

New Jersey.

$\{111\}$ often striated parallel to their intersection, and $\{111\}$ may be rounded off or step-like toward the termination. Common forms: $s\, c\, m\, a$.

P h y s. Cleavage $\{001\}$ perfect, $\{110\}$ indistinct. Fracture conchoidal. Brittle. H. 2½–3. G. 2.75–2.85; 2.81 (calc.). Luster vitreous to slightly waxy, pearly on the $\{001\}$ cleavage. Usually gray or yellowish in color, sometimes colorless, or reddish due to included iron oxide. Streak white. Transparent to translucent. Taste slightly saline. Whitens in water owing to leaching of sodium sulfate and deposition of gypsum. Shows translation gliding.[4]

O p t.[5] In transmitted light, colorless.

Orientation		n	
X		1.515	Biaxial negative $(-)$.
$Y \wedge c$	$12°$	1.535	$2V\ 7°,\ 2E\ 10°$.
$Z\ \ b$		1.536	$r > v$, strong; also horizontal dispersion.

The indices of refraction, optical orientation, $2V$ and dispersion vary markedly with temperature.[6] The optical character and the position of the optical indicatrix remain sensibly constant between $0°$ and $100°$. The optic plane at ordinary temperatures is $\perp \{010\}$ with horizontal dispersion, and $r > v$. With increasing temperature the optic angle decreases to zero and then opens again in the plane $\{010\}$, with inclined dispersion and $r < v$. The temperature of uniaxiality ($2V = 0°$) varies with wavelength: $17.8°$ (blue), $35.7°$ (green), $45.8°$ (yellow), $58.2°$ (red). The dispersion is very strong giving abnormal interference figures in white light.

C h e m. Sulfate of sodium and calcium, $Na_2Ca(SO_4)_2$. Na_2O 22.29, CaO 20.16, SO_3 57.55, total 100.00. Analyses agree closely with the composition required by the formula.[7]

Tests. B.B. decrepitates, turns white, and fuses at $1\frac{1}{2}$ to a white enamel. Completely soluble in HCl.

O c c u r. A widespread constituent of salt deposits both of oceanic and lacustrine origin. It is also found as isolated crystals embedded in clastic sediments formed under arid conditions, in cavities in basic extrusive rocks, as a fumarole deposit and with nitrate deposits in extremely arid regions. A few of the many occurrences are mentioned below.

First recognized as an independent species as crystals in halite at Villarubia near Ocana, in the province of Toledo, and since found in several other localities in Spain. It occurs as brick-red crystals of resinous appearance with polyhalite and halite at Varangéville near Nancy, and elsewhere in the Triassic salt deposits of Lorraine, France. At Douglashall near Westerregeln with halite in salty clay layers and at various other localities in the German Triassic potash deposits, often as fine crystals. In masses several meters thick with anhydrite at Hallstatt, and elsewhere in the Triassic salt deposits of upper Austria. Found with thenardite and sassolite as a fumarole deposit at Vulcano in the Lipari Islands. With halite and polyhalite in Permian salt deposits at Kairovka, near Sterlitamak in Bashkiria, and elsewhere in southeastern Russia. Crystallized on cracks in rock salt at the Mayo Mine in the Punjab Salt Range, India. With thenardite and bloedite in the Taltal nitrate region of the Atacama desert and elsewhere with the nitrate deposits of northern Chile. In the United States very sparingly with gypsum in cavities in basalt at West Paterson, New Jersey. Its former presence together with anhydrite, prehnite, and zeolites in basalt cavities at Paterson, Great Notch, and other places in this region is indicated by abundant molds and casts.[9] Similarly its former presence in Triassic red shales is indicated by calcite-filled casts at Blackwell's Mills, New Jersey.[10] With polyhalite, halite, and anhydrite in Permian salt deposits penetrated by deep borings in Ector and Upton

Counties, Texas, and in Eddy County, New Mexico. With thenardite, mirabilite, and halite in the salt beds of the Verde Valley, Yavapai County, Arizona; as single crystals embedded in clays under the surface salt crusts of Saline Valley, Inyo County, and rather abundant in the Searles Lake salt deposits, San Bernardino County, California. In Canada with gypsum in upper Silurian beds at Gypsumville, Manitoba.

Alter. In oceanic salt deposits glauberite has replaced halite or polyhalite.[8] Molds and casts of prehnite and quartz after glauberite have been found in great number in the New Jersey basalt localities. Pseudomorphs of both calcite and opal after glauberite have been reported. Stable in dry atmosphere.

Artif. The double salt $Na_2Ca(SO_4)_2$ is not crystallized from fusion in the system Na_2SO_4-$CaSO_4$.[11] From aqueous solutions of Na_2SO_4 and $CaSO_4$ it may crystallize at equilibrium above about 30°. At 35° its field of crystallization is bounded by the gypsum and thenardite fields, at 50° and 75° by the anhydrite and thenardite fields.[12]

Name. Glauberite was named in allusion to the fact that it contained a considerable quantity of the salt (Na_2SO_4) formerly called Glauber's salt after the German chemist Johann Rudolf Glauber (1604–1668).

CIEMPOZUELITE. Ciempozuelita *Areitio* (*R. soc. española hist. nat., Anales*, **2**, 393, 1873). Efflorescences in the Consuelo mine near Ciempozuelos, Spain, having a composition close to $Na_6Ca(SO_4)_4$. Perhaps a mixture with glauberite.

Ref.

1. Rounded from $a:b:c = 1.21998:1:1.02749$; β 67°49'07" by Zepharovich, *Ak. Wien, Ber.*, **69**, 16 (1874), on Westerregeln crystals.
2. Goldschmidt (**4**, 46, 1918); Schaller, *U. S. Geol. Sur., Bull. 832*, 55 (1932); Schaller and Henderson, *U. S. Geol. Sur., Bull. 833*, 27 (1932); Pardillo, *Ac. cienc. arts, Barcelona, Mem.*, **25**, 1 (1934) in *Jb. Min.*, I, 344 (1935). Many of the rare or uncertain forms which follow were observed on pseudomorphs after glauberite or are vicinal.

010	$\bar{3}$02	114	113	449	8.8.17	447	223	445	221
023	$\bar{2}$01	3.3.11	338	6.6.13	9.9.19	335	557	667	661
021	118	5.5.17	337	7.7.15	10.10.21	558	334	778	

3. Pardillo (1934) by rotation method on Chincon, Spain, crystals.
4. Wetzel, *Zs. pr. Geol.*, **32**, 119 (1924).
5. Larsen and Berman (156, 1934); also Godlevsky, *Mém. soc. russe min.*, **68**, 576 (1939).
6. See Laspeyres, *Zs. Kr.*, **1**, 529 (1877); Kraus, *Zs. Kr.*, **52**, 321 (1913); and Kraus and Peck, *Michigan Ac. Sc., Ann. Rept.*, **19**, 95 (1917).
7. For a tabulation of analyses, see Hintze (**1** [3], 3721 (1929); also Godlevsky (1939).
8. See also Schaller and Henderson, *U. S. Geol. Sur., Bull. 833*, 27 (1932), who also report that glauberite replaces anhydrite in the Texas-New Mexico Permian salt deposits.
9. Schaller, *U. S. Geol. Sur., Bull. 832*, 55 (1932).
10. Hawkins, *Am. J. Sc.*, [5], **16**, 361 (1928), reports calcite casts in red shales.
11. Müller, *Jb. Min., Beil.-Bd.*, **30**, 32 (1910).
12. Hill and Wills, *Am. Chem. Soc., J.*, **60**, 1647 (1938). The solution at 35° in equilibrium with gypsum and glauberite contains 22.65 weight per cent Na_2SO_4 and 0.206 weight per cent $CaSO_4$.

28.4.3 LANGBEINITE GROUP

28.4.3.1 LANGBEINITE [$K_2Mg_2(SO_4)_3$]. Zuckschwerdt (Zs. anorg. Chem., 356, 1891).

Cryst. Isometric; tetartoidal [1]—2 3.

Forms: [2]

a 001	o 111	e' 012	$'e$ 102	n 112
d 011	$-o$ $\bar{1}11$	$'f$ 103	$-j$ $\bar{1}22$	$-n$ $\bar{1}12$	209(?)

Structure cell.[3] Space group probably $P2_13$. a_0 9.96 kX. Cell contents $K_8Mg_8(SO_4)_{12}$.

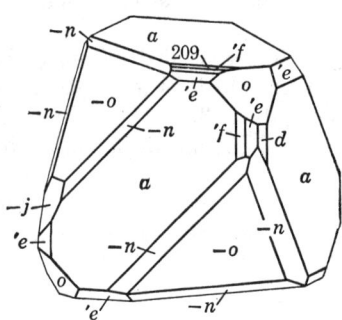

Habit. Rarely in crystals. As nodular masses or disseminated grains in salt deposits; also bedded.

Phys. No cleavage. Fracture conchoidal. Brittle. H. $3\frac{1}{2}$–4. G. 2.83; 2.77 (calc.). Luster vitreous, bright. Colorless and transparent, occasionally yellowish, rose, reddish, violet, greenish, or gray. Often contains inclusions of halite, anhydrite, or other substances. Piezoelectric.[5] M.P. 930°.[4]

Opt.[6] In transmitted light, colorless. Does not show circular polarization.

	Li	Na	Tl	
n (Hall, Tyrol)	1.5323	1.5347	1.5370	Isotropic
n (Solvayhall)	1.5281	1.5329	1.5344	

Chem. A sulfate of potassium and magnesium, $K_2Mg_2(SO_4)_3$. It has been shown that Mg can be replaced by Ca in any proportion,[7] but such replacement has not been found in natural materials. A small bromine content has been reported.[8]

Anal.[9]

	K_2O	MgO	SO_3	Rem.	Total	G.
1.	22.70	19.43	57.87		100.00	2.762
2.	22.23	19.08	57.27	1.25	99.83	2.84
3.	22.72	19.91	57.99		100.62	2.825
4.	22.37	19.15	57.44	0.78	99.74	2.858

1. $K_2Mg_2(SO_4)_3$. 2. Mayo mine, Punjab Salt Range.[10] Rem. is NaCl 0.41, H_2O 0.84. 3. Hall, Tyrol.[11] 4. Lea County, New Mexico.[12] Rem. is Na_2O 0.48, ign. loss 0.25, insol. 0.05.

Tests. Only very slowly soluble in water.[13] Easily fusible. Turns white on slight heating.

Occur. Known only as a constituent of oceanic salt deposits. First noted in borings at Wilhelmshall near Halberstadt and later found in larger amount at Solvayhall near Bernburg, where it occurs interbedded with carnallite (below) and "hartsalz" (above), and elsewhere in the north German potash deposits. As large nodules or rounded masses embedded in halite at Hall, Tyrol, and at Hallstatt, Upper Austria. At Stebnik near Drohobycz, Galicia. In the Mayo mine, Punjab Salt Range, India.

In the United States it was first noted in cores of borings for potash salts in Lea and Eddy Counties, New Mexico, mostly associated with halite and sylvite. Partly in crystals up to 2 cm. in diameter. Some beds contain up to 90 per cent langbeinite. Now mined on a large scale from a bed 5 to 7 feet thick at a depth of 755 feet, 15 miles east of Carlsbad, New Mexico.

Alter. At the surface langbeinite alters to the hydrated sulfates, picromerite, and epsomite.[14] Pseudomorphs of kainite after langbeinite have been observed in the New Mexico deposits.[15]

Artif. Formed by heating the mixed anhydrous sulfates of potassium and magnesium to 80°.[16] Crystallizes from aqueous solution above 61° at which temperature the solution is also in equilibrium with $MgSO_4 \cdot 6H_2O$ (*hexahydrite*) and $K_2Mg(SO_4)_2 \cdot 4H_2O$ (*leonite*).[17] From a solution containing the ions Na, K, Mg, SO_4, Cl, and saturated with NaCl, corresponding to the brines of the oceanic salt deposits, it may crystallize above 37°.[18]

Name. In honor of A. Langbein of Leopoldshall.

Ref.
1. Luedecke, *Zs. Kr.*, **29**, 255 (1898).
2. Luedecke (1898) and Sachs, *Ak. Berlin, Ber.*, 376 (1902).
3. Gossner and Koch, *Zs. Kr.*, **80**, 455 (1931).
4. Nacken, *Ges. Wiss. Göttingen, Nachr.*, 602 (1907), who gives temperature-composition diagram for the system $MgSO_4$-K_2SO_4.
5. Bond, *Bell Sys. Tech. J.*, **22**, 145 (1943), reports langbeinite to be piezoelectric. This effect was not detected by Luedecke (1898) or by Gossner and Koch (1931).
6. Luedecke (1898) and Görgey, *Min. Mitt.*, **28**, 334 (1909).
7. Ramsdell, *Am. Min.*, **20**, 569 (1935).
8. Winkler, *Zs. angew. Chem.*, **30**, 95 (1917), found 0.016 per cent Br in langbeinite from Leinetal, Germany.
9. For additional analyses see Hintze (**1** [3], 3729, 1929).
10. Mallet, *Min. Mag.*, **12**, 159 (1900).
11. Görgey (1909).
12. Fahey in Schaller and Henderson, *U. S. Geol. Sur., Bull. 833*, 41 (1932).
13. The slower rate of solution in water has been used in separating langbeinite from associated halite and other salts either in preparing it for analyses or for commercial utilization.
14. Görgey (1909).
15. Schaller and Henderson (1932).
16. Ide, *Kali*, **29**, 83 (1935)—*Min. Abs.*, **6**, 351 (1936).
17. *Int. Crit. Tables*, **4**, 352 (1928).
18. As determined by van't Hoff and co-workers and summarized in Doelter (**4** [2], 93, 1926).

28.4.3.2 **Manganolangbeinite** [$K_2Mn_2(SO_4)_3$]. Zambonini and Carobbi (*Acc. Napoli, Rend.* [3], **30**, 123, 1924).

Isometric, tetartoidal—2 3. As small tetrahedra; also $\{210\}$ and $\{211\}$ in artificial material. Space group [1] $P2_13$, with a_0 10.014 Å and cell contents $K_8Mn_8(SO_4)_{12}$. G. 3.02. Color rose-red. Isotropic with n 1.572. A sulfate of potassium and divalent manganese, $K_2Mn_2(SO_4)_3$. The natural material has not been analyzed. Found with thenardite, halite, sylvite, and aphthitalite in a cavern in lava on Vesuvius, Italy. Artificial crystals [1] have been obtained from fusion or water solutions of K_2SO_4 and $2MnSO_4$.

Ref.
1. Bellanca, *Acc. Linc., Att., Cl. sc.*, [8], **2**, 451 (1947).

29 HYDRATED ACID AND NORMAL SULFATES

TYPE 1. HYDRATED ACID SULFATES

29.1.1 RHOMBOCLASE [$HFe(SO_4)_2 \cdot 4H_2O$]. Rhomboklas *Krenner* (*Ak. Értes.*, Budapest, **2**, 96, 1891; *Földt. Közl.*, **37**, 204, 1907). Saures Ferrisulfate *Scharizer* (*Zs. Kr.*, **35**, 345, 1901).

Cryst.[1] Orthorhombic; dipyramidal—$2/m\ 2/m\ 2/m$.

$a:b:c = 0.5577:1:0.9370;\qquad p_0:q_0:r_0 = 1.6801:0.9370:1$

$q_1:r_1:p_1 = 0.5577:0.5952:1;\qquad r_2:p_2:q_2 = 1.0672:1.7930:1$

Forms:[2]

	ϕ	$\rho = C$	ϕ_1	$\rho_1 = A$	ϕ_2	$\rho_2 = B$
c 001	.....	0°00′	0°00′	90°00′	90°00′	90°00′
b 010	0°00′	90 00	90 00	90 00	.....	0 00
s 120	41 52½	90 00	90 00	48 07½	0 00	41 52½
m 110	60 51	90 00	90 00	29 09	0 00	60 51
v 113	60 51	32 40	17 20½	61 52½	60 45	74 45½
w 223	60 51	52 03	31 59½	46 28½	41 45½	67 24½
p 111	60 51	62 32	43 08	39 12	30 45½	64 23½
q 221	60 51	75 26	61 55	32 18	16 34½	61 52½

Less common and rare:

	d 011	t 041	n 118	r 114	u 241	

Habit. Thin tabular {001}. Common forms $b\ c\ m\ p$. Also as stalactites with a radiating bladed structure.

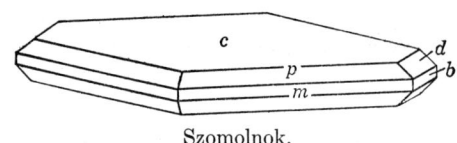

Szomolnok.

Phys. Cleavage {001} perfect, {110} good. Fracture conchoidal to fibrous; cleavage folia are flexible. H. 2. G. 2.23.[2] Luster subvitreous to pearly. Colorless to white, gray or pale yellow, or tinted greenish or bluish due to admixture. Transparent.

Opt. In transmitted light, colorless.

ORIENTATION	n(Na; Chile)[2]	n (artif.)[3]	
X c	1.534	1.533	Biaxial positive (+).
Y a	1.553	1.555	$2V\ 27°$ (Na).[2]
Z b	1.638	1.635	

Chem. A hydrated acid ferric sulfate, $HFe(SO_4)_2 \cdot 4H_2O$. Analysis gave:

	CaO	CuO	FeO	Fe_2O_3	SO_3	H_2O	Rem.	Total
1.				24.87	49.88	25.25		100.00
2.	0.10	0.03	0.35	24.54	49.27	[25.54]	0.15	[99.98]

1. $HFe(SO_4)_2 \cdot 4H_2O$. 2. Szomolnok.[4] Rem is MgO plus Na_2O and Al_2O_3 trace. The analysis is given as cited with water by difference.

Tests. B.B. decrepitates and fuses to a dark brown bead. Slowly soluble in water, easily soluble in acids.

Occur. At Szomolnok (Schmöllnitz), Czechoslovakia, associated with szomolnokite, copiapite, and other sulfates as an alteration of pyritic ore. From the Esperanza mine, Pasco, Peru, as incrustations in abandoned workings with chalcanthite, roemerite, epsomite. Also at Alcaparrosa, near Cerritos Bayos, Chile, with szomolnokite and roemerite.

Artif.[5] Rhomboclase is a stable phase in the system Fe_2O_3-SO_3-H_2O at temperatures up to 140°.

Name. In allusion to the basal cleavage and crystal shape.

Ref.

1. Bandy, *Am. Min.*, **23**, 740 (1938), on Alcaparrosa crystals of better quality than those measured by Krenner, *Cbl. Min.*, 265 (1928), from Czechoslovakia.
2. Bandy (1938).
3. Posnjak and Merwin, *J. Am. Chem. Soc.*, **44**, 1983 (1922).
4. Loczka analysis in Krenner (1928).
5. Posnjak and Merwin (1922). See also Scharizer, *Zs. Kr.*, **35**, 345 (1901); **43**, 113 (1907); **56**, 353 (1921); **65**, 335 (1927), who gives dehydration data.

29.1.2 **MINASRAGRITE** [$(VO)_2H_2(SO_4)_3 \cdot 15H_2O$]. *Schaller* (*J. Washington Ac. Sc.*, **5**, 7, 1915; **7**, 501, 1917).

Cryst.[1] Monoclinic; prismatic—$2/m$.

$a:b:c = 0.7196:1:0.6656;\quad \beta\ 110°57';\quad p_0:q_0:r_0 = 0.9250:0.6216:1$

$r_2:p_2:q_2 = 1.6088:1.4880:1;\quad \mu\ 69°03';\quad p_0'\ 0.9904,\ q_0'\ 0.6656,\ x_0'\ 0.3829$

Forms:

	ϕ	ρ	ϕ_2	$\rho_2 = B$	C	A
c 001	90°00′	20°57′	69°03′	90°00′	.....	69°03′
m 110	56 06	90 00	0 00	56 06	72°44′	33 54
d 011	29 54½	37 31	69 03	58 08	31 52	72 19
h $\bar{1}$01	−90 00	31 16½	121 16½	90 00	52 13½	121 16½
o $\bar{1}$11	−42 23	42 01½	110 57	60 22	66 44	116 49½

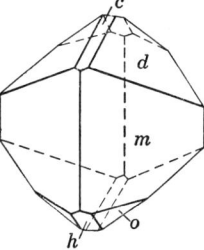

Habit. As a delicate efflorescence composed of minute crystals; also in granular aggregates, small mammillary masses or in spherulites.

Phys. Cleavage lacking.[2] H. and G. not known. Luster vitreous. Color blue.

Opt. In transmitted light, blue.

ORIENTATION	n^3	n^4	PLEOCHROISM	
X b	1.515	1.518 ± 0.003	Deep blue	Biaxial negative $(-)$.
Y	1.525	1.530 ± 0.003	Blue	$2V$ medium large.
$Z \wedge c$ small	1.545	1.542 ± 0.003	Colorless	

Chem. An acid hydrated vanadyl sulfate, $(VO)_2H_2(SO_4)_3 \cdot 15H_2O$.
Anal.

	1	2
V_2O_4	23.89	24.64
SO_3	34.59	33.17
H_2O	41.52	42.19
Total	100.00	100.00

1. $(VO)_2H_2(SO_4)_3 \cdot 15H_2O$. 2. Minasragra. Recalculated to 100 after deducting 66.16 per cent insol. and a few per cent of admixed melanterite, morenosite, and gypsum.

Tests. Easily soluble in cold water. Readily fusible in C.T.

Occur. Found as an efflorescence on patronite at Minasragra near Cerro de Pasco, Peru. Forms similarly on museum specimens of patronite.
Name. From the locality.

Ref.
1. Palache, *Am. Min.*, **19**, 197 (1934).
2. A reported cleavage on {010} was not verified by Palache (1934).
3. Schaller (1917).
4. Larsen (110, 1921) on type material.

TYPE 2. $A_2(XO_4) \cdot xH_2O$

29.2.1 **LECONTITE** [$Na(NH_4,K)(SO_4) \cdot 2H_2O$]. Taylor (*Am. J. Sc.*, **26**, 273, 1858).

Cryst.[1] Orthorhombic; dipyramidal—$2/m\ 2/m\ 2/m$.

$$a:b:c = 0.7848:1:1.5317; \qquad p_0:q_0:r_0 = 1.9517:1.5317:1$$

$$q_1:r_1:p_1 = 0.7848:0.5124:1; \qquad r_2:p_2:q_2 = 0.6529:1.2742:1$$

Forms:

	ϕ	$\rho = C$	ϕ_1	$\rho_1 = A$	ϕ_2	$\rho_2 = B$
c 001	....	0°00′	0°00′	90°00′	90°00′	90°00′
g 120	32°15′	90 00	90 00	57 45	0 00	32 15
m 110	51 52½	90 00	90 00	38 07½	0 00	51 52½
d 104	90 00	26 00½	0 00	63 59½	63 59½	90 00
e 101	90 00	62 52	0 00	27 08	27 08	90 00

Habit. As narrow prismatic crystals up to an inch in length, also very short and broad.

Phys. H. 2–2½. Luster vitreous. Colorless and transparent. Taste saline and rather bitter. Stable in air.

Opt.[2] In transmitted light, colorless.

ORIENTATION	n (Honduras)	
X	1.440 ± 0.003	Biaxial negative $(-)$.
Y	1.452 ± 0.003	$2V\ 40° \pm 1°$ (meas.).
Z	1.453 ± 0.003	$r < v$, rather strong.

Chem. A hydrated sulfate of sodium and ammonium, $Na(NH_4,K)(SO_4) \cdot 2H_2O$. A small amount of K substitutes for NH_4 in the only reported analysis of natural material.

	K_2O	$(NH_4)_2O$	Na_2O	SO_3	H_2O	Insol.	Total
1.	2.75	13.33	17.68	45.68	20.56		100.00
2.	2.67	12.94	17.56	44.97	19.45	2.41	100.00

1. $Na(NH_4,K)(SO_4) \cdot 2H_2O$ with $K:(NH_4) = 1:8.8$. 2. Honduras. With P_2O_5 tr.

Tests. Soluble in water.

Occur. From the cave of Las Piedras, near Comayagua, Honduras, embedded in bat guano.

Artif.[3] $Na(NH_4)(SO_4) \cdot 2H_2O$ crystallizes from a water solution of sodium sulfate and excess ammonium sulfate.

Name. After the American entomologist John L. LeConte (1825–1883), who discovered the mineral.

Ref.

1. J. D. Dana in Taylor (1858) on natural crystals (analysis 2) of very poor quality. Lecontite probably is identical with artificial $Na(NH_4)(SO_4) \cdot 2H_2O$, described crystallographically by Mitscherlich, *Ann. Phys.*, **58**, 568 (1843), and Lang, *Ak. Wien. Sitzber.*, **45**, 108 (1862).
2. Larsen (99, 1921).
3. Dawson, *J. Chem. Soc. London*, **113**, 675 (1918).

29.2.2 **MIRABILITE** [$Na_2SO_4 \cdot 10H_2O$]. Sal mirabile *Glauber* (Tractatus de natura salium, 1658). Natürliches Wundersalz, Glaubersalz *Germ.* Glauber Salt. Sulfate of Soda. Sel de Glauber, Soude sulfatée *Fr.* Alcali minerale vitriolatum *Bergmann* (1782). Gediegen Glaubersalz (from Saidschitz and Sedlitz) *Reuss* (*Crell's Ann.*, **2**, 18, 1791) = Natürliches Bittersalz pt. *Lenz* (**1**, 489, 1794) = Reussin *Karsten* (40, 1800). Exanthalose pt. *Beudant* (**2**, 475, 1832). Mirabilite *Haidinger* (488, 1845).

Cryst. Monoclinic; prismatic—$2/m$.

$a:b:c = 1.1096:1:1.2388;$ $\beta \; 107°45';$ [1] $p_0:q_0:r_0 = 1.1164:1.1798:1$

$r_2:p_2:q_2 = 0.8476:0.9463:1;$ $\mu \; 72°15';$ $p_0' \; 1.1722, \; q_0' \; 1.2388, \; x_0' \; 0.3201$

Forms:[2]

	ϕ	ρ	ϕ_2	$\rho_2 = B$	C	A
c 001	90°00′	17°45′	72°15′	90°00′	.....	72°15′
b 010	0 00	90 00	.....	0 00	90°00′	90 00
a 100	90 00	90 00	0 00	90 00	72 15	.....
m 110	49 59	90 00	0 00	49 59	76 30	40 01
μ 011	14 29½	51 59½	72 15	40 17	49 43	78 38
v 021	7 21½	68 11	72 15	22 58	67 02	83 10
r $\bar{1}$01	−90 00	40 26	130 26	90 00	58 11	130 26
d 111	50 18	62 43½	33 49½	55 24½	49 50½	46 51
n $\bar{1}$11	−34 31½	56 22½	130 26	46 41	67 27	118 05

Less common:

w 102   l $\bar{1}$02   H $\bar{2}$03   ε 112   u 221   y $\bar{1}$12

Additional forms on artificial crystals:

f 120   h 052   k 103   i $\bar{1}$04   q 101

Habit.[3] Short prismatic [001] and somewhat resembling borax in habit; commonly prismatic to acicular [010]; also thin tabular {100} or {001}, or lath-like due to flattening on {100} and elongation parallel [010] (artificial crystals). Massive, as efflorescent crusts or masses of interlocking fibers; granular; stalactitic.

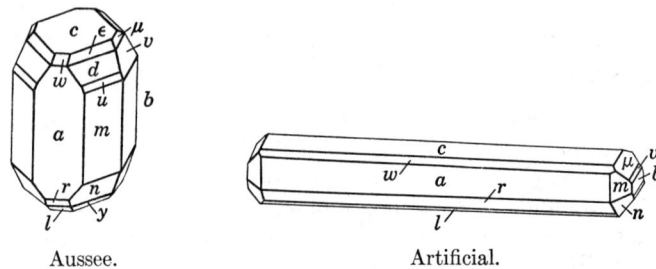

Aussee.　　　　　　　　　Artificial.

Twinning. Rare: (1) By interpenetration on {001};[4] (2) on {100}.

P h y s. Cleavage {100} perfect; also {001}, {010} and {011}.[5] Fracture conchoidal. H. $1\frac{1}{2}$–2. G. 1.490.[6] Luster vitreous. Colorless and transparent to white and opaque. Streak white. Taste cool, then feebly saline and bitter.

O p t. In transmitted light, colorless.

Orientation	n [7]	n(Na) [5]	
X b	1.394 ± 0.003	1.396	Biaxial negative (−).
Y	1.396 ± 0.003	1.4103	$2V$ 75°56′ (Na; meas.),
$Z \wedge c$ +31° (Li)	1.398 ± 0.003	1.419 (calc.)	$2E$ 120°23′.
			$r < v$, strong, crossed.

{010} sections show abnormal interference colors; other sections give sharp extinction.

C h e m. Sodium sulfate decahydrate, $Na_2SO_4 \cdot 10H_2O$.
Anal.[8]

	1	2	3	4
Na_2O	19.24	19.19	18.67	19.02
SO_3	24.85	24.77	25.16	25.37
H_2O	55.91	56.25	55.28	54.84
Rem.		0.31		0.77
Total	100.00	100.52	99.11	100.00

1. $Na_2SO_4 \cdot 10H_2O$. 2. Lacul-Sarat, Roumania.[9] Rem. is Cl 0.24, Mg 0.07. 3. Kirkby Thore, Westmoreland, England.[10] 4. Isle Royale mine, Michigan.[11] Rem. is K_2O 0.77. Recalculated to 100 after deducting 5.69 per cent insoluble.

Tests. On gentle heating in C.T. partially fuses with separation of the anhydrous salt, and affords much water. Loses its water in dry air and falls to a white powder. Very soluble in water (5.0 g. $Na_2SO_4 \cdot 10H_2O$ in 100 g. H_2O at 0°, 50.0 g. at 32.38° which is the transition point to Na_2SO_4 in pure solution).

O c c u r. Found as a deposit from saline lakes, playas, and springs, and as an efflorescence on clayey, alkalic soils or in caves and covered spots in recent lavas. The solubility of mirabilite decreases markedly with decreasing temperature, and the mineral is often deposited from saline waters during the winter season, as at Great Salt Lake, Utah, and the Karabugas Gulf, Caspian Sea, and as a deposition from sea spray on rocks along the shore (Antarctica). Commonly associated with gypsum, halite, thenardite, trona, glauberite, aphthitalite, bloedite, epsomite.

Mirabilite occurs [12] in the salt deposits of Austria at Hallein, Hallstadt and Ischl in Salzburg and at Aussee in Styria. Abundantly in the hot springs at Carlsbad and as an efflorescence with epsomite in the region of Sedlitz, Saidschitz, and Bilin, in Bohemia. In Spain at Miranda de Ebro, Prov. Alava; in Italy at Montedoro, Prov. Girgenti, Sicily, and as a hydration product of mixed salts in fumaroles on Vesuvius; as a winter deposit from some of the saline lakes in the Yeniseisk, Astrakhan, and Kuban regions, U.S.S.R., and similarly from Lacul-Sarat, Roumania. From brine springs in the Kamysh-Kurgan salt region, Tadzhikstan, U.S.S.R.

In the United States, mirabilite occurs widely in the arid regions of the west and southwest. In California in a large deposit at Soda Lake on the Carrizo Plain, San Luis Obispo County, and dissolved in large amounts in the waters of Owens Lake, Mono Lake, and Searles Lake. In Colorado dissolved in the water of ponds and soda springs, and as a deposit therefrom, about 12 miles south of Denver in Jefferson County. In Nevada found on alkali flats in the Buena Vista Valley, the salt residues here and at some other localities in Nevada representing residues left by the evaporation of former Lake Lahontan; as a deposit from alkaline spring waters at Wabuska, Mason Valley, Lyon County, at Goodsprings, Clark County, and elsewhere; in the soda lakes near Ragtown. In alkali flats and admixed with the gypsum sands of the Tularosa Basin west of Valmont, Dona Ana County, New Mexico. From small alkali lakes in Okanogan County, Washington; as deposits along the south shore of Great Salt Lake, Utah, and as a deposit from the lake waters during the winter. Found abundantly as a deposit from saline lakes in Wyoming, notably the Downey Lakes and the Union Pacific Lakes in Albany County, in Natrona County at Sodium, at the Gill Lakes, and in mud basins impregnated with salts near Independence Rock. Found in Canada, notably in alkali lakes in western Saskatchewan.

A r t i f.[13] Mirabilite is the stable phase in the system Na_2SO_4-H_2O below 32.38° and thenardite above; the transition temperature is lowered by the presence of other salts.

N a m e. The artificial salt was discovered by J. R. Glauber (1603–1668), a German chemist, about the middle of the seventeenth century, while he was operating with sulfuric acid and common salt; the name *sal mirabile* was his own expression of surprise at its formation.

Ref.

1. Rosický, *Zs. Kr.*, **45**, 473 (1908), on artificial crystals.
2. Goldschmidt (**4**, 49, 1918); $h\{\overline{2}03\}$ from Pardillo, *Bol. r. soc. española hist. nat.*, **15**, 153 (1915)—*Min. Abs.*, **1**, 78 (1920).
3. See Haidinger (**2**, 31, 1825); Hausmann (**2**, 1182, 1847); Zepharovich, *Zs. Kr.*, **3**, 100 (1879); Wells, *U. S. Geol. Sur.*, *Bull. 717*, 2 (1923); and Rosický (1908).
4. Pardillo (1915).

5. Rosický (1908).
6. Rosický (1908) by suspension method; Petterson, *Nova Acta Upsala* (1874) in Hintze (**1** [3B], 4266, 1929), gave 1.485–1.492. Values in the range 1.46–1.47, averaging 1.467, have been reported by Zambonini (171, 1935), Görgey, *Min. Mitt.*, **29**, 206 (1910), and several earlier workers.
7. Larsen (111, 1921) on recrystallized material from Carrizo Plains, California, and confirmed by Schaller, *U. S. Geol. Sur., Bull. 833*, 81 (1932); Larsen and Berman (148, 1934) give nX 1.393, nY 1.395, nZ 1.397, $2V$ 76°; Peck, *Am. Min.*, **2**, 62 (1917), gives nY 1.437 which is close to the value $nY \sim 1.44$ of Miller, *Trans. Cambridge Phil. Soc.*, **7**, 215 (1842); compare also data of Rosický (1908). Some of these observations probably were made on partially dehydrated material.
8. For earlier analyses see Hintze (1929).
9. Poni, *Ann. sc. univ. Jassy*, **1**, 15 (1900).
10. Best analysis in Trechmann, *Min. Mag.*, **13**, 73 (1901).
11. Peck (1917).
12. On the occurrence see further in Wells (1923) and Hintze (1923).
13. Doelter (4 [2], 35, 1926).

TYPE 3. $A_2B(XO_4)_2 \cdot xH_2O$

29.3.1   **S Y N G E N I T E** [$K_2Ca(SO_4)_2 \cdot H_2O$].   Zepharovich (*Lotos*, **22**, 137, 1872). Kaluszite Rumpf (*Min. Mitt.*, **2**, 117, 1872).

C r y s t.[1] Monoclinic; prismatic—$2/m$.

$a:b:c = 1.3691:1:0.8747;$ $\beta\ 104°05';$ $p_0:q_0:r_0 = 0.6389:0.8484:1$

$r_2:p_2:q_2 = 1.1789:0.7530:1;$ $\mu\ 75°55';$ $p_0'\ 0.6587, q_0'\ 0.8747, x_0'\ 0.2509$

Forms:[2]

	ϕ	ρ	ϕ_2	$\rho_2 = B$	C	A
c 001	90°00′	14°05′	75°55′	90°00′	.....	75°55′
b 010	0 00	90 00	....	0 00	90°00′	90 00
a 100	90 00	90 00	0 00	90 00	75 55	.....
m 110	36 59	90 00	0 00	36 59	80 22	53 01
λ 210	56 25	90 00	0 00	56 25	78 18	33 35
ϑ 310	66 07½	90 00	0 00	66 07½	77 09	23 52½
q 011	16 00½	42 18	75 55	49 41½	40 18½	79 18½
r 101	90 00	42 17½	47 42½	90 00	28 12½	47 42½
u $\bar{1}$01	−90 00	22 11	112 11	90 00	36 16	112 11
ω $\bar{1}$11	−24 59½	43 59	112 11	50 59½	51 12½	107 04

Less common:

g 120	k 610	d 504	g $\bar{7}$04	t $\bar{1}$12	w 124	e $\bar{2}$11
y 10.3.0	l 710	j $\bar{3}$04	v $\bar{2}$01	π $\bar{2}$21	i 411	

Structure cell.[3]   Space group $P2_1/m$. a_0 9.70 kX, b_0 7.15, c_0 6.20; β 104°; $a_0:b_0:c_0 = 1.356:1:0.867$. Cell contents $K_4Ca_2(SO_4)_4 \cdot 2H_2O$.

Habit. In crystalline crusts and lamellar aggregates. Individual crystals, sometimes attaining dimensions of several centimeters, tabular {100} to prismatic [001] and often rich in forms. Faces in the zone [001] striated vertically. Artificial crystals prismatic to acicular [001].

SYNGENITE

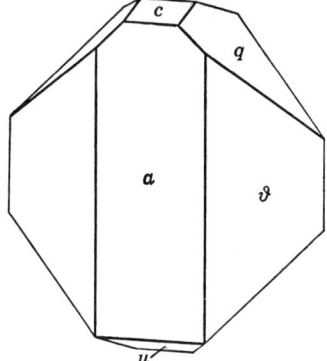

Twinning. Contact twins on {100} common.

Phys. Cleavage {110} and {100} perfect, {010} distinct.[4] Fracture conchoidal. H. $2\frac{1}{2}$. G. 2.603 (Kalusz), 2.579 (Sondershausen); 2.597 (calc.). Luster vitreous, colorless or faintly yellow due to inclusions; milky white. Transparent to translucent. Does not alter in dry air.

Opt.[5] In transmitted light, colorless.

ORIENTATION		n(Na)	
$X \wedge c$	$-2°17'$	1.5010	Biaxial negative $(-)$.
Y		1.5166	$28°18'$ Na (meas.).
Z	b	1.5176	$r < v$, very strong.

The optic plane is $\perp$ {010} at ordinary temperatures. With increase of temperature $2V$ decreases, becoming zero in yellow light at 158°, and at higher temperatures opens out again in the plane {010}. The temperature of uniaxiality varies with the wavelength: 127° (red), 158° (yellow), 172° (green), $177\frac{1}{2}°$ (blue).

Chem. A hydrated sulfate of potassium and calcium, $K_2Ca(SO_4)_2 \cdot H_2O$.

Anal.

	K_2O	CaO	SO_3	H_2O	Rem.	Total	G.
1.	28.68	17.08	48.75	5.49		100.00	2.597
2.	28.03	16.97	49.04	5.81	0.64	100.49	2.603
3.	28.63	17.85	48.49	5.45		100.42	
4.	28.80	17.20	48.40	5.41		99.81	2.579

1. $K_2Ca(SO_4)_2 \cdot H_2O$. 2. Kalusz, Galicia.[6] Rem. is MgO. 3. Kalusz.[7] 4. Sondershausen, Germany.[8]

Tests. B.B. easily fusible. In C.T. decrepitates violently and gives off water. In water dissolves in part with separation of gypsum.

Occur. A rare constituent of oceanic salt deposits. Also known as a product of volcanic action. First noted in druses in halite at Kalusz, Galicia. Found with aphthitalite and mirabilite at Stebnik, Galicia. With halite in crystalline crusts deposited by brines accumulated in the lower parts of the potash mine Glück Auf near Sondershausen, Germany. In cavities in a block ejected during the April, 1906, eruption of Vesuvius. As an incrustation on lava in the crater of Haleakala, Maui, Hawaiian Islands.

Alter. Loses water between 220° and 250°, becoming turbid.[8]

Artif. In the system K_2SO_4-$CaSO_4$-H_2O the compound $K_2Ca(SO_4)_2 \cdot H_2O$ crystallizes between 0° and 100° from solutions in which K_2SO_4 greatly predominates.[9] The Rb, Cs, and (NH_4) analogues of syngenite have also been prepared.[10]

Name. From συγγενής, *related*, because of its chemical resemblance to polyhalite.

Ref.
1. Elements of Laszkiewicz, *Arch. Min. Soc. Warsaw*, **3**, 61 (1928), who reviews the closely agreeing elements of earlier observers.
2. Laszkiewicz (1928) and Goldschmidt (**8**, 111, 1922).
3. Laszkiewicz, *Arch. Min. Soc. Warsaw*, **12**, 8 (1936), on crystals from Kalusz using rotation and Weissenberg methods.
4. Mügge, *Jb. Min.*, I, 266 (1895), and Schreiber, *Jb. Min., Beil.-Bd.*, **37**, 247 (1914).
5. Schreiber (1914) who also summarizes previous observations and reports in detail on pronounced changes of optical properties with varying temperature as well as changes of morphological constants with temperature. See also Buttgenbach, *Soc. géol. Belgique, Ann.*, **49**, 176 (1926).
6. Völker in Vrba, *Lotos*, **22**, 211 (1872), and Zepharovich, *Ak. Wien, Ber.*, **67** (1) 128 (1873).
7. Ullik in Rumpf (1872).
8. Schreiber (1914).
9. Hill, *Ann. Chem. Soc., J.*, **56**, 1071 (1934). Schreiber (1914) has summarized the numerous earlier syntheses.
10. D'Ans, *Ber. deutsch. chemische Gesellschaft*, **39**, 3326 (1906), **41**, 1776 (1908).

29.3.2 **Koktaite** [$(NH_4)_2Ca(SO_4)_2 \cdot H_2O$]. Sekanina (*Acta Ac. Sc. Nat. Moravo-Silesiacae*, **20**, no. 1, 1948).

Monoclinic, and apparently isostructural with syngenite. Crystals acicular or fibro-lamellar. Observed forms: $\{100\}$, $\{110\}$, $\{001\}$, $\{011\}$, $\{101\}$; $(100) \wedge (110)$ $54°$. Twins on $\{100\}$ are frequent. No cleavage. G. 2.09. Colorless to white. Optically biaxial negative $(-)$ with nX 1.524, nY 1.532, nZ 1.536; $Y = b$, $Z' \wedge c$ on $\{110\}$ $2°$; $2V$ $72°$. Identical with artificial ammonium-syngenite, $(NH_4)_2Ca(SO_4)_2 \cdot H_2O$. Soluble in water with the precipitation of gypsum. Found as pseudomorphs after gypsum with mascagnite and ammonia alum on the waste-heaps of a lignite mine at Žeravice near Kyjov in southeastern Moravia.

29.3.3 **KROEHNKITE** [$Na_2Cu(SO_4)_2 \cdot 2H_2O$]. Krönnkite *Domeyko* (5th app., 33, 1876; 250, 1879). Krömkite. Kröhnkite. Salvadorite *Herz* (*Zs. Kr.*, **26**, 16, 1896).

Cryst.[1] Monoclinic; prismatic—$2/m$.

$a:b:c = 0.4586:1:0.4357;$ $\quad \beta\ 108°30';$ $\quad p_0:q_0:r_0 = 0.9500:0.4132:1$

$r_2:p_2:q_2 = 2.4202:2.2993:1;$ $\quad \mu\ 71°30';$ $\quad p_0'\ 1.0018, q_0'\ 0.4357, x_0'\ 0.3346$

KROEHNKITE

Forms:[2]

	ϕ	ρ	ϕ_2	$\rho_2 = B$	C	A
c 001	90°00'	18°30'	71°30'	90°00'	.....	71°30'
b 010	0 00	90 00	....	0 00	90°00'	90 00
a 100	90 00	90 00	0 00	90 00	71 30	.....
k 130	37 27½	90 00	0 00	37 27½	78 52½	52 32½
h 120	48 58½	90 00	0 00	48 58½	76 09	41 01½
m 110	66 29½	90 00	0 00	66 29½	73 05	23 30½
j 210	77 44	90 00	0 00	77 44	71 56	12 16
q 011	37 31½	28 47	71 30	67 33	22 27	72 57
s 021	21 00½	43 01½	71 30	50 26	39 34	75 50½
v 201	90 00	66 50½	23 09½	90 00	48 20	23 09½
w 111	71 56½	54 34	36 48½	75 22½	37 17½	39 13½
e $\bar{1}$11	−56 51½	38 33	123 42½	70 05	54 49	121 27
n $\bar{1}$32	−14 16½	33 59½	99 26½	57 11½	42 03½	97 55½
d $\bar{1}$21	−37 26½	47 39½	123 42½	54 04	60 15	116 42
f $\bar{1}$31	−27 02½	55 43½	123 42½	42 36½	65 29½	112 04

Less common and rare:

α 0.1.10 y 032 δ $\bar{1}$12 g 132 l 451 π $\bar{3}$11
β 015 ϵ 031 r $\bar{2}$21 x 121 o 9.10.1 p $\bar{2}$11
γ 012 u 102 i $\bar{1}$23 z 231 λ 671

Structure cell.[9] Space group $P2_1/c$. a_0 5.78 kX, b_0 12.58, c_0 5.48; β 108°30'; $a_0:b_0:c_0 = 0.459:1:0.436$. Cell contents $Na_4Cu_2(SO_4)_4 \cdot 4H_2O$.

Habit. Short prismatic [001]; also octahedral with large {110} and {011} and somewhat elongated [100]. Common forms: $m\ q\ b\ s\ e$. As prismatic or fibrous aggregates and crusts; massive, granular.

Twinning.[3] On {101}, common; twins sometimes heart-shaped.

Phys. Cleavage {010} perfect, {$\bar{1}$01} very imperfect. Fracture conchoidal. H. 2½–3. G. 2.90; 2.95 (calc.). Luster vitreous. Color sky-blue to pale blue, inclining toward greenish blue, on exposure. Transparent.

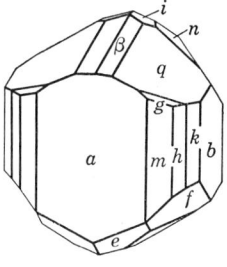

Chuquicamata.

Opt.[4] In transmitted light, pale blue to colorless.

Orientation	n(Na) (anal. 2)	
$X \wedge c$ +48°	1.544	Biaxial negative (−).
$Y\ b$	1.578	$2V$ 78°42' (meas.)
Z	1.601	$r < v$, weak; inclined.

Chem. A hydrated sulfate of sodium and copper, $Na_2Cu(SO_4)_2 \cdot 2H_2O$. Analysis gave:[5]

	Na_2O	CuO	SO_3	H_2O	Total
1.	18.36	23.56	47.41	10.67	100.00
2.	18.89	23.25	47.60	10.72	100.46

1. $Na_2Cu(SO_4)_2 \cdot 2H_2O$. 2. Chuquicamata.[6]

Tests. B.B. easily fusible. C.T. affords acid water at a low temperature. Easily soluble in water to an acid solution.

O c c u r. A secondary mineral found abundantly at Chuquicamata in Antofagasta, Chile, associated with atacamite, bloedite, chalcanthite, antlerite, and natrochalcite. Also from Chile from the Salvador mine, Quetena, near Calama (*salvadorite*) [8] with tamarugite, from Collahurasi, and from El Cobre de Mejillones and the Incahuasi district in Atacama.

A l t e r. Becomes green and opaque on exposure.

A r t i f.[7] Crystallizes from a solution of the component sulfates above 17°.

N a m e. After B. Kröhnke, who first analyzed the mineral.

Ref.
1. Elements of Palache, *Am. J. Sc.*, **237**, 447 (1939), on Chuquicamata crystals; orientation of Darapsky, *Jb. Min.*, I, 192 (1889). Transformation: Palache and Warren, *Am. J. Sc.*, **26**, 342 (1908), *Zs. Kr.*, **45**, 534 (1908), to Darapsky (1889): $\overline{1}01/010/001$.
2. Palache (1939) and Gordon, *Ac. Sc. Philadelphia, Not. Nat.*, no. 72 (1941). Vicinal: ψ $\overline{7}.13.1$, ω $\overline{1}3.28.2$.
3. Gordon (1941).
4. Merwin in Palache and Warren (1908).
5. Earlier analyses in Hintze (**1** [3B], 4455, 1929).
6. Warren analysis in Palache and Warren (1908).
7. See Benrath, *Zs. anorg. Chem.*, **179**, 369 (1929); Caven and Johnston, *J. Chem. Soc. London*, **129**, 2358 (1927); and Koppel, *Zs. phys. Chem.*, **42**, 1 (1902).
8. Made identical with kroehnkite by Gordon (1941); earlier suggested to be identical with pisanite by Bandy, *Am. Min.*, **23**, 718 (1938).
9. Richmond in Palache (1939) by Weissenberg method on Chuquicamata crystal.

29.3.4   **L O E W E I T E** [$Na_4Mg_2(SO_4)_4 \cdot 5H_2O$]. *Haidinger* (*Böhm. Ges., Abh.*, [5], **4**, 663, 1847).

Crystallization uncertain, perhaps trigonal.[1] Found naturally only as anhedral grains embedded in other salts.

P h y s. Fracture conchoidal.[2] H. $2\frac{1}{2}$–3 (Ischl).[3] G. 2.374 (Ischl),[3] 2.423 (Stassfurt).[4] Luster vitreous. Colorless; also reddish yellow due to included iron oxide. Transparent. Soluble in water. Taste slightly bitter.

O p t.[5] In transmitted light, colorless.

	n(Na)	
O	1.490	Uniaxial negative (−).
E	1.471	

C h e m. A hydrated sulfate of sodium and magnesium, $Na_4Mg_2(SO_4)_4 \cdot 5H_2O$.

Anal.

	1	2	3	4	5
Na_2O	20.17	18.97	18.58	19.1	18.1
K_2O				0.7	2.9
MgO	13.11	12.78	14.31	13.6	12.8
SO_3	52.07	52.35	52.35	52.2	51.3
H_2O	14.65	14.45	14.80	13.9	13.7
Rem.		0.66		0.5	1.0
Total	100.00	99.21	100.22	100.0	99.8
G.		2.376			

1. $Na_4Mg_2(SO_4)_4 \cdot 5H_2O$. 2. Ischl, Upper Austria.[3] Rem. is $(Fe,Al)_2O_3$. 3. Ischl.[8] 4,5. Stassfurt.[4] Rem. is NaCl.

BLOEDITE 447

Occur. Known only in salt deposits of oceanic origin. In crystalline masses with coarse anhydrite at Ischl and Hallstatt, Upper Austria. With vanthoffite in bloedite at Hall, Tyrol. With langbeinite and aphthitalite in the Stassfurt area, Germany.

Alter. Becomes dull and coated with a white crust in moist air.

Artif. $Na_4Mg_2(SO_4)_4 \cdot 5H_2O$ may crystallize from a solution containing only its constituent ions between 59.5°, at which temperature the solution is also saturated with $MgSO_4 \cdot 6H_2O$ (*hexahydrite*) and $Na_2Mg(SO_4) \cdot 4H_2O$ (*bloedite*), and at 180° or more, the invariant point at which the solution is saturated with $Na_4Mg_2(SO_4)_4 \cdot 5H_2O$ (*loeweite*), $MgSO_4 \cdot H_2O$ (*kieserite*) and $Na_6Mg(SO_4)_4$ (*vanthoffite*).[6] From a solution containing the ions Na, K, Mg, SO_4, Cl, and saturated with NaCl, corresponding to the brines of the oceanic salt deposits, it may crystallize in the interval 43° to 100°.[7]

Name. After Alexander Loewe (1808–?), German chemist.

CHILE-LOEWEITE. Wetzel (*Chem. Erde*, **3**, 375, 1928). A provisional name for a material supposedly differing somewhat from loeweite in composition. Found mixed with other salts at Taltal in the Chilean desert. Apparently trigonal, with nO 1.470, nE 1.434.

CHROM-LOEWEITE. Wetzel (*Chem. Erde*, **3**, 375, 1928). A supposed chromian variety of loeweite from Taltal. nO 1.496, nE 1.448.

Ref.

1. Originally described by Haidinger (1847) as tetragonal. Artificial crystals were stated by Görgey, *Min. Mitt.*, **29**, 198 (1910) to be trigonal, $a:c = 1:0.7017$, but no crystal measurements were reported. The identity of the two materials was supposedly established from other properties.
2. Very indistinct (tetragonal) pyramidal and prismatic cleavages and an indistinct basal cleavage were reported by Haidinger (1847) on Perneck (Ischl) material but this lacks verification.
3. Haidinger (1847).
4. Kubierschky, *Ak. Berlin, Ber.*, 404 (1903).
5. Görgey, *Min. Mitt.*, **28**, 343 (1909) on material from Hall, Tyrol.
6. Blasdale and Robson, *J. Am. Chem. Soc.*, **50**, 35 (1928).
7. As determined by van't Hoff and co-workers and summarized in Doelter (4 [2], 93, 1926).
8. von Hauer, *Geol. Reichsanst., Jb.*, 605 (1856).

BLOEDITE GROUP

29.3.5.1 **BLOEDITE** [$Na_2Mg(SO_4)_2 \cdot 4H_2O$]. Blödit *John* (*Chemische Schriften*, **6**, 240, 1821). Astrakanit *Rose* (**2**, 270, 1842). Astrakhanite. Simonyit *Tschermak* (*Ak. Wien, Ber.*, **60** [1], 718, 1869). Warthit *Quenstedt* (645, 1877). Natronkalisimonyit *Koechlin* (*Min. Mitt.*, **21**, 356, 1902). Blödite

Cryst.[1] Monoclinic; prismatic—$2/m$.

$a:b:c = 1.3492:1:0.6717$;[2] $\beta\ 100°48\frac{1}{2}'$; $p_0:q_0:r_0 = 0.4979:0.6598:1$

$r_2:p_2:q_2 = 1.5156:0.7546:1$; $\mu\ 79°11\frac{1}{2}'$; $p_0'\ 0.5068,\ q_0'\ 0.6717,\ x_0'\ 0.1909$

448 SULFATES

Forms:[3]

	ϕ	ρ	ϕ_2	$\rho_2 = B$	C	A
c 001	90°00′	10°48½′	79°11½′	90°00′	.....	79°11½′
a 100	90 00	90 00	0 00	90 00	79°11½′	.....
ν 120	20 40	90 00	0 00	20 40	86 12½	69 20
m 110	37 02	90 00	0 00	37 02	83 31	52 58
n 210	56 28	90 00	0 00	56 28	81 00½	33 32
λ 310	66 10	90 00	0 00	66 10	80 07½	23 50
d 011	15 52	34 55½	79 11½	56 35	33 25	80 59½
e 021	8 05½	53 36½	79 11½	37 09½	52 50½	83 30
q $\bar{2}$01	−90 00	39 26½	129 26½	90 00	50 15	129 26½
p 111	46 05½	44 05	55 06	61 09	36 54½	59 55½
o 121	27 26½	56 33	55 06	42 13½	52 09	67 23
s $\bar{2}$11	−50 46	46 43½	129 26½	62 35	55 25	124 20

Less common:

b 010	τ 450	r $\bar{1}$01	u $\bar{1}$11	z 131	v $\bar{2}$12	t $\bar{3}$11
μ 130	l 320	w $\bar{1}$12	y $\bar{2}$21	f $\bar{1}$44	x $\bar{1}$21	

Structure cell.[4] Space group $P2_1/a$. a_0 11.04 kX, b_0 8.15, c_0 5.49; β 100°41′; $a_0:b_0:c_0$ = 1.355:1:0.674. Cell contents $Na_8Mg_4(SO_4)_8 \cdot 16H_2O$.

Habit. Crystals short prismatic [001], often highly modified. Also massive granular or compact. Common forms: $m\ n\ d\ c\ p\ s$.

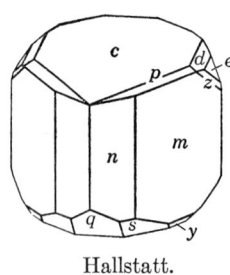

Hallstatt.

Phys. Fracture conchoidal. Brittle. H. 2½–3. G. 2.25 ± 0.03; 2.274 (calc.). Colorless, sometimes bluish green or reddish due to minute inclusions. Luster vitreous. Transparent. Taste faint, saline, and bitter.

Opt.[5] In transmitted light, colorless.

Orientation		n	
$X \wedge c$	37°	1.483	Biaxial negative (−).
Y	b	1.486	$2V$ 71° (meas.).
Z		1.487	

Chem. A hydrated sulfate of sodium and magnesium, $Na_2Mg(SO_4)_2 \cdot 4H_2O$. Experiments[6] show that K can substitute for Na up to about K:Na = 1:30 in artificial material (see also analysis 5). The chemically analogous potassium salt, leonite, is not isostructural with bloedite.

Anal.[7]

	1	2	3	4	5	6
Na_2O	18.53	18.00	18.86	18.53	18.14	18.26
K_2O					0.43	
MgO	12.06	12.09	12.65	11.97	11.94	11.93
SO_3	47.87	47.61	47.17	47.82	48.09	48.11
H_2O	21.54	21.49	21.82	21.54	21.52	21.37
Rem.		0.39		0.07	0.07	
Total	100.00	99.58	100.50	99.93	100.19	99.67
G.	2.274	2.251	2.244			

1. $Na_2Mg(SO_4)_2 \cdot 4H_2O$. 2. Ischl, Upper Austria. Rem. is Cl 0.31, Fe_2O_3 0.08. Especially pure material from original locality.[8] 3. Christina mine, Hallstatt, Upper Austria (*simonyite*).[9] 4. Varcha mine. Salt Range, Punjab. Rem. is NaCl 0.07.[10] 5. Potassian bloedite. Kalusz, Eastern Galicia (*natronkalisimonyite*). Rem. is insol. 0.07.[11] 6. Soda Lake, Carrizo Plain, San Luis Obispo County, California.[12]

Tests. Easily soluble in cold water.

Occur. In salt deposits of oceanic origin with halite, kainite, carnallite, polyhalite, and other salts. Probably not an original constituent of these deposits but formed by metamorphism, often in coarse crystals embedded in other salts or lining cracks. Also in lacustrine salt deposits mostly with halite, thenardite, and mirabilite, sometimes in crystals of great size. With nitrates in extremely arid regions (Atacama).

In Germany in the potash mines near Stassfurt. In Upper Austria in the salt mines at Ischl and at Hallstatt (*simonyite*). At Kalusz, Eastern Galicia, and at Monte Sambuco in the province of Caltanissetta, Sicily. In foot-thick layers (*astrakanite*) with halite in the salt lakes near Korduansk east of the mouth of the Volga. With halite, thenardite, and mirabilite in gypsiferous sandy clays near Uzun-su, Transcaspia. With halite in the Mayo (*warthite*) and Varcha mines in the Salt Range, Punjab. With darapskite in the nitrate region of the Atacama Desert [13] and with mirabilite and thenardite at Chuquicamata,[14] Chile. In the United States at the Laguna Salina in Estancia Valley, Torrance County, New Mexico, and at Soda Lake, Carrizo Plain, San Luis Obispo County, California.[15]

Alter. Bloedite was sometimes reported in the older literature as crumbling or losing water in dry air, whereas other observers found it remained unchanged.[16] This difference is probably due to the effect of other unstable minerals such as epsomite or mirabilite mixed with some specimens. Clean crystals of bloedite remain unchanged on exposure to air.

Artif. $Na_2Mg(SO_4)_2 \cdot 4H_2O$ crystallizes from solutions containing only its constituent ions between 20.6°, at which temperature the solution is also in equilibrium with $MgSO_4 \cdot 7H_2O$ (*epsomite*) and $Na_2SO_4 \cdot 10H_2O$ (*mirabilite*), and 71°, at which temperature the solution is also in equilibrium with $MgNa_6(SO_4)_4$ (*vanthoffite*) and $Mg_2Na_4(SO_4)_4 \cdot 5H_2O$ (*loeweite*).[17] From aqueous solutions containing the ions Na, K, Mg, Cl, and SO_4 and saturated with NaCl, corresponding to the brines of the oceanic salt deposits, $Na_2Mg(SO_4)_2 \cdot 4H_2O$ may crystallize in the temperature interval 4.5° to 59.5°.[18]

Name. After the German chemist Carl August Bloede (1773–1820).

Ref.

1. Symmetry from structure cell, Lauro, *Per. Min.*, **11**, 39 (1940).
2. Angles of Koechlin, *Ann. Naturhist. Hofmuseum, Vienna*, **15**, 103 (1900), on crystals from Hallstatt, Upper Austria.
3. Goldschmidt (**1**, 207, 1913).
4. Lauro (1940) on crystals from Monte Sambuco, Sicily.
5. Orientation, Laszkiewicz, *Arch. Min. Soc. Warsaw*, **5**, 79 (1929); indices, Schaller, *Am. Min.*, **17**, 530 (1932).
6. van't Hoff and Barschall, *Ak. Berlin, Ber.*, 367 (1903).
7. For additional analyses see Hintze (1 [3B], 4468, 1929).
8. v. Hauer, *Geol. Reichsanst., Wien., Jb.*, 605 (1856).
9. Tschermak, *Ak. Wien., Ber.*, **60** [1], 718 (1869).
10. Mallet, *Min. Mag.*, **11**, 311 (1897).
11. Koechlin, *Min. Mitt.*, **21**, 356 (1902).
12. Schaller, *Washington Ac. Sc., J.*, **3**, 75 (1913).

13. Dietze, Zs. Kr., **19**, 445 (1891).
14. Bandy, Am. Min., **23**, 719 (1938).
15. For additional localities see Hintze (1929).
16. This was made the basis for distinguishing the supposed new species, simonyite, by Tschermak (1869). The observations on dehydration have been summarized by Jaeger, Min. Mitt., **22**, 103 (1903).
17. Data on the system Na-Mg-SO$_4$-H$_2$O have been contributed by many investigators and are summarized in Int. Crit. Tables, **4**, 349–351 (1928) and in Doelter (4 [2], 49–53, 1926). See also Nikolaev, C. r. ac. sc. U.R.S.S., **46**, 322 (1945).
18. As determined by van't Hoff and co-workers and summarized in Doelter (4 [2], 93, 1926).

29.3.5.2 **L E O N I T E** [K$_2$Mg(SO$_4$)$_2$·4H$_2$O]. Kalium-Astrakanit *van der Heide* (*Ber. deutsche chemische Gesellschaft*, **26**, 414, 1893). Kaliastrakanit, *Naupert* and *Wense* (*Ber. deutsche chemische Gesellschaft*, **26**, 873, 1893). Kalium-Blödit, Leonit, *Tenne* (*Zs. deutsch. geol. Ges.*, **48**, 632, 1896).

C r y s t.[1] Monoclinic; prismatic—$2/m$.

$a:b:c = 1.2322:1:1.0376;\quad \beta\,95°09';\quad p_0:q_0:r_0 = 0.8421:1.0334:1$

$r_2:p_2:q_2 = 0.9677:0.8148:1;\quad \mu\,84°51';\quad p_0'\,0.8455,\,q_0'\,1.0376,\,x_0'\,0.0901$

Forms:

		ϕ	ρ	ϕ_2	$\rho_2 = B$	C	A
c	001	90°00′	5°09′	84°51′	90°00′	.....	84°51′
b	010	0 00	90 00	....	0 00	90°00′	90 00
a	100	90 00	90 00	0 00	90 00	84 51	.....
m	110	39 10½	90 00	0 00	39 10½	86 45	50 49½
o	310	67 45	90 00	0 00	67 45	85 14	22 15
M	021	2 29	64 17½	84 51	25 49	64 11	87 45½
w	101	90 00	43 05½	46 54½	90 00	37 55½	46 54½
d	201	90 00	60 41½	29 18½	90 00	55 31½	29 18½
δ	2̄01	−90 00	58 00½	148 00½	90 00	63 09½	148 00½
p	111	42 02½	54 24½	46 54½	52 51	51 02½	57 00½
ξ	1̄12	−32 40	31 38½	108 24	63 47½	34 40	106 27
π	1̄11	−36 03½	52 04½	127 04	50 22½	55 13	117 40
q	311	68 26½	70 30	20 50½	69 44	65 43½	28 46

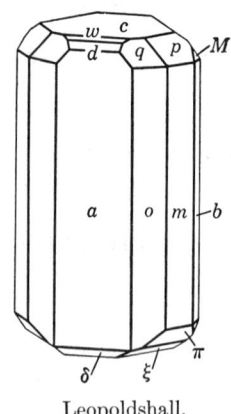

Leopoldshall.

Habit. Crystals tabular {100} and elongated [001]. Usually anhedral, intergrown with other salt minerals.

Twinning. Artificial crystals show twinning on {100}.[2] Lamellar twinning is observed under the microscope.[3]

P h y s. Conchoidal fracture. H. 2½–3. G. 2.201.[4] Colorless, also yellowish. Luster waxy to vitreous. Transparent. Taste bitter.

O p t.[5] In transmitted light, colorless.

Orientation	n	
X	1.479	Biaxial positive (+).
Y b	1.482	2V near 90°.
Z ∧ a small	1.487	

C h e m. A hydrated sulfate of potassium and magnesium, $K_2Mg(SO_4)_2 \cdot 4H_2O$. In artificial crystals the K may be replaced by Na to nearly one quarter [6] (see *bloedite*) but natural materials correspond closely to the ideal formula. Leonite is isostructural with a number of compounds of the type $AB_2(SO_4)_2 \cdot 4H_2O$, in which A = K, Rb, Cs, NH_4, Tl and B = Mg, Mn, Zn, Cd, Cu, Ni, Co, Fe.

Anal.[7]

	K_2O	MgO	SO_4	H_2O	Rem.	Total
1.	25.69	10.99	43.67	19.65		100.00
2.	25.62	10.37	43.76	19.57	0.50	99.82

1. $K_2Mg(SO_4)_2 \cdot 4H_2O$.   2. Leopoldshall, Germany.[8] Rem. is KCl.

Tests. In C.T. loses water and fuses. Easily soluble in water.

O c c u r. Known only as a secondary or "metamorphic" mineral in salt deposits of oceanic origin. Found sparingly at Westerregeln, Leopoldshall, Neu Stassfurt, and elsewhere in the German potash districts. In the United States, found with kainite, polyhalite, halite, and sylvite from the cores of wells in Eddy and Lea Counties, New Mexico.[9]

A l t e r. Becomes coated with a white crust in air, presumably due to hydration forming the hexahydrate, picromerite.

A r t i f. $K_2Mg(SO_4)_2 \cdot 4H_2O$ crystallizes from solutions containing only its constituent ions between 41°, at which temperature the solution is also in equilibrium with $MgSO_4 \cdot 7H_2O$ (*epsomite*) and $K_2Mg(SO_4)_2 \cdot 6H_2O$ (*picromerite*) and 89°, at which temperature the solution is also in equilibrium with $K_2Mg(SO_4)_3$ (*langbeinite*) and K_2SO_4.[10] From aqueous solutions containing the ions Na, K, Mg, Cl, and SO_4 and saturated with NaCl, $K_2Mg(SO_4)_2 \cdot 4H_2O$ may crystallize between 18° and $61\frac{1}{2}°$.[11]

N a m e. In honor of Leo Strippelmann, director of the salt works at Westerregeln, Germany.

Ref.

1. Elements recalculated by LaForge (priv. comm.) from the measurements of Tenne (1896) and Strandmark, *Zs. Kr.*, **36**, 461 (1902), with a and c interchanged to conform with the structure cell of the isostructural compound $K_2Mn(SO_4)_2 \cdot 4H_2O$, as determined by Anspach, *Zs. Kr.*, **101**, 39 (1939). Form list from Strandmark (1902). Transformation: Tenne and Strandmark to new, 001/0$\overline{1}$0/100.
2. Strandmark (1902).
3. Görgey, *Min. Mitt.*, **29**, 207 (1910); see also Schaller and Henderson, *U. S. Geol. Sur., Bull. 833*, 44 (1932).
4. Görgey (1910).
5. Indices, Schaller and Henderson (1932). Orientation, Görgey (1910).
6. van't Hoff and Barschall, *Ak. Berlin, Ber.*, 367 (1903).
7. For analyses of material from Westerregeln see Naupert and Wense (1893) and from New Mexico see Schaller and Henderson (1932).
8. Strandmark (1902).
9. Schaller and Henderson (1932).
10. Data from various sources summarized in *Int. Crit. Tables*, **4**, 352 (1928).
11. As determined by van't Hoff and co-workers and summarized in Doelter (4 [2], 93, 1926).

29.3.6 Wattevilleite. *Singer* (Inaug. Diss., Würzburg, 18, 1879).

Perhaps orthorhombic or monoclinic. As aggregates of very minute acicular or hair-like crystals; in part twins. G. 1.81. Luster silky. Color snow-white. Taste first sweet, then astringent.

O p t.[1] In transmitted light, colorless.

nX	1.435 ± 0.003	Biaxial negative $(-)$.
nY	1.455 ± 0.003	$2V\ 48° \pm 3°$ (meas.).
nZ	1.459 ± 0.003	Dispersion not perceptible.

The extinction of the fibers is not uniform; X appears parallel to the elongation in some and normal in others. Some fibers normal to X show a large extinction angle.

C h e m. A hydrated sulfate of sodium and calcium of uncertain formula, perhaps $Na_2Ca(SO_4)_2 \cdot 4H_2O$. Analysis gave (after deducting 33.69 per cent hygroscopic water): K_2O 4.74, Na_2O 10.46, CaO 16.87, MgO 2.49, CoO 1.30, NiO 1.05, FeO 0.88, Al_2O_3 0.24, SO_3 44.01, H_2O 17.73, total 99.77.

Tests. B.B. swells up and fuses with difficulty to a white blebby enamel. Very soluble in water, affording an acid solution from which crystals of gypsum separate on standing.

Found associated with other sulfates in pyritic lignite in the Einigkeit mine on the Bauersberg, near Bischofsheim, Bavaria. Needs verification.

Ref.

1. Larsen (156, 1921). The optical properties are close to those of kalinite.

29.3.7 PICROMERITE GROUP

MONOCLINIC; PRISMATIC—$2/m$

	$a:b:c$	β
Picromerite, $K_2Mg(SO_4)_2 \cdot 6H_2O$	$0.7413:1:0.4993$	$104°48'$
Cyanochroite, $K_2Cu(SO_4)_2 \cdot 6H_2O$	$0.7490:1:0.5088$	$104\ 28$
Boussingaultite, $(NH_4)_2Mg(SO_4)_2 \cdot 6H_2O$	$0.7400:1:0.4918$	$107\ 06$

The artificial equivalents of these rare minerals have been the subject of careful crystallographic study. They belong to an extensive series of isostructural compounds, often referred to as "Tutton's salts," of the formula $A_2B(XO_4)_2 \cdot 6H_2O$ where A = K, Rb, Cs, NH_4, Tl, B = Mg, Cu, Co, Zn, Mn, etc., and X = S or Se. The crystal structure [1] is pseudo-cubic on $\{\bar{2}01\}$ and resembles that of the alums.

The occurrence of these relatively soluble minerals at the original localities, fumaroles in Italy in each case, is not well-authenticated. Picromerite also occurs in the oceanic salt deposits (in which it has commonly been referred to as *schoenite*). Cyanochroite remains somewhat ill-defined.

Ref.

1. Hofmann, *Zs. Kr.*, **78**, 279 (1931).

PICROMERITE

29.3.7.1 PICROMERITE [$K_2Mg(SO_4)_2 \cdot 6H_2O$]. Picromeride *Scacchi* (*Mem. sullo Incendio Vesuv.*, 191, 1855). Schoenite *Reichardt* (*Jb. Min.*, 340, 1866).

Cryst.[1] Monoclinic; prismatic—$2/m$

$a:b:c = 0.7413:1:0.4993$; $\beta\ 104°48'$; $p_0:q_0:r_0 = 0.6735:0.4827:1$

$r_2:p_2:q_2 = 2.0715:1.3953:1$; $\mu\ 75°12'$; $p_0'\ 0.6967,\ q_0'\ 0.4993,\ x_0'\ 0.2642$

Forms:

	ϕ	ρ	ϕ_2	$\rho_2 = B$	C	A
$c\ 001$	90°00'	14°48'	75°12'	90°00'	.....	75°12'
$b\ 010$	0 00	90 00	.....	0 00	90°00'	90 00
$a\ 100$	90 00	90 00	0 00	90 00	75 12	.....
$m\ 110$	54 22½	90 00	0 00	54 22½	78 01	35 37½
$q\ 011$	27 53	29 27½	75 12	64 14	25 46	76 42
$e\ \overline{2}01$	−90 00	48 28½	138 28½	90 00	63 16½	138 28½
$u\ \overline{1}11$	−40 54	33 27	113 23½	65 23	44 23½	111 09

Less common:

$s\ 130$ $n\ 120$ $\mu\ 230$ $o\ 111$

Structure cell.[2] Space group $P2_1/a$. $a_0\ 9.04\ kX$, $b_0\ 12.24$, $c_0\ 6.095$; [$\beta\ 104°48'$]; $a_0:b_0:c_0 = 0.739:1:0.498$. Cell contents $K_4Mg_2(SO_4)_4 \cdot 12H_2O$.

Habit. Crystals short prismatic [001]. As crusts on other salts; also massive. In bedded deposits.

Phys. Cleavage {$\overline{2}01$} perfect.[3] H. 2½. G. 2.028 (artif.),[3] 2.039 (calc.). Luster vitreous. Colorless or white; also reddish, yellowish, or gray due to admixture. Transparent. Taste bitter.

Opt.[3] In transmitted light, colorless.

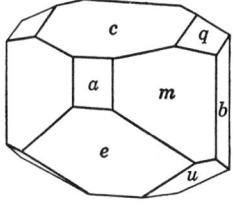

Aschersleben.

ORIENTATION	n(Na)	
$X \wedge a$ −1°	1.4607	Biaxial positive (+).
$Y\ \ \ b$	1.4629	$2V\ 47°54'$ (Na; meas.).
Z	1.4755	$r > v$, weak.

Chem. A hydrated sulfate of potassium and magnesium, $K_2Mg(SO_4)_2 \cdot 6H_2O$. The water is wholly lost at about 133°.[5]

Anal.[4]

	K_2O	MgO	SO_3	H_2O	Rem.	Total
1.	23.39	10.01	39.76	26.84		100.00
2.	23.28	10.40	39.74	26.87	0.28	100.57

1. $K_2Mg(SO_4)_2 \cdot 6H_2O$. 2. Leopoldshall, Germany. Rem. is Cl.[5]

Tests. Easily fusible. Soluble in cold water.

Occur. Reported very sparingly from fumarole deposits at Vesuvius. In the kainite zones of some oceanic salt deposits. With halite and anhydrite at Leopoldshall, and in beds up to a meter thick at Aschersleben, Germany. With epsomite at Kalusz, Eastern Galicia.

Artif. $K_2Mg(SO_4)_2 \cdot 6H_2O$ crystallizes from a solution containing only its constituent ions between −5°, at which temperature the solution is also in equilibrium with $MgSO_4 \cdot 12H_2O$ and ice, and 47.5°, at which temperature the solution is also in equilibrium with K_2SO_4 and $K_2Mg(SO_4)_2 \cdot 4H_2O$ (*leonite*).[6] The salt also crystallizes at temperatures below 26° from aqueous solutions containing the ions Na, K, Mg, Cl, and SO_4 and saturated with NaCl.[7]

Name. The name picromeride, from πικρός, *bitter*, and μέρος, *part*, in allusion to the magnesium present, was originally applied to material crystallized from an aqueous solution of mixed salts from Vesuvian fumaroles. The mineral itself, however, has since been identified at that locality. The name schoenite was first used for material from Leopoldshall and has been widely used for the mineral in the oceanic salt deposits.

Ref.
1. Axial elements of Tutton, *Zs. Kr.*, **21**, 491 (1893), on artificial crystals.
2. Hofmann, *Zs. Kr.*, **78**, 279 (1931), by the rotation method on artificial crystals.
3. Tutton, *Zs. Kr.*, **27**, 113 (1897), on artificial crystals.
4. For other analyses see Dana (949, 1892).
5. Reichardt, *Jb. Min.*, 340 (1866).
6. Data on the system $K-Mg-SO_4-H_2O$ have been obtained by several investigations and are summarized in *Int. Crit. Tables*, **4**, 352 (1928).
7. As found by the investigations of van't Hoff and co-workers which are summarized in Doelter (4 [2], 93, 1926).

29.3.7.2 **C Y A N O C H R O I T E** $[K_2Cu(SO_4)_2 \cdot 6H_2O]$. Cianocroma *Scacchi* (Memoria sullo Incendio Vesuviano, 191, 1855). Cyanochroite *Dana* (649, 1868).

Cryst.[1] Monoclinic; prismatic—$2/m$.

$a:b:c = 0.7490:1:0.5088;$ $\beta\ 104°28';$ $p_0:q_0:r_0 = 0.6793:0.4927:1$

$r_2:p_2:q_2 = 2.0298:1.3788:1;$ $\mu\ 75°32';$ $p_0'\ 0.7015, q_0'\ 0.5088, x_0'\ 0.2580$

Forms:[2] (artificial)

c 001	a 100	m 110	t 130	η $\overline{1}01$	n $\overline{1}11$
b 010	l 210	s 120	o 011	e $\overline{2}01$	μ $\overline{1}21$

Habit. In crystalline crusts; crystals tabular $\{001\}$.

Phys. Cleavage $\{\overline{2}01\}$, perfect. G. 2.224 (artif.).[3] Color greenish blue. Luster vitreous. Transparent. Water soluble.

Opt.[3]

Orientation		n(Na)	
$X \wedge c$	$+18°33'$	1.4836	Biaxial positive (+).
Y	b	1.4864	$2V$ 46°32' (meas., Na).
Z		1.5020	$r < v$, strong.

Chem. A hydrated sulfate of potassium and copper, $K_2Cu(SO_4)_2 \cdot 6H_2O$. CuO 18.00, K_2O 21.31, SO_3 36.23, H_2O 24.46, total 100.00. Natural crystals have not been analyzed.

Occur. Presumed to have occurred in crystalline crusts formed in fumaroles of Vesuvius during the eruptions of 1855 and 1906 and obtained in crystals by evaporation of water solutions of these crusts.[4]

Artif. By evaporation of an equimolar water solution of K_2SO_4 and $CuSO_4$.

Name. Cyanochroite from κύανος, *blue*, and χρόα, *color*, in allusion to the color. The original name cianocroma was modified by Dana [5] to avoid ambiguity, the material containing no chromium.

Ref.
1. Natural crystals have not been measured. Elements of Tutton, *Zs. Kr.*, **21**, 546 (1893). Scacchi (1855) reported closely similar values on crystals obtained by recrystallization of a solution of crusts of the mineral.
2. Goldschmidt (**2**, 200, 1913).
3. Tutton, *Zs. Kr.*, **27**, 191 (1897), on artificial crystals.
4. Scacchi (1855) and Lacroix, *Bull. soc. min.*, **30**, 223 (1907).
5. Dana (1868).

29.3.7.3 BOUSSINGAULTITE [$(NH_4)_2Mg(SO_4)_2 \cdot 6H_2O$]. Bechi (*C.R.*, **58**, 583, 1864). Cerbolit Popp (*Ann. Chem., Suppl. Bd.*, **8**, 1, 1872).

Cryst.[1] Monoclinic; prismatic—$2/m$.

$a:b:c = 0.7400:1:0.4918;\quad \beta\ 107°06';\quad p_0:q_0:r_0 = 0.6646:0.4701:1$

$r_2:p_2:q_2 = 2.1275:1.4139:1;\quad \mu\ 72°54';\quad p_0'\ 0.6953,\ q_0'\ 0.4918,\ x_0'\ 0.3076$

Forms:[8]

c 001	l 130	m 110	r 201	o 111	n 121
b 010	k 120	q 011	s $\bar{2}$01	p $\bar{1}$11	

Structure cell.[2] Space group $P2_1/a$. a_0 9.28 kX, b_0 12.57, c_0 6.20; [β 107°06']; $a_0:b_0:c_0 = 0.738:1:0.493$. Cell contents $(NH_4)_4Mg_2(SO_4)_4 \cdot 12H_2O$.

Habit. Crystals short prismatic [001] with {001} prominent. Usually massive, as crusts or stalactites.

Phys. Cleavage {$\bar{2}$01} perfect. H. 2. G. 1.722 (artif.),[3] 1.722 (calc.). Colorless to yellowish pink. Transparent. Water soluble.

Opt. In transmitted light, colorless.

Orientation	n[4]	n(Na)[9]	
X	1.470	1.4716	Biaxial positive (+).
Y b	1.472	1.4730	2V 51°11' (meas., Na).[9]
Z ∧ c ~ 12°	1.479	1.4786	$r > v$, perceptible.

Chem. A hydrated sulfate of ammonium and magnesium, $(NH_4)_2Mg(SO_4)_2 \cdot 6H_2O$. The water is largely lost at 150°.

Anal.

	$(NH_4)_2O$	MgO	SO_3	H_2O	Rem.	Total	G.
1.	14.44	11.18	44.40	29.98		100.00	1.722
2.	10.86	11.54	43.49	31.48	0.94	98.31	

1. $(NH_4)_2Mg(SO_4)_2 \cdot 6H_2O$. 2. South Mountain, Ventura County, California.[5] Rem. is K_2O 0.22, Na_2O 0.60, Fe_2O_3 0.08, Al_2O_3 0.04.

Occur. First found in the boric acid fumaroles of Travale, Tuscany. In the United States it has been found in the vicinity of The Geysers in Sonoma County, California, and as stalactites and incrustations formed by heated gases escaping through crevices in sandstone and shale at South Mountain, Ventura County, California. Found in small colorless to yellowish pink crystals in cavities after a fire in an anthracite refuse bank near Mahanoy City, Schuylkill County, Pennsylvania.

A r t i f. Formed in crystals by evaporation of an equi-molal solution of $(NH_4)_2SO_4$ and $MgSO_4$ between 0° and 100°, the investigated range,[6] or more.

N a m e. After the French chemist Jean-Baptiste Boussingault (1802–1887). The name cerbolite, sometimes considered to be a synonym of boussingaultite, was originally intended [7] as a group name for double salts of this type.

Ref.

1. Natural crystals have not been measured. Elements after Tutton, Zs. Kr., **41**, 328 (1905), on artificial crystals.
2. Hofmann, Zs. Kr., **78**, 279 (1931), by rotation and oscillation methods on artificial crystals.
3. Cleavage and G. on artificial crystals, Tutton (1905).
4. Larsen and Shannon, Am. Min., **5**, 127 (1920), on natural crystals from South Mountain, California.
5. Shannon analysis in Larsen and Shannon (1920). This is the only complete analysis so far reported. For a partial analysis of material from Sonoma County, California, see Goldsmith, Proc. Ac. Sc., Philadelphia, **28**, 264 (1873).
6. Benrath and Thiemann, Zs. anorg. Chem., **208**, 179 (1932), who give a crystallization diagram for the system NH_4-Mg-SO_4-H_2O for temperatures 0° to 100° C.
7. Popp (1872).
8. Goldschmidt (**1**, 227, 1913).
9. Tutton (1905) by prism method on artificial crystals.

TYPE 4. $A_m B_n (XO_4 \cdot)_p x H_2 O$, WITH $(m+n):p < 3:2$ AND $>1:1$

29.4.1 F E R R I N A T R I T E [$Na_3Fe(SO_4)_3 \cdot 3H_2O$]. Ferronatrite *Mackintosh* (Am. J. Sc., [3], **38**, 244, 1889). Gordaite *Frenzel* (Min. Mitt., **11**, 218, 1890). Ferrinatrite *Scharizer* (Zs. Kr., **41**, 210, 1906). Leucoglaucite *Ungemach* (Bull. soc. min., **58**, 203, 1935).

C r y s t.[1] Hexagonal—P; rhombohedral—$\bar{3}$ (?).

$a:c = 1:0.5563;$ $\alpha\ 111°05';$ $p_0:r_0 = 0.6424:1;$ $\lambda\ 55°49'$

Forms:

	ϕ	$\rho = C$	A_1	A_2
c 0001	.....	0°00'	90°00'	90°00'
m 10$\bar{1}$0	30°00'	90 00	30 00	90 00
a 11$\bar{2}$0	0 00	90 00	60 00	60 00
s 10$\bar{1}$2	30 00	17 48½	74 38½	90 00
r 10$\bar{1}$1	30 00	32 43	62 05½	90 00
η 01$\bar{1}\bar{1}$	−30 00	32 43	90 00	62 05½

Less common:

u 11$\bar{2}$1 υ 21$\bar{3}$1

Habit. Short prismatic [0001]. As isolated crystals or in stellate groups; also in fibrous aggregates or cleavable masses; cryptocrystalline.

P h y s. Cleavage {10$\bar{1}$0} perfect, {11$\bar{2}$0} less perfect.[2] Brittle. Splintery fracture. H. 2½. G. 2.55–2.61. Luster vitreous. Color grayish

FERRINATRITE

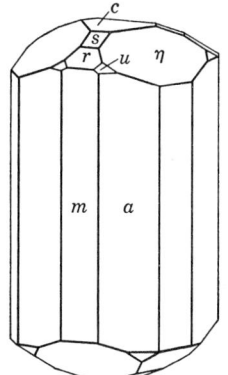

Mina la Compania, Chile.

white, whitish green, bluish green, pale amethystine. Small crystals colorless. Transparent.

O p t. In transmitted light, colorless.

	n(Na) [3]	
O	1.558	Uniaxial positive (+).
E	1.613	

C h e m. A hydrated sulfate of sodium and ferric iron, $Na_3Fe(SO_4)_3 \cdot 3H_2O$.

Anal.

	1	2	3	4	5	6	7
Na_2O	19.91	18.34	19.95	20.22	20.03	20.06	19.92
K_2O		0.40	tr.		0.39		0.40
Fe_2O_3	17.10	17.23	17.30	17.69	17.47	16.91	19.00
SO_3	51.42	50.25	51.30	50.85	51.00	51.29	49.75
H_2O	11.57	11.14	11.39	11.90	11.62	11.50	15.13
Rem.		2.43					0.30
Total	100.00	99.79	99.94	100.66	100.51	99.76	104.50
G.		2.547–2.578	2.61				2.57

1. $Na_3Fe(SO_4)_3 \cdot 3H_2O$. 2. Chile.[4] Rem. is Al_2O_3 0.43, insol. 2.00. 3. Mina la Compania, Chile.[5] 4. Sierra Gorda, Chile.[6] (*gordaite*). 5. Alcaparrosa, Cerritos Bayos, Chile.[7] 6. Artificial ferrinatrite.[8] 7. Chuquicamata, Chile.[9] Rem. is insol. Original sum given as 100.54.

O c c u r. A secondary mineral mostly occurring with other iron sulfates in arid regions; also as a fumarole deposit. Found with aphthitalite and alum in a fumarole deposit at Vesuvius. Occurs at various points in the Atacama Desert, Chile; at Mina la Compania, south of Sierra Gorda (probably the type locality), in crystals and cleavable aggregates with metavoltine, copiapite, and coquimbite; at Alcaparrosa, Cerritos Bayos, in white, cryptocrystalline aggregates with sideronatrite and copiapite; at Chuquicamata with metasideronatrite and metavoltine; at the Mina Salvadora near Quetena with tamarugite and sideronatrite and at Tierra Amarilla near Quetena (*leucoglaucite*) [10] with copiapite, quenstedtite, and roemerite.

Alter. Loses nearly all its water at or near 100° C.[11] On standing in moist air ferrinatrite changes to sideronatrite.[8]

Artif. Formed by the action of concentrated sulfuric acid on sideronatrite.[12]

Name. The original name ferronatrite, from the content of iron and sodium, was changed to ferrinatrite for consistency with the trivalent state of the iron in the mineral.

Ref.

1. Elements and forms from Gordon, *Not. Nat., Ac. Sc. Philadelphia*, no. 103 (1942). Earlier measurements were reported by Arzruni and Frenzel, *Zs. Kr.*, **18**, 596 (1891). The lattice mode is indicated as hexagonal primitive by Gordon's form list; see also Ungemach, *Bull. soc. min.*, **58**, 203 (1935), whose Laue photograph along [0001] indicates the point symmetry to be $\bar{3}$ or 3. The morphology indicates a center of symmetry and the class thus appears to be $\bar{3}$.
2. Arzruni and Frenzel (1891). Bandy, *Am. Min.*, **23**, 731 (1938), mentions only $\{10\bar{1}0\}$ perfect cleavage; Gordon (1942) mentions only $\{11\bar{2}0\}$ perfect cleavage.
3. Genth and Penfield, *Am. J. Sc.*, **40**, 202 (1890), on Mina la Compania crystals by the prism method. Concordant data were obtained by Gordon (1942) on this and on Quetena and Alcaparrosa material.
4. Mackintosh (1889).
5. Genth and Penfield (1890).
6. Arzruni and Frenzel (1891).
7. Pittman analysis in Gordon (1942).
8. Scharizer, *Zs. Kr.*, **41**, 209 (1906).
9. Henderson analysis in Bandy (1938).
10. The reported analysis of this mineral by Ungemach (1935), made on a 50-mg. sample, approximates to the formula $HFe(SO_4)_2 \cdot 2H_2O$ but in other respects the mineral is identical with ferrinatrite and is here so considered.
11. Genth and Penfield (1890) and Arzruni and Frenzel (1891) report slight water loss at 100°, whereas Mackintosh (1889) reports almost complete dehydration at 110°. Scharizer (1906) found that artificial material showed no water loss at 70° but almost complete dehydration at 100°.
12. Scharizer (1906); see also Mellor (**14**, 346, 1935).

29.4.2 **POLYHALITE** $[K_2Ca_2Mg(SO_4)_4 \cdot 2H_2O]$. Stromeyer (*J. Chemie u. Phys.*, **29**, 389, 1820). Mamanite Goebel (*Ac. sc. St. Pétersbourg, Bull.*, **9**, 1, 1866). Krugite pt. Precht (*Ber. deutsche chemische Gesellschaft*, **14**, 2138, 1881).

Cryst.[1] Triclinic; pinacoidal—$\bar{1}$.

$a:b:c = 0.7176:1:0.4657;$ $\alpha\ 91°39',\ \beta\ 90°06\tfrac{1}{2}',\ \gamma\ 91°53'$

$p_0:q_0:r_0 = 0.6490:0.4660:1;$ $\lambda\ 88°20\tfrac{1}{2}',\ \mu\ 89°50\tfrac{1}{2}',\ \nu\ 88°06\tfrac{1}{2}'$

$p_0'\ 0.6493$ $q_0'\ 0.4661$ $x_0'\ 0.0019$ $y_0'\ 0.0289$

Forms:

		φ	ρ	A	B	C
b	010	0°00'	90°00'	88°06½'	.....	88°20½'
a	100	88 06½	90 00	.....	88°06½'	89 50½
M	1$\bar{1}$0	124 25½	90 00	36 19	124 25½	90 51
o	1$\bar{2}$0	144 32	90 00	52 38½	144 32	91 17
y	$\bar{1}$01	−89 19½	32 54½	89 38	89 50½	33 02
z	$\bar{1}$11	−53 47½	38 43½	119 29½	68 18½	37 51½
d	1$\bar{3}$1	154 13½	56 15½	70 19½	138 29½	57 42½
l	$\bar{3}$13	−75 51½	33 43	122 14½	82 12½	33 27
n	$\bar{1}$31	−24 43	57 08	109 01	40 16½	55 40½
t	$\bar{1}$51	−15 28	67 36	102 32½	26 59½	66 02

POLYHALITE

Less common:
m 110   w 1̄40   h 021   x 101   e 1̄11   δ 131   ξ 2̄12   τ 1̄51   g 3̄31
v 1̄30   γ 011   s 01̄1   ε 111   u 1̄11   λ 3̄13   ν 1̄31   f 31̄1   i 5̄11

Habit. Crystals small and rare; tabular {010} or elongated [001] and showing many forms. Common forms: *b a z y t M l o d n*. Usually massive, also fibrous to foliated.

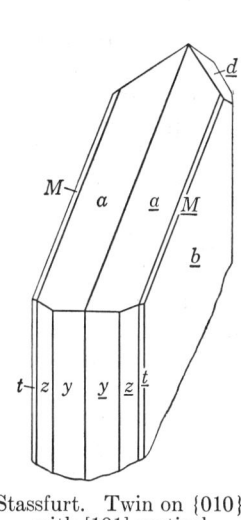

Stassfurt. Twin on {010} with [101] vertical.

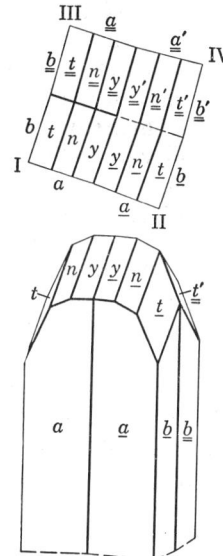

Twinning. On {010} and {100}, very common and often polysynthetic.
Phys. Cleavage {101̄} perfect. Parting {010} in foliated varieties. H. $3\frac{1}{2}$. G. 2.78. Luster vitreous to resinous. Colorless and transparent; white or gray; often salmon-pink to brick-red due to finely divided inclusions of iron oxide.
Opt.[2] In transmitted light, colorless.

ORIENTATION	n (Texas)	n (Ischl)	
X	1.547	1.548	Biaxial negative (−).
Y	1.560	1.562	
Z	1.567	1.567	
2V	62°	~70°	

Polysynthetic twinning is conspicuous under the microscope.

Chem. A hydrated sulfate of potassium, calcium and magnesium, $K_2Ca_2Mg(SO_4)_4 \cdot 2H_2O$.
Anal.[3]

	K_2O	CaO	MgO	SO_3	H_2O	Rem.	Total
1.	15.62	18.60	6.69	53.11	5.98		100.00
2.	15.66	18.75	6.48	52.40	6.07	0.27	99.63
3.	15.37	18.54	6.16	52.60	6.50	0.59	99.76
4.	14.41	18.89	6.60	52.42	6.34	0.49	99.15

1 $K_2Ca_2Mg(SO_4)_4 \cdot 2H_2O$. 2. Brick red. McDowell Well, Glasscock County, Texas.[4] Rem. is insol. 3. Salmon colored, McDowell Well.[4] Rem. is Na_2O 0.20, insol. 0.39. 4. Maman, Iran.[5] Rem. is Na_2O 0.07, Fe_2O_3 0.20, Cl 0.10, insol. 0.12.

Tests. Fusible at $1\frac{1}{2}$. Decomposed by water with the separation of gypsum or of gypsum and syngenite.

O c c u r. A widespread constituent of oceanic salt deposits, mostly associated with halite and anhydrite. Very rare as a volcanic product. First recognized as a distinct mineral in the salt mines at Ischl, Upper Austria. Also found at Hallstatt, Upper Austria, and at Hall, Tyrol. The lowest part of the potash deposits at Stassfurt consists of fine-grained halite 15 meters thick through which is dispersed polyhalite attaining 7 per cent by weight of the rock. Elsewhere in the German potash deposits polyhalite occurs in many places and in association with carnallite, kieserite, langbeinite, and vanthoffite. At Varangéville and elsewhere in Lorraine, France, it occurs with glauberite and anhydrite. At Stebnik, Galicia, in anhydrite with halite. At Vesuvius polyhalite has been found with halite and sylvite in a crystalline crust on lava. It has been identified in a boring at Aislaby, near Whitby, England, at a depth of 4500 feet. Nodules of polyhalite occur in the halite beds at Maman, northwestern Iran. Polyhalite, mostly associated with halite and sylvite, is an important constituent of salt deposits penetrated by borings in the region of Saratov along the lower Volga, in western Kazakhstan, and in Bashkirian, U.S.S.R. At certain horizons polyhalite constitutes up to 85 per cent of the salts. In the United States polyhalite has been found at many points in the Texas-New Mexico potash region as blebs in halite, with anhydrite, and in beds up to eight feet thick. It has also been identified at several horizons in salt penetrated by a deep boring in Grand County, Utah.

A l t e r. Dehydrated by heating at 320° with the formation of a mixture of $K_2CaMg(SO_4)_3$ and $CaSO_4$ (*anhydrite*).[6]

A r t i f. Can be formed by warming a compressed mixture of K_2SO_4, $MgSO_4$, and gypsum.[7] Crystallizes from an aqueous solution of its constituents or from a solution approximating the composition of oceanic brines over a wide range of temperatures.[8]

N a m e. From πολύς, *many*, and ἅλς, *salt*, in allusion to its composition.

MAMANITE Goebel (*Ac. sc. Leningrad, Bull.*, **9**, 1, 1866) has been shown [9] to be polyhalite.

KRUGITE Precht (*Ber. deut. chem. Ges.*, **14**, 2138, 1881) has been shown [10] to be polyhalite admixed with anhydrite.

Ref.

1. Setting of Peacock, *Am. Min.*, **23**, 38 (1938); form list of Görgey, *Min. Mitt.*, **33**, 48 (1915), on crystals from Stassfurt. Transformation: Görgey to Peacock, $\frac{1}{2}01/010/\frac{1}{2}00$.
2. Larsen and Berman, 159 (1934), and Görgey, *Min. Mitt.*, **29**, 209 (1910).
3. For a tabulation of older analyses see Hintze (**1** [3], 4485, 1930).
4. Schaller and Henderson, *U. S. Geol. Sur., Bull. 833*, 73 (1932).
5. Ladame, *Schweiz. min. Mitt.*, **22**, 242 (1942).
6. Ramsdell, *Am. Min.*, **20**, 569 (1935), identified the dehydration products by x-ray diffraction. Lepeshkov and Bodaleva, *Ac. sc. Leningrad, C.r.*, **27**, 978 (1940), established the temperature by differential thermal analysis.
7. Ide, *Kali*, **29**, 83 (1935)—*Min. Abs.*, **6**, 351 (1936).

8. The lowest temperature at which polyhalite will form in these systems has not been established precisely but lies below 25° C. in both cases. Data on these systems derived from various sources are summarized in *Int. Crit. Tables*, **4**, 279 and 349 (1928). For equilibria at 100° C. see Conley, Gabriel, and Partridge, *J. Phys. Chem.*, **42**, 587 (1938).
9. van't Hoff and Voerman, *Ak. Berlin, Ber.*, 984 (1904).
10. Görgey (1910).

29.4.3   **L E I G H T O N I T E** [$K_2Ca_2Cu(SO_4)_4 \cdot 2H_2O$]. *Palache* (*Am. Min.*, **23**, 34, 1938).

C r y s t.[1] Triclinic (pseudo-orthorhombic).

$a:b:c = 0.7043:1:0.4578;$ $p_0:q_0:r_0 = 0.6500:0.4578:1$

$q_1:r_1:p_1 = 0.7043:1.5385:1;$ $r_2:p_2:q_2 = 2.1844:1.4198:1$

Forms:

	ϕ	$\rho = C$	ϕ_1	$\rho_1 = A$	ϕ_2	$\rho_2 = B$
b 010	0°00'	90°00'	90°00'	90°00'	....	0°00'
a 100	90 00	90 00	....	0 00	0°00'	90 00
m 110	54 50½	90 00	90 00	35 09½	0 00	54 50½
e 101	90 00	33 01½	0 00	56 58½	56 58½	90 00
p 111	54 50½	38 29	24 36	59 25	56 58½	69 00
q 131	25 19½	56 39	53 56½	69 04	56 58½	40 58½

Less common: n 130   d 031

Habit. Blades or laths elongated [001] and flattened {100}; rarely equant. Some crystals have an hour-glass appearance due to curved

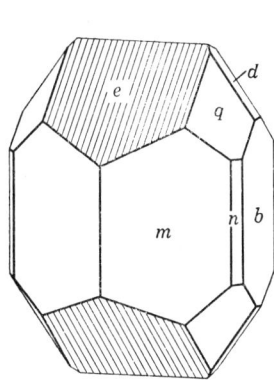 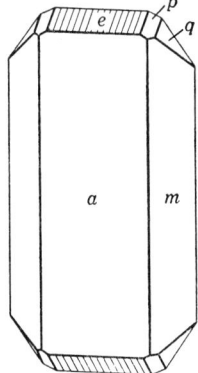

surfaces cutting away more or less of the front edges of the prism. Common forms: *a m e q*. Also as cross-fiber veinlets.

Twinning. The crystals show mimetic orthorhombic symmetry due to microscopic, repeated lamellar twinning on (100) and (010) of a sub-rectangular triclinic structure. The morphological description, above, is based on the pseudo-orthorhombic elements.

Phys. No cleavage. H. 3. G. 2.95. Luster vitreous. Color pale watery blue to greenish blue. Transparent to translucent.
Opt. In transmitted light, pale blue.

Orientation	n(Na)	
X near b	1.578 ± 0.002	Biaxial negative (−).
Y near c	1.587 ± 0.002	$2V$ 86°09′ (calc.).
Z near a	1.595 ± 0.002	$r > v$, fairly strong.

(010) sections show twin lamellae parallel [001] with extinction $Y \wedge [001] = 3°$; in (001) sections the twin lamellae are parallel [010] with extinction $X \wedge [010] = 3°$ to $5°$.

Chem. A hydrated sulfate of potassium, calcium, and copper, $K_2Ca_2Cu(SO_4)_4 \cdot 2H_2O$. Probably isostructural with polyhalite.[4]
Anal.

	1	2	3	4
K_2O	14.68	13.8	13.62	13.93
Na_2O		tr.	0.98	0.56
CaO	17.45	17.3	17.50	18.41
CuO	12.39	12.2	11.24	11.97
SO_3	49.87	48.8	50.75	49.33
H_2O at 105°	5.61	0.2	5.98	5.71
H_2O at 900°		6.8		
Insol.		0.5		
Total	100.00	99.6	100.07	99.91

1. $K_2Ca_2Cu(SO_4)_4 \cdot 2H_2O$. 2. Chuquicamata.[2] Cross-fiber material. 3,4. Chuquicamata;[3] 3, cross-fiber material and 4, crystals.

Occur. At Chuquicamata, Chile, as cross-fiber veinlets and as crystals in open crevices associated chiefly with atacamite and kroehnkite.

Name. After Tomas Leighton, Professor of Mineralogy at the University of Santiago, Chile.

Ref.

1. Palache (1938). Peacock, *Am. Min.*, **23**, 38 (1938), shows that polyhalite and leightonite are very closely related geometrically.
2. Carter analysis in Palache (1938).
3. Gonyer analysis in Palache (1938).
4. Suggested by the geometrical and chemical similarities (Peacock (1938)) and the near identity of their x-ray powder photographs (Frondel, *priv. comm.*, 1947).

TYPE 5. $AB(XO_4)_2 \cdot xH_2O$

29.5.1 **KRAUSITE** [$KFe(SO_4)_2 \cdot H_2O$]. Foshag (*Am. Min.*, **16**, 352, 1931).

Cryst.[1] Monoclinic.

$a:b:c = 1.5424:1:1.7639$; $\beta\ 102°47\frac{1}{2}′$; $p_0:q_0:r_0 = 1.1436:1.7201:1$

$r_2:p_2:q_2 = 0.5813:0.6648:1$; $\mu\ 77°12\frac{1}{2}′$; $p_0′\ 1.1727,\ q_0′\ 1.7639,\ x_0′\ 0.2271$

KRAUSITE

Forms:

		φ	ρ	φ₂	ρ₂ = B	C	A
c	001	90°00'	12°47½'	77°12½'	90°00'	.....	77°12½'
a	100	90 00	90 00	0 00	90 00	77°12½'	.....
m	110	33 37	90 00	0 00	33 37	82 57½	56 22
l	210	53 03½	90 00	0 00	53 03½	79 48½	36 56½
f	1̄02	−90 00	19 45½	109 45½	90 00	32 33	109 45½
d	1̄01	−90 00	43 24	133 24	90 00	56 11½	133 24
s	1̄12	−22 10	43 36	109 45½	50 18½	49 34	105 05
r	2̄11	−50 13	70 03½	154 44	53 01½	80 03½	136 15

Habit. Usually short prismatic [001]; also equant to tabular {001}. The larger crystals are rough and opaque, the smaller ones often brilliant and clear.

P h y s. Cleavage {001} perfect, {100} good. H. 2½. G. 2.840. Luster vitreous. Color pale lemon-yellow. Streak white. Transparent.

O p t. In transmitted light, colorless to pale yellow.

ORIENTATION	n	PLEOCHROISM	
X	1.588	Colorless	Biaxial positive (+).
Y ∧ c −35°	1.650	Pale yellow	2V large.
Z b	1.722	Pale yellow	

C h e m. A hydrated sulfate of potassium and ferric iron, $KFe(SO_4)_2 \cdot H_2O$.

Anal.

	Na₂O	K₂O	CaO	FeO	Fe₂O₃	SO₃	H₂O	SiO₂	Insol.	Total	G.
1.		15.45			26.18	52.47	5.90			100.00	
2.	0.64	14.71	0.12	0.24	24.94	51.05	5.59	2.19	0.92	100.40	2.840

1. $KFe(SO_4)_2 \cdot H_2O$. 2. Borate near Yermo, San Bernardino County, California.

Tests. Slowly decomposed by water. Slowly soluble in HCl. B.B. decrepitates strongly and reduces to a black scoriaceous mass. C.T. decrepitates, turns ocherous yellow, then brown and finally fuses to a dull vesicular mass. Yields acid water.

O c c u r. Found with alunite, coquimbite, and other sulfates at Borate near Yermo, in the Calico Hills, San Bernardino County, California. Also in drusy crusts in limestone with voltaite, halotrichite, and other iron minerals at the Santa Maria mine, Velardeña, Durango, Mexico.

N a m e. After Edward H. Kraus (1875-), American mineralogist, of the University of Michigan.

Ref.

1. Forms from Foshag, *Am. Min.*, **16**, 352 (1931). Elements recomputed from Foshag's observations by La Forge, priv. comm. (1935).

29.5.2 **VOLTAITE** [$(K,Fe'')_3Fe'''(SO_4)_3 \cdot 4H_2O$ (?)]. *Breislak* (*Essai min. sur la solfatara de Pozzuoli*, Naples, 155, 1792). Voltaite *Scacchi* (17, 1841; *J. pr. Chem.*, **28**, 487, 1843; *Zs. deutsch. geol. Ges.*, **4**, 164, 1852). Pettkoite *Paulinyi* (*Jb. Min.*, 457, 1867).

Cryst. Isometric.[1]

Forms:

      a 001 o 111 d 011 q 112 r 332

Structure cell.[2] Space group uncertain. a_0 27.33 kX. Cell contents uncertain.

Habit. Cubical, octahedral, or, rarely, dodecahedral and in combinations of these. Also massive, granular.

Phys. No cleavage. Fracture conchoidal. H. 3. G. 2.7 ± 0.1. Luster resinous. Color greenish black to black, also dark oil-green. Opaque; on the edges or in thin splinters translucent and greenish. Streak grayish green.

Opt. In transmitted light, oil-green to pale green.

	Cyprus [6]	Arizona [7]	California [4]	Schemnitz [5]	Chile [3]	Turkey [16]
n	1.593	1.594	1.596	1.600	1.602	1.602

	Arizona [4]	Chile [5]	Alcaparrosa [8]
n	1.604	1.605	1.608

Voltaite is sometimes isotropic but often is weakly birefringent showing a sectoral distribution of uniaxial or biaxial parts or an isotropic core and an outer anisotropic zone.[9] The birefringence probably is anomalous and due, as in the isometric alums, to strain arising in compositional variation.

Chem. A hydrated sulfate of ferric iron, ferrous iron, and potassium. The formula is uncertain,[10] particularly as to the role of Mg and divalent Fe, and to water content, and has been given as $KFe''Fe'''(SO_4)_3 \cdot 4H_2O$ or $(K_2Fe'')_3Fe'''(SO_4)_3 \cdot 4H_2O$. The relatively high content of trivalent cations in analysis 3 may be due to partial oxidation of divalent iron; in this analysis Al substitutes to a large extent for Fe'''. The H_2O begins to be lost at about 200°.

Anal.

	1	2	3	4	5	6
K_2O	9.22	5.99	4.73	3.65	4.52	2.37
Na_2O			0.50			1.62
MgO			0.48	3.00	1.55	7.35
FeO	14.06	16.59	14.07	14.47	8.82	5.24
Al_2O_3		2.45	1.58	2.14	6.06	3.72
Fe_2O_3	15.62	11.84	13.47	11.38	14.34	13.85
SO_3	47.00	48.33	46.78	47.02	47.83	49.12
H_2O	14.10	15.34	15.70	16.44	16.13	16.60
Rem.			2.32	1.12		
Total	100.00	100.54	99.63	99.22	99.25	99.87
G.		2.695			2.75	2.6

1. $KFe''Fe'''(SO_4)_3 \cdot 4H_2O$. 2. Artificial.[11] 3. Szomolnok.[12] Rem. is ZnO 1.69, CuO 0.55, NiO 0.08. 4. Huelva, Spain.[13] Rem. is CuO 0.22, CaO 0.90. 5. Jerome, Arizona.[4] 6. Persia.[14]

VOLTAITE 465

Tests. Decomposed by water to an acid solution and leaving a citron-yellow precipitate. Soluble in acids.

Occur. Found originally in the solfatara of Pozzuoli near Naples, Italy, with halotrichite; also in Italy with alunogen and metavoltine in fumaroles of the Atrio del Cavallo on Vesuvius and in the sulfur cave of Miseno near Naples. With metavoltine and roemerite at Madeni Zakh, Persia. In Czechoslovakia at Szomolnok, Schemnitz, and with melanterite at Kremnitz. At the Rammelsberg mine near Goslar in the Harz, Germany. In the pyrite mines of Huelva, Spain, with coquimbite and melanterite. From the Skouriotissa mine, Cyprus. In Chile found with coquimbite at Chuquicamata and Quetena and with pickeringite at Alcaparrosa. Reported from Cerro Potosi, Bolivia, with sideronatrite. In the United States voltaite occurs in the fire zone at the United Verde mine at Jerome, Arizona. In California at the Redington mine, Napa County; with melanterite from the Mount Lassen area, Shasta County; in the sulfate deposit near Borate, San Bernardino County; sparingly at The Geysers, Sonoma County. In the Copper Queen mine, Bisbee, Arizona, with coquimbite, rhomboclase, kornelite, roemerite.

Alter. Alters on exposure to yellow, powdery coquimbite (?).[4]

Artif.[15] By crystallization of a solution of the constituent salts containing an excess of sulfuric acid together with some aluminum sulfate. The pure iron or aluminum compounds are unstable. The following analogues of ordinary voltaite ($KFe''Fe'''$) have been synthesized: $KMgFe'''$, $KMn''Fe'''$, $KCo''Fe'''$, $TlFe''Fe'''$, $TlMgFe'''$, $TlCdFe'''$, $RbCdFe'''$, $(NH_4)MgFe'''$, $KZnFe'''$, $RbMgFe'''$, and others.

Name. After the Italian physicist Alessandro Volta (1745–1827).

Ref.

1. Classed as isometric on the basis of crystal form and the occasional isotropic character of both natural and artificial material; see also the x-ray studies of Gossner and Arm, *Zs. Kr.*, **72**, 202 (1929), Gossner and Besslein, *Cbl. Min.*, 358 (1934). The crystals are often birefringent and have been interpreted as pseudo-isometric twins of tetragonal individuals by Blaas, *Ak. Wien, Sitzber.*, **87**, 143 (1883); *Min. Mitt.*, **3**, 499 (1881).
2. Gossner and Arm (1929); Gossner and Besslein (1934).
3. Larsen (155, 1921).
4. Anderson, *Am. Min.*, **12**, 287 (1927).
5. Ulrich, *Časopis pro min. a geol.*, 1923—*Jb. Min.*, I, 330 (1927).
6. Mélon and Donnay, *Soc. géol. Belgique, Bull.*, **59**, B162 (1936).
7. Lausen, *Am. Min.*, **13**, 226 (1928).
8. Bandy, *Am. Min.*, **23**, 749 (1938).
9. For descriptions see Gossner and Arm (1929), Ulrich (1923), and especially Gossner and Bäuerlein, *Jb. Min., Beil.-Bd.*, **66**, 1 (1933).
10. See discussion in Scharizer, *Zs. Kr.*, **54**, 127 (1914), **75**, 82 (1930); Gossner and Arm (1929); Gossner and Besslein (1934); Gossner, *Zbl. Min.*, 262 (1936); Gossner and Fell, *Ber.*, **65**, 393 (1932).
11. Gossner and Arm (1929).
12. Scharizer (1914).
13. Serrano analysis in Collins, *Min. Mag.*, **20**, 32 (1923).
14. Blaas (1883).
15. See Mellor (**14**, 352, 1935).
16. Schröder, *Min. Abs.*, **10**, 123 (1947).

29.5.3 TAMARUGITE GROUP

29.5.3.1 TAMARUGITE [$NaAl(SO_4)_2 \cdot 6H_2O$]. Alumbre nativo (?) *Domeyko* (*Min.*, 2d App. to 3rd Ed., 30, 1883). Natronalaun (?) *Cleve* (*Kongl. Svenska Vet.-Ak. Handl.*, **9**, 31, 1870). Tamarugite *Schulze* (*Verh. Ver. Santiago*, **2**, 56, 1889). Lapparentite *Ungemach* (*Bull. soc. min.*, **58**, 209, 1935).

Cryst.[1] Monoclinic; prismatic—$2/m$.

$a:b:c = 0.2918:1:0.2415;\quad \beta\,94°49\tfrac{1}{2}';\quad p_0:q_0:r_0 = 0.8277:0.2407:1$

$r_2:p_2:q_2 = 4.1547:3.4387:1;\quad \mu\,85°10\tfrac{1}{2}';\quad p_0'\,0.8306,\ q_0'\,0.2415,\ x_0'\,0.0844$

Forms:[2]

	ϕ	ρ	ϕ_2	$\rho_2 = B$	C	A
c 001	90°00′	4°49½′	85°10½′	90°00′	.....	85°10½′
b 010	0 00	90 00	....	0 00	90°00′	90 00
a 100	90 00	90 00	0 00	90 00	85 10½	.....
g 120	59 49	90 00	0 00	59 49	85 50	30 11
γ 230	66 26	90 00	0 00	66 26	85 34½	23 34
m 110	73 47	90 00	0 00	73 47	85 22	16 13
e 011	19 15½	14 21	85 10½	76 28	13 32	85 18½
f 021	9 54½	26 07½	85 10½	64 17½	25 42½	85 39½
i 031	6 38½	36 07	85 10½	54 10	35 50	86 05½
j 041	4 59½	44 07½	85 10½	46 05	43 55	86 31½
p 111	75 12½	43 25½	17 32½	79 53½	38 46½	48 21
o $\bar{1}11$	−72 04	38 06½	126 44	79 02½	42 43	125 57½
r 121	62 10	45 58½	47 32½	70 23	41 45½	50 31
u 141	43 26½	53 04½	47 32½	54 31	49 50½	56 39

Less common and rare:

ϵ 1.12.0 α 150 h 210 θ 172 τ 171 λ 1.14.1 w $\bar{1}31$
δ 180 l 140 k 051 v 151 π 181 t $\bar{1}21$ x $\bar{1}41$
β 160 n 130 s 131 σ 161 ω 1.10.1 μ $\bar{2}52$

Habit. Crystals tabular {010} or short prismatic [001]. Also as fibrous or fine-granular masses.

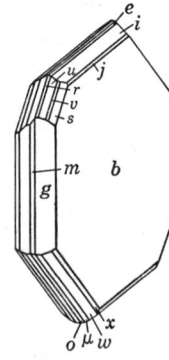

Phys. Cleavage {010} perfect but not easy. H. ~3. G. 2.07. Luster vitreous. Colorless and transparent. Taste sweetish and astringent.

TAMARUGITE 467

O p t. In transmitted light, colorless.

ORIENTATION		n(Na)[3]	n (Tierra Amarilla)[4]	
$X \wedge c$	$+4°$ to $5°$	1.484 ± 0.001	1.490	Biaxial positive (+).
Y	b	1.486 ± 0.001	1.492	$2V \sim 60°$.
Z		1.497 ± 0.001	1.504	

Polysynthetic twinning has been reported.[5]

C h e m. A hydrated sulfate of sodium and aluminum, $NaAl(SO_4)_2 \cdot 6H_2O$.

Anal.[10]

	1	2	3	4	5
Na_2O	8.85	9.04	9.25	8.64	9.11
Al_2O_3	14.56	14.48	14.50	14.66	15.07
SO_3	45.72	45.66	45.64	45.48	45.44
H_2O	30.87	30.86	31.11	31.40	30.21
Rem.		0.32		0.48	
Total	100.00	100.36	100.50	100.66	99.83
G.		2.04		2.05	

1. $NaAl(SO_4)_2 \cdot 6H_2O$. 2. Cerro Pintados.[6] Rem. is CaO 0.20, Cl 0.12. 3. St. Bartholomew, West Indies.[7] 4. Miseno, Italy.[8] Rem. is Cl 0.48, CaO tr. 5. Alcaparrosa.[9] H_2O includes H_2O at 104° 15.10.

Tests. Soluble in water.

O c c u r. A secondary mineral formed principally under arid conditions by the oxidation of sulfides in aluminous and alkali-rich environments. Associated with pickeringite, coquimbite, quenstedtite, gypsum, sideronatrite, mendozite, kroehnkite, halite. Originally from Cerro Pintados, about 80 km. southeast of Iquique, Tarapacá, Chile, in the Tamarugal Pampa of the Atacama desert; also from Chile from Alcaparrosa near Cerritos Bayos, Quetena, Chuquicamata, la Compania about 8 km. south of Sierra Gorda, and Tierra Amarilla about 15 km. southeast of Copiapó (*lapparentite*). On the Island of St. Bartholomew in the West Indies. From the Skouriotissa mine, Cyprus.[11] With alums in the sulfur cave at Miseno near Naples, Italy. From the Gascogne River, Australia.[12] Found in the United States near Eureka, St. Louis County, and at Fulton, Calloway County, in Missouri, as an alteration of mendozite.[13]

A l t e r. Mendozite and soda alum dehydrate, on exposure or slight heating, to tamarugite.

N a m e. From the locality in the Tamarugal Pampa, Chile.

LAPPARENTITE. *Ungemach* (*Bull. soc. min.*, **58**, 209, 1935). Shown to be identical with tamarugite.[14]

Ref.

1. Elements of Ungemach, *Bull. soc. min.*, **58**, 209 (1935), on Tierra Amarilla crystals.
2. Ungemach (1935), Bandy, *Am. Min.*, **23**, 740 (1938), and Gordon, *Ac. Sc. Philadelphia, Not. Nat.*, no. 57 (1940).
3. Gordon (1940) on Cerro Pintados, Alcaparrosa, and Quetena material in agreement with the data of Larsen (141, 1921).
4. Bandy (1938) on type lapparentite (partly dehydrated ?).

5. Larsen (1921).
6. Schulze (1889).
7. Cleve (1870).
8. Zambonini, *Acc. Napoli, Rend.*, **46**, 321 (1907).
9. Collins and Pitman analysis in Gordon (1940).
10. Additional analyses are cited by Gordon (1940). The analysis of Ungemach (1935) on lapparentite = tamarugite is erroneous.
11. Mélon and Donnay, *Soc. géol. Belgique, Bull.*, **59**, 162 (1936).
12. Simpson, *J. Roy. Soc. Western Australia*, **9**, 62 (1923).
13. Keller, *Am. Min.*, **20**, 537 (1935).
14. Gordon (1940).

29.5.3.2 **A M A R I L L I T E** [$NaFe(SO_4)_2 \cdot 6H_2O$]. *Ungemach* (*C.R.*, **197**, 1132, 1933; *Bull. soc. min.*, **58**, 200, 1935).

C r y s t.[1] Monoclinic; prismatic—$2/m$.

$a:b:c = 0.7757:1:1.1482$; $\quad \beta\, 95°37'$; $\quad p_0:q_0:r_0 = 1.4803:1.1427:1$

$r_2:p_2:q_2 = 0.8751:1.2954:1$; $\quad \mu\, 84°23'$; $\quad p_0'\, 1.4874, q_0'\, 1.1482, x_0'\, 0.0983$

Forms:

		ϕ	ρ	ϕ_2	$\rho_2 = B$	C	A
c	001	90°00'	5°37'	84°23'	90°00'	.....	84°23'
b	010	0 00	90 00	.....	0 00	90°00'	90 00
a	100	90 00	90 00	0 00	90 00	84 23	.....
m	110	52 20	90 00	0 00	52 20	85 33½	37 40
d	021	2 27	66 29½	84 23	23 38	66 22	87 45½
g	$\bar{1}01$	−90 00	54 15	144 15	90 00	59 52	144 15
p	112	55 43	45 32½	49 54	66 17½	41 00	53 51½
q	111	54 05½	62 56½	32 14	58 31	58 27	43 50
x	$\bar{1}12$	−48 20½	40 49	122 50	64 15	45 08½	119 14
y	$\bar{1}11$	−50 25½	60 58½	144 15	56 08½	65 21½	132 22½

Less common:
$\quad r\, 665 \qquad s\, 332 \qquad t\, 221 \qquad u\, 331 \qquad w\, \bar{1}13 \qquad z\, \bar{2}21$

Habit. Equant; thick-tabular {001}; rarely prismatic [001]. Extended ramified groups of parallel individuals. Common forms: $c\ a\ m\ g\ q$.

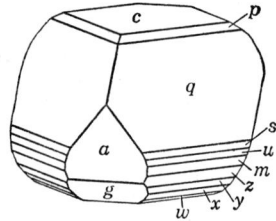

Tierra Amarilla.

P h y s. Cleavage {110} good but not easily developed. Fracture conchoidal. H. 2½–3. G. 2.19. Luster brilliant, vitreous inclining to adamantine. Color pale yellow with a greenish tint. Transparent. Taste astringent.

MENDOZITE 469

O p t.[2]

ORIENTATION	n	
$X \wedge c \quad -39°$	1.532 ± 0.002	Biaxial positive (+).
$Y \quad b$	1.555 ± 0.003	$2V$ large.
$Z \wedge c \quad +51$	1.591 ± 0.002	$r < v$.

C h e m. A hydrated sulfate of sodium and trivalent iron, $NaFe(SO_4)_2 \cdot 6H_2O$. Analysis gave:

	Na_2O	Fe_2O_3	SO_3	H_2O	Total
1.	8.18	21.06	42.24	28.52	100.00
2.	7.14	21.39	43.59	28.45	100.57

1. $NaFe(SO_4)_2 \cdot 6H_2O$. 2. Tierra Amarilla.

Tests. Soluble in water.

O c c u r. Found as veinlets in massive coquimbite at Tierra Amarilla near Copiapó, Chile.

Ref.
1. Ungemach (1935), with change of form letters. Uncertain: 031, $\bar{1}14$.
2. Berman, priv. comm. (1936).

29.5.4 MENDOZITE GROUP

MONOCLINIC; PRISMATIC—2/m

	$a:b:c$	β
Mendozite, $NaAl(SO_4)_2 \cdot 11H_2O$	$2.5060:1:0.9125$	$109°1'$
Kalinite, $KAl(SO_4)_2 \cdot 11H_2O$		

The names mendozite and kalinite are here used to designate the fibrous, monoclinic alums as distinct from the isometric species soda alum and potash alum, with which minerals they appear to have been generally confused. The natural occurrences of mendozite and kalinite, however, are not well-established.

29.5.4.1 M E N D O Z I T E [$NaAl(SO_4)_2 \cdot 11H_2O$]. Native Soda-Alum *Thomson* (*Ann. Lyceum Nat. Hist., New York*, **3**, 19, 1828). Natronalun (?) *Huot* (**2**, 448, 1841). Subsesquisulfate of Alumina (?) *Thomson* (*Phil. Mag.*, **22**, 192, 1843). Mendozite pt. *Dana* (653, 1868). Soda Alum pt. Natronalaun, Natrumalaun pt. *Germ.*

C r y s t.[1] Monoclinic; prismatic—$2/m$.

$a:b:c = 2.5060:1:0.9125; \quad \beta\ 109°01'; \quad p_0:q_0:r_0 = 0.3641:0.8627:1$

$r_2:p_2:q_2 = 1.1592:0.4221:1; \quad \mu\ 70°59'; \quad p_0'\ 0.3852,\ q_0'\ 0.9125,\ x_0'\ 0.3447$

Forms:

	ϕ	ρ	ϕ_2	$\rho_2 = B$	C	A
c 001	90°00'	19°01'	70°59'	90°00'	.....	70°59'
b 010	0 00	90 00	.....	0 00	90°00'	90 00
a 100	90 00	90 00	0 00	90 00	70 59	.....
m 110	22 53	90 00	0 00	22 53	82 43	67 07
q 011	20 41½	44 17½	70 59	75 43	14 17	49 13
y $\bar{2}11$	−25 00½	45 12	113 03½	49 59	55 21½	107 27½

Less common:

	a $\bar{3}10$	s 302	x 211

470   SULFATES

Habit. Artificial crystals are prismatic [001] with large {100}; also pseudo-rhombohedral with {100} and {$\bar{2}11$}. As fibrous masses.

Phys. Cleavage {100} good, {001} and {010} indistinct.[2] H. $\sim$3. G. 1.730,[3] 1.765.[4] Colorless and transparent; white. On exposure becomes white and turbid (*tamarugite*).

Opt. In transmitted light, colorless.

Orientation	n (artificial)[5]	
X b	1.449 ± 0.003	Biaxial negative ($-$).
$Y \wedge c$ 30°	1.461 ± 0.003	$2V$ 56° $\pm$ 1° (meas.).
Z	1.463 ± 0.003	Distinct crossed dispersion.[3]

Chem. A hydrated sulfate of sodium and aluminum, $NaAl(SO_4)_2 \cdot 11H_2O$. The water content is uncertain and may be $12H_2O$.

Anal.

	1	2	3
Na_2O	7.04	7.96	7.26
Al_2O_3	11.58	12.00	11.27
SO_3	36.37	37.70	34.73
H_2O	45.01	41.96	[46.74]
Total	100.00	99.62	[100.00]
G.		1.88	

1. $NaAl(SO_4)_2 \cdot 11H_2O$. 2. Mendozite? Mendoza, Argentina.[6] 3. Mendozite? Shimané, Japan.[7]

Tests. Soluble in water.

Occur. and **Name.** The name mendozite is here applied to the monoclinic compound $NaAl(SO_4)_2 \cdot 11H_2O$ or $12H_2O$ well-known as an artificial salt but not yet identified with certainty in nature. The name soda alum is restricted to the isometric compound $NaAl(SO_4)_2 \cdot 12H_2O$. Most reported occurrences of natural NaAl alum cannot be referred with certainty to either one of these species for lack of crystallographic or optical evidence. The materials from St. Juan near Mendoza, Argentina, and Shimané, Idzumo, Japan (analyses 2 and 3), are tentatively placed with mendozite on the basis of a fibrous habit—a not very reliable criterion—and relatively low water content ($\sim 11H_2O$). Birefringent NaAl alums occur at Quetena, Alcaparrosa, and Chuquicamata, Chile, and may represent mendozite.[8] An efflorescence in a road cut near Eureka, St. Louis County, Missouri, has optical properties indicative of mendozite, and a similar mineral occurs at Fulton, Callaway County, Missouri.[9] Other occurrences which cannot be definitely placed with either mendozite or soda alum include Pritchard Ranch near St. Helena in Napa County, near Hidden Springs in San Bernardino County, and near Volcano in Amador County, all in California; and Box Elder County, Utah.

Alter. On exposure dehydrates to tamarugite, $NaAl(SO_4)_2 \cdot 6H_2O$.

Artif.[2] As crystals by placing a layer of absolute alcohol over a saturated solution of NaAl alum at 8°–12°; also from a supersaturated solution of NaAl alum over 20°.

POTASH ALUM

Ref.
1. Soret, *Arch. sc. phys. nat. Genève,* **11,** 63 (1884), on artificial crystals with $\sim$11.5H$_2$O. Surganov, *Ac. sc. St. Pétersbourg, Bull.,* **3,** 1057 (1909), gave comparable measurements on artificial crystals with $\sim$11H$_2$O.
2. Soret (1884) and Surganov (1909).
3. Soret (1884).
4. Surganov (1909).
5. Larsen (108, 1921). Bandy, *Am. Min.,* **23,** 722 (1938), gives nY 1.466, nZ 1.472, $X = b$, $Z \wedge$ twin plane 40° for twinned mendozite (?) from Chuquicamata, Chile. Keller, *Am. Min.,* **20,** 537 (1935), gives nX 1.436, nY 1.455, nZ 1.459, $2V$ 40°–50°, $r > v$, $X = b$, for artificial mendozite (?).
6. Thomson, *Ann. Lyceum Nat. Hist.,* New York, **3,** 19 (1828).
7. Mori analysis in Divers, *Chem. News,* **44,** 218 (1881).
8. Bandy (1938).
9. Keller (1935).

29.5.4.2 Kalinite [KAl(SO$_4$)$_2\cdot$11H$_2$O (?)]. Potash Alum, Native Alum, Kalialaun, Kalinischer Alumsulphat, pt.? *older authors.* Alumen pt.? *Weisbach* (9, 1875). Kalinite pt.? *Dana* (652, 1868). Potassalumite *Winchell* (259, 1931).

The name kalinite is here applied to a fibrous and birefringent type of KAl alum distinct from isometric potash alum. Probably monoclinic and related to mendozite. Not known artificially. Unanalyzed materials from San Bernardino County, California, and Mount Wingen, Australia, appear from optical evidence [1] to belong here, and perhaps other occurrences are included among material hitherto ascribed without crystallographic or optical evidence to potash alum. A biaxial KAl alum occurs with jarosite at Quetena, Chile.[2]

	nX	nY	nZ	$2V$	$Y \wedge c$	Sign
Australia	1.430	1.452	1.458	52° $\pm$ 1°	13°	–
California	1.429		1.456	$\sim$0°	$\sim$0°	–

Ref.
1. Larsen (94, 1921). The seemingly uniaxial material from California and similar uniaxial types of NaAl alum (see Larsen, p. 108) may perhaps represent types of potash alum and soda alum with strong anomalous birefringence or be distinct species. See also Winchell (259, 1931).
2. Bandy, *Am. Min.,* **23,** 721 (1938).

29.5.5 ALUM GROUP

ISOMETRIC; DIPLOIDAL—$2/m\ \bar{3}$

	a_0
Potash alum, KAl(SO$_4$)$_2\cdot$12H$_2$O	12.133 kX
Ammonia alum, (NH$_4$)Al(SO$_4$)$_2\cdot$12H$_2$O	12.215
Soda alum, NaAl(SO$_4$)$_2\cdot$12H$_2$O	12.19

The alums comprise a large and important group of compounds conforming to the general formula $A'B'''(SO_4)_2\cdot12H_2O$, where A' is Na, K, Rb, Cs, (NH$_4$), etc., and B''' is Al, Fe, Cr, Mn, Ga, etc. Of these, only the KAl, NH$_4$Al, and NaAl alums are known to occur in nature. The first two mentioned, potash alum and ammonia alum, have structures [1] of the α-type and form a complete solid-solution series in artificial preparations at least. Soda alum has a structure of the γ-type and does not accept significant amounts of K or (NH_4) in substitution for Na. Alums of the β-type, such as CsAl alum, in which the A ion is of large size, do not occur in nature. All the alums are highly soluble and occur as surface evaporation products in arid regions or in sheltered places.

Ref.
1. Lipson and Beevers, *Proc. Roy. Soc. London*, **148A**, 664 (1935); Lipson, *Proc. Roy. Soc. London*, **151A**, 347 (1935).

29.5.5.1 **P O T A S H   A L U M** [$KAl(SO_4)_2 \cdot 12H_2O$]. Native Alum pt. Kalialaun, Kalinischer Alumsulphat pt. *Germ.* Alumen pt. *Weisbach* (**9**, 1875). Kalinite pt. *Dana* (652, 1868).

C r y s t. Isometric; diploidal—$2/m\ \overline{3}$.

Forms:

| a 001 | d 011 | o 111 | e 012 | n 112 | p 122 |

Structure cell.[1] Space group $Pa3$. a_0 12.133 ± 0.002 kX. Cell contents $K_4Al_4(SO_4)_8 \cdot 48H_2O$.

Habit. Octahedral from pure water solution, cubical from alkaline solutions; the habit varies accompanying the adsorption of dyes or other substances present in the crystallizing solution.[2] Natural potash alum is usually massive with a columnar or granular structure; also stalactitic, and as mealy coatings.

Twinning. On {111}, very rare.

Etch figures.[7] Etching effects conform to diploidal symmetry.

P h y s. Fracture conchoidal. Traces of cleavage on {111}. H. 2–$2\tfrac{1}{2}$.[3] G. 1.757; 1.753 (calc.). Luster vitreous. Colorless and transparent; white. Taste sweetish and astringent.

O p t. In transmitted light, colorless. Isotropic.

$$nD^5\ 1.4562 \pm 0.0001$$

λ	B	C	D	E	F	G
n^4	1.4530	1.4540	1.4564	1.4593	1.4618	1.4661

Potash alum together with ammonia alum and other alums frequently shows weak anomalous birefringence.[6] The crystals usually appear as pyramidal composites of uniaxial individuals arranged with their optic axes perpendicular to the external crystal faces. The birefringence appears to be due to internal strain arising from compositional variation; it also may be produced by mechanical deformation.

C h e m. A hydrated sulfate of potassium and aluminum, $KAl(SO_4)_2 \cdot 12H_2O$. (NH_4) can substitute for K (analysis 2), and a complete series to ammonia alum has been obtained in artificial material. Na does not substitute for K in artificial preparations (but see analysis 2).

Anal.

	K_2O	Na_2O	$(NH_4)_2O$	Al_2O_3	Fe_2O_3	SO_3	H_2O	Rem.	Total	G.
1.	9.93			10.75		33.75	45.57		100.00	1.753
2.	5.75	1.35	2.42	10.40	0.80	34.00	45.37	0.77	100.86	

1. $KAl(SO_4)_2 \cdot 12H_2O$. 2. Vesuvius.[8] Rem. is Cr_2O_3 0.16, Mn_2O_3 0.01, CaO 0.36, Cl 0.24; Cs, Rb. tr.

POTASH ALUM

Tests. Soluble in water (11.4 g. in water at 20°). Melts in its water of crystallization at ~91°. On rapid heating to below red heat intumesces and forms a bulky, porous mass of the anhydrous sulfate ("burnt alum").[10]

O c c u r. Potash alum, together with the other natural alums, commonly occurs as an efflorescence or crevice filling in argillaceous rocks and brown coals containing disseminated pyrite or marcasite. The mineral is derived by the action of sulfuric acid formed during the weathering of the sulfides on alkali-rich aluminous silicates. Similarly by the action of sulfurous vapors or solutions on feldspathic or leucite-rich rocks in fumaroles or solfataras. Associated minerals include alunogen, pickeringite, epsomite, melanterite, gypsum, sulfur. Many localities are known including several that have been of commercial importance. Found in Italy as products of volcanic activity on Vesuvius, in the Grotte Faraglione on Vulcano, and in the solfatara of Miseno. At numerous brown-coal deposits in Germany, as at Duttweiler near Saarbrücken and Arzberg in the Fichtelgebirge. In France at Decazeville and elsewhere in brown-coal deposits in the Plateau Central. Also at numerous places in Galicia and Aragon in Spain. In England at Whitby, Yorkshire, and at Hurlet and Campsie near Glasgow, Scotland. In Chile at Chuquicamata. Found in the United States in Alum Cave, Sevier County, Tennessee. At numerous places in California, including the Sulfur Bank cinnabar mine, Lake County, The Geysers in Sonoma County, and the Nortonville coal mine in Contra Costa County. Near Silver Peak in Esmeralda County, Nevada.

A r t i f.[9] As crystals, which can be grown to very large sizes, from a water solution of the component salts.

H i s t o r y. The *alumen* of Pliny and later ages included a number of naturally occurring efflorescent sulfates. KAl alum—the "alum" of commerce and popular usage—was not clearly recognized as a separate entity until the analysis of Vauquelin in 1797. The massive, fibrous forms of the mineral continued to be confused with other minerals of similar appearance or constitution in the various descriptive mineralogies of the early nineteenth century. Alum was an important article of trade in the fifteenth century and earlier as a mordant in dyeing. The substance was then largely obtained by the calcination and lixivation of alunite—alumstone—obtained from various places in the Near East under Turkish control and after about 1460 from the alunite deposit at Tolfa in Italy. Alum was manufactured in competition with Roman alum in the seventeenth century and later in Yorkshire, England, by the roasting and lixivation of pyritic, aluminous schists. This method was supplanted by the extraction of leucite rocks or calcined aluminous shales with sulfuric acid. Ammonia alum, using by-product ammonia from the coal-gas industry, has been used in place of the potassium salt.

Ref.

1. Lipson and Beevers, *Proc. Roy. Soc. London*, **148A**, 664 (1935) on artificial crystals.
2. For literature see Groth (**2**, 565, 1908) and Frondel, *Am. Min.*, **25**, 91 (1940).

On the growth rate of various faces see Spangenberg, *Zs. Kr.*, **61**, 189 (1925); *Jb. Min., Beil.-Bd.*, **57**, 1197 (1928), and Valeton, *Zs. Kr.*, **56**, 434 (1921).
 3. On the variation of hardness with direction see Pfaff, *Sitzber. bayer. Ak. Wiss.*, 255 (1884).
 4. Indices of Soret, *Arch. sc. phys. nat. Genève*, **12**, 376 (1884) on artificial crystals.
 5. Wendekamm, *Zs. Kr.*, **85**, 169 (1933).
 6. For description and interpretation see Brauns, *Die opt. Anom. der Kristalle*, Leipzig, 1891, and *Jb. Min.*, II, 102 (1883), I, 96 (1885); Klocke, *Jb. Min.*, I, 56 (1880), II, 267 (1881); Beckenkamp, *Zs. Kr.*, **51**, 492 (1913); Pockels, *Jb. Min., Beil..Bd.*, **8**, 217 (1892).
 7. See Bauhans, *Verh. Nat. Ver. Heidelberg*, **12**, 319 (1913); Klocke, *Zs. Kr.*, **2**, 126 (1878); Wulff, *Zs. Kr.*, **5**, 81 (1881); Friedel, *C.R.*, **179**, 796 (1924).
 8. Alfani, *Per. Min.*, **4**, 395 (1933).
 9. On the system K_2SO_4-$Al_2(SO_4)_3$-H_2O see Britton, *J. Chem. Soc. London*, **121**, 982 (1922).
 10. For dehydration data see Spangenberg and Baldermann-Fiola, *Jb. Min., Monatsh.*, Abt. A, 113 (1949).

29.5.5.2 **S O D A   A L U M** [$NaAl(SO_4)_2 \cdot 12H_2O$]. Soda Alum pt., Natronalaun pt. *older authors.* Mendozite pt. *Dana* (653, 1868). Sodalumite *Winchell* (259, 1931).

C r y s t. Isometric; diploidal—$2/m\,\overline{3}$.

Forms:
(artificial) o 111 a 001

Structure cell.[1] Space group $Pa3$. a_0 12.19 ± 0.02 kX. Cell contents $Na_4Al_4(SO_4)_8 \cdot 48H_2O$. Soda alum is not isostructural with potash alum.

Habit. Artificial crystals are octahedral.

P h y s. Fracture conchoidal. H. ~3. G. 1.67. Luster vitreous. Colorless and transparent.

O p t.[2] In transmitted light, colorless. Isotropic.

λ	B(686mμ)	C(656mμ)	D(589mμ)	E(527mμ)	F(486mμ)	G(431mμ)
n	1.4356	1.4365	1.4388	1.4418	1.4441	1.4480

C h e m. A hydrated sulfate of sodium and aluminum, $NaAl(SO_4)_2 \cdot 12H_2O$. Analyses of natural material known to belong to this species are lacking. K does not substitute for Na to a significant extent.[4]

Tests. Soluble in water (110 g. of the anhydrous salt in 100 ml. of water at 15°). Fuses in its water of crystallization at about 63°. Loses $6H_2O$ at about 50°, forming tamarugite.

O c c u r. The name soda alum is here applied to the isometric compound $NaAl(SO_4)_2 \cdot 12H_2O$. A number of occurrences of NaAl alum in nature have been reported (see *mendozite*), but none can be referred with certainty to the present species.

A r t i f.[3] As crystals from a water solution of the component salts. Soda alum crystallizes with more difficulty than potash alum.

Ref.
 1. Lipson, *Proc. Roy. Soc. London*, **151A**, 347 (1935).
 2. Soret, *Arch. sc. phys. nat. Genève*, **13**, 9 (1885), on artificial crystals.
 3. Dobbins and Addleston, *J. Phys. Chem.*, **39**, 637 (1935), and Mellor (**5**, 342, 1924).
 4. Krickmeyer, *Zs. phys. Chem.*, **21**, 78 (1896).

Ammonia Alum

29.5.5.3 AMMONIA ALUM [(NH$_4$)Al(SO$_4$)$_2$·12H$_2$O]. Ammonalun *Beudant* (2, 497, 1832). Tschermigit *von Kobell* (44, 1853). Ammoniakalaun *Germ.* Čermíkite *Czech.*

Cryst. Isometric; diploidal—$2/m\,\bar{3}$.

Forms:
  a 001   d 011   o 111

Structure cell.[1] Space group $Pa3$. a_0 12.215 kX ± 0.002. Cell contents (NH$_4$)$_4$Al$_4$(SO$_4$)$_8$·48H$_2$O. Isostructural with potash alum.

Habit. Artificial crystals are octahedral from pure water solution. Also as fibrous or columnar masses or as efflorescences.

Phys. Fracture conchoidal. H. 1½.[2] G. 1.645; 1.641 (calc.). Luster of crystals vitreous, of fibrous masses silky. Colorless and transparent to white. Taste sweetish and astringent.

Opt.[9] In transmitted light, colorless. Isotropic, but may show anomalous birefringence due to strain arising in mechanical deformation or compositional variation (see potash alum).

λ	B(686mμ)	C(656mμ)	D(589mμ)
n,(NH$_4$)Al(SO$_4$)$_2$·12H$_2$O	1.4560	1.4569	1.4594
n,(NH$_4$,K)Al(SO$_4$)$_2$·12H$_2$O with NH$_4$:K = 635:365	1.4553	1.4563	1.4586

λ	E(527mμ)	F(486mμ)	G(431mμ)
n,(NH$_4$)Al(SO$_4$)$_2$·12H$_2$O	1.4623	1.4648	1.4692
n,(NH$_4$,K)Al(SO$_4$)$_2$·12H$_2$O with NH$_4$:K = 635:365	1.4617	1.4642	1.4685

Chem. A hydrated sulfate of ammonium and aluminum, (NH$_4$)Al(SO$_4$)$_2$.12H$_2$O. Natural material conforms closely to the formula. K can substitute for (NH$_4$), and a complete series to potash alum has been obtained in artificial preparations.[3]

Anal.[4]

	1	2	3	4
(NH$_4$)$_2$O	5.75	4.78	4.46	5.23
Al$_2$O$_3$	11.24	11.86	11.59	11.57
SO$_3$	35.32	35.78	35.61	35.11
H$_2$O	47.69	46.84	48.11	47.82
Rem		1.19	tr.	0.40
Total	100.00	100.45	99.77	100.13
G.				1.645

1. (NH$_4$)Al(SO$_4$)$_2$·12H$_2$O. 2. Utah.[5] Rem. is Na$_2$O 0.48, K$_2$O 0.25, CaO 0.10, MgO 0.28, Cl 0.04, Fe$_2$O$_3$ tr., insol. 0.04. 3. Tokod, Hungary.[6] Rem. is CaO, MgO, Fe$_2$O$_3$. (NH$_4$)$_2$O is NH$_3$. Total in original given as 99.97. 4. Wamsutter, Wyoming.[7] Rem. is Na$_2$O 0.21, MgO 0.13, insol. 0 06; K$_2$O, CaO, Cl, Fe$_2$O$_3$ tr.

Tests. Soluble in water (19.2 g. in 100 ml. water at 25°). Melts at ∼93° in its water of crystallization.

Occur. Found chiefly in lignite or brown-coal beds and bituminous shales. Also as a deposit in fumaroles and in burning coal seams or the waste-heaps of coal mines. From Bohemia of Czechoslovakia in brown coal at Tschermig east of Kaaden, and at Dux and in the Niedergeorgental at Brüx. In Eocene lignite at Tokod, Com. Eszteryom, Hungary. In the solfatara of Pozzuoli at Naples, Italy, and in the crater of Etna. With sulfur and mascagnite on Nyamlagira volcano, Belgian Congo. In the United States with gypsum and ammoniojarosite in bituminous shale near Wamsutter, Wyoming, and with epsomite and ammoniojarosite in lignitic shale in southern Utah. At The Geysers, Sonoma County, California, with boussingaultite, mascagnite, and voltaite, and at Sulfur Bank, Lake County. In the Tucumcari region, Quay County, New Mexico. With mascagnite, salammoniac, alunogen, pickeringite, letovicite, and sulfur on burning waste-heaps in the coal basin of Kladno, Bohemia.[8]

A r t i f. As crystals from a water solution of the component salts.

Ref.

1. Lipson and Beevers, *Proc. Roy. Soc. London*, **148A**, 664 (1935).
2. On variation in hardness with direction, see Pfaff, *Sitzber. bayer. Ak. Wiss.*, 255 (1884).
3. Krickmeyer, *Zs. phys. Chem.*, **21**, 79 (1896), and Klug and Alexander, *J. Am. Chem. Soc.*, **62**, 1492 (1940).
4. For additional analyses see Hintze (1 [3B], 4498, 1930).
5. Shannon, *Proc. U. S. Nat. Mus.*, **74**, art. 13 (1929).
6. Emszt analysis in Liffa and Emszt, *Földt. Közl.*, **51–52**, 45, 105 (1923).
7. Erickson, *Washington Ac. Sc., J.*, **12**, 49 (1922).
8. Rost, *Ac. Sc. Bohême, Bull.*, 1937.
9. Indices of Soret, *Arch. sc. phys. nat. Genève*, **12**, 573, 569 (1884). See also Wendekamm, *Zs. Kr.*, **85**, 169 (1933).

TYPE 6. $A(XO_4) \cdot xH_2O$

29.6.1 **Bassanite** [$2CaSO_4 \cdot H_2O$ (?)]. *Zambonini* (*Min. Vesuviana*, 327, 1910).

Color white. As pseudomorphs after gypsum, composed of microscopic needles in parallel arrangement. G. 2.69-2.76. The needles show parallel extinction with positive elongation and weak birefringence. Analysis gave: CaO 40.65, SO_3 58.50, H_2O 0.60, total 99.75. On standing in moist air the water content increases to 2.80 per cent. Converted to anhydrite when heated to redness. Found in cavities of leucite-tephrite blocks thrown out during the April, 1906, eruption of Vesuvius. Also found with gibbsite in fumaroles of the eruption of 1911. Named after Professor Francesco Bassani of the University of Naples.

The properties of bassanite suggest that it is identical with $2CaSO_4 \cdot H_2O$ (soluble-anhydrite; see under gypsum). The white, chalky coating on some natural gypsum crystals may be the same substance. Artificial[1] $2CaSO_4 \cdot H_2O$ is hexagonal-R, with a_0 6.82 kX, c_0 6.24; nO 1.558, nE 1.586; G. 2.7. The water is zeolitic and is held in channels in the crystal structure; it is nearly or entirely lost at about 130°.

Ref.

1. Caspari, *Proc. Roy. Soc. London*, **155A**, 41 (1936), and Gaubert, *C.R.*, **197**, 72 (1933). See also Berg and Sveshnikova, *Bull. ac. sc. U.R.S.S., Cl. sc. chim.*, no. 1, 19 (1946).

29.6.2 KIESERITE GROUP

MONOCLINIC; PRISMATIC—$2/m$

	$a:b:c$	β
Kieserite, $MgSO_4 \cdot H_2O$	0.8960:1:0.9779	$116°5\frac{1}{2}'$
Szomolnokite, $FeSO_4 \cdot H_2O$	0.9344:1:1.0078	116 14
Szmikite, $MnSO_4 \cdot H_2O$		

X-ray powder diffraction study [1] and other evidence indicate that these species are isostructural. All are water-soluble; kieserite occurs abundantly in oceanic salt deposits, whereas szomolnokite and szmikite are rare species formed as efflorescences during the oxidation of sulfides.

Ref.

1. Hammel, *Ann. chim.*, **11**, 247 (1939).

29.6.2.1 K I E S E R I T E [$MgSO_4 \cdot H_2O$]. Reichardt (*Nova Acta Leopold. Ac.*, **27**, 634, 1860). Martinsite *Kenngott* (23, 1859) [not the martinsite of *Karsten* (*Ak. Berlin, Ber.*, 245, 1845)].

C r y s t.[1] Monoclinic; prismatic—$2/m$.

$a:b:c = 0.8960:1:0.9779;$ $\beta\ 116°05\frac{1}{2}';$ $p_0:q_0:r_0 = 1.0914:0.8782:1$

$r_2:p_2:q_2 = 1.1386:1.2427:1;$ $\mu\ 63°54\frac{1}{2}';$ $p_0'\ 1.2153,\ q_0'\ 0.9779,\ x_0'\ 0.4897$

Forms:

	ϕ	ρ	ϕ_2	$\rho_2 = B$	C	A
m 110	$51°10\frac{1}{2}'$	$90°00'$	$0°00'$	$51°10\frac{1}{2}'$	$69°57\frac{1}{2}'$	$38°49\frac{1}{2}'$
u 011	26 36	47 $33\frac{1}{2}$	63 $54\frac{1}{2}$	48 $42\frac{1}{2}$	41 $17\frac{1}{2}$	70 42
v 111	60 10	63 02	30 $23\frac{1}{2}$	63 $40\frac{1}{2}$	41 $38\frac{1}{2}$	39 22
x $\bar{1}12$	$-13\ 33\frac{1}{2}$	26 42	96 $43\frac{1}{2}$	64 06	40 53	96 03
p $\bar{1}11$	$-36\ 34\frac{1}{2}$	50 $36\frac{1}{2}$	125 58	51 38	68 $26\frac{1}{2}$	117 25

Less common:
c 001 w 021 h 447 Y 331 s $\overline{14}.14.3$ b 010 t $\bar{1}01$ z 221 y $\bar{3}34$

Structure cell.[2] Space group $C2/n$. a_0 6.89 kX, b_0 7.69, c_0 7.52; $\beta\ 116°05\frac{1}{2}';$ $a_0:b_0:c_0 = 0.8960:1:0.9779$. Cell contents $Mg_4(SO_4)_4 \cdot 4H_2O$.

Habit. Mostly massive, coarse to fine-granular, and intergrown with other salts. The rare crystals are mostly bipyramidal $\{\bar{1}11\}$, $\{110\}$, and sometimes are highly modified.

Twinning.[3] Contact twins on $\{001\}$ rare. Polysynthetic twinning, with twin axis [110] (?), is commonly seen in grains under the microscope.

P h y s. Cleavage $\{110\}$ and $\{111\}$ perfect, $\{\bar{1}11\}$, $\{\bar{1}01\}$, and $\{011\}$ imperfect. Friable to firm. H. $3\frac{1}{2}$. G. 2.571 (meas.).[4] 2.571 (calc.). Luster vitreous. Colorless, grayish white or yellowish. Translucent.

O p t.[5] In transmitted light, colorless.

Orientation	n	
X	1.520	Biaxial positive (+).
Y b	1.533	$2V$ (Na) 55° (meas.).
$Z \wedge c\ -76\frac{1}{2}°$	1.584	$r > v$, moderate.

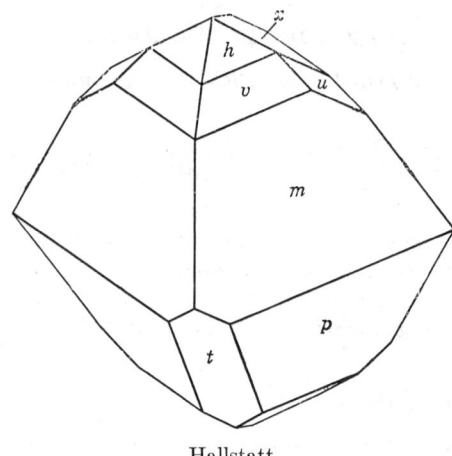

Hallstatt.

Chem. Magnesium sulfate monohydrate, $MgSO_4 \cdot H_2O$. Analyses [6] correspond closely to the composition required by the formula. MgO 29.13, SO_3 57.85, H_2O 13.02, total 100.00. The water is lost at about 350°.[7]

Tests. Dissolves in water very slowly. B.B. easily fusible.

Occur. Kieserite is known chiefly in salt deposits of oceanic origin in which it normally occurs in the upper portions interbedded with halite or associated with carnallite and other potash salts. First recognized at Stassfurt, where it was found in halite, and since found at many points in the north German potash deposits. Found in good crystals at the Hildesia mine, Hildesheim and at Wathlingen, Hannover, and with boracite, sulfoborite, and celestite at Westerregeln. Good crystals also at Hallstatt, Upper Austria, in halite. At Kalusz, Galicia, associated with carnallite. With halite at Monte Sambuco, province of Caltanisetta, Sicily. With halite in places making up 99 per cent of salt deposits penetrated by borings at Ozinky, southwest of Saratov, Russia. With halite and sylvite in the Mayo mine, Punjab Salt Range, India. Found with gypsum in 1919 on leached lavas on the floor of Kilauea crater, Hawaii. In the United States found at many points in the Permian salt deposits of west Texas and New Mexico, mostly with polyhalite, anhydrite, and halite. Found in small amount in salt deposits penetrated by deep borings in Grand County, Utah.

Alter. Alterations of kieserite to leonite and, where in anhydrite, to polyhalite have been noted;[8] also found as pseudomorphs after sylvite and langbeinite. In moist air kieserite becomes coated over with epsomite.

Artif. Formed by crystallization of an aqueous solution above 69° at which temperature the mono- and hexahydrate are in equilibrium with a saturated solution containing 37.1 per cent $MgSO_4$.[9] From solutions containing the ions K, Na, Ca, Mg, SO_4, and Cl, approximating the brines formed by evaporation of sea water, kieserite may form at temperatures as low as $17\frac{1}{2}°C$.[10]

N a m e. In honor of D. G. Kieser, President of the Jena Academy.

Ref.
1. Orientation and elements from structure cell. Crystals first measured by Tschermak, *Ak. Wien, Ber.*, [1], **63**, 314 (1871), whose setting was earlier followed. Transformation: Tschermak to structure cell, $100/0\overline{1}0/\frac{1}{2}0\frac{1}{2}$. Forms from Goldschmidt (5, 27, 1918) and Grandinger, *Cbl. Min.*, 49 (1917).
2. Weinert, *Jb. Min., Beil.-Bd.*, **A75**, 297 (1939) on crystals from Wathlingen, Hannover.
3. Bücking, *Ak. Berlin, Ber.*, **28**, 533 (1899); Tschermak (1871); Grandinger (1917); Schaller and Henderson, *U. S. Geol. Sur., Bull. 833*, 40 (1933), describe the polysynthetic twinning; Grandinger gives the composition plane as approximately ($\overline{4}43$).
4. Average of measurements by Tschermak (1871) and Görgey, *Min. Mitt.*, **29**, 205 (1910), on material from Hallstatt.
5. Indices, average of values of Schaller and Henderson (1933), Görgey (1910), and Lück, *Kali*, **4**, 540 (1910), *Jb. Min.*, **2**, 172 (1911).
6. For summary see Hintze (1 [3], 4356, 1929).
7. Lepeshkov and Bodaleva, *Ac. sc., Leningrad*, **27**, 978 (1940), and Lück (1910).
8. Schaller and Henderson (1933).
9. Seidell, *Solubility of Inorg. Comp.*, **1**, 985 (1940).
10. van't Hoff and co-workers summarized in Doelter (4 [2], 93, 1926).

29.6.2.2 **S Z O M O L N O K I T E** [$FeSO_4 \cdot H_2O$]. *Krenner (Ak. Magyar, Értes.,* **2**, 96, 1891—*Cbl. Min.*, 268, 1928). Ferropallidite *Scharizer (Zs. Kr.,* **37**, 547, 1903).

C r y s t.[1] Monoclinic; prismatic—$2/m$.

$a:b:c = 0.9344:1:1.0078;$ $\beta\ 116°14';$ $p_0:q_0:r_0 = 1.0786:0.9040:1$

$r_2:p_2:q_2 = 1.1062:1.1931:1;$ $\mu\ 63°46';$ $p_0'\ 1.2024, q_0'\ 1.0078, x_0'\ 0.4928$

Forms:

	ϕ	ρ	ϕ_2	$\rho_2 = B$	C	A
m 110	50°02′	90°00′	0°00′	50°02′	70°12′	39°58′
u 011	26 03½	48 17	63 46	47 53	42 07	70 51½
v 111	59 16	63 06½	30 32	62 53	41 53	39 56½
z 221	55 49½	74 11	19 02½	56 40½	53 35	37 50
$p\ \overline{1}11$	−35 09	50 57	125 21½	50 35	68 26½	116 33½
$d\ \overline{2}21$	−62 12½	65 10½	152 23½	64 58	88 45	143 24½

Less common:
b 010 h 043 g 441 k 133 $f\ \overline{1}31$

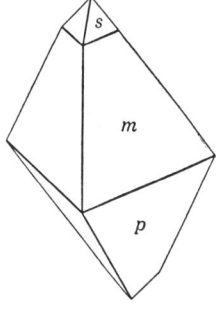
Szomolnok.

Habit. Bipyramidal $\{\overline{1}11\}$ and $\{110\}$; commonly distorted, tabular parallel to $\{\overline{1}11\}$; parallel growth common. Also globular to stalactitic. Transparent when unoxidized and light in color.

Twinning. Common (Chile);[2] twin law undetermined.

P h y s. Fracture conchoidal to uneven. Brittle. H. 2½. G. 3.03–3.07. Luster vitreous. Color sulfur-yellow, yellowish or reddish brown; also blue or colorless.

O p t.[3]

Orientation		n	
$X \wedge c$	$-26°$	1.591	Biaxial positive (+).
Y	b	1.623	$2V\ 80°$.
Z		1.663	$r > v$, strong.

C h e m. Ferrous sulfate monohydrate, $FeSO_4 \cdot H_2O$.
Anal.

	FeO	SO$_3$	H$_2$O	Rem.	Total	G.
1.	42.28	47.12	10.60		100.00	
2.	39.42	47.96	10.36	3.33	101.07	3.035
3.	40.94	46.66	10.33	1.79	99.72	
4.	38.94	45.77	10.45	4.99	100.15	

1. $FeSO_4 \cdot H_2O$. 2. Szomolnok, Slovakia.[4] Rem. is $(Na,K)_2O$ 0.31, CaO 0.07, CuO 1.20, ZnO 0.14, MgO 0.25, Fe_2O_3 1.36. 3. Chile [5] (*ferropallidite*). Rem. is Fe_2O_3 0.92, insol. 0.87. 4. Tintic Standard mine, Dividend, Utah.[6] Rem. is PbO 0.09, CuO 0.20, ZnO 1.18, Fe_2O_3 0.59, P_2O_5 0.27, insol. 2.66.

Tests. Dissolves in water very slowly, yielding a brown solution. In C.T. gives off acid water and leaves a black residue.

O c c u r. Commonly associated with pyrite and secondary sulfates. Deposited from solution of high acid content and high concentration. First recognized at Szomolnok (Schmölnitz, Smolnik), Slovakia, where it was found with rhomboclase. With fibroferrite at Valachov hill near Skrivan, western Bohemia. Rare at Quetena, Chile, with halotrichite. At Alcaparrosa, Chile, with roemerite and rhomboclase. With copiapite and other iron sulfates at the Santa Elena mine, La Alcaparrosa, prov. San Juan, Argentina. In the United States at the Tintic Standard mine, Dividend, Utah.

A l t e r. Alters to roemerite or rhomboclase.

A r t i f. May be made by dehydrating the heptahydrate at 100° C.[7] Not formed from an aqueous solution of ferrous sulfate but only from solutions containing a large concentration of sulfuric acid. At 25° the hepta- and monohydrate are in equilibrium with a solution containing 27.78 per cent H_2SO_4 and 10.70 per cent $FeSO_4$ by weight.[8]

Ref.

1. Elements and form list from Bandy, *Am. Min.*, **23**, 669 (1938). Orientation to conform to new orientation of kieserite with which szomolnokite is isostructural. Transformation Bandy to new: $100/0\bar{1}0/\tfrac{1}{2}0\tfrac{1}{2}$. For powder diffraction data and cell constants based on assumed orthogonal axes for artificial $FeSO_4 \cdot H_2O$ see Hammel, *Ann. chim.*, **11**, 247 (1939).
2. Bandy (1938).
3. Bandy (1938) on crystals from Chile.
4. Krenner (1928).
5. Scharizer (1903).
6. Schaller, *U. S. Geol. Sur., Bull. 878*, 125 (1937).
7. Florentin, *Soc. chim., Bull.*, [4], **13**, 362 (1913).
8. Cameron, *J. Phys. Chem.*, **34**, 692 (1930), which see for equilibrium data on the systems $FeSO_4$-H_2O and $FeSO_4$-H_2SO_4-H_2O superseding results of certain earlier investigations.

29.6.2.3 S Z M I K I T E [MnSO$_4 \cdot$H$_2$O]. *Schroeckinger (Geol. Reichsanst., Verhl., 115, 1877).*

Probably monoclinic.[1] Stalactitic masses with a botryoidal surface. Fracture splintery to earthy. H. 1$\frac{1}{2}$. G. 3.15. Color dirty white to reddish, rose-red.

O p t.[2]

Orientation		n	
X		1.562 ± 0.003	Biaxial positive (+).
Y		1.595 ± 0.003	$2V$ near $90°$.
Z	b	1.632 ± 0.003	

C h e m. Manganous sulfate monohydrate, MnSO$_4 \cdot$H$_2$O.

Anal.

	MnO	SO$_3$	H$_2$O	Total	G.
1.	41.97	47.37	10.66	100.00	
2.	41.70	47.27	11.05	100.02	3.15

1. MnSO$_4 \cdot$H$_2$O.   2. Felsöbánya, Roumania.[3]

O c c u r. At Felsöbánya, Roumania, as an efflorescence.

A r t i f. Can be formed by dehydration of any of the higher hydrates at 100° C. or by hydration of the anhydrous salt by atmospheric moisture at ordinary temperature and humidity.[4] The monohydrate of manganous sulfate is the hydrate in equilibrium with saturated aqueous solution above 24$\frac{1}{2}$°C. At this temperature the solution, containing 39.3 per cent MnSO$_4$, is also in equilibrium with MnSO$_4 \cdot$5H$_2$O. The monohydrate shows retrograde solubility, the saturated solution at its boiling point, 100.7° C., containing but 26.1 per cent MnSO$_4$.[5]

N a m e. After Ignaz Szmik, a Hungarian mining official at Felsöbánya.

Ref.
1. The x-ray powder diffraction pattern of artificial MnSO$_4 \cdot$H$_2$O corresponds closely to those of kieserite and szomolnokite—Hammel, *Ann. chim.*, **11**, 247 (1939).
2. Larsen and Glenn, *Am. J. Sc.*, **50**, 233 (1920), on artificial crystals, except orientation which was determined on natural material.
3. Average of two almost identical analyses by Schrauf and by Dietrich in Schroeckinger (1877).
4. Krepelka and Rejha, *Coll. Czech. Chem. Comm.*, **3**, 517 (1931).
5. Krepelka and Rejha, *Coll. Czech. Chem. Comm.*, **5**, 67 (1933).

Gypsum

Gypsum, CaSO$_4 \cdot$2H$_2$O, has a layer structure [1] in which double sheets of (SO$_4$)$^=$ ions coordinated by Ca are interlaminated parallel {010} by sheets of H$_2$O molecules, these representing the plane of perfect cleavage. The direction of minimal thermal expansion and the slow optical vibration direction are very close in position to chains of tightly packed Ca-O ions extending in {010}. Gypsum is isostructural with the members of the Brushite Group, and all are here described in the same crystallographic orientation. The mineral is one of the very few natural products that do not show any significant variation in chemical composition.

Ref.
1. Wooster, *Zs. Kr.*, **94**, 375 (1936).

29.6.3 GYPSUM [CaSO$_4$·2H$_2$O].

Γύψος [mostly burnt gypsum] *Herodotus, Plato, Theophrastus.* Σεληνίτης, Ἀφροσέληνον *Dioscorides* (**5**, 152, 159). Lapis specularis (principal part), Gypsum (burnt gypsum only) *Pliny.* Lapis specularis, Gypsum, σεληνίτης, *Germ.* Gips, Fraueneis, *Ital.* Lumen de Scaiola [Scagliola] *Agricola* (Foss., 251; Interpr., 465, 1546). Glacies Mariæ, Marienglas [selenite], Gips, Gypsum, Alabastrum [fine-grained gypsum], Selenites [crystals] *Wallerius* (50, 1747). Marmor fugax *Linnaeus* (1736). Gypsum, Terra calcarea acido vitrioli saturata, Alabaster, Selenites, *Cronstedt* (18, 1758). Montmartrite *Delamétherie* (**2**, 380, 1812). Gips, Gyps, Fraueneis *Werner.* Geso *Ital.* Yeso *Span.* Sulphate of Lime, Alabaster, Plaster Stone. Gypsite. Chaux sulfatée, Albâtre *Fr.* Satin Spar.

Perhaps in part Ἀλαβαστρίτης *Theophrastus, Pliny.*

Cryst.[1] Monoclinic; prismatic—$2/m$.

$a:b:c = 0.3722:1:0.4124$; $\beta\ 113°50\frac{1}{2}'$; $p_0:q_0:r_0 = 1.1080:0.3772:1$

$r_2:p_2:q_2 = 2.6511:2.9374:1$; $\mu\ 66°09\frac{1}{2}'$; $p_0'\ 1.2114$, $q_0'\ 0.4124$, $x_0'\ 0.4419$

Forms:[2]

	ϕ	ρ	ϕ_2	$\rho_2 = B$	C	A
c 001	90°00'	23°50½'	66°09½'	90°00'	....	66°09½'
b 010	0 00	90 00	....	0 00	90°00'	90 00
a 100	90 00	90 00	0 00	90 00	66 09½	....
r 180	20 09½	90 00	0 00	20 09½	81 59½	69 50½
k 160	26 05	90 00	0 00	26 05	79 45½	63 55
i 150	30 26	90 00	0 00	30 26	78 11	59 34
h 140	36 17½	90 00	0 00	36 17½	76 09½	53 42½
g 130	44 24	90 00	0 00	44 24	73 34½	45 36
M 120	55 45	90 00	0 00	55 45	70 29	34 15
n 011	46 58½	31 09	66 09½	69 20	20 40	67 46½
x 021	28 11	43 06	66 09½	52 58	37 02	71 10½
s 031	19 39½	52 43½	66 09½	41 28	48 32	74 28½
e $\overline{1}$03	90 00	2 11	87 49	90 00	21 39½	87 49
G $\overline{1}$02	−90 00	9 18	99 18	90 00	33 08½	99 18
d $\overline{1}$01	−90 00	37 34½	127 34½	90 00	61 25	127 34½
w $\overline{1}$13	15 29½	8 07	87 49	82 10½	22 58	87 50½
l $\overline{1}$11	−61 48½	41 07½	127 34½	71 54	62 57	125 25½
u $\overline{1}$33	5 16½	22 30	87 49	67 36	30 46	87 59
v $\overline{1}$22	−21 39½	23 55½	99 18	67 51½	39 09	98 36½

Less common:

| η 270 | ψ 340 | α 110 | z 320 | λ $\overline{2}$03 | σ $\overline{1}$34 | γ $\overline{3}$46 | y $\overline{1}$31 |

Structure cell.[3] Space group C_{2h}^6. a_0 5.67 kX, b_0 15.15, c_0 6.28; β 113°50'; $a_0:b_0:c_0 = 0.372:1:0.415$. Cell contents Ca$_4$(SO$_4$)$_4$·8H$_2$O.

Habit. Untwinned crystals very commonly simple in habit: thin to thick tabular {010} with {$\overline{1}$11}, {120} and often also {$\overline{1}$03}; usually slightly elongated [001] or by extension of {$\overline{1}$11}; {120} and {010} often coarsely striated [001], and {010} sometimes showing step-wise depressions or rough. Also prismatic [001] with a nearly equant cross section and either short and stout or elongated into pencil-like or acicular forms; the prism zone occasionally rich in forms and striated or grooved [001]. The

GYPSUM

crystals also lenticular by rounding of {$\bar{1}11$} and {$\bar{1}03$} or with rounded or irregular terminations. Simple crystals frequently have warped or curved surfaces, and long-prismatic crystals occur bent into irregular or hoop-shaped forms or are twisted about [001]. Lenticular crystals, with large {$\bar{1}11$}, {011} and subordinate {010}, {120}, often occur intergrown as rosette-like aggregates. Also granular massive or rock-forming; foliated;

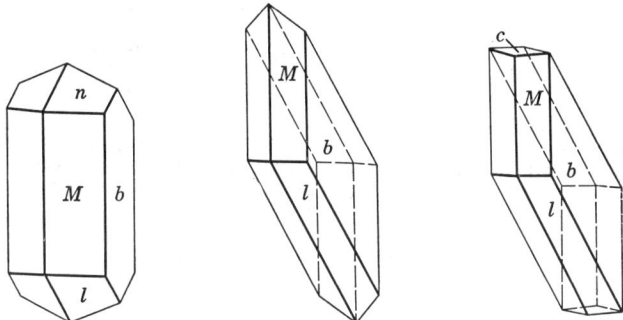

pulverulent; as distorted flower- or cauliflower-like growths on the walls of caves (helictites); [9] as fibrous veinlets or concretions.[8] Gypsum crystals have been found up to five feet in length.

Twinning. (1) Very common on {100} as contact twins (the familiar swallow-tail twins), with a re-entrant angle formed ordinarily by {$\bar{1}11$}; also as cruciform penetration twins. (2) On {$\bar{1}01$} as contact twins, sometimes butterfly- or heart-shaped or elongated along {$\bar{1}11$}; uncommon; the twinned individuals situated on the same or opposite sides of the twin plane. (3) On {$\bar{2}09$}.[4]

Etching. The etch figures,[5] solution bodies,[6] and dehydration figures [7] conform to prismatic $(2/m)$ symmetry.

Phys. Cleavage {010} eminent, easily yielding thin polished foliae; {100} distinct, giving a surface with conchoidal fracture; {011}, giving a fibrous fracture {001}. Cleavage fragments have a rhombic form with plane angles of 66° and 114°. Easily flexible but not elastic. H. 2, varying somewhat with direction.[10] G. 2.317 ± 0.005; 2.30 (calc.). Luster subvitreous; {010} cleavages often pearly. Colorless and transparent; also white, gray, yellowish or brownish when massive. Streak white. Readily undergoes translation gliding [11] with T{010}, t[001], which also may be produced by torsion about [001] or bending {010} about [010]. The crystals sometimes contain clay, sand, or other impurities symmetrically distributed along internal edge or face loci or show differential fluorescence in face loci.[12] Cleavage plates may show asterism when held against a source of light.[13] Not piezoelectric.

Opt. In transmitted light, colorless and transparent. Biaxial positive (+). At room temperature dispersion $r > v$ and strongly inclined; over 91°, $r < v$ and horizontal.

INDICES AT DIFFERENT TEMPERATURES AND WAVELENGTHS [14]

λ	nX		nY		nZ	
	12°	105°	12°	105°	12°	105°
Li(671)	1.5178	1.5154	1.5201	1.5158	1.5270	1.5243
Na(589)	1.5207	1.5184	1.5230	1.5188	1.5299	1.5274
Tl(535)	1.5231	1.5209	1.5255	1.5213	1.5325	1.5300

2E AT DIFFERENT TEMPERATURES AND WAVELENGTHS [14]

	11.5°	48°	75°
Li	99°16′	74°26′	52°14′
Na	100 36	75 40	54 18
Tl	100 34	75 23	54 22

$2V$ for Na at room temperature about 58°.

Optical orientation: $Y = b$ and $Z \wedge c = +51°52′$ (for Na at 18.5°). $2V$ decreases with increasing temperature and the mineral becomes uniaxial at about 91° (for Na), the optic plane opening out perpendicular to {010} at higher temperatures.[15]

Chem. Calcium sulfate dihydrate, $Ca(SO_4) \cdot 2H_2O$. CaO 32.57, SO_3 46.50, H_2O 20.93, total 100.00. Reported analyses are only of technical interest. Trace amounts of Ba and Sr apparently can substitute for Ca.[16]

Dehydration.[17] Four phases have been reported in the system $CaSO_4$-H_2O. These include $CaSO_4 \cdot 2H_2O$ (*gypsum*), $2CaSO_4 \cdot H_2O$, γ-$CaSO_4$ (*"soluble-anhydrite"*), and β-$CaSO_4$ (*anhydrite*). The hemihydrate is rhombohedral in crystallization. Soluble anhydrite has been variously reported to be identical with the hemihydrate and to contain zeolitic water or to be a distinct phase and perhaps itself polymorphous (γ-$CaSO_4$ and an orthorhombic or triclinic phase). The hemihydrate also has been said to be polymorphous. The stability relations between the several phases are not well-defined. In water, gypsum converts to anhydrite above about 42°, and the reverse change takes place at lower temperatures. Gypsum converts to the metastable hemihydrate in water at about 98°, the transition temperature being lowered by the presence of dissolved NaCl or $CaCl_2$. Heated in air, gypsum slowly converts to the hemihydrate at about 70° or less, depending on the moisture content of the air, and rapidly at about 90° or over (the product also being termed by some as soluble-anhydrite). Commercial plaster of Paris consists principally of the hemihydrate and is made by heating gypsum at 190°–200°; it often contains anhydrite or undecomposed gypsum. Heating gypsum at much higher temperatures produces anhydrite (dead-burned plaster).

Tests. B.B. fuses at $2\frac{1}{2}$–3. In C.T. loses water and becomes opaque, white. Slightly soluble in water (0.208 g. $CaSO_4$ in 100 cc. of H_2O solution at 25°). Soluble in HCl.

Var. *Crystallized*, or *Selenite*. In distinct crystals or broad folia, and transparent.

Fibrous, or *Satin spar*. Coherent aggregates with a parallel fibrous structure, and translucent with the pearly opalescence of moonstone.

Massive, or *Alabaster*. Fine-grained, either white or delicately shaded, and when pure and translucent valuable as an ornamental stone. Often dull-colored and impure with clay, iron oxide, calcite, or anhydrite.

Occur. Gypsum is the commonest of the sulfate minerals. It forms extensive bedded sedimentary deposits associated with limestones, red shales and sandstone, marl, and clay, and although not confined to rocks of any particular age occurs especially in Permian and Triassic formations.

GYPSUM 485

In the concentration of oceanic brines in closed basins, gypsum is ordinarily deposited first and is followed, as the salt concentration increases, by the deposition of anhydrite and then of rock salt; this may be followed in an ideal sequence, as at Stassfurt, by the crystallization of polyhalite, kieserite, and then carnallite. The bedded anhydrite of these deposits may be later altered by hydration to gypsum on a geologic scale. Gypsum deposits of considerable size occur in saline lakes and salt pans. Gypsum also is found in deposits of native sulfur, as in Sicily, and is produced around fumaroles and solfataras in volcanic regions by the action of sulfureous vapors and waters on calcic minerals. Also as an efflorescence or incrustation on certain soils or in limestone caves, as concretionary aggregates in recent clays and marls, as the cap rock of salt domes, and in the gossan of mineral deposits where sulfate solutions formed by the oxidation of pyrite have reacted with limestone. A few localities that are of interest because of the quality of their specimens are mentioned beyond.[18]

In Italy in the sulfur mines of Sicily, as at Rocalmuto, Girgenti, and Cianciana, with celestite and aragonite; on Vesuvius; as alabaster at Castellina, Chiusdino, and Volterra in Tuscany. From the salt mines of Bex, Vaud, Switzerland. In France, common in the Tertiary clays and marls of the Paris basin (whence the name plaster of Paris) notably in Montmartre, and in the Triassic of Dép. Landes, Pyrenées. At many localities in the salt and potash deposits of northern Germany; also at Eisleben and near-by localities in Saxony, and in Württemberg. In Austria at Ischl in Salzburg, Aussee in Styria and Hall near Innsbruck in Tyrol. At the salt deposits of Wieliczka and Bochnia and the sulfur deposit of Swoszowice in Krakow, Poland. In England in the clay at Shotover Hill near Oxford; in Derbyshire. As very large crystals at Naica, Chihuahua, Mexico.[19] Gypsum occurs in the United States in commercial deposits in western New York, Michigan, Iowa, Kansas, New Mexico, Colorado, California, Utah, and other states. Fine crystals have been found in the drusy cavities of the Lockport dolomite in Niagara County, New York. Also at Ellsworth, Mahoning County, Ohio; as large crystals from South Wash, Wayne County, Utah; in the Bad Lands of southwestern South Dakota. In Canada, extensive deposits occur in New Brunswick, especially at Hillsborough, Albert County, and at many points in Nova Scotia.

Alter. Anhydrite readily hydrates to form gypsum at ordinary temperatures, and bedded anhydrite deposits may so alter on a geologic scale. Pseudomorphs of other minerals after gypsum crystals are common, especially of chalcedony, opal, or quartz (Paris basin); also of calcite, aragonite, celestite, malachite, anhydrite. Gypsum has been found as pseudomorphs after calcite, anhydrite, and halite and as the petrifying material of fossils.

Artif. Readily obtained in distinct crystals and twins by reaction of soluble calcium salts and sulfates, although not in dilute solutions at high acid concentrations.

Name.[20] Gypsum from γίψος, applied especially to the calcined mineral or *plaster*. Selenite from σελήνη, *moon*, in allusion to the moon-like white reflections of the mineral or to the quality of the light transmitted by semi-pellucid gypsum slabs or cleavages used as windows.[21]

Ref.

1. Setting of Terpstra, *Zs. Kr.*, **97**, 229 (1937), corresponding to the cell of minimum translations in the structural lattice. Transformation: from the B-centered cell of Des Cloizeaux, *Bull. soc. min.*, **9**, 175 (1886), Dana (933, 1892), and Goldschmidt (**4**, 93, 1918) to Terpstra, $\frac{1}{2}0\frac{1}{2}/0\overline{1}0/001$. The same setting is here taken for the isostructural compounds brushite and pharmacolite. Angles of Des Cloizeaux (1886). On the variation in angle with temperature see Beckenkamp, *Zs. Kr.*, **6**, 450 (1882).
2. Dana (1892) and Goldschmidt (1918). 56 rare and uncertain forms are listed by Buttgenbach, *Bull. soc. min.*, **53**, 47 (1930) and Goldschmidt (in the old setting). Also Γ {0$\overline{2}$5} (Tokody, *Ann. Mus. Nat. Hungar.*, *Min. Geol. Pal.*, **32**, 12 (1939)).
3. Cell dimensions of Gossner, *Fortschr. Min.*, **21**, 34 (1937) and *Zs. Kr.*, **96**, 488 (1937), in the new setting. Space group of Wooster, *Zs. Kr.*, **94**, 375 (1936). See also de Jong and Bouman, *Zs. Kr.*, **100**, 275 (1938). On variation in cell dimensions see Nagy, *Zs. Phys.*, **51**, 410 (1928).
4. Parsons, *Univ. Toronto Stud.*, *Geol. Ser.*, no. 32, 25 (1932).
5. Grengg, *Min. Mitt.*, **33**, 201 (1915); Rosický, *Ak. Česká, Roz.*, Cl. 2, **25**, no. 13 (1916); Viola, *Zs. Kr.*, **28**, 573 (1897); Baumhauer, *Ak. München, Sitzber.*, 169 (1875). Natural etch and growth accessories have been often described, cf. Himmel, *Cbl. Min.*, 342 (1927).
6. Gross, *Zs. Kr.*, **57**, 145 (1922).
7. Grengg, *Zs. Kr.*, **55**, 1 (1915); Gaudefroy, *Bull. soc. min.*, **42**, 284 (1919).
8. Matsuura, *Jap. J. Geol. Geog.*, **4**, 65 (1927); Richardson, *Min. Mag.*, **19**, 77 (1920).
9. Huff, *J. Geol.*, **48**, 641 (1940).
10. Auerbach, *Ann. Phys.*, **58**, 357 (1896); Mügge, *Jb. Min.*, I, 50 (1884).
11. Mügge, *Jb. Min.*, II, 14 (1883), I, 90 (1898); also another doubtful translation direction. On pressure and percussion figures on {010} see Reuss, *Ann. Phys.*, **136**, 135 (1869), *Ak. Berlin, Sitzber.*, 259 (1883); Mügge (1884).
12. Josten, *Cbl. Min.*, 432 (1932).
13. Josten (1932) and Rutherford, *Univ. Toronto Stud.*, *Geol. Ser.*, no. 44, **71** (1940).
14. Tutton, *Zs. Kr.*, **46**, 135 (1909); Hutchinson and Tutton, *Zs. Kr.*, **52**, 223 (1913).
15. Hutchinson and Tutton (1913); Kraus and Young, *Cbl. Min.*, 356 (1914); Berek, *Jb. Min.*, Beil.-Bd., **33**, 583 (1912); Berger, et al., *Ak. Wiss. Leipzig, Ber.*, **81**, 171 (1929).
16. Carobbi, *Ann. R. Osservat. Vesuviano*, [3], **2**, 125 (1925); Miropolsky and Borovick, *C. r. ac. sc. U.R.S.S.*, **38**, 33 (1943).
17. Mellor (3, 767, 1923); Ramsdell and Partridge, *Am. Min.*, **14**, 59 (1929); Tourtsev, *Bull. ac. sc. U.R.S.S.*, Ser. Geol., no. 4, 180 (1939); Beljankin and Feodotiev, *Trav. inst. pétrog. ac. sc. U.R.S.S.*, no. 6, 453 (1934); Gaubert, *C. R.*, **197**, 72 (1933); Gallitelli, *Per. Min.*, **4**, 132 (1933); Weiser, et al., *J. Am. Chem. Soc.*, **58**, 1261 (1936); Büssem and Gallitelli, *Zs. Kr.*, **96**, 376 (1937); Hill, *J. Am. Chem. Soc.*, **59**, 2242 (1937); Posnjak, *Am. J. Sc.*, **35**, 247 (1939); Caspari, *Proc. Roy. Soc. London*, **155A**, 41 (1936); Berg and Sveshnikova, *Bull. ac. sc. U.R.S.S.*, Cl. sc. chim., no. 1, 19 (1946); Beljankin and Lapin, *C. r. ac. sc. U.R.S.S.*, **51**, 535 (1946).
18. An extensive list of localities is given in Hintze (**1** [3B], 4274, 1929) and deposits of economic importance are described by Dammer and Tietze, Die nutzbaren mineralien, Stuttgart, 2 ed., 1927.
19. Foshag, *Am. Min.*, **12**, 252 (1927).
20. For a history of the nomenclature see Dana (936, 1892) and Bromehead, *Min. Mag.*, **26**, 325 (1943).
21. Cf. Bromehead (1943).

29.6.4 **Ilesite** [$(Mn,Zn,Fe)SO_4 \cdot 4H_2O$ (?)]. *Wuensch* (*Mining Index*, Leadville, Colo., Nov. 5, 1881); *Iles* (*Am. Chem. J.*, **3**, 420, 1881).

Monoclinic? In loosely adherent crystalline aggregates, prismatic. Color clear green, becoming white on partial dehydration. G. 2.25 (artif. $MnSO_4 \cdot 4H_2O$).[1] Transparent.

CHALCANTHITE 487

O p t.

n (artif. $MnSO_4 \cdot 4H_2O$) [2]

X	1.511 ± 0.003	Biaxial negative $(-)$.
Y	1.519 ± 0.003	$2V$ moderate.
Z	1.521 ± 0.003	

C h e m. Probably manganous sulfate tetrahydrate, $(Mn,Zn,Fe)SO_4 \cdot 4H_2O$. Zn and Fe" substitute for Mn", with $Fe:Zn:Mn = 1:1.3:5.4$ in the only reported analysis. The water content is uncertain. Analysis gave:

	FeO	MnO	ZnO	SO_3	H_2O	Total
1.	4.15	22.11	6.11	35.59	32.04	100.00
2.	4.18	22.31	5.97	36.07	31.60	100.13

1. $(Mn,Zn,Fe)SO_4 \cdot 4H_2O$ with $Fe:Zn:Mn = 1:1.3:5.4$. 2. Hall Valley, Colorado.[3]

Tests. Soluble in water (68.8 g. $MnSO_4$ per 100 g. H_2O at 40°).

O c c u r. Found as an oxidation product in sulfide veins in several mines at the head of Hall Valley, Park County, Colorado. The material may be an alteration product of a higher hydrate.

A r t i f.[4] The tetrahydrate (monoclinic) of manganous sulfate is obtained by crystallization of the water solution at about 30° to 40°; the pentahydrate (triclinic) is formed at about 15° and the heptahydrate (monoclinic) below about 9°.

N a m e. After M. W. Iles of Denver.

Ref.

1. Topsøe, *Bibl. Univ., Arch. sc. phys. nat. Genève*, **45**, 223 (1872).
2. Larsen and Glenn, *Am. J. Sc.*, **50**, 225 (1920).
3. Wuensch (1881); also Iles (1881).
4. Mellor (**12**, 402, 1932); Larsen and Glenn (1920).

29.6.5 CHALCANTHITE GROUP

TRICLINIC; PINACOIDAL—$\bar{1}$

	a_0	b_0	c_0	α	β	γ
Chalcanthite, $CuSO_4 \cdot 5H_2O$	6.110 kX	10.673	5.95	97°35'	107°10'	77°33'
Siderotil, $FeSO_4 \cdot 5H_2O$						
Pentahydrite, $MgSO_4 \cdot 5H_2O$						

Chalcanthite and the not well-established minerals pentahydrite and siderotil are isostructural with a number of artificial salts variously including the pentahydrated sulfates and selenates of Mn, Co, Cu, Zn, and the chromate and molybdate of Mg. In the structure [1] of chalcanthite, the Cu ions are coordinated octahedrally by four H_2O molecules and two O ions belonging to (SO_4) groups; the fifth molecule of water, relatively loosely held, is coordinated only by oxygen ions and other water molecules.

Ref.

1. Beevers and Lipson, *Proc. Roy. Soc. London*, **146A**, 570 (1934).

29.6.5.1 **CHALCANTHITE** [$CuSO_4 \cdot 5H_2O$]. Χαλκανθον, Chalcanthum pt. *Dioscorides, Pliny*. Atramentum sutorium pt. *Pliny*. Atramentum coeruleum *Agricola, Gesner*. Vitriolum cupri, Vitriolum Cypri, Vitriolum veneris *Wallerius, Cronstedt*. Sulphate of Copper, Blue vitriol, Copper vitriol, Cyprian vitriol, Blue Stone. Kupfervitriol *Germ*. Copper sulfate pentahydrate. Couperose bleue, Cuivre sulfaté *Fr*. Vitriolo di Rame, *Ital*. Cyanose *Beudant* (**2**, 486, 1832). Chalkanthit *von Kobell* (31, 1853). Vitriolo azul *Span*.

C r y s t.[1] Triclinic; pinacoidal—$\bar{1}$.

$a:b:c = 0.5715:1:0.5575; \quad \alpha\ 97°44', \beta\ 107°26', \gamma\ 77°20'$

$p_0:q_0:r_0 = 0.9907:0.5452:1; \quad \lambda\ 85°45\tfrac{1}{2}', \mu\ 73°46\tfrac{1}{2}', \nu\ 100°54\tfrac{1}{2}'$

$p_0'\ 1.0413,\ q_0'\ 0.5730,\ x_0'\ 0.3139,\ y_0'\ 0.0778$

Forms:[2]

		φ	ρ	A	B	C
c	001	76°05½′	17°55½′	73°46½′	85°45½′	....
b	010	0 00	90 00	100 54½	......	85°45½′
a	100	100 54½	90 00	......	100 54½	73 46½
π	130	33 54	90 00	67 01	33 54	76 49
λ	120	47 08½	90 00	53 46½	47 08½	74 22½
m	110	69 49	90 00	31 05½	69 49	72 11
ν	210	85 00½	90 00	15 54½	85 00½	72 18
f	3$\bar{1}$0	110 47	90 00	9 52½	110 47	75 20½
n	2$\bar{1}$0	115 19	90 00	14 24	115 19	76 12½
M	1$\bar{1}$0	126 59	90 00	26 04½	126 59	78 48½
l	1$\bar{2}$0	142 42	90 00	41 47	142 42	82 59
N	1$\bar{4}$0	157 41	90 00	56 46½	157 41	87 25
k	011	25 45	35 51	81 22½	58 10	27 35½
τ	021	14 23	51 38	87 16½	40 35	45 10½
q	0$\bar{1}$1	147 38	30 23	69 45½	115 17½	29 32
t	0$\bar{2}$1	163 37½	48 04	70 03½	135 32½	49 47½
g	0$\bar{3}$1	169 10½	59 06	71 28	147 26	61 40½
d	201	97 38	67 12½	23 00½	97 03	51 21
ϕ	$\bar{1}$01	− 68 48	37 14	126 32½	77 21½	52 45½
p	111	71 15	54 40½	44 50½	74 47½	36 50½
x	$\bar{1}$12	−119 10½	12 44	99 42½	96 10	30 23½
ω	$\bar{1}\bar{1}$1	−112 49	37 33	120 27½	103 40	55 19
o	$\bar{1}$11	− 39 53	47 51½	125 04	55 19½	57 25
e	121	52 28	59 19	55 13	58 24½	43 23½
h	$\bar{1}$22	−153 33	23 53½	96 14	111 16	37 53½
Ψ	131	39 56½	64 20½	64 03½	46 17	50 30
S	141	31 36	68 35½	70 48	37 32½	57 01½
r	1$\bar{2}$1	133 26	61 29	42 11½	127 10	53 07
σ	$\bar{1}$21	− 26 30½	57 48	120 56½	40 47	63 14½
ξ	$\bar{1}\bar{2}$1	−140 52½	48 26½	110 43	125 29	63 26½
ζ	$\bar{1}$31	−153 52	58 07½	102 53	139 40½	70 28½

CHALCANTHITE

Structure cell.[3] Space group $P\bar{1}$. a_0 6.110 kX, b_0 10.673, c_0 5.95; α 97°35', β 107°10', γ 77°33'; $a_0:b_0:c_0 = 0.5725:1:0.5575$. Cell contents $Cu_2(SO_4)_2 \cdot 10H_2O$.

Habit. Crystals short prismatic [001], less commonly thick tabular $\{\bar{1}\bar{1}1\}$. Common forms: a π λ m M ω. Also stalactitic or reniform; as cross-fiber veinlets; massive, granular.

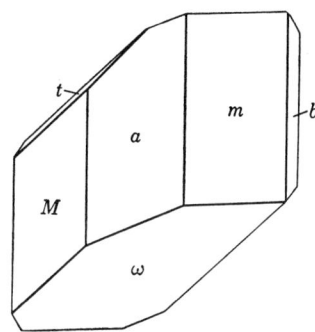 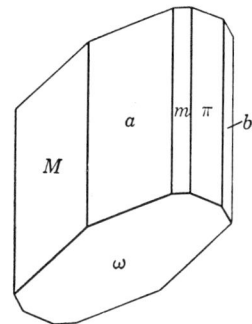

Twinning.[4] Rarely as cruciform intergrowths of two individuals with the edges [001] and [100] on one crystal respectively parallel to [100] and [001] on the other, and the planes $\{010\}$ in common.

Etching. Etch figures and dehydration figures conform to pinacoidal symmetry.[15]

P h y s. Cleavage $\{1\bar{1}0\}$ imperfect, $\{110\}$ in traces. Fracture conchoidal. H. $2\frac{1}{2}$. G. 2.286 (artif.);[5] 2.282 (calc.). Luster vitreous. Color berlin-blue to sky-blue, of different shades; sometimes a' little greenish. Streak colorless. Subtransparent to translucent. Taste metallic and nauseous.

O p t. In transmitted light, colorless to pale blue.

	ORIENTATION[6]		$n(Na)$[7]	
	ϕ	ρ		
X	169°	76°	1.5140	Biaxial negative $(-)$.
Y	76°	78°	1.5368	$2V$ 56°02' (meas.).[7]
Z	$-57°$	18°	1.5434	$r < v$.

C h e m. Copper sulfate pentahydrate, $CuSO_4 \cdot 5H_2O$. In artificial material,[8] Mg substitutes for Cu up to about Mg:Cu = 1:20.9; Mn substitutes up to about Mn:Cu = 1:10.2; Zn substitutes up to about Zn:Cu = 1:11.5; and Fe'' substitutes up to about Fe:Cu = 1:18.5. In natural material, Fe'' is probably present in significant amounts and other divalent metals in lesser amounts, but the extent of solid solution and the variation in optical and other properties therewith are not clearly established.

Anal.[9]

	1	2	3	4
CuO	31.87	31.14	30.5	31.40
FeO			0.5	
SO_3	32.06	32.06	32.0	32.13
H_2O	36.07	35.70	[36.4]	35.90
Rem.		0.81	0.6	0.76
Total	100.00	99.71	[100.0]	100.19

1. $CuSO_4 \cdot 5H_2O$. 2. Leona Heights, California.[10] Rem. is SiO_2. 3. Recsk, Hungary [11] Rem. is CaO. 4. Luishia, Katanga.[12] Rem. is Fe_2O_3.

Tests. On heating loses water and at a higher temperature, SO_3, but does not fuse. Soluble in water (22.7 g. $CuSO_4$ in 100 g. H_2O at 25°).

O c c u r. A secondary mineral found with other hydrated sulfates of copper or iron, less frequently nickel, cobalt, and magnesium, in the oxidized zone of sulfidic copper ores. Commonly associated minerals include melanterite, fibroferrite, rhomboclase, epsomite, goslarite, pickeringite, retgersite, morenosite, brochantite, gypsum. Deposits of commercial size have been found in arid regions. The salt is commonly present dissolved in mine waters, from which it may crystallize as stalactites and crusts on the timbers and walls of the mine workings; the copper may be recovered from such waters by plating-out on metallic iron (cement-copper). Chalcanthite is widespread in nature, and only a selected few localities can be listed here.

In Germany recognized in early times at Rammelsberg and Goslar in the Harz and known to the ancients from deposits on the island of Cyprus. At Herrengrund, Bohemia, and well-crystallized at Zaječar, Serbia. As a fumarole deposit on Vesuvius. From the Ting Tang mine, Gwennap, and other mines in Cornwall. Abundantly as stalactites in mine workings and in solution in mine water at Rio Tinto, Spain, and similarly in mines in County Wicklow, Ireland. Chalcanthite occurs abundantly in copper deposits in the arid Pacific coast regions of South America, and at some places has constituted an important ore mineral. Notable localities here include Chuquicamata, Quetena, and Copaquire in Chile.

In the United States formerly abundant in the Ducktown district, Polk County, Tennessee; also at the Kings Mountain mine, Cleveland County, North Carolina. With morenosite and retgersite at the Gap Nickel mine, Lancaster County, Pennsylvania. Formerly a minor ore in the oxidized zone at Butte, Montana. At the Bluestone mine, Yerington district, Lyon County, Nevada, and formerly shipped in large quantities to the Comstock Lode for use in the reduction of silver ores. In Colorado at numerous localities in Clear Creek County. Widespread in Arizona, notably at the United Verde mine, Yavapai County, and in the Clifton-Morenci district, Greenlee County. In California, especially at the Alma pyrite mine, Leona Heights, Alameda County, and in Trinity County at the Island Mountain mine and in the New River district.

Alter.[13] In a dry atmosphere at ordinary temperature chalcanthite dehydrates surficially to an opaque greenish white powdery aggregate of $CuSO_4 \cdot H_2O$. On heating in the open air the pentahydrate loses $2H_2O$ at 105°, $2H_2O$ more at 117°, and forms the anhydrous salt at 258°. Melts incongruently under pressure at about 100° with the formation of $CuSO_4 \cdot 3H_2O$. Chalcanthite forms by dehydration of boothite, $CuSO_4 \cdot 7H_2O$.

Artif.[14] Easily obtained in large crystals by slow evaporation of a water solution; $CuSO_4 \cdot 3H_2O$ is obtained from solution above approximately 108°.

Name. From *chalcanthum*, its old name, meaning *flowers of copper*.

Ref.

1. Elements of Tutton (301, 1911; **1**, 282, 1922) from his classical study of the artificial salt, transformed to the orientation of Kupffer, *Ann. Phys.*, **8**, 217 (1826), as derived from Kupffer's fundamental angles by Boeris, *Riv. min.*, **33**, 5 (1907). Kupffer's data have been variously recomputed and presented in the literature. Tutton (1911), Barth and Tunell, *Am. Min.*, **18**, 187 (1933), and Beevers and Lipson, *Proc. Roy. Soc. London*, **146A**, 570 (1934), used the orientation of Barker in Groth (**2**, 419, 1908). Dana (944, 1892) used another position. Goldschmidt (**5**, 105, 1918) adopted another lattice. Transformations: Dana to Kupffer, $\overline{1}00/0\overline{1}0/001$; Goldschmidt to Kupffer, $\overline{1}00/011/00\overline{1}$; Barker to Kupffer, $\overline{1}00/010/00\overline{1}$.
2. Goldschmidt (1918); Ungemach, *Bull. soc. min.*, **58**, 214 (1935); and Bandy, *Am. Min.*, **23**, 718 (1938). Uncertain forms: $3\overline{4}0$, $1\overline{4}0$, $1\overline{5}0$.
3. Barth and Tunell (1933) on artificial crystals; their elements, in the orientation of Tutton, have been computed from $\alpha^*, \beta^*, \gamma^*$ and transformed to the new position. See also Gossner and Brückl, *Zs. Kr.*, **69**, 422 (1929); and for the crystal structure, Beevers and Lipson (1934).
4. Boeris (1907; *Att. soc. ital. sc. nat. Milan*, **44**, 73, 1905); see also Haas, *Bull. soc. min.*, **43**, 228 (1920).
5. Retgers, *Zs. phys. Chem.*, **3**, 289 (1889); very closely agreeing values have been reported by others.
6. Optical orientation of Barth and Tunell (1933) referred to the adopted orientation and confirming the orientation of Pape, *Ann. phys.*, **6**, 35 (1874).
7. Lavenir, *Bull. soc. min.*, **14**, 115 (1891), by prism method on artificial crystals.
8. See Retgers, *Zs. phys. Chem.*, **15**, 563 (1894) and **16**, 583 (1895); Stortenbeker, *Zs. phys. Chem.*, **17**, 647 (1895) and **34**, 111 (1900); Collins, *Min. Mag.*, **20**, 32 (1923).
9. For earlier analyses, on material of very doubtful purity, see Doelter (**4** [2], 280, 1927).
10. Schaller, *Univ. Calif., Dept. Geol. Bull.*, **3**, 191 (1903).
11. Vavrinecz, *Magyar Chem. Foly.*, **35**, 4 (1929).
12. Mélon, *Bull. soc. min.*, **51**, 194 (1928).
13. Mellor (**3**, 243, 1923) and Rinne, *Jb. Min.*, I, 7 (1899).
14. Mellor (1923); Etard, *C.R.*, **104**, 1614 (1887); Posnjak and Tunell, *Am. J. Sc.*, **18**, 1 (1929).
15. Walker, *Am. J. Sc.*, **5**, 176 (1898); Gaudefroy, *Bull. soc. min.*, **35**, 129 (1914); Blasius, *Zs. Kr.*, **10**, 238 (1885).

29.6.5.2 **Siderotil** [$FeSO_4 \cdot 5H_2O$ (?)]. Siderotyl Schrauf (*Jb. geol. Reichsanst. Wien*, **41**, 380, 1891).

Found as fibrous crusts and divergent groups of needle-like crystals. White to yellowish; pale greenish white. Crystallographic [4] and physical data are lacking.

Opt. In transmitted light, colorless.

ORIENTATION	n (California) [1]	
X	1.528 ± 0.003	Biaxial negative (−).
Y	1.537 ± 0.003	2V rather large.
Z	1.545 ± 0.003	$r > v$.

Chem. Ferrous sulfate pentahydrate, $FeSO_4 \cdot 5H_2O$ (?). The water content is uncertain, and the natural material may be the tetrahydrate. Analysis gave:

	FeO	SO$_3$	H$_2$O	Total
1.	29.70	33.10	37.20	100.00
2.	30.0	34.3	[34.0]	[98.3]

1. $FeSO_4 \cdot 5H_2O$. 2. Idria. On a few milligram weight sample. Fe_2O_3 31.7 per cent as residue after ignition.

Occur. Originally found with melanterite at Idria, Gorizia, Italy. Also reported as an alteration of melanterite from an unstated locality [1] in California, and from the Mt. Diablo mercury mine, Contra Costa County, California.[2]

Artif.[3] The existence of $FeSO_4 \cdot 5H_2O$ as an artificial compound is doubtful; the tetrahydrate is ordinarily formed by dehydration of the heptahydrate or by crystallization from water at temperatures (56° to 64°) above those affording the heptahydrate.

Name. From σίδηρος, *iron*, and τίλος, *fiber*, in allusion to its composition and structure.

Ref.

1. Larsen (134, 1921). Dehydrated artificial $FeSO_4 \cdot 7H_2O$ is said to have identical optical properties.
2. Ross, *U. S. Geol. Sur., Bull. 922B*, 44 (1940).
3. Mellor (**14**, 249, 1935).
4. On the morphology of the artificial salt see Marignac, *Ann. mines*, **9**, 11 (1856).

29.6.5.3 Pentahydrite [$MgSO_4 \cdot 5H_2O$]. Epsomite *Hobbs* (*Am. Geol.*, **36**, 184, 1905). Double Sulphate of Copper and Magnesium *Keller* (*Proc. Am. Phil. Soc.*, **47**, 81, 1908). Pentahydrite *Frondel* (priv. comm., 1948). Magnesium sulfate pentahydrate.

Minerals probably identical with artificial triclinic $MgSO_4 \cdot 5H_2O$ and isostructural with chalcanthite have been described from several localities. An "epsomite" found with alunogen at Cripple Creek, Colorado, corresponds in composition with the pentahydrate (anal. 2) but is not otherwise described.[1] A light blue mineral found with chalcanthite at Copaquire, Tarapacá Province, Chile,[2] apparently is a cuprian variety with Cu:Mg ∼ 1:1.5 (anal. 3). A granular pseudomorphous dehydration product of epsomite from The Geysers, Sonoma County, California,[3] is essentially pure $MgSO_4 \cdot 5H_2O$ (anal. 4). A mineral found with pickeringite as a deposit on mine timbers in the Comstock Lode, Nevada,[4] is close to (Mg,Cu,Zn,Fe,Mn)$SO_4 \cdot 5H_2O$ with Mg:Cu:Zn:Fe:Mn = 55:26:16:4.5:1 (anal. 5). The latter mineral is massive with a fine-granular or slightly platy texture; color light greenish blue; optically negative with nX 1.495, nY 1.512, nZ 1.518, 2V 55°, dispersion $r < v$.

Anal.

	1	2	3	4	5
MgO	19.15	19.35	11.39	17.91	9.40
CuO			12.43		9.00
ZnO					5.60
FeO			1.01	0.23	1.36
MnO			0.32	0.14	0.30
NiO			0.06	0.11	
SO_3	38.07	38.51	35.70	38.13	35.07
H_2O	42.78	42.03	38.38	42.97	[39.07]
Rem.				0.13	0.20
Total	100.00	99.89	99.29	99.62	[100.00]

1. $MgSO_4 \cdot 5H_2O$. 2. Cripple Creek, Colorado.[1] 3. Copaquire, Chile.[2] 4. Sonoma County, California[3] Rem. is CaO. 5. Comstock Lode, Nevada.[4] Rem. is insol.

A r t i f.[5] $MgSO_4 \cdot 5H_2O$ is triclinic pinacoidal, with $a:b:c = 0.621:1:0.5605$; $\alpha\ 98°30'$, $\beta\ 109°00'$, $\gamma\ 75°05'$. Observed forms: b 010, a 100, m 110, $u\ 1\bar{1}0$, q 011, $x\ 0\bar{1}1$, $w\ \bar{1}11$, $\xi\ \bar{1}21$. Crystals are elongated [001] with a and w prominent. No cleavage. G. 1.718. Optically negative (−), with nY 1.491, $2V$ 45°08′, $r < v$. Obtained [6] in crystals together with the hexahydrate and tetrahydrate by evaporation over H_2SO_4 of a solution of magnesium sulfate containing added H_2SO_4 or magnesium chloride; also reported by dehydration over H_2SO_4 of the heptahydrate.

Ref.

1. Hobbs (1905).
2. Keller (1908).
3. Allen and Day, *Carnegie Inst. Washington Publ.*, **378**, 42 (1927).
4. Milton and Johnston, *Econ. Geol.*, **33**, 749 (1938).
5. Wyrouboff, *Bull. soc. min.*, **12**, 371 (1889).
6. Mellor (**4**, 523, 1923) and Wyrouboff (1889).

COBALT-CHALCANTHITE *Larsen* and *Glenn* (*Am. J. Sc.*, **50**, 225, 1920).

A name given to triclinic (?) $CoSO_4 \cdot 5H_2O$ formed by the partial dehydration of bieberite or of artificial $CoSO_4 \cdot 7H_2O$ or $6H_2O$. Biaxial negative (−), $2V$ medium, with nX 1.531, nY 1.549, nZ 1.552; faintly pleochroic with X eosine-pink and Z pale rose-pink; dispersion not strong. There is no evidence that this compound has formed directly in nature.

29.6.6 HEXAHYDRITE GROUP

MONOCLINIC; PRISMATIC—2/m

	$a:b:c$	β
Hexahydrite, $MgSO_4 \cdot 6H_2O$	1.4018:1:3.3890	98°14′
Bianchite, $(Zn,Fe)SO_4 \cdot 6H_2O$	1.3847:1:3.3516	98 12 [$ZnSO_4 \cdot 6H_2O$]

The minerals of this group, not yet analyzed structurally, are isostructural with the monoclinic artificial hexahydrated sulfates and selenates of Mg, Co, Ni, and Zn. Several of these compounds have tetragonal polymorphs isostructural with retgersite. The pure Fe″ member is not known artificially, although Fe″ substitutes for Zn to a considerable extent in bianchite.

29.6.6.1 HEXAHYDRITE [MgSO$_4$·6H$_2$O]. *Johnston (Geol. Sur. Canada, Sum. Rep.*, 1910, 256, 1911).

Cryst.[1] Monoclinic; prismatic—$2/m$.

$a:b:c = 1.4018:1:3.3890;$ $\beta\,98°14';$ $p_0:q_0:r_0 = 2.4176:3.3541:1$

$r_2:p_2:q_2 = 0.2981:0.7208:1;$ $\mu\,81°46';$ $p_0'\,2.4428,\,q_0'\,3.3890,\,x_0'\,0.1447$

Forms:

	ϕ	ρ	ϕ_2	$\rho_2 = B$	C	A
c 001	90°00'	8°14'	81°46'	90°00'	.....	81°46'
a 100	90 00	90 00	0 00	90 00	81°46'	.....
m 110	35 47	90 00	0 00	35 47	85 12	54 13
σ $\bar{1}$04	−90 00	24 59	114 59	90 00	33 13	114 59
ρ $\bar{1}$02	−90 00	47 07	137 07	90 00	55 21	137 07
o 112	38 52½	65 19½	36 12½	44 58½	60 20½	55 13½
ξ $\bar{1}$11	−34 08½	76 16½	156 29	36 29	80 59	123 02½
w $\bar{1}$12	−32 26	63 31½	137 07	40 56	68 08	118 41½
y $\bar{1}$14	−28 48½	44 02	114 59	52 28½	48 26	109 34½

Structure cell.[2] Space group $C2/c$. a_0 10.04 kX, b_0 7.15, c_0 24.34; β 98°34'; $a_0:b_0:c_0 = 1.404:1:3.404$. Cell contents Mg$_8$(SO$_4$)$_8$·48H$_2$O.

Habit. Coarse columnar to delicately fibrous. Rarely in good crystals, and then usually thick tabular {001}.

Twinning.[3] (a) On {001}. (b) On {110}.

Phys. Cleavage {100}, perfect. Fracture conchoidal. G. 1.757;[4] 1.745 (calc.). Colorless to white, sometimes pale greenish. Luster pearly to vitreous. Transparent; usually white and opaque. Taste bitter, salty.

Opt.[3] In transmitted light, colorless.

Orientation		n	
$X \wedge c$	−25°	1.426	Biaxial negative (−).
Y	b	1.453	$2V$ 38° (meas.).
Z		1.456	

Chem. A hydrated sulfate of magnesium, MgSO$_4$·6H$_2$O.
Anal.

	1	2	3	4
MgO	17.64	17.15	17.88	17.28
SO$_3$	35.04	34.52	34.64	34.27
H$_2$O	47.32	46.42	47.32	48.57
Rem.		1.78	0.13	
Total	100.00	99.87	99.97	100.12
G.		1.757	1.71	1.756

1 MgSO$_4$·6H$_2$O 2. Bonaparte River, British Columbia.[4] Rem. is insol. 3. Near Onoville, Okanogan County, Washington.[5] Rem. is (Al,Fe)$_2$O$_3$ 0.10, insol. 0.03. 4. Kelčany, Moravia.[6]

Occur. Found sparingly as a dehydration product of epsomite, sometimes as pseudomorphs. Also, rarely, as a direct deposit in salt lakes. First found at a locality on the Bonaparte River, British Columbia. As an efflorescence on epsomite near Oroville, Washington. Pseudomorphous

BIANCHITE

after epsomite at Kelčany, Moravia, and Kladno, Bohemia. In spear-shaped crystals with halite in the Saki salt lakes, Crimea.[3]

Alter. Reported to rehydrate to epsomite in most air.[7]

Artif. Crystallizes from an aqueous solution between 48° and 69°. At lower temperatures the heptahydrate, epsomite, and at higher the monohydrate, kieserite, are formed. The hexahydrate has been reported as a metastable crystalline phase in contact with solution from 0° to 48° and 69° to 100°.[8]

Name. Named in allusion to the water content.

Ref.

1. Elements and symmetry from structure cell. Form list from Dolivo-Dobrovolsky, *Mem. soc. russe min.*, [2], **58**, 3 (1929). Transformation: Dolivo-Dobrovolsky to new elements, 100/010/002. See also Groth (**2**, 422, 1908).
2. Ide, *Naturwiss.*, **26**, 411 (1938), on artificial material.
3. Dolivo-Dobrovolsky (1929).
4. Johnston (1911).
5. Walker and Parsons, *Univ. Toronto Stud., Geol. Ser.*, **24**, 21 (1927).
6. Kokta, *Publ. Fac. Sc. Univ. Masaryk*, Brno, 166 (1935).
7. Walker and Parsons (1927).
8. Seidell, *Solubilities of Inorg. Comp.*, New York, **1**, 985 (1940).

29.6.6.2 **BIANCHITE** [$ZnSO_4 \cdot 6H_2O$]. Andreatta (*Acc. Linc., Rend.*, [6], **11**, 760, 1930).

Cryst.[1] Monoclinic; prismatic—$2/m$.

$$ZnSO_4 \cdot 6H_2O$$

$a:b:c = 1.3847:1:3.3516;\quad \beta\,98°12';\quad p_0:q_0:r_0 = 2.4205:3.3173:1$

$r_2:p_2:q_2 = 0.3014:0.7296:1;\quad \mu\,81°48';\quad p_0'\,2.4455,\,q_0'\,3.3516,\,x_0'\,0.1441$

$$(Zn,Fe)SO_4 \cdot 6H_2O \text{ with } Zn:Fe = 2:1$$

$a:b:c = 1.3788:1:3.3324;\quad \beta\,98°30';\quad p_0:q_0:r_0 = 2.4169:3.2958:1$

$r_2:p_2:q_2 = 0.3034:0.6333:1;\quad \mu\,81°30';\quad p_0'\,2.4437,\,q_0'\,3.3324,\,x_0'\,0.1495$

Forms:[2] (Artificial crystals):

| c 001 | a 100 | m 110 | σ 10$\bar{1}$ | ξ 11$\bar{1}$ | w 11$\bar{2}$ | ζ 11$\bar{3}$ | y 11$\bar{4}$ | r 102 |

Habit. As crusts of indistinct crystals. Artificial crystals are tabular {001} with {110} and {112} prominent.

Twinning. On {001}, common in artificial material.

Phys. H. $\sim 2\frac{1}{2}$. G. 2.07 (for $ZnSO_4 \cdot 6H_2O$),[3] 2.031 (for $(Zn,Fe)SO_4 \cdot 6H_2O$ with $Zn:Fe = 2:1$). Color white, becoming yellowish on oxidation of the iron. Luster vitreous. Transparent.

Opt. In transmitted light, colorless.

ORIENTATION n (for $Zn:Fe \sim 2:1$)

$X \wedge c$	$-26°$	1.465	Biaxial negative $(-)$.
Y	b	1.494	$2V$ 10° (meas.).
Z		1.495	

Chem. Zinc sulfate hexahydrate, $ZnSO_4 \cdot 6H_2O$. The name bianchite was originally given to a particular ferroan variety (with Zn:Fe = 2:1 in analysis 2) but the name is here extended to all regions of composition in which Zn predominates over Fe'' and other substituting elements.

Anal.

	1	2	3
ZnO	30.19	20.01	15.37
FeO		8.84	13.57
SO_3	29.70	30.13	30.24
H_2O	40.11	39.92	40.82
Rem.		1.02	
Total	100.00	99.92	100.00

1. $ZnSO_4 \cdot 6H_2O$. 2. Raibl, Venezia Giulia.[4] Rem. is insol. 3. $(Zn,Fe)SO_4 \cdot 6H_2O$ with Zn:Fe = 1:1.

Tests. Easily soluble in water.

Occur. Found with goslarite, melanterite, gypsum, and hydrozincite on the walls of mines at Raibl, Predil, in the Julian Alps.

Artif. Crystals containing Fe in substitution for Zn can be obtained up to at least Zn:Fe $\sim$ 2:1 by crystallization from mixed solutions of iron and zinc sulfates containing free sulfuric acid.

Name. After Angelo Bianchi (1892–), Italian mineralogist.

Ref.

1. Elements of $ZnSO_4 \cdot 6H_2O$ from the angles of Wyrouboff, *Bull. soc. min.*, **12**, 377 (1889), in the unit of the structure cell found for the isostructural mineral hexahydrite. Data for $(Zn,Fe)SO_4 \cdot 6H_2O$ from Andreatta, *Acc. Linc., Rend.*, [6], **16**, 62 (1932). Transformation: Wyrouboff and Andreatta to new, 100/010/002.
2. Andreatta (1932) and Wyrouboff (1889). Rare or doubtful: k 10.0.18, f 778, g 1.1.$\overline{10}$, h 7.7.20.
3. Thorpe and Watts, *J. Chem. Soc. London*, **37**, 102 (1880).
4. Andreatta (1930).

RETGERSITE

TETRAGONAL; TRAPEZOHEDRAL—4 2 2

Retgersite, $NiSO_4 \cdot 6H_2O$, is isostructural with the tetragonal polymorphs of the artificial hexahydrated selenates of Ni and Zn. It has an unstable monoclinic polymorph, known only artificially, that is isostructural with the members of the Hexahydrite Group. The structure [1] is based on (SO_4) groups together with octahedral coordination-groups of water molecules about the Ni ions, as in the structure of morenosite, $NiSO_4 \cdot 7H_2O$.

Ref.

1. Beevers and Lipson, *Zs. Kr.*, **83**, 123 (1932).

29.6.7 RETGERSITE [$NiSO_4 \cdot 6H_2O$]. Frondel and Palache (Am. Min., **34**, 188, 1949). α-$NiSO_4 \cdot 6H_2O$. Blue $NiSO_4 \cdot 6H_2O$.

Cryst.[1] Tetragonal; trapezohedral—4 2 2.

$$a:c = 1:2.7038; \quad p_0:r_0 = 2.7038:1$$

Forms:[2]

	ϕ	ρ	A	$\overline{M}$
c 001	.....	0°00′	90°00′	90°00′
m 110	45°00′	90 00	45 00	90 00
h 013	0 00	42 01½	90 00	61 44½
i 012	0 00	53 30½	90 00	55 21½
j 011	0 00	69 42	90 00	48 27½
o 113	45 00	51 53	56 12	90 00
p 112	45 00	62 23½	51 12	90 00

Structure cell.[3] Space group $P4_12_1$ or $P4_32_1$. a_0 6.776 ± 0.003 kX, c_0 18.249 ± 0.009; $a_0:c_0 = 1:2.693$. Cell contents $Ni_4(SO_4)_4 \cdot 24H_2O$.

Habit. Artificial crystals are thick tabular {001} grading to short prismatic [001]. The natural material usually occurs as fibrous crusts or veinlets, the fibers elongated [001], or as twisted spire- or horn-like tufts; rarely as single crystals (Minasragra).

Etch figures.[4] Conformable to tetragonal-trapezohedral symmetry.

Phys. Cleavage {001} perfect; also traces of a cleavage {110} in crushed grains under the microscope. Fracture subconchoidal to uneven. Brittle. H. 2½. G. 2.04 (Nevada), 2.07 (artif.);[5] 2.070 (calc.). Luster vitreous. Color deep emerald-green with a tinge of blue. Streak greenish white. Optical rotatory power,[6] $\alpha_D = 1.50°$ per mm. Taste slightly bitter and metallic.

Minasragra.

Opt.[7] In transmitted light, pale green.

λ	C	D	F	G	
nO	1.5078	1.5109	1.5173	1.5228	Uniaxial negative (−).
nE	1.4844	1.4873	1.4930		

Chem. Nickel sulfate hexahydrate, $NiSO_4 \cdot 6H_2O$. Mg probably can substitute for Ni to a limited extent.

Anal.

	1	2
NiO	28.42	26.87
MgO		0.65
FeO		0.63
SO_3	30.46	30.32
H_2O	41.12	[41.53]
Total	100.00	[100.00]
G.	2.07	2.04

1. $NiSO_4 \cdot 6H_2O$. 2. Cottonwood Canyon, Nevada.[10]

Tests. Soluble in water (30.2 g. NiSO$_4$ in 100 g. H$_2$O at 33°). Other tests as with morenosite.

Occur. A secondary mineral formed by the oxidation of nickel-bearing minerals and formed as an efflorescence under essentially atmospheric conditions. Found with annabergite as an alteration product of niccolite in Cottonwood Canyon, Churchill County, Nevada.[8] With morenosite and minasragrite on patronite at Minasragra, Peru, and with ferroan chalcanthite at the Gap Nickel mine, Lancaster County, Pennsylvania. Also from Lichtenberg, near Bayreuth, Bavaria, and from Lobenstein, Thuringia, in Germany.

Artif.[9] Easily obtained as large crystals by crystallization from water solution in the range from 31.5° to 53.3°, or at ordinary temperatures from solutions containing a sufficient excess of free H$_2$SO$_4$.

Name. After Jan Willem Retgers (1856–1896), Dutch chemical crystallographer.

Ref.

1. Angles and unit of Scacchi, *Acc. Napoli, Att.*, **1**, no. 11 (1863) on artificial material in the orientation of the structure cell. Transformation: Scacchi to new, 110/$\bar{1}$10/001.
2. Mitscherlich, *Ann. Phys.*, **12**, 146 (1828), Rammelsberg (92, 1855), Brooke, *Ann. Phil.*, **22**, 437 (1823), on artificial crystals.
3. Space group and structure analysis by Beevers and Lipson, *Zs. Kr.*, **83**, 123 (1932), cell dimensions of Borghijs, *Natuurwet. Tijdschr.*, **19**, 115 (1937), by powder method, all on artificial material.
4. Blasius, *Zs. Kr.*, **10**, 227 (1885); Baumhauer (57, 1894); Borghijs (1937).
5. Mrose, priv. comm. (1948), on artificial crystal by microbalance, confirming the earlier values of 2.074 and 2.064 by Topsøe, *Arch. sc. phys. nat. Genève*, [2], **45**, 223 (1872), and Gossner, *Ber. deutsch. chem. Ges.*, **40**, 2374 (1907).
6. Borghijs (1937); see also Beevers and Lipson (1932) and Rudnick, Slack, and O'Connor, *Phys. Rev.*, **58**, 1003 (1940).
7. Topsøe and Christiansen, *Ann. chim. phys.*, **1**, [5], 63 (1874), by prism method on artificial material. On the optical rotation see Mathieu, *C.R.*, **222**, 223 (1946).
8. Ferguson, *Univ. Nevada Bull.*, **33**, no. 5 (1939).
9. Seidell, *Solubilities of Inorg. Compounds*, New York, 1940, 1347, and Mellor (**15**, 453, 1936).
10. Hallowell analysis in Frondel and Palache (1948).

29.6.8 MELANTERITE GROUP

MONOCLINIC; PRISMATIC—2/m

	$a:b:c$	β
Melanterite, FeSO$_4\cdot$7H$_2$O	1.1828:1:1.5427	104°16'
Pisanite, (Fe,Cu)SO$_4\cdot$7H$_2$O		
Kirovite, (Fe,Mg)SO$_4\cdot$7H$_2$O		
Boothite, CuSO$_4\cdot$7H$_2$O	1.1622:1:1.5000	105 36
Bieberite, CoSO$_4\cdot$7H$_2$O	1.1815:1:1.5325	104 40
Mallardite, MnSO$_4\cdot$7H$_2$O	1.2245:1:1.5727	104 51
Zinc-melanterite, (Zn,Cu)SO$_4\cdot$7H$_2$O		

The members of the Melanterite Group, all well-known as artificial compounds, are isostructural with the artificial heptahydrated selenates of Fe and Co and with the monoclinic polymorph of MgSO$_4\cdot$7H$_2$O. These compounds enter into partial or complete series, best known in artificial

MELANTERITE SERIES

preparations, by mutual substitution of their cations. The species names pisanite and kirovite are here employed for parts of the series extending from melanterite respectively toward boothite and monoclinic $MgSO_4 \cdot 7H_2O$, the latter not known as a mineral. The members of the group are water-soluble and are found in arid or sheltered places in nature as oxidation products of sulfides.

The crystal structure of the Melanterite Group is not known.

MELANTERITE SERIES

29.6.8.1 **MELANTERITE** [$FeSO_4 \cdot 7H_2O$]. Μελαντηρία, Χάλκανθον, etc. *Dioscorides*. Chalcanthum, Atramentum sutorium, etc. *Pliny*. Melanteria, Atramentum sutorium viride *Agricola*. Vitriolum pt. *Albertus Magnus*. Atramentum viride *Gesner*. Vitriolum viride, ferri, martis *Wallerius*. Green Vitriol. Copperas. Sulphate of Iron. Eisenvitriol *Germ*. Fer sulfaté *Fr*. Mélantérie *Beudant* (**2**, 482, 1832). Melantherite *Chapman* (14, 1843). Alcaparrosa verde *S. Amer.*, *Raimondi* (212, 1878). Vitrolo verde *Span*. Melanterit *Haidinger* (489, 1850).

Luckite *Carnot* (*Bull. soc. min.*, **2**, 168, 1879). Bourbolite pt. *Lefort* (*C.R.*, **55**, 919, 1862). Sommairite (*Brit. Mus. Nat. Hist., Students Guide to Coll. of Min.*, 1897; *Coll. de Min. du Mus. d'Histoire Nat., Paris, Guide du Visiteur*, 2nd ed., 32, 1900; *Min. Mag.*, **14**, 410, 1907). Calingastite *Angelelli* and *Trelles* (*Bol. obras sanit. de la nacion*, Buenos Aires, nos. 8–10, 40, 1938).

29.6.8.2 **PISANITE** [$(Fe,Cu)SO_4 \cdot 7H_2O$]. Pisanit *Kenngott* (10, 1860 [1859]). Cyanoferrite *Adam* (66, 1865). Cuproferrite *Des Cloizeaux* (157, 1867). Kupfereisenvitriol *Germ*. Salvadorite *Herz* (*Zs. Kr.*, **26**, 16, 1896).

29.6.8.3 **KIROVITE** [$(Fe,Mg)SO_4 \cdot 7H_2O$]. Vertushkov (*Bull. ac. sc. U.R.S.S., Sér. geol.*, no. 1, 109, 1939); Cuprokirovite *Vertushkov* (1939). Jarošit, Kuprojarošit *Kokta* (*Sbornik Klubu Přiro. Brně*, **19**, 75, 1937—*Min. Abs.*, **7**, 316, 1939).

Cryst.[1] Monoclinic; prismatic—$2/m$.

$$FeSO_4 \cdot 7H_2O$$

$a:b:c = 1.1828:1:1.5427; \quad \beta\ 104°16'; \quad p_0:q_0:r_0 = 1.3043:1.4951:1$

$r_2:p_2:q_2 = 0.6688:0.8724:1; \quad \mu\ 75°44'; \quad p_0'\ 1.3458, q_0'\ 1.5427, x_0'\ 0.2543$

Forms:[2]

		ϕ	ρ	ϕ_2	$\rho_2 = B$	C	A
c	001	90°00'	14°16'	75°44'	90°00'	.....	75°44'
b	010	0 00	90 00	.....	0 00	90°00'	90 00
a	100	90 00	90 00	0 00	90 00	75 44	.....
m	110	41 06	90 00	0 00	41 06	80 40½	48 54
o	011	9 21½	57 24	75 44	33 46½	56 13½	82 07½
w	103	90 00	34 43	55 17	90 00	20 27	55 17
v	101	90 00	58 00	32 00	90 00	43 44	32 00
s	$\overline{1}05$	−90 00	0 51	90 51	90 00	15 07	90 51
t	$\overline{1}01$	−90 00	47 26½	137 26½	90 00	61 42½	137 26½
r	111	46 03	65 46½	32 00	50 44	55 53½	48 58
γ	$\overline{1}21$	−19 26½	73 00½	137 26½	25 36	78 11	108 33½

Structure cell.[3] Space group $F2/d$. a_0 15.34 kX, b_0 12.98, c_0 20.02 [$FeSO_4 \cdot 7H_2O$]; $a_0:b_0:c_0 = 1.183:1:1.543$; [$\beta$ 104°16']. Cell contents $(Fe,Cu)_{16}(SO_4)_{16} \cdot 112H_2O$.

Habit. Equant to short prismatic [001] with large {110} and {001}; also thick tabular {010} or {10$\bar{1}$}, or octahedral due to equal development of {110}, {001} and {$\bar{1}$01}. Cuprian melanterite and pisanite are often simple in habit and pseudo-rhombohedral with {110} and {001} and short

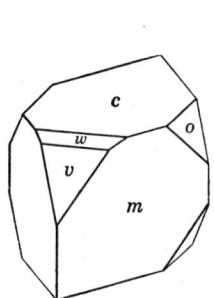
Melanterite. Artificial.

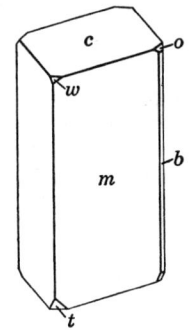
Pisanite. Ducktown.

to long prismatic [001]. Usually in stalactitic or concretionary forms; as fibrous to capillary aggregates and crusts; massive, pulverulent.

P h y s. Cleavage {001} perfect, {110} distinct. Fracture conchoidal. Brittle. H. 2. G. 1.898 ($FeSO_4 \cdot 7H_2O$);[4] 1.84 (calc.). Luster vitreous. Color green of various shades, passing into greenish blue and blue with increasing substitution of Cu for Fe. Also greenish white, especially in fibrous and massive types. Streak colorless. Subtransparent to translucent. Taste sweetish, astringent, and metallic. On exposure to dry air becomes yellowish white and opaque.

O p t. In transmitted light, colorless to pale green.

Orientation	Melanterite n(Na) (pure artif.)[5]	Kirovite n (anal. 9)[16]	Pisanite n (Fe:Cu~1.7:1)[6]	
X	1.4713	1.467	1.472	Biaxial positive (+).
Y	1.4782		1.479	$r > v$, weak, inclined.
Z	1.4856	1.476	1.487	
$Z \wedge c$	61°	12°		
2V	85°27' (meas.)	Large	Very large	

C h e m. Ferrous sulfate heptahydrate, $FeSO_4 \cdot 7H_2O$. Cu substitutes for Fe, and a series extends from the iron end-member up to about Cu:Fe = 1.89:1 [CuO 18.3 per cent];[7] the maximum copper content varies somewhat with the temperature of crystallization. The species name melanterite is here applied arbitrarily to material in this series with Fe:Cu > 5:1 [CuO 4.75 per cent], or (Fe,Mg, etc.):Cu > 5:1, and the name pisanite to material with a copper content above this ratio. Pisanite apparently is isostructural with boothite, $CuSO_4 \cdot 7H_2O$, but is separated from that species by a gap in the solid solution series. Mg substitutes for Fe and a partial series extends up to about Fe:Mg = 1:1.38 in artificial material.[20] The species name kirovite is here applied to natural material in this series with Mg:(Fe,Cu) > 1:5 [MgO 2.46 per cent]; in analysis 9, Mg:Fe = 1.04:1. A number of other divalent metals may substitute [20] in part or entirety for Fe in artificial monoclinic $FeSO_4 \cdot 7H_2O$, including Zn, Co, Ni, and Mn. In natural material, Zn:Fe = 1:2.26 in analysis 6, and Mn:Fe = 1:11.2 in analysis 7. Several substituting cations may be present simultaneously; the cuprian kirovite of analysis 11 has Cu:Mg:Fe = 1:2.9:3.8. The Al_2O_3 reported in many analyses is probably due to admixture, and the Fe_2O_3 to admixture or partial oxidation. Melanterite and pisanite dehydrate to the pentahydrate (*siderotil*) or to lower hydrates at room temperature, depending on the relative humidity.[21] Melanterite is dimorphous with tauriscite.

Anal.[8]

	1	2	3	4	5	6	7
MgO		1.87		2.16			0.2
FeO	25.84	22.27	24.15	20.35	25.23	17.74	21.7
CuO			0.14	0.04	2.21		
ZnO						8.92	
MnO				0.05			1.9
SO_3	28.80	29.19	28.81	31.15	31.14	28.85	26.3
H_2O	45.36	45.79	[43.20]	43.90	41.42	44.21	[42.2]
Rem.			3.70	1.96			7.7
Total	100.00	99.12	[100.00]	99.61	100.00	99.72	[100.0]
G.	1.84		1.87			1.95	

1. $FeSO_4 \cdot 7H_2O$. 2. Melanterite. Falun, Sweden.[9] 3. Melanterite Recsz, Hungary.[10] Rem. is insol. 3.05, Al_2O_3 0.65. 4. Magnesian melanterite. Alsósajó, Hungary.[11] Rem. is K_2O 0.07, Na_2O 0.08, CaO 0.13, NiO tr., Al_2O_3 0.63, Fe_2O_3 0.83, insol. 0.22. 5 Cuprian melanterite. Zmeinogorsky mine, Altai, U.S.S.R.[12] 6. Zincian melanterite. Laurium, Greece.[13] 7. Manganoan melanterite. Lucky Boy mine, Utah (*luckite*).[14] Rem. is CaO 0.5, insol. 7.2.

	8	9	10	11	12	13	14
MgO		5.55	7.45	3.36	4.29		0.97
FeO	17.10	12.75	18.48	15.18	16.37	11.00	8.51
CuO	0.04	0.30	3.18	4.40	9.17	13.09	18.81
ZnO		0.50	0.38				
MnO		0.18	0.05	tr.			
SO_3	30.13	30.51	30.11	29.93	29.00	28.52	27.93
H_2O	47.30	46.68	44.50	46.50	45.46	[44.92]	44.75
Rem.		1.42	0.51			1.50	
Total	100.12	99.79	100.57	100.30	100.00	[100.00]	100.00
G.	1.818	1.76	1.81	1.868		1.95	

8. Kirovite. Szomolnok, Czechoslovakia (*jarošite*).[15] 9. Kirovite. Kirovgrad, Urals.[16] Rem. is CaO tr., Al_2O_3 1.42, Fe_2O_3 tr. 10. Cuprian kirovite. Kirovgrad, Urals (*cuprokirovite*).[16] Rem. is CaO tr., Al_2O_3 0.29, Fe_2O_3 0.22. 11. Cuprian kirovite. Szomolnok, Czechoslovakia (*cuprojarošite*).[15] 12. Pisanite. Leona Heights, California.[17] 13. Pisanite. Ravensthorpe, Western Australia.[18] NiO 0.16, CoO 0 57, CaO 0.27, Al_2O_3 0.39, SiO_2 0.11. 14. Pisanite. Quetena, Chile (*salvadorite*).[19]

Tests. In C.T. yields water and on strong heating SO_2 and SO_3. On charcoal turns at first brown, then red, and finally black, becoming magnetic. Soluble in water (26.6 g. $FeSO_4$ per 100 g. H_2O at 20°).

Var.

MELANTERITE

Cuprian. Pisanite, pt., *some authors.* Contains Cu in substitution for Fe, and grades into pisanite. Color green to bluish green.

Magnesian. Contains Mg in substitution for Fe, with diminished indices of refraction and gravity, and grades into kirovite.

Zincian. Sommairite (applied to the material of analysis 6—see Spencer, *Min. Mag.,* 14, 410, 1907). Calingastite *Angelelli* and *Trelles* (1938). Contains Zn in substitution for Fe, with Zn:Fe = 1:2.4 in material from Argentina [22] and 1:2.26 in analysis 6.

PISANITE

Magnesian. Contains Mg in substitution for (Fe,Cu)—see analyses 10, 11, 13.

KIROVITE

Cuprian. Cuprokirovite *Vertushkov* (1939). Cuprojarošite *Kokta* (1937). Contains Cu in substitution for (Fe,Mg). Color pale blue.

O c c u r. Melanterite, pisanite, and kirovite are secondary minerals formed by the oxidation of pyrite, marcasite, and cupriferous pyrite ores. All these species are readily soluble in water, and occur typically as efflorescences on the walls and timbers of mine workings in pyritic ore deposits, in the oxidized zone of such deposits especially in arid regions, and in sheltered crevices in pyritic sedimentary or metamorphic rocks; also in coal and lignite deposits as an alteration especially of marcasite. Associated minerals include epsomite, chalcanthite, pickeringite, halotrichite, alunogen, alums, gypsum, and, as further oxidation products, copiapite, botryogen, fibroferrite, and other basic ferric sulfates. Among the more important localities of melanterite may be mentioned [23] the long-known occurrences at Rammelsberg near Goslar in the Harz, Bodenmais in Bavaria, and Rio Tinto in Spain. Also at Sain-Bel, near Lyon, Rhône, France, and Falun, Sweden. In the Blyava district in the southern Urals, U.S.S.R. Kirovite and a cuprian variety occurs at the Kalata mine, Kirovgrad, in the Ural Mountains, U.S.S.R. Also at Szomolnok, Czechoslovakia (*jarošite, cuprojarošite*); at Idria, Gorizia, Italy; with pisanite in the Alma mine, Leona Heights, Alameda County, California. Pisanite was originally described from a copper mine in the interior of Turkey, the exact locality unknown; it also occurs at Lading in Carinthia, at Recsz, Hungary, at the Quetena mine near Calama, Chile (*salvadorite*),[25] and at localities for melanterite in general. In the United States, melanterite, pisanite, or both species occur at numerous localities. In California in the Alma mine near Leona Heights, Alameda County, and in the copper deposits of Shasta and Trinity counties. At Ducktown, Tennessee. In the Butte district, Montana, and at Bingham Canyon, Utah. At the Comstock Lode, Lincoln County, Nevada. Manganoan melanterite occurs in the Lucky Boy mine, Butterfield Canyon, Salt Lake County, Utah

MELANTERITE SERIES 503

(*luckite*). A nickelian variety (2.5 per cent NiO) occurs in the Mt. Diablo district, Contra Costa County, California.

Alter. By dehydration to siderotil, and by oxidation ultimately to copiapite and other ferric sulfates. The alteration of pyrite into green vitriol, puzzling to early chemists, was shown by Lavoisier in 1777 to be due to the absorption of oxygen and the formation of sulfate—one of the observations primarily responsible for the downfall of the phlogiston theory.

Artif.[20] Melanterite crystallizes from water solution at temperatures up to 56°; lower hydrates are obtained from solutions containing sufficient free sulfuric acid. Cuprian melanterite and pisanite are obtained from solutions containing added copper sulfate; and magnesian and other compositional variants may be similarly obtained by appropriate additions.

Name.[24] Melanterite from μελαντηρια, copperas. Pisanite after the French mineral chemist and mineral dealer, Félix Pisani (1831–1920). *Copperas*, ultimately from the Latin *cuprum*, *Cyprium* [*Cyprian*-], *copper*, may derive from the formation of the compound as a by-product in obtaining copper from *aqua cuprosa* or *cuperosa*, *copper-water*, by the cementation process. Kirovite after S. M. Kirov. Jarošite after Zdeněk Jaroš, Czech mineralogist.

Ref.

1. Elements of Zepharovich, *Ak. Wien, Sitzber.*, **79**, 183 (1879), on pure artificial $FeSO_4 \cdot 7H_2O$ crystals. The angles vary appreciably with substitution of other cations for Fe; on variation with Cu content see Vavrinecz, *Zs. Kr.*, **66**, 167 (1927); Mg content, Zepharovich (1879) and Vertushkov (1939). Zepharovich (1879) gives $a:b:c$ = 1.1799:1:1.5434, β 104°26′ for Fe:Mg = 3:2; Vavrinecz (1927) gives $a:b:c$ = 1.1694:1:1.5589, β 105°06′, for Fe:Cu = 5.9:4.1 (pisanite).

2. Goldschmidt (**3**, 111, 1916); Hey, *Min. Mag.*, **22**, 413 (1930); Ungemach, *Bull. soc. min.*, **58**, 159 (1935); Machatschki, *Zbl. Min.*, 53 (1935). Also, for cuprian melanterite and pisanite Goldschmidt (**6**, 157, 1920, Vavrinecz (1927), Schaller, *U. S. Geol. Sur., Bull.* 610, 161 (1916). Melanterite and pisanite forms are here grouped together. Less common and rare:

β 150	ω 0.1.12	904	$\bar{9}08$	q 221	$D\,\bar{2}21$
120	e 013	502	$\bar{2}01$	5.5.22	$\bar{12}.12.1$
f 320	102	301	$\bar{8}01$	$\rho\,\bar{1}12$	121
k 530	203	$\bar{1}04$	115	$E\,\bar{3}35$	x 161
h 210	302	$g\,\bar{2}05$	α 112	$\bar{9}98$	$\delta\,\bar{2}11$
i 810	201	$\bar{1}02$	332	$p\,\bar{1}11$	$y\,\bar{2}31$

3. Westenbrink, *Rec. trav. chim. Pays-Bas*, **46**, 105 (1927); see also Ness, *Naturwiss.*, **28**, 78 (1940) on artificial $FeSO_4 \cdot 7H_2O$.
4. Retgers, *Zs. phys. Chem.*, **3**, 534 (1889); almost identical values are given by others.
5. Erofejeff, *Ak. Wien, Sitzber.*, **56**, 63 (1867).
6. Larsen (120, 1921).
7. See Collins, *Min. Mag.*, **20**, 32 (1923).
8. For additional analyses see Smythe, *The Vasculum*, Newcastle-on-Tyne, **19**, 12 (1933)—*Min. Abs.*, **5**, 283 (1933); Hey (1930); Milton and Johnston, *Econ. Geol.*, **33**, 749 (1938); Niggli and Faesy, *Zs. Kr.*, **59**, 258 (1924); Vavrinecz (1927); Eckel, *Am. Min.*, **18**, 449 (1933); Mélon, *Ann. soc. géol. Belgique*, **67**, B56 (1944).
9. Edgren, *Geol. För. Förh.*, **23**, 329 (1901).

10. Vavrinecz, *Magyar Chem. Foly.*, **35**, 1 (1929).
11. Zsivny, *Zs. Kr.*, **66**, 651 (1928).
12. Pilipenko, *Bull. Imp. Tomsk. Univ.*, no. 63 (1915)—*Min. Abs.*, **2**, 113 (1923).
13. Michel, *Bull. soc. min.*, **17**, 204 (1894).
14. Carnot (1879).
15. Kokta (1937).
16. Vertushkov (1939).
17. Schaller, *Univ. California, Dept. Geol., Bull.*, **3**, 191 (1903).
18. Simpson, *J. Roy. Soc. Western Australia*, **23**, 17 (1937).
19. Herz (1896).
20. See Retgers and Mellor (**14**, 253, 1935).
21. Mellor (1935) and Eckel (1933). Blasius, *Zs. Kr.*, **10**, 231 (1885), describes the dehydration ellipsoids on the crystals.
22. Angelelli and Trelles (1938) and Gordon, *Ac. Sc., Philadelphia, Not. Nat.*, no. 89 (1941), with ZnO 8.42, FeO 16.67, CuO 1.29 per cent.
23. A lengthy list of localities is given by Hintze (**1** [3B], 4362, 1929).
24. The early synonomy of the monoclinic vitriols is discussed in Dana (941, 1892) and Mellor (**14**, 242, 1935).
25. Taken as identical with pisanite by most authors, but indicated by Gordon, *Ac. Sc. Philadelphia, Not. Nat.*, **72** (1941), to be identical with kroehnkite.

29.6.8.4 **BOOTHITE** [$CuSO_4 \cdot 7H_2O$]. Schaller (*Univ. California Dept. Geol. Bull.*, **3**, 207, 1903).

Cryst. Monoclinic.

$a:b:c = 1.1622:1:1.5000;$ $\quad \beta\, 105°36';$ $\quad p_0:q_0:r_0 = 1.2907:1.4447:1$

$r_2:p_2:q_2 = 0.6922:0.8933:1;$ $\quad \mu\, 74°24';$ $\quad p_0'\, 1.3400,\, q_0'\, 1.5000,\, x_0'\, 0.2792$

Forms:

		ϕ	ρ	ϕ_2	$\rho_2 = B$	C	A
c	001	90°00′	15°36′	74°24′	90°00′	....	74°24′
a	100	90 00	90 00	....	0 00	90°00′	90 00
m	110	41 46½	90 00	0 00	41 46½	79 41	48 13½
t	$\bar{1}01$	−90 00	46 41½	136 41½	90 00	62 17½	136 41½
z	$\bar{3}01$	−90 00	75 02	165 02	90 00	90 38	165 02
π	$\bar{1}12$	−27 31½	40 13½	111 20½	55 04	49 04	107 21½
e	$\bar{1}11$	−35 16	61 26½	136 41½	44 11	71 05½	120 28½
σ	$\bar{1}21$	−19 28½	72 33	136 41½	25 55	78 16½	108 32½

Habit. Usually massive with a crystalline structure or fibrous, rarely in crystals.

Phys. Cleavage {001} imperfect. H. 2–2½. G. ~2.1. Luster vitreous, in fibrous material silky or pearly. Color blue, slightly paler than chalcanthite. Transparent to translucent.

Opt. In transmitted light, pale blue.

Orientation	n [1]	
X near c	1.47	Biaxial, sign uncertain.
Y b	1.48	$2V$ large.
Z	1.49	

Chem. Copper sulfate heptahydrate, $CuSO_4 \cdot 7H_2O$. Boothite apparently is separated by a gap in the solid solution series from the isostructural species pisanite,[2] $(Cu,Fe)SO_4 \cdot 7H_2O$.

BIEBERITE

Anal.

	1	2	3	4	5
CuO	27.85	27.83	28.53	27.21	28.0
FeO		tr.	0.28	0.84	
MgO		tr.	tr.	0.66	
SO_3	28.02	28.37	28.65	28.37	28.1
H_2O-	44.13	36.64	43.76	38.28	37.8
H_2O+		7.42		5.11	6.1
Total	100.00	100.26	101.22	100.47	100.0

1. $CuSO_4 \cdot 7H_2O$. 2. Leona Heights.[3] Fibrous material. Recalculated after deducting 3.70 insol. 3. Leona Heights.[3] Massive material. Recalculated after deducting 3.54 insol. 4. Campo Seco.[4] Recalculated after deducting 3.96 insol. 5. Sain-Bel, France.[5]

Tests. In C.T. loses water, whitens but does not fuse and on long ignition turns black. Easily soluble in cold water.

O c c u r. Originally found at the Alma pyrite mine near Leona Heights, Alameda County, California, associated with chalcanthite, melanterite, and pisanite; also at a copper mine near Campo Seco, Calaveras County. Noted from a pyrite mine at Sain-Bel, near Lyon, Rhône, France.[5]

A l t e r. Dehydrates to chalcanthite on exposure in dry air. On heating, loses $6H_2O$ at 105° and the remainder at a higher temperature.

A r t i f.[6] Obtained at 15°–20° by nucleating a supersaturated solution of copper sulfate with a monoclinic crystal of $FeSO_4 \cdot 7H_2O$. Probably a stable deposit from copper sulfate solution below 0°.

N a m e. After Edward Booth.

Ref.

1. Larsen and Berman (150, 1934).
2. Should this series be found to be complete, the name boothite may be restricted to an arbitrary range of substitution of Fe" for Cu, perhaps up to Fe:Cu = 1:2. The nomenclature of the series with the monoclinic $FeSO_4 \cdot 7H_2O$ then would be: melanterite from Fe 100 up to Cu:Fe = 1:5, pisanite from this ratio up to Fe:Cu = 1:2, and boothite from this ratio up to Cu 100. On the ranges of mutual substitution of Cu, Fe", Zn, Mg, Mn, Co in the artificial monoclinic heptahydrated vitriols see literature in Doelter (4 [2], 259, 293, etc., 1927).
3. Schaller (1903).
4. Schaller, *Am. J. Sc.*, **17**, 192 (1904).
5. Lacroix (4, 227, 1910).
6. de Boisbaudran, *C.R.*, **65**, 1249 (1867).

CUPROMAGNESITE. Scacchi (*Acc. Napoli, Rend.*, 211, 1872; *Acc. Napoli, Atti*, **6**, 58, 1874). Found as bluish green crusts on lava of 1872 at Vesuvius. Crystals obtained from a water extract of the crusts are monoclinic and have the composition $(Cu,Mg)SO_4 \cdot 7H_2O$.

29.6.8.5 **B I E B E R I T E** [$CoSO_4 \cdot 7H_2O$]. Cobalt Vitriol *Sage* (*J. Phys.*, **39**, 53, 1791). Kobaltvitriol *Kopp* (*Allg. J. Chem.*, **6**, 157, 1808). Red Vitriol. Sulphate of Cobalt. Rhodhalose *Beudant* (**2**, 481, 1832). Bieberit *Haidinger* (489, 1845).

C r y s t.[1] Monoclinic; prismatic—$2/m$.

$a:b:c = 1.1815:1:1.5325;$ $\beta\ 104°40';$ $p_0:q_0:r_0 = 1.2971:1.4826:1$

$r_2:p_2:q_2 = 0.6745:0.8749:1,$ $\mu\ 75°20';$ $p_0'\ 1.3408,\ q_0'\ 1.5325,\ x_0'\ 0.2617$

Forms: [2]

		ϕ	ρ	ϕ_2	$\rho_2 = B$	C	A
c	001	90°00′	14°40′	75°20′	90°00′	.....	75°20′
b	010	0 00	90 00	.....	0 00	90°00′	90 00
m	110	41 11	90 00	0 00	41 11	80 24	48 49
e	013	27 07½	29 51½	75 20	63 42	26 18	76 53
o	011	9 41½	57 15	75 20	34 00	56 00	81 52
f	103	90 00	35 19½	54 40½	90 00	20 39½	54 40½
v	101	90 00	58 02	31 58	90 00	43 62	31 58
t	1̄01	−90 00	47 10½	137 10½	90 00	61 50½	137 10½
p	111	24 49	59 22	76 40½	38 39	54 14	68 50
n	121	13 01	72 22	54 40½	21 48	69 40	77 36
ν	1̄21	−19 23½	72 53½	137 10½	25 38½	78 29½	108 30½

Structure cell.[3] Space group $F2/d$. a_0 15.45 kX, b_0 13.08, c_0 20.04; [β 104°40′]; $a_0:b_0:c_0 = 1.181:1:1.532$. Cell contents $Co_{16}(SO_4)_{16}\cdot 112H_2O$.

Habit. Artificial crystals in form very like melanterite. As crusts and stalactites.

P h y s. Cleavage {001} perfect, {110} fair. H. ~2. G. 1.96; 1.83 (calc.). Luster vitreous. Color rose-red or flesh-red. Subtransparent. Dehydrates readily to the hexahydrate at room temperature, becoming opaque and mealy.

O p t. In transmitted light, colorless to pale rose.

Orientation		n(Na) (artif.) [4]	
X		1.4748	Biaxial positive (+).
Y	b	1.4820	$2V$ 88°01′.
Z ∧ c	+29°18′	1.4885	

C h e m. Cobalt sulfate heptahydrate, $CoSO_4\cdot 7H_2O$. In artificial material, Fe″ substitutes for Co and a complete series extends to $FeSO_4\cdot 7H_2O$; Cu substitutes for Co in part. In natural material, small amounts of Cu and Mg occur in substitution for Co, with Mg:Co = 1:2.8 in analysis 3.

Anal.

	1	2	2
MgO		0.88	3.86
CuO		0.30	
CoO	26.66	23.30	19.91
SO$_3$	28.48	28.81	29.05
H$_2$O	44.86	45.22	46.83
Total	100.00	98.51	99.65

1. $CoSO_4\cdot 7H_2O$. 2. Glücksstern mine, Siegen.[5] 3. Bieber, Hesse.[6]

Tests. In C.T. affords water and at higher temperatures SO_3. Soluble in water (27.2 g. $CoSO_4$ in 100 g. saturated solution at 25°).

O c c u r. A secondary mineral formed by the oxidation of sulfide and arsenide ores containing cobalt. Bieberite is relatively soluble and is much less common than melanterite and pisanite. Associated minerals include erythrite, annabergite, pharmacolite. Found as an efflorescence in old mine workings at Bieber in Hesse, Germany; at Siegen, Westphalia, in the cobalt veins; from Leogang in Salzburg, Austria. At Chalanches, Isère, France. From Tres Puntos, near Copiapó, Chile; and from Lomagundi, Southern Rhodesia. At Island Mountain, Trinity County, California.

A r t i f.[7] Obtained as crystals from water solution at temperatures below 40°.

N a m e. From the locality at Bieber.

Ref.

1. Marignac, *Mém. soc. phys. nat. Genève*, **14**, 245 (1855) on artificial crystals.
2. Goldschmidt (**1**, 194, 1913).
3. Westenbrink, *Proc. Ac. Sc. Amsterdam*, **29**, 1223 (1926).
4. Porter, *Festschr. V. Goldschmidt*, Heidelberg, 210, 1928. See also Larsen and Glenn, *Am. J. Sc.*, **50**, 225 (1920).
5. Schnabel in Rammelsberg (266, 1860).
6. Winkelblech, *Ann. Chem.*, **13**, 265 (1845).
7. Mellor (**14**, 761, 1935).

29.6.8.6   **M A L L A R D I T E** [$MnSO_4 \cdot 7H_2O$]. *Carnot* (*Bull. soc. min.*, **2**, 117, 1879).

C r y s t.[1] Monoclinic; prismatic—$2/m$.

$a:b:c = 1.2245:1:1.5727$; $\beta\ 104°51'$; $p_0:q_0:r_0 = 1.2844:1.5202:1$

$r_2:p_2:q_2 = 0.6578:0.8449:1$; $\mu\ 75°09'$; $p_0'\ 1.3287, q_0'\ 1.5727, x_0'\ 0.2652$

Forms:

	ϕ	ρ	ϕ_2	$\rho_2 = B$	C	A
c 001	90°00′	14°51′	75°09′	90°00′	.....	75°09′
m 110	40 11½	90 00	0 00	40 11½	80°29′	49 48½
q 011	9 34½	57 54½	75 09	33 20½	56 39½	81 54
s 103	90 00	35 18	54 42	90 00	20 27	54 42
r 101	90 00	57 54	32 06	90 00	43 03	32 06
p $\overline{1}$01	−90 00	46 46	136 46	90 00	61 37	136 46

Habit. Artificial crystals are tabular {001}. As fibrous masses and crusts.

P h y s. Cleavage {001} good, probably also on {110}. H. ∼2. G. 1.846 (artif.).[1] Luster vitreous. Color pale rose. Dehydrates rapidly at room temperature, becoming opaque and mealy.

O p t.[2] In transmitted light, colorless. Biaxial positive (+), with $Y = b$ and $Z \wedge c = +43°$. Indices of refraction not known.

C h e m. Manganese sulfate heptahydrate, $MnSO_4 \cdot 7H_2O$. Cu and Fe substitute in part for Mn, in artificial material.[5] Analysis gave:

	CaO	MgO	MnO	SO_3	H_2O	Insol.	Total
1.			25.60	28.89	45.51		100.00
2.	0.7	0.6	23.6	29.0	44.5	1.6	100.0

1. $MnSO_4 \cdot 7H_2O$.   2. Lucky Boy mine, Utah.[3]

Tests. In C.T. loses water, and on strong heating affords SO_3 and leaves a brown residue. Soluble in water (53.5 g. $MnSO_4 \cdot 7H_2O$ in 100 parts H_2O).

Occur. Found with manganoan melanterite as an oxidation product at the Lucky Boy mine, Butterfield Canyon, Salt Lake County, Utah. A cuprian and zincian variety is said to occur in the Bayard area of the Central District, Grant County, New Mexico.

Artif.[4] Obtained as crystals from water solution at temperatures below 9°. Lower hydrates are obtained at higher temperatures.

Name. After Ernest Mallard (1833–1894), French crystallographer.

Ref.

1. Elements and forms of Gunther, Inaug. Diss., Jena, 1908, cited in *Zs. Kr.*, **50**, 91 (1912), on artificial crystals.
2. Mallard in Carnot (1879). See also Larsen and Glenn, *Am. J. Sc.*, **50**, 225 (1920).
3. Carnot (1879).
4. Mellor (**12**, 401, 1932) and Gunther (1908).
5. Stortenbecker, *Zs. phys. Chem.*, **16**, 577 (1895), and Retgers, *Zs. phys. Chem.*, **16**, 590 (1895).

ZINC-MELANTERITE [$(Zn,Cu)SO_4 \cdot 7H_2O$]. Zinc-copper melanterite *Larsen and Glenn* (*Am. J. Sc.*, **50**, 225, 1920). Zinkboothit *Doelter* (4 [2], 296, 1927).

Apparently monoclinic and isostructural with the members of the melanterite group. Massive, columnar; under the microscope seen to consist of rod-like crystals. H. ~2. G. 2.02. Luster vitreous. Color pale greenish-blue. Dehydrates readily at room temperature to the pentahydrate (?).[3]

Opt. In transmitted light, very pale blue-green.

Orientation	n	
X	1.479	Biaxial positive (+).
$Y \wedge$ elong. ~34°	1.483	$2V$ near 90°.
Z	1.488	Dispersion slight.

Chem. Heptahydrated sulfate of zinc and copper, $(Zn,Cu,Fe)SO_4 \cdot 7H_2O$. In the only reported analysis, $Zn:Cu:Fe = 100:98:19$. Zinc-melanterite apparently represents the monoclinic polymorph of cuprian goslarite, and is related to zincian pisanite and zincian boothite. The name is here restricted to regions of composition in which $Zn >$ (Cu,Fe, etc.). The pure end-member, monoclinic $ZnSO_4 \cdot 7H_2O$, is not stable in artificial systems, but intermediate members of a series toward boothite have been synthesized.[1]

Anal.

	CuO	ZnO	FeO	SO_3	H_2O	Insol.	Total
1.	13.88	14.19		27.93	44.00		100.00
2.	12.37	12.89	2.14	28.78	42.61	1.11	99.90

1. $(Zn,Cu)SO_4 \cdot 7H_2O$ with $Zn:Cu = 1:1$. 2. Vulcan, Colorado.[2]

Tests. B.B. fuses with intumescence to a white froth which turns black on continued heating. Readily soluble in water.

Occur. Found in considerable amounts as an oxidation product of pyrite-chalcopyrite-sphalerite ore at the Good Hope and Vulcan mines at Vulcan, Gunnison County, Colorado.

Name. In allusion to its composition and membership in the melanterite group.

Ref.

1. Stortenbecker, *Zs. phys. Chem.*, **22**, 60 (1897), and Hollmann, *Zs. phys. Chem.*, **54**, 98 (1906).
2. Glenn analysis in Larsen and Glenn (1920).
3. This material is biaxial negative (−), with moderate $2V$ and nX 1.513, nY 1.533, nZ 1.540—Larsen and Glenn (1920).

29.6.9 EPSOMITE GROUP

ORTHORHOMBIC; DISPHENOIDAL—2 2 2

	a_0	b_0	c_0
Epsomite, $MgSO_4 \cdot 7H_2O$	11.94 kX	12.03	6.865
Goslarite, $ZnSO_4 \cdot 7H_2O$	11.85	12.09	6.83
Morenosite, $NiSO_4 \cdot 7H_2O$	11.8	12.0	6.80
Tauriscite, $FeSO_4 \cdot 7H_2O$	....	....	....

The structure [1] of the members of this group contains tetrahedral (SO_4) groups together with six water molecules in octahedral coordination about the metal ions. A seventh, relatively loosely held molecule of H_2O is linked to oxygen ions of the (SO_4) groups and to other water molecules; it is easily lost on dehydration and yields the structurally related hexahydrates of the $NiSO_4 \cdot 6H_2O$ (*retgersite*) [2] or $MgSO_4 \cdot 6H_2O$ (*hexahydrite*) types.

The members of the group enter into extensive series through the substitution of the divalent cations. Complete series extend between the Mg-Ni and Mg-Zn members. Only limited amounts of Cu, Fe, and Mn substitute for (Mg,Zn,Ni) and the heptahydrates of these elements crystallize in the monoclinic Melanterite Group. The existence of the ferrous iron member of the Epsomite Group, tauriscite, is very uncertain. The members of the group are water-soluble and occur chiefly as efflorescences in arid or sheltered places.

Ref.
1. Beevers and Schwartz, *Zs. Kr.*, **91**, 157 (1935).
2. Beevers and Lipson, *Zs. Kr.*, **83**, 123 (1932).

29.6.9.1 **EPSOMITE** [$MgSO_4 \cdot 7H_2O$]. Epsom Salt. Sal nativum catharticum *Hermann* (*De Sale nativo cathartico in fodinis Hungariae recens invento*, Posonii, 1721). Sal neutrum acidulare, Sal Anglicanum *Wallerius* (184, 1747). Sel d'Epsom (*French transl.* Wallerius, **1**, 339, 1753). Halotrichum *Scopoli* (*De Hydrarg. Idriense Tent.*, Venet. 1761 [Klaproth's *Beitr.*, **3**, 104, 1802], *Princip. Min.*, 1772). Magnesia vitriolata (Sal Anglicus, Epsomensis, Seidlitzensis, Seydschützensis, Sal amarus, Sal catharticum, etc.) *Bergmann* (1782). Bittersalz *Werner*. Gletschersalz. Haarsalz pt. Bitter Salts. Epsomite *Beudant* (445, 1824). Reichardtit *Krause* (*Zs. Nat. Halle*, **44**, 554, 1874; *Arch. Pharm.*, **5**, 423, 1874; **6**, 41, 1875). Fauserite *Breithaupt* (*B. H. Ztg.*, **24**, 301, 1865). Seelandite *Brunlechner* (*Jb. Nat. Land.-Mus. Klagenfurt*, **22**, 192, 1893). Ferroepsomite *Vertushkov* (*Bull. ac. sc.*, *U.R.S.S.*, *Sér. géol.*, 109, 1939).

C r y s t.[1] Orthorhombic; disphenoidal—2 2 2.

$a:b:c = 0.9901:1:0.5709; \quad p_0:q_0:r_0 = 0.5766:0.5709:1$

$q_1:r_1:p_1 = 0.9901:1.7343:1; \quad r_2:p_2:q_2 = 1.7516:1.0100:1$

Forms: [2]

		ϕ	$\rho = C$	ϕ_1	$\rho_1 = A$	ϕ_2	$\rho_2 = B$
b	010	0°00′	90°00′	90°00′	90°00′	.....	0°00′
a	100	90 00	90 00	.....	0 00	0°00′	90 00
f	120	26 47½	90 00	90 00	63 12½	0 00	26 47½
m	110	45 17	90 00	90 00	44 43	0 00	45 17
g	210	63 39½	90 00	90 00	26 20½	0 00	63 39½
l	310	71 44	90 00	90 00	18 16	0 00	71 44
v	011	0 00	29 43½	29 43½	90 00	90 00	60 16½
r	021	0 00	48 47½	48 47½	90 00	90 00	41 12½
n	101	90 00	29 58	0 00	60 02	60 02	90 00
x	201	90 00	49 04	0 00	40 56	40 56	90 00
z	111	45 17	39 03½	29 43½	63 24	60 02	63 41
t	121	26 47½	51 59	48 47½	69 12	60 02	45 19
s	211	63 39½	52 09	29 43½	44 57½	40 56	69 29½

Structure cell.[3] Space group $P2_12_12_1$. a_0 11.94 kX, b_0 12.03, c_0 6.865; $a_0:b_0:c_0 = 0.9925:1:0.5707$. Cell contents $Mg_4(SO_4)_4 \cdot 28H_2O$.

Habit. Artificial crystals short prismatic [001] to equant; common forms *m b z n*. The habit varies with the presence of cosolutes.[4] The

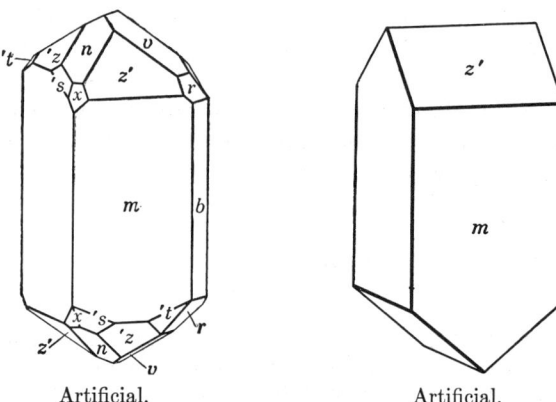

Artificial.   Artificial.

natural material is rarely well-crystallized. Usually as fibrous to hair-like or acicular crusts, the fibers elongated [001]; also as woolly efflorescences; as botryoidal or reniform masses; stalactitic.

Twinning.[4] On {110}, rare.

Etch figures.[5] Conformable to disphenoidal symmetry.

P h y s. Cleavage {010} perfect, {101} distinct. Fracture conchoidal. H. 2–2½. G. 1.677 ± 0.002 (artif.),[6] 1.650 (calc.). Luster vitreous; of fibrous types silky to earthy. Colorless and transparent in pure single crystals, massive aggregates white and translucent; also pink (cobaltian) or greenish (nickelian). Weakly diamagnetic. Optical rotatory power [7] along the optic axis, 1.98° per mm. for $\lambda = 579m\mu$. Translation gliding: [8] $T\{110\}$, $t[1\bar{1}0]$?; $T\{100\}$, $t[0\bar{1}0]$?; $T\{011\}$, $t[011]$?; $T\{101\}$, $t[\bar{1}01]$? Epsomite and the zinc and nickel analogues form oriented overgrowths [9] upon muscovite, with epsomite {100} or {110} parallel muscovite {001} and epsomite [001] parallel muscovite [120], [310], [150], [1$\bar{2}$0], [3$\bar{1}$0], [1$\bar{5}$0]. Effloresces in dry air. Taste bitter and saline to metallic.

O p t.[13] In transmitted light, colorless.

Orientation		$n(B)$	$n(D)$	$n(F)$	$n(G)$	
X	a	1.4299	1.4325	1.4378	1.4400	Biaxial negative ($-$).
Y	c	1.4523	1.4554	1.4611	1.4623	$r < v$, weak.
Z	b	1.4572	1.4609	1.4666	1.4694	
2V (calc.)		49°57'	51°35'	51°11'	50°25'	

The indices of refraction vary with increasing substitution of Ni or Zn for Mg, increasing linearly toward the values for pure morenosite or goslarite, respectively.[10] Optical data are lacking for the partial series with Mn, Fe, Cu, Co.

C h e m. Magnesium sulfate heptahydrate, $MgSO_4 \cdot 7H_2O$. Epsomite forms partial or complete solid-solution series with a number of other divalent metals [11] in artificial material. Complete series exist to $NiSO_4 \cdot 7H_2O$ and $ZnSO_4 \cdot 7H_2O$. Fe'' substitutes for Mg up to a maximum of Fe'':Mg $\sim$ 1:5. Mn'' substitutes for Mg up to a maximum of Mn:Mg $\sim$ 4:10. Cu and Co can also substitute to a limited extent for Mg. The name epsomite is here applied to those regions of composition in which the atomic per cent of Mg is greater than that of any individual substituting element, regardless of the total summation of the substituting elements.

In dry air at ordinary temperature epsomite loses up to $1H_2O$ (the change being reversible) and affords hexahydrite; in vacuum or at higher temperatures various lower hydrates are formed.[12]

Anal.[14]

	1	2	3	4	5	6	7	8	9
MgO	16.36	15.24	16.26	17.24	13.36	12.76	5.15	14.06	11.40
ZnO		0.53		0.17	1.37	2.82			
FeO		0.049		tr.	0.64	0.015		4.42	7.77
CoO		0.11		1.06		0.068			
NiO		tr.		0.07	0.13	tr.			
MnO		0.40		0.20	2.63	4.28	19.61		
SO_3	32.48	32.33	32.41	34.15	31.23	31.54	34.49	31.80	31.39
H_2O	51.16	51.10	51.32	47.44	[48.94]	48.41	42.66	49.49	49.40
Rem.		tr.	0.113	0.20	1.70	tr.			tr.
Total	100.00	99.759	100.103	100.53	[100.00]	99.893	101.91	99.77	99.96
G.		1.678	1.68	1.68–1.69		1.70		1.888	1.681

1. $MgSO_4 \cdot 7H_2O$. 2. Herrengrund, Czechoslovakia.[15] Rem. is insol. 3. Ashcroft, British Columbia.[16] Rem. is $(Na,K)_2O$ 0.055, $(Al,Fe)_2O_3$ 0.05, Cl 0.003, insol. 0.005. 4. Neusohl, Czechoslovakia.[15] Rem. is insol. 5. New Brancepeth, Scotland.[17] Rem. is insol. 6. Schemnitz, Hungary.[15] Rem. is insol. 7. Herrengrund (*fauserite*).[18] Probably a mixture of $(Mg,Mn)SO_4 \cdot 7H_2O$ and $(Mn,Mg)SO_4 \cdot 5H_2O$. 8. Idria, Italy.[19] 9. Lafayette, Montgomery County, Pennsylvania.[20] Rem. is CuO.

Var. *Nickelian.* A complete series exists to morenosite, $NiSO_4 \cdot 7H_2O$. Color pale green, increasing in depth with increasing substitution of Ni.

Manganoan. Fauserite *Breithaupt* (*B. H. Ztg.*, **24**, 301, 1865). Contains Mn in substitution for Mg up to a maximum of about Mn:Mg = 2:5.

Material with a much higher content of Mn has been reported (anal. 7) but probably is a mixture.[21] Color pink.

Zincian. Only small amounts of Zn have been found in substitution for Mg in natural material, with Zn:(Mg,Mn,Co) $\sim$ 1:11 in analysis 6, but a complete series to $ZnSO_4 \cdot 7H_2O$ (*goslarite*) can be obtained synthetically.

Tests. Heated in the C.T. melts in its own water of crystallization and at higher temperatures affords much acid water. B.B. fuses and finally affords an infusible alkaline mass of $MgSO_4$. Very soluble in water (36.4 g. $MgSO_4$ in 100 g. H_2O at 25°).

O c c u r. Commonly found as crusts or delicate fibrous efflorescences on the walls of mine workings in coal or metal deposits; on the walls or floors of limestone caves; as efflorescences in sheltered places on outcrops of dolomitic or calcareous sedimentary rocks and gypsum, or on magnesian igneous and metamorphic rocks particularly in the presence of pyrite or pyrrhotite. In the oxidized zone of pyritic sulfide deposits in arid regions. Common in the waters of mineral springs; also dissolved in salt lakes and as a deposit therefrom during the dry summer season. Also found in oceanic salt deposits as an alteration product of kieserite or formed by the action of magnesium-rich solutions on anhydrite, gypsum, or polyhalite. Often associated with melanterite and other iron sulfates, gypsum, halotrichite, pickeringite, alums, alunogen, mirabilite. Only a few of the many known localities can be cited here.

In mineral waters and as a deposit therefrom in the long-known occurrences at Epsom, Surrey, England, and at Sedlitz and Saidschitz, Bohemia. In silky fibers (hair salt) at the mercury deposit of Idria, Gorizia (Carniola), Italy. In small amounts in the fumarole deposits on Vesuvius. From mine workings at Herrengrund, Neusohl, and Hodrusbánya, Czechoslovakia, and with mascagnite, salammoniac, copiapite, etc., from burning pyritic coal heaps at Kladno. A massive variety (*reichardite*) occurs in thin layers with carnallite at Leopoldshall, Anhalt, and at Stassfurt, Saxony, Germany. In Spain at Calatayud, Zaragoza, and numerous other localities.[22] In France in a coal mine at Peychagnard, Isère, and elsewhere.[23] In the United States, on the walls and mingled with the earth of the floors in limestone caves in Kentucky, Tennessee, and Indiana. As a deposit from salt lakes in Albany, Carbon, and Natrona counties, notably near Wilcox Station, Albany County, Wyoming. Near Manti, Sanpete County, Utah. In the Carlsbad potash district, Eddy County, New Mexico, as a hydration product of kieserite and langbeinite. Found widely in Nevada in saline surface deposits in playas; also in mine workings in the Comstock Lode, Virginia City.[24] At the pyrite mines of Leona Heights, Alameda County, California, and at many other places in this state. Abundantly with mirabilite in salt lakes on Kruger Mountain near Oroville, Washington, some crystals being several feet long.[25] In Canada notably as lake deposits near Ashcroft, British Columbia.[26]

A l t e r. Dehydrates easily on exposure at ordinary conditions to hexahydrite, $MgSO_4 \cdot 6H_2O$, or at higher temperatures to lower hydrates.

GOSLARITE

A r t i f.[27] Epsomite crystallizes from pure water solutions at temperatures below about 50°; the hexahydrate forms over this point but may be obtained in metastable equilibrium at lower temperatures.

N a m e. From the occurrence at Epsom, England.

FAUSERITE. *Breithaupt (B. H. Ztg.*, **24**, 301, 1865). Orthorhombic, with angles and habit close to epsomite. Color reddish and yellowish white. Supposedly $(Mn,Mg)SO_4 \cdot 7H_2O$ with $Mn:Mg \sim 2:1$ in analysis 7. Shown [21] to be probably a mixture of $(Mg,Mn)SO_4 \cdot 7H_2O$ with $Mn:Mg \sim 1:2$ and $(Mn,Mg)SO_4 \cdot 5H_2O$.

Ref.

1. Brooke, *Ann. Phil.*, **22**, 40 (1823), on artificial crystals; see also Miller (546, 1852); Rammelsberg, *Ann. Phys.*, **91**, 324 (1854); Mahl, *Jb. Min., Beil.-Bd.*, **59**, 273 (1929).
2. Goldschmidt (**3**, 151, 1916). Also {221}, Zirkl, *Min. Mitt.*, **1**, 185 (1948).
3. Barnes and Hunter, *Nature*, **130**, 96 (1932), confirming the earlier work of Cardoso, *Zs. Kr.*, **63**, 19 (1926), and Westenbrink, *Proc. Ac. Sc. Amsterdam*, **29**, 1223 (1926), on artificial crystals.
4. Johnsen, *Jb. Min., Beil.-Bd.*, **23**, 315 (1907).
5. Blasius, *Zs. Kr.*, **10**, 227 (1885), and Gaudefroy, *C.R.*, **157**, 61 (1913).
6. Range of six best values reported in the literature.
7. Longchambon, *Bull. soc. min.*, **45**, 238 (1922).
8. Johnsen, *Cbl. Min.*, 33, 1915.
9. Royer, *C.R.*, **194**, 1088 (1932).
10. Dufet, *Bull. soc. min.*, **3**, 180 (1880), **12**, 22 (1889), **1**, 58 (1878); Hutton, *Am. Min.*, **32**, 553 (1947); Porter, *Zs. Kr.*, **75**, 288 (1930).
11. Cf. Groth (**2**, 401, 1908); Retgers, *Zs. phys. Chem.*, **3**, 534 (1889), **16**, 580 (1895); Hey, *Min. Mag.*, **22**, 510 (1931).
12. Johnsen, *Cbl. Min.*, 289, 1915; Mellor (**4**, 329, 1923); Merwin, *Zs. Kr.*, **55**, 113 (1915).
13. Indices of Borel, *Arch. sc. phys. nat. Genève* [3], **34**, 230 (1895), by prism method on artificial material.
14. For additional analyses see Hintze (**1** [3B], 4346, 1929); Hey (1931); Zsivny, *Ann. Hist.-Nat. Mus. Nat. Hungar.*, **13**, 577 (1915); Kokta, *Příroda*, Brno, **32**, 381 (1930); Shannon, *Proc. U. S. Nat. Mus.*, **74**, Art. 13 (1929); Vavrinecz, *Magyar Chem. Foly.*, **38**, 140 (1932); Mélon, *Ann. soc. géol. Belgique*, **67**, B56 (1944).
15. Hey (1931).
16. Walker, *Univ. Toronto Stud., Geol. Ser.*, 44, 1921.
17. Smythe, *The Vasculum*, Newcastle-on-Tyne, **19**, 12 (1933)—*Min. Abs.*, **5**, 283 (1933).
18. Mollnár in Breithaupt, *B. H. Ztg.*, **24**, 301 (1865).
19. Onorato, *Acc. Linc., Atti.*, [6], **2**, 204 (1925), who gives additional analyses, one with FeO 4.68 per cent.
20. Eyermann, cited in *Zs. Kr.*, **54**, 100 (1914).
21. See Hey (1931).
22. Tenne and Calderón (**2**, 235, 1910).
23. Lacroix (**4**, 212, 1910).
24. Milton and Johnston, *Econ. Geol.*, **33**, 749 (1938).
25. Jenkins, *Am. J. Sc.*, **46**, 638 (1918).
26. Walker, *Univ. Toronto Stud., Geol. Ser.*, no. 12, 43 (1921).
27. Mellor (**4**, 322, 1923); on the conditions of formation in the presence of other salts, see also Doelter (**4** [2], 46, 1929); Benrath and Neumann, *Zs. anorg. Chem.*, **242**, 70 (1939).

29.6.9.2 **G O S L A R I T E** [$ZnSO_4 \cdot 7H_2O$]. Atramentum sutorium, candidum, potissimum reperitur Goselariæ, translucidem, crystalli instar *Agricola* (213, 1546). Atramentum album fossile durum Goslarianum *Gesner* (13, 1565). Vitriolum Zinci album nativum, Galizensten, Hvit Viktril *Wallerius* (157, 1747). Zinc Vitriol, White Vitriol, White Copperas, Sulphate of Zinc. Zinc sulfatée, Couperose blanche *Fr.*

Gallizinite *Beudant* (446, 1824). Galiznite. Goslarite *Haidinger* (490, 1847). Ferro-Goslarite *Wheeler* (*Am. J. Sc.*, **41**, 212, 1891). Cuprogoslarite *Rogers* (*Kansas Univ. Quart.*, **8**, 105, 1899).

C r y s t.[1] Orthorhombic; disphenoidal—2 2 2.

$$a:b:c = 0.9804:1:0.5631; \quad p_0:q_0:r_0 = 0.5744:0.5631:1$$

$$q_1:r_1:p_1 = 0.9804:1.7411:1; \quad r_2:p_2:q_2 = 1.7759:1.0200:1$$

Forms:[1]

		ϕ	$\rho = C$	ϕ_1	$\rho_1 = A$	ϕ_2	$\rho_2 = B$
b	010	0°00′	90°00′	90°00′	90°00′	.....	0°00′
a	100	90 00	90 00	.....	0 00	0°00′	90 00
f	120	27 01½	90 00	90 00	62 58½	0 00	27 01½
m	110	45 34	90 00	90 00	44 26	0 00	45 34
v	011	0 00	29 23	29 23	90 00	90 00	60 37
r	021	0 00	48 24	48 24	90 00	90 00	41 36
n	101	90 00	29 52	0 00	60 08	60 08	90 00
x	201	90 00	48 57½	0 00	41 02½	41 02½	90 00
z	111	45 34	38 48½	29 23	63 24½	60 07½	63 58½
t	121	27 01½	51 39½	48 24	69 07½	60 07½	45 40½
s	211	63 53½	51 59½	29 23	44 58½	41 02½	69 42½

Structure cell.[2] Space group $P2_12_12_1$. a_0 11.85 kX, b_0 12.09, c_0 6.83; $a_0:b_0:c_0 = 0.982:1:0.566$. Cell contents $Zn_4(SO_4)_4 \cdot 28H_2O$.

Habit. Artificial crystals stout prismatic [001]. Usually as efflorescent crusts or stalactitic or stalagmitic masses with a fibrous structure; massive, granular, or fibrous.

P h y s. Cleavage {010} perfect. Brittle. H. 2–2½. G. 1.978 (crystals)[3] 1.94 (calc.). Luster vitreous; silky in fibrous material. Colorless and transparent in pure crystals, inclining toward brown, green, or blue in material containing Fe, Mn, Cu in solid solution; massive material white. Optical rotation[4] per mm., 2.41° for $\lambda = 579$, 4.05° for λ 436. Strongly diamagnetic. Dehydrates easily to a dull, white powder. Taste astringent, metallic, and nauseous.

Oriented growths.[5] Upon muscovite with goslarite {100} [001] or {010} [001] parallel muscovite {001} [120], [310], [150], [1$\bar{2}$0], [3$\bar{1}$0], or [1$\bar{5}$0].

O p t. In transmitted light, colorless.

Orientation	n(Na) (artif.)[6]	n (Comstock Lode, anal. 6)[7]	n (Comstock Lode, anal. 5)[7]	
X b	1.4568	1.463	1.447	Biaxial negative (−).
Y c	1.4801	1.475		$r > v$, very weak.
Z a	1.4844	1.480	1.470	
2V	46°10′ (meas.)	mod.	small	

C h e m. Zinc sulfate heptahydrate, $ZnSO_4 \cdot 7H_2O$. A complete series extends to $MgSO_4 \cdot 7H_2O$ in artificial material,[6] with Mg:Zn = 46:54 in analysis 5. Cu substitutes for Zn to an apparent maximum of Cu:Zn = 1:47 in artificial material,[8] but with Cu:Zn = 1:3.5 in analysis 9 and Cu:(Zn,Mg) = 1:5.7 in analysis 6. Fe″ substitutes for Zn to an apparent maximum of Fe:Zn = 1:8.65 in artificial material,[9] but with Fe:(Zn,etc.)

= 1:3.36 in analysis 7 and 1:8.05 in analysis 6. Mn″ substitutes for Zn with Mn:Zn = 1:2.9 in analysis 8, but this occurrence needs verification. Goslarite dehydrates to the hexahydrate at about 33° and to the monohydrate at about 100°. Some analyses of natural material approximate to $6H_2O$, due to partial dehydration.

Anal.[10]

	1	2	3	4	5	6	7	8	9
MgO		0.81	1.58	4.19	6.90	4.86	1.78		
CuO					0.19	4.27	0.91		6.68
ZnO	28.30	23.12	22.91	18.80	14.05	11.77	19.88	21.7	23.83
FeO		1.12	0.95	1.88		2.85	6.41		0.13
NiO		0.06	0.22						
MnO		1.24	0.54		0.20	0.30		6.5	
SO_3	27.84	27.29	27.76	28.32	30.00	29.41	26.80 }	71.74	[27.02]
H_2O	43.86	42.42	41.84	46.18	[48.19]	[46.11]	43.70 }		41.76
Rem.			3.65	4.50	0.35	0.47	0.43		
Total	100.00	99.71	100.30	99.72	[100.00]	[100.00]	99.48	99.94	[99.42]

1. $ZnSO_4 \cdot 7H_2O$. 2,3. Ushaw Moor, Durham, England.[11] Rem. is insol. 4. Magnesian goslarite. Raibl.[12] Rem. is insol. 5. Magnesian goslarite. Comstock Lode, Nevada.[13] Rem. is $(Al,Fe)_2O_3$ 0.31, insol. 0.16. Contains about 2 per cent pickeringite. 6. Cuprian and magnesian goslarite. Comstock Lode.[13] Rem. is $(Al,Fe)_2O_3$ 0.30, insol. 0.13. 7. Ferroan goslarite. Freiberg, Saxony.[14] 8. Manganoan goslarite. Rammelsberg.[15] 9. Cuprian goslarite. Galena, Kansas (*cuprogoslarite*).[16]

Tests. In C.T. yields water. Soluble in water (53.8 g. $ZnSO_4$ per 100 g. H_2O).

Var. Magnesian. A complete series exists to the magnesium analogue, epsomite, with accompanying linear variation in the indices of refraction and gravity.

Ferroan. Ferro-Goslarite Wheeler (*Am. J. Sc.*, **41**, 212, 1891). Contains Fe″ in substitution for Zn. Color green to brown.

Cuprian. Cuprogoslarite Rogers (*Kansas Univ. Quart.*, **8**, 105, 1899). Contains Cu″ in substitution for Zn. Color bluish.

Occur. Goslarite is a secondary mineral, generally formed by the alteration of sphalerite, and occurs especially as an efflorescence on the walls of mine passages. Associated with epsomite, melanterite, pickeringite, chalcanthite, gypsum. Long known from the Rammelsberg mine near Goslar in the Harz and Schemnitz in Bohemia; also from Freiberg and Altenberg, Saxony. In France from the pyrite mine at Sain-Bel near Lyon, Rhône. At Falun, Kopparberg, Sweden. In the Almagrera mine, Tharsis, Huelva province, Spain. In Italy at Capanne Vecchie, Elba; from Raibl, Venezia Giulia. In the Esperanza mine, Cerro de Pasco, Peru, and at La Alcaparrosa, San Juan province, Argentina. In the United States, at the Gagnon mine, Butte, Montana; at Bingham Canyon, Utah; in mines on the Comstock Lode, Nevada. In Arizona in the Globe district, Gila County, the Clifton-Morenci district, Greenlee County, and the Silver Bell district, Pima County. In New Mexico, especially in the Central district, Grant County. At Island Mountain, Trinity County, California. A ferroan variety at Webb City, Jasper County, Missouri (*ferro-goslarite*), and a cupran variety at Galena, Cherokee County, Kansas (*cuprogoslarite*).

Artif. Easily obtained as crystals by evaporation of the water solution below 38°.

Name. From the occurrence at Goslar in the Harz.

Ref.

1. Goldschmidt (**9,** 119, 1923) on artificial crystals. On the variation in angles with Mg content see Dufet, *Bull. soc. min.*, **12,** 22 (1889).
2. Westenbrink, *Proc. Ac. Sc. Amsterdam*, **29,** 1223 (1926)—Strukturber. (I, 388, 1931).
3. Average of ten best separate determinations reported in the literature; range 1.953–2.036.
4. Longchambon, *Bull. soc. min.*, **45,** 239 (1922).
5. Royer, *C.R.*, **194,** 620, 1088 (1932), artificially, from water solution.
6. Dufet, *Bull. soc. min.*, **3,** 187 (1880) and Doelter (4 [2], 257, 1927). See also Porter, *Zs. Kr.*, **75,** 288 (1930).
7. Milton and Johnston, *Econ. Geol.*, **33,** 749 (1938).
8. Stortenbecker, *Zs. phys. Chem.*, **22,** 60 (1897).
9. Wyrouboff, *Bull. soc. min.*, **3,** 73 (1880).
10. For additional analyses see Morgante, *Atti Accad. Ven.-Trent.-Istr.*, **22,** 49 (1932); Collins, *Min. Mag.*, **20,** 32 (1923); Syrokomsky cited in *Min. Abs.*, **2,** 136 (1923); Lacroix (**4,** 219, 1910); Hintze (1 [3B], 4353, 1929); Doelter (4 [2], 254, 1927).
11. Smythe, *The Vasculum*, Newcastle-on-Tyne, **19,** 12 (1923)—*Min. Abs.*, **5,** 283 (1933).
12. Morgante (1932).
13. Milton analysis in Milton and Johnston (1938).
14. Frenzel, *Jb. Min.*, 675, 1875.
15. Hausmann, *Hercyn. Archiv.*, **3,** 573 (1808).
16. Rogers (1899); orthorhombic symmetry is not proved, and this material may belong in a monoclinic solid solution series with $CuSO_4 \cdot 7H_2O$—see Stortenbecker (1897).

29.6.9.3 **MORENOSITE** [$NiSO_4 \cdot 7H_2O$]. Nickel-Viktril, Vitriolum ferrum et nicolum continens ("of a deep green color, with Kupfernickel, in Cobalt mines") *Cronstedt* (114, 1758). Niccolum vitriolatum (interdum e mineris sulphuratis fotiscentibus genitum) *Bergmann* (50, 1782). Sulfato de niquel (from Galicia) *Casares* (1849; see Alcibar in *Rev. min.*, 305, 1850). Sulfato de nickel, Morenosita *Casares* (*Rev. min.*, 176, 1851). Nickel Vitriol *Hunt* (in Dana, 679, 1850). Pyromelin *von Kobell* (*Gel. Anz. Münch.*, **35,** 215, 1852; *J. pr. Chem.*, **58,** 44, 1853). Nickel-epsomite.

Cryst.[1] Orthorhombic; disphenoidal—2 2 2.

$$a:b:c = 0.9817:1:0.5656; \quad p_0:q_0:r_0 = 0.5761:0.5656:1$$

$$q_1:r_1:p_1 = 0.9817:1.7357:1; \quad r_2:p_2:q_2 = 1.7680:1.0186:1$$

Forms:[2]

		ϕ	$\rho = C$	ϕ_1	$\rho_1 = A$	ϕ_2	$\rho_2 = B$
b	010	0°00′	90°00′	90°00′	90°00′	.....	0°00′
f	120	26 59½	90 00	90 00	63 00½	0°00′	26 59½
m	110	45 31½	90 00	90 00	44 28½	0 00	45 31½
v	011	0 00	29 29½	29 29½	90 00	90 00	60 30½
r	021	0 00	48 31	48 31	90 00	90 00	41 29
n	101	90 00	29 57	0 00	60 03	60 03	90 00
x	201	90 00	49 02½	0 00	40 57½	40 57½	90 00
z	111	45 31½	38 55	29 29½	63 22	60 03	63 53½
w	1$\bar{1}$1	134 38½	38 55	29 29½	63 22	60 03	116 38
t	121	26 59½	51 46	48 31	69 07	60 03	45 34½
s	211	63 51½	52 04½	29 29½	44 55	40 57½	69 39½
ξ	2$\bar{1}$1	116 08½	52 04½	29 29½	44 55	40 57½	110 20½

MORENOSITE 517

Structure cell. [3] Space group $P2_12_12_1$. a_0 11.8 kX, b_0 12.0, c_0 6.80; $a_0:b_0:c_0 = 0.983:1:0.564$. Cell contents $Ni_4(SO_4)_4 \cdot 28H_2O$.

Habit. Artificial crystals are short prismatic [001] with m, z, w, b, and n prominent. The natural material occurs only as efflorescent crusts of indistinct crystals and fibers; stalactitic.

Etch figures.[6] Conformable with disphenoidal symmetry.

P h y s. Cleavage {010} distinct. Fracture conchoidal. H. 2–2½. G. 1.953;[4] 1.93 (calc.). Luster vitreous. Color apple-green to greenish white. Streak white, faintly greenish. Transparent when unaltered. Optical rotatory power[7] along the optic axis, 6.1° per mm. for λ = 579mμ. Weakly paramagnetic. Taste metallic astringent.

O p t. In transmitted light, green.

ORIENTATION		n(Na)[5]	n (Minasragra)[9]	
X	a	1.4693	1.470	Biaxial negative (−).
Y	c	1.4893	1.493	2V 41°54′ (meas.).
Z	b	1.4923	1.500	$r > v$, weak.

The indices decrease linearly to the values for epsomite, $MgSO_4 \cdot 7H_2O$, as Mg substitutes for Ni.[14]

C h e m. Nickel sulfate heptahydrate, $NiSO_4 \cdot 7H_2O$. A complete series exists to epsomite, $MgSO_4 \cdot 7H_2O$, and intermediate members of the series[20] have been found in nature (anals. 4, 5; see also *epsomite*). In artificial material, Fe substitutes[8] for Mg up to at least Fe:Ni = 1:5. V has been found spectrographically.[9] Cu can substitute for Ni to about Cu:Ni = 2:98 in artificial material.[19] Part at least of the water content over $6H_2O$ probably can be lost without breakdown of the structure.

Anal.

	1	2	3	4	5	6
NiO	26.59	26.76	27.16	17.13	18.5	14.16
MgO		0.25		5.80	6.5	7.65
SO$_3$	28.51	28.54	29.58	29.87	28.7	30.36
H$_2$O	44.90	44.43	42.49	47.20	46.5	47.83
Rem.		0.27	0.31			
Total	100.00	100.00	99.79	100.00	100.2	100.00

1. $NiSO_4 \cdot 7H_2O$. 2. Riechelsdorf, Hesse-Nassau.[10] Rem. is As_2O_5. Not certainly morenosite. 3. Valtournanche, Piedmont.[11] Rem. is Al_2O_3 0.10, Fe_2O_3 0.21, Cu tr. Water-soluble portion of a sample containing 5.35 per cent insoluble. 4. Val Malenco, Lombardy.[12] Recalculated to 100 after deducting 11.32 per cent insoluble. 5. Zermatt, Switzerland.[13] Not certainly morenosite. 6. $(Ni,Mg)SO_4 \cdot 7H_2O$ with Ni:Mg = 1:1.

Var. *Magnesian.* Contains Mg in substitution for Ni, the species name morenosite being applied to all compositions in the series up to Ni:Mg = 1:1. The specific gravity and indices of refraction[14] vary linearly between morenosite and the pure magnesium analogue, epsomite. The green color becomes paler as Mg increases.

Tests. In O.T. affords much acid water, swells up, and hardens finally becoming yellow and opaque ($NiSO_4$). Easily soluble in water (41.2 g. $NiSO_4$ per 100 g. H_2O at 25°).

Occur. A secondary mineral formed by oxidation of nickel-bearing sulfides and deposited as a coating under essentially atmospheric conditions. Often associated with annabergite. Numerous localities have been reported but in many instances it is not certain which of the several known hydrates of nickel sulfate is represented (see also *retgersite*). Material definitely of the species here described has been found at Joachimsthal, Bohemia,[15] in the Val Malenco, Lombardy, Italy, at Cap Hortegal, Galicia, Spain (the original locality), and with retgersite at Minasragra, Peru. Localities of less certain authenticity include the following: Busserailles in the Valtournanche, Piedmont, Italy; Zermatt, Valais, Switzerland; in cobalt mines in Helsingland, Sweden; in nickel deposits at numerous localities [16] in the Rhine provinces and at Lichtenberg near Bayreuth, Bavaria (*pyromelin*); from Judet, Haute-Garonne, France, and at Mouzaïa, Algeria; in the San Miguel district, La Mar Province, Peru. In the United States, said to occur at the Gap Nickel Mine, Lancaster County, Pennsylvania. In California near Julian, San Diego County, and from the Phoenix cinnabar mine, Napa County. In Canada abundantly in the Sudbury District and at the Wallace mine in the Algoma District, Ontario.

Alter. Dehydrates easily to retgersite.

Artif.[17] The orthorhombic heptahydrate, morenosite, crystallizes below 31.5° from pure water solution although an unstable monoclinic hydrate may be obtained in this region.[18] The tetragonal hexahydrate (*retgersite*) crystallizes above 31.5° to 53.3° above which point a monoclinic hexahydrate is stable. The tetragonal hexahydrate or a lower hydrate may form at ordinary temperatures from solutions containing sulfuric acid.

Name. After Señor Moreno, of Spain. Pyromeline, from πῦρ, -ος, *fire*, and μήλινος, *yellow*, because the mineral turns yellow (the color of the anhydrous sulfate) when heated.

Ref.

1. Marignac, *Arch. sc. phys. nat. Genève*, [1], **14,** 238 (1855), on artificial crystals.
2. Goldschmidt (**6,** 63, 1920) and Marignac (1855), on artificial crystals.
3. Beevers and Schwartz, *Zs. Kr.*, **91,** 157 (1935), by rotation method on artificial material with analysis of structure.
4. Gossner, *Ber.*, **40,** 2374 (1907), and comparable values have been reported by others.
5. Dufet, *Bull. soc. min.*, **1,** 58 (1878), by prism method on artificial crystals. Slightly lower values were reported by Topsøe and Christiansen, *Ann. Phys., Ergzbd.*, **6,** 549 (1874).
6. Blasius, *Zs. Kr.*, **10,** 238 (1885).
7. Longchambon, *Bull. soc. min.*, **45,** 239 (1922).
8. Retgers, *Zs. phys. Chem.*, **16,** 577 (1895), and Wyrouboff, *Bull. soc. min.*, **3,** 73 (1880).
9. Frondel and Palache, *Am. Min.*, **34,** 188 (1949).
10. Fulda, *Ann. chem. Pharm.*, **131,** 213 (1864); another analysis by Koerner, *idem*. The As_2O_5 may be due to admixed annabergite.
11. Cavinato, *Per. Min.*, **9,** 141 (1938).
12. Cavinato, *Acc. Linc., Att.*, [6], **25,** 399 (1937).
13. Pisani, *Bull. soc. min.*, **15,** 48 (1892).
14. Hutton, *Am. Min.*, **32,** 553 (1947).
15. Ulrich, *Časopis Mus. Česk.*, **95,** 123 (1921).

16. Laspeyres, *Verh. nat. Verein, Bonn*, 143, 375 (1893).
17. See Seidell, *Solubilities of Inorg. Compounds*, New York, 1347 (1940).
18. Federov, *Bull. ac. sc. St. Pétersbourg*, 18, 15 (1903).
19. Fock, *Zs. Kr.*, 28, 386 (1897).
20. Dufet, *C.R.*, 86, 880 (1878); *Bull. soc. min.*, 1, 58 (1878).

29.6.9.4 **Tauriscite** [$FeSO_4 \cdot 7H_2O$ (?)]. Tauriszit *Volger (Jb. Min.*, 152, 1855).

As well-formed crystals up to a centimeter in size with the forms and angles of epsomite. Luster vitreous. Color green (?). Transparent. Stated to be $FeSO_4 \cdot 7H_2O$, but analyses are lacking. Found with melanterite and potash alum at Wingdälle, Canton Uri (Pagus Tauriscorum of the Romans), Switzerland. Also said to occur at Szomolnok, Czechoslovakia.[1] Fe can substitute for Mg to a limited extent in the orthorhombic polymorph of $MgSO_4 \cdot 7H_2O$ (*epsomite*), but the pure iron analogue of which tauriscite is supposedly representative has not been synthesized. The existence of the mineral needs confirmation.

Ref.
1. Krenner, *Földt. Közl.*, 17, 556 (1887).

TYPE 7. $A_2B(XO_4)_4 \cdot xH_2O$

29.7.1 **R A N S O M I T E** [$Cu(Fe,Al)_2(SO_4)_4 \cdot 7H_2O$]. Lausen (*Am. Min.*, 13, 221, 1928).

C r y s t. Orthorhombic.[1]

$$a:b:c = 1.7407:1:0.5168$$

Forms:

b 010	a 100	m 110	n 210	e 053	d 301	o 111	s 211	t 311

Habit. As crusts and radiating tufts of needle-like crystals elongated [001].

P h y s. Cleavage {001}, perfect.[2] H. $2\frac{1}{2}$. G. 2.632. Color bright sky-blue. Luster vitreous, pearly on cleavage. Transparent.

O p t. In transmitted light, pale blue.

nX 1.631 ± 0.005 Biaxial positive (+).
nY 1.643 ± 0.005
nZ 1.695 ± 0.005

C h e m. A hydrated sulfate of copper, ferric iron, and aluminum, $Cu(Fe,Al)_2(SO_4)_4 \cdot 7H_2O$. Al substitutes for Fe''' with $Al:Fe = 1:9.4$ in the only reported analysis.

Anal.

	CuO	Fe_2O_3	Al_2O_3	SO_3	H_2O	Total	G.
1.	11.70	21.22	1.44	47.09	18.55	100.00	
2.	11.29	22.57	1.52	46.30	18.82	100.50	2.632

1. $Cu(Fe,Al)_2(SO_4)_4 \cdot 7H_2O$, with $Al:Fe = 1:9.4$. 2. United Verde mine, Jerome, Arizona.

O c c u r. Formed with other hydrated sulfates as the result of a fire in the United Verde mine, Jerome, Arizona.

Name. After Frederick Leslie Ransome (1868-1935), American mining geologist.

Ref.

1. For angle table and crystal drawing see Lausen, *Am. Min.*, **13**, 221 (1928). Crystal imperfections were such as to leave an uncertainty of several degrees in some angles.
2. Described as either pinacoidal or prismatic by Lausen (1928).

PHILLIPITE. Domeyko (*Min. Chili*, 5th App., 38, 1876; 3rd Ed., 248, 1879). Compact, granular, or with a fibrous structure. Luster vitreous. Color azure-blue. Translucent. Taste astringent. Analysis gave: MgO 0.85, CuO 14.39, Fe_2O_3 9.80, Al_2O_3 tr., SO_3 28.96, H_2O 43.72, iron subsulfate 2.28, total 100.00. This approximates to $Cu_3Fe_2'''(SO_4)_6 \cdot 40H_2O$. Found as an alteration product of chalcopyrite at copper mines in the Cordilleras of Condes, Santiago province, Chile.

29.7.2 **ROEMERITE** [$Fe''Fe_2'''(SO_4)_4 \cdot 14H_2O$]. Grailich (*Ak. Wien., Ber.*, **28**, 272, 1858). Bückingit Linck (*Jb. Min.*, I, 213, 1888). Louderbackite Lausen (*Am. Min.*, **13**, 220, 1928).

Cryst.[1] Triclinic; pinacoidal—$\bar{1}$.

$a:b:c = 0.4214:1:0.4174;$ $\quad \alpha\ 91°17',\ \beta\ 100°30',\ \gamma\ 85°31'$

$p_0:q_0:r_0 = 0.9931:0.4116:1;$ $\quad \lambda\ 89°31\tfrac{1}{2}',\ \mu\ 79°34',\ \nu\ 94°19\tfrac{1}{2}'$

$p_0'\ 1.0101,\ q_0'\ 0.4186;$ $\quad x_0'\ 0.1853,\ y_0'\ 0.0084$

Forms:

		φ	ρ	A	B	C
c	001	87°24½	10°30'	79°34'	89°31½'	.....
b	010	0 00	90 00	94 19½	.....	89°31½'
a	100	94 19½	90 00	.....	94 19½	79 34
j	130	40 29	90 00	53 50	40 29	82 51
k	120	52 55	90 00	41 24	52 55	81 21½
m	110	71 13	90 00	23 06½	71 13	79 55
M	1$\bar{1}$0	116 10	90 00	21 50½	116 10	80 47½
K	1$\bar{2}$0	132 12½	90 00	37 53	132 12½	82 34
W	0$\bar{1}$1	155 42½	24 14	78 40	111 58	22 26½
r	$\bar{1}$11	−58 31½	43 56½	128 08	68 45½	52 54

Less common:

i 140	L 2$\bar{3}$0	I 1$\bar{4}$0	x 021	Y 0$\bar{3}$1	D $\bar{1}$01	t $\bar{1}$11	s $\bar{1}$21
N 2$\bar{1}$0	J 1$\bar{3}$0	w 011	X 0$\bar{2}$1	d 101	p 111	q 121	u $\bar{1}$21

Habit. Cuboidal, with prominent development of the zone [001]; thick tabular {010}. Incrusting crystal aggregates; granular; stalactitic.

Phys. Cleavage {010}, perfect, also {001}, less good. Fracture uneven. H. 3–3½. G. 2.174. Color rust-brown to yellow, also reported to be violet-brown. Luster oily to vitreous. Translucent. Taste saline, astringent.

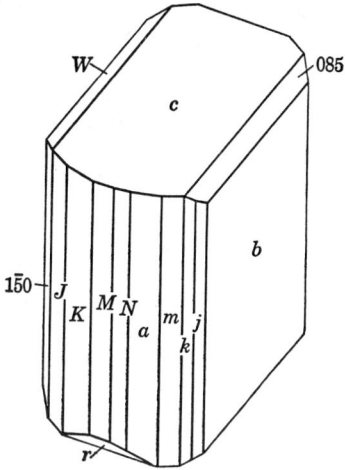

Island Mountain.

O p t.[2] In transmitted light, yellow-brown.

	n (Atacama)	n (Tierra Amarilla)	PLEOCHROISM	
X	1.524 ± 0.003	1.519 ± 0.003	Reddish yellow	Biaxial negative $(-)$.
Y	1.571 ± 0.003	1.570 ± 0.003	Pale yellow	$r > v$, very strong.
Z	1.583 ± 0.003	1.580 ± 0.003	Yellow-brown	Crossed dispersion
$2V$	$51° \pm 2°$ (meas.)	$45° \pm 3°$		also reported.

C h e m. A hydrated sulfate of ferrous and ferric iron, $Fe''Fe_2'''(SO_4)_4 \cdot 14H_2O$. The water content is rather variable. Small amounts of Zn sometimes substitute for Fe''.

Anal.[11]

	1	2	3	4	5	6	7
ZnO		1.97	3.06				
FeO	8.94	6.26	5.80	8.71	6.94	7.01	8.78
Fe_2O_3	19.86	20.63	19.77	20.11	20.60	20.84	19.55
SO_3	39.83	41.54	39.71	39.79	38.30	39.34	38.40
H_2O	31.37	28.00	31.17	30.99	33.40	31.33	30.98
Rem.		1.08	0.29	0.16		3.43	2.21
Total	100.00	99.48	99.80	99.76	99.24	101.95	99.92
G.		2.174				2.185	

1. $Fe''Fe_2'''(SO_4)_4 \cdot 14H_2O$. 2. Rammelsberg, Harz. Rem. is CaO 0.58, insol. 0.50. Material contaminated with about 1½ per cent gypsum.[3] 3. Rammelsberg. Rem. is MgO 0.25, insol. 0.04.[4] 4. Tierra Amarilla, Chile. Rem. is insol.[4] 5. Island Mountain, Trinity County, California.[5] 6. United Verde mine, Jerome, Arizona. Rem. is Na_2O 0.88, and Al_2O_3 2.55.[6] 7. Pfaffenreuth, Bavaria.[10] Rem. is Al_2O_3 1.45, MgO 0.14, CaO 0.62. Average of two analyses.

O c c u r. Found with copiapite and other iron sulfates formed from the oxidation of pyrite.[7] With copiapite at Rammelsberg, Harz; at the pyrite mine, Bayerland, near Waldsassen, Bavaria; and in the Tmavýdůl coal mine, near Rtyně, Bohemia. With melanterite at the Blyava sulfate deposits, south Urals, and with voltaite at Madeni Zakh, Persia. In Chile, at Tierra Amarilla, partly in crystals in halotrichite, and abundantly at Alcaparrosa with szomolnokite. In the United States with coquimbite and other sulfates formed by the alteration of pyrrhotite at Island Mountain, Trinity County, California. Formed, as the result of a mine fire, with copiapite on massive pyrite at the United Verde mine, Jerome, Arizona (*louderbackite*).[8] With coquimbite and other iron sulfates in crusts on walls of the Copper Queen mine, Bisbee, Arizona.

A r t i f. It has been claimed that roemerite can be formed by the interaction of melanterite and rhomboclase in moist air, and that it may be crystallized from a solution of ferrous and ferric sulfates with a substantial excess of sulfuric acid.[9]

N a m e. After the German geologist Friedrich Adolph Roemer (1809–1869). Louderbackite was named after the American geologist George D. Louderback (1874–   ), of the University of California.

Ref.

1. Elements and angle table of Wolfe, *Am. Min.*, **22**, 736 (1937). Other settings were used by earlier investigators. The setting of Linck, *Zs. Kr.*, **15**, 22 (1889) was used by Dana (1892) and Ungemach, *Bull. soc. min.*, **58**, 160 (1935). Transformation: Linck to Wolfe, $010/0\bar{1}\bar{1}/\bar{1}00$; Goldschmidt (7, 125, 1922) and Landon, *Am. Min.*, **12**, 279 (1927), to Wolfe (1937), $0\bar{1}0/\bar{1}10/00\bar{1}$. All forms listed were reported by Ungemach on crystals from Tierra Amarilla, Chile. Forms included in the angle table have been reported by three or more observers from various localities.
2. Indices and $2V$ from Larsen (127, 1921). Pleochroism as reported by Anderson, in Landon (1927). Orientation imperfectly known.
3. Tschermak analysis in Grailich (1858).
4. Scharizer, *Zs. Kr.*, **37**, 529 (1902).
5. Landon (1927).
6. Buehrer analysis in Lausen, *Am. Min.*, **13**, 220 (1928).
7. Conditions for the formation of roemerite and associated sulfates have been discussed by Merwin and Posnjak, *Am. Min.*, **22**, 567 (1937).
8. Lausen (1928). X-ray powder patterns of louderbackite and roemerite from the type localities are identical (Pearl, priv. comm., 1947). The differences in the optical properties may be ascribed to slight variation in composition.
9. Scharizer (1902). Identification of the synthetic material as roemerite is based largely on analysis. Some differences in optical properties were noted, and no crystals large enough for measurement were obtained.
10. Gossner and Drexler, *Zbl. Min.*, 1935A, 267.
11. For additional analyses see Hintze (**1** [3B], 4527, 1930).

29.7.3 HALOTRICHITE GROUP

MONOCLINIC; SPHENOIDAL—2

	a_0	b_0	c_0	β
Halotrichite, $Fe''Al_2(SO_4)_4 \cdot 22H_2O$	20.47 kX	24.24	6.167	$\sim 101°$
Pickeringite, $MgAl_2(SO_4)_4 \cdot 22H_2O$	20.8	24.2	6.17	~ 95
Apjohnite, $Mn''Al_2(SO_4)_4 \cdot 22H_2O$				
Dietrichite, $ZnAl_2(SO_4)_4 \cdot 22H_2O$				
Bilinite, $Fe''Fe_2'''(SO_4)_4 \cdot 22H_2O$				
Redingtonite, $(Fe'',Mg,Ni)(Cr,Al)_2(SO_4)_2 \cdot 22H_2O$ (?)				

The crystal structure of the Halotrichite Group is not known. A complete series by mutual substitution of the divalent cation extends between halotrichite and pickeringite, and at least a partial series extends from these species toward apjohnite. Substitution between the trivalent cations Fe''', Al, and Cr is much less marked. The members of the group are fibrous in habit, water-soluble, and occur principally as efflorescences in sheltered places.

29.7.3.1 PICKERINGITE [MgAl$_2$(SO$_4$)$_4$·22H$_2$O]. *Hayes* (Am. J. Sc., **46**, 360, 1844). Magnesia Alum. Magnesia-Alaun, Talkerde-Alaun *Germ*. Alumbre. Haarsalz pt., Federalaun pt. Bosjemanite *Dana* (654, 1868). Bushmanite, Buschmanite, Boschjesmanite. Manganese Alum pt. Mangano-magnesian Alum. Magnesiumapjohnit, Eisenpickeringit, Manganpickeringit *Meixner* and *Pillewizer* (Zbl. Min., 263, 1937). Keramohalite pt. Ferropickeringite.

Picroallumogene *Roster* (Boll. com. geol. Ital., 302, 1876). Pikroalumogen, Picralluminite. Sonomaite *Goldsmith* (Proc. Ac. Sc., Philadelphia, 263, 1876). Dumreicherit *Doelter* (Vulk. Gesteine u. Min. der Capverde Inseln, 93, 1882). Stüvenite ? *Darapsky* (Verh. Ver. Santiago, 107, 1886; Jb. Min., I, 125, 1887). Sesqui-Magnesiaalaun *Darapsky* (Jb. Min., I, 131, 1887). Aromite *Darapsky* (Jb. Min., I, 49, 1890).

29.7.3.2 HALOTRICHITE [Fe''Al$_2$(SO$_4$)$_4$·22H$_2$O]. Haarsalz pt., Federalaun pt., Bergbutter pt., Eisenalaun pt. *older authors*. Federalaun vom Freyenwalde *Klaproth* (Beitr. Kr. Min., **3**, 102, 1802). Eisenalaun *Germ*. Iron Alum. Halotrichit *Glocker* (691, 1839). Hversalt *Forchhammer* (Jahresber. Chem. Min., **23**, 263, 1843). Halotrichine *Scacchi* (Mem. geol. camp. Nap., 84, 1849).

Masrite *Richmond* and *Off* (J. Chem. Soc. London, **61**, 491, 1892).

C r y s t. Monoclinic; sphenoidal—2.

PICKERINGITE[1]

$a:b:c = 0.8655:1:0.2551;\quad \beta\, 96°33\tfrac{1}{2}';\quad p_0:q_0:r_0 = 0.2948:0.2534:1$

$r_2:p_2:q_2 = 3.9459:1.1638:1;\quad \mu\, 83°26\tfrac{1}{2}';\quad p_0'\, 0.2967,\, q_0'\, 0.2551,\, x_0'\, 0.1150$

Forms:

Left	Right	ϕ	ρ	ϕ_2	$\rho_2 = B$	C	A
'b	b 010	0°00'	90°00'	....	0°00'	90°00'	90°00'
'm	m 110	49 18½	90 00	0°00'	49 18½	85 02½	40 41½
'y	y 031	8 32½	37 44	83 26½	52 45½	37 14½	84 47
	d 101	90 00	22 22½	67 37½	90 00	15 49	67 37½
'p	p 111	58 13	25 50½	67 37½	76 43½	20 32½	68 15
'q	q 221	54 14½	41 07½	54 41	67 24	35 58	57 45
'P	P $\bar{1}$11	−35 27½	17 23½	100 18	75 54½	21 50	99 59
'g	g 131	28 16½	40 59½	67 37½	54 43	38 15½	71 53½
'i	i 211	70 11½	36 58½	54 41	78 14½	30 52½	55 32
'j	j 231	42 47½	46 12	54 41	58 01	41 57½	60 38½
'r	r 311	75 45½	46 02½	44 51	79 48	39 42½	45 45½
't	t 321	63 05	48 25½	44 51	70 12½	42 39	48 10
'v	v 421	66 44	52 15½	40 07	71 48	46 17	43 24½

HALOTRICHITE[2]

$a:b:c = 0.845:1:0.254;\quad \beta\, 100°36';\quad p_0:q_0:r_0 = 0.301:0.250:1$

$r_2:p_2:q_2 = 4.005:1.204:1;\quad \mu\, 79°24';\quad p_0'\, 0.306,\, q_0'\, 0.254,\, x_0'\, 0.187$

Forms:

	φ	ρ	φ₂	ρ₂ = B	C	A
b 010	0°00′	90°00′	....	0°00′	90°00′	90°00′
l 250	25 43	90 00	0°00′	25 43	85 25½	64 17
m 110	50 17	90 00	0 00	50 17	80 52	39 43

Structure cell.[3] Space group $P2$.

	a_0	b_0	c_0	β	$a_0:b_0:c_0$
Pickeringite (anal. 2)	20.8 kX	24.2	6.17	~95°	0.860:1:0.255
Halotrichite (anal. 12)	20.47 kX	24.24	6.167	~101°	0.845:1:0.254

Cell contents $(Mg,Fe)_4Al_8(SO_4)_{16} \cdot 88H_2O$.

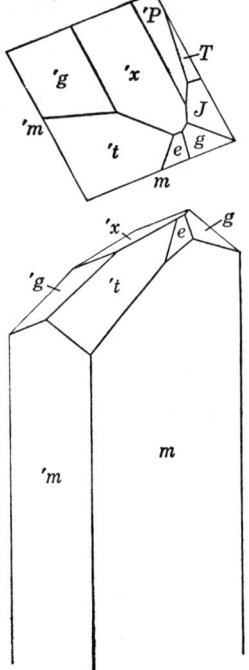

Pickeringite. Quetena.

Habit. Acicular [001]; distinctly terminated crystals are very rare. As radial or matted aggregates of acicular or hair-like crystals; tufted, spheroidal, as an incrustation or efflorescence; asbestiform. Thicker crystals are sometimes observed to be hollow or to contain pores filled with liquid or gas.

Phys. Cleavage {010} poor. Fracture conchoidal. Brittle. H. 1½. G.: pickeringite end-member 1.73–1.79, 1.84 (calc.); halotrichite end-member 1.89, 1.95 (calc.). Luster vitreous. Pickeringite colorless to pure white, sometimes tinted yellowish or reddish; halotrichite colorless to white, also yellowish or greenish. Taste astringent.

Opt.[4] In transmitted light, colorless.

Orientation		n(Na) pickeringite (anal. 2)	n manganoan pickeringite (South Africa)	n(Na) halotrichite (anal. 12)	
X		1.475	1.478	1.480	Biaxial negative (−).
Y	b	1.480	1.482	1.486	
Z		1.483	1.482	1.490	
Z ∧ c		36°	~29°	38°	
2V		60°	small	35°	

Chem. A hydrated sulfate of aluminum, magnesium, and divalent iron, $(Mg,Fe)Al_2(SO_4)_4 \cdot 22H_2O$. A doubtless complete series extends between Mg and Fe'', although most analyses are close to the pure endmembers. The names pickeringite and halotrichite are here applied to the halves of the series with Mg > Fe and Fe > Mg, respectively. Mn'' substitutes for (Mg,Fe) and at least a partial series extends towards apjohnite. Fe''' substitutes for Al in the series, usually only in small amounts but with Fe:Al = 1:3.7 in analysis 4 (see *bilinite*). Small amounts of Co and Cu also substitute for (Mg,Fe). The water content has been established [5] as $22H_2O$, but most analyses depart more or less therefrom. Some analyses apparently have been made on material admixed with alunogen, alums, or other fibrous sulfates.

Anal. [6]

	1	2	3	4	5	6	7
MgO	4.69	4.35	4.16	4.06	3.69	4.28	3.95
FeO				0.23		tr.	1.65
MnO			1.02	tr.	2.17	1.98	
Al_2O_3	11.87	12.30	11.64	9.52	11.52	11.52	11.07
Fe_2O_3			0.24	4.04			0.23
SO_3	37.29	37.84	36.86	36.62	36.77	35.44	35.28
H_2O	46.15	44.66	46.10	45.42	45.74	44.82	47.00
Rem.		0.59	0.41	0.70	0.20	2.10	0.12
Total	100.00	99.74	100.43	100.59	100.09	100.14	99.30
G.	1.84						1.79

1. $MgAl_2(SO_4)_4 \cdot 22H_2O$. 2. Pickeringite. Quetena, Chile.[7] Rem. is CaO 0.09, insol. 0.50. 3. Pickeringite. Portland, Conn.[5] Rem. is CuO 0.17, NiO tr., insol. 0.24. 4. Ferrian pickeringite. Opálbánya, Bohemia.[8] Rem. is CaO. 5. Manganoan pickeringite. Bushman River, South Africa.[9] (*bosjemanite*). Rem. is KCl. 6. Manganoan pickeringite. Terlan, Tyrol.[10] Rem. is insol. H_2O includes H_2O+ 15.18, H_2O- 29.64. 7. Ferroan pickeringite. Kalata mine, Urals, U.S.S.R.[11] Rem. is CuO.

	8	9	10	11	12	13
MgO	2.64	2.84	0.85			
FeO	3.95	7.48	6.86	4.23	7.28	8.07
MnO	0.24	0.21	0.47	2.56		
Al_2O_3	10.99	9.07	10.54	10.62	10.26	11.45
Fe_2O_3	0.68	0.50	1.12	1.63	0.65	
SO_3	35.61	34.66	36.42	36.78	37.28	35.97
H_2O	46.18	[45.24]	43.63	[40.35]	43.33	44.51
Rem.			0.37	3.83	1.59	
Total	100.29	[100.00]	100.26	[100.00]	100.39	100.00
G.		1.87			1.895	1.95

8. Ferroan pickeringite. Dienten, Austria.[10] H_2O includes H_2O+ 18.24, H_2O- 27.94. 9. Magnesian halotrichite. Franquenies, Belgium.[12] 10. Halotrichite. Roccalumera, Sicily.[13] Rem. is insol. 11. Manganoan halotrichite (*masrite*). Egypt.[14] Rem. is CoO 1.02, insol. 2.61, "masrium oxide" 0.20. 12. Halotrichite. Chuquicamata, Chile.[7] Rem. is CuO 0.66, insol. 0.93. 13. $Fe''Al_2(SO_4)_4 \cdot 22H_2O$.

Tests. B.B. fuses easily in its water of crystallization; the M.P., 95°–105°, is slightly lower than that of alunogen, $\sim$113°. Completely soluble in water to an acid solution.

Var. Manganoan. Bosjemanite *Dana* (654, 1868). Bushmanite. Mangano-magnesium Alum. Magnesiumapjohnite, Manganpickeringite *Meixner* and *Pillewizer* (*Zbl. Min.*, 263, 1937). Masrite *Richmond* and *Off*

(*J. Chem. Soc. London*, **61**, 491, 1892). Mn" substitutes for (Mg,Fe) and at least a partial series extends from pickeringite and halotrichite toward apjohnite. The manganoan varieties of these species (for which special given names are superfluous) tend toward pink or rose tints of color and the extinction angle, $Z \wedge c$, apparently decreases with increasing content of Mn.

Occur. Pickeringite and halotrichite are commonly formed as the products of weathering of pyritic and aluminous sedimentary or other rocks and accumulate as efflorescences in sheltered places. Also in the gossan of pyritic ore deposits, especially in arid regions, and in fumaroles in volcanic regions. Often observed as a recent deposit in mine workings, especially in pyritic lignite or coal seams. Associated minerals include kalinite and other alums, alunogen, melanterite, copiapite, gypsum, epsomite. Several of these species, especially alunogen, occur as fibrous or acicular efflorescences not easily distinguished on casual examination from pickeringite and halotrichite, and these have been loosely grouped, especially in the older literature, under the name of hair salts (*Haarsalz*). A few of the many known occurrences follow.

Pickeringite occurs on alum schist at Wetzelstein near Saalfeld and near Lehesten, Thuringia, Germany. A manganoan variety at Terlano in the south Tyrol and a ferroan variety at Dienten in Salzburg, Austria. From the iron mines of Elba, Italy. At several places in Bohemia of Czechoslovakia, notably Valachov Hill near Skřivan; at Opálbánya in the Vörösvágás opal district (ferrian). A highly ferroan variety at Aix-les-Bains, Savoy, France. Manganoan pickeringite also occurs in a cave near the Bushman (Boschjesman) River in Cape Province, South Africa, in the Maderaner Thal, Uri, Switzerland, and at Alum Point, Salt Lake County, Utah, U.S.A. Pickeringite occurs abundantly at numerous places in the arid regions along the western coast of South America; as in the Cerros Pintados near Iquique, Tarapacá, and at Quetena and Chuquicamata, in Chile. In the United States near Tucumcari, Quay County, and near Las Vegas, San Miguel County, in New Mexico; at The Geysers, Sonoma County, California. Near Fallon, Churchill County, Nevada.[16] In Canada on shale on the banks of the Meander River, Newport, Nova Scotia.

Halotrichite occurs in Germany at Mörsfeld east of Obermoschell in the Bavarian Pfalz, at Reichenbach in Saxony, and on brown coal at the Wohlforth mine, Gusterhain, Hesse-Nassau; at Zuckmantel, Silesia; in the Idria mines. At the solfatara of Pozzuoli near Naples. From Björkbakkagård, Finland, and from the solfatara of Krisuvig, Iceland. From Fahlun, Sweden. A manganoan and cobaltian variety has been described from Egypt (*masrite*). At Urumiya, Persia. In France in the mines of Huelgoet and Poullaouen, Brittany. At numerous places in South America, as at Tierra Amarilla near Copiapó and Quetena, Alcaparrosa and Chuquicamata, Chile. In the United States abundant in the Alum Mountain district on the Gila River in Grant County, New Mexico. In California in the sulfate deposit close to the borax mines in the Calico Hills, San Bernardino County, and around hot springs on Lassen Peak, Shasta County. In Canada at the Glace Bay coal mines, Nova Scotia.

A r t i f.[15] Halotrichite is a stable phase in the system $FeSO_4$-Al_2-$(SO_4)_3$-H_2O. Pickeringite (?) has been obtained from a dilute H_2SO_4 solution of Mg and Al sulfates; alunogen is first deposited from a water solution of the salt itself.

N a m e. Pickeringite after John Pickering (1777–1846). Halotrichite from the Latin form, *halotrichum*, which in turn is from the older German name Haarsalz.

Ref.
1. Elements of Bandy, *Am. Min.*, **23**, 724 (1938), on crystals from Quetena, Chile (analysis 2). Sphenoidal symmetry recognized by Gordon, *Ac. Sc. Philadelphia, Not. Nat.*, no. 101 (1942). Forms of Gordon (1942), who lists 79 less common and rare forms. Also the following forms from Bandy (1938) based on presumed holohedral symmetry, the right or left forms not distinguished: n 210, w 011, f 121.
2. Bandy (1938) on crystals from Chuquicamata, Chile (analysis 12); right and left forms not distinguished.
3. Bandy (1938) by Weissenberg method on analyzed material; his space group, $P2/m$, based on presumed holohedral symmetry, is here made $P2$.
4. Bandy (1938) and Larsen (41, 1921); see also Meixner and Pillewizer, *Zbl. Min.*, 263 (1937). Many further measurements are cited in the literature.
5. Schairer and Lawson, *Am. J. Sc.*, **11**, 301 (1926); see also Uhlig, *Cbl. Min.*, 723 (1912).
6. For numerous additional analyses see Hintze (1 [3B], 4505, 4512, 1930) and Doelter (4 [2], 523, 545, 1927); also Meixner and Pillewizer (1937); Van Tassel, *Mus. Belgique, Bull.*, **20**, no. 16 (1944); Rutherford, *Am. Min.*, **17**, 401 (1932); Tokody, *Földt. Közl.*, **62**, 187 (1933); Vavrinecz, *Magyar Chem. Foly.*, **35**, 1 (1929); Kokta, *Příroda*, Brno, **32**, 381 (1930); Kokta, *Publ. Fac. Sc. Univ. Masaryk*, no. 166 (1932)—*Min. Abs.*, **5**, 488 (1934); Gordon (1942).
7. Gonyer analysis in Bandy (1938).
8. Zsivny, *Zs. Kr.*, **55**, 629 (1920).
9. Stromeyer, *Ann. Phys.*, **31**, 137 (1834).
10. Meixner and Pillewizer (1937).
11. Vertushkov, *C. r. ac. sc. U.R.S.S.*, **30**, 334 (1941).
12. Van Tassel (1944).
13. Lo Sardo, *Per. Min.*, **8**, 281 (1937).
14. Richmond and Off, *J. Chem. Soc. London*, **61**, 491 (1892).
15. Occleshaw, *J. Chem. Soc. London*, **127**, 2598 (1925); Knauer, *Ann. Chemie*, **14**, 261 (1835); Uhlig (1912); Wirth, *Zs. angew. Chem.*, **26**, 81 (1913).
16. Hewett, *U. S. Geol. Sur., Bull.* 750E (1924).

29.7.3.3 **A P J O H N I T E** [$Mn''Al_2(SO_4)_4 \cdot 22H_2O$]. Manganese Alum *Apjohn* (*Phil. Mag.*, **12**, 103, 1838). Manganalaun. Apjohnit *Glocker* (298, 1847).

Monoclinic. As masses and crusts composed of fibrous or acicular crystals; also asbestiform. H. $1\frac{1}{2}$. G. 1.78. Luster of aggregates silky. Colorless to white, also tinted rose, pale green or yellow. Optical data are lacking.

C h e m. A hydrated sulfate of aluminum and manganese, $Mn''Al_2(SO_4)_4 \cdot 22H_2O$. Probably complete series extend to pickeringite and halotrichite through manganoan varieties of those species. The name apjohnite is applied to material with Mn > (Mg,Fe).

Anal.

	1	2	3	4
FeO				0.39
MgO		0.36		0.30
MnO	7.97	6.60	7.44	8.73
Al_2O_3	11.46	10.65	10.47	10.03
SO_3	36.00	33.51	35.90	35.47
H_2O	44.57	48.15	46.99	44.78
Rem.			1.54	0.37
Total	100.00	99.27	102.34	100.07

1. $Mn''Al_2(SO_4)_4 \cdot 22H_2O$. 2. Delagoa Bay, Africa.[1] 3. Delagoa Bay.[2] Rem. is $(NH_4)_2O$. 4. Alum Cave, Tennessee.[3] Rem. is $(Co,Ni)O$ 0.29, CuO 0.02, insol. 0.06.

Tests. Soluble in water.

Occur. Originally from Delagoa Bay, Portuguese East Africa. Also from Alum Cave, Sevier County, Tennessee, where large masses of the substance have collected under an overhanging cliff at the headwaters of Little Pigeon creek. An unanalyzed mineral from Szomolnok in Czechoslovakia with G. 1.787, mean index of refraction 1.467 and $Z \wedge c = 27°-28°$ has been referred to apjohnite.[5] Other occurrences classed here in the past are properly manganoan varieties of pickeringite or halotrichite, which see.

Artif.[4] Found in the system $MnSO_4$-$Al_2(SO_4)_3$-H_2O.

Name. After James Apjohn (1796–?), Professor of Chemistry and Mineralogy in Trinity College, Dublin.

Ref.

1. Apjohn (1838); see also Kane, *Ann. Phys.*, **44**, 471 (1838).
2. Ludwig, *Arch. Pharm.*, **143**, 97 (1870).
3. Brown, *Am. Chem. J.*, **6**, 97 (1884).
4. Caven and Mitchell, *J. Chem. Soc. London*, **127**, 527 (1925).
5. Kokta, *Sborník Klubu Příroda.*, **19**, 75 (1937)—*Min. Abs.*, **7**, 316 (1939).

29.7.3.4 DIETRICHITE [$(Zn,Fe,Mn)Al_2(SO_4)_4 \cdot 22H_2O$]. von Schroeckinger (*Verhl. geol. Bundesanst. Wien*, 189, 1878).

In fine-fibrous, tufted aggregates as an efflorescence or incrusting. Monoclinic. H. 2. Color dirty white to brownish yellow. Luster silky.

Chem. A hydrated sulfate of aluminum, zinc, and divalent iron and manganese, $(Zn,Fe,Mn)Al_2(SO_4)_4 \cdot 22H_2O$. In the only reported analysis, $Zn:Fe:Mn = 1.86:1.77:1$.

	MgO	FeO	MnO	ZnO	Al_2O_3	SO_3	H_2O	Total
1.		3.07	1.69	3.66	11.41	35.83	44.34	100.00
2.	0.33	3.11	1.74	3.70	10.92	35.94	44.38	100.12

1. $(Zn,Fe,Mn)Al_2(SO_4)_4 \cdot 22H_2O$ with $Zn:Fe:Mn = 1.86:1.77:1$. 2. Felsöbánya.

Opt.[1] In transmitted light, colorless.

ORIENTATION		n	
X	b	1.475 ± 0.003	Biaxial positive (+).
Y		1.480 ± 0.003	$2V$ large.
$Z \wedge c$	$\sim 29°$	1.488 ± 0.003	

Tests. Soluble in water.

O c c u r. Found as a recent efflorescence in mine workings at Felsöbánya, Hungary.
A r t i f. The pure zinc salt has been reported to crystallize from a solution of its components.[2]
N a m e. After Dr. Dietrich of Příbram, Bohemia, who analyzed the mineral.

Ref.
1. Larsen (68, 1921).
2. Kane, *Trans. Irish Ac.*, **17**, 423 (1837).

29.7.3.5 Bilinite [Fe″Fe$_2$‴(SO$_4$)$_4$·22H$_2$O]. Šebor (*Sborník Klubu přírodověd.*, no. 2, 1913—*Jb. Min.*, I, 395, 1914).

Radial-fibrous aggregates. H. ~ 2. G. 1.875. Color white to yellowish. Mean index of refraction ~ 1.500; inclined extinction up to 35°–39°.
C h e m. A hydrated sulfate of divalent and trivalent iron, Fe″Fe$_2$‴(SO$_4$)$_4$·22H$_2$O. Analyses gave:

	Na$_2$O	MgO	FeO	Fe$_2$O$_3$	SO$_3$	H$_2$O	C	Total
1.			7.58	16.84	33.78	41.80		100.00
2.	0.90	0.04	5.86	15.95	32.80	39.82	2.11	97.48
3.	0.29	0.13	6.93	15.88	34.87	41.77		99.87

1. Fe″Fe$_2$‴(SO$_4$)$_4$·22H$_2$O. 2,3. Schwaz, Bohemia; 2, outer parts; 3, inner parts.

O c c u r. An alteration of iron sulfide in lignite at Schwaz near Bilin, Bohemia.

Bilinite appears from its composition and optical properties to be the ferric iron analogue of halotrichite.

29.7.3.6 Redingtonite. Becker (*U. S. Geol. Sur.*, Mon. *13*, 279, 1888); Melville and Lindgren (*U. S. Geol. Sur.*, Bull. *61*, 23, 1890).

Massive with a fine parallel-fibrous structure. G. 1.761. Luster silky. Color white, purple on fractures across the fiber length. Optically shows inclined extinction up to 38°. Birefringence feeble. Analysis gave: MgO 1.850, MnO tr., NiO 1.001, FeO 4.579, Fe$_2$O$_3$ 0.186, Cr$_2$O$_3$ 7.512, Al$_2$O$_3$ 5.136, SO$_3$ 35.352, H$_2$O+ 14.340, H$_2$O− 27.083, insol. 3.457, total 100.506. Soluble in water. Found in the Redington mercury mine, Knoxville, Napa County, California. Redingtonite may prove on further examination to be a chromium member of the halotrichite group, (Fe″,Mg,Ni)(Cr,Al)$_2$(SO$_4$)$_4$·22H$_2$O.

SCLEROSPATHITE. Schlerospathite *Petterd* (*Pap. Proc. Roy. Soc. Tasmania*, **27**, 1902; 157, 1910). Sklerospathite.
As compact felted masses, tough under the hammer, with a spicular surface. Color white. Luster silky. Analysis gave: Fe$_2$O$_3$ 14.0, Cr$_2$O$_3$ 10.64, SO$_3$ 27.20, ign. loss 39.19, insol. 10.77, total 101.80. Soluble in water. B.B. swells to a brown mass. Found with another unidentified sulfate of Fe and Cr at the Salisbury mine, Blue Tier, near Beaconsfield, Tasmania. Needs verification.

TYPE 8. $A_2(XO_4)_3 \cdot xH_2O$

29.8.1 LAUSENITE [$Fe_2(SO_4)_3 \cdot 6H_2O$]. Butler (Am. Min., 13, 594, 1928). Rogersite Lausen (Am. Min., 13, 203, 1928).

Monoclinic.[1] Lumpy aggregates of minute fibers. Silky luster. White.
O p t. In transmitted light, colorless. Fibers elongated parallel to c.

ORIENTATION		n	
$X \wedge c$	$27°$	1.598 ± 0.005	Biaxial negative ($-$).
Y		1.628 ± 0.005	$2V$ large.
Z		1.654 ± 0.005	

C h e m. A hydrated sulfate of ferric iron, $Fe_2(SO_4)_3 \cdot 6H_2O$.
Anal.

	Na_2O	K_2O	Fe_2O_3	Al_2O_3	SO_3	H_2O	Total
1.			31.45		47.27	21.28	100.00
2.	1.23	0.06	28.07	1.40	47.90	20.64	99.30

1. $Fe_2(SO_4)_3 \cdot 6H_2O$. 2. United Verde mine, Jerome, Arizona.

O c c u r. Formed with copiapite and other sulfates as the result of a fire in the United Verde mine, Jerome, Arizona.
A r t i f. The artificial compound, $Fe_2(SO_4)_3 \cdot 6H_2O$, crystallizes readily and is stable in contact with its saturated solution from somewhat above 50° to about 150°.[2]
N a m e. Originally named rogersite after the American mineralogist Austin Flint Rogers (1877–) of Stanford University. Because the name rogersite was preoccupied, it was renamed lausenite after Carl Lausen, mining engineer, who first described the species.

Ref.
1. Lausen (1928), on optical grounds.
2. Posnjak and Merwin, Am. Chem. Soc. J., 44, 1965 (1922) report the artificial compound, $Fe_2(SO_4)_3 \cdot 6H_2O$, to be monoclinic, as slender colorless laths elongated [001] and flattened {010}; $X \wedge c = 22°$ for red and 26° for blue; $Y = b$; nX 1.605, nY 1.635, nZ 1.657.

29.8.2 KORNELITE [$Fe_2(SO_4)_3 \cdot 7H_2O$]. Krenner (Ak. Magyar, Értes., 22, 131, 1888; Mat. Termés. Ért., 42, 1, 1926).

C r y s t.[1] Monoclinic; prismatic—$2/m$.

$a:b:c = 0.7073:1:0.5419$; $\beta\ 97°05'$; $p_0:q_0:r_0 = 0.7662:0.5378:1$

$r_2:p_2:q_2 = 1.8596:1.4247:1$; $\mu\ 82°55'$; $p_0'\ 0.7721$, $q_0'\ 0.5419$, $x_0'\ 0.1243$

KORNELITE

Forms:

	ϕ	ρ	ϕ_2	$\rho_2 = B$	C	A
b 010	0°00′	90°00′	.....	0°00′	90°00′	90°00′
a 100	90 00	90 00	0°00′	90 00	82 55	.....
m 110	54 56	90 00	0 00	54 56	84 12½	35 04
l 210	70 40	90 00	0 00	70 40	83 20	19 20
e 011	12 55	29 04½	82 55	61 44	28 16	83 46
g 102	90 00	27 02	62 58	90 00	19 57	62 58
i 101	90 00	37 40½	52 19½	90 00	30 35½	52 19½
o 1̄11	−50 05	40 11	122 56	65 32½	45 48	119 40

Also:

f 350   k 530   d 2̄03   t 122   u 212   r 1̄22   s 2̄12

Habit. Lath-like {010} and acicular [001]. As crusts, tufted aggregates or globular masses with a radial-fibrous structure.

Twinning. On {100}, polysynthetic.

Phys. Cleavage {010}. G. 2.306. Luster of fibrous masses, silky. Color pale rose-pink to violet.

Opt.

Orientation	n (artificial)[2]	n[3]	
$X \wedge c$ 20° ± 2°	1.572 ± 0.003	1.567	Biaxial positive (+).
Y	1.586 ± 0.003	1.581	2V 49°–62°.
Z b	1.640 ± 0.003	1.638	$r > v$, perceptible.

Chem. A hydrated ferric sulfate, $Fe_2(SO_4)_3 \cdot 7H_2O$. The water content is somewhat uncertain and may be higher than $7H_2O$.

Anal.

	1	2	3
Fe_2O_3	30.36	29.23	30.17
SO_3	45.66	43.15	44.55
H_2O	23.98	25.96	24.92
Rem.		1.48	0.29
Total	100.00	99.82	99.93

1. $Fe_2(SO_4)_3 \cdot 7H_2O$. 2. Tintic Standard mine, Utah.[4] Rem. is insol. 0.86, sol. residue 0.62. 3. Szomolnok, Hungary.[5] Rem. is CaO 0.06, Na_2O 0.11, K_2O 0.09, $(NH_4)_2O$ 0.03.

Tests. Soluble in water.

Occur. A secondary mineral, originally found with voltaite and coquimbite in the pyrite mine of Szomolnok, County Szepes, Hungary. Also at the Tintic Standard mine, Dividend, Utah, and with coquimbite, roemerite, voltaite, rhomboclase as an efflorescence in workings of the Copper Queen mine, Bisbee, Arizona.[6]

Artif.[2] Kornelite is a stable phase below 80° in the system Fe_2O_3-SO_3-H_2O.

Name. After Kornel Hlavacsek.

Ref.

1. Schaller, priv. comm. (1949), and *Am. Min.*, **16,** 116 (1931), on crystals from Utah (analysis 3). Vicinal: 531, 951, 12.6.1, 20.10.1, 841, 14.7.1, 14.7.2, 12.5.1, 14.4.1, 29.6.1, 13.6.1.

2. Posnjak and Merwin, *J. Am. Chem. Soc.*, **44**, 1965 (1922).
3. Larsen and Berman (107, 1934), on Utah material.
4. Schaller, *U. S. Geol. Sur.*, *Bull. 878*, 123 (1937).
5. Loczka, *Mat. Termés. Ért.*, **42**, 6 (1926).
6. Merwin and Posnjak, *Am. Min.*, **22**, 567 (1937).

29.8.3 **C O Q U I M B I T E** [$Fe_2(SO_4)_3 \cdot 9H_2O$]. Neutrales schwefelsaures Eisenoxyd Rose (*Ann. Phys.*, **27**, 310, 1833). White Copperas. Coquimbite *Breithaupt* (100, 1841). Blakeite *Dana* (447, 1850).

C r y s t.[1] Hexagonal—P; 3 m, 3 2 or $\bar{3}$ m.

$$a:c = 1:1.5643; \quad p_0:r_0 = 1.8063:1$$

Forms:

	ϕ	ρ	M	A_2
c 0001	....	0°00′	90°00′	90°00′
m 10$\bar{1}$0	30°00′	90 00	60 00	90 00
a 11$\bar{2}$0	0 00	90 00	90 00	60 00
p 10$\bar{1}$3	30 00	31 03	75 03	90 00
q 10$\bar{1}$2	30 00	42 05	70 25	90 00
r 10$\bar{1}$1	30 00	61 02	64 03½	90 00
e 11$\bar{2}$2	0 00	57 24½	90 00	65 05
f 11$\bar{2}$1	0 00	72 16½	90 00	61 33½
s 31$\bar{4}$4	16 06	58 26½	76 20	78 11½

Structure cell.[2] Lattice hexagonal. a_0 10.85 kX, c_0 17.03; $a_0:c_0 = 1:1.563$. Cell contents $Fe_8(SO_4)_{12} \cdot 36H_2O$.

Habit. Short prismatic [0001] with {10$\bar{1}$0} and {11$\bar{2}$0}; pyramidal {10$\bar{1}$1}. Also massive, granular.

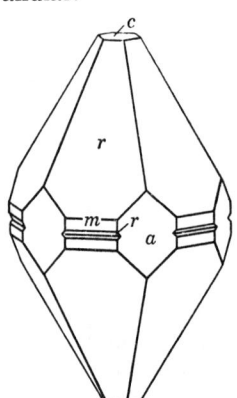

Tierra Amarilla.

Oriented growths. With paracoquimbite in parallel position; as lamellar intergrowths on {0001} or as scepter-like overgrowths.

P h y s. Cleavage {10$\bar{1}$1} imperfect, {10$\bar{1}$0} difficult. H. 2½. G. 2.11 ± 0.01; 2.137 (calc.). Luster vitreous. Color pale violet to deep amethystine; also yellowish or greenish. Transparent. Taste astringent.

COQUIMBITE

Opt. Uniaxial positive (+).

	n (Arizona)[3] anal. 4	n(Na) (Chile)[4]	n(Na) (Chile)[5]	n(Li) (Chile)[5]
O	1.536	1.5455	1.5519	1.5469
E	1.572	1.5547	1.5575	1.5508

The identity of the particular Chilean material for which indices are given is uncertain and may be coquimbite or paracoquimbite; the Arizona material probably is coquimbite.

Chem. A hydrated sulfate of ferric iron, $Fe_2(SO_4)_3 \cdot 9H_2O$. Dimorphous with paracoquimbite. Al substitutes for Fe''' up to at least $Al:Fe''' = 1:1.49$ (analysis 4). The water is largely lost at 93°, and the remainder at 161°.[6]

Anal.[10]

	1	2	3	4	5
Fe_2O_3	28.41	28.71	28.94	14.69	25.84
Al_2O_3			tr.	6.93	2.25
SO_3	42.74	42.31	42.37	44.05	44.19
H_2O	28.85	28.82	28.48	31.72	27.25
Rem.		0.40		2.13	0.31
Total	100.00	100.24	99.79	99.52	99.84
G.				2.066	

1. $Fe_2(SO_4)_3 \cdot 9H_2O$. 2. Alcaparrosa.[7] Rem. is insol. 3. Tierra Amarilla.[8] 4. Jerome, Arizona.[3] Rem. is Na_2O (due to admixture?). 5. Huelva, Spain.[9] Rem. is CaO 0.15, MgO 0.16.

Tests. Soluble in cold water; if the water solution is heated hydrous ferric oxide is precipitated. Soluble in acids.

Occur. Found abundantly at localities in Chile, including Tierra Amarilla, Quetena, Chuquicamata and Alcaparrosa, where it is associated with paracoquimbite, voltaite, szomolnokite, roemerite, copiapite, and other secondary sulfates. From Skouriotissa, Cyprus, and the Concepción mine, Huelva, Spain. Reported from Rammelsberg in the Harz, Germany. Found in the United States in the fire zone in the United Verde mine, Jerome, Arizona; also from Bisbee, Arizona. In California in the Redington mercury mine, Napa County, in the sulfate deposit near Borate, San Bernardino County, and elsewhere. It is uncertain in most instances whether the above occurrences, other than Chilean, refer to coquimbite or paracoquimbite.

Alter. Alters in dry air to a white powder.

Artif.[11] As crystals from solutions of ferric sulfate containing an excess of sulfuric acid.

Name. From the [supposed] occurrence in the province of Coquimbo, Chile.

Ref.

1. Elements and forms of Ungemach, *Bull. soc. min.*, **58**, 165 (1935), on Tierra Amarilla crystals (analysis 3), who showed that the mineral as earlier known included two separate species, one hexagonal (coquimbite proper) and the other rhombohedral (paracoquimbite) which are related crystallographically. For earlier morphological descriptions see Linck, *Zs. Kr.*, **15**, 7 (1889), and Arzruni, *Zs. Kr.*, **3**, 516 (1879).

534 SULFATES

2. Hocart in Ungemach (1935).
3. Lausen, *Am. Min.*, **13**, 215 (1928).
4. Arzruni (1879).
5. Linck (1889).
6. For dehydration data see Scharizer, *Zs. Kr.*, **65**, 339 (1927).
7. Henderson analysis in Bandy, *Am. Min.*, **23**, 729 (1938).
8. Ungemach (1935).
9. Serrano analysis in Collins, *Min. Mag.*, **20**, 32 (1923).
10. For additional analyses see Hintze (**1** [3B], 4401, 1929) and Dana (956, 1892).
11. Mellor (**14**, 307, 1935).

29.8.4 **PARACOQUIMBITE** [$Fe_2(SO_4)_3 \cdot 9H_2O$]. *Ungemach (Bull. soc. min.*, **58**, 165, 1935; *C.R.*, **197**, 1132, 1933). [Not paracoquimbite Klvana (Böhm. Ges., Ber., 272, [1881] 1882) = slavikite].

Cryst. Hexagonal—R; rhombohedral—$\bar{3}$.

$a:c = 1:4.6928;$ $\alpha\ 34°54';$ $p_0:r_0 = 5.4118:1;$ $\lambda\ 116°47'$

Forms:

			ϕ	$\rho = C$	A_1	A_2
c	0001	111	.....	0°00'	90°00'	90°00'
a	$11\bar{2}0$	$10\bar{1}$	0°00'	90 00	60 00	60 00
r	$10\bar{1}4$	211	30 00	53 34	45 50	90 00
t	$10\bar{1}1$	100	30 00	79 32½	31 36½	90 00
S	$01\bar{1}2$	110	−30 00	69 44½	90 00	35 40
Q	$01\bar{1}5$	221	−30 00	47 18	90 00	50 28
e	$11\bar{2}6$	321	0 00	83 54½	60 11	60 11

Also:

$n\ 1.0.\bar{1}.10$ $O\ 01\bar{1}8$ $p\ 10\bar{1}7$ $V\ 0.5.\bar{5}.22$ $W\ 02\bar{2}7$ $f\ 11\bar{2}3$

Structure cell.[1] Lattice rhombohedral. $a_0\ 10.90\ kX$, $c_0\ 51.15$; $a_0:c_0 = 1:4.693$. Cell contents $Fe_{24}(SO_4)_{36} \cdot 108H_2O$.

Habit. Rhombohedral, often with large $\{01\bar{1}2\}$; also equant, pseudo-cubic, with large $\{0001\}$ and $\{01\bar{1}2\}$, or prismatic [0001]. Massive, granular.

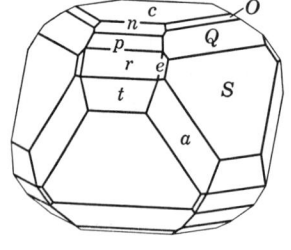

Oriented growths. With coquimbite, in parallel position; contact surface $\{0001\}$. As scepter-like overgrowths.

Twinning. On $\{0001\}$, common.

Phys. Cleavage $\{01\bar{1}2\}$ and $\{10\bar{1}4\}$, imperfect. H. 2½. G. 2.11 ± 0.01; 2.115 (calc.). Luster vitreous. Color pale violet. Transparent. Optical data are lacking. Taste astringent.

Chem. A hydrated ferric sulfate, $Fe_2(SO_4)_3 \cdot 9H_2O$. Dimorphous with coquimbite. 44.88 per cent of the water is lost at 105°. Analysis gave:

	Fe_2O_3	Al_2O_3	SO_3	H_2O	Total
1.	28.41		42.74	28.85	100.00
2.	29.79	tr.	41.38	28.69	99.86

1. $Fe_2(SO_4)_3 \cdot 9H_2O$. 2. Tierra Amarilla.

Occur. Found with coquimbite at Tierra Amarilla and Alcaparrosa and with coquimbite, roemerite, szomolnokite, quenstedtite, and voltaite at Quetena, in Chile. Other occurrences probably have been confused with coquimbite.

Ref.
1. Hocart in Ungemach (1935).

29.8.5 **QUENSTEDTITE** [$Fe_2(SO_4)_3 \cdot 10H_2O$]. *Linck* (Zs. Kr., **15**, 11, 1889).

Cryst.[1] Triclinic; pinacoidal—$\bar{1}$.

$a:b:c = 0.2621:1:0.2776;$ $\alpha\ 94°10',\ \beta\ 101°44\frac{1}{2}',\ \gamma\ 96°18\frac{1}{2}'$

$p_0:q_0:r_0 = 1.0628:0.2734:1;$ $\lambda\ 84°24',$ $\mu\ 77°41',\ \nu\ 82°40'$

$p_0'\ 1.0906,\ q_0'\ 0.2806,\ x_0'\ 0.2079,\ y_0'\ 0.1002$

Forms:[2]

		φ	ρ	A	B	C
b	010	0°00'	90°00'	82°40'	.....	84°24'
a	100	82 40	90 00	.....	82°40'	77 41
k	130	47 47½	90 00	34 52½	47 47½	77 33
l	120	57 04½	90 00	25 35½	57 04½	77 06½
m	110	68 47	90 00	13 52½	68 47	77 03
M	1$\bar{1}$0	97 27	90 00	14 47	97 27	79 09
L	1$\bar{2}$0	111 18½	90 00	28 39	111 18½	81 11½
w	011	28 37½	23 27½	76 29	69 33	14 50
v	021	17 26½	34 44	76 11	57 04½	27 19½
x	031	12 26½	43 58½	76 25	47 18½	37 05
W	0$\bar{1}$1	130 57½	15 23½	79 50	100 01	15 37½
V	0$\bar{2}$1	155 44	26 49½	82 27	114 17½	29 54
X	0$\bar{3}$1	164 20½	37 36½	84 56	125 59	41 35½
Y	0$\bar{4}$1	168 30½	46 12½	87 00	135 01½	50 38
s	$\bar{1}$21	−59 08½	45 31	124 06	68 32	53 31½
f	$\bar{1}$41	−38 53½	54 18½	64 51	50 47½	58 12
S	$\bar{1}$2$\bar{1}$	−124 29	46 40½	130 20½	114 19½	59 32½

Less common:

| c 001 | u 032 | P 1$\bar{1}$1 | T 1$\bar{3}$1 | j 140 | y 041 | r 111 | D 1$\bar{4}$1 |
| K 1$\bar{3}$0 | Z 0$\bar{5}$1 | o 121 | E $\bar{1}$31 | I 1$\bar{4}$0 | p 111 | t 131 | |

Habit. Aggregates of minute crystals which are commonly tabular {010} or short prismatic [100], sometimes highly modified.

Twinning. On {010} common.

Phys. Cleavage {010} perfect, {100} good. H. 2½. G. 2.147. Color pale violet to reddish violet. Transparent.

Opt.[3] In transmitted light, colorless to pale rose.

	Orientation		$n(Na)$	
	φ	ρ		
X	−43°	45°	1.547	Biaxial positive (+).
Y	128°	43°	1.566	$2V$ 70° (meas.).
Z	−138°	88°	1.594	$r < v$, strong; horizontal.

Extinction on (010) against [001] = 30°.

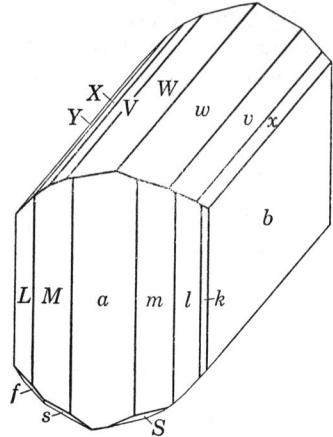

Chem. A hydrated ferric sulfate, $Fe_2(SO_4)_3 \cdot 10H_2O$.
Anal.

	Fe_2O_3	SO_3	H_2O	Rem.	Total	G.
1.	27.53	41.41	31.06		100.00	
2.	27.66	39.83	31.35	0.40	99.24	2.11
3.	27.88	39.37	31.69		98.94	2.147

1. $Fe_2(SO_4)_3 \cdot 10H_2O$. 2. Tierra Amarilla, Chile.[4] Rem. is CaO 0.40, Al_2O_3 tr., MgO tr. 3. Tierra Amarilla.[5] The SO_3 is low owing to analytical accident.

Tests. Readily soluble in water; on warming, the solution deposits a basic salt.

Occur. Found with coquimbite and copiapite at Tierra Amarilla near Copiapó, and at Alcaparrosa near Calama, Chile.

Alter. Changes to coquimbite, $Fe_2(SO_4)_3 \cdot 9H_2O$, by loss of water.[6] Loses about 7 molecules of water at 105° and is almost completely dehydrated at 220°.[7]

Artif. Has not been found as a stable phase in any recent studies of the system Fe_2O_3-SO_3-H_2O.[8]

Name. After the German mineralogist Friedrich August Quenstedt (1809–1889).

Ref.

1. Originally described as monoclinic. Re-examination of the type material by Ungemach, *Bull. soc. min.*, **58**, 97 (1935), showed it to be triclinic. The B angles of {100}, {110}, {120}, etc., are very nearly the supplements of the B angles of {1$\bar{1}$0}, {1$\bar{2}$0}, {1$\bar{3}$0}, etc., giving the zone [100] a monoclinic prismatic appearance.
2. Ungemach (1935). {032} from Bandy, *Am. Min.*, **23**, 669 (1938).
3. Bandy (1938) on crystals from Alcaparrosa, Chile, which by goniometric measurement were shown to correspond to the material of Linck and Ungemach. Larsen's (61, 1921) observations were made on copiapite from Montpelier, Iowa, which had been erroneously described as quenstedtite by Kuntze, *Am. Geol.*, **23**, 119 (1899).
4. Linck (1889).
5. Ungemach (1935).
6. Bandy (1938).
7. Dehydration has been studied by Linck (1889) and Ungemach (1935) but their observations beginning at or near 100° fail to show indication of changes to the nine-hydrate, coquimbite, or the seven- or six-hydrates known as artificial substances.
8. Summarized by Seidell, *Solubilities of Inorganic Compounds*, 541 (1940).

ALUNOGEN

29.8.6 A L U N O G E N [$Al_2(SO_4)_3 \cdot 18H_2O$]. Haarsalz pt. *early authors.* Hydrotrisulfate d'alumine *Beudant* (449, 1824). Davite (?) *Mill* (*Quart. J.*, **25**, 382, 1828). Alunogène *Beudant* (**2**, 488, 1832). Solfatarite pt. *Shepard* (188, 1835). Keramohalit *Glocker* (689, 1839). Saldanite *Huot* (**2**, 451, 1841). Stypterit *Glocker* (297, 1847). Halotrichit pt. *Hausmann* (**2**, 1174, 1847) [not Halotrichit *Glocker*]. Schwefelsaures Thonerde. Haarsalz *Rammelsberg* (269, 1875). Alunogenite.

C r y s t.[1] Triclinic; pinacoidal—$\bar{1}$.

$a:b:c = 0.8355:1:0.6752;$ $\alpha\ 89°58',\ \beta\ 97°26',\ \gamma\ 91°52'$

$p_0:q_0:r_0 = 0.8086:0.6699:1;$ $\lambda\ 89°48',\ \mu\ 82°34',\ \nu\ 88°08'$

$p_0'\ 0.8155,\ q_0'\ 0.6756,\ x_0'\ 0.1305,\ y_0'\ 0.0035$

Forms:

		ϕ	ρ	A	B	C
b	010	0°00'	90°00'	88°08'	.....	89°48'
a	100	88 08	90 00	.....	88°08'	82 34
l	230	38 05½	90 00	50 02½	38 05½	85 16½
o	530	62 04½	90 00	26 03½	62 04½	83 21½
g	410	76 31	90 00	11 37	76 31	82 44½
i	610	80 18½	90 00	7 49½	80 18½	82 39½
M	1$\bar{1}$0	128 31½	90 00	40 23½	128 31½	84 19½
Ψ	043	8 12½	42 25	83 13½	48 07	41 41
γ	053	6 35½	48 40	83 39½	41 36½	48 12½
x	021	5 30	53 41½	84 04	36 20	53 28
p	111	53 16	49 43	51 15	62 51	43 48½
r	353	39 17	56 12	56 51	49 58½	51 33
s	373	30 29	61 47½	61 52	40 35	58 03½
t	131	24 41½	66 10	65 51½	33 47	63 04½
σ	$\bar{1}$31	−161 32	65 10	108 23	149 24½	67 53½

Habit. Crystals small and rare; prismatic [001] or {010} with a six-sided outline about [010]. Usually in delicate fibrous masses or crusts; as an efflorescence; also massive, fibrous.

Twinning.[2] On {010}.

P h y s. Cleavage {010} perfect, also (?) {100} and {$\bar{3}$13}.[3] H. 1½–2. G. 1.77 (natural), 1.78 (artif.).[2] Luster vitreous to silky. Single-crystals are colorless and transparent, the aggregates white or tinged yellow or reddish by impurities. Taste acid and sharp, like alum.

O p t. In transmitted light, colorless.

ORIENTATION	n (New Mexico)[4]	n (Bolivia)[5]	n (Francisco de Vergara)[6]	
$X \sim b$	1.460	1.475	1.459	Biaxial positive (+).
Y	1.461	1.478	1.463	
$Z \wedge c\ \ 42°$	1.470	1.485	1.473	
2V	31°			

The indices of refraction vary with the water content.[4] The New Mexico material cited above has 15.5 H_2O; on heating to 75° the water content was about 12.5 H_2O and the indices were nX 1.483, nY 1.484, nZ 1.496. Over about 90° the material becomes isotropic with $n \sim 1.50$ which increases up to 1.54 at 290°. The mineral does not materially re-hydrate on exposure.

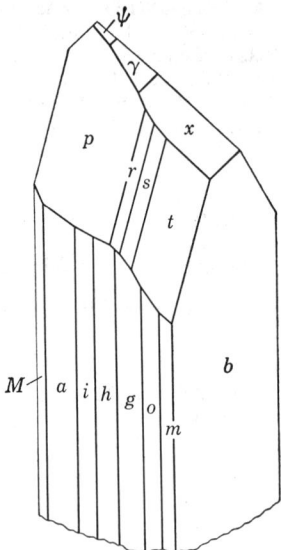

Francisco de Vergara, Chile.

Chem. A hydrous sulfate of aluminum, $Al_2(SO_4)_3 \cdot nH_2O$. In fully hydrated material n appears to be 18 and can vary from 18 down to at least 12.5, depending on the temperature and humidity apparently without destruction of the crystal structure.[11] In most natural material n is about 15 or 16. Fe''' apparently can substitute in small amounts for Al. The K, Mg, and Fe'' reported in many analyses are probably due to admixture, as of potash alum, epsomite, halotrichite, pickeringite.

Anal.[12]

	1	2	3	4	5	6
CaO					0.84	
MgO		nil			0.31	
Al_2O_3	14.90	16.17	17.33	16.08	15.25	16.59
Fe_2O_3				0.33		
SO_3	35.09	37.35	41.04	38.25	38.51	37.74
H_2O	50.01	46.10	41.44	45.66	[44.99]	44.64
Rem.		0.01			0.10	0.94
Total	100.00	99.63	99.81	100.32	[100.00]	99.91
G.		1.77		1.718	1.772	

1. $Al_2(SO_4)_3 \cdot 18H_2O$. 2. Francisco de Vergara, Chile.[7] Rem. is Na_2O. Original total given as 99.62. 3. Francisco de Vergara[8] (*meta-alunogen*). 4. Valachov, Czechoslovakia.[9] 5. Opálbánya, Bohemia.[10] Rem. is insol. 6. Pintado Canyon, New Mexico.[11] Rem. is insol.

Tests. In the C.T. yields water and at a higher temperature sulfuric acid. Easily soluble in water (27 g. $Al_2(SO_4)_3$ in 100 g. sat soln. at 25°). Alunogen melts at about 114°, higher than the members of the halotrichite group.

Occur. Alunogen occurs principally as an efflorescence or crevice filling in coal formations, shales or slates containing pyrite or marcasite or in the gossan or adjoining wall rock of pyritic ore deposits especially in arid regions; the mineral is formed by the action of sulfate solutions resulting from the oxidation of the sulfides on aluminous minerals. Associated species include halotrichite, pickeringite, epsomite, alums, gypsum, melanterite, and other iron sulfates. Also found with sulfur and gypsum as a product of fumarolic or solfataric action on volcanoes. The following are among the many known localities. In Bohemia at Kolosoruk in brown coal, at Rudain near Koenigsberg, on the pyritic shales of Valachov hill near Skřivaň, and in the opal mines of Cervenica (Opálbánya). As a fumarole deposit on Vesuvius, at the solfatara of Pozzuoli near Naples, and in the sulfur caves of Miseno, Italy. Also as a fumarolic incrustation in the Avacha volcano, Kamchatka,[13] and on Mt. Pelée, Martinique. In France at the pyrite mine of Sain-Bel, Rhône, and in the coal mines of Ronchamp, Haute-Saône. At numerous places in Chile and Peru, as in the Cerros Pintados near Iquique, and at Francisco de Vergara (*metalunogen*), Antofagasta, in Chile. Found as an efflorescence at numerous places in the United States. Abundantly at Smoky Mountain, Jackson County, North Carolina. On weathered shale around Ithaca, New York. Formed under fumarolic conditions in a mine fire at the United Verde mine, Jerome, Arizona. Extensive deposits in the Alum Mountain district, Grant County, New Mexico, on cliffs as crusts up to three or four feet thick with a fluted outer surface and a porous cellular structure; also in this state in Pintado Canyon in the Pastura district, Guadalupe County, in Sandoval County as a thick bed, and at other localities. Found in California around the hot springs in the Mount Lassen area, Shasta County, at The Geysers near Cloverdale in Sonoma County, and elsewhere. In Canada, found in British Columbia near Vernon with epsomite in veins, as crusts near Savona and as fibrous masses near Spatsum in the Ashcroft Mining Division, and on Blair Creek in the Kamloops Mining Division; in Nova Scotia on a shale heap at the Scotia mine, Spring Hill, Cumberland County.

A r t i f.[14] Obtained as crystals from solution in HCl; lower hydrates have been reported but are of uncertain validity. Numerous hydrated basic sulfates of aluminum have been reported.

N a m e. From *alun* and γενναν, *to make alum*.

MILLOSEVICHITE. Panichi (*Acc. Linc., Att.*, **22**, I, 303, 1913). Triclinic. As granular masses which show small clear crystals on their surfaces. Luster vitreous. Color violet-blue. Hygroscopic and alters easily, the color changing to gray. Stated to be a normal iron-aluminum sulfate, but no analysis is given. Found in the Alum Grotto on the Island of Vulcano, Lipari Islands, Italy. Named after the Italian mineralogist F. Millosevich (1875–1942), of the University of Rome. Probably a ferrian alunogen.

META-ALUNOGEN. Gordon (*Ac. Sc. Philadelphia, Not. Nat.*, no. 101, 1942). A name given to a white substance with a waxy or pearly luster formed by the natural dehydration of alunogen at Francisco de Vergara, Chile. The analysis (no. 3, above) corresponds to $Al_2(SO_4)_3 \cdot 13\frac{1}{2}H_2O$. Biaxial positive (+) with nX 1.469, nY 1.473, nZ 1.491, and $2E$ large. The distinction from alunogen is based on the optical orientation. The measured crystals of colorless alunogen from the locality with $16H_2O$

are said to have Z perpendicular to the trace of the perfect cleavage, presumably {010}, and meta-alunogen with $13\frac{1}{2}H_2O$ to have X perpendicular to this cleavage. The latter orientation, however, is that ordinarily reported for alunogen including the measured and analyzed crystals from Opálbánya with $16H_2O$. The separate identity of alunogen and meta-alunogen thus is uncertain. If meta-alunogen is a distinct phase, and not merely alunogen that has lost zeolitic (?) water without accompanying structural change, then much of what has been termed alunogen in the past and here so accepted may represent this species.

Ref.

1. Gordon, *Ac. Sc. Philadelphia, Not. Nat.*, no. 101 (1942), on crystals from Francisco de Vergara, Chile (analysis 2). Gordon's angles differ appreciably from those earlier reported by Hlawatsch, *Festschr. V. Goldschmidt*, 154 (1928), on small crystals of poor quality from Opálbánya. The water content of the crystals from the two occurrences is identical, but the optical orientation apparently differs (see *meta-alunogen*). Less common and rare forms:

c 001	O 5$\bar{3}$0	δ 0$\bar{1}$3	R 3$\bar{5}$3		x $\bar{3}$53
k 120	Y 3$\bar{2}$0	ϵ 0$\bar{2}$3	S 3$\bar{7}$3		r $\bar{1}$21
m 110	I 7$\bar{9}$0	W 0$\bar{1}$1	e $\bar{3}$13		ξ $\bar{3}$73
n 320	k 3$\bar{5}$0	ζ 0$\bar{4}$3	ϕ $\bar{3}$43		z $\bar{2}$01
f 310	L 4.$\bar{11}$.0	η 0$\bar{5}$3	γ $\bar{3}$53		ω $\bar{6}$13
h 920	J 1$\bar{3}$0	X 0$\bar{2}$1	τ $\bar{1}$21		χ $\bar{6}$23
F 3$\bar{1}$0	α 013	d 323	ρ $\bar{3}.\bar{10}.3$		π $\bar{6}$43
Z 5$\bar{2}$0	β 023	P 1$\bar{1}$1	θ $\bar{3}$13		λ $\bar{6}$13
N 2$\bar{1}$0	q 011	U 3$\bar{4}$3	w $\bar{1}$11		$\bar{6}$23

2. Hlawatsch (1928).
3. Perfect {010} cleavage is reported by Gordon (1942) and others; Hlawatsch (1928), however, cites only {100} and {$\bar{3}$13}.
4. Larsen and Steiger, *Am. J. Sc.*, **15**, 1 (1928).
5. Larsen (38, 1921).
6. Gordon (1942).
7. Pitman analysis in Gordon (1942).
8. Collins analysis in Gordon (1942).
9. Jirkovsky, cited in *Jb. Min.*, I, 106 (1928).
10. Doht analysis in Hlawatsch (1928).
11. Steiger analysis in Larsen and Steiger (1928).
12. For numerous additional analyses see Lausen, *Am. Min.*, **13**, 208 (1928); Hintze (1 [3B], 4404, 1929); Petrov, *Trav. inst. pétrog. ac. sc. U.R.S.S.*, no. 3, 89 (1933).
13. Zavaritsky, *Trans. Centr. Geol. & Prosp. Inst., Leningrad*, no. 35 (1935).
14. See Wirth and Bakke, *Zs. anorg. Chem.*, **79**, 360 (1913), Mellor (**5**, 333, 1924), Seidell, *Sol. of Inorg. Comp.*, 3rd ed., **1**, 97 (1940), Henry and King, *J. Am. Chem. Soc.*, **71**, 1142 (1949), Bassett and Goodwin, *J. Chem. Soc. London*, 2239 (1949). On the dehydration see also Kremann and Hüttinger, *Jb. geol. Reichsanst. Wien*, **58**, 637 (1908).

30 ANHYDROUS SULFATES CONTAINING HYDROXYL OR HALOGEN

TYPE 1. $A_m(XO_4)_p Z_q$, WITH $m:p > 2:1$

30.1.1 B R O C H A N T I T E [$Cu_4(SO_4)(OH)_6$]. *Lévy (Ann. Phil.,* **8**, 241, 1824).
Königine *Lévy (Ann. Phil.,* **11**, 194, 1826). Brongniartine *Huot* (**1**, 331, 1841).
Krisuvigite *Forchhammer (Skand. Nat. Stockholm,* 1842, Arsb. 192, 1843). Waringtonite *Maskelyne (Chem. News,* **10**, 263, 1864; *Phil. Mag.,* **29**, 475, 1865). Warringtonite.

C r y s t.[1] Monoclinic; prismatic—$2/m$.

$a:b:c = 1.3283:1:0.6135;\quad \beta\ 103°21';\quad p_0:q_0:r_0 = 0.4619:0.5969:1$

$r_2:p_2:q_2 = 1.6753:0.7738:1;\quad \mu\ 76°39';\quad p_0'\ 0.4747,\ q_0'\ 0.6135,\ x_0'\ 0.2373$

Forms:[2]

	ϕ	ρ	ϕ_2	$\rho_2 = B$	C	A	Orth.
c 001	90°00'	13°21'	76°39'	90°00'	.....	76°39'	e 0$\bar{1}$2
b 010	0 00	90 00	.....	0 00	90°00'	90 00	a 100
a 100	90 00	90 00	0 00	90 00	76 39	.....	b 0$\bar{1}$0
l 120	21 09	90 00	0 00	21 09	85 13½	68 51	*2$\bar{1}$0
m 110	37 44	90 00	0 00	37 44	81 52½	52 16	h, m 1$\bar{1}$0
d 210	57 08	90 00	0 00	57 08	78 49	32 52	d 1$\bar{2}$0
p 011	21 09	33 20½	76 39	59 10	30 50	78 34	p 2$\bar{1}$2
r 031	7 21	61 41	76 39	29 11	60 49	83 32	*6$\bar{1}$2
y 201	90 00	49 53	40 07	90 00	36 32	40 07	*0$\bar{5}$2
P 111	49 15	43 13½	54 33	63 27	34 01½	58 44½	*2$\bar{3}$2
π $\bar{1}$11	−21 09	33 20½	103 21	59 10	39 54½	101 26	p 212
V $\bar{1}$22	0 00	31 32	90 00	58 28	33 58	90 00	v 101
α $\bar{1}$62	0 00	61 29	90 00	28 31	62 19½	90 00	*301
B 211	62 40	53 11	40 07	68 26	41 39	44 40	*2$\bar{5}$2
β $\bar{2}$11	−49 15	43 13½	125 27	63 27	53 54	121 15½	*232

Less common and rare:

E 140	q 021	γ 702	ξ $\bar{2}$01	A $\bar{1}$62	ψ $\bar{1}$31
n 430	z 104	δ $\bar{1}$02	θ $\bar{3}$01	x $\bar{1}$42	Δ $\bar{1}\bar{1}$.4.4
F 410	i 102	ϵ $\bar{1}$01	ν $\bar{4}$01	t $\bar{2}$52	Φ $\bar{3}$11
o 012	u 304	χ $\bar{7}$04	Σ $\bar{3}$31	ω $\bar{2}$12	

Structure cell.[3] Space group $P2/a$. $a_0\ 13.05\ kX$, $b_0\ 9.83$, $c_0\ 6.01$; $\beta\ 103°22'$; $a_0:b_0:c_0 = 1.328:1:0.611$. Cell contents $Cu_{16}(SO_4)_4(OH)_{24}$.

Habit. Stout prismatic to acicular [001], but sometimes elongated [010] or, more rarely, [100]; also tabular {001}. Common forms: c a m d p π y v B. As loosely coherent aggregates of acicular crystals; in groups and drusy crusts; massive, granular.

Twinning.[4] On {100} with composition surface {100}; common; the twins often symmetrical and pseudo-orthorhombic in appearance.

Phys. Cleavage {100} perfect. Fracture uneven to conchoidal. H. $3\frac{1}{2}$–4. G. 3.97; 4.09 (calc.). Luster vitreous, on the cleavage somewhat

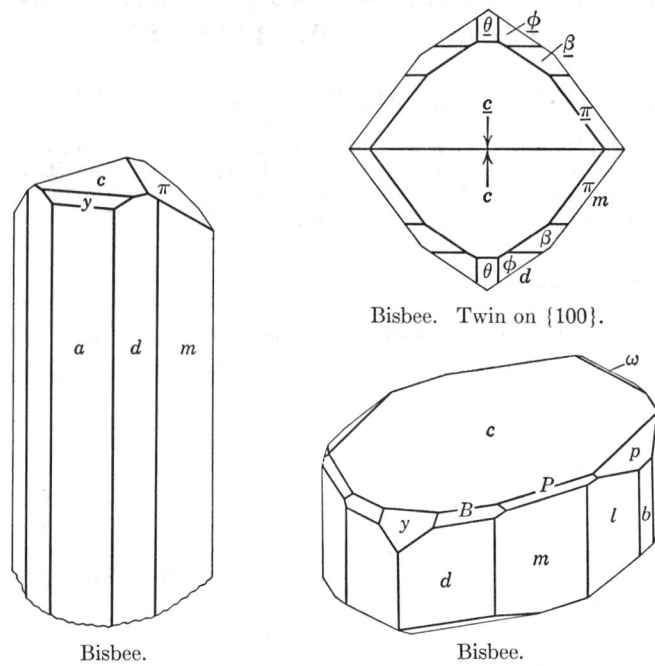

Bisbee. Twin on {100}.

Bisbee. Bisbee.

pearly. Color emerald-green to blackish green, also light green. Streak pale green. Transparent to translucent.

Opt.[5] In transmitted light, bluish green.

Orientation	n (Na) (anal. 5)	
$X \sim a$	1.728	Biaxial negative (−).
$Y\ \ b$	1.771	$2V\ 77° \pm 2°$.
$Z \sim c$	1.800	$r < v$, medium.
		Slightly pleochroic in bluish green.

Chem. A basic sulfate of copper, $Cu_4(SO_4)(OH)_6$.

Anal.[6]

	1	2	3	4	5
CuO	70.36	69.11	70.29	70.26	69.04
SO_3	17.70	17.07	17.54	18.32	17.61
H_2O	11.94	13.81	11.96	11.03	12.30
Rem.				tr.	0.95
Total	100.00	99.99	99.79	99.61	99.90
G.		4.09	3.88		

1. $Cu_4(SO_4)(OH)_6$. 2. Katanga.[7] 3. Collahurasi, Chile.[8] Average of two. 4. Tsumeb.[9] Rem. is FeO. 5. Frisco, Utah.[10] Rem. is insol.

Tests. B.B. fuses. In C.T. yields water which at higher temperatures is acid. Soluble in acids.

O c c u r. Brochantite is a secondary mineral found in the oxidized zone of copper deposits especially in arid regions. The mineral forms under conditions of relatively low acidity. Associated minerals include malachite, azurite, tenorite, cuprite, linarite, caledonite, cerussite, atacamite, chrysocolla, cyanotrichite, limonite. Numerous localities are known. Found in Russia at Gumeshevsk southwest of Ekaterinburg in the Urals, and at Nizhne-Tagilsk. From Rézbánya, Roumania. Near Nassau on the Lahn in Hesse-Nassau, Germany. From Rosas and Sa Duchessa, Sardinia, Italy. At Rio Tinto, Spain, and from Cap Garonne, Var, France. Found in various localities in Cornwall, and at Roughten Gill, Cumberland, England. At Krisuvig, Iceland. In South West Africa at Tsumeb, and at Katanga in the Belgian Congo; from Aïn-Barbar, Constantine, Algeria. Found abundantly at numerous localities in the desert regions of Chile, notably Chuquicamata, Paposo, and Potrerillos in Antofagasta, Challacollo in Atacama, and Collahurasi in Tarapaca. In New South Wales, Australia, at Broken Hill; New Caledonia.

In the United States, brochantite occurs abundantly with malachite in the Bisbee district, Cochise County, and in the Clifton-Morenci district, Greenlee County, Arizona; also from the Mammoth mine at Tiger, in Pinal County, and the United Verde mine in Yavapai County. From the Monarch mine, Chaffee County, Colorado. In Utah in the Tintic district, Juab County, from Frisco, Beaver County, the Apex mine at St. George in Washington County, and Dry Canyon in Toole County. In the Organ district, Dona Ana County, New Mexico. In Idaho in the Seven Devils district, Adams County, and the Alder Creek district, Custer County. In California at the Cerro Gordo mine in Inyo County.

A l t e r. Observed as pseudomorphs after malachite and azurite; also found altered to chrysocolla.

A r t i f.[11] Obtained as crystals in the system $CuO-SO_3-H_2O$.

N a m e. After A. J. M. Brochant de Villiers (1772–1840), French geologist and mineralogist.

Ref.

1. Elements of Palache, *Am. Min.*, **24**, 463 (1939). Originally described as orthorhombic by Lévy (1824) and so accepted by Dana (925, 1892), and, questionably, by Goldschmidt (**1**, 234, 1913), but shown to be monoclinic, with pseudo-orthorhombic symmetry due to twinning, by Palache (1939). Transformation: orthorhombic to monoclinic position, $0\bar{1}\tfrac{1}{2}/100/001$.
2. Palache (1939).
3. Richmond in Palache (1939) by Weissenberg method. Comparable data with confirmation of monoclinic symmetry are given by Lauro, *Per. Min.*, **12**, 419 (1941).
4. For a detailed description of the twinning, with numerous crystal drawings, see Palache (1939).
5. Indices of Posnjak and Tunell, *Am. J. Sc.*, **18**, 1 (1929), orientation of Palache (1937).
6. For additional analyses see Hintze (**1** [3B], 4220, 1929); Cavinato, *Per. Min.*, **10**, 259 (1939); Vavrinecz, *Magyar Chem. Foly.*, **35**, 1 (1929).
7. Schoep and Buysse, *Bull. soc. belge géol.*, **33**, 72 (1923).

8. Ford, *Am. J. Sc.*, **30**, 16 (1910).
9. Biehl, Inaug. Diss. Münster, 38 (1919).
10. Posnjak and Tunell (1929).
11. See Posnjak and Tunell (1929); Weiser, Milligan, and Cook, *J. Am. Chem. Soc.*, **64**, 503 (1942); Ciminali, *Rend. soc. min. ital.*, **1**, 34 (1941); Binder, *Ann. chim.*, **5**, 337 (1936).

30.1.2 **A N T L E R I T E** [$Cu_3(SO_4)(OH)_4$]. *Hillebrand (U. S. Geol. Sur., Bull. 55, 54, 1889).* Stelznerit *Arzruni* and *Thaddéeff (Zs. Kr., 31, 232, 1899).* Heterobrochantite *Buttgenbach (Soc. géol. Belgique, Bull., 49, 164, 1926).*

C r y s t.[1] Orthorhombic; dipyramidal—$2/m\ 2/m\ 2/m$.

$a:b:c = 0.6867:1:0.5027;\qquad p_0:q_0:r_0 = 0.7321:0.5027:1$

$q_1:r_1:p_1 = 0.6867:1.3659:1;\qquad r_2:p_2:q_2 = 1.9891:1.4562:1$

Forms:[2]

	ϕ	$\rho = C$	ϕ_1	$\rho_1 = A$	ϕ_2	$\rho_2 = B$
c 001	.....	0°00′	0°00′	90°00′	90°00′	90°00′
b 010	0°00′	90 00	90 00	90 00	.....	0 00
a 100	90 00	90 00	.....	0 00	0 00	90 00
f 130	25 53½	90 00	90 00	64 06½	0 00	25 53½
e 120	36 03½	90 00	90 00	53 56½	0 00	36 03½
m 110	55 31½	90 00	90 00	34 28½	0 00	55 31½
o 011	0 00	26 41½	26 41½	90 00	90 00	63 18½
k 201	90 00	55 40	0 00	34 20	34 20	90 00
r 111	55 31½	41 36½	26 41½	56 48½	53 47½	67 55
s 122	36 03½	31 52½	26 41½	71 53½	69 53½	64 43½
x 233	44 09	35 01	26 41½	66 26½	63 59	65 41
z 121	36 03½	51 12	45 09½	62 41½	53 47½	50 57
A 131	25 53½	59 11	56 27½	67 58½	53 47½	39 24½
C 211	71 03	57 08½	26 41½	37 23½	34 20	74 10
F 311	77 06½	66 04	26 41½	27 00	24 29	78 14

Less common:

| h 140 | g 560 | q 403 | v 123 | w 142 | B 151 | E 522 |
| i 230 | n 013 | u 221 | t 133 | y 313 | D 231 | |

Structure cell.[3] Space group *Pnam.* a_0 8.22 kX, b_0 11.97, c_0 6.02; $a_0:b_0:c_0 = 0.687:1:0.503$. Cell contents $Cu_{12}(SO_4)_4(OH)_{16}$.

Habit. Usually thick tabular {010}, also equant or short prismatic [001]. As cross-fiber veinlets or friable interlaced aggregates of acicular or fibrous crystals; felt-like; granular.

P h y s. Cleavage {010} perfect, {100} poor. Brittle. H. 3½. G. 3.88; 3.93 (calc.). Luster vitreous. Color emerald-green to blackish green, also light green; the color is very like brochantite and atacamite. Streak paler green. Translucent.

O p t.[4]

Orientation	n	Pleochroism	
X b	1.726	Yellow-green	Biaxial positive (+).
Y a	1.738	Blue green	$2V$ 53°.
Z c	1.789	Green	

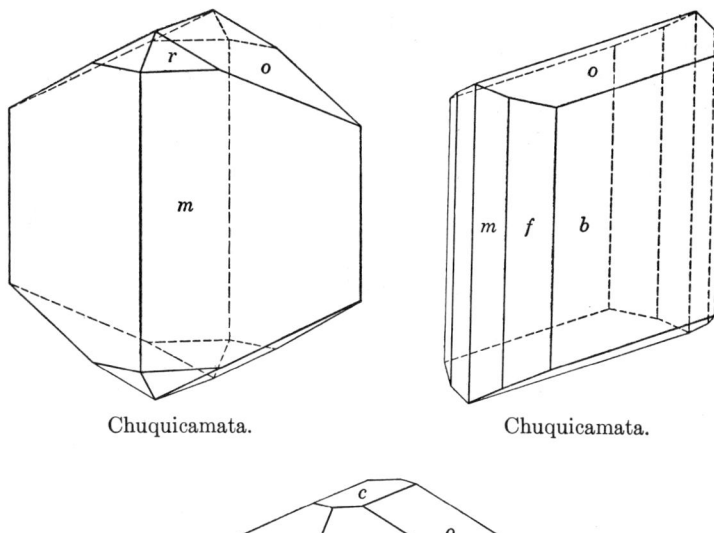

Chuquicamata. Chuquicamata.

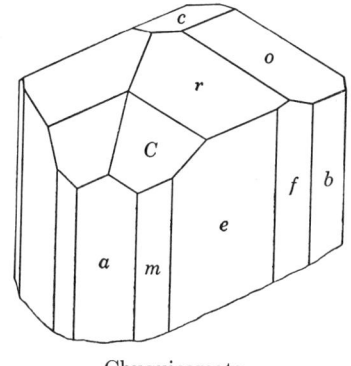

Chuquicamata.

Chem. A basic sulfate of copper, $Cu_3(SO_4)(OH)_4$.
Anal.[7]

	1	2	3
CuO	67.28	67.08	66.34
SO$_3$	22.57	22.40	22.32
H$_2$O	10.15	10.22	10.52
Rem.		0.84	0.88
Total	100.00	100.54	100.06
G.	3.93	3.884	

1. $Cu_3(SO_4)(OH)_4$. 2. Remolinos, Chile.[5] Rem. is insol. 0.44, Fe$_2$O$_3$ 0.34, CaO 0.06.
3. Chuquicamata, Chile.[6] Rem. is insol.

Tests. B.B. fusible. Soluble in dilute H$_2$SO$_4$.

Occur. Antlerite is a secondary mineral found in the oxidized zone of copper deposits in arid regions. Associated with atacamite, brochantite, chalcanthite, kroehnkite, natrochalcite, linarite, gypsum. Originally from the Antler mine in the Hualpai Mountains, Mohave County, Arizona; also from Bisbee, Cochise County. From the Northern Light mine, Warsuk range, near Black Mountain, Nevada, and from Kennecott, Alaska. In the Sierra Mojada, Coahuila, Mexico. From Remolinos, Vallinar, Chile (*stelznerite*). Antlerite is the principal ore mineral of copper in the oxidized zone of the great mine at Chuquicamata, Chile.[9] With linarite in Kazakhstan, U.S.S.R.

A r t i f. The conditions of formation of antlerite in the system CuO-SO_3-H_2O have been investigated.[8] With decreasing concentration of SO_3, chalcanthite, antlerite and brochantite are stable in sequence.

A l t e r. Cuprite has been found as pseudomorphs after fibrous antlerite.

Ref.

1. Orientation and angles of Palache, *Am. Min.*, **24**, 293 (1939); angles averaged from new measurements on Bisbee and Chuquicamata crystals and older measurements of Ungemach, *Bull. soc. min.*, **47**, 124 (1924), on Chuquicamata crystals. Transformation: Ungemach (1924) to Palache (1939), 001/010/100.
2. Palache (1939). Uncertain: 160, 570, 580, 015, 255, 161.
3. Richmond in Palache (1939) by Weissenberg method.
4. Posnjak and Tunell, *Am. J. Sc.*, **18**, 1 (1929), on analyzed artificial crystals.
5. Thaddéeff and Arzruni, *Zs. Kr.*, **31**, 232 (1899).
6. Audrieth and Martens, *Am. Min.*, **10**, 161 (1925).
7. See also Hintze (1 [3B], 4224, 1929).
8. See Posnjak and Tunell (1929); Binder, *Ann. chim.*, **5**, 337 (1936); Guillot and Geneslay, *Bull. soc. chim.*, **4**, 125 (1937).
9. See Jarrell, *Econ. Geol.*, **39**, 251 (1944).

30.1.3 **Caracolite.** *Websky* (*Ak. Berlin, Sitzber.*, 1045, 1886).

As a crystalline incrustation. The imperfect crystals range in size up to a millimeter or so and are seemingly hexagonal with a hexagonal pyramid and prism and large base. Measured angles (mean): cp 50°08′, pp' 37°44′, mm' 60°36′. Optical evidence and sutures on the pyramid edges indicate that the crystals are pseudo-hexagonal trillings, analogous to aragonite.[1] Cleavage not noted. H. $4\frac{1}{2}$. G. ~ 5.1. Luster vitreous. Colorless and transparent; also grayish or stained green.

O p t.[2] Colorless.

Orientation	n	
X	1.743 ± 0.005	Biaxial negative $(-)$.
Y	1.754 ± 0.005	$2V$ very large.
Z	1.764 ± 0.005	$r > v$, rather strong.

Shows complex polysynthetic twinning with rather large extinction angles.

C h e m. A basic chloride-sulfate of sodium and lead of unknown formula. A partial analysis on very impure material gave:

Na_2O	Pb	Cu	FeO	ZnO	SO_3	Cl	Insol.	Total
Present	50.88	2.51	0.33	0.29	16.70	10.18	1.84	82.73

Tests. B.B. fuses easily to a white to brown enamel. In C.T. fuses to a greenish brown mass and affords water.[3] Fragments are partly decomposed by water becoming white and opaque and losing sodium chloride to the solution. Partly decomposed by HNO_3, completely by ammonium acetate or hot KOH solution.

Schairerite

Occur. From the Mina Beatriz, Sierra Gorda, Atacama, between Caracoles and the Bay of Mejillones, Chile. Associated with percylite, bindheimite, anglesite, and galena.

Ref.
1. Regarded as orthorhombic by Websky (1886), with $a:b:c = 0.5843:1:0.4213$, and twinned on (110). See also Fletcher, *Min. Mag.*, **8**, 172 (1889), who observed (hexagonal notation) $\{0001\}$, $\{10\bar{1}1\}$, $\{10\bar{1}0\}$, and rare $\{11\bar{2}0\}$ and $\{11\bar{2}2\}$. Goldschmidt (**2**, 104, 1913) made the crystals questionably hexagonal with $a:c = 1:0.6264$.
2. Larsen (52, 1921); the optical data here given for daviesite also refer to caracolite.
3. Anhydrous according to Sandberger, *Jb. Min.*, II, 75 (1887).

30·1·4 Chlorothionite $[K_2Cu(SO_4)Cl_2]$. *Scacchi* (*Acc. Napoli, Rend.*, 203, 1872).

Orthorhombic; dipyramidal—$2/m\ 2/m\ 2/m$. Found as bright blue crystalline incrustations in fumaroles of the eruption of 1872 of Vesuvius, Italy. Composition $K_2Cu(SO_4)Cl_2$. Analysis gave:

	K	Cu	Cl	SO$_4$	Loss	Total
1.	25.33	20.59	22.97	31.11		100.00
2.	26.29	19.56	20.04	32.99	1.12	100.00

1. $K_2Cu(SO_4)Cl_2$. 2. Vesuvius.

Artificial crystals [1] have G. 2.67, H. $2\frac{1}{2}$, and are optically biaxial positive (+) with moderately large $2V$ and $r > v$. Space group $Pnma$, with a_0 6.105 kX, b_0 7.697, c_0 16.132; cell contents $K_8Cu_4(SO_4)_4Cl_8$. Water-soluble.

Ref.
1. Bellanca, *Per. Min.*, **15**, 33 (1946); Zambonini (329, 1910); Goldschmidt (**2**, 151, 1913).

30.1.5 SCHAIRERITE $[Na_3(SO_4)(F,Cl)]$. *Foshag* (*Am. Min.*, **16**, 133, 1931).

Cryst.[1] Hexagonal (scalenohedral—$\bar{3}\ 2/m$ (?)).

$a:c = 1:2.7468$; $\alpha\ 55°01\frac{1}{2}'$; $p_0:r_0 = 3.1717:1$; $\lambda\ 110°82'$

Forms:[2]

			ϕ	ρ	A_1	A_2
c	0001	111	.....	0°00'	90°00'	90°00'
m	$10\bar{1}0$	$2\bar{1}\bar{1}$	30°00'	90 00	30 00	90 00
r	$10\bar{1}1$	100	30 00	72 30	34 19	90 00
e	$01\bar{1}2$	110	−30 00	57 46	90 00	42 54

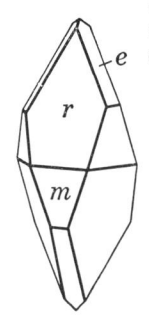

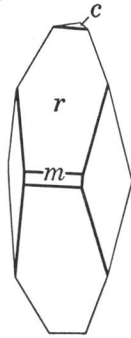

Twin on $\{0001\}$ (?).

Structure cell.[3] Apparently hexagonal—P with a_0 12.12 kX, c_0 19.19; $a_0:c_0 = 1:1.583$.

Habit. Rhombohedral, with $\{03\bar{3}2\}$ prominent. Artificial crystals are tabular $\{0001\}$.

Twinning. On $\{0001\}$, common.

Phys. Fracture conchoidal.[4] Brittle. H. $3\frac{1}{2}$. G. 2.612; 2.67 (calc.). Luster vitreous. Colorless. Transparent.

Opt. In transmitted light, colorless.

	n (Searles Lake)	n(Na) (artif. $Na_3(SO_4)F$) [5]	
O	1.440	1.436	Uniaxial positive (+).
E	1.445	1.439	

Chem. Sodium sulfate-fluoride, $Na_3(SO_4)(F,Cl)$. In the natural material of analysis 2, Cl substitutes for F with $F:Cl \sim 4:1$. This series is limited, however, and the compound $Na_6(SO_4)_2FCl$ is isometric.

Anal.

	1	2
Na	36.86	35.77
K		0.13
SO_4	51.32	50.01
F	8.22	8.08
Cl	3.60	3.44
Rem.		1.55
Total	100.00	98.98

1. $Na_3(SO_4)(F,Cl)$ with $F:Cl = 4:1$. 2. Searles Lake.[6] Rem. is CaO 0.30, $(Fe,Al)_2O_3$ 0.15, insol. 0.20, ign. loss 0.90.

Tests. Easily fusible to a white porcelaneous mass. Slowly but completely soluble in water.

Occur. Found in a well sample from the central salt crust of Searles Lake, San Bernardino County, California. Associated with gaylussite, tychite, pirssonite, thenardite, trona, hanksite, calcite.

Artif.[5] Obtained as crystals in the system Na_2SO_4-NaF-NaCl-H_2O. The artificial crystals apparently are pseudo-hexagonal sectoral intergrowths of individuals of lower symmetry which invert[7] to the hexagonal form at *ca.* 105° when heated.

Name. After John F. Schairer (1904–), physical chemist, of the Geophysical Laboratory, Washington.

Ref.

1. Elements modified after Foshag (1931).
2. Foshag (1931). Schairer in Foote and Schairer, *J. Am. Chem. Soc.*, **52**, 4202 (1930), gives {10$\bar{1}$3} and {02$\bar{2}$3} on artificial crystals.
3. Frondel, *Am. Min.*, **25**, 352 (1940), by Weissenberg method on natural crystal.
4. Cleavage {0001} is given by Foote and Schairer (1930) but is not verified by Foshag (1931).
5. Foote and Schairer (1930).
6. Foshag (1931).
7. Wolters, *Jb. Min., Beil.-Bd.*, **30**, 55 (1910).

30.1.6 SULFOHALITE [$Na_6ClF(SO_4)_2$]. *Hidden* and *Mackintosh* (*Am. J. Sc.*, **36**, 463, 1888). Sulphohalite.

Cryst.[1] Isometric; hexoctahedral—$4/m\,\bar{3}\,2/m$ (?).

Forms:

a 001   o 111   d 011

Structure cell. Space group uncertain, probably $Fm3m$ or $F\bar{4}3$. a_0 10.15 kX,[2] 10.08 kX.[3] Cell contents $Na_{24}Cl_4F_4(SO_4)_8$.

SULFOHALITE

Habit. Usually dodecahedral; also octahedral, or in combinations of a, o, and d.

Phys. Fracture conchoidal. H. $3\frac{1}{2}$. G. 2.500;[4] 2.48 (calc. for a_0 10.08). Colorless to pale greenish yellow or gray. Luster weak, vitreous to greasy. Transparent. Taste weakly saline.

Opt. In transmitted light, colorless. Isotropic. Index of refraction:[5] 1.455.

Chem. A chloride-fluoride-sulfate of sodium, $Na_6ClF(SO_4)_2$. The Cl and F appear to be structurally distinct and not isomorphous.

Anal.[6]

	1	2	3	4
Na_2O	32.24	32.37	32.50	31.56
Na	11.96	11.60	11.35	13.41
Cl	9.23	9.10	9.19	9.20
F	4.94	4.71	[4.71]	4.95
SO_3	41.63	41.79	42.00	40.76
Rem.		0.25	0.25	0.65
Total	100.00	99.82	[100.00]	100.53
G.	2.48	2.500	2.5	

1. $Na_6ClF(SO_4)_2$. 2. Searles Lake.[7] Rem. is K_2O 0.10, ign. loss 0.15. 3. Searles Lake.[8] Rem. is ign. loss at 200°C. 4. Otjiwalundo, S. W. Africa.[9] Rem. is H_2O 0.15, insol. 0.50.

Tests. Easily fusible. Slowly soluble in cold water.

Occur. Found associated with hanksite in drill cuttings from the saline beds of Searles (Borax) Lake, San Bernardino County, California. Crystals have been found up to 3 cm. in size. Later recognized as an abundant constituent of the Otjiwalundo salt pan, 250 kilometers north of Otavi, South West Africa, where it occurs with trona, thenardite, and pirssonite.

Artif. Not known artificially.[10]

Name. In allusion to the composition (given originally in lack of knowledge of the F content of the mineral).

Ref.

1. The crystal class and space group is not rigorously fixed by the available evidence —cf. Watanabé, *Proc. Imp. Ac. Tokyo*, **10**, 575 (1934), and Pabst, *Zs. Kr.*, **89**, 514 (1934). Morphological evidence of tetrahedral symmetry is mentioned by Hidden and Mackintosh, *Am. J. Sc.*, **41**, 438 (1891).
2. Watanabé (1934) by powder method.
3. Pabst (1934) by powder method.
4. An average of closely agreeing values obtained by the suspension method by Penfield, *Am. J. Sc.*, **9**, 425 (1900); Hidden and Mackintosh (1888) give 2.489, and Foshag, *Am. J. Sc.*, **49**, 76 (1920), 2.43.
5. Foshag (1920); Larsen in Gale and Hicks, *Am. J. Sc.*, **38**, 273 (1914); Foshag, *Am. Min.*, **18**, 431 (1933).
6. In the original analysis by Mackintosh in Hidden and Mackintosh (1888) the F was overlooked.
7. Penfield (1900).
8. Hicks in Gale and Hicks (1914).
9. Foshag (1933).
10. Unsuccessful attempts to synthesize the compound were made by van't Hoff and Saunders, *Ak. Berlin, Sitzber.*, **1**, 387 (1898), and de Schulten (priv. comm. in Penfield (1900)), but these were based on the erroneous original formula of Hidden and Mackintosh (1888), $Na_8Cl_2(SO_4)_3$.

TYPE 2. $A_2(XO_4)Z_q$

30.2.1 LANARKITE [$Pb_2(SO_4)O$]. Sulphato-carbonate of Lead *Brooke* (*Edinburgh N. Phil. J.*, **3**, 117, 1820). Prismatisches schwefelkohlensaures Blei *Leonhard* (253, 1826). Plomb sulfatocarbonaté *Dufrénoy* (**3**, 255, 1856). Lanarkite *Beudant* (**2**, 366, 1832). Dioxylith *Breithaupt* (**2**, 180, 1841). Kohlenvitriol-bleispath, Halbvitriolblei, Bleisulfocarbonate *Germ.*

C r y s t.[1] Monoclinic; prismatic—$2/m$.

$a:b:c = 2.4149:1:1.2424$; $\beta\ 115°48'$; $p_0:q_0:r_0 = 0.5145:1.1186:1$

$r_2:p_2:q_2 = 0.8940:0.4600:1$; $\mu\ 64°12'$; $p_0'\ 0.5714, q_0'\ 1.2424, x_0'\ 0.4834$

Forms:[2]

		ϕ	ρ	ϕ_2	$\rho_2 = B$	C	A
c	001	90°00'	25°48'	64°12'	90°00'	.....	64°12'
m	110	24 42	90 00	0 00	24 42	79°33'	65 18
d	103	90 00	33 58½	56 01½	90 00	8 10½	56 01½
o	205	90 00	35 27	54 33	90 00	9 39	54 33
f	201	90 00	58 24½	31 35½	90 00	32 36½	31 35½
E	$\bar{4}03$	−90 00	15 34	105 34	90 00	41 22	105 34
F	$\bar{2}01$	−90 00	33 24	123 24	90 00	59 12	123 24
U	$\bar{4}01$	−90 00	60 58½	150 58½	90 00	86 46½	150 58½
R	$\bar{8}01$	−90 00	76 15	166 15	90 00	102 03	166 15
P	$\bar{1}11$	−4 03	51 14½	95 03	38 56½	57 20	93 09½

Structure cell.[3] Space group $C2/m$. $a_0\ 13.73\ kX$, $b_0\ 5.68$, $c_0\ 7.07$; $\beta\ 116°13'$; $a_0:b_0:c_0 = 2.417:1:1.245$. Cell contents $Pb_8(SO_4)_4O_4$.

Habit. Crystals elongated [010]. Also massive.

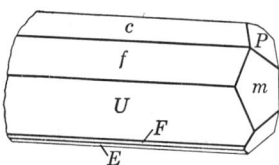

Twinning.[5] Polysynthetic, in the zone [010]; rare.

P h y s. Cleavage $\{\bar{2}01\}$, perfect, $\{\bar{4}01\}$ and $\{201\}$, imperfect; also $\{010\}$.[6] Thin laminae flexible. H. 2–2½. G. 6.92;[4] 7.00 (calc.). Luster adamantine inclining to resinous, on cleavage surfaces pearly. Color gray to greenish white, pale yellow. Streak white. Transparent to translucent. Fluoresces yellow in ultraviolet radiation and in x-rays.

O p t.[7] In transmitted light, colorless.

Orientation	n(Na)	
X	1.928 ± 0.003	Biaxial negative (−).
$Y\ b$	2.007 ± 0.003	$2V\ 60° \pm 2°$.
$Z \wedge c\ \ 30°$	2.036 ± 0.003	$r > v$, strong, inclined.

C h e m. An oxysulfate of lead, $Pb_2(SO_4)O$. Early thought to be a carbonate-sulfate (see synonymy) but the CO_2 later was shown [8] to be due to admixture.

Anal.[10]

	1	2	3
PbO	84.79	84.63	84.1
SO_3	15.21	15.37	15.2
Ign. loss			0.5
Total	100.00	100.00	99.8
G.		6.8	

1. $Pb_2(SO_4)O$. 2. Leadhills.[8] Recalculated to 100 from original sum of 98.66 after deducting 0.83 ign. loss. 3. Leadhills.[9]

Tests. B.B. fuses with difficulty (M.P. 977°). Soluble in KOH and in warm HNO_3; in cold HNO_3 leaves a deposit of $PbSO_4$.

O c c u r. Found sparingly in the Susanna mine, Leadhills, Lanarkshire, Scotland, associated with leadhillite, cerussite, and caledonite. Also from Tanne in the Harz Mountains, and in the Herrensegen mine in the Kinzigtal, Black Forest, Germany. From Bieberwier northwest of Innsbruck in the Tyrol, Austria. At the Laquorre mine near Aulus, Ariège, France. Reported from Challocolla, Atacama, Chile.

A l t e r. To cerussite, leadhillite and anglesite.

A r t i f.[11] Obtained in crystals by the slow inter-diffusion of sodium sulfate and alkalic lead acetate or lead formate solutions; also found in the fusion system PbO-$PbSO_4$.

N a m e. From the original locality in Lanarkshire, Scotland.

Ref.

1. Angles of Schrauf, *Zs. Kr.*, **1**, 31 (1877), on Leadhills material; unit and orientation of Richmond and Wolfe, *Am. Min.*, **23**, 399 (1938), conforming to the base-centered structure cell. Transformation: from the pseudo-rectangular cell ($\beta = 91°49'$) of Schrauf to new, $\overline{1}0\tfrac{1}{2}/0\tfrac{1}{2}0/\overline{1}0\tfrac{1}{4}$; Goldschmidt (**2**, 281, 1890) to new, $\overline{1}0\tfrac{2}{3}/0\tfrac{1}{3}0/\overline{1}0\tfrac{1}{6}$; new to Schrauf, $\overline{1}02/030/\overline{1}02$; new to Goldschmidt, $\overline{1}04/030/\overline{1}02$.
2. Schrauf (1877); Richmond and Wolfe (1938). Rare or doubtful: 100, 101, $\overline{1}01$, 401, 601, $\overline{9}.10.3$. Discarded: $\overline{6}.2.57$, $\overline{234}.4.57$, $\overline{75}.4.18$.
3. Richmond and Wolfe (1938) by Weissenberg method on Leadhills crystal.
4. Richmond and Wolfe (1938) by flotation method on Leadhills crystal; earlier reported values range down to 6.3. de Schulten, *Bull. soc. min.*, **21**, 142 (1898), gave 6.923 for artificial material.
5. Observed optically by Lacroix (**4**, 146, 1910) on Laquorre crystals.
6. Reported by Lacroix (1910) but not noted by other observers.
7. Richmond and Wolfe (1938); see also Larsen (97, 1921) and Lacroix (1910).
8. Pisani, *C.R.*, **76**, 114 (1873).
9. Collie, *J. Chem. Soc. London*, **55**, 91 (1889).
10. Additional analyses are cited by Dana (923, 1892).
11. de Schulten (1898); Barford, *Overs. Dansk. Vidensk. Selskr. Forh.*, no. 3, 13 (1869); Jaeger and Germs, *Zs. anorg. Chem.*, **119**, 151 (1921).

30.2.2 **D O L E R O P H A N I T E** [$Cu_2(SO_4)O$]. Dolerophano Scacchi (*Acc. Napoli, Att.*, **5**, 26, 1873). Dolerophan.

C r y s t.[1] Monoclinic; prismatic—$2/m$.

$a:b:c = 1.4842:1:1.2089$; $\beta\ 122°18\tfrac{1}{2}'$; $p_0:q_0:r_0 = 0.8145:1.0217:1$

$r_2:p_2:q_2 = 0.9787:0.8127:1$; $\mu\ 57°41\tfrac{1}{2}'$; $p_0'\ 0.9637, q_0'\ 1.2089, x_0'\ 0.6324$

Forms:[2]

	ϕ	ρ	ϕ_2	$\rho_2 = B$	C	A
c 001	90°00′	32°18½′	57°41½′	90°00′	.....	57°41½′
b 010	0 00	90 00	.....	0 00	90°00′	90 00
a 100	90 00	90 00	0 00	90 00	57 41½	.....
m 110	38 33½	90 00	0 00	38 33½	70 32½	51 26½
d $\bar{1}$01	−90 00	18 20	108 20	90 00	50 38½	108 20
B $\bar{4}$03	−90 00	33 07½	123 07½	90 00	65 26	123 07½
e $\bar{2}$01	−90 00	52 19½	142 19½	90 00	84 38	142 19½
f $\bar{4}$01	−90 00	72 45½	162 45½	90 00	105 04	162 45½
r $\bar{1}$12	13 59	31 55	81 26½	59 08	38 13	82 39½
s $\bar{1}$11	−15 19½	51 25	108 20	41 04	65 22½	101 55½
n $\bar{1}$33	14 26	51 18	72 43	40 54	50 46½	78 47

Less common:
o 410 w 011 μ $\bar{3}$04 j $\bar{4}$05 l 201 h $\bar{8}$03 y 601 v 111 p $\bar{3}$14 q $\bar{3}$12

Structure cell.[3] Space group $C2/m$. a_0 9.39 kX, b_0 6.30, c_0 7.62; β 122°41½′; $a_0:b_0:c_0 = 1.490:1:1.209$. Cell contents $Cu_8(SO_4)_4O_4$.

Habit. Crystals small, usually elongated [010] and more or less tabular on $\{\bar{1}01\}$. Faces in the zone [010] often striated [010], other faces smooth and lustrous. Translucent to opaque.

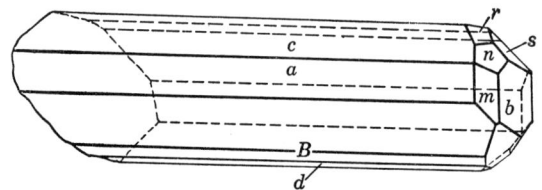

P h y s. Cleavage $\{\bar{1}01\}$ perfect. H. 3. G. 4.17;[4] 4.16 (calc.). Color chestnut-brown to dark brown and nearly black. Streak yellowish brown.

O p t.[5] In transmitted light, yellow-brown.

ORIENTATION	n(Na)	PLEOCHROISM	
X	1.715	Deep brown	Biaxial positive (+).
Y b	1.820	Brownish yellow	$2V$ 85°.
Z ∧ c −10°	1.880	Lemon-yellow	$r > v$, very strong; crossed.

C h e m. An oxysulfate of copper, $Cu_2(SO_4)O$.

Anal.

	1	2	3	4
CuO	66.53	62.27	65.20	65.95
SO$_3$	33.47	36.07	33.49	34.43
Rem.		1.22	[1.31]	
Total	100.00	99.56	[100.00]	100.38

1. $Cu_2(SO_4)O$. 2,3. Vesuvius.[6] Rem. is insol. 4. Artificial.[7]

Tests. B.B. fuses, leaving a black scoriaceous residue. Easily soluble in HNO_3. Slowly decomposed by cold water, giving a blue solution and residue, and by moist air.

O c c u r. Associated with chalcocyanite, euchlorin, and eriochalcite as a sublimate formed during the eruption of Vesuvius in October, 1868.

LINARITE 553

Artif.[8] Obtained as a powder by ignition of $CuSO_4$ at 650°. Also observed in the floor of a copper reverbatory furnace.[7]

Name. From δολερός, *fallacious*, and φαίνεσθαι, *to appear*, because its appearance does not suggest its composition.

Ref.

1. Angles of Zambonini (161, 1935); orientation and unit of the x-ray cell of Richmond and Wolfe, *Am. Min.*, **25**, 606 (1940), and identical with that of Goldschmidt (*Index*, 511, 1886). Transformations: Scacchi (1873) to Goldschmidt (1886), $\overline{4}00/040/\overline{3}03$; Dana (924, 1892) to Goldschmidt, $00\overline{4}/040/\overline{3}03$; Strandmark, *Zs. Kr.*, **36**, 457 (1902), and Zambonini (1935) to Goldschmidt, $\overline{1}0\overline{1}/0\overline{1}0/001$.
2. Richmond and Wolfe (1940). Uncertain forms: $\overline{10}.0.1$, $\overline{13}.0.1$, $\overline{14}.0.1$, $\overline{8}83$, $\overline{2}69$.
3. Richmond in Richmond and Wolfe (1940) by Weissenberg method on a Vesuvius crystal.
4. Richmond in Richmond and Wolfe (1940) on Vesuvius material.
5. Berman in Richmond and Wolfe (1940).
6. Scacchi (1873).
7. Strandmark (1902).
8. Binder, *Ann. chim.*, **5** [11], 337 (1936).

30.2.3 LINARITE $[PbCu(SO_4)(OH)_2]$. Crystallized Blue Carbonate of Copper *Sowerby* (3, 5, fig. 203, 1809). Cupreous Sulfate of Lead *Brooke* (*Ann. Phil.*, **4**, 117, 1822). Linarite *Glocker* (618, 1839); *Phillips-Alger* (552, 1844). Cupreous Anglesite. Plomb sulfaté cuprifère *Levy* (**2**, 455, 1837). Bleilasur *Breithaupt* (1823). Kupferbleispat, Kupferbleivitriol *Germ.*

Cryst.[1] Monoclinic; prismatic—$2/m$.

$a:b:c = 1.7161:1:0.8296;\quad \beta\ 102°37\frac{1}{2}';\quad p_0:q_0:r_0 = 0.4834:0.8095:1$

$r_2:p_2:q_2 = 1.2353:0.5972:1;\quad \mu\ 77°22\frac{1}{2}';\quad p_0'\ 0.4956,\ q_0'\ 0.8296,\ x_0'\ 0.2240$

Forms:[2]

		ϕ	ρ	ϕ_2	$\rho_2 = B$	C	A
c	001	90°00′	12°37½′	77°22½′	90°00′	.....	77°22½′
b	010	0 00	90 00	.....	0 00	90°00′	90 00
a	100	90 00	90 00	0 00	90 00	77 22½	.....
m	110	30 50½	90 00	0 00	30 50½	83 34	59 09½
l	210	50 03½	90 00	0 00	50 03½	80 21	39 56½
w	012	28 22	25 14½	77 22½	67 58	22 02	78 18½
v	011	15 06½	40 40½	77 22½	51 00½	38 59½	80 13
y	101	90 00	35 44½	54 15½	90 00	23 07	54 15½
d	$\overline{1}08$	90 00	9 12½	80 47½	90 00	3 25	80 47½
o	$\overline{2}03$	−90 00	6 04½	96 04½	90 00	18 42	96 04½
t	$\overline{5}06$	−90 00	10 42	100 42	90 00	23 19½	100 42
s	$\overline{1}01$	−90 00	15 11½	105 11½	90 00	27 49	105 11½
x	$\overline{3}02$	−90 00	27 27	117 27	90 00	30 04½	117 27
u	$\overline{2}01$	−90 00	37 29½	127 29½	90 00	50 07	127 29½
p	$\overline{7}01$	−90 00	72 52½	162 52½	90 00	85 30	162 52½
e	$\overline{1}11$	−18 07½	41 07	105 11½	51 19	46 20	101 48½
n	$\overline{2}21$	−24 49	61 19	127 29½	37 13½	67 11	111 36½
g	$\overline{2}11$	−42 45½	48 29½	127 29½	56 39	57 37	120 33½

Less common or rare:

η 501	$\overline{8}09$	$\pi\ \overline{7}03$	$q\ \overline{1}12$	311	$\overline{12}.2.11$	$\overline{7}16$
H 14.0.1	$\overline{9}.0.10$	$\beta\ \overline{12}.0.5$	718	$\alpha\ \overline{13}.1.13$	$\overline{11}.1.10$	$\overline{22}.1.14$
$\overline{3}04$	$\rho\ \overline{39}.0.20$	111	211	$\delta\ \overline{9}19$	$\overline{8}17$	$\overline{5}23$

Structure cell.[3] Space group $P2_1/m$. a_0 9.66 kX, b_0 5.64, c_0 4.67; β 102°48′; $a_0:b_0:c_0 = 1.714:1:0.828$. Cell contents $Pb_2Cu_2(SO_4)_2(OH)_4$.

Habit. Crystals elongated [010], and often tabular $\{\bar{1}01\}$ or $\{001\}$ rarely $\{100\}$. Crystals either singly or in groups; as crusts or confused aggregates of prismatic crystals.

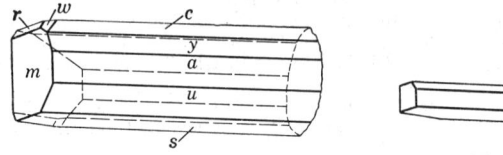
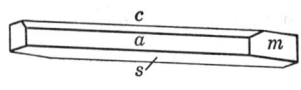

Mammoth mine, Arizona.   New Caledonia.

Twinning. On $\{100\}$ fairly common; also reported on $\{001\}$.

P h y s. Cleavage $\{100\}$ perfect, $\{001\}$ imperfect. Fracture conchoidal. Brittle. H. $2\frac{1}{2}$. G. 5.35;[4] 5.33 (calc.). Luster vitreous to sub-adamantine. Color deep azure-blue. Streak pale blue. Translucent.

O p t.[5] In transmitted light, deep blue.

Orientation	n(Na) (Sardinia)	Pleochroism[6]	
$X \wedge c \sim -24°$	1.809	Pale blue	Biaxial negative (−).
Y	1.838	Clear blue	$2V$ 80°(Na).
Z b	1.859	Prussian blue	$r < v$, strong.

C h e m. A basic sulfate of lead and copper, $PbCu(SO_4)(OH)_2$. The H_2O is lost at 190°.[9]

Anal.[11]

	1	2	3	4	5
CuO	19.85	19.08	19.88	19.61	26.43
PbO	55.68	55.52	75.17	55.91	47.15
SO$_3$	19.97	19.94		19.93	20.70
H$_2$O	4.50	5.35	4.73	4.44	5.00
Rem.		0.18		0.30	1.22
Total	100.00	100.07	99.78	100.19	100.50
G.	5.33		5.23		

1. $PbCu(SO_4)(OH)_2$. 2. Arakawa, Japan.[7] Rem. is Fe_2O_3. 3. Slocan, B. C.[8] 4. Arenas, Sardinia.[9] Rem. is ZnO 0.04, MgO 0.26. 5. Linarite? Kazakhstan.[10] Rem. is insol.

Tests. In C.T. decrepitates, loses water and darkens in color. Easily fusible on charcoal. Soluble in dilute nitric acid with separation of lead sulfate.

O c c u r. Widespread in small amounts as a secondary mineral in the oxidized zone of copper and lead deposits. Associated with cerussite, malachite, brochantite, caledonite, aurichalcite, anglesite, leadhillite, hemimorphite, chrysocolla. Only a few of the known localities can be mentioned here. Originally from Linares, Jaen province, Spain. At Rézbánya, Transylvania, and at Lölling near Hüttenberg in Carinthia, Austria. In Germany in the Schapbachtal and at Badenweiler in the Black Forest, and from Nassau on the Lahn in Hesse-Nassau. In Russia at Nerchinsk in Transbaikalia, near Beresovsk in the Urals, and from several localities in

Kazakhstan. From San Giovanni and Arenas in Sardinia. In Cumberland, England, at Red Gill, Roughten Gill and near Keswick, as fine crystals sometimes an inch in length; also at Leadhills, Lanarkshire, and elsewhere in Scotland. Abundant at Tsumeb, South West Africa. In South America in the Ortiz mine, Catamarca province, Argentina, and from localities in Chile and Peru. At Broken Hill, New South Wales. From Arakawa, Ugo province, and Kamioka, Hida province, in Japan. In the United States, a notable occurrence in the Mammoth mine, Tiger, Arizona. In the Tintic district and at Eureka, Utah. At Park City, Utah, and Butte, Montana. In the Bay Horse district, Custer County, and the Coeur d'Alene district, Shoshone County, Idaho. Fine specimens came from the Cerro Gordo mines, Inyo County, California. In Canada from Beaver Mountain in the Slocan district and from the Atlin district in British Columbia.

Alter. Found altered to antlerite, and to cerussite with malachite.

Name. From the locality in Spain at Linares.

Ref.

1. Koksharov (**5**, 206, 1866), *Bull. ac. imp. Sc.*, *St. Pétersburg*, **13**, 472 (1869), from an extended study of Cumberland crystals. Transformation: Goldschmidt (**5**, 158, 1918) to Koksharov, $10\overline{1}/0\overline{1}0/001$.
2. Goldschmidt (1918). See also Shannon, *U. S. Nat. Mus. Bull.*, **131**, 455 (1926).
3. Berry, priv. comm. (1946), by Weissenberg method on Red Gill crystals.
4. Frondel, priv. comm. (1949), on Mammoth crystal.
5. Brugnatelli, *Zs. Kr.*, **28**, 307 (1897), and Fenoglio, *Per. Min.*, **3**, 4 (1932).
6. Lacroix (**4**, 152, 1910).
7. Tamura cited by Wada (78, 1904).
8. Johnston, *Geol. Sur. Canada*, *Summ. Rpt.*, 260 (1910).
9. Fenoglio (1932).
10. Chukhrov, *C. r. ac. sc. U.R.S.S.*, **22**, 257 (1939); **15**, 95 (1937).
11. For earlier analyses, see Hintze (**1** [3B], 4205, 1929).

30.2.4 ALUNITE GROUP

HEXAGONAL—R; DITRIGONAL-PYRAMIDAL—3M

	c_0/a_0	a_0	c_0
Alunite, $KAl_3(SO_4)_2(OH)_6$	2.493	6.96 kX	17.35
Natroalunite, $NaAl_3(SO_4)_2(OH)_6$			
Jarosite, $KFe_3(SO_4)_2(OH)_6$	2.361	7.20	17.00
Ammoniojarosite, $NH_4Fe_3(SO_4)_2(OH)_6$	2.361	7.20	17.00
Natrojarosite, $NaFe_3(SO_4)_2(OH)_6$	2.270	7.18	16.30
Argentojarosite, $AgFe_3(SO_4)_2(OH)_6$	2.271	7.22	16.40
Carphosiderite, $(H_2O)Fe_3(SO_4)_2[(OH)_5 \cdot H_2O]$	2.360	7.16	16.70
Beaverite, $Pb(Cu,Fe,Al)_3(SO_4)_2(OH)_6$	2.351	7.20	16.94
Plumbojarosite, $PbFe_6(SO_4)_4(OH)_{12}$	4.667	7.20	33.60

The members of the Alunite Group are isostructural and conform to the general formula $A'B'''(SO_4)_2(OH)_6$ where A = Na, K, Pb, Rb, NH$_4$, Ag, or H$_2$O and B = Al or Fe'''. The minerals of the Plumbogummite and Beudantite Groups have a very similar structure. In the Alunite structure-type,[1] the A atom is in 12-coordination between six O and six (OH) ions and the B atom is in six-coordination between two O and four

(OH) ions. The A position is very tolerant with respect to ionic size. In plumbojarosite half of the available A positions are occupied by Pb'' ions, and the others remain vacant to provide valence compensation. In the members of the Beudantite Group there is a 1:1 replacement of the $(SO_4)^=$ tetrahedra by $(AsO_4)^\equiv$ or $(PO_4)^\equiv$ tetrahedra, the A positions in this case being completely filled by the divalent cations Ca, Sr, or Pb. In the Plumbogummite Group, the cerium member florencite continues this series; here all the $(SO_4)^=$ positions in alunite are occupied by $(PO_4)^\equiv$. The members of the Plumbogummite Group containing the divalent A cations Pb, Ca, Ba, or Sr have a neutral H_2O molecule in place of an $(OH)^-$ in order to effect valence compensation. In carphosiderite the A positions probably contain H_2O, valence compensation again being effected by a substitution of H_2O for $(OH)^-$.

Crystals of the members of the Alunite Group are small, imperfect, and rare. They are either tabular $\{0001\}$ or pseudo-cubic on a rhombohedron designated $\{10\bar{1}1\}$ in the old morphological setting or $\{01\bar{1}2\}$ in the unit of the structure cell (for plumbojarosite $\{10\bar{1}4\}$). As in the Beudantite and Plumbogummite Groups, the iron members are optically negative, and the aluminum members optically positive. Biaxial optical anomalies are sometimes observed, especially in the minerals of the Beudantite Group. All these minerals have $\{0001\}$ cleavage. A considerable range of serial substitution is observed among the A cations, notably between jarosite and natrojarosite. The Fe''' and Al in the B positions do not ordinarily form appreciable series although a complete series may extend between alunite and jarosite. A considerable range of compositional variation also is found in crandallite, gorceixite, and deltaite in the Plumbogummite Group.

Ref.
1. Hendricks, *Am. Min.*, **22**, 773 (1937). See also Lemmon, *Am. Min.*, **22**, 939 (1937), and McConnell, *Am. J. Sc.*, **240**, 649 (1942), for a discussion of the group relations.

ALUNITE SERIES

30.2.4.1 **A L U N I T E** [$KAl_3(SO_4)_2(OH)_6$]. Alumen de Tolpha *Gesner* (13, 1565). Romersk Alunsten *Wallerius* (163, 1747). Alaunstein. Alumstone *Jameson* (**1**, 319, 1804). Aluminilite *Delamétherie* (**2**, 113, 1797). Alum de Rome, pt. *Haüy* (**2**, 119, 1801). Pierre alumineuse de la Tolfa *Fr.* Alunite *Cordier* (*Mém. mus. d'hist. nat.*, **6**, 204, 1820); *Beudant* (449, 1824). Kalioalunit *Doelter* (4 [2], 497, 1927). Loewigite *Mitscherlich* (*J. pr. Chem.*, **83**, 474, 1861). Ignatiewite *Flug* (*Russ. Ges. Min., Verh.*, [2], **23**, 116, 1887). Newtonite *Brackett* and *Williams* (*Am. J. Sc.*, **42**, 11, 1891). Kauaiite *Goldsmith* (*Proc. Ac. Nat. Sc. Philadelphia*, 105, 1894). Calafatite *Calderon* (**2**, 205, 1910).

30.2.4.2 **N A T R O A L U N I T E** [$NaAl_3(SO_4)_2(OH)_6$]. Almeriite *Calderon* (**2**, 203, 1910). Natroalunite *Schaller* (*Am. J. Sc.*, **32**, 359, 1911). Alkanasul *Westman* (*Bol. minero soc. nac. mineria, Santiago de Chile*, **43**, 433, 1931). Alunite pt. *some authors*. Soda-alunite *Laudermilk* (*Am. Min.*, **20**, 57, 1935). Natronalunit *Doelter* (4 [2], 495, 1927).

Cryst.[1] Hexagonal—R; ditrigonal-pyramidal—$3\ m$.

$a:c = 1:2.493;\qquad \alpha\ 59°14';\qquad p_0:r_0 = 2.878:1;\qquad \lambda\ 109°47'$

Alunite

Forms:[2]

			ϕ	$\rho = C$	A_1	A_2
c	0001	111	.....	0°00′	90°00′	90°00′
m	$10\bar{1}0$	$2\bar{1}\bar{1}$	30°00′	90 00	30 00	90 00
a	$11\bar{2}0$	$10\bar{1}$	0 00	90 00	60 00	60 00
e	$10\bar{1}1$	100	30 00	70 50½	35 06½	90 00
s	$01\bar{1}2$	110	−30 00	55 12½	90 00	44 40

Less common:

$x\ 03\bar{3}8$ 11.11.2 $y\ 03\bar{3}7$ 10.10.1 $z\ 03\bar{3}5$ $88\bar{1}$

Structure cell.[3] Space group $R3m$. a_{rh} 7.04, α 59°14′; a_0 6.96 kX, c_0 17.35 (*alunite*); $a_0:c_0 = 1:2.493$. Cell contents $(K,Na)Al_3(SO_4)_2(OH)_6$ in the rhombohedral unit.

Habit. Rarely in small crystals, as druses in massive alunite or as aggregates; tabular {0001} or terminated by flat vicinal rhombohedra; rhombohedral, often pseudo-cubic {$01\bar{1}2$}; lenticular due to distortion or vicinal

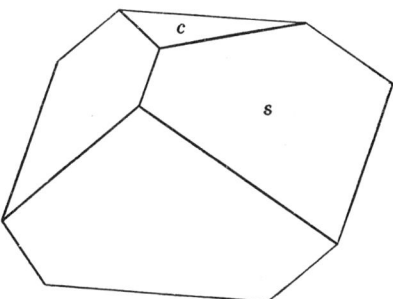

development. Massive, granular to dense, as rocky masses often admixed with other minerals especially quartz, kaolin, halloysite, or diaspore; fibrous to columnar.

Phys. Cleavage {0001} distinct, {$01\bar{1}2$} in traces. Fracture of crystals conchoidal; of dense masses splintery-uneven to conchoidal. Brittle. H. 3½–4; the higher apparent hardness of some massive types is due to admixed quartz. G. 2.6–2.9 (massive); 2.82 (calc., *alunite*). Luster vitreous, somewhat pearly on {0001}. Color white when pure; also grayish, yellowish, reddish, reddish brown. Streak white. Transparent to subtranslucent. Strongly pyroelectric.[4]

Opt.[5] In transmitted light, colorless.

	n(Na) (alunite, Tolfa)	
O	1.572	Uniaxial positive (+).
E	1.592	

Chem. A basic sulfate of aluminum, potassium, and sodium, $(K,Na)\cdot Al_3(SO_4)_2(OH)_6$. Na substitutes for K at least up to Na:K = 7:4 (analysis 5); the name natroalunite is given to that part of the series with Na > K, and alunite to material with K > Na. Na is ordinarily present in alunite in amounts of 1 per cent Na_2O or less. Fe''' apparently substitutes for Al in only very small amounts (see *jarosite*). Many of the reported analyses of alunite were made on impure material.

Anal.[6]

	K_2O	Na_2O	Al_2O_3	SO_3	H_2O	Rem.	Total	G.
1.	11.37		36.92	38.66	13.05		100.00	2.82
2.	10.02		39.65	35.50	14.83		100.00	2.752
3.	10.46	0.33	37.18	38.34	12.90	0.89	100.10	
4.	4.48	2.78	38.05	38.50	11.92	3.82	99.55	2.78
5.	4.26	4.41	39.03	38.93	13.35	0.50	100.48	2.83

1. $KAl_3(SO_4)_2(OH)_6$. 2. Alunite. Tolfa, Vecchia, Italy.[7] 3. Alunite. Coarse granular near Marysvale, Piute County, Utah.[8] Rem. is P_2O_5 0.58, H_2O- 0.09, SiO_2 0.22. 4. Sodian alunite. Tres Cerritos, Mariposa County, California.[9] Rem. is CaO 0.55, Fe_2O_3 0.23, TiO_2 0.40, SiO_2 2.64. 5. Natroalunite. National Belle mine, Ouray County, Colorado.[10] Rem. is insol.

Tests. B.B. infusible. Insoluble in water and practically insoluble in hydrochloric or nitric acids. Slowly soluble in dilute sulfuric acid. After ignition readily soluble in nitric acid.

O c c u r. A widespread rock-making mineral in rocks that have been "alunitized," a process that is mostly a phase of solfataric action. In some places formed by the action of sulfuric acid, resulting from the oxidation of pyrite in aluminous rocks.[11] The process of alunitization is often paralleled by kaolinization and silicification.

Used as a raw material in the manufacture of alum since early times, alunite has been mined for this purpose or as a source of potash in many countries. The well-known deposits at Tolfa, near Civita Vecchia, Italy, are veins in trachyte and have been the basis of alum manufacture there since the middle of the fifteenth century. Large deposits in altered tuffs near Beregszasz in Hungary have also been used for alum manufacture. Also in extensive deposits near Almeria, Spain, and at Zaglik in Transcaucasia. At Fanshan in southwestern China. Lentil-shaped crystals of alunite occur embedded in gypsum near Hadji-Kân in eastern Turkmenistan.[12] Commercially important deposits occur near Bullah Delah, in New South Wales.

In the United States alunite is widespread in altered or mineralized volcanic rocks of the west. Large bodies of alunite occur at the head of South River, Mineral County, and at Red Mountain, Hinsdale County, in Colorado. It has been found in small crystals with quartz and kaolin in a solfatarized andesite in the Rosita Hills, Custer County, Colorado.[13] It is a characteristic mineral of the ore bodies in altered volcanic rocks of the Goldfield district and occurs with cinnabar in the opalized rhyolites of Meiklejohn Mountain, near Beatty, Nye County, Nevada. It is ubiquitous in the altered and kaolinized volcanic rocks about the hot springs of Lassen Volcanic National Park, California. The large and well-known deposits of alunite near Marysvale, Piute County, Utah, are associated with kaolinization of folded early Tertiary volcanic rocks.[18] They have been mined intermittently for several decades. Alunite with specularite

ALUNITE

and quartz makes up a large body of fine-grained altered gneissoid schist at Hickey's Pond near Placentia Bay, Newfoundland.

Natroalunite is relatively rare. Among the known occurrences are Sugarloaf Butte near Quartzite, Arizona, and the Funeral Range Mountains near Death Valley, California. In Colorado at the National Belle mine on Red Mountain near Silverton in Hinsdale County, and in the Rosita Hills and at Knickerbocker Hill in Custer County; also at Calico Peak in the Rico Mountains. From the desert strip on Molokai, Hawaiian Islands, and from Ukusu, Izu province, Japan. From Kalgoorlie and Kanowna, Western Australia; and from Salamanca, Chile. Strontian natroalunite (SrO_2 2.81 per cent) occurs with andalusite on White Mountain, Mono County, California.[17]

Alter. Suffers no change on heating to 500°. Above this temperature it loses water and changes to $KAl(SO_4)_2$ with which must be mixed excess $Al_2(SO_4)_3$. Above 800° this material is transformed, with loss of SO_3, to Al_2O_3 and K_2SO_4, finally changing at 1400° with loss of most of the K_2O and all SO_3 to a phase having the composition $K_2O \cdot 10Al_2O_3$.[14]

Artif. Alunite has been crystallized by heating a solution of alum and aluminum sulfate in a sealed tube at 230°.[15]

Name. The name alunite, a contraction of aluminilite, was first applied to the material from Tolfa, Italy, to which also were applied all the older synonyms.

LOEWIGITE. *Mitscherlich (J. pr. Chem.*, **83**, 474, 1861). Löwigite. Differs from alunite only in having a slightly higher water content. Found at Tolfa, Italy; Beregszasz, Hungary; and at Tabrze in upper Silesia.

IGNATIEWITE. *Flug (Russ. Ges. Min., Vh.*, **23** [2], 116, 1887—*Zs. Kr.*, **13**, 306, 1887). An impure hydrated sulfate of potassium and aluminum. Found in the district of Bakhmuf, Ekaterinoslav, Russia.

NEWTONITE. *Brackett and Williams (Am. J. Sc.*, **42**, 11, 1891). Described as a rhombohedral kaolin mineral but later shown to be alunite.[16]

KAUAIITE. *Goldsmith (Proc. Ac. Sc. Philadelphia*, 105, 1894). A white, chalky basic sulfate of potassium and aluminum found on the island of Kauai, Hawaiian Islands. Probably fine-grained alunite.

CALAFATITE. *Calderon* (**2**, 205, 1910). Differs from alunite only in having a slightly higher water content. Found near Almeria, Spain.

Ref.

1. Symmetry and elements (*alunite*) from structure cell. The morphological elements, $a:c = 1:1.252$, of Breithaupt in Zippe, *Geol. Reichsanst.*, 3 [4], 25 (1852) probably are less accurate. The variation in ratio with substitution of Na for K is not known. Transformation: Breithaupt to new, $0\overline{1}00/00\overline{1}0/\overline{1}000/0002$.
2. Forms abridged from Breithaupt (1852) and Jereméjev, *Mém. soc. russe min.*, **18** [2], 221 (1883). See also Kagaya, *J. Jap. Assoc. Min. Petr. Ec. Geol.*, **16**, 276 (1936).
3. Hendricks, *Am. Min.*, **22**, 773 (1937), on alunite from Rosita Hills, Colorado.
4. Hendricks (1937).
5. Michel-Lévy and Lacroix, *Les Minéraux des Roches*, 140 (1888).
6. For additional analyses see Hintze (**1** [3], 4192, 1929); Doelter (4 [2], 495, 497, 1927); Kawano, *J. Jap. Assoc. Min. Petr. Ec. Geol.*, **17**, 31 (1937); Laudermilk, *Am. Min.*, **20**, 57 (1935).
7. Cordier (1820).
8. Schaller, *U. S. Geol. Sur., Bull. 610*, 150 (1916).

9. Turner, *Am. J. Sc.*, **5**, 424 (1898).
10. Hurlburt, *Am. J. Sc.*, **48**, 130 (1894).
11. For a summary of the modes of formation and occurrence of alunite see Butler and Gale, *U. S. Geol. Sur., Bull. 511* (1912).
12. Jereméjev (1883).
13. For analysis see Cross, *Am. J. Sc.*, **41**, 466 (1891).
14. According to Fink, Van Horn, and Pazour, *J. Ind. Eng. Chem.*, **23**, 1248 (1931), who identified the phases formed by means of x-ray powder patterns. See also Kashkai, *C. r. ac. sc. U.R.S.S.*, **24**, 931 (1939) and Bayliss, et al., *Australian J. Sc. Res.* **1**, 343 (1948).
15. Mitscherlich, *J. pr. Chem.*, **83**, 471 (1861).
16. Foshag, *Am. Min.*, **11**, 33 (1926).
17. Lemmon, *Am. Min.*, **22**, 939 (1937), with G. 2.90, nO 1.587, nE 1.608.
18. Willard and Proctor, *Econ. Geol.*, **41**, 619 (1946).

ALUMIAN. Breithaupt (*B. H. Ztg.*, **17**, 53, 1858).
Microscopic rhombohedral (?) crystals with indistinct cleavage; also massive. White. H. 2–3. G. 2.70–2.78. Luster of crystals vitreous. Subtranslucent. Uniaxial positive [1] (+) with nO 1.583, nE 1.602. Contains 37–38 per cent Al_2O_3 with (SO_4) but no water, and the formula $Al_2(SO_4)_2O$ has been suggested. From the Sierra Almagrera, Spain. Probably identical with natroalunite or alunite.[2]

Ref.

1. Larsen (38, 1921).
2. Hlawatsch, *Festschr. V. Goldschmidt*, 162 (1928), obtained a potassium flame test on probably authentic material.

30.2.4.3 **J A R O S I T E** [$KFe_3(SO_4)_2(OH)_6$]. Gelbeisenerz Rammelsberg (*Ann. Phys.*, **43**, 132, 1838). Misy Haidinger (512, 1845). Vitriolgelb Hausmann (1205, 1847). Jarosit Breithaupt (*B. H. Ztg.*, **6**, 68, 1852). Pastréit Bergemann (*Nat. Ver. Rheinland*, **27**, 17, 1866). Kolosorukite Weisbach (42, 1875). [Not jarošit of Kokta (*Sbornik Klubu Přiro. Brně*, **19**, 75, 1937)]. Moronolite Shepard (4, 1857; suppl. app.).

C r y s t.[1] Hexagonal—R; ditrigonal-pyramidal—3 m.

$a:c = 1:2.3611;$ $\alpha\; 61°38';$ $p_0:r_0 = 2.7264:1;$ $\lambda\; 108°48'$

Forms:

	ϕ		$\rho = C$	A_1	A_2
$c\; 0001$	111	. . . . .	0°00'	90°00'	90°00'
$s\; 10\bar{1}1$	100	30°00'	69 51½	35 36	90 00
$r\; 01\bar{1}2$	110	−30 00	53 44	90 00	45 42½

Structure cell.[2] Space group $R3m$. a_{rh} 7.34, $\alpha\; 61°38'$; a_0 7.20 kX, c_0 17.00; $a_0:c_0 = 1:2.361$. Cell contents $KFe_3(SO_4)_2(OH)_6$ in the rhombohedral unit.

Habit. Crystals minute and indistinct, usually either pseudo-cubic $\{01\bar{1}2\}$ or tabular $\{0001\}$; sometimes distorted or rounded by vicinal development.

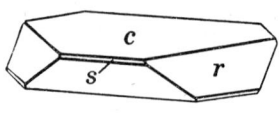

As crusts or coatings of microscopic crystals; granular massive; fibrous; as nodular, tuberose or coralloidal masses and crusts; in small concretionary forms. Also pulverulent; earthy.

Phys. Cleavage $\{0001\}$ distinct. Fracture uneven to conchoidal. Brittle. H. 2½–3½. G. 2.91–3.26 (meas.); 3.25 (calc.). Luster subadamantine to vitreous on faces, resinous on fracture. Color ocherous,

JAROSITE

amber yellow to dark brown. Streak pale yellow, sometimes glistening. Translucent. Strongly pyroelectric.[2]

O p t.[3]

	n	PLEOCHROISM	
E (X)	1.715 ± 0.003	Colorless	Uniaxial negative (−).
(Y)	1.817	Reddish brown	Usually anomalously biaxial,
O (Z)	1.820	Reddish brown	with very small $2V$, and sectoral development.

C h e m. A basic sulfate of potassium and ferric iron, $KFe_3(SO_4)_2$-$(OH)_6$. Na substitutes for K and a series toward natrojarosite extends to at least Na:K = 1:2.4 (analysis 5). Al ordinarily is present in substitution for Fe''' in only very small amount, but the material of analysis 5 has Al:Fe $\sim$ 1:1 and is midway in a series toward alunite.

Anal.[4]

	K_2O	Na_2O	Fe_2O_3	Al_2O_3	SO_3	H_2O	Rem.	Total	G.
1.	9.41		47.83		31.97	10.79		100.00	3.25
2.	7.13	0.84	51.10		28.57	10.56	2.40	100.60	3.144
3.	9.05	0.33	51.16		28.93	10.24	0.29	100.00	3.163
4.	7.75	0.55	48.63		32.02	10.61	0.29	99.85	3.02
5.	6.00	1.68	28.73	18.90	29.47	10.57	4.65	100.00	2.91

1. $KFe_3(SO_4)_2(OH)_6$. 2. Iron Arrow mine, Chaffee County, Colorado.[5] Rem. is SiO_2. 3. Mammoth mine, Tintic district, Utah.[6] Rem. is SiO_2. 4. Saint Félix-de-Pallières, France [7] (*pastreite*). Rem. is insol. 5. Kopěc near Vodolka, north of Prague.[8] Rem. is CaO 0.84, MgO 1.55, insol. 2.26.

Tests. B.B. darkens in color and gives off SO_2 at red heat. In C.T. affords acid water. Insoluble in water. Soluble in hydrochloric acid.

O c c u r. A secondary mineral widespread as crusts and coatings on ferruginous ores and in cracks of adjoining rocks. Often unrecognized when mixed with limonite.[9] The older names *gelbeisenerz* and *misy* were, for the most part, used somewhat loosely and referred also to natrojarosite, carphosiderite, and perhaps other minerals. Originally described from the Barranco Jaroso, Sierra Almagrera, Spain. On brown coal near Luschitz between Bilin and Kolosoruk, in vein quartz at Schlaggenwald and at many other places in Bohemia. On limonite at the Thekla mine near Hartmannsgrün and the Frischglück mine near Schwarzenberg, Saxony. On altered mine dumps (*pastreite*) at Saint Félix-de-Pallières (Gard) and with barite in cracks in Triassic arkose at Blanot near Mâcon, France. On limonite at Laurium, Greece. Disseminated with quartz in a tuff near the Carrasco mine, as mammillary crusts in cavities at Chocaya, Potosi, at Huanuni, Cerro de Llallagua and elsewhere in Bolivia. Common at Chuquicamata and also found at Quetena and Alcaparrosa, in northern Chile. In pyrite deposits in the Urals, Kazakhstan and the Altai, U.S.S.R.[12]

In the United States as minute brilliant crystals in seams and cavities of siliceous limonite and hematite in the Iron Arrow mine, Chaffee County, Colorado. As brownish yellow crusts on cellular quartz in the Vulture mine, Maricopa County, Arizona. As a pulverulent lining of cavities in siliceous limonite at the Mammoth mine, Tintic district, Utah. In hexagonal scales on massive hematite in the Shattuck-Arizona mine, Bisbee, Arizona. As golden brown thick tabular crystals on cherty rock near a limestone-quartzite contact on Cherry Creek in the Mackay district, Custer County, Idaho. In the Pioche district, Lincoln County, Nevada, jarosite is most common in veins in limestone and is particularly abundant in the May Day vein, Bristol area.[10] In brecciated ferruginous sandstone, with rockbridgeite, one mile southeast of Midvale, Virginia, and at many other localities in the United States.

A r t i f. Jarosite has been synthesized at 110° C. by crystallization from an acid solution of K_2SO_4 and excess $Fe_2(SO_4)_3$.[11]

N a m e. From the original locality in the Jaroso ravine in the Sierra Almagrera, Spain.

Ref.

1. Symmetry and axial elements from structure cell of Hendricks, *Am. Min.*, **22**, 773 (1937). With this choice of axial elements c is about double the values usually given from goniometric measurements and obtained on crystals of poor quality. Koenig, *Am. Chem. J.*, **2**, 375 (1880) reported $a:c = 1:1.2492$. Transformation: Dana (974, 1892) to new setting, 0100/00$\bar{1}$0/1000/0002. Forms of Penfield, *Am. J. Sc.*, **39**, 73 (1890).
2. Hendricks (1937) on crystals from Meadow Valley mine, Pioche, Nevada.
3. Larsen (92, 1921) on crystals from the Mammoth mine, Tintic district, Utah. The biaxial character is anomalous. Gordon, *Ac. Sc. Philadelphia, Proc.*, **77**, 1 (1925), and Azema, *Bull. soc. min.*, **33**, 130 (1910), report it to be uniaxial, whereas Schaller, *U. S. Geol. Sur., Bull. 610*, 137 (1916), describes crystals from the Shattuck-Arizona mine, Bisbee, Arizona, as being partly uniaxial and partly divided into biaxial sectors. See also natrojarosite.
4. For additional analyses see Hintze (1 [3], 4200, 1929).
5. Koenig, *Ac. Nat. Sc. Philadelphia, Proc.*, 331 (1880).
6. Genth, *Am. J. Sc.*, **39**, 73 (1890).
7. Azema (1910).
8. Jirkovsky, *Časopis Národ. Musea, Praha*, **101**, 151 (1927).
9. See Blanchard, *Am. Min.*, **29**, 111 (1944), also Locke, *Leached Outcrops as Guides to Copper Ore*, 107, Baltimore (1926), who discusses the position of jarosite among the associated secondary minerals.
10. Westgate and Knopf, *U. S. Geol. Sur., Prof. Pap. 171*, 47 (1932).
11. Fairchild, *Am. Min.*, **18**, 543 (1933).
12. Breshenkov, *C. r. ac. sc. U.R.S.S.*, **52**, 329 (1946).

30.2.4.4   A M M O N I O J A R O S I T E [$(NH_4)Fe_3(SO_4)_2(OH)_6$]. Shannon (*Am. Min.*, **12**, 424, 1927; *U. S. Nat. Mus., Proc.*, **74** [13], 1929).

C r y s t.[1] Hexagonal—R; ditrigonal-pyramidal—$3\ m$.

Structure cell.[2] Space group $R3m$. a_{rh} 7.028, α 61°38′; a_0 7.20 kX, c_0 17.00; $a_0:c_0 = 1:2.361$. Cell contents $(NH_4)Fe_3(SO_4)_2(OH)_6$ in the rhombohedral unit.

Habit. Small lumps and irregular flattened nodules composed of microscopic tabular grains, some with hexagonal outline.

P h y s. G. 3.112 (calc.). Color light ocherous yellow. Luster dull and waxy to earthy.

O p t. In transmitted light, pale yellow to nearly colorless.

nO	1.800 ± 0.005	Uniaxial negative ($-$).
nE	1.750 ± 0.005	

NATROJAROSITE

Chem. A basic sulfate of ammonium and ferric iron, $(NH_4)Fe_3(SO_4)_2(OH)_6$.

Anal.

	$(NH_4)_2O$	K_2O	Fe_2O_3	SO_3	H_2O	Rem.	Total	G.
1.	5.43		49.93	33.37	11.27		100.00	3.112
2.	4.23	1.56	49.30	34.49	9.86	1.18	100.62	

1. $(NH_4)Fe_3(SO_4)_2(OH)_6$. 2. Utah. Rem. is Na_2O 0.22, CaO 0.05, MgO 0.13, Al_2O_3 0.02, insol. 0.76.

Occur. Found with ammonia alum, epsomite, and jarosite in a black lignitic shale on the west side of the Kaibab fault in southern Utah, the exact locality not known. Reported to occur mixed with natrojarosite in pyritic shales at Valachov, Bohemia.[3]

Artif. Reported formed artificially.[3]

Name. In allusion to its ammonium content, analogous to the names of other members of the jarosite group.

Ref.
1. Symmetry from structure cell.
2. Hendricks, *Am. Min.*, **22**, 773 (1937). The assignment of the space group is determined by the observation of pyroelectricity in other members of the jarosite group.
3. Jirkovsky, *Vest. Král. České Spol. Nauk.* Class 2, [36] (1932).

30.2.4.5 **NATROJAROSITE** [$NaFe_3(SO_4)_2(OH)_6$]. Penfield and Hillebrand (*Am. J. Sc.*, **14**, 211, 1902). Jarosite pt. *older authors*. Natronjarosite *wrong orthog.*

Cryst.[1] Hexagonal—R; ditrigonal-pyramidal—$3\,m$.

$a:c = 1:2.2702;\quad \alpha\ 63°23';\quad p_0:r_0 = 2.6214:1;\quad \lambda\ 108°42'$

Forms:

	ϕ	ρ	$\rho = C$	A_1	A_2
$c\ 0001$	111	....	0°00'	90°00'	90°00'
$e\ 10\bar{1}1$	100	30°00'	69 07	54 01	90 00
$s\ 01\bar{1}2$	110	−30 00	52 39½	90 00	46 29

Structure cell.[2] Space group $R3m$. a_{rh} 6.835, α 63°23'; a_0 7.18 kX, c_0 16.30; $a_0:c_0 = 1:2.270$. Cell contents $NaFe_3(SO_4)_2(OH)_6$ in the rhombohedral unit.

Habit. Earthy masses or crusts composed of minute scaly crystals; pulverulent. Crystals pseudo-cubic $\{01\bar{1}2\}$ or flattened $\{0001\}$ with an hexagonal outline.

Phys. Cleavage $\{0001\}$ perfect. Fracture conchoidal. Brittle. H. 3. G. 3.18;[3] 3.29 calc. Color yellow, golden brown, cinnamon-brown. Luster vitreous. Transparent to translucent.

Opt.[4]

	n	DICHROISM	
O	1.832 ± 0.005	Pale yellowish	Uniaxial negative (−).
E	1.750 ± 0.005	Nearly colorless	May show anomalous birefringence, in sectors.

Chem. A basic sulfate of sodium and ferric iron, $NaFe_3(SO_4)_2(OH)_6$. K substitutes for Na, and a series toward the isostructural mineral jarosite exists up to at least K:Na = 1:3.2 (analysis 3).

Anal.[5]

	1	2	3	4	5	6
K_2O		0.35	2.02	0.68	2.28	0.15
Na_2O	6.40	6.03	4.20	6.32	4.28	5.51
Fe_2O_3	49.42	50.98	50.94	49.86	48.23	48.08
SO_3	33.04	30.96	30.43	32.30	33.71	34.65
H_2O	11.14	11.15	11.73	10.93	10.76	11.83
Rem.		0.47	0.69	0.36	0.99	
Total	100.00	99.94	100.01	100.45	100.25	100.22
G.	3.29	3.18		3.11		

1. $NaFe_3(SO_4)_2(OH)_6$. 2. Soda Springs Valley, Esmeralda County, Nevada.[6] Rem. is CaO 0.04, As_2O_5 0.20, SiO_2 0.23. 3. Mina San Toy, Santa Eulalia, Chihuahua, Mexico.[7] Rem. is insol. 4. Kundip, West Australia.[8] Rem. is FeO 0.16, P_2O_5 0.20. 5. Georgia Sunset Claim, near Kingman, Arizona.[9] Rem. is CaO 0.05, MgO 0.05, FeO 0.58, Al_2O_3 0.09, insol. 0.22. 6. Chuquicamata, Chile.[10]

Tests. Slowly soluble in hydrochloric acid. In C.T. unchanged at moderate temperature.

Occur. Found with gypsum, alunite, jarosite, and other sulfates formed upon the oxidation of pyrite. The most common product of the alteration of pyritic shales at Valachov, western Bohemia, and at numerous other places in Czechoslovakia. In cracks in Eocene clay at Refsnaes, Zealand, Denmark. With gypsum in caverns in alunitic rock near Modum, Norway. At Cap Calamita on Elba, as thin coatings on limonite. Very common in the Maikain deposits in the region of Pavlodar, Kazakhstan, U.S.S.R. With chalcanthite, kroehnkite, and sulfur at Chuquicamata, Chile. With mimetite and gypsum in the Mina San Toy, Santa Eulalia, Chihuahua, Mexico. In the United States coating or filling cavities in quartzite at the Buxton mine, Lawrence County, South Dakota; on the east side of Soda Springs Valley, Esmeralda County, Nevada (the type locality); at the Georgia Sunset Claim, near Kingman, Arizona. From Kundip in the Philips River goldfield, Western Australia.

Alter. On heating, natrojarosite loses its water in the interval between 230° and 384°.[11] Pseudomorphs of limonite after natrojarosite have been reported [12] in old slags at Wiesloch, Baden, Germany.

Name. In allusion to the fact that it contains sodium rather than potassium as does jarosite. Before the name natrojarosite was coined both minerals were designated by jarosite or its synonyms.

Ref.

1. Axial elements from structure cell. Transformation: Penfield to new setting, $0\bar{1}00/00\bar{1}0/\bar{1}000/0002$.
2. Hendricks, *Am. Min.*, **22**, 773 (1937). The assignment of the space group is determined by the observation of pyroelectricity on the related mineral jarosite.
3. Penfield and Hillebrand (1902) at 30° on material from Esmeralda County, Nevada.
4. Larsen (1921) on material from the type locality, Esmeralda County, Nevada. Cesàro, *Ac. Belgique, Bull.*, 138 (1905), found natrojarosite from South Dakota to be

biaxial. Bandy, *Am. Min.*, **23**, 669 (1938), found crystals from Chuquicamata, Chile, to be divided into three biaxial sectors, each optically negative with small $2V$, and interpreted the material as orthorhombic, pseudo-rhombohedral due to twinning.
  5. For additional analyses see Doelter (**4** [2], 589, 1927) and Hintze (**1** [3B], 4202, 1929).
  6. Penfield and Hillebrand (1902).
  7. Koenig, *Ac. Sc. Philadelphia, J.*, **15**, 425 (1912).
  8. Simpson and Brown, *J. Roy. Soc. Western Australia*, **1**, 45 (1916).
  9. Shannon and Gonyer, *Washington Ac. Sc., J.*, **17**, 536 (1927).
  10. Gonyer in Bandy, *Am. Min.*, **23**, 669 (1938).
  11. Jirkovsky, *Vestnik Král. Čes. Spol. Nauk*, Class 2, [36] (1932).
  12. Himmel and Mueller, *Geol. Rundsch.*, **23a**, 7 (1933), who refer to the pseudo-morphed mineral as jarosite but cite angles indicating that it was probably natro-jarosite.

30.2.4.6 **A R G E N T O J A R O S I T E** [$AgFe_3(SO_4)_2(OH)_6$]. *Schaller (Washington Ac. Sc., J.*, **13**, 233, 1923) and *Schempp* (*Am. J. Sc.*, **6**, 73, 1923). Argento-jarosite.

C r y s t. Hexagonal—R; ditrigonal-pyramidal—$3\ m$.[1]

Structure cell.[2] Space group $R3m$. a_{rh} 6.87, α 63°21'; a_0 7.22 kX, c_0 16.40; $a_0:c_0 = 1:2.271$. Cell contents $AgFe_3(SO_4)_2(OH)_6$ in the rhombohedral unit.

Habit. Very fine-grained masses and coatings composed of micaceous crystals or scales flattened {0001} with hexagonal outline.

P h y s. Cleavage [3] {0001}. G. 3.66; 3.81 (calc.). Color yellow to brown. Brilliant luster.

O p t.[3]

	n	Dichroism	
O	1.882	Yellow	Uniaxial negative (−).
E	1.785	Pale yellow	

C h e m. A basic sulfate of silver and ferric iron, $AgFe_3(SO_4)_2(OH)_6$.

Anal.

	Ag_2O	PbO	Fe_2O_3	SO_3	H_2O	Rem.	Total	G.
1.	20.35		42.05	28.11	9.49		100.00	3.81
2.	18.00	0.61	43.07	28.15	9.81	0.71	100.35	3.66

  1. $AgFe_3(SO_4)_2(OH)_6$. 2. Tintic Standard mine, Dividend, Utah.[4] Rem. is K_2O 0.42, insol. 0.29.

O c c u r. A secondary mineral, found with anglesite, barite, and quartz at the Tintic Standard mine, Dividend, Utah.

A r t i f. Argentojarosite has been synthesized from an acid solution at 110° C.[5]

N a m e. In allusion to its content of silver in place of the potassium in jarosite.

Ref.

  1. Symmetry from structure cell.
  2. Hendricks, *Am. Min.*, **22**, 773 (1937), on artificial material. The choice of space group is based on the observation of pyroelectricity in the related minerals alunite and jarosite.
  3. Larsen and Berman (91, 1934).
  4. Analysis by Schaller, *U. S. Geol. Sur., Bull. 878*, 121 (1937). For other analyses of material from the same locality see Schempp (1923) and Schaller (1937).
  5. Fairchild, *Am. Min.*, **18**, 543 (1933).

30.2.4.7   CARPHOSIDERITE [$(H_2O)Fe_3(SO_4)_2\{(OH)_5H_2O\}$]. Karphosiderite *Breithaupt* (*J. Chemie u. Phys.*, **50**, 314, 1827). Cyprusite *Reinsch* (*Roy. Soc. London, Proc.*, **33**, 119, 1882). Utahite *Damour* (*Bull. soc. min.*, **7**, 128, 1884). Borgströmite *Saxén* (*Medd. Geol. Fören., Helsingfors*, 1921, 20—*Comm. géol. Finlande, Bull.*, **65**, 50, 1923). Borgstroemite.

Cryst.[1] Hexagonal—R; ditrigonal-pyramidal—$3\,m$.

$a{:}c = 1{:}2.3603;$ $\alpha\ 61°39';$ $p_0{:}r_0 = 2.7254{:}1;$ $\lambda\ 108°47'$

Forms:

	ϕ	$\rho = C$	A_1	A_2
c 0001	111	.....	0°00'	90°00'	90°00'
m 10$\bar{1}$0	2$\bar{1}\bar{1}$	30°00'	90 00	30 00	90 00
s 01$\bar{1}$2	110	−30 00	53 43½	90 00	45 43

Structure cell.[2] Space group $R3m$. a_{rh} 6.99, $\alpha\ 61°39';\ a_0$ 7.16 kX, c_0 16.70; $a_0{:}c_0 = 1{:}2.360$. Cell contents $(H_2O)Fe_3(SO_4)_2\{(OH)_5H_2O\}$ in the rhombohedral unit.

Habit. Earthy or minutely scaly crusts and aggregates; reniform. Microscopic crystals are hexagonal plates or rhombohedrons.

Phys. Cleavage {0001} doubtful.[3] H. 4–4½.[4] G. 2.496–2.905;[5] 3.17 (calc.). Luster dull to resinous or glistening. Color golden yellow, dark straw-yellow.

Opt.[6]

	n	Dichroism	
O	1.816	Deep yellow	Uniaxial negative (−).
E	1.728	Pale yellow	

Chem. A hydrated basic sulfate of ferric iron, $(H_2O)Fe_3(SO_4)_2\{(OH)_5H_2O\}$. Agreement of analyses with theoretical composition is not good. Analysis 3 indicates a slight substitution of Al for Fe'''.

Anal.

	1	2	3	4	5
Fe_2O_3	49.83	40.00	49.68	58.82	51.22
SO_3	33.31	25.52	35.34	28.45	29.07
H_2O	16.86	14.67	11.06	9.35	19.74
Rem.		23.81	3.89	3.19	
Total	100.00	104.00	99.97	99.81	100.03
G.	3.17	2.78			

1. $(H_2O)Fe_3(SO_4)_2\{(OH)_5H_2O\}$. 2. Greenland.[7] Rem. is sand 14.78, gypsum 9.03. 3. Near Kynussa, Cyprus[8] (*cyprusite*). Rem. is Al_2O_3. 4. Eureka Hill mine, Tintic district, Utah[9] (*utahite*). Rem. is As_2O_5. 5. Otravaara, Finland[10] (*borgströmite*).

Tests. Insoluble in water. Soluble in hydrochloric acid.

Occur. As an incrustation in localities where pyrite is weathered. Mixed with quartz, limonite, and other jarosites as a constituent of the gossan. In crusts one centimeter thick in the Upernivik District, Langø, Greenland (the original *carphosiderite*); reported from other localities in Greenland. In large amounts at the ancient copper mines near Kynussa, Cyprus (*cyprusite*). With limonite as an alteration of pyrite in the Otravaara region, Finland (*borgströmite*). In the United States at the Eureka Hill mine, Tintic district, Utah (*utahite*).

A r t i f. Can be crystallized from acid solutions of ferric sulfate up to 170° C.[11]

N a m e. Carphosiderite from κάρφος, *straw*, and σίδηρος, *iron*, in allusion to the composition. Borgströmite after J. L. H. Borgström, Finnish chemist. The name carphosiderite has been but little used. It has been shown [12] by powder diffraction patterns that carphosiderite, borgströmite, cyprusite, and utahite from the type localities are all closely similar to artificial $(H_2O)KFe_3(SO_4)_2\{(OH)_5H_2O\}$. Some materials called carphosiderite have later been shown [15] to be jarosite. On the other hand some so-called jarosites may be carphosiderite.

The following substances probably are identical with carphosiderite.

APATELITE. *Meillet* (*Ann. mines*, [4], **3**, 808, 1843).

A hydrated ferric sulfate found at Meudon, Seine, and elsewhere in France. Later reported to contain a large proportion of Al_2O_3.[13] Another analysis on material supposedly equivalent to the original apatelite showed no aluminum.[14]

RAIMONDITE. *Breithaupt* (*B. H. Ztg.*, **25**, 149, 1866).

A basic ferric sulfate from a tin mine in Bolivia whose properties and composition are very close to carphosiderite. Scaly hexagonal plates. Perfect basal cleavage. H. 4. G. 3.19–3.22. Luster pearly. Ocher to honey-yellow. Analysis gave: Fe_2O_3 46.58, SO_3 35.54, H_2O 17.88, total 100.00.

PLANOFERRITE. *Darapsky* (*Zs. Kr.*, **29**, 213, 1898).

A hydrated ferric sulfate from near Antofagasta, Chile. H. 3. Minute brittle yellow plates with basal cleavage. Apparently orthorhombic. Analysis gave: Fe_2O_3 31.20, SO_3 15.57, H_2O 51.82, insol. 1.41, total 100.00.

Ref.

1. Symmetry and elements from structure cell. Forms from Arzruni, *Bull. soc. min.*, **7**, 126 (1884), on utahite. Transformation: Arzruni to new setting, 0100/0010/1000/0002.
2. Hendricks, *Am. Min.*, **22**, 773 (1937). On artificial material the choice of space group was based on the observation of pyroelectricity in the related minerals, alunite and jarosite.
3. The original description makes no mention of cleavage. Bøggild (186, 1905) records cleavage in one direction. For utahite $\{10\bar{1}0\}$ is reported a possible cleavage by Arzruni (1884). The artificial material is reported to lack cleavage, Posnjak and Merwin, *Am. Chem. Soc., J.*, **44**, 1977 (1922).
4. Bøggild (1905) on material from the type locality; Breithaupt (1827) gave $5\frac{1}{4}$ to $5\frac{3}{4}$.
5. Range of values reported by Breithaupt (1827) and Bøggild (1905).
6. Posnjak and Merwin (1922) on artificial material. Larsen (65, 1921) gives nO 1.83, nE 1.72 for cyprusite.
7. Pisani, *C.R.*, **58**, 242 (1864).
8. Deby, *Roy. Mic. Soc., J.*, [2], **4**, 1, 186 (1884).
9. Damour (1884).
10. Saxén (1923).
11. Posnjak and Merwin (1922).
12. For carphosiderite and borgströmite, Hendricks (1937); for cyprusite, *ASTM Card* II-1294; for utahite, priv. comm. Pabst (1948).
13. Lacroix, **4**, 246 (1910).
14. Magne, *Bull. soc. min.*, **65**, 39 (1942), who proposes the name pseudo-apatelite for the aluminous material described by Lacroix.
15. Lacroix, *Bull. soc. min.*, **10**, 142 (1887); **4**, 145 (1910).

30.2.4.8 B E A V E R I T E [Pb(Cu,Fe,Al)$_3$(SO$_4$)$_2$(OH)$_6$]. *Butler* and *Schaller* (*Wash. Ac. Sc., J.*, **1**, 26, 1911—*Am. J. Sc.*, **32**, 418, 1911).

C r y s t.[1] Hexagonal—*R*; ditrigonal-pyramidal—3 *m*.
Structure cell.[2] Space group $R3m$. a_{rh} 7.012, α 61°49′; a_0 7.203 A, c_0 16.94; a_0:c_0 = 1:2.351. Cell contents Pb(Cu,Fe,Al)$_3$(SO$_4$)$_2$(OH)$_6$ in the rhombohedral unit.
Habit. Earthy and friable masses composed of microscopic hexagonal plates.
P h y s. G. 4.36;[3] 4.31 (calc.). Color canary-yellow.
O p t.[3] Uniaxial negative. nO variable, 1.85 ± 0.02. Birefringence strong.
C h e m. A basic sulfate of lead, copper, ferric iron, and aluminum, Pb(Cu,Fe,Al)$_3$(SO$_4$)$_2$(OH)$_6$. The single analysis shows Cu:Fe:Al close to 2:3:1.[4]

Anal.

	PbO	CuO	Fe$_2$O$_3$	Al$_2$O$_3$	SO$_3$	H$_2$O	Total
1.	33.71	12.02	18.08	3.85	24.18	8.16	100.00
2.	32.50	10.74	19.13	4.03	23.60	10.00	100.00

1. Pb(Cu,Fe,Al)$_3$(SO$_4$)$_2$(OH)$_6$ with Cu:Fe:Al = 2:3:1. 2. Beaver County, Utah.[5] After deducting 10.05 insol. and recalculating to 100.

Tests. Insoluble in water. Soluble in hydrochloric acid.

O c c u r. With plumbojarosite in the oxidized parts of lead-copper ores in arid regions. First described from the Horn Silver mine, near Frisco, Utah, a similar material has also been noted at the Alta Consolidated mine in the Alta district and elsewhere in Utah. Beaverite constitutes about 20 per cent of a mixture with plumbojarosite at the Boss mine, Yellow Pine district, Nevada.

A l t e r.[5] Practically no water is lost when heated to 250°, but completely dehydrated at 590°.

N a m e. From the county in which first recognized.

Ref.

1. Crystals of beaverite have not been measured. X-ray powder diffraction patterns show that it is rhombohedral and has cell dimensions close to those of other minerals in the alunite group. For this reason it is assigned to the ditrigonal-pyramidal class.
2. Pabst, priv. comm. (1948), by powder method.
3. Larsen, 45 and 201 (1921).
4. If the Pb position is fully occupied one-third of the trivalent atoms in the threefold position must be replaced by divalent atoms to keep charges balanced.
5. Butler and Schaller (1911).

30.2.4.9 P L U M B O J A R O S I T E [PbFe$_6$(SO$_4$)$_4$(OH)$_{12}$]. *Hillebrand* and *Penfield* (*Am. J. Sc.*, **14**, 211, 1902). Vegasite *Knopf* (*Washington Ac. Sc., J.*, **5**, 497, 1919).

C r y s t.[1] Hexagonal—*R*; hexagonal scalenohedral—$\bar{3}\,2/m$.

a:c = 1:4.6667; α 35°05′; p_0:r_0 = 5.3886:1; λ 116°45′

Forms:

		φ	ρ = C	A_1	A_2
c	0001	111	.....	0°00'	90°00'	90°00'
s	$10\bar{1}4$	211	30°00'	53 25	45 56½	90 00
e	$01\bar{1}2$	110	−30 00	69 38	90 00	35 43

Structure cell.[2] Space group $R\bar{3}m$. a_{rh} 11.95, α 35°05'; a_0 7.20 kX, c_0 33.60; $a_0:c_0 = 1:4.667$. Cell contents $PbFe_6(SO_4)_4(OH)_{12}$ in the rhombohedral unit.

Habit. Crusts, lumps, or compact masses composed of microscopic hexagonal plates; also pulverulent, earthy to ocherous.

P h y s. Cleavage $\{10\bar{1}4\}$, fair.[3] H. soft (talc-like feel). G. 3.665; 3.71 (calc.). Color golden brown to dark brown. Luster dull to glistening or silky.

O p t.[4]

	n	Dichroism	
O	1.875	Yellow-brown	Uniaxial negative (−).
E	1.786	Nearly colorless	

C h e m. A basic sulfate of lead and ferric iron, $PbFe_6(SO_4)_4(OH)_{12}$.
Anal.

	PbO	Fe_2O_3	SO_3	H_2O	Rem.	Total	G.
1.	19.74	42.37	28.33	9.56		100.00	3.71
2.	19.84	42.37	27.06	9.56	1.32	100.15	3.665
3.	18.46	42.87	27.67	10.14	1.23	100.37	
4.	18.32	42.11	27.59	9.16	3.07	100.25	3.60

1. $PbFe_6(SO_4)_4(OH)_{12}$. 2. Cooks Peak, Luna County, New Mexico.[5] Rem. is K_2O 0.17, Na_2O 0.21, CaO 0.05, CuO 0.27, MgO 0.01, Al_2O_3 0.10, SiO_2 0.51. 3. American Fork district, Utah County, Utah.[6] Rem. is K_2O 0.15, Na_2O 0.52, CaO 0.06, CuO 0.10, insol. 0.40. 4. Beaver County, Utah.[7] Rem. is $(K,Na)_2O$ 0.13, ZnO 0.30, insol. 2.64.

Tests. Only very slowly soluble in acids.

O c c u r. Widespread as a secondary mineral in the oxidized zone of lead mines in arid regions. Plumbojarosite is an important constituent of the ore in the lead mines at Bolkardag, southern Anatolia, Turkey. It occurs with anglesite, jarosite, and cassiterite as an alteration product of teallite in the Carguaicollo tin deposits, Dept. Potosi, Bolivia. First reported from the Cooks Peak district, Luna County, New Mexico, and since has been noted in a great many localities in western United States. With jarosite and limonite at the San Jose and Three Brothers mines, Grant County, New Mexico. In the American Fork and Tintic districts and numerous other localities in Utah. Abundant in small ocherous masses in quartz at the Boss mine, Clark County, Nevada. Sparingly in the lead-silver veins of the Kaufman and Weaver mine, Lemhi County, Idaho. Abundant in brown oxide ore in the Tombstone district, Cochise County, Arizona. An alteration product of galena at the Barr mine, Picher district, Oklahoma.

N a m e. In allusion to its content of lead in place of the potassium of jarosite.

VEGASITE. *Knopf* (*Washington Ac. Sc., J.*, **5**, 497, 1915). An ocherous material occurring in lumps at a prospect near the Boss mine, Clark County, Nevada. The analysis suggests a contaminated plumbojarosite. Reported to be uniaxial positive.

Ref.
1. Symmetry and axial ratio from structure cell. Transformation: Hillebrand and Penfield to new, $0\bar{1}00/00\bar{1}0/\bar{1}000/0004$. Forms of Hillebrand and Penfield (1902). The agreement between the morphological ratio and the structure cell is only approximate owing to the minute size and poor quality of the measured crystals.
2. Hendricks, *Am. Min.*, **22**, 773 (1937), on material from the type locality.
3. Hillebrand and Wright, *Am. J. Sc.*, **30**, 191 (1910). The {0001} cleavage characteristic of other members of the group is not reported for plumbojarosite.
4. Larsen (121, 1921) on analyzed material from the type locality. Some plates showed hexagonal segments with small axial angle.
5. Hillebrand and Penfield (1902).
6. Hillebrand and Wright (1910).
7. Butler and Schaller, *Am. J. Sc.*, **32**, 422 (1911).

The following basic iron sulfates are of uncertain identity.

PLAGIOCITRITE. *Sandberger; Singer* (Inaug. Diss., Würzburg, 13, 1879). Microscopic prismatic crystals, apparently monoclinic or triclinic. Color lemon-yellow. G. 1.88 (?). Alters in air. Soluble in water to an acid solution which deposits hydrous iron oxide on warming. Decomposes on exposure, becoming orange-yellow. Analysis gave, after deducting 9.85 per cent hygroscopic water: Na_2O 4.04, K_2O 4.23, MgO 1.19, CaO 0.43, CoO 0.58, NiO 0.97, FeO 1.64, Al_2O_3 14.37, Fe_2O_3 7.95, SO_3 35.44, H_2O 29.42, total 100.26. Found with other secondary iron sulfates in pyritic tuff at the Bauersberg near Bischofsheim, Bavaria, Germany.

CLINOPHAEITE. Klinophäit *Sandberger; Singer* (Inaug. Diss., Würzburg, 16, 1879). In microscopic crystals, probably monoclinic, with $c\{001\}$, $m\{110\}$, $d\{101\}$; prismatic angles 85° and 95°. G.2.979. Color blackish green. Vitreous luster. Analysis gave, after deducting 7.88 per cent hygroscopic water: Na_2O 6.35, K_2O 21.79, CaO 0.77, MgO 1.88, (Ni,Co)O 0.76, FeO 6.06, Al_2O_3 4.04, Fe_2O_3 9.48, SO_3 37.01, H_2O 14.72, total 102.86. Soluble with difficulty in water to an acid solution which deposits hydrous iron oxide on boiling. B.B. fusible with intumescence to a black magnetic residue. From the Bauersberg near Bischofsheim, Bavaria.

CLINOCROCITE. Klinocrocit *Sandberger; Singer* (Inaug. Diss., Würzburg, 9, 1879). As microscopic, apparently monoclinic crystals. Color deep saffron-yellow. Qualitative tests indicate a hydrous sulfate of Al, Fe''', Na, and K. Found with clinophaeite near Bischofsheim, Bavaria.

UNNAMED FERRIC SULFATES. *Mackintosh* (*Am. J. Sc.*, **38**, 243, 245, 1889). Brief descriptions are given of ill-defined iron sulfates from an unstated locality in Chile. Substance A occurs as pulverulent orange flakes arranged in parallel tabular layers with copiapite and amarantite. The analysis is close to $Fe_2(SO_4)_2(OH)_2 \cdot 3H_2O$ and the substance may be a dehydration product of amarantite. About 0.3 per cent H_2O is lost at 110°. Substance B is white and pulverulent. The analysis approximates to $Fe_4''Fe_2'''(SO_4)_6(OH)_2 \cdot 18H_2O$, which is close to copiapite. About 9.6 per cent H_2O are lost at 110°. Substances C and D are white and pulverulent.

	Na_2O	FeO	Fe_2O_3	Al_2O_3	SO_3	H_2O	Total
A.			41.22		41.24	[17.54]	[100.00]
B.	0.58	22.51	12.16		38.00	[26.75]	[100.00]
C.	4.42	30.81	5.64	0.65	47.90	[10.58]	[100.00]
D.	0.33	35.05	5.14		45.61	[13.87]	[100.00]

RUBRITE. *Darapsky* (*Jb. Min.*, I, 65, 1890). In indistinct lamellar crystals of a deep red color, penetrated by white nodules and clear zones. Analysis gave: CaO 4.10, MgO 5.62, Al_2O_3 3.01, Fe_2O_3 18.22, SO_3 41.15, H_2O 27.64, total 99.84. After deducting epsomite and gypsum, this is close to $Fe'''(SO_4)(OH) \cdot H_2O$. From the neighborhood of the Rio Loa, Chile.

EUCHLORIN

30.2.5 **E U C H L O R I N.** *Euclorina A. Scacchi* (1869); *E. Scacchi* (*Acc. Napoli, Rend.*, 126, 1884). Euchlorite. Euchlorine.

C r y s t.[1] Orthorhombic.

$a:b:c = 0.7616:1:1.8755;$ $\quad p_0:q_0:r_0 = 2.4626:1.8755:1$

$q_1:r_1:p_1 = 0.7616:0.4061:1;$ $\quad r_2:p_2:q_2 = 0.5332:1.3130:1$

Forms:[2]

	ϕ	$\rho = C$	ϕ_1	$\rho_1 = A$	ϕ_2	$\rho_2 = B$
c 001	.....	0°00′	0°00′	90°00′	90°00′	90°00′
b 010	0°00′	90 00	90 00	90 00	.....	0 00
p 120	33 17	90 00	90 00	56 43	0 00	33 17
d 013	0 00	32 01	32 01	90 00	90 00	57 59
e 011	0 00	61 56	61 56	90 00	90 00	28 04
n 103	90 00	39 23	0 00	50 37	50 37	90 00
m 101	90 00	67 54	0 00	22 06	22 06	90 00

Habit. Crystals rectangular tablets {001} and slightly elongated [100]. As an incrustation.

P h y s. Cleavage in two directions.[3] Color emerald-green.

O p t.[3] In transmitted light, emerald green.

ORIENTATION	n	PLEOCHROISM	
X	1.580	Pale grass green	Biaxial positive (+).
Y	1.605	Grass-green	$2V$ moderately large.
Z	1.644	Bright yellow-green	

C h e m. A sulfate of copper, potassium, and sodium of uncertain formula. Analyses on material of questionable purity gave:

	Na_2O	K_2O	CuO	SO_3	Total
1.	7.28	7.33	41.14	44.25	100.00
2.	6.03	8.32	42.28	43.37	100.00
3.	6.48	8.04	41.50	43.98	100.00
4.	5.48	10.34	37.87	42.96	96.65

1, 2, 3. Vesuvius.[4] 4. Vesuvius.[5] Undet., probably H_2O.

Tests. Partly soluble in water.

O c c u r. Found with dolerophanite, eriochalcite, chalcocyanite, and melanothallite in fumaroles of the eruption of 1868 of Vesuvius.

N a m e. From εὔχλωρος, *pale green*, in allusion to the color.

Ref.

1. E. Scacchi (1884) and Goldschmidt (**3**, 156, 1916).
2. Scacchi (1884) and Zambonini (167, 1910).
3. Frondel, *Min. Mag.*, **29**, 34 (1950), on material not known with certainty to be euchlorin.
4. Scacchi (1884).
5. Rammelsberg (87, 1886).

31 HYDRATED SULFATES CONTAINING HYDROXYL OR HALOGEN

TYPE 1. $A_m B_n (XO_4)_p Z_q \cdot xH_2O$, WITH $(m+n):p > 4:1$

CONNELLITE GROUP

31.1.1.1 **CONNELLITE** [$Cu_{19}(SO_4)Cl_4(OH)_{32} \cdot 3H_2O$?]. Copper ore of an azure-blue color, composed of needle crystals *Rashleigh* (*Brit. Min.*, **2**, 13, Pl. 12, figs. 1, 6, 1802). Sulphato-chloride of Copper *Connell* (*Rpt. British Assoc.*, 1847). Connellite *Dana* (523, 1850). Footeite *Koenig* (*Proc. Ac. Nat. Sc. Philadelphia*, 291, 1891; *Zs. Kr.*, **19**, 601, 1891). Ceruleofibrite *Holden* (*Am. Min.*, **7**, 80, 1922).

31.1.1.2 **BUTTGENBACHITE**   [$Cu_{19}(NO_3)_2Cl_4(OH)_{32} \cdot 3H_2O$?] *Schoep* (*C.R.*, **181**, 421, 1925; *Bull. soc. chim. Belgique*, **34**, 313, 1925).

Cryst. Hexagonal—P; dihexagonal-dipyramidal—$6/m\ 2/m\ 2/m$.

Connellite.[1] $a:c = 1:0.6696$; $p_0:r_0 = 0.7732:1$

Buttgenbachite.[2] $a:c = 1:0.6694$; $p_0:r_0 = 0.7731:1$

Forms:

	ϕ	$\rho = C$	M	A_2
$c\ 0001$	.....	0°00′	90°00′	90°00′
$m\ 10\bar{1}0$	30°00′	90 00	60 00	90 00
$a\ 11\bar{2}0$	0 00	90 00	90 00	60 00
$n\ 13\bar{4}0$	18 26	90 00	71 34	78 26
$p\ 20\bar{2}1$	30 00	57 06½	65 10½	90 00
$d\ 22\bar{4}1$	0 00	69 31½	90 00	62 04
$w\ 22.4.\overline{26}.3$ (?)	21 39	80 53½	68 38	81 45½

Structure cell.[3]   Space group probably $C6/mmc$.
Connellite: $a_0\ 13.574\ kX$ $c_0\ 9.07$ $a_0:c_0 = 1:0.668$
Buttgenbachite: $a_0\ 13.526$ $c_0\ 9.13$ $a_0:c_0 = 1:0.675$
Cell contents uncertain.

Habit. Acicular [0001] and striated [0001]. In radiated groups of needles and as felted aggregates.

Phys. H. 3. G. 3.36 (*connellite*), 3.33 (*buttgenbachite*). Luster vitreous. Color fine azure-blue. Streak pale greenish blue. Translucent.

Opt. In transmitted light, blue in color and not pleochroic. Uniaxial positive (+).

	nO	nE
Connellite, Bisbee [4]	1.730	1.754
Connellite, Bisbee [5]	1.724	1.746
Connellite, Bisbee [6]	1.738	1.752
Connellite, Sardinia [7]	1.735	1.758
Buttgenbachite, Likasi [8]	1.738	1.752

Chem. A hydrated basic chloride-sulfate-nitrate of copper, in which (SO_4) and (NO_3) apparently substitute mutually. The formula and the structural interpretation of the substitution are uncertain. The following formulas for the end-members of the series represent the analyses relatively closely:

Connellite $Cu_{19}(SO_4)Cl_4(OH)_{32} \cdot 3H_2O$
Buttgenbachite $Cu_{19}(NO_3)_2Cl_4(OH)_{32} \cdot 3H_2O$

Anal.

	1	2	3	4	5	6
CuO	73.96	73.38	73.41	75.96	71.56	72.97
SO_3	3.92	3.15	3.84	3.43	nil	
Cl	6.94	6.82	7.05	6.37	6.02	6.84
N_2O_5		0.72	0.30		5.40	5.21
H_2O	16.75	17.13	16.81	16.07	17.34	16.52
	101.57	101.20	101.41	101.83	100.32	101.54
O = Cl	1.57	1.53	1.59	1.42	1.28	1.54
Total	100.00	99.67	99.82	100.41	99.04	100.00

1. $Cu_{19}(SO_4)Cl_4(OH)_{32} \cdot 3H_2O$. 2. Connellite. Czar mine, Bisbee.[4] 3. Connellite. Grand Central mine, Tintic.[4] 4. Connellite. Calumet and Arizona mine, Bisbee.[5] H_2O includes 0.25 below 220°, 12.06 at 220° to 260°, 2.10 at 260°–300°, 1.66 above 300°. 5. Buttgenbachite. Likasi, Belgian Congo.[9] 6. $Cu_{19}(NO_3)_2Cl_4(OH)_{32} \cdot 3H_2O$.

Tests. B.B. easily fusible. In C.T. affords much acid water. Soluble in acids and in ammonium hydroxide, insoluble in water.

O c c u r. Connellite was originally from Cornwall, England, from the Wheal Providence; also from Wheal Damsel, Wheal Unity and the Carharrack mine at St. Day, the Marke Valley mine, and in the Camborne district. The mineral is here associated, as at other localities, with cuprite, malachite, azurite, chalcophyllite, and spangolite. Connellite also occurs at Arenas, Sardinia, and Mouzaïa, Algeria. In the United States at several mines at Bisbee, Cochise County, and at Ajo, Pima County, in Arizona. Also in the Grand Central mine in the Tintic district, Utah. Buttgenbachite is known only from Likasi in the Belgian Congo, Africa.

N a m e. Connellite after Arthur Connell, who first examined the mineral. Buttgenbachite after the Belgian mineralogist Henri Buttgenbach (1874–).

Ref.

1. Elements of Penfield, *Am. J. Sc.*, **40,** 82 (1890), on analyzed connellite crystals from Camborne, Cornwall, with the c-axis halved, in agreement with the structure cell. Discordant results, due to the small size and poor quality of the available crystals, have been reported by Maskelyne, *Phil. Mag.*, **25,** 39 (1863), Pelloux, *Ann. mus. civico stor. nat. Genova*, **5,** 205 (1912), and Palache and Merwin, *Am. J. Sc.*, **28,** 537 (1909).
2. Elements and forms of Palache, priv. comm. (1940), on Likasi crystals. Angle table calculated to connellite elements. See also Buttgenbach, *Soc. géol. Belgique, Ann.*, **50,** B 35 (1926) for approximate measurements.
3. Weichel, priv. comm. (1948), by Weissenberg method on Bisbee connellite and Likasi buttgenbachite.
4. Ford and Bradley, *Am. J. Sc.*, **39,** 670 (1915).
5. Palache and Merwin (1909).
6. Weichel (1948).
7. Pelloux (1912).
8. Weichel (1948). Buttgenbach (1926) states the mineral to be negative with $nO > 1.74$ and scarcely appreciable birefringence.
9. Schoep (1925). The occurrence of intermediate members of the series at Likasi is indicated by Buttgenbach (1926).

31.1.2 Glaucocerinite. Glaucokerinit *Dittler* and *Koechlin* (*Cbl. Min.*, 13, 1932).

As warty masses with a radial fibrous structure and a concentric banding of color. H. 1; wax-like. G. 2.749. Color sky-blue to turquois-blue, inwardly passing to white; also greenish, gray, or brownish due to impurities.

O p t. Fibers give parallel extinction. Birefringence high. nZ 1.542. Elongation positive.

C h e m. A hydrated basic sulfate of copper, zinc, and aluminum of uncertain formula, perhaps $Zn_{13}Cu_7Al_8(SO_4)_2(OH)_{60} \cdot 4H_2O$. An analysis [1] on a 50-mg. air-dried sample gave: CuO 19.26, ZnO 37.95, Al_2O_3 15.40, SO_3 5.79, H_2O- 16.31, H_2O below 500° 5.69, total 100.40.

O c c u r. Found with smithsonite, adamite, gypsum, malachite on old specimens in the Vienna Museum from Laurium, Greece.

N a m e. From γλαυκός, *blue*, and κήρινος, *wax-like*.

Ref.
1. Freh and Dittler analysis in Dittler and Koechlin (1932).

31.1.3 M O O R E I T E $[(Mg,Zn,Mn)_8(SO_4)(OH)_{14} \cdot 4H_2O]$. *Bauer* and *Berman* (*Am. Min.*, 14, 165, 1929).

C r y s t.[1] Monoclinic; prismatic—$2/m$.

$a:b:c = 0.551:1:0.961;$ $\beta\ 122°23';$ $p_0:q_0:r_0 = 1.743:0.812:1$

$r_2:p_2:q_2 = 1.232:2.148:1;$ $\mu\ 57°37';$ $p_0'\ 2.064,\ q_0'\ 0.961,\ x_0'\ 0.634$

Forms:

		ϕ	ρ	ϕ_2	$\rho_2 = B$	C	A
c	001	90°00′	32°23′	57°37′	90°00′	....	57°37′
b	010	0 00	90 00	....	0 00	90°00′	90 00
a	100	90 00	90 00	0 00	90 00	57 37	....
g	011	33 25	49 01½	57 37	50 56	39 04	65 26
d	101	90 00	69 40	20 20	90 00	37 17	20 20
f	$\bar{1}02$	−90 00	21 42	111 42	90 00	54 05	111 42
e	$\bar{1}01$	−90 00	55 02	145 02	90 00	87 25	145 02
v	$\bar{1}15$	49 01	16 20	77 31½	79 24	22 28½	77 46½
t	$\bar{1}13$	− 9 34	17 59½	93 05½	72 16	39 08	92 56½
r	$\bar{1}11$	−56 06	59 52	145 02	61 09½	87 44½	135 52½
p	121	54 32	73 12	20 20	56 15½	48 34½	38 46

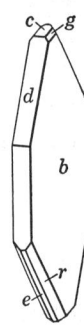

Structure cell.[2] Space group $P2_1/m$. a_0 11.18 kX, b_0 20.25, c_0 19.52; [β 122°23′]; $a_0:b_0:c_0 = 0.552:1:0.964$. Cell contents $(Mg,Mn,Zn)_{104}(SO_4)_{13}(OH)_{182} \cdot 52H_2O$.

Habit. Tabular to platy $\{010\}$. Often as subparallel aggregates grouped on $\{010\}$.

P h y s. Cleavage $\{010\}$ perfect. H. 3. G. 2.47; 2.54 (calc.). Luster vitreous. Colorless. Transparent.

TORREYITE 575

Opt. In transmitted light, colorless.

ORIENTATION		n	
X	b	1.533	Biaxial negative (−).
Y		1.545	$2V \sim 50°$.
$Z \wedge c$	$44°$	1.547	$r > v$, perceptible.

Chem. A hydrated basic sulfate of magnesium, manganese, and zinc, $(Mg,Zn,Mn)_8(SO_4)(OH)_{14} \cdot 4H_2O$, with $Mg:Zn:Mn \sim 4:2:1$ in the only reported analysis.

Anal.

	1	2
MgO	25.26	25.38
ZnO	25.50	24.58
MnO	11.11	11.93
SO$_3$	10.97	10.99
H$_2$O	27.16	27.12
Total	100.00	100.00
G.	2.54	2.47

1. $(Mg,Zn,Mn)_8(SO_4)(OH)_{14} \cdot 4H_2O$ with $Mg:Zn:Mn = 4:2:1$. 2. Sterling Hill, New Jersey.[3] Average of two recalculated to 100 after deducting CaCO$_3$ 0.89 and SiO$_2$ 0.06. B$_2$O$_3$ shown to be present, probably as admixed fluoborite.

Tests. Soluble in acids.

Occur. At Sterling Hill, Sussex County, New Jersey, with rhodochrosite, zincite, torreyite, fluoborite, and altered pyrochroite in veinlets cutting massive calcite-franklinite-willemite ore.

Name. After Gideon E. Moore (1842–1895), chemist, who early investigated the minerals of Sterling Hill and Franklin, New Jersey.

Ref.
1. Data of Bauer and Berman (1929) recomputed by LaForge, priv. comm. (1936).
2. Prewitt-Hopkins, *Am. Min.*, **34**, 589 (1949), by Weissenberg method on the type material.
3. Bauer analysis in Bauer and Berman (1929).

31.1.4 **TORREYITE** [$(Mg,Mn,Zn)_7(SO_4)(OH)_{12} \cdot 4H_2O$]. Delta-mooreite Bauer and Berman (*Am. Min.*, **14**, 165, 1929). Torreyite *Prewitt-Hopkins* (*Am. Min.*, **34**, 589, 1949).

Massive, granular to foliated. Cleavage {010} good. H. 3. G. 2.665. Luster vitreous to somewhat pearly. Color bluish white. Translucent.

Twinning.[1] Under the microscope shows intricate polysynthetic twinning with twin plane in the zone [010].

Opt. In transmitted light, colorless.

ORIENTATION		n	
X	b	1.570	Biaxial negative (−).
Y		1.584	$2V \sim 40°$.
Z		1.585	

Chem. A hydrated basic sulfate of magnesium, manganese, and zinc, $(Mg,Mn,Zn)_7(SO_4)(OH)_{12} \cdot 4H_2O$, with $Mg:Mn:Zn \sim 5:3:4$ in the only reported analysis.

Anal.

	1	2
MgO	17.00	17.27
ZnO	27.45	26.30
MnO	17.94	17.98
SO_3	11.57	11.64
H_2O	26.04	26.39
SiO_2		0.08
Total	100.00	99.66

1. $(Mg,Mn,Zn)_7(SO_4)(OH)_{12} \cdot 4H_2O$ with $Mg:Mn:Zn = 5:3:4$. 2. Sterling Hill, N. J.[2] B_2O_3 present, probably as admixed fluoborite.

Tests. Soluble in acids.

Occur. At Sterling Hill, Sussex County, New Jersey, with mooreite, fluoborite, and altered pyrochroite in veinlets cutting calcite-franklinite-willemite ore.

Name. After John Torrey (1796–1873), American naturalist, who early (1822) studied the Franklin minerals. Originally named delta-mooreite, from a supposed varietal relation to mooreite, but later shown [1] to be a distinct species.

Ref.
1. Prewitt-Hopkins (1949).
2. Bauer analysis in Bauer and Berman (1929).

31.1.5 **SPANGOLITE** [$Cu_6Al(SO_4)(OH)_{12}Cl \cdot 3H_2O$]. *Penfield (Am. J. Sc., 39, 370, 1890).*

Cryst.[1] Hexagonal—R; ditrigonal-pyramidal—$3m$.

$a:c = 1:1.7414;$ [2] $\alpha\ 75°17';$ $p_0:r_0 = 2.0108:1;$ $\lambda\ 101°41\frac{1}{2}'$

Forms: [3]

Lower	Upper		ϕ	$\rho = C$	A_1	A_2
$\bar{c}$	c 0001	111	....	0°00'	90°00'	90°00'
	m 10$\bar{1}$0	2$\bar{1}\bar{1}$	30°00'	90 00	30 00	90 00
	$-m$ 01$\bar{1}$0	11$\bar{2}$	$-30\ 00$	90 00	90 00	30 00
	a 11$\bar{2}$0	10$\bar{1}$	0 00	90 00	60 00	60 00
$\bar{n}$	n 10$\bar{1}$3	522	30 00	33 50	61 10$\frac{1}{2}$	90 00
$\bar{o}$	o 10$\bar{1}$2	411	30 00	45 09$\frac{1}{2}$	52 01	90 00
p	p 10$\bar{1}$1	100	30 00	63 33$\frac{1}{2}$	39 09$\frac{1}{2}$	90 00
$\bar{y}$	y 20$\bar{2}$1	5$\bar{1}\bar{1}$	30 00	76 02	32 49	90 00
$-\bar{n}$	$-n$ 01$\bar{1}$3	441	$-30\ 00$	33 50	90 00	61 10$\frac{1}{2}$
$-\bar{o}$	$-o$ 01$\bar{1}$2	110	$-30\ 00$	45 09$\frac{1}{2}$	90 00	52 01
$-\bar{p}$	$-p$ 01$\bar{1}$1	22$\bar{1}$	$-30\ 00$	63 33$\frac{1}{2}$	90 00	39 09$\frac{1}{2}$
$-\bar{y}$	$-y$ 02$\bar{2}$1	11$\bar{1}$	$-30\ 00$	76 02	90 00	32 49

Less common and rare:

k 10$\bar{1}$4, $\bar{k}$, $-k$, $-\bar{k}$ l 60$\bar{6}$7, $\bar{l}$, $-l$, $-\bar{l}$ z 30$\bar{3}$1, $\bar{z}$, $-z$, $-\bar{z}$
r 30$\bar{3}$4, $\bar{r}$, $-r$, $-\bar{r}$ x 30$\bar{3}$2, $\bar{x}$, $-x$, $-\bar{x}$

Structure cell.[4] Space group $C3c$. a_0 8.245 Å, c_0 14.34; $a_0:c_0 = 1:1.739$. Cell contents $Cu_{12}Al_2(SO_4)_2(OH)_{24}Cl_2 \cdot 6H_2O$.

Spangolite

Habit. Often holohedral in aspect and either short prismatic [0001] or, more commonly, tabular {0001}. Less commonly hemimorphic with large {000$\bar{1}$} and tapering toward the antilogous pole. Prism and trigonal pyramid faces horizontally striated, with series of trigonal pyramids and prisms in oscillatory combination. The geometrically equivalent positive

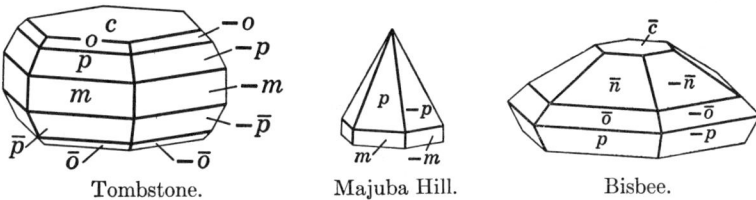

Tombstone.   Majuba Hill.   Bisbee.

and negative trigonal pyramids apparently are always equally developed giving the crystal a hexagonal appearance.

Twinning.[5] On {0001}, rare, the acute (antilogous) poles of the two individuals joined together.

Etch figures. Pits produced on {0001} and {000$\bar{1}$} vary with the kind and concentration of solvent used but conform to ditrigonal-pyramidal symmetry.

P h y s. Cleavage {0001} perfect; also on $p\{10\bar{1}1\}$, $\bar{p}$, $-p$ and $-\bar{p}$, distinct. Fracture conchoidal. Brittle. H. 2 on {0001}, about 3 on the inclined faces. G. 3.14 (Tombstone);[6] 3.14 (calc.). Luster vitreous. Color dark green, also emerald-green to bluish green. Streak pale green. Pyroelectric.

O p t. In transmitted light, pale green.

	n (Tombstone)[7]	n (Majuba Hill)[8]	n (Tintic)[8]	
O	1.694	1.681 ± 0.002	1.686 ± 0.002	Uniaxial negative ($-$).
E		1.627 ± 0.002	1.638 ± 0.002	Weakly dichroic, with O green and E bluish green.

C h e m. A hydrated basic sulfate of copper and aluminum, $Cu_6Al(SO_4)(OH)_{12}Cl \cdot 3H_2O$. Small amounts of OH apparently can substitute for Cl.

Anal.

	1	2
CuO	59.82	59.51
Al_2O_3	6.39	6.60
SO_3	10.03	10.11
Cl	4.44	4.11
H_2O	20.32	20.41
	101.00	100.74
O = Cl	1.00	0.92
Total	100.00	99.82
G.	3.14	3.14

1. $Cu_6Al(SO_4)(OH)_{12} \cdot 3H_2O$. 2. Region of Tombstone, Arizona.[9] Average of four analyses, three partial.

Tests. Fuses at about 3 to a black slaggy mass. In C.T. yields abundant water with an acid reaction. Insoluble in water but easily soluble in acids.

Occur. A secondary mineral found associated with cuprite, chrysocolla, azurite, malachite, connellite and, less commonly, liroconite, brochantite, clinoclase. Originally found in the region of Tombstone, Cochise County, Arizona, but the exact locality is not known. Later found in the Copper Queen and Czar mines at Bisbee, Cochise County, and in the Metcalf mine in the Clifton-Morenci district, Greenlee County, Arizona. In the Grand Central mine, Tintic district, Utah, and at Majuba Hill, Pershing County, Nevada. At St. Day, Cornwall, and from Arenas, south of Iglesias, Sardinia.

Name. After Mr. Norman Spang of Etna, Allegheny County, Pennsylvania, who supplied the original specimen for study.

Ref.
 1. Originally made hexagonal-scalenohedral by Penfield (1890) but later shown by Frondel, *Am. Min.*, **34**, 181 (1949), to be ditrigonal-pyramidal.
 2. Angles and unit of Penfield (1890) on Tombstone crystals in the orientation of the structure cell. Transformation: Penfield to new, $\frac{1}{3}\bar{2}00/\frac{1}{3}\bar{1}00/i/000\frac{1}{2}$.
 3. Penfield (1890) and Frondel (1949).
 4. Frondel (1949) by Weissenberg method on Tintic and Bisbee crystals.
 5. Miers, *Min. Mag.*, **10**, 273 (1894), on Cornwall crystals.
 6. Penfield (1890). Values also have been reported of 3.14 (Bisbee), 3.117 (Sardinia) and 3.07 (Cornwall)—see Frondel (1949), Pelloux, *Ann. Mus. Civico Stor. Nat. Genova*, **4**, [3a], 194 (1909), and Miers (1894).
 7. Glass cited in Frondel (1949). Penfield's values of nO 1.694 and nE 1.641 appear to be in error.
 8. Frondel (1949). The variation in indices of refraction and apparently also in specific gravity may be due to mutual substitution of OH and Cl.
 9. Penfield (1890).

31.1.6 CYANOTRICHITE [$Cu_4Al_2(SO_4)(OH)_{12} \cdot 2H_2O$]. Kupfersammeterz, Kupfersammterz *Werner* (62, 1808). Velvet Copper Ore *Jameson* (**3**, 153, 1816). Sammeterz *Breithaupt* (168, 1823; 320, 1832). Cuivre velouté *Fr.* Cyanotrichit *Glocker* (587, 1839). Lettsomite *Percy* (*Phil. Mag.*, **36**, 100, 1850).

Orthorhombic. As plush- or wool-like aggregates and coatings of minute acicular crystals, also radial-fibrous or tufted. G. 2.95 (Cap Garonne [1]), 2.737 (Morenci [2]), 2.76.[4] Luster silky. Color sky-blue or azure-blue. Streak pale blue.

Opt.

ORIENTATION	n (Arizona [3])	n (Mednorudyansk) [4]	PLEOCHROISM	
$X \perp$ laths	1.588	1.591	Nearly colorless	Biaxial positive (+).
Y	1.617	1.620	Pale blue	$2V$ 82°.
Z elong.	1.655	1.654	Bright blue	$r < v$, large.

Chem. A hydrated basic sulfate of copper and aluminum, $Cu_4Al_2(SO_4)(OH)_{12} \cdot 2H_2O$. Analysis gave: [5]

	CuO	CaO	Al_2O_3	Fe_2O_3	SO_3	H_2O	Insol.	Total
1.	49.39		15.82		12.42	22.37		100.00
2.	47.50	0.11	15.59	0.43	12.19	23.20	1.46	100.48
3.	48.22		15.72		13.22	23.14	0.30	100.60

1. $Cu_4Al_2(SO_4)(OH)_{12} \cdot 2H_2O$. 2. Grand Canyon, Arizona.[6] 3. Mednorudyansk, U.S.S.R.[4]

Tests. Soluble in acids.

ZINCALUMINITE 579

Occur. A secondary mineral found sparingly in copper deposits. With azurite, malachite, and an aluminum sulfate at Moldawa in the Banat, Roumania. From Laurium, Greece. At Cap Garonne, Var, France, as coatings on sandstone associated with azurite, tyrolite, adamite, brochantite, olivenite. Reported from Rio, Elba, and Traversella, Italy. From the Springbok mine, Namaqualand, Africa,[7] and Leadhills, Scotland.[7] In Russia from Mednorudyansk near Nizhne Tagilsk and Kounrad north of Lake Balkash, and in the Berkara Mountains, Kazakhstan.[8] In the United States found in Arizona in the Morenci district and with brochantite in the Grandview mine, Grand Canyon, Coconino County. From the American Eagle mine, Tintic district, Utah. In Nevada in the Mason Park district near Yerington and with spangolite and chalcophyllite at Majuba Hill, in Nevada.[7]

Name. From κυάνος, blue, and θρίξ, hair. Lettsomite after the English mineralogist William G. Lettsom (1805–1887).

Ref.

1. Lacroix (**4**, 265, 1910).
2. Genth, *Am. J. Sc.*, **40**, 118 (1890).
3. Palache and Vassar, *Am. Min.*, **11**, 213 (1926). Larsen (65, 1921) and Sumin, *Bull. ac. sc. U.R.S.S.*, *Ser. géol.*, no. 6, 77 (1941), give identical data on Tintic and Kounrad material.
4. Sumin (1941).
5. For additional analyses see Dana (963, 1892).
6. Vassar analysis in Palache and Vassar (1926).
7. Occurrence verified by x-ray powder comparison with Moldawa and Grand Canyon material—Frondel, priv. comm. (1947).
8. Chukhrov, *C. r. ac. sc. U.R.S.S.*, **46**, 370 (1945), with analysis.

31.1.7 Zincaluminite [$Zn_6Al_6(SO_4)_2(OH)_{26} \cdot 5H_2O$]. *Bertrand* and *Damour* (*Bull. soc. min.*, **4**, 135, 136, 1881).

In tufts and crusts of minute crystals, forming very thin hexagonal plates; possibly orthorhombic [1] with a prism angle of nearly 60°.

Phys. Observations on cleavage are lacking. H. $2\frac{1}{2}$–3. G. 2.26. Color white to bluish white and pale blue.

Opt.[2] In transmitted light, colorless. Sometimes biaxial, with small 2V.

nO 1.534 ± 0.003   Uniaxial negative (−).
nE 1.514 ± 0.003

Chem. A basic hydrated sulfate of zinc and aluminum, $Zn_6Al_6(SO_4)_2(OH)_{26} \cdot 5H_2O$. A small amount of Cu apparently substitutes for Zn.

Anal.

	1	2
CuO		1.85
ZnO	38.19	34.69
Al_2O_3	23.92	25.48
SO_3	12.52	12.94
H_2O	25.37	25.04
Total	100.00	100.00

1. $Zn_6Al_6(SO_4)_2(OH)_{26} \cdot 5H_2O$. 2. Laurium, Greece.[3] Recalculated to 100 after deducting a small amount of gangue.

Tests. Heated C.T. affords an abundance of slightly alkaline water. Soluble in acids and alkalies.

O c c u r. Associated with smithsonite, serpierite, and other secondary minerals at the zinc mines of Laurium, Greece.

N a m e. In allusion to the composition.

Ref.
1. Suggested by measurements of the prism angle of minute crystals under the microscope by Bertrand (1881).
2. Larsen (159, 1921).
3. Damour (1881).

LAMPROPHANITE. Lamprophan *Igelström* (*Ak. Stockholm, Öfv.*, **23**, 93, 1866). In thin cleavable folia. H. 3. G. 3.07. Luster pearly. Color and streak white. Analysis gave: $(Na,K)_2O$ 14.02, CaO 24.65, MgO 5.26, MnO 7.90, FeO tr., PbO 28.00, SO_3 11.17, H_2O 8.35, total 99.35. Not wholly soluble in acids. From Långban, Sweden. Needs verification.

31.1.8 Woodwardite. Church (*Chem. News*, **13**, 85, 113, 1866; *J. Chem. Soc. London*, **19**, 130, 1866).

As minute, botryoidal concretions with a fine-fibrous structure; spherulitic. Color greenish blue to turquois-blue. Translucent to almost transparent. G. 2.38. Optically [1] biaxial positive (+) with parallel extinction. The reported indices of refraction vary widely, probably due to differences in water content: nX 1.552, nY 1.555, nZ 1.565 (Klausen);[2] nX 1.571, nZ 1.576 (Cornwall).[1] A hydrated basic sulfate of copper and aluminum. Formula probably $Cu_4Al_2(SO_4)(OH)_{12} \cdot 2–4H_2O$. The reported analyses[3] are unsatisfactory. Originally found in Cornwall, England; also from Klausen, in the Trentino, Italy. Woodwardite has been regarded as impure langite or as a mixture of cyanotrichite and gibbsite but has been shown to be a distinct species.[1] Named after S. P. Woodward.

Ref.
1. See Meixner, *Zbl. Min.*, 238 (1940).
2. Larsen (98, 1921) [langite, no. 3, from Klausen].
3. Summarized by Meixner (1940).

31.1.9 CHALCOALUMITE $[CuAl_4(SO_4)(OH)_{12} \cdot 3H_2O]$ Larsen and *Vassar* (*Am. Min.*, **10**, 79, 1925).

Probably triclinic. In rather porous botryoidal crusts composed mostly of minute matted fibers; the outer layers of the crusts are vitreous, somewhat darker colored and composed of relatively large lath-like crystals. The best crystals have a terminal edge at about 60° to the length, and some are equilateral triangles. The laths are sometimes twinned with the composition face parallel to the long edge and nearly normal to the flat face; these exhibit re-entrant angles, with symmetrical faces and extinction.

P h y s. Probably several perfect cleavages. Brittle. H. $2\frac{1}{2}$. G. 2.29. Luster dull to vitreous. Color turquois-green to pale blue and bluish gray. Streak white.

URANOPILITE 581

Opt. In transmitted light, colorless.

nX	1.523	Biaxial positive (+).
nY	1.525	$2V$ small.
nZ	1.532	$r > v$, strong.

Laths lying on the flat face do not extinguish in white light and have negative elongation; Y (?) is inclined about 32° to the length and lies in the acute angle of the inclined termination. On edge, the laths extinguish at about 40° with negative elongation. Matted fibers appear to have parallel extinction, with positive elongation, and show abnormal blue interference colors.

Chem. A basic hydrated sulfate of copper and aluminum, $CuAl_4(SO_4)(OH)_{12} \cdot 3H_2O$.

Anal.

	1	2	3
Na_2O		0.50	
K_2O			
CaO		0.01	[1.01]
MgO		0.05	
CuO	15.14	14.78	14.56
Al_2O_3	38.78	38.71	38.88
SO_3	15.23	14.80	14.67
H_2O-	30.85	28.95	30.60
H_2O+		1.90	
Insol.		0.09	0.28
Total	100.00	99.79	[100.00]

1. $CuAl_4(SO_4)(OH)_{12} \cdot 3H_2O$. 2,3. Bisbee, Arizona.[1]

Tests. Fusible at 5. Slowly soluble in cold dilute acids.

Occur. Associated with malachite and azurite as thin crusts upon limonite stalactites at the Copper Queen mine, Bisbee, Cochise County, Arizona. Observed altered to gibbsite.

Name. In allusion to the composition.

Ref.

1. Vassar analysis in Larsen and Vassar (1925).

31.1.10 **URANOPILITE** $[(UO_2)_6(SO_4)(OH)_{10} \cdot 12H_2O]$. *Weisbach* (*Jb. Min.*, II, 258, 1882). Basisches Uransulphat *Dauber* (*Ann. Phys.*, **92**, 237, 1854). Uranocher pt.

Probably monoclinic. As velvety incrustations and globular or reniform masses composed of microscopic needles or laths, elongated [001] and flattened {010}. Color bright yellow, sometimes lemon-yellow or golden yellow. Luster silky. Cleavage {010} perfect. G. 3.7–4.0. Fluoresces bright yellowish green in ultraviolet light.

Opt. In transmitted light, colorless to pale yellow. Not perceptibly pleochroic.

	n		
Orientation	(Joachimsthal)[1]	n(Na)[2]	
$X \sim b$	1.621	1.623	Biaxial positive (+).
$Y \wedge c\ +18°$	1.623	1.625	$2V$ rather large (Na), 0°
Z	1.631	1.634	for some wavelengths.
			$r < v$, extreme; also $r > v$.

Chem. A hydrated basic uranium sulfate, $(UO_2)_6(SO_4)(OH)_{10} \cdot nH_2O$. n is close to 11 or 12H_2O in fully hydrated material. In a dry atmosphere at ordinary temperatures or on heating to 60°–70° the water content decreases [6] to 5H_2O. The Ca and Fe''' reported in the analyses are probably due to admixed gypsum and limonite.

Anal.

	1	2	3	4	5	6	7	8
CaO		2.08	1.96		1.63		tr.	0.60
Fe$_2$O$_3$					tr.		tr.	1.17
UO$_3$	81.63	77.17	77.46	79.9	79.89	80.10	80.78	80.19
SO$_3$	3.81	3.18	4.56	4.0	4.24	4.88	4.18	4.15
H$_2$O	14.56	16.59	14.69	14.3	13.98	14.16	14.32	13.81
Rem.		0.39	1.33		0.08	0.33		
Total	100.00	99.41	100.00	98.2	99.82	99.47	99.28	99.92

1. $(UO_2)_6(SO_4)(OH)_{10} \cdot 12H_2O$. 2,3. Johanngeorgenstadt.[3] Rem. is SiO$_2$. 4. Joachimsthal.[4] 5. St. Just, Cornwall.[5] Rem. is PbO. 6. Chinkolobwe, Katanga.[7] Rem. is insol. H$_2$O includes H$_2$O+ 3.92. 7. Příbram, Bohemia.[5] 8. Joachimsthal.[5]

Occur. A secondary mineral found associated with gypsum and meta-uranopilite on altering uraninite. From Johanngeorgenstadt, Joachimsthal, and Příbram in Bohemia and from St. Just, Cornwall, England. Also from Chinkolobwe, Katanga, in the Belgian Congo.

Artif. Numerous hydrated uranyl sulfates have been reported but none have been shown to be identical with uranopilite or meta-uranopilite.

Name. From *uranium* and πιλος, *felt*, alluding to the composition and structure.

Ref.
1. Larsen (150, 1921).
2. Nováček, *Soc. sc. Bohême, Mém.*, no. 7 (1935), average of numerous measurements.
3. Schulze analysis in Weisbach (1882).
4. Dauber (1854).
5. Nováček (1935), who cites two additional analyses.
6. For dehydration data see Nováček, *Věst. Král. České Spol. Nauk*, no. 17 (1942).
7. Mélon in Buttgenbach, *Mém. inst. roy. colon. Belge*, **6**, 449 (1935).

31.1.11 META-URANOPILITE [$(UO_2)_6(SO_4)(OH)_{10} \cdot 5H_2O$]. Beta-uranopilite, β-uranopilite Nováček (*Mém. soc. sc. Bohême*, no. 7, 1935). Meta-uranopilite Frondel (priv. comm., 1948).

As needles and laths. Color yellow, grayish, brown, or green. Fluoresces yellowish green in ultraviolet light.

Opt. Pleochroism not perceptible.

Orientation		n	
X	b	1.72	Biaxial negative (−).
Y	c (elong.)	1.76	
Z	a	1.76	

Chem. A hydrated basic sulfate of uranium, $(UO_2)_6(SO_4)(OH)_{10} \cdot 5H_2O$. Analysis gave:[1]

	UO$_3$	Fe$_2$O$_3$	SO$_3$	H$_2$O	Total
1.	86.84		4.05	9.11	100.00
2.	82.40	2.03	4.17	9.40	98.00

1. $(UO_2)_6(SO_4)(OH)_{10} \cdot 5H_2O$. 2. Joachimsthal.

Occur. From Joachimsthal, Bohemia, and probably from other localities of uranopilite.

Ref.
1. For dehydration data on both uranopilite and meta-uranopilite see Nováček, *Věst. Král. České Spol. Nauk*, no. 17 (1942).

TYPE 2. $A_4(XO_4)Z_q \cdot xH_2O$

31.2.1 Klebelsbergite. Zsivny (*Mat. Termés. Ért.*, **46**, 19, 25, 1929).

Monoclinic.[1] As tufts of tiny acicular crystals elongated [001] and flattened {010}. Observed forms: {001}, {010}, {110}, {101}, and {$\bar{1}$01}; β 91°48'. Brittle. Color dark sulfur-yellow. Transparent. H. and G. not known.

Opt. In transmitted light, colorless to yellow.

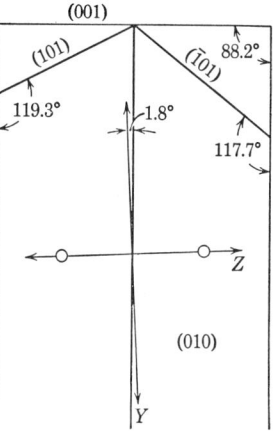

Orientation	n	
X b	>1.74	Biaxial negative (−).
Y ∧ c 1°48'		Not pleochroic.
Z		$r < v$.

Chem. Shown by microchemical tests to be essentially a basic antimony sulfate containing also a little H_2O, Fe, Mg, Na, K, and traces of Bi and P_2O_5. Heated in C.T. turns reddish brown, then white and finally fuses giving off acid water. Slowly soluble in cold concentrated HCl.

Found in the interstices of columnar aggregates of stibnite from Felsöbánya, Roumania. Named after Kuno Klebelsberg, Hungarian educator.

Ref.
1. Plane angles measured on a microscope stage are cited by Zsivny (1929) but are too few and poor to establish elements.

31.2.2 LANGITE [$Cu_4(SO_4)(OH)_6 \cdot H_2O$?]. A new British mineral *Maskelyne* (*Phil. Mag.*, **27**, 316, 1864). Langite *Maskelyne* in Pisani (*C.R.*, **59**, 633, 1864); *Maskelyne* (*Phil. Mag.*, **29**, 473, 1865).

Cryst.[1] Orthorhombic.

$a:b:c = 0.5347:1:0.6346;$ $p_0:q_0:r_0 = 1.1868:0.6346:1$

$q_1:r_1:p_1 = 0.5347:0.8426:1;$ $r_2:p_2:q_2 = 1.5758:1.8702:1$

Forms:

	ϕ	$\rho = C$	ϕ_1	$\rho_1 = A$	ϕ_2	$\rho_2 = B$
c 001	.....	0°00'	0°00'	90°00'	90°00'	90°00'
b 010	0°00'	90 00	90 00	90 00	.....	0 00
a 100	90 00	90 00	.....	0 00	0 00	90 00
m 110	32 24	90 00	90 00	57 36	0 00	32 24
f 021	0 00	51 46	51 46	90 00	90 00	38 14

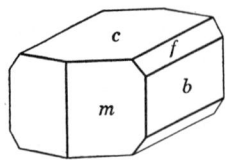

Habit. Crystals small, equant or elongated [100]. As laths or scales, forming fibro-lamellar crusts; also earthy.

Twinning. On {110}, common, in part repeated as aggregates flattened {001} and resembling aragonite, or star-shaped.

P h y s. Cleavage {001} and {010}. H. $2\frac{1}{2}$–3. G. 3.48–3.50 (Cornwall).[2] Luster of crystals vitreous, of crusts somewhat silky. Color fine blue to greenish blue. Translucent.

O p t.[3]

ORIENTATION	n (Cornwall)[4]	PLEOCHROISM	
X	c	1.654	Light yellowish green	Biaxial negative (−).
Y	b	1.713	Blue-green	$2E$ 66°–104°.
Z	a	1.722	Sky-blue	

C h e m. A hydrated basic sulfate of copper, probably $Cu_4(SO_4)(OH)_6 \cdot H_2O$.

Anal.[5]

	1	2	3	4	5
CuO	67.66	65.82	65.92	67.48	66.94
SO$_3$	15.32	16.42	16.77	16.79	14.59
H$_2$O	17.02	18.32	16.19	15.73	16.47
Rem.			1.12		
Total	100.00	100.56	100.00	100.00	98.00

1. $Cu_4(SO_4)(OH)_6 \cdot H_2O$. 2. Cornwall.[2] 3. Cornwall.[6] Rem. is CaO 0.83, MgO 0.29. 4. Cornwall.[7] 5. Andacollo, Chile.[8] Color green. Brochantite?

Tests. On heating loses water and changes in color through bright green and olive-green to black. Insoluble in water but easily soluble in acids or in ammonia.

O c c u r. A secondary mineral resulting from the oxidation of copper sulfides and found associated with gypsum and basic copper sulfates. Originally found at St. Blazey and St. Just in Cornwall, England, in minute twinned crystals and as blue crusts, partly earthy, sometimes associated with connellite or intimately admixed with gypsum. Langite has been reported from other localities but the identification in most instances is uncertain, and there is little doubt but that various minerals of similar appearance or chemical constitution have been confused under the name. Probably authentic occurrences[9] are at Eschach, Styria, Austria, and at Virneberg near Rheinbreitbach, Germany. Reported from Herrengrund north of Neusohl, Czechoslovakia, and at Viel-Salm, Belgium. Langite (perhaps brochantite) is also reported from the El Cobre mines and Andacolla in Chile, and from Mexico.

A r t i f. Langite has been reported[10] in the system CuO-SO_3-H_2O. Cu_2SO_5 (*dolerophanite*) formed by ignition of $CuSO_4$ at 650° affords langite when treated with cold water and antlerite with hot water. Also reported[11] by precipitation of copper sulfate solution with potassium hydroxide.

N a m e. After Victor von Lang (1838–1921), physicist and physical crystallographer, long Professor of Physics in the University of Vienna.

FELSÖBÁNYAITE

Ref.
1. Angles of Maskelyne (1865), orientation of Dana (961, 1892). Transformations: Maskelyne to Dana, 010/100/001; Dana to Goldschmidt (5, 116, 1918), 020/001/100. See also Brezina, *Zs. Kr.*, **3**, 374 (1879).
2. Maskelyne (1864).
3. The optical data of Larsen (97, 1921) on three reputed specimens of langite all disagree, and probably none are of langite. See also Meixner, *Zbl. Min.*, 11 (1941) and Koritnig, *Zbl. Min.*, 154 (1941).
4. Meixner (1941).
5. For additional analyses of material referable with some doubt to this species see Hintze (1 [3B], 4385, 1929).
6. Pisani (1864).
7. Church, *J. Chem. Soc. London*, **18**, 87 (1865).
8. Field, *Phil. Mag.*, **24**, 123 (1862).
9. See Meixner (1941) and Koritnig (1941).
10. Binder, *Ann. chimie*, **5** [11], 337 (1936); see also Posnjak and Tunell, *Am. J. Sc.* **18**, 1 (1929), who state langite to be a mixture.
11. Field (1862).

31.2.3 Felsöbányaite [$Al_4(SO_4)(OH)_{10} \cdot 5H_2O$ (?)]. Felsöbanyite *Kenngott* (*Ak. Wien, Sitzber.*, **10**, 294, 1853), *Haidinger* (*Ak. Wien, Sitzber.*, **12**, 183, 1854). Felsobanyite, Felsöbanyite.

Probably orthorhombic. As globular, radial aggregates of lamellar crystals; also as intergrown lath-like crystals tabular {001} and elongated [100] with a terminal angle [1] of 66°24'.

Phys. Good cleavages {010} and {100}, and also {001} (?). H. $1\frac{1}{2}$. G. 2.33. Color yellow to white. Luster pearly on cleavage surfaces.

Opt.[2] In transmitted light, colorless.

Orientation		n	
X	a	1.516 ± 0.003	Biaxial positive (+).
Y	b	1.518 ± 0.003	2V 48° ± 2° (meas.), 2E 77°.
Z	c	1.533 ± 0.005	$r > v$.

Chem. A hydrated basic sulfate of aluminum, perhaps $Al_4(SO_4)(OH)_{10} \cdot 5H_2O$.

Anal.

	1	2
Al_2O_3	43.93	45.63
SO_3	17.25	16.47
H_2O	38.82	37.27
Total	100.00	99.37

1. $Al_4(SO_4)(OH)_{10} \cdot 5H_2O$. 2. Felsöbánya.[3] Average of two analyses.

Tests. B.B. swells, exfoliates, and turns white, but does not fuse.

Occur. At Felsöbánya, Hungary, with marcasite, stibnite, and barite.
Name. From the locality.

Ref.
1. Krenner, *Cbl. Min.*, 138 (1928A), measured under the microscope.
2. Larsen (159, 1921); Krenner (1928) gives $Y = b$.
3. von Hauer, *Ak. Wien, Sitzber.*, **12**, 188 (1854).

31.2.4 Basaluminite [$Al_4(SO_4)(OH)_{10} \cdot 5H_2O$]. Bannister and Hollingworth (*Nature*, **162**, 565, 1948); Hollingworth and Bannister (*Min. Mag.*, **29**, 1, 1950).

Perhaps hexagonal. White compact masses with a conchoidal fracture. Not plastic. G. $\sim$2.12. Microcrystalline with a mean index of refraction of 1.515–1.519 (Na); elongation negative. The strongest line of the x-ray powder pattern is 9.4 Å. Analysis gave (average of two after deducting allophane and Fe_2O_3): Al_2O_3 42.85, SO_3 17.15, H_2O [40.00], total [100.00]. This is close to $Al_4(SO_4)(OH)_{10} \cdot 5H_2O$. On heating, $5H_2O$ are lost at about 150°. Found as coatings on joint surfaces and as veinlets in workings in ironstone (siderite) at Irchester, Northamptonshire, and Clifton Hill, Brighton, Sussex, England; also from Épernay, Marne, France. Associated with hydrobasaluminite, allophane, gypsum, and aragonite.

31.2.5 Hydrobasaluminite [$Al_4(SO_4)(OH)_{10} \cdot 36H_2O$?]. Bannister and Hollingworth (*Nature*, **162**, 565, 1948); Hollingworth and Bannister (*Min. Mag.*, **29**, 1, 1950).

White plastic clay-like masses. The strongest x-ray powder line is 12.6 Å. Composition approximately $Al_4(SO_4)(OH)_{10} \cdot 36H_2O$. The mineral dehydrates on exposure to form powdery basaluminite and does not regain its plastic character when wetted. Found with basaluminite at Irchester, Northamptonshire, England.

WINEBERGITE [$Al_4(SO_4)(OH)_{10} \cdot 7H_2O$ (?)]. Gümbel (*Ostbayer. Grenzgeb.*, 260, 1868; in Roth, **1**, 239, 1879). Doughtyite Headden (*Colorado Sc. Soc. Proc.*, **8**, 55, 1905).

Names given to probably identical ill-defined materials found as mine sludges or precipitates from spring water. Color white. Apparently a hydrated basic sulfate of aluminum, $Al_4(SO_4)(OH)_{10} \cdot 7H_2O$.

Anal.

	1	2	3
MgO		0.78	tr.
FeO		2.60	
Fe_2O_3			0.45
Al_2O_3	40.76	40.80	39.51
SO_3	16.01	15.61	15.00
H_2O	43.23	40.21	41.80
Rem.			3.91
Total	100.00	100.00	100.67

1. $Al_4(SO_4)(OH)_{10} \cdot 7H_2O$. 2. Bodenmais (?).[1] 3. Colorado.[2] Rem. is SiO_2 1.91, sand 1.56, ZnO 0.44.

Occur. Originally from Bodenmais and with an iron sulfate from Lowmühl near Passau, Bavaria (*winebergite*). Also as a deposit formed by the reaction of acid and alkaline waters at the Doughty Springs, Delta County, Colorado (*doughtyite*).

Ref.

1. Gümbel (1868).
2. Headden (1905).

PARALUMINITE. Steinberg (*J. pr. Chem.*, **32**, 495, 1844). Pissophan *Breithaupt* (101, 1832); Pissophanite *Dana* (661, 1868); Garnsdorfite *Brooke* and *Miller* (544, 1852). Eisenparaluminite, Ferri-paraluminite *Pilipenko* (*Wiss. Schr. Univ. Saratow*, **6**, 167, 1927).

Massive, usually in fibrous reniform or nodular masses in clay, like aluminite; also reported as a slimy precipitate and as an efflorescence of acicular crystals in stonework.[1] Color white to pale yellow. The name paraluminite originally was applied to a substance from near Halle, Saxony, close in composition to $2Al_2O_3 \cdot SO_3 \cdot 15H_2O$, and other ill-defined substances, some at least of very doubtful purity, approximating to this ratio later were placed therewith.[2] Ferri-paraluminite, from Mt. Sokolova, near Saratov on the Danube, U.S.S.R. (anals. 6,7), and the horn-like pissophanite from Garnsdorf, Thuringia (anal. 8) may be varieties in which Fe''' substitutes for Al or perhaps are mixtures of an aluminum sulfate or hydrous oxide with an iron sulfate.

	1	2	3	4	5	6	7	8
Al_2O_3	36.79	36.17	35.96	36.54	43.00	36.20	21.73	35.2
Fe_2O_3						2.57	17.46	9.8
SO_3	14.45	14.54	14.04	15.56	13.37	14.11	14.44	12.6
H_2O	48.76	49.03	50.00	46.89	43.63	46.90	42.47	41.7
Rem.							3.57	0.7
Total	100.00	99.74	100.00	98.99	100.00	99.78	99.67	100.0
G.						1.85	2.13	1.98

1. $2Al_2O_3 \cdot SO_3 \cdot 15H_2O$. 2,3. Halle.[3] 4. Halle.[4] 5. Huelgoet, Brittany.[5] 6,7. Mt. Sokolova, Saratov (*ferri-paraluminite*).[6] Rem. in anal. 7 is CaO 1.76, FeO 0.74, MgO 1.07 with Li, K, and organic matter present. Anal. 6 is average of two after deducting 4.3 and 5.5 per cent quartz; anal. 7 has 23.24 per cent insol. deducted. 8. Garnsdorf, Thuringia (*pissophanite*); average of two analyses;[7] rem. is insol.

A specimen from Halle labeled paraluminite but not known definitely to be authentic has been described optically (col. 1), and two later occurrences have been identified as such on this basis (cols. 2,3).

	n (Halle) [8]	n (in concrete) [1]	n (on stonework) [1]
X	1.462 ± 0.003	1.463 ± 0.003	1.463 ± 0.005
Y	1.470 ± 0.003	1.471	
Z	1.471 ± 0.003	1.471	1.470
Sign	—	—	
$2V$	Small to 0°	Small	Small to 0°
Elong.	X	X	X

Ref.

1. Hutton, *New Zealand J. Sc. Tech.*, **26**, 242 (1945).
2. For additional analyses see Dana (661, 1868).
3. Schmidt and Martens in Steinberg, *J. pr. Chem.*, **32**, 495 (1845).
4. Dieck, *Zs. Nat. Halle*, **8**, 265 (1857).
5. Berthier cited in Dana (1868).
6. Pilipenko, *Wiss. Schr. Univ. Saratow*, **6**, 167 (1927)—*Jb. Min.*, I, 297 (1928).
7. Erdmann, *J. Chemie u. Phys.*, **62**, 104 (1831); see also Cornu, *Cbl. Min.*, 329 (1909).
8. Larsen (118, 1921).

31.2.6 **Glockerite.** Vitriolocker *Berzelius* (*Afhandl. Fys. Kem. Min.*, **5**, 157, 1816). Fer sous-sulfaté terreux *Berzelius* (1819). Vitriol Ocher. Pittizite *Beudant* (447, 1824). Glockerit *Naumann* (254, 1855). Hydroglockerite *Greenly* (*Mem. Geol. Sur. Gr. Britain*, **2**, 832, 1919).

Stalactitic or incrusting, also earthy. Fracture conchoidal to earthy. Brittle. G. variable, low. Luster resinous or dull. Color brown to ocher-yellow, brownish black to pitch-black; dull green. Streak ocher-yellow to brown. Opaque to translucent.

Opt.[1] Material from Zuckmantel, Silesia, is composed of minute intertwined red fibers, with $nX \sim 1.76$, $nZ \sim 1.81$.

Chem. A hydrated basic sulfate of ferric iron. The formula is sometimes given as $Fe_4'''(SO_4)(OH)_{10} \cdot 1-3H_2O$, but the analyses vary widely and the substance is best regarded as an impure hardened gel-mass of indefinite composition.[7]

Anal.[6]

	Fe_2O_3	SO_3	H_2O	Rem.	Total
1.	60.73	17.83	21.00		99.56
2.	57.85	13.44	26.36	[2.35]	[100.00]
3.	68.75	9.80	15.52	5.93	100.00
4.	57.01	13.97	28.83		99.81

1. Germantown, Pennsylvania.[2] 2. Parys Mount, Anglesey (*hydroglockerite*).[3] Rem. is SiO_2 2.14, Al_2O_3, etc., 0.21. 3. Rammelsberg, Harz.[4] Rem. is ZnO 1.29, CuO 0.50, gangue 4.14. 5. Gap mine, Lancaster County, Pennsylvania.[5]

Occur. A secondary mineral, in part of recent formation, resulting from the oxidation of iron sulfides. As stalactites up to several feet in length from Obergrund near Zuckmantel, Silesia, and compact and earthy from the Rammelsberg mine near Goslar in the Harz, Germany. From Falun, Sweden, and Modum, Norway. As efflorescences in pyrite mines in the Urals, U.S.S.R.[8] At Parys Mount, Anglesey, England (*hydroglockerite*). In the United States at Germantown and at the Gap nickel mine, Lancaster County, Pennsylvania. Ill-defined, hydrous ferric-sulfate sinters have been reported from several other localities.

Name. After the German mineralogist Ernst Friedrich Glocker (1783–1858).

Ref.

1. Larsen (80, 1921).
2. Eyerman cited in Gordon (147, 1922).
3. Church, *Min. Mag.*, **11**, 13 (1895).
4. Jordan, *J. pr. Chem.*, **9**, 95 (1836).
5. Roepper analysis in Genth cited by Gordon (1922).
6. See also Dana (970, 1892); Hintze (**1** [3B], 4441, 1929); Vertushkov, *Mém. soc. russe min.*, **69**, 485 (1940); Erdmann, *J. Chemie u. Phys.*, **62**, 104 (1831).
7. See also Posnjak and Merwin, *J. Am. Chem. Soc.*, **44**, 1965 (1922).
8. Vertushkov (1940).

TYPE 3. $A_m B_n (XO_4)_p Z_q \cdot xH_2O$, WITH $(m + n):p$ FROM 5:2 TO 3:1

31.3.1 Kamarezite. Busz (*Sitzber. niederrhein. Ges., Bonn*, **50**, 83, 1893; *Jb. Min.*, I, 115, 1895).

Orthorhombic (?). In minute crystals elongated [001] and flattened {010} with domical terminations.[1] Vertically striated. Cleavage {100} perfect. H. 3. G. 3.98. Color grass-green. Transparent. Biaxial with $Y = b$ and $Bx_a = a$; $2V$ large.

Chem. A hydrated basic sulfate of copper, perhaps $Cu_3(SO_4)(OH)_4 \cdot 6H_2O$.

Anal.

	1	2
CuO	51.57	51.50
FeO		0.69
SO$_3$	17.29	17.52
H$_2$O	31.14	[30.29]
Total	100.00	[100.00]

1. Cu$_3$(SO$_4$)(OH)$_4 \cdot$6H$_2$O.   2. Kamareza, Greece.[2]

Heated in C.T. decrepitates and gives off acid water, mostly at a high temperature. Easily soluble in acids and in ammonia, but insoluble in water. Found at Kamareza, near Laurium, Greece. Needs verification.

Ref.
1. The angles for the domes are given as (101) ∧ (100) ∼ 59°, (201) ∧ (100) ∼ 40°, from measurements on a microscope stage, and afford $a:c = 1:0.60$. Observed forms: {100}, {010}, {101}, {201}.
2. Klingemann analysis in Busz (1895); average of four partial analyses.

31.3.2   **E T T R I N G I T E** [Ca$_6$Al$_2$(SO$_4$)$_3$(OH)$_{12} \cdot$26H$_2$O]. *Lehmann (Jb. Min.,* 273, 1874).

C r y s t.[1]   Hexagonal—P.

$$a:c = 1:1.9084; \quad p_0:r_0 = 2.2036:1$$

Forms:[2]

		φ	ρ = C	M	A$_2$
c	0001	.....	0°00′	90°00′	90°00′
m	10$\bar{1}$0	30°00′	90 00	60 00	90 00
a	11$\bar{2}$0	0 00	90 00	90 00	60 00
f	10$\bar{1}$4	30 00	28 51	76 02½	90 00
e	10$\bar{1}$2	30 00	47 46½	68 16	90 00
r	10$\bar{1}$1	30 00	65 35½	62 55	90 00

Structure cell.[3] Space group $C6/mmc$. a_0 11.24 ± 0.02 kX, c_0 21.45 ± 0.05; $a_0:c_0 = 1:1.908$. Cell contents Ca$_{12}$Al$_4$(SO$_4$)$_6$(OH)$_{24} \cdot$52H$_2$O.

Habit. As small crystals prismatic [0001], usually without terminal faces.

P h y s. Cleavage {10$\bar{1}$0}, perfect. H. 2–2½. G. 1.77;[4] 1.76 (calc.). Colorless and transparent, becoming white and opaque on partial dehydration.

O p t. In transmitted light, colorless.

	n (Scawt Hill)[5]	n (artif.)[6]	
O	1.4655 ± 0.0005	1.464	Uniaxial negative (−).
E	1.4618 ± 0.0005	1.458	

As the mineral dehydrates the indices of refraction increase, E relatively rapidly, and the sign changes from negative to positive.[6]

C h e m. A hydrated basic sulfate of calcium and aluminum, Ca$_6$Al$_2$-(SO$_4$)$_3$(OH)$_{12} \cdot$26H$_2$O. About three-fourths of the water is lost at 110° and apparently is zeolitic; the remainder is lost at red heat.[7]

Anal.

	1	2	3	4
CaO	26.81	26.6	27.27	26.31
Al_2O_3	8.12	7.0	7.76	9.72
SO_3	19.14	18.8	16.64	18.54
H_2O	45.93	46.3	45.82	45.41
Rem.		0.8		
Total	100.00	99.5	97.49	99.98
G.		1.772	1.750	

1. $Ca_6Al_2(SO_4)_3(OH)_{12} \cdot 26H_2O$. 2. Scawt Hill, Antrim.[8] Rem. is CO_2. 3. Ettringen.[9] 4. Tombstone, Arizona.[10] Recalculated after deducting SiO_2 1.90 as a hydrous Ca,Al silicate of stated composition. Not certainly ettringite.

Tests. Fusible to a white enamel. Partly decomposed by water, giving an alkaline solution. Easily soluble in dilute acids.

Occur. Originally found in cavities of metamorphosed limestone inclusions in a leucite-nepheline-tephrite near Ettringen, between Mayen and the Laacher See, Rhine Province, Germany. With afwillite, portlandite, and hydrocalumite in a contact zone between limestone and dolerite at Scawt Hill, near Larne, County Antrim, Ireland. As an alteration of Ca, Al silicates in the Lucky Cuss mine, Tombstone, Cochise County, Arizona.

Artif.[11] Found in the system $CaO-Al_2O_3-CaSO_4-H_2O$. The compound $Ca_4Al_2(SO_4)(OH)_{12} \cdot 6H_2O$ is formed at higher temperatures.

Name. From the locality.

Ref.

1. Elements of the x-ray cell of Bannister, *Min. Mag.*, **24**, 324 (1936), the *c*-axis of the morphological cell of Lehmann (1874) and Brauns (173, 1922) being doubled. Crystal class uncertain.
2. $\{11\bar{2}0\}$ from Bannister (1936); $\{10\bar{1}1\}$ from Brauns (173, 1922); others from Lehmann (1874).
3. Bannister (1936) by Laue and rotation methods on a Scawt Hill crystal. Holohedry assumed.
4. Bannister (1936) on Scawt Hill material; other reported values range between 1.75 and 1.79.
5. Bannister (1936) by prism method; Brauns (1922) gave very close values on Ettringen crystals. Larsen and Berman (78, 1934) give nO 1.488 and nE 1.474 on probably partly dehydrated material.
6. Lerch, Ashton, and Bogue, *Bur. Standards J. Research*, **2**, 715 (1929).
7. See Bannister (1936) for discussion.
8. Hey analysis in Bannister (1936).
9. Lehmann (1874).
10. Moses, *Am. J. Sc.*, **45**, 489 (1893).
11. Lerch, *et al.* (1929) and Jones, *Trans. Faraday Soc.*, **35**, 1484 (1939).

31.3.3 **DEVILLITE** [$Cu_4Ca(SO_4)_2(OH)_6 \cdot 3H_2O$]. Devilline *Pisani* (*C.R.*, **59**, 813, 1864) = Lyellite *Maskelyne* (*Chem. News*, **10**, 263, 1864). Herrengrundite *Brezina* (*Zs. Kr.*, **3**, 359, 1879; **13**, 670, 1887). Urvölgyite *Szabó* (*Min. Mitt.*, **2**, 311, 1879; *Lit. Ber. Ungarn*, **3**, 510, 1879).

Cryst.[1] Monoclinic.

$a:b:c = 1.8161:1:1.4002;$ $\beta\, 91°10';$ $p_0:q_0:r_0 = 0.7710:1.3999:1$

$r_2:p_2:q_2 = 0.7143:0.5507:1;$ $\mu\, 88°50';$ $p_0'\, 0.7712,\ q_0'\, 1.4002,\ x_0'\, 0.0204$

DEVILLITE

Forms: [2]

	ϕ	ρ	ϕ_2	$\rho_2 = B$	C	A
c 001	90°00'	1°10'	88°50'	90°00'	.....	88°50'
m 110	54 28	90 00	0 00	54 28	89°03'	35 32
ϵ 101	90 00	38 22	51 38	90 00	37 12	51 38
e $\bar{1}$01	−90 00	36 54	126 54	90 00	38 04	126 54
q $\bar{2}$21	−40 12½	67 01	146 41½	45 19½	67 46	126 28

Habit. In thin six-sided plates flattened {001} and striated [010]. Usually as interlaced crusts or as rosettes of plate- or lath-like crystals.
Twinning. On {010}.
P h y s. Cleavage {001} perfect, on {110} and {101} or {10$\bar{1}$} distinct. Not very brittle. H. 2½. G. 3.13. Luster vitreous, on {001} sometimes pearly. Color dark emerald-green to verdigris-green and bluish green. Streak light green. Transparent.
O p t.[3] In transmitted light, green.

Orientation	n (Herrengrund)	Pleochroism	
$X \sim c$	1.585 ± 0.003	Very pale green	Biaxial negative (−).
Y	1.649 ± 0.003	Venice-green	$2V$ 39° ± 2° (meas.).
Z b	1.660 ± 0.003	Turquois-green	$r < v$, marked.

C h e m. A hydrated basic sulfate of copper and calcium, $Cu_4Ca(SO_4)_2(OH)_6 \cdot 3H_2O$.
Anal.[4]

	1	2	3	4
CaO	8.73	8.59	8.17	9.14
CuO	49.53	49.52	49.96	48.42
SO$_3$	24.92	24.62	24.59	24.28
H$_2$O	16.82	16.73	17.76	16.52
Rem.		0.14		0.96
Total	100.00	99.60	100.48	99.32

1. $Cu_4Ca(SO_4)_2(OH)_6 \cdot 3H_2O$. 2. Herrengrund.[5] Rem. is FeO. 3. Herrengrund.[6] 4. Uspensky mine, Kazakhstan.[7] Rem. is (Al,Fe)$_2$O$_3$ 0.34, SiO$_2$ 0.62, MgO tr.

Tests. In C.T. affords water and at higher temperatures acid fumes. Insoluble in water and concentrated H$_2$SO$_4$ but completely soluble in HNO$_3$.

O c c u r. Originally from Cornwall, England. Found at Herrengrund (Urvölgy) north of Neusohl, Czechoslovakia, as a secondary mineral associated with gypsum, azurite, and malachite (*herrengrundite*).[8] Probably from Schwaz, Tyrol. Also on the walls and timbers of old mine workings in the Uspensky mine, Kazakhstan, U.S.S.R. From the Ecton copper mine, Montgomery County, Pennsylvania, U.S.A.

N a m e. After the French chemist H. E. Sainte-Claire Deville (1818–1881).

Ref.

1. Goldschmidt (**4**, 131, 1918) from angles of Brezina (1879) on crystals of very poor quality from Herrengrund; see also Goldschmidt (**2**, 150, 1890). The crystallographic data of Brezina and of Szabó (1879) are not in good agreement. Transformation: Dana (962, 1892) to Goldschmidt (1918), 100/010/00½.

2. Goldschmidt (1918); also β 980, γ 540, ζ 450, n 350, θ 120, x 250, δ 10.0.7, d $\overline{10}.0.7$. Few of the forms can be regarded as certain.
3. Larsen (84, 1921); Larsen and Berman (176, 1934).
4. An additional analysis, cited by Brezina (1879), is low in CaO (2.05 per cent) and probably is erroneous. The Ca was regarded by Brezina as due to admixed gypsum.
5. Schenek analysis in Szabó (1879).
6. Winkler analysis in Weisbach (1886).
7. Chukhrov and Senderova, *C. r. ac. sc. U.R.S.S.*, **23**, 165 (1939).
8. On the identity of devillite and herrengrundite see Meixner, *Zbl. Min.*, 244 (1940).

31.3.4   Arnimite [$Cu_5(SO_4)_2(OH)_6 \cdot 3H_2O$?].   *Weisbach (Jb. Berg.-hütt. Sachsen, 86, 1886).*

Forms a bright green incrustation consisting of short acicular or scaly crystals.

Chem. A hydrated basic sulfate of copper, $Cu_5(SO_4)_2(OH)_6 \cdot 3H_2O$, according to the only analysis (made on material containing gypsum).

	CaO	CuO	(Fe,Al)$_2$O$_3$	SO$_3$	H$_2$O	Total
1.		59.73		24.04	16.23	100.00
2.	0.56	56.81	0.35	24.43	[17.85]	[100.00]

1. $Cu_5(SO_4)_2(OH)_6 \cdot 3H_2O$.   2. Planitz.[1]

Occur. On porcelaneous jasper in the coal region at Planitz near Zwickau, Germany.

Arnimite has been considered to be a calcium-free variety of devillite. The x-ray powder pattern and optical properties, however, differ from those of devillite and the mineral appears to be a distinct species.[2]

Ref.
1. Winkler analysis in Weisbach (1886).
2. Frondel, priv. comm. (1949), found an authentic specimen to be sensibly orthorhombic with nX 1.720 $\pm$ 0.003. The five darkest powder lines are: d 4.84, I 10; 3.60, 9; 2.68, 8; 2.57, 8; 2.13, 8.

31.3.5   SERPIERITE [$(Cu,Zn,Ca)_5(SO_4)_2(OH)_6 \cdot 3H_2O$ (?)].   *Des Cloizeaux (Bull. soc. min., 4, 89, 1881).*

Cryst.[1] Orthorhombic.

$$a:b:c = 0.8586:1:1.3637; \qquad p_0:q_0:r_0 = 1.5883:1.3637:1$$

$$q_1:r_1:p_1 = 0.8586:0.6296:1; \qquad r_2:p_2:q_2 = 0.7333:1.1647:1$$

Forms:

	ϕ	$\rho = C$	ϕ_1	$\rho_1 = A$	ϕ_2	$\rho_2 = B$
c 001	.....	0°00'	0°00'	90°00'	90°00'	90°00'
m 110	49°21'	90 00	90 00	40 39	0 00	49 21
d 034	0 00	45 38½	45 38½	90 00	90 00	44 21½
e 011	0 00	53 45	53 45	90 00	90 00	36 15
p 111	49 21	64 28	53 45	46 47½	32 11½	54 00

Habit. As crusts or tufted aggregates of tiny lath-like crystals elongated [100] and flattened {001}; as botryoidal masses with a satiny surface.

SERPIERITE

Phys. Cleavage {001}, perfect.[2] G. 2.52.[3] Luster vitreous, on the cleavage pearly. Color sky-blue. Transparent.

Opt.[4] In transmitted light, greenish blue.

ORIENTATION		n	PLEOCHROISM	
X	c	1.584 ± 0.003	Nearly colorless	Biaxial negative (−).
Y	a	1.642 ± 0.003	Deep greenish blue	$2V\ 35° \pm 2°$ (meas.).
Z	b	1.647 ± 0.003	Deep greenish blue	$r > v$, strong.

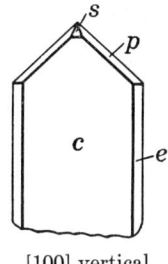

[100] vertical.

Chem. A hydrated basic sulfate of copper, zinc, and calcium, of uncertain formula, perhaps $(Cu,Zn,Ca)_5(SO_4)_2(OH)_6 \cdot 3H_2O$.

Anal.

	1	2	3
CaO	5.57	8.00	n.d.
CuO	43.49	36.12	34.77
ZnO	9.82	13.95	15.46
SO$_3$	24.55	24.29	23.54
H$_2$O	16.57	16.75	16.75
Total	100.00	99.11	90.52

1. $(Cu,Zn,Ca)_5(SO_4)_2(OH)_6 \cdot 3H_2O$ with Cu:Zn:Ca = 0.77:0.17:0.14. 2,3. Laurium.[3]

Tests. Easily soluble in acids.

Occur. Found with smithsonite at Laurium, Greece. Also with smithsonite at Ross Island, Killarney, Ireland.[5] Found with linarite and cyanotrichite at Akchagyl, Kazakhstan, U.S.S.R.[6]

Name. After J. B. Serpieri, an Italian engineer, active in the development of the Laurium mines.

Ref.

1. Des Cloizeaux (1881) and Goldschmidt (**8,** 37, 1922). Uncertain forms: f 043, g 053, h 081, s 203.
2. Not reported by Des Cloizeaux (1881) but observed by Larsen (133, 1921) on Laurium material and by Russell, *Min. Mag.*, **21,** 386 (1927), on crystals from Ireland.
3. Frenzel, *Min. Mitt.*, **14,** 121 (1894).
4. Des Cloizeaux (1881); indices from Larsen (1921).
5. Russell (1927).
6. Chukhrov, *C. r. ac. sc. U.R.S.S.*, **45,** 370 (1945).

PULSZKYITE. Krenner in Tokody (*Schweiz. min. Mitt.*, **28,** 707, 1948).
A partly described mineral found as green, hexagonal plates {0001} and {10$\bar{1}$0}. Optically, uniaxial negative (−). Chemically, a sulfate of copper and zinc. From Urvölgy, Hungary [Špana Dolina, Slovakia], with devillite. Named after Franz Pulszky (1814–1897), a director of the Hungarian National Museum.

TYPE 4. $(AB)_z(XO_4)Z_q \cdot xH_2O$

31.4.1 K A I N I T E [KMg(SO$_4$)Cl·3H$_2$O]. Zincken (Jb. Min., 310, 1865).

C r y s t.[1] Monoclinic; prismatic—$2/m$.

$a:b:c = 1.1727:1:0.6093$; $\beta\,94°54'$; $p_0:q_0:r_0 = 0.5196:0.6071:1$

$r_2:p_2:q_2 = 1.6473:0.8559:1$; $\mu\,85°06'$; $p_0'\,0.5215,\ q_0'\,0.6093,\ x_0'\,0.0857$

Forms:[2]

	ϕ	ρ	ϕ_2	$\rho_2 = B$	C	A
c 001	90°00′	4°54′	85°06′	90°00′	.....	85°06′
b 010	0 00	90 00	.....	0 00	90°00′	90 00
a 100	90 00	90 00	0 00	90 00	85 06	.....
m 021	3 49	52 11½	85 06	37 58	52 02	86 59½
r 201	90 00	46 12½	43 47½	90 00	41 18½	43 47½
o 421	59 22	68 23	24 43½	61 43½	64 11½	36 52½
ω $\overline{4}21$	−57 16	67 11½	153 26½	60 06	71 20½	140 50½
x $\overline{4}61$	−28 41	76 30½	153 26½	31 27½	78 54	117 49½

Less common: d 110 s 011 h 061 i $\overline{2}01$ e 16.6.3
                α 025 f 083 n 101 v 221 β $\overline{4}23$
                l 023 g 031 t 401 w 423

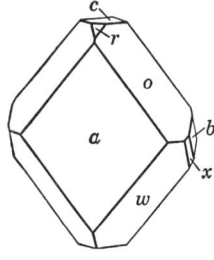

Stassfurt.

Structure cell.[15] Space group $C2/m$. a_0 19.05 Å, b_0 16.24, c_0 9.86; β 94°55′; $a_0:b_0:c_0 = 1.173:1:0.607$. Cell contents K$_{16}Mg_{16}$(SO$_4$)$_{16}Cl_{16}$·48H$_2$O.

Habit. Granular massive and interbedded with other salts. As crystalline coatings and clusters in vugs or cracks in which single crystals may have dimensions of several centimeters. Thick tabular {100} to equidimensional, often rich in forms.

P h y s. Cleavage {001} perfect.[3] Fracture smooth to splintery. Brittle. H. 2½–3. G. 2.15; 2.24 (calc.). Luster vitreous. Colorless; also, especially in massive types, gray, blue, violet,[5] yellowish, or reddish due to inclusions of foreign matter. Good crystals commonly are quite transparent. Taste salty and bitter.

O p t.[4] In transmitted light, colorless.

Orientation	n (New Mexico)	
X	1.494	Biaxial negative (−).
Y b	1.505	$2V \sim 90°$.
$Z \wedge c$ 13°	1.516	$r > v$, very weak.

Blue or violet material may be perceptibly pleochroic with X violet, Y blue, Z yellowish.[6]

C h e m. A hydrated sulfate and chloride of potassium and magnesium, KMg(SO$_4$)Cl·3H$_2$O. Analyses agree closely with the calculated composition. A small content of Br and I has been found.[7]

KAINITE

Anal.[8]

	K	MgO	SO$_4$	Cl	H$_2$O	Rem.	Total
1.	15.70	16.19	32.16	14.24	21.71		100.00
2.	13.30	16.72	32.50	13.51	20.95	2.77	99.75

1. KMgSO$_4$Cl·3H$_2$O. 2. Eddy County, New Mexico.[9] Rem. is Na 1.03, insol. 1.74.

Occur. Found only in potash deposits of oceanic origin. Mostly as a product of metamorphism attending deformation of the potash beds or due to the effects of meteoric waters which in places have produced a kainite capping.[10] First noted at Stassfurt and since found at many points in the middle and north German potash deposits, often as a principal constituent of large salt bodies with sylvite, halite, carnallite, kieserite, and other salt minerals. Besides the Stassfurt region where it has been mined in many places, it is also important in the Werra-Fulda region. Blue or violet crystals have been found in the potash mines at Asse near Wolfenbüttel and Hansa-Silberberg near Hannover. At Kalusz, Galicia, it occurs interbedded with other salts and also as crystals in druses in rock salt. At Holyn near Kalusz it is associated with schoenite. Kainite deposits have been found at Osinki in the region of Saratov on the lower Volga. In the United States it has been found with sylvite and langbeinite in the Permian potash beds in Eddy County, New Mexico, where it was first noted in drill cores.

Alter. Decomposed by water to epsomite and sylvite. On heating over sulfuric acid at 60° for 20 days it does not lose water.[11]

Artif. KMgSO$_4$Cl·3H$_2$O crystallizes from MgCl$_2$-rich solutions in the system K-Mg-SO$_4$-Cl-H$_2$O between 13°, at which temperature the solution is also in equilibrium with epsomite, carnallite, and sylvite, and 85° at which temperature the solution is also in equilibrium with langbeinite, kieserite, and sylvite.[12] The temperature of incongruent melting (85°) is greatly elevated by pressure.[13] From solutions containing Na in addition to the ions of the compound and usually considered closely equivalent to the brines from which oceanic salts have formed, kainite can crystallize over the range 12° to 83°.[14]

Name. From καινος, *recent*, alluding to its recent (secondary) formation.

Ref.

1. Angles of Groth, *Ann. Phys.*, **137**, 442 (1869); unit and orientation of the structure cell. Transformation: Groth to new, 002/010/½00.
2. Goldschmidt (**5**, 1, 1918).
3. Noted by Groth (1869) and Bücking, *Zs. Kr.*, **15**, 569 (1889). Groth also noted {110} distinct and {010} indistinct; Bücking did not observe {110} cleavage but reported {111} distinct and {1̄11} imperfect.
4. Indices of Schaller and Henderson, *U. S. Geol. Sur.*, *Bull. 833*, 38 (1932). Orientation of Groth (1869); Zepharovich (1882) found $Z \wedge c = 15°38'$.
5. Leonhardt and Kühn, *Cbl. Min.*, 193 (1935), found violet material to be decolorized at 150°.
6. Baumgärtel, *Cbl. Min.*, 449 (1905).
7. Boeke, *Zs. Kr.*, **45**, 388 (1908). Leonhardt and Kühn (1935) found H$_2$S in blue kainite.

8. For a tabulation of older analyses see Hintze (1 [3B], 4551, 1930).
9. Schaller and Henderson (1932).
10. For a discussion of the conditions of formation see Meyerhoffer, *Zs. anorg. Chem.*, **34**, 145 (1903), and Schlesinger, Zorkin, and Petuchova, *Ac. sc., Leningrad, C.r.*, **27**, 466 (1940). For the processes involved in its formation from the principal constituents of the German potash deposits, carnallite, kieserite, and halite, see especially Rózsa, *Cbl. Min.*, 505 (1916). Its occurrence in New Mexico suggests that it may have formed there by the action of solutions on langbeinite and sylvite, Schaller and Henderson (1932).
11. Schlesinger, Zorkin, and Petuchova (1940).
12. Data on system summarized in *Int. Crit. Tab.*, **4**, 284 (1928). See also Schlesinger, Zorkin, and Petuchova (1940), who crystallized kainite from a solution of carnallite, kieserite, and sylvite.
13. See Geller, *Zs. Kr.*, **60**, 414 (1924).
14. From the researches of van't Hoff and his co-workers, summarized in Doelter (4 [2], 93, 1926).
15. Evans, priv. comm. (1949), by precession method on Stassfurt crystals.

ANHYDROKAINITE. *Jänecke* (*Kali*, **7**, 140, 1913). Basaltkainit *Rózsa* (*Cbl. Min.*, 505, 1916).

$KMg(SO_4)Cl$ produced by the dehydration of kainite by the intrusion of basalt into salt-deposits in northern Germany.

31.4.2 **U N G E M A C H I T E** [$K_3Na_9Fe(SO_4)_6(OH)_3 \cdot 9H_2O$]. *Peacock* and *Bandy* (*Am. Min.*, **23**, 314, 1938).

C r y s t. Hexagonal—R; rhombohedral—$\bar{3}$.

$a{:}c = 1{:}2.2966;$ $\alpha\ 62°51\tfrac{1}{2}';$ $p_0{:}r_0 = 2.6519{:}1;$ $\lambda\ 108°15\tfrac{1}{2}'$

Forms: [1]

			ϕ	$\rho = C$	A_1	A_2
c	0001	111	.....	0°00'	90°00'	90°00'
a	11$\bar{2}$0	10$\bar{1}$	0°00'	90 00	60 00	60 00
g	10$\bar{1}$7	322	30 00	20 45	72 08	90 00
h	10$\bar{1}$4	211	30 00	33 32½	61 24½	90 00
i	20$\bar{2}$5	311	30 00	46 41½	50 56½	90 00
r	10$\bar{1}$1	100	30 00	69 20½	35 52	90 00
H	01$\bar{1}$5	221	−30 00	27 56½	90 00	66 03½
K	02$\bar{2}$7	331	−30 00	37 09	90 00	58 28
M	01$\bar{1}$2	110	−30 00	52 58½	90 00	46 15½
N	02$\bar{2}$1	11$\bar{1}$	−30 00	79 19½	90 00	31 40½
p	11$\bar{2}$3	210	0 00	56 51	65 15	65 15
P	2$\bar{1}\bar{1}$3	201	60 00	56 51	33 09	114 45
t	2$\bar{1}\bar{3}$1	20$\bar{1}$	10 53½	81 53	49 36	71 05½
U	$\bar{1}$3$\bar{2}$2	56$\bar{3}$	−49 06½	74 05½	108 21	19 12
V	12$\bar{3}$2	21$\bar{1}$	−10 53½	74 05½	71 39	50 59

Structure cell. Space group $R\bar{3}$. $a_0\ 10.84 \pm 0.02\ kX$, $c_0\ 24.82 \pm 0.05$; $a_0{:}c_0 = 1{:}2.290$. $a_{rh}\ 10.37$; $\alpha\ 62°59\tfrac{1}{2}'$. Cell contents $K_3Na_9Fe(SO_4)_6(OH)_3 \cdot 9H_2O$ in the rhombohedral unit.

Habit. Thick tabular {0001}. Common forms $r\ c\ M\ p\ h$.

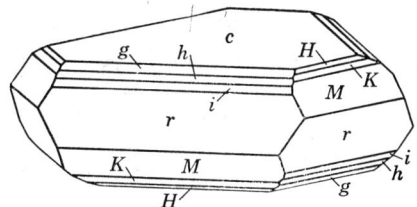

Phys. Cleavage {0001} perfect. Fracture irregular. Brittle. H. $2\frac{1}{2}$. G. 2.287; 2.29 (calc.). Luster vitreous. Colorless to pale yellow. Transparent.

Opt. In transmitted light, colorless.

	$n(\text{Na})$	
O	1.502 ± 0.002	Uniaxial negative $(-)$.
E	1.449 ± 0.002	

Chem. A hydrated basic sulfate of potassium, sodium, and ferric iron, $K_3Na_9Fe(SO_4)_6(OH)_3 \cdot 9H_2O$.[2]

Anal.

	1	2
K_2O	12.08	11.63
Na_2O	23.85	22.15
Fe_2O_3	6.83	7.88
SO_3	41.07	41.23
H_2O	16.17	17.11
Total	100.00	100.00
G.	2.29	2.287

1. $K_3Na_9Fe(SO_4)_6(OH)_3 \cdot 9H_2O$. 2. Chuquicamata.[3] Recalculated to 100 after deducting 2.07 insol. Has N_2O_5 tr.

Tests. Easily soluble in dilute HCl.

Occur. With sideronatrite and clino-ungemachite as vein fillings and in cavities in jarosite and metasideronatrite at Chuquicamata, Chile.

Name. After Henri-Léon Ungemach (1880–1936), Belgian crystallographer.

Ref.

1. Rare forms: $D\ 0.1.\bar{1}.11$, $m\ 10\bar{1}0$, $L\ 01\bar{1}3$, $j\ 40\bar{4}7$, $E\ 01\bar{1}8$, $q\ 21\bar{3}4$, $e\ 1.0.\bar{1}.10$, $F\ 0.2.\bar{2}.13$, $W\ 12\bar{3}5$. Accessory forms: $b\ 1.0.\bar{1}.13$, $d\ 2.0.\bar{2}.23$, $f\ 2.0.\bar{2}.17$, $k\ 4.0.\bar{4}.19$, $G\ 0.4.\bar{4}.23$, $I\ 0.4.\bar{4}.17$, $J\ 0.3.\bar{3}.12$, $s\ 31\bar{4}5$.
2. Frondel, priv. comm. (1947).
3. Gonyer analysis in Peacock and Bandy (1938).

31.4.3 Clino-ungemachite. *Peacock* and *Bandy* (*Am. Min.*, **23**, 314, 1938).

Cryst.[1] Monoclinic (pseudo-rhombohedral); prismatic—$2/m$.

$a:b:c = 1.6327:1:1.7308$; $\beta\ 110°40'$; $p_0:q_0:r_0 = 1.0601:1.6194:1$

$r_2:p_2:q_2 = 0.6175:0.6546:1$; $\mu\ 69°20'$; $p_0'\ 1.1330$, $q_0'\ 1.7308$, $x_0'\ 0.3772$

Forms:

		φ	ρ	φ2	ρ2 = B	C	A
c	001	90°00′	20°40′	69°20′	90°00′	.....	69°20′
a	100	90 00	90 00	0 00	90 00	69°20′	.....
m	110	33 12½	90 00	0 00	33 12½	78 53	56 47½
d	012	23 33	43 21	69 20	51 00	39 00	74 05
e	011	12 17½	60 33	69 20	31 41½	58 18½	79 19
f	101	90 00	56 29½	33 30½	90 00	35 49½	33 30½
g	$\bar{1}02$	−90 00	10 43	100 43	90 00	31 23	100 43
h	$\bar{1}01$	−90 00	37 05	127 05	90 00	57 45	127 05
p	111	41 06½	66 28½	33 30½	46 18	54 06½	52 55½
o	$\bar{1}11$	−23 35½	62 06	127 05	35 55	71 45½	110 42½

Less common and rare:
b 010 l 331 q $\bar{3}31$ s $\bar{1}31$ u $\bar{3}13$ w 533 y $\bar{3}11$
k 113 n $\bar{1}13$ r $\bar{3}15$ t $\overline{10}.1.10$ v $\bar{1}31$ x 211 z 811

Habit. Thick tabular {001}. Common forms: c a g p o. Crystals are pseudo-rhombohedral and indistinguishable in appearance from ungemachite.

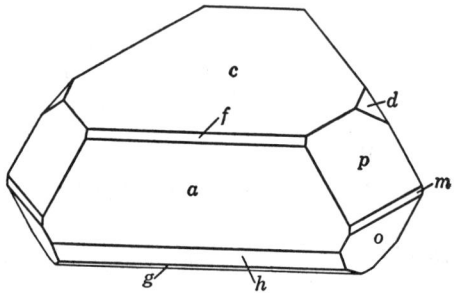

Occur. Clino-ungemachite was found intimately associated, in very small proportion, with ungemachite at Chuquicamata, Chile. The mineral was distinguished from the rhombohedral species only by goniometric measurements. Only six minute crystals were found. The physical and chemical properties are presumably close to those of ungemachite, but the available material does not permit further study.

Ref.
1. The cell selected is all-face-centered (on morphological evidence) and is preferred over the conventional base-centered cell in order to preserve the striking morphological resemblance to ungemachite. The symmetry class is not established beyond question.

31.4.4 ZIPPEITE [$(UO_2)_2(SO_4)(OH)_2 \cdot nH_2O$]. Uranblüthe ? *Zippe* (*Verhl. Ges. Mus. Böhm.*, 81, 1824). Basisches schwefelsaures Uranoxyd ? *John* (*Chem. Unters.*, **5**, 1821; *Jb. Min.*, 299, 1845). Zippeit *Haidinger* (510, 1845). Uranoker, pt. ? *Vogl* (*Min. Joachimsthal*, 124, 1857). Dauberite ? *Adam* (64, 1869).

Probably orthorhombic. As fragile crusts and reniform or spheroidal aggregates composed of microscopic crystals; also pulverulent to earthy. The crystals are flattened {010} and usually are spindle- or lens-shaped, sometimes rhomboidal or lath-like. Luster of the aggregates dull to silky. Probably a perfect cleavage on {010}. Color orange-yellow. The mineral closely resembles uranopilite in external characters. Fluoresces green.

O p t. Biaxial negative (−). The acute bisectrix, X, is perpendicular to the plates, {010}. Pleochroic, with X nearly colorless, Y pale yellow to orange-yellow, Z deep yellow to orange-yellow. The wide variation observed in the indices of refraction is unexplained.

Locality	Anal.	nX	nY	nZ	$Z \wedge c$	$2V$
Joachimsthal [1]	2	1.575	1.615	1.646		
Joachimsthal [1]	4	1.616	1.677	1.700		
Joachimsthal [2]		1.620	1.680	1.720	$\sim$32°	Large.
Joachimsthal [2]		1.630	1.70	1.720		Rather large.
Fruita, Utah [2]		1.630	1.689	1.739	41°	Large.
Joachimsthal [1]	3	1.636	1.694	1.732		
Gilpin County, Colorado [2]		1.660	1.710	1.760	$\sim$40°	

C h e m. A hydrated basic sulfate of uranium, probably $(UO_2)_2(SO_4)(OH)_2 \cdot 4H_2O$. The Ca reported in most analyses is believed due to admixed gypsum. It is not certain if all of the analyses cited actually refer to this species.

Anal.

	1	2	3	4	5	6	7
CaO		1.88	4.13	3.58		2.62	3.03
UO_3	77.08	71.98	73.47	74.76	70.94	66.05	58.48
Fe_2O_3		1.17			0.41	0.86	2.46
SO_3	10.79	10.02	10.19	10.15	7.12	10.17	10.22
H_2O	12.13	13.95	11.32	11.37	20.88	20.06	20.58
Rem.					0.24		4.02
Total	100.00	99.00	99.11	99.86	99.59	99.76	98.79

1. $(UO_2)_2(SO_4)(OH)_2 \cdot 4H_2O$. 2,3,4. Joachimsthal.[1] Microchemical. 5,6,7. Joachimsthal.[3] Rem. in no. 5 is CuO; in no. 7, PbO 2.21, MnO 0.35, SiO_2 1.46.

O c c u r. Zippeite is a secondary mineral, in part probably of recent formation, found associated with gypsum, uranopilite, and limonite on altering uraninite. Known from Joachimsthal, Bohemia. Minerals probably identical with zippeite have been reported [4] from Příbram and Schlaggenwald in Bohemia, from Cornwall, England, and in the United States from Gilpin County, Colorado, with johannite, and from Fruita, Wayne County, Utah,[5] in asphaltic sandstone. Also reported from Great Bear Lake, Canada.

A r t i f. Obtained as a crystalline precipitate by slowly adding very dilute NH_4OH to a solution of uranyl sulfate until a precipitate just begins to form and then allowing to stand.

N a m e. After Franz X. M. Zippe (1791–1863), Austrian mineralogist. The early synonymy of zippeite is uncertain.

Ref.
1. Nováček, *Soc. sc. Bohême, Mém.*, Cl. 2, no. 7 (1935).
2. Larsen (159, 1921).
3. Lindacker analysis in Vogl (1857).
4. See Nováček (1935).
5. Hess, *U. S. Geol. Sur., Bull. 750D* (1924).

URANOCHALCITE. Uranochalzit *Breithaupt* (173, 1841). Urangrün *Hartmann.*
VOGLIANITE. Basisch-schwefelsaures Uranoxidoxidul *Vogl* (*Min. Joachimsthal,* 1857). Voglianite *Dana* (668, 1868).
URACONITE. Uranocker pt. *Vogl* (*Min. Joachimsthal,* 1857). Uraconise? *Beudant* (**2**, 672, 1832). Uraconite *Dana* (668, 1868).
MEDJIDITE. *Smith* (*Am. J. Sc.,* **5**, 337, 1848).

The above-mentioned names have been proposed for ill-defined substances supposedly uranium sulfates. Color yellow to green, also brown (*medjidite*). Found as nodular or botryoidal crusts and earthy to scaly or acicular coatings at Joachimsthal, Bohemia; also massive from near Adrianople, Turkey (*medjidite*). The reported chemical analyses [1] probably are not reliable. Specimens reputedly of these minerals have proved on examination [2] to be cuprosklodowskite, liebigite, or other known species. An unidentified uranium sulfate (?) from Gilpin County, Colorado,[3] consists of a lemon-yellow powder composed of minute fibers and laths. Probably orthorhombic. Biaxial positive (+), with nX 1.75, nY 1.79, nZ 1.85; $2V$ medium; $r < v$, strong. Z = elongation, and X is perpendicular to the flattening. Other uranium sulfates loosely termed uraconite have been reported from the Rathgeb mine near San Andreas, Calaveras County, California, and from Pennsylvania at Avondale, Crozer's quarry at Chester and Leiperville in Delaware County, and Fairmount Park in Philadelphia. Also from Canada at Snowdon, Haliburton County, and Madoc, Hastings County, in Ontario. The cited names can hardly be applied to new occurrences of uranium sulfates for lack of a definitive redescription of the original materials.

Ref.

1. Summarized by Dana (978, 1892) and Nováček, *Soc. sc. Bohême, Mém.,* Cl. 2, no. 7 (1935).
2. Nováček (1935) and Frondel, priv. comm. (1948).
3. Larsen (149, 1921); called uraconite (?).

31.4.5 A L U M I N I T E [$Al_2(SO_4)(OH)_4 \cdot 7H_2O$]. Reine Thonerde (from Halle) *Werner* (176, 1780). Native Argill *Kirwan* (**1**, 175, 1784). Aluminit *Haberle* (*Das Mineralreich,* etc., 1807; Karsten, 48, 1808). Hallite *Delamétherie* (**2**, 1812). Websterite *Brongniart* (in Haüy, **2**, 125, 1822). Hydrosulfate d'alumine, Websterite *Beudant* (449, 1824).

As reniform or nodular masses composed of tiny needles or fibers; also spherulitic; as veinlets. The needles have been variously described [1] from microscopical study as having oblique or orthogonal terminations, and the optical extinction as inclined or parallel. The mineral may be monoclinic, perhaps elongated [010], or orthorhombic, but it is not certain if all the observations have been made on the same substance.

Fracture of the aggregates, earthy. Friable. H. 1–2. G. 1.66–1.82. Luster dull, earthy. Color white. Opaque.

O p t.[2] In transmitted light, colorless.

ORIENTATION	n (Newhaven)	n (Green River)	
X elong.	1.459	1.460	Biaxial positive (+).
Y	1.464		
Z	1.470	1.470	
$2V$	~90°	Large	

An apparent dehydration product of aluminite is biaxial positive (+) with large $2V$ and a large extinction angle; nY 1.50 to 1.52.

C h e m. A hydrated basic aluminum sulfate, $Al_2(SO_4)(OH)_4 \cdot 7H_2O$.

Anal.[3]

	1	2	3	4
Al_2O_3	29.62	30.08	31.34	30.54
SO_3	23.27	23.63	22.63	21.20
H_2O	47.11	46.44	45.02	46.53
Rem.			0.60	1.89
Total	100.00	100.15	99.59	100.16

1. $Al_2(SO_4)(OH)_4 \cdot 7H_2O$. 2. Punjab Salt Range.[4] 3. Mt. Sokolova, Saratov, U.S.S.R.[5] Rem. is CaO 0.34, insol. 0.26. 4. Gaal, Saxony.[7] Rem. is CaO 0.86, MgO 0.05, SiO_2 0.98.

Tests. In C.T. affords much water which at higher temperatures becomes acid. B.B. infusible. Soluble in acids.

Occur. As a concretionary deposit in Recent or Tertiary clays, marls, or lignites and formed by the action of sulfate solutions derived by the alteration of marcasite or pyrite on aluminous silicates. Long known from the neighborhood of Halle, Saxony. In Bohemia at Mühlhausen near Kralup and at Miletic and Velvary. In France at Épernay, Marne, and at Auteuil near Paris. With halite and sylvite on Vesuvius. In volcanic tuff near Zaglik, Azerbaijan, and from Mt. Sokolova near Saratov on the Volga, U.S.S.R. In chalk at New Haven, Sussex, England (*websterite*). In the Salt Range, Punjab, India. In the United States as a coating on limestone at Joplin, Missouri, and at the Green River crossing of the Denver and Rio Grande Railroad, Utah.

Artif.[6] Numerous basic aluminum sulfates have been reported, but their relation to the natural substances is uncertain.

Ref.

1. Fischer, *Zs. Kr.*, **4**, 375 (1880); Zahálka, *Böhm. Ges., Sitzber.*, no. 23 (1911); Larsen (**38**, 1921); Lacroix (**4**, 243, 1910); Carobbi, *Per. Min.*, **3**, 204 (1932).
2. Larsen (1921).
3. Earlier analyses are listed by Hintze (**1** [3B], 4432, 1929) and Doelter (**4** [2], 387, 1927).
4. Oldham, *Rec. Geol. Sur. India*, **30**, 110 (1897).
5. Pilipenko, *Wiss. Schr. Univ. Saratow*, **6**, 167 (1927)—*Jb. Min.*, I, 298 (1928).
6. Mellor (**5**, 336, 1924).
7. Kashkai, *C. r. ac. sc. U.R.S.S.*, **24**, 931 (1939).

LAPPARENTITE. Rost (*Intern. ac. sc. Bohême, Bull.*, preprint, 7 pp., 1937).

A mineral thought to be identical with the original lapparentite (= *tamarugite*) but apparently a distinct species. Orthorhombic? As chalky white masses. G. 1.892. Crystals are rhomboidal tablets with internal angles of 82° and 98°. Optically biaxial positive (+) with symmetrical extinction; Z is perpendicular to the flattening, and Y bisects the acute angle of the tablets. nX 1.461, nY 1.470, nZ 1.484. Composition close to $Al_2(SO_4)_2(OH)_2 \cdot 9H_2O$ from the analysis: Na_2O 1.09, MgO 0.52, Al_2O_3 22.07, SO_3 35.51, H_2O 40.69, total 99.88. Found with ammonia alum and copiapite on burning waste-heaps of the Schoeller coal mine at Libušín in the Kladno district, Bohemia. A sulfate with similar optical properties occurs at Wackerdorf, Germany.

WERTHEMANITE. Raimondi (*Min. Pérou*, Paris, 244, 1878).

As white, pulverulent masses. G. 2.80. Gives an argillaceous odor and adheres to the tongue. Analysis gave Fe_2O_3 1.25, Al_2O_3 45.00, SO_3 34.50, H_2O 19.25, total 100.00. This is close to $Al_2(SO_4)(OH)_4 \cdot H_2O$. B.B. infusible; slowly soluble in warm sulfuric acid. Found in a bed of clay near Chachapoyas, Santa Lucia, Peru. Needs verification.

TYPE 5. $A_3(XO_4)_2Z_q \cdot xH_2O$

31.5.1 NATROCHALCITE [$NaCu_2(SO_4)_2(OH) \cdot H_2O$]. Palache and Warren (Am. J. Sc., **26**, 342, 1908; Zs. Kr., **45**, 534, 1908).

Cryst.[1] Monoclinic; prismatic—$2/m$.

$a:b:c = 1.4239:1:1.2140$; $\beta\ 118°42\frac{1}{2}'$; $p_0:q_0:r_0 = 0.8526:1.0648:1$

$r_2:p_2:q_2 = 0.9392:0.8007:1$; $\mu\ 61°17\frac{1}{2}'$; $p_0'\ 0.9721$, $q_0'\ 1.2140$, $x_0'\ 0.5477$

Forms:[2]

	ϕ	ρ	ϕ_2	$\rho_2 = B$	C	A
c 001	90°00'	28°42½'	61°17½'	90°00'	.....	61°17½'
b 010	0 00	90 00	.....	0 00	90°00'	90 00
a 100	90 00	90 00	0 00	90 00	61 17½	.....
m 110	38 41	90 00	0 00	38 41	72 31½	51 19
v 112	59 35	50 10	44 03	67 07	28 22	48 32
p 111	51 23	62 47½	33 20½	56 17½	42 42½	45 59
u 221	45 45	73 58	21 52	47 53	55 02½	46 30
w 331	43 34	78 45	16 06	44 42½	60 16½	47 28½
q $\bar{1}11$	−19 16	52 08	112 59½	41 49½	65 35½	105 06
x $\bar{2}21$	−29 54½	70 21	144 24	35 17	86 01½	118 00½

Less common and rare: $k\ \bar{2}01 \quad z\ \bar{7}12$

Structure cell.[3] Space group $C2/m$. $a_0\ 8.74\ kX$, $b_0\ 6.15$, $c_0\ 6.53$; $\beta\ 118°42\frac{1}{2}'$; $a_0:b_0:c_0 = 1.422:1:1.211$. Cell contents $Na_2Cu_4(SO_4)_4(OH)_2 \cdot 2H_2O$.

Habit. Pyramidal {111}; {110} is usually well-developed, but the faces of the other forms are small and often not present in their full complement. Also as cross-fiber veinlets.

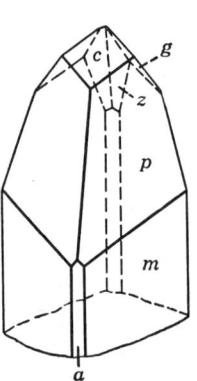

Chuquicamata.　　　　Chuquicamata.

Phys. Cleavage {001} perfect. H. 4½. G. 3.49 ± 0.02;[4] 3.54 (calc.). Color bright emerald-green. Luster vitreous. Streak greenish white. Transparent.

METASIDERONATRITE

Opt. In transmitted light, green.

ORIENTATION		n(Na) [5]	
$X \wedge c$	$-12°$	1.649	Biaxial positive (+).
Y	b	1.655	$2V\ 36°48'$ (Na, meas.).
Z		1.714	$r < v$, strong; dispersion of bisectrices slight, inclined.

Chem.[6] A hydrated basic sulfate of sodium and copper, $NaCu_2(SO_4)_2(OH) \cdot H_2O$.

Anal.

	1	2	3
Na_2O	8.22	8.44	7.98
CuO	42.18	41.95	42.01
SO_3	42.44	42.10	42.42
H_2O	7.16	7.70	7.71
Rem.		0.75	
Total	100.00	100.94	100.12
G.	3.54		3.49

1. $NaCu_2(SO_4)_2(OH) \cdot H_2O$. 2. Chuquicamata.[7] Rem. is Cl 0.05 (due to admixed atacamite), insol. 0.70 per cent. 3. Chuquicamata.[8]

Tests. B.B. decrepitates and fuses easily (about 1) to a black bead. In C.T. fuses to a dark enamel and affords acid water. Slowly soluble in water, easily soluble in acids.

Occur. Found only at Chuquicamata near Calama, Antofagasta province, Chile, associated with kroehnkite, antlerite, bloedite, atacamite, brochantite, chalcanthite, and gypsum.

Name. In allusion to the composition.

Ref.

1. Palache and Warren (1908).
2. Palache and Warren (1908), Palache, *Am. J. Sc.*, **237**, 451 (1939).
3. Richmond in Palache (1939) by Weissenberg method.
4. Palache (1939), correcting the earlier reported value of 2.83.
5. Merwin in Palache and Warren (1908).
6. The formula was erroneously given in the original paper as $Na_2Cu_4(SO_4)_3(OH)_2 \cdot 2H_2O$.
7. Warren analyst in Palache and Warren (1908).
8. Gonyer analyst in Palache (1939).

31.5.2 **METASIDERONATRITE** [$Na_4Fe_2(SO_4)_4(OH)_2 \cdot 3H_2O$]. *Bandy* (*Am. Min.*, **23**, 733, 1938).

Cryst. Orthorhombic; dipyramidal—$2/m\ 2/m\ 2/m$.

$a:b:c = 0.4571:1:0.1187;$ $\quad p_0:q_0:r_0 = 0.2579:0.1187:1$

$q_1:r_1:p_1 = 0.4571:3.8598:1;$ $\quad r_2:p_2:q_2 = 8.4246:2.1877:1$

Forms:

	ϕ	$\rho = C$	ϕ_1	$\rho_1 = A$	ϕ_2	$\rho_2 = B$
b 010	0°00'	90°00'	90°00'	90°00'	.....	0°00'
m 110	65 26	90 00	90 00	24 34	0°00'	65 26
e 011	0 00	6 46	6 46	90 00	90 00	83 14

Habit. Prismatic [001] crystals rare; usually in coarse to fine crystalline aggregates.

Phys. Cleavage {100} and {010} perfect, {001} eminent. Fracture fibrous. H. $2\frac{1}{2}$. G. 2.46. Luster silky. Color golden yellow to straw-yellow.

Opt. In transmitted light, yellow in color.

Orientation		n	Pleochroism	
X	a	1.543	Colorless	Biaxial positive (+).
Y	b	1.575	Light yellow	$2V$ 60°.
Z	c	1.634	Brownish yellow	$r > v$, strong.

Chem. A basic hydrated sulfate of sodium and ferric iron, $Na_4Fe_2(SO_4)_4(OH)_2 \cdot 3H_2O$.

Anal.

	Na_2O	Fe_2O_3	SO_3	H_2O	Rem.	Total	G.
1.	18.34	23.62	47.38	10.66		100.00	
2.	17.56	22.90	48.66	9.75	0.86	99.73	2.46

1. $Na_4Fe_2(SO_4)_4(OH)_2 \cdot 3H_2O$. 2. Chuquicamata.[1] Rem. is K_2O 0.26, insol. 0.60.

Tests. Insoluble in cold water; soluble in boiling water with decomposition. Soluble in dilute acids. In C.T. yields abundant acid water.

Occur. At Chuquicamata, Chile, intimately associated with metavoltine, less commonly with ferrinatrite, alums and other secondary sulfates.

Artif. Metasideronatrite can be produced from sideronatrite by dehydration over sulfuric acid [2] and may be converted to ferrinatrite by treatment with a solution of sodium sulfate and sulfuric acid.

Name. In allusion to the similarity to sideronatrite, from which metasideronatrite differs in composition only in having a lower water content. There is, however, no evidence that metasideronatrite has formed in nature from sideronatrite.

BARTHOLOMITE. *Cleve (Ak. Stockholm, Handl.,* **9** [12], 31, 1870). A yellow hydrated basic sulfate of sodium and ferric acid found as nodules composed of small needles as an alteration of pyrite on the island of St. Bartholomew, West Indies. Analysis of impure material gave: Na_2O 17.08, MgO 0.63, NaCl 2.88, Fe_2O_3 22.71, SO_3 44.75, H_2O 8.08, insol. 3.56, total 99.69. Probably identical with metasideronatrite or sideronatrite.

Ref.
1. Henderson analysis in Bandy (1938).
2. Bandy (1938). See also Scharizer, *Zs. Kr.,* **41**, 223 (1905), who studied the dehydration of artificial sideronatrite.

31.5.3 SIDERONATRITE [$Na_2Fe(SO_4)_2(OH) \cdot 3H_2O$]. *Raimondi* (233, 1878). Urusite *Frenzel (Min. Mitt.,* **2**, 133, 1880).

Cryst. Orthorhombic (?).[1]

Habit. Minute needles elongated [001].[2] As nodular masses or fibrous crusts, also earthy to pulverulent.

Phys. Cleavage {100} perfect.[2] H. $1\frac{1}{2}$–$2\frac{1}{2}$. G. 2.15–2.35. Color lemon-yellow, pale orange to straw-yellow, yellow brown. Streak pale yellow to yellowish white.

O p t.[2] In transmitted light, pale yellow to nearly colorless.

ORIENTATION		n	PLEOCHROISM	
X	a	1.508 ± 0.003	Nearly colorless	Biaxial positive (+).
Y	b	1.525 ± 0.003	Very pale amber-yellow	$2V\ 58° \pm 5°$ (calc.).
Z	c	1.586 ± 0.003	Pale amber-yellow	$r > v$, strong.

C h e m. A hydrated basic sulfate of sodium and ferric iron, $Na_2Fe(SO_4)_2(OH)\cdot 3H_2O$.

Anal.

	Na_2O	Fe_2O_3	SO_3	H_2O	Rem.	Total	G.
1.	16.99	21.87	43.87	17.27		100.00	
2.	15.59	21.60	43.26	15.35	4.26	100.06	2.153
3.	16.50	21.28	42.08	19.80		99.66	2.22
4.	16.28	22.58	45.15	16.03		100.04	2.31
5.	16.39	21.77	44.22	17.07		99.45	2.355

1. $Na_2Fe(SO_4)_2(OH)\cdot 3H_2O$. 2. San Simon mine, Huantajaya, Chile.[3] Rem. is NaCl 1.06, insol. 3.20. 3. Island of Cheleken, Caspian Sea.[4] (*urusite*). 4. Sierra Gorda, Chile.[5] 5. Mina de la Compania, near Sierra Gorda, Chile.[6]

Tests. Insoluble in cold water, decomposed by boiling water with separation of ferric oxide.[7]

O c c u r. Found with other secondary sulfates in very arid regions. First described from the San Simon mine, Huantajaya, province of Tarapacá, Chile. With ferrinatrite at the Mina de la Compania near Sierra Gorda, Province of Tocapilla, Chile. With voltaite at Potosi, Bolivia. Under a foot-thick layer of melanterite on the Urus plateau, Cheleken Island, Caspian Sea.

A l t e r. Changes to ferrinatrite by contact with concentrated sulfuric acid,[8] and to metasideronatrite by partial dehydration. On heating, all but 2.36 per cent of water is lost at 125°; the remainder is held to relatively high temperature.[9]

A r t i f. Can be crystallized from a solution of ferric sulfate and sodium sulfate.[10]

N a m e. From σίδηρος, *iron*, and *natrium*, *sodium*, from the bases which it contains.

Ref.

1. Crystal system inferred from optical properties. No crystal measurements have been reported—see Frenzel (1880).
2. Larsen (134, 1921) on material from Sierra Gorda, Chile.
3. Raimondi (1878).
4. Frenzel (1880).
5. Frenzel, *Min. Mitt.*, **11**, 214 (1890).
6. Genth and Penfield, *Am. J. Sc.*, **40**, 201 (1890).
7. According to both Raimondi (1878) and Frenzel (1880). Genth and Penfield (1890) report that it is decomposed by cold water into an insoluble basic ferric sulfate.
8. Scharizer, *Zs. Kr.*, **41**, 215 (1906).
9. Scharizer (1906). This behavior is in harmony with formula here used.
10. Scharizer (1906). See also Mellor, **14**, 345 (1935).

31.5.4 JOHANNITE [$Cu(UO_2)_2(SO_4)_2(OH)_2 \cdot 6H_2O$]. Uranvitriol *John (Chem. Unters.*, **5**, 254, 1821). Johannit *Haidinger (Edinburgh J. Sc.*, **4**, 306, 1830; *Abhl. Böhm. Ges. Prag.*, 1830). Sulfate vert d'urane *Beudant*. Gilpinite *Larsen* and *Brown (Am. Min.*, **2**, 75, 1917).

Cryst.[1] Triclinic; pinacoidal—$\bar{1}$.

$a:b:c = 0.9182:1:0.3799; \quad \alpha\ 90°54\frac{1}{2}',\ \beta\ 90°38',\ \gamma\ 110°37'$

$p_0:q_0:r_0 = 0.4419:0.4059:1; \quad \lambda\ 88°48',\ \mu\ 88°59',\ \nu\ 69°22'$

$p_0'\ 0.4420,\ q_0'\ 0.4060; \quad x_0'\ 0.0110,\ y_0'\ 0.0210$

Forms:[2]

		φ	ρ	A	B	C
b	010	0°00'	90°00'	69°22'	.....	88°48'
a	100	69 22	90 00	.....	69°22'	88 59
m	110	36 21½	90 00	33 00½	36 21½	88 39½
N	1$\bar{2}$0	147 46½	90 00	78 24½	147 46½	90 41
M	1$\bar{1}$0	121 10½	90 00	51 48½	121 10½	90 05
p	111	36 04½	35 47½	60 44	69 51½	34 27
Q	1$\bar{1}$1	118 21½	25 45½	73 26	101 55	25 48½
q	$\bar{1}$11	−56 02	25 53½	104 39	75 52½	25 46½
P	$\bar{1}\bar{1}$1	−143 20	33 59½	118 03½	116 38½	35 20
R	$\bar{3}$11	−124 43½	56 14½	143 45	118 16	56 27

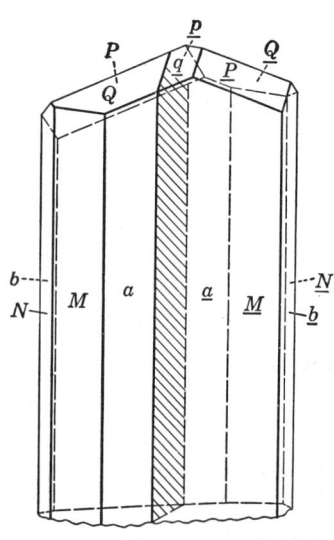

Joachimsthal. Twin on [001].

Structure cell.[6] Dimensions of the F-centered cell corresponding to the pseudo-monoclinic morphological unit: a_0 16.51 kX, b_0 17.98, c_0 6.83 [α 90°54½', β 90°38', γ 110°37']. Cell contents $Cu_4(UO_2)_8(SO_4)_8(OH)_8 \cdot 24H_2O$.

Habit. Prismatic [001] and thick tabular {100}. Common forms: $b\ a\ M\ N\ P\ Q$. As subparallel or drusy aggregates; also as coatings and small spheroidal aggregates of lath-like fibers.

Twinning. Twin axis [001], composition plane {010} or near thereto; both simple and lamellar, common.

Phys. Cleavage {100} good. Not brittle. H. 2–2½. G. 3.32; 3.27 (calc.). Luster vitreous. Color emerald-green to apple-green. Streak paler. Transparent to translucent. Taste bitter.

O p t. Strongly pleochroic, with X colorless, Y pale yellow, Z greenish yellow or canary-yellow. Biaxial.

	ORIENTATION [2]		n (Joachimsthal) [3]	n (Colorado) [3]
	ϕ	ρ		
X	$-101°$	$85°$	1.572	1.577
Y	37	8	1.595	1.597
Z	169	85	1.614	1.616
$2V$ (Na)			$\sim 90°$	$\sim 90°$
Sign			$-$	$+$
Dispersion			$r > v$, strong	$r < v$, strong

C h e m. A hydrated basic sulfate of copper and uranium, $Cu(UO_2)_2(SO_4)_2(OH)_2 \cdot 6H_2O$. Analysis gave: [4]

	CuO	UO_3	SO_3	H_2O	Total
1.	8.48	61.00	17.07	13.45	100.00
2.	8.07	61.34	16.59	13.84	99.84

1. $Cu(UO_2)_2(SO_4)_2(OH)_2 \cdot 6H_2O$. 2. Joachimsthal.[5] Microchemical on 35 mg. No H_2O-, H_2O+ at 210°.

Tests. In C.T. highly heated gives off water and SO_2 and turns brown and finally black. Decomposed by water, soluble in acids.

O c c u r. A secondary mineral, found at Joachimsthal and Johanngeorgenstadt, Bohemia. From Cornwall, England. Also from the Central City district, Gilpin County, Colorado (*gilpinite*). Formed by the alteration of uraninite and associated with gypsum.

N a m e. After the Archduke Johann (1782–1859) of Austria, founder of the Styrian Landesmuseum in Graz.

Ref.

1. Elements and forms of Peacock, *Zs. Kr.*, **90**, 112 (1935), on crystals from Joachimsthal, who showed the symmetry was triclinic and not monoclinic as earlier given by Ježek, *Bull. int. ac. sc. Prague*, **70**, 358 (1916), and Haidinger, *Ann. Phys.*, **20**, 472 (1830). Transformation: Ježek to Peacock, $\overline{2}01/20\overline{1}/010$; Goldschmidt (**4**, 216, 1918) to P., $\overline{1}01/10\overline{1}/0\overline{1}0$. Uncertain form: $\{5\overline{3}0\}$.
2. Peacock (1935).
3. Larsen and Berman, *Am. Min.*, **11**, 1 (1926), who give additional data.
4. Earlier, divergent analyses on small samples are cited in Peacock (1935) and Doelter (**4** [2], 648, 1927).
5. Nováček analysis in Peacock (1935), also in Nováček, *Soc. sc. Bohême, Mém.*, no. 7, Cl. 2 (1935) who discusses the earlier analyses.
6. Hurlbut, *Am. Min.*, **35**, 531 (1950), by the Weissenberg method on the crystals studied by Peacock (1935). Transformation: from the face-centered to primitive cell, $\frac{1}{2}0\overline{1}/0\frac{1}{2}\frac{1}{2}/001$. Elements of the P-cell: $a:b:c = 0.9382:1:0.7143$, $\alpha\ 108°50'$, $\beta\ 112°03'$, $\gamma\ 64°52'$. The morphological cell angles are employed.

31.5.5 Vernadskite $[Cu_4(SO_4)_3(OH)_2 \cdot 4H_2O]$. Vernadskijte, Wernadskyite Zambonini (337, 1910).

Aggregates of minute birefringent crystals. H. $3\frac{1}{2}$. G. > 3.3. Color green. Soluble in HCl. Formula $Cu_4(SO_4)_3(OH)_2 \cdot 4H_2O$ from the analysis: CuO 49.15, SO_3 37.01, H_2O [13.84], total [100.00]. Found as an alteration product of dolerophanite on Vesuvius. Named after the Russian geochemist Vladimir Ivanovich Vernadsky (1863–1945).

TYPE 6. $A(XO_4)Z_q \cdot xH_2O$

31.6.1 Metahohmannite [$Fe_2(SO_4)_2(OH)_2 \cdot 3H_2O$]. Bandy (Am. Min., **23**, 748, 1938).

Pulverulent masses formed by the partial dehydration of hohmannite. Color orange. Biaxial positive (+), with nX 1.709 (pale yellow), nY 1.718 (reddish yellow), nZ 1.734 (reddish brown) [Chuquicamata]. Composition $Fe_2(SO_4)_2(OH)_2 \cdot 3H_2O$ from the analyses:

	Fe_2O_3	SO_3	H_2O	Total
1.	40.75	40.86	18.39	100.00
2.	39.25	39.61	20.29	99.15
3.	41.22	41.24	[17.54]	[100.00]

1. $Fe_2(SO_4)_2(OH)_2 \cdot 3H_2O$. 2. Alcaparrosa.[1] Original total given as 99.25. 3. Chile.[2]

Found as an alteration product of hohmannite at Chuquicamata, Quetena, and Alcaparrosa, Chile. The alteration takes place very rapidly in dry air at room temperature.

Ref.
1. Ungemach, Bull. soc. min., **58**, 97 (1935).
2. Mackintosh, Am. J. Sc., **38**, 243 (1889). The identity of this material with metahohmannite is not entirely certain.

31.6.2 BUTLERITE [$Fe(SO_4)(OH) \cdot 2H_2O$]. Lausen (Am. Min., **13**, 211, 1928).

Cryst.[1] Monoclinic (?), with orthorhombic pseudo-symmetry.

$a:b:c = 0.8752:1:0.7897;$ $\beta\ 108°35';$ $p_0:q_0:r_0 = 0.9023:0.7485:1$

$r_2:p_2:q_2 = 1.3360:1.2054:1;$ $\mu\ 71°25';$ $p_0'\ 0.9519,\ q_0'\ 0.7897,\ x_0'\ 0.3362$

Forms:[2]

		ϕ	ρ	ϕ_2	$\rho_2 = B$	C	A
c	001	90°00′	18°35′	71°25′	90°00′	.....	71°25′
a	100	90 00	90 00	0 00	90 00	71°25′	.....
m	110	50 19	90 00	0 00	50 19	75 48	39 41
d	011	23 03½	40 38½	71 25	53 11	36 49	75 13
h	$\bar{1}$01	−90 00	31 37	121 37	90 00	50 12	121 37
u	$\bar{1}$11	−37 56½	45 02½	121 37	56 05	57 55	115 47½
v	$\bar{1}$21	−21 18	59 28	121 37	36 38	67 33	108 14

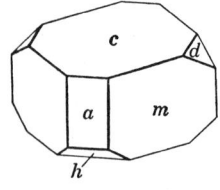

Argentina.

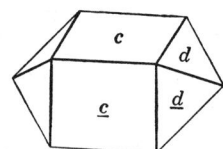

Argentina. Twin on {$\bar{1}$05}.

Habit. Usually tabular {001} or {100} with a tendency for elongation [010]; also octahedral (Arizona).

BUTLERITE 609

Twinning.[3] On $\{\bar{1}05\}$, very common.
Oriented growths. With parabutlerite (which see).
P h y s. Cleavage $\{100\}$ perfect.[4] H. $2\frac{1}{2}$. G. 2.55. Luster vitreous. Color deep orange. Streak pale yellow.
O p t. In transmitted light, pale yellow-orange.

ORIENTATION[5]	n (Argentina)[5]	n (Arizona)[6]	n (artificial)[7]	PLEOCHROISM[8]	
$X \wedge c$ $-18°$	1.593	1.604	1.588	Colorless	Biaxial. $2V$
Y	1.665	1.674	1.678	Faint yellow	large.
Z b	1.741	1.731	1.749	Light yellow	
Opt.	+	−	−		

C h e m. A hydrated basic ferric sulfate, $Fe(SO_4)(OH) \cdot 2H_2O$. Dimorphous with parabutlerite. Analysis gave:

	Na_2O	FeO	Fe_2O_3	Al_2O_3	SO_3	H_2O	Total
1.			38.96		39.06	21.98	100.00
2.	2.73	0.41	36.31	0.55	38.63	22.83	101.46

1. $Fe(SO_4)(OH) \cdot 2H_2O$. 2. Arizona.[9] The Na_2O and FeO are probably due to admixture.

O c c u r. Originally from the United Verde copper mine at Jerome, Arizona, where the mineral formed under fumarolic conditions with copiapite and other ferric sulfates in burning pyritic ore. At the Santa Elena mine, La Alcaparrosa, Barreal Dept., Argentina, with fibroferrite, zincian melanterite, parabutlerite, and other iron sulfates. Reported from Chuquicamata, Chile.

A r t i f.[7] Stable below 100° in the system $Fe_2O_3\text{-}SO_3\text{-}H_2O$.
N a m e. After Gurdon Montague Butler (1881–), mining geologist, of the University of Arizona.

Ref.

1. The true symmetry of butlerite is uncertain. The type crystals from Arizona were described as orthorhombic by Lausen (1928). The Argentina crystals were made out as monoclinic by Gordon, *Ac. Sc. Philadelphia, Not. Nat.*, no. 89 (1941), in accordance with the observations of Posnjak and Merwin, *J. Am. Chem. Soc.*, **44**, 1978 (1912), on artificial material apparently isostructural with butlerite rather than parabutlerite. Bandy, *Am. Min.*, **23**, 742 (1938), has, however, observed in Arizona crystals a twin plane in $\{100\}$ of Lausen's orientation, with the optical orientation in each segment such that the most probable symmetry is triclinic. Lausen's material may be an inversion pseudomorph, and there is some uncertainty as to the mutual identity of the several substances described.
2. Elements and forms of Gordon (1941) on Argentina crystals.
3. Gordon (1941). See also Bandy (1938).
4. Gordon (1941). A second, imperfect cleavage of unknown orientation is mentioned by Lausen (1928).
5. Gordon (1941) on Argentina material.
6. Lausen (1928).
7. Posnjak and Merwin (1912).
8. Gordon (1941) and Posnjak and Merwin (1912); Lausen (1928) gives X pale brownish yellow and Z pale canary-yellow.
9. Buehrer analysis in Lausen (1928).

31.6.3 PARABUTLERITE [Fe(SO$_4$)(OH)·2H$_2$O]. *Bandy (Am. Min.,* **23,** 742, 1938).

Cryst.[1] Orthorhombic; dipyramidal—$2/m\,2/m\,2/m$.

$a:b:c = 0.7310:1:0.7218;\qquad p_0:q_0:r_0 = 0.9874:0.7218:1$

$q_1:r_1:p_1 = 0.7310:1.0128:1;\qquad r_2:p_2:q_2 = 1.3855:1.3680:1$

Forms:[2]

		ϕ	$\rho = C$	ϕ_1	$\rho_1 = A$	ϕ_2	$\rho_2 = B$
l	140	18°53′	90°00′	90°00′	71°07′	0°00′	18°53′
m	110	53 50	90 00	90 00	36 10	0 00	53 50
n	540	59 41	90 00	90 00	30 19	0 00	59 41
s	012	0 00	19 50½	19 50½	90 00	90 00	70 09½
f	034	0 00	28 25½	28 25½	90 00	90 00	61 34½
e	011	0 00	35 49	35 49	90 00	90 00	54 11
g	032	0 00	47 16½	47 16½	90 00	90 00	42 43½
d	101	90 00	44 38	0 00	45 22	45 22	90 00
p	111	53 50	50 44	35 49	51 19	45 22	62 49
r	212	69 55½	46 26	19 50½	47 07	45 22	75 36

Less common and rare:
    o 021 t 205 j 155 i 4.5.10 k 255 x 414 z 232

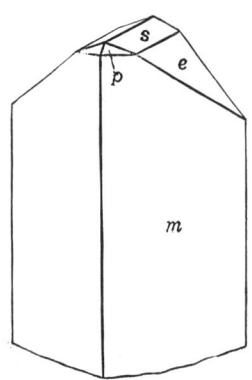

Habit. Prismatic [001]. Common forms: $m\,e\,r\,s\,p$. Prism zone striated [001].

Twinning. On {142}, rare.[3]

Oriented growths. Intergrowths with butlerite have been described,[3] with {011} of parabutlerite nearly parallel to {001} and {100} of butlerite.

Phys. Cleavage {110} poor. Fracture conchoidal. Brittle. H. 2½. G. 2.55. Luster vitreous. Color light orange to light orange-brown. Transparent.

Opt. In transmitted light, pale yellow in color.

ORIEN-TATION		n (Na; Alcaparrosa, Chile)[4]		n (Argentina)[3]		
X	b	1.598	Pale yellow	1.589	Colorless	Biaxial positive (+).
Y	c	1.663	Greenish yellow	1.660	Faint yellow	$r > v$, moderate.
Z	a	1.737	Brownish yellow	1.750	Light yellow	
2V (meas.)		87°		43½°		

Chem. A hydrated basic sulfate of ferric iron, Fe(SO$_4$)(OH)·2H$_2$O. Dimorphous with butlerite.

Anal.

	1	2
Fe$_2$O$_3$	38.96	39.21
SO$_3$	39.06	39.15
H$_2$O	21.98	22.00
Insol.		0.23
Total	100.00	100.59

1. Fe(SO$_4$)(OH)·2H$_2$O. 2. Alcaparrosa, Chile.[5]

AMARANTITE

Tests. In C.T. yields acid water. Insoluble in cold or hot water; soluble in dilute acids.

Occur. Found at the mine Alcaparrosa near Cerritos Bayos, Antofagasta province, Chile, as a bed about 8 inches thick above copiapite and formed by the alteration of that species; also found at Quetena and at Chuquicamata in Antofagasta province with copiapite and jarosite. At the Santa Elena mine, La Alcaparrosa, San Juan province, Argentina, with butlerite, zincian melanterite, fibroferrite, and other sulfates in the oxidized zone of pyritic veins.

Alter. Observed as pseudomorphs after copiapite.

Ref.
1. Bandy (1938). Gordon, *Ac. Sc. Philadelphia, Not. Nat.*, no. 89 (1941) gives another orientation; transformatino: Gordon to Bandy, 010/$\overline{1}$00/001.
2. Bandy (1938) and Gordon (1941).
3. Gordon (1941).
4. Bandy (1938).
5. Henderson analysis in Bandy (1938).

31.6.4 **A M A R A N T I T E** [$Fe(SO_4)(OH) \cdot 3H_2O$]. Frenzel (*Min. Mitt.*, **9**, 398, 1887). Paposite (?) *Darapsky* (*Bol. soc. min.*, Santiago, Chile, No. 92, 735, 1887; *Jb. Min.*, I, 23, 1889).

Cryst.[1] Triclinic; pinacoidal—$\overline{1}$.

$a:b:c = 0.7692:1:0.5738;$ $\alpha\ 95°38\frac{1}{2}',\ \beta\ 90°23\frac{1}{2}',\ \gamma\ 97°13'$

$p_0:q_0:r_0 = 0.7484:0.5784:1;$ $\lambda\ 84°16',\ \mu\ 88°53',\ \nu\ 82°42\frac{1}{2}'$

$p_0'\ 0.7522,\ q_0'\ 0.5813,\ x_0'\ 0.0069,\ y_0'\ 0.1004$

Forms:[2]

		ϕ	ρ	A	B	C
c	001	3°56'	5°45'	88°53'	84°16'	.....
b	010	0 00	90 00	82 42½	.....	84°16'
a	100	82 42½	90 00	.....	82 42½	88 53
m	110	47 47½	90 00	34 55	47 47½	85 51½
M	1$\overline{1}$0	123 04½	90 00	40 22	123 04½	92 47½
L	1$\overline{2}$0	145 02½	90 00	62 20	145 02½	94 28
J	1$\overline{4}$0	161 30	90 00	78 47½	161 30	95 18½
E	0$\overline{1}$1	179 10½	25 41	92 48½	115 41	31 25
x	101	75 25	37 53	52 28½	81 06½	36 24½
p	111	44 05½	47 15½	54 59	58 10	42 59½
P	1$\overline{1}$1	117 06½	40 13½	57 48	107 07	42 45½
ʃ	1$\overline{2}$2	138 47	29 58	73 48½	112 02½	34 13½
n	121	29 00	57 13½	60 09½	42 39½	52 03½
N	1$\overline{2}$1	142 05	50 47	66 45½	127 41	55 09½
Y	1$\overline{3}$1	154 03½	59 51	73 57	141 02½	64 52
ω	2$\overline{1}$1	100 56½	56 46½	37 23½	99 08½	57 39½
σ	$\overline{2}$12	−68 12	38 31½	122 58½	76 37½	37 06
φ	$\overline{1}$21	−32 20	54 06½	110 03½	46 48½	49 33½

Less common and rare:

d 180	t 510	h 012	v 201	Δ $\bar{1}12$	τ $3\bar{1}1$	
g 160	R $3\bar{1}0$	e 011	U $\bar{1}02$	Π $\bar{1}11$	ν $4\bar{1}1$	
i 150	Q $2\bar{1}0$	f 021	X $\bar{1}01$	z 144	λ $\bar{1}31$	
j 140	D $3\bar{2}0$	w 031	V $\bar{2}01$	y 131	α $\bar{1}41$	
k 130	C $3\bar{4}0$	H $0\bar{1}2$	θ $\bar{3}01$	ϵ $1\bar{3}3$	ξ $\bar{2}11$	
l 120	B $2\bar{3}0$	Σ $0\bar{3}2$	ψ $\bar{4}01$	T $2\bar{1}2$	γ $\bar{3}11$	
o 430	K $1\bar{3}0$	F $0\bar{2}1$	η $2\bar{2}1$	O $2\bar{3}2$	Υ $\bar{1}22$	
q 210	A $2\bar{7}0$	W $0\bar{3}1$	δ $\bar{1}12$	Z $\bar{1}41$	Φ $\bar{1}21$	
r 310	I $1\bar{5}0$	Ω $0\bar{4}1$	π $\bar{1}11$	S $1\bar{5}1$	Ξ $\bar{2}11$	
s 720	G $1\bar{6}0$	u $\bar{1}02$	β $\bar{2}21$	μ $3\bar{2}2$	Γ $\bar{3}11$	

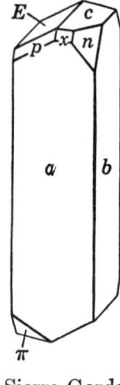

Sierra Gorda.

Habit. Elongated [001], with large {100} and {010} and affording a nearly square cross section or flattened {100}. Striated [001]. Common forms: $c\ b\ a\ E\ x\ P$. In radiating or matted aggregates of acicular crystals; columnar or bladed.

Phys. Cleavage {010} and {100} perfect. Brittle. H. $2\frac{1}{2}$. G. 2.189,[3] 2.286.[4] Luster vitreous. Color amaranth-red to brownish red and orange-red. Streak lemon-yellow. Transparent.

Opt.[5]

Orientation	ϕ	ρ	n(Na)	Pleochroism	
X	82°	72°	1.516	Colorless	Biaxial negative ($-$).
Y	178°	68°	1.598	Pale yellow	2V 30°.
Z	$-44°$	29°	1.621	Reddish brown	$r < v$, horizontal.

Chem. A hydrated basic sulfate of ferric iron, $Fe(SO_4)(OH) \cdot 3H_2O$. On heating, the water is lost in three stages.[12]

Anal.[10]

	1	2	3	4	5	6
Fe_2O_3	35.81	37.26	37.46	35.92	37.31	35.64
SO_3	35.91	35.58	35.46	36.18	37.55	35.98
H_2O	28.28	27.62	28.29	28.13	25.14	28.23
Rem.			0.70			
Total	100.00	100.46	101.91	100.23	100.00	99.85
G.		2.286			2.188	

1. $Fe(SO_4)(OH) \cdot 3H_2O$. 2. Sierra Gorda.[8] 3. La Compania.[6] Rem. is Na_2O 0.59, K_2O 0.11. 4. Paposa, Chile (*paposite?*).[9] 5. Tierra Amarilla.[3] H_2O includes H_2O 23.50 at 220°, 1.64 by diff. at red heat. 6. Chile.[7]

Tests. Decomposed by cold water with the formation of an insoluble basic salt. Soluble in HCl.

Occur. A secondary mineral, found associated with hohmannite, fibroferrite, chalcanthite, copiapite, sideronatrite, coquimbite. Found in Chile at Tierra Amarilla near Copiapó, La Compania near Sierra Gorda, and Quetena, Alcaparrosa, and Chuquicamata. Paposite was from the Union mine in the Reventon district near Paposa, Chile. With copiapite from the Santa Maria Mountains in Riverside County, California.

Artif.[11] Obtained in the system Fe_2O_3-SO_3-H_2O.

Name. From ἀμάραντος, *amaranth*, on account of its color.

Ref.

1. Elements of Penfield, *Am. J. Sc.*, **40**, 199 (1890) on crystals from La Compania near Sierra Gorda, Chile.
2. Penfield (1890); Ungemach, *Bull. soc. min.*, **58**, 115 (1935); Bandy, *Am. Min.*, **23**, 747 (1938). Uncertain: 230, 320, 410, 2$\bar{5}$0, 1$\bar{9}$0, 0$\bar{5}$1, 301, 4$\bar{1}$4, $\bar{1}$31.
3. Ungemach (1935).
4. Penfield (1890).
5. Bandy (1938).
6. Genth, *Am. J. Sc.*, **40**, 199 (1890).
7. Scharizer, *Zs. Kr.*, **65**, 347 (1927).
8. Frenzel (1887).
9. Gräbner analysis in Frenzel, *Min. Mitt.*, **11**, 223 (1890).
10. See Dana (967, 968, 1892) for other analyses.
11. Scharizer in Doelter (4 [2], 574, 1927); see also Posnjak and Merwin, *J. Am. Chem. Soc.*, **44**, 1965 (1922).
12. For dehydration data see Scharizer (1927).

31.6.5 **H O H M A N N I T E** [$Fe_2(SO_4)_2(OH)_2 \cdot 7H_2O$]. Frenzel (*Min. Mitt.*, **9**, 397, 1887). Castanite Darapsky (*Jb. Min.*, II, 267, 1890). [Not castanite Bandy (*Am. Min.*, **17**, 534, 1931) = amarantite.]

C r y s t.[1] Triclinic; pinacoidal—$\bar{1}$.

$a:b:c = 0.7283:1:0.8936;\quad \alpha\ 90°10\tfrac{1}{2}',\ \beta\ 91°10',\ \gamma\ 101°13'$

$p_0:q_0:r_0 = 1.2526:0.9081:1;\quad \lambda\ 89°35\tfrac{1}{2}',\ \mu\ 88°46\tfrac{1}{2}',\ \nu\ 78°46\tfrac{1}{2}'$

$p_0'\ 1.2528,\ q_0'\ 0.9083,\ x_0'\ 0.0203,\ y_0'\ 0.0072$

Forms:

		φ	ρ	A	B	C
c	001	70°35'	1°14'	88°46½'	89°35½'	.....
b	010	0 00	90 00	78 46½	.....	89°35½'
M	110	46 50½	90 00	31 56	46 50½	88 52
m	1$\bar{1}$0	118 24	90 00	39 37½	118 24	89 10
e	0$\bar{1}$1	178 42½	42 02	96 04½	132 01	42 26
s	1$\bar{1}$2	117 07	35 29½	62 54½	105 20½	34 39½

Habit. Crystals short prismatic [001], the faces generally rounded and dull or striated. Usually as granular aggregates of subhedral crystals from which well-formed crystals may project into cavities.

P h y s. Cleavage {010} perfect, {110} and {1$\bar{1}$0} less perfect. H. 3. G. 2.2. Luster vitreous, brilliant. Color chestnut-brown to burnt orange and light amaranth-red. Powder orange-yellow. Transparent to translucent. On exposure rapidly dehydrates and crumbles to metahohmannite.

O p t.

ORIENTATION	n (Knoxville)[2]	n(Na) (Chuquicamata)[3]	PLEOCHROISM	
X	1.553	1.559	Very pale yellow	Biaxial negative (−).
Y	1.643	1.643	Pale greenish yellow	2V 40°.
Z	1.657	1.655	Dark greenish yellow	r > v, extreme.

In (010) sections a slightly off-centered optic axis figure is obtained and the extinction is $Y' \wedge c = 22°\text{--}23°$.

614   SULFATES

Chem. A hydrated basic sulfate of ferric iron, $Fe_2(SO_4)_2(OH)_2 \cdot 7H_2O$. $4H_2O$ is lost on heating at 27°, and the last of the remainder at 360°.
Anal.

	1	2	3	4	5	6
Fe_2O_3	34.42	33.92	34.47	35.58	37.03	36.86
SO_3	34.52	33.80	35.11	33.84	35.76	36.85
H_2O	31.06	30.76	30.31	30.08	27.71	26.34
Rem.		1.15	0.22			0.53
Total	100.00	99.63	100.11	99.50	100.50	100.58
G.		2.118	2.2		2.17	

1. $Fe_2(SO_4)_2(OH)_2 \cdot 7H_2O$. 2. Sierra Gorda, Chile (*castanite*).[4] Rem. is barite 1.15, Al_2O_3 tr. 3. Knoxville, California.[5] Rem. is insol. Original total given as 100.21. 4. Sierra Gorda.[6] 5. Sierra Gorda.[7] 6. Chile.[8] Rem. is insol.

Tests. In C.T. turns dark in color and affords acid water. Decomposed by hot water but practically insoluble in cold. Easily soluble in HCl. B.B. a fragment carefully heated turns first pale orange, then grayish brown, next dark red, and finally black and fuses on thin edges.

Occur. A secondary mineral originally found with copiapite, amarantite, and sideronatrite at Sierra Gorda near Copiapó, Chile (also *castanite*). With chalcanthite, picromerite, amarantite, fibroferrite, copiapite at Chuquicamata and Quetena, Chile. With sulfur and cinnabar at the Redington mine, Knoxville, Napa County, California.
Alter. To metahohmannite.
Name. After Thomas Hohmann, mining engineer of Valparaiso, Chile, who discovered the mineral. Castanite from κάστανα, chestnut, in allusion to the color. Castanite has been considered to be identical with amarantite [9] but was later shown [10] to be in all probability the same as hohmannite.

Ref.
1. Elements recalculated by La Forge, priv. comm. (1935) from the measurements of Rogers, *Am. Min.*, **16**, 396 (1931), on crystals from Knoxville, California.
2. Rogers (1931).
3. Bandy, *Am. Min.*, **23**, 746 (1938).
4. Darapsky (1890).
5. Shepard analysis in Rogers (1931).
6. Frenzel (1887).
7. Frenzel, *Min. Mitt.*, **11**, 215 (1890).
8. Darapsky, *Jb. Min.*, I, 56 (1890).
9. Hintze (**1** [3B], 4426, 1929).
10. Ungemach, *Bull. soc. min.*, **58**, 97 (1935), and Bandy (1938).

31.6.6   **FIBROFERRITE** [$Fe(SO_4)(OH) \cdot 5H_2O$ (?)]. Rose (*Ann. Phys.*, **27**, 316, 1833). Fibroferrite *Prideaux* (*Phil. Mag.*, **18**, 397, 1841). Stypticit *Hausmann* (**2**, 1202, 1847). Copiapite *Smith* (*Am. J. Sc.*, **18**, 375, 1854).

Cryst.[1] Orthorhombic.

$$a:b:c = 0.5604:1:0.2162; \qquad p_0:q_0:r_0 = 0.3858:0.2162:1$$

$$q_1:r_1:p_1 = 0.5604:2.5920:1; \qquad r_2:p_2:q_2 = 4.6253:1.7844:1$$

FIBROFERRITE

Forms:

		ϕ	$\rho = C$	ϕ_1	$\rho_1 = A$	ϕ_2	$\rho_2 = B$
b	010	0°00′	90°00′	90°00′	90°00′	.....	0°00′
a	100	90 00	90 00	.....	0 00	0°00′	90 00
m	110	60 44	90 00	90 00	29 16	0 00	60 44
d	101	90 00	21 06	0 00	68 54	68 54	90 00
g	401	90 00	57 03½	0 00	32 56½	32 56½	90 00
t	131	30 44½	37 02½	32 58	72 04	68 54	58 49
ε	161	16 34	53 32½	52 22½	76 41	68 54	39 14½
u	532	71 25	45 30	17 58	47 28	46 02	76 51½

Habit. As fine-fibrous crusts and masses; botryoidal; radial-fibrous. Fibers elongated [001].

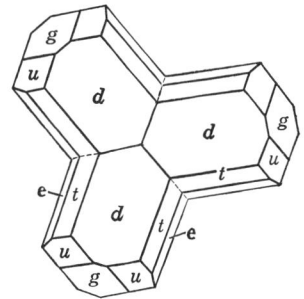

Argentina. Trilling projected on {001}.

Twinning. On {110}, repeated, as triplets elongated [001].

P h y s. Cleavage {001} perfect. H. ~2½. G. 1.84–2.1, 2.52 (Valachov). Luster of aggregates silky, pearly. Color pale yellow or straw-yellow to nearly white, also greenish gray and yellowish green to pale green.

O p t. In transmitted light, colorless to faint yellow.

ORIEN-TATION	n (Quatsino)[2]	n (Arizona)[3]	n (Chuquicamata)[4]	PLEOCHROISM	
X	1.518	1.533	1.513	Colorless	Biaxial positive
Y	1.518	1.534	1.535	Colorless	(+).
Z c	1.561	1.575	1.571	Pale amber-yellow	$2V$ sensibly 0°.

The variation in the indices of refraction probably is due to variation in water content of the mineral.[5]

C h e m. A hydrated basic sulfate of ferric iron, probably $Fe(SO_4)(OH) \cdot 5H_2O$ for fully hydrated material. The water is largely lost below 100° and the content at room temperature probably varies with the humidity;[6] the reported analyses mostly afford values between 3 and $5H_2O$. Some analyses cite small and irrational amounts of divalent iron, and the mineral as found may be the more or less oxidized equivalent of an original ferrous-ferric sulfate. Al substitutes for Fe''' with Al:Fe = 1:1.7 in analysis 5.

Anal.[7]

	1	2	3	4	5
FeO			1.50	0.28	0.31
Fe_2O_3	30.83	30.66	30.58	32.68	20.38
SO_3	30.91	30.66	32.03	32.48	34.28
H_2O	38.26	37.44	36.40	33.20	37.09
Rem.		1.24		1.08	8.24
Total	100.00	100.00	100.51	99.72	100.30
G.		1.92	2.09	1.901	2.52

1. $Fe(SO_4)(OH) \cdot 5H_2O$. 2. La Alcaparrosa, Argentina.[8] Rem. is S 0.40, insol. 0.84. 3. Capo Calamita, Elba.[9] 4. Quatsino, British Columbia.[10] Rem. is insol. 5. Valachov Hill, Bohemia.[11] Rem. is Al_2O_3 7.60, CaO 0.34, SiO_2 0.30, MgO tr.

Tests. Decomposed by water, yielding an acid solution and a brown precipitate of hydrous iron oxide.

O c c u r. A secondary mineral formed commonly by the oxidation of pyrite. Associated with copiapite, botryogen, szomolnokite, melanterite, epsomite, jarosite, gypsum. Originally from Tierra Amarilla near Copiapó, Chile; also from Quetena and with amarantite and hohmannite at Chuquicamata. In veins up to 3 meters in width at the Santa Elena mine, La Alcaparrosa, San Juan, Argentina. In Bohemia on Valachov Hill near Skřivaň and as a weathering of pyritic slates in the Praha district. From Skouriotissa, Cyprus. In the Blyava sulfate deposit in the southern Urals, U.S.S.R. In Italy at Cetine near Rosia, Sienna, and at Capo Calamita and Vigneria on Elba. From the mines of Paillières, Dept. Gard, France, and from Pöham, Salzburg, Austria. From Yetar Spring near Chidlows, Western Australia. In the United States from Red Cliff and Cimarron (?), Colorado. From Genette Mountain, Arizona. In California in the pyrrhotite deposit at Island Mountain, Trinity County; with krausite, coquimbite, and other sulfates in the Calico Hills near Borate, San Bernardino County, and in the Redington mine at Knoxville, Napa County. In Canada from Quatsino, British Columbia, in bog iron ore.

N a m e. From the iron content and fibrous mode of aggregation.

Ref.

1. Elements and forms of Gordon, *Ac. Sc. Philadelphia, Not. Nat.*, no. 102 (1942), on minute crystals from La Alcaparrosa, Argentina.
2. Walker, *Univ. Toronto Stud., Geol. Ser.*, no. 14, 87 (1922).
3. Larsen (75, 1921).
4. Bandy, *Am. Min.*, **23**, 748 (1938); nX perhaps a misprint for 1.531.
5. Additional optical data are given by Ulrich, *Časopis Národ. Musea, Praha*, **109**, 79 (1935); Meixner, *Zbl. Min.*, 110 (1939); Prider, *J. Roy. Soc. Western Australia*, **28**, 79 (1943); Jirkovsky, *Časopis Národ. Musea, Praha*, **104**, 16 (1930); and others.
6. For dehydration data see Walker (1922) and Scharizer, *Zs. Kr.*, **65**, 335 (1927).
7. For additional analyses see Hintze (1 [3B], 4428, 1929); Scharizer (1927); Headden, *Proc. Colorado Sc. Soc.*, **8**, 60 (1905).
8. Angelelli and Chaudet, *Rev. min., Soc. argentine min. geol.*, **8**, 46 (1937).
9. Manasse, *Proc. att. soc. toscana*, **27** (1911).
10. Todd analysis in Walker (1922).
11. Jirkovsky (1930).

BOTRYOGEN

31.6.7 BOTRYOGEN [MgFe'''(SO$_4$)$_2$(OH).7H$_2$O]. Rother Eisen-Vitriol *Berzelius* (*Afhandl. Fys. Kem. Min.*, **4**, 307, 1815). Red Iron Vitriol. Fer sulfaté rouge *Fr.* Botryogen *Haidinger* (*Ann. Phys.*, **12**, 491, 1828). Neoplase pt. *Beudant* (**2**, 483, 1832). Botryt *Glocker* (300, 1847).
Quetenite *Frenzel* (*Min. Mitt.*, **11**, 217, 1890). Rubrite *Darapsky* (*Jb. Min.*, I, 65, 1890). Kubeite *Darapsky* (*Jb. Min.*, I, 163, 1898). Cubeïte. Palacheite *Eakle* (*Univ. Calif. Dept. Geol., Bull.*, **3**, 231, 1903).

Cryst.[1] Monoclinic; prismatic—$2/m$.

$a:b:c = 0.5893:1:0.3996;$ $\beta\,100°01';$ $p_0:q_0:r_0 = 0.6781:0.3935:1$

$r_2:p_2:q_2 = 1.4747:1.7232:1;$ $\mu\,79°59';$ $p_0'\,0.6886,\,q_0'\,0.3996,\,x_0'\,0.1766$

Forms:[2]

		φ	ρ	φ₂	ρ₂ = B	C	A
b	010	0°00′	90°00′	....	0°00′	90°00′	90°00′
a	100	90 00	90 00	0°00′	90 00	79 59	.....
k	130	29 52½	90 00	0 00	29 52½	85 02	60 07½
l	120	40 45	90 00	0 00	40 45	83 29	49 15
m	110	59 52½	90 00	0 00	59 52½	81 21	30 07½
p	011	23 50½	23 36	79 59	68 31	21 29	80 41
s	021	12 27½	39 18	79 59	51 48	38 12	82 08½
d	101	90 00	40 54	49 06	90 00	30 53	49 06
o	$\bar{1}$01	−90 00	27 04	117 04	90 00	37 05	117 04
r	111	65 14	43 39	49 06	73 11½	34 45½	51 11½
n	$\bar{1}$11	−51 58½	32 58½	117 04	70 25	41 16½	115 23
u	131	35 51	55 56	49 06	47 49	50 30½	60 58½
v	$\bar{1}$21	−32 35½	43 29½	117 04	54 33½	49 27½	111 45½

Less common:
e 270 *f* 250   *h* 350 *t* 450   *i* 210 *F* 310   *P* 031 *z* 221   *D* 121 *w* 141   *q* $\bar{1}$31 *x* $\bar{1}$71   *y* $\bar{2}$11

Habit. Prismatic [001]; smaller crystals usually long prismatic, large crystals short prismatic with {$\bar{1}$01} large and often striated [100]. Vertical zone striated [001]. Common forms: *o b m l n*. Mostly as reniform, botryoidal, or globular aggregates with a radiating structure and crystalline surface.

Phys. Cleavage {010} perfect, {110} good. Fracture conchoidal, interrupted. Brittle. H. 2–2½. G. 2.14. Luster vitreous. Color light to dark orange-red. Streak ocher-yellow. Transparent to translucent.

Opt.[2]

Orientation	n (Quetena, anal. 2)	Pleochroism	
X b	1.523	Colorless to pale brown	Biaxial positive (+).
Y	1.530	Cinnamon-brown	2V 42°.
Z ∧ c +12°	1.582	Golden yellow	r > v, strong.

Chem. A hydrated basic sulfate of magnesium and trivalent iron, MgFe'''(SO$_4$)$_2$(OH)·7H$_2$O. Fe'', Zn, and Mn'' substitute at least in part for Mg, with Fe''':Mg = 1:1.55 in analysis 4 and Zn:Mg = 1:2.9 in analysis 5. Much of the water is lost at 100° or below.[10]

Anal.[8]

	1	2	3	4	5	6
MgO	9.64	9.40	7.63	3.59	6.65	9.35
FeO			2.23	4.12	0.53	
Fe_2O_3	19.28	19.73	16.38	20.50	18.73	19.51
SO_3	38.55	38.45	36.94	40.95	37.64	38.37
H_2O	32.53	32.00	33.99	30.82	31.04	32.28
Rem.			2.83		5.09	
Total	100.00	100.58	100.00	99.98	99.68	99.51
G.				2.138		

1. $MgFe'''(SO_4)_2(OH) \cdot 7H_2O$. 2. Quetena, Chile.[3] Original total 100.18. 3. Falun, Sweden.[4] Rem. is CaO 0.90, MnO 1.93. Average of two. 4. Persia.[5] 5. Falun, Sweden.[6] Rem. is ZnO 4.82, insol. 0.27. 6. Knoxville, California (*palacheite*).[7]

Tests. B.B. intumesces and gives off water, leaving a reddish yellow earthy residue. Partly soluble in boiling water, leaving an ocherous residue. Soluble in HCl.

Occur. A secondary mineral found in the sulfate capping of pyritic ore deposits, especially in arid regions. Associated with copiapite, amarantite, hohmannite, coquimbite, voltaite, pickeringite, epsomite, gypsum, chalcanthite. Found abundantly in Chile at Chuquicamata, Quetena (*quetenite*)[9] and Alcaparrosa; also from the Rio Loa (*rubrite, kubeite*).[9] At the Santa Elena mine, La Alcaparrosa, San Juan, Argentina. From the copper mine of Falun, Sweden, and the region of Madeni Zakh, Persia. Reported from Rammelsberg in the Harz, Germany, and Mine en Charbes, Val de Villé, Alsace. In the United States from the Redington mine, Knoxville, Napa County, California (*palacheite*);[9] also at the Palisades mine near Calistoga, Napa County. From Cornwall, Lebanon County, Pennsylvania.

Name. From βότρυς, *a bunch of grapes*, and γεννᾶν, *to bear*, in allusion to the botryoidal and stalactitic masses found originally at Falun. Palacheite after Charles Palache (1869–   ), American mineralogist.

Ref.

1. Elements and angles of Berman in Bandy, *Am. Min.*, **23**, 749 (1938) on quetenite = botryogen from Quetena. Concordant angles were obtained by Eakle, *Univ. California Dept. Geol., Bull.*, **3**, 321 (1903) on palacheite = botryogen from Knoxville; transformation: Eakle to new, 10$\overline{1}$/010/001.
2. Bandy (1938).
3. Henderson analysis in Bandy (1938).
4. Hockauf, *Zs. Kr.*, **12**, 251 (1886).
5. Blaas, *Ak. Wien, Sitzber.*, **87**, 161 (1883).
6. Cleve, *Upsala Univ. Åarskrift*, **22** (1862).
7. Eakle (1903).
8. For a tabulation of other analyses of botryogen and synonyms see Bandy (1938).
9. On the identity of these substances with botryogen see Bandy (1938) and Eakle, *Am. J. Sc.*, **16**, 379 (1903), on palacheite.
10. For dehydration data see Eakle (1903).

IDRIZITE. Schrauf (*Jb. Geol. Reichsanst. Wien*, **41**, 379, 1892).

Compact; crystalline. H. 3. G. 1.829. Color yellow-gray. Analysis gave: MgO 4.51, FeO and MnO 3.10, Fe_2O_3 8.70, Al_2O_3 8.59, SO_3 33.94, H_2O 40.80, total 99.64. The composition is rather similar to that of botryogen, although with more H_2O, and idrizite may be the aluminum analogue of that species, $(Mg,Fe,Mn)(Al,Fe''')(SO_4)_2$-$(OH) \cdot 7H_2O$ with Al:Fe = 1.55:1. Insoluble in water but soluble in HCl. From the Idria mercury mine in Carniola.

31.6.8 GUILDITE [$(Cu,Fe'')_3(Fe''',Al)_4(SO_4)_7(OH)_4 \cdot 15H_2O$]. *Lausen* (*Am. Min.*, **13**, 203, 1928).

Cryst. Monoclinic.[1]

$$a:b:c = 0.037:1:1.407; \quad \beta\ 105°17'$$

Forms:

c 001	a 100	m 110	d 540	k 011	l 101	h $\bar{1}$01

Habit. Short prismatic, pseudo-cubic crystals up to half a centimeter in width.

Phys. Cleavage {001} and {100}, perfect. Fracture conchoidal. Brittle. H. $2\frac{1}{2}$. G. 2.72. Luster vitreous. Color brown. Streak pale, canary-yellow. Translucent, cleavage fragments transparent.

Opt.

	n	Pleochroism	
X	1.623 ± 0.005	Pale yellow	Biaxial positive (+).
Y	1.630 ± 0.005	Pale yellow	$2V$ small.
Z	1.684 ± 0.005	Greenish yellow	

Chem. A basic hydrated sulfate of copper and ferric iron, $(Cu,Fe'')_3(Fe''',Al)_4(SO_4)_7(OH)_4 \cdot 15H_2O$. Small amounts of Fe'' and Al substitute for Cu and Fe''', respectively.

Anal.

	Na_2O	CuO	FeO	Fe_2O_3	Al_2O_3	SO_3	H_2O	Total	G.
1.		15.35	1.48	19.31	2.18	39.88	21.80	100.00	
2.	1.23	15.78	1.49	19.12	2.11	39.68	22.15	101.56	2.725

1. $(Cu,Fe'')_3(Fe''',Al)_4(SO_4)_7(OH)_4 \cdot 15H_2O$, with $Cu:Fe'' = 9:1$ and $Fe''':Al = 6:1$.
2. United Verde mine, Jerome, Arizona.

Occur. Formed with coquimbite and other sulfates as the result of a fire at the United Verde mine, Jerome, Arizona.

Name. After Frank Nelson Guild (1870–1939) of the University of Arizona.

Ref.

1. For angle table and crystal drawing see Lausen, *Am. Min.*, **13**, 203 (1928). Crystal imperfections were such as to leave an uncertainty of several degrees in some angles.

31.6.9 METAVOLTINE [$(K,Na,Fe'')_5Fe_3'''(SO_4)_6(OH)_2 \cdot 9H_2O$ (?)]. *Blaas* (*Ak. Wien, Sitzber.*, **87**, 155, 1883). α-Metavoltine *Scharizer* (*Zs. Kr.*, **65**, 1, 1927). Metavoltaite.

Cryst.[1] Hexagonal.

$$a:c = 1:0.9468; \quad p_0:r_0 = 1.0932:1$$

Forms:

	ϕ	$\rho = C$	M	A_2
c 0001	.....	0°00'	90°00'	90°00'
a 10$\bar{1}$0	30°00'	90 00	60 00	90 00
u 10$\bar{1}$3	30 00	20 01$\frac{1}{2}$	80 08$\frac{1}{2}$	90 00
v 10$\bar{1}$2	30 00	28 39$\frac{1}{2}$	76 07$\frac{1}{2}$	90 00
x 50$\bar{5}$7	30 00	37 59	72 04$\frac{1}{2}$	90 00
p 10$\bar{1}$1	30 00	47 33	68 21	90 00
o 30$\bar{3}$2	30 00	58 37$\frac{1}{2}$	64 44	90 00
l 20$\bar{2}$1	30 00	65 25$\frac{1}{2}$	62 57$\frac{1}{2}$	90 00
t 40$\bar{4}$1	30 00	77 07	60 49$\frac{1}{2}$	90 00

Structure cell.[2] a_0 19.43 kX, c_0 18.60; $a_0:c_0 = 1:0.957$. Cell contents $K_{40}Fe_{24}'''(SO_4)_{48}(OH)_{16} \cdot 72H_2O$.

Habit. Tabular {0001} with a hexagonal outline; usually minute in size. As granular or scaly aggregates.

Phys. Cleavage {0001} perfect. H. $2\frac{1}{2}$. G. 2.5, 2.396 (artificial).[2] Luster resinous. Color yellowish brown and orange-brown to greenish brown. Translucent.

Opt. Uniaxial negative $(-)$.

	n (La Compania)[3]	n (Chuquicamata)[4]	n (Chile)[5]	n (Peru)[5]
O	1.595	1.589–1.590	1.588	1.591
E	1.581	1.572–1.574	1.578	1.573

DICHROISM:

O	Dark yellow	Brown	Dark yellow	Deep orange-yellow
E	Light yellow	Pale greenish yellow	Light yellow	Very pale yellow

Chem. A hydrated basic sulfate essentially of ferric iron and potassium, probably $(K,Na,Fe'')_5Fe_3'''(SO_4)_6(OH)_2 \cdot 9H_2O$. Na and apparently also Fe'' substitute for K, with Na > K in some material. The water content is uncertain.[6] A monoclinic or triclinic polymorph (β-metavoltine) is known artificially.[14]

Anal.

	1	2	3	4	5	6	7
K_2O	20.74	21.44	20.13	21.03	9.87	4.69	4.82
Na_2O			0.35		4.65	8.15	8.89
FeO					2.92		2.83
Fe_2O_3	21.10	21.54	21.67	21.49	21.20	23.31	22.55
SO_3	42.30	42.86	43.23	42.98	46.90	45.42	44.22
H_2O	15.86	14.49	14.48	14.35	14.58	17.83	17.23
Total	100.00	100.33	99.86	99.85	100.12	99.40	100.54
G.					2.5		

1. $K_5Fe_3'''(SO_4)_6(OH)_2 \cdot 9H_2O$. 2. Artificial.[2] 3. Artificial.[7] Original total given as 100.00. 4. Vesuvius.[8] On material recrystallized from water solution. 5. Persia.[9] 6. Chuquicamata.[10] Original total given as 99.56. 7. La Compania.[11]

Tests. Partially soluble in water, the solution giving a reddish precipitate on heating. Slowly decomposed by dilute acids.

Occur. Originally found with voltaite and botryogen near Madeni Zakh, Persia, as an alteration of pyritic trachyte. Also from the Grotta dello Zolfo at Miseno and with alums and pickeringite at the solfatara of Pozzuoli, near Naples, and in fumaroles on Vesuvius, Italy; from caves near Porto di Levante, Vulcano, Lipari Islands, and Milo in the Cyclades.[12] Found in Chile with metasideronatrite, alunogen, copiapite, chalcanthite at Chuquicamata; with coquimbite and alunogen at Quetena; with copiapite and pickeringite at Alcaparrosa; and with ferrinatrite, coquimbite, and copiapite at La Compania. In the United States reported from near Borate in the Calico Hills, San Bernardino County, California.

A r t i f.[13] By evaporation preferably at 70°–80° of a water solution of ferric sulfate and potassium sulfate in excess, and from a water solution of KFe alum. β-metavoltine is obtained from relatively acid solutions.

N a m e. From μετά, *with*, and *voltine*, because found associated with voltaite at the original locality.

Ref.
1. Elements and forms of Gordon, *Ac. Sc. Philadelphia, Not. Nat.*, no. 64 (1940), on crystals from La Compania, Chile (analysis 7). Gordon's elements are comparable to those of the structure cell of Gossner and Arm, *Zs. Kr.*, **72**, 205 (1929), on artificial crystal with the composition $K_5Fe_3'''(SO_4)_6(OH)_2 \cdot 8H_2O$ (analysis 2). The crystallography given by Bandy, *Am. Min.*, **23**, 735 (1938), for metavoltine (?) from Chuquicamata (analysis 6) does not agree with this data.
2. Gossner and Arm (1929).
3. Gordon (1940).
4. Bandy (1938).
5. Larsen (109, 1921).
6. For dehydration data see Scharizer, *Zs. Kr.*, **58**, 420 (1923); **65**, 1 (1927). About half of the water is lost at 100° and below.
7. Scharizer (1923).
8. Zambonini, *Acc. Napoli, Att.*, **13**, 18 (1906).
9. Blaas (1883).
10. Henderson analysis in Bandy (1938).
11. Pitman and Collins analysis in Gordon (1940).
12. On the Italian occurrences see Lacroix, *Bull. soc. min.*, **30**, 33 (1907); Zambonini, *Acc. Napoli, Rend.*, 158 (1908); Aquilar, *Soc. nat. Napoli, Boll.*, **25**, 28 (1911).
13. Scharizer (1923, 1927), Gossner and Arm (1929); see also Zambonini (1908) and Mellor (**14**, 341, 1935) on the composition of the artificial compound ("Maus' salt").
14. See Gossner, *Zbl. Min.*, 262 (1936), and Scharizer (1927).

31.6.10 **S L A V I K I T E** [$(Na,K)_2Fe_{10}'''(OH)_6(SO_4)_{13} \cdot 63H_2O$ (?)]. *Jirkovsky* and *Ulrich* (*Vešt. Stát. Geol. Úst. Českoslov. Rep.*, **2**, 345, 1926). Paracoquimbite *Klvana* (*Böhm. Ges., Ber.*, **272**, [1881] 1882). [Not paracoquimbite *Ungemach* (*Bull. soc. min.*, **58**, 165, 1935).]

C r y s t.[1] Hexagonal—R (?).

$a:c = 1:1.389$; $\alpha\ 85°00\tfrac{1}{2}'$; $p_0:r_0 = 1.604:1$; $\lambda\ 94°35\tfrac{1}{2}'$

Forms:

			φ	ρ = C	A_1	A_2
c	0001	111	....	0°00′	90°00′	90°00′
r	10$\bar{1}$1	100	30°00′	58 03½	42 42	90 00
h	50$\bar{5}$2	4$\bar{1}\bar{1}$	30 00	32 41	62 07	90 00
N	02$\bar{2}$1	11$\bar{1}$	−30 00	38 44	90 00	57 11½
H	05$\bar{5}$2	77$\bar{8}$	−30 00	32 41	90 00	62 07

Habit. Minute crystals tabular {0001} with prominent {10$\bar{1}$1}. Fine granular; scaly; incrusting.

P h y s. Observations on cleavage and hardness are lacking. G. 1.905 (Bohemia), 1.99 (Argentina).[2] Luster vitreous. Color greenish yellow.

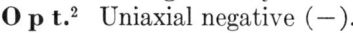

O p t.[2] Uniaxial negative (−).

	n (Argentina)	n (Austria)	n(Na) (Bohemia)	DICHROISM
O	1.533	1.537	1.530 ± 0.001	Lemon-yellow
E	1.497	1.498	1.506 ± 0.002	Almost colorless

Chem. A hydrated basic sulfate of trivalent iron containing some sodium and potassium. Formula [3] perhaps $(Na,K)_2Fe_{10}'''(OH)_6(SO_4)_{13} \cdot 63H_2O$.

Anal.

	Na_2O	K_2O	Fe_2O_3	SO_3	H_2O	Rem.	Total	G.
1.	1.60	0.61	25.78	33.61	38.40		100.00	
2.	1.63	0.57	20.08	34.06	38.76	4.82	99.92	1.905
3.	0.29	....	23.93	34.62	37.21	4.40	100.45	1.99

1. $(Na,K)_2Fe_{10}'''(OH)_6(SO_4)_{13} \cdot 63H_2O$ with $K:Na = 1:4$. 2. Valachov Hill, Bohemia.[4] H_2O includes H_2O- 3.10. Rem. is CaO 0.01, Al_2O_3 4.29, insol. 0.52. 3. La Alcaparrosa, Argentina.[5] Rem. is MgO.

Occur. Found with other sulfates as a product of the oxidation of pyrite. First found at Valachov Hill, near Skřivaň, central Bohemia, where it occurs with halotrichite, pickeringite, and gypsum among the weathering products of pyritic shales. On phyllitic slate partly as crusts on pickeringite at Troja near Prague (*paracoquimbite*).[6] With fibroferrite, copiapite, melanterite, and epsomite at Pöham, near Radstadt, Salzburg, Austria. In a vein with fibroferrite in Santa Elena mine, La Alcaparrosa, San Juan province, Argentina.

Alter. Loses about 3 per cent water on heating to 110°.

Name. After the Czech mineralogist František Slavík, Professor of Mineralogy in the Charles University, Prague.

Ref.

1. Elements and forms of Gordon, *Ac. Sc. Philadelphia, Not. Nat.*, **89** (1941). The morphology suggests the hexagonal scalenohedral class, but the symmetry and lattice type have not been determined.
2. Data on material from Bohemia, Jirkovsky, and Ulrich (1926); Argentina, Gordon (1941); Austria, Meixner, *Cbl. Min.*, **110** (1939).
3. The formula here given is from Jirkovsky and Ulrich (1926). Analyses of others show less sodium. Meixner (1939) reported some sodium, amount undetermined, in slavikite containing 21.4 per cent Fe_2O_3. Rost, *Ak. Česká Roz.*, **50**, [9] (1940)—*Min. Abs.*, **9**, 204 (1946), found alkalies in impure slavikite and attributed them to admixed natrojarosite; he assigned the formula $Fe_2O_3 \cdot 2SO_3 \cdot 9H_2O$.
4. Jirkovsky and Ulrich (1926).
5. Gordon (1941).
6. Shown to be identical with slavikite by Rost, *Ak. Česká Roz.*, **51**, [8] (1941)—*Min. Abs.*, **9**, 204 (1946).

FRANQUENITE. Van Tassel (*Bull. mus. hist. nat. Belg.*, **20**, no. 16, 1944).

A yellow efflorescence composed of minute hexagonal scales. G. 1.87–1.94. Optically uniaxial negative (−), with nO 1.531, nE 1.494. Analysis gave: MgO 4.23, FeO 1.00, Fe_2O_3 14.84, Al_2O_3 5.51, SO_3 32.77, H_2O 40.69, total 99.04. This is close to $(Mg,Fe'')(Fe''',Al)_3(SO_4)_4(OH)_3 \cdot 21H_2O$. 28.2 per cent H_2O is lost at 110°. Slowly soluble in cold water. Found with fibroferrite (?) ferroan pickeringite, melanterite, and gypsum on shale at Mousty, Franquenies, Belgium. Further investigation may prove this substance to be identical with slavikite.

COPIAPITE GROUP

31.6.11.1 COPIAPITE [(Fe,Mg)Fe$_4'''$(SO$_4$)$_6$(OH)$_2 \cdot$20H$_2$O]. Μισυ *Dioscorides*. Misy [from Cyprus] *Pliny* (**34,** 31). Misy *Germ.* Gelb Atrament [from the Harz, etc.] *Agricola* (*Nat. Foss.*, 213, 457; *Interpr.*, 466, 1546). Misy, Gul Atrament Sten, Lapis atramentarius flavus *Wallerius* (159, 1747). Misy *Hausmann* (1061, 1813; 1203, 1847); *Dana* (964, 1892). Gelbeisenerz *Breithaupt* (97, 238, 1823; 223, 1832). Yellow Copperas. Basisches schwefelsaures Eisenoxyd *Rose* (*Ann. Phys.*, **27**, 314, 1833). Copiapite *Haidinger* (489, 1845A). Xanthosiderit pt. *Glocker* (65, 1847). Ihlëite *Schrauf* (252, 1877). Janosite *Böckh* and *Emszt* (*Földt. Közl.*, **35**, 139, 1905). Pseudocopiapite *Ungemach* (*Bull. soc. min.*, **58**, 152, 1935) Ferrocopiapite, Ferricopiapite *Berry* (*Am. Min.*, **23,** pt. 2, no. 12, 3, 1938; *Univ. Toronto Stud., Geol. Ser.*, no. 51, 21, 1947). Aluminocopiapite *Berry* (*Univ. Toronto Stud., Geol. Ser.*, no. 51, 21, 1947).

31.6.11.2 MAGNESIOCOPIAPITE [(Mg,Fe)Fe$_4'''$(SO$_4$)$_6$(OH)$_2 \cdot$20H$_2$O]. *Berry* (*Am. Min.*, **23,** pt. 2, no. 12, 3, 1938; *Univ. Toronto Stud., Geol. Ser.*, no. 51, 21, 1947). Knoxvillite *Becker* (*U. S. Geol. Sur., Mon. 13,* 279, 1888); *Melville* and *Lindgren* (*U. S. Geol. Sur., Bull. 61,* 24, 1890).

31.6.11.3 CUPROCOPIAPITE [CuFe$_4'''$(SO$_4$)$_6$(OH)$_2 \cdot$20H$_2$O]. *Bandy* (*Am. Min.*, **23,** 737, 1938).

Cryst.[1] Triclinic; pinacoidal—$\bar{1}$.

$a:b:c = 0.4005:1:0.3971;$ $\alpha\ 93°58\frac{1}{2}',\ \beta\ 102°08',\ \gamma\ 98°50'$

$p_0:q_0:r_0 = 1.0010:0.3929:1;$ $\lambda\ 83°58',\ \mu\ 77°03\frac{1}{2}',\ \nu\ 80°04\frac{1}{2}'$

$p_0'\ 1.0301,\ q_0'\ 0.4043,\ x_0'\ 0.2161,\ y_0'\ 0.1081$

Forms:[2]

		ϕ	ρ	A	B	C	Z
c	001	63°25$\frac{1}{2}$'	13°35'	77°03$\frac{1}{2}$'	83°58'	.....	77°48$\frac{1}{2}$'
b	010	0 00	90 00	80 04$\frac{1}{2}$	.....	83°58'	0 00
a	100	80 04$\frac{1}{2}$	90 00	.....	80 04$\frac{1}{2}$	77 03$\frac{1}{2}$	0 00
$M2$	1$\bar{2}$0	121 53	90 00	41 48$\frac{1}{2}$	121 53	82 56$\frac{1}{2}$	0 00
$d\frac{1}{2}$	012	34 51$\frac{1}{2}$	20 42$\frac{1}{2}$	75 34$\frac{1}{2}$	73 08	10 50	77 48$\frac{1}{2}$
d	011	22 52	29 04$\frac{1}{2}$	74 44$\frac{1}{2}$	63 24	20 34	77 48$\frac{1}{2}$
$d2$	021	13 16	43 17	74 20	48 08$\frac{1}{2}$	35 49$\frac{1}{2}$	77 48$\frac{1}{2}$
$d3$	031	9 17$\frac{1}{2}$	53 14	74 42$\frac{1}{2}$	37 45$\frac{1}{2}$	46 12$\frac{1}{2}$	77 48$\frac{1}{2}$
e	0$\bar{1}$1	143 53	20 08	81 15$\frac{1}{2}$	106 09	22 11	77 48$\frac{1}{2}$
f	101	76 56	51 38$\frac{1}{2}$	38 28	79 47$\frac{1}{2}$	38 35$\frac{1}{2}$	39 05$\frac{1}{2}$
$g\frac{1}{2}$	$\bar{1}$02	− 86 12	16 16$\frac{1}{2}$	105 47$\frac{1}{2}$	88 56	28 44$\frac{1}{2}$	106 14$\frac{1}{2}$
g	$\bar{1}$01	− 94 58	38 43	128 32$\frac{1}{2}$	93 06$\frac{1}{2}$	51 29$\frac{1}{2}$	128 36$\frac{1}{2}$
$g2$	$\bar{2}$01	− 97 45	61 21$\frac{1}{2}$	151 17	96 48	74 13$\frac{1}{2}$	151 08
s	$\bar{1}$11	− 67 15	40 53$\frac{1}{2}$	123 26	75 20	50 36$\frac{1}{2}$	128 36$\frac{1}{2}$
r	$\bar{1}$1$\bar{1}$	−120 40$\frac{1}{2}$	42 52$\frac{1}{2}$	129 31	110 18$\frac{1}{2}$	56 26	128 36$\frac{1}{2}$

Other common forms:

m	110	M	1$\bar{1}$0	$p3$	131	$v3$	$\bar{2}$31	$r\frac{1}{2}$	$\bar{2}$12	$w3$	$\bar{2}$31
$M\frac{1}{2}$	2$\bar{1}$0	$M3$	1$\bar{3}$0	$s2$	$\bar{1}$21	$u\frac{3}{2}$	$\bar{1}$32	$r2$	$\bar{1}$2$\bar{1}$	$w6$	$\bar{2}$61

Structure cell.[3] Space group $P\bar{1}$. a_0 7.33 kX, b_0 18.15, c_0 7.27; α 93°51', β 101°30', γ 99°23'; $a_0:b_0:c_0 = 0.4037:1:0.4005$. Cell contents (Fe,Mg,Cu)(Fe''',Al)$_4$(SO$_4$)$_6$(OH)$_2 \cdot$20H$_2$O ideally.

Habit. Crystals tabular {010}, sometimes highly modified. Usually in loose aggregations of minute scales; granular or incrusting.

Twinning.[4] Contact twins with twin axis [101] and composition plane {010}, not common.

Phys. Cleavage {010} perfect, {$\bar{1}$01} imperfect. H. $2\frac{1}{2}$–3. G. 2.08–2.17, 2.147 (Tierra Amarilla [5]), 2.154 (Chuquicamata [6]); the variation in G. with composition is not clearly defined. Luster pearly on {010}. Color sulfur-yellow to orange- or golden yellow; when massive often greenish yellow to olive-green. Transparent to translucent.

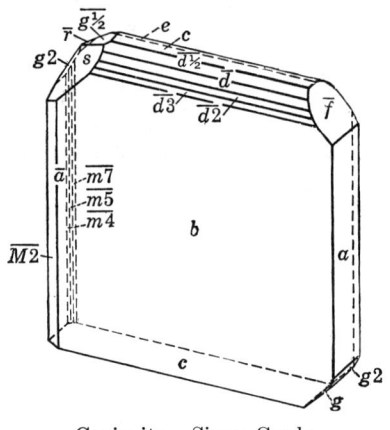

Copiapite. Sierra Gorda.

Opt.[7] Copiapite is optically positive (+). X is sensibly normal to the plane of perfect cleavage and platy development, {010}. Y and Z are practically coincident with the nearly rectangular diagonals [$\bar{1}$01] and [101], respectively, of the pseudo-orthorhombic plates. The optic sign, orientation, and absorption of copiapite are nearly constant, whereas the indices of refraction vary markedly with the chemical composition. The absorption formula is $X \doteq Z > Y$, with X commonly greenish yellow to yellow, Y yellow to colorless, Z sulfur-yellow to yellow-green. In cuprocopiapite X and Z are distinctly green. Dispersion $r > v$, strong.

	Anal.	nX	nY	nZ	$2V$
Copiapite, oxidized [8]	2	1.531	1.546	1.597	52° (meas.)
Copiapite, oxidized [9]	4	1.530	1.550	1.592	69° (meas.)
Copiapite [10]	5	1.509	1.532	1.577	73° (calc.)
Copiapite ("pseudocopiapite") [7]		1.510	1.531	1.578	56° (meas.)
Copiapite, aluminian [7]	3	1.525	1.543	1.590	Mod.
Magnesiocopiapite [9]	9	1.510	1.535	1.575	Mod.
Magnesiocopiapite ("knoxvillite") [9]		1.507	1.529	1.576	67° (calc.)
Cuprocopiapite [7]	10	1.558	1.575	1.620	63° (meas.)

Chem. A hydrated basic sulfate of trivalent iron with divalent iron, magnesium, copper, or zinc. The ideal formula [11] may be written $AB_4(SO_4)_6(OH)_2 \cdot 20H_2O$, where A is ordinarily Fe'', Mg, Cu, Zn and B is Fe''', Al. A complete series exists in the A position in the case of Fe'' and Mg, these being the divalent cations ordinarily present, and probably also with Cu and Zn as well. The names copiapite, magnesiocopiapite, and cuprocopiapite are given to the regions of composition with Fe'' > (Mg, Cu, etc.), Mg > (Fe, Cu, etc.) and Cu > (Fe, Mg, etc.), respectively. Material with Zn > (Fe, Mg, etc.) has been analyzed,[12] but a species name is not here given in lack of substantiating evidence. Small amounts of other cations, including Mn'', Ca, Ni, and alkalies, apparently also substitute in the A position. The Fe'' of the ideal composition may oxidize in part or entirety to Fe'''. Al is commonly present and substitutes for Fe''' in the B position with Al:Fe = 1:4.1 in analysis 3. Cr has been reported but may be due to admixture.

Anal.[13]

	1	2	3	4	5	6
FeO	5.75		4.01	0.48	4.06	
CuO			1.00			
Fe_2O_3	25.55	30.75	23.21	30.51	26.10	31.92
Al_2O_3			3.62	tr.	tr.	
SO_3	38.43	39.28	38.44	38.87	38.37	38.89
H_2O	30.27	29.97	27.76	28.54	30.68	[29.19]
Rem.			0.68	2.55	0.71	
Total	100.00	100.00	98.72	100.95	99.92	[100.00]

1. Fe''Fe$_4$'''(SO$_4$)$_6$(OH)$_2$·20H$_2$O. 2. Copiapite. Artificial.[14] 3. Cuprian copiapite. Island Mountain, California.[17] Rem. is (Na,K)$_2$O 0.16, insol. 0.52. 4. Copiapite. Tierra Amarilla.[10] Rem. is insol. 5. Copiapite. Capo d'Arco, Elba.[10] Rem. is insol. 6. Copiapite. Sierra Gorda.[15]

	7	8	9	10	11	12
MgO	tr.	3.75	3.26		0.85	
FeO	0.94	0.27	0.52		0.45	0.44
CuO				5.72		
Fe_2O_3	21.91	24.46	27.44	27.66	26.19	27.28
Al_2O_3	4.15			1.47		1.72
SO_3	38.87	38.48	39.47	41.62	38.22	39.83
H_2O	33.33	32.39	27.84	23.51	31.33	29.92
Rem.	0.61	0.83	1.16	0.21	2.87	0.55
Total	99.81	100.18	99.69	100.19	99.91	99.74

7. Aluminian copiapite. Vigneria, Elba.[10] Rem. is insol. 8. Magnesiocopiapite. Falun, Sweden.[16] Rem. is MnO 0.16, ZnO 0.58, insol. 0.09. 9. Magnesiocopiapite. Blythe, California.[17] Rem. is insol. 10. Cuprocopiapite. Chuquicamata.[18] Rem. is insol. 11. Zincian copiapite. Rammelsberg, Harz.[19] Rem. is ZnO 2.47, insol. 0.40. 12. Copiapite. Chuquicamata.[18] Rem. is insol.

Tests. Easily soluble in water to a yellowish solution, with an acid reaction; the solution becomes turbid on heating. In C.T. affords much water which at higher temperatures is acid.

Var. *Aluminian.* Aluminocopiapite Berry (*Univ. Toronto Stud., Geol. Ser.*, no. 51, 21, 1947). Contains Al in substitution for Fe'''.

Ferrian. Oxidized copiapite. Ferricopiapite Berry (*Am. Min.*, **23**, pt. 2, no. 12, 3, 1938; *Univ. Toronto Stud., Geol. Ser.*, no. 51, 21, 1947). Contains Fe''' over the requirements of the formula, owing to oxidation of ferrous iron or to the substitution of Fe''' in the A position. The indices of refraction are relatively high. Analyses 2, 4, 6, 7.

Occur. The members of the copiapite series are secondary minerals associated with melanterite, alunogen, fibroferrite, halotrichite, and other sulfates formed by the oxidation of sulfides, especially pyrite. Copiapite occurs as loosely coherent masses of sulfur-yellow crystals at Goslar and Rammelsberg in the Harz, Germany. Greenish yellow and pulverulent in graphitic slates at Vashegy, Czechoslovakia (*janosite*[20]). With botryogen at Falun, Sweden. At Vigneria, Capo Calamita and Capo d'Arco, Elba.

Common in coal mines in France; in large masses in the coal near Commentry in the Plateau Central. In northern Chile found at Tierra Amarilla near Copiapó and common at Alcaparrosa, Quetena and Chuquicamata (*cuprocopiapite* pt.) in the northeastern part of the Atacama desert. In the United States found in California at the Redington mercury mine at Knoxville, Napa County (*knoxvillite* [21]), at the Island Mountain copper mine in Trinity County, with amarantite near Blythe in the Santa Maria Mountains, Riverside County, at the Sulfur Bank hot springs in Lake County, at the Alma mine, Leona Heights, Alameda County, and elsewhere. Found in Nevada notably in the mine workings of the Comstock Lode, and near Steamboat Springs. In Arizona in the Copper Queen mine, Bisbee, and in the fire zone of the United Verde mine at Jerome.

A l t e r. Begins to lose water on heating to 70° or less. Dehydration is complete or nearly so below 300°.[22]

A r t i f. In the system Fe_2O_3-SO_3-H_2O, copiapite may crystallize from solutions of nearly neutral composition at temperatures up to about 90°. At 50° copiapite is in equilibrium together with kornelite on the acid side of its stability range and with $Fe(SO_4)(OH) \cdot 2H_2O$ on the basic side. At lower temperatures its stability range meets that of coquimbite on the acid side.[23]

N a m e. From the city of Copiapó, Chile, near which the mineral is found. The Grecian μίσυ, *misy*, from the Island of Cyprus, and the misy of Agricola and other early writers from the Harz and other localities is possibly for the most part copiapite but a positive identification is not possible.

Ref.

1. Considered monoclinic by Linck, *Zs. Kr.*, **15**, 1 (1889), the first to measure crystals of copiapite, it has also often been reported to be orthorhombic on account of its misleading optical orientation. Shown to be triclinic by Ungemach, *Bull. soc. min.*, **58**, 97 (1935), whose angles, on crystals from Sierra Gorda (analysis 6), are here used. Setting and form list of Palache, Peacock, and Berry, *Univ. Toronto Stud., Geol. Ser.*, no. 50, 9 (1946); transformation: Ungemach (1935) to Palache, *et al.*, $\frac{1}{2}0\frac{1}{6}/\frac{1}{2}\frac{1}{2}0/\frac{1}{2}0\frac{\bar{1}}{6}$. Ungemach's angles are close to those obtained by Palache on Chuquicamata crystals (analysis 12). Morphological data are lacking on magnesiocopiapite; some approximate measurements on cuprocopiapite are given by Palache, *et al.* (1946). The name pseudocopiapite was proposed by Ungemach (1935) for crystals from Tierra Amarilla having slightly different elements. Berry, *Univ. Toronto Stud., Geol. Ser.*, no. 41, 7 (1938), confirmed the existence of such material, with the elements of Ungemach in the new setting as $a:b:c = 0.3938:1:0.3951$, α 91°18½′, β 102°08′, γ 98°59′. The difference in elements is due presumably to variation in composition, but analyses are lacking. The properties of pseudocopiapite lie within the observed or expected range of those of copiapite, and the name is not useful as a species or varietal designation.

2. For a full list of observed forms, totaling 308, see Palache, *et al.* (1946). Less common forms appearing on the figure, which shows a crystal lacking holohedral development, are $M4$ $\{1\bar{4}0\}$, $M5$ $\{1\bar{5}0\}$, $M7$ $\{1\bar{7}0\}$.

3. Peacock in Palache, *et al.* (1946) by Weissenberg method on Chuquicamata crystals (analysis 12). Berry (1938) obtained on Tierra Amarilla pseudocopiapite a_0 7.26 kX, b_0 18.67, c_0 7.45; α 90°58′, β 104°35′, γ 98°36′; $a_0:b_0:c_0 = 0.389:1:0.399$.

4. Ungemach (1935) in the new setting.
5. Berry (1938).
6. Berman in Palache, *et al.* (1946).

7. Berry, *Univ. Toronto Stud.*, *Geol. Ser.*, no. 51, 21 (1947), who summarizes the optical data.
8. Posnjak and Merwin, *J. Am. Chem. Soc.*, **44**, 1965 (1922).
9. Larsen (61, 1921).
10. Manasse, *Soc. toscana sc. nat.*, *Proc. verb.*, **20**, 1 (1911).
11. See the discussion of Berry (1947). Berry places Fe''' and Al in part in the A position of the formula, as isomorphous substitutions during initial growth, in the case of members of the series too low in Fe'' or other divalent metals and correspondingly too high in Fe''' in the B position for the requirements of the formula. Such ferrian material also may be interpreted as due to oxidation of Fe'' originally present, with Al restricted to the B position.
12. See analysis 11 and Berry (1947), analyses 37, 40.
13. For numerous additional analyses see Berry (1947).
14. Posnjak and Merwin, *J. Am. Chem. Soc.*, **44**, 1965 (1922).
15. Ungemach (1935).
16. Mauzelius analysis in Sjögren, *Geol. För. Förh.*, **17**, 268 (1895).
17. Foshag analysis in Bandy, *Am. Min.*, **23**, 737 (1938).
18. Gonyer analysis in Bandy (1938).
19. Scharizer, *Zs. Kr.*, **52**, 372 (1913).
20. Shown to be identical with copiapite—see Toborffy, *Zs. Kr.*, **43**, 369 (1907).
21. See Berry (1947) on the identity with copiapite.
22. On the dehydration of copiapite see Linck (1889), Darapsky, *Jb. Min.*, I, 49 (1890), and Scharizer, *Zs. Kr.*, **65**, 335 (1927).
23. Posnjak and Merwin (1922); Merwin and Posnjak, *Am. Min.*, **22**, 567 (1937).

32 COMPOUND SULFATES

TYPE 1. MISCELLANEOUS

32.1.1 H A N K S I T E [$Na_{22}K(SO_4)_9(CO_3)_2Cl$]. Hidden (Am. J. Sc., 30, 133, 1885).

Cryst. Hexagonal; dipyramidal—$6/m$.

$$a:c = 1:2.025;\ ^1 \qquad p_0:r_0 = 2.338:1$$

Forms: [2]

	ϕ	$\rho = C$	M	A_2
c 0001	.....	0°00′	90°00′	90°00′
m 10$\bar{1}$0	30°00′	90 00	60 00	90 00
p 20$\bar{2}$5	30 00	43 05	70 02	90 00
o 10$\bar{1}$2	30 00	49 27	67 40	90 00
s 10$\bar{1}$1	30 00	66 51	62 38	90 00

Structure cell.[3] Space group $C6_3/m$. a_0 10.46 kX, c_0 21.18; $a_0:c_0$ = 1:2.025. Cell contents $Na_{44}K_2(SO_4)_{18}(CO_3)_4Cl_2$.

Habit. Large euhedral crystals up to 3 inches in size, often in interpenetrating groups. In hexagonal prisms, usually short prismatic [0001]

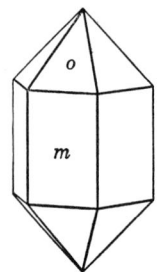
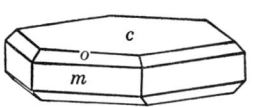

or tabular {0001}; also as quartzoids with large {10$\bar{1}$2}. {10$\bar{1}$0} striated horizontally.

Phys. Cleavage {0001}, good. Fracture uneven. Brittle. H. 3-3$\frac{1}{2}$. G. 2.562;[4] 2.57 (calc.). Luster vitreous to dull. Colorless, sometimes faintly yellowish or gray due to inclusions of clay particles. Streak white. Transparent to translucent. Taste saline. Weakly fluorescent in pale yellow in long-wave ultraviolet radiation. Sometimes falls to pieces or deliquesces when mixed with other salts.

Opt.[5] In transmitted light, colorless.

nO	1.481	Uniaxial negative (−).
nE	1.461	

C h e m. A chloride-sulfate of sodium and potassium, $Na_{22}K(SO_4)_9(CO_3)_2Cl$. The K and Cl content, not recognized in the original description, has been shown [6] to be essential.

Anal.

	1	2	3
Na_2O	43.57	40.86	43.61
K	2.50	2.33	2.39
Cl	2.27	2.13	2.28
CO_2	5.62	5.42	5.63
SO_3	46.04	43.59	45.78
Rem.		5.73	0.12
Total	100.00	100.06	99.81
G.	2.57		2.545

1. $Na_{22}K(SO_4)_9(CO_3)_2Cl$. 2. Searles Lake.[7] Rem. is insol. 4.41, ign. loss 1.32. 3. Searles Lake.[8] Rem. is insol. Very pure crystal.

Tests. Easily fusible. Easily soluble in water. Effervesces weakly in dilute acids.

O c c u r. Hanksite is abundant in the salt deposits of Searles Lake, San Bernardino County, California.[7] There it is found in drill cores from various horizons, embedded in mud or associated with halite, borax, trona, and aphthitalite. Also reported from the sinks of Death Valley, Inyo County, California.[7]

A r t i f. Hanksite is one of the solid phases obtained by crystallization in the system [9] $NaCl-KCl-Na_2SO_4-K_2SO_4-Na_2CO_3-K_2CO_3-H_2O$ at 20° C. Here it forms in equilibrium with halite, thenardite, aphthitalite, and burkeite. The supposed compound $4Na_2SO_4 \cdot Na_2CO_3$ has been said [10] to be identical with hanksite, but later work has shown [11] this material to be a member of a solid-solution series with burkeite. On heating, hanksite breaks down to a mixture of KCl and $9Na_2SO_4 \cdot 2Na_2CO_3$, the latter a hexagonal polymorph of the burkeite series.

N a m e. After Henry G. Hanks (1826–1907), a former State Mineralogist of California.

Ref.

1. Ratio of the structure cell of Ramsdell, *Am. Min.*, **24**, 109 (1939). The original morphological cell of Hidden (1885), apparently based on measurements to only half-degrees, has c halved.
2. Goldschmidt (**4**, 109, 1918).
3. Ramsdell (1939) by Weissenberg method on natural material. Similar values were reported by Gossner and Koch, *Zs. Kr.*, **80**, 416 (1931).
4. Hidden (1885).
5. Pabst, priv. comm. (1948), on Searles Lake material. The [identical] indices reported by de Schulten, *C.R.*, **123**, 1325 (1896), were not obtained on hanksite.
6. Ramsdell (1939) and Pratt, *Am. J. Sc.*, **2**, 135 (1896).
7. Dana and Penfield, *Am. J. Sc.*, **30**, 136 (1885).
8. Pratt (1896).
9. See Teeple, *The Industrial Devel. of Searles Lake Brines*, New York, 102 (1929).
10. de Schulten (1896); also Wegscheider, *Min. Mitt.*, **39**, 316 (1928).
11. Ramsdell (1939).

32.1.2 CALEDONITE [$Cu_2Pb_5(SO_4)_3(CO_3)(OH)_6$]. Cupreous Sulfatocarbonate of Lead Brooke (Edinburgh Phil. J., **3**, 117, 1820). Plomb sulfato-carbonaté cuprifère Dufrénoy (**3**, 260, 1847). Kupferhaltiges schwefelkohlensaures Blei Leonhard (254, 1826). Prismatischer Kupferbleispat Breithaupt (53, 1823). Halblasurblei Hausmann (1847). Calédonite Beudant (**2**, 367, 1832).

Cryst.[1] Orthorhombic; dipyramidal—$2/m\ 2/m\ 2/m$.

$a:b:c = 0.3555:1:0.3263;$ $p_0:q_0:r_0 = 0.9180:0.3263:1$

$q_1:r_1:p_1 = 0.3555:1.0893:1;$ $r_2:p_2:q_2 = 3.0647:2.8134:1$

Forms:[2]

	ϕ	$\rho = C$	ϕ_1	$\rho_1 = A$	ϕ_2	$\rho_2 = B$	G.&L.	B.&M.	Sch.
b 010	0°00′	90°00′	90°00′	90°00′	.....	0°00′	010	001	001
a 100	90 00	90 00	.....	0 00	0°00′	90 00	100	010	100
ψ 160	25 07½	90 00	90 00	64 52½	0 00	25 07½	...	013	$\bar{1}03$
f 140	35 07	90 00	90 00	54 53	0 00	35 07	...	012	±102
κ 130	43 09½	90 00	90 00	46 50½	0 00	43 09½	...	023	...
e 120	54 35½	90 00	90 00	35 24½	0 00	54 35½	110	011	+101
m 110	70 26	90 00	90 00	29 34	0 00	70 26	...	021	$\bar{2}01$
x 011	0 00	18 04½	18 04½	90 00	90 00	71 55½	012	201	...
d 021	0 00	33 07½	33 07½	90 00	90 00	56 52½	...	101	...
α 102	90 00	24 39½	0 00	65 20½	65 20½	90 00	...	...	...
ν 101	90 00	42 33	0 00	47 27	47 27	90 00	101	110	110
β 201	90 00	61 25½	0 00	28 34½	28 34½	90 00	...	...	...
t 111	70 26	44 15	18 04½	48 53½	47 27	76 29	212	221	±221
r 121	54 35½	48 24	33 07½	52 27	47 27	64 19½	111	111	±111
s 131	43 09½	53 18½	44 23½	56 44	47 27	54 12	232	223	±223
y 141	35 07	57 55½	52 32½	60 49½	47 27	46 07½	...	...	...
τ 161	25 07½	65 11	62 56½	67 20	47 27	34 44	...	113	...

Structure cell.[3] Space group $Pnmm$. $a_0\ 7.14\ kX$, $b_0\ 20.06$, $c_0\ 6.55$; $a_0:b_0:c_0 = 0.356:1:0.326$. Cell contents $Cu_4Pb_{10}(SO_4)_6(CO_3)_2(OH)_{12}$.

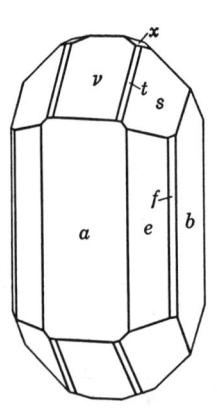

Las Cruces, New Mexico.

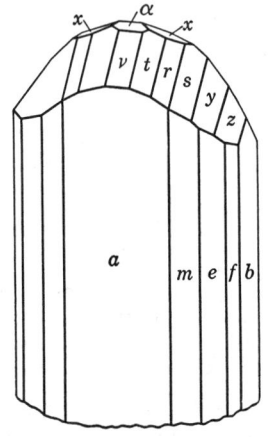

Talisman mine, Utah.

CALEDONITE 631

Habit. Elongated [001]; crystals usually small and in divergent groups. Often striated and with vicinal faces in the zone [001]. Common forms: $b\ a\ e\ x\ v\ t\ r\ s$. As coatings; rarely massive.

P h y s. Cleavage {010} perfect, {100} and {101} incomplete. Fracture uneven. Rather brittle. H. $2\frac{1}{2}$–3. G. 5.76 ± 0.01,[4] 5.6 (Mammoth mine, Arizona);[13] 5.68 (calc.). Luster resinous. Color deep verdigris-green or bluish green. Streak greenish white. Translucent.

O p t. In transmitted light, pale bluish green.

ORIENTATION[5]		n (Inyo Co., Calif.)[6]	
X	c	1.818 ± 0.003	Biaxial negative (−).
Y	a	1.866 ± 0.003	$2V \sim 85°$.
Z	b	1.909 ± 0.003	$r < v$, barely perceptible.
			Weakly pleochroic.

C h e m. A basic carbonate-sulfate of lead and copper, $Cu_2Pb_5(SO_4)_3(CO_3)(OH)_6$. The CO_2 was long regarded as due to admixed cerussite but has been shown[7] to be an essential constituent.

Anal.[8]

	1	2	3	4
CuO	9.86	9.73	10.7	9.07
PbO	69.17	69.18	67.7	66.69
SO_3	14.89	14.15	15.6	12.43
CO_2	2.73	3.16	1.9	3.29
H_2O	3.35	3.78	3.5	3.34
Rem.				0.20
Total	100.00	100.00	99.4	95.02
G.				6.13

1. $Cu_2Pb_5(SO_4)_3(CO_3)(OH)_6$. 2. Challacollo, Chile.[9] Recalculated to 100 after deducting insol. 2.31. 3. Leadhills, Scotland.[10] 4. Toroku mine, Japan.[12] Rem. is insol. and As_2O_5 tr. Average of two. Low total due to loss of $PbSO_4$ during analysis.

Tests. In C.T. decrepitates, turns dark and opaque, and affords water. B.B. swells and easily fuses to a black bead. Soluble with effervescence in nitric acid, leaving a deposit of lead sulfate.

O c c u r. A secondary mineral found rather widely although in small amounts in the oxidized zone of copper-lead deposits. Associated with cerussite, anglesite, linarite, leadhillite, malachite, azurite, and less frequently, lanarkite, hydrocerussite, mimetite, smithsonite, wulfenite, etc. Not all the known localities can be listed here. At Rezbánya, Transylvania; in the Preobrashensky mine, Beresov, Urals, U.S.S.R. At Mala Calzetta near Iglesias, Sardinia. In the Nuissière mine, Chenelette, Dept. Rhône, France. At Leadhills, Lanarkshire, and at Wanlockhead, Dumfriesshire, in Scotland; at Red Gill, Cumberland, England. From Challacollo, in the Sierra Gorda, Atacama, Chile. At Tsumeb, South West Africa.

In the United States in the Mammoth mine, Tiger, Arizona. In the Organ district, Dona Ana County, New Mexico. Reported from several localities in Idaho.[11] In California at Cerro Gordo, Inyo County, and in the Blind Spring Hill district, Mono County. From the Talisman mine,

Beaver Creek, Utah. At Butte, Montana. In Canada from Beaver Mountain, Slocan, British Columbia.

Alter. Observed altered to cerussite.

Name. From *Caledonia*, because found in Scotland.

Ref.

1. Elements of Goldschmidt (398, 1897), representing the average of four earlier observers, transformed to the position of Greg and Lettsom (403, 1858) but with b doubled to conform to the structure cell of Richmond in Palache, *Am. Min.*, **24**, 441 (1939). Transformations: Greg and Lettsom to new, 100/020/001; Schrauf, *Ak. Wien, Sitzber.*, **64**, 179 (1871), to new, 100/002/010; Brooke and Miller (561, 1852) and Goldschmidt (1897) to new, 010/002/100. The monoclinic interpretation of Schrauf (1871) and Jeremejev, *Vh. russ. min. Ges.*, **17**, 213 (1883) is not supported by the optical and x-ray evidence. See also Yosimura, *J. Fac. Sc. Hokkaidô Univ.*, 4 [4], 453 (1939).
2. Palache (1939; and priv. comm., 1941). Rare and uncertain forms:

001	1.24.0	1.10.0	10.1.10	171	142	241
1.48.0	1.20.0	180	787	181	211	
1.40.0	1.16.0	021	3.10.3	191	221	
1.32.0	1.12.0	201	151	1.10.1	231	

3. Richmond in Palache (1939) by Weissenberg method.
4. Palache and Richmond (1939).
5. Berman in Palache (1939). The orientation $X = a$ (or $X = b$ before transformation from Brooke and Miller's position) given by Dana (925, 1892) and Larsen and Berman (203, 1934) is in error.
6. Larsen (52, 1921). Des Cloizeaux (205, 1867) gave for red $2V$ 82°37′ and nY 1.846, for blue $2V$ 83°03′ and nY 1.864.
7. Berg, *Min. Mitt.*, **20**, 390 (1901).
8. For other analyses see Hintze (1 [3B], 4257, 1929).
9. Liebert analysis in Berg (1901).
10. Collie, *J. Chem. Soc., London*, **55**, 9 (1889).
11. Shannon (453, 1926).
12. Yosimura (1939).
13. Mrose, priv. comm. (1948), by microbalance.

32.1.3 Wherryite [$Pb_4Cu(CO_3)(SO_4)_2(OH,Cl)_2O$ (?)]. *Fahey, Daggett,* and *Gordon* (*Am. Min.*, **35**, 93, 1950).

Massive, fine-granular. Color light green. G. 6.45. Optically biaxial negative $(-)$, with nX 1.942, nY 2.010, nZ 2.024 (all ± 0.005; for Na), and $2V$ 50° (calc.). A basic carbonate-sulfate of lead and copper, apparently $Pb_4Cu(CO_3)(SO_4)_2(OH,Cl)_2O$. Analysis gave (F not determined):[1]

PbO	CuO	SO_3	CO_2	Cl	H_2O-	H_2O+	Insol.	Total	O=Cl
72.9	7.3	13.0	3.1	0.9	nil	1.2	2.2	100.4	0.2

Tests. In C.T. turns black without decrepitation and melts. Slowly soluble in cold HCl or HNO_3.

Occur. Found at the Mammoth mine, at Tiger, Pinal County, Arizona, as a secondary mineral associated with leadhillite, paralaurionite, diaboleite, chrysocolla, cerussite.

Named after Edgar T. Wherry (1885–), American mineralogist and plant ecologist.

Ref.

1. Fahey analysis in Fahey, *et al.* (1950).

BURKEITE

32.1.4 BURKEITE [$Na_6(CO_3)(SO_4)_2$]. *Foshag* (*Am. Min.*, **20**, 50, 1935).

Cryst.[1] Orthorhombic.

$$a:b = 0.574:1$$

Habit. Tabular {100}, with narrow faces of {010}, {110}, and {001}; the crystal faces are rough and uneven. As reticulated aggregates of platy crystals; also in pea-shaped nodules and as irregular warty masses. Massive.

Twinning. Common, as X-shaped penetration twins with twin plane {110} and $b \wedge \underline{b}$ ~59°42'. The re-entrant angles are partially filled with small, rough crystals apparently in random position.

Phys. Fracture conchoidal. Brittle. H. $3\frac{1}{2}$. G. 2.57. Luster vitreous to greasy, resembling that of cryolite. Color white to pale buff or grayish. Transparent.

Opt. In transmitted light, colorless.

ORIENTATION		n	n (artif.)	
X	c	1.448	1.450	Biaxial negative (−).
Y	a	1.489	1.490	$2V$ 34° (meas.).
Z	b	1.493	1.492	$r > v$, distinct.

Chem. A carbonate-sulfate of sodium, $Na_6(CO_3)(SO_4)_2$. SO_4 and CO_3 can substitute mutually over a wide range in artificial material.[4]

Anal.

	1	2
Na_2O	47.67	47.89
CO_2	11.28	11.72
SO_3	41.05	39.96
Rem.		0.17
Total	100.00	99.74
G.		2.57

1. $Na_6(CO_3)(SO_4)_2$. 2. Searles Lake. Rem. is insol. 0.04, Cl 0.09, H_2O 0.04.

Tests. B.B. fuses easily and quietly to a clear bead. Completely soluble in cold water.

Occur. Found with gaylussite, trona, sulfohalite, borax, northupite, thenardite, calcite, tychite, halite, and rarely hanksite in drill cuttings from a clay stratum in Searles (Borax) Lake, San Bernardino County, California.

Artif. Occurs in the system Na_2SO_4-Na_2CO_3-H_2O above about 25°, and in related systems.[2] Obtained in large amounts during the refining of Searles Lake brines.

Name. After W. E. Burke, chemical engineer and director of research of the American Potash and Chemical Corporation, who discovered [3] the artificial salt.

Ref.

1. Symmetry inferred from optical properties and the habit of rough crystals. Partial axial ratio from approximate measurements on twinned crystals.

2. See Teeple, *Industrial Devel. of Searles Lake Brines*, Am. Chem. Soc. Monog., no. 49, New York (1929).
3. Burke, *J. Ind. Eng. Chem.*, **13,** 249 (1921).
4. Schroeder, Berk, Partridge, and Gabriel, *J. Am. Chem. Soc.*, **58,** 846 (1936); Kracek and Ksanda, *J. Phys. Chem.*, **34,** 1741 (1930).

UNNAMED MINERAL. Schaller (*Am. Min.*, **25,** 213, 1940).

A scaly white mineral resembling alunogen in appearance and properties, found with halotrichite, siderotil, and szomolnokite in the Tintic Standard mine, Dividend, Utah. Soluble in cold water. Does not lose H_2O on heating at 105°. Composition said to be $2Al_2O_3 \cdot 4SO_3 \cdot P_2O_5 \cdot 24H_2O$.

SELENATES AND TELLURATES
SELENITES AND TELLURITES

CLASS 33. Selenates and tellurates
Type 1. Miscellaneous
- 33.1.1 Ahlfeldite — $Ni(SeO_4) \cdot 6H_2O$ (?)
- 33.1.2 Teineite — $Cu(Te,SO_4) \cdot 2H_2O$
- 33.1.3 Montanite — $(BiO)_2(TeO_4) \cdot 2H_2O$

CLASS 34. Selenites and tellurites
Type 1. $A(XO_3) \cdot xH_2O$
- 34.1.1 Chalcomenite — $CuSeO_3 \cdot 2H_2O$
- 34.1.2 Cobaltomenite — $CoSeO_3 \cdot 2H_2O$ (?)
- 34.1.3 Kerstenite — $PbSeO_3 \cdot 2H_2O$ (?)

Type 2. $A_2(XO_3)_3 \cdot xH_2O$
- 34.2.1 Emmonsite — $Fe_2(TeO_3)_3 \cdot 2H_2O$ (?)
- 34.2.2 Mackayite — $Fe_2(TeO_3)_3 \cdot xH_2O$ (?)
- 34.2.3 Blakeite — (ferric tellurite)

33 SELENATES AND TELLURATES

TYPE 1. MISCELLANEOUS

33.1.1 Ahlfeldite. Herzenberg (*Zbl. Min.*, 189, 1935; *Inst. Boliviano de Ing. de Minas y Geol., Publ. Técu*, La Paz, no. 5, 1945; *Mineria Boliviana*, **2**, no. 25, 1945).

Greenish to yellowish crystals, with a reddish brown coating. Probably triclinic. Qualitative tests indicate the mineral to be a hydrous selenite of nickel distinct from cobaltomenite. Found as an alteration product of penroseite at the Pacajake mine, 30 km. from Colquechaca, Bolivia. Named after Friedrich Ahlfeld, German mining engineer and mineralogist.

33.1.2 TEINEITE $[Cu_{13}(SO_4)_3(TeO_4)_{10} \cdot 26H_2O]$. Yosimura (*J. Fac. Sc. Hokkaidô Univ.*, [4], **4**, 465, 1939).

Cryst. Orthorhombic; dipyramidal—$2/m\ 2/m\ 2/m$.

$a:b:c = 0.7051:1:0.7860;\qquad p_0:q_0:r_0 = 1.1147:0.7860:1$

$q_1:r_1:p_1 = 0.7051:0.8971:1;\qquad r_2:p_2:q_2 = 1.2723:1.4182:1$

Forms:

	ϕ	$\rho = C$	ϕ_1	$\rho_1 = A$	ϕ_2	$\rho_2 = B$
b 010	0°00′	90°00′	90°00′	90°00′	.....	0°00′
m 110	54 48½	90 00	90 00	35 11¾	0°00′	54 48½
e 011	0 00	38 10	38 10	90 00	90 00	51 50
f 073	0 00	61 24	61 24	90 00	90 00	28 36

Habit. Prismatic [001], sometimes also flattened {010}. As crusts and intersecting aggregates.

Phys. Cleavage {010}, good, {001} and {100}, poor. Brittle. H. $2\frac{1}{2}$. G. 3.80. Color deep sky-blue, inclining to cobalt-blue or bluish gray when altered. Streak bluish white.

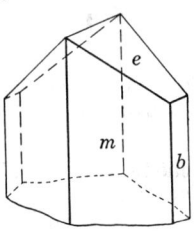

Opt. In transmitted light, bright blue to greenish blue.

Orientation		n	Pleochroism	
X	a	1.767	Greenish blue	Biaxial negative ($-$).
Y	b	1.782	Blue	2V 36°.
Z	c	1.791	Indigo-blue	Absorption $Z > Y > X$.

Chem. A hydrated sulfate-tellurate of copper, $Cu_{13}(SO_4)_3(TeO_4)_{10} \cdot 26H_2O$, or perhaps $Cu[(Te,S)O_4] \cdot 2H_2O$ with S:Te = 1:3.33 (analysis 2).

Anal.

	1	2
CuO	29.56	29.54
SO$_3$	6.86	6.96
TeO$_3$	50.19	50.63
H$_2$O	13.39	12.87
Total	100.00	100.00
G.		3.80

1. $Cu_{13}(SO_4)_3(TeO_4)_{10} \cdot 26H_2O$. 2. Teine mine. Recalculated to 100 after deducting insol. 6.1 per cent.

Tests. B.B. fuses at 2 to a black bead. In C.T. loses water without decomposition of the tellurate. Dissolves in HCl to a greenish yellow solution and in HNO$_3$ to a blue solution; a precipitate of TeO$_2$ may form which dissolves on heating.

Occur. Found with tellurite as an oxidation product of tetrahedrite and native tellurium in the Takinosawa vein of the Teine mine, Hokkaidô, Japan.

Alter. To malachite and azurite.

33.1.3 Montanite. *Genth (Am. J. Sc., 45, 318, 1868)*

Soft and earthy to compact. Luster dull to waxy. Color yellowish, greenish to white. Opaque.

Opt.[1] Material from Highland, Montana, appears as fibers which tend to lie on a face nearly normal to X. Biaxial negative ($-$), with 2V small and $r < v$, extreme. Shows abnormal green interference colors. nY 2.09 ± 0.03, with birefringence 0.01.

Chem. A tellurate of bismuth, perhaps $(BiO)_2(TeO_4) \cdot 2H_2O$.

Anal.

	1	2	3	4
Bi_2O_3	68.78	66.78	71.90	50.68
Fe_2O_3		0.56	0.32	14.38
TeO_3	25.90	26.83	23.90	27.65
H_2O	5.32	[5.44]	[2.80]	6.16
Rem.		0.39	1.08	1.00
Total	100.00	[100.00]	[100.00]	99.87

1. $(BiO)_2(TeO_4) \cdot 2H_2O$. 2. Highland, Montana.[2] Rem. is PbO. 3. Highland.[2] Rem. is CuO. 4. Norongo, New South Wales.[3] Rem. is gangue. Contains admixed limonite.

O c c u r. Found as an alteration of tetradymite at Highland, Montana, and in Davidson County, North Carolina. In New South Wales with tetradymite at Norongo near Captain's Flat, in part reddish brown and admixed with limonite as pseudomorphs after pyrite (analysis 4); also with bismutite and tetradymite at Nanima near Yass. Said to occur in the Organ district, Dona Ana County, New Mexico.

Ref.
1. Larsen (112, 1921).
2. Genth (1868), who cites a third analysis.
3. Mingaye, *Australasian Assoc. Adv. Sc.*, **1,** 116 (1888).

34 SELENITES AND TELLURITES

TYPE 1. $AXO_3 \cdot xH_2O$

34.1.1 CHALCOMENITE [$CuSeO_3 \cdot 2H_2O$]. Chalcoménite *Des Cloizeaux* and *Damour* (*Bull. soc. min.*, **4**, 51, 164, 1881).

Cryst.[1] Orthorhombic; disphenoidal—2 2 2.

$$a:b:c = 0.7325:1:0.8077; \qquad p_0:q_0:r_0 = 1.1026:0.8077:1$$

$$q_1:r_1:p_1 = 0.7325:0.9070:1; \qquad r_2:p_2:q_2 = 1.2381:1.3651:1$$

Forms:[2]

	ϕ	$\rho = C$	ϕ_1	$\rho_1 = A$	ϕ_2	$\rho_2 = B$
c 001	.....	0°00′	0°00′	90°00′	90°00′	90°00′
b 010	0°00′	90 00	90 00	90 00	.....	0 00
n 120	34 19	90 00	90 00	55 41	0 00	34 19
m 110	53 46½	90 00	90 00	36 13½	0 00	53 46½
β 012	0 00	21 59½	21 59½	90 00	90 00	68 00½
q 011	0 00	38 55½	38 55½	90 00	90 00	51 04½
t 021	0 00	58 14½	58 14½	90 00	90 00	31 45½
d 101	90 00	47 47½	0 00	42 12½	42 12½	90 00
o 111	53 46½	53 48½	38 55	49 22½	42 12½	61 31
p $\bar{1}11$	−53 46½	53 48½	−126 13½	49 22½	137 47½	61 31
δ 141	18 50½	73 40½	72 48	71 56½	42 12½	24 44

Structure cell.[3] Space group $P2_12_12_1$. a_0 6.56 kX, b_0 9.10, c_0 7.36; $a_0:b_0:c_0 = 0.731:1:0.809$. Cell contents $Cu_4(SeO_3)_4 \cdot 8H_2O$.

Habit. Minute crystals, acicular [001]. Prismatic faces often striated or rounded.

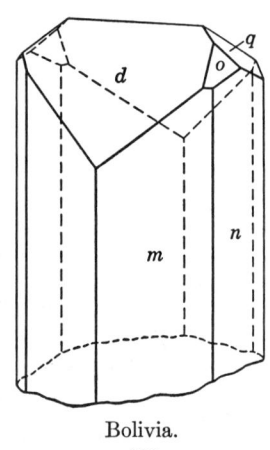

Bolivia.

P h y s. Cleavage none. H. 2–2½. G. 3.35 (Hiaco);[3] 3.40 (calc.). Luster vitreous. Color bright blue. Transparent.
O p t. In transmitted light, pale blue.

ORIEN-TATION	n(Na) (Hiaco)[3]	n (Cerro de Cacheuta)[5]	PLEOCHROISM	
X a	1.712 ± 0.002	1.710 ± 0.003	Pale blue	Biaxial negative $(-)$.
Y c	1.732 ± 0.002	1.731 ± 0.003	Darker blue	
Z b	1.732 ± 0.002	1.732 ± 0.003	Darker blue	
$2V$	$\sim 0°$	$34°$ (Li)		
Disp.	$r < v$, strong	$r > v$, strong		

C h e m. Copper selenite dihydrate, $Cu(SeO_3) \cdot 2H_2O$. Analysis gave:[4]

	CuO	SeO$_2$	H$_2$O	Total
1.	35.12	48.98	15.90	100.00
2.	35.40	48.12	15.30	98.82

1. $Cu(SeO_3) \cdot 2H_2O$. 2. Cerro de Cacheuta.[4]

Tests. B.B. fusible. In C.T. affords water and a sublimate of SeO$_2$. Soluble in acids.

O c c u r. With azurite and cerussite as an oxidation product of primary selenides of copper and lead in veins in the Cerro de Cacheuta, Mendoza, Argentina. Also from the Sierra de Umango and Sierra Famatina, La Rioja, Argentina. From the Hiaco mine, Pacajake, about 30 km. from Colquechaca, Bolivia, associated with cobaltomenite and molybdomenite as an alteration of penroseite.

A r t i f.[6] In crystals by heating the precipitated salt with water in a closed tube.

N a m e. From χαλκός, *copper*, and μήνη (for σελήνη), the *moon*.

Ref.
1. Palache, *Am. Min.*, **22**, 790, 1123 (1937), on Hiaco mine crystals. Transformations: Des Cloizeaux and Damour (1881) and Dana (980, 1892) to new, $10\overline{2}/\overline{1}04/010$; Goldschmidt (**2**, 132, 1913) to new, $20\overline{1}/\overline{2}0\overline{2}/020$; Friedel and Sarasin, *Bull. soc. min.*, **4**, 176, 225 (1881), and Groth (**2**, 300, 1908) to new, 100/001/010.
2. Palache (1937) and Ungemach in Palache (1937).
3. Berman in Palache (1937).
4. Damour analysis in Des Cloizeaux and Damour (1881). Also a partial analysis by Gonyer in Palache (1937) of Hiaco crystals with CuO:SeO$_2 \sim 1:1$.
5. Larsen (55, 1921).
6. Friedel and Sarasin (1881).

34.1.2 Cobaltomenite. Bertrand (*Bull. soc. min.*, **5**, 90, 1882).

In minute rose-red monoclinic crystals, resembling erythrite. Biaxial negative $(-)$. The axial plane is parallel to the elongation of the crystals, and the acute bisectrix is normal thereto and inclined at a large angle to a cleavage. Contains cobalt and is perhaps a cobalt selenite. Found with chalcomenite and molybdomenite as an alteration of selenides in the Cerro de Cacheuta, Mendoza, Argentina. Several artificial cobalt selenites are known,[1] including $CoSeO_3$ and $CoSeO_3 \cdot 2H_2O$.

Ref.
1. Mellor (**10**, 840, 1930).

34.1.3 Kerstenite. Selenichtsaures Bleioxyd *Kersten* (*Ann. Phys.*, **46**, 277, 1839). Selenbleispath, Bleiselenit *Germ.* Kerstenite *Dana* (669, 1868). Glasbachit *Adam* (52, 1869). Molybdomenite *Bertrand* (*Bull. soc. min.*, **5**, 90, 1882). Selenite de plomb.

Minerals that can be regarded as lead selenite or selenate have been reported, but their true nature remains uncertain. Kerstenite was found as botryoidal masses with a fibrous fracture as an alteration of zorgite at the Friedrichsglück mine near Hildburghausen, Thuringia, Germany. Color sulfur-yellow, with a greasy to vitreous luster. Cleavage distinct in one direction. H. 3–4. Contains some copper in addition to lead and selenious oxide. Molybdomenite was found as very thin, apparently orthorhombic scales. Color white, with pearly luster. Cleavable in two directions. Biaxial positive (+), with the axial plane normal to intersection of two cleavages and X perpendicular to the best cleavage and plane of flattening. Some varieties contain copper in addition to lead and selenium. Found with chalcomenite in the Cerro de Cacheuta, Mendoza, Argentina. Several basic and normal lead selenites have been synthesized,[1] including $Pb_2Se_2O_7$ and $PbSeO_3$. Kerstenite was named after K. M. Kersten (1803–1850), Professor of Chemistry in the Freiberg School of Mines. Molybdomenite from μόλυβδος, *lead*, and μήνη, *moon*.

Ref.
1. Mellor (**10**, 833, 1930).

TYPE 2. $A_2(XO_3)_3 \cdot xH_2O$

34.2.1 E M M O N I T E [$Fe_2(TeO_3)_3 \cdot 2H_2O$ (?)]. *Hillebrand* (*Proc. Colorado Sc. Soc.*, **2**, pt. 1, 20, 1885).
Durdenite *Dana* and *Wells* (*Am. J. Sc.*, **40**, *78, 1890*).

C r y s t. Monoclinic (?).[1]
Habit. As fibrous crusts with a small botryoidal surface, grading into isolated globular aggregates, and as compact microcrystalline masses; also as radial or lichen-like groups of rough acicular crystals, and as druses of thin scales.
Twinning. Observed in microscopic crystals from Goldfield, Nevada.[2]
P h y s. Cleavage {010} perfect, with two unequal cleavages inclined thereto and making plane angles of 85° and 95° in {010}.[3] H. about 5. G. 4.52 (Silver City, New Mexico),[2] 4.53 (Cripple Creek).[3] Luster vitreous. Color yellowish green. Transparent in small grains.
O p t. In transmitted light, pale yellow-green in color.

ORIENTATION	n (Cripple Creek)[4]	n (Goldfield)[2]	
X	1.962 ± 0.002	1.962 ± 0.004	Biaxial negative (−).
Y b	2.09 ± 0.02		$r > v$, strong.
Z	2.10 ± 0.02	2.12 ± 0.02	Weakly pleochroic, with absorption
$2V$	20° ±	small	$Z > X$.

X is nearly perpendicular to a cleavage trace on {010}.

EMMONSITE 641

Chem. A hydrated ferric tellurite, perhaps $Fe_2(TeO_3)_3 \cdot 2H_2O$. Se substitutes in part for Te (anal. 8). On heating, emmonsite fuses, loses water, and leaves $Fe_2(TeO_3)_3$.[2]

Anal.

	1	2	3	4	5	6	7	8	9
Fe_2O_3	23.67	[20.02]	[20.10]	[21.30]	[20.30]	22.67	22.81	22.79	25.41
Fe		14.00	14.06	14.90	14.20				
TeO_2	70.99	[74.80]	[74.01]	[73.90]	[74.00]	70.83	71.80	70.20	62.34
Te		59.77	59.15	59.05	59.14				
SeO_2									2.12
H_2O+	5.34	3.28				} 4.68	4.82	0.21	} 10.13
H_2O-									
Rem.					2.50	0.34	0.58	0.54	
Total	100.00	[98.10]	[94.11]	[95.20]	[96.80]	98.52	100.01	93.74	100.00
G.						4.53			

1. $Fe_2(TeO_3)_3 \cdot 2H_2O$. 2, 3, 4, 5. Tombstone, Arizona.[5] Oxide per cent calculated from determination as element. Rem. in anal. 5 is ZnO 1.94, CaO 0.56. 6, 7, 8. Cripple Creek, Colorado.[6] Rem. in anal. 6 is P_2O_5 0.34; anal. 7, Al_2O_3 0.58; anal. 8, Al_2O_3 0.54. 9. Honduras (*durdenite*).[7] Recalc. to 100 after deduction of 23.89 insol.

Tests. Melts easily to a deep red liquid, loses water, and leaves an anhydrous residue. Easily soluble in strong acid.

Occur. A secondary mineral, formed by the oxidation of tellurides and native tellurium. Associated with tellurite, finely divided gold, limonite, quartz, and other tellurites of iron. Originally found with cerussite near Tombstone, Arizona, the exact locality unknown. Later found at Cripple Creek, Teller County, Colorado; also near Silver City, New Mexico; and with mackayite and blakeite at Goldfield, Nevada. At the El Plomo mine in the Ojojoma district, Tegucigalpa, Honduras (*durdenite*).

Artif. Anhydrous ferric tellurite has been prepared, but artificial hydrates are not known.[2]

Name. After Samuel Franklin Emmons (1841–1911), economic geologist, for many years on the U. S. Geological Survey. Durdenite after Henry S. Durden, of San Francisco, from whom the specimens were received.

DURDENITE. Dana and *Wells* (*Am. J. Sc.*, **40**, 78, 1890).
Found as small mammillary crusts in oxidized vein material containing native tellurium and selen-tellurium in the El Plomo mine, Tegucigalpa, Honduras. Supposedly $Fe_2(TeO_3)_3 \cdot 4H_2O$ (anal. 9). Shown to be identical with emmonsite.[2]

Ref.

1. Inferred from the cleavage and the optical data of Cross in Hillebrand (1885) and Schaller in Hillebrand, *Am. J. Sc.*, **18**, 433 (1904). Morphological and x-ray data are lacking.
2. Frondel and Pough, *Am. Min.*, **29**, 211 (1944).
3. Cross and Schaller in Hillebrand (1885; 1904).
4. Larsen (71, 1921; 208, 1934); the value for nX is from Frondel and Pough (1944). Larsen gives 1.95. The optical data given by Larsen (70, 1921) for durdenite refer to an unidentified mineral.
5. Hillebrand (1885).
6. Hillebrand (1904).
7. Wells in Dana and Wells, *Am. J. Sc.*, **40**, 78 (1890).
8. Hillebrand (1904) after correcting for impurities.

34.2.2 MACKAYITE [$Fe_2(TeO_3)_3 \cdot xH_2O$?]. *Frondel* and *Pough* (*Am Min.*, **29**, 211, 1944).

Cryst.[1] Tetragonal; ditetragonal dipyramidal—$4/m\ 2/m\ 2/m$ (?).

$$a:c = 1:1.259; \quad p_0:r_0 = 1.259:1$$

Forms:

	ϕ	ρ	A	M
a 010	0°00′	90°00′	90°00′	45°00′
m 110	45 00	90 00	45 00	90 00
g 012	0 00	32 00	90 00	68 00
f 112	45 00	41 41	61 57	90 00

Structure cell.[2] Space group probably $I4/acd$. a_0 11.70 ± 0.02 kX, c_0 14.95 ± 0.02; $a_0:c_0 = 1:1.278$.

Habit. Short prismatic [001] to pyramidal {112}.

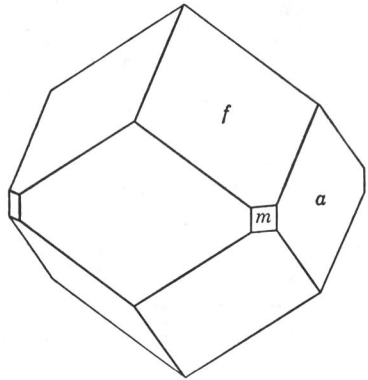

Phys. Fracture subconchoidal. Brittle. H. $4\frac{1}{2}$. G. 4.86.[3] Luster vitreous. Color light peridot-green to olive- and brownish green. Streak light green. Transparent.

Opt. In transmitted light, greenish yellow to yellowish green in color.

	n(Li)	DICHROISM	
O	2.19 ± 0.02	Green	Uniaxial positive (+).
E	2.21 ± 0.02	Yellowish green	

Chem. A hydrated ferric tellurite of unknown formula, perhaps $Fe_2(TeO_3)_3 \cdot xH_2O$. On heating, mackayite fuses, loses water, and yields anhydrous $Fe_2(TeO_3)_3$.

Occur. Found sparingly in the McGinnity shaft and the Sheets-Ish lease of the Mohawk mine at Goldfield, Nevada. Associated with emmonsite, blakeite, tellurite, alunite, barite, and quartz in the oxidized portion of a vein containing native tellurium and pyrite.

Name. After John W. Mackay (1831–1902), financier and mine operator on the Comstock Lode, Nevada, who endowed the School of Mines of the University of Nevada.

BLAKEITE

Ref.

1. The morphological ratio is used in preference to the x-ray values. Crystal class indicated by the probable space group, $I4/acd$.
2. Frondel in Frondel and Pough (1944) by the Weissenberg method.
3. The value is uncertain owing to the small powdered sample (10 mg.) available.

34.2.3 Blakeite. *Frondel* and *Pough* (*Am. Min.*, **29,** 211, 1944). [Not blakeite of J. D. Dana (447, 1850) = coquimbite.]

Found only massive, as microcrystalline crusts. The x-ray powder pattern is distinct from that of the other iron tellurites. Friable, and harsh to the touch. H. > 2. G. > 3.1. Color deep reddish brown to deep brown. Streak yellowish brown. Luster dull.

O p t. In transmitted light, golden yellow to yellowish brown and brown in color. Isotropic or nearly so. $n(\text{Li}): 2.16 \pm 0.02$.

C h e m. An anhydrous ferric tellurite of unknown formula. On heating, blakeite breaks down into a mixture of ferric tellurite and ferric tellurate (?).

O c c u r. Found very sparingly at Goldfield, Nevada, associated with mackayite and emmonsite in the oxidized zone of a vein containing native tellurium and pyrite. A possibly similar mineral has been reported from Tombstone, Arizona,[1] and Cripple Creek, Colorado.[2]

N a m e. After William P. Blake (1826–1910), geologist and mineralogist who worked in the American southwest.

Ref.

1. Hillebrand, *Proc. Colorado Sc. Soc.*, **2,** 20 (1885).
2. Knight, *Proc. Colorado Sc. Soc.*, **5,** 66 (1894–96).

CHROMATES

The chromates are anisodesmic compounds containing discrete $(CrO_4)^=$ groups and are very similar to sulfates, with which they may form substitutional series. The natural chromates are rare and few in number. They comprise principally a group of relatively insoluble chromates and compound-chromates of lead found as secondary minerals in ore deposits. The alkali chromates and the iodate-chromate, dietzeite, are relatively soluble compounds found in the arid-region *caliche* deposits of Chile.

CLASS 35. Anhydrous normal chromates
    Type 1. $A_2(XO_4)$
        35.1.1 Tarapacaite $K_2(CrO_4)$
    Type 2. $A_2(X_2O_7)$
        35.2.1 Lopezite $K_2(Cr_2O_7)$
    Type 3. $A(XO_4)$
        35.3.1 Crocoite $Pb(CrO_4)$
        35.3.2 Phoenicochroite $Pb_3(CrO_4)_2O$ (?)

CLASS 36. Compound chromates
    Type 1. Miscellaneous
        36.1.1 Vauquelinite $Pb_5(CrO_4)_2(PO_4)_2$ (?)
        36.1.2 Beresovite $Pb_6(CrO_4)_3(CO_3)O_2$
        Dietzeite   [in iodates]

35 ANHYDROUS NORMAL CHROMATES

TYPE 1. $A_2(XO_4)$

35.1.1 TARAPACAITE [K_2CrO_4]. *Raimondi* (274, 1878).

Cryst.[1] Orthorhombic; dipyramidal—$2/m\ 2/m\ 2/m$.

$a:b:c = 0.5694:1:0.7298$; $p_0:q_0:r_0 = 1.2817:0.7298:1$

$q_1:r_1:p_1 = 0.5694:0.7802:1$; $r_2:p_2:q_2 = 1.3702:1.7652:1$

Forms:

	ϕ	$\rho = C$	ϕ_1	$\rho_1 = A$	ϕ_2	$\rho_2 = B$
c 001	....	0°00′	0°00′	90°00′	90°00′	90°00′
b 010	0°00′	90 00	90 00	90 00	....	0 00
m 110	60 20½	90 00	90 00	29 39½	0 00	60 20½
q 011	0 00	36 07½	36 07½	90 00	90 00	53 52½
Q 021	0 00	55 35	55 35	90 00	90 00	34 25
o 111	60 20½	55 51½	36 07	44 00	37 57½	65 49½

Structure cell.[2] Space group $Pncm$. a_0 5.92 kX, b_0 10.40, c_0 7.61; $a_0:b_0:c_0 = 0.571:1:0.732$. Cell contents $K_8(CrO_4)_4$. Isostructural with mascagnite.

Habit. Thick tabular {001}.

Twinning. Pseudo-hexagonal twins on {110} analogous to aragonite have been observed on both natural [3] and artificial crystals.

P h y s. Cleavage {001} and {010} distinct. G. 2.74 (artif.); [4] 2.735 (calc.). Transparent. Color bright canary-yellow.

O p t.[5]

Orientation		n	
X	b	1.687	Biaxial negative $(-)$.
Y	a	1.722	$2V$ 52°.
Z	c	1.731	$r > v$, weak.

C h e m. Potassium chromate, K_2CrO_4. Complete analyses are lacking.[6]

Tests. Easily soluble in water.

O c c u r. From nitrate deposits in the Tocopilla pampa associated with lopezite and dietzeite, and from other nitrate deposits, especially in the so-called *caliche azufrado*, in Atacama, Tarapacá, and Antofagasta provinces, Chile.

A r t i f. A well-known artificial product, easily obtained in crystals from water solution.

N a m e. From the locality in Tarapacá province, Chile.

Ref.

1. Mitscherlich, *Ann. Phys.*, **18**, 169 (1830), quoted in Brendler, *Zs. Kr.*, **58**, 445 (1923). Forms also noted in Brendler.
2. Zachariasen and Ziegler, *Zs. Kr.*, **80**, 164 (1931), by Laue and oscillation methods on artificial crystals.
3. Brendler (1923).
4. Gossner, *Zs. Kr.*, **39**, 166 (1904), on artificial crystals by the suspension method.
5. Larsen and Berman (192, 1934) on artificial crystals.
6. The natural material was proved by Brendler (1923) by chemical tests and morphology to be identical with the artificial salt.

TYPE 2. $A_2(X_2O_7)$

35.2.1 **L O P E Z I T E** [$K_2(Cr_2O_7)$]. Bandy (*Am. Min.*, **22**, 929, 1937).

C r y s t.[1] Morphological data are lacking for the natural material. Artificial crystals are triclinic; pinacoidal—$\bar{1}$.

$a:b:c = 0.5575:1:0.5511;$ $\alpha\ 98°00',\ \beta\ 90°51',\ \gamma\ 96°13'$

$p_0:q_0:r_0 = 0.9847:0.5543:1;$ $\lambda\ 81°51\frac{1}{2}',\ \mu\ 88°15\frac{1}{2}',\ \nu\ 83°36'$

$p_0'\ 0.9948,\ q_0'\ 0.5600,\ x_0'\ 0.0148,\ y_0'\ 0.1431$

Common forms:	Less common:	Rare or uncertain:			
c 001	m 110	ν 150	v 041	w 0$\bar{5}$2	o 1$\bar{1}$1
b 010	M 1$\bar{1}$0	x 2$\bar{1}$0	κ 0$\bar{1}$2	r 101	π $\bar{1}$11
a 100	t 011	l 021	k 0$\bar{2}$1	ρ $\bar{1}$01	ω $\bar{1}$11
q 0$\bar{1}$1					

Structure cell.[2] Space group $P\bar{1}$; a_0 7.50 kX, b_0 13.40, c_0 7.38; α 98°00′, β 90°51′, γ 96°13′; $a_0:b_0:c_0 = 0.560:1:0.551$; cell contents $K_8Cr_8O_{28}$.

Habit. Found as ball-like aggregates about a millimeter in diameter. Artificial crystals are short prismatic [001].

Twinning. Observed on artificial crystals.[3]

P h y s. For artificial crystals:[3] cleavage {010} perfect, {100} and {001} distinct. H. 2½. G. 2.69; 2.660 (calc.). Transparent. Color orange-red to aurora-red.

O p t.

	Orientation[4]		n (lopezite)	n(Na) (artificial)[4]	
	ϕ	ρ			
X	−177°33′	68°08′	1.714	1.7202	Biaxial positive (+).
Y	96°57′	88°36′	1.732	1.7380	$r > v$, medium.
Z	3°09′	21°56′	1.805	1.8197	
$2V$			50°	51°33′	

C h e m. Shown to be potassium dichromate, $K_2Cr_2O_7$, by microchemical tests and by the optical identity with the artificial substance. Analyses are lacking.

Tests. Easily soluble in water to an orange solution.

O c c u r. Found with tarapacaite, dietzeite, and ulexite in vugs in massive nitrate rock at the Oficina Maria Elena, near Tocopilla, and at the Oficina Rosario on the Iquique pampa, Chile.

A r t i f. A well-known artificial product, easily obtained in crystals from water solution.

N a m e. After Emiliano Lopez of Iquique, Chile, a mineral collector associated with the nitrate mining industry in Chile.

Ref.

1. Angles and unit of Schabus, *Ak. Wien., Ber.*, **5**, 369 (1850) [Groth, **2**, 586, 1908]; orientation of Wolfe, priv. comm. (1946). Transformation: Schabus to Wolfe, 100/-0$\bar{1}$0/001.

2. Gossner and Mussgnug, *Zs. Kr.*, **72**, 476 (1930), by the rotation method on artificial crystals. Transformation: G. and M. to Wolfe, 100/001/0$\bar{1}$0. G. and M. γ should be 89°09′, not 90°51′.

3. Cf. Groth (**2**, 586, 1908). Twin plane {010}; twin axes [001] and [101]; all have {010} as composition plane.

4. Dufet, *Bull. soc. min.*, **13**, 346 (1890). Transformation: Dufet to Wolfe, 100/-001/010. ø and ρ values of optical orientation derived from angles given by Dufet.

TYPE 3. $A(XO_4)$

35.3.1 **C R O C O I T E** [Pb(CrO$_4$)]. Red lead-ore from Beresov *Lomonosov* (*Grundlagen der Metallurgie*, **1**, 44, 1763). Nova minera Plumbi *Lehmann* (*Nov. Comm. Ac. Petrop.*, 1766). Minerai de plomb rouge *Pallas* (*Reise durch versch. Prov. russ. Reichs*, **2**, 235, 1770). Minera Plumbi rubra *Wallerius* (**2**, 309, 1775). Rotbleierz, Rothes Bleierz *Werner* (296, 1774). Plombe rouge de Sibérie *Macquart* (*J. phys.*, **36**, 389, 1789); *Vauquelin* (*J. phys.*, **45**, 393, 1794; **46**, 152, 311, 1798). Plomb chromaté *Haüy* (**3**, 357, 1801). Chromate of lead. Chromsaures Blei, Bleichromat, Chrombleispath *Germ*. Kallochrom *Hausmann* (1084, 1813). Crocoise *Beudant* (**2**, 669, 1832). Krokoisit *von Kobell* (282, 1838). Bleiischer Chromspath *Breithaupt* (**2**, 262, 1841). Lehmannite *Brooke* and *Miller* (557, 1852). Beresofite *Shepard* (121, 1844).

Cryst.[1] Monoclinic; prismatic—$2/m$.

$a:b:c = 0.9602:1:0.9171$; $\beta\ 102°33'$; $p_0:q_0:r_0 = 0.9551:0.8952:1$

$r_2:p_2:q_2 = 1.1171:1.0669:1$; $\mu\ 77°27'$; $p_0'\ 0.9785,\ q_0'\ 0.9171,\ x_0'\ 0.2226$

Forms:[2]

		ϕ	ρ	ϕ_2	$\rho_2 = B$	C	A
c	001	90°00′	12°33′	77°27′	90°00′	.....	77°27′
b	010	0 00	90 00	.....	0 00	90°00′	90 00
a	100	90 00	90 00	0 00	90 00	77 27	.....
f	120	28 05	90 00	0 00	28 05	84 08	61 55
m	110	46 51	90 00	0 00	46 51	80 52½	43 09
T	530	60 39	90 00	0 00	60 39	79 05	29 21
d	210	64 53½	90 00	0 00	64 53½	79 39	25 06½
w	012	25 53½	27 00½	77 27	65 53	24 07	78 33½
z	011	13 38½	43 20½	77 27	48 10	41 50	80 41
y	021	6 55	61 34½	77 27	29 11	60 49	83 55
h	101	90 00	50 13	39 47	90 00	37 40	39 47
k	$\bar{1}01$	−90 00	37 05	127 05	90 00	49 38	127 05
t	111	52 38	56 30½	39 47	59 35½	46 56½	48 29
λ	$\bar{1}12$	−30 10½	27 56½	104 56	66 06	35 47½	103 37½
v	$\bar{1}11$	−39 29½	49 55	127 05	53 48½	58 29	119 07½

Less common:

| α 310 | ζ 350 | ρ 502 | x 801 | $l\ \bar{4}01$ | $\vartheta\ \bar{6}01$ | s 441 | $\beta\ \bar{3}12$ | $A\ \bar{5}11$ |
| g 320 | e 201 | n 401 | $\chi\ \bar{3}01$ | $\epsilon\ \bar{5}01$ | π 221 | $u\ \bar{2}11$ | $\phi\ \bar{3}11$ | |

Structure cell.[3] Space group $P2_1/a$. $a_0\ 7.10\ kX,\ b_0\ 7.40,\ c_0\ 6.80$; $\beta\ 102°27'$; $a_0:b_0:c_0 = 0.959:1:0.919$. Cell contents $Pb_4(CrO_4)_4$.

Habit. Crystals usually prismatic [001]; also elongated parallel to $[\bar{1}01]$; sometimes octahedral, with $\{111\}\ \{\bar{1}11\}$, or acute rhombohedral with $\{110\}$

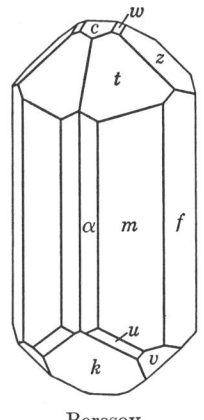
Beresov.

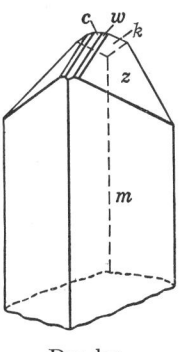
Dundas.

$\{\bar{h}0l\}$. Faces mostly smooth and brilliant; $\{110\}$ usually striated [001], and the steep orthodomes rounded or distorted. The crystals frequently are cavernous or hollow. Also massive, imperfectly columnar to granular.

Phys. Cleavage {110} rather distinct, {001} and {100} indistinct. Fracture small conchoidal to uneven. Sectile. H. $2\frac{1}{2}$–3. G. 5.99 ± 0.03;[4] 6.06 (Dundas);[12] 6.12 (artif.);[5] 5.97 (calc.). Luster adamantine to vitreous. Color hyacinth-red; also deep orange-red, orange, yellow. Streak orange-yellow. Transparent.

Opt. In transmitted light, orange-red.

Orientation[6]	n(Li)[7]	Pleochroism	
X	2.29 ± 0.02	Orange-red	Biaxial positive (+).
Y b	2.36 ± 0.02	Orange-red	$2V$ 57°(Li),[7] 54°03′(Na).[6]
$Z \wedge c$ +5$\frac{1}{2}$°	2.66 ± 0.02	Blood-red	$r > v$, very strong, inclined.

Chem. Lead chromate, $PbCrO_4$. PbO 69.06, CrO_3 30.94, total 100.00. The reported analyses [8] conform closely to the formula given. On isomorphous substitution in artificial material, see further under **Artif.** The element chromium was discovered in specimens of this mineral from Beresov by Vauquelin in 1797.

Tests. In C.T. decrepitates, blackens, but recovers its original color on cooling. B.B. fuses at 1$\frac{1}{2}$.

Occur. A secondary mineral, found associated with pyromorphite, cerussite, and, less frequently, wulfenite, vanadinite, vauquelinite, descloizite in the gossan of mineral deposits. First recognized in the Beresov district, near Ekaterinburg (Sverdlovsk), in the Urals, U.S.S.R.; also from Mursinsk, north of Ekaterinburg, and at Nizhne-Tagilsk. From Rézbánya, Roumania, and the Penchalonga mine, in the Umtali district, Rhodesia. In good crystals from Goyabeira, near Congonhas do Campo, Minas Geraes, Brazil. From Labo in North Camarines province, Luzon, Philippine Islands. A notable occurrence in the Adelaide, West Comet, Dundas Extension, and other mines in the Dundas district, Tasmania, where the mineral occurs in large amounts and as fine crystals associated with chromian cerussite, dundasite, massicot, wad; also at the Heazlewood and Whyte River mines, Tasmania.

In the United States, crocoite has been found in California associated with wulfenite at the Darwin mines, Inyo County, and at the El Dorado mine, near Indio, Riverside County. Found in Arizona at mines in the Vulture Mountains, Maricopa County, associated with other secondary chromium and vanadium minerals, and at the Mammoth mine, Pinal County, with wulfenite and vanadinite.

Artif. $Pb(CrO_4)$ is known in three modifications.[9] The monoclinic form, crocoite, is stable at ordinary temperatures and can form mixed crystals containing considerable amounts of (SO_4). A tetragonal form can be obtained from fusion and is stable only at high temperatures or as mixed crystals with (MoO_4). An unstable orthorhombic form, more orange in tone than crocoite, can take up (SO_4) in solid solution and is stable when the sulfate is in considerable excess. The artificial monoclinic form also can take up (MoO_4) in solid solution.[11] Well-formed crocoite crystals have been variously obtained [10] by inter-diffusion of solutions of lead nitrate and potassium chromate, by cooling hot solutions of $PbCrO_4$ in HCl or HNO_3, and by the slow dropwise addition of $K_2Cr_2O_7$ into a solution of lead nitrate in dilute HCl.

N a m e. From κρόκος, *saffron*, in allusion to the color.

Ref.
1. Dauber, *Ak. Wien, Ber.*, **42**, 19 (1860).
2. Goldschmidt (*Atlas*, **7**, 130, 1922).

Rare forms:

470	331	6.10.9	812	11.10.1	$\overline{13}$.1.5	$\overline{7}$11
850	$\overline{2}$23	435	711	12.4.1	$\overline{6}$12	$\overline{9}$11
10.3.0	2.1.10	352	841	$\overline{3}$28	$\overline{4}$11	$\overline{9}$31
032	154	412	911	$\overline{2}$65	$\overline{5}$21	$\overline{18}$.4.1
$\overline{4}$03	123	953	11.1.1	$\overline{8}$.7.10	$\overline{6}$21	

Uncertain forms:

380	601	332	10.3.4	13.5.1	$\overline{4}$85	$\overline{15}$.7.5
230	$\overline{8}$03	$\overline{6}$65	532	$\overline{1}$.20.18	$\overline{10}$.9.10	$\overline{13}$.1.4
340	$\overline{7}$02	3.4.12	12.9.4	$\overline{3}$.8.12	$\overline{11}$.10.10	$\overline{7}$22
450	112	598	652	$\overline{2}$36	$\overline{3}$62	
085	554	452	16.5.4	$\overline{1}$23	$\overline{13}$.8.6	
501	443	783	11.3.1	$\overline{3}$48	$\overline{7}$13	
$\overline{9}$22	$\overline{12}$.3.2	$\overline{4}$16	$\overline{7}$43	$\overline{21}$.3.5	$\overline{11}$.3.4	
$\overline{9}$32	$\overline{13}$.5.2	$\overline{4}$56	$\overline{5}$12	$\overline{17}$.5.4	$\overline{12}$.1.4	
$\overline{15}$.2.3	$\overline{11}$.5.1	$\overline{11}$.10.16	$\overline{5}$32	$\overline{18}$.3.4	$\overline{12}$.5.4	

3. Brill, *Zs. Kr.*, **77**, 506 (1931), by rotation and Weissenberg methods. See also Gliszczynski, *Zs. Kr.*, **101**, 1 (1939).
4. Range of best-reported values.
5. de Schulten, *Bull. soc. min.*, **27**, 135 (1904).
6. Des Cloizeaux, *Bull. soc. min.*, **5**, 103 (1882).
7. Larsen (63, 1921) by immersion method.
8. Hintze (**1** [3B], 4030, 1929).
9. Wagner, Haug, and Zipfel, *Zs. anorg. Chem.*, **208**, 249 (1932). See also Jaeger and Germs, *Zs. anorg. Chem.*, **119**, 154 (1921).
10. Cf. Mellor (**11**, 291, 1931).
11. Schultze, *Ann. Chem. Pharm.*, **126**, 52 (1863).
12. Mrose, priv. comm. (1946), by microbalance on single crystal.

35.3.2 P H O E N I C O C H R O I T E [$Pb_3(CrO_4)_2O$?]. Melanochroit *Hermann* (*Ann. Phys.*, **28**, 162, 1833). Phœnikochroit *Glocker* (612, 1839). Subsesquichromate of Lead *Thomson* (1836). Phönicit *Haidinger* (504, 1845). Phönizit. Phenicochroite.

Orthorhombic ? Found as imperfect tabular crystals, often reticularly intergrown. Also massive, and as thin coatings.

P h y s. Cleavage in one direction, perfect. H. 3–3½. G. 5.75. Color between cochineal- and hyacinth-red; becomes lemon-yellow on exposure. Streak brick-red. Luster resinous or adamantine, glimmering. Transparent on thin edges.

O p t.[1]

	n(Li)	
X	2.34 ± 0.02	Biaxial positive (+).
Y	2.38 ± 0.02	2V medium.
Z	2.65 ± 0.02	$r > v$, strong.

C h e m. A basic lead chromate, perhaps $Pb_3(CrO_4)_2O$.

Anal.

	1	2
PbO	77.00	76.69
CrO_3	23.00	23.31
Total	100.00	100.00
G.		5.75

1. $Pb_3(CrO_4)_2O$. 2. Beresov, Urals.[2]

Tests. Fuses easily on charcoal to a dark mass, which is crystalline when cold. Soluble in HCl with separation of lead chloride.

O c c u r. Originally found at Beresov, in the Urals, U.S.S.R., associated with crocoite, vauquelinite, and pyromorphite in altered quartz-galena veins in limestone. Also said [3] to occur in the gossan of the Adelaide Proprietary mine, Dundas, Tasmania.

A r t i f. Several basic lead chromates have been prepared,[4] but their relation to the natural substance is not known.

N a m e. From φοίνικος, *deep red*, and χρόα, *color*, in allusion to the color.

Ref.
1. Larsen (119, 1921).
2. Hermann (1833).
3. Petterd (114, 1910).
4. Cf. Mellor (**11**, 303, 1931).

36 COMPOUND CHROMATES

TYPE 1. MISCELLANEOUS

36.1.1 V A U Q U E L I N I T E. Vauqueline *Berzelius* (*Afhandl. Fys. Kem. Min.*, **6**, 246, 1818). Vauquelinite *Berzelius* (202, 1819). Plomb chromé *Haüy* (**3**, 363, 1822). Chromate of Lead and Copper *Phillips* (369, 1837). Laxmannite *Nordenskiöld* (*Ak. Stockholm, Öfv.*, **24**, 655, 1867; *Ann. Phys.*, **137**, 299, 1869). Chromphosphorkupferbleispath *John* (*Jb. Min.*, 67, 1845). Phosphorchromit *Hermann* (*J. pr. Chem.*, **1**, 196, 1870).

C r y s t. Monoclinic; prismatic—$2/m$.

$a:b:c = 0.747:1:1.407;$ $\quad \beta\ 110°10'$ [1] $\quad p_0:q_0:r_0 = 1.884:1.321:1$

$r_2:p_2:q_2 = 0.757:1.426:1;$ $\quad \mu\ 69°50';$ $\quad p_0'\ 2.007,\ q_0'\ 1.407,\ x_0'\ 0.367$

Forms: [2]

	φ	ρ	φ₂	$ρ_2 = B$	C	A
c 001	90°00'	20°10'	69°50'	90°00'	....	69°50'
a 100	90 00	90 00	0 00	90 00	69°50'	....
m 110	54 58	90 00	0 00	54 58	73 36	35 02
h $\bar{1}01$	−90 00	58 37½	148 37½	90 00	78 47½	148 37½

Less common or rare:

s 410	f 120	e 102	n $\bar{1}02$
z 320	d 011	x 304	p $\bar{3}04$

VAUQUELINITE

Habit. Crystals usually minute, often wedge-shaped; irregularly aggregated, and in mammillary fibrous forms. Also reniform or botryoidal; granular; compact.

Twinning. On $\{102\}$.[3]

Phys. Fracture uneven. Brittle. H. $2\frac{1}{2}$–3. G. 6.02.[4] Luster adamantine to resinous. Color green to brown; apple-green, siskin-green, olive-green, ocher-brown, liver-brown; sometimes nearly black. Streak greenish or brownish. Transparent in thin pieces.

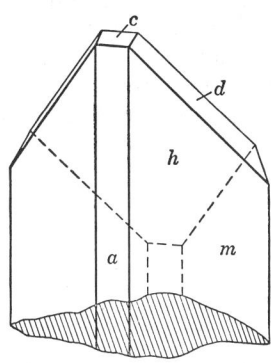

Beresov.

Opt.[5]

	n	Pleochroism	
X	2.11 ± 0.02	Pale green	Biaxial negative (−).
Y	2.22 ± 0.02	Pale brown	$2V$ near zero.
Z	2.22 ± 0.02	Pale brown	X = elongation (c?).

Chem. A chromate-phosphate of lead and copper of uncertain formula. In analysis 8, $(Pb,Cu):(Cr,P):O \sim 4:3:12$. It is not certain whether all the reported analyses represent the same mineral.

Anal.

	1	2	3	4	5	6	7	8
PbO	61.26	61.06	68.33	70.60	62.70	61.09	62.59	62.06
CuO	12.43	10.85	7.36	4.57	9.58	11.91	12.19	10.31
CrO_3	15.26	16.76	10.13	15.80	11.95	26.79	21.46	17.44
P_2O_5	8.05	8.57	9.94	9.78	9.23		3.55	8.66
Fe_2O_3	1.09	1.28	2.80				0.70	0.50
H_2O	1.31	0.90	1.16		3.00			1.12
Total	99.40	99.42	99.72	100.75	96.46	[99.79]	100.49	100.09
G.	5.77		5.80		6.06			

1,2. Beresov (*laxmannite*).[6] 3. Beresov (*phosphorchromite*).[7] 4. Beresov (*phosphorchromite*).[8] 5. Beresov (*vauquelinite*).[9] 6,7,8. Beresov.[10] Total of anal. 6 given in original as 99.94.

Tests. B.B. on charcoal slightly intumesces and fuses to a gray submetallic globule, yielding at the same time small globules of metal. Partly soluble in HNO_3.

Occur. Originally found at Beresov, in the Urals, U.S.S.R., associated with crocoite, pyromorphite, and mimetite. Found as twinned crystals supposedly from Pontigibaud, Puy-de-Dôme, France.[11] The following localities also have been reported but in lack of adequate descriptions must be regarded as uncertain. From Leadhills, Lanarkshire, and the High Pirn mine, Wanlock Dod, Dumfriesshire, Scotland.[12] With crocoite at Congonhas do Campo, Minas Geraes, Brazil, and from Colombia. In Tasmania [15] with phoenicochroite at the Adelaide Proprietary mine, Dundas, and near George's Bay. In the United States [13] from the Pequa mines (?), Lancaster County, Pennsylvania; at an old copper mine at Sparta, near Sing Sing, Westchester County, New York; [14] from mines in the Vulture Mountains, Maricopa County, Arizona, with crocoite.

Alter. Reported as an alteration of pyromorphite [16] and as an incrustation pseudomorph after dolomite.

Name. After Louis Nicolas Vauquelin (1763–1829), French chemist and discoverer of chromium, who first drew attention to this species. Laxmannite after Eric Laxmann (1737–1796), Russian mineralogist and chemist, who early drew attention to the minerals of Beresov.

Ref.

1. Orientation and unit of Nordenskiöld, *Ak. Stockholm, Öfv.*, **24**, 655, 1867; angles average of: Nordenskiöld (1867), Koksharov, **8**, 345 (1878), Des Cloizeaux, *Bull. soc. min.*, **5**, 53 (1882); Chirva, *Annals Leningrad State Univ., Ser. Geol.*, **1**, 19 (1935) [*Min. Abs.*, **7**, 460 (1940)]. Transformation: Des Cloizeaux to Nordenskiöld, $\frac{1}{2}0\bar{1}/0\frac{1}{2}0/00\frac{1}{3}$.
2. Goldschmidt (**9**, 54, 1923). Transformation: Goldschmidt to Nordenskiöld, $\frac{1}{2}00/010/001$. Uncertain forms: 370, 940, $\bar{2}.9.14 = \bar{1}46?$, $\bar{4}3.15.4 = 931?$.
3. Haidinger, *Edinburgh J. Sc.*, **7**, 213 (1827), on crystals supposedly from Pontigibaud, France; see also Lacroix (**4**, 448, 1910) and Des Cloizeaux (1882).
4. Average of 6.06 Nikolajev in Koksharov (1878) and 5.986 Haidinger (1827).
5. Larsen (153, 1921).
6. Nordenskiöld, *Ann. Phys.*, **137**, 299 (1869).
7. Hermann, *J. pr. Chem.*, **1**, 447 (1870).
8. Pisani, *Bull. soc. min.*, **3**, 196 (1880).
9. Nikolajev in Koksharov (**8**, 353, 1878).
10. Chirva, *Annals Leningrad State Univ., Ser. Geol.*, **1**, 19 (1935)—*Min. Abs.*, **7**, 460 (1939).
11. Cf. Lacroix (**4**, 448, 1910).
12. Davies, *Min. Mag.*, **1**, 112 (1877).
13. Localities mentioned in Dana (916, 1892).
14. Cf. Torrey, *Annals. Lyc. Nat. Hist. New York*, **4**, 76 (1837).
15. Petterd (196, 1910).
16. Blum (197, 1863).

FORNACITE. Lacroix (*Bull. soc. min.*, **38**, 198, 1915; **39**, 84, 1916). An ill-defined mineral, found as confused groups of small prismatic crystals that appear to be monoclinic in crystallization. No cleavage. Hardness and gravity not given. Color deep olive-green. In transmitted light, small grains are golden yellow in color. Biaxial positive (+), with $2V$ rather large, strong dispersion and strong birefringence. Chemical tests indicate that the mineral is a basic chromate-arsenate of lead and copper, analogous to vauquelinite. Completely soluble in HCl. Found with dioptase at Djoué, in the French Congo, Africa. Named after the colonial governor, Lucien Fourneau. Needs verification.

JOSSAITE. Breithaupt (*B. H. Ztg.*, **17**, 54, 1858). An ill-defined material, possibly a chromate of lead and zinc, found with vauquelinite at Beresov, Urals, U.S.S.R. Orthorhombic, with a prism angle of 62°–70° and traces of prismatic cleavage. Color orange-yellow. H. 3; G. 5.2 (?).

36.1.2 **Beresovite.** *Samoilov (Soc. nat. Moscou, Bull.,* **11,** no. 2, 290, 1897—*Zs. Kr.,* **31,** 519, 1899).

An ill-defined mineral, found as small lamellae with a perfect cleavage. G. 6.69. Color deep red. Pleochroic. Analysis gave (average of two): PbO 79.30, CrO_3 17.94, CO_2 2.46, total 99.70, corresponding to $Pb_6(CrO_4)_3(CO_3)O_2$. Found associated with cerussite and crocoite at Beresov, in the Urals, U.S.S.R. Later reported [1] from the Magnet mine, Russell County, Tasmania, in part altered to crocoite and to massicot. Needs verification.

Ref.

1. Petterd (25, 1910).

PHOSPHATES, ARSENATES, AND VANADATES

The phosphates, arsenates, and vanadates are anisodesmic oxysalts in which the anionic units of structure are tetrahedral $(XO_4)^{\equiv}$ groups, in which X is P, As, or V. There is commonly a partial or complete substitutional series between P and As and between As and V, less commonly between P and V. Substitutional series between these oxysalts and sulfates, molybdates, tungstates, and chromates, all of which possess tetrahedral $(XO_4)^{=}$ anions, generally are lacking. In only a few instances have sulfates and phosphates been found to be isostructural, as in the case of brushite and gypsum.

CLASS 37. Anhydrous acid phosphates, etc.

Type 1. Miscellaneous
37.1.1 Monetite $CaH(PO_4)$
37.1.2 Schultenite $PbH(AsO_4)$

CLASS 38. Anhydrous normal phosphates, etc.

Type 1. $AB(XO_4)$
38.1.1 Triphylite group
38.1.1.1 Triphylite $LiFe(PO_4)$
38.1.1.2 Lithiophilite $LiMn(PO_4)$
38.1.1.3 Hühnerkobelite $(Na,Ca)Fe(PO_4)$
38.1.1.4 Varulite $(Na,Ca)Mn(PO_4)$
38.1.2 Natrophilite $NaMn(PO_4)$
38.1.3 Sicklerite series
38.1.3.1 Ferri-sicklerite $(Li,Fe''',Mn'')(PO_4)$
38.1.3.2 Sicklerite $(Li,Mn'',Fe''')(PO_4)$
38.1.4 Alluaudite series
38.1.4.1 Alluaudite $(Na,Fe''',Mn'')(PO_4)$
38.1.4.2 Mangan-alluaudite $(Na,Mn'',Fe''')(PO_4)$
38.1.5 Heterosite series
38.1.5.1 Heterosite $(Fe''',Mn''')(PO_4)$
38.1.5.2 Purpurite $(Mn''',Fe''')(PO_4)$
38.1.6 Beryllonite $NaBe(PO_4)$
38.1.7 Arrojadite $Na_2(Fe'',Mn'')_5(PO_4)_4$

Type 2. $A_3B_2(XO_4)_3$
38.2.1 Berzeliite series
38.2.1.1 Berzeliite $(Mg,Mn)_2(Ca,Na)_3(AsO_4)_3$
38.2.1.2 Manganberzeliite $(Mn,Mg)_2(Ca,Na)_3(AsO_4)_3$
38.2.2 Caryinite $(Ca,Pb,Na)_5(Mn,Mg)_4(AsO_4)_5$ (?)

Type 3. $A_3(XO_4)_2$
38.3.1 Whitlockite $Ca_3(PO_4)_2$
38.3.2 Graftonite $(Fe,Mn,Ca)_3(PO_4)_2$

CLASSIFICATION

Type 4. $A(XO_4)$
- 38.4.1   Xenotime $Y(PO_4)$
- 38.4.2   Monazite $(Ce,La,Di)(PO_4)$
- 38.4.3   Berlinite $Al(PO_4)$
- 38.4.4   Rooseveltite $Bi(AsO_4)$

CLASS 39.   Hydrated acid phosphates, etc.

Type 1. $(AB)_m H_n(XO_4)_p \cdot xH_2O$, with $m + n:p > 2:1$
- 39.1.1   Stercorite $Na(NH_4)H(PO_4) \cdot 4H_2O$
- 39.1.2   Hannayite $Mg_3(NH_4)_2H_4(PO_4)_4 \cdot 8H_2O$
- 39.1.3   Hureaulite $Mn_5H_2(PO_4)_4 \cdot 4H_2O$

Type 2. $AH(XO_4) \cdot xH_2O$
- 39.2.1   Brushite group
  - 39.2.1.1   Brushite $CaH(PO_4) \cdot 2H_2O$
  - 39.2.1.2   Pharmacolite $CaH(AsO_4) \cdot 2H_2O$
- 39.2.2   Haidingerite $CaH(AsO_4) \cdot H_2O$
- 39.2.3   Newberyite $MgH(PO_4) \cdot 3H_2O$
- 39.2.4   Forbesite $(Ni,Co)H(AsO_4) \cdot 3\tfrac{1}{2}H_2O$ (?)
- 39.2.5   Roesslerite group
  - 39.2.5.1   Roesslerite $MgH(AsO_4) \cdot 7H_2O$
  - 39.2.5.2   Phosphorroesslerite $MgH(PO_4) \cdot 7H_2O$

CLASS 40.   Hydrated normal phosphates, etc.

Type 1. $AB(XO_4) \cdot xH_2O$
- 40.1.1   Struvite $(NH_4)Mg(PO_4) \cdot 6H_2O$

Type 2. $AB_2(XO_4)_2 \cdot xH_2O$
- 40.2.1   Dickinsonite $Na_6(Mn,Fe,Ca)_{14}H_2(PO_4)_{12} \cdot H_2O$
- 40.2.2   Fillowite $Na_6(Mn,Fe,Ca)_{14}H_2(PO_4)_{12} \cdot H_2O$ (?)
- 40.2.3   Fairfieldite group
  - 40.2.3.1   Fairfieldite $Ca_2(Mn,Fe)(PO_4)_2 \cdot 2H_2O$
  - 40.2.3.2   Collinsite $Ca_2(Mg,Fe)(PO_4)_2 \cdot 2H_2O$
- 40.2.4   Roselite group
  - 40.2.4.1   Roselite $Ca_2(Co,Mg)(AsO_4)_2 \cdot 2H_2O$
  - 40.2.4.2   Brandtite $Ca_2Mn(AsO_4)_2 \cdot 2H_2O$
- 40.2.5   Reddingite series
  - 40.2.5.1   Reddingite $(Mn,Fe)_3(PO_4)_2 \cdot 3H_2O$
  - 40.2.5.2   Phosphoferrite $(Fe,Mn)_3(PO_4)_2 \cdot 3H_2O$
- 40.2.6   Landesite $Fe_6Mn_{20}(PO_4)_{16} \cdot 27H_2O$ (?)
- 40.2.7   Stewartite   Mn phosphate
- 40.2.8   Salmonsite   Mn,Fe phosphate
- 40.2.9   Anapaite $Ca_2Fe(PO_4)_2 \cdot 4H_2O$
- 40.2.10   Parahopeite $Zn_3(PO_4)_2 \cdot 4H_2O$
- 40.2.11   Hopeite $Zn_3(PO_4)_2 \cdot 4H_2O$
- 40.2.12   Phosphophyllite $Zn_2(Fe,Mn)(PO_4)_2 \cdot 4H_2O$
- 40.2.13   Trichalcite $Cu_3(AsO_4)_2 \cdot 5H_2O$ (?)
- 40.2.14   Picropharmacolite $(Ca,Mg)_3(AsO_4)_2 \cdot 6H_2O$ (?)
- 40.2.15   Vivianite group
  - 40.2.15.1   Vivianite $Fe_3(PO_4)_2 \cdot 8H_2O$
  - 40.2.15.2   Erythrite $Co_3(AsO_4)_2 \cdot 8H_2O$
  - 40.2.15.4   Koettigite $Zn_3(AsO_4)_2 \cdot 8H_2O$
  - 40.2.15.5   Symplesite $Fe_3(AsO_4)_2 \cdot 8H_2O$

- 40.2.16   Bobierrite $Mg_3(PO_4)_2 \cdot 8H_2O$
- 40.2.17   Hoernesite $Mg_3(AsO_4)_2 \cdot 8H_2O$

PHOSPHATES, ARSENATES, VANADATES

Type 3. $A(XO_4) \cdot xH_2O$

40.3.1	Variscite group	
40.3.1.1	Variscite	$Al(PO_4) \cdot 2H_2O$
40.3.1.2	Strengite	$Fe(PO_4) \cdot 2H_2O$
40.3.1.3	Scorodite	$Fe(AsO_4) \cdot 2H_2O$
40.3.1.4	Mansfieldite	$Al(AsO_4) \cdot 2H_2O$
40.3.2	Metavariscite group	
40.3.2.1	Metavariscite	$Al(PO_4) \cdot 2H_2O$
40.3.2.2	Metastrengite	$Fe(PO_4) \cdot 2H_2O$
40.3.3	Weinschenkite	$(Y,Er)(PO_4) \cdot 2H_2O$
40.3.4	Churchite	$(Ce,Ca)(PO_4) \cdot 2H_2O$
40.3.5	Rhabdophane	$(Ce,Y)(PO_4) \cdot H_2O$

CLASS 41. Anhydrous phosphates, etc., containing hydroxyl or halogen

Type 1. $A_m(XO_4)_pZ_q$, with $m:p > 4:1$

41.1.1	Sahlinite	$Pb_{14}(AsO_4)_2O_9Cl$
41.1.2	Holdenite	$(Mn'',Ca)_4(Zn,Mg,Fe'')_2(AsO_4)(OH)_5O_2$ (?)
41.1.3	Hematolite	$(Mn'',Mg)_4Al(AsO_4)(OH)_8$
41.1.4	Chlorophoenicite group	
41.1.4.1	Chlorophoenicite	$(Zn,Mn)_5(AsO_4)(OH)_7$
41.1.4.2	Magnesium-chlorophoenicite	$Mg_5(AsO_4)(OH)_7$
41.1.5	Synadelphite	$(Mn,Mg,Ca,Pb)_4(AsO_4)(OH)_5$

Type 2. $A_7(XO_4)_2Z_q$

41.2.1	Lacroixite	Na,Ca,Al fluo-phosphate
41.2.2	Morinite	Na,Ca,Al fluo-phosphate
41.2.3	Ježekite	$Na_4CaAl_2(PO_4)_2(OH)_2F_2O$ (?)
41.2.4	Allactite	$Mn_7(AsO_4)_2(OH)_8$

Type 3. $A_3(XO_4)Z_q$

41.3.1	Clinoclase	$Cu_3(AsO_4)(OH)_3$
41.3.2	Cornetite	$Cu_3(PO_4)(OH)_3$
41.3.3	Georgiadesite	$Pb_3(AsO_4)Cl_3$
41.3.4	Atelestite	$Bi_3(AsO_4)O_2(OH)_2$ (?)
41.3.5	Flinkite	$Mn_2''Mn''''(AsO_4)(OH)_4$
41.3.6	Retzian	

Type 4. $(AB)_5(XO_4)_2Z_q$

41.4.1	Walpurgite	$Bi_4(UO_2)(AsO_4)O_4 \cdot 3H_2O$
41.4.2	Erinite	$Cu_5(AsO_4)_2(OH)_4$
41.4.3	Pseudomalachite	$Cu_5(PO_4)_2(OH)_4 \cdot H_2O$ (?)
41.4.4	Arsenoclasite	$Mn_5(AsO_4)_2(OH)_4$
41.4.5	Andrewsite	$(Cu,Fe'')_3Fe_6'''(PO_4)_4(OH)_{12}$
41.4.6	Laubmannite	$Fe_3''Fe_6'''(PO_4)_4(OH)_{12}$

Type 5. $AB(XO_4)Z_q$

41.5.1	Adelite group	
41.5.1.1	Adelite	$CaMg(AsO_4)(OH,F)$
41.5.1.2	Conichalcite	$CaCu(AsO_4)(OH)$
41.5.1.3	Austinite	$CaZn(AsO_4)(OH)$
41.5.1.4	Duftite	$PbCu(AsO_4)(OH)$
41.5.2	Descloizite group	
41.5.2.1	Descloizite	$ZnPb(VO_4)(OH)$
41.5.2.2	Mottramite	$CuPb(VO_4)(OH)$
41.5.2.3	Pyrobelonite	$MnPb(VO_4)(OH)$
41.5.2.4	Calciovolborthite	$CuCa(VO_4)(OH)$

Classification

Type 5. $AB(XO_4)Z_q$—Continued

41.5.2.5	Turanite	$Cu_2(VO_4)(OH)$ (?)
41.5.3	Volborthite	Cu vanadate
41.5.4	Herderite series	
41.5.4.1	Herderite	$CaBe(PO_4)(F,OH)$
41.5.4.2	Hydroxyl-herderite	$CaBe(PO_4)(OH,F)$
41.5.5	Amblygonite series	
41.5.5.1	Amblygonite	$(Li,Na)Al(PO_4)(F,OH)$
41.5.5.2	Montebrasite	$(Li,Na)Al(PO_4)(OH,F)$
41.5.5.3	Natromontebrasite	$(Na,Li)Al(PO_4)(F,OH)$
41.5.6	Tilasite	$CaMg(AsO_4)F$
41.5.7	Durangite	$NaAl(AsO_4)F$
41.5.8	Plumbogummite group	
41.5.8.1	Plumbogummite	$PbAl_3(PO_4)_2(OH)_5H_2O$
41.5.8.2	Gorceixite	$BaAl_3(PO_4)_2(OH)_5H_2O$
41.5.8.3	Goyazite	$SrAl_3(PO_4)_2(OH)_5H_2O$
41.5.8.4	Crandallite	$CaAl_3(PO_4)_2(OH)_5H_2O$
41.5.8.5	Deltaite	$Ca(Al_2Ca)(PO_4)_2(OH)_4H_2O$
41.5.8.6	Florencite	$CeAl_3(PO_4)_2(OH)_6$
41.5.8.7	Dussertite	$BaFe_3(AsO_4)_2(OH)_5H_2O$
41.5.9	Chenevixite	$Cu_2Fe_2(AsO_4)_2(OH)_4 \cdot H_2O$ (?)
41.5.10	Brazilianite	$NaAl_3(PO_4)_2(OH)_4$
41.5.11	Griphite	$(Na,Ca,Fe,Al)_3Mn_2(PO_4)_{2.5}(OH,F)_2$
41.5.12	Arseniopleite	Mn basic arsenate

Type 6. $A_2(XO_4)Z_q$

41.6.1	Wagnerite	$Mg_2(PO_4)F$
41.6.2	Triplite	$(Mn'',Fe'')(PO_4)F$
41.6.3	Triploidite group	
41.6.3.1	Triploidite	$(Mn'',Fe'')_2(PO_4)(OH)$
41.6.3.2	Wolfeite	$(Fe'',Mn'')_2(PO_4)(OH)$
41.6.3.3	Sarkinite	$Mn_2(AsO_4)(OH)$
41.6.4	Sarcopside	$(Fe,Mn,Ca)_7(PO_4)_4F_2$ (?)
41.6.5	Olivenite group	
41.6.5.1	Olivenite	$Cu_2(AsO_4)(OH)$
41.6.5.2	Libethenite	$Cu_2(PO_4)(OH)$
41.6.5.3	Adamite	$Zn_2(AsO_4)(OH)$
41.6.6	Frondelite series	
41.6.6.1	Frondelite	$Mn''Fe_4'''(PO_4)_3(OH)_5$
41.6.6.2	Rockbridgeite	$Fe''Fe_4'''(PO_4)_3(OH)_5$
41.6.7	Tarbuttite	$Zn_2(PO_4)(OH)$
41.6.8	Augelite	$Al_2(PO_4)(OH)_3$
41.6.9	Dufrenite	$Fe''Fe_4'''(PO_4)_3(OH)_5 \cdot 2H_2O$ (?)
41.6.10	Dewindtite	$Pb_3(UO_2)_5(PO_4)_4(OH)_4 \cdot 10H_2O$
41.6.11	Phosphuranylite	Ca uranyl phosphate

Type 7. $A_5(XO_4)_3Z_q$

41.7.1	Apatite series	
41.7.1.1	Fluorapatite	$Ca_5(PO_4)_3F$
41.7.1.2	Chlorapatite	$Ca_5(PO_4)_3Cl$
41.7.1.3	Hydroxylapatite	$Ca_5(PO_4)_3(OH)$
41.7.1.4	Carbonate-apatite	$\sim Ca_{10}(PO_4)_6(CO_3) \cdot H_2O$
41.7.2	Pyromorphite series	
41.7.2.1	Pyromorphite	$Pb_5(PO_4)_3Cl$
41.7.2.2	Mimetite	$Pb_5(AsO_4)_3Cl$
41.7.2.3	Vanadinite	$Pb_5(VO_4)_3Cl$
41.7.3	Svabite series	

Phosphates, Arsenates, Vanadates

Type 7. $A_5(XO_4)_3Z_q$—Continued

41.7.3.1	Svabite	$Ca_5(AsO_4)_3(F,OH)$
41.7.3.2	Hedyphane	$(Ca,Pb)_5(AsO_4)_3Cl$

41.7.4	Dehrnite	$(Ca,Na,K)_5(PO_4)_3(OH)$
41.7.5	Lewistonite	$(Ca,K,Na)_5(PO_4)_3(OH)$
41.7.6	Fermorite	$(Ca,Sr)_5(P,AsO_4)_3(F,OH)$
41.7.7	Wilkeite	$Ca_5(P,S,Si,CO_4)_3(OH)$
41.7.8	Ellestadite	$Ca_5(Si,S,P)_3(Cl,F,OH)$
41.7.9	Tavistockite	$Ca_3Al_2(PO_4)_2(OH)_6$
41.7.10	Arsenobismite	Bi basic arsenate

Type 8. $(AB)_3(XO_4)_2Z_q$

41.8.1	Lazulite series	
41.8.1.1	Lazulite	$(Mg,Fe'')Al_2(PO_4)_2(OH)_2$
41.8.1.2	Scorzalite	$(Fe'',Mg)Al_2(PO_4)_2(OH)_2$
41.8.2	Souzalite	$(Mg,Fe'')_3(Al,Fe''')_4(PO_4)_4(OH)_6 \cdot 2H_2O$
41.8.3	Carminite	$PbFe_2(AsO_4)_2(OH)_2$
41.8.4	Parsonsite	$Pb_2(UO_2)(PO_4)_2 \cdot 2H_2O$

Class 42. Hydrated phosphates, etc., containing hydroxyl or halogen

Type 1. $(AB)_m(XO_4)_pZ_q \cdot xH_2O$, with $m:p > 3:1$

42.1.1	Borickite	$CaFe_5(PO_4)_2(OH)_{11} \cdot 3H_2O$

Type 2. $(AB)_3(XO_4)Z_q \cdot xH_2O$

42.2.1	Veszelyite	$(Cu,Zn)_3(As,PO_4)(OH)_3 \cdot 2H_2O$
42.2.2	Tsumebite	$Pb_2Cu(PO_4)(OH)_3 \cdot 3H_2O$
42.2.3	Hemafibrite	$Mn_3(AsO_4)(OH)_3 \cdot H_2O$
42.2.4	Freirinite	$Na_3Cu_3(AsO_4)_2(OH)_3 \cdot H_2O$
42.2.5	Liroconite	$Cu_2Al(AsO_4)(OH)_4 \cdot 4H_2O$
42.2.6	Evansite	$Al_3(PO_4)(OH)_6 \cdot 6H_2O$
42.2.7	Liskeardite	$(Al,Fe)_3(AsO_4)(OH)_6 \cdot 5H_2O$

Type 3. $(AB)_5(XO_4)_2Z_q \cdot xH_2O$

42.3.1	Cornwallite	$Cu_5(AsO_4)_2(OH)_4 \cdot H_2O$
42.3.2	Tyrolite	$Cu_5Ca(AsO_4)_2(CO_3)(OH)_4 \cdot 6H_2O$ (?)
42.3.3	Akrochordite	$MgMn_4(AsO_4)_2(OH)_4 \cdot 4H_2O$ (?)
42.3.4	Ceruleite	$CuAl_4(AsO_4)_2(OH)_8 \cdot 4H_2O$
42.3.5	Renardite	$Pb(UO_2)_4(PO_4)_2(OH)_4 \cdot 7H_2O$
42.3.6	Dumontite	$Pb_2(UO_2)_3(PO_4)_2(OH)_4 \cdot 3H_2O$

Type 4. $A_2(XO_4)Z_q \cdot xH_2O$

42.4.1	Bayldonite	$(Cu,Pb)_2(AsO_4)(OH)$ (?)
42.4.2	Leucochalcite	$Cu_2(AsO_4)(OH) \cdot H_2O$
42.4.3	Tagilite	$Cu_2(PO_4)(OH) \cdot H_2O$
42.4.4	Spencerite	$Zn_4(PO_4)_2(OH)_2 \cdot 3H_2O$
42.4.5	Isoclasite	$Ca_2(PO_4)(OH) \cdot 2H_2O$
42.4.6	Euchroite	$Cu_2(AsO_4)(OH) \cdot 3H_2O$
42.4.7	Delvauxite	$Fe_2(PO_4)(OH)_3 \cdot xH_2O$ (?)

Type 5. $(AB)_m(XO_4)_pZ_q \cdot xH_2O$, with $m:p = 2:1$

42.5.1	Leucophosphite	$K_2(Fe,Al)_7(PO_4)_4(OH)_{11} \cdot 6H_2O$
42.5.2	Childrenite series	
42.5.2.1	Childrenite	$(Fe'',Mn'')Al(PO_4)(OH)_2 \cdot H_2O$
42.5.2.2	Eosphorite	$(Mn'',Fe'')Al(PO_4)(OH)_2 \cdot H_2O$
42.5.3	Davisonite	$Ca_3Al(PO_4)_2(OH)_3 \cdot H_2O$ (?)

Classification

Type 5. $(AB)_m(XO_4)_p Z_q \cdot xH_2O$, with $m:p = 2:1$—*Continued*

42.5.4	Wardite	$Na_4CaAl_{12}(PO_4)_8(OH)_9 \cdot 3H_2O$
42.5.5	Millisite	$(Na,K)CaAl_6(PO_4)_4(OH)_9 \cdot 3H_2O$
42.5.6	Lehiite	$(Na,K)_2Ca_5Al_8(PO_4)_8(OH)_{12} \cdot 6H_2O$ (?)
42.5.7	Mixite	$Cu_{11}Bi(AsO_4)_5(OH)_{10} \cdot 6H_2O$ (?)

Type 6. $(AB)_m(XO_4)_p Z_q \cdot xH_2O$, with $m:p = 7:4$

42.6.1	Sampleite	$NaCaCu_5(PO_4)_4Cl \cdot 5H_2O$
42.6.2	Turquois group	
42.6.2.1	Turquois	$CuAl_6(PO_4)_4(OH)_8 \cdot 4H_2O$
42.6.2.2	Chalcosiderite	$CuFe_6(PO_4)_4(OH)_8 \cdot 4H_2O$
42.6.3	Ludlamite	$(Fe'',Mg,Mn)_3(PO_4)_2 \cdot 4H_2O$
42.6.4	Arseniosiderite	$Ca_3Fe_4(AsO_4)_4(OH)_4 \cdot 4H_2O$ (?)
42.6.5	Eguëite	$CaFe_{14}'''(PO_4)_{10}(OH)_{14} \cdot 21H_2O$ (?)
42.6.6	Mitridatite	Ca,Fe phosphate
42.6.7	Richellite	Ca,Fe phosphate
42.6.8	Englishite	$K_2Ca_4Al_8(PO_4)_8(OH)_{10} \cdot 9H_2O$

Type 7. $A_3(XO_4)_2 Z_q \cdot xH_2O$

42.7.1	Legrandite	$Zn_{14}(AsO_4)_9(OH) \cdot 12H_2O$
42.7.2	Beraunite	$Fe''Fe_4'''(PO_4)_3(OH)_5 \cdot 3H_2O$ (?)
42.7.3	Coeruleolactite	$Al_3(PO_4)_2(OH)_3$
42.7.4	Wavellite	$Al_3(PO_4)_2(OH)_3 \cdot 5H_2O$
42.7.5	Sterrettite	$Al_6(PO_4)_4(OH)_6 \cdot 5H_2O$
42.7.6	Troegerite	$(UO_2)_3(AsO_4)_2 \cdot 12H_2O$

Type 8. $(AB)_m(XO_4)_p Z_q \cdot xH_2O$, with $m:p = 3:2$

42.8.1	Bermanite	$(Mn,Mg)_5(Mn,Fe)_8(PO_4)_8(OH)_{10} \cdot 15H_2O$ (?)
42.8.2	Roscherite	$(Ca,Mn,Fe)_2Al(PO_4)_2(OH) \cdot 2H_2O$
42.8.3	Minyulite	$KAl_2(PO_4)_2(OH) \cdot 3\tfrac{1}{2}H_2O$ (?)
42.8.4	Tinticite	$Fe_3'''(PO_4)_3 \cdot 3\tfrac{1}{2}H_2O$
42.8.5	Metavauxite	$FeAl_2(PO_4)_2(OH)_2 \cdot 8H_2O$
42.8.6	Paravauxite	$FeAl_2(PO_4)_2(OH)_2 \cdot 8H_2O$
42.8.7	Vauxite	$FeAl_2(PO_4)_2(OH)_2 \cdot 7H_2O$
42.8.8	Gordonite	$MgAl_2(PO_4)_2(OH)_2 \cdot 8H_2O$
42.8.9	Calcioferrite	$Ca_3Fe_3(PO_4)_4(OH)_3 \cdot 8H_2O$ (?)
42.8.10	Xanthoxenite	$Ca_2Fe(PO_4)_2(OH) \cdot 1\tfrac{1}{2}H_2O$
42.8.11	Montgomeryite	$Ca_4Al_5(PO_4)_6(OH)_5 \cdot 11H_2O$
42.8.12	Overite	$Ca_3Al_8(PO_4)_8(OH)_6 \cdot 15H_2O$
42.8.13	Torbernite group	
42.8.13.1	Torbernite	$Cu(UO_2)_2(PO_4)_2 \cdot 8–12H_2O$
42.8.13.2	Autunite	$Ca(UO_2)_2(PO_4)_2 \cdot 10–12H_2O$
42.8.13.3	Uranocircite	$Ba(UO_2)_2(PO_4)_2 \cdot 8H_2O$
42.8.13.4	Saléeite	$Mg(UO_2)_2(PO_4)_2 \cdot 10H_2O$
42.8.13.5	Zeunerite	$Cu(UO_2)_2(AsO_4)_2 \cdot 10–16H_2O$
42.8.13.6	Uranospinite	$Ca(UO_2)_2(AsO_4)_2 \cdot 8H_2O$
42.8.14	Metatorbernite group	
42.8.14.1	Metatorbernite	$Cu(UO_2)_2(PO_4)_2 \cdot 8H_2O$
	Meta-autunite	$Ca(UO_2)_2(PO_4)_2 \cdot 2–6H_2O$
42.8.14.2	Metazeunerite	$Cu(UO_2)_2(AsO_4)_2 \cdot 8H_2O$
42.8.15	Bassetite	Fe'' uranyl phosphate

Type 9. $(AB)_m(XO_4)_p Z_q \cdot xH_2O$, with $m:p < 3:2$

42.9.1	Pharmacosiderite	$Fe_3(AsO_4)_2(OH)_3 \cdot 5H_2O$
42.9.2	Cacoxenite	$Fe_4(PO_4)_3(OH)_3 \cdot 12H_2O$
42.9.3	Vashegyite	$Al_4(PO_4)_3(OH)_3 \cdot nH_2O$ (?)
42.9.4	Taranakite	$K_2Al_6(PO_4)_6(OH)_2 \cdot 18H_2O$ (?)

660   Phosphates, Arsenates, Vanadates

Class 43. Compound phosphates, etc.

Type 1. $AB(XO_4)Z_q$

43.1.1		Beudantite group	
	43.1.1.1	Beudantite	$PbFe_3(AsO_4)(SO_4)(OH)_6$
	43.1.1.2	Corkite	$PbFe_3(PO_4)(SO_4)(OH)_6$
	43.1.1.3	Hinsdalite	$(Pb,Sr)Al_3(PO_4)(SO_4)(OH)_6$
	43.1.1.4	Svanbergite	$SrAl_3(PO_4)(SO_4)(OH)_6$
	43.1.1.5	Woodhouseite	$CaAl_3(PO_4)(SO_4)(OH)_6$
43.1.2		Lindackerite	$Cu_6Ni_3(AsO_4)_4(SO_4)(OH)_4 \cdot 5H_2O$ (?)

Type 2. Miscellaneous

43.2.1	Chalcophyllite	$Cu_{18}Al_2(AsO_4)_3(SO_4)_3(OH)_{27} \cdot 33H_2O$
43.2.2	Ardealite	$Ca_2H(PO_4)(SO_4) \cdot 4H_2O$
43.2.3	Kribergite	$Al_4(PO_4)_2(SO_4)_2(OH)_2 \cdot 8H_2O$ (?)
43.2.4	Diadochite	$Fe_2(PO_4)(SO_4)(OH) \cdot 5H_2O$
43.2.5	Sarmientite	$Fe_2(AsO_4)(SO_4)(OH) \cdot 5H_2O$
43.2.6	Pitticite	Fe sulfate-arsenate
43.2.7	Kolbeckite	Ca,Be,Al silicate-phosphate
	Vauquelinite	[with chromates]
	Lunebergite	[with borates]
	Cahnite	[with borates]
	Seamanite	[with borates]
	Bradleyite	[with carbonates]

37 ANHYDROUS ACID PHOSPHATES, ETC.

TYPE 1. MISCELLANEOUS

37.1.1 **M O N E T I T E** [$CaH(PO_4)$]. *Shepard* (*Am. J. Sc.*, **23**, 400, 1882). Glaubapatite pt. *Shepard* (*Am. J. Sc.*, **22**, 96, 1856).

C r y s t.[1] Triclinic; pinacoidal—$\bar{1}$.

$a:b:c = 0.8244:1:0.6467;$ $\alpha\ 89°43',\ \beta\ 95°03',\ \gamma\ 94°22'$

$p_0:q_0:r_0 = 0.7867:0.6461:1;$ $\lambda\ 89°54',\ \mu\ 84°57\frac{1}{2}',\ \nu\ 85°38\frac{1}{2}'$

$p_0'\ 0.7898,\ q_0'\ 0.6486,\ x_0'\ 0.0883,\ y_0'\ 0.0018$

Forms:[2]

		ϕ	ρ	A	B	C
c	001	88°50′	5°03′	84°57½′	89°54′	.....
b	010	0 00	90 00	85 38½	.....	89°54′
m	110	47 46½	90 00	37 52	47 46½	86 12
e	101	85 34½	41 18	48 49½	87 34½	36 00
E	$\bar{1}01$	−84 45½	35 04½	124 31	86 59½	39 33½
x	121	65 53	43 49	49 20	73 34	39 12½
s	321	58 45	70 46	32 38½	60 40½	66 25½
t	$\bar{1}21$	−27 11	56 30	108 57	41 52	59 26½

Habit. As massive aggregates of minute crystals; also as crusts or stalactites with rough crystalline surfaces. Crystals small with rough faces, often arranged in interpenetrating groups; usually flattened {010} with a rhombohedral outline.

Phys. Indistinctly cleavable in three directions.[4] Fracture uneven. Brittle. H. $3\frac{1}{2}$. G. 2.929 (artif.).[3] Luster vitreous. Color pale yellowish white; the pure artificial salt is white. Translucent.

Opt. In transmitted light, colorless.

	n(Na)[5]	
X	1.587	Biaxial positive (+).
Y	~1.615	$2V$ large.
Z	1.640	$r > v$, weak.

In {010} sections an extinction direction makes an angle of 23° with the edge with {10$\bar{1}$}, and in {001} sections, an angle of 30° with the edge with {010}.

Chem. Calcium acid phosphate, $CaH(PO_4)$. CaO 41.21, P_2O_5 52.18, H_2O 6.61, total 100.00. Analyses of pure natural material are lacking.[6]

Occur. From the Islands of Moneta and Mona, in the Caribbean Sea about 40 miles from the port of Mayaguez, Puerto Rico. The mineral occurs with gypsum in a phosphate-rock deposit in limestone underlying a bed of bird guano. Also found with whitlockite on Los Monges Islands in the Caribbean Sea off the Gulf of Maracaibo, Venezuela (*glaubapatite* pt.) [7] and with apatite and newberyite on Ascension Island in the South Atlantic. In a cave deposit of phosphates at Gunong Jerneh, Malaya.

Alter. Finely divided monetite takes up water from the atmosphere and forms brushite.

Artif.[8] Obtained in crystals by heating precipitated $CaHPO_4 \cdot 2H_2O$ with water in a closed tube at 150°, by slow inter-diffusion of Na_2HPO_4 and $CaCl_2$ solutions and in numerous other ways.

Name. From the locality.

Ref.

1. Angles and unit of de Schulten, *Bull. soc. min.*, **27**, 120 (1904), on artificial crystals. Transformation: de Schulten to new orientation, 001/010/$\bar{1}$00.
2. de Schulten (1904).
3. de Schulten, *Bull. soc. min.*, **24**, 332 (1901).
4. Larsen and Berman (111, 1934); Dana in Shepard (1882) mentions only an apparent cleavage {001}.
5. Hill and Hendricks, *Ind. Eng. Chem.*, **38**, 440 (1936).
6. Analyses of mixtures with gypsum are cited by Shepard (1882) and Willbourn, *J. Malayan Branch Roy. Asiatic Soc.*, **3**, Pt. 3, 57 (1925).
7. Shown to be identical with monetite by Frondel, *Am. Min.*, **28**, 215 (1943).
8. Mellor (**3**, 881, 1923); de Schulten, *Bull. soc. min.*, **24**, 323 (1901); **26**, 15 (1903); Hill and Hendricks (1936).

37.1.2 **SCHULTENITE.** [$PbH(AsO_4)$]. Spencer (*Min. Mag.*, **21**, 149, 1926).

Cryst.[1] Monoclinic; prismatic—$2/m$.

$a:b:c = 0.8649:1:0.7201;\quad \beta\ 95°23\frac{1}{2}';\quad p_0:q_0:r_0 = 0.8326:0.7169:1$

$r_2:p_2:q_2 = 1.3949:1.1614:1;\quad \mu\ 84°36\frac{1}{2}';\quad p_0'\ 0.8363,\ q_0'\ 0.7201,\ x_0'\ 0.0944$

Forms:

	φ	ρ	φ2	ρ2 = B	C	A
c 001	90°00′	5°23½′	84°36½′	90°00′	.....	84°36½′
b 010	0 00	90 00	.....	0 00	90°00′	90 00
n 140	16 11½	90 00	0 00	16 11½	88 30	73 48½
l 130	21 09½	90 00	0 00	21 09½	88 03½	68 50½
k 120	30 08½	90 00	0 00	30 08½	87 18	59 51½
m 110	49 16	90 00	0 00	49 16	83 44½	40 44
h 210	66 42½	90 00	0 00	66 42½	77 24	23 17½
e 011	7 28	35 59½	84 36½	54 21½	35 38½	85 37½
p 111	52 09½	49 34½	47 10	62 09½	45 24½	53 03
q 1̄11	−45 43	45 53	126 26½	59 55	49 51	120 55½
u 121	32 46	59 43½	47 10	43 26	56 55	62 08
s 2̄11	−65 22½	59 56½	147 31½	68 51½	64 52	141 53½

Less common: w 323   v 232   r 1̄22   t 2̄21

Habit. Flattened {010}; usually rhomb-shaped and resembling gypsum. {010} striated [001], {001} striated [100].

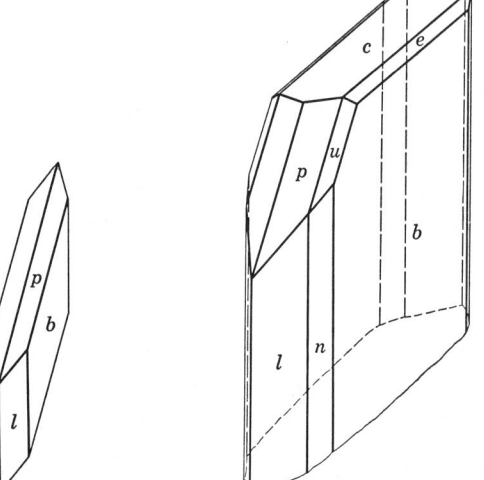

Phys. Cleavage {010} good. Brittle. H. 2½. G. 5.943, 6.06 ± 0.01 (artif.).[2] Luster brilliant, vitreous to almost adamantine. Colorless and transparent. Streak white.

Opt. In transmitted light, colorless.

Orientation	n(Na)[3]	
X b	1.8903 (calc.)	Biaxial positive (+).
Y ∧ c −24°	1.9097	2V 58°14′ (Na, meas.).
		2E 136°38′.
Z ∧ c +66°	1.9765	Dispersion strong.

Chem. Lead hydrogen arsenate, $PbH(AsO_4)$. On heating breaks down to $Pb_2As_2O_7$.

Anal.

	1	2
PbO	64.30	63.97
As_2O_5	33.11	32.18
H_2O	2.59	2.88
Total	100.00	99.03

1. $PbH(AsO_4)$. 2. Tsumeb.[4]

Occur. From Tsumeb, South West Africa, associated with anglesite, bayldonite, and bayldonite pseudomorphs after mimetite (?) and azurite (?). Anglesite also was observed forming pseudomorphs probably after schultenite.

Artif.[5] Obtained in crystals by cooling a boiling solution of the salt in dilute HNO_3, and in other ways. $PbHAsO_4$ is the commonest constituent of the "lead arsenate" of commerce and is widely used as an insecticide.

Ref.

1. Spencer (1926). The elements, originally computed from only three angles, are here recomputed from all the measured angles. Transformation: de Schulten, *Bull. soc. min.*, **27**, 113 (1904), and Groth (**2**, 822, 1908) to Spencer, 001/010/100.
2. From values 6.076 and 6.053 of de Schulten (1904) and McDonnell and Smith, *J. Am. Chem. Soc.*, **38**, 2030 (1916).
3. By prism method.
4. Mountain analysis in Spencer (1926).
5. Mellor (**9**, 193, 1929).

38 ANHYDROUS NORMAL PHOSPHATES, ETC.

TYPE 1. $AB(XO_4)$

38.1.1 TRIPHYLITE GROUP

ORTHORHOMBIC; DIPYRAMIDAL—$2/m\ 2/m\ 2/m$

	a_0	b_0	c_0
Triphylite series			
Triphylite, $LiFe''(PO_4)$	$6.00\ kX$	10.34	4.67 [Fe:Mn = 3:1]
Lithiophilite, $LiMn''(PO_4)$			
Hühnerkobelite series			
Hühnerkobelite, $(Na,Ca)Fe''(PO_4)$			
Varulite, $(Na,Ca)Mn''(PO_4)$			
Natrophilite, $NaMn''(PO_4)$	6.32	10.52	4.97

A complete series by mutual substitution of Fe'' and Mn'' extends between triphylite and lithiophilite and probably between hühnerkobelite and varulite. The Triphylite Series and Hühnerkobelite Series are isostructural according to the evidence of x-ray powder diffraction photographs. The structure of triphylite-lithiophilite, and presumably also of hühnerkobelite-varulite, is very similar to that of the silicates of the olivine type.[1] The x-ray powder photograph of natrophilite differs considerably from those of the members of the Triphylite Group. Natrophilite has the same space group and cell dimensions as the Triphylite Group and the two differ structurally[2] only in that Na and Mn in the former are distributed statistically over $4(a) + 4(c)$, whereas Li or Na and (Mn,Fe) occupy these positions separately in the latter. There is a partial analogy in this to the relation between forsterite, $MgMg(SiO_4)$, and monticellite, $CaMg(SiO_4)$. Natrophilite can also be described as the disordered equivalent of varulite.

The members of the Triphylite Group enter into an interesting sequence of alterations whereby the alkali ions are leached from the structure with a concomitant oxidation first of Fe'' and then of Mn'' to effect valence compensation (see further under **Alter.** in the description of triphylite-lithiophilite). It may be noted that triphylite-lithiophilite are common minerals, whereas hühnerkobelite-varulite are uncommon and natrophilite is known only from a single occurrence. All are high-temperature minerals found in granite pegmatites. Other iron-manganese phosphates that occur similarly are triplite, $(Fe'',Mn'')_2(PO_4)F$, triploidite, $(Mn'',Fe'')_2(PO_4)(OH)$, wolfeite, $(Fe'',Mn'')_2(PO_4)(OH)$, sarcopside, $(Fe'',Mn'')_7(PO_4)_4F_2$, arrojadite, $Na_2(Fe'',Mn'')_5(PO_4)_4$, and graftonite, $(Fe'',Mn'',Ca)_3(PO_4)_2$.

Ref.

1. Gossner and Strunz, *Zs. Kr.*, **83**, 415 (1932).
2. Byström, *Ark. Kemi*, **17**, 1 (1943).

TRIPHYLITE-LITHIOPHILITE SERIES

38.1.1.1 **T R I P H Y L I T E** [Li(Fe″,Mn″)(PO$_4$)]. Triphylin *Fuchs* (*J. pr. Chem.*, **3**, 98, 1834; **5**, 319, 1835). Tetraphylin *Berzelius* (*Jahresber.*, **15**, 211, 1836). Perowskyn *Nordenskiöld* (*Ann. Phys.*, **36**, 473, 1835).

38.1.1.2 **L I T H I O P H I L I T E** [Li(Mn″,Fe″)(PO$_4$)]. *Brush* and *Dana* (*Am. J. Sc.*, **16**, 118, 1878; **18**, 45, 1879).

C r y s t.[1] Orthorhombic; dipyramidal—$2/m\ 2/m\ 2/m$ (?).

$a:b:c = 0.5823:1:0.4541$; $p_0:q_0:r_0 = 0.7798:0.4541:1$

$q_1:r_1:p_1 = 0.5823:1.2823:1$; $r_2:p_2:q_2 = 2.2022:1.7173:1$

Forms:[4]

	ϕ	$\rho = C$	ϕ_1	$\rho_1 = A$	ϕ_2	$\rho_2 = B$
b 010	0°00′	90°00′	90°00′	90°00′	.....	0°00′
a 100	90 00	90 00	.....	0 00	0°00′	90 00
n 130	29 47	90 00	90 00	60 13	0 00	29 47
ε 120	40 39	90 00	90 00	49 21	0 00	40 39
d 011	0 00	24 25½	24 25½	90 00	90 00	65 34½
l 021	0 00	42 14½	42 14½	90 00	90 00	47 45½
v 203	90 00	27 28	0 00	62 32	62 32	90 00
e 101	90 00	37 57	0 00	52 03	52 03	90 00
w 201	90 00	57 16½	0 00	32 43½	32 43½	90 00

Structure cell. Space group $Pmnb$ if holohedral.

	Fe:Mn	a_0	b_0	c_0	$a_0:b_0:c_0$
Varuträsk[2]	~1.3:1(?)	6.038kX	10.374	4.711	0.582:1:0.454
Hagendorf[3]	~3 :1(?)	6.00 kX	10.34	4.67	0.580:1:0.452

Cell contents Li$_4$(Fe″, Mn″)$_4$(PO$_4$)$_4$.

Habit. Crystals rare, and then usually coarse with uneven surfaces; stout prismatic [100]. Commonly massive, cleavable to compact. An-

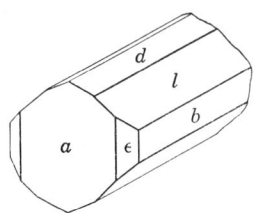

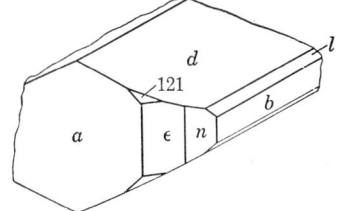

Triphylite. Norwich, Mass.   Triphylite. Newport, N. H.

hedral or subhedral single-crystals of triphylite 10 to 15 feet in greatest dimension have been found in pegmatites.

P h y s. Cleavage {100} nearly perfect, {010} imperfect, {011} interrupted. Fracture uneven to subconchoidal. H. 4–5. G. 3.34 (Mn endmember) to 3.58 (Fe end-member), 3.50 (for Fe:Mn = 1:1);[5] the G. does not vary linearly with the Fe:Mn ratio but decreases relatively rapidly toward the Mn end of the series. Luster vitreous to subresinous. Color bluish gray or greenish gray in triphylite, becoming clove-brown, yellowish brown to honey-yellow and salmon in lithiophilite. Exposed surfaces often

are brownish or dark gray to nearly black due to alteration. Streak colorless to grayish white. Transparent to translucent.
Oriented growths. With graftonite (which see).
O p t.[6] In transmitted light, colorless to pale yellow, pink, etc.

Mn:Fe	9:1	3:2	2:3	3:7	>1:4
nX(Na)	1.669	1.680	1.689	1.694	
nY	1.673	1.681	1.689	1.695	
nZ	1.682	1.688	1.695	1.700	>1.71
$2V$	~65°	~48°	0°	~55°	0°+
Sign	+	+	+	+	−
X	c	c		a	c
Y	a	a		c	b
Z	b	b	b	b	a

The indices of refraction increase with increase in content of Fe″. The index in the direction of [001] at the Mn end of the series changes from the least (X) to the greatest (Z) with increasing Fe, the series passing through a uniaxial condition twice. The dispersion is strong, $r < v$. When $2V$ is sensibly 0° for intermediate wavelengths, anomalous interference colors are obtained in white light in which the extreme colors of the spectrum, red and blue, are dispersed in planes at right angles to each other. Members of the series are not pleochroic or are weakly pleochroic, with some lithiophilite having X deep pink, Y pale greenish yellow, Z pale pink. The indices of refraction are markedly lowered by the substitution of Mg for (Fe,Mn); the material of analysis 2 has nX 1.675 (c), nY 1.684 (a), nZ 1.685 (b) with $2V$ 25°.

C h e m. A phosphate of lithium and divalent iron and manganese, Li(Fe″,Mn″)(PO$_4$). Fe″ and Mn″ substitute mutually, and a doubtless complete series extends between the essentially pure end-members. The names triphylite and lithiophilite are applied to the halves of the series with Fe > Mn and Mn > Fe, respectively. Mg substitutes for (Fe,Mn), with Mg:(Fe,Mn) = 1:2.8 in analysis 2 but is usually present in amounts up to only about 2 weight per cent MgO. Ca similarly is present in substitution for (Fe,Mn) in small amounts, although highly calcian material, with Ca:(Fe,Mn) = 1:3.1 (analysis 8) has been reported. Only small amounts of Na have been reported in substitution for Li. The Fe$_2$O$_3$ reported in some analyses is due to oxidation of Fe″.

Anal.[7]

	1	2	3	4	5	6	7	8	9	10
Li$_2$O	9.47	8.36	6.95	8.59	8.86	8.36	9.23	5.51	9.11	9.53
Na$_2$O		1.05	2.71		0.15	0.12	0.24	1.48	0.24	
FeO	45.54	32.93	31.09	35.06	29.13	21.70	12.57	7.45	2.94	
MnO		3.11	8.49	11.40	15.96	21.13	32.65	30.53	42.58	45.22
CaO		nil	6.13		0.31	0.65	tr.	9.70	tr.	
MgO		7.38			0.42	0.23		nil		
P$_2$O$_5$	44.99	46.03	41.08	44.43	44.87	46.41	45.05	43.43	44.95	45.25
H$_2$O		0.77	0.42		0.48	0.99	0.30	1.81	0.65	
Rem.		3.97	3.26	0.60	0.28	0.64	0.17	0.50	0.16	
Total	100.00	103.60	100.13	100.08	100.46	100.23	100.21	100.41	100.63	100.00
G.		3.44					3.482			

1. LiFe″(PO$_4$). 2. Triphylite. Newport, New Hampshire.[8] Rem. is K$_2$O 0.45, SiO$_2$ 0.33, Al$_2$O$_3$ 3.19, H$_2$O includes H$_2$O− 0.20. 3. Triphylite. Sukula, Finland.[9] Rem. is Fe$_2$O$_3$ 2.76, insol. 0.50, F tr., K$_2$O tr. 4. Hagendorf, Bavaria.[10] Rem. is insol. 5. Triphylite. Center Strafford, N. H.[11] Rem. is K$_2$O 0.07, insol. 0.21. 6. Triphylite. Pointe du Bois, Manitoba.[12] Rem. is K$_2$O 0.44, insol. 0.20. 7. Lithiophilite. Branchville, Conn.[13] Rem. is insol. 8. Lithiophilite. Wodgina, Western Australia.[14] Rem. is K$_2$O 0.17, insol. 0.16. H$_2$O includes H$_2$O− 0.07, Cl 0.07, F 0.10. 9. Lithiophilite. Buckfield, Maine.[13] Rem. is insol. 10. LiMn″(PO$_4$).

TRIPHYLITE-LITHIOPHILITE

Tests. B.B. fuses at $1\frac{1}{2}$. In C.T. sometimes decrepitates, darkens in color, and gives off traces of water. Soluble in acids.

Occur. Triphylite and lithiophilite occur as primary minerals in granite pegmatites, especially those of the so-called complex type containing lithia and phosphate minerals. Commonly associated species include triploidite-wolfeite, eosphorite, spodumene, albite, beryl, amblygonite, graftonite. Triphylite is found in Bavaria, Germany, at Hühnerkobel near Rabenstein (probably the correct locality for material bearing old labels marked Bodenmais, Zwiesel, or Rabenstein), and also at Hagendorf near Pleystein in the Oberpfalz. In Finland from Kietyö near Tammela and at Sukula. At Varuträsk, about 20 km. southeast of Boliden in northern Sweden. In France at La Vilate near Chanteloube and at Huréaux, Haute-Vienne, for the most part deeply altered. In the United States, triphylite occurs abundantly at numerous pegmatites in the central region of New Hampshire, notably at the Palermo and near-by mines at North Groton, at the Ruggles mine near Grafton Center, at Alexandria, at the Smith mine near Newport (magnesian), and at Center Strafford. In Massachusetts at Huntington (formerly Norwich); in Maine at Peru, Newry, Lords Hill in Stoneham and Black Mountain. In South Dakota in the pegmatites of the Keystone district, and at Pala in San Diego County, California. At Pointe du Bois, Manitoba, Canada. Lithiophilite occurs at Mangualde, Portugal, and sparingly at Varuträsk, Sweden. Calcian lithiophilite occurs at Wodgina and Yandeyarra in Western Australia. In the United States found in Connecticut at Branchville (the original locality) and at Portland. In Maine at Buckfield, Poland, and Norway in Oxford County. Found in large amounts in the pegmatites of the Custer district, South Dakota. At Pala, San Diego County, California. In Canada in the Yellowknife-Beaulieu area, Northwest Territories. In the Karibib area, South West Africa, and in the pegmatite area south of the Orange River in Namaqualand, South Africa.

Alter.[15] Triphylite-lithiophilite are relatively unstable chemically and readily undergo alteration. This is of two contrasting types, depending on the environmental conditions. In one type, crystals of these species formed early during the crystallization of the pegmatite are altered or re-worked by later hydrothermal solutions with the formation of triploidite, wolfeite, reddingite, phosphoferrite, eosphorite, fairfieldite, dickinsonite, fillowite, rhodochrosite, siderite. The second type of alteration takes place when the crystals are attacked by meteoric waters in the zone of weathering. This alteration involves first the oxidation of Fe'' to Fe''' (the Mn with its higher oxidation potential remaining unchanged) giving rise to the minerals of the sicklerite series, and then the oxidation of Mn'' to Mn''' to give the minerals of the heterosite series. During oxidation Li is leached concomitantly to effect valence compensation. These changes apparently are not accompanied by marked changes in crystal structure and the oxidation products, commonly arranged in concentric zones around unaltered cores of triphylite-lithiophilite, are oriented single-crystal pseudomorphs after the original crystals. Continued weathering may ultimately form brown to black, friable pseudomorphs composed of hydrous iron and manganese oxides. This alteration sequence usually is accompanied by the direct crystallization of other secondary iron and manganese phosphates, all ultimately derived by the decomposition of the triphylite and lithiophilite. Here may be mentioned hureaulite, stewartite, strengite, metastrengite, dufrenite, frondelite, rockbridgeite, xanthoxenite, cacoxenite, beraunite, and especially vivianite.

Hühnerkobelite and varulite similarly go through a stepwise oxidation of Fe'' and Mn'' with concomitant leaching of Na and in the final stage of alteration form heterosite and purpurite. The sequence of alteration and the nomenclature is tabulated below:

Li Series		Na Series	
Triphylite	Lithiophilite	Hühnerkobelite	Varulite
$Li(Fe'',Mn'')(PO_4)$	$Li(Mn'',Fe'')(PO_4)$	$Na(Fe'',Mn'')(PO_4)$	$Na(Mn'',Fe'')(PO_4)$
Ferri-sicklerite	Sicklerite	Alluaudite	Mangan-alluaudite
$(Li,Fe''',Mn'')(PO_4)$	$(Li,Mn'',Fe''')(PO_4)$	$(Na,Fe''',Mn'')(PO_4)$	$(Na,Mn'',Fe''')(PO_4)$
Heterosite	Purpurite	Heterosite	Purpurite
$(Fe''',Mn''')(PO_4)$	$(Mn''',Fe''')(PO_4)$	$(Fe''',Mn''')(PO_4)$	$(Mn''',Fe''')(PO_4)$

The individuality of the various members of these alteration sequences has been a matter of much confusion, especially in the older literature; and additional work is desirable. Among the older names that have been proposed for these minerals, or mixtures thereof, and now recognized only as synonyms, may be mentioned pseudotriplite, neopurpurite and melanchlor pt. (all = heterosite), lemnäsite (= alluaudite), pseudoheterosite (= sicklerite), headdenite and soda-triphylite (= arrojadite).

A r t i f.[16] Crystals of $LiMn''(PO_4)$ have been obtained by heating LiCl and $MnCl_2$ in a stream of $POCl_3$ and water vapor at 850°, and by melting together LiCl, $Li_3(PO_4)$, and $Mn_3(PO_4)_2$.

N a m e. Triphylite from τρίς, *threefold*, and φυλή, *family*, in allusion to containing three cations. Lithiophilite from *lithium* and φιλός, *friend*.

Ref.

1. Elements of the structure cell of Björling and Westgren, *Geol. För. Förh.*, **60**, 67 (1938), on Varuträsk triphylite (with $Fe:Mn \sim 1.3:1$ (?)), reoriented to make $c < a < b$. Transformations: B. and W. (1938), Dana (756, 1892), Tschermak, *Ak. Wien, Sitzber.*, **47**, 282 (1863), to new, 001/010/100; Goldschmidt (9, 6, 1923) to new: 001/0½0/100. The reported morphological elements (see below) are unsatisfactory because of the poor quality of the measured crystals; and the variation in angles with the Fe:Mn ratio is not established.

Locality	a	b	c	FeO	MnO	MgO	Ref.
Newport, N. H.	0.5346	1	0.4425	32.93	3.11	7.38	Palache, priv. comm.
Manitoba	0.5806	1	0.4536	21.70	21.13	0.23	Walker, ref. 12
Norwich	0.5265	1	0.4348	$(Fe:Mn \sim 1.2:1$ (?))			Tschermak (1863)
Artificial	0.5834	1	0.4522	$LiMn(PO_4)$			Zambonini and Malossi, ref. 16
———	0.5526	1	0.4355	……			Lacroix (**4**, 361, 1910)

2. Björling and Westgren (1938).
3. Gossner and Strunz, Zs. Kr., **83,** 415 (1932).
4. Dana (756, 1892) and Goldschmidt (1923). Chapman, Am. Min., **28,** 90 (1943), doubtfully reports {121}. All the reported forms are on triphylite; also {010}, {011}, {120}, {101} on artificial $LiMn''(PO_4)$.
5. Values taken from the graphical summary of Chapman (1943).
6. Summarized from the data of Penfield and Pratt, Am. J. Sc., **50,** 387 (1895), and Chapman (1943).
7. For summaries of analyses see Mason, Geol. För. Förh., **63,** 117 (1941), and Hintze (**1** [4A], 226, 1922); also Eriksson, Ark. Kemi, **23,** no. 8 (1946).
8. Lindgren analysis in Chapman (1943).
9. Mäkinen, Bull. comm. géol. Finlande, no. 35, 96 (1913).
10. Laubmann and Steinmetz, Zs. Kr., **55,** 561 (1920).
11. Gonyer analysis in Switzer, Am. Min., **23,** 811 (1938).
12. Oswald in Walker, Univ. Toronto Stud., Geol. Ser., no. 30, 10 (1931).
13. Vassar in Landes, Am. Min., **10,** 383 (1925).
14. Berggren analysis in Quensel, Geol. För. Förh., **59,** 77 (1937).
15. For a summary discussion see Mason (1941), Geol. För. Förh., **64,** 335 (1942); also Lindberg, Am. Min., **35,** 59 (1950).
16. Zambonini and Malossi, Zs. Kr., **80,** 442 (1931).

38.1.1.3 **HÜHNERKOBELITE** $[(Na,Ca)(Fe'',Mn'')_2(PO_4)_2]$. Arrojadite Mason (Geol. För. Förh., **64,** 335, 1942); Eriksson (Ark. Kemi, **23A,** no. 8, 1946). Hühnerkobelite Lindberg (Am. Min., **35,** 59, 1950).

38.1.1.4 **VARULITE** $[(Na,Ca)(Mn'',Fe'')_2(PO_4)_2]$. Quensel (Geol. För. Förh., **59,** 77, 1937).

Probably orthorhombic. Massive, granular. Cleavage {100} good, {010} distinct.[1] H. 5. G. 3.5–3.6. Color dull olive-green, becoming yellowish or brownish on alteration. Luster vitreous.

Opt. Biaxial positive (+).

ORIENTATION	Hühnerkobelite, Norrö [2]	Hühnerkobelite, Hühnerkobel [3]	Varulite, Skrumpetorp [4]	Varulite, Varuträsk [5]
nX	1.718	1.754	1.708	1.720
nY b				
nZ	1.731	>1.77	1.722	1.732
$2V$			large	70° (Na)
Disp.				$r > v$

Pleochroic with X yellowish green and Z grass-green.

Chem. A phosphate of sodium, calcium, divalent iron, and divalent manganese, essentially $(Na,Ca)(Fe'',Mn'')_2(PO_4)_2$. The names hühnerkobelite and varulite are applied to the parts of the series with Fe > Mn and Mn > Fe, respectively. The Fe'' in the material so far analyzed is in part oxidized to Fe''' and indicates a passage to the related minerals alluaudite and mangan-alluaudite or is due to admixture of these species.

Anal.	1	2	3	4	5	6
Li_2O		0.36	0.25		1.65	0.88
Na_2O	4.66	3.73	9.67	9.72	8.08	7.12
K_2O		0.05	0.06	tr.	0.12	0.12
CaO	8.44	9.70	1.35	3.60	2.30	4.86
MgO		0.68	2.55	0.09	0.13	
FeO	10.81	7.09	18.15	12.01	11.03	7.52
MnO	21.35	6.44	9.79	21.06	25.31	25.30
Fe_2O_3	12.02	26.49	14.45	6.44	5.32	8.35
P_2O_5	42.72	39.40	41.93	44.60	44.93	42.80
H_2O+		4.49	0.85	1.52	0.64	0.75
H_2O-		0.24	0.20	0.14	0.18	0.14
Insol.		1.88	0.50	0.44	0.28	1.80
Total	100.00	100.55	99.75	99.70	99.97	100.06
G.			3.55	3.61		3.581

1. $(Na,Ca)(Fe'',Fe''',Mn'')_2(PO_4)_2$ with $Na:Ca = 1:1$ and $Fe'':Fe''':Mn = 1:1:2$.
2. Hühnerkobelite. Hühnerkobel, Bavaria.[6] Contains some alluaudite. 3. Hühnerkobelite. Norrö, Sweden.[7] 4. Varulite. Skrumpetorp, Sweden.[8] Total includes F 0.08.
5. Varulite. Varuträsk, Sweden.[9] With Cl tr. 6. Varulite. Varuträsk.[10] Total includes F 0.06 and Al_2O_3 0.36.

Occur. A primary mineral found in granite pegmatites; also as a hydrothermal alteration product of triphylite. Varulite was originally from Varuträsk, Sweden, and also occurs at Skrumpetorp, Östergötland, Sweden. Hühnerkobelite occurs at Norrö on the island of Ranö in the Stockholm archipelago, Sweden, and with alluaudite and eosphorite at Hühnerkobel, Bavaria, Germany.

Alter. To alluaudite and mangan-alluaudite by oxidation of the iron.

Name. Varulite from the locality at Varuträsk. The iron-rich matrial from Hühnerkobel and Norrö was originally described under the name arrojadite in the belief that this species was the iron analogue of varulite. Later comparison [11] of type arrojadite with this material proved their separate identity and the new name hühnerkobelite, from the locality in Bavaria, was then introduced for the present species.

Ref.
1. The reported cleavages (Quensel, 1937) are here set analogous to triphylite.
2. Eriksson (1946).
3. Mason (1942); these data are out of line with the others and may be in error.
4. Mason, *Geol. För. Förh.*, **62**, 373 (1940).
5. Quensel (1937).
6. Berggren analysis in Mason (1942).
7. Berggren analysis in Eriksson (1946).
8. Berggren analysis in Mason (1940).
9. Berggren analysis in Quensel, *Geol. För. Förh.*, **62**, 297 (1940); an additional analysis is here cited.
10. Berggren analysis in Quensel (1937).
11. Lindberg, *Am. Min.*, **35**, 59 (1950).

38.1.2 NATROPHILITE [$NaMn(PO_4)$]. Brush and Dana (*Am. J. Sc.*, **39**, 205, 1890).

Cryst. Orthorhombic.

$$a:b:c = 0.601:1:0.472; \qquad p_0:q_0:r_0 = 0.785:0.472:1$$

$$q_1:r_1:p_1 = 0.601:1.273:1; \qquad r_2:p_2:q_2 = 2.119:1.664:1$$

NATROPHILITE

Forms: [1]

	ϕ	$\rho = C$	ϕ_1	$\rho_1 = A$	ϕ_2	$\rho_2 = B$
a 100	90°00′	90°00′	.....	0°00′	0°00′	90°00′
e 120	39 44½	90 00	90°00′	50 15½	0 00	39 44½
d 011	0 00	25 16	25 16	90 00	90 00	64 44
l 021	0 00	43 21	43 21	90 00	90 00	46 39

Structure cell.[2] Space group $Pmna$ if holohedral. a_0 6.32 Å, b_0 10.52, c_0 4.97; $a_0:b_0:c_0 = 0.601:1:0.472$. Cell contents $Na_4Mn_4(PO_4)_4$. The crystal structure of natrophilite is related to but not identical with that of triphylite-lithiophilite.

Habit. Crystals rare; indistinct, and resembling triphylite. Commonly as cleavable masses or granular.

P h y s. Cleavage {100} good, {010} indistinct, {021} interrupted. Fracture conchoidal. H. 4½–5. G. 3.41; 3.47 (calc.). Luster bright resinous to nearly adamantine, somewhat pearly on the {100} cleavage. Color deep wine-yellow. Transparent to translucent.

O p t.[3]

Orientation		n	
X	c	1.671 ± 0.003	Biaxial positive (+).
Y	a	1.674 ± 0.003	$2V$ 75° ± 5°.
Z	b	1.684 ± 0.003	$r < v$, strong.

C h e m. A phosphate of sodium and manganese, $NaMn(PO_4)$. Fe'' substitutes for Mn'', with Fe:Mn $\sim$1:13 in the only reported analysis.

	Li_2O	Na_2O	FeO	MnO	P_2O_5	H_2O	Insol	Total
1.		17.92		41.01	41.07			100.00
2.	0.19	16.79	3.06	38.19	41.03	0.43	0.81	100.50

1. $NaMn(PO_4)$. 2. Branchville.[4]

Tests. B.B. fuses very easily. Soluble in acids.

O c c u r. Found sparingly with lithiophilite, triploidite, eosphorite, hureaulite, fairfieldite, dickinsonite in granite pegmatite at Branchville, Fairfield County, Connecticut.

A l t e r. To an unidentified fine-fibrous, yellowish mineral with a silky luster.

N a m e. From *natrium* and φίλος, *a friend*, because it contains much sodium.

Ref.

1. Brush and Dana (1890), from approximate measurements. Elements of the structure cell.
2. Byström, *Ark. Kemi*, **17B**, no. 4 (1943), by Weissenberg method on Branchville material. Transformation: Byström to new, 010/100/001; Brush and Dana (1890) to new, 001/010/100.
3. Larsen (115, 1921).
4. Wells analysis in Brush and Dana (1890).

SICKLERITE SERIES

38.1.3.1 FERRI-SICKLERITE [(Li,Fe''',Mn'')(PO$_4$)]. *Quensel* (*Geol. För. Förh.*, **59**, 77, 1937). Fe-sicklerite.
38.1.3.2 SICKLERITE [(Li,Mn'',Fe''')(PO$_4$)]. Pseudohétérosite *Lacroix* (**4**, 469, 1910). Sicklerite *Schaller* (*Washington Ac. Sc., J.*, **2**, 143, 1912). Mangansicklerite. Mn-sicklerite.

Cryst.[1] Orthorhombic.

$$a:b:c = 0.588:1:0.475; \quad p_0:q_0:r_0 = 0.808:0.475:1$$

$$q_1:r_1:p_1 = 0.588:1.238:1; \quad r_2:p_2:q_2 = 2.105:1.701:1$$

Structure cell.[2] Orthorhombic. a_0 5.939 kX, b_0 10.086, c_0 4.787; $a_0:b_0:c_0 = 0.588:1:0.475$. Cell contents (Li,Mn'',Fe''')$_4$(PO$_4$)$_4$. The cell dimensions increase with increasing content of Mn.

Habit. Found only massive as an alteration product of triphylite-lithiophilite, usually as rims around partly altered masses or crystals of the latter species and in crystallographic continuity therewith.

Phys. Cleavage {100} good. H. ~4. G. 3.2–3.4. Color yellowish brown to dark brown. Streak and powder light yellowish brown to brown and reddish brown, not purple as in heterosite-purpurite. Subtranslucent to opaque.

Opt.[3]

Orientation		n (sicklerite, Pala)	Pleochroism	
X	a	1.715	Deep reddish	Biaxial negative (−).
Y		1.735	Paler reddish	2V medium large.
Z		1.745	Very pale reddish	$r > v$, very strong.

Chem. A phosphate of lithium, divalent manganese, and trivalent iron, essentially (Li,Mn'',Fe''')(PO$_4$). The ratio of Mn to Fe is dependent on the ratio of these mutually substituting cations in the original triphylite-lithiophilite from which the present species have been derived by oxidation of the Fe''. The names ferri-sicklerite and sicklerite are applied to the halves of the series with Fe > Mn and Mn > Fe and are analogous to triphylite and lithiophilite, respectively. The mechanism of the valence compensation accompanying the oxidation may involve partial leaching of Li according to the formula Li$_{1-y}$(Mn$_{1-x}$'',Fe$_{x-y}$'',Fe$_y$''')(PO$_4$). Complete oxidation of both Fe and Mn results in the total loss of Li and the formation of heterosite-purpurite, (Mn''',Fe''')(PO$_4$).

Anal.

	1	2	3	4	5
Li_2O	3.26	3.72	4.89	3.80	3.94
Na_2O	0.88	0.81	1.17		1.22
FeO	0.22	0.59	nil		nil
MnO	15.20	19.13	26.14	33.60	29.61
CaO	1.88	1.36	0.84	0.20	3.44
MgO	1.70	0.11	0.59		nil
Fe_2O_3	29.08	27.20	14.58	11.26	4.72
Mn_2O_3		nil	4.89	2.10	8.19
P_2O_5	44.64	44.80	44.45	43.10	44.83
H_2O	1.23	1.02	1.76	1.71	4.13
Rem.	2.42	1.66	0.45	4.18	0.40
Total	100.51	100.40	99.76	99.95	100.48
G.	3.27	3.39		3.45	

1. Ferri-sicklerite. Lövlunden, Sweden.[4] Rem. is insol. H_2O includes H_2O- 0.79.
2. Ferri-sicklerite. Varuträsk, Sweden.[4] Rem. is insol. H_2O includes H_2O- 0.51.
3. Sicklerite. Eräjärvi, Finland.[5] Rem. is K_2O 0.13, insol. 0.32. H_2O includes H_2O- 1.00. 4. Sicklerite. Pala, California.[6] Rem. is insol. 5. Sicklerite. Wodgina, Western Australia.[5] Rem. is K_2O 0.08, insol. 0.32. H_2O includes H_2O- 0.11.

Tests. Easily fusible. Soluble in acids.

O c c u r.[7] Ferri-sicklerite and sicklerite are secondary minerals formed by the alteration of triphylite and lithiophilite in the zone of weathering. The first described member of the series was the sicklerite from Pala, San Diego County, California, where it occurred as an alteration product of lithiophilite and was associated with hureaulite, stewartite, salmonsite, and metastrengite. Sicklerite also has been found at Varuträsk in Sweden, Eräjärvi (near Oriväsi) and Tammela in Finland, and at Wodgina, Western Australia. Ferri-sicklerite has been found at Varuträsk, Sweden, and Tammela, Finland; in Bavaria at Hühnerkobel and Hagendorf; in France in the pegmatite district in Haute-Vienne, Plateau Central. Probably from the Euriowie Range of New South Wales, Australia. In the United States, ferri-sicklerite has been identified in pegmatites at Center Strafford, Rochester and North Groton in New Hampshire and at Peru and Stoneham in Maine. Sicklerite and ferri-sicklerite doubtless occur in many other pegmatites in New England and other regions that carry lithiophilite and triphylite.

A l t e r. To heterosite and purpurite by oxidation of the Mn'' and leaching of the Li.

N a m e. Sicklerite after the Sickler family, formerly of Pala, California.

Ref.

1. Elements and symmetry of the structure cell of Björling and Westgren, *Geol. För. Förh.*, **60**, 67 (1938), on Varuträsk ferri-sicklerite (analysis 2 ?).
2. Björling and Westgren (1938); transformation from their orientation to new: 001/010/100.
3. Schaller (1912) and Dana (1892; 3rd App., 72, 1915).
4. Berggren analysis in Quensel, *Geol. För. Förh.*, **59**, 77 (1937).
5. Berggren analysis in Mason, *Geol. För. Förh.*, **63**, 117 (1941).
6. Schaller (1912).
7. See also the summary of Mason (1941).

ALLUAUDITE SERIES

38.1.4.1   A L L U A U D I T E [(Na,Fe''',Mn'')(PO$_4$)]. *Damour* (*Ann. mines*, **13**, 341, 1848). Fe-alluaudite *Quensel* (*Geol. För. Förh.*, **59**, 77, 1937). Ferri-alluaudite.
38.1.4.2   M A N G A N - A L L U A U D I T E [(Na,Mn'',Fe''')(PO$_4$)]. Mn-alluaudite *Quensel* (*Geol. För. Förh.*, **59**, 77, 1937). Lemnäsite *Pehrman* (*Acta Ac. Aboensis, Math. Phys.*, **12**, no. 6, 1939).

Massive, as an alteration product of varulite-hühnerkobelite; also compact granular, and in radiating fibrous or globular aggregates. Cleavage {100} and {010} good.[1] H. 5–5$\frac{1}{2}$. G. 3.4–3.5. Color dirty yellow to brownish yellow, also dull greenish black; outwardly brownish black or black due to alteration, but washing in dilute acid restores the yellow color. Streak and powder dirty yellow. Subtranslucent to opaque.

O p t.[2] In transmitted light, yellow to yellow-green and not pleochroic.

Orientation	n (Lövlunden)
X	1.760
Y	
Z	1.775

C h e m. A phosphate of sodium, divalent manganese, and trivalent iron, essentially (Na,Fe''',Mn'')(PO$_4$). The ratio of Fe to Mn is dependent on the ratio of these cations in the parent varulite-hühnerkobelite, from which these minerals are derived by oxidation of the Fe'' (see also *sicklerite* and *ferri-sicklerite*). The names alluaudite and mangan-alluaudite are given to material with Fe > Mn and Mn > Fe, respectively. The H$_2$O reported in the analyses probably is non-essential. Oxidation of the Mn'' apparently is accompanied by leaching of Na and the formation of heterosite or purpurite.[3]

Anal.

	1	2	3	4	5	6
Li$_2$O			0.07		0.02	0.28
K$_2$O					0.42	0.13
Na$_2$O	7.85	5.47	8.19	7.20	8.84	6.94
CaO			2.70		4.59	2.16
MgO			0.11	1.17	0.53	
FeO			3.19		1.92	1.98
MnO	17.97	23.08	20.66	28.17	27.87	26.24
Fe$_2$O$_3$	20.23	25.62	20.56	19.83	12.76	16.44
Mn$_2$O$_3$		1.06				
Al$_2$O$_3$					1.08	
P$_2$O$_5$	53.95	41.25	42.52	46.70	40.94	42.76
H$_2$O+		2.65	0.96	0.66	1.48	1.60
H$_2$O−			0.17	0.24	0.13	0.88
Insol.			0.60	1.06	1.24		0.54
Total	100.00	99.73	100.19	106.01	100.58	99.95
G.		3.468	3.58	3.576		

1. (Na,Fe''',Mn'')(PO$_4$) with Na:Fe:Mn = 1:1:1.   2. Alluaudite.   Chanteloube, France.[4]   3. Alluaudite.   Sukula, Finland.[5]   4. Mangan-alluaudite.   Varuträsk, Sweden.[6]   With F 0.80 included in the total. Original total given as 100.03.   5. Mangan-alluaudite (*lemnäsite*).   Lemnäs, Finland.[7] With Al$_2$O$_3$ 1.08.   6. Mangan-alluaudite.   Varuträsk.[8]

Tests. B.B. easily fusible. Soluble in acids.

O c c u r. Alluaudite and mangan-alluaudite are secondary minerals. Alluaudite was originally found in pegmatite at Chanteloube, France; also from Sukula, Finland. Mangan-alluaudite occurs at Varuträsk, Sweden, and at Lemnäs, Finland.
A l t e r. To heterosite and purpurite.
N a m e. After F. Alluaud, who discovered the Chanteloube material.
Ref.
1. Damour (1848) mentions three rectangular cleavages.
2. Quensel, *Geol. För. Förh.*, **59**, 77 (1937).
3. See Mason, *Geol. För. Förh.*, **63**, 117 (1941); Quensel (1937) and *Geol. För. Förh.*, **62**, 300 (1940).
4. Damour (1848).
5. Berggren analysis in Mason, *Geol. För. Förh.*, **62**, 369 (1940).
6. Blix analysis in Quensel (1937).
7. Pehrman (1939).
8. Berggren analysis in Quensel (1940).

HETEROSITE SERIES

38.1.5.1 H E T E R O S I T E [(Fe''', Mn''')(PO_4)]. Heteposite [error for heterosite] *Alluaud* in *Vauquelin (Ann. chim. phys.*, **30**, 294, 1825). Heterosite, Heterozite *Alluaud (Ann. sc. nat.*, **8**, 346, 1826). Purpurite *Graton* and *Schaller (Am. J. Sc.*, **20**, 146, 1905). Ferripurpurite *Schaller (Am. J. Sc.*, **24**, 152, 1907). Pseudotriplite *Blum* (537, 1845). Melanchlor *Fuchs (J. pr. Chem.*, **17**, 160, 1839). Neopurpurite *de Jesus (Com. Serv. Geol. Portugal*, **19**, 65, 1933). Na-heterosite *Quensel (Geol. För. Förh.*, **59**, 77, 1937).
38.1.5.2 P U R P U R I T E [Mn''',Fe''')(PO_4)]. Manganipurpurite *Schaller (Am. J. Sc.*, **24**, 152, 1907). Purpurite *Schaller (U. S. Geol. Sur., Bull. 490*, 72, 1910). Na-purpurite *Quensel (Geol. För. Förh.*, **59**, 77, 1937).

C r y s t.[1] Orthorhombic.

$$a:b:c = 0.492:1:0.601; \quad p_0:q_0:r_0 = 1.223:0.601:1$$

$$q_1:r_1:p_1 = 0.492:0.818:1; \quad r_2:p_2:q_2 = 1.664:2.034:1$$

Structure cell.[2] a_0 5.819 kX, b_0 9.680, c_0 4.760; $a_0:b_0:c_0 = 0.492:1:0.601$. Cell contents (Fe''',Mn''')$_4$(PO_4)$_4$.

P h y s. Cleavage {100} good, {010} imperfect; the cleavage surfaces sometimes are curved or crinkled. Fracture uneven. Brittle. H. 4–4½. G. 3.2–3.4. Luster on fresh fracture surfaces satiny. Color deep rose to reddish purple. Streak and powder paler in tint. Often outwardly altered and then dull or earthy in luster and dark brown to brownish black in color; brief washing in dilute acid restores the true color. Subtranslucent to opaque.

O p t.[3]

ORIENTATION	Purpurite?, Maine	Purpurite?, Peru, Maine	Heterosite, Limoges	Heterosite, La Vilate
nX a		1.85 ± 0.02	1.86 ± 0.01	
nY	1.92 ± 0.02	1.86 ± 0.02	1.89 ± 0.01	1.84–1.87
nZ		1.92 ± 0.02	1.91 ± 0.01	
$2V$	38°	Mod.	Large	Large

Strongly pleochroic, with X greenish gray, gray to rose-red, Y and Z deep blood-red to purplish red. Absorption $Z = Y > X$ or $Z > Y \gg X$. Dispersion very strong and optic axis sections may show anomalous green interference colors.

Chem. A phosphate of iron and manganese in their trivalent states, $(Fe''',Mn''')(PO_4)$. The ratio of Fe to Mn is dependent on the ratio of these cations obtaining in the original triphylite or lithiophilite from which the present species have been derived by oxidation. The names heterosite and purpurite are applied to the parts of the series with Fe > Mn and Mn > Fe, respectively. The Ca and Mg present in small amounts was originally present in substitution for (Fe'',Mn''). H_2O is present in small and variable amounts and appears to be non-essential.[4]

Anal.[5]

	1	2	3	4	5	6	7
Li_2O		tr.		0.11	0.46	0.83	tr.
Na_2O		0.84		1.07	1.12	0.72	tr.
CaO		1.48		1.28	1.70	1.16	1.37
MgO				0.04	0.17	nil	tr.
MnO					3.60	1.53	
Mn_2O_3	26.25	29.25	29.35	24.45	20.42	15.14	12.08
Fe_2O_3	26.55	15.89	24.60	28.20	27.44	33.42	38.36
P_2O_5	47.20	47.30	41.60	38.72	41.93	43.79	43.45
H_2O+		5.26	7.62	3.74	1.65	2.33	4.82
H_2O-				2.33	1.67	1.08	
Insol.		0.52		0.27	0.30		0.19
Total	100.00	100.54	103.17	100.21	100.46	100.00	100.27
G.					3.409	3.398	3.40

1. $(Fe''',Mn''')(PO_4)$ with Fe:Mn = 1:1. 2. Purpurite. Kings Mountain, North Carolina.[6] 3. Purpurite. Chanteloube, France.[7] 4. Heterosite *(neopurpurite)*. Mangualde, Portugal.[8] 5. Heterosite. Varuträsk, Sweden.[9] 6. Heterosite. Erongo, South West Africa.[9] 7. Heterosite. Hill City, South Dakota.[6]

Tests. B.B. easily fusible. In C.T. readily gives off a little water and turns yellowish brown. Easily soluble in HCl.

Occur. Heterosite and purpurite are secondary minerals formed by oxidation of triphylite and lithiophilite through the intermediate stage of ferri-sicklerite and sicklerite. Heterosite has been found in Haute-Vienne, France, near Limoges, at La Vilate near Chanteloube and at Huréaux. Also from Varuträsk, Sweden; the Hühnerkobel pegmatite near Rabenstein, Bavaria (*pseudotriplite* [10]); and Mangualde, Portugal (*neopurpurite* [10]). Also from Erongo, South West Africa, and the Euriowie Range in New South Wales. In the United States from Hill City, Pennington County, South Dakota; Huntington (Norwich), Massachusetts; Branchville, Connecticut; Palermo mine, North Groton, New Hampshire; Newry, Oxford County, Maine. Purpurite occurs at Chanteloube, France, at Wodgina, Western Australia, and in the United States at the Faires mine, Kings Mountain, Gaston County, North Carolina, and at Pala and Rincon, San Diego County, California. Also in the Custer district, Pennington County, South Dakota, and doubtless at many localities of lithiophilite.

Alter. To a black, pitchy material containing Fe_2O_3 23.2, Mn_2O_3 29.1, P_2O_5 31.2, H_2O 14.1 (Varuträsk);[9] an apparently similar mineral occurs in the Custer district, South Dakota,[11] and elsewhere.

N a m e. Heterosite from ἕτερος, *another*, probably because it is a second manganese-containing mineral from the same locality. Purpurite from the Latin *purpura*, in allusion to the color.

Ref.
1. Elements and symmetry of the x-ray structure cell of Björling and Westgren, *Geol. För. Förh.*, **60,** 67 (1938).
2. Björling and Westgren (1938) on heterosite from Varuträsk (analysis 5?); transformation: B. and W. to new, 001/010/100.
3. Larsen (84, 124, 1921).
4. Mason, *Geol. För. Förh.*, **63,** 117 (1941), from dehydration and x-ray study.
5. Additional analyses are tabulated by Mason (1941).
6. Schaller, *U. S. Geol. Sur., Bull. 490*, 72 (1911).
7. Pisani in Lacroix (**4,** 470, 1910).
8. de Jesus (1933).
9. Berggren in Quensel, *Geol. För. Förh.*, **59,** 88 (1937).
10. See Mason (1941) on the identity with heterosite.
11. Fisher, *South Dakota Geol. Sur., Rpt.*, 50 (1945).

38.1.6   B E R Y L L O N I T E [$NaBe(PO_4)$]. *Dana (Am. J. Sc.*, **36,** 290, 1888); *Dana and Wells (Am. J. Sc.*, 37, 23, 1889).

C r y s t.[1] Monoclinic prismatic $(2/m)$, with marked orthorhombic pseudo-symmetry.

$a:b:c = 1.0426:1:1.8215$; $\beta\ 90°00'$; $p_0:q_0:r_0 = 1.7471:1.8215:1$

$q_1:r_1:p_1 = 1.0426:0.5724:1$; $\mu\ 90°00'$; $r_2:p_2:q_2 = 0.5490:0.9591:1$

Forms:[2]

	ψ	$\rho = C$	ϕ_1	$\rho_1 = A$	ϕ_2	$\rho_2 = B$
c 001	.....	0°00'	0°00'	90°00'	90°00'	90°00'
b 010	0°00'	90 00	90 00	90 00	.....	0 00
a 100	90 00	90 00	.....	0 00	0 00	90 00
d 120	25 37½	90 00	90 00	64 22½	0 00	25 37½
m 110	43 48½	90 00	90 00	46 11½	0 00	43 48½
f 210	62 28	90 00	90 00	27 32	0 00	62 28
κ 014	0 00	24 29	24 29	90 00	90 00	65 31
ϑ 013	0 00	31 16	31 16	90 00	90 00	58 44
η 012	0 00	42 19½	42 19½	90 00	90 00	47 40½
ε 011	0 00	61 14	61 14	90 00	90 00	28 46
ζ 032	0 00	69 54	69 54	90 00	90 00	20 06
γ 021	0 00	74 39	74 39	90 00	90 00	15 21
l 102	90 00	41 08½	0 00	48 51½	48 51½	90 00
k 203	90 00	49 21	0 00	40 39	40 39	90 00
e 101	90 00	60 13	0 00	60 13	29 47	90 00
i 201	90 00	74 02	0 00	74 02	15 58	90 00
w 112	43 48½	51 36½	42 19½	57 08½	48 51½	55 33
v 111	43 48½	68 23	61 14	49 56½	29 47	47 51½
ψ 121	25 37½	76 06	74 39	65 11	29 47	28 55½
t 213	62 28	52 43	31 16	45 07½	40 39	68 25
s 212	62 28	63 05½	42 19½	37 45	29 47	65 39½
r 211	62 28	75 45½	61 14	30 44½	15 58	63 23

***Structure cell.*[3]** Space group $P2_1/c$. a_0 8.13 kX, b_0 7.76, c_0 14.17; β 90°00′; $a_0:b_0:c_0 = 1.0477:1:1.8260$. Cell contents $Na_{12}Be_{12}(PO_4)_{12}$.

Habit. Tabular {010} to short prismatic [010]; the crystals are often highly complex, particularly in the zones [100] and [010]. Faces in the

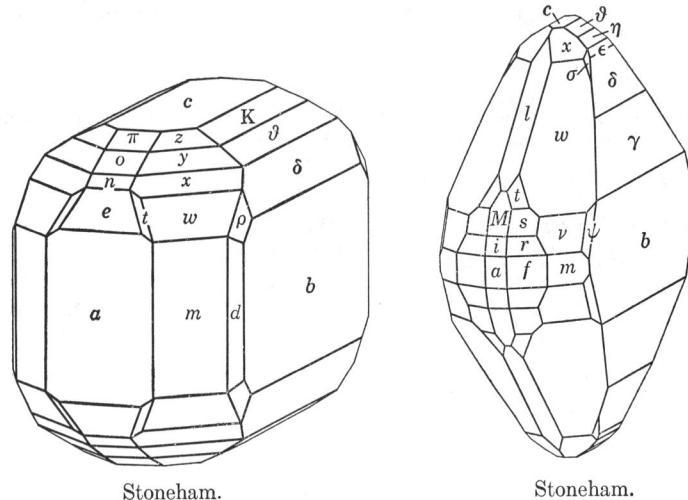

Stoneham.   Stoneham.

zone [010] near {100} may be united in oscillatory combination. The crystal faces frequently are dull or roughened or in some cases delicately etched.

***Twinning.*[4]** Twin plane {101}, both as contact and penetration twins; sometimes repeated, or in pseudo-hexagonal stellate forms ($a \wedge \underline{a} \sim 120°$); also polysynthetic.

P h y s. Cleavage {010} perfect, {100} good but interrupted, {101} indistinct,[5] {001} in traces. Fracture conchoidal. Brittle. H. $5\frac{1}{2}$–6. G. 2.81;[6] 2.81 (calc.). Luster vitreous, brilliant; on {010} sometimes pearly. Colorless to snow-white or pale yellowish. Transparent to translucent. The crystals often show a columnar structure due to the presence of hollow canals and fluid cavities arranged parallel to [010].

O p t.[9] In transmitted light, colorless.

ORIENTATION		n(Li)	n(Na)	n(Tl)	
X	b	1.5492	1.5520	1.5544	Biaxial negative (−).
Y	a	1.5550	1.5579	1.5604	$r < v$, weak.
Z	c	1.560	1.561	1.564	
2V (meas.)		67°51′	67°56′	67°57′	

C h e m. A phosphate of sodium and beryllium, $NaBe(PO_4)$.

Anal.

	1	2	3
Na_2O	24.41	23.64	23.28
BeO	19.70	19.84	19.12
P_2O_5	55.89	55.86	55.40
Ign. loss		0.08	0.52
Rem.			1.83
Total	100.00	99.42	100.15

1. $NaBe(PO_4)$. 2. Stoneham.[7] 3. Newry.[8] Rem. is CaO 0.40, K_2O 0.92, Li_2O 0.07, Al_2O_3 0.21, Fe_2O_3 0.07, insol. 0.16.

Tests. B.B. decrepitates and fuses at about 3 to a cloudy glass. Slowly soluble in acids.

O c c u r. Originally found in the disintegrated outcrop of a pegmatite at the base of McKean Mountain near Stoneham, Oxford County, Maine. Here associated with smoky quartz crystals, mica, apatite, triplite, beryl, cassiterite, orthoclase, and albite. Later found with herderite, eosphorite, albite, and tourmaline in pegmatite at Newry, Maine.

A l t e r. To herderite, the alteration first developing a fibration parallel [010] of the beryllonite and finally leaving a cavernous pseudomorph.

A r t i f.[10] Obtained in hexagonal tablets or prisms by slow cooling of a fusion of BeO and NaCl in sodium pyrophosphate or of BeO in sodium metaphosphate.

Ref.

1. Angles and unit of Dana and Wells (1889) in the orientation of the structure cell of Gossner and Besslein, *Cbl. Min.*, 144 (1934). Transformation: D. and W. to G. and B., $100/00\bar{1}/010$. The symmetry of beryllonite, originally taken as orthorhombic, was shown to be monoclinic by the x-ray study of Gossner and Besslein (1934); the angle table and physical properties are here given as of an orthorhombic crystal. On the structural resemblance to trimerite see Gossner and Besslein (1934) and Strunz, *Zs. Kr.*, **98**, 76 (1937).

2. Dana and Wells (1889). Less common and rare:

μ 016	β 031	p 106	n 103	g 401	y 114	u 221	σ 123	ρ 132
λ 015	A 041	π 105	j 302	ω 116	x 113	τ 136	χ 122	T 412
ξ 023	q 1.0.12	o 104	h 301	z 115	$\varnothing$ 223	Q 124	Δ 313	R 411

3. Gossner and Besslein (1934).
4. Dana and Wells (1889); Gossner and Besslein (1934).
5. Dana (759, 1892), apparently correcting a cleavage on {103} given by Dana and Wells (1889).
6. Berman in Palache and Shannon, *Am. Min.*, **13**, 392 (1928), on Newry and Stoneham material.
7. Wells analysis in Dana and Wells (1889).
8. Shannon analysis in Palache and Shannon (1928).
9. Dana and Wells (1889). The values for nZ are probably low; Gaubert, *Bull. soc. min.*, **30**, 108 (1907), gives $nZ(Na) = 1.5618$.
10. Ouvrard, *C.R.*, **110**, 1334 (1890).

38.1.7 **A R R O J A D I T E** $[Na_2(Fe'',Mn'')_5(PO_4)_4]$. Unnamed phosphate *Headden* (*Am. J. Sc.*, **41**, 416, 1891). Arrojadita *Guimarães* (*Publ. da Inspect. de Obras Contra as Seccas, Rio de Janeiro*, no. 58, 1925; *Bol. fac. fil. Univ. São Paulo*, no. 30 (Min. no. 5, 3–16) 1942). Soda-triphylite *Ziegler* (*South Dakota School of Mines, Bull.*, **10**, 1914). Headdenite *Quensel* (*Geol. För. Förh.*, **59**, 77, 1937).

Cryst.[1] Monoclinic; prismatic—$2/m$ (?).
Structure cell.[2] Space group $C2/m$ if holohedral. a_0 16.60 Å, b_0 10.02, c_0 23.99; β 93°37'; $a_0:b_0:c_0 = 1.656:1:2.389$. Cell contents $Na_{24}(Fe'',Mn'')_{60}(PO_4)_{48}$.
Habit. As large cleavable masses.
Phys. Cleavage {001} good, {201} poor. Fracture uneven to subconchoidal. H. 5. G. 3.55. Luster vitreous to somewhat greasy. Color dark green. Translucent.
Opt.[2]

ORIENTATION	n (Serra Branca)	n (Nickel Plate)	PLEOCHROISM	
X b	1.662	1.664	Colorless	Biaxial negative ($-$).
$Y \wedge c$ $+21\frac{1}{2}$	1.668	1.670	Pale green	$r < v$
Z	1.672	1.675	Pale yellow-green	
$2V$	80°	86°		

Chem. Essentially a phosphate of sodium, divalent iron, and manganese, $Na_2(Fe'',Mn'')_5(PO_4)_4$. Significant amounts of K and Ca substitute for Na; Mg, Al are present in substitution for (Fe'',Mn''). The Fe_2O_3 reported in some material is due to oxidation of Fe''. The role of the small amounts of H_2O and F present is uncertain.
Anal.

	1	2	3
Na_2O	8.79	6.40	4.67
K_2O		1.74	1.45
Li_2O		0.09	tr.
CaO		2.46	5.69
MgO		1.04	1.85
FeO	50.95	28.22	19.84
MnO		15.78	12.33
P_2O_5	40.26	40.00	34.32
H_2O		0.91	5.40
Rem.		3.57	14.57
Total	100.00	99.87	100.12
G.		3.55	

1. $Na_2Fe_5''(PO_4)_4$. 2. Nickel Plate mine (*headdenite*).[3] Rem. is Al_2O_3 2.66, F 0.80, insol. 0.11. F = O = 0.34 deducted. 3. Serra Branca.[4] Rem. is Fe_2O_3 12.39, SnO_2 1.52, insol. 0.66. H_2O includes H_2O- 0.44. Analysis made on slightly altered material.

Tests. Soluble in dilute acids.

Occur. Originally found as large masses in pegmatite at the Nickel Plate mine in the Keystone district, Pennington County, South Dakota, associated with graftonite, cassiterite, spodumene, beryl, and muscovite. Also from the Etta mine in the same district. Later found (and named *arrojadite*) in pegmatite at Serra Branca, Picuhy, Parahyba, Brazil. So-called arrojadite from Hühnerkobel, Bavaria, and Norrö, Sweden, is identical not with this species but with hühnerkobelite.
Name. Arrojadite after the Brazilian geologist Miguel Arrojado Lisbôa. Headdenite after William P. Headden (1850–1932), American mineralogist. The name headdenite was given to the mineral from South

Dakota early described but not named by Headden, in lack of knowledge of the identity [5] of this mineral with the arrojadite from Brazil.

Ref.
1. Elements of the structure cell of Lindberg, *Am. Min.*, **35**, 59 (1950), on the material of Headden, *Am. J. Sc.*, **41**, 416 (1891), from the Nickel Plate mine (analysis 2).
2. Lindberg (1950).
3. Lindberg (1950); also an earlier, inferior analysis on the same material by Headden (1891).
4. Guimarães (1942).
5. Established by x-ray and optical study of type material by Lindberg (1950).

TYPE 2. $A_3B_2(XO_4)_3$

BERZELIITE SERIES

	a_0
Berzeliite, $(Mg,Mn)_2(Ca,Na)_3(AsO_4)_3$	$\sim$12.3 kX
Manganberzeliite, $(Mn,Mg)_2(Ca,Na)_3(AsO_4)_3$	$\sim$12.5

The Berzeliite Series is isometric hexoctahedral and has a crystal structure [1] similar to that of garnet, $A_2B_3(SiO_4)_3$. It also may be noted that the hydrated phosphate griphite,[2] $(Na,Ca,Fe,Al)_3Mn_2(PO_4)_{3-x}(OH)_{4x}$ where $x \sim 0.5$, has a structure similar to garnet, as do the artificial compounds [3] $Na_3Al_2(PO_4)_3$ and $NaCa_2Mg_2(PO_4)_3$.

Ref.
1. Bubeck and Machatschki, *Zs. Kr.*, **90**, 44 (1935).
2. McConnell, *Am. Min.*, **27**, 452 (1942).
3. Thilo, *Naturwiss.*, **29**, 329 (1941).

38.2.1.1 **BERZELIITE** [$(Mg,Mn)_2(Ca,Na)_3(AsO_4)_3$]. Berzeliit *Kühn* (*Ann. Chem. Pharm.*, **34**, 211, 1840). Magnesian Pharmacolite *Dana* (239, 1844). Chaux arseniatée anhydre *Dufrénoy*. Berzelit *Haidinger* (495, 1895). Kühnite *Brooke* and *Miller* (481, 1851). Pseudoberzeliite *Lindgren* (*Geol. För. Förh.*, **7**, 291, 1884). Dubbelbrytande Berzeliite *Lindgren* (*Geol. För. Förh.*, **5**, 552, 1881). Soda Berzeliite *Sjögren* (*Bull. Geol. Inst. Upsala*, **2**, 92, 1895). Natron-berzeliite *Bäckström* (*Zs. Kr.*, **26**, 102, 1896). Magnesium-Berzeliit *Bubeck* (*Geol. För. Förh.*, **56**, 526, 1934). Mg-Berzeliite. Pyrrho-arsenite pt.

38.2.1.2 **MANGANBERZELIITE** [$(Mn,Mg)_2(Ca,Na)_3(AsO_4)_3$]. Pyrrhoarsenite *Igelström* (*Bull. soc. min.*, **9**, 218, 1886). Pyrrharsenite. Chloroarsenian? *Igelström* (*Geol. För. Förh.*, **15**, 471, 1893). Manganberzeliit *Igelström* (*Zs. Kr.*, **23**, 592, 1894). Mn-Berzeliite.

Cryst. Isometric; hexoctahedral—$4/m\,\bar{3}\,2/m$.

Forms: [1]

| a 001 | d 011 | e 012 | n 112 |

Structure cell.[2] Space group $Ia3d$. a_0 12.35 kX (2.3 per cent MnO), 12.46 (19.4 per cent MnO). Cell contents $(Mg,Mn)_{16}(Ca,Na)_{24}(AsO_4)_{24}$. Berzeliite is isotypic with garnet.[4]

Habit. Rarely in trapezohedrons with small modifying faces. Usually massive or as rounded grains.

Phosphates, Arsenates, Vanadates

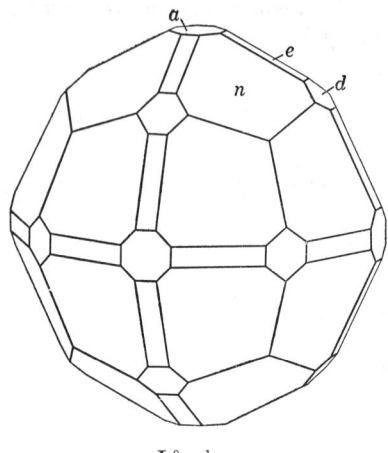

Långban.

P h y s. Cleavage none.[3] Fracture subconchoidal to uneven. Brittle. H. $4\frac{1}{2}$–5. G. ~4.08 (Mg end-member) to ~4.46 (Mn end-member).[5] Luster resinous. Color yellow or honey-yellow to orange-yellow and yellowish red; the depth of the red tint increases with the content of Mn. Streak nearly white to orange-yellow. Transparent to translucent. M.P. 1180° (2.3 per cent MnO), 1090° (19.4 per cent MnO).

O p t. In transmitted light, colorless to orange. Isotropic. Sometimes shows weak anomalous birefringence.

	Berzeliite end-member [5]	Manganberzeliite end-member [5]
n(670 mμ)	1.702 ± 0.003	1.781 ± 0.003
n(535 mμ)	1.710 ± 0.003	1.790 ± 0.003

C h e m. An arsenate of magnesium, divalent manganese, calcium and sodium, $(Mg,Mn)_2(Ca,Na)_3(AsO_4)_3$. Mn and Mg substitute mutually and form a probably complete series; the names berzeliite and manganberzeliite are applied to the halves of the series with Mg > Mn and Mn > Mg, respectively. Na, overlooked in the older analyses, is present in rather constant amount (Na:Ca ~2:5) in substitution for Ca. Sb substitutes for As up to at least Sb:As = 1:10.5 (analysis 9).

Anal.[6]

	1	2	3	4	5	6	7	8	9
Na$_2$O	4.64	4.25	4.49	4.29	4.21	5.05	4.20		
K$_2$O		0.19	0.10	0.15	0.06	0.09			
CaO	21.00	20.08	20.39	18.91	19.08	18.34	18.97	18.35	20.21
MgO	14.10	12.86	12.28	8.68	1.88	0.72		3.50	9.20
MnO		3.05	2.61	8.84	18.66	21.41	22.40	19.18	10.82
FeO		0.33			0.87	0.38		tr.	
PbO		0.30		0.68	0.14				
As$_2$O$_5$	60.26	57.57	57.29	57.03	53.90	52.90	54.43	50.92	53.23
Sb$_2$O$_5$						tr.		2.60	6.54
H$_2$O		0.22	0.81	0.85	0.44	0.40			
Rem.		1.06	2.11	0.93		0.24		5.23	
Total	100.00	99.91	100.08	100.36	99.24	99.53	100.00	99.78	100.00
G.		4.02	4.13	4.29		4.21			

1. $Mg_2(Ca,Na)_3(AsO_4)_3$ with Na:Ca = 2:5. 2. Berzeliite. Långban.[7] Rem. is SiO$_2$ 0.91, P$_2$O$_5$ 0.15. 3. Berzeliite. Långban.[8] Rem. is P$_2$O$_5$ 0.46, Al$_2$O$_3$ 1.13, Fe$_2$O$_3$ 0.04, insol. 0.48. 4. Berzeliite. Långban.[8] Rem. is Al$_2$O$_3$ 0.39, Fe$_2$O$_3$ 0.08, insol. 0.46. Original total given as 100.66. 5. Manganberzeliite. Långban.[8] Original total given as 100.31. 6. Manganberzeliite. Långban.[9] Rem. is V$_2$O$_5$, Cl tr. 7. Mn$_2$ (Ca, Na)$_3$ (AsO$_4$)$_3$ with Na:Ca = 2:5. 8. Antimonian manganberzeliite. Sjö mine [10] (*pyrrhoarsenite*). Rem. is CO$_2$ 1.27, BaSO$_4$ 3.96 (insol.). 9. Antimonian berzeliite. Sjö mine (*pyrrhoarsenite*).[11] Recalculated to 100 after deduction of dolomite and insol.

CARYINITE

Tests. B.B. fuses easily to a black bead if rich in Mn, slightly less readily to a gray or brown bead if rich in Mg. Easily soluble in HNO_3 or HCl.

Var. *Antimonian.* Pyrrhoarsenite pt. *Igelström* (1886). Contains Sb in substitution for As, in both berzeliite (anal. 9) and manganberzeliite (anal. 8). Color yellow to deep yellowish red. Optical data and cell dimensions are lacking.

Occur. Originally from Långban in Wermland, Sweden, in limestone skarn associated with hausmannite, manganophyllite, caryinite, and other rare minerals of the locality. Also from the Moss mine near Nordmark in Wermland, and the Sjö mine in Grythytte parish, Örebro, in Sweden; the mineral occurs in limestone skarn or in secondary veinlets therein associated with hausmannite, rhodonite, tephroite, manganosite, pyrochroite, barite. Pseudoberzeliite, originally from Långban, and pyrrharsenite from the Sjö mine apparently are identical with berzeliite.[12]

Name. After the Swedish chemist Jöns Jacob Berzelius (1779–1848).

Ref.

1. Goldschmidt (**1,** 191, 1913).
2. Bubeck and Machatschki, *Zs. Kr.,* **90,** 44 (1935), and Bubeck, *Geol. För. Förh.,* **56,** 525 (1934).
3. An unidentified cleavage is reported by Igelström, *Bull. soc. min.,* **9,** 218 (1886), on pyrrhoarsenite.
4. See Bubeck and Machatschki (1935); McConnell, *Am. Min.,* **27,** 452 (1942); Machatschki, *Zs. Kr.,* **73,** 123 (1930); **74,** 230 (1930).
5. Estimated from data of Landergren, *Geol. För. Förh.,* **52,** 123 (1930).
6. For additional analyses see Hintze (1 [4A], 214, 1922) and Doelter (3 [1], 635, 639, 1914).
7. Mauzelius analysis in Landergren (1930).
8. Almström analysis in Landergren (1930).
9. Mauzelius analysis in Sjögren, *Bull. Geol. Inst. Upsala,* **2,** 92 (1895).
10. Högbom, *Geol. För. Förh.,* **9,** 397 (1887).
11. Igelström, *Jb. Min.,* I, 49 (1889).
12. See discussion in Hintze (1922).

38.2.2 **CARYINITE** [$(Ca,Pb,Na)_5(Mn,Mg)_4(AsO_4)_5$ (?)]. Karyinit, Karyinit *Lundström* (*Geol. För. Förh.,* **2,** 178, 223, 1874). Caryinite *Dana* (754, 1892).

Orthorhombic.[1] Fine-granular and as cleavable masses. Distinct cleavage on $\{110\}$ and $\{010\}$; (010) ∧ (110) ~49°15′. H. 4. G. 4.29. Luster greasy. Color nut-brown to yellowish brown. Subtranslucent.

Opt.[2] In transmitted light, pale yellow-brown.

ORIENTATION		n	
X	c	1.776 ± 0.005	Biaxial positive (+).
Y	a	1.780 ± 0.005	$2V\ 41° \pm 3°$.
Z	b	1.805 ± 0.005	$r > v$, slight.
			Not pleochroic.

The indices of refraction and the optic angle are somewhat variable.

Chem. An arsenate principally of calcium, manganese, lead, and sodium. The formula is uncertain, perhaps $(Ca,Pb,Na)_5(Mn,Mg)_4(AsO_4)_5$

or $(Mn,Ca,Na,Mg,Pb)_3(AsO_4)_2$. Caryinite is close to manganberzeliite in composition, but contains lead. Analyses gave:

	Na_2O	CaO	MgO	PbO	MnO	FeO	As_2O_5	Rem.	Total	G.
1.	5.33	12.52	3.19	9.47	19.21		50.28		100.00	
2.	5.16	12.12	3.09	9.21	18.66	0.54	49.78	2.12	100.68	4.29
3.		12.80	4.72	11.70	17.61	0.60	52.49	0.08	[100.00]	4.25

1. $(Ca,Pb,Na)_5(Mn,Mg)_4(AsO_4)_5$ with Ca:Pb:Na = 5.3:1:4.0 and Mn:Mg = 3.4:1.
2. Långban.[3] Rem. is BaO 1.03, K_2O 0.37, P_2O_5 0.19, H_2O 0.53, Cl tr. 3. Långban.[4] Rem. is Cl. Analysis recalculated to 100 after deducting CO_2 3.86 as $CaCO_3$ and insol. 0.65; the analyzed sample also contained admixed berzeliite.

Tests. B.B. fusible to a black bead. Soluble in nitric acid.

O c c u r. Found only at Långban, Wermland, Sweden, associated with berzeliite, calcite, hedyphane, and hausmannite in veinlets in skarn containing rhodonite and schefferite. The caryinite alters to berzeliite.

N a m e. From κἄρῠϊνος, *nut-brown*, alluding to its color.

Ref.
1. Indicated by the optical study of Sjögren, *Bull. Geol. Inst. Upsala*, **2**, 87 (1895).
2. Indices and 2V of Larsen (53, 1921); orientation of Sjögren (1895).
3. Mauzelius analysis in Sjögren (1895).
4. Lundström (1874).

CHLOROARSENIAN. Igelström (*Geol. För. Förh.*, **15**, 471, 1893; *Zs. Kr.*, **22**, 468, 1893).
A name casually applied to an anhydrous manganese arsenate (?) found with basiliite at the Sjö mine, Örebro, Sweden. Cleavable grains or crystals with a yellowish green color and vitreous luster.

TYPE 3. $A_3(XO_4)_2$

38.3.1 W H I T L O C K I T E $[Ca_3(PO_4)_2]$. Frondel (*Am. Min.*, **26**, 145, 1941). Zeugite *Julien* (*Am. J. Sc.*, **40**, 371, 1865). Pyrophosphorite *Shepard* (*Am. J. Sc.*, **15**, 49, 1878). Martinite *Kloos* (*Samml. geol. Reichsmus. Leiden*, **1**; *Jb. Min.*, I, 41, 1888).

C r y s t.[1] Hexagonal—R; scalenohedral—$\bar{3}\ 2/m$.

$a:c = 1:3.5473;$ $\alpha\ 44°40';$ $p_0:r_0 = 4.0961:1;$ $\lambda\ 114°33\frac{1}{2}'$

Forms:[1]

		ϕ	$\rho = C$	A_1	A_2
c	0001	111	.....	0°00'	90°00'	90°00'
a	$11\bar{2}0$	$10\bar{1}$	0°00'	90 00	60 00	60 00
k	$1.0.\bar{1}.10$	433	30 00	22 16½	70 50	90 00
f	$10\bar{1}4$	211	30 00	45 41	51 43	90 00
j	$20\bar{2}3$	711	30 00	69 53	35 35½	90 00
h	$01\bar{1}8$	332	−30 00	27 07	90 00	66 45
g	$01\bar{1}3$	441	−30 00	53 47	90 00	45 40½
e	$01\bar{1}2$	330	−30 00	63 58½	90 00	38 54

WHITLOCKITE

Structure cell. [2] Space group probably $R\bar{3}c$. a_{rh} 13.67, α 44°21'; a_0 10.32 kX, c_0 36.9; $a_0:c_0 = 1:3.576$. Cell contents in the rhombohedral unit, $Ca_{21}(PO_4)_{14}$.

Habit. Crystals rhombohedral in development, rarely tabular {0001}. Also coarse-granular to earthy. Common forms: $e\ c\ f\ a$.

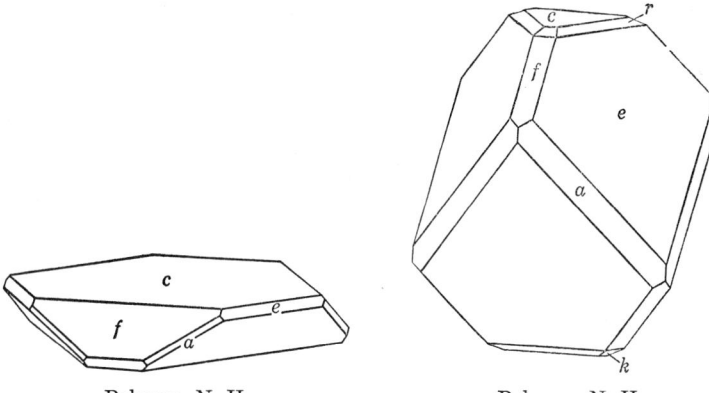

Palermo, N. H.   Palermo, N. H.

P h y s. No cleavage. Fracture uneven to subconchoidal. H. 5. G. 3.12; 3.15 (calc.). Luster vitreous inclining to subresinous. Colorless to white, gray or yellowish. Transparent to translucent.

O p t. In transmitted light, colorless.

	n (Palermo, anal. 3)	n (Curaçao, contains CO_2) [3]	
O	1.629	1.607	Uniaxial negative (−).
E	1.626	1.604	

C h e m. Tricalcium phosphate, $Ca_3(PO_4)_2$. Small amounts of Mn'', Fe'', and Mg substitute for Ca, with Mn:Fe:Mg:Ca = 1:3.2:4.4:78 in analysis 2. Small amounts of (CO_3) substitute for (PO_4) in some types,[3] analogous to carbonate-apatite. Whitlockite is dimorphous, but the alpha polymorph [4] is not known to occur in nature.

Anal. [8]

	1	2	3	4
CaO	54.22	46.84	46.90	48.15
MgO		1.91	2.53	3.53
FeO		2.34	1.91	
MnO		0.76		0.01
Fe_2O_3		0.20	1.73	0.05
P_2O_5	45.78	45.94	45.68	45.87
H_2O		0.66	0.48	2.00
Rem.		1.29	0.57	0.10
Total	100.00	99.94	99.80	99.71
G.	3.19	3.09	3.12	2.96

1. $Ca_3(PO_4)_2$. 2. Palermo mine, New Hampshire.[5] Rem. is Na_2O 0.59, K_2O 0.01, insol. 0.34, CO_2 0.14, F 0.07, Cl 0.03, Al_2O_3 0.11. 3. Palermo mine, New Hampshire.[6] Rem. is F 0.06, Cl tr., insol. 0.51. 4. Sebdou, Algeria.[7] Rem. is SiO_2.

***Var.** Carbonation.* Carbonate-whitlockite. Zeugite *Julien (Am. J. Sc.,* **40,** 371, 1865). Martinite *Kloos (Samml. geol. Reichsmus. Leiden,* **1;** *Jb. Min.,* I, 41, 1888).
Contains small amounts of CO_2, with relatively low indices of refraction. Found in phosphate-rock deposits in the West Indies.[3]

Tests. Infusible. Easily soluble in dilute acids.

O c c u r. Originally found associated with siderite, quartz, apatite, ludlamite, fairfieldite, xanthoxenite, and triphylite as a hydrothermal mineral in pegmatite at the Palermo mine near North Groton, New Hampshire. Also found in phosphate-rock deposits [3] in the West Indies on Table Mountain, Curaçao *(martinite* [9]), on the island of Sombrero near the Virgin Islands *(zeugite* [9]), on the Los Monges keys in the Gulf of Maracaibo, and from an unnamed locality *(pyrophosphorite* [9]). Also as a cave deposit at Sebdou, Oran, Algeria.

A r t i f. Observed in certain slags.[4]

N a m e. After Herbert P. Whitlock (1868–1948), American mineralogist and Curator of Mineralogy at the American Museum of Natural History.

Ref.
1. Frondel (1941), *Am. Min.,* **34,** 692 (1949), on crystals from Palermo mine, New Hampshire.
2. Frondel (1941) by Weissenberg method on the Palermo crystals of analysis 3.
3. Frondel, *Am. Min.,* **28,** 227 (1943).
4. See Trömel, *Mitt. Kaiser Wilhelm Inst. Eisenforsch.,* **14,** 25 (1932), Trömel and Möller, *Zs. anorg. Chem.,* **206,** 227 (1932), Bredig, Franck, and Füldner, *Zs. Elektrochem.,* **38,** 158 (1932), *ibid.,* **39,** 959 (1933).
5. Peck analysis in Frondel (1949).
6. Gonyer analysis in Frondel (1941).
7. Bennett analysis in Bannister, *Min. Mag.,* **28,** 29 (1947).
8. Unsatisfactory analyses of minerals *(zeugite, martinite,* and *pyrophosphorite)* later shown to be identical with whitlockite were given by Julien (1865), Kloos (1888), and Shepard (1878).
9. Shown to be identical with whitlockite by Frondel (1943).

38.3.2 **G R A F T O N I T E** [$(Fe,Mn,Ca)_3(PO_4)_2$]. Penfield *(Am. J. Sc.,* **9,** 20, 1900). Repossite *Grill (Per. Min.,* **6,** 19, 1935).

C r y s t.[1] Monoclinic; prismatic—$2/m$.

$$a:b:c = 0.766:1:0.533; \quad \beta\ 99°12'$$

Structure cell.[2] Space group $P2_1/c$. a_0 8.87 Å, b_0 11.57, c_0 6.17; $\beta\ 99°12' \pm 15'$; $a_0:b_0:c_0 = 0.766:1:0.533$. Cell contents $(Fe,Mn,Ca)_{12}(PO_4)_8$.

Habit. Stout prismatic, as rough composite crystals. Commonly massive, cleavable.

Oriented growths.[3] Graftonite often occurs in coarsely laminated intergrowths with triphylite, the two minerals mutually oriented with graftonite {010} [100] parallel triphylite {102} [010].

P h y s. Cleavage {010} good and {100} fair,[4] but not easily observed. Fracture uneven to subconchoidal. H. 5. G. 3.67–3.79 varying with the Fe:Mn:Ca ratio; 3.775 meas., 3.72 calc. (Nickel Plate mine, South Dakota; anal. 5). Luster vitreous to slightly resinous. Color salmon-pink to reddish brown when fresh, but usually dark brown from alteration. Streak white to faintly pinkish.

O p t.[7] In transmitted light, nearly colorless.

ORIENTATION		n (Rice mine, N. H., anal. 7)[3]	n (Grafton, N. H., anal. 2)[5]	n (Brissago anal. 6)[6]	n (Nickel Plate, S. D., anal. 5)[3]	
X	b	1.695	1.700	1.705	1.709	Biaxial positive (+).
Y		1.699	1.705	1.708	1.714	Faintly pleochroic,
$Z \wedge c$	$-36°$	1.719	1.724	1.722	1.736	with X and Y col-
2V			small	43°	60°	orless, Z pink.
Disp.			$r > v$		$r < v$, strong	

The indices of refraction decrease with increasing content of Ca and increase with increasing content of Fe and Mn, more rapidly with Fe.

C h e m. A phosphate of calcium and divalent iron and manganese, $(Fe,Mn,Ca)_3(PO_4)_2$. A limited isomorphous series extends between Fe, Mn, and Ca, with Fe:Mn:Ca = 4.6:4.3:1 in analysis 3, 2.6:1.5:1 in analysis 2, and 1.76:1.0:1 in analysis 7. Graftonite is not isostructural with whitlockite, caryinite, or fillowite.[9]

Anal.

	1	2	3	4	5	6	7
CaO	9.50	9.23	4.71	4.50	6.00	7.95	12.80
MgO		0.40			0.10	0.53	n.d.
FeO	31.51	30.65	27.78	32.33	30.70	32.58	28.84
MnO	18.13	17.62	25.48	23.32	21.81	15.65	15.96
P_2O_5	40.86	41.20	40.03	38.94	39.66	40.81	41.65
H_2O		0.75	0.60	0.54	0.60	tr.	n.d.
Rem.		0.33	1.22		0.89	2.60	
Total	100.00	100.18	99.82	99.63	99.76	100.12	99.25
G.		3.672	3.796	3.76		3.71	

1. $(Fe,Mn,Ca)_3(PO_4)_2$ with Fe:Mn:Ca = 2.59:1.51:1. 2. Grafton, N. H.[10] Rem. is $(Li,Na)_2O$. 3. Greenwood, Maine.[11] Rem. is K_2O 0.05, Na_2O 0.16, Li_2O 0.37, $CaCO_3$ 0.46, insol. 0.18. 4. Olgiasca, Italy.[12] 5. Nickel Plate mine, South Dakota.[8] Rem. is insol. 0.16, Al_2O_3 0.20, Li_2O 0.05, Na_2O 0.28, F 0.20. 6. Brissago, Switzerland.[13] Rem. is Al_2O_3. 7. Rice mine, New Hampshire.[8] Partial analysis.

Tests. B.B. fusible at about 2 to a black magnetic globule. Easily soluble in acids.

O c c u r. Found as a primary mineral in complex granite pegmatites, often in laminated intergrowths with triphylite. The triphylite usually is dark colored due to partial oxidation or may be more or less completely altered to ferri-sicklerite or heterosite. Originally found on Melvin Mountain about five miles west of Grafton, Grafton County, New Hampshire. Also found at the Palermo mine [14] and the near-by Rice mine near North Groton, Grafton County, and at Center Strafford [15] in New Hampshire. At Greenwood, Maine, and with arrojadite in the Nickel Plate mine,

Pennington County, South Dakota. A mineral found in pegmatite at Olgiasca, Lake Como, Italy (*repossite*) is identical [16] with graftonite. At Brissago, Tessin, Switzerland. In the pegmatites of the Kondakovo district,[17] Eastern Siberia, U.S.S.R.

Name. Graftonite from the locality. Repossite after Emilio Repossi, Italian geologist.

Ref.

1. Elements of the structure cell of Lindberg, *Am. Min.*, **35**, 59 (1950). These elements cannot be satisfactorily correlated with the morphological elements of Penfield (1900) obtained from contact measurements on rough crystals from Grafton. Penfield gave $a:b:c = 0.886:1:0.582$, β 114°00′, with the forms b {010}, a {100}, n {130}, l {120}, m {110}, d {011}, e {021}, p {111}. Penfield's cell probably is approximately related to the structure cell by the transformation 101/010/001.
2. Lindberg (1950) by the Weissenberg method on the material of analysis 5.
3. Penfield (1900) and Berman, *Am. Min.*, **12**, 170 (1927); face symbols in Penfield's cell.
4. Lindberg (1950) on Nickel Plate mine, South Dakota, material.
5. Larsen (81, 1921).
6. Parker, de Quervain and Weber, *Schweiz. min. Mitt.*, **9**, 302 (1939).
7. Optical data also are given by Berman (1927); Grill (1935); Switzer, *Am. Min.*, **23**, 814 (1937); Glass and Fahey, *Am. Min.*, **22**, 1035 (1937). Extinction angle of Lindberg (1950) on Nickel Plate material in orientation of the structure cell.
8. Lindberg (1950).
9. Frondel, *Am. Min.*, **26**, 145 (1941).
10. Penfield (1900), who cites another analysis by Ford on impure material.
11. Fahey analysis in Glass and Fahey (1937).
12. Gallitelli analysis in Grill (1935).
13. Jakob analysis in Parker, *et al.* (1939).
14. Berman (1927).
15. Switzer (1937).
16. Strunz, *Zbl. Min.*, 248 (1939).
17. Tatarinov, *Micas of the U.S.S.R.*, Leningrad, 1937—*Min. Abs.*, **9**, 30 (1946).

TYPE 4. $A(XO_4)$

XENOTIME

The tetragonal species xenotime, YPO_4, has the same structure [1] as zircon, $ZrSiO_4$, as do the artificial compounds [2] YVO_4 and $CaCrO_4$. The compounds $PbCrO_4$ and $SrCrO_4$, with larger cations, crystallize with the structure of monazite, $CePO_4$. In xenotime small amounts of Si^4 apparently can substitute for P^5, valence compensation probably being effected by a concomitant substitution of Zr^4 and U^4 for Y^3.

Ref.

1. Vegard, *Phil. Mag.*, **33**, 395 (1917), and Strunz, *Zs. Kr.*, **94**, 60 (1936).
2. Clouse, *Zs Kr.*, **76**, 285 (1930); **83**, 161 (1932).

38.4.1 **XENOTIME** [$Y(PO_4)$]. Phosphorsyrad Ytterjord *Berzelius* (*Ak. Stockholm, Handl.*, **2**, 334, 1824). Phosphorsaure Yttererde *Germ.* Phosphate of Yttria. Xenotime *Beudant* (**2**, 552, 1832). Ytterspath *Glocker* (959, 1831). Castelnaudite *Damour* (*L'Institut*, 78, 1853). Hussakite *Kraus* and *Reitinger* (*Zs. Kr.*, **34**, 268, 1901).

XENOTIME

Cryst.[1] Tetragonal; ditetragonal-dipyramidal—$4/m\ 2/m\ 2/m$.

$a:c = 1:0.8757;\quad p_0:r_0 = 0.8757:1$

Forms:

	ϕ	ρ	A	$\overline{M}$
c 001	.....	0°00′	90°00′	90°00′
m 010	0°00′	90 00	90 00	45 00
a 110	45 00	90 00	45 00	90 00
z 011	0 00	41 12½	90 00	62 14
x 031	0 00	69 09½	90 00	48 38
e 112	45 00	31 46	68 08½	90 00
f 111	45 00	51 05	56 37½	90 00
τ 121	26 34	62 57	66 32	73 38½

Structure cell.[2] Space group $I4/amd$. a_0 6.88 kX, c_0 6.03; $a_0:c_0 = 1:0.876$. Cell contents $Y_4(PO_4)_4$. Isostructural with zircon.

Twinning. On {111}, rare.

Oriented growths. Frequently found in parallel growth with zircon.

Habit. Short to long prismatic [001]; also equant, pyramidal. The crystals often closely resemble zircon in habit. Common forms: $m\ z\ a\ e$. As rudely radial aggregates of coarse crystals; in rosettes.

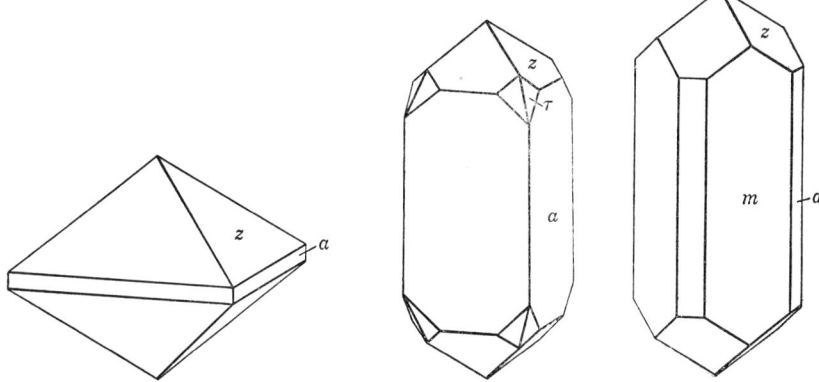

Phys. Cleavage {100} complete. Fraction uneven to splintery. Brittle. H. 4–5. G. 4.4–5.1; 4.25 (calc. for $Y(PO_4)$). Luster vitreous inclining toward resinous. Color commonly yellowish brown to reddish brown, also hair-brown, flesh-red, grayish white, wine-yellow, pale yellow, greenish. Streak pale brown, yellowish or reddish. Translucent to opaque. Moderately paramagnetic.

Opt.[3] In transmitted light, colorless to very pale yellowish green, yellow, or yellowish brown.

	n(Na)	n(Na; New Zealand)	
O	1.721	1.720	Uniaxial positive (+).
E	1.816	1.827	

Xenotime may be weakly dichroic with O pink, yellow, or yellowish brown, and E brownish yellow, grayish brown, or greenish.

Chem. Essentially yttrium phosphate, $Y(PO_4)$. Yttrium is sometimes substituted for in considerable part by erbium and, in lesser amount, by other rare earths, such as cerium, lanthanum, scandium. Th, U, Zr, Be, and Ca also substitute for Y in small amounts. (SO_4) apparently can substitute for (PO_4) in amounts up to 1 per cent or so SO_3; reported highly sulfatian xenotime (*hussakite*) has been discredited [9] as due to erroneous analyses. (SiO_4) is often present in amounts up to several per cent, presumably in substitution for (PO_4) but in some instances at least due to admixed zircon.

Anal.[4]

	1	2	3	4	5	6
$(Y,Er)_2O_3$	61.40	57.78	56.81	51.82	64.97	63.25
La_2O_3			0.93	2.14		
Al_2O_3			0.77	4.80		
Fe_2O_3			0.65		0.09	
ThO_2			tr.	2.47		
UO_2			4.13	3.17		
ZrO_2			1.95	1.90		
SiO_2			3.46	4.32	0.57	
P_2O_5	38.60	32.11	30.31	25.38	33.42	35.99
Rem.		0.18	0.84	3.65	0.83	0.63
Total	100.00	100.07	99.85	99.65	99.88	99.87
G.		5.106	4.68	4.46	4.577	

1. $Y(PO_4)$. 2. El Paso County, Colorado.[5] Rem. is ign. loss. Atomic weight of rare earths is 118. 3. Brindletown, North Carolina.[6] Rem. is CaO 0.21, F 0.06, H_2O 0.57. Atomic weight is 106. 4. Iisaka, Japan.[7] Rem. is FeO 1.10, MnO 1.72, MgO 0.02, CaO 0.61, H_2O 0.20. La_2O_3 includes Ce_2O_3. 5. South Mountain, North Carolina.[7] Rem. is MgO 0.01, CaO 0.05, Al_2O_3 0.02, SO_3 0.75. P_2O_5 in the original given as 34.42. 6. Dattas, Brazil.[10] Rem. is SO_3 0.11, insol. 0.52.

Tests. B.B. infusible but turns grayish white; when moistened with sulfuric acid colors the flame bluish green. Very slightly attacked or not attacked by acids, depending on the composition, but decomposed by fused sodium carbonate, potassium acid sulfate, and, with more difficulty, by borax or salt of phosphorus.

Occur. Widely distributed as a minor accessory mineral in acidic and alkalic igneous rocks and, in larger crystals, in pegmatites associated therewith. Also in mica- and quartz-rich gneissic metamorphic rocks and in quartz veinlets or pegmatitic patches contained in such rocks. In veins of the Alpine type and, rarely, in metamorphosed calcareous rocks. Commonly associated minerals include monazite, zircon, rutile, anatase, magnetite, ilmenite, hematite, sillimanite, alkalic feldspars, and, less commonly, fergusonite and other columbate-tantalates. A common detrital mineral. A few of the more prominent localities follow.

Found frequently in the granite pegmatites of southern Norway associated with gadolinite, polycrase-euxenite, allanite, zircon, yttrotantalite, thorite: well-known localities [8] here include Hitterö in Vest-Agder; at Tvedestrand and at Garta and other places near Arendal in Aust-Agder; at Raade near Moss in Olstfold, and at Kragerö, Telemark. At Ytterby northeast of Stockholm, Sweden. In Switzerland found on Mt. Fibia southwest of St. Gotthard, Ticino, in the Binnenthal, Valais, and in the

Tavetschtal, Graubünden. In Silesia in granite at Königshain near Görlitz and with monazite at Schreiberhau. At Pisek and Schüttenhofen, Bohemia. In the alluvial deposits of Kiravoravo, Madagascar; as an accessory mineral in granites of Cape Province, South Africa. A xenotime-like mineral containing much cerium and apparently orthorhombic occurs in apatite-magnetite rock in the Manbhum district, Bihar and Orissa province, India.[12] In river sands in North Westland, New Zealand. Xenotime occurs widely distributed in granitic rocks and in alluvial deposits in Brazil, as in Minas Geraes at Dattas (*hussakite*)[9] near Diamantina and at Pomba; also in the diamond washings of Bahia.

In the United States, found widely as a detrital mineral in the placer gold deposits of Polk, McDowell, and Burke Counties and elsewhere in North Carolina, and similarly in Georgia and Alabama. In several minor occurrences in schist on Manhattan Island, New York City, New York. With fluocerite and bastnaesite in Cheyenne Canyon, Pikes Peak district, El Paso County, Colorado.

A r t i f.[11] In crystals by fusion of precipitated yttrium phosphate in potassium sulfate, borax, or salt of phosphorus; also by fusion of yttrium oxide with potassium pyrophosphate.

N a m e. From κενός, *vain*, and τιμή, *honor*, recalling the fact that the yttrium in it had been mistaken for a new element.

Ref.

1. Elements and forms of Goldschmidt (**9**, 102, 1923); rare or doubtful: 015, 047, 3.24.10. The orientation of Dana (748, 1892) and Hintze (**1** [4A], 240, 1922) corresponds to the compound, face-centered unit.
2. Vegard, *Phil. Mag.*, **4**, 511 (1927), by powder method on unanalyzed crystal.
3. Hutton, *Am. Min.*, **32**, 141 (1947); Kraus and Reitinger, *Zs. Kr.*, **34**, 268 (1901).
4. For additional analyses see Hintze (1922).
5. Penfield, *Am. J. Sc.*, **45**, 398 (1893).
6. Eakins analysis in Hidden, *Am. J. Sc.*, **46**, 255 (1893).
7. Tschernik, *Verh. min. Ges. St. Petersburg*, **45**, 425 (1907)—*Zs Kr.*, **47**, 291 (1910).
8. Brögger, *Vidensk.-Selsk. Skr.*, Oslo, *Math.-Nat. Kl.*, **1**, no. 6, 6 (1906).
9. Hussak, *Cbl. Min.*, 533 (1907); Brögger, *Nytt Mag.*, **42**, 1 (1904).
10. Florence analysis in Hussak (1907).
11. Radominski, *C.R.*, **80**, 306 (1875); Duboin, *C.R.*, **107**, 622 (1888); Florence, *Jb. Min.*, II, 139 (1898).
12. Tipper, *Rec. Geol. Sur. India*, **51**, 31 (1920).

38.4.2 M O N A Z I T E [(Ce,La,Y,Th)(PO$_4$)]. Monazit *Breithaupt* (*J. Chemie u. Phys.*, **55**, 301, 1829). Monacite *bad orthogr.* Mengite *Brooke* (*Phil. Mag.*, **10**, 139, 1831). Edwardsite *Shepard* (*Am. J. Sc.*, **32**, 162, 1837). Eremite *Shepard* (*Am. J. Sc.*, **32**, 341, 1837). Monazitoid *Hermann* (*J. pr. Chem.*, **40**, 21, 1847). Urdit *Forbes* and *Dahll* (*Nytt Mag.*, **8**, 227, 1855). Turnerite *Lévy* (*Ann. Phil.*, **5**, 241, 1823). Kryptolith *Wöhler* (*Gel. Anz. Gött.*, **19**, 1846; *Ann. Phys.*, **67**, 424, 1846). Cryptolite. Phosphocerite *Watts* (*J. Chem. Soc. London*, **2**, 131, 1849). Kårarfveit *Radominski* (*C.R.*, **78**, 764, 1874). Korarfveite.

C r y s t.[1] Monoclinic; prismatic—$2/m$.

$a:b:c = 0.9693:1:0.9256;$ $\beta\ 103°40';$ $p_0:q_0:r_0 = 0.9549:0.8994:1$

$r_2:p_2:q_2 = 1.1119:1.0617:1;$ $\mu\ 76°20';$ $p_0'\ 0.9827,\ q_0'\ 0.9256,\ x_0'\ 0.2432$

Forms:[2]

		φ	ρ	φ₂	ρ₂ = B	C	A
c	001	90°00′	13°40′	76°20′	90°00′	.....	76°20′
b	010	0 00	90 00	.....	0 00	90°00′	90 00
a	100	90 00	90 00	0 00	90 00	76 20	.....
n	120	27 57½	90 00	0 00	27 57½	83 38½	62 02½
m	110	46 43	90 00	0 00	46 43	80 06	43 17
l	210	64 47	90 00	0 00	64 47	77 39½	25 13
y	310	72 34	90 00	0 00	72 34	76 58½	17 26
g	012	27 43½	27 36	76 20	65 47	24 13	77 33
e	011	14 43½	43 44½	76 20	48 02	41 58	79 53
u	021	7 29	61 49½	76 20	29 04½	60 55½	83 24½
w	101	90 00	50 47½	39 12½	90 00	37 07½	39 12½
x	$\bar{1}01$	−90 00	36 29	126 29	90 00	50 09	126 29
r	111	52 57	56 56	39 12½	59 40	46 31	48 01½
d	112	−28 12	27 42	103 56	65 48½	36 11	102 41½
v	$\bar{1}11$	−38 37½	49 50	126 29	53 20½	59 04	118 29½
s	121	33 31	65 45	39 12½	40 31	58 48	59 46½
o	$\bar{1}21$	−21 46½	63 21½	126 29	33 53½	69 03½	109 22
i	$\bar{2}11$	−61 44½	62 54½	149 51½	65 04½	75 06	141 39
z	311	−71 07	70 43½	159 43	72 12½	83 42	153 16

Structure cell.[3] Space group $P2_1/n$. a_0 6.782 kX, b_0 6.993, c_0 6.445; β 76°22′; $a_0:b_0:c_0 = 0.970:1:0.923$. Cell contents $(Ce,La,Y,Th)_4(PO_4)_4$.

Habit. Crystals commonly small but sometimes large and coarse. Often flattened {100} or elongated [010]; sometimes prismatic by extension of {$\bar{1}11$}, equant, or wedge-shaped by the large development of {100} and {$\bar{1}11$}. Common forms:[4] $a\ v\ m\ x\ e\ w\ b\ u\ z\ r$. The faces are often rough, striated or uneven.

Twinning. (1) Twin plane {100}, common; sometimes cruciform. (2) Twin plane {001}, lamellar; rare. Twinning also has been doubtfully reported[5] on {201} and {$\bar{9}02$}.

Phys. Cleavage {100} distinct and {010} difficult, also indistinct cleavage sometimes observed on {110}, {101} and {011}; well-marked

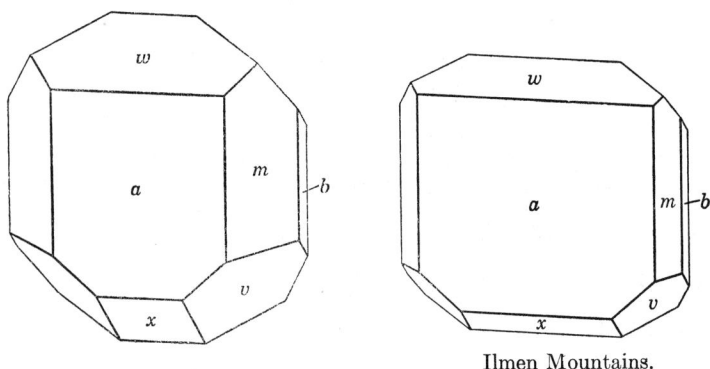

Ilmen Mountains.

MONAZITE

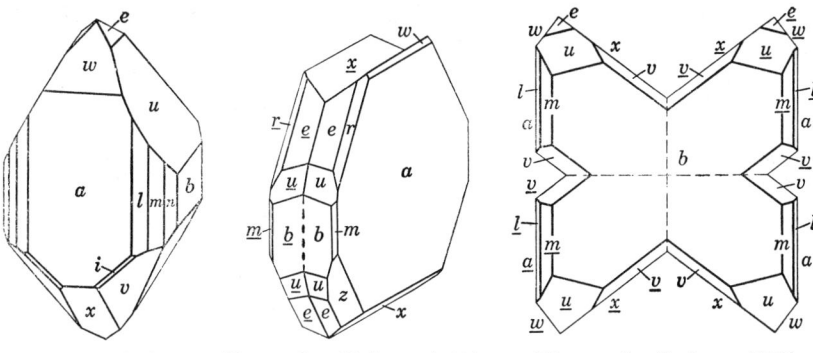

New York.   Tavetsch. Twin on {100}.   Binnental. Twin on {100}.

parting is often present on {001}, rarely on {$\bar{1}11$}. The ease and quality of the {100} and {010} cleavage varies apparently due to alteration of the crystals; the {010} cleavage is perfect in most monazite from Alpine localities. Fracture conchoidal to uneven. Brittle. H. 5–5½. G. 4.6–5.4, mostly 5.0 to 5.2; the G. increases with content of Th. Luster variable, usually resinous or waxy, but inclining toward vitreous or adamantine. Color yellowish or reddish brown to brown; also shades of yellow, greenish, nearly white. Subtransparent to subtranslucent. Streak white or faintly colored. Moderately paramagnetic.

Opt.[8] In transmitted light, yellowish brown or yellow to colorless.

ORIENTATION		n(Na) (Dattas, Brazil)[6]	n(Na) (Madagascar)[7]	n (Bolivia) anal. 11	
X	b	1.7902	1.8004	1.785	Biaxial positive (+).
Y		1.7912	1.8008	1.787	$r < v$, sometimes $r > v$;
Z ∧ c	2° to 6°	1.8441	1.8494	1.840	horizontal dispersion,
2V		15°33′	11°00′		weak.
					Pleochroism faint or not perceptible.

Chem. A phosphate essentially of the cerium metals,[9] (Ce,La)(PO$_4$). The La earths consist principally of La and ordinarily are present in about a 1:1 ratio with Ce. Th is usually present in substitution for (Ce,La). The amount of Th present ordinarily ranges from a few weight per cent ThO$_2$ to 10 or 12 per cent, but a series probably extends up to at least 30 per cent ThO$_2$; thorium-free material is rare. The Y earths substitute in small amounts for (Ce,La) together with minor amounts of Ca and, less abundantly, Mg, Fe″, Fe‴ (perhaps due to oxidation), Al, Zr, Mn, Be, Sn. U has been occasionally reported[10] in amounts up to about 1 weight per cent U$_3$O$_8$. He is generally present.[11] Si is often reported in amounts up to several per cent SiO$_2$. The Si has been ascribed to alteration or to admixed thorite[23] but may in part be due to substitution of Si for P. Monazite has been employed for the determination of geological age by the radioactive decay method, involving the determination of the Pb:Th + U ratio, but is less reliable for this purpose than is uraninite.[12]

Anal.[13]

	1	2	3	4	5	6
CaO		0.16	0.41	0.35		0.52
MgO				0.02		0.27
Ce_2O_3	34.99	30.38	27.37	22.63	31.85	21.08
$(La,Nd)_2O_3$	34.74	29.60	30.13	34.63	27.90	31.27
Y_2O_3		1.33	2.14	4.66	2.93	3.53
Fe_2O_3		1.50	0.81	0.08	0.42	0.66
Al_2O_3		0.10	0.17	0.10	0.21	0.80
ThO_2		6.19	10.29	7.32	9.15	11.08
SiO_2		0.85	1.03	1.54		2.98
P_2O_5	30.27	29.70	27.67	27.89	27.45	27.52
H_2O				0.40		0.56
Ign. loss		0.33	0.20		0.74	
Rem.				0.65		0.42
Total	100.00	100.14	100.22	100.27	100.65	100.69
G.			5.23	5.27	5.11	5.17

1. $(Ce,La)PO_4$ with Ce:La = 1:1. 2. Nsan Oban, Nigeria.[14] 3. Ratnapura, India.[14]
4. Dickens Township, Nipissing dist., Ontario.[15] Rem. is PbO 0.33, U_3O_8 0.32, C tr.
5. Miandrarivo, Madagascar.[16] 6. Ishikawa, Japan.[17] Rem. is UO_3 0.42.

	7	8	9	10	11
CaO	0.39	0.20	0.83	0.37	0.34
MgO			0.09	tr.	0.22
Ce_2O_3	31.63	31.90	22.00	32.29	31.41
$(La,Nd)_2O_3$	29.68	28.46	32.72	41.83	33.19
Y_2O_3	2.86		1.15		5.08
Fe_2O_3	0.68	1.50	0.44	tr.	
Al_2O_3		0.17	1.20		
ThO_2	5.65	10.22	12.00		nil
SiO_2	1.22	0.90	1.56	0.95	0.27
P_2O_5	26.81	26.82	27.22	24.90	29.29
H_2O		0.46	0.63		
Ign. loss	0.40				
Rem.	0.84		0.80		
Total	100.16	100.63	100.64	100.34	99.80
G.			5.16		5.173

7. Impilaks, Finland.[18] Rem. is SnO_2 0.84. 8. Monazite sand. Travancore.[14] 9. Gaya district, Bihar, India.[19] Rem. is U_3O_8 0.2677, PbO 0.5331, Mn tr., H_2O includes H_2O+ 0.48, H_2O- 0.15. 10. Shinkolobwe, Katanga.[20] Ce_2O_3 reported as CeO_2. Contains about 0.2 ThO_2. 11. Llallagua, Bolivia.[21]

Tests. B.B. turns gray but does not fuse. In C.T. unchanged. Slowly decomposed by acids, or by fusion with potassium bisulfate or sodium carbonate.

Occur. Monazite is widely disseminated as an accessory mineral in granitic igneous rocks and gneissic metamorphic rocks, and the detrital sands in regions of such rocks may contain monazite in commercial quantities. Monazite also occurs as relatively large crystals in pegmatites associated with granitic or syenitic igneous rocks; associated minerals include zircon, xenotime, gadolinite, samarskite, fergusonite, magnetite, apatite, columbite, ilmenite. Monazite occurs at numerous localities in Norway, especially in the granite-pegmatite districts in the southern part of the country. Well-known localities here include Arendal and various near-by points; Dillingö and Raade near Moss in Olstfold; Tvedestrand and Risör in Aust-Agder; Hvalö and Nöterö on Kristiana Fiord; at Hitterö. In Sweden at Lilla Holma near Stromstad and at Kårarfvet near Falun; in Finland in the pegmatites of Sordavala and Impilaks parishes on Lake Ladoga. Near Miask in the Ilmen Mountains, U.S.S.R.; in the gold sands of the Sanarka River in the southern Urals. From near St. Christophe near Bourg d'Oisans, Isère, France (probably the original locality of *turnerite*); also found in veins of the Alpine type associated with anatase, ilmenite, rutile, sphene, albite, etc., in the Swiss Alps as in the Binnenthal, Valais, and in the Val Tavetsch and Val Cornera, Grisons. At Säulenkopf near Pregatten in the Tirolean Alps; at Pisek and Schüttenhofen in Bohemia; and in the sanidine-bombs of the Laacher See, Germany. Monazite occurs, often in crystals of exceptional size and quality in the potassic pegmatites of Madagascar as at Mt. Volhambohitra and elsewhere in the Ankazobe district, and at Ampangabe and Ambatofotsikely; also in the black sands of the Manajary basin. In Africa in sands in the Embabaan district, Swaziland, South Africa. With wolframite in the tin deposit at Mt. Bischoff, Tasmania; in the tin placers of the Moolyella district, Western Australia, with wolframite in the Torrington district, New South Wales, and elsewhere in Australia. Monazite occurs in commercial amounts in various places in India, notably Travancore, and on the west coast of Ceylon. In tin deposits on Banka and Billiton and in the Malay States. Important detrital deposits have been found in Minas Geraes, Espirito Santo, Rio de Janeiro, and Bahia provinces, Brazil. From Llallagua, Bolivia.

In the United States, considerable amounts of monazite occur in the sands of Burke, Polk, Lincoln, McDowell, Cleveland, and Rutherford Counties in North Carolina; also in this state in veins in schist at Milholland's Mill and at Stony Point in Alexander County, and in pegmatite at Mars Hill, Madison County. In Spartanburg, Cherokee, and Laurens Counties and elsewhere in South Carolina. In beach sands in Florida. With microlite in pegmatite at Amelia Court House, Virginia. In New York, as crystals with sillimanite, near Yorktown Heights, Westchester County, and in several minor occurrences in schist in the northern part of Manhattan Island, New York City. From Boothwyn, Delaware County, Pennsylvania. In Connecticut notably at Yantic Falls, Norwich. In goldplacer sands near Centerville, Boise County, Idaho. As large crystals in pegmatite in the Petaca district, Rio Arriba County, New Mexico. Near Nuevo and elsewhere in Riverside County, California. In pegmatite at Villeneuve, Ottawa County, Quebec, Canada.

Alter. Found surficially altered to a yellowish or reddish brown, opaque material of unknown nature.

Artif.[22] In crystals by fusion of admixed phosphate and chloride of cerium.

Name. From μονάζειν, *to be solitary*, in allusion to its rare occurrence in the first-known localities.

Ref.
1. Elements of E. S. Dana (749, 1892), *Am. J. Sc.*, **24**, 247 (1882), on unanalyzed crystals from Milholland's Mill, Alexander County, North Carolina. The angles vary measurably for crystals from different localities, owing to variation in composition—for tabulations see Dana (1882, 1892) and Goldschmidt (**2**, 399, 1890).
2. Parker, *Am. Min.*, **22**, 572 (1937); Goldschmidt (**6**, 51, 1920). Rare or doubtful:

μ 130	δ 106	χ 11.0.1	π 221	ϑ $\bar{1}22$
α 014	ψ 105	ρ $\bar{1}03$	λ 212	ν $\bar{5}66$
κ 013	S 102	Φ $\bar{3}02$	p 211	η $\bar{1}32$
0.10.11	h 305	E $\bar{2}03$	Δ 122	283
β 043	11.0.10	ϵ $\bar{4}03$	U $\bar{1}13$	Σ $\bar{5}16$
γ 041	q 701	σ $\bar{3}01$	Γ $\bar{3}13$	τ $\bar{5}17$

3. Gliszczynski, *Zs. Kr.*, **101**, 1 (1939), on unanalyzed crystals from Perdatsch, Switzerland. Parrish, *Am. Min.*, **24**, 651 (1939), gives a_0 6.76 kX, b_0 7.00, c_0 6.42 on the material of analysis 11; G 5.17 (meas.), 5.06 (calc.).
4. Parker (1937).
5. Schetelig, *Norsk geol. Tidsskr.*, **2**, no. 9 (1913).
6. Busz, *Jb Min., Beil.-Bd.*, **39**, 492 (1914).
7. Sabot, Inaug. Diss. Ghent (1914)—*Jb. Min.*, 138 (1920).
8. For additional measurements of the indices of refraction and other optical data see summaries in Hintze (1\[4A], 298, 1922); also Gordon, *Ac. Sc. Philadelphia, Not. Nat.*, no. 2 (1939) and Hutton, *Am. Min.*, **32**, 141 (1947).
9 For spectrographic studies of the rare-earth content of monazite see Bearth, *Schweiz min. Mitt.*, **14**, 442 (1934), Thoreau, Breckpot, and Vaes, *Ac. roy. Belgique, Bull.*, 22 [5], 1111 (1936), Sahama and Vähätalo, *C. r. soc. géol. Finlande*, no. 14, 50 (1941).
10. For summary see Holmes, *Proc. Roy. Soc. Edinburgh*, **63**, 115 (1948).
11. On the He content of monazite see literature in Hintze (1922, p. 309).
12. For summary see Holmes (1948).
13. For additional analyses see Doelter (3 [1], 546, 1914); Hintze (**1** [4A], 368, 1923); Mawson, *Trans. Roy. Soc. S. Australia*, **68**, 334 (1944); Zemel, *J. Appl. Chem , Leningrad*, **9**, 1969 (1936); Niggli and Faesy, *Zs Kr.*, **62**, 557 (1925).
14. Johnstone, *J. Soc. Chem. Ind.*, **33**, 55 (1914).
15. Ellsworth, *Am. Min.*, **17**, 19 (1932).
16. Pisani analysis in Lacroix (**1**, 351, 1922).
17. Shibata and Kimura, *J. Chem. Soc. Japan*, **42**, 957 (1921).
18. Ramsay and Zilliacus, *Öfv Finska Vet.-Soc. Förh.*, **39** (1897).
19. Sarkar, *Proc. Indian Ac. Sc.*, **13**, Sect. A, 245 (1941).
20. Thoreau, Breckpot, and Vaes, *Ac roy. Belgique, Bull.*, [5], **22**, 1111 (1936).
21. Gordon, *Ac. Sc. Philadelphia, Not. Nat.*, no. 2, (1939)
22. Radominsky, *C.R.*, **80**, 309 (1895).
23. See discussion in Hintze (1922) and Doelter (1914).

BERLINITE

Berlinite, $AlPO_4$, is isostructural [1] with the artificial compound $AlAsO_4$ and with quartz, $SiSiO_4$. Berlinite shows the same thermal inversions [2] as SiO_2, although at slightly lower temperatures.

Ref.
1. Strunz, *Zs. Kr.*, **103**, 228 (1941).
2 Beck, *J. Am. Ceramic Soc.*, **32**, 147 (1949).

38.4.3 BERLINITE [$Al(PO_4)$]. Blomstrand (*Ak. Stockholm, Öfv.*, **25**, 198, 1868; *J. pr. Chem.*, **105**, 338, 1868).

C r y s t.[1] Hexagonal—R; trigonal-trapezohedral—3 2. Artificial crystals are very similar in habit to quartz, with $\{10\bar{1}1\}$, $\{01\bar{1}1\}$, and $\{10\bar{1}0\}$. Natural material is massive, granular.

Structure cell.[2] Space group $C3_12 = C3_22$. a_{rh} 4.62, α 64°26' (artif. AlPO$_4$); a_0 4.93 kX, c_0 10.94; $a_0:c_0 = 1:2.219$. Cell contents Al$_3$(PO$_4$)$_3$. Berlinite is isostructural with Al(AsO$_4$) and isotypic with α-SiO$_2$, quartz.

P h y s. No cleavage. Fracture conchoidal. H. $\sim 6\frac{1}{2}$. G. 2.64 (natural), 2.56 (artif.);[3] 2.62 (calc.). Luster vitreous. Colorless to grayish or pale rose. Transparent to translucent. Resembles quartz. Inverts at 583° to a polymorph isotypic with β-SiO$_2$.

O p t. In transmitted light, colorless.

n(Na), artif.[4]

O	1.524	Uniaxial positive (+).
E	1.530	

C h e m. Aluminum orthophosphate, AlPO$_4$. Al$_2$O$_3$ 41.80, P$_2$O$_5$ 58.20, total 100.00. The only reported analysis on natural material apparently was made on a sample containing capillary, non-essential water or an admixed hydrated aluminum phosphate: Al$_2$O$_3$ 40.27, Fe$_2$O$_3$ 0.26, P$_2$O$_5$ 54.84, H$_2$O 4.14, total 99.51.

Tests. B.B. whitens but does not fuse. Easily soluble in alkalies, but practically insoluble in cold acids.

O c c u r. Found with augelite, attacolite, and other phosphates at the Westanå iron mine, near Näsum, Kristianstad, Sweden.

A r t i f. Obtained as crystals from solution in phosphoric acid. Berlinite or the tridymite-like or cristobalite-like polymorphs thereof are obtained by heating hydrated aluminum phosphates.[5]

N a m e. After N. J. Berlin (1812– ?), pharmacologist, of the University of Lund.

Ref.

1. The identity of the natural material with artificial AlPO$_4$ was established by Strunz, *Zs. Kr.*, **103**, 228 (1941).
2. Huttenlocher, *Zs. Kr.*, **90**, 508 (1935); Strunz (1941) gives a_0 4.92 kX, c_0 10.91 for natural material.
3. Mrose, priv. comm. (1949).
4. Huttenlocher (1935).
5. Manly, *Am. Min.*, **35**, 108 (1950).

38.4.4 Rooseveltite [BiAsO$_4$]. *Rooseveltita* Herzenberg (*Facultad. Nac. de Ing., Univ. Téc. Oruro, Bolivia, Bol. Téc.*, no. 1, 1946).

As thin botryoidal crusts. Brittle. H. 4–4$\frac{1}{2}$. G. 6.86. Color gray. Luster adamantine. Anisotropic with very high indices of refraction. Composition BiAsO$_4$, from the analysis Bi$_2$O$_3$ 67.2, As$_2$O$_5$ 33.2, total 100.4. B. B. fusible. Soluble in acids. Found in wood-tin veinlets in rhyolitic and dacitic lava flows at Santiaguilo, Macha, Potosi, Bolivia. Monoclinic and isostructural with monazite.[1] Named after Franklin D. Roosevelt (1882–1945), thirty-second President of the United States.

Ref.

1. Frondel, priv. comm. (1950) on type material.

39 HYDRATED ACID PHOSPHATES, ETC.

TYPE 1. $(AB)_m H_n (XO_4)_p \cdot x H_2 O$, WITH $m + n : p > 2:1$

39.1.1 STERCORITE [$Na(NH_4)H(PO_4) \cdot 4H_2O$]. *Herapath* (*Quart. J. Chem. Soc. London*, **2**, 70, 1850). Microcosmic Salt. Native Salt of Phosphorus. Phosphorsalz *Germ.* Estercorrita *Span.*

C r y s t.[1] Triclinic; pseudo-monoclinic.

$a:b:c = 2.908:1:1.859$; $\beta\ 98°30'$; $p_0:q_0:r_0 = 0.639:1.839:1$

$r_2:p_2:q_2 = 0.544:0.348:1$; $\mu\ 81°30'$; $p_0'\ 0.646$, $q_0'\ 1.859$, $x_0'\ 0.150$

Forms:[2] On artificial crystals:

c 001	m 110	r 101	ρ $\bar{1}$01	x 112
a 100	n 310	s 201	σ $\bar{2}$01	ξ 11$\bar{2}$

Habit. Artificial crystals short prismatic [001]. The natural material occurs as crystalline masses and nodules.

Twinning.[3] Artificial crystals generally simulate monoclinic symmetry due to repeated twinning of triclinic individuals on {010}. The faces ordinarily are even and lustrous without re-entrant angles.

P h y s. No cleavage. H. 2. G. 1.574 (artif.),[3] 1.615 (natural).[4] Luster vitreous. Colorless (artificial); white to yellowish and brownish (natural). Transparent. M.P. 79°.

O p t.[3] In transmitted light, colorless.

Orientation	n	
X	1.439	Biaxial positive (+).
Y	1.442	$2V\ 35°34'$.
Z	1.469	$r > v$, rather strong.

The optic plane is nearly normal to {010}, and Z is nearly normal to {001}. On {001}, $X' \wedge b = 9°35'$; on {100}, $X' \wedge a = 1°20'$ with strong dispersion. {010} sections may show two sets of lamellar twinning at about 90°.

C h e m. A hydrated acid phosphate of sodium and ammonium, $Na(NH_4)H(PO_4) \cdot 4H_2O$. The reported analyses of natural material do not conform closely to the cited formula.

Anal.

	1	2	3
Na_2O	14.82	15.75	14.50
$(NH_4)_2O$	12.46	7.68	8.48
P_2O_5	33.96	34.33	34.54
H_2O	38.76	42.24	42.48
Total	100.00	100.00	100.00

1. $Na(NH_4)H(PO_4) \cdot 4H_2O$. 2. Ichaboe Island.[4] 3. Guañape Islands.[5]

Tests. Easily fusible. Soluble in water (17 grams in 100 cc. cold water). On heating, the ammonia is lost at about 200°, and at higher temperatures the water is driven off leaving molten sodium hexametaphosphate—a flux used in blowpipe analysis.

Occur. Found in a guano deposit on the Island of Ichaboe, South West Africa. Also from the Guañape Islands, off the coast of Peru.

Artif. Obtained as crystals by evaporation of a water solution of the salt, preferably in the range from $+10°$ to $-6°$ (below which point another phase appears).

Name. From *stercus*, dung.

Ref.

1. Elements of Schaschek, *Min. Mitt.*, **32**, 402 (1914), on mimetically twinned, pseudo-monoclinic artificial crystals. The chemically analogous salt, $Na(NH_4)H(AsO_4) \cdot 4H_2O$, is monoclinic and not twinned with $a:b:c = 2.8723:1:1.8589$, $\beta\ 98°59'$ (Schaschek, 1914).
2. Schaschek (1914) and Mitscherlich, *Ann. Chem., Phys.* **19**, 399 (1821).
3. Schaschek (1914).
4. Herapath (1850).
5. Raimondi (28, 1878).

DITTMARITE. MacIvor (*Chem. News*, **55**, 215, 1887; **85**, 181, 1902).

In small transparent orthorhombic (?) crystals. Analysis gave: MgO 25.67, $(NH_4)_2O$ 3.94, MnO 0.08, FeO 0.38, P_2O_5 46.51, H_2O 23.42, total 100.00. This is close to the formula $Mg_5(NH_4)H_4(PO_4)_5 \cdot 8H_2O$. The crystals lose water at $100°$–$105°$ and become turbid. Found with struvite, newberyite, hannayite, and schertelite in bat guano of the Skipton Caves near Ballarat, Victoria, Australia.

SCHERTELITE. Muellerite *MacIvor* (*Chem. News*, **55**, 215, 1887; **85**, 181, 1902) [not müllerite *Zambonini* (*Zs. Kr.*, **32**, 157, 1899)]. Schertelite [erron. written Schertalite] *MacIvor* (*Chem. News*, **85**, 217, 1902).

As small indistinct flat crystals. Analysis gave: FeO 0.20, MnO 0.05, MgO 12.17, $(NH_4)_2O$ 16.15, P_2O_5 43.88, H_2O 27.55, total 100.00. This is very close to $Mg(NH_4)_2H_2(PO_4)_2 \cdot 4H_2O$. The water is lost at about $120°$, and magnesium metaphosphate is formed on ignition. Found sparingly with struvite and newberyite in the drier layers of the bat guano deposit in the Skipton Caves near Ballarat, Victoria. Named after Professor Arnulf Schertel (1841-1902) of the Freiberg Mining Academy.

39.1.2 **HANNAYITE** $[Mg_3(NH_4)_2H_4(PO_4)_4 \cdot 8H_2O]$. Ulrich and vom Rath (Ber. Niederrhein. Ges., **35**, January 7, 11, 1878; **36**, January 13, 5, 1879).

Cryst.[1] Triclinic; pinacoidal—$\bar{1}$ (?).

$a:b:c = 0.6990:1:0.9743;$ $\alpha\ 122°31', \beta\ 126°46', \gamma\ 54°09'$

$p_0:q_0:r_0 = 1.4497:0.9627:1;$ $\lambda\ 73°15', \mu\ 65°28', \nu\ 112°58'$

$p_0'\ 1.8898,\ q_0'\ 1.2549,\ x_0'\ 0.7471,\ y_0'\ 0.3757$

Forms:[2]

		ϕ	ρ	A	B	C
c	001	63°18′	39°54′	65°28′	73°15′	.....
a	100	112 58	90 00	.....	112 58	65°28′
l	130	29 53½	90 00	83 04½	29 53½	58 02½
m	110	73 26	90 00	39 32	73 26	50 50½
M	1$\bar{1}$0	138 52	90 00	25 54	138 52	80 48
ω	$\bar{1}$33	165 13	33 13½	70 24	121 59½	55 18½

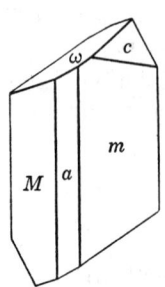

Habit. In small and slender crystals, elongated [001] and striated parallel thereto.

Phys. Cleavage {001} complete; {110}, {1$\bar{1}$0} and {130} poor. H. low. G. 1.893. Color yellowish.

Opt.[3] In transmitted light, colorless.

ORIENTATION	n	
$X \perp \{001\}$ (?)	1.555 ± 0.003	Biaxial negative (−).
$Y \wedge c \sim 33°$ (?)	1.572 ± 0.003	$2V\ 42° \pm 1°$ (meas.).
Z	1.575 ± 0.003	Dispersion not perceptible.

Chem. A hydrated acid phosphate of magnesium and ammonium, $Mg_3(NH_4)_2H_4(PO_4)_4 \cdot 8H_2O$. Loses 21 per cent H_2O between 100° and 115° and the remainder at a high heat.

Anal.

	1	2	3
MgO	18.98	18.90	18.36
$(NH_4)_2O$	8.17	8.09	8.10
P_2O_5	44.58	45.70	44.63
H_2O	28.27	28.20	28.51
Rem.			0.40
Total	100.00	100.89	100.00

1. $Mg_3(NH_4)_2H_4(PO_4)_4 \cdot 8H_2O$. 2. Skipton Caves.[4] Average of two. 3. Skipton Caves.[5] Rem. is FeO 0.31, MnO 0.087.

Tests. On heating in a platinum crucible loses water and ammonia, and the residue is fusible B.B. Before ignition easily soluble in acids.

Occur. Found with struvite, brushite, newberyite, dittmarite, bobierrite (?), and schertelite in bat guano of the Skipton Caves, southwest of Ballarat, Victoria, Australia.

Name. After J. B. Hannay (1855– ?), chemist, of the University of Manchester.

Ref.

1. Goldschmidt (**4**, 110, 1918); see also vom Rath (1878).
2. Goldschmidt (1918).
3. Larsen (**83**, 1921). It is not certain if this material, which can be interpreted as monoclinic with $Y = b$, actually is identical with hannayite.
4. MacIvor analysis in vom Rath (1879).
5. MacIvor, *Chem. News*, **85**, 181 (1902).

39.1.3 HUREAULITE [$Mn_5H_2(PO_4)_4 \cdot 4H_2O$]. Alluaud (*Ann. sc. nat.*, **8**, 334, 1826; cited in Vauquelin, *Ann. chim. phys.*, **30**, 302, 1825). Palaite Schaller (*J. Washington Ac. Sc.*, **2**, 143, 1912). Pseudopalaite de Jesus (*Com. Serv. Geol. Portugal*, **19**, 65, 1933 [1936]).

Cryst.[1] Monoclinic; prismatic–$2/m$.

$a:b:c = 1.9307:1:1.0470;$ $\quad \beta\ 96°40';$ $\quad p_0:q_0:r_0 = 0.5423:1.0400:1$

$r_2:p_2:q_2 = 0.9615:0.5215:1;$ $\quad \mu\ 83°20';$ $\quad p_0'\ 0.5460,\ q_0'\ 1.0470,\ x_0'\ 0.1169$

HUREAULITE

Forms:[2]

	ϕ	ρ	ϕ_2	$\rho_2 = B$	C	A
c 001	90°00'	6°40'	83°20'	90°00'	.....	83°20'
b 010	0 00	90 00	.....	0 00	90°00'	90 00
a 100	90 00	90 00	0 00	90 00	83 20	.....
m 110	27 32½	90 00	0 00	27 32½	86 55	62 27½
α $\bar{2}01$	−90 00	44 17	134 17	90 00	50 57	134 17
p 113	40 34½	24 40½	73 21½	71 31	20 56	74 14½
δ 112	36 40½	33 08	68 42	64 00	29 36	70 57
g 111	32 20	51 06	56 31½	48 53	47 46½	65 24
ϵ $\bar{1}11$	−22 17	48 32	113 13½	46 06	51 20	106 30
k $\bar{5}12$	−67 14½	53 32	141 18	71 52	59 43	137 52

Structure cell.[3] Space group probably $P2/c$. a_0 17.42 Å, b_0 9.12, c_0 9.50; [β 96°40']; $a_0:b_0:c_0 = 1.908:1:1.040$. Cell contents $Mn_{20}H_8(PO_4)_{16} \cdot 16H_2O$.

Habit. Crystals short prismatic [001], sometimes also tabular {100}; equant. The crystals isolated or grouped, occasionally fascicled as in stilbite. Also massive, compact, scaly, or imperfectly fibrous.

P h y s. Cleavage {100}, good. H. 3½. G. 3.19 ± 0.01 (Vilate), 3.15 (Branchville); 3.23 (calc.). Luster vitreous, somewhat greasy, bright. Color orange, orange-red, brownish orange, reddish to yellowish brown, violet-rose, pale rose, red, amber; also gray to nearly colorless. Streak nearly white. Transparent to translucent.

O p t.[4]

Orientation	n (Branchville)	Pleochroism	
X b	1.647	Colorless	Biaxial negative (−).
Y	1.654	Yellow to pale rose	$r < v$, very strong;
$Z \wedge c$ 75°	1.660	Reddish yellow to reddish brown	crossed, strong. $2V$ 75°.

The variation in optical properties accompanying variation in the Mn:Fe ratio is not clearly established.

C h e m. A hydrated acid phosphate of manganese, $(Mn,Fe'')_5H_2(PO_4)_4 \cdot 4H_2O$. Fe'' substitutes for Mn'', and a series extends from at least Fe:Mn = 1:3.0 (Huréaux, anal. 4) down to Fe:Mn = 1:9.4 (Branchville, anal. 2) and probably to the pure manganese compound.

Anal.[8]

	1	2	3	4
FeO		4.56	7.86	11.10
MnO	48.66	42.29	41.67	32.85
P_2O_5	38.98	38.36	38.00	38.00
H_2O	12.36	12.20	11.98	18.00
Insol.		2.70	0.38	
Total	100.00	100.11	99.89	99.95
G.		3.23	3.149	

1. $Mn_5H_2(PO_4)_4 \cdot 4H_2O$. 2. Branchville, Connecticut.[5] Average of two. Insol. includes CaO 0.94. 3. La Vilate, France.[6] 4. Huréaux, France.[7]

Tests. B.B. fuses to an orange- or reddish yellow bead. Affords H_2O in the C.T. Easily soluble in acids.

O c c u r. Originally found with rockbridgeite, cacoxenite, and vivianite in cavities in heterosite formed by the alteration of triphylite in pegmatite at Huréaux in St. Sylvestre and at Vilate near Chanteloube, both north of Limoges, Haute Vienne, France. From Mangualde, Beira, Portugal (*pseudopalaite*).[11] Said to occur with sarcopside at Michelsdorf, Silesia. In the United States found with fairfieldite, dickinsonite, reddingite, and eosphorite as a hydrothermal alteration of lithiophilite in pegmatite at Branchville, Connecticut. With stewartite and metastrengite in altered lithiophilite at Pala, San Diego County, California (*palaite* pt.).[10] Also with dickinsonite and lithiophilite at the Strickland quarry, Portland, Connecticut, and with metastrengite, strengite, and heterosite in weathered triphylite at the Palermo mine, North Groton, New Hampshire.

A r t i f. Obtained as crystals [9] by neutralization with ammonia of a solution of manganese phosphate. The Cd analogue also has been prepared.

Ref.
1. Elements of Murdoch, *Am Min.*, **28**, 19 (1943), on Pala crystals. Transformation: Dana (832, 1892) to M , 100/010/002; Goldschmidt (**4**, 178, 1918) to M., $\overline{1}01/010/001$.
2. Murdoch (1943), who adds 9 new forms, 4 doubtful, to the summary of Goldschmidt (1918). Doubtful: 611, 12.1.1, 232, $\overline{6}13$.
3. McCullough in Murdoch (1943) by rotation method.
4. Larsen and Berman (177, 1934).
5. Wells analysis in Brush and Dana, *Am. J. Sc.*, **39**, 210 (1890).
6. Damour, *Ann. chim phys.*, **53**, 293 (1858).
7. Dufrénoy, *Ann. chim. phys.*, **41**, 338 (1829).
8. For additional analyses see Hintze (**1** [3B], 824, 1931); de Jesus (1933); Schaller (1912).
9. de Schulten, *Bull. soc. min.*, **27**, 123 (1904), who gives crystallographic and optical data.
10. Considered identical with hureaulite by Schaller, priv comm (1946).
11. On the identity with hureaulite see Mason, *Geol. För. Förh.*, **63**, 175 (1941).

WENZELITE [$(Mn,Mg,Fe)H(PO_4) \cdot 2H_2O$?]. Wentzelite *Müllbauer* (*Zs. Kr.*, **61**, 333, 1925).

C r y s t. Monoclinic; prismatic—$2/m$.

$a:b:c = 2.3239:1:2.8513$; $\beta\ 133°38'$; $p_0:q_0:r_0 = 1.2269:2.0637:1$

$r_2:p_2:q_2 = 0.4846:0.5945:1$; $\mu\ 46°22'$; $p_0'\ 1.6952$; $q_0'\ 2.8513$; $x_0'\ 0.9534$

Forms:

		ϕ	ρ	ϕ_2	$\rho_2 = B$	C	A
c	001	90°00'	43°38'	46°22'	90°00'	.....	46°22'
a	100	90 00	90 00	0 00	90 00	46°22'	.....
m	110	30 47½	90 00	0 00	30 47½	65 43	59 12½
ρ	$\overline{1}01$	−90 00	36 34	126 34	90 00	80 12	126 34
τ	$\overline{1}13$	−22 09	45 51	111 13½	48 21	71 29½	105 41½

Found as rosettes of small, intergrown crystals. Color light red to flesh-red.

O p t.[1] In transmitted light, colorless.

ORIENTATION		n	
X	b	1.646	Biaxial negative (−).
Y		1.655	$2V$ medium large.
Z		1.657	$r < v$, perceptible.

BRUSHITE

Chem. A hydrated acid phosphate of manganese, magnesium and iron of uncertain formula,[2] perhaps $(Mn,Mg,Fe)H(PO_4)\cdot 2H_2O$.

Anal.

	1	2	3
MnO	22.09	21.13	23.85
MgO	6.73	6.83	6.78
FeO	6.00	6.01	6.07
P_2O_5	39.88	39.37	38.85
H_2O	25.30	23.66	23.39
Total	100.00	97.00	98.94

1. $(Mn,Mg,Fe)H(PO_4)\cdot 2H_2O$, with $Mn:Mg:Fe = 56:30:15$. 2,3. Hagendorf.

Occur. Found with vivianite and rockbridgeite in pegmatite at Hagendorf, Bavaria, Germany.

Name. After Hieronymus Wenzel [not Wentzel] who discovered the phosphate minerals at Pleystein.

Probably identical with hureaulite.

Ref.
1. Larsen and Berman (177, 1934).
2. See Steinmetz, *Jb. Min.*, I, 54 (1926).

BALDAUFITE. *Müllbauer* (*Zs. Kr.*, **61**, 334, 1925).

Monoclinic. Small striated crystals prismatic [001], closely resembling wenzelite in both habit and angles; also massive. $a:b:c = 2.21:1:1.84$; β 133°18′; with $a\{100\}$, $c\{001\}$, $m\{110\}$, $p\{\bar{1}01\}$. Optically negative, with $nX = b = 1.652$, $nY = 1.657$, $nZ = 1.662$, $2V$ large.[1] Composition perhaps $H_2(Fe,Mn,Ca,Mg)_5(PO_4)_4\cdot 5H_2O$, with $Fe:Mn:Ca:Mg \sim 10:2:2:1$. Analysis gave:[2] CaO 6.23, MgO 2.73, MnO 9.26, FeO 27.13, P_2O_5 39.05, H_2O 15.60, total 100.00. Found with dufrenite in a phosphate-bearing pegmatite at Hagendorf near Pleystein, Bavaria. Needs verification, but specimens apparently are not extant. Named after Richard Baldauf (1848–1931), German mining engineer and mineralogist.

Ref.
1. Larsen and Berman (178, 1934).
2. Average of two analyses, both probably originally recalculated to 100 after deducting insoluble.

TYPE 2. $AH(XO_4)\cdot xH_2O$

39.2.1 BRUSHITE GROUP

MONOCLINIC

	a_0	b_0	c_0	β
Brushite, $CaHPO_4\cdot 2H_2O$	5.88 kX	15.15	6.37	117°28′
Pharmacolite, $CaH(AsO_4)\cdot 2H_2O$	6.00	15.40	6.29	114 47
Ardealite, $Ca_2H(PO_4)(SO_4)\cdot 4H_2O$	5.67	14.64	6.28	$\sim$113
Weinschenkite, $(Y,Er)PO_4\cdot 2H_2O$	5.46	15.12	6.28	113 24
Churchite, $(Ce,Ca)PO_4\cdot 2H_2O$				

The minerals of this group are related in formula and lattice dimensions. They have been given here a crystallographic setting that corresponds to the new setting of gypsum with which they are presumably isostructural. The limited data on ardealite[1] indicates it to be the link between the group and gypsum. No crystallographic information is available on churchite except that it is probably monoclinic and has one perfect cleavage which probably corresponds to the perfect {010} cleavage of the other members of this group.

Substitution of anions has not been observed in this group,[2] ardealite being a phase of fixed, 1:1, phosphate-sulfate ratio. Substitution of cations has been noted to any considerable extent only in the yttrium and cerium phosphates in this group.

Ref.
1. Halla, *Zs. Kr.*, **80**, 349 (1931).
2. O'Daniel, *Fortschr. Min.*, **23**, cviii (1939).

39.2.1.1 BRUSHITE [CaHPO$_4 \cdot$2H$_2$O]. *Moore* (*Am. J. Sc.*, **39**, 43, 1865). Epiglaubite ? *Shepard* (*Am. J. Sc.*, **22**, 96, 1856). Metabrushite *Julien* (*Am. J. Sc.*, **40**, 369, 1865). Stoffertite *Klein* (*Ak. Berlin, Ber.*, 720, 1901).

C r y s t.[1] Monoclinic; sphenoidal—2.

$a:b:c = 0.388:1:0.420;$ $\quad \beta\ 117°28';$ $\quad p_0:q_0:r_0 = 1.083:0.373:1$

$r_2:p_2:q_2 = 2.683:2.905:1;$ $\quad \mu\ 62°32';$ $\quad p_0'\ 1.220,\ q_0'\ 0.420,\ x_0'\ 0.520$

Forms:[2]

		ϕ	ρ	ϕ_2	$\rho_2 = B$	C	B
c	001	90°00'	27°28'	62°32'	90°00'	.....	62°32'
b	010	0 00	90 00	.....	0 00	90°00'	90 00
f	120	55 27	90 00	0 00	55 27	67 40½	34 33
d	$\bar{1}$01	−90 00	35 00	125 00	90 00	62 28	125 00
l	$\bar{1}$11	−59 02½	39 14	125 00	71 01	64 05	122 50½
P	1$\bar{1}$1	103 34½	60 48½	29 53½	101 49½	34 29½	31 56½

Structure cell.[3] Space group probably $A2$. a_0 5.88 kX, b_0 15.15, c_0 6.37; β 117°28'; $a_0:b_0:c_0 = 0.388:1:0.420$. Cell contents Ca$_4H_4$(PO$_4$)$_4 \cdot$8H$_2$O.

Habit. Efflorescences and cavity linings of minute crystals, needle-like or prismatic to tabular {010}. Some habits resemble gypsum. Also earthy or powdery; foliated.

P h y s. Cleavage {010} and {001} perfect. H. 2½. G. 2.328 (natural crystals),[4] 2.326 (artificial crystals);[5] 2.257 (calc.). Luster vitreous, pearly on cleavage. Colorless to pale yellow. Transparent to translucent. Piezoelectric.[5]

O p t.[6] In transmitted light, colorless.

Orientation	n(Na)	
$X \wedge c$ $-30°$	1.539	Biaxial positive (+).
Y	1.546	$2V$ 86° (meas.).
Z b	1.551	$r > v$; crossed dispersion noticeable.

C h e m. A hydrated acid phosphate of calcium, CaHPO$_4 \cdot$2H$_2$O. Some analyses show departures in water content.

BRUSHITE

Anal.[7]

	CaO	P$_2$O$_5$	H$_2$O	Rem.	Total	G.
1.	32.58	41.25	26.17		100.00	2.257
2.	32.69	41.41	26.36		100.46	2.208
3.	32.98	42.72	23.33	1.36	100.39	2.29–2.36
4.	30.83	37.96	30.88	0.49	100.16	2.28

1. CaHPO$_4$·2H$_2$O. 2. Aves Island, West Indies.[8] 3. Sombrero Island, West Indies (*metabrushite*).[9] H$_2$O includes organic matter. Rem. is MgO 0.52, (Al,Fe)$_2$O$_3$ 0.79, SO$_3$ 0.05. 4. Mona Island, Puerto Rico (*stoffertite*).[10] H$_2$O is ign. loss. Rem. is SO$_3$.

Tests. B.B. exfoliates and fuses. In C.T. yields acid water. Readily soluble in hydrochloric acid.

O c c u r. Widespread in small amounts in insular and continental phosphate deposits. Frequently occurs as an incrustation upon ancient human and animal bones and has been observed in human urinary calculi. First noted in guano from Aves Island in the Caribbean Sea. Also found on Sombrero Island (*metabrushite*) and Mona Island (*stoffertite*) in the West Indies. Crystals up to 2 cm. long in cavities in phosphorite at Quercy, near Limoges, in powdery form in the Minerva Cave in the Cevennes Mountains, Hérault, and on human skeletons both in prehistoric bone deposits and in ancient coffins at several points in France. Similar occurrences have been noted in Germany, Austria, and several localities in Czechoslovakia. In a cave near Oran, Algeria, it is found in such quantity that it has been worked commercially.

A l t e r. Loses water readily even at ordinary temperature, eventually turning to CaHPO$_4$, monetite.[11] Water loss at ordinary temperature may occur even at relatively high humidity, but all water of crystallization is expelled only with difficulty at 100° C.[12] The name metabrushite was applied to partly dehydrated material. Pseudomorphs of carbonate-hydroxyl-apatite have been reported by several observers[13] and have been formed artificially.[14]

A r t i f. Formed by precipitation from a calcium chloride solution by sodium acid phosphate. May be grown in crystals by slow evaporation of an acetic acid solution of calcium phosphate.[15] Data on the system CaO-H$_2$O-P$_2$O$_5$ are incomplete, but it is known that CaHPO$_4$ crystallizes only as the dihydrate above 0°. The upper limit of formation of this dihydrate is near 80° C.[16]

N a m e. After the American mineralogist George Jarvis Brush (1831–1912), Professor in Yale University.

Ref.

1. Usually listed as prismatic, $2/m$, but the observation of piezoelectricity is not in harmony with this. Elements from structure cell. Transformation: from earlier setting of Dana (828, 1892) to new (gypsum), 103/0$\overline{3}$0/103.

2. Form list from Lacroix, **4**, 491 (1910), includes only forms established on natural crystals by goniometric methods. Additional forms have been found on artificial crystals or reported on the basis of microscopic observation on natural crystals—see Chudoba, *Cbl. Min.*, 291 (1928), and Koehler, *Min. Mitt.*, **37**, 93 (1926).

3. Space group according to Terpstra, *Zs. Kr.*, **97**, 229 (1937). Structure cell with a_0 and c_0 the smallest translations in {010} in the orientation adopted for gypsum by De Jong and Bouman, *Zs. Kr.*, **100**, 275 (1938). See also under gypsum. Dimensions

calculated from the cell of Hill and Hendricks, *Ind. Eng. Chem.*, **28**, 441 (1936). Transformation: Hill and Hendricks cell to structure cell, $10\bar{1}/0\bar{1}0/102$. It is a coincidence that c_0 of Hill and Hendricks and of the structure cell are both 6.37, there being two translations in {010} nearly identical in length. See also Terpstra (1937), who, on the basis of less accurate measurements, gives 6.3 and 6.4 for these two translations.

4. Sekanina, *Publ. Fac. Sc. Univ. Masaryk, Brno* [220] (1935) on Bitov, Moravia, crystals.
5. Terpstra (1937).
6. Klein, *Ak. Berlin, Sitzber.*, 720 (1901). See also Sekanina (1935); Hintze (**1** [4B], 788, 1931); Van Tassel, *Bull. Mus. Hist. Nat. Belgique*, **22**, no. 17 (1944); Mélon and Dallemagne, *Bull. soc. géol. Belgique*, **69**, B19 (1946).
7. For additional analyses see Hintze (1931).
8. Moore (1865), mean of two closely agreeing analyses.
9. Julien (1865).
10. Finkener anal. in Klein (1901).
11. See Frondel, *Am. Min.*, **28**, 215 (1943). For thermal curves see Sekanina (1935).
12. Hill and Hendricks (1936).
13. Lacroix (**4**, 492, 1910) and Frondel (1943).
14. Bassett, *J. Chem. Soc London*, **111**, 620 (1917).
15. de Schulten, *Bull. soc. min.*, **26**, 11 (1903).
16. Bassett, *Zs anorg. Chem.*, **53**, 34 (1907). Demontovich and Serubina, *Biochem. Zs.*, **163**, 464 (1925), have determined the solubility product of $CaHPO_4 \cdot 2H_2O$ in many solutions. Hill and Hendricks (1936) have published solubility curves for $CaHPO_4 \cdot 2H_2O$ in water containing $CaHPO_4$ and H_3PO_4 at 21°, 25° and 50.7°.

39.2.1.2 PHARMACOLITE [$CaH(AsO_4) \cdot 2H_2O$]. Arseniksaurer Kalk *Selb* (*Allg. J. Chem.*, **4**, 537, 1800). Pharmacolith *Karsten* (75, 1800). Arsenikblüthe *Werner*. Arsenicite *Beudant* (**2**, 593, 1832).

C r y s t.[1] Monoclinic.

$a:b:c = 0.3883:1:0.4082;\quad \beta\ 114°32';\quad p_0:q_0:r_0 = 1.0512:0.3713:1$

$r_2:p_2:q_2 = 2.6929:2.8309:1;\quad \mu\ 65°28';\quad p_0'\ 1.1556,\ q_0'\ 0.4082,\ x_0'\ 0.4564$

Forms:

	ϕ	ρ	ϕ_2	$\rho_2 = B$	C	A
b 010	0°00′	90°00′	....	90°00′	90°00′	0°00′
m 110	70 32½	90 00	0°00′	70 32½	66 57	19 27½
x 011	48 11½	31 29	65 28	69 37½	20 22½	67 05½
s $\bar{1}12$	−30 44½	13 21½	96 55½	78 33	33 16½	96 47

Less common:
a 100 π 232 M $\bar{1}32$ d $\bar{2}32$

Structure cell.[2] a_0 6.00 kX, b_0 15.40, c_0 6.29; β 114°47′; $a_0:b_0:c_0 = 0.389:1:0.408$. Cell contents $Ca_4H_4(AsO_4)_4 \cdot 8H_2O$.

Habit. As delicate silky fibers or acicular clusters; also botryoidal and stalactitic. Small crystals rare, flattened on {010} or needle-like [001].

Phys. Cleavage {010} perfect. Fracture uneven. Flexible in thin laminae. H. 2–2½. G. 2.53–2.73; 2.710 ± 0.010 (artif.);[3] 2.703 (calc.). Luster vitreous, pearly on cleavage. Colorless, white or grayish. Good crystals are transparent.

Opt.[4] In transmitted light, colorless.

Orientation	n(Na)	
$X \wedge c$ −29°	1.583	Biaxial negative (−).
Y	1.589	$2V$ (Na) 79°24′ (meas.).
Z b	1.594	$r > v$.

Pharmacolite

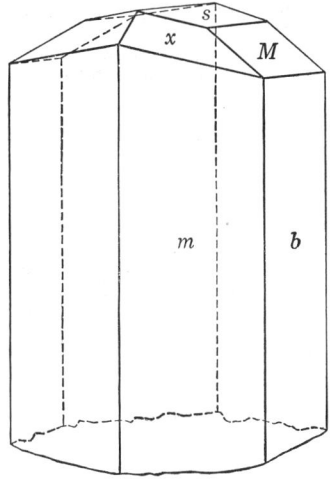

Joachimsthal.

Chem.[5] A hydrated acid arsenate of calcium, $CaH(AsO_4) \cdot 2H_2O$.

	CaO	As$_2$O$_5$	H$_2$O	Rem.	Total	G.
1.	25.96	53.20	20.84		100.00	2.703
2.	23.90	50.54	23.80	1.85	100.09	2.535

1. $CaH(AsO_4) \cdot 2H_2O$. 2. Sainte-Marie-aux-Mines, Alsace.[6] Rem. is MgO 0.30, Fe$_2$O$_3$ 0.35, P$_2$O$_5$ 0.30, SiO$_2$ 0.70.

Tests. B.B. in O.F. fuses with intumescence to a white enamel. Readily soluble in acids but insoluble in water.

Occur. Found sparingly with other arsenates, especially haidingerite, in the oxidized parts of arsenical ores. First recognized at Wittichen in the Black Forest, Germany, where it occurs with haidingerite, roesslerite, and wapplerite. Also found at Andreasberg in the Harz and several other German localities. As excellent but very small crystals at Joachimsthal, Bohemia, and at Sainte-Marie-aux-Mines (Markirch), Alsace. In the United States with haidingerite at the White Caps mine, Manhattan, Nevada, and reported with erythrite at the O.K. mine, San Gabriel Canyon, Los Angeles County, California.

Alter. Dehydrates rapidly at 60°.[7]

Artif. Has been obtained by precipitation by sodium acid arsenate from a solution of calcium chloride in the presence of excess hydrochloric acid [8] and in various other ways. In the system CaO-As$_2$O$_5$-H$_2$O formed at 17° and 40° from solution of a pH range 6.8–6.0 and 6.4–6.0, the monohydrate forming at lower pH. Not formed at 60° or higher.[9]

Name. From φάρμακον, *poison*, in allusion to the arsenic content.

Ref.

1. Orientation to conform to structure cell in setting adopted for gypsum, which see. Elements recalculated from Des Cloizeaux in Lacroix, **4**, 498 (1910). Transformation:

Des Cloizeaux and Goldschmidt (**6**, 136, 1920) to new orientation, $\bar{1}0\bar{3}/030/200$. Forms from Goldschmidt (1920).
2. Cell of smallest dimensions calculated from cell measured by Gossner, *Fortschr. Min.*, **21**, 34 (1937), who gave no space group information. Transformation: Gossner to new, $\bar{1}0\tfrac{1}{2}/0\bar{1}0/001$.
3. Guerin, *Ann. chim*, **16**, 101 (1941).
4. Dufet, *Bull soc. min.*, **11**, 187 (1888), on artificial crystals; also Des Cloizeaux, *Bull. soc min.*, **11**, 192 (1888).
5. For a tabulation of other analyses see Hintze (**1** [4], 784, 1931).
6. Jannettaz, *Bull. soc. min*, **11**, 212 (1888).
7. Guerin (1941), presumably to the monohydrate, haidingerite.
8. de Schulten, *Bull soc min.*, **26**, 18 (1903).
9. Guerin (1941) who studied equilibria in the system at 17°, 40°, 60°, and 90°. He showed that a higher metastable hydrate of $CaHAsO_4$ is formed first and changes to the dihydrate or monohydrate in contact with solution, depending on the pH.

39.2.2 HAIDINGERITE [$CaH(AsO_4) \cdot H_2O$]. Turner (*Edinburgh J. Sc.*, **6**, 317, 1827). (Not the haidingerite of Berthier, *Ann. chim. phys.*, **35**, 351, 1827).

Cryst.[1] Orthorhombic; dipyramidal—$2/m\ 2/m\ 2/m$.

$$a:b:c = 0.8391:1:0.4990; \quad p_0:q_0:r_0 = 0.5947:0.8391:1$$

$$q_1:r_1:p_1 = 0.8391:1.6816:1; \quad r_2:p_2:q_2 = 2.0040:1.1918:1$$

Forms:

	ϕ	$\rho = C$	ϕ_1	$\rho_1 = A$	ϕ_2	$\rho_2 = B$
b 010	0°00′	90°00′	90°00′	90°00′	.....	0°00′
a 100	90 00	90 00	.....	0 00	0°00′	90 00
m 110	35 19½	90 00	90 00	54 40½	0 00	35 19½
t 011	0 00	40 00	40 00	90 00	90 00	50 00
g 102	90 00	16 33½	0 00	73 26½	73 26½	90 00
k 201	90 00	49 56½	0 00	40 03½	40 03½	90 00

Less common:

i 401   n 542   s 421

Twinning. $\{110\}$, rare.[2]

Habit. Fine-grained botryoidal to fibrous coatings. Rarely in crystals, equi-dimensional to short prismatic [001] or [100].

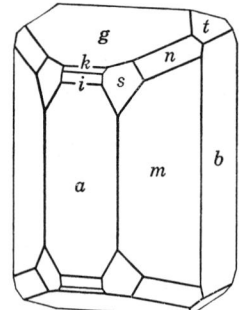
Joachimsthal.

Phys. Cleavage $\{010\}$ perfect. Sectile; thin laminae slightly flexible. H. 2–2½. G. 2.848, 2.962 ± 0.008 (artif.).[3] Luster vitreous, inclining to adamantine, pearly on cleavage. Colorless, white. Transparent in small crystals.

Opt.[4] In transmitted light, colorless.

Orientation		n (Joachimsthal)	
X	b	1.590 ± 0.003	Biaxial positive (+).
Y	a	1.602 ± 0.003	2V 58° ± 3° (meas.).
Z	c	1.638 ± 0.003	$r > v$, weak.

Chem. A hydrated acid arsenate of calcium, $CaH(AsO_4) \cdot H_2O$.

Anal.

	CaO	As_2O_5	H_2O	Total	G.
1.	28.32	58.03	13.65	100.00	
2.	28.39	57.52	14.32	100.23	2.848

1. $CaH(AsO_4) \cdot H_2O$.   2. Bohemia.[5]

Tests. Like pharmacolite.

Occur. Found with pharmacolite and other arsenates as an oxidation product of arsenical ores. First recognized on specimens from Joachimsthal, Bohemia. Also near Wittichen in the Black Forest, Germany, in twinned crystals. Reported from Schneeberg and Johanngeorgenstadt, Saxony. In the United States with pharmacolite and pitticite at the White Caps mine, Manhattan, Nevada.

Alter. On heating loses water in several stages, being completely dehydrated to the pyroarsenate at 200° C.[3]

Artif. May be precipitated together with pharmacolite from a solution of calcium chloride by sodium acid arsenate in the presence of hydrochloric acid at concentrations greater than those required to precipitate pharmacolite alone.[6] The Ba and Sr analogues also are known. In the system $CaO\text{-}As_2O_5\text{-}H_2O$ it appears at temperatures to 60° C. or higher over a pH range about 6.0 to 2.4. At lesser acid concentrations and low temperatures pharmacolite forms; at higher acid concentrations, $CaHAsO_4$.[3]

Name. After the Austrian mineralogist Wilhelm Haidinger (1795–1871).

Ref.

1. Form list and elements from the original and only measurement of natural crystals, Haidinger, *Edinburgh J. Sc.*, **3**, 302 (1825). Artificial crystals measured by de Schulten, *Bull. soc. min.*, **26**, 20 (1903) give closely similar results Transformation: Goldschmidt (4, 105, 1918) to Haidinger, $100/010/00\frac{1}{2}$.
2. Sandberger, 407 (1885).
3. Guerin, *Ann chim.*, **16**, 101 (1941).
4. Indices and $2V$ from Larsen, 82 (1921), on crystals from Joachimsthal, Bohemia; orientation from Des Cloizeaux, *Bull. soc. min.*, **11**, 195 (1888).
5. Turner, *Edinburgh J. Sc.*, **3**, 308 (1925).
6. de Schulten, *Bull. soc. min.*, **26**, 18 (1903).

39.2.3   **NEWBERYITE** [$MgH(PO_4) \cdot 3H_2O$].   vom Rath (*Ber. Niederrhein. Ges.*, **36**, 5, 1879).

Cryst.[1]   Orthorhombic; dipyramidal—$2/m\ 2/m\ 2/m$.

$a:b:c = 0.9548:1:0.9360;\quad p_0:q_0:r_0 = 0.9803:0.9360:1$

$q_1:r_1:p_1 = 0.9548:1.0201:1;\quad r_2:p_2:q_2 = 1.0684:1.0473:1$

Forms:

	ϕ	$\rho = C$	ϕ_1	$\rho_1 = A$	ϕ_2	$\rho_2 = B$
c 001	....	0°00′	0°00′	90°00′	90°00′	90°00′
b 010	0°00′	90 00	90 00	90 00	....	0 00
a 100	90 00	90 00	....	0 00	0 00	90 00
f 021	0 00	61 53½	61 53½	90 00	90 00	28 06½
e 102	90 00	26 06½	0 00	63 53½	63 53½	90 00
o 111	46 19½	53 35	43 06½	54 24½	45 34	56 14½

Less common:

m 110	n 750	l 210	d 101	p 112	r 211
t 430	v 320	g 011	q 302	h 223	s 722

PHOSPHATES, ARSENATES, VANADATES

Habit. Crystals equi-dimensional, short prismatic [001] or tabular {100}; also tabular {010} or dipyramidal {111} in artificial crystals.

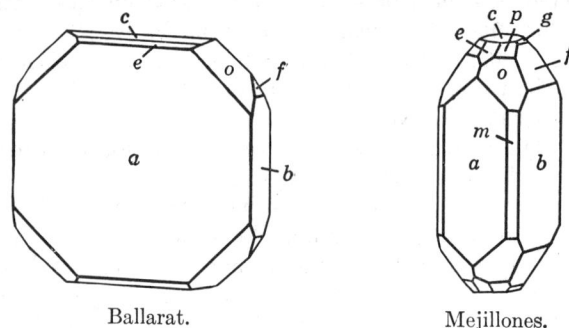

Ballarat. Mejillones.

P h y s. Cleavage {010} perfect, {001} imperfect. H. 3–3½. G. 2.10 (Mejillones), 2.123 (artif.).[2] Luster vitreous. Colorless. Transparent.
O p t.[3] In transmitted light, colorless.

Orientation		n	
X	a	1.514 ± 0.003	Biaxial positive (+).
Y	b	1.517 ± 0.003	$2V\ 44°46'$.
Z	c	1.533 ± 0.003	$r < v$, perceptible.

C h e m. A hydrated acid phosphate of magnesium, $MgH(PO)_4 \cdot 3H_2O$.
Anal.

	MgO	P_2O_5	H_2O	Rem.	Total
1.	23.12	40.72	36.16		100.00
2.	22.37	40.73	[35.84]	1.06	[100.00]

1. $MgH(PO_4) \cdot 3H_2O$. 2. Skipton Caves, near Ballarat, Victoria.[4] Rem. is Fe_2O_3 0.85, Mn_2O_3 0.21.

Tests. Scarcely soluble in cold water. Readily soluble in dilute hydrochloric acid.

O c c u r. Recognized as a new species in Skipton Caves, near Ballarat, Victoria, where it occurs in large (inch-square) crystals with hannayite in bat guano. Found as small reticulated crystals resembling a zeolite in cavities in basalt in a lava tunnel filled with bat guano on the Island of Reunion. With collophane on basalt from the roof of a cavern on Ascension Island, South Atlantic. In clear, highly modified crystals on cracks in guano, Mejillones, Chile. An analysis of impure magnesium phosphate in a mammoth tusk found in frozen swamp muck near Dawson, Yukon territory, Canada, has been interpreted [9] as representing a mixture of newberyite, struvite, and magnesite.

A l t e r. Some varieties are reported to become cloudy on exposure but this may be due to included impurities. At 130° dehydrates rapidly [5] to $MgHPO_4$ which on ignition changes further to the pyrophosphate.

A r t i f. May be formed by dehydration of $MgHPO_4 \cdot 7H_2O$ at ordinary temperatures.[5] Crystallized from a solution of $MgHPO_4 \cdot 7H_2O$ in 25 per cent acetic acid evaporated on a steam bath and in a variety of other ways between room temperature and 150° C.[6] Can be crystallized together with struvite at ordinary temperature.[7] In the system $MgO-P_2O_5-H_2O$ at 25° the trihydrate is the only stable hydrate of $MgHPO_4$.[8]

N a m e. After J. Cosmo Newbery of Melbourne.

Ref.
1. Elements and form list of Schmidt, *Zs. Kr.*, **7**, 26 (1882).
2. de Schulten, *Bull. soc. min.*, **26**, 24 (1903).
3. Indices from Larsen, 115 (1921), orientation and $2V$ from Des Cloizeaux, *Bull. soc. min.*, **2**, 82 (1879), both on material from Skipton Cave. See also Schmidt (1882); Des Cloizeaux, *Bull. soc min.*, **2**, 82 (1879); Richards, *Am. Min.*, **13**, 397 (1928).
4. MacIvor, *Chem. News*, **55**, 216 (1887).
5. Hägele and Machatschki, *Cbl. Min.*, 297 (1939).
6. de Schulten (1903).
7. de Schulten, *Bull. soc. min.*, **26**, 95 (1903).
8. Cameron and Bell, *J. Phys Chem.*, **11**, 313 (1907).
9. Hoffman, *Am. J. Sc.*, **11**, 149 (1901).

39.2.4   Forbesite. Hydrous Bibasic Arsenate of Nickel and Cobalt *Forbes* (*Phil. Mag.*, **25**, 103, 1863). Forbesit *Kenngott* (46, 1868).

Radial fibrous crusts. H. $2\frac{1}{2}$. G. 3.13.[1] Color grayish white. Luster dull to silky or resinous. A hydrated arsenate of nickel and cobalt of uncertain formula, perhaps $H(Ni,Co)(AsO_4) \cdot 3\frac{1}{2}H_2O$, with $Ni:Co = 2:1$.

Anal.

	1	2
NiO	19.02	19.71
CoO	9.54	9.24
As_2O_5	43.91	44.05
H_2O	27.53	26.98
Total	100.00	99.98

1. $H(Ni,Co)(AsO_4) \cdot 3\frac{1}{2}H_2O$, with $Ni:Co = 2:1$.   2. Chile.

Tests. B.B. fusible with difficulty in the reducing flame to a magnetic bead. In C.T. affords water, becoming darker in color.

O c c u r. As an alteration of chloanthite (?) in a decomposed basic igneous rock at a mine about 20 miles east of Flamenco, in the Atacama desert, Chile.

A r t i f.[2] Possibly isostructural with the artificial salt $MnH(PO_4) \cdot 3H_2O$.

N a m e. After David Forbes (1825–1876), English chemist and geologist.

Needs verification.

Ref.
1. Highest of the values 3.134, 3.069, 3.054 reported by Forbes (1863).
2. Mellor (**12**, 450, 1932).

39.2.5 ROESSLERITE GROUP

39.2.5.1 ROESSLERITE [MgH(AsO$_4$)·7H$_2$O]. *Blum* (*Wett. Ges. Hanau*, **32**, 1861). Wapplerite pt. *Frenzel* (*Min. Mitt.*, 279, 1874).

C r y s t.[1] Monoclinic; prismatic—$2/m$.

$a:b:c = 0.4473:1:0.2598$; $\beta\,94°26'$; $p_0:q_0:r_0 = 0.5808:0.2590:1$

$r_2:p_2:q_2 = 3.8607:2.2423:1$; $\mu\,85°34'$; $p_0'\,0.5826$, $q_0'\,0.2598$, $x_0'\,0.0775$

F o r m s: (on artificial crystals)[2] $b\,010$, $m\,110$, $l\,120$, $d\,011$, $t\,031$, $w\,111$, $o\,\bar{1}11$, $N\,131$; (on "wapplerite" = roesslerite ?)[3] $a\,100$, $n\,210$, $O\,411$, $p\,211$, $W\,\bar{4}11$, $\pi\,\bar{2}11$, $g\,231$, $e\,251$, $\gamma\,\bar{2}31$, also b, m, l, d, t, w, o, N.

Habit. In fine-grained crusts, sometimes hair-like. Rarely in good crystals, small, short prismatic [001]. Artificial crystals are tabular {010}.

P h y s. Cleavage {111} imperfect. H. 2–3. G. 1.943 (artif.).[4] Luster vitreous to dull. Colorless or white, sometimes faintly colored by inclusions. Transparent.

O p t.[5] In transmitted light, colorless.

Orientation		n	
$X \wedge c$	14°	1.525 ± 0.005	Biaxial positive (+).
Y		1.53 ± 0.01	$2V$ small.
Z	b	1.550 ± 0.005	

C h e m. A hydrated acid arsenate of magnesium, MgH(AsO$_4$)·7H$_2$O.
Anal.

	MgO	As$_2$O$_5$	H$_2$O	Total
1.	13.88	39.58	46.54	100.00
2.	14.22	40.16	45.62	100.00

1. MgH(AsO$_4$)·7H$_2$O. 2. Bieber, near Hanau, Germany.[6] Water determined by difference includes trace of CoO.

Tests. Readily soluble in dilute hydrochloric acid; slightly soluble in water.

O c c u r. A secondary mineral found with other arsenates in the oxidized zones of arsenical ore deposits. First noted at Bieber, near Hanau, Germany, where it occurs with pharmacolite and erythrite. At the Sophie mine near Wittichen and more abundantly with haidingerite and erythrite at the Wolfgang mine, near Alpirsbach in the Black Forest. At Joachimsthal, Bohemia, in weathered (partly dehydrated) crystals with pharmacolite and haidingerite; also in crystals described as wapplerite.

A l t e r. Natural crystals lose water on exposure to air becoming white and dull.[7] Artificial crystals are reported to be stable in air.[2]

Artif.[4] Artificial roesslerite is precipitated together with MgNH$_4$-AsO$_4$·6H$_2$O (the arsenate analogue of struvite) from an acid solution of disodium arsenate and ammonium sulfate by a solution of magnesium sulfate. Under slightly higher acid concentration roesslerite alone is formed. From a chloride solution of magnesium and calcium roesslerite can be crystallized together with pharmacolite, duplicating the paragenesis at Joachimsthal.

Name. After Karl Rössler of Hanau, Germany.

WAPPLERITE. *Frenzel (Min. Mitt.,* 279, 1874). Originally described as a hydrated arsenate of calcium and magnesium. Probably a mixture of pharmacolite and roesslerite.[3] Crystal measurements of the supposedly triclinic wapplerite were probably made on roesslerite.[4]

Ref.

1. Elements for artificial crystals of Haushofer, *Zs Kr*, **7**, 259 (1882). Measurements on presumed weathered roesslerite, Tschermak, *Ak. Wien, Ber.*, **56**, 824 (1867), gave different elements. Crystals of so-called wapplerite, measured by Schrauf, *Zs. Kr.*, **4**, 277 (1880), and reported as triclinic with γ and α departing only $10\frac{1}{2}'$ and $14'$, respectively, from $90°$ were considered by de Schulten, *Bull. soc min.*, **26**, 99, 95 (1903), to have been roesslerite. If this is accepted, the elements of natural roesslerite become $a:b:c = 0.4504:1:0.2616$, $\beta\ 95°20'$.
2. Haushofer (1882)
3. Schrauf (1880).
4. de Schulten (1903).
5. Larsen, 156 (1921), on fibers from Joachimsthal, Bohemia, referred to as wapplerite. Orientation of Haushofer (1882).
6. Analysis by Delffs in Blum (1861).
7. Blum (1861) who reported an exfoliation of the natural crystals accompanying dehydration and Tschermak (1867) who found the partly dehydrated crystals to have a composition corresponding to $MgHAsO_4 \cdot 4H_2O$. Such a hydrate has not been reported by others.
8. de Schulten (1903); Friedrich and Robitsch, *Cbl. Min.*, 142 (1939).

39.2.5.2. PHOSPHORROESSLERITE $[MgH(PO_4) \cdot 7H_2O]$. *Friedrich* and *Robitsch (Cbl. Min.,* 142, 1939).

Cryst.[1] Monoclinic; prismatic—$2/m$.

$a:b:c = 0.4455:1:0.2602$; $\beta\ 94°56'$; $p_0:q_0:r_0 = 0.5841:0.2592:1$

$r_2:p_2:q_2 = 3.8575:2.2530:1$; $\mu\ 85°04'$; $p_0'\ 0.5862, q_0'\ 0.2602,\ x_0'\ 0.0863$

Forms:

	ϕ	ρ	ϕ_2	$\rho_2 = B$	C	A
b 010	0°00'	90°00'	....	0°00'	90°00'	90°00'
p 110	66 04	90 00	0°00'	66 04	85 29½	23 56
r 011	41 58	42 02	85 04	78 51½	11 08½	50 08

Less common:
a 100 l 140 g 120 w 111 o 111

Structure cell.[2] Space group $A2/a$. $a_0\ 11.35\ kX$, $b_0\ 25.36$, $c_0\ 6.60$; $\beta\ 95°$; $a_0:b_0:c_0 = 0.4475:1:0.2605$. Cell contents $Mg_8H_8(PO_4)_8 \cdot 56H_2O$.

Habit. Crystals equi-dimensional or short prismatic [100], also skeletal or crusted.

Phys. Fracture conchoidal. H. $2\frac{1}{2}$. G. 1.725 ± 0.004, 1.728 (artif.);[3] 1.717 (calc.). Luster vitreous when fresh, dull due to water loss. Rarely clear, mostly yellowish due to included iron oxides and organic matter.

Opt. In transmitted light, colorless.

Orientation		n	
X	b	1.477	Biaxial negative $(-)$.
Y		1.485	$2V$ 38°10′ (meas.).
$Z \wedge c$	$+6\frac{1}{2}°$	1.486	$r > v$.

Chem. A hydrated acid phosphate of magnesium, $MgH(PO_4) \cdot 7H_2O$. Isostructural and probably isomorphous with roesslerite.

Anal.

	MgO	P_2O_5	H_2O	Insol.	Total	G.
1.	16.36	28.81	54.83		100.00	1.707
2.	16.28	28.07	54.51	0.08	98.94	1.725

1. $MgH(PO_4) \cdot 7H_2O$. 2. Schellgaden, Austria.

Occur. Found in cold (10° C.), wet muck in abandoned mine workings near Schellgaden, province of Salzburg, Austria.

Alter. Natural crystals are more or less dehydrated to pseudomorphs of a lower hydrate in the dry parts of the mine or lose water in dry air. Artificial crystals are stable in contact with mother liquor at room temperature but on exposure to air at this temperature become altered to the trihydrate [4] within a few days.

Artif. Crystallizes in needles [010] with the forms {010} {120}, {140}, and {1̄11} when a magnesium sulfate solution is precipitated by phosphate ions at room temperature.[5] In the system $MgO-P_2O_5-H_2O$, $MgHPO_4 \cdot 7H_2O$ does not appear at 25° C.[6]

Name. From the related mineral roesslerite, of which it is the phosphate analogue.

Ref.

1. Friedrich and Robitsch (1939). Elements agree closely with those derived from measurements of artificial crystals by Haushofer, *Zs Kr.*, **7**, 259 (1882), as corrected by Hägele and Machatschki, *Cbl. Min.*, 297 (1939) [$a:b:c = 0.4451:1:0.2577$, β 94°18′].
2. Hägele and Machatschki (1939).
3. de Schulten, *Bull. soc. min.*, **26**, 99 (1903).
4. Hägele and Machatschki (1939) who established the identity of the alteration product by x-ray examination.
5. Hägele and Machatschki (1939).
6. Cameron and Bell, *J. Phys. Chem*, **11**, 363 (1907).

40 HYDRATED NORMAL PHOSPHATES, ETC.

TYPE 1. $AB(XO_4) \cdot xH_2O$

40.1.1 S T R U V I T E [$Mg(NH_4)(PO_4) \cdot 6H_2O$]. Struveit *Ulex* (*Ak. Stockholm, Öfv.*, **3**, 32, 1845; *Ann. Phys.*, **58**, 99, 1846; **66**, 41, 1848). Guanite *Teschemacher* (*Phil. Mag.*, **28**, 546, 1846). "Triple-phosphate" *Med.*

C r y s t.[1] Orthorhombic; pyramidal—*m m* 2.

$a:b:c = 0.8827:1:1.6102;$ $p_0:q_0:r_0 = 1.8242:1.6102:1$

$q_1:r_1:p_1 = 0.8827:0.5482:1;$ $r_2:p_2:q_2 = 0.6210:1.1329:1$

Forms:[2]

		ϕ	$\rho = C$	ϕ_1	$\rho_1 = A$	ϕ_2	$\rho_2 = B$
c	001	.....	0°00′	0°00′	90°00′	90°00′	90°00′
a	100	90°00′	90 00	.....	0 00	0 00	90 00
m	110	48 34	90 00	90 00	41 26	0 00	48 34
w	013	0 00	28 13½	28 13½	90 00	90 00	61 46½
s	011	0 00	58 09½	58 09½	90 00	90 00	31 50½
q	102	90 00	42 22	0 00	47 38	47 38	90 00
h	101	90 00	61 16	0 00	28 44	28 44	90 00
k	201	90 00	74 40½	0 00	15 19½	15 19½	90 00
t	111	48 34	67 39½	58 09½	46 06	28 44	52 15½

Structure cell.[3] Space group $Pmc2$. a_0 6.09 kX, b_0 6.97, c_0 11.18; $a_0:b_0:c_0 = 0.874:1:1.604$. Cell contents $Mg_2(NH_4)_2(PO_4)_2 \cdot 12H_2O$.

Habit. Very variable: equant; wedge-shaped, due to large and unequally developed {101} and {10$\bar{1}$}, or coffin-shaped when viewed along

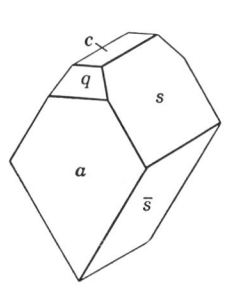
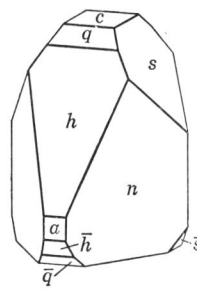
Ballarat.
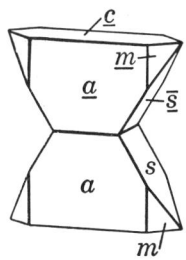
Braunschweig.

[100]; short prismatic [001], [010], or [100]; thick tabular {100}. The distribution of the pyramidal or domical faces is often irregular. Usually found as distinct crystals, ranging up to one inch or so in size, with a marked hemimorphic habit.

Twinning. On {001}, the acute poles of [001] united in the composition plane, {001}, with deep re-entrant angles.

P h y s. Cleavage {001} good, {100} poor. Brittle. Fracture subconchoidal to uneven. H. 2. G. 1.711;[4] 1.705 (calc.). Luster vitreous. Colorless; sometimes slightly yellowish or brown due to impurities, and becoming white and chalky on dehydration. Transparent to translucent. Tasteless, being but slightly soluble. Pyroelectric.[5] Piezoelectric. Etch pits conform to pyramidal symmetry.[6]

O p t. In transmitted light, colorless.

ORIENTATION		n(Na) [7]	
X	b	1.495	Biaxial positive (+).
Y	c	1.496	$2V$ 37°22′ (meas.).
Z	a	1.504	$r < v$, strong.

C h e m. Magnesium ammonium phosphate hexahydrate, $Mg(NH_4)(PO_4) \cdot 6H_2O$. Small amounts of Mn'' and Fe'' substitute for Mg.

Anal.[8]

	1	2	3	4	5
MgO	16.43	16.57	16.07	13.15	13.46
MnO		tr.	0.16	2.01	1.12
FeO		0.95	0.81	2.22	3.06
$(NH_4)_2O$	10.61 ⎫	54.49	10.57 ⎫	53.64	53.76
H_2O	44.04 ⎭		43.57 ⎭		
P_2O_5	28.92	28.81	28.82	28.05	28.56
Total	100.00	100.82	100.00	99.07	99.96

1. $Mg(NH_4)(PO_4) \cdot 6H_2O$. 2. Skipton Caves, Victoria.[9] 3. Skipton Caves.[10] 4. Hamburg, Germany.[11] Nearly colorless. 5. Hamburg.[11] Amber colored.

Tests. In C.T. loses water and ammonia and turns opaque. Easily soluble in acids, very slightly soluble in water (0.05 g. in 100 ml. H_2O at 20°).

O c c u r. Formed by reaction of magnesian solutions with ammoniacal solutions of phosphates. Found in deposits of guano or dung, in putrescent matter [12] or canned foods,[13] or formed by bacterial action on various organic compounds.[14] Also a constituent of human urinary sediments, and as bladder or kidney concretions in alkaline and infected urines.[15] First found in nature under an old church in Hamburg, Germany, in a bed of peat underlying deposits of organic matter; also in a deposit of cattle dung at Homburg v.d. Höhe.[16] From the Liim Fjord, Denmark, in a sediment rich in organic remains.[17] In guano at Saldanha Bay, Cape Province, South Africa, and in bat guano of the Skipton Caves southwest of Ballarat, Victoria, Australia. With newberyite in the tooth of a mammoth from Quartz Creek near Dawson, Yukon.[18] With brushite in caves on the Island of Reunion.

A r t i f.[19] A well-known laboratory product, formed as a precipitate in the analytical determination of magnesium or phosphorus. Obtained in large crystals by slow reaction of magnesium sulfate solution with an acid solution of ammonium phosphate.

DICKINSONITE

A l t e r. Becomes white and pulverulent on standing in a dry, warm atmosphere.

N a m e. After H. C. G. Struve (1772–1851), of the Russian diplomatic service.

Ref.
1. Angles of Sadebeck, *Min. Mitt.*, 113 (1877), in the unit and orientation of the structure cell. Transformation: Sadebeck (1877) and Dana (806, 1892) to new, $0\frac{1}{2}0/\overline{1}00/001$. The present position was earlier used by Goldschmidt (3, 167, 1890) and Hausmann (1847). Transformation: Goldschmidt (8, 95, 1922) to new, $\overline{1}00/001/010$.
2. Goldschmidt (1922) Less common and rare: b 010, x 057, μ 031, i 105, β 801, r 113, n 120. Upper and lower forms not here distinguished
3. Wolfe, priv. comm (1946), by Weissenberg method.
4. de Schulten, *Bull. soc. min.*, **26**, 95 (1903).
5. Kalkowsky, *Zs. Kr.*, **11**, 1 (1886).
6 Sadebeck (1877).
7. Bøggild, *Medd. Dansk. Geol. For.*, **13**, 25 (1907).
8. For additional analyses see Hintze (1 [4A], 1201, 1933).
9. Pittman, *Contrib. Min. Victoria*, 56 (1870)
10. MacIvor, *Chem. News*, **55**, 215 (1887).
11. Ulex, *Jb Min.*, 51 (1851).
12. Porter, *Am. Min.*, **9**, 93 (1924); Arzruni, *Zs Kr.*, **18**, 60 (1890).
13. Palache, *Am. Min.*, **8**, 72 (1923); Ayres, *Am Min.*, **27**, 387 (1942).
14. Robinson, *Proc. Cambridge Phil. Soc.*, **6**, 360 (1889).
15. Frondel and Prien, *J. Urology*, **57**, 970 (1947).
16. Kalkowsky (1886).
17. Bøggild (1917).
18. Hoffmann, *Am. J. Sc.*, **11**, 149 (1901).
19. Mellor (4, 384, 1923).

TYPE 2. $AB_2(XO_4)_2 \cdot xH_2O$

40.2.1 D I C K I N S O N I T E $[H_2Na_6(Mn,Fe,Ca,Mg)_{14}(PO_4)_{12} \cdot H_2O]$. Brush and Dana (*Am. J. Sc.*, **16**, 114, 1878).

C r y s t.[1] Monoclinic; prismatic—$2/m$.

$a:b:c = 1.6784:1:2.4814$; $\beta\ 104°41'$; $p_0:q_0:r_0 = 1.4784:2.4004:1$

$r_2:p_2:q_2 = 0.4166:0.6159:1$; $\mu\ 75°19'$; $p_0'\ 1.5284,\ q_0'\ 2.4814,\ x_0'\ 0.2620$

Forms:

	ϕ	ρ	ϕ_2	$\rho_2 = B$	C	A
c 001	90°00'	14°41'	75°19'	90°00'	.....	75°19'
b 010	0 00	90 00	.....	0 00	90°00'	90 00
m 110	31 38	90 00	0 00	31 38	82 21½	58 22
d 102	90 00	45 44½	44 15½	90 00	31 03½	44 15½
g 302	90 00	68 37½	21 22½	90 00	53 56½	21 22½
j 401	90 00	81 05	8 55	90 00	66 24	8 55
$D\ \overline{1}02$	−90 00	26 40	116 40	90 00	41 21	116 40
$E\ \overline{3}04$	−90 00	41 29	131 29	90 00	56 10	131 29
$F\ \overline{1}01$	−90 00	51 42	141 42	90 00	66 23	141 42
$H\ \overline{2}01$	−90 00	70 18½	160 18½	90 00	84 59½	160 18½
p 111	35 48½	71 54	29 11	39 34½	63 48	56 12½
$P\ \overline{1}11$	−27 02½	70 15½	141 42	33 02	77 23	115 20

Structure cell.[2] Space group $C2/c$. a_0 16.70 kX, b_0 9.95, c_0 24.69; [β 104°41′]; $a_0:b_0:c_0 = 1.695:1:2.507$. Cell contents $H_8Na_{24}(Mn,Fe)_{56}$-$(PO_4)_{48}\cdot 4H_2O$.

Habit. Tabular {001} and often pseudo-rhombohedral in habit; triangular striations on {001}. Commonly foliated to micaceous; also curved lamellar, radiated, or stellated; as disseminated scales.

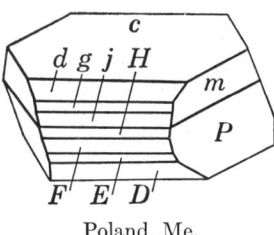

Poland, Me.

Phys. Cleavage {001} perfect, almost micaceous. Fracture uneven. Very brittle. H. $3\frac{1}{2}$–4. G. 3.41 (Branchville),[3] 3.38 (Poland, Maine);[3] 3.42 (calc.). Luster vitreous, on the cleavage face somewhat pearly. Color oil- to olive-green, grass-green (especially in massive varieties); also yellowish green to brownish green. Streak nearly white. Transparent to translucent.

Opt. In transmitted light, green.

ORIENTATION	n (Maine)[7]	n (Branchville)[4]	PLEOCHROISM	
X b	1.648 ± 0.003	1.658 ± 0.003	Pale olive-green	Biaxial positive
$Y \wedge c \sim +15°$	1.655 ± 0.003	1.662 ± 0.003	Paler olive-green	(+).
Z	1.662 ± 0.003	1.671 ± 0.003	Very pale yellowish green	$r > v$, rather strong.
$2V$	Near 90°	Moderate		

Chem. A hydrated acid phosphate of sodium and manganese, $H_2Na_6(Mn,Fe,Ca,Mg)_{14}(PO_4)_{12}\cdot H_2O$.[2] Divalent iron, calcium, and magnesium substitute for Mn, with Ca:Mg:Fe:Mn = 1:1.17:4.86:12.7 in analysis 4.

Anal.

	1	2	3	4
Li_2O		0.17	0.22	0.20
Na_2O	9.14	7.46	7.37	7.41
K_2O		1.52	1.80	1.73
CaO	1.96	2.15	2.09	2.01
MgO	1.65			1.67
FeO	12.18	13.25	12.96	12.33
MnO	31.42	31.58	31.83	31.83
P_2O_5	41.88	39.57	40.89	40.78
H_2O	1.77	1.65	1.63	1.82
Insol.		2.58	0.82	1.00
Total	100.00	99.93	99.61	100.78
G.	3.42			3.266

1. $H_2Na_6(Mn,Fe,Ca,Mg)_{14}(PO_4)_{12}\cdot H_2O$, with Ca:Mg:Fe:Mn = 1:1.17:4.86:12.7. 2,3. Branchville.[5] 4. Poland, Maine.[6]

Tests. B.B. easily fusible. In C.T. yields a little water, the first portions neutral and the last faintly acid. Soluble in acids.

Occur. In pegmatite at Branchville, Fairfield County, Connecticut, associated with eosphorite, triploidite, lithiophilite, rhodochrosite, reddingite, fairfieldite. Later found in pegmatite at the Berry quarry, near Poland, Maine, associated as at Branchville. Reported from Portland, Connecticut, with lithiophilite.

FILLOWITE

Alter. To fairfieldite (Maine).

Name. After the Rev. William Dickinson, formerly of Redding, Connecticut, who early collected and drew attention to the rare minerals of the locality.

Ref.

1. Angles of Wolfe, *Am Min.*, **26**, 339 (1941), on material from Poland, Maine, in the orientation and unit of the structure cell. Transformation: Brush and Dana (1878) to new, $\overline{1}00/010/102$. The original Branchville crystals of Brush and Dana were of inferior quality, and the correspondence of their angles with the present angles is not close.
2. Wolfe (1941).
3. Wolfe (1941); slightly lower values have been reported by Brush and Dana (1878) and by Berman and Gonyer, *Am. Min.*, **15**, 375 (1930).
4. Larsen (68, 1921)
5. Wells analyses in Brush and Dana, *Am. J. Sc.*, **39**, 213 (1890) An earlier analysis on impure material was reported by Penfield in Brush and Dana (1878).
6. Gonyer analysis in Berman and Gonyer (1930).
7. Berman and Gonyer (1930).

40.2.2 **FILLOWITE** [$H_2Na_6(Mn,Fe,Ca)_{14}(PO_4)_{12} \cdot H_2O$ (?)]. *Brush* and *Dana* (*Am. J. Sc.*, **17**, 363, 1879; *Zs. Kr.*, **3**, 578, 1879).

Cryst.[1] Monoclinic.

$a:b:c = 1.7303:1:1.4190;$ $\beta\ 90°09';$ $p_0:q_0:r_0 = 0.8201:1.4190:1$

$r_2:p_2:q_2 = 0.7047:0.5779:1;$ $\mu\ 89°51';$ $p_0'\ 0.8201, q_0'\ 1.4190, x_0'\ 0.0026$

Forms:

	ϕ	ρ	ϕ_2	$\rho_2 = B$	C	A
c 001	90°00'	0°09'	89°51'	90°00'	.....	89°51'
d 201	90 00	58 40	31 20	90 00	53°31'	31 20
$p\ \overline{1}11$	−29 57	58 35½	129 16	42 18½	58 40	115 13

Habit. Pseudo-rhombohedral. In granular crystalline masses.

Phys. Cleavage {001} nearly perfect. Fracture uneven. Brittle. H. 4½. G. 3.43. Luster subresinous to greasy. Color wax-yellow, yellowish to reddish brown, colorless. Streak white. Transparent to translucent.

Opt.[2] In transmitted light, colorless to yellow.

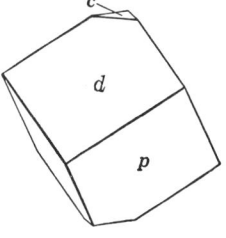

Orientation	n	
$X \sim c$	1.672 ± 0.003	Biaxial positive (+).
Y	1.672 ± 0.003	$2V$ small.
Z	1.676 ± 0.003	$r < v$ (strong?).

Chem. A hydrated acid phosphate of sodium, manganese, divalent iron and calcium, $H_2Na_6(Mn,Fe,Ca)_{14}(PO_4)_{12} \cdot H_2O$. Ca:Fe:Mn = 1: 1.77:7.6 in analysis 2. The exact formula, however, is uncertain and as given differs from that of dickinsonite only in the ratio of the divalent cations.

Anal.

	1	2	3
Na_2O	9.08	5.74	5.44
Li_2O		0.06	0.07
CaO	3.69	4.08	3.63
FeO	8.38	9.33	9.69
MnO	35.52	39.42	39.58
P_2O_5	41.57	39.10	39.68
H_2O	1.76	1.66	1.58
Quartz		0.88	1.02
Total	100.00	100.27	100.69

1. $H_2Na_6(Mn,Fe,Ca)_{14}(PO_4)_{12} \cdot H_2O$ with Ca:Fe:Mn = 1:1.77:7.6. 2. Branchville.[3] Average of two. 3. Branchville.[4]

Tests. B.B. fuses at $1\frac{1}{2}$ with intumescence to a black feebly magnetic mass, coloring the flame momentarily pale green, then intensely yellow. In C.T. yields a little neutral water. Soluble in acids.

O c c u r. In pegmatite at Branchville, Fairfield County, Connecticut, associated with triploidite, fairfieldite, and reddingite.

N a m e. After A. N. Fillow of Branchville, Connecticut, who originally owned and operated the Branchville pegmatite and first collected the rare manganese phosphates for which the locality is known.

Ref.

1. Elements of Brush and Dana (1879). On the morphological resemblance to dickinsonite see Dana (810, 1892).
2. Larsen (75, 1921).
3. Penfield analysis in Brush and Dana (1879).
4. Wells analysis in Brush and Dana, *Am. J. Sc.*, **39**, 215 (1890).

40.2.3 FAIRFIELDITE GROUP

40.2.3.1 **F A I R F I E L D I T E** [$Ca_2(Mn,Fe)(PO_4)_2 \cdot 2H_2O$]. *Brush* and *Dana* (*Am. J. Sc.*, **17**, 359, 1879). Leucomanganit *Sandberger* (*Jb. Min.*, 370, 1879; I, 185, 1885).

C r y s t.[1] Triclinic; pinacoidal—$\bar{1}$.

$a:b:c = 0.8791:1:0.8331;$ $\quad \alpha\ 102°05',\ \beta\ 108°42\frac{1}{2}',\ \gamma\ 90°05'$

$p_0:q_0:r_0 = 0.9266:0.7891:1;$ $\quad \lambda\ 77°12',\ \mu\ 70°50',\ \nu\ 85°45'$

$p_0'\ 1.0032,\ q_0'\ 0.8543,\ x_0'\ 0.3386,\ y_0'\ 0.2399$

Forms:[2]

	ϕ	ρ	A	B	C	M
c 001	54°41'	22°32'	70°50'	77°12'	.....	83°39$\frac{1}{2}$'
b 010	0 00	90 00	85 45	.....	77°12'	127 56$\frac{1}{2}$
a 100	85 45	90 00	.....	85 45	70 50	42 11$\frac{1}{2}$
M 1$\bar{1}$0	127 56$\frac{1}{2}$	90 00	42 11$\frac{1}{2}$	127 56$\frac{1}{2}$	83 39$\frac{1}{2}$	.....
P $\bar{1}$11	−136 08$\frac{1}{2}$	43 41$\frac{1}{2}$	120 56$\frac{1}{2}$	119 52	65 55$\frac{1}{2}$	94 04

Structure cell.[3] Space group $P\bar{1}$. $a_0\ 5.77\ kX,\ b_0\ 6.56,\ c_0\ 5.47;\ \alpha\ 102°05',\ \beta\ 108°42\frac{1}{2}',\ \gamma\ 90°05\frac{1}{2}';\ a_0:b_0:c_0 = 0.8791:1:0.8331$. Cell contents $Ca_2(Mn,Fe)(PO_4)_2 \cdot 2H_2O$.

FAIRFIELDITE

Habit. Crystals prismatic to equant; often composite. Usually in foliated to lamellar crystalline aggregates and sometimes resembling gypsum; occasionally curved, foliated, or fibrous; in radiating masses.

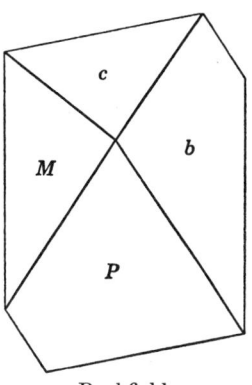

Buckfield.

P h y s. Cleavage {001} perfect, {010} good, {1$\bar{1}$0} distinct. Fracture uneven. Brittle. H. 3$\frac{1}{2}$. G. 3.08 (Branchville);[4] 3.09 (calc.). Luster pearly to sub-adamantine; on the perfect {001} cleavage brilliant, resembling gypsum. Color white; also greenish white, pale straw-yellow, salmon-yellow. Streak white. Transparent.

O p t. In transmitted light, colorless.

	ORIENTATION[8]		n (Branchville)[5]	n (Poland,[6] anal. 5)	n (Buckfield,[7] anal. 6)	
	ϕ	ρ				
X	120°	60°	1.636 ± 0.003	1.640 ± 0.002	1.633	Biaxial positive
Y	−102°	36°	1.644 ± 0.003	1.650 ± 0.002	1.641	(+).
Z	18°	69°	1.654 ± 0.003	1.660 ± 0.002	1.652	$r > v$, moderate.
2V			Very large	86° ± 1° (meas.)		

C h e m. A hydrated phosphate of calcium and divalent manganese, $Ca_2(Mn,Fe)(PO_4)_2 \cdot 2H_2O$. Fe'' substitutes for Mn'', with Fe:Mn = 1:1.8 in analysis 3.

Anal.

	1	2	3	4	5	6
Na$_2$O		0.73	0.30		0.41	
K$_2$O		0.13				
CaO	31.03	28.85	30.76	30.02	30.85	29.77
FeO	7.10	5.62	7.00	3.42	4.75	1.00
MnO	12.62	15.55	12.40	17.40	14.82	19.68
P$_2$O$_5$	39.28	38.39	39.62	[37.69]	39.55	37.79
H$_2$O	9.97	9.98	9.67	9.81	9.70	9.94
Insol.		1.31	0.55	1.66	0.50	1.07
Total	100.00	100.56	100.30	[100.00]	100.58	99.25
G.	3.09	3.15		3.07	3.016	

1. $Ca_2(Mn,Fe)(PO_4)_2 \cdot 2H_2O$, with Fe:Mn = 1:1.8. 2,3. Branchville.[9] No. 2 clear and lustrous; 3, friable and duller. 4. Branchville.[10] 5. Poland, Maine.[11] 6. Buckfield, Maine.[12]

Tests. B.B. glows, blackens, and fuses at a little over 4 to a dark yellowish brown mass. In C.T. gives off a little neutral water; turns first yellow, then dark brown, and becomes magnetic. Soluble in acids.

O c c u r. Originally from Branchville, Fairfield County, Connecticut, in pegmatite associated with eosphorite, triploidite, dickinsonite, reddingite, fillowite, and lithiophilite. With dufrenite and altered triphylite in pegmatite at Hühnerkobel near Rabenstein, Bavaria (*leucomanganite*). With rhodochrosite, eosphorite, reddingite, and other manganese phosphates at the Berry quarry, Poland, Maine; also at Buckfield, Maine.

A l t e r. Oxidizes to white or greenish products;[13] also found as an alteration of dickinsonite and as pseudomorphs after rhodochrosite.

Name. From the county in which the original locality is situated.

Ref.
1. Elements of Wolfe, *Am Min.*, **25**, 748 (1940), based on the structure cell. The original Branchville crystals of Brush and Dana (1879) and those of later description from other localties are of too poor quality to yield satisfactory morphological data. The transformation from the approximate elements of B. and D. to the new elements is problematical; {111} of B. and D. is probably the new {1̄11}, with old {010} and {100} = new {001} and {010}.
2. Wolfe (1940) on Buckfield crystals. Additional forms were reported by Brush and Dana (1879), but their transformed equivalents are not known.
3. Wolfe (1940) by Weissenberg method on Buckfield crystals.
4. Wolfe (1940); Brush and Dana (1879); *Am. J. Sc.*, **39**, 212 (1890) give 3.15 and 3.07.
5. Larsen (74, 1921).
6. Berman and Gonyer, *Am. Min.*, **15**, 375 (1930).
7. Landes, *Am. Min.*, **10**, 386 (1925).
8 Wolfe (1940).
9. Penfield analyses in Brush and Dana (1879).
10. Wells analysis in Brush and Dana (1890).
11 Gonyer analysis in Berman and Gonyer (1930).
12. Vassar analysis in Landes (1925).
13 Landes (1925).

40.2.3.2 COLLINSITE [$Ca_2(Mg,Fe)(PO_4)_2 \cdot 2H_2O$]. Poitevin (*Canada Dept. Mines, Geol. Sur., Bull. 46*, 1, 1927).

Cryst. Triclinic; pinacoidal—$\bar{1}$ (?).

Structure cell.[1] Space group $P\bar{1}$. a_0 5.70 kX, b_0 6.72, c_0 5.38; α 96°48$\frac{1}{2}$′, β 107°16$\frac{1}{2}$′, γ 104°32′; $a_0:b_0:c_0 = 0.8479:1:0.8002$. Cell contents $Ca_2(Mg,Fe)(PO_4)_2 \cdot 2H_2O$.

Habit. Layers composed of radial blades or lath-like fibers elongated [100].

Phys. Cleavage[2] {001} and {010}, fair. H. 3$\frac{1}{2}$. G. 2.99; 3.04 (calc.). Luster of aggregates silky. Color light brown. Translucent.

Opt. In transmitted light, pale yellow-brown to colorless.

Orientation	n	
X	1.632	Biaxial positive (+).
Y	1.642	$2V$ 80°.
Z	1.657	

Chem. A hydrated phosphate of calcium and magnesium, $Ca_2(Mg,Fe'')(PO_4)_2 \cdot 2H_2O$. Fe'' substitutes for Mg, with Fe:Mg = 1:2.28 in analysis 3.

Anal.

	1	2	3
CaO	32.98	32.18	32.03
MgO	8.24	6.34	9.31
FeO	6.44	6.86	7.31
P_2O_5	41.74	39.83	41.13
H_2O	10.60	12.28	9.69
Rem.		2.37	0.37
Total	100.00	99.86	99.84
G.		2.95	2.99

1. $Ca_2(Mg,Fe)(PO_4)_2 \cdot 2H_2O$ with Fe:Mg = 1:2.28. 2. François Lake, British Columbia.[3] Rem. is Fe_2O_3 0.80, Mn_2O_3 0.36, F 0.27, CO_2 0.23. 3. François Lake, B.C.[4] Rem. is insol.

ROSELITE

Tests. B.B. fuses with intumescence to a brownish slag. Easily soluble in acids.

Occur. Found with dark brown, fibrous carbonatian fluorapatite at François Lake, British Columbia, Canada. The minerals occur as concentric, alternate crusts with a radial fibrous structure on fragments of andesite in a vein-like deposit. Asphaltum is associated.

Name. After William H. Collins (1878–1937), former Director of the Geological Survey of Canada.

Ref.

1. Wolfe, *Am. Min.*, **25**, 746 (1940).
2. Four cleavages were noted by Poitevin (1927), but only two, at an angle of ~76°, were verified by Wolfe (1940).
3. Thompson analysis in Poitevin (1927).
4. Gonyer analysis in Wolfe (1940).

40.2.4 ROSELITE GROUP

40.2.4.1 ROSELITE [$(Ca,Co)_2(Co,Mg)(AsO_4)_2 \cdot 2H_2O$]. Lévy (*Ann. Phil.*, **8**, 439, 1824; *Edinburgh J. Sc.*, **2**, 177, 1825).

Cryst.[1] Monoclinic; prismatic—$2/m$.

$a:b:c = 0.4390:1:0.4398$;[2] $\beta\ 100°53'$; $p_0:q_0:r_0 = 1.0018:0.4319:1$

$r_2:p_2:q_2 = 2.3155:2.3198:1$; $\mu\ 79°07'$; $p_0'\ 1.0202$, $q_0'\ 0.4398$, $x_0'\ 0.1923$

Forms:[3]

		ϕ	ρ	ϕ_2	$\rho_2 = B$	C	A
c	001	90°00′	10°53′	79°07′	90°00′	....	79°07′
b	010	0 00	90 00	....	0 00	90°00′	90 00
a	100	90 00	90 00	0 00	90 00	79 07	....
n	140	30 06½	90 00	0 00	30 06½	84 34	59 53½
k	3.10.0	34 50	90 00	0 00	34 50	83 48½	55 10
l	130	37 42½	90 00	0 00	37 42½	83 22	52 17½
j	120	49 14	90 00	0 00	49 14	81 46½	40 46
m	110	66 41	90 00	0 00	66 41	80 01	23 19
o	012	41 10	16 17	79 07	77 49	12 11	79 22
p	122	57 57	39 39	54 55	70 12½	30 53	57 15½
t	254	51 57	41 44	54 55	65 46½	33 42½	58 23½
r	$\bar{1}38$	21 27	10 03	86 17½	80 39	11 46	86 20½
s	$\bar{1}44$	− 8 07	23 57	93 35½	66 18	27 33	93 17½
q	$\bar{1}22$	−35 51	28 29	107 38	67 15½	35 52	106 13

Structure cell.[4] Space group $P2_1/c$. $a_0\ 5.60\ kX$, $b_0\ 12.80$, $c_0\ 5.60$; $\beta\ 100°45'$; $a_0:b_0:c_0 = 0.4374:1:0.4374$. Cell contents $Ca_4(Co,Mg)_2(AsO_4)_4 \cdot 4H_2O$.

Habit. Short prismatic [001] or, rarely, thick tabular $\{001\}$; the terminations are usually simple and of monoclinic or orthorhombic aspect with large $\{122\}$ and $\{\bar{1}22\}$. Common forms: $c\ b\ a\ m\ n\ p$. Also in druses of interlocking crystals and as spherical aggregates.

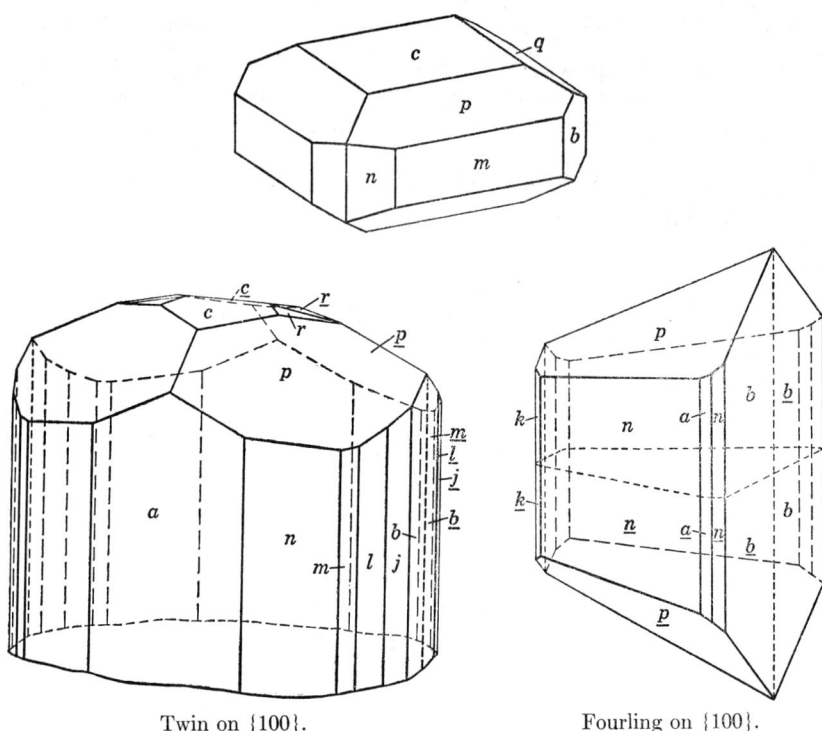

Twin on {100}.   Fourling on {100}.

Twinning.[5] On {100}, very common, with {100} as composition plane; also as fourlings.

P h y s. Cleavage {010} perfect and easy. H. $3\frac{1}{2}$. G. 3.50–3.74, increasing with Co content; 3.65 (calc.).[7] Luster vitreous. Color pink or pale rose to dark rose, darkest in material relatively high in Co. Transparent to translucent.

O p t.[6] In transmitted light, rose in color. The indices of refraction and optical orientation vary with the amount of Co in substitution for Ca, Mg. The darker colored crystals often show marked color zoning, due to variation in composition, with accompanying variation in the optical properties.

ORIENTATION	n(Na) (light rose)	PLEOCHROISM	
$X \wedge c \sim 0°$ to $4°$	1.694 ± 0.003	Pale rose	Biaxial positive (+).
Y b	1.704 ± 0.003	Paler rose	Dispersion, $r < v$,
$Z \wedge c \sim 90°$ to $+91°$	1.719 ± 0.003	Palest rose	marked.
$2V$	$75°$		
	(dark rose)		
X b	1.725 ± 0.003	Dark rose	Dispersion, $r < v$,
$Y \wedge c +12°$ to $+20°$	1.728 ± 0.003	Pale rose	weak.
$Z \wedge c -78°$ to $-70°$	1.735 ± 0.003	Paler rose	
$2V$	$60°$		

BRANDTITE

Chem. A hydrated arsenate of calcium, cobalt, and magnesium, $(Ca,Co)_2(Co,Mg)(AsO_4)_2 \cdot 2H_2O$. Co, Ca, and Mg substitute mutually. Roselite is isostructural with brandtite, $Ca_2Mn(AsO_4)_2 \cdot 2H_2O$.

Anal.

	1	2	3	4	5	6
CaO	25.75	25.17	24.93	23.72	21.9	19.2
MgO	4.63	4.22	3.95	4.67	4.3	4.8
CoO	8.60	10.03	10.56	12.45	[12.1]	[15.9]
As_2O_5	52.75	52.41	52.93	49.96	50.9	49.6
H_2O	8.27	8.22	8.35	9.69	10.8	10.5
Total	100.00	100.05	100.72	100.49	[100.0]	[100.0]
G.		3.56		3.46	3.506	3.738

1. $Ca_2(Co,Mg)(AsO_4)_2 \cdot 2H_2O$ with Co:Mg = 1:1. 2,3. Daniel mine, Schneeberg.[8]
4. Schneeberg.[9] 5. Daniel mine, Schneeberg.[10] Analysis on 24.5 mg. 6. Rappold mine, Schneeberg.[10] Analysis on 37.6 mg.

Tests. B.B. easily fusible. On slight heating splinters and turns dark blue, becoming rose again on cooling. Easily soluble in acids.

Occur. First found about 1800 in the Rappold mine and later (1874) in the Daniel mine at Schneeberg, Saxony. The mineral occurs with chalcedony and drusy quartz in cavities in quartzose vein material. Reported from Schapbach, Bavaria.[11]

Name. After Gustav Rose (1798–1873), Professor of Mineralogy in the University of Berlin.

Ref.

1. Originally described as orthorhombic by Lévy (1824). Later made triclinic and pseudo-orthorhombic by Schrauf, *Min. Mitt.*, 137 (1874), and so accepted by Dana (810, 1892) and Goldschmidt (7, 127, 1922), but shown to be monoclinic by Peacock, *Am. Min.*, 21, 589 (1936), and, by x-ray study, by Wolfe, *Am. Min.*, 25, 750 (1940).
2. Angles and orientation of Peacock (1936) in the unit of the structure cell. Transformation: Peacock to new, $\bar{1}00/010/001$; Schrauf (1874) to Peacock, $0\bar{1}4/300/030$; Dana to Peacock, $\bar{1}04/0\bar{3}0/300$.
3. Peacock (1936). Schrauf's form list contains fictitious forms due to wrongly assumed orthorhombic pseudo-symmetry.
4. Wolfe (1940) by Weissenberg method.
5. Of Schrauf's seven twin laws only this one is confirmed by Peacock (1936); 180° rotation about [001], 180° rotation about [301], and reflection on $\{\bar{1}03\}$ are further theoretically probable twin operations. See also Wolfe (1940).
6. Peacock (1936); see also Larsen (128, 1921).
7. For Co:Mg = 1:1. Wolfe (1940) obtained G. 3.695 on a dark rose-colored crystal.
8. Winkler analysis in Weisbach, *Jb. Min.*, 407 (1877).
9. Winkler, *Zs. pr. Chem.*, **10**, 190 (1874).
10. Schrauf (1874).
11. Sandberger (1885).

40.2.4.2 **BRANDTITE** [$Ca_2Mn(AsO_4)_2 \cdot 2H_2O$]. *Nordenskiöld* (*Ak. Stockholm, Öfv.*, **45**, 418, 1888).

Cryst.[1] Monoclinic; prismatic—$2/m$.

$a:b:c = 0.4360:1:0.4475;$ [2] $\beta\ 99°36\frac{1}{2}';$ $p_0:q_0:r_0 = 1.0264:0.4412:1$

$r_2:p_2:q_2 = 2.2665:2.3262:1;$ $\mu\ 80°23\frac{1}{2}';$ $p_0'\ 1.0410,\ q_0'\ 0.4475,\ x_0'\ 0.1693$

Forms: [3]

	φ	ρ	φ₂	ρ₂ = B	C	A
c 001	90°00′	9°36½′	80°23½′	90°00′	.....	80°23½′
b 010	0 00	90 00	.....	0 00	90°00′	90 00
a 100	90 00	90 00	0 00	90 00	80 23½	.....
n 140	30 11	90 00	0 00	30 11	85 11	59 49
l 130	37 47½	90 00	0 00	37 47½	84 07½	52 12½
j 120	49 19	90 00	0 00	49 19	82 43½	40 41
h 340	60 11	90 00	0 00	60 11	81 40½	29 49
m 110	66 44½	90 00	0 00	66 44½	81 10½	23 15½
u 112	72 01½	35 57	55 24	79 34	26 57	56 03½
p 122	57 01½	39 25½	55 24	69 47	31 44	57 48½
v 142	37 37½	48 29½	55 24	53 37	43 08	62 48
q $\bar{1}22$	−38 07½	29 38	109 21	67 06½	36 17	107 46½

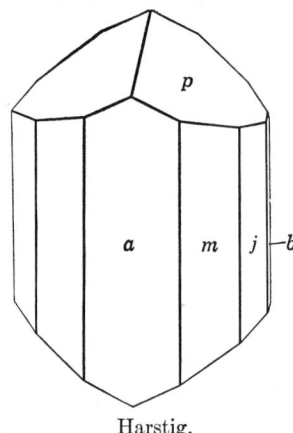

Harstig.

Structure cell.[4] Space group $P2_1/c$. a_0 5.65 kX, b_0 12.80, c_0 5.65; β 99°30′; $a_0:b_0:c_0 = 0.4412:1:0.4412$. Cell contents $Ca_4Mn_2(AsO_4)_4 \cdot 4H_2O$.

Habit. Stout prismatic [001]. Common forms: b a m n p. {100} striated [001]. Crystals often united in radiated groups; also in rounded or reniform masses with a radial fibrous structure.

Twinning. On {100}, common.

P h y s. Cleavage {010}, perfect. H. 3½.[5] G. 3.67; 3.70 (calc.). Luster vitreous. Colorless to white. Transparent to translucent.

O p t. In transmitted light, colorless.

Orientation [6]		n [6]	n [7]	
X	b	1.709 ± 0.003	1.707	Biaxial positive (+).
Y ∧ c	6°	1.711 ± 0.003		2V 23° ± 1° (meas.).
Z ∧ c	84°	1.724 ± 0.003	1.729	r < v, rather strong.

C h e m. A hydrated arsenate of calcium and manganese, $Ca_2Mn(AsO_4)_2 \cdot 2H_2O$. Brandtite is isostructural with roselite.

Anal.

	CaO	MgO	PbO	MnO	As₂O₅	H₂O	Rem.	Total
1.	24.98			15.80	51.19	8.03		100.00
2.	25.07	0.90	0.96	14.03	50.48	8.09	0.18	99.71

1. $Ca_2Mn(AsO_4)_2 \cdot 2H_2O$. 2. Harstig, Sweden.[8] Rem. is FeO 0.05, P₂O₅ 0.05, Cl 0.04, insol. 0.04.

Tests. Easily fusible to a brown bead. Soluble in dilute acids.

O c c u r. Found first at the Harstig mine at Pajsberg near Persberg, Vermland, Sweden. Associated with barite, calcite, sarkinite, flinkite, caryopilite, galena, and native lead. Also from Sterling Hill, New Jersey.[9]

N a m e. After the Swedish chemist George Brandt (1694–1768).

Ref.
1. Originally described as triclinic by Nordenskiöld in Lindström, *Geol. För. Förh.*, **13**, 123 (1891), but shown to be monoclinic by Aminoff, *Geol. För. Förh.*, **41**, 161 (1919), and by x-ray study, by Wolfe, *Am. Min.*, **25**, 750 (1940).
2. Angles and orientation of Aminoff (1919) in the unit of the structure cell. Transformation: Aminoff to new, $\frac{1}{2}00/010/001$; from the triclinic interpretation of Dana (811, 1892) and Goldschmidt (1, 228, 1913) to Aminoff (1919), $\overline{1}04/0\overline{3}0/300$.
3. Forms of Aminoff (1919) relettered to correspond to roselite. Uncertain forms: λ {343} Dana = {$\overline{3}86$}; Dana's X {$4\overline{3}3$} has a false symbol and probably is v {142}.
4. Wolfe (1940) by Weissenberg method.
5. Wolfe (1940), correcting the earlier value of 5–5½.
6. Larsen (50, 1921).
7. Aminoff (1919), who gives $Y = b$.
8. Lindström (1891).
9. Gaines, priv. comm. (1949).

Reddingite Series

40.2.5.1   **R E D D I N G I T E** [(Mn,Fe)$_3$(PO$_4$)$_2 \cdot$3H$_2$O].   Brush and Dana (*Am. J. Sc.*, **16**, 120, 1878; *ibid.*, **17**, 365, 1879).
40.2.5.2   **P H O S P H O F E R R I T E** [(Fe,Mn)$_3$(PO$_4$)$_2 \cdot$3H$_2$O].   Laubmann and Steinmetz (*Zs. Kr.*, **55**, 569, 1920) = Reddingite Steinmetz (*Zs. Kr.*, **64**, 405, 1926).

C r y s t.[1] Orthorhombic; dipyramidal—$2/m\ 2/m\ 2/m$.

$$a:b:c = 0.9428:1:0.8620; \qquad p_0:q_0:r_0 = 0.9143:0.8620:1$$

$$q_1:r_1:p_1 = 0.9428:1.0937:1; \qquad r_2:p_2:q_2 = 1.1601:1.0607:1$$

Forms:[2]

	ϕ	$\rho = C$	ϕ_1	$\rho_1 = A$	ϕ_2	$\rho_2 = B$
c 001	....	0°00'	0°00'	90°00'	90°00'	90°00'
b 010	0°00'	90 00	90 00	90 00	....	0 00
a 100	90 00	90 00	....	0 00	0 00	90 00
k 110	46 41	90 00	90 00	43 19	0 00	46 41
w 021	0 00	59 53	59 53	90 00	90 00	30 07
S 103	90 00	16 57	0 00	73 03	73 03	90 00
d 101	90 00	42 26	0 00	47 34	47 34	90 00
f 201	90 00	61 19½	0 00	28 40½	28 40½	90 00
g 301	90 00	69 58	0 00	20 02	20 02	90 00
p 111	46 41	51 29	40 45½	55 18	47 34	57 32
t 122	27 56½	44 18	40 45½	70 54	65 26	51 54½
u 212	64 45½	45 18½	23 19	49 59	47 34	72 21½
z 322	57 51	58 18½	40 45½	43 54½	36 06	63 04½
y 211	64 45½	63 41	40 45½	35 50	28 40½	67 32

Less common:

       R 102 r 112 s 223 v 233 x 311

Structure cell.[3] Space group $Pmna$. a_0 9.52 kX, b_0 10.06, c_0 8.70 (reddingite, anal. 5); $a_0:b_0:c_0 = 0.9463:1:0.8648$. Cell contents (Mn,Fe)$_{12}$(PO$_4$)$_8 \cdot$12H$_2$O.

Habit. Octahedral with large {111}, or tabular {010}; the crystals often in parallel grouping. Also massive, granular; coarsely fibrous.

P h y s. Cleavage {010}, poor. Fracture uneven. Brittle. H. 3–3½. G. 3.0–3.2;[4] 3.24 (calc. for Mn:Fe = 3:1). Luster vitreous to subresinous.

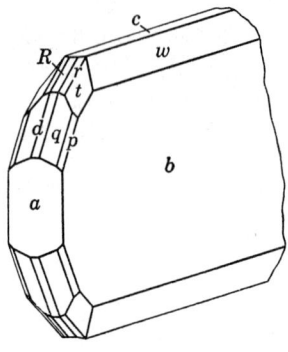
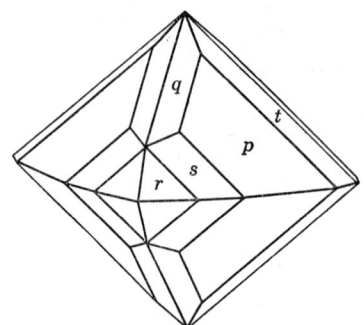

Phosphoferrite.   Reddingite. Branchville.

Color pinkish white or pale rose-pink to yellowish white and colorless (*reddingite*); pale green (*phosphoferrite*); sometimes dark reddish brown from alteration. Translucent to transparent.

O p t. In transmitted light, colorless or faintly tinted pink, etc. The indices of refraction increase with increasing content of iron in the series.

ORIEN-TATION	Reddingite Buckfield [5]	Reddingite Branch-ville [6]	Reddingite Poland [7]	Phospho-ferrite Hagendorf [16]	
nX a	1.643	1.651	1.655	1.672	Biaxial positive (+).
nY b	1.648	1.656	1.662	1.680	$r > v$, distinct.
nZ c	1.674	1.683	1.683	1.700	PLEOCHROIC:
$2V$		41° (meas.)	65° (meas.)	68° (meas.)	X Colorless
Fe:Mn	1:22		1:3	2.7:1	Y Pinkish brown
					Z Pale yellow (Buckfield)

C h e m. A hydrated phosphate of divalent manganese and iron, $(Mn,Fe)_3(PO_4)_2 \cdot 3H_2O$. Mn'' and Fe'' substitute mutually, and a series extends from at least Fe:Mn = 1:22 (anal. 2) to at least Fe:Mn = 2.7:1 (anal. 7). The species names reddingite and phosphoferrite are applied to the parts of the series with Mn > Fe and Fe > Mn, respectively.

Anal.

	1	2	3	4	5	6	7	8
Na_2O			0.31		tr.			
CaO		0.73	0.78	0.71	0.15	0.63	1.20	
FeO		2.19	5.43	7.89	12.68	17.13	37.52	52.38
MnO	52.05	48.15	46.29	43.22	38.36	34.51	13.63	
P_2O_5	34.73	34.54	34.52	35.16	34.52	34.90	34.39	34.49
H_2O	13.22	13.14	13.08	12.27	13.16	13.18	13.32	13.13
Rem.		0.65			1.40	0.13		
Total	100.00	99.40	100.41	99.25	100.27	100.48	100.06	100.00
G.			3.10		3.136	3.204	3.10	

1. $Mn_3(PO_4)_2 \cdot 3H_2O$. 2. Reddingite. Buckfield, Maine.[8] Rem. is MgO 0.35, insol. 0.30. 3. Reddingite. Branchville, Connecticut.[9] Average of two recalculated after deducting 12 per cent quartz. Has Li_2O tr. 4. Reddingite. Branchville.[10] Recalculated after deducting 4.4 per cent quartz. 5. Reddingite. Poland, Maine.[11] Rem. is Fe_2O_3 0.95, insol. 0.45, K_2O tr. 6. Reddingite. Branchville, Conn.[12] Rem. is quartz. 7. Phosphoferrite. Hagendorf, Bavaria.[13] Average of two. 8. $Fe_3(PO_4)_2 \cdot 3H_2O$.

Tests. B.B. fuses easily to a blackish brown non-magnetic globule. In C.T. yields water; first whitens, then turns yellow and finally brown but does not become magnetic. Soluble in acids.

Occur. Reddingite was found originally at Branchville, Fairfield County, Connecticut, associated with lithiophilite and other manganese phosphates in pegmatite. Also found similarly at Poland and especially at Buckfield, in Maine. Phosphoferrite has been found only at Hagendorf near Pleystein, Oberpfalz, Bavaria, with triploidite, ludlamite, and vivianite in pegmatite.

Artif.[14] A hydrated manganese phosphate perhaps identical with reddingite has been obtained by reaction of manganous carbonate with excess phosphoric acid.

Name. Reddingite from Redding township, in which the Branchville locality is situated. Phosphoferrite [15] in allusion to the composition.

Ref.

1. Elements re-computed by Palache, priv. comm. (1936), from all measured angles of Steinmetz, *Zs Kr.*, **64**, 405 (1926), on Hagendorf crystals of phosphoferrite (anal. 7); unit of Brush and Dana (1878), *Am. J. Sc.*, **39**, 211 (1890), on Branchville reddingite in the orientation of the structure cell of Wolfe, *Am. Min.*, **25**, 752 (1940). Transformation: B. and D. to new, $00\bar{1}/010/100$. Data on the variation of angles with composition in the reddingite-phosphoferrite series are lacking; the original elements of B. and D. on the Branchville reddingite are inaccurate, owing to the poor quality of the measured crystals.
2. Steinmetz (1926) on Hagendorf phosphoferrite and confirmed by Palache (1936); Brush and Dana (1878) give b, f, p, r, s, t, q for Branchville reddingite.
3. Wolfe (1940) by Weissenberg method on Poland reddingite (anal. 5).
4. The G. and the nature of its variation with composition in this series is not well-defined. Wolfe (1940) gives 3.22 for Poland material with Mn:Fe = 3:1.
5. Landes, *Am. Min.*, **10**, 387 (1925).
6. Larsen (126, 1921).
7. Berman and Gonyer, *Am. Min.*, **15**, 379 (1930).
8 Vassar analysis in Landes (1925).
9. Wells analyses in Brush and Dana (1878).
10. Wells analysis in Brush and Dana (1879).
11 Gonyer analysis in Berman and Gonyer (1930).
12. Wells analysis in Brush and Dana (1890).
13. Steinmetz (1926).
14. Debray, *Ann chim. phys.*, **61**, 433 (1861).
15. This name, originally proposed by Laubmann and Steinmetz (1920) was later discarded by Steinmetz (1925) when the relation to reddingite was recognized but is here retained for the iron-rich part of the series with reddingite.
16. Mrose, priv comm. (1948).

40.2.6 **LANDESITE** [$Fe_6Mn_{20}(PO_4)_{16} \cdot 27H_2O$ (?)]. *Berman* and *Gonyer* (*Am. Min.*, **15**, 384, 1930).

In rough, pseudomorphous (?) octahedral-like crystals with one good cleavage and a poor cleavage at right angles to the first. H. 3–3½. G. 3.026. Color brown.

Opt. In transmitted light, yellowish brown.

Orientation	n	Pleochroism	
X ($\perp$ inferior cleavage)	1.720	Dark brown	Biaxial negative ($-$).
Y	1.728	Light brown	$2V$ large.
Z ($\perp$ best cleavage)	1.735	Yellow	

Chem. A hydrated phosphate of trivalent iron and divalent manganese of uncertain formula, perhaps $Fe_6Mn_{20}(PO_4)_{16} \cdot 27H_2O$. Small amounts of Ca and Mg apparently substitute for Mn.

Anal.

	1	2
CaO		1.39
MgO		3.07
MnO	40.33	33.65
Fe_2O_3	13.61	13.91
Mn_2O_3		2.69
P_2O_5	32.26	31.94
H_2O	13.80	13.60
Insol.		0.13
Total	100.00	100.38
G.		3.026

1. $Fe_6Mn_{20}(PO_4)_{16} \cdot 27H_2O$. 2. Poland, Maine.[1] FeO not determined.

Occur. Found with rhodochrosite, eosphorite, fairfieldite, lithiophilite, and apatite as an alteration product of reddingite in a granite pegmatite at the Berry Quarry, Poland, Maine.

Name. After Kenneth K. Landes (1899–), Professor of Geology at the University of Michigan, who has done much work on the pegmatites of Maine.

Ref.
1. Gonyer analysis in Berman and Gonyer (1930).

40.2.7 Stewartite. *Schaller (Washington Ac. Sc., J., 2, 143, 1912).*

Probably triclinic. As minute crystals and tufts of fibers. Color brownish yellow. G. 2.94. Optically[1] biaxial negative (−) with $2V$ large. Extinction inclined on all crystal edges. Indices of refraction and pleochroism: nX 1.63 (colorless), nY 1.658 (pale yellow), nZ 1.66 (yellow); $r < v$, strong.

In composition essentially a hydrated phosphate of manganese. Originally found with palaite and hureaulite as an alteration product of lithiophilite at the Stewart mine, Pala, San Diego County, California. Also reported[2] from the Palermo and Fletcher pegmatites, North Groton, New Hampshire, and from Newry, Maine. A mineral[3] found with hureaulite at Vilate near Chanteloube, Haute-Vienne, France, apparently is identical with stewartite.

Ref.
1. Mrose, priv. comm. (1949), on probably authentic material from Palermo, New Hampshire. Principal powder lines (Fe/Mn): 10.06, 3.46, 5.82, 4.41, 2.81.
2. Mrose (1949).
3. Lacroix (**4**, 506, 1910) ["Mineral A"].

40.2.8 Salmonsite $[Mn_9Fe_2(PO_4)_8 \cdot 14H_2O \; (?)]$. *Schaller (Washington Ac. Sc., J., 2, 143, 1912).*

Probably orthorhombic. In fibrous masses with two pinacoidal cleavages. H. 4. G. 2.88. Color buff.

ANAPAITE

Opt. In transmitted light, yellow.

	n	PLEOCHROISM	
X	1.655 ± 0.005	Nearly colorless	Biaxial positive (+).
Y	1.66 ± 0.01	Yellow	2V, very large.
Z elong.	1.670 ± 0.005	Orange-yellow	$r > v$, strong.

Chem. A hydrated phosphate of trivalent iron and divalent manganese, probably $Mn_9Fe_2(PO_4)_8 \cdot 14H_2O$. Analysis gave:

	CaO	MnO	FeO	Fe_2O_3	P_2O_5	H_2O+	H_2O-	Insol.	Total
1.		39.45		9.87	35.09	15.59			100.00
2.	1.06	37.74	0.13	9.53	34.86	15.30	0.43	1.40	100.45

1. $Mn_9Fe_2(PO_4)_8 \cdot 14H_2O$. 2. Stewart mine.

Occur. Found as an alteration product of hureaulite in pegmatite at the Stewart mine, Pala, San Diego County, California.

Name. After Frank A. Salmon, formerly of Pala. Needs verification.

40.2.9 **ANAPAITE** [$Ca_2Fe(PO_4)_2 \cdot 4H_2O$]. *Sachs* (*Ak. Berlin, Ber.*, **18**, 1902). Tamanit *Popoff* (*Zs. Kr.*, **37**, 267, 1902; *Ac. Sc. St. Pétersbourg, Mus. geol., Trav.*, **4**, 49, 1910). Anapäite. Messelit *Muthmann* (*Zs. Kr.*, **17**, 93, 1890).

Cryst.[1] Triclinic; pinacoidal—$\bar{1}$.

$a:b:c = 0.9401:1:0.8575; \quad \alpha\ 101°34\tfrac{1}{2}',\ \beta\ 104°05\tfrac{1}{2}',\ \gamma\ 71°03\tfrac{1}{2}'$

$p_0:q_0:r_0 = 0.9447:0.8793:1; \quad \lambda\ 82°23',\ \mu\ 78°54',\ \nu\ 106°52'$

$p_0'\ 0.9827,\ q_0'\ 0.9147,\ x_0'\ 0.2511,\ y_0'\ 0.1379$

Forms:[2]

		ϕ	ρ	A	B	C
c	001	61°13½'	15°59'	78°54'	82°23'	.....
b	010	0 00	90 00	106 52	......	82°23'
a	100	106 52	90 00	......	106 52	78 54
W	0$\bar{1}$1	162 05	39 13½	68 51	127 00½	60 36½
r	101	97 02½	42 26	48 20	94 44½	30 34
s	$\bar{1}$01	−58 28	33 35	122 21	73 11	43 27
P	1$\bar{1}$1	131 42½	57 56	39 44	124 19	53 57½
o	$\bar{1}$11	−125 30	40 15½	113 14½	112 02½	56 09½

Structure cell.[3] Space group $P\bar{1}$. a_0 6.41 kX, b_0 6.88, c_0 5.86; [α 101°34½', β 104°05½', γ 71°03½']; $a_0:b_0:c_0 = 0.9317:1:0.8512$. Cell contents $Ca_2Fe(PO_4)_2 \cdot 4H_2O$.

Habit. Tabular {110}. Common forms: $c\ b\ a\ W\ r\ s$. Also as rosette-like aggregates or crusts of subparallel crystals.

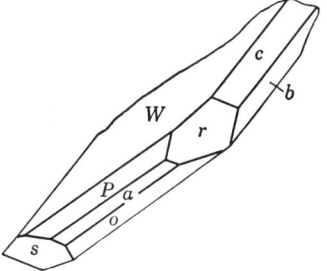

Phys. Cleavage {001} perfect, {010} distinct. H. 3½. G. 2.81;[4] 2.80 (calc.). Luster vitreous. Color green to greenish white. Streak white. Transparent.

O p t.[5] In transmitted light, colorless to pale green.

ORIENTATION

	ϕ	ρ	n	
X	−119°	81°	1.602 ± 0.003	Biaxial positive (+).
Y	147°	70°	1.613 ± 0.003	$2V$ 54° ± 2°.
Z	−6°	21°	1.649 ± 0.003	$r > v$, perceptible.

C h e m. A hydrated phosphate of calcium and divalent iron, $Ca_2Fe(PO_4)_2 \cdot 4H_2O$.

Anal.

	1	2	3	4	5
CaO	28.18	27.77	27.72	28.32	26.35
FeO	18.05	18.07	20.00	17.49	18.66
P_2O_5	35.67	35.51	34.50	34.36	35.31
H_2O	18.10	18.47	18.33	18.64	[18.48]
Rem.				1.46	1.20
Total	100.00	99.82	100.55	100.27	[100.00]
G.	2.80	2.81	2.812	2.85	

1. $Ca_2Fe(PO_4)_2 \cdot 4H_2O$. 2. Taman.[6] 3. Taman.[7] Average of two. 4. Taman.[8] Rem. is Fe_2O_3 0.84, CO_2 0.62. Average of several. 5. Taman.[9] Rem. is MgO 0.81, Fe_2O_3 0.39.

Tests. On heating, loses acid water above 120°. Easily soluble in HCl and HNO_3.

O c c u r. Found in crevices in oolitic iron ore at Sheljesny Rog near Anapa on the Taman peninsula on the Black Sea, U.S.S.R. Also found in nodular concretions in Miocene clay at Prats-Sampsor, Spain,[12] and in well cores in Kings County, California.

Alter. To an ill-defined hydrous phosphate of calcium and ferric iron (see *mitridatite*).[10]

Name. From the locality.

MESSELITE. *Muthmann* (*Zs. Kr.*, **17**, 93, 1890). A substance close in composition to $Ca_2Fe(PO_4)_2 \cdot 2\frac{1}{2}H_2O$ found as rudely stellate groups of indistinct triclinic crystals in a bituminous clay rock at Messel in Hesse, Germany. Later shown [11] to be anapaite altered in part to collinsite.

Ref.

1. Angles of Palache, *Zs. Kr.*, **86**, 280 (1933); unit and orientation of Wolfe, *Am. Min.*, **25**, 788 (1940) conforming with the structure cell. Transformation: Palache to new, 001/010/$\overline{1}$00. Palache's orientation differs from those of Sachs (1902) and Popoff (1902). Transformations: Sachs to Palache, 112/1$\overline{1}$0/00$\overline{2}$; Popoff to Palache, 101/020/-$\overline{1}$01.
2. Palache (1933).
3. Wolfe (1940) by Weissenberg method.
4. Identical values of Sachs (1902) and Popoff (1902).
5. Wolfe (1940) and Larsen (39, 1921).
6. Sachs (1902).
7. Popoff (1910).
8. Loczka, *Zs. Kr.*, **37**, 438 (1902).
9. Chukhrov, *Trans. Lomonosoff Inst*, Ac. Sc., U S.S.R., no. 7, 273 (1936).
10. Chukhrov, *Trans. Lomonosoff Inst.*, Ac. Sc., U.S.S.R., no. 10, 131 (1937).
11. Wolfe (1940).
12. Pardillo, *Treballs Mus. Cienc. Nat. Barcelona*, **9**, no. 1 (1924).

40.2.10 PARAHOPEITE [$Zn_3(PO_4)_2 \cdot 4H_2O$]. *Spencer (Min. Mag.,* **15,** 18, 1908).

Cryst.[1] Triclinic; pinacoidal—$\bar{1}$.

$a:b:c = 0.7747:1:0.7017;$ $\alpha\ 93°17\frac{1}{2}',\ \beta\ 91°55',\ \gamma\ 91°19'$

$p_0:q_0:r_0 = 0.9096:0.7012:1;$ $\lambda\ 86°39\frac{1}{2}',\ \mu\ 88°00',\ \nu\ 88°34\frac{1}{2}'$

$p_0'\ 0.9117,\ q_0'\ 0.7028,\ x_0'\ 0.0335,\ y_0'\ 0.0584$

Forms:[2]

		φ	ρ	A	B	C
c	001	29°50½'	3°51'	88°00'	86°39½'	.....
b	010	0 00	90 00	88 34½	.....	86°39½'
a	100	88 34½	90 00	.....	88 34½	88 00
m	110	51 29	90 00	37 05½	51 29	86 25½
M	1̄10	126 44	90 00	38 09½	126 44	90 28
d	011	2 31	37 18½	87 36½	52 44	33 55½
e	032	1 43½	48 04	87 39½	41 58	44 41½
f	021	1 18½	55 40½	87 44½	34 21	52 18½
g	01̄2	173 28½	16 26	88 33½	106 19½	19 40
D	01̄1	177 01½	32 50	89 09½	122 47	36 07½
p	111	50 19	50 50	52 29½	60 19½	47 14½
u	322	60 24½	58 10	41 30	65 12	54 52
P	11̄1	123 20½	48 31	52 34	114 19	48 52
r	1̄11	−49 55½	48 55½	124 22½	60 58	48 21
R	1̄1̄1	−127 14	47 47½	53 04½	116 37½	51 21½

Less common and rare:

k 160	K 1̄60	q 131	x 1̄43	s 1̄21	S 1̄2̄1
l 120	L 1̄20	v 423	w 1̄12	z 5̄12	Q 1̄31
n 310	h 203	t 231	y 1̄62	j 2̄71	

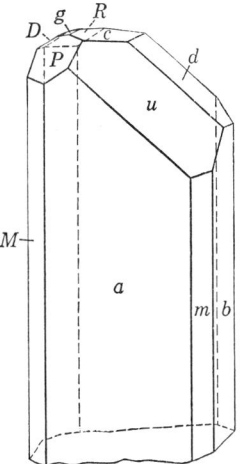

Structure cell.[3] Space group $P\bar{1}$. $a_0\ 5.755\ kX$, $b_0\ 7.535,\ c_0\ 5.292;\ [\alpha\ 93°17\frac{1}{2}',\ \beta\ 91°55',\ \gamma\ 91°19'];$ $a_0:b_0:c_0 = 0.7638:1:0.7023.$ Cell contents $Zn_3(PO_4)_2 \cdot 4H_2O.$

Habit. Elongated [001] and tabular {100}, the crystals often as sub-parallel aggregates or as fan-like or tufted groups.

Twinning.[2] On {100}, polysynthetic, common.

Phys. Cleavage {010} perfect. H. $3\frac{3}{4}$. G. 3.31; 3.304 (calc.). Luster vitreous, on the cleavage somewhat pearly. Colorless and transparent.

O p t.[4] In transmitted light, colorless.

ORIENTATION	n	
X near a	1.614 ± 0.003	Biaxial positive (+).
$Y \wedge c$ on $\{100\}$ $30°$	1.625 ± 0.003	$2V$ nearly $90°$.
Z	1.637 ± 0.003	$r < v$, perceptible.

There is no noticeable change in the optical properties on heating up to about 163°, at which point the crystals commence to become white and opaque.

C h e m. Hydrated zinc orthophosphate, $Zn_3(PO_4)_2 \cdot 4H_2O$. Dimorphous with hopeite. On heating, no water is lost at 139° and almost all at 233°.

Anal.

	1	2	3
ZnO	53.28	53.0	54.69
P_2O_5	31.00	31.6	30.46
H_2O	15.72	15.6	15.31
Total	100.00	100.2	100.46
G.	3.30	3.31	3.236

1. $Zn_3(PO_4)_2 \cdot 4H_2O$. 2. Rhodesia.[5] A duplicate determination of H_2O (ignition) gave 15.8 per cent. 3. Salmo, British Columbia.[7]

O c c u r. A secondary mineral, found with hemimorphite, tarbuttite, pyromorphite, limonite at the Broken Hill mines, northwest Rhodesia. Also from the Hudson Bay mine, Salmo, Nelson district, British Columbia, with hopeite, spencerite, and hemimorphite.

A r t i f.[6] By reaction of hot solutions of zinc chloride and sodium ammonium phosphate.

N a m e. In allusion to the polymorphic relation to hopeite.

Ref.

1. Unit and orientation of Ledoux, Walker, and Wheatley, *Min. Mag.*, **18**, 101 (1919), but with elements re-computed by La Forge, priv. comm. (1936), using all but a few of their measured angles. The new elements are more consistent with the majority of the measured angles and are closer to the x-ray values.
2. Ledoux, *et al.* (1919).
3. Wolfe, *Am. Min.*, **25**, 788 (1940), by Weissenberg method.
4. Indices of Larsen (117, 1921).
5. Spencer (1908).
6. Mrose, priv. comm. (1948).
7. Walker, *Univ. Toronto Stud., Geol. Ser.*, no. 10, 16 (1918).

40.2.11 **H O P E I T E** [$Zn_3(PO_4)_2 \cdot 4H_2O$]. Brewster (*Edinburgh Phil. J.*, **6**, 184, 1822; *Trans. Roy. Soc. Edinburgh*, **10**, 107, 1826 (1823)). Prismatoidischer Zinkphyllit Breithaupt (38, 1832). Stilbite duovigésimale *Haüy* (cf. Des Cloizeaux, *Bull. soc. min.*, **2**, 133, 1879). α-Hopeite, β-Hopeite Spencer (*Min. Mag.*, **15**, 1, 1908).

C r y s t.[1] Orthorhombic; dipyramidal—$2/m\ 2/m\ 2/m$.

$a:b:c = 0.5761:1:0.2741;$[2] $p_0:q_0:r_0 = 0.4758:0.2741:1$

$q_1:r_1:p_1 = 0.5761:2.1018:1;$ $r_2:p_2:q_2 = 3.6483:1.7358:1$

HOPEITE

Forms:[3]

		ϕ	$\rho = C$	ϕ_1	$\rho_1 = A$	ϕ_2	$\rho_2 = B$
c	001	.....	0°00′	0°00′	90°00′	90°00′	90°00′
b	010	0°00′	90 00	90 00	90 00	.....	0 00
a	100	90 00	90 00	.....	0 00	0 00	90 00
s	230	49 10	90 00	90 00	40 50	0 00	49 10
m	110	60 03½	90 00	90 00	29 56½	0 00	60 03½
u	011	0 00	15 19½	15 19½	90 00	90 00	74 40½
e	031	0 00	39 26	39 26	90 00	90 00	50 34
f	101	90 00	25 26½	0 00	64 33½	64 33½	90 00
g	201	90 00	43 35	0 00	46 25	46 25	90 00
t	111	60 03½	28 46½	15 19½	65 21	64 33½	76 06
r	131	30 03½	43 32	39 26	69 49½	64 33½	53 24½
α	141	23 27½	50 05	47 38	72 13½	64 33½	45 17
ϑ	211	73 56	44 43	15 19½	47 27½	46 25	78 46

Less common:
y 190   n 130   o 250   μ 540   k 210   h 102   ε 161   β 321
x 290   z 7.18.0   j 560   w 320   d 061   q 121   ζ 171   γ 421

Structure cell.[4] Space group $Pnma$. a_0 10.64 kX, b_0 18.32, c_0 5.03; $a_0:b_0:c_0 = 0.5808:1:0.2745$. Cell contents $Zn_{12}(PO_4)_8 \cdot 16H_2O$.

Habit. Tabular {010} to prismatic [001], the crystals occurring singly or as tufted or divergent aggregates and crusts. The development of the

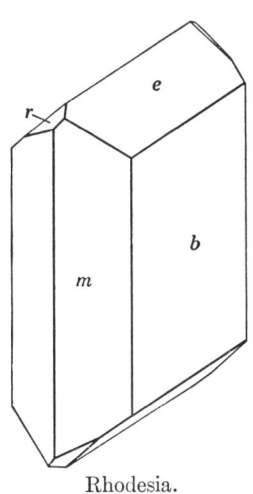
Rhodesia.

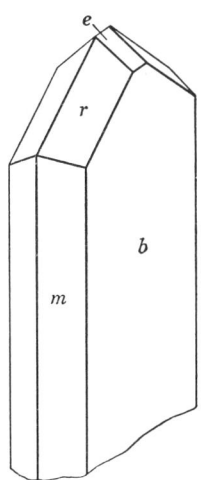

faces often is irregular and the crystals may simulate disphenoidal or hemimorphic symmetry. Also in reniform masses; compact.

Etching. Etch figures conform to holohedral symmetry.[5]

Twinning. Not observed.[6]

Phys. Cleavage {010} perfect, {100} good, {001} poor. Fracture uneven. Brittle. H. 3¼. G. 3.05 ± 0.05;[7] 3.08 (calc.). Luster vitreous;

on the {010} cleavage, pearly. Colorless to grayish white and pale yellow. Transparent to translucent. Streak white.

Opt. In transmitted light, colorless.

ORIENTATION	n (Rhodesia) [8]	n (Rhodesia) [9]	
X a	1.589	1.574 ± 0.003	Biaxial negative (−).
Y c	1.598	1.582 ± 0.003	$r < v$, perceptible.
Z b	1.599	1.582 ± 0.003	
$2V$	37°	Small	

Sections on {100} show a marked zonal growth structure, the optic angle, birefringence and probably the indices of refraction varying in successive zones (see α- and β-hopeite, beyond).

Chem. Hydrated zinc orthophosphate, $Zn_3(PO_4)_2 \cdot 4H_2O$. Dimorphous with parahopeite. On heating,[5] about half of the water is lost at temperatures up to about 135° and the remainder at about 240°.

Anal.

	1	2	3
ZnO	53.28	52.1	51.9
P_2O_5	31.00	31.8	[31.9]
H_2O	15.72	16.1	16.2
Total	100.00	100.0	[100.00]
G.	3.08	3.04	3.03

1. $Zn_3(PO_4)_2 \cdot 4H_2O$. 2. Broken Hill mine, Rhodesia (α-*hopeite*).[5] 3. Broken Hill mine, Rhodesia (β-*hopeite*).[5]

Tests. In C.T. gives off much water and turns white (both hot and cold) and opaque with a porcelaneous aspect. B.B. fuses readily to a colorless bead. Easily soluble in dilute HCl.

Occur. Originally found about 1820 sparingly associated with hemimorphite in the zinc mine of Altenberg (Vieille Montagne) in the Moresnet district between Aix-la-Chapelle and Liége in the Rheinland, Germany. Later found abundantly in bone-breccia in a limestone cave at the Broken Hill mine, northwestern Rhodesia,[5] associated with tarbuttite, hemimorphite, smithsonite, vanadinite. With spencerite in the Hudson Bay mine, Salmo, Nelson district, British Columbia.[10]

Artif.[11] Obtained in crystals by reaction of phosphate solutions with zinc sulfate solution, or from a solution of zinc phosphate in acetic acid.

Name. After Thomas Charles Hope (1766–1844), Professor of Chemistry at the University of Edinburgh.

α-HOPEITE, β-HOPEITE. Spencer (*Min. Mag.*, **15**, 1, 1908).
Hopeite crystals from Rhodesia show a zonal growth structure, the material of adjacent zones differing somewhat in properties although apparently of identical composition (anals. 2,3). α-Hopeite has a higher specific gravity and birefringence, with $X = a$, $Y = c$ and $2E \sim 58\frac{1}{2}°$. β-Hopeite has $X = a$, sometimes with $Y = c$ and $2E \sim 32\frac{1}{2}°$ or sometimes $Y = b$ and $2E \sim 20°$. On heating, α-hopeite breaks down at 105° and β-hopeite at 139°. The x-ray diffraction patterns of the two materials are identical, and the differences in properties and thermal behavior may be due to a slight variation in water content.

HOPEITE 737

Ref.
1. Crystal class established as holohedral by x-ray evidence, etch figures, and lack of piezoelectricity (see Wolfe, *Am. Min.*, **25**, 795 (1940), and Spencer, *Min. Mag.*, **15**, 1 (1908)). Lower symmetry was suggested by Ungemach, *Bull. soc. min.*, **33**, 132 (1910), on morphological grounds, but this appears to be due to fortuitous irregularities in the relative development of the crystal faces.
2. Angles of Spencer (1908) in the unit and orientation of the structure cell of Wolfe (1940). Spencer's elements are close to the x-ray elements and to the mean of the morphological elements reported by Walker, *J. Washington Ac. Sc.*, **6**, 685 (1916), Buttgenbach, *Soc. géol. Belgique, Ann.*, **42**, M93 (1921) and *Ac. Belgique, Bull.*, no. 5, 567 (1909), Ungemach, *Bull. soc. min.*, **33**, 132 (1910), and earlier workers (see Spencer (1908)). Transformation: Spencer to new, 010/300/001; Ungemach (1910) to new, 010/001/100.
3. Spencer (1908) and Ungemach (1910); 7 18.0 and 11.9.0 from Walker (1916). Rare or doubtful forms: 025, 11.9.0, i 340, v 430, δ 151, λ 191, ξ 3.11.3, η 3.10.3.
4. Wolfe (1940) by Weissenberg method.
5. Spencer (1908).
6. Seemingly regular zonal groupings have been described by Buttgenbach, *Soc. géol. Belgique, Ann.*, **42**, 93 (1919).
7. Range found by Spencer (1908) on Rhodesian crystals.
8. Wolfe (1940).
9. Larsen (87, 1921).
10. Walker (1916), *Univ. Toronto Stud., Geol. Ser.*, no. 10 (1918).
11. See Spencer (1908), Mellor (4, 658, 1923), and Eberly, Gross, and Crowell, *J. Am. Chem. Soc.*, **42**, 1433 (1920).

HIBBENITE [$Zn_7(PO_4)_4(OH)_2 \cdot 7H_2O$]. Phillips (*Am. J. Sc.*, **42**, 275, 1916).

As imperfect orthorhombic crystals, thick tabular on {010}, closely resembling hopeite in both habit and angles. Forms: $a\{100\}$, $b\{010\}$, $s\{230\}$, $e\{031\}$, and $r\{131\}$, with $a:b:c = 0.548:1:0.276$.[1]

Phys. Cleavage {010} perfect, {100} good, {001} imperfect. Brittle. H.$3\frac{3}{4}$. G. 3.21. Luster vitreous inclining toward pearly on the cleavages. Color pale yellow, almost white. Translucent.

Opt.[2] In transmitted light, colorless.

ORIENTATION		n	
X	b	1.582 ± 0.003	Biaxial negative $(-)$.
Y	c	1.592 ± 0.003	$2V$ $54° \pm 1°$ (meas.).
Z	a	1.593 ± 0.003	$r < v$, perceptible.

Chem. A hydrated basic zinc phosphate, $Zn_7(PO_4)_4(OH)_2 \cdot 7H_2O$. Part of the water is lost at ordinary conditions and part is retained to red heat.

Anal.

	1	2
ZnO	56.53	57.55
P_2O_5	28.17	28.83
H_2O	15.30	13.71
Total	100.00	100.09

1. $Zn_7(PO_4)_4(OH)_2 \cdot 7H_2O$. 2. Salmo, British Columbia. Average of two analyses on air-dried sample. A fresh sample gave H_2O 14.8 per cent.

Tests. B.B. turns yellow when hot and fuses easily. In C.T. decrepitates strongly and yields much water. Readily soluble in dilute acids.

Occur. Found with spencerite, salmoite, and hemimorphite at the Hudson Bay mine, Salmo, Nelson district, British Columbia.

Name. After John Grier Hibben, former president of Princeton University.

The status of hibbenite is uncertain, and the existing data may have been obtained on a mixture of hopeite and spencerite.[3]

Ref.
1. Elements recalculated from the fundamental angles of Phillips (1916) and given in the unit and orientation of hopeite. Transformation: Phillips to new, 010/300/001. s {120} and p {111} as originally given are inconsistent with the angles cited; p is here taken as {111} = {131}, and s then becomes {230}.
2. Larsen (85, 1921).
3. Non-type specimens labeled hibbenite from the Harvard collection and Royal Ontario Museum gave the x-ray powder pattern of spencerite (Frondel, priv. comm., 1948).

40.2.12 PHOSPHOPHYLLITE [$Zn_2(Fe,Mn)(PO_4)_2 \cdot 4H_2O$]. Laubmann and Steinmetz (Zs. Kr., 55, 566, 1920).

C r y s t.[1] Monoclinic; prismatic—$2/m$.

$a:b:c = 2.0265:1:2.0792$; $\beta\ 120°25'$; $p_0:q_0:r_0 = 1.0260:1.7930:1$

$r_2:p_2:q_2 = 0.5577:0.5722:1$; $\mu\ 59°35'$; $p_0'\ 1.1898, q_0'\ 2.0792, x_0'\ 0.5871$

Forms:[2]

	ϕ	ρ	ϕ_2	$\rho_2 = B$	C	A
c 001	90°00'	30°25'	59°35'	90°00'	.....	59°35'
a 100	90 00	90 00	0 00	90 00	30°25'	.....
g 120	15 58	90 00	0 00	15 58	76 16½	74 02
m 110	29 47	90 00	0 00	29 47	64 38	60 13
k 210	48 51	90 00	0 00	48 51	49 30½	41 09
x 013	40 16	42 15	59 35	59 08	47 45	64 35
π 011	15 46	65 10	59 35	29 09	65 09	75 44
τ 101	90 00	60 38	29 22	90 00	30 13	29 22
α $\bar{1}02$	−90 00	0 27	90 27	90 00	30 52	90 27
m $\bar{3}04$	−90 00	16 58½	106 58½	90 00	46 83½	106 58½
s $\bar{3}02$	−90 00	50 08	140 08	90 00	80 33	140 08
ω 111	40 31	69 55½	29 22	44 26	52 46½	52 24
λ 211	54 58½	74 34	18 38	56 25	51 00½	37 52
o $\bar{2}11$	−40 46	69 59	150 50½	44 38	90 53	127 51
p $\bar{3}11$	−55 07	74 37	161 28	56 32	99 53½	142 16½

Structure cell.[3] Space group $P2_1/c$. a_0 10.23 kX, b_0 5.08, c_0 10.49; β 120°15'; $a_0:b_0:c_0 = 2.014:1:2.065$. Cell contents $Zn_4(Fe,Mn)_2(PO_4)_4 \cdot 8H_2O$.

Habit. Thick tabular {100}. Common forms: $a\ m\ \pi\ \alpha\ \omega\ p$.

Twinning. (1) On {100}, common, sometimes polysynthetic. (2) On {$\bar{1}02$}, rare.[4]

P h y s. Cleavage {100} perfect, {010} distinct, {$\bar{1}02$} distinct. Brittle. H. 3–3½. G. 3.08,[5] 3.13;[6] 3.14 (calc.). Luster vitreous. Colorless to pale bluish green. Transparent.

Hagendorf.

Fluoresces violet in short-wave ultraviolet radiation.

O p t.[7] In transmitted light, colorless.

Orientation	n(Na)	
X	1.595	Biaxial negative (−).
$Y \wedge c\ +50°$	1.614	$2V \sim 45°$.
$Z\ \ b$	1.616	$r > v$, perceptible.

TRICHALCITE 739

Chem. A hydrated phosphate of zinc, iron, and manganese, $Zn_2(Fe,Mn)(PO_4)_2 \cdot 4H_2O$.

Anal.

	1	2
ZnO	36.30	34.26
FeO	11.35	12.24
MnO	4.62	4.96
P_2O_5	31.66	32.51
H_2O	16.07	16.52
Total	100.00	100.49

1. $Zn_2(Fe,Mn)(PO_4)_2 \cdot 4H_2O$ with Fe:Mn = 17:7. 2. Hagendorf.[8]

Tests. B.B. easily fusible. In C.T. decrepitates, turns gray and loses water. Soluble in acids.

Occur. A secondary mineral derived from sphalerite and iron-manganese phosphates in pegmatite at Hagendorf near Pleystein in the Oberpfalz, Bavaria, Germany. Associated with triplite, triphylite, sphalerite, apatite, vivianite, rockbridgeite, strengite, fairfieldite, metastrengite.

Name. In allusion to its being a phosphate with a perfect cleavage.

Ref.

1. Unit, orientation, and angles of Palache and Berman, *Am. Min.*, **12**, 180 (1927). Transformation: Laubmann and Steinmetz (1920) to P. and B., 101/010/200. On the dimensional relations to hopeite see Wolfe, *Am. Min.*, **25**, 792 (1940).
2. Laubmann and Steinmetz (1920), Steinmetz, *Zs. Kr.*, **64**, 405 (1926), and Kleber, *Jb. Min., Beil.-Bd.*, **70A**, 203 (1935), who discusses the form development in detail.
3. Wolfe (1940) by Weissenberg method. See also Strunz, *Naturwiss.*, **34** (1942) who discusses the structural relation to metastrengite.
4. See Wolfe (1940) and Kleber (1935).
5. Laubmann and Steinmetz (1920) and Kleber (1935).
6. Wolfe (1940).
7. Kleber (1935).
8. Steinmetz (1926).

40.2.13 **TRICHALCITE** [$Cu_3(AsO_4)_2 \cdot 5H_2O$?]. Trichalcit *Hermann* (*J. pr. Chem.*, **73**, 212, 1858).

Cryst.[1] Orthorhombic.

$a:b:c = 0.384:1:0.207;\qquad p_0:q_0:r_0 = 0.539:0.207:1$

$q_1:r_1:p_1 = 0.384:1.855:1;\qquad r_2:p_2:q_2 = 4.831:2.604:1$

Structure cell.[2] Space group not known. a_0 10.34 kX, b_0 26.9, c_0 5.57; $a_0:b_0:c_0 = 0.384:1:0.207$.

Habit. Minute pseudo-hexagonal twinned aggregates flattened on $\{010\}$ and bounded laterally by $\{100\}$. In radiated groups, columnar; in dendritic forms.

Twinning.[3] On $\{101\}$, repeated, to give pseudo-hexagonal aggregates.

Phys. There is a well-defined cleavage in the zone [001]. H. $2\frac{1}{2}$. G. not known. Luster vitreous to pearly, of aggregates sometimes silky. Color verdigris-green to blue green.

O p t. In transmitted light, pale bluish green and not pleochroic.

Orientation		n (Turginsk) [4]	
X	b	1.67 ± 0.01	Biaxial negative $(-)$.
Y	elong.	1.686 ± 0.003	$2V$ large.
Z		1.698 ± 0.003	Shows sectoral twinning.

C h e m. A hydrated arsenate of copper. The formula is uncertain, perhaps $Cu_3(AsO_4)_2 \cdot 5H_2O$.
Anal.

	CuO	As$_2$O$_5$	P$_2$O$_5$	H$_2$O	Total
1.	42.71	41.13		16.16	100.00
2.	44.19	38.73	0.67	16.41	100.00

1. $Cu_3(AsO_4)_2 \cdot 5H_2O$. 2. Turginsk, Urals.

Tests. In C.T. decrepitates, yields much water, and turns dark brown. B.B. fusible. Easily soluble in cold HCl.

O c c u r. Originally from the Turginsk copper mine near Bogolovsk or from Beresov in the Ural Mountains, U.S.S.R. Also as oxidation crusts on chalcopyrite-arsenopyrite ore in the Liberal King mine, Shoshone County, Utah. Said to occur at Schneeberg, Saxony.

N a m e. From τρίς, *three*, and χαλκός, *copper*.

Ref.
1. Symmetry and axial ratio from the structure cell of Wolfe, *Am. Min.*, **25**, 799 (1940), on Turginsk material. Transformation: from the hypothetical orientation of Shannon, *Proc. U. S. Nat. Mus.*, **62**, Art. 9 (1922), for Idaho crystals, 010/001/100.
2. Wolfe (1940) by Weissenberg method.
3. Shannon (1922).
4. Larsen (144, 1921). Comparable data are given by Shannon (1922) for the Idaho material.

40.2.14 PICROPHARMACOLITE [(Ca,Mg)$_3$(AsO$_4$)$_2 \cdot$6H$_2$O (?)]. Stromeyer (*Ann. Phys.*, **61**, 185, 1819).

Probably monoclinic. In aggregates of small spherical, botryoidal forms with radiating foliated structure; also as silky fibers or minute acicular crystals. Under the microscope appears as rectangular prisms elongated [001] with perfect cleavages {100} and {010}.[1] Color white. Luster feeble pearly. Opaque. G. 2.58 (Joplin).[2]

O p t. In transmitted light, colorless.

Orientation		n (Riechelsdorf) [1]	
$X \wedge c$	$\sim 37°$	1.631	Biaxial positive $(+)$.
Y	b	1.632	$2V$ $40° \pm 2°$ (meas.).
Z		1.640	$r < v$, rather strong.

C h e m. A hydrated arsenate of calcium and magnesium. The formula is uncertain, perhaps (Ca,Mg)$_3$(AsO$_4$)$_2 \cdot$6H$_2$O. About 1 H$_2$O is lost at room temperature over H$_2$SO$_4$.[2] In the reported analyses, Ca is slightly low and H$_2$O high for the formula suggested, and the substance may be an acid salt.

Anal.

	1	2	3	4
CaO	28.53	24.65	25.77	22.40
MgO	3.73	3.22	3.73	6.60
As_2O_5	46.07	46.97	46.93	47.48
H_2O-	21.67	23.98	24.01	11.60
H_2O+				11.44
Rem.		1.00		0.38
Total	100.00	99.82	100.44	99.90
G.				2.583

1. $(Ca,Mg)_3(AsO_4)_2 \cdot 6H_2O$ with $Ca:Mg = 5.5:1$. 2. Riechelsdorf.[3] Rem. is CoO.
3. Freiberg.[4] 4. Joplin, Missouri.[2] Rem. is MnO_2 0.21 and insol. 0.17.

O c c u r. From Riechelsdorf and as a recent efflorescence in mine workings at Freiberg, in Saxony. Also coating dolomite at Joplin, Missouri.

N a m e. From the content of magnesia and the chemical similarity to pharmacolite.

Ref.
1. Larsen (120, 1921) and Frenzel, *Jb. Min.*, 786 (1873). The reported optical properties indicate that the species is distinct from pharmacolite.
2. Genth, *Am J. Sc.*, **40**, 204 (1890).
3. Stromeyer (1819).
4. Frenzel (1873).

40.2.15 VIVIANITE GROUP

MONOCLINIC; PRISMATIC—2/m

	a_0	b_0	c_0	β
Vivianite, $Fe_3(PO_4)_2 \cdot 8H_2O$	10.039 kX	13.38	4.687	104°18′
Annabergite, $Ni_3(AsO_4)_2 \cdot 8H_2O$	10.122	13.28	4.698	104 45
Erythrite, $Co_3(AsO_4)_2 \cdot 8H_2O$	10.184	13.34	4.73	105 01
Koettigite, $Zn_3(AsO_4)_2 \cdot 8H_2O$	10.11	13.31	4.70	103 50
Bobierrite, $Mg_3(PO_4)_2 \cdot 8H_2O$	9.946	27.65	4.639	104 01
Hoernesite, $Mg_3(AsO_4)_2 \cdot 8H_2O$				

Symplesite, $Fe_3(AsO_4)_2 \cdot 8H_2O$

The members of the Vivianite Group proper comprise the four species first named above. Bobierrite has a similar formula and cell, but with b_0 doubled, and appears to be apart from this Group although closely related thereto. It is not known whether hoernesite is to be classed with bobierrite or vivianite. Symplesite has been interpreted variously as triclinic or as monoclinic and isostructural [1] with vivianite. The structure [2] of vivianite is built up of (PO_4) tetrahedra and of octahedral groupings of H_2O and O about the Fe ions; these are linked together into sheets extending along {010}. The sheets are held together by tetrahedral groupings of H_2O which represent the plane of perfect {010} cleavage.

Vivianite does not enter into a wide range of cationic substitutions, but a complete series extends between erythrite and annabergite and possibly between these species and koettigite. Vivianite is of special interest because of the optical changes accompanying the gradual oxidation of the ferrous iron.

Ref.
1. Cf. Mori and Shoji, Acta Cryst., **3**, 1 (1950), who find a_0 10.25 Å, b_0 13.48, c_0 4.71, β 103°50′, on Kiura, Japan, crystals [added in press].
2. Mori and Shoji (1950).

40.2.15.1   **V I V I A N I T E** [Fe$_3''$(PO$_4$)$_2 \cdot$8H$_2$O]. Bloa Järnjord, Naturligit Berlinerblätt, Calx Martis phlogisto juncta, etc. *Cronstedt* (182, 1758). Cæruleum Berolinense nativum *von Born* (**1**, 136, 1772). Ocre martiale bleue, Bleu de Prusse natif *de Lisle* (**3**, 295, 1783). Natürliche Berlinblau, Phosphorsaurer Eisen *Klaproth* (*Crell's Ann.*, **1**, 390, 1784). Eisenblau, Blaueisenerde *Germ.* Vivianit [from Cornwall] *Werner* (41, 1817); Breithaupt, Hoffmann, **4b**, 146, 1817). Phosphate of Iron. Blue Iron Earth. Fer phosphaté, Fer azuré *Fr.* Eisenglimmer *Mohs* (212, 1824). Eisen-Phyllit *Breithaupt* (26, 1823). Glaukosiderit *Glocker* (857, 1831). Mullicite *Thomson* (**1**, 452, 1836). Anglarite *Berthier* (*Ann. mines*, **12**, 303, 1837) = Angelardite *Lacroix* (**4**, 522, 1910). Odontolite pt.
Paravivianite *Popoff* (*Cbl. Min.*, 112, 1906; *Bull. Ac. Sc. St. Petersburg*, **1**, 127, 1907). Alpha-kertschenite, Beta-kertschenite, Oxykertschenite *Popoff* (*Trav. mus. Géol., Ac. St. Pétersburg*, **4**, 99, 1910—*Zs. Kr.*, **52**, 606, 1913). Bosphorite *Dvoichenko* (*Mem. Crimean Soc. Sc. & Nat.*, **4**, 113, 1914—*Zs. Kr.*, **61**, 586, 1925). Kerchenite. α-, β-kertschenite.

C r y s t.[1]   Monoclinic; prismatic—$2/m$.

$a:b:c = 0.7498:1:0.3509;$ $\beta\ 104°26';$ $p_0:q_0:r_0 = 0.4679:0.3398:1$

$r_2:p_2:q_2 = 2.9431:1.3772:1;$ $\mu\ 75°34';$ $p_0'\ 0.4832, q_0'\ 0.3509, x_0'\ 0.2574$

Forms:[2]

	ϕ	ρ	ϕ_2	$\rho_2 = B$	C	A
c 001	90°00′	14°26′	75°34′	90°00′	.....	75°34′
b 010	0 00	90 00	.....	0 00	90°00′	90 00
a 100	90 00	90 00	0 00	90 00	75 34	.....
m 110	54 01	90 00	0 00	54 01	78 22	35 59
y 310	76 23½	90 00	0 00	76 23½	75 59	13 36½
e 021	20 08½	36 46½	75 34	55 48	34 12	78 08
o $\overline{2}$03	−90 00	3 42	93 42	90 00	18 08	93 42
w $\overline{2}$01	−90 00	35 20	125 20	90 00	49 46	125 20
z 111	64 39	39 20	53 28½	74 15½	26 54½	55 03
x 221	54 01	50 03½	45 58½	63 13½	39 11½	51 39
v $\overline{2}$21	−45 18	44 55½	125 20	60 12½	55 54½	120 08

Less common:

h 250	203	301	t $\overline{4}$01	$\overline{6}$21
g 011	k 101	d 801	l $\overline{8}$01	
f 043	n 201	$\overline{3}$01	r $\overline{1}$11	

Structure cell.[3] Space group $C2/m$. a_0 10.039 kX, b_0 13.388, c_0 4.687; β 104°18′; $a_0:b_0:c_0 = 0.7499:1:0.3501$. Cell contents Fe$_6$(PO$_4$)$_4\cdot$16H$_2$O.

Habit. Usually prismatic [001], sometimes also flattened {010}, or, rarely, {100}; also equant or tabular {010}. The crystals often are rounded by vicinal development into blade-like or lanceolate shapes. In stellate groups. As reniform, globular, or tubular masses or concretions, or incrusting, with a divergent bladed or fibrous structure. Also earthy; pulverulent.

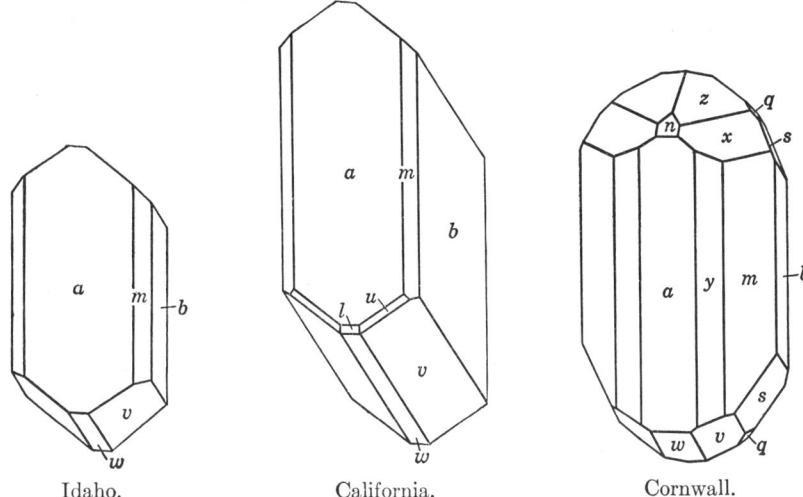

Idaho. California. Cornwall.

Twinning.[4] Lamellae apparently due to twinning have been observed on {010} and correspond approximately to (304).

Phys. Cleavage {010} perfect and easy; in traces on {$\bar{1}$06} and {100}. Fracture fibrous nearly perpendicular to [001].[5] Flexible in thin {010} laminae; sectile. Shows translation gliding with T(010), t[001]. H. $1\frac{1}{2}$–2. G. 2.68 ± 0.01; 2.71 (calc.). Luster vitreous, on {010} pearly; also dull and earthy. Colorless and transparent when fresh and unaltered, but rapidly becoming pale blue or greenish blue due to oxidation [6] and deepening on further exposure to dark blue or dark greenish blue, indigo-blue, or bluish black. Streak and powder colorless to bluish white but soon changing to dark blue or brown. Transparent when colorless or weakly tinted, translucent in deeply colored crystals or transparent only in thin flakes.

Opt.[7]

Orientation		n(Na) (Litošice)	n (Rodna Vecche)	n (black; Leadville)	
X	b	1.5788	1.5816	1.616	Biaxial positive (+).
Y		1.6024	1.6042	1.656	$r < v$ weak;
$Z \wedge c$	$+28\frac{1}{2}°$	1.6294	1.6365	1.675	horizontal.
$2V$		$83\frac{1}{2}°$	$80°56'$	$63\frac{1}{2}°$	
Pleochroism					
X		Blue	Deep blue	Indigo	
Y		Pale yellowish green	Pale bluish green	Green-yellow	
Z		Pale yellowish green	Pale yellowish green	Yellow-olive	

With increasing oxidation, the refractive indices increase, the birefringence decreases and the pleochroism on {010} becomes stronger.

Chem. Octahydrated ferrous phosphate, $Fe_3(PO_4)_2 \cdot 8H_2O$. Small amounts of Mn'', Mg, and Ca sometimes substitute for Fe'', with $Ca:Mn:Mg:Fe'' = 1:4.1:5.6:5.4$ in analysis 3. Vivianite usually contains more or less ferric iron due to oxidation. The oxidation can extend over a wide range without apparent change in crystal structure but with accompanying variation in color and optical properties (see *Oxidized vivianite*, below). The mechanism of the oxidation is not established but may involve a concomitant conversion of (H_2O) to $(OH) + H$.

Anal.[14]

	1	2	3	4	5	6	7	8	9
CaO			0.48	0.85	0.02	tr.	0.11	0.49	0.79
MgO			1.92	0.57			0.09	1.54	1.22
MnO			2.01		0.25		0.08	1.99	2.57
FeO	42.96	44.10	39.12	37.20	32.64	32.70	23.47	9.50	
Fe_2O_3				4.27	9.43	12.04	20.32	32.89	41.82
P_2O_5	28.31	27.17	27.01	27.25	29.99	28.51	28.25	28.19	28.04
H_2O	28.73	27.95	28.75	27.55	27.70	25.43	27.38	25.04	24.98
Rem.		0.10		1.73	0.12	0.97			
Total	100.00	99.32	99.29	99.42	100.15	99.65	99.70	99.64	99.42
G.			2.66		2.693			2.65	2.65

1. $Fe_3(PO_4)_2 \cdot 8H_2O$. 2. Cantwell's Bridge, Delaware.[8] Colorless. Rem. is SiO_2. 3. Taman peninsula, U.S.S.R. (*paravivianite*).[9] 4. Adamov, Moravia.[10] H_2O+ 13.18. H_2O- 14.37. Rem. is insol. 1.42, Al_2O_3 0.31. 5. Plant City, Florida.[11] Pale green. Rem. is SiO_2 0.12, TiO_2 tr., H_2O+ 15.84, H_2O- 11.86. 6. Rovečin, Moravia.[12] Rem. is insol. and Al_2O_3 tr. 7. Kertsch peninsula (*beta-kertschenite*).[13] 8. Kertsch peninsula (*alpha-kertschenite*).[13] Average of two. 9. Kertsch peninsula (*oxykertschenite*).[13]

Tests. In C.T. yields neutral water, whitens, and exfoliates. B.B. fusible to a gray, magnetic bead. M.P. 1114°. Easily soluble in acids. Darkens in color in H_2O_2.

Var. *Oxidized Vivianite.* Alpha-kertschenite, Beta-kertschenite, Oxy-kertschenite *Popoff* (1910). Bosphorite *Dvoichenko* (1914). Kertschenite.

Material high in ferric iron produced by partial, continuous oxidation of the original ferrous compound. Color dark blue or indigo-blue to nearly black, with relatively high indices of refraction and strong pleochroism. Streak dark-colored. With complete conversion of Fe'', fragile, cleavable pseudomorphs may be obtained that are brown in color, dull in luster, and give very diffuse x-ray patterns. The extent of oxidation that can be tolerated without breakdown in structure is not known.

Occur. Vivianite occurs relatively well-crystallized as a secondary mineral in the gossan of metallic ore deposits, and as a weathering product of primary iron-manganese phosphates in pegmatites. Widespread in clays, usually in concretionary shapes, and in recent alluvial or other sedimentary deposits where the mineral often is associated with bone, decayed wood, or other organic remains; in glauconitic sediments and pebble-phosphate rock. Also in lignite and peat, forest soils, and bog-iron ores. As the coloring matter of fossil bone or teeth (*odontolite*) and then often mistaken for turquois.

VIVIANITE

Found in Cornwall, England, at St. Agnes and elsewhere with pyrrhotite in the tin veins. In France at places where coal measures have been on fire, as at Commentry in Allier and at Cransac in Aveyron. In Germany in limonite ores in the Amberg-Auerbach region, Bavaria, and as an alteration of triphylite in the pegmatites around Hagendorf, Bavaria, and Marienbad, Bohemia. With siderite in the brown coal of Mecklenburg. In Bohemia notably at Litošice, Rovečin, and Valdic. From Rodna Vecche (Oradna), Roumania. With siderite impregnations in the moors of Drenthe province, Holland. In Russia, especially in sedimentary iron-ores on the Kerch and Taman peninsulas on the Black Sea (*paravivianite, kertschenite*). Crystals over 10 cm. in size occur in the tin veins at Llallagua, Bolivia; also at Tasna and Tatasi. Large crystals also occur at the Ashio mines, Shimotsuke province, Japan. From the Wannon river, Victoria, Australia.

In the United States, vivianite occurs abundantly at Allentown and Shrewsbury, Monmouth County, and at Mullica Hill, Gloucester County, in New Jersey. In green-sand at Middletown, New Castle County, Delaware. At Plant City, Florida, in phosphate rock. Abundantly in fine specimens at Leadville, Lake County, Colorado. Incrusting a fossil tusk in a gold placer in Clearwater County, Idaho. In California near Camptonville, Yuba County, and in diatomite in a Tertiary Lake bed near Burney, Shasta County. In Canada in bog-iron ore at Côte St. Charles, Vaudreuil County, Quebec.

A r t i f.[15] Obtained as colorless crystals by heating precipitated ferrous phosphate in a solution of sodium phosphate, by standing a solution of ferrous ammonium phosphate in hydrofluoric acid, and in other ways.

N a m e. After J. G. Vivian, an English mineralogist who discovered the mineral in Cornwall.

Ref.

1. Angles and orientation of vom Rath, *Ann. Phys.*, **136**, 405 (1869); unit of the structure cell of Barth, *Am. Min.*, **22**, 325 (1937). Transformation: vom Rath (1869), Dana (814, 1892), Goldschmidt (**3**, 273, 1890; **9**, 57, 1923) to new, 100/010/00½. As noted by Wolfe, *Am. Min.*, **25**, 738 (1940), the structural cell with smallest translations is body-centered, the transformation from Dana (1892) being 10$\bar{1}$/010/00½; the conventional C-centered cell is here taken.

2. Goldschmidt (**9**, 57, 1923); also Ulrich, *Rozpr. České Ak.*, **23**, no. 33 (1925); Gordon, *Proc. Ac. Sc. Philadelphia*, **76**, 335 (1924). Rare and uncertain: 320, 540, 160, 28.0.1, 501, 702, A 209, μ $\bar{1}$01, $\bar{3}$01, γ $\bar{7}$02, $\bar{5}$01, s $\bar{2}$61, p $\bar{1}$51, u $\bar{8}$21, q $\bar{1}$31, $\bar{3}$.2 3, α 8 3 3, β 3.5.7, ψ 2.16.3, i $\bar{1}$6.6.3.

3 Barth (1937) by Weissenberg method on Monserrat, Bolivia, material. Takané and Ômori, *J. Jap. Assoc. Min. Petr. Econ. Geol.*, **16**, 234 (1936), and Yamaguti, *Proc. Physico-Math. Soc. Japan*, **18** [3], 372 (1936), give comparable data

4. Ulrich (1925).

5. See Mügge, *Jb. Min.*, I, 71 (1898), I, 53 (1884); Buttgenbach, *Soc. géol. Belgique, Mém.*, **40**, 3 (1913); Kuhara, *Mem. Coll. Eng. Kyoto*, **2**, 71 (1918).

6. On the change in color and optical properties accompanying oxidation, see Watson, *Am. Min.*, **3**, 159 (1918); Ulrich, *Rozpr. České Ak.*, **31**, no. 10 (1922), **33**, no 33 (1925); Zieleniewski, *Arch. Min. Soc. Warsaw*, **15**, 51 (1945).

7. Ulrich (1922; 1925; *Zs. Kr.*, **64**, 143 (1926)) and Zieleniewski (1945). Slightly different data are reported by others, the divergences due to variation in extent of

oxidation or to unrecognized isomorphous substitution (Mg, Mn, etc.). See also Gaubert, *Bull soc. min.*, **27**, 212 (1904); Rosický, *Böhm. Ak., Abh*, no. 28, 17 (1908); Larsen and Berman (110, 1934). On the axial dispersion see Petrow, *Jb. Min., Beil.-Bd.*, **37**, 457 (1914), and on dispersion of nZ-nY see Lincio, *Heidelberg Ak. Wiss., Math.-Nat. Kl., Abt. A, Abh.*, 15 (1914).
8. Fisher, *Am. J. Sc.*, **9**, 84 (1850).
9. Popoff (1906).
10. Pelíšek, *Příroda*, **28**, 279 (1935).
11. Gooch analysis in Watson and Gooch, *Washington Ac. Sc., J.*, **8**, 82 (1918).
12. Kovář, *Böhm. Ak., Abh.*, no. 9 (1898).
13. Popoff (1910).
14. For additional analyses see Hintze (1 [3B], 1235, 1933).
15. Mellor (**14**, 392, 1935).

40.2.15.2 **E R Y T H R I T E** [$(Co,Ni)_3(AsO_4)_2 \cdot 8H_2O$]. Kobold-Blüthe *Brückmann* (161, 1727). Kobolt Blomma, Flos Cobalti [the crystals], Koboltbeslag [impure earthy], Cobalti minera colore rubro, etc. *Wallerius* (234, 1747). Koboltblüte, Koboltbeschlag, Ochra Cobalti rubra *Cronstedt* (212, 1758). Kobaltblüthe *Germ.* Cobalt Bloom, Red Cobalt, Cobalt Ocher. Cobaltum Acido arsenico mineralisatum *Bergmann* (134, 1782; **2**, 444, 1780 [the first analysis]). Arsenate of Cobalt. Cobalt arseniaté *Fr.* Erythrine *Beudant* (**2**, 596, 1832). Rhodoise *Huot* (**1**, 313, 1841).

40.2.15.3 **A N N A B E R G I T E** [$(Ni,Co)_3(AsO_4)_2 \cdot 8H_2O$]. Ochra Niccoli, Niccolum calciforme *Cronstedt* (218, 1758). Nickelocker. Nickelblüthe. Nickel Bloom. Nickel Ocher; Nickel Green; Arsenate of Nickel, Nickel Arseniaté *Fr.* Annabergite *Brooke* and *Miller* (503, 1852). Dudgeonite *Heddle* (*Min. Mag.*, **8**, 200, 1889). Cabrerite *Dana* (561, 1868). Wasserhaltige Nickeloxyd-Magnesia *Ferber* (*B. H. Ztg.*, **22**, 306, 1863).

C r y s t. Monoclinic; prismatic—$2/m$.

ERYTHRITE [1]

$a:b:c = 0.7633:1:0.3546;\quad \beta\, 105°01';\quad p_0:q_0:r_0 = 0.4646:0.3425:1$

$r_2:p_2:q_2 = 2.9198:1.3564:1;\quad \mu\, 74°59';\quad p_0'\, 0.4810,\, q_0'\, 0.3546,\, x_0'\, 0.2683$

Forms: [2]

	ϕ	ρ	ϕ_2	$\rho_2 = B$	C	A
b 010	0°00'	90°00'	....	0°00'	90°00'	90°00'
a 100	90 00	90 00	0°00'	90 00	74 59	....
l 340	45 29½	90 00	0 00	45 29½	79 21	44 30½
m 110	53 35½	90 00	0 00	53 35½	77 58	36 24½
s 320	63 49½	90 00	0 00	63 49½	76 33	26 10½
q 301	90 00	59 42	30 18	90 00	44 41	30 18
o $\bar{2}$03	−90 00	3 00	93 00	90 00	18 01	93 00
w $\bar{2}$01	−90 00	34 45	124 45	90 00	49 46	124 45
p 221	60 02½	54 51	39 06½	65 54	42 21	44 54
v $\bar{2}$21	−44 22	44 46½	124 45	59 46	56 04½	119 30½

ERYTHRITE–ANNABERGITE

ANNABERGITE [4]

$a:b:c = 0.7619:1:0.3537;$ $\beta\ 104°45';$ $p_0:q_0:r_0 = 0.4642:0.3420:1$

$r_2:p_2:q_2 = 2.9236:1.3572:1;$ $\mu\ 75°15';$ $p_0'\ 0.4801,\ q_0'\ 0.3537,\ x_0'\ 0.2633$

Forms: [5]

	ϕ	ρ	ϕ_2	$\rho_2 = B$	C	A
c 001	90°00'	14°45'	75°15'	90°00'	.....	75°15'
b 010	0 00	90 00	.....	0 00	90°00'	90 00
a 100	90 00	90 00	0 00	90 00	75 15	.....
m 110	53 57	90 00	0 00	53 37	78 10½	36 23
w $\bar{2}$01	−90 00	34 52½	124 52½	90 00	49 37½	124 52½
r 111	64 33½	39 28	53 22½	74 09	26 47	54 58½
p 221	60 06½	54 50	39 06	65 57½	42 29½	44 52
v $\bar{2}$21	−44 34½	44 48	124 52½	59 52	55 55½	119 38

Structure cell.[3] Space group $C2/m$.

	a_0	b_0	c_0	β	$a_0:b_0:c_0$
Erythrite	10.184 kX	13.340	4.730	105°01'	0.7633:1:0.3546
Annabergite	10.122	13.284	4.698	104°45'	0.7619:1:0.3537

Cell contents $(Co,Ni)_6(AsO_4)_4 \cdot 16H_2O$.

Habit. Prismatic to acicular [001] and flattened {010}. The crystals are deeply striated or furrowed [001], also striated on {010} parallel {$h0l$} or {$\bar{h}0l$}. Crystals of erythrite and especially of annabergite are rare and of poor quality. Erythrite often as radial or stellate groups; in globular or reniform shapes with a drusy surface and columnar or coarse-fibrous structure; also earthy and pulverulent. Annabergite usually as fine-crystalline coatings or earthy.

Phys. Cleavage {010} perfect, also {100} and {$\bar{1}$02} indistinct. Flexible in thin {010} laminae. Translation gliding [6] with T{010}, t[001]. Sectile. H. 1½–2½, least on {010}. G. erythrite, 3.06 (Schneeberg[7]), 3.178 (artif.[8]), 3.18 (calc.); annabergite, 3.07 (Laurium [7]), 3.0 (artif.[8]), 3.23 (calc.). Luster weakly adamantine, pearly on {010}; also dull and earthy. Color of erythrite crimson-red and peach-red; the color becomes paler with increasing content of Ni and is still pale rose or pale pink at Co:Ni ∼1:1, then becoming white or gray, pale green, and fine apple-green in highly niccolian material. Streak paler than the color. Transparent to translucent in crystals. Single crystals may show color-banding or be tipped by material of a different color.

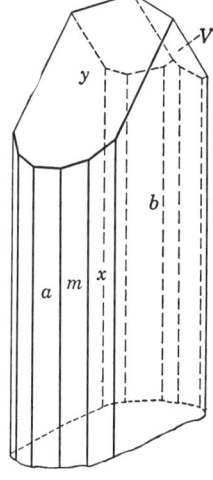

Erythrite. Cobalt, Ontario.

Opt.[9]

ORIEN-TATION	Erythrite	Erythrite Schneeberg	Erythrite Mont-Chemin	Anna-bergite	
X	b	1.629	1.622	1.622	Biaxial positive
	1.626				(+); also
Y	1.661	1.663	1.660	1.658	negative (−).
Z	1.699	1.701	>1.681	1.687	$r > v$.
$Z \wedge c$	+31°	+30° ± 1°	$31\frac{1}{2}°$	+36°	
2V	90°±	Very large		84°	

PLEOCHROISM

X	Pale pinkish	Pale pinkish	Very pale rose
Y	Very pale violet	Pale violet	Pale violet-rose
Z	Red	Red	Deep rose

Chem. Octahydrated arsenate of cobalt and nickel, $(Co,Ni)_3(AsO_4)_2 \cdot 8H_2O$. Ni and Co substitute mutually to form a complete series. The names erythrite and annabergite are applied to the halves of the series with Co > Ni and Ni > Co, respectively. Significant amounts of Ca, Zn, Fe″, and Mg also substitute for (Co,Ni)—see **Var.**, below. (PO_4) has not been reported in substitution for (AsO_4).

Anal.[10]

	1	2	3	4	5	6	7
CaO					0.42		
MgO						0.50	
FeO		4.01	1.49	3.04	3.51	1.13	
ZnO							8.54
CoO	37.54	33.42	34.11	30.36	23.75	16.33	11.97
NiO			0.52	3.71	11.26	17.37	17.23
As$_2$O$_5$	38.39	38.30	37.86	38.10	36.42	38.13	34.63
H$_2$O	24.07	24.08	24.19	[24.79]	23.52	22.30	22.55
Rem.			2.17		0.86	4.00	5.06
Total	100.00	99.81	100.34	[100.00]	99.74	100.03	99.98
G.	3.18						

1. $Co_3(AsO_4)_2 \cdot 8H_2O$. 2. Erythrite. Rappold mine, Schneeberg.[11] 3. Erythrite, O'Brien mine, Cobalt, Ontario.[12] Rem. is SiO$_2$ 1.28, CuO 0.89. Color deep peach-red. 4. Erythrite, St. Anton mine, Wittichen, Baden[13] 5. Erythrite, Joachimsthal, Bohemia.[14] Rem. is SO$_3$. 6. Annabergite. Gowganda, Ontario.[12] Color pale pink. Rem. is SiO$_2$ 2.34 and cobaltite 1.66. 7. Annabergite. Arburese, Sardinia.[15] Rem. is Fe$_2$O$_3$ 0.83., insol. 4.23.

	8	9	10	11	12	13	14
CaO		0.84		3.51	9.32		
MgO	2.47	1.12	0.21	3.74		6.16	
FeO		0.30	0.56			1.10	
CoO	7.23	6.43	3.40	0.50	0.76	tr.	
NiO	28.40	29.30	33.82	32.64	25.01	26.97	37.46
As$_2$O$_5$	35.18	38.31	37.25	36.64	39.33	40.45	38.44
H$_2$O	23.12	24.04	24.83	23.94	25.01	25.26	24.10
Rem.	3.60		0.41				
Total	100.00	100.34	100.48	100.97	99.43	99.94	100.00
G.			2.907			3.01	

8. Annabergite. Arburese, Sardinia.[15] Rem. is insol. Color pale green. 9. Annabergite. Cobalt, Ontario.[12] Color white. 10. Annabergite. Cobalt, Ontario.[12] Color faint green. Rem. is SiO$_2$. 11. Annabergite. Silver Cliff, Colorado.[16] 12. Annabergite. Pibble mine, Scotland (*dudgeonite*).[17] 13 Annabergite. Laurium, Greece (*cabrerite*).[18] 14. $Ni_3(AsO_4)_2 \cdot 8H_2O$.

Tests. B.B. easily fusible. In C.T. yields water at a low heat and turns bluish (*erythrite*) or darkens in color (*annabergite*). Soluble in acids.

Var. Calcian. Dudgeonite *Heddle* (*Min. Mag.*, **8**, 200, 1889). Contains Ca in substitution for (Ni, Co), with Ca:(Ni, Co) = 1:2.1 in calcian annabergite from Scotland (anal. 12); white and pulverulent. A highly calcian erythrite with Ca:(Co, Ni) = 1:2.7 has been reported from Schneeberg.[19] Material with Ca > (Co,Ni) is not known.

Zincian. Contains Zn in substitution for (Ni, Co), with Zn:(Ni, Co) = 1:3.7 in analysis 7, and is part of a series toward koettigite. Color rose.

Ferroan. Contains Fe'' in substitution for (Co, Ni) with Fe'':(Co, Ni) = 1:8.0 in the ferroan erythrite of analysis 2. This series apparently is limited. The chemically analogous compound $Fe_3''(AsO_4)_2 \cdot 8H_2O$, symplesite, is triclinic and not isostructural with the minerals at hand.

Magnesian. Cabrerite *Dana* (561, 1868); Wasserhaltige Nickeloxyd-Magnesia *Ferber* (*B. H. Ztg.*, **22**, 306, 1863). Contains Mg in substitution for (Ni, Co) with Mg:(Ni, Co) ~1:1.4 in the original magnesian annabergite from Spain [20] and 1:2.46 in the material of analysis 13 from Laurium. The analyses of the Laurium material, at least, may represent the average composition of zoned crystals consisting of dark green, nickel-rich material and pale green nickel-poor parts (perhaps *nickelian hoernesite*) with relatively low indices of refraction.[21]

Occur. Erythrite and annabergite are secondary minerals usually formed by the oxidation of the arsenides of cobalt and nickel. Associated with symplesite, morenosite, retgersite, malachite, adamite, scorodite. Erythrite occurs in Germany as fine specimens at Schneeberg, Saxony; also from Saalfeld and Glücksbrunn near Schweina, Thuringia; and from Wittichen in the Black Forest, Baden, and Riechelsdorf in Hesse-Nassau. At Joachimsthal, Běloves and Příbram in Bohemia. From Schladming in Styria and Leogang in Salzburg, Austria. From Mont-Chemin and the Turtmannthal in Wallis, Switzerland. Earthy varieties occur at Chalanches near Allemont, Isère, and at Markirch in Alsace, France. From Tunaberg, Sweden. At various localities in Cornwall and at Alston Moor, Cumberland, England. From several localities in Chile, and from the Veta Rica mine, Coahuila, Mexico. Found in the United States in Nevada in the northeast part of Churchill County and in Cottonwood Canyon in the Stillwater Range; in the Black Hills, Yavapai County, Arizona; with annabergite in the Black Hawk district, Grant County, New Mexico. In California at the Bishop mine, east of Long Lake, Inyo County, in mines in San Gabriel Canyon, Los Angeles County, and elsewhere. Common in the Blackbird district, Lemhi County, Idaho. In Canada found abundantly in the Cobalt area, Timiskaming District, Ontario.

Annabergite is less common than erythrite. It occurs at Annaberg and Schneeberg, Saxony, and Riechelsdorf, Hesse-Nassau; at Mohrungen near Sangershausen in the Harz. At Leogang, Salzburg, Austria, and Dobsina, Gömör, Hungary. From the Arburese mine, Gonnosfanadiga, Sardinia. Magnesian varieties occur at Laurium, Greece, and in the Sierra Cabrera

Almeria province, Spain (*cabrerite*). Annabergite also is found at Chalanches, Isère, Spain, and the Pibble mine near Creetown, Kirkudbrightshire, Scotland (*calcian*). Found in the United States at Silver Cliff, Custer County, Colorado; abundant with retgersite at the Lovelock mine, Cottonwood Canyon, Humboldt County, Nevada. In Canada notably at Gowganda and Cobalt in the Temiskaming.

A r t i f.[22] Obtained as crystals by slow precipitation of a $NiSO_4$ or $CoSO_4$ solution with Na_2HAsO_4, and in other ways.

N a m e. Erythrite from ἐρυθρός, *red*. Annabergite from the locality in Saxony.

Ref.

1. Elements of the C-centered structure cell of Barth, *Am. Min.*, **22**, 325 (1937), on unanalyzed erythrite from Schneeberg. The morphological measurements on erythrite are relatively inaccurate—see Shannon, Green, *Trans. Canad. Inst.*, *Toronto*, **8**, 443 (1910); Lincio, *Sitzber. Heidelberg Ak. Wiss.*, *Math -phys. Kl., Abh.*, 15 (1914), Brezina, *Min. Mitt.*, 19 (1872), Barth (1937). Transformation: Dana (817, 1892) and Goldschmidt (5, 33, 1918) to new, 100/010/00$\frac{1}{2}$.
2. Goldschmidt (1918) and Barth (1937). Rare or uncertain for erythrite: N 610, x 350, n 201, y 102, r 111, q 613.
3. Barth (1937) by Weissenberg method on unanalyzed specimens of erythrite and annabergite from Schneeberg and Laurium.
4. Elements of the C-centered structure cell of Barth (1937). The morphological elements of Barth on Laurium crystals are less accurate.
5. Barth (1937).
6. Mügge, *Jb. Min.*, I, 74 (1898). On percussion figures on {010} see Mügge, *Jb. Min.*, I, 53 (1884), and Lincio (1914).
7. Wolfe, *Am. Min.*, **25**, 804 (1940).
8. de Schulten, *C.R.*, **136**, 1444 (1904).
9. Larsen (40, 72, 1921); Larsen and Berman (120, 178, 1934); Friedlaender, *Verh. Schweiz. naturfor. Ges.*, 112 (1942); Ulrich, *Zs. Kr.*, **64**, 143 (1926). All data on unanalyzed material.
10. For additional analyses see Hintze (1 [3B], 1261, 1264, 1933) and Sinkler, *Econ. Geol.*, **37**, 136 (1942).
11. Kersten, *Ann. Phys.*, **60**, 151 (1843).
12. Todd analysis in Walker and Parsons, *Univ. Toronto Stud.*, *Geol Ser.*, no. 17, 13 (1924).
13. Petersen, *Ann. Phys.*, **134**, 86 (1868).
14. Lindacker analyses in Vogl, *Gangverhält. und Min. Joachimsthal*, *Teplitz*, 1856.
15. Rossetti (1942).
16. Genth, *Proc. Am. Phil. Soc.*, **23**, 46 (1885).
17. Heddle, *Min. Mag.*, **8**, 200 (1889).
18. Sachs, *Cbl. Min.*, 198 (1906).
19. See the early and perhaps erroneous analysis by Kersten (1843) with CaO 8.00 per cent.
20. See analyses cited in Dana (819, 1892).
21. Such zoning was observed in crystals here examined by Frondel, priv. comm. (1949). The optical data of Larsen (51, 1921) for cabrerite indicate that his specimen was annabergite (see also Larsen, p. 40), as also was the cabrerite specimen examined by Barth (1937).
22. de Schulten, *Bull. soc. min.*, **26**, 87 (1903); *C.R.*, **136**, 1444 (1903); Ducru, *C.R.*, **131**, 675, 702 (1900).

LAVENDULAN. Breithaupt (*J. pr. Chem.*, **10**, 505, 1837). Lavendulite. [Not lavendulan Goldsmith (*Proc. Ac. Sc. Philadelphia*, 192, 1877) = freirinite.]

As thin botryoidal crusts. Color lavender-blue. Luster vitreous to waxy. H. $2\frac{1}{2}$–3. G. 3.014. Under the microscope [1] composed of minute radiated fibers or plates. Slightly pleochroic in shades of blue. Biaxial with nY 1.715 ± 0.005, nZ 1.725 ± 0.003. Z is inclined to the fiber length. The fibers show zones of varying composition with accompanying variation in indices of refraction. Analyses are lacking; affords qualitative tests for Cu, Co, Ni, and As. Found with erythrite at Joachimsthal, Bohemia. Lavendulan may be a cuprian erythrite or perhaps the copper analogue of erythrite.

Ref.
1. Foshag, *Am. Min.*, **9**, 29 (1924).

40.2.15.4   **KOETTIGITE**   [$Zn_3(AsO_4)_2 \cdot 8H_2O$].   Zinkarseniat *Köttig* (*J. pr. Chem.*, **48**, 183, 1849). Köttigite *Dana* (487, 1850).

C r y s t.[1]   Monoclinic; prismatic—$2/m$.

$a:b:c = 0.7593:1:0.3531$; $\beta\ 104°30'$; $p_0:q_0:r_0 = 0.4650:0.3419:1$

$r_2:p_2:q_2 = 2.9253:1.3603:1$; $\mu\ 75°30'$; $p_0'\ 0.4803, q_0'\ 0.3531, x_0'\ 0.2586$

Forms:

	ϕ	ρ	ϕ_2	$\rho_2 = B$	C	A
b 010	0°00'	90°00'	....	0°00'	90°00'	90°00'
a 100	90 00	90 00	0°00'	90 00	75 30	....
m 110	53 41	90 00	0 00	53 41	78 21½	36 19
n 201	90 00	50 38½	39 21½	90 00	36 08½	39 21½
v $\bar{2}21$	−44 49½	44 52½	125 04	59 58½	55 50½	119 50

Structure cell.[2] Space group $C2/m$. a_0 10.11 kX, b_0 13.31, c_0 4.70; β 103°50'; $a_0:b_0:c_0 = 0.7593:1:0.3531$. Cell contents $Zn_6(AsO_4)_4 \cdot 16H_2O$.

Habit. Prismatic [001] and flattened {010}. Massive, or in crusts with a crystalline surface and fibrous structure.

P h y s. Cleavage {010} perfect. H. 2½–3. G. 3.33;[3] 3.32 (calc.). Luster of fracture silky. Color light carmine and peach-blossom-red. Streak reddish white. Translucent.

O p t.[4] In transmitted light, pale rose.

Orientation n(Na) (Schneeberg)
X b 1.622   Biaxial positive (+).
Y   1.638 $2V$ 74°.
Z ∧ c 37°   1.671

C h e m. Octahydrated arsenate of zinc, $(Zn,Co,Ni)_3(AsO_4)_2 \cdot 8H_2O$. Co and Ni substitute for Zn with Ni:Co:Zn = 1:3.4:14.0 in the only reported analysis (see also zincian *annabergite*).

Anal.

	CoO	NiO	ZnO	As_2O_5	H_2O	Total
1.			39.50	37.19	23.31	100.00
2.	6.91	2.00	30.52	[37.17]	23.40	[100.00]

1. $Zn_3(AsO_4)_2 \cdot 8H_2O$. 2. Schneeberg.[5] With CaO tr.

Tests. B.B. easily fusible. In C.T. loses water at a low temperature and turns smalt-blue. Soluble in acids.

Occur. A secondary mineral derived by the alteration of smaltite and sphalerite at the Daniel mine, Schneeberg, Saxony, Germany.

Artif.[6] Obtained by aging of the gelatinous precipitate obtained from mixed solutions of $ZnSO_4$ and Na_2HAsO_4. Cobaltian koettigite also has been synthesized.

Name. After Otto Köttig (1824– ?), a chemist of Schneeberg, Saxony.

Ref.
1. Forms of Wolfe, *Am. Min.*, **25**, 804 (1940), on crystals from Schneeberg; elements of the structure cell.
2. Wolfe (1940) by Weissenberg method.
3. Wolfe (1940).
4. Wolfe (1940); the data do not agree with those of Larsen (96, 1921).
5. Köttig (1849).
6. de Schulten, *Bull. soc. min.*, **26**, 91 (1903).

40.2.15.5 SYMPLESITE [$Fe_3(AsO_4)_2 \cdot 8H_2O$]. Breithaupt (*J. pr. Chem.*, **10**, 501, 1837).

Cryst.[1] Triclinic; pinacoidal—$\bar{1}$.

$a:b:c = 0.8320:1:0.5027;$ $\alpha\ 99°55',\ \beta\ 97°22\frac{1}{2}',\ \gamma\ 105°57\frac{1}{2}'$

$p_0:q_0:r_0 = 0.6190:0.5185:1;$ $\lambda\ 77°26',\ \mu\ 79°18\frac{1}{2}',\ \nu\ 72°18'$

$p_0'\ 0.6928,\ q_0'\ 0.5803,\ x_0'\ 0.0439,\ y_0'\ 0.0243$

Forms:

		ϕ	ρ	A	B	C
c	001	61°02'	2°52½'	79°18½'	77°26'	.....
m	110	39 50½	90 00	32 27½	39 50½	87°19½'
Q	9$\bar{1}$0	102 29	90 00	30 11	102 29	87 51
M	1$\bar{1}$0	119 15½	90 00	46 57½	119 15½	88 29½
r	1$\bar{1}$3	165 03½	45 40	91 58½	133 43	46 17½

Structure cell.[2] Space group $P\bar{1}$. $a_0\ 7.85\ kX,\ b_0\ 9.39,\ c_0\ 4.71;\ \alpha\ 99°55',\ \beta\ 97°22\frac{1}{2}',\ \gamma\ 105°57\frac{1}{2}';\ a_0:b_0:c_0 = 0.832:1:0.503$. Cell contents $Fe_3(AsO_4)_2 \cdot 8H_2O$.

Habit. In spherical aggregates with a coarsely fibrous, radial structure. Also in small imperfect crystals elongated [001] and sometimes flattened $\{1\bar{1}0\}$.

Twinning.[3] On $\{1\bar{1}0\}$.

Phys. Cleavage $\{1\bar{1}0\}$ perfect. Fracture uneven. Brittle. H. $\sim 2\frac{1}{2}$. G. 3.01 (Lobenstein),[4] 3.02 (calc.). Luster vitreous, of cleavage planes pearly. Color light green to leek-green, inclining toward greenish black and deep indigo-blue on partial oxidation. Transparent when fresh. Streak bluish white. Translation gliding[6] with $T\{1\bar{1}0\}$, $t[001]$.

Opt.[5]

ORIENTATION	n (Lobenstein)	PLEOCHROISM	
$X \perp (1\bar{1}0)$	1.635 ± 0.005	Deep blue	Biaxial negative $(-)$.
Y	1.668 ± 0.003	Nearly colorless	$2V\ 86\frac{1}{2}°$ (meas.).
$Z \wedge c\ \ 31\frac{1}{2}°$	1.702 ± 0.003	Yellowish	$r > v$, strong.

With increasing alteration, the extinction angle increases up to about 50° and the absorption colors change from blue to green or green to brown and decrease in intensity.

BOBIERRITE

Chem. Hydrated ferrous arsenate, $Fe_3(AsO_4)_2 \cdot 8H_2O$. The Fe'' may oxidize more or less completely to Fe''' with accompanying changes in color and optical properties,[6] analogous to vivianite. Analysis gave:

	FeO	As_2O_5	H_2O	Total	G.
1.	36.56	38.99	24.45	100.00	3.02
2.	34.73	37.84	[27.43]	[100.00]	2.964

1. $Fe_3(AsO_4)_2 \cdot 8H_2O$. 2. Hüttenberg, Carinthia.[7] Recalculated after deducting 7.7 per cent quartz.

Tests. B.B. infusible but becomes black and magnetic. In C.T. yields much water; at a high temperature some arsenious acid sublimes, imparting an acid reaction to the water, and leaves a black, magnetic residue. Soluble in acids. In strong KOH solution or in 30 per cent H_2O_2 turns reddish brown.

Occur. A secondary mineral found associated with pharmacosiderite, scorodite, erythrite, annabergite, limonite. Found in Germany at Lobenstein and Saubach in Voigtland and with roselite at Neustädtel near Schneeberg, Saxony. At Pisek, Bohemia, and at Hüttenberg, Carinthia, Austria. With quartz, in cavities in hornstone at Felsöbánya, Roumania. From Pizzo Cipolla near Mandanici, Messina province, Italy. Also from the Magnet mine, Tasmania. Reported from Tagish lake, Yukon, Canada.

Name. From σύν and πλησιάζειν, *to bring together*, in allusion to its relations to other minerals.

Ref.
1. Krenner, *Termés. Füzetek*, **10**, 83, 108 (1886), *Zs. Kr.*, **13**, 70 (1887), described imperfect crystals from Felsöbánya as monoclinic, $a:b:c = 0.7806:1:0.6812$, $\beta\ 107°17'$, with {001}, {010}, {100}, {110}, {013}. The x-ray study of Wolfe, *Am. Min.*, **25**, 801 (1940), indicates the mineral to be triclinic. Krenner's morphological elements transformed to the structure cell, $a:b:c = 0.8604:1:0.4972$, are not in satisfactory agreement with the x-ray elements here accepted for the species. Transformation: Krenner (1886) to Wolfe (1940), $\frac{2}{5}\bar{1}0/\frac{2}{5}\bar{1}0/00\frac{1}{2}$.
2. Wolfe (1940) by Weissenberg method on twinned Lobenstein fibers.
3. Identified by Wolfe (1940) by x-ray study.
4. Wolfe (1940).
5. Larsen (140, 1921).
6. Heide, *Zs. Kr.*, **67**, 81 (1928).
7. Bořický, *Vh. min. Ges.*, **3**, 98 (1868).

FERRISYMPLESITE. Walker and Parsons (*Univ Toronto Stud., Geol. Ser.*, no. 17, 16, 1924). Small irregular masses of resinous luster and deep amber-brown color. Under the microscope strongly birefringent with a fibrous structure and parallel extinction. In composition apparently a hydrated ferric arsenate and thought to be the oxidized equivalent of symplesite. Found intimately admixed with erythrite and annabergite in the Hudson Bay mine, Cobalt, Ontario.

40.2.16 BOBIERRITE [$Mg_3(PO_4)_2 \cdot 8H_2O$]. Phosphate de Magnésie tribasique et hydraté *Bobierre* (*Les Mondes*, 691, April, 1868). Bobierrite *Dana* (795, 1868). Hautefeuillite *Michel* (*Bull. soc. min.*, **16**, 40, 1893).

Cryst.[1] Monoclinic.

$a:b:c = 0.3596:1:0.1668$; $\beta\ 104°01'$; $p_0:q_0:r_0 = 0.4638:0.1618:1$

$r_2:p_2:q_2 = 6.1792:2.8662:1$; $\mu\ 75°59'$; $p_0'\ 0.4781, q_0'\ 0.1668, x_0'\ 0.2496$

Structure cell.[2] Space group $P2_1/c$. a_0 9.946 kX, b_0 27.654, c_0 4.639; β 104°01′; $a_0:b_0:c_0 = 0.3596:1:0.1668$. Cell contents $Mg_{12}(PO_4)_8 \cdot 32H_2O$.

Habit. Minute acicular or fibrous crystals elongated [001] and flattened {010}. As crystalline agglomerations, looking like white spots in the guano in which it is embedded; also massive and as rosette-like flattened aggregates; lamellar.

Phys. Cleavage {010} perfect. H. $2-2\frac{1}{2}$. G. 2.195 (artif.);[3] 2.17 (calc.). Luster weakly vitreous. Colorless and transparent, also white.

Opt.[4] In transmitted light, colorless.

Orientation	n (Minnesota)	n (New Zealand) anal. 4	
X	1.510	1.5468	Biaxial positive (+).
Y b	1.520	1.5533	$r < v$, perceptible.
Z ∧ c 29°	1.543	1.5820	
2V	71° ± 3°	57° ± 3°	

Chem. Octahydrated phosphate of magnesium, $Mg_3(PO_4)_2 \cdot 8H_2O$. The Ca reported in analysis 3 probably is due to admixed apatite. Fe″ and Mn substitute for Mg, with Mn:Fe:Mg = 1:5.0:9.8 in analysis 4.

Anal.

	CaO	FeO	MgO	P_2O_5	H_2O	Rem.	Total
1.			29.72	34.87	35.41		100.00
2.			29.97	34.59	35.38		99.94
3.	5.71		25.12	34.52	34.27		99.62
4.		15.10	16.40	31.45	30.75	6.30	100.00

1. $Mg_3(PO_4)_2 \cdot 8H_2O$. 2. Mejillones, Chile.[5] 3. Bamle, Sweden (*hautefeuillite*).[6] 4. Ferroan bobierrite. New Zealand.[8] Rem. is Fe_2O_3 0.20, MnO 2.95, Na_2O 0.04, CO_2 tr., insol. 0.30, organic matter 2.81. H_2O includes $H_2O - 2.45$.

Tests. B.B. exfoliates and fuses. In C.T. loses water at a low temperature. Easily soluble in acids.

Occur. Found in guano on the island of Mejillones, Chile. With apatite at Odegärden near Bamle, Norway (*hautefeuillite*). In cavities in a fossil elephant tusk in Pleistocene gravel near Edgerton, Pipestone County, Minnesota.[7] As a fossil enterolith found in New Zealand.[8]

Artif.[3] Obtained in crystals by slow precipitation of a $MgSO_4$ solution by a solution of Na_2HPO_4 and $NaHCO_3$.

Name. After the French agricultural chemist P. A. Bobierre (1823–1881), who first described the mineral.

Ref.
1. Elements of the structure cell of Barth, *Am. Min.*, **22**, 325 (1937), on Mejillones fibers. The morphological data of Bobierre (1868), Michel (1893) on hautefeuillite (= bobierrite), and de Schulten, *Bull. soc. min.*, **26**, 81 (1903), on artificial crystals are unsatisfactory. Bobierrite apparently is structurally distinct from the vivianite group although closely related thereto.
2. Barth (1937) by the Weissenberg method.
3. de Schulten (1903).
4. Gruner and Stauffer, *Am. Min.*, **28**, 339 (1943). See also Barth (1937) and Larsen (49, 1921).
5. Lacroix, *C.R.*, **106**, 631 (1888).
6. Michel (1893).
7. Gruner and Stauffer (1943).
8. Hutton, *New Zealand J. Sc. Tech.*, Sect. B, **23**, 9 (1941).

HOERNESITE

40.2.17 HOERNESITE [$Mg_3(AsO_4)_2 \cdot 8H_2O$]. Hörnesit *Haidinger* (*Jb. geol. Reichsanst. Wien*, **11**, 41, 1860; *Ak. Wien, Sitzber.*, **40**, 18, 1860).

C r y s t.[1] Monoclinic; prismatic—$2/m$.

$a:b:c = 0.7676:1:0.3591; \quad \beta\ 104°25'; \quad p_0:q_0:r_0 = 0.4678:0.3478:1$

$r_2:p_2:q_2 = 2.8753:1.3451:1; \quad \mu\ 75°35'; \quad p_0'\ 0.4830, q_0'\ 0.3591, x_0'\ 0.2571$

Forms:

	ϕ	ρ	ϕ_2	$\rho_2 = B$	C	A
b 010	0°00′	90°00′	....	0°00′	90°00′	90°00′
m 110	53 22	90 00	0°00′	53 22	78 28½	36 38
i 450	47 06	90 00	0 00	47 06	79 29½	42 54
n 201	90 00	50 44	39 19	90 00	36 19	39 19
w $\bar{2}$01	−90 00	35 20	125 20	90 00	49 45	125 20
v $\bar{2}$21	−44 37½	45 15½	125 20	59 38	56 07	119 56

Habit. Prismatic [001] and flattened {010}. Also columnar; radial-foliated.

P h y s. Cleavage {010} perfect, {100} poor. Flexible. H. 1. G. 2.73 (Joachimsthal),[2] 2.57 (Fiano),[3] 2.609 (artif.).[4] Luster pearly on cleavage. Color white. Transparent.

O p t.[5] In transmitted light, colorless.

ORIENTATION	n (Joachimsthal)	
X b	1.563	Biaxial positive (+).
Y	1.571	$2V$ 60°.
$Z \wedge c$ 31°	1.596	

C h e m. Octahydrated arsenate of magnesium, $Mg_3(AsO_4)_2 \cdot 8H_2O$.
Anal.

	1	2	3
MgO	24.44	25.54	24.28
As_2O_5	46.44	46.33	46.30
H_2O	29.12	29.07	29.32
Total	100.00	100.94	99.90

1. $Mg_3(AsO_4)_2 \cdot 8H_2O$. 2. Banat, Hungary.[6] 3. Fiano, Italy.[3] With tr. of CaO, FeO, MnO.

Tests. B.B. easily fusible. Soluble in acids.

O c c u r. Originally found in the vicinity of Oravicza or Csiklova in the Banat, Roumania; also with nagyagite from Nagyág, Roumania, and from Joachimsthal, Bohemia. With nocerite, fluorite, and hydromagnesite in blocks of metamorphosed limestone in tuff at Fiano near Naples, Italy.

N a m e. After M. Hoernes (1815–1868), curator of the Imperial Mineral Cabinet in Vienna.

Ref.

1. Forms and angles of Zambonini, *Com. geol. ital., Mem.*, **7**, 127 (1919), on minute Fiano crystals. The c-axis of Zambonini's unit is halved to conform to vivianite; it is

not known, however, whether hoernesite is isostructural with vivianite or bobierrite. See Wolfe, *Am. Min.*, **25**, 806 (1940), for inconclusive x-ray data.
2. Wolfe (1940).
3. Zambonini (1919).
4. de Schulten, *Bull. soc. min.*, **26**, 81 (1903).
5. Larsen (86, 1921).
6. von Hauer, *Jb. Geol. Reichsanst. Wien*, **11**, 10 (1860).

TYPE 3. $A(XO_4) \cdot xH_2O$

40.3.1 VARISCITE GROUP

ORTHORHOMBIC; DIPYRAMIDAL—2/m 2/m 2/m

	a_0	b_0	c_0
Variscite, $AlPO_4 \cdot 2H_2O$	9.85 kX	9.55	8.50
Strengite, $FePO_4 \cdot 2H_2O$	10.05	9.80	8.65
Mansfieldite, $AlAsO_4 \cdot 2H_2O$			
Scorodite, $FeAsO_4 \cdot 2H_2O$	10.30	10.31	8.92

The members of the Variscite Group are isostructural, but the details of their crystal structure are not yet known. The phosphate members of the group are isodimorphous [1] with the members of the monoclinic Metavariscite Group. A complete substitutional series involving Fe''' and Al extends between the phosphate members variscite and strengite and probably between the arsenate members mansfieldite and scorodite. Substitution between (AsO_4) and (PO_4) apparently is limited to a small range at the arsenate end of the series.

Ref.
1. See Strunz and Sztrókay, *Zbl. Min.*, 272 (1939).

VARISCITE-STRENGITE SERIES

40.3.1.1 VARISCITE [$Al'''(PO_4) \cdot 2H_2O$]. Peganite *Breithaupt* (*J. Chemie u. Phys.*, **60**, 308, 1830). Variscite *Breithaupt* (*J. pr. Chem.*, **10**, 506, 1837). Sphaerite. Sphärit *Zepharovich* (*Ak. Wien, Sitzber.*, **56**, 24, 1867). Redondite *Shepard* (*Am. J. Sc.*, **47**, 428, 1869). Barrandite pt., *var. authors.* Utahlite *Kunz* (*U. S. Geol. Sur., 16th Ann. Rpt.*, Pt. 4, 602, 1895). Amatrice pt. *Sterrett* (*U.S. Geol. Sur., Min. Res.*, Pt. 2, 832, 1908). Chlor-utahlite *Sterrett* (*U. S. Geol. Sur., Min. Res.*, Pt. 2, 853, 1909). Gelvariscite, Uhligite *Cornu* (*Zs. Chem. Ind. Kolloide*, **4**, 17, 1909). Lucinite *Schaller* (*U. S. Geol. Sur., Bull. 610*, 56, 1916). Alpha-variscite *Ulrich* (*Rozpr. České Ak., Cl. 2*, **39**, no. 17, 1930).

40.3.1.2 STRENGITE [$Fe'''(PO_4) \cdot 2H_2O$]. Barrandite *Zepharovich* (*Ak. Wien, Sitzber.*, **56**, 20, 1867). Strengite *Nies* (*Jb. Min.*, 8, 1877).

Cryst. Orthorhombic; dipyramidal—$2/m\ 2/m\ 2/m$.

VARISCITE,[1] $Al(PO_4) \cdot 2H_2O$

$a:b:c = 1.0217:1:0.8918;\qquad p_0:q_0:r_0 = 0.8729:0.8918:1$

$q_1:r_1:p_1 = 1.0217:1.1457:1;\qquad r_2:p_2:q_2 = 1.1213:0.9788:1$

VARISCITE-STRENGITE

Forms: [2]

	ϕ	$\rho = C$	ϕ_1	$\rho_1 = A$	ϕ_2	$\rho_2 = B$
c 001	.....	0°00′	0°00′	90°00′	90°00′	90°00′
b 010	0°00′	90 00	90 00	90 00	.....	0 00
e 120	26 04½	90 00	90 00	63 55½	0 00	26 04½
d 201	90 00	60 11½	0 00	29 48½	29 48½	90 00
p 111	44 23	51 17½	41 43½	56 55	48 53	56 06½
i 121	26 04½	63 16	60 46½	66 53	48 53	36 39½
r 131	18 04	70 26½	69 30½	73 00½	48 53	26 23½
s 211	62 56½	62 58½	41 43½	37 30½	29 48½	66 05½

STRENGITE,[3] $Fe(PO_4) \cdot 2H_2O$

$$a:b:c = 1.0244:1:0.8858; \quad p_0:q_0:r_0 = 0.8647:0.8858:1$$

$$q_1:r_1:p_1 = 1.0244:1.1565:1; \quad r_2:p_2:q_2 = 1.1289:0.9762:1$$

Forms: [4]

	ϕ	$\rho = C$	ϕ_1	$\rho_1 = A$	ϕ_2	$\rho_2 = B$
c 001	.....	0°00′	0°00′	90°00′	90°00′	90°00′
b 010	0°00′	90 00	90 00	90 00	.....	0 00
a 100	90 00	90 00	.....	0 00	0 00	90 00
d 120	26 01	90 00	90 00	63 59	0 00	26 01
m 110	44 18½	90 00	90 00	45 41½	0 00	44 18½
r 101	90 00	40 51	0 00	49 09	49 09	90 00
h 201	90 00	59 57½	0 00	30 02½	30 02½	90 00
e 012	0 00	23 53½	23 53½	90 00	90 00	66 06½
ρ 011	0 00	41 32	41 32	90 00	90 00	48 28
p 111	44 18½	51 04	41 32	57 05	49 09	56 10½
s 121	26 01	63 06	60 33½	66 58	49 09	36 44
i 211	62 52½	62 46	41 32	37 41	30 02½	66 05½

Structure cell. Space group $Pcab$.

	a_0	b_0	c_0	$a_0:b_0:c_0$
Variscite (Lucin; anal. 1?) [5]	9.85 kX	9.55	8.50	1.031:1:0.890
Strengite (Pleystein; anal. 12?) [5]	10.05 kX	9.80	8.65	1.026:1:0.883

Cell contents $(Al''',Fe)_8(PO_4)_8 \cdot 16H_2O$.

Habit. Crystals of variscite are rare; these are octahedral {111}, with {001} and other forms as modifying faces only. Variscite ordinarily occurs as fine-grained masses, nodules, veinlets, or crusts; also chalcedonic to opaline. Strengite crystals are more variable in habit: octahedral {111}; thick to thin tabular {001}; stout prismatic [100] or [010]. Strengite generally occurs as spherical and botryoidal aggregates with radial fibrous structure and a drusy surface, and as crusts.

Twinning. On {201}, rare.[6]

Phys. Cleavage {010} good, {001} poor. Massive variscite has an uneven to splintery fracture, tending toward conchoidal in very fine-grained or glassy types. H. 3½ to 4½, highest in variscite. G. 2.57 (variscite end-member, crystals), 2.61 (calc. for $AlPO_4 \cdot 2H_2O$), microcrystalline and massive variscite has a lower G., mostly 2.2 to 2.5; 2.87 (strengite end-member, crystals), 2.90 (calc. for $FePO_4 \cdot 2H_2O$). Luster vitreous (crystals) to faintly waxy in dense variscite. Color of variscite end-member pale green to emerald green, also bluish green to colorless; of strengite end-member peach-blossom-red, carmine, violet, also nearly colorless. Streak white. Transparent to translucent.

Opt. In transmitted light, colorless to pale green, pink, etc.

ORIEN-TATION		Variscite end-member, anal.[7]	Strengite end-member [8]	
nX	a	1.563	1.707	Biaxial negative (−)
nY	c	1.588	1.719	(variscite) or positive (+)
nZ	b	1.594	1.741	(strengite).
				$r < v$, perceptible (variscite) to strong (strengite).
				$2V$ moderate, sometimes small in strengite.

Optical data have been reported [9] for numerous unanalyzed members of the series. The indices reported for the essentially pure variscite end-member vary widely, owing probably to the presence of non-essential water in cryptocrystalline types or to partial loss of essential water by standing at room temperature. A similar, unexplained variation is found among unanalyzed specimens of the presumed strengite end-member. A selection of the reported measurements is given below.

	Variscite, Messbach, Saxony [10]	Variscite, Langenstriegis, Saxony [10]	Variscite, Langenstriegis, Saxony [10]	Ferrian variscite, Třenice [11]	Aluminian strengite, Černovice [12]	Strengite, Pala [10]	Strengite, Giessen [10]
nX	1.550	1.554	1.562	1.566	1.650	1.697	1.708
nY	1.565	1.571	1.583	1.584	[1.660]	1.714	1.708
nZ	1.570	1.576	1.587	1.593	1.680	1.772	1.745

Chem. A hydrated phosphate of aluminum and iron, $(Al''',Fe)(PO_4)\cdot 2H_2O$. A complete series extends between Fe''' and Al'''. The species names variscite and strengite are applied to material with $Al > Fe$ and $Fe > Al$, respectively. (AsO_4) does not substitute for (PO_4) in significant amounts in these species, although small amounts of (PO_4) substitute for (AsO_4) in the isostructural phase scorodite (which see). Most analyses of variscite and strengite are close to the Al or Fe ends of the series. Part of the water is lost below 110°, and all is driven off at 180° in variscite.[22]

Anal.[32]

	1	2	3	4	5	6
CaO			0.65	0.22	0.80	0.85
Al_2O_3	32.26	32.40	31.63	28.83	30.08	24.93
Fe_2O_3		0.06	tr.	2.27	3.42	6.50
Cr_2O_3		0.18		0.73		
P_2O_5	44.94	44.73	44.99	41.98	41.50	41.15
H_2O	22.80	22.68	23.20	25.98	24.50	23.47
Rem.		0.32	0.01	0.80		3.23
Total	100.00	100.37	100.48	100.81	100.30	100.13
G.	2.61	2.53	2.57			2.60

1. $Al'''(PO_4)\cdot 2H_2O$. 2. Variscite. Lucin, Utah (*lucinite*).[10] Rem. is V_2O_3. 3. Variscite. Langenstriegis (*peganite*).[13] Rem. is insol. 4. Variscite. Železnik, Czechoslovakia.[14] Rem. is Na_2O 0.07, K_2O 0.12, MgO 0.10, CuO 0.09, FeO 0.16, insol. 0.26. 5. Variscite. Connétable, French Guiana.[15] 6. Variscite. Třenice, Czechoslovakia.[11] Rem. is insol.

	7	8	9	10	11	12	13
CaO		0.57		1.20	0.83		
Al_2O_3	20.45	16.60	12.50	8.15			
Fe_2O_3	12.66	14.40	26.08	30.00	41.42	43.40	42.72
Cr_2O_3	0.73						
P_2O_5	42.19	43.20	38.93	36.29	37.60	38.24	38.00
H_2O	22.95	24.00	20.61	21.60	19.11	18.89	19.28
Rem.	1.37	1.60	1.04	3.38	1.26		
Total	100.35	100.37	99.16	100.62	100.22	100.53	100.00
G.	2.44						2.90

7. Ferrian variscite. Ninghanboun Hills, Western Australia.[16] Rem. is Na_2O tr., K_2O 0.96, $(NH_4)_2O$ tr., TiO_2 0.21, FeO 0.20, H_2O includes H_2O- 1.97. 8. Ferrian variscite. Redonda Island (*redondite*).[17] Rem. is insol. 9. Aluminian strengite. Černovice, Bohemia (*barrandite*).[18] Rem. is SiO_2. 10. Aluminian strengite. Manhattan, Nevada.[19] Rem. is MgO 1.36, FeO 0.90, insol. 1.12, H_2O includes H_2O at 100° 1.88. 11. Strengite. Mangualde, Portugal.[20] Rem. is MgO 0.10, Na_2O 0.55, SiO_2 0.10, Mn_2O_3 0.51, H_2O includes H_2O- 1.17. 12. Strengite. Pleystein, Bavaria.[21] 13. $Fe'''(PO_4) \cdot 2H_2O$.

Tests. B.B. the high-Al members are infusible but the high-Fe members fuse easily to a black, lustrous bead. Coarsely crystalline variscite end-member is insoluble in HCl but becomes soluble after heating to about 140°; the fine-grained or chalcedonic types are more or less soluble in HCl before heating. Soluble in alkalies. On slight heating the color of green variscite may become lavender to purplish red.[22] Strengite end-member is soluble in HCl but not in HNO_3.

Var. Ferrian variscite. Redondite *Shepard* (*Am. J. Sc.*, **47**, 428, 1869). Contains Fe in substitution for Al, with accompanying presumed linear variation in gravity, indices of refraction and unit cell dimensions toward the iron end-member.

Aluminian strengite. Barrandite *Zepharovich* (*Ak. Wien, Sitzber.*, **56**, 20, 1867). Contains Al in substitution for Fe, with accompanying presumed linear variation in properties toward the aluminum end-member.

Occur. The relatively highly aluminous members of the series (*variscite*) are typically deposited under surface or near-surface conditions in cavities or breccias by the action of phosphatic meteoric waters on aluminous rocks. Associated minerals include wavellite, crandallite, metavariscite, microcrystalline apatite, chalcedony, limonite. The iron-rich members of the series (*strengite*) also are surface or near-surface products formed by the alteration of iron-containing phosphates, such as triphylite in pegmatite or dufrenite, or are found in limonite ore deposits and gossans. Associated minerals include beraunite, cacoxenite, vivianite, metastrengite, frondelite-rockbridgeite, dufrenite, microcrystalline apatite. A majority of the known localities for these species are cited below.

Variscite occurs at Messbach and at Langenstriegis east of Frankenberg (*peganite*, pt.) [23] in Saxony, Germany. In Austria at Mixnitz and at Brandberg near Leoben in Styria. In Bohemia of Czechoslovakia at Železnik, Třenice, and Zaječow near St. Benigna (*sphaerite*, pt.).[24] A variscite-like mineral has been reported [25] from Mura-Panda in Katanga, Belgian Congo. In the United States variscite occurs in a large deposit near Fairfield, Utah County, Utah,[26] where the mineral forms nodules in a brecciated zone in phosphatic sediments and is associated with a variety of rare phosphates of Ca, Mg, Al, and alkalies. Also in Utah at Amatrice Hill near Stockton and at Lucin in Box Elder County. From Montgomery, Garland County, Arkansas; [27] at Moore's Mill, Cumberland County, Pennsylvania. In a number of small deposits in Nevada, mostly in Esmeralda County. Ferrian variscite occurs associated with apatite in insular rock-phosphate deposits,[28] as on the island of Redonda (*redondite*) [29] in the Antilles, Gran Roque Island in Venezuela, de la Perle Island near Martinique, and in the Pacific on Clipperton Atoll and on Malpelo Island, Colombia; in these occurrences the mineral has been formed by the action of phosphatic solutions derived from guano on aluminous igneous rocks. Also at Ninghanboun Hills, Western Australia.

Strengite occurs in Germany in the Eleonore iron mine near Giessen, and near Waldgirmes, Thuringia; also well-crystallized as an alteration of iron phosphates in pegmatite at Pleystein and Hagendorf, Bavaria, and at Mangualde, Beira province, Portugal. In veinlets in magnetite ore at Kirunavaara, Sweden. In the United States with rockbridgeite near Midvale, Rockbridge County, Virginia; at Moore's Mill, Cumberland County, and the Noblis mine, Bearville, Lancaster County, Pennsylvania; as an oxidation product of triphylite in the Palermo pegmatite, North Groton, New Hampshire. Aluminian strengite occurs at Manhattan, Nye County, Nevada; on the Island of Connétable, French Guiana; at Černovice near Příbram, Bohemia (*barrandite*).[29] A manganian (?) variety is found in pegmatite at Pala, San Diego County, California.

A l t e r . Variscite is readily altered by alkaline solutions containing alkaline earths or alkalies and affords crandallite, wardite, and chemically related species.[30]

A r t i f.[31] Strengite (metastrengite?) has been obtained as crystals by heating a phosphoric acid solution of ferric chloride in a closed tube at about 185°. Strengite and variscite can be obtained as fine-grained precipitates by addition of NaOH solution to acid solutions containing Fe or Al chlorides and dihydrogen phosphates.

N a m e . Variscite from *Variscia*, the ancient name of the Voigtland district in Germany where the mineral was first found (at Messbach near Plauen in Saxon Voigtland). Strengite after J. A. Streng (1830–1897), German mineralogist, of the University of Giessen.

Ref.

1. Angles and unit of Schaller, *U. S. Geol. Sur., Bull. 610*, 56 (1916), on lucinite (= *variscite*) from Lucin, Utah (analysis 2), in the orientation of the structure cell. This orientation is selected to best show the dimensional and other relations existing between the isodimorphous variscite and metavariscite groups—see Strunz and Sztrókay, *Zbl. Min.*, **272** (1939), and McConnell, *Am Min.*, **25**, 719 (1940). Transformation: Schaller to new, 010/001/100. Kolbeck, *Min. Mitt.*, **52**, 363 (1941), obtained $a:b:c$ = 1.030:1:0.8955 on analyzed crystals from Sosa, Saxony, with {111} and {001}. The morphological elements and the dimensions of the structure cell are presumed to vary linearly with Al/Fe ratio in the series.

2. Schaller (1916).

3. Angles and unit of Himmel and Schroeder, *Zbl. Min.*, 289 (1935), on Pleystein crystals (analysis 12 ?) in the orientation of the structure cell. Transformation: H. and

S. to new, 010/001/100. See also Laubmann and Steinmetz, *Zs Kr.*, **55**, 542 (1920), and Goldschmidt (**3**, 89, 1922).
4. Himmel and Schroeder (1935) and Goldschmidt (1922). Rare or doubtful: t 021, f 023, g 058, q 223, x 3.10.4, π 12.10.15, k 304.
5. McConnell, *Am. Min.*, **25**, 719 (1940). Almost identical measurements are given for strengite end-member by Kokkoros, *Prakt. Ac. Athénes*, **13**, 337 (1938)—*Min. Abs.*, **8**, 140 (1941).
6. Himmel and Schroeder (1935) in strengite.
7. Schaller (1916) on crystals from Lucin.
8. Frondel, priv. comm. (1949), on unanalyzed Pleystein crystals probably representing strengite end-member.
9. Schaller (1916); Gordon, *Proc. Ac. Sc., Philadelphia*, **77**, 6 (1925); Larsen (**44**, 118, 138, 152, 1921); Ulrich, *Rozpr. České Ak.*, Cl. 2, **39**, no. 17 (1930).
10. Schaller (1916)   The strengite from Pala may be metastrengite.
11. Ulrich, with Fe:Al = 1:6 in analysis 6.
12. Gordon (1925).
13. Moschetti, *Acc. Torino, Att.*, **53**, 1062 (1918).
14. Loczka analysis in Ulrich, *Rozpr. České Ak.*, *Cl. 2*, **31**, no. 10 (1922).
15. Pisani analysis in Lacroix (**4**, 480, 1910).
16. Simpson, *J. Roy. Soc. Western Australia*, **18**, 61 (1932).
17. Shepard, *Am. J. Sc.*, **47**, 428 (1869).
18. Bořický analysis in Zepharovich, *Ak. Wien, Sitzber*, **66**, 22 (1867).
19. Shannon, *Am. Min.*, **8**, 182 (1923).
20. de Jesus, *Com. Serv. Geol. Portugal*, **19**, 65 (1935).
21. Laubmann and Steinmetz, *Zs. Kr.*, **55**, 546 (1920).
22. The changes in color and optical properties on heating are described by Schaller, *Proc. U. S. Nat. Mus.*, **41**, 413 (1912); *Zs. Kr.*, **50**, 321 (1912).
23. Shown to be identical with variscite by Moschetti (1918) and Larsen and Schaller (1925).
24. Shown to be identical with variscite by Pearl, priv. comm. (1947).
25. Schoep, *Soc. géol. Belgique, Bull.*, **37**, 89 (1927), with G. 2.92, nX 1.578, $nY \sim$ 1.590, nZ 1.599, $2V$ 32°   Analysis Fe$_2$O$_3$ 3.98, Al$_2$O$_3$ 22.97, P$_2$O$_5$ 44.33, H$_2$O 21.22, insol. 0.90, total 93.40.
26. See Larsen, *Am. Min*, **27**, 281 (1942), for a description of the mineralogy and paragenesis.
27. Described by Chester, *Am. J. Sc.*, **15**, 207 (1878), and Schaller (1916) and shown identical with variscite by x-ray study (Frondel, priv. comm., 1949).
28. See Lacroix (**4**, 481, 1910) and McConnell, *Geol. Soc. Am. Bull.*, **54**, 707 (1943).
29. Shown to be identical with variscite by Frondel, priv. comm. (1949), by x-ray and optical study.
30. See Larsen (1942).
31. de Schulten, *C.R.*, **100**, 15 (1885), Mellor (**5**, 362, 1924), Cole and Jackson, *J. Phys. Coll. Chem.*, **54**, 128 (1950).
32. For additional analyses see Hintze (**1** [4A], 1283, 1288, 1933); Ulrich (1930); Machatschki, *Cbl. Min.*, 321 (1929); Pelloux, *Ann. Mus. Genova*, **5**, 470 (1912); Tschirwinsky, *Ann. géol. min. Russie*, **7**, 28 (1904).

The following are ill-defined substances that may be identical with variscite-strengite or wavellite.

MEYERSITE. Elschner (*Koll.-Zs.*, **31**, 94, 1922). Found as agate-like, metacolloidal masses in cavities in lava on Necker Island, Hawaii. Analysis gave: AlPO$_4$ 66.33, FePO$_4$ 2.52, H$_2$O 26.10, total 94.95. Darker-colored zones in the colloform crusts have up to 10 per cent FePO$_4$. Named after H. H. Meyers of the Mellon Institute, Pittsburgh. Probably ferrian variscite.

CALLAINITE. Callaina? *Pliny* (**37**, 33). Turquois pt. Callais *Damour* (*C.R.*, **59**, 936, 1864). Callainite *Dana* (825, 1892). Callaïnit *Germ.* A name given to the substance of Neolithic beads found at various places in Brittany,[1] originally in the dolmens at Mané-er-H'roeck in Lockmariaquer. Massive, wax-like. Color apple-green to emerald-green, spotted or veined with whitish or bluish. The mean index of refraction (1.576) and the physical properties are similar to those of microcrystalline variscite.[1] Analysis gave (Mané-er-H'roeck):[2] CaO 0.70, Al$_2$O$_3$ 29.57, Fe$_2$O$_3$ 1.82.

Cr_2O_3 tr., P_2O_5 42.58, H_2O 23.62, insol. 2.10, total 100.39, corresponding to $Al(PO_4)\cdot 2\frac{1}{2}H_2O$. Named from the supposed identity of the material with the *callais* of Pliny. Probably variscite.

GLOBOSITE. Breithaupt (*B. H. Ztg.*, **24**, 321, 1865). Small globular concretions found with limonite at the Arme Hilfe mine near Hirschberg, Thuringia. Also reported from Schneeberg, Saxony. H. 5–5½. G. 2.83. Color wax-yellow to yellowish gray. Luster greasy adamantine. Streak white. Analysis gave: CaO 2.40, MgO 2.40, CuO 0.48, Fe_2O_3 40.86, P_2O_5 28.89, As_2O_5 tr., SiO_2 0.24, H_2O and F 23.94, total 99.21. The analysis is probably erroneous. Perhaps identical with strengite.

PLANERITE. Hermann (*Soc. nat. Moscou, Bull.*, **35**, [2], 240, 1862). Subcrystalline, botryoidal crusts. H. 5. G. 2.65. Color verdigris-green, passing to olive-green on exposure. Luster dull. Translucent on edges, with a birefringence [3] of 0.017. Analysis gave: FeO 3.52, CuO 3.72, Al_2O_3 37.48, P_2O_5 33.94, H_2O 20.93, total 99.59. B.B. decrepitates, yielding much neutral water. Only slightly attacked by acids but soluble in boiling NaOH solution. Found at the copper mines of Gumeshevsk, Ural Mountains, U.S.S.R., and named after the director of the mines. A metacolloidal material from near Jakubeny in Bukovina has been ascribed [4] to planerite, and the possible identity of coeruleolactite and fischerite with planerite has been suggested, but such views are speculative in lack of rigorous descriptive criteria newly obtained on type material.

FISCHERITE. Hermann (*J. pr. Chem.*, **33**, 285, 1844).

Orthorhombic; pseudo-hexagonal.[5] $a:b:c = 0.5937:1:?$ Forms: c 001, b 010, m 110, g 120. Crystals small, often six-sided prisms with m and b; as acicular crystals grouped in druses in radiating form, as scales, and in radial fibrous crusts. H. 5. G. 2.46. Luster vitreous. Color grass-green to olive-green and verdigris-green. Translucent.

O p t.[6]

	ORIENTATION	n(Na)(Urals)	
X	b		Biaxial positive (+).
Y	a	1.557	$2V$ 61°51′ (Na), 62°05′ (red).
Z	c		$r > v$.

C h e m. A hydrated basic aluminum phosphate. The only reported analysis [2] does not differ greatly in ratios from those of wavellite: CuO 0.80, Fe_2O_3 1.20, Al_2O_3 38.47, P_2O_5 29.03, H_2O 27.50, gangue 3.00, total 100.00.

B.B. becomes white and clouded and yields much water but no fluorine. Soluble in H_2SO_4. Found in crevices of a ferruginous sandstone and clay slate at Nizhne Tagilsk in the Urals, U.S.S.R. Named after G. Fischer (1771–1853), Professor of Natural History in the University of Moscow. A supposedly identical mineral has been reported [7] as occurring in botryoidal enamel-like crusts at Roman-Gladna, Transylvania (*gelfischerite*).[8] The material from Russia probably is identical with wavellite.[9]

RICHMONDITE. Kenngott (*Viertel-jahrschr. nat. Ges. Zürich*, **11**, 225, 1866). Gibbsite Hermann (*J. pr. Chem.*, **40**, 32, 1847; **47**, 1, 1849). A substance found with true gibbsite, $Al(OH)_3$, in a limonite deposit at Richmond, Berkshire County, Massachusetts. Analysis gave:[10] Al_2O_3 26.66, P_2O_5 37.62, H_2O 35.72, total 100.00. This is close to $AlPO_4\cdot 4H_2O$.

Ref.

1. Lacroix (**4**, 480, 1910).
2. Damour (1864).
3. Lacroix, *Bull. soc. min.*, **9**, 4 (1886).
4. See Leitmeier, *Zs. Kr.*, **55**, 353 (1916), and Grosspietsch, *Verh. geol. Reichsanst. Wien*, 149 (1919).
5. Koksharov (**1**, 31, 1853) from $mm''' = 61°28'$; also $gg' = 80°08'$. Compare wavellite and the Arkansas variscite of Chester, *Am. J. Sc.*, **15**, 207 (1878).
6. Des Cloizeaux cited in Koksharov (**7**, 23, 1875).

SCORODITE–MANSFIELDITE

7. See Slávik, *Ac. sc. Bohême, Bull.*, **22**, 32 (1918); Larsen, *Am. Min.*, **2**, 31 (1917); Larsen (75, 1921).
8. Cornu, *Zs. Chem. Ind. Kolloide*, **4**, 17 (1909).
9. Wherry, *Am. Min.*, **2**, 32 (1917).
10. Hermann, *J. pr. Chem.*, **40**, 32 (1847).

KONINCKITE. Cesàro (*Soc. géol. Belgique, Ann. Mem.*, **11**, 247, 1883–84). Orthorhombic (?). In small spherical aggregates of radiating needles; in one instance terminated by an oblique plane. Cleavage transverse to the elongation. H. $3\frac{1}{2}$. G. 2.3. Luster vitreous. Color and streak yellow. Transparent.

Opt.[1] In transmitted light, pale yellow to colorless.

ORIENTATION		n	
X	elong.	1.645	Biaxial.
Y			$2V$ not very large.
Z		1.656	In-part nearly isotropic with $n \sim 1.65$.

Chem. A hydrated ferric phosphate, apparently $FePO_4 \cdot 3H_2O$. Al substitutes in small amount for Fe''', with Al:Fe = 1:4.8 (anal. 2).

Anal.

	1	2
Fe_2O_3	33.05	34.4
Al_2O_3	4.40	4.6
P_2O_5	35.51	34.8
H_2O	27.04	[26.2]
Total	100.00	[100.00]

1. $(Fe,Al)PO_4 \cdot 3H_2O$ with Al:Fe = 1:4.8. 2. Richelle, Belgium. Average of two.

Tests. B.B. easily fusible. Easily soluble in hot HCl or HNO_3.
Occur. Found with richellite at Richelle near Visé, Belgium.
Name. After the Belgian geologist L. G. de Koninck (1809–1887).
Needs further study.

Ref.

1. Cesàro, *Ac. Belgique, Mém.*, **53**, 1 (1897), and Larsen (97, 1921).

40.3.1.3 **SCORODITE** [$Fe'''(AsO_4) \cdot 2H_2O$]. Cupreous Arsenate of Iron. Cupromartial Arsenate *Bournon* (*Phil. Trans.*, 191, 1801). Martial Arsenate of Copper. Cuivre arseniaté ferrifére *Haüy* (91, 1809). Scorodit *Breithaupt* (*Hoffmann's Handbuch*, **4**, 2, 182, 1817). Scorodite and Néoctèse *Beudant* (**2**, 605, 607, 1832); Des Cloizeaux (*Ann. chim. phys.*, **10**, 402, 1844). Arseniksinter, Eisensinter *Hermann* (*Bull. Soc. Imp. Nat. Moscou*, **1**, 254, 1845). Kobalt-scorodit *Lippmann* (von Hornberg, *Zool. min. Ver. Regensberg*, **11**, 172, 1857). Jogynaite ? *Nordenskiöld* (in Koksharov, *Bull. ac. sc. St. Pétersbourg*, **19**, 571, 1873; *Min. Russ.*, **6**, 320, 1870). Lossenite *Milch* (*Zs. Kr.*, **24**, 100, 1895). Loaisite *Codazzi* (*Cbl. Min.*, 183, 1908). Aluminoscorodite *Ito* and *Sakurai* (Wada's *Min. of Japan*, 3rd ed., **1**, 1947).

40.3.1.4 **MANSFIELDITE** [$Al'''(AsO_4) \cdot 2H_2O$]. *Allen* and *Fahey* (*Am. Min.*, **33**, 122, 1948).

Cryst. Orthorhombic; dipyramidal—$2/m\ 2/m\ 2/m$.

SCORODITE,[1] $Fe(AsO_4) \cdot 2H_2O$

$a:b:c = 1.0404:1:0.9030;\qquad p_0:q_0:r_0 = 0.8679:0.9030:1$

$q_1:r_1:p_1 = 1.0404:1.1522:1;\qquad r_2:p_2:q_2 = 1.1074:0.9612:1$

764   PHOSPHATES, ARSENATES, VANADATES

Forms: [2]

		ϕ	$\rho = C$	ϕ_1	$\rho_1 = A$	ϕ_2	$\rho_2 = B$
c	001	....	0°00′	0°00′	90°00′	90°00′	90°00′
b	010	0°00′	90 00	90 00	90 00	....	0 00
a	100	90 00	90 00	....	0 00	0 00	90 00
e	120	25 37	90 00	90 00	64 23	0 00	25 37
m	110	43 52	90 00	90 00	55 08	0 00	43 52
M	012	0 00	24 18	24 18	90 00	90 00	65 42
h	011	0 00	42 05	42 05	90 00	90 00	47 55
o	021	0 00	61 01½	61 01½	90 00	90 00	28 58½
n	101	90 00	40 57½	0 00	49 02½	49 02½	90 00
d	201	90 00	60 03	0 00	29 57	29 57	90 00
p	111	43 52	51 23	42 05	57 13	49 02½	55 42½
i	121	25 40	63 28½	61 01½	67 12	49 02½	36 15
s	211	62 31	62 55½	42 05	37 49	29 57	65 44

Less common:
  M· 320   L 210   q 102   k 403   g 301   w 401   l 112

Structure cell.[3] Space group $Pcab$. a_0 10.30 kX, b_0 10.01, c_0 8.92; $a_0:b_0:c_0 = 1.029:1:0.891$ (scorodite end-member?). Cell contents $(Fe,Al)_8 (AsO_4)_8 \cdot 16H_2O$.

Habit. Scorodite crystals are usually pyramidal {111}; also tabular {001} or prismatic [010]. Common forms: $p\ d\ c\ n\ i$. The crystals often

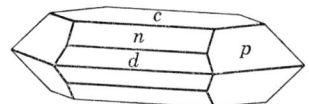

Scorodite. Djebel Debar, Algeria.

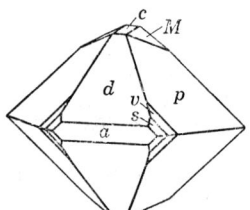

Scorodite. Blagodatnoi, Ural Mts.

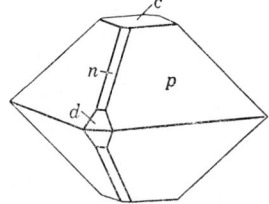

Scorodite. Djebel Debar, Algeria.

are aggregated into crusts or irregular groups. Also massive, crystalline, or porous and sinter-like; dense to earthy. Mansfieldite occurs as porous, cellular masses and crusts of spherulitic fibers.

P h y s. Cleavage {201} imperfect, {001} and {100} in traces. Fracture subconchoidal. H. $3\frac{1}{2}$–4. G. 3.28 (scorodite end-member, anal. 2), 3.31 (calc.); 3.03 (mansfieldite end-member, anal. 7). Luster of scorodite crystals strongly vitreous to subadamantine and subresinous. Color of scorodite crystals pale leek-green or grayish green to liver-brown, sometimes nearly colorless, or bluish, violet, yellow; earthy material is pale green to pale grayish or brownish green. Mansfieldite is white to pale gray. Transparent in crystals.

O p t. In transmitted light, colorless to faint green or greenish brown.

ORIENTATION		Scorodite, Durango (anal. 2) [4]	Aluminian scorodite, Oregon (anal. 6) [5]	Mansfieldite, Oregon (anal. 7) [5]	
nX	a	1.784	1.741	1.622	Biaxial positive (+).
nY	c	1.795	1.744	1.624	$r > v$, strong.
nZ	b	1.814	1.768	1.642	Scorodite is faintly
2V		~75°	~40°	~30°	pleochroic, with absorption $Z > X$, Y.

The indices of refraction, birefringence and 2V decrease with decreasing Fe/Al ratio. Divergent optical properties have been reported for other scorodites:

	Phosphatian scorodite, Idaho (anal. 5) [6]	Scorodite, Japan (anal. 4) [7]	Scorodite, Cornwall [6]	Scorodite, Japan [6]
nX	1.738	1.771(Na)	1.810	1.888
nY	1.742	1.805	1.880	1.895
nZ	1.765	1.820	1.925	1.915
2V	med.		~90°	62° (calc.)
Disp.	$r > v$, strong		none	$r > v$, strong

C h e m. A hydrated arsenate of ferric iron and aluminum, (Fe,Al)(AsO$_4$)·2H$_2$O. Fe''' and Al''' substitute mutually and a complete series [8] probably extends from scorodite, with Fe > Al, to mansfieldite, with Al > Fe. (PO$_4$) substitutes in small amounts for (AsO$_4$) in scorodite, with P:As = 1:5.7 in analysis 5, and a partial series at least extends towards strengite. Scorodite loses all its water between 220° and 250°.[7]

Anal.[12]

	1	2	3	4	5	6	7	8
Fe$_2$O$_3$	34.60	34.79	34.13	34.33	34.02	25.72	0.88	
Al$_2$O$_3$		none	none		tr.	5.76	23.30	25.24
As$_2$O$_5$	49.79	49.52	48.42	49.85	44.40	48.88	56.43	56.91
P$_2$O$_5$		none	none		4.80	1.72	0.59	
H$_2$O+	15.61	15.44	15.73	15.84	12.25	15.86	17.77	17.85
H$_2$O−			0.23		5.08	1.18		
Rem.		0.38	1.65		0.32	1.00	1.03	
Total	100.00	100.13	100.16	100.02	100.87	100.12	100.00	100.00
G.		3.278	3.413			3.135	3.031	

1. Fe'''(AsO$_4$)·2H$_2$O. 2. Scorodite. Durango, Mexico.[9] Rem. is TiO$_2$ 0.02, Sb$_2$O$_5$ 0.06, SiO$_2$ 0.30. 3. Scorodite. Gold Hill, Utah.[10] Rem. is MgO 0.01, CaO 0.38, FeO 0.84, insol. 0.42. 4. Scorodite. Kiura mine, Bungo, Japan.[7] 5. Phosphatian scorodite. Black Pine, Idaho.[11] Rem. is Cr$_2$O$_3$. 6. Aluminian scorodite. Hobart Butte, Oregon.[9] Rem. is TiO$_2$ 0.06, SiO$_2$ 0.20, Sb$_2$O$_5$ 0.74. 7. Mansfieldite. Hobart Butte, Oregon.[9] Rem. is TiO$_2$ 0.91, Sb$_2$O$_5$ 0.12. Recalculated after deduction of 30.68 per cent impurities.

Tests. Scorodite fuses B.B. to a lustrous, gray magnetic mass. In C.T. loses water and turns yellow. Soluble in acids; in strong alkalies decomposed with separation of hydrous iron oxide.

Var. *Aluminian* scorodite. Contains Al''' in substitution for Fe''', with accompanying linear decrease in the indices of refraction, $2V$, birefringence, and gravity.

Phosphatian scorodite. Contains (PO_4) in substitution for (AsO_4), with decreased indices of refraction.

O c c u r. Scorodite is typically a secondary mineral formed in gossans by the oxidation of arsenopyrite or other arsenic-containing minerals. Associated with pharmacosiderite, beudantite, vivianite, limonite, gypsum, iron-vitriols, chalcedonic quartz, clay. Also found as a primary hydrothermal deposit (Saubach, Voigtland, Germany).[16] Only a few of the known localities can be listed.

In Germany at Dernbach northwest of Montabaur, Hesse-Nassau; at Saubach in the Voigtland; on the Graul near Schwartzenberg, Saxony. In Austria at Lölling, Carinthia. At Schlaggenwald and Schönfeld, Bohemia. From the Adun-Chilon Mountains, south of Nerchinsk, Transbaikalia, U.S.S.R. In France at Vaulry, northwest of Limoges, and elsewhere in Haute-Vienne; at Morbihan, Brittany. From Djebel Debar, Constantine, Algeria, and Gonnosfanadiga, Sardinia. From Laurium, Greece (*lossenite*, pt.). From St. Stephens and elsewhere in Cornwall; from the Virtuous Lady mine near Tavistock, Devonshire. Near Ouro Preto, Minas Geraes, Brazil. At Kiura, Bungo, Japan. In the United States, scorodite occurs near Carmel, Putnam County, New York. At the Canton mine and elsewhere in Cherokee County, Georgia. In the Gold Hill mine, Toole County, Utah, with adamite, austinite, and hemimorphite, and with various copper arsenates in the Mammoth mine and elsewhere in the Tintic district, Utah. In the Monte Cristo district, Snohomish County, Washington, and at Black Pine, Cassia County, Idaho. As thin coatings and stains on the siliceous sinter of the geysers and hot springs of the Yellowstone region, Wyoming;[13] similarly at Steamboat Springs, Nevada. At numerous minor localities in California, Nevada, Colorado, and other western states. In Canada with erythrite in the Nipissing mine, Temiskaming district, Ontario.

Mansfieldite occurs with scorodite, realgar, and kaolinite at Hobart Butte, Lane County, Oregon.

A r t i f.[14] Obtained as crystals by heating metallic iron or precipitated ferric arsenate with H_2O and As_2O_5 in a closed tube at temperatures up to 150° C.

A l t e r. To limonite.

N a m e. Scorodite from σκορόδιον, *garlic-like*, alluding to its odor when heated. Mansfieldite after George R. Mansfield (1875–   ), geologist, of the U. S. Geological Survey.

LOSSENITE. *Milch* (*Zs. Kr.*, **24**, 100, 1895). Found at Laurium, Greece. Shown[15] to be a mixture of scorodite and beudantite.

Ref.
1. Elements and unit of Goldschmidt (**8**, 58, 1922) in an orientation corresponding to strengite and variscite. Transformation: G. to new, 010/001/100. The reported angles for scorodite vary widely, owing doubtless to variation in composition—for literature see Goldschmidt (**3**, 135, 1890; 1922), Hintze (**1** [4A], 1289, 1933), and ref. 2.
2. Goldschmidt (1922); Chirva, *Bull. ac. sc., U.R.S.S.* [6], 19, 731 (1925); Piazza, *Acc. Linc., Rend.*, [6], **6**, 70 (1927). Numerous vicinal faces are listed by Ito and Shiga, *Min. Mag.*, **23**, 130 (1932). Rare and uncertain: Z 520, R 805, D 323, V 453, P 714, α 722, j 421, Q 405, t 325, y 423, X 513, A 724, β 411.
3. Kokkoros, *Prakt. Ac Athènes*, **13**, 337 (1938) on Callington, Cornwall, scorodite crystals. McConnell, *Am. Min.*, **25**, 719 (1940) and Strunz, *Zs. Kr.*, **99**, 513 (1938) give comparable values.
4. Allen and Fahey, *Am. Min.*, **33**, 122 (1948); comparable values on analyzed, essentially pure scorodite end-member are given by Foshag, *Am. Min.*, **22**, 482 (1937), and Foshag, Berman, and Doggett, *Am. Min.*, **15**, 390 (1930).
5. Allen and Fahey (1948).
6. Larsen (132, 1921).
7. Ito and Shiga (1932).
8. Analyses are lacking for material between mansfieldite end-member and the aluminian scorodite of analysis 6, but some intermediate members are indicated by the optical data of Allen and Fahey (1948), and a complete series is suggested by analogy to strengite-variscite.
9. Fahey analysis in Allen and Fahey (1948).
10. Foshag analysis in Foshag, *et al.* (1930).
11. Shannon (432, 1926) cited in Larsen (132, 1921).
12. For additional analyses see Hintze (1933); Foshag (1937); Codazzi, *Cbl. Min.*, 183 (1908); Rossetti, *Per. Min.*, **12**, 433 (1941).
13. Hague, *Am. J. Sc.*, **34**, 171 (1887).
14. Mellor (**9**, 224, 1929).
15. Pearl, priv. comm. (1947).
16. Heide, *Zs. Kr.*, **67**, 33 (1928).

40.3.2 METAVARISCITE GROUP

MONOCLINIC; PRISMATIC—$2/m$

	a_0	b_0	c_0	β
Metavariscite, $AlPO_4 \cdot 2H_2O$	5.15 kX	9.45	8.45	$\sim 90°$
Metastrengite, $FePO_4 \cdot 2H_2O$	5.28	9.75	8.71	$90°36'$

Metavariscite and metastrengite are polymorphs of the orthorhombic species variscite and strengite. Monoclinic polymorphs of mansfieldite and scorodite are not known. A complete series probably extends between metavariscite and metastrengite.

40.3.2.1 METAVARISCITE. [$Al(PO_4) \cdot 2H_2O$]. Variscite *Schaller* (*U. S. Geol. Sur., Bull. 509*, 48, 1912; *610*, 69, 1916). Metavariscite *Larsen* and *Schaller* (*Am. Min.*, **10**, 23, 1925).

Cryst.[1] Monoclinic; prismatic—$2/m$.

$a:b:c = 0.5460:1:0.8944;$ $\beta\ 90°00';$ $p_0:q_0:r_0 = 1.6381:0.8944:1$

$r_2:p_2:q_2 = 1.1180:1.8315:1;$ $\mu\ 90°00';$ $p_0'\ 1.6381,\ q_0'\ 0.8944,\ x_0'\ 0.0000$

PHOSPHATES, ARSENATES, VANADATES

Forms: [2]

	φ	ρ	φ₂	ρ₂ = B	C	A
c 001	90°00′	0°00′	90°00′	90°00′	.....	90°00′
b 010	0 00	90 00	.....	0 00	90°00′	90 00
a 100	90 00	90 00	0 00	90 00	90 00	.....
l 130	31 24½	90 00	0 00	31 24½	58 36	58 35½
m 110	61 22	90 00	0 00	61 22	28 38	28 38
h 011	0 00	41 48½	90 00	48 11½	41 48½	90 00
d 021	0 00	60 47½	90 00	29 12½	60 47½	90 00
j 052	0 00	65 54½	90 00	24 05½	65 54½	90 00
w 031	0 00	69 33½	90 00	20 26½	69 33½	90 00
p 1̄22	−42 29	50 29½	129 19½	55 13½	50 32½	121 24

Less common and rare:

    f 025   q 012   u 043   e 101

Structure cell.[5] Space group probably $P2_1/m$. a_0 5.15 kX, b_0 9.45, c_0 8.45; [β ~90°]; $a_0:b_0:c_0$ = 0.545:1:0.894. Cell contents $Al_4(PO_4)_4 \cdot 8H_2O$.

Habit. Crystals thin to thick tabular {010} and either equant or slightly elongated [001] or [100] to give lath-like shapes; rarely long prismatic

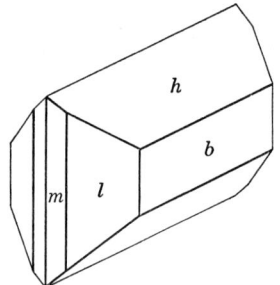 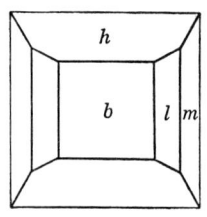

Lucin. Orthographic projection on {010}.

[100]. The form development is markedly pseudo-orthorhombic. Common forms: b m l h. Also as loosely coherent granular masses.

Twinning. On {102}, as contact twins.

Phys. Cleavage {010}.[3] H. ~3½.[3] G. 2.54; 2.53 (calc.). Color pale green. Streak white. Luster vitreous. Crystals transparent, masses translucent.

Opt. In transmitted light, colorless.

Orientation	n (crystals)[4]	n (small grains)[4]	
X		1.551	1.551	Biaxial positive (+).
Y b	1.558	1.561	2V 55°.
Z		1.582	1.585	r < v, perceptible.

Chem. A hydrated aluminum phosphate, $Al(PO_4) \cdot 2H_2O$. Dimorphous with variscite.[4] Fe‴ probably can substitute for Al, and a series may extend to metastrengite (which see).

Anal.

	1	2
Al_2O_3	32.26	32.40
P_2O_5	44.93	44.73
H_2O	22.81	22.68
Rem.		0.56
Total	100.00	100.37
G.	2.53	2.54

1. $Al(PO_4) \cdot 2H_2O$. 2. Lucin.[6] Rem. is V_2O_3 0.32, Cr_2O_3 0.18, Fe_2O_3 0.06.

Tests. As with variscite.

Occur. As tiny crystals associated with crystals of variscite in cavities in massive variscite nodules on Utahlite Hill, five miles northeast of Lucin, Box Elder County, Utah. Also found massive with variscite at Candelaria, Nevada. In phosphatized andesite on Malpelo Island, Colombia, in the Pacific Ocean.[7]

Alter. To variscite (?).

Name. In allusion to the dimorphous relation to variscite.

Ref.

1. Originally described as orthorhombic (under the name variscite) by Schaller (1912, 1916) but shown by x-ray study, McConnell, *Am. Min.*, **25**, 719 (1940), to be isostructural with metastrengite and monoclinic as earlier suggested by Strunz and Sztrókay, *Zbl. Min.*, **272** (1939), and Ulrich, *Rozpr. České Ak.*, *Cl. 2*, **39**, no. 17 (1930). Elements of Schaller with the unit and orientation changed to conform to the structure cell. Transformation: Schaller to new, $00\frac{1}{2}/010/\overline{1}00$.
2. Schaller (1916); it is uncertain whether certain forms given in orthorhombic position by Schaller are positive or negative forms in the monoclinic interpretation.
3. Frondel, priv. comm. (1948).
4. Larsen and Schaller (1925).
5. McConnell, *Am. Min.*, **25**, 719 (1940).
6. Schaller (p. 65, 1916).
7. McConnell, *Geol. Soc. Am., Bull.*, **54**, 707 (1943).

40.3.2.2   METASTRENGITE [$FePO_4 \cdot 2H_2O$]. Phosphosiderite *Bruhns* and *Busz* (*Zs. Kr.*, **17**, 555, 1890). Clinobarrandite *McConnell* (*Am. Min.*, **25**, 719, 1940). Vilateite (?) *Lacroix* (**4**, 477, 1910).

Cryst.[1] Monoclinic; prismatic—$2/m$.

$a:b:c = 0.5449:1:0.8968$; $\beta\ 90°36'$; $p_0:q_0:r_0 = 1.6458:0.8968:1$

$r_2:p_2:q_2 = 1.1151:1.8353:1$; $\mu\ 89°24'$; $p_0'\ 1.6459, q_0'\ 0.8968, x_0'\ 0.0105$

Forms:[2]

		ϕ	ρ	ϕ_2	$\rho_2 = B$	C	A
c	001	90°00′	0°36′	89°24′	90°00′	.....	89°24′
b	010	0 00	90 00	.....	0 00	90°00′	90 00
a	100	90 00	90 00	0 00	90 00	89 24	.....
m	110	61 25	90 00	0 00	61 25	89 28½	28 35
n	210	74 45½	90 00	0 00	74 45½	89 25½	15 14½
o	410	82 14½	90 00	0 00	82 14½	89 24½	7 45½
h	011	6 40½	42 05	89 24	48 16	41 44	85 32
f	103	90 00	29 12½	60 47½	90 00	28 36½	60 47½
e	101	90 00	58 53	31 07	90 00	58 17	31 07
d	111	61 34	62 02	31 07	65 08	61 30½	39 02½
r	$\overline{1}11$	−61 15½	61 48	148 33½	64 55½	61 16½	140 36

Rare or doubtful:

p 710	g 034	t 041	$\overline{5}04$	117	i 771	$\overline{1}12$	

Structure cell.[3] Space group probably $P2_1/n$. a_0 5.28 kX, b_0 9.75, c_0 8.71; [β 90°36′]; $a_0:b_0:c_0 = 0.542:1:0.893$. Cell contents $Fe_4(PO_4)_4 \cdot 8H_2O$.

Habit. Crystals tabular {010} or stout prismatic [001]. Also as botryoidal or reniform masses and crusts with a radial-fibrous structure.

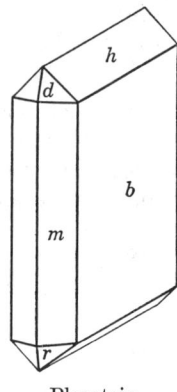

Pleystein.

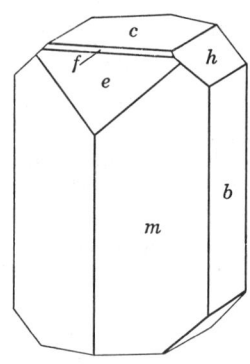

Sardinia.

Twinning. On {101}, frequently as interpenetration twins.

Phys. Cleavage {010} good, {001} indistinct. Fracture uneven. H. 3½–4. G. 2.76; 2.76 (calc.). Luster vitreous to subresinous. Color of crystals peach-blossom-red or reddish violet, rarely moss-green; aggregates bright rose-red to nearly colorless. Transparent to translucent.

Opt. In transmitted light, rose to colorless.

Orientation [4]	n(Na) [4]	Pleochroism	
$X \wedge c \sim 4°$	1.692	Pale rose	Biaxial negative (−).
Y b	1.725	Carmine-red	$2V$ 62° (Na).
Z	1.738	Colorless	$r > v$, very strong.

Chem. A hydrated ferric phosphate, $FePO_4 \cdot 2H_2O$. Dimorphous with strengite. Al apparently substitutes for Fe′′′ in large amount in some material,[11] and a complete series may extend to metavariscite analogous to the strengite-variscite series. Mn may substitute for Fe (*vilateite*).[13]

Anal.

	1	2	3
Fe_2O_3	42.73	44.30	44.38
P_2O_5	37.99	38.85	37.71
H_2O	19.28	17.26	17.31
Total	100.00	100.41	99.40
G.	2.76	2.76	

1. $FePO_4 \cdot 2H_2O$. 2. Kalterborn mine, Siegen.[5] 3. Kreuzberg, Bavaria.[6] Average of two.

Tests. B.B. easily fusible to a black magnetic bead. In C.T. turns yellow and opaque and loses water. Soluble completely in HCl, nearly insoluble in HNO_3.

Occur. Originally found in the Kalterborn mine, near Eiserfeld, Siegen, Germany, in limonite iron ore. Found in Bavaria [7] with pharmacosiderite in pegmatite at the Kreuzberg in Pleystein, also in pegmatite between Wildenau and Plössberg, and with strengite and dufrenite (?) in pegmatite at Hagendorf. In the St. Giovannedu mine, Gonnesa, Sardinia.[4] At Kirunavaara, Sweden,[8] with strengite in magnetite ore. In the United States, with aluminian strengite at Manhattan, Nevada (*clinobarrandite*); also in pegmatite at the Palermo and Fletcher quarries near North Groton, New Hampshire, and at Pala, San Diego County, California.[9] In phosphatized andesite on Malpelo Island, Colombia, in the Pacific Ocean.[12] A manganian variety (*vilateite*)[13] occurs with heterosite and hureaulite in pegmatite at La Vilate near Chanteloube, France.

Artif.[10] Microscopic rose-red monoclinic crystals perhaps identical with metastrengite have been obtained by heating a solution of ferric chloride with phosphoric acid in a closed tube at 180°–190°.

Name. Phosphosiderite from phosphorus and σίδηρος, *iron*, in allusion to the composition. The name metastrengite is here used parallel to metavariscite to reveal the dimorphous relation to strengite.

Ref.

1. Originally described as orthorhombic, and so accepted by Dana (823, 1892) and Goldschmidt (6, 152, 1920), but shown to be monoclinic by morphological, optical, and etch study by De Angelis, *Ann. Mus. Civ. Stor. Nat. Genova*, **52**, 138 (1926), and by x-ray methods by McConnell, *Am. Min.*, **24**, 636 (1939). Elements of De Angelis; see also Koechlin, *Cbl. Min.*, 290 (1934), and Laubmann and Steinmetz, *Zs. Kr.*, **55**, 523 (1920).
2. Goldschmidt (1920) and De Angelis (1926). It is uncertain whether $\{034\}$, $\{041\}$, and $\{771\}$ reported in orthorhombic position by Bruhns and Busz (1890) are positive or negative in the monoclinic interpretation.
3. Strunz and Sztrókay, *Zbl. Min.*, 272 (1939); comparable values are given by McConnell (1939). Space group of McConnell, *Am. Min.*, **25**, 719 (1940).
4. De Angelis (1926).
5. Bruhns and Busz (1890).
6. Laubmann and Steinmetz (1920).
7. Müllbauer, *Zs. Kr.*, **61**, 318 (1925), and Laubmann and Steinmetz (1920).
8. Koechlin (1934).
9. Mrose, priv comm. (1949).
10. de Schulten, *C.R.*, **100**, 1522 (1885).
11. Inferred from the presence of a monoclinic mineral (*clinobarrandite*) isostructural with metavaricite in the aluminian strengite ("barrandite" of Shannon, *Am. Min.*, **8**, 182, 1923) from Manhattan, Nevada, as shown by McConnell (1940).
12. McConnell, *Bull. Geol. Soc. Am.*, **54**, 707 (1943).
13. Presumed but not proved to be identical with metastrengite. Described by Lacroix (1910) as monoclinic, with $a:b:c = 1.6958:1:0.8886$, β 90°33′, and forms $a\{001\}$, $b\{010\}$, $m\{110\}$, $d\{201\}$, $n\{011\}$, $\{\bar{3}11\}$, $\{\bar{3}41\}$; G 2.745. The elements are close to those of metastrengite with the a-axis tripled. An analysis is lacking, but Mn is present in significant amounts in addition to Fe, (PO_4), and H_2O. This mineral is identical with the so-called Type 1 hureaulite from La Vilate described by Des Cloizeaux, *Ann. mines*, **53**, 293 (1858).

40.3.3 WEINSCHENKITE [$(Y,Er)(PO_4) \cdot 2H_2O$]. Laubmann (*Geognost. Jahreshefte, Geol. Landes. München*, **35**, 193, 1923). [Not weinschenkite *Murgoci* (*C.R.*, **175**, 372, 426, 1922)].

Cryst.[1] Monoclinic; prismatic—$2/m$ (?).

$a:b:c = 0.361:1:0.415$; β 129°24′; $p_0:q_0:r_0 = 1.150:3.207:1$

$r_2:p_2:q_2 = 3.119:3.585:1$; μ 50°36′; p_0' 1.488, q_0' 0.415, x_0' 0.821

Forms:[2]

	ϕ	ρ	ϕ_2	$\rho_2 = B$	C	A
b 010	0°00′	90°00′	....	0°00′	90°00′	90°00′
a 100	90 00	90 00	0°00′	90 00	50 36	.....
m 110	74 25	90 00	0 00	74 25	52 19	15 35
u $\bar{1}$01	−90 00	33 42	123 42	90 00	73 06	123 42

Structure cell.[3] Space group $P2/a$ (?). a_0 5.46 kX, b_0 15.12, c_0 6.28; β 113°24′; $a_0:b_0:c_0 = 0.361:1:0.415$. Cell contents $(Y,Er)_4(PO_4)_4 \cdot 8H_2O$. Weinschenkite is isostructural with gypsum.

Habit. Found as crusts and radial-fibrous spherulites and rosettes composed of lath-like crystals elongated [001] and flattened {010}.

P h y s. Cleavage {$\bar{1}$01}. G.[4] 3.05 (calc. for pure artificial $YPO_4 \cdot 2H_2O$), increasing with increasing substitution of Er, Yb, Dy, Sm, or Ce for Y; 3.26 meas., 3.27 (calc. for material from Virginia with Y:(Er,Yb, etc.) ∼4:1). Luster silky. Colorless to snow-white. Thermoluminescent on gentle heating.

O p t. In transmitted light, colorless.

Orientation	n (Bavaria)[5]	n (Virginia)[4]	
X b	1.600	1.605	Biaxial positive (+).
Y	1.608	1.612	$2V$ medium small.
Z	1.645	1.645	Marked dispersion of
$Z \wedge c$	∼30°	∼35°	bisectrices (Virginia).

C h e m. Hydrated yttrium phosphate, $Y(PO_4) \cdot 2H_2O$, containing small amounts [6] of Er, Yb, Dy, Ce, etc., in substitution for Y. On heating, loses about $1\frac{1}{2}H_2O$ at roughly 180°, the remainder being held to much higher temperatures.[4]

Anal.

	1	2
Y_2O_3	51.34	52.90
P_2O_5	32.28	31.15
H_2O	16.38	15.96
Insol.		0.23
Total	100.00	100.24

1. $Y(PO_4) \cdot 2H_2O$. 2. Bavaria.[7] The Y_2O_3 contains small amounts of other rare earths; average atomic weight 100.05.

Tests. B.B. infusible. Soluble in hot acids, insoluble in alkalies.

O c c u r. Originally found associated with wavellite, crandallite, cacoxenite, beraunite, and dufrenite in a limonite deposit at the Nitzelbuch mine, Auerbach, Bavaria. Also found in a deposit of manganiferous limonite at the Kelly Bank mine near Vesuvius, Rockbridge County, Virginia. At both occurrences the mineral was deposited by meteoric waters, and at the Virginia occurrence at least the rare earths may be of biochemical origin.

A r t i f.[8] Obtained as a fine-grained precipitate by reaction of yttrium nitrate and tribasic sodium phosphate solutions.

Name. After Ernst Weinschenk (1865–1921), Professor of Petrography at the University of Munich.

Ref.

1. Elements of the x-ray cell of Strunz, *Naturwiss.*, **30**, 64 (1942), on Bavarian material. Transformation: Strunz to new setting (gypsum), 10$\bar{1}$/0$\bar{1}$0/001.
2. Laubmann (1923).
3. Strunz (1942).
4. Milton, Murata, and Knechtel, *Am. Min.*, **29**, 92 (1944).
5. Larsen and Berman (111, 1934).
6. See spectrographic studies of Henrich, *J. pr. Chem.*, **142**, 1 (1935), and of Murata in Milton, et al. (1944).
7. Henrich (1935).
8. Mellor (5, 684, 1924) and Milton, et al. (1944).

40.3.4 **CHURCHITE** [(Ce,Ca)(PO$_4$)·2H$_2$O]. *Church (Chem. News,* **12**, 121, 1865); *Williams (Chem. News,* **12**, 183, 1865).

Monoclinic (?).[1] As fan-shaped groups of minute columnar crystals.

Phys. One perfect cleavage. Conchoidal fracture. H. 3. G. 3.14. Luster vitreous; pearly on cleavage. Color smoke-gray, tinged with flesh-red. Transparent.

Opt.[2] In transmitted light, colorless.

nX	1.620 ± 0.003	Sensibly uniaxial
nY	1.620	positive (+).
nZ ⊥ plates	1.654	

Chem. A hydrated phosphate of cerium in which calcium replaces cerium to a limited extent, (Ce,Ca)(PO$_4$)·2H$_2$O.

Anal.

	CaO	Ce$_2$O$_3$	P$_2$O$_5$	H$_2$O	Total	G.
1.		60.52	26.19	13.29	100.00	
2.	5.42	51.87	28.48	14.93	100.70	3.14

1. Ce(PO$_4$)·2H$_2$O. 2. Cornwall.[3] The cerium includes some didymium.

Occur. As coatings on a siliceous rock from a copper lode in Cornwall, England.

Artif. CePO$_4$·2H$_2$O crystallizes in platy, elongated crystals from the precipitate obtained from a cerium nitrate solution by sodium phosphate.[4]

Name. After the English chemist Arthur Herbert Church (1834–1915).

Ref.

1. Considered probably monoclinic by Church, *J. Chem. Soc. London*, **18**, 259 (1865). Larsen (58, 1921) suggested orthorhombic symmetry on optical grounds. Zambonini, *Bull. soc. min.*, **38**, 206 (1915), found artificial cerium phosphate dihydrate to be probably monoclinic.
2. Larsen (1921).
3. Church (1865).
4. Zambonini (1915). The angle between the edges of the plates is 116°, close to the angle β, 113°50′, of gypsum.

40.3.5 RHABDOPHANE [(Ce,Y,La,Di)(PO$_4$)·H$_2$O]. *Lettsom* (*Zs. Kr.*, **3**, 191, 1878; *Cryst. Soc., Proc.*, 105, 1882); *Lecoq de Boisbaudran* (*C.R.*, **86**, 1028, 1878). Scovillite *Brush* and *Penfield* (*Am. J. Sc.*, **25**, 459, 1883; *Am. J. Sc.*, **27**, 200, 1884). Skovillit.

Cryst. Tetragonal or hexagonal (?).[1]

Habit. Incrustations, botryoidal, globular, or stalactitic with radiated fibrous structure.

Phys. Fracture uneven. H. $3\frac{1}{2}$. G. 3.94–4.01. Luster greasy. Color brown, pinkish, or yellowish white. Translucent.

Opt.[2] Colorless fibers with positive elongation.

$$nO \quad 1.654 \pm 0.003 \quad \text{Uniaxial positive } (+).$$
$$nE \quad 1.703 \pm 0.003$$

Chem. A hydrated phosphate of cerium, yttrium, and rare-earth elements, (Ce,Y,La,Di)(PO$_4$)·H$_2$O. The proportions of the elements occurring with cerium have not been fully determined.

Anal.

	Ce$_2$O$_3$	(Ce,Y)$_2$O$_3$	(Y,Er)$_2$O$_3$	(La,Di)$_2$O$_3$	P$_2$O$_5$	H$_2$O	Rem.	Total	G.
1.	64.83				28.05	7.12		100.00	
2.		61.69			24.64	7.50	5.69	99.52	
3.			8.51	55.17	24.94	7.37	3.84	99.83	3.94–4.01

1. Ce(PO$_4$)·H$_2$O. 2. Cornwall.[3] Rem. is (Al,Fe)$_2$O$_3$ 1.93, SiO$_2$ 3.76. Spectroscopic examination showed Ce, Y, Di and La but no Er. 3. Salisbury, Connecticut (*scovillite*).[4] Rem. is Fe$_2$O$_3$ 0.25, CO$_2$ 3.59.

Tests. B.B. infusible. Easily soluble in hydrochloric acid.

Occur. First identified on specimens from Cornwall. In the United States as thin incrustations in the limonite deposit at Salisbury, Connecticut.

Alter. The water is largely retained on heating to 200° or more.[5]

Name. From ῥάβδος, *a rod*, and φαίνεσθαι, *to appear*, in allusion to the characteristic bands shown in its spectrum.

Ref.

1. Indicated by the optical properties.
2. Larsen (126, 1921) on material from Salisbury, Connecticut. Bertrand, *Bull. soc. min.*, **3**, 58 (1880), concluded from microscopic examination of spherulites of rhabdophane from Cornwall that the material is uniaxial positive.
3. Hartley, *J. Chem. Soc. London*, **41**, 210 (1882).
4. Brush and Penfield (1883).
5. Hartley, *J. Chem. Soc London*, **45**, 167 (1884).

41 ANHYDROUS PHOSPHATES, ETC., CONTAINING HYDROXYL OR HALOGEN

TYPE 1. $A_m(XO_4)_p Z_q$, WITH $m:p > 4:1$

41.1.1 S A H L I N I T E [$Pb_{14}(AsO_4)_2O_9Cl_4$]. *Aminoff* (*Geol. För. Förh.*, **56**, 493, 1934).

Monoclinic. In aggregates of small thin scales. Cleavage {010}, perfect. H. 2–3. G. 7.95. Color pale sulfur-yellow.
O p t. Biaxial negative $(-)$, with very high indices of refraction. $X = b$. $2E = 96\frac{1}{2}°$.
C h e m. A basic chloride-arsenate of lead, $Pb_{14}(AsO_4)_2O_9Cl_4$.
Anal.

	1	2
CaO		0.46
PbO	90.20	89.33
As_2O_5	6.64	6.57
Cl	4.09	4.05
CO_2		0.43
H_2O		0.10
	100.93	100.94
O = Cl	0.93	0.91
Total	100.00	100.03

1. $Pb_{14}(AsO_4)_2O_9Cl_4$. 2. Långban.[1] Average of one complete and one partial analysis.

O c c u r. At Långban, Sweden, in a dolomite rock impregnated with hausmannite.
N a m e. After Carl Sahlin.

Ref.
1. Blix analysis in Aminoff (1934).

41.1.2 H O L D E N I T E [$(Mn,Ca)_4(Zn,Mg,Fe)_2(AsO_4)(OH)_5O_2$ (?)]. *Palache* and *Shannon* (*Am. Min.*, **12**, 144, 1927).

C r y s t.[1] Orthorhombic; dipyramidal—$2/m\ 2/m\ 2/m$.

$a:b:c = 0.3811:1:0.2755;\qquad p_0:q_0:r_0 = 0.7230:0.2755:1$

$q_1:r_1:p_1 = 0.3811:1.3811:1;\qquad r_2:p_2:q_2 = 3.6298:2.6240:1$

Forms:

	ϕ	$\rho = C$	ϕ_1	$\rho_1 = A$	ϕ_2	$\rho_2 = B$
b 010	0°00′	90°00′	90°00′	90°00′	.....	0°00′
a 100	90 00	90 00	....	0 00	0°00′	90 00
n 130	41 10½	90 00	90 00	48 49½	0 00	41 10½
m 110	69 08½	90 00	90 00	20 51½	0 00	69 08½
e 011	0 00	15 24	15 24	90 00	90 00	74 36
p 111	69 08½	37 44	15 24	55 07½	54 08	77 25
s 131	41 10½	47 40½	39 34½	60 52	54 08	56 11
q 211	79 13	55 48½	15 24	35 39	34 40	81 05½
t 251	46 23½	63 24	54 01½	49 39	34 40	51 55½
r 311	82 45½	65 25½	15 24	25 33½	24 45	83 25½

Less common and rare:
c 001   l 120   f 031   d 102   x 182   w 151   u 7.16.2

Structure cell.[2] Space group $Bmam$. a_0 11.97 kX, b_0 31.15, c_0 8.58; $a_0:b_0:c_0 = 0.3842:1:0.2754$. Cell contents $(Mn,Ca)_8(Zn,Mg,Fe)_4(AsO_4)_2(OH)_{10}O_4$ (?).

Habit. Equant to thick tabular {100}, with a and u relatively large.

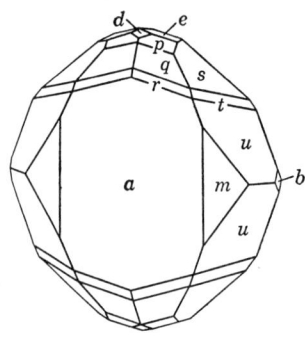

Phys. Cleavage {010}, poor. Fracture subconchoidal. H. 4. G. 4.11;[3] 4.27 (calc. for analysis 3). Color clear pink to yellowish red and deep red. Luster vitreous. Translucent.

Opt.[4] In transmitted light, pink.

Orientation		n	
X	c	1.769	Biaxial positive (+).
Y	b	1.770	2V 30°20′ (meas.).
Z	a	1.785	$r > v$, perceptible.

Chem. A hydrated basic arsenate of manganese and zinc of uncertain formula, probably $(Mn,Ca)_4(Zn,Mg,Fe)_2(AsO_4)(OH)_5O_2$. Small amounts of Ca, Fe″, and Mg substitute for Mn and Zn, with Ca:Mn = 2:23 and Fe:Mg:Zn = 2:3:25 in analysis 3.

Anal.

	1	2	3
CaO		3.03	2.68
MnO	46.78	44.08	41.92
ZnO	26.84	22.90	25.12
MgO		1.36	1.61
FeO		1.62	2.00
As_2O_5	18.95	19.40	19.32
H_2O	7.43	7.61	7.35
Total	100.00	100.00	[100.00]
G.		4.27	4.11

1. $Mn_4Zn_2(AsO_4)(OH)_5O_2$. 2. $(Mn,Ca)_4(Zn,Mg,Fe)_2(AsO_4)(OH)_5O_2$ with Mn:Ca = 46:4 and Zn:Mg:Fe = 25:3:2. 3. Franklin, New Jersey.[5] Analysis recalculated to 100 after deduction of CO_2[1.09] as $CaCO_3$ and SiO_2 2.01 as willemite.

HEMATOLITE

O c c u r. Known only as a single specimen from Franklin, New Jersey. The crystals occur with barite, galena, pyrochroite, and fibrous willemite in a veinlet cutting massive franklinite ore.

N a m e. After Albert Fairchild Holden (1866–1913), mining engineer and mineral collector, in whose collection the mineral was first noticed.

Ref.
1. The ratio $a:b$ originally given by Palache and Shannon (1927) as $0.3802:1$ is here recalculated from the fundamental angles as $0.3811:1$.
2. Prewitt-Hopkins, *Am. Min.*, **34**, 589 (1949), by Weissenberg method on type material.
3. Prewitt-Hopkins (1949) by microbalance; 4.07 was given originally.
4. Berman in Palache and Shannon (1927).
5. Shannon analysis in Palache and Shannon (1927).

41.1.3 HEMATOLITE [$(Mn'',Mg)_4Al(AsO_4)(OH)_8$]. Aimatolith *Igelström* (*Ak. Stockholm, Öfv.*, **41**, 85, 1884; *Geol. För. Förh.*, **7**, 211, 1884). Hämatolith. Diadelphit *Sjögren* (*Geol. För. Förh.*, **7**, 233, 369, 1884; *Zs. Kr.*, **10**, 130, 1885).

C r y s t.[1] Hexagonal—R; rhombohedral—$\bar{3}$ (?).

$a:c = 1:4.4425;$ $\alpha\ 36°40\frac{1}{2}';$ $p_0:r_0 = 5.1298:1;$ $\lambda\ 116°25\frac{1}{2}'$

Forms:[2]

		ϕ	$\rho = C$	A_1	A_2
c 0001	111	....	0°00'	90°00'	90°00'
B 0.1.$\bar{1}$.29	10.10.9	−30°00'	10 02	90 00	81 19½
Q 0.2.$\bar{2}$.13	553	−30 00	38 17	90 00	57 33
P 01$\bar{1}$5	221	−30 00	45 44	90 00	51 40½
S 02$\bar{2}$5	771	−30 00	64 01	90 00	38 52½
T 01$\bar{1}$2	110	−30 00	68 42	90 00	36 13

Structure cell.[3] Space group probably $R\bar{3}$. a_{rh} 13.07, α 36°53'; a_0 8.27 Å, c_0 36.51; $a_0:c_0 = 1:4.415$. Cell contents in the rhombohedral unit, $(Mn'',Mg)_{12}Al_3(AsO_4)_3(OH)_{24}$.

Habit. Thick tabular $\{0001\}$ with large $\{01\bar{1}5\}$, or rhombohedral with $\{01\bar{1}5\}$ alone. Rhombohedral faces striated horizontally.

P h y s. Cleavage $\{0001\}$ perfect. Fracture uneven. Brittle. H. 3½. G. 3.49;[4] 3.48 (calc.). Luster vitreous, on cleavage surfaces pearly to submetallic. Color brownish red, garnet-red or blood-red to almost black. Streak light chocolate-brown to reddish brown. Translucent to nearly opaque.

O p t. In transmitted light, reddish orange to yellowish brown in color. Crystals are sometimes divided into differently oriented, anomalously biaxial sectors.[7]

	n [5]	n (red) [6]	n (blue) [6]	
O	1.733 ± 0.003	1.723	1.740	Uniaxial negative (−).
E	1.714 ± 0.003			Not pleochroic.

Chem. A basic arsenate of aluminum, divalent manganese, and magnesium, $(Mn'',Mg)_4Al(AsO_4)(OH)_8$. A small amount of Fe''', and probably also Mn''', substitutes for Al. $Mn:Mg \sim 5:1$ in the reported analyses.

Anal.[10]

	1	2	3	4
CaO		0.76	0.66	0.71
MnO	47.69	49.67	46.86	50.98
MgO	5.31	7.14	6.66	5.38
Al_2O_3	10.07	8.33	6.39 ⎱	8.61
Fe_2O_3		1.46	1.01 ⎰	
As_2O_5	22.70	19.09	21.55	22.54
H_2O	14.23	14.09	13.93	14.02
Rem.			0.64	
Total	100.00	100.54	97.70	102.24

1 $(Mn,Mg)_4Al(AsO_4)(OH)_8$. 2. Nordmark, Sweden [8] 3. Nordmark.[8] Rem. is insol. 4. Nordmark.[9]

Tests. B.B. infusible. In C.T. decrepitates, gives off water, and turns black. Easily soluble in acids.

Occur. As crystals in thin veinlets or cavities in crystalline limestone at the Moss mine, Nordmark, Sweden. Associated with magnetite, jacobsite, barite, fluorite.

Alter. To manganite.

Name. From αἷμα, *blood*, in allusion to its color. Diadelphite from δις-, *twice*, and ἀδελφός, *brother*, because of its close association with allactite.

Ref.

1. Elements of Sjögren, *Zs. Kr.*, **10**, 131 (1885); unit of the structure cell of Berry, *Am. Min.*, **33**, 489 (1948). Transformation: Sjögren to new, $0\overline{1}00/00\overline{1}0/\overline{1}000/0005$. Sjögren's angles are used in preference to those of Lorenzen, *Öfv. Vet. Ak. Stockholm*, no. 4, 95 (1884), Sjögren, *Geol. För. Förh.*, **7**, 234 (1884), and Berry (1948).
2. Sjögren (1885) and Berry (1948). The form $\{70\overline{7}3\}$ of Lorenzen (1884) is here simplified to $\{01\overline{1}2\}$ after transformation.
3. Berry (1948) by Weissenberg and precession methods.
4. Berry (1948); Sjögren (1885) gave 3.3–3 4.
5. Larsen (83, 1921).
6. Sjögren (1885) by prism method.
7. Bertrand, *Bull. soc. min.*, **7**, 124 (1884).
8. Lundström analysis in Sjögren (1885).
9. Sjögren (1885).
10. Additional analyses are cited by Sjögren (1885) and Igelström (1884) and are listed by Hintze (1 [4B], 1175, 1933).

CHLOROPHOENICITE GROUP

41.1.4.1 CHLOROPHOENICITE $[(Mn,Zn)_5(AsO_4)(OH)_7]$. *Foshag* and *Gage* (*J. Washington Ac. Sc.*, **14**, 362, 1924).

Cryst.[1] Monoclinic; prismatic—$2/m$.

$a:b:c = 2.3233:1:2.1685$; $\beta\ 105°34'$; $p_0:q_0:r_0 = 0.9334:2.0890:1$

$r_2:p_2:q_2 = 0.4787:0.4468:1$; $\mu\ 74°26'$; $p_0'\ 0.9689,\ q_0'\ 2.1685,\ x_0'\ 0.2786$

CHLOROPHOENICITE

Forms:

	φ	ρ	φ₂	ρ₂ = B	C	A
c 001	90°00′	15°34′	-74°26′	90°00′	.....	74°26′
a 100	90 00	90 00	0 00	90 00	74°26′	.....
s 106	90 00	23 45½	66 14½	90 00	8 11½	66 14½
r 102	90 00	37 20½	52 39½	90 00	21 06½	52 39½
k 1̄04	90 00	2 05	92 05	90 00	13 29	92 05
h 2̄03	-90 00	20 10	110 10	90 00	35 44	110 10
p 111	29 54½	68 12½	38 43	36 24	61 11½	62 25

Habit. Long prismatic [010]. The crystals are deeply striated [010], with etched and dull terminal faces. {100} relatively smooth; {h0l} faces uneven or warped.

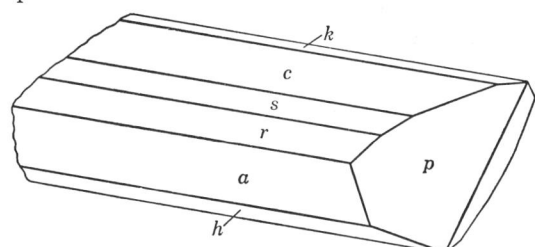

P h y s. Cleavage {100} good. Brittle. H. 3½. G. 3.46.[2] Luster vitreous, on cleavage surfaces pearly. Color light grayish green in natural light, pink to light purplish red in strong artificial light. Translucent.

O p t.[3] In transmitted light, nearly colorless.

ORIENTATION		n	
X		1.682	Biaxial negative (−).
Y	b	1.690	2V 83° ± 2° (meas.).
Z		1.697	r > v, strong.

C h e m. A basic arsenate of zinc and manganese, probably $(Mn,Zn)_5(AsO_4)(OH)_7$. Small amounts of Ca, Mg, and Fe″ substitute for (Mn,Zn), with Mg:Ca:Zn:Mn = 1:1.8:11.0:14.6 in the only reported analysis.

Anal.

	1	2
CaO	3.27	3.36
MgO	1.31	1.34
ZnO	29.04	29.72
FeO		0.48
MnO	33.59	34.46
As₂O₅	21.17	19.24
H₂O	11.62	11.60
Total	100.00	100.20

1. $(Mn,Zn,Ca,Mg)_5(AsO_4)(OH)_7$, with Mg:Ca:Zn:Mn = 1:1.8:11.0:14.6. 2. Franklin, New Jersey.[4]

Tests. B.B. fusible with difficulty; in C.T. turns black and evolves abundant water. Soluble in acids.

PHOSPHATES, ARSENATES, VANADATES

Occur. Found with leucophoenicite, tephroite, gageite, pyrochroite, willemite, and calcite on the surface of cracks in massive franklinite-willemite ore at Franklin, New Jersey. Also found with calcite and barite in secondary veinlets in franklinite ore at Sterling Hill, New Jersey.

Name. From χλωρός, *green*, and φοινικος, *purple-red*, in allusion to the color changes in natural and artificial light.

Ref.
1. Elements recalculated by Palache, priv. comm. (1946), from the measurements of Foshag, Berman, and Gage, *Proc. U. S. Nat. Mus.*, **70**, Art. 20 (1927), whose elements and angle table are incorrect. The original angles are variable, owing to the poor quality of the crystals, and the present elements are not accurate.
2. Palache, *U. S. Geol. Sur., Prof. Pap. 180*, 122 (1935).
3. Foshag, Berman, and Gage (1927).
4. Foshag analysis in Foshag, Berman, and Gage (1927).

41.1.4.2 MAGNESIUM-CHLOROPHOENICITE [$(Mg,Mn)_5(AsO_4)(OH)_7$]. *Palache (U. S. Geol. Sur., Prof. Pap. 180, 124, 1935).*

Monoclinic. As rosettes and radial aggregates of fibers. The fibers have a perfect lengthwise cleavage. G. 3.37. Colorless to white.

Opt. In transmitted light, colorless.

Orientation		n	
X		1.669	Biaxial positive (+).
Y	elong.	1.672	$r < v$, strong.
Z		1.677	

Chem. A basic arsenate of magnesium and manganese, probably $(Mg,Mn)_5(AsO_4)(OH)_7$, with $Mn:Mg = 1:4$ in the only analysis. Magnesium-chlorophoenicite is isostructural with chlorophoenicite.[2]

	ZnO	MgO	MnO	Fe_2O_3	As_2O_5	SiO_2	$CaCO_3$	H_2O	Total
1.		39.31	17.29		28.04			15.36	100.00
2.		39.64	18.05		28.00			14.31	100.00
3.	8.90	20.95	15.57	3.85	21.16	3.36	6.29	10.81	99.89

1. $(Mg,Mn)_5(AsO_4)(OH)_7$ with $Mn:Mg = 1:4$. 2. Analysis 3 recalculated to 100 after deducting 12.49 willemite, 6.29 calcite, 5.55 franklinite. 3. Franklin, New Jersey.[1]

Occur. Found with zincite and carbonates in a veinlet in franklinite ore at Franklin, New Jersey.

Ref.
1. Gonyer analysis in Palache (1935).
2. Frondel, priv. comm. (1949), by x-ray powder study.

41.1.5 SYNADELPHITE [$(Mn,Mg,Ca,Pb)_4(AsO_4)(OH)_5$]. Synadelphite *Sjögren (Geol. För. Förh., 7, 220, 382, 1884).* Allodelphite *Quensel* and *von Eckermann (Geol. För. Förh., 52, 639, 1930).* Plumbosynadelphite *Hurlbut (Am. Min., 22, 526, 1937).*

Cryst.[1] Triclinic; pseudo-orthorhombic.

$$a:b:c = 0.5333:1:0.5851; \quad p_0:q_0:r_0 = 1.0970:0.5851:1$$

$$q_1:r_1:p_1 = 0.5333:0.9598:1; \quad r_2:p_2:q_2 = 1.7094:1.8751:1$$

SYNADELPHITE

Forms:[2]

	ϕ	$\rho = C$	ϕ_1	$\rho_1 = A$	ϕ_2	$\rho_2 = B$
b 010	0°00′	90°00′	90°00′	90°00′	....	0°00′
m 110	61 56	90 00	90 00	28 04	0°00′	61 56
d 011	0 00	30 03	30 30	90 00	90 00	59 30
p 121	43 13	58 02	49 30	54 30	42 12	51 51

Structure cell.[3] Sensibly orthorhombic. a_0 9.91 kX, b_0 18.70, c_0 10.65; $a_0:b_0:c_0 = 0.532:1:0.5695$. Cell contents $(Mn,Mg,Ca,Pb)_{40}(AsO_4)_{10}(OH)_{50}$.

Habit and twinning. Short prismatic [001], with {110} striated [001]. The seemingly single individuals, however, are found on optical study[3] to be fourlings twinned by reflection on (100) and (010) of a pseudo-orthogonal triclinic lattice. Also massive, as reniform crusts.

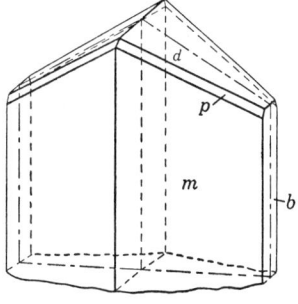

P h y s. Cleavage {010}, imperfect. Fracture uneven to conchoidal. Brittle. H. $4\frac{1}{2}$. G. 3.57 (colorless),[4] 3.79 or more in red, plumbian material;[4] 3.56 (calc. for material of anal. 5). Crystals commonly are zoned in color: the inner parts colorless, whereas the outer parts, relatively high in lead, are red, red-brown to dark brown or almost black in color. Luster of the colorless material vitreous, of plumbian material adamantine to sub-metallic. Transparent in colorless material, translucent to opaque in plumbian varieties.

O p t.[5] In transmitted light, colorless to pale brown or reddish.
Anal.

	Orientation (anal. 5)		n,Na (anal. 5)	n,Na (anal. 6)	
	ϕ	ρ			
X	−8°	86°	1.750	1.851	Biaxial positive (+).
Y	−172°	86°	1.751	1.864	$r > v$.
Z	42°	7°	1.761	1.894	
2V (meas.)			37°	40°	
2V (calc.)			35°15′	67°37′	

	Pleochroism		
	X	Colorless	Light brown
	Y	Colorless	Brown
	Z	Light brown	Dark red-brown

C h e m. A basic arsenate essentially of divalent manganese. Small amounts of Mg, Pb, and Ca substitute for Mn″, with Ca:Mg:Pb:Mn = 1:6.5:6.5:33.5 in analysis 6. The formula[6] probably is $(Mn,Mg,Ca,Pb)_4(AsO_4)(OH)_5$. Si apparently substitutes for As in small amounts (anal. 5, 6); the large amount of Si reported in analysis 4, however, has been ascribed to analytical error.

Anal.	1	2	3	4	5	6
Na_2O				0.53	0.62	0.59
K_2O				0.74	0.79	0.70
CaO		3.76	0.28	1.10	1.55	1.25
MgO		2.19		6.22	4.62	5.89
PbO				0.39		3.24
MnO	63.95	35.71	56.43	50.30	53.10	52.27
Mn_2O_3		11.79				
Al_2O_3		6.16		1.50	1.41	
Fe_2O_3		1.23		0.98	0.86	0.48
As_2O_5	25.90	29.31	32.43	21.91	26.89	26.18
SiO_2				6.23	1.45	0.63
H_2O	10.15	11.39	11.33	8.82	8.52	8.74
Rem.			0.36	0.77		
Total	100.00	101.54	100.83	99.49	99.81	99.97
G.				3.57	3.57	3.79

1. $Mn_4(AsO_4)(OH)_5$. 2. Nordmark, Sweden.[7] 3. Nordmark.[8] Rem. is FeO 0.17, insol. 0.19. 4. Långban, Sweden.[9] Rem. is As_2O_3 0.62, Sb_2O_5 0.15. 5. Långban, Sweden.[10] Colorless interior of crystals. 6. Långban.[10] Red outer zone of crystals.

Tests. B.B. fuses at about 2 to a black slaggy bead. In C.T. affords water and turns black. Easily soluble in acids.

O c c u r. Originally from the Moss mine, Nordmark, Sweden, associated with pyrochroite, manganite, jacobsite, calcite, allactite, hemafibrite, and hematolite in drusy cavities in manganese ore. Later described (*allodelphite, plumbosynadelphite*) from Långban, Sweden, with manganophyllite and hausmannite along joint planes in dolomite.

A l t e r. Oxidizes readily on exposure to air.

N a m e. Synadelphite from σύν, *with*, and ἀδελφός, *brother*, referring to the fact that it occurs associated with several chemically similar minerals. Allodelphite, named from ἄλλος, *other*, and ἀδελφός, *brother*, in allusion to its close relationship to synadelphite, has been shown [3] to be identical with that species. Plumbosynadelphite is a superfluous varietal designation for plumbian synadelphite.

Ref.

1. Hurlbut, *Am. Min.*, **22**, 526 (1937), from new measurements on Långban crystals, in the orientation and unit of the structure cell. Transformation: Sjögren (1884) to Hurlbut, 001/200/010; Quensel and von Eckermann (1930) to Hurlbut, 010/200/001. The results of morphological and x-ray study are consistent with orthorhombic symmetry, but optical study indicates that the crystals actually are interpenetration twins of sensibly orthogonal triclinic individuals. The symmetry was earlier regarded as tetragonal and, later, as monoclinic by Sjögren (1884) and as orthorhombic by Hamberg, *Geol. För. Förh.*, **11**, 222 (1889).
2. Sjögren (1884) and Hurlbut (1937).
3. Hurlbut (1937).
4. Hurlbut (1937) and Quensel and von Eckermann (1930), the latter also reporting 3.91 on another sample.
5. Hurlbut (1937), correcting the earlier measurements of Quensel and von Eckermann (1930) on zoned crystals. 2V ranges down to 0° in some material.
6. For discussion see Machatschki, *Geol. För. Förh.*, **53**, 187 (1931), and Hurlbut (1937).
7. A. Sjögren analysis in Sjögren, *Zs Kr.*, **10**, 143 (1885).
8. Blix analysis in Quensel and von Eckermann (1930).
9. Almström analysis in Quensel and von Eckermann (1930).
10. Gonyer analysis in Hurlbut (1937).

TYPE 2. $A_7(XO_4)_2Z_q$

41.2.1 Lacroixite. Slavík (*Bull. soc. min.*, **37**, 157, 1914; *Ac. sc. Bohême, Bull.*, no. 4, 19, 1914).

C r y s t.[1] Probably monoclinic; pseudo-orthorhombic.

$$a:b:c = 0.815:1:1.600$$

Forms:[2]

		ϕ	ρ
a	010	0°00′	90°00′
b	100	90 00	90 00
m	110	50 49	90 00
p	111	50 49	68 27
q	131	22 14½	79 05

Found only in fragmentary crystals. Indistinct cleavage parallel to (111) and (1$\bar{1}$1) only, indicating monoclinic symmetry. H. 4½. G. 3.126. Luster vitreous to resinous. Color pale yellow to pale green, at times almost white.

O p t.[3] In transmitted light, colorless.

nX	1.545	Biaxial negative (−).
nY	1.554	2V near 90°.
nZ	1.565	Dispersion slight.
		Extinction angle large (?).

C h e m. A fluo-phosphate of sodium, calcium, and aluminum of unknown formula. Partly altered material contains a considerable amount of Mn″, which spectrographic examination shows is lacking in the fresh material. Analysis gave:[4] Na_2O 14.92, CaO 19.46, MnO 8.43, Al_2O_3 18.87, P_2O_5 28.83, F 6.53, ign. loss 5.46, SiO_2 0.95, total 100.70 (less O = F = 2.75).

Tests. Easily soluble in HCl and H_2SO_4.

O c c u r. Found with ježekite, apatite, childrenite, roscherite, and tourmaline in druses in granite at Greifenstein near Ehrenfriedersdorf, Saxony, Germany.

N a m e. After Alfred Lacroix (1863–1948), French mineralogist.

Ref.

1. Ratio recalculated from the averaged measured angles of Slavík (1914), *Ac. sc. Bohême*, **20**, 372 (1915), on crystals of poor quality.
2. Slavík (1915); given in orthorhombic interpretation.
3. Larsen and Berman (159, 1934).
4. Jilek analysis in Slavík (1914).

41.2.2 Morinite. Lacroix (*Bull. soc. min.*, **14**, 187, 1891).

Originally described as occurring both as tiny prismatic monoclinic crystals, pale rose in color, and as lamellar masses. The crystals were later shown[1] to be identical with ježekite. The lamellar mineral, described below, may represent a distinct species but its status is doubtful.[2] H. ~4.

G. 2.95. Easily cleavable. Luster vitreous to pearly. Color wine-red. Streak white. Translucent. Chemically a basic fluo-phosphate of sodium, calcium, and aluminum. The reported analysis of the lamellar mineral corresponds to the formula $Na_2Ca_3Al_3H(PO_4)_4F_6 \cdot 8H_2O$, which differs considerably from that of ježekite. Analysis gave:[3] Na_2O 5.10, CaO 13.55, Al_2O_3 17.50, P_2O_5 32.95, F 13.00, H_2O+ 17.60, H_2O at 120° 0.20, SiO_2 1.50, total 101.40.

Found with wardite, wavellite, fibrous apatite,[4] and cassiterite as an alteration of amblygonite at Montebras, Creuze, Plateau Central, France. Named after Mr. Morineau, director of the tin mine at Montebras.

Ref.

1. Slavík, *Bull. soc. min.*, **37**, 152 (1914).
2. A rose-colored lamellar fragment labeled morinite and supplied by Slavík proved on x-ray and optical study to be identical with ježekite (Frondel, priv. comm., 1947).
3. Carnot and Lacroix, *Bull. soc. min.*, **31**, 149 (1908).
4. Frondel, priv. comm. (1947).

41.2.3 **J E Ž E K I T E** [$Na_4CaAl_2(PO_4)_2(OH)_2F_2O$ (?)]. Slavík (*Bull. soc. min.*, **37**, 152, 1914; *Ac. sc. Bohême, Bull.*, no. 4, 19, 1914). Morinite pt. Lacroix (*Bull. soc. min.*, **14**, 187, 1891).

C r y s t.[1] Monoclinic; prismatic—$2/m$.

$a:b:c = 0.8959:1:1.0241$; $\beta\, 105°31\tfrac{1}{2}'$; $p_0:q_0:r_0 = 1.1431:0.9867:1$

$r_2:p_2:q_2 = 1.0134:1.1585:1$; $\mu\, 74°28\tfrac{1}{2}'$; $p_0'\, 1.1864$, $q_0'\, 1.0241$, $x_0'\, 0.2778$

Forms:[1]

	ϕ	ρ	ϕ_1	ρ_2	C	A
c 001	90°00′	15°31½′	74°28½′	90°00′	.....	74°28½′
b 010	0 00	90 00	.....	0 00	90°00′	90 00
a 100	90 00	90 00	0 00	90 00	74 28½	.....
m 110	49 12	90 00	0 00	49 12	78 18½	40 48
f 012	28 29	30 13½	74 28½	63 44½	26 15½	76 06½
e 011	15 10½	46 42	74 28½	45 23	44 37	79 01
t 104	90 00	29 52½	60 07½	90 00	14 21	60 07½
s 102	90 00	41 03½	48 56½	90 00	25 32	48 56½
u $\overline{1}$01	−90 00	42 15½	132 15½	90 00	57 47	132 15½

Habit. Prismatic [001] and tabular {100}. The terminal faces are often rounded, and the vertical faces are striated [001]. Also as radial fibrous or columnar crusts or aggregates.

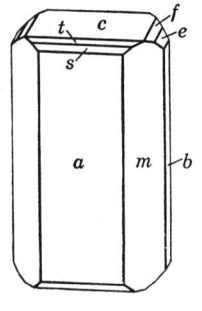

P h y s. Cleavage {100}, perfect, {001} imperfect. H. 4–4½. G. 2.94. Luster vitreous, tending toward pearly on the {100} cleavage surfaces. Colorless to white, also pale rose. Streak white. Translucent.

O p t.[2] In transmitted light, colorless to pale rose.

Orientation		n	
$X \wedge a$	30°	1.551	Biaxial negative (−).
Y	b	1.561	$2V$ medium small.
Z		1.562	$r < v$, perceptible.

Chem. A basic fluophosphate of sodium, calcium, and aluminum, probably $Na_4CaAl_2(PO_4)_2(OH)_2F_2O$.
Anal.

	1	2
Na	20.58	18.71
Li		0.86
CaO	12.55	13.50
Al_2O_3	22.81	21.92
P_2O_5	31.76	30.30
F	8.50	8.15
OH	3.80	7.26
Total	100.00	100.70

1. $Na_4CaAl_2(PO_4)_2(OH)_2F_2O$. 2. Greifenstein.[3] With Fe_2O_3 tr.

Tests. B.B. infusible. Incompletely decomposed by cold H_2SO_4 or aqua regia, completely so in hot H_2SO_4.

Occur. Found with roscherite in druses in granite at Greifenstein near Ehrenfriedersdorf, Saxony. The prismatic crystallized type of morinite (which see) from Montebras, Creuze, France, is identical with ježekite.
Name. After Bohuslav Ježek, Bohemian mineralogist.

Ref.
1. Slavík, *Bull. soc. min.*, **37**, 152 (1914).
2. Larsen and Berman (160, 1934) on type material.
3. Skarnitzl analysis in Slavík (1914).

41.2.4 **A L L A C T I T E** [$Mn_7(AsO_4)_2(OH)_8$]. Allaktit *Sjögren* (*Geol. För. Förh.*, **7**, 109, 1884; *Ak. Stockholm, Öfv.*, **41**, no. 3, 29, 1884).

Cryst.[1] Monoclinic; prismatic—$2/m$.

$a:b:c = 0.9007:1:0.4494;$ $\beta\ 114°04';$ $p_0:q_0:r_0 = 0.4989:0.4103:1$

$r_2:p_2:q_2 = 2.4372:1.2159:1;$ $\mu\ 65°56';$ $p_0'\ 0.5464,\ q_0'\ 0.4494,\ x_0'\ 0.4466$

Forms:[2]

		ϕ	ρ	ϕ_2	$\rho_2 = B$	C	A
c	001	90°00′	24°04′	65°56′	90°00′	....	65°56′
b	010	0 00	90 00	....	0 00	90°00′	90 00
a	100	90 00	90 00	0 00	90 00	65 56	....
t	120	31 18	90 00	0 00	31 18	77 46	58 42
o	110	50 34	90 00	0 00	50 34	71 38½	39 26
f	210	67 39	90 00	0 00	67 39	67 50½	22 21
y	011	44 49½	32 21	65 56	67 42	22 18	67 50
e	$\bar{2}01$	−90 00	32 52	112 52	90 00	56 56	112 52
v	111	65 39	47 24	45 12	72 20	26 56	47 53
q	$\bar{1}11$	−12 31	24 43	95 42	65 55	37 35	95 12
A	$\bar{3}21$	−53 00	56 12½	140 01	48 25	76 43	120 00½

Less common and rare:
  n 430 l 830 u 021 w 201 x $\bar{4}01$ i $\bar{2}21$ d $\bar{2}31$ S $\bar{2}51$

Structure cell.[3] Space group $P2_1/a$. a_0 11.01 kX, b_0 12.10, c_0 5.498; [β 114°04′]; $a_0:b_0:c_0 = 0.9097:1:0.4544$. Cell contents $Mn_{14}(AsO_4)_4$-$(OH)_{16}$.

Habit. Elongated [010], either as slender prisms or tabular {100}. Sometimes as rosette-like aggregates.

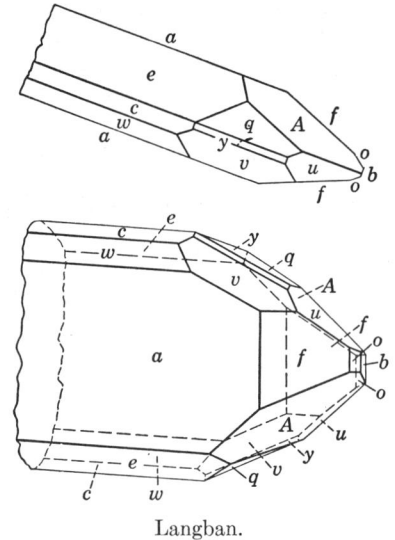

Langban.

P h y s. Cleavage {001} distinct.[4] Fracture uneven. Brittle. H. $4\frac{1}{2}$. G. 3.83; 3.94 (calc.). Luster vitreous, somewhat greasy on fracture surfaces. Color dark to light purplish red and brownish red. Streak grayish to faint brown. Translucent.

O p t.

λ	Långban[5]						Långban[6]	PLEO-
	486mμ	527	540	589	656	687	Na	CHROISM
nX	1.763				1.755		1.755	Blood-red
	(calc.)				(calc.)			
nY				1.7788	1.7732	1.7704	1.772	Pale yellow
nZ	1.7937	1.7869	1.7841				1.774	Sea-green

Biaxial negative ($-$). $2V$ almost 0° for Na. Dispersion $r > v$, strong. Optical orientation:[5] for wavelengths > 573mμ, $Y = b$, and $X \wedge c = 51°17'$; uniaxial at 573mμ; for wavelengths < 573mμ the axial plane is perpendicular to {010}.

C h e m. A basic arsenate of divalent manganese, $Mn_7(AsO_4)_2(OH)_8$. Small amounts of Ca and Mg substitute for Mn″.

Anal.[7]

	1	2	3	4
CaO		0.57	0.10	2.01
MgO		1.92		1.34
FeO			0.49 }	58.64
MnO	62.19	60.16	62.23 }	
As_2O_5	28.78	28.10	28.19	29.10
H_2O	9.03	9.08	8.99	8.97
Total	100.00	99.83	100.00	100.06

1. $Mn_7(AsO_4)(OH)$. 2. Långban, Sweden.[8] 3. Moss mine, Sweden.[9] 4. Långban.[10]

Tests. B.B. fusible with much difficulty. In C.T. loses water and blackens. Easily soluble in HCl.

O c c u r. Originally found in the Moss mine, Nordmark, Sweden, associated with fluorite, synadelphite, hematolite, hausmannite, pyrochroite; also at the Brattfors mine in the Nordmark district with catoptrite. Later found at Långban, Sweden, associated with calcite, barite, pyrochroite, fluorite, hausmannite, and native lead in veinlets and crevices. In the United States found at Franklin and Sterling Hill, New Jersey, with calcite, franklinite, fluorite, and willemite in small veinlets in the massive ore.

N a m e. From ἀλλάκτειν, *to change*, in allusion to the pleochroism.

Ref.
 1. Elements of Palache and Berman, *Kungl. Svenska Vetenskap., Handl.*, [3], **11**, no. 4, 32 (1933), on Långban crystals (analysis 2). Transformation: Aminoff, *Geol. För. Förh.*, **43**, 24 (1921), to P. and B., $\bar{1}01/0\bar{1}0/001$; Sjögren to P. and B., $\bar{1}01/0\frac{3}{4}0/001$; Goldschmidt (33, 1897) to P. and B., $\bar{1}01/0\frac{3}{4}0/100$.
 2. Palache and Berman (1933); Aminoff (1921); Sjögren (1884).
 3. Hurlbut, priv. comm. (1949), by the Weissenberg method on the crystals of Palache and Berman (1933).
 4. Doubtfully reported on {100} and {010} by Aminoff (1921) and Sjögren (1884).
 5. Aminoff (1921) on Långban crystals, who also gives $2E$ as a function of wavelength.
 6. Palache and Berman (1933).
 7. Additional analyses tabulated by Hintze (1 [4B], 1184, 1933).
 8. Gonyer in Palache and Berman (1933).
 9. Sjögren (1884).
 10 Lundström analysis in Sjögren, *Ak. Stockholm, Öfv.*, **44**, 109 (1887).

TYPE 3. $A_3(XO_4)Z_q$

41.3.1 C L I N O C L A S E [$Cu_3AsO_4(OH)_3$]. Strahliges Olivenerz *Karsten* (*Ges. nat. Freunde Berlin, N. Schr.*, **3**, 288, 1801). Arseniate of copper, 4th species, *Bournon* (*Phil. Trans.*, **91**, 181, 1801). Strahlenerz *Karsten* (64, 97, 1808). Cuivre arseniaté. *Haüy* (3, 504, 1822). Strahlenkupfer *Hausmann* (1050, 1813). Strahlerz *Werner*. Klinoklas *Breithaupt* (1830). Siderochalcit *Glocker* (840, 1831). Aphanèse *Beudant* (**2**, 602, 1832). Aphanesite *Shepard* (**2**, 35, 1835). Abichit *Bernhardi* (in *Glocker* [579, 1839]). Clinoclasite *Dana* (570, 1868).

C r y s t. Monoclinic; prismatic—$2/m$.[1]

$a:b:c = 1.9109:1:1.1223;$ $\beta\ 99°22';$ $p_0:q_0:r_0 = 0.5873:1.1073:1$

$r_2:p_2:q_2 = 0.9031:0.5304:1;$ $\mu\ 80°38';$ $p_0'\ 0.5953,\ q_0'\ 1.1223,\ x_0'\ 0.1650$

Forms: [2]

	φ	ρ	φ₂	ρ₂ = B	C	A
c 001	90°00′	9°22′	80°38′	90°00′	.....	80°38′
a 100	90 00	90 00	0 00	90 00	80°38′	.....
m 110	27 56½	90 00	0 00	27 56½	85 37½	62 03½
d 301	90 00	65 51½	27 08½	90 00	53 29½	27 08½
f 302̄	−90 00	36 03	126 03	90 00	45 25	126 03
g 2̄01	−90 00	45 43½	135 43½	90 00	55 05½	135 43½
h 5̄02	−90 00	52 55½	142 55½	90 00	62 17½	142 55½
o 111	34 07	53 35	52 45	48 13½	48 46	63 10
q 1̄11	−20 58½	50 14½	113 17	44 07½	54 06½	105 58½

Structure cell.[3] Space group $P2_1/a$. a_0 12.36 kX, b_0 6.45, c_0 7.23; β 99°30′; $a_0:b_0:c_0 = 1.916:1:1.121$. Cell contents $Cu_{12}(AsO_4)_4(OH)_{12}$.

Habit. Crystals elongated [010] and tabular {001}; also rhombohedral in aspect, or elongated [001] with {100} prominent. As isolated crystals or grouped into rosettes; as crusts or coatings, sometimes densely aggregated with a fibrous structure.

P h y s. Cleavage {001}, perfect. Fracture uneven. Rather brittle. H. 2½–3. G. 4.38;[4] 4.42 (calc.). Luster vitreous, pearly on cleavage.

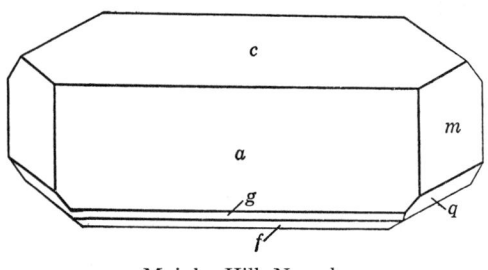

Majuba Hill, Nevada.

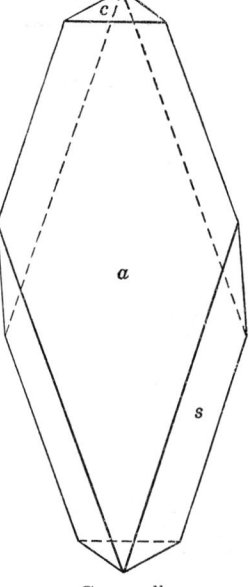

Cornwall.

Color dark greenish black to greenish blue. Streak bluish green. Transparent to translucent.

O p t.[5] Blue-green in transmitted light.

Orientation	n	Pleochroism	
X	1.756	Pale blue-green	Biaxial negative (−).
Y b	1.874	Light blue-green	2V 50° (meas.).
Z near a	1.896	Benzol-green	$r < v$, very strong.

C h e m. A basic arsenate of copper, $Cu_3(AsO_4)(OH)_3$. Small amounts of P substitute for As, with P:As = 1:11 in analysis 2.

Anal.[11]

	CuO	ZnO	Fe_2O_3	As_2O_5	P_2O_5	H_2O	Rem.	Total	G.
1.	62.71			30.19		7.10		100.00	
2.	62.80		0.49	27.08	1.50	7.57		99.44	4.312
3.	62.34	0.06	0.12	29.59	0.05	7.73	0.06	99.95	

1. $Cu_3AsO_4(OH)_3$. 2. Cornwall.[6] 3. Mammoth mine, Tintic district, Utah.[7] Rem. is $0.06\ SiO_2$.

Tests. Same as for olivenite.

Occur. Associated with olivenite and other secondary copper minerals. In Cornwall, at Ting Tang mine, Wheal Unity and Wheal Gorland; also at Bedford United mines, near Tavistock, Devonshire. Near Saida in Saxony, and at Markirch in the Vosges. In the United States at Mammoth mine, Tintic district, Juab County, Utah. At Majuba Hill, Pershing County, Nevada.[8]

Artif. A compound corresponding to the given formula has been prepared by the prolonged action of hot arsenic acid on copper.[9]

Name. From κλίνειν, *to incline*, and κλαν, *to break*, in allusion to the oblique basal cleavage.[10] The terms *cupreous arsenate of iron* and *cuivre arseniaté ferrifère*, often given in older reference works as synonymous, are inappropriate and rest on faulty reading of Bournon (1801).

Ref.
1. Symmetry and axial elements from Palache and Berry, *Am. Min.*, **31**, 243 (1946).
2. Palache and Berry (1946). Rare and doubtful: $s\{552\}$, $k\{\overline{4}01\}$, $\{972\}$, $u\{761\}$.
3. Berry (1946) by Weissenberg method on material from Cornwall. Material from Nevada and Utah gives closely similar values.
4. Washington in Hillebrand and Washington, *Am. J. Sc.*, **35**, 305 (1888), on Utah material.
5. Indices and 2V, Hurlbut in Palache and Berry (1946) on material from Majuba Hill and Tintic. Orientation, pleochroism, and dispersion, Larsen and Berman (204, 1934). See also Des Cloizeaux (**2**, 417, 1874).
6. Damour, *Ann. chim. phys.*, **13**, 412 (1845).
7. Hillebrand anal. in Hillebrand and Washington (1888).
8. Gianella, *Am. Min.*, **31**, 259 (1946).
9. Coloriano, *Bull. soc. chim.*, **45**, 707 (1886).
10. Breithaupt (1830).
11. For additional analyses see Palache and Berry (1946) and Hintze (**1** [4B], 1106, 1933).

41.3.2 CORNETITE $[Cu_3(PO_4)(OH)_3]$. *Cesàro* (*Ann. soc. géol. Belgique, Bull.*, **39**, 241, 1912). Cornetite *Buttgenbach* (452, 1917).

Cryst.[1] Orthorhombic; dipyramidal—$2/m\ 2/m\ 2/m$.

$a:b:c = 0.7703:1:0.5074;\quad p_0:q_0:r_0 = 0.6587:0.5074:1$

$q_1:r_1:p_1 = 0.7703:1.5181:1;\quad r_2:p_2:q_2 = 1.9708:1.2982:1$

Forms:[2]

		ϕ	$\rho = C$	ϕ_1	$\rho_1 = A$	ϕ_2	$\rho_2 = B$
d	210	68°56'	90°00'	90°00'	21°04'	0°00'	68°56'
M	021	0 00	45 25½	45 25½	90 00	90 00	44 34½
r	221	52 23½	58 59	45 25½	47 14½	37 12	58 28
v	121	32 59	50 25½	45 25½	65 11	56 37½	49 43

Structure cell.[3] Space group *Pbca*. a_0 10.86 kX, b_0 14.07, c_0 7.10; $a_0:b_0:c_0 = 0.772:1:0.505$. Cell contents $Cu_{24}(PO_4)_8(OH)_{24}$.

Habit. Short prismatic [001] to equant. {210} usually rounded. As crusts of minute crystals.

Twinning. Reported on {$h0l$}.[4]

P h y s. No cleavage. H. ~$4\frac{1}{2}$. G. 4.10; 4.10 (calc.). Luster vitreous. Color deep blue or peacock-blue, to greenish blue. Transparent to translucent.

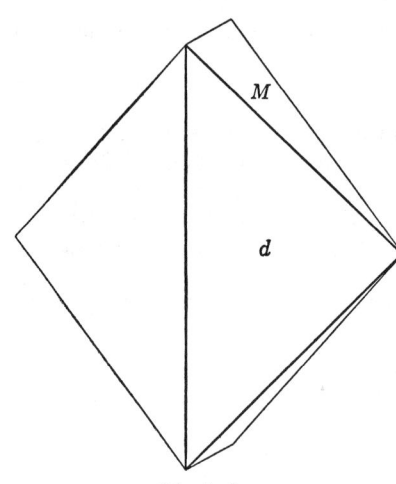
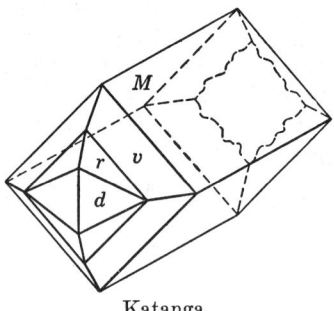

Rhodesia. Katanga.

O p t.[4] In transmitted light, greenish blue. May show color zoning in shades of greenish blue.

Orientation		n (green)	
X	b	1.765	Biaxial negative (−).
Y	a	1.81	$2V$ ~33°.
Z	c	1.82	Not pleochroic.
			$r < v$ strong.

C h e m. A basic phosphate of copper, $Cu_3(PO_4)(OH)_3$. The material from Katanga contains a small amount of Co in substitution for Cu.

Anal.

	CuO	P_2O_5	H_2O	Total	G.
1.	70.89	21.08	8.03	100.00	4.10
2.	70.78	19.81	9.41	100.00	4.10

1. $Cu_3(PO_4)(OH)_3$. 2. Bwana M'Kubwa.[4] Recalculated to 100 after deducting 4.03 insol. and 0.53 Fe_2O_3. Water given as 0.18 hygroscopic and 8.77 ign. loss.

Tests. Soluble in cold HCl.

O c c u r. Originally from the L'Étoile du Congo mine near Elizabethville, Katanga, in the Belgian Congo, associated with heterogenite. Also from Bwana M'Kubwa, northern Rhodesia, with wad, native copper, malachite. In the United States in the Blue Jay and Empire-Nevada mines, Yerington, Nevada.

N a m e. After Jules Cornet (1865-1929), Belgian geologist.

Ref.
1. Angles of Hutchinson and MacGregor, *Min. Mag.*, **19**, 225 (1921), in the orientation and unit of the structure cell of Berry, *Am. Min.*, **31**, 190 (1946). Transformation: H. and M. to Berry, 001/100/0½0; Ungemach, *Ann. Soc. géol. Belg.*, **52**, 75 (1929) to Berry, 001/200/010. See also Cesàro (1912) and Hutchinson and MacGregor, *Nature*, **92**, 364 (1913).
2. Hutchinson and MacGregor (1921). Ungemach (1929) gives also 110, 100, 102, 010, 001, 321, 421 [transformed].
3. Berry (1946) by Weissenberg method on Bwana M'Kubwa crystal.
4. Larsen and Berman (201, 1934).

41.3.3 GEORGIADESITE [Pb₃(AsO₄)Cl₃]. *Lacroix* and *de Schulten* (*C.R.*, **145**, 783, 1907; *Bull. soc. min.*, **31**, 86, 1908). Georgiasdésite.

C r y s t.[1] Monoclinic.

$a:b:c = 1.7675:1:1.5438;$ $\beta\ 102°33\tfrac{1}{2}';$ $p_0:q_0:r_0 = 0.8735:1.5069:1$

$r_2:p_2:q_2 = 0.6636:0.5797:1;$ $\mu\ 77°26\tfrac{1}{2}';$ $p_0'\ 0.8949,\ q_0'\ 1.5438,\ x_0'\ 0.2228$

Forms:

	ϕ	ρ	ϕ_2	$\rho_2 = B$	C	A
c 001	90°00'	12°33½'	77°26½'	90°00'	.....	77°26½'
a 100	90 00	90 00	0 00	90 00	77°26½'	.....
m 110	30 06	90 00	0 00	30 06	83 44½	59 54
x 011	8 12½	57 20	77 26½	33 34	56 26	83 05½
E 102	90 00	33 50	56 10	90 00	21 16½	56 10
y 111	35 54½	62 19	41 49	44 10½	55 30	58 43
z 312	63 45	60 11	32 34½	67 26	49 07	38 54½

Habit and twinning. As stubby pseudo-hexagonal tablets with a platy composite structure parallel [100], apparently due to lamellar twinning on {100} and {1̄04}. {001} grooved and striated [010].

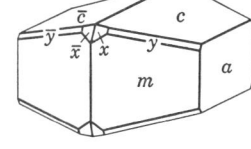

P h y s. Cleavage is lacking. H. 3½. G. 7.1. Luster resinous. Color white or brownish yellow.
O p t.[2] In transmitted light, colorless.

Orientation	n	
X		2.17	Biaxial positive (+).
Y	b	2.17	2V very large.
Z	c	2.18	$r < v$, strong.

C h e m. A chloride-arsenate of lead, Pb₃(AsO₄)Cl₃.
Anal.

	1	2
PbO	38.62	38.86
Pb	35.85	36.38
As₂O₅	13.26	12.49
Cl	12.27	12.47
Total	100.00	100.20

1. Pb₃(AsO₄)Cl₃. 2. Laurium.

Tests. Easily fusible. In C.T. yields a sublimate of PbCl₂. Soluble in HNO₃.

792 PHOSPHATES, ARSENATES, VANADATES

Occur. From Laurium, Greece, in altered lead slag associated with laurionite, fiedlerite, and matlockite. Only a single specimen is known.

Name. After Mr. Georgiadès, a director of the mine at Laurium.

Ref.
1. Angles of Lacroix and de Schulten (1908) recalculated by Palache, priv. comm. (1935), to a monoclinic interpretation. Originally considered to be orthorhombic and pseudo-hexagonal ($m \wedge m' = 60°12'$) with the elements $a:b:c = 0.5770:1:0.2228$ and the forms $\{010\}$, $\{110\}$, $\{011\}$, $\{0.11.4\}$, $\{451\}$, $\{16.5.4\}$, and $\{4.15.2\}$. Palache's simplifying monoclinic interpretation, here followed, considers the crystals to be polysynthetic lamellar twins. Transformation: old to new, $01\bar{1}/100/004$ (approx.).
2. Larsen and Berman (144, 1934); it is not certain that the data actually refer to georgiadesite.

41.3.4   ATELESTITE [$Bi_3(AsO_4)O_2(OH)_2$ (?)]. *Breithaupt* (307, 1832). Rhagite *Weisbach (Jb. Min.*, 302, 870, 1874). Atalesite.

Cryst.[1] Monoclinic; prismatic—$2/m$.

$a:b:c = 0.9334:1:1.5051$; $\beta\ 109°17'$; $p_0:q_0:r_0 = 1.6125:1.4207:1$

$r_2:p_2:q_2 = 0.7039:1.1350:1$; $\mu\ 70°43'$; $p_0'\ 1.7083,\ q_0'\ 1.5051,\ x_0'\ 0.3499$

Forms:[2]

		ϕ	ρ	ϕ_2	$\rho_2 = B$	C	A
c	001	90°00′	19°17′	70°43′	90°00′	....	70°43′
a	100	90 00	90 00	0 00	90 00	70°48′	....
m	110	48 37	90 00	0 00	48 37	75 39	41 23
e	011	13 05	57 05½	70 43	35 08½	54 51½	79 02½
d	101	90 00	64 05	25 55	90 00	44 48	25 55
g	$\bar{1}01$	−90 00	53 38½	143 38½	90 00	72 55½	143 38½
o	111	53 49½	68 35	25 55	56 40	53 38½	41 17
q	313	76 18	64 44	25 55	77 38	46 07½	28 31½

Less common or rare: b 010 l 310

Habit. Minute crystals tabular $\{100\}$, with $a\ g\ d\ q$ usually prominent. $q\{313\}$ somewhat rounded, other faces smooth and brilliant. Also as spherical or mammillary crystalline aggregates, smooth on the surface.

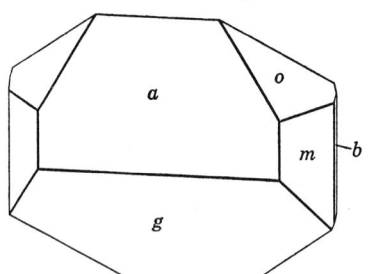

Phys. Cleavage $\{001\}$ indistinct. Fracture subconchoidal. H. $4\frac{1}{2}$–5. G. 6.82.[3] Luster resinous to adamantine. Color sulfur-yellow (crystals) to yellowish green or wax-yellow. Transparent to translucent.

Opt.[4] In transmitted light, pale yellow to colorless.

	n (Neuhilfe mine)	n (Weisser Hirsch mine)	
X	2.14 ± 0.01	~2.10	Biaxial positive (+).
Y	2.15 ± 0.01		$2V\ 44° \pm 2°$ (Na, meas.);
Z	2.18 ± 0.01	~2.15	$2E\ 107°$.
			$r < v$, rather strong.

Chem. A basic arsenate of bismuth, probably $Bi_3(AsO_4)O_2(OH)_2$ (analysis 2).

Anal.

	1	2	3
Bi_2O_3	84.02	82.41	72.76
As_2O_5	13.81	14.12	14.20
H_2O	2.17	1.92	4.62
Rem.		0.51	6.85
Total	100.00	98.96	98.43
G.		6.4	6.82

1. $Bi_3(AsO_4)O_2(OH)_2$. 2. Neuhilfe mine, Schneeberg [5] (crystals). Rem. is Fe_2O_3 0.51. Original total given as 98.99. 3. Weisser Hirsch mine, Schneeberg [6] (mammillary). Rem. is CoO 1.47, CaO 0.50, $(Fe,Al)_2O_3$ 1.62, gangue 3.26.

Tests. B.B. fusible. In C.T. crumbles to a yellow powder and loses water. Easily soluble in HCl, with difficulty in HNO_3.

Occur. Found with bismutite, eulytite, and quartz at the Neuhilfe mine, Schneeberg, Saxony. Later described (rhagite) [7] from the Weisser Hirsch mine, at Neustädtl near Schneeberg, associated with walpurgite, torbernite, and bismutite.

Name. Atelestite from ἀτελής, *incomplete*, presumably because its composition was unknown when first described. Rhagite from ῥάξ (ῥαγος), *grape*, in allusion to the color and botryoidal grouping.

Ref.

1. Busz, *Zs. Kr.*, **15**, 625 (1889), on Neuhilfe mine crystals. See also Rath, *Ann. Phys.*, **136**, 422 (1869), and Goldschmidt (**1**, 265, 1886) who used other positions.
2. Goldschmidt (**1**, 120, 1913).
3. Winkler, *J. pr. Chem.*, **10**, 190 (1874), on fibrous material from the Weisser Hirsch mine (*rhagite*).
4. Larsen (42, 1921) and Frondel, priv. comm. (1947).
5. Busz (1889).
6. Winkler (1874).
7. Shown to be identical with atelestite by Frondel, *Am. Min.*, **28**, 538 (1943).

41.3.5 **FLINKITE** [$Mn_2''Mn'''(AsO_4)(OH)_4$ (?)]. Hamberg (*Geol. För. Förh.*, **11**, 212, 1889).

Cryst.[1] Orthorhombic.

$a:b:c = 0.4131:1:0.7386; \quad p_0:q_0:r_0 = 1.7879:0.7386:1$

$q_1:r_1:p_1 = 0.4131:0.5593:1; \quad r_2:p_2:q_2 = 1.3539:2.4207:1$

Forms:[2]

	ϕ	$\rho = C$	ϕ_1	$\rho_1 = A$	ϕ_2	$\rho_2 = B$
c 001	....	0°00′	0°00′	90°00′	90°00′	90°00′
b 010	0°00′	90 00	90 00	90 00	....	0 00
m 110	67 33	90 00	90 00	22 27	0 00	67 33
e 101	90 00	60 47	0 00	29 13	29 13	90 00
k 111	67 33	62 40	36 26½	34 48½	29 13	70 10½

Habit. Thin tabular {001} and slightly elongated [010]. The crystals often are rounded in the zone [100], and {001} is striated [100]. Isolated crystals rare, usually as feather-like aggregates.

P h y s. Brittle. H. $4\frac{1}{2}$. G. 3.87. Luster vitreous to somewhat greasy. Color greenish brown to dark green. Transparent.

O p t.[3] In transmitted light, brownish yellowish green.

ORIENTATION		n	PLEOCHROISM	
X	b	1.783 ± 0.003	Pale brownish green	Biaxial positive (+).
Y	c	1.801 ± 0.003	Yellow-green	$2V$ large.
Z	a	1.834 ± 0.003	Orange-brown	$r > v$, weak.

C h e m. A basic arsenate of divalent and trivalent manganese, probably $Mn_2''Mn'''(AsO_4)(OH)_4$. Small amounts of Mg and Ca substitute for Mn'', Fe''' substitutes for Mn''', and Sb substitutes for As with Sb:As = 1:16.4 in the only reported analysis.

Anal.

	1	2
CaO		0.4
MgO		1.7
MnO	38.16	35.8
Mn_2O_3	21.24	20.2
$(Fe,Al)_2O_3$		1.5
As_2O_5	30.91	29.1
Sb_2O_5		2.5
H_2O	9.69	9.9
Total	100.00	101.1

1. $Mn_2''Mn'''(AsO_4)(OH)_4$. 2. Harstig mine.

Tests. B.B. easily fusible. Soluble in HCl.

O c c u r. From the Harstig mine at Pajsberg near Persberg, Sweden, in veinlets in magnetite ore associated with sarkinite, brandtite, caryopilite, nadorite, native lead, manganoan calcite, barite.

N a m e. After the Swedish mineralogist and collector Gustav Flink (1849–1931).

Ref.
1. Hamberg (1889).
2. Hamberg (1889). Doubtful: $p\{0.1\,10\}$, $m\{014\}$, $n\{027\}$.
3. Larsen, 75 (1921).

41.3.6 **R E T Z I A N.** Sjögren (*Bull. Geol. Inst. Upsala*, **2**, 54, 1894; *Geol. För. Förh.*, **19**, 1897).

C r y s t.[1] Orthorhombic.

$$a:b:c = 0.441:1:0.727; \quad p_0:q_0:r_0 = 1.647:0.727:1$$

$$q_1:r_1:p_1 = 0.441:0.607:1; \quad r_2:p_2:q_2 = 1.376:2.266:1$$

Forms:

	ϕ	$\rho = C$	ϕ_1	$\rho_1 = A$	ϕ_2	$\rho_2 = B$
c 001	.....	0°00′	0°00′	90°00′	90°00′	90°00′
b 010	0°00′	90 00	90 00	90 00	.....	0 00
n 130	37 03½	90 00	90 00	52 56½	0 00	37 03½
m 110	66 11	90 00	90 00	23 49	0 00	66 11
d 101	90 00	58 44	0 00	31 16	31 16	90 00

RETZIAN

Habit. Prismatic [001], sometimes also tabular {010}.
Twinning.[2] Doubtfully observed on {150}.
Phys. Fracture conchoidal to uneven. H. 4. G. 4.15. Luster vitreous to greasy. Color dark chocolate-brown to chestnut-brown; the

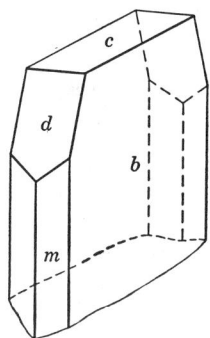

face loci sometimes are differentially pigmented. Streak light brown. Subtranslucent.

Opt.[3] In transmitted light, brown.

ORIENTATION		n	PLEOCHROISM	
X	c	1.777 ± 0.005	Nearly colorless to orange-yellow	Biaxial positive (+). 2V large.
Y	b	1.788 ± 0.005	Yellowish brown to dark brown	$r < v$, weak. Absorption
Z	a	1.800 ± 0.005	Red-brown to crimson	$Z > Y > X$.

Chem. A basic arsenate of calcium, rare earths, and divalent manganese of uncertain formula. Analysis gave:[4]

CaO	MgO	Rare earths	MnO	FeO	PbO	As$_2$O$_5$	SiO$_2$	Insol.	H$_2$O	Total
19.2	2.7	10.3	30.2	1.7	0.2	24.4	0.5	4.3	8.4	101.9

Tests. B.B. almost infusible. In C.T. yields water. Soluble in acids.

Occur. Found sparingly with jacobsite in cavities in a dolomite rock containing braunite at the Moss mine, Nordmark, Sweden.

Name. After the Swedish naturalist Anders Jahan Retzius (1742–1821).

Ref.
1. Sjögren (1894) on crystals of poor quality Retzian is rather similar geometrically to flinkite.
2. Sjögren (1894).
3. Larsen (126, 1921) and Sjögren (1894).
4. Mauzelius analysis in Sjögren (1894) on 0 08 gram.

AZOVSKITE. *Efremov* (*Trans. Lomonossov Inst. Ac. Sc. U.S.S.R.*, **10**, 151, 1937). Asovskite.

Massive, cryptocrystalline, as nodules, shells, and reticulated veinlets. Fracture flat conchoidal. Brittle. H. ~4. G. 2.5. Color dark brown, streak brown. Luster pitchy.

O p t. In transmitted light, brown-red. Slightly birefringent, with mean refractive index about 1.758.

C h e m. A basic ferric phosphate, essentially $Fe_3(PO_4)(OH)_6 \cdot 3H_2O$. The role of the minor constituents reported in the analysis is uncertain and may be due to admixture.

Anal.

	1	2
Fe_2O_3	57.22	52.73
Mn_2O_3		1.68
P_2O_5	16.96	15.90
H_2O	25.82	23.15
Rem.		5.63
Total	100.00	99.09

1. $Fe_3(PO_4)(OH)_6 \cdot 3H_2O$. 2. Taman, Russia. Rem. is CaO 2.84, SiO_2 2.64, CO_2 0.15, MgO tr. H_2O includes H_2O- 11.28, H_2O+ 11.87.

Tests. Easily soluble in HCl and HNO_3, but nearly insoluble in hot concentrated H_2SO_4.

Occur. Found in sedimentary iron ores in the Zheleznaya ravine on the Taman peninsula, Sea of Azov, U.S.S.R.

Needs verification.

TYPE 4. $(AB)_5(XO_4)_2Z_q$

41.4.1 WALPURGITE [$Bi_4(UO_2)(AsO_4)_2O_4 \cdot 3H_2O$ (?)]. Walpurgin *Weisbach* (*Jb. Min.*, 870, 1871; 1, 1877).

Cryst.[1] Triclinic.

$a:b:c = 0.638:1:0.525;$ $\alpha\ 101°40', \beta\ 110°49', \gamma\ 88°17'$

$p_0:q_0:r_0 = 0.748:0.488:1;$ $\lambda\ 78°10', \mu\ 69°10', \nu\ 87°22'$

$p_0'\ 0.824,\ q_0'\ 0.537;\ x_0'\ 0.380,\ y_0'\ 0.224$

Forms:

	ϕ	ρ	A	B	C
b 010	0°00′	90°00′	87°22′	.....	78°10′
m 110	55 04	90 00	32 18	55°04′	66 16
M 1$\bar{1}$0	121 15	90 00	33 53	121 15	78 58
x $\bar{1}\bar{1}$1	−129 27	29 29	113 31	129 27	53 10

Less common:

a 100   d 0$\bar{1}$1   o 111   z $\bar{2}$21   γ $\bar{6}$76   e 021   f $\bar{1}$01   p $\bar{1}$11   y $\bar{1}$21   w $\bar{1}\bar{2}$1

Structure cell.[2] Space group probably $P\bar{1}$. a_0 7.13 Å, b_0 10.44, c_0 5.49; $\alpha\ 101°40', \beta\ 110°49', \gamma\ 88°17'$; $a_0:b_0:c_0 = 0.638:1:0.525$. Cell contents $Bi_4(UO_2)(AsO_4)_2O_4 \cdot 3H_2O$ (?).

Habit. Lath-like [001] and tabular {010} with an oblique termination. As subparallel aggregates and radial groupings.

Twinning. Twin plane and composition surface {010}, common, the twinned aggregates pseudomonoclinic and resembling gypsum.

Phys. Cleavage {010} perfect. H. $3\frac{1}{2}$. G. 6.69 (calc.).[3] Luster adamantine to greasy. Color wax-yellow or straw-yellow. Transparent to translucent. The crystals generally are cloudy, owing to microscopic inclusions and have a transparent zone at the free termination.

O p t.[4] In transmitted light, colorless to pale yellow. Not pleochroic.

ORIENTATION	n (Joachimsthal)	n (Schneeberg)	
X	1.90 ± 0.03	1.871 ± 0.005	Biaxial negative $(-)$.
Y	2.00 ± 0.03	1.975 ± 0.005	Dispersion slight.
Z	2.05 ± 0.03	2.005 ± 0.005	
$2V$	Medium large	$52°$ (calc.)	

The rhomboidal laths have plane angles of about $66°$ and $114°$. X is nearly normal to the plates, and Y' makes an angle of about $12°$ with the long edge in the obtuse angle. Plates turned on edge show twinning parallel to the flat face with an extinction angle of about $8°$. The crystals are zoned, probably with variation in the As:P ratio, and the indices of refraction vary therein by at least 0.03. The clear borders have slightly higher indices of refraction.

C h e m. A hydrated arsenate of bismuth and uranium. The formula is uncertain, perhaps $Bi_4(UO_2)(AsO_4)_2O_4 \cdot 3H_2O$ (from analysis 2). A considerable amount of P substitutes for As in the material of analysis 3. Analyses 3, 4, and 5 were made on cloudy material.

Anal.

	1	2	3	4	5
Bi_2O_3	62.05	61.8	61.87	61.43	59.34
UO_3	19.05	18.7	16.16	20.29	20.54
P_2O_5		0.9	5.88		
As_2O_5	15.30	14.1	12.45	11.88	13.03
H_2O	3.60	3.7	3.42	4.32	4.65
Total	100.00	99.2	99.78	97.92	97.56
G.			5.95		

1. $Bi_4(UO_2)(AsO_4)_2O_4 \cdot 3H_2O$. 2. Schneeberg.[5] Microchemical analysis on clear terminations. 3. Schneeberg.[6] 4,5. Schneeberg.[7]

O c c u r. A secondary mineral found with troegerite, zeunerite, uranosphaerite, torbernite, and uranospinite in the Walpurgis vein of the Weisser Hirsch mine at Neustädtl, Schneeberg, Saxony. Also at Joachimsthal, Bohemia.

Ref.

1. Elements of the x-ray cell of Evans, *Am. Min.*, **35**, 1021 (1950). The morphology is described by Weisbach (1877) and Fischer, *Zs. Kr*, **106**, 25 (1946). Transformations: Weisbach to new, $\bar{x}0\bar{1}/0\bar{x}\bar{1}/001$ (x undetermined); Fischer to new, $0\bar{1}\bar{1}/302/001$.
2. Evans (1950) by precession method.
3. Measured values of 5.76 and 5.95 have been reported but are probably low due to inclusions.
4. Larsen (155, 1921).
5. Fischer, *Jb. Min., Monatsh., Abt. A*, 44 (1948).
6. Gonyer analysis in Evans (1950).
7. Winkler, *J. pr. Chem.*, **7**, 6 (1873).

MERRILLITE [$Na_2Ca_3(PO_4)_2O$ (?)]. Monticellite-like mineral *Merrill* (*Proc. Nat. Ac. Sc.*, **1**, 302, 1917; *Am. J. Sc.*, **43**, 322, 1917). Merrillite *Wherry* (*Am. Min.*, **2**, 119, 1917).

As anhedral grains disseminated in very small amounts in stony meteorites. Very brittle. Sometimes exhibits an indistinct hexagonal cleavage. G. 3.10. Colorless. Optically[1] uniaxial negative $(-)$ with nO 1.623, nE 1.620. Chemical analyses[2] of very small, impure samples obtained from the Allegan and New Concord stones indicate the composition to be $Na_2Ca_3(PO_4)_2O$. Also reported from the Kimble County, Texas,[3] Bjurbole, Homestead, Waconda, New Concord, and other meteorites.

Named after George P. Merrill (1854–1929), Curator of Geology in the U. S. National Museum.

Ref.
1. Shannon and Larsen, *Am. J. Sc.*, **9**, 250 (1925).
2. Shannon in Shannon and Larsen (1925).
3. Barnes, *Univ. Texas Publ.*, *3945*, 623 (1940).

41.4.2 E R I N I T E [$Cu_5(OH)_4(AsO_4)_2$]. Haidinger (*Ann. Phys.*, **14**, 228, 1828; *Phil. Mag.*, **4**, 154, 1828). [Not erinite Des Cloizeaux (*Ann. chim.*, **13**, 420, 1845) and others = chalcophyllite.]

Orthorhombic (?). In mammillated crusts and groups, concentric in structure and fibrous, and rough or satiny from the terminations of very minute crystals; also as vitreous crusts.

P h y s. Cleavage in one direction. Fracture conchoidal to uneven. H. $4\frac{1}{2}$. G. 4.04.[1] Luster almost dull, slightly resinous. Color dark green to emerald-green, slightly inclining to grass-green. Subtranslucent to nearly opaque.

O p t. In transmitted light, green.

Orientation	n (Tintic, Utah)[2]	
X	1.820 ± 0.005	Biaxial negative ($-$).
Y elongation	1.86 ± 0.01	$2V$ moderate.
$Z \perp$ cleavage	1.88 ± 0.01	$r < v$, moderate.

C h e m. A basic arsenate of copper, probably $Cu_5(OH)_4(AsO_4)_2$. Small amounts of Zn and Ca apparently substitute for Cu in the material of analyses 3 and 4.

Anal.[3]

	1	2	3	4
CuO	59.94	59.44	57.67	57.51
ZnO			1.06	0.59
CaO			0.32	0.51
As_2O_5	34.63	33.78	33.53	31.91
P_2O_5			0.10	
$H_2O - 280°$	5.43	5.01	1.14	3.22
$H_2O + 280°$			6.08	5.93
Rem.		1.77	0.14	0.20
Total	100.00	100.00	100.04	99.87

1. $Cu_5(OH)_4(AsO_4)_2$. 2. Cornwall.[4] Rem. is Al_2O_3. 3,4. Tintic, Utah.[5] Rem. is Fe_2O_3, with MgO tr. Analysis 3 recalculated after deducting 3.90 per cent insoluble. Both analyses on air-dried material.

Tests. B.B. easily fusible. Soluble in nitric acid.

O c c u r. A secondary mineral formed by the oxidation of arsenical copper ores. Found typically with olivenite; also with liroconite, clinoclase, linarite, azurite, chrysocolla. The mineral was first described as from County Limerick, Ireland. Reported from Germany at Markirch in the Vosges and at Freudenstadt in the Black Forest. From Glen Gonner, Leadhills, Scotland. Found in the United States in the American Eagle, Mammoth, and Centennial Eureka mines in the Tintic district, Utah.

N a m e. This emerald-green mineral was named erinite by Haidinger because "It unites, what is rarely the case with mineral names, the comparatively trite and prosaical allusion to the native country, with the poetical recollection of the characteristic verdure of the Emerald Isle"; but the specimens probably originated in Cornwall.[6]

Ref.

1. Haidinger (1828).
2. Larsen (72, 1921).
3. Two additional analyses, both incomplete, of Tintic material are cited by Hillebrand and Washington, *Am. J. Sc.*, **35**, 299 (1888).
4. Turner, *Phil. Mag.*, **4**, 155 (1828).
5. Hillebrand analyses in Hillebrand and Washington (1888).
6. Greg and Lettsom (321, 1858).

41.4.3 PSEUDOMALACHITE [$Cu_5(PO_4)_2(OH)_4 \cdot H_2O$?]. Kupfergrün pt. *Wallerius*. Phosphorsaures Kupfer pt. *Karsten, Klaproth* (*Ges. nat. Freunde Berlin, N. Schr.*, **3**, 304, 1801). Phosphorkupfer *Karsten* (64, 97, 1808). Phosphor-kupfererz *Werner*. Cuivre phosphaté *Haüy* (92, 1809). Phosphate of Copper. Pseudomalachite *Hausmann* (1035, 1813). Phosphorocalcit *Glocker* (847, 1831). Ypoléime *Beudant* (**2**, 570, 1832). Ehlit, Prasin, Prasin-chalzit *Breithaupt* (45, 49, 1832). Thrombolite; Thrombolith *Breithaupt* (*J. pr. Chem.*, **15**, 321, 1838). Lunnit *Bernhardi* in Glocker (578, 1839). Kupfer-diaspore *Kühn* (*Ann. Chemie*, **51**, 125, 1844). Dihydrite *Hermann* (*J. pr. Chem.*, **37**, 178, 1846).

Cryst.[1] Monoclinic; prismatic—$2/m$.

$a:b:c = 2.9634:1:0.7787;\quad \beta\,91°01';\quad p_0:q_0:r_0 = 0.2628:0.7786:1$

$r_2:p_2:q_2 = 1.2844:0.3375:1;\quad \mu\,88°59';\quad p_0'\,0.2625,\,q_0'\,0.7787,\,x_0'\,0.0177$

Forms:[2]

c 001	m 110	χ 0$\bar{8}$5	z 301	W $\overline{10}.0.1$	h 432	Ω $\overline{3}11$
b 010	n 540	q 101	τ $\bar{2}01$	f 332	X $\overline{4}32$	
a 100	l 430	t 201	ς $\bar{3}01$	d 885	Γ $\overline{10}.8.5$	

Structure cell.[3] Space group $P2_1/a$.

	a_0	b_0	c_0	β	$a_0:b_0:c_0$
Rheinbreitbach	17.00±0.1 kX	5.75±0.03	4.48±0.03	90°±1°	2.95:1:0.779
Rheinbreitbach	17.01±0.05 kX	5.74±0.02	4.47±0.02	91°01'±15'	2.96:1:0.779
Ehl	17.07±0.05 kX	5.75±0.02	4.48±0.02	90°57'±15'	2.97:1:0.779

Cell contents $Cu_{10}(PO_4)_4(OH)_8 \cdot 2H_2O$ (?).

Habit. Single crystals are rare; prismatic [001], usually with uneven faces and small in size. The crystals often are united in subparallel aggregates with a drusy surface or in hemispherical forms; also reniform, botryoidal, or massive with a radial fibrous structure and concentric banding, the fibers elongated [010]; foliated; microcrystalline or dense; colloform.

Twinning. On {100}.

Phys. Cleavage {010} distinct. Fracture splintery. H. $4\frac{1}{2}$–5. G. 4.35 ± 0.05 (crystals), 4.0–4.3 (fibrous types), 4.36 (calc.). Luster vitreous. Color dark emerald-green to dark green and blackish green (crystals); fibrous material is often somewhat lighter in color, also bluish green. Streak paler than the color. Translucent to subtranslucent.

Opt. In transmitted light, green to bluish green.

ORIENTATION	n (Virneberg)[4]	n (Rheinbreitbach)[4]	n (Ehl)[4]	n (Portugal)[5]
X	1.80	1.789	1.785	1.719
Y	1.86	1.835	1.850	1.763
Z b	1.88	1.845	1.862	1.805
2V	46°	~50°	~50°	~90°
X ∧ c	23°	21°	21°	22°

Usually optically biaxial negative (−) with $r < v$, strong; sometimes optically positive (+) with $r > v$. Pleochroism not perceptible or weak, with X bluish green to pale green, Y yellowish green, Z deep bluish green to blue-green. The significance of the variation in indices of refraction is not known.

Chem. A hydrated basic phosphate of copper. The reported analyses are somewhat divergent but usually are close to $Cu_5(PO_4)_2(OH)_4 \cdot H_2O$ or $Cu_5(PO_4)_2(OH)_4$.

Anal.[6]

	1	2	3	4	5	6	7	8
CuO	66.99	70.89	71.44	69.98	66.97	69.02	69.25	68.13
P_2O_5	23.91	21.08	20.00	21.04	22.07	23.23	23.86	22.73
H_2O	9.10	8.03	8.20	7.65	7.59	8.09	6.76	8.51
Rem.				1.92	3.31		0.19	0.48
Total	100.00	100.00	99.64	100.59	99.94	100.34	100.06	99.85
G.			3.58	4.102			4.309	4.25

1. $Cu_5(PO_4)_2(OH)_4 \cdot H_2O$. 2. $Cu_3(PO_4)(OH)_3$. 3. Las Coste mine, Tarn, France.[7] 4. Katanga, Belgian Congo.[8] Rem. is Al_2O_3 1.84, SiO_2 0.08. 5. Ehl, Germany.[9] Rem. is FeO 0.30, SiO_2 3.01. Chrysocolla admixed. Fibrous. 6. Nizhne Tagilsk.[9] 7. Rheinbreitbach, Germany.[9] Rem. is FeO. 8. Cornwall.[10] Rem. is SiO_2.

Tests. B.B. fuses at about 2. In C.T. yields water and turns black. Soluble in acids.

Occur. A secondary mineral, found associated with malachite, chrysocolla, tenorite, pyromorphite, chalcedony, limonite in the oxidized zone of copper deposits. Among the notable localities are those in Germany at Ehl near Linz and at Virneberg near Rheinbreitbach in the Rhineland, and at Hirschberg in Thuringia. In Czechoslovakia at Libethen near Neusohl, and at Rezbánya (*thrombolite*, pt.)[11] and Moldawa in Roumania. As large malachite-like masses at Nizhne Tagilsk in Urals, U.S.S.R. In England in small amounts at Liskeard and elsewhere in Cornwall. From Viel-Salm, Belgium, and at the Las Coste mine, Alban-la-Fraysse, Dept. Tarn, France. In Africa at Mindouli in the French Congo and at Katanga in the Belgian Congo. From the Collier Bay district in Western Australia. In the United States found in Pennsylvania at the Wheatley mines in Chester County and at the Ecton and Perkiomen mines in Montgomery County. Also from Cabarrus County, North Carolina, and near Phillipsburg, Montana.

Name. Pseudomalachite from ψευδής, *false*, and *malachite* because not malachite although resembling it. The name lunnite, after the Rev. F. Lunn who early (1821) analyzed the Rheinbreitbach material, has been used as a group name to include the supposedly chemically distinct minerals ehlite, pseudomalachite (*phosphorocalcite*), and dihydrite—the latter crystallized and the two former massive and fibrous. Ehlite also has been classed separately. These substances are now known to be identical and following Dana are grouped under the original name pseudomalachite.

Ref.
1. Monoclinic symmetry established by the x-ray Weissenberg study of Berry, priv. comm. (1946), on crystals from Ehl and Rheinbreitbach The morphological development is holohedral. The elements here given are of the structure cell. Pseudomalachite (including *lunnite*, *ehlite*, and *dihydrite*, all of which are identical with fibrous *pseudomalachite* as shown by the x-ray study of Berry (1946) and Frondel, priv. comm. (1942)) has been variously described as monoclinic or triclinic (see Hintze (**1** [4B], 1093, 1933), Goldschmidt (**5**, 173, 1918), Barth and Berman, *Chem. Erde*, **5**, 22 (1930)). The elements of the structure cell are close to the triclinic elements of Schrauf, *Zs. Kr.*, **4**, 1 (1880) [$a:b:c = 2.8252:1:1.5339$; $\alpha\ 89°29\frac{1}{2}'$, $\beta\ 91°00\frac{1}{2}'$, $\gamma\ 90°39\frac{1}{2}'$] but with c halved
2. Schrauf (1880), Goldschmidt (1918), with c halved. Transformation: Goldschmidt, Schrauf to new, 100/010/00$\frac{1}{2}$.
3. Berry (1946) by Weissenberg method.
4. Barth and Berman (1930).
5. Larsen (68, 1921).
6. Numerous additional analyses are listed by Hintze (**1** [4B], 880, 1101, 1933).
7. Lacroix (**4**, 433, 1910).
8. Cesàro and Bellière, *Soc. géol Belgique, Ann.*, **45**, 172 (1922).
9. Schrauf, *Zs Kr*, **4**, 1 (1880)
10. Heddle, *Phil. Mag.*, **10**, 39 (1855).
11. A mixture largely of pseudomalachite, as found by x-ray study—Frondel, priv. comm. (1949).

41.4.4 **A R S E N O C L A S I T E**[$Mn_5(AsO_4)_2(OH)_4$]. Arsenoklasit *Aminoff* (*Kungl. Svenska Vetenskap. Handl.*, **9**, no. 5, 52, 1931).

C r y s t.[1] Orthorhombic.
Structure cell.[2] $a_0\ 9.19\ kX$, $b_0\ 18.01$, $c_0\ 5.79$; $a_0:b_0:c_0 = 0.510:1:0.313$. Cell contents $Mn_{20}(AsO_4)_8(OH)_{16}$.
Habit. Massive, granular.
P h y s. Cleavage {010} perfect. H. 5–6. G. 4.16; 4.27 (calc.). Color red.
O p t.[3]

Orientation		n(Na)	
X	b	1.787	Biaxial negative (−).
Y	a	1.810	2V 53°26',
Z	c	1.816	2E 108°57'.

C h e m. A basic arsenate of divalent manganese, $Mn_5(AsO_4)_2(OH)_4$.
Anal.

	1	2
MnO	57.29	55.01
As_2O_5	36.92	36.96
H_2O	5.79	5.90
Rem.		1.55
Total	100.00	99.42
G.	4.27	4.16

1. $Mn_5(AsO_4)_2(OH)_4$. 2. Långban. Rem. is MgO 0.87, FeO tr., BaO 0.11, CuO 0.57; H_2O includes H_2O- 0.04.

Occur. Occurs at Långban, Sweden, with sarkinite (which it closely resembles) and adelite along fissures in dolomite which have been impregnated with hausmannite.

Name. From ἀρσενικόν, arsenic, and κλάσις, cleavage.

Ref.
1. Symmetry indicated by x-ray Laue study.
2. By rotation and Laue methods; possible space groups $Pmmm$, Pmm, Pmn, $P222$ to $P2_12_12_1$.
3. Aminoff (1931) by prism method, with original value for nX (1.808) corrected by Spencer, Min Abs., **4**, 496 (1931).

41.4.5 ANDREWSITE [(Cu,Fe″)$_3$Fe$_6$‴(PO$_4$)$_4$(OH)$_{12}$]. Maskelyne (Chem. News, **24**, 99, 1871; J. Chem. Soc. London, **28**, 586, 1875).

Orthorhombic (?). Botryoidal aggregates with a radial fibrous structure. Cleavable in two directions parallel to the fiber length. H. 4. G. 3.475. Luster somewhat silky. Color dark green to bluish green.

Opt.[1] In transmitted light, green.

Orientation	n	Pleochroism	
X	1.813	Pale yellowish green	Biaxial positive (+).
Y	1.820	Emerald-green	$2V$ large.
Z	1.830	Yellow to olive-green	$r < v$, extreme.

Chem. A basic phosphate of copper and ferric iron, (Cu,Fe″)$_3$Fe$_6$‴(PO$_4$)$_4$(OH)$_{12}$. Fe″ substitutes for Cu to about Fe:Cu = 1:1.38 in the only reported analysis, and a series may extend to the isostructural species laubmannite.[1]

Anal.

	1	2	3
CaO		0.09	0.09
FeO	8.23	7.11	8.58
MnO		0.60	0.60
CuO	12.58	10.86	10.86
Fe$_2$O$_3$	43.55	44.64	43.01
Al$_2$O$_3$		0.92	0.92
P$_2$O$_5$	25.81	26.09	26.09
H$_2$O	9.83	8.79	8.79
SiO$_2$		0.49	0.49
Total	100.00	99.59	99.43
G.		3.475	

1. (Cu,Fe″)$_3$Fe$_6$‴(PO$_4$)$_4$(OH)$_{12}$ with Cu:Fe = 1.38:1. 2. Cornwall.[2] 3. Analysis 2 with enough Fe$_2$O$_3$ converted to FeO to make RO:R$_2$O$_3$ = 1:1.

Occur. Found in the West Phoenix mine, near Liskeard, Cornwall, England, associated with limonite, "dufrenite," chalcosiderite, and cuprite. The fibrous aggregates grade inwardly to a brown, hydrated iron phosphate.[3]

Name. After Thomas Andrews (1813–1885), English chemist.

Ref.
1. Frondel, Am. Min., **34**, 534 (1949).
2. Flight analysis in Maskelyne (1875).
3. An analysis is cited by Maskelyne (1875).

41.4.6 LAUBMANNITE [$Fe_3''Fe_6'''(PO_4)_4(OH)_{12}$]. Frondel (Am. Min., 34, 513, 1949).

Orthorhombic (?). As crusts with a parallel-fibrous structure. Cleavable in one and perhaps two directions parallel to the fiber length. Brittle. H. $3\frac{1}{2}$–4. G. 3.33. Luster vitreous to silky in aggregates. Color grayish green and greenish brown to brown; unoxidized material probably is dark green. Subtranslucent to opaque.

O p t. In transmitted light, green to brown in color.

Orientation	n (Arkansas)	Pleochroism	
X	1.840	Pale buff	Biaxial positive (+).
Y	1.847	Greenish brown to olive-green	$r < v$, extreme. $2V$ moderate, variable.
Z	1.892	Reddish brown to olive-brown	Absorption $Z > Y > X$.

C h e m. A basic phosphate of ferrous and ferric iron, $Fe_3''Fe_6'''(PO_4)_4(OH)_{12}$. Small amounts of Ca and Mn'' substitute for Fe'' in the material of analysis 3. In its unoxidized state laubmannite is the ferrous iron analogue of andrewsite; the ferrous iron is largely oxidized to the ferric state in the material as found, as with dufrenite and frondelite-rockbridgeite.

Anal.

	1	2	3	4
MgO		0.01	0.01	
CaO		1.14	1.14	
FeO	19.83	2.07	15.47	19.74
MnO		2.40	2.40	
Fe_2O_3	44.09	57.88	42.99	44.95
Al_2O_3		0.05	0.05	
P_2O_5	26.13	25.95	25.95	26.21
H_2O-		0.44	0.44	0.67
H_2O+	9.95	10.06	10.06	8.40
Total	100.00	[100.00]	98.51	99.97
G.		3.33		

1. $Fe_3''Fe_6'''(PO_4)_4(OH)_{12}$. 2. Shady, Polk County, Arkansas.[1] Recalculated to 100 after deducting quartz. 3. Analysis 2 with enough Fe_2O_3 converted to FeO to make $RO:R_2O_3 = 1:1$. 4. Laubmannite (?). Nitzelbuch mine, Bavaria.[2]

O c c u r. From Shady, Polk County, Arkansas, with limonite. A fibrous, dark green mineral from the Nitzelbuch mine, Amberg-Auerbach district, Bavaria, Germany, apparently is identical with laubmannite (analysis 4).

N a m e. After the German mineralogist Heinrich Laubmann.

Ref.
1. Hallowell analysis in Frondel (1949).
2. Spengel analysis in Laubmann, *Geognost. Jahreshefte, Geol. Landes. München*, **35**, 193 (1923).

TYPE 5. $AB(XO_4)Z_q$

41.5.1 ADELITE GROUP AND 41.5.2 DESCLOIZITE GROUP

ORTHORHOMBIC

	a_0	b_0	c_0	Space group
Adelite Group				
Adelite, $CaMg(AsO_4)(OH,F)$	7.47 kX	8.94	5.88	$P2_12_12_1$
Conichalcite, $CaCu(AsO_4)(OH)$	7.42	9.20	5.85	$P2_12_12_1$ (?)
Austinite, $CaZn(AsO_4)(OH)$	7.43	9.00	5.90	$P2_12_12_1$
Duftite, $PbCu(AsO_4)(OH)$	7.50	9.12	5.90	$P2_12_12_1$ (?)
Descloizite Group				
Descloizite, $ZnPb(VO_4)(OH)$	7.56	9.39	6.05	$Pnam$
Mottramite, $CuPb(VO_4)(OH)$				$Pnam$
Pyrobelonite, $MnPb(VO_4)(OH)$	7.84	9.45	6.09	$Pnam$
Calciovolborthite, $CuCa(VO_4)(OH)$				
Turanite, $Cu_2(VO_4)(OH)$ (?)				

The members of the Adelite Group are isostructural and possess orthorhombic disphenoidal (2 2 2) symmetry. All are rare in occurrence and little is known of their compositional variation. Zn substitutes for Cu in conichalcite, and at least a partial series extends toward austinite. Apparently only small amounts of P can substitute for As, and the phosphate analogues of these species are not known. The members of the Libethenite Group are related in formula type, $AA(PO_4)(OH)$, and probably in structure, but are orthorhombic dipyramidal ($2/m\ 2/m\ 2/m$) and have the space group $Pnnm$. Small amounts of F can substitute for (OH) in adelite at least; with a higher content of F in this species, polymorphism intervenes with the formation of the monoclinic species tilasite, $CaMg(AsO_4)F$.

In the Descloizite Group, mottramite and descloizite form a complete series by mutual substitution of Cu and Zn. Pyrobelonite is a member of the group, but the position of calciovolborthite and especially of turanite is uncertain. The symmetry of the Descloizite Group is orthorhombic dipyramidal ($2/m\ 2/m\ 2/m$) and differs from the symmetry of the Adelite Group. The two groups, however, are closely allied in crystal structures as shown by the dimensional relations and by the similarity in x-ray powder diffraction patterns. As substitutes for V in the Descloizite Group to a significant, although apparently limited, extent, and P is present only in very small amounts.

41.5.1.1 **A D E L I T E** [$CaMg(AsO_4)(OH,F)$]. *Sjögren* (*Geol. För. Förh.*, **13**, 781, 1891; *Bull. Geol. Inst. Upsala*, **1**, 56, 1893).

C r y s t.[1] Orthorhombic; disphenoidal—2 2 2.

$a:b:c = 0.8294:1:0.6650;$ $p_0:q_0:r_0 = 0.8018:0.6650:1$

$q_1:r_1:p_1 = 0.8294:1.2472:1;$ $r_2:p_2:q_2 = 1.5038:1.2057:1$

Adelite

Forms:[2]

	ϕ	$\rho = C$	ϕ_1	$\rho_1 = A$	ϕ_2	$\rho_2 = B$
l 120	31°05′	90°00′	90°00′	58°55′	0°00′	31°05′
m 110	50 19½	90 00	90 00	39 40½	0 00	50 19½
d 011	0 00	33 37½	33 37½	90 00	90 00	56 22½
f 021	0 00	53 03½	53 03½	90 00	90 00	36 56½
e 031	0 00	63 22½	63 22½	90 00	90 00	26 37½
o 101	90 00	38 43½	0 00	51 16½	51 16½	90 00
p 1$\bar{1}$1	129 40½	46 10	−33 37½	56 16½	51 16½	117 25½

Structure cell.[3] Space group $P2_12_12_1$. a_0 7.47 kX, b_0 8.94, c_0 5.88; $a_0:b_0:c_0 = 0.836:1:0.658$. Cell contents $Ca_4Mg_4(AsO_4)_4(OH)_4$.

Habit. Crystals elongated [100]. Usually massive.

Phys. Fracture uneven to conchoidal. H. 5. G. 3.73 ± 0.03; 3.79 (calc.). Luster resinous. Colorless to gray, bluish gray, yellowish gray, yellow, light green. Transparent.

Opt. In transmitted light, colorless.

Orientation[2]		n (Jacobsberg)[4]	n(Na) (Långban)[2]	
X	a	1.712		Biaxial positive (+).
Y	c	1.721	1.707	$r < v$, perceptible.
Z	b	1.731		
2V		∼90°	68°36′ (meas.)	

Chem. A basic arsenate of calcium and magnesium, $CaMg(AsO_4)(OH)$. F substitutes for OH with F:OH = 1:2.44 in analysis 5. The extent of this substitution in the orthorhombic structure appears to be limited, and material high in fluorine is monoclinic (*tilasite*). Small amounts of Pb and Mn″ substitute for Ca and Mg, respectively. P substitutes for As with P:As = 1:9.6 in analysis 5.

Anal.

	1	2	3	4	5	6
CaO	25.45	25.43	23.13	24.04	23.43	25.52
BaO		tr.		0.23		0.81
PbO		0.39	2.41	2.79	0.17	
MgO	18.30	17.05	19.25	17.90	17.10	18.98
MnO		1.64	1.27	0.48	4.38	1.69
P_2O_5					3.01	
As_2O_5	52.17	50.04	48.52	50.28	47.40	49.73
Cl		0.24	tr.	tr.	0.31	
F					1.44	
H_2O	4.08	4.25	3.99	3.90	3.33	[2.36]
Rem.		0.56	1.97	0.40		0.91
	100.00	99.60	100.54	100.02	100.57	[100.00]
Cl,F = O		0.05			0.62	
Total		99.55			99.95	
G.		3.71	3.72	3.76	3.70	

1. $CaMg(AsO_4)(OH)$. 2. Kittel mine, Nordmark.[5] Rem. is $(Fe,Al)_2O_3$ 0.30, Cu 0.26. 3. Jacobsberg.[5] Rem. is SiO_2 1.88, FeO 0.09. 4. Långban.[5] Rem. is FeO 0.08, CuO 0.32. 5. Långban.[6] Original total given as 99.89. 6. Långban[7] (*adelite?*). Rem. is $(Fe,Al)_2O_3$ 0.83, ZnO(?) 0.08.

Tests. B.B. fuses easily to a gray enamel. Soluble in dilute acids.

Occur. Found in the manganese ore deposits of Vermland, Sweden: at Långban with sarkinite, arsenoclasite, braunite, hedyphane; at Jacobsberg, and at the Kittel mine, Nordmark, with hausmannite, magnetite, and native copper. A mineral perhaps identical with adelite or tilasite occurs at the Moss mine, Nordmark (analysis 6).

Name. From ἄδηλος, indistinct.

Ref.
1. The original crystallographic description of Sjögren (1893) was made by mistake on a pyroxene as shown by Aminoff, *Kungl. Svenska Vetenskap., Handl.*, [3], **11**, no. 4, 24 (1933). The elements and unit are those of Aminoff on Långban crystals (analysis 5) with a and b interchanged in accordance with the convention $c < a < b$. Transformation: Aminoff to new, 010/$\bar{1}$00/001.
2. Aminoff (1933)
3. Hägele, *Jb. Min., Beil.-Bd.*, **75**, 101 (1939), on Långban crystal; see also Aminoff (1933).
4. Larsen (35, 1921); the optic orientation given in Sjögren (1893) and Larsen and Berman (131, 1934) is in error.
5. Mauzelius analysis in Sjögren (1893, 1891).
6. Blix analysis in Aminoff (1933).
7. Lundström analysis in Sjögren (1893).

41.5.1.2 CONICHALCITE [CaCu(AsO$_4$)(OH)].

Konichalcit *Breithaupt* and *Fritzsche* (*Ann. Phys.*, **77**, 139, 1849). Barthite *Henglein* and *Meigen* (*Cbl. Min.*, 353, 1914). Staszycyt, Staszicite *Morozewicz* (*Bull. int. ac. sc. Cracovie, Cl. Sc., Ser. A*, 4, 1918). Higginsite *Palache* and *Shannon* (*Am. Min.*, **5**, 155, 1920).

Cryst.[1] Orthorhombic; disphenoidal—2 2 2.

$$a:b:c = 0.7940:1:0.6242; \quad p_0:q_0:r_0 = 0.7861:0.6242:1$$

$$q_1:r_1:p_1 = 0.7940:1.2720:1; \quad r_2:p_2:q_2 = 1.6021:1.2594:1$$

Forms:[2]

	ϕ	$\rho = C$	ϕ_1	$\rho_1 = A$	ϕ_2	$\rho_2 = B$
c 001		0°00′	0°00′	90°00′	90°00′	90°00′
m 110	51°33′	90 00	90 00	38 27	0 00	51 33
C 023	0 00	22 35½	22 35½	90 00	90 00	67 24½
M 011	0 00	31 58½	31 58½	90 00	90 00	58 01½
g 021	0 00	51 18½	51 18½	90 00	90 00	38 41½
j 041	0 00	68 56½	68 56½	90 00	90 00	21 03½
y 203	90 00	27 39½	0 00	62 20½	62 20½	90 00
e 201	90 00	57 32½	0 00	32 27½	32 27½	90 00
o′ 111	51 33	45 06½	31 58½	56 18	51 49½	63 51½
p′ 221	22 46½	53 33	51 18½	71 51½	62 20½	42 07½
′o 1$\bar{1}$1	128 27	45 06½	−31 58½	56 18	51 49½	116 08½
′p 2$\bar{2}$1	157 13½	53 33	−51 18½	71 51½	62 20½	137 52½
r′ 243	32 12	44 31½	39 46	68 03½	62 24	53 36
′r 2$\bar{4}$3	147 48	44 31½	−39 46	68 03½	62 24	126 24

Less common or rare:
B 012   z 101   A′ 647   s′ 825   x′ 623   ′q 6$\bar{4}$7   ′S 8$\bar{2}$5   ′t 6$\bar{2}$3

Structure cell.[3] Space group uncertain, perhaps $P2_12_12_1$. a_0 7.42 kX, b_0 9.20, c_0 5.85; $a_0:b_0:c_0 = 0.806:1:0.636$. Cell contents Ca$_4Cu_4$(AsO$_4$)$_4(OH)_4$.

Habit. Crystals are equant to short prismatic [010]. Usually as botryoidal to reniform crusts and masses with a radial fibrous structure.
P h y s. Cleavage lacking. Fracture uneven. Brittle. H. $4\frac{1}{2}$. G. 4.33 (crystals, Bisbee),[4] 4.29 (calc.); the G. of the fibrous types ranges down to about 4.1. Luster vitreous to somewhat greasy. Color grass-green to yellowish green, pistachio-green, or emerald-green. Streak green. Subtranslucent.

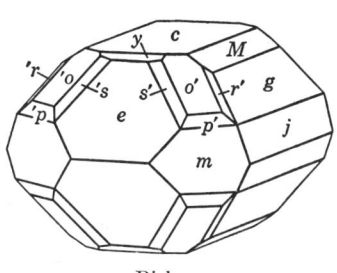

Bisbee.

O p t. In transmitted light, pale green to yellowish green in color. The optical constants vary considerably as the composition varies, but the relationships are not clearly defined. The indices may vary within single crystals or between different parts of crusted aggregates.

ORIENTATION	n (Bisbee, anal. 2)[4]	n (Guchab)[5]	n (Tintic, anal. 4)[6]	n (Bristol)[7]
X c	1.800	1.780	1.778	1.730
Y b	1.831	1.795		
Z a	1.846	1.815	1.801	1.771
Sign	–	+	+	–
Elong.			+	+
$2V$	~90°	~90°	~0°	~25°
PLEOCHROISM				
X	Green		Nearly colorless	
Y	Yellow-green		Pale greenish	
Z	Blue-green		Pale bluish	
Dispersion	$r > v$, rather strong	$r < v$, moderate	None	$r > v$, very strong

C h e m. A basic arsenate of calcium and copper, $CaCu(AsO_4)(OH)$. Zn substitutes for Cu with Zn:Cu = 1:3.7 in analysis 6[8] and a partial series extends towards austinite. Small amounts of Mg have been reported presumably in substitution for Cu. P substitutes for As up to at least P:As = 1:2.15 (analysis 3),[9] and V also substitutes for As with V:As = 1:16.6 in analysis 2. The botryoidal fibrous material usually contains a small amount of water in excess of the formula given.

Anal.

	1	2	3	4	5	6
CaO	21.61	21.67	21.36	19.79	23.10	20.80
MgO				0.54	1.90	0.27
CuO	30.65	29.83	31.76	28.68	31.55	26.45
ZnO				2.86		7.30
V_2O_5		2.05	1.78			
As_2O_5	44.27	42.90	30.68	39.94	36.40	38.77
P_2O_5			8.81	0.14	1.30	
H_2O	3.47	3.55	5.61	5.52	5.15	5.56
Rem.				2.53	0.40	0.91
Total	100.00	100.00	100.00	[100.00]	99.80	100.06
G.		4.33			4.15	4.227

1. $CaCu(AsO_4)(OH)$. 2. Higgins mine, Bisbee, Arizona [10] (*higginsite*). Recalculated to 100 after deducting Fe_2O_3 0.48, MnO 2.84, gangue 0.86, present as admixed wad and limonite. 3. Hinojosa de Cordoba, Spain.[11] 4. American Eagle mine, Tintic.[12] Rem. is Ag 0.30, Fe_2O_3 0.36, CO_2 [0.97], gangue 0.90. 5. Maja-Tass, Siberia.[13] Rem. is Fe_2O_3. 6. Miedzianka, Poland [14] (*staszicite*). Rem. is FeO 0.63, MnO 0.14, SiO_2 0.14.

Tests. In C.T. decrepitates, turns black, and gives off a little neutral water. B.B. fuses at about 3. Readily soluble in HCl or HNO_3.

Occur. A secondary mineral found in the oxidized zone of copper deposits associated with limonite, austinite, olivenite, clinoclase, libethenite, brochantite, malachite, azurite, jarosite. Originally from Hinojosa de Cordoba, Andalusia, Spain. At Maja-Tass, province of Akmolinsk, western Siberia, U.S.S.R., and at Graul near Schwarzenberg, Saxony, Germany. As an alteration of tennantite at Miedzianka near Kielce, Poland (*staszicite*). From the copper mine at Guchab near Otavi, South West Africa (*barthite*). At Collahuasi, Tarapacá province, Chile. In the United States, a notable occurrence in the American Eagle and other mines in the Tintic district, Juab County, Utah, where the mineral occurs with chenevixite and other arsenates as an alteration of enargite. A well-crystallized occurrence in the Higgins mine at Bisbee, Cochise County, Arizona (*higginsite*); [15] also reported from Globe Hills, Gila County, Arizona. In Nevada with chrysocolla and tenorite at the Bristol mine, Lincoln County; at the Simon mine and the Calavada mine, Mineral County; at the Empire-Nevada mine, Yerington, Lyon County; at Good Springs, Clark County. Reported from Caruther's mine, Trans-Pecos, Texas.

Name. Conichalcite from κονία, *powder*, and χαλκος, *lime*. The name conichalcite clearly has priority over the later names higginsite, staszicite, and barthite (the latter a zoned parallel intergrowth of conichalcite and austinite) and is here retained for the species.

BARTHITE. *Henglein* and *Meigen* (*Cbl. Min.*, 353, 1914).

Originally described as a basic arsenate of zinc and copper but later shown [16] to contain also calcium and to be identical with conichalcite-austinite. The crystals are zoned: [17] the inner pale green to colorless zones are austinite, perhaps containing some Cu in substitution for Zn, and the thin outer grass-green zones are essentially conichalcite. Right- and left-handed crystals sometimes are twinned together on {001}.

Ref.

1. Angles of Palache and Shannon (1920) on Higgins mine crystals; unit of the structure cell with $c < a < b$. Palache and Shannon to new, 001/010/$\bar{1}$00. Disphenoidal symmetry established by Fisher, *Zs. Kr.*, **105**, 268 (1944), on Guchab crystals (*barthite*); the Higgins mine crystals (*higginsite*) have a dipyramidal appearance here interpreted as due to the equal development of right and left sphenoids.
2. Palache and Shannon (1920).
3. Richmond, *Am. Min.*, **25**, 441 (1940), by Weissenberg method on type crystals of higginsite; nearly identical cell dimensions are given by Strunz, *Zs. Kr.*, **101**, 496 (1939).
4. Larsen and Berman (202, 1934).
5. Larsen (44, 1921) on outer deep-green part of a zoned crystal (see *barthite*).
6. Larsen (60, 1921).
7. Gillson, *Am. Min.*, **11**, 109 (1926), who gives data on material from additional localities.
8. The identity of the material of this analysis with conichalcite has not been demonstrated by rigorous means.
9. A partial analysis with P_2O_5 9.10 has also been reported by Fritzsche, *Ann. Phys.*, **77**, 180 (1849), on material from the same locality, Hinojosa de Cordoba.
10. Shannon analysis in Palache and Shannon (1920).
11. Fritzsche (1849).
12. Hillebrand, *Proc. Colorado Sc. Soc.*, **1**, 114 (1883).

13. Michel, *Bull. soc. min.*, **32**, 50 (1909).
14. Morozewicz (1918).
15. Shown to be identical with conichalcite by Strunz (1939) and on the type specimen by Frondel, priv. comm. (1948).
16. Fisher (1944), who describes the morphology of the zoned crystals. The x-ray powder pattern of the zoned aggregate is identical with that of austinite but with a slightly smaller cell—Frondel, priv. comm. (1947).
17. See Fisher (1944) and Larsen (44, 1921).

41.5.1.3 **A U S T I N I T E** [$CaZn(AsO_4)(OH)$]. *Staples* (*Am. Min.*, **20**, 112, 1935). Brickerit Ahlfeld and Mosebach (*Zbl. Min.*, 226, 287, 1936); Ahlfeld (*Jb. Min., Beil.-Bd.*, **66A**, 41, 1932).

C r y s t.[1] Orthorhombic; disphenoidal—2 2 2.

$$a:b:c = 0.832:1:0.657; \quad p_0:q_0:r_0 = 0.790:0.657:1$$

$$q_1:r_1:p_1 = 0.832:1.266:1; \quad r_2:p_2:q_2 = 1.522:1.202:1$$

Forms:

	ϕ	$\rho = C$	ϕ_1	$\rho_1 = A$	ϕ_2	$\rho_2 = B$
b 010	0°00′	90°00′	90°00′	90°00′	.....	0°00′
m 110	50 14½	90 00	90 00	39 45½	0°00′	50 14½
q 011	0 00	33 18½	33 18½	90 00	90 00	56 41½
p 111	50 14½	45 46	33 18½	56 34½	51 42	62 43½
ρ 1̄11	129 45½	45 46	33 18½	56 34½	51 42	117 16½

Structure cell.[2] Space group $P2_12_12_1$. a_0 7.43 kX, b_0 9.00, c_0 5.90; $a_0:b_0:c_0 = 0.826:1:0.656$. Cell contents $Ca_4Zn_4(AsO_4)_4(OH)_4$.

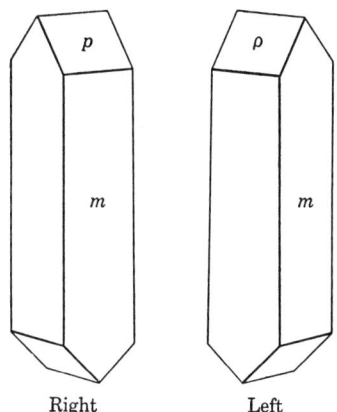

Right   Left

Habit. In minute bladed or acicular crystals elongated [100], sometimes with sceptre-like terminations. Etch figures in dilute HCl distinguish right- and left-handed crystals. Also as radially fibrous crusts and nodules.

P h y s. Cleavage {011}, good. Brittle. H. 4–4½. G. 4.13;[3] 4.37 (calc.). Colorless to white and pale yellowish white. Luster sub-adamantine (crystals) to silky (fibrous aggregates).

810   PHOSPHATES, ARSENATES, VANADATES

Opt. In transmitted light, colorless.

ORIENTATION		$n(Na)$ [4]	
X	a	1.759 ± 0.003	Biaxial positive (+).
Y	c	1.763 ± 0.003	$2V \sim 47°$.
Z	b	1.783 ± 0.003	

Chem. A basic arsenate of calcium and zinc, $CaZn(AsO_4)(OH)$.

Anal.

	1	2	3
CaO	21.46	19.2	21.12
ZnO	31.13	32.5	30.51
As_2O_5	43.96	42.7	42.48
H_2O+	3.45	3.6	3.47
H_2O-			0.12
Rem.		2.5	2.45
Total	100.00	100.5	100.15
G.		4.12	4.13

1. $CaZn(AsO_4)(OH)$. 2. Gold Hill, Utah.[5] Rem. is P_2O_5 0.1, insol. 2.4. 3. Lomitos, Bolivia.[6] Rem. is Sb_2O_5 0.53, Fe_2O_3 0.69, MnO 0.12, insol. 1.11.

Occur. Found with adamite, limonite, and quartz in oxidized ore at Gold Hill, Tooele County, Utah. Later described (*brickerite*) [7] as fibrous veinlets with calcite and chalcedony in the Lilli mine near Lomitos, Sica Sica province, Bolivia. Reported with barthite from Guchab, South West Africa.

Alter. Turns dull white and becomes carbonated.

Name. After Austin F. Rogers (1877–), American mineralogist.

Ref.

1. Ratio of Staples (1935) on Utah crystals in orientation of x-ray cell of Richmond, *Am. Min.*, **25**, 441 (1940); transformation: Staples to new, 001/010/100.
2. Richmond (1940) by Weissenberg method on Utah crystals.
3. Ahlfeld and Mosebach (1936); Staples (1935) gave approx. 4.12.
4. Staples (1935); Ahlfeld and Mosebach (1936) give nX 1.752, nY 1.756, nZ 1.779, $2V$ 41° on the Bolivian material.
5. Ellestad analysis in Staples (1935).
6. Brendler, *Am. Min.*, **23**, 347 (1938); an earlier, unsatisfactory analysis of Bolivian material was reported by Barrande-Hesse in Ahlfeld and Mosebach (1936).
7. Shown to be identical with austinite by Brendler (1938) and Ahlfeld and Mosebach (1936).

41.5.1.4   **DUFTITE** [$PbCu(AsO_4)(OH)$].   Pufahl (*Cbl. Min.*, 289, 1920).

Cryst.[1] Orthorhombic; disphenoidal—2 2 2(?).

$a:b:c = 0.822:1:0.647;$ $p_0:q_0:r_0 = 0.7871:0.647:1$

$q_1:r_1:p_1 = 0.822:1.2705:1;$ $r_2:p_2:q_2 = 1.5456:1.2165:1$

Forms:

		ϕ	$\rho = C$	ϕ_1	$\rho_1 = A$	ϕ_2	$\rho_2 = B$
n	130	22°04½′	90°00′	90°00′	67°55½′	0°00′	22°04½′
m	110	50 35	90 00	90 00	39 25	0 00	50 35
d	011	0 00	32 54	32 54	90 00	90 00	57 06

Structure cell.[2] Space group uncertain. a_0 7.50 kX, b_0 9.12, c_0 5.90; $a_0:b_0:c_0 = 0.822:1:0.647$. Cell contents $Pb_4Cu_4(AsO_4)_4(OH)_4$.
Habit. In aggregates and crusts of tiny crystals elongated [001] with curved and rough faces.
P h y s. H. 3. G. 6.98 (calc.).[3] Luster vitreous on fracture surfaces, dull on crystal faces. Color bright olive-green to gray-green. Streak pale green to white. Subtranslucent.
O p t.[4] In transmitted light, pale apple-green.

nX	2.03 ± 0.01	Biaxial negative (−).
nY	2.06 ± 0.01	$2V$ large.
nZ	2.08 ± 0.01	$r > v$, perceptible.

C h e m. A basic arsenate of lead and copper, $PbCu(AsO_4)(OH)$.
Anal.

	1	2
CaO		0.75
PbO	52.31	50.10
CuO	18.65	19.32
ZnO		0.46
As_2O_5	26.93	26.01
SiO_2		0.44
H_2O	2.11	2.73
Total	100.00	99.81

1. $PbCu(AsO_4)(OH)$. 2. Tsumeb. H_2O includes H_2O- 0.08.

O c c u r. Found with azurite and yellowish films of a bauxite-like material at Tsumeb, South West Africa.
N a m e. After G. Duft, a director of the mines at Tsumeb.

Ref.
1. Forms of Berman, priv. comm. (1932); ratio of the structure cell of Richmond, *Am. Min.*, **25**, 458 (1940). Berman's approximate morphological measurements gave $a:b:c = 0.869:1:0.623$. Disphenoidal symmetry assumed from analogy to austinite.
2. Richmond (1940) by Weissenberg method.
3. The measured G., 6.19, of Pufahl (1920) appears to be in error.
4. Barth and Berman, *Chem. Erde*, **5**, 30 (1930).

DESCLOIZITE SERIES

41.5.2.1 **D E S C L O I Z I T E** [(Zn,Cu)Pb(VO$_4$)(OH)]. Dechenite *Bergemann* (*Ann. Phys.*, **80**, 393, 1850). Aræoxene *von Kobell* (*J. pr. Chem.*, **50**, 496, 1850). Descloizite *Damour* (*Ann. chim. phys.*, **41**, 72, 78, 1854). Eusynchite *Fischer* and *Nessler* (*Ber. Verh. nat. Ges. Freiburg*, **1**, 33, 1854); Rhombischer Vanadit *Zippe* (*Ak. Wien, Ber.*, **44**, [1], 197, 1861); *Tschermak* (*ibid.*, **44**, [2], 157, 1861); *Schrauf* (*Ann. Phys.*, **116**, 355, 1862). Tritochorit *Frenzel* (*Min. Mitt.*, **3**, 506, 1880; **4**, 97, 1881). Cuprodescloizite *Rammelsberg* (*Ak. Berlin, Ber.*, 1215, 1883). La Ramarita *Miguel Velasquez de Leon* (*Naturaleza*, **7**, 65, 1884); Ramirite. Schaffnerit *Vigener* (*Ber. Niederrhein. Ges.*, 87, 1884).

41.5.2.2 **M O T T R A M I T E** [(Cu,Zn)Pb(VO$_4$)(OH)]. Chileite *Domeyko* (*Ann. mines*, **14**, 145, 1848). Cuprovanadite *Adam* (33, 1869). Mottramite *Roscoe* (*Proc. Roy. Soc. London*, **25**, 111, 1876). Psittacinite *Genth* (*Proc. Am. Phil. Soc.*, **14**, 229, 1876). Vesbine *Scacchi* (*Acc. Napoli, Att.*, **8**, 1, 1879).

PHOSPHATES, ARSENATES, VANADATES

Cryst.[1] Orthorhombic; dipyramidal—$2/m\ 2/m\ 2/m$.

$a:b:c = 0.8022:1:0.6448;$ $\quad p_0:q_0:r_0 = 0.8038:0.6448:1$

$q_1:r_1:p_1 = 0.8022:1.2441:1;$ $\quad r_2:p_2:q_2 = 1.5509:1.2466:1$

Forms:[2]

		ϕ	$\rho = C$	ϕ_1	$\rho_1 = A$	ϕ_2	$\rho_2 = B$
c	001	.....	0°00′	0°00′	90°00′	90°00′	90°00′
b	010	0°00′	90 00	90 00	90 00	.....	0 00
a	100	90 00	90 00	.....	0 00	0 00	90 00
v	120	31 56	90 00	90 00	58 04	0 00	31 56
m	110	51 16	90 00	90 00	38 44	0 00	51 16
d	210	68 08½	90 00	90 00	21 51½	0 00	68 08½
M	011	0 00	32 49	32 49	90 00	90 00	57 11
l	031	0 00	62 40	62 40	90 00	90 00	27 20
r	101	90 00	38 47½	0 00	51 12½	51 12½	90 00
e	201	90 00	58 07	0 00	51 53	51 53	90 00
o	111	51 16	45 51½	32 49	55 57½	51 12½	63 19
p	121	31 56	56 39	52 12½	63 46½	51 53	44 51

Less common and rare:

s 013 y 021 ϵ 112 i 146 x 131 ω 431
z 012 f 102 k 168 q 287 h 231 t 10.1.1

Structure cell.[3] Space group $Pnam$. a_0 7.56 kX, b_0 9.39, c_0 6.05; $a_0:b_0:c_0 = 0.805:1:0.644$. Cell contents $(Zn,Cu)_4Pb_4(VO_4)_4(OH)_4$.

Habit. Variable: often pyramidal $\{111\}$; prismatic [001]; rarely tabular $\{100\}$ or short prismatic [100]. The crystal faces usually are uneven or rough, and subparallel growth is frequent. Common forms: $m\ o\ d\ a\ b$. Commonly as drusy crusts of intergrown crystals; also stalactitic, or massive with a coarse fibrous structure and mammillary or botryoidal surface; sometimes massive granular, compact to friable.

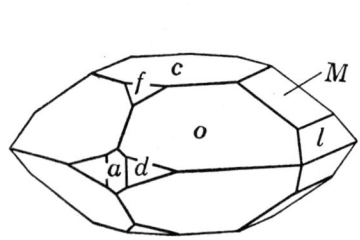
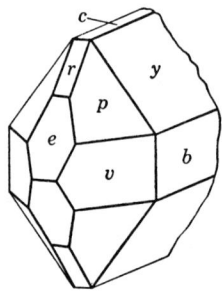

Descloizite. Lake Valley, New Mexico.   Descloizite. Otavi.

Phys. Cleavage none. Fracture small conchoidal to uneven. Brittle. H. 3–3½ on fracture surfaces, somewhat harder on the external faces. G. ~6.2 (descloizite end-member) to ~5.9 (mottramite end-member), decreasing with substitution of As for V. Luster greasy. Color variable, usually brownish red to blackish brown but including shades from orange-

DESCLOIZITE–MOTTRAMITE 813

red to deep reddish brown and nearly black; also grass-green to olive-green and siskin-green, especially in mottramite. Streak orange to brownish red or yellowish. Transparent to nearly opaque. Crystals often show zonal growth with varying color and optical properties.

Opt.

ORIEN-TATION	n (descloizite, Mammoth, Ariz.) [4]	n (mottramite, Bisbee, Ariz.) [4]	n (mottramite, Oruro, Bol.) [4]	
X c	2.185 ± 0.01	2.17 ± 0.02	2.21	Biaxial negative (−).
Y b	2.265 ± 0.01	2.26 ± 0.02	2.31	$r > v$, strong; sometimes $r < v$ and optically positive (+).
Z a	2.35 ± 0.01	2.32 ± 0.02	2.33	
$2V$	~90°	~73°	~45°	

Weakly to strongly pleochroic, with X and Y canary-yellow to greenish yellow and Z brownish yellow. Fibers have negative elongation.

Chem. A basic vanadate of lead, copper, and zinc, $(Zn,Cu)Pb(VO_4)(OH)$. A complete series exists between Zn and Cu; the names descloizite and mottramite are applied to the halves of the series with Zn > Cu and Cu > Zn, respectively. Mn substitutes for (Zn,Cu) up to at least Mn:(Zn,Cu) = 1:2.8 (analysis 4); [5] both Mn and Fe″ are ordinarily present in amounts less than 1 weight per cent. As substitutes for V up to at least As:V = 1:2.9 (analysis 2); [6] P sometimes substitutes for V in amounts up to a few tenths of a per cent. Non-essential, adsorbed or capillary H_2O is often present over the requirements of the formula,[7] especially in massive fibrous material. Traces of Cl are present in substitution for (OH) or are due to admixed vanadinite.

Anal.[8]

	1	2	3	4	5	6	7	8	9
PbO	55.47	54.10	52.95	53.36	55.30	55.78	54.50	54.56	55.64
ZnO	19.21	19.46	12.45	13.15	10.08	8.86	5.12	2.62	0.31
CuO	0.56	0.61	8.51	1.21	9.86	10.29	13.90	16.84	17.05
FeO	tr.	tr.		0.56		0.72	0.53	0.58	
MnO	tr.	tr.	tr.	4.56		0.01	tr.	tr.	
V_2O_5	22.76	16.26	19.01	23.05	22.53	22.73	22.65	22.73	21.21
As_2O_5	nil	7.13	3.84	0.11		nil	nil	nil	1.33
H_2O+ 350°	2.17	2.29 ⎱	2.65	2.27	2.23	2.24	2.09	2.29 ⎱	3.57
H_2O- 350°	0.02	0.08 ⎰				0.12	0.58	0.40 ⎰	
Rem.	0.02	0.27	0.55	0.86		0.11	0.89		0.91
Total	100.21	100.20	99.96	99.13	100.00	100.86	100.26	100.02	100.02
G.	6.26	6.24				6.18	5.96	5.80	

1. Descloizite. Abenab, South West Africa.[9] Rem. is SiO_2. 2. Descloizite. Niederschlettenbach, Bavaria.[9] Rem. is SiO_2 (*dechenite*). 3. Descloizite. Nogales, Arizona.[10] Rem. is Fe_2O_3 0.20, insol. 0.35. 4. Descloizite. Argentina.[11] Rem. is Cl 0.08, insol. 0.78. 5. $(Zn,Cu)Pb(VO_4)(OH)$ with Zn:Cu = 1:1. 6. Mottramite. Grootfontein, South West Africa.[9] Rem. is SiO_2. 7. Mottramite. Otavi, South West Africa.[9] Rem. is SiO_2. 8. Mottramite. Tsumeb, South West Africa.[9] 9. Mottramite. Bisbee, Arizona.[12] Rem. is P_2O_5 0.24, CrO_3 0.50, insol. 0.17. Average of three.

Tests. B.B. easily fusible. In C.T. decrepitates and affords little water. Easily soluble in acids.

Occur. A secondary mineral found principally in the oxidized zone of ore deposits where it is chiefly associated with vanadinite, also with pyromorphite, mimetite, cerussite. Also locally as a near-surface deposit in sandstones as a concentration by meteoric waters. Both descloizite and mottramite often occur at the same locality. From Obir, Carinthia, Austria. At Bena de Podru, Sassaro, Sardinia (mottramite). As large deposits (both descloizite and mottramite) at Abenab, Grootfontein, Tsumeb, Olifantsfontein, Uitsab, Guchab, and elsewhere in the Otavi region, South West Africa;[15] the minerals occur in sandy pockets in limestone and dolomite. Also in Africa at Broken Hill in Rhodesia, in the Katanga district, Belgian Congo, and at Djebba, Tunis, and Saïda, Oran, Algeria. In England on sandstone at Mottram St. Andrew, Cheshire (the original locality of mottramite); and from Pim Hill near Shrewsbury, Shropshire. In Germany from Niederschlettenbach, Lauterthal, Rhenish Bavaria (*dechenite*)[13] and from a near-by locality at Dahn in the Lauterthal (*arœoxene*);[13] from Hofsgrund near Freiburg, Baden (*eusynchite*).[13] As a coating on lava on Vesuvius, Italy (*vesbine*).[14] Cuprian descloizite occurs as crusts and reniform masses in San Luis Potosi, Mexico (*ramirite, tritochorite, cuprodescloizite*).[13] Descloizite occurs in South America in the Sierra Córdoba, Argentina; mottramite occurs between Caxata and Yaco in Inquisivi province, Bolivia, and at Mina Grande near Arqueros, Chile (*chileite*).[13] In the United States known from numerous minor localities. Mottramite occurs in the Silver Star district, Montana (*psittacinite*);[13] also in Arizona in the Tombstone and Bisbee districts in Bisbee County, near Nogales in Santa Cruz County, and with descloizite in the Mammoth mine in Pinal County. In New Mexico notable occurrences of descloizite in the Georgetown district, Grant County, and in the Caballos Mountains, Hillsboro and Lake Valley districts in Sierra County (in part mottramite).

Alter. Found as pseudomorphs after vanadinite.

Name. After the French mineralogist Alfred L. O. L. Des Cloizeaux (1817–1897).

Ref.

1. Angles of Diefenbach, *Zs. Kr.*, **74,** 155 (1930), on descloizite with ∼3 per cent CuO and ∼17 per cent ZnO from Otavi. Comparable ratios are afforded by goniometric and x-ray measurement of descloizite from Lake Valley, New Mexico, with ∼1 per cent CuO and ∼17.4 per cent ZnO (see Bannister, *Min. Mag.*, **23,** 376 (1933) and Diefenbach (1930)). The variation in angle with Cu:Zn ratio in the series is not established. Transformation: Diefenbach (1930) *et al.* to new, 001/010/100.
2. Diefenbach (1930) and Goldschmidt (**3,** 32, 1916). {012} from Meixner, *Jb. Min.*, I, 79 (1937).
3. Bannister (1933) by rotation method on Lake Valley material. See also Hägele, *Jb. Min., Beil.-Bd.*, **75,** 101 (1939).
4. Larsen (67, 64, 1921). See also Bannister (1933).
5. Impure material with MnO 5.32 has been analyzed by Damour, *Ann. chim. phys.*, **41,** 72 (1854).
6. A questionable analysis with As_2O_5 10.52 was reported by Bergemann, *Jb. Min.*, 399 (1857).
7. See discussion in Bannister (1933).
8. For tabulations of additional analyses see Diefenbach (1930); Hintze (1 [4A], 666, 1931); Dittler and Hueber, *Min. Mitt.*, **41,** 173 (1931).
9. Hey analysis in Bannister (1933).
10. Headden, *Proc. Colorado Sc. Soc.*, **7,** 141 (1903).
11. Fresenius analysis in Guild, *Zs. Kr.*, **49,** 322 (1911).

12. Wells, *Am. J. Sc.*, **36**, 636 (1913).
13. On the identity with descloizite-mottramite see Bannister (1933); Strunz, *Naturwiss.*, **27**, 423 (1930) and *Zs. Kr.*, **101**, 496 (1939).
14. Zambonini and Carobbi, *Am. Min.*, **12**, 1 (1927), show this to be identical with mottramite.
15. For description see Diefenbach (1930) and Schwellnus, *Trans. Geol. Soc. South Africa*, **48**, 49 (1946).

HÜGELITE. *Dürrfeld* (*Zs. Kr.*, **51**, 278, 1913). Huegelite.
Monoclinic ? Crystals prismatic [001], with {110}, {001} and {011}; (110) ∧ (1$\bar{1}$0) ~ 46°02′, (001) ∧ (110) ~62°47′, with $a:b:c$ = 0.49:1:0.38 and β 119°48′. G. ~5.1. Luster greasy to adamantine. Color orange-yellow to yellowish brown. Streak pale yellow. Biaxial positive (+) with extreme dispersion; $2V$ small (for red), large with Y = elong. (for blue); nY 1.915 ± 0.005. Qualitative tests indicate the mineral to be a vanadate of lead and zinc; PbO 32.59 per cent. Found at Reichenbach near Lahr, Baden, Germany. Needs verification.

41.5.2.3   **P Y R O B E L O N I T E** [MnPb(VO$_4$)(OH)]. *Flink* (*Geol. För. Förh.*, **41**, 433, 1919).

C r y s t. Orthorhombic; dipyramidal—$2/m\ 2/m\ 2/m$.

$a:b:c$ = 0.8040:1:0.6509; $p_0:q_0:r_0$ = 0.8096:0.6509:1

$q_1:r_1:p_1$ = 0.8040:1.2352:1; $r_2:p_2:q_2$ = 1.5363:1.2438:1

Forms:

	ϕ	$\rho = C$	ϕ_1	$\rho_1 = A$	ϕ_2	$\rho_2 = B$
c 001	.....	0°00′	0°00′	90°00′	90°00′	90°00′
a 100	90°00′	90 00	.....	0 00	0 00	90 00
n 120	31 52½	90 00	90 00	58 07½	0 00	31 52½
m 110	51 12	90 00	90 00	38 48	0 00	51 12
d 011	0 00	33 03½	33 03½	90 00	90 00	56 56½
f 031	0 00	62 53	62 53	90 00	90 00	27 07
e 201	90 00	58 18	0 00	31 42	31 42	90 00
p 111	51 12	46 05½	33 03½	55 50½	51 00½	63 10
o 221	51 12	64 18	52 28	45 23½	31 42	55 37½

Structure cell.[1] Space group *Pnam*. a_0 7.84 kX, b_0 9.45, c_0 6.09; $a_0:b_0:c_0$ = 0.802:1:0.644. Cell contents Mn$_4$Pb$_4$(VO$_4$)$_4$(OH)$_4$. Isostructural with descloizite-mottramite.

Habit. Acicular [001].

P h y s. Cleavage not observed. Fracture conchoidal. Brittle. H. 3½. G. 5.377; 5.39 (calc.). Color fire-red, resembling proustite. Luster adamantine. Streak orange-yellow or reddish. Transparent.

O p t. Pleochroic in red-brown with absorption $Y > X, Z$.

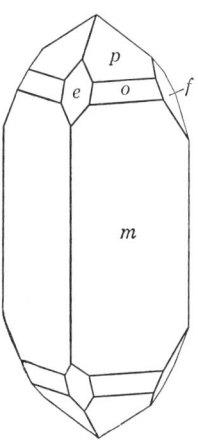

ORIENTATION		n^2	
X	a	2.32	Biaxial negative (−).
Y	c	2.36	$2V$ 29°.
Z	b	2.37	$r > v$, easily perceptible.

Chem. A basic vanadate of manganese and lead, $MnPb(VO_4)(OH)$. In the reported analysis, cited below, Mn:Pb ~5:3. Mn may in part be in substitution for Pb, the formula then being $Mn(Pb,Mn)(VO_4)(OH)$, but the departure from the 1:1 ratio of the formula first given probably is due to analytical error.

	CaO	MgO	MnO	PbO	FeO	V_2O_5	P_2O_5	SiO_2	H_2O	Total
1.			18.00	56.64		23.08			2.28	100.00
2.	0.79	0.60	25.01	48.82	0.47	20.03	0.05	0.21	[3.02]	[100.00]

1. $MnPb(VO_4)(OH)$. 2. Långban.[3] The analysis, here given as originally cited, sums only to 99.00.

Occur. Found at Långban, Sweden, associated with hausmannite, manganite, pyrochroite, barite, calcite.

Name. From πῦρ, *fire*, and βελόνη, *needle*, in allusion to color and habit.

Ref.

1. Richmond, *Am. Min.*, **25**, 460 (1940), by Weissenberg method. See also Strunz, *Zs Kr.*, **101**, 496 (1939), who gives comparable data.
2. Larsen and Berman (211, 1934).
3. Mauzelius analysis in Flink (1919).

VOLBORTHITE AND CALCIOVOLBORTHITE

The identity and mutual relations of the basic vanadates of copper and calcium variously described as volborthite, calciovolborthite, tangeite, and uzbekite are not clearly established. An x-ray study [1] indicates that the so-called volborthite of Genth [2] from Woskressensk, Perm, Russia, is identical with the uzbekite from Ferghana described optically by Barth and Berman [3] and with the mineral from the carnotite deposits of the Colorado-Utah region commonly identified as calciovolborthite. This mineral is tentatively described here as volborthite. The calciovolborthite (*kalkvolborthit*) of Credner [4] from Friedrichsroda, Germany, and some of the material labeled volborthite from Perm, Russia, is distinct from the above mineral and is tentatively described here as calciovolborthite. Turanite is found to be distinct from both these species and is related to conichalcite and mottramite. Tangeite appears to be identical with calciovolborthite.[5] It also has been suggested that calciovolborthite is only a calcian variety of volborthite; and that volborthite is isostructural with descloizite-mottramite.[5]

Ref.

1. Frondel, priv. comm. (1948).
2. Genth, *Proc. Am. Phil. Soc.*, **17**, 122 (1877).
3. Barth and Berman, *Chem. Erde*, **5**, 30 (1930).
4. Credner, *Ann. Phys.*, **74**, 546 (1848).
5. Strunz, *Zs. Kr.*, **101**, 496 (1939).

41.5.2.4 Calciovolborthite [CuCa(VO$_4$)(OH)]. Kalk-volborthit *Rammelsberg*. Calcio-volborthite *D'Achiardi* (I Metalli, **2**, 492, 1883). Calcvolborthite *Adam* (33, 1869). Tangeite *Nenadkevič* and *Volkov* (*C.r. ac. sc. U.R.S.S.*, 43, 1926). Tanguéite. Turkestan-volborthite (?) *Antipov* (*Gornyi Zhurnal, St. Petersburg*, **4**, 255, 1908).

Orthorhombic ? As scaly aggregates; also fibrous to dense. Perfect cleavage in one direction. H. $3\frac{1}{2}$. G. uncertain. Luster vitreous to pearly on the cleavage. Color greenish yellow and siskin-green to dark green and olive-green. Subtranslucent.

O p t. In transmitted light, pale green.

	n (tangeite, Tyuya Muyun) [1]	PLEOCHROISM	
X	2.00	Brown	Biaxial negative ($-$).
Y	2.01	Brown	$2V$ large.
Z	2.02	Green	$r > v$, strong.

C h e m. A basic vanadate of copper and calcium, probably CuCa(VO$_4$)(OH). Cu replaces Ca in part in some of the material referred to this species (analysis 2).

Anal.

	1	2	3	4	5
CuO	33.77	44.15	26.92	32.70	33.29
CaO	23.80	12.28	19.89	22.40	22.69
V$_2$O$_5$	38.60	36.58	38.53	37.65	38.45
H$_2$O	3.83	4.62	4.71	4.47	3.90
SiO$_2$			2.45	0.98	0.20
Rem.		1.00	7.39	1.63	0.60
Total	100.00	98.63	99.89	99.83	99.13

1. CuCa(VO$_4$)(OH). 2. Friedrichroda, Thuringia.[2] Rem. is MgO 0.50, MnO 0.40, insol. 0.10. 3. Tyuya Muyun (*tangeite;* dense).[3] Rem. is U$_3$O$_8$ 3.67, (Fe,Al)$_2$O$_3$ 2.08, SO$_3$ 0.40, MoO$_3$ tr., BaO 1.24. 4. Tyuya Muyun (*tangeite*, fibrous).[3] Rem. is Al$_2$O$_3$ 0.61, Fe$_2$O$_3$ 1.02. 5. Tyuya Muyun (*tangeite*).[3] Rem. is Fe$_2$O$_3$ 0.10, Al$_2$O$_3$ 0.50, U$_3$O$_8$ tr.

Tests. B.B. easily fusible. Soluble in acids.

O c c u r. A secondary mineral found in sandstone and in the oxidized zone of deposits containing vanadium minerals. Originally from Friedrichroda in Thuringia, Germany, with crednerite and manganese oxides. Also at Tyuya Muyun in the Ferghana district of Turkestan, U.S.S.R. (*tangeite*), and in the Perm district, Urals. Minerals resembling calciovolborthite and volborthite but of uncertain identity have been reported from a number of localities: Potechino in the Minusinsk district, Jenissei, Siberia, with allophane; from Tendu and Luiswichi in the Belgian Congo and Tsumeb in South West Africa. In the United States in small amounts with carnotite and tyuyamunite in sandstone in southwestern Colorado and adjoining parts of Utah, as at Naturita, Paradox Valley, and Uravan in Montrose County, near Telluride in San Miguel County, and Garo in Park County, in Colorado. Also from near Tombstone in Cochise County, Arizona, and in California from the Mammoth copper mine on Grindstone Creek in Glenn County, and at Camp Signal in San Bernardino County.

N a m e. Tangeite from the occurrence in the Tange ravine in Tyuya Muyun.

Ref.

1. Barth and Berman, *Chem. Erde*, **5**, 30 (1930). Larsen (154, 1921) gives optical data on material from Glenn County, California, and Perm, Urals, that cannot be stated with certainty to be volborthite or calciovolborthite.
2. Credner, *Ann. Phys.*, **74**, 546 (1848).
3. Nenadkevič and Volkov, *C. r. ac. sc. U.R.S.S* , 43 (1926).

UNNAMED MINERAL. Hillebrand (*Am. J. Sc.*, **8**, 201, 1924). Color olive-green or deep green. A vanadate of copper and barium and possibly also containing calcium but lacking uranium. Nearly or quite uniaxial, optically negative (−), with weak birefringence and a mean (?) index of refraction of 2.03–2.04. Found admixed with carnotite and tyuyamunite in the Paradox Valley, Montrose County, Colorado.

UNNAMED MINERAL. Hillebrand and Merwin (*Am. J. Sc.*, **35**, 441, 1913).

Rosettes and patches of minute reticulated scales. Color greenish yellow. Birefringent, with a mean index of refraction about 1.92. A hydrated arsenate-vanadate of copper and calcium probably related to calciovolborthite. Analysis gave (after deducting 13.5 per cent insoluble): CaO 15.3, BaO 2.3, MgO 0.5, $(Na,K)_2O$ 0.2, CuO 37.1, Fe_2O_3 0.5, oxides of Mn, Co and Al 3.2, As_2O_5 17.2, V_2O_5 16.0, P_2O_5 0.8, H_2O+ 4.3, H_2O- 1.0, CO_2 0.9, SiO_2 0.7, total 100.0. Found associated with volborthite as coatings on sandstone at Richardson, Grand County, Utah.

41.5.2.5 **Turanite.** *Nenadkevič* (*Bull. Ac. St. Petersburg*, **3**, 185, 1909).

Probably orthorhombic. As reniform crusts and spherical concretions with a radial fibrous structure. Color olive-green. H. 5. Optically [1] biaxial negative (−), with $2V$ medium, $r > v$ strong, and positive elongation. Indices of refraction and pleochroism: nX 2.00 (brown), nY 2.01 (brown), nZ 2.02 (green). Composition stated to be $Cu_5(VO_4)_2(OH)_4$. Found with various vanadium and uranium minerals in cavities in limestone at Tyuya Muyun, Ferghana, Turkestan, U.S.S.R. Turanite appears to be distinct from volborthite and calciovolborthite and to be related to mottramite.[2]

Ref.

1. Larsen and Berman (207, 1934).
2. Frondel, priv. comm. (1948) by x-ray powder study of probably authentic material.

41.5.3 **Volborthite.** Volborthite ? *Hess* (*Bull. Ac. St. Petersburg*, **4**, 21, 1838; *J. pr. Chem.*, **14**, 52, 1838). Knauffite *Planer* (*Arch. Wiss. Kunde*, **8**, 135, 1849). Vanadate of Copper. Vanadinsaures Kupfer. Volborthite *Genth* (*Proc. Am. Phil. Soc.*, **17**, 122, 1877). Uzbekite *Fersman* (*Priroda*, Leningrad, no. 7–9, col. 238, 1925). Usbekit *Kurbatov* (*Cbl. Min.*, 345, 1926A). α-Uzbekite, β-Uzbekite *Kurbatov* and *Kargin* (*C. r. ac. sc. U.R.S.S.*, no. 75–80, 1927). Siberian volborthite.

Monoclinic ? As scaly, spongy, or fibrous crusts and as rosette-like aggregates; also reticulated. Sometimes as scales with a triangular or hexagonal outline.

P h y s. Perfect cleavage in one direction. H. $3\frac{1}{2}$. G. uncertain, 3.5–3.8. Luster vitreous to pearly on the cleavage. Color dark olive-green to green and yellowish green. Subtranslucent.

VOLBORTHITE

O p t. In transmitted light, green to greenish yellow. Faintly pleochroic. Biaxial.

ORIENTATION	n (uzbekite)[1]	n (Utah)[2]
X	2.01	2.01
Y	2.04	2.05
Z	2.07	2.10 (calc.)
$2V$	Large	$\begin{cases} 68° \text{ (Li)} \\ 83° \text{ (Na)} \end{cases}$
Sign	–	$\begin{cases} + \text{ (red)} \\ - \text{ (violet)} \end{cases}$
Disp.	$r < v$	$r > v$ (inclined).

C h e m. Essentially a hydrated vanadate of copper. The formula is uncertain, perhaps $Cu_3(VO_4)_2 \cdot 3H_2O$. Small amounts of Ca and Ba apparently substitute for Cu.

Anal.

	CuO	CaO	BaO	V_2O_5	H_2O+	H_2O-	SiO_2	Rem.	Total
1.	50.29			38.32	11.39				100.00
2.	44.69	0.31		37.71	12.82	0.53	1.17	2.30	99.53
3.	48.4	3.9	2.7	30.6	6.4	1.8	0.6	5.6	[100.00]
4.	34.04	4.29	4.29	13.62	[33.15]		1.38	9.23	[100.00]

1. $Cu_3(VO_4)_2 \cdot 3H_2O$. 2. Uzbekistana district, Ferghana (*uzbekite*).[3] Rem. is NiO 0.90, (Al,Fe)$_2$O$_3$ 1.40. 3. Richardson, Utah.[2] Rem. is P$_2$O$_5$ 0.3, (K,Na)$_2$O 0.7, MgO 0.3, CO$_2$ 2.4, As$_2$O$_5$ 1.1, Fe$_2$O$_3$ 0.8. Recalculated after deducting 30.6 per cent insol. 4. Woskressensk, Perm, Russia.[4] Rem. is MgO 3.01, Al$_2$O$_3$ 4.45, Fe$_2$O$_3$ 1.77. Erroneous analysis on impure material.

Tests. B.B. easily fusible. Soluble in acids.

O c c u r. A secondary mineral, originally found at Syssersk and Nizhne Tagilsk in the Ural Mountains and also at Woskressensk in Permian sandstone in Perm, Russia. In the Uzbekistana district in Ferghana, Russia (*uzbekite*). With carnotite in sandstone at Richardson in the canyon of the Grand River, Grand County, Utah. A similar mineral, perhaps calciovolborthite (which see), has been reported from a number of other localities.

A r t i f.[5] Both anhydrous and hydrated copper vanadates have been reported but their relation to volborthite has not been established.

N a m e. After Alexander von Volborth (1800–1876), Russian paleontologist who first noticed the mineral.

Ref.

1. Barth and Berman, *Chem. Erde*, **5**, 30 (1930).
2. Hillebrand and Merwin, *Am. J Sc.*, **35**, 441 (1913), on material termed calciovolborthite (?).
3. Kurbatov and Kargin, *C. r. ac sc. U R.S.S.*, no. 75–80 (1927), who cite additional analyses on impure material. See also Fersman, *Abh. pr. Geol. Bergwirtsch.*, **19**, 36 (1930).
4. Genth, *Proc. Am. Phil. Soc.*, **17**, 122 (1877).
5. Mellor (**9**, 767, 1929).

41.5.4 HERDERITE SERIES

MONOCLINIC; PRISMATIC—$2/m$

	$a:b:c$	β
Herderite, $CaBe(PO_4)F$		
Hydroxyl-herderite, $CaBe(PO_4)(OH)$	0.6307:1:1.2822	90°06'

A probably complete series extends between herderite and hydroxylherderite by mutual substitution of (OH) and F. These species have the same crystal structure [1] as the borosilicate datolite, $CaB(SiO_4)(OH)$. The structures probably are based on a three-dimensional linkage of (BeO_4) and (PO_4) or (BO_4) and (SiO_4) tetrahedra, respectively.

Ref.

1. Strunz, *Zs. Kr.*, **93**, 146 (1936).

41.5.4.1 **H E R D E R I T E** [$CaBe(PO_4)(F,OH)$]. *Haidinger* (*Phil. Mag.*, **4**, 1, 1828). Allogonite *Breithaupt* (23, 1830; 78, 1832). Glucinite *Hidden* (*Am. J. Sc.*, **27**, 135, 1884). Fluor-herderite.

41.5.4.2 **H Y D R O X Y L - H E R D E R I T E** [$CaBe(PO_4)(OH,F)$]. Hydro-herderite, Hydro-fluorherderite *Penfield* (*Am. J. Sc.*, **47**, 329, 1894).

C r y s t.[1] Monoclinic; prismatic—$2/m$.

$a:b:c = 0.6307:1:1.2822;$ $\beta\ 90°06';$ $p_0:q_0:r_0 = 2.0330:1.2822:1$

$r_2:p_2:q_2 = 1.5598:1.5855:1;$ $\mu\ 89°54';$ $p_0'\ 2.0331, q_0'\ 1.2822, x_0'\ 0.0017$

Forms:[2]

		ϕ	ρ	ϕ_2	$\rho_2 = B$	C	A
c	001	90°00'	0°06'	89°54'	90°00'	.....	89°54'
b	010	0 00	90 00	.....	0 00	90°00'	90 00
a	100	90 00	90 00	0 00	90 00	89 54	.....
l	120	38 24½	90 00	0 00	38 24½	89 56½	51 35½
m	110	57 45½	90 00	0 00	57 45½	89 55	32 14½
t	012	0 09½	32 40	89 54	57 20	32 40	89 55
v	011	0 04½	52 03	89 54	37 57	52 03	89 56½
s	021	0 02½	68 42	89 54	21 18	68 42	89 58½
e	102	90 00	45 31	44 29	90 00	45 25	44 29
p	113	57 49½	38 45	55 48½	70 32	38 40	58 00½
q	112	57 48½	50 16½	44 29	65 48½	50 11	49 24
n	111	57 47	67 25½	26 10½	60 30½	67 20½	38 37½
$q: \bar{1}12$		−57 43	50 12	135 25½	65 46½	50 17	130 30½
$n: \bar{1}11$		−57 44½	67 24	153 47½	60 28½	67 29	141 19½
$r: \bar{1}23$		−38 20	47 27½	124 03½	54 41½	47 31½	117 11½

Less common and rare:

μ 130	$\alpha\ \bar{1}04$	$d:\bar{1}01$	k 126	$z:\bar{1}24$	$x\ \bar{1}22$
u 013	$S\ \bar{1}03$	r 116	w 144	$z\ \bar{1}34$	$i\ \bar{2}43$
d 103	$e:\bar{1}02$	$p:\bar{1}13$	γ 121	$h\ \bar{2}14$	$P\ \bar{1}31$

HERDERITE–HYDROXYL-HERDERITE

Structure cell.[3] Space group $P2_1/c$. a_0 4.80 kX, b_0 7.68, c_0 9.80; β 90°06′; $a_0:b_0:c_0$ = 0.625:1:1.276. Cell contents $Ca_4Be_4(PO_4)_4(OH,F)_4$.

Habit. Stout prismatic [100] or [001]; also thick tabular {001}. In aspect usually pseudo-orthorhombic, rarely monoclinic; sometimes pseudo-hexagonal [001]. Common forms: $c\ b\ m\ t\ v\ e\ q\ n\ n$:. Also as botryoidal or spheroidal aggregates with a radial fibrous structure.

Twinning. (1) Twin plane {001}. (2) Twin plane {100}. Most crystals are twinned although some show no outward signs thereof.

Phys. Cleavage {110}, interrupted. Fracture subconchoidal. H. 5–5½. The G. increases with F content; 2.95 (hydroxyl-herderite) to 3.01 (hydroxyl-herderite with OH:F = 1.5:1); 2.94 (calc. for hydroxyl-herderite). Luster vitreous or subvitreous. Colorless to pale yellow or greenish white. Transparent to translucent.

Opt. In transmitted light, colorless.

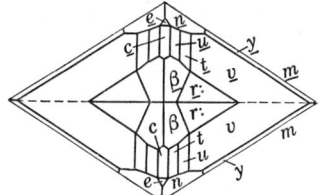

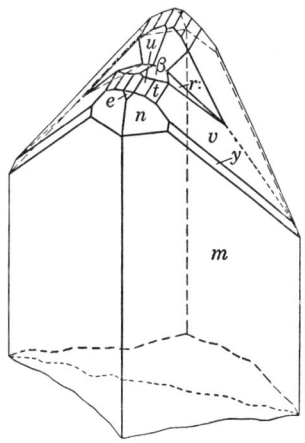

Herderite. Topsham, Me. Twin on {100}.

ORIENTATION	Hydroxyl-herderite n(Na), Paris,[4] anal. 2.	Herderite (?) n, Topsham [5]	Herderite (?) n(Na), Stoneham [6]	
$X \wedge c \sim +87°$		1.591	1.592	Biaxial negative (−).
$Y\ b$	1.632	1.611	1.612	$r > v$, inclined.
$Z \wedge c \sim -3°$		1.619	1.621	
$2V$ [7]	72°	75° ± 5°	67°	

Chem. A basic phosphate of calcium and beryllium, $CaBe(PO_4)(OH,F)$. (OH) and F substitute mutually, and a probably complete series extends between the (OH) and F end-members. The names hydroxyl-herderite and herderite are applied to material with (OH) > F and F > (OH), respectively.

Anal.[8]

	1	2	3	4	5	6	7
CaO	34.82	34.04	32.24	33.67	33.65	33.21	34.39
BeO	15.53	16.13	16.50	15.51	15.04	15.76	15.34
P_2O_5	44.06	44.05	39.74	43.74	43.43	44.31	43.53
H_2O	5.59	5.85	7.97	3.70	0.61		
F		tr.	0.87	5.27	8.93	11.32	11.65
Rem.		0.44	3.44		0.46		
	100.00	100.51	100.76	101.89	102.12	104.60	104.91
F = O			0.37	2.22	3.76	4.77	4.91
Total	100.00	100.51	100.39	99.67	98.36	99.83	100.00
G.			2.952	2.851	3.01		

1. CaBe(PO$_4$)(OH). 2. Hydroxyl-herderite. Paris, Maine.[9] Rem. is insol. 3. Hydroxyl-herderite. Newry, Maine.[10] Rem. is MnO 0.16, MgO 0.76, Al$_2$O$_3$ 0.50, insol. 2.02. Fibrous material. 4. Hydroxyl-herderite. Stoneham, Maine.[11] 5. Herderite. Stoneham, Maine.[12] Rem. is MnO 0.11, Fe$_2$O$_3$ 0.15, Al$_2$O$_3$ 0.20. The F determination is said to be probably low, and the H$_2$O reported may actually be F. 6. Herderite. Stoneham.[14] 7. CaBe(PO$_4$)F.

Tests. B.B. becomes white and opaque and fuses with difficulty. Strongly heated in C.T. affords acid water. Soluble in acids.

Occur. Found typically as a pegmatite mineral formed during the later, hydrothermal, stages of deposition. The original herderite is known only from a few specimens found prior to 1825 at the tin mines of Ehrenfriedersdorf, Saxony. Also found in pegmatite at Epprechtstein and at Reinersreuth in the Fichtelgebirge, Bavaria, Germany. In pegmatite at Mursinsk, Ural Mountains, Russia. In the United States, herderite occurs at Stoneham, Oxford County, Maine, in pegmatite. Hydroxyl-herderite is found at Hebron, Paris, and Newry, in Oxford County, Maine. Herderite or hydroxyl-herderite, as yet unanalyzed, occurs at Auburn and Poland in Androscoggin County, and at Buckfield and Greenwood in Oxford County, in Maine. Also in New Hampshire at the Palermo mine and the Fletcher mine, near North Groton, Grafton County.

Alter. Observed as an alteration product of beryllonite (Newry, Maine).[13]

Name. After S. A. W. von Herder (1776–1838), mining official in Freiberg, Saxony.

Ref.

1. Angles and orientation of Penfield, *Am. J. Sc.*, **47**, 329 (1894), on hydroxyl-herderite (analysis 2) in the unit of Yatsevitch, *Am. Min.*, **20**, 426 (1935), corresponding to the structure cell. The elements apparently vary only slightly with the OH:F ratio—see Ford, *Zs. Kr.*, **50**, 97 (1912). Dana, *Am. J. Sc.*, **27**, 229 (1884), 760 (1892), gave herderite as orthorhombic; the monoclinic symmetry was first established by Penfield. Transformations: Penfield to new, 100/010/003; Goldschmidt (**4**, 127, 1918) to new, 100/010/002. Herderite is isotypic with datolite.

2. Yatsevitch (1935) and Goldschmidt (1918). Uncertain forms: f 209, E 803, β $\bar{1}$15, g 114, O 443, y 133, G $\bar{3}$49, H 135.

3. Strunz, *Zs. Kr*, **93**, 146 (1936), by rotation method on Topsham material (hydroxyl-herderite ?).

4. Penfield (1894).
5. Yatsevitch (1935).
6. Bertrand in Des Cloizeaux, *Bull. soc. min.*, **9**, 141 (1886).
7. For additional measurements of the optic angle see Penfield (1894) and Des Cloizeaux, *Bull. soc. min.*, **7**, 130 (1884).
8. For additional analyses see Hintze (**1** [4A], 688, 1931).
9. Wells analysis in Penfield (1894).
10. Shannon analysis in Palache and Shannon, *Am. Min*, **13**, 394 (1928).
11. Penfield and Harper, *Am. J. Sc.*, **32**, 107 (1886).
12. Genth, *Am. Phil. Soc.*, **21**, 694 (1884).
13. Palache and Shannon (1928).
14. Mackintosh, *Am. J. Sc.*, **27**, 135 (1884).

AMBLYGONITE SERIES

41.5.5.1 A M B L Y G O N I T E [(Li,Na)Al(PO$_4$)(F,OH)]. Amblygonite *Breithaupt* (*Breithaupt's Hoffmann*, **4B**, 159, 1817). Montebrasite *Des Cloizeaux* (*C.R.*, **73**, 306, 1247, 1871). Hebronite *von Kobell* (*Ak. München, Sitzber.*, 284, 1872).
41.5.5.2 M O N T E B R A S I T E [(Li,Na)Al(PO$_4$)(OH,F)]. *Des Cloizeaux* (*Ann. chim. phys.*, **27**, 385, 1872). Montebrazit *Germ.*
41.5.5.3 N A T R O M O N T E B R A S I T E [(Na,Li)Al(PO$_4$)(OH,F)]. Natramblygonite *Schaller* (*Am. J. Sc.*, **31**, 48, 1911) = Fremontite *Schaller* (*J. Washington Ac. Sc.*, **4**, 354, 1914; *U. S. Geol. Sur., Bull. 610*, 141, 1916) = Natronamblygonit *Germ.* = Natromontebrasite *Gonnard* (*Bull. soc. min.*, **36**, 120, 1913).

C r y s t.[1] Triclinic; pinacoidal—$\bar{1}$.

$a:b:c = 0.7255:1:0.7028;$ $\alpha\ 111°59\frac{1}{2}',\ \beta\ 97°46\frac{1}{2}',\ \gamma\ 68°16\frac{1}{2}'$

$p_0:q_0:r_0 = 0.9669:0.7495:1;$ $\lambda\ 69°22',\ \mu\ 90°13',\ \nu\ 110°21'$

$p_0'\ 1.0427,\ q_0'\ 0.8083,\ x_0'\ 0.1366,\ y_0'\ 0.3800$

Forms:[2]

		ϕ	ρ	A	B	C
c	001	19°46$\frac{1}{2}'$	21°59'	90°13'	69°22'	.....
b	010	0 00	90 00	110 21	......	69°22'
a	100	110 21	90 00	.....	110 21	90 13
w	110	65 30	90 00	44 51	65 30	74 51
W	1$\bar{1}$0	140 08	90 00	29 47$\frac{1}{2}$	140 08	100 54$\frac{1}{2}$
q	021	3 55	63 27	104 40	26 49	42 33
d	0$\bar{1}$1	162 19	24 12$\frac{1}{2}$	75 22	112 59$\frac{1}{2}$	43 37$\frac{1}{2}$
u	0$\bar{2}$1	173 42	51 12$\frac{1}{2}$	69 32$\frac{1}{2}$	140 46$\frac{1}{2}$	71 24$\frac{1}{2}$
r	111	53 27$\frac{1}{2}$	54 12$\frac{1}{2}$	63 42	61 07	37 20$\frac{1}{2}$
t	$\bar{1}$1	− 94 28$\frac{1}{2}$	40 09	125 49$\frac{1}{2}$	92 53	52 26$\frac{1}{2}$
v	$\bar{3}$32	−102 14$\frac{1}{2}$	53 41$\frac{1}{2}$	132 46$\frac{1}{2}$	99 50$\frac{1}{2}$	67 05$\frac{1}{2}$
x	$\bar{2}$31	−125 58	66 00$\frac{1}{2}$	120 26$\frac{1}{2}$	122 27	84 35

Less common and rare:
Z 120 o 011 $f\ \overline{2}23$ $s\ \overline{1}21$ $g\ \overline{1}22$ $k\ \overline{3}52$ $i\ \overline{5}72$

Structure cell.[3] Space group $P\bar{1}$. $a_0\ 5.18\ kX,\ b_0\ 7.11,\ c_0\ 5.03;\ \alpha\ 112°02\frac{1}{2}'$, $\beta\ 97°49\frac{1}{2}',\ \gamma\ 68°07\frac{1}{2}';\ a_0:b_0:c_0 = 0.729:1:0.709$. Cell contents Li$_2Al_2$(PO$_4$)$_2(OH,F)_2$.

Habit. Small crystals are equant to short prismatic [010]. The crystals ordinarily are rough, especially when large. Also in large cleavable masses; columnar; compact.

Twinning. (1) on {$\bar{1}$11}, common, with composition plane {$\bar{1}$11}. The twins usually are tabular parallel to {$\bar{1}$11} and the twinned individuals of about equal size; also tabular {110} and the twinned individuals of very unequal size. (2) On {111}, rare; lamellar.

P h y s. Cleavage {100} perfect, {110} good, {0$\bar{1}$1} distinct, {001} imperfect.[4] Fracture uneven to subconchoidal. Brittle. H. 5$\frac{1}{2}$–6. G. $\sim$ 3.11 (amblygonite end-member) to $\sim$2.98 (montebrasite end-member); 3.04–3.1 (natromontebrasite). Luster vitreous to greasy; pearly on well-developed cleavages. Color usually white to milky or creamy white; also yellowish, beige, salmon-pink, greenish, bluish, gray; rarely colorless and water-clear. Transparent to translucent.

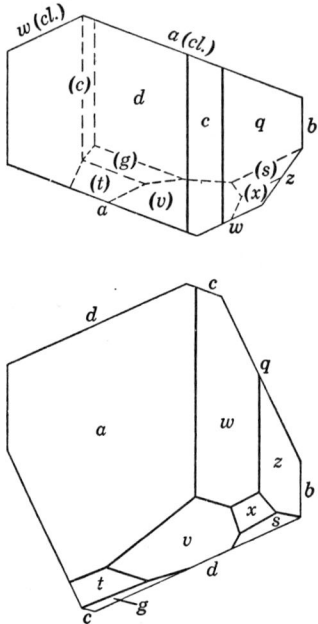

Amblygonite. Newry, Me.

O p t. In transmitted light, colorless. Biaxial.

ORIENTATION [5]		Amblygonite n(Na) (Chursdorf) [6]	Amblygonite n(Na) (Utö) [6]	Montebrasite n(Na) (Karibib) [7]	Montebrasite n (Kimito) [8]	Natromontebrasite n (Colorado) [9]
	ϕ ρ					
X	19° 69°	1.5783	1.5910	1.594	1.611	1.594
Y	−78° 72°		1.6046	1.608	1.619	1.603
Z	156° 28°	1.5983	1.6125	1.616	1.633	1.615
2V				75° ± 2°	73°22′	Very large
G.		3.101	3.065	3.085	3.002	3.04
Analysis no.		.....	5	6	10	11

The variation in optical properties [10] with composition in the amblygonite series is not clearly defined. The indices of refraction and the optic angle decrease with increasing content of F and of Na. Montebrasite end-member is optically positive (+) with $r < v$; 2V passes through 90° at about 60 per cent (OH); and amblygonite is optically negative (−) with $r > v$. The dispersion of the axial angle is weak or not perceptible; the dispersion of the indicatrix is both crossed and inclined and varies in degree with composition. Polysynthetic twinning frequently is observed.

C h e m. A basic phosphate of aluminum, sodium, and lithium, (Li,Na)Al(PO$_4$)(F,OH). Most analyses show a slight excess of water,[11] and the formula has been written (Li,Na)$_4$Al$_4$(PO$_4$)$_4$(F,OH)$_4 \cdot$H$_2$O. Na and Li substitute mutually; Li ordinarily is much in excess of Na, and only one instance has been found of material with Na > Li (analysis 11, with Na:Li = 1.69:1). F and OH also substitute mutually to form a probably complete series. The following nomenclature is used for the series:

	Li > Na	Na > Li
F > OH	Amblygonite	(Not known)
OH > F	Montebrasite	Natromontebrasite

Anal.[12]

	1	2	3	4	5	6
Li_2O	10.10	9.60	6.70	8.97	7.84	9.65
Na_2O		0.59	5.30	2.04	2.71	0.43
Al_2O_3	34.46	37.60	35.50	33.55	33.64	35.19
P_2O_5	48.00	46.85	45.91	48.24	43.40	47.54
F	12.85	10.40	9.00	11.26	11.10	5.40
H_2O		0.14	0.70	1.75	2.18	3.30
Rem.			0.50	0.13	3.50	0.43
	105.41	105.18	103.61	105.94	104.37	101.94
F = O	5.41	4.38	3.79	4.74	4.66	2.27
Total	100.00	100.80	99.82	101.20	99.71	99.67
G.					3.065	3.085

1. $LiAl(PO_4)F$. 2. Amblygonite. Montebras, France.[13] 3. Amblygonite. Montebras, France.[24] Rem. is CaO. H_2O is ign. loss. 4. Amblygonite. Penig, Saxony.[14] Rem. is Mn_2O_3. 5. Amblygonite. Utö, Sweden.[15] Rem. is SiO_2. 6. Montebrasite. Karibib, South West Africa.[16] Rem. is K_2O 0.10, MgO 0.33. H_2O includes $H_2O - 0.04$.

	7	8	9	10	11	12	13
Li_2O	9.24	8.80	9.14	8.85	3.21	10.24	
Na_2O	0.66	1.41	2.17	0.83	11.23		19.14
Al_2O_3	33.90	35.80	34.39	33.87	33.59	34.94	31.47
P_2O_5	47.44	46.70	47.79		44.35	48.65	43.83
F	5.45	2.27	1.80	0.57	5.63		
H_2O	5.05	4.54	5.29	5.99	4.78	6.17	5.56
Rem.		0.48	0.14	1.59	0.14		
	101.74	100.00	100.72		102.93	100.00	100.00
F = O	2.29		0.76		2.37		
Total	99.45	100.00	99.96		100.56	100.00	100.00
G.	3.032	2.89	3.002	3.002	3.04		

7. Montebrasite. Hebron, Maine.[14] 8. Montebrasite. Mogi das Cruzes, Brazil.[18] Rem. is SiO_2. 9. Montebrasite. Varuträsk, Sweden.[17] Rem. is Fe_2O_3 0.04, insol. 0.10. H_2O includes $H_2O - 0.09$. 10. Montebrasite. Kimito, Finland.[19] Rem. is SiO_2 0.40, Fe_2O_3 0.13, MnO 0.02, MgO 0.33, CaO 0.53, K_2O 0.18. H_2O includes $H_2O - 0.15$. 11. Natromontebrasite. Fremont County, Colorado.[20] Rem. is K_2O. 12. $LiAl(PO_4)(OH)$. 13. $NaAl(PO_4)(OH)$.

Tests. B.B. fuses easily with intumescence and cools to an opaque white bead. In C.T. affords some water and at a high temperature HF. Soluble with difficulty in acids. Amblygonite and montebrasite color a flame red (Li), natromontebrasite yellow (Na).

Occur. The members of the amblygonite series occur chiefly in granite pegmatites of the lithium- and phosphate-rich type, often in crystals of enormous size. The principal associated minerals include spodumene, lithiophilite-triphylite, apatite, lepidolite, petalite, pollucite, and tourmaline. Also found with cassiterite, topaz, and mica in high-temperature tin veins and in greisen. Amblygonite and montebrasite are not separately distinguished in all the localities mentioned here. There often is a considerable variation in composition among different specimens from the same locality.

Amblygonite occurs at Chursdorf and Arnsdorf near Penig and with cassiterite at Geyer, in Saxony, Germany. At Königswart and Punau near Marienbad, in Bohemia; natromontebrasite (?) occurs at Jeclov near Jihlava in Moravia of Czechoslovakia. Near San Piero, Elba. In Spain montebrasite occurs in a tin deposit at Cáceres in Province Estramadura; also near Silleda in Pontevedra. Both montebrasite and amblygonite occur at Montebras, Creuse, France (the original locality). In Sweden, amblygonite occurs on the island of Utö near Stockholm, and montebrasite is found in large masses in the pegmatite at Varuträsk. Montebrasite occurs with petalite in the lepidolite pegmatites of the Karibib district, South West Africa. In pegmatite at Mogi das Cruzes, São Paulo, Brazil (montebrasite). In Australia at Ravensthorpe and Ubini, Western Australia (montebrasite). In the United States amblygonite and montebrasite occur as giant crystals in pegmatites in the Custer district and elsewhere in the Black Hills, South Dakota; masses weighing hundreds of tons have been mined at the Peerless and Bob Ingersoll mines near Keystone and at the Giant-Volney mine near Tinton. Natromontebrasite occurs in pegmatite in Eight Mile Park, Fremont County, Colorado.[21] In California, found at Pala, Mesa Grande, and elsewhere in San Diego County; also at the Fano mine, Coahuila Mountain, Riverside County, and at Turtle Mountain, San Bernardino County. With lepidolite and microlite at the Harding pegmatite, Taos County, New Mexico. From Mitchell Wash, Yavapai County, Arizona. Small amounts occur in pegmatite at the Ruggles mine and the Palermo mine, Grafton County, New Hampshire. As transparent crystals, the finest known, in the New Pit at Newry, Oxford County, Maine; also in Maine at Hebron, Buckfield, Peru, Mt. Mica near Paris, Auburn, Poland, Greenwood, and other places and usually corresponding in composition to montebrasite. Sparingly at Branchville and Portland, Connecticut (montebrasite). In Canada found in pegmatites in the Yellowknife-Beaulieu area, Northwest Territories, and in the Oiseaux River district in southeastern Manitoba. Near New Ross, Lunenberg County, Nova Scotia (montebrasite).

Alter. To dense mixtures of kaolin and mica, often as rims surrounding rounded nodules of unaltered material (Varuträsk; Greenwood, Maine).[22] Also to turquois, wavellite, wardite, and morinite.[23]

Name. Amblygonite from ἀμβλύς, *blunt*, and γόνυ, *angle*. Montebrasite after the locality.

Ref.

1. Elements of Palache, Richmond, and Wolfe, *Am. Min.*, **28**, 39 (1943), on Hebron crystals probably near analysis 7 in composition (*montebrasite*). Transformations: Des Cloizeaux, *C.R.*, **57**, 357 (1863), to Palache, *et al.*, 002/Ī12/200; Dana (545, 1873) to Palache, *et al.* 001/011/100.

2. Palache, Richmond, and Wolfe (1943).
3. Richmond in Palache, et al (1943) by Weissenberg method on Hebron material (near analysis 7). Tengnér, Geol. För. Förh., **62**, 332 (1940), and Richmond, Am. Min., **25**, 473 (1940), give comparable data. See also Strunz, Zbl. Min., 248 (1939).
4. Not all observers agree on the existence or relative perfection of the cleavages—see Palache, et al. (1943).
5. Nel, Am. Min., **31**, 51 (1946), on material of analysis 6 (see indices); Palache et al. (1943) give slightly divergent data on unanalyzed material from Hebron (probably near analysis 7) but no indices. The orientation doubtless varies with composition.
6. Backlund, Geol. För. Förh., **40**, 757 (1918).
7. Nel (1946), who also gives indices for F (486 mμ) and C (656 mμ).
8. Pehrman, Acta Ac. Aboensis, Math Phys., **15**, no. 2 (1945).
9. Larsen (76, 1921).
10. See summaries of Winchell, Am. Min., **11**, 246 (1926), and Backlund (1918). Additional optical data are given by Quensel, Geol. För. Förh., **59**, 455 (1937), and Sekanina, Spisy přiro. fak. Masarykovy Univ., no. 180 (1933), and others.
11. See discussion in Penfield, Am. J. Sc., **18**, 295 (1879).
12. For additional analyses see Hintze (1 [4A], 626, 1924); Quensel (1937).
13. Pisani, Ann. mines, **29**, 82 (1873).
14. Penfield (1879).
15. Sahlbom analysis in Backlund (1918).
16. Liebenberg analysis in Nel (1946).
17. Blix analysis in Quensel (1937).
18. Cunha and Costa, Notas Prelim. Div. Geol. Min. Brasil, no. 22 (1941).
19. Pehrman (1945).
20. Schaller, Am. J. Sc, **31**, 48 (1911).
21. See Heinrich, Am. Min., **33**, 561 (1948).
22. See Quensel (1937) and Landes, Am. Min., **10**, 403, 410 (1925).
23. See Lacroix (**4**, 416, 1910).
24. von Kobell, Ak. München, Sitzber., 284 (1872).

DURANGITE AND TILASITE

MONOCLINIC

	a_0	b_0	c_0	β	Space group
Durangite, NaAl(AsO$_4$)F	7.30 kX	8.46	6.53	119°22′	$C2/c$
Tilasite, CaMg(AsO$_4$)F	7.56	8.95	6.66	121 00	Cc (?)

Durangite and tilasite are closely related in crystal structure, although they do not fall into the same space group or crystal class. Both species also are related in structure [1] to the silicate sphene, CaTi(SiO$_4$)O, and contain isolated (PO$_4$) or (SiO$_4$) groups.

Ref.

1. Strunz, Zs Kr., **96**, 7 (1937), Zbl. Min., 59 (1938), and Kokkoros, Zs. Kr., **99**, 38 (1938). Durangite is here given in a setting analogous to tilasite and sphene

41.5.6 **T I L A S I T E** [CaMg(AsO$_4$)F]. Fluor-adelite, Tilasite Sjögren (Geol. För. Förh., **17**, 291, 1895).

C r y s t.[1] Monoclinic; domatic—m.

$a:b:c = 0.8391:1:0.7503;$ $\beta\,121°00';$ $p_0:q_0:r_0 = 0.8942:0.6431:1$

$r_2:p_2:q_2 = 1.5550:1.3905:1;$ $\mu\,59°00';$ $p_0'\,1.0433, q_0'\,0.7503, x_0'\,0.6009$

828   PHOSPHATES, ARSENATES, VANADATES

Forms: [2]

Lower	Upper		φ	ρ	φ2	ρ2 = B	C	A
	b	010	0°00′	90°00′	.....	0°00′	90°00′	90°00′
$\overline{M}$	M	011	38 41½	43 52	59°00′	57 15	32 45	64 19½
$\bar{e}$		$\overline{1}$01	−90 00	23 52	113 52	90 00	54 52	113 52
$\bar{p}$	p	111	65 28½	61 02½	31 18½	68 42	34 25	37 15
$\bar{x}$	x	$\overline{1}$11	−30 31½	41 03½	113 52	55 32½	61 40	109 29
$\bar{r}$	r	133	51 39½	50 25	46 30½	61 26½	30 57½	52 48½
$\bar{u}$		$\overline{2}$13	−20 43½	14 58	95 24½	76 01½	38 39	95 14½
$\bar{o}$		$\overline{1}$31	−63 02	26 24	113 52	78 22	55 41½	113 20½
$\bar{y}$	y	$\overline{2}$11	−63 12½	59 00	146 03½	67 16	87 17	139 55
$\bar{v}$		$\overline{3}$11	−73 28½	69 14½	158 25½	74 34½	80 54	153 41½

Less common or rare:

$\bar{c}$ 001   h 120   g 210 $\bar{A}$ $\overline{1}$22   z $\overline{2}$51   δ $\overline{5}$61

Structure cell.[3] Space group probably Cc. a_0 7.56 kX, b_0 8.95, c_0 6.66; [β 121°00′]; $a_0:b_0:c_0 = 0.839:1:0.750$. Cell contents $Ca_4Mg_4(AsO_4)_4F_4$. Tilasite resembles sphene in structure.

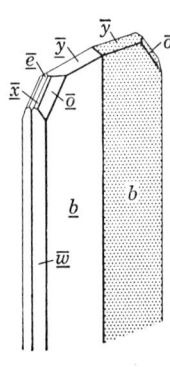

Twin on {001}. [100] vertical.

Habit. Elongated [100] and sometimes also flattened on {010}, the crystals occasionally in subparallel groups. Also massive.

Twinning. On {001}, common, as symmetrical contact twins.

Phys. Cleavage {10$\overline{1}$} good, also parting on {1$\overline{3}$3}, {10$\overline{2}$}, and {0$\overline{1}$1}. Brittle. H. 5. G. 3.77 ± 0.02;[4] 3.78 (calc.). Luster resinous, on cleavage surfaces vitreous. Color gray to violet-gray (Långban), olive-green to apple-green (India). Translucent. Piezoelectric.[5]

Opt.[6] In transmitted light, colorless or pale green.

ORIENTATION	n (India)	
X ∧ c ∼ 30°	1.640	Biaxial negative (−).
Y	1.660	2V 82°44′ (meas.).
Z b	1.675	r < v, slight.

Chem. A fluor-arsenate of calcium and magnesium, $CaMg(AsO_4)F$. Small amounts of OH substitute for F (see adelite).

Anal.

	1	2	3	4	5
CaO	25.22	25.69	25.27	25.32	25.68
FeO		0.21		0.14	0.55
MgO	18.13	18.12	17.32	18.22	18.34
P_2O_5		0.10		tr.	0.43
As_2O_5	51.70	51.13	48.33	50.91	50.35
F	8.55	7.45	7.95	8.24	7.18
H_2O		0.43	0.07	0.28	0.73
Rem.		0.13	4.21	0.47	0.11
	103.60	103.26	103.15	103.58	103.37
F = O	3.60	3.04	3.15	3.47	3.02
Total	100.00	100.22	100.00	100.11	100.35
G.	3.78	3.76	3.79	3.78	3.77

DURANGITE

1. CaMg(AsO$_4$)F. 2. Långban.[7] Rem. is insol. 0.08, MnO 0.05, Cl tr. 3. Långban.[7] Rem. is MnO 0.15, CO$_2$ 0.83, BaO 0.62, insol. 2.61. 4. Långban.[8] Rem. is MnO 0.16, Na$_2$O 0.29, Cl 0.02. 5. Kajlidongri, India.[9] Average of two. Rem. is SrO 0.06, insol. 0.05.

Tests. B.B. easily fusible. Readily soluble in HCl or HNO$_3$.

Occur. At Långban, Vermland, Sweden, as grains or veinlets with berzeliite and barite in dolomitic limestone that carries hausmannite, in part as crystals of varied habit associated with allactite, pyroaurite, pyrochroite, dixenite, hematite, calcite in drusy cavities. Also found at Kajlidongri, Jhabua State, Central Provinces, India, with quartz, barite, spessartite, braunite.

Name. After the Swedish mining engineer Daniel Tilas (1712–1772).

Ref.
1. Angles and unit of Smith and Prior, *Min. Mag.*, **16**, 84 (1911), re-oriented with interchange of a and c in conformance with the structure cell. Transformation: Smith and Prior to new, 001/0$\bar{1}$0/100. See also Aminoff, *Geol. För Förh.*, **45**, 144 (1923).
2. Aminoff (1923) and Smith and Prior (1911).
3. Strunz, *Zs. Kr.*, **96**, 7 (1937); *Cbl. Min*, 59 (1938), by rotation and powder methods on Långban crystal; the space group assumes domatic symmetry   See also Aminoff, *Kungl. Svenska Vetenskap. Ak. Handl.*, **11**, no 4, 28 (1933).
4. From the value 3.75 reported by Strunz (1937) and those cited under **Anal.**
5. Observed by Smith and Prior (1911) on India crystals but not confirmed by Strunz (1937) on Långban material.
6. Smith in Smith and Prior (1911).
7. Mauzelius analysis in Aminoff (1923), who cites two additional partial analyses, one with H$_2$O 0.95.
8. Mauzelius analysis in Sjögren (1895).
9. Prior analysis in Smith and Prior (1911).

41.5.7   **DURANGITE** [NaAlF(AsO$_4$)].   Brush (*Am. J. Sc.*, **48**, 179, 1869).

Cryst.[1] Monoclinic; prismatic—$2/m$.

$a:b:c = 0.773:1:0.840;$ $\beta\ 115°46';$ $p_0:q_0:r_0 = 1.087:0.757:1$

$r_2:p_2:q_2 = 1.322:1.436:1;$ $\mu\ 64°14';$ $p_0'\ 1.207,\ q_0'\ 0.840,\ x_0'\ 0.483$

Forms:[2]

	ϕ	ρ	ϕ_2	$\rho_2 = B$	C	A
b 010	0°00'	90°00'	.....	0°00'	90°00'	90°00'
a 100	90 00	90 00	0°00'	90 00	64 14	.....
m 110	55 10	90 00	0 00	55 10	69 06	34 50
e 021	29 53	44 05½	64 14	52 53½	37 06½	69 43
p 111	63 34	62 04½	30 37	66 50	40 02	37 42
q $\bar{1}$12	−16 03	23 36½	96 53½	67 22	39 00½	96 21½
π $\bar{1}$11	−40 46	48 01½	125 55	55 44	66 55	119 02½

Structure cell.[3] Space group $C2/c$. a_0 6.53 kX, b_0 8.46, c_0 7.00; $\beta\ 115°00';\ a_0:b_0:c_0 = 0.772:1:0.827$ [natural]; a_0 6.69 kX, b_0 8.66, c_0 7.27; $\beta\ 115°46';\ a_0:b_0:c_0 = 0.773:1:0.840$ [artificial].   Cell contents Na$_4$Al$_4$F$_4$(AsO$_4$)$_4$. Isostructural with sphene.

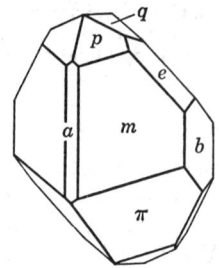

Habit. In crystals, the habit oblique pyramidal with {I11} and {110} predominating; faces usually dull and rough. Artificial crystals [7] are prismatic [101], rarely tabular.

Twinning. Interpenetration twins on {001} are common in artificial crystals.

P h y s. Cleavage {110}, distinct. Fracture uneven. Brittle. H. 5. G. 3.94–4.07, highest in dark-colored, ferrian material; 3.616 (calc. for $NaAlF(AsO_4)$). Luster vitreous. Color orange-red, light and dark; green (artificial material). Streak cream-yellow. Translucent.

O p t.[4] In transmitted light, orange-yellow.

Orientation	n (Durango)	Pleochroism	
$X \wedge c$ −25°	1.634 ± 0.003	Orange-yellow	Biaxial negative (−).
Y	1.673 ± 0.003	Pale orange-yellow	$2V$ 45° (meas.).
Z b	1.685 ± 0.003	Nearly colorless	$r > v$, perceptible, horizontal distinct.

C h e m. A fluo-arsenate of sodium and aluminum, $NaAlF(AsO_4)$. Small amounts of Li substitute for Na, and Fe''' substitutes for Al with Fe:Al = 1:2.91 in analysis 4.

Anal.

	1	2	3	4
Li_2O		0.81	0.70	0.65
Na_2O	14.91	11.66	11.86	13.06
Al_2O_3	24.52	20.68	20.09	17.19
MnO		1.30	1.28	
Mn_2O_3				2.08
Fe_2O_3		4.78	5.06	9.23
As_2O_5	55.28	55.10	53.22	53.11
F	9.14	n.d.	n.d.	7.67
Total	103.85			102.99
O = F	3.85			3.23
Total	100.00			99.76
G.		3.94		4.07

1. $NaAlF(AsO_4)$. 2,3. Durango.[5] Light-colored crystals. 4. Durango.[6] Dark-colored crystals.

Tests. B.B. fuses at 2. In C.T. blackens at a moderate temperature, regaining its color on cooling; at a higher temperature fuses to a yellow glass and gives a faint white volatile sublimate, etching the tube slightly. Soluble in H_2SO_4.

O c c u r. Found with cassiterite, hematite, and topaz in the state of Durango, Mexico, at the Barranca tin mine, 18 miles northeast of Coneto and about 90 miles northeast of the city of Durango. Also with amblygonite and cassiterite in a pegmatite near Lake Ramsay, New Ross, Lunenberg County, Nova Scotia.

A r t i f.[7] Obtained in crystals by the reaction of powdered cryolite with syrupy ortho-arsenic acid in a closed tube at 200°.

N a m e. From the locality.

Ref.
1. Unit and orientation of Des Cloizeaux, *Ann. chim. phys.*, **4**, 401 (1875); ratio of the structure cell of Machatschki, *Zs. Kr*, **103**, 221 (1941), on synthetic material; transformation: M to D., 001/010/101.
2. Goldschmidt (**3**, 77, 1916); Dana (780, 1892).
3. Machatschki (1941) and Kokkoros, *Zs Kr.*, **99**, 38 (1938).
4. Machatschki (1941) and Des Cloizeaux (1875); indices from Larsen (70, 1921).
5. Brush (1869).
6. Hawes analysis in Brush, *Am J Sc.*, **11**, 464 (1876).
7. Machatschki (1941).

41.5.8 PLUMBOGUMMITE GROUP

HEXAGONAL—R

	c_0/a_0	a_0	c_0
Plumbogummite, $PbAl_3(PO_4)_2(OH)_5H_2O$	.....		
Gorceixite, $BaAl_3(PO_4)_2(OH)_5H_2O$	.....		
Goyazite, $SrAl_3(PO_4)_2(OH)_5H_2O$	2.368	6.97 kX	16.51
Crandallite, $CaAl_3(PO_4)_2(OH)_5H_2O$	.....		
Deltaite, $Ca(Al_2Ca)(PO_4)_2(OH)_5H_2O$	2.307	6.98	16.10
Florencite, $CeAl_3(PO_4)_2(OH)_6$	1.1588		
Dussertite, $BaFe_3(AsO_4)_2(OH)_5H_2O$	.....		

The members of this group are isostructural with the minerals of the Alunite Group (which see) and the Beudantite Group. The dimensions of the unit cell of florencite are not known; presumably the old morphological unit, here retained, should be doubled to conform with that of goyazite and the Alunite Group.

41.5.8.1 P L U M B O G U M M I T E [$PbAl_3(PO_4)_2(OH)_5H_2O$]. Plomb rouge en stalactites—tantot en globules *de Lisle* (*Demeste Lettres Min.*, **2**, 399, 1779; *Crist.*, **3**, 399, 1783). Sel acide-phosphorique-martial *de Laumont* (*J. phys.*, **28**, 385, 1786). Plomb-gomme *de Laumont*. Aluminiate de Plomb avec eau de combinaison *Berzelius* (*Nouv. Min.*, 283, 1819). Bleigummi, Blei-aluminat, etc., *Berzelius* (*J. Chemie u. Phys.*, **27**, 65, 1819). Native Aluminiate of Lead *Smithson* (*Ann. Phil.*, **14**, 31, 1819 [citing S. Tennant as having first analyzed the mineral]). Plomb hydro-alumineux *Haüy* (**3**, 410, 1822). Gummispath *Breithaupt* (**56**, 1832). Plombgomme *Beudant* (**2**, 1832). Plumbo-gummite *Shepard* (**2**, 113, 1835). Plumbo-resinite *Dana* (230, 1837). Bleigummi, Gummibleispath, Bleihydroaluminat *Germ.* Hitchcockite *Shepard* (*Rep. Canton Mine, Ga.*, 1856; 401, 1857). Bischofit *Fischer* (*Jb. Min.*, 466, 1862). Schadeite *Lazarevič* (*Koll.-Zs.*, **4**, 306, 1909).

C r y s t.[1] Hexagonal—*R* (?).

Habit. As botryoidal, reniform, stalactitic, or globular crusts or masses, often with a concentric structure; also compact massive. The mineral may resemble drops or coatings of gum. Microscopically radial fibrous or spherulitic. Rarely as minute crystals with an hexagonal outline.

P h y s. Fracture of masses uneven to subconchoidal. Brittle. H. $4\frac{1}{2}$–5. G. 4.014;[2] 4.08 (calc.). Color grayish white, yellowish gray, or yellow to yellowish or reddish brown; also greenish or bluish. Luster dull to resinous, gum-like. Streak colorless to white. Translucent.

O p t.

	n (Canton, Ga.)[3]	
O	1.653 ± 0.01	Uniaxial positive (+).
E	1.675 ± 0.01	In part sensibly isotropic.

Chem. A hydrated basic phosphate of lead and aluminum, $PbAl_3(PO_4)_2(OH)_5H_2O$. The analyses do not agree well with this formula, taken from analogy to other members of the plumbogummite group; all show excess water, and some show varying amounts of CO_2.

Anal.[4]

	1	2	3	4	5
PbO	38.40	34.36	37.03	38.91	35.66
Al_2O_3	26.31	29.48	28.74	20.98	25.11
P_2O_5	24.44	17.58	18.64	19.14	22.30
CO_2		2.77	3.12	4.66	
H_2O	10.85	14.71	12.73	15.44	16.19
Rem.		0.82		1.12	1.45
Total	100.00	99.72	100.26	100.25	100.71

1. $PbAl_3(PO_4)_2(OH)_5H_2O$. 2. Canton mine, Georgia (*hitchcockite*).[5] Rem. is insol. 3. Roughten Gill, Cumberland.[5] 4. Huelgoat, Brittany.[5] Rem. is SO_3 0.96, Cl 0.16. 5. Near Diamantina, Brazil.[6] Rem. is CaO 0.62, CeO 0.16, SiO_2 0.67.

Tests. B.B. swells up but is imperfectly fused. In C.T. decrepitates and affords much water. Soluble in hot acids. Some specimens on heating turn black and afford an organic odor.

Occur. A secondary mineral often found with pyromorphite in lead deposits. In France at Huelgoat, Brittany; at Nussière near Beaujeu with pyromorphite, cerussite, anglesite, and wulfenite, and at Montchonay, Dept. Rhône. With pyromorphite and mimetite at Roughten Gill and Dry Gill near Caldbeck, Cumberland. As rolled fragments (favas) with a fibrous structure in the diamantiferous sands in the region around Diamantina, Minas Geraes, Brazil. In the United States with marcasite at the Canton mine, Georgia, and reported from Mine la Motte, Missouri.

Alter. Found as pseudomorphs after pyromorphite and barite (Cumberland).

Name.[7] From *plumbum*, lead, and *gummi*, gum. Schadeite after Heinrich Schade of the University of Kiel. Hitchcockite after Edward Hitchcock (1793–1867), American geologist.

Ref.

1. Measurable crystals have not been found but the optical properties, composition, and x-ray powder data of Pabst, priv. comm. (1948), indicate that plumbogummite belongs to the goyazite group and hence is rhombohedral. See also Prior, *Min Mag.*, **12**, 249 (1900).
2. Genth, *Am. J. Sc.*, **23**, 424 (1857), on Canton, Georgia, material. The values 4.88 and 6.421 of Dufrénoy, *Ann. mines*, **8**, 243 (1835), on Nussière material and of Breithaupt (**2**, 506, 1841) are incompatible with the composition.
3. Larsen (121, 1921).
4. For older analyses see Dana (855, 1892) and Hintze (**1** [4B], 1158, 1933).
5. Hartley, *Min. Mag.*, **12**, 223 (1900).
6. Florence analysis in Hussak, *Min Mitt*, **25**, 341 (1906).
7. For a history of the nomenclature see Hintze (1933). Wherry, *Proc. U. S. Nat. Mus.*, **51**, 83 (1916), discusses a proposed nomenclature for the goyazite group.

FERRAZITE. Lee and de Moraes (*Am. J. Sc.*, **48**, 353, 1919).

Found as discoidal pebbles (favas) of granular structure in the diamond sands of the Diamantina district, Minas Geraes, Brazil. G. 3.0–3.3. Color dark yellowish white, resembling old ivory. A hydrated phosphate of Pb, Ba, and Al. Analysis of a sample assumed to contain kaolin gave: BaO 8.87, CaO tr., PbO 45.63, Al_2O_3 3.48, SiO_2 2.44, P_2O_5 26.24, H_2O 14.20, total 100.86. Named after J. B. de Araujo Ferraz of the Brazilian Geological Survey.

41.5.8.2 G O R C E I X I T E [$BaAl_3(PO_4)_2(OH)_5H_2O$ (?)]. *Hussak (Min. Mitt.*, **25**, 336, 1906).

Hexagonal—R.[1] In grains and pebbles which are in part microcrystalline. Porcelaneous fracture. H. 6. G. 3.036–3.185. Color brown, sometimes mottled. Luster vitreous to dull.

O p t.[2] Colorless in thin sections. Uniaxial positive (+). Mean index of refraction 1.625; birefringence low.

C h e m. A hydrated basic phosphate of barium and aluminum of uncertain formula, perhaps $BaAl_3(PO_4)_2(OH)_5H_2O$. Ca and Ce substitute for Ba to some extent.

Anal.

	BaO	CaO	CeO	Al_2O_3	P_2O_5	H_2O	Rem.	Total	G.
1.	29.98			29.96	27.75	12.31		100.00	
2.	15.42	3.55	1.55	35.00	22.74	14.62	6.32	99.20	3.101
3.	15.30	2.24	2.35	35.20	21.47	14.73	8.92	100.21	3.098
4.	11.88		7.00	37.96	22.39	15.05	6.29	100.57	3.185
5.	26.02			34.59	24.10	15.29		100.00	

1. $BaAl_3(PO_4)_2(OH)_5H_2O$. 2,3. Rio Abaeto, Minas Geraes, Brazil.[3] Rem. (2) is Fe_2O_3 4.10, TiO_2 0.67, SiO_2 1.55. Rem. (3) is Fe_2O_3 1.67, TiO_2 0.75, SiO_2 6.50. 4. Somabula, Southern Rhodesia.[4] Rem. is MgO 1.28, Fe_2O_3 3.76, SiO_2 1.25. Ce reported as CeO_2. 5. $BaAl_4(PO_4)_2(OH)_8H_2O$.

O c c u r. As favas in the diamantiferous sands of Brazil and Africa. In Brazil near Curralinho and Dattas in the vicinity of Diamantina and very abundantly in the Rio Abaete, Rio Bagagem and Douradinhos, Minas Geraes; Rio Paranahijba and Verissimo in the southern part of Goyaz; Patrocinio de Sapucahy and Rio Canoas in São Paulo. In Africa near Dompim, Bonsa River, Gold Coast; Oiyi District, Sierra Leone; and in Triassic gravels at Somabula, Southern Rhodesia. At Issineru, Mazaruni district, British Guiana.

N a m e. After Henrique Gorceix (1842–1919), first director of the School of Mines at Ouro Preto, Brazil.

GERAESITE. *Farrington (Bull. Geol. Soc. Am.*, **23**, 728, 1912). Described as a hydrous barium aluminum phosphate more acidic than gorceixite from the favas accompanying diamonds in river sands in Brazil.[5]

Ref.

1. No crystal measurements have been reported. System from indexed powder pattern, Pabst, priv. comm. (1948).
2. Optical character from Hussak (1906); refractive index from Gaubert, *Bull. soc. min.*, **30**, 108 (1907).
3. Hussak (1906).
4. MacGregor, *Bull. Imp. Inst., London*, **39**, 399 (1941). For analyses of phosphates in diamantiferous gravels that may be in part gorceixite, see Farrington, *Am. J. Sc.*, **41**, 357 (1916), and Beard, *Bull. Imp. Inst., London*, **39**, 160 (1941).
5. In a later discussion of the favas, giving analyses, Farrington (1916) did not use the name geraesite.

41.5.8.3 G O Y A Z I T E [SrAl$_3$(PO$_4$)$_2$(OH)$_5$H$_2$O]. *Damour* (*Bull. soc. min.*, **7**, 204, 1884). Hamlinite *Hidden* and *Penfield* (*Am. J. Sc.*, **39**, 511, 1890). Bowmanite *Solly* (*Min. Mag.*, **14**, 72, 1905).

C r y s t.[1] Hexagonal—R; scalenohedral—$\bar{3}\ 2/m$.

$a:c = 1:2.3901;$ $\alpha\ 61°05';$ $p_0:r_0 = 2.7599:1;$ $\lambda\ 109°01'$

Forms:

		ϕ	$\rho = C$	A_1	A_2
c 0001	111	....	0°00'	90°00'	90°00'
r 10$\bar{1}$2	411	30°00'	54 04	45 28½	90 00
n 20$\bar{2}$1	5$\bar{1}$1	30 00	79 44	31 33	90 00
f 01$\bar{1}$1	22$\bar{1}$	−30 00	70 05	90 00	35 29½

Structure cell.[2] Space group $R\bar{3}m$. a_{rh} 6.82, α 61°28'. a_0 6.97 kX, c_0 16.51; $a_0:c_0 = 1:2.3688$. Cell contents SrAl$_3$(PO$_4$)$_2$(OH)$_5$H$_2$O in the rhombohedral unit.

Habit. Small rhombohedral crystals, pseudo-cubic {10$\bar{1}$2} or tabular {0001}. Rhombohedral faces often striated horizontally. Also as pebbles or rounded grains.

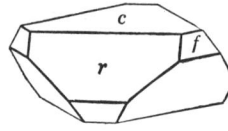

P h y s. Cleavage {0001} perfect. H. 4½–5. G. 3.26; 3.29 (calc.). Colorless, pink, or honey-yellow. Luster greasy to resinous, on {0001} pearly. Transparent.

O p t.[3] In transmitted light, colorless or faintly tinted.

	n(Na) (Brazil)	n (Colorado)	n (Ukraine)	
O	1.6294	1.620 ± 0.005	1.635	Uniaxial positive (+).
E	1.6387	1.630 ± 0.005	1.641	

Sometimes shows anomalous biaxial character, basal plates being divided into sectors, with $2V$ up to 30°. Zonal growth also may be shown. Thick grains may be weakly pleochroic with O light pink, E yellowish or greenish (Ukraine).

C h e m. A hydrated basic phosphate of strontium and aluminum, SrAl$_3$(PO$_4$)$_2$(OH)$_5$H$_2$O.[4] Ba substitutes for Sr with Ba:Sr = 1:6.8 in analysis 3, and F is present in some material in substitution for (OH).

Anal.

	BaO	SrO	Al$_2$O$_3$	P$_2$O$_5$	H$_2$O	Rem.	Total	G.
1.		22.45	33.12	30.77	13.66		100.00	3.29
2.		17.33	50.66	14.87	16.67		99.53	3.26
3.	4.00	18.43	32.30	28.92	12.00	4.53	100.18	3.159–3.283
4.	0.15	19.8	35.6	24.8	15.7	3.2	99.25	3.219–3.266

1. SrAl$_3$(PO$_4$)$_2$(OH)$_5$H$_2$O. 2. Minas Geraes, Brazil.[5] 3. Oxford County, Maine[6] (*hamlinite*). Rem. is Na$_2$O 0.40, K$_2$O 0.34, Fe$_2$O$_3$ 0.90, SiO$_2$ 0.96, F 1.93. Total after deducting 0.81 O for F is 99.37. 4. Lengenbach, Binnental, Valais, Switzerland (*bowmanite*).[7] Rem. is K$_2$O 0.4, SiO$_2$ 2.8. The figure for H$_2$O is ign. loss.

Tests. Fuses at 4. In C.T. gives abundant water.[8] Slowly soluble in acids.

O c c u r. In the diamond-bearing sands of Serra de Congonhas, near Diamantina, Minas Geraes, Brazil. With pyrite in crevices of dolomite at Lengenbach, Binnental, Valais, Switzerland (*bowmanite*).[8] In implanted crystals, partly on anhydrite, in the Simplon tunnel, Switzerland. Associated with barite and quartz and with pyrite and clayey inclusions in the Carboniferous limestones of the Belaya Kalitva region, Donetz basin, and in the breccias of the Romny and Issachki salt domes, Ukraine. In the United States (*hamlinite*)[8] with bertrandite, apatite, and herderite on feldspar and mica in a pegmatite on Lords Hill near Stoneham, Oxford County, Maine; also at Paris, Greenwood, and Buckfield in Oxford County; in the Palermo pegmatite, North Groton, New Hampshire. At the Eagle Rock mine, Boulder County, Colorado.

Name. From the province of Goyaz, in which were found the principal diamond deposits of Brazil. Hamlinite after A. C. Hamlin, a collector of Maine minerals and gems. Bowmanite after H. L. Bowman (1874–1942), Professor of Mineralogy at Oxford.

Ref.

1. Axial ratio and forms of Ježek, *Bull Ac Sc. Bohême*, **13** (1908), with *c* doubled to conform to structure cell. Transformation: Ježek to new, 1000/0100/0010/0002.
2. Gossner, *Zs. Kr.*, **96**, 488 (1937), by rotation method on material of unstated locality.
3. Ježek (1908); Pitkoskaya, *Ac sc. Leningrad, C. r.*, **25**, 502 (1939); Weber, *Jb. Min., Beil.-Bd.*, **57A**, 563 (1928); Larsen (82, 1921).
4. The formula here written conforms to the type for other minerals in this and related groups. It may also be written as a pyrophosphate, but the obvious structural similarity to alunite makes this inappropriate. Owing to poor agreement of analyses with required composition, other formulas have been proposed.
5. Damour (1884) SrO originally reported as CaO, but Gramont, *Bull soc min.*, **40**, 26 (1917), showed by spectrographic examination that goyazite contains much Sr and little Ca.
6. Penfield, *Am. J. Sc ,* **4**, 313 (1897).
7. Summary of several partial analyses by Bowman, *Min Mag.*, **14**, 389 (1907), as given by Bader, *Schweiz. min. Mitt ,* **14**, 319 (1934).
8. On the identity with goyazite see Bowman, *Min. Mag.*, **14**, 389 (1907); and Schaller, *Am. J. Sc.*, **32**, 359 (1911), **43**, 165 (1917).

41.5.8.4 **C R A N D A L L I T E** [$CaAl_3(PO_4)_2(OH)_5H_2O$]. Kalkwavellit *Kosmann* (*Zs. deutsche geol. Ges.*, **21**, 799, 1869). Crandallite *Loughlin* and *Schaller* (*Am. J. Sc.*, **43**, 69, 1917). Pseudowavellite *Laubmann* (*Ber. deutsche chemische Gesellschaft*, **55B**, 3016, 1922; *Geognost. Jahreshefte, München*, **35**, 203, 1922).

C r y s t.[1] Hexagonal—*R*.

Habit. Rarely as minute trigonal prisms terminated by {0001}, or as rosettes of fibers elongated at right angles to [0001]. Usually massive, as nodular aggregates with concentric layers ordinarily composed of subparallel or matted fibers but sometimes fine-granular and grading into dense agate-like or chalcedonic forms; also as banded spherules with a radial fibrous structure.

Phys. Cleavage {0001}, perfect. H. 5. G. 2.78 (Bolivia),[2] 2.92 (Fairfield, Utah).[3] Luster of crystals vitreous, of massive material grading from vitreous in chalcedonic types to dull and chalky. Color yellow, varying to yellowish white, white, and gray.

Opt. In transmitted light, colorless. Uniaxial positive (+).

	n (Silver City, Utah; anal. 3)	n (Fairfield, Utah: fine fibrous; anal. 5)	n (Fairfield: vitreous granular; anal. 6)	n (Fairfield: white chalky; anal. 7)	n (Llallagua; anal. 8)
O	1.60	1.618	1.622	1.619	1.618
E	1.59	1.623	1.631	1.627	1.620

Chem. A hydrated basic phosphate of calcium and aluminum, $CaAl_3(PO_4)_2(OH)_5H_2O$. This formula corresponds to the ideal alunite-type structure. Analyses show marked departures. Sr, Ba, and rare earths may substitute for Ca (analyses 3 and 4). Fe''' may substitute for Al (analysis 4). Some Ca may enter into positions indicated by the formula as Al positions, giving transitions towards deltaite, which see. All analyses show high H_2O and low P_2O_5, indicating substitution of hydroxyl for (PO_4). This substitution may lead to a composition, expressed by the formula $CaAl_3(PO_4)_{3/2}(OH)_2(OH)_5H_2O$, very close to analysis 8.

	BaO	SrO	CaO	Al_2O_3	P_2O_5	H_2O-	H_2O+	Rem.	Total	G.
1.			13.55	36.93	34.29		15.23		100.00	
2.			11.04	37.52	25.24	1.00	17.90	7.70	100.40	
3.		2.21	7.50	38.71	27.09	1.29	18.86	4.74	100.40	
4.	0.67		16.86	28.18	30.10		18.76	6.81	101.38	
5.			15.30	34.15	32.23		17.50	2.06	101.24	2.92
6.			14.74	35.92	30.53		17.39	1.18	99.76	2.88
7.			10.4	37.5	32.6		17.6	1.9	100.0	
8.			13.7	38.8	28.1		20.0		100.6	2.78

1. $CaAl_3(PO_4)_2(OH)_5H_2O$. 2. Dehrn, Hesse-Nassau, Germany (*kalkwavellit*).[5] Rem. is MgO 0.24, SiO_2 4.92, CO_3 2.54. 3. Brooklyn mine, near Silver City, Utah.[6] After deduction of 35.13 insoluble. Rem. is MgO 0.94, SO_3 3.80. 4. Amberg, Bavaria.[7] Rem. is rare earths 1.02, Fe_2O_3 5.79. 5. Fairfield, Utah.[8] Rem. is MgO 2.06. 6. Fairfield, Utah.[9] Rem. is Na_2O 0.33, K_2O 0.13, MgO 0 72. 7. Fairfield, Utah.[9] Rem. is Na_2O 1.4, K_2O 0.5. 8. Llallagua, Bolivia.[10]

Tests. Fuses at $2\frac{1}{2}$ to a white enamel. Difficultly soluble in acids both before and after ignition.

Occur. First noted in fibrous aggregates in cavities in the phosphate deposits of Dehrn and Ahlbach, Hesse-Nassau, Germany (*kalkwavellite*).[11] White radiating incrustations on wavellite and limonite in an iron ore deposit at Amberg, Bavaria (*pseudowavellite*). As microcrystalline coatings on cassiterite and quartz, Llallagua, Bolivia. In the United States, in cavities in a quartz-barite aggregate at the Brooklyn mine, near Silver City, Tintic district, Juab County, Utah. Later found abundantly as an alteration product and replacement of variscite nodules in a brecciated zone in limestone five miles west of Fairfield, Utah County, Utah.[12] The crandallite is the first-formed alteration product and is closely associated with deltaite, millisite, wardite, and lehiite; this was followed by the deposition in open cavities in the massive variscite and crandallite of relatively well-crystallized gordonite, englishite, montgomeryite, overite, and sterrettite; and finally by the deposition in small amounts of members of the apatite group. Alunite and small amounts of calcite are other associated minerals. Crandallite also occurs with variscite at the Amatrice Hill deposit, located about 25 miles northwest of Fairfield, and probably also in the variscite deposits at Lucin, Utah, and in Esmeralda County, Nevada. As an alteration product in pegmatite near Harney City, South Dakota.

Alter. Observed as pseudomorphs after gordonite.

Artif. A substance apparently identical with crandallite-deltaite is formed by the action of solutions containing $Ca(OH)_2$ and $NaOH$ on fragments of variscite.[13]

Name. After M. L. Crandall. The older name kalkwavellite is dropped as misleading.[14] Available evidence indicates that crandallite and pseudowavellite are best considered a single species with some variation of composition, the name crandallite having priority.

Ref.

1. Crystal system from indexed powder patterns of crandallite from Utah and Bolivia, Pabst, priv. comm (1948). See also Larsen, *Am. Min.*, **27**, 288 (1942).
2. Gordon, *Proc. Ac. Sc. Philadelphia*, **96**, 336 (1944).
3. Larsen and Shannon, *Am. Min.*, **15**, 307 (1930)*B*.
4. For source of data on refractive indices see references with analyses. The original crandallite from Silver City was reported by Larsen and Shannon, *Am. Min* , **15**, 303 (1930)*A*, to be biaxial with moderate $2V$.
5. Shannon analysis in Larsen and Shannon (1930)*A*. For an older analysis of the material from Dehrn see Kosmann (1869).
6. Schaller analysis in Loughlin and Schaller (1917).
7. Kieffer analysis in Laubmann (1922).
8. Shannon analysis in Larsen and Shannon (1930)*B*.
9. Gonyer analysis in Larsen and Shannon (1930)*B*.
10. Gordon (1944).
11. Material from the type locality for kalkwavellite has been shown to be identical with crandallite, Larsen and Shannon (1930)*A*.
12. The mineralogy and paragenesis of this occurrence are described in detail by Larsen, *Am. Min.*, **27**, 288 (1942), who employed the name pseudowavellite.
13 Larsen (1942).
14. Larsen and Shannon (1930)*A*.

41.5.8.5 **DELTAITE** $[Ca(Al_2,Ca)(PO_4)_2(OH)_4H_2O]$. *Larsen* and *Shannon* (*Am. Min.*, **15**, 321, 1930).

Cryst.[1] Hexagonal—R; ditrigonal-pyramidal—3 m.

Structure cell.[2] Space group $R3m$. a_{rh} 6.71, α 62°40′, a_0 6.98 kX, c_0 16.10; $a_0:c_0 = 1:2.307$. Cell contents $Ca_2Al_2(PO_4)_2(OH)_4H_2O$ in the rhombohedral unit.

Habit. As minute crystals, variously trigonal prisms terminated by {0001} or by steep rhombohedra, or as nearly square rhombohedra (?) modified by {0001}. Also as cherty or chalky crusts composed of matted fibers.

Phys. Cleavage {0001} (?)[3] H. 5. G. $\sim$2.95; 2.89 (calc.). Color gray to lavender and pale blue; also canary-yellow.

Opt. In transmitted light, colorless.

	n (crystals)	n (gray, cherty)	
O	1.641 ± 0.003	1.630	Uniaxial positive (+).
E	1.650 ± 0.003	1.640	

Chem. A hydrated basic phosphate of calcium and aluminum, $Ca(Al_2,Ca)(PO_4)_2(OH)_4H_2O$. Deltaite is isostructural with crandallite and forms at least a partial series toward that species.[2]

Anal.

	1	2	3	4
CaO	27.35	24.0	22.95	24.37
MgO		0.9		3.40
Al_2O_3	24.86	28.0	29.36	26.87
P_2O_5	34.61	33.4	32.84	27.60
H_2O+	13.18	13.1	14.22	14.29
Rem.		0.6	0.04	2.06
Total	100.00	100.0	99.41	98.59

1. $Ca(Al_2,Ca)(PO_4)_2(OH)_4H_2O$. 2. Fairfield, Utah.[4] Spherulitic material. Rem. is Na_2O 0.5, K_2O 0.1. Recalculated to 100 after deduction of 20 per cent crandallite. 3. Fairfield, Utah.[4] Gray cherty crust. Rem. is H_2O- 0.04. 4. Fairfield, Utah.[4] Chalky layer; not certainly deltaite. Rem. is insol. 0.60, Na_2O 0.33, K_2O 1.13. Contains several per cent of CO_2.

Occur. Found intergrown with crandallite in variscite nodules from Fairfield, Utah County, Utah.

Name. In allusion to the characteristic triangular cross section of the crystals, like the Greek letter delta.

Ref.

1. Crystal class from morphology and x-ray Weissenberg study by Larsen, *Am. Min.*, **27**, 288 (1942).
2. Larsen (1942).
3. Presumed from the isostructural relation to crandallite.
4. Gonyer analysis in Larsen and Shannon (1930).

41.5.8.6 FLORENCITE $[CeAl_3(PO_4)_2(OH)_6]$. *Hussak* and *Prior* (*Min. Mag.*, **12**, 244, 1900). Stiepelmannite *Ramdohr* and *Thilo* (*Zbl. Min.*, 1, 1940).

Cryst.[1] Hexagonal—R; scalenohedral—$\bar{3}\ 2/m$.

$a:c = 1:1.1588;$ $\alpha\ 53°13\frac{1}{2}';$ $p_0:r_0 = 1.3381:1;$ $\lambda\ 87°51'$

Forms:

		ϕ	$\rho = C$	A_1	A_2
c 0001	111	....	0°00'	90°00'	90°00'
m $10\bar{1}0$	$2\bar{1}\bar{1}$	30°00'	90 00	30 00	90 00
r $10\bar{1}1$	100	30 00	53 13½	46 04½	90 00
f $02\bar{2}1$	$11\bar{1}$	−30 00	69 30½	90 00	35 47

Habit. As small rhombohedral crystals, either $\{02\bar{2}1\}$ predominant or $\{10\bar{1}1\}$ and then pseudo-cubical.

Phys. Cleavage $\{0001\}$ good, $\{11\bar{2}0\}$ in traces. Fracture splintery to subconchoidal. H. 5–6. G. 3.586 (Brazil), 3.67–3.71 (Africa). Color clear pale yellow. Luster greasy to resinous. Transparent.

Opt. In transmitted light, colorless.

	n (Brazil)[2]	n (Africa)[3]	
O	1.680 ± 0.01	1.695	Uniaxial positive (+).
E		1.705	

C h e m. A basic phosphate of the cerium earths and aluminum, $CeAl_3(PO_4)_2(OH)_6$. Some Ca and Y may substitute for the Ce earths. The distribution of the several rare earths in florencite has not been fully determined. F is probably present in some material.

Anal.

	Ce_2O_3	Al_2O_3	P_2O_5	H_2O	Rem.	Total	G.
1.	31.99	29.80	27.68	10.53		100.00	
2.	28.00	32.28	25.61	10.87	2.55	99.31	3.586
3.	29.25	30.83	[26.94]	11.07	1.91	[100.00]	3.695

1. $CeAl_3(PO_4)_2(OH)_6$. 2. Near Diamantina, Brazil.[4] Rem. is CaO 1.31, Fe_2O_3 0.76, SiO_2 0.48. 3. Klein Spitzkopje, S. W. Africa (*stiepelmannite*).[3] Rem. is CaO 0.50, ZrO_2 1.12, SiO_2 0.29.

Tests. B.B. infusible. Partly soluble in HCl. In C.T. gives off acid water, which is completely lost only at a high temperature.

O c c u r. A rare constituent of the sands of Tripuhy, near Ouro Preto, Minas Geraes, Brazil, associated with cinnabar, monazite, xenotime, etc.; in crystals and rounded grains in diamond-bearing sand from Matta dos Creoulos on the Rio Jequetinhonha near Diamantina, and as an accessory constituent of mica schists of Morro do Caixambú, near Ouro Preto. With microcline and fluorite in a pegmatite at Klein Spitzkopje, South West Africa (*stiepelmannite*).[5]

N a m e. After Dr. W. Florence, who made a preliminary chemical examination of the mineral.

Ref.

1. Elements calculated from the interfacial angles of Hussak and Prior (1900) as discussed by Dana (App. II, 42, 1914). The x-ray structure cell of Ramdohr and Thilo, *Zbl. Min.*, 1 (1940), is not here accepted.
2. Larsen (76, 1921).
3. Ramdohr and Thilo (1940).
4. Hussak and Prior (1900).
5. Originally made a distinct species on the supposition that it contained Y and Yb but later found to contain Ce earths essentially—see Ygberg, *Ark. Kemi*, **20**, 13 (1945).

41.5.8.7   D U S S E R T I T E [$BaFe_3(AsO_4)_2(OH)_5H_2O$]. Barthoux (*C.R.*, **180**, 299, 1925).

Probably hexagonal—R.[1] As minute crystals flattened {0001} and grouped as rosettes or as crusts.

P h y s. Cleavage not observed.[2] H. $3\frac{1}{2}$. G. 3.75. Color green to pistachio-green.

O p t. In transmitted light, yellowish green.

	n (Algeria)[3]	n (Mapimi)[4]	
O	1.87 ± 0.01	1.870	Uniaxial negative $(-)$.
E	1.85 ± 0.01	1.845	Sometimes abnormally biaxial with $2V$ 15° to 20°.

C h e m. A hydrated basic arsenate of barium[5] and ferric iron, $BaFe_3(AsO_4)_2(OH)_5H_2O$. Loss of water: at 120°, 0.3%; at 700°, 9.3%; at 1000°, 9.8%.

Anal.

	1	2
BaO	22.37	20.93
Fe_2O_3	34.93	34.57
As_2O_5	33.51	31.23
H_2O	9.19	9.30
Rem.		4.18
Total	100.00	100.21

1. $BaFe_3(AsO_4)_2(OH)_5H_2O$. 2. Djebel Debar, Constantine.[6] Rem. is CaO 0.08, insol. 4.10.

Occur. On lamellar or spongy quartz at Djebel Debar, northeast of Hammam Meskhoutine, Constantine, Algeria. With arseniosiderite and carminite at Mapimi, Durango, Mexico.

Name. After D. Dussert, French mining engineer.

Ref.

1. Inferred from the optical properties, the apparently rhombohedral habit of the Mapimi and Algeria crystals, and the structural resemblance to jarosite indicated by the x-ray powder study of McConnell, *Am. J. Sc.*, **240**, 651 (1942).
2. {0001} cleavage cited in Larsen and Berman (91, 1934).
3. Barth and Berman, *Chem. Erde*, **5**, 36 (1930).
4. Foshag, *Am. Min.*, **22**, 479 (1937).
5. Originally thought on the basis of an erroneous analysis to be a calcium-iron arsenate; later shown by Berman, in McConnell (1942), to contain barium instead of calcium.
6. Gonyer analysis in McConnell (1942).

41.5.9 **CHENEVIXITE** $[Cu_2Fe_2(AsO_4)_2(OH)_4 \cdot H_2O (?)]$. *Pisani* (C.R., **62**, 690, 1866).

Massive; compact earthy to opaline; as acicular to lath-like crystals perhaps pseudomorphous (Tintic). Fracture earthy to subconchoidal. Brittle. H. $3\frac{1}{2}$–$4\frac{1}{2}$. G. 3.93. Color greenish yellow to olive-green and dark green. Streak greenish yellow.

Opt.[1] Microcrystalline with a mean index of refraction of about 1.88; birefringence in some of the material rather strong (Tintic).

Chem. A basic hydrated (?) arsenate of copper and ferric iron of uncertain formula, perhaps $Cu_2Fe_2(AsO_4)_2(OH)_4 \cdot H_2O$ (anals. 3,4). P apparently substitutes in part for As in the Cornwall material (anal. 2); and possibly Al for Fe.

Anal.

	1	2	3	4	5
CaO		0.34	0.55	0.44	1.06
CuO	26.40	31.70	26.88	26.31	20.72
Fe_2O_3	26.49	25.10	26.94	27.37	24.10
Al_2O_3			1.17	0.66	2.28
As_2O_5	38.14	32.20	34.62	35.14	38.44
P_2O_5		2.30			tr.
H_2O	8.97	8.66	9.25	9.33	13.33
Rem.			0.94	0.56	0.07
Total	100.00	100.30	100.35	99.81	100.00

1. $Cu_2Fe_2(AsO_4)_2(OH)_4 \cdot H_2O$. 2. Cornwall.[2] 10.3 per cent quartz deducted. 3. American Eagle mine, Tintic, Utah.[3] Rem. is MgO 0.23, quartz 0.71. 4. American Eagle mine, Tintic.[4] Rem. is MgO 0.16, quartz 0.40. 5. Broken Hill, New South Wales.[5] Recalc. to 100 after deducting Bi_2S_3 0.06, SiO_2 0.30, gangue 7.64. Rem. is MgO.

Tests. B.B. on charcoal fuses easily, gives off arsenical fumes, and leaves a black magnetic scoria with grains of copper. In C.T. usually decrepitates and yields water. Easily soluble in acids.

Occur. Originally from Cornwall, England, impregnating quartz-rich vein material. From the American Eagle and other mines in the Tintic district, Utah,[6] with olivenite. From the Consols mine, Broken Hill, and reported from the Ardlethan tin field, New South Wales. Said to occur with olivenite at Tsumeb, South West Africa and with scorodite at Klein Spitzkopje, South West Africa. From Chuquicamata, Chile.[6]

Name. After Richard Chenevix (1774–1830), French chemist, who early (1801) analyzed an arsenate of copper and iron from Cornwall.

Ref.
1. Larsen (56, 1921).
2. Pisani (1866); see also Dana (583, 1868).
3. Mackenzie, *Min. Mag.*, 6, 181 (1885).
4. Genth, *Proc. Colorado Sc. Soc.*, 1, 115 (1884).
5. Mingaye anal. in Smith, *New South Wales Dept Mines, Min. Res.*, no. 34, 80 (1926).
6. Shown by x-ray powder study to be identical with the Cornwall material—Frondel, priv. comm. (1947).

41.5.10 BRAZILIANITE [$NaAl_3(PO_4)_2(OH)_4$]. Pough and Henderson (*Am. Min.*, 30, 572, 1945) [not brazilianite of Mawe (*Descr. Cat. Mins.*, London, 54, 1818)].

Cryst. Monoclinic; prismatic—$2/m$.

$a:b:c = 1.1056:1:0.6992$; $\beta\ 97°22'$; $p_0:q_0:r_0 = 0.6324:0.6934:1$

$r_2:p_2:q_2 = 1.4421:0.9122:1$; $\mu\ 82°38'$; $p_0'\ 0.6377, q_0'\ 0.6992, x_0'\ 0.1293$

Forms:[1]

		ϕ	ρ	ϕ_2	$\rho_2 = B$	C	A
c	001	90°00'	7°22'	82°38'	90°00'	.....	82°38'
b	010	0 00	90 00	.....	0 00	90°00'	90 00
a	100	90 00	90 00	0 00	90 00	82 38	.....
m	110	42 22	90 00	0 00	42 22	85 02½	47 38
n	011	10 28½	35 25	82 38	55 15½	34 44½	83 57
x	$\bar{1}$01	−90 00	26 57	116 57	90 00	34 19	116 57
v	$\bar{3}$01	−90 00	60 43½	150 43½	90 00	68 05½	150 43½
o	111	47 39	46 04	52 30½	60 58½	41 00½	57 51
g	$\bar{1}$11	−36 01½	40 50½	116 57	58 04	45 30	112 37
s	211	63 32½	57 29½	35 27	67 55½	50 58	40 59
q	$\bar{1}$21	−19 58½	56 06	116 57	38 44	58 53	106 28½

Structure cell.[2] Space group $P2_1/n$. a_0 11.19 Å, b_0 10.08, c_0 7.06; [$\beta\ 97°22'$]; $a_0:b_0:c_0 = 1.110:1:0.700$. Cell contents $Na_2Al_6(PO_4)_4(OH)_8$.

Habit.[3] Equant to short prismatic [001], with the prism zone striated [001]; also elongated [100] with a four-sided appearance due to the relatively large development of n and g and the near suppression of the prism zone. Common forms: n g b m v x. Also globular with a radial fibrous structure.

Phys. Cleavage {010} good. Brittle. Fracture conchoidal. H. $5\frac{1}{2}$. G. 2.983 ± 0.005;[4] 3.025 (calc.). Color chartreuse yellow to pale yellow. Luster vitreous. Streak colorless. Transparent.

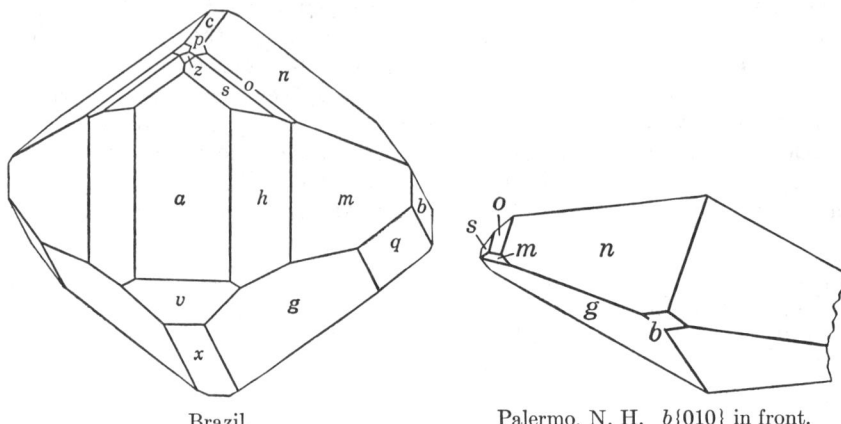

Brazil.   Palermo, N. H. b{010} in front.

Opt. In transmitted light, colorless.

Orientation	n(Na) (Brazil)[6]	n(Na) (Palermo)[6]	
$X \wedge c$ $-20°$	1.602 ± 0.001	1.602 ± 0.001	Biaxial positive (+).
Y b	1.609 ± 0.001	1.609 ± 0.001	$r < v$, faint.
Z	1.621 ± 0.001	1.623 ± 0.001	
2V (calc.)	75°	71°	

Chem. A basic phosphate of sodium and aluminum, $NaAl_3(PO_4)_2(OH)_4$. F is lacking in significant amounts.[5]

Anal.

	1	2	3
Na_2O	8.56	8.29	8.42
K_2O		0.20	0.37
Al_2O_3	42.25	42.85	43.82
Fe_2O_3		0.03	
TiO_2		0.05	
P_2O_5	39.23	38.79	37.97
H_2O+	9.96	9.91	9.65
H_2O-		0.04	
Total	100.00	100.16	100.23
G.	3.025	2.985	2.980

1. $NaAl_3(PO_4)_2(OH)_4$. 2. Palermo mine, New Hampshire.[7] Includes MnO trace; F and Li_2O none. 3. Conselheira Pena, Minas Geraes, Brazil.[8] Includes Cl trace; F none.

Tests. On heating B.B., turns white, expands slightly, and fuses with difficulty. The yellow color is lost at a low temperature. Slowly decomposed by hydrofluoric acid and by hot sulfuric acid.

Occur. Found associated[9] with muscovite, albite, apatite, and tourmaline in cavities in a pegmatite near Conselheira Pena, Minas Geraes, Brazil. Later found with whitlockite and apatite as a hydrothermal

GRIPHITE   843

mineral in cavities in granite pegmatite at the Palermo (No. 1) mine, near North Groton, Grafton County, New Hampshire. Also in pegmatite at the Smith mine near Newport, New Hampshire.

N a m e. After the country in which it was first found.

Ref.
1. Angles recalculated from the elements of Pough and Henderson (1945). Concordant measurements are reported by Frondel and Lindberg, *Am. Min.*, **33**, 135 (1948), on crystals from the Palermo mine, New Hampshire. Less common or rare forms: β 340, i 210, h 310, j 610, z 101, w $\overline{2}01$, p 113.
2. Hurlbut and Weichel, *Am. Min.*, **31**, 507 (1946), by Weissenberg method on material from Brazil.
3. Growth accessories on the principal faces are described by Pough and Henderson (1945).
4. Average of values 2.985 ± 0.005 and 2.980 ± 0.005 reported on crystals from Conselheira Pena and Palermo, respectively, by Frondel and Lindberg (1948).
5. Spectrographic amounts of Ca, Si, Cu, Mg, Mn, V, Ba, and other elements have been reported by Harrison in Frondel and Lindberg (1948).
6. Lindberg in Frondel and Lindberg (1948) on the analyzed samples.
7. Lindberg analysis in Frondel and Lindberg (1948).
8. Henderson analysis in Pough and Henderson (1945).
9. The occurrence and association has been described by Pecora and Fahey, *Am. Min.*, **34**, 83 (1949).

41.5.11   **G R I P H I T E.** *Headden (Am. J. Sc., **41**, 415, 1891).*

Massive, sometimes reniform.

Structure cell.[1] Isometric, body centered, with a_0 12.26 kX. Cell contents $(Na,Al,Ca,Fe)_{24}Mn_{16}(PO_4)_{20}(OH)_{16}$. The crystal structure is similar to that of garnet but probably of lower symmetry. Sometimes metamict but recrystallizes on heating to about 500°.

P h y s. No cleavage. Fracture uneven to conchoidal. Brittle. H. $5\frac{1}{2}$. G. 3.40;[2] 3.40 (calc.). Luster resinous to vitreous. Color dark brown to brownish black. Translucent.

O p t.[3] In transmitted light, yellowish brown to brown. Isotropic, with variable index of refraction, 1.63–1.66.

C h e m. A basic phosphate of manganese, calcium, aluminum, sodium, and iron. The formula is of the garnet type, $A_2B_3(ZO_4)_3$, and may be written $(Na,Ca,Fe,Al)_3Mn_2(PO_4)_{2.5}(OH,F)_2$.

Anal.

	1	2	3
Na_2O	5.52	5.25	3.47
CaO	7.47	6.72	5.98
FeO	4.00	1.97	10.67
MnO	29.64	29.13	18.12
Al_2O_3	10.13	8.74	9.44
Fe_2O_3		2.36	6.95
P_2O_5	38.52	39.68	39.15
Cl	0.11	0.25	0.12
F	tr.	2.35	3.03
H_2O	4.29	3.67	2.62
Rem.	0.61	0.82	0.65
Total	100.29	99.89	100.20
G.	3.40		3.43

1. Riverton Lode, near Harney City, South Dakota.[4] Rem. is Li_2O tr., K_2O 0.30, MgO 0.15, insol. 0.16. Average of two complete and three partial analyses. Probably

contains some CO_2. 2. Near Rapid City, South Dakota.[5] Rem. is Li_2O 0.13, K_2O tr., MgO tr., SiO_2 0.43, CO_2 0.26. O = F = 1.05 deducted from total. 3. Mt. Ida, Australia.[6] Rem. is K_2O 0.18, Li_2O 0.47.

Tests. Easily fusible. Readily soluble in acids.

Occur. Found in masses ranging upwards of 50 pounds in weight in granite pegmatite at the Riverton Lode, near Harney City, Pennington County, South Dakota. The masses are somewhat darker in color toward the surface owing to partial oxidation. A mineral originally referred to triplite but doubtless identical with griphite was later reported from a pegmatite near Rapid City, South Dakota (analysis 2). From Mt. Ida, Northern Territory, Australia.

Name. From γρίφος, *puzzle*, in allusion to the chemical composition.

Ref.
1. McConnell, *Am. Min.*, **27**, 452 (1942), by x-ray powder method on type material of Headden.
2. Headden (1891) verified by McConnell (1942).
3. Larsen (81, 1921).
4. Headden (1891).
5. Eakins, *U S. Geol. Sur., Bull. 60*, 135 (1890).
6. Dwiggins and Yelmgren analysis in Jaffe, *Am. Min.*, **31**, 404 (1946).

41.5.12   Arseniopleite. *Igelström (Bull. soc. min.,* **11**, 39, 1888; *Jb. Min.,* II, 117, 1888).

Hexagonal—R (?). Massive, granular. Cleavage rhombohedral; fracture conchoidal. H. 3–4. Color brownish red to cherry red. Streak yellowish brown. Opaque in the mass.

Opt.[1] In transmitted light, light apricot-orange. Non-pleochroic.

nO 1.794 ± 0.003   Uniaxial positive (+).
nE 1.803 ± 0.003   Sometimes biaxial with small $2V$.

Chem. A basic arsenate of divalent and trivalent manganese of uncertain formula, perhaps $(Mn'',Ca,Pb,Mg)_9(Mn''',Fe)_2(AsO_4)_6(OH)_6$. The presence of both valences of Mn and their relative amounts, however, is not clearly established.

CaO	MgO	PbO	MnO *	Fe_2O_3	Sb_2O_5	As_2O_5	H_2O	Cl	Total
8.11	3.10	4.48	28.25	3.68	tr.	44.98	5.67	tr.	98.27

* Regarded as MnO 21.25, Mn_2O_3 7.80.

Tests. B.B. fuses easily to a black non-magnetic bead. In C.T. affords water. Easily soluble in dilute HCl or HNO_3.

Occur. As thin veins or nodules with rhodonite in crystallized limestone at the Sjö mine, Grythytte parish, Örebro, Sweden.

Name. From the Latin *arsenicum* and Greek πλείων, *more*, because it adds to the number of related minerals already described.

Ref.
1. Larsen (41, 1921); also Bertrand in Igelström (1888) who notes the uniaxial character.

ATTACOLITE. *Blomstrand* (*Ak. Stockholm, Öfv.*, **25**, 201, 1868).
Massive, indistinctly crystalline. Color pale red. G. 3.09. Analysis gave (average of two after deduction of 8.6 per cent SiO_2): Na_2O 0.45, CaO 13.19, MgO 0.33, MnO 8.02, Fe_2O_3 3.98, Al_2O_3 29.75, P_2O_5 36.06, H_2O 6.90, total 98.68. This corresponds roughly to $(Ca,Mn)_3Al_4(PO_4)_4(OH)_6$. B.B. fuses easily, with intumescence. Incompletely decomposed by acids. Found rarely with berlinite and lazulite at the iron mine of Westanå in Kristianstad, Sweden. Evidence of species validity is lacking. A specimen of uncertain authenticity was biaxial positive[1] with nX 1.655, nY 1.664, nZ 1.675, $2V$ about $40°$, $r < v$, medium.

Ref.

1. Mrose, priv. comm. (1948).

CIRROLITE. Kirrolith *Blomstrand* (*Ak. Stockholm, Öfv.*, **25**, 202, 1868). Cirrolite *Dana* (799, 1892). Cirrolith.
Compact, without a trace of cleavage. H. 5–6. G. 3.08. Color pale yellow. Analysis gave (after deducting 4.60 insol.) the following figures, which approximate to the formula $Ca_3Al_2(PO_4)_3(OH)_3$:

CaO	MgO	PbO	MnO	FeO	Al_2O_3	P_2O_5	H_2O	Total
29.37	0.21	0.11	2.24	0.91	20.54	41.17	5.06	99.61

B.B. very easily fusible to a white enamel. Slowly decomposed by HCl. Found at the iron mine at Westanå, Scania, Sweden. Evidence of species validity is lacking.

PLEURASITE *Igelström* (*Geol. För. Förh.*, **11**, 391, 1889; *Jb. Min.*, I, 253, 1890). Pleurastite.
In opaque, bluish black masses. Streak black with a faint tinge of red. Fracture conchoidal. H. 4. Luster submetallic. Very weakly magnetic. Not analyzed: contains As, a little Sb, Mn, and divalent Fe. On charcoal fuses to a black magnetic bead. In C.T. affords much water. Easily soluble in acids. Found as bands a centimeter or less in thickness adjacent to arseniopleite (and hence named from πλευρά, side) at the Sjö mine, Örebro, Sweden. Needs verification.

SJOGRUVITE. *Igelström* (*Geol. För. Förh.*, **14**, 309, 1892; *Zs. Kr.*, **22**, 471, 1893).
Massive; cleavable. Color and streak yellow. Darkens in color on exposure to air. Magnetic. Thin splinters are blood-red by transmitted light. A hydrated arsenate of manganese and ferric iron with minor lead and calcium. Analysis gave:

CaO	PbO	MnO	Fe_2O_3	As_2O_5	H_2O	Total
3.61	1.74	27.26	11.29	49.46	6.81	100.17

Easily soluble in cold HCl without evolution of chlorine. B.B. fuses to a black bead. Found with jacobsite at the Sjö mine, Örebro, Sweden. Perhaps a ferrian variety of arseniopleite. Needs verification.

TYPE 6. $A_2(XO_4)Z_q$

41.6.1 **W A G N E R I T E** [$Mg_2(PO_4)F$]. Phosphorsaurer Talk, Wagnerit *Fuchs* (*J. Chemie u. Phys.*, **33**, 269, 1821). Magnésie phosphatée *Fr.* Pleuroklas *Breithaupt* (50, 193, 1823). Kjerulfin *von Kobell* (*J. pr. Chem.*, **7**, 272, 1873). Crifiolite *Scacchi* (*Acc. Napoli*, **1**, no. 5, 1886). Cryphiolite. Kryphiolith.

C r y s t.[1] Monoclinic; prismatic—$2/m$.

$a:b:c = 0.9569:1:0.7527$; $\beta\ 108°07'$; $p_0:q_0:r_0 = 0.7866:0.7154:1$

$r_2:p_2:q_2 = 1.3978:1.0996:1$; $\mu\ 71°53'$; $p_0'\ 0.8276$, $q_0'\ 0.7527$, $x_0'\ 0.3272$

Forms:[2]

		ϕ	ρ	ϕ_2	$\rho_2 = B$	C	A
c	001	90°00′	18°07′	71°53′	90°00′	.....	71°53′
b	010	0 00	90 00	.....	0 00	90°00′	90 00
a	100	90 00	90 00	0 00	90 00	71 53	.....
M	120	28 48	90 00	0 00	28 48	81 23	61 12
m	110	47 43	90 00	0 00	47 43	76 42	42 17
h	320	58 46	90 00	0 00	58 46	74 35	31 14
t	012	41 00½	26 30½	71 53	70 19	19 41	72 58½
r	011	23 29½	39 22½	71 53	54 25½	35 34½	75 21
e	021	12 16	57 00½	71 53	34 57	55 03	79 44½
w	$\bar{1}$01	−90 00	26 35	116 35	90 00	44 42	116 35
s	111	56 54	54 02½	40 53½	63 46	39 44½	47 18
n	$\bar{1}$12	−12 57½	21 07	94 57	69 27	30 31	94 38
z	$\bar{1}$11	−33 37	42 06½	116 35	56 03½	53 52	111 47½
i	$\bar{1}$22	− 6 34	37 09	94 57	53 08	42 36	93 57½

Structure Cell.[3] Space group $P2_1/a$. a_0 11.90 kX, b_0 12.51, c_0 9.63; β 108°07′; $a_0:b_0:c_0 = 0.951:1:0.770$. Cell contents $Mg_{32}(PO_4)_{16}F_{16}$.

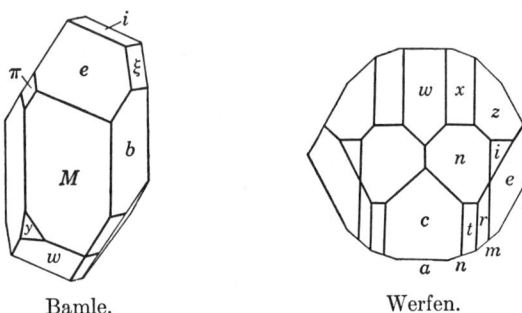

Bamle.   Werfen.

Habit. Prismatic [001], the prismatic planes vertically striated or vicinally rounded; the crystals sometimes are large and coarse. Rarely tabular {100}. Also massive.

Phys. Cleavage {100} and {120} imperfect, {001} in traces. Fracture uneven or subconchoidal to splintery. H. 5–5½. G. 3.15, slightly less in opaque, altered (?) or calcian material; 3.15 (calc.). Luster vitreous to slightly resinous. Color yellow, of different shades; often grayish, also flesh-red, greenish. Translucent; nearly opaque in altered material.

Opt.[4] In transmitted light, colorless.

Orientation		n(Na)	
X		1.5678	Biaxial positive (+).
Y	b	1.5719	$2V$ 28°24½′, $2E$ 45°22½′.
$Z \wedge c$	−21½°	1.5824	$r > v$, weak, inclined.

Chem. Magnesium fluophosphate, $Mg_2(PO_4)F$. Ca and Fe″ substitute in part for Mg, with Ca:Mg = 1:13.3 in analysis 3, and Fe″:

WAGNERITE 847

(Mg,Ca) = 1:27.5 in analysis 2. Highly calcian material has been reported and is due perhaps to alteration (analysis 5).

Anal.[5]

	1	2	3	4	5
CaO		2.53	4.81		13.45
MgO	49.59	47.56	46.01	48.33	37.57
FeO		4.32		0.95	0.58
P_2O_5	43.65	40.18	42.35	43.43	41.42
F	11.68	9.51	5.06	11.48	10.02
Rem.			4.23	0.70	1.55
	104.92	104.10	102.46	104.89	104.59
F = O	4.92	4.00	2.13	4.83	4.22
Total	100.00	100.10	100.33	100.06	100.37
G.	3.15	3.153	3.10		

1. $Mg_2(PO_4)F$. 2. Werfen, Germany.[6] 3. Bamle, Norway (*kjerulfine*).[8] Rem. is $(Fe,Al)_2O_3$ 0.65, $(Na,K)_2O$ 1.54, insol. 2.04. 4. Werfen.[7] Rem. is MnO. Fresh crystals. 5. Werfen.[7] Altered crystals. Rem. is MnO 0.54, SiO_2 0.62, CO_2 0.39.

Tests. B.B. fuses with difficulty to a greenish gray glass. Soluble in acids.

Occur. Found in quartz veins traversing clay slate at Höllgraben near Werfen and Radelgraben near Bischofshofen in Salzburg, Austria, associated with ferroan magnesite, lazulite, and chlorite. Also at Havredal, Oedegården and Porsgrund, near Bamle, Telemark, Norway (*kjerulfine*).[9] With apatite in lava of 1872 on Vesuvius, Italy (*cryphiolite*).[10] A mineral perhaps identical with wagnerite has been reported [11] from pegmatite near Mangualde, Beira province, Portugal.

Alter. To apatite.[12]

Artif.[13] Wagnerite and the corresponding calcium compound, together with the chloride and bromide analogues thereof, have been obtained by fusion of the metal halide with ammonium phosphate or the corresponding metal phosphate. The reported barium and strontium analogues of wagnerite have not been verified.[14]

Name. After F. M. von Wagner (1768–1851), mining official in Munich.

Ref.

1. Goldschmidt (**9**, 63, 1923), corresponding to the unit and orientation of the structure cell; angles of Brooke and Miller (489, 1852). Dana (775, 1892), Hintze (**1** [4A], 688, 1931), and Hegemann and Steinmetz, *Cbl. Min.*, **45** (1927), use a unit with b halved; the angles of the latter authors afford $a:b:c = 0.9341:1:0.7595$ in the present unit, β 108°17′.
2. Goldschmidt (1923), Hegemann and Steinmetz (1927). Rare and uncertain:

γ 140	λ 230	l 210	π 101	q $\bar{3}01$	v 122	x $\bar{2}12$
ζ 250	9.11.0	k 054	Y 201	o 221	d 314	ψ $\bar{2}61$
ν 470	μ 890	f 032	y $\bar{2}01$	u $\bar{2}21$	ξ $\bar{1}62$	

3. Kraus and Mussgnug, *Naturwiss.*, **26**, 801 (1938). Richmond, *Am. Min.*, **25**, 470 (1940), gives comparable data.
4. Hegemann and Steinmetz (1927). Brögger, *Zs. Kr.*, **3**, 476 (1879), found so-called kjerulfine from Bamle to be optically negative (−) with $2V$ 37°49′ (Na).

5. For additional analyses see Hintze (1 [4A], 694, 1931).
6. Gonyer analysis in Richmond (1940).
7. Hegemann and Steinmetz (1927).
8. Frederici analysis in Bauer, *Jb. Min.*, II, 77 (1880)
9. Shown to be identical with wagnerite by Pisani, *Bull. soc. min.*, **2**, 43 (1879), Bauer (1880), Brögger (1879), and others.
10. Shown to be identical with wagnerite by Zambonini (200, 1910).
11. de Jesus, *Com. Serv. Geol. Portugal*, **19**, 65 (1936).
12. Pisani (1879); see also Hegemann and Steinmetz (1927).
13. Mellor (4, 388, 1923).
14. Klement and Dihn, *Zs. anorg. Chem.*, **240**, 31 (1938).

SPODIOSITE. Tiberg (*Geol. För. Förh.*, **1**, 84, 1872).

Orthorhombic (?) As crystals flattened {010} and elongated [001]. Principal forms:[1] a{100}, b{010}, c{001}, p{111}, m{110}; (110) ∧ (010) 48°11.4′, (001) ∧ (111) 67°10.2′, affording $a:b:c$ = 0.894:1:1.584. Cleavage {010} distinct, {001} indistinct. H. 5. G. 2.94. Color ash-gray inclining to brown. Streak white. Luster dull porcelain-like, but vitreous.

O p t.[2] The principal vibration directions are not perpendicular to the cleavages. Fragments lying on one cleavage give an angle of Z' to the trace of the other cleavage of about 35°.

	n	
X	1.663	Biaxial positive (+).
Y	1.674	$2V$ 69° (calc.).
Z	1.699	$r > v$, rather strong.

C h e m. A calcium fluo-phosphate. The formula is uncertain, perhaps $Ca_2(PO_4)F$ analogous to wagnerite.[3] The two reported analyses are unsatisfactory.

	CaO	MgO	MnO	Fe_2O_3	P_2O_5	SiO_2	CO_2	F	H_2O	Rem.	Total
1.	49.81	2.27	0.55	1.24	32.20	1.15	3.90	[4.71]	2.70	1.47	[100.00]
2.	45.84	8.56		2.38	29.62	8.74		2.94	3.76		101.84

1. Nyttsta Kran mine, Sweden.[4] Rem. is Al_2O_3 1.11, As_2O_5 0.24, Cl 0.12. 2. Nordmark, Sweden.[5] Fe_2O_3 includes Al_2O_3.

Tests. B.B. thinnest splinters fuse to a white enamel. Soluble in acids.

O c c u r. Found with serpentine, chondrodite, magnetite, and calcite at the Nyttsta Kran mine, north of Filipstad in Wermland, Sweden; also at Nordmark, Sweden. Named from σποδιος, *ash grey*.

Spodiosite is ill-defined and of uncertain validity as a species. Some specimens at least have been found [6] to be pseudomorphs of apatite after an unknown mineral. An artificial compound [7] with the composition $Ca_2(PO_4)Cl$ has properties close to those attributed to spodiosite.

Ref.

1. Nordenskiöld, *Geol. För. Förh*, **15**, 460 (1893), on Nordmark crystals; see also Tiberg (1872) for approximate measurements (mb = 48°) on Nyttsta Kran mine crystals.
2. Larsen (136, 1921), on material from the Ladn mine, Freystad (?), Wermland, Sweden, according to his original notes (unpublished).
3. Sjögren, *Geol. För. Förh.*, **7**, 666 (1885).
4. Lundström analysis in Tiberg (1872)
5. Nordenskiöld (1893).
6. Strunz, *Naturwiss.*, **27**, 423 (1939).
7. Cameron and McCaughey, *J. Phys. Chem.*, **15**, 463 (1911). Orthorhombic tablets with mb ~49°; cleavage {001}; G. 3.04; $nX = a$ = 1.649, $nY = b$ = 1.665, $nZ = c$ = 1.670.

TRIPLITE

41.6.2   T R I P L I T E [(Mn'',Fe'',Mg,Ca)$_2$(PO$_4$)(F,OH)]. Phosphate natif de fer melangé de manganèse [from Limoges] *Vauquelin* (*J. mines*, **11**, 295, 1802; *Ann. Chem.*, **41**, 242, 1802). Eisenpecherz *Werner* (1808). Manganèse phosphaté *Lucas* (**1**, 169, 1806). Phosphormangan *Karsten* (72, 1808). Manganèse phosphaté ferrifère *Haüy* (1809). Triplit *Hausmann* (1079, 1813). Eisenapatit *Fuchs* (*J. pr. Chem.*, **18**, 499, 1839). Zwiselite *Breithaupt* (**2**, 299, 1841). Phosphate of Iron and Manganese. Zwieselite *Glocker* (244, 1847). Talktriplit *L. J. Igelström* (*Ak. Stockholm, Öfv.*, **39**, 86, 1882). Metatriplite pt. (?) *de Jesus* (*Com. Serv. Geol. Portugal*, **19**, 65, 1936).

C r y s t.[1] Monoclinic; prismatic—$2/m$ (?).

$a:b:c = 1.836:1:1.531;$ $\beta\ 105°53';$ $p_0:q_0:r_0 = 0.834:1.472:1$

$r_2:p_2:q_2 = 0.679:0.566:1;$ $\mu\ 74°07';$ $p_0'\ 0.867,\ q_0'\ 1.531,\ x_0'\ 0.285$

Forms:[2]

	ϕ	ρ	ϕ_2	$\rho_2 = B$	C	A
c 001	90°00'	15°53'	74°07'	90°00'	.....	74°07'
a 100	90 00	90 00	0 00	90 00	74°07'	.....
m 110	29 31½	90 00	0 00	29 31½	82 15	60 28½
e 011	10 32½	57 17½	74 07	34 11	55 49	81 08½
π 101	90 00	49 02½	40 57½	90 00	33 09½	40 57½
w $\bar{1}$01	−90 00	30 12	120 12	90 00	46 05	120 12
v 112	43 11	46 23½	54 18	58 08	36 58	60 17½
u $\bar{2}$11	−43 25½	64 37½	145 23½	48 59½	75 59	128 23½

Structure cell. Space group $I2/m$. $a_0\ 11.90\ kX$, $b_0\ 6.48$, $c_0\ 9.92$; $\beta\ 105°53'$; $a_0:b_0:c_0 = 1.836:1:1.531$ (Mica Lode, Colorado [3]); $a_0\ 12.03\ kX$, $b_0\ 6.46$, $c_0\ 10.03$; $\beta\ 105°42'$; $a_0:b_0:c_0 = 1.862:1:1.553$ (Bagdad, Arizona [4]). Cell contents (Fe,Mn)$_{16}$(PO$_4$)$_8$F$_8$.

Habit. Crystals rough and incompletely developed. Usually massive.

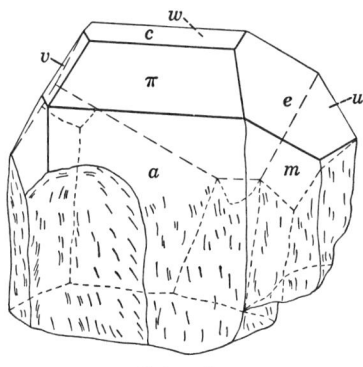

Colorado.

P h y s. Cleavage {001} good, {010} fair, {100} poor; the cleavages may not be seen easily in altered material.[5] Fracture uneven to subconchoidal. H. 5–5½, decreasing with alteration. G. variable due to isomorphous substitution and to alteration, usually 3.5 to 3.9; 3.94 (calc.).[6]

Luster vitreous to resinous. Color usually dark brown or chestnut-brown, also reddish brown to flesh-red and salmon-pink in highly manganoan material and becoming brownish black to black on alteration. Streak white to brown. Subtranslucent to opaque.

O p t.[7] In transmitted light, pale brownish yellow to dark reddish brown.

ORIENTATION	Mica Lode, Colorado [9]	7U7 Ranch, Arizona (anal. 8) [10]	Mt. Loma, Arizona (anal. 4) [10]	Varuträsk (anal. 6) [11]
nX	1.643	1.651	1.662	1.673
nY b	1.647	1.653	1.673	1.681
nZ	1.668	1.665	1.684	1.691
$Z \wedge c$	$-41°$	(see ref. 8)		
$2V$	25°	28°	88°(−)	76°

ORIENTATION	Finland (anal. 9) [12]	Zwiesel, Bavaria (anal. 10) [12]	
nX	1.684	1.696	Biaxial positive (+).
nY b	1.693	1.704	$r > v$, moderate to strong.
nZ	1.703	1.713	Pleochroism usually distinct in shades of yellow brown or reddish brown with absorption $X > Z, Y$.
$Z \wedge c$			
$2V$	87½° (calc.)	87° (calc.)	

The indices of refraction decrease with increasing content of Mg and of Mn.

C h e m. A fluo-phosphate chiefly of divalent iron and manganese, with smaller amounts of calcium and magnesium, $(Mn'',Fe'',Mg,Ca)_2(PO_4)(F,OH)$. (OH) substitutes in small part for F, but the limit is uncertain because of the probable presence of non-essential water due to alteration. Fe and Mn substitute mutually over a range of at least Fe:Mn = 1:35 (analysis 2) to Fe:Mn = 2.2:1 (analysis 10). Mg substitutes for (Fe,Mn,Ca) up to at least Mg:(Fe,Mn,Ca) = 1:2.36 (analysis 8), and Ca substitutes for (Fe,Mn) up to Ca:(Fe,Mn) = 1:8.7 (analysis 6). Triplite is not isostructural with triploidite-wolfeite, $(Fe,Mn)_2(PO_4)(OH)$.

Anal.[13]

	1	2	3	4	5
CaO		2.86	3.18	2.17	2.65
MgO		1.21	0.58	0.31	
FeO	31.97	1.68	4.95	6.68	21.90
MnO	31.56	57.63	52.40	53.77	37.35
P$_2$O$_5$	31.58	31.84	32.81	32.20	33.09
F	8.45	7.77	9.09	7.58	7.83
H$_2$O			0.35		0.45
Rem.			1.17	0.40	0.22
	103.56	102.99	104.53	103.11	103.49
F = O	3.56	3.27	3.83	3.19	3.30
Total	100.00	99.72	100.70	99.92	100.19
G.		3.79	3.58	3.84	3.838

1. $(Fe'',Mn'')_2(PO_4)F$ with Fe:Mn = 1:1. 2. Aurum, Nevada.[14] 3. Chatham, Connecticut.[15] Rem. is insol. 4. Mt. Loma, Bagdad, Arizona.[16] Rem. is Fe$_2$O$_3$. 5. Lemnäs, Finland.[17] Rem. is Na$_2$O 0.22, K$_2$O tr.; H$_2$O includes H$_2$O+ 0.38.

	6	7	8	9	10	11
CaO	4.68	3.48	2.48	0.50	1.69	14.91
MgO	tr.	tr.	11.87		0.80	17.42
FeO	19.96	23.54	11.68	38.52	41.96	16.12
MnO	32.07	34.84	34.55	24.20	18.40	14.86
P_2O_5	30.63	31.50	33.32	32.65	32.85	32.82
F	6.72	6.41	8.02	6.02	7.21	n.d.
H_2O	0.75	1.58	0.75	0.62		
Rem.	7.82	0.88	0.52			
	102.63	102.23	103.19	102.51	102.91	96.13
F = O	2.83	2.69	3.38	2.53	3.03	
Total	99.80	99.54	99.81	99.98	99.88	
G.	3.84	3.87		3.927	3.97	

6. Varuträsk, Sweden.[18] Rem. is Fe_2O_3 0.78, Li_2O 0.10, K_2O 0.33, Na_2O 0.39, insol. 6.22; H_2O includes H_2O+ 0.48. 7. La Rioja Province, Argentina.[19] Rem. is insol. 0.28, Na_2O 0.27, K_2O 0.33, Cl tr. 8. 7U7 Ranch, Bagdad, Arizona.[16] Rem. is Na_2O 0.52. 9. Kimito, Finland.[19] 10. Zwiesel, Bavaria.[19] 11. Horrsjöberg, Sweden (*talktriplite*).[20]

Var. *Magnesian.* Talktriplite *Igelström* (*Ak. Stockholm, Öfv.*, **39**, 86, 1882; *Bull. soc. min.*, **5**, 301, 1883). Contains Mg (and usually also Ca) in substitution for (Fe,Mn), with Ca:Mg:Fe:Mn = 1:6.6:3.7:11 in analysis 8 and Ca:Mg:Fe:Mn = 1:1.6:0.79:0.84 in analysis 11; the species identity of the latter material, however, is not certain. Increase in Mg (and Ca) content is accompanied by a decrease in indices of refraction and unit cell dimensions.

Tests. Easily fusible to a steel-gray, magnetic globule. Soluble in acids.

Occur. A primary mineral in granite-pegmatites of the complex, phosphate-rich type. Associated with triploidite-wolfeite, triphylite-lithiophilite, apatite, tourmaline, etc. Also in a high temperature vein (?) with wolframite, scheelite, and sulfides (Aurum, Nevada). Not all the reported occurrences can be mentioned here. Originally found at Limoges near Chanteloube, Haute-Vienne, France; also at La Vilate and elsewhere in this region.[21] At numerous localities in Bavaria,[22] notably Hagendorf, the Kreuzberg, Pleystein, and Hühnerkobel near Rabenstein (*zwieselite*); in Moravia [23] at Cyrillhof and Wien, and in the neighborhood of Marienbad; at Schlaggenwald, Bohemia. In Finland at Sukula and at Lemnäs, Kimito parish. In Sweden at Horrsjöberg in Wermland (*talktriplite*) with lazulite, apatite, rutile, kyanite; at Skrumpetorp and at Lake Lilla Elysjö, Östergötland; very rarely at Varuträsk. In Africa in the Karibib area, South West Africa, and at Lomagundi, Southern Rhodesia. At Mangualde, Beira province, Portugal. In Argentina in the Sierra de Zapata, Catamarca province, and in Dept. Calamuchita, Cordoba. In the United States, triplite occurs at Stoneham, Hebron, and Mt. Apatite near Auburn, Maine; at Chatham, Branchville, and the Strickland quarry, Portland, Connecticut. At Amelia, Virginia. At the 7U7 Ranch and Mt. Loma near Bagdad in the Eureka district, Yavapai County, Arizona. In California at Pala,

San Diego County, and with hübnerite at Camp Signal, near Goffs. Near Aurum, Reagan district, White Pine County, Nevada. In Colorado in the Mica Lode and School Section pegmatites, Fremont County, and in Dead Man's Gulch, near Colorado Springs, El Paso County.

A r t i f. An artificial compound of the composition has been reported.[25]

A l t e r. Triplite may alter under hydrothermal conditions to triploidite-wolfeite or alluaudite, and on weathering to vivianite or dufrenite-like minerals or to wad.[24] Triplite, unlike triphylite-lithiophilite, apparently does not alter to heterosite-purpurite. The mineral is commonly found more or less altered by superficial oxidation and hydration.

N a m e. From τριπλόος, *three-fold*, probably in allusion to its three cleavages. Zwieselite from the locality at Rabenstein, near Zwiesel, Bavaria.

Ref.

1. Unit and orientation of Pehrman, *Acta Ac. Aboensis, Math. Physica,* **12,** no. 6 (1939), corresponding to the structure cell; elements of the structure cell of Wolfe and Heinrich, *Am. Min.,* **32,** 518 (1947). Triplite is closely related dimensionally to wagnerite.
2. Pehrman (1939); Wolfe and Heinrich (1947); Laubmann and Steinmetz, *Zs. Kr.,* **55,** 559 (1920). Form letters of Pehrman. Less common: h 310, i $\bar{1}$12, ϵ $\bar{1}$21, q $\bar{3}$01, $\bar{3}$23.
3. Wolfe in Wolfe and Heinrich (1947) by Weissenberg method on unanalyzed crystals; space group of Wolfe.
4. Richmond, *Am. Min.,* **25,** 468 (1940), by Weissenberg method on material of analysis 4.
5. Pehrman (1939) cites cleavages on {$\bar{3}$01}, {001}, and {010} but not on {100} in Lemnäs material.
6. Calculated from cell dimensions of the analyzed Bagdad, Arizona, material; meas. G. = 3.84.
7. For additional optical data see Otto, *Min. Mitt.,* **47,** 98 (1935); Wolfe and Heinrich (1947); Hess and Hunt, *Am. J Sc.,* **36,** 51 (1913); Henderson, *Am. Min.,* **18,** 104 (1933); Shannon, *Proc. U S Nat. Mus.,* **58,** 444 (1920); Pehrman (1939).
8. The optical orientation is given as $X = b$ and $Z \wedge a = 22°$ (7U7 Ranch) or $Y = b$ and $Z \wedge a = 42°$ (Mt. Loma), the latter material being optically negative. Des Cloizeaux (180, 1867) gave $Y = b$ and $Z \wedge c = 42°10'$ (red), 41°53' (yellow) on Chanteloube material; Pehrman (1939) gives $X \wedge c = 44°$.
9. Wolfe and Heinrich (1947).
10. Hurlbut, *Am. Min* , **21,** 656 (1936).
11. Mason, *Geol För. Förh.,* **63,** 285 (1941).
12. Otto (1935).
13. For additional analyses see Hintze (1 [4A], 700, 1931) and Otto (1939).
14. Hess and Hunt (1913).
15. Shannon (1920).
16. Gonyer analysis in Hurlbut (1936).
17. Pehrman (1939).
18. Mason (1941).
19. Otto (1939).
20. Igelström (1882).
21. Lacroix (**4,** 431, 1910).
22. See Laubmann and Steinmetz, *Zs Kr.,* **55,** 523 (1920); Müllbauer, *Zs. Kr.,* **61,** 318 (1925); and Mason, *Geol. För. Förh.,* **64,** 335 (1942).
23. Kovář and Slavík, *Verh. geol. Reichsanst. Wien,* **50,** 347 (1900); Sellner, *Zs. Kr.,* **59,** 504 (1924).
24. An analysis of altered triplite is given by de Jesus (1936); see also Kovář and Slavík (1900).
25. Deville and Caron, *C.R.,* **47,** 985 (1858); *Ann. chim. phys* , **67,** 454 (1863).

41.6.3 TRIPLOIDITE GROUP

MONOCLINIC; PRISMATIC—2/m

	a_0	b_0	c_0	β
Triploidite, $(Mn'',Fe'')_2(PO_4)(OH)$	12.26 Å	13.38	9.90	108°04'
Wolfeite, $(Fe'',Mn'')_2(PO_4)(OH)$	12.12	13.16	9.73	108 18
Sarkinite, $Mn_2(AsO_4)(OH)$	12.65	13.51	10.15	108 44

Triploidite and wolfeite form a probably complete series by substitution of Mn'' and Fe''. Sarkinite is isostructural with these species, but evidence is lacking of a series between them. F does not substitute in significant amounts for (OH), and it may be noted that triplite, $(Fe,Mn)_2(PO_4)F$, is not isostructural with wolfeite.

41.6.3.1 **TRIPLOIDITE** [$(Mn'',Fe'')_2(PO_4)(OH)$]. *Brush* and *Dana* (*Am. J. Sc.*, **16**, 42, 1878).
41.6.3.2 **WOLFEITE** [$(Fe'',Mn'')_2(PO_4)(OH)$]. *Frondel* (*Am. Min.*, **34**, 692, 1949).

Cryst.[1] Monoclinic; prismatic—2/m.

$a:b:c = 0.9286:1:0.7463$; β 108°14'; $p_0:q_0:r_0 = 0.8037:0.7088:1$

$r_2:p_2:q_2 = 1.4108:1.1338:1$; μ 71°46'; p_0' 0.8462, q_0' 0.7463, x_0' 0.3294

Forms:[2]

		ϕ	ρ	ϕ_2	$\rho_2 = B$	C	A
c	001	90°00'	18°14'	71°46'	90°00'	.....	71°46'
b	010	0 00	90 00	....	0 00	90°00'	90 00
a	100	90 00	90 00	0 00	90 00	71 46	.....
n	120	29 33	90 00	0 00	29 33	81 07½	60 27
m	110	48 35½	90 00	0 00	48 35½	76 26	13 34
e	021	12 26½	56 48½	71 46	35 12	54 48	79 36½
ζ	203	90 00	41 47	48 13	90 00	23 33	48 13
p	$\bar{2}21$	−42 24	63 40½	143 44	48 33½	76 35	127 11

Structure cell. Space group $P2_1/a$.

	Triploidite, Branchville[3]	Wolfeite, Palermo[4]	Wolfeite, Hagendorf[5]
Fe:Mn	1:3.30	3.39:1	>3:1
a_0	12.26 Å	12.20 Å	12.12 Å
b_0	13.38	13.17	13.16
c_0	9.90	9.79	9.73
β	108°04'	108°00'	108°18'
$a_0:b_0:c_0$	0.916:1:0.740	0.926:1:0.743	0.921:1:0.739

Cell contents $(Mn'',Fe'')_{32}(PO_4)_{16}(OH)_{16}$.

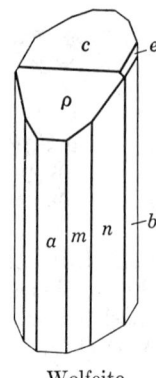

Wolfeite.
Hagendorf.

Habit. Crystals prismatic [001], the prism zone striated vertically. Commonly as parallel-fibrous to columnar aggregates; also divergent or confusedly fibrous; rarely granular.

Phys. Cleavage {010} good, {120} fair, {010} poor, {110} very poor.[4] Fracture uneven to subconchoidal. H. $4\frac{1}{2}$–5. G. 3.66 (est. for pure Mn-end of series) to 3.83 (est. for pure Fe-end of series).[4] Luster vitreous to greasy adamantine. Color pinkish, wine-yellow to yellowish brown in material high in Mn grading toward reddish brown and dark clove-brown in material high in Fe; rarely green. Streak nearly white. Transparent to translucent.

Opt. In transmitted light, pale pink to pale brown.

ORIEN-TATION	Triploidite, Branchville [6]	Wolfeite, Palermo [4]	Wolfeite, Hagendorf [5]	
nX b	1.725 ± 0.003	1.741 ± 0.003	1.748 ± 0.003	Biaxial positive (+).
nY	1.723 ± 0.003	1.742 ± 0.003	1.749 ± 0.003	$2V$ moderate.
nZ	1.730 ± 0.003	1.746 ± 0.003	1.753 ± 0.003	$r > v$, very strong; dispersion of bisectrices marked.
$Z \wedge c$	$-4°$			
Fe:Mn	1:3.30	3.39:1	>3:1	

Faintly pleochroic in thick grains with absorption $Z > X, Y$.

Chem. A basic phosphate of divalent manganese and iron, $(Mn'',Fe'')_2(PO_4)(OH)$. Mn and Fe can mutually substitute, and a series exists from Fe:Mn = 1:3.30 (anal. 2) to Fe:Mn = 3.39:1 (anal. 7) at least and probably to the pure iron end-member. The names triploidite and wolfeite are applied to those parts of the series with Mn > Fe and Fe > Mn, respectively. Small amounts of Mg and Ca substitute for (Mn'',Fe''), with Mg:(Fe'',Mn'') = 1:6.92 in analysis 4 and Ca:(Fe'',Mn'') = 1:22.9 in analysis 5. The Fe_2O_3 reported in some analyses is due to partial oxidation of FeO. Triploidite-wolfeite is isostructural with sarkinite,[7] $Mn_2(AsO_4)(OH)$.

Anal.

	1	2	3	4	5	6	7	8
CaO		0.33	(~0.90)	0.56	2.00	1.27	0.19	
MgO				4.74		0.40	2.28	
MnO	63.95	48.45	43.46	29.85	21.73	17.92	13.12	
FeO		14.88	18.87	24.31	36.44	33.37	44.44	64.24
Fe_2O_3					4.26	0.22	7.78	0.70
P_2O_5	31.99	32.11	32.62	30.89	34.20	32.44	32.90	31.73
H_2O	4.06	4.08	4.14	4.20	4.50	4.48	3.78	4.03
F		none		tr.	0.09	0.88	0.51	
Rem.			1.09	0.94	1.27	2.47	2.19	
Total	100.00	99.85	100.18	99.75	100.45	101.01	100.11	100.00
G.		3.697			3.78		3.79	
Fe:Mn		1:3.30	1:2.39	1:1.07	1.66:1	2.22:1	3.39:1	

1. $Mn_2''(PO_4)(OH)$.   2. Triploidite, Branchville, Connecticut.[8]   3. Triploidite, Branchville.[8] Rem. is quartz, CaO estimated.   4. Triploidite, Wien, Moravia.[9] Fe_2O_3

due to oxidation of FeO, and calculated as such in the Fe:Mn ratio.   Rem. is CO_2 0.59, SiO_2 0.35.   5. Wolfeite, Skrumpetorp, Sweden.[10]   Rem. is Na_2O 0.93, K_2O tr., H_2O – 0.19, insol. 0.15.   6. Wolfeite, Cyrillhof, Moravia.[9]   Rem. is $(Na,K)_2O$ 0.17, insol. 2.30. Fe_2O_3 due to oxidation of FeO and calculated as such in the Fe:Mn ratio.   7. Wolfeite, Palermo mine, New Hampshire.[11]   Rem. is Na_2O 0.14, K_2O 0.05, Li_2O 0.56, insol. 1.44.   8. $Fe_2''(PO_4)(OH)$.

Tests. B.B. fuses quietly. In C.T. gives off neutral water, turns black and becomes magnetic. Soluble in acids.

O c c u r.[4] Triploidite and wolfeite occur in granite pegmatite closely associated with triplite, lithiophilite, and triphylite and may form by the hydrothermal alteration of these species. Triploidite was originally found with eosphorite, dickinsonite, and rhodochrosite in lithiophilite at Branchville, Fairfield County, Connecticut; also found as an alteration of triplite at Wien, Moravia (Czechoslovakia). Wolfeite occurs as an alteration of triphylite at the Palermo pegmatite, North Groton, New Hampshire. Also found with triplite at Skrumpetorp, Sweden; at Cyrillhof, Moravia (Czechoslovakia); and at Hagendorf, Bavaria.

N a m e. Triploidite from *triplite* and εἶδος, *form*, in allusion to its resemblance to triplite in physical characters and composition. Wolfeite after Caleb Wroe Wolfe (1908–), American crystallographer.

Ref.

1. Elements of E. S. Dana in Dana (779, 1892) but with *b* doubled to conform to the structure cell of Kokkoros, *Zbl. Min.*, 278 (1938), as confirmed by Richmond, *Am. Min.*, **25**, 469 (1940), and Frondel, *Am. Min.*, **34**, 692 (1949). Dana's elements are on triploidite from Branchville with Fe:Mn = 1:3 30 (anal. 2). Müllbauer, *Zs. Kr.*, **61**, 318 (1925), obtained 0.9258:1:0.74425, β 108°18′, on wolfeite from Hagendorf with Fe:Mn > 3:1 (see Frondel, 1949). Transformation, Dana (1892) to Kokkoros (1938): $\frac{1}{2}00/010/00\frac{1}{2}$.
2. Dana (1892) and Müllbauer (1925).
3. Richmond (1940).
4. Frondel (1949).
5. Kokkoros (1938).
6. Larsen, 145 (1921).
7. Hägele, *Zbl Min.*, 267 (1938).
8. Penfield analysis in Brush and Dana (1878).
9. Herles analysis in Kovář and Slavík, *Verh geol. Reichsanst. Wien*, **50**, 347 (1900).
10. Berggren analysis in Mason, *Geol. För. Förh.*, **62**, 373 (1941).
11. Peck analysis in Frondel (1949).

41.6.3.3   **S A R K I N I T E** [$Mn_2(AsO_4)(OH)$]. Kondroarsenit *Igelström* (*Ak. Stockholm, Öfv.*, **22**, 3, 1865). Chondrarsenite *Dana* (562, 1868). Xanthoarsénite *Igelström* (*Bull. soc. min.*, **7**, 237, 1884). Xantharsenite *Dana* (796, 1892). Polyarsenite *Igelström* (*Ak. Stockholm, Öfv.*, **42**, 257, 1885; *Bull. soc. min.*, **8**, 369, 1885). Sarkinite *Sjögren* (*Geol. För. Förh.*, **7**, 724, 1885).

C r y s t.[1] Monoclinic prismatic—$2/m$.

$a:b:c = 0.9322:1:0.7577;$ $\beta\ 108°12\frac{1}{2}';$ $p_0:q_0:r_0 = 0.8128:0.7198:1$

$r_2:p_2:q_2 = 1.3198:1.1293:1;$ $\mu\ 71°47\frac{1}{2}';$ $p_0'\ 0.8557,\ q_0'\ 0.7577,\ x_0'\ 0.3289$

Forms:[2]

		φ	ρ	φ2	ρ2 = B	C	A
c	001	90°00′	18°12½′	71°47½′	90°00′	.....	71°47½′
b	010	0 00	90 00	.....	0 00	90°00′	90 00
a	100	90 00	90 00	0 00	90 00	71 47½	.....
M	120	29 30½	90 00	0 00	29 30½	81 09	60 29½
n	210	66 07	90 00	0 00	66 07	73 24	23 53
o	021	12 14½	57 11	71 47½	34 47	55 13	79 44
d	101	90 00	40 33½	49 26½	90 00	22 21	49 26½
f	201	90 00	59 42	30 18	90 00	41 29½	30 18
C	1̄01	−90 00	27 47	117 47	90 00	45 59½	117 47
k	121	38 01	62 32	40 10	45 39	52 26½	56 52½
p	1̄41	− 9 51½	71 59½	117 47	20 27	75 57	99 22½

Structure cell.[3] Space group $C2_1/a$. a_0 12.65 kX, b_0 13.51, c_0 10.15; β 108°44′; $a_0:b_0:c_0 = 0.936:1:0.751$. Cell contents $Mn_{32}(AsO_4)_{16}(OH)_{16}$.

Habit. Crystals usually thick tabular {100} and somewhat elongated [010]; also short prismatic [010], or shortened along this axis and tabular.

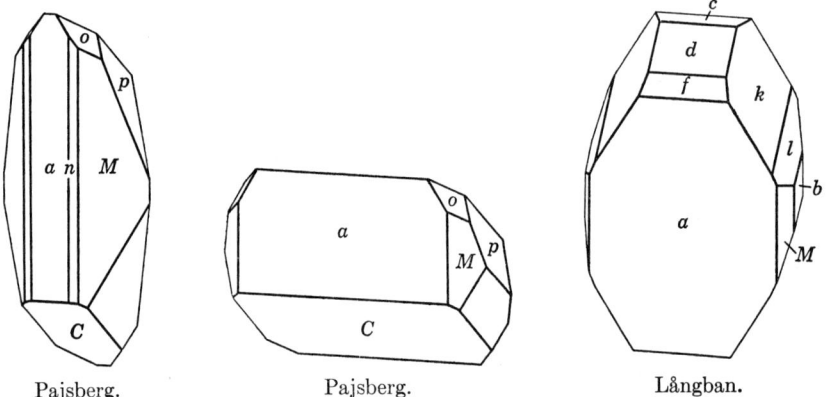

Pajsberg.　　　　Pajsberg.　　　　Långban.

{100} and {001} often uneven and {001} striated [100]. Sometimes grouped in rudely spherical forms; granular.

Phys. Cleavage {100}, distinct. Fracture subconchoidal to uneven. H. 4–5. G. 4.08–4.18, 4.18 (Långban);[4] 4.04 (calc.). Luster greasy. Color flesh-red to dark blood-red, rose-red, reddish yellow to yellow. Streak rose-red to yellow.

Opt.[5] In transmitted light, pale rose to yellow.

Orientation		n	
$X \wedge c$	−54°	1.793	Biaxial negative (−).
Y	b	1.807	$2V$ 83°.
Z		1.809	Weakly pleochroic, with absorption $X > Z > Y$. Dispersion not observed.

Chem. A basic arsenate of manganese, $Mn_2(AsO_4)(OH)$. Small amounts of Ca, Mg, and Fe substitute for Mn, with a reported[6] maximum of Mg:Ca:Mn = 1:1.7:14.3. Sb substitutes for As, with Sb:As = 1: 42.3 in analysis 5 and probably in a slightly greater amount in the material of analysis 6. Sarkinite is isostructural with triploidite, but only a very small amount of P has been reported in substitution for As (anal. 3).

Anal.

	1	2	3	4	5	6
CaO		1.09	1.40	1.99	2.85	1.93
MgO		0.29	0.98	0.28	0.77	6.08
FeO		0.15	0.13		tr.	3.11
MnO	53.38	50.60	51.60	50.91	49.88	43.60
As_2O_5	43.23	42.55	41.60	41.02	39.23	33.26
H_2O	3.39	3.44	3.06	3.33	3.15	[12.02]
Rem.		1.54	1.60	2.73	4.88	
Total	100.00	99.66	100.37	100.26	100.76	[100.00]
G.	4.04	4.178	4.15	4.117	4.08	

1. $Mn_2(AsO_4)(OH)$. 2. Långban, Sweden.[7] Rem. is Na_2O 1.30, Al_2O_3 0.10, insol. 0.14. 3. Pajsberg, Sweden.[6] Rem. is PbO 0.25, P_2O_5 0.21, CO_2 0.76, insol. 0.38. 4. Pajsberg (*chondrarsenite*).[8] Rem. is CO_2 1.80, insol. 0.93. 5. Sjö mine, Sweden[9] (*polyarsenite*). Rem. is Sb_2O_5 1.37, CO_2 3.51. 6. Långban (*xantharsenite*).[10] As_2O_5 perhaps contains about 3 per cent Sb_2O_5.

Tests. B.B. fuses easily to black, non-magnetic bead. In C.T. decrepitates, turns brown, and affords water. Easily soluble in dilute acids.

Occur. From the Harstig mine, Pajsberg, Sweden, in cavities associated with native lead, bementite, brandtite, calcite, and barite; the crystals are either elongated [010] or are somewhat shortened along this axis (the latter type earlier described under the name *chondrarsenite*[11]). Also described practically simultaneously with the Pajsberg sarkinite from the Sjö mine in Grythytte parish, Örebro, Sweden (*polyarsenite*),[12] with hematostibite, magnetite, jacobsite, and tephroite. At Långban, Sweden, associated with hausmannite and hedyphane (*xanthoarsenite*, pt.).

Name. From σάρκινος, *made of flesh*, in allusion to the blood-red color and greasy luster.

Ref.

1. Angles of Flink and Hamberg, *Geol. För. Förh.*, **10**, 380 (1888) on Pajsberg crystals; unit and orientation of the structure cell of Hägele, *Zbl. Min*., 267 (1938). Transformation: Flink and Hamberg (1888), Dana (779, 1892), and Goldschmidt (**8**, 9, 1922) to new, $10\bar{1}/0\bar{2}0/001$.
2. Flink and Hamberg (1888) and Flink, *Geol. För. Förh.*, **46**, 661 (1924). Less common and doubtful: r 7.10.0, l 131, q $\bar{1}21$, g $\bar{5}41$, h $\bar{9}41$.
3. Hägele (1938) by rotation and Weissenberg methods on Pajsberg material. Richmond, *Am. Min.*, **25**, 466 (1940), gives comparable data but with b_0 halved.
4. Flink, *Geol För. Förh.*, **46**, 661 (1924), on the essentially pure compound (anal. 2); the lower reported values are probably due to substitution of Ca and Mg for Mn.
5. Larsen and Berman (201, 1934) on material probably from Långban. Larsen (130, 1921) for Harstig material gives nX 1.780, nY 1.793, nZ 1.802 with $Z \wedge c = \pm 43°$.
6. Igelström (1865).
7. Almström analysis in Flink (1924).
8. Mauzelius analysis in Sjögren, *Geol. För. Förh.*, **28**, 401 (1906).
9. Söderbaum analysis in Igelström (1885).
10. Igelström (1884).
11. See Sjögren (1906).
12. See Flink and Hamberg (1888).

XANTHARSENITE. Xanthoarsenite *Igelström* (*Bull. soc. min.*, **7**, 237, 1884). Xantharsenite *Dana* (796, 1892).
Massive, granular. Brittle. Color sulfur-yellow. No cleavage. Optically biaxial.
Composition essentially $(Mn,Mg,Fe,Ca)_3(AsO_4)(OH)_3 \cdot H_2O$.

CaO	MgO	FeO	MnO	As_2O_5*	H_2O	Total
1.93	6.08	3.11	43.60	33.26	[12.02]	[100.00]

*Contains 1 to 3 per cent Sb_2O_5.

Found with hausmannite, magnetite, and hematite at the Sjö mine, Grythytte parish, Örebro, Sweden. Needs verification.

41.6.4 Sarcopside $[(Fe,Mn,Ca)_7(PO_4)_4F_2 \ (?)]$. Sarkopsid *Websky* (*Zs. deutsche geol. Ges.*, **20**, 245, 1868).

Probably monoclinic. As irregular masses with a fibrous structure, sometimes in distorted six-sided plates (Silesia).

P h y s. Distinct cleavage approximately perpendicular to the fibers and another less distinct cleavage parallel to the fiber direction (= [001]?). Fracture splintery to fibrous. H. 4. G. 3.73 (Silesia),[1] 3.64 (New Hampshire).[2] Luster silky and glistening. Color flesh-red to reddish brown, changing on exposure to dark brown, blue, lavender, or green. Translucent.

O p t.[3] In transmitted light, flesh-red.

ORIENTATION		n (New Hampshire)	
$X \wedge c$	45°	1.675	Biaxial negative (−).
Y		1.728	$2V$ small.
Z	b	1.730	$r > v$, perceptible.

C h e m. A fluo-phosphate of divalent iron, manganese, and calcium of uncertain formula, perhaps $(Fe'',Mn'',Ca)_7(PO_4)_4F_2$. The trivalent iron reported in the analyses is due probably to partial oxidation.

Anal.

	1	2	3	4	5
CaO	4.93	3.40	4.38	3.40	4.70
MgO			0.68		0.73
MnO	11.69	20.57	10.83	20.57	11.64
FeO	44.98	30.53	39.87	38.48	44.51
Fe_2O_3		8.83	1.70		
P_2O_5	35.64	34.73	33.26	34.73	35.72
F	4.77	n.d.	4.35	n.d.	4.67
H_2O		[1.94]	1.53		
Rem.			3.22		
	102.01	[100.00]	99.82		101.97
O = F	2.01		1.83		1.97
Total	100.00		97.99		100.00

1. $(Fe,Mn,Ca)_7(PO_4)_4F_2$ with Fe:Mn:Ca = 57:15:8. 2. Silesia.[4] Average of three. 3. New Hampshire.[2] Average of two. Rem. is insol. 4. Analysis 2 with Fe_2O_3 recalculated to FeO. 5. Analysis 3 recalculated after deducting insol. and H_2O and converting Fe_2O_3 to FeO.

Tests. Like triplite.

OLIVENITE 859

Occur. Originally from a pegmatite near Michelsdorf in the Eulengebirge, Silesia, Germany, with hureaulite and vivianite. Also in a small pegmatitic lens in gneiss on the Gingrass farm near Deering, New Hampshire.[2]

Alter. Surficially to vivianite and unidentified iron-manganese phosphates.

Name. From σάρξ, *flesh*, and ὄψις, *view*, in allusion to its color.

Ref.
1. Websky (1868); highest value of range 3.69–3.73.
2. Holden, *Am. Min.*, **5**, 99 (1920); **9**, 205 (1924).
3. Larsen and Berman (193, 1934).
4. Websky (1868).

41.6.5 OLIVENITE GROUP

ORTHORHOMBIC; DIPYRAMIDAL—$2/m\ 2/m\ 2/m$

	a_0	b_0	c_0
Olivenite, $Cu_2(AsO_4)(OH)$	$8.16\,kX$	8.54	5.86
Libethenite, $Cu_2(PO_4)(OH)$	8.08	8.43	5.90
Adamite, $Zn_2(AsO_4)(OH)$	8.30	8.51	6.04

The crystal structure [1] of the members of the Olivenite Group is similar to that of andalusite, $Al_2(SiO_4)O$. A partial series extends between olivenite and libethenite by mutual substitution of As and P, and a partial series extends from adamite toward olivenite by substitution of Cu for Zn. All the members of the group are secondary minerals formed under essentially atmospheric conditions, whereas andalusite is formed at high temperatures.

Ref.
1. Heritsch, *Zs. Kr.*, **102**, 1 (1940), and Strunz, *Zs. Kr.*, **94**, 60 (1936).

41.6.5.1 **OLIVENITE** [$Cu_2(AsO_4)(OH)$]. Arseniksaures Kupfererz (from Cornwall) *Klaproth* (*Schrft. Ges. Nat. Freunde Berlin*, **7**, 160, 1786). Olivenerz (from Cornwall) *Werner* (*Bergm. J.*, 382, 385, 1789). Olive Copper Ore *Kirwan* (**2**, 151, 1796). Olive-green Copper Ore *Rashleigh* (**1**, pl. 11, fig. 2, 1797; **2**, pl. 6, 1802). Holzkupfererz. Wood-Copper. Wood-Arsenate. Cuivre arseniaté en octaèdre aigus *Bournon* (*Phil. Trans.*, 177, 1801). Pharmacochalzit pt. *Hausmann* (**3**, 1042, 1813); Olivenkupfer (*idem*, 1045); Pharmacolzit (*idem*, 1025, 1847). Olivenite pt. *Jameson* (**2**, 335, 1820); *Leonhard* (283, 1821).

Cryst.[1] Orthorhombic; dipyramidal—$2/m\ 2/m\ 2/m$ (?).

$a:b:c = 0.9485:1:0.6810;$ [2] $p_0:q_0:r_0 = 0.7180:0.6810:1$

$q_1:r_1:p_1 = 0.9485:1.3928:1;$ $r_2:p_2:q_2 = 1.4684:1.0543:1$

Forms: [3]

	ϕ	$\rho = C$	ϕ_1	$\rho_1 = A$	ϕ_2	$\rho_2 = B$
b 010	0°00′	90°00′	90°00′	90°00′	.....	0°00′
a 100	90 00	90 00	.....	0 00	0°00′	90 00
m 110	46 31	90 00	90 00	43 29	0 00	46 31
r 012	0 00	18 48	18 48	90 00	90 00	71 12
e 011	0 00	34 15½	34 15½	90 00	90 00	55 44½
v 101	90 00	35 40½	0 00	54 19½	54 19½	90 00
p 111	46 31	44 42	34 15½	59 19	54 19½	61 03

Less common and rare:

    c 001   f 013   d 025   s 034   l 043

Structure cell.[4] Space group $Pnnm$ (?). a_0 8.16 kX, b_0 8.54, c_0 5.86; $a_0:b_0:c_0 = 0.955:1:0.686$. Cell contents $Cu_8(AsO_4)_4(OH)_4$.

Habit. Variable; often elongated [100], also short prismatic to acicular [001], less frequently tabular on {011}, {100}, or {001}. Often in globular and reniform shapes with a fibrous structure, the fibers straight and divergent and rarely irregular; curved lamellar; massive, granular to earthy; nodular.

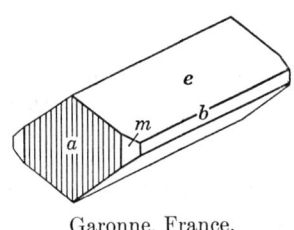
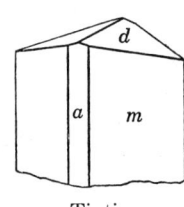
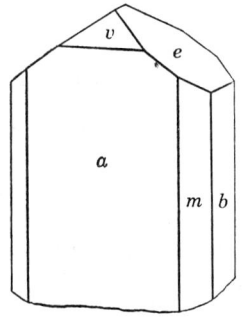

Garonne, France.   Tintic.   Tintic.

Twinning. Reported on optical grounds in Cornwall crystals.[7]

Phys. Cleavage {011} and {110} indistinct. Fracture conchoidal to irregular. H. 3. G. 4.46 (Tintic; Tsumeb),[14] 3.9–4.4 (Cornwall, phosphatic);[15] 4.57 (calc.). Luster adamantine to vitreous; of some fibrous varieties pearly to silky. Color various shades of olive-green; also greenish brown and brown, rarely straw-yellow, grayish green, or grayish white. Streak olive-green to brown. Translucent to opaque.

Opt. In transmitted light, pale green.

ORIENTATION		n (Chuquicamata)[8]	n (Tintic, anal. 4)[5]	n (Cornwall)[5]	
X		1.780	1.772 ± 0.005	1.747 ± 0.005	Biaxial positive, sometimes negative.
Y	c	1.820	1.810 ± 0.005	1.788 ± 0.005	$2V \sim 90°$.
Z		1.865	1.863 ± 0.005	1.829 ± 0.005	$r < v$, strong (opt. +); also $r > v$, strong (opt. −).

Some Cornwall crystals show a spiral interference figure apparently due to twinning.[7] Sometimes weakly pleochroic[6] in green and yellow with absorption $Y > X, Z$.

OLIVENITE

Chem. A basic arsenate of copper, $Cu_2(AsO_4)(OH)$. P substitutes for As up to about P:As = 1:3.5 (analysis 3), and a partial series thus extends toward libethenite. Small amounts of Fe apparently substitute for Cu, with Fe:Cu = 1:17 in analysis 5.

Anal.[9]

	1	2	3	4	5
FeO			tr.		2.78
CuO	56.22	56.65	56.38	55.40	52.67
P_2O_5			5.96	0.06	
As_2O_5	40.60	39.80	33.50	40.05	39.76
H_2O	3.18	3.55	4.16	3.39	4.09
Rem.				0.81	
Total	100.00	100.00	100.00	99.71	99.30

1. $Cu_2(AsO_4)(OH)$. 2. Cornwall.[10] 3. Cornwall.[11] 4. Eagle mine, Tintic, Utah.[12] Rem. is CaO 0.16, insol. 0.40, Fe_2O_3 0.25. 5. Tsumeb, South West Africa.[13]

Tests. B.B. easily fusible. Soluble in acids and in ammonia.

Occur. A secondary mineral found in the oxidized zone of ore deposits. Associated with adamite, malachite, azurite, limonite, scorodite. Not all the known localities can be mentioned here. Originally from Cornwall, England, where it occurs at a number of mines both as crystals and fibrous aggregates; also at Alston Moor, Cumberland, and Tavistock, Devonshire. In Czechoslovakia at Bêloves near Nachod. With bayldonite at Tsumeb, South West Africa. Found in Chile at Collahuasi and Copiapó and with chenevixite (?) at Chuquicamata. In the United States olivenite is of frequent occurrence at the American Eagle, Mammoth, and other mines in the Tintic district, Utah, where it is associated with clinoclase, tyrolite, conichalcite, chenevixite. At Majuba Hill, Pershing County, Nevada, with cornetite; in the Blackbird district, Lemhi County, Idaho.

Name. In allusion to the color.

Ref.

1. The crystal class is uncertain. The crystals ordinarily conform to dipyramidal symmetry and Heritsch, *Zs. Kr.*, **98**, 351 (1937); **99**, 466 (1938), has assigned the space group *Pnnm*. Richmond, *Am. Min.*, **25**, 453 (1940), however, uniquely determined the space group $P2_12_12_1$ on Tintic material; these crystals exhibited pyramidal morphological symmetry, with [100] polar, and gave no evidence of disphenoidal symmetry as required by his space group.
2. Goldschmidt (**6**, 90, 1920).
3. Goldschmidt (1920); also Shannon (425, 1926) and Richmond (1940).
4. Cell dimensions of Richmond (1940) on Tintic crystal by Weissenberg method; space group of Heritsch (1937; 1938).
5. Larsen (116, 1921). The relatively low indices of the Cornwall material may be due to a substitution of P for As.
6. Hillebrand and Washington, *Am. J. Sc.*, **35**, 298 (1888), and Winchell (132, 1933).
7. See Schaller, *U. S. Geol. Sur., Bull. 679*, 116 (1921), and Larsen (116, 1921).
8. Jarrell, *Am. Min.*, **24**, 632 (1939).
9. For additional analyses see Dana (785, 1892; 564, 1868).
10. Richardson in Thomson (**1**, 614, 1836).
11. Hermann, *J. pr. Chem.*, **33**, 291 (1844).
12. Hillebrand, *Proc. Colorado Sc. Soc.*, **1**, 113 (1884).
13. Biehl, Diss. Münster (1919)—*Zs. Kr.*, **62**, 328 (1925).
14. Mrose, priv. comm. (1947), by microbalance.
15. Range of values reported by Hermann (1844); Damour, *Ann. chim. phys.*, **13**, 412 (1845); Bournon, *Trans. Roy. Soc. London*, 177 (1801).

41.6.5.2 LIBETHENITE [$Cu_2(PO_4)(OH)$]. Olivenerz, pt. *Werner*. Phosphorkupfererz, pt. Phosphate of Copper, pt. Cuivre phosphaté, pt. *Haüy*. Octaedrisches Phosphorkupfer *Leonhard* (143, 1821). Blättricher Olivenmalachite pt. *Hausmann* (1036, 1813). Libethenite *Breithaupt* (267, 1823). Aphérèse *Beudant* (**2,** 569, 1832). Pseudo-libethenite *Rammelsberg* (344, 1860).

Cryst.[1] Orthorhombic; dipyramidal—$2/m\ 2/m\ 2/m$.

$$a:b:c = 0.9606:1:0.7024; \quad p_0:q_0:r_0 = 0.7312:0.7024:1$$

$$q_1:r_1:p_1 = 0.9606:1.3676:1; \quad r_2:p_2:q_2 = 1.4237:1.0410:1$$

Forms:[2]

	ϕ	$\rho = C$	ϕ_1	$\rho_1 = A$	ϕ_2	$\rho_2 = B$
b 010	0°00′	90°00′	90°00′	90°00′	....	0°00′
a 100	90 00	90 00	....	0 00	0°00′	90 00
m 110	46 09	90 00	90 00	43 51	0 00	46 09
h 540	52 27½	90 00	90 00	37 32½	0 00	52 27½
α 320	57 22	90 00	90 00	32 38	0 00	57 22
t 210	64 20½	90 00	90 00	25 39½	0 00	64 20½
δ 310	72 14½	90 00	90 00	17 45½	0 00	72 14½
e 011	0 00	35 05	35 05	90 00	90 00	54 55
d 101	90 00	36 10½	0 00	53 49½	53 49½	60 27
s 111	46 09	45 23½	35 05	59 06½	53 49½	60 27

Structure cell.[3] Space group *Pnnm*. a_0 8.08 kX, b_0 8.43, c_0 5.90; $a_0:b_0:c_0 = 0.958:1:0.700$. Cell contents $Cu_8(PO_4)_4(OH)_4$.

Habit. Short prismatic [001], or slightly elongated [100]; also equant. Common forms *m e s*. The crystals ordinarily are of poor quality, with a

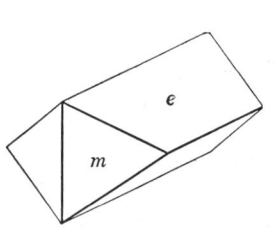
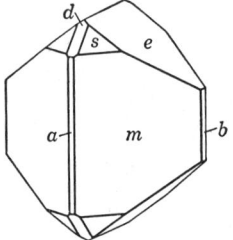

composite structure. {110} vertically grooved or striated, and {011} striated parallel the edge with {111}.

Etching. Conforms to holohedral symmetry.[4]

Twinning. Not observed.[5]

Phys. Cleavage {100} and {010} very indistinct.[6] Fracture conchoidal to uneven. H. 4. G. 3.97;[12] 3.93 (calc.). Luster vitreous on crystal faces to somewhat greasy on fracture surfaces. Color light to dark olive-green, also deep green and blackish green. Translucent.

LIBETHENITE 863

Opt. In transmitted light, bluish green to pale green.

ORIENTATION		n (Libethen)[7]	n (Cornwall)[7]	
X	b	1.701 ± 0.003	1.704 ± 0.003	Biaxial negative (−).
Y	c	1.743 ± 0.003	1.747 ± 0.003	$2V$ nearly 90°.
Z	a	1.787 ± 0.003	1.790 ± 0.003	$r > v$, strong.

Sometimes faintly pleochroic with X pale blue with a yellowish cast, and Z pale blue with a greenish cast (Libethen).[7]

Chem. A basic phosphate of copper, $Cu_2(PO_4)(OH)$. Small amounts of As have been reported in substitution for P, with As:P = 1:19 in analysis 2, and a partial series extends between libethenite and olivenite. No substitution has been observed for Cu (compare adamite).

Anal.[8]

	1	2	3	4
CuO	66.54	66.29	66.42	66.98
P_2O_5	29.69	26.46	29.31	28.89
As_2O_5		2.30		
H_2O	3.77	4.04	3.74	4.04
Total	100.00	99.09	99.47	99.91

1. $Cu_2(PO_4)(OH)$. 2. Libethen.[9] 3. Coquimbo, Chile.[10] 4. Loanda, Congo.[11]

Tests. B.B. fuses to a black crystalline bead. Easily soluble in acids and ammonia.

Occur. A secondary mineral, found in the oxidized zone of ore deposits associated with malachite, azurite, limonite, pyromorphite. Originally from Libethen, near Neusohl, Roumania. Later found at a small number of additional localities: at the Wheal Phoenix, Cornwall; in France at Montebras, Plateau Central, and at Ardillats, Rhône; in Russia at Nizhne Tagilsk and Wissokaya Gora, in the Urals; at Mindouli and Likasi in the Belgian Congo. From Chuquicamata, Chile, and the Mercedes mine, Coquimbo, Chile. In the United States found at Yerington, Lyon County, Nevada; in the Coronado vein, Clifton-Morenci district, Greenlee County, and the Castle Dome mine, Gila County, Arizona; in the Santa Rita district, Grant County, New Mexico; at the Perkiomen mine, Montgomery County, Pennsylvania.

Artif.[13] Formed as crystals by heating precipitated cupric phosphate with water and preferably excess phosphoric acid in a closed tube at temperatures up to *ca.* 200°.

Name. From the locality at Libethen.

Ref.

1. Elements of Melczer, *Zs. Kr.*, **39**, 288 (1904), recalculated by La Forge, priv. comm. (1936). Shown not to be piezoelectric by Strunz, *Zs. Kr.*, **94**, 60 (1936).
2. Goldschmidt (**5**, 149, 1918). Rare: β{650}, l{410}.
3. Strunz (1936) by rotation method on a Cornwall crystal. See also Richmond, *Am. Min.*, **25**, 452 (1940), and Heritsch, *Zs. Kr.*, **102**, 1 (1939).
4. Melczer (1904).
5. Not observed optically or macroscopically by Melczer (1904) and others; Schrauf, *Zs. Kr.*, **4**, 19 (1880), postulated microscopic mimetic twinning to support his monoclinic interpretation of the morphology.
6. Reported by Des Cloizeaux, *Ann. mines*, **14**, 343 (1858), but not later described.
7. Larsen (100, 1921). Des Cloizeaux (1858) gives $2V$ 81°38′, nY 1.739 for red; $2V$ 81°08′, nY 1.743 for yellow; $2V$ 80°20′, nY 1.755 for blue.

8. For other analyses see Dana (786, 1892; 563, 1868).
9. Bergemann, *Ann. Phys.*, **104**, 190 (1890).
10. Field, *Chem. Gaz.*, **17**, 400 (1859).
11. Mueller, *J. Chem. Soc., London*, **11**, 202 (1858).
12. Frondel, priv. comm. (1947), on a Cornwall crystal by the microbalance. Strunz (1936) obtained 3.928 by suspension method.
13. Friedel and Sarasin, *Bull. soc. min.*, **2**, 157 (1879), and Debray, *C.R.*, **52**, 44 (1861).

41.6.5.3 **A D A M I T E** [$Zn_2(OH)(AsO_4)$]. Adamine *Friedel* (*C.R.*, **62**, 692, 1866). Adamite *Dana* (565, 1868). Cuproadamite, Cobaltoadamite *Lacroix* (**4**, 424, 1910).

C r y s t.[1] Orthorhombic; dipyramidal—$2/m\ 2/m\ 2/m$.

$a:b:c = 0.9769:1:0.7086;$ $p_0:q_0:r_0 = 0.7254:0.7086:1$

$q_1:r_1:p_1 = 0.9769:1.3786:1;$ $r_2:p_2:q_2 = 1.4112:1.0236:1$

Forms:[2]

		ϕ	$\rho = C$	ϕ_1	$\rho_1 = A$	ϕ_2	$\rho_2 = B$
b	010	0°00′	90°00′	90°00′	90°00′	....	0°00′
t	120	27 06½	90 00	90 00	62 53½	0°00′	27 06½
m	110	45 40½	90 00	90 00	44 19½	0 00	45 40½
h	210	63 58	90 00	90 00	26 02	0 00	63 58
n	530	59 37½	90 00	90 00	30 22½	0 00	59 37½
l	011	0 00	35 19½	35 19½	90 00	90 00	54 40½
d	101	90 00	35 57½	0 00	54 02½	54 02½	90 00
o	111	45 40½	45 24	35 19½	59 23	54 02½	60 09½
x	212	63 58	38 55	19 30½	55 38½	54 02½	74 00
i	121	27 06½	57 52	54 47½	67 18½	54 02½	41 05

Less common and rare:

c 001	j 320	s 350	f 506	α 102	E 013	T 231	z 311
a 100	ε 750	w 250	g 607	e 205	F 053	p 123	y 412
k 410	7.10.0	K 140	r 203	β 103	G 081	v 421	u 1.10.1

Structure cell. Space group[3] *Pnnm*. a_0 8.30 Å, b_0 8.51, c_0 6.04 (Mapimi,[4] analysis 2), $a_0:b_0:c_0 = 0.9753:1:0.7055$; a_0 8.32 kX, b_0 8.54, c_0 6.08 (Thasos),[5] $a_0:b_0:c_0 = 0.974:1:0.712$. Cell contents $Zn_8(AsO_4)_4(OH)_4$.

Habit. Variable: often elongated [010], also elongated [001] like olivenite, rarely elongated [100]; sometimes tabular {101} or equant. Com-

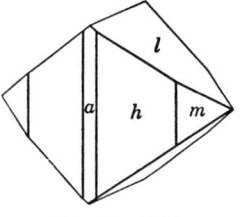

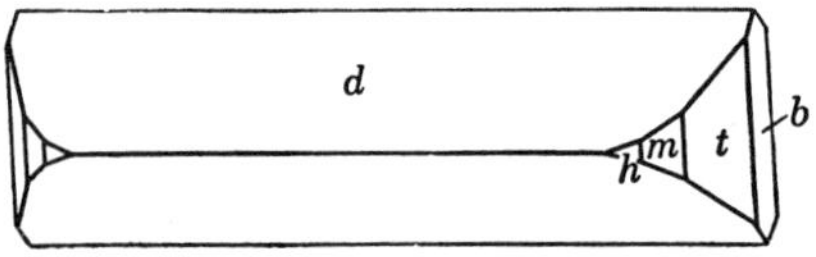

Mapimi.

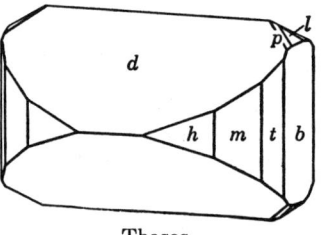

Chloride Cliff.   Thasos.

ADAMITE

mon forms: $d\ m\ l\ h\ b\ t$. Crystals often merged together in crusts or as rudely radial aggregates.

P h y s. Cleavage {101} good, {010} poor. Fracture uneven to subconchoidal. H. $3\frac{1}{2}$. G. 4.32–4.48, 4.43 (Mapimi, anal. 2); [6] 4.435 (calc.). Luster vitreous. Color variable: light yellow, honey-yellow, brownish yellow, pale green to green, rarely colorless or white; also bright green in cuproan material and violet to rose in cobaltian material. Transparent to translucent. Some specimens fluoresce lemon-yellow in ultraviolet radiation.

O p t. In transmitted light, colorless or faintly tinted in yellow, green, rose, etc. The reported [7] indices of refraction vary widely presumably owing to variation in composition, but the relationships are not clearly established.

ORIEN-TATION	n (Mapimi) [8]	n (Tsumeb; cuprian) [8]	n (Tsumeb; cobaltian) [8]	n (Laurium) [9]	n [10]	n (Balkash) [11]
$X\ \ a$	1.722	1.742	1.722	1.708	1.708	1.710
$Y\ \ c$	1.742	1.768	1.738	1.734	1.744	1.741
$Z\ \ b$	1.763	1.773	1.761	1.758	1.773	1.763
$2V$ (obs.)	88° ± 2°	15° ± 5°	88° ± 2°	87° ± 5°	90°±	86°±
$2V$ (calc.)	89°37′	23°28′	88°52′	86°28′		78°54′
Sign	+	−	+	−		−
DISPERSION	$r < v$, strong	$r > v$, strong	$r < v$, strong	$r > v$, strong	$r > v$, strong	
PLEOCHROISM						
X		Nearly colorless	Pale rose			Pink
Y		Bluish green	Pale rose			Pale rose
Z		Yellow-green	Pale purple			Colorless

C h e m. A basic arsenate of zinc, $Zn_2(AsO_4)(OH)$. Cu substitutes for Zn up to at least Cu:Zn = 1:1.33 (analysis 8), and a partial series thus extends towards olivenite. Co substitutes for Zn, with Co:Zn = 1:8.8 in analysis 7, and small amounts of Fe″ apparently also substitute for Zn. *Anal.*[12]

	1	2	3	4	5	6	7	8	9
CuO					0.64		1.75	23.45	tr.
CoO						3.92	5.16	0.52	
FeO			tr.	1.48	0.18				
ZnO	56.77	56.78	56.98	54.32	55.97	52.50	49.11	31.85	54.90
As_2O_5	40.09	38.96	39.80	39.95	40.17	38.50	39.24	39.85	39.80
H_2O	3.14	3.53	[3.22]	4.55	4.01	3.57	4.25	3.68	3.45
Rem.		0.26		tr.			tr.	0.87	1.19
Total	100.00	99.53	[100.00]	100.30	100.97	98.49	99.51	100.22	99.34
G.	4.435	4.43	4.484	4.338		4.35	4.35		4.319

1. $Zn_2(AsO_4)(OH)$. 2. Mapimi, Mexico.[14] Rem. is SiO_2. 3. Thasos, Turkey.[13] 4. Chañarcillo, Chile.[15] Rem. is MnO. 5. Laurium, Greece.[18] 6. Cap Garonne, France (cobaltian).[16] 7. Cap Garonne, France (cobaltian and cuprian).[17] Rem. is Fe_2O_3. 8. Cap Garonne, France (cuprian).[16] Rem. is CaO. 9. Balkhash region, U.S.S.R.[11] Rem. is MnO 0.11, rare earths 0.39, SiO_2 0.69.

Tests. In C.T. decrepitates feebly, yields a little water, and becomes white and porcelaneous. Easily soluble in dilute acids.

O c c u r. A secondary mineral, found in the oxidized zone of ore deposits containing primary zinc- and arsenic-rich minerals. Often associated with smithsonite, calcite, malachite, azurite, limonite, olivenite, hemimorphite, quartz. Originally from Chañarcillo, Chile, with limonite, native silver, and cerargyrite. At Cap Garonne, near Hyères, Dept. Var, France, together with cobaltian and cuproan varieties (*cobalto-adamite, cuproadamite*) incrusting sandstone. At Reichenbach near Lahr in the Black Forest, Baden, Germany, and at Mt. Valerio near Campiglia Marittima, Tuscany, Italy. A notable occurrence at the ancient zinc mines of Laurium, Greece, filling drusy cavities in cellular smithsonite; also on the Island of Thasos, Turkey. In Africa at Aïn Achour, near Guelma, Constantine, and elsewhere in Algeria; in fine specimens at Tsumeb, near Otavi, South West Africa, in part cuproan and cobaltian. In large, well-crystallized specimens from the Ojuela mine, Mapimi, Durango, Mexico, associated with scorodite, mimetite, hemimorphite, calcite, limonite.

In the United States, found with smithsonite at the Simon mine, Cedar Mountain, Mineral County, Nevada; at Chloride Cliff, Inyo County, California. In Utah at the Iron Blossom mine, Tintic district, Juab County, and with austinite at Gold Hill, Toole County.

A r t i f.[19] Obtained in crystals by the slow reaction of hot solutions of zinc sulfate and disodium acid arsenate.

N a m e. After Gilbert-Joseph Adam (1795–1881), French mineralogist, who supplied the first specimens for examination.

Ref.

1. The elements reported for the species by different investigators vary widely, owing to the poor quality of the measured crystals and in part to compositional variation; the values given are the average of eight sets tabulated by Mrose, *Am. Min.*, **33**, 449 (1948).
2. Ungemach, *Bull. soc. min.*, **44**, 122 (1922). {7.10.0} and {750} from Kukharenko, *Mem. soc. russe min.*, **68** [2], 589 (1939).
3. Space group determined independently by Strunz, *Zs. Kr.*, **94**, 60 (1936); Kokkoros, *Zs. Kr.*, **96**, 417 (1937); and Mrose (1948).
4. Mrose (1948); nearly identical values are given by Kokkoros (1937) on Laurium material.
5. Strunz (1936).
6. The wide variation in the numerous determinations of G.—see summary in Mrose (1948)—presumably is due to compositional variation, but the relation is uncertain; the highest values apparently represent the essentially pure compound.
7. Summarized in Mrose (1948).
8. Mrose (1948).
9. Larsen (35, 1921).
10. Larsen and Berman (195, 1934).
11. Kukharenko (1939).
12. For additional analyses see Hintze (**1** [4A], 649, 1931).
13. Rosický, *Zs. Kr.*, **48**, 656 (1910).
14. McLean analysis in Mrose (1948).
15. Friedel, *C.R.*, **62**, 692 (1866).
16. Pisani, *C.R.*, **70**, 1001 (1870).
17. Damour, *C.R.*, **67**, 1124 (1868).
18. Friedel, *Bull. soc. min.*, **1**, 31 (1878).
19. de Schulten, *Bull. soc. min.*, **26**, 91 (1903).

FRONDELITE SERIES

41.6.6.1 **F R O N D E L I T E** [(Mn'',Fe'')Fe$_4$'''(PO$_4$)$_3$(OH)$_5$]. *Lindberg* (*Am. Min.*, 34, 541, 1949).
41.6.6.2 **R O C K B R I D G E I T E** [(Fe'',Mn'')Fe$_4$'''(PO$_4$)$_3$(OH)$_5$]. *Frondel* (*Am. Min.*, 34, 513, 1949). Dufrenite pt., *earlier authors*.

C r y s t.[1] Orthorhombic.

$$a:b:c = 0.8166:1:0.3063$$

Structure cell.[2] Space group probably $B22_1$ or $B22_12$. a_0 13.89 Å, b_0 17.01, c_0 5.21; $a_0:b_0:c_0 = 0.8166:1:0.3063$. Cell contents (Mn'',Fe'')$_4$Fe$_{16}$'''(PO$_4$)$_{12}$(OH)$_{20}$.

Habit. Botryoidal masses and crusts with a radial fibrous or fine-columnar structure. The surface of the crusts sometimes is drusy and composed of indistinct crystals with rounded or exfoliated terminations. The fibers are elongated [001].

P h y s. Cleavage {100} excellent, {010} good, {001} fair. Brittle. Fracture uneven. H. 4½, less in altered material. G. 3.3–3.49, variable due to porosity and in part to oxidation and variation in Mn/Fe ratio. Luster vitreous to dull. Color dark green or olive-green to greenish black in fresh material becoming brownish green to dark reddish brown on oxidation. Subtranslucent. Radial aggregates often show concentric color banding.

O p t. In transmitted light, bluish green to olive-brown and yellow-brown in color.

ORIENTATION	n (Frondelite, anal. 2)[3]	n (Rockbridgeite, anal. 4)[4]	n (Rockbridgeite, anal. 7?)[4]	n (oxidized rockbridgeite, anal. 3)[3]
X c	1.860	1.875	1.873	1.915
Y	1.880	1.880	1.880	1.927
Z	1.893	1.897	1.895	1.939
PLEOCHROISM				
X	Pale yellow-brown	Pale yellow-brown	Pale brown	Pale yellow-brown
Y	Orange-brown	Bluish green	Bluish green	Orange-brown
Z	Orange-brown	Dark bluish green	Dark bluish green	Orange-brown
SIGN	−	+	+	
2V	mod.	mod.	mod.	large
DISPERSION	$r > v$	$r < v$		$r > v$ (if +)

The indices of refraction increase with increasing oxidation of Fe'' to Fe''' and with increase in the Fe/Mn ratio. Absorption ordinarily $Z > Y > X$, sometimes $Z > X > Y$, with bluish green tints predominating in fresh material and olive, brown, or yellow tints in oxidized material.

C h e m. A basic phosphate of trivalent iron and of divalent manganese and iron, probably (Mn'',Fe'')Fe$_4$'''(PO$_4$)$_3$(OH)$_5$. Divalent manganese and iron substitute mutually and a probably complete series extends between the manganese and iron end-members. The names frondelite and rockbridgeite are applied to the halves of the series with Mn > Fe and Fe > Mn, respectively. Small amounts of Mg and Ca substitute for (Mn'',Fe''), and Al and perhaps Mn''' substitute for Fe'''. The material as found is usually more or less oxidized [5] by conversion of Fe'' to Fe'''.

Anal.

	1	2	3	4	5	6	7	8	9
MnO	10.94	7.74	4.10	3.73	2.84	2.24	0.24	0.403	
FeO		nil	nil	5.51	2.66	0.99	6.06	6.144	11.07
MgO		0.20	tr.		tr.		2.16	0.762	
CaO		0.02			tr.			1.124	
Fe_2O_3	49.27	48.85	55.20		55.00	55.84	50.89	50.845	49.19
Al_2O_3		1.31	0.35		tr.		0.29	0.212	
P_2O_5	32.84	31.28	31.67		30.43	32.86	31.66	31.761	32.80
H_2O	6.95	7.52	6.98		8.06	7.96	8.35	8.531	6.94
Rem.		3.17	0.87		[1.01]			0.20	0.115
Total	100.00	100.09	99.17		[100.00]	99.89	99.85	99.897	100.00
G.		3.453	3.490			3.33	3.454	3.382	

1. $Mn''Fe_4'''(PO_4)_3(OH)_5$ (frondelite). 2. Frondelite. Sapucaia, Brazil.[3] Rem. is Na_2O 0.98, K_2O 0.12, Mn_2O_3 1.75, insol. 0.32. 3. Ferroan frondelite. Fletcher mine, New Hampshire.[3] Rem. is Na_2O 0.23, Mn_2O_3 0.32, ZnO 0.16, insol. 0.16. Oxidized. 4. Manganoan rockbridgeite. Fletcher mine, New Hampshire.[3] 5. Manganoan rockbridgeite. Polk County, Arkansas.[6] Rem. is largely SiO_2. 6. Manganoan rockbridgeite. Palermo mine, New Hampshire.[7] 7. Rockbridgeite. Rockbridge County, Virginia.[8] Rem. is SiO_2. 8. Rockbridgeite. Rockbridge County, Virginia.[9] Rem. is insol. 9. $Fe''Fe_4'''(PO_4)_3(OH)_5$ (rockbridgeite).

Tests. Easily fusible to a magnetic globule. In C.T. yields water. Soluble in HCl but not in HNO_3 or H_2SO_4.

Occur. The members of the series are secondary minerals found in limonite deposits or as alteration products of triphylite or other manganese-iron phosphates in pegmatite. Associated minerals include vivianite, strengite, metastrengite, heterosite, goethite. Frondelite occurs in pegmatite at Sapucaia, Conselheira Pena, Minas Geraes, Brazil, and a ferroan variety occurs with metastrengite at the Fletcher mine, North Groton, New Hampshire, U. S. A. Rockbridgeite occurs in the United States in a limonite deposit on South Mountain near Midvale, Rockbridge County, Virginia, in limonite concretions at Greenbelt, Maryland, and with limonite in Polk County, Arkansas. Manganoan varieties occur as an alteration of triphylite in pegmatite at the Palermo and Fletcher mines near North Groton, Grafton County, New Hampshire. Rockbridgeite also occurs in Germany near Herdorf, Westphalia, at Ullersreuth, Saxony, and in pegmatite at Hagendorf and at the Kreuzberg, Pleystein in Bavaria. In pegmatite at Chanteloube, France. Other occurrences probably have been earlier described under the name dufrenite.

Alter. See under dufrenite.

Name. Frondelite after Clifford Frondel (1907–), American mineralogist. Rockbridgeite after the locality in Rockbridge County, Virginia. The materials from Virginia and certain of the other localities had earlier been placed with dufrenite but were later shown [4] to be a distinct species.

Ref.

1. Symmetry from x-ray Weissenberg and optical study of Lindberg (1949) on fibers. Elements of the structure cell.
2. Lindberg (1949) on material of analysis 2 (frondelite). Also, for rockbridgeite (from powder photographs): a_0 13.73Å, b_0 16.82, c_0 5.18 (anal. 7 ?); a_0 13.76, b_0 16.94, c_0 5.19 (anal. 4); a_0 13.72Å, b_0 16.94, c_0 5.19 (oxidized equivalent of preceding, anal. 3).
3. Lindberg (1949).
4. Frondel (1949).
5. See discussion of rockbridgeite and dufrenite in Frondel (1949).
6. Hallowell analysis in Frondel (1949).
7. Gonyer analysis in Frondel (1949).
8. Massie, *Chem. News*, **42,** 181 (1880).
9. Campbell, *Am. J. Sc.*, **22,** 65 (1881).

41.6.7 **TARBUTTITE** [$Zn_2(PO_4)(OH)$]. *Spencer* (*Nature*, **76,** 215, 1907; *Min. Mag.*, **15,** 22, 1908).

Cryst.[1] Triclinic; pinacoidal—$\bar{1}$.

$$a{:}b{:}c = 0.6296{:}1{:}0.5971; \quad \alpha\ 89°37\tfrac{1}{2}', \beta\ 91°28\tfrac{1}{2}', \gamma\ 107°41'$$

$$p_0{:}q_0{:}r_0 = 0.9954{:}0.6265{:}1; \quad \lambda\ 89°55\tfrac{1}{2}', \mu\ 88°34\tfrac{1}{2}', \nu\ 72°19'$$

$$p_0'\ 0.9957,\ q_0'\ 0.6267,\ x_0'\ 0.0258,\ y_0'\ 0.0014$$

Forms:[2]

	ϕ	ρ	A	B	C
c 001	87°00′	1°28½′	88°34½′	89°55½′	.....
b 010	0 00	90 00	72 19	.....	89°55½′
a 100	72 19	90 00	.....	72 19	88 34½
m 110	45 35½	90 00	26 43½	45 35½	88 53½
k $2\bar{1}0$	90 39½	90 00	18 20½	90 39½	88 31½
M $1\bar{1}0$	108 52	90 00	36 33	108 52	88 37½
f 101	72 41	45 35	44 25	77 43½	44 16
G 1.6.23	20 35	10 47½	83 20½	79 54½	10 17
s 121	32 02	61 27	47 56	41 53	60 35½
Z $1\bar{2}1$	134 15½	53 41	67 44	124 13	52 41½
W ı11	−70 41	44 21½	124 27	76 37½	45 44
l ı21	−44 06	52 59	110 48½	55 01½	53 57½
E $\overline{34}7$	−141 56½	31 42½	115 44½	114 26½	32 42
u $\overline{11}1$	135 09	52 37	134 49	124 17	53 43½

Structure cell.[3] Space group $P\bar{1}$. a_0 8.097 kX, b_0 12.91, c_0 7.688; α 89°34½′, β 91°37½′, γ 107°47′; $a_0{:}b_0{:}c_0 = 0.6271{:}1{:}0.5957$. Cell contents $Zn_{16}(PO_4)_8(OH)_8$.

Habit. Equant to short prismatic [001]. Usually in sheaf-like aggregates or as crusts, the individual crystals rounded and deeply striated. Common forms: *c G s Z W l E u*.

Phys. Cleavage {010} perfect. Fracture uneven. H. 3¾. G. 4.12; 4.19 (calc.). Luster vitreous, on the cleavage pearly. Sometimes colorless and transparent, usually pale shades of yellow, brown, red, or green and translucent. Streak white.

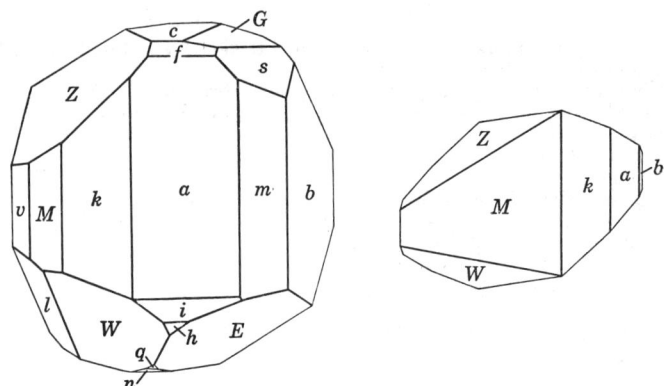

O p t.[4] In transmitted light, colorless.

ORIENTATION

	φ	ρ	n(Na)	
X	7°	58°	1.660 ± 0.003	Biaxial negative (−).
Y	159°	25°	1.705 ± 0.003	2V 50° (meas.).
Z	−86°	80°	1.713 ± 0.003	Dispersion of the bisectrices strong.

C h e m. A basic zinc phosphate, $Zn_2(PO_4)(OH)$.

Anal.

	1	2
ZnO	67.05	66.6
P_2O_5	29.24	29.2
H_2O	3.71	3.8
Total	100.00	99.6
G.	4.19	4.12

1. $Zn_2(PO_4)(OH)$. 2. Broken Hill mine, Rhodesia.[5]

Tests. B.B. fuses readily to a clear, yellow bead which on cooling crystallizes to an opaque, dark gray bead. In C.T. at a high temperature decrepitates slightly and gives off a small amount of water. The material when hot is of a bright yellow color; on cooling, the pieces are white and opaque with a porcelaneous appearance.

O c c u r. At the zinc mines of Broken Hill in northwestern Rhodesia associated with pyromorphite, hopeite, descloizite, vanadinite, hemimorphite, cerussite, smithsonite, limonite. Formed in the upper part of the zone of oxidation by the action of phosphatic solutions on earlier formed hemimorphite and cerussite.

A l t e r. Observed as pseudomorphs after smithsonite, descloizite and hemimorphite (?). Also found altered to parahopeite.

N a m e. After Percy Coventry Tarbutt, a director of the Broken Hill Exploration Company, who collected some of the first specimens.

Ref.

1. Unit, orientation, and angles of Richmond, *Am. Min.*, **23**, 881 (1938), on crystals superior in quality to those of Spencer (1908) and Rosický, *Bull. ac. sc. Bohême*, **18**, no. 35 (1913). Transformation: Spencer to new, $\overline{1}10/212/\overline{1}10$. On geometrical relations to libethenite see Richmond, *Am. Min.*, **25**, 475 (1940).

2. Form list as revised by Richmond (1938). Less common and rare: $v\ 2\bar{3}0$, $q\ \bar{1}02$, $i\ \bar{3}01$, $H\ 8.13.\bar{1}0$, $B\ 412$, $F\ 1\bar{2}3$, $r\ \bar{3}11$, $n\ \bar{1}03$, $h\ \bar{1}01$, $N\ \bar{1}12$, $t\ 141$, $P\ 1\bar{2}5$, $p\ \bar{5}95$, $o\ 1\bar{2}3$.
3. Richmond (1938) by Weissenberg method.
4. Richmond (1938). See also Buttgenbach and Mélon, *Ac. roy. Belgique, Bull.,* Cl. sc., [5], **17**, 892 (1931), and Larsen (141, 1921).
5. Spencer (1908).

41.6.8 A U G E L I T E [$Al_2(PO_4)(OH)_3$]. Blomstrand (*Ak. Stockholm, Öfv.,* **25**, 199, 1868).

C r y s t.[1] Monoclinic; prismatic—$2/m$.

$a:b:c = 1.6419:1:0.6354;\quad \beta\ 112°26\tfrac{1}{2}';\quad p_0:q_0:r_0 = 0.3870:0.5873:1$

$r_2:p_2:q_2 = 1.7027:0.6589:1;\quad \mu\ 67°33\tfrac{1}{2}';\quad p_0'\ 0.4187,\ q_0'\ 0.6354,\ x_0'\ 0.4130$

Forms:[2]

		ϕ	ρ	ϕ_2	$\rho_2 = B$	C	A
c	001	90°00′	22°26$\tfrac{1}{2}$′	67°33$\tfrac{1}{2}$′	90°00′	....	67°33$\tfrac{1}{2}$′
b	010	0 00	90 00	....	0 00	90°00′	90 00
a	100	90 00	90 00	0 00	90 00	67 33$\tfrac{1}{2}$	.....
j	130	12 23$\tfrac{1}{2}$	90 00	0 00	12 23$\tfrac{1}{2}$	85 18	77 36$\tfrac{1}{2}$
m	110	33 23	90 00	0 00	33 23	77 52$\tfrac{1}{2}$	56 37
t	310	63 10	90 00	0 00	63 10	70 05	26 50
g	910	80 25$\tfrac{1}{2}$	90 00	0 00	80 25$\tfrac{1}{2}$	67 53$\tfrac{1}{2}$	9 34$\tfrac{1}{2}$
r	021	18 00	53 11$\tfrac{1}{2}$	67 33$\tfrac{1}{2}$	40 24$\tfrac{1}{2}$	49 35$\tfrac{1}{2}$	75 40$\tfrac{1}{2}$
f	201	90 00	51 21	38 39	90 00	28 54$\tfrac{1}{2}$	38 39
x	$\bar{2}01$	−90 00	23 00	113 00	90 00	45 26$\tfrac{1}{2}$	113 00
n	111	52 37$\tfrac{1}{2}$	46 18$\tfrac{1}{2}$	50 15	63 58	30 55$\tfrac{1}{2}$	54 56
d	332	47 31$\tfrac{1}{2}$	54 41	43 51	56 34	40 10$\tfrac{1}{2}$	53 00
o	$\bar{1}11$	− 0 31	32 26	90 19$\tfrac{1}{2}$	57 34	38 54	90 16$\tfrac{1}{2}$

Less common:

k 120 l 230 s 210 p 221 q 441

Structure cell.[3] Space group $C2/m$. $a_0\ 13.10\ kX$, $b_0\ 7.96$, $c_0\ 5.06$ (all ± 0.02); $\beta\ 112°27' \pm 20'$; $a_0:b_0:c_0 = 1.646:1:0.636$. Cell contents $Al_8(PO_4)_4(OH)_{12}$.

Habit. Thick tabular {001}; less commonly prismatic to acicular [001] or as thin triangular plates flattened {110}. Dominant forms: $m\ c\ x\ o$. {110} striated [001]; {$\bar{2}01$} usually rough; {001} sometimes replaced by vicinal faces in the zones cx or cm. Also massive.

P h y s. Cleavage {110} perfect, {$\bar{2}01$} good, {001} and {$\bar{1}01$} imperfect. Fracture uneven. Brittle. H. $4\tfrac{1}{2}$–5. G. 2.696;[4] 2.704 (calc.). Colorless to white; also yellowish to pale rose. Luster vitreous, on the {110} cleavage pearly. Streak white. Transparent.

Laws, Calif.

O p t.[5] In transmitted light, colorless.

ORIENTATION		n(Na)	
X	b	1.5736	Biaxial positive (+).
$Y \wedge c$	$-56°$	1.5759	$2V$ 50°49′ (meas.),
$Z \wedge c$	$+34°$	1.5877	47°56′ (calc.).
			No dispersion observed.

C h e m. A basic phosphate of aluminum, $Al_2(PO_4)(OH)_3$.

Anal.

	1	2	3	4
CaO		1.09	0.11	0.90
MnO		0.31		
Al_2O_3	50.98	49.15	51.40	50.28
Fe_2O_3		0.89		
P_2O_5	35.51	35.04	34.60	35.33
H_2O	13.51	12.85	13.77	13.93
Total	100.00	99.33	99.88	100.44
G.	2.704	2.77	2.696	

1. $Al_2(PO_4)(OH)_3$. 2. Westanå, Sweden.[6] Admixed quartz deducted. 3,4. Machacamarca, Bolivia.[7] CaO probably adventitious.

Tests. B.B. becomes opaque but does not fuse. In C.T. yields water. Slowly soluble in hot concentrated HCl.

O c c u r. Originally from the Westanå iron mine near Näsum, west of Carlsham, Kristianstad, Sweden, where it occurs massive with berlinite, attacolite, lazulite. Found well-crystallized at Machacamarca near Potosi, Bolivia, associated with bournonite and pyrite. Also from Bolivia with pyrite and jamesonite (?) in the Carmen vein at Tatasi and Portugalete in Potosi province, and with arsenopyrite, stannite, pyrite, and quartz at Oruro.[8] In the United States at White Mountain, Mono County, California, associated with lazulite, rutile, pyrophyllite, barite in a deposit of andalusite.[9] With amblygonite at Mbale, Uganda.[10]

N a m e. From αὐγή, luster.

Ref.

1. Angles of Spencer in Prior and Spencer, *Min. Mag.*, **11**, 16 (1895); unit of the x-ray cell of Peacock and Moddle, *Min. Mag.*, **26**, 105 (1941), with c of Spencer halved. Holohedry indicated by lack of piezoelectric or pyroelectric response—Prior and Spencer (1895), Peacock and Moddle (1941), Frondel, priv. comm. (1947)—although etch effects suggest the hemihedral class (Prior and Spencer, 1895) and some crystals appear hemimorphic (Pough in Peacock and Moddle, 1941).
2. Peacock and Moddle (1941). Doubtful: 530, 510, 102, $\overline{16}.0.1$, 112, 223, 445, 443, 12.12.5, 661, $\overline{1}13$, $\overline{2}41$.
3. Peacock and Moddle (1941) by Weissenberg method on a California crystal.
4. Prior and Spencer (1895) by pycnometer at 22°.
5. Prior and Spencer (1895) by prism method on Machacamarca crystal; orientation verified by Peacock and Moddle (1941).
6. Blomstrand (1868).
7. Prior analysis in Prior and Spencer (1895).
8. Spencer, *Min. Mag.*, **12**, 1 (1898); **14**, 308 (1907).
9. Lemmon, *Am. Min.*, **20**, 664 (1935).
10. Roberts, *Bull. Imp. Inst. London*, **46**, 342 (1948).

BOLIVARITE. Navarro and Barea (*Bol. soc. espan. hist. nat.*, **21**, 326, 1921). Found as cryptocrystalline, pale greenish yellow crusts. Brittle with splintery-conchoidal fracture. H. $2\frac{1}{2}$. G. 2.05. Feebly birefringent. Luster vitreous. Analysis gave Al_2O_3 44.07, P_2O_5 34.93, H_2O 20.60, total 99.60, which is close to the formula $Al_2(PO_4)(OH)_3 \cdot H_2O$. Found in crevices in granite near Pontevedra, Spain. Needs verification.

AMPHITHALITE. Amfihalit *Igelström* (*Ak. Stockholm, Öfv.*, **23**, 93, 1866; *B. H. Ztg.*, **25**, 309, 1866). Massive. H. 6. Color milk-white. Analysis gave CaO 5.76, FeO and MnO tr., MgO 1.55, Al_2O_3 48.50, P_2O_5 30.06, H_2O 12.47, total 98.34. B.B. infusible. Insoluble in acids. Found with lazulite, rutile, and cyanite at Horrsjöberg, Wermland, Sweden. Needs verification.

41.6.9 D U F R E N I T E [$Fe''Fe_4'''(PO_4)_3(OH)_5 \cdot 2H_2O$]. Strahlstein pt. *Jordan* (*Min. Reisebem.*, 243, 1803). Grüneisenstein, strahlicher, ochrichter, pt. *Ullmann* (152, 1814). Grüneisenerde, pt. *Werner*. Fer oxidé, terreaux, jaune verdâtre, pt. *Haüy* (**4**, 106, 1822). Grün-Eisenstein (?) *Karsten* (*Arch. Berg. u. Hütten.*, **15**, 243, 1827). Dufrénite *Brongniart* (20, 1833). Green Iron Ore, pt. Kraurit *Breithaupt* (152, 1841). Dufrenite, pt., of most authors. Dufrenite *Frondel* (*Am. Min.*, **34**, 513, 1949).

Probably monoclinic, pseudo-orthorhombic.[1] Rarely as indistinct, rounded crystals in subparallel or sheaf-like aggregates. Ordinarily as botryoidal masses or crusts with a radial fibrous structure, the surface of the crusts sometimes formed of drusy crystals with rounded or exfoliated terminations.

P h y s. Two cleavages, one perfect, are present parallel to the fiber direction, with traces of a third cleavage at right angles to the other two. Brittle. H. $3\frac{1}{2}$–$4\frac{1}{2}$. G. variable, 3.1–3.34, owing to the fibrous character of the mineral and to variation in the degree of alteration.[2] Luster vitreous to silky. Color dark green or olive-green grading to greenish black in fresh material and becoming olive-brown to reddish brown with increasing oxidation. Translucent to nearly opaque.

O p t.[3] In transmitted light, bluish green to reddish brown or yellow.

	Rock Run, Alabama	Siegen, Westphalia	Wheal Phoenix, crystals	Hirschberg, Westphalia
nX	1.820	1.842	1.837	1.832
nY	1.830	1.850	1.845	1.837
nZ	1.925	1.875	1.895	1.890
PLEOCHROISM				
X	Deep blue	Pale brown	Pale yellow-brown	Deep bluish green
Y	Buff	Brown	Pale brown	Pale yellow-brown
Z	Deep red-brown	Dark brown	Red-brown	Deep olive-brown
$2V$	Small	Small	Small	Very small
DISPERSION	$r < v$	$r < v$	$r > v$	$r < v$

Biaxial positive (+). Dispersion extreme. Absorption usually $Z > Y > X$, sometimes $Z > X > Y$. Oxidation of Fe'' to Fe''' is accompanied by an increase in the indices of refraction and by a change in the absorption color of X from blue or bluish green to brown and yellow-brown. The values for nX and nZ cited were measured statistically and probably represent minimum and maximum values in unequally oxidized, inhomogeneous samples and hence may not afford a measure of the true birefringence.

Chem. A hydrated basic phosphate of both divalent and trivalent iron.[2] Formula uncertain, probably $Fe''Fe_4'''(PO_4)_3(OH)_5 \cdot 2H_2O$. Small amounts of Ca and Cu substitute for Fe'', with Ca:Fe = 1:3.8 in analysis 2 and Cu:Fe = 1:10 in analysis 3. Al substitutes for Fe''' in some material probably belonging to this species (analysis 6). Dufrenite as found is ordinarily more or less oxidized by conversion of Fe'' to Fe''' (analyses 3, 4, 5).

A dufrenite-like mineral containing much Mn'' in place of Fe'' also has been analyzed.[4]

Anal.[5]

	1	2	3	4	5	6
CaO		1.68	1.50			0.64
MgO		0.17	tr.			
FeO	10.49	6.80		2.2	1.53	8.34
Fe_2O_3	46.61	47.03	55.63	56.5	60.20	40.15
Al_2O_3		0.87				4.44
P_2O_5	31.07	31.10	30.26	31.8	31.82	31.26
H_2O	11.83	11.47	10.62	9.1	8.03	14.39
Rem.		0.43	1.48	0.1		1.32
Total	100.00	99.55	99.49	99.7	101.58	100.54
G.		3.08	3.233		3.39	

1. $Fe''Fe_4'''(PO_4)_3(OH)_5 \cdot 2H_2O$. 2. Wheal Phoenix, Cornwall.[6] Fibrous. Rem. is SiO_2. 3. Wheal Phoenix.[7] Crystals. Rem. is SiO_2 0.53, CuO 0.95. 4. Rock Run, Alabama.[8] Rem. is SiO_2. 5. Dufrenite (?). Waldgirmes, Hesse.[9] 6. Dufrenite (?) Dandarragan, Western Australia.[10] Rem. is CO_2 0.24, SiO_2 0.48, C 0.40, MnO 0.20.

Tests. B.B. easily fusible. Soluble in dilute acids.

Occur. Dufrenite is a secondary mineral found with limonite in the gossan of veins and in iron ore deposits. Only the following localities, among the many reported [11] for so-called dufrenite, have been definitely shown to belong to this species. In Germany at Hirschberg and Leobenstein in Thuringia; at Ullersreuth and Hauptmannsgrün near Reichenbach, Saxony; at Siegen, Westphalia; and probably at the Rothläufchen mine, Waldgirmes, Hesse. At the Wheal Phoenix, Cornwall, England, in part as distinct crystals. In the United States at Rock Run, Cherokee County, Alabama.

Alter.[3] Dufrenite (and the related species frondelite and rockbridgeite) alter readily, first by a more or less complete oxidation of the ferrous iron present and finally by leaching of the P_2O_5. In the first stage of alteration the color changes from greenish black to olive-brown and reddish brown with an accompanying increase in the indices of refraction but with retention of other properties; in the final stage the crystal structure breaks down leaving an ocherous pseudomorph of hydrous iron oxide (goethite).

Name. After the French mineralogist P. A. Dufrénoy (1792–1857). Kraurite is from κραῦρος, *harsh, dry*. The name dufrenite as hitherto applied included several different basic iron phosphates and has been redefined [2] to apply to the particular mineral here described.

Ref.

1. Imperfect, seemingly orthorhombic crystals have been described by Miers in Kinch and Butler, *Min. Mag.*, **7**, 65 (1886), from Wheal Phoenix (analysis 3) and by Streng, *Jb. Min.*, I, 110 (1881), from Waldgirmes, Hesse (analysis 5). The latter material, not certainly identical with dufrenite, had the forms {100}, {010}, {011}, {110}, and {120} with $a:b:c = 0.873:1:0.426$. The mineral apparently is pseudo-orthorhombic since some material shows crossed dispersion.
2. Frondel, *Am. Min.*, **34**, 513 (1949).
3. For discussion and additional data see Frondel (1949).
4. Schaller in Clarke, *U. S. Geol. Sur.*, *Bull. 419*, 303 (1910).
5. Additional analyses of minerals earlier referred to dufrenite but not known definitely to belong to this species are cited by Frondel (1949) and Hintze (**1** [4B], 1122, 1933).
6. Kinch, *Min. Mag.*, **8**, 112 (1888).
7. Kinch and Butler (1886).
8. Wells analysis cited in Frondel (1949).
9. Streng (1881).
10. Simpson, *J. Nat. Hist. Soc. Western Australia*, **4**, 3 (1911).
11. See Hintze (1933).

41.6.10 **DEWINDTITE** [$Pb_3(UO_2)_5(PO_4)_4(OH)_4 \cdot 10H_2O$]. Schoep (*C.R.*, **174**, 623, 1922; *Bull. soc. min.*, **48**, 77, 1925). Stasite Schoep (*C.R.*, **174**, 875, 1922; *Bull. soc. belge géol.*, **33**, 190, 1923).

Orthorhombic. As microscopic rectangular tablets flattened {100} and terminated by {001}; striated [001]. Also pulverulent or compact. G. 5.03.[1] Color canary-yellow. Cleavage {100}. Fluoresces green in ultraviolet light.

Opt.[3] Not pleochroic.

Orientation		n	
X	b	1.762	Biaxial positive (+).
Y	c	1.763	$r < v$.
Z	a		2E large.

Chem. A hydrated basic phosphate of lead and uranium, $Pb_3(UO_2)_5(PO_4)_4(OH)_4 \cdot 10H_2O$. The water is largely or entirely lost below 300°.[2] Analyses gave:[2]

	PbO	UO_3	P_2O_5	H_2O	Rem.	Total
1.	25.75	55.02	10.93	8.30		100.00
2.	24.85	54.80	10.14	7.93		97.72
3.	26.20	55.77	10.62	6.71	0.40	99.70

1. $Pb_3(UO_2)_5(PO_4)_4(OH)_4 \cdot 10H_2O$.   2. Kasolo.   Crystals.   3. Kasolo (*stasite*). H_2O is ign. loss, rem. is insol.

Tests. B.B. easily fusible. In C.T. loses water and turns orange. Soluble in acids.

Occur. Found associated with torbernite and other secondary uranium minerals at Kasolo, Katanga in the Belgian Congo. Also at Wölsendorf, Bavaria.

Name. After Jean Dewindt, a Belgian geologist.

Ref.

1. Schoep (1922) on stasite (= dewindtite).
2. Additional analyses are given by Schoep (1922, 1925).
3. Schoep (1925) gives $nY - nX$ as 0.004 by direct measurement.

41.6.11 Phosphuranylite. *Genth* (*Am. Chem. J.*, **1**, 92, 1879).

Tetragonal or pseudo-tetragonal. As earthy or scaly coatings and crusts; also as microscopic rectangular plates and laths. Cleavage {001} perfect but not easily observed. H. $\sim 2\frac{1}{2}$. Color deep yellow to golden yellow.

O p t.[1] In transmitted light, deep yellow to golden yellow in color and distinctly pleochroic. Sometimes sensibly uniaxial negative (−) but ordinarily biaxial negative (−) with Z perpendicular to the flattening. $2V$ is variable and ranges up to about 35° but usually is 5° to 20°. In biaxial material, $r > v$ strong. The indices of refraction vary over a wide range, probably as a function of the water content of the mineral.

Orientation	n (Flat Rock mine, North Carolina)	n (Ruggles mine, New Hampshire)	n (Rosmaneira, Portugal)	Pleochroism
X or E	1.690	1.668	1.660	Colorless to pale yellow
Y	1.718	1.710	1.700	Golden yellow
Z or O	1.718	1.710	1.701	Golden yellow

C h e m. A hydrated phosphate of calcium and uranium of uncertain formula. Phosphuranylite appears to be related structurally to dewindtite.

Tests. Easily soluble in acids.

O c c u r. A secondary mineral found associated with autunite and, less commonly, uranophane, beta-uranophane, hydrated uranium oxides, and opal. The mineral is found typically in the weathered zone of granite pegmatites that carry uraninite. Originally found at the Flat Rock pegmatite and the Buchanan pegmatite in Mitchell County, North Carolina. Later found [1] at the Ruggles pegmatite near Grafton Center and the Palermo pegmatite near North Groton, New Hampshire; from Newry, Oxford County, Maine; Bedford, Westchester County, New York; Branchville, Fairfield County, Connecticut. In Portugal at Rosmaneira and Carrasca in Sabugal and at Urgeirica, Cannas de Senhorim. From Wölsendorf, Bavaria, Germany. In South America at Pamplonita, Santander do Norte, Colombia, and at Memoes near Equador, Rio Grande do Norte, Brazil.

N a m e. In allusion to the composition, a phosphate of uranium. The original material was considered by Genth to be a uranyl phosphate, $(UO_2)_3(PO_4)_2 \cdot 6H_2O$, on the basis of an erroneous analysis.

Ref.

1. Frondel, priv. comm. (1949).

TYPE 7. $A_5(XO_4)_3Z_q$

APATITE GROUP

HEXAGONAL—P; HEXAGONAL-DIPYRAMIDAL—6/m

	$a:c$
Apatite Series	
Fluorapatite, $Ca_5(PO_4)_3F$	1:0.7346
Chlorapatite, $Ca_5(PO_4)_3Cl$	1:0.70
Hydroxylapatite, $Ca_5(PO_4)_3(OH)$	1:0.737 (x-ray)
Carbonate-apatite, $\sim Ca_{10}(PO_4)_6(CO_3) \cdot H_2O$	
Pyromorphite Series	
Pyromorphite, $Pb_5(PO_4)_3Cl$	1:0.7632
Mimetite, $Pb_5(AsO_4)_3Cl$	1:0.7224
Vanadinite, $Pb_5(VO_4)_3Cl$	1:0.7122
Svabite Series	
Svabite, $Ca_5(AsO_4)_3(F,OH)$	1:0.7109
Hedyphane, $(Ca,Pb)_5(AsO_4)_3Cl$	1:0.7052
Dehrnite, $(Ca,Na,K)_5(PO_4)_3(OH)$	1:0.737 (x-ray)
Lewistonite, $(Ca,K,Na)_5(PO_4)_3(OH)$	1:0.737 (x-ray)
Fermorite, $(Ca,Sr)_5(P,AsO_4)_3(F,OH)$	1:0.7292 (x-ray)
Wilkeite, $Ca_5(P,S,Si,CO_4)_3(OH)$	1:0.73 (x-ray)
Ellestadite, $Ca_5(Si,S,PO_4)_3(Cl,F,OH)$	1:0.725 (x-ray)

The structure of the members of the Apatite Group has been described by Náray-Szabó,[1] Mehmel,[2] and others. The members of the group fall into two main series, the Apatite Series and the Pyromorphite Series. The Svabite Series can be considered as intermediate between these; dehrnite, wilkeite and related species, grouped together at the bottom of the tabulation, are essentially substitutional variants of the Apatite Series. The members of the Pyromorphite Series are composed of relatively large atoms, other than P, and the cell volume of this series is about one-fifth larger than that of the Apatite Series.

The members of the Apatite Group enter into a wide range of substitutional solid solutions. The dominant cation always is either Ca or Pb. Sr, rare earths (principally Ce) and to a less extent Na (and K) substitute in part for Ca in the Apatite Series, and Ca substitutes in small part for Pb in the Pyromorphite Series. Fermorite is essentially a highly strontian (and arsenatian) variety of fluorapatite. It is interesting to note that Pb is lacking in significant amounts in substitution for Ca in the Apatite Series and that Ca substitutes to only a limited extent, and ordinarily is nearly lacking, in the Pyromorphite Series. This probably is due in part to environmental conditions, but a degree of structural control also is apparent. The substitution of the larger cations is favored in the expanded structures containing (AsO_4) and Cl; Na and rare earths, on the other hand, are present principally in fluorapatite and hydroxylapatite. Mg and Ba do not occur in significant amounts in the structure-type because of the relatively large dissimilarities in size with the Ca ion. Wagnerite is roughly the magnesium analogue of apatite in geochemical regards. Structural control also is apparent in the substitutions involving the (XO_4) anion. Thus Si, S, and C substitute for P, although to only a limited extent, in the Apatite Series but are lacking in the Pyromorphite and Svabite Series. Large monovalent cations such as K probably do not substitute for Pb in the arsenate and vanadate members because valence compensation cannot be effected by a coupled substitution of Si or C for the X atom, analogous to the coupled substitution of (Na,Ca) and (P,Si,C) in apatite. The Pyromorphite Series, containing Pb and Cl, also is marked by more or less complete series between V, As, and P, whereas As and especially V, on the other hand, are present in only very small amounts or lacking in the Apatite Series. With regard to the halogen and (OH) component, it may be noted that chlorapatite is relatively rare and formed at high temperatures, a consequence presumably of the critical fit of the F or (OH) ions in their oxygen coordination; the displacement of (OH) by F by reaction of hydroxylapatite with fluorine-containing solutions— an important biochemical reaction—and the formation of fluorapatite instead of chlorapatite or hydroxylapatite in the extensive phosphate-rock deposits of marine origin is a consequence of the particularly favorable fit of F in the structure. The non-existence of the (OH) and F analogues of the members of the Pyromorphite Series is a further illustration of these dimensional relations.

Ref.
1. Náray-Szabó, *Zs. Kr.*, **75**, 387 (1930).
2. Mehmel, *Zs. Kr.*, **75**, 323 (1930); *Zs. phys. Chem.*, **15**, 223 (1931).

41.7.1 APATITE SERIES

General synonyms. From Spain: Chrysolite ordinaire *de Lisle* (1772; **2**, 271, 1783) = Spargelgrüne Steinkrystalle aus Spanien nähern Apatit *Werner* (*Bergm. J.*, 74, 1790) = Spargelstein *Werner;* Asparagus Stone; Pierre d'Asperge *Fr.;* Asparagolithe *Abildgaard* (*Ann. chim.*, **32**, 195, 1800); Chaux phosphatée *Vauquelin* (*Ann. chim.*, **26**, 123, 1798); Phosphate of Lime. From Saxony: Aquamarin *Brünnich* (in his ed. of *Cronstedt*, Copenhagen, 1770). Améthiste basaltine *Sage* (**1**, 231, 1777); *de Lisle* (**2**, 254, 1783) = Apatit *Werner* (*Gerhard's Grundr.*, 281, 1786; *Bergm. J.*, 576, 1788 and 378, 1789); Phosphorsaurer-Kalk *Klaproth* (*Bergm. J.*, 294, 1788); Sächsischer Beryll, Augustit *Trommsdorf* (*J. de pharm.*, **8**, 153, 1800). Moroxit [Norway] *Abildgaard* (*Moll's Berg.-Hütten., Jb.*, **2**, 432, 1798). Lazur-apatit [Russia] *Nordenskiöld* (*Bull. soc. nat. Moscou*, 30, 217, 224, 1857). Apatit [all vars. except moroxite] *Karsten* (36, 1800); Apatite [incl. moroxite] *Haüy* (**2**, 1801). Massive: La Pierre Phosphorique [from Spain] *Davila* = Phosphate calcaire *Proust* (*J. phys.*, **32**, 241, 1788); *Pelletier* (*Ann. chim.*, **7**, 1790) = Phosphorite *Kirwan* (**1**, 129, 1794), *Karsten* (52, 1808); Osteolith *Bromeis* (*Ann. Chem. Pharm.*, **79**, 1851) = Bone-phosphate.

Talkapatit *Hermann* (*J. pr. Chem.*, **31**, 101, 1844). Manganapatit *Siewert* (*Zs. Nat. Halle*, **10**, 339, 1874). Cupro-apatit (?) *Adam* (45, 1869). Epiphosphorit *Breithaupt* (*B. H. Ztg.*, **25**, 194, 1866). Pyroguanite, Glaubapatite pt. *Shepard* (*Am. J. Sc.*, **22**, 96, 1856). Sombrerite *Phipson* (*J. Chem. Soc. London*, **15**, 277, 1862). Estramadurite *Roscoe* and *Schorlemmer* (*Treat. on Chem.*, **1**, 459, 1877). Kietyöite *Nordenskiöld* (*Min. Finland*, 154, 1863). Odontolite, pt. Floridite *Cox* (*Proc. Am. Assoc.*, **39**, 260, 1891). Ciplyte (?) *Ortlier* (*Bull. soc. min.*, **13**, 160, 1890).

Apatite Series

41.7.1.1 FLUORAPATITE [$Ca_5(PO_4)_3F$]. Eupyrchroite *Emmons* (*Rep. Geol. New York*, 252, 1838). Francolite *Brooke* and *Henry* (*Phil. Mag.*, **36**, 134, 1850). Fluorapatit *Rammelsberg* (353, 1860); Fluor-apatite. Staffelite *Stein* (*Jb. Min.*, 716, 1866). Voelckerite *Rogers* (*Am. J. Sc.*, **33**, 475, 1912); Oxyapatite *Rogers* (*ibid.*). Nauruite *Elschner* (*Corallogene Phosphat-Insel Austral Oceanien*, Lubeck, 1913; *Cbl. Min.*, 543, 1914). Sulfatapatit *Brauns* (*Jb. Min., Beil.-Bd.*, **41**, 60, 1916). Fluormanganapatit *Laubmann* and *Steinmetz* (*Zs. Kr.*, **55**, 563, 1920). Cerapatite *Fersman* (*Jb. Min., Beil.-Bd.*, **55**, 40, 1928); Cerium-apatite. Fluorcollophane, Fluocollophanite. Mangualdite *de Jesus* (*Com. Serv. Geol. Portugal*, **19**, 142, 1933). Saamite *Volkova* and *Melentiev* (*C.r. ac. sc., U.R.S.S.*, **25**, 120, 1939). Manganvoelckerite *Quensel* (*Geol. För. Förh.*, **59**, 257, 1937). Mangan-fluorapatite *Mason* (*Geol. För. Förh.*, **63**, 279, 1941). Manganapatite pt. *some authors*.

41.7.1.2 CHLORAPATITE [$Ca_5(PO_4)_3Cl$]. Chlorapatit *Rammelsberg* (353, 1860). Chlor-apatite.

41.7.1.3 HYDROXYLAPATITE [$Ca_5(PO_4)_3(OH)$]. Hydro-apatite *Damour* (*Ann. mines*, **10**, 65, 1856). Pyroclasite *Shepard* (*Am. J. Sc.*, **22**, 96, 1856). Ornithite *Julien* (*Am. J. Sc.*, **40**, 371, 1865). Hydrated oxygen apatite [artif.] *Warrington* (*J. Chem. Soc. London*, **26**, 983, 1873). Monite *Shepard* (*Am. J. Sc.*, **23**, 400, 1882). Hydroxyapatite *Schaller* (*U. S. Geol. Sur., Bull. 509*, 89, 1912). Hydroxylapatit *Burri, Jakob, Parker*, and *Strunz* (*Schweiz. min. Mitt.*, **15**, 327, 1935). Manganhydroxyapatite *Mason* (*Geol. För. Förh.*, **63**, 279, 1941.

41.7.1.4 CARBONATE-APATITE [$\sim Ca_{10}(PO_4)_6(CO_3) \cdot H_2O$]. Pseudoapatite (?) *Breithaupt* (*Glocker's Min. Jahresh.*, 217, 1837). Kollophan *Sandberger* (*Jb. Min.*, 308, 1870) = Collophane *Rogers* (*Am. J. Sc.*, **3**, 269, 1922) = Collophanite *Dana* (808, 1892). Dahllite *Brögger* and *Bäckström* (*Ak. Stockholm, Öfv.*, **45**, 493, 1888). Carbapatit *Chirvinsky* (*Ann. Géol. min. Russie*, **8**, 251, 256, 1906). Podolite *Chirvinsky* (*Cbl. Min.*, 279, 1907). Quercyite, Quercyite-α, Quercyite-β *Lacroix* (**4**, 579, 1910). Carbonatapatit *Brauns* (*Jb. Min., Beil.-Bd.*, **41**, 60, 1916). Carbonate-apatite. Carbonapatit. Kurskite *Chirvinsky* (*Jb. Min.*, II, 61, 71, 1911; *Mat. for Knowl. of Nat. Prod. Forces of Russia*, Petrograd, no. 30, 1919). Grodnolite *Morozewicz* (*Bull. soc. min.*, **47**, 46, 1924). α-Dahllite, β-Dahllite, γ-Dahllite.

Cryst.[1] Hexagonal—P; hexagonal-dipyramidal—$6/m$.

$$a:c = 1:0.7346; \quad p_0:r_0 = 0.8482:1$$

Forms:[2]

	ϕ	$\rho = C$	M	A_2
$c\ 0001$	.....	$0°00'$	$90°00'$	$90°00'$
$m\ 10\bar{1}0$	$30°00'$	$90\ 00$	$60\ 00$	$90\ 00$
$a\ 11\bar{2}0$	$0\ 00$	$90\ 00$	$60\ 00$	$60\ 00$
$k\ 41\bar{5}0$	$19\ 06\frac{1}{2}$	$90\ 00$	$70\ 54\frac{1}{2}$	$79\ 06\frac{1}{2}$
$r\ 10\bar{1}2$	$30\ 00$	$22\ 59$	$78\ 44\frac{1}{2}$	$90\ 00$
$x\ 10\bar{1}1$	$30\ 00$	$40\ 18\frac{1}{2}$	$71\ 08$	$90\ 00$
$y\ 20\bar{2}1$	$30\ 00$	$59\ 29$	$64\ 29$	$90\ 00$
$v\ 11\bar{2}2$	$0\ 00$	$36\ 18$	$90\ 00$	$72\ 47$
$s\ 11\bar{2}1$	$0\ 00$	$55\ 45\frac{1}{2}$	$90\ 00$	$65\ 35$
$\mu\ 21\bar{3}1$	$10\ 53\frac{1}{2}$	$65\ 59$	$80\ 03\frac{1}{2}$	$72\ 36$
$o\ 31\bar{4}2$	$16\ 06$	$56\ 49$	$76\ 34\frac{1}{2}$	$78\ 24$
$n\ 31\bar{4}1$	$16\ 06$	$71\ 53\frac{1}{2}$	$74\ 43$	$76\ 48$

Less common:

$h\ 21\bar{3}0$	$\zeta\ 5.0.\bar{5}.12$	$\alpha\ 30\bar{3}2$	$\pi\ 40\bar{4}1$	$\psi\ 11\bar{2}4$	$i\ 21\bar{3}2$
$\tau\ 10\bar{1}6$	$\eta\ 30\bar{3}5$	$w\ 70\bar{7}3$	$\chi\ 1.1.\bar{2}.12$	$d\ 22\bar{4}1$	$\omega\ 7.3.\bar{10}.3$
$\sigma\ 10\bar{1}3$	$\epsilon\ 30\bar{3}4$	$z\ 30\bar{3}1$	$\phi\ 11\bar{2}6$	$q\ 43\bar{7}1$	$\rho\ 41\bar{5}1$

Phosphates, Arsenates, Vanadates

Structure cell.[3] Space group $C6_3/m$.

Species	Anal. no.	a_0	c_0	$a_0:c_0$	Ref.
Fluorapatite	2	9.36 kX	6.88	1:0.735	5
Fluorapatite (manganoan)	4	9.35	6.83	1:0.731	4
Fluorapatite (manganoan)	MnO 10.3%	9.33	6.80	1:0.729	5
Fluorapatite (carbonatian)	6	9.34	6.89	1:0.738	9
Chlorapatite	8	9.52	6.85	1:0.719	6
Hydroxylapatite	Artif.	9.40	6.93	1:0.737	6
Hydroxylapatite (fluorian)	10	9.42	6.935	1:0.736	7
Hydroxylapatite (manganoan)	11	9.54	6.73	1:0.705	4
Carbonate-apatite	12	9.41	6.88	1:0.731	8

Cell contents $Ca_{10}(PO_4)_6(F,Cl,OH)_2$.

Habit. Crystals short to long prismatic [0001], especially in the apatite of crystalline limestones and of igneous rocks, with $\{10\bar{1}0\}$ and $\{10\bar{1}1\}$

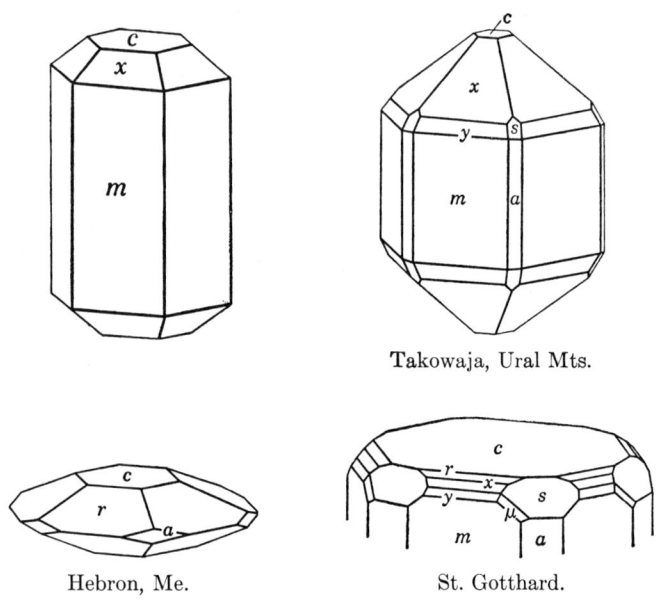

Takowaja, Ural Mts.

Hebron, Me.   St. Gotthard.

dominant; also thick tabular {0001}, commonly in the crystals of hydrothermal origin in pegmatites and veins, with $\{10\bar{1}0\}$, relatively large {0001}, and often also $\{10\bar{1}1\}$ or low pyramids. The tabular crystals sometimes are complex; the third-order forms ordinarily present only as small modifying faces. Also massive, coarse granular to compact; sometimes globular or reniform with a subfibrous, scaly, or imperfectly columnar structure or as fibrous crusts (especially in carbonate-apatite); stalactitic; earthy; oolitic. Sometimes rock-forming, in bedded deposits which may be of great extent; as nodular concretions in clays and shales; conglomeratic.

Twinning.[10] Twin plane $\{11\bar{2}1\}$, as contact twins; rare.[10] (2) Twin plane $\{10\bar{1}3\}$; rare. Reported with twin planes $\{10\bar{1}0\}$ and $\{11\bar{2}3\}$.

APATITE SERIES 881

Etching.[11] Etch pits and solution bodies conform to hexagonal dipyramidal ($6/m$) symmetry.

Oriented growths. Microscopic needle-like inclusions of rutile in apatite have been described,[12] the c-axes of the two species parallel. Monazite apparently also occurs as oriented inclusions. Zoned or overgrown crystals of different members of the apatite series are sometimes found, especially of carbonate-apatite enclosing a fluorapatite, and the accessory apatite of igneous rocks very commonly has a relatively deeply colored and pleochroic core [13] which may be crowded with minute inclusions.

P h y s. Cleavage $\{0001\}$ indistinct and $\{10\bar{1}0\}$ in traces, but varying in ease and quality. Fracture conchoidal to uneven. Brittle. H. 5, varying with direction [14] as in fluorite and in part closely approaching that species. G. of crystals usually 3.1–3.2 for fluorapatite and chlorapatite; 3.18 (artif., F),[15] 3.17 (artif., Cl); [15] pure hydroxylapatite and carbonate-apatite somewhat lower, 2.9–3.1. Luster vitreous, inclining to subresinous. Color usually sea-green, asparagus-green, bluish green, grayish green; also violet-blue, violet, amethystine; sometimes colorless, pale greenish white, gray, brown, flesh-red, rose-red, green, clear blue; also in manganoan varieties dark green to deep blue-green. The violet color is lost on heating.[16] Transparent to opaque. Streak white. Occasionally shows a bluish opalescence. Much apatite is fluorescent [17] in ultraviolet light, cathode rays or x-rays; also phosphorescent and sometimes strongly thermoluminescent. Not piezoelectric.

O p t. In transmitted light, colorless or faintly tinted. Uniaxial negative $(-)$. The members of the apatite group are at times abnormally biaxial with $2V$ up to $20°$ and may show a sixfold sectoral structure [45] in basal sections with often a central uniaxial core. Colored crystals are weakly to strongly dichroic, with absorption $E > O$.

Body color	Violet	Pale green	Yellow	Blue
O	Deep violet	Pale yellow	Yellowish brown	Sky-blue
E	Reddish violet	Pale bluish green	Dark green	Greenish blue

The indices of refraction in the apatite group are highest for chlorapatite and decrease both by the substitution of F or OH for Cl and by the substitution of C for P. The indices [18] are increased, however, by the substitution of Mn for Ca. Carbonate-apatite also has a higher birefringence.

Species	Anal. no.	nO	nE	Ref.
Chlorapatite	Artificial	1.6684	1.6675	19
Chlorapatite (fluorian)	7	1.658	1.653	20
Hydroxylapatite	9	1.651 (Na)	1.644	21
Hydroxylapatite (fluorian)	10	1.6452 (Na)	1.6413	7
Hydroxylapatite (manganoan)	11	1.661 (Na)	1.657	4
Fluorapatite	Artificial	1.6325 (Na)	1.630	15
Fluorapatite	3	1.6357 (Na)	1.6328	23
Fluorapatite (manganoan)	4	1.6459 (Na)	1.6411	24
Fluorapatite (carbonatian)	6	1.629	1.624	9
Fluorapatite (carbonatian)	15	1.625–1.627	1.620–1.622	26
Fluorapatite (carbonatian)	14	1.627–1.630	1.614–1.617	25
Carbonate-apatite	12	1.628± (Na)	1.619±	8
Carbonate-apatite	13	1.603	1.598	27

INDICES AT VARIOUS WAVELENGTHS (FLUORAPATITE)

λ	Rotenkopf, Zillertal [28]		Knappenwand, Sulzbachtal [28]	
	nO	nE	nO	nE
688	1.6425	1.6382	1.6356	1.6326
588	1.6460	1.6416	1.6388	1.6357
502	1.6517	1.6470	1.6445	1.6413
447	1.6574	1.6525	1.6497	1.6462

The dense varieties of carbonatian hydroxyl-fluorapatite which comprise the bulk of continental and insular phosphate rock and fossil bone, and including all collophane, are usually quite isotropic due to aggregate polarization in a mass composed of submicroscopic crystallites and afford weak and diffuse x-ray diffraction patterns.[29] Sometimes faintly birefringent with extinction either undulose or perpendicular to the banding in specimens with colloform structure. The index of refraction varies widely, reflecting the large and variable content of non-essential water, and ranges from roughly 1.565 to 1.625 with most specimens about 1.59–1.61.

Chem. Fluorapatite, chlorapatite, and hydroxylapatite conform to the formula $Ca_5(PO_4)_3(F,Cl,OH)$. In this formula, F, Cl, and OH can substitute mutually to form complete series to the essentially pure end-members. The species fluorapatite comprises the range of composition with F > Cl or OH, chlorapatite has Cl > F or OH, and hydroxylapatite has OH > F or Cl. CO_2 is sometimes found in small amounts as an essential constituent, giving rise to carbonatian varieties (see below). The mechanism of this type of compositional variation has been the subject of much investigation.[30] It apparently involves the substitution of (CO_3OH) groups for (PO_4), and the general formulation may be written as $Ca_5(P_{1-x}C_xO_{4-x}OH_x)_3(F,Cl,OH)$. The substitution of C for P also can be compensated electrostatically by a coupled substitution of alkalies for Ca (see dehrnite); a substitution of C for Ca also has been advocated. Relatively highly carbonatian types are distinguished by the species name carbonate-apatite. The range of composition to which this name should apply hitherto has not been sharply defined; it is here used for material in which CO_3 > F, Cl or OH in molecular per cent. Some analyses of carbonate-apatite approach the formula $Ca_{10}(PO_4)_6(CO_3) \cdot H_2O$.

Other types of compositional variation also are known in apatite. Mn" can substitute for Ca up to at least Mn:Ca = 1:5.7 (~10 weight per cent MnO), especially in the fluorapatite of granite pegmatites. Rare earths, principally Ce, are frequently present[31] in substitution for Ca, especially in the apatite of alkalic igneous rocks, but only rarely in appreciable amounts (analysis 5). Sr sometimes substitutes for Ca, with Sr:Ca = 1:6.8 in analysis 5, but generally is absent. Mg occurs rarely and only in very small amounts in apatite. Alkalies are present only in traces in ordinary apatite, although they may be present in large amounts in the isostructural and probably isomorphous silicate-phosphates (see ellestadite). (SiO_4) and (SO_4) occur in apatite in substitution[32] for (PO_4) although ordinarily absent or present only in traces (see, however, wilkeite and ellestadite). The substitution of Si^4 for P^5 is compensated electrostatically by a concomitant substitution of S^6 for P^5 or of Na' for Ca". (AsO_4) can substitute for (PO_4), and a series probably extends to svabite (and fermorite); As usually is absent. A sulfidic apatite in which S'' apparently substitutes for 2(OH) has been reported in urinary calculi.[33] Small amounts of Al, Cr''', Cr^6, and Fe''' apparently can enter apatite in solid solution.[49] The existence of a compound $Ca_{10}(PO_4)_6O$, the so-called voelckerite or oxyapatite, has been claimed or postulated. Structural considerations and the evidence of artificial systems, however, indicate that such a compound does not exist.[34] The seeming content of oxygen in some apatites, indicated by the high summation of the analyses, may be due to analytical error in the determination of OH, which in part at least is said to be held up to 1400°, or in the determination of F. In many of the older analyses F was probably overlooked or measured erroneously.

Anal.[35]

	1	2	3	4	5	6	7
CaO	55.38	55.16	55.59	50.31	42.38	53.94	52.40
MnO		0.12		5.32	0.05		1.52
P_2O_5	42.06	41.30	41.95	41.50	38.33	38.13	40.98
F	1.25	3.67	3.02	3.41	3.73	3.71	1.15
Cl	2.33	0.09			nil		3.74
H_2O	0.56	0.01		0.28	0.28	0.47	0.06
CO_2		0.50	pres.			3.40	
Rem.		1.29	0.33	0.65	16.90	2.27	0.75
	101.58	102.14	100.89	101.47	101.67	101.92	100.60
O = F,Cl	1.58	1.56		1.56	1.57	1.56	
Total	100.00	100.58	100.89	99.91	100.10	100.36	
G.		3.176	3.206	3.22	3.355	~3.15	

1. $Ca_5(PO_4)_3(F,Cl,OH)$ with F:Cl:OH = 1:1:1. 2. Fluorapatite. Faraday Township, Ontario.[36] Rem is FeO 0.14, Fe_2O_3 0.63, Al_2O_3 0.24, insol. 0.28. 3. Fluorapatite. Sunk, Styria.[23] Rem. is FeO 0.08, MgO tr., insol. 0.02, ign. loss 0.23. 4. Manganoan fluorapatite. Varuträsk, Sweden.[24] Rem. is FeO 0.26, MgO 0.04, insol. 0.35; H_2O includes H_2O- 0.03; alkalies nil. 5. Strontian fluorapatite. Khibina tundra, U.S.S.R. (*saamite*).[37] Rem. is SiO_2 1.55, rare earths 3.22, Fe_2O_3 0.12, Al_2O_3 0.34, SrO 11.42, MgO 0.05, Na_2O 0.13, K_2O 0.07, As and Zr tr. 6. Carbonatian fluorapatite. Wheal Franco, Devonshire (*francolite*).[9] Rem. is MgO 0.10, Fe_2O_3 0.34, insol. 1.83. 7. Fluorian chlorapatite. Kurokura, Japan.[20] Rem. is FeO 0.21, MgO 0.54.

	8	9	10	11	12	13	14	15
CaO	52.97	55.84	55.47	44.14	53.16	56.47	54.84	51.30
MnO		0.07	0.06	7.50				tr.
P_2O_5	40.50	42.05	42.19	40.26	38.57	35.75	35.01	34.63
F	0.17	0.16	1.01	nil	0.19		5.60	3.66
Cl	4.13	tr.			0.02		0.03	nil
H_2O		1.86	1.73	2.87	1.20	4.44	1.67	2.20
CO_2	nil			nil	4.46	3.36	4.43	3.20
Rem.	2.43	0.25	0.60	4.32	2.28	0.67	1.08	6.93
	101.20	100.23	101.06	99.09	99.88	100.69	102.66	101.92
O = F,Cl	0.98	0.07	0.43		0.08		2.36	1.54
Total	100.22	100.16	100.63	99.09	99.80	100.69	100.30	100.38
G.	3.25	3.21	3.21	3.19	~2.93		3.116	~3.15

8. Chlorapatite. Kragerö, Norway.[6] Rem. is MgO 0.29, Na_2O 0.22, K_2O 0.10, SiO_2 1.16, Fe_2O_3 0.18, ign. loss at 1000° 0.48. 9. Hydroxylapatite. Georgia.[21] Rem. is MgO 0.10, insol. 0.15. 10. Fluorian hydroxylapatite. Hospenthal, Switzerland.[7] Rem. is insol. 11. Manganoan hydroxylapatite. Varuträsk, Sweden.[22] Rem. is FeO 3.07, SrO 0.18, Na_2O 0.58, insol. 0.49, H_2O- 0.31. 12. Carbonate-apatite. Mouillac, France.[8] Rem. is SO_3 0.05, SiO_2 0.40, Al_2O_3 0.44, Na_2O 0.77, K_2O 0.28, Fe_2O_3 0.34,

H_2O $-110°$ 0.48, H_2O $110°-300°$ 0.21, H_2O $+300°$ 0.51. 13. Carbonate-apatite. St. Paul's Rocks, Atlantic Ocean.[27] Rem. is MgO 0.30, SiO_2 0.01, SO_3 0.36. H_2O- 0.53. 14. Carbonatian fluorapatite. Richtersveld, South Africa.[25] Rem. is MgO 0.53, Na_2O 0.20, Fe_2O_3 0.25, insol. 0.10. H_2O $-105°$ 0.16, H_2O $105°-300°$ 0.62, H_2O $+300°$ 0.89. 15. Carbonatian fluorapatite. Milburn, New Zealand.[26] Rem. is MgO 0.41, Na_2O 0.46, K_2O 0.17, SiO_2 2.27, Al_2O_3 1.18, Fe_2O_3 2.11, FeO 0.27, TiO_2 0.06. H_2O- 1.02.

Var.

FLUORAPATITE

Chlorian. Chlor-fluorapatite. The term chlorian may be applied to the range of composition of fluorapatite in the binary series toward chlorapatite in which Cl is present in considerable amounts although still subordinate to F in atomic per cent; also similarly applicable to regions of the ternary series between F, Cl, and OH in which $F \geq Cl \gg OH$. Highly chlorian fluorapatite is rare (see analysis 7); the entrance of Cl is accompanied by an increase in G. and in indices of refraction.

Hydroxylian. Hydroxyl-fluorapatite. Voelckerite pt. Used to designate regions of composition in fluorapatite in which OH is present in relatively large amounts, i.e. $F \geq OH \gg Cl$. Hydroxylian fluorapatite and fluorian hydroxylapatite (with $OH \geq F \gg Cl$), and often containing some CO_2, constitute the great bulk of the crystallized apatites and the massive phosphate rocks.

Carbonatian. Eupyrchroite *Emmons* (1838). Francolite *Brooke* and *Henry* (1850). Nauruite *Elschner* (1913). Staffelite *Stein* (1866). Fluorcollophane pt. Carbonate-apatite pt. Contains considerable amounts of CO_2. The substitution is accompanied by a slight increase in the indices of refraction and in the birefringence.

Manganoan. Mangan-fluorapatite. Manganapatite pt. Fluormanganapatite *Laubmann* and *Steinmetz* (1920). Contains Mn'' in substitution for Ca at least up to Mn:Ca = 1:5.7. Color often dark blue-green, with increased indices of refraction and G.

Strontian. Saamite *Volkova* and *Melentiev* (1939). Contains Sr in substitution for Ca (analysis 5), especially in the apatite of alkalic igneous rocks. Rare earths often also are present.

CHLORAPATITE

Relatively highly fluorian, hydroxylian, or carbonatian varieties of chlorapatite are rare or unknown.

HYDROXYLAPATITE

Fluorian. Fluor-hydroxylapatite. See hydroxylian fluorapatite.

Chlorian. Cl is either absent or present in only very small amounts in hydroxylapatite, and a very large gap exists in the binary series toward chlorapatite.

Manganoan. Mangan-hydroxyapatite. Manganapatite pt. Contains Mn'' in substitution for Ca up to at least Mn:Ca = 1:7.7 (analysis 11).

Carbonatian. Collophane pt. Carbonate-apatite pt. Grades into carbonatian fluorapatite (which see) and carbonate-apatite.

COLLOPHANE

The name collophane is here used as a generic designation for the massive, cryptocrystalline types of apatite such as constitute the bulk of phosphate rock and fossil bone. Used in this sense the term is analogous to limonite, wad, gummite, and bauxite. Collophane in its physical appearance is often opaline or horn-like, with a dense, layered or colloform structure; sometimes concretionary, nodular, spherulitic, or pulverulent. Color grayish white, yellowish, or brown; luster weakly vitreous to subresinous or dull. The H. (3–4) and G. (2.5–2.9) are relatively low; see further under O p t. The identification of collophane as a particular species of the apatite group can be effected by x-ray and chemical examination; it commonly is a carbonatian, intermediate member of the hydroxylapatite-fluorapatite series.

Tests. B.B. fusible on the edges with difficulty. Soluble in HCl or HNO_3, the carnatian varieties and carbonate-apatite dissolving with slight effervescence.

O c c u r. The members of the apatite group are among the commonest and most widely distributed of all minerals and are by far the most abundant of the minerals containing phosphorus. Fluorapatite and fluorian hydroxylapatite together with carbonatian varieties of these are the important members of the group, whereas essentially pure chlorapatite, carbonate-apatite, and hydroxylapatite are rare and restricted in occurrence. The fluorine-containing types occur in almost all igneous rocks [38] as an early-formed accessory mineral, usually in microscopic crystals, and may occur as extremely large bodies as late-magmatic segregations from alkalic igneous rocks (Kola peninsula). Also found crystallized in the pegmatitic facies of both acidic and basic types of igneous rocks; in magnetite deposits; and in hydrothermal veins, especially those formed at relatively high temperatures, and in veins of the Alpine type. Apatite is very common in both regionally and contact metamorphosed rocks, especially in the crystalline limestones where it is associated with sphene, zircon, pyroxene, amphibole, spinel, vesuvianite, phlogopite; also in talc and chlorite schists; as a contact metamorphic mineral. Also as extensive bedded deposits of marine origin, and in residual or detrital deposits derived therefrom; and as replacements of limestone or coral by phosphatic solutions derived by the leaching of guano deposits.[46]

The following localities for the fluorine-rich types of apatite may be mentioned for the fine specimens that they have afforded. In Russia on the Takowaja River, east of Ekaterinburg, Ural Mountains, with emerald and chrysoberyl in mica schist; at Miask in the Ilmen Mountains. Enormous deposits occur in alkalic igneous rocks in the Khibina tundra, Kola peninsula.[39] At Schlaggenwald and Pisek in Bohemia, Czechoslovakia. In Germany in pegmatite druses at Epprechtstein and Waldstein in the Fichtel Mountains, Bavaria; in the sanidinite of the Laacher See, Rhein Province (sulfatian pt.); in the tin veins of Ehrenfriedersdorf, Saxony.

In Austria, especially on the Knappenwand in the Untersulzbachtal, Salzburg, and in the Zillertal, Tyrol. In Switzerland at Gletsch on the Rhone glacier; in veins at various places on the St. Gotthard and Aar massifs. In Italy notably at Biella in Piedmont in syenite, and on Elba in pegmatite. Apatite occurs in Spain at Jumilla, Murcia, in andesite-tuff, and at Cáceres and Logrosán, Estremadura, in veins. Found in the tin veins of Villeder, France, and similarly in Cornwall, England; also at the Wheal Franco (carbonatian; *francolite*) at Tavistock, Devonshire. In Sweden at Gellivara in veins in gabbroic rock and at Nordmark in magnetite ores. Fine yellow crystals are found in the magnetite deposit at Cerro Mercado, Durango, Mexico. In the United States, fluorapatite occurs in fine crystals and sometimes of a deep violet color in pegmatite druses in New England, notably at Mt. Apatite, Auburn, Maine. Also in the magnetite deposits of the Adirondack Mountains, New York; in the crystalline limestone of Sussex County, New Jersey, as at Franklin. In Canada found as fine crystals, sometimes weighing over 500 pounds, in Grenville marble as irregular veins or contact deposits; the chief localities are in Ottawa County, Quebec, in the townships of Buckingham, Templeton, Portland, Hull, and Wakefield, and in central Ontario at North Burgess in Lanark County, Eganville, and Sebastopol in Renfrew County, and Bedford in Frontenac County.

Chlorapatite is found chiefly in veins in gabbroic rocks in southeastern Norway in the region between Lillesard and Bamle, especially at Oedegaarden; associated minerals include rutile, ilmenite, sphene, magnetite, pyrrhotite, hornblende, wernerite. Chlorapatite also occurs in some meteorites. Hydroxylapatite occurs in talc-schist in the Hospenthal, Uri, Switzerland; Rossa in the Val Devero, Ossola, Italy, with sphene in a diallage-serpentine rock;[40] and near Holly Springs, Cherokee County, Georgia, in talc-schist. Carbonate-apatite occurs extensively as nodules in phosphate rock near the Ushitsa river in Podolia (*podolite*) and similarly in Kursk (*kurskite*), in Russia, and near Cody in the Beartooth Mountains, Wyoming, U.S.[48] Also as crusts upon chlorapatite at Oedegaarden, Norway (*dahllite*); at Mouillac, France; in crevices in sodalite rock at Kangerdluarsuk, Greenland;[47] in a guano-derived deposit on St. Paul's Rocks in the Atlantic; as zonal overgrowths upon fluorapatite at Magnet Cove, Arkansas,[41] and similarly (or as fibrous crusts) in pegmatite at various New England localities, as at Palermo, New Hampshire, and Black Mountain, Greenwood, and Newry in Maine. The apatite of phosphate-rock deposits and of fossil bone and shell is an intermediate member of the fluorapatite-hydroxylapatite series and generally contains a significant amount of CO_2. The manganoan varieties of apatite are exclusively found in pegmatites, as at Branchville, Connecticut, and Varuträsk, Sweden.

Alter. Pseudomorphs have been found of serpentine, kaolin, and turquois after apatite, and of apatite after pyromorphite (?), prosopite, and brushite. Apatite occurs rarely as the petrifying material of fossil wood.[42]

A r t i f.[43] Fluorapatite and chlorapatite have been obtained in crystals by fusion of $Ca_3(PO_4)_2$ or of hydroxylapatite with CaF_2 or $CaCl_2$; also chlorapatite by heating $CaCl_2$, $(NH_4)_2H(PO_4)$, and NH_4Cl in a closed tube, by fusion of $Ca_3(PO_4)_2$ with a large excess of NaCl, by heating $Ca_3(PO_4)_2$ with NaCl and H_2O in a closed tube at 250°, and in other ways. Hydroxylapatite [44] is afforded by precipitation of solutions of calcium salts by ammoniacal solutions of phosphates; the precipitate may contain excess adsorbed Ca or (PO_4). Also formed by prolonged treatment of $Ca_3(PO_4)_2$ or $CaHPO_4$ in hot water; by heating $4CaO \cdot P_2O_5$ (hilgenstockite) at about 1100° in the presence of water vapor or by boiling in water.

A number of compounds of the apatite type not known in nature also have been synthesized. These include $Sr_5(PO_4)_3(OH)$, $Pb_5(PO_4)_3(OH)$, $Ba_5(PO_4)_3(OH)$, $Ca_9Na(PO_4)_6Cl$, $3Ca_3(PO_4)_2 \cdot 2H_2O$, $Ca_5(AsO_4)_3Cl$, $Na_6Ca_4(SO_4)_6F_2$, $Ca_8La_2(PO_4)_6O_2$, $Ca_{10}Si_3S_3O_{24}F_2$, $Ca_{10}Si_2P_2S_2O_{24}F_2$, etc.

N a m e. Apatite from ἀπατᾶν, *to deceive*, the mineral having been early confused with aquamarine, amethyst, olivine, fluorite, and other species. The mineral was first called apatite in 1788 by Werner although earlier recognized by some as a distinct species. Its constitution as a calcium phosphate, and its similarity in composition to calcined bone was established by the analyses of Proust and of Klaproth in 1788, although the content of F and Cl was not recognized. The *spargelstein* or *asparagolith*, *asparagus-stone*, was a yellowish-green, prismatic variety originally from Murcia, Spain. Francolite was named after its occurrence at the Wheal Franco near Tavistock in Devonshire, England. Dahllite after the brothers T. and J. Dahll, Norwegian geologists. Voelckerite after the English agricultural chemist J. A. Voelcker (1822–1884), who first proposed the existence of an oxyapatite.

Ref.

1. Angles of Koksharov (**2**, 39, 1854; **5**, 86, 1866) on fluorapatite from Ehrenfriedersdorf, Saxony. Virtually identical values have been reported by others for essentially pure fluorapatite—see summaries of Wolff and Palache, *Zs. Kr.*, **36**, 438 (1902), and Hausen, *Acta Ac. Aboensis, math. phys. Kl.*, **5**, no. 3 (1929). Essentially pure hydroxylapatite has almost identical elements; c decreases, however, with substitution of Cl for (F,OH), and essentially pure chlorapatite has $c \sim 0.70$. The morphological elements of carbonate-apatite are not known. On the variation in angle with temperature see Eissner, Inaug. Diss. Leipzig (1913).

2. Goldschmidt (**1**, 73, 1913); also Parker, *Zs. Kr.*, **64**, 224 (1926); Gennaro, *Acc. Torino, Att.*, **66**, 433 (1931); Whitlock, *Am. Mus. Nov.*, No. 190 (1925); Bianchi, *Atti soc. ital. sci. nat.*, **58**, 306 (1919); Hawkins, *Am. Min.*, **7**, 27 (1922); Niggli and Faesy, *Zs. Kr.*, **62**, 154 (1925); Mieleitner, *Zs. Kr.*, **56**, 90 (1921). The reported forms are principally for fluorapatite, and the variation in morphology with composition is not clearly established. On vicinal development see Kalb, *Zs. Kr.*, **74**, 469 (1930); Lorenz, *Ber. Ak. Leipzig*, **73**, 249 (1921); Karnojitsky, *Vh. Min. Ges. St. Petersburg*, **33**, 65 (1895).

Rare and doubtful:

$10.9.\overline{19}.0$	$\lambda\ 4.0.\overline{4}.21$	$\beta\ 50\overline{5}7$	$O\ 12.0.\overline{12}.5$	$S\ 32\overline{5}2$	$U\ 71\overline{8}1$
$5\overline{4}90$	$f\ 10\overline{1}4$	$g\ 70\overline{7}9$	$D\ 13.0.\overline{13}.4$	$M\ 9.4.\overline{13}.4$	$V\ 81\overline{9}1$
$l\ \ \ 43\overline{7}0$	$20\overline{2}7$	$\delta\ 40\overline{4}5$	$80\overline{8}1$	$M_1\ 52\overline{7}2$	$Y\ 10.1.\overline{11}.1$
$W\ 52\overline{7}0$	$e\ 7.0.\overline{7}.11$	$j\ 70\overline{7}8$	$3.1.\overline{4}.280$	$N\ 72\overline{9}2$	$6.\overline{6}.\overline{12}.1$
$2.0.\overline{2}.19$	$20\overline{2}3$	$\gamma\ 80\overline{8}9$	$31\overline{4}6$	$T\ \ 51\overline{6}1$	$10\overline{1}9$

3. On the structure of apatite see Hendricks, Jefferson, and Mosley, Zs. Kr., **81**, 352 (1932); Gruner and McConnell, Zs. Kr., **97**, 208 (1937); Mehmel, Zs. Kr., **75**, 323 (1930) and Zs. phys. Chem., **15**, 223 (1931); Náray-Szabó, Zs. Kr., **75**, 387 (1930); Hendricks and Hill, Science, **96**, 255 (1942); Bale, Am. J. Roentg. Rad. Therapy, **43**, 735 (1940); Beevers and McIntyre, Min. Mag., **27**, 254 (1945).
4. Mason, Geol. För. Förh., **63**, 279 (1941).
5. McConnell, Am. Min., **23**, 1 (1938).
6. Hendricks, Jefferson, and Mosley (1932).
7. Burri, Jakob, Parker, and Strunz, Schweiz. Min. Mitt., **15**, 327 (1935).
8. McConnell, Am. J. Sc., **36**, 296 (1938).
9. Sandell, Hey, and McConnell, Min. Mag., **25**, 395 (1939).
10. Lacroix (**4**, 387, 1910); Hidden and Washington, Am. J. Sc., **33**, 501 (1887); Washington, J. Geol., **3**, 25 (1895); Himmelbauer, Ak. Wien, Sitzber., **138**, 251 (1929); Heddle (**2**, 158, 1901).
11. Lorenz, Ak. Leipzig, Sitzber., math.-phys. Kl., **73**, 267 (1921); Baumhauer, Ak. Berlin, Sitzber., **42**, 863 (1887) and **45**, 447 (1899); Dürrfeld, Zs. Kr., **50**, 590 (1912); Honess (98, 1927); Royer, C.R., **202**, 1346 (1936); Ichikawa, Am. J. Sc., **14**, 231 (1927).
12. Barthoux, Bull. soc. min., **48**, 225 (1925).
13. See Baker, Am. Min., **26**, 382 (1941); Niggli and Faesy, Zs. Kr., **62**, 166 (1925).
14. For measurements of the vectorial hardness see Pöschl, Härte der fest. Körper, **55** (1909), and Holmquist, Geol. För. Förh., **38**, 501 (1916); the variation with composition is not established.
15. Nacken, Cbl. Min., 547 (1912).
16. Wolff and Palache (1902). On the color see also Hoffmann, Chem. Erde, **11**, 552 (1938).
17. On luminescence in apatite see Köhler and Haberlandt, Chem. Erde, **9**, 88(1934); Ulrich, Min. Abs., **7**, 529 (1940); Iwase, Sc. Pap. Inst. Phys. Chem. Res. Tokyo, **27**, no. 567, 1 (1935).
18. Otto, Min. Mitt., **47**, 89 (1935).
19. Borgström, Finska Kem. Medd., no. 2, 51 (1932). Nacken (1912) gives nO 1.6667, birefringence nearly zero.
20. Harada, J. Fac. Sc. Hokkaidô Univ., **4**, 11 (1938).
21. Mitchell, Faust, Hendricks, and Reynolds, Am. Min., **28**, 356 (1943).
22. Mason, Geol. För. Förh., **63**, 279 (1941).
23. Grosspietsch, Zs. Kr., **54**, 461 (1915).
24. Quensel, Geol. För. Förh., **59**, 257 (1937).
25. de Villiers, Am. J. Sc., **240**, 443 (1942).
26. Hutton and Seelye, Trans. Roy. Soc. New Zealand, **72**, pt. 3, 191 (1942).
27. Washington, Am. Min., **14**, 369 (1929).
28. Baumhauer, Zs. Kr., **45**, 555 (1908).
29. Frondel, Am. Min., **28**, 215 (1943); Frondel and Prien, Science, **95**, 431 (1942); Rogers, Bull. Geol. Soc. Am., **35**, 535 (1924); Hendricks, et al., Indust. Eng. Chem., **23**, 1413 (1931).
30. See Gruner and McConnell (1937); McConnell (1938a, 1938b); de Villiers (1942); Hendricks and Hill (1942); McConnell and Gruner, Am. Min., **25**, 157 (1940); Borneman-Starinkevitch, C. r. ac. sc. U.R.S.S., **19**, 253 (1938); McConnell, ibid., **22**, 87 (1939); Belov, ibid., **22**, 89 (1939); McConnell, ibid., **25**, 46 (1939); Sandell, et al. (1939).
31. See Volkova and Melentiev, C. r. ac. sc. U.R.S.S., **25**, 120 (1939); Borneman-Starinkevitch, ibid., 39 (1924); Fersman, ibid., 42 (1924); Zambonini, Zs. Kr., **58**, 226 (1923); Sahama, Bull. Comm. Géol. Finlande, no. 126, 50 (1941); Bellucci and Grassi, Gazz. chim. ital., **49**, 232 (1919).
32. On the substitution of S and Si for P in apatite see Dihn and Klement, Zs. Elektrochem., **48**, 331 (1942), and McConnell, Am. Min., **22**, 977 (1937).
33. Hudson, Cherkes, and Buchwald, J. Urology, **53**, 654 (1945), and Frondel, priv. comm. (1946).
34. Rogers, Am. J. Sc., **33**, 475 (1912); Bredig, Franck and Fuldner, Zs. Elektrochem., **39**, 959 (1933); Trömel and Möller, Zs. anorg. Chem., **206**, 227 (1932).
35. Additional analyses are tabulated by Hintze (**1** [4A], 562, 1924); also Gordon, Proc. Ac. Sc. Philadelphia, **96**, 333 (1944); Volodchenkova and Melentiev, C. r. ac. sc. U.R.S.S., **39**, 34 (1943); Thewlis et al., Trans. Faraday Soc., **35**, 358 (1939); Dittler and Hueber, Ann. Mus. Wien, **46**, 185 (1932), Alfani, Per. Min., **3**, 220 (1932), Deans, Min. Mag., **25**, 135 (1938); Simpson, J. Roy. Soc. Western Australia, **15**, 99 (1929); ibid., **20**, 47 (1933); Otto (1935); Mason, Geol. För. Förh., **63**, 383 (1941); Shannon and

Larsen, *Am. J. Sc.,* **9**, 250 (1925); Ramdohr, *Jb. preuss. geol. Landesanst.,* **40**, 284 (1919); Afanasev, *C. r. ac. sc. U.R.S.S.,* **62**, 677 (1948). On trace elements see Jolibois and Hébert, *C.R.,* **222**, 569 (1946).
36. Dadson, *Univ. Toronto Stud., Geol. Ser.,* no. 35, 51 (1935).
37. Volkova and Melentiev (1939).
38. Kind, *Chem. Erde,* **12**, 50 (1938).
39. Antonov, *State Chem. Tech. Publ., Leningrad,* **7**, 196 pp. (1934); Granigg, *Zs. pr. Geol.,* **41**, 1 (1933).
40. Bianchi, *Atti Soc. Ital., Soc. Nat.,* **458**, 306 (1919).
41. McConnell and Gruner, *Am. Min.,* **25**, 157 (1940).
42. See Simpson, *Min. Mag.,* **19**, 29 (1920).
43. Mellor (**3**, 896, 1923); Nacken, *Cbl. Min.,* 545 (1912); Bredig, *et al., Zs. Elektrochem.,* **38**, 158 (1932); Eisenberger, Lehrman, and Turner, *Chem. Rev.,* **26**, 257 (1940).
44. Schleede, *et al., Zs. Elektrochem.,* **38**, 633 (1932); Trömel and Möller, *Zs. anorg. Chem.,* **206**, 227 (1932); Trömel, *Zs. phys. Chem.,* **158**, 422 (1932); Klement, *Zs. anorg. Chem.,* **242**, 215 (1939); Fouretier, *C.R.,* **205**, 413 (1937); Müller, *Helvet. Chim. Acta,* **30**, 2069 (1947).
45. On the optical anomalies see Deans (1938); Wolff and Palache (1902); Brauns, *Jb. Min., Beil.-Bd.,* **41**, 73 (1916); Viola, *Boll. com. geol. Ital.,* **6**, 106 (1905); Lacroix (**4**, 558, 1910); Chirvinsky, *Cbl. Min.,* 279 (1907).
46. On the mineralogy of phosphate-rock deposits see Frondel (1943); Hendricks, *et al.* (1931); Chirvinsky (1919); Lacroix (**4**, 561, 1910); Mansfield, *Am. J. Sc.,* **238**, 863 (1940); McConnell, *Bull. Geol. Soc. Am.,* **54**, 707 (1943). For analyses see Doelter (**3**, [1], 333, 354, 1914).
47. Bøggild, *Zs. Kr.,* **55**, 417 (1920).
48. McConnell, *Am. Min.,* **20**, 693 (1935).
49. Pieruccini, *Att. soc. tosc.,* **54**, (1947); Minguzzi, *Per. Min.,* **12**, 343 (1941).

41.7.2 PYROMORPHITE SERIES

41.7.2.1 **PYROMORPHITE** [$Pb_5(PO_4,AsO_4)_3Cl$]. Grön Blyspat, Minera plumbi viridis pt. *Wallerius* (296, 1748). Mine de Plomb verte *Wallerius* (**1**, 536, 1853). Grünbleierz, Braunbleierz *Schultze (Dresden Mag.,* **2**, 70, 1761; **2**, 467, 1765). Grün Bleyerz, Phosphorsaurehaltig [from Zschopau] *Klaproth (Crell's Ann.,* **1**, 394, 1784). Green Lead Ore, Brown Lead Ore. Phosphate of Lead. Phosphorsaures Blei, Phosphorblei, Buntbleierz *Germ.* Plomb phosphaté *Fr.* Polychrom, Pyromorphit *Hausmann* (1089, 1090, 1813). Traubenblei *Hausmann* (1093, 1813). Polysphærit *Breithaupt* (54, 1832). Nussièrite *Danhauser, Barruel (Ann. chim. phys.,* **62**, 217, 1836). Miesit *Breithaupt* (285, 1841). Cherokine *Shepard* (407, 1857; *Am. J. Sc.,* **24**, 38, 1857). Plumbeine, Sexangulit *Breithaupt (B. H. Ztg.,* **22**, 36, 44, 1863); Blaubleierz. Collieite *Brown (Trans. Dumfriesshire & Galloway, Nat. Hist. Antiq. Soc.,* **13**, 72, 1927).

41.7.2.2 **MIMETITE** [$Pb_5(AsO_4,PO_4)_3Cl$]. Minera plumbi Viridis pt., Plumbum arsenico mineralisatum *Wallerius* (296, 1748). Plomb vert arsenical [from Andalusia] *Proust (J. phys.,* **30**, 394, 1787); [from Roziers] *Fourcroy (Mem. Ac. Sc., Paris,* 1789). Arsenikalisches Bleyerz *Lenz* (2, 24, 1794). Grünbleierz pt., Buntbleierz pt., Flockenerz, Traubenblei pt., Arsensaures Blei *Germ.* Arsenate of Lead, Green Lead Ore pt. Plomb arseniaté *Fr.* Pyromorphite pt. *Mohs.* Mimetèse *Beudant* (**2**, 594, 1832). Mimetene *Shepard* (1835). Mimetesit *Breithaupt* (289, 1841). Mimetit *Haidinger* (1845), *Glocker* (254, 1847). Kampylit *Breithaupt* (**2**, 291, 1841); Campylite. Endlichite, pt.

C r y s t.[1] Hexagonal—*P*; hexagonal-dipyramidal—$6/m$.

PYROMORPHITE [2]

$$a{:}c = 1{:}0.7632; \qquad p_0{:}r_0 = 0.8501{:}1$$

Forms:[3]

	ϕ	$\rho = C$	M	A_2
c 0001	.....	0°00'	90°00'	90°00'
m 10$\bar{1}$0	30°00'	90 00	60 00	90 00
a 11$\bar{2}$0	0 00	90 00	90 00	60 00
x 10$\bar{1}$1	30 00	40 22	71 06½	90 00
y 20$\bar{2}$1	30 00	59 32	64 28	90 00
π 40$\bar{4}$1	30 00	73 36½	61 20	90 00
S 11$\bar{2}$1	0 00	55 49	90 00	65 34

Less common:

	ϕ	$\rho = C$	M	A_2
h 21$\bar{3}$0	ϵ 30$\bar{3}$4	P 80$\bar{8}$1	R 90$\bar{9}$1	u 21$\bar{3}$1

Mimetite [4]

$a{:}c = 1{:}0.7224; \quad p_0{:}r_0 = 0.8342{:}1$

Forms:[5]

	ϕ	$\rho = C$	M	A_2
c 0001		0°00'	90°00'	90°00'
m 10$\bar{1}$0	30°00'	90 00	60 00	90 00
d 11$\bar{2}$0	0 00	90 00	90 00	60 00
x 10$\bar{1}$1	30 00	39 50	71 19	90 00
s 11$\bar{2}$1	0 00	55 19	90 00	65 43½
u 21$\bar{3}$1	10 53½	65 37½	80 05½	72 39

Less common:

h 21$\bar{3}$0	r 10$\bar{1}$2	α 30$\bar{3}$2	y 20$\bar{2}$1	z 30$\bar{3}$1

Structure cell.[6] Space group $C6_3/m$.

	a_0	c_0	$a_0{:}c_0$
Pyromorphite	9.95 kX	7.31	1:0.735
Mimetite	10.24 kX	7.43	1:0.725

Cell contents $Pb_{10}(PO_4,AsO_4)_6Cl_2$.

Habit. Crystals prismatic [0001] and generally simple in habit with {10$\bar{1}$0}{0001} or these and {10$\bar{1}$1}; also equant, rarely tabular {0001} or pyramidal. Often in rounded barrel-shaped forms; also spindle-shaped,

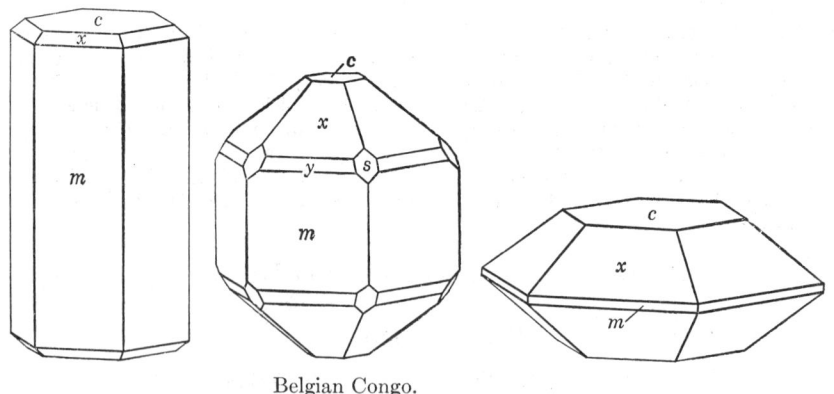

Belgian Congo.

Pyromorphite–Mimetite

or with cavernous basal terminations. As branching groups of prismatic crystals in nearly parallel position, tapering down to a point. Mimetite sometimes acicular. Often globular, reniform, botryoidal, or wart-like, with usually a subcolumnar structure; granular.

Etch figures.[9] Conform to hexagonal-dipyramidal ($6/m$) symmetry.
Twinning.[10] On $\{11\bar{2}2\}$, very rare.
P h y s. Cleavage $\{10\bar{1}1\}$ in traces. Fracture uneven to subconchoidal. Brittle. H. $3\frac{1}{2}$–4. G. 7.04 ± 0.04 (pyromorphite end-member; 7.14 calc.) to 7.24 (mimetite end-member; 7.28 calc.);[7] the G. decreases to 5.9 or less in varieties containing Ca in substitution for Pb. Luster resinous to sub-adamantine. Color of pyromorphite green, yellow, and brown of various shades, also wax-yellow, orange-yellow, orange-red, brownish red, grayish; very rarely colorless and transparent. Color of mimetite usually pale yellow to yellowish brown, orange-yellow, white, or colorless. Streak white or nearly so. Ordinarily subtransparent to translucent. Piezoelectric.[8] Crystals often show concentric zones of growth,[11] in which the P:As ratio varies with the outer zones higher in As.

O p t.[12] In transmitted light, colorless or faintly tinted and then weakly pleochroic with absorption $E > O$.

λ	Pyromorphite (end-member)		Mimetite (end-member)	
	nO	nE	nO	nE
404.66	2.144	2.131	2.263	2.239
589.3	2.058	2.048	2.147	2.128
690.75	2.041	2.030	2.124	2.106

The indices of refraction are markedly lowered from the values given by the substitution of Ca for Pb.

The members of the series, especially when high in As, generally are biaxial negative ($-$) and show a division of basal sections into six triangular areas, sometimes with a uniaxial negative ($-$) core.[13] Some crystals are wholly uniaxial. In the biaxial parts, the optic plane usually is parallel to a side of the hexagon with $X = c$, dispersion $r < v$, and $2V$ variable up to about 42°. $2V$ apparently decreases with increasing content of P.

C h e m. A chloride-phosphate-arsenate of lead, $Pb_5(PO_4,AsO_4)_3Cl$. P and As substitute mutually, and a complete series extends between the essentially pure end-members. The names pyromorphite and mimetite are applied to the halves of the series with $P > As$ and $As > P$, respectively. Ca frequently substitutes for Pb, with Ca:Pb = 1:2.1 in analysis 10 and 1:1.1 in the impure material of analysis 7. A complete series apparently does not extend in natural material between calcian varieties of pyromorphite-mimetite and chlorapatite; with increase of Ca beyond Pb:Ca = 1:1 mimetite passes into hedyphane. F and (OH) apparently substitute for Cl in pyromorphite-mimetite only in trace amounts; these constituents, however, ordinarily are not sought for in analyses. Fe'' has been reported in amounts up to about 1 weight per cent FeO. Cr is sometimes present in very small amounts. V is present in substitution for (As,P) (analysis 10), and more highly vanadian material has been reported (see under ***Var.***). Vanadinite forms a series toward mimetite up to about V:As = 1:1 and a very limited series toward pyromorphite. Rare earths, Sr,

Ba, Mn have been reported in trace amounts.[14] CO_2 is sometimes present in traces.[35]

Anal.[15]

	1	2	3	4	5	6	7
PbO	79.81	81.21	81.19	81.33	80.13	79.29	52.64
CaO		0.45	0.06		0.56		12.30
P_2O_5	7.61	15.77	16.32	16.11	15.65	15.56	19.80
As_2O_5	10.61	tr.	0.01	0.13	0.90	2.34	4.06
Cl	2.54	2.64	2.53	2.71	2.59	2.85	1.95
Rem.		0.48	0.20		0.51		9.64
	100.57	100.55	100.31	100.28	100.34	100.04	100.39
O = Cl	0.57	0.60	0.57	0.61	0.59	0.64	0.44
Total	100.00	99.95	99.74	99.67	99.75	99.40	99.95
G.		7.01–7.05	7.02				

	8	9	10	11	12	13	14
PbO	77.29	77.86	68.46	71.32	66.77	75.07	74.58
CaO			8.31	3.46	2.81		
P_2O_5	11.31	8.31	5.36	3.44	1.93	0.29	0.14
As_2O_5	8.98	11.51	12.06	19.65	25.99	22.89	23.17
Cl	2.31	1.72	2.47	2.57	2.32	2.30	2.39
Rem.			2.90				
	99.89	99.40	99.56	100.44	99.82	100.55	100.28
O = Cl	0.52	0.39	0.56	0.58	0.50	0.52	0.54
Total	99.37	99.01	99.00	99.86	99.32	100.03	99.74
G.				6.65			7.32

1. $Pb_5(PO_4,AsO_4)_3Cl$, with P:As = 1:1. 2. Pyromorphite. Ems, Germany.[16] Average of three. Rem. is Fe_2O_3 0.02, Al_2O_3 0.02, H_2O 0.14, insol. 0.30. 3. Pyromorphite. Braubach, Germany.[16] Rem. is Fe_2O_3 0.01, Al_2O_3 0.02, H_2O 0.08, insol. 0.09. 4. Pyromorphite. Wissen, Germany.[17] 5. Pyromorphite. Moyie, British Columbia.[18] Rem. is FeO 0.46, insol. 0.05. 6. Pyromorphite. Zschopau, Germany.[19] 7. Calcian pyromorphite (*nussièrite*). Nussières, France.[20] Rem. is FeO 2.44, SiO_2 7.20. 8. Pyromorphite. Roughten Gill, Cumberland.[19] 9. Pyromorphite. Roziers, Pontigibaud, France.[21] 10. Mimetite (calcian and vanadian). Mina Grande, Arqueros, Chile.[22] Rem. is CuO 0.96, V_2O_5 1.94. 11. Mimetite. Villevieille, France.[23] 12. Mimetite. Tsumeb, Africa.[24] 13. Mimetite. Bena e Padru, Sardinia.[25] 14. Mimetite. Phoenixville, Pennsylvania.[26]

Var. Calcian. Polysphærit *Breithaupt* (1832), Nussièrite *Danhauser, Barruel* (1836), Miesit *Breithaupt* (1841). Contains Ca in substitution for Pb, with attendant lowering of the specific gravity and indices of refraction. Color usually brown to yellowish gray and nearly white. Analyses 7, 10, 11, 12.

Vanadian. Collieite *Brown* (1927). A black botryoidal mineral from Leadhills, Scotland, containing about 4.1 per cent V_2O_5 apparently is a vanadian and calcian variety of pyromorphite;[27] G. 6.9–7.0. Vanadian material also has been reported from Chile (anal. 10).

Curved Crystals. Campylite. Pseudo-campylite *Dana* (770, 1892). Crystals of pyromorphite-mimetite distorted into barrel-shaped forms have been distinguished by the name campylite (from καμπύλος, curved). Originally from Drygill, Cumberland (phosphatian mimetite); also from Roziers and Villevieille in France and from other localities. The curved habit is not restricted to material of any particular composition.

Tests. B.B. fuses easily (1½). On charcoal fuses without reduction (pyromorphite) to a globule which on cooling assumes a crystalline polyhedral form. In C.T. gives a white sublimate of lead chloride. Soluble in HNO_3 and KOH. Pyromorphite is slightly soluble in carbonated water.[28]

O c c u r. Pyromorphite and mimetite are secondary minerals found in the oxidized zone of lead deposits. Associated species include cerussite and limonite principally, also smithsonite, hemimorphite, anglesite, malachite, leadhillite, caledonite, vanadinite, wulfenite, descloizite, mottramite, bayldonite. A few of the many known localities are given below. Pyromorphite occurs in Germany at Friedrichsegen near Ems and at Dernbach in Hesse-Nassau, Badenweiler in Baden, Bernkastel on the Moselle in the Rhine Provinces, at Zschopau and Schneeberg in Saxony. In Bohemia of Czechoslovakia at Příbram, Mies, and Bleistadt. From Beresovsk near Ekaterinburg in the Urals, U.S.S.R. At Sarrabus and Gennamari on Sardinia, Italy. In France, especially at Huelgoat and Poullaouen in Brittany; at Pontigibaud, Puy-de-Dôme, Nussière in Rhône, and Croix-aux-Mines, Vosges. In England from various localities in Cornwall, from Roughten Gill and Caldbeck Fells in Cumberland, and in Scotland at Leadhills. From Mindouli in the Belgian Congo and at Djebel Mahseur, Algeria, in Africa. At Broken Hill, New South Wales. In Mexico at the Ojuela mine, Mapimi. In the United States, pyromorphite occurs in Pennsylvania as fine specimens at the Wheatley mines, Phoenixville, Chester County, and the Ecton and Perkiomen mines in Montgomery County. From the Canton mine, Cherokee County, Georgia. At the Southampton lead mine, Hampden County, Massachusetts. In many mines in the Coeur d'Alene district, Shoshone County, Idaho.[29] In New Mexico in the Caballos Mountains district, Sierra County, and in the Central district, Grant County. At Leadville, Colorado. In Canada at Moyie, British Columbia.

Mimetite is less common than pyromorphite. It occurs in Germany at Badenweiler in Baden, at Borstein in Hesse, and notably at Johanngeorgenstadt in Saxony; also at Bräunsdorf, Saxony, and Lampersdorf, Silesia. At Příbram, Bohemia. In the Nerchinsk district, Transbaikalia, U.S.S.R. In France in the Pontigibaud district, Puy-de-Dôme. Found at many places in Cornwall, England; also at Roughten Gill and Drygill in Cumberland and at Leadhills and Wanlockhead in Scotland. At Långban, Sweden. At Broken Hill, New South Wales. In Algeria in the zinc deposits of Sidi Rouman, Bou-Thaleb, Djebel Grouz, and elsewhere; abundant with bayldonite and cerussite at Tsumeb, South West Africa. In Mexico at Santa Eulalia, Chihuahua, and from the Ojuela mine, Mapimi, and Mina del Diablo, Durango. As an impregnation of dacite tuff at Lomitos, Bolivia.[30] In the United States, mimetite was found at Phoenixville, Pennsylvania; at the Mammoth mine, Pinal County, and the Vulture district in Yavapai County, Arizona; in the Cerro Gordo district, Inyo County, California; in the Tintic district, Utah; in the Eureka district, Nevada.

Alter. Pseudomorphs of pyromorphite after galena and cerussite are common. Galena often occurs as more or less complete pseudomorphs after pyromorphite or forms thin films on the surface of pyromorphite crystals (*Blaubleierz, Plumbeine*); the two minerals are mutually oriented, with galena $\{001\}$ [001] parallel pyromorphite $\{0001\}$ $(11\bar{2}0)$ $(10\bar{1}0)$ [0001].[31] Pseudomorphs also have been found of apatite after pyromorphite, and incrustation pseudomorphs of limonite, chalcedony, hemimorphite, and vauquelinite after pyromorphite. Mimetite has been found at Durango, Mexico, as pseudomorphs after an unknown mineral, perhaps anglesite;[32] also altered to bayldonite (Tsumeb).

Artif.[33] Obtained as large crystals by cooling a melt of lead phosphate or lead arsenate with excess lead chloride. Intermediate members of the series can be obtained by appropriate admixture.[34] Also by heating these reagents with water vapor in a closed tube at lower temperatures, or by heating ammonium arsenate or phosphate with lead chloride and water in a closed tube. Mimetite also has been obtained in crystals by heating $PbHAsO_4$ in alkali chloride solution or by treating a solution of $PbHAsO_4$ in HCl with lead acetate. Pyromorphite is formed when $PbCl_2$ solution is added to a weak HCl solution containing suspended $Ca_3(PO_4)_2$. The F and I analogues of pyromorphite also have been reported synthesized.

Name. Pyromorphite from πῦρ, *fire*, and μορφή, *form*, because the globule produced by melting a fragment assumes a crystalline shape on cooling. Mimetite from μιμητής, *an imitator*, from its resemblance to pyromorphite.

Ref.

1. The x-ray diffraction effects accord with the Laue symmetry $6/m$, and the etch symmetry together with the characteristic hexagonal-dipyramidal morphological development of the isostructural species vanadinite and apatite indicate the point symmetry to be $6/m$. However, some crystals at least are piezoelectric, as noted below, and apparently have inverted to a markedly pseudo-hexagonal, orthorhombic modification.

2. Elements of Haidinger (**2**, 134, 1825) on Breisgau crystals. For a discussion of the variation in angles with the P:As ratio see Amadori, *Ist. Lombardo, Rend.*, [2], **49**, 137 (1916).

3. Dana (770, 1892), Hintze (**1** [4A], 572, 1924), Goldschmidt (**7**, 4, 1922). Uncertain: $20\bar{2}9$, 15.0.15.14.

4. Elements of Haidinger (**2**, 135, 1825) on Johanngeorgenstadt crystals.

5. Hintze (**1** [4A], 595, 1924); Goldschmidt (**6**, 45, 1920); Wherry, *Proc. U. S., Nat. Mus.*, **54**, 373 (1918); Anderson, *Rec. Australian Mus.*, **13**, 1 (1920). Uncertain: $70\bar{7}1$.

6. Hendricks, Jefferson, and Mosley, *Zs. Kr.*, **81**, 352 (1932), on mimetite with P_2O_5 1.62, As_2O_5 21.72, CaO 1.27 per cent from Santa Eulalia, Mexico, and pyromorphite from Hunter district, Idaho. See also Mehmel, *Zs. phys. Chem.*, **15A**, 223 (1931); Zambonini and Ferrari, *Acc. Linc., Atti*, [6], **7**, 283 (1928); Aminoff and Parsons, *Geol. För. Förh.*, **49**, 438 (1927).

7. Lietz, *Zs. Kr.*, **77**, 437 (1931).

8. Crystals of pyromorphite from Ems and of mimetite from Tsumeb are measurably piezoelectric under both hydrostatic and directed pressure and are optically biaxial, presumably due to inversion (Frondel, priv. comm., 1947). If the crystals are mimetic orthorhombic twins, as indicated by the optical properties, the point symmetry is then $mm2$. Apatite is not piezoelectric when tested similarly.

9. Baumhauer, *Jb. Min.*, 411 (1876).

10. Goldschmidt and Schroeder, *Zs. Kr.*, **51**, 362 (1912); also reported on $\{20\bar{2}1\}$ by Klein, *Cbl. Min.*, 748 (1902).

11. See Drescher, *Cbl. Min.*, 257 (1926); Schouten, *Econ. Geol.*, **29**, 611 (1934).

12. Lietz (1931), who gives numerous other prism measurements in part on analyzed material. The data here cited, on Ems pyromorphite and Wheal Alfred, Cornwall, mimetite represent the essentially pure end-members of the series. See also Bowman, *Min. Mag.*, **13**, 324 (1903).
13. See Bertrand, *Bull. soc. min.*, **4**, 35 (1881); Jannettaz, *Bull. soc. min.*, **4**, 39 (1881); Bowman (1903); Jannettaz and Michel, *Bull. soc. min.*, **4**, 196 (1881); Drescher (1926). Bertrand gives $X = a$.
14. Carobbi, *Acc. Napoli, Rend.*, [3], **32**, 54 (1926).
15. For numerous additional analyses see Hintze (**1** [4A], 590, 603, 1924); also Igelström, *Geol. För. Förh.*, **22**, 229 (1865); Chirva, *Trav. inst. Lomonossov, ac. sc. U.R.S.S.*, no. 5, 86 (1935); Anderson (1920); Mélon, *Soc. géol. Belgique, Ann.*, **66**, B56 (1943).
16. Lietz (1931), who cites other analyses.
17. Haege, *Min. Siegerland.*, 36 (1888).
18. Bowles, *Am. J. Sc.*, **24**, 40 (1909).
19. Jannettaz and Michel (1881).
20. Barruel, *Ann. chim. phys.*, **62**, 217 (1836).
21. Bertrand analysis in Lacroix (**4**, 407, 1910).
22. Domeyko, *Ann. mines*, **14**, 115 (1848).
23. Damour, *Bull. soc. min.*, **6**, 84 (1883).
24. Biehl, Inaug. Diss. Münster, 40 (1919), who gives another analysis.
25. Serra, *Acc. Linc., Rend.*, **18**, 361 (1909).
26. Smith, *Am. J. Sc.*, **20**, 248 (1885).
27. Collie, *J. Chem. Soc. London*, **55**, 94 (1889).
28. Bischoff (**3**, 742, 1866).
29. Shannon (418, 1926).
30. Ahlfeld, *Jb. Min., Beil.-Bd.*, **66**, 41 (1932).
31. See Mügge, *Jb. Min., Beil.-Bd.*, **16**, 350 (1903).
32. Genth and vom Rath, *Proc. Am. Phil. Soc.*, **24**, 33 (1887).
33. Hintze (**1** [4A], 578, 598, 1924); Mellor (**7**, 883, 1927; **9**, 260, 1929); McDonnell and Smith, *Am. J. Sc.*, **42**, 139 (1916); Jolibois and Chaudron, *C.R.*, **192**, 1650 (1931).
34. Michel, *Bull. soc. min.*, **10**, 133 (1887); Amadori (1916).
35. Seaman, priv. comm. (1949), found a faint trace of CO_2 in pyromorphite from 2 of 18 localities tested and none in mimetite (9 localities) or vanadinite (3 localities).

BELLITE. Petterd (*Notes on Tasmanian Minerals*, priv. publ., Tasmania, 1904; see also *Petterd* (23, 1910)).

A supposed arsenate-chromate of lead found with crocoite and mimetite at the Magnet mine, Russell County, Tasmania. Occurs as velvety coatings and delicate tufts of minute hexagonal crystals. H. $2\frac{1}{2}$. G. ~5.5. Color bright crimson-red, also bright yellow to orange. Optically[1] negative $(-)$, with nO 2.16, nE 2.14; faintly pleochroic in pale pink with absorption $O > E$. Analysis gave PbO 61.68, Al_2O_3 0.01, SiO_2 7.59, CrO_3 22.61, V_2O_5 0.11, P_2O_5 0.05, As_2O_5 6.55, Cl 0.52, SO_3 0.05, total 99.17. Named after Mr. W. R. Bell, of Tasmania. Identical with or near mimetite.[2]

Ref.

1. Larsen (45, 1921).
2. Palache, priv. comm. (1939), measured a minute crystal and found the angles to be close to mimetite. Chemical tests indicated little or no Cr. The optical properties also are close to mimetite. The reported chemical analysis is presumed to be erroneous or to represent a mixture.

41.7.2.3 **V A N A D I N I T E** [$Pb_5(VO_4)_3Cl$]. Plomb brun, Braunbleierz of Zimapan, *early authors*. Chromate de plomb brun [from Descotil's analysis] *Brongniart* (**2**, 204, 1807). Vanadinbleierz *Rose* (*Ann. Phys.*, **29**, 455, 1833). Vanadinit *von Kobell* (283, 1838). Vanadate of Lead. Vanadin-spath, Vanadinbleispath, Vanadinsaures Blei *Germ.* Plomo pardo *Domeyko*. Endlichite *Genth* and *vom Rath* (*Proc. Am. Phil. Soc.*, **22**, 367, 1885). Cuprovanadite *Yanishevsky* (*Trans. Geol. Prospect. Serv. U.S.S.R.*, fasc., 109, 19, 1931).

C r y s t.[1] Hexagonal—P; hexagonal-dipyramidal—$6/m$.

$$a:c = 1:0.7122; \quad p_0:r_0 = 0.8223:1$$

Forms: [2]

	ϕ	$\rho = C$	M	A_2
c 0001	.....	0°00'	90°00'	90°00'
m 10$\bar{1}$0	30°00'	90 00	60 00	90 00
a 11$\bar{2}$0	0 00	90 00	90 00	60 00
h 21$\bar{3}$0	10 53½	90 00	79 06½	70 53½
x 10$\bar{1}$1	30 00	39 26	71 29	90 00
y 20$\bar{2}$1	30 00	58 42	64 42½	90 00
s 11$\bar{2}$1	0 00	54 55½	90 00	65 50½
u 21$\bar{3}$1	10 53½	65 19	80 07	72 42

Less common:

d 53$\bar{8}$0	k 30$\bar{3}$4	β 50$\bar{5}$3	v 11$\bar{2}$2	l 32$\bar{5}$2
g 31$\bar{4}$0	o 70$\bar{7}$6	A 70$\bar{7}$4	η 33$\bar{6}$2	t 52$\bar{7}$2
H 72$\bar{9}$0	δ 50$\bar{5}$4	q 50$\bar{5}$2	Σ 22$\bar{4}$1	p 41$\bar{5}$2
f 51$\bar{6}$0	γ 40$\bar{4}$3	z 30$\bar{3}$1	ξ 41$\bar{5}$4	
σ 10$\bar{1}$3	70$\bar{7}$5	π 40$\bar{4}$1	i 21$\bar{3}$2	
r 10$\bar{1}$2	α 30$\bar{3}$2	N 50$\bar{5}$1	e 32$\bar{5}$3	

Structure cell.[3] Space group $C6_3/m$. a_0 10.31 kX, c_0 7.34; $a_0:c_0$ = 1:0.712. Cell contents $Pb_{10}(VO_4)_6Cl_2$. Isostructural with pyromorphite-mimetite.

Habit. Crystals short to long prismatic [0001], usually with smooth faces and sharp edges; also acicular to hair-like. Sometimes cavernous,

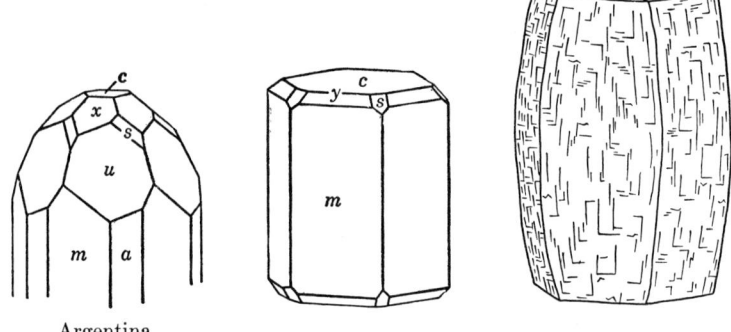

Argentina.

the crystals hollow prisms; also in rounded forms and in parallel groupings like pyromorphite-mimetite; skeletalized; in implanted globules. Common forms: $m\ c\ x\ y\ a\ u$.

Phys. Fracture uneven to conchoidal. Brittle. H. 2¾ to 3. G. 6.88 for the essentially pure mineral, 6.92 (calc.), decreasing with substitution of Ca for Pb; reported values vary from ~6.5 to 7.1. Luster subresinous to sub-adamantine. Color orange-red, ruby red, brownish red in all transitions; also reddish brown, brown, brownish yellow, yellow, pale straw-yellow. Streak white or yellowish. Subtransparent to nearly opaque. The crystals sometimes show concentric zones of varying composition.

VANADINITE 897

Opt. In transmitted light, colorless or weakly tinted and then pleochroic with absorption $E < O$. Uniaxial negative $(-)$.

	Vanadinite, Obir, Carinthia [4]		Arsenian vanadinite, Hillsboro, N. M.[5]			
λ	nO	nE	nO	nE	nO	nE
435.83	2.628	2.505				
589.3	2.416	2.350	2.25	2.20	2.358	2.311
690.75	2.370	2.313				

The indices of refraction are lowered by substitution of As or P for V or of Ca for Pb.

Chem. A chloride-vanadate of lead, $Pb_5(VO_4)_3Cl$. Both P and As substitute for V, up to at least $P:V = 1:4.7$ (anal. 4) and $As:V = 1:1$ (anal. 8). Small amounts of Ca, Zn, and Cu substitute for Pb (see analyses).[12]

Anal.[13]

	1	2	3	4	5	6	7	8
CaO		0.07		3.25				0.34
PbO	78.80	78.07	74.58	74.22	77.31	77.49	74.79	73.48
V_2O_5	19.26	19.10	19.49	[17.68]	17.66	16.98	17.34	10.98
As_2O_5		tr.			2.60	3.06	5.67	13.52
P_2O_5		0.30		2.93	0.35	0.29		tr.
Cl	2.50	2.66	2.44	2.48	2.54	2.41	2.61	2.45
Rem.			3.43			0.48	0.54	
	100.56	100.20	99.94	[100.56]	100.46	100.71	100.95	100.77
O = Cl	0.56	0.60	0.55	0.56	0.57	0.54	0.59	0.55
Total	100.00	99.60	99.39	[100.00]	99.89	100.17	100.36	100.22
G.					6.88		6.57	6.864

1. $Pb_5(VO_4)_3Cl$. 2. Obir, Carinthia.[6] 3. Sierra Cordoba, Argentina.[7] Rem. is ZnO 2.91, H_2O 0.52. 4. Wanlockhead, Scotland.[8] 5. Hillsboro, New Mexico.[9] 6. Oracle, Arizona.[10] Rem. is Fe_2O_3. 7. Yuma County, Arizona.[11] Rem. is insol. 8. Lake Valley, New Mexico (*endlichite*).[14]

Tests. B.B. easily fusible. In C.T. decrepitates and gives a faint sublimate of lead chloride. Easily soluble in HNO_3 to a yellow solution; soluble in HCl to a green solution with deposition of lead chloride.

Var. Arsenian. Endlichite *Genth* and *vom Rath* (1885). Contains As in substitution for V, with $As:V = 1:1$ in the material from Hillsboro, New Mexico. Color yellow of various shades, with diminished indices of refraction.

Occur. Vanadinite is a secondary mineral found chiefly in the oxidized zone of lead deposits. Associated with pyromorphite, mimetite, descloizite, mottramite, wulfenite, cerussite, anglesite, limonite. The vanadium content of the mineral is derived by the alteration of vanadiferous sulfides and silicates of the gangue and wall rocks. Occurrences to be noted are at Beresovsk near Ekaterinburg in the Urals, U.S.S.R.; at Obir near Eisenkappel in Carinthia, Austria; at Bena e Padru at Sassari, Sardinia. In Scotland at Wanlockhead, Dumfries, with smithsonite; also at Leadhills. Large crystals are found at Djebel Mahseur near Oudjda, Morocco;[15] also at Saïda, Algeria, and Djebba, Tunis. From Otteskoop in the Marico district, Transvaal, and Djoué in the French Congo, Africa.

At various localities in the Sierra de Córdoba, Argentina. Vanadinite was originally found at Zimapan, Hidalgo, Mexico; also in the Santa Eulalia district, Chihuahua.

In the United States, vanadinite occurs chiefly in the mining regions of New Mexico and Arizona. In Arizona, magnificent specimens have been found in the Old Yuma mine in Pima County, and the Red Cloud mine in Yuma County (an occurrence first noted by Silliman in 1881); [16] also from the Globe district in Gila County and the Mammoth mine near Tiger (Schultz) in Pinal County. In New Mexico, the Hillsboro and Lake Valley districts in Sierra County have afforded beautiful specimens (in part highly arsenian); also in notable specimens from the Caballos Mountains district in Sierra County, and the Central and Georgetown districts in Grant County. In California at Camp Signal near Goffs in San Bernardino County, and the El Dorado mine near Indio in Riverside County.

A r t i f.[17] By slowly cooling a fusion of lead vanadate with excess lead chloride at about 900°, and by heating ammonium vanadate, lead chloride, and excess ammonium chloride in a closed tube at about 180°. Compounds containing Ca, Ba, or Sr in place of Pb and Br in place of Cl have been synthesized.

A l t e r. Descloizite has been found as pseudomorphs after vanadinite.

N a m e. The mineral now called vanadinite was first found by A. M. del Rio (1764–1849), Professor in the School of Mines of Mexico, at Zimapan. An analysis made in 1801 shows 14.8 per cent of the oxide of a new metal, called erythronium, which later (1804) was considered to be chromium and the mineral a chromate of lead. Wöhler showed that del Rio's mineral was a vanadate shortly after the discovery of vanadium by Sefström in 1830 in iron ore from Taberg, Sweden.

Ref.

1. Elements of Vrba, *Zs. Kr.*, **4,** 353 (1880), on Obir, Carinthia, crystals probably essentially free from As, P, and Ca. The ratio $a:c$ apparently increases with increase in content of As, the material from Lake Valley, New Mexico, with As:V = 1:1 (analysis 8) having $a:c$ = 1:0.7495 (Penfield, *Am. J. Sc.*, **32,** 441, 1886).
2. Goldschmidt (**9,** 49, 1923).
3. Hendricks, Jefferson, and Mosley, *Zs. Kr.*, **81,** 352 (1932), on Yuma County, Arizona, crystals with As_2O_5 3.17, V_2O_5 16.20, CaO 0.26 per cent.
4. Lietz, *Zs. Kr.*, **77,** 454 (1931), who gives data for material from other localities.
5. Larsen (**71,** 1921) and Bowman, *Min. Mag.*, **13,** 324 (1903).
6. Lietz (1931).
7. Döring, *Bol. ac. cienc. Córdoba*, **5,** 498 (1883).
8. Frenzel, *Min. Mitt.*, **3,** 505 (1881).
9. Jannasch analysis in Goldschmidt, *Zs. Kr.*, **32,** 561 (1900).
10. Genth, *Proc. Am. Phil. Soc.*, **22,** 365 (1885).
11. Harwood (1910) cited by Lietze (1931).
12. Yanishevsky, *Trans. Geol. Prospect. Serv. U.S.S.R., fasc.*, 109, 19 (1931)—*Min. Mag.*, **24,** 607 (1937)—reports 1.55 CuO in material from Kazakhstan.
13. For additional analyses see Hintze (**1** [4A], 618, 1924); also Comucci, *Acc. Linc., Att.*, [6], **3,** 335 (1926); Lietze (1931); Wagner and Marchand, *Trans. Geol. Soc. S. Africa*, **23,** 59 (1921).
14. Genth and vom Rath (1885) who cite another analysis.
15. Barthoux, *Bull. soc. min.*, **47,** 36 (1924).
16. Silliman, *Am. J. Sc.*, **22,** 198 (1881).
17. Mellor (**9,** 809, 1929), Lietz (1931).

41.7.3 SVABITE SERIES

41.7.3.1 S V A B I T E [(Ca,Pb)$_5$(AsO$_4$,PO$_4$)$_3$(F,Cl,OH)]. *Sjögren (Geol. För. Förh., 13, 789, 1891; Bull. Geol. Inst. Upsala, 1, 50, 1892).*

Cryst.[1] Hexagonal—P; hexagonal-dipyramidal—$6/m$.

$$a:c = 1:0.7109; \quad p_0:r_0 = 0.8209:1$$

Forms:

	ϕ	$\rho = C$	M	A_2
c 0001	....	0°00′	90°00′	90°00′
m 10$\bar{1}$0	30°00′	90 00	60 00	90 00
h 21$\bar{3}$0	10 53½	90 00	79 06½	70 53½
x 10$\bar{1}$1	30 00	39 23	71 27½	90 00
v 11$\bar{2}$2	0 00	35 24½	90 00	73 09½
s 11$\bar{2}$1	0 00	54 53	90 00	65 51½
e 21$\bar{3}$2	10 53½	47 21½	82 00½	76 04

Less common:

r 10$\bar{1}$2 y 20$\bar{2}$1 p 41$\bar{5}$2

Habit. Short prismatic [0001]. Also massive.

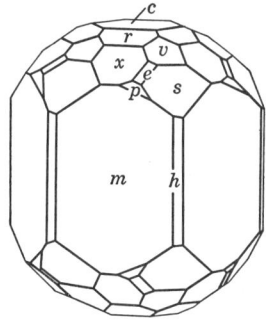

Phys. Cleavage {10$\bar{1}$0} indistinct. Brittle. H. 4–5. G. 3.5–3.8, least in material low in Pb and high in P. Luster vitreous to subresinous. Colorless and transparent; also yellowish white to gray and grayish green and translucent.

Opt.[2] In transmitted light, colorless.

	Svabite As end-member	Svabite As:P = 1:1	Svabite, Harstig	
nO	~1.707	~1.672	1.706 ± 0.003	Uniaxial negative (−).
nE			1.698 ± 0.003	

Chem. Essentially a fluoride-arsenate of calcium, Ca$_5$(AsO$_4$)$_3$F. P substitutes for As up to at least P:As = 1:2 (analysis 6). Pb substitutes for Ca up to at least Pb:Ca = 1:38 (analysis 3), and both Mg and Mn″ are sometimes also present in minor amounts. Small amounts of both

PHOSPHATES, ARSENATES, VANADATES

Cl and (OH) substitute for F. With increase of P svabite forms a series toward fluorapatite, and with increase of Pb and Cl a series toward hedyphane and mimetite.

Anal.

	1	2	3	4	5	6
CaO	44.08	42.07	37.22	42.5	42.68	45.89
MgO		0.52	3.90	0.7	0.56	0.84
MnO		0.26	0.19			1.23
FeO		0.08	0.14			
PbO		3.02	4.52	tr.		0.51
P_2O_5		0.38	tr.	0.9	2.72	12.54
As_2O_5	54.19	51.05	50.92	52.2	50.60	35.24
Cl		0.12	0.08	0.1	0.54	
F	2.99	1.99	2.80	not det.	not det.	1.41
H_2O		0.25	0.33	1.0	2.28	1.32
Rem.		1.55	1.24	2.1	3.50	1.83
	101.26	101.29	101.34	99.5	102.88	100.81
O = F,Cl	1.26	0.87	1.20		0.12	0.59
Total	100.00	100.42	100.14	99.5	102.76	100.22
G.		3.77	3.82	3.52	3.695	3.54

1. $Ca_5(AsO_4)_3F$. 2,3. Jakobsberg, Sweden.[3] Rem. in 2 is Na_2O 0.56, K_2O 0.30, SO_3 0.69; in 3, Na_2O 0.39, K_2O 0.28, SO_3 0.57. 4. Harstig, Sweden.[3] F not det. Rem. is insol. 5. Långban, Sweden.[4] Rem. is Na_2O 1.24, K_2O 0.37, Sb_2O_5 0.55, Al_2O_3 0.34, insol. 1.00. H_2O is ign. loss. 6. Franklin, New Jersey.[5] Rem. is ZnO 1.54, CO_2 tr., insol. 0.29.

Tests. B.B. fuses with difficulty to a black slag. Soluble in dilute acids.

O c c u r. Found in Sweden at the Harstig mine, Pajsberg, with schefferite, garnet, brandtite, and sarkinite; massive at Jakobsberg in the Nordmark district with hausmannite; at Långban with hematite, mica, and garnet. In the United States, found as rough hexagonal prisms with rounded terminations embedded in franklinite ore at Franklin, New Jersey.

N a m e. After Anton Svab (1703–1768), Swedish mining official.

Ref.

1. Elements and forms of Flink, *Geol. För. Förh.*, **47**, 127 (1925), on Långban crystals (analysis 5). Sjögren (1892; also *Geol. För. Förh*, **17**, 313, 1895) obtained 1:0.7143 on Harstig crystals (analysis 4).
2. Harstig data from Larsen (140, 1921); other data from approximate index-composition curve of Bauer and Berman, *Am. Min.*, **15**, 347 (1930).
3. Mauzelius analyses in Sjögren (1892).
4. Almström analysis in Flink, *Geol. För. Förh.*, **47**, 127 (1925).
5. Bauer analysis in Bauer and Berman (1930); a partial analysis is cited with As_2O_5 16.2, P_2O_5 28.4 (*arsenian apatite*).

41.7.3.2 H E D Y P H A N E [$(Ca,Pb)_5(AsO_4)_3Cl$]. Breithaupt (*J. Chemie u. Phys.*, **60**, 310, 1830).

C r y s t.[1] Hexagonal—P; hexagonal-dipyramidal—$6/m$.

$$a:c = 1:0.7052; \qquad p_0:r_0 = 0.8143:1$$

HEDYPHANE

Forms:[2]

		φ	ρ = C	M	A_2
c	0001		0°00′	90°00′	90°00′
m	10$\bar{1}$0	30°00′	90 00	60 00	90 00
a	11$\bar{2}$0	0 00	90 00	90 00	60 00
r	10$\bar{1}$2	30 00	22 09	79 08	90 00
x	10$\bar{1}$1	30 00	39 09½	71 36	90 00
y	20$\bar{2}$1	30 00	58 27	64 47	90 00
v	11$\bar{2}$2	0 00	35 11½	90 00	73 15
s	11$\bar{2}$1	0 00	54 40	90 00	65 55½

Less common:

α 30$\bar{3}$2   g 70$\bar{7}$4   p 41$\bar{5}$2

Habit. Prismatic [0001], also pyramidal to thick tabular {0001}. Massive.

Phys. Cleavage {10$\bar{1}$1}. Fracture subconchoidal. Brittle. H. 4½. G. 5.82. Luster bright, greasy to resinous. Color white to yellowish white and buff, also bluish. Translucent.

Opt. In transmitted light, colorless.

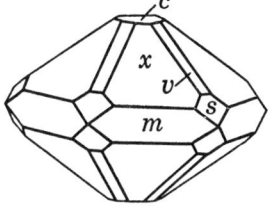

Franklin, N. J.

	Franklin, N. J.[3]	Franklin, N. J.[4] (anal. 3)	
nO	1.958(Na)	2.03	Uniaxial
nE	1.948	2.01	
Sign	+	−	

Chem. A chloride-arsenate of calcium and lead, $(Ca,Pb)_5(AsO_4)_3Cl$. Ca and Pb substitute mutually with Ca:Pb close to 1:1 in the modern analyses. With increase in Pb over the 1:1 ratio, hedyphane passes into mimetite. The name hedyphane is here applied to the half of this series with Ca > Pb. As so defined, hedyphane is essentially the Cl analogue of svabite.

Anal.

	1	2	3	4	5
CaO	13.10	7.85	14.98	14.09	10.50
BaO		8.27			
MgO		0.25	0.10		
PbO	52.13	50.89	52.77	51.03	57.45
P_2O_5		0.55		n.d.	3.19
As_2O_5	32.21	29.01	29.94	22.78	28.51
Cl	3.31	3.14	2.98	2.66	3.06
Rem.		0.32	0.76		
	100.75	100.28	101.53	90.56	102.71
O = Cl	0.75	0.71	0.67		0.69
Total	100.00	99.57	100.86		102.02
G.		5.82		5.49	

1. $(Ca,Pb)_5(AsO_4)_3Cl$ with Ca:Pb = 1:1. 2. Långban, Sweden.[5] Rem. is $(Na,K)_2O$ 0.24, Fe_2O_3 0.08. 3. Franklin, New Jersey.[4] Rem. is (Mn,Fe)O 0.28, ZnO 0.23, H_2O 0.08, insol. 0.17. Original total cited as 100.85. 4. Långban.[6] 5. Långban.[7] Accepting the accuracy of the analysis, this material is properly classed as calcian mimetite.

Tests. B.B. fusible with difficulty. Soluble in HNO_3.

Occur. Originally from Långban, Sweden, where it in part occurs with barite and barylite in veinlets in hematite. Also from the Harstig mine, Pajsberg, Sweden, with tephroite and calcite in veinlets cutting magnetite. In the United States found at Franklin, New Jersey, associated with calcite, willemite, native copper, native lead, hancockite, and rhodonite in veinlets cutting massive franklinite ore.

Name. From ἡδυ-φανής, *beautifully bright*, in allusion to the high luster.

Ref.
1. Elements of Aminoff, *Geol. För. Förh.*, **45**, 124 (1923), on unanalyzed Långban crystals, Sjögren, *Geol. För. Förh.*, **14**, 250 (1892), *Bull. Geol. Inst. Upsala*, **1**, 1 (1892) obtained 1:0.7063 on Harstig crystals.
2. Aminoff (1923), Sjögren (1892).
3. Palache and Berman, *Am. Min.*, **12**, 180 (1927); a check here made on their material proves it to be optically positive (+).
4. Foshag and Gage, *Am. Min.*, **10**, 351 (1925).
5. Lundström, *Geol. För. Förh.*, **4**, 266 (1879).
6. Kersten, *J. Chem. Phys.*, **62**, 22 (1831), as recalculated by Rammelsberg (337, 1875).
7. Michaelson, *J. Chem. Phys.*, **62**, 22 (1831).

PLEONECTITE. Pleonektit *Igelström* (*Geol. För. Förh.*, **11**, 210, 1889; *Jb. Min.*, II, 40, 1889).

Found as embedded grains with arseniopleite, hausmannite, rhodonite, and calcite at the Sjö mine, Örebro, Sweden. Indistinct cleavage. H. 4. Luster resinous. Grayish white in color. Not analyzed; qualitative tests suggest that the mineral is identical with hedyphane or mimetite.

41.7.4 DEHRNITE [(Ca,Na,K)$_5$(PO$_4$)$_3$(OH)]. Larsen and Shannon (*Am. Min.*, **15**, 303, 1930).

Hexagonal; isostructural with apatite.[1]

Structure cell.[1]

	a_0	c_0	$a_0:c_0$
Dehrn, Germany	9.31 kX	6.87	1:0.7379
Fairfield, Utah	9.35	6.89	1:0.7369

Habit. As botryoidal crusts composed of fibrous to bladed crystals; also as minute hexagonal prisms in subparallel or sheaf-like groups.

Phys. Cleavage {0001} perfect. H. ~5. G. 3.04–3.09. Colorless to pale green, greenish white, or grayish. Luster weakly vitreous.

Opt.[2] Optically negative (−). The crystals sometimes are uniaxial but commonly are biaxial or have a uniaxial core and a border zone that is divided into six biaxial segments. The biaxial parts have relatively low indices of refraction with small $2V$; $X = c$ and the axial plane makes an angle of about 12° with the prism faces.

	n (Dehrn,	n (Utah, core)	n (Utah, border)	n (Utah)
E or X	1.614	1.633	1.585	1.610
Y				1.619
O or Z	1.622	1.640	1.600	1.620
2V	0°	0°	small	small

Chem. The name dehrnite is applied to members of the apatite group that contain significant amounts of alkalies in substitution for Ca, with

Na > K. Material with K > Na is classed with lewistonite. Valence compensation for the substitution of Na,K for Ca presumably is effected by the coupled substitution of C for P or of OH for O; the formula may be written $(Ca,Na,K)_5(PO_4,CO_3)_3(OH)$. Analyses gave (see also analysis 2 under lewistonite):

	Na_2O	K_2O	CaO	MgO	Al_2O_3	P_2O_5	CO_2	H_2O	Insol.	Total	G.
1.	7.11	1.20	50.88		tr.	37.12	1.49	1.68	0.12	99.60	3.04
2.	4.36	5.90	47.65	0.85	1.00	35.68	3.30	1.91		100.65	3.09
3.	3.70	0.30	54.86		0.65	36.66	1.22	3.02		100.41	3.07

1. Dehrn, Germany.[3] No F or Cl. H_2O includes $H_2O - 0.16$. 2. Fairfield, Utah.[4] Zoned crystals. 3. Fairfield, Utah.[4] Crusts.

Tests. B.B. fuses readily to a white enamel. Soluble in acids.

O c c u r. Originally found with crandallite in brecciated phosphate rock at Dehrn, Nassau, Germany. Also found at Fairfield, Utah. Here dehrnite occurs with wardite, deltaite, lehiite, montgomeryite, gordonite, millisite, lewistonite, crandallite, and other phosphates in more or less altered nodules of variscite.[5] The dehrnite occurs with other apatite-like minerals as yet unidentified (nO 1.610–1.635) as balls, tufts, and sheaf-like or subparallel aggregates of fibers or acicular hexagonal prisms lining open crevices in the crandallite-variscite masses as the last-formed of the various minerals.

Ref.

1. McConnell, Am. Min., **23**, 1 (1938).
2. Larsen and Shannon (1930a); Am. Min., **15**, 324 (1930b).
3. Shannon in Larsen and Shannon (1930a).
4. Shannon in Larsen and Shannon (1930b).
5. Larsen, Am. Min., **27**, 297, 364 (1942).

41.7.5   L E W I S T O N I T E [$(Ca,K,Na)_5(PO_4)_3(OH)$]. *Larsen* and *Shannon* (Am. Min., **15**, 326, 1930).

Hexagonal; isostructural with apatite.[1]

Structure cell.[1] a_0 9.35 kX, c_0 6.89; $a_0:c_0 = 1:0.7369$.

Habit. As stout hexagonal prisms either isolated or as subparallel or sheaf-like aggregates; also as powdery crusts and as spherules with a fibrous structure.

P h y s. Cleavage {0001} perfect, probably also {10$\bar{1}$0} perfect. H. $\sim$5. G. 3.08. Colorless to pale green and white.

O p t. Optically negative ($-$). Sometimes uniaxial but commonly biaxial, the crystals showing a uniaxial core and a border zone that is divided into six biaxial segments. The biaxial parts have moderate $2V$; $X = c$ and the axial plane makes an angle of about 16° with the prism faces.

	n (anal. 2)	n (core)	n (biaxial; anal. 1)
E or X	1.611		1.613
Y			1.623
O or Z	1.621	1.60	1.624
$2V$	0°	0°	42°

Chem. The name lewistonite is applied to members of the apatite group that contain significant amounts of alkalies in substitution for Ca, with K > Na. Material with Na > K is classed with dehrnite. The formula may be written $(Ca,K,Na)_5(PO_4)_3(OH)$. The mechanism of the valence compensation may involve the coupled substitution of Al for Ca or of OH for O. Analyses gave:

	Na$_2$O	K$_2$O	CaO	MgO	Al$_2$O$_3$	P$_2$O$_5$	H$_2$O	Rem.	Total	G.
1.	0.47	3.71	41.41	7.10	3.67	32.31	8.60	1.72	98.99	3.06
2.	4.34	1.36	46.78		2.53	37.92	7.69		100.62	3.08

1. Fairfield, Utah.[2] Biaxial. Contains about 7 per cent of crandallite and 3 per cent of deltaite as impurities. Rem. is SiO$_2$ 1.12, insol. 0.60. 2. Fairfield.[2] Uniaxial fibers. The material of this analysis was originally classed as lewistonite on the basis of the high water content and low alkali content, but the ratio of Na:K would place the mineral with dehrnite.

Occur. Found with dehrnite (which see), deltaite, crandallite, and other phosphates in crevices in variscite nodules at Fairfield, Utah.

Name. From the town of Lewiston, Utah, located near the Fairfield deposit.

Ref.

1. McConnell, *Am. Min.*, **23**, 1 (1938).
2. Shannon analysis in Larsen and Shannon (1930).

41.7.6 **FERMORITE** $[(Ca,Sr)_5(P,AsO_4)_3(F,OH)]$. Smith and Prior (*Min. Mag.*, **16**, 84, 1911). Strontianapatite.

Hexagonal; hexagonal-dipyramidal $(6/m)$. Isostructural with apatite. Massive, granular.

Structure cell.[1] Space group $C6_3/m$. a_0 9.60 kX, c_0 7.00; $a_0:c_0$ = 1:0.7292.

Phys. Fracture even. H. 5. G. 3.518. Luster greasy. Color pale pinkish-white. Streak white. Translucent.

Opt. Uniaxial negative $(-)$. Birefringence weak, with a mean index of refraction of about 1.660.

Chem. An apatite-type phosphate-arsenate of calcium and strontium, $(Ca,Sr)_5(P,AsO_4)_3(F,OH)$. The only reported analysis has Sr:Ca = 1:8.2 and As:P = 1:1.3. Analysis gave:[2] CaO 44.34, SrO 9.93, P$_2$O$_5$ 20.11, As$_2$O$_5$ 25.23, F 0.83, H$_2$O tr., insol. 0.08, less O = F = 0.35, total 100.17. F determination probably low.

Tests. B.B. infusible. Soluble in acids.

Occur. Found at Sitapár, Chindwára district, Central Provinces, India, in veinlets in manganese ore consisting of braunite, hollandite, and pyrolusite.

Name. After Lewis Leigh Fermor, formerly director of the Geological Survey of India.

Ref.

1. McConnell, *Am. Min.*, **23**, 1 (1938).
2. Prior analysis in Smith and Prior (1911).

41.7.7 WILKEITE [$Ca_5(P,S,Si,CO_4)_3(OH)$]. *Eakle* and *Rogers* (*Am. J. Sc.*, **37**, 362, 1914).

Hexagonal, hexagonal-dipyramidal ($6/m$); isostructural with apatite. Crystals indistinct, with rounded edges and corners; also massive granular. Forms: $\{10\bar{1}0\}$, $\{11\bar{2}0\}$, $\{31\bar{4}0\}$, $\{10\bar{1}1\}$, and unidentified third-order pyramids; $a:c \sim 1:0.730$.

Structure cell.[1] Space group $C6_3/m$.

Anal. no.	a_0	c_0	$a_0:c_0$
2	9.48 kX	6.91	1:0.7289
3	9.40 kX	6.89	1:0.7330

Phys. Cleavage $\{0001\}$ imperfect. Very brittle. H. ~ 5. G. 3.1. Luster subresinous on fracture surfaces. Color pale rose-red, also yellow. Translucent.

Opt.[1] Uniaxial negative ($-$).

	n (anal. 2)	n (anal. 3)
O	1.650 ± 0.002	1.640 ± 0.002
E	1.646	1.636

Chem. An apatite-like calcium sulfate-silicate-phosphate. Formula essentially $Ca_5(P,S,Si,CO_4)_3(OH)$, with P:S:Si = 2:1:1 in the original analysis. These ratios, however, vary somewhat in different specimens, and wilkeite apparently is an intermediate member of a series extending from apatite to ellestadite.

Anal.

	CaO	MnO	P_2O_5	SO_3	SiO_2	CO_2	F	Cl	H_2O	Total	G.
1.	54.44	0.77	20.85	12.28	9.62	2.10			tr.	100.06	3.234
2.			14.4		11.2	pres.	0.9	0.8			3.12
3.			32.2		2.9	pres.	2.1	0.4			3.16

1. Crestmore.[2] 2,3. Crestmore.[3]

Tests. B.B. fusible with difficulty on thin edges. Easily soluble in HCl or HNO_3, leaving part of the silica as a flocculent residue.

Occur. Found with diopside, vesuvianite, garnet, crestmoreite, and blue calcite in contact metamorphosed marble at Crestmore, Riverside County, California. The sulfate-apatite from the sanidinites of the Laacher See, Rheinland, Germany,[4] may be the same mineral. Also from Kyshtym in the Urals, U.S.S.R.[5]

Alter. To crestmoreite.

Name. After R. M. Wilke, mineral dealer and collector of Palo Alto, California.

Ref.

1. McConnell, *Am. Min.*, **22**, 977 (1937), who graphically summarizes the variation in G. and indices of refraction in the series.
2. Eakle analysis in Eakle and Rogers (1914).
3. Ellestad analyses in McConnell (1937).
4. See Brauns, *Jb. Min., Beil.-Bd.*, **41**, 60 (1916).
5. Borneman-Starinkevitch, *C. r. ac. sc. U.R.S.S.*, **19**, 253 (1938).

41.7.8 ELLESTADITE [$Ca_5(Si,S,P,CO_4)_3(Cl,F,OH)$]. *McConnell* (*Am. Min.*, **22**, 977, 1937).

Hexagonal; hexagonal-dipyramidal ($6/m$); isostructural with apatite. Massive, granular.

Structure cell. Space group $C6_3/m$. a_0 9.53 kX, c_0 6.91; $a_0:c_0 = 1:0.7251$.

Phys. Cleavage {0001} and {10$\bar{1}$0} indistinct. H. ~5. G. 3.068. Color pale rose.

Opt. Uniaxial negative (−).

nO	1.655 ± 0.002
nE	1.650 ± 0.002

Chem. An apatite-like calcium sulfate-silicate, containing (SO_4) and (SiO_4) in place of (PO_4). Ellestadite apparently is a member of a series extending through wilkeite to apatite. Britholite and abukumalite also are similar [2] but contain much Ce or V in place of Ca and lack S. Formula essentially $Ca_5(Si,S,P,CO_4)_3(Cl,F,OH)$, with Si:S:P ~2.9:2.6:0.4 in the only reported analysis: [1] CaO 55.18, MgO 0.47, MnO 0.01, Fe_2O_3 0.22, Al_2O_3 0.13, P_2O_5 3.06, SO_3 20.69, SiO_2 17.31, CO_2 0.61, Cl 1.64, F 0.57, H_2O+ 0.53, H_2O- 0.10, less O = Cl, F = 0.61, total 99.91.

Occur. Found as veinlets associated with wollastonite, vesuvianite, diopside, okenite, wilkeite, and blue calcite in the contact metamorphosed marble at Crestmore, Riverside County, California.

Name. After R. B. Ellestad, American analytical chemist.

Ref.
1. Ellestad analysis in McConnell (1937).
2. Machatschki, *Zbl. Min.*, 161 (1939).

41.7.9 TAVISTOCKITE [$Ca_3Al_2(PO_4)_2(OH)_6$]. Hydrated Calciumaluminic Phosphate (?) *Church* (*J. Chem. Soc. London*, **18**, 263, 1865). Tavistockite *Dana* (582, 1868). Bialite *Buttgenbach* (*Soc. géol. Belgique, Ann., Publ. Congo Belge*, **51**, C117, 1929).

Orthorhombic. In microscopic acicular crystals, sometimes aggregated into spherulitic or irregular stellate groups, constituting a white pearly powder. The crystals are elongated [001] and flattened {100} with {110} and are terminated by an obtuse pyramid; the plane angle (010) ∧ (011) is about 58° (Katanga).[1] Cleavage {100}, perfect. Color white. Luster pearly, brilliant. Transparent.

Opt. In transmitted light, colorless.

Orientation		n (Tavistock)[2]	n (Katanga)[1]	
X	b	1.522 ± 0.003	1.525	Optically positive (+).
Y	a	1.530 ± 0.003		$2V$ 74° (calc.).
Z	c	1.544 ± 0.003	~1.544	Dispersion not perceptible.

Chem. A basic phosphate of calcium and aluminum, $Ca_3Al_2(PO_4)_2(OH)_6$. Qualitative tests suggest that the material from Katanga is a magnesian variety.

Anal.

	1	2
CaO	36.08	36.27
Al_2O_3	21.86	22.40
P_2O_5	30.47	30.36
H_2O	11.59	12.00
Total	100.00	101.03

1. $Ca_3Al_2(PO_4)_2(OH)_6$. 2. Tavistock.[3]

Tests. B.B. infusible. Soluble with difficulty in acids.

Occur. At Tavistock, Devonshire, England, in cavities with quartz, childrenite, pyrite, and chalcopyrite. Later found at Mushishimano, Katanga, Belgian Congo, on a compact brown phosphatic rock.

Name. From the locality. Bialite after Lucian Bia (?–1892), an explorer of the Belgian Congo.

Ref.

1. Buttgenbach (1929).
2. Larsen (142, 1921).
3. Church (1865).

41.7.10 Arsenobismite. Arseno-Bismite *Means* (*Am. J. Sc.*, **41**, 127, 1916).

As friable or ocherous microcrystalline masses. Color yellowish brown to yellowish green. Optically nearly or quite isotropic with a mean (?) index of refraction above 1.86. Specific gravity ~5.7. Chemically a hydrated bismuth arsenate, of unknown formula. Analysis of the acid-soluble part (50.08 per cent) of a mixture with barite and limonite gave:[1] BaO 31.90, CuO 0.12, CaO 0.62, MgO tr., Al_2O_3 0.44, Fe_2O_3 3.88, Bi_2O_3 28.17, PbO 1.12, Sb_2O_3 1.26, P_2O_5 0.04, As_2O_5 10.59, SO_3 17.06, SiO_2 1.42, H_2O+ 1.43, H_2O- 1.09, total 99.14. Found in considerable amounts in oxidized ore in the Mammoth mine, Tintic district, Utah. Also found with bindheimite and bismutite at Tazna, Bolivia.[2] Both the Tintic and Bolivian minerals are associated with another hydrated bismuth arsenate which appears to be a distinct species:[3] color dead-white to yellow and orange; pulverulent to dense, very fine-grained masses, birefringent, with indices of refraction over 2.04. Also found with pucherite at Schneeberg Saxony.

Ref.

1. Wells analysis in Means (1916).
2. The validity of the species and the Bolivian occurrence was established by x-ray study of type material by Frondel, *Am. Min.*, **28**, 536 (1943).
3. Frondel (1943).

TYPE 8. $(AB)_3(XO_4)_2Z_q$

41.8.1 LAZULITE SERIES

41.8.1.1 **L A Z U L I T E** [$(Mg,Fe'')Al_2(PO_4)_2(OH)_2$]. Himmelblau Fossil von Steiermark *Widenmann* (*Bergm. J.*, 346, 1791). Smalteblaue Fossil von Vorau *Stütz* (*Schrift. Ges. nat. Freunde Berlin*, **9**, 352, 1791). Natürliche Smalt; Berlinerblau, Eisenblau [= vivianite]; Bergblau [= chrysocolla]; Unächter Lasurstein *Stütz* (*Einricht. Nat. Wien*, 49, 1793). Lazulit *Klaproth* (*Schrift. Ges. nat. Freunde Berlin*, **10**, 90, 1792; *Beitr. chem. Kenntn. Min.*, **1**, 197, 1795). Dichter blauer Feldspath [from Krieglach, Styria] *Klaproth* (*Beitr.*, **1**, 14, 1795). Lazulith *Klaproth* (*Beitr.*, **4**, 279, 1807). Blue Spar. Blue Feldspar. Wahrscheinlich neue Foss. aus Salzburg., Siderit von *Moll* (*Berg.-Hütten. Jb.*, **4**, 71, 1799); Mollit *Haberle* (Handbuch, 1804); = Lazulith *Mohs* (**1**, 427, 1804). Blauspath *Werner*. Voraulite *Delamétherie* (1812). Azurite *Jameson* (**2**, 542, 1805). Phosphorsäure Thonerde, etc. *Fuchs* (*J. Chemie u. Phys.*, **24**, 373, 1818). Klaprothite *Beudant* (464, 1824). Klaprothine *Beudant* (**2**, 576, 1832). Calcium-lazulite *Watson* (*J. Washington Ac. Sc.*, **11**, 386, 1921). Berkeyite *Kerr* (*Jeweler's Circ.*, **92**, 67, 1926).

41.8.1.2 **S C O R Z A L I T E** [$(Fe'',Mg)Al_2(PO_4)_2(OH)_2$]. *Pecora* and *Fahey* (*Am. Min.*, **34**, 83, 1949).

C r y s t.[1] Monoclinic; prismatic—$2/m$.

$a:b:c = 0.9750:1:1.6483;$ $\quad \beta\, 90°46';$ $\quad p_0:q_0:r_0 = 1.6906:1.6482:1$

$r_2:p_2:q_2 = 0.6067:1.0234:1;$ $\quad \mu\, 89°14';$ $\quad p_0'\, 1.6907,\, q_0'\, 1.6483,\, x_0'\, 0.0134$

Forms:[2]

	ϕ	ρ	ϕ_2	$\rho_2 = B$	C	A
c 001	90°00'	0°46'	89°14'	90°00'	.....	89°14'
b 010	0 00	90 00	....	0 00	90°00'	90 00
a 100	90 00	90 00	0 00	90 00	89 14	.....
m 110	45 43½	90 00	0 00	45 43½	89 27	44 16½
d 011	0 28	58 45½	89 14	31 15	58 45	89 36
y 103	90 00	29 59	60 01	90 00	29 13	60 01
t 101	90 00	59 35	30 25	90 00	58 49	30 25
s $\bar{1}01$	−90 00	59 12	149 12	90 00	59 58	149 12
z 112	46 10	49 58	49 20½	57 59	49 24½	56 28
p 111	45 57	67 08	30 24	50 10	66 34	48 31½
e $\bar{1}11$	−45 30	66 58	149 12	49 50	55 46½	131 01½

Structure cell.[3] Space group $P2_1/n$.

	a_0	b_0	c_0	β
Lazulite, Werfen	7.12 Å	7.24	7.10	118°55'
Lazulite, Graves Mountain, Georgia	7.14	7.27	7.16	119°18'
Lazulite, Churchill River, Canada	7.16	7.25	7.14	118°47'
Scorzalite, Corrego Frio	7.15	7.32	7.14	119°00'

Cell contents $(Mg,Fe'')_2Al_4(PO_4)_4(OH)_4$.

Habit. Crystals usually acute pyramidal, with large {111} and {1̄11} and small {101}; also tabular on (1̄11) or (101). Massive, compact to granular.

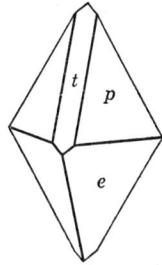

Lazulite. Graves Mountain.

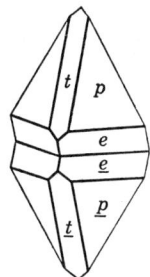

Lazulite. Graves Mountain. Twin on {100}.

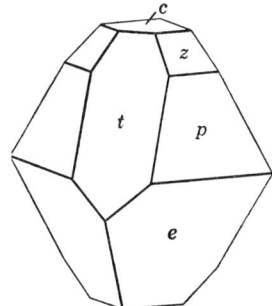

Lazulite. Werfen.

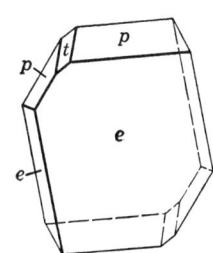

Lazulite. Graves Mountain.

Twinning. (1) On {100} common, sometimes lamellar or polysynthetic, with composition surface {001} or {100}; usually with a re-entrant angle. (2) On {223}, rare. (3) Reported [4] on {210}, {021}, and perhaps other planes.

P h y s. Cleavage {110} indistinct to good, {101} indistinct. Fracture uneven to splintery. Brittle. H. $5\frac{1}{2}$–6. G. 3.08 (extrapolated for Mg end-member; 3.14 calc.) to 3.38 (extrapolated for Fe end-member; 3.39 calc.). Luster vitreous. Color azure-blue, also sky-blue, bluish white; also deep blue or bluish green. Streak white. Subtranslucent to opaque; sometimes transparent and of gem quality.

O p t.[5] In transmitted light, strongly pleochroic with X colorless, Y blue, Z darker blue. Absorption $Z > Y > X$.

ORIENTATION	n (Mg end)	n (Fe:Mg=1:1)	n (Fe:Mg=4:1)	
$X \wedge c$ $9\frac{1}{2}°$	1.604	1.626	1.639	Biaxial negative (−).
Y b	1.626	1.654	1.670	$r < v$, perceptible.
Z	1.637	1.663	1.680	
$2V$	69°42′ calc.		58°17′ calc.	

With increase in content of Fe″, the indices of refraction and birefringence increase linearly and $2V$ decreases slightly.

Chem. A basic phosphate of aluminum, magnesium, and ferrous iron, $(Mg,Fe)Al_2(PO_4)_2(OH)_2$. A probably complete isomorphous series [6] extends between Fe″ and Mg. The names lazulite and scorzalite are applied to the halves of the series with Mg > Fe and Fe > Mg, respectively. Material high in Fe″ is relatively rare. Fe‴, presumably in substitution for Al, is often present in amounts less than 1 weight per cent Fe_2O_3. The SiO_2 and CaO frequently reported in analyses are ascribed to admixture, and the validity of a reported highly calcian variety is very doubtful.[10]

Anal.[7]

	1	2	3	4	5	6	7	8
MgO	13.34	11.97	10.38	10.02	7.98	4.23	2.93	
FeO		2.80	3.95	5.22	8.94	14.74	17.06	21.53
MnO		nil	nil	0.03	0.14	0.11	0.10	
CaO		0.08	0.06	0.06	0.13	0.02	0.03	
Al_2O_3	33.73	32.55	32.49	32.37	31.53	30.87	30.80	30.54
Fe_2O_3		0.49	0.60	0.75	0.89	0.54	0.13	
TiO_2		0.16	0.20	0.22	0.23	0.10	0.10	
P_2O_5	46.97	46.12	45.79	45.42	44.31	42.90	42.67	42.53
H_2O+	5.96	5.90	6.48	5.83	6.27	5.86	6.10	5.40
Total	100.00	100.07	99.95	99.92	100.42	99.54	99.92	100.00
G.		3.118	3.122	3.160	3.190	3.268	3.327	

1. $MgAl_2(PO_4)_2(OH)_2$. 2. Lazulite. Dattas, Brazil.[8] 3. Lazulite. Graves Mountain, Georgia.[8] 4. Lazulite. Clubbs Mountain, North Carolina.[8] 5. Lazulite. Breyfogle Canyon, California.[8] 6. Scorzalite. Corrego Frio, Brazil.[8] Total includes ZnO 0.17. 7. Scorzalite. Victory, South Dakota.[8] 8. $Fe''Al_2(PO_4)_2(OH)_2$.

Tests. B.B. infusible. Slowly soluble in hot acids.

Occur. The members of the lazulite-scorzalite series occur associated with quartz, rutile, corundum, pyrophyllite, cyanite, andalusite, sillimanite, dumortierite, garnet, muscovite. The chief occurrences are (1) as disseminated grains or masses in aluminous, high-grade metamorphic rocks, especially in quartz-rich facies thereof or in quartzites, (2) in quartz veins or dikes in such rocks, and then sometimes well-crystallized, and (3) in granite pegmatites. Lazulite occurs as fine crystals with wagnerite in quartz-carbonate veins in clay slate at various localities near Werfen, Salzburg, Austria;[9] also at Vorau near Graz. From near Ratten (or Fischbach) and Krieglach in the Mürztal, Styria, Austria. In Switzerland near Zermatt, Valais. With svanbergite at Horrsjöberg near Nȳ as crystals up to six inches in length and with berlinite at Westanå, Sweden. From the Sierra do Roberedo, Portugal. Found at a number of localities in Madagascar, notably on Mt. Bity in quartz veins in itabirite, at Ranomainty southeast of Betafo and in the neighborhood of Jamalezo. In Brazil at Dattas, Minas Geraes, in quartz veins in itacolumite. With sericite as pseudomorphs after orthoclase phenocrysts in altered quartz-porphyry in the Real Socavon, Potosi, Bolivia. In the United States, lazulite occurs well-crystallized with kyanite and rutile in quartzite on Graves Mountain, Lincoln County, Georgia; also at Clubbs Mountain and Crowders Mountain in Gaston County, North Carolina, with corundum.

In veinlets in quartzite at Chittenden north of Rutland, Vermont. In California near Markleeville, Alpine County, with andalusite and rutile in garnetiferous quartzite, in a vein in schist at Breyfogle Canyon near Death Valley, Inyo County, and notably in the andalusite mine near Mocalno in the White Mountains, Mono County. In the South Canyon district, Dona Ana County, New Mexico.

Scorzalite is known from the Corrego Frio pegmatite, Minas Geraes, Brazil, where it occurs with souzalite and brazilianite.[11] Also from the Victory pegmatite near Custer, Custer County, South Dakota.

A l t e r. To gibbsite (?).

N a m e. Lazulite from the older German name, *lazurstein*, blue stone. Scorzalite after E. P. Scorza, mineralogist of the Departamento nacional da produção mineral, Brazil.

Ref.

1. Angles of Prüfer, *Naturwiss. Abh. Wien*, **1**, 169 (1847) on lazulite from Werfen; setting of Prüfer (1847), Dana (798, 1892), Goldschmidt (**5**, 131, 1916), Hintze (1 [4B], 1126, 1933). The nearly orthogonal, pseudo-hexagonal morphological cell is retained here. Transformation: from the morphological cell (space group $B2_1/a$) to the simplest structure cell (space group $P2_1/n$), $\frac{1}{2}0\frac{1}{2}/010/001$.
2. Dana (1892), Goldschmidt (1916).
3. Berry, *Am. Min.*, **33**, 750 (1948), by Weissenberg method. Pecora and Fahey, *Am. Min.*, **35**, 1 (1950), give powder spacing data for members of the series.
4. Baier, *Zbl. Min.*, 145 (1941).
5. Optical data summarized by Pecora and Fahey (1950); indices from their graph.
6. See Pecora and Fahey (1950).
7. For additional analyses see summary in Pecora and Fahey (1950); also Hintze (1 [3B], 1132, 1933).
8. Fahey analysis in Pecora and Fahey (1950).
9. See Meixner, *Berg. -u. hütten. Jb., Montan. Hochsch. Leoben*, **85**, 1, 33 (1937).
10. See Watson, *J. Washington Ac. Sc.*, **11**, 386 (1921), Meixner (1937), Pecora and Fahey (1950).
11. Pecora and Fahey, *Am. Min.*, **34**, 83, 684 (1949).

TROLLEITE. Blomstrand (*Ak. Stockholm, Öfv.*, **25**, 199, 1868).

Massive, lamellar. Indistinct cleavage[1] in two directions at an angle of 110°52'. Fracture even to conchoidal. H. 5½–6. G. 3.10. Color pale green. Luster vitreous. Biaxial,[1] with optic plane parallel to the best of the two cleavages. $2E \sim 81°44'$. Dispersion $r > v$, weak. Analysis gave CaO 0.97, Fe_2O_3 2.75, Al_2O_3 43.26, P_2O_5 46.72, H_2O 6.23, total 99.93. This is close to $Al_4(PO_4)_3(OH)_3$. Before B.B. turns white but does not fuse. Scarcely attacked by acids. Found with berlinite and other phosphates at the iron mine of Westanå in Kristianstad, Sweden. Named after the Swedish chemist H. G. Trolle-Wachtmeister (1782–1871). Evidence of species validity is lacking.

Ref.

1. Des Cloizeaux (**2**, 453, 1893).

41.8.2 **S O U Z A L I T E** [$(Mg,Fe'')_3(Al,Fe''')_4(PO_4)_4(OH)_6 \cdot 2H_2O$]. *Pecora* and *Fahey* (*Am. Min.*, **34**, 83, 1949).

Probably monoclinic. As coarse fibrous masses. Cleavable in two directions approximately at right angles, one of good and the other of fair quality. H. 5½–6. G. 3.087. Color green. Shows polysynthetic twinning with the composition plane parallel to the good cleavage direction.

Opt.

ORIENTATION	n	PLEOCHROISM	
$X \perp$ clv.	1.618	Green	Biaxial negative $(-)$.
Y	1.642	Blue	$2V\ 68°$ (calc.).
Z elong.	1.652	Yellow	$r > v$, extreme.

The twin lamellae have a maximum extinction of $12°$.

Chem. A hydrated basic phosphate of aluminum, trivalent and divalent iron and magnesium, $(Mg,Fe'')_3(Al,Fe''')_4(PO_4)_4(OH)_6 \cdot 2H_2O$. Fe''' substitutes for Al, and Fe'' for Mg, with $Fe''':Al = 1:15.4$ and $Fe'':Mg = 1:1.49$ in the only reported analysis.

	CaO	MgO	FeO	SnO	MnO	Al_2O_3	Fe_2O_3	TiO_2	P_2O_5	H_2O+	Total
1.		9.73	11.64			25.73	2.62		38.17	12.11	100.00
2.	0.02	9.62	11.49	0.04	0.31	26.07	2.65	0.07	37.70	12.04	100.01

1. $(Mg,Fe'')_3(Al,Fe''')_4(PO_4)_4(OH)_6 \cdot 2H_2O$, with $Fe'':Mg = 1:1.49$ and $Fe''':Al = 1:15.4$. 2. Corrego Frio, Brazil.[1]

Tests. B.B. infusible. Soluble with difficulty in HCl.

Occur. Found as a hydrothermal alteration product of scorzalite in the Corrego Frio pegmatite, north of Divino, Minas Geraes, Brazil.

Name. After A. J. A. de Souza, former Director of the Departamento nacional da produção mineral, Brazil.

Ref.

1. Fahey analysis in Pecora and Fahey (1949).

41.8.3 **CARMINITE** $[PbFe_2(AsO_4)_2(OH)_2]$. Carminspath *Sandberger* (*Ann. Phys.*, **80**, 391, 1850; **103**, 345, 1858). Karminspat. Carmine Spar. Carminite *Dana* (410, 1854).

Cryst.[1] Orthorhombic; dipyramidal—$2/m\ 2/m\ 2/m$.

$$a:b:c = 0.7373:1:0.4663; \qquad p_0:q_0:r_0 = 0.6324:0.4663:1$$

$$q_1:r_1:p_1 = 0.7373:1.5812:1; \qquad r_2:p_2:q_2 = 2.1445:1.3562:1$$

Forms:[1]

	ϕ	$\rho = C$	ϕ_1	$\rho_1 = A$	ϕ_2	$\rho_2 = B$
o 010	0°00′	90°00′	90°00′	90°00′	.....	0°00′
m 110	53 36	90 00	90 00	36 24	0°00′	53 36
e 011	0 00	25 00	25 00	90 00	90 00	57 16

Habit. Crystals lath-like, flattened on $\{010\}$ and elongated $[001]$ (Mapimi). In tufted or satiny aggregates of fine needles; also massive with a radiated structure or in radially fibrous spherical aggregates.

Phys. Cleavage $\{110\}$ (?), distinct. Brittle. H. $3\frac{1}{2}$. G. 4.10,[2] 5.22 (anal. 4). Luster vitreous, pearly on the cleavage. Color carmine-red, also tile-red to reddish brown. Streak reddish yellow. Translucent.

O p t. In transmitted light, red.

ORIENTATION	n (Mapimi)[3]	n (Cornwall)[4]	PLEOCHROISM	
X c	2.070	2.05	Pale yellowish red	Biaxial positive (+).
Y a (?)	2.070	2.05	Dark carmine-red	$2V$ medium (Na).
Z b (?)	2.080	2.06	Dark carmine-red	$r < v$, strong.

C h e m. A basic arsenate of lead and ferric iron, $PbFe_2(AsO_4)_2(OH)_2$ (analyses 2, 3).[5]

Anal.

	1	2	3	4
CaO		0.44	0.06	
PbO	35.38	37.30	36.57	38.14
Fe_2O_3	25.33	23.43	23.81	23.07
Al_2O_3		0.96	0.43	
As_2O_5	36.44	33.98	34.49	33.22
H_2O	2.85	3.00	3.13	3.71
Rem.		0.85	0.62	1.65
Total	100.00	99.96	99.11	99.79
G.				5.22

1. $PbFe_2(AsO_4)_2(OH)_2$. 2. Mapimi, Mexico.[3] Rem. is MgO 0.06, FeO 0.21, insol. 0.58; H_2O includes H_2O- 0.10. 3. Colorado.[3] Rem. is MgO 0.02, FeO 0.14, insol. 0.46. 4. Western Australia.[6] Rem. is P_2O_5 0.30, SO_3 1.35.

Tests. B.B. fuses easily with intumescence to a black non-magnetic bead. In C.T. fuses, turns dark brown, and gives a little water. Slowly soluble in HCl with separation of $PbCl_2$ and completely soluble in HNO_3.

O c c u r. Originally from the Luise iron mine near Horhausen, Rhine Province, Germany, with beudantite; also from Ems, Hesse-Nassau. At the Hingston Down Consols mine, Calstock, Cornwall, England, with scorodite, mimetite, pharmacosiderite. At Mapimi, Durango, Mexico, with scorodite, dussertite, arseniosiderite, anglesite, cerussite. From Wyloo, Ashburton district, Western Australia. In the United States from an unstated locality in Colorado (anal. 3) and reported from Eureka, Nevada, and the Tintic District, Utah.

N a m e. In allusion to the color.

Ref.

1. Foshag, *Am. Min.*, **22**, 479 (1937), on poor crystals from Mapimi.
2. Sandberger (1858) on Horhausen material.
3. Foshag (1937).
4. Larsen and Berman (143, 1934), who give $r > v$, extreme.
5. Foshag (1937). Earlier considered to be $Pb_3Fe_{10}(AsO_4)_{12}$ on the basis of an approximate analysis by Müller in Sandberger (1858) on Horhausen material.
6. Le Mesurier, *J. Roy. Soc. Western Australia*, **25**, 137 (1939).

41.8.4 **P A R S O N S I T E** $[Pb_2(UO_2)(PO_4)_2 \cdot 2H_2O]$. Schoep (*C.R.*, **176**, 171, 1923).

Monoclinic.[1] As crusts and powdery aggregates of minute lath-like crystals. The crystals are elongated [001] and flattened {010} with {001}, {101}, and {100}; (001) $\wedge$ (100) $\sim 81°$, (101) $\wedge$ (100) $\sim 132°$.

P h y s. Cleavage not observed. H. $2\frac{1}{2}$–3.[2] G. 5.37 (Ruggles mine [2]). Luster sub-adamantine. Color pale citron-yellow. Transparent to translucent. Not fluorescent in ultraviolet light.

O p t.[2] In transmitted light, pale yellow. Not pleochroic.

ORIENTATION	n (Kasolo)[4]	n (Ruggles)	n (Wölsendorf)	
X	1.85	~1.870	~1.795	Biaxial negative $(-)$.
Y b				
Z	1.862	~1.890	~1.815	
$Z \wedge c$ (usually)	12°	6°–23°	2°–36°	

The optical orientation varies and sometimes is $Y = b$ with $X \wedge c = 6°$–$23°$. The indices of refraction also vary with nX up to about 1.88, and nZ sometimes below 1.81. $Z \wedge c$ usually is 12°–14°.

C h e m. A hydrated phosphate of lead and uranium, $Pb_2(UO_2)(PO_4)_2 \cdot 2H_2O$. The water content is uncertain and probably varies with atmospheric conditions.

Anal.

	1	2	3
PbO	49.03	47.43	44.71
UO_3	31.42	34.68	29.67
P_2O_5	15.59	14.46	15.08
H_2O	3.96	3.43	1.56
Rem.			8.25
Total	100.00	100.00	99.27
G.		5.37	6.23

1. $Pb_2(UO_2)(PO_4)_2 \cdot 2H_2O$. 2. Ruggles mine, New Hampshire.[3] Recalculated to 100 after deducting 5.64 per cent quartz. 3. Kasolo, Belgian Congo.[5] Rem. is CaO 0.63, CuO 0.25, Al_2O_3 1.23, TeO_3 3.01, MoO_3 0.43, CO_2 1.19, insol. 1.51. Original total given as 99.47.

Tests. B.B. fusible. In C.T. yields water and turns deeper yellow. Soluble in acids.

O c c u r. Originally found at Kasolo, Katanga, in the Belgian Congo, associated with torbernite, kasolite, and dewindtite. Also from the Ruggles mine near Grafton Center, New Hampshire,[2] with autunite and phosphuranylite, and from Wölsendorf, Bavaria, Germany, with fluorite and uranocircite.

N a m e. After Arthur L. Parsons (1873–), formerly Professor of Mineralogy at the University of Toronto.

Ref.

1. Frondel, *Am. Min.*, **35**, 245 (1950) from x-ray and optical evidence; c_0 6.8 kX.
2. Frondel (1950).
3. Gonyer analysis in Frondel (1950).
4. Billiet, *Bull. soc. min.*, **49**, 136 (1926).
5. Schoep (1923).

42 HYDRATED PHOSPHATES, ETC., CONTAINING HYDROXYL OR HALOGEN

TYPE 1. $(AB)_m(XO_4)_pZ_q \cdot xH_2O$, WITH $m:p > 3:1$

42.1.1 Borickite. Delvauxite [from Leoben] von Hauer (Jb. geol. Reichsanst., Wien, **5**, 68, 1854); [from Nenačovic] Bořicky (Nat. Zs. Lotos, March, 1867). Borickite Dana (588, 1868). Boryckite, Borickyt.

Compact reniform masses, in part opaline. H. $3\frac{1}{2}$. G. $\sim$2.70.[1] Color and streak reddish brown. Luster weak waxy. Opaque. Isotropic,[2] with n variable 1.57–1.67.

The name originally was applied to material from Leoben (anal. 2), and later to a similar mineral from Nenačovic (anal. 3), which approximates in composition to $CaFe_5(PO_4)_2(OH)_{11} \cdot 3H_2O$. Borickite also has been grouped with type foucherite and delvauxite (and various chemically similar substances ascribed thereto), but all are ill-defined and their mutual relations are uncertain.

Anal.

	1	2	3
CaO	7.47	8.16	7.29
MgO			0.41
Fe$_2$O$_3$	53.20	52.29	52.99
P$_2$O$_5$	18.92	20.49	19.35
H$_2$O	20.41	19.06	19.96
Total	100.00	100.00	100.00

1. $CaFe_5(PO_4)_2(OH)_{11} \cdot 3H_2O$. 2. Leoben, Styria.[3] Average of two. 3. Nenačovic, Bohemia.[4] Recalculated after deduction of impurities.

Tests. B.B. fuses easily to a black mass. Soluble in HCl.

Occur. In gossan at Nenačovic, south of Kladno, Bohemia. At Leoben, Styria, Austria.

Named after Emanuel Bořicky (1840–1881), Czech petrographer.

Ref.

1. Vála and Helmhacker, Arch. naturw. Landes. Böhmen, **2**, 381 (1874)—Jb. Min., 317 (1875).
2. Larsen (49, 1921).
3. von Hauer (1854).
4. Bořicky (1867).

FOUCHERITE. Berthier (Ann. mines, **9**, 519, 1836). Fuchérite Leymérie (**2**, 340, 1867). Fouchérite Lacroix (**4**, 535, 1910).

Compact, brownish red nodules. H. $3\frac{1}{2}$. G. 2.7. The reported analysis is close to $Ca(Fe,Al)_4(PO_4)_2(OH)_8 \cdot 7H_2O$.

	1	2
CaO	8.02	7.71
Al_2O_3	4.17	4.50
Fe_2O_3	39.16	38.50
P_2O_5	20.30	19.50
H_2O	28.35	28.50
Total	100.00	98.71

1. $Ca(Fe,Al)_4(PO_4)_2(OH)_8 \cdot 7H_2O$ with Fe:Al = 6:1. 2. Fouchères.[1]

B.B. fuses to a black magnetic mass. Easily soluble in HCl. Found at Fouchères, Aube, Champagne, France, in Neocomien sandstone. Minerals found as colloform crusts at Litošice (n 1.648) and the Hrbek mine, St. Benigna, Czechoslovakia, have been ascribed [2] to this species. Foucherite is close to borickite in composition, and the identity of the two substances, both ill-defined, has been urged but is not proved.

Ref.

1. Pisani in Lacroix (1910). Also an earlier, incorrect analysis by Berthier, *Ann. mines*, **9**, 519 (1836).
2. Ulrich, *Rozpr. České Ak.*, **31**, no. 10 (1922); Slavík, *Bull. int. ac. Bohême*, **22**, 32 (1918).

TYPE 2. $(AB)_3(XO_4)Z_q \cdot xH_2O$

42.2.1 VESZELYITE $[(Cu,Zn)_3(PO_4)(OH)_3 \cdot 2H_2O]$. Schrauf (*Anz. Ak. Wien*, **11**, 135, 1874; *Zs. Kr.*, **4**, 31, 1879). Arakawaite Wakabayashi and Komada (*J. Geol. Soc. Tokyo*, **28**, 191, 1921). Kipushite Buttgenbach (*Ac. roy. Belgique, Bull., Cl. Sc.*, 905, 1926).

C r y s t.[1] Monoclinic; prismatic—$2/m$.

$a:b:c = 0.9542:1:0.7288$; $\beta\ 103°23'$; $p_0:q_0:r_0 = 0.7638:0.7090:1$

$r_2:p_2:q_2 = 1.4104:1.0773:1$; $\mu\ 76°37'$; $p_0'\ 0.7851, q_0'\ 0.7288, x_0'\ 0.2379$

Forms:[2]

	ϕ	ρ	ϕ_2	$\rho_2 = B$	C	A
c 001	90°00'	13°23'	76°37'	90°00'		76°37'
m 110	47 08	90 00	0 00	47 08	80°14'	42 52
e 011	18 04$\frac{1}{2}$	37 28$\frac{1}{2}$	76 37	54 40	35 20	79 07
i 111	54 32	51 28$\frac{1}{2}$	44 21	63 00	41 07	50 25
σ 121	35 04	60 41	44 21	44 28	53 41	59 56$\frac{1}{2}$

Structure cell.[3] Space group $P2_1/a$. $a_0\ 9.84$ Å, $b_0\ 10.17$, $c_0\ 7.48$; $\beta\ 103°25'$; $a_0:b_0:c_0 = 0.9675:1:0.7355$. Cell contents $(Cu,Zn)_{12}(PO_4)_4(OH)_{12} \cdot 8H_2O$.

Habit. Short prismatic [001] and thick tabular {100}; also equant or octahedral with large {100} and {011}. As granular aggregates of indistinct crystals.

P h y s. Cleavage {001} and {110}. G. 3.4 ± 0.1; 3.42 (calc. for anal. 1). Luster vitreous. Color greenish blue to dark blue. Translucent.

VESZELYITE 917

O p t.[6] In transmitted light, greenish blue. Faintly pleochroic, with Z blue and X greenish blue.

ORIENTATION	n (Arakawa mine)[4]	n (Moravicza)[5]	
X	1.618	1.640	Biaxial positive (+).
Y b	1.622	1.658	$r < v$, weak to strong.
$Z \wedge c$	1.658	1.695	
$-35°$ to $-43°$			
$2V$	$38\frac{1}{2}°$	$71° \pm 5°$	

C h e m. A hydrated basic phosphate of copper and zinc, $(Cu,Zn)_3$-$(PO_4)(OH)_3 \cdot 2H_2O$. Zn and Cu substitute mutually, with Cu:Zn ~3:2 in the reported analyses. (AsO_4) substitutes for (PO_4) with As:P = 1:1.03 in analysis 3. The water is entirely lost below red heat.[7]

Anal.

	1	2	3	4	5
CuO	38.20	37.82	37.34	40.44	35.99
ZnO	26.05	26.69	25.20	23.64	28.94
P_2O_5	18.93	18.43	9.01	19.01	19.90
H_2O	16.82	16.87	17.05	16.22	14.31
Rem.		0.11	10.41		0.65
Total	100.00	99.92	99.01	99.31	99.79
G.		3.34	3.53	3.09	>3.37

1. $(Cu,Zn)_3(PO_4)(OH)_3 \cdot 2H_2O$ with Cu:Zn = 3:2. 2. Moravicza.[8] Rem. is PbO 0.05, FeO 0.06, NiO tr. 3. Moravicza.[9] Rem. is As_2O_5 10.41. 4. Arakawa mine, Japan (*arakawaite*).[4] 5. Kipushi, Belgian Congo.[10] Rem. is SiO_2.

Tests. Soluble in acids.

O c c u r. Originally found at Moravicza (Vaskö) in the Banat, Roumania. In the Arakawa mine, Ugo province, Japan (*arakawaite*).[11] In Africa in the Broken Hill mine, northern Rhodesia,[12] and at Kipushi, Katanga, in the Belgian Congo (*kipushite*).[11] Associated with secondary copper minerals.

N a m e. After the Hungarian mining engineer A. Veszelyi, who discovered the mineral.

Ref.

1. Angles and unit of Zsivny, *Mat. termés. Ért.*, **48**, 331 (1931), *Zs. Kr.*, **82**, 87 (1932), on Moravicza crystals (analysis 2); orientation of the structure cell. Transformation: Zsivny (1931) and Schrauf II (1874) to new, $001/0\bar{1}0/\bar{1}00$. Roughly comparable angles were reported by Buttgenbach, *Ac. roy. Belgique, Bull., Cl. Sc.*, **18**, 43 (1932) and (1926), on "kipushite" and by Wakabayashi and Komada (1921) and Ito, *Zs. Kr.*, **65**, 305 (1927), on "arakawaite"; the differences may be due to variation in the Cu:Zn ratio (analyses 4, 5).
2. Zsivny (1931). Rare or doubtful: b 010, δ 102, o 113, Q $\bar{1}15$, t 345, n 425, r 495, p 625, s 352. Also rounded, vicinal faces in the zone [010], giving an elliptical outline.
3. Berry, *Am. Min.*, **33**, 750 (1948), by Weissenberg method on Moravicza crystals.
4. Wakabayashi and Komada (1921).
5. Larsen (153, 1921).
6. The optical data and their apparent variation with Cu:Zn ratio are discussed by Zsivny (1931, 1932).
7. Zsivny (1931, 1932) gives dehydration data.
8. Zsivny (1931, 1932).
9. Schrauf (1874).
10. Bolsius analysis in Buttgenbach (1932).

11. On the identity of arakawaite and kipushite with veszelyite see Buttgenbach, *Bull. ac. roy. Belgique, Cl. Sc.*, 448 (1941), and Zsivny (1931, 1932).
12. Mennell, *Min. Mag.*, **19**, 69 (1920).

42.2.2   **T S U M E B I T E** [$Pb_2Cu(PO_4)(OH)_3 \cdot 3H_2O$]. *Busz* (*Festschr. med.-naturwiss. Ges. Münster*, 1912; *Zs. Kr.*, **51**, 526, 1912). Preslite *Rosický* (*Zs. Kr.*, **51**, 521, 1912). Arsentsumebite (*Bull. soc. min.*, **58**, 4, 1935).

C r y s t.[1] Monoclinic.

$a:b:c = 0.6546:1:0.6745;$ $\beta\, 94°22';$ $p_0:q_0:r_0 = 1.0304:0.6725:1$

$r_2:p_2:q_2 = 1.4869:1.5321:1;$ $\mu\, 85°38';$ $p_0'\, 1.0334,\ q_0'\, 0.6745,\ x_0'\, 0.0764$

Forms:

	ϕ	ρ	ϕ_2	$\rho_2 = B$	C	A
a 100	90°00′	90°00′	0°00′	90°00′	85°38′	.....
m 110	56 52	90 00	0 00	56 52	86 20½	33°08′
u 011	6 27½	34 10	85 38	56 04½	33 55½	86 22½
v 021	3 14½	53 29½	85 38	36 37½	53 22½	87 23½
d 101	90 00	47 58½	42 01½	90 00	43 36½	42 01½
f $\bar{1}04$	−90 00	10 18½	100 18½	90 00	14 40½	100 18½
r 241	38 27½	73 49	25 01	41 14	71 08	53 19

Less common:
        k 140 l 120 n 320 e 201

Habit. Crusts of intergrown crystals. Always twinned, with twin plane {$\bar{1}22$}, sometimes as trillings or more complex groups. Often thick

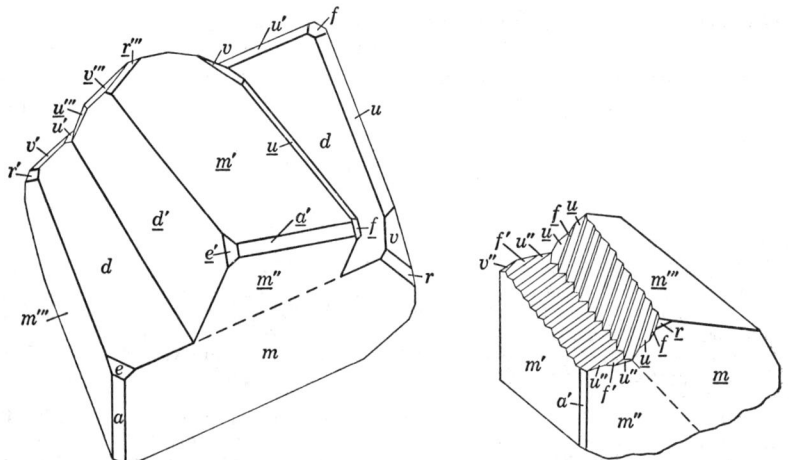

tabular on {101} or {110}. {011} and {$\bar{1}04$} usually in oscillatory combination, sometimes giving a zig-zag suture in twins. {100} narrow and bright, {110} uneven or composite.

P h y s. Fracture uneven. Brittle. H. 3½. G. 6.13. Luster vitreous, brilliant. Color emerald-green. Transparent.

O p t.[2] In transmitted light, green in color with faint pleochroism and absorption $Z > X$.

Orientation	n	
X	1.885 ± 0.005	Biaxial positive (+).
Y	1.920 ± 0.005	$2V$ near $90°$.
Z	1.956 ± 0.005	$r < v$, strong.

C h e m. A hydrated basic phosphate of lead and copper, $Pb_2Cu(PO_4)(OH)_3 \cdot 3H_2O$.

Anal.

	1	2	3	4
PbO	65.84	63.77	65.09	64.71
CuO	11.73	11.79	11.97	12.13
P_2O_5	10.47	12.01	10.26	10.62
H_2O	11.96	12.33	[12.68]	12.09
Rem.				0.78
Total	100.00	99.90	[100.00]	100.33
G.		6.133	6.09	

1. $Pb_2Cu(PO_4)(OH)_3 \cdot 3H_2O$. 2. Tsumeb.[3] 3. Tsumeb.[4] 4. Tsumeb.[5] Rem. is ZnO 0.54 and CO_2 0.24. Original sum 99.79.

Tests. B.B. fusible. In C.T. loses water and blackens. Easily soluble in HCl, slowly in HNO_3.

O c c u r. A secondary mineral, found with smithsonite, cerussite, and azurite at Tsumeb, South West Africa. Also from Morenci, Arizona.

Ref.

1. Elements and forms of LaForge, *Am. Min.*, **23**, 772 (1938). $o\{210\}$ uncertain.
2. Larsen (147, 1921).
3. Rüsberg and Dubigk analysis in Busz (1912).
4. Frejka analysis in Rosický (1912).
5. Gonyer analysis in LaForge (1938).

42.2.3 **H E M A F I B R I T E** $[Mn_3(AsO_4)(OH)_3 \cdot H_2O]$. Aimafibrit *Igelström* (*Öfv. Ak. Stockholm*, **41**, 86, 1884). Hämafibrite.

C r y s t.[1] Orthorhombic.

$a:b:c = 0.9504:1:1.0931;$ $p_0:q_0:r_0 = 1.1501:1.0931:1$

$q_1:r_1:p_1 = 0.9504:0.8695:1;$ $r_2:p_2:q_2 = 0.9148:1.0522:1$

Forms:[2]

	ϕ	$\rho = C$	ϕ_1	$\rho_1 = A$	ϕ_2	$\rho_2 = B$
a 100	90°00′	90°00′		0°00′	0°00′	90°00′
l 120	34 28½	90 00	90°00′	55 31½	0 00	34 28½
p 111	53 56	61 42	47 33	44 37½	33 40½	58 47

Habit. Crystals prismatic [001]. Usually as radially fibrous spherical aggregates, the fibers elongated [001].

P h y s. Cleavage $\{100\}$, good, $\{120\}$ poor. Fracture uneven. Brittle. H. 3. G. 3.65.[3] Luster vitreous on crystal faces, greasy on fracture surfaces. Color brownish red to garnet-red, becoming brownish black or black on exposure. Streak light brick-red. Transparent to translucent.

O p t.[4] In transmitted light, red-brown.

ORIENTATION		n	
X	a	1.87 ± 0.01	Biaxial positive (+).
Y	b	1.88 ± 0.01	2V moderate.
Z	c	1.93 ± 0.01	Not pleochroic.
			$r > v$.

C h e m. A hydrated basic arsenate of divalent manganese, $Mn_3(AsO_4)(OH)_3 \cdot H_2O$. The water content is uncertain.[5]

Anal.

	1	2	3
MgO			0.41
FeO		0.79	0.25
MnO	57.09	57.94	58.02
As_2O_5	30.83	30.76	30.88
H_2O	12.08	12.01	12.01
Total	100.00	101.50	101.57

1. $Mn_3(AsO_4)(OH)_3 \cdot H_2O$. 2,3. Nordmark.[6]

Tests. B.B. easily fusible to a black bead which on prolonged heating turns brown. Easily soluble in acids. In C.T. affords water and turns black.

O c c u r. At the Moss mine, Nordmark, Sweden, with carbonates and secondary manganese oxides.

A l t e r. To unidentified hydrous manganese oxides.

N a m e. From αἷμα, *blood-red*, and *fiber*, in allusion to the color and structure.

Ref.
1. Angles of Sjögren, *Zs. Kr.*, **10**, 126 (1885), unit and orientation of Machatschki, *Geol. För. Förh.*, **53**, 191 (1931), to show relation to synadelphite.
2. Sjögren (1885).
3. Highest value of the range 3.50–3.65 reported by Sjögren (1885).
4. Larsen (83, 1921) and Sjögren (1885).
5. The water expressed as H_2O in the formula is regarded by Machatschki (1931) as secondary.
6. A. Sjögren and Lundström analyses in Sjögren (1885); an earlier analysis by Igelström (1884), containing much Fe, Mg, Ca, appears to be in error.

ELFSTORPITE. Igelström (*Geol. För. Förh.*, **15**, 472, 1893; *Zs. Kr.*, **22**, 468, 1893).
As grains or orthorhombic (?) crystals with one cleavage. Brittle. H. 4. Color and streak whitish gray. Inferred to be a hydrated arsenate of divalent manganese. With basiliite and tephroite from the Sjö mine, Örebro, Sweden.

42.2.4 F R E I R I N I T E $[Na_3Cu_3(AsO_4)_2(OH)_3 \cdot H_2O]$. Foshag (*Am. Min.*, **9**, 30, 1924). Lavendulan Goldsmith (*Proc. Ac. Sc. Philadelphia*, 192, 1877) [not lavendulan Breithaupt, *J. pr. Chem.*, **10**, 505, 1837].

Probably tetragonal. In satiny aggregates of fine flakes. Cleavage {001} good, {110} imperfect.

O p t. In transmitted light, greenish blue.

	n	DICHROISM	
O	1.748	Deep greenish blue	Uniaxial negative (−).
E	1.645	Light greenish blue	

LIROCONITE

C h e m. A basic hydrated arsenate of sodium and copper, $Na_3(Cu,Ca)_3(AsO_4)_2(OH)_3 \cdot H_2O$. The water content is uncertain. Ca substitutes for Cu with Ca:Cu = 1:3.4 in analysis 2.

Anal.

	1	2
Na_2O	15.33	14.36
CaO		6.16
CuO	39.35	29.62
As_2O_5	37.89	38.80
H_2O	7.43	9.17
Rem.		1.34
Total	100.00	99.45

1. $Na_3Cu_3(AsO_4)_2(OH)_3 \cdot H_2O$. 2. San Juan, Chile. Rem. is Fe_2O_3 0.76, insol. 0.58.

Tests. Fuses easily with intumescence to a black mass. Easily soluble in HCl.

O c c u r. With erythrite, cuprite, malachite, and cobaltian wad at the Blanca mine, San Juan, Dept. Freirina, Chile.

N a m e. From the locality.

42.2.5 **LIROCONITE** $[Cu_2Al(AsO_4)(OH)_4 \cdot 4H_2O]$. Octahedral Arseniate of Copper (from Cornwall) *Bournon* (*Phil. Trans.*, 174, 1801; *Rashleigh's Brit. Min.*, **2**, Pl. 2, 5, 11, 1802). Linsenerz *Werner* (1803; *Ludwig's Min.*, **2**, 215, 1804); *Karsten* (64, 1808). Linsenkupfer *Hausmann* (1051, 1813). Lirokon-malachite pt. *Mohs* (180, 1822). Chalcophacit *Glocker* (859, 1831). Liroconite *Beudant* (**2**, 600, 1832). Lenticular Copper. Cuivre Arseniaté Octaèdre obtus *Haüy*.

C r y s t.[1] Monoclinic; prismatic—$2/m$.

$a:b:c = 1.6809:1:1.3190$; $\quad \beta\; 91°27'$; $\quad p_0:q_0:r_0 = 0.7847:1.3186:1$

$r_2:p_2:q_2 = 0.7584:0.5951:1$; $\quad \mu\; 88°33'$; $\quad p_0'\; 0.7850$, $q_0'\; 1.3190$, $x_0'\; 0.0253$

Forms:[2]

	ϕ	ρ	ϕ_2	$p_2 = B$	C	A
c 001	90°00'	1°27'	88°33'	90°00'	....	88°33'
b 010	0 00	90 00	....	0 00	90°00'	90 00
a 100	90 00	90 00	0 00	90 00	88 33	....
m 110	30 45½	90 00	0 00	30 45½	89 15½	59 14½
e 011	1 06	52 50	88 33	37 10½	52 49½	89 07½

Structure cell.[3] Space group $I2/a$. a_0 12.67 kX, b_0 7.55, c_0 9.86; $\beta\; 91°23'$; $a_0:b_0:c_0 = 1.679:1:1.306$. Cell contents $Cu_8Al_4(AsO_4)_4(OH)_{16} \cdot 16H_2O$.

Habit. Crystals thin or lenticular [001] with a flat octahedral appearance, the faces of m and e striated parallel their intersection edges.

P h y s. Cleavage {110} and {011} indistinct. Fracture uneven to conchoidal. H. 2–2½. G. 2.9–3.0, varying with the P:As ratio; 2.95, 3.01 (Cornwall[4]); 2.97 (calc. for P:As = 1:4). Luster vitreous, inclining to resinous. Color and streak sky-blue to verdigris-green. Transparent to translucent.

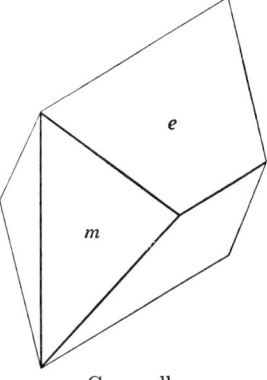

Cornwall.

O p t.[5] In transmitted light, pale blue to pale bluish green.

Orientation		n	
X		1.612 ± 0.003	Biaxial negative $(-)$.
Y	b	1.652 ± 0.003	$2V\ 72° \pm 5°$ (calc.).
$Z \wedge a$	$+25°$	1.675 ± 0.003	$r < v$, moderate.
			Not pleochroic.

C h e m. A hydrated basic arsenate of copper and aluminum, $Cu_2Al(AsO_4)(OH)_4 \cdot 4H_2O$. (PO_4) substitutes for (AsO_4) up to about $P:As = 1:3.8$ (anal. 2).

Anal.[6]

	1	2	3	4
CuO	36.74	36.38	37.18	36.73
Al_2O_3	11.77	10.85	9.68	
Fe_2O_3		0.98		
As_2O_5	26.53	23.05	22.22	23.85
P_2O_5		3.73	3.49	1.02
H_2O	24.96	25.01	25.49	
Total	100.00	100.00	98.06	
G.		2.985	2.964	2.97

1. $Cu_2Al(AsO_4)(OH)_4 \cdot 4H_2O$. 2. Cornwall.[7] 3. Cornwall.[8] 4. Cornwall.[9] Also with H_2O in vacuo 6.40 and at 100° 9.85 per cent.

Tests. B.B. turns olive-green and fuses easily to a dark gray glass. Soluble in acids.

O c c u r. A secondary mineral found in the oxidized zone of copper deposits associated with cuprite, malachite, olivenite, chalcophyllite, limonite. Found at several mines in the neighborhood of Redruth and St. Day, Cornwall; and in Devonshire. At Herrengrund, Czechoslovakia. In Germany at Schwarzemberg, Sagisdorf, and Mortelgrunde near Saida in Saxony; also at Ullersreuth, Saxony, and at Hirschberg, Thuringia. From the Preobrashensky mines in the Urals, U.S.S.R. Said to occur at the Cerro Gordo mine, Inyo County, California, with linarite and caledonite.

N a m e. From λειρός, *pale*, and κονία, *powder*, in allusion to the color of its streak. Linsenerz refers to the flat-lenticular habit of the crystals.

Ref.

1. Goldschmidt (**5,** 163, 1918), whose unit and orientation corresponds to the structure cell. Transformation: from the original elements of Des Cloizeaux (**2,** 71, 1859), later accepted by Dana (853, 1892), to the structure cell, $001/0\bar{1}0/100$. Point symmetry established by Berry, priv. comm. (1946).
2. Goldschmidt (1918).
3. Berry (1946) by Weissenberg method on unanalyzed Cornwall crystals.
4. Berry (1946) and Mrose, priv. comm. (1948), on unanalyzed material; earlier reported values range from 2.88 to 2.99.
5. Larsen (101, 1921) and Des Cloizeaux (1859).
6. For earlier analyses see Hintze (**1** [4B], 953, 1931).
7. Hermann, *J. pr. Chem.*, **33,** 296 (1844).
8. Damour, *Ann. chim. phys.*, **13,** 414 (1845).
9. Church, *Min. Mag.*, **11,** 1 (1895).

EVANSITE

42.2.6 EVANSITE [$Al_3(PO_4)(OH)_6 \cdot 6H_2O$]. Forbes (Phil. Mag., **28**, 341, 1864). Bernonite (?) Adam (73, 1869).

Massive. As opaline, botryoidal, or reniform coatings sometimes with a concentric, colloform structure; also stalactitic.

Phys. Fracture conchoidal. Very brittle. H. 3-4. G. variable, 1.8-2.2, usually about 1.9. Luster vitreous, inclining to resinous or waxy. Colorless to milky white, sometimes tinged with blue, green, or yellow; varieties containing much Fe_2O_3 are brown, reddish brown, or red. Streak white or weakly tinted. Transparent to translucent.

Opt. In transmitted light, colorless to brown. Isotropic; in part birefringent. Index of refraction [1] variable in part:

	Szirk (milky)	Alsô-Sajô (milky)	Nizná-Slava (milky)	Nizná-Slava (colorless)	Goldburg (brown)
n	~1.485	1.445 to 1.465	~1.455	~1.485	~1.485

Chem. A basic hydrated phosphate of aluminum, $Al_3(PO_4)(OH)_6 \cdot 6H_2O$. About half of the water is lost at 110°, and most of the remainder below 300°.[2] The Fe''' reported in some analyses presumably substitutes for Al if not due to admixture.

Anal.

	1	2	3	4	5	6	7
CaO			0.41	0.23		4.32	1.03
MgO			tr.	tr.		tr.	0.75
Fe_2O_3			0.87	1.92		5.49	
Al_2O_3	39.60	39.31	39.22	38.05	40.19	34.48	38.33
P_2O_5	18.40	19.05	18.02	18.48	18.11	19.14	[21.70]
H_2O	42.00	39.95	41.47	41.29	41.27	36.96	38.19
Insol.		1.41	0.15	0.44			
Total	100.00	99.72	100.14	100.41	99.57	100.39	[100.00]
G.			1.874	1.937	1.939	1.98	

1. $Al_3(PO_4)(OH)_6 \cdot 6H_2O$. 2. Zeleznik, Czechoslovakia.[3] 3,4. Gross-Tresny, Bohemia.[4] No. 3 white, 4 yellow. 5. Zeehan district, Tasmania.[5] 6. Goldburg, Idaho.[2] Brown. 7. Columbiana, Alabama.[2] Light yellow. H_2O by ign. loss.

Tests. B.B. infusible. In C.T. yields neutral water, decrepitates and falls to a powder. Easily soluble in acids.

Occur. A secondary mineral often found with limonite and allophane. Originally from Mt. Zeleznik near Szirk in Com. Gömör, Czechoslovakia. In Hungary with variscite at Vashegy, in fine specimens at Alsô-Sajô, and at Verespaták and Offenbanya. In a graphite deposit at Gross-Tresny, Moravia; and at Litošice and Nizná-Slava in Bohemia. In France at Epernay, Dept. Marne. At several localities in Spain,[6] notably Teis near Vigo. In fissures in shale at Yoredale Rocks, near Macclesfield, East Cheshire, England. Occurs in graphitic gneiss in the Vatoinandry district, eastern Madagascar. At Mt. Zeehan, Tasmania. In the United States, evansite has been found at Columbiana and Coalville in the Coosa coal field of Alabama, and at Goldburg, Custer County, Idaho.

Name. After Brooke Evans of Birmingham, England, who brought the first specimens from Hungary in 1855.

Ref.
1. Frondel, priv. comm. (1948).
2. Schaller, *Am. J. Sc.*, **24**, 155 (1907).
3. Forbes (1864).
4. Kovár, cited in *Zs. Kr.*, **31**, 524 (1899).
5. Smith, *Proc. Roy. Soc. New South Wales*, **27**, 382 (1893).
6. Iglesias, *Bol. soc. espan. hist. nat.*, **27**, 319 (1927).

ROSIÉRÉSITE. *Lacroix* (*Min. France*, **4**, 532, 1910).

As opaline stalactitic masses with a concentric structure. G. 2.2. Color greenish yellow, yellow, and light brown. Isotropic. Index of refraction about 1.50. A hydrous phosphate of aluminum with lead and copper. Analysis gave:[1] CuO 3.0, PbO 10.0, Al_2O_3 23.0, P_2O_5 25.5, As_2O_5 tr., H_2O 38.0, total 99.5. B.B. does not fuse and turns black. Easily soluble in HNO_3. Found as a recent deposit in abandoned workings of the copper mine at Rosiérés, near Carmaux, Tarn, France. Needs verification.

Ref.
1. Berthier, *Ann. mines*, **19**, 669 (1841).

42.2.7   **Liskeardite.** *Maskelyne* (*Nature*, **18**, 426, 1878).

Orthorhombic (?). Massive, as thin incrusting layers with a radial fibrous structure. Soft. Color white, inclining toward greenish, bluish, or brownish white.

O p t.[1] In transmitted light, colorless.

ORIENTATION	n	
X	1.661	Biaxial positive (+).
Y	1.675	$2V$ nearly 90°.
Z elongation	1.689	

The indices of refraction vary about ±0.01, and the values given are near the average.

C h e m. A hydrated basic arsenate of aluminum and ferric iron, perhaps $(Al,Fe)_3(AsO_4)(OH)_6 \cdot 5H_2O$, with Fe:Al = 1:6 in the only reported analysis.

Anal.

	1	2
Fe_2O_3	8.06	7.64
Al_2O_3	30.89	28.23
As_2O_5	27.08	26.96
H_2O	33.97	34.05
Rem.		2.86
Total	100.00	99.74

1. $(Al,Fe)_3(AsO_4)(OH)_6 \cdot 5H_2O$, with Fe:Al = 1:6.   2. Liskeard.[2] Rem. is CuO 1.03, CaO 0.72, SO_3 1.11; H_2O includes 4.35 lost over H_2SO_4, 10.96 at 100°, 5.55 at 120°, 8.22 at 190°, 4.97 by ignition with PbO.

O c c u r. Found coating scorodite, arsenopyrite, chalcopyrite, pyrite, quartz at Liskeard, Cornwall. A perhaps similar mineral occurs[3] as fibrolamellar aggregates of monoclinic (?) crystals at Cap Garonne, near Hyères, Var, France; G. 3.011, color greenish white, cleavage parallel the elongation with the acute bisectrix (X) perpendicular thereto.

Ref.
1. Larsen (101, 1921).
2. Flight, *J. Chem. Soc. London*, **43**, 140 (1883).
3. Lacroix, *Bull. soc. min.*, **24**, 27 (1901).

TYPE 3. $(AB)_5(XO_4)_2Z_q \cdot xH_2O$

42.3.1 CORNWALLITE [$Cu_5(AsO_4)_2(OH)_4 \cdot H_2O$]. *Zippe* (*Böhm. Ges., Abh.*, [5], **4**, 649, 1847).

As small botryoidal crusts with a radial fibrous structure, somewhat resembling malachite. Fracture conchoidal. Not very brittle. H. $4\frac{1}{2}$. G. 4.166. Color verdigris-green to blackish green.

O p t.[1] In transmitted light, emerald-green.

	n	
X	~1.81	Biaxial positive (+).
Y	1.815	$2V$ small.
Z	~1.85	Elongation negative, in part positive.

C h e m. A hydrated basic arsenate of divalent copper, $Cu_5(AsO_4)_2(OH)_4 \cdot H_2O$. There is some substitution of phosphorus for arsenic, with As:P ~7:1 in analysis 2. Cornwallite may be isostructural with pseudomalachite and, further, may be identical with erinite.

Anal.

	CuO	As_2O_5	P_2O_5	H_2O	Total	G.
1.	59.31	29.98	2.65	8.06	100.00	
2.	59.95	30.47	2.71	8.23	101.36	4.17

1. $Cu_5[(As,P)O_4]_2(OH)_4 \cdot H_2O$, with the ratio As:P = 7:1.   2. Cornwall.[2]

Tests. F. 2 to $2\frac{1}{2}$ to a black glass.

O c c u r. With olivenite and tenorite from unspecified localities in Cornwall, England.

Ref.

1. Larsen, *U. S. Geol. Sur., Bull. 679*, 63 (1921).
2. Church, *J. Chem. Soc. London*, **21**, 276 (1868).

42.3.2 TYROLITE [$Cu_5Ca(AsO_4)_2(CO_3)(OH)_4 \cdot 6H_2O$ (?)]. Kupferschaum *Werner* (Hoffmann, **3**, 180, 1816; 19, 50, 1817). Kupaphrite *Shepard* (**1**, 294, 1835). Tirolit *Haidinger* (509, 1845).

C r y s t.[1] Orthorhombic.

Structure cell.[7] Space group *Pmma*. a_0 10.50 Å, b_0 54.71, c_0 5.59; $a_0:b_0:c_0 = 0.1919:1:0.1022$. Cell contents uncertain.

Habit. Rarely as lath-like crystals or scales flattened on {010} and elongated [100] or [001]. Usually as fan-shaped and closely foliated aggregates; also as crusts and reniform masses with a radiated foliaceous structure and a drusy surface; divergent fibrous.

P h y s. Cleavage {001} perfect. Thin laminae are flexible. Sectile. H. ~2. G. 3.0–3.2, 3.25 (Tintic). Luster vitreous, on {001} pearly. Color pale apple-green to verdigris-green, inclining to sky-blue. Streak a little paler. Translucent.

O p t. In transmitted light, pale green.

ORIENTATION		n (Tintic)[2]	PLEOCHROISM	
X	b	1.694 ± 0.003	Pale grass-green	Biaxial negative $(-)$.
Y	c	1.726 ± 0.003	Pale yellowish green	$2V\ 36° \pm 3°$ (meas.). $r > v$, strong.
Z	a	1.730 ± 0.003	Pale grass-green	

C h e m. A hydrated basic arsenate of copper and calcium of uncertain formula, perhaps $Cu_5Ca(AsO_4)_2(CO_3)(OH)_4 \cdot 6H_2O$ (analysis 4)[5] or $Cu_9Ca_2(AsO_4)_4(OH)_{10} \cdot 10H_2O$. In material from Tintic, Utah, (CO_3) is absent and (SO_4) is present (analyses 2, 3). About half of the water is lost at $100°$.

Anal.[6]

	1	2	3	4
CaO	6.43	6.78	9.10	6.44
CuO	45.63	45.08	42.60	46.24
As$_2$O$_5$	26.36	28.52	27.87	27.07
SO$_3$		2.23	2.45	
CO$_2$	5.05			5.05
H$_2$O	16.53	17.21	16.23	15.68
Rem.		0.24	0.97	
Total	100.00	100.06	99.22	100.48
G.			3.27	

1. $Cu_5Ca(AsO_4)_2(CO_3)(OH)_4 \cdot 6H_2O$. 2. Mammoth mine, Tintic, Utah.[3] Rem. is Fe$_2$O$_3$ 0.08, insol. 0.16. 3. Mammoth mine.[4] Rem. is (Fe,Al)$_2$O$_3$. 4. Falkenstein, Tyrol.[5]

Tests. B.B. easily fusible to a steel-gray globule. In C.T. decrepitates and yields acid water. Soluble in acids and in ammonia.

O c c u r. A secondary mineral, found associated with erythrite, malachite, azurite, chrysocolla, aurichalcite, cuprite, limonite, and other oxidation products in copper deposits. Found in Czechoslovakia at Libethen, at Pojnik near Neusohl and at other localities. At Herrengrund in Hungary. In Germany at Saalfeld, Thuringia, and Bieber and Riechelsdorf, Hesse; at Schneeberg, Saxony. In the Tyrol, Austria, at numerous localities especially at Kogel near Brixlegg, on the Falkenstein near Schwaz, and at Rattenberg. From Nerchinsk, Transbaikalia, U.S.S.R. In France at Cap Garonne, Var; in Italy at Campiglia, and in Spain at Linares in the Sierra Morena. From Matlock, Derbyshire, England. In the United States found with chalcophyllite and conichalcite in the Tintic district, Utah.

N a m e. From its occurrence in the Tyrol.

Ref.

1. E. S. Dana in Hillebrand and Dana, *Am. J. Sc.*, **39**, 273 (1890), gives rough measurements on ill-formed microscopic crystals.
2. Larsen (147, 1921).
3. Hillebrand analysis in Hillebrand and Dana (1890).
4. Pearce analysis in Hillebrand and Washington, *Am. J. Sc.*, **35**, 301 (1888).
5. Church, *Min. Mag.*, **11**, 5 (1895).
6. For earlier analyses see Hintze (**1** [4B], 1078, 1931).
7. Berry, *Am. Min.*, **33**, 193 (1948) [abstract].

42.3.3 AKROCHORDITE [MgMn$_4$(AsO$_4$)$_2$(OH)$_4$·4H$_2$O (?)]. Acrochordite Flink (Geol. För. Förh., 44, 773, 1922).

Monoclinic.[1] In wart-like or spherical aggregates of minute crystals. Cleavage in two mutually perpendicular directions. H. $3\frac{1}{2}$. G. 3.194. Color red-brown with yellow tint. Translucent.

O p t.[2]

Orientation		n	
X	b	1.672	Biaxial positive (+).
$Y \wedge c$	$\sim 45° \pm$	1.676	$2V$ medium large.
Z		1.683	$r < v$, fairly strong.

C h e m. A hydrated basic arsenate of magnesium and divalent manganese, probably (Mn,Mg)$_5$(AsO$_4$)$_2$(OH)$_4$·4H$_2$O, with Mg:Mn = 1:3.2 (analysis 2).

Anal.

	1	2
MgO	6.09	6.94
MnO	42.88	38.98
As$_2$O$_5$	34.72	33.51
P$_2$O$_5$		0.42
H$_2$O	16.31	16.78
Rem.		3.68
Total	100.00	100.31

1. MgMn$_4$(AsO$_4$)$_2$(OH)$_4$·4H$_2$O.   2. Långban, Sweden.[3]   Rem. is Na$_2$O 1.18, K$_2$O 0.55, CaO 0.99, FeO 0.46, Mn$_2$O$_3$ 0.50. Analysis made on a small sample which may have contained admixed pyrochroite.

Tests. On heating loses water and becomes gray-black in color. Easily soluble in dilute H$_2$SO$_4$, giving a purple solution.

O c c u r. Found in small amounts with pyrochroite and barite in hausmannite ore from the "Japan" workings at Långban, Sweden.

N a m e. From ἀκροχορδών, a wart, in allusion to the form of aggregation.

Ref.
1. Indicated by the optical study of Quensel in Flink (1922) and of Barth and Berman, Chem. Erde, 5, 41 (1930).
2. Barth and Berman (1930).
3. Almström analysis in Flink (1922); see also Almström, Geol. För. Förh., 45, 117 (1923).

42.3.4 Ceruleite [CuAl$_4$(AsO$_4$)$_2$(OH)$_8$·4H$_2$O]. Céruléite Dufet (Bull. soc. min., 23, 147, 1900). Coerulite.

Massive, compact and clay-like but composed of extremely minute rod-like crystals. G. 2.803. Color turquois-blue. Chemically, a hydrated basic arsenate of copper and aluminum, CuAl$_4$(AsO$_4$)$_2$(OH)$_8$·4H$_2$O. About 1.45 per cent water is lost at 180°, the remainder at high temperatures.

Anal.

	1	2
CuO	12.10	11.80
Al$_2$O$_3$	31.01	31.26
As$_2$O$_5$	34.96	34.56
H$_2$O	21.93	22.32
Total	100.00	99.94

1. CuAl$_4$(AsO$_4$)$_2$(OH)$_8$·4H$_2$O.   2. Chile.

Tests. Easily soluble in acids.

O c c u r. From the gold mine Emma Luisa near Huanaco, Taltal province, Chile. Named in allusion to its color.

42.3.5 **R E N A R D I T E** [$Pb(UO_2)_4(PO_4)_2(OH)_4 \cdot 7H_2O$]. Schoep (Bull. soc. min., 51, 247, 1928).

C r y s t.[1] Orthorhombic. $a:c:b = 1.209:1:?$

Forms: a 100 b 010 d 101

Habit. Drusy crusts of minute crystals flattened {100} and slightly elongated [001].

P h y s. Cleavage {100} perfect. G. >4. Color yellow. Transparent.

O p t.

Orientation		n	Pleochroism	
X	a	1.715 ± 0.003	Yellow	Biaxial negative (−).
Y	c	1.736 ± 0.003	Yellow	$2V\ 41°01'$ (calc.).
Z	b	1.739 ± 0.003	Colorless	$r > v$.

C h e m. A hydrated basic phosphate of lead and uranium, $Pb(UO_2)_4(PO_4)_2(OH)_4 \cdot 7H_2O$. Analysis gave:

	PbO	UO_3	P_2O_5	H_2O	Rem.	Total
1.	13.16	68.92	8.37	9.55		100.00
2.	12.26	64.82	8.15	8.74	6.53	100.50
3	13.05	68.98	8.67	9.30		100.00

1. $Pb(UO_2)_4(PO_4)_2(OH)_4 \cdot 7H_2O$. 2. Kasolo. Rem. is $(Fe,Al)_2O_3$ 3.68, MoO_3 0.74, insol. 2.11. H_2O includes H_2O- 4.97. 3. Analysis 2 recalculated to 100 after deducting remainder.

Tests. Fuses to a black scoriaceous mass. In C.T. yields water and turns brown. Soluble in acids.

O c c u r. With dewindtite, dumontite, and torbernite at the Kasolo mine, Katanga, Belgian Congo.

N a m e. After A. F. Renard (1842–1903) of the University of Ghent.

Ref.

1. See also Schoep, Ann. mus. Congo belge, 1 (1930), and Buttgenbach, Les Min. de Belgique et du Congo belge, 457 (1947). $a:c$ from the measured ρ angle of {101}, 39°47'.

42.3.6 **D U M O N T I T E** [$Pb_2(UO_2)_3(PO_4)_2(OH)_4 \cdot 3H_2O$]. Schoep (C.R., 179, 693, 1924).

C r y s t.[1] Orthorhombic. $a:b:c = ?:1:1.327$.

Forms: a 100 c 001 b 010 d 011 e 013

As small crystals elongated [001] and flattened {010}. Color and streak ocher-yellow. Fluoresces greenish.

O p t.

Orientation		n^2	Pleochroism	
X	a	1.88	Pale yellow	Biaxial positive (+).
Y	c			$r < v$.
Z	b	1.89	Deep yellow	$2V$ large.

Chem. A hydrated basic phosphate of lead and uranium, $Pb_2(UO_2)_3$-$(PO_4)_2(OH)_4 \cdot 3H_2O$. The water is largely or entirely lost below 300°. Analysis gave:

	PbO	UO$_3$	P$_2$O$_5$	H$_2$O	Rem.	Total
1.	29.05	55.85	9.24	5.86		100.00
2.	27.19	56.49	8.65	5.78	1.01	99.12

1. $Pb_2(UO_2)_3(PO_4)_2(OH)_4 \cdot 3H_2O$. 2. Chinkolobwe. Rem. is TeO_3.

Tests. In C.T. loses water and turns orange. Soluble in acids.

Occur. With torbernite at Chinkolobwe, Katanga, in the Belgian Congo.

Name. After André Dumont (1809–1857), Belgian geologist.

Ref.
1. From measurements under the microscope, with (001) ∧ (011) ~53°, (001) ∧ (013) 23°30′ by Schoep (1924), *Ann. mus. Congo belge*, pt. 2, **1**, 32 (1930).
2. Billiet, *Bull. soc. min.*, **49**, 136 (1926).

TYPE 4. $A_2(XO_4)Z_q \cdot xH_2O$

42.4.1 **BAYLDONITE** [(Cu,Pb)$_2$(AsO$_4$)(OH) (?)]. Church (*J. Chem. Soc. London*, **18**, 265, 1865). Parabayldonite, Cuproplumbite *Biehl* (Inaug. Diss. Münster, 1919—*Min. Abs.*, **1**, 202, 1921; *Zs. Kr.*, **62**, 341, 1925).

Probably monoclinic. In minute mammillary concretions with a fibrous structure and a drusy surface; also massive, fine-granular to powdery, and as crusts. H. $4\frac{1}{2}$. G. 5.5. Luster strongly resinous. Color siskin-green to apple-green and yellow-green. Subtranslucent.

Opt.[1]

ORIENTATION	n (Cornwall)	
X b	1.95 ± 0.01	Biaxial positive (+).
Y ∧ elong. ~45°		2V large.
Z	1.99 ± 0.01	$r < v$.

Chem. A basic arsenate of copper and lead. Formula uncertain, perhaps (Cu,Pb)$_2$(AsO$_4$)(OH). Cu and Pb substitute mutually over a considerable range, with Pb:Cu = 1:2.87 in analysis 2 and Pb:Cu = 1:1.08 in analysis 4. Fe″ apparently substitutes for (Cu,Pb) with Fe:(Cu,Pb) = 1:5.0 in analysis 4.

Anal.

	1	2	3	4	5	6
CaO					0.81	0.21
FeO			4.52	5.75	2.38	1.89
CuO	33.03	30.88	23.09	16.51	22.68	29.25
PbO	32.29	30.13	37.76	42.93	43.52	36.44
As$_2$O$_5$	32.16	31.76	30.31	29.09	25.60	28.81
H$_2$O	2.52	4.58	5.15	4.21	3.10	3.27
Rem.		2.65		0.78	2.09	0.46
Total	100.00	[100.00]	100.83	99.27	100.18	100.33
G.		5.35	5.50	5.21	5.44	5.50

1. (Cu,Pb)$_2$(AsO$_4$)(OH) with Pb:Cu = 1:2.87. 2. Cornwall.[2] Rem. is Fe$_2$O$_3$, CaO and loss. 3,4. Tsumeb.[3] Rem. is SiO$_2$. 5. Tsumeb (*parabayldonite*).[3] Rem. is Sb$_2$O$_5$. 6. Tsumeb (*cuproplumbite*).[3] Rem. is Sb$_2$O$_5$.

Tests. B.B. easily fusible. In C.T. loses water and turns black. Soluble with difficulty in HCl.

Occur. A secondary mineral found with mimetite, olivenite, and other arsenates in the oxidized zone of the copper deposit at Tsumeb, South West Africa (*parabayldonite, cuproplumbite,* in part).[4] Originally from copper mines near St. Day, Cornwall, England. Reported from Diou, Allier, France.[5]

Alter. From mimetite (Tsumeb).

Name. After John Bayldon.

Ref.
1. Larsen (45, 1921).
2. Church (1865).
3. Biehl (1919), who cites additional analysis.
4. These substances are probably, but not shown certainly, identical with bayldonite.
5. Lacroix (4, 513, 1910).

42.4.2 **LEUCOCHALCITE** [$Cu_2(AsO_4)(OH) \cdot H_2O$ (?)]. Sandberger, Petersen (*Jb. Min.*, I, 263, 1881). Leukochalcit *Germ.*

Orthorhombic (?). As a delicate coating of needle-like crystals. Color white to greenish white. Luster silky.

Opt.[1] In transmitted light, colorless.

nX	1.79 ± 0.01	Biaxial positive (+).
nY elong.	1.807 ± 0.003	$2V$ large.
nZ	1.84 ± 0.01	$r < v$, strong.
		Parallel extinction.

Chem. A hydrated basic arsenate of copper, perhaps $Cu_2(AsO_4)(OH) \cdot H_2O$. A small amount of P apparently substitutes for As, and Ca and Mg may be in substitution for Cu, but the purity of the analyzed material is doubtful. Analysis gave: CaO 1.56, MgO 2.28, CuO 47.10, P_2O_5 1.60, As_2O_5 [37.89], H_2O (and CO_2 trace) 9.57, total [100.00].

Tests. In C.T. first turns green, then black and loses water. On charcoal fusible to a black glass.

Occur. With malachite and calcite at the Wilhelmine mine near Schöllkrippen in the Spessart, Bavaria, Germany. A mineral referred to leucochalcite occurs as pale green fibrous masses with dioptase at Mindouli in the Belgian Congo.[2]

Name. From λευκός, *white,* and χαλκός, *copper,* in allusion to the color and composition.

Ref.
1. Larsen (100, 1921).
2. Lacroix, *Bull. soc. min.*, **31,** 255 (1908); the mineral differs in that Z = elongation.

42.4.3 **TAGILITE** [$Cu_2(PO_4)(OH)\cdot H_2O$]. Tagilith Hermann (J. pr. Chem., **37**, 184, 1846).

Orthorhombic (?). As porous, earthy coatings and reniform concretions with a fine-fibrous structure. H. 3. G. ~3.5. Color emerald-green to mountain-green.

O p t.[1] In transmitted light, bluish green.

ORIEN- TATION		n (Nizhne Tagilsk)	n (Moravicza) tagilite (?)	
X	elong.	1.69 ± 0.01	1.685 ± 0.005	Biaxial negative (−).
Y		1.84 ± 0.01	1.820 ± 0.005	$2V$ small.
Z		1.85 ± 0.01	1.820 ± 0.005	

C h e m. A hydrated basic copper phosphate, $Cu_2(PO_4)(OH)\cdot H_2O$.
Anal.

	1	2	3	4
CuO	61.89	61.29	62.38	61.70
Fe_2O_3		1.50		
P_2O_5	27.60	26.44	26.91	27.42
H_2O	10.51	10.77	10.71	10.25
Total	100.00	100.00	100.00	99.37

1. $Cu_2(PO_4)(OH)\cdot H_2O$. 2,3. Nizhne Tagilsk. No. 2 contains limonite. 3. Tagilite (?). Coquimbo, Chile.[2]

Tests. Like libethenite.

O c c u r. Originally found with libethenite and pseudomalachite at Nizhne Tagilsk in the Ural Mountains, U.S.S.R. Also identified by optical means from Moravicza, Hungary, and on chemical grounds from the Mercedes mine, near Coquimbo, Chile. Unanalyzed monoclinic crystals resembling liroconite in habit, with distinct {010} cleavage and G. ~4.08, found at the Arme Hilfe mine, Ullersreuth, Saxony, Germany, have been referred[3] to tagilite.

Ref.
1. Larsen (140, 1921).
2. Field, *Gazz. chim. Ital.*, **17**, 225 (1859).
3. Breithaupt, *B. H. Zs.*, **24**, 301 (1865).

42.4.4 **SPENCERITE** [$Zn_4(PO_4)_2(OH)_2\cdot 3H_2O$]. Walker (Min. Mag., **18**, 76, 1916).

C r y s t.[1] Monoclinic; prismatic—$2/m$ (?).

$a:b:c = 1.0125:1:1.0643;\quad \beta\ 116°47';\quad p_0:q_0:r_0 = 1.0512:0.9501:1$

$r_2:p_2:q_2 = 1.0525:1.1063:1;\quad \mu\ 63°13';\quad p_0'\ 1.1775,\ q_0'\ 1.0643,\ x_0'\ 0.5048$

Forms: [2]

		ϕ	ρ	φ_2	$\rho_2 = B$	C	A
c	001	90°00′	26°47′	63°13′	90°00′	.....	63°13′
b	010	0 00	90 00		0 00	90°00′	90 00
a	100	90 00	90 00	0 00	90 00	63 13	.....
t	120	28 57	90 00	0 00	28 57	77 24	61 03
m	110	47 53½	90 00	0 00	47 53½	70 28	42 06½
l	520	70 07½	90 00	0 00	70 07½	64 55½	19 52½
f	021	13 20½	65 26	63 13	27 45½	62 14½	77 53
k	$\bar{1}02$	90 00	47 33½	42 26½	90 00	20 46½	42 26½
g	$\bar{1}01$	90 00	59 16½	30 43½	90 00	32 29½	30 43½
q	$\bar{2}21$	53 20½	74 20	19 16½	54 54½	31 54	39 26
x	121	38 19	69 46	30 43½	42 36	55 11	54 25½
z	$\bar{2}41$	33 53½	78 58	19 16½	35 26	53 01	56 49
s	346	− 6 45	35 32½	94 48	54 44½	45 55½	93 55
y	$\bar{1}21$	−17 32½	65 52	123 55½	29 31	76 03	105 57½

Less common and rare:

  n 230 e 023 i $\bar{1}04$ j $\bar{3}04$ h $\bar{2}01$ p $\bar{1}11$

Habit. Crystals tabular {100} and elongated [001], the termination usually blunt and rectangular with {$\bar{1}02$}, {001}, and {$\bar{1}01$} but sometimes lance-like with large {$\bar{1}21$} and {$\bar{2}21$}. {$h0l$} striated [010] due to twinning. Commonly massive, stalactitic, with a columnar to platy structure.

Twinning. Twin plane and composition surface {100}, polysynthetic.

Etch figures. Pits produced by HNO_3 on {100} show a vertical plane of symmetry.

P h y s. Cleavage {100} perfect, {010} good, {001} distinct. H. 3. G. 3.14. Color pure white. Luster pearly on the cleavages to vitreous.

O p t.[3] In transmitted light, colorless.

Orientation	n	
X near a	1.586	Biaxial negative (−).
Y	1.602	$2V$ 49°, $2E$ 83° (meas.).
Z b	1.606	$r > v$, moderate.

Sections across {100} show polysynthetic twin lamellae with a maximum extinction of about 6°.

C h e m. A hydrated basic zinc phosphate, $Zn_4(PO_4)_2(OH)_2 \cdot 3H_2O$. On heating,[4] $3H_2O$ are lost below 140° and the remainder is lost only at about 500°.

Anal.

	1	2	3
ZnO	60.33	60.18	60.39
P_2O_5	26.31	26.19	26.13
H_2O+	13.36	3.50	13.44
H_2O-		9.81	
Total	100.00	99.68	99.96
G.		3.145	3.123

1. $Zn_4(PO_4)_2(OH)_2 \cdot 3H_2O$. 2. Salmo, British Columbia.[2] 3. Average of two analyses. 3. Salmo, British Columbia.[5]

ISOCLASITE

Tests. Heated in C.T. decrepitates and yields much water, the residue yellow while hot and white when cold. B.B. easily fusible. Readily soluble in acids.

O c c u r. Found with hemimorphite, hopeite, salmoite, and a zinc-rich clay (*sauconite?*) in a cave in oxidized zinc ore at the Hudson Bay mine, Salmo, Nelson mining district, British Columbia.

N a m e. After Leonard James Spencer (1870–   ), British mineralogist and editor, formerly Keeper of Minerals in the British Museum of Natural History.

Ref.
1. Walker, *Washington Ac. Sc., J.,* **7**, 456 (1917); *Univ. Toronto Stud., Geol. Ser.,* no. 10, 5 (1918). Transformation: Walker to new, 100/010/001.
2. Walker (1916).
3. Larsen and Berman (167, 1934).
4. Walker and Parsons, *Univ. Toronto Stud., Geol. Ser.,* 58 (1921), with optical study of the dehydration products.
5. Phillips, *Am. J. Sc.,* **42**, 275 (1916).

SALMOITE. Larsen (*U. S. Geol. Sur., Bull. 679,* 135, 1921).

A colorless mineral found as lens-shaped crystals and grains admixed with spencerite and hopeite at the Hudson Bay mine, Salmo, British Columbia. Optically negative (−), with nX 1.645, nY 1.683, nZ 1.695 (all ± 0.003), $2V$ about 60°, $r > v$ perceptible. No cleavage. Probably a basic zinc phosphate.[1]

Ref.
1. Phillips, *Am. J. Sc.,* **42**, 275 (1916). Larsen's specimen was obtained from Phillips.

42.4.5   I S O C L A S I T E [$Ca_2(PO_4)(OH) \cdot 2H_2O$]. Isoklas *Sandberger (J. pr. Chem.,* **2**, 125, 1870; *Jb. Min.,* 306, 1870). Isoclasite *Dana* (835, 1892).

Monoclinic.[1] In dull, minute crystals, prismatic [001] and as cotton-like fibers. Cleavage {010}, distinct. H. $1\frac{1}{2}$. G. 2.92. Colorless to snow-white. Luster vitreous to pearly on cleavage surfaces.

O p t.[2] In transmitted light, colorless.

ORIENTATION	n	
X b	1.565 ± 0.003	Biaxial positive (+).
Y	1.568 ± 0.003	$2V$ med.
$Z \wedge c$ small	1.580 ± 0.003	

C h c m. A basic hydrated calcium phosphate, $Ca_2(PO_4)(OH) \cdot 2H_2O$.

Anal.

	1	2
CaO	49.15	49.51
P_2O_5	31.11	29.90
H_2O+	19.74	18.53
H_2O-		2.06
Total	100.00	100.00

1. $Ca_2(PO_4)(OH) \cdot 2H_2O$.   2. Joachimsthal.[3]

Tests. B.B. loses neutral water and fuses. Easily soluble in acids.

O c c u r. With chalcedony and dolomite at Joachimsthal, Bohemia.

A l t e r. White altered crystals gave on analysis: Na_2O 9.80, CaO 1.00, MgO 17.30, $(Al,Fe)_2O_3$ 0.36, P_2O_5 34.00, H_2O+ 9.22, H_2O- 24.26, insol. 0.18, total 96.12.

Name. From ἴσος, *equal*, and κλάσις, *fracture*, in allusion to the cleavage.

Ref.
1. Symmetry inferred from the optical properties. Approximate measurements with a contact goniometer on altered crystals gave 110 ∧ 1̄10 = 136°50′ and 001 ∧ 110 = 110°.
2. Larsen (92, 1921).
3. Köttnitz analysis in Sandberger (1870).

42.4.6 EUCHROITE [$Cu_2(AsO_4)(OH) \cdot 3H_2O$]. Breithaupt (172, 266, 1823).

Cryst.[1] Orthorhombic; disphenoidal—2 2 2.

$a:b:c = 0.9635:1:0.5866;$[2] $\quad p_0:q_0:r_0 = 0.6088:0.5866:1$

$q_1:r_1:p_1 = 0.9635:1.6425:1; \quad r_2:p_2:q_2 = 1.7047:1.0379:1$

Forms:[3]

		φ	ρ = C	φ₁	ρ₁ = A	φ₂	ρ₂ = B
b	010	0°00′	90°00′	90°00′	90°00′	.....	0°00′
a	100	90 00	90 00		0 00	0°00′	90 00
m	110	46 04	90 00	90 00	43 56	0 00	46 04
e	011	0 00	30 24	30 24	90 00	90 00	69 36
d	021	0 00	49 33½	49 33½	90 00	90 00	40 26½
M	101	90 00	31 20	0 00	58 40	58 40	90 00
s	302	90 00	42 24	0 00	47 36	47 36	90 00
l	201	90 00	50 36	0 00	39 24	39 24	90 00

Structure cell.[4] Space group $P2_12_12_1$. a_0 10.05 kX, b_0 10.50, c_0 6.11; $a_0:b_0:c_0 = 0.957:1:0.582$. Cell contents $Cu_8(AsO_4)_4(OH)_4 \cdot 12H_2O$.

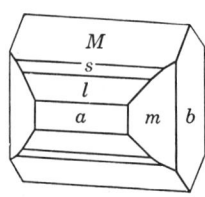

Habit. Short prismatic [010] to equant; rarely thick tabular {100}. Faces *m s l* striated [010]. The crystals ordinarily are holohedral in appearance.

Phys. Cleavage {101} and {110} in traces. Fracture uneven to subconchoidal. Rather brittle. H. 3½–4. G. 3.44 (Libethen),[5] 3.45 (calc.). Luster vitreous. Color bright emerald- or leek-green. Transparent to translucent.

Opt. In transmitted light, bright bluish green.

Orientation[9]		n (Libethen)[6]	
X	c	1.695 ± 0.003	Biaxial positive (+).
Y	a	1.698 ± 0.003	2V 29° ± 1° (meas.).[6]
Z	b	1.733 ± 0.003	r > v, moderate.
			Weakly pleochroic.

Chem. A hydrated basic arsenate of copper, $Cu_2(AsO_4)(OH) \cdot 3H_2O$. A small amount of ($PO_4$) substitutes for ($AsO_4$) in analysis 3.

Anal.[7]

	CuO	As₂O₅	P₂O₅	H₂O	Total	G.
1.	47.21	34.09		18.70	100.00	3.45
2.	47.26	30.90	1.48	19.28	98.92	3.42

1. $Cu_2(AsO_4)(OH) \cdot 3H_2O$. 2. Libethen.[8] H₂O includes H₂O − 3.12 and H₂O + 16.16.

Tests. B.B. fusible. In C.T. loses water and turns yellow-green. Soluble in acids.

Occur. From Libethen near Neusohl, Hungary, with olivenite as crystals lining crevices in mica schist.

Name. From εὔχροια, beautiful color.

Ref.
1. Disphenoidal symmetry established by Berry, priv. comm. (1946).
2. Angles and unit of Haidinger, *Ann. Phys.*, **5**, 165 (1825) and *Ed. J. Sc.*, **2**, 133 (1825) in the orientation of the structure cell. Transformation: Haidinger to new, 010/001/100.
3. Goldschmidt (**3**, 157, 1916).
4. Berry (1946) by Weissenberg method on Libethen crystals.
5. Frondel, priv. comm. (1946); earlier reported values range from 3.34 to 3.42.
6. Larsen (**73**, 1921). See also Des Cloizeaux (**2**, 30, 1859).
7. For other analyses see Dana (566, 1868) and Hintze (**1** [4B], 876, 1931).
8. Church, *Min. Mag.*, **11**, 1 (1895).
9. Grailich and Lang, *Ak. Wien, Sitzber.*, **27**, 47 (1857).

42.4.7 Delvauxite. Delvauxene *Dumont* (*L'Institut*, 121, 1839); *Delvaux* (*Ac. Belgique, Bull.*, 147, 1838). Delvauxite *Haidinger* (512, 1845). Picite *Nies* (*Jb. Min.*, 176, 1877; *Ber. Oberhess. Ges.*, **19**, 112, 1880); *Streng* (*Jb. Min.*, I, 116, 1881); Picites resinaceus (?) *Breithaupt* (**3**, 897, 1847).

A name given originally to an ill-defined hydrous ferric phosphate, approximately $Fe_4(PO_4)_2(OH)_6 \cdot nH_2O$, from Berneau, near Visé, in Liége, Belgium (anals. 1–3). Color yellowish brown to brownish black or reddish. G. 1.85. The name was later applied with more or less certainty to materials from other localities: Litošice, Bohemia,[1] as isotropic concretionary masses of a chestnut-brown color, G. ∼2.00, H. $2\frac{1}{2}$, n ∼1.726 (anal. 4); Železník, Bohemia,[1] with n(Na) 1.716; Pisek, Bohemia,[2] as small botryoidal, red-brown crusts with a yellow streak, conchoidal fracture and weak waxy luster, G. 2.79; Payerbach, near Grillenberg, lower Austria, G. 1.845 (anal. 5). Picite apparently is identical; it was found originally at the Eleonore mine on the Dünsberg near Giessen, Nassau, Germany, as stalactitic or botryoidal gel-like coatings. H. 3–4. G. 2.83. Color dark brown, streak yellow, luster vitreous to greasy (anal. 6). Other localities for so-called picite include Hrbek near St. Benigna and Trubin, Bohemia,[4] and near Waldgirmes, Nassau;[10] reported as nodules (G. 2.38–2.40; n 1.677–1.680) in the oolitic iron ores of the Kerch peninsula, U.S.S.R.[11]

Anal.

	1	2	3	4	5	6
CaO				2.31		
Fe_2O_3	34.20	36.62	40.44	42.57	34.23	46.50
P_2O_5	16.04	16.57	18.20	20.85	17.38	24.47
H_2O	49.76	46.81	41.13	[23.16]	∼50.00	28.03
Rem.				11.11		1.00
Total	100.00	100.00	99.77	[100.00]		100.00

1,2. Visé, Belgium.[5] Recalculated after deduction of ca. 10.5 per cent calcite. The water is largely lost below 100°.[6] 3. Visé.[7] 4. Litošice, Bohemia.[8] Rem. is SO_3 0.75, Al_2O_3 6.69, SiO_2 3.67, CaO tr., MgO tr. 5. Payerbach, Austria.[3] 6. Eleonore mine, Nassau.[9] Rem. is Al_2O_3. Recalculated to 100 after deduction of 2.10 insol.

Named after J. S. P. J. Delvaux de Feuffe (1782–?), Belgian chemist, who first analyzed the mineral.

Ref.
1. Ulrich, *Rozpr. České Ak.*, **31**, no. 10 (1922).
2. Vrba, *Zs. Kr.*, **15**, 207 (1889).
3. Dittler, *Zs. Chem. Ind. Kolloide*, **4**, 300 (1909).
4. Ulrich (1922); Slavík and Ulrich, *Jb. Min.*, I, 33, 1926; Slavík, *Bull. Intern. Ac. Bohême*, **22**, 32 (1918).
5. Dumont (1839).
6. Church, *Chem. News*, **10**, 157 (1864); and von Hauer, *Jb. geol. Reichsanst.*, *Wien*, **5**, 68 (1854).
7. Delvaux (1838).
8. Veselý, *Rozpr. České Ak.*, **31**, no. 9 (1922).
9. Nies (1880).
10. Streng (1881).
11. Sidorenko, *C. r. ac. sc. U.R.S.S.*, **43**, 264 (1944).

TYPE 5. $(AB)_m(XO_4)_p Z_q \cdot xH_2O$, WITH $m:p = 2:1$

42.5.1 Leucophosphite. Simpson (*J. Roy. Soc. Western Australia*, **18**, 69, 1931–32).

As fine-grained chalk-like masses. Friable. G. between 2.30 and 2.65. Color white to greenish. Birefringent. Analysis gave [1] (after deducting quartz 52.75, rutile 0.48, chromite 1.07, carbonaceous material 1.03 per cent): Na_2O 0.13, K_2O 7.88, $(NH_4)_2O$ 0.09, MgO 0.73, MnO 0.22, CaO tr., Fe_2O_3 32.82, Al_2O_3 12.73, P_2O_5 26.69, CO_2 0.17, H_2O+ 12.28, H_2O- 6.59, total 100.33. Formula approximately $K_2(Fe,Al)_7(PO_4)_4(OH)_{11} \cdot 6H_2O$. Soluble in HCl, insoluble in water. Found with variscite, chalcedony, and opal in veinlets in serpentine on the shore of Weelhamby Lake, Ninghanboun Hills, Western Australia. Apparently formed by the action of solutions derived from bird guano upon serpentine. Named in allusion to its color and composition.

Ref.
1. Murray analysis in Simpson (1931–32).

42.5.2 CHILDRENITE SERIES

42.5.2.1 **C H I L D R E N I T E** [$(Fe'',Mn'')Al(PO_4)(OH)_2 \cdot H_2O$]. Brooke (*Quart. J. Sc. Lit. Arts*, **16**, 274, 1823).

42.5.2.2 **E O S P H O R I T E** [$(Mn'',Fe'')Al(PO_4)(OH)_2 \cdot H_2O$]. Brush and Dana (*Am. J. Sc.*, **16**, 35, 1878).

C r y s t.[1] Orthorhombic; dipyramidal—$2/m\ 2/m\ 2/m$.

CHILDRENITE (Fe:Mn $\sim$9:1)

$$a:b:c = 0.7780:1:0.5258; \qquad p_0:q_0:r_0 = 0.6758:0.5258:1$$

$$q_1:r_1:p_1 = 0.7780:1.4798:1; \qquad r_2:p_2:q_2 = 1.9020:1.2853:1$$

EOSPHORITE (Fe:Mn $\sim$1:4)

$$a:b:c = 0.7745:1:0.5139; \qquad p_0:q_0:r_0 = 0.6635:0.5139:1$$

$$q_1:r_1:p_1 = 0.7745:1.5072:1; \qquad r_2:p_2:q_2 = 1.9459:1.2911:1$$

Forms: [2]

	Childrenite	Eosphorite	ϕ	$\rho = C$	ϕ_1	$\rho_1 = A$	ϕ_2	$\rho_2 = B$
b 010	*	*	0°00'	90°00'	90°00'	90°00'	.....	0°00'
a 100	*	*	90 00	90 00	.....	0 00	0°00'	90 00
u 140	*		17 53½	90 00	90 00	72 06½	0 00	17 53½
g 120	*		32 50½	90 00	90 00	57 09½	0 00	32 50½
m 110	*	*	52 14	90 00	90 00	37 46	0 00	52 14
n 320		*	62 41	90 00	90 00	27 19	0 00	62 41
p 111	*	*	52 14	40 00½	27 12	59 27½	56 26	66 48½
q 232		*	40 43	45 29½	37 51	62 17	56 26	57 17
s 121	*	*	32 50½	50 44½	45 47½	65 10½	56 26	49 25
r 131	*		23 17	59 13	57 02	70 09	56 26	37 54

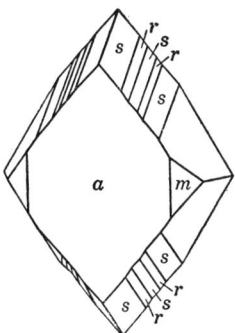

Childrenite. Cornwall.

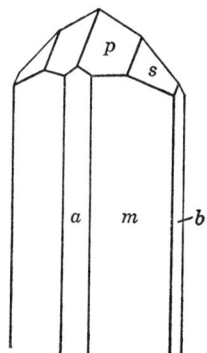

Eosphorite. Poland, Me.

Structure cell.[3] Space group $Bba2$. a_0 10.38 Å, b_0 13.36, c_0 6.911; $a_0:b_0:c_0 = 0.776:1:0.517$ [Fe:Mn $\sim$9:1]. Cell contents $(Fe'',Mn)_8Al_8(PO_4)_8(OH)_{16} \cdot 8H_2O$.

Habit. Childrenite equant or pyramidal to short prismatic [001] and thick tabular {010}; also platy {100}. Eosphorite short to long prismatic [001]. The prism zone is striated [001]. Often in radial groupings of distinct crystals and grading into botryoidal masses or crusts with a coarse fibrous structure; rarely massive.

Phys. Cleavage {100}, poor. Fracture subconchoidal to uneven. H. 5. G. 3.25 ± 0.03 (*childrenite* end-member) to 3.06 ± 0.02 (*eosphorite* end-member);[4] 3.18 (calc. for *childrenite*). Luster vitreous to somewhat resinous. Color brown and yellowish brown in childrenite to pink or rose-red in eosphorite. Streak white. Transparent to translucent.

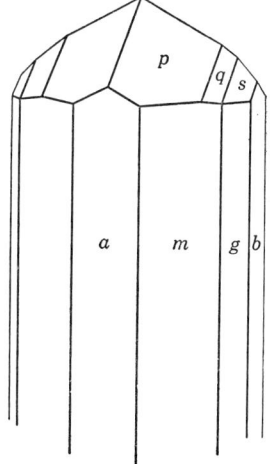

Eosphorite. Branchville.

Opt.[4] In transmitted light, colorless.

ORIENTATION	n (childrenite end-member)	n (eosphorite end-member)	PLEOCHROISM	
X b	1.649	1.628	Yellow	Biaxial negative (−).
Y a	1.683	1.648	Pink	$r < v$, strong.
Z c	1.691	1.657	Pale pink to	
2V		~50°	colorless	

Chem. A hydrated basic phosphate of aluminum with divalent iron and manganese, $(Fe'',Mn'')Al(PO_4)(OH)_2 \cdot H_2O$. Fe'' and Mn substitute mutually, and a complete series probably extends between the two species, but the reported analyses are close to the ends of the series. The observed limits of substitution are $Fe:Mn = 3.0:1$ in childrenite (analysis 4) and $Fe:Mn = 1:3.2$ in eosphorite (analysis 5). Small amounts of Ca and Mg substitute for (Fe,Mn) in some analyses. The Fe''' reported in some analyses is apparently due to oxidation of Fe''.

Anal.[5]

	1	2	3	4	5	6	7	8
CaO		0.65			0.54	0.84		
FeO	31.26	28.56	21.69	23.45	7.40	3.74	1.38	
MnO		3.11	4.21	7.74	23.51	27.65	29.94	30.98
Al_2O_3	22.18	21.43	21.11	15.85	22.19	20.51	22.37	22.26
P_2O_5	30.88	29.92	30.70	30.65	31.05	30.38	29.89	31.02
H_2O	15.68	15.80	16.37	17.10	15.60	15.59	15.34	15.74
Rem.		0.37	5.36	4.54	0.33	1.82	0.90	
Total	100.00	99.84	99.44	99.33	100.62	100.53	99.82	100.00
G.			3.05	3.22	3.134	3.067		

1. $Fe''Al(PO_4)(OH)_2 \cdot H_2O$. 2. Childrenite. Crinnis mine, Cornwall.[6] Rem. is Na_2O. 3. Childrenite? Llallagua, Bolivia.[7] Rem. is SiO_2 2.63, Fe_2O_3 2.15, MgO 0.58. 4. Childrenite. Tavistock.[8] Rem. is MgO 1.03, Fe_2O_3 3.51. 5. Eosphorite. Branchville, Conn.[9] Rem. is Na_2O. Average of two. 6. Eosphorite. Newry, Maine.[10] Rem. is BeO 0.27, MgO 0.36, Fe_2O_3 0.56, F 0.18, insol. 0.45. 7. Eosphorite. Buckfield, Maine.[11] Rem. is insol., F tr. 8. $Mn''Al(PO_4)(OH)_2 \cdot H_2O$.

Tests. B.B. swells up and fuses to a black magnetic bead. In C.T. affords neutral water. Soluble in acids.

Occur. Childrenite was described originally from a specimen purchased at Tavistock in Devonshire, England, and said to have come from an excavation for a canal. Later found at the Wheal Crebor and the George and Charlotte mine, both near Tavistock, associated with siderite, quartz, pyrite, and apatite in hydrothermal veins. Also from the Crinnis mine at St. Austell, Cornwall, and with zinnwaldite, tourmaline, apatite, and kaolin in pegmatitic granite at Greifenstein near Ehrenfriedersdorf, Saxony, Germany. From Llallagua, Bolivia,[12] with paravauxite. Eosphorite was first described from Branchville, Fairfield County, Connecticut, United States, associated with rhodochrosite, lithiophilite, triploidite, dickinsonite in granite pegmatite. Also associated with rhodochrosite, apatite, and manganese phosphates in pegmatite at Newry, Poland, Buckfield, Mt. Mica, Black Mountain, and Hebron, all in Maine, and at the Palermo mine near North Groton, New Hampshire. From Hagendorf near Pleystein, Bavaria, Germany.

N a m e. Childrenite after J. G. Children (1777–1852), English chemist and mineralogist. Eosphorite from ἐωσφορος, *dawn-bearing*, in allusion to the pink color.

Ref.
1. Eosphorite elements computed by LaForge, priv. comm. (1937), from an average of all angles reported by Drugman, *Min. Mag.*, **17**, 193 (1915), on analyzed crystals from Poland, Maine, and by Brush and Dana (1878) on Branchville crystals (analysis 5). Childrenite elements on crystals of poor quality from the George and Charlotte mine by Miller (519, 1852).
2. Goldschmidt (**2**, 136, 1913); Drugman (1915); Slavík, *Bull. Ac. Sc. Bohême*, **14**, no. 4 (1914). The childrenite forms are here calculated to the eosphorite ratio. Also 301, uncertain, on childrenite.
3. Barnes, *Am. Min.*, **34**, 12 (1949), on Tavistock crystals by precession method.
4. G. and indices by graphical extrapolation of published values: Otto, *Min. Mitt.*, **47**, 89 (1935); Drugman (1915); Palache and Shannon, *Am. Min.*, **13**, 392 (1928); Landes, *Am. Min.*, **10**, 384 (1925); Larsen (57, 72, 1921); Mason, *Geol. För. Förh.*, **64**, 335 (1942).
5. For additional analyses see Hintze (1 [4B], 930, 933, 1931).
6. Otto (1935).
7. Gordon, *Proc. Ac. Sc. Philadelphia*, **96**, 349 (1944).
8. Church, *J. Chem. Soc. London*, **26**, 103 (1873).
9. Penfield, *Am. J. Sc.*, **16**, 40 (1878).
10. Shannon analysis in Palache and Shannon (1928).
11. Vassar analysis in Landes (1925).
12. This material is not comparable with other members of the series in G. and indices of refraction; see also analysis 3.

42.5.3 **D A V I S O N I T E** [$Ca_3Al(PO_4)_2(OH)_3 \cdot H_2O$ (?)]. Dennisonite Larsen and Shannon (*Am. Min.*, **15**, 322, 1930).

As botryoidal or spherulitic crusts composed of stout fibers with cross sections that appear to be hexagonal.

P h y s. Cleavage basal, perfect. H. $4\frac{1}{2}$. G. 2.85. Color white.
O p t. In transmitted light, colorless.

nO 1.601 Uniaxial negative (−).
nE 1.591

C h e m. A hydrated basic phosphate of calcium and aluminum, perhaps $Ca_3Al(PO_4)_2(OH)_3 \cdot H_2O$. Alkalies, not determined in the single analysis of the Fairfield material, may be present in small amounts as essential constituents.

Anal.

	1	2	3
CaO	36.80	40.30	41.70
MgO		1.38	5.92
Al_2O_3	22.30	11.82	15.15
P_2O_5	31.05	33.10	30.33
H_2O+	9.85	11.60	6.67
Cl			0.22
Total	100.00	98.20	99.99

1. $Ca_3Al(PO_4)_2(OH)_3 \cdot H_2O$. 2. Fairfield, Utah.[1] Recalculated to original sum after deduction of a few per cent of admixed crandallite. Alkalies not determined; analysis on 0.1 gram sample. 3. Davisonite (?), Diamantina, Brazil.[2] Recalculated after deducting 28.56 per cent insol.

Occur. At Fairfield, Utah County, Utah, lining cavities in crandallite and deltaite in variscite nodules. A mineral referred [3] to davisonite has been found at Damasio, near Diamantina, Minas Geraes, Brazil (analysis 3).

Name. After J. M. Davison (1840–1915), of the University of Rochester, who first described wardite from the Fairfield locality. The name was written erroneously as Dennison by Larsen.

Ref.
1. Shannon analysis in Larsen and Shannon (1930).
2. de Andrade analysis in Guimarães, *Monog. Serv. Geol. Minas Geraes*, no. 2 (1934).
3. Guimarães (1934); the indices of refraction given (nO 1.633, nE 1.628, for Na) are considerably higher than those of davisonite.

42.5.4 **WARDITE** [$Na_4CaAl_{12}(PO_4)_8(OH)_{18} \cdot 6H_2O$]. *Davison* (Am. J. Sc., **2**, 154, 1896). Soumansite *Lacroix* (**4**, 541, 1910).

Cryst.[1] Tetragonal (pyramidal—4 (?)).

$$a:c = 1:2.682; \quad p_0:r_0 = 2.682:1$$

Forms:

	ϕ	ρ	A	M
c 001	....	0°00′	90°00′	90°00′
a 010	0°00′	90 00	90 00	45 00
t 012	0 00	53 17	90 00	55 28
u 011	0 00	69 33	90 00	48 30½
b 114	45 00	43 28½	60 53½	90 00

Structure cell.[2] Space group $P4$, or $P4_3$. a_0 7.04 ± 0.02 kX, c_0 18.88 ± 0.02; $a_0:c_0 = 1:2.682$. Cell contents $Na_4CaAl_{12}(PO_4)_8(OH)_{18} \cdot 6H_2O$.

Habit. Pyramidal {102} or {114}, with {001} usually present as a small brilliant face. {102} is striated horizontally, and both it and the other forms aside from {001} are generally uneven and give poor reflections. Also as granular aggregates and crusts, as subparallel aggregates of coarse fibers, and as radially fibrous and concentrically banded spherulites.

Phys. Cleavage {001} perfect. H. 5. G. 2.87 (Montebras),[3] 2.81 (Fairfield);[4] 2.87 (calc.). Color blue-green to pale green, grading to colorless. Luster vitreous. Transparent.

Opt. In transmitted light, colorless. Basal sections of crystals from Montebras (*soumansite*) are divided into four anomalously biaxial sectors with $Y = [101]$, $Z = [001]$, and $2E$ (+) ranging up to 70°.

	n (Montebras)[5]	n (Fairfield; anal. 2)[5]	
O	1.586	1.590 ± 0.005	Uniaxial positive (+).
E	1.595	1.599 ± 0.005	

Chem. A hydrated basic phosphate of sodium, calcium and aluminum, $Na_4CaAl_{12}(PO_4)_8(OH)_{18} \cdot 6H_2O$. Wardite appears to be related structurally to millisite.

Anal.[8]

	1	2	3
Na_2O	7.61	5.98	7.0
K_2O		0.24	0.6
CaO	3.44		3.0
Al_2O_3	37.53	[38.25]	36.6
P_2O_5	34.84	34.46	34.9
H_2O	16.58	17.87	17.9
Rem.		3.20	
Total	100.00	[100.00]	100.0
G.	2.87	2.77	2.81

1. $Na_4CaAl_{12}(PO_4)_8(OH)_{18} \cdot 6H_2O$. 2. Fairfield, Utah.[6] Analysis probably made on an impure sample. Rem. is MgO 2.40, CuO 0.04, FeO 0.76. 3. Fairfield, Utah.[7] Analysis recalculated after deduction of 8 per cent millisite.

Tests. Fuses at 3 with intumescence to a blebby glass. Difficultly but completely soluble in acids.

Occur. In variscite nodules at Fairfield, Utah County, Utah, associated with millisite and crandallite. The wardite is often intergrown with millisite, or occurs as layers alternating with that mineral. Wardite also occurs as an alteration product of amblygonite at Montebras in Soumans, Dept. Creuse, France (*soumansite*), and has been reported from the variscite deposits at Amatrice Hill and Lucin, Utah. As large crystals in pegmatite at Beryl Mountain near West Andover, New Hampshire.[10]

Name. After Henry A. Ward (1834–1906), American naturalist, collector, and dealer.

SOUMANSITE. Lacroix (4, 541, 1910). Shown to be identical with wardite.[9]

Ref.

1. Elements of the x-ray cell of Larsen, *Am. Min.*, **27**, 292 (1942). Forms of Pough, *Am. Mus. Nov.*, no. 932 (1937). Transformation: Pough to new, 100/010/002. The x-ray ratio is taken in preference to the morphological ratio ($a:c = 1:2.6234$) based on crystals of very poor quality. The identity of some forms is uncertain; {13.0.24}, and {9.0.14} of Pough are taken as {102} and {203}. Also doubtful: 302, 227, 138, 4.9.24, 3.5.15.
2. Larsen (1942) by Weissenberg method on Fairfield crystal.
3. Lacroix (1910).
4. Larsen and Shannon, *Am. Min.*, **15**, 315 (1930); Davison (1896) gave 2.77.
5. Larsen and Shannon (1930).
6. Davison (1896).
7. Shannon analysis in Larsen and Shannon (1930).
8. A partial analysis on 40 mg. by Pisani cited by Lacroix (1910) of Montebras material gave Al_2O_3 36.5, P_2O_5 31.5 per cent.
9. Larsen and Shannon (1930), verified by an x-ray rotation photograph on [001] by Frondel, priv. comm. (1947), on type material.
10. Frondel, priv. comm. (1949).

42.5.5 **MILLISITE** [$(Na,K)CaAl_6(PO_4)_4(OH)_9 \cdot 3H_2O$]. *Larsen* and *Shannon* (*Am Min.*, **15,** 329, 1930).

As chalcedonic crusts or spherules with a finely fibrous structure. Probably tetragonal.[1]

Phys. Cleavage {001} (?).[1] H. $5\frac{1}{2}$. G. 2.83; 2.87 (calc.).[2] Color white to light gray.

Opt. In transmitted light, colorless.

nX 1.584 Biaxial positive (+).
nY 1.598 $2V$ moderate.
nZ 1.602 Elongation negative.

Chem. A hydrated basic phosphate of sodium, calcium and aluminum, $(Na,K)CaAl_6(PO_4)_4(OH)_9 \cdot 3H_2O$. In the sample analyzed, $Na:K = 5:2$. Millisite apparently is isostructural with wardite, and is related simply in composition to that species.

Anal.

	1	2
Na_2O	2.71	2.8
K_2O	1.65	1.7
CaO	6.87	7.1
MgO		0.2
Al_2O_3	37.45	36.3
P_2O_5	34.77	33.8
H_2O	16.55	18.0
Total	100.00	99.9

1. $(Na,K)CaAl_6(PO_4)_4(OH)_9 \cdot 3H_2O$, with $Na:K = 5:2$. 2. Fairfield, Utah.[3] Average of two analyses recalculated after deduction of 5 per cent wardite and 1 per cent davisonite.

Tests. Fuses at $3\frac{1}{2}$ with slight intumescence to a blebby glass.

Occur. Found interlayered with green wardite in variscite nodules at Fairfield, Utah County, Utah.

Name. After F. T. Millis, of Lehi, Utah, who found the first specimens.

Ref.

1. Inferred from the x-ray powder diffraction study of Larsen, *Am. Min.*, **27**, 294 (1942), which indicates the mineral to be isostructural with wardite.
2. Calculated from the measured cell volume of wardite and analysis 2.
3. Shannon analysis in Larsen and Shannon (1930).

42.5.6 **LEHIITE** [$(Na,K)_2Ca_5Al_8(PO_4)_8(OH)_{12} \cdot 6H_2O$ (?)]. Larsen and Shannon (*Am. Min.*, **15**, 329, 1930).

Found as layers and crusts composed of subparallel aggregates of fibers. H. $5\frac{1}{2}$. G. 2.89. Color white to gray.

Opt. In transmitted light, colorless.

nX 1.600 Biaxial negative ($-$).
nY 1.615 $2V$ very large.
nZ 1.629 Fibers show a large extinction angle.

Very finely fibrous material from Fairfield described later [1] has approximately parallel extinction with nX parallel elongation $= 1.605$, $n \perp$ elongation $= 1.620$.

Chem. A hydrated basic phosphate of sodium, potassium, calcium, and aluminum of uncertain formula, perhaps $(Na,K)_2Ca_5Al_8(PO_4)_8(OH)_{12} \cdot 6H_2O$. In the only reported analysis, $Na:K \sim 2:1$.

Anal.

	1	2
Na_2O	2.68	3.08
K_2O	2.03	2.25
CaO	18.15	18.10
Al_2O_3	26.39	27.79
P_2O_5	36.76	34.64
H_2O	13.99	14.19
Insol.		0.58
Total	100.00	100.63

1. $(Na,K)_2Ca_5Al_8(PO_4)_8(OH)_{12} \cdot 6H_2O$, with $Na:K = 2:1$. 2. Fairfield, Utah.[2] Analysis on 0.2 gram.

O c c u r. At Fairfield, near Lehi, Utah County, Utah, associated with wardite, millisite, crandallite, and occasionally gordonite in the outer shells of altered variscite nodules.

N a m e. From the occurrence near Lehi, Utah.

Ref.
1. Larsen, *Am. Min.*, **27**, 294 (1942).
2. Gonyer analysis in Larsen and Shannon (1915).

42.5.7 **M I X I T E** $[Cu_{11}Bi(AsO_4)_5(OH)_{10} \cdot 6H_2O]$. *Schrauf* (*Zs. Kr.*, **4**, 278, 1880). Chlorotile *Frenzel* (*Min. Mitt.*, 42, 1875).

C r y s t.[1] Hexagonal—*P*.

$$a:c = 1:0.431; \qquad p_0:r_0 = 0.497:1$$

Structure cell.[2] Laue group $6/m$; space group uncertain. a_0 13.84 ± 0.10 Å, c_0 5.96 ± 0.05, $a_0:c_0 = 1:0.431$. Cell contents $Cu_{11}Bi(AsO_4)_5(OH)_{10} \cdot 6H_2O$.

Habit. In hair-like crystals, elongated [0001] and deeply striated vertically, which appear six-sided under the microscope. As tufted groups or incrusting; also compact, in reniform or spherical masses with a partial concentric fibrous structure; as cross-fiber veinlets.

P h y s. H. 3–4. G. 3.79 (Utah);[3] 3.77 (calc.). Luster dull (on aggregates) to brilliant (on single crystals). Color emerald-green to bluish green, pale green, or whitish. Streak somewhat lighter than the color.

O p t.[4] In transmitted light, pale green to colorless.

	n (Tintic district)	n (Mammoth mine, Tintic)	PLEOCHROISM	
O	1.743 ± 0.003 (non-pleochroic)	1.730 ± 0.003	Nearly colorless	Uniaxial positive (+).
E	1.830 ± 0.003	1.810 ± 0.003	Bright green	

C h e m.[5] A hydrated basic arsenate of copper and bismuth, perhaps $Cu_{11}Bi(AsO_4)_5(OH)_{10} \cdot 6H_2O$. Zn, Fe″, and Ca may replace Cu, and P may replace As to a limited extent. About one-half the water content is lost on heating to 175°.[8]

Anal. [10]

	1	2	3
CuO	46.53	44.23	43.89
ZnO			2.70
FeO		1.52	
CaO		0.83	0.26
Bi_2O_3	12.39	12.25	11.18
As_2O_5	30.54	29.51	28.79
P_2O_5		1.05	0.06
H_2O	10.54	11.06	11.04
Rem.			1.39
Total	100.00	100.45	99.31
G.	3.77		3.79

1. $Cu_{11}Bi(AsO_4)_5(OH)_{10} \cdot 6H_2O$. 2. Joachimsthal, Bohemia.[6] 3. Tintic district, Utah.[7] Rem. is Fe_2O_3 0.97, SiO_2 0.42.

Occur. First described from Joachimsthal, Bohemia, where it occurs with bismutite, smaltite, and native bismuth. Found with erythrite in clefts in barite at the St. Anton mine near Wittichen, and at Freudenstadt and Bulach, in Baden. Also from Schneeberg, Graul, and Neustädtl in Saxony, and Neubulach, Württemberg. In the United States at the Mammoth mine and elsewhere in the Tintic district, Utah. With bismutite at El Carmen mine, Durango, Mexico.

Name. In honor of A. Mixa, a mining official at Joachimsthal.

CHLOROTILE. *Frenzel* (*Min. Mitt.*, 42, 1875). Originally described as a hydrated cupric arsenate from Schneeberg and Zinnwald, Saxony, but shown [9] to be identical with mixite.

Ref.

1. Crystal system and axial ratio from structure cell of Pabst, priv. comm. (1948).
2. Pabst (1948) on crystals from Tintic district, Utah, by Weissenberg method.
3. Hillebrand and Washington, *Am. J. Sc.*, **35**, 298 (1888), on crystals from Tintic district, Utah. The value 2.66 of Schrauf (1880) is erroneous.
4. Larsen, 111 (1921).
5. The formula originally assigned by Schrauf (1880) differs only slightly from that here given which is based on cell counts from the only two available complete analyses. It may be presumed that copper and its substituents occupy a twelvefold position or combination of positions and some groups of (OH) ions occupy (AsO_4) positions as indicated in the formula $Cu_{11}Bi(AsO_4)_5(OH)_4(OH)_6 \cdot 6H_2O$.
6. Schrauf (1880).
7. Hillebrand analyst in Hillebrand and Washington (1888).
8. Schrauf (1880).
9. Gordon, *Ac. Sc. Philadelphia, Proc.*, **77**, 1 (1925), and confirmed by x-ray examination by Frondel, priv. comm. (1947).
10. For additional analyses see Schrauf (1880) and Hillebrand and Washington (1888).

KEHOEITE *Headden* (*Am. J. Sc.*, **46**, 22, 1893).

White, chalky masses. The washed powder has G. 2.34. Isotropic,[1] with n ranging from about 1.52 to 1.54. Analysis gave MgO 0.08, CaO 2.70, ZnO 11.64, Al_2O_3 24.84, Fe_2O_3 0.78, P_2O_5 26.76, SO_3 0.50, Cl tr., H_2O 31.06, insol. 1.76, total 100.12 (original total given as 100.02). This approximates to $Al_8(Zn,Ca)_3(PO_4)_6(OH)_{12} \cdot 21H_2O$ with Zn:Ca ~3:1. On heating, 14.2 weight per cent is lost at 110°, 3.34 per cent additional at 120°, the remainder of the water only at red heat. Infusible. Insoluble in water but easily soluble in dilute acids or concentrated KOH solution. Found as seams and bunches in the sphalerite-galena-pyrite ore of the Merritt mine, Galena, Lawrence County, South Dakota. Named after Mr. Henry Kehoe, who first observed the mineral.

Ref.

1. Larsen (95, 1921).

TYPE 6. $(AB)_m(XO_4)_p Z_q \cdot xH_2O$, WITH $m:p = 7:4$

42.6.1   S A M P L E I T E [$NaCaCu_5(PO_4)_4Cl \cdot 5H_2O$]. *Hurlbut* (*Am. Min.*, **27**, 586, 1942).

C r y s t.[1]   Orthorhombic; dipyramidal—$2/m\ 2/m\ 2/m$.

$a:b:c = 0.2526:1:0.2513;$ $p_0:q_0:r_0 = 0.9949:0.2513:1$

$q_1:r_1:p_1 = 0.2526:1.0052:1;$ $r_2:p_2:q_2 = 3.9793:3.9588:1$

Forms:

	ϕ	$\rho = C$	ϕ_1	$\rho_1 = A$	ϕ_2	$\rho_2 = B$
c 001	....	0°00′	0°00′	90°00′	90°00′	90°00′
b 010	0°00′	90 00	90 00	90 00	....	0 00
m 110	75 49½	90 00	90 00	14 10½	0 00	75 49½
d 150	38 22½	90 00	90 00	51 37½	0 00	38 22½
r 021	0 00	26 41	26 41	90 00	90 00	63 19
p 111	75 49½	45 44½	14 06½	46 01½	45 12½	79 54

Structure cell.[2]   Orthorhombic. a_0 9.70 kX, b_0 38.40, c_0 9.65; $a_0:b_0:c_0 = 0.2526:1:0.2513$. Cell contents $Na_8Ca_8Cu_{40}(PO_4)_{32}Cl_8 \cdot 40H_2O$.

Habit. In very thin, lath-like crystals flattened {010} and elongated [001].

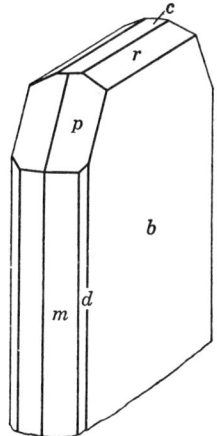

P h y s. Cleavage {010} perfect, {100} and {001} good. H. ~4. G. 3.20; 3.26 (calc.). Luster pearly on {010}. Color light blue to blue-green. Transparent.

O p t. In transmitted light, blue.

Orientation		n(Na)	Pleochroism	
X	b	1.629 ± 0.001	Deep blue	Biaxial negative (−).
Y	a	1.677 ± 0.001	Light blue	2V 22°34′ (calc.).
Z	c	1.679 ± 0.001	Colorless	$r > v$.

Chem. A hydrated chloride-phosphate of sodium, calcium, and copper, $(Na,K)CaCu_5(PO_4)_4Cl \cdot 5H_2O$.

Anal.

	1	2
Na_2O	3.50	3.11
K_2O		1.49
CaO	6.33	5.83
MgO		0.52
CuO	44.89	44.12
Cl	3.99	4.00
P_2O_5	32.03	32.10
H_2O	10.16	9.74
Total	100.90	100.91
$O = Cl$	0.90	0.91
	100.00	100.00
G.	3.26	3.20

1. $NaCaCu_5(PO_4)_4Cl \cdot 5H_2O$. 2. Chuquicamata, Chile.[3]

Tests. Fuses at 2 to a black globule. Easily soluble in acids.

Occur. Found in highly oxidized ore at Chuquicamata, Chile, associated with gypsum, atacamite, libethenite, jarosite, gypsum, and limonite.

Name. After Mat Sample, for many years superintendent of the mine at Chuquicamata.

Ref.

1. Only approximate morphological measurements could be obtained, owing to the poor quality of the crystals, and the angle table is computed from the elements of the structure cell.
2. From c-axis Weissenberg photographs.
3. Gonyer analysis in Hurlbut (1942).

42.6.2 TURQUOIS GROUP

TRICLINIC; PINACOIDAL—$\bar{1}$

	a_0	b_0	c_0	α	β	γ
Turquois, $CuAl_6(PO_4)_4(OH)_8 \cdot 4H_2O$	7.47 kX	9.93	7.67	111°39′	115°23′	69°26′
Chalcosiderite, $CuFe_6(PO_4)_4(OH)_8 \cdot 4H_2O$	7.66	10.18	7.88	112 29	115 18	69 00

Turquois and chalcosiderite form a probably complete series by mutual substitution of Fe''' and Al, although most analyses are very close to the Al end of the series. Both species form under essentially atmospheric conditions, turquois in altered aluminous rocks and the rare species chalcosiderite principally in the ferruginous gossan of ore deposits.

42.6.2.1 **T U R Q U O I S** [$CuAl_6(PO_4)_4(OH)_8 \cdot 4H_2O$]. Callais ? Callaina *Pliny* (37, 56, 33). Firuzegi *Persian*. Turques, Turquois pt., of the sixteenth century and later. Türkis pt. *Germ.* Turchesa *Ital.* Turquoise *Fr.* Turquoise *Tavernier* (Voy. en Turquie, en Persie, etc., Paris, 1678). Turchine *Bocconi* (Museo di Fisica, etc., 278, 1697).

TURQUOIS-CHALCOSIDERITE

Orientalischer Türkis *Demetrius Agaphi, N. Nordenskiöld (Beitr.*, **5**, 261), *Pallas (ibid.*, 265). Turquois orientale, Calaite, Agaphite, Johnite *Fischer (Mem. soc. nat. Moscou*, **1**, 1806; *Onomasticon Min. Mus. Imp. Moscou*, 1811; *Essai sur la Turquoise, Moscou*, 1816—abstr. in *Ann. Phil.*, **14**, 406, 1819); *John (Mem. soc. nat. Moscou*, **1**, 1806; *J. Chemie u. Phys.*, **3**, 93, 1807). Hydrargillite pt. *Hausmann* (444, 1813). Turquoise de vielle roche (in distinction from Odontolite, or Turquoise de nouvelle roche, called also Occidental Turquoise). Kallait, Kalait *Germ.* Turchesia *Ital.* Turquesa *Span.* Chalchuite *Blake (Am. J. Sc.*, **25**, 197, 1883). Ferri-turquois *Fairbanks (The Mineralogist, Portland, Oregon*, **10**, 44, 1942). Rashleighite *Russell (Min. Mag.*, **28**, 353, 1948).

42.6.2.2 **C H A L C O S I D E R I T E** [$CuFe_6(PO_4)_4(OH)_8 \cdot 4H_2O$]. Chalkosiderit *Ullmann* (323, 1824). *Maskelyne (J. Chem. Soc. London*, **28**, 586, 1875). Alumo-chalcosiderite *Jahn* and *Gruner (Mitt. Vogtländ. Ges. Naturfor.*, **1**, no. 8, 1, 1933).

C r y s t.[1] Triclinic; pinacoidal—$\bar{1}$.

TURQUOIS [2]

$a:b:c = 0.7523:1:0.7703;$ $\alpha\ 111°39',\ \beta\ 115°23',\ \gamma\ 69°26'$

$p_0:q_0:r_0 = 1.0192:0.7453:1;$ $\lambda\ 75°02',\ \mu\ 69°54',\ \nu\ 103°17'$

$p_0'\ 1.1677,\ q_0'\ 0.8539,\ x_0'\ 0.4744,\ y_0'\ 0.2959$

Forms:

		ϕ	ρ	A	B	C
c	001	58°03'	29°13'	69°54'	75°02'	.....
b	010	0 00	90 00	103 17	.....	75°02'
a	100	103 17	90 00	.....	103 17	69 54
m	110	62 44	90 00	40 33	62 44	60 53
M	1$\bar{1}$0	134 38	90 00	31 21	134 38	83 29

CHALCOSIDERITE [3]

$a:b:c = 0.7525:1:0.7741;$ $\alpha\ 112°29',\ \beta\ 115°18',\ \gamma\ 69°00'$

$p_0:q_0:r_0 = 1.0182:0.7496:1;$ $\lambda\ 74°15',\ \mu\ 70°20',\ \nu\ 103°29'$

$p_0'\ 1.1700,\ q_0'\ 0.8614,\ x_0'\ 0.4725,\ y_0'\ 0.3119$

Forms:

		ϕ	ρ	A	B	C
c	001	56°34'	29°31'	70°20'	75°15'	.....
b	010	0 00	90 00	103 29	.....	74°15'
a	100	103 29	90 00	.....	103 29	70 20
m	110	62 39	90 00	40 50	62 39	60 40
M	1$\bar{1}$0	134 55	90 00	31 26	134 55	84 17
d	2$\bar{3}$0	143 59	90 00	40 30	143 59	88 44
	5$\bar{9}$0	148 02	90 00	44 33	148 02	90 43
	3$\bar{7}$0	153 30	90 00	50 01	153 30	93 25
g	1$\bar{3}$0	158 17	90 00	54 48	158 17	95 45
u	$\bar{1}\bar{1}$1	−112 47	35 49	118 09	103 06	65 01

Structure cell.[4] Space group $P\bar{1}$.

	a_0	b_0	c_0	α	β	γ	$a_0:b_0:c_0$
Turquois	7.47 kX	9.93	7.67	111°39'	115°23'	69°26'	0.7523:1:0.7703
Chalcosiderite	7.66	10.18	7.88	112°29'	115°18'	69°00'	0.7525:1:0.7741

Cell contents $Cu(Al,Fe)_6(PO_4)_4(OH)_8 \cdot 4H_2O$.

Habit. Turquois crystals are rare: short prismatic [001] with large faces of c {001}, b {010}, and M {1$\bar{1}$0}. Usually massive, dense, and

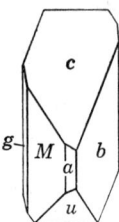
Chalcosiderite. Cornwall.

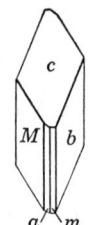
Turquois.

cryptocrystalline to fine-granular; also as veinlets or crusts and in stalactitic or concretionary shapes. Chalcosiderite occurs as crusts or sheaf-like groups of distinct crystals; short prismatic [001] with large {001}, {010}, {1$\bar{1}$0}, and {$\bar{1}\bar{1}$1}.

Phys. Cleavage {001} perfect, {010} good. Fracture of massive turquois, small conchoidal to smooth. H. 5–6 (turquois), 4½ (chalcosiderite). G. turquois, 2.6–2.8 (massive), 2.84 (crystals),[5] 2.91 (calc.); chalcosiderite, 3.22 (crystals),[5] 3.26 (calc.). Luster of crystals vitreous, of massive turquois somewhat waxy, feeble. Color of massive turquois sky-blue, bluish green to apple-green, greenish gray and subtranslucent to opaque, crystals bright blue and transparent; chalcosiderite is light siskin-green and transparent. The color of much turquois is altered by treatment with acids or ammonia or by exposure to a dry atmosphere. Streak white or greenish to pale green.

Opt.[5]

	Turquois			Chalcosiderite		
	ϕ	ρ	n(Na)	ϕ	ρ	n(Na)
X	−30°	60°	1.61	107°	57°	1.775
Y	63°	83°	1.62	−143°	68°	1.840
Z	163°	32°	1.65	−32°	40°	1.844
Sign	+			−		
2V		40° ± 2°			22° ± 2°	
Disp.		$r < v$, strong			$r > v$, very strong; crossed	

Both turquois and chalcosiderite are weakly pleochroic in thick grains, with X colorless and Z pale blue or pale green.

Chem. A hydrated basic phosphate of ferric iron and aluminum, $Cu(Al,Fe)_6(PO_4)_4(OH)_8 \cdot 4H_2O$. The names turquois and chalcosiderite are applied to material with Al > Fe''' and Fe''' > Al, respectively. Most analyses of turquois are close to the Al end of the series. Only two analyses of chalcosiderite have been reported. Fe″ substitutes for Cu in turquois, with Fe″:Cu = 1:1.26 in analysis 5. Analyses of turquois usually show somewhat more water than required by the above formula.

Anal.[8]

	1	2	3	4	5	6	7	8	9
CuO	9.78	8.57	9.00	7.70	7.40	7.87	8.15	6.82	8.06
FeO				0.72	5.32				
Al_2O_3	37.60	35.03	36.50	33.42	44.82	20.84	4.45	10.45	
Fe_2O_3		1.44	0.21	4.37		21.29	42.81	34.26	48.56
P_2O_5	34.90	34.18	34.13	34.41	30.38	28.60	29.93	33.82	28.77
H_2O	17.72	19.38	20.12	19.35	11.86	16.45	15.00	13.70	14.61
Rem.		0.93			0.22	4.36	0.61	0.87	
Total	100.00	99.53	99.96	99.97	100.00	99.41	100.95	99.92	100.00
G.	2.91	2.791	2.84	2.76		3.02		3.0	

1. $CuAl_6(PO_4)_4(OH)_8 \cdot 4H_2O$. 2. Turquois. Lincoln County, Nevada.[6] Rem. is insol. 3. Turquois. Virginia.[7] Recalculated after deducting 12.57 per cent quartz. 4. Turquois. Los Cerillos Mountains.[8] Recalculated after deducting 2.24 per cent insoluble. 5. Turquois. Columbus district, Nevada.[9] 6. Turquois (ferrian). St. Austell, Cornwall (*rashleighite*).[18] Rem. is As_2O_5 2.11, SiO_2 2.25. 7. Chalcosiderite. Wheal Phoenix, Cornwall.[10] Rem. is As_2O_5. 8. Chalcosiderite (aluminian). Schneckenstein, Germany [12] (*alumo-chalcosiderite*). Rem. is CaO. 9. $CuFe_6(PO_4)_4(OH)_8 \cdot 4H_2O$.

Tests. Turquois turns brown but does not fuse before the blowpipe, and the flame is not tinted or only weakly tinted green. Imitation turquois usually fuses easily and colors the flame a distinct green. In C.T. decrepitates, yields water, and turns brown or black (CuO). Soluble with difficulty in HCl, readily after moderate ignition.

Occur. Turquois is a secondary mineral found with limonite, chalcedony, and kaolin and formed by the action of surface waters, usually in arid regions, on aluminous igneous or sedimentary rocks. The turquois found in igneous flow rocks such as trachyte is probably ultimately derived by the alteration of accessory apatite and disseminated copper sulfides. Turquois of fine quality has been obtained since times of antiquity from a deposit in Persia on the southern slopes of the Ali-Mirsa-Kuh Mountains, northwest of the village of Madèn near Nishâpûr, Khorasan; the mineral occurs in irregular patches and narrow seams in a brecciated trachyte and in the surrounding clay slate. From the Sinai Peninsula in the Wadi Maghara, Egypt, with limonite in seams in porphyry.[13] A greenish blue variety comes from Karkaralinsk in Semipalatinsk, Siberia; also in the Kara-tjube Mountains, south of Samarkand, Turkestan. At Got near Angolola, Abyssinia. A well-known German locality is at Jordansmühl, Silesia; also found at Ölsnitz and Messbach in Vogtland, Saxony. With wavellite at Montebras, Creuse, France, as an alteration of amblygonite in pegmatite. In the United States found near Lynch Station, Campbell County, Virginia, as tiny crystals composing crusts or spherical groups. Large deposits early mined by the Indians [14] and Mexicans and later extensively worked occur in altered trachyte about 20 miles southwest of Santa Fe in the Los Cerillos Mountains, New Mexico; [15] also in this state at the Azure mine near Silver City in the Burro Mountains district and the Eureka district in Grant County, and at numerous minor occurrences.

In Colorado near Leadville, Lake County, at the Hall mine near Villagrove, Saguache County, and elsewhere.[16] In Arizona at Courtland, Cochise County, at Ithaca Peak in the Mineral Park district, Mohave County, the Castle Dome mine in the Miami district in Gila County, and at other places. Found at many places in Nevada, notably Columbus, in Esmeralda County. In California in the Turquois Mountains, in the Solo district, and at the Gove mine near Cottonwood siding on the Santa Fe Railroad, San Bernardino County.

Chalcosiderite occurs with dufrenite, andrewsite, and goethite as a secondary mineral in gossan at the Wheal Phoenix mine, Cornwall, England. Also found with pharmacosiderite at Schneckenstein, Saxony (*alumochalcosiderite*), and with dufrenite (?) at Siegen, Westphalia. In the United States from Bisbee, Arizona.

Artif. Obtained[11] by reaction of malachite, hydrous aluminum oxide, and phosphoric acid at 100°. Imitation turquois has been made by compressing or melting together precipitated aluminum phosphate and copper phosphate, also by staining chalcedony or other materials, and as a glass. The imitations can be distinguished with certainty by x-ray powder diffraction photographs, and the melting point and flame tests also are useful. Natural turquois of inferior color is often artificially treated to give it the tint desired.

Alter. Turquois has frequently been found as pseudomorphs after orthoclase; also after apatite, bone, and teeth. The mineral does not lose essential water below about 180°–200° (although small amounts of nonessential, capillary water may be lost, sometimes with accompanying color change) but apparently is completely dehydrated below 600°.

Name. Turquois means *Turkish* and is probably so called because it was originally brought to Europe from Persia through Turkey. The names Oriental Turquois and Occidental Turquois have been variously used to refer to the source of the material or to its quality, that of the East being superior. *Bone-turquois*, also called *fossil-turquois* or *odontolite* and including much of what is termed occidental turquois, is fossil bone or tooth and consists of microcrystalline apatite colored by a phosphate of iron (vivianite); a hydrochloric acid solution thereof does not give a blue color when treated with ammonia as does true turquois. The *Callais* and *Callaina* of Pliny and the *chalchihuitl* prized by the Aztecs of Mexico are considered to have been turquois.[17]

Ref.

1. The triclinic symmetry of turquois and its isostructural relation to chalcosiderite was first established by Schaller, *Am. J. Sc.*, **33**, 35 (1912), *Zs. Kr.*, **50**, 120 (1912). The crystallography and inter-relations of the two species were later definitively redescribed by Graham, *Univ. Toronto Stud., Geol. Ser.*, no. 52, 39 (1947).

2. Unit, orientation, and angles of the structure cell of Graham (1947) on the type material of Schaller (1912) from Virginia (analysis 3). Transformation: Schaller (1912) to structure cell, $\frac{1}{2}\frac{1}{2}\frac{1}{2}/\frac{1}{2}\frac{1}{2}\frac{1}{2}/001$; structure cell to Schaller, $110/\overline{1}1\overline{1}/001$. Forms of Schaller (1912) verified by Graham (1947).

3. Unit, orientation, and angle of the structural cell of Graham (1947) on Wheal Phoenix crystals. Transformation: Maskelyne (1875) to structure cell, as for turquois. Forms of Maskelyne.

4. Graham (1947) by Weissenberg method.
5. Orientation of Graham (1947), indices of Larsen (1921).
6. Penfield, *Am. J. Sc.*, **10**, 346 (1900).
7. Schaller, *Am. J. Sc.*, **33**, 35 (1912).
8. For additional analyses see Hintze (**1** [3B], 941, 1931); Pearl, *Gemmologist, London*, **14**, 62 (1945); Davy, *Trans. Roy. Geol. Soc. Cornwall*, **16**, 43 (1929).
9. Carnot, *C.R.*, **118**, 995 (1894).
10. Flight analysis in Maskelyne (1875).
11. Hoffmann, *Zbl. Min.*, 429 (1927), and Mayer, *Chem. Erde*, **9**, 311 (1935).
12. Jahn and Gruner, *Mitt. Vogtland. Ges. Naturfor.*, **1**, no. 8, 1 (1933).
13. Davy (1929).
14. On the early mining and use of turquois in North and South America see Ball, *Bull. Bur. Am. Ethnology, Smithsonian Inst.*, no. 128 (1941), and Pogue, *The Turquois*, Washington, 1915 (*Mem. Nat. Ac. Sc.*, **12**, Pt. 2).
15. See Northrop, *Univ. New Mexico, Bull.*, no. **379**, 313 (1942).
16. Pearl, *Econ. Geol.*, **36**, 335 (1941).
17. For the early history and synonomy of turquois see Dana (844, 1892), Hintze (1931), and Pogue (1915).
18. Smythe analysis in Russell, *Min. Mag.*, **28**, 353 (1948).

HARBORTITE. Brandt (*Chem. Erde*, **7**, 383, 1932).

Spherulites and octahedral masses, the latter perhaps pseudomorphs. Color white to brown. H. 5–5½. G. 2.80 (spherulites). Optically biaxial with weak birefringence and a mean index of refraction of about 1.61; fibers show parallel extinction and negative elongation. A hydrated aluminum phosphate approximating to $Al_3(PO_4)_2(OH)_3 \cdot 3H_2O$.

Anal.[1]

	Na_2O	CaO	Fe_2O_3	Al_2O_3	P_2O_5	TiO_2	SiO_2	H_2O	Total
1.	3.7	1.2	3.8	33.4	35.7	1.0	2.8	18.0	99.6
2.	4.5	1.2	5.1	32.7	34.1	1.2	0.8	18.6	98.2
3.	4.7	1.6	4.5	32.7	35.4	1.3	0.6	18.6	99.4

1, 2. White spherulites. 3. White octahedra.

Found with dufrenite (?) in phosphatized laterite near the coast of northern Brazil in Maranhao. Named after Erich Harbort (1879–1929) of Berlin-Charlottenberg. Needs verification.[2]

Ref.

1. Nine additional analyses are cited.
2. The x-ray powder pattern resembles that of turquois (Jung, *Chem. Erde*, **9**, 320 (1935)).

HENWOODITE. Collins (*Min. Mag.*, **1**, 11, 1876).

Botryoidal masses with a faint indication of a radial fibrous structure. Fracture conchoidal. H. 4–4½. G. 2.67. Color turquois-blue. Streak white with bluish green tinge.

Chem. A hydrated basic phosphate of copper and aluminum of uncertain formula. Analysis gave (in part an average of two):

CuO	CaO	Fe_2O_3	Al_2O_3	P_2O_5	SiO_2	H_2O	Undet.	Total
7.10	0.54	2.74	18.24	48.94	1.37	17.10	[3.97]	[100.00]

Tests.[1] In C.T. decrepitates slightly, turns brownish green, and gives off water. B.B. infusible. Soluble with difficulty in warm HNO_3.

Occur. Found with limonite and chalcosiderite at the West Phoenix mine, Cornwall, England. Named after William J. Henwood (1805–1875), a mining engineer of Cornwall.

Ref.

1. Foster, *Min. Mag.*, **1**, 8 (1876), and Ross, *Chem. News*, Oct. 13 (1876)—*Min. Mag.*, **1**, 59 (1876).

42.6.3 L U D L A M I T E [(Fe'',Mg,Mn)$_3$(PO$_4$)$_2$·4H$_2$O]. *Maskelyne* and *Field* (*Phil. Mag.*, **3**, 52, 135, 525, 1877). Lehnerite *Müllbauer* (*Zs. Kr.*, **61**, 318, 1925).

C r y s t.[1] Monoclinic; prismatic—$2/m$.

$a:b:c = 2.2527:1:1.9820$; $\beta\ 100°33'$; $p_0:q_0:r_0 = 0.8798:1.9845:1$

$r_2:p_2:q_2 = 0.5132:0.4515:1$; $\mu\ 79°27'$; $p_0'\ 0.8950$, $q_0'\ 1.9820$, $x_0'\ 0.1862$

Forms:

		ϕ	ρ	ϕ_2	$\rho_2 = B$	C	A
c	001	90°00′	10°33′	79°27′	90°00′	.....	79°27′
a	100	90 00	90 00	0 00	90 00	79°27′	.....
m	110	24 18	90 00	0 00	24 18	85 15½	65 42
l	011	5 22	63 19½	79 27	27 10	62 50	85 12½
t	201	90 00	63 09½	26 50½	90 00	52 36½	26 50½
d	$\bar{1}$01	−90 00	35 19½	125 19½	90 00	45 52½	125 19½
k	$\bar{2}$01	−90 00	58 03½	148 03½	90 00	68 36½	148 03½
r	112	32 36	49 38	57 39	50 04	44 36½	65 46
p	111	28 37	66 06½	42 46	36 37	61 25½	64 02
q	1$\bar{1}$1	−19 40	64 35½	125 19	31 44	68 31½	107 42½

Structure cell.[2] Space group $P2_1/a$. $a_0\ 10.48 \pm 0.06\ kX$, $b_0\ 4.63 \pm 0.05$, $c_0\ 9.16 \pm 0.06$; $\beta\ 100°36'$; $a_0:b_0:c_0 = 2.26:1:1.98$. Cell contents (Fe'',Mg,Mn)$_6$(PO$_4$)$_4$·8H$_2$O.

Habit. Tabular {001}, sometimes in parallel aggregates. Also massive, granular.

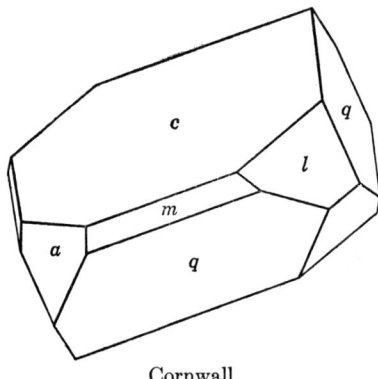

Cornwall.

P h y s. Cleavage {001} perfect, {100} indistinct. H. 3½. G. 3.12–3.19; 3.24 (calc.). Luster vitreous. Color bright green to apple-green. Streak pale greenish white. Translucent.

O p t. In transmitted light, nearly colorless.

Orientation		n (Cornwall)[4]	n (Hagendorf)[3]	n (Palermo)[2]	
X		1.653	1.650	1.650	Biaxial positive (+).
Y	b	1.675	1.669	1.667	$r > v$, perceptible.
$Z \wedge c$	−67°	1.697	1.689	1.688	
$2V$		82°	Large		

Chem. A hydrated ferrous phosphate, $(Fe'',Mg,Mn)_3(PO_4)_2 \cdot 4H_2O$. Mg and Mn″ substitute in small amount for Fe″, with Mn:Mg:Fe ~4:5:64 in analysis 3.

Anal.

	1	2	3	4
MgO			2.21	0.96
MnO			3.10	
FeO	50.17	52.76	45.91	49.22
P_2O_5	33.06	30.11	33.87	32.95
H_2O	16.77	16.98	14.91	16.12
Insol.				0.57
Total	100.00	99.85	100.00	99.82
G.	3.24	3.12	[3.19]	3.14

1. $Fe_3(PO_4)_2 \cdot 4H_2O$. 2. Cornwall.[5] 3. Hagendorf.[6] Recalculated to 100 after deducting Al and insol. 4. Palermo mine, New Hampshire.[7] Original total given as 99.85.

Tests. In C.T. decrepitates, turns dark blue, and gives off water. Soluble in acids.

Occur. Originally from the Wheal Jane mine near Truro, Cornwall, England. Reported from Stösgen near Linz, in the Rheinland, Germany. With phosphoferrite, triploidite, triplite, triphylite, apatite in pegmatite at Hagendorf, Bavaria (*lehnerite*).[8] With fairfieldite, vivianite, siderite, whitlockite as an alteration of triphylite at the Palermo pegmatite, North Groton, New Hampshire, United States. In the Blackbird district, Lemhi County, Idaho, with vivianite.[9]

Name. After Mr. Ludlam of London. Lehnerite after F. Lehner, mineral collector, of Pleystein, Bavaria.

Ref.

1. Elements and forms of Maskelyne and Field (1877), also in Goldschmidt (5, 171, 1918).
2. Wolfe, *Am. Min.*, 34, 94 (1949).
3. Berman, *Am. Min.*, 10, 428 (1925).
4. Larsen and Berman (124, 1934).
5. Flight analysis in Maskelyne and Field (1877).
6. Müllbauer (1925), *Jb. Min.*, I, 54 (1926); G. from Berman (1925).
7. Gonyer analysis in Wolfe (1949).
8. Shown to be identical with ludlamite by Berman (1925) and Wolfe (1949).
9. Glass and Vhay, *Am. Min.*, 34, 335 (1949).

42.6.4 **ARSENIOSIDERITE** [$Ca_3Fe_4(AsO_4)_4(OH)_4 \cdot 4H_2O$ (?)]. Arseniosiderite *Dufrénoy* (*Ann. mines*, 2, 343, 1842; *C.R.*, 16, 22, 1843). [Not arseniosiderite *Glocker* (321, 1839) = löllingite.] Arsenokrokite, Arsenocrocites *Glocker* (226, 1847). Mazapilite *Koenig* (*Proc. Ac. Sc. Philadelphia*, 192, 1888; *Zs. Kr.*, 17, 85, 1889). Yukonite *Tyrrell* and *Graham* (*Trans. Roy. Soc. Canada*, 7, Sect. IV, 3, 1913).

Hexagonal or tetragonal (?) [1] Massive: as radial fibrous aggregates, the fibers flattened {001} and easily separable; also as granular pseudomorphs.

Phys. Cleavage {001}. H. $4\frac{1}{2}$ in granular types, decreasing to $1\frac{1}{2}$ or so in fibrous material. G. 3.60.[2] Luster submetallic to silky. Color golden yellow to yellowish brown and reddish brown (fibrous); also black or brownish black (dense granular). Streak ocher-yellow. Opaque except in small grains.

Opt. In transmitted light, reddish brown to brownish yellow.

	n (Romanêche)[3]	n (Mazapil)[3]	DICHROISM	
O	1.870 ± 0.005	1.898 ± 0.005	Reddish brown	Uniaxial negative $(-)$.
E	1.792 ± 0.005	1.815 ± 0.005	Nearly colorless	

Chem. A hydrated basic arsenate of calcium and iron, probably $Ca_3Fe_4(AsO_4)_4(OH)_4 \cdot 4H_2O$.
Anal.

	1	2	3	4	5	6
CaO	15.94	12.53	15.53	14.82	14.44	10.00
Fe_2O_3	30.26	39.37	35.75	30.53	32.71	35.72
As_2O_5	43.56	38.74	39.86	43.60	42.67	34.06
H_2O	10.24	9.36	7.87	9.83	9.34	20.28
Rem.			0.65	0.39	1.41	
Total	100.00	100.00	99.66	99.17	100.57	100.06
		3.88	3.36	3.58		2.86

1. $Ca_3Fe_4(AsO_4)_4(OH)_4 \cdot 4H_2O$. 2. Romanêche, France.[4] 3. Romanêche, France.[5] Average of two. Rem. is MgO 0.18, K_2O 0.47. 4. Mazapil, Mexico[6] (*mazapilite*). Rem. is Sb_2O_5 0.25, P_2O_5 0.14. 5. Mapimi, Mexico.[7] Rem. is PbO 0.28, FeO 0.12, MgO 0.61, insol. 0.40. 6. Yukon[10] (*yukonite*).

Tests. B.B. turns reddish and fuses to a black, magnetic mass. In C.T. affords neutral water. Easily soluble in hot acids.

Occur. Originally from the manganese deposit at Romanêche, Saône et Loire, France, with quartz, hematite, goethite, romanechite, and psilomelane. Reported in Germany at Schneeberg, Saxony, with erythrite and roselite; at Auerbach, Hesse; with pharmacosiderite, at Bulach, Württemberg; and with erythrite at Wittichen, Baden. As felted masses with scorodite, symplesite, pitticite, and pharmacosiderite on loellingite at Hüttenberg, Carinthia, Austria. Found in Mexico at the Jesus Maria mine, Mazapil district, Zacatecas, with pharmacolite, chrysocolla, calcite, as pseudomorphs after scorodite (*mazapilite*);[8] and at the Ojuela mine, Mapimi, Durango, with carminite as pseudomorphs after scorodite. At Franklin, New Jersey; and said to occur in the Tintic district, Utah. A gel-like, concretionary mineral from Tagish lake, Yukon, Canada (*yukonite*)[10] apparently belongs here; it is brownish black in color, isotropic, with H. 2–3, extremely brittle with a conchoidal fracture, and flies apart in water.

Alter. Found as pseudomorphs after siderite[9] and scorodite.

Name. From *arsenic* and σίδηρος, *iron*.

Ref.

1. Inferred from optical properties. The so-thought orthorhombic crystals of mazapilite (= arseniosiderite) are pseudomorphs after scorodite—Foshag, *Am. Min.*, **22**, 483 (1937).
2. Mrose, priv. comm. (1947), by microbalance on Mazapil material. Koenig (1888) gave 3.58 on Mazapil material; on Romanêche material, Dufrénoy, *Ann. mines*, **2**, 346 (1842), gave 3.52, and Rammelsberg, *Ann. Phys.*, **68**, 508 (1846), gave the apparently erroneous value 3.88.
3. Larsen (42, 1921).

4. Rammelsberg (1846); also other analyses from this locality by Dufrénoy (1842) and Barthoux, *Bull. soc. min.*, **48**, 104 (1925). See further Lacroix, *Bull. soc. min.*, **9**, 3 (1886).
5. Church, *J. Chem. Soc. London*, **26**, 102 (1873).
6. Koenig (1888).
7. Foshag (1937).
8. Shown to be identical with arseniosiderite by Larsen, *Am. Min.*, **3**, 12 (1918), and Foshag (1937).
9. Sandberger, *Jb. Min.*, I, 251 (1886).
10. Tyrrell and Graham (1913).

42.6.5 **Eguëite.** *Garde (C.R., 148, 1618, 1909). Eguéïite Lacroix (4, 536, 1910).*

As small nodules with a fibrous-lamellar structure and showing tiny monoclinic (?) crystals in cavities. Very friable. G. 2.60. Color brownish yellow. Streak yellow. Luster vitreous to slightly greasy. Isotropic, with occasional birefringent spots. Mean index of refraction, n 1.65. A hydrous phosphate of ferric iron with minor calcium and aluminum. Analysis gave:[1] CaO 2.28, Al_2O_3 1.50, Fe_2O_3 44.20, P_2O_5 30.30, H_2O 20.47, insol. 0.75, total 99.50. This approximates to $CaFe_{14}(PO_4)_{10}(OH)_{14} \cdot 21H_2O$. In C.T. blackens and gives much water. On charcoal fuses at 1 with intumescence to a black globule. Soluble in cold HCl. Eguëite was found embedded in a clay impregnated with trona and thenardite at Koukourdei (Bir Salado) near Hangara in the Egai (Eguéï) region of the Chad territory, Sudan, Africa.

Ref.

1. Pisani analysis in Garde (1909) as corrected by Lacroix (1910).

42.6.6 **Mitridatite.** *Popov (1910). Mitridatite Dvoichenko (Zap. Krym. Obshch. Est., 4, 114, 1914); Chukhrov (Trans. Lomonossov Inst. Ac. Sc. U.S.S.R., Ser. Min., no. 10, 139, 1937); Sidorenko (C. r. ac. sc. U.R.S.S., 48, 53, 1945). Calcium ferri-phosphate Efremov (Mem. soc. russe min., 65, 225, 1936). Mithridatite.*

A name given to ill-defined calcium ferri-phosphates found at various localities on the Kerch and Taman peninsulas, Russia. Massive, as nodules, crusts, and veinlets. Structure earthy and pulverulent or friable to dense and gum-like. H. $2\frac{1}{2}$ in dense material. Color greenish yellow to green and dark green; also brownish green, yellowish brown to brownish black. Luster dull and earthy to resinous. Isotropic.

Chem. Essentially a basic phosphate of calcium and iron containing much water, in part doubtless nonessential. It is not certain if the reported analyses represent one or several different minerals. The CO_2 reported apparently is due to admixed calcite, and the SO_3 to gypsum or possibly a sulfate-phosphate. The analyses for the most part approach the ratio $CaO:Fe_2O_3:P_2O_5 = 1:1:1$, giving the formula $CaFe_2(PO_4)_2$-$(OH)_2 \cdot nH_2O$ (compare *calcioferrite* and *xanthoxenite*).

Anal.

	1	2	3	4	5	6
CaO	12.43	16.05	12.28	15.85	10.69	12.70
MgO	tr.	0.96	1.04	0.98	tr.	
MnO					0.27	0.78
FeO	0.29	0.52	1.86	0.62	0.97	3.08
Fe_2O_3	37.81	36.69	31.57	36.19	43.27	33.68
Mn_2O_3	2.34	0.51	0.53	0.94		
P_2O_5	30.09	30.61	22.16	29.84	29.88	22.98
CO_2	5.28	0.65	2.06	n.d.		n.d.
SO_3		0.08	0.47	0.58		
H_2O+	11.83	8.82	7.19	8.51	14.92	26.78
H_2O-		4.26	7.57	6.33		
Rem.		0.32	12.61	0.44		
Total	100.07	99.47	99.34	100.28	100.00	100.00

1. Kamysh-Burun, Kerch.[1] 2. Kamysh-Burun.[2] Nodules. Rem. is insol. 3. Kamysh-Burun.[2] Veinlets. Rem. is insol. 4. Kamysh-Burun.[2] In shells. Rem. is insol. 5. Kamysh-Burun.[3] Recalc. to 100 after deducting 4.92 insol. and 1.29 CO_2 as $CaCO_3$. 6. Zheleznaya ravine, Taman.[3] Recalc. to 100 after deducting 15.24 insol.

	7	8	9	10
CaO	8.20	9.25	12.36	4.30
MgO	0.62	1.03	0.73	0.58
MnO	0.12			
FeO	0.53		0.78	
Fe_2O_3	28.96	34.73	31.54	35.67
Mn_2O_3		0.25	0.88	0.25
P_2O_5	22.70	17.15	24.48	7.87
SO_3		4.99	2.66	0.25
H_2O+	13.46	26.80	7.38	4.39
H_2O-	20.89		11.55	11.15
Rem.	4.07	6.49	8.19	35.08
Total	99.55	100.69	100.55	99.54

7. Zheleznaya ravine.[4] Rem. is Al_2O_3 2.65, SiO_2 1.42. 8. Novo-Karantiny mine, Kerch.[4] Brown, gel-like. Rem. is insol. 9. Novo-Karantiny mine.[4] Green, earthy. Rem. is insol. 10. Novo-Karantiny mine.[4] Green, earthy. Rem. is insol.

Tests. B.B. turns brown-red but does not fuse. Soluble in hot acids.

Occur. A metacolloid found in oolitic sedimentary iron ores on the Taman and Kerch peninsulas, southern Russia. In part formed by the alteration of anapaite (anals. 8–10) and vivianite.

Ref.

1. Popov (1910) [exact reference not known].
2. Chukhrov (1937).
3. Sidorenko (1945).
4. Chukhrov, *Trans. Lomonossov Inst. Ac. Sc., U.S.S.R.* **10**, 131 (1937).

42.6.7 **Richellite.** Cesàro and Desprets (*Soc. géol. Belgique, Ann., Mém.,* **10**, 36, 1883); Cesàro (*ibid.,* **11**, 257, 1884).

Massive; compact or foliated, also as radially fibrous globules. H. 2–3. G. ~2. Color and streak reddish to yellowish brown. Luster greasy to horn-like.

C h e m. A basic hydrous phosphate of calcium and iron of uncertain formula, perhaps $Ca_3Fe_{10}(PO_4)_8(OH)_{12} \cdot nH_2O$.
Anal.

	1	2	3	4
CaO	5.76	5.53	6.18	7.19
Al_2O_3	1.81	1.79	2.82	3.64
Fe_2O_3	28.71		29.63	29.67
P_2O_5	28.78	28.55	27.23	25.49
HF	6.11		1.22	0.96
H_2O-	23.33		6.90	9.47
H_2O+	6.10		25.64	23.63
Total	100.60		99.62	100.05

1, 2. Visé, Belgium.[1] H_2O- at 100°, H_2O+ at red heat. 3. Visé. Dense type.[2] H_2O- hygroscopic. 4. Visé. Foliated type.[2] H_2O- hygroscopic.

Tests. B.B. fuses easily to a weakly magnetic enamel. Readily soluble in acids.

O c c u r. At Richelle near Visé in Liége, Belgium, associated with halloysite, allophane, and koninckite.

N a m e. From the locality.

Ref.
1. Cesàro and Desprets (1883).
2. Cesàro (1884).

42.6.8 **E N G L I S H I T E** [$K_2Ca_4Al_8(PO_4)_8(OH)_{10} \cdot 9H_2O$]. *Larsen* and *Shannon* (*Am. Min.*, 15, 328, 1930).

Probably monoclinic.[1] Occurs as layers and aggregates of subparallel plates.

P h y s. Cleavage {001} perfect, micaceous. The cleavage plates are curved and composite. Flexible. H. ~3. G. ~2.65. Colorless and transparent. Luster vitreous, on cleavage surfaces pearly.

O p t. In transmitted light, colorless.

ORIENTATION	n	
$X\ c$	1.570 ± 0.005	Biaxial negative $(-)$.
Y near a		$2V$ small.
Z near b	1.572 ± 0.005	

C h e m. A hydrated basic phosphate of potassium, calcium, and aluminum of uncertain formula, perhaps $K_2Ca_4Al_8(PO_4)_8(OH)_{10} \cdot 9H_2O$.
Anal.

	1	2
Na_2O		1.6
K_2O	6.09	5.4
CaO	14.50	14.1
Al_2O_3	26.42	24.7
P_2O_5	36.71	37.8
H_2O+	16.28	16.5
Total	100.00	100.1

1. $K_2Ca_4Al_8(PO_4)_8(OH)_{10} \cdot 9H_2O$. 2. Fairfield, Utah.[2] Analysis, on 180 mg., recalculated after deduction of 7 per cent variscite and 2 per cent wardite.

Phosphates, Arsenates, Vanadates

Occur. Found with montgomeryite, wardite, millisite, and crandallite in variscite nodules at Fairfield, Utah County, Utah.
Name. After George L. English (1864–1944), American mineral dealer and collector.

Ref.
1. From x-ray Laue photograph of a cleavage flake by Larsen, *Am. Min.*, **27**, 296 (1942).
2. Shannon analysis in Larsen and Shannon (1930).

TYPE 7. $A_3(XO_4)_2Z_q \cdot xH_2O$

42.7.1 LEGRANDITE [$Zn_{14}(OH)(AsO_4)_9 \cdot 12H_2O$]. *Drugman* and *Hey* (*Min. Mag.*, **23**, 175, 1932).

Cryst.[1] Monoclinic.

$a:b:c = 1.6076:1:1.2886;$ $\beta\,104°25';$ $p_0:q_0:r_0 = 0.8016:1.2480:1$

$r_2:p_2:q_2 = 0.8013:0.6422:1;$ $\mu\,75°35';$ $p_0'\,0.8276,\,q_0'\,1.2886,\,x_0'\,0.2571$

Forms:

	ϕ	ρ	ϕ_2	$\rho_2 = B$	C	A
c 001	0°00'	14°25'	75°35'	90°00'	.....	75°35'
a 100	90 00	90 00	0 00	90 00	75°35'	.....
m 110	32 43	90 00	0 00	32 43	82 17	57 17
p $\bar{1}11$	−23 53	35 38½	119 42½	41 47	61 25½	109 17

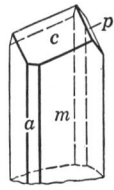

Structure cell.[2] a_0 12.70 kX, b_0 7.90, c_0 10.18; β 75°35'; $a_0:b_0:c_0 = 1.607:1:1.288$. Cell contents $Zn_{14}(OH)(AsO_4)_9 \cdot 12H_2O$.
Habit. Radiating aggregates of crystals prismatic [001].
Phys. Cleavage {100}, fair.[3] H. ~5³. G. 4.01 ± 0.05; 4.00 (calc.). Colorless to canary-yellow. Transparent.
Opt. In transmitted light, colorless to pale yellow.

Orientation	n(Na)[4]	n[3]	Pleochroism	
X b	1.675 ± 0.005	1.792	Colorless to yellow	Biaxial positive (+).
Y	1.690 ± 0.005	1.709	Colorless to yellow	$2V$ 36°±.
$Z \wedge c$ +36° to 40°	1.735 ± 0.005	1.741	Yellow	$r < v$, distinct.

Chem. A hydrated basic zinc arsenate, $Zn_{14}(OH)(AsO_4)_9 \cdot 12H_2O$.
Anal.

	1	2
ZnO	47.49	46.68
MnO		0.05
Fe_2O_3		2.14
As_2O_5	43.11	42.02
H_2O	9.40	9.36
Total	100.00	100.25
G.		4.01

1. $Zn_{14}(OH)(AsO_4)_9 \cdot 12H_2O$.   2. Lampazos, Mexico.[5] Total iron determined as Fe_2O_3.

BERAUNITE

Occur. Found with siderite, mimetite (?), and pyrite on massive sphalerite from the Flor de Peña mine, Lampazos, Nuevo Leon, Mexico.

Name. After Mr. Legrand, a Belgian mine manager, who collected the only known specimen.

Ref.
1. Only approximate morphological measurements could be obtained, owing to the poor quality of the crystals and the elements are here computed from the x-ray data.
2. Bannister in Drugman and Hey (1932).
3. Berman, priv. comm. (1933).
4. Drugman and Hey (1932).
5. Hey analysis in Drugman and Hey (1932).

42.7.2 **B E R A U N I T E** [Fe″Fe₄‴(PO$_4$)$_3$(OH)$_5$·3H$_2$O (?)]. *Breithaupt* (156, 1841; *B. H. Ztg.*, 402, 1853). Eleonorite *Nies* (*Ber. Oberhess. Ges.*, **19**, 111, 1880); *Streng* (*Jb. Min.*, I, 102, 1881). Dufreniberaunite *Wherry* (*Proc. U. S. Nat. Mus.*, **47**, 509, 1914).

Cryst.[1] Monoclinic.

$a:b:c = 2.7550:1:4.0157$; $\beta\ 131°27'$; $p_0:q_0:r_0 = 1.4576:3.0099:1$

$r_2:p_2:q_2 = 0.3322:0.4843:1$; $\mu\ 48°33'$; $p_0'\ 1.9447,\ q_0'\ 4.0157,\ x_0'\ 0.8832$

Forms:

	ϕ	ρ	ϕ_2	$\rho_2 = B$	C	A
c 001	90°00′	41°27′	131°27′	90°00′	.....	48°33′
a 100	90 00	90 00	0 00	90 00	48°33′	.....
f 111	35 09	78 29½	19 28½	36 45½	58 28	55 39
p $\bar{1}$11	−14 48½	76 28	136 42½	19 57½	89 22½	104 23

Habit. Tabular {100} and more or less elongated [010]; {100} striated [010]. Crystals rare and small. Usually as radiated foliated globules and

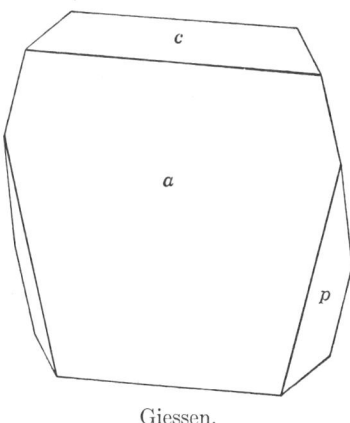

Giessen.

crusts, or as radial-fibrous aggregates; as discoidal concretions with a coarse-fibrous structure.

Twinning. On {100}, sometimes as interpenetration twins.

Phys. Cleavage {100} good. H. $3\frac{1}{2}$-4. G. 2.8–2.99, 3.08 (Middletown).[3] Luster vitreous, inclining to pearly on cleavage surfaces. Color reddish brown to dark hyacinth-red and blood-red; rarely dark greenish brown or dull green. Streak yellow to olive-drab. Translucent.

Opt. Biaxial.

Orientation	n (Giessen)[2]	n (Middle-town)[3] anal. 2	n (Palermo mine, N. H.)[9] dull green
X	1.775	1.775	1.707
$Y \wedge c\ \ 1\frac{1}{2}°$ to 5°	1.786		1.735
$Z\ \ b$	1.815	1.820	1.738

Pleochroism			
X	Pale flesh	Palest yellow	Blue-green
Y	Pale flesh	Pale yellow	Pale olive-green
Z	Carnelian-red	Reddish brown	Olive-green
2V	Med. large	Large	~25°
Disp.	$r > v$, marked	$r > v$, marked	$r < v$, marked
Sign	+	+	−

Chem. A hydrated basic phosphate of both ferrous and ferric iron. The formula[3] is uncertain, perhaps $Fe''Fe_4'''(PO_4)_3(OH)_5 \cdot 3H_2O$. The mineral as ordinarily found is completely oxidized, and the composition then is very close to $Fe_5'''(PO_4)_3(OH)_6 \cdot 2\text{--}3H_2O$.

Anal.[8]

	1	2	3	4	5	6
FeO		1.92				
Fe_2O_3	55.43	54.41	55.98	52.05	49.60	54.11
Al_2O_3		0.02			4.50	0.43
P_2O_5	29.56	30.17	28.99	31.78	30.93	27.24
H_2O	15.01	13.45	14.41	16.56	14.81	15.60
Total	100.00	99.97	99.38	100.39	99.84	99.99
G.		3.08				2.95

1. $Fe_5'''(PO_4)_3(OH)_6 \cdot 3H_2O$. 2. Middletown, New Jersey.[4] 3. St. Benigna, Bohemia.[5] 4. Waldgirmes, Hesse.[6] 5. Sevier County, Arkansas.[7] 6. Dandaragan, Western Australia.[10] Total includes Na_2O 0.58, K_2O 0.12, insol. 1.91.

Tests. B.B. easily fusible to a black bead. Easily soluble in HCl.

Occur. Found in secondary iron ore deposits and as an alteration of triphylite or other primary phosphates in pegmatites. Associated minerals include vivianite, dufrenite, rockbridgeite, strengite, cacoxenite, wavellite, limonite. Originally from the Hrbek mine at St. Benigna near Beraun, Bohemia. Also from the Eleonore iron mine on the Dünsberg near Giessen, Hesse, Germany, and at the Rothläufchen mine near Waldgirmes in the same region. At Scheibenberg, Saxony; at the Nitzelbuch mine in the Amberg iron district, and in pegmatite at Hühnerkobel, Wildenau, and the Kreuzberg, Pleystein, Bavaria. From Kirunavaara, Sweden. At the iron and manganese mine of Roury Glen, County Cork, Ireland. Reported from the Kerch peninsula, U.S.S.R. In the United States from Middletown, New Jersey; Hellertown, Northampton County, and Moore's Mill, Cumberland County, Pennsylvania. From Sevier County, Arkansas. In altered triphylite in pegmatite at the Palermo mine, North Groton, New Hampshire.

Alter. To limonite; also found as an alteration product of vivianite.

Ref.
1. Streng, *Jb. Min.*, I, 102 (1881), on crystals from Waldgirmes; see also Bořický, *Ak. Wien, Sitzber.*, 56 (1867), on St. Benigna crystals.
2. Larsen (46, 1921).
3. Frondel, *Am. Min.*, **34**, 536 (1949).
4. Hallowell analysis in Frondel (1949).
5. Bořický (1867).
6. Streng (1881).
7. Koenig, *Zs. Kr.*, **17**, 85 (1890).
8. For additional analyses see Hintze (**1** [3B], 898, 1931), and Wherry, *Proc. U. S. Nat. Mus.*, **47**, 501 (1914).
9. Mrose, priv. comm. (1949).
10. Le Mesurier, *J. Roy. Soc. Western Australia*, **27**, 133 (1943).

42.7.3 Coeruleolactite. Cœruleolactin Petersen (*Jb. Min.*, 353, 1871).

As cryptocrystalline to microcrystalline or fibrous veinlets, crusts, and botryoidal aggregates. Fracture uneven to conchoidal. H. 5. G. 2.57 (Nassau); 2.69 (Pennsylvania). Color milk-white passing to light blue. Streak white.

Opt. Nearly or quite uniaxial. Optically positive (+). Elongation positive (+).

n (Pennsylvania) [1]
O 1.580 ± 0.005
E 1.588 ± 0.005

Chem. A hydrated basic phosphate of aluminum of uncertain formula, perhaps $Al_3(PO_4)_2(OH)_3 \cdot 4H_2O$. The CaO and CuO reported may be essential constituents.

Anal.

	1	2	3
CaO		2.41	
CuO		1.40	4.25
Al_2O_3	38.81	35.11	38.27
Fe_2O_3		0.93	
P_2O_5	36.04	36.33	36.31
H_2O	25.15	21.23	21.70
Rem.		2.02	0.54
Total	100.00	99.43	101.07
G.		2.57	2.696

1. $Al_3(PO_4)_2(OH)_3 \cdot 4H_2O$. 2. Rindsberg mine, Nassau.[2] Rem. is SiO_2 1.82, ZnO tr., MgO 0.20, F tr. The CaO, MgO, CuO and SiO_2 is ascribed by the analyst to admixture. 3. Chester County, Pennsylvania.[3] Rem. is insol.

Tests. B.B. decrepitates but infusible. Soluble in acids and in alkalies.

Occur. Originally found in limonite at the Rindsberg mine, near Katzenellnbogen, Nassau, Germany. An apparently identical mineral [4] occurs with wavellite at General Trimble's mine, East Whiteland, Chester County, Pennsylvania.

Name. From *coeruleus*, blue, and *lac, lactis*, milk, in allusion to its color.

Ref.
1. Larsen (59, 1921).
2. Petersen (1871).
3. Genth, *Min. Rept. Penn.*, 143 (1875).
4. The x-ray powder pattern of this mineral differs from that of variscite, metavariscite, and wavellite (Frondel, priv. comm., 1949).

42.7.4 W A V E L L I T E [$Al_3(OH)_3(PO_4)_2 \cdot 5H_2O$]. *Babbington* (*Phil. Trans.*, 162, 1805). Hydrargillite *Davy* (*Phil. Trans.*, 155, 162, 1805). Devonite *Thomson; von Moll* (*Jahrb. Efem.*, **5**, 148, 1809). Strahliger Hydrargillit [= columnar var. of diaspore] *Hausmann* (443, 1813). Lasionit *Fuchs* (*J. Chem. Phys.*, **62**, 379, 1831). Thonerdephosphat *Germ.* Alumine phosphatée *Fr.* Subphosphate of Alumina. Kapnicit *Kenngott* (*Uebers.*, 1855, 1856).

C r y s t. Orthorhombic.[1]

$a:b:c = 0.5577:1:0.4061;$ $p_0:q_0:r_0 = 0.7282:0.4061:1$

$q_1:r_1:p_1 = 0.5577:1.3733:1;$ $r_2:p_2:q_2 = 2.4624:1.7931:1$

Forms:[2]

	ϕ	$\rho = C$	ϕ_1	$\rho_1 = A$	ϕ_2	$\rho_2 = B$
c 001	.....	0°00′	0°00′	90°00′	90°00′	90°00′
b 010	0°00′	90 00	90 00	90 00	....	0 00
a 100	90 00	90 00	.....	0 00	0 00	90 00
n 340	53 22	90 00	90 00	36 38	0 00	53 22
m 110	60 51	90 00	90 00	29 09	0 00	60 51
l 430	67 18	90 00	90 00	22 42	0 00	67 18
N 210	74 25	90 00	90 00	15 35	0 00	74 25
h 310	79 28	90 00	90 00	10 32	0 00	79 28
p 101	90 00	36 03½	0 00	53 56½	53 56½	90 00
s 111	60 51	39 49	22 06	55 59½	53 56½	71 49½
o 121	41 52½	47 29½	39 05	60 31½	53 56½	56 42½

Structure cell.[3] Space group $Pcmn$. a_0 9.60 kX, b_0 17.31, c_0 6.98; $a_0:b_0:c_0 = 0.555:1:0.403$. Cell contents $Al_{12}(OH)_{12}(PO_4)_8 \cdot 20H_2O$.

Habit. Crystals rare; stout to long prismatic [001], with {110} striated [001]. Usually as hemispherical or globular aggregates with a radial fibrous or stellate structure; also as crusts or stalactitic; rarely as chalcedonic or opaline masses.

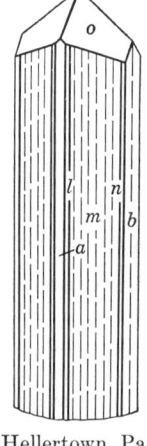

Hellertown, Pa.

P h y s. Cleavage {110} perfect, {101} good, {010} distinct.[4] Fracture uneven to subconchoidal. Brittle. H. 3¼–4. G. 2.36;[5] 2.37 (calc.). Luster vitreous

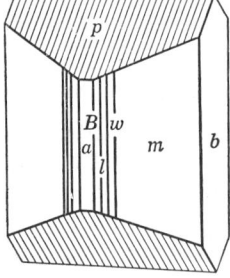

Llallagua.

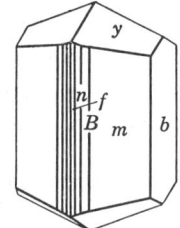

Trimble's mine, Pa.

inclining toward pearly and resinous. Color greenish white and green to yellow, also yellowish brown, brown, brownish black, blue, white, colorless. Streak white. Translucent.

O p t.[6] In transmitted light, colorless.

ORIENTATION	n (Černovice, anal. 3)[7]	n (Hellertown, anal. 2)[8]	n (Třenice)[9]	
X b	1.535	1.525	1.520	Biaxial positive (+).
Y a	1.543	1.535	1.526	$2V \sim 71°$.
Z c	1.561	1.550	1.545	$r > v$, weak.

Weakly pleochroic, with X greenish and Z yellowish; absorption $X > Z$.

The indices apparently increase with increasing substitution of Fe''' for Al and of F for OH.

C h e m. A hydrated basic phosphate of aluminum, $Al_3(OH)_3(PO_4)_2 \cdot 5H_2O$. F often substitutes for OH in small amount, with $F:OH = 1:20$ in analysis 4. Very small amounts of Fe''' substitute for Al with $Fe:Al = 1:15$ in analysis 3. Fe'', Ca, Mg, and Cr''' apparently also may be minor essential constituents (analysis 3). The water is largely lost below 200°.[7]

Anal.[10]

	1	2	3	4	5	6
Al_2O_3	37.11	36.5	31.01	37.44	35.89	35.77
Fe_2O_3			3.18	0.64		0.30
P_2O_5	34.47	33.4	32.72	33.40	34.16	33.65
F		0.8	0.60	2.79	n.d.	2.05
H_2O	28.42	28.6	27.92	26.45	28.46	27.47
Rem.		1.1	5.17		1.39	1.40
	100.00	100.4	100.60	100.72	99.90	100.64
F = O		0.3	0.25	1.17		0.86
Total	100.00	100.1	100.35	99.55		99.78
G.		2.325	2.358			2.36

1. $Al_3(OH)_3(PO_4)_2 \cdot 5H_2O$. 2. Hellertown, Pennsylvania.[11] Rem. is SiO_2. F determination probably low. 3. Černovice, Bohemia.[7] Rem. is CuO 0.06, MgO 0.32, CaO 0.68, FeO 2.63, Cr_2O_3 0.52, SiO_2 0.96, H_2O includes H_2O- 1.59. 4. Clonmel, Ireland.[12] 5. Amberg-Auerbach, Bavaria.[13] Rem. contains F and Al. 6. Llallagua, Bolivia.[14] Rem. is CaO 0.80, SiO_2 0.56, MgO 0.04. H_2O includes H_2O- 1.00.

Tests. B.B. infusible. In C.T. becomes opaque and affords much water with an acid reaction (F). Easily soluble in acids.

O c c u r. A secondary mineral, widespread in small amounts as a deposit in crevices in aluminous, low-grade metamorphic rocks; in limonite and phosphate-rock deposits; also rarely as a late-formed mineral in hydrothermal veins (Bolivia). Among the more important localities are the following: Černovice, Zbirow, Zeleznik, and neighboring localities in Bohemia. Kapnik, Roumania. With amblygonite in the tin veins of Montebras, Creuse, France. In Germany at Langenstriegis and Frankenberg in Saxony; at Waldgirmes and in the phosphate deposits of the Lahn-Dill region, Hesse, Nassau; at Amberg-Auerbach in the Oberpfalz, Bavaria. Wavellite was originally discovered in a clay slate near Barnstable, Devonshire, England; also in Cornwall. From Ireland at Clonmel in County Tipperary and from Tracton and Kinsale in County Cork. Very abundantly in the tin veins of Llallagua, Bolivia,[14] with apatite, vauxite, paravauxite, marcasite; also at Oruro and Potosi.

In the United States, found in Pennsylvania at General Trimble's mine (East Whiteland), Chester County; Moore's Mill, Cumberland County; Hellertown, Northampton County. In Arkansas near Hot Springs, Garland County; at Magnet Cove, Hot Spring County in the novaculite beds; in the Silver City region and neighboring districts in Montgomery County. Near Coal City, St. Clair County, Alabama. In phosphate-rock deposits at Dunellen, Marion County, Florida.

Name. After William Wavell (?–1829), physician, of Horwood Parish, Devonshire, who discovered the mineral.

Ref.

1. The angular measurements reported by different observers vary considerably owing to the generally poor quality of the crystals and probably also the isomorphous substitution of F = OH and Fe''' = Al. The elements given are the average of those reported by Gordon, *Proc. Ac. Sc. Philadelphia*, **74**, 113 (1922); Wherry, *Proc. U. S. Nat. Mus.*, **54**, 379 (1919); Cesàro, *Zs. Kr.*, **31**, 90 (1899); and Ungemach, *Bull. soc. min.*, **35**, 536 (1912). As a group these vary from the early measurements of Senff, *Ann. Phys.*, **18**, 474 (1830) [$a:b:c = 0.50489:1:0.37514$]. The structure cell of Jansen, *Zs. Kr.*, **85**, 239 (1933), is in close accord with the ratio of Senff with c doubled but needs verification on analyzed material and is not here accepted.

2. Gordon (1922) and Shannon, *Proc. U. S. Nat. Mus.*, **62**, art. 9 (1922). Less common, rare and doubtful forms:

w 650	B 10.7.0	z 850	k 510	q 13.1.0	y 122
R 970	i 320	f 520	t 710	g 270	r 5.11.6

3. Gordon, priv. comm. (1949), on Llallagua crystals. Caglioti, *Atti V° congr. naz. chim.*, pt. 1, 310 (1936)—*Min. Abs.*, **7**, 88 (1938), gives a_0 9.7 kX, b_0 17.4, c_0 7.07. See also Jansen (1933).

4. Orlov, *Zs. Kr.*, **77**, 317 (1931), notes also {101} and {100}; Senff (1830) gives {101} and {010} only; Des Cloizeaux (**2**, 455, 1874) gives {110} and {010}; Gordon (1922) gives {110} only.

5. Other reported values range down to 2.32.

6. Additional optical data are given by Orlov (1931); Shannon (1922); Krenner, *Cbl. Min.*, 163 (1930); Des Cloizeaux, *Ann. chim. phys.*, **27**, 405 (1872); Ulrich, *Věst. Stát. Geol. Úst. Československ. Rep.*, **6**, 108 (1930).

7. Orlov (1931).
8. Wherry (1919).
9. Slavík, *Ac. sc. Bohême, Bull.*, **22**, 32 (1918).
10. For additional analyses see Hintze (**1** [4B], 908, 1931).
11. Wynkoop analysis in Wherry (1919).
12. Carnot, *C.R.*, **118**, 995 (1894).
13. Laubmann, *Geognost. Jahresh.*, **35**, 193 (1922).
14. Gordon, *Proc. Ac. Sc. Philadelphia*, **96**, 279 (1944).

ZEPHAROVICHITE. Bořický (*Ak. Wien, Sitzber.*, **59**, 593, 1869).
Crystalline to compact, horn-like in aspect. Fracture conchoidal. H. $5\frac{1}{2}$. G. 2.37. Color greenish, yellowish, or grayish white. Translucent. Analyses of impure material gave:

	CaO	MgO	Fe_2O_3	Al_2O_3	P_2O_5	H_2O	SiO_2	Total
1.	1.07	0.41		29.77	35.56	26.70	5.46	98.97
2.	0.54	tr.		28.44	37.46	26.57	6.05	99.06
3.	1.38		0.86	29.60	37.80	28.98	0.46	99.08

Found with cacoxenite, limonite, strengite, and picite at Třenice near Cerhovic southwest of Beraun in Bohemia of Czechoslovakia. Named after the Austrian mineralogist Victor L. von Zepharovich (1830–1890). Zepharovichite probably is a mixture consisting principally of microcrystalline wavellite.[1]

Ref.

1. Pearl, priv. comm. (1947).

42.7.5   S T E R R E T T I T E ($Al_6(PO_4)_4(OH)_6 \cdot 5H_2O$]. Eggonite *Schrauf* (*Zs. Kr.,* **3**, 352, 1879). Sterrettite *Larsen* and *Montgomery* (*Am. Min.,* **25**, 513, 1940).

C r y s t.[1] Orthorhombic; disphenoidal—2 2 2.

$a:b:c = 0.8662:1:0.5325; \quad p_0:q_0:r_0 = 0.6147:0.5325:1$

$q_1:r_1:p_1 = 0.8663:1.6268:1; \quad r_2:p_2:q_2 = 1.8779:1.1545:1$

Forms:[2]

	ϕ	$\rho = C$	ϕ_1	$\rho_1 = A$	ϕ_2	$\rho_2 = B$
b 010	0°00′	90°00′	90°00′	90°00′	....	0°00′
a 100	90 00	90 00	....	0 00	0°00′	90 00
m 110	49 06	90 00	90 00	40 54	0 00	49 06
w 011	0 00	28 02	28 02	90 00	90 00	61 58
d 101	90 00	31 34½	0 00	58 25½	58 25½	90 00

Less common: v 031   451

Structure cell.[3] Space group $P2_12_12_1$. a_0 8.90 kX, b_0 10.20, c_0 5.43 (all ± 0.02); $a_0:b_0:c_0 = 0.872:1:0.532$. Cell contents $Al_{12}(PO_4)_8(OH)_{12} \cdot 10H_2O$.

Habit. Prismatic [100]. {100} may be striated, with diagonal sutures suggestive of twinning.

Twinning.[4] Apparently twinned about one or more of the (polar) crystal axes with {001} as the principal composition plane and perhaps {031} as a subordinate composition plane.

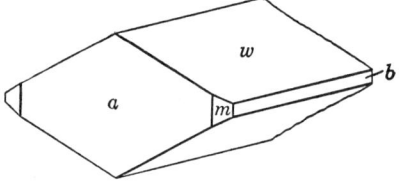

P h y s. Cleavage {110} fair, {100} and {001} poor. H. 5. G. 2.44;[5] 2.47 (calc.). Luster vitreous. Colorless, sometimes yellowish. Transparent.

O p t.[4] In transmitted light, colorless.

Orientation		n	
X	a	1.572	Biaxial negative (−).
Y	b	1.590	2V 60°.
Z	c	1.601	$r > v$, barely perceptible.

C h e m. A hydrated basic phosphate of aluminum, $Al_6(PO_4)_4(OH)_6 \cdot 5H_2O$.

Anal.

	1	2
Al_2O_3	41.67	39.07
P_2O_5	38.69	40.10
H_2O	19.64	20.36
Total	100.00	99.53

1. $Al_6(PO_4)_4(OH)_6 \cdot 5H_2O$. 2. Fairfield, Utah.[6] CaO and MgO absent, alkalies not determined.

Tests. Heated in C.T. fuses readily, giving off water and leaving a dark infusible mass. Almost insoluble in hot acids.

O c c u r. Found originally on specimens of silver ore from Felsöbánya, Roumania (first given as Altenberg, Moresnet, Belgium) and erroneously described as a cadmium silicate (*eggonite*).[7] Later found with wardite in porous masses of crandallite formed by the alteration of variscite nodules at Fairfield, Utah County, Utah.

N a m e. After Douglas B. Sterrett, a geologist of the U. S. Geological Survey, who early described the variscite deposits of Utah and Nevada.

Ref.
1. Larsen and Montgomery (1940) on Fairfield crystals. Concordant but less accurate elements were obtained by Krenner, *Cbl. Min.*, **27** (1929) on Felsöbánya material.
2. Larsen and Montgomery (1940). {451} from Krenner (1929).
3. Larsen in Larsen and Montgomery (1940) by Weissenberg method on Fairfield crystal.
4. Larsen and Montgomery (1940).
5. Bannister, *Min. Mag.*, **26**, 131 (1941), on Felsöbánya crystals; Larsen and Montgomery's (1940) value, 2.36, is low.
6. ʳGonyer analysis in Larsen and Montgomery (1940).
7. Later shown by Krenner (1929) to be a hydrated aluminum phosphate and then proved by Bannister (1941) to be identical with sterrettite.

42.7.6 **T R O E G E R I T E** $[(UO_2)_3(AsO_4)_2 \cdot 12H_2O]$. Trögerite *Weisbach* (*Jb. Min.*, 870, 1871).

C r y s t.[1] Tetragonal (?).

$$a:c = 1:2.16; \quad p_0:r_0 = 2.16:1$$

Forms:

	ϕ	ρ	A	M
c 001	.....	0°00′	90°00′	90°00′
a 010	0°00′	90 00	90 00	45 00
l 120	26 34	90 00	63 26	71 34
e 012	0 00	47 12	90 00	58 44½
d 011	0 00	65 09½	90 00	50 05
i 021	0 00	76 58	90 00	46 27½
p 111	45 00	71 52½	47 46½	90 00

Less common: h 032 u 331

Habit. Thin tabular {001}. The nonpinacoidal forms often are asymmetrically developed, and the crystals then simulate monoclinic symmetry if {001} in the tetragonal setting is taken as {010}. {001} often composite and uneven, the other faces small and horizontally striated. As subparallel aggregates.

P h y s. Cleavage {001} perfect, micaceous; also {100} good, and probably also cleavable on {011}. H. 2–3. Color lemon-yellow. Luster pearly on {001}.

Oriented growths. Zeunerite crystals are very commonly found overgrowing troegerite in parallel position.

O p t.[2] In transmitted light, pale lemon-yellow. Not pleochroic.

Orientation		n	n	
X	c	1.585 ± 0.005	1.580	Uniaxial or biaxial,
Y		1.630 ± 0.005		negative $(-)$.
$Z \wedge a$	12°–14°	1.630 ± 0.005	1.624	
$2V$		Very small	0°	

C h e m. A hydrated arsenate of uranium, apparently $(UO_2)_3(AsO_4)_2 \cdot 12H_2O$. The reported analysis may be erroneous and the composition may be $H_2(UO_2)_2(AsO_4)_2 \cdot 10H_2O$ (analogous to artificial hydrogen-autunite).
Anal.

	UO_3	As_2O_5	H_2O	Total
1.	65.81	17.62	16.57	100.00
2.	63.76	19.64	14.81	98.21

1. $(UO_2)_3(AsO_4)_2 \cdot 12H_2O$.   2. Schneeberg.[3]

Tests. In C.T. loses water and turns golden brown. Soluble in acids.

O c c u r. Found with walpurgite and zeunerite in the Walpurgis vein of the Weisser Hirsch mine at Neustädtl, Schneeberg, Saxony. Reported from the Bald Mountain district, Pennington County, South Dakota.

N a m e. After R. Tröger, mining official at Schneeberg.

Ref.

1. Interpreted as probably tetragonal by Goldschmidt, *Zs. Kr.*, **31**, 468 (1899), on geometrical grounds, although the optical properties indicate monoclinic symmetry. In the monoclinic setting, $a:b:c = 0.463:1:0.463$, β 90°; transformation: tetragonal to monoclinic, 100/001/010. Earlier made monoclinic by Schrauf (1871), *Min. Mitt.*, 185 (1872), in another setting intended to show a supposed analogy to gypsum.
2. Orientation of Goldschmidt (1899). Indices of Larsen (145, 1921) on Weisser Hirsch crystals associated with zeunerite (biaxial data); neither this nor the uniaxial material is known with certainty to be troegerite. The water content doubtless is variable, and the optical properties may vary therewith as with autunite.
3. Winkler, *J. pr. Chem.*, **7**, 7 (1873).

TYPE 8. $(AB)_m(XO_4)_pZ_q \cdot xH_2O$, WITH $m:p = 3:2$

42.8.1 **B E R M A N I T E** $[(Mn,Mg)_5(Mn,Fe)_8(PO_4)_8(OH)_{10} \cdot 15H_2O \,(?)]$. Hurlbut (*Am. Min.*, **21**, 656, 1936).

C r y s t. Orthorhombic; dipyramidal—$2/m\ 2/m\ 2/m$.

$a:b:c = 0.6890:1:2.2018;\qquad p_0:q_0:r_0 = 3.1956:2.2018:1$

$q_1:r_1:p_1 = 0.6890:0.3038:1;\qquad r_2:p_2:q_2 = 0.4542:1.4514:1$

Forms:

	ϕ	$\rho = C$	ϕ_1	$\rho_1 = A$	ϕ_2	$\rho_2 = B$
c 001	....	0°00′	0°00′	90°00′	90°00′	90°00′
b 010	0°00′	90 00	90 00	90 00	....	0 00
m 110	55 26	90 00	90 00	34 34	0 00	55 26
r 116	55 26	32 53½	20 09	63 26	61 57½	72 03
n 113	55 26	52 17½	36 16½	49 21	43 11½	63 19½
o 112	55 26	62 44	47 45	42 57	32 02½	59 43
p 111	55 26	75 33	65 34½	37 07	17 22½	56 40½

Structure cell.[1] $a_0\ 6.25\ kX,\ b_0\ 8.92,\ c_0\ 19.61;\ a_0:b_0:c_0 = 0.701:1:2.198$. Cell contents uncertain.

Habit. Tabular {001}. In fan-shaped or rosette-like aggregates of crystals and as lamellar masses.

P h y s. Cleavage {001} perfect, {110} imperfect. H. $3\frac{1}{2}$. G. 2.84; 2.83 (calc.). Luster vitreous to slightly resinous. Color reddish brown, darkening on exposure.

O p t. In transmitted light, reddish yellow to brownish red.

Orientation		n	Pleochroism	
X	c	1.687 ± 0.003	Light red	Biaxial negative $(-)$.
Y	b	1.725 ± 0.003	Pale yellow	$2V$ $74°$
Z	a	1.748 ± 0.003	Deep red	$r < v$.

C h e m. Essentially a hydrated basic phosphate of divalent and trivalent manganese. Small amounts of Fe''' substitute for Mn''', and of Mg, Ca, Na for Mn''. The only available analysis [2] affords $(Mn,Mg,Ca,Na)_5$-$(Mn,Fe)_8(PO_4)_8(OH)_{10} \cdot 15H_2O$ with $Mn'':Mg:(Ca,Na) = 19:6:2$ and $Mn''':Fe''' = 9:1$.

Na_2O	CaO	MgO	MnO	Mn_2O_3	Fe_2O_3	P_2O_5	H_2O	Total
0.32	0.72	2.39	13.79	28.76	3.03	31.39	19.33	99.73

Tests. B.B. swells, separates into scales and then fuses easily to a globule. In C.T. swells and affords abundant neutral water. Soluble in nitric acid.

O c c u r. Found as a secondary mineral in narrow veinlets and in drusy cavities in triplite in a pegmatite on the 7U7 Ranch near the Bagdad copper mine, 25 miles west of Hillside, Arizona. Several unidentified phosphates occur with the bermanite.

N a m e. After Harry Berman (1902–1944), American mineralogist.

Ref.
1. By rotation method.
2. Gonyer analysis in Hurlbut (1936).

42.8.2 R O S C H E R I T E $[(Ca,Mn,Fe)_2Al(PO_4)_2(OH) \cdot 2H_2O]$. *Slavík (Ak. Ceská, Bull.—Bull. intern. ac. sc. Bohême,* No. 4, 1914).

C r y s t.[1] Monoclinic.

$a:b:c = 0.94:1:0.88;$ $\beta\ 99°50';$ $p_0:q_0:r_0 = 0.94:0.87:1$

$r_2:p_2:q_2 = 1.15:1.08:1;$ $\mu\ 80°10';$ $p_0'\ 0.95,\ q_0'\ 0.88,\ x_0'\ 0.17$

Forms:

		ϕ	ρ	ϕ_2	$\rho_2 = B$	C	A
c	001	90°00′	9°50′	80°10′	90°00′	.....	80°10′
b	010	0 00	90 00	.....	0 00	90°00′	90 00
a	100	90 00	90 00	0 00	90 00	80 10	.....
m	110	47 11	90 00	0 00	47 11	82 48	42 48
d	$\bar{1}01$	−90 00	37 50	127 50	90 00	47 40	127 50

Habit. Short prismatic [001], with an eight- or six-sided cross section; also thin tabular {001} and elongated [010], with a rectangular outline due to large {100} and {010} and subordinate {110}. Also as vermicular aggregates of thin plates.

P h y s. Cleavage {001} good, {010} distinct. H. $4\frac{1}{2}$. G. 2.92. Color dark brown to olive-green.

O p t.[2] In transmitted light, yellowish green to brown.

Greifenstein.

Orientation		n	Pleochroism	
X	b	1.624	Yellow to olive-green	Biaxial negative $(-)$.
$Y \wedge c$	$-15°$	1.639	Yellow-brown, greenish	$2V$ large. Crossed dispersion;
Z		1.643	Chestnut-brown	$r > v$, very strong. Shows abnormal interference colors.

C h e m. A hydrated basic phosphate of aluminum, calcium, and divalent manganese and iron: $(Ca,Mn,Fe)_2Al(PO_4)_2(OH) \cdot 2H_2O$. In the only reported analysis, $Ca:Mn:Fe = 10:10:7$.

Anal.

	1	2
CaO	11.25	11.48
MnO	14.23	14.47
FeO	10.09	10.13
Al_2O_3	13.80	13.74
P_2O_5	38.44	38.01
H_2O	12.19	12.17
Total	100.00	100.00
G.		2.916

1. $(Ca,Mn,Fe)_2Al(PO_4)_2(OH) \cdot 2H_2O$ with $Ca:Mn:Fe = 10:10:7$. 2. Greifenstein.[3] Recalculated to 100 after deducting 4.58 per cent insol.

Tests. In C.T. affords water. B.B. fuses to a black magnetic slag. Soluble in acids.

O c c u r. Found with ježekite, lacroixite, childrenite, apatite, and tourmaline in drusy cavities of a granite at Greifenstein, near Ehrenfriedersdorf, Saxony. In the United States from pegmatite at Newry and Black Mountain, Maine.

N a m e. After Walter Roscher, a mineral collector of Ehrenfriedersdorf.

Ref.
1. Elements based on approximate measurements of poor crystals.
2. Larsen and Berman (174, 1934), and Slavík (1914).
3. Preis analysis in Slavík (1914).

42.8.3 MINYULITE [KAl$_2$(PO$_4$)$_2$(OH)·3$\frac{1}{2}$H$_2$O (?)]. *Simpson* and *Le Mesurier* (*J. Roy. Soc. Western Australia*, **19**, 13, 1933).

Orthorhombic (?). Radial aggregates of needles with much the appearance of wavellite.

Phys. Probably a cleavage parallel to the elongation. Brittle. H. 3$\frac{1}{2}$. G. 2.45. Colorless to white. Luster silky. Transparent.

Opt. In transmitted light, colorless.

Orientation	n	
X elong.	1.531	Biaxial positive (+).
Y	1.534	
Z	1.538	

Chem. A hydrated basic phosphate of potassium and aluminum, probably KAl$_2$(PO$_4$)$_2$(OH)·3$\frac{1}{2}$H$_2$O.

Anal.

	1	2
K$_2$O	12.97	12.30
Na$_2$O		0.45
Al$_2$O$_3$	28.08	29.98
P$_2$O$_5$	39.10	35.58
H$_2$O + 200°	19.85	2.79
H$_2$O − 200°		17.84
Total	100.00	98.94

1. KAl$_2$(PO$_4$)$_2$(OH)·3$\frac{1}{2}$H$_2$O.   2. Western Australia.[1] With Fe$_2$O$_3$ and F in traces.

Tests. In C.T. decrepitates and yields much acid water. Fusible to a white bead. Soluble in hot concentrated acids and in warm dilute NaOH.

Occur. Near Minyulo Well, Dandaragan, Western Australia, with dufrenite in the altered outcrops of glauconitic phosphate beds of Cretaceous age.

Ref.

1. Le Mesurier analysis.

42.8.4 Tinticite [Fe$_3'''$(PO$_4$)$_2$(OH)$_3$·3$\frac{1}{2}$H$_2$O]. *Stringham* (*Am. Min.*, **31**, 395, 1946).

Dense, earthy to porcelaneous masses composed of submicroscopic orthorhombic (?) crystals. Color creamy white with a yellowish green tint. H. ∼2$\frac{1}{2}$. G. ∼2.8. Mean index of refraction about 1.745. A hydrated basic phosphate of trivalent iron, Fe$_3$(PO$_4$)$_2$(OH)$_3$·3$\frac{1}{2}$H$_2$O. Analyses gave:[1]

	Na$_2$O	K$_2$O	CaO	MgO	Fe$_2$O$_3$	Al$_2$O$_3$	TiO$_2$
1.	0.45	0.32	0.36	0.24	48.84	0.18	0.04
2.					49.28		

	SiO$_2$	P$_2$O$_5$	SO$_3$	H$_2$O+	H$_2$O−	Total
1.	0.19	28.40	1.07	18.42	1.32	98.83
2.		30.21	0.71	17.21	1.72	99.13

METAVAUXITE 971

Found with jarosite and limonite in a limestone cave near the Tintic Standard mine, Tintic District, Utah. Formed by the reaction of phosphatic solutions derived from bat guano with iron from the alteration of pyrite.

Ref.
1. Analyses by Peck [1] and Hamm [2].

42.8.5 **M E T A V A U X I T E** [$FeAl_2(PO_4)_2(OH)_2 \cdot 8H_2O$]. *Gordon* (*Am. Min.*, **12**, 264, 1927; *Proc. Ac. Sc. Philadelphia*, **96**, 348, 1944).

C r y s t.[1] Monoclinic; prismatic—$2/m$.

$a:b:c = 1.0671:1:0.7272$; $\beta\, 98°02'$; $p_0:q_0:r_0 = 0.6815:0.7201:1$

$r_2:p_2:q_2 = 1.3888:0.9464:1$; $\mu\, 81°58'$; $p_0'\, 0.6882,\, q_0'\, 0.7272,\, x_0'\, 0.1411$

Forms:

		ϕ	ρ	ϕ_2	$\rho_2 = B$	C	A
a	100	90°00'	90°00'	0°00'	90°00'	81°58'	......
m	110	43 25½	90 00	0 00	43 25½	84 29	46°34½'
n	210	62 09	90 00	0 00	62 09	82 54	27 51
j	$\bar{1}01$	−90 00	28 14½	118 14½	90 00	36 16½	118 14½
r	111	48 45	47 48	50 20	60 45½	42 01	56 09
v	$\bar{1}11$	−36 27	42 07	118 14½	57 21½	47 15	113 29
p	$\bar{2}11$	−59 31	55 06	141 00½	65 25	62 06½	134 58½

Less common:
o 120 l 320 k 310 s 211

Structure cell.[2] Space group $P2_1/c$. $a_0\, 10.21\, kX$, $b_0\, 9.57$, $c_0\, 6.93$; $\beta\, 98°02'$; $a_0:b_0:c_0 = 1.067:1:0.724$. Cell contents $Fe_2Al_4(PO_4)_4(OH)_4 \cdot 16H_2O$.

Habit. Prismatic to acicular [001], as subparallel to radial aggregates.

P h y s. Brittle. H. 3. G. 2.345; 2.35 (calc.). Luster vitreous, inclining to silky in fibrous aggregates. Colorless, white, or pale green. Transparent to translucent.

O p t. In transmitted light, colorless.

ORIENTATION		n(Na)	
X	b	1.550 ± 0.001	Biaxial positive (+).
Y		1.561 ± 0.001	$2V$ large.
$Z \wedge c$	∼−17°	1.577 ± 0.001	

Llallagua. Back view.

C h e m. A hydrated basic phosphate of aluminum and divalent iron, $FeAl_2(PO_4)_2(OH)_2 \cdot 8H_2O$. Very small amounts of Ca and Mg substitute for Fe.

Anal.

	1	2	3
CaO		0.27	0.80
MgO		0.55	0.57
FeO	15.03	1.457	17.00
Fe_2O_3		0.35	tr.
Al_2O_3	21.33	21.01	17.38
SiO_2		0.05	0.60
P_2O_5	29.71	29.55	28.53
H_2O+	33.93	14.00	14.50
H_2O-		19.22	20.50
Total	100.00	99.57	99.88
G.		2.345	

1. $FeAl_2(PO_4)_2(OH)_2 \cdot 8H_2O$. 2. Llallagua.[5] 3. Llallagua.[3]

Occur. Found with vauxite, paravauxite, and wavellite as a supergene mineral at Llallagua, Bolivia. Also found at Tasna, near Urjuni, Bolivia.[4]

Name. In allusion to the chemical relationship to vauxite.

Ref.

1. Angles and forms of Gordon, *Proc. Ac. Sc. Philadelphia*, **96**, 348 (1944); orientation and unit of the structure cell of Nuffield, priv. comm. (1948). Transformation: Gordon to new, $\overline{1}0\overline{1}/0\overline{1}0/001$.
2. Nuffield (1948) by Weissenberg method.
3. Shannon analysis in Gordon (1944).
4. Peacock, *Zs. Kr.*, **86**, 203 (1933).
5. Gordon (1944).

42.8.6 PARAVAUXITE $[FeAl_2(PO_4)_2(OH)_2 \cdot 8H_2O]$. Gordon (*Science*, **56**, 50, 1922; *Proc. Ac. Sc., Philadelphia*, **75**, 261, 1923).

Cryst.[1] Triclinic; pinacoidal—$\overline{1}$.

$a:b:c = 0.4963:1:0.6613;$ $\alpha\ 106°45',\ \beta\ 110°43',\ \gamma\ 72°06\frac{1}{2}'$

$p_0:q_0:r_0 = 1.3409:0.6500:1;$ $\lambda\ 78°22',\ \mu\ 73°05',\ \nu\ 103°15'$

$p_0'\ 1.4636,\ q_0'\ 0.7095,\ x_0'\ 0.3781,\ y_0'\ 0.2201$

Forms:[2]

		ϕ	ρ	A	B'	C
c	001	59°47½′	23°38′	73°05′	78°22′	.....
b	010	0 00	90 00	103 15	.....	78°22′
a	100	103 15	90 00	.....	103 15	73 05
m	110	75 17½	90 00	27 57½	75 17½	67 17
M	1$\overline{1}$0	126 15½	90 00	23 00½	126 15½	80 47½
k	0$\overline{1}$1	142 03½	31 35	65 54½	114 24	36 02
z	0$\overline{5}$4	150 26½	37 28½	65 35	121 57	43 35
K	$\overline{1}$11	−39 36	58 39½	132 54	48 51	65 07
s	$\overline{2}$11	−94 12	68 01½	152 12½	93 54	89 29½

Less common:

t 1$\overline{2}$0   f 011   e $\overline{1}$01   h $\overline{1}$12   g $\overline{1}$23   H $\overline{1}$32

PARAVAUXITE

Structure cell.[3] a_0 5.23 kX, b_0 10.52, c_0 6.955; α 107°16½', β 111°24', γ 72°29'; $a_0:b_0:c_0$ = 0.4971:1:0.6611. Cell contents $FeAl_2(PO_4)_2(OH)_2 \cdot 8H_2O$.

Habit. Short prismatic [001] to thick tabular {010}. Occasionally as subparallel or radial aggregates. Common forms: *b a m c k z s*.

Phys. Cleavage {010} perfect. Fracture conchoidal. Brittle. H. 3. G. 2.36; 2.38 (calc.). Luster vitreous, on {010} cleavages pearly. Colorless to pale greenish white. Streak white. Transparent to translucent.

Opt. In transmitted light, colorless.

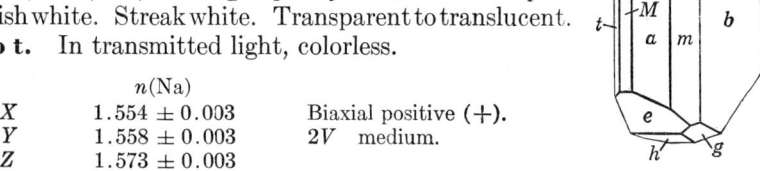

	n(Na)	
X	1.554 ± 0.003	Biaxial positive (+).
Y	1.558 ± 0.003	2V medium.
Z	1.573 ± 0.003	

Chem. A hydrated basic phosphate of aluminum and divalent iron, $FeAl_2(PO_4)_2(OH)_2 \cdot 8H_2O$. Very small amounts of Ca and Mg substitute for Fe.

Anal.

	1	2	3	4
CaO		0.02		0.28
MgO		0.56		0.22
FeO	15.03	13.71	13.59	13.60
Fe_2O_3		0.47	0.60	1.52
Al_2O_3	21.33	21.26	21.48	21.00
SiO_2		0.07	0.02	1.20
P_2O_5	29.71	29.81	29.58	27.64
H_2O+	33.93	17.77	18.07	18.99
H_2O-		16.38	16.08	16.06
Total	100.00	100.05	99.42	100.51
G.	2.38	2.358		

1. $FeAl_2(PO_4)_2(OH)_2 \cdot 8H_2O$. 2. Llallagua.[4] 3. Llallagua.[5] 4. Llallagua.[6]

Occur. At Llallagua, Bolivia, where it occurs in the tin veins with vauxite, metavauxite, and wavellite and is later formed than these.

Name. In allusion to the chemical relationship to vauxite.

Ref.

1. Angles of Gordon, *Proc. Ac. Sc. Philadelphia*, **96**, 345 (1944); orientation and unit of the structure cell of Nuffield, priv. comm. (1948). Transformation: Gordon to new, 100/110/001.
2. Gordon (1923, 1944). Doubtful forms:

1.$\overline{12}$.0	1$\overline{5}$0	3.$\overline{11}$.0	4.$\overline{13}$.0	4.$\overline{11}$.0	2$\overline{5}$0	9.$\overline{16}$.0
1$\overline{6}$0	3.$\overline{14}$.0	2$\overline{7}$0	5.$\overline{16}$.0	3$\overline{8}$0	5.$\overline{12}$.0	
2.$\overline{11}$.0	4.$\overline{15}$.0	3.$\overline{10}$.0	6.$\overline{17}$.0	5.$\overline{13}$.0	8.$\overline{17}$.0	

3. Nuffield (1948) by Weissenberg method.
4. Gordon (1944).
5. Hallowell analysis in Gordon (1944).
6. Shannon analysis in Gordon (1944).

42.8.7 V A U X I T E [$Fe''Al_2(PO_4)_2(OH)_2 \cdot 7H_2O$]. Gordon (Science, **56**, 50, 1922; Proc. Ac. Sc. Philadelphia, **75**, 261, 1923).

C r y s t.[1] Triclinic; pinacoidal—$\bar{1}$.

$a:b:c = 0.7878:1:0.5247; \quad \alpha\ 98°34',\ \beta\ 91°21\tfrac{1}{2}',\ \gamma\ 109°09'$

$p_0:q_0:r_0 = 0.6972:0.5553:1; \quad \lambda\ 80°27',\ \mu\ 85°32\tfrac{1}{2}',\ \nu\ 70°24'$

$p_0'\ 0.7072,\ q_0'\ 0.5632,\ x_0'\ 0.0240,\ y_0'\ 0.1683$

Forms:[2]

		ϕ	ρ	A	B	C
b	010	0°00′	90°00′	70°24′	……	80°27′
M	$1\bar{1}0$	116 04½	90 00	45 40½	116°04½′	92 57½
h	$1\bar{4}0$	161 42½	90 00	91 18½	161 42½	98 38
d	101	59 34	38 40½	52 08	71 32½	33 24½
u	$\bar{1}01$	−83 52½	32 51	119 15½	86 41	33 43
s	$1\bar{1}1$	102 52	35 18	60 49½	97 24	37 11½
e	$\bar{1}\bar{1}1$	−134 33	42 01½	127 22	118 00½	49 58½

Structure cell.[7] Space group $P\bar{1}$. a_0 9.07 kX, b_0 11.55, c_0 6.14; α 98°52′, β 92°22′, γ 107°42½′; $a_0:b_0:c_0 = 0.7853:1:0.5316$. Cell contents $Fe_2Al_4(PO_4)_4(OH)_4 \cdot 14H_2O$.

Habit. Minute crystals tabular {010} and elongated [001], or, less commonly, [101]. Order of form importance: $b\ M\ e\ d\ s\ u\ h$. As subparallel to radial aggregates and nodules.

Twinning.[3] Twin plane {010}, with composition plane {010}.

P h y s. Cleavage none. Brittle. H. 3½. G. 2.39; 2.52 (calc.). Luster vitreous. Color sky-blue to venetian blue, becoming greenish on exposure. Streak white. Transparent.

O p t.[2] In transmitted light, pale blue.

	n(Na)	Pleochroism	
X	1.551 ± 0.003	Colorless	Biaxial positive (+).
Y	1.555 ± 0.003	Blue	$2V$ 32°.
Z	1.562 ± 0.003	Colorless	$r > v$, marked.
			Z is about $\perp$ to {010}.

C h e m. A hydrated basic phosphate of aluminum and divalent iron, $Fe''Al_2(PO_4)_2(OH)_2 \cdot 7H_2O$. Very small amounts of Ca and Mg substitute for Fe''.

Anal.[4]

	1	2	3
CaO		0.77	0.08
MgO		0.28	0.28
FeO	15.62	15.54	14.96
Fe_2O_3		0.60	tr.
Al_2O_3	22.17	21.42	23.80
P_2O_5	30.87	30.52	30.40
H_2O+		22.92	23.37
H_2O-	31.34	8.49	7.22
Rem.			0.20
Total	100.00	100.54	100.31
G.		2.389	

1. $Fe''Al_2(PO_4)_2(OH)_2 \cdot 7H_2O$. 2. Llallagua.[5] 3. Llallagua.[6] Rem. is SiO_2.

Occur. Found associated with wavellite and paravauxite in the tin veins of Llallagua, Bolivia. The vauxite, paravauxite, metavauxite, and childrenite are secondary minerals, derived by the solution of apatite in supergene waters, and are usually deposited upon wavellite.

Name. After George Vaux, Jr. (1863–1927), a private mineral collector of Bryn Mawr, Pennsylvania.

Ref.
1. Elements recalculated from the new angles of Gordon, *Proc. Ac. Sc. Philadelphia*, **46**, 344 (1944); unit and orientation of the structure cell of Nuffield, priv. comm. (1948). Transformation: Gordon to Nuffield, $\overline{2}01/230/00\overline{1}$.
2. Gordon (1944).
3. Gordon (1923, 1944).
4. An earlier analysis is cited by Gordon (1923).
5. Gordon (1944).
6. Shannon cited in Gordon (1944).
7. Nuffield (1948) by Weissenberg method.

42.8.8   **GORDONITE** [$MgAl_2(PO_4)_2(OH)_2 \cdot 8H_2O$]. *Larsen* and *Shannon* (*Am. Min.*, **15**, 331, 1930).

Cryst.[1] Triclinic; pinacoidal—$\overline{1}$.

$a:b:c = 0.4990:1:0.6635;$ $\alpha\ 107°25',\ \beta\ 111°04',\ \gamma\ 72°22'$ [2]

$p_0:q_0:r_0 = 1.3315:0.6498:1;$ $\lambda\ 77°38',\ \mu\ 72°48\tfrac{1}{2}',\ \nu\ 102°40\tfrac{1}{2}'$

$p_0'\ 1.4607,\ q_0'\ 0.7129,\ x_0'\ 0.3852,\ y_0'\ 0.2350$

Forms:[2]

		ϕ	ρ	A	B	C	Z
c	001	58°37'	24°17'	72°48½'	77°38'	.....	68°56'
b	010	0 00	90 00	102 40½	.....	77°38'	0 00
a	100	102 40½	90 00	.....	102 40½	72 48½	0 00
m	110	74 29	90 00	28 11½	74 29	66 41½	0 00
M	1$\overline{1}$0	125 57	90 00	23 16½	125 57	80 53	0 00
y	011	22 07	45 39½	83 16	48 30½	29 07½	68 56
k	0$\overline{1}$1	141 08	31 32½	65 49	114 02	36 24	68 56
s	$\overline{2}$11	−86 13	67 58	156 19	86 29½	88 15½	157 55

Less common:

$j\ 5\overline{1}0$ $i\ 3\overline{1}0$ $l\ 3\overline{2}0$ $t\ 1\overline{2}0$ $f\ 1\overline{5}0$ $x\ 013$
$p\ 4\overline{1}0$ $n\ 2\overline{1}0$ $r\ 2\overline{3}0$ $d\ 1\overline{3}0$ $e\ 1\overline{7}0$ $w\ 0\overline{5}6$

Structure cell.[1] Space group $P\overline{1}$. $a_0\ 5.235\ kX,\ b_0\ 10.49,\ c_0\ 6.96$; $\alpha\ 107°25',\ \beta\ 111°04',\ \gamma\ 72°22';\ a_0:b_0:c_0 = 0.4990:1:0.6635$. Cell contents $MgAl_2(PO_4)_2(OH)_2 \cdot 8H_2O$. Gordonite is isostructural with paravauxite.

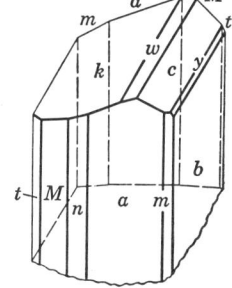

Habit. Usually in bundles, radiating outward in sheaf-like aggregates with all individuals of one group similarly terminated. Crystals are prismatic [001] to platy {010}; strongly striated [001], less marked [100]. Dominant forms: $b\ a\ M\ k\ c$. {001} is sometimes lacking, the crystals being terminated entirely by {0$\overline{1}$1}; rarely doubly terminated.

Phys. Cleavage {010} perfect, {100} fair, {001} poor.[3] Fracture conchoidal. Brittle. H. 3½. G.

2.23;[4] 2.22 (calc.). Luster vitreous, pearly on {010}. Color commonly smoky white to colorless; rare crystals are pale pink at their tips, or are pale green. Streak white. Transparent.

O p t.[5] In transmitted light, colorless.

ORIENTATION	n	
X about $\perp$ {010}	1.534	Biaxial positive (+).
Y	1.543	$2V$ 73°.
Z	1.558	$r > v$, easily perceptible.

C h e m. A hydrated basic phosphate of magnesium and aluminum, $MgAl_2(PO_4)_2(OH)_2 \cdot 8H_2O$.

Anal.

	MgO	Al_2O_3	P_2O_5	H_2O+	H_2O-	Total
1.	9.03	22.84	31.80	36.33		100.00
2.	10.01	20.68	32.80	16.80	18.20	98.49

1. $MgAl_2(PO_4)_2(OH)_2 \cdot 8H_2O$. 2. Fairfield, Utah.[5]

Tests. Fusibility 3. In the C.T. yields neutral water. Soluble in acids.

O c c u r. Found with variscite in phosphate nodules at Fairfield, Utah County, Utah. Usually in cavities between successive shells of variscite and crandallite, but frequently attached to wardite crusts.

N a m e. After Samuel G. Gordon (1897–), American mineralogist.

Ref.

1. Unit and orientation of Pough, *Am. Min.*, **22**, 625 (1937), from a goniometric study of poor-quality crystals which gave the ratios: $a:b:c = 0.5192:1:0.6942$; α 109°27′, β 110°57½′, γ 71°40½′. Transformation: Larsen and Shannon (1930) to Pough, 100/110/001. The axial ratios here given are calculated from the averaged cell constants of Larsen, *Am. Min.*, **27**, 295 (1942): a_0 5.25 kX, b_0 10.49, c_0 6.97; α 107°20′, β 111°12′, γ 72°12′, and Nuffield, priv. comm. (1948): a_0 5.22, b_0 10.49, c_0 6.95; α 107°29½′, β 110°55½′, γ 72°32½′.
2. Larsen and Shannon (1930); Pough (1937). Z gives the phi angle of the form when (010) is made polar and the azimuth of (100) is set at 0°. Uncertain: 510, $4\bar{7}0$, 140, $\bar{1}01$.
3. Pough (1937). The perfect cleavage, {110} in the structural setting, given by Larsen and Shannon (1930) is probably {100}.
4. Larsen (1942), the average of 7 determinations with the Berman microbalance on clear crystals. Larsen and Shannon (1930) gave 2.26.
5. Larsen and Shannon (1930).

42.8.9 Calcioferrite $[Ca_3Fe_3(PO_4)_4(OH)_3 \cdot 8H_2O$ (?)]. Calcoferrit *Blum* (*Jb. Min.*, 287, 1858). Calcioferrite *Dana* (578, 1868).

Possibly hexagonal. In foliated nodular or reniform masses.

P h y s. Cleavage perfect parallel to the foliation; also traces of another at right angles to the perfect one and traces of a third cleavage oblique to the same. Brittle. H. 2½. G. 2.53. Color sulfur-yellow, greenish yellow to siskin-green and yellowish white. Streak sulfur-yellow to white. Luster pearly on cleavage surfaces. Thin laminae translucent.

O p t.[2] In transmitted light, yellow. Uniaxial negative (−); in part very fine-grained and nearly isotropic. nO 1.57–1.58, variable.

C h e m. A hydrated basic phosphate essentially of calcium and iron of uncertain formula, perhaps $Ca_3Fe_3(PO_4)_4(OH)_3 \cdot 8H_2O$.

Anal.

	1	2
CaO	19.50	14.81
MgO		2.65
Al_2O_3		2.90
Fe_2O_3	27.76	24.34
P_2O_5	32.90	34.01
H_2O	19.84	20.56
Total	100.00	99.27

1. $Ca_3Fe_3(PO_4)_4(OH)_3 \cdot 8H_2O$. 2. Battenberg, Bavaria.[1]

Tests. Fuses easily to a black magnetic globule. Easily decomposed by HCl.

O c c u r. In nodules in a bed of Tertiary clay at Battenberg in Rhenish Bavaria. The exterior of the nodules is yellowish or reddish brown impure calcioferrite.

N a m e. In allusion to the composition.

Ref.
1. Reissig in Rammelsberg (316, 1875).
2. Larsen (52, 1921).

42.8.10 X A N T H O X E N I T E $[Ca_2Fe(PO_4)_2(OH) \cdot 1\frac{1}{2}H_2O]$. *Laubmann* and *Steinmetz* (*Zs. Kr.*, **55**, 579, 1920).

Monoclinic or triclinic. As masses and crusts composed of indistinct platy or lath-like crystals.[1]

P h y s. One perfect cleavage. H. $\sim 2\frac{1}{2}$.[2] G. 2.8 (Hühnerkobel), 2.97 (Palermo).[2] Luster dull to waxy. Color pale yellow to brownish yellow.

O p t. In transmitted light, pale yellow to pale lemon-yellow in color.

	n (Palermo)[2]	
X	1.704 ± 0.003	Biaxial negative (−).
Y	1.715 ± 0.003	$2V$ large.
Z	1.724 ± 0.003	$r < v$, strong.
		Weakly pleochroic.

C h e m. A hydrated basic phosphate of calcium and trivalent iron, $Ca_2Fe(PO_4)_2(OH) \cdot 1\frac{1}{2}H_2O$. Small amounts of Mn'' and Mg substitute for Ca, with Ca:Mn:Mg = 19.7:2.84:1 in the only reported analysis.[3]

Anal.

	CaO	MgO	MnO	Fe_2O_3	P_2O_5	H_2O+	H_2O-	Rem.	Total
1.	30.31			21.58	38.37	9.74			100.00
2.	24.99	0.91	4.55	21.68	37.62	9.13	0.86	0.79	100.53

1. $Ca_2Fe(PO_4)_2(OH) \cdot 1\frac{1}{2}H_2O$. 2. Palermo mine, New Hampshire.[4] Rem. is insol. 0.78, Al_2O_3 0.01.

O c c u r. Originally from the Hühnerkobel pegmatite near Rabenstein, Bavaria, associated with cacoxenite and dufrenite (?) as an alteration product of triphylite. Found abundantly with whitlockite, eosphorite-childrenite, rockbridgeite, and other phosphates in a triphylite-rich pegmatite at the Palermo mine, North Groton, New Hampshire.

N a m e. In allusion to its color and the originally supposed chemical resemblance to cacoxenite.

Ref.
1. It is not entirely certain if the original xanthoxenite from Hühnerkobel described by Laubmann and Steinmetz (1920) is identical with the so-called xanthoxenite from Palermo described by Frondel, *Am. Min.*, **34**, 692 (1949), and on which material the present description is largely based. The Hühnerkobel mineral occurs as indistinct gypsum-like crystals with a clinodome making an angle with [001] of about 41°. Cleavage {010}. Biaxial negative (−), with $Z \wedge c = +36°$.
2. Frondel (1949).
3. The original Hühnerkobel mineral was said by Laubmann and Steinmetz (1920) on the basis of qualitative tests and a partial analysis to be a hydrous ferric phosphate with minor Mg, Ca, Mn, Al, P_2O_5 32.61 per cent, H_2O 16.10 per cent (ignition).
4. Hallowell analysis in Frondel (1949).

42.8.11   **MONTGOMERYITE** [$Ca_4Al_5(PO_4)_6(OH)_5 \cdot 11H_2O$].   Larsen (*Am. Min.*, **25**, 315, 1940).

C r y s t.   Monoclinic; prismatic—$2/m$.

$a:b:c = 0.4145:1:0.2580;$ $\beta\, 91°34';$ $p_0:q_0:r_0 = 0.6224:0.2579:1$

$r_2:p_2:q_2 = 3.8775:2.4133:1;$ $\mu\, 88°26';$ $p_0'\, 0.6224, q_0'\, 0.2579, x_0'\, 0.0275$

Forms: [1]

	ϕ	ρ	ϕ_2	$\rho_2 = B$	C	A
a 100	90°00'	90°00'	0°00'	90°00'	88°26'	.....
b 010	0 00	90 00	....	0 00	90 00	90°00'
f 170	19 01	90 00	0 00	19 01	89 29½	70 59
g 150	25 46	90 00	0 00	25 46	89 19	64 14
h 290	28 12	90 00	0 00	28 12	89 15½	61 48
i 140	31 06	90 00	0 00	31 06	89 11½	58 54
j 270	34 35	90 00	0 00	34 35	89 06½	55 25
k 130	38 48½	90 00	0 00	38 48½	89 01	51 11½
m 110	67 29½	90 00	0 00	67 29½	88 33	22 30½
q 151	26 44½	55 18½	56 59	42 45½	54 37	68 17½

Structure cell.[2]   Space group $C\, 2/c$. a_0 9.99 ± 0.02 kX, b_0 24.10 ± 0.02, c_0 6.25 ± 0.05; $\beta\, 91°28'$; $a_0:b_0:c_0 = 0.4145:1:0.2593$. Cell contents $Ca_8Al_{10}(PO_4)_{12}(OH)_{10}\cdot 22H_2O$.

Habit.   Laths flattened {010} and elongated [001]. Striated parallel [001] with numerous vicinal prism forms. The crystals often occur in subparallel growths in contact on {010}. Also massive, as subparallel aggregates of coarse plates.

P h y s.   Cleavage {010} perfect; {100} poor. H. 4. G. 2.53 ± 0.05; 2.51 (calc.). Color usually deep green, rarely pale green to colorless. Luster vitreous.

O p t.   In transmitted light, colorless.

Orientation	$n(Na)$	Pleochroism	
$X \wedge c$ +60°	1.572 ± 0.002	Colorless to pale green	Biaxial negative (−).
Y	1.578 ± 0.002	Colorless	$2V\, 75° \pm 10°$.
$Z\, b$	1.582 ± 0.002	Colorless	$r < v$, perceptible.

C h e m.   A basic hydrated phosphate of calcium and aluminum, $Ca_4Al_5(PO_4)_6(OH)_5 \cdot 11H_2O$.

Anal.

	1	2	3
CaO	19.53	19.07	18.89
Al_2O_3	22.19	21.32	21.56
P_2O_5	37.10	37.70	37.63
H_2O	21.18	21.65	21.71
Total	100.00	99.74	99.79

1. $Ca_4Al_5(PO_4)_6(OH)_5 \cdot 11H_2O$. 2, 3. Fairfield, Utah.[3] Sample 2 about 99 per cent pure, sample 3 about 95 per cent pure.

Occur. In variscite nodules from Fairfield, Utah County, Utah, associated with crandallite, englishite, wardite, millisite, gordonite, and members of the apatite group.

Name. After Arthur Montgomery (1909–), who with Edwin Over mined the Fairfield deposit and first recognized the present species as new.

Ref.

1. Less common and doubtful: 190, l 120, 350, 230, 340, 450, x 021, y 041, p 111, P $\bar{1}$11, R $\bar{1}$31, Q $\bar{1}$51.
2. Weissenberg method by Larsen (1940).
3. Gonyer analysis in Larsen (1940).

42.8.12 **OVERITE** [$Ca_3Al_8(PO_4)_8(OH)_6 \cdot 15H_2O$]. Larsen (*Am. Min.*, **25**, 315, 1940).

Cryst. Orthorhombic; dipyramidal—$2/m\ 2/m\ 2/m$.

$a:b:c = 0.7864:1:0.3795;$ $p_0:q_0:r_0 = 0.4826:0.3795:1$

$q_1:r_1:p_1 = 0.7864:2.0721:1;$ $r_2:p_2:q_2 = 2.6350:1.2717:1$

Forms:[1]

	ϕ	$\rho = C$	ϕ_1	$\rho_1 = A$	ϕ_2	$\rho_2 = B$
b 010	0°00′	90°00′	90°00′	90°00′	....	0°00′
a 100	90 00	90 00	.....	0 00	0°00′	90 00
f 130	22 58½	90 00	90 00	67 01½	0 00	22 58½
m 110	51 49	90 00	90 00	38 11	0 00	51 49
y 021	0 00	37 12	37 12	90 00	90 00	52 48
q 121	32 27	41 58	37 12	68 58½	64 14½	55 38½

Structure cell.[2] Space group *Bmam*. a_0 14.75 kX, b_0 18.74, c_0 7.12; $a_0:b_0:c_0 = 0.7871:1:0.3799$. Cell contents $Ca_6Al_{16}(PO_4)_{16}(OH)_{12} \cdot 30H_2O$.

Habit. Flattened on {010} and elongated [001] with a platy to lath-like shape. Often as subparallel growths on {010}. Also massive, as subparallel aggregates of coarse plates.

Phys. Cleavage {010} perfect, {100} poor. H. 3½–4. G. 2.53; 2.47 (calc.). Color light apple-green to colorless. Luster vitreous.

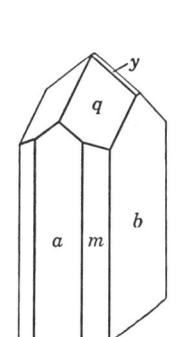
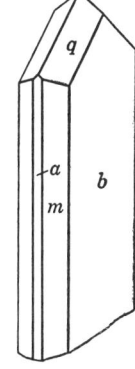

Opt. In transmitted light, colorless.

ORIENTATION		n(Na)	
X	c	1.568 ± 0.002	Biaxial negative $(-)$.
Y	a	1.574 ± 0.002	$2V\ 75° \pm 10°$.
Z	b	1.580 ± 0.002	$r > v$, weak.

Chem. A basic hydrated phosphate of aluminum and calcium, $Ca_3Al_8(PO_4)_8(OH)_6 \cdot 15H_2O$.

Anal.

	1	2
CaO	11.67	11.62
Al_2O_3	28.11	27.99
P_2O_5	38.08	37.91
H_2O	22.14	22.04
Insol.		0.11
Total	100.00	99.67

1. $Ca_3Al_8(PO_4)_8(OH)_6 \cdot 15H_2O$.   2. Fairfield, Utah.[3]

Tests. B.B. fuses at 2 with intumescence and leaves a white chalky mass. C.T. gives off abundant neutral water. Easily soluble in hot nitric acid.

Occur. Found in cavities in variscite nodules at Fairfield, Utah County, Utah, associated with crandallite, deltaite (?), and an apatite-like mineral.

Name. After Edwin Over (1905–), a mineral collector of Colorado Springs, Colorado, who with Arthur Montgomery worked the Fairfield locality and recognized the present mineral as a new species.

Ref.

1. Less common and doubtful forms: e 150, g 250, h 120, i 350, j 430, k 320, l 310, n 410.
2. Weissenberg method by Larsen (1940).
3. Gonyer analysis in Larsen (1940).

42.8.13 AND 42.8.14 TORBERNITE AND METATORBERNITE GROUPS

TETRAGONAL; DITETRAGONAL DIPYRAMIDAL—$4/m\ 2/m\ 2/m$

	a_0	c_0
Torbernite Group		
Torbernite, $Cu(UO_2)_2(PO_4)_2 \cdot 8\text{–}12H_2O$	$7.05\ kX$	20.5
Autunite, $Ca(UO_2)_2(PO_4)_2 \cdot 10\text{–}12H_2O$	6.989	20.63
Uranocircite, $Ba(UO_2)_2(PO_4)_2 \cdot 8H_2O$ (?)		
Saléeite, $Mg(UO_2)_2(PO_4)_2 \cdot 10H_2O$	6.980	19.813
Zeunerite, $Cu(UO_2)_2(AsO_4)_2 \cdot 10\text{–}16H_2O$		
Uranospinite, $Ca(UO_2)_2(AsO_4)_2 \cdot 8H_2O$		
Metatorbernite Group		
Metatorbernite, $Cu(UO_2)_2(PO_4)_2 \cdot 8H_2O$	$6.95\ kX$	8.60
Meta-autunite, $Ca(UO_2)_2(PO_4)_2 \cdot 2\text{–}6H_2O$	6.98	8.42
Metazeunerite, $Cu(UO_2)_2(AsO_4)_2 \cdot 8H_2O$	7.13	8.83

The members of the Torbernite and Metatorbernite Groups conform to the formula $A(UO_2)_2(XO_4)_2 \cdot nH_2O$, where A is Cu, Ca, Ba, or Mg and

X is P or As. Both groups have closely related structures [1] of the layer type. In autunite, with which the other members of the Torbernite Group are isostructural, there are layers parallel {001} built of (PO_4) tetrahedra together with U^6 ions in distorted octahedral coordination with O, two O ions being more closely associated with the U^6 to give linear uranyl (UO_2) groups. The Ca or other A ions together with the H_2O molecules occupy relatively open cavities between the vertically stacked layers. The autunite-type structure is stable for values of nH_2O over about 8 and ranging zeolitically up to a maximum of 12 (?). The meta-autunite (or meta-autunite-I) type of structure [1] is stable for values of nH_2O ranging from 6–8 down to 5 (metazeunerite) or $2\frac{1}{2}$ (meta-autunite). The transition temperature varies with the nature of the A and X atoms and with the vapor pressure, but in general it is near room temperature at low humidities and below 100° from water solution.[3] In the meta-autunite structure, the {001} layers containing the oxygen-coordinated U^6 and P atoms are identical with those of autunite but are moved laterally relative to one another by a distance of $a/2$, giving a more close-packed vertical stacking and reducing the c period by one-half. The water content of this structure varies zeolitically between limits. Isostructural compounds with the formula $A_2(UO_2)_2(XO_4)_2 \cdot nH_2O$, where A is Na, K, NH_4, or H, have been prepared artificially; these substances undergo reversible base exchange [2] with Ca or other divalent cations. At somewhat higher temperatures a lower hydrate is formed (meta-autunite-II, etc.), not known in nature, which appears to have a different type of structure. The transition between the meta-I and meta-II hydrates is not reversible. The transition between the meta-I and the fully hydrated structures is reversible in the case of autunite at least, although not readily or not at all in torbernite and zeunerite. Because of these relations and the occurrence of the transition at or near room temperatures there is considerable uncertainty whether the substances as represented by museum specimens are of primary formation or are secondary dehydration products.

Ref.

1. Beintema, *Rec. trav. chim. Pays-Bas*, **57**, 155 (1938).
2. Fairchild, *Am. Min.*, **14**, 265 (1929), and Beintema (1938).
3. Hallimond, *Min. Mag.*, **17**, 326 (1916); **19**, 43 (1920); Beintema (1938); Weiss-Frondel, *Am. Min.*, **36**, 249 (1951).

42.8.13.1 **TORBERNITE** [$Cu(UO_2)_2(PO_4)_2 \cdot 8-12H_2O$]. Mica viridis cryst. [from Joachimsthal] *von Born* (**1**, 42, 1772). Grüner Glimmer [from Saxony] *Werner* (Cronstedt, 217, 1780). Torberit *Werner, Karsten* (43, 1793) [later spelled *torbernite*, as in Ludwig's *Werner* (**1**, 308, 1803)]. Chalkolith *Werner* (*Bergm. J.*, 376, 1789); Urankalk durch Kupfer gefärbt, Uranites spathosus pt. *Klaproth* (*Schrift. Ges. Nat. Berlin*, **9**, 273, 1789; Beitr., **2**, 217, 1797). Uranglimmer *Werner* (1800; *Ludwig's Werner*, **1**, 55, 1803). Urane oxydé *Haüy* (**4**, 319, 1801). Uranite *Aikin* (1814). Uran-Mica *Jameson* (1820). Uranphyllit *Breithaupt* (1820). Phosphate of Uranium containing Phosphate of Copper *R. Phillips* (*Ann. Phil.*, **5**, 57, 1823). Phosphate of Uranium and Copper *Berzelius* (*Jahresber.*, 1823). Kupfer-Uranit *Germ.* Copper-Uranite. Torberite *Brooke* and *Miller* (517, 1852). Cuprouranit *Breithaupt* (*Berg.-H. Ztg.*, **24**, 302, 1865).

982 PHOSPHATES, ARSENATES, VANADATES

C r y s t.[1] Tetragonal; ditetragonal-dipyramidal—$4/m\ 2/m\ 2/m$.

$a:c = 1:2.974;\quad p_0:r_0 = 2.974:1$

Forms:[2]

		ϕ	ρ	A	M
c	001	.....	0°00′	90°00′	90°00′
a	010	0°00′	90 00	90 00	45 00
m	110	45 00	90 00	45 00	90 00
o	013	0 00	44 45	90 00	60 08½
e	011	0 00	71 25	90 00	47 55
l	112	45 00	64 34	50 19	90 00
p	111	45 00	76 37½	46 32	90 00

Habit. Thin to thick tabular {001}, usually square in outline; rarely pyramidal. Subparallel growth is frequent. Also in foliated, micaceous, or scaly aggregates.

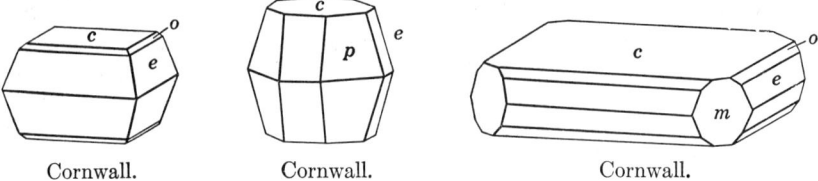

Cornwall.　　　　Cornwall.　　　　Cornwall.

Structure cell.[13] Space group $I4/mmm$. a_0 7.05 kX, c_0 20.5; $a_0:c_0 = 1:2.90$. Cell contents $Cu_2(UO_2)_4(PO_4)_4 \cdot 16\text{-}24H_2O$.

Twinning. On {110}, rare.

Oriented growths. Parallel growths with autunite are common; also with uranospinite, zeunerite, and bassetite.

P h y s. Cleavage {001} perfect, micaceous; {100} indistinct. Laminae rather brittle, more so than autunite. H. 2–2½. G. 3.22.[3] Luster vitreous to sub-adamantine, pearly on {001}. Color emerald- and grass-green, sometimes leek-green, apple-green, or siskin-green. Streak paler than the color. Transparent to translucent. Not fluorescent in ultraviolet light.[11]

O p t.

	n (Cornwall)[4]	DICHROISM	
O	1.592	Sky-blue	Uniaxial negative (−).
E	1.582	Green	

The indices of refraction and the gravity vary with water content in the range from 8 to 12 H_2O, but correlated data are lacking. Most of the published optical data on so-called torbernite are of uncertain affiliation and may belong to metatorbernite (which see).

C h e m. A hydrated phosphate of copper and uranium, $Cu(UO_2)_2(PO_4)_2 \cdot 8\text{-}12H_2O$. Small amounts of As sometimes substitute for P, with As:P = 1:6.8 in analysis 4. Pb apparently substitutes for Cu in the material of analysis 5, with Pb:Cu = 1:6.8. There is no evidence of an isomorphous series with autunite. V, Ba, Mn, and other elements have been found in spectrographic amounts.[12] The water content of torbernite varies with humidity and temperature between 8 and $12H_2O$. $4H_2O$ are lost at 55° to 75° or at room temperature over a desiccant irreversibly (?) forming metatorbernite, $Cu(UO_2)_2(PO_4)_2 \cdot 8H_2O$. Torbernite is the phase deposited from aqueous solutions up to about 75°; above this temperature metatorbernite would be formed by direct crystallization.[5]

Anal.[6]

	1	2	3	4	5	6
CuO	7.88	8.49	7.73	8.13	7.15	8.58
PbO					2.95	0.11
UO_3	56.65	61.00	57.03	60.71	60.35	60.54
P_2O_5	14.06	15.14	14.50	13.54	14.42	14.40
As_2O_5				3.24		
H_2O	21.41	15.37	20.30	15.36	14.31	14.62
Rem.			0.59		1.69	0.76
Total	100.00	100.00	100.15	100.98	100.87	99.01
G.					~3.9	3.68

1. $Cu(UO_2)_2(PO_4)_2 \cdot 12H_2O$. 2. $Cu(UO_2)_2(PO_4)_2 \cdot 8H_2O$. 3. Leupoldsdorf, Bavaria.[7] Rem. is SiO_2. 4. Cornwall (*metatorbernite?*).[8] 5. Katanga (*metatorbernite?*).[9] Rem. is $(Fe,Al)_2O_3$ 0.88, Te 0.63, insol. 0.18, MoO_3 tr. Mean of 8 partial analyses. 6. Tincroft mine, Redruth (*metatorbernite?*).[10] Rem. is $(Fe,Al)_2O_3$ 0.61, insol. 0.15. Mean of 6 partial analyses.

Tests. B.B. easily fusible to a black mass. Soluble in acids.

O c c u r. Torbernite is a secondary mineral found associated with autunite and other secondary uranium minerals as an oxidation product of uraninite, especially in the gossan of veins carrying both copper sulfides and uraninite (Cornwall). The distinction between torbernite and metatorbernite has not been made for most of the reported localities. Among the more important localities for its occurrence are Joachimsthal, Zinnwald, and Schlaggenwald in Bohemia, and at Johanngeorgenstadt and Schneeberg in Saxony, Germany. Found widely and sometimes as very fine specimens in Cornwall, England, notably at mines in the neighborhood of Redruth, St. Austel, St. Just, and St. Agnes. At numerous minor localities in France. In Spain in Cacéres province, Estramadura, and in northwestern Portugal in the neighborhood of Vizeu and Sabugal. Found abundantly at Kasolo, Chinkolobwe, and elsewhere in the Katanga district, Belgian Congo, Africa. At Mt. Painter in the Flinders Range, South Australia. In the United States found sparingly at a small number of places. As an alteration of uraninite in pegmatite at Haddam Neck, Connecticut, and in a few other pegmatites in the New England states; autunite is much more common in these occurrences. In South Dakota at the Hannibal mine near Terry, Lawrence County, and near Keystone in Pennington County. In Utah in the La Sal Mountains in San Juan County, and probably widespread in trace amounts in the carnotite-sandstones of the plateau region of Colorado and Utah. Occurs in New Mexico in the White Signal district in Grant County and the San Lorenzo district in Socorro County.

N a m e. After Torbern Bergmann (1735–1784), Swedish chemist.

Ref.
1. Angles of Hallimond, *Min. Mag.*, **17**, 326 (1916), unit and orientation of Hallimond (1916), Schrauf, *Min. Mitt.*, 181 (1872), Dana (856, 1892). Transformation: Goldschmidt (5, 101, 1918) to present cell, 100/010/002. The numerous reported measurements (see Hallimond) are not in good agreement, owing to the poor quality of the crystals; some of the earlier observations may have been made on zeunerite, metatorbernite, or other related species. Morphologically torbernite is tetragonal holohedral, but the mineral has been considered to be monoclinic or orthorhombic on the basis of the frequently observed optically biaxial character, from an apparent inequality in ease of cleavage on vertical planes, and from monosymmetric etch pits produced on (001)—see Walker, *Am. J. Sc.*, **6**, 41 (1898). The biaxial character, however, apparently is a strain effect due to variation in water content or in composition and most material is uniaxial. The etch symmetry sometimes is tetragonal and depends on the etching conditions—see Rinne, *Cbl. Min.*, 618 (1901).
2. Hallimond (1916); other forms have been reported but are of doubtful validity—see Goldschmidt (1918).
3. Hallimond, *Min. Mag.*, **19**, 43 (1920); earlier reported values for torbernite, in part higher, doubtless included metatorbernite.
4. Cited by Larsen (195, 1921). Millosevich, *Acc. Linc., Rend.*, **21**, 594 (1912), gave nO 1.590, nE 1.581 (Na) for Gunnis Lake mine, Cornwall, crystals.
5. See Hallimond, *Min. Mag.*, **17**, 326 (1916), **19**, 43 (1920); Rinne, *Cbl. Min.*, 618 (1901); Buchholz, *Cbl. Min.*, 362 (1903).
6. Additional analyses are cited by Hintze (1 [4B], 993, 1931). It is uncertain whether some of these analyses belong to torbernite or to metatorbernite for lack of distinguishing data.
7. Henrich, *J. pr. Chem.*, **96**, 73 (1917).
8. Winkler, *J. pr. Chem.*, **7**, 10 (1873).
9. Steinkuhler, *Bull. soc. chim. Belgique*, **32**, 270 (1923).
10. Steinkuhler, *Bull. soc. chim. Belgique*, **32**, 253 (1923).
11. See Meixner, *Chem. Erde*, **12**, 433 (1940).
12. Forjaz, *C.R.*, **164**, 102 (1917).
13. Goldsztaub, *Bull. soc. min.*, **55**, 7 (1932), on crystals from Lachaux, Puy-de-Dôme, France.

FRITZSCHEITE. Breithaupt (*B. H. Ztg.*, **24**, 302, 1865).

Supposed from qualitative chemical tests to be a manganese uranium phosphate belonging in the torbernite group. Found as rectangular (or nearly so) plates with a perfect basal cleavage and a vitreous to pearly luster. H. 2–3. G. 3.50 (?). Color reddish brown to hyacinth-red. Transparent. Said to occur at Autun, France, at Neuhammer near Neudeck in Bohemia, and at Johanngeorgenstadt and Steinig in Saxony, Germany.

42.8.13.2 **A U T U N I T E** [$Ca(UO_2)_2(PO_4)_2 \cdot 10$–$12H_2O$]. A variety of Uranglimmer, Urankalk, or Chalcolite [Torbernite] prior to 1819. Sel à base de chaux, où l'oxide d'urane joue le rôle d'acide *Berzelius* (295, 1819). Uranit *Berzelius* (**4**, 46, 1823). Kalk-Uranit, Kalk-Uranglimmer *Germ.* Lime-Uranite. Autunite *Brooke* and *Miller* 519, 1852). Calcouranit *Breithaupt* (*B. H. Ztg.*, **24**, 302, 1865).

C r y s t.[1] Tetragonal; ditetragonal-dipyramidal—$4/m\ 2/m\ 2/m$.

$$a:c = 1:2.952; \quad p_0:r_0 = 2.952:1$$

Forms:[2]

	φ	ρ	A	M
c 001	....	0°00′	90°00′	90°00′
a 010	0°00′	90 00	90 00	45 00
m 110	45 00	90 00	45 00	90 00
l 120	26 34	90 00	63 26	71 34
d 011	0 00	71 17	90 00	47 57½
p 111	45 00	76 32	46 33½	90 00
q 112	26 34	64 24	50 23	90 00

AUTUNITE

Structure Cell.[3] Space group $I4/mmm$. a_0 6.989 kX, c_0 20.63; $a_0:c_0 =$ 1:2.952. Cell contents $Ca_2(UO_2)_4(PO_4)_4 \cdot 20-24H_2O$.

Habit. Thin tabular {001}, closely approaching torbernite in form and angles. The crystals usually deviate slightly in angle from tetragonal symmetry apparently due to malformation. Subparallel growth common. Also as foliated or scaly aggregates; as thick crusts with a serrated surface, composed of crystals standing on edge.

Twinning. On {110}.

Oriented growths. With torbernite, common.

P h y s. Cleavage {001} perfect, {100} indistinct. Not brittle. H. $2-2\frac{1}{2}$. G. 3.1–3.2, varying with the water content (?); 3.14 (calc. for $10\frac{1}{2}H_2O$). Luster vitreous, pearly on {001}. Color lemon-yellow to sulfur-yellow, sometimes greenish yellow to pale green. Streak yellowish. Transparent to translucent. Strongly fluorescent in yellowish green in ultra-violet light;[12] the dehydration product meta-autunite I is less strongly fluorescent.

O p t.

	ORIENTATION	n (Autun)[5]	n (Maryland)[6]	n[13]	DICHROISM
(E)	X c	1.553	1.555		Colorless to pale yellow
	Y	1.575	1.575	1.58–1.59	Yellow to dark yellow
(O)	Z	1.577	1.578	1.59–1.60	Yellow to dark yellow

Natural autunite sometimes is uniaxial negative but usually is of an anomalous biaxial negative character. The biaxiality is dependent on the water content of the crystals,[4] and $2V$ decreases with decreasing water content. In the biaxial material, $2V$ ranges from $0°$ to about $53°$ and usually is $10°–30°$ with $r > v$, strong; $Z = c$ and Y is parallel to one of the diagonals of the square tablets. The tetragonal dehydration product meta-autunite I also shows a sectoral, biaxial character with $2V$ decreasing to $0°$ with decreasing water content; $Z = c$ and Y is perpendicular to the edges of the tablets. The indices of refraction in both hydrates increase with decreasing water content. Some of the reported descriptions of autunite probably refer to meta-autunite I.

C h e m. A hydrated phosphate of calcium and uranium, $Ca(UO_2)_2(PO_4)_2 \cdot 10-12H_2O$. Small amounts of Ba and Mg occur in substitution for Ca (analyses 3, 6). Cu apparently does not substitute for Ca to give a series with torbernite. V, Pb, and other elements have been reported in spectrographic amounts.[14] The water content apparently varies between 10 and $12H_2O$ in autunite. On drying or slight heating autunite passes reversibly to meta-autunite I (a tetragonal phase with $6\frac{1}{2}-2\frac{1}{2}H_2O$ and a_0 6.98 kX, c_0 8.42) and this on heating to $\sim 80°$ passes irreversibly to meta-autunite II (an orthorhombic phase with $0-6H_2O$ and a_0 6.45 kX, b_0 6.97, c_0 8.65).[3] Neither the meta-I nor the meta-II hydrate occurs as a primary deposit in nature, although most museum specimens of autunite if exposed to a warm, dry atmosphere rapidly alter to meta-autunite I (*metakalkuranite*, Germ.).

Autunite readily undergoes base exchange; H substitutes for Ca in acid solutions to give a non-isostructural compound, and Na can substitute for Ca in strong brine solutions to give a non-isostructural sodium-autunite.[3]

Anal.[11]

	1	2	3	4	5	6
CaO	5.69	5.31	6.09	6.01	6.06	6.56
UO_3	58.00	60.84	61.70	60.62	60.95	58.85
P_2O_5	14.39	13.40	15.21	14.52	14.61	14.80
H_2O	21.92	20.33	14.90	18.47	18.32	19.60
Rem.			1.57			0.26
Total	100.00	99.88	99.47	99.62	99.94	100.07
G.						3.198

1. $Ca(UO_2)_2(PO_4)_2 \cdot 12H_2O$. 2. Autun, France.[7] 3. Autun, France.[8] Rem. is BaO. Meta-autunite? 4. Leupoldsdorf, Bavaria.[9] 9. Beira Alta province, Portugal.[9] 6. Mt. Painter, South Australia.[10] Rem. is MgO.

Tests. B.B. fusible. Soluble in acids.

Occur. Autunite is a secondary mineral found chiefly in the zone of oxidation and weathering of hydrothermal veins and pegmatites, and is derived by the alteration of uraninite or other uranium-containing minerals. Among the more notable occurrences are Saint-Symphorien and other localities near Autun in Saône-et-Loire, France. In Portugal at Sabugal southeast of Guarda and at Vizeu. In Germany in the Johanngeorgenstadt district and at Falkenstein in Saxony and in the pegmatites of the Hagendorf and Pleystein area in Bavaria. In Cornwall, England, at Redruth, St. Austel, and other localities in the oxidized zone of the uraninite-bearing veins. In the Katanga district, Belgian Congo, Africa. At Mt. Painter in the Flinders Range, South Australia. In a peaty alluvium south of Antsirabe, Madagascar. In the United States found at a large number of minor localities. In New England found in pegmatite at numerous points, especially the Ruggles mine near Grafton Center, New Hampshire. In New York in pegmatite at Bedford, Westchester County. In the pegmatites near Keystone in the Black Hills, Utah. In North Carolina in pegmatite at Spruce Pine, Penland, the Flat Rock mine, and elsewhere in Mitchell County. From the White Signal district, Grant County, New Mexico.

Artif.[3] By reaction of solutions of monocalcium phosphate and uranyl nitrate at low temperatures.

Alter. To phosphuranylite.

Ref.

1. Tetragonal symmetry established by the x-ray study of Beintema, *Rec. trav. chim. Pays-Bas*, **57**, 155 (1938), on Limoges crystals and artificial material. Elements of the x-ray cell of Beintema. Autunite has been considered to be orthorhombic or monoclinic with marked tetragonal pseudosymmetry on the basis of morphological measurements (on crystals of very poor quality) and the biaxial character—see especially Des Cloizeaux, *Ann. mines*, **11**, 261 (1854), **14**, 339 (1858), who gave $a:b:c = 0.9875:1:2.8517$; Brezina, *Zs. Kr.*, **3**, 273 (1879), who gave $a:b:c = 0.3463:1:0.3525$ and $\beta\ 90°30'$; and Hallimond, *Min. Mag.*, **17**, 221 (1916).
2. Goldschmidt (5, 5, 1918) and Dana (857, 1892), in the tetragonal interpretation.
3. Beintema (1938).
4. Beintema (1938), who also gives optical data on meta-autunite I and meta-autunite II (not known in nature).
5. Michel-Lévy and Lacroix, *Les Minéraux des Roches*, 157 (1888).
6. Shannon, *Am. Min.*, **11**, 35 (1926).

7. Church, *J. Chem. Soc. London*, **28**, 109 (1875).
8. Berzelius, *Ann. Phys.*, **1**, 379 (1844).
9. Ewald cited by Henrich, *Ber.*, **55**, 1212 (1922).
10. Greig cited by Smith, *Rec. Australian Mus.*, **15**, 69 (1926).
11. For additional analyses see Hintze (1 [4B], 982, 1931).
12. See Iimori and Iwase, *Sc. Pap. Inst. Phys. Chem. Res. Tokyo*, **34**, 372 (1938), and Meixner, *Chem. Erde*, **12**, 433 (1940).
13. Meixner (1940).
14. Forjaz, *C.R.*, **164**, 102 (1917).

42.8.13.3 **URANOCIRCITE** [$Ba(UO_2)_2(PO_4)_2 \cdot 8H_2O$]. *Weisbach (Berg.-u. hütten. Jb., Abh.*, 48, 1877; *Jb. Min.*, 185, 404, 1877). [Earlier confused with autunite.]

Tetragonal? In crystals resembling autunite. Cleavage {001} perfect, {100} and {010} distinct. H. 2-2½. G. 3.53. Luster pearly on {001}. Color yellow-green. Transparent to translucent. Found in parallel growths with autunite and torbernite; also with barium chloride (artificially).[1]

O p t. In transmitted light, pale canary-yellow. Faintly pleochroic.

ORIENTATION	n (Falkenstein)[2]	n (artif.)[3]	PLEOCHROISM	
X c	1.610 ± 0.003	1.604	Nearly colorless	Biaxial negative (−).
Y	1.623 ± 0.003		Pale canary-yellow	Shows sets of twin lamellae parallel {100} and {010}.
Z	1.623 ± 0.003	1.613	Pale canary-yellow	
2V	Small	0°		

On heating, the biaxial material becomes uniaxial at 100°–150° accompanying loss of water to the dihydrate.

C h e m. A hydrated phosphate of barium and uranium, $Ba(UO_2)_2(PO_4)_2 \cdot 8H_2O$. $6H_2O$ is lost over H_2SO_4 or by heating below 100°.[4] Analysis gave:

	BaO	UO_3	P_2O_5	H_2O	Total
1.	15.16	56.56	14.04	14.24	100.00
2.	14.57	56.86	15.06	13.99	100.48

1. $Ba(UO_2)_2(PO_4)_2 \cdot 8H_2O$. 2. Falkenstein.[5]

Tests. Soluble in acids. B.B. fusible.

O c c u r. Originally from Bergen near Falkenstein in Voigtland, Saxony, as a secondary mineral in quartz veins; also from Schneeberg in Saxony. With parsonsite and other secondary uranium minerals in the fluorite deposit at Wölsendorf, Bavaria. Reported from Srdnia Gora in the Banat, Hungary, and from Rosmaneira, Spain.

A r t i f.[6] Crystals containing $10H_2O$ and isostructural with autunite have been prepared by reaction of phosphoric acid with a mixed solution of barium chloride and uranyl nitrate. On heating, a lower hydrate with $2-6H_2O$ and isostructural with meta-autunite is formed; a_0 6.952 kX, c_0 8.515.

N a m e. From *uranium*, and κίρκος, *a falcon*, because it is an uranium mineral from Falkenstein.

Ref.
1. Gaubert, *Bull. soc. min.*, **27,** 222 (1904).
2. Larsen (149, 1921).
3. Fairchild, *Am. Min.*, **14,** 265 (1929), on material made from autunite by base exchange.
4. For dehydration data see Church, *Min. Mag.*, **1,** 234 (1877), and Gaubert, *Bull. soc. min.*, **27,** 222 (1904).
5. Winkler analysis in Weisbach (1877). See also Church (1877) and Fairchild (1929) for partial analyses.
6. Beintema, *Rec. trav. chim. Pays-Bas*, **57,** 155 (1938).

42.8.13.4 **S A L É E I T E** [$Mg(UO_2)_2(PO_4)_2 \cdot 10H_2O$]. *Thoreau* and *Vaes* (*Soc. géol. Belgique, Bull.*, **42,** 96, 1932). Saléite.

C r y s t.[1] Tetragonal; ditetragonal-dipyramidal—$4/m\,2/m\,2/m$ (?).

$$a:c = 1:2.839; \qquad p_0:r_0 = 2.839:1$$

Forms:[2]

	ϕ	ρ	A	M
c 001	.....	0°00'	90°00'	90°00'
a 010	0°00'	90 00	90 00	45 00
d 120	26 34	90 00	63 26	71 34
e 012	0 00	54 50	90 00	54 41

Structure cell.[2] Space group $I4/mmm$ if holohedral. a_0 6.980 A, c_0 19.813; $a_0:c_0 = 1:2.839$. Cell contents $Mg_2(UO_2)_4(PO_4)_4 \cdot 20H_2O$.
Habit. Tablets flattened on {001}.

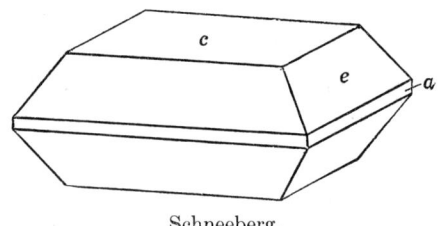

Schneeberg.

Oriented growths. Upon torbernite, with parallel axes.
P h y s. Cleavage {001} perfect; also on {010} and {110}, indistinct. H. 2–3. G. 3.27; 3.27 (calc.; both values on Schneeberg material of analysis 4). Color yellow to lemon-yellow. Transparent; translucent when partly dehydrated.
O p t.

ORIENTATION	n (Schneeberg [3])		n (Katanga [4])	
E (or X)	1.559 ± 0.002	(colorless)	1.559	Uniaxial negative (−),
Y			1.570	sometimes anomalously biaxial.
O (or Z)	1.574 ± 0.002	(pale greenish yellow)	1.574	

The Schneeberg material (with 10 H_2O) is uniaxial. The Katanga crystals are biaxial with $2V$ 61° and $r > v$ strong and show a division into two or four sectors with $Z = c$ and Y in each sector perpendicular to the edge of the plate.

ZEUNERITE 989

Chem. A hydrated phosphate of magnesium and uranium, $Mg(UO_2)_2(PO_4)_2 \cdot 10H_2O$. In the material of analysis 4, As substitutes for P with $As:P = 1:4.5$. The Schneeberg mineral with $10H_2O$ is fully hydrated and is isostructural[3] with autunite, $Ca(UO_2)_2(PO_4)_2 \cdot 10H_2O$. The Katanga material, with $8H_2O$ and biaxial, sectoral crystals, probably is isostructural with the meta-I hydrate of autunite and should then be separated as a species under the name meta-saléeite.

Anal.

	1	2	3	4
MgO	4.49	4.31	5.01	4.31
UO_3	63.67	61.23	64.07	60.32
P_2O_5	15.80	15.19	14.44	11.98
As_2O_5				4.50
H_2O	16.04	19.27	16.48	[18.89]
Total	100.00	100.00	[100.00]	[100.00]
G.			<3.3	3.27

1. $Mg(UO_2)_2(PO_4)_2 \cdot 8H_2O$ 2. $Mg(UO_2)_2(PO_4)_2 \cdot 10H_2O$. 3. Katanga.[5] Recalculated to 100 after deduction of 2.79 per cent insol. 4. Schneeberg, Saxony.[6]

Occur. A secondary mineral, originally found with torbernite and dewindtite at Chinkolobwe, Katanga, in the Belgian Congo. Also from Schneeberg, Saxony, with uranophane and zeunerite.

Name. After A. Salée.

Ref.

1. Tetragonal symmetry established by Mrose, *Am. Min.*, **35**, 525 (1950). Earlier classed as orthorhombic, pseudo-tetragonal by Thoreau and Vaes (1932) and as monoclinic by Schoep, *Med. Kl. Wetens. Kon. Vlaam. Ac.*, 65 (1939). Ratio of the structure cell.
2. Mrose (1950) on Schneeberg crystals.
3. Mrose (1950).
4. Schoep (1939).
5. Mollet analysis in Thoreau and Vaes (1932).
6. Gonyer analysis in Mrose (1950).

42.8.13.5 **Zeunerite** $[Cu(UO_2)_2(AsO_4)_2 \cdot 10\text{-}16H_2O]$.

All specimens of natural so-called zeunerite that have been examined by x-ray means have proved to be the meta-I hydrate with 5 to 8 H_2O and to be isostructural with meta-autunite and metatorbernite. Fully hydrated material with 10–16 H_2O and isostructural with autunite and torbernite has not yet been shown to occur in nature. In order to maintain parallelism of nomenclature, the name metazeunerite is here given to the species and the name zeunerite proper is held in reserve status for natural occurrences of the 10–16 H_2O hydrate that later may be demonstrated.

Artificial[1] $Cu(UO_2)_2(AsO_4)_2 \cdot 10\text{-}16H_2O$ is tetragonal. Optically uniaxial negative $(-)$ with nO 1.602 to 1.610. The water content varies zeolitically within the range of stability of the phase, and the indices of refraction vary therewith. Heated to 65° C. in air, this phase breaks down to metazeunerite with a water content varying zeolitically between 5 and 8 H_2O.

Ref.
1. Weiss-Frondel, *Am. Min.*, **36**, 249 (1951).

URANOSPATHITE. *Hallimond (Min. Mag.*, **17**, 230, 1915). Uranspat *Germ.*
Apparently orthorhombic, pseudo-tetragonal. As fan-like groups of rectangular plates and laths flattened {001} and striated parallel to the elongation [010]. The corners of the plates are often modified equally by edges approximately at 45°. Cleavage {001} perfect, {100} good, {010} (?) fibrous. G. 2.50. Color yellow to pale green. Sometimes twinned on {110}, in cruciform groups.

Opt.

ORIENTATION		n(Na)	PLEOCHROISM	
X	c	[~1.49]	Pale yellow	Biaxial negative (−).
Y	a	1.510	Deep yellow	$2V$ 69°(Na).
Z	b	1.521	Deep yellow	

A hydrated uranyl compound, perhaps an arsenate or phosphate related to the autunite group, but the full composition is not known. Loses H_2O in a desiccator at room temperature and becomes uniaxial. Soluble in acids. In oriented growths with bassetite, the latter having {010} [001] parallel uranospathite {001} [010]. Found at Redruth, Cornwall, England. Named from *uranium* and σπάθη, *a broad blade*, in allusion to the crystal habit.

42.8.13.6   **U R A N O S P I N I T E**   [$Ca(UO_2)_2(AsO_4)_2 \cdot 8H_2O$]. *Weisbach (Jb. Min.*, 315, 1873).

C r y s t.[1] Tetragonal.

$$a:c = 1:1.4561; \quad p_0:r_0 = 1.4561:1$$

Forms:

	ϕ	ρ	A	$\overline{M}$
c 001	....	0°00′	90°00′	90°00′
q 015	0°00′	16 14	90 00	78 36
d 011	0 00	55 31	90 00	54 21
r 021	0 00	71 03	90 00	48 01

Habit. Thin tabular {001}.

P h y s. Cleavage {001} perfect, {100} distinct. H. 2–3. G. 3.45. Color lemon-yellow to siskin-green. Luster pearly on {010}.

O p t.[2] Usually anomalously biaxial negative (−) but sometimes uniaxial negative; artificial material is uniaxial. In biaxial material $X = c$ with $r > v$, moderate. The crystals often are zoned with uniaxial and biaxial parts and may show a parallel core of zeunerite.

ORIENTATION	n (Schneeberg)	n (Schneeberg)	n (Schneeberg)	PLEOCHROISM
X or E	1.560	1.56	1.55	Nearly colorless
Y	1.582		1.567	Pale yellow
Z or O	1.587	1.586	1.572	Pale yellow
2V	46°	0°	62°	

C h e m. A hydrated arsenate of calcium and uranium, $Ca(UO_2)_2(AsO_4)_2 \cdot nH_2O$. The water content of the natural material of analysis 2 is ~$10H_2O$. The artificial material of analysis 3 has ~$8H_2O$. The water content (and optical properties) doubtless vary at ordinary temperatures with the humidity, as with torbernite.

Anal.

	1	2	3
CaO	5.40	5.47	5.62
UO_3	55.11	59.18	59.01
As_2O_5	22.14	19.37	23.01
H_2O	17.35	16.19	14.27
Total	100.00	100.21	101.91

1. $Ca(UO_2)_2(AsO_4)_2 \cdot 10H_2O$.   2. Schneeberg.[3]   3. Artificial.[3]

Occur. A secondary mineral found associated with zeunerite, uranocircite, troegerite, and walpurgite in the Weisser Hirsch mine at Neustädtl near Schneeberg, Saxony, Germany. Reported [4] from near Pahreah, Kane County, Utah.

Artif.[3] Obtained in crystals by adding a solution of uranyl nitrate to a solution of lime in excess arsenic acid.

Name. From *uranium* and σπίνος, *siskin*, in allusion to its composition and color.

Ref.
1. Elements, forms and symmetry of Goldschmidt (**9**, 41, 1923). Originally considered to be orthorhombic pseudo-tetragonal by Weisbach (1873) on the ground of the optically biaxial character.
2. Larsen (151, 1921); Goldschmidt, *Zs. Kr.*, **31**, 468 (1899).
3. Winkler, *J. pr. Chem.*, **7**, 11 (1873).
4. Butler, *U. S. Geol. Sur., Prof. Paper 111*, 115 (1920).

42.8.14 METATORBERNITE GROUP

42.8.14.1 METATORBERNITE $[Cu(UO_2)_2(PO_4)_2 \cdot 8H_2O]$. Metakupferuranit *Rinne* (*Cbl. Min.*, 618, 1901). Metatorbernite *Hallimond* (*Min. Mag.*, **17**, 326, 1916). Meta-chalcolite. Meta-torbernite I.

Cryst. Tetragonal;[1] ditetragonal-dipyramidal—$4/m\ 2/m\ 2/m$.

$$a{:}c = 1{:}1.14; \quad p_0{:}r_0 = 1.14{:}1$$

Forms:

		ϕ	ρ	A	M
c	001	....	0°00′	90°00′	90°00′
d	011	0°00′	48 44½	90 00	57 53½

Structure cell.[2] Space group $P4/nmm$. a_0 6.95 ± 0.03 kX, c_0 8.60 ± 0.04; $a_0{:}c_0 = 1{:}1.238$. Cell contents $Cu(UO_2)_2(PO_4)_2 \cdot 8H_2O$.

Habit. Tablets on {001}; often in rosettes or sheaf-like aggregates of irregularly curved and composite crystals.

Phys. Cleavage {001} perfect. Rather brittle. H. $2\frac{1}{2}$. G. 3.5–3.7, 3.70 (Gunnis Lake);[3] 3.76 (calc.). Luster vitreous to sub-adamantine, pearly on {001}. Color pale green to dark green. Transparent to translucent. Not fluorescent.[4]

Opt.[6]

λ	640mμ	515mμ	440mμ	White light [7]
nO	1.618 }	1.633	1.649	1.610–1.628
nE	1.622 }		1.646	
Sign	+	Isotropic	−	+

Optically uniaxial; sometimes biaxial due to strain with small $2V$. Pleochroic, with O sky-blue and E green. Metatorbernite and to a less extent torbernite show anomalous interference colors.

Chem. A hydrated phosphate of copper and uranium, $Cu(UO_2)_2$-$(PO_4)_2 \cdot 8H_2O$. Metatorbernite is stable from solution over about 75° (see torbernite). On heating,[8] $4H_2O$ are lost at about 100° forming orthorhombic metatorbernite II, and at about 125° an orthorhombic monohydrate is formed which is stable to a higher temperature. It is not certain if the following analyses refer to metatorbernite or torbernite:

	CuO	UO_3	P_2O_5	As_2O_5	H_2O	SiO_2	Total	G.
1.	8.48	61.01	15.14		15.37		100.00	
2.	11.39	58.77	13.43	2.83	13.55		99.97	3.7
3.	8.50	59.67	14.00		15.00	0.40	97.57	

1. $Cu(UO_2)_2(PO_4)_2 \cdot 8H_2O$. 2. Lurisia, Piedmont.[9] 3. Gunnis Lake mine, Cornwall.[10]

Tests. B.B. easily fusible. Soluble in acids.

Occur. A secondary or perhaps a low temperature (> 75°) hydrothermal mineral. Considered to be represented as a product of direct crystallization in nature (by reason of the absence of cracks and turbidity found in artificial material made by dehydration of torbernite) by the crystals from the Gunnis Lake mine, Calstock, Cornwall, England. Material identified as metatorbernite from the water content, optical properties, G., or x-ray measurements has been reported from numerous localities, but it is not certain if this has formed directly or is a product of alteration of torbernite. Among these localities may be mentioned Lurisia in Piedmont, Italy; Wheal Basset in Cornwall; the Katanga district in the Belgian Congo; Temple Mountain[5] in the San Rafael Swell, Utah, and Spruce Pine in North Carolina in the United States; and Wilberforce, Ontario, Canada.

Artif.[11] Metatorbernite (perhaps torbernite) has been obtained by treating artificial sodium autunite in a hot solution of copper chloride.

Ref.

1. Angles and forms of Hallimond, *Min. Mag.*, **19**, 43 (1920), on Gunnis Lake mine, Cornwall, crystals, unit of the structure cell. Transformation: Hallimond to new, 100/010/00½. Morphological data on material not certainly known to be metatorbernite are given by Pelloux, *Atti Soc. Ligustica Sc. Lett. Genova*, **13**, 137 (1934), Buttgenbach, *Soc. géol. Belgique, Ann.*, **47** (annex), 31C (1924).
2. Berry, priv. comm. (1946), on Cornwall material by Weissenberg method.
3. Hallimond (1920).
4. R. H. Lund, priv. comm. (1949), on Gunnis Lake crystals.
5. Hess, *U. S. Geol. Sur., Bull. 750D*, 70 (1924).
6. Dispersion data of Bowen, *Am. J. Sc.*, **48**, 195 (1919); indices estimated from curves.
7. Range of values reported by Pelloux (1934) and Stočes, *Rozpr. Česká Ak.*, **27**, no. 27 (1918) for metatorbernite (?).
8. See Rinne, *Cbl. Min.*, 618 (1901), and Buchholz, *Cbl. Min.*, 362 (1903).
9. Bruna analysis in Pelloux (1934).
10. Pisani, *C.R.*, **52**, 817 (1861).
11. Fairchild, *Am. Min.*, **14**, 265 (1929).

42.8.14.2 M E T A Z E U N E R I T E [$Cu(UO_2)_2(AsO_4)_2 \cdot 8H_2O$]. Zeunerit *Weisbach* (*Jb. Min.*, 207, 1872; 315, 1873; *Berg.-u. hütten.*, *Jb.*, **42**, 53, 1877). Kupferuranit, Kupfer-Uranglimmer, pt.

C r y s t.[1] Tetragonal; ditetragonal-dipyramidal—$4/m\ 2/m\ 2/m$.

$$a:c = 1:1.250; \quad p_0:r_0 = 1.250:1$$

Forms:[2]

	ϕ	ρ	A	M
c 001	....	0°00′	90°00′	90°00′
a 010	0°00′	90 00	90 00	45 00
n 013	0 00	22 37	90 00	74 13
f 043	0 00	59 02	90 00	52 40½
P 021	0 00	68 12	90 00	48 58
i 041	0 00	78 41½	90 00	46 06

Doubtful:

| m 110 | g 012 | s 023 | y 011 | v 111 |

Structure cell.[3] Space group $P4/nmm$. a_0 7.13 Å, c_0 8.83; $a_0:c_0 = 1:1.238$. Cell contents $Cu(UO_2)_2(AsO_4)_2 \cdot 8H_2O$.

Habit. Tabular {001} and resembling torbernite; also acute pyramidal. Forms other than {001} ordinarily are rough and striated horizontally.

Oriented growths. Upon troegerite[4] and uranospinite,[5] with parallel axes.

P h y s. Cleavage {001} perfect, {100} distinct. Fracture uneven. Brittle. H. 2–2½. G. 3.64;[5] 3.79 (calc.). Luster vitreous, pearly on {001}. Color grass-green to emerald-green. Fluoresces yellow-green in both long- and short-wave ultraviolet light.

O p t. In transmitted light, green in color and weakly dichroic. Uniaxial negative (−).

	n (Schneeberg)[6]	n (Tintic)[7]	n (Schneeberg)[6]	DICHROISM
O	1.643	1.647	1.651	Grass-green
E	1.623	1.630	1.635	Pale green

The indices of refraction vary continuously with the content of zeolitic water within the limits of stability of the phase. Artificial material[7] with $8H_2O$ has nO 1.645–1.648 and with $5H_2O$ has nO 1.654.

C h e m. A hydrated arsenate of copper and uranium, $Cu(UO_2)_2(AsO_4)_2 \cdot 8H_2O$. ($PO_4$) doubtless can substitute for (AsO_4), but analytical evidence of this in natural material is lacking. The water content can vary zeolitically over a range from $8H_2O$ to $5H_2O$. On heating to 110°, metazeunerite breaks down to a hydrate with $2\frac{1}{2}H_2O$ (?) isostructural with artificial meta-autunite-II.

Anal.

	CuO	UO_3	As_2O_5	H_2O	Total
1.	7.76	55.78	22.41	14.05	100.00
2.	7.49	55.86	20.94	15.68	99.97

1. $Cu(UO_2)_2(AsO_4)_2 \cdot 8H_2O$. 2. Schneeberg.[8]

Tests. B.B. fusible. Easily soluble in acids.

Occur. A secondary mineral found associated with torbernite, troegerite, uranospinite, walpurgite, erythrite, mimetite, etc. Originally from the Weisser Hirsch mine near Schneeberg, Saxony. Also at Joachimsthal and Zinnwald, Bohemia, and at Wittichen in the Black Forest, Bavaria. With chalcophyllite, olivenite, and pharmacosiderite at Cap Garonne, Var, France. At Wheal Garland, Cornwall, England, and at Cala Mästra on the island of Monte Cristo, Italy. In the United States found with secondary copper arsenates in the Centennial Eureka mine, Tintic, Utah, and at Majuba Hill, Nevada.

Artif.[8] Obtained as tiny crystals by adding uranyl nitrate solution to a hot solution of precipitated copper carbonate in arsenic acid. When precipitated at room temperature or below, a hydrate is obtained which contains $\sim 16 H_2O$ when freshly prepared but which loses water zeolitically down to $\sim 10 H_2O$ on standing in dry air (see zeunerite). This phase breaks down to metazeunerite on heating to 65° in air.[7] Metazeunerite apparently does not re-convert to zeunerite when immersed in water.

Name. After Gustav A. Zeuner (1828– ?), civil engineer and Director of the School of Mines at Freiberg.

Ref.

1. Elements of Schrauf, *Min. Mitt.*, **2**, 181 (1872) on Schneeberg crystals. The crystals described by Ježek, *Rozpr. České Ak.*, **31**, Cl. 2, no. 15 (1922), have relatively low G. and indices of refraction and may be torbernite.
2. Goldschmidt (**9**, 106, 1923).
3. Weiss-Frondel, *Am. Min.*, **36**, 249 (1951), by Weissenberg method on unanalyzed Tintic crystals.
4. Goldschmidt, *Zs. Kr.*, **31**, 468 (1899).
5. Weiss-Frondel (1951); the value probably is low.
6. Larsen (151, 158, 1921).
7. Weiss-Frondel (1951) who summarizes the available data.
8. Winkler, *J. pr. Chem.*, **7**, 115 (1873).

42.8.15 **BASSETITE** *Hallimond* (*Min. Mag.*, **17**, 221, 1915).

Cryst. Monoclinic (pseudo-orthorhombic).

$a:b:c = 0.3472:1:0.3456;\quad \beta\ 90°43';\quad p_0:q_0:r_0 = 0.9951:0.3456:1$

$r_2:p_2:q_2 = 2.8935:2.8805:1;\quad \mu\ 89°17';\quad p_0'\ 0.9951,\ q_0'\ 0.3456,\ x_0'\ 0.0125$

Forms:

		ϕ	ρ	ϕ_2	$\rho_2 = B$	C	A
b	010	0°00′	90°00′	....	0°00′	90°00′	90°00′
n	120	55 13	90 00	0°00′	55 13	89 24½	34 47
m	110	70 51	90 00	0 00	70 51	89 19½	19 09
d	011	2 04½	19 04½	89 17	70 56	19 04	89 19½
f	$\bar{1}01$	−90 00	44 30	134 30	90 00	45 13	134 30
p	111	71 04	46 48½	44 47	76 19	46 48½	46 24
o	121	55 33	50 42½	44 47	64 02½	50 42	50 20½
r	$\bar{1}21$	−54 52½	50 13½	134 30	65 45½	50 14	128 57
s	$\bar{1}41$	−35 24½	59 28½	134 30	45 24	59 28½	119 56½

Habit. As fan-like or chessboard-like groupings of thin tablets with an almost rectangular outline and flattened on {010}.

Twinning. Often twinned, with two or more individuals arranged approximately at 90° with {010} in common. [001] of one individual is parallel to [100] of the other, with the contact surface (010) in some instances and (110) against (011) in others.

P h y s. Cleavage {010} perfect, also {100} and {001} distinct. G. 3.10. Color yellow. Transparent. Not fluorescent.

O p t.

Orientation		n(Na)	Pleochroism	
X	b	[∼1.56]	Pale yellow	Biaxial negative (−).
Y		1.574	Deep yellow	$2V$ ∼62°.
$Z \wedge c$	−4°	1.580	Deep yellow	

Between crossed nicols some plates are divided into four quadrants by rather broad dark bands, due to compensation between overlapping parts of adjacent twinned crystals. The bands are approximately parallel to the extinction directions and are not parallel to the edges of the plates.

C h e m. Analyses are lacking. Apparently a hydrated uranium phosphate containing divalent iron.[1] Loses H_2O in a desiccator at room temperature; the plates then break up into four sectors with the axial planes at right angles and the extinction angle increases to about 20°. Soluble in acids.

Found at the Basset group of mines, Redruth, Cornwall, England. In part as oriented growths with uranospathite and torbernite. Bassetite may be the iron analogue of autunite or meta-autunite.

Ref.

1. Said by Hallimond (1915) to be $Ca(UO_2)_2(PO_4)_2 \cdot 8(?)H_2O$, from a supposed identity with material earlier analyzed by Church, *J. Chem. Soc. London*, **12**, 109 (1875). Microchemical tests by Meixner, *Chem. Erde*, **12**, 433 (1940), *Naturwiss.*, **27**, 454 (1939), on bassetite (?) indicate the absence of Ca and the presence of Fe.

TYPE 9. $(AB)_m(XO_4)_pZ_q \cdot xH_2O$, WITH $m:p < 3:2$

42.9.1 P H A R M A C O S I D E R I T E [$Fe_3(AsO_4)_2(OH)_3 \cdot 5H_2O$]. Fer minéralisé par l'acide arsenique ? *Proust* (*Ann. Chem.*, **1**, 195, 1790). Arsenicated Iron Ore *Kirwan* (**2**, 189, 1796). Olivenerz, Arseniksaures Eisen im Würfeln kryst. [from Carharrack] *Klaproth* (*Ges. nat. Freunde Berlin, Schr.*, **1**, 161, 1786; *Beitr.*, **3**, 194, 1802). Würfelerz, var. of Olivenerz *Lenz* (**2**, 18, 151, 1794). Würfelerz *Karsten* (66, 1808). Cube Ore. Pharmakosiderit *Hausmann* (1065, 1813).

C r y s t. Isometric; hextetrahedral—$\bar{4}\,3\,m$.

Forms:[1]

	a 001	d 011	o 111	$-o\,\bar{1}11$	u 122

Structure cell.[2] Space group $P\bar{4}3m$. a_0 7.94 kX. Cell contents $1\frac{1}{2}[Fe_3(AsO_4)_2(OH)_3 \cdot 5H_2O]$.

Habit. Commonly in cubes with faces striated diagonally or replaced by a vicinal trapezohedron near {1.1.40}; also tetrahedral; rarely granular or earthy.

Phys. Cleavage {001} imperfect to good. Fracture uneven. Rather sectile. H. $2\frac{1}{2}$. G. 2.797 (Cornwall); [10] 2.90 (calc.). Luster adamantine to greasy. Color olive-green, passing into honey-yellow, yellowish brown, and dark brown; also hyacinth-red, brownish red, grass-green, or emerald-green. Transparent to translucent. Weakly piezoelectric [9] and pyroelectric.

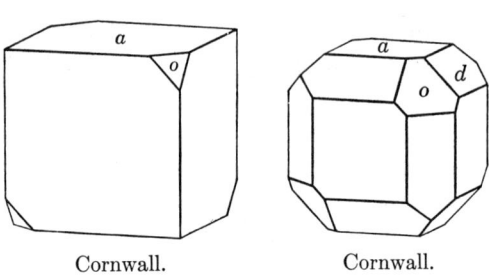

Cornwall. Cornwall.

Opt. In transmitted light, green, yellow-brown, etc. Apparently always shows optical anomalies [4] of tensional origin, the crystals composed of six biaxial sectors based on the faces of {001}. The optical orientation of the sectors varies in different specimens, apparently as a function of composition, and also with temperature. In (001) sections, crystals from Vaulry have the optic plane parallel [001] in the border zones, and parallel [011] in the central sector with the acute bisectrix X perpendicular (001). Lamellar twinning may be present parallel to the sides of the border sectors. Variously biaxial positive (+) with $r < v$ strong or biaxial negative (−) with $r > v$ strong; $2V$ usually large. Birefringence very weak. The indices of refraction vary with the color and ultimately with the chemical composition:

	Saubach, pale yellow [5]	Saubach, greenish [5]	Cornwall [6]	Cornwall [7]
n or nY (Na)	1.687	1.704	1.693	1.676

Chem. A hydrated basic ferric arsenate, $Fe_3(AsO_4)_2(OH)_3 \cdot 5H_2O$. Small amounts of P may substitute for As, with P:As = 1:11 in analysis 2. On heating,[5] $5H_2O$ of apparently zeolitic nature is lost up to 233°; the water of constitution is lost completely at 390°. The crystals take up K and (NH_4) ions from solution and turn brown or reddish brown; these bases may be exchanged by subsequent treatment with acids, which restores the original green color.[5] The small and variable amount of K in natural crystals is of this nature.[3]

Anal.

	1	2	3
K_2O		[1.51]	4.54
Fe_2O_3	40.84	39.29	37.58
As_2O_5	39.19	37.53	37.16
P_2O_5		2.04	1.20
H_2O	19.97	19.63	18.85
Total	100.00	[100.00]	99.33
G.		2.798	

1. $Fe_3(AsO_4)_2(OH)_3 \cdot 5H_2O$. 2, 3. Cornwall.[8]

Tests. B.B. easily fusible with sprouting to a magnetic bead. In C.T. loses neutral water. Soluble in HCl. Crystals placed in ammonia turn red, and on re-immersion in dilute HCl the original green color is restored.

O c c u r. Found both as a hydrothermal deposit (Saubach),[5] and, more commonly, as an oxidation product of arsenopyrite and other arsenic-rich minerals. In the latter occurrences, pharmacosiderite is associated with scorodite, erythrite, beudantite, symplesite, arseniosiderite, pitticite, limonite. A few of the many known localities follow. In Germany at Bösenbrunn, Pöhla, and near Schwarzenberg in Saxony; at Kahl and Aschaffenberg in the Spessart, Bavaria; at Saubach in Voigtland with scorodite in altered quartz porphyry; at Bulach in Württemberg. From Königsberg southwest of Schemnitz in Czechoslovakia. In France at Vaulry and at Puy-les-Vignes near Saint-Léonard, Haute-Vienne. Found at numerous localities in Cornwall, England, including mines at St. Day, Liskeard, Redruth, and Calstock. At Capo Bianco, Elba, and Mouzaïa in Algiers. In the United States, found with scorodite and various copper arsenates derived from enargite in the Tintic district, Utah.

A l t e r. To limonite and to wad; also as pseudomorphs after siderite.

A r t i f.[2] The aluminum analogue of pharmacosiderite (*alumopharmacosiderite*) has been obtained by reaction of aluminum sulfate and potassium monohydrogen arsenate solutions in a closed tube at 250°.

N a m e. From φάρμακον, *poison* (in allusion to the arsenic present), and σίδηρος, *iron*.

Ref.

1. Goldschmidt (**6**, 137, 1920).
2. Hägele and Machatschki, *Fortschr. Min.*, **21**, 77 (1937).
3. See also Zemann, *Min. Mitt.*, **1**, ser. 3, 1 (1948), who proposes the formula $KFe_4(OH)_4(AsO_4)_3 \cdot 6\text{-}7H_2O$.
4. See Lacroix (**4**, 520, 1910); Hocart, *Bull. soc. min.*, **57**, 5 (1934); Brauns (349, 1891); Larsen (118, 1921); and especially Heide, *Zs. Kr.*, **67**, 60 (1928).
5. Heide (1928).
6. Larsen (1921).
7. Gaubert, *Bull. soc. min.*, **30**, 108 (1907).
8. Hartley, *Min. Mag.*, **12**, 152 (1899).
9. Frondel, priv. comm. (1949), by oscilloscope test.
10. Average of closely agreeing values of Hartley (1899) and Mrose, priv. comm. (1949).

42.9.2 **C A C O X E N I T E** [$Fe_4(PO_4)_3(OH)_3 \cdot 12H_2O$]. Kakoxen Steinmann (*Vortr. Böhm. Ges.*, Prague, 1825). Cacoxene.

C r y s t.[1] Hexagonal. $a:c = 1:1.325$.

Forms:

c 0001 a 11$\bar{2}$0 o 11$\bar{2}$2 (?)

Structure cell.[2] Hexagonal. a_0 7.92 Å, c_0 10.5; $a_0:c_0 = 1:1.325$. Cell contents $Fe_4(PO_4)_3(OH)_3 \cdot 12H_2O$.

Habit. Tiny crystals acicular [0001], sometimes with a hexagonal cross section and indistinct pyramidal faces. As tufted or radial aggregates or fibrous coatings; spherulitic.

P h y s. Cleavage not observed. H. 3–4. G. 2.2–2.4, 2.26 (Hellertown);[3] 2.26 (calc.). Luster silky. Color yellow to brownish yellow, golden yellow, reddish yellow, rarely greenish.

O p t. In transmitted light, yellow.

	n (Hellertown)[3]	n (Trenič)[4]	n (Tonopah, Nev.)[4]	DICHROISM	
O	1.575	1.580	1.585	Pale yellow	Uniaxial positive (+).
E	1.635	1.640	1.656	Canary-yellow to orange-yellow	

C h e m. A hydrated basic phosphate of ferric iron, $Fe_4(PO_4)_3(OH)_3 \cdot 12H_2O$. All the reported analyses are not in good agreement with this formula. In some material small amounts of Al substitute for Fe'''. The green color sometimes observed suggests that the mineral may contain ferrous iron in its original state, analogous to dufrenite.

Anal.[8]

	1	2	3	4	5
Fe_2O_3	41.18	40.50	40.37	41.46	48.57
Al_2O_3			2.89		
P_2O_5	27.46	26.78	26.18	25.71	19.76
H_2O	31.36	32.43	30.59	32.81	31.80
Rem.			0.14		
Total	100.00	99.71	100.17	99.98	100.13
G.		2.26			

1. $Fe_4(PO_4)_3(OH)_3 \cdot 12H_2O$. 2. Hellertown, Pennsylvania.[3] 3. Eleonore mine, Giessen, Germany.[5] Rem. is insol. 4. St. Benigna, Bohemia.[6] 5. Eleonore mine, Giessen.[7]

Tests. B.B. turns brown and fuses to a black, magnetic bead. Easily soluble in acids.

O c c u r. A secondary mineral found with dufrenite, rockbridgeite, beraunite, strengite, wavellite, and limonite. In Germany at the Eleonore mine near Giessen and the Rothlaüfchen mine at Waldgirmes in the Rhine provinces; in Bavaria in limonite ore in the Amberg-Auerbach district, at Weilburg and as an alteration of iron-manganese phosphates in pegmatite at Hagendorf and Rabenstein. In Bohemia at the Hrbek mine at St. Benigna and at Trenič, Zbiroff, and Čerhovic. At Rochefort-en-Terre, Morbihan, France. In magnetite ore at Kirunavaara, Sweden. Found in the United States at Hellertown, Northampton County, Moore's Mill, Cumberland County, and Noble's mine, Lancaster County, in Pennsylvania. On earthy hematite at Antwerp, New York. At Tonopah, Nevada.

N a m e. From κᾰκός and ξένος, *a bad guest*, because the phosphorus content injures the quality of the iron made from the limonite ore in which the mineral occurs.

Ref.

1. Elements of structure cell; forms from approximate measurements on microscopic crystals by Gordon, priv. comm. (1949). Luquer, *Am. J. Sc.*, **46**, 154 (1893), also notes pyramidal forms and {0001} on crystals from Lobenstein.
2. Gordon (1949) by rotation method.
3. Gordon (1949).

4. Larsen (51, 1921). Inclined extinction due presumably to subparallel growth is sometimes observed.
5. Niess, *Jb. Min.*, I, 101 (1881).
6. von Hauer, *Jb. Min.*, 191 (1854).
7. Church, *Min. Mag.*, **11**, 1 (1895).
8. For additional analyses see Hintze (1 [4B], 915, 1931).

42.9.3 **Vashegyite** [$Al_4(PO_4)_3(OH)_3 \cdot nH_2O$ (?)]. *Zimányi (Zs. Kr.*, **47**, 53, 1909; *Mat. Termés. Ért.*, **27**, 64, 1909).

Massive, compact to porous; microcrystalline with a fibrous structure. H. 2–3. G. 1.96 (Vashegy). Luster dull. Color white, also pale green to yellow and brownish. Opaque to subtranslucent.

Opt.[1] Appears as minute fibers with positive elongation and birefringence about 0.02, $nY = 1.48 \pm 0.01$ (Vashegy).

Chem. A hydrous basic aluminum phosphate. Formula uncertain, perhaps $Al_4(PO_4)_3(OH)_3 \cdot nH_2O$.

Anal.

	1	2	3
Al_2O_3	30.07	28.33	26.48
Fe_2O_3		1.19	1.28
P_2O_5	31.40	31.32	26.55
H_2O	38.53	38.97	44.68
Rem.		0.57	1.04
Total	100.00	100.38	100.03
G.		1.964	2.0

1. $Al_4(PO_4)_3(OH)_3 \cdot 13H_2O$. 2. Vashegy.[2] Rem. is K_2O 0.16, Na_2O 0.05, CO_2 0.12, insol. 0.24. 3. Vashegyite? Plauen.[3] Rem. is CuO.

Tests. B.B. infusible. Easily soluble in acids.

Occur. From an iron mine at Vashegy near Szirk, Comitat Gömör, Hungary, with variscite and evansite. Minerals referred to vashegyite have been found at Thiergarten near Plauen in Voigtland, Germany (analysis 3), and in Bohemia at Litošice and Zeleznik.[4] A similar mineral also occurs with variscite in veinlets in altered slate near Manhattan, Nye County, Nevada.[5]

Ref.

1. Larsen (153, 1921).
2. Loczka analysis in Zimányi (1909).
3. Jahn and Gruner, *Mitt. Vogtland. Ges. Naturfor.*, **1**, no. 8, 20 (1933); fine fibrous with positive elongation and mean index of refraction 1.505.
4. Ulrich, *Rozpr. Česká Ak.*, **31**, 2 (1922); largely isotropic with $n \sim 1.504$.
5. Wherry, *Washington Ac. Sc., J.*, **6**, 105 (1916); Clinton, *Am. Min.*, **14**, 434 (1929). The material from this locality also is known under the trade names of trainite and sabalite (see Schaller, *U. S. Geol. Sur., Min. Res.*, Pt. II, 163, 1918).

42.9.4 **TARANAKITE** [$K_2Al_6(PO_4)_6(OH)_2 \cdot 18H_2O$ (?)]. *Hector* and *Skey* (*Rpts. of Jurors, New Zealand Exped.*, 423, 1865); *Cox* (*Trans. New Zealand Inst.*, **15**, 385, 1882). Minervite *Gautier* (*Ann. mines*, **5**, 23, 1894; *C.R.*, **116**, 928, 1022, 1171, 1271, 1893). Palmerite *Casorio* (*Acc. Georgofili, Att.*, **1**, July 3, 1904).

Massive, clay-like, pulverulent to compact. Color white; also gray or yellowish white. Very soft; unctuous to the touch. G. 2.15.[1] Under the microscope exhibits minute birefringent particles and lath-like crystals. Mean index of refraction,[2] 1.49–1.51.

Chem. A hydrated basic phosphate of potassium and aluminum of uncertain formula, perhaps $K_2Al_6(PO_4)_6(OH)_2 \cdot 18H_2O$. The variation between the reported analyses doubtless is due to admixture. Rb is present in spectrographic amounts in material from Minerva Grotto.[6] About half of the water is lost at 100°, and the remainder at red heat.

Anal.[9]

	1	2	3	4
K_2O	8.06	4.20	8.28	8.04
$(NH_4)_2O$			0.52	0.90
CaO		0.55	1.40	tr.
Fe_2O_3			0.83	1.17
Al_2O_3	26.18	21.43	18.59	22.89
P_2O_5	36.46	35.05	37.28	37.10
H_2O	29.30	33.06	28.20	29.16
Rem.		5.71	4.68	0.38
Total	100.00	100.00	99.78	99.64
G.[1]	2.15	2.15		2.15
n [1]		1.502	1.50	1.507

1. $K_2Al_6(PO_4)_6(OH)_2 \cdot 18H_2O$. 2. Sugarloaves, New Zealand (*taranakite*).[3] Rem. is insol. 0.80, Cl 0.46, SO_3 tr., FeO 4.45. Contains admixed vashegyite. 3. Minerva Grotto, France (*minervite*).[4] Rem. is MgO 0.33, insol. 4.35, traces of F, Cl, SO_3. 4. Monte Alburno, Italy (*palmerite*).[5] Rem. is insol. 0.36, Na_2O 0.02, MgO tr.

Tests. B.B. fuses readily. Easily soluble in acids.

Occur. Found as deposits in caves or locally along sea coasts as a product of the reaction of phosphatic solutions derived from bat or seabird guano on clays or aluminous rocks. From bird colonies at the Sugarloaves, Taranaki, New Zealand (*taranakite*), and on Islas Leones, Patagonia. In caves at Grotte de Minerve, Aude valley, Hérault, France (*minervite*); Monte Alburno, Salerno, Italy (*palmerite*); Misserghin, Oran, Algeria; Jenolan, New South Wales; the island of Rèunion, Indian Ocean.

Name. Taranakite and minervite from the localities; palmerite after Prof. Paride Palmeri. Minervite was early considered [7] identical with palmerite, and both minerals were later shown [8] to be identical with taranakite.

Ref.

1. Bannister in Bannister and Hutchinson, *Min. Mag.*, **28**, 31 (1947).
2. Range of values reported by Barth and Berman, *Chem. Erde*, **5**, 114 (1912), and Bannister, ref. 1.
3. Hector and Skey (1865).
4. Carnot, *Ann. Mines*, **8**, 319 (1895).
5. Casoria (1904).
6. Hutchinson in Bannister and Hutchinson (1947).
7. Lacroix, *Bull. soc. min.*, **33**, 36 (1910); **35**, 114 (1912).
8. Bannister and Hutchinson (1947).
9. See also an analysis by Pisani in Lacroix (1910) of material from Réunion very low in K_2O possibly belonging to this species.

43 COMPOUND PHOSPHATES, ETC.

TYPE 1. $AB(XO_4)Z_q$

43.1.1 BEUDANTITE GROUP

HEXAGONAL—R; SCALENOHEDRAL—$\bar{3}\,2/m$

	c/a	a_0	c_0
Beudantite, $PbFe_3(AsO_4)(SO_4)(OH)_6$			
Corkite, $PbFe_3(PO_4)(SO_4)(OH)_6$	1.1842		
Hinsdalite, $(Pb,Sr)Al_3(PO_4)(SO_4)(OH)_6$	1.2677		
Svanbergite, $SrAl_3(PO_4)(SO_4)(OH)_6$	2.439	6.99 Å	16.75
Woodhouseite, $CaAl_3(PO_4)(SO_4)(OH)_6$	2.340	6.96	16.27

The members of this group are isostructural with the Plumbogummite Group and the Alunite Group (which see). The cell dimensions of corkite, hinsdalite, and beudantite are not known. The old morphological unit of these species has been retained, although presumably it should be doubled to conform with the structure cell of svanbergite and the Alunite Group. A small range of substitutional solid solution is found among the divalent cations, and in some analyses there is a slight departure from a 1:1 ratio between (SO_4) and (PO_4). An anomalous biaxial optical character is common among the members of this group.

43.1.1.1 BEUDANTITE $[PbFe_3(AsO_4)(SO_4)(OH)_6]$. *Lévy (Ann. Phil.,* **11,** 194, 1826) [not Beudantina Covelli *(Acc. Napoli, Att.,* **4,** 17, 1839)].

Cryst.[1] Hexagonal—R.
Habit. Rhombohedral crystals, often pseudo-cubic.
Phys. Cleavage {0001} easy. H. $3\frac{1}{2}$–$4\frac{1}{2}$. G. 4–4.3. Color black, dark green, or brown. Streak grayish yellow or greenish. Luster vitreous to resinous. Transparent to translucent.
Opt. Uniaxial negative (−). Optical anomalies are common. Crystals from Laurium show three biaxial sectors about a uniaxial core.[2]
Chem. A basic sulfate-arsenate of lead and ferric iron, $PbFe_3(AsO_4)(SO_4)(OH)_6$. Minor amounts of Al substitute for Fe''' in analysis 2. P substitutes for As in small amounts, and at least a partial series extends toward corkite.
Anal.[3]

	PbO	Fe_2O_3	As_2O_5	SO_3	H_2O	Rem.	Total
1.	31.35	33.68	16.14	11.24	7.59		100.00
2.	30.65	33.09	13.36	12.30	7.95	2.63	99.98
3.	24.80	39.57	10.34	14.82	7.76	2.71	100.00

1. $PbFe_3(AsO_4)(SO_4)(OH)_6$. 2. Laurium, Greece.[4] Rem. is Al_2O_3. 3. Belvedere mine, Mt. McGrath, Western Australia.[5] Rem. is Na_2O 0.34, K_2O 0.50, Al_2O_3 0.44, SiO_2 1.43.

Tests. Fuses to a gray slag before the blowpipe. Soluble in hydrochloric acid.

Occur. With pharmacosiderite and hematite at Horhausen in the Rhineland, Germany. In cavities in vein quartz at Blond near Vaulry, Haute-Vienne, France. In clusters of pure crystals with scorodite at Laurium, Greece. In the oxidized ore of the Belvedere gold and lead mine, near Mt. McGrath, Western Australia.

Name. After the French mineralogist François Sulpice Beudant (1787–1850). At the time the name was given to the material from Horhausen only incomplete chemical tests were reported. Thereafter, the name was applied to similar material from County Cork, Ireland and from Dernbach near Montabaun, Hesse-Nassau,[6] but chemical tests [7] soon showed that the Horhausen material was essentially a sulfate-arsenate, whereas that from the other two localities was essentially a sulfate-phosphate. Later the names corkite and dernbachite were suggested [8] for the varieties from Cork and Dernbach, and finally the name corkite was adopted for the sulfate-phosphate.[9] In the older literature the name beudantite covers both substances.

MIRIQUIDITE. *Frenzel* (*Jb. Min.*, 673, 1874). Found with pyromorphite at Schneeberg, Saxony. Brittle. H. 4. Blackish brown in color with an ocherous streak. Vitreous. Translucent. Gives tests for Pb, Fe, As, P, and H_2O and fuses before the blowpipe. Crystals are rhombohedral. This substance may be beudantite or corkite.

Ref.

1. Crystallographic measurements on beudantite date to the time before this species was distinguished from corkite. The best measurements appear to have been made on what is now recognized as corkite. The axial ratios of the two minerals do not differ greatly.
2. See Lacroix, *Bull. soc. min.*, **38**, 35 (1915).
3. For a compilation of older analyses see Hintze (**1** [4A], 729, 1931).
4. Lacroix (1915).
5. Simpson, *J. Roy. Soc. Western Australia*, **24**, 110 (1938). Designated beudantite-plumbojarosite by the author, this material shows a departure from the ideal beudantite composition towards that of the Jarosite Group.
6. Dauber, *Ann. Phys.*, **100**, 579 (1857).
7. Percy, *Phil. Mag.*, **37**, 161 (1850), Rammelsberg, *Ann. Phys.*, **100**, 581 (1857), Sandberger, *Ann. Phys.*, **100**, 611 (1857).
8. Adam (1869).
9. Lacroix (1910).

43.1.1.2   C O R K I T E [$PbFe_3(PO_4)(SO_4)(OH)_6$]. *Adam* (49, 1869). Dernbachite, Bieirosite *Adam* (49, 1869). Korkit *Germ.*

Cryst.[1] Hexagonal—R.

$a:c = 1:1.1842;$ $\alpha\ 91°16';$ $p_0:r_0 = 1.3674:1;$ $\lambda\ 88°42'$

Forms:

c 0001	111	o 50$\bar{5}$1	11.$\bar{4}$.$\bar{4}$	e 02$\bar{2}$1	11$\bar{1}$	r 04$\bar{4}$1	55$\bar{7}$
s 10$\bar{1}$1	100	q 01$\bar{1}$1	22$\bar{1}$	w 05$\bar{5}$2	77$\bar{8}$	v 05$\bar{5}$1	22$\bar{3}$

CORKITE

Habit. Crystals rhombohedral, generally pseudo-cubic $\{10\bar{1}1\}$, similar to beudantite.

Phys. Cleavage $\{0001\}$ perfect. H. $3\frac{1}{2}$–$4\frac{1}{2}$. G. 4.295. Color dark green, yellowish green to pale yellow. Luster vitreous to resinous.

Opt. Uniaxial negative $(-)$. Mostly showing anomalous biaxial character and lamellar structure.[2] The mean index of refraction is reported as 1.96 ± 0.01 (Dernbach)[3] and 1.930 ± 0.01 (Utah); birefringence weak.

Chem. A basic sulfate-phosphate of lead and ferric iron, $PbFe_3(PO_4)(SO_4)(OH)_6$. Analyses show some substitution of Cu for Pb and some preponderance of (SO_4) over (PO_4).

Anal.

	PbO	CuO	Fe$_2$O$_3$	P$_2$O$_5$	SO$_3$	H$_2$O	Rem.	Total	G.
1.	33.41		35.89	10.63	11.98	8.09		100.00	
2.	24.05	2.45	40.69	8.97	13.76	9.77	0.24	99.93	4.295
3.	32.33	1.35	34.61	9.35	12.72	8.45	0.56	99.37	

1. $PbFe_3(PO_4)(SO_4)(OH)_6$. 2. Glendore iron mine, Cork, Ireland.[4] Rem. is As_2O_5.
3. Probably from Dernbach, near Montabaur, Hesse-Nassau.[5] Rem. is insol.

Tests. B.B. fusible with difficulty. Easily soluble in warm hydrochloric acid.

Occur. With limonite on chert in the Glendore iron mine, County Cork, Ireland. At Dernbach near Montabaur, Hesse-Nassau, with pyromorphite and limonite. Coating clefts in cellular limonite in the iron deposits near Ljubija, Bosnia. Small dark crystals in a vein near St. Antonio di Gennemari, Sardinia. Corkite is reported to be abundant in the neighborhood of Maikain, Kazakhstan, associated with silver halides connected with weathering of galena in cavities in quartz and barite.[6] In the United States at the Harrington-Hickory and Wild Bull mines, Beaver County, Utah.[7]

Name. From the county of Cork, Ireland.[8]

Ref.

1. Axial elements from Dauber, *Ann. Phys.*, **100**, 579 (1857), on crystals from Dernbach, which he referred to as beudantite but which were shown to be the sulfate-phosphate now called corkite.
2. Miers, *Min. Mag.*, **12**, 242 (1900).
3. Larsen (1921) reported the optical properties of material from Dernbach for beudantite, but all reports of chemical examination of the mineral from this locality indicate that it is corkite.
4. Rammelsberg, *Ann. Phys.*, **100**, 581 (1857).
5. Hartley, *Min. Mag.*, **12**, 234 (1900). The material analyzed was taken from the collections of the British Museum of Natural History. Though labeled as being from Cork it seemed from the characteristics of the specimen that it was probably from Dernbach. An earlier analysis of corkite from Dernbach was reported by Sandberger, *Ann. Phys.*, **100**, 611 (1857).
6. Chukhrov, *Ac. sc. Leningrad, C.r.*, **27**, 246 (1940).
7. Butler and Schaller, *Am. J. Sc.*, **32**, 418 (1911).
8. The name corkite as well as dernbachite and bieirosite were originally intended to designate varieties of beudantite. See also the discussion of the name beudantite.

43.1.1.3 **H I N S D A L I T E** [(Pb,Sr)Al$_3$(PO$_4$)(SO$_4$)(OH)$_6$]. *Larsen* and *Schaller* (*Washington Ac. Sc., J.*, **1**, 25, 1911; *Am. J. Sc.*, **32**, 251, 1911; *Zs. Kr.*, **50**, 106, 1911).

C r y s t.[1] Hexagonal—*R*.

$a:c = 1:1.2677$; $\alpha\, 88°40'$; $p_0:r_0 = 1.4638:1$; $\lambda\, 91°18'$

Forms:

$c\, 0001$ $s\, 10\bar{1}1$

Habit. Crystals are pseudo-cubic $\{10\bar{1}1\}$ or tabular $\{0001\}$, with rough or dull faces. Also massive, granular.

P h y s. Cleavage $\{0001\}$ perfect. H. $4\frac{1}{2}$. G. 3.65. Nearly colorless, with a greenish cast. Luster vitreous to greasy.

O p t. The optical anomalies [2] common in the beudantite group are conspicuous. A uniaxial core may be surrounded by six biaxial segments in which the planes of the optic axes are arranged radially to the core at about 60° to each other. The indices and the optic angles ($2E\, 0°$ to $40°$) of the biaxial segment are variable due to zonal growth of the crystals with compositional variation.

	n [3]	
O	1.671	Uniaxial positive (+).
E	1.689	

C h e m. A basic sulfate-phosphate of lead and aluminum, (Pb,Sr)Al$_3$(PO$_4$)(SO$_4$)(OH)$_6$. Sr substitutes for Pb with Pb:Sr $\sim$5:1 in the reported analysis.

Anal.

	PbO	SrO	Al$_2$O$_3$	P$_2$O$_5$	SO$_3$	H$_2$O	Total	G.
1.	33.14	3.08	27.24	12.65	14.26	9.63	100.00	
2.	31.75	3.11	26.47	14.50	14.13	10.25	100.21	3.65

1. (Pb,Sr)Al$_3$(PO$_4$)(SO$_4$)(OH)$_6$ with Pb:Sr = 5:1.   2. Golden Fleece mine, near Lake City, Hinsdale County, Colorado.

Tests. Infusible. Practically insoluble in acids, but decomposed by fusion in sodium carbonate.

O c c u r. Found abundantly in a vein with quartz, barite, rhodochrosite, and sulfides cutting volcanic rocks in the Golden Fleece mine, near Lake City, Hinsdale County, Colorado.

N a m e. For Hinsdale County, Colorado, where the mineral was discovered.

Ref.

1. Axial ratio from Larsen and Schaller (1911). For several related minerals the determination of the structure cell has led to a doubling of the axial ratio.
2. Larsen and Schaller (1911).
3. Indices for the uniaxial material as given by Larsen and Berman (1934). In the biaxial parts nZ varies from 1.678 to 1.700.

43.1.1.4 **S V A N B E R G I T E** [SrAl$_3$(PO$_4$)(SO$_4$)(OH)$_6$]. *Igelström (Ak. Stockholm, Öfv.*, **11**, 156, 1854). Harttite *Hussak (Min. Mitt.*, **25**, 339, 1906).

C r y s t.[1] Hexagonal—R; scalenohedral—$\bar{3}\,2/m$ (?).

$a:c = 1:2.3565;$ $\alpha\ 61°44';$ $p_0:r_0 = 2.7210:1;$ $\lambda\ 108°45'$

Forms:

			ϕ	$\rho = C$	A_1	A_2
c	0001	111	.....	0°00'	90°00'	90°00'
e	10$\bar{1}$1	100	30°00'	69 49	35 57½	90 00
s	01$\bar{1}$2	110	−30 00	53 41	90 00	45 35
d	02$\bar{2}$1	11$\bar{1}$	−30 00	79 35	90 00	31 36

Less common:
m 10$\bar{1}$0 2$\bar{1}\bar{1}$ t 10$\bar{1}$7 322   u 10$\bar{1}$4 211   p 01$\bar{1}$5 221   o 05$\bar{5}$2 77$\bar{8}$

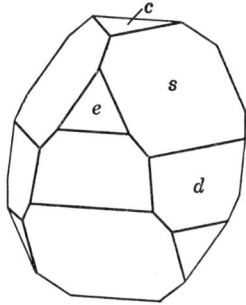

Structure cell.[2] Space group $R\bar{3}m$. a_{rh} 6.91 ± 0.05, $\alpha\ 60°58\frac{1}{2}'$; a_0 6.99 ± 0.05 Å, c_0 16.75 ± 0.05; $c_0:a_0 = 2.3963:1$. Cell contents SrAl$_3$(PO$_4$)(SO$_4$)(OH)$_6$ in the rhombohedral unit.

Habit. Rhombohedral, sometimes pseudocubic {10$\bar{1}$2}. Also granular.

P h y s. Cleavage {0001} distinct. H. 5. G. 3.22; 3.24 (calc.). Colorless to yellow, rose, or reddish brown. Luster vitreous to adamantine. Translucent.

O p t. In transmitted light, colorless. The optical anomalies common in the beudantite group are reported to be absent in svanbergite.[3]

	n(Na) (Horrsjöberg)[3]	n (Hawthorne)[4]	
O	1.631	1.635 ± 0.002	Uniaxial positive (+).
E	1.646	1.649 ± 0.002	

C h e m. A basic sulfate-phosphate of strontium and aluminum, SrAl$_3$(PO$_4$)(SO$_4$)(OH)$_6$. The few good analyses show some substitution of Ca for Sr and preponderance of (PO$_4$) over (SO$_4$). Slight substitution of Pb for Sr has also been reported.

Anal.

	SrO	CaO	Al$_2$O$_3$	P$_2$O$_5$	SO$_3$	H$_2$O	Rem.	Total	G.
1.	22.44		33.13	15.38	17.34	11.71		100.00	3.24
2.	17.99	1.75	32.68	17.55	15.66	11.59	2.81	100.03	3.20
3.	16.80	2.14	33.66	21.17	11.53	12.53	2.44	100.27	3.21
4.	12.84	3.25	36.91	16.70	17.34	12.51	0.24	99.79	3.22

1. SrAl$_3$(PO$_4$)(SO$_4$)(OH)$_6$. 2. Horrsjöberg, Wermland, Sweden.[5] Rem. is (Na,K)$_2$O 0.40, MgO 0.44, Fe$_2$O$_3$ 0.55, SiO$_2$ 1.42. 3. Paraguassu, Bahia, Brazil (*harttite*).[6] Rem. is CeO 1.02, TiO$_2$ 1.42. On the original through misprint the figure for CaO is given as 2.80. 4. Near Hawthorne, Mineral County, Nevada.[7] Rem. is Fe$_2$O$_3$.

Tests. B.B. fusible with difficulty. Insoluble in acids. Yields acid water abundantly in closed tube.

Occur. Originally found at Horrsjöberg, Wermland, Sweden, as euhedral crystals in quartzite with kyanite, pyrophyllite, and other aluminous minerals. At the abandoned Westanå iron mine in Skåne, Sweden, in quartz with hematite. As one of the so-called "phosphate favas" in the diamond gravels of the Rio São Jose, near Paraguassu, Bahia, Brazil (*harttite*). With kaolinite, diaspore, and pyrophyllite close to a granite-quartzite contact at Chizeuil, near Chalmoux, Saône et Loire, France.[8] In the United States, with pyrophyllite in the Dover andalusite mine, 12 miles northwest of Hawthorne, Mineral County, Nevada.

N a m e. After the Swedish chemist Lars Fredrik Svanberg (1805–1878), professor at the University of Uppsala. Harttite, at first assigned a different formula, has been shown to correspond better to the formula of svanbergite.[9]

TIKHVINITE. *Ansheless* and *Vlodavetz* (*Mem. soc. russe min.*, (2) **56**, 53, 1927). Described as a new basic sulfate-phosphate of strontium and aluminum. Found as cavity fillings in bauxite in the Tikhvin district, Russia. It is close to svanbergite in composition.

MUNKFORSSITE. *Igelström* (*Zs. Kr.*, **27**, 601, 1897). Described from Dicksberg, Ransäter, Wermland, Sweden, as a mineral allied to svanbergite but containing little water. Of doubtful validity.[3]

MUNKRUDITE. *Igelström* (*Zs. Kr.*, **28**, 311, 1897). The name was applied to an unanalyzed mineral supposedly related to svanbergite, from near Dicksberg, Ransäter, Wermland, Sweden. Its validity is most doubtful.[3]

Ref.

1. Axial elements and form list from Switzer, *Am. Min.*, **34**, 104 (1949). Symmetry from structure cell. Setting to correspond to rhombohedral cell. Transformation: from original setting of Dauber, *Ann. Phys.*, **100**, 579 (1857), to new, $0\bar{1}00/00\bar{1}0/\bar{1}000/-0002$. The variation in axial ratio with the Sr:Ca ratio is not known.
2. By rotation and Weissenberg method on crystals from Nevada, Switzer (1949). Ygberg, *Ark. Kemi*, **20**, (4) (1945), found a_0 6.96, c_0 16.8 on material from Horrsjöberg, Sweden. For a discussion of space group see Switzer (1949) and Pabst, *Am. Min.*, **32**, 16 (1947).
3. Ygberg (1945).
4. Switzer (1949).
5. Blix analysis in Ygberg (1945), who also gives a tabulation of older faulty analyses of material from Horrsjöberg and from Westanå.
6. Hussak (1906).
7. Gonyer analysis in Switzer (1949).
8. Lacroix, *Bull. soc. min.*, **41**, 14 (1918).
9. Schaller, *Am. J. Sc.*, **32**, 363 (1911).

43.1.1.5 W O O D H O U S E I T E [$CaAl_3(PO_4)(SO_4)(OH)_6$]. *Lemmon* (*Am. Min.*, **22**, 939, 1937).

C r y s t.[1] Hexagonal—R; scalenohedral—$\bar{3}\ 2/m$.

$a:c = 1:2.340;\quad \alpha\ 62°01';\quad p_0:r_0 = 2.703:1;\quad \lambda\ 108°38'$

Forms:

		ϕ	$\rho = C$	A_1	A_2
c	0001	111	.....	0°00'	90°00'	90°00'
e	$10\bar{1}1$	100	30°00'	69 42	35 41	90 00
s	$01\bar{1}2$	110	−30 00	53 30	90 00	45 52½
d	$02\bar{2}1$	$11\bar{1}$	−30 00	79 31	90 00	31 37
o	$05\bar{5}2$	$77\bar{8}$	−30 00	81 35	90 00	31 03

Structure cell.[2] Space group $R\bar{3}m$. a_{rh} 6.75, α 62°04'; a_0 6.961 kX, c_0 16.27; $a_0:c_0$ = 1:2.338. Cell contents $CaAl_3(PO_4)(SO_4)(OH)_6$ in the rhombohedral unit.

Habit. Small crystals, mostly pseudo-cubic $\{10\bar{1}2\}$, also tabular $\{0001\}$. The faces often are curved, striated, or rough.

P h y s. Cleavage $\{0001\}$ excellent. H. $4\frac{1}{2}$. G. 3.012; 3.001 (calc.). Colorless to flesh-colored or white. Luster vitreous, pearly on $\{0001\}$. Translucent or transparent.

O p t. Basal sections of larger crystals show six radial biaxial segments, with $2V$ up to 20°, as do several other members of the beudantite group. May show zonal banding.

   nO 1.636 ± 0.003 Uniaxial positive (+).
   nE 1.647 ± 0.003

C h e m. A basic sulfate-phosphate of calcium and aluminum, $CaAl_3(PO_4)(SO_4)(OH)_6$. Ba substitutes for Ca to a minor extent.

Anal.

	BaO	CaO	Al$_2$O$_3$	P$_2$O$_5$	SO$_3$	H$_2$O	Rem.	Total	G.
1.		13.54	36.93	17.15	19.33	13.05		100.00	3.001
2.	1.00	12.31	36.63	18.13	17.59	13.45	0.76	99.87	3.012

1. $CaAl_3(PO_4)(SO_4)(OH)_6$. 2. White Mountain, Mono County, California. Rem. is Na$_2$O 0.08, K$_2$O 0.02, SrO 0.25, MgO 0.11, SiO$_2$ 0.30.

Tests. Soluble in acid after water has been driven off by heating in closed tube.

O c c u r. Lining vugs in quartz veins cutting an andalusite deposit on White Mountain, Mono County, California. Commonly associated with topaz, augelite, lazulite, and pyrophyllite, all of which formed previously to it.

N a m e. After Mr. C. D. Woodhouse, mineral collector, of Santa Barbara, California.

Ref.

1. Reoriented to conform to rhombohedral lattice mode. Transformation: Lemmon to new setting, $0\bar{1}00/00\bar{1}0/\bar{1}000/0002$.
2. Pabst, *Am. Min.*, **32**, 16 (1947).

43.1.2 **Lindackerite** [$Cu_6Ni_3(AsO_4)_4(SO_4)(OH)_4 \cdot 5H_2O$]. *Haidinger; Vogl (Geol. Reichsanst., Jb., 4, 552, 1853).*

Monoclinic.[1] As minute laths grouped in rosettes or crusts.

P h y s. Cleavage $\{010\}$ perfect. H. $2-2\frac{1}{2}$. G. 2.0–2.5. Luster vitreous. Color apple-green. Streak pale green to white.

O p t.[2] In transmitted light, colorless to pale green.

Orientation	n	
$X \wedge$ elong. 26°	1.629 ± 0.003	Biaxial positive (+).
Y b	1.662	$2V$ 73° $\pm$ 5° (calc.).
Z	1.727	$r < v$, strong.

Chem. A hydrated basic arsenate and sulfate of copper and nickel, probably $Cu_6Ni_3(AsO_4)_4(SO_4)(OH)_4 \cdot 5H_2O$.

Anal.

	CuO	NiO	As$_2$O$_5$	SO$_3$	H$_2$O	Rem.	Total
1.	34.91	16.39	33.62	5.86	9.22		100.00
2.	36.34	16.15	28.58	6.44	9.32	2.90	99.73

1. $Cu_6Ni_3(AsO_4)_4(SO_4)(OH)_4 \cdot 5H_2O$. 2. Joachimsthal, Bohemia. Rem. is FeO.

O c c u r. From Joachimsthal, Bohemia, where it was associated with erythrite, pitticite, annabergite, and other secondary minerals.

N a m e.[3] After the Austrian chemist Joseph Lindacker who made the analysis.

Needs further study.

Ref.
1. Originally described as orthorhombic, but the optical properties indicate that it is probably monoclinic.
2. Larsen, 101 (1921).
3. The name was proposed by Haidinger in an editorial note following the description by Vogl (1893).

TYPE 2. MISCELLANEOUS

43.2.1 **C H A L C O P H Y L L I T E** [$Cu_{18}Al_2(AsO_4)_3(SO_4)_3(OH)_{27} \cdot 33H_2O$]. Cuivre arseniaté lamelliforme *Haüy* (3, 509, 1801); *Vauquelin* (*J. mines*, **10**, 562, 1801). Blättriges Olivenerz, Kupferglimmer *Karsten* (in Hoffmann, **1**, 543, 1801; *Ludwig's Werner*, 180, 1803). Copper Mica *Jameson* (1820). Kupferphyllit *Breithaupt* (42, 1832). Erinite *Beudant* (**2**, 598, 1832); *Des Cloizeaux* (*Ann. chim. phys.*, **13**, 420, 1845). Chalkophyllit *Breithaupt* (149, 1847). Tamarite *Brooke* and *Miller* (512, 1852). Métachalcophyllite *Gaubert* (*Bull. soc. min.*, **27**, 224, 1904).

C r y s t.[1] Hexagonal—R; scalenohedral—$\bar{3}\ 2/m$ (?).

$a:c = 1:5.342;$ $\alpha\ 30°59';$ $p_0:r_0 = 6.168:1;$ $\lambda\ 116°89\frac{1}{2}'$

Forms:[2]

		ϕ	$\rho = C$	A_1	A_2
c 0001	111	.....	0°00'	90°00'	90°00'
m 10$\bar{1}$0	2$\bar{1}\bar{1}$	30°00'	90 00	30 00	90 00
e 10$\bar{1}$4	211	30 00	57 02	43 24	90 00
y 10$\bar{1}$1	100	30 00	80 47$\frac{1}{2}$	31 15$\frac{1}{2}$	90 00
n 70$\bar{7}$4	6$\bar{1}\bar{1}$	30 00	84 42$\frac{1}{2}$	30 25	90 00
w 0.1.$\bar{1}$.12	13.13.11	−30 00	78 59$\frac{1}{2}$	90 00	31 46$\frac{1}{2}$
x 01$\bar{1}$8	332	−30 00	37 38	90 00	58 04$\frac{1}{2}$
d 01$\bar{1}$6	774	−30 00	45 47$\frac{1}{2}$	90 00	51 37$\frac{1}{2}$
s 01$\bar{1}$5	221	−30 00	50 58	90 00	47 43
v 01$\bar{1}$4	552	−30 00	57 02	90 00	43 24
t 01$\bar{1}$3	441	−30 00	64 04	90 00	38 51
r 01$\bar{1}$2	110	−30 00	72 02	90 00	34 32

Structure cell.[3] Space group $R\bar{3}m$ (?). $a_{rh}\ 20.49$, $\alpha\ 30°40'$; $a_0\ 10.75\ kX$, $c_0\ 57.40$; $a_0:c_0 = 1:5.341$. Cell contents $Cu_{18}Al_2(AsO_4)_3(SO_4)_3(OH)_{27} \cdot 33H_2O$.

CHALCOPHYLLITE

Habit. Tabular {0001}, in six-sided crystals of rhombohedral aspect. {0001} sometimes triangularly striated. Also foliated massive; in druses; as rosette-like aggregates.

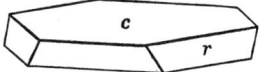 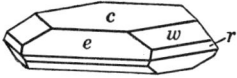

Phys. Cleavage {0001} perfect, micaceous, {10$\bar{1}$1} in traces. Not brittle. H. 2. G. 2.67 in fully hydrated material, decreasing with loss of water; 2.64 (calc.). Luster vitreous to sub-adamantine, becoming pearly especially on {0001} on partial dehydration. Color emerald-green; also grass-green to bluish green. Streak paler than the color. Transparent to translucent.

Opt.[4] In transmitted light, green.

	n (fully hydrated)	n (dried at 110°)	DICHROISM	
O	1.618	1.680	Bluish green	Uniaxial negative (−).
E	1.552	1.618	Almost colorless	

The water content varies at room temperature, depending on the relative humidity, and the indices of refraction vary markedly therewith.

Chem. A hydrated basic sulfate-arsenate of copper and aluminum. The formula is uncertain, perhaps $Cu_{18}Al_2(AsO_4)_3(SO_4)_3(OH)_{27} \cdot 33H_2O$. The (SO_4) was overlooked in early analyses.[5] Small amounts of (PO_4) sometimes substitute for (AsO_4). The ratio of (AsO_4) to (SO_4) apparently varies and is in excess of 1:1 in the reported analyses.[6] On heating,[7] about 14 per cent H_2O is lost at 110° (which may be regained by standing in a moist atmosphere), and the remainder near red heat.

Anal.

	1	2	3	4
CuO	48.44	45.93	46.54	46.56
Al_2O_3	3.45	4.74	3.49	3.48
As_2O_5	11.66	14.46	13.23	13.26
P_2O_5			0.67	0.64
SO_3	8.12	7.04	6.67	7.98
H_2O+	28.33	10.33	14.04	28.21
H_2O-		17.93	14.40	
SiO_2			1.33	
Total	100.00	100.43	100.37	100.13

1. $Cu_{18}Al_2(AsO_4)_3(SO_4)_3(OH)_{27} \cdot 33H_2O$. 2. Cornwall.[8] Average of two. 3. Teniente mine, Rancagua, Chile.[9] 4. Cornwall.[10]

Tests. In C.T. decrepitates, yields much water, and leaves a residue of olive-green, isotropic flakes. B.B. easily fusible. Soluble in acids and in ammonia.

Occur. A secondary mineral found with cuprite, tenorite, malachite, azurite, connellite, pharmacosiderite, clinoclase in the oxidized zone of copper deposits. Found in Germany at Schmiedeberg and at Mordelgrund near Saida, Saxony. In Hungary at Herrengrund and Moldava. From

Schwaz, in the Tyrol, Austria, and at Cap Garonne, Var, France. From Nizhni Tagilsk in the Urals, U.S.S.R. Found at Redruth, St. Day, Liskeard, and other localities in Cornwall, England. In the Teniente mine near Rancagua, Chile. In the United States, chalcophyllite occurs at Bisbee, Arizona, and in the Tintic District, Utah. Also from Sodaville, Mineral County, and with spangolite and cyanotrichite at Majuba Hill, Nevada.

Alter. To chrysocolla.

Name. From χαλκός, *copper*, and φύλλον, *a leaf*, alluding to the micaceous structure.

Ref.
1. Angles of Palache and Merwin, *Am. J. Sc.*, **28**, 537 (1909), on Bisbee crystals, unit of the structure cell. Transformation: P. and M. and Dana (840, 1892) to new, $0\overline{1}00/00\overline{1}0/\overline{1}000/0002$.
2. Goldschmidt (5, 68, 1918) and Piazza, *Per. Min.*, **2**, 104 (1931).
3. Berry (priv. comm., 1946) on re-hydrated crystals by Weissenberg method. The partly dehydrated mineral, as ordinarily found, has a marked pseudo-cell with $c' = c/2$ and $a' = a$ and a quite different sequence of intensities in (000*l*).
4. Indices of Shannon, *Am. J. Sc.*, **7**, 31 (1924); see also Gaubert, *Bull. soc. min.*, **27**, 223 (1904).
5. Cited in Dana (840, 1892).
6. For discussion see Piazza (1931).
7. Thermal data in Shannon (1924), Piazza (1931), Gaubert (1904), Hartley, *Min. Mag.*, **12**, 120 (1898).
8. Hartley (1898).
9. Shannon (1924).
10. Piazza (1931).

43.2.2 ARDEALITE $[Ca_2H(PO_4)(SO_4)\cdot 4H_2O]$. Schadler (*Cbl. Min.*, 40, 1932).

As very fine-grained powdery masses. Apparently monoclinic and isostructural [1] with gypsum and brushite. Color light yellow. G. 2.300; 2.38 (calc.).

Structure cell.[2] a_0 5.67 kX, b_0 14.64, c_0 6.28; [β 113°50′]; $a_0:b_0:c_0 = 0.387:1:0.429$. Cell contents $Ca_4H_2(PO_4)_2(SO_4)_2\cdot 8H_2O$.

Chem. A hydrated acid sulfate-phosphate of calcium, $Ca_2H(PO_4)(SO_4)\cdot 4H_2O$.

Anal.

	1	2	3
CaO	32.59	31.61	32.13
SO$_3$	23.24	21.25	23.08
P$_2$O$_5$	20.62	21.85	23.53
H$_2$O	23.55	25.14	n.d.
Insol.		0.39	
Total	100.00	100.24	

1. $Ca_2H(PO_4)(SO_4)\cdot 4H_2O$. 2. Cioclovina cave, Transylvania. 3. Transylvania.[3]

Occur. Found intimately admixed with minute crystals of brushite and gypsum in a phosphate deposit in the limestone cave of Cioclovina, Transylvania.

Name. From Ardeal, the old Roumanian name for Transylvania.

Ref.
1. Halla, *Zs. Kr.*, **80**, 349 (1931), by powder method on natural material, whose work indicates that the substance is not a mechanical mixture of brushite and gypsum.
2. Halla (1931). Transformation: Halla to new (gypsum) cell, $\bar{1}0\frac{1}{2}/0\bar{1}0/001$. β taken from gypsum.
3. Rist analysis in Hill and Hendricks, *Ind. Eng. Chem.*, **28**, 440 (1936).

43.2.3   **Kribergite.**   *Du Rietz (Geol. För. Förh.*, **67**, 78, 1945).

White, compact, and chalk-like showing a felted lamellar to spherulitic structure under the microscope. G. 1.92. Mean index of refraction 1.484, birefringence about 0.002. Analysis gave:[1] MgO 0.02, Al_2O_3 38.45, V 0.003, SO_3 13.05, P_2O_5 31.27, H_2O 17.22, total 100.013. This approximates to $Al_{16}(PO_4)_8(SO_4)_3(OH)_{18} \cdot 10H_2O$. Found filling crevices in cupriferous pyrite at the Kristineberg mine, Västerbotten, Sweden.

Ref.
1. Helger analysis in Du Rietz (1945).

43.2.4   **D I A D O C H I T E** [$Fe_2(PO_4)(SO_4)(OH) \cdot 5H_2O$]. Diadochit *Breithaupt (J. pr. Chem.*, **10**, 503, 1837). Phosphoreisensinter *Rammelsberg*. Destinezite *Forir* and *Jorissen (Bull. soc. belge géol.*, **7**, 117, 1881). Orthodiadochite. Eisensinter, pt. Geldiadochit *Cornu (Cbl. Min.*, 330, 1909; *Zs. Chem. Ind. Kolloide*, **4**, 17, 1909).

Probably triclinic. Microcrystalline, in reniform, nodular, or earthy masses, in part composed of microscopic six-sided plates of various habits

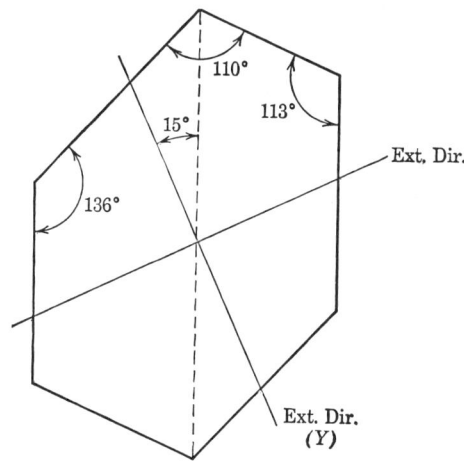

Visé.   Commonest type of plate as seen under microscope.

(see figures);[1] also gel-like and amorphous, in botryoidal, stalactitic, or colloform crusts or masses.

P h y s.   Fracture earthy or uneven to conchoidal. Pulverulent to brittle. H. variable, 3 to 4 in the glassy types, less in earthy material. The G. varies widely with the state of aggregation and water content; ordinarily 2.0–2.4. Color yellow to yellowish brown and brown; also reddish brown, greenish yellow, pale greenish, or yellowish white. Dull and opaque in earthy types; nearly transparent with a waxy, horn-like or subvitreous luster in gel masses.

Opt. In transmitted light, pale yellow to yellowish brown. The glassy types are isotropic [2] with n variable, 1.60–1.61.

	n (crystals; Visé) [2]	n (crystals; Cyprus) [3]	n (crystals; Chvaletice) [4]	
X	1.615 ± 0.005	~1.618		Biaxial positive (+).
Y	1.625 ± 0.005	~1.618	1.6380 (Y ')	2V small.
Z	1.665 ± 0.005	1.670	1.6658 (Z ')	$r > v$, rather strong.

The crystals from Visé have X nearly normal to the flattening, and Z' makes an angle of about 16° to the long edge; on edge the angle of Z' to the elongation is about 14°; [2] see also the figures.[5]

Chem. A hydrated basic sulfate-phosphate of iron, $Fe_2(PO_4)(SO_4)(OH) \cdot 5H_2O$. The gel types contain a considerable amount of non-essential water above the requirements of this formula. Sarmientite apparently is the arsenate analogue of diadochite, and the two minerals may be isostructural. It is not certain if all the reported [6] analyses actually refer to this species, and some doubtless represent mixtures.

Anal.

	1	2	3	4	5	6	7
CaO			0.19	0.30		0.17	
Fe_2O_3	38.97	37.60	37.66	36.63	35.09	32.91	39.69
P_2O_5	17.32	16.76	16.50	16.70	16.07	19.17	14.82
SO_3	19.53	18.85	19.32	13.37	17.25	10.85	15.14
H_2O	24.18	25.65	26.47	32.43	30.12	34.45	30.35
Rem.		1.40	0.04	0.45	1.60	2.22	
Total	100.00	100.26	100.18	99.88	100.13	99.77	100.00

1. $Fe_2(PO_4)(SO_4)(OH) \cdot 5H_2O$. 2. Visé, Belgium [7] (*destinezite*). H_2O includes 0.30 hygroscop. H_2O; rem. is insol. 3. Chvaletice, Bohemia [8] (*destinezite*). Rem. is insol. 4. Peychagnard, France [9] (*diadochite*). Rem. is As_2O_5. 5. Hředl, Bohemia [10] (*diadochite*). H_2O includes 2.00 hygroscop. H_2O; rem. is aluminum silicate. 6. Feengrotte, Saalfeld, Thuringia [11] (*diadochite*). H_2O includes 14.67 ign. loss, 19.78 H_2O at 100°; rem. is Al_2O_3. 7. Ansbach, Thuringia [12] (*diadochite*).

Tests. B.B. fusible to a black mass. C.T. affords much acid water and turns brownish red. Easily soluble in acids. Little or no sulfate is lost on extraction with water.[13]

Occur. A near-surface secondary mineral, found in the gossan of pyritic deposits and sometimes formed by the action of sulfate solutions on earlier formed secondary phosphates such as delvauxite [14] or pyromorphite; also as a recent deposit in mine workings. Associated with limonite, delvauxite, foucherite, borickite, vashegyite, and other hydrous aluminum phosphates, vivianite, pitticite, melanterite. From alum-slate at Ansbach near Gräfenthal, Saalfeld, Thuringia (the original locality for diadochite); also found similarly at other localities in the same region,[14] notably the Garnsdorf mine (the so-called Feengrotte) near Saalfeld. In France at Huelgoat, Finistère, and in the anthracite mine of Peychagnard,

Isère. At numerous localities in Bohemia, notably at Litošice, Chvaletice, Nučic, and Hředl near Rakonitz. From Leoben, Styria, and at Vashegy, Gömör Com., and Eisenbach in Hungary. In Belgium at Visé, Argenteau (the original locality for destinezite), and at Védrin. Reported from numerous other localities, including the New Idria mine, San Benito County, California.[16]

Name. Diadochite from διάδοχος, *a successor:* the name may be considered to mark the fact that it is sometimes formed by the alteration of earlier-formed phosphates. Destinezite after M. Destinez. The name destinezite was originally applied to a mineral now known to be only a relatively coarsely crystalline variety of diadochite (a metacolloid in its original occurrences); diadochite has priority and is here used as the species designation.[15]

Ref.

1. Cesàro, *Mém. soc. belge géol.*, **12**, 173 (1885); *Mém. ac. sc. belge*, **53**, 1 (1897), who draws attention to certain geometrical similarities to gypsum. A supposed specimen of diadochite from Styria gave an x-ray pattern resembling that of jarosite; McConnell, *Am. J. Sc.*, **240**, 656 (1942).
2. Larsen (67, 1921).
3. Mélon and Donnay, *Bull. soc. belge géol.*, **59**, B162 (1936).
4. Ulrich, *Rozpr. Česká Ak.*, **31**, no. 10 (1922).
5. Cesàro (1885, 1897), who with Lacroix (4, 597, 1910) gives $X \wedge c\ 39°$ in crystals of gypsum-like habit. See also Slavík, *Bull. intern. ac. Bohême*, **22**, 32 (1918).
6. For additional analyses see Hintze (1 [4B], 750, 755, 1073, 1931).
7. Cesàro (1885).
8. Veselý, *Bull. intern. ac. Bohême*, **23**, 177 (1923).
9. Carnot, *Bull. soc. min.*, **3**, 40 (1880).
10. Feistmantel, *Lotos*, **23**, 33 (1873).
11. von Wichdorff, 1911, cited in Hintze (1931, p. 755).
12. Rammelsberg (331, 1875).
13. Veselý (1933); see also Rammelsberg (5, Suppl., 1843) and Ulrich, *Rozpr. Česká Ak.*, **31**, no. 10 (1922).
14. See literature cited in Hintze (1931).
15. Following Dana (867, 1892).
16. Rogers, *Am. Min.*, **23**, 178 (1938).

43.2.5 SARMIENTITE [$Fe_2(AsO_4)(SO_4)(OH) \cdot 5H_2O$]. Angelelli and Gordon (*Ac. Sc. Philadelphia, Not. Nat.*, no. 92, 1941; *Publ. Inst. de Fisiog. y Geol., Univ. Nac. del Litoral, Argentina*, no. XII, 1942).

Cryst.[1] Monoclinic; prismatic—$2/m$ (?).

$a:b:c = 0.3415:1:0.5242;\quad \beta\ 97°39';\quad p_0:q_0:r_0 = 1.5345:1.5195:1$

$r_2:p_2:q_2 = 1.9248:2.9546:1;\quad \mu\ 82°21';\quad p_0'\ 1.5488, q_0'\ 0.5242, x_0'\ 0.1343$

Forms:

	ϕ	ρ	ϕ_2	$\rho_2 = B$	C	A
b 010	0°00'	90°00'	.....	0°00'	90°00'	90°00'
m 110	71 17½	90 00	0°00'	71 17½	82 45½	18 42½
w 011	14 22	28 25	82 21	62 33	27 27	83 13

Habit. As nodular masses composed of microscopic crystals prismatic [001] and flattened {010}.

Phys. Data on the cleavage and hardness are lacking. G. 2.58. Color pale yellow-orange.

Opt.

Orientation		n	
X		1.628	Biaxial positive (+).
Y	b	1.635	$2V$ 38° (calc.).
$Z \wedge c$	12°	1.698	

Chem. A hydrated basic sulfate-arsenate of ferric iron, $Fe_2(AsO_4)(SO_4)(OH) \cdot 5H_2O$.

Anal.

	1	2
CaO		0.27
Fe_2O_3	35.20	36.57
As_2O_5	25.32	22.68
SO_3	17.64	18.28
H_2O	21.84	22.86
Total	100.00	100.66

1. $Fe_2(AsO_4)(SO_4)(OH) \cdot 5H_2O$. 2. Santa Elena mine, Alcaparrosa, Argentina.[2]

Occur. Found as nodules embedded in fibroferrite together with copiapite, botryogen, szomolnokite, zincian melanterite, slavikite, gypsum, and epsomite at the Santa Elena mine, La Alcaparrosa, between San Juan and Calingasta, Department Barreal, Argentina. The iron sulfates occur in the gossan of pyritic sulfide veins in diabase.

Name. After Domingo Faustino Sarmiento (1811–1888), Argentinian educator and statesman.

Ref.

1. From approximate measurements on a crystal about 25 microns in length. Possibly triclinic.
2. Collins analysis in Angelelli and Gordon (1941).

43.2.6 **Pitticite.** Eisenbranderz *Torber*. Eisenpecherz *Karsten* [not Werner] (66, 98, 1808). Fer oxydé, resinite *Haüy* (98, 1809). Pittizit *Hausmann* (285, 1813). Eisensinter *Werner* (Hoffmann, **3**, *b*, 302, 1816; **4**, *b*, 141, 1817). Arseneisensinter *Germ.* Pitch Iron Ore. Diarsenate of Iron. Sideretine *Beudant* (2, 609, 1832) [not pittizite *Beudant* (**2**, 484, 1832)]. Pitticit *Breithaupt* (1022, 1847).

Massive: reniform to botryoidal or stalactitic; as opaline crusts; also earthy and as slimy coatings.

H. and G. variable depending on the state of aggregation and water content; H. mostly 2–3, G. ordinarily 2.2–2.5. Opaline masses are very brittle. Fracture earthy to conchoidal and splintery. Luster dull to vitreous, sometimes greasy. Color yellowish and reddish brown, brownish red to brownish black, also yellow, gray, almost white. Streak yellow to white. Opaque in earthy material to transparent in opaline types. May show deep red internal reflections from cracks.

Opt. In transmitted light, pale yellowish to reddish brown. Isotropic; $n \sim 1.616$ (Manhattan, Nevada),[1] 1.635 (Freiberg, Saxony).[2]

KOLBECKITE

Chem. A hydrous sulfate-arsenate of ferric iron of uncertain formula. The reported analyses [3] may vary widely, in part at least owing to admixture. Some are not far from $Fe_2(AsO_4)(SO_4)(OH) \cdot nH_2O$, and these may represent the arsenate analogue of diadochite.

Anal.

	1	2	3	4	5
Fe_2O_3	34.85	33.73	34.01	32.54	54.66
As_2O_5	26.70	26.06	14.54	33.99	24.67
SO_3	13.91	10.04	17.97	7.28	5.20
H_2O	24.54	29.25	29.64	24.92	15.47
Rem.			4.24		
Total	100.00	99.08	100.40	98.73	100.00

1. Schwartzenberg, Saxony.[4]   2. Freiberg, Saxony.[5]   3. Manhattan, Nevada.[1] Rem. is Al_2O_3 3.11, FeO 0.72, insol. 0.41, MgO and Ca tr.   4. Redruth, Cornwall.[6]   5. Stieglitzstollen, Gastein, Austria.[7]

Tests. B.B. fusible to a black mass, with loss of As_2O_5. In C.T. affords much acid water. Easily soluble in acids, and decomposed by strong alkalies leaving a residue of iron oxide.

Occur. A near-surface secondary mineral, often formed by the oxidation of arsenopyrite, realgar, or orpiment, and found associated with limonite, haidingerite, pharmacolite, erythrite, pharmacosiderite, diadochite, scorodite, melanterite, gypsum, alunite. Also as a recent deposit from mine and spring waters. Numerous poorly authenticated occurrences have been reported.[3] Found in Saxony in old mines at Freiberg, Schneeberg, and at Stamm Asser at Graul near Schwartzenberg. In the alum-slate workings near Saalfeld, Thuringia; at Joachimsthal and Příbram, Bohemia; at Felsöbánya and Offenbánya, Roumania. From Redruth, Cornwall. As an exudation on mine walls at Manhattan, Nevada. Substances doubtfully referred to pitticite have been described from the Clarissa mine, Tintic district, Utah,[8] and from Djebel Debar, Constantine [9] and Tomsk, Russia [10] (the two latter containing calcium).

Name. From πιττά, *pitch*, because early called pitch-ore. The name is best considered as a generic term for hydrous ferric sulfate-arsenates whose species identity is unknown.

Ref.

1. Foshag and Clinton, *Am. Min.*, **12**, 290 (1927).
2. Larsen (121, 1921).
3. Cited in Hintze (1 [4B], 1067, 1931).
4. Rammelsberg (384, 1860).
5. Stromeyer, *Ann. Phys.*, **61**, 181 (1843).
6. Church, *Chem. News*, **24**, 136 (1871).
7. Rammelsberg (355, 1875).
8. Genth, *Am. J. Sc.*, **40**, 205 (1890).
9. Barthoux, *Bull. soc. min.*, **48**, 106 (1925).
10. Pilipenko, *Bull. Univ. Tomsk*, **28**, 1 (1907).

43.2.7   Kolbeckite. *Edelmann (Berg.-u. hütten. Sachsen, Jb., **100**, 74A, 1926).*

Probably monoclinic, with {110}, {011}, and {001} and twinned on {100} to give an orthorhombic aspect. Crystals short prismatic [001].

Cleavage {010} distinct. Fracture conchoidal. Brittle. H. $3\frac{1}{2}$–4. G. 2.39. Luster vitreous to pearly. Color bright blue to blue-gray. Strongly pleochroic.

Chem. Apparently a hydrated silicate-phosphate of beryllium, aluminum, and calcium. An average of two microchemical analyses gave:[1] CaO 3.22, BeO 8.74, Fe_2O_3 0.29, Al_2O_3 21.35, P_2O_5 33.80, SiO_2 9.25, H_2O 23.45, total 100.10.

Tests. Decomposed by acids.

Occur. Found as a few small crystals in a quartz-wolframite vein in the Sadisdorf copper mine near Schmiedeberg, Saxony, Germany.

Name. After Friedrich Kolbeck (1860–1943), German mineralogist, of the Mining Academy at Freiberg.

Ref.

1. Thurnwald and Benedetti-Pichler, *Mikrochem.*, **11**, 200 (1932).

ANTIMONATES; ANTIMONITES AND ARSENITES

CLASS 44. Antimonates
Type 1. $A_2X_2O_6(O,OH,F)$
  44.1.1 Bindheimite group
    44.1.1.1 Bindheimite $Pb_2Sb_2O_6(O,OH)$
    44.1.1.2 Romeite $Ca_2Sb_2O_6(O,OH,F)$
  44.1.2 Monimolite $(Pb,Ca)_3Sb_2O_8$ (?)
  44.1.3 Tripuhyite $Fe_2Sb_2O_7$ (?)
  44.1.4 Flajolotite $FeSbO_4 \cdot \tfrac{3}{4}H_2O$ (?)
Type 2. Miscellaneous
  44.2.1 Derbylite $Fe_6Ti_6Sb_2O_{23}$ (?)
  44.2.2 Swedenborgite $NaBe_4SbO_7$
  44.2.3 Catoptrite Mn,Al antimonate-silicate

CLASS 45. Acid and normal antimonites and arsenites
Type 1. Miscellaneous
  45.1.1 Armangite $Mn_3(AsO_3)_2$
  45.1.2 Trigonite $MnPb_3H(AsO_3)_3$
  45.1.3 Trippkeite $CuAs_2O_4$
  45.1.4 Schafarzikite $FeSb_2O_4$

CLASS 46. Basic or halogen-containing antimonites, arsenites
Type 1. Miscellaneous
  46.1.1 Ecdemite $Pb_6As_2O_7Cl_4$ (?)
  46.1.2 Heliophyllite $Pb_6As_2O_7Cl_4$ (?)
  46.1.3 Finnemanite $Pb_5(AsO_3)_3Cl$
  46.1.4 Nadorite $PbSbO_2Cl$
  46.1.5 Melanostibian $(Mn,Fe)_6(SbO_3)_2O_3$

44 ANTIMONATES

TYPE 1. $A_2X_2O_6(O,OH,F)$

44.1.1 BINDHEIMITE GROUP

ISOMETRIC; HEXOCTAHEDRAL—$4/m\,\bar{3}\,2/m$

	a_0
Bindheimite, $Pb_2Sb_2O_6(O,OH)$	10.4 kX
Romeite, $Ca_2Sb_2O_6(O,OH,F)$	10.25
Monimolite, $(Pb,Ca)_3Sb_2O_8$ (?)	

Bindheimite and romeite have crystal structures [1] of the pyrochlore type, $A_2B_2O_6(O,OH,F)$, and are probably best classed as multiple oxides rather than among anisodesmic compounds. The ill-defined hydroromeite may be a link between romeite and the pyrochlore-type substance stibiconite, $Sb_6O_6(OH)$. The formula of monimolite is not well-established, but it is similar in habit and occurrence to romeite and may be intermediate between it and bindheimite. A number of supposedly distinct species have been found to be only minor compositional variants of romeite, and all have here been reduced to a varietal status. These include atopite, schneebergite, lewisite, mauzeliite, and weslienite.

Ref.
1. Natta and Baccaredda, *Zs. Kr.*, **85**, 271 (1933); Machatschki, *Chem. Erde*, **7**, 56 (1932).

44.1.1.1 **BINDHEIMITE** [$Pb_2Sb_2O_6(O,OH)$]. Bleiniere *Karsten* (77, 1800). Stibiogalenites *Glocker* (257, 1847). Bleinierite *Nicol* (319, 1849). Moffrasite *Leymerie* (**2**, 283, 1859). Blumit *Fischer* (*Jb. Min.*, 559, 1863) [not the blumit = huebnerite of *Liebe* (*Jb. Min.*, 652, 1863)]. Bindheimite *Dana* (591, 1868). Pfaffite *Adam* (37, 1869). Antimoniate of Lead. Antimonbleispath, Antimonsaures Bleioxyd *Germ.*

Cryst.[1] Isometric; hexoctahedral—$4/m\,\overline{3}\,2/m$.
Structure cell.[2] Space group $Fd3m$. a_0 10.4 kX. Cell contents $Pb_{16}Sb_{16}O_{56}(H_2O)_8$.
Habit. As cryptocrystalline masses, dense to earthy, or as incrustations. Also as nodular to reniform masses with concentric layering; sometimes opaline. Pseudomorphous.
Phys. Fracture earthy to conchoidal. H. $4–4\frac{1}{2}$. G. 4.6–5.6; 9.14 (calc.). Luster resinous, dull, or earthy. Color yellow, brown, reddish brown, gray, white, greenish. Streak white to yellowish. Opaque to translucent.
Opt. In transmitted light, colorless to shades of yellow or brown. Isotropic, but often contains microcrystalline birefringent material due to admixture. Index of refraction[3] variable, n = 1.84–1.87.
Chem. A pyroantimonate of lead, $Pb_2Sb_2O_6(O,OH)$. Ca may substitute for Pb, and at least a partial series extends toward the isostructural mineral romeite. The water content is variable,[4] and dehydration does not change the structure. Most analyses have been made on impure material, and their agreement with the formula is poor.

Anal.[5]

	PbO	CaO	Sb_2O_5	H_2O	Rem.	Total	G.	a_0
1.	56.65		41.06	2.29		100.00	9.14	10.4
2.	57.32		41.53	1.15		[100.00]		
3.	44.12		40.35	5.31	8.85	98.63		
4.	41.02	4.49	34.85	3.86	14.33	98.55	5.45	10.37
5.	51.80	tr.	27.59	3.00	17.43	99.82	5.13	10.42

1. $Pb_2Sb_2O_7 \cdot H_2O$. 2. Hypotheek mine, Coeur d'Alene district, Idaho.[6] Recalculated after deduction of limonite 4.49, cerussite 3.64, and quartz 2.28 per cent. 3. Wamsley mine, Mineral County, Nevada.[7] Rem. is Fe_2O_3 5.22, insol. 3.63. 4. Camerata Cornello, Italy.[8] Rem. is CuO 3.75, Fe_2O_3 1.97, Al_2O_3 1.16, CO_2 3.83, insol. 3.62. 5. Gorno, Italy.[8] Rem. is CuO 2.20, MnO 1.38, Fe_2O_3 4.84, CO_2 5.36, insol. 3.65.

Tests. In C.T. affords water and turns darker in color. Dissolves leaving a residue of antimonic oxide in HNO_3 and of lead chloride in HCl.

BINDHEIMITE

Occur. A widespread mineral in the oxidized zones of antimonial lead ores, generally intimately mixed with other cryptocrystalline secondary minerals and often pseudomorphous after various primary lead-antimony ores.[9] Only a few of the known localities are recorded here. First recognized at Nerchinsk, Siberia. Occurs with malachite in altered bournonite at Waitschach near Hüttenberg, Carinthia (formerly often referred to as *wölchite* [10]) and as an ocherous alteration of bournonite at Oberzeiring, Styria, in Austria. At Camerata Cornello, Val Brembana and at Gorno, Val Seriana, Lombardy, mixed with secondary carbonates. Associated with jamesonite and nadorite at the Bodannon antimony mine, St. Endellion, Cornwall. With nadorite and cerussite at Hamman N'bail, in the province of Constantine, Algeria. Abundant and in part rich in silver at Broken Hill, New South Wales. Associated with bournonite and pseudomorphous after zinkenite at Machacamarca, Potosi, Bolivia. In the United States bindheimite occurs abundantly in lead mining districts of the west. Solid masses of bindheimite or quartz and bindheimite form veins in the Montezuma mine, Arabia district, Humboldt County, and numerous other localities in Nevada. It is common in all the oxidized ores in the Coeur d'Alene district, Shoshone County, Idaho.

Alter. Bindheimite may alter to massicot [11] and has itself been reported as pseudomorphs after jamesonite,[12] tetrahedrite and galena,[13] bournonite,[14] and zinkenite.[15]

Artif. $Pb_2Sb_2O_7$, the isostructural anhydrous analogue of bindheimite has been prepared by precipitation from a neutral solution of potassium pyroantimonate by lead acetate [16] and by sintering lead acetate and antimonic acid at low red heat and purifying the product by washing with boiling water.[17]

Name. After Johann Jacob Bindheim (1750–1825), German chemist, who made the first analysis of bindheimite on material from Nerchinsk, Siberia.[18]

Ref.

1. Crystal system and symmetry from structure cell.
2. Natta and Baccaredda, *Zs. Kr.*, **85**, 271 (1933), and Hägele, *Cbl. Min.*, 45 (1937), from powder photographs. a_0 varies from 10.37 to 10.43 kX for bindheimite from various localities, the lower values being associated with substitution of Ca for Pb. The amount of water shown in the cell contents corresponds to filling of an 8-fold position in the cell. Most analyses show more than this water content. The artificial anhydrous analogue $Pb_2Sb_2O_7$ has a_0 10.44 kX, according to Natta and Baccaredda (1933) and 10.68, according to Baccaredda, *Gazz. chim. ital.*, **66**, 539 (1936).
3. Larsen, 47 (1921), on unanalyzed material from the Eureka district, Nevada, and from Fresno County, California.
4. Materials intermediate in composition between bindheimite and hydroromeite have been described by Natta and Baccaredda (1937), who also report that stibiconite is isostructural with bindheimite.
5. For additional analyses see Hintze (1 {4B}, 838, 1931); Le Mesurier, *J. Roy. Soc. Western Australia*, **25**, 137 (1939).
6. Shannon, 438 (1926).
7. Shannon, *Econ. Geol.*, **15**, 88 (1920).
8. Natta and Baccaredda (1933). G. of analyses 3 and 4 after dehydration, 6.36 and 6.84, respectively. Both analyses were made on contaminated material. The carbonate content could be entirely removed without altering the x-ray powder pattern.

9. See Shannon (1920) for a discussion of bindheimite as an ore mineral.
10. Meixner, *Cbl. Min.*, **38** (1937).
11. Lacroix, **4**, 514 (1910).
12. Knopf, *U. S. Geol. Sur.*, *Bull.* **660**, 249 (1918).
13. Shannon (1920).
14. Meixner (1937).
15. Prior and Spencer, *Min. Mag.*, **11**, 23 (1895).
16. Natta and Baccaredda (1933).
17. Baccaredda, *Gazz. chim. ital.*, **66**, 539 (1936). The $Pb_2Sb_2O_7$ obtained by sintering has G. (meas.) 6.72, G. (calc.) 8.40.
18. Bindheim, *Ges. nat. Freunde Berlin, Schr.*, **10**, 374 (1792).

44.1.1.2 **R O M E I T E** [$(Ca,Fe,Mn,Na)_2(Sb,Ti)_2O_6(O,OH,F)$]. Roméine *Damour* (*Ann. mines*, **20**, 247, 1841; **3**, 179, 1853). Atopite *Nordenskiöld* (*Geol. För. Förh.*, **3**, 376, 1877). Schneebergite *Brezina* (*Geol. Reichsanst., Verh.*, 313, 1880). Mauzeliite *Sjögren* (*Geol. För. Förh.*, **17**, 313, 1895). Lewisite *Hussak* and *Prior* (*Min. Mag.*, **11**, 80, 1895). Weslienite *Flink* (*Geol. För. Förh.*, **45**, 567, 1923).

C r y s t.[1] Isometric; hexoctahedral—$4/m\,\bar{3}\,2/m$.

Forms:[2]

| a 001 | d 011 | o 111 | m 113 | n 112 | g 133 |

Structure cell.[3] Space group $Fd3m$.

a_0	Locality
10.245 kX	St. Marcel, Piedmont
10.27	Ouro Preto, Brazil (titanian romeite, "lewisite")
10.28	Minas Geraes, Brazil (sodian romeite, "atopite")
10.31	Långban, Sweden (sodian romeite, "atopite")
10.28	Långban, Sweden ("weslienite," anal. 4)
10.29	Schneeberg, Saxony (ferroan romeite, "schneebergite")

Cell contents $Ca_{16}Sb_{16}O_{56}$. Romeite is isostructural with pyrochlore.

Habit. Small octahedral crystals, often embedded or in groups. Also massive.

Twinning. On {111}, not common.

P h y s. Cleavage {111} imperfect. Fracture splintery to uneven. Variously reported as brittle or not brittle. H. $5\frac{1}{2}$–$6\frac{1}{2}$. G. 4.7–5.4, varying with composition; 5.326 (calc. for $Ca_2Sb_2O_7$ and $a_0 = 10.26\,kX$). Luster vitreous to greasy or sub-adamantine. Color usually pale yellow to yellowish brown; also reddish brown, dark brown. Streak nearly colorless to pale yellow. Transparent to subtranslucent.

O p t.[4] Isotropic. Some varieties show weak to moderate anomalous birefringence, which may be distributed in sectors or zones. The indices of refraction vary markedly with composition.

	n	Birefringence
Manganoan romeite (anal. 2)	1.87 ± 0.01	Weak to moderate
Sodian romeite (anal. 3)	1.83 ± 0.01	Weak
Ferroan romeite (anal. 5)	2.09	Low
Miguel Burnier (different crystals)	1.817–1.854	Isotropic

C h e m. Essentially a pyroantimonate of calcium with the ideal formula $Ca_2Sb_2O_7$. A variety of isomorphous substitutions of both simple and coupled types may occur in the structure.[5] These include principally the substitution of Na, Mn'', Fe'', and Pb for Ca, Ti for Sb, and OH and F for O. The general formula may be written $(Ca,Na,Fe,Mn,Pb)_2(Sb,Ti)_2O_6(O,OH,F)$. The limits of the several substitutions are not well-defined.

Anal.[6]

	1	2	3	4	5	6	7
Na_2O		0.81	5.08	2.03	0.10	0.99	2.70
CaO	25.74	15.81	14.81	19.01	17.42	15.93	17.97
PbO				0.51			6.79
MnO		6.27	2.62	0.43		0.38	1.27
FeO		1.12	1.29	2.92	8.51	4.55	0.79
TiO_2				0.30		11.35	7.93
Sb_2O_5	74.26	74.72	74.72	72.17	72.48	67.52	59.25
F				3.50			[3.63]
H_2O		1.39	1.12	0.66	1.67		0.87
Rem.				0.52	0.30		0.33
Total	100.00	100.12	99.64	102.05	100.48	100.72	[101.53]
F = O				1.48			1.53
				100.57			100.00
G.	5.326			4.967	5.41	4.950	5.11

1. $Ca_2Sb_2O_7$. 2. Manganoan romeite, St. Marcel, Piedmont, Italy.[7] 3. Sodian romeite. Miguel Burnier, Minas Geraes, Brazil.[7] 4. Långban, Sweden [8] (*weslienite*). Rem. is K_2O 0.17, MgO 0.35. 5. Ferroan romeite. Schneeberg, Tyrol [7] (*schneebergite*). Rem. is insol. 6. Titanian romeite. Tripuhy, Minas Geraes, Brazil [9] (*lewisite*). 7. Plumbian romeite. Jakobsberg mine, Sweden [10] (*mauzeliite*). Rem. is K_2O 0.22, MgO 0.11.

Tests. B.B. infusible (romeite). Lewisite and mauzeliite are fusible. Unchanged in C.T. and O.T. Insoluble in acids. Decomposed by fusion with sodium carbonate.

Var. 1. *Sodian.* Atopite *Nordenskiöld* (Geol. För. Förh., **3**, 376, 1877). Contains substantial amounts of Na in substitution for Ca, with relatively low index of refraction and specific gravity.

2. *Ferroan.* Schneebergite *Brezina* (Geol. Reichsanst., Verh., 313, 1880). Contains Fe'' in substitution for Ca, with Fe:Ca = 1:2.6 in analysis 5.

3. *Manganoan.* Atopite pt. Contains Mn'' in substitution for Ca, with Na sometimes also present in considerable amounts.

4. *Titanian.* Lewisite *Hussak* and *Prior* (Min. Mag., **11**, 80, 1895). Contains Ti in substitution for Sb, with Ti:Sb = 1:2.9 in analysis 6 and 1:3.7 in analysis 7.

5. *Plumbian.* Mauzeliite *Sjögren* (Geol. För. Förh., **17**, 313, 1895). Contains Pb in substitution for Ca with Ti also present in substitution for Sb in analysis 7.

6. *Fluorian.* Weslienite *Flink* (Geol. För. Förh., **45**, 567, 1923). Contains a considerable amount of F in substitution for O. F probably has been overlooked in other analyses of romeite.

Occur. The original romeite was found with epidote and manganese oxides at the Saint Marcel mine, Piedmont, Italy. Also found in Vermland, Sweden, in veinlets of hedyphane traversing rhodonite (*atopite*) [11] or hematite and manganophyllite (*weslienite*) [11] at Långban, and with calcite, hausmannite, and other manganese minerals at Jakobsberg (*mauzeliite*).[11]

With calcite, magnetite, chalcopyrite, and sphalerite at Schneeberg in the Austrian Tyrol (*schneebergite*).[11] In Minas Geraes, Brazil, with cinnabar in eluvial sands at Tripuhy near Ouro Preto (*lewisite*) [11] and in manganese ores at Miguel Burnier. Said to occur with nadorite and bindheimite at Hamman N'bail, Constantine, Algeria.[12]

Alter. The titanian romeite from Tripuhy is surficially altered to a pulverulent, sulfur-yellow material.

Artif.[13] $Ca_2Sb_2O_7$ has been prepared by sintering $Ca(NO_3)_2 \cdot 4H_2O$ and H_3SbO_4 at low red heat and purifying the product by washing with boiling water. The isostructural compounds $Cd_2Sb_2O_7$, $SbSb_2O_6(OH)$, and $Na_2Sb_2O_6$ also have been prepared.

Name. Romeite, after the French crystallographer J. B. L. Romé de Lisle (1736–1790). Lewisite, after W. J. Lewis (1847–1926), formerly Professor of Mineralogy in Cambridge University. Mauzeliite, after the Swedish mineral chemist R. Mauzelius (1864–1921). Weslienite, after J. G. H. Weslien, manager of the Långban mines. Atopite from ατοπος, *unusual*, on account of its rarity.

Ref.

1. The original romeite was reported to be tetragonal by Dufrénoy in Damour (1841). All later morphological and x-ray studies of members of the romeite series show that the species is isometric.
2. Form list from Goldschmidt (**1**, 121 [atopite], 1913) and Schaller, *U. S. Geol. Sur., Bull. 610*, 81, 95 (1916). {112} from Pelloux, *Mus. civ. storia nat. Genova, Ann.*, [3], **6**, 22 (1913).
3. For x-ray data see Machatschki, *Chem. Erde*, **7**, 56 (1932); Aminoff, *Ak. Stockholm, Handl.*, [3], **11**, 14 (1933); Zedlitz, *Zs. Kr.*, **81**, 253 (1932); Machatschki and Zedlitz, *Zs. Kr.*, **82**, 72 (1932); Machatschki, *Zs. Kr.*, **73**, 159 (1930).
4. Indices of Larsen (128, 130, 1921) and Rose, *Cbl. Min.*, 268 (1919). On the dispersion of sodian and manganoan romeite ("atopite") from Brazil see Rose (1919). On the optical anomalies see Larsen (1921); Bertrand, *Bull. soc. min.*, **4**, 237 (1881); and Schaller (1916).
5. For a discussion of the variations in composition in structures of this type see Pabst, *Am. Min.*, **24**, 575 (1939) and Machatschki (1932).
6. For additional analyses see Damour (1853); Nordenskiöld (1877); and Hussak, *Cbl. Min.*, 240 (1905) [discussed in Schaller (1916)].
7. Schaller (1916).
8. Aminoff (1933).
9. Hussak and Prior, *Min. Mag.*, **11**, 80 (1895).
10. Sjögren, *Geol. För. Förh.*, **17**, 313 (1895).
11. On the identity of these supposedly distinct species with romeite see Hintze (*Erg.-Bd.*, 571, 1937), Schaller (1916), and the references under 3, above.
12. Lacroix (4, 360, 1910).
13. Baccaredda, *Gazz. chim. ital.*, **66**, 539 (1936); Dihlström and Westgren, *Zs. anorg. Chem.*, **235**, 153 (1937); Schrewelius, *Zs. anorg. Chem.*, **238**, 241 (1938).

HYDROROMEITE. Natta and Baccaredda (*Zs. Kr.*, **85**, 271, 1933).

Calcium-bearing antimony ochers found as yellow to brown pseudomorphs after stibnite from Spain. One from Villafranca, Galicia, with G. 3.50 and H. $3\frac{1}{2}$ gave on analysis CaO 19.19, Fe_2O_3 1.98, Sb_2O_5 62.90, H_2O 13.48, CO_2 3.33, total 100.88. Another, from Higueras, Cordoba, with G. 3.66 and H. 5 gave CaO 14.38, Fe_2O_3 1.50, Sb_2O_5 70.01, H_2O 12.27, total 98.16. Water is lost gradually on heating, dehydration being complete at about 400°. X-ray powder patterns are identical with those of romeite both before and after dehydration; $a_0 = 10.25$ kX. A similar mineral from Grosseto, Italy,[1] approximates to the formula $(Sb''',Ca)Sb_4O_{11}(OH)_2$. Analyses of white (and yellow) material gave: CaO 6.94 (8.93), Sb_2O_5 64.91 (67.39), Sb_2O_3 12.71 (10.10), SiO_2 3.59 (1.55), Al_2O_3 0.01 (0.01), Fe_2O_3 0.70 (0.50), H_2O 10.61 (11.10), total 99.47 (99.58); G. 4.34 (4.08); a_0 10.24Å.

Ref.
1. Fornaseri, *Per. Min.*, **15**, 47 (1946).

44.1.2 M O N I M O L I T E [$(Pb,Ca)_3Sb_2O_8$?]. *Igelström (Ak. Stockholm, Öfv.*, **22**, 227, 1865).

C r y s t.[1] Isometric; probably hexoctahedral—$4/m\,\bar{3}\,2/m$.

Forms:

a 001 d 011 o 111 m 113

Habit.[2] Euhedral crystals of two types: I, octahedral with m, relatively common; II, cubic with o and d, rare.

Phys. Cleavage {111} indistinct. Fracture small conchoidal, splintery. H. $4\frac{1}{2}$–6. G. 5.9–7.3. Luster greasy to adamantine. Color yellow, gray-green, dark brown. Streak (I) straw-yellow, (II) cinnamon-brown. Translucent on thin edges. Isotropic, or weakly birefringent.

Chem. An antimonate of lead, containing also substantial amounts of calcium and ferrous iron. Type I approximates to $(Pb,Ca)_3Sb_2O_8$ with some excess FeO (anal. 2). Type II is close to $(Pb,Fe)_3Sb_2O_8$ (anal. 3). A pyrochlore-type structure would require a formula such as $(Pb,Ca,Fe)_2Sb_2O_7$.

Anal.

	PbO	CaO	MnO	FeO	Sb_2O_5	Rem.	Total	H.	G.
1.	42.40	7.59	6.20		40.29	3.25	99.73	$4\frac{1}{2}$–5	5.94
2.	42.74	9.70	0.41	5.38	40.51	1.10	99.84	5	6.58
3.	55.33		1.16	5.57	38.18		100.24	6	7.29

1. Harstig mine.[3] Rem. is MgO. 2. Harstig mine.[4] Type I. Rem. is MgO 0.56, Na_2O 0.54. Original sum given as 99.24. 3. Harstig mine.[4] Type II.

Occur. With tephroite in calcite veins in ore at the Harstig iron and manganese mine, Pajsberg, Vermland, Sweden. Reported from Långban, Sweden.[5]

Name. From μόνιμος, *stable*, because decomposed chemically with great difficulty.

Ref.
1. Considered tetragonal by Nordenskiöld, *Ak. Stockholm, Öfv.*, **27**, 550 (1870), and Igelström (1865), but shown to be isometric by Flink, *Ak. Stockholm, Bihang*, **12**, Pt. 2, [2], 35 (1887), who reported the forms listed. Monimolite is hexoctahedral if isostructural with romeite and pyrochlore.
2. Flink (1887).
3. Igelström (1865).
4. Flink (1887).
5. Nordenskiöld, *Geol. För. Förh.*, **3**, 379 (1877), on the basis of a museum specimen of uncertain authenticity.

44.1.3 **T R I P U H Y I T E** [$Fe_2Sb_2O_7$ (?)]. *Hussak* and *Prior* (*Min. Mag.*, **11**, 302, 1897). Jujuyite *Ahlfeld* (*Econ. Geol.*, **43**, 598, 1948).

As microcrystalline aggregates. G. 5.82. Color dull greenish yellow to dark brown. Streak canary-yellow. Translucent.

O p t.[1] Small grains in transmitted light are bright canary-yellow in color. Non-pleochroic.

nX	2.19		Biaxial positive (+).
nY	2.20		$2V$ small.
nZ	2.33		$r < v$ very strong.

C h e m. An antimonate of iron, perhaps $Fe'''SbO_4$ or $Fe_2''Sb_2O_7$.
Anal.

	FeO	Sb_2O_5	Rem.	Total	G.
1.	30.75	69.25		100.00	
2.	27.70	66.68	5.62	100.00	5.82

1. $Fe_2Sb_2O_7$. 2. Tripuhy, Brazil. Rem. is CaO 0.82, Al_2O_3 1.40, TiO_2 0.86, SiO_2 1.35, undet. 1.19. The state of oxidation of the iron was not determined in the analysis.

Tests. Infusible. Insoluble in acids.

O c c u r. Found as fragments in the cinnabar-bearing gravels of Tripuhy near Ouro Preto, Minas Geraes, Brazil, associated with lewisite, monazite, rutile, kyanite, magnetite, and other heavy detrital minerals. Also with opaline silica as small veins in dacite lava near Doncellas, Jujuy, Argentina (*jujuyite*).[2]

N a m e. From the original locality.

Ref.

1. Larsen, 145 (1921).
2. Shown to be identical with tripuhyite by Frondel, priv. comm. (1950).

44.1.4 **Flajolotite.** *Lacroix* (**4**, 509, 1910).

Compact or earthy, nodular. Color lemon-yellow. Formula perhaps $4FeSbO_4 \cdot 3H_2O$.

Anal.

	Fe_2O_3	Al_2O_3	SiO_2	Sb_2O_5	H_2O	Total
1.	31.30			63.41	5.29	100.00
2.	31.4			63.5	5.1	100.0
3.	28.0	6.2	7.7	53.8	5.0	100.7

1. $4FeSbO_4 \cdot 3H_2O$. 2. Algeria.[1] 3. Algeria.[2]

Insoluble in acids. Found with calcite, celestite, and nadorite at Hamman N'bail, south of Djebel Nador, Constantine, Algeria.

Ref.

1. Flajolot, *Ann. mines*, **20**, 28 (1871).
2. Pisani in Lacroix (1910).

The following ill-defined materials require further study to establish any of them as mineral species.

AMMIOLITE. *Dana* (534, 1850). Antimonite de Mercure *Domeyko* (*Ann. mines*, **6**, 183, 1844). Antimoniato de cobre con cinabrio terroso *Domeyko* (129, 1860). A deep red to scarlet, earthy powder. Regarded as a mixture of cinnabar with an antimonate of copper. Found as an alteration product of mercurian tetrahedrite at mines in Chile.[1]

AREQUIPITE. *Raimondi* (167, 1878). A supposed silico-antimonate of lead found with cerussite and chrysocolla at the Victoria mine near Tibaya, Arequipa province, Peru. Compact, wax-like with conchoidal fracture. H. ~6. Color honey-yellow. Perhaps a mixture largely of bindheimite.

CORONGUITE. *Raimondi* (88, 1878). Earthy, pulverulent, sometimes slightly lamellar. H. $2\frac{1}{2}$–3. G. 5.05. Color grayish yellow outwardly, blackish with resinous luster internally. An antimonate of lead and silver, if homogeneous. Found at mines in the Corongo district, Pallasca province, and at Pasacancha, Pomabamba province, Peru.

BARCENITE. *Mallet* (*Am. J. Sc.*, **16**, 306, 1878). Fine-granular, compact, or porous, in part columnar and pseudomorphous after livingstonite. H. $5\frac{1}{2}$. G. 5.34. Luster dull to slightly resinous. Color dark gray, nearly black. From Huitzuco, Guerrero, Mexico. Supposedly an antimonate of mercury.

TAZNITE. *Domeyko* (*C.R.*, **85**, 977, 1877). Earthy, imperfectly fibrous. Color yellow. A supposed arsenate-antimonate of bismuth containing H_2O and some Cl from Tazna and Choroloque, Bolivia. Specimens perhaps of this substance have been found to be mixtures of kaolin with bismuth arsenate [2] or of bismutite with bismuth arsenate or a bindheimite-like substance.[3]

Ref.

1. See also Rivot, *Ann. mines*, **6**, 556 (1854), and Field, *J. Chem. Soc. London*, **12**, 27 (1860).
2. Bannister, *Min. Mag.*, **24**, 55 (1935).
3. Frondel, priv. comm. (1949).

TYPE 2. MISCELLANEOUS

MAGNETOSTIBIAN. *Igelström* (*Zs. Kr.*, **23**, 212, 1894). As embedded grains and granular aggregates. Luster metallic. Color and streak black. Magnetic. Apparently an extremely basic antimonate of iron and manganese. Analysis gave (after deducting 68.6 per cent admixed impurities): FeO 17.16, MnO 59.11, Fe_2O_3 12.36, As_2O_5 1.54, Sb_2O_5 9.83, = 100.00. Soluble in HCl without evolution of chlorine. Found with basiliite, calcite, and tephroite at the Sjö mine, Örebro, Sweden. Evidence of species validity is lacking.

BASILIITE. *Igelström* (*Geol. För. Förh.*, **14**, 307, 1892; *Zs. Kr.*, **22**, 470, 1893). Massive, foliated. Luster metallic. Color steel-blue. Streak dark brown. Opaque; in thin splinters transparent and blood-red in color. Non-magnetic. A very basic hydrated antimonate of trivalent manganese. Analysis gave: Mn_2O_3 70.01, Fe_2O_3 1.91, Sb_2O_5 13.09, H_2O 15.00 = 100.01.

In C.T. affords water and turns black and then red-brown. Easily soluble in warm HCl with evolution of chlorine. Alters on exposure to air and acquires a yellow to brown ocherous coating. Found with hausmannite and calcite at the Sjö mine, Örebro, Sweden. Named after the alchemist Basil Valentine, who early wrote on the properties of antimony. Needs verification.

44.2.1 **DERBYLITE** [$Fe_6'''Ti_6Sb_2O_{23}$ (?)]. *Hussak* and *Prior* (*Min. Mag.*, **11**, 85, 176, 1897).

Cryst. Orthorhombic; dipyramidal—$2/m\ 2/m\ 2/m$.

$a:b:c = 0.9661:1:0.5502; \quad p_0:q_0:r_0 = 0.5695:0.5502:1$

$q_1:r_1:p_1 = 0.9661:1.7559:1; \quad r_2:p_2:q_2 = 1.8175:1.0351:1$

Forms:

	ϕ	$\rho = C$	ϕ_1	$\rho_1 = A$	ϕ_2	$\rho_2 = B$
c 001	.....	0°00′	0°00′	90°00′	90°00′	90°00′
a 100	90°00′	90 00	.....	0 00	0 00	90 00
m 110	45 59½	90 00	90 00	44 00½	0 00	45 59½

Habit. Prismatic [001]. m smooth, a rough, c rare and uneven.

Twinning. On {011}; commonly as cruciform twins crossing at 57°38½′, rarely as trillings.

P h y s. No cleavage. Fracture conchoidal. Very brittle. H. 5. G. 4.53. Luster resinous. Color pitch black. Opaque.

O p t.[1] In small grains translucent with a dark brown color. Not pleochroic.

ORIENTATION	n(Li)	
X	2.45	Biaxial positive (+).
Y	2.45	$2V$ nearly 0°.
Z	2.51	

C h e m. An antimonate of titanium and divalent iron. The formula is uncertain, probably $Fe_6''Ti_6Sb_2O_{23}$. Analysis gave:

	Na$_2$O	K$_2$O	CaO	FeO	Al$_2$O$_3$	TiO$_2$	Sb$_2$O$_5$	SiO$_2$	Ign.	Total
1.				34.93		38.85	26.22			100.00
2.	0.76	0.28	0.32	32.10	3.17	34.56	24.19	3.50	0.50	99.38
3.				35.33		38.04	26.63			100.00

1. $Fe_6''Ti_6Sb_2O_{23}$. 2. Tripuhy.[2] 3. Analysis 2 recalculated after deduction of extraneous (?) oxides.

Tests. Infusible. Insoluble in acids. Decomposed by fusion in KHSO$_4$.

O c c u r. Found in the cinnabar-bearing gravels of Tripuhy near Ouro Preto, Minas Geraes, Brazil. Associated with lewisite, tripuhyite, xenotime, monazite, zircon, rutile, hematite, muscovite.

N a m e. After the American geologist Orville A. Derby (1851–1915), formerly Director of the Geological Survey of Brazil.

Ref.
1. Larsen and Berman (147, 1934).
2. Prior analysis in Hussak and Prior (1897).

FERROSTIBIAN. *Igelström* (*Geol. För. Förh.*, **11**, 389, 1889; *Jb. Min.*, I, 250, 1890).

In monoclinic (?) crystals with {100}, {010}, and {001}, the last face rectangular. Cleavage in two or three directions. H. 4. Luster submetallic. Color black to grayish black. Streak brownish black tending to red. Opaque; in thin sections blood-red. Weakly magnetic. Apparently a very basic hydrated antimonate of divalent iron and manganese. Analysis gave: FeO 22.60, MnO 46.97, (Mg,Ca)CO$_3$ 2.14, Sb$_2$O$_5$ 14.80, SiO$_2$ 2.24, H$_2$O 10.34 = 99.09.

B.B. fuses on thin edges to a black magnetic glass. Dissolves only imperfectly in acids. Found intergrown with rhodonite at the Sjö mine, Grythytte parish, Örebro, Sweden. Needs verification.

RHODOARSENIAN. *Igelström* (*Zs. Kr.*, **22**, 469, 1893).

Found as small rose-red spherules embedded in arseniopleite at the Sjö mine, Örebro, Sweden. H. 4. Luster vitreous. Transparent. Analysis gave (after deducting CaCO$_3$): CaO 21.53, MgO 5.37, MnO 49.28, As$_2$O$_5$ 12.17, H$_2$O 11.65, Pb and Cl tr. = 100. Regarded as the arsenic analogue of ferrostibian. Needs verification.

MANGANOSTIBITE. Manganostibiite *Igelström* (*Geol. För. Förh.*, **7**, 210, 1884; *Bull. soc. min.*, **7**, 120, 1884).

Monoclinic (?). In embedded grains, sometimes with crystal outlines. Fracture uneven. Color black. Streak chocolate-brown. Luster greasy. Opaque.

O p t.[1] In transmitted light, dark reddish brown and nearly opaque.

ORIENTATION		n	PLEOCHROISM	
X	b	1.92 ± 0.02	Reddish brown	Biaxial negative $(-)$.
Y		1.95 ± 0.02		$2V$ small.
$Z \wedge c$	very large	1.96 ± 0.02	Nearly opaque	

As nearly opaque rods and fibers with positive elongation.

C h e m. A highly basic manganese antimonate-arsenate, perhaps $(Mn,Fe,Mg,Ca)_{10}(Sb,AsO_4)_2O_7$, with $As:Sb = 1:2.3$. Analysis gave: CaO 4.62, MgO 3.00, FeO 5.00, MnO 55.77, As_2O_5 7.44, Sb_2O_5 24.09, total 99.92.

Tests. B.B. infusible. Completely soluble in HCl, partly so in HNO_3.

O c c u r. Found very rarely with hausmannite in the Moss mine, Nordmark, Sweden.

N a m e. In allusion to the composition.

Needs verification.

Ref.

1. Larsen (104, 1921) and Larsen and Berman (206, 1934); it is not certain that the data refer to manganostibite.

HEMATOSTIBITE. Hämatostibiite *Igelström* (*Bull. soc. min.*, **8**, 143, 1885). Hematostibiite.

Orthorhombic (?). As embedded grains and as lamellar masses. Perfect cleavage in one direction. H. 5. Color black. Streak brown. Luster submetallic on cleavage surfaces. Opaque. Biaxial negative $(-)$, with $X \perp$ cleavage and $2V$ very small.[1] Blood-red by transmitted light. Chemically a highly basic antimonate of divalent manganese and iron. Analysis gave (after deducting much calcite and tephroite): MnO 51.7, FeO 9.5, $(Mg,Ca)O$ 1.6, Sb_2O_5 37.2 = 100.0.

Tests. Infusible on charcoal. Easily soluble in HCl.

O c c u r. At the Sjö mine, Grythytte parish, Örebro, Sweden, with tephroite, calcite, barite, jacobsite, and chondrarsenite (?).

N a m e. From αἷμα, blood, and *stibium*, antimony, in allusion to its color and composition.

Needs study; possibly identical with manganostibite or catoptrite.

Ref.

1. Bertrand in Igelström (1885).

44.2.2 **S W E D E N B O R G I T E** [$NaBe_4SbO_7$]. *Aminoff* (*Zs. Kr.*, **60**, 262, 1924).

C r y s t.[1] Hexagonal—P; dihexagonal-pyramidal—$6\ m\ m$.

$$a:c = 1:1.6309;\ ^2 \quad p_0:r_0 = 1.8832:1$$

Forms:[3]

		ϕ	$\rho = C$	M	A_2
c	0001	.....	0°00'	90°00'	90°00'
m	$10\bar{1}0$	30°00'	90 00	60 00	90 00
q	$10\bar{1}4$	30 00	25 12½	77 42	90 00
o	$10\bar{1}3$	30 00	32 07	74 35	90 00
x	$10\bar{1}2$	30 00	43 16½	69 57½	90 00
p	$10\bar{1}1$	30 00	62 02	63 47½	90 00
y	$20\bar{2}1$	30 00	75 08	61 06	90 00

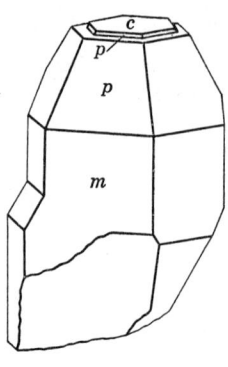

Structure cell.[4] Space group $C6mc$. a_0 5.42 kX, c_0 8.80. $a_0:c_0 = 1:1.624$. Cell contents $Na_2Be_8Sb_2O_{14}$.

Habit. Short prismatic [0001] with $c\ p\ m$ dominant. c and p sometimes are unequally developed at opposite terminations of [0001].

Twinning. Observed, but not described.[5]

Etch Figures. Pits produced by molten sodium carbonate on $\{10\bar{1}0\}$ and $\{10\bar{1}1\}$ show vertical planes of symmetry, but definite results could not be obtained on $\{0001\}$.

Phys. Cleavage $\{0001\}$, distinct. Fracture subconchoidal. H. ~8. G. 4.285; 4.316 (calc.). Colorless and transparent; also pale wine-yellow to honey-yellow.

Opt. In transmitted light, colorless.

λ	n (486mμ)	n (589mμ)	n (687mμ)	
O	1.7822	1.7724	1.7684	Uniaxial negative $(-)$.
E	1.7791	1.7700	1.7658	

Chem. An antimonate of sodium and beryllium, $NaBe_4SbO_7$. Small amounts of P and of Ca and Mg apparently substitute for Sb and Na, respectively.

Anal.

	1	2	3
Na_2O	10.58	8.50	[8.49]
CaO		0.94	0.90
MgO		0.52	0.65
BeO	34.18	34.72	34.92
Sb_2O_5	55.24	54.17	54.41
P_2O_5		0.23	0.63
Rem.		0.60	
Total	100.00	99.68	[100.00]
G.	4.316	4.285	

1. $NaBe_4SbO_7$. 2. Långban.[6] Rem. is K_2O 0.21, H_2O 0.39. Average of three partial analyses on material dried at 120°. 3. Långban.[7] Sample heated and acid washed before analysis.

Tests. Insoluble in acids. Decomposed by fused Na_2CO_3 or $KHSO_4$.

Occur. At Långban, Sweden, as crystals up to 8 mm. in size with calcite, hematite, bromellite, richterite, and manganophyllite in skarn.

Name. After the Swedish philosopher Emanuel Swedenborg (1688–1772).

Ref.

1. The polar nature of [0001] is suggested by morphological evidence but is not proved.
2. Aminoff (1924); identical elements $(a:c = 1:1.6308 \pm 0.0003)$ were obtained morphologically by Pauling, Klug, and Winchell, *Am. Min.*, **20**, 492 (1935). On the morphological resemblance to corundum see Gossner, *Cbl. Min.*, 289 (1925).
3. Aminoff (1924).
4. Space group of Aminoff, *Svenska Vet.-Ak. Stockholm, Handl.* [3], **11**, no. 4, 3 (1933), and Pauling, *et al.* (1935). Cell dimensions of Aminoff by powder method; Pauling *et al.* give a_0 5.47 kX, c_0 8.92, with $a_0:c_0 = 1:1.631$ and G. 4.18 (calc.).

5. Pauling, et al. (1935).
6. Almström analysis in Aminoff (1924); the Al_2O_3 reported in his analysis is here converted to BeO.
7. Blix analysis in Aminoff (1933); Sb_2O_5 given as 55.41 in original appears to be a misprint for 54.41.

44.2.3 CATOPTRITE.   *Flink (Geol. För. Förh.*, 39, 426, 1917). Katoptrite.

Cryst.[1] Monoclinic; prismatic—$2/m$.

$a:b:c = 0.3981:1:0.2450;$ $\beta\,101°13';$ $p_0:q_0:r_0 = 0.6154:0.2403:1$

$r_2:p_2:q_2 = 4.1611:2.2508:1;$ $\mu\,78°47';$ $p_0'\,0.6274,\ q_0'\,0.2450,\ x_0'\,0.1983$

Forms:[2]

		ϕ	ρ	ϕ_2	$\rho_2 = B$	C	A
c	001	90°00'	11°13'	78°47'	90°00'	....	78°47'
b	010	0 00	90 00	....	0 00	90°00'	90 00
a	100	90 00	90 00	0 00	90 00	78 47	....
n	140	32 38	90 00	0 00	32 38	83 58½	57 22
M	120	52 01	90 00	0 00	52 01	81 11	37 59
m	110	68 39½	90 00	0 00	68 39½	79 33½	21 20½
d	011	38 59½	17 29½	78 47	76 29½	13 30½	79 06
e	031	15 06	37 16½	78 47	54 13	35 47	80 55
r	111	73 28½	40 44	50 27½	79 18½	30 07½	51 16½
o	$\bar{1}11$	−60 16½	26 17½	113 13½	77 19	36 25½	112 37½
p	$\bar{1}31$	−30 17	40 23½	113 13½	55 58½	46 53	109 04½

Habit. Crystals usually tabular {010} and somewhat elongated [001]; also elongated [101], or equant. As minute grains to masses a centimeter or so in size, the larger masses seldom with crystal faces. Common forms: $b\ a\ o\ p\ c$.

Phys. Cleavage {100}, perfect, micaceous. Brittle. H. 5½. G. 4.5. Luster metallic, brilliant. Color iron-black to jet-black. Opaque; in thin flakes translucent and fire-red in color.

Opt.[3] In transmitted light, red in color.

Orientation	n	
$X \sim \perp \{100\}$	1.92	Biaxial negative (−).
$Y\ \ b$	1.95	$2V$ small.
$Z \wedge c\ \ \sim\!-3°$	1.95	$r > v$ strong, inclined. Strongly pleochroic in red-brown to red-yellow.

Chem. A silico-antimonate of aluminum and divalent manganese, with some substitution of Mg and Fe'' for Mn'' and of Fe''' for Al. The reported analysis affords the ratio $14(Mn,Mg,Fe)O \cdot 2(Al,Fe)_2O_3 \cdot Sb_2O_5 \cdot 2SiO_2$, but the formulation is uncertain. Analysis gave (average of four partial analyses):[4] CaO 0.58, MgO 3.06, FeO 2.44, MnO 52.61, Al_2O_3 9.50, Fe_2O_3 3.58, Sb_2O_5 20.76, SiO_2 7.75, H_2O 0.11, total 100.39.

Occur. At the Brattfors mine, near Nordmark, Sweden, with magnetite in crystalline limestone.

Name. From κάτοπτρον, *mirror*, in allusion to the perfectly reflecting cleavage surfaces.

Ref.
1. Orientation and angles of Flink (1917) recalculated by La Forge, priv. comm. (1935); unit of Palache, priv. comm. (1936). Transformation: Flink to Palache, $\frac{1}{2}00/010/00\frac{1}{2}$.
2. Flink (1917). Uncertain: $q\{\bar{1}71\}$.
3. Larsen and Berman (206, 1934).
4. Mauzelius analysis in Flink (1917).

LAMPROSTIBIAN. *Igelström* (*Geol. För. Förh.*, **15**, 471, 1893; *Zs. Kr.*, **22**, 467, 1893).

In foliated or scaly masses. Brittle. H. 4. Luster brilliant. Color lead-gray. Streak red. Opaque, except in very thin splinters and then blood-red in color. Not magnetic. Inferred to be an anhydrous antimonate of divalent iron and manganese. Soluble with difficulty in hot concentrated HCl without evolution of Cl. Found with tephroite and calcite at the Sjö mine, Örebro, Sweden. Needs verification. Named from λαμπρός, *shining*, and *stibium*, antimony.

CHONDROSTIBIAN. *Igelström* (*Geol. För. Förh.*, **15**, 343, 1893; *Zs. Kr.*, **22**, 43, 1893).

Color yellowish red to dark brownish red. Transparent in small pieces. Feebly magnetic. An analysis on material containing 51 per cent of impurities indicates the mineral to be essentially a hydrous antimonate of trivalent iron and manganese. Found with tephroite and dolomite as grains embedded in barite at the Sjö mine, Örebro, Sweden.

STIBIATIL. *Igelström* (*Geol. För. Förh.*, **11**, 391, 1889; *Jb. Min.*, I, 254, 1890).

In prismatic monoclinic (?) crystals with rectangular and rhombic cross section. H. 5–5$\frac{1}{2}$. Luster metallic. Color iron-black. Streak iron-black, inclining toward brown. Opaque. Chemically, an antimonate of iron and manganese. An approximate analysis gave: FeO 26, Mn_2O_3 44, Sb_2O_5 (and H_2O) 30 = 100.

In C.T. affords water. B.B. fuses on thin edges to a black magnetic bead. Incompletely soluble in acids. Found at the Sjö mine, Grythytte parish, Örebro, Sweden. Needs verification.

45 ACID AND NORMAL ANTIMONITES AND ARSENITES

TYPE 1. MISCELLANEOUS

45.1.1 A R M A N G I T E [$Mn_3(AsO_3)_2$]. *Aminoff* and *Mauzelius* (*Geol. För. Förh.*, **42**, 301, 1920).

C r y s t.[1] Hexagonal—P; hexagonal-scalenohedral—$\bar{3}\,2/m$ (?).

$$a{:}c = 1{:}0.6558;\,{}^2 \quad p_0{:}r_0 = 0.7572{:}1$$

Forms:[3]

	ϕ	ρ	M	A_2
c 0001	....	0°00′	90°00′	90°00′
m 10$\bar{1}$0	30°00′	90 00	60 00	90 00
f 10$\bar{1}$1	30 00	37 08	72 26	90 00
p 20$\bar{2}$1	30 00	56 33½	65 20½	90 00

Structure cell.[4] Space group $R3m$, $R32$ or $R\bar{3}m$ (?). a_0 13.44 kX, c_0 8.72; $a_0{:}c_0 = 1{:}0.6596$. Cell contents Mn_{27}-$(AsO_3)_{18}$ in the hexagonal unit.

Habit. Short prismatic [0001].
Twinning. On {02$\bar{2}$1}, lamellar.
P h y s. Cleavage {0001}, fair. H. ~4. G. 4.43;[5] 4.47 (calc.). Color black. Streak brown. Transparent in thin splinters.
O p t.[6] In transmitted light, yellow to brown.

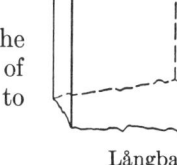

Långban.

| nO | 2.01 | Uniaxial negative (−). |
| nE | 1.99 | Not pleochroic. |

C h e m. Manganese arsenite, $Mn_3(AsO_3)_2$. The CO_2 together with the Ca, Mg, and a proportion of the Mn reported in the analysis is probably due to an admixed carbonate.

Anal.

	MnO	FeO	PbO	CaO	MgO	As_2O_3	Sb_2O_3	CO_2	H_2O	Insol.	Total
1.	51.82					48.18					100.00
2.	45.06	2.19	0.32	2.83	0.49	42.92	0.40	5.08	0.71	0.20	100.00
3.	47.06	2.49	0.36			48.82	0.46		0.81		100.00

1. $Mn_3(AsO_3)_2$. 2. Långban.[7] FeO includes some Fe_2O_3 present perhaps as admixed limonite. Average of three analyses. 3. Analysis 2 recalculated after deduction of insol., CO_2, CaO, MgO and MnO 3.69 as $(Ca,Mn,Mg)CO_3$.

Tests. In C.T. turns gray, loses a little water, and at higher temperature fuses into a grayish black, porous glass. Easily soluble in HCl, separating As_2O_3 which dissolves on dilution and warming.

Occur. At Långban, Sweden, in calcite-barite veinlets with hematite and unidentified minerals.

Name. In allusion to the chemical composition, from *ar*senic and *man*ganese.

Ref.
1. Crystal class as stated by Aminoff and Mauzelius (1920) but the form development as illustrated is trigonal not rhombohedral. The possible space groups found by Aminoff, *Svenska Vet. Ak. Stockholm, Handl.*, [3], **11**, no. 4, 19 (1933), are, however, rhombohedral.
2. Angles and orientation of Aminoff and Mauzelius (1920); unit of structure cell of Aminoff (1933). Transformation: old to new, $1000/0100/0010/000\frac{1}{2}$.
3. Aminoff and Mauzelius (1920).
4. Aminoff (1933).
5. As calculated from original determination of 4.23 on material containing 12 per cent admixed carbonate.
6. Larsen and Berman (91, 1934).
7. Mauzelius analysis in Aminoff and Mauzelius (1920).

45.1.2   **TRIGONITE**   [$MnPb_3H(AsO_3)_3$].   *Flink* (*Geol. För. Förh.*, **42**, 436, 1920).

Cryst.[1] Monoclinic; domatic—*m*.

$a:b:c = 1.0740:1:1.6590;$ $\beta\ 91°31';$ $p_0:q_0:r_0 = 1.5447:1.6584:1$

$r_2:p_2:q_2 = 0.6030:0.9314:1;$ $\mu\ 88°29';$ $p_0'\ 1.5452,\ q_0'\ 1.6590,\ x_0'\ 0.0265$

Forms:[2]

Lower	Upper	ϕ	ρ	ϕ_2	$\rho_2 = B$	C	A
$\bar{c}$	c 001	90°00′	1°31′	88°29′	90°00′	.....	88°29′
	b 010	0 00	90 00	.....	0 00	90°00′	90 00
	a 100	90 00	90 00	0 00	90 00	88 29	.....
	$-a$ $\bar{1}00$	-90 00	90 00	180 00	90 00	91 31	180 00
	m 110	42 58	90 00	0 00	42 58	88 58	47 02
	$-m$ $\bar{1}10$	-42 58	90 00	180 00	42 58	91 02	132 58
	l 210	61 46$\frac{1}{2}$	90 00	0 00	61 46$\frac{1}{2}$	88 40	28 13$\frac{1}{2}$
	h 014	3 39$\frac{1}{2}$	22 34	88 29	67 29	22 31	88 36
	g 012	1 50	39 41$\frac{1}{2}$	88 29	50 20	39 40	88 50
$\bar{f}$	f 011	0 55	58 55$\frac{1}{2}$	88 29	31 05$\frac{1}{2}$	58 54$\frac{1}{2}$	89 13
$\bar{p}$	p 101	90 00	57 32	32 28	90 00	56 01	32 28
$\bar{s}$	$-s$ $\bar{1}01$	-90 00	56 38	146 38	90 00	58 09	146 38
	o 111	43 27$\frac{1}{2}$	66 22	32 28	48 19	65 19$\frac{1}{2}$	50 56$\frac{1}{2}$
$\bar{n}$	$\bar{1}11$	-42 28$\frac{1}{2}$	66 02	146 38	47 37$\frac{1}{2}$	66 22$\frac{1}{2}$	128 06

Structure cell.[3] Space group Pn. $a_0\ 7.25 \pm 0.04\ kX$, $b_0\ 6.80 \pm 0.04$, $c_0\ 11.07 \pm 0.04$; $\beta\ 91°49' \pm 15'$; $a_0:b_0:c_0 = 1.066:1:1.628$. Cell contents uncertain, probably $Mn_2Pb_6H_2(AsO_3)_6$ for brown material with G. 6.33 (calc.).

Trigonite 1033

Habit. Domatic. Thick tabular {010} with a three-sided outline due to the large development of c, $\bar{s}$, $\bar{p}$; also complexly modified. As subparallel growths.

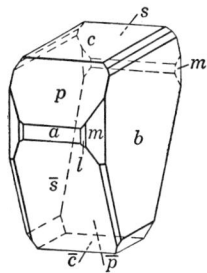
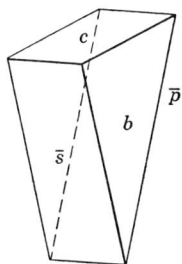

Phys. Cleavage {010} perfect, {101} good. Fracture uneven. H. 2–3. G. variable,[4] 6.1–7.1. Luster vitreous to adamantine. Color variable, sulfur-yellow to yellowish brown and dark brown. Translucent.

Opt.[5] In transmitted light, pale yellow to brownish yellow and not pleochroic.

Orientation		n	
X		2.07	Biaxial negative (−).
Y	b	2.10	$2V$ very large.
Z		2.12	$r < v$, perceptible.
			Ext. angle on $c = 45°$.

Chem. An acid arsenite of lead and manganese, $MnPb_3H(AsO_3)_3$. The composition apparently varies due to isomorphous substitution but the nature of the variation is not known. Analysis gave:

	CaO	MgO	FeO	MnO	PbO	As$_2$O$_3$	H$_2$O	Cl	Insol.	Total
1.				6.78	64.00	28.36	0.86			100.00
2.	0.23	0.11	0.15	6.79	63.40	28.83	0.81	tr.	0.13	100.45

1. $MnPb_3H(AsO_3)_3$. 2. Långban.[6]

Tests. B.B. easily fusible. Soluble in dilute acids.

Occur. At Långban, Wermland, Sweden, associated with native lead and dixenite in crevices in dolomite-hausmannite ore.

Name. From τρίγωνον, *a triangle*, in reference to the shape of the crystals.

Ref.
1. Flink (1920).
2. Flink (1920); the large basal pedion is made upper. The form indices for o and n are reversed in Flink's tabulation, p. 439 (see figures and text, p. 442).
3. Wolfe and Frondel, priv. comm. (1946), by Weissenberg method.
4. The G. and color varies accompanying a compositional variation of an unknown nature. Flink (1920) gave G. 8.28.
5. Larsen and Berman (209, 1934).
6. Mauzelius analysis in Flink (1920).

45.1.3 TRIPPKEITE. vom Rath (Ber. niederrhein. Ges., 209, 1880).

C r y s t.[1] Tetragonal; ditetragonal-dipyramidal—$4/m\ 2/m\ 2/m$.

$a:c = 1:0.6477; \quad p_0:r_0 = 0.6477:1$

Forms:[2]

	ϕ	ρ	A	M
c 001	....	0°00′	90°00′	90°00′
a 010	0°00′	90 00	90 00	45 00
m 110	45 00	90 00	45 00	90 00
u 011	0 00	32 56	90 00	67 23
o 021	0 00	52 20	90 00	55 58
y 122	26 34	35 54	74 48	79 19
x 121	26 34	55 22	68 24	74 55

Less common:

e 061 z 232

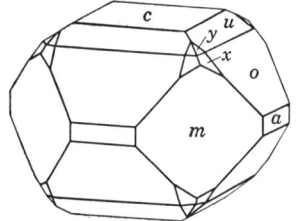

Structure cell.[3] Lattice primitive tetragonal. a_0 8.55 ± 0.02 kX, c_0 5.65 ± 0.02; $a_0:c_0 = 1:0.661$.

Habit. Short prismatic [001], the crystals often somewhat bent. Dominant forms: $o\ a\ c$.

Phys. Cleavage {100} perfect, {110} good. Cleaving tends to break the crystals into shreds or fibers parallel [001] which are easily bent. Soft. G.[6] 4.8 ± 0.5. Brilliant luster. Color greenish blue.

O p t.[4] In transmitted light, pale blue-green.

nO 1.90 ± 0.01   Uniaxial positive (+).
nE 2.12 ± 0.01   Not perceptibly pleochroic.

Chem. An arsenite of copper[5] of unknown formula. No quantitative analysis has been made.

Tests. Easily soluble in HCl or HNO_3.

Occur. Found with olivenite in copper deposits near Copiapó, Chile.

Name. After the Polish mineralogist Paul Trippke (?–1880), the discoverer of the mineral.

Ref.

1. Symmetry from morphology. The setting here adopted is from Goldschmidt, **3**, 239 (1891), and agrees with the structure cell. Transformation: vom Rath (1880) to Goldschmidt (1891), $\frac{1}{2}\bar{1}0/\frac{1}{2}\frac{1}{2}0/00\frac{1}{2}$.
2. The form z reported by vom Rath as {24.5.20} is very close to {514} which on transformation becomes {232}.
3. Pabst, priv. comm. (1948), by powder method.
4. Larsen, 145 (1921).
5. Damour, *Bull. soc. min.*, **3**, 178 (1880), on the basis of qualitative tests. The formula $5CuO \cdot 2As_2O_3$ based on the supposed isomorphism of trippkeite with schafarzikite is not well-established. See also ref. 5 under schafarzikite.
6. Pabst (1948) by microbalance.

45.1.4 **Schafarzikite** [$Fe_5Sb_4O_{11}$]. *Krenner* (*Zs. Kr.*, **56**, 198, 1921).

Cryst. Tetragonal; ditetragonal-dipyramidal—$4/m\ 2/m\ 2/m$.

Forms:[2] $a:c = 1:0.9538;$[1] $p_0:r_0 = 0.9538:1$

	φ	ρ	A	M
c 001	.....	0°00′	90°00′	90°00′
a 100	0°00′	90 00	90 00	45 00
m 110	45 00	90 00	45 00	90 00
e 012	0 00	25 30	90 00	72 17
v 021	0 00	62 20	90 00	51 13½
l 031	0 00	70 44	90 00	48 07½
p 111	45 00	53 27	55 23	90 00
r 132	18 26	56 27	74 43	68 07

Habit. Prismatic [001]. Dominant forms: *a e r v c*. {100} striated parallel [001].

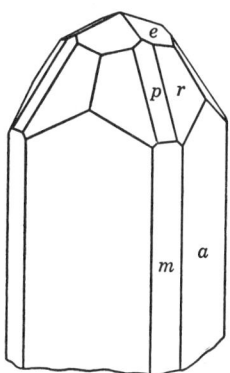

Phys.[3] Cleavage {110} perfect, {100} very good, {001} trace. H. $3\frac{1}{2}$. G. 4.3. Luster metallic. Color red to red-brown. Streak brown. Opaque; yellow in thin sheets.

Opt.[3]

	DICHROISM	
nE	Brownish yellow	Uniaxial positive (+).
$nO > 1.74$	Straw-yellow	Birefringence weak.

Chem.[5] An antimonite of ferrous iron, $Fe_5Sb_4O_{11}$.
Anal.

	FeO	Sb_2O_3	Total	G.
1.	38.40	61.60	100.00	
2.	38.2	61.3	99.5	4.3

1. $Fe_5Sb_4O_{11}$. 2. Pernek, Czechoslovakia.[4]

Occur. Found very sparingly with kermesite, valentinite, and senarmontite on stibnite in the antimony mine at Pernek, near Malaczka, Slovakia.

Name. After the Hungarian mineralogist Ferenc Schafarzik (1854–1927).

Ref.
1. Tokody, *Zs. Kr.*, **62**, 123 (1925). Krenner (1921) gives $a:c = 1:0.9787$ which Tokody corrects to $1:0.9792$. Both Krenner and Tokody supposed schafarzikite to be isomorphous with trippkeite. The form development of the two minerals, however, shows little similarity. In the absence of x-ray data for schafarzikite and of a quantitative chemical analysis for trippkeite, their supposed isomorphism is not well-established. Hence a new setting, as for trippkeite, has not been adopted for schafarzikite.
2. Tokody (1925); Krenner (1921). Tokody's $e\{101\}$ should be $e\{102\}$.
3. Tokody (1925).
4. Hueber, *Cbl. Min.*, 337 (1932). Microanalysis made on 4.308 mg.
5. Zemann, priv. comm. (1950) added in press, finds schafarzikite to be $FeSb_2O_4$, with a_0 8.59 kX, c_0 5.92, and isostructural with trippkeite ($CuAs_2O_4$).

46 BASIC OR HALOGEN-CONTAINING ANTIMONITES, ARSENITES

TYPE 1. MISCELLANEOUS

46.1.1 ECDEMITE. Ekdemit A. E. Nordenskiöld (*Geol. För. Förh.*, **3**, 379, 1877).

Tetragonal. As coatings of small tabular crystals and as coarsely foliated masses. Cleavage $\{001\}$ distinct. H. $2\frac{1}{2}$–3. G. 7.14. Luster vitreous on cleavage surfaces, on fracture surfaces greasy. Color greenish yellow to yellow. Translucent.

Opt.[1] Uniaxial negative $(-)$.

	n(Li) (Långban)
O	2.32 ± 0.02
E	2.25 ± 0.02

Basal sections are in part biaxial and may contain systems of biaxial lamellae intimately intergrown (see *heliophyllite*).

Chem. An oxychloride of lead and arsenic. The formula is uncertain,[2] perhaps $Pb_6As_2O_7Cl_4$. Analysis gave:

	PbO	As$_2$O$_3$	Cl	O = Cl	Total	G.
1.	81.32	12.01	8.61	1.94	100.00	
2.	83.45	10.60	8.00	1.81	100.24	7.14

1. $Pb_6As_2O_7Cl_4$. 2. Långban.

Tests. Easily fusible to a yellow mass, with a sublimate of lead chloride. Soluble in nitric acid or warm hydrochloric acid and in alkalies.

Occur. Found as crystals and masses associated with heliophyllite in manganoan calcite at Långban, Wermland, Sweden. Ecdemite apparently also constitutes the uniaxial parts of the intergrowths with heliophyllite found at the Harstig mine, Pajsberg, at Jakobsberg and Långban, all in Wermland, Sweden. Reported as bright orange incrustations on wulfenite from the Mammoth mine, Pinal County, Arizona.

Artif.[3] Not obtained from melts containing $PbCl_2$, PbO, and As_2O_3.
Name. From ἔκδημος, *unusual*, in allusion to the composition.

Ref.
1. Larsen (71, 1921).
2. See Sillén and Melander, *Zs. Kr.*, **103**, 420 (1941), and Hamberg, *Geol. För. Förh.*, **11**, 229 (1889).
3. Sillén and Melander (1941).

46.1.2   **HELIOPHYLLITE.**   Flink (*Ak. Stockholm, Öfv.*, **45**, 574, 1888).

Orthorhombic, pseudo-tetragonal. In acute pyramidal crystals, with (001) ∧ (111) ~53° [Pajsberg][1] and the inclined faces strongly striated horizontally; also tabular {001} with {011}, {101}, and (001) ∧ (011) 65°36′ and (011) ∧ (01$\bar{1}$) 78°48′ [Långban].[2] Commonly massive, coarsely foliated or granular.

Phys. Cleavage {011} nearly perfect. H. ~2. G. 6.89 (Harstig). Luster vitreous on cleavage planes, greasy on fracture surfaces. Color yellow to greenish yellow. Translucent.

Opt. Biaxial negative (−) with low birefringence. 2V large. Strong dispersion, $r < v$. The axial plane is perpendicular {001}. Not pleochroic.

Basal sections of heliophyllite are in part uniaxial and in part biaxial;[3] the former may be ecdemite. The biaxial parts appear as systems of lamellae crossing at right angles in parquet fashion; they are sometimes also arranged in the direction of a ditetragonal prism (210). The lamellae extinguish at 45° to the rectangular outlines of the crystal, and the axial plane is parallel to their length. The lamellae remain unchanged up to at least 200°.

Chem. An oxychloride of lead and arsenic. The formula is uncertain, perhaps $Pb_6As_2O_7Cl_4$. Apparently dimorphous with ecdemite.

Anal.

	1	2	3	4
CaO			0.08	0.11
(Fe,Mn)O		0.54	0.07	0.16
PbO	81.32	80.70	81.03	80.99
As_2O_3	12.01	11.69	10.85	10.49
Sb_2O_3			0.56	1.38
Cl	8.61	8.00	8.05	7.96
	101.94	100.93	100.64	101.09
O = Cl	1.94	1.81	1.82	1.80
Total	100.00	99.12	98.82	99.29
G.		6.886		

1. $Pb_6As_2O_7Cl_4$. 2. Harstig mine, Sweden.[4] 3. Harstig mine (Type I).[1] 4. Harstig mine (Type II).[1]

Tests. As with ecdemite.

Occur. Found intergrown with ecdemite. At Jakobsberg, with inesite at the Harstig mine near Pajsberg, and at Långban (in part as individual crystals), in Wermland, Sweden.

Alter. To a fine-grained, gray material.

Name. From ἥλιος, *sun*, and φίλλον, *leaf*, in allusion to the color and structure.

Both heliophyllite and ecdemite are ill-defined and need further study, particularly as to their mutual relations.

Ref.
1. Hamberg, *Geol. För. Förh.*, **11**, 229 (1889).
2. Brögger quoted by Flink (1888) on apparently orthorhombic crystals found associated with ecdemite by Nordenskiöld, *Geol. För. Förh.*, **3**, 379 (1877); the crystals are regarded as twinned fourlings. Sillén and Melander, *Zs. Kr.*, **103**, 420 (1941), found material from Harstig to be biaxial but with tetragonal Laue symmetry ($4/mmm$) and the cell dimensions $a_0 = b_0 = 10.8$ kX, c_0 25.6. See also Strunz, *Naturwiss.*, **30**, 89 (1942).
3. See Hamberg (1889) and Sillén and Melander (1941).
4. Flink (1888).

46.1.3 FINNEMANITE [$Pb_5(AsO_3)_3Cl$]. *Aminoff* (*Geol. För. Förh.*, **45**, 160, 1923).

C r y s t.[1] Hexagonal—P; hexagonal-dipyramidal—$6\ m$ or hexagonal pyramidal—6.

$$a:c = 1:0.6880; \quad p_0:r_0 = 0.7945:1$$

Forms:[2]

	ϕ	$\rho = C$	M	A_2
c 0001	.....	0°00′	90°00′	90°00′
m 10$\bar{1}$0	30°00′	90 00	60 00	90 00
p 10$\bar{1}$1	30 00	38 28	71 52½	90 00

Structure cell.[3] Space group not determined. a_0 10.21 kX, c_0 6.975; $a_0:c_0 = 1:0.683$. Cell contents $Pb_{10}(AsO_3)_6Cl_2$.

Habit. As crusts of small crystals prismatic [0001].

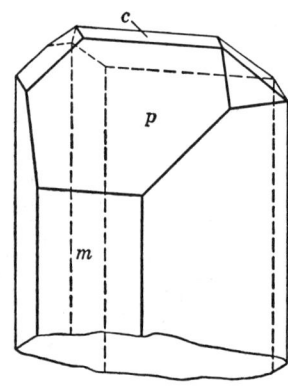

P h y s. Cleavage {10$\bar{1}$1}, distinct. H. 2½. G. 7.265;[4] 7.55 (calc.). Luster sub-adamantine. Color gray to black, in thin flakes somewhat olive-green. Translucent to opaque.

O p t. In transmitted light, olive-green in color.

	n (486mμ)	n (589mμ)	n (687mμ)	
O	2.3634	2.2949	2.2651	Uniaxial negative (−).
E	2.3449	2.2847	2.2594	

Chem. A chlorarsenite of lead, $Pb_5(AsO_3)_3Cl$. Finnemanite apparently is isostructural with the analogous arsenate, mimetite.

Anal.

	1	2
Na_2O		0.24
K_2O		0.44
CaO		0.39
PbO	77.49	76.83
As_2O_3	20.61	20.54
Cl	2.46	2.42
Rem.		tr.
	100.56	100.86
$O = Cl$	0.56	0.55
Total	100.00	100.31

1. $Pb_5(AsO_3)_3Cl$. 2. Långban.[5] Rem. is FeO tr., Sb_2O_3 tr.

Occur. Found lining crevices in granular hematite in the Hindenberg shaft at Långban, Sweden.

Name. After K. J. Finneman, of Långban, who first noted the mineral.

Ref.

1. Crystal classes indicated by Laue study and morphology, Aminoff and Parsons, Geol. För. Förh., **49**, 438 (1927).
2. Aminoff (1923). Rare or uncertain: $11\bar{2}0$, $20\bar{2}1$.
3. Aminoff and Parsons (1927) by rotation method.
4. Flink in Aminoff (1923).
5. Almström analysis in Aminoff (1923).

46.1.4 **NADORITE** [$PbSbO_2Cl$]. *Flajolot* (C.R., **71**, 237, 406, 1870). Ochrolith *Flink* (Ak. Stockholm, Öfv., **46**, 5, 1889). Ochrolite.

Cryst.[1] Orthorhombic; dipyramidal—$2/m\ 2/m\ 2/m$.

$a:b:c = 0.4589:1:0.4450;\qquad p_0:q_0:r_0 = 0.9697:0.4450:1$

$q_1:r_1:p_1 = 0.4589:1.0312:1;\qquad r_2:p_2:q_2 = 2.2472:2.1791:1$

Forms:[2]

		ϕ	$\rho = C$	ϕ_1	$\rho_1 = A$	ϕ_2	$\rho_2 = B$
c	001	....	0°00′	0°00′	90°00′	90°00′	90°00′
b	010	0°00′	90 00	90 00	90 00	....	0 00
a	100	90 00	90 00	....	0 00	0 00	90 00
δ	1.11.0	11 12½	90 00	90 00	78 47½	0 00	11 12½
ε	170	17 17½	90 00	90 00	72 42½	0 00	17 17½
d	160	19 57½	90 00	90 00	70 02½	0 00	19 57½
e	150	23 33	90 00	90 00	66 27	0 00	23 33
η	130	35 59½	90 00	90 00	54 00½	0 00	35 59½
m	110	65 21	90 00	90 00	24 39	0 00	65 21
r	011	0 00	23 59½	23 59½	90 00	90 00	66 00½
q	021	0 00	41 40	41 40	90 00	90 00	48 20
p	041	0 00	60 40½	60 40½	90 00	90 00	29 19½
l	101	90 00	44 07	0 00	45 53	45 53	90 00
s	131	35 59½	58 47	53 10	59 49½	45 53	46 13

Structure cell.[3] Space group $Cmcm$. a_0 5.59 kX, b_0 12.20, c_0 5.43; $a_0:b_0:c_0 = 0.458:1:0.445$. Cell contents $Pb_4Sb_4O_8Cl_4$.

Habit. Tabular {010} or prismatic [100]; also lenticular [010] with a square or octagonal outline. Sometimes in subparallel or divergent groups.

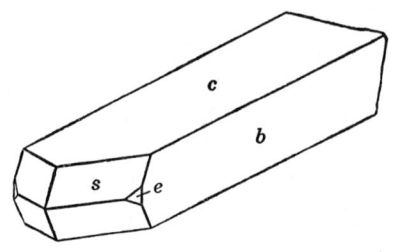 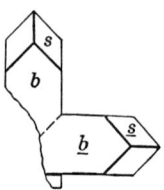

Twin on {101}.

Twinning. On {101}, the twin units nearly at right angles (91°45′).
Phys. Cleavage {010} perfect. H. $3\frac{1}{2}$–4. G. 7.02; 7.06 (calc.). Luster resinous to adamantine. Color smoky brown, also brownish yellow to yellow. Streak yellowish white. Translucent.

Opt.[4]

ORIENTATION		n(Li) (Djebel Nador)	n(Li) (Harstig mine)	
X	b	2.30 ± 0.01	2.30	Biaxial positive (+).
Y	c	2.35 ± 0.01	2.34	$2V$ very large.
Z	a	2.40 ± 0.01	2.36	$r > v$, strong.

Twinned units at 45° are seen on cleavage flakes.

Chem. An oxychloride of lead and antimony, $PbSbO_2Cl$. Pb 52.27, Sb 30.71, O 8.07, Cl 8.95, total 100.00. The reported [5] analyses are in close agreement with the formula given.

Occur. Originally found with smithsonite and bindheimite at Djebel Nador, Constantine, Algeria. As an alteration of jamesonite at the Bodannon mine, St. Endellion, Cornwall. With mimetite and hematite at the Harstig mine near Pajsberg, Wermland, Sweden (*ochrolite*). From Långban, Sweden.

Alter. To cerussite.

Artif.[6] Obtained in crystals by fusion of PbCl with Sb_2O_3 or of PbO with SbOCl.

OCHROLITE. Ochrolith Flink (*Ak. Stockholm, Öfv.*, **46**, 5, 1889). An orthorhombic, yellow mineral from the Harstig mine, Sweden. Supposedly $Pb_6Sb_2O_7Cl_4$, but the original analysis, made on a very small sample, was largely in error. Shown [7] by x-ray study to be identical with nadorite.

Ref.

1. Angles of Cesàro, *Bull. soc. min.*, **11**, 44 (1888), unit of the structure cell of Sillén and Melander, *Zs. Kr.*, **103**, 420 (1941) re-oriented to make $c < a < b$. Transformations: Cesàro (1888) to new, $00\frac{1}{3}/100/0\frac{1}{3}0$; Sillén and Melander to new, 100/001/010; Goldschmidt (**6**, 66, 1920) to new, $00\frac{1}{3}/010/\frac{1}{2}00$.
2. Goldschmidt (1920) and Barthoux, *Bull. soc. min.*, **48**, 246 (1925). Rare or uncertain: 1.15.0, 8.45.0, 0.15.1, 2.45.2, 12.111.17, 4.51.5.

3. Sillén and Melander (1941) on Djebel Nador material by Weissenberg method.
4. Larsen (113, 1921); Russell, *Min. Mag.*, **21**, 272 (1927); Larsen and Berman (211, 1934).
5. Cited in Dana (863, 1892).
6. Cesàro (1888); Sillén and Melander (1941); Bolfa, Pastant, and Roubault, *C.R.*, **228**, 1739 (1949).
7. Sillén and Melander (1941) on type material.

46.1.5 **Melanostibian.** *Igelström (Zs. Kr.*, **21**, 246, 1893).

As foliated masses, also as tiny striated crystals showing a prism and pyramid and perhaps tetragonal or orthorhombic. Sometimes twinned. Cleavage both prismatic and basal, distinct. H. 4. Color black. Streak cherry-red. Opaque.

C h e m. An antimonite of divalent iron and manganese, essentially $(Mn,Fe)_6(SbO_3)_2O_3$. Analysis gave: CaO 1.97, MgO 1.03, FeO 27.30, MnO 29.62, Sb_2O_3 37.50, H_2O 1.06, total 98.48. Weakly magnetic after ignition. Soluble in HCl.

O c c u r. Found in veinlets in dolomite at the Sjö mine, Grythytte parish, Örebro, Sweden. Named in allusion to the color and antimony content. Needs verification.

VANADIUM OXYSALTS

Vanadium is very similar in the chemistry of its oxysalts to both phosphorus and arsenic. In the orthovanadates, V occurs in tetrahedral coordination with oxygen to give anisodesmic compounds with discrete $(VO_4)^{\equiv}$ groups. These compounds generally are isostructural with phosphates and arsenates of the same formula type, and V can substitute serially therein to a greater or less extent for As and P. In this work, vanadates of this nature have been described with the phosphates and arsenates. A number of other natural vanadium oxysalts are known, however, in which the nature of the anionic configuration is unknown. Some of these apparently contain polynuclear complexes, and others contain V in several different valences. These minerals are here gathered together for convenience of reference into a single class and are arranged arbitrarily in decreasing ratio of cations to V.

CLASS 47. Vanadium Oxysalts
Type 1. Miscellaneous

47.1.1	Carnotite	$K_2(UO_2)_2(VO_4)_2 \cdot 3H_2O$
47.1.2	Tyuyamunite	$Ca(UO_2)_2(VO_4)_2 \cdot nH_2O$
47.1.3	Sengierite	$Cu(UO_2)(VO_4)(OH) \cdot 4\text{--}5H_2O$ (?)
47.1.4	Ferghanite	$U_3(VO_4)_2 \cdot 6H_2O$
47.1.5	Kolovratite	Ni vanadate ?
47.1.6	Fervanite	$Fe_4V_4O_{16} \cdot 5H_2O$
47.1.7	Steigerite	$Al_2(VO_4)_2 \cdot 6\frac{1}{2}H_2O$
47.1.8	Pucherite	$Bi(VO_4)$
47.1.9	Brackebuschite	$Pb_4MnFe(VO_4)_4 \cdot 2H_2O$
47.1.10	Pintadoite	$Ca_2V_2O_7 \cdot 9H_2O$
47.1.11	Rossite	$CaV_2O_6 \cdot 4H_2O$
47.1.12	Metarossite	$CaV_2O_6 \cdot 2H_2O$
47.1.13	Pascoite	$Ca_2V_6O_{17} \cdot 11H_2O$
47.1.14	Uvanite	$U_2V_6O_{21} \cdot 15H_2O$
47.1.15	Sincosite	$Ca(VO)_2(PO_4)_2 \cdot 5H_2O$
47.1.16	Rauvite	$CaU_2V_{12}O_{36} \cdot 20H_2O$
47.1.17	Melanovanadite	$Ca_2V_4^4V_6^5O_{25}$
47.1.18	Hewettite	$CaV_6O_{16} \cdot 9H_2O$
47.1.19	Metahewettite	$CaV_6O_{16} \cdot 9H_2O$
47.1.20	Fernandinite	$CaO \cdot V_2O_4 \cdot 5V_2O_5 \cdot 14H_2O$

47 VANADIUM OXYSALTS

TYPE 1. MISCELLANEOUS

CARNOTITE AND TYUYAMUNITE

	a_0	b_0	c_0	β
Carnotite, $K_2(UO_2)_2(VO_4)_2 \cdot 3H_2O$	6.59 Å	8.40	10.43	104°12'
Tyuyamunite, $Ca(UO_2)_2(VO_4)_2 \cdot nH_2O$	19.41	8.26	10.40	90

Synthetic $K(UO_2)(VO_4)$ appears to be the anhydrous prototype of one of the several minerals at present included under the name carnotite. The synthetic compound has a layer structure [1] consisting of sheets of the composition $(UO_2VO_4)_n{}^-$ interlaminated parallel $\{001\}$ with sheets of K ions. Linear uranyl ions, $(UO_2)^{++}$, are present as in autunite and all other known oxysalts containing U^6; in carnotite these are coordinated with (VO_4) groups. The layer structure offers an explanation of the ready base exchange and variable water content of carnotite. Tyuyamunite probably has a related layer structure with the metal-oxygen sheets bonded together by Ca ions; it may be likened to the brittle micas.

Ref.
1. Sundberg and Sillén, *Ark. Kemi*, **1**, no. 42, 337 (1949).

47.1.1 CARNOTITE [$K_2(UO_2)_2(VO_4)_2 \cdot 3H_2O$]. Friedel and Cumenge (*C.R.*, **128**, 532, 1899; *Bull. soc. min.*, **22**, 26, 1899).

Orthorhombic or monoclinic? As a powder or loosely coherent microcrystalline aggregates, sometimes compact; disseminated; rarely as crusts of imperfect platy crystals flattened $\{001\}$. Microscopic crystals [1] are flattened $\{001\}$ and rhomboidal $\{110\}$, also lath-like [010] with $\{100\}$ and $\{110\}$ or $\{120\}$; (110) $\wedge$ ($1\bar{1}0$) $\sim78°$.

Structure cell.[8] Artificial anhydrous $K(UO_2)(VO_4)$ is monoclinic with the space group $P2_1/c$. a_0 6.590Å, b_0 8.403, c_0 10.430; β 104°12'; $a_0:b_0:c_0 = 0.784:1:1.242$. Cell contents $K_4(UO_2)_4(VO_4)_4$. G. 5.03 (calc.). X-ray powder study indicates that this substance is isostructural with some natural (hydrated) material referred to carnotite.

Phys. Cleavage $\{001\}$ perfect. Luster dull or earthy, also pearly or silky when coarsely crystalline. Color bright yellow to lemon-yellow, also greenish yellow.

Opt.

ORIENTATION	n (Long Park, Colorado) [2]	n(Na) (anal. 2) [3]	PLEOCHROISM	
X c	1.750		Nearly colorless	Biaxial negative $(-)$.
Y b	1.925	2.06	Canary-yellow	
Z a	1.950	2.08	Canary-yellow	
2V	$\sim40°$	$\sim50°$		

The indices of refraction vary with the water content, increasing as the amount of water present decreases.

Chem. A hydrated vanadate of potassium and uranium, $K_2(UO_2)_2$-$(VO_4)_2 \cdot nH_2O$. The water content is partly zeolitic and varies with the humidity at ordinary temperature;[4] n is probably 3 in fully hydrated material and can range down at least to $\sim$1 H_2O. Small amounts of Ca, Ba, Mg, Fe''', Na, etc., are reported in most analyses of carnotite, due to isomorphous substitution or, more likely, to admixture.

Anal.[6]

	Na_2O	K_2O	CaO	MgO	Fe_2O_3	UO_3	V_2O_5	H_2O	Rem.	Total
1.		10.44				63.42	20.16	5.98		100.00
2.	0.35	9.58	0.64	0.22	0.04	65.62	21.12	1.35	0.48	99.40
3.	0.16	10.00	0.66	0.30	0.55	62.26	20.57	4.90	0.37	99.77

1. $K_2(UO_2)_2(VO_4)_2 \cdot 3H_2O$. 2. Cane Springs Pass, Utah.[5] Rem. is Al_2O_3 0.16, insol. 0.32, with tr. of CuO, PbO, P_2O_5. Air dried. 3. Temple Mountain, Utah.[4] Dried at 16 mm. H_2O. Rem. is CuO 0.07, SO_3 0.26, insol. 0.04.

Tests. B.B. infusible (a distinction from tyuyamunite). Easily soluble in acids.

Occur. It is not certain if all the occurrences that have been referred to this species actually are carnotite. In the United States found principally in the plateau region of southwestern Colorado and in adjoining districts of Utah, New Mexico, and Arizona. The mineral occurs chiefly in cross-bedded sandstones of Triassic or Jurassic age either disseminated or locally as small, relatively pure masses especially around petrified or carbonized tree trunks or other vegetal matter. The carnotite is of secondary origin and has been formed by the action of meteoric waters on pre-existing uranium and vanadium minerals. Associated species include tyuyamunite principally, also volborthite, calciovolborthite, rossite, hewettite, vanoxite, pintadoite, uvanite, rauvite, asphaltite. The Colorado localities include Paradox Valley and Roc and La Sal creeks in Montrose County, along the Rio Dolores southwest of Gateway in Mesa County, near Placerville and Cedar in San Miguel County, and near Meeker in Rio Blanco County. In Utah notably on San Rafael Swell west of Greenriver, Emery County, in the Henry Mountains in Garfield County, in the La Sal Mountains, and near Richardson and Thompson's in Grand County and north of Monticello in San Juan County. Occurrences also are known in Arizona, Nevada, New Mexico. Carnotite occurs in the Pottsville conglomerate near Mauch Chunk, Carbon County, Pennsylvania. At Radium Hill near Olary, South Australia, as alteration films on a uranium-containing mixture of ilmenite, davidite, magnetite, rutile. In red sandstone from Katanga in the Belgian Congo.

Artif.[7] Minute orthorhombic yellow scales with the composition $K(UO_2)(VO_4)$ have been obtained by fusion of potassium metavanadate and ammonium pyrouranate.

Alter.[4] Carnotite can be converted to tyuyamunite by treatment with calcium bicarbonate solution.

Name. After Marie-Adolphe Carnot (1839–1920), French mining engineer and chemist.

TYUYAMUNITE

Ref.
1. Crook and Blake, *Min. Mag.*, **15**, 271 (1924), and Larsen (52, 1921).
2. Larsen (1921).
3. Hess and Foshag, *Proc. U. S. Nat. Mus.*, **72**, Art. 12 (1927), on material dried over concentrated H_2SO_4 (1.32 per cent H_2O); when rehydrated at 19 mm. v.p. of H_2O (1.72 per cent H_2O) the indices were nY 2.04, nZ 2.06.
4. Hillebrand, *Am. J. Sc.*, **8**, 201 (1924).
5. Hess and Foshag (1927).
6. Additional analyses on material of inferior purity or of uncontrolled water content are listed by Hintze (1 [4B], 1002, 1931) and Doelter (3 [I], 844, 1918).
7. Canneri and Pestelli, *Gazz. chim. ital.*, **54**, 641 (1924).
8 Sundberg and Sillén, *Ark. Kemi*, **1**, no. 42, 337 (1949), with description of structure.

47.1.2 T Y U Y A M U N I T E [$Ca(UO_2)_2(VO_4)_2 \cdot nH_2O$]. *Nenadkevič* (*Ac. Sc. St. Petersburg, Bull.*, [6], **6**, 945, 1912). Calciocarnotite *Hillebrand* (*Am. J. Sc.*, **35**, 440, 1913). Tuyamunite. Tujamunite. Tjuiamunite. Tjujamunite. Tïujamunite. Tyuyamuyunite.

C r y s t.[1] Orthorhombic.

$$a:b:c = 1.303:1:2.337; \quad p_0:q_0:r_0 = 1.794:2.337:1$$

$$q_1:r_1:p_1 = 2.337:0.558:1; \quad r_2:p_2:q_2 = 0.428:0.767:1$$

Forms:[2]

		ϕ	$\rho = C$	ϕ_1	$\rho_1 = A$	ϕ_2	$\rho_2 = B$
c	001	....	0°00'	0°00'	90°00'	90°00'	90°00'
b	010	0°00'	90 00	90 00	90 00	....	0 00
a	100	90 00	90 00	....	0 00	0 00	90 00
l	120	21 00	90 00	90 00	69 00	0 00	21 00
m	110	37 30½	90 00	90 00	52 29½	0 00	37 30½
d	101	90 00	60 52	0 00	29 08	29 08	90 00
p	111	37 30½	71 15	66 50	54 47	29 08	41 18½

Structure cell.[13] Orthorhombic. a_0 10.40 Å, b_0 8.26, c_0 19.41; $a_0:b_0:c_0 = 1.26:1:2.35$.

Habit. As scales and laths flattened {001} and elongated [100]; as radial aggregates. The crystal faces are dull and sometimes curved. Commonly massive, compact to cryptocrystalline; also pulverulent.

P h y s. Cleavage {001} perfect, micaceous; {010} and {100} distinct. Not brittle; prone to cake on grinding. H. ~2. G. variable, increasing with decreasing water content, 3.67–4.35 (Colorado);[3] also 3.3–3.4 (Tyuya Muyun). Luster of crystals adamantine, pearly on {001}; massive material waxy. Color canary-yellow, also lemon-yellow to greenish yellow due to exposure or to admixture of copper salts. Translucent to opaque.

O p t. In transmitted light, colorless to pale yellow. Faintly pleochroic, with X nearly colorless, Y pale canary-yellow, Z canary-yellow. Dispersion $r < v$.

ORIEN-TATION [7]		n (Red Creek, Utah) [4]	n (Henry Mountains, Utah) [5]	n (Paradox Valley) [6]	n (Paradox Valley) [6]
X	c	1.670	1.72 ± 0.01	1.75–1.80 (calc.)	1.78 (calc.)
Y	b	1.870	1.868	1.927–1.932	1.895
Z	a	1.895	1.953	1.965–1.968	1.92
2V		36°	48°	~48°	

On dehydration the optical orientation, sign, and 2V remain unchanged, but the indices of refraction increase. The birefringence progressively decreases when Ca is leached from the mineral by acids.

Chem. A hydrous basic vanadate of uranium and calcium, $Ca(UO_2)_2(VO_4)_2 \cdot nH_2O$. The water content is partly zeolitic and varies markedly with the humidity at ordinary temperature;[8] n ordinarily is 9 or 10 but can range continuously down at least to $\sim 4H_2O$ on strong drying. Analyses of tyuyamunite, carnotite, and similar substances should be made on samples whose water content refers to known conditions, such as the vapor pressure of H_2O (~ 16 mm.) afforded by an H_2SO_4 solution of G. 1.05 at 20°.

Small amounts of Ba, Mg, Pb, Cu, and K are reported in many analyses of tyuyamunite, but it is not known if these are due to isomorphous substitution or to admixture.

Anal.[12]

	Na_2O	K_2O	CaO	BaO	MgO	CuO	UO_3	V_2O_5	H_2O	Rem.	Total
1.			5.66				57.78	18.37	18.19		100.00
2.			5.99				63.09	21.00	7.04		97.12
3.			6.75		0.08		57.29	18.49	15.37	1.25	99.23
4.	0.21	0.18	5.01		0.09	0.10	58.34	18.72	17.35		100.00
5.	0.02	0.18	5.63	0.54	0.09	0.04	55.91	18.76	18.83		100.00
6.		0.5	5.9				57.7	19.5	16.3		99.9

1. $Ca(UO_2)_2(VO_4)_2 \cdot 6H_2O$. 2. Tyuya Muyun.[9] Dried over P_2O_5. CaO contains some SrO. With Tl tr. 3. Tyuya Muyun.[10] Air dried. Rem. is PbO 0.14, CO_2 0.91, insol. 0.20. 4. Tyuya Muyun.[3] Recalculated to 100 after deducting 3.39 insol. and Fe_2O_3. Li_2O in Na_2O. 5. Calamity Creek, Mesa County, Colorado.[3] Recalculated to 100 after deducting 1.78 insol. and Fe_2O_3. 6. Henry Mountains, Utah.[11] Recalculated to 100 after deducting 9.61 insol. and Fe_2O_3.

Tests. B.B. fuses easily to a dark liquid (a distinction from carnotite). Insoluble in dilute acetic acid, soluble in HCl, HNO_3, H_2SO_4.

Occur. Originally found at Tyuya Muyun (the name of a hill traversed by the gorge of the Aravan River and resembling a camel's hump) about 62 km. southeast of Fedchenko in Ferghana, Turkestan, U.S.S.R. The tyuyamunite occurs in veinlets or disseminated in Carboniferous limestone, or fills karst caverns in the limestone, and is associated with malachite, ferghanite, turanite, barite, calcite, and several ill-defined vanadium minerals. Tyuyamunite occurs widely in the carnotite districts of western Colorado and eastern Utah. In Colorado notably at Paradox Valley, Montrose County, and Thompson's, Grand County; in Utah at Richardson on the northwest side of the La Sal Mountains, in the Henry Mountains, Garfield County, and on Red Creek, Browns Park, Uintah County. The tyuyamunite occurs with carnotite, with which it is easily confused, as fine-grained aggregates associated with fossil wood in sandstone. Also from the Bisbee district, Arizona.

Artif. Formed by treating carnotite with a solution of calcium bicarbonate; the reverse change can be effected by treating tyuyamunite with potassium-mercuric iodide solution. The interchange of Ca and K is accompanied by variation in the water content, and the two substances appear to be structurally distinct.

Sengierite 1047

Ref.

1. Elements of Dolivo-Dobrovolsky, *Mem. soc. russe min.*, [2], **54**, 359 (1925), on minute crystals of poor quality from Tyuya Muyun.
2. Dolivo-Dobrovolsky (1925); Merwin in Hillebrand, *Am. J. Sc.*, **8**, 201 (1924); Chirvinsky, *Min. Mag.*, **20**, 287 (1925).
3. Hillebrand (1924).
4. Larsen (148, 1921), who gives $nY = 1.87 \pm 0.01$ and $2V$ mod. for Tyuya Muyun crystals.
5. Ross in Hess, *U. S. Geol. Sur.*, *Bull.* **750D**, 73 (1924).
6. Merwin in Hillebrand (1924).
7. For further optical data see Chlopin, *C. r. ac. sc. Russie*, **73** (1925), and Dolivo-Dobrovolsky (1925).
8. For dehydration data see Hillebrand (1924), Chlopin (1925).
9. Nenadkevič (1912).
10. Chlopin (1925).
11. Schaller analysis in Hess (1924).
12. For additional analyses see Hintze (**1** [4B], 1005, 1931).
13. Weiss-Frondel, priv. comm. (1949).

47.1.3 **SENGIERITE** [$Cu(UO_2)(VO_4)(OH) \cdot 4$–$5H_2O$ (?)]. Vaes and Kerr (*Am. Min.*, **34**, 109, 1949).

Cryst. Orthorhombic; dipyramidal—$2/m\ 2/m\ 2/m$.

$a:b:c = 0.762:1:0.739;\qquad p_0:q_0:r_0 = 0.970:0.739:1$

$q_1:r_1:p_1 = 0.762:1.031:1;\qquad r_2:p_2:q_2 = 1.353:1.312:1$

Forms:

	ϕ	$\rho = C$	ϕ_1	$\rho_1 = A$	ϕ_2	$\rho_2 = B$
c 001	....	0°00′	0°00′	90°00′	90°00′	90°00′
b 010	0°00′	90 00	90 00	90 00	....	0 00
m 110	52 42	90 00	90 00	37 18	0 00	52 42
d 011	0 00	36 28	36 28	90 00	90 00	53 32
p 111	52 42	50 39	36 28	52 02½	45 52½	62 03½

Doubtful:

331   441

Habit. Thin plates flattened {001}, with a six-sided outline. As coatings.

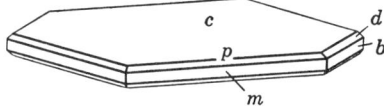

Phys. Cleavage {001} perfect. Brittle. H. 2½. G. ∼4. Luster vitreous. Color yellowish green. Streak light green. Transparent.

Opt.

Orientation		n	Pleochroism	
X	c	1.77	Bluish green	Biaxial negative (−).
Y	b	1.94	Olive-green	$2V \sim 38°$.
Z	a	1.97	Yellowish green to colorless	$r < v$, strong.

Chem. A hydrated basic vanadate of copper and uranium, probably $Cu(UO_2)(VO_4)(OH) \cdot 4$–$5H_2O$. On heating, sengierite loses 3.94 per cent

H_2O at 100°, 10.56 per cent at 300° (regained in part on exposure for a loss of 6.48 per cent, or $4H_2O$), and the remainder on ignition. Analysis gave:

	CuO	UO_3	$(Fe,Al)_2O_3$	V_2O_5	H_2O	Total
1.	14.55	52.33		16.64	16.48	100.00
2.	14.82	47.45	1.43	15.96	15.77	95.43

1. $Cu_2(UO_2)_2(VO_4)_2(OH)_2 \cdot 10H_2O$. 2. Katanga.

Tests. In C.T. loses water, assumes a bronze shade with a metallic luster, and becomes opaque. Soluble in acids.

O c c u r. A secondary mineral, found with volborthite, vandenbrandite, malachite, and black oxides of cobalt, etc., in the Elizabethville-Jadotville area of Katanga in the Belgian Congo.

N a m e. After Edgard Sengier, an official of the Union Minière du Haut Katanga.

UNNAMED MINERAL. *Chernik (Bull. ac. sc. Russie,* [6], **16**, 505, 1922). Color dark green. H. 4. G. 4.46. Analysis gave: K_2O 0.80, BaO 0.84, CuO 7.23, U_3O_8 38.27, V_2O_5 38.28, H_2O 7.80, total 93.22. Found as a crystalline crust associated with tyuyamunite at Tyuya Muyun, Ferghana, Turkestan, U.S.S.R.

47.1.4 F E R G H A N I T E $[U_3(VO_4)_2 \cdot 6H_2O]$. *Antipov (Gornyi Zhurnal, St. Petersburgh,* **4**, 259, 1908—*Jb. Min.,* II, 38, 1909).

Orthorhombic ? As scales flattened {001}. Fragments have a six-sided outline with plane interior angles of 106° (*mm'* ?) and 127° (*bm* ?). Cleavage {001} perfect; also a second cleavage {100} (?) perpendicular to the optic plane. H. $\sim 2\frac{1}{2}$. G. 3.31. Luster waxy. Color sulfur-yellow. Translucent. Optically biaxial, with 2V large, and $X = c$, $Y = b$, $Z = a$. The indices of refraction and birefringence are low. Not pleochroic.

C h e m. Essentially a hydrated vanadate of uranium, $U_3(VO_4)_2 \cdot 6H_2O$. Analysis gave:

	Li_2O	UO	V_2O_5	H_2O	Total
1.		72.44	17.31	10.25	100.00
2.	1.22	69.30	17.60	10.75	98.87

1. $U_3(VO_4)_2 \cdot 6H_2O$. 2. Turkestan. Another analysis gave 3.49 Li_2O.

Found with other uranium minerals at Tyuya Muyun in the Ferghana district, Turkestan, U.S.S.R. Ferghanite may represent a leached or weathered product of tyuyamunite.[1]

Ref.

1. See Chirvinsky, *Min. Mag.,* **20**, 294 (1925); Nenadkevič, *Bull. ac. sc. St. Petersburg,* **6**, 945 (1912); Hillebrand, *Am. J. Sc.,* **8**, 201 (1924).

47.1.5 Kolovratite. *Vernadsky (C. r. ac. sc. Russie,* 37, 1922).

As incrustations and botryoidal crusts. Color yellow to greenish yellow. A vanadate of nickel of uncertain composition. Partial analyses of ore samples gave:[1]

	1	2	3
NiO	6.50	8.38	12.22
V_2O_5	11.55	5.94	6.20

Reported to be widely distributed in quartz schists and carbonaceous slates in Ferghana, Russian Turkestan, notably at Kara-chagyr and Uch-Kurgan. Named after L. S. Kolovrat-Chervinsky, a Russian radiologist. Several different vanadates of nickel are known artificially.[2]

Ref.
1. Preobrazhensky analysis in Chirvinsky, *Min. Mag.*, **20**, 290 (1925).
2. Mellor (**9**, 791, 1929).

47.1.6 **FERVANITE** [$Fe_4V_4O_{16} \cdot 5H_2O$]. Hess and Henderson (*Am. Min.*, **16**, 273, 1931).

Probably monoclinic. In parallel-fibrous aggregates. Cleavage is not apparent. Color golden brown. Luster brilliant.

Opt.

nX	2.186 ± 0.005	Biaxial negative ($-$).
nY	2.222 ± 0.005	$2V$ very small.
nZ	2.224 ± 0.005	

Extinction is slightly inclined to the fiber length.

Chem. A hydrated ferric vanadate, $Fe_4V_4O_{16} \cdot 5H_2O$.

Anal.

	1	2
Fe_2O_3	41.30	41.89
V_2O_5	47.05	46.10
H_2O-	11.65	12.01
Total	100.00	100.00

1. $Fe_4V_4O_{16} \cdot 5H_2O$. 2. Gypsum Valley, San Miguel County, Colorado.[1] Recalculated to 100 after deduction of 9.40 per cent insoluble, 7.34 per cent gypsum.

Tests. Apparently insoluble in water.

Occur. Found at Polar Mesa in the La Sal Mountains, Grand County, Utah, and at Gypsum Valley, San Miguel County, Colorado. Also at numerous other localities in the uranium-vanadium district of southwestern Colorado and southeastern Utah. Associated with gypsum, metahewettite, carnotite, and various black vanadium minerals. Other materials apparently composed largely of ferric vanadates different from fervanite have been described from Quisque, Peru, and Gypsum Valley, Colorado.[2]

Name. In allusion to the composition, iron (*ferrum*) and *vanadium*.

Ref.
1. Henderson analysis in Hess and Henderson (1931).
2. See discussion of analyses by Hillebrand, *Trans. Am. Inst. Min. Engr.*, **40**, 294 (1909), and Schaller (unpb., 1926) in Hess and Henderson (1931).

47.1.7 **Steigerite** [$Al_2(VO_4)_2 \cdot 6\frac{1}{2}H_2O$]. Henderson (*Am. Min.*, **20**, 769, 1935).

Occurs as canary-yellow pulverulent coatings that under the microscope are variously composed of cryptocrystalline fibrous material resembling chalcedony, gum-like masses, and occasionally flat plates. Mean index of refraction 1.710 ± 0.005. Compact aggregates have a rather waxy luster.

C h e m. A hydrated aluminum vanadate, $Al_2(VO_4)_2 \cdot 6\frac{1}{2}H_2O$. The water content is uncertain. The x-ray powder diffraction pattern is unlike that of fervanite.

Anal.

	1	2
Al_2O_3	25.42	25.14
Fe_2O_3		1.50
V_2O_5	45.37	44.44
H_2O-	29.21	8.08
H_2O+		21.04
Total	100.00	100.20

1. $Al_2(VO_4)_2 \cdot 6\frac{1}{2}H_2O$. 2. San Miguel County, Colorado.

Tests. Easily soluble in dilute mineral acids to a deep cherry-red solution. Insoluble in water.

O c c u r. Found with fervanite and gypsum coating fractures in and around nodular concretions of corvusite in sandstone along the north wall of Gypsum Valley, San Miguel County, Colorado.

A r t i f. An aluminum vanadate chemically identical with steigerite but giving only few and faint x-ray diffractions has been prepared by reaction of aluminum sulfate solution with a soluble calcium vanadate made by boiling hewettite or metahewettite in saturated lime water.

N a m e. After George Steiger (1869–1944), former Chief Chemist of the U. S. Geological Survey.

47.1.8   **P U C H E R I T E** [$BiVO_4$]. *Frenzel (J. pr. Chem.,* **4,** 227, 361, 1871).

C r y s t.[1] Orthorhombic; dipyramidal—$2/m\ 2/m\ 2/m$.

$a:b:c = 1.0654:1:2.3357;\quad p_0:q_0:r_0 = 2.1923:2.3357:1$

$q_1:r_1:p_1 = 1.0654:0.4561:1;\quad r_2:p_2:q_2 = 0.4281:0.9386:1$

Forms:[2]

	ϕ	$\rho = C$	ϕ_1	$\rho_1 = A$	ϕ_2	$\rho_2 = B$
c 001	....	0°00′	0°00′	90°00′	90°00′	90°00′
a 100	90°00′	90 00	....	0 00	0 00	90 00
k 210	61 57½	90 00	90 00	28 02½	0 00	61 57½
w 012	0 00	49 25½	49 25½	90 00	90 00	40 34½
x 011	0 00	66 49½	66 49½	90 00	90 00	23 10½
n 112	43 11	58 01½	49 25½	54 31	42 22½	51 48
q 111	43 11	72 41	66 49½	49 12½	24 31	45 53½
r 212	61 57½	68 04½	49 25½	35 02½	24 31	64 08½
s 522	66 55	80 28½	66 49½	24 52½	10 20½	67 15½

Structure cell.[3] a_0 5.38 kX, b_0 5.04, c_0 11.98; $a_0:b_0:c_0 = 1.069:1:2.379$. Cell contents $Bi_4(VO_4)_4$.

PUCHERITE

Habit. Crystals small, often with curved faces and a marked composite structure. Usually tabular {001}, also acicular. {111} striated parallel edge {001} {111}. Also massive; ocherous, or as coatings of minute crystals.

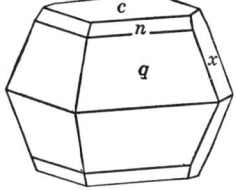

Minas Geraes.

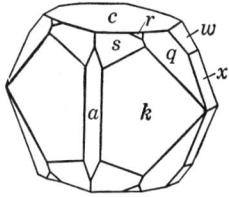

Schneeberg.

Phys. Cleavage {001}, perfect. Fracture subconchoidal. Brittle. H. 4. G. 6.57 (calc.). Luster vitreous to adamantine. Color dark brownish red to reddish brown and yellowish brown. Streak yellow. Opaque to transparent.

Opt.[4] In transmitted light, yellow-brown.

ORIENTATION		n(Li)	
X	c	2.41 ± 0.02	Biaxial negative (−).
Y	a	2.50 ± 0.02	$2V$ 19° ± 5° (meas.).
Z	b	2.51 ± 0.02	$r < v$, extreme.

Chem. Bismuth vanadate, $BiVO_4$. Small amounts of P and As substitute for V, with P:As:V = 1:1.7:12.9 in analysis 3.

Anal.[7]

	1	2	3	4
Bi_2O_3	71.93	73.39	73.16	70.88
P_2O_5		tr.	1.34	
As_2O_5			3.66	
V_2O_5	28.07	27.31	22.19	27.65
Rem.				1.47
Total	100.00	100.70	100.35	100.00

1. $BiVO_4$. 2, 3. Schneeberg, Saxony.[5] 4. Pala Chief mine, San Diego County, California.[6] Yellow, ocherous material. Recalculated to 100 after deducting insol. 7.37. Rem. is H_2O.

Tests. In C.T. decrepitates. B.B. on charcoal fuses. Soluble in HCl with evolution of Cl to a deep red solution, which on dilution becomes green and deposits a yellow basic chloride.

Occur. A secondary mineral found associated with bismutite and beyerite. Originally found in the oxidized portion of a Bi-Ag-U-Cu vein in the Pucher shaft of the Wolfgang mine at Schneeberg, Saxony. Reported from the Arme Hilfe mine at Ullersreuth, Thuringia, and from Sosa near Eibenstock, Saxony. In pegmatites in Madagascar with native bismuth at Mt. Bity and with bismuthinite at Ampangabe. From São José at Brejauba, Minas Geraes, Brazil. As ocherous alteration products of native bismuth in pegmatite at the Stewart and Pala Chief mines near Pala, San Diego County, California.

A r t i f.[8] Obtained in crystals by the evaporation of a mixed solution of bismuth nitrate and vanadium chloride.

Ref.
1. Angles of Websky, *Min. Mitt.*, 245 (1872), on Schneeberg crystals, unit of the structural cell of de Jong and Lange, *Am. Min.*, **21**, 809 (1936). On the morphological resemblance to brookite and columbite see Websky (1872) and Brögger, *Vidensk. Skr. Oslo, Mat. nat. Cl. 1*, no. 6, 76 (1906). Transformations: Websky I (1872) and Dana (755, 1892) to new, 200/010/001; Goldschmidt (**6**, 179, 1920) to new, 200/010/002; Görgey, in Doelter (**3** [1], 843, 1918), to new, 100/010/00$\frac{5}{2}$; Brögger (1906) to new, 100/001/010; Websky II (1872) to new, 010/001/200.
2. Goldschmidt (1920); $n\{112\}$ from Shannon, *Proc. U. S. Nat. Mus.*, **62**, Art. 9 (1923).
3. de Jong and Lange (1936).
4. Larsen (123, 1921) and Cesàro, *Ac. roy. Belgique, Bull.*, 142 (1905).
5. Frenzel (1871); *Jb. Min.*, 514 (1872).
6. Schaller, *Zs. Kr.*, **49**, 228 (1911).
7. For additional analyses see Frenzel (1871) and Schaller (1911).
8. Frenzel, *Jb. Min.*, 680 (1875).

47.1.9 B R A C K E B U S C H I T E [$Pb_4MnFe(VO_4)_4 \cdot 2H_2O$]. Doering (quoted by Rammelsberg, *Zs. deutsch. geol. Ges.*, **32**, 711, 1880); *Bol. ac. cienc. Córdoba*, **5**, 501, 1883.

C r y s t.[1] Monoclinic.
Structure cell.[2] Space group $P2_1/m$ (if holohedral). a_0 8.92 Å, b_0 6.16, c_0 7.69; β 111°47'; $a_0:b_0:c_0 = 1.448:1:1.248$. Cell contents $Pb_4(Mn,Fe)_2(VO_4)_4 \cdot 2H_2O$.

Habit. Acicular [010] with $\{100\}$ and $\{001\}$ but without terminal faces; the crystals are striated [010] and are sometimes flattened parallel to the elongation. As tufts or groups of crystals; also dendritic or botryoidal.

P h y s. G. 6.05;[2] 6.07 (calc.). Luster submetallic. Color dark brown to black. Streak yellow. Translucent to nearly opaque.

O p t.[3] In transmitted light, reddish brown in color.

	n(Li)	PLEOCHROISM	
X	2.28 ± 0.02	Nearly colorless	Biaxial positive (+).
Y	2.36 ± 0.02	Dark reddish brown	$2V$ large.
Z	2.48 ± 0.02	Clear reddish brown	$r > v$, strong.

C h e m. A hydrated vanadate of lead, manganese, and iron, $Pb_4MnFe(VO_4)_4 \cdot 2H_2O$.

Anal.

	1	2
PbO	62.20	61.00
FeO	5.01	4.65
MnO	4.94	4.77
ZnO		1.29
CuO		0.42
V_2O_5	25.34	25.32
P_2O_5		0.18
H_2O	2.51	2.03
Total	100.00	99.66

1. $Pb_4MnFe(VO_4)_4 \cdot 2H_2O$. 2. Sierra de Córdoba, Argentina.[4]

ROSSITE 1053

Occur. Found with descloizite and vanadinite in the oxidized zone of lead-zinc veins in the western part of the Sierra de Córdoba, Córdoba province, Argentina.

Name. After Ludwig Brackebusch (?-1906), Professor of Mineralogy in the University of Córdoba, Argentina.

Ref.
1. Symmetry established by x-ray study of Berry, *Am. Min.*, **33**, 489 (1948). Morphological measurements are lacking.
2. Berry (1948).
3. Larsen and Berman (146, 1934).
4. Doering (1880, 1883).

47.1.10 Pintadoite [$Ca_2V_2O_7 \cdot 9H_2O$]. *Hess* and *Schaller* (*J. Washington Ac. Sc.*, **4**, 576, 1914).

As a thin efflorescence, which when dissolved in water and crystallized forms twinned lath-shaped crystals. Color light to dark green. Slightly pleochroic in yellow-green, with moderate to high birefringence.

Chem. A hydrated calcium vanadate, $Ca_2V_2O_7 \cdot 9H_2O$.
Anal.

	1	2
CaO	24.59	22.6
V_2O_5	39.87	42.4
H_2O	35.54	35.0
Total	100.00	100.0

1. $Ca_2V_2O_7 \cdot 9H_2O$. 2. Frisco No. 2 Claim, Cañon Pintado, Utah.[1] Recalculated to 100 after deducting unstated amounts of gypsum and insoluble material.

Tests. Slowly soluble in cold water.

Occur. Found as efflorescences on protected outcroppings of the McElmo sandstone in the Cañon Pintado, San Juan County, Utah, and elsewhere in the uranium-vanadium field of southeastern Utah.

Name. From the locality in the Cañon Pintado.

Ref.
1. Schaller analysis in Hess and Schaller (1914).

47.1.11 ROSSITE [$CaV_2O_6 \cdot 4H_2O$]. *Foshag* and *Hess* (*Proc. U. S. Nat. Mus.*, **72**, 1, 1927).

Cryst.[1] Triclinic. $\alpha\ 98°17'$, $\beta\ 120°12'$, $\gamma\ 89°34'$.

Forms:

	ϕ	ρ
c 001	72°54'	31°31'
a 100	85 38	90 00
m 110	53 48	90 00
y $\bar{1}01$	−86 38	44 13

Habit. Artificial crystals are usually elongated [001] and more or less flattened on {010}; rarely tabular {001}, or lath-like or needle-like [001]. The natural material occurs as glassy lumps surrounded by flaky alteration rims of metarossite.

Twinning. On {100}.

Phys. Cleavage {010}, good. Brittle. H. 2–3. G. 2.45. Luster vitreous to somewhat pearly. Color yellow (Martius yellow to pinard-yellow). Transparent.

Opt.[2] In transmitted light, yellow.

Orientation		n (artif.)	
X		1.710	Biaxial; negative (?).
$Y \wedge b$	$\sim 45°$	1.770	$2V$ large.
$Z \sim c$		1.840	Dispersion very strong.

Chem. A hydrated calcium vanadate, $CaV_2O_6 \cdot 4H_2O$. Analysis gave:

	CaO	MgO	V_2O_5	H_2O	Insol.	Total
1.	18.09		58.67	23.24		100.00
2.	18.00	0.14	58.00	22.90	1.60	100.64

1. $CaV_2O_6 \cdot 4H_2O$. 2. Bull Pen Canyon, Colorado.[2]

Tests. B.B. fuses easily. In C.T. fuses easily and loses H_2O. Slowly soluble in water. The water is largely or entirely lost at about 120°.

Occur. Found as small glassy kernels embedded in metarossite filling small veinlets in sandstone containing carnotite and gypsum at Bull Pen Canyon, San Miguel County, Colorado.

Artif. As crystals from a water solution of the mineral or of metarossite.

Alter. Dehydrates to metarossite at room temperature.

Name. After Clarence S. Ross (1880–), geologist and mineralogist, of the U. S. Geological Survey.

Ref.
1. Elements and forms of Foshag in Foshag and Hess (1927) on artificial crystals; pyramidal forms are lacking. The angle β is given in the original as 97°24′. The angles cited in the table are averaged direct measurements.
2. Foshag in Foshag and Hess (1927).

47.1.12 METAROSSITE [$CaV_2O_6 \cdot 2H_2O$]. *Foshag* and *Hess* (*Proc. U. S. Nat. Mus.*, **72**, 9, 1927).

Found as soft and friable, platy to flaky masses and veinlets. Color light yellow. Luster dull, pearly. H. and G. not known.

Opt.[1] In transmitted light, pale yellow to colorless.

nX	1.840	Biaxial positive (+).
nY	>1.85	$2V$ large.
nZ	>1.85	Dispersion strong.

Chem. A hydrated calcium vanadate, $CaV_2O_6 \cdot 2H_2O$.

Anal.[1]

	1	2	3
CaO	20.47	20.04	19.60
MgO		0.10	0.13
V_2O_5	66.38	64.08	64.20
H_2O	13.15	13.56	14.08
Insol.		2.72	2.48
Total	100.00	100.50	100.49

1. $CaV_2O_6 \cdot 2H_2O$. 2, 3. San Miguel County, Colorado.

PASCOITE

Tests. The blowpipe and chemical tests are identical with those of rossite. Soluble in water.

Occur. Found as veinlets in carnotite-bearing sandstone in Bull Pen Canyon, San Miguel County, Colorado. The mineral is a dehydration product of rossite and sometimes encloses glassy, unaltered kernels of that species.

Artif. Rossite crystallizes from a water solution of metarossite, but the crystals gradually pass over to metarossite on exposure to dry air.

Name. In allusion to its relation to rossite.

Ref.
1. Foshag in Foshag and Hess (1927).

47.1.13 **PASCOITE** [$Ca_2V_6O_{17} \cdot 11H_2O$]. *Hillebrand, Merwin,* and *Wright* (*Zs. Kr.*, **54**, 226, 1915; *Proc. Am. Phil. Soc.*, **53**, 31, 1914).

Cryst.[1] Triclinic.

$a:b:c = 0.9144:1:1.1189;$ $\alpha\ 93°28',\ \beta\ 95°02',\ \gamma\ 104°42'$

$p_0:q_0:r_0 = 1.2627:1.1523:1;$ $\lambda\ 85°04\frac{1}{2}',\ \mu\ 83°52\frac{1}{2}',\ \nu\ 74°54'$

$p_0'\ 1.2723,\ q_0'\ 1.1610,\ x_0'\ 0.0880,\ y_0'\ 0.0865$

Forms:

	ϕ	ρ	A	B	C
b 010	0°00'	90°00'	74°54'	......	85°04½'
m 110	39 27½	90 00	35 26½	39°27½'	83 00½
M 1̄10	124 02	90 00	49 08	124 02	88 36½
p 111	39 49	64 03½	42 37	46 19	57 03½
q 1̄11	−141 01	61 03½	135 08	132 52	68 03
r 131	156 45½	73 19	82 11½	151 39½	70 53

Rare or doubtful: 5̄20   1̄41   1̄11

Structure cell.[2] $a_0\ 8.875\ kX,\ b_0\ 9.706,\ c_0\ 10.86;\ \alpha\ 93°28',\ \beta\ 95°02',\ \gamma\ 104°42';\ a_0:b_0:c_0 = 0.9144:1:1.1189.$

Habit. As granular crusts, rarely showing minute lath-like crystals with oblique terminations. {1̄10} striated [001], and {111} striated [10̄1].

Phys. Cleavage {010} distinct. Fracture conchoidal. H. $\sim 2\frac{1}{2}$. G. 1.87; 1.88 (calc.). Luster vitreous to sub-adamantine. Color dark red-orange to yellow-orange. Streak cadmium-yellow. Translucent. On partial dehydration turns dirty yellow in color.

Opt.

	n	PLEOCHROISM	
X	1.775 ± 0.005	Light cadmium-yellow	Biaxial negative (−).
Y	1.815 ± 0.005	Cadmium-yellow	$2V\ 50\frac{1}{2}°$ (Na), 56° (Li).
Z	1.825 ± 0.005	Orange	Crossed dispersion, strong Absorption $Z > Y > X$.

The optic plane is perpendicular {010}.

Chem. A hydrated vanadate of calcium, $Ca_2V_6O_{17} \cdot 11H_2O$. The water content is uncertain. About two-thirds of the water is lost over a strong desiccant at room temperature or by heating to 100°. Analysis gave:

	CaO	MoO$_3$	V$_2$O$_5$	H$_2$O+ 100°	H$_2$O− 100°	Undet.	Total
1.	13.10		63.75	23.15			100.00
2.	12.6	0.3	64.6	7.8	13.8	0.9	100.00

1. Ca$_2$V$_6$O$_{17}$·11H$_2$O. 2. Minasragra.[3]

Tests. Easily fusible to a deep red liquid. Soluble in water.

Occur. Found as a recent efflorescence on tunnel walls in the vanadium deposit at Minasragra, Pasco province, Peru. In the United States from Paradox Valley, San Miguel County, Colorado, and from Grand County, Utah, with carnotite in sandstone.

Ref.

1. Morphology from Berman, priv. comm. (1942), on Minasragra crystals. Elements of the structure cell.
2. Berman (1942) by Weissenberg method on Minasragra crystals.
3. Hillebrand analysis in Hillebrand, Merwin, and Wright (1915).

47.1.14   U V A N I T E [U$_2$V$_6$O$_{21}$·15H$_2$O]. *Hess* and *Schaller (Washington Ac. Sc., J.,* **4,** 576, 1914).

Probably orthorhombic. As minutely crystalline masses and coatings. Two pinacoidal cleavages.[1] Color brownish yellow.

Opt.[1] In transmitted light, pale brown.

	n	PLEOCHROISM	
X	1.817	Light brown	Biaxial positive (+).
Y	1.879	Dark brown	2V 52°.
Z	2.057	Greenish yellow	

Chem. A hydrated uranium vanadate, U$_2$V$_6$O$_{21}$·15H$_2$O (?).

Anal.

	1	2
CaO		1.73
UO$_3$	41.22	39.60
V$_2$O$_5$	39.31	37.70
H$_2$O	19.47	18.28
Rem.		1.69
Total	100.00	99.00

1. U$_2$V$_6$O$_{21}$·15H$_2$O. 2. Temple Rock, Utah.[2] Rem. is K$_2$O 0.30, MgO 0.04, P$_2$O$_5$ 0.06, As$_2$O$_5$ 0.05, insol. 1.24.

Tests. Insoluble in water, but soluble in ammonium carbonate solution.

Occur. Associated with carnotite, rauvite, hewettite, metatorbernite, hyalite, and gypsum in asphaltic sandstone (Shinarump) at Temple Rock, 45 miles southwest of Greenriver, Emery County, Utah.

Name. In allusion to the presence of uranium and vanadium in the mineral.

Ref.

1. Cited as priv. comm. from authors by Ford (App. III, 81, 1915 to Dana, 1892).
2. Schaller analysis in Hess and Schaller (1914).

47.1.15 SINCOSITE [$CaV_2O_2(PO_4)_2 \cdot 5H_2O$]. *Schaller (Washington Ac. Sc., J., 12, 195, 1922; Am. J. Sc., 8, 462, 1924).*

Cryst.[1] Tetragonal. Crystals are rectangular scales or thin plates with {001}, {100}, and, in traces, {110} (?). The crystals are striated parallel [100] on {001} and are often composed of many superposed plates in nearly parallel position. Usually aggregated into compact, scaly, or irregular masses; also as crudely radial aggregates and as small nodules admixed with gypsum and carbonaceous matter.

Twinning. On {110}, rare.

Phys. Cleavage {001} good, {100} and {110} poor. Brittle. H. low. G. ~2.84. Color leek-green, varying toward olive-green, brownish green, or yellowish green, according to the state of aggregation and degree of alteration. Luster vitreous, inclining toward a brassy, submetallic luster in altered crystals. Streak green. Transparent in small grains.

Opt. In transmitted light, variously light green, olive-green, brownish green, yellowish green, or bluish green. Relatively unaltered material is uniaxial negative ($-$), with nO ~1.680, nE ~1.655; both values vary slightly in different crystals. In part biaxial negative ($-$), apparently due to partial dehydration, with areas and bands of both uniaxial and biaxial material present in most crystals. $2E$ varies from about 10° to a maximum of about 83°, sometimes varying within single crystals with the smallest value in the marginal parts and the largest in the center. Dispersion $r > v$, faintly perceptible at small axial angles and increasing to nearly extreme at large angles. Indices for material with $2E = 83°$: nX 1.675, nY 1.690 (calc.), nZ 1.693. Both the uniaxial and the biaxial material is strongly pleochroic, with $E = X$ nearly colorless to pale yellow and $O = Z$ gray-green.

Chem. A hydrated phosphate of calcium and tetravalent vanadium, $CaV_2O_2(PO_4)_2 \cdot 5H_2O$. The water content is uncertain. The material alters on exposure to a lower hydrate, with possibly a partial oxidation of V^4 to V^5 in the brassy submetallic material. Sincosite may be isostructural with autunite or meta-autunite.

Anal.

	1	2	3
CaO	12.35	12.1	13.3
V_2O_4	36.54	36.3	37.8
V_2O_5		0.0	
P_2O_5	31.27	31.7	31.1
H_2O	19.84	19.9	[17.1]
Insol.		0.3	0.7
Total	100.00	100.3	[100.0]

1. $CaV_2O_2(PO_4)_2 \cdot 5H_2O$. 2. Sincos, Peru. On green crystals from a single specimen.
3. Sincos, Peru. On admixed green, olive, and brassy material from several specimens.

Tests. In C.T. changes in color from green to dark brown and black, decrepitates, loses water, and sinters to a dark-colored mass. A faint sublimate of Se is generally obtained. Readily soluble in dilute acids to a blue solution. Insoluble in water.

Occur. Found in a black carbonaceous shale of Cretaceous age near Sincos, Department of Junin, Peru. Sincos is about 100 miles east of Lima and the same distance southeast of Minasragra. The sincosite occurs in small veinlets, usually parallel to the bedding planes, and as irregular masses and small gypsum-bearing nodules in particular shale beds.

Name. From the locality.

Ref.
1. Tetragonal symmetry is assigned on the basis of the uniaxial optical character of the unaltered material.

47.1.16 R A U V I T E [$CaU_2V_{12}O_{36} \cdot 20H_2O$]. *Hess (Eng. Min. J.-Press*, 114, 274, 1922; *U. S. Geol. Sur., Bull. 750D*, 68, 1924).

As dense, slickensided masses, botryoidal crusts, and filmy coatings. Color purplish to bluish black. Streak yellowish brown to olive. Under the microscope minutely crystalline, with a mean index of refraction of about 1.88.

Chem. A hydrated vanadate of calcium and uranium of uncertain formula, perhaps $CaU_2V_{12}O_{36} \cdot 20H_2O$. On standing, fresh samples dehydrate partially and become cracked.

Anal.

	1	2
CaO	2.72	2.7
UO_3	27.72	28.1
V_2O_4	8.04	2.8
V_2O_5	44.06	49.0
H_2O	17.46	17.5
Total	100.00	100.1

1. $CaU_2V_{12}O_{36} \cdot 20H_2O$. 2. Temple Rock, Utah.[1] Analysis recalculated after deducting 27.06 per cent impurities.

Occur. Found filling cracks and impregnating sandstone (Shinarump) at Temple Rock and near-by Flat Top, San Rafael Swell, Emery County, Utah. Associated with carnotite, uvanite, hewettite, gypsum, and metatorbernite.

Name. From the chemical symbols of the constituent elements, Ra, U, V.

Ref.
1. Schaller analysis in Hess (1924).

47.1.17 M E L A N O V A N A D I T E [$Ca_2V_4^4V_6^5O_{25}$]. *Lindgren (Proc. Nat. Ac. Sc.*, 7, 249, 1921).

Cryst.[1] Monoclinic; prismatic—$2/m$.

$a:b:c = 0.4737:1:0.5815;\quad \beta\,91°22\tfrac{1}{2}';\quad p_0:q_0:r_0 = 1.2276:0.5813:1$

$r_2:p_2:q_2 = 1.7202:2.1116:1;\quad \mu\,88°37\tfrac{1}{2}';\quad p_0'\,1.2279,\,q_0'\,0.5815,\,x_0'\,0.0240$

MELANOVANADITE

Forms:

	ϕ	ρ	ϕ_2	$\rho_2 = B$	C	A
b 010	0°00′	90°00′	......	0°00′	90°00′	90°00′
h 230	54 36½	90 00	0°00′	54 36½	88 52½	35 23½
l 530 *	74 08½	90 00	0 00	74 08½	88 40½	15 51½
w $\bar{1}$01	−90 00	12 51½	102 51½	90 00	14 24	102 51½
g 012	4 43	16 16	88 37½	73 47½	16 12½	88 41
d 032	1 34½	41 06½	88 37½	48 54½	41 05½	88 58
p 111	65 05	54 04½	38 37	70 03	52 50	42 44½
s $\bar{1}$21	−45 59½	59 08½	140 17	53 23	58 09½	128 08

* Uncertain, perhaps {210}.

Habit. Velvety, divergent bunches of crystals elongated [001], the prism faces usually rounded or striated.

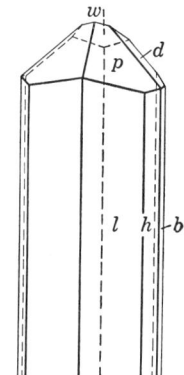

Phys. Cleavage {010} perfect. Brittle. H. 2½. G. 3.477. Luster almost submetallic. Color black. Streak dark reddish brown. Opaque.

Opt.[2] Very thin flakes are translucent with a dark reddish brown color and distinct pleochroism.

ORIENTATION		n	PLEOCHROISM	
X		1.73	Light reddish brown	Biaxial negative (−).
$Y \wedge c$	15°	1.96	Deep reddish brown	$2V$ medium.
Z	b	1.98	Dark reddish brown	

Chem. A calcium vanadyl vanadate, $Ca_2V_4^4V_6^5O_{25}$. The original mineral apparently is anhydrous,[3] but on standing in a damp atmosphere takes up loosely held water to at least 15 per cent H_2O.

Anal.

	1	2	3
CaO	11.33	9.89	10.65
V_2O_4	33.53	33.34	33.48
V_2O_5	55.14	52.61	49.38
Rem.		3.82	7.29
Total	100.00	99.66	100.80

1. $Ca_2V_4V_6O_{25}$. 2. Minasragra.[4] Rem. is MgO 0.27, $(Al,Fe)_2O_3$ 1.89, SiO_2 1.66.
3. Minasragra.[4] On material dried at 105° after standing in a damp atmosphere. Rem. is H_2O 5.90, $(Al,Fe)_2O_3$ 1.39.

1060 VANADIUM OXYSALTS

Tests. B.B. fuses very easily to a brown liquid. Easily soluble in acids.

O c c u r. Found with pascoite, pyrite, native copper, gypsum, and "patronite" in crevices in black shale at Minasragra near Cerro de Pasco, Peru. Alters to pascoite.

N a m e. In allusion to the color and composition.

Ref.
1. Palache in Lindgren, Hamilton, and Palache, *Am. J. Sc.*, **3**, 195 (1922).
2. Larsen and Berman (206, 1934).
3. See discussion in Lindgren, *et al.* (1922).
4. Hamilton analysis in Lindgren, *et al.* (1922).

47.1.18 H E W E T T I T E [$CaV_6O_{16} \cdot 9H_2O$]. *Hillebrand, Merwin*, and *Wright* (*Proc. Am. Phil. Soc.*, **53**, 31, 1914; *Zs. Kr.*, **54**, 209, 1914).

Probably orthorhombic. Found as nodular aggregates and coatings of fibers or microscopic needles. G. 2.55 (with $9H_2O$), 2.618 (on air-dried material). Color deep red. Luster somewhat silky.

O p t. In transmitted light, red to orange.

ORIENTATION	n(Li)	PLEOCHROISM	
X	1.77	Light orange-yellow	Biaxial negative (−).
Y	2.18	Light orange-yellow	2V medium.
Z c	2.35–2.4	Dark red	

C h e m. A hydrated calcium vanadate, $CaV_6O_{16} \cdot 9H_2O$ or perhaps $CaV_6O_{15}(OH)_2 \cdot 8H_2O$. Hewettite is dimorphous with metahewettite. A small amount of Mo is present in the material of analysis 2, presumably in substitution for V. The water content varies reversibly over a small range with atmospheric changes in humidity and temperature.[1]

Anal.

	1	2	3
Na_2O		0.15	
CaO	7.34	7.38	7.15
V_2O_4		1.21	0.43
V_2O_5	71.43	68.19	71.20
MoO_3		1.56	0.99
H_2O	21.23	21.33	21.04
Rem.		0.28	
Total	100.00	100.10	100.81

1. $CaV_6O_{16} \cdot 9H_2O$. 2. Minasragra, Peru.[2] Rem. is Fe_2O_3 0.11, insol. 0.17; MgO and K_2O none. Analysis made on material in equilibrium with water vapor at 21.8 mm. and 25°. 3. Minasragra, Peru.[4]

Tests. On heating loses water, changing in color through various shades of brown to bronze, but apparently without any change in structure until complete hydration. The dehydrated sample on exposure to moisture regains only part of the original amount of water and does not change color. Easily fusible to a dark red liquid. Slightly soluble in water.

O c c u r. Found in the oxidized zone of the vanadium deposit at Minasragra near Cerro de Pasco, Peru, as an alteration product of patronite. Also found rarely in the carnotite-bearing sandstone of Paradox

METAHEWETTITE 1061

Valley, Montrose County, Colorado. Alaite[3] from Ferghana, Russian Turkestan, may be identical with hewettite.

Name. After D. Foster Hewett (1881–), geologist of the U. S. Geological Survey.

Ref.
1. A partial study of the dehydration phenomena is given in the original paper.
2. Hillebrand analysis.
3. Described in Volume I, page 603, of this work as $V_2O_5 \cdot H_2O$ (?).
4. Schaller, *U. S. Geol. Sur., Bull. 878*, 118 (1937).

47.1.19 METAHEWETTITE [$CaV_6O_{16} \cdot 9H_2O$]. *Hillebrand, Merwin,* and *Wright (Proc. Am. Phil. Soc.,* **53**, 31, 1914; *Zs. Kr.,* **54**, 209, 1914).

Probably orthorhombic. As pulverulent masses, composed of microscopic tablets or laths, and as parallel- or radial-fibrous to bladed aggregates or coatings. The tablets are flattened on {001} and are bounded laterally by {010} and {110}, with a prism angle of 57°. G. 2.51 (material with $9H_2O$), 2.94 (material with $3H_2O$). Color deep red. Powder maroon to brownish red. Luster dull to somewhat silky.

Opt. In transmitted light, red to orange.

Orientation		n(Li)	Pleochroism	
X	c	1.70	Light orange-yellow	Biaxial negative (−).
Y	b	2.10	Deep red	$2V$ 52° (calc.).
Z	a	~2.23	Deeper red	

Chem. A hydrated calcium vanadate, $CaV_6O_{16} \cdot 9H_2O$. Dimorphous with hewettite. The water content varies reversibly over a small range with atmospheric changes in humidity and temperature.[1]

Anal.

	1	2
CaO	7.34	7.25
V_2O_3		0.35
V_2O_5	71.43	70.01
MoO_3		0.13
H_2O	21.23	21.30
Insol.		0.80
Rem.		0.39
Total	100.00	100.23

1. $CaV_6O_{16} \cdot 9H_2O$. 2. Thompson's, Utah.[2] Rem. is MgO 0.03, K_2O 0.09, Na_2O 0.08, Fe_2O_3 0.19. Analysis made on material in equilibrium with water vapor at 21.8 mm. and 25°.

Tests. On heating loses water and becomes progressively darker red until the last molecule of water begins to be lost, then becoming yellow-brown. Easily fusible to a dark red liquid. Slightly soluble in water.

Occur. Found as an impregnation in sandstone, generally coating the sandstone grains but sometimes filling cavities and crevices. Found in Paradox Valley, Montrose County, Colorado, at Thompson's, Grand County, Utah, and at other places in southwestern Colorado and southeastern Utah. Often immediately associated with gypsum, native selenium, and an unidentified hydrous silicate of vanadium, aluminum, and potassium.

N a m e.   In allusion to its relation to hewettite.

Ref.
1. A partial study of the dehydration phenomena is given in the original paper.
2. Hillebrand analysis.

47.1.20   **F E R N A N D I N I T E** [$CaO \cdot V_2O_4 \cdot 5V_2O_5 \cdot 14H_2O$]. *Schaller (J. Washington Ac. Sc.*, **5,** 7, 1915; Dana, App. III, 29, 1915).

Massive, cryptocrystalline to fibrous; rarely in rectangular plates. Color dull green.

O p t.   In transmitted light, light green to dark olive-green to brownish green and nearly opaque. Birefringence strong. Mean index of refraction [1] about 2.05. Not pleochroic.

C h e m.   A hydrated calcium vanadyl vanadate, probably $CaO \cdot V_2O_4 \cdot 5V_2O_5 \cdot 14H_2O$.

Anal.

	1	2
CaO	4.05	3.83
MoO_3		1.58
V_2O_4	11.99	11.63
V_2O_5	65.73	63.33
H_2O	18.23	18.07
Rem.		1.56
Total	100.00	100.00

1. $CaO \cdot V_2O_4 \cdot 5V_2O_5 \cdot 14H_2O$.   2. Minasragra. Rem. is K_2O 0.59, MgO 0.07, Fe_2O_3 0.90. Recalculated to 100 after deducting insol. 12.18 per cent.

Tests.   Easily soluble in acids to a green solution; sufficiently soluble in cold water to give a yellow solution.

O c c u r.   Found in the vanadium deposit at Minasragra, near Cerro de Pasco, Peru.

N a m e.   After Eulagio E. Fernandini, a former owner of the deposit.

Ref.
1. Larsen (74, 1921).

MOLYBDATES AND TUNGSTATES

The molybdates and tungstates are anisodesmic oxysalts that contain distorted $(XO_4)^=$ groups, in which X is Mo or W. The distortion is of the nature of a compression along a twofold axis of the idealized tetrahedral group, and the anion is intermediate in shape between the symmetrical coordination tetrahedra of the sulfates and phosphates and the planar, quadrilateral coordination groups such as in $K_2(PtCl_4)$. Because of the distorted shape and relatively large size of the anion in molybdates and tungstates there are no substitutional series between them and sulfates. Partial or complete anionic series are found between isostructural molybdates and tungstates, as in the Scheelite and Wulfenite Groups.

CLASS 48. Normal anhydrous molybdates and tungstates
Type 1. $A(XO_4)$
48.1.1   Wolframite group
48.1.1.1   Huebnerite $Mn(WO_4)$
48.1.1.2   Wolframite $(Fe,Mn)(WO_4)$
48.1.1.3   Ferberite $Fe(WO_4)$
48.1.2   Sanmartinite $(Zn,Fe)(WO_4)$
48.1.3   Scheelite group
48.1.3.1   Scheelite $Ca(WO_4)$
48.1.3.2   Powellite $Ca(MoO_4)$
48.1.4   Wulfenite group
48.1.4.1   Wulfenite $Pb(MoO_4)$
48.1.4.2   Stolzite $Pb(WO_4)$
48.1.5   Raspite $Pb(WO_4)$

CLASS 49. Basic and hydrated molybdates and tungstates
Type 1. Miscellaneous
49.1.1   Cuprotungstite $Cu_2(WO_4)(OH)_2$
49.1.2   Koechlinite $(BiO)_2(MoO_4)$
49.1.3   Ferritungstite $Fe_2(WO_4)(OH)_4 \cdot 4H_2O$ (?)
49.1.4   Lindgrenite $Cu_3(MoO_4)_2(OH)_2$
49.1.5   Ferrimolybdite $Fe_2(MoO_4)_3 \cdot 8H_2O$ (?)
49.1.6   Thorotungstite Th,Al tungstate (?)
49.1.7   Anthoinite $Al(WO_4)(OH) \cdot H_2O$

48 NORMAL ANHYDROUS MOLYBDATES AND TUNGSTATES

TYPE 1. $A(XO_4)$

48.1.1 WOLFRAMITE SERIES

48.1.1.1 H U E B N E R I T E [MnWO$_4$]. *Riotte* (*Reese River Reveille*, Austin, Nevada, 1865); *Credner* (*B. H. Ztg.*, **24**, 370, 1865). Manganowolframit *Weisbach* (40, 1875). Megabasit *Breithaupt* (*B. H. Ztg.*, **11**, 189, 1852). Blumit *Breithaupt*, *Liebe* (*Jb. Min.*, 652, 1863). Manganese tungstate. Hübnerite.

48.1.1.2 W O L F R A M I T E [(Fe,Mn)WO$_4$]. Lupi Spuma, Lapis niger ex quo conflatur candidum plumbum (= tin) *Agricola* (*De Natura Fossilium*, 255, 1546). Wolfram *Henckel* (171, 1725). Volfram, Wolffram, Ferrum arsenico mineralisatum, Spuma lupi *Wallerius* (268, 1747; **1**, 484, 1753). Wolf, Wolfart, Wolfort, Wolfrig, Wolffert, Wolfrath, Woolferam (tin miners of Saxony and Bohemia). Tungstate ferrugineux, Tungstate manganèsié, Écume de loup *Fr.* Licafro *Ital.* Lupus jovis, molybdaenum *Linnaeus* (*Natursystem des Mineralreichs*, Gmelin, **3**, 78, 1778). Cal, Call, Gal, Mock-lead (Cornish miners) *Pryce* (317, 321, 324, 1778). Wolfram = Tungstic acid, iron, and manganese *J. J.* and *F. de Elhuyar* (*A Chemical Examination of Wolfram and Examination of a New Metal Which Enters into its Composition*, London, 1785). Wolframum, Manganesia, parva cum portoine martis et jovis mixta *Cronstedt* (Magellan's ed., 208, 866, 1788, London), see also *Cronstedt* (107, 1758). Tungstate of Iron and Manganese, Scheelate of Iron and Manganese. Scheeleisenerz. Schéelin ferruginé *Haüy* (**4**, 314, 1801; **4**, 366, 1822). Prismatisches Scheel-Erz *Mohs*. Prismatic scheelium ore *Haidinger* (**2**, 387, 1825). Wolframit *Breithaupt* (227, 1832). Wolfram, Eisenscheel *Beudant* (**2**, 659, 1832). Wolfram *Dana* (373, 1837).

48.1.1.3 F E R B E R I T E [FeWO$_4$]. Ferberit *Liebe* (*Jb. Min.*, 641, 1863), attributing the name to Breithaupt. Ferrowolframit *Weisbach* (43, 1875). Reinit *Fritsch* (*Zs. Nat. Halle*, **3**, 864, 1878); *Luedecke* (*Jb. Min.*, 286, 1879). Iron tungstate. Ferrotungstat.

C r y s t.[1] Monoclinic; prismatic—$2/m$.

$a:b:c = 0.8255:1:0.8664;$ $\beta\ 90°28';$ [2] $p_0:q_0:r_0 = 1.0495:0.8664:1$

$r_2:p_2:q_2 = 1.1542:1.2114:1;$ $\mu\ 89°32';$ $p_0'\ 1.0496,\ q_0'\ 0.8664,\ x_0'\ 0.0081$

Forms: [3]

		ϕ	ρ	ϕ_2	$\rho_2 = B$	C	A
c	001	90°00'	0°28'	89°32'	90°00'	.....	89°32'
b	010	0 00	90 00	.....	0 00	90°00'	90 00
a	100	90 00	90 00	0 00	90 00	89 32	.....
r	120	31 12	90 00	0 00	31 12	89 45½	58 48
m	110	50 27½	90 00	0 00	50 27½	89 38½	39 32½
l	210	67 34	90 00	0 00	67 34	89 34	22 26
h	310	74 37	90 00	0 00	74 37	89 33	15 23
f	011	0 32	40 54½	89 32	49 05½	40 55½	89 39
t	102	90 00	28 03	61 57	90 00	27 35	61 57
y	$\bar{1}02$	−90 00	27 19½	117 19½	90 00	27 47½	117 19½
Δ	112	50 53½	34 28½	61 57	69 04½	34 07	63 56½
ω	111	50 40½	53 49	43 24	59 14	53 27½	51 21½
e	$\bar{1}12$	−50 01½	33 59½	117 19½	68 57	34 21	115 22
o	$\bar{1}11$	−50 14½	53 34	136 10	59 02	53 55½	128 12½
π	121	31 24	63 46½	43 24	40 02	63 32	62 08
s	$\bar{1}21$	−31 00½	63 41	136 10	39 48	63 55	117 30

Less common, rare, or doubtful forms:

T 150	M 510	g 103	z 113	τ 321	Ω 10.5.7
P 140	N 11.2.0	χ 101	A 337	U 541	
S 7.11.0	j 610	H 904	ψ $\bar{5}$52	W 572	
J 320	L 710	γ $\bar{1}$.0.11	B 123	V $\bar{1}$22	
R 15.7.0	n 810	δ $\bar{3}$04	E 5.9.14	α $\bar{1}$32	
C 940	η 013	λ $\bar{1}$01	p 214	d $\bar{2}$11	
F 520	φ 095	β $\bar{4}$03	D 313	k $\bar{6}$14	
q 830	w 021	ζ $\bar{5}$02	X 754	π 012	
G 720	u 104	Γ $\bar{2}$01	κ 211	I 403	

Structure cell.[4] Space group $P2/c$. Cell dimensions:

	a_0	b_0	c_0	$a_0:b_0:c_0$	β
MnWO$_4$ (art.)	4.84 kX	5.76	4.97	0.841:1:0.863	90°53′
Huebnerite	4.82	5.76	4.97	0.837:1:0.863	90 53
Wolframite	4.78	5.73	4.98	0.835:1:0.869	90 26
Ferberite	4.71	5.69	4.95	0.828:1:0.870	90 00
FeWO$_4$ (art.)	4.70	5.69	4.93	0.825:1:0.866	90 00

(Data obtained on huebnerite from Silverton, Colorado, containing 0–5% FeO, ca. 20% MnO; wolframite from Zinnwald, Bohemia, containing 5–15% FeO, ca. 14% MnO; ferberite from Colorado containing 15–25% FeO, 2–8% MnO.)

Contains (Mn,Fe)$_2$(WO$_4$)$_2$ in the unit cell.

Habit. Huebnerite is commonly prismatic to long prismatic [001], less often short prismatic, and flattened or tabular {100}. Striated or furrowed

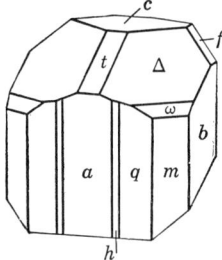

Wolframite. Bolivia.

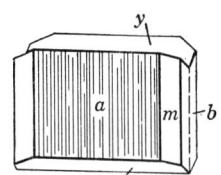

Wolframite. Bolivia.
Twin on {100}.

[001]. Common forms: *b n a q h t f c*. Often in large groups of parallel or subparallel crystals; less often in semi-spherical radiating groups of crystals.

Wolframite is commonly short-prismatic [001], less frequently long-prismatic, and somewhat flattened or tabular {100}; rarely elongated [010] and flattened {001}; rarely tabular {102}, {$\bar{1}$02}, or {010}; sometimes equant. Common forms: *a l m t f c b*. Usually striated [001]; rarely {001} striated [010]. In large groups of subparallel crystals; also lamellar or massive granular; rarely in braided intergrowths of acicular crystals.

Ferberite is commonly elongated [010] and somewhat flattened {100}. Usually striated [001] with {001} often striated [010]. Less often short prismatic [001] and flattened {100}. Common forms: *c b a m l t f*, the crystals often having a wedge-shaped appearance. Also in groups of bladed crystals, or massive.

The faces of $t\{102\}$, $y\{\bar{1}02\}$, and $c\{001\}$ in crystals of this series generally are rough, striated, or uneven; t and y commonly are indistinguishable.

Twinning. (1) Twin plane $\{100\}$, common in all members of the series.[5] Usually as simple contact twins with composition face (100) or, rarely, (001); also as interpenetration twins, sometimes simulating the Carlsbad twins in orthoclase; very rarely lamellar. (2) Twin plane $\{023\}$, common in wolframite and ferberite.[6] Usually as simple contact twins, rarely as repeated twins or interpenetrating.

Phys. Cleavage $\{010\}$ perfect; parting sometimes observed $\{100\}$, $\{102\}$. Crystals often show a well-developed zonal growth structure with more or less distinct parting parallel to the external faces. Fracture uneven. Brittle. H. 4–4½, greatest in the high-iron members; the hardness on $\{001\}$ is greater than on $\{010\}$.[14] G. increases with increasing content of Fe from 7.12 (huebnerite[7]) to 7.51 (ferberite[8]); 7.25 (calc. for $MnWO_4$), 7.60 (calc. for $FeWO_4$). Luster submetallic to metallic-adamantine, tending toward resinous in huebnerite. Color: huebnerite is yellowish brown to reddish brown, rarely brownish black; wolframite is dark grayish or brownish black to iron-black; ferberite is black. Color banding sometimes present, due to variation in the Fe:Mn ratio. Sometimes tarnished iridescent. Streak: huebnerite is yellow to reddish brown or greenish gray; wolframite reddish brown to brownish black and black; ferberite brownish black to black. Huebnerite is transparent; the transparency decreases with increasing content of Fe, and ferberite is nearly or quite opaque. The high-iron members of the series are weakly magnetic.

Opt. In transmitted light, yellow, orange, olive-green, red, reddish brown, or brown. Members of the series high in Fe are transparent only in very small grains. Pleochroism is often perceptible,[12] especially in material high in Mn, with absorption $Z > Y > X$. The refractive indices increase with increase in Fe, both in the infrared[37] and in the visible regions.

Orientation	1	2	3
λ	Li		
X b		2.17 ± 0.01	2.20 ± 0.02
Y		2.22 ± 0.01	
$Z \wedge c$ 17° to 21°		2.32 ± 0.01	2.30 ± 0.02
in the acute angle [9]			
Sign		+	+
$2V$	75°	73° ± 5° (calc.)	large

1. Artificial $MnWO_4$.[10] 2. Huebnerite, South Homestake mine, White Oaks, New Mexico (0.55 per cent FeO).[11] 3. Huebnerite, Pony, Montana.[11]

Orientation	4	5	6
λ	Li	Li	Li
X b	2.26 ± 0.02	2.31 ± 0.03	
Y	2.32 ± 0.02		2.40 ± 0.03
$Z \wedge c$ 17° to 21°	2.42 ± 0.02	2.46 ± 0.03	
in the acute angle [9]			
Sign	+		
$2V$	78°36′ (calc.)		

WOLFRAMITE SERIES

4. Wolframite, Cornwall, England.[11] 5. Wolframite, Mariposa County, California.[11]
6. Ferberite, Nugget claim, Rollinsville, Gilpin County, Colorado.[11]

PLEOCHROISM [12]

	Silverton, Colorado (Huebnerite, 2.91 FeO)	Pelagatos mine, Conchucos, Peru (Huebnerite, G. 7.09, in sections of varying thickness)			Vizeu, Portugal (Wolframite, 5.87 FeO)
X	Green	Bright yellow	Orange-red	Orange-red	Light red-brown
Y	Yellowish brown	Greenish yellow	Orange-red	Bright red	Red-brown
Z		Olive-green	Brick-red	Dark red	Dark red-brown, nearly black

In polished section,[13] gray to white in color, similar to sphalerite, with deep brownish red to blood-red internal reflections (distinctly brighter in huebnerite). Weak reflection pleochroism. Distinctly anisotropic. Reflection percentages: green 17, orange 15, red 14 (wolframite).

Chem. A tungstate of manganese and iron, $(Mn,Fe)WO_4$. A complete isomorphous series exists between Mn'' and Fe''. The species names applied [15] to particular ranges of composition are tabulated below. The

Name	Atomic per cent substitution		Corresponding weight percentages	
	Fe	Mn	FeO	MnO
Huebnerite	0 to 20	100 to 80	0 to 4.8	23.4 to 18.7
Wolframite	20 to 80	80 to 20	4.8 to 18.9	18.7 to 4.7
Ferberite	80 to 100	20 to 0	18.9 to 23.6	4.7 to 0

departure from $(Mn,Fe):W = 1:1$ shown by some analyses of ferberite and other members of the series apparently is due to analytical error. Ca is often reported in small amounts and probably is due to microscopic inclusions of scheelite.[15] Cb and Ta are sometimes reported in amounts up to 2.2 weight per cent $(Cb,Ta)_2O_5$ (anal. 7), but it is not known whether these elements are essential constituents of the mineral, presumably in substitution for W, or whether, as appears in some instances at least,[15] they are due to inclusions of columbite-tantalite. Sc, In, Ti, V, Mo, Al have been found [16] in spectrographic amounts.

Anal.[17]

	1	2	3	4	5
MnO	23.42	23.40	20.54	20.29	12.55
FeO		0.24	3.01	5.33	10.81
CaO		0.13			0.80
MgO					0.12
WO_3	76.58	75.58	75.12	74.78	74.84
$(Cb,Ta)_2O_5$		0.05			0.26
Rem.		0.62			1.00
Total	100.00	100.02	99.71	100.40	100.38
G.	7.25	7.177			7.272

1. $MnWO_4$. 2. Huebnerite. Uncompahgre district, Ouray County, Colorado.[18] Rem. is SiO_2. 3. Huebnerite. Lawrence County, South Dakota.[19] 4. Wolframite. Bodiosa, Vizeu, Portugal.[20] 5. Wolframite. Cornwall, England.[21] Rem. is SiO_2 0.30, Fe_2O_3 0.70.

	6	7	8	9	10
MnO	8.31	3.37	1.41	0.19	
FeO	15.75	18.18	22.54	24.37	23.65
CaO		0.24			
MgO			0.18	tr.	
WO_3	75.45	73.74	75.07	75.21	76.35
$(Cb,Ta)_2O_5$	tr.	2.20			
Rem.	0.23	2.06	0.88		
Total	99.74	99.79	100.08	99.77	100.00
G.	7.10	7.162	7.283	7.079	7.60

6. Wolframite. Burnt Hill Brook, York County, New Brunswick, Canada.[22] Rem. is SiO_2 0.23. Average of three concordant analyses. 7. Ferberite. Cave Creek, Maricopa County, Arizona.[21] Rem. is CuO 1.34, SiO_2 0.72. The $(Cb,Ta)_2O_5$ had Ta_2O_5 1.50, Cb_2O_5 0.70. 8. Ferberite. Riddarhyttan, Sweden.[23] Rem. is SiO_2 0.88, MoO_3 tr. 9. Ferberite (reinite). Kurasawa, Kai province, Japan.[24] 10. $FeWO_4$.

Tests. Wolframite and ferberite are easily fusible (2½–3) to a globule which has a crystalline surface on cooling; huebnerite is less fusible. Fused wolframite is highly magnetic, but pure huebnerite is not. The members of the series are more or less decomposed by prolonged treatment with hot concentrated sulfuric or hydrochloric acid; also decomposed by aqua regia with the separation of tungstic oxide.

Occur.[25] Members of the wolframite series constitute the principal ore of tungsten. The principal types of occurrence may be broadly classed as follows: (1) in greisen, quartz-rich veins, or pegmatitic veins immediately associated with granitic intrusive rocks. Many of the deposits are of so-called pneumatolytic origin. Associated minerals include cassiterite, arsenopyrite, specular hematite, tourmaline, lithia mica, topaz, and, less abundantly, beryl, fluorite, danburite, molybdenite, scheelite, and native bismuth. Wolframite rarely occurs in normal granite pegmatites, such as those of New England. (2) In high-temperature hydrothermal veins, often genetically associated with deposits of the preceding type. Associated minerals include various sulfides, particularly pyrrhotite, pyrite, chalcopyrite, arsenopyrite, and bismuthinite, together with cassiterite, scheelite, native bismuth, molybdenite, hematite, magnetite, tourmaline, topaz, albite, apatite, and danburite. In this and the preceding type of deposit, wall-rock alteration often is marked, with development of tourmaline, mica, topaz, fluorite, quartz, and chlorite. (3) In mesothermal veins, usually not immediately associated with deposits of the preceding type. Here sulfides are relatively abundant, and cassiterite may or may not be present. Other associated minerals include scheelite, native bismuth, bismuthinite, fine-grained quartz, siderite, and, less frequently, barite, fluorite, tourmaline, and adularia. Wolframite and ferberite sometimes are almost the only metallic minerals present. (4) Huebnerite and less frequently wolframite sometimes occur in very minor amounts in epithermal veins, as at Tonopah, Nevada, and Cripple Creek, Colorado. (5) Wolframite occasionally is found in small amounts in contact-metamorphic deposits adjacent to granitic intrusives, where it occurs with scheelite, diopside, axinite, and other silicates. (6) In alluvial or eluvial deposits.

Only a few of the more important mining districts and localities of specimen interest can be mentioned. The principal commercial occurrences

WOLFRAMITE SERIES 1069

of ferberite are in the United States, especially in Boulder County, Colorado. Small amounts or specimens have been obtained in Germany at Ehrenfriedersdorf, Saxony, and at Ölsnitz, Vogtland; in France in Haute-Vienne; in the cryolite deposit at Ivigtut, Greenland; at Calacalani, Colquiri, Bolivia; in New South Wales and Queensland. Ferberite was originally found in quartz veins near Aquilas in the Sierra Almagrera, in southern Spain. Huebnerite is less important than wolframite or ferberite as an ore mineral. It has been mined in the United States, in Peru at Morococha and in the Conchucos district, and in New South Wales and Queensland. Also found, among other localities, at Schlaggenwald, Bohemia (*megabasite*), and in France at Valcroze, Lozère, Plateau Central, and at Adervielle, Hautes-Pyrénées.

Wolframite is widespread and often is an important ore of tungsten. The major part of the production has come from China, the United States, Portugal, Cornwall, Bolivia, the Malayan peninsula and Burma, and Australia. Among the more important or interesting localities can be mentioned Schneeberg, Zinnwald, Altenberg, and Annaberg in Saxony; Neudorf and Harzgerode in the Harz, Germany; Schlaggenwald and Zinnwald, Bohemia. In France in Brittany and in Dept. Haute-Vienne, Plateau Central. In Russia in the Nerchinsk district and elsewhere in Transbaikalia. Important deposits of wolframite have long been mined in Cornwall, England, where the mineral occurs [26] with tourmaline in the cassiterite veins. Found in Beira Beixa province, Portugal, and in Orense and Córdoba provinces in northwestern Spain. In Africa in the Mutue-Fides-Stavoren district in the Transvaal. Numerous important deposits occur in the Malay States and Burma, notably the Tavoy and Mawchi district, and in Australia. Here may be cited the Deepwater, Emmaville, Barrier Range, and Ardlethan districts in New South Wales [28] and the Heberton district in Queensland; also in Tasmania. Wolframite is also mined in Japan, Burma, Siam, and Korea. The world's richest and most extensive deposits of wolframite occur in the Nanling Range in southern China.[36] The mineral here is found chiefly in hypothermal veins associated with cassiterite, molybdenite, arsenopyrite, and bismuthinite; also with cassiterite in greisen and with scheelite, molybdenite, and cassiterite in pegmatites. Wolframite also occurs at numerous localities in Bolivia,[27] including Chicote Grande, Llallagua, Tasna, and Chicote; and in Argentina in the Sierra de Córdoba and the Sierra de San Louis.

In the United States, members of the wolframite series occur in commercial amounts in many of the western states. The principal producing district has been in Boulder County and adjacent parts of northern Gilpin County in Colorado.[29] Here ferberite, sometimes beautifully crystallized, occurs as the principal metallic mineral in mesothermal veins in sheeted and crushed zones in granitic and gneissic rocks associated with minor amounts of fine-grained quartz and sporadic hematite, magnetite, scheelite, pyrite and other sulfides, adularia, and tellurides. Other localities in Colorado include the Uncompahgre district in Ouray County (huebnerite) and near Silverton and Gladstone in San Juan County (huebnerite). In Idaho found in the Blue Wing district, Lemhi County (huebnerite) associated with molybdenite and other sulfides and argentian tetrahedrite; also near Soldier Mountain, Camas County (ferberite). Small amounts of huebnerite were mined at Butte, Montana. In South Dakota, wolframite and small amounts of huebnerite and ferberite have been mined from veins and pegmatites near Hill City, Oreville, and Keystone, Pennington County, and from near Deadwood, Custer County. In New Mexico, found at Copper Mountain, in the Picuris district, Taos County (wolframite), in the Elizabethtown district, Colfax County (ferberite), and in the Nogal and White Oaks district, Lincoln County (huebnerite). Huebnerite was first described from the Erie and Enterprize veins near Ellsworth, in the Mammoth district, southwest of Austin, Nye County, Nevada; also from the Regan district, north of Osceola, Nevada. Members of the wolframite series are mined at numerous localities in Arizona,[30] among which may be cited the Cave Creek district, Maricopa County (ferberite), the Boriana mine, Hualpai Mountains, Mohave County, where wolframite occurs with scheelite and chalcopyrite in quartz veins, the Eureka district in Yavapai County, and the Little Dragoon Mountains in Cochise County, where quartz veins contain huebnerite, scheelite, and fluorite.

A l t e r. The members of the series are relatively stable chemically, and often occur as alluvial or eluvial concentrates, but occasionally are found decomposed to iron and manganese oxides, ferritungstite, hydrotungstite, or tungstite. Wolframite and ferberite have been found as remarkably perfect pseudomorphs after scheelite (Trumbull, Connecticut,[31] and *reinite*, from Kimbosan, Japan). Scheelite, usually later-formed than wolframite, often is observed more or less completely replacing wolframite. Pseudomorphs also have been found of quartz, hematite, marcasite, and kaolin after wolframite and of quartz after huebnerite.

A r t i f. Members of the series have been obtained in good crystals by fusion [32] of $NaWO_4$, $NaCl$, and $MnCl_2$ or $FeCl_2$; also by strongly heating a mixture of WO_3 and Fe_2O_3 in a stream of HCl gas, with ferberite, tungstite, and magnetite crystallizing toward the cold end of the tube.[33] Artificial $MgWO_4$, $ZnWO_4$, $NiWO_4$, and $CoWO_4$ are isostructural with $(Mn,Fe)WO_4$.[34]

N a m e. The origin of the name wolframite is not known. According to Agricola, wolfram is derived from *volf*, *wolf*, and *rahm*, *ram*, *froth*, *cream*, *soot* (*Latin*, Lupi spuma). The name perhaps alludes to an objectionable scum or product formed during the smelting of tin ores containing tungsten or, since *lupus* and *wolf* were alchemists' terms for antimony, can be taken to mean that the mineral was mistaken for an antimony ore. The name may also have derived from *Wolfrig* or similar terms used by early Saxon miners, in allusion to the action of the mineral when present in tin ores to decrease—eat away or devour—the yield of tin during concentration or smelting. Pryce said that the mineral was regarded as objectionable by the early Cornish tin miners because, being of equal gravity, it could not be mechanically separated from cassiterite; magnetic methods of separation are presently employed.

Huebnerite after Adolph Hübner, metallurgist, of Freiberg, Saxony. Megabasite from μέγας, *greater, more*, and βάσις, *base*, because it was thought to contain more basic ingredients than wolframite. Ferberite after Rudolph Ferber of Gera, Germany. Reinite after Johannes Justus Rein (1835–?), German geographer and traveler, who brought the first specimens from Japan. Blumite [not blumite, Fischer, = bindheimite] after the German mineralogist Johann Reinhard Blum (1802–1883), Professor of Mineralogy at Heidelberg.

In the early literature and in some modern publications of a technological nature the name wolframite is applied to all members of the series.

REINITE. *Fritsch (Zs. Nat. Halle*, **3**, 864, 1878), *Luedecke (Jb. Min.*, 286, 1879). A name given to a supposed tetragonal modification of $FeWO_4$. Shown [35] to be wolframite pseudomorphous after scheelite.

Ref.

1. Early regarded as orthorhombic by Haüy (**4**, 366, 1822); Rose, *Ann. Phys.*, **64**, 175 (1845); Miller (473, 1852); and others. The monoclinic symmetry was first noted by Des Cloizeaux, *Ann. chim. phys.*, **28**, 163 (1850).
2. Angles and ratio of Goldschmidt (Winkeltabellen, 366, 1897), the average of earlier measurements by Krenner, *Min. Mitt.*, **5**, 9 (1875), Des Cloizeaux, *Ann. chim. phys.*, **28**, 163 (1850), **19**, 168 (1870), and Seligmann, *Zs. Kr.*, **11**, 347 (1886). Lévy (**3**, 362, 1837) gives elements in a different setting, transformation Lévy to Goldschmidt 101/010/002. There is no substantial variation in angles with composition—cf. Schaller in Hess and Schaller, *U. S. Geol. Sur., Bull. 583*, 40 (1914). On the morphological resemblance to columbite see Brögger, *Zs. Kr.*, **45**, 87 (1908), and Rose (1845).
3. Form list from Dana (982, 1892); Goldschmidt (**9**, 86, 1923); Hess and Schaller (1914); Koch, *Ann. Hist.-Nat. Mus. Nat. Hungarici*, **22**, 146 (1925)—*Min. Abs.*, **3**, 101 (1926); Jahn, *Vogtländ. Ges. Naturfor. Mitt.*, no. 3, 1 (1926)—*Min. Abs.*, **3**, 154 (1926); Fisher, *Am. Min.*, **15**, 104 (1930); Franzenau and Tokody, *Math. Naturwiss. Ber. aus Ungarn*, **38**, 236 (1931)—*Min. Abs.*, **5**, 135 (1934); and Gordon, *Ac. Sc. Philadelphia, Proc.*, **76**, 335 (1924), **96**, 279 (1944).
4. Broch, *Norsk. Ak., Oslo, Mat.-Nat. Kl., Skrifter*, no. 8 (1929), by powder method. The reported variation in axial ratio is greater than that found by morphological methods. See also Bunin, Klimov, and Umansky, *J. Phys. Chem., Ac. Sc. U.S.S.R.*, **14**, 844 (1940), who give a_0 4.78 kX, b_0 5.73, c_0 4.98.
5. Cf. Dufrénoy (**2**, 636, 1856); *Atlas*, Pl. 74, 1856); Breithaupt (**3**, Tafel 16, 1847); Des Cloizeaux (1850), Greg and Lettsom (352, 1858), Schneiderhöhn and Ramdohr (**2**, 607, 1931).
6. Cf. Hess and Schaller (1914); Jahn (1926); Bøggild, *Medd. Grønland.*, no. 32, 179 (1905); Kerndt, *J. pr. Chem.*, **42**, 81 (1847); Rose (1845), Maskelyne (369, 1895), Naumann (**2**, Tafel 32, 1830). Twin lamellae of uncertain nature are described by Purgold, *Isis*, Dresden, 73 (1883).
7. Obtained by microbalance on unanalyzed crystals from Cerro de Pasco, Peru; crystals from Silverton, Colorado, gave 7.17 (Mrose, priv. comm., 1946). Hillebrand, *U. S. Geol. Sur., Bull. 20*, 96 (1885), gave 7.177 for analyzed material from Ouray, Colorado, with 0.24 FeO. Values much below 7.12 for huebnerite have been reported in the literature but are of uncertain validity.
8. Kerndt (1847) on ferberite (Mn:Fe ~1:4) from Chanteloube, Monte Video, Ehrenfriedersdorf, and Nertschinsk. Chernik, *Zs. Kr.*, **31**, 513 (1899), gives 7.53 for nearly pure ferberite from Batum.
9. Des Cloizeaux (cited in Hintze (**1** [3B], 4112, 1929)); Duparc, *Min. Mitt.*, **38**, 116 (1925); Genth and Penfield, *Am. J. Sc.*, **43**, 184 (1892). Groth and Arzruni, *Ann. Phys.*, **149**, 235 (1873), place the acute bisectrix in the acute angle. Bravo, *Arch. assoc. peruana Prog. Cienc.*, **1**, 141 (1921)—*Zs. Kr.*, **59**, 487 (1924), reported an extinction angle of 3° on [001] in (100) sections.

10. Groth and Arzruni (1873).
11. Larsen (157, 1921).
12. Cf. Tronquoy, *Bull. soc. min.*, **36,** 113 (1913), Larsen (1921), Duparc (1925), Genth and Penfield (1892).
13. Schneiderhöhn and Ramdohr (**2,** 606, 1931) and Short (167, 168, 1940).
14. Tronquoy (1913).
15. Cf. Hess and Schaller (1914, p. 20, 36, 37) and Van Horn, *Am. Min.*, **15,** 461 (1930). On the Cb, Ta content see also Parga, *Anal. soc. españ. fis. quim.*, **28,** 905 (1930).
16. Cf. Eberhard in Hess and Schaller (1914, p. 18), Doelter (**4** [2], 846, 1928), and Pereira-Ferraz, *C.R.*, **173,** 1170 (1921).
17. For additional analyses see Hess and Schaller (1914) and Doelter (**4** [2], 825, 1928); also Ahlfeld, *Chem. Erde*, **7,** 121 (1932); Parga (1930); Angelelli and Chaudet, *Rev. Min., Soc. Argentine Min. Geol.*, **10,** 74 (1939); Ellsworth and Jolliffe, *Univ. Toronto Stud., Geol. Ser.*, no. 40, 71 (1937); Lacroix, *Bull. serv. Geol. Indochine*, **20,** 1933—*Min. Abs.*, **6,** 21 (1935).
18. Hillebrand, *U. S. Geol. Sur., Bull. 20*, 96 (1885).
19. Headden, *Proc. Colorado Sc. Soc.*, **8,** 175 (1906).
20. Duparc, *Min. Mitt.*, **38,** 117 (1925).
21. Wherry, *Proc. U. S. Nat. Mus.*, **47,** 501 (1914).
22. Swanson, *Univ. Toronto Stud., Geol. Ser.*, no. 20, 28 (1925).
23. Assarson in Geijer, *Geol. För. Förh.*, **45,** 434 (1923).
24. Kodera anal. in Wada (*Minerals of Japan*, 76, 1904).
25. See also Hess, *U. S. Geol. Sur., Bull. 652* (1915); Hintze (**1** [3B], 4103, 4117, 4143, 1929); Lindgren (1933; p. 596, 659, 761); Kerr, *Geol. Soc. Am., Mem.*, **15** (1946); Li and Wang (*Tungsten*, New York, 1943).
26. Cf. Dewey and Bromehead (*Mem. Geol. Sur., 1916, Spec. Rpt.*, Tungsten and Manganese Ores).
27. Cf. Ahlfeld and Reyes (1938; p. 63).
28. Cf. Smith, *Mineralogy of New South Wales, Dept. of Mines, Min. Res.*, no. 34, 1926, p. 132.
29. Hess and Schaller (1914); Loomis, *Econ. Geol.*, **32,** 952 (1937); Lovering, *Econ. Geol.*, **36,** 229 (1941).
30. Wilson, *Arizona Bureau of Mines, Geol. Ser., Bull.* no. 148 (1941).
31. Warren, *Am. J. Sc.*, **11,** 369 (1901); also Hintze (**1** [3B], 4067, 4113, 1929).
32. Geuther and Forsberg, *Ann. Chem. Pharm.*, **120,** 270 (1861); Michel, *Bull. soc. min.*, **2,** 142 (1879).
33. Debray, *C.R.*, **55,** 287 (1862); **64,** 603 (1866).
34. Machatschki, *Zs. Kr.*, **67,** 163 (1928), and Broch (1929).
35. Cf. Wada (1904, p. 76) and Jimbo, *Beitr. Min. Japan*, no. 5, 256 (1915).
36. Ke-Chin Hsu, *Econ. Geol.*, **38,** 431 (1943); Juan, *Econ. Geol.*, **41,** 399 (1946).
37. Bailly, *Soc. géol. Belgique, Ann.*, **65,** B133 (1942).

UNNAMED MINERAL. MacGregor (*Min. Mag.*, **27,** 157, 1945). An apparently monoclinic manganese tantalate found as radiating, twinned needles in greisen at Bikita, Southern Rhodesia.

48.1.2 **SANMARTINITE** [(Zn,Fe,Ca)WO$_4$]. *Angelelli* and *Gordon* (*Ac. Sc. Philadelphia, Not. Nat.*, no. 205, April, 1948).

Cryst. Monoclinic; prismatic—$2/m$.

$a:b:c = 0.8004:1:0.8499;$ $\beta\,90°28';$ $p_0:q_0:r_0 = 1.0618:0.8499:1$

$r_2:p_2:q_2 = 1.1766:1.2493:1;$ $\mu\,89°32';$ $p_0'\,1.0618, q_0'\,0.8499, x_0'\,0.0082$

Forms:

	ϕ	ρ	ϕ_2	$\rho_2 = B$	C	A
b 010	0°00'	90°00'	.....	0°00'	90°00'	90°00'
a 100	90 00	90 00	0°00'	90 00	85 29	.....
m 110	51 19½	90 00	0 00	51 19½	89 38	38 40½
t 102	90 00	28 19½	61 40½	90 00	27 51½	61 40½
Δ 112	51 45	34 28	61 40½	69 29½	34 06	63 36½

SCHEELITE GROUP

Structure cell.[1] Space group $P2/c$. a_0 4.712 kX, b_0 5.738, c_0 4.958; [β 90°28′]; $a_0:b_0:c_0 = 0.8212:1:0.8641$. Cell contents $(Zn,Fe,Ca)_2(WO_4)_2$. Pure artificial[2] $ZnWO_4$ has a_0 4.68 kX, b_0 5.73, c_0 4.95; β 90°30′; G. 7.79 (calc.).

Habit. Fine-granular masses, in part as microscopic crystals tabular {001} and as reticular aggregates.

Phys. Cleavage {010}, perfect. Luster resinous. Color dark brown to brownish black in masses; microscopic crystals are reddish brown with red internal reflections. G. 6.70;[6] 7.44 (calc. from anal. 1).

Chem. Essentially zinc tungstate, $(Zn,Fe,Ca,Mn)WO_4$, with small amounts of Fe″, Ca, and Mn″ in substitution for Zn.

Anal.

	1	2	3	4
CaO		1.48	1.54	6.32
ZnO	25.98	18.18	15.74	11.70
FeO		7.24	8.28	7.74
MnO		1.73	1.00	0.74
WO_3	74.02	72.62	71.70	71.20
Insol.		0.24	1.10	1.80
Total	100.00	101.49	99.36	99.50

1. $ZnWO_4$. 2. San Martin, Argentina.[3] Original total given as 101.25. 3, 4. San Martin. Preliminary analyses.[4]

Occur. Found with willemite as an alteration of scheelite in a quartz vein in Los Cerrillos, near San Martin, Department of San Martin, San Luis Province, Argentina.

Artif.[5] Obtained by heating the pure oxides in stoichiometric proportions at 800°.

Ref.

1. By powder method; space group from presumed isostructural relation to wolframite.
2. Broch, *Norsk. Ak. Skr.*, no. 8 (1929).
3. Hallowell analysis in Angelelli and Gordon (1948).
4. Lab. of Direc. de Minas y Geol., Buenos Aires.
5. Mellor (**11**, 788, 1931).
6. Weiss in Angelelli and Gordon (1948) by pycnometer method.

SCHEELITE AND WULFENITE GROUPS

TETRAGONAL

	$a_0:c_0$	a_0	c_0
Scheelite Group			
Scheelite, $CaWO_4$	1:2.165	5.246 kX	11.349
Powellite, $CaMoO_4$	1:2.187	5.23	11.44
Wulfenite Group			
Stolzite, $PbWO_4$	1:2.207	5.452	12.031
Wulfenite, $PbMoO_4$	1:2.236	5.401	12.079

The scheelite type of structure is possessed by a number of oxysalts of the AXO_4 type with large X ions, usually I, W, or Mo, and ordinarily with relatively large cations although the structure is quite tolerant in this regard. Scheelite and powellite appear to conform to tetragonal dipyramidal (4/m) symmetry. Wulfenite, however, apparently lacks a horizontal plane of symmetry, and its point symmetry is tetragonal pyramidal (4), although its structure [1] and that of the rare mineral stolzite is very similar to, if not identical with, that of scheelite. Raspite is a monoclinic polymorph of stolzite. Probably complete series involving the mutual substitution of W and Mo extend between both scheelite-powellite and wulfenite-stolzite. Wulfenite, stolzite, and, peculiarly, powellite are secondary minerals, whereas scheelite is typically a high-temperature hydrothermal or skarn mineral although sometimes of low temperature or secondary origin.

Ref.

1. See Sillén and Nylander, *Ark. Kemi*, **17A**, no. 4 (1943).

48.1.3.1 S C H E E L I T E [CaWO$_4$]. Tennspat, Lapides stanniferi spathecei "lik en huit spat" (from Bohemia) *Wallerius* (303, 1747). Not Tungsten von Bastnaes [= Cerite] *Cronstedt* (*Ak. Stockholm, Handl.*, 1751; 183, 1758). Stannum spathosum subdiaphanum album *Linnaeus* (1768). Tungsten (= Tungstic acid and lime) *Scheele* (*Ak. Stockholm, Handl.*, 1781). Schwerstein *Werner* (*Bergm. J.*, 386, 1789), *Karsten* (26, 1791). Scheelerz *Karsten* (56, 1800; 74, 1808). Tungstate of Lime. Tungstein. Scheelin calcaire *Haüy* (**4**, 372, 1801). Scheelspath *Breithaupt* (23, 1820). Scheelit *Leonhard* (594, 1821). Calciumwolframite. Calcium tungstate.
 Trimontite *Iwase* (1877) in *Harada* (*J. Fac. Sc. Hokkaido Univ.*, **3**, [4], 357, 1936); *Wada* (1904, p. 75). Seyrigite *Lacroix* (*C.R.*, **210**, 273, 1940). Molybdoscheelite *Strunz* (Mineralogische Tabellen, 144, 1941).

C r y s t.[1] Tetragonal; dipyramidal—$4/m$.

$$a\!:\!c = 1\!:\!2.1717; \quad p_0\!:\!r_0 = 2.1717\!:\!1$$

Forms: [2]

	ϕ	ρ	A	M
c 001	.....	0°00′	90°00′	90°00′
β 013	0°00′	35 54	90 00	65 30
p 011	0 00	65 16½	90 00	50 02½
o 114	45 00	37 31	64 29½	90 00
e 112	45 00	56 55½	53 40	90 00
g 134	18 26	59 47	74 08½	67 16
h 123	26 34	58 17½	67 38½	74 23½
s 121	26 34	78 22	64 01½	71 57½

Less common:
 f 014 v 012 d 1.1.10 k 235 δ 132 t 354

Structure cell.[3] Space group $I4_1/a$. a_0 5.246 kX, c_0 11.349; $a_0\!:\!c_0 = 1\!:\!2.165$. Cell contents Ca$_4$(WO$_4$)$_4$.

Habit. Octahedral with {011} or {112} predominant; sometimes tabular {001}. {001} usually rough; {112} often diagonally striated, usually parallel to [$\bar{3}$11], the intersection with {121}. Also massive, granular; columnar.

Twinning.[4] On {110}, common, as penetration more often than contact twins. Composition plane usually (110); also (001).

Scheelite 1075

Etching.[5] Etch figures conform to tetragonal-dipyramidal symmetry, except with optically active solvents.

Oriented growths.[27] Upon wolframite, with scheelite {001} [110] parallel wolframite {010} [001].

P h y s. Cleavage {101} distinct, {112} interrupted, {001} usually indistinct.[6] Fracture uneven to subconchoidal. H. $4\frac{1}{2}$–5. G. 6.10 ± 0.02, 6.09 (calc.); the gravity decreases with increasing substitution of Mo, with G. ~5.9 for material with MoO_3 ca. 8.0 per cent and G. = 5.48 for MoO_3

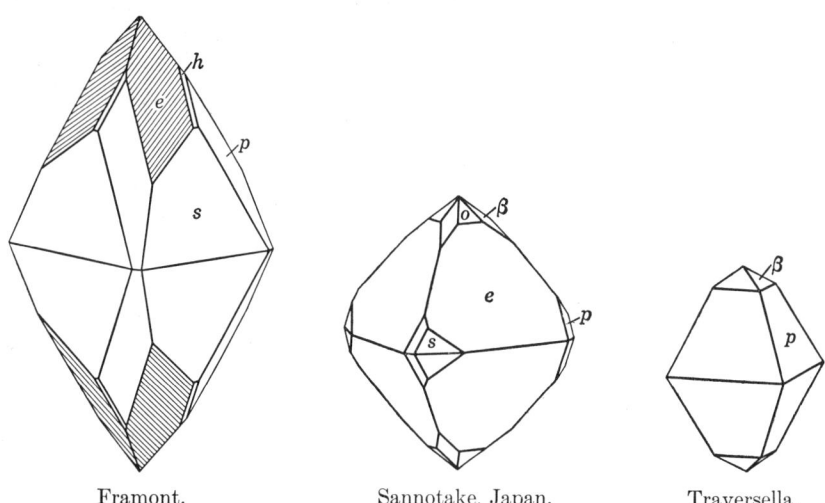

Framont.   Sannotake, Japan.   Traversella.

24.0 per cent (anal. 9). Luster vitreous, inclining to adamantine. Colorless to white, usually yellowish white, pale yellow or brownish; also greenish, gray, reddish, orange-yellow. Material with little or no Mo is usually white or nearly so. Sometimes deep green due to surficial alteration to cuprotungstite. Streak white. Transparent. Fluoresces bright bluish white in x-rays, short ultraviolet radiation, and cathode-rays. The fluorescence inclines to white and yellowish white with increasing content of Mo.[25] Thermoluminescent.

O p t. In transmitted light, colorless. Sometimes exhibits weak anomalous birefringence. The indices presumably decrease with increasing content of Mo, but data are lacking. Uniaxial positive (+). Indices:[7]

λ	667	C	D	570	533	E	475
nE	1.9263	1.9281	1.9365	1.9375	1.9442	1.9368	1.9525
nO	1.9107	1.9124	1.9200	1.9208	1.9273	1.9298	1.9344

C h e m. Calcium tungstate, $CaWO_4$. Mo can substitute for W, and a partial isomorphous series extends towards powellite, $CaMoO_4$, to at least Mo:W = 1:1.38 (anal. 9). Scheelite ordinarily contains small amounts of Mo but this element was not sought for in most old and some modern analyses. Growth zones with varying W:Mo ratio and color are often present. Small amounts of rare earths [8] and Cb, Ta [9] are sometimes present in substitution for Ca and W, respectively. Traces of F and Cl have been reported; [10] also Si (apparently due to admixture in some instances at least), Fe,[11] and Bi.[11]

Anal.[12]

	1	2	3	4	5	6	7	8	9
CaO	19.47	19.49	19.57	19.43	19.23	18.33	19.73	20.33	[22.43]
MgO		tr.	0.05	tr.	0.55	1.67			
MoO$_3$		0.07	0.58	0.72	1.47	2.46	3.15	8.23	24.01
WO$_3$	80.53	80.17	79.51	79.68	78.75	77.35	77.03	71.08	[53.56]
Rem.			0.25						
Total	100.00	99.73	99.96	99.83	100.00	99.81	99.91	99.64	[100.00]
G.		6.02						5.88	5.484

1. CaWO$_4$. 2. Schwartzenberg, Saxony.[13] Outer brownish layer over white nucleus containing only a trace of MoO$_3$. Original sum given as 99.83. 3. Traversella.[14] Yellow crystal. Rem. is Cb$_2$O$_5$ 0.08, Ta$_2$O$_5$ 0.02, rare earths 0.14, H$_2$O 0.01, BaO and SrO tr. 4, 5, 6, 7. Traversella.[15] Color: 4, orange-yellow; 5, greenish brown; 6, reddish brown; 7, colorless. 8. Zinnwald, Saxony.[16] 9. Ambondrombe, Madagascar (*seyrigite*).[17]

Var. Molybdian. Moliboscheelite *Strunz* (Mineralogische Tabellen, 144, 1941). Seyrigite *Lacroix* (*C.R.*, **210**, 273, 1940). Contains Mo in substitution for W up to at least 24 weight per cent MoO$_3$ (Mo:W = 1:1.38). Material with more than 2 or 3 weight per cent of MoO$_3$ is very rare.

Tests. B.B. fusible with difficulty (5) to a semi-transparent glass. Decomposed by HCl or HNO$_3$, leaving a yellow powder of hydrous tungstic oxide which is soluble in ammonia.

O c c u r. Scheelite is widespread and often is an important ore of tungsten. Like wolframite, with which it often is associated, scheelite is typically a high-temperature mineral. The principal types of occurrence include: (1) Contact metamorphic deposits formed adjacent to granitic intrusives in limestone. Associated minerals include garnet, diopside, tremolite, hornblende, epidote, wollastonite, vesuvianite, sphene, axinite, molybdenite, fluorite, with minor amounts of sulfides, principally pyrite and chalcopyrite, and rarely wolframite. The contact-metamorphic rock is commonly referred to as tactite. The principal economic deposits of scheelite are of this type. (2) High-temperature quartz-rich hydrothermal veins and greisen, usually immediately associated with granitic intrusives and containing important amounts of wolframite and cassiterite. Other associated minerals include tourmaline, apatite, topaz, fluorite, mica, albite, arsenopyrite, pyrite, pyrrhotite, molybdenite, chalcopyrite, native bismuth, and bismuthinite. The scheelite is subordinate in amount to and later-formed than wolframite and often is only an accessory mineral. (3) In pegmatites, usually only as an accessory constituent but sometimes in important amounts (Oreana, Nevada). (4) In relatively small amounts in hydrothermal veins formed at moderate to low temperatures (Leadville, Colorado; Bolivia; Atolia, California). A few of the more important or interesting localities are listed below.

In the wolframite-cassiterite veins of Zinnwald, Altenberg and Schwarzenberg in Saxony, similarly at Neudorf and Harzgerode in the Harz and at Tirpersdorf, Vogtland. Also found in the tin veins of Zinnwald and Schlaggenwald in Bohemia. In veins with calcite and epidote in gneiss in the Unter-Sulzbachthal and elsewhere in the Austrian Alps. Specimens also have been obtained from veins of the Alpine type in the Haslithal and elsewhere in the Swiss Alps. In Italy found with tourmaline, fluorite, and chalcopyrite in veins in diorite at Predazzo in the Tyrol; at Traversella in Piedmont, often in fine crystals, and at Gerrei, Sardinia. In France, especially at Meymac, Corrèze, in part altered to tungstite (cf. *meymacite*), and at Framont in the Vosges, Alsace. Scheelite is a frequent, although minor, constituent of the cassiterite veins in Cornwall, England; also found with wolframite at Carrock Fells, Cumberland. In Russia, scheelite is found in contact metamorphic deposits in the Nuratinsky Range, Uzbek, in the Kuznetsk Ala-Tan region, Siberia, in the Krasnoyarsk district, Khakass, and at Boevka in Transbaikalia. Deposits of scheelite are found in Burma and on the Malayan peninsula, especially at Kinta, where it occurs with fluorite in a contact zone. Also found variously in Japan, and as a minor associate of wolframite in the tungsten district of southern China. At Hillgrove, New South Wales, with stibnite, and at Mt. Ramsay, Tasmania. Scheelite is found in small amount along with wolframite and cassiterite in sulfide-rich veins in northern Bolivia.[18] Important deposits occur in northeastern Brazil.[26]

In the United States, scheelite is found in commercial amounts in many of the western states. One of the largest economic deposits is near Mill City, in the Humboldt Range, Nevada, where the mineral occurs [19] in a contact metamorphic deposit. Also mined at numerous other localities in Nevada, notably at Silver Dyke, near Mina, with albite and minor sulfides in fissure veins, at Oreana [20] with beryl, oligoclase, phlogopite, fluorite in pegmatite; in the Paradise Range. In Arizona [21] with wolframite in vein deposits at the Boriana mine, Hualpai Mountains, Mohave County, and in veins in the Huachuca Mountains, and Little Dragoon Mountains, in Cochise County, in contact deposits in the Helvetia district and elsewhere in Pima County. In New Mexico in minor amounts in Hidalgo County and elsewhere. Important deposits of scheelite are found in California, notably near Bishop, Inyo County, in epithermal veins at Atolia, San Bernardino County, and at Randsburg, Kern County. Found in Idaho at Yellow Pine, and in the Coeur d'Alene district, Shoshone County. In Utah in the Cottonwood-American Fork districts and in the Mineral Range, Beaver County. In Connecticut at Long Hill, Trumbull, and at Lane's mine, west of Monroe.

In Canada, found in the Bridge River district, British Columbia, with stibnite and ferroan dolomite in low-temperature veins in serpentine, and in the Cariboo and Nelson districts, British Columbia. Also in quartz veins in Halifax County, Nova Scotia, and in Marlow township, Frontenac County, Quebec.

Alter. In the zone of weathering, scheelite alters to tungstite, hydrotungstite, or, in the presence of copper-bearing solutions, to cuprotungstite. Wolframite sometimes is found as perfect pseudomorphs after scheelite, but scheelite more often is found more or less completely replacing wolframite,[22] in both cases apparently as a result of hypogene processes. Pseudomorphs also have been found of quartz, tungstite (?), kaolin, and bismutite after scheelite.

Artif.[23] Obtained in crystals from a melt of Na_2WO_4 with $CaCl_2$ and NaCl, by heating precipitated $CaWO_4$ with Ca in a stream of HCl gas, by adding dropwise a solution of Na_2WO_4 to a water solution of $CaCl_2$ made acid with HCl, and as a sublimate by heating a mixture of $CaWO_4$ with NaCl and KCl. $CaWO_4$ forms a partial series with $BaWO_4$ in artificial material.[24] Also obtained as single crystals from fusion and by the Verneuil process.

Name. Named in honor of the Swedish chemist Karl Wilhelm Scheele (1742–1786), who proved the existence of tungstic oxide in scheelite in 1781. The isolation of metallic tungsten (from the Swedish *tung*, heavy or *ponderous*, and *sten*, stone) was first reported by the Spanish chemists J. J. and F. de Elhuyar in 1783, but it may have been earlier accomplished by Scheele.

SEYRIGITE. Lacroix (*C.R.*, **210**, 273, 1940). A molybdian scheelite, with MoO_3 24.01 per cent (Mo:W = 1:1.38), found as large crystals associated with phlogopite, diopside, sphene, and calcite in pegmatite at Ambondrombe, southern Madagascar. Color golden yellow. G. 5.484.

Ref.

1. Evidence of less than dipyramidal (4/m) symmetry is lacking—compare wulfenite. Angles of Dauber, *Ann. Phys.*, **107**, 272 (1859), unit of the structure cell. Transformation: Dauber (1859), Dana (985, 1892), Goldschmidt (**8**, 12, 1922) to new, 1$\bar{1}$0/110/002.

2. Goldschmidt (1922). No satisfactory separation of right- and left-handed forms has been made. Rare or uncertain:

010	120	085	3.3.10	332	3.5.12	11.13.24	10.11.11
110	018	115	113	156	154	7.8.13	
170	027	9.9.32	5.5.14	166	255	233	
130	038	227	7.7.16	125	358	344	

3. Space group and crystal structure by Dickinson, *J. Am. Chem. Soc.*, **42**, 85 (1920); Vegard, *Phil. Mag.*, **1**, 1151 (1926); Sillén and Nylander, *Ark. Kemi*, **17A**, no. 4 (1943). Cell dimensions from Aanerud, *Norsk. Ak., Skr.*, no. 13 (1931), on artificial material by the powder method.

4. Cf. Bauer, *Württemberg. Naturwiss. Jahreshefte*, 154 (1874).

5. Traube, *Jb. Min., Beil.-Bd.*, **10**, 457 (1895–96); Honess (1927); Royer, *C.R.*, **202**, 1346 (1936).

6. Cf. Colomba, *Acc. Linc., Rend.*, **15**, 281 (1906), on crystals from Traversella with good cleavage {001}.

7. Zambonini, *C.R.*, **162**, 835 (1916), by prism method on artificial crystals. See also Harada, *J. Fac. Sc. Hokkaido Univ.*, **4**, [2], 279 (1934).

8. Cf. de Rohden, *C.R.*, **159**, 318 (1914); Servigne, *C.R.*, **210**, 440 (1940); Carobbi, *Gazz. chim. ital.*, **54**, 59 (1924); Marsh, *J. Chem. Soc. London*, 577 (1943). On artificial mixed crystals with Ce see Zambonini, *Zs. Kr.*, **58**, 226 (1923).

9. Cf. Carnot, *C.R.*, **79**, 637 (1874), who found 0.4 Ta_2O_5 in material from Meymac, France; also Carobbi (1924), anal. 3.

POWELLITE

10. Breithaupt, J. Chemie u. Phys., **54**, 130 (1851), and Plattner-Richter, Probierkunst m.d. Lothrohr, 185 (1878).
11. Lindroth and Mauzelius, Geol. För. Förh., **44**, 110 (1922).
12. For additional analyses see Hintze (**1** [3B], 4083, 1929); Doelter (4 [2], 814 (1928); also Lacroix, Bull. serv. Geol. Indochine, **20** (1933)—Min. Abs., **6**, 21 (1935).
13. Traube, Jb. Min., Beil.-Bd., **7**, 238 (1890).
14. Carobbi (1924).
15. Colomba, Acc. Linc., Rend., **15**, 281 (1906).
16. Traube, Jb. Min., Beil.-Bd., **7**, 232 (1890).
17. Lacroix, C.R., **210**, 273 (1940).
18. Cf. Ahlfeld-Reyes (1938, p. 62).
19. Kerr, Univ. Nevada Bull., **28**, no. 2 (1934).
20. Kerr, Econ. Geol., **33**, 390 (1938).
21. Cf. Wilson, Arizona Bur. Mines, Geol. Ser., Bull. *148* (1941).
22. Cf. Hintze (**1** [3B], 4067, 4113, 1929); Petterd (1910, p. 200); Wada (1904, p. 76).
23. Cf. Mellor (**11**, 783, 1931).
24. Aanerud (1931).
25. Cf. Greenwood, Econ. Geol., **28**, 56 (1943).
26. Johnston and Vasconcellos, Econ. Geol., **40**, 34 (1945).
27. Ramdohr, Heidelberger Beitr. zur Min., **1**, 105 (1949).

48.1.3.2   **P O W E L L I T E** [Ca(Mo,WO$_4$)]. Melville (Am. J. Sc., **41**, 138, 1891).

C r y s t.   Tetragonal; dipyramidal—$4/m$.

$$a:c = 1:2.1762; \quad p_0:r_0 = 2.1762:1 \quad {}^1$$

Forms: [2]

	ϕ	ρ	A	$\overline{M}$
c 001	....	0°00′	90°00′	90°00′
p 011	0°00′	65 19	90 00	50 01
e 112	45 00	56 59	53 38	90 00
r 156	11 18½	61 36	80 04	60 48
h 123	26 34	58 21	67 37½	74 23
S 211	63 26	78 23	28 49	108 02½

Less common:
b 013 x 129 u 127 U 217 f 136 F 316 j 4.7.11 W 323

Structure cell. [3] Space group $I4_1/a$. a_0 5.23 kX, c_0 11.44; $a_0:c_0$ = 1:2.187. Cell contents Ca$_4$(MoO$_4$)$_4$. Isostructural with scheelite.

Habit.   Usually pyramidal; also thin tabular {001}. e sometimes striated parallel to intersection with vertical plane {110}; diagonal stria-

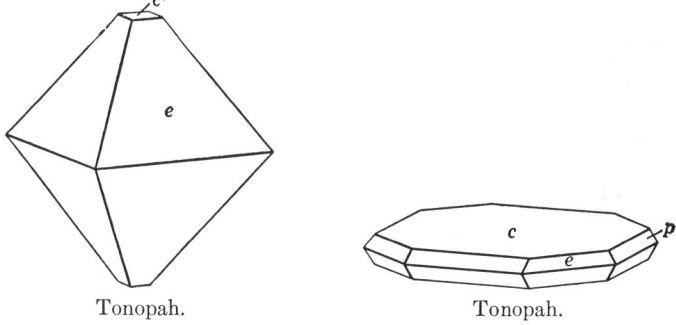

Tonopah.   Tonopah.

tions parallel to [13$\overline{1}$] observed on e and W. Tetragonal-dipyramidal symmetry appears only in the zones of [$\overline{1}$11]. Also massive, with a foliated structure pseudomorphous after molybdenite, or pulverulent to ocherous. As crusts of merged crystals.

Phys. Indistinct cleavages reported [4] on {112}, {011}, and {001}. Fracture uneven. H. $3\frac{1}{2}$–4. G. 4.23 ± 0.02 [5] for $CaMoO_4$, 4.22 (calc.); the G. increases markedly with increase of W, with G. 4.526 in material containing Mo:W ~9.2:1 (anal. 5), and 4.61 in artificial material [6] with Mo:W ~10:1. Color straw-yellow, brown, greenish yellow, pale greenish blue; also dirty white to gray, blue, and nearly black (deep blue in transmitted light). Luster sub-adamantine on crystal faces, greasy on fracture surfaces; foliated pseudomorphous material often has a pearly luster. Transparent. Fluoresces creamy yellow to golden yellow in ultraviolet light.

Opt. Uniaxial positive (+).

	n [7]	n (667) [8]	n (570) [8]	n (533) [8]
O	1.967 ± 0.005	1.959	1.974	1.982
E	1.978 ± 0.005	1.967	1.984	1.993

Bluish green crystals from Houghton County, Michigan, are dichroic with O blue, E green.[13]

Chem. Calcium molybdate, $CaMoO_4$. W substitutes for Mo up to about Mo:W = 9:1,[9] and a complete series apparently does not exist to scheelite, $CaWO_4$, in natural material. Rare earths substitute for Ca in artificial material.[10]

Anal.[10]

	1	2	3	4	5
CaO	28.48	27.65	28.11	27.30	25.55
MgO		tr.		0.16	0.16
MoO_3	71.52	69.84	71.67	67.84	58.58
WO_3			nil	1.65	10.28
Rem.		2.53	0.34	2.48	4.90
Total	100.00	100.02	100.12	99.43	99.47
G.	4.22		4.22	4.349	4.526

1. $CaMoO_4$. 2. Kasa Keskin, Turkey. Rem. is Fe_2O_3 0.54, H_2O 0.08, quartz 1.91.[11] 3. Western Altai, Siberia. Rem. is CuO.[12] The copper is probably present as admixed malachite. 4. Houghton County, Michigan. Rem. is Fe_2O_3 0.96, SiO_2 1.52. Another partial analysis gave WO_3 4.50.[13] 5. Houghton County, Michigan. Rem. is SiO_2 3.25, Fe_2O_3 1.65.[14]

Tests. Fuses at about 5 to a gray mass. Decomposed by HCl and HNO_3.

Occur. A secondary mineral, often formed by the alteration of molybdenite. Most of the known localities are mentioned below. In the U.S.S.R., found with molybdenite and scheelite in contact metamorphic deposits in the Nuratinsky Range, Uzbek, and in the Minusinsk district, Yeniseisk, Urals; in the Byelousovsky mine, western Altai, and in the Nizhne-Borsinsky placer, Transbaikalia. From Hüseyin beyobasi, Kasa Keskin, Vilayet Ankara, Turkey. In garnet skarn at Azegour, Morocco. In the United States, found originally with copper ores and garnet on the Peacock claim, Seven Devils district, Adams County, Utah; also in the OK mine, San Francisco district, Beaver County and in the Clifton tungsten district, Tooele County, Utah. In the South Hecla and Isle Royale mines, Houghton County, Michigan, associated with native copper and epidote. In Nevada, in vugs in altered rhyolite in the Tonopah Divide mine, near Tonopah, and near Oak Springs, Nye County; also reported [15] in contact metamorphic deposits with scheelite at Tungsten near Mill City, Luning, the White Pine district, and at other localities. At Baringer Hill, Llano County, Texas. Found in California in the Pine Creek tungsten mine, west of Bishop, Inyo County, and in the El Paso Mountains northwest of Randsburg, Kern County. Reported from the Helvetia district, Pima County, and from the Cerbat Range, Mohave County, Arizona. From Iron Mountain, Sierra Cuchillo, Sierra County, New Mexico.

A l t e r. Alters to ferrimolybdite, and frequently found as pseudomorphs after molybdenite.

A r t i f.[16] Obtained in crystals by the fusion of sodium molybdate with calcium chloride and sodium chloride, or by heating a mixture of calcium molybdate with sodium chloride at 1200°.

N a m e. After John Wesley Powell (1834–1902), American explorer and geologist.

Ref.

1. Angles of Pough, *Am. Min.*, **22**, 57 (1937), unit of structure cell; body-centered orientation. Transformation: Pough to new, $1\bar{2}0/\frac{1}{2}\frac{1}{2}0/001$.
2. Pough (1937). Rare or uncertain: 010, 1.7.28, 235, 5.6.11.
3. Vegard, *Phil. Mag.*, **1**, 1151 (1926). Maucher, *Zs. angew. Min.*, **1**, 103 (1938), gives a_0 5.21 kX, c_0 11.40, for material of anal. 2.
4. Palache, *Am. J. Sc.*, **7**, 367 (1899), reported {112}, and Koenig and Hubbard, *Am. J. Sc.*, **46**, 356 (1893), reported {011} and {001} on crystals from Houghton County, Michigan.
5. Range of best reported values.
6. Michel, *Bull. soc. min.*, **17**, 612 (1894).
7. Larsen (122, 1921) by immersion method on tungstenian type material of Melville (1891), anal. 5.
8. Zambonini, *Zs. Kr.*, **58**, 248 (1923), by prism method on artificial $CaMoO_4$.
9. Cf. anal. 5 on natural material and the analysis by Michel (1894) on artificial tungstenian material.
10. For additional analyses see Hintze (1 [3B], 4036, 1929).
11. Maucher (1938).
12. Pilipenko, *Bull. Tomsk. Univ.*, no. 63 (1915)—*Min. Abs.*, **2**, 109 (1923).
13. Koenig and Hubbard (1893).
14. Melville (1891).
15. Cf. Gianella, *Univ. Nevada Bull.*, **35**, no. 6, 80 (1941).
16. Hiortdahl, *Zs. Kr.*, **12**, 411 (1887); Michel (1894); Zambonini (1923).

48.1.4.1 **W U L F E N I T E** [$PbMoO_4$]. Plumbum spatosum flavo-rubrum (from Annaberg) *von Born* (**1**, 90, 1772). Kärntherischer Bleispath *Jacquin* (*Misc. Austriaca*, **2**, 1781, Vienna); *Wulfen* (*Abhandl. vom Kärnthner Bleispath*, 1875, Vienna). Plomb jaune *de Lisle* (**3**, 387, 1783). Gelbbleierz *Werner* (*Bergm. J.*, 384, 1789). Yellow Leadspar, Molybdenated Lead Ore *Kirwan* (**2**, 212, 1796). Plomb molybdaté *Haüy* (**3**, 353, 1801). Molybdate of Lead. Molybdänbleispath, Bleimolybdat, Molybdänbleierz *Germ*. Mélinose *Beudant* (**2**, 664, 1832). Wulfenit *Haidinger* (504, 1841). Chromowulfenite *Schrauf* (*Ak. Wien, Sitzber.*, **63**, [1], 184, 1871). Chrommolybdänbleierz, Chrommolybdänbleispath.

Eosite *Schrauf* (*Ak. Wien, Sitzber.*, **63**, 176, 1871). Achrematite pt. ? *Mallet* (*J. Chem. Soc. London*, **28**, 1141, 1875). Chillagite *Ullman* (*Roy. Soc. New South Wales, J.*, **46**, 186, 1912).

Cryst.[1] Tetragonal; pyramidal—4.

$$a:c = 1:2.2308;{}^{2} \quad p_0:r_0 = 2.2308:1$$

Forms:[3]

	ϕ	ρ	A	$\overline{M}$
c 001	.....	0°00′	90°00′	90°00′
a 010	0°00′	90 00	90 00	45 00
m 110	45 00	90 00	45 00	90 00
μ 170	8 08	90 00	81 52	53 08
f 150	11 18½	90 00	78 41½	56 18½
k 130	18 26	90 00	71 34	63 26
g 120	26 34	90 00	63 26	71 34
s 013	0 00	36 38	90 00	65 02½
n 011	0 00	65 51½	90 00	49 49
d 021	0 00	77 22	90 00	46 22½
t 116	45 00	27 44	70 47	90 00
z 115	45 00	32 15	67 50	90 00
u 114	45 00	38 16	64 02	90 00
y 113	45 00	46 26½	59 10½	90 00
e 112	45 00	58 07½	58 14	90 00
φ 3.4.75	36 52	8 27½	84 55	88 48½

Less common or rare:

ν 1.11.0   l 3.11.0   ρ 017   Y 014   r 032   σ 3.3.10   ε 111   χ 121
δ 140   Σ 019   Λ 015   Γ 043   X 118   q 334   π 123

Structure cell.[4] Space group uncertain. a_0 5.401 kX, c_0 12.079; $a_0:c_0 = 1:2.236$. Cell contents $Pb_4(MoO_4)_4$.

Habit. Crystals commonly square tabular {001}, sometimes extremely thin, or with a flat vicinal pyramid replacing {001}; less frequently octa-

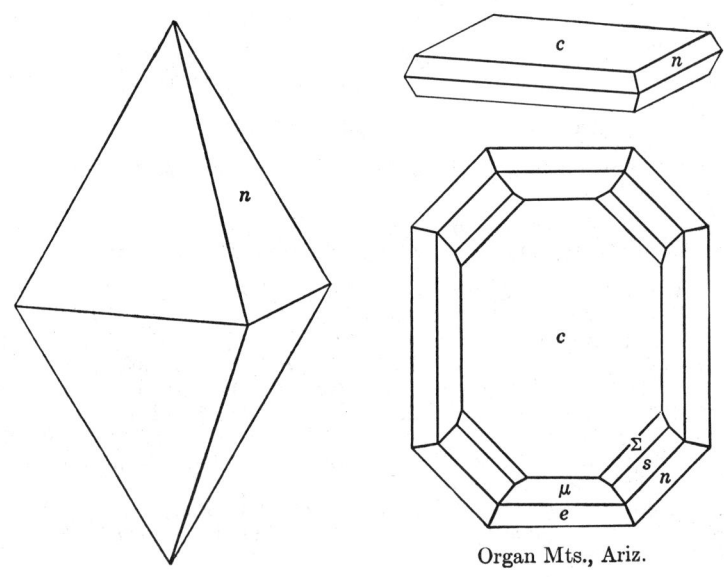

Organ Mts., Ariz.

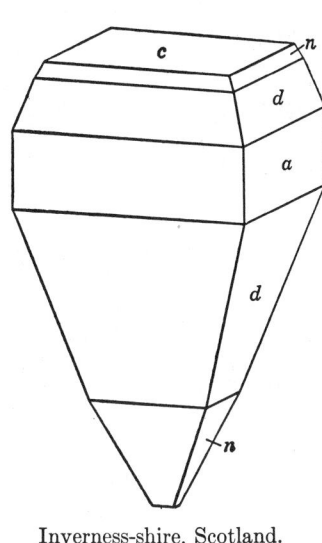 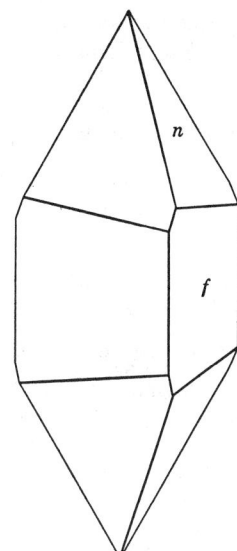

Inverness-shire, Scotland.

hedral in habit; infrequently prismatic [001]; cuboidal. Also massive, coarse to fine granular. Commonest forms: *c n a s u e*.

Twinning. Reported [5] on {001} on the assumption that [001] is polar; otherwise interpreted as parallel growths.

Etching. Etch figures conform to tetragonal-dipyramidal symmetry.[6]

P h y s. Cleavage {011} distinct, {001} and {013} indistinct. Fracture subconchoidal to uneven. Not very brittle. H. $2\frac{3}{4}$–3. G. 6.5–7.0,[7] decreasing with the content of Ca and increasing with W (7.5 with W:Mo ~1:1 in anal. 6); 6.78 ± 0.01;[8] 6.88 (calc.). Luster resinous to adamantine. Color orange-yellow to wax-yellow; also yellowish gray, grayish white, siskin- and olive-green, brown, reddish brown; also orange to bright red, due to chromium.[9] Streak white. Transparent. M.P. 1065°.[10]

O p t. Uniaxial negative (−). Sometimes anomalously biaxial,[11] with 2V up to 8°. Not optically active.[14] Weakly pleochroic in orange and yellow tints.[12] Indices:[13]

Red Cloud mine, Yuma County, Arizona (orange-red crystal)

λ	686.7	656.3	589.3	527.0	517.5
nO	2.3620	2.3724	2.4053	2.4542	2.4611
nE	2.2558	2.2635	2.2826	2.3131	2.3162

C h e m. Lead molybdate, $PbMoO_4$. W substitutes for Mo in the material of anal. 6 with W:Mo ~1:1, indicating at least a partial series toward the isostructural compound stolzite, $PbWO_4$, but W is lacking in most reported analyses. Ca substitutes for Pb to at least Ca:Pb = 1:1.7 (anal. 4), indicating at least a partial series toward the isostructural compound powellite, $Ca(Mo,W)O_4$. The role of the Fe_2O_3 reported in some analyses is uncertain. Small amounts of V [15] and traces of Cr [16] and As [17] also have been reported, presumably in substitution for Mo. In artificial material,[18] W, Cr, and S substitute for Mo.

Anal.[19]

	1	2	3	4	5	6
PbO	60.79	60.91	42.47	47.00	60.30	54.25
CaO			2.88	6.88		
MoO_3	39.21	38.87	50.47	46.12	37.47	17.52
WO_3						28.22
V_2O_5					1.28	
Rem.			4.65			
Total	100.00	99.78	100.47	100.00	99.05	99.99
G.		6.88				7.5

1. $PbMoO_4$. 2. Oudida, Morocco.[20] 3. Zinnwald, Bohemia.[21] Rem. is Fe_2O_3 3.09, MnO 0.04, MgO 1.52; MoO_3 contains undet. amount of WO_3. 4. Calcian wulfenite. Chile.[22] 5. Wheatley mine, Chester County, Pennsylvania.[23] 6. Tungstenian wulfenite (*chillagite*). Chillagoe, Queensland.[24]

Var. *Calcian.* Contains Ca in substitution for Pb up to at least Ca:Pb = 1:1.7 (anal. 4). The substitution of Ca is accompanied by a decrease in gravity, cell dimensions, and axial ratio.[25] Material with *ca.* 1 per cent CaO has G. ~6.7.

Vanadian. Eosite (?) *Schrauf* (*Ak. Wien, Sitzber.*, **63**, 176, 1871). Contains V in substitution for Mo (anal. 5). Color red. The so-called eosite from Leadhills, Scotland, apparently is a highly vanadian wulfenite.

Tungstenian. Chillagite *Ullman* (*Roy. Soc. New South Wales, J.*, **46**, 186, 1912). Material from Chillagoe, Queensland (anal. 6) contains W in substitution for Mo from about 3:5 to slightly over 1:1 in different samples;[24] color straw- to ocher-yellow; a_0 5.43 kX, c_0 12.08; G. 7.5. Material with W:Mo > 1:1 is properly termed molybdian stolzite.

Tests. B.B. decrepitates and fuses below 2. Decomposed by evaporation in HCl with separation of lead chloride and molybdic oxide; soluble in concentrated H_2SO_4 and alkalies; decomposed by HNO_3 with separation of molybdic oxide.

Occur. A secondary mineral, formed in the oxidized zone of deposits of lead- and molybdenum-containing minerals. Usually associated with pyromorphite, vanadinite, mimetite, cerussite, limonite, wad, calcite, and altered or corroded galena; less frequently found with hemimorphite, descloizite, anglesite, crocoite. Wulfenite is the next most common molybdenum mineral after molybdenite. Only a few of the many known localities can be mentioned here.

In Germany, found at Berggiesshübel and Zinnwald in Saxony, at Hausbaden and in the Schapbachtal, Baden; in Austria at Annaberg, and at Bleiberg,[33] Rudnik, Mies, and elsewhere in Carinthia. In fine crystals, sometimes apparently hemimorphic, at Příbram, Bohemia. With barite at Sarrabus and Gennemari, Sardinia. In North Africa at Oudida and Jebel Mahser, Morocco, and at Sidi Rouman, near Tebessa, Algeria. In Australia, in the Broken Hill district, New South Wales; a tungstenian variety (*chillagite*) in the Christmas Gift mine, Chillagoe, Queensland.

Fine specimens have been obtained in Mexico at Mapimi, Durango, the Sierra de los Lamentos, Chihuahua, and Magdalena, Sonora. In the United States, wulfenite is found abundantly at numerous localities in the western and southwestern parts of the country. A famous occurrence in the Red Cloud and Hamburg mines, in the Trigo Mountains, Yuma County, Arizona, where brilliant orange-red tablets up to 2 inches on an edge have been found; also in notable specimens from the Mammoth mine, Pinal County, from the Tombstone district, Cochise County, and from the Patagonia Mountains, Santa Cruz County. In New Mexico, found especially in the Central district, Grant County, and near Las Cruces in the Organ Mountains, Dona Ana County. In Nevada, at Eureka, Eureka County, and in the Lucin district, Elko County. In large crystals, sometimes paper-thin, at the Tacoma mine and elsewhere in the Lucin district, Box Elder County, Utah. Formerly found in small amounts at Loudville, near Southampton, Hampshire County, Massachusetts, and in the Wheatley mines, Phoenixville, Chester County, Pennsylvania.

A l t e r. Wulfenite has been said to alter to descloizite, ilsemannite, and vanadinite. Also found as incrustation pseudomorphs after calcite, cerussite, pyromorphite, and mimetite. Quartz occurs as incrustation pseudomorphs after wulfenite.

A r t i f. Obtained in crystals by fusion of ammonium molybdate or sodium molybdate with $PbCl_2$ and NaCl,[26] and by prolonged digestion of powdered $PbMoO_4$ in alkaline solutions.[27] Found (with Pb_2MoO_5) in the system $PbO-MoO_3$.[28]

N a m e. After the Austrian mineralogist Franz Xaver Wülfen (Wulffen) (1728–1805), Jesuit, who in 1785 wrote an extended monograph on the lead ores of Carinthia.

EOSITE. *Schrauf (Ak. Wien, Sitzber.,* **63,** 176, 1871). Found associated with cerussite and pyromorphite at Leadhills, Scotland, as minute, imperfectly developed tetragonal octahedrons. Color deep aurora-red. Qualitative analysis indicated the mineral to be a lead molybdate-vanadate. Probably a vanadian wulfenite.[29] Named in allusion to the color.

CHILLAGITE. *Ullman (Roy. Soc. New South Wales, J.,* **46,** 186, 1912). Found as tabular tetragonal [30] crystals associated with cerussite in gossan in the Christmas Gift mine, Chillagoe, Queensland, Australia. Analyses [31] show a range of PbO 53.9 to 54.7, WO_3 21.1 to 29.5, MoO_3 16.3 to 22.2, corresponding to $Pb(Mo,W)O_4$. A tungstenian wulfenite.[32]

ACHREMATITE. *Mallet (J. Chem. Soc. London,* **28,** 1141, 1875). An ill-defined, fine-grained massive material found admixed with limonite at Guanaceré, Chihuahua, Mexico. Color pale sulfur-yellow to orange and red. Luster resinous to adamantine. G. 5.96–6.18. Analyses indicate the material to be a chlor-arsenate-molybdate of lead, but the homogeneity of the material is doubtful. Perhaps a mixture of mimetite and wulfenite. Named from $ἀχρήματος$, *useless*, in allusion to the fact that it was mistaken for a silver ore.

Ref.

1. Pyramidal (4) symmetry indicated by the morphology, although not always clearly shown—see Bach, *Jb. Min., Beil.-Bd.,* **54,** 380 (1926); Koning, *Proc. Kon. Nederl. Ak. Wet.,* **51,** 390 (1948); Russell, *Min. Mag.,* **27,** 151 (1946). Dipyramidal

($4/m$) symmetry is suggested by the etch-pit symmetry (see ref. 6), the apparent absence of piezoelectricity, and the apparent presence of a space-group glide plane on (001).

2. Angles of Koksharov (**8**, 394, 1878), orientation and unit of the structure cell. Transformation: Koksharov (1878), Goldschmidt (**8**, 12, 1922), Hintze (**1** [3B], 4059, 1929), and Dana (989, 1892) to new, $1\bar{1}0/\frac{1}{2}\frac{1}{2}0/001$. The angular measurements reported by different observers vary widely.

3. Bach (1926). Distinction of right and left forms is impossible without recognition of top and bottom; no adequate treatment of the merohedral development has been made, and all forms are here listed as *upper right*. $\phi\{3.4.75\}$ is vicinal and variable in position but frequently present. For additional vicinal and uncertain forms see Barthoux, *Bull. soc. min.*, **47**, 36 (1924); Comucci, *Acc. Linc., Att.*, **3**, 335 (1926); Shannon, *U. S. Nat. Mus. Bull.*, **131**, 474 (1926); Padurova, *Mem. soc. russe min.*, **58**, 109 (1929); Barić—*Min. Abs.*, **7**, 360 (1939). Uncertain and vicinal:

1.13.0	0.1.12	029	1.1.32	117	447	2.3.62	187
230	0.2.17	0.6.13	1.1.24	3.3.16	772	9.10.248	132
0.1.44	018	012	1.1.18	3.3.14	1.17.36	9.11.100	4.19.2
0.1.16	017	1.1.528	1.1.16	5.5.22	1.3.32	5.21.44	

4. Vegard, *Phil. Mag.*, **1**, 1151 (1926); Vegard and Refsum, *Norsk Ak., Skr., Mat.-Nat. Kl.*, no. 2 (1927); space group given as $I4_1/a$. See also ref. 32, and Sillén and Nylander, *Ark. Kemi*, **17A**, no. 4 (1943).

5. Bach (1926); Hlawatsch, *Ann. Mus., Wien*, **38**, 15 (1925).

6. Traube, *Jb. Min., Beil.-Bd.*, **10**, 457 (1896); Honess (92, 1927); see also Royer, *C.R.*, **202**, 1346 (1936).

7. Usual range of reported values.

8. Mrose, priv. comm. (1946), by microbalance on crystals from six localities.

9. Haberlandt and Schroll, *Experientia*, **6**, 89 (1950).

10. Jaeger and Germs, *Zs. anorg. Chem.*, **119**, 158 (1921).

11. Cf. de Gramont, *Bull. soc. min.*, **16**, 127 (1893).

12. Frondel, priv. comm. (1946).

13. Ites, *Preisschr. Göttingen* (1903), corrected by Ehringhaus, *Jb. Min., Beil.-Bd.*, **43**, 566 (1920). See also Baumhauer, *Zs. Kr.*, **47**, 7 (1910).

14. Johnsen, *Cbl. Min.*, 712 (1908).

15. Cf. anal. 5; also Wöhler, *Ann. Chem. Pharm.*, **102**, 383 (1857), and Schrauf, *Ak. Wien, Sitzber.*, **63**, 176 (1871), on eosite (= *vanadian wulfenite*).

16. Cf. Schrauf (1871); Groth, *Zs. Kr.*, **7**, 592 (1882); Haberlandt and Schroll (1950).

17. Regnard, *Bull. soc. min.*, **5**, 2 (1882).

18. Jaeger and Germs (1921); see also Zambonini, *Zs. Kr.*, **58**, 226 (1923), on isomorphism with rare earths.

19. For additional analyses see Doelter (4 [2], 785, 1927).

20. Comucci, *Acc. Linc., Att., Cl. Sc.*, [6], **3**, 335 (1926).

21. Jung, *Jb. Min., Beil.-Bd.*, **64A**, 197 (1931).

22. Domeyko anal. in Rammelsberg (283, 1875).

23. J. L. Smith, *Am. J. Sc.*, **20**, 245 (1855).

24. Ullman, *Roy. Soc. New South Wales, J.*, **46**, 186 (1912); see also Mingaye, *Rec. Geol. Sur. New South Wales*, **9**, 171 (1916), for additional analyses.

25. Zepharovich, *Zs. Kr.*, **8**, 583 (1883).

26. Manross, *Ann. Chem. Pharm.*, **82**, 358 (1852); Schultze, *ibid.*, **49**, 126 (1863).

27. Dittler, *Zs. Kr.*, **53**, 158 (1913); Cesàro, *Ac. roy. Belgique, Bull.*, 327 (1905).

28. Jaeger and Germs (1921).

29. Schrauf (1871) gives $c = 1.378$ (= 1.949 in new position) which is not close to wulfenite, but the quality of the crystals measured was very poor.

30. Smith and Cotton, *Roy. Soc. New South Wales, J.*, **46**, 207 (1912), list 23 forms, but the measurements are of little value owing to the very poor quality of the crystals.

31. Cf. Mingaye (1916).

32. X-ray study of original material by Quodling and Cohen, *Roy. Soc. New South Wales, J.*, **71**, 543 (1938), proves the mineral to be isostructural with stolzite and wulfenite with intermediate cell dimensions (a_0 5.43 kX, c_0 12.08).

33. Schroll, *Min. Mitt.*, [3], **1**, 325 (1949).

48.1.4.2   S T O L Z I T E [PbWO$_4$]. Scheel-Bleispath *Breithaupt* (14, 1820). Tungstate of lead. Bleiwolframat, Bleischeelat, Wolframbleierz, Scheelbleispat, Scheelsaures Blei *Germ*. Scheeltine *Beudant* (**2**, 662, 1832). Stolzit *Haidinger* (504, 1845).

C r y s t.[1]   Tetragonal; pyramidal—4 (?).

$a:c = 1:2.2070;$ $p_0:r_0 = 2.2070:1$

Forms:[2]

		φ	ρ	A	$\overline{M}$
c	001	.....	0°00′	90°00′	90°00′
a	010	0°00′	90 00	90 00	45 00
v	012	0 00	47 49	90 00	58 24
n	011	0 00	65 37½	90 00	49 54
o	021	0 00	77 14	90 00	46 24
u	114	45 00	37 58	64 13	90 00
e	112	45 00	57 21	53 27½	90 00
ε	111	45 00	72 14	47 40½	90 00
π	213	63 26	58 42½	40 09½	74 19½

Less common:

f 150   N 210   δ 217   ρ 516   y 134   z 312   s 311   A 321

Structure cell.[3] Space group uncertain. a_0 5.452 kX, c_0 12.031; $a_0:c_0 = 1:2.207$. Cell contents Pb$_4$(WO$_4$)$_4$.

Habit. Usually dipyramidal, with {111}, {101}, or other pyramidal forms; also thick tabular {001}; infrequently prismatic [001]. Prism faces horizontally striated; pyramidal faces often striated parallel to the edge {h0l}{hhl} or diagonally striated parallel to the edges with third-order pyramids. Most common forms: *n e c*.

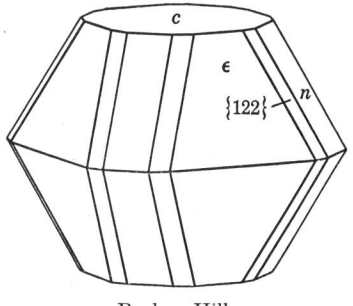

Broken Hill.

Phys. Cleavage {001} imperfect; {011} indistinct. Fracture conchoidal to uneven. Brittle. H. 2½–3. G. 7.9–8.3,[4] 8.34 (Broken Hill;[5] Nigeria[6]); 8.40 (calc.). Luster resinous, sub-adamantine. Color reddish brown, brown, fawn, yellowish gray, straw-yellow; also green, yellow-red, red. Streak uncolored. Transparent in thin pieces. M.P. 1123°.

Opt.

	n (Broken Hill)[7]	n (Broken Hill) (Na)[8]	
O	2.27 ± 0.01	2.2685	Uniaxial negative (−).
E	2.19 ± 0.01	2.182	

Chem. Lead tungstate, PbWO$_4$. Dimorphous with raspite.[9] The few reported analyses do not reveal substitution of Mo for W and show only a very slight substitution of Ca (and Mn, Mg ?) for Pb. A tungstenian wulfenite, Pb(Mo,W)O$_4$, with Mo:W ∼1:1 is known, however, and at least a partial series extends from wulfenite. In artificial material [10] Mo, Cr, and S substitute for W.

Anal.[11]

	1	2	3
PbO	49.04	47.44	47.78
CaO			0.92
WO_3	50.96	51.34	50.92
Rem.		0.78	0.49
Total	100.00	99.56	100.11
G.	8.40		8.30

1. $PbWO_4$. 2. Broken Hill, New South Wales. Rem. is MnO 0.78 and MgO tr.[12] 3. Marianna de Itacolumy, Brazil. Rem. is $(Fe,Al)_2O_3$ 0.30, MgO 0.19.[13]

Tests. Fuses at 2 to a crystalline, metallic bead. Decomposed by HCl with separation of yellow tungstic acid.

Occur. A secondary mineral, found associated with limonite, wad, vanadinite, mimetite, wulfenite, cerussite in the oxidized zone of ore deposits containing primary tungsten minerals. Much less common than wulfenite. Originally found in the tin veins of Zinnwald, Saxony. Also found in Europe at Berggiesshübel, Saxony; Bleiberg, Carinthia; Bena e Padru, Sardinia. With cerussite at the Force Craig lead mine, near Keswick, Cumberland, England. A notable occurrence in the Proprietary mine, Broken Hill, New South Wales, where crystals up to an inch in size were found implanted on wad. From Marianna de Itacolumy, near Ouro Preto, Minas Geraes, Brazil. Found in tin placers at Abuja, northern Nigeria, Africa.

In the United States, stolzite has been found at the Manhan lead mine, Southampton, Massachusetts; in scheelite deposits in the Huachuca Mountains, south of Tombstone, and the Primos mine, near Dragoon, in Arizona; in the Grouse Creek Mountains, near Lucin, Utah; with pyromorphite at the Wheatley lead mines, Chester County, Pennsylvania. In Canada with scheelite and ferritungstite (?) in the Cariboo district, British Columbia.

Artif.[14] Obtained in small crystals by sublimation and from fusion.

Name. After Dr. Stolz, of Teplitz, Bohemia, who first drew attention to the species.

Ref.

1. Pyramidal (4) symmetry assigned from isomorphous relation to wulfenite; no evidence of lower than dipyramidal (4/m) symmetry is shown by the morphology. Angles of Hlawatsch, *Zs. Kr.*, **29**, 130 (1897), unit and orientation of the structure cell. Transformation: Hlawatsch (1897), Dana (989, 1892), Goldschmidt (**8**, 87, 1922) to new, $\frac{1}{2}\bar{1}0/\frac{1}{2}\frac{1}{2}0/001$.

2. Goldschmidt (1922); Palache, *Am. Min.*, **26**, 429 (1941), Glastonbury and Semmens, *Roy. Soc. South Australia, Trans. and Proc.*, **53**, 258 (1929); Hodge-Smith, *Rec. Australian Mus.*, **19**, 165 (1934). All forms are given as in the *upper-right* octant. Rare or uncertain:

110	019	041	113	1.9.10	174
130	017	1.1.18	338	178	121
320	013	116	1.17.18	137	

3. Cell dimensions of Aanerud, *Norsk. Ak., Skr., Mat.-Nat. Kl.*, no. 13 (1931), by powder method on artificial material; see also Vegard and Refsum, *ibid.*, no. 2 (1927).
4. Usual range of reported values.
5. Mrose (1946) by microbalance on single crystal.
6. Berman (1941) by microbalance on single crystal.
7. Larsen (138, 1921) by immersion.
8. Hlawatsch (1898) by prism method.
9. Jaeger and Germs, *Zs. anorg. Chem.*, **119**, 145 (1921), found α-$PbWO_4$ stable up to 877° and β-$PbWO_4$ (*stolzite ?*) stable from 877° to 1123°, the melting point. See also Jung, *Jb. Min., Beil.-Bd.*, **64A**, 197 (1931).
10. Cf. Jaeger and Germs (1921); also Zambonini, *Gaz. chim. ital.*, **50**, Pt. 2, 128 (1920), on the relation to $Bi_2(WO_4)_3$.
11. For additional analyses see Hintze (**1** [3B], 4094, 1929).
12. Treadwell anal. in Hlawatsch (1898).
13. Florence, *Cbl. Min.*, 725 (1903).
14. Traube, *Cbl. Min.*, 682 (1901); Manross, *Ann. Chem. Pharm.*, **81**, 243 (1851), **82**, 357 (1852); Michel, *Bull. soc. min.*, **2**, 142 (1897); Zambonini (1920).

48.1.5   **R A S P I T E** [$PbWO_4$].   Hlawatsch (*Ann. Mus. Wien*, **12**, 38, 1897; *Zs. Kr.*, **29**, 137, 1898; **31**, 8, 1899).

C r y s t.[1]   Monoclinic; prismatic—$2/m$.

$a:b:c = 1.3450:1:1.1147$; $\beta\,107°37'$; $p_0:q_0:r_0 = 0.8288:1.0624:1$

$r_2:p_2:q_2 = 0.9412:0.7801:1$; $\mu\,72°23'$; $p_0'\,0.8696, q_0'\,1.1147, x_0'\,0.3175$

Forms:[2]

	ϕ	ρ	ϕ_2	$\rho_2 = B$	C	A
c 001	90°00′	17°37′	72°23′	90°00′	.....	72°23′
b 010	0 00	90 00	.....	0 00	90°00′	90 00
a 100	90 00	90 00	0 00	90 00	72 23	.....
m 110	37 57½	90 00	0 00	37 15	79 16½	52 02½
d 011	15 54	49 13	72 23	43 16	46 44	78 02
f $\bar{1}$02	−90 00	6 41½	96 41½	90 00	24 18½	96 41½
e $\bar{1}$01	−90 00	28 54	118 54	90 00	46 31	18 54
p 122	34 01	53 22	53 03	48 18½	45 12	63 19½

Habit. Usually tabular {100}, with this form striated parallel [010], and somewhat elongated parallel [010]; also elongated [100] or thin tabular {$\bar{1}$01}. Common forms: *a d c b*.

Twinning. (1) On {100}, common.[3] (2) On {$\bar{1}$02}.[4]

Minas Geraes.

P h y s. Cleavage {100}, perfect. H. $2\frac{1}{2}$–3. G. 8.46.[5] Color yellowish brown, light yellow; gray. Luster adamantine.

O p t.[6]

ORIENTATION		n	
X		2.27 ± 0.02	Biaxial positive (+).
Y	b	2.27 ± 0.02	$2V$ nearly zero.
$Z \wedge c$	$\sim$30°	2.30 ± 0.02	

C h e m. Lead tungstate, $PbWO_4$. Dimorphous with stolzite.[7]

Anal.

	1	2
PbO	49.03	49.06
WO_3	50.97	48.32
Rem.		1.43
Total	100.00	98.81

1. $PbWO_4$. 2. Broken Hill, New South Wales. Rem. is Fe_2O_3 and MnO.[8]

Tests. Fusible at 2½ to 3. Decomposed by HCl with separation of yellow tungstic acid.

Occur. Found originally in the Proprietary mine, Broken Hill, New South Wales, associated with stolzite in manganiferous, oxidized vein material. Also found as small crystals in the gold placers of Sumidouro, Minas Geraes, Brazil, associated with stolzite and scheelite. Reported from tin veins on the Cerro Estaño, east of Guanajuato, Mexico.

Artif. $PbWO_4$ has been prepared in various ways,[9] but the relation of these products to raspite is not known.

Name. After Mr. Rasp, the discoverer of the Broken Hill mines.

Ref.

1. Hlawatsch, *Zs. Kr.*, **42**, 587 (1907).
2. Hlawatsch (1907). Vicinal {1.12.12}.
3. Hlawatsch (1898); Hussak, *Cbl. Min.*, 723 (1903).
4. Hlawatsch, *Cbl. Min.*, 423 (1905).
5. Florence in Hussak (1903) by hydrostatic balance on ¼-gram loose crystals from Brazil (given 8.465).
6. Larsen (125, 1921) on material from Broken Hill.
7. Jaeger and Germs, *Zs. anorg. Chem.*, **119**, 145 (1921), found α-$PbWO_4$ stable up to 877° and β-$PbWO_4$ stable from 877° to 1123°, the melting point; the relation of these forms to stolzite and raspite is not known. See also Jung, *Jb. Min., Beil.-Bd.*, **64A**, 197 (1931).
8. Treadwell anal. in Hlawatsch (1898), on 0.1331 gram.
9. Cf. Mellor (**11**, 792, 1921); see also ref. 7.

49 BASIC AND HYDRATED MOLYBDATES AND TUNGSTATES

TYPE 1. MISCELLANEOUS

49.1.1 CUPROTUNGSTITE [$Cu_2(WO_4)(OH)_2$]. Cuproscheelite *Whitney* (*Proc. California Ac. Sc.*, **3**, 287, 1866). Tungstate de cuivre *Domeyko* (*Ann. mines*, **16**, 537, 1869). Cuprotungstite *Adam* (32, 1869).

Found as microcrystalline masses and crusts, friable to compact. Hardness and gravity unknown. Color pistachio-green, passing to olive-green, leek-green, and emerald green. Luster vitreous; also waxy to earthy. Streak greenish gray to greenish yellow. Strongly birefringent,[1] with a mean index of refraction of 2.15 ± 0.02.

Chem. A basic copper tungstate, probably $Cu_2(WO_4)(OH)_2$.[2] The large amounts of Ca reported in early analyses [3] are no doubt due to admixed scheelite, but Ca possibly may substitute for Cu in small part.

Anal.

	1	2	3
CuO	38.89	32.66	32.68
CaO		4.12	2.89
MgO		0.67	0.45
WO_3	56.71	55.36	59.04
H_2O	4.40	7.19	4.94
Total	100.00	100.00	100.00

1. $Cu_2(WO_4)(OH)_2$. 2, 3. Cave Creek, Arizona. Each an average of two analyses recalculated to 100 after deduction of 76 and 40 per cent gangue, respectively.[2]

Tests. Easily soluble in acids. B.B. fusible to a black glass.

Occur. A secondary mineral formed by the alteration of scheelite. The description given above is based largely on material from Cave Creek, Maricopa County, Arizona. The identity of this mineral with the material from other reported localities is very probable but not proved. Found originally in the vicinity of La Paz, Lower California, Mexico (*cuproscheelite*). From the copper mines of Llamuco near Santiago, Chile (*cuprotungstite*). Reported from Villa Salto, Sardinia, and from Sorpresa, Montoro, Spain. With wolframite and scheelite at the Mitake mine, Prov. Kai, and at Ofuku, Prov. Nagato, Japan. In the Mutue-Fides-Stavoren district, Transvaal, Africa. At the San Lorenzo mine, Arizpe district, Sonora, Mexico. In various mines in Georgiana County, New South Wales. In the United States found in the San Andres Mountains, Socorro County, New Mexico; at the Green Monster mine, near White River, Kern County, California.

Alter. Found as more or less complete pseudomorphs after scheelite. The green cuprotungstite incrusts or veins the outer portions of the scheelite crystals and grades inwardly through more or less altered zones (*cuproscheelite*) into unaltered scheelite.

Name. In allusion to the composition. The names cuproscheelite and cuprotungstite were earlier applied by Dana [4] and others to copper-calcium and copper tungstates supposedly analogous in composition and crystal form to scheelite. The original cuproscheelite of Whitney was later found,[5] however, to be a mixture in varying degree of a basic copper tungstate, for which the name cuprotungstite is retained, with scheelite. The analyzed material of Domeyko [3] from Chile doubtless also is such a mixture.

Ref.
1. Larsen (64, 1921) on material from Cave Creek, Arizona.
2. Schaller, *Am. Min.*, **17**, 234 (1932).
3. Cf. Hintze (1 [3B], 4087, 1929).
4. Dana (988, 1892).
5. Hess, *U. S. Geol. Sur., Bull. 652*, 32 (1917).

49.1.2 **KOECHLINITE** [$(BiO)_2(MoO_4)$]. *Schaller (U. S. Geol. Sur., Bull. 610, 10, 1916).*

Cryst.[1] Orthorhombic; dipyramidal—$2/m\ 2/m\ 2/m$.

$a:b:c = 0.3419:1:0.3410; \quad p_0:q_0:r_0 = 0.9974:0.3410:1$

$q_1:r_1:p_1 = 0.3419:1.0026:1; \quad r_2:p_2:q_2 = 2.9325:2.9248:1$

Forms:[2]

	ϕ	$\rho = C$	ϕ_1	$\rho_1 = A$	ϕ_2	$\rho_2 = B$
c 001	....	0°00′	0°00′	90°00′	90°00′	90°00′
b 010	0°00′	90 00	90 00	90 00	....	0 00
n 021	0 00	34 17½	34 17½	90 00	90 00	55 42½
h 041	0 00	53 45	53 45	90 00	90 00	36 15
p 131	44 16½	55 00½	45 39	55 07	45 04½	54 05
s 151	30 19½	63 09	59 36½	63 13½	45 04½	39 38

Structure cell.[3] Space group $Cmc(a\ ?)$. $a_0\ 5.48\ kX \pm 0.02$, $b_0\ 16.16 \pm 0.04$, $c_0\ 5.48 \pm 0.02$; $a_0:b_0:c_0 = 0.339:1:0.339$. Markedly pseudo-tetragonal. Cell contents $(BiO)_8(MoO_4)_4$.

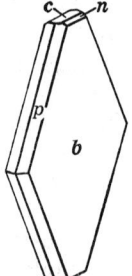

Schneeberg.

Habit. Thin, square to rectangular plates and laths flattened parallel {010} and striated parallel [010]. Common forms *b p n*. Also massive, earthy.

Twinning. Twin plane {101}, both as contact and penetration twins. Composition face {101} or {010} in contact twins.

Oriented growths. Parallel growths with an unidentified orthorhombic mineral have been noted.

Phys. Cleavage {010} perfect; also an imperfect cleavage {0kl}. Very brittle. Hardness and specific gravity not known; calculated G. 8.29. Color greenish yellow. Streak pale greenish yellow. Transparent.

O p t.[4] In transmitted light, greenish yellow.

ORIENTATION		n(Li)	
X	c	2.52	Biaxial negative (−).
Y	a	2.61	$2V$ large.
Z	b	2.67	$r < v$, rather strong.
			Weakly pleochroic.

C h e m. A bismuth molybdate, $(BiO)_2(MoO_4)$.
Anal.

	1	2
Bi_2O_3	76.51	77.1
MoO_3	23.49	22.4
H_2O		0.2
Total	100.00	99.7

1. $(BiO)_2(MoO_4)$. 2. Schneeberg. Average of three partial analyses after deduction of quartz.

Tests. On heating in the C.T. the mineral becomes brown and fuses readily to a dark brown mass which on cooling changes to a very pale yellow and finally becomes white. Easily soluble in HCl, less readily in HNO_3.

O c c u r. Originally found in the Daniel mine, Schneeberg, Saxony, associated with several unidentified minerals in cavities in vein material composed of quartz, native bismuth, and smaltite. Later found[5] as an alteration product of tetradymite (?) at the Dunallan mine, Coolgardie, Western Australia, and as white to yellow earthy masses associated with bismoclite at Bygoo, New South Wales.

N a m e. After Rudolf Koechlin (1862–1939), Austrian mineralogist and curator of the mineral collection in the Hof-Museum, Vienna.

Ref.
1. Angles of Schaller (1916); unit and orientation of the structure cell. Transformation: Schaller to new, 001/300/010.
2. Schaller (1916). Rare or uncertain: 011, 0.12.5, 031, 061, 292, 133, 296.
3. Wolfe and Frondel, priv. comm. (1946), by Weissenberg method.
4. Larsen and Berman (212, 1934).
5. Frondel, *Am. Min.*, **28**, 536 (1943).

49.1.3 Ferritungstite. Schaller (*J. Washington Ac. Sc.*, **1**, 24, 1911; *Am. J. Sc.*, **32**, 161, 1911; *Zs. Kr.*, **50**, 112, 1911; *U. S. Geol. Sur., Bull.* 509, 83, 1912).

Found as earthy coatings composed of microscopic hexagonal plates. Color pale yellow to brownish yellow. Optically[1] negative (−), nO 1.80, nE 1.72. G. 5.57 (Argentina[2]).

C h e m. A hydrated basic tungstate of iron, probably $Fe_2(WO_4)(OH)_4 \cdot 4H_2O$.
Anal.

	1	2	3
Fe_2O_3	31.95	32.3	32.5
WO_3	46.42	45.1	42.6
H_2O	21.63	22.6	24.9
Total	100.00	100.0	100.0

1. $Fe_2(WO_4)(OH)_4 \cdot 4H_2O$. 2, 3. Washington. Recalculated to 100 after deduction of 14.7 and 16.0 per cent insoluble.

Tests. Decomposed by acids with separation of yellow tungstic oxide.

Occur. Found as an alteration product of wolframite at the Germania mine, Deer Trail district, in northeastern Washington. Also reported as an alteration of wolframite in the Cerro Liquinaste, Jujuy, Argentina.[2]

Name. In allusion to the composition.

Ref.
1. Larsen (74, 1921).
2. Kittl, *Rev. min., Soc. Argentina Min. Geol.*, **10**, 78 (1939).

49.1.4 LINDGRENITE [$Cu_3(MoO_4)_2(OH)_2$]. Palache (*Am. Min.*, **20**, 484, 1935).

Cryst.[1] Monoclinic; prismatic—$2/m$.

$a:b:c = 0.4001:1:0.3852$; $\beta\,98°23'$; $p_0:q_0:r_0 = 0.9628:0.3811:1$

$r_2:p_2:q_2 = 2.6241:2.5263:1$; $\mu\,81°37'$; $p_0'\,0.9732, q_0'\,0.3852, x_0'\,0.1474$

Forms:

	ϕ	ρ	ϕ_2	$\rho_2 = B$	C	A
b 010	0°00′	90°00′	....	0°00′	90°00′	90°00′
o 110	68 24½	90 00	0°00′	68 24½	82 12½	21 35½
C 101	90 00	48 15½	41 44½	90 00	39 52½	41 44½
A $\bar{1}01$	−90 00	39 33	129 33	90 00	47 56	129 33
h 111	71 02	49 50½	41 44½	75 37	41 58½	43 43
q 211	79 34½	64 50½	25 31½	80 34½	56 36½	27 06½

Less common and rare:

a 100 t 130 s 120 p 011 r 021 n $\bar{1}11$ M $\bar{1}21$ k $\bar{1}41$

Structure cell.[4] Space group $P2_1/n$. a_0 5.613 Å, b_0 14.03, c_0 5.405; $\beta\,98°23'$; $a_0:b_0:c_0 = 0.4001:1:0.3852$. Cell contents $Cu_6(MoO_4)_4(OH)_4$.

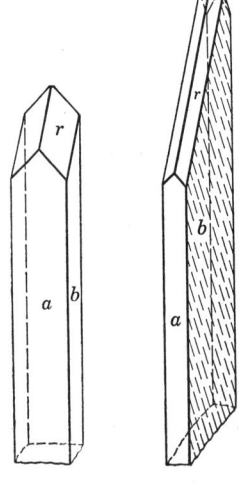

Habit. Tabular or platy {010}; rarely acicular [101]; sometimes doubly terminated. {010} striated parallel [001]. Also massive, platy.

Phys. Cleavage {010} perfect; {101} and {100}, very poor. H. 4½. G. 4.26; 4.29 (calc.). Color green; yellowish green in thin plates. Transparent.

Opt. In transmitted light, pale green to yellow-green.

Orientation	n	
$X \wedge c$ +9°	1.930 ± 0.003	Biaxial negative (−).
Y	2.002 ± 0.003	$2V$ 71°.
Z b	2.020 ± 0.003	$r > v$.

Chem. A basic copper molybdate, $Cu_3(MoO_4)_2(OH)_2$.

Anal.

	1	2
CuO	43.97	42.8
MoO$_3$	53.71	53.7
H$_2$O	3.32	3.5
Total	100.00	100.0
G.		4.26

1. Cu$_3$(MoO$_4$)$_2$(OH)$_2$. 2. Chuquicamata, Chile.[2] Recalculated to 100 after deduction of 1.43 Fe$_2$O$_3$ and 3.34 SiO$_2$ as impurities.

Tests. Easily fusible. In C.T. darkens quickly, gives a slight brownish coating near the assay, and melts. Easily soluble in HCl and HNO$_3$.

O c c u r. A secondary mineral found associated with antlerite, limonite, hematite, and quartz at Chuquicamata, Chile.

A r t i f. Basic copper molybdates have been reported,[3] but their relation to lindgrenite is not known.

N a m e. After Waldemar Lindgren (1860–1939), American mining geologist and teacher.

Ref.

1. Elements of the structure cell of Barnes, *Am. Min.*, **34**, 163 (1949); **34**, 611 (1949). Transformation: Palache (1935) to new, $\bar{1}0\frac{1}{2}/0\bar{1}0/\frac{1}{2}0\frac{1}{2}$. The original cell was B-centered.
2. Gonyer analysis in Palache (1935).
3. Cf. Mellor (**11**, 558, 1931).
4. Barnes (1949) by precession method.

49.1.5 **F E R R I M O L Y B D I T E** [Fe$_2$(MoO$_4$)$_3$·8H$_2$O ?]. Wasserbleiocker *Karsten* (54, 79, 1800). Molybdänocker *Karsten* (70, 1808). Molybdänoxyd *Hausmann* (**1**, 336, 1813). Molybdena or Molybdic Ocher. Molybic Acid. Molybdänocker, Molybdänoxyd, Molybdänsäure *Germ*. Molybdine *Greg* and *Lettsom* (348, 1858; earlier cited in Dana (144, 1854)). Molybdite *Breithaupt* (B. H. Ztg., **17**, 125, 1858). Ferrimolybdite *Pilipenko* (*Festschr. Vernadsky, Beilage Mater. Kenntnis geol. Baues Russ. Reichs, Moscow*, 1914—*Jb. Min.*, I, 191, 1915).

Massive, as fibrous crusts and tufted or radial fibrous aggregates; also subfibrous, and as an earthy powder or coating. Probably orthorhombic.[1]

P h y s. Very soft. G. 2.99 (earthy, Colorado [2]). Color canary-yellow; also straw-yellow, greenish yellow. Luster silky; earthy. Streak pale yellow.

O p t.[3] In transmitted light, colorless to canary-yellow. Usually appears as minute needles. The indices vary in different specimens presumably due to variation in water content.

PLEOCHROISM

nX	1.78	1.74	1.720	Clear and nearly colorless	Biaxial positive (+).
nY	1.79	1.75	1.733	Clear and nearly colorless	$r < v$, marked.
nZ	2.04	1.95	1.935	Dirty gray to canary-yellow	Z = elongation.
$2V$	small	small	28°(meas.)		

C h e m. A hydrated ferric molybdate, probably Fe$_2$(MoO$_4$)$_3$·8H$_2$O. The water content is uncertain; most of the water is lost at temperatures below 125° and dehydration is complete at 250° C.[4]

Anal.[9]

	1	2	3	4	5
Fe_2O_3	21.70	21.87	21.84	18.47	20.30
MoO_3	58.71	59.85	60.80	61.57	59.51
H_2O	19.59	18.28	17.36	19.96	20.19
Total	100.00	100.00	100.00	100.00	100.00
G.					2.99

1. $Fe_2(MoO_4)_3 \cdot 8H_2O$. 2. Westmoreland, N. H. Recalc. to 100 after deduction of 4.66 insol.[5] 3. Santa Rita Mountains, Arizona. Average of two anals. after deducting 4.86 and 2.66 insol. and recalc. to 100.[6] 4. Yeniseisk, Siberia. Recalc. to 100 after deduction of 5.4 insol.[7] 5. Hortense, Colorado. Recalc. to 100 after deduction of 22 per cent quartz and molybdenite.[8]

Tests. Easily soluble in HCl, and slowly soluble in ammonia with separation of ferric hydroxide. On heating in an open crucible,[5] the yellow mineral darkens and becomes dark gray, appearing almost black with a slight olive tint, then becomes light yellow again, and on further heating changes to deep orange. On heating for some time at a higher temperature the mineral, on cooling, becomes a permanent dark green. By further heating, all the molybdenum is volatilized and dark red ferric oxide remains.

Occur. A secondary mineral, commonly formed by the alteration of molybdenite. Associated with limonite. Only a few of the occurrences [10] known or presumed to be of this species can be mentioned here. In the U.S.S.R. from the Alexjevsky mine in the Minusinsk district, Yeniseisk, Urals, and from Pitkäranta, Lake Ladoga. From Bivongi, Calabria, Italy. In Australia, found in molybdenite veins and pipes at Kingsgate, Bald Knob, Whipstick, Wilson's Downfall, and Wunglebong in New South Wales; from Mulgine, Western Australia; from the Hampshire mine, Tasmania. In the United States, ferrimolybdite has been found at Westmoreland, New Hampshire; Granite Hill, Cuttingsville, Vermont; Vinalhaven, Maine; and at Chester, Delaware County, Pennsylvania. At Pioneer Mills, Cabarrus County, North Carolina, and in Heard County, Georgia. Found with molybdenite at numerous localities in the western United States. Near Alta, in Little Cottonwood Canyon, Salt Lake County, Utah; in the Red River district, Taos County, and the El Porvenir district, San Miguel County, in New Mexico. In Arizona, near Madera Canyon, Santa Rita Mountains, Pinal County, at Kingman, Mohave County, and in the Copper Creek district, Galiuro Mountain, Pinal County. In several mines at Tonopah, Nye County, Nevada. Found in California near Nevada City, Nevada County, at Red Mountain in the Ritter Range, Madera County, and at Cameron and Silverado Creek, Mono County.

Artif.[11] Obtained as a yellow powder by reaction of ferric chloride and ammonium molybdate in acetic acid solution of pH 3.5. The precipitate obtained in solutions of less acidity may contain admixed hydrous molybdic oxide or a basic salt, $Fe_2(MoO_4)_2(OH)_2 \cdot 6H_2O$. The basic salt possibly occurs in nature.

Name. The mineral long-known as molybdite was shown by Schaller in 1907 to be a hydrated ferric molybdate and not molybdic oxide as supposed. The name ferrimolybdite was later proposed for the species, in allusion to the composition, and the name molybdite, in lack of proved natural occurrences of MoO_3, is here relegated to the synonymy.

Ref.

1. Suggested by the optical properties. The properties and orthorhombic morphology given in Dana (201, 1892) are based on artificial crystals of MoO_3.
2. Schaller, *Zs. Kr.*, **44**, 13 (1908), by pycnometer, after correcting for 22 per cent impurities. Simpson, *J. Roy. Soc. Western Australia*, **12**, 57 (1926), also gives 2.99 for material from Mulgine.
3. Larsen and Berman (133, 135, 138, 1934), Larsen (112, 1921).
4. For partial dehydration data see Schaller, *Am. J. Sc.*, **23**, 297 (1907), *Zs. Kr.*, **44**, 13 (1908), and Carobbi, *Bull. Soc. Nat. Napoli*, **41**, 169 (1930).
5. Schaller (1907).
6. Guild, *Am. J. Sc.*, **23**, 455 (1907).
7. Pilipenko (1914).
8. Schaller (1908).
9. For additional analyses see Simpson (1926); Smith, *Rec. Australian Mus.*, **14**, 101 (1923), Carobbi, *Acc. Napoli, Rend.*, **33**, 53 (1927); Schaller (1907).
10. Cf. molybdite and ferrimolybdite in Hintze (1 [2A], 1259, 1915; Ergzbd., 166, 1936).
11. Carobbi (1930) and Schaller (1907).

49.1.6 Thorotungstite. Scrivenor and Shenton (*Am. J. Sc.*, **13**, 487, 1927). Found as minute, apparently orthorhombic crystals with a lath-like habit.[1] The crystals show a transverse cleavage and a longitudinal cleavage not so well-marked. Color yellow. G. 5.55. Optically negative (−), with indices over 1.74 and strong birefringence. Analysis gave [2] CaO 1.02, MgO trace, rare earths (almost all Ce) 1.77, Al_2O_3 4.31, Fe_2O_3 1.35, ThO_2 16.00, ZrO_2 1.96, SiO_2 (as quartz) 0.48, WO_3 69.69, ign. loss (H_2O) 4.18, total 100.76. Formula uncertain; essentially a basic tungstate or oxide with (Al,Fe):(Th,Ca,Ce,Zr):W ~1:1:3. Found as an alteration product of wolframite or scheelite in an eluvial cassiterite deposit at Pulai in the Kinta district, Perak, Malay States. Needs investigation.

Ref.

1. Rough sketches of the crystals as seen under the microscope, given by Scrivenor and Shenton (1927), show laths flattened on a pinacoid with prism and dome faces.
2. Shenton anal. in Scrivenor and Shenton (1927).

49.1.7 A N T H O I N I T E [$Al(WO_4)(OH) \cdot H_2O$]. Varlamoff (*Soc. géol. Belgique, Bull.*, **70**, B153, 1947).

White chalky masses resembling kaolin, also as indistinct pseudomorphous (?) crystals. H. 1 and G. ~4.6 (of the fine-grained masses). Apparently isotropic.

C h e m. A hydrated basic tungstate of aluminum, $Al(WO_4)(OH) \cdot H_2O$.

Anal.

	1	2	3
CaO		0.25	
Al_2O_3	16.45	16.41	15.74
Fe_2O_3		0.69	
WO_3	74.83	73.23	73.32
H_2O+	8.72	8.70	9.42
SiO_2		0.73	0.69
Total	100.00	100.01	99.17

1. $Al(WO_4)(OH) \cdot H_2O$. 2. Belgian Congo. Average of two.[1] 3. Belgian Congo.[2] H_2O includes $H_2O - 0.22$.

Tests. Slowly attacked by HCl and HNO_3, easily soluble in strong KOH solution.

Occur. Found in placer concentrates containing cassiterite and wolframite at Mt. Misobo, Kalima district, Maniema, Belgian Congo; also associated with ferberite in quartz veins. Also found at Ruanda, Kifuruwe region, in the Belgian Congo.

Alter. Apparently replaced by ferberite in the vein occurrences.

Name. After Raymond Anthoine, Belgian mining engineer.

Ref.

1. Gastelier analysis in Varlamoff (1947).
2. Analysis by Laboratoire de Panda (Congo Belge) in Varlamoff, who cites additional analyses.

PATERAITE. *Haidinger (Jb. Geol. Bundesanst.,* **7,** 196, 1856). A supposed cobalt molybdate, found admixed with pyrite and bismuthinite in the Elias mine, Joachimsthal, Bohemia. Massive; color black, but probably due to sulfidic impurities. The only available analysis[1] is on very impure material. Easily fusible, and soluble in acids. Needs verification. Named after A. Patera, who first examined the mineral.

Ref.

1. Laube, *Jb. Geol. Bundesanst.,* **14,** 303 (1864); also cited in Dana (991, 1892).

ORGANIC COMPOUNDS

The only natural organic compounds considered in the present volume are the salts between inorganic cations (including ammonium) and organic acids. These include the oxalates together with the mellates, citrates, sulfocyanates, and acetates. The very numerous natural oxygen-free and oxygenated hydrocarbons will be treated in Volume III of this work.

CLASS 50. Salts of organic acids
Type 1. Oxalates
50.1.1 Whewellite $CaC_2O_4 \cdot H_2O$
50.1.2 Weddellite $CaC_2O_4 \cdot 2H_2O$
50.1.3 Humboldtine $FeC_2O_4 \cdot 2H_2O$
50.1.4 Oxammite $(NH_4)_2C_2O_4 \cdot H_2O$

Type 2. Mellates, Citrates, Sulfocyanates, Acetates
50.2.1 Mellite $Al_2C_{12}O_{12} \cdot 18H_2O$
50.2.2 Earlandite $Ca_3(C_6H_5O_7)_2 \cdot 4H_2O$
50.2.3 Julienite $Na_2Co(SCN)_4 \cdot 8H_2O$
50.2.4 Calclacite $CaCl(C_2H_3O_2) \cdot 5H_2O$

50 SALTS OF ORGANIC ACIDS

TYPE 1. OXALATES

50.1.1 **WHEWELLITE** [$Ca(C_2O_4) \cdot H_2O$]. Oxalate of Lime *Brooke* (*Phil. Mag.*, **16**, 449, 1840). Oxacalcite *Shepard* (111, 1844). Whewellite *Brooke* and *Miller* (623, 1852). Kohlenspath *Frenzel* (*Min. Mitt.*, **11**, 83, 1889). Thierschite *Liebig* (*Ann. Chemie*, **86**, 113, 1853).

Cryst.[1] Monoclinic; prismatic—$2/m$.

$a:b:c = 0.8696:1:1.3695$; $\beta\ 107°18\frac{1}{2}'$; $p_0:q_0:r_0 = 1.5749:1.3075:1$

$r_2:p_2:q_2 = 0.7648:1.2045:1$; $\mu\ 72°41\frac{1}{2}'$; $p_0'\ 1.6496,\ q_0'\ 1.3695,\ x_0'\ 0.3116$

Forms:[2]

		ϕ	ρ	ϕ_2	$p_2 = B$	C	A
c	001	90°00′	17°18½′	72°41½′	90°00′	.....	72°41½′
b	010	0 00	90 00		0 00	90°00′	90 00
u	120	31 03½	90 00	0 00	31 03½	81 10	58 56½
m	110	50 18	90 00	0 00	50 18	76 46	39 42
z	014	42 18½	24 50½	72 41½	71 54	18 06	73 34½
y	012	24 28	36 57½	72 41½	56 49½	33 10½	75 35
x	011	12 49	54 33	72 41½	37 24½	52 35½	79 35½
k	102	90 00	48 39	41 21	90 00	31 20½	41 21
e	1̄01	−90 00	53 13½	143 13½	90 00	70 32	143 13½
f	112	58 55½	52 59½	41 21	65 39½	38 54½	46 50½
s	1̄32	−14 01½	64 43	117 10	28 41	69 58½	102 39½

1100 ORGANIC COMPOUNDS

Habit and twinning. Only as crystals. Simple crystals are equant or short prismatic [001] and the faces are usually irregularly developed. Twins with twin plane $e\{\bar{1}01\}$ are very common; these are either heart-shaped or prismatic and of pseudo-orthorhombic appearance and may be with or without re-entrant angles. Common forms: $c\ x\ b\ m\ u\ s\ e$.

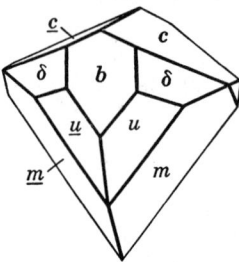
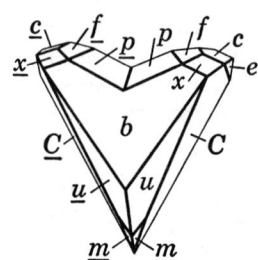

Schlan, Bohemia. Twin on $\{\bar{1}01\}$.   Burgk, Saxony.

Etching.[3] Conforms to holohedral monoclinic symmetry.

P h y s. Cleavage $\{\bar{1}01\}$ good, $\{010\}$ imperfect, $\{001\}$ and $\{110\}$ indistinct. Fracture conchoidal. Brittle. H. $2\frac{1}{2}$–3. G. 2.23. Luster vitreous, on $\{010\}$ somewhat pearly. Colorless and transparent; sometimes yellowish or brownish.

O p t.[4] In transmitted light, colorless.

		n (Burgk)[5]		
ORIENTATION	Li	Na	Tl	
$X\ \ b$	1.4878	1.4909	1.4939	Biaxial positive (+).
Y	1.5513	1.5554	1.5599	$r < v$
$Z \wedge c\ \ +30°$	1.6450	1.6502	1.6567	
$2V$ (meas.)	83°45′	83°55′	84°15′	

The extinction angle on $\{010\}$ of twins to the twin boundary is $6\frac{1}{2}°$.

C h e m. Calcium oxalate monohydrate, $Ca(C_2O_4)\cdot H_2O$.
Anal.

	1	2	3	4
CaO	38.38	38.46	38.83	38.36
C_2O_3	49.28	49.65	49.38	50.28
H_2O	12.34	12.14	12.31	11.36
Total	100.00	100.25	100.52	100.00
G.		2.222	2.226	2.22

1. $Ca(C_2O_4)\cdot H_2O$. 2. Pchery, Bohemia.[6] 3. Brüx, Bohemia.[7] 4. Maikop, Caucasus.[8] Recalculated to 100 after deducting 6.68 per cent $CaCO_3$.

Tests. B.B. decomposes[11] to CaO. Soluble in acids but insoluble in water.

O c c u r. Whewellite ordinarily is of organic origin and occurs in association with vegetal remains, but it has been found as a primary hydrothermal deposit in ore veins. The substance also occurs as microscopic crystals in living plant cells and as calculi or as a sediment in the human urinary tract. Among the natural occurrences may be mentioned

Burgk near Dresden, Saxony, where crystals up to several inches in size occur with calcite in the footwall of a coal seam. Also found in Saxony in coal at Zwickau and with calcite and native silver in a vein at Freiberg. Found with lignite or coal at various points in Czechoslovakia, notably Kladno, Pchery, and at Bruch near Dux where lignite beds are covered by flows of basaltic rocks. Also from veins at Recsk and Kapnik in Hungary; and from the Maikop district in the northern Caucasus, U.S.S.R. In calcite veins in shale on the Yareg River, southern Timan, U.S.S.R.[10] With tetrahedrite, carbonates, and quartz in veins at the Saint-Sylvestre mine, Urbeis, Alsace.[9] The locality of the original crystals described by Brooke is not known.

A r t i f. Obtained as microscopic crystals by reaction of calcium salts with oxalates in water solution.

N a m e. After William Whewell (1794–1866), English natural scientist and philosopher.

Ref.

1. Elements of Miller, *Phil. Mag.*, **16**, 450 (1840) and (623, 1852). See also Becke, *Min. Mitt.*, **26**, 391 (1907), who found $a:b:c = 0.8628:1:1.3677$, $\beta\ 107°00'$, on good crystals from Brüx.
2. Goldschmidt (**9**, 70, 1923). The form frequency and combinations have been discussed by Kolbeck, Goldschmidt, and Schröder, *Beitr. Kr. Min.*, **1**, 199 (1918); Kašpar, *Věstnik. Král. Čes. Spol. Nauk*, Cl. 2, no. 34 (1932), and *Knihov. Stat. Geol. Úst. Česk. Repub.*, **20**, 1 (1939); Frey, *Schweiz. min. Mitt.*, **4**, 16 (1924); Ježek, *Ac. sc. Bohême, Bull.*, **14**, 1 (1909); and others. Rare, vicinal, and doubtful forms:

a 100	n 230	α 031	P 111	χ 454	q $\overline{3}19$
F 190	D 340	e: 041	ϵ $\overline{1}14$	H 575	V $\overline{3}58$
v_1 170	n_1 790	ν 205	ϕ $\overline{1}12$	ω 232	o $\overline{3}16$
Σ 160	650	N 203	G $\overline{1}11$	γ 121	δ $\overline{1}21$
ϑ 150	A 540	μ_1 507	Y 459	T 423	U $\overline{3}12$
ζ 290	r 210	μ 101	L 458	J 342	π $\overline{2}11$
v 140	310	τ $\overline{1}05$	Z 7.9.14	p:231	σ $\overline{3}21$
l_1 270	r_1 410	t $\overline{1}03$	K 346	B 341	ρ $\overline{11}.8.3$
l 130	w 016	κ $\overline{1}02$	S 10.5.16	Δ $\overline{2}38$	g $\overline{4}31$
250	013	h 1.1.10	M 545	Q $\overline{3}.7.17$	C $\overline{6}51$
470	i 032	j 119	W 878	R $\overline{12}.9.40$	ξ $\overline{14}.26.1$
350	ψ 0.11.5	λ 118	X 9.10.9	p $\overline{2}16$	

3. Kašpar (1939).
4. For summary see Wherry, *Washington Ac. Sc., J.*, **12**, 196 (1922).
5. Ježek, *Ac. sc. Bohême, Bull.*, **16**, 1 (1911), by refractometer method at 24°–28°.
6. Millbauer analysis in Slavík, *Rozpr. böhm. Ak. Prag*, **17**, no. 38 (1908).
7. Plzák analysis in Ježek, *Ac. sc. Bohême, Bull.*, **13**, 1 (1908).
8. Vassoyevicz and Razumovsky, *Mem. soc. russe min.*, **57**, 275 (1928).
9. Ungemach, *Bull. soc. min.*, **32**, 20 (1909).
10. Kalyuzhnyi, *C. r. ac. sc. U.R.S.S.*, **59**, 1631 (1948).
11. Dehydration and infrared absorption data are given by Pobeguin, *et al.*, *C.R.*, **216**, 500, 808 (1943).

50.1.2   **W E D D E L L I T E** [$CaC_2O_4 \cdot 2H_2O$]. Hydrated Calcium Oxalate Bannister and Hey (*Discovery Reports*, **13**, 60, 1936). Weddellite Frondel and Prien (*Science*, **95**, 431, 1942).

C r y s t.[1] Tetragonal; tetragonal-dipyramidal—$4/m$.

$$a:c = 1:0.591; \quad p_0:r_0 = 0.591:1$$

Forms: [2]

	φ	ρ	A	M
r 011	0°00'	30°35'	90°00'	68°55'

Structure cell.[3] Space group $I4/m$. a_0 12.40 ± 0.02 kX, c_0 7.37 ± 0.02; $a_0:c_0 = 1:0.594$. Cell contents $Ca_8(C_2O_4)_8 \cdot 16H_2O$.

Habit. Pyramidal {101}; sometimes aggregated into groups.

P h y s. No cleavage. Fracture subconchoidal. H. about 4. G. 1.94 ± 0.01;[4] 1.91 (calc.). Colorless to white; also yellowish brown to brown, due to included organic matter. Transparent.

O p t. In transmitted light, colorless. Indices:[5]

nO 1.523 ± 0.003   Uniaxial positive (+).
nE 1.544 ± 0.003

C h e m. Calcium oxalate dihydrate, $CaC_2O_4 \cdot 2H_2O$. There has been some uncertainty as to the water content,[6] whether $2H_2O$ or $3H_2O$. Analyses of natural material are lacking.

Tests. Decomposes to $CaCO_3$ at about 270° C. Insoluble in water.

O c c u r. Found as tiny isolated crystals in bottom muds of the central Weddell Sea, Antarctica. Weddellite and whewellite are the principal constituents of the oxalate type of urinary calculi in humans.[8]

A r t i f. Obtained in crystals by reaction of soluble calcium salts with oxalates under the proper conditions.[7]

N a m e. From the locality.

Ref.

1. Crystal class given by Bannister and Hey (1936), but without evidence.
2. Forms other than {101} were not observed on natural crystals or among several hundred urinary calculi containing weddellite and examined by Prien and Frondel, *J. Urology*, **57,** 949 (1947). Tetragonal crystals with a prismatic habit have been described by Nakano, *J. Biochem.*, **2,** 437 (1923), and Hammarsten, *C. r. des trav. lab. Carlsberg*, **17,** 1 (1929), but these may represent another hydrate.
3. Bannister in Bannister and Hey (1936) by rotation and oscillation methods.
4. Bannister and Hey (1936) by suspension method on artificial crystals; corrected for impurities.
5. Prien and Frondel (1947) and Bannister and Hey (1936) on both artificial and natural material, by immersion method.
6. Cf. Bannister and Hey (1936); also Pobeguin, *C.R.*, **216,** 500 (1943), and Lecompte, Pobeguin, and Wyart, *C.R.*, **216,** 808 (1943), Hammarsten (1929).
7. Cf. Hammarsten (1929) and Pobeguin (1943).
8. Cf. Prien and Frondel (1947).

50.1.3 **H U M B O L D T I N E** [$Fe(C_2O_4) \cdot 2H_2O$]. Faser Resin (Honigsteinsaures Eisen ?) *Breithaupt* (75, 1820). Humboldtine, Oxalsaures Eisen *de Rivero* (*Ann. chim. phys.*, **18,** 207, 1821). Eisen-Resin *Breithaupt* (*Ann. Phys.*, **70,** 426, 1822). Oxalit *Breithaupt* (1823). Humboldtit *Leonhard* (789, 1826).

C r y s t.[1] Orthorhombic.

$a:b:c = 0.7730:1:1.1039;$ $p_0:q_0:r_0 = 1.4282:1.1039:1$

$q_1:r_1:p_1 = 0.7730:0.7002:1;$ $r_2:p_2:q_2 = 0.9059:1.2938:1$

Forms:

	ϕ	$\rho = C$	ϕ_1	$\rho_1 = A$	ϕ_2	$\rho_2 = B$
c 001	.....	0°00′	0°00′	90°00′	90°00′	90°00′
a 100	90°00′	90 00	.....	0 00	0 00	90 00
m 110	52 18	90 00	90 00	37 42	0 00	52 18
d 101	90 00	55 00	0 00	35 00	35 00	90 00

Habit. Distinct crystals are rare; as prisms elongated [001] or as plates. Usually capillary, or in botryoidal or incrusting forms with a fibrous structure; also fine-granular to earthy and compact.

P h y s. Cleavage {110} perfect, {100} and {010} imperfect. H. $1\frac{1}{2}$–2. G. 2.28. Luster dull to resinous. Color yellow to amber-yellow. Transparent in crystals.

O p t.[2]

Orientation		n	Pleochroism	
X	a	~1.494	Very pale yellowish green	Biaxial positive (+).
Y	b	1.561	Pale greenish yellow	2V large.
Z	c	1.692	Intense yellow	

C h e m. Ferrous oxalate dihydrate, $Fe(C_2O_4) \cdot 2H_2O$. Analysis gave:[3]

	FeO	MgO	C_2O_3	H_2O	Total	G.
1.	39.94		40.03	20.03	100.00	
2.	40.72	tr.	40.18	[19.10]	[100.00]	2.28

1. $Fe(C_2O_4) \cdot 2H_2O$ 2. Capo d'Arco, Elba.[2]

Tests. In C.T. loses water, turns black, and becomes magnetic. Heated in air turns greenish brown and then reddish brown. Soluble in acids.

O c c u r. Found in Czechoslovakia with gypsum in brown coal at Kollosoruk southwest of Bilin; also with ammonia alum at Tschermig and in brown coal at Luschitz. In Germany found at Grossalmerode in Hesse-Nassau and in coal at Potschappel near Dresden. In Canada at Kettle Point, Bosanquet township, Ontario.

A r t i f.[2] Obtained as a yellow powder by reaction of oxalic acid or ammonium oxalate with ferrous chloride or ferrous ammonium oxalate solutions.

N a m e. After Alexander von Humboldt (1769–1859).

Ref.

1. Elements and forms of Manasse, *Acc. Linc., Rend.*, **19**, 138 (1910), *Mem. Soc. Tosc.*, **28**, 118 (1912), on crystals from Capo d'Arco, Elba (analysis 2).
2. Manasse (1910).
3. Older analyses are cited by Hintze (**1** [4B], 1346, 1933).

50.1.4 **O X A M M I T E** [$(NH_4)_2C_2O_4 \cdot H_2O$]. Shepard (*Rural Carolinian*, **1**, 471, May 1870). Guañapite *Raimondi* (30, 33, 1878).

Orthorhombic; disphenoidal—2 2 2. Found as lamellar masses, pulverulent or rarely as distinct crystals. Artificial crystals are sphenoidal in habit, with $a:b:c = 0.7799:1:0.3700$.[1] Cleavage {001} distinct. H. $2\frac{1}{2}$. G. ~1.5. Colorless (artificial) to yellowish white. Transparent.

O p t.[2] In transmitted light, colorless.

ORIENTATION n(Na) (artificial)
X	c	1.438	Biaxial negative (−).
Y	a	1.547	2V 62°.
Z	b	1.595	$r < v$, distinct.

C h e m. Ammonium oxalate monohydrate,[3] $(NH_4)_2C_2O_4 \cdot H_2O$. Analysis of natural material [4] (recalculated after deducting 5.5 per cent organic matter): NH_4 21.95, C_2O_4 53.30, H_2O 24.75, total 100.00.

O c c u r. Found with mascagnite in the guano deposits of the Guañape Islands, Peru.

A r t i f. Obtained as crystals by evaporation of the water solution.

Ref.
1. See summary in Groth (**3**, 150, 1910).
2. Brio, *Ak. Wien, Sitzber.*, **55**, 870 (1867).
3. Identity of the natural material with artificial $(NH_4)_2C_2O_4 \cdot H_2O$ established by x-ray and optical study by Frondel, priv. comm. (1948).
4. Tanner, *Chem. News*, **32**, 162 (1875).

UNNAMED OXALATE. Lacroix (*Bull. soc. min.*, **9**, 51, 1886).
Apparently an oxalate of sodium and ammonium, found with stercorite as small masses consisting of cleavable, micaceous laminae in a guano deposit in Peru. Crystallization probably orthorhombic. Optically, biaxial negative (−) with $2E$ about 15°, $r < v$, weak, and X perpendicular to the laminae.

TYPE 2. MELLATES, CITRATES, SULFOCYANATES, ACETATES

50.2.1 M E L L I T E $[Al_2C_{12}O_{12} \cdot 18H_2O]$. Honigstein [from Thuringia] *Werner* (*Bergm. J.*, **1**, 380, 395, 1789). Honigstein *Karsten* (*Mus. Lesk.*, **2**, Pt. 1, 335, 1789). Succin transparent en cristaux octaèdres, Pierre de miel *von Born* (**2**, 90, 1790). Mellites *Gmelin* (*Linn. Syst.*, **3**, 282, 1793). Mellilite *Kirwan* (**2**, 68, 1796). Mellite *Haüy* (**4**, 445, 1822). Honigstein, Melilithus, = Honigsteinsäure (acidum melilithicum) + Alaunerde + Wasser *Klaproth* (*Ak. Berlin*, 1799; *Beitr.*, **3**, 114, 1801). Xylocryptite (?) *Becquerel* (*J. Phys.*, **89**, 237, 1819).

C r y s t.[1] Tetragonal; trapezohedral—4 2 2 (?).

$$a:c = 1.0554:1; \quad p_0:r_0 = 1.0554:1$$

Forms:[2]

	ϕ	ρ	A	M
c 001	....	0°00′	90°00′	90°00′
a 010	0°00′	90 00	90 00	45 00
m 110	45 00	90 00	45 00	90 00
r 011	0 00	46 32½	90 00	59 07
e 112	45 00	36 44	64 59	90 00

Structure cell.[3] Space group probably $P4_12$ or $P4_32$. a_0 22.0 kX, c_0 23.3; $a_0:c_0 = 1:1.055$. Cell contents $Al_{32}C_{192}O_{192} \cdot 288H_2O$.

Habit. Crystals small and rare; prismatic [001] or pyramidal {011}. Also massive, fine-granular, as coatings or nodular.

P h y s. Cleavage {011} indistinct. Fracture conchoidal. Slightly sectile. H. 2–2½. G. 1.64; 1.65 (calc.). Luster resinous, inclining to vitreous. Color honey-yellow, often reddish or brownish; rarely white. Streak white. Transparent to translucent. Pyroelectric.[6] Fluoresces blue in short-wave ultraviolet light.

O p t.[4] In transmitted light, colorless to pale yellow. Weakly pleochroic with nO yellowish brown and nE yellow. Sometimes abnormally biaxial with small $2V$ and $X = c$. Uniaxial negative $(-)$.

λ	687mμ	589	527	397
nO	1.5345	1.5393	1.5435	1.5611
nE	1.5079	1.5110	1.5146	1.5277

C h e m. Hydrated aluminum mellate, $Al_2C_{12}O_{12} \cdot 18H_2O$. Al_2O_3 14.27, C 20.17, O 20.16, H_2O 45.40, total 100.00. The reported analyses [5] are close to the formula given.

Tests. Whitens in the flame of a candle but does not take fire. Insoluble in water and alcohol but soluble in HNO_3. On heating, $12H_2O$ are lost at about 100°, and decomposition sets in at about 300°.

O c c u r. Found as a secondary deposit in crevices in brown coal and lignite. In Germany found at Artern in Thuringia and near Bitterfeld in Saxony; at Luschitz near Bilin in Bohemia, and near Walchow and Boskowitz in Moravia. At Malowka in the Bogoroditsk district, Tula, U.S.S.R. With gypsum in lignitic clay in the Paris basin, Seine, France.

A r t i f.[7] Obtained in crystals by diffusion of aluminum chloride solution into a solution of an alkali salt of mellitic acid.

N a m e. Mellite from μέλι, *honey*, in allusion to the color.

Ref.

1. Angles of Dauber, *Ann. Phys.*, **94**, 410 (1855); orientation of the x-ray cell of Barth and Ksanda, *Am. Min.*, **18**, 8 (1933). Transformation: Dauber to B. and K., 110/110/001.
2. Goldschmidt (**6**, 28, 1920).
3. Barth and Ksanda (1933).
4. Indices of Schrauf, *Ak. Wien, Sitzber.*, 41 (1860). On the optical anomalies see Des Cloizeaux (15, 1867).
5. For summary see Hintze (**1** [4B], 1346, 1933) and Doelter (4 [3], 798, 1930).
6. Hankel, *Ann. Phys.*, **18**, 422 (1883).
7. Friedel and Balsohn, *Bull. soc. min.*, **4**, 26 (1881); Friedel and Crafts, *ibid.*, **3**, 189 (1880).

50.2.2 **E A R L A N D I T E** [$Ca_3(C_6H_5O_7)_2 \cdot 4H_2O$]. Bannister (*Discovery Reports*, **13**, 67, 1936).

As warty, fine-grained nodules. Color white to pale yellow. G. 1.95. Mean refractive index 1.56.

C h e m. Calcium citrate tetrahydrate, $Ca_3(C_6H_5O_7)_2 \cdot 4H_2O$.

Anal.

	1	2	3	4
CaO	29.49	28.63	31.6	29.01
H	3.18	3.48		
C	25.26	24.01		
O	42.07	[43.88]		
Total	100.00	[100.00]		

1. $Ca_3(C_6H_5O_7) \cdot 4H_2O$. 2. Weddell Sea.[1] 3, 4. Weddell Sea.[2]

Occur. Found in oceanic bottom samples from the Weddell Sea, Antarctica (71°22′S, 16°34′W, 1410 fathoms).
Artif.[3] By reaction of soluble calcium salts and citrates.
Name. After Arthur Earland, oceanographer.

Ref.
1. Perrins microanalysis in Bannister (1936).
2. Hey microanalysis in Bannister (1936).
3. The identity of the natural and the artificial substances proved by Bannister (1936) by x-ray study.

50.2.3 JULIENITE [$Na_2Co(SCN)_4 \cdot 8H_2O$]. Schoep (Natuurw. Tijdschr., Antwerp, 10, 58, 1928).

Cryst.[1] Tetragonal.

$$a:c = 1:0.6030; \quad p_0:r_0 = 0.6030:1$$

Forms:

	ϕ	ρ	A	$\overline{M}$
a 010	0°00′	90°00′	90°00′	45°00′
k 5.13.0	21 02½	90 00	68 57½	66 02½
l 140	14 02	90 00	75 58	59 02
m 110	45 00	90 00	45 00	90 00
o 021	0 00	50 20	90 00	57 01½

Structure cell.[2] Tetragonal. a_0 9.22 kX, c_0 5.56; $a_0:c_0 = 1:0.6030$. Cell contents $Na_2Co(SCN)_4 \cdot 8H_2O$.

Habit. Thin crusts of minute needles elongated [001]. Acicular [001] (artif.).

Phys. Cleavable [001].[2] G. 1.648 (artif.),[2] 1.680 (calc.). Color blue.

Opt.

	Katanga[3]	Artif.[4]	
nO	1.556	1.556	Uniaxial positive (+).
nE	1.645	1.642	

Chem. A hydrated sodium cobalt thiocyanate,[5] $Na_2Co(SCN)_4 \cdot 8H_2O$.

Anal.

	Na	Co	SCN	H$_2$O	Total
1.	9.56	12.24	48.26	29.94	100.00
2.	10.35	12.5	47.6	29.87	100.32

1. $Na_2Co(SCN)_4 \cdot 8H_2O$. 2. Katanga.[4]

Tests. Soluble in water or alcohol yielding a rose-colored solution.

Occur. Found with cobaltian wad (?) in a white talc schist at Chamibumba, near Kambove, Katanga, Belgian Congo.

Artif.[6] Obtained as crystals from a water solution containing two mols NaSCN and one mol $Co(SCN)_2$; also from a water solution of the mineral.

Name. After Henry Julien (?–1920).

Ref.

1. Natural crystals have not been measured. Artificial crystals were used by Schoep, *Natuurw. Tijdschr.*, **13**, 147 (1931), to establish the ratio 1:1.2059. The powder and rotation x-ray study gives 1:0.6030; transformation: Schoep to Schoep and Billiet, 200/020/001.
2. Schoep and Billiet (1935).
3. Schoep (1928).
4. Cuvelier, *Natuurw. Tijdschr.*, **15**, 17 (1933).
5. Originally described erroneously as the Co analogue of buttgenbachite.
6. Cuvelier and de Sweemer, *Ann. serv. des mines, Com. spéc. du Katanga*, **3**, 67 (1933).

PIGOTITE. Johnston (*Phil. Mag.*, **17**, 382, 1840).

A reddish brown or brown, resinous coating formed on granite on the coast of Cornwall by the action of decomposing plant slime.[1] A salt of aluminum and an organic acid, approximately $4Al_2O_3 \cdot C_{12}H_{10}O_8 \cdot 27H_2O$. In C.T. yields much water and empyreumatic products, leaving a black mass. Heated to redness slowly burns away leaving a gray ash.

Ref.

1. Creighton, *Geol. Mag.*, **1** [IV], 223 (1894).

50.2.4   **C A L C L A C I T E** [$CaCl(C_2H_3O_2) \cdot 5H_2O$]. van Tassel (*Mus. Belgique, Bull.*, **21**, 1, 1945).

As hair-like efflorescences. Color white. G. 1.5. Optically biaxial positive (+), with $2V$ 80° and nX 1.468, nY 1.484, nZ 1.515; Z = elongation. Chemically a hydrated chloride-acetate of calcium, $CaCl(C_2H_3O_2) \cdot 5H_2O$. Analysis gave:

	Ca	Cl	($C_2H_3O_2$)	H_2O	Total
1.	17.84	15.78	26.28	40.10	100.00
2.	17.6	15.4	25.5	39.5	98.0

1. $CaCl(C_2H_3O_2) \cdot 5H_2O$. 2. Efflorescence on a museum specimen of calcareous schist.

Found as recent formation on specimens of calcareous rocks and fossils kept in oak museum cases (from which the acetic acid was derived). Found similarly, admixed with other efflorescent salts, on pottery kept in museum cases.[1]

A r t i f.[2] An artificial salt of the same composition is obtained by crystallization of an equimolar solution of calcium chloride and calcium acetate.

N a m e. From the composition, *cal*cium, *Cl*, and *a*cetate.

Ref.

1. Gettens and Frondel, priv. comm. (1948).
2. Fritsche, *Ann. Phys.*, **28**, 121 (1833); Dubsky and Tesarik, *Coll. trav. chim. czechoslov.*, **1**, 571 (1929).

INDEX

Names of numbered and accepted species and also folios of pages on which main description appears are printed in bold-face type.

Abichit, 787
Achrematite, 1081, 1085
Adamite, 859, **864**
Adelite, 804
α-AgI, 19
β-AgI, 22
Agaphite, 947
Ahlfeldite, 635
Akrochordite, 927
Alabaster, 482, 484
Alaunstein, 556
Alkanasul, 556
Allactite, 785
Allodelphite, 780
Allogonite, 820
Allomorphit, 408
Alluaudite, 674
Almagrerite, 428
Almeraite, 92, 94
Almeriite, 556
Alpha-ascharite, 375
Alpha-boracite, 378
Alpha-Dahllite, 879
Alpha-Hopeite, 734, 736
Alpha-kertschenite, 742, 744
Alpha-metavoltine, 619
Alpha-Usbekite, 818
Alpha-variscite, 756
Alstonite, 182, 218
Alum, 471
Alumian, 560
Aluminilite, 556
Aluminite, 600
Aluminocopiapite, 623, 625
Alumino-scorodite, 763
Alumo-chalcosiderite, 947
Alumohydrocalcite, 280
Alumopharmacolite, 997
Alunite, 555, **556**
Alunogen, 537
Alunogène, 537
Amarantite, 611
Amarillite, 468
Amatrice, 756
Ambatoarinite, 293
Amblygonite, 823
Ammiolite, 1024
Ammonia Alum, 471, **475**
Ammoniakalaun, 475
Ammonia-niter, 305
Ammoniak-salpeter, 305

Ammonioborite, 366
Ammoniojarosite, 555, 562
Amphithalite, 873
Anapaite, 731
Ancylite, 291
Andersonite, 239
Andrewsite, 802
Angelardite, 742
Anglarite, 742
Anglesite, 407, 420
Angleso-barite, 408, 411
Anhydrite, 424
Anhydrokainite, 596
Ankerite, 207, 208
Ankylit, 291
Annabergite, 741, 746
Anthoinite, 1097
Antimonbleispath, 1018
Antlerite, 544
Antofagastite, 44
Antozonit, 29
Apatelite, 567
Apatite, 877, 878
Aphanëse, 787
Aphérèse, 862
Aphrit, 142, 153
Aphthalose, 400
Aphthitalite, 400
Apjohnite, 522, **527**
Araeoxene, 811
Aragonite, 182
Arakawaite, 916
Arcanite, 398, **399**, 400
Ardealite, 703, 1010
Arequipite, 1025
Argentine, 142, 144, 153
Argentojarosite, 555, 565
Argentopercylite, 78
Argyroceratite, 11
Arksutite, 123
Armangite, 1031
Arnimite, 592
Aromite, 523
Arrojadite, 669, **679**
Arseneisensinter, 1014
Arsenicite, 706
Arsenikblüthe, 706
Arseniksinter, 763
Arseniopleite, 844
Arseniosiderite, 953
Arsenobismite, 907
Arsenoclasite, 801

1109

INDEX

Arsenocrocites, 953
Arsenokrokite, 953
Arsentsumebite, 918
Artinite, 263
Arzrunite, 130
Ascharite, 375
Asparagus Stone, 878
Astrakhanite, 447
Atacamite, 69
Atelestite, 792
Atelite, 74
Atlaserz, 253
Atopite, 1020, 1021
Attacolite, 845
Augelite, 871
Augustit, 878
Aurichalcite, 249
Austinite, 804, 809
Autunite, 980, 984
Avogadrite, 97
Azovskite, 795
Azurite, 264, 908

Baeumlerite, 91
Bakerite, 363
Baldaufite, 703
Bandylite, 373
Bararite, 103, 106
Barcenite, 1025
Baricalcite, 142, 154
Barite, 407, 408
Barium-Parisite, 285
Barolite, 194
Baroselenite, 408
Barrandite, 756, 759
Barthite, 806, 808
Bartholomite, 604
Barystrontianite, 196
Barytes, 408
Barytine, 408
Barytite, 408
Barytoanglesite, 420
Barytocalcite, 218, 220
Barytocelestite, 408, 415
Barytocölestin, 408, 411, 415, 417
Barytsalpeter, 305
Basaluminite, 586
Basicérine, 289
Basiliite, 1025
Basobismutite, 259, 262
Bassanite, 476
Bassetite, 994
Bastnaesite, 289
Baudisserite, 162, 271
Bayldonite, 929
Bayleyite, 237
Beaverite, 555, 568
Bechilite, 347, 365
Beiyinite, 291
Bellingerite, 313
Bellite, 895
Belonesite, 37, 39
Beraunite, 959

Beresofite, 646
Beresovite, 653
Bergblau, 264
Berggrün, 252
Berglasur, 264
Bergmehl, 142
Bergmilch, 227
Berkeyite, 908
Berlinite, 696
Bermanite, 967
Bernonite, 923
Beryllonite, 677
Berzeliite, 681
Berzélite, 57
Beta-ascharite, 375
Beta-boracite, 378
Beta-Dahllite, 879
Beta-Hopeite, 734, 736
Beta-Kertschenite, 742, 744
Beta-uranopilite, 582
Beta-Usbekite, 818
Beudantite, 1001
Beyerite, 258, 281
Bialite, 906
Bianchite, 493, 495
Bieberite, 498, 505
Bieirosite, 1002
Bilinite, 522, 529
Bindheimite, 1017, 1018
Bischofit, 831
Bischofite, 46
Bismoclite, 59, 60
Bismuthite, 259
Bismutite, 259
Bismutosphärit, 259, 261
Bitter Salts, 509
Bitterspath, 208
Black-band Ore, 166
Blakeite, 532, 643
Blättriges Olivenerz, 1008
Blauspath, 908
Bleigummi, 831
Bleihornerz, 256
Bleikerat, 256
Bleilasur, 553
Bleimalachite, 252
Bleiniere, 1018
Bleinierite, 1018
Bleispath, 200
Bloedite, 447
Blue Iron Earth, 742
Blue John, 29
Blue Stone, 488
Blue Vitriol, 488
Blumit, 1018, 1064
Bobierrite, 741, 753
Boksputite, 259, 261
Boleite, 78
Bolivarite, 872
Bonamite, 176
Bone-phosphate, 878
Bone-turquois, 950
Boothite, 498, 504

Index

Boracite, 378
Borax, **339**
Bordosite, 25, 27
Borgströmite, 566
Borickite, 915
Borocalcite, 345, 347, 365
Boromagnesite, 375
Boronatrocalcite, 345
Borspar, 349
Bosjemanite, 523, 525
Bosphorite, 742, 744
Botallackite, 76
Botryogen, 617
Botryt, 617
Bouglisite, 420
Bourbolite, 499
Boussingaultite, 452, 455
Bowmanite, 834
Brackebuschite, 1052
Bradleyite, 295
Brandtite, 725
Braunbleierz, 889
Brazilianite, 841
Breunnerite, 162, 164
Brickerite, 809
Brochantite, 541
Bromargyrite, 11
Bromchlorsilber, 11
Bromit, 11
Bromlite, 218
Bromsilber, 11
Bromyrite, 3, 11
Brongniartin, 431
Brongniartine, 541
Brossit, 208, 213
Brown Spar, 164, 166, 208
Bruiachite, 29
Brushite, 703, 704
Bückingit, 520
Buetschliite, 231
Buntbleierz, 889
Buratite, 249
Burkeite, 633
Bushmanite, 523, 525
Butlerite, 608
Buttgenbachite, 572

Cabrerite, 746, 749
Cacoxenite, 997
Cadmiumzinkspath, 176
Cadwaladerite, 77
Cahnite, 386
Calafatite, 556, 559
Calaite, 947
Calamine, 176
Calcareobarite, 408, 411
Calcimangite, 142, 153
Calcio-ancylite, 291
Calciocarnotite, 1045
Calciocelestite, 415, 417
Calcioferrite, 976
Calciostrontianit, 196
Calciovolborthite, 804, 816, 817

Calcite, 141, **142**
μ-Calcite, 181
Calcium-lazulite, 908
Calclacite, 1107
Calcnitre, 306
Calcouranit, 984
Calc Spar, 142
Caledonite, 630
Caliche, 300
Calingastite, 499, 502
Callaina, 946
Callainite, 761
Calomel, 25
Calstronbarite, 408
Campylite, 889, 892
Camsellite, 375
Capnit, 176, 179
Capreite, 142
Caracolite, 546
Carbapatit, 879
Carbocérine, 241
Carbonapatit, 879
Carbonate-apatite, 877, **879**
Carbonate-whitlockite, 686
Carbonyttrine, 275
Carminite, 912
Carnallite, 92
Carnotite, 1043
Carphosiderite, 555, **566**
Caryinite, 683
Castanite, 613
Castelnaudite, 688
Catoptrite, 1029
Cawk, 408
Cegamit, 247
Celestite, 407, 415
Celestobarite, 408, 411
Cerapatite, 879
Cerargyrite, 3, 11
Cerasite, 57
Cerbolit, 455
Cerium-apatite, 879
Ceruleite, 927
Ceruleofibrite, 572
Céruse, 200
Cerussa, 200
Cerussite, 182, **200**
Chalcanthite, 487, 488
Chalchuite, 947
Chalcoalumite, 580
Chalcocyanite, 429
Chalcomenite, 638
Chalcophacite, 921
Chalcophyllite, 1008
Chalcosiderite, 946, **947**
Chalkolith, 981
Chalybite, 166
Chenevixite, 840
Cherokine, 889
Chessylite, 264
Childrenite, 936
Chileite, 811
Chile-Loeweite, 447

INDEX

Chilisalpeter, 300
Chillagite, 1081, 1085
Chimborazite, 183
Chiolite, 110, 123
Chloraluminite, 50
Chlorapatite, 877, 879
Chlorargyrite, 11
Chlorarsenian, 681, 684
Chlor-fluorapatite, 884
Chlormanganokalite, 109
Chlornatrokalite, 7, 9
Chlorobromite, 11
Chlorocalcite, 91
Chlorochalcit, 69
Chloromagnesite, 39, 41
Chlorophane, 29
Chlorophoenicite, 778
Chlorothionite, 547
Chlorotile, 943, 944
Chloroxiphite, 84
Chlorsilber, 11
Chlorsilberspath, 11
Chlor-Spath, 57
Chlor-utahlite, 756
Chodneffite, 123
Chondrarsenite, 855
Chondrostibian, 1030
Chrombleispath, 646
Chrom-Loeweite, 447
Chrommolybdänbleierz, 1081
Chromowulfenite, 1081
Chubutite, 56
Churchillite, 57
Churchite, 703, 773
Ciempozuelite, 433
Ciplyte, 878
Cirrolite, 845
Clay Iron Ore, 166
Clay Ironstone, 166
Clinobarrandite, 769
Clinoclase, 787
Clinocrocite, 570
Clinophaeite, 570
Clino-ungemachite, 597
Cobalt Bloom, 746
Cobalt-Chalcanthite, 493
Cobaltoadamite, 864
Cobaltocalcite, 141, 142, 154, 175
Cobaltomenite, 639
Cobaltosphärosiderit, 166, 169
Cobalt Vitriol, 505
Coccinite, 42
Codazzite, 208
Coeruleolactite, 961
Colemanite, 349
Collbranite, 321
Collieite, 889, 892
Collinsite, 722
Collophane, 879, 885
Conchite, 183, 191
Conichalcite, 804, 806
Conite, 208
Connellite, 572

Copiapite, 623
Copperas, 499
Copper Mica, 1008
Copper-Uranite, 981
Coquimbite, 532
Cordylite, 285
Corkite, 1001, **1002**
Cornetite, 789
Cornwallite, 925
Coronguite, 1025
Cotunnite, 42
Crandallite, 831, 835
Creedite, 129
Crocoise, 646
Crocoite, 646
Cromfordite, 256
Cryolite, 110
Cryolithionite, 99
Cryphiolite, 845
Cryptohalite, 103, 104, 106
Cryptolite, 691
Cryptomorphite, 341, 345, 347
Ctypeite, 183
Cube Ore, 995
Cumengite, 79
Cuproadamite, 864
Cupro-apatit, 878
Cuprocopiapite, 623
Cuprodescloizite, 811
Cuproferrite, 499
Cuprogoslarite, 514, 515
Cupro-iodargyrite, 22
Cuprokirovite, 499, 502
Cupromagnesite, 505
Cuproplumbite, 929
Cuproscheelite, 1091
Cuprotungstite, 1091
Cuprouranit, 981
Cuprovanadite, 811, 895
Cuprozincite, 252
Cyanochroite, 452, 454
Cyanoferrite, 499
Cyanose, 488
Cyanotrichite, 578
Cyprusite, 566

Dahllite, 879
Dakeite, 236
Darapskite, 309
Dauberite, 598
Daubréeite, 59, 60
Daviesite, 58
Davisonite, 939
Davite, 537
Dawsonite, 276
Dechenite, 811
Dehrnite, 877, 902
Deltaite, 831, 837
Delta-mooreite, 175
Delvauxite, 915, 935
Dennisonite, 939
Derbylite, 1025
Derbyshire Spar, 29

INDEX 1113

Dernbachite, 1002
Descloizite, 804, **811**
Destinezite, 1011
Devilline, 590
Devillite, 590
Devonite, 962
Dewindtite, 875
Diaboleite, 82
Diadelphit, 777
Diadochite, 1011
Dialogite, 171
Dickinsonite, 717
Diderichite, 275
Dietrichite, 522, **528**
Dietzeite, 318
Dihydrite, 799
Dioxylith, 550
Dittmarite, 699
Dogtooth Spar, 142
Dolerophanite, 551
Dolomite, 207, **208**
Doppelspat, 142
Doughtyite, 586
Douglasite, 100
Dréelite, 408
Drewite, 142
Dry-bone, 176
Dudgeonite, 746, 749
Dufreniberaunite, 959
Dufrenite, 867, **873**
Duftite, 804, **810**
Dumontite, 928
Dumreicherite, 523
Dundasite, 279
Durangite, 827, **829**
Durdenite, 640, 641
Dussertite, 831, **839**

Earlandite, 1105
Ecdemite, 1036
Edwardsite, 691
Eggonite, 965
Eglestonite, 51
Eguëite, 955
Ehlit, 799
Eichwaldite, 330
Eisenalaun, 523
Eisenapatit, 849
Eisenblüthe, 183
Eisenboracit, 378, 380
Eisenbranderz, 1014
Eisendolomit, 208
Eisenglimmer, 742
Eisenkalkankylit, 291
Eisenparaluminite, 586
Eisenpecherz, 849, 1014
Eisen-Phyllit, 742
Eisenpickeringit, 523
Eisen-Resin, 1102
Eisensinter, 763, 1011, 1014
Eisenspath, 166
Eisenstassfurtit, 378, 380
Eisenvitriol, 499

Eisstein, 110
Elatolite, 142, 151, 158
Eleonorite, 959
Elfstorpite, 920
Ellestadite, 877, **906**
Elpasolite, 110, **114**
Embolite, 11, 12, 13
Emerald Nickel, 245
Emmonite, 196
Emmonsite, 640
Enceladite, 326
Endlichite, 889, 895, 897
Englishite, 957
Eosite, 1081, 1085
Eosphorite, 936
Epiglaubite, 704
Epiphosphorit, 878
Epsomite, 509
Epsom Salt, 509
Erbsenstein, 142, 153, 183, 188
Eremeyevite, 330
Eremite, 691
Erinite, 798, 1008
Eriochalcite, 44
Erythrite, 741, **746**
Erythrocalcite, 44
Erythrosiderite, 101
Erzbergit, 183
Eschwegite, 415
Estramadurite, 878
Ettringite, 589
Euchlorin, 570
Euchroite, 934
Eupyrchroite, 879, 884
Eusynchite, 811
Evansite, 923
Evigtokite, 119
Exanthalose, 439

Fairchildite, 222
Fairfieldite, 720
Faserkalk, 183
Faser Resin, 1102
Fauserite, 509, 511, 513
Fe-alluaudite, 674
Federalaun, 523
Felsöbányaite, 585
Ferberite, 1064
Ferghanite, 1048
Fermorite, 877, **904**
Fernandinite, 1062
Ferrazite, 832
Ferri-alluaudite, 674
Ferricopiapite, 623, 625
Ferrimolybdite, 1095
Ferrinatrite, 456
Ferri-paraluminite, 586
Ferripurpurite, 675
Ferri-sicklerite, 672
Ferrisymplesite, 753
Ferritungstite, 1093
Ferri-turquois, 947
Ferrocalcite, 142, 154

Ferrocopiapite, 623
Ferrodolomite, 208
Ferroepsomite, 509
Ferro-Goslarite, 514, 515
Ferroludwigite, 322
Ferronatrite, 456
Ferropallidite, 479
Ferropickeringite, 523
Ferrostibian, 1026
Ferrowolframite, 1064
Ferruccite, 97, 98
Fervanite, 1049
Fe-sicklerite, 672
Fibroferrite, 614
Fiedlerite, 67
Fillowite, 719
Finnemanite, 1038
Firuzegi, 946
Fischerite, 762
Flajolotite, 1024
Flinkite, 793
Flockenerz, 889
Florencite, 831, 838
Floridite, 878
Flos Ferri, 183, 188
Fluellite, 110, 124
Fluoborite, 369
Fluocerine, 289
Fluocerite, 48
Fluor-adelite, 827
Fluorapatite, 877, 879
Fluorbaryt, 29
Fluor-collophane, 884
Fluorcollophane, 879
Fluor-herderite, 820
Fluorite, 28, 29
Fluormanganapatit, 879, 884
Fluor Spar, 29
Flusscerit, 48
Flussyttrocalcit, 29, 33
Flutherite, 240
Fontainebleu limestone, 142, 153
Footeite, 572
Forbesite, 711
Fornacite, 652
Foucherite, 915
Francolite, 879, 884
Franklandite, 345, 347
Franquenite, 622
Freirinite, 920
Fremontite, 823
Fritzscheite, 984
Frondelite, 867

Gajite, 208, 264
Galenoceratite, 256
Gallizinite, 514
Gamma-Dahllite, 879
Garnsdorfite, 586
Gaylussite, 234
Gearksutite, 110, **119**
Gekrösstein, 424
Gelbbleierz, 1081

Gelbeisenerz, 560, 623
Geldiadochite, 1011
Gelvariscite, 756
Georgiadesite, 791
Geraesite, 833
Gerhardtite, 308
Gesso, 482
Gibbsite, 762
Gilpinite, 606
Ginorite, 364
Giobertite, 162
Giorgiosite, 274
Gips, 482
Glasbachit, 640
Glaserite, 399, 400
Glaubapatite, 660, 878
Glauberite, 431
Glauber Salt, 439
Glaucocerinite, 574
Glaukosiderit, 742
Glendonite, 160
Globosite, 762
Glockerite, 587
Glucinite, 820
Gorceixite, 831, 833
Gordaite, 456
Gordonite, 975
Goslarite, 509, **515**
Goyazite, 831, 834
Graftonite, 686
Green Lead Ore, 889
Green Vitriol, 499
Greinerite, 208, 213
Griphite, 843
Grodnolite, 879
Grünbleierz, 889
Grüneisenstein, 873
Grüner glimmer, 981
Guañapite, 1103
Guanite, 715
Guildite, 619
Gummibleispath, 831
Gunnisonite, 29
Gurhofian, 208
Gypsum, 481, 482

Haarsalz, 509, 523, 537
Hagemannite, 116, 117
Haidingerite, 708
Halblasurblei, 630
Halite, 3, 4
β-Halite, 7
Hallite, 600
Halochalzit, 69
Halotrichite, 522, 523, 537
Hamartite, 289
Hambergite, 370
Hamlinite, 834
Hanksite, 628
Hannayite, 699
Harbortite, 951
Harttite, 1005
Hautefeuillite, 753

Hayesine, 345, 347, 365
Headdenite, 679
Heavy Spar, 408
Hebronite, 823
Hedyphane, 877, 900
Heintzite, 367
Heliophyllite, 1037
Hemafibrite, 919
Hematolite, 777
Hematostibite, 1027
Henwoodite, 951
Hepatit, 408
Herderite, 820
Herrengrundite, 590
Herrerite, 176, 179
Heterobrochantite, 544
Heterosite, 675
Hewettite, 1060
Hexahydrite, 493, 494
Hibbenite, 737
Hibbertite, 271
Hieratite, 103
Higginsite, 806
Hilgardite, 382
Hinsdalite, 1001, 1004
Hintzeite, 367
Hislopite, 142
Hitchcockite, 831
Hoernesite, 741, 755
Hoevelit, 7
Hohmannite, 613
Hokutolite, 408, 411
Holdenite, 775
Honigstein, 1104
Hopeite, 734
Hornblei, 256
Hornerz, 11
Horn Mercury, 25
Hornsilber, 11
Howlite, 362
Huantajayite, 4, 6
Huebnerite, 1064
Huelvite, 171
Hügelite, 815
Hühnerkobelite, 669
Hulsite, 321, **326**
Humboldtine, 1102
Hureaulite, 700
Hussakite, 688
Huyssenite, 378, 382
Hydrargillite, 947, 962
Hydrargyrit, 25
Hydro-apatite, 879
Hydrobasaluminite, 586
Hydrobismutite, 259, 262
Hydroboracite, 353
Hydroborocalcite, 345
Hydrocalcite, 227
Hydrocerit, 241
Hydrocerussite, 270
Hydroconite, 228
Hydrocyanite, 429
Hydrodolomite, 271, 273

Hydrofluocerite, 289
Hydro-fluorherderite, 820
Hydrogiobertite, 271, 273
Hydroglockerite, 587
Hydrohalite, 15
Hydro-herderite, 820
Hydrolanthanit, 241
Hydromagnesite, 271
Hydromagnocalcit, 271
Hydromanganocalcit, 271
Hydromelanothallite, 77
Hydronickelmagnesite, 271
Hydrophilite, 41, 91
Hydroromeite, 1022
Hydroxyapatite, 879
Hydroxylapatite, 877, 879
Hydroxyl-fluorapatite, 884
Hydroxyl-Herderite, 820
Hydrozincite, 247

Idrizite, 618
Iglésiasite, 200
Iglit, 183
Ignatiewite, 556, 559
Ihlëite, 623
Ilesite, 486
Inderborite, 355
Inderite, 360
Inyoite, 358
Iodargyrit, 22
Iodembolite, 11, 12, 13
Iodit, 22
Iodobromite, 11, 13
Iodyrite, 22
Iron alum, 523
Iron-boracite, 378, 380
Iron-dolomite, 208
Isoclasite, 933

Janosite, 623
Jarlite, 110, 118
Jarošit, 499
Jarosite, 555, 560
Jarrowite, 160
Jeremejevite, 330
Jezekite, 784
Jogynaite, 763
Johachidolite, 384
Johannite, 606
Johnite, 947
Jossaite, 652
Jujuyite, 1024
Julienite, 1106
Junckérite, 166
Justit, 86

Kainite, 594
Kakoxen, 997
Kalialaun, 471
Kaliastrakanit, 450
Kaliborite, 367
Kalicinite, 136
Kalicit, 136

Kalinite, 469, **471,** 472
Kalinitrat, 303
Kalioalunite, 556
Kalisalpeter, 303
Kalium-Astrakanit, 450
Kalium-Blödit, 450
Kalkmagnesit, 271
Kalksalpeter, 306
Kalkspath, 142
Kalk-Uranglimmer, 984
Kalk-Uranit, 984
Kalk-volborthit, 817
Kalkwavellit, 835
Kallait, 947
Kallochrom, 646
Kaluszite, 442
Kamarezite, 588
Kampylit, 889, 892
Kapnicit, 962
Kårafveit, 691
Karminspath, 912
Karphosiderite, 566
Karstenit, 424
Karyinite, 683
Kauaiite, 556, 559
Kehoeite, 944
Kempite, 73
Keramohalite, 523, 537
Kerargyrite, 11
Kerasine, 57, 256
Kerat, 11
Kerchenite, 742
Kernite, 335
Kerstenite, 640
Kieserite, 477
Kietyöite, 878
Kipushite, 916
Kirovite, 498, **499**
Kirrolith, 845
Kischtimite, 289
Kischtim-Parisit, 289
Kjerulfin, 845
Klaprothine, 908
Klebelsbergite, 583
Kleinite, 87
Klinocrocit, 570
Klinoklas, 787
Klinophäit, 570
Knauffite, 818
Knoxvillite, 623
Kobalt-oligonspat, 166, 169
Kobalt-scorodite, 763
Koboltblüthe, 746
Kobaltspath, 175
Koechlinite, 1092
Koenenite, 86
Koettigite, 741, **751**
Kohlengalmei, 176
Kohlenspath, 1099
Koktaite, 444
Kolbeckite, 1015
Kollophan, 879
Kolosorukite, 560

Kolovratite, 1048
Kondroarsenit, 855
Königine, 541
Koninckite, 763
Kordylit, 285
Kornelite, 530
Kotoite, 328
Kramerite, 343
Kraurite, 873
Krausite, 462
Kremersite, 101
Kreuzbergite, 124, 125
Kribergite, 1011
Krisuvigite, 541
Kroehnkite, 444
Krokoisit, 646
Krugite, 458, 460
Kryolith, 110
Kryphiolith, 845
Kryptolith, 691
Ktypeïte, 183, 191
Kubeite, 617
Kühnite, 681
Kupaphrite, 925
Kupferbleispat, 553
Kupfer-diaspore, 799
Kupfereisenvitriol, 499
Kupferglimmer, 1008
Kupferhornerz, 69
Kupferlasur, 264
Kupferphyllit, 1008
Kupfersammterz, 578
Kupferschaum, 925
Kupfer-Uranglimmer, 993
Kupfer-Uranit, 981
Kupferuranit, 993
Kupfervitriol, 488
Kupferzinkblüthe, 249
Kuprojarošit, 499
Kurnakovite, 360
Kurskite, 879
Kutnahorite, 207, **217**

Lacroixite, 171, 783
Lamprophanite, 580
Lamprostibian, 1030
Lanarkite, 550
Lancasterite, 271
Landesite, 729
Langbeinite, 434
Langite, 583
Lansfordite, 228
Lanthanite, 241
Lapparentite, 466, 467, 601
Larderellite, 365
Lasionite, 962
Lasur, 264
Laubmannite, 803
Laurionite, 62
Lausenite, 530
Lautarite, 312
Lavendulan, 750, 920
Lawrencite, 39, 40

INDEX 1117

Laxmannite, 650
Lazulite, 908
Lazur-apatit, 878
Leadhillite, 295
Lechedor, 6
Lecontite, 438
Leedsite, 408
Leesbergite, 208
Legrandite, 958
Lehiite, 942
Lehmannite, 646
Lehnerite, 952
Leightonite, 461
Lemnäsite, 674
Leonite, 450
Leopoldit, **7**
Letovicite, 397
Lettsomite, 578
Leucochalcite, 930
Leucoglaucite, 456
Leucomanganit, 720
Leucophosphite, 936
Lewisite, 1020, 1021
Lewistonite, 877, 903
Libethenite, 859, 862
Liebigite, 240
Lime-malachite, 253
Lime-Uranite, 984
Linarite, 553
Lindackerite, 1007
Lindgrenite, 1094
Linsenerz, 921
Linsenkupfer, 921
Liparit, 29
Liroconite, 921
Liskeardite, 924
Lithiophilite, 664, 665
Loaisite, 763
Loeweite, 446
Loewigite, 556, 559
Lopezite, 645
Lorettoite, 56
Loseyite, 244
Lossenite, 763, 766
Louderbackite, 520
Lublinite, 142, 144
Lucinite, 756
Luckite, 499
Lucullan, 208
Ludlamite, 952
Ludwigite, 321
Lueneburgite, 385
Lunnite, 799
Lyellite, 590

Mackayite, 642
Magnesia Alum, 523
Magnesiasalpeter, 307
Magnesinitre, 307
Magnesiocopiapite, 623
Magnesio-dolomite, 208
Magnesioludwigite, 321
Magnesiosussexite, 375

Magnesite, 141, 162
Magnesiumapjohnite, 523, 525
Magnesium-berzeliite, 681
Magnesium-chlorophoenicite, 780
Magnetostibian, 1025
Makite, 407
Malachite, 252
Malladrite, 103, 105
Mallardite, 498, 507
Mamanite, 458, 460
Mangan-alluaudite, 674
Mangan-ankerite, 208
Manganapatite, 878, 879, 884
Manganberzeliite, 681
Mangandolomite, 208, 213, 217
Manganese Alum, 527
Mangan-fluorapatite, 879, 884
Mangan-hydroxyapatite, 879, 884
Manganipurpurite, 675
Mangankalkankylit, 291
Manganocalcite, 142, 153, 171, 173
Manganolangbeinite, 435
Manganosiderite, 166, 171, 173
Manganosphärit, 166, 169
Manganostibite, 1027
Manganowolframite, 1064
Manganpickeringit, 523, 525
Mangan-sicklerite, 672
Manganspath, 171
Manganvoelckerite, 879
Mangualdite, 879
Mansfieldite, 756, 763
Marcylite, 69
Marionite, 247
Marmor, 142, 208
Marshite, 18, 20
Martinite, 684, 686
Martinsite, 4, 477
Mascagnite, 398
Masrite, 523, 525
Matlockite, 59
Mauzeliite, 1020, 1021
Maxite, 295
Mazapilite, 953
Medjidite, 600
Megabasite, 1064
Megabromite, 11, 13
Melanchlor, 675
Melanochroit, 649
Melanostibian, 1041
Melanothallite, 44, 77
Melanovanadite, 1058
Melanterite, 498, 499
Mélinose, 1081
Mellilite, 1104
Mellite, 1104
Mellonite, 96
Mendipite, 56, 57
Mendozite, 469, 474
Mengite, 691
Mercallite, 395
Merkur-Hornerz, 25
Merkur-Kerat, 25

Merkurspath, 25
Merrillite, **797**
Mesitin, 162, 164
Mesitite, 162, 164
Messelite, 731, 732
Messingblüthe, 249
Messingite, 249
Meta-alunogen, 539
Meta-autunite, 980, **985**
Meta-autunite I, 981, 985
Meta-autunite II, 981, 985
Metabrushite, 704
Meta-chalcolite, 991
Métachalcophyllite, 1008
Metahewettite, 1061
Metahohmannite, 608
Metahydroboracite, 355
Meta-jarlite, 118, 119
Metakalkuranite, 985
Metakupferuranit, 991
Metanhydrit, 424
Metanocerite, 86
Metarossite, 1054
Metasideronatrite, 603
Metastrengite, 767, **769**
Metathenardite, 407
Metatorbernite, 980, **991**
Meta-torbernite I, 991
Metatriplite, 849
Meta-uranopilite, 582
Metavariscite, 767
Metavauxite, 971
Metavoltine, 619
Metazeunerite, 980, **993**
Meyerhofferite, 356
Meyersite, 761
Michel-Lévyte, 408
Microbromite, 11, 12
Microcosmic Salt, 698
Miemit, 208
Miersite, 18, **19**
Miesit, 889, 892
Millisite, 941
Millosevichite, 539
Mimetene, 889
Mimetèse, 889
Mimetite, 877, **889**
Minasragrite, 437
Minervite, 999
Minyulite, 970
Mirabilite, 439
Miriquidite, 1002
Misenite, 396
Misy, 560, 623
Mitridatite, 955
Mitscherlichite, 100
Mixite, 943
Mn-alluaudite, 674
Mn-Berzeliite, 681
Mn-sicklerite, 672
Moffrasite, 1018
Mohavite, 337
Mollit, 908

Molybdanbleispath, 1081
Molybdänocker, 1095
Molybdänoxyd, 1095
Molybdenbleierz, 1081
Molybdine, 1095
Molybdite, 1095
Molybdomenite, 640
Molybdoscheelite, 1074, 1076
Molysite, 47
Monazite, 691
Monazitoid, 691
Monetite, 660
Monheimite, 176, 178
Monimolite, 1017, **1023**
Monite, 879
Monkforssite, 1006
Montanite, 636
Montebrasite, 823
Montgomeryite, 978
Montmartrite, 482
Mooreite, 574
Morenosite, 509, **516**
Morinite, 783, 784
Moronolite, 560
Moroxit, 878
Mosesite, 89
Mossottite, 183, 189
Mottramite, 804, **811**
Muellerite, 699
Mullicite, 742
Munkrudite, 1006
Muriazit, 424
Musite, 282
Mysorin, 253

Nadelstein, 182
Nadorite, 1039
Nahcolite, 134
Na-heterosite, 675
Nantokite, 18
Na-purpurite, 675
Natramblygonite, 823
Natrikalite, 4
Natrit, 230
Natroalunite, 555, **556**
Natrocalcite, 234
Natrochalcite, 602
Natrojarosite, 555, **563**
Natromontebrasite, 823
Natron, 230
Natronalaun, 469
Natronamblygonite, 823
Natron-berzeliite, 681
Natronitrite, 300
Natronkalisimonyit, 447
Natronsalpeter, 300
Natrophilite, 664, **670**
Nauruite, 879, 884
Neocolemanite, 349, 352
Néoctèse, 763
Neoplase, 617
Neopurpurite, 675
Neotyp, 142, 154

Nesquehonite, 225
Newberyite, **709**
Newtonite, 556, 559
Nicholsonite, 183, 189
Nickel Bloom, 746
Nickelblüthe, 746
Nickel Smaragd, 245
Nickel Vitriol, 516
Nipholith, 123
Niter, 303
Nitrammite, 305
Nitratin, 300
Nitratite, 300
Nitre, 303
Nitrobarite, 305
Nitrocalcite, 306
Nitroglauberite, 311
Nitromagnesite, 307
Nitrum, 303
Nocerite, 85
Nordenskiöldine, 332
Normal-ankerite, 208, 213
Normal-dolomite, 208
Normal-parankerit, 208, 213
Normannite, 259
Northupite, 278
Nussierite, 889, 892

Oborite, 291
Occidental Turquois, 947
Ochrolite, 1039, 1040
Octahedral borax, 337
Odontolite, 742, 878, 950
Oligonite, 166, 169
Oligonspath, 166, 169
Olivenerz, 859, 862, 995
Olivenite, 859
Olivenkupfer, 859
Oriental Turquois, 947
Ornithite, 879
Orthobromite, 11, 13
Orthodiadochite, 1011
Oserskit, 183
Osteolith, 878
Ostwaldit, 11
Otavite, 141, **181**
Overite, 979
Oxacalcite, 1099
Oxalit, 1102
Oxammite, 1103
Oxyapatite, 879
Oxykertschenite, 742, 744

Pachnolite, 110, 114
Paigeite, 321, 322
Palacheite, 617
Palaite, 700
Palmerite, 999
Palmierite, 403
Pandermite, 341, 343
Paposite, 611
Parabayldonite, 929
Parabutlerite, 610

Paracoquimbite, **534**, 621
Paragearksutite, 119
Parahilgardite, 383
Parahopeite, 733
Paralaurionite, 64
Paraluminite, 586
Parankerit, 208, 213
Parasit, 378
Paratacamite, 74
Paraurichalcite-I, 251
Paraurichalcite-II, 252
Paravauxite, 972
Paravivianite, 742
Parisite, **282**, 287
Parsonsite, 913
Pascoite, 1055
Pastréit, 560
Patagosite, 142
Pateraite, 1098
Paternoite, 363
Pearl Spar, 208
Peganite, 756
Pelagosite, 183
Pelosiderite, 166
Pencatite, 271
Penfieldite, 66
Pennite, 271
Pentahydrite, 487, 492
Pentahydrocalcite, 228
Percylite, 81
Perlspath, 208
Perowskyn, 665
Pettkoite, 464
Pfaffite, 1018
Pharmacochalzit, 859
Pharmacolite, 703, 706
Pharmacolzit, 859
Pharmacosiderite, 995
Phillipite, 520
Phoenicochroite, 649
Phönicit, 649
Phosgenite, 256
Phosgen-spath, 256
Phosphocerite, 691
Phosphoferrite, 727
Phosphophyllite, 738
Phosphorchromit, 650
Phosphoreisensinter, 1011
Phosphorite, 878
Phosphorkupfererz, 799, 862
Phosphorocalcit, 799
Phosphorroesslerite, 713
Phosphosiderite, 769
Phosphuranylite, 876
Picite, 935
Pickeringite, 522, 523
Picralluminite, 523
Picroallumogene, 523
Picromerite, 452, 453
Picropharmacolite, 740
Pigotite, 1107
Pinakiolite, 321, 324
Pinnoite, 334

Pinolite, 162
Pintadoite, 1053
Pirssonite, 232
Pisanite, 498, **499**
Pisolite, 142, 153, 183, 188
Pissophanite, 586
Pistomesite, 162, 164
Pitch Iron Ore, 1014
Pitticite, 1014
Pittizite, 587
Plagiocitrite, 570
Planerite, 762
Planoferrite, 567
Plaster of Paris, 484
Pleonectite, 902
Pleurasite, 845
Pleuroklas, 845
Pleysteinite, 124, 125
Plombgomme, 831
Plumbeine, 889
Plumbiodite, 317
Plumbo-aragonite, 183, 188
Plumbocalcite, 142, 154
Plumbodolomite, 208, 213
Plumbogummite, 831
Plumbojarosite, 555, **568**
Plumbonacrite, 270
Plumbo-resinite, 831
Plumbosynadelphite, 780
Podolite, 879
Polyarsenite, 855
Polyhalite, 458
Polysphaerit, 889, 892
Ponite, 171, 173
Potash Alum, 471, **472**
Potassalumite, 471
Potassium Fluoride, 28
Powellite, 1073, **1079**
Prasin, 799
Prasin-chalzit, 799
Predazzite, 271
Preslite, 918
Priceite, 341
Probertite, 343
Proidonite, 100
Prosopite, 110, **121**
Protocalcite, 142
Prunnerite, 142
Pseudo-apatelite, 567
Pseudo-apatite, 879
Pseudoberzeliite, 681
Pseudoboleite, 80
Pseudo-campylite, 892
Pseudocopiapite, 623
Pseudocotunnite, 96
Pseudogaylussite, 160
Pseudoheterosite, 672
Pseudo-libethenite, 862
Pseudomalachite, 799
Pseudomendipite, 57
Pseudonocerite, 29
Pseudopalaite, 700
Pseudotriplite, 675

Pseudowavellite, 835
Psimythit, 295
Psittacinite, 811
Pucherite, 1050
Pulszkyite, 593
Purpurite, 675
Pyrobelonite, 804, **815**
Pyroclasite, 879
Pyroconite, 114
Pyroguanite, 878
Pyromelin, 516
Pyromorphite, 877, **889**
Pyrophosphorite, 684
Pyrosmaragd, 29
Pyrotechnite, 404
Pyrrhoarsenite, 681, 683

Quenstedtite, 535
Quercyite, 879
Quercyite-α, 879
Quercyite-β, 879
Quetenite, 617

Rafaëlit, 64
Raimondite, 567
Ralstonite, 110, **126**
Ramirite, 811
Ransomite, 519
Rashleighite, 947
Rasorite, 335
Raspite, 1089
Ratofkite, 29
Rauvite, 1058
Reddingite, 727
Redingtonite, 522, 529
Red Iron Vitriol, 617
Redondite, 756, 759
Red Vitriol, 505
Reichardtit, 509
Reichit, 142
Reinite, 1064, 1071
Remolinite, 69
Renardite, 928
Repossite, 686
Retgersite, 496, **497**
Retzian, 794
Reussin, 439
Rhabdophane, 774
Rhagite, 792
Rhodhalose, 505
Rhodizite, 329
Rhodoarsenian, 1026
Rhodochrosite, 141, **171**
Rhodoise, 746
Rhomboclase, 436
Rhomb Spar, 208
Richellite, 956
Richmondite, 762
Ridolphite, 208
Rinneite, 107
Risséite, 249
Rockbridgeite, 867
Rock Salt, 4

Roemerite, 520
Roesslerite, 712
Rogersite, 530
Rohwand, 208
Roméine, 1020
Romeite, 1017, **1020**
Rooseveltite, **697**
Rosasite, **251**
Roscherite, **968**
Roselite, **723**
Rosiéresite, 924
Rossite, **1053**
Rotbleierz, 646
Roweite, **377**
Rubrite, 570, 617
Rutherfordine, **274**

Saamite, 879, 884
Sahlinite, **775**
Salammoniac, **15**
Saldanite, 537
Saléeite, 980, **988**
Salesite, **315**
Sal gemma, 4
Salmiak, 15
Salmoite, 933
Salmonsite, 730
Salpeter, 303
Saltspar, 4
Salvadorite, 444, 499
Salzkupfererz, 69
Sampleite, **945**
Sanmartinite, **1072**
Sarcopside, **858**
Sardinian, 420
Sarkinite, 853, **855**
Sarmientite, **1013**
Satin Spar, 482, 484
Scacchite, **39, 40**
Schadeite, 831
Schafarzikite, **1035**
Schaffnerite, 811
Schairerite, **547**
Schätzellit, **7**
Schaumkalk, 191
Schaumspath, 142
Scheelerz, 1074
Scheelite, 1073, **1074**
Scheelspath, 1074
Schertelite, 699
Schieferspath, 142, 153
Schneebergite, 1020, 1021
Schoenite, 453
Schoharite, 408
Schokoladenstein, 171
Schroeckingerite, **236**
Schuilingite, 252
Schultenite, **661**
Schützit, 415
Schwartzembergite, **317**
Schwerspath, 408
Schwerstein, 1074
Sclerospathite, 529

Scorodite, 756, **763**
Scorzalite, **908**
Scovillite, 774
Seamanite, **388**
Sebkhainite, 92
Seelandite, 509
Selenbleispath, 640
Selenite, 482, 484
Sellaite, 28, **37**
Sengierite, **1047**
Serpierite, **592**
Sesqui-Magnesia-Alaun, 523
Seyrigite, 1074, 1076, 1078
Sexangulit, 889
Sharpite, 275
Shortite, **222**
Siberian volborthite, 818
Sicilianite, 415
Sicklerite, **672**
Sideretine, 1014
Siderite, 141, **166**
Siderochalcit, 787
Siderodot, 166
Sideronatrite, **604**
Sideroplesite, 166, 169
Siderose, 166
Siderotil, 487, **491**
Silberhornerz, 11
Silber-Kerat, 11
Silicoborocalcite, 362
Simonyit, 447
Sincosite, **1057**
Sjögruvite, 845
Slavikite, **621**
Smaragdochalcit, 69
Smithsonite, 141, **176**
Soda, 230
Soda Alum, 469, 471, **474**
Soda-alunite, 556
Soda Berzeliite, 681
Sodalumite, 474
Soda-Niter, **300**
Soda-triphylite, 679
Sodium-autunite, 985
Solfatarite, 537
Soluble-anhydrite, 476, 484
Sombrerite, 878
Sommairite, 499, 502
Sonomaite, 523
Soumansite, 940, 941
Souzalite, **911**
Spangolite, **576**
Spargelstein, 878
Spartaite, 142, 153
Spatheisenstein, 166
Spathic Iron, 166
Spencerite, **931**
Sphaerite, 756
Sphaerocobaltite, 175
Spherosiderite, 166
Spodiosite, 848
Sprudelstein, 183
Staffelite, 879, 884

1122   INDEX

Stagmatite, 47
Stasite, 875
Stassfurtit, 378
Staszicite, 806
Steigerite, 1049
Steinsalz, 4
Stelznerit, 544
Stercorite, 698
Sterrettite, 965
Stewartite, 730
Stibiatil, 1030
Stibiogalenites, 1018
Stiepelmannite, 838
Stinkfluss, 29
Stoffertite, 704
Stolzite, 1073, 1087
Strahlbaryt, 408
Strahlenerz, 787
Strahlenkupfer, 787
Strengite, 756
Stromnite, 196
Strontianapatite, 904
Strontianite, 182, 196
Strontianocalcite, 142, 154
Struvite, 715
Studtite, 275
Stüvenite, 523
Stypterit, 537
Stypticit, 614
Subhydrocalcite, 227
Sulfatapatit, 879
Sulfoborite, 387
Sulfohalite, 548
Susannite, 298
Sussexite, 375
Svabite, 877, 899
Svanbergite, 1001, **1005**
Swartzite, 238
Swedenborgite, **1027**
Sylvine, 7
Sylvite, 3, 7
Symplesite, 741, **752**
Synadelphite, 780
Synchisite, 287
Syngenite, 442
Szaibelyite, 375
Szmikite, 477, 481
Szomolnokite, 477, **479**

Tachyhydrite, 95
Tagilite, 931
Talkapatit, 878
Talktriplit, 849
Tallingite, 76
Tamanit, 731
Tamarite, 1008
Tamarugite, 466
Tangeite, 816, 817
Tanguéite, 817
Taranakite, 999
Tarapacaite, 644
Taraspite, 208
Tarbuttite, 869

Tarnowitzite, 183, 188
Tauriscite, 509, 519
Tautoklin, 208, 213
Tavistockite, 906
Taylorite, 398, **400**
Taznite, 1025
Teepleite, 372
Teineite, 635
Tengerite, 275
Terlinguaite, 52
Teruelite, 208
Teschemacherite, 137
Tetraphylin, 665
Texasite, 245
Tharandit, 208
Thenardite, 404
Thermokalite, 134
Thermonatrite, 224
Thierschite, 1099
Thinolite, 160
Thomäit, 166
Thomsenolite, 110, 116
Thoneisenstein, 166
Thorotungstite, 1097
Thrombolite, 799
Tikhvinite, 1006
Tilasite, 827
Tincal, 337
Tincalconite, 337
Tinkalzit, 345
Tinticite, 970
Tirolit, 925
Tocornalite, 25
Torberite, 981
Torbernite, 980, 981
Torrensite, 171
Torreyite, 575
Traubenblei, 889
Trichalcite, 739
Trigonite, 1032
Trihydrocalcite, **227**
Trimontite, 1074
Tripe Stone, 424
Triphylin, 665
Triphylite, 664, **665**
Triplite, 849
Triploidite, 853
Trippkeite, 1034
Tripuhyite, 1024
Tritochorit, 811
Troegerite, 966
Trolleite, 911
Trona, 138
Trudellite, 131
Tschermigit, 475
Tsumebite, 918
Turanite, 804, 816, **818**
Turchesia, 947
Turkestan-volborthite, 817
Turkey-fat ore, 179
Türkis, 946
Turnerite, 691
Turpeth, 25

Index

Turquois, 946
Tychite, 294
Tyrolite, 925
Tysonite, 48
Tyuyamunite, 1043, **1045**

Uhligite, 756
Ulexite, 345
Ungemachite, 596
Unnamed minerals, 127, 237, 570, 634, 818, 1048, 1072, 1104
Uraconise, 600
Uraconite, 600
Uranblüthe, 598
Uranglimmer, 981
Urangrün, 600
Uranite, 981, 984
Uran-mica, 981
Uranochalcite, 600
Uranocher, 581, 598
Uranocircite, 980, **987**
Uranopilite, 581
Uranospathite, 990
Uranospinite, 980, **990**
Uranothallite, 240
Uranphyllit, 981
Uranvitriol, 606
Urao, 138
Urdit, 691
Urusite, 604
Urvölgyite, 590
Usbekite, 818
Utahite, 566
Utahlite, 756
Uvanite, 1056
Uzbekite, 816, 818

Vanadinbleispath, 895
Vanadinite, 877, **895**
Vanthoffite, 430
Variscite, 756
Varulite, 664, **669**
Vashegyite, 999
Vaterite, 181
Vaterite-A, 142
Vaterite-B, 181
Vauquelinite, 650
Vauxite, 974
Veatchite, 348
Vegasite, 568, 570
Vernadskite, 607
Vesbine, 811
Veszelyite, 916
Viellaurite, 171
Vilateite, 769
Villiaumite, 3, **10**
Vitriolbleierz, 420
Vitriolgelb, 560
Vitriolocker, 587
Vivianite, 741, **742**
Voelckerite, 879
Voglianite, 600
Voglite, 237

Volborthite, 816, **818**
Voltaite, 464
Vonsenite, 322
Voraulite, 908
Vulpinit, 424

Wagnerite, 845
Walmstedtite, 162, 164
Walpurgite, 796
Waltherite, 262
Wapplerite, 712, 713
Wardite, 940
Waringtonite, 541
Warthit, 447
Warwickite, 326
Wasserbleiocker, 1095
Wattevilleite, 452
Wavellite, 962
Weberite, 110, **127**
Websterite, 600
Weddellite, 1101
Weibyeite, 293
Weinschenkite, 703, **771**
Weisbachit, 420
Weissbleierz, 200
Wenzelite, 702
Werthemanite, 601
Weslienite, 1020, 1021
Wherryite, 632
Whewellite, 1099
White Copperas, 513, 532
White Lead Ore, 200
White Vitriol, 513
Whitlockite, 684
Wilkeite, 877, **905**
Winebergite, 586
Winkworthite, 362
Wiserite, 245
Wismuthspath, 259
Witherite, 182, **194**
Wölchite, 1019
Wolfeite, 853
Wolframite, 1064
Wood-Copper, 859
Woodhouseite, 1001, **1006**
Woodwardite, 580
Wulfenite, 1073, **1081**
Würfelerz, 995
Würfelgyps, 424
Würfelspath, 424

Xantharsenite, 855, 858
Xanthosiderit, 623
Xanthoxenite, 977
Xenotime, 688
Xylocryptite, 1104

Yeremeyevite, 330
Ypoléime, 799
Ytterspath, 275, 688
Yttrium Calcium Fluoride, 37
Yttrocalcite, 29, 33
Yttrocererite, 29, 33

Yttrocerite, 29, 33
Yttrofluorite, 32
Yttroflusspath, 29, 33
Yttroparisite, 285
Yukonite, 953

Zamboninite, 37, 39
Zaratite, 245
Zeiringit, 183, 189
Zepharovichite, 964
Zeugite, 684, 686
Zeunerite, 980, **989**, 993
Zincaluminite, 579
Zinc-melanterite, 498, 508
Zincocalcite, 142, 154

Zinconise, 247
Zincorhodochrosite, 171
Zinc Vitriol, 513
Zinkazurit, 264
Zink-Bleispath, 200
Zinkblüthe, 247
Zinkboothit, 508
Zinkeisenspath, 176, 178
Zinkosite, 428
Zinkspath, 176
Zippeite, 598
Zirklerite, 87
Zölestin, 415
Zootinsalz, 300
Zwieselite, 849